Element	Symbol	Atomic Number	Average Atomic Mass[a]
Actinium	Ac	89	[227]
Aluminum	Al	13	26.982
Americium	Am	95	[243]
Antimony	Sb	51	121.76
Argon	Ar	18	39.948
Arsenic	As	33	74.922
Astatine	At	85	[210]
Barium	Ba	56	137.33
Berkelium	Bk	97	[247]
Beryllium	Be	4	9.0122
Bismuth	Bi	83	208.98
Bohrium	Bh	107	[270]
Boron	B	5	10.811
Bromine	Br	35	79.904
Cadmium	Cd	48	112.41
Calcium	Ca	20	40.078
Californium	Cf	98	[251]
Carbon	C	6	12.011
Cerium	Ce	58	140.12
Cesium	Cs	55	132.91
Chlorine	Cl	17	35.453
Chromium	Cr	24	51.996
Cobalt	Co	27	58.933
Copernicium	Cn	112	[285]
Copper	Cu	29	63.546
Curium	Cm	96	[247]
Darmstadtium	Ds	110	[281]
Dubnium	Db	105	[268]
Dysprosium	Dy	66	162.50
Einsteinium	Es	99	[252]
Erbium	Er	68	167.26
Europium	Eu	63	151.96
Fermium	Fm	100	[257]
Flerovium	Fl	114	[289]
Fluorine	F	9	18.998
Francium	Fr	87	[223]
Gadolinium	Gd	64	157.25
Gallium	Ga	31	69.723
Germanium	Ge	32	72.63
Gold	Au	79	196.97

Element	Symbol	Atomic Number	Average Atomic Mass[a]
Hafnium	Hf	72	178.49
Hassium	Hs	108	[277]
Helium	He	2	4.0026
Holmium	Ho	67	164.93
Hydrogen	H	1	1.0079
Indium	In	49	114.82
Iodine	I	53	126.90
Iridium	Ir	77	192.22
Iron	Fe	26	55.845
Krypton	Kr	36	83.798
Lanthanum	La	57	138.91
Lawrencium	Lr	103	[262]
Lead	Pb	82	207.2
Lithium	Li	3	6.941
Livermorium	Lv	116	[293]
Lutetium	Lu	71	174.97
Magnesium	Mg	12	24.305
Manganese	Mn	25	54.938
Meitnerium	Mt	109	[276]
Mendelevium	Md	101	[258]
Mercury	Hg	80	200.59
Molybdenum	Mo	42	95.96
Moscovium[b]	Mc	115	[288]
Neodymium	Nd	60	144.24
Neon	Ne	10	20.180
Neptunium	Np	93	[237]
Nickel	Ni	28	58.693
Nihonium[b]	Nh	113	[284]
Niobium	Nb	41	92.906
Nitrogen	N	7	14.007
Nobelium	No	102	[259]
Oganesson[b]	Og	118	[294]
Osmium	Os	76	190.23
Oxygen	O	8	15.999
Palladium	Pd	46	106.42
Phosphorus	P	15	30.974
Platinum	Pt	78	195.08
Plutonium	Pu	94	[244]
Polonium	Po	84	[209]

Element	Symbol	Atomic Number	Average Atomic Mass[a]
Potassium	K	19	39.098
Praseodymium	Pr	59	140.91
Promethium	Pm	61	[145]
Protactinium	Pa	91	231.04
Radium	Ra	88	[226]
Radon	Rn	86	[222]
Rhenium	Re	75	186.21
Rhodium	Rh	45	102.91
Roentgenium	Rg	111	[280]
Rubidium	Rb	37	85.468
Ruthenium	Ru	44	101.07
Rutherfordium	Rf	104	[265]
Samarium	Sm	62	150.36
Scandium	Sc	21	44.956
Seaborgium	Sg	106	[271]
Selenium	Se	34	78.96
Silicon	Si	14	28.086
Silver	Ag	47	107.87
Sodium	Na	11	22.990
Strontium	Sr	38	87.62
Sulfur	S	16	32.065
Tantalum	Ta	73	180.95
Technetium	Tc	43	[98]
Tellurium	Te	52	127.60
Tennessine[b]	Ts	117	[294]
Terbium	Tb	65	158.93
Thallium	Tl	81	204.38
Thorium	Th	90	232.04
Thulium	Tm	69	168.93
Tin	Sn	50	118.71
Titanium	Ti	22	47.867
Tungsten	W	74	183.84
Uranium	U	92	238.03
Vanadium	V	23	50.942
Xenon	Xe	54	131.29
Ytterbium	Yb	70	173.05
Yttrium	Y	39	88.906
Zinc	Zn	30	65.38
Zirconium	Zr	40	91.224

[a]Average atomic mass values for most elements are from *Pure Appl. Chem.* (2011) 83, 359. Those for B, C, Cl, H, Li, N, O, Si, S, and Tl are from *Pure Appl. Chem.* (2009) 81, 2131, and are within the ranges cited in the first reference. Atomic masses in brackets are the mass numbers of the longest-lived isotopes of elements with no stable isotopes.
[b]Names and symbols for these elements were recommended by the International Union for Pure and Applied Chemistry (IUPAC) in June 2016.

SECOND EDITION

Chemistry

An Atoms-Focused Approach

Thomas R. Gilbert
NORTHEASTERN UNIVERSITY

Rein V. Kirss
NORTHEASTERN UNIVERSITY

Natalie Foster
LEHIGH UNIVERSITY

Stacey Lowery Bretz
MIAMI UNIVERSITY

W. W. NORTON & COMPANY
NEW YORK • LONDON

W. W. Norton & Company has been independent since its founding in 1923, when William Warder Norton and Mary D. Herter Norton first published lectures delivered at the People's Institute, the adult education division of New York City's Cooper Union. The firm soon expanded its program beyond the Institute, publishing books by celebrated academics from America and abroad. By mid-century, the two major pillars of Norton's publishing program—trade books and college texts—were firmly established. In the 1950s, the Norton family transferred control of the company to its employees, and today—with a staff of four hundred and a comparable number of trade, college, and professional titles published each year—W. W. Norton & Company stands as the largest and oldest publishing house owned wholly by its employees.

Printed in Canada

Editor: Erik Fahlgren
Developmental Editor: John Murdzek
Project Editor: Diane Cipollone
Assistant Editor: Arielle Holstein
Production Manager: Eric Pier-Hocking
Managing Editor, College: Marian Johnson
Managing Editor, College Digital Media: Kim Yi
Media Editor: Christopher Rapp
Associate Media Editor: Julia Sammaritano
Media Project Editor: Marcus Van Harpen
Media Editorial Assistants: Tori Reuter and Doris Chiu
Ebook Production Manager: Mateus Teixeira
Marketing Manager, Chemistry: Stacy Loyal
Associate Design Director: Hope Miller Goodell
Photo Editor: Aga Millhouse
Permissions Manager: Megan Schindel
Composition: Graphic World
Illustrations: Imagineering—Toronto, ON
Manufacturing: Transcontinental Interglobe

Library of Congress Cataloging-in-Publication Data
Names: Gilbert, Thomas R. | Kirss, Rein V. | Foster, Natalie. | Bretz, Stacey Lowery, 1967-
Title: Chemistry : an atoms-focused approach / Thomas R. Gilbert, Northeastern University, Rein V. Kirss, Northeastern University, Natalie Foster, Lehigh University, Stacey Lowery Bretz, Miami University.
Description: Second edition. | New York : W.W. Norton & Company, Inc., [2018] | Includes index.
Identifiers: LCCN 2016049892 | **ISBN 9780393284218 (hardcover)**
Subjects: LCSH: Chemistry.
Classification: LCC QD33.2 .G54 2018 | DDC 540—dc23 LC record available at https://lccn.loc.gov/2016049892

W. W. Norton & Company, Inc., 500 Fifth Avenue, New York, NY 10110

www.wwnorton.com

W. W. Norton & Company Ltd., 15 Carlisle Street, London W1D 3BS

1 2 3 4 5 6 7 8 9 0

Brief Contents

Contents

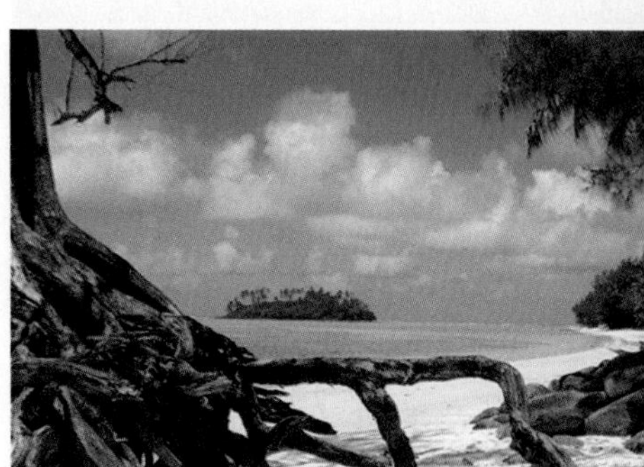

Why does black ironwood sink in seawater? *(Chapter 1)*

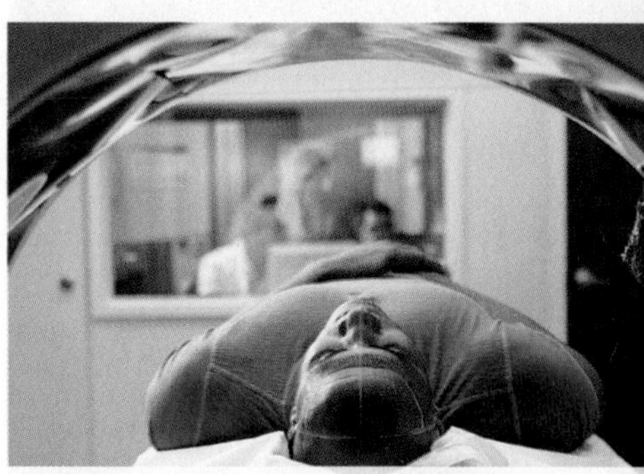

How do MRI machines work? *(Chapter 2)*

What is responsible for the shimmering, colorful display known as an aurora? *(Chapter 3)*

How does lightning produce ozone? *(Chapter 4)*

What molecule is an active ingredient in cough syrup? *(Chapter 5)*

Why are controlled fires often seen on oil rigs? *(Chapter 6)*

Why is this river green? *(Chapter 7)*

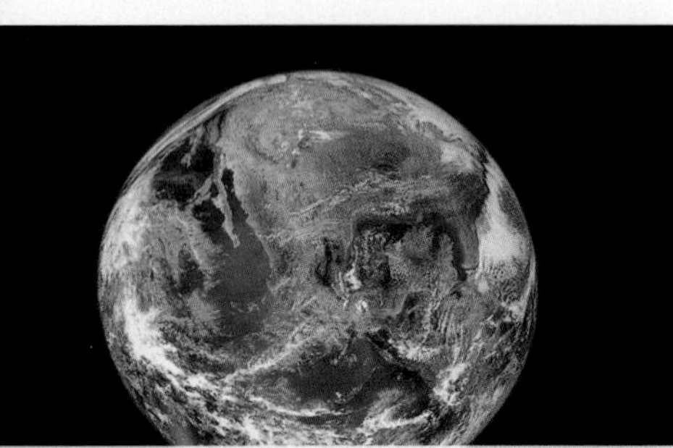
What processes control the composition of seawater? *(Chapter 8)*

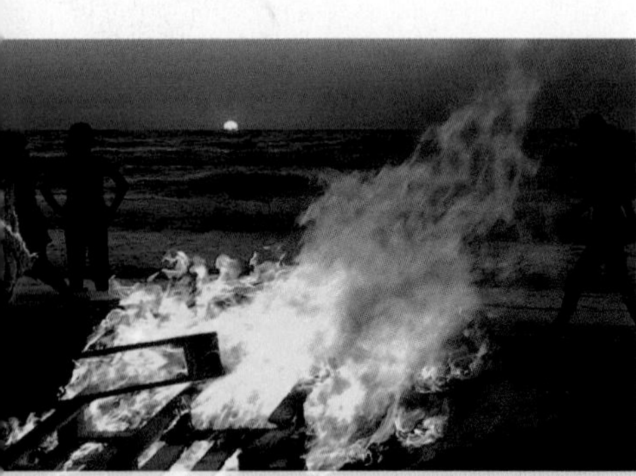
What reactions occur when wood burns? *(Chapter 9)*

What allows hot-air balloons to fly? *(Chapter 10)*

How does this sailboat turn seawater into drinking water? *(Chapter 11)*

What caused this ship to rust? *(Chapter 12)*

What causes smog? *(Chapter 13)*

How is chemical equilibrium manipulated to produce the ammonia needed to fertilize crops? *(Chapter 14)*

What is responsible for the color of hydrangeas? *(Chapter 15)*

How do increasing CO_2 levels threaten coral reefs? *(Chapter 16)*

How do we power cars that do not rely on gasoline? *(Chapter 17)*

Why are skyscrapers built from steel? *(Chapter 18)*

Why is Kevlar so strong? *(Chapter 19)*

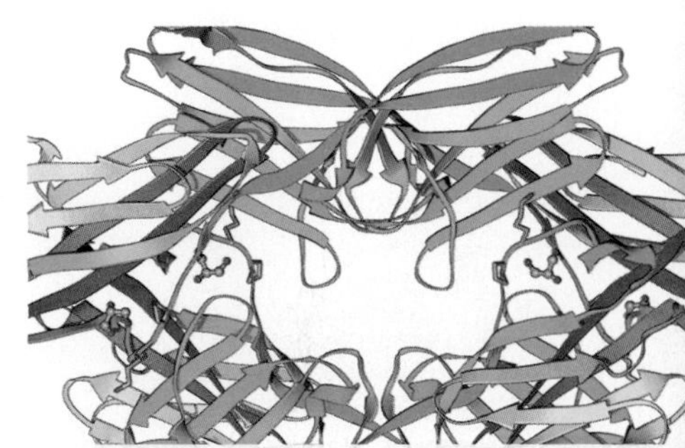

How large can a biomolecule be? *(Chapter 20)*

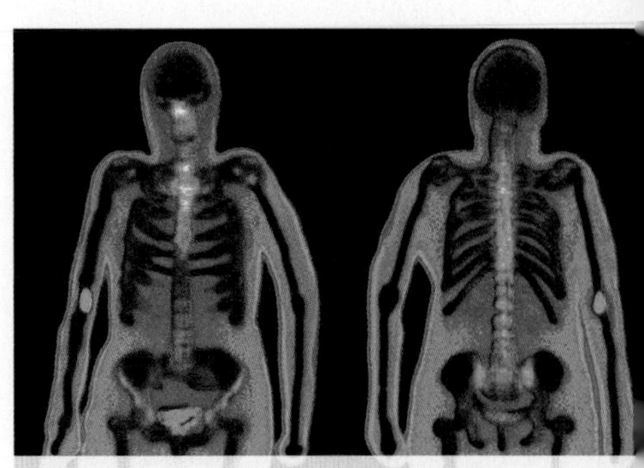

How are radioactive nuclei used in diagnostic medicine? *(Chapter 21)*

What are the crystals in hard cheeses made of? *(Chapter 22)*

What makes aquamarine crystals blue? *(Chapter 23)*

Applications

ChemTours

About the **Authors**

Thomas R. Gilbert has a BS in chemistry from Clarkson and a PhD in analytical chemistry from MIT. After 10 years with the Research Department of the New England Aquarium in Boston, he joined the faculty of Northeastern University, where he is currently associate professor of chemistry and chemical biology. His research interests are in chemical and science education. He teaches general chemistry and science education courses and conducts professional development workshops for K–12 teachers. He has won Northeastern's Excellence in Teaching Award and Outstanding Teacher of First-Year Engineering Students Award. He is a fellow of the American Chemical Society and in 2012 was elected to the ACS Board of Directors.

Rein V. Kirss received both a BS in chemistry and a BA in history as well as an MA in chemistry from SUNY Buffalo. He received his PhD in inorganic chemistry from the University of Wisconsin, Madison, where the seeds for this textbook were undoubtedly planted. After two years of postdoctoral study at the University of Rochester, he spent a year at Advanced Technology Materials, Inc., before returning to academics at Northeastern University in 1989. He is an associate professor of chemistry with an active research interest in organometallic chemistry.

Natalie Foster is emeritus professor of chemistry at Lehigh University in Bethlehem, Pennsylvania. She received a BS in chemistry from Muhlenberg College and MS, DA, and PhD degrees from Lehigh University. Her research interests included studying poly(vinyl alcohol) gels by NMR as part of a larger interest in porphyrins and phthalocyanines as candidate contrast enhancement agents for MRI. She taught both semesters of the introductory chemistry class to engineering, biology, and other nonchemistry majors and a spectral analysis course at the graduate level. She is the recipient of the Christian R. and Mary F. Lindback Foundation Award for distinguished teaching and a Fellow of the American Chemical Society.

Stacey Lowery Bretz is a University Distinguished Professor in the Department of Chemistry and Biochemistry at Miami University in Oxford, Ohio. She earned her BA in chemistry from Cornell University, MS from Pennsylvania State University, and a PhD in chemistry education research (CER) from Cornell University. She then spent one year at the University of California, Berkeley, as a post-doc in the Department of Chemistry. Her research expertise includes the development of assessments to characterize chemistry misconceptions and measure learning in the chemistry laboratory. Of particular interest is method development with regard to the use of multiple representations (particulate, symbolic, and macroscopic) to generate cognitive dissonance, including protocols for establishing the reliability and validity of these measures. She is a fellow of both the American Chemical Society and the American Association for the Advancement of Science. She was the recipient of the E. Phillips Knox Award for Undergraduate Teaching in 2009 and the Distinguished Teaching Award for Excellence in Graduate Instruction and Mentoring in 2013, Miami University's highest teaching awards.

Preface

Dear Student,

They say you can't judge a book by its cover. Still, you may be wondering why we chose to put peeling wallpaper on the cover of a chemistry book. Actually, the cover photo is not wallpaper but the bark of a Pacific Madrone tree, *Arbutus menziesii*. The illustration shows a molecular view of the cellulose that is a principal component of tree's trunk, including its peeling bark and the heartwood beneath it.

Our cover illustrates a central message of this book: the properties of substances are directly linked to their atomic and molecular structures. In our book we start with the smallest particles of matter and assemble them into more elaborate structures: from subatomic particles to single atoms to monatomic ions and polyatomic ions, and from atoms to small molecules to bigger ones to truly gigantic polymers. By constructing this layered particulate view of matter, we hope our book helps you visualize the properties of substances and the changes they undergo during chemical reactions.

With that in mind, we begin each chapter with a **Particulate Review** and **Particulate Preview** on the very first page. The goal of these tools is to prepare you for the material in the chapter. The Particulate Review assesses important prior knowledge that you need to interpret particulate images in the chapter. The Particulate Preview asks you to expand your prior knowledge and to speculate about the new concepts you will see in the chapter. It is also designed to focus your reading by asking you to look out for key terms and concepts.

As you develop your ability to visualize atoms and molecules, you will find that you don't have to resort to memorizing formulas and reactions as a strategy for surviving general chemistry. Instead, you will be able to understand why elements combine to form compounds with particular formulas and why substances react with each other the way they do.

PARTICULATE **REVIEW**

Phase Changes and Energy

In Chapter 9, we explore the energy changes that accompany both physical and chemical changes. Particulate representations of the three phases of water are shown here.

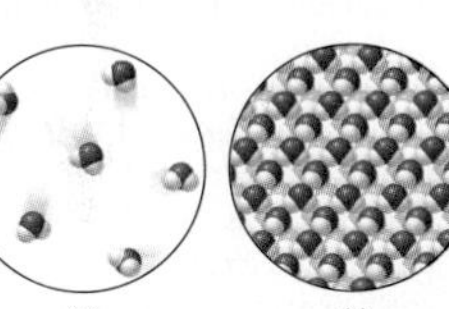

(a) (b) (c)

- Which representation depicts the solid phase of water? The liquid? The gaseous?
- Is energy added or released during the physical change from (a) to (b)? What intermolecular forces are involved?
- Describe the energy changes that accompany the physical changes from (a) to (c) and from (c) to (a).

(Review Section 1.4 and Section 6.2 if you need help answering these questions.)

(Answers to Particulate Review questions are in the back of the book.)

PARTICULATE **PREVIEW**

Breaking Bonds and Energy Changes

Calcium chloride, shown in the accompanying figure, is used to melt ice on sidewalks. As you read Chapter 9, look for ideas that will help you understand the energy changes that accompany the breaking and forming of bonds.

- What kind of bonds must be broken for calcium chloride to dissolve in water? Is energy absorbed or released in order to break these bonds?
- Which color spheres represent the chloride ions? Label the polar covalent bonds in water using δ+ and δ−.
- What intermolecular interactions form as the salt dissolves? Is energy absorbed or released as these attractions form?

Context

While our primary goal is for you to be able to interpret and even predict the physical and chemical properties of substances based on their atomic and molecular structures, we would also like you to understand how chemistry is linked to other scientific disciplines. We illustrate these connections using contexts drawn from fields such as biology, medicine, environmental science, materials science, and engineering. We hope that this approach helps you better understand how scientists apply the principles of chemistry to treat and cure diseases, to make more efficient use of natural resources, and to minimize the impact of human activity on our planet and its people.

Problem-Solving Strategies

Another major goal of our book is to help you improve your problem-solving skills. To do this, you first need to recognize the connections between the information provided in a problem and the answer you are asked to find. Sometimes the hardest part of solving a problem is distinguishing between information that is relevant and information that is not. Once you are clear on where you are starting and where you are going, planning for and carrying out a solution become much easier.

To help you hone your problem-solving skills, we have developed a framework that we introduce in Chapter 1. It is a four-step approach we call COAST, which is our acronym for (1) **C**ollect and **O**rganize, (2) **A**nalyze, (3) **S**olve, and (4) **T**hink About It. We use these four steps in *every* Sample Exercise and in the solutions to *odd-numbered* problems in the Student's Solutions Manual. They are also used in the hints and feedback embedded in the Smartwork5 online homework program. To summarize the four steps:

Collect and Organize helps you understand where to begin to solve the problem. In this step we often rephrase the problem and the answer that is sought, and we identify the relevant information that is provided in the problem statement or available elsewhere in the book.

Analyze is where we map out a strategy for solving the problem. As part of that strategy we often estimate what a reasonable answer might be.

Solve applies our analysis of the problem from the second step to the information and relations from the first step to actually solve the problem. We walk you through each step in the solution so that you can follow the logic and the math.

Think About It reminds us that an answer is not the last step in solving a problem. We should check the accuracy of the solution and think about the value of a quantitative answer. Is

SAMPLE EXERCISE 9.8 Calculating ΔH°_{rxn} Using Hess's Law **LO5**

One reason furnaces and hot-water heaters fueled by natural gas need to be vented is that incomplete combustion can produce toxic carbon monoxide:

Equation A: $2\ CH_4(g) + 3\ O_2(g) \rightarrow 2\ CO(g) + 4\ H_2O(g) \quad \Delta H^\circ_A = ?$

Use thermochemical equations B and C to calculate ΔH°_A:

Equation B: $CH_4(g) + 2\ O_2(g) \rightarrow CO_2(g) + 2\ H_2O(g) \quad \Delta H^\circ_B = -802\ kJ$

Equation C: $2\ CO(g) + O_2(g) \rightarrow 2\ CO_2(g) \quad \Delta H^\circ_C = -566\ kJ$

Collect and Organize We are given two equations (B and C) with thermochemical data and a third (A) for which we are asked to find ΔH°. All the reactants and products in equation A are present in B and/or C.

Analyze We can manipulate equations B and C algebraically so that they sum to give the equation for which ΔH° is unknown. Then we can calculate the unknown value by applying Hess's law. Methane is a reactant in A and B, so we will use B in the direction written. CO is a product in A but a reactant in C, so we have to reverse C to get CO on the product side. Reversing C means that we must change the sign of ΔH°_C. If the coefficients in B and the reverse of C do not allow us to sum the two equations to obtain equation A, we will need to multiply one or both by appropriate factors.

Solve Comparing equation B as written and the reverse of C:

(B) $CH_4(g) + 2\ O_2(g) \rightarrow CO_2(g) + 2\ H_2O(g) \quad \Delta H^\circ_B = -802\ kJ$

(C, reversed) $2\ CO_2(g) \rightarrow 2\ CO(g) + O_2(g) \quad -\Delta H^\circ_C = +566\ kJ$

with equation A, we find that the coefficient of CH_4 is 2 in A but only 1 in B, so we need to multiply all the terms in B by 2, including ΔH°_B:

(2B) $2\ CH_4(g) + 4\ O_2(g) \rightarrow 2\ CO_2(g) + 4\ H_2O(g) \quad 2\ \Delta H^\circ_B = -1604\ kJ$

When we sum C (reversed) and 2B, the CO_2 terms cancel out and we obtain equation A:

(C, reversed)	$\cancel{2\ CO_2(g)} \rightarrow 2\ CO(g) + \cancel{O_2(g)}$	$-\Delta H^\circ_C = +566\ kJ$
+ (2B)	$2\ CH_4(g) + \overset{3}{\cancel{4}}\ O_2(g) \rightarrow \cancel{2\ CO_2(g)} + 4\ H_2O(g)$	$2\ \Delta H^\circ_B = -1604\ kJ$
(A)	$2\ CH_4(g) + 3\ O_2(g) \rightarrow 2\ CO(g) + 4\ H_2O(g)$	$\Delta H^\circ_A = -1038\ kJ$

Think About It Our calculation shows that incomplete combustion of two moles of methane is less exothermic ($\Delta H^\circ_A = -1038$ kJ) than their complete combustion ($2\ \Delta H^\circ_B = -1604$ kJ), which makes sense because the CO produced in incomplete combustion reacts exothermically with more O_2 to form CO_2. In fact, the value of ΔH°_C for the reaction $2\ CO(g) + O_2(g) \rightarrow 2\ CO_2(g)$ is the difference between −1604 kJ and −1038 kJ.

Practice Exercise It does not matter how you assemble the equations in a Hess's law problem. Show that reactions A and C can be summed to give reaction B and result in the same value for ΔH°_B.

it realistic? Are the units correct? Is the number of significant figures appropriate? Does it agree with our estimate from the Analyze step?

Suggestion: Some Sample Exercises that are based on simple concepts and single-step solutions are streamlined by combining Collect, Organize, and Analyze steps, but the essential COAST features are always maintained.

Many students use the **Sample Exercises** more than any other part of the book. Sample Exercises take the concepts being discussed and illustrate how to apply them to solve problems. We think that repeated application of the COAST framework will help you refine your problem-solving skills, and we hope that the approach will become habit-forming for you. When you finish a Sample Exercise, you'll find a **Practice Exercise** to try on your own. The next few pages describe how to use the tools built into each chapter to gain a conceptual understanding of chemistry and to connect the microscopic structure of substances to their observable physical and chemical properties.

Chapter Structure

As mentioned earlier, each chapter begins with the Particulate Review and Particulate Preview to help you prepare for the material ahead.

If you are trying to decide what is most important in a chapter, check the **Learning Outcomes** listed on the first page. Whether you are reading the chapter from first page to last or reviewing it for an exam, the Learning Outcomes should help you focus on the key information you need and the skills you should develop. You will also see which Learning Outcomes are linked to which Sample Exercises in the chapter.

Learning Outcomes

LO1 Distinguish between isolated, closed and open thermodynamic systems and between endothermic and exothermic processes
Sample Exercise 9.1

LO2 Relate changes in the internal energies of thermodynamic systems to heat flows and work done
Sample Exercises 9.2, 9.3

LO3 Calculate the heat gained or lost during changes in temperature and physical state
Sample Exercises 9.4, 9.5

LO4 Use calorimetry data to calculate enthalpies of reaction and heat capacities of calorimeters
Sample Exercises 9.6, 9.7

LO5 Calculate enthalpies of reaction using Hess's law and enthalpies of formation
Sample Exercises 9.8, 9.9, 9.10

LO6 Estimate enthalpies of reaction using average bond energies
Sample Exercise 9.11

LO7 Estimate enthalpies of solution and lattice energies using the Born–Haber cycle and Hess's law
Sample Exercise 9.12

LO8 Calculate and compare fuel values and fuel densities
Sample Exercises 9.13, 9.14

As you study each chapter, you will find **key terms** in boldface in the text and in a running glossary in the margin. We have deliberately duplicated these definitions so that you can continue reading without interruption but quickly find them when doing homework or studying. All key terms are also defined in the Glossary in the back of the book.

Many concepts are related to others described earlier in the book. We point out these relationships with **Connection** icons in the margins. We hope they enable you to draw your own connections between major themes covered in the book.

CONNECTION In Chapter 1, we defined *energy* as the ability to do work. We also introduced the *law of conservation of energy* and the concept that energy cannot be created or destroyed but can be changed from one form of energy to another.

CHEMTOUR
Bond Polarity and Polar Moelcules

To help you develop your own microscale view of matter, we use **molecular art** to enhance photos and figures, and to illustrate what is happening at the atomic and molecular levels.

If you're looking for additional help visualizing a concept, we have about 100 **ChemTours**, denoted by the ChemTour icon, available online at https://digital.wwnorton.com/atoms2. ChemTours demonstrate dynamic processes and help you visualize events at the molecular level. Many of the ChemTours allow you to manipulate variables and observe the resulting changes.

Concept Tests are short, conceptual questions that serve as self-checks by asking you to stop and answer questions related to what you just read. We designed them to help you see for yourself whether you have grasped a key concept and can apply it. We have an average of one Concept Test per section and many have visual components. We provide the answers to all Concept Tests in the back of the book.

CONCEPT **TEST**

Suppose two identical pots of water are heated on a stove until the water inside them begins to boil. Both pots are then removed from the stove. One of the two is covered with a tight lid; the other is not, and both are allowed to cool.

a. What type of thermodynamic system—open, closed, or isolated—describes each of the cooling pots?

b. Which pot cools faster? Why?

(Answers to Concept Tests are in the back of the book.)

At the end of each chapter is a special Sample Exercise that draws on several key concepts from the chapter and occasionally others from preceding chapters to solve a problem that is framed in the context of a real-world scenario or incident. We call these **Integrated Sample Exercises**. You may find them more challenging than most exercises that precede them in each chapter, but please invest your time in working through them because they represent authentic exercises that will enhance your problem-solving skills.

Also at the end of each chapter are a thematic **Summary** and a **Problem-Solving Summary**. The first is a brief synopsis of the chapter, organized by learning outcomes. Key figures provide visual cues as you review. The Problem-Solving Summary is unique to this general chemistry book—it outlines the different types of problems you should be able to solve, where to find examples of them in the Sample Exercises, and it reminds you of key concepts and equations.

PROBLEM-SOLVING SUMMARY

Type of Problem	Concepts and Equations		Sample Exercises
Identifying endothermic and exothermic processes	During an endothermic process, heat flows into the system from its surroundings ($q > 0$). During an exothermic process, heat flows out from the system into its surroundings ($q < 0$).		**9.1**
Calculating *P–V* work	$w = -P\Delta V$		**9.2**
Relating ΔE, *q*, and *w*	$\Delta E = q + w = q - P\Delta V$	(9.3, 9.4)	**9.3**
Calculating heat transfer (*q*) associated with a change of temperature or state of a substance	Heating either an object: $q = C_p\Delta T$ or a mass (*m*) of a pure substance: $q = mc_p\Delta T$ or a quantity of a pure substance in moles (*n*): $q = nc_{P,n}\Delta T$ Melting a solid at its melting point: $q = n\Delta H_{fus}$ Vaporizing a liquid at its boiling point: $q = n\Delta H_{vap}$	 (9.8) (9.9) (9.10) (9.11) (9.12)	**9.4, 9.5**
Calculating C	q = −q = −C ΔT		**9.6, 9.7**

Following the summaries are groups of questions and problems. The first group consists of **Visual Problems.** In many of them, you are asked to interpret a molecular view of a sample or a graph of experimental data. The last Visual Problem in each chapter contains a **Visual Problem Matrix.** This grid consists of nine images followed by a series of questions that will test your ability to identify the similarities and differences among the macroscopic, particulate, and symbolic images.

Concept Review Questions and Problems come next, arranged by topic in the same order as they appear in the chapter. Concept Reviews are qualitative and often ask you to explain why or how something happens. Problems are paired and can be quantitative, conceptual, or a combination of both. **Contextual problems** have a title that describes the context in which the problem is placed. Finally, **Additional Problems** can come from any section or combination of sections in the chapter. Some of them incorporate concepts from previous chapters. Problems marked with an asterisk (*) are more challenging and often take multiple steps to solve.

We want you to have confidence in using the answers in the back of the book as well as the Student's Solutions Manual, so we used a rigorous triple-check accuracy program for this book. Each end-of-chapter question or problem was solved independently by the Solutions Manual author, Karen Brewer, and by two additional chemical educators. Karen compared her solutions to those from the two reviewers and resolved any discrepancies. This process is designed to ensure clearly written problems and accurate answers in the appendices and Solutions Manual.

9.8. Use representations [A] through [I] in Figure P9.8 to answer questions a–f.
   a. Match two of the particulate images to the phase change for liquid nitrogen in [B].
   b. Match two of the particulate images to the phase change for dry ice (solid CO_2) in [H].
   c. Which, if any, of the photos correspond to [D]? Are these endothermic or exothermic?
   d. Which, if any, of the photos correspond to [F]? Are these endothermic or exothermic?
   e. What bonds break when the solid ammonium nitrate in [E] dissolves in water to activate the cold pack?
   f. Which particulate images show an element or compound in its standard state?

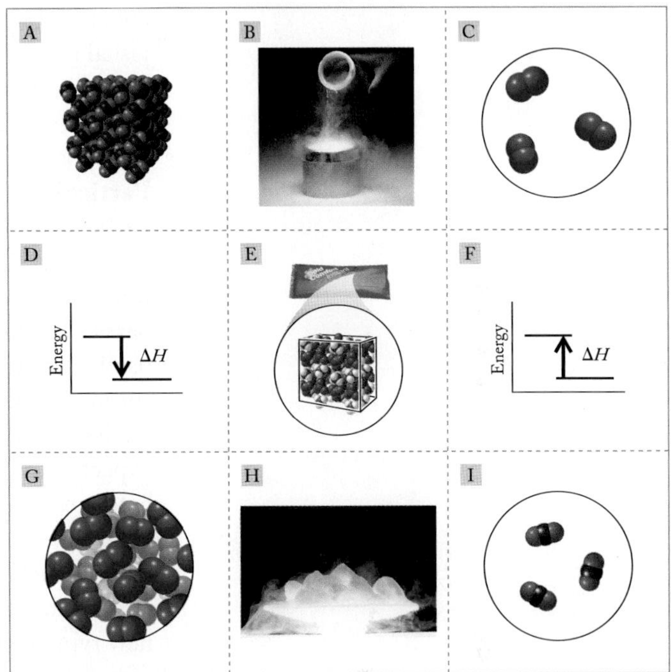

FIGURE P9.8

Dear Instructor,

This book takes an atoms-focused approach to teaching chemistry. Consequently, the sequence of chapters in the book and the sequence of topic in many of the chapters are not the same as in most general chemistry textbooks. For example, we devote the early chapters to providing an in-depth view of the particulate nature of matter including the structure of atoms and molecules and how the properties of substances link directly to those structures.

After two chapters on the nature of chemical bonding, molecular shape, and theories to explain both, we build on those topics as we explore the intermolecular forces that strongly influence the form and function of molecules, particularly those of biological importance.

Once this theoretical foundation has been laid, we examine chemical reactivity and the energetics of chemical reactions. Most general chemistry books don't complete their coverage of chemistry and energy until late in the book. We finish the job in Chapter 12, which means that students already understand the roles of energy and entropy in chemical reactions before they encounter chemical kinetics and the question of how they happen. The kinetics chapter is followed by several on chemical equilibrium, which introduce the phenomenon in terms of what happens when reactions proceed to a measureable extent in both forward and reverse directions and how interactions between and within particles influence the contacts that drive chemical changes.

Changes in the Second Edition

As authors of a textbook, we are very often asked: "Why is a second edition necessary? Has the science changed that much since the first edition?" Although chemistry is a vigorous and dynamic field, most basic concepts presented in an introductory course have not changed dramatically. However, two areas tightly intertwined in this text—pedagogy and context—have changed significantly, and those areas are the drivers of this new edition. Here are some of the most noteworthy changes we made throughout this edition:

- We welcome Stacey Lowery Bretz as our new co-author. Stacey is a chemistry education researcher and her insights and expertise about accurate visual representations to support consistent pedagogy as well as about student misconceptions and effective ways to address them are evident throughout the book.
- The most obvious examples are the new **Particulate Review** and **Particulate Preview** questions at the beginning of each chapter. The Review is a diagnostic element highlighting important prior knowledge students must draw upon to successfully interpret molecular (particulate) images in the chapter. The Review consists of a few questions based on particulate art. The Preview consists of a short series of questions about a particulate image that ask students to extend their prior knowledge and speculate about material in the chapter. The goal of the Preview is to direct students as they read, making reading more interactive. Students are not expected to know the correct answers to the questions posed in the Preview before they start the chapter but are to use them as a guide while reading. Overviews of each Particulate Review and Preview section can be found in the Instructor's Resource Manual and the lecture PowerPoints.
- In addition to the Particulate Review and Preview feature, Stacey authored a new type of visual problem: the **Visual Problem Matrix**. The matrix consists of macroscopic, particulate, and symbolic images in a grid, followed by a series of questions asking students to identify commonalities and differences across the images. Versions of all of these new problems are in the lecture PowerPoint slides to use in group activities and lecture quizzes. They are also available in Smartwork5 as individual problems and in pre-made assignments to use before or after class.
- We evaluated each Sample Exercise and streamlined many of those based on simple concepts and single-step solutions by combining the Collect and Organize and Analyze steps. We revised other Sample Exercises throughout the book based on reviewer and user feedback.
- The treatment of how to evaluate the precision and accuracy of experimental values in Chapter 1 has been expanded to include more rigorous treatment of the variability in data sets and in the identification of outliers.
- We have expanded our coverage of aqueous equilibrium by adding a second chapter that doubles the number of Sample Exercises and includes Concept Tests that focus on the molecules and ions present during titrations and in buffers.
- We took the advice of reviewers and now have two descriptive chemistry chapters at the end of the book. These chapters focus on main group chemistry and transition metals, both within the context of biological and medical applications.

- We have revised or replaced at least 10% of the end-of-chapter problems. We incorporated feedback from users and reviewers to address areas where we needed more problems or additional problems of varying difficulty.
- A new version of Smartwork, Smartwork5, offers more than 3600 problems in a sophisticated and user-friendly platform. Four hundred new problems were designed to support the new visualization pedagogy. In addition to being tablet compatible, Smartwork5 integrates with the most common campus learning management systems.

The nearly 100 ChemTours have been updated to better support lecture, lab, and independent student learning. The ChemTours include images, animations, and audio that demonstrate dynamic processes and help students visualize and understand chemistry at the molecular level. Forty of the ChemTours now contain greater interactivity and are assignable in Smartwork5. The ChemTours are linked directly from the ebook and are now in HTML5, which means they are tablet compatible.

Teaching and Learning Resources

Smarkwork5 Online Homework For General Chemistry

digital.wwnorton.com/atoms2

Smartwork5 is the most intuitive online tutorial and homework management system available for general chemistry. The many question types, including graded molecule drawing, math and chemical equations, ranking tasks, and interactive figures, help students develop and apply their understanding of fundamental concepts in chemistry.

Every problem in Smartwork5 includes response-specific feedback and general hints using the steps in COAST. Links to the ebook version of *Chemistry: An Atoms-Focused Approach*, Second Edition, take students to the specific place in the text where the concept is explained. All problems in Smartwork5 use the same language and notation as the textbook.

Smartwork5 also features Tutorial Problems. If students ask for help in a Tutorial Problem, the system breaks the problem down into smaller steps, coaching them with hints, answer-specific feedback, and probing questions within each step. At any point in a Tutorial, a student can return to and answer the original problem.

Assigning, editing, and administering homework within Smartwork5 is easy. Smartwork5 allows the instructor to search for problems using both the text's Learning Objectives and Bloom's taxonomy. Instructors can use pre-made assignment sets provided by Norton authors, modify those assignments, or create their own. Instructors can also make changes in the problems at the question level. All instructors have access to our WYSIWYG (What You See Is What You Get) authoring tools—the same ones Norton authors use. Those intuitive tools make it easy to modify existing problems or to develop new content that meets the specific needs of your course.

Wherever possible, Smartwork5 makes use of algorithmic variables so that students see slightly different versions of the same problem. Assignments are graded automatically, and Smartwork5 includes sophisticated yet flexible tools for managing class data. Instructors can use the class activity report to assess students' performance on specific problems within an assignment. Instructors can also review individual students' work on problems.

Smartwork5 for *Chemistry: An Atoms-Focused Approach*, Second Edition, features the following problem types:

- End-of-Chapter Problems. These problems, which use algorithmic variables when appropriate, all have hints and answer-specific feedback to coach students through mastering single- and multi-concept problems based on chapter content. They make use of all of Smartwork5's answer-entry tools.
- ChemTour Problems. Forty ChemTours now contain greater interactivity and are assignable in Smartwork5.
- Visual and Graphing Problems. These problems challenge students to identify chemical phenomena and to interpret graphs. They use Smartwork5's Drag-and-Drop and Hotspot functionality.
- Reaction Visualization Problems. Based on both static art and videos of simulated reactions, these problems are designed to help students visualize what happens at the atomic level—and why it happens.
- Ranking Task Problems. These problems ask students to make comparative judgments between items in a set.
- Nomenclature Problems. New matching and multiple-choice problems help students master course vocabulary.
- Multistep Tutorials. These problems offer students who demonstrate a need for help a series of linked, step-by-step subproblems to work. They are based on the Concept Review problems at the end of each chapter.
- Math Review Problems. These problems can be used by students for practice or by instructors to diagnose the mathematical ability of their students.

Ebook

digital.wwnorton.com/atoms2

An affordable and convenient alternative to the print text, the Norton Ebook lets students access the entire book and much more: they can search, highlight, and take notes with ease. The Norton Ebook allows instructors to share their notes with students. And the ebook can be viewed on most devices—laptop, tablet, even a public computer—and will stay synced between devices.

The online version of *Chemistry: An Atoms-Focused Approach*, Second Edition, also provides students with one-click access to the nearly 100 ChemTour animations.

The online ebook is available bundled with the print text and Smartwork5 at no extra cost, or it may be purchased bundled with Smartwork5 access.

Norton also offers a downloadable PDF version of the ebook.

Student's Solutions Manual

by Karen Brewer, Hamilton University

The Student's Solutions Manual provides students with fully worked solutions to select end-of-chapter problems using the **COAST** four-step method (**C**ollect and **O**rganize, **A**nalyze, **S**olve, and **T**hink About It). The Student's Solutions Manual contains several pieces of art for each chapter, designed to help students visualize ways to approach problems. This artwork is also used in the hints and feedback within Smartwork.

Clickers in Action: Increasing Student Participation in General Chemistry

by Margaret Asirvatham, University of Colorado, Boulder

This instructor-oriented resource provides information on implementing clickers in general chemistry courses. *Clickers in Action* contains more than 250 class-tested, lecture-ready questions, with histograms showing student responses, as well as insights and suggestions for implementation. Question types include macroscopic observation, symbolic representation, and atomic/molecular views of processes.

Test Bank

by Daniel E. Autrey, Fayetteville State University

Norton uses an innovative, evidence-based model to deliver high-quality and pedagogically effective quizzes and testing materials. Each chapter of the Test Bank is structured around an expanded list of student learning objectives and evaluates student knowledge on six distinct levels based on Bloom's Taxonomy: Remembering, Understanding, Applying, Analyzing, Evaluating, and Creating.

Questions are further classified by section and difficulty, making it easy to construct tests and quizzes that are meaningful and diagnostic, according to each instructor's needs. More than 2500 questions are divided into multiple choice and short answer.

The Test Bank is available with ExamView Test Generator software, allowing instructors to effortlessly create, administer, and manage assessments. The convenient and intuitive test-making wizard makes it easy to create customized exams with no software learning curve. Other key features include the ability to create paper exams with algorithmically generated variables and export files directly to Blackboard, Canvas, Desire2Learn, and Moodle.

Instructor's Solutions Manual

by Karen Brewer, Hamilton University

The Instructor's Solutions Manual provides instructors with fully worked solutions to every end-of-chapter Concept Review and Problem. Each solution uses the **COAST** four-step method (**C**ollect and **O**rganize, **A**nalyze, **S**olve, and **T**hink About It).

Instructor's Resource Manual

by Anthony Fernandez, Merrimack College

This complete resource manual for instructors has been revised to correspond to changes made in the Second Edition. Each chapter begins with a brief overview of the text chapter followed by suggestions for integrating the contexts featured in the book into a lecture, summaries of the textbook's Particulate Review and Preview sections, suggested sample lecture outlines, alternate contexts to use with each chapter, and instructor notes for suggested activities from the *ChemConnections* and *Calculations in Chemistry,* Second Edition, workbooks. Suggested ChemTours and laboratory exercises round out each chapter.

Instructor's Resource Disc

This helpful classroom presentation tool features the following:

- Stepwise animations and classroom response questions are included. Developed by Jeffrey Macedone of Brigham Young University and his team, these animations, which use native PowerPoint functionality and textbook art, help instructors to walk students through nearly 100 chemical concepts and processes. Where appropriate, the slides contain two types of questions for students to answer in class: questions that ask them to predict what will happen next and why, and questions that ask them to apply knowledge gained from watching the animation. Self-contained notes help instructors adapt these materials to their own classrooms.
- Lecture PowerPoint slides (authored by Cynthia Lamberty, Cloud County Community College) include a suggested classroom-lecture script in an accompanying Word file. Each chapter opens with a set of multiple-choice questions based on the textbook's Particulate Review and Preview section and concludes with another set of questions based on the textbook's Visual Problems matrix.
- All ChemTours are included.
- *Clickers in Action* clicker questions for each chapter provide instructors with class-tested questions they can integrate into their course.
- Labeled and unlabeled photographs, drawn figures, and tables from the text are available in PowerPoint and JPEG.

Downloadable Instructor's Resources

digital.wwnorton.com/atoms2
This password-protected site for instructors includes the following:

- Stepwise animations and classroom response questions are included. Developed by Jeffrey Macedone of Brigham Young University and his team, these animations, which use native PowerPoint functionality and textbook art, help instructors to walk students through nearly 100 chemical concepts and processes. Where appropriate, the slides contain two types of questions for students to answer in class: questions that ask them to predict what will happen next and why, and questions that ask them to apply knowledge gained from watching the animation. Self-contained notes help instructors adapt these materials to their own classrooms.
- Lecture PowerPoints are available.
- All ChemTours are included.
- Test bank is available in PDF, Word RTF, and *ExamView* Assessment Suite formats.
- Solutions Manual is offered in PDF and Word, so that instructors may edit solutions.
- All end-of-chapter questions and problems are available in Word along with the key equations.
- Labeled and unlabeled photographs, drawn figures, and tables from the text are available in PowerPoint and JPEG.
- *Clickers in Action* clicker questions are included.

- Course cartridges: Available for the most common learning management systems, course cartridges include access to the ChemTours and StepWise animations, links to the ebook and Smartwork5.

Acknowledgments

Our thanks begin with our publisher, W. W. Norton, for supporting us in writing a book that is written the way we much prefer to teach general chemistry. We especially wish to acknowledge the hard work and dedication of our editor/motivator/taskmaster, Erik Fahlgren. Erik has been an indefatigable source of guidance, perspective, persuasion, and inspiration to all of us.

We are pleased to acknowledge the contributions of an outstanding developmental editor, John Murdzek. John's clear understanding and expertise in science, along with his wry wit, have helped us improve the presentation of core concepts and applied content of the book.

Diane Cipollone is our project editor who crossed t's and dotted i's to make sure each page was attractive and easy to navigate. Assistant editor Arielle Holstein is like a lighthouse in the fog: reliable, competent, and unfailingly patient in managing the constant flood of questions, information, and schedule updates. Thanks as well to Aga Millhouse and Rona Tuccillo for finding just the right photo again and again; production manager Eric Pier-Hocking for his work behind the scenes; Julia Sammaritano for managing the print ancillaries; Chris Rapp for his creative skill in the creation of digital media that enhance effective communication of content and ideas; and Stacy Loyal for her unwavering support and steadfast commitment to getting this book in the hands of potential users ("Serve that ace!"). The entire Norton team is staffed by skilled, dedicated professionals who are delightful colleagues to work with and, as a bonus, to relax with, as the occasion allows.

Many reviewers, listed here, contributed to the development and production of this book. We owe an extra special thanks to Karen Brewer for her dedicated work on the Solutions Manuals and for her invaluable suggestions on how to improve the inventory and organization of problems and concept questions at the end of each chapter. She, along with Timothy Brewer (Eastern Michigan University) and Timothy W. Chapp (Allegheny College), comprised the triple-check accuracy team who helped ensure the quality of the back-of-book answers and Solutions Manuals. Finally, we wish to acknowledge the care and thoroughness of Drew Brodeur, Hill Harman, Julie Henderleiter, Amy Johnson, Brian Leskiw, Richard Lord, Marc Knecht, Thomas McGrath, Anne-Marie Nickel, Jason Ritchie, Thomas Sorensen, Uma Swamy, Rebecca Weber, and Amanda Wilmsmeyer for checking the accuracy of the myriad facts that frame the contexts and the science in the pages that follow.

Thomas R. Gilbert
Rein V. Kirss
Natalie Foster
Stacey Lowery Bretz

Second Edition Reviewers:

Kevin Alliston, Wichita State University
Daniel Autrey, Fayetteville State University
Nathan Barrows, Grand Valley State University
Chris Bender, The University of South Carolina Upstate
Mary Ellen Biggin, Augustana College
Randy A. Booth, Colorado State University
Simon Bott, University of Houston
John C. Branca, Wichita State University
Jonathan Breitzer, Fayetteville State University
Drew Brodeur, Worcester Polytechnic Institute
Jasmine Bryant, University of Washington
Jerry Burns, Pellissippi State Community College
Andrea Carroll, University of Washington
Christina Chant, Saint Michael's College
Ramesh Chinnasamy, New Mexico State University
Travis Clark, Wright State University
David Cleary, Gonzaga University
Keying Ding, Middle Tennessee State University
John DiVincenzo, Middle Tennessee State University
Stephen Drucker, University of Wisconsin, Eau Claire
Sheryl Ann Dykstra, Pennsylvania State University
Mark Eberhart, Colorado School of Mines
Jack Eichler, University of California, Riverside
Michael Evans, Georgia Institute of Technology
Renee Falconer, Colorado School of Mines
Hua-Jun Fan, Prairie View A&M University
Max Fontus, Prairie View A&M University
Carol Fortney, University of Pittsburgh
Matthew Gerner, University of Arkansas
Peter Golden, Sandhills Community College
Maojun Gong, Wichita State University
Benjamin Hafensteiner, University of Rochester
Hill Harman, University of California, Riverside
Roger Harrison, Brigham Young University
Julie Henderleiter, Grand Valley State University
Amanda Holton, University of California, Irvine
Amy Johnson, Eastern Michigan University
Crisjoe Joseph, University of California, Santa Barbara
Marc Knecht, University of Miami
Colleen Knight, College of Coastal Georgia
Ava Kreider-Mueller, Clemson University
John Krenos, Rutgers University
Maria Krisch, Trinity College
Brian Leskiw, Youngstown State University
Joseph Lodmell, College of Coastal Georgia
Richard Lord, Grand Valley State University
Sudha Madhugiri, Collin College, Preston Ridge
Anna Victoria Martinez-Saltzberg, San Francisco State University
Jason Matthews, Florida State College at Jacksonville
Thomas McGrath, Baylor University
Alice Mignerey, University of Maryland
Tod Miller, Augustana College
Stephanie Myers, Augusta University
Anne-Marie Nickel, Milwaukee School of Engineering
Chad Rezsnyak, Tennessee Tech University
Dawn Richardson, Collin College, Preston Ridge
Hope Rindal, Western Washington University
Jason Ritchie, The University of Mississippi
Lary Sanders, Wright State University
Margaret Scheuermann, Western Washington University
Allan Scruggs, Gonzaga University
Thomas Sorensen, University of Wisconsin, Milwaukee
John Stubbs, The University of New England
Uma Swamy, Florida International University
Lucas Tucker, Siena College
Gabriele Varani, University of Washington
Rebecca Weber, University of North Texas
Karen Wesenberg-Ward, Montana Tech of the University of Montana
Amanda Wilmsmeyer, Augustana College
Eric Zuckerman, Augusta University

First Edition Reviewers:

Ioan Andricioaei, University of California, Irvine
Merritt Andrus, Brigham Young University
David Arnett, Northwestern College
Christopher Babayco, Columbia College
Carey Bagdassarian, University of Wisconsin, Madison
Craig Bayse, Old Dominion University
Vladimir Benin, University of Dayton
Philip Bevilacqua, Pennsylvania State University
Robert Blake, Glendale Community College
David Boatright, University of West Georgia
Petia Bobadova-Parvanova, Rockhurst University
Stephanie Boussert, DePaul University
Jasmine Bryant, University of Washington
Michael Bukowski, Pennsylvania State University
Charles Burns, Wake Technical Community College
Jon Camden, University of Tennessee at Knoxville
Tara Carpenter, University of Maryland, Baltimore County
David Carter, Angelo State University
Allison Caster, Colorado School of Mines
Colleen Craig, University of Washington
Gary Crosson, University of Dayton
Guy Dadson, Fullerton College
David Dearden, Brigham Young University
Danilo DeLaCruz, Southeast Missouri State University
Anthony Diaz, Central Washington University
Greg Domski, Augustana College
Jacqueline Drak, Bellevue Community College
Michael Ducey, Missouri Western State University
Lisa Dysleski, Colorado State University
Amina El-Ashmawy, Collin College
Doug English, Wichita State University
Jim Farrar, University of Rochester
MD Abul Fazal, College of Saint Benedict & Saint John's University

Anthony Fernandez, Merrimack College
Lee Friedman, University of Maryland
Arthur Glasfeld, Reed College
Daniel Groh, Grand Valley State University
Megan Grunert, Western Michigan University
Margaret Haak, Oregon State University
Tracy Hamilton, University of Alabama at Birmingham
David Hanson, Stony Brook University
Roger Harrison, Brigham Young University
David Henderson, Trinity College
Carl Hoeger, University of California, San Diego
Adam Jacoby, Southeast Missouri State University
James Jeitler, Marietta College
Christina Johnson, University of California, San Diego
Maria Kolber, University of Colorado
Regis Komperda, Wright State University
Jeffrey Kovac, University of Tennessee at Knoxville
Jeremy Kua, University of California, San Diego
Robin Lammi, Winthrop University
Annie Lee, Rockhurst University
Willem Leenstra, University of Vermont
Ted Lorance, Vanguard University
Charity Lovitt, Bellevue Community College
Suzanne Lunsford, Wright State University
Jeffrey Macedone, Brigham Young University
Douglas Magde, University of California, San Diego
Rita Maher, Richland College
Heather McKechney, Monroe Community College
Anna McKenna, College of Saint Benedict & Saint John's University
Claude Mertzenich, Luther College
Gellert Mezei, Western Michigan University
Katie Mitchell-Koch, Emporia State University
Stephanie Morris, Pellissippi State Community College
Nancy Mullins, Florida State College at Jacksonville
Joseph Nguyen, Mount Mercy University
Sherine Obare, Western Michigan University
Edith Osborne, Angelo State University
Ruben Parra, DePaul University
Robert Parson, University of Colorado
Brad Parsons, Creighton University
James Patterson, Brigham Young University
Garry Pennycuff, Pellissippi State Community College
Thomas Pentecost, Grand Valley State University
Sandra Peszek, DePaul University
John Pollard, University of Arizona
Gretchen Potts, University of Tennessee at Chattanooga
William Quintana, New Mexico State University
Cathrine Reck, Indiana University, Bloomington
Alan Richardson, Oregon State University
Dawn Richardson, Collin College, Preston Ridge
James Roach, Emporia State University
Jill Robinson, Indiana University
Perminder Sandhu, Bellevue Community College
James Silliman, Texas A&M University, Corpus Christi
Joseph Simard, University of New England
Kim Simons, Emporia State University
Sergei Smirnov, New Mexico State University
Justin Stace, Belmont University
Alyssa Thomas, Utica College
Jess Vickery, SUNY Adirondack
Wayne Wesolowski, University of Arizona
Thao Yang, University of Wisconsin, Eau Claire

Chemistry

An Atoms-Focused Approach

1

Matter and Energy

An Atomic Perspective

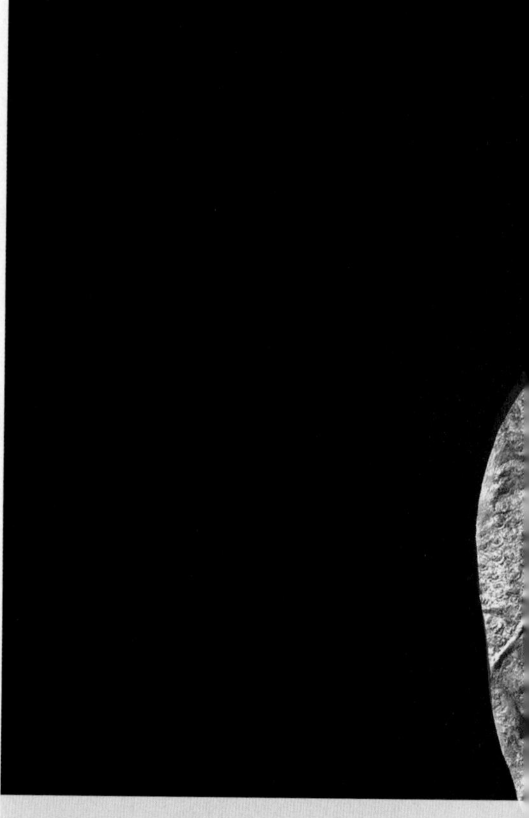

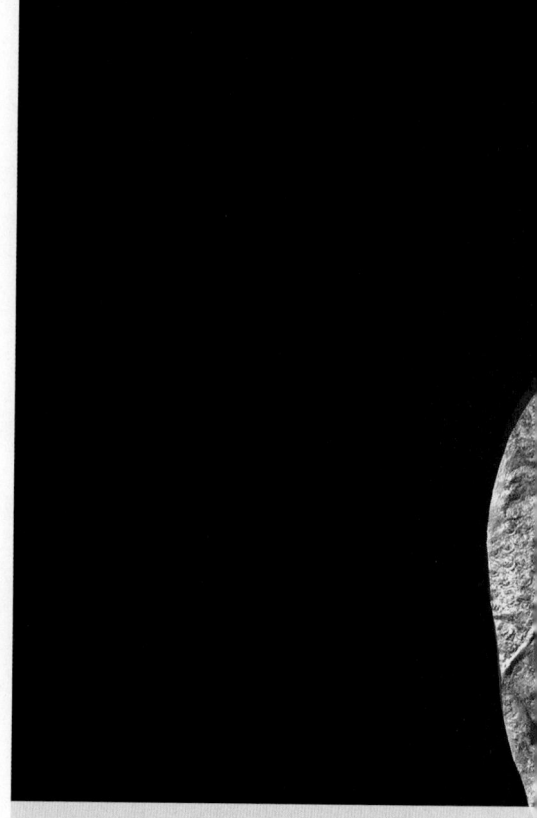

BRONZE AGE BATTLE GEAR This Greek shield decoration from the 6th century BCE is made of bronze, which is a mixture of copper and tin atoms. Tin atoms create irregularities in the layers of copper atoms in bronze. As a result, the layers do not pass each other as easily, making bronze objects harder and less easily deformed than copper objects.

PARTICULATE **REVIEW**

Solids, Liquids, and Gases

In Chapter 1, we explore the particulate nature of matter. Chemists use colored spheres to represent atoms of different elements. Liquid nitrogen (an element) can be used to make ice cream while dry ice (solid carbon dioxide) is used to keep ice cream cold on a hot day.

- Which representation depicts liquid nitrogen?
- Which representation depicts dry ice?
- Which representation depicts carbon dioxide vapor?

(Answers to Particulate Review questions are in the back of the book.)

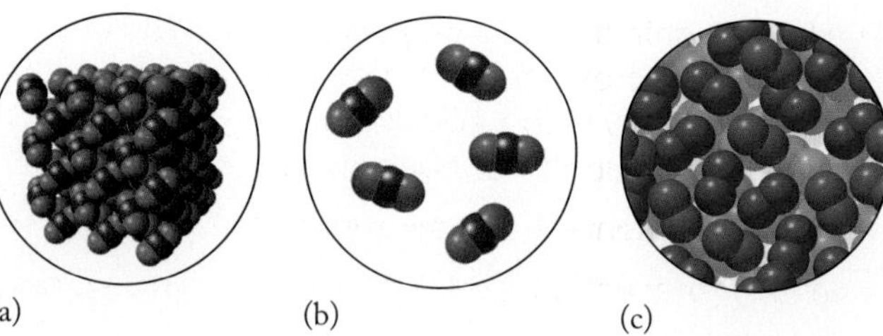

PARTICULATE **PREVIEW**

Elements versus Compounds

The bronze shield on this page is a mixture of copper and tin atoms. Some of the representations shown depict a molecule made of two atoms or an array made from two ions. As you read Chapter 1, look for ideas that will help you answer these questions:

- Which representation depicts molecules of a compound?
- Which representation depicts molecules of an element?
- Which representation depicts a compound consisting of an array of ions?
- Which representation depicts an element consisting of an array of atoms?

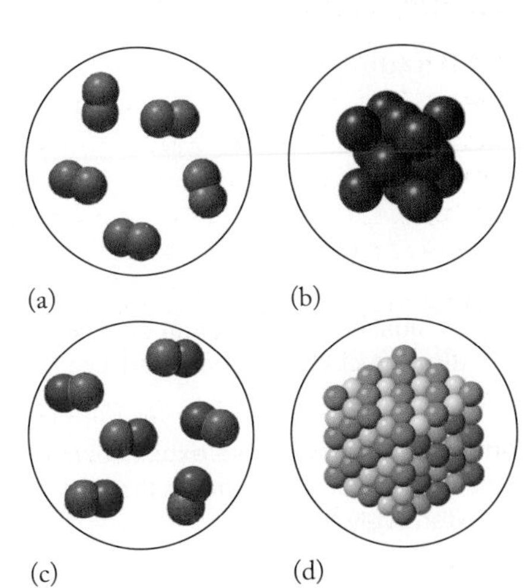

Learning Outcomes

LO1 Describe the scientific method

LO2 Apply the COAST approach to solving problems
Sample Exercises 1.1–1.12

LO3 Distinguish between the classes of matter and between the physical and chemical properties of pure substances
Sample Exercises 1.1–1.3

LO4 Describe the states of matter and how their physical properties can be explained by the particulate nature of matter
Sample Exercise 1.4

LO5 Distinguish between heat, work, potential energy, and kinetic energy, and describe the law of conservation of energy

LO6 Use molecular formulas and molecular models to describe the elemental composition and three-dimensional arrangement of the atoms in compounds

LO7 Distinguish between exact and uncertain values and express uncertain values with the appropriate number of significant figures
Sample Exercises 1.5, 1.6

LO8 Accurately convert values from one set of units to another
Sample Exercises 1.7–1.10

LO9 Express the results of experiments in ways that accurately convey their certainty
Sample Exercises 1.11, 1.12

1.1 Exploring the Particulate Nature of Matter

Atoms and Atomism

The chapter-opening photo shows a Greek shield decoration from the 6th century BCE. It's made of bronze, which is a blend of copper and tin. For thousands of years ancient craftsmen produced bronze using furnaces blazing with mixtures of fuel, such as wood or charcoal, and chunks of metal-containing minerals. When the minerals in the furnace contained copper and lesser amounts of tin, the bronze that was produced could be fashioned into tools and weapons that were much stronger and more durable than those made of copper alone.

To ancient metalworkers, turning minerals into metals was more art than science. They knew how to build and operate metal-producing furnaces, called smelters, but they had little understanding of the chemical changes that, for example, converted copper minerals into copper metal. Today we know what those changes are, and we can explain why mixtures of metals such as bronze are much stronger than their parent metals, because *we know the structures of these materials at the atomic level.*

We know, for example, that the atoms in copper metal are arranged in ordered, tightly packed layers, as shown in the opening photo. Copper wire or foil is easily bent because the layers of copper atoms can slide past each other when subjected to an external force. When slightly larger atoms of tin are also present as shown in the magnified view in the opening photo, the resulting imperfections inhibit the layers of copper atoms from sliding past each other. An object made of bronze, therefore, is much harder to bend than if it were made of pure copper. As a result, Bronze Age tools and weapons held their shape better, stayed sharper longer, and, in the case of shields and body armor, provided better protection for warriors in battle.

In this chapter we begin an exploration of how the properties of materials are linked to their atomic-level structure. As we do, we need to acknowledge the Greek philosophers of the late Bronze Age who espoused *atomism*, a belief

atom the smallest particle of an element that retains the chemical characteristics of the element.

scientific theory a concise explanation of widely observed phenomena that has been extensively tested.

element a substance that cannot be separated into simpler substances by any chemical process.

that all forms of matter are composed of extremely tiny, indestructible building blocks called **atoms**. Atomism is an example of a natural philosophy; it is not a **scientific theory**. The difference between the two is that while both seek to explain natural phenomena, scientific theories are concise explanations of natural phenomena based on observation and experimentation, and they are testable. An important quality of a valid scientific theory is that it accurately predicts the results of future experiments and can even serve as a guide to designing those experiments. The ancient Greeks did not have the technology to test whether matter really is made of atoms—but we do.

Consider the images in Figure 1.1. On the bottom is a photograph of silicon (Si) wafers, the material used today to make computer chips and photovoltaic cells. The magnified view above it is a photomicrograph of a silicon wafer produced by an instrument called a scanning tunneling microscope (STM).[1] The fuzzy spheres are individual atoms of silicon, the smallest representative particles of silicon. If you could grind a sample of pure silicon into the finest dust imaginable, the tiniest particle of the dust you could obtain that still had the properties of silicon would be an atom of silicon.

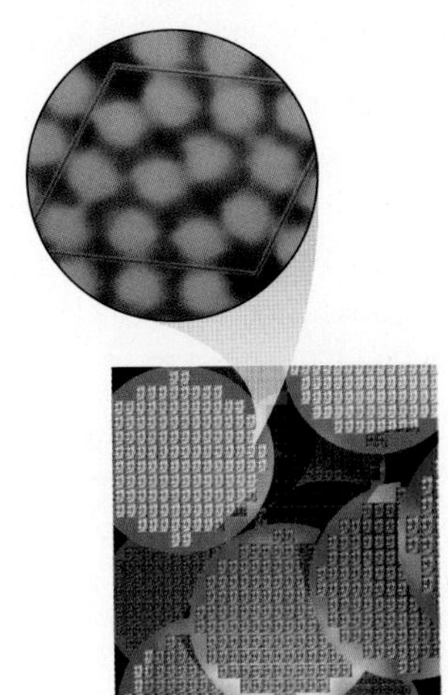

FIGURE 1.1 Silicon wafers are widely used to make computer chips and photovoltaic cells for solar panels. Since the 1980s, scientists have been able to image individual atoms using an instrument called a scanning tunneling microscope (STM). In the STM image (top), the irregular shapes are individual silicon atoms. The radius of each atom is 117 picometers (pm), or 117 trillionths of a meter. Atoms are the tiniest particles of silicon that still retain the chemical characteristics of silicon.

Atomic Theory: The Scientific Method in Action

Scanning tunneling microscopes have been used to image atoms since the early 1980s, but the scientific theory that matter was composed of atoms evolved two centuries earlier during a time when chemists in France and England made enormous advances in our understanding of the composition of matter. Among them was French chemist Antoine Lavoisier (1743–1794), who published the first modern chemistry textbook in 1789. It contained a list of substances that he believed could not be separated into simpler substances. Today we call such "simple" substances **elements** (Figure 1.2). The silicon in Figure 1.1 is an element, as are copper and tin. The periodic table of the elements inside the front cover of this textbook contains over 100 others.

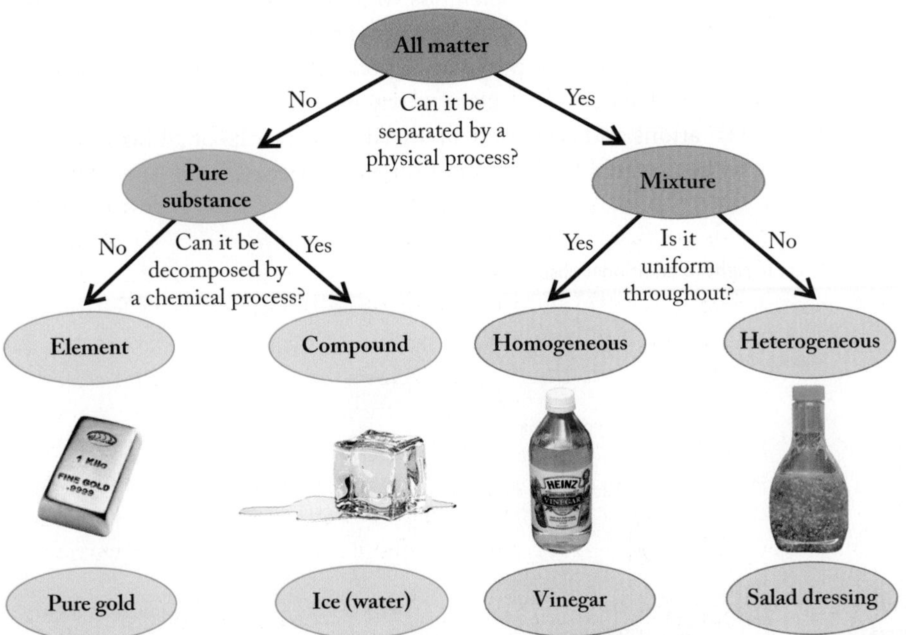

FIGURE 1.2 Matter is classified as shown in this diagram. The two principal categories are pure substances and mixtures. A substance may be a compound (such as water) or an element (such as gold). When the substances making up a mixture are distributed uniformly, as they are in vinegar (a mixture of acetic acid and water), the mixture is homogeneous. When the substances making up a mixture are not distributed uniformly, as in salad dressing, the mixture is heterogeneous.

[1]German physicist Gerd Binnig (b. 1947) and Swiss physicist Heinrich Rohrer (1933–2013) shared the 1986 Nobel Prize in Physics for their development of scanning tunneling microscopy.

Lavoisier and other scientists conducted experiments that examined the patterns in how elements combined with other elements to form **compounds**. These experiments followed a systematic approach to investigating and understanding natural phenomena known as the **scientific method** (Figure 1.3). When such investigations reveal consistent patterns and relationships, they may be used to formulate concise descriptions of fundamental scientific truths. These descriptions are known as **scientific laws**.

When the French chemist Joseph Louis Proust (1754–1826) studied the composition of compounds containing different metals and oxygen, he concluded that these compounds always contained the same proportions of their component elements. His **law of definite proportions** applies to all compounds. An equivalent law, known as the **law of constant composition**, states that a compound always has the same *elemental composition* by mass no matter what its source. Thus, the composition of pure water is always the same: 11.2% by mass hydrogen and 88.8% by mass oxygen.

When Proust published his law of definite proportions, some of the leading chemists of the time refused to believe it. Their own experiments seemed to show, for example, that the compound that tin formed with oxygen had variable tin content. These scientists did not realize that their samples were actually mixtures of two different compounds with different compositions, which Proust was able to demonstrate. Still, acceptance of Proust's law required more than corroborating results from other scientists; it also needed to be explained by a scientific theory. That is, there needed to be a convincing argument that explained *why* the composition of a compound was always the same.

Scientific laws and theories complement each other in that scientific laws describe natural phenomena and relationships, and scientific theories explain *why* these phenomena and relationships are always observed. Scientific theories usually start out as tentative explanations of why a set of experimental results was obtained or why a particular phenomenon is consistently observed. Such a tentative explanation is called a **hypothesis** (Figure 1.3). An important feature of a hypothesis is that it can be tested through additional observations and experiments. A hypothesis also enables scientists to accurately predict the likely outcomes of future observations and experiments. Further testing and observation might support a hypothesis or disprove it, or perhaps require that it be modified. A hypothesis that withstands the tests of many experiments, accurately explaining further observations and accurately predicting the results of additional experimentation, may be elevated to the rank of scientific theory.

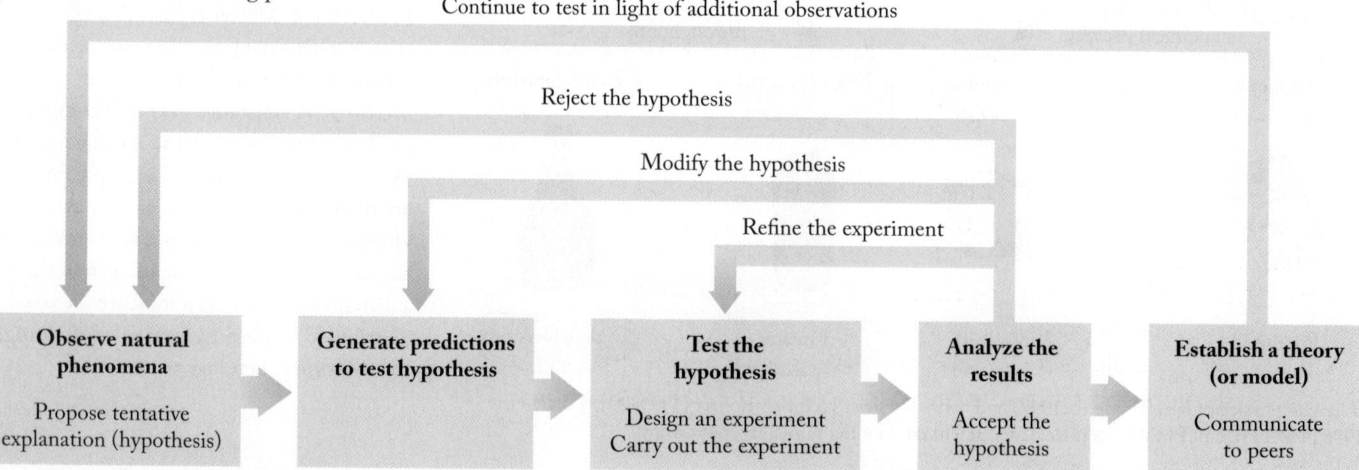

FIGURE 1.3 In the scientific method, observations lead to a tentative explanation, or hypothesis, which leads to more observations and testing, which may lead to the formulation of a succinct, comprehensive explanation called a theory. This process is rarely linear: it often involves looping back, because the results of one test lead to additional tests and a revised hypothesis. Science, when done right, is a dynamic and self-correcting process.

A scientific theory explaining Proust's law of definite proportions was proposed by John Dalton (1766–1844) in 1803. Whereas Proust studied the composition of the solid compounds formed by metals and oxygen, Dalton's own research focused on the composition and behavior of gases. Dalton observed that when two elements combine to form gaseous compounds, they may form two or more different compounds with different compositions. Similarly, Proust had discovered that tin (Sn) and oxygen (O) combined to form one compound that was 88.1% by mass Sn and 11.9% O and a second compound that was 78.8% Sn and 21.2% O. Dalton noted that the ratio of oxygen to tin in the second compound,

$$\frac{21.2\%\ \text{O}}{78.8\%\ \text{Sn}} = 0.269$$

was very close to twice of what it was in the first compound,

$$\frac{11.9\%\ \text{O}}{88.1\%\ \text{Sn}} = 0.135$$

Similar results were obtained with other sets of compounds formed by pairs of elements. Sometimes their compositions would differ by a factor of 2, as with oxygen and tin (and with oxygen and carbon), and sometimes their compositions differed by other factors, but in all cases they differed *by ratios of small whole numbers*. This pattern led Dalton to formulate the **law of multiple proportions**: when two elements combine to make two (or more) compounds, the ratio of the masses of one of the elements, which combine with a given mass of the second element, is always a ratio of small whole numbers. For example, 15 grams of oxygen combines with 10 grams of sulfur under one set of reaction conditions, whereas only 10 grams of oxygen combines with 10 grams of sulfur to form a different compound under a different set of reaction conditions. The ratio of the two masses of oxygen,

$$\frac{15\ \cancel{\text{g oxygen}}}{10\ \cancel{\text{g oxygen}}} = \frac{3}{2}$$

is indeed a ratio of two small whole numbers and is consistent with Dalton's law of multiple proportions.

To explain the laws of definite proportions and multiple proportions, Dalton proposed the scientific theory that *elements are composed of atoms*. Thus, Proust's compound with the O:Sn ratio of 0.135 contains one atom of oxygen for each atom of tin, whereas his compound with twice that O:Sn ratio (0.269) contains *two* atoms of O per atom of Sn. These atomic ratios are reflected in the **chemical formulas** of the two compounds: SnO and SnO_2, in which the subscripts after the symbols represent the relative number of atoms of each element in the substance. The absence of a subscript means the formula contains one atom of the preceding element. Similarly, the two compounds that sulfur and oxygen form have an oxygen ratio of 3:2 because their chemical formulas are SO_3 and SO_2, respectively.

Since the early 1800s, scientists have learned much more about the atomic, and even subatomic, structure of the matter that makes up our world and the universe that surrounds us. Although the laws developed two centuries ago are still useful, Dalton's atomic theory, like many theories, has undergone revisions as new discoveries have been made. Dalton assumed, for example, that all of the atoms of a particular element were the same. We will see in Chapter 2 that atoms have internal components and structures, only some of which are the same for all the atoms of a given element. Atoms can differ in other ways, too, that the scientists of 1800 could not have observed or even imagined.

compound a substance composed of characteristic proportions of two or more elements chemically bonded together.

scientific method an approach to acquiring knowledge based on the observation of phenomena, the development of a testable hypothesis, and additional experiments that test the validity of the hypothesis.

scientific law a concise and generally applicable statement of a fundamental scientific principle.

law of definite proportions the principle that compounds always contain the same proportions of their component elements.

law of constant composition the principle that all samples of a particular compound have the same elemental composition.

hypothesis a tentative and testable explanation for an observation or a series of observations.

law of multiple proportions the principle that, when two masses of one element react with a given mass of another element to form two different compounds, the two masses of the first element have a ratio of two small whole numbers.

chemical formula a notation for representing the elemental composition of a pure substance using the symbols of the elements; subscripts indicate the relative number of atoms of each element in the substance.

1.2 COAST: A Framework for Solving Problems

Throughout the rest of this chapter and this book, you will find Sample Exercises designed to help you better understand chemical concepts and develop your problem-solving skills. Each Sample Exercise follows a systematic approach to problem solving that we encourage you to apply to the Practice Exercises that follow each Sample Exercise and to the end-of-chapter problems. We use the acronym COAST (**C**ollect and **O**rganize, **A**nalyze, **S**olve, and **T**hink about the answer) to represent the four steps in our approach to problem solving. As you read about it here and use it later, keep in mind that COAST is merely a *framework* for solving problems, not a recipe. Use it as a guide to help you develop your own approach to solve each problem.

Collect and Organize First, start by sorting through the information given in the problem and identifying other relevant information. These actions help you understand the problem, including the fundamental chemical principles on which it is based. As part of the collecting and organizing process,

- Identify the key concept of the problem.
- Identify and define the key terms used to express that concept. You may find it useful to restate the problem in your own words.
- Sort through the information given in the problem, separating what is pertinent from what is not.
- Assemble any supplemental information that may be needed, including equations, definitions, and constants.

Analyze The next step is to analyze the information you have collected to determine how to relate it to the answer you seek. Sometimes it is easier to work backward to create the relationships: consider the nature of the answer first, and think about how you might get to it from the information provided in the problem and other sources. If the problem is quantitative and requires a numerical answer, frequently the units of the initial values and the final answer will help you identify how they are connected and which equation (or equations) may be useful.

For some problems, drawing a sketch based on molecular models or an experimental setup may help you visualize how the starting points and final answer are connected. You should also look at the numbers involved and estimate your answer. Having an order-of-magnitude ("ballpark") estimate of your final answer before entering numbers into your calculator provides a check on the accuracy of your calculated answer.

Some Sample Exercises test your understanding of a single concept or their solution involves only a single step. In these exercises we may combine the Collect and Organize and the Analyze steps.

Solve For most conceptual questions, the solution flows directly from your analysis of the problem. To solve quantitative problems, you need to insert the starting values and the appropriate constants into the relevant equations or conversion factors and calculate the answer. In the Solve step, make sure that units are consistent and cancel as needed and that the certainty of the quantitative information is reflected in how many *significant figures* (see Section 1.7) you used.

Think About It Finally, you need to think about your result. Does your answer make sense based on your own experience and what you have just learned? Is the value for a quantitative answer reasonable—is it close to your estimate from the

Analyze step? Are the units correct and the number of significant figures appropriate? Then ask yourself how confident you are that you could solve another problem, perhaps drawn from another context but based on the same chemical concept. You may also think about how this problem relates to other observations you may have made about matter in your daily life.

The COAST approach should help you solve problems in a logical way and avoid certain pitfalls, such as grabbing an equation that seems to have the right variables and simply plugging numbers into it or resorting to trial and error. As you study the steps in each Sample Exercise, try to answer these questions about each step:

- **What** is done in this step?
- **How** is it done?
- **Why** is it done?

After answering the questions, you will be ready to solve the Practice Exercises and end-of-chapter problems in a systematic way.

matter anything that has mass and occupies space.

mass (*m*) the property that defines the quantity of matter in an object.

chemistry the study of the composition, structure, and properties of matter and of the energy consumed or given off when matter undergoes a change.

intensive property a property that is independent of the amount of substance present.

extensive property a property that varies with the amount of substance present.

physical property a property of a substance that can be observed without changing the substance into another substance.

density (*d*) the ratio of the mass (*m*) of an object to its volume (*V*).

1.3 Classes and Properties of Matter

All things that are physically real—from the air we breathe to the ground we walk on—are forms of matter. Scientists define **matter** as everything in the universe that has **mass (*m*)** and occupies space. **Chemistry** is the study of the composition, structure, and properties of matter and the changes it undergoes.

Matter is classified based on its composition, as shown in Figure 1.2. The simplest forms of matter—elements and compounds—are pure substances with compositions that do not change unless they are involved in a chemical reaction. They also have distinctive properties. Pure gold, for example (Figure 1.4a), has a distinctive color; it's soft for a metal, it's malleable (it can be hammered into very thin sheets called gold leaf), it's ductile (it can be drawn into thin wires), and it melts at 1064°C. These properties, which characterize a pure substance but are independent of the amount of the substance in a sample, are called **intensive properties**. Other properties, such as the particular length, width, mass, and volume of an ingot of gold, are called **extensive properties** because they depend on how much of the substance is present in a particular sample.

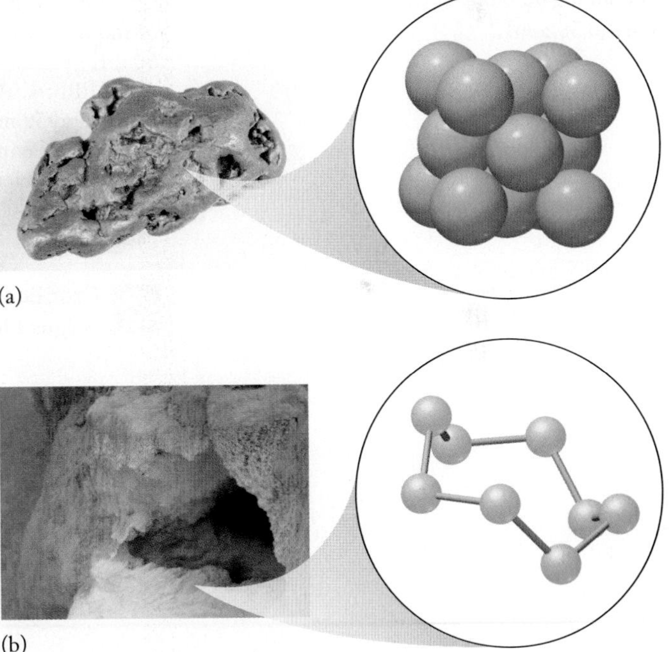

FIGURE 1.4 (a) Gold and (b) sulfur are among the few elements that may occur in nature uncombined with other elements.

The properties of substances are either *physical* or *chemical*. **Physical properties**, such as those described for gold in the previous paragraph, can be observed or measured without changing the substance into another substance. Another physical property is **density (*d*)**, which is the ratio of the mass (*m*) of a substance or object to its volume (*V*):

$$d = \frac{m}{V} \tag{1.1}$$

CONCEPT TEST

Which of the following properties of a sample of pure iron are intensive? (a) mass, (b) density, (c) volume, (d) hardness

(Answers to Concept Tests are in the back of the book.)

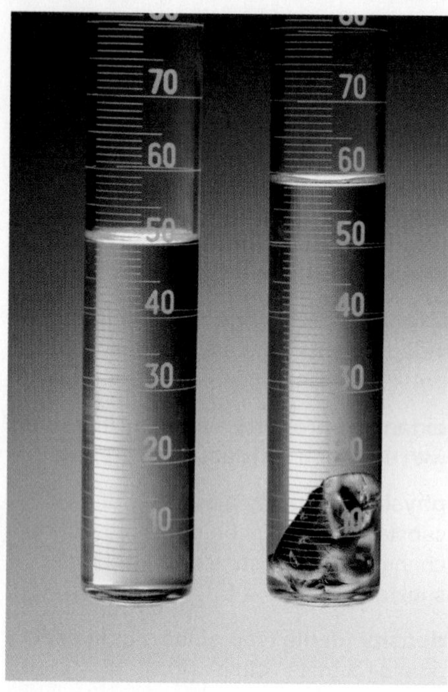

FIGURE 1.5 When a nugget believed to be pure gold is placed in a graduated cylinder containing 50.0 mL of water, the volume rises to 58.5 mL.

SAMPLE EXERCISE 1.1 Calculating Density **LO2**

Suppose you are asked to determine whether or not a nugget could be gold based on its density. You weigh the nugget and find that it has a mass of 164 grams (g). Then you carefully add it to a graduated cylinder that contains 50.0 milliliters (mL) of water (Figure 1.5) and observe that the water level in the cylinder rises to 58.5 mL. What is the density of the nugget in grams per cubic centimeter? Could it be gold?

Collect and Organize Density is the ratio of the mass of an object to its volume (Equation 1.1). We are given the mass of the gold nugget and the volume of the water it displaces. Appendix A3.2 lists the density of pure gold as 19.3 g/cm^3. We know from the table of units inside the back cover of this book that 1 mL = 1 cm^3 (cubic centimeter).

Analyze The volume of the nugget is equal to the volume of water it displaces, which is the difference between 58.5 and 50.0 mL.

Solve The density of the nugget is

$$d = \frac{m}{V} = \frac{164\ \text{g}}{(58.5 - 50.0)\ \text{mL}} = \frac{164\ \text{g}}{8.5\ \text{mL}} = 19\ \frac{\text{g}}{\text{mL}} = 19\ \frac{\text{g}}{\text{cm}^3}$$

The calculated density is close to the density of pure gold near room temperature. Therefore, the nugget *could* be gold, and it very likely is gold, given its appearance.

Think About It A careful look at the images in Figure 1.5 shows that the meniscus of the water in the graduated cylinder before the nugget was added just touches the 50 mL graduation, so we can estimate that the volume is 50.0 mL to the nearest tenth of a milliliter. After the nugget is added, the meniscus is about halfway between the 58 and 59 mL graduations, so we can estimate that the total volume is then 58.5 mL. Having estimates of the beginning and final volumes to the nearest tenth of a milliliter lets us estimate the difference between them to the nearest tenth, or 8.5 mL. We discuss measured values and the uncertainties in them in greater detail in Section 1.7.

Practice Exercise A silver-colored metallic cube is 2.54 cm on a side and weighs 146 g. Which of the following metals is it most likely to be?

Metal	Density (g/cm^3)
Aluminum	2.7
Cadmium	8.7
Copper	9.0
Iron	7.9
Lead	11.3
Nickel	8.9
Zinc	7.1

(Answers to Practice Exercises are in the back of the book.)

Gold is one of the few elements that occur in nature uncombined with other elements. The tendencies of most elements to react with other substances—that is, to participate in **chemical reactions** in which they are transformed into compounds with different chemical identities and properties—represent the

chemical properties of the elements. Chemical properties include whether or not a particular element reacts with another element or with a particular compound. They also include how rapidly the reactions take place and what products are formed.

Throughout this book we will see many examples of how the properties of substances are linked to the behavior of the particles that make them up, including the ways in which these particles interact with the particles that make up other substances. As we saw in Section 1.1, these particles may be atoms, but they may also be groups of atoms that are held together in a characteristic pattern by forces called **chemical bonds**. Many of these groups of atoms are neutral **molecules**, meaning that they have no net electrical charge. However, some atoms or molecules acquire a net positive or negative electrical charge, which means that they are monatomic (single-atom) or molecular **ions**. We will discuss how and why ions form in Chapter 2, but for now the key point is that the particles that make up elements and compounds may be atoms, molecules, or ions.

The physical and chemical properties of compounds are different from those of the elements of which they are composed. For example, water is a liquid at room temperature, whereas hydrogen and oxygen, the elements that combine to form water, are gases. Water is a liquid at room temperature because molecules of H_2O are much more strongly attracted to each other than are molecules of hydrogen (H_2) or molecules of oxygen (O_2). Liquid water expands when it freezes at 0°C, but liquid hydrogen and oxygen contract when they freeze at −259°C and −219°C, respectively. Oxygen supports combustion reactions, and hydrogen is highly flammable, but water neither supports combustion nor is flammable. Indeed, it is widely used to extinguish fires.

chemical reaction the conversion of one or more substances into one or more different substances.

chemical property a property of a substance that can be observed only by reacting the substance to form another substance.

chemical bond a force that holds two atoms or ions in a molecule or a compound together.

molecule a collection of atoms chemically bonded together.

ion an atom or molecule that has a net positive or negative charge.

SAMPLE EXERCISE 1.2 Distinguishing Physical and Chemical Properties — LO3

Which of the following properties of gold are chemical and which are physical?

a. Gold metal, which is insoluble in water, can be made soluble by reacting it with a mixture of nitric and hydrochloric acids known as aqua regia.
b. Gold melts at 1064°C.
c. Gold can be hammered into sheets so thin that light passes through them.
d. Gold metal can be recovered from gold ore by treating the ore with a solution containing cyanide, which reacts with and dissolves gold.

Collect, Organize, and Analyze Chemical properties describe how a substance reacts with other substances, whereas physical properties can be observed or measured without changing one substance into another. Properties (a) and (d) describe reactions that chemically change gold metal into compounds of gold that dissolve in water. Properties (b) and (c) describe processes in which elemental gold remains elemental gold. When it melts, gold changes its physical state from solid to liquid but not its chemical identity. When it is hammered flat, it is still solid, elemental gold.

Solve Properties (a) and (d) are chemical properties, whereas (b) and (c) are physical properties.

Think About It When possible, rely on your experiences and observations. Gold jewelry does not dissolve in water, so dissolving gold metal requires a change in its chemical identity: it can no longer be elemental gold. On the other hand, physical processes such as melting do not alter the chemical identity of the gold. Gold can be melted and then cooled to produce solid gold again.

 **Practice Exercise** Which of the following properties of water are chemical and which are physical?

a. Water normally freezes at 0.0°C.
b. Water and carbon dioxide combine during photosynthesis, forming sugar and oxygen.
c. A cork floats on water, but a piece of copper sinks.
d. During digestion, starch reacts with water to form sugar.

(Answers to Practice Exercises are in the back of the book.)

Most of the matter in nature exists as mixtures of elements and compounds. **Mixtures** are composed of two or more pure substances and are classified as either *homogeneous* or *heterogeneous* (Figure 1.2). The substances in a **homogeneous mixture** are distributed uniformly, and the composition and appearance of the mixture are uniform throughout. Homogeneous mixtures are also called **solutions**, a term that scientists apply to homogeneous mixtures of gases and solids as well as liquids. In contrast, the substances in a **heterogeneous mixture** are not distributed uniformly and contain distinct regions of different composition. They include mixtures of **immiscible liquids**, which, like the oil and water in salad dressing, do not dissolve in each other, forming a single solution. Nearly all the forms of matter we encounter, including the air we breathe and the food and drink we consume, are mixtures.

Separating Mixtures

The substances in mixtures can be separated from one another based on differences in their physical properties. The differences are closely linked to how strongly the particles in each substance interact with each other and with the particles of the other substances in the mixture. For example, the water in

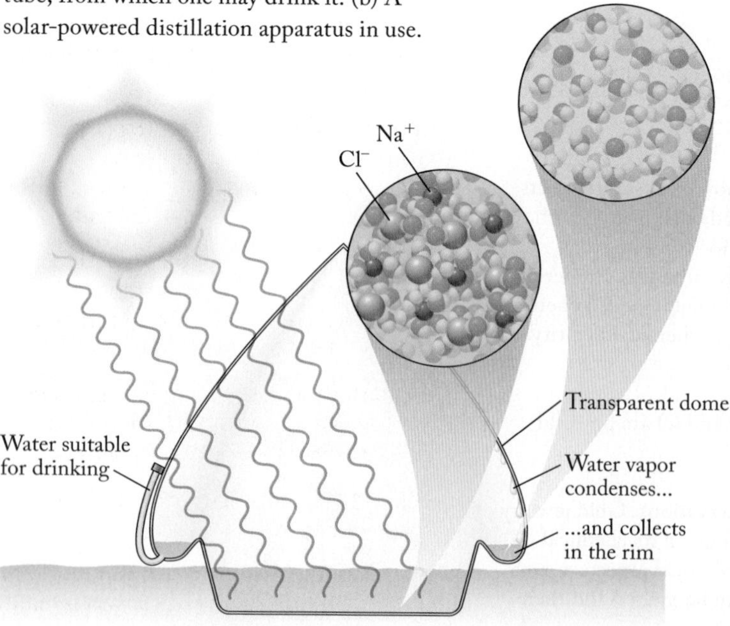

(a)

(b)

FIGURE 1.6 (a) In a solar-powered distillation apparatus used in survival gear to provide fresh water from seawater, sunlight passes through the transparent dome and heats a pool of seawater. Water vapor rises from the pool, contacts the inside of the transparent dome (which is relatively cool), and condenses. The distilled water collects in the depression around the rim and then passes into the attached tube, from which one may drink it. (b) A solar-powered distillation apparatus in use.

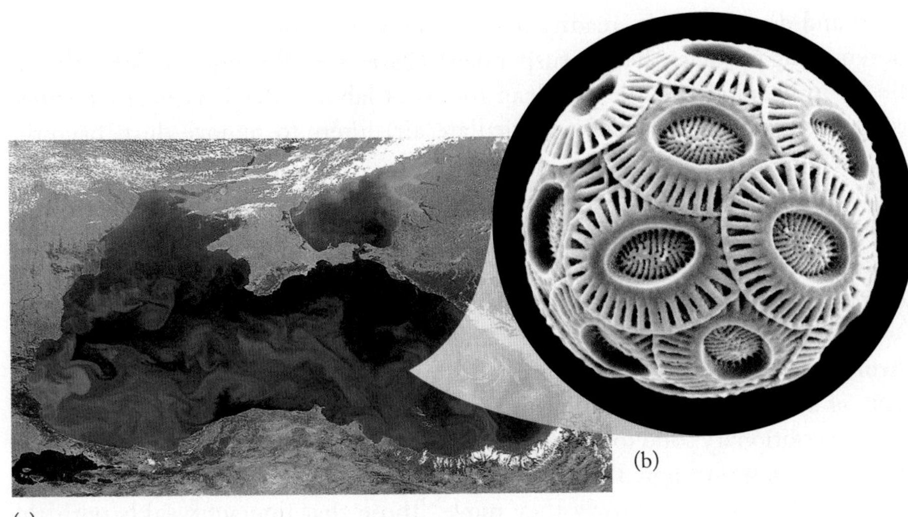

FIGURE 1.7 (a) Very high concentrations (called "blooms") of phytoplankton known as *coccolithophores* in the Black Sea were photographed by a NASA satellite. During intense blooms, coccolithophores turn the color of the sea a milky aquamarine. The color comes from chlorophyll and other pigments in the phytoplankton; the milkiness comes from sunlight scattering off the organisms' textured exterior shells. (b) A single coccolithophore cell.

seawater can be made drinkable by separating it from the salts (mostly sodium chloride, NaCl) that are dissolved in it. One process for doing so is **distillation** (Figure 1.6), in which a component of the mixture (water in this case) is separated by evaporation, and the resulting vapor is recovered by condensation. Distillation works as a separation technique whenever the components of a mixture have different **volatilities**—that is, one or more of them vaporize more readily than the others. The volatility of a substance is inversely proportional to the strength of the interactions between its particles: as the strength of the interactions increases, the probability that particles of the substance will have enough energy to break away from adjacent particles in the liquid phase and become particles of vapor decreases.

In addition to dissolved salts, seawater may contain single-celled plants called phytoplankton (Figure 1.7). Scientists who study these microscopic algae separate them from seawater by **filtration**, which works in this case because the phytoplankton cells, which are many micrometers in diameter, are larger than the pores in the filter (Figure 1.8). The cells are trapped on the filter while the

mixture a combination of pure substances in variable proportions in which the individual substances retain their chemical identities and can be separated from one another by a physical process.

homogeneous mixture a mixture in which the components are distributed uniformly throughout and the composition and appearance is uniform.

solution another name for a *homogeneous mixture*. Solutions are often liquids, but they may also be solids or gases.

heterogeneous mixture a mixture in which the components are not distributed uniformly, so that the mixture contains distinct regions of different compositions.

immiscible liquids combinations of liquids that do not mix with, or dissolve in, each other.

distillation a process using evaporation and condensation to separate a mixture of substances with different volatilities.

volatility a measure of how readily a substance vaporizes.

filtration a process for separating solid particles from a liquid or gaseous sample by passing the sample through a porous material that retains the solid particles.

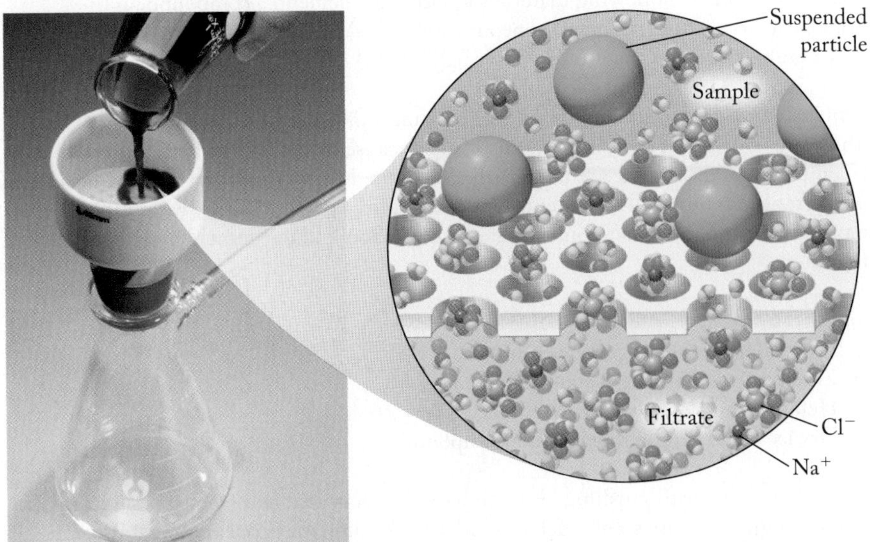

FIGURE 1.8 Particles suspended in a liquid, such as a culture of phytoplankton in seawater, can be separated from the liquid by filtration. The suspended particles and filter pores are many thousands of times larger than water molecules.

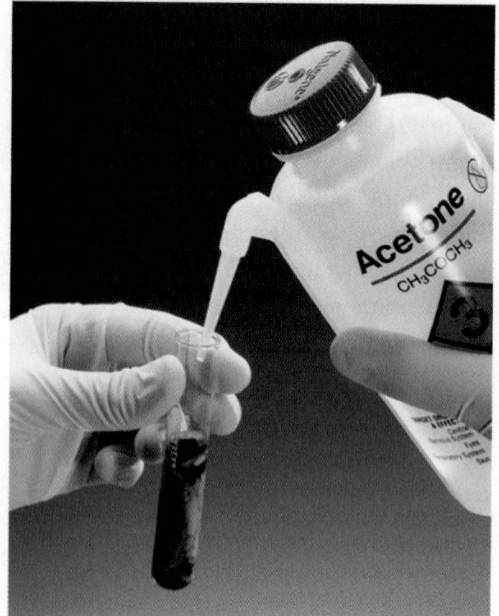

(a)

(b)

FIGURE 1.9 (a) Chlorophyll and other pigments can be extracted from a phytoplankton sample trapped on a filter using the solvent acetone. (b) Dissolved pigments can be separated from one another using chromatography, producing a characteristic pattern of colored bands.

water and dissolved salts readily pass through. Filtration is a useful technique for removing particles that are suspended in gases as well as liquids. For example, the air in hospitals and in the clean rooms of laboratories is typically purified using HEPA (high-efficiency particulate air) filters to remove dust, bacteria, and even viruses.

The components of the phytoplankton caught on the filter in Figure 1.8 can be further separated by soaking the wet filter in acetone (Figure 1.9a), which dissolves some of the compounds present inside the phytoplankton cells. These compounds include chlorophyll and other pigments that have distinctive colors. Once dissolved in acetone, the pigments can be separated from each other using **chromatography** (Figure 1.9b). In a chromatographic separation, the components of a liquid or gaseous mixture are distributed between two phases: one of them is a stationary solid (or liquid-coated solid), and the other is a moving liquid or gas. The more strongly the particles in the mobile phase interact with the stationary solid, the more slowly they move. Those that interact weakly with the stationary solid move more rapidly.

CONCEPT TEST

Which physical process—distillation, filtration, or chromatography—would you use to perform each of the following separations?

a. Removing particles of rust from drinking water.
b. Separating the different coloring agents in a sample of ink.
c. Separating volatile compounds normally found in natural gas that have dissolved in a sample of crude oil.

(Answers to Concept Tests are in the back of the book.)

SAMPLE EXERCISE 1.3 Distinguishing Between Different Classes of Matter — LO3

Classify each of the following materials as either an element, a compound, a homogeneous mixture, or heterogeneous mixture: (a) fruit salad, (b) filtered air in a scuba tank, (c) helium gas inside a party balloon, (d) dry ice (solid carbon dioxide).

Collect, Organize, and Analyze Fruit salad is a mixture of different kinds of fruit. The pieces of fruit are not mixed uniformly because of their large size. Filtered air is a mixture of gases that are uniformly distributed throughout the inner volume of the tank. Helium is an element. Dry ice is solid carbon dioxide, which, given its name, is a compound that contains atoms of carbon and oxygen chemically bonded together.

Solve

a. Fruit salad is a heterogeneous mixture.
b. Filtered air is a homogeneous mixture.
c. Helium is an element.
d. Dry ice (solid carbon dioxide) is a compound.

Think About It Distinguishing between heterogeneous and homogeneous mixtures can be challenging because some, such as murky river water, may appear to be uniformly mixed, but if the water is allowed to stand in a glass, solid particles may settle to the bottom, demonstrating that it is actually a heterogeneous mixture.

Practice Exercise Which of the following are pure substances? (a) oxygen gas, (b) human bones, (c) a brick wall, (d) a wooden baseball bat

(Answers to Practice Exercises are in the back of the book.)

chromatography a process involving a stationary and a mobile phase for separating a mixture of substances based on their different affinities for the two phases.

solid a phase of matter that has a definite shape and volume.

liquid a phase of matter that occupies a definite volume but flows to assume the shape of its container.

gas a phase of matter that has neither definite volume nor definite shape, and that expands to fill its container; also called *vapor*.

1.4 The States of Matter

The forms of matter described in Section 1.3 can exist in three phases or physical states: solid, liquid, or gas. Their characteristic properties are as follows:

- A **solid** has a definite volume and shape.
- A **liquid** has a definite volume but not a definite shape. Instead, it takes the shape of its container.
- A **gas** (or *vapor*) has neither a definite volume nor a definite shape. Rather, it expands to occupy the entire volume and shape of its container. Unlike solids and liquids, gases are highly compressible, which means they can be squeezed into smaller volumes.

Consider the three states of water shown with photographs and particle-view magnifications in Figure 1.10. All three phases contain the same particles: molecules made up of two atoms of hydrogen chemically bonded to a central atom of oxygen. In solid ice (Figure 1.10a), each water molecule is surrounded by four others and locked in place in a hexagonal array of molecules that extends in all three spatial dimensions. Molecules in the array may vibrate a little, depending on their temperature, but they are *not* free to move past the molecules that surround them. Thus, ice is rigid at both the molecular and macroscopic levels. The molecules in liquid water (Figure 1.10b), on the other hand, are more randomly ordered and can flow past one another. They are still in close proximity to each other, but their nearest neighbors change over time. There is little space between the molecules of water in ice and liquid

(a) Solid

(b) Liquid

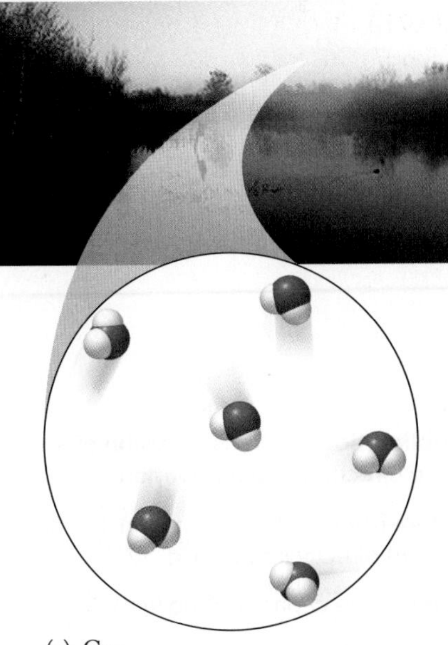

(c) Gas

FIGURE 1.10 The three states of water. (a) In the solid state, each water molecule in ice is held in place in a rigid, three-dimensional array. (b) In the liquid state, the molecules are close together but free to tumble over one another. (c) In the gas state, the molecules are far apart, largely independent of one another, and move freely. Water vapor is invisible, but we can see clouds and fog that form when atmospheric water vapor condenses forming tiny drops of liquid water.

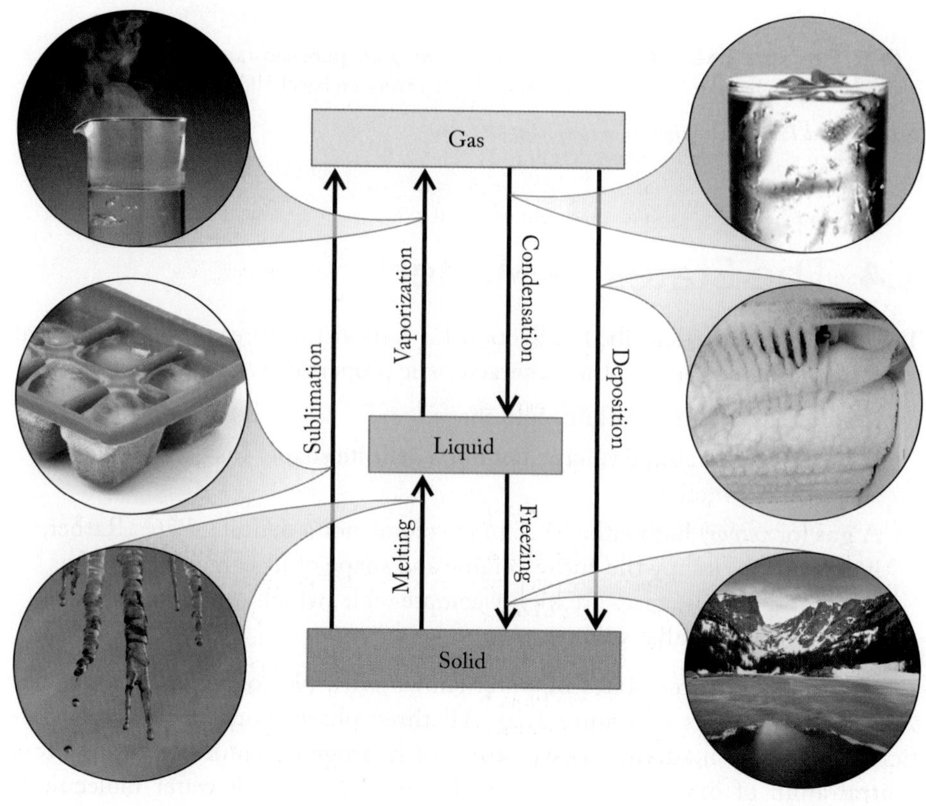

FIGURE 1.11 Matter may change from one state to another when energy is either added or removed. Arrows pointing upward represent transformations that require adding energy, whereas arrows pointing downward represent transformations that release energy.

water, but molecules of water vapor (Figure 1.10c) are widely separated. The volume the particles occupy is negligible relative to the volume occupied by the vapor itself. Most of the volume is made up of nothing—just empty space between the particles of gas. This empty space between particles explains why gases are so compressible.

The state of a substance may change if it is heated up or cooled down. Consider, for example, the changes that water undergoes in each of the processes in Figure 1.11. The upward-pointing arrows on the left side of the figure mean that heat is absorbed when (1) an icicle *melts*, (2) ice cubes in a frost-free freezer slowly become water vapor in a process called **sublimation**, and (3) bubbles of water vapor form in boiling water as the liquid undergoes *vaporization*. The downward-pointing arrows on the right side mean that heat is released when (1) water vapor *condenses* as drops of liquid water on a glass holding a cold drink on a hot day, (2) liquid water *freezes* on the surface of a pond in early winter, and (3) water vapor forms solid ice directly in a process called **deposition**. In Section 1.5, we examine the impacts of these transfers of heat on the behavior of the particles that make up solids, liquids, and gases.

CONCEPT TEST

Ice cream vendors often use dry ice (solid carbon dioxide, CO_2) to keep their ice cream frozen. Over time, the dry ice disappears as solid CO_2 turns into CO_2 gas.

a. What is the name of this change in physical state?

b. What is the name of the reverse process in which dry ice is produced from CO_2 gas?

(Answers to Concept Tests are in the back of the book.)

sublimation the transformation of a solid directly into a gas (vapor).

deposition the transformation of a gas (vapor) directly into a solid.

energy the capacity to do work (w).

work (w) the exertion of a force (F) through a distance (d): $w = F \times d$.

SAMPLE EXERCISE 1.4 Distinguishing Between Particulate Views of the Different States of Matter LO4

Which physical state is represented in each box of Figure 1.12? (The particles could be atoms or molecules.) What change of state is indicated by each arrow? What would the changes of state be if both arrows pointed in the opposite direction?

(a)

(b)

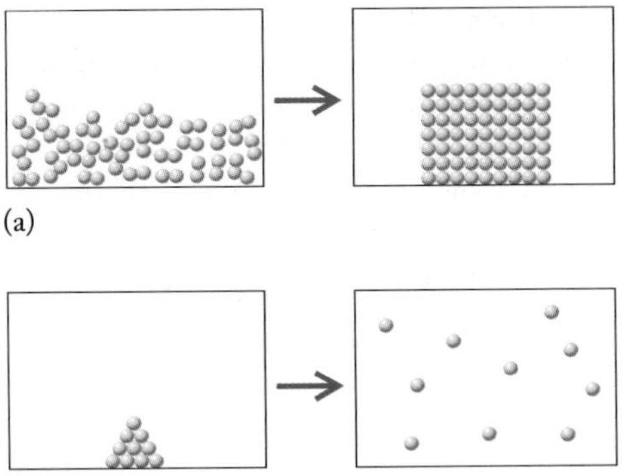

FIGURE 1.12

Collect, Organize, and Analyze We need to look for patterns among the particles in each box:

- An ordered arrangement of particles that does not fill the box or match the shape of the box represents a solid.
- A less-ordered arrangement that partially fills the box and conforms to its shape represents a liquid.
- A dispersed array of particles distributed throughout the box represents a gas.

Solve The particles in the left box of Figure 1.12(a) partially fill the box and adopt the shape of the container, so they represent a liquid. The particles in the box on the right are ordered and form a shape that does *not* conform to that of the box, so they represent a solid. The arrow represents a liquid turning into a solid, so the physical process is freezing. An arrow in the opposite direction would represent melting. The particles in the box on the left in Figure 1.12(b) represent a solid because they are ordered and do not adopt the shape of the box. Those to the right are dispersed throughout the box, so they represent a gas. The arrow represents a solid turning into a gas, so the physical process is sublimation. The reverse process—a gas becoming a solid—would be deposition.

Think About It The particles in the liquid and solid in Figure 1.12(a) are very close together, whereas the gas particles shown in the right panel of Figure 1.12(b) are widely separated and have much more freedom to move about.

Practice Exercise (a) What physical state is represented by the particles in each box of Figure 1.13, and which change of state is represented? (b) Which change of state would be represented if the arrow pointed in the opposite direction?

(Answers to Practice Exercises are in the back of the book.)

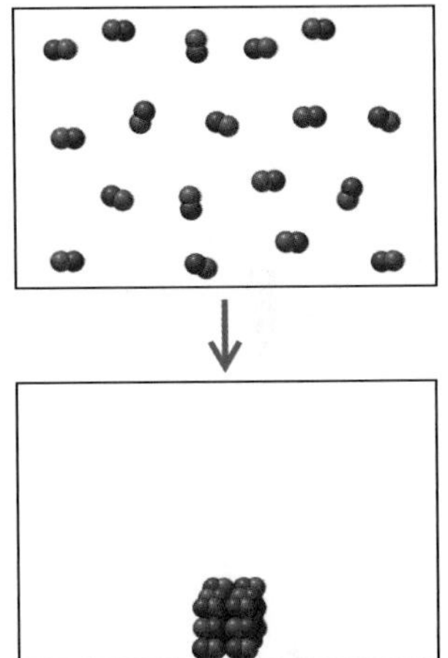

FIGURE 1.13

1.5 Forms of Energy

In Section 1.4 we noted that phase changes are accompanied by the absorption or release of heat. These heat transfers either increase or decrease the energy content of the gas, liquid, or solid undergoing the phase change. Thus, the upward-pointing arrows in Figure 1.11 represent increases in the energy of the molecules of water in ice as it melts or sublimes, or in liquid water as it vaporizes. The downward-pointing arrows represent decreases in the energy of water vapor as it condenses to liquid water or is deposited as ice, or as liquid water freezes. In this section we examine several different forms of energy and how and why they change during physical and chemical changes.

Let's begin by defining what we mean by heat and energy. In the physical sciences, **energy** is the capacity to do work, and **work** (w) is the exertion of a force (F) through a distance (d):

$$w = F \times d \tag{1.2}$$

FIGURE 1.14 Olympic sprinters converting chemical energy into kinetic energy.

For example, the sprinters in Figure 1.14 are rapidly expending energy and doing work as their muscles move their bodies through a distance of 100 m in about 11 s. They were not moving at all in their starting blocks, so the runners initially had to expend energy to accelerate to top speeds exceeding 10 m/s.

The energy inside sprinters' bodies (and yours) is derived from chemical reactions fueled by glucose (blood sugar). The chemical energy stored in glucose is an example of **potential energy (PE)**, which is the energy stored in an object because of its position or composition. When the energy in glucose is released during vigorous exercise, such as running the 100-m dash, some of it is transformed into **kinetic energy (KE)**, which is the energy of motion. The amount of kinetic energy in a moving object is the product of the mass (m) of the object and its speed (u):

$$\text{KE} = \frac{1}{2}mu^2 \tag{1.3}$$

Equation 1.3 confirms that a heavy object has more kinetic energy than a lighter one moving at the same speed. Similarly, two objects having the same mass but traveling at different speeds have different kinetic energies. If one of them is moving twice as fast as the other, then its kinetic energy is 2^2 or four times the kinetic energy of the other.

Other portions of the chemical energy released in a sprinter's body do other types of work, such as pumping blood through the circulation system. Some of it is also given off as heat. **Heat** is the transfer of energy that takes place because of a difference in temperatures. Heat is spontaneously transferred from a warm object, such as the skin of a sprinter, to a cooler one, such as the sprinter's surroundings. The dispersion of the energy released by biochemical processes in the body—some of it from doing work of various kinds and some of it released as heat—happens in accordance with the **law of conservation of energy**, which states that energy cannot be created or destroyed, but it can be converted from one form to another.

The particles that make up all forms of matter also have kinetic energy that depends on their temperature and physical state. For example, the tiny molecules of oxygen in the air you are breathing are moving around at supersonic speeds. The microscopic views of the three states of water in Figure 1.10 show that the water molecules in ice are locked in place and have limited kinetic energy; they have more when they can tumble over each other in liquid water, and they have a lot more when they are free to move as individual gas molecules in water vapor.

CONCEPT **TEST**

If the speed of a vehicle increases by 22%, by what factor does its kinetic energy increase?

(Answers to Concept Tests are in the back of the book.)

FIGURE 1.15 Molecular views of (a) carbon dioxide gas and (b) a mixture of three gases: hydrogen (pairs of white spheres), oxygen (pairs of red spheres), and nitrogen (pairs of blue spheres).

1.6 Formulas and Models

As we noted in Section 1.4, a molecule is a collection of atoms held together in a characteristic pattern and proportion by chemical bonds. Figure 1.10 provided several molecular views of water molecules. Figure 1.15(a) depicts molecules of carbon dioxide, another common molecular compound. Each is composed of a central carbon atom, represented by a black sphere, which is bonded to two atoms of oxygen, the red spheres. (The atomic color palette inside the back cover of this book shows the standard colors used to represent atoms of the elements we study most often.) Some pure elements also exist as molecules. The molecular views of hydrogen, nitrogen, and oxygen gas in Figure 1.15(b) show that these elements exist as *diatomic* (two-atom) molecules: H_2, N_2, and O_2, respectively. Fluorine, chlorine, bromine, and iodine also form diatomic molecules: F_2, Cl_2, Br_2, and I_2, respectively.

The chemical formulas of molecular compounds are also called *molecular* formulas. The element symbols and subscripts in a **molecular formula** indicate how many atoms of each element are present in one molecule of the compound. The molecular formulas of acetone, a common solvent, and acetic acid, the ingredient that gives vinegar its distinctive aroma and taste, are given in Figure 1.16(a). These two formulas provide information about the number of atoms of carbon, hydrogen, and oxygen in a molecule of each compound, but they do not tell us how the atoms are bonded together, nor do they tell us anything about the shape of the molecules. One way to show how they are connected is to draw a **structural formula** (Figure 1.16b), which includes the chemical bonds between atoms.

Sometimes the structures of molecules are represented using *condensed* structural formulas (Figure 1.16c) in which the symbols of elements appear in a pattern that shows how the atoms are arranged relative to one another. These structures may also omit common structural components, such as C—H bonds or C atoms bonded to other C atoms. For example, the structural formula of acetic acid shows that three of the four hydrogen atoms are bonded to the same carbon atom. The same information is conveyed in the condensed structural formula by grouping three H atoms before the first C atom in the formula (H_3C–).

Ball-and-stick models (Figure 1.16d) provide three-dimensional representations of molecules. They use balls to represent atoms and sticks to represent chemical bonds. Although ball-and-stick models accurately show the angles between

potential energy (PE) the energy stored in an object because of its position or composition.

kinetic energy (KE) the energy of an object in motion due to its mass (*m*) and its speed (*u*).

heat the transfer of energy from one object or place to another due to differences in the temperatures of the objects or places.

law of conservation of energy the principle that energy cannot be created or destroyed but can be changed from one form to another.

molecular formula a chemical formula that shows how many atoms of each element are in one molecule of a pure substance.

structural formula a representation of a molecule that uses short lines between the symbols of elements to show chemical bonds between atoms.

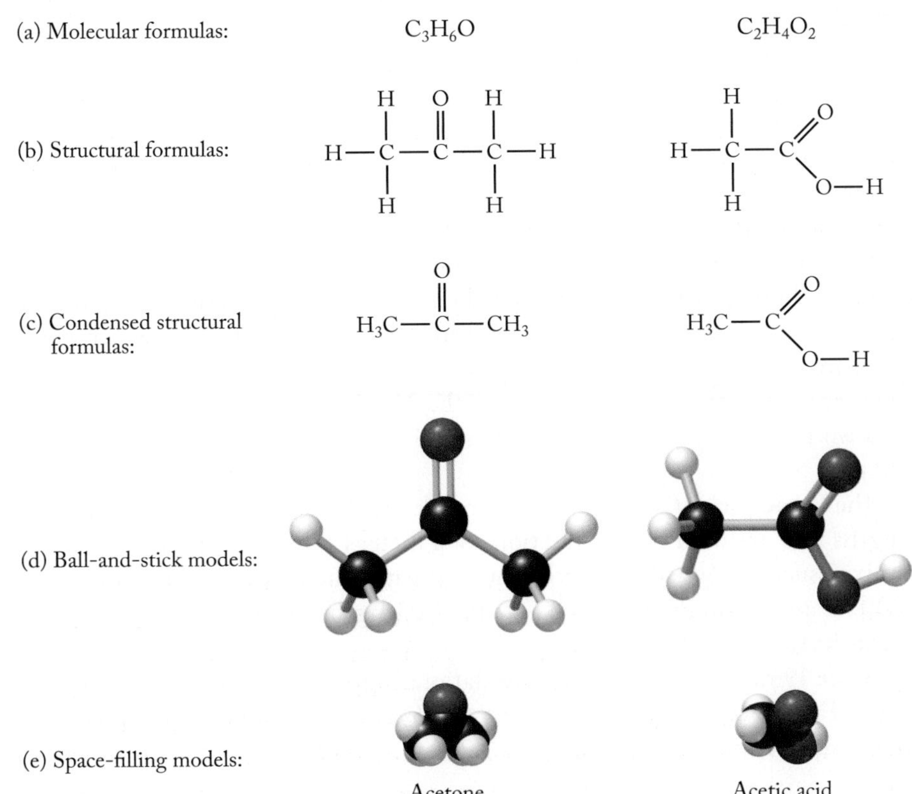

FIGURE 1.16 Five ways to represent the arrangement of atoms in molecules of acetone and acetic acid: (a) molecular formulas; (b) structural formulas; (c) condensed structural formulas; (d) ball-and-stick models, where white spheres represent hydrogen atoms, black spheres represent carbon atoms, and red spheres represent oxygen atoms; (e) space-filling models.

FIGURE 1.17 Space-filling model of methanol.

the bonds in a molecule, the sticks make the atoms seem far apart when they actually overlap each other. *Space-filling* models (Figure 1.16e) more accurately show how the atoms are arranged in a molecule and its overall three-dimensional shape, but it is sometimes hard to see all the atoms and the angles between the bonds, especially in molecules with many atoms.

CONCEPT TEST

Figure 1.17 shows the space-filling model of methanol (also called methyl alcohol and wood alcohol). What is its molecular formula?

(Answers to Concept Tests are in the back of the book.)

FIGURE 1.18 Crystals of sodium chloride consist of ordered three-dimensional arrays of Na^+ and Cl^- ions.

Not all compounds are molecular. Some consist of positively and negatively charged ions that are attracted to one another because they have opposite electrical charges. (Particles with the same charge repel each other.) One of the most common **ionic compounds**, and the principle component of table salt, is sodium chloride (NaCl). It consists of ordered three-dimensional arrays (Figure 1.18), or crystals, of sodium (Na^+) ions and chloride (Cl^-) ions, where the superscripts in their symbols indicate the electrical charge on the ions. The lack of subscripts in its chemical formula tells us that the ratio of sodium ions to chloride ions in a crystal of NaCl is 1:1. There are no "molecules" of NaCl, so we never refer to NaCl as a molecular formula. Rather, NaCl represents the simplest whole-number ratio of the ions in its three-dimensional structure. This ratio is referred to as the **empirical formula** of a compound. The term *empirical* is used in science to describe information that is obtained through experimentation. In this case, if we were to analyze the chemical composition of sodium chloride, we would discover that the ratio of sodium ions to chloride ions is 1:1, which would lead to the conclusion that the empirical formula of the compound is NaCl.

1.7 Expressing Experimental Results

Advances in scientific inquiry in the late 18th century, including those which led to the atomic theory of matter, created a heightened awareness of the need for accurate measurements and an international system of units for expressing the results of those measurements. In 1791 French scientists proposed a standard unit of length, which they called the **meter (m)** after the Greek *metron,* which means "measure." They based the length of the meter on 1/10,000,000 of the distance along an imaginary line running from the North Pole to the equator. By 1794 hard work by teams of surveyors had established the length of the meter that is still used today.

The French scientists also settled on a decimal-based system for designating lengths that are multiples or fractions of a meter (Table 1.1). They chose Greek prefixes such as *kilo-* for lengths much greater than 1 meter (1 kilometer = 1000 m) and Latin prefixes such as *centi-* for lengths much smaller than a meter (1 centimeter = 0.01 m).

Since 1960 scientists have by international agreement used a modern version of the French *metric* system of units: the *Système International d'Unités*, commonly referred to as **SI units**. Table 1.2 lists six SI base units; many others are

ionic compound a compound that consists of a characteristic ratio of positive and negative ions.

empirical formula a chemical formula in which the subscripts represent the simplest whole-number ratio of the atoms or ions in a compound.

meter (m) the standard unit of length, equivalent to 39.37 inches.

SI units a set of base and derived units used worldwide to express distances and quantities of matter and energy.

joule (J) the SI unit of energy, equivalent to 1 kg · $(m/s)^2$.

TABLE 1.1 Commonly Used Prefixes for SI Units

PREFIX		VALUE		
Name	Symbol	Numerical	Exponential	Example
zetta	Z	1,000,000,000,000,000,000,000	10^{21}	$1\ \text{Zm} = 10^{21}\text{m}$
exa	E	1,000,000,000,000,000,000	10^{18}	$1\ \text{Em} = 10^{18}\text{m}$
peta	P	1,000,000,000,000,000	10^{15}	$1\ \text{Pm} = 10^{15}\text{m}$
tera	T	1,000,000,000,000	10^{12}	$1\ \text{Tm} = 10^{12}\text{m}$
giga	G	1,000,000,000	10^{9}	$1\ \text{Gm} = 10^{9}\text{m}$
mega	M	1,000,000	10^{6}	$1\ \text{Mm} = 10^{6}\text{m}$
kilo	k	1000	10^{3}	$1\ \text{km} = 10^{3}\text{m}$
hecto	h	100	10^{2}	$1\ \text{hm} = 10^{2}\text{m}$
deka	da	10	10^{1}	$1\ \text{dam} = 10\ \text{m}$
deci	d	0.1	10^{-1}	$1\ \text{dm} = 10^{-1}\text{m}$
centi	c	0.01	10^{-2}	$1\ \text{cm} = 10^{-2}\text{m}$
milli	m	0.001	10^{-3}	$1\ \text{mm} = 10^{-3}\text{m}$
micro	μ	0.000001	10^{-6}	$1\ \mu\text{m} = 10^{-6}\text{m}$
nano	n	0.000000001	10^{-9}	$1\ \text{nm} = 10^{-9}\text{m}$
pico	p	0.000000000001	10^{-12}	$1\ \text{pm} = 10^{-12}\text{m}$
femto	f	0.000000000000001	10^{-15}	$1\ \text{fm} = 10^{-15}\text{m}$
atto	a	0.000000000000000001	10^{-18}	$1\ \text{am} = 10^{-18}\text{m}$
zepto	z	0.000000000000000000001	10^{-21}	$1\ \text{zm} = 10^{-21}\text{m}$

derived from them. For example, a common SI unit for volume, the cubic meter (m^3), is derived from the base unit for length, the meter. A common SI unit for speed, meters per second (m/s), is derived from the base units for length and time. The SI unit for energy is the **joule (J)**, which is equivalent to $1\ \text{kg} \cdot (\text{m/s})^2$. This is consistent with the relationship introduced in Equation 1.3 (Section 1.5) between kinetic energy and mass and speed: $\text{KE} = \frac{1}{2}mu^2$. When mass ($m$) is expressed in kilograms and speed (u) is in meters per second, then the units of KE are $\text{kg} \cdot (\text{m/s})^2$, or joules.

Table 1.3 lists some of the SI units and their equivalents in the U.S. customary system of units. They include the volume corresponding to 1 cubic decimeter (a cube 1/10 meter on a side), which we call a liter (L). Since there are exactly 10 decimeters in a meter, there are exactly $10^3\ \text{dm}^3$, or 10^3 L, in 1 cubic meter ($1\ \text{m}^3$). Another important exact unit equivalence in Table 1.3 is 2.54 cm = 1 in.

The temperature (T) scales in Table 1.3 also deserve our attention because temperature has been the most frequently measured quantity since the first thermometers were developed nearly 500 years ago. Modern thermometers based on the thermal expansion of liquids such as mercury and alcohol were introduced in the early 18th century by German scientist (and glassblower) Daniel Gabriel Fahrenheit (1686–1736). He later developed a temperature scale with a zero point corresponding to the freezing point of a concentrated salt solution and an upper value (100°) corresponding to the average internal temperature of the human body. The Fahrenheit scale, still widely used in the United States, is a slightly modified version of the one based on those two reference temperatures. Later

TABLE 1.2 Six SI Base Units

Quantity or Dimension	Unit Name	Unit Abbreviation
Mass	kilogram	kg
Length	meter	m
Temperature	kelvin	K
Time	second	s
Electric current	ampere	A
Quantity of a substance	mole	mol

absolute zero (0 K) zero point on Kelvin temperature scale; theoretically the lowest temperature possible.

precision the extent to which repeated measurements of the same variable agree.

TABLE 1.3 Conversion Factors for SI and Other Commonly Used Units

Quantity or Dimension	Equivalent Units
Mass (*m*)	1 kg = 2.205 pounds (lb) = 35.27 ounces (oz)
Length (distance)	1 m = 39.37 inches (in) = 3.281 feet (ft) = 1.094 yards (yd) 1 km = 0.6214 mile (mi) 1 in = 2.54 cm (exactly)
Volume (*V*)	1 m^3 = 35.31 ft^3 = 1000 liters (L) (exactly) 1 L = 0.2642 gallon (gal) = 1.057 quarts (qt)
Temperature (*T*)	*T*(K) = *T*(°C) + 273.15 *T*(°C) = (5/9)[*T*(°F) − 32]

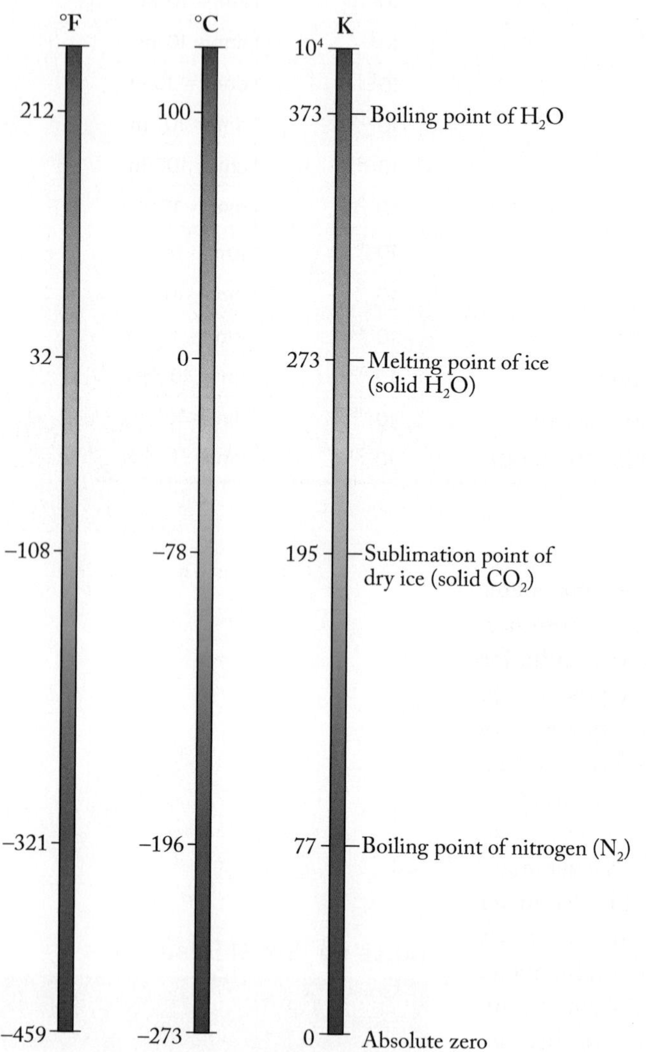

FIGURE 1.19 Three temperature scales are commonly used today, although the Fahrenheit scale is rarely used in scientific work.

in the 18th century, Swedish astronomer Anders Celsius (1701–1744) proposed an alternative temperature scale based on the freezing point (0°C) and average boiling point at sea level (100°C) of pure water. In 1848 British scientist William Thomson (later Lord Kelvin, 1824–1907) proposed another temperature scale that was not based on the physical properties of any substance but rather on the notion that there is a lower limit to temperature, called **absolute zero (0 K)**. That temperature is the zero point of the Kelvin scale named in his honor. Figure 1.19 compares these three temperature scales.

To conduct science today, the length of the meter and the dimensions of the other SI units must be known or defined by quantities that are much more constant and precisely known than, for example, the distance from the North Pole to the equator. Two such quantities are the speed of light (*c*) and time. In 1983, 1 m was redefined as the distance traveled in 1/299,792,458 of a second in a vacuum by the light emitted from a helium–neon laser. This modern definition of the meter is consistent with the one adopted in France in 1794.

All scientific measurements are limited in how well we can know the results. Nobody is perfect, and no analytical method is perfect either. Therefore, measured quantities and values derived from measured quantities always have within them some degree of uncertainty. On the other hand, some quantities are known exactly because they can be determined by counting, such as the number of steps in a ladder or the number of eggs in a carton. There is no uncertainty in such a value.

SAMPLE EXERCISE 1.5 Distinguishing Exact from Uncertain Values **LO7**

Which of the following quantities associated with the Washington Monument in Washington, DC (Figure 1.20), are exact values and which are inexact?

a. The monument is made of 36,941 white marble blocks.
b. The monument is 169 m tall.
c. There are 893 stair steps to the top.

d. The mass of the aluminum apex is 2.8 kg.
e. The area of the foundation is 1487 m^2.

Collect and Organize We must distinguish between values based on an exact number, such as 12 eggs in a dozen, from those that are not exact numbers, such as the mass of an egg.

Analyze One way to distinguish exact from inexact values is to answer the question, "Which values represent quantities that can be counted?" Exact values can be counted.

Solve The number of marble blocks (a) and the number of stairs (c) are quantities we can count, so they are exact numbers. The other three quantities are based on measurements of length (b), mass (d), and area (e), so they are inexact.

Think About It The "you can count them" property of exact numbers was used in this exercise, which assumes that there is no uncertainty in a counted value. Can you think of a type of counting whose certainty is sometimes challenged?

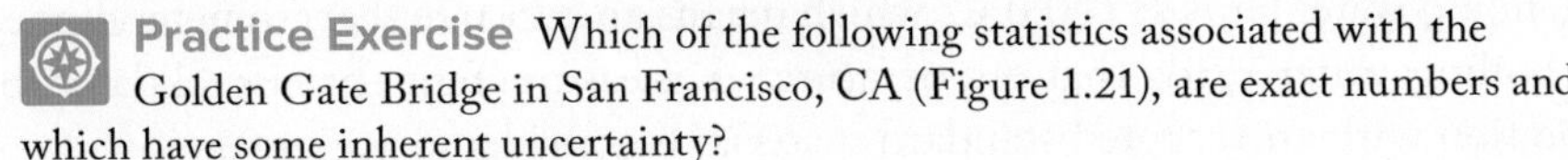

Practice Exercise Which of the following statistics associated with the Golden Gate Bridge in San Francisco, CA (Figure 1.21), are exact numbers and which have some inherent uncertainty?

a. The roadway is six lanes wide.
b. The width of the bridge is 27.4 m.
c. The bridge has a mass of 381 million kg.
d. The length of the bridge is 2740 m.
e. The "FasTrack" toll for a car traveling south is $6.25 (as of July 1, 2015).

(Answers to Practice Exercises are in the back of the book.)

FIGURE 1.20 The Washington Monument.

FIGURE 1.21 The Golden Gate Bridge.

Precision and Accuracy

Two terms—*precision* and *accuracy*—are used to describe how well we know a measured quantity or a value calculated from a measured quantity. **Precision** indicates how repeatable a measurement is. Suppose we used the balance on the top in Figure 1.22 to determine the mass of a penny over and over again, and suppose the reading on the balance was always 2.53 g. These results tell us that the mass of the penny is precisely 2.53 g. We can also say that we are certain of its mass to the nearest 0.01 g.

Now suppose we used the balance on the bottom in Figure 1.22 to determine the mass of the same penny five more times and obtained the following values:

Measurement	Mass (g)
1	2.5270
2	2.5271
3	2.5272
4	2.5271
5	2.5271

The small variability in the last decimal place is not unusual when using a balance that can report masses to the nearest 0.0001 g. These results are quite consistent with one another, so we can say that the balance is precise. Many other factors—particles of dust landing on the balance, vibrations of the laboratory bench, or the transfer of moisture from our fingers to the penny as we handled it—could have produced a change in mass of 0.0001 g or more.

One way to express the precision of these results is to cite the range between the highest and lowest values—in this example, 2.5270 to 2.5272. Range can also

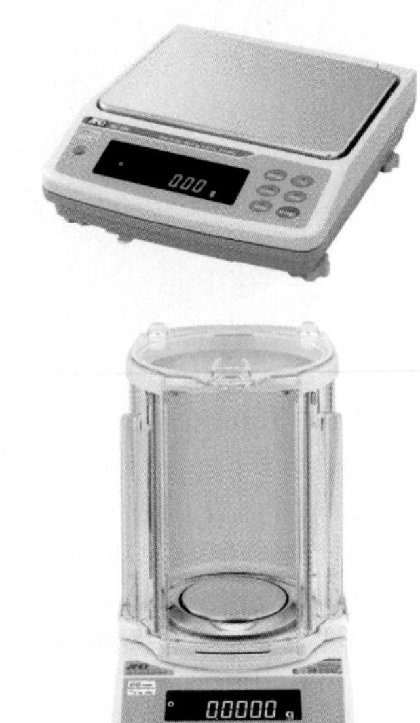

FIGURE 1.22 The mass of a penny can be measured to the nearest 0.01 g with the balance on the top and to the nearest 0.0001 g with the balance on the bottom.

(a)

(b)

(c)

FIGURE 1.23 (a) Three dart throws meant to hit the center of the target are both accurate and precise. (b) Three throws meant to hit the center of the target are precise but not accurate. (c) This set of three throws is neither precise nor accurate.

accuracy agreement between an experimental value and the true value.

significant figures all the certain digits in a measured value plus one estimated digit. The greater the number of significant figures, the greater the certainty with which the value is known.

be expressed using the average value (2.5271 g) and the range above (0.0001) and below (0.0001) the average that includes all the observed results. A convenient way to express the observed range in this case is 2.5271 ± 0.0001, where the symbol ± means "plus or minus" the value that follows it.

While precision relates to the agreement among repeated measurements, **accuracy** reflects how close the measured value is to the true value. Suppose the true mass of our penny is 2.5267 g. That means the average result obtained with the bottom balance in Figure 1.22—2.5271 g—is 0.0004 g too high. Thus, the measurements made on this balance may be precise to within 0.0001 g, but they are not accurate to within 0.0001 g of the true value. A way to visualize the difference between accuracy and precision is presented in Figure 1.23.

How can we be sure that the results of measurements are accurate? The accuracy of a balance can be checked by weighing objects of known mass. A thermometer can be calibrated by measuring the temperature at which a substance changes state. For example, an ice-water bath should have a temperature of 0.0°C. At sea level, liquid water boils at 100.0°C, which means an accurate thermometer dipped into boiling water reads that temperature. A measurement that is validated by calibration with an accepted standard material is considered accurate.

Significant Figures

Let's now learn a way to express how well we know the results of measurements or calculated values derived from measurements. Our approach makes use of **significant figures**. The number of significant figures used to report a value indicates how certain we are of the value. For example, if the two balances shown in Figure 1.22 are working properly, then we can use the one on the top to weigh a penny to the nearest 0.01 g, obtaining a mass value of 2.53 g. The balance on the bottom can measure the mass of the same penny to the nearest 0.0001 g, or 2.5271 g. The mass obtained with the top balance has three significant figures: the 2, 5, and 3 are considered *significant* because we are confident in their values. The mass obtained using the bottom balance has five significant figures (2, 5, 2, 7, and 1). The mass of the penny can be determined with greater certainty with the balance on the bottom.

Now suppose an aspirin tablet is placed on the balance on the bottom in Figure 1.22, and the display reads 0.0810 g. How many significant figures are there in this value? You might be tempted to say five because five digits are displayed. However, the first two zeros are *not* considered significant because they serve only to determine the location of the decimal point. Those two zeros function the way exponents do when we express values using scientific notation. Expressing 0.0810 g using scientific notation means moving the decimal point two places to the right so that it is between the first and second nonzero digits and then adding the exponent 10^{-2} to convey how many places the decimal was moved (two) and that it moved to the right (the negative sign): 8.10×10^{-2} g. Only the three digits in the decimal part indicate how precisely we know the value; the "−2" in the exponent does not. (See Appendix 1 for a review of how to express values using scientific notation.)

Why is the rightmost zero in 0.0810 g significant? The balance, if operating correctly, can measure masses to the nearest 0.0001 g, so we may assume that the last digit is significant. If we dropped the zero and recorded a value of only 0.081 g, we would be implying that we only knew the value to the nearest 0.001 g, which is not the case. The following guidelines will help you determine which zeros (highlighted in green) are significant and which are not:

1. Zeros at the beginning of a value, as in 0.0592, are never significant: they just set the decimal place.

2. Zeros after a decimal point and after a nonzero digit, as in 3.00×10^8, are always significant.
3. Zeros at the end of a value that contains no decimal point, as in 96,500, may or may not be significant.[2] They may be there only to set the decimal place. We can use scientific notation to indicate whether or not these terminal zeros are significant. If they serve only to set the decimal point, then the value in scientific notation is 9.65×10^4, where the positive 4 exponent means the decimal point was moved four places to the left from its original location.
4. Zeros between nonzero digits, as in 101.3, are always significant.

CHEMTOUR Significant Figures

CHEMTOUR Scientific Notation

CONCEPT TEST

How many significant figures are there in the values used as examples in guidelines 1, 2, and 4?

(Answers to Concept Tests are in the back of the book.)

Significant Figures in Calculations

Now let's consider how significant figures are used to express the results of calculations involving measured quantities. The significant figure rules should be used only at *the end of a calculation*, never on intermediate results. This policy helps avoid errors that may happen when we round off intermediate values.

Rounding off means that we drop the *insignificant digits* (all digits to the right of the last significant digit) and then either increase the value of the last significant digit by 1 or leave it unchanged, depending on the value of the first insignificant digit. For example, if we needed to round 76.4523 to three significant figures, then 5 is the first of the insignificant digits that must be dropped. If this first insignificant digit had been greater than 5, we would have rounded up (thus, 76.46 would be rounded to 76.5). If it had been less than 5, we would have rounded down (thus, 76.44 would be rounded to 76.4). Because it is 5, we check whether there are any nonzero digits to the right of it. When there are (as in this case), we round up. Therefore, 76.4521 is rounded to 76.5.

Had there been no nonzero digits to the right of the 5 (for example, in 76.45 or 76.450), then we would have needed a tie-breaking rule for rounding down or up. A good rule to follow is to round to the nearest even number. Thus, rounding 76.45 (or 76.450) to three significant figures makes both values 76.4, because the 4 in the tenths place is the nearest even number. However, 76.55 (or 76.550) would be rounded to 76.6, because the 6 in the tenths place is the nearest even number.

To see how significant figures impact calculations involving measured values, let's return to the density calculation from Sample Exercise 1.1. Recall that we determined the density of a gold nugget using the volume of water it displaced, which was the difference between an initial value of 58.0 mL and a final value of 58.5 mL. Taking the difference of the two values, we found that the volume of our nugget was

$$\begin{array}{r} 58.5\ \text{mL} \\ -50.0\ \text{mL} \\ \hline 8.5\ \text{mL} \end{array}$$

[2]Some books add decimal points after terminal zeros (e.g., 1000.) to indicate that the zeros are significant. We do not follow this practice, in part because it does not work for values in which only some of the terminal zeros are significant.

weak-link rule the rule that the result of a calculation is known only as well as the least well-known value used in the calculation.

The difference has only two significant figures, even though the initial and final values had three, because the initial and final volumes are only known to the nearest tenth of a milliliter. Therefore, we can only know the difference between them to the nearest tenth of a milliliter. In general, *when measured numbers are added or subtracted, the result has the same number of digits to the right of the decimal as the measured number with the fewest digits to the right of the decimal.* This rule is an example of a more general one that applies to all calculations involving uncertain values: *we can know the result of a calculation only as well as we know the least well-known value that went into the calculation.* This is called the **weak-link rule**. It applies to adding and subtracting uncertain values, as just described, but it's applied differently for multiplication and division, as we see next by recalculating the density value from Sample Exercise 1.1.

Recall that the density of the metal sample was calculated by dividing its mass (164 g) by the difference between two volume values:

$$d = \frac{m}{V} = \frac{164\ \text{g}}{(58.5 - 50.0)\ \text{mL}} = \frac{164\ \text{g}}{8.5\ \text{mL}} = 19.3\ \frac{\text{g}}{\text{mL}}$$

As noted previously, we know the value of the denominator to only two significant figures. The weak-link rule applied to division states that the significant figures in a quotient can be no greater than the number of significant figures in the value used in the calculation with the lesser number of significant figures. In this example, the numerator has three, but the denominator has only two, so the density value should have only two, which means rounding off 19.3 to 19 g/mL. In general, *when measured numbers are multiplied or divided, the result has the same number of significant figures as the value used to calculate it with the smallest number of significant figures.*

SAMPLE EXERCISE 1.6 Using Significant Figures in Calculations LO7

Suppose we add a penny with a mass of 2.5271 g to 49 other pennies with a combined mass of 124.01 g. What is the combined mass of the 50 pennies?

Collect and Organize We are asked to calculate the combined mass of 49 pennies that were weighed to the nearest 0.01 g and a single penny that was weighed to the nearest 0.0001 g. To express the results of this calculation, we must follow the weak-link rule: we can know a combination of measured values only as well as we know the least well-known measured value.

Analyze We are summing two values in this example, so the weak-link value is the one with the fewer digits to the right of its decimal point, 124.01 g, which is the combined mass of the 49 pennies.

Solve Adding the mass of the 50th penny to the mass of the other 49, we have

$$\begin{array}{r} 124.01\ \text{g} \\ +\ \ 2.5271\ \text{g} \\ \hline 126.5371\ \text{g} \end{array} = 126.54\ \text{g}$$

Think About It We can know the value of the sum to only the nearest 0.01 g, so we round 126.5371 to 126.54. We round the "3" up to 4 because the first digit to be dropped is the "7" highlighted in red.

Practice Exercise According to the rules of golf, a golf ball cannot weigh more than 45.97 g and it must be at least 4.267 cm in diameter. The volume of a sphere is $(4/3)\pi r^3$, where r is the radius (half the diameter).

a. What is the maximum density of a golf ball, expressed in g/cm^3 to the appropriate number of significant figures?
b. Is such a golf ball more dense or less dense than water? (You may need to look up the density of water—unless you are a golfer, in which case you probably already know the answer.)

(Answers to Practice Exercises are in the back of the book.)

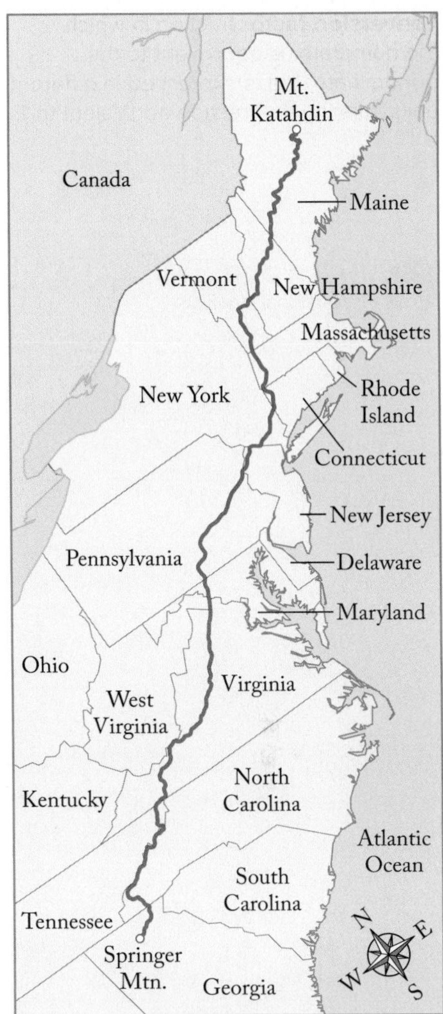

FIGURE 1.24 Map of the Appalachian Trail.

CONCEPT TEST

In the summer of 2015 American ultramarathoner Scott Jurek set a record by hiking the Appalachian Trail (Figure 1.24) from its southern terminus at Springer Mountain, Georgia, to its northern terminus on the peak of Mt. Katahdin, Maine, in 46 days, 8 hours, and 7 minutes. According to the Appalachian Trail Conservancy, the trail is 2175 miles long.

a. What was Mr. Jurek's average speed in miles/day?
b. Which do you think is the weak link in calculating his average speed: the actual distance hiked or the time that it took?

(Answers to Concept Tests are in the back of the book.)

1.8 Unit Conversions and Dimensional Analysis

On July 23, 1983, Air Canada Flight 143, a Boeing 767 that had been in service only a few months, was near the halfway point on a flight from Montreal, Quebec, to Edmonton, Alberta, when it ran out of fuel. Without power, the plane became a very large, very heavy glider. Fortunately, the pilot of Flight 143 was also an experienced glider pilot, and he was able to make a successful emergency landing at an abandoned airbase in Gimli, Manitoba. There were no serious injuries, even though the nose gear collapsed during the landing (Figure 1.25). The plane was repaired and flown out of Gimli and back into service, where it remained until its retirement in 2008. During its 25 years of operation, the plane became widely known in aviation circles as the *Gimli Glider*.

Why did Flight 143 run out of fuel? There were several reasons, including faulty communications between ground and flight crews about a malfunctioning fuel gauge. However, the most immediate reason was an error in calculating how much fuel the plane needed to fly from Montreal to Edmonton. The error occurred at a time when Canada and its national airline were in the process of converting from a unit system called the *Imperial System*, in which masses are expressed in ounces and pounds (as in the United States today), to the SI system, in which the base unit of mass is the kilogram. The Boeing 767 that landed in Gimli was among the first in the Air Canada fleet to use the SI system, so its fuel capacity and consumption were expressed in kilograms. The plane was supposed to take off from Montreal with 22,300 kg of fuel in its tanks, but it left with less than half that. Let's explore why.

To make sure the plane had enough fuel, but not too much (to avoid transporting excess weight), the refueling crew in Montreal measured the volume of fuel

FIGURE 1.25 Air Canada Flight 143 after landing safely in Gimli, Manitoba.

conversion factor fraction in which the numerator is equivalent to the denominator but is expressed in different units, making the fraction equivalent to 1.

CHEMTOUR Dimensional Analysis

already in its tanks and found that they contained 7700 L. Next, they converted this volume of jet fuel into an equivalent mass of fuel because they knew the plane needed 22,300 kg of fuel to fly to Edmonton. To convert volume in liters to mass in kilograms, they should have multiplied 7700 L by the density of jet fuel, which is 0.80 kg/L:

$$7700 \cancel{\text{L}} \times \frac{0.80 \text{ kg}}{1 \cancel{\text{L}}} = 6200 \text{ kg}$$

Note how the volume units cancel out and we obtain 6200 kg for the mass of 7700 L of fuel. In this calculation density is used as a **conversion factor**, which is a ratio of equivalent quantities expressed using the initial and desired final units of measure. This equivalency means that multiplying a quantity by a conversion factor is like multiplying by 1: the quantity does not change, but its units do. Equation 1.4 illustrates how any conversion factor (highlighted in blue) can be used to convert the initial units on a quantity to the desired units:

$$\cancel{\text{Initial units}} \times \frac{\text{desired units}}{\cancel{\text{initial units}}} = \text{desired units} \tag{1.4}$$

The Montreal refueling crew next had to calculate the mass of jet fuel to be added by subtracting the mass of fuel onboard from that needed for the flight:

$$\begin{array}{r} 22{,}300 \text{ kg needed} \\ -6200 \text{ kg onboard} \\ \hline 16{,}100 \text{ kg to be added} \end{array}$$

Then they had to calculate *the volume* of jet fuel to be added because jet fuel is pumped into airplanes the way gasoline is pumped into automobiles: with a metering system based on volume. The refueling crew calculated the volume of jet fuel in liters corresponding to the mass of fuel needed. In this calculation the reciprocal of the density serves as the conversion factor so that mass units cancel out:

$$16{,}100 \cancel{\text{kg}} \times \frac{1 \text{ L}}{0.80 \cancel{\text{kg}}} = 20{,}100 \text{ L} \approx 20{,}000 \text{ L}$$

Had this volume of fuel been added, Flight 143 would have made it to Edmonton. Unfortunately, the refueling crew added only 4900 L of fuel! Why? Because they used the wrong density value to convert volume to mass and again when they converted mass to volume. The value they used was 1.77 instead of 0.80. They had used 1.77 for many years because it is the density of jet fuel expressed in pounds per liter, not kilograms per liter. Using the wrong conversion factor meant they overestimated the mass of fuel already onboard the airplane (they multiplied by 1.77 instead of 0.80), and then they underestimated how much fuel to add by using the reciprocal of 1.77 instead of 0.80 to calculate the volume of fuel to pump into the plane's tanks. As a result, Flight 143 left Montreal with less than half as much fuel as it should have carried.

The flight of the *Gimli Glider* is not the only disaster (or near disaster) caused by mixing up units. In 1999 the *Mars Climate Orbiter* space probe crashed on the surface of Mars instead of orbiting it because the programmers who wrote the software that controlled the spacecraft used one set of units to express the thrust of its rocket engines and the engineers who designed the engines used another. These examples demonstrate that we must take great care to use the correct units when expressing the quantities of substances. Our goal here in Section 1.8 is to help you become more skilled at converting measured and calculated values from one set of units to another. You will have an opportunity to develop this skill in the sample and practice exercises that follow.

SAMPLE EXERCISE 1.7 Converting Units LO8

The temperature of interstellar space is 2.73 K. What is this temperature on the Celsius and Fahrenheit scales?

Collect and Organize We are asked to convert a temperature from Kelvin to degrees Celsius and degrees Fahrenheit. Table 1.3 contains equations that relate Kelvin and Celsius temperatures, and Celsius and Fahrenheit temperatures.

Analyze We will first convert 2.73 K into an equivalent Celsius temperature and then calculate an equivalent Fahrenheit temperature. We can estimate that the value in degrees Celsius should be close to absolute zero (about −273°C). Because a Fahrenheit degree is about half the size of a Celsius degree, the temperature on the Fahrenheit scale should be a little less than twice the absolute value of the temperature on the Celsius scale (or around −500°F).

Solve
A. Converting 2.73 K to degrees Celsius:

$$T(\text{K}) = T(^\circ\text{C}) + 273.15$$

or

$$T(^\circ\text{C}) = T(\text{K}) - 273.15$$
$$= 2.73 - 273.15 = -270.42^\circ\text{C}$$

Converting this Celsius temperature to Fahrenheit:

$$T(^\circ\text{C}) = (5/9)[T(^\circ\text{F}) - 32]$$

or

$$T(^\circ\text{F}) = (9/5)T(^\circ\text{C}) + 32$$
$$= 9/5(-270.42) + 32 = -454.76^\circ\text{F}$$

Think About It Converting Kelvin to Celsius temperatures is a matter of subtracting their zero points, which differ by 273.15. The sizes of the degrees on both scales are the same. However, the Celsius and Fahrenheit scales differ by both their zero points and the sizes of their degrees, so the math involved requires two steps. The calculated Celsius value of −270.42°C is only a few degrees above absolute zero, as we estimated. The Fahrenheit value is within 10% of our estimate, so it is reasonable, too.

Practice Exercise The temperature of the moon's surface varies from −233°C at night to 123°C during the day. What are these temperatures on the Kelvin and Fahrenheit scales?

(Answers to Practice Exercises are in the back of the book.)

SAMPLE EXERCISE 1.8 Calculating the Kinetic Energy of a Moving Object LO8

One of tennis star Serena Williams's serves was clocked at 207.0 km/h (128.6 mph) during the 2013 Australian Open. If the mass of the tennis ball she served was 58.0 g, what was its kinetic energy in joules?

Collect and Organize We know the mass (m) and speed (u) of a served tennis ball and are asked to calculate its kinetic energy (KE). These variables are related by Equation 1.3: $\text{KE} = \frac{1}{2}mu^2$. One joule of energy is equivalent to 1 kg · $(\text{m/s})^2$.

Analyze Before using Equation 1.3, we need to convert the speed given from 207.0 km/h to m/s and to convert the mass of the tennis ball given from 58.0 g to kg.

Solve The mass of the tennis ball: $58.0 \text{ g} \times \dfrac{1 \text{ kg}}{1000 \text{ g}} = 0.0580 \text{ kg}$

The speed of the ball:

$$\frac{207.0\ \cancel{\text{km}}}{\cancel{\text{h}}} \times \frac{1000\ \text{m}}{\cancel{\text{km}}} \times \frac{\cancel{\text{h}}}{60\ \cancel{\text{min}}} \times \frac{\cancel{\text{min}}}{60\ \text{s}} = 57.50\ \text{m/s}$$

The kinetic energy of the ball:

$$\text{KE} = \tfrac{1}{2}mu^2 = \tfrac{1}{2}(0.0580\ \text{kg})(57.50\ \text{m/s})^2 = 95.9\ \text{kg} \cdot (\text{m/s})^2 = 95.9\ \text{J}$$

Think About It The calculated value is expressed with three significant figures because there are three significant figures in the mass of the ball and four in its velocity. Therefore, the mass value is the weak link in this calculation. There are only two digits in the time conversion factors, but they are exact numbers and do not affect how well we know the final calculated value.

Practice Exercise On September 25, 2010, Aroldis Chapman of the Cincinnati Reds set a Major League Baseball record by throwing a pitch that was clocked at 105.1 mph. If the mass of the baseball he threw was 146 g, what was its kinetic energy in joules?

(Answers to Practice Exercises are in the back of the book.)

SAMPLE EXERCISE 1.9 Converting Customary U.S. Units and SI Units I — LO8

Natural gas is the most widely used fuel in the United States for generating electricity and for heating homes and hot water. It is also used in industry as a starting material (also called a "feedstock") for making fertilizer, fabrics, and plastics. In 2016 U.S. industries consumed natural gas at the rate of 22.5 billion cubic feet per day. Use the exact conversion factor, 1 in = 2.54 cm, to convert this rate to cubic meters of natural gas per second.

Collect and Organize We need to convert a rate expressed in billions of cubic feet (ft^3) per day into cubic meters (m^3) per second. One billion units = 10^9 units. Also 1 foot = 12 inches (exactly), and 1 m = 10^2 cm (exactly).

Analyze To use the foregoing distance equivalents, we will need to cube both sides of each equation to obtain volume equivalents. Doing so means cubing everything on both sides of each equation—that is, the coefficients as well as the units:

$$(1\ \text{ft})^3 = (12\ \text{in})^3 \text{ or } 1\ \text{ft}^3 = 1728\ \text{in}^3$$
$$\text{and } (1\ \text{in})^3 = (2.54\ \text{cm})^3 \text{ or } 1\ \text{in}^3 = 16.387\ \text{cm}^3$$
$$\text{and } (1\ \text{m})^3 = (10^2\ \text{cm})^3 \text{ or } 1\ \text{m}^3 = 10^6\ \text{cm}^3$$

One meter is a little longer than 1 yard or 3 feet; 1 cubic meter should be a bit more than 3^3 or about 30 cubic feet. There are 24 h × 60 min/h × 60 s/min or about 10^5 seconds in a day. Therefore, the calculated rate value should be about $(20 \times 10^9)/(30 \times 10^5)$ or about 7×10^3.

Solve Let's handle the volume and time conversions separately and then combine them:

$$22.5 \times 10^9\ \cancel{\text{ft}^3} \times \frac{1728\ \cancel{\text{in}^3}}{1\ \cancel{\text{ft}^3}} \times \frac{16.387\ \cancel{\text{cm}^3}}{1\ \cancel{\text{in}^3}} \times \frac{1\ \text{m}^3}{10^6\ \cancel{\text{cm}^3}} = 6.3713 \times 10^8\ \text{m}^3$$

$$1\ \cancel{\text{d}} \times \frac{24\ \cancel{\text{h}}}{1\ \cancel{\text{d}}} \times \frac{60\ \cancel{\text{min}}}{1\ \cancel{\text{h}}} \times \frac{60\ \text{s}}{1\ \cancel{\text{min}}} = 86{,}400\ \text{s}$$

Taking the quotient of the calculated volume and time values:

$$\text{rate} = \frac{6.3713 \times 10^8\ \text{m}^3}{86{,}400\ \text{s}} = 7.37 \times 10^3\ \text{m}^3/\text{s}$$

Think About It The calculated value is reassuringly close to our estimate. It is expressed with three significant figures because there are that many in the initial rate value. All of the conversion factors used are based on exact values and do not affect how well we know the calculated rate.

Practice Exercise Suppose it takes exactly one minute for the heart of an adult male at rest to recirculate all the blood (5.6 L) in his body. How rapidly does his heart pump his blood? Express your answer in cm^3/s and in^3/s.

(Answers to Practice Exercises are in the back of the book.)

SAMPLE EXERCISE 1.10 Converting Customary U.S. Units and SI Units II — LO8

A pediatrician prescribes the antibiotic amoxicillin to treat a 2-year-old child suffering from an ear infection. The therapeutic dose is 75 mg of the drug per kilogram of body mass of the patient per day. The drug is administered twice each day in the form of a flavored liquid that contains 125 mg of amoxicillin per milliliter. If the child weighs 28 lb, how many milliliters of the liquid should be administered in each dose?

Collect and Organize We need to calculate the volume of liquid amoxicillin in milliliters to be administered twice each day to provide a daily dose of 75 mg/kg patient. We know the patient's mass (28 lb) and the concentration of liquid amoxicillin (125 mg/mL). We know from Table 1.3 that 1 kg = 2.205 lb.

Analyze Let's start with the mass of the patient: 28 lb. Dosage is expressed in mg amoxicillin/kg patient, so we need to convert 28 lb to kilograms and then multiply that patient mass by the daily dosage in mg amoxicillin/kg patient to calculate the mg amoxicillin to be given each day. To convert this mass to an equivalent volume of the amoxicillin solution, we invert the concentration of the liquid from "mg amoxicillin/mL" to "mL/mg amoxicillin" so that the mass units cancel out. Finally, we need to divide the daily volume in two because the drug is administered twice a day. The resulting conversion factors are

$$\frac{1\ \text{kg}}{2.205\ \text{lb}} \quad \frac{75\ \text{mg}}{\text{kg} \cdot \text{day}} \quad \frac{1\ \text{mL}}{125\ \text{mg}} \quad \frac{1\ \text{day}}{2\ \text{doses}}$$

To estimate the result of our calculation, we note that a kilogram is a little more than two pounds, so a 28-lb child weighs a little less than 14 kg—say, 12 kg. The product of 12 kg × 75 mg amoxicillin/kg patient is about 1000 mg amoxicillin. Dividing this value by the concentration of the liquid (125 mg/mL) gives a volume of about 8 mL per day, or 4 mL twice a day.

Solve

$$28\ \cancel{\text{lb}} \times \frac{1\ \cancel{\text{kg}}}{2.205\ \cancel{\text{lb}}} \times \frac{75\ \cancel{\text{mg}}}{\cancel{\text{kg}} \cdot \cancel{\text{day}}} \times \frac{1\ \text{mL}}{125\ \cancel{\text{mg}}} \times \frac{1\ \cancel{\text{day}}}{2\ \text{doses}} = 3.8\ \frac{\text{mL}}{\text{dose}}$$

Think About It The answer nearly matches our estimate. We rounded the result to two significant figures because the child's weight and the therapeutic daily dose both have only two significant figures. The number of doses per day (two) is an exact number and has no uncertainty.

Practice Exercise A student planning a party has \$20 to spend on her favorite soft drink. It is on sale at store A for \$1.29 for a 2-L bottle (plus 10-cent deposit); at store B the price of a 12 pack of 12 fl oz cans is \$2.99 (plus a 5-cent deposit per can). At which store can she buy the most of her favorite soft drink for no more than \$20? (There are 29.57 mL in exactly 1 U.S. fl oz, which is a unit of volume, not mass.)

(Answers to Practice Exercises are in the back of the book.)

1.9 Assessing and Expressing Precision and Accuracy

Recall from Section 1.7 that experimental results are considered accurate when they have been validated by checking them with an accepted standard material of known composition. In many laboratories these accuracy checks are made by periodically analyzing *control samples*, which contain known quantities of the substance(s) of interest. Control samples are used in clinical chemistry labs to ensure the accuracy of routine analyses of blood serum and urine samples. One substance that is included in comprehensive blood tests is creatinine, a product of protein metabolism that doctors use to monitor patients' kidney function. Suppose that a control sample known to contain 0.681 mg creatinine per deciliter of sample is analyzed after every 10 patient samples, and that the results of five control sample analyses are 0.685, 0.676, 0.669, 0.688, and 0.692 mg/dL. How precise (repeatable) are these results, and do they mean that the analyses of the control sample (and those subsequently conducted on the real samples) are accurate?

A widely used method for expressing the precision of data such as these involves calculating the average of all the values and then calculating how much each value deviates from the average. To calculate the average, we sum the five control sample results and divide by the number of values (5):

$$\frac{(0.685 + 0.676 + 0.669 + 0.688 + 0.692)}{5}\ \text{mg/dL} = \frac{3.410}{5}\ \text{mg/dL} = 0.682\ \text{mg/dL}$$

We can write a general equation to represent this calculation for any number of measurements n of any parameter x. Individual results are identified by the generic symbol x_i, where i can be any integer from 1 to n, the number of measurements made. The capital Greek sigma (Σ) with an i subscript, $\Sigma_i(x_i)$, represents the *sum* of all the individual x_i values. This kind of average is called the **mean** or **arithmetic mean**, and it is represented by the symbol $\bar{x}$ in Equation 1.5:

$$\bar{x} = \frac{\Sigma_i(x_i)}{n} \qquad (1.5)$$

To evaluate the precision of the five analyses of the control sample, we first calculate how much each of the values, x_i, deviates from the mean value $\bar{x}$: $(x_i - \bar{x})$. To average these deviations, we first square each deviation: $(x_i - \bar{x})^2$. Then we sum all of these squared values, divide the sum by $(n - 1)$, and finally take the square root of the quotient. The results of these calculations are shown in Table 1.4,

TABLE 1.4 Calculating the Standard Deviation of the Control Sample Creatinine Values (mg/dL)

x_i	$x_i - \bar{x}$	$(x_i - \bar{x})^2$	s
0.685	0.003	0.000009	
0.676	−0.006	0.000036	$s = \sqrt{\frac{\Sigma_i(x_i - \bar{x})^2}{n - 1}}$
0.669	−0.013	0.000169	
0.688	0.006	0.000036	$= \sqrt{\frac{0.000350}{5 - 1}}$
0.692	0.010	0.000100	$= 0.00935$
	$\Sigma_i(x_i - \bar{x})^2 = 0.00035$		

where the final result is called the **standard deviation** (s) of the control sample values. Equation 1.6 puts the steps for calculating s in equation form:

$$s = \sqrt{\frac{\Sigma_i(x_i - \bar{x})^2}{n - 1}} \tag{1.6}$$

The standard deviation is sometimes reported with the mean value using a ± sign to indicate the uncertainty in the mean value. In the case of the five control samples, this expression is 0.682 ± 0.009 mg/dL, where the calculated s value is rounded off to match the last decimal place of the mean. The smaller the value of s, the more tightly clustered the measurements are around the mean and the more precise the data are.

Now that we have evaluated the precision of the control sample data, we should evaluate how accurate they are—that is, how well they agree with the actual creatinine value of the control sample, 0.681 mg/dL. The mean of the five values is within 0.001 mg/dL of the actual value, so the results certainly seem accurate, but there is also a way to quantitatively express the certainty that they are accurate. It involves calculating the **confidence interval**, which is a range of values around a calculated mean (0.682 mg/dL in our control sample analyses) that probably contains the true mean value, μ. Once we have calculated the size of this range, we can determine whether the actual creatinine value is within it. If it is, then the analyses are considered accurate.

To calculate the confidence interval, we use a statistical tool called the t-distribution and Equation 1.7:

$$\mu = \bar{x} \pm \frac{ts}{\sqrt{n}} \tag{1.7}$$

A table of t values is located in Appendix 1. Table 1.5 contains a portion of it. The values are arranged based on two parameters: the number of values in a set of data (actually, $n - 1$), and the confidence level we wish to use in our decision making. A commonly used confidence level in chemical analysis is 95%. Using it means that the chances are 95% that the range we calculate using Equation 1.7 will contain the true mean value (in this case, the amount of the creatinine in our control sample). The 95% t value for $n - 1 = 5 - 1 = 4$ in Table 1.5 is 2.776, so using the mean ($\bar{x}$) and standard deviation (s) values calculated previously gives us the following value for the true mean value (μ):

$$\mu = \bar{x} \pm \frac{ts}{\sqrt{n}} = \left(0.682 \pm \frac{2.776 \times 0.0094}{\sqrt{5}}\right) \text{mg/dL} = (0.682 \pm 0.012) \text{ mg/dL}$$

Thus, we can say with 95% certainty that the true mean of our control sample data is between 0.670 and 0.694 mg/dL. Because this range includes the actual creatinine concentration (0.681 mg/dL) of the control sample, we can infer with 95% confidence that these five control analyses (and the analyses of the patients' samples) are accurate.

mean (arithmetic mean) an average calculated by summing all of the values in a series and then dividing the sum by the number of values.

standard deviation (s) a measure of the amount of variation, or dispersion, in a set of related values.

confidence interval a range of values that has a specified probability of containing the true value of a measurement.

TABLE 1.5 Values of t

	CONFIDENCE LEVEL (%)		
$(n - 1)$	90	95	99
3	2.353	3.182	4.541
4	2.132	2.776	3.747
5	2.015	2.571	3.365
10	1.812	2.228	2.764
20	1.725	2.086	2.528
∞	1.645	1.960	2.326

CONCEPT **TEST**

Instead of calculating a standard deviation to express the variability in the data as we did in Table 1.4, we could have calculated a simple average deviation based on the mean of the absolute values of the deviations in the second column. What is the average deviation of the data? How does it differ from the standard deviation value? Suggest one reason for this difference.

(Answers to Concept Tests are in the back of the book.)

SAMPLE EXERCISE 1.11 Evaluating the Precision of Analytical Results **LO9**

A group of students collects a sample of water from a river near their campus and divides it up among five other groups of students. All six groups independently determine the concentration of dissolved oxygen in the sample. Their results are as follows: 9.2, 8.6, 9.0, 9.3, 9.1, and 8.9 mg O_2/L. Calculate the mean, standard deviation, and 95% confidence interval of these results.

Collect, Organize, and Analyze Equations 1.5, 1.6, and 1.7 can be used to calculate the mean, standard deviation, and 95% confidence interval of the results of the six analyses. The t value (Table 1.5) to use in Equation 1.7 for $n - 1 = 5$ is 2.571. Mean and standard deviation functions are also included in many programmable calculators and in computer spreadsheet applications such as Microsoft Excel.

Solve

a. Using Equation 1.5 to calculate the mean:

$$\bar{x} = \frac{\Sigma_i(x_i)}{n} = \left(\frac{9.2 + 8.6 + 9.0 + 9.3 + 9.1 + 8.9}{6}\right) \text{mg/L} = \frac{54.1}{6} = 9.02 \text{ mg/L}$$

b. We use a data table like Table 1.4 with Equation 1.6 to calculate the standard deviation (s) value:

x_i	$x_i - \bar{x}$	$(x_i - \bar{x})^2$	s
9.2	0.18	0.0324	
8.6	−0.42	0.1764	$s = \sqrt{\frac{\Sigma_i(x_i - \bar{x})^2}{n - 1}}$
9.0	−0.02	0.0004	
9.3	0.28	0.0784	$= \sqrt{\frac{0.3084}{6 - 1}}$
9.1	0.08	0.0064	$= 0.248$
8.9	−0.12	0.0144	
	$\Sigma_i(x_i - \bar{x})^2 = 0.3084$		

c. Using Equation 1.7 to calculate the 95% confidence interval:

$$\mu = \bar{x} \pm \frac{ts}{\sqrt{n}} = \left(9.02 \pm \frac{2.571 \times 0.248}{\sqrt{6}}\right) \text{mg/L} = (9.02 \pm 0.26) \text{ mg/L}$$

Scientists often round off $\frac{ts}{\sqrt{n}}$ values to only one digit and round off the mean value to the same number of digits after the decimal point. These steps give expressions that better reflect the variability in the data and the uncertainty in the mean. In the preceding example 9.02 ± 0.26 mg/L becomes 9.0 ± 0.3 mg/L.

Think About It Did you notice that the six data points used in these calculations each contained two significant figures (each was known to the nearest tenth of a milligram of oxygen per liter), yet we initially expressed the mean with three significant figures (to the hundredths place)? This happened because we knew the sum of the six results (54.1) to three significant figures, and divided this value by an exact number (6 values). Therefore, the quotient could be reported with three significant figures: 9.02. This increase in significant figures—and in the students' confidence in knowing the actual concentration of dissolved oxygen in the river—illustrates the importance of replicating analyses: doing so gives us more certainty about the true value of an experimental value than a single determination of it.

outlier a data point that is distant from the other observations.

Grubbs' test a statistical test used to detect an outlier in a set of data.

Practice Exercise Analyses of a sample of Dead Sea water produced the following results for the concentration of sodium ions: 35.9, 36.5, 36.3, 36.8, and 36.4 mg/L. What are the mean, standard deviation, and 95% confidence interval of these results?

(Answers to Practice Exercises are in the back of the book.)

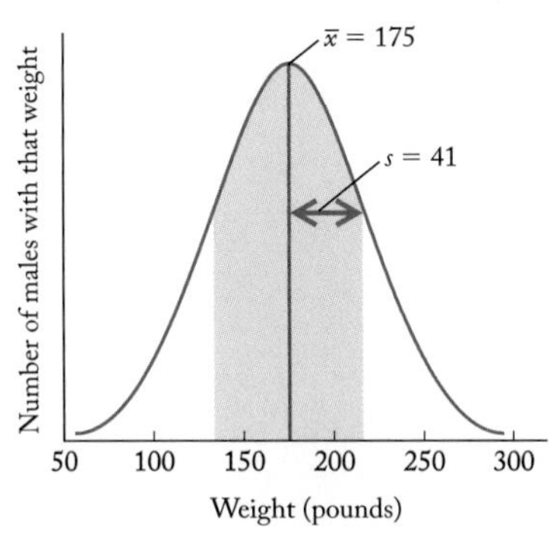

FIGURE 1.26 The distribution of body weight values among 19-year-old American males is an example of a normal distribution.

An assumption built into calculating means, standard deviations, and confidence intervals is that the variability in the data is random. Random means that data points are as likely to be above the mean value as below it and that there is a greater probability of values lying close to the mean than far away from it. This kind of distribution is called a *normal* distribution. Large numbers of such data produce a distribution profile called a bell curve, like the one shown in Figure 1.26, which is based on a study conducted by the U.S. Centers for Disease Control and Prevention. The data in Figure 1.26 are for the weights of 19-year-old American males and have a mean of 175 lb and a standard deviation of 41 lb (79 ± 19 kg). In randomly distributed data, 68% of the values—represented by the area under the curve highlighted in pale red—are within one standard deviation of the mean.

Sometimes a set of data contains an **outlier,** that is, an individual value that is much farther away from the mean than any of the other values. You may be tempted to simply ignore such a value, but unless there is a valid reason for doing so, such as accidentally leaving out a step in an analytical procedure, it is unethical to disregard a value just because it is unexpected or not similar to the others.

Suppose a college freshman discovers a long-forgotten piggy bank full of pennies and, being short of money, decides to pack them into rolls of 50 to cash them in at a bank. To avoid having to count hundreds of pennies, the student decides to use a balance in a general chemistry lab that can weigh up to 300 g to the nearest 0.001 g. The student individually weighs 10 different pennies from the piggy bank to determine the average mass of a penny, intending to multiply the average by 50 to calculate how many to weigh out for each roll. The results of the 10 measurements are listed in Table 1.6 from lowest to highest values.

TABLE 1.6 Masses in Grams of 10 Circulated Pennies

2.486
2.495
2.500
2.502
2.502
2.505
2.506
2.507
2.515
3.107

The results reveal that nine of the ten masses are quite close to 2.5 g, but the tenth is considerably heavier. Was there an error in the measurement, or is there something unusual about that tenth penny? To answer questions such as these—and the broader one of whether an unusually high or low value is statistically different enough from the others in a set of data to be labeled an outlier—we analyze the data using *Grubbs' test of a single outlier*, or **Grubbs' test.** In this test, the absolute difference between the suspected outlier and the mean of a set of data is divided by the standard deviation of the data set. The result is a statistical parameter that has the symbol *Z*:

$$Z = \frac{|x_i - \bar{x}|}{s} \tag{1.8}$$

If this calculated *Z* value is greater than the reference *Z* value (Table 1.7) for a given number of data points and a particular confidence level—usually 95%—then the suspect data point is determined to be an outlier and can be discarded.

TABLE 1.7 Reference *Z* Values for Grubbs' Test

	CONFIDENCE LEVEL (%)	
n	95	99
3	1.155	1.155
4	1.481	1.496
5	1.715	1.764
6	1.887	1.973
7	2.020	2.139
8	2.126	2.274
9	2.215	2.387
10	2.290	2.482
11	2.355	2.564
12	2.412	2.636

To apply Grubbs' test to the mass of the tenth and heaviest penny, we first calculate the mean and standard deviation of the masses of all 10 pennies. These values are 2.562 ± 0.191 g. We use them in Equation 1.8 to calculate Z:

$$Z = \frac{|x_i - \bar{x}|}{s} = \frac{|3.107 - 2.562|}{0.191} = 2.85$$

Next, we check the reference Z values in Table 1.7 for $n = 10$ data points, and we find that our calculated Z value is greater than both 2.290 and 2.482—the Z values above which we can conclude with 95% and 99% confidence that the 3.107 g data point is an outlier. Stated another way, the probability that this data point *is not* an outlier is less than 1%. Note that Grubbs' test can be used *only once* to identify *only one* outlier in a set of data.

The tenth penny probably weighed much more than the others because U.S. pennies minted since 1983 weigh 2.50 g new, but those minted before 1983 weighed 3.11 g new. The older pennies are 95% copper and 5% zinc, whereas the newer pennies are 97.5% zinc and are coated with a thin layer of copper. (Copper is about 25% more dense than zinc.) Thus, a piggy bank containing hundreds of pennies will likely have a few that are heavier than most of the others.

SAMPLE EXERCISE 1.12 Testing Whether a Data Point Should Be Considered an Outlier — LO9

Use Grubbs' test to determine whether or not the lowest value in the set of sodium ion concentration data from Practice Exercise 1.11 should be considered an outlier at the 95% confidence level.

Collect, Organize, and Analyze We want to determine whether the lowest value in the following set of five results is an outlier: 35.8, 36.6, 36.3, 36.8, and 36.4 mg/L. Grubbs' test (Equation 1.8) is used to determine whether a data point should be deemed an outlier. If the resulting Z value is equal to or greater than the reference Z values for $n = 5$ from Table 1.7, the suspect value is an outlier.

Solve Using the statistics functions of a programmable calculator, we find that the mean and standard deviation of the data set are 36.38 ± 0.38 mg/L. We calculate the value of Z using Equation 1.8:

$$Z = \frac{|x_i - \bar{x}|}{s} = \frac{|35.8 - 36.38|}{0.38} = 1.5$$

This calculated Z value is less than 1.715, which is the reference Z value for $n = 5$ at the 95% confidence level (Table 1.7). Therefore, the lowest value is *not* an outlier, and it should be included with the other four values in any analysis of the data.

Think About It Although the lowest value may have appeared to be considerably lower than the others in the data set, Grubbs' test tells us that it is not *significantly* lower at the 95% confidence level.

Practice Exercise Duplicate determinations of the cholesterol concentration in a blood serum sample produce the following results: 181 and 215 mg/dL. The patient's doctor is concerned about the difference between the two results and the fact that values above 200 mg/dL are considered "borderline high," so she orders a third cholesterol test. The result is 185 mg/dL. Should the doctor consider the 215 mg/dL value an outlier and discard it?

(Answers to Practice Exercises are in the back of the book.)

SAMPLE EXERCISE 1.13 Integrating Concepts: Searching for Cheaper Gas

When would it be worthwhile making a special trip across town (or out of the country) to buy gas at a station where it is cheaper than nearby?

Suppose a driver lives in Niagara Falls, Ontario, on a day not long ago when the price of gasoline at a local station was 1.06 Canadian dollars (CAD) per liter. On the same day the price of gasoline at a station in Niagara Falls, New York (4.0 km away), was 2.15 U.S. dollars (USD) per gallon, and the currency exchange rate was 1.00 USD = 1.30 CAD. How much money in Canadian dollars would the driver save by filling up with 60.0 liters of gasoline at the American station? In calculating the driver's savings, consider that his car can go 12.0 km on one liter of gasoline (about 26 miles per gallon), and that he has to make a round trip, which includes crossing the Rainbow Bridge over the Niagara River. The toll on the bridge was 4.50 CAD for vehicles crossing to the Canadian side (there was no toll the other way).

Collect and Organize The problem gives the gasoline prices at a nearby station and one that is 4.0 km away, and asks how much money would be saved by driving to the distant station to purchase 60.0 liters of gasoline. The prices are given in different currencies and volume units: liters and U.S. gallons. According to Table 1.3, there are 0.2642 gal / L. We are also given the distance to the American station in km, the fuel efficiency of the car in km/L and a one-way bridge toll in CAD.

Analyze The problem contains many pieces of information, and we need to sort them out. A logical starting point is the different prices of gasoline. We have to assume that gas is cheaper at the American station; otherwise, why make the trip? To compare prices we need to express both using the same set of units. Converting the price at the American station into Canadian dollars per liter makes sense because (1) the car's fuel tank capacity is given in liters, (2) its fuel efficiency is expressed in km/L, and (3) the distance to the American station is given in km. We can use the distance and fuel efficiency values to calculate the volume of gas consumed driving to and from the American gas station.

To estimate the savings, we start with the approximate difference in the cost of 60.0 liters of gasoline locally and at the American station. Locally the cost is about (60 L × 1 CAD/L) ≈ 60 CAD. The cost at the American station is about (60 L × 1/4 gal / L × 2 USD/gal × 1.3 CAD / USD) ≈ 40 CAD, which represents a savings of about (60 – 40) = 20 CAD to fill up. The cost of the gasoline consumed is about (8 ~~km~~ × 1 ~~L~~ / 12 ~~km~~ × 1 CAD/~~L~~) = ~ 1 CAD, and the bridge toll is 4.5 CAD, so the overall savings should be about (20 – 5.5) ≈ 15 CAD.

Solve

1. Cost of buying gasoline at the local station:

$$60.0\ \cancel{L} \times 1.06\ \text{CAD}/\cancel{L} \times = 63.60\ \text{CAD}$$

2. Cost of buying gasoline at the American station:

$$60.0\ \cancel{L} \times 0.2642\ \cancel{gal} / \cancel{L} \times 2.15\ \text{USD}/\cancel{gal} \times 1.30\ \text{CAD}/\ \cancel{USD} = 44.31\ \text{CAD}$$

3. Difference in fill-up costs (63.60 − 44.31) = 19.29 CAD
4. Cost of gasoline to drive to and from the American station:

$$2 \times 4.0\ \cancel{km} \times 1\ \cancel{L} / 12.0\ \cancel{km} \times 1.06\ \text{CAD}/\cancel{L}) = 0.71\ \text{CAD}$$

5. Net savings factoring in the bridge toll:

$$19.29 - (0.71 + 4.50) = 14.08\ \text{CAD}$$

Think About It The calculated savings are about what we estimated. Would you be willing to drive to another country to save about $14 on a tankful of gasoline? How about driving the same distance within your own country to save only half as much? What additional considerations besides those in this Sample Exercise would impact your decision?

SUMMARY

LO1 The **scientific method** starts with observations of natural phenomena and/or the results of laboratory experiments; next, a tentative explanation, or **hypothesis**, is developed that explains the observations and results; then the hypothesis is tested through further experimentation before a **scientific theory** is formulated that explains all the results and observations available. A **scientific law** is a comprehensive, succinct description of a phenomenon or process. Dalton's atomic theory explains Proust's **law of definite proportions** and Dalton's **law of multiple proportions**. (Section 1.1)

LO2 The COAST framework used in this book to solve problems has four components: **C**ollect and **O**rganize information and ideas, **A**nalyze the information to determine how it can be used to obtain the answer, **S**olve the problem (often the math-intensive step), and **T**hink about the answer. (Section 1.2)

LO3 The principal classes of matter are **mixtures** and **pure substances.** Pure substances may be either **elements** or **compounds** (elements chemically combined together). A **chemical formula** indicates the proportion of elements in a substance. The properties of a substance are either **intensive properties**, which are independent of quantity, or **extensive properties**, which are related to the quantity of the substance. The **physical properties** of a substance can be observed without changing the substance into another one, whereas the **chemical properties** of a substance (such as flammability) can be observed only through chemical reactions involving the substance. The **density (*d*)** of an object or substance is the ratio of its mass to its volume. Mixtures may be **homogeneous** (such mixtures are also called **solutions**) or **heterogeneous**, and they can be separated by **physical processes** such as **distillation**, **filtration**, and **chromatography**. Distillation separates substances of differing **volatility**. (Section 1.3)

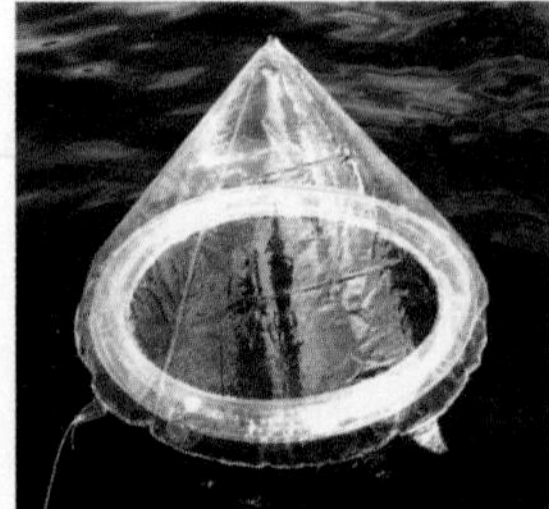

LO4 The states (or phases) of matter include **solid**, in which the particles have an ordered structure; **liquid**, in which the particles are free to move past each other; and **gas** (or *vapor*), in which the particles have the most freedom and completely fill their container. Familiar phase changes include melting, freezing, vaporization, and condensation. The transformation of a solid directly into a gas is **sublimation**; the reverse process is **deposition**. (Section 1.4)

LO5 **Energy** can be defined as the ability to do **work**. **Heat** is the flow of energy due to a difference in temperature. **Potential energy (PE)** is the energy in an object due to its position or composition. **Kinetic energy (KE)** is the energy of motion. According to the **law of conservation of energy**, energy cannot be created or destroyed. (Section 1.5)

LO6 The composition of molecular compounds is described by their molecular formulas. The three-dimensional arrangements of their atoms may be represented by structural formulas, ball-and-stick models, and space-filling models. (Section 1.6)

LO7 Measured quantities and values derived from them are inherently uncertain. Exact values include those derived from counting objects or those that are defined, such as 60 seconds in a minute. The appropriate number of **significant figures** is used to express the certainty in the result of a measurement or calculation. The **precision** of any set of measurements indicates how repeatable the measurement is, whereas the **accuracy** of a measurement indicates how close to the true value the measured value is. (Section 1.7)

LO8 The International System of Units (SI), in which the **meter (m)** is the standard unit of length, evolved from the metric system and is widely used in science to express the results of measurements. Prefixes naming powers of 10 are used with SI base units to express quantities much larger or much smaller than the base units. Dimensional analysis uses **conversion factors** (fractions in which the numerators and denominators have different units but represent the same quantity) to convert a value from one unit into another unit. (Section 1.8)

LO9 The average value and variability in repeated measurements or analyses are determined by calculating the **arithmetic mean** ($\overline{x}$), **standard deviation** (s), and **confidence interval**. An **outlier** in a data set may be identified based on the results of Grubbs' test. (Section 1.9)

PARTICULATE **PREVIEW WRAP-UP**

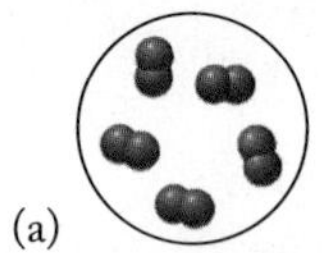

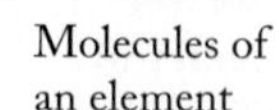

(a) Molecules of an element

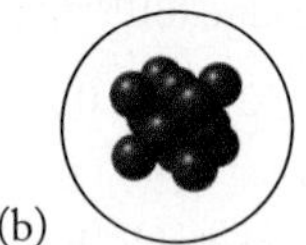

(b) Atoms of an element

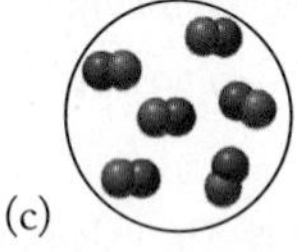

(c) Molecules of a compound

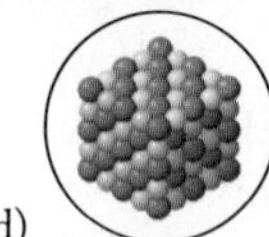

(d) Ionic compound

PROBLEM-SOLVING SUMMARY

Type of Problem	Concepts and Equations	Sample Exercises
Calculating density	$d = \frac{m}{V}$	**1.1**
Distinguishing exact from uncertain values	Quantities that can be counted are exact. Measured quantities or conversion factors that are not exact values are inherently uncertain.	**1.5**
Using significant figures in calculations	Apply the weak-link rule: the number of significant figures in a calculated quantity involving multiplication or division can be no greater than the number of significant figures in the least certain value used to calculate it. When adding or subtracting values, the number of digits after the decimal point in the sum or difference can be no greater than in the value that has the fewest digits after its decimal point.	**1.6**
Converting between temperature scales	$T(\text{K}) = T(°\text{C}) + 273.15$ $T(°\text{C}) = (5/9)[T(°\text{F}) - 32]$	**1.7**
Calculating kinetic energy	$\text{KE} = \frac{1}{2}mu^2$	**1.8**
Converting units using dimensional analysis	Converting values from one set of units to another involves multiplication by one or more conversion factors, which are set up so that the original units cancel out.	**1.9–1.10**

Type of Problem	Concepts and Equations	Sample Exercises
Calculating the mean ($\bar{x}$), standard deviation (s), and confidence interval of a set of data	$\bar{x} = \frac{\sum_i (x_i)}{n}$ $s = \sqrt{\frac{\sum_i (x_i - \bar{x})^2}{n - 1}}$ $\mu = \bar{x} \pm \frac{ts}{\sqrt{n}}$	**1.11**
Testing whether a suspect data point (x_i) is an outlier	Use Grubbs' test: calculate the value of Z using Equation 1.8: $Z = \frac{\lvert x_i - \bar{x} \rvert}{s}$ and compare it to the appropriate reference value in Table 1.7. If the calculated Z is greater, the suspect data point is an outlier.	**1.12**

VISUAL PROBLEMS

(Answers to boldface end-of-chapter questions and problems are in the back of the book.)

1.1. For each image in Figure P1.1, identify what class of pure substance is depicted (an element or compound) and identify the physical state(s).

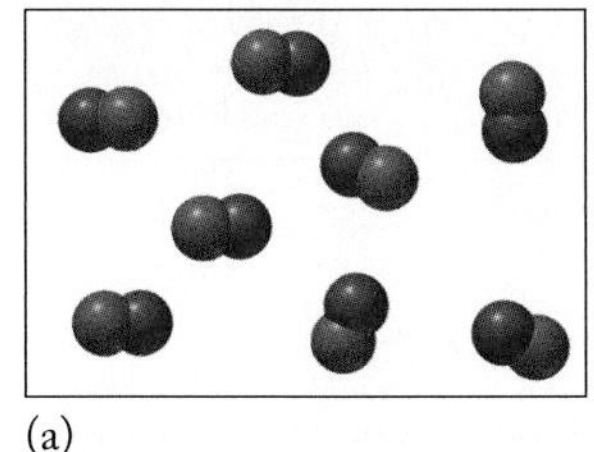

(a) (b)

FIGURE P1.1

1.2. For each image in Figure P1.2, identify what class of matter is depicted (an element, a compound, a mixture of elements, or a mixture of compounds) and identify the physical state.

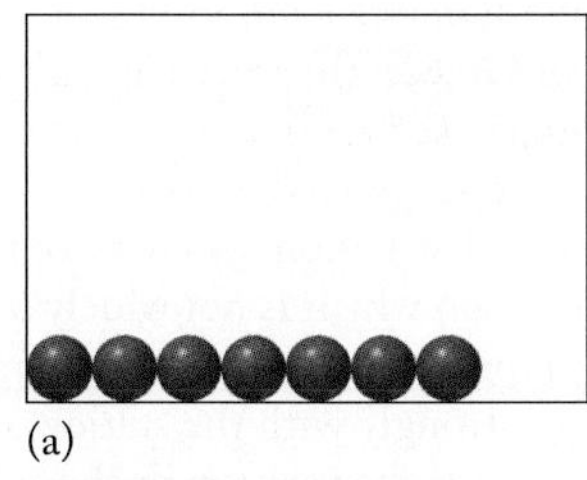

(a)

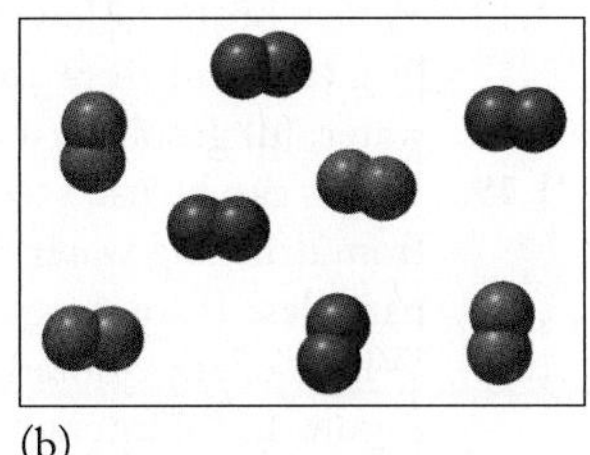

(b)

FIGURE P1.2

1.3. Which of the following statements best describes the change depicted in Figure P1.3?

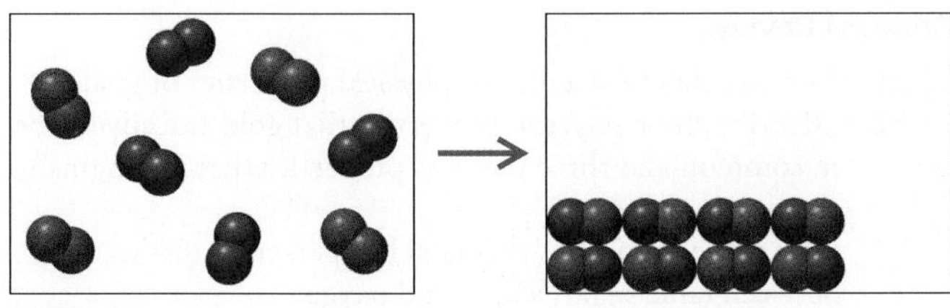

FIGURE P1.3

a. A mixture of two gaseous elements undergoes a chemical reaction, forming a gaseous compound.
b. A mixture of two gaseous elements undergoes a chemical reaction, forming a solid compound.
c. A mixture of two gaseous elements undergoes deposition.
d. A mixture of two gaseous elements condenses.

1.4. Which of the following statements best describes the change depicted in Figure P1.4?
a. A mixture of two gaseous elements is cooled to a temperature at which one of them condenses.
b. A mixture of two gaseous compounds is heated to a temperature at which one of them decomposes.
c. A mixture of two gaseous elements undergoes deposition.
d. A mixture of two gaseous elements reacts to form two compounds, one of which is a liquid.

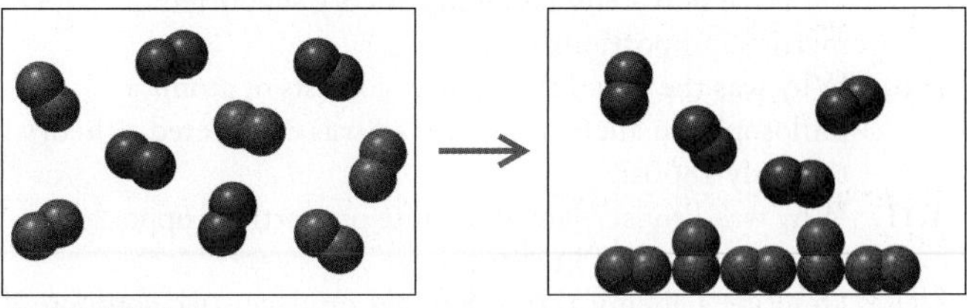

FIGURE P1.4

1.5. A space-filling model of formic acid is shown in Figure P1.5. What is the molecular formula of formic acid?

FIGURE P1.5

1.6. Use representations [A] through [I] in Figure P1.6 to answer questions a–f.
   a. Which molecule contains the most atoms?
   b. Which compound contains the most elements?
   c. Which representation depicts a solid solution?
   d. Which representation depicts a homogeneous mixture?
   e. Which pure substances are compounds?
   f. Which pure substances are elements?

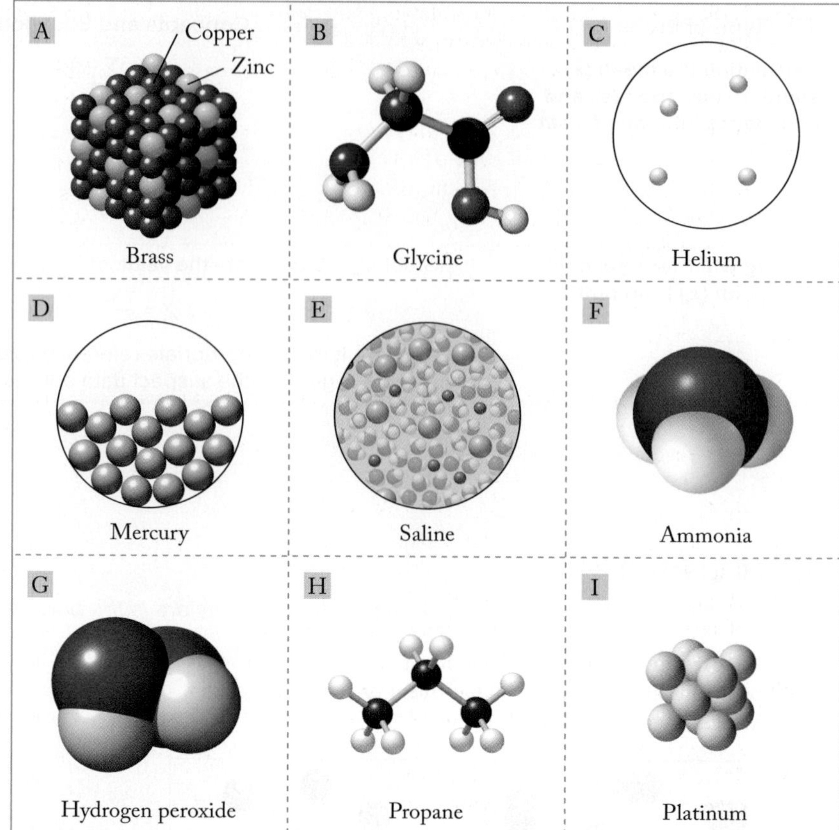

FIGURE P1.6

QUESTIONS AND PROBLEMS

Atomic Theory: The Scientific Method in Action

Concept Review

1.7. How does a hypothesis differ from a scientific theory?

1.8. How does a hypothesis become a theory?

1.9. Describe how Dalton's atomic theory supported his law of multiple proportions.

1.10. Why was the belief that matter consists of atoms a philosophy in ancient Greece but was considered a theory in the early 1800s?

1.11. Why was Proust's law of definite proportions opposed by many scientists of his time?

*1.12 Describe a chemical reaction that produces two compounds whose compositions illustrate Dalton's law of multiple proportions.

1.13. Describe how a scientific theory differs from the meaning of *theory* as it is used in normal conversation.

1.14. Can a theory be proven?

Classes of Matter

Concept Review

1.15. Which of the following foods is a heterogeneous mixture? (a) bottled water; (b) a Snickers bar; (c) grape juice; (d) an uncooked hamburger

1.16. Which of the following foods is a homogeneous mixture? (a) freshly brewed coffee; (b) vinegar; (c) a slice of white bread; (d) a slice of ham

1.17. Which of the following foods is a heterogeneous mixture? (a) apple juice; (b) cooking oil; (c) solid butter; (d) orange juice; (e) tomato juice

1.18. Which of the following is a homogeneous mixture? (a) a bronze sword from ancient Greece; (b) sweat; (c) Nile River water; (d) gasoline; (e) compressed air in a scuba tank

***1.19.** Filters can be used to remove suspended particles of soil from drinking water. Would distillation also remove these particles? If so, suggest a reason why it is not widely used.

1.20. Which of the colored compounds in the photograph in Figure 1.9(b) interact more strongly with the stationary phase of the separation: those at the very top or those below them? Assume the liquid phase migrates upward.

Properties of Matter

Concept Review

1.21. List one chemical and four physical properties of gold.

1.22. Describe three physical properties that gold and silver have in common and three physical properties that distinguish them.

1.23. Give three properties that enable a person to distinguish between table sugar, water, and oxygen.

1.24. Give three properties that enable a person to distinguish between table salt, sand, and copper.

1.25. Indicate whether each of the following properties is a physical or chemical property of sodium (Na):
a. Its density is greater than that of kerosene and less than that of water.
b. It has a lower melting point than most other metals.
c. It is an excellent conductor of heat and electricity.
d. It is soft and can be easily cut with a knife.
e. Freshly cut sodium is shiny, but it rapidly tarnishes when it comes in contact with air.
f. It reacts very vigorously with water, releasing hydrogen gas (H_2).

1.26. Indicate whether each of the following is a physical or chemical property of hydrogen gas (H_2):
a. At room temperature, its density is less than that of any other gas.
b. It reacts vigorously with oxygen (O_2) to form water.
c. Liquefied H_2 boils at −253°C.
d. H_2 gas does not conduct electricity.

1.27. Can an extensive property be used to identify a substance? Explain why or why not.

1.28. Which of the following are intensive properties of a sample of a substance? (a) freezing point; (b) heat content; (c) temperature

***1.29.** Is the capacity of carbon dioxide to extinguish fires linked to its chemical properties, its physical properties, or both? Explain your answer.

1.30. The stainless steel used to make kitchen knives and many other tools gets its name from its capacity to resist corrosion and, therefore, *stain less*. Is this a chemical or physical property of stainless steel?

States of Matter

Concept Review

1.31. In what ways are the arrangements of water molecules in ice and liquid water similar and in what ways are they different?

1.32. What occupies the space between the particles that make up a gas?

1.33. Substances have characteristic *triple points*, unique combinations of temperature and pressure at which substances can simultaneously exist as solids, liquids, and gases. In which of these three states do the particles of a substance at its triple point have the greatest motion and in which state do they have the least motion?

1.34. A pot of water on a stove is heated to a rapid boil. Identify the gas inside the bubbles that forms in the boiling water.

1.35. A brief winter storm leaves a dusting of snow on the ground. During the sunny but very cold day after the storm, the snow disappears even though the air temperature never gets above freezing. If the snow didn't melt, where did it go?

1.36. Equal masses of water undergo condensation, deposition, evaporation, and sublimation.
a. Which of the processes is accompanied by the *release* of the greatest amount of energy?
b. In which of the processes is the greatest amount of energy *absorbed*?

Forms of Energy

Concept Review

1.37. How are energy and work related?

1.38. Explain the difference between potential energy and kinetic energy.

1.39. Which of the following statements about heat are true?
a. Heat is the transfer of energy from a warmer place to a cooler one.
b. Heat flows faster from a full container of hot coffee than a half-full container of coffee at the same temperature.
c. A cup of hot coffee loses heat faster than the same cup full of warm coffee.

1.40. Describe three examples of energy transfer that happen when you speak on a cell phone to a friend.

Problems

1.41. A subcompact car with a mass of 1400 kg and a loaded dump truck with a mass of 18,000 kg are traveling at the same speed. How many times more kinetic energy does the dump truck have than the car?

1.42. **Speed of Baseball Pitches** Major League Baseball pitchers throw a pitch called a *changeup*, which looks like a fastball leaving the pitcher's hand but has less speed than a typical fastball. How much more kinetic energy does a 92 mph fastball have than a 78 mph changeup? Express your answer as a percentage of the kinetic energy of the changeup.

Making Measurements and Expressing the Results; Unit Conversions and Dimensional Analysis

Concept Review

1.43. Describe in general terms how the SI and U.S. customary systems of units differ.

1.44. Suggest two reasons why SI units are not more widely used in the United States.

1.45. Both the Fahrenheit and Celsius scales are based on reference temperatures that are 100 degrees apart. Suggest a reason why the Celsius scale is preferred by scientists.

1.46. In what way are the Celsius and Kelvin scales similar and in what way are they different?

1.47. What is meant by an *absolute* temperature scale?

1.48. Can a temperature in °C ever have the same value in °F?

Problems

Note: Some physical properties of the elements are listed in Appendix 3.

1.49. **Olympic Mile** An Olympic "mile" is actually 1500 m. What percentage is an Olympic mile of a U.S. mile (5280 ft)?

1.50. A sport-utility vehicle has an average mileage rating of 18 miles per gallon. How many gallons of gasoline are needed for a 389-mile trip?

1.51. A single strand of natural silk may be as long as 4.0×10^3 m. What is this length in miles?

1.52. The speed of light in a vacuum is 2.998×10^8 m/s. What is the speed of light in km/h?

1.53. If a wheelchair-marathon racer moving at 13.1 miles per hour expends energy at a rate of 665 Calories per hour, how much energy in Calories would be required to complete a marathon race (26.2 miles) at that pace?

1.54. **Boston Marathon** To qualify to run in the 2016 Boston Marathon, a distance of 26.2 miles, an 18-year-old woman had to have completed another marathon in 3 hours and 35 minutes or less. Translate this qualifying time and distance into average speeds expressed in (a) miles per hour and (b) meters per second.

1.55. **Nearest Star** At a distance of 4.3 light-years, Proxima Centauri is the nearest star to our solar system. What is the distance to Proxima Centauri in kilometers? (The speed of light in space is 2.998×10^8 m/s.)

1.56. **Sports Car** The Porsche Boxster Spyder (on the left in Figure P1.56) is powered by a 320-horsepower gasoline engine. The electric motor in the Tesla Roadster (on the right in Figure P1.56) is rated at 215 kilowatts. Which is the more powerful sports car? (1 horsepower = 745.7 watts.)

FIGURE P1.56

***1.57.** The level of water in an Olympic-size swimming pool (50.0 m long, 25.0 m wide, and about 2 m deep) needs to be lowered 3.0 cm. If water is pumped out at a rate of 5.2 L per second, how long will it take to lower the water level 3.0 cm?

*1.58. The price of a popular soft drink is \$1.00 for 24 fluid ounces (fl oz) or \$0.75 for 0.50 L. Which is a better buy? (1 qt = 32 fl oz.)

1.59. Suppose a runner completes a 10K (10.0 km) road race in 41 minutes and 23 seconds. What is the runner's average speed in meters per second?

1.60. **Kentucky Derby Record** In 1973 a horse named Secretariat ran the fastest Kentucky Derby in history, taking 1 minute and 59.4 seconds to run 1.25 miles. What was Secretariat's average speed in (a) miles per hour and (b) meters per second?

1.61. What is the mass of a magnesium block that measures 2.5 cm × 3.5 cm × 1.5 cm?

1.62. What is the mass of an osmium block that measures 6.5 cm × 9.0 cm × 3.25 cm? Do you think you could lift it with one hand?

1.63. A chemist needs 35.0 g of concentrated sulfuric acid for an experiment. The density of concentrated sulfuric acid at room temperature is 1.84 g/mL. What volume of the acid is required?

1.64. What is the mass of 65.0 mL of ethanol? (Its density at room temperature is 0.789 g/mL.)

1.65. A brand new silver U.S. dollar weighs 0.934 oz. Express this mass in grams and kilograms. (1 oz = 28.35 g.)

1.66. A U.S. dime weighs 2.5 g. What is the U.S. dollar value of exactly 1 kg of dimes?

1.67. What volume of gold would be equal in mass to a piece of copper with a volume of 125 cm^3?

*1.68. A small hot-air balloon is filled with 1.00×10^6 L of air at a temperature at which the density of air is 1.18 g/L. As the air in the balloon is heated further, it expands and 9×10^4 L escapes out the open bottom of the balloon. What is the density of the heated air remaining inside the balloon?

1.69. What is the volume of 1.00 kg of mercury?

1.70. A student wonders whether a piece of jewelry is made of pure silver. She determines that its mass is 3.17 g. Then she drops it into a 10-mL graduated cylinder partially filled with water and determines that its volume is 0.3 mL. Could the jewelry be made of pure silver?

***1.71.** The average density of Earth is 5.5 g/cm^3. The mass of Venus is 81.5% of Earth's mass, and the volume of Venus is 88% of Earth's volume. What is the density of Venus?

1.72. Earth has a mass of 6.0×10^{27} g and an average density of 5.5 g/cm^3.
a. What is the volume of Earth in cubic kilometers?
*b. Geologists sometimes express the "natural" density of Earth after doing a calculation that corrects for gravitational squeezing (compression of the core because of high pressure). Should the natural density be more or less than 5.5 g/cm^3?

***1.73.** **Utility Boats for the Navy** A plastic material called high-density polyethylene (HDPE) was once evaluated for use in impact-resistant hulls of small utility boats for the U.S. Navy. A cube of this material measures 1.20×10^{-2} m on a side and has a mass of 1.70×10^{-3} kg. Seawater at the surface of the ocean has a density of 1.03 g/cm^3. Will this cube float on water?

1.74. **Dimensions of the Sun** The sun is a sphere with an estimated mass of 2×10^{30} kg. If the radius of the sun is 7.0×10^5 km, what is the average density of the sun in units of grams per cubic centimeter? The volume of a sphere is $\frac{4}{3}\pi r^3$.

1.75. The Golden Jubilee diamond (Figure P1.75) has a mass of 545.67 carats (1 carat = 0.200 g). What is the mass of the diamond in (a) grams and (b) ounces? (1 pound = 16 ounces.)

FIGURE P1.75

1.76. The density of diamond is 3.51 g/cm^3. What is the volume of the Golden Jubilee diamond in Figure P1.75?

1.77. Which of the following numbers have just three significant figures? (a) 7.02; (b) 6.452; (c) 302; (d) 6.02×10^{23}; (e) 12.77; (f) 3.43

1.78. Which of the following numbers have four significant figures? (a) 0.0592; (b) 0.08206; (c) 8.314; (d) 273.15; (e) 5.091×10^3; (f) 9.490

1.79. Perform each of the following calculations and express the answer with the correct number of significant figures:

a. $3.15 \times 2255 / 7.7 =$
b. $(6.7399 \times 10^{-18}) \times (1.0135 \times 10^3) / (52.67 + 0.144) =$
c. $(4.7 + 58.69)/(6.022 \times 10^{23} \times 6.864) =$
d. $(76.2 - 60.0)/[43.53 \times (9.988 \times 10^4)] =$

1.80. Perform each of the following calculations, and express the answer with the correct number of significant figures:

a. $[(12 \times 60.0) + 55.3]/(5.000 \times 10^3) =$
b. $3.1416 \times (2.031)^2 \times 3.75 \times 8.00 =$
c. The number of cubic centimeters in 389 cubic inches
d. The average (mean) of 8.7, 8.5, 8.5, 8.9, and 8.8

1.81. Liquid helium boils at 4.2 K. What is the boiling point of helium in degrees Celsius?

1.82. Liquid hydrogen boils at −253°C. What is the boiling point of H_2 on the Kelvin scale?

1.83. **Topical Anesthetic** Ethyl chloride acts as a mild topical anesthetic because it chills the skin when sprayed on it. It dulls the pain of injury and is sometimes used to make removing splinters easier. The boiling point of ethyl chloride is 12.3°C. What is its boiling point on the Fahrenheit and Kelvin scales?

1.84. **Dry Ice** The temperature of the dry ice (solid carbon dioxide) in ice cream vending carts is −78°C. What is this temperature on the Fahrenheit and Kelvin scales?

1.85. **Record Low** The lowest temperature measured on Earth is −128.6°F, recorded at Vostok, Antarctica, in July 1983. What is this temperature on the Celsius and Kelvin scales?

1.86. **Record High** The highest temperature ever recorded in the United States is 134°F at Greenland Ranch, Death Valley, California, on July 13, 1913. What is this temperature on the Celsius and Kelvin scales?

1.87. **Critical Temperature** The discovery of "high-temperature" superconducting materials in the mid-1980s spurred a race to prepare the material with the highest superconducting temperature. The critical temperatures (T_c)—the temperatures at which the material becomes superconducting—of $YBa_2Cu_3O_7$, Nb_3Ge, and $HgBa_2CaCu_2O_6$ are 93.0 K, −250.0°C, and −231.1°F, respectively. Convert these temperatures into a single temperature scale, and determine which superconductor has the highest T_c value.

1.88. The boiling point of O_2 is −183°C, whereas the boiling point of N_2 is 77 K. As air is cooled, which gas condenses first?

Assessing and Expressing Precision and Accuracy

Concept Review

1.89. How many suspect data points can be identified in a data set using Grubbs' test?

1.90. Which confidence interval is the largest for a given value of n: 50%, 90%, or 95%?

1.91. The concentration of ammonia in an aquarium tank is determined each day for a week. Which of these measures of the variability in the results of these analyses is greater: (a) mean ± standard deviation, or (b) 95% confidence interval? Explain your selection.

1.92. If the results of Grubbs' test indicate that a suspect data point is not an outlier at the 95% confidence level, could it be one at the 99% confidence level?

Problems

***1.93.** The widths of copper lines in printed circuit boards must be close to a design value. Three manufacturers were asked to prepare circuit boards with copper lines that are 0.500 μm (micrometers) wide (1 μm = 1×10^{-6} m). Each manufacturer's quality control department reported the following line widths on five sample circuit boards (given in micrometers):

Manufacturer 1	Manufacturer 2	Manufacturer 3
0.512	0.514	0.500
0.508	0.513	0.501
0.516	0.514	0.502
0.504	0.514	0.502
0.513	0.512	0.501

a. What is the mean and standard deviation of the data provided by each manufacturer?
b. For which of the three sets of data does the 95% confidence interval include 0.500 μm?
c. Which of the data sets fit the description "precise and accurate," and which is "precise but not accurate"?

1.94. **Diabetes Test** Glucose concentrations in the blood above 110 mg/dL can be an early indication of several medical conditions, including diabetes. Suppose analyses of a series of blood samples from a patient at risk of diabetes produce the following results: 106, 99, 109, 108, and 105 mg/dL.

a. What are the mean and the standard deviation of the data?
b. Patients with blood glucose levels above 120 mg/dL are considered diabetic. Is this value within the 95% confidence interval of these data?

1.95. Use Grubbs' test to decide whether the value 3.41 should be considered an outlier in the following data set from the analyses of portions of the same sample conducted by six groups of students: 3.15, 3.03, 3.09, 3.11, 3.12, and 3.41.

1.96. Use Grubbs' test to decide whether any one of the values in the following set of replicate measurements should be considered an outlier: 61, 75, 64, 65, 64, and 66.

Additional Problems

*1.97. **Agricultural Runoff** A farmer applies 1.50 metric tons of a fertilizer that contains 10% nitrogen to his fields each year (1 metric ton = 1000 kg). Fifteen percent of the fertilizer washes into a stream that runs through the farm. If the stream flows at an average rate of 1.4 cubic meters per minute, what is the additional concentration of nitrogen (expressed in milligrams of nitrogen per liter) in the stream water due to the farmer's yearly application of fertilizer?

1.98. Your laboratory instructor has given you two shiny, light-gray metal cylinders (A and B). Your assignment is to determine which one is made of aluminum (d = 2.699 g/mL) and which one is made of titanium (d = 4.54 g/mL). The mass of each cylinder was determined on a balance to five significant figures. The volume of each was determined by immersing it in a partially filled graduated cylinder as shown in Figure P1.98.

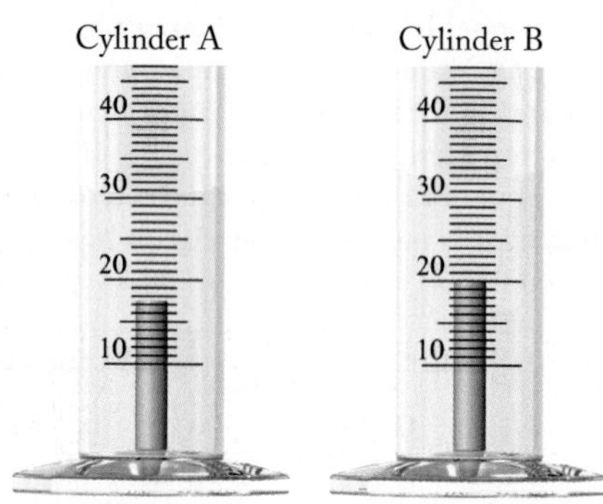

FIGURE P1.98

The initial volume of water was 25.0 mL in each graduated cylinder. The following data were collected:

	Mass (g)	Height (cm)	Diameter (cm)
Cylinder A	15.560	5.1	1.2
Cylinder B	35.536	5.9	1.3

a. Calculate the volume of each cylinder using the dimensions of the cylinder only.
b. Calculate the volume from the water displacement method.
c. Which volume measurement allows for the greater number of significant figures in the calculated densities?
d. Express the density of each cylinder to the appropriate number of significant figures.

*1.99. Table salt contains 1.54 g of chlorine (as chloride ions) for every 1.00 g of sodium ions. Which of the following mixtures would react to produce NaCl with no sodium or chlorine left over?
a. 11.0 g of sodium and 17.0 g of chlorine
b. 6.5 g of sodium and 10.0 g of chlorine
c. 6.5 g of sodium and 12.0 g of chlorine
d. 6.5 g of sodium and 8.0 g of chlorine

*1.100. The wood of the black ironwood tree (*Krugiodendron ferreum*, Figure P1.100), which grows in the West Indies and coastal areas of South Florida, is so dense that it sinks in seawater. Does it sink in fresh water, too? Explain your answer.

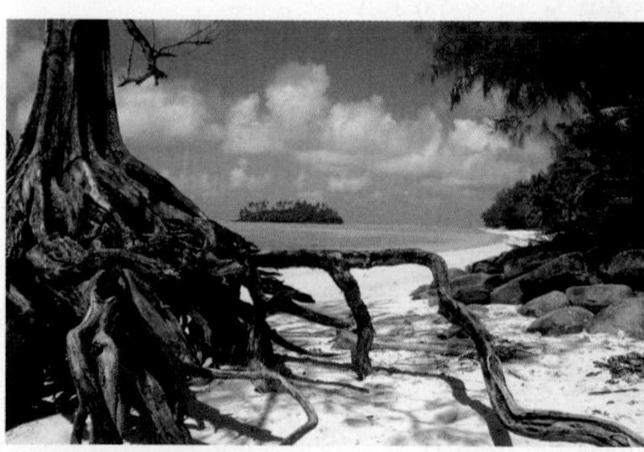

FIGURE P1.100

1.101. Manufacturers of trail mix have to control the distribution of ingredients in their products. Deviations of more than 2% from specifications may cause production delays and supply problems. A favorite trail mix is supposed to contain 67% peanuts and 33% raisins. Bags of trail mix were sampled from the production line on different days with the following results:

Day	Number of Peanuts	Number of Raisins
1	50	32
11	56	26
21	48	34
31	52	30

On which day(s) did the product meet the specification of 65% to 69% peanuts?

*1.102. Gasoline and water are immiscible. Regular-grade (87 octane) gasoline has a lower density (0.73 g/mL) than water (1.00 g/mL). A 100-mL graduated cylinder with an inside diameter of 3.2 cm contains 34.0 g of gasoline and 34.0 g of water. What is the combined height of the two liquid layers in the cylinder? The volume of a cylinder is $\pi r^2 h$, where r is the radius and h is the height.

1.103. **Drug Overdose** In 1999 a drug overdose incident occurred when a prescription that called for a patient to receive 0.5 *grain* of the powerful sedative phenobarbital each day was misread, and the patient was given 0.5 *gram* of the drug. Actually, four intravenous injections of 130 mg each were administered each day for 3 days. How many times as much phenobarbital was administered with respect to the prescribed amount? (1 grain = 64.79891 mg.)

*1.104. **Mercury in Dental Fillings** The controversy over human exposure to mercury from dental fillings (Figure P1.104) is linked to concerns that mercury may volatilize from fillings made of a combination of silver and mercury and may then be breathed into the lungs and absorbed into the blood. In 1995 the U.S. Environmental Protection Agency (EPA) set a safe exposure level for mercury vapor in air of 0.3 $\mu g/m^3$. Typically, an adult breathes in 0.5 L of air 15 times per minute.

a. What rate of volatilization of mercury (in μg/minute) from dental fillings would create an exposure level of 0.3 μg Hg/m^3 in the air entering the lungs of an adult?
b. The safe exposure level to inhaled mercury vapor adopted by Health Canada is only 0.06 $\mu g/m^3$. What rate of volatilization of mercury (in μg/minute) from dental fillings would create this exposure level in air entering the lungs of a child who breathes in 0.35 L of air 18 times per minute?

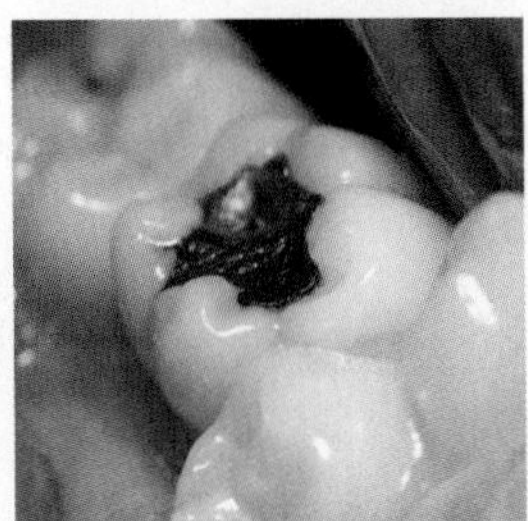

FIGURE P1.104

***1.105.** The digital thermometers used in a hospital are evaluated by immersing them in an ice-water bath at 0.0°C and then in boiling water at 100.0°C. The following results were obtained for three thermometers (A, B, and C):

	Measured Temperature (°C)	
Thermometer	Ice Water	Boiling Water
A	−0.8	99.4
B	0.2	99.8
C	0.4	101.0

a. Which of the three thermometers, if any, would detect an increase of 0.1°C in the temperature of a patient?
b. Which of the three thermometers, if any, would accurately give a reading of 36.8°C (under-the-tongue temperature) for a patient without a fever?

1.106. **Deepwater Horizon Oil Spill** According to the U.S. government, 4.9 billion barrels of crude oil flowed into the Gulf of Mexico following the explosion that destroyed the Deepwater Horizon drilling rig in April 2010. Express this volume of crude oil in liters and in cubic kilometers. (1 barrel of oil = 42 gallons.)

1.107. ***New Horizons*** As the *New Horizons* spacecraft approached Pluto (Figure P1.107) in June 2015, it was traveling away from the sun at a speed of 14.51 km/s. What was this velocity in miles per hour?

FIGURE P1.107

1.108. **Sodium in Candy Bars** Three different analytical techniques were used to determine the quantity of sodium in a Mars Milky Way candy bar. Each technique was used to analyze five portions of the same candy bar, with the following results (expressed in milligrams of sodium per candy bar):

mg of Na: Technique 1	mg of Na: Technique 2	mg of Na: Technique 3
109	110	114
111	115	115
110	120	116
109	116	115
110	113	115

The actual quantity of sodium in the candy bar was 115 mg. Which techniques would you describe as precise, which as accurate, and which as both? What is the range of the values (the difference between the highest and lowest measurements) for each technique?

smartwork5

If your instructor uses Smartwork5, log in at digital.wwnorton.com/atoms2.

2

Atoms, Ions, and Molecules

The Building Blocks of Matter

MAGNETIC RESONANCE IMAGING Magnetic resonance imaging (MRI) is used extensively to diagnose injuries and diseases in soft tissue. MRI signals are produced by the protons in the nuclei of the hydrogen atoms in molecules of H_2O.

PARTICULATE **REVIEW**

Atoms and Molecules

In Chapter 1, we learned that all substances, be they elements or compounds, are composed of atoms. In many substances these atoms form ions or are bonded together in molecules. In this chapter and the next we explore the structure of atoms to better understand why they form the ions and molecules they do.

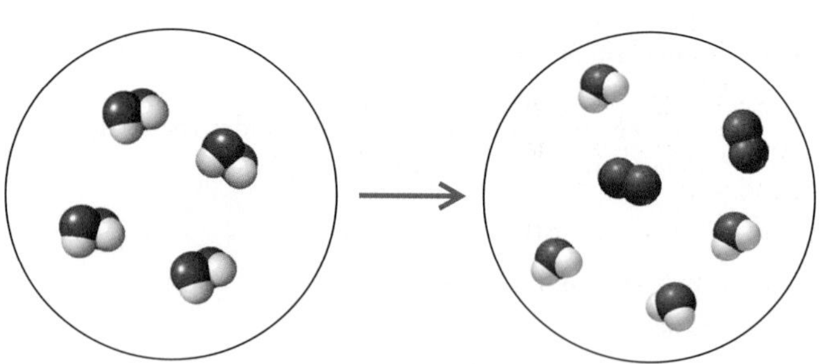

The decomposition of hydrogen peroxide produces water and oxygen, as shown here.

- How many atoms are there in one molecule of hydrogen peroxide?
- How many oxygen molecules are depicted here?
- What is the ratio of hydrogen atoms to oxygen atoms in hydrogen peroxide?
- What is the ratio of hydrogen atoms to oxygen atoms in water?

(Review Sections 1.1, 1.3, and 1.4 if you need help.)

(Answers to Particulate Review questions are in the back of the book.)

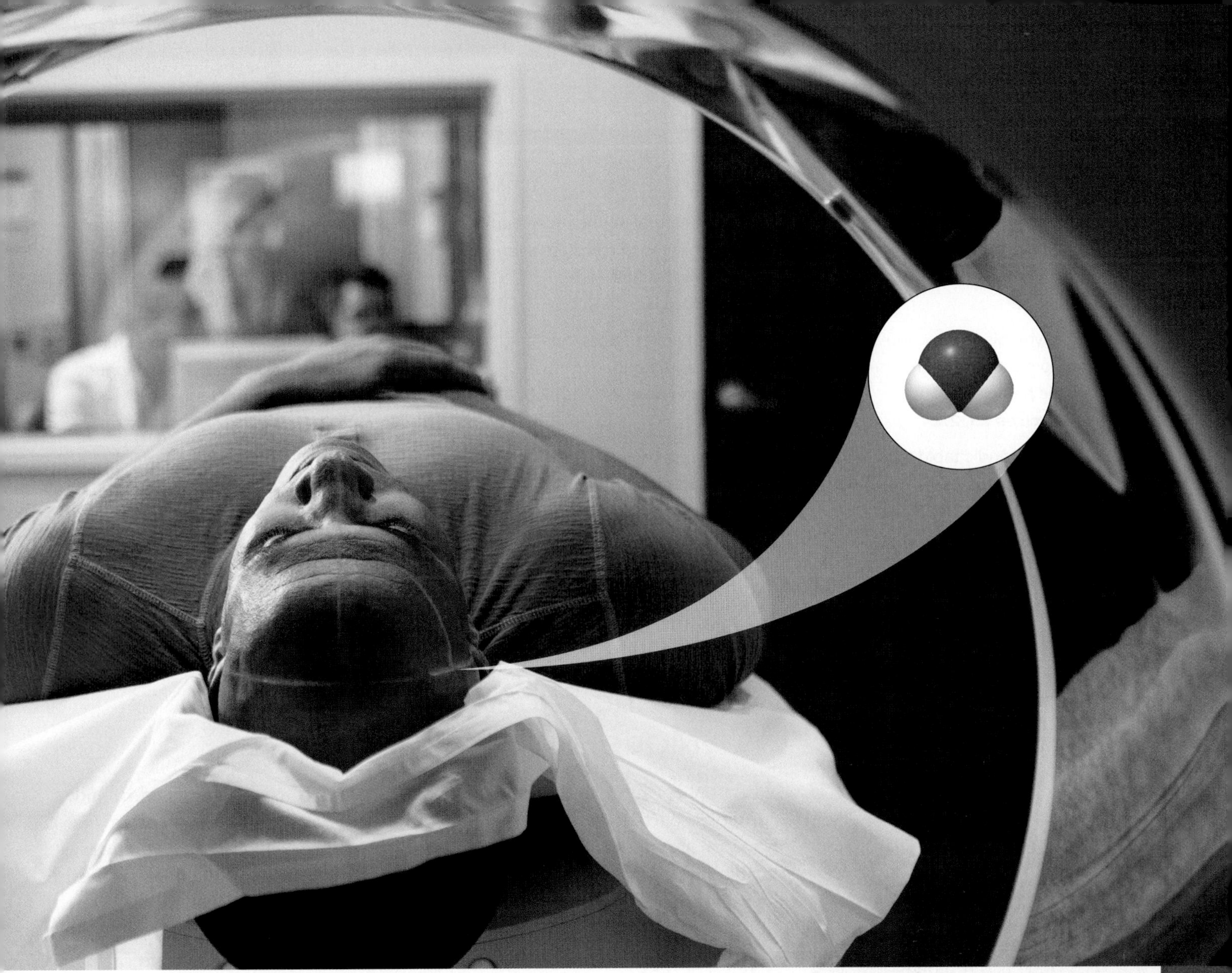

PARTICULATE **PREVIEW**

Nuclei and Nucleons

Three nuclei are depicted here. Assume all the particles that make up each nucleus are visible. As you read Chapter 2, look for ideas that will help you answer these questions:

- What is the mass number of each nucleus?
- What is the atomic number of each nucleus?
- Which two nuclei are isotopes of each other?

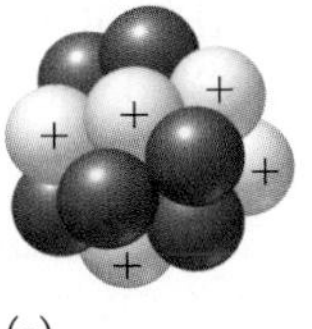

(a)

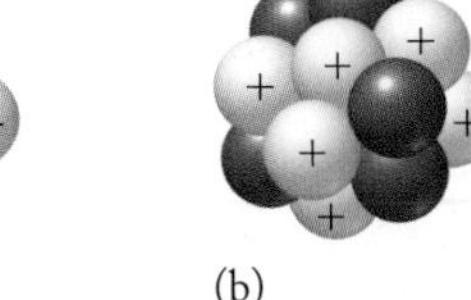

(b)

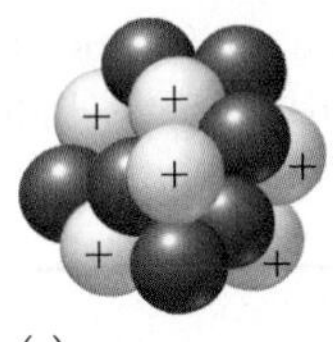

(c)

Learning Outcomes

LO1 Explain how the experiments of Thomson, Millikan, and Rutherford contributed to our understanding of atomic structure

LO2 Write symbols to describe atoms, their monatomic ions, and the three types of subatomic particles
Sample Exercises 2.1, 2.2

LO3 Use the periodic table to explain and predict the properties of elements
Sample Exercise 2.3

LO4 Use natural abundances and isotopic masses to calculate average atomic mass values
Sample Exercises 2.4, 2.5

LO5 Interconvert mass, number of particles, and number of moles
Sample Exercises 2.6, 2.7, 2.8, 2.9

LO6 Use chemical formulas and average atomic masses to calculate molecular masses and molar masses
Sample Exercise 2.10

LO7 Determine the molecular mass of a molecule from its mass spectrum
Sample Exercise 2.11

LO8 Determine the isotopic abundance of an element from its mass spectrum or from the mass spectrum of a compound containing the element
Sample Exercise 2.12

2.1 When Projectiles Bounced Off Tissue Paper: The Rutherford Model of Atomic Structure

Philosophers in ancient Greece proposed that matter was composed of indestructible particles called atoms. For over 2000 years the concept of an *indestructible* atom also meant an *indivisible* atom. However, a series of discoveries made in the 1890s proved that atoms are not indivisible. Instead, they are made of even tinier **subatomic particles**: neutrons, protons, and electrons.

FIGURE 2.1 In 1897 J. J. Thomson discovered electrons when he studied how gases conduct electricity. The research earned him the 1906 Nobel Prize in Physics.

CHEMTOUR
Cathode Ray Tube

subatomic particles the neutrons, protons, and electrons in an atom.

cathode rays streams of electrons emitted by the cathode in a partially evacuated tube.

electron a subatomic particle that has a relative charge of 1− and negligible mass.

Electrons

Discovery of the first subatomic particle came in 1897 in the laboratory of English physicist Joseph John (J. J.) Thomson (1856–1940, Figure 2.1). During the 1890s, Thomson's research included investigations of how gases conduct electricity. In this work he used a device called a cathode-ray tube, or CRT (Figure 2.2), which consists of a glass tube from which most of the air has been removed. Electrodes within the tube are attached to the poles of a high-voltage power supply. An electrode called the cathode is connected to the negative terminal of the power supply, and an electrode called the anode is connected to the positive terminal. When these connections are made, electricity travels the length of the glass tube in the form of a beam of **cathode rays** that flows from the cathode toward the anode, passing through a hole cut in the center of the anode. Cathode rays are invisible to the naked eye, but when the end of the tube is coated with a phosphorescent material, a glowing spot appears where the beam hits the coating.

Thomson discovered that cathode-ray beams could be deflected by magnetic (Figure 2.2a) and electric (Figure 2.2b) fields. This behavior revealed that cathode rays were not rays of energy, but rather charged particles of matter. The directions of the deflections told him that their charges were negative. By adjusting the strengths of the electric and magnetic fields, Thomson balanced out the deflections (Figure 2.2c) so that the particles passed straight through the tube. From the strengths of the opposing electric and magnetic fields, he was able to calculate the mass-to-charge ratio of the particles. He also observed that the deflection pattern and the calculated mass-to-charge ratio were always the same no matter what cathode material he used to generate the beam of particles. This observation convinced Thomson that these particles, which he called *corpuscles* but which we now know as **electrons**, were fundamental particles that occur in all forms of matter.

In 1909, American physicist Robert Millikan (1868–1953) advanced Thomson's work by determining the charge of an electron and, indirectly, its mass. Figure 2.3 illustrates Millikan's experimental apparatus. It consisted of two chambers filled with air (that is, mostly N_2 and O_2). A fine spray of oil drops produced in the top chamber fell through a hole into the bottom chamber. Highly energetic X-rays also passed through the bottom chamber, colliding with the N_2 and O_2 molecules.

The interaction of X-rays with N_2 and O_2 molecules resulted in the loss of an electron, yielding a molecular *ion* with a positive charge:

$$N_2 \rightarrow N_2^+ + e^- \quad (2.1)$$

The superscripts in Equation 2.1 indicate the electrical charges on the nitrogen ion (1+) and the electron (1−).

Particles of matter can also gain electrons. As Millikan's oil drops fell through the bottom chamber, they collided with and absorbed free electrons, thereby acquiring a negative charge. He observed the rate of fall of the negatively charged oil drops with a microscope in the side of the bottom chamber (Figure 2.3). Because like charges repel one another and opposite charges attract, Millikan could adjust the drops' rate of fall using charged metal plates located above and below the bottom chamber, thus creating a vertical electric field within it. From the strength of the electric field and the rate of fall of the drops, he calculated the charges on them. Millikan discovered that the charge on each drop was always a

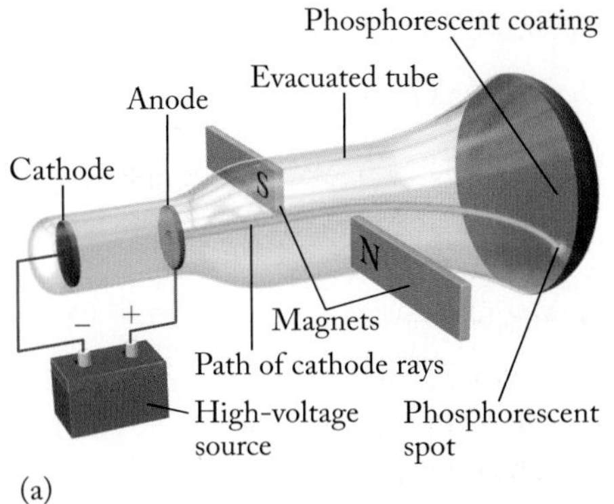

(a)

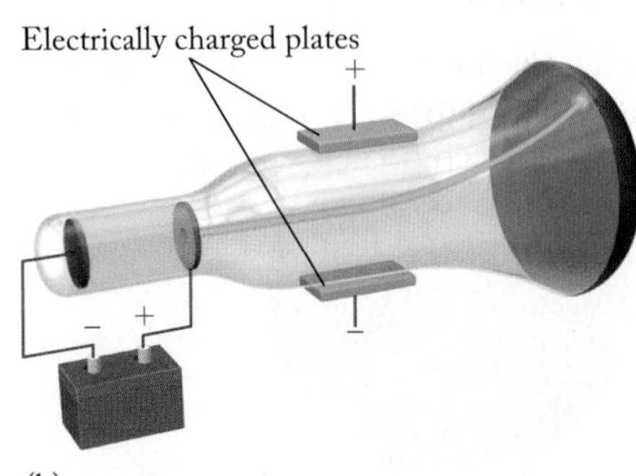

(b)

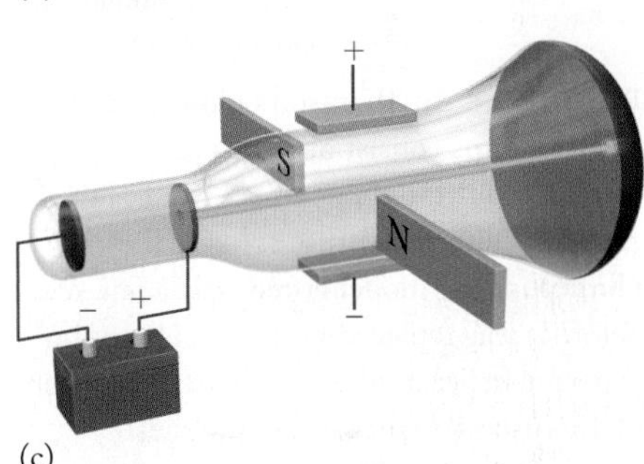

(c)

FIGURE 2.2 A cathode ray is generated when electricity is passed through a tube from which most of the air has been removed. Though invisible to the unaided eye, the path of the rays shown in green can be inferred by the bright spot it makes on a phosphorescent material coated on the end of the tube. (a) Cathode rays are deflected by magnetic fields and (b) by electric fields; (c) electric and magnetic fields tuned to balance out the deflections.

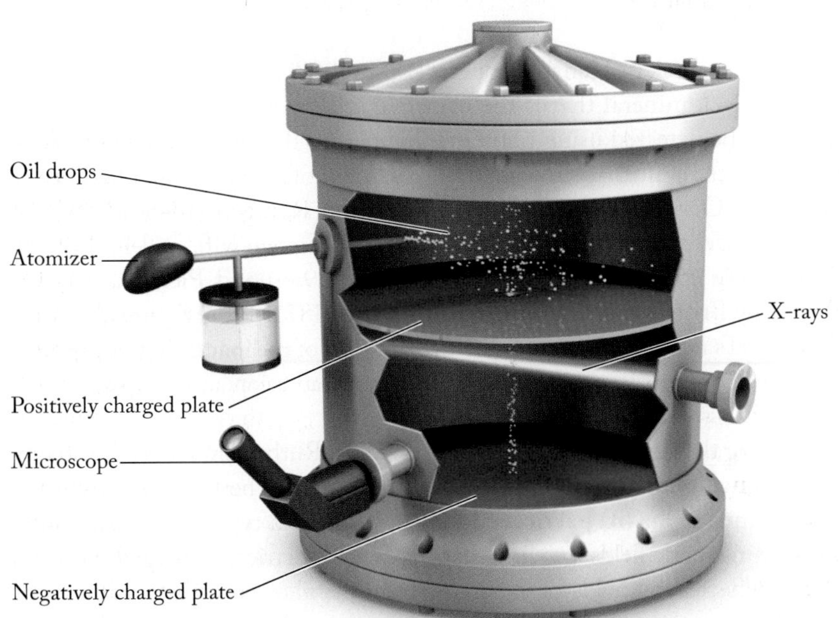

FIGURE 2.3 In Millikan's oil-drop experiment, X-rays ionized the air in the lower chamber, producing electrons that were absorbed by tiny drops of oil falling through the chamber. The descent of the electrically charged drops could be slowed, stopped, or even reversed by applying an electric field with charged plates above and below the chamber. A microscope allowed Millikan to follow the descent of the oil drops.

CONNECTION An ion is an atom or molecule with an electrical charge (Section 1.3).

whole-number multiple of a minimum value. He concluded that this minimum value must be the charge on one electron. Millikan's experiments and calculation yielded a value for the charge of an electron (e) that was within 1% of the modern value. Knowing Thomson's value of the electron's mass-to-charge ratio, Millikan was able to calculate the electron's mass (m_e). We can reproduce Millikan's calculation using modern values of an electron's charge (-1.60218×10^{-19} C, where C is the abbreviation for *coulomb*, the SI unit of charge) and a mass-to-charge ratio, -5.6856×10^{-12} kg/C:

$$m_e = e \times (m_e/e) = (-1.60218 \times 10^{-19}\ \cancel{\text{C}}) \times (-5.6856 \times 10^{-12}\ \text{kg}/\cancel{\text{C}})$$
$$= 9.10938 \times 10^{-31}\ \text{kg}$$

CONCEPT TEST

By adjusting the electrical charges on the top and bottom plates of the lower chamber of his apparatus, Millikan was able to slow, stop, or even reverse the fall of oil drops. Which of the two plates must have been positively charged?

(Answers to Concept Tests are in the back of the book.)

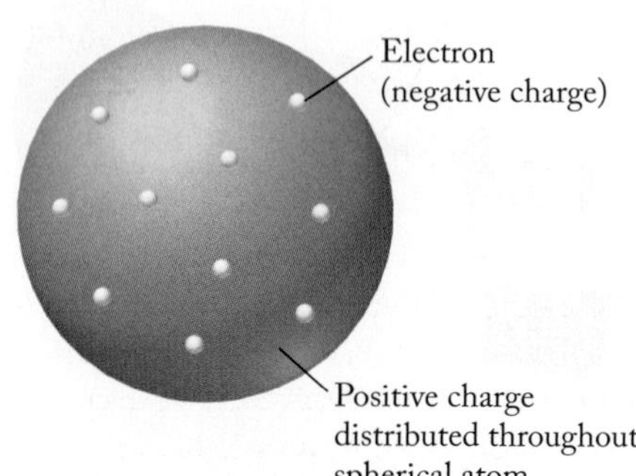

FIGURE 2.4 In Thomson's plum-pudding model, atoms consist of electrons distributed throughout a massive, positively charged, but very diffuse sphere. The plum-pudding model lasted only a few years before it was replaced by a model based on experiments carried out under the direction of Thomson's former student, Ernest Rutherford.

Scientists knew that matter was electrically neutral, so the discovery of the electron raised the possibility that other positively charged subatomic particles existed, too. At the time scientists didn't know how the electrons and positive charges were arranged inside atoms. Thomson proposed a model of the atom (Figure 2.4) in which electrons were distributed throughout the atom like raisins in an English plum pudding (or blueberries in a muffin). Thomson's "plum-pudding" model lasted only a few years. Its demise was linked to another scientific discovery of the 1890s: radioactivity.

Radioactivity

In 1896 French physicist Henri Becquerel (1852–1908) discovered that pitchblende, a brownish-black mineral that is the principal source of uranium, produces radiation that can be detected using photographic plates. Becquerel and his contemporaries initially thought that this radiation consisted of X-rays, which had just been discovered by German scientist Wilhelm Conrad Röntgen (1845–1923).[1] Additional experiments by Becquerel, by the Polish and French wife-and-husband team of Marie Curie (born Marie Skłodowska, 1867–1934) and Pierre Curie (1859–1906), and by British scientist Ernest Rutherford (1871–1937, Figure 2.5) showed that Becquerel's radiation was actually several types of **radioactivity**, a term used to describe the spontaneous emission of high-energy radiation and particles by radioactive materials such as pitchblende.

FIGURE 2.5 Ernest Rutherford was born in New Zealand. He was awarded a scholarship in 1894 that enabled him to go to Trinity College in Cambridge, England, where he was a research assistant in the laboratory of J. J. Thomson. His contributions included characterizing the properties of α and β particles. By 1907 he was a professor at the University of Manchester, where his gold-foil experiments led to our modern view of atomic structure. He received the Nobel Prize in Chemistry in 1908.

In studying the particles emitted by pitchblende, Rutherford found that one type, which he named **beta (β) particles**, penetrated materials better than another type, which he named **alpha (α) particles**. He knew that both types of particles could be deflected by magnetic fields, proving that they were electrically charged. How much they were deflected by fields of different strengths allowed him to calculate their mass-to-charge ratios. He found that this ratio for β particles exactly matched

[1]Röntgen discovered X-rays in experiments with a cathode-ray tube much like the apparatus used by J. J. Thomson. After completely encasing the tube in a black carton, Röntgen discovered that invisible rays escaped the carton and were detected by a photographic plate. Because he knew so little about the rays, he called them X-rays.

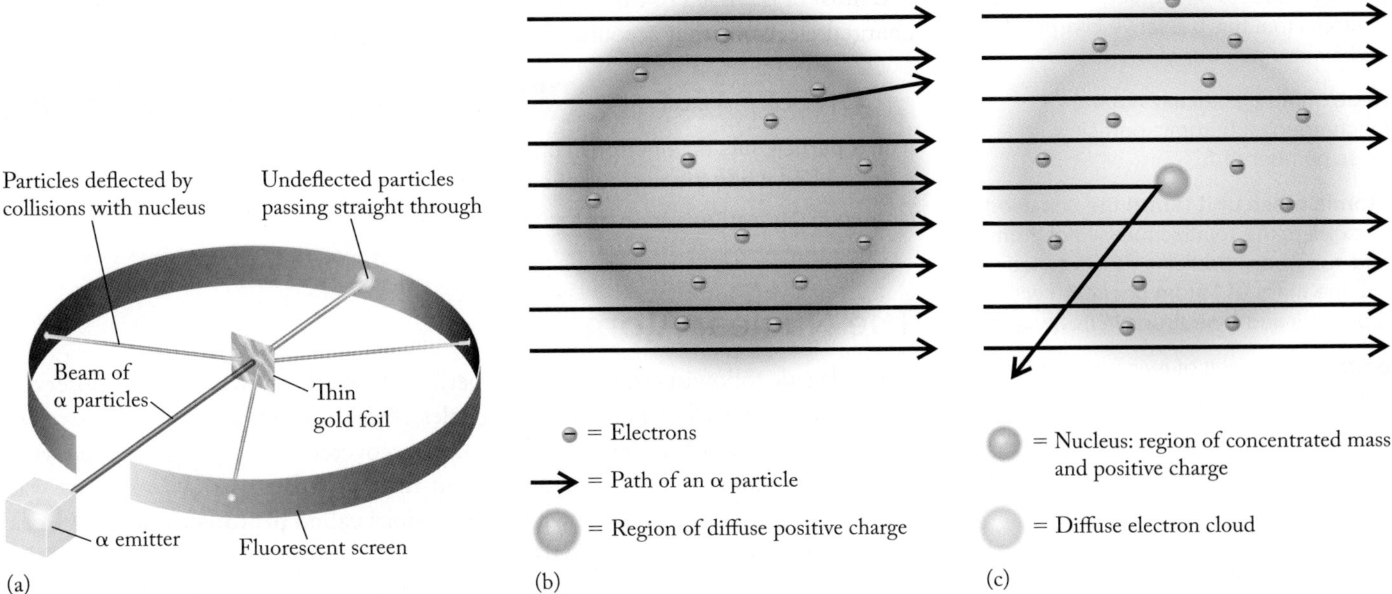

FIGURE 2.6 (a) The design of Rutherford's gold-foil experiments. (b) If Thomson's plum-pudding model had been correct, most of the α particles would have passed straight through the thin gold foil, though a few might have been deflected slightly. (c) In fact, most did pass straight through, but a few were scattered widely as shown in (a). This unexpected result led to the theory that an atom has a small, positively charged nucleus that contains most of the mass of the atom.

the mass-to-charge ratio of the electron that had been determined previously by Thomson, which suggested that β particles were simply high-energy electrons.

The direction in which α particles are deflected in a magnetic field is opposite that of β particles, which means α particles are positively charged. If the β particle (an electron) is assigned a relative charge of 1−, the corresponding charge of an α particle is 2+. Rutherford also discovered that α particles are about 10,000 times more massive than β particles.

In 1909 Rutherford directed two of his students at Manchester University—Hans Geiger (1882–1945, for whom the Geiger counter was named) and Ernest Marsden (1889–1970)—to test the plum-pudding model by bombarding a thin foil of gold with a beam of α particles (Figure 2.6a). If the plum-pudding model was correct, then most of the particles would pass straight through the diffuse spheres of positive charge that made up the gold atoms, though a few might interact with the electrons (the "raisins" embedded in the pudding) enough to be deflected slightly (Figure 2.6b).

Geiger and Marsden did observe that most of the α particles passed straight through the gold, as Rutherford expected. However, about 1 in every 8000 particles was deflected by an average angle of 90° (Figure 2.6c), and a very few bounced almost straight back at their source. Rutherford later described his amazement at the result: "It was almost as incredible as if you had fired a 15-inch shell[2] at a piece of tissue paper and it came back and hit you."

Thomson's plum-pudding model could not explain such large angles of deflection. Rutherford concluded that the deflections occurred because a tiny fraction of the α particles encountered small regions of high positive charge and large mass. Based on the relative numbers of α particles that were deflected in this way, Rutherford determined that the diameter of a gold atom was over 10,000 times greater than the diameter of the region of positive charge at its center. Rutherford's model of the atom is the basis for our current understanding of atomic structure. It assumes that an atom consists of a tiny **nucleus** that contains the positive charge

radioactivity the spontaneous emission of high-energy radiation and particles by materials.

beta (β) particle a radioactive emission that is a high-energy electron.

alpha (α) particle a radioactive emission with a charge of 2+ and a mass equivalent to that of a helium nucleus.

nucleus (of an atom) the positively charged center of an atom that contains nearly all the atom's mass.

[2]The most widely used "heavy" guns on British battleships in the early 20th century fired shells with a diameter of 15 inches (38 cm).

proton a subatomic particle in the nuclei of atoms that has a relative charge of 1+ and a mass number of 1.

neutron an electrically neutral (uncharged) subatomic particle with a mass number of 1.

atomic mass unit (amu) the unit used to express the relative masses of atoms and subatomic particles that is exactly 1/12 the mass of 1 atom of carbon with 6 protons and 6 neutrons in its nucleus.

dalton (Da) a unit of mass equal to 1 atomic mass unit.

and most of the mass of the atom and is surrounded by a diffuse cloud of negatively charged electrons that accounts for most of the volume of the atom.

CONCEPT **TEST**

If an α particle hits an electron in an atom of gold, why doesn't it bounce back the way it does when it hits the nucleus of a gold atom?

(Answers to Concept Tests are in the back of the book.)

The Nuclear Atom

In the decade following the gold-foil experiments, Rutherford and others observed that bombarding elements with α particles sometimes changed, or *transmuted*, the elements into other elements. They also discovered that hydrogen nuclei were frequently produced during transmutation reactions. By 1920 a consensus was growing that hydrogen nuclei, which Rutherford called **protons** (from the Greek *protos*, meaning "first"), were part of all nuclei. For example, to account for the mass and charge of an α particle, Rutherford assumed that it was made of four protons, two of which had combined with two electrons to form two electrically neutral particles, which he called **neutrons**. Repeated attempts to produce neutrons by neutralizing protons with electrons were unsuccessful. However, in 1932 one of Rutherford's former students, James Chadwick (1891–1974), was the first to successfully detect and characterize free neutrons. With the discovery of neutrons, the current model of atomic structure was complete, as illustrated by the gold atom in Figure 2.7.

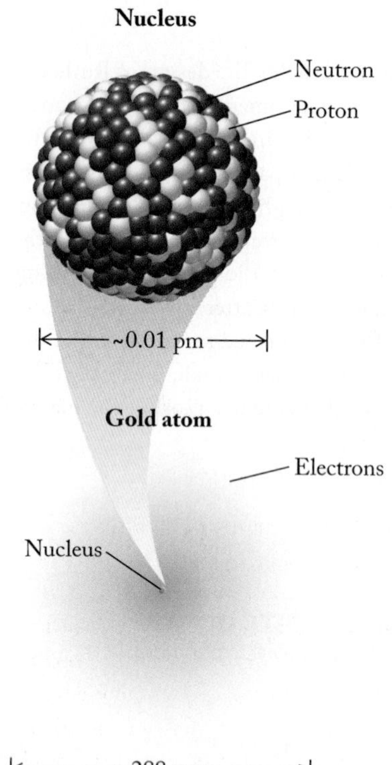

FIGURE 2.7 The modern view of Rutherford's model of the gold atom includes a nucleus that is about 1/10,000 the overall size of the atom. The nucleus would be too small to see if drawn to scale in the left drawing.

Table 2.1 summarizes the properties of neutrons, protons, and electrons. Their masses are expressed in kilograms (the SI standard unit of mass) and in **atomic mass units (amu)**. One amu is exactly 1/12 the mass of a carbon atom that has 6 protons and 6 neutrons in its nucleus. Atomic mass units are also called **daltons (Da)** or *unified atomic mass units* (u). The term *dalton* honors English chemist John Dalton, who published the first table of atomic masses in 1803. Dividing the mass in kilograms of any of the particles in Table 2.1 by its mass in amu shows that 1 amu = 1.66054×10^{-27} kg. We use this equality in several calculations later in this chapter. Note that this value and the mass values in Table 2.1 are expressed using six significant figures, which is more than enough for our calculations.

According to the data in Table 2.1, the masses of neutrons and protons are similar and are much greater than the mass of an electron. The relative masses of these particles are reflected in their *mass numbers*, which are not real mass values, but rather a way of counting the number of subatomic particles in an atom. Neutrons and protons, which make up nearly all of an atom's mass, each have mass numbers of 1, whereas the mass number of an electron is 0, reflecting its minimal contribution to the mass of an atom.

Rutherford's pioneering work with atomic nuclei represented only the beginning of decades of research on their composition and properties. For example, in

TABLE 2.1 Properties of Subatomic Particles

Particle	Symbol	Mass (amu)	Mass Number	Mass (kg)	Charge (relative value)	Charge (C)
Neutron	${}_{0}^{1}n$	1.00866	1	1.67493×10^{-27}	0	0
Proton	${}_{1}^{1}p$	1.00728	1	1.67262×10^{-27}	1+	$+1.60218 \times 10^{-19}$
Electron	${}_{-1}^{0}e$ or ${}_{-1}^{0}\beta$	5.48580×10^{-4}	0	9.10938×10^{-31}	1−	-1.60218×10^{-19}

the 1930s it was discovered that when atoms with nuclei that have odd numbers of protons and/or neutrons are placed inside magnetic fields, the nuclei can absorb and re-emit tiny quantities of energy. This discovery led to the development of a technique for the medical imaging of soft tissue known as magnetic resonance imaging or MRI (see the chapter-opening figure).

isotopes atoms of an element containing the same number of protons but different numbers of neutrons.

nuclide a specific isotope of an element.

atomic number (*Z*) the number of protons in the nucleus of an atom.

nucleon a proton or neutron in a nucleus.

mass number (*A*) the number of nucleons in an atom.

2.2 Nuclides and Their Symbols

As Thomson continued to experiment with cathode-ray tubes, he modified them so that he could study the beams of positively charged particles that flowed from the anode toward the cathode as cathode rays (electrons) flowed in the opposite direction. A combination of electric and magnetic fields surrounding the tubes of these *positive-ray analyzers* deflected positively charged ions according to their charges and masses. If all the ions in a particular beam had the same charge, say 1+, and different masses, the ions with the greatest mass would be deflected the least, and those with the smallest mass would be deflected the most.

In 1912 Thomson and his research assistant, Francis W. Aston (1877–1945), observed that when they passed an electric current through a tube that contained small quantities of neon gas, two bright patches formed on a photographic plate, as shown in Figure 2.8. They assumed that the patches were produced by neon atoms that had lost electrons in the tube, forming positively charged Ne^+ ions. The presence of two patches meant that Ne^+ ions and their parent Ne atoms had two different masses. The one with lower mass produced the brighter patch, indicating that more of its ions hit the plate and that it must be the more abundant of the two.

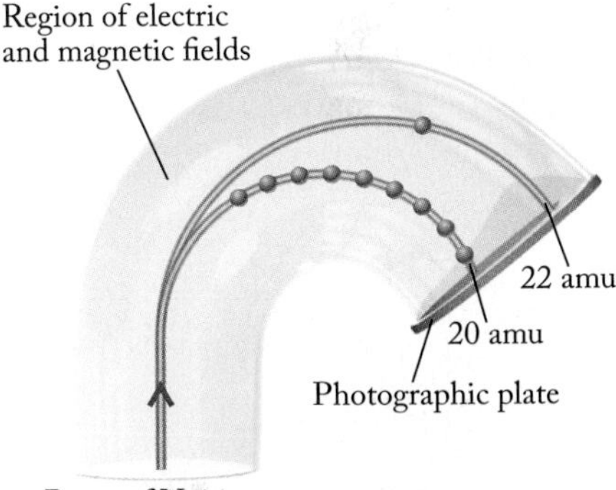

FIGURE 2.8 Aston's positive-ray analyzer. A beam of Ne^+ ions passing through electric and magnetic fields separates into two beams. Ions with a mass of about 20 amu—90% of the total—are deflected more than the other 10%, which have a mass of about 22 amu. Aston's positive-ray analyzer was the forerunner of the modern mass spectrometer.

Today we know that the Ne^+ ions detected by Thomson and Aston came from two different **isotopes** of neon. Isotopes are atoms of the same element that have the same number of protons (10 for neon) in their nuclei but different numbers of neutrons. The lighter Ne isotope has 10 neutrons per nucleus, giving it a total mass of about 20 amu. Atoms of the less abundant isotope have 12 neutrons in their nuclei, giving them a mass of about 22 amu. The term **nuclide** is used to refer to any atom of any element that has a particular number of neutrons in its nucleus.

Since the time of John Dalton, scientists had defined an element as *matter composed of identical atoms, all of which have the same mass*. The work of Thomson, Aston, Chadwick, and others meant this definition had to be modified: henceforth an element was defined as *matter composed of atoms all having the same number of protons in their nuclei*. This number of protons is called the **atomic number (*Z*)** of the element. The total number of **nucleons** (neutrons and protons) in the nucleus of an atom defines its **mass number (*A*)**. The isotopes of a given element thus all have the same atomic number *Z* but different mass numbers *A*.

The symbol we use to represent a particular nuclide has the generic form

$$^{A}_{Z}\mathrm{X}$$

where X represents the one- or two-letter symbol for the element, *Z* is the element's atomic number, and *A* is the mass number of the particular nuclide. For example, two of the isotopes of oxygen (O) and two of the isotopes of lead (Pb) are written

$$^{16}_{8}\mathrm{O} \quad ^{18}_{8}\mathrm{O} \quad ^{206}_{82}\mathrm{Pb} \quad ^{208}_{82}\mathrm{Pb}$$

Because *Z* and X provide the same information—each by itself identifies the element—the subscript *Z* may be omitted, so that the nuclide symbol may be simplified to

$$^{A}\mathrm{X}$$

The name of a nuclide may also be spelled out as the name of the element followed by the mass number of the nuclide. For example, the names of the two isotopes of neon detected by Thomson and Aston may be written *neon-20* and *neon-22* or abbreviated as *Ne-20* and *Ne-22*, respectively.

The $^{A}_{Z}X$ symbol can also be used to represent subatomic particles, as shown in Table 2.1. As with nuclides, the superscripts of the symbols are the particles' mass numbers. The subscripts, however, represent the relative charges of the particles: 0 for a neutron, 1 for a proton, and 1− for an electron. This notation for the subscripts makes sense because an element's atomic number Z is simply the number of protons in the nucleus of an atom of that element, which is the positive charge of that nucleus.

SAMPLE EXERCISE 2.1 Relating the Symbols of Nuclides to the Composition of Their Nuclei **LO2**

Several nuclides of gaseous elements are useful for the MRI of pulmonary (lung) function. Write symbols in the form $^{A}_{Z}X$ for the nuclides that have (a) 2 protons and 1 neutron, (b) 36 protons and 47 neutrons, and (c) 54 protons and 75 neutrons.

Collect, Organize, and Analyze We are given the number of protons in the nucleus of an atom of element X, which defines its atomic number (Z) and also defines its one- or two-letter symbol. The sum of the number of nucleons (protons plus neutrons) is the mass number (A) of the nuclide. We are asked to write symbols in the form $^{A}_{Z}X$.

Solve

a. This nuclide has 2 protons, so $Z = 2$, which is the atomic number of helium. Two protons plus 1 neutron gives a mass number of 3. Therefore, this nuclide is helium-3, or $^{3}_{2}He$.
b. This nuclide has 36 protons, so $Z = 36$, which is the atomic number of krypton. Thirty-six protons plus 47 neutrons gives a mass number of 83. This isotope is krypton-83, or $^{83}_{36}Kr$.
c. This nuclide has 54 protons, so $Z = 54$, which is the atomic number of xenon. The mass number is $54 + 75 = 129$, so this isotope is xenon-129, or $^{129}_{54}Xe$.

Think About It In working through this exercise, did you use the periodic table of the elements inside the front cover to identify the symbol of the element once you knew its atomic number? It's a straightforward search because the elements in the table are arranged in order of increasing atomic number. Only these isotopes of helium, krypton, and xenon are useful for MRI.

Practice Exercise Use the $^{A}_{Z}X$ format to write the symbols of the nuclides whose atoms each have (a) 26 protons and 30 neutrons, (b) 7 protons and 8 neutrons, (c) 17 protons and 20 neutrons, and (d) 19 protons and 20 neutrons.

(Answers to Practice Exercises are in the back of the book.)

SAMPLE EXERCISE 2.2 Identifying Ionic Nuclides and Writing Their Isotopic Symbols **LO2**

In March 2011, an earthquake and tsunami off the coast of northeastern Japan crippled nuclear reactors at a power station in Fukushima, Japan. The resulting explosions and fires released radiation into the atmosphere and ocean that included two ionic nuclides $^{134}Cs^{+}$ and $^{131}I^{-}$ and two other single-atom ions, labeled (c) and (d) in the table, with the following numbers of protons, neutrons, and electrons in each of their nuclei:

	Symbol	Protons	Neutrons	Electrons
(a)	$^{134}Cs^{+}$			
(b)	$^{131}I^{-}$			
(c)		55	82	54
(d)		94	145	90

Complete the table by determining the numbers of protons, neutrons, and electrons in each $^{134}Cs^{+}$ ion and $^{131}I^{-}$ ion and by writing the $^{A}X^{Q}$ symbols of ionic nuclides (c) and (d), where Q is the charge of each ion.

Collect, Organize, and Analyze We are to relate the numbers of neutrons, protons, and electrons and the $^{A}X^{Q}$ symbols for a total of four radioactive single-atom ions. The number of protons in the nucleus of an atom or single-atom ion defines its atomic number, which in turn defines what element it is (X). The sum of the number of protons plus neutrons is the mass number (A), and the charge (Q) is the difference between the number of protons and the number of electrons. Q has a negative value in ions with more electrons than protons and a positive value in ions with more protons than electrons.

Solve

a. Cs is the symbol of cesium, which has an atomic number of 55. Therefore, there are 55 protons in the nucleus of a $^{134}Cs^{+}$ ion, and there are $134 - 55 = 79$ neutrons. The ion has a 1+ charge, which means it has one fewer electron than protons, or 54 electrons.

b. I is the symbol of iodine, whose atomic number is 53. Therefore, there are 53 protons in the nucleus of a $^{131}I^{-}$ ion, and there are $131 - 53 = 78$ neutrons. The charge on the ion is 1−, so it has one more electron than protons, or $53 + 1 = 54$.

c. This ion has 55 protons, which makes its parent element cesium (Cs). Its mass number (A) is $55 + 82 = 137$, and its charge is $55 - 54 = 1+$. The charge symbol of an ion with a single positive charge is simply +, so the symbol of the ion is $^{137}Cs^{+}$.

d. This ion has 94 protons, which makes its parent element plutonium (Pu). Its mass number (A) is $94 + 145 = 239$, and its charge is $94 - 90 = 4+$. The symbol of the ion is $^{239}Pu^{4+}$.

Use these results to fill in the table:

	Symbol	Protons	Neutrons	Electrons
(a)	$^{134}Cs^{+}$	55	79	54
(b)	$^{131}I^{-}$	53	78	54
(c)	$^{137}Cs^{+}$	55	82	54
(d)	$^{239}Pu^{4+}$	94	145	90

Think About It Each of the ions in this Sample Exercise has an *unequal* number of protons and electrons. In contrast, each atom in the previous Sample Exercise had an *equal* number of protons and electrons and was electrically neutral.

Practice Exercise Complete the following table by adding the appropriate numbers of protons, neutrons, and electrons, or writing the $^{A}X^{Q}$ symbol related to each of these four ions present in seawater.

	Symbol	Protons	Neutrons	Electrons
(a)	$^{40}Ca^{2+}$			
(b)	$^{79}Br^{-}$			
(c)		17	18	18
(d)		11	12	10

(Answers to Practice Exercises are in the back of the book.)

2.3 Navigating the Periodic Table

Long before chemists knew about subatomic particles and the concept of atomic numbers, they knew that groups of elements such as Li, Na, and K, or F, Cl, and Br had similar properties and that, when the elements were arranged by increasing atomic mass, there were repeating patterns of similar properties. This *periodicity* in the properties of the elements inspired several 19th-century scientists to create tables of the elements in which the elements were arranged in patterns based on similarities in their chemical properties.

By far the most successful of these scientists was Russian chemist Dmitri Mendeleev (1834–1907). In 1872 he published a table (Figure 2.9a) that is widely considered the forerunner of the modern **periodic table of the elements** (inside the front cover and Figure 2.9b). In addition to organizing all the elements that were known at the time, Mendeleev also left empty cells in his table for as yet unknown elements so that he could align the known elements with others having similar chemical properties in the same columns. Based on the locations of the empty cells, Mendeleev was able to predict the chemical properties of the missing elements. These predictions greatly facilitated the subsequent discovery of these elements by other scientists. Note that Mendeleev arranged the elements in his periodic table in order of increasing atomic mass, but in the modern periodic table the elements appear in order of their atomic numbers.

CONCEPT **TEST**

Suggest a reason why the elements in Mendeleev's version of the periodic table are in order of atomic mass and not atomic number.

(Answers to Concept Tests are in the back of the book.)

Group Number →

Row	I	II	III	IV	V	VI	VII	VIII
1	1 H							
2	7 Li	9.4 Be	11 B	12 C	14 N	16 O	19 F	
3	23 Na	24 Mg	27.3 Al	28 Si	31 P	32 S	35.5 Cl	
4	39 K	40 Ca	44 ?	48 Ti	51 V	52 Cr	55 Mn	56 Fe, 59 Co, 59 Ni
5	63 Cu	65 Zn	68 ?	72 ?	75 As	78 Se	80 Br	
6	85 Rb	87 Sr	88 ?Yt	90 Zr	94 Nb	96 Mo	100 ?	104 Ru, 104 Rh, 106 Pd
7	108 Ag	112 Cd	113 In	118 Sn	122 Sb	125 Te	127 J	
8	133 Cs	137 Ba	138 ?Di	140 ?Ce				
9								
10			178 ?Er	180 ?La	182 Ta	184 W		195 Os, 197 Ir, 198 Pt
11	199 Au	200 Hg	204 Tl	207 Pb	208 Bi			
12				231 Th		240 U		

(a)

Atomic number / Symbol for element

Period	1	2	3	4	5	6	7	8	9	10	11	12	13	14	15	16	17	18
1	1 H																	2 He
2	3 Li	4 Be											5 B	6 C	7 N	8 O	9 F	10 Ne
3	11 Na	12 Mg											13 Al	14 Si	15 P	16 S	17 Cl	18 Ar
4	19 K	20 Ca	21 Sc	22 Ti	23 V	24 Cr	25 Mn	26 Fe	27 Co	28 Ni	29 Cu	30 Zn	31 Ga	32 Ge	33 As	34 Se	35 Br	36 Kr
5	37 Rb	38 Sr	39 Y	40 Zr	41 Nb	42 Mo	43 Tc	44 Ru	45 Rh	46 Pd	47 Ag	48 Cd	49 In	50 Sn	51 Sb	52 Te	53 I	54 Xe
6	55 Cs	56 Ba	57 La	72 Hf	73 Ta	74 W	75 Re	76 Os	77 Ir	78 Pt	79 Au	80 Hg	81 Tl	82 Pb	83 Bi	84 Po	85 At	86 Rn
7	87 Fr	88 Ra	89 Ac	104 Rf	105 Db	106 Sg	107 Bh	108 Hs	109 Mt	110 Ds	111 Rg	112 Cn	113 Nh	114 Fl	115 Mc	116 Lv	117 Ts	118 Og

6 Lanthanides	58 Ce	59 Pr	60 Nd	61 Pm	62 Sm	63 Eu	64 Gd	65 Tb	66 Dy	67 Ho	68 Er	69 Tm	70 Yb	71 Lu
7 Actinides	90 Th	91 Pa	92 U	93 Np	94 Pu	95 Am	96 Cm	97 Bk	98 Cf	99 Es	100 Fm	101 Md	102 No	103 Lr

(b)

FIGURE 2.9 Two views of the periodic table. (a) Mendeleev organized his periodic table around similar properties and atomic masses. He assigned three elements with similar properties to group VIII in rows 4, 6, and 10. As a result, his rows 4 and 5 together contain spaces for 18 elements, corresponding to the 18 groups in the modern periodic table. (b) In the modern table, elements are arranged in order of atomic number (Z). Those in tan are metals, those in blue are nonmetals, and those in green are metalloids (also called semimetals).

The elements in modern periodic tables are arranged in seven horizontal rows (**periods**) and 18 columns (**groups** or **families**). The periods are numbered at the far left of each row, whereas the group numbers appear at the top of each column. The periodic table inside the front cover has a second set of column headings consisting of a number followed by the letter A or B. These secondary headings, though not used in this text, were widely used in earlier versions of the periodic table.

Atoms and molecules may lose or gain more than one electron (Figure 2.10). Those that lose two, three, or four electrons, for example, also become positively charged ions, called **cations**, with corresponding charges of 2+, 3+, and 4+. When atoms or molecules gain electrons, they form negatively charged ions, called **anions**, with charges equivalent to the number of electrons gained: 1− from gaining one electron, 2− from gaining two electrons, and so on. Actually, *all* the cations of group 1 and group 2 elements have charges of 1+ and 2+, respectively.

Group	Charges of the Group's Most Common Monatomic Cations
1	1+
2	2+
3	3+
4	4+
13	3+

The charges on the common single-atom, or *monatomic*, anions shown in Figure 2.10 of group 17 elements are always 1−, those in group 16 are always 2−, and those in group 15 are always 3−. Note how in all three groups the charges of the anions are equal to the group number minus 18 (the total number of groups in the table).

The first row of the periodic table contains only two elements—hydrogen and helium—and the second and third rows each contain only eight. Starting with the fourth row, all 18 columns are full. Actually, the sixth and seventh rows contain more elements than there is space for in an 18-column array. The additional elements appear in the two separate rows at the bottom of the main table. Elements in the row with atomic numbers from 58 to 71 are called the lanthanides (after element 57, lanthanum) and those with atomic numbers between 90 and 103 are called actinides (after element 89, actinium). All the isotopes of the elements with atomic numbers above 83, including all of the actinide elements, are radioactive. These **radionuclides** spontaneously emit high-energy radiation and particles and are transformed into other nuclides. Elements with $Z > 94$ are not found in nature and elements 93 and 94 occur in only extremely small (trace) amounts in minerals that also contain uranium ($Z = 92$).

Several of the groups of elements have names in addition to numbers—names that are based on chemical properties common to the elements in that group (Figure 2.11a). For example, the elements of group 17 are called **halogens**. The word *halogen* is derived from the Greek word for "salt former." Chlorine is

periodic table of the elements a chart of the elements in order of their atomic numbers and in a pattern based on their physical and chemical properties.

period (of elements) all the elements in a row of the periodic table.

group or **family** (of elements) all the elements in a column of the periodic table.

cation a positively charged ion.

anion a negatively charged ion.

radionuclide a radioactive (unstable) nuclide.

halogen an element in group 17 of the periodic table.

CONNECTION The structure of NaCl as an ordered three-dimensional array of sodium (Na^+) ions and chloride (Cl^-) ions is described in Figure 1.18.

FIGURE 2.10 Periodic trends in the charges of common monatomic ions.

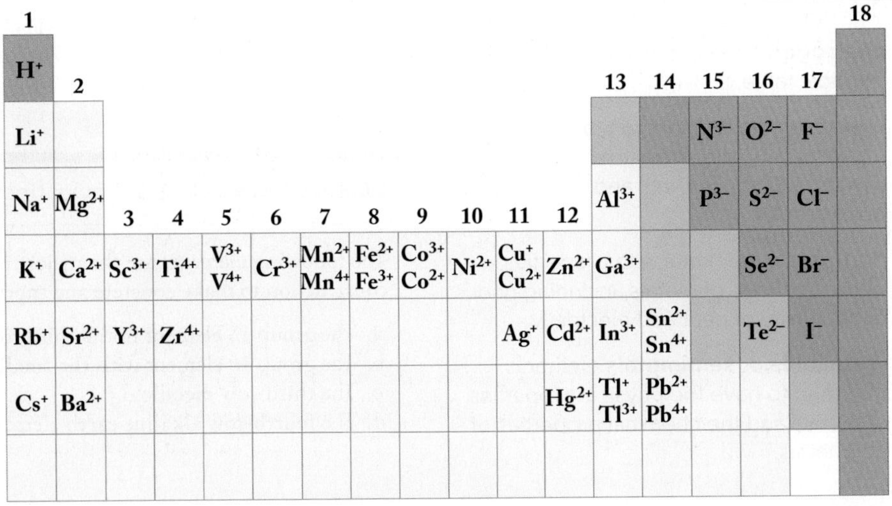

1	2	3	4	5	6	7	8	9	10	11	12	13	14	15	16	17	18
H^+																	
Li^+														N^{3-}	O^{2-}	F^-	
Na^+	Mg^{2+}											Al^{3+}		P^{3-}	S^{2-}	Cl^-	
K^+	Ca^{2+}	Sc^{3+}	Ti^{4+}	V^{3+} V^{4+}	Cr^{3+}	Mn^{2+} Mn^{4+}	Fe^{2+} Fe^{3+}	Co^{3+} Co^{2+}	Ni^{2+}	Cu^+ Cu^{2+}	Zn^{2+}	Ga^{3+}			Se^{2-}	Br^-	
Rb^+	Sr^{2+}	Y^{3+}	Zr^{4+}							Ag^+	Cd^{2+}	In^{3+}	Sn^{2+} Sn^{4+}		Te^{2-}	I^-	
Cs^+	Ba^{2+}										Hg^{2+}	Tl^+ Tl^{3+}	Pb^{2+} Pb^{4+}				

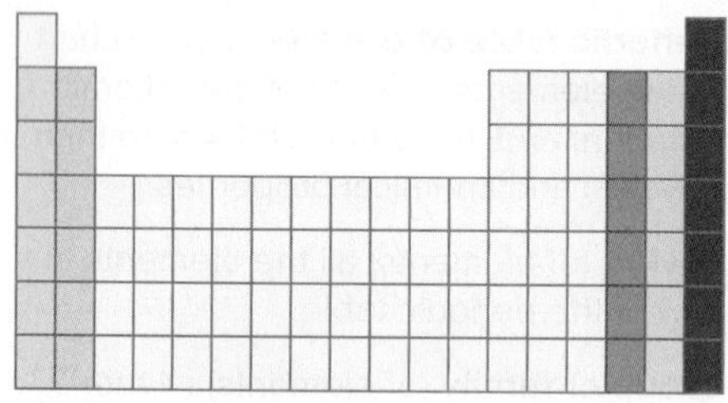

(a)

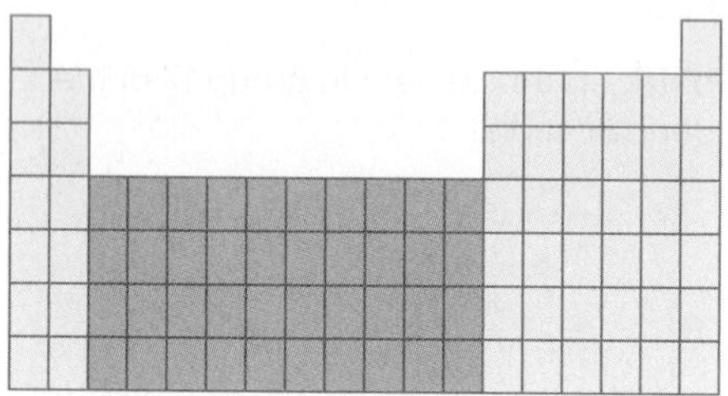

(b)

FIGURE 2.11 (a) The commonly used names of groups 1, 2, 17, and 18 of the periodic table. (b) The *main group* (or *representative*) elements are in groups 1, 2, and 13–18. They are separated by the *transition metals* in groups 3–12.

a typical halogen. It forms 1:1 binary (two-element) ionic compounds (compounds that consist of positive and negative ions) with the group 1 elements; a familiar example is sodium chloride, NaCl (the principal ingredient in table salt). The ratio is 1:1 because all the group 1 elements form cations that have a 1+ charge, and all the group 17 elements form 1− anions (as we saw in Figure 2.10). Equal numbers of these anions and cations mean that their charges cancel out, as they must in all ionic compounds so that, overall, they are neutral substances. For their part, the group 1 elements (except for hydrogen) are called **alkali metals**.

Group 2 elements are called **alkaline earth metals**. They form ionic compounds with halogens in which the ratio of cations to anions is 1:2 because all the group 2 cations have 2+ charges, so a neutral compound must have twice as many anions with 1− charges as alkaline earth cations. A common example of such a compound is calcium chloride, $CaCl_2$, which is widely used in regions with cold winters to melt ice and snow from exterior steps and sidewalks. Alkaline earth metals form ionic compounds with the group 16 elements in a 1:1 cation:anion ratio because the group 16 elements form 2− anions. The group 16 elements are sometimes referred to as the **chalcogens**, a name derived from the association of O, S, Se, and Te with copper (*chalco*) in minerals.

The elements in the periodic table can be categorized very broadly as metals, nonmetals, and metalloids (or semimetals), as shown in Figure 2.9(b). **Metals** (tan cells) tend to conduct heat and electricity well, they're malleable (capable of being shaped by hammering) and ductile (capable of being drawn out into wires), and all but mercury (Hg) are shiny solids at room temperature. **Nonmetals** (blue cells) are poor conductors of heat and electricity. Most are gases at room temperature; the solids among them tend to be brittle (I_2), and bromine (Br_2) is a liquid at room temperature. **Metalloids** or **semimetals** (green cells) are so named because they tend to have the physical properties of metals but the chemical properties of nonmetals.

Groups 1, 2, and 13 through 18 are referred to collectively as **main group elements** or **representative elements** (Figure 2.11b). They include the most abundant elements in the solar system and many of the most abundant on Earth. They are the "A" elements in the former group numbering system. The elements in groups 3 through 12 are called **transition metals**; they are the former "B" elements. All of them except mercury exhibit the classic physical properties of metals: they are hard, shiny, ductile, malleable solids and excellent conductors of heat and electricity. The group 18 elements are called **noble gases** because for the most part they do not interact with other elements. Becquerel's α particle is actually the nucleus of a helium atom, the first element in the group.

alkali metal an element in group 1 of the periodic table.

alkaline earth metal an element in group 2 of the periodic table.

chalcogen an element in group 16 of the periodic table.

metals elements that are typically shiny, malleable, ductile solids that conduct heat and electricity well and tend to form positive ions.

nonmetals elements with properties opposite those of metals, including poor conductivity of heat and electricity.

metalloids or **semimetals** elements that tend to have the physical properties of metals and the chemical properties of nonmetals.

SAMPLE EXERCISE 2.3 Navigating the Periodic Table LO3

The following elements are the major components of Portland cement, which is used in construction to make concrete and mortar. Which elements are they?

a. The group 13 element in the third row (period)
b. The group 16 element with the smallest atomic number
c. The third-row metalloid
d. The fourth-row alkaline earth element

Collect, Organize, and Analyze We are asked to identify elements based on the locations of their symbols in the periodic table. In parts a, c, and d, we are given the row number; in b we are not provided the row number directly, but we know that the element has the lowest atomic number of all the elements in its group. Information about group locations comes from the group number in a and b, the type of element in c, and the group name in d.

Solve (a) The cell address is group 13, row 3: Al, aluminum; (b) the element at the top of group 16 is O, oxygen; (c) the only metalloid in row 3 is in group 14: Si, silicon; (d) all the alkaline earths, including the one in row 4, are in group 2: Ca, calcium.

Think About It Each element has a unique location in the periodic table that is linked to both its atomic number and its chemical properties. Chemists usually describe element locations in terms of their group number or name (the column they're in) and their row number.

 **Practice Exercise** What are the symbol and name of each of the following elements?

a. The metalloid in group 15 closest in mass to the noble gas krypton
b. A representative element in the third row that is an alkaline earth metal
c. A transition metal in the sixth row and in the same group as zinc (Zn)

(Answers to Practice Exercises are in the back of the book.)

main group elements or **representative elements** the elements in groups 1, 2, and 13 through 18 of the periodic table.

transition metals the elements in groups 3 through 12 of the periodic table.

noble gases the elements in group 18 of the periodic table.

average atomic mass the weighted average of the masses of all isotopes of an element, calculated by multiplying the natural abundance of each isotope by its mass in atomic mass units and then summing the products.

natural abundance the proportion of a particular isotope, usually expressed as a percentage, relative to all the isotopes of that element in a natural sample.

2.4 The Masses of Atoms, Ions, and Molecules

At the center of each of the cells in the periodic table inside the front cover is the symbol of an element. The number above the symbol is the element's atomic number Z; the number below the symbol is the element's **average atomic mass**. More precisely, the number below the symbol is the *weighted* average of the masses of all the isotopes of the element.

To understand the meaning of a weighted average, let's consider the masses and **natural abundances** of the three isotopes of neon in the table to the right. Natural abundances are usually expressed as percentages. Thus, 90.4838% of all neon atoms are neon-20, 9.2465% are neon-22, and only 0.2696% are neon-21. The natural abundance of neon-21 is so small that Thomson and Aston could not detect it with their positive-ray analyzer. However, modern versions of their instrument, called *mass spectrometers*, are capable of ionizing atoms and molecules and separating and detecting their ions with much more sensitivity. These instruments can provide precise natural abundance values for even low-abundance isotopes, as illustrated by the neon data.

Isotope	Mass (amu)	Natural Abundance (%)
Neon-20	19.9924	90.4838
Neon-21	20.9940	0.2696
Neon-22	21.9914	9.2465

To calculate the average atomic mass of neon we multiply the mass of each of its isotopes by its natural abundance (in the language of mathematics, we *weight* the isotope's mass using natural abundance as the *weighting factor*) and then sum the three weighted masses. To simplify the calculation we first convert the percent abundance values into their decimal equivalents:

$$\begin{aligned} \text{Average atomic mass of neon} = \quad & 19.9924 \text{ amu} \times 0.904838 = 18.08988 \text{ amu} \\ & + 20.9940 \text{ amu} \times 0.002696 = 0.05660 \text{ amu} \\ & + 21.9914 \text{ amu} \times 0.092465 = 2.03344 \text{ amu} \\ \hline & \qquad\qquad\qquad\qquad\quad 20.17992 \text{ amu} \end{aligned}$$

In this calculation and similar ones later in the chapter, we carry one more digit (highlighted in red) than we are allowed by significant-figure rules for multiplication to avoid rounding errors. Rounding the sum to the appropriate number of significant figures gives us an average atomic mass of 20.1799 amu. No single atom of neon has this average atomic mass; instead, every atom of neon in the universe has the mass of one of the three neon isotopes.

This method for calculating average atomic mass works for all elements. The general equation for doing the calculation is

$$m_X = a_1m_1 + a_2m_2 + a_3m_3 + \cdots \tag{2.2}$$

where m_X is the average atomic mass of element X, which has isotopes with masses $m_1, m_2, m_3, \ldots,$ for which the natural abundances expressed in decimal form are $a_1, a_2, a_3, \ldots.$

SAMPLE EXERCISE 2.4 Calculating Average Atomic Mass **LO4**

We briefly discussed the application of ^{83}Kr to MRI in Sample Exercise 2.1. There are six stable isotopes of krypton with the following natural abundances:

Symbol	Mass (amu)	Natural Abundance (%)
^{78}Kr	77.920	0.35
^{80}Kr	79.916	2.25
^{82}Kr	81.913	11.60
^{83}Kr	82.914	11.50
^{84}Kr	83.912	57.00
^{86}Kr	85.911	17.30

Use these data to calculate the average atomic mass of krypton.

Collect, Organize, and Analyze We know the masses and natural abundances of each of the six isotopes of krypton and need to calculate the average atomic mass. Using Equation 2.2, we multiply the mass of each isotope by its natural abundance expressed as a decimal and then add the products together. In this calculation we will carry one more digit (highlighted in red) than allowed by the significant-figure rules in Chapter 1 to avoid rounding errors.

Solve

$$\begin{aligned}\text{Average atomic mass} = {} & 77.920\text{ amu} \times 0.0035 = 0.273\text{ amu}\\ & + 79.916\text{ amu} \times 0.0225 = 1.798\text{ amu}\\ & + 81.913\text{ amu} \times 0.1160 = 9.502\text{ amu}\\ & + 82.914\text{ amu} \times 0.1150 = 9.535\text{ amu}\\ & + 83.912\text{ amu} \times 0.5700 = 47.830\text{ amu}\\ & + 85.911\text{ amu} \times 0.1730 = 14.863\text{ amu}\\ & \hline \qquad\qquad\qquad\qquad\qquad\qquad 83.801\text{ amu}\end{aligned}$$

Rounding this sum to the nearest 0.01 amu gives us 83.80 amu.

Think About It The six values of natural abundances expressed as decimals should add up to 1.0000, and they do. Sometimes this is not the case (check the neon abundances in the previous calculation). Uncertainties in the last decimal place may be due to uncertainties in measured values or in rounding them off.

Practice Exercise Xenon-129 is one of nine stable isotopes of xenon and is used to evaluate lung function by magnetic resonance imaging. The following table includes all the stable isotopes of Xe.

Symbol	Mass (amu)	Natural Abundance (%)
^{124}Xe	123.906	0.10
^{126}Xe	125.904	0.09
^{128}Xe	127.904	1.91
^{129}Xe	128.905	26.40
^{130}Xe	129.904	4.10
^{131}Xe	130.905	21.20
^{132}Xe	131.905	26.90
^{134}Xe	133.905	10.40
^{136}Xe	135.907	8.90

Calculate the average atomic mass of xenon by including (a) all nine isotopes and (b) only the seven most abundant isotopes.

(Answers to Practice Exercises are in the back of the book.)

SAMPLE EXERCISE 2.5 Calculating Natural Isotopic Abundances LO4

Silver (Ag) has two stable isotopes: ^{107}Ag, 106.90 amu and ^{109}Ag, 108.90 amu. The average atomic mass of silver is 107.87 amu. What is the natural abundance of each isotope?

Collect and Organize We know the masses of each of two isotopes of silver and the average atomic mass of silver. We are asked to calculate the natural abundance of each.

Analyze Unlike Sample Exercise 2.4, we can't substitute directly into Equation 2.2 because we do not know the coefficients a_1 and a_2. However, if we let $x = a_1$ (the natural abundance of one of the isotopes expressed as a decimal), then the natural abundance of the other isotope $(a_2) = 1 - x$. The average mass (107.87 amu) is about halfway between the exact masses of the two isotopes. This suggests that the two isotopes are present in nearly equal proportions, so the values of a_1 and a_2 should be nearly the same.

Solve Substituting x and $1 - x$ into Equation 2.2 gives us an equation that we can solve for x:

$$m_{\text{Ag}} = x m_{\text{Ag-107}} + (1 - x) m_{\text{Ag-109}}$$

$$107.89 \text{ amu} = x(106.90 \text{ amu}) + (1 - x)(108.90 \text{ amu})$$

$$x = 0.5050 \qquad (1 - x) = 0.4950$$

Therefore, the percent natural abundances are the following:

$$^{107}\text{Ag}: 0.5050 \times 100 = 50.5\% \text{ and } ^{109}\text{Ag}: 0.4950 \times 100 = 49.5\%$$

Think About It We predicted that the natural abundances of ^{107}Ag and ^{109}Ag would be nearly the same, and the calculated abundances are consistent with our prediction. This approach only works for elements that have just two isotopes.

Practice Exercise Iridium has two stable isotopes: iridium-191 (190.96 amu) and iridium-193 (192.96 amu). The average atomic mass of iridium is 192.22 amu. What is the natural abundance of each isotope?

(Answers to Practice Exercises are in the back of the book.)

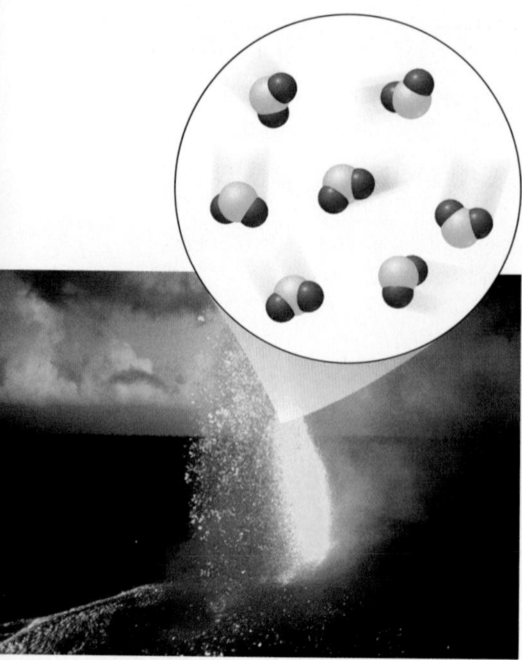

FIGURE 2.12 Volcanic eruptions, such as this one of Kilauea in Hawaii in March 2011, can introduce thousands of metric tons (1 ton = 1000 kg) of SO_2 into the atmosphere each day.

When we apply the concept of particle mass to molecular compounds, the particles involved are not individual atoms but rather groups of atoms chemically bonded together in molecules. Just as an atom of an element has an average atomic mass, a molecule of a molecular compound has a **molecular mass** that is the sum of the average atomic masses of the atoms in it. Like average atomic masses, molecular masses are expressed in amu. To calculate the molecular mass of a compound, we simply add up the average atomic masses of the atoms in each of its molecules. For example, the molecular mass of sulfur dioxide (SO_2), which is one of the gases released during volcanic eruptions (Figure 2.12), is the sum of the average atomic mass of one atom of sulfur and two atoms of oxygen:

$$\left(\frac{1\ \cancel{\text{atom S}}}{\text{molecule SO}_2} \times \frac{32.065\ \text{amu}}{1\ \cancel{\text{atom S}}}\right) + \left(\frac{2\ \cancel{\text{atoms O}}}{\text{molecule SO}_2} \times \frac{15.999\ \text{amu}}{1\ \cancel{\text{atom O}}}\right)$$

$$= \frac{64.063\ \text{amu}}{\text{molecule SO}_2}$$

Recall that there are no molecules in ionic compounds such as sodium chloride (NaCl). Instead ionic compounds consist of three-dimensional arrays of positive and negative ions. The lack of molecules means that ionic compounds do not have *molecular* masses. However, the formulas of ionic compounds define quantities called **formula units**, the smallest electrically neutral unit in an ionic compound. A formula unit has a particular **formula mass**: the sum of the average atomic masses of the cations and anions that make up a neutral formula unit. Thus, the formula unit of NaCl consists of one Na^+ ion and one Cl^- ion. The formula of calcium chloride is $CaCl_2$, so its formula unit consists of one calcium ion, Ca^{2+}, and *two* chloride ions, Cl^-, and its formula mass is the combined mass of 1 Ca^{2+} ion and 2 Cl^- ions:

$$\left(\frac{1\ \cancel{\text{Ca}^{2+}\ \text{ion}}}{\text{formula unit}} \times \frac{40.078\ \text{amu}}{1\ \cancel{\text{Ca}^{2+}\ \text{ion}}}\right) + \left(\frac{2\ \cancel{\text{Cl}^-\ \text{ion}}}{\text{formula unit}} \times \frac{35.453\ \text{amu}}{1\ \cancel{\text{Cl}^-\ \text{ion}}}\right)$$

$$= \frac{110.984\ \text{amu}}{\text{formula unit}}$$

Note that we used the average atomic masses of calcium and chlorine from the periodic table for the masses of the Ca^{2+} and Cl^- ions in this calculation. We can do that because losing or gaining an electron or two has very little impact on the mass of an atom; nearly all of its mass is concentrated in its nucleus. Moreover, the masses of the electrons lost when atoms form cations are balanced by the masses gained when atoms form anions, so the overall mass of a formula unit is exactly the same as the sum of the average atomic masses of the elements in it.

2.5 Moles and Molar Masses

Individual atoms are very tiny particles, which is why we need very powerful instruments, such as scanning tunneling microscopes (Figure 1.12), to see them. Atoms also have very little mass. The average mass of a gold atom—196.97 amu—is heavy for an atom, but it wouldn't even register on a conventional scale. What, then, is 196.97 amu in grams? If 1 amu equals 1.66054×10^{-27} kg, then the mass of an average gold atom is only

$$196.97\ \cancel{\text{amu}}\ \text{Au} \times \frac{1.66054 \times 10^{-27}\ \cancel{\text{kg}}}{1\ \cancel{\text{amu}}} \times \frac{1000\ \text{g}}{1\ \cancel{\text{kg}}} = 3.2708 \times 10^{-22}\ \text{g Au}$$

In our macroscopic (visible) world, chemists usually work with quantities of substances that they can see, transfer from one container to another, and weigh on balances. Inevitably these quantities of substances contain enormous numbers of atoms, ions, or molecules.

To deal with such very large numbers, chemists need a unit that relates macroscopic quantities of substances, such as their masses expressed in grams, to the number of particles they contain. That unit is the **mole (mol)**, the SI base unit for expressing quantities of substances (see Table 1.2).

One mole of a substance is defined as the quantity of the substance that contains the same number of particles as the number of carbon atoms in exactly 12 g of the isotope carbon-12. This number, to five significant figures, is 6.0221×10^{23}. This very large value is called **Avogadro's number (N_A)** after the Italian scientist Amedeo Avogadro (1776–1856), whose research enabled other scientists to accurately determine the atomic masses of the elements.

To put a number of this magnitude in perspective, it would take 9.4 trillion computer flash drives, each capable of storing 64 gigabytes (6.4×10^{10} bytes) of data, to store a mole of bytes. In a different analogy, some estimates put the number of cells in the human body at 37 trillion (37×10^{13}). Avogadro's number corresponds to the number of cells in 16 billion people, or more than twice Earth's current population of just under 8 billion.

Dividing the number of particles in a sample by Avogadro's number yields the number of moles of those particles. For example, U.S. pennies minted before 1962 each contained about 2.4×10^{22} atoms of Cu. That number is equivalent to 4.0×10^{-2} moles of copper:

$$2.4 \times 10^{22} \text{ \sout{atoms} Cu} \times \frac{1 \text{ mol}}{6.0221 \times 10^{23} \text{ \sout{atoms}}} = 4.0 \times 10^{-2} \text{ mol Cu}$$

On the other hand, multiplying a number of moles by Avogadro's number gives us the number of particles in that many moles. For example, silicon wafers used in solar panels are typically 10–20 cm on a side and between 200 and 300 μm thick. A silicon wafer with dimensions 10.0 cm by 10.0 cm by 2.00×10^{-2} cm contains 0.166 mole of silicon. The number of silicon atoms in such a wafer is

$$0.166 \text{ \sout{mol} Si} \times \frac{6.0221 \times 10^{23} \text{ atoms}}{1 \text{ \sout{mol}}} = 1.00 \times 10^{23} \text{ atoms Si}$$

These atom-to-mole conversions are illustrated in Figure 2.13, and the quantities of some common elements equivalent to 1 mole of each are shown in Figure 2.14. While it is sometimes useful to know the number of atoms of an element in a sample, most of the time we'll focus on how many *moles* of an element are present.

CONCEPT TEST

How does a unit of measure such as 500 facial tissues in a box relate to the concept of the mole?

(Answers to Concept Tests are in the back of the book.)

molecular mass the mass in amu of one molecule of a molecular compound.

formula unit the smallest electrically neutral unit of an ionic compound.

formula mass the mass in amu of one formula unit of an ionic compound.

mole (mol) an amount of a substance that contains Avogadro's number ($N_A = 6.0221 \times 10^{23}$) of particles (atoms, ions, molecules, or formula units).

Avogadro's number (N_A) the number of carbon atoms in exactly 12 grams of carbon-12; $N_A = 6.0221 \times 10^{23}$. It is the number of particles in 1 mole.

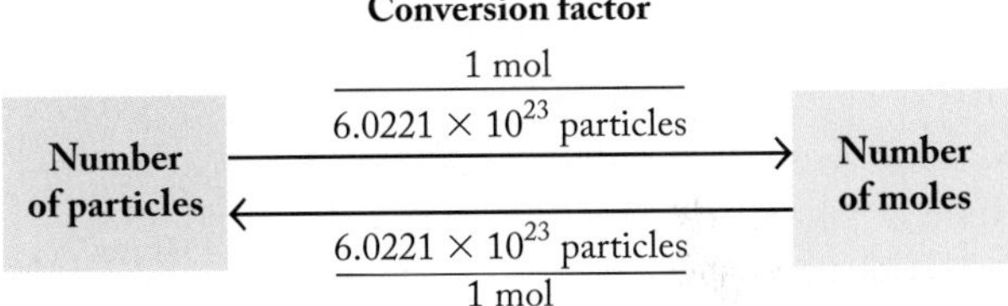

FIGURE 2.13 Converting between a number of particles and an equivalent number of moles (or vice versa) is a matter of dividing (or multiplying) by Avogadro's number.

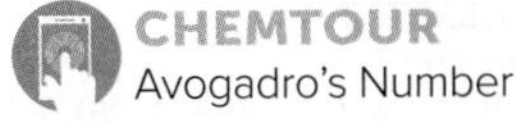

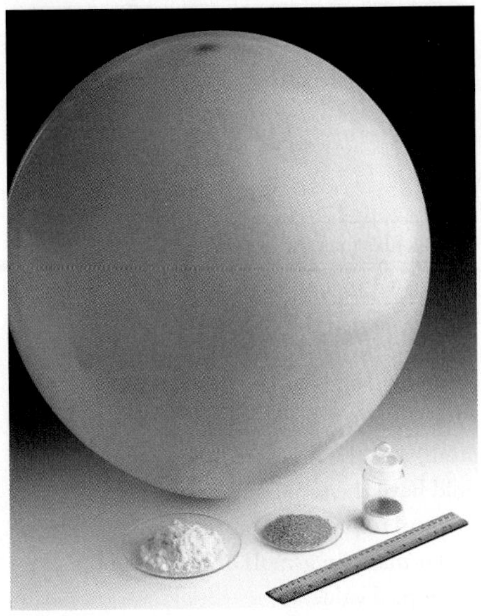

FIGURE 2.14 The quantities shown represent 1 mole of each element: 4.0026 grams of helium gas fill the balloon and, left to right in front, 32.065 grams of solid sulfur, 63.546 grams of copper metal, and 200.59 grams of liquid mercury.

SAMPLE EXERCISE 2.6 Converting Number of Moles into Number of Particles **LO5**

The silicon used to make computer chips has to be extremely pure. For example, it must contain less than 3×10^{-10} moles of phosphorus (a common impurity in silicon) per mole of silicon. What is this level of impurity expressed in atoms of phosphorus per mole of silicon?

Collect and Organize The problem states the maximum number of moles of phosphorus allowed per mole of silicon and asks us to calculate the equivalent number of atoms of P per mole of Si. Avogadro's number defines the number of atoms in 1 mole of an element: 6.0221×10^{23}.

Analyze We can convert a number of moles into an equivalent number of atoms by multiplying by Avogadro's number because its denominator contains our initial unit (mol) and its numerator contains the units we seek (atoms). We can get a rough estimate of the correct answer by rounding each starting value to the nearest power of ten and combining exponents: $10^{24} \times 10^{-10} = 10^{14}$.

Solve

$$\frac{3 \times 10^{-10}\ \cancel{\text{mol P}}}{1\ \text{mol Si}} \times \frac{6.0221 \times 10^{23}\ \text{atoms P}}{1\ \cancel{\text{mol P}}} = \frac{2 \times 10^{14}\ \text{atoms P}}{\text{mol Si}}$$

Think About It The result reveals that even a very small number of moles of an impurity translates to a very large number of atoms. The exponent of our calculated value matches our estimate, so our answer is reasonable.

Practice Exercise If 1.0 mL of seawater contains about 2.5×10^{-14} moles of dissolved gold, how many atoms of gold are in the seawater?

(Answers to Practice Exercises are in the back of the book.)

79 Au 196.97	
Atomic mass of Au	196.97 amu/atom
Mass of 1 mol of Au	196.97 g
Molar mass of Au	196.97 g/mol

(a)

Molecular mass of SO_3	80.062 amu/molecule
Mass of 1 mol of SO_3	80.062 g
Molar mass of SO_3	80.062 g/mol

(b)

FIGURE 2.15 (a) The atomic mass (in amu/atom) and the molar mass (in g/mol) of gold have the same numerical value. (b) The molecular mass (in amu/molecule) and the molar mass (in g/mol) of SO_3 have the same numerical value.

molar mass ($\mathcal{M}$) the mass of 1 mole of a substance.

Molar Mass

The mole provides an important link between the atomic mass values in the periodic table and measurable masses of elements and compounds. To see how this link works, let's convert the average mass of an atom of gold, 196.97 amu, into an equivalent mass expressed in grams per mole of gold:

$$\frac{196.97\ \cancel{\text{amu}}\ \text{Au}}{1\ \cancel{\text{atom}}\ \text{Au}} \times \frac{6.0221 \times 10^{23}\ \cancel{\text{atoms}}}{1\ \text{mol}} \times \frac{1.66054 \times 10^{-27}\ \cancel{\text{kg}}}{1\ \cancel{\text{amu}}} \times \frac{1000\ \text{g}}{1\ \cancel{\text{kg}}} = \frac{196.97\ \text{g Au}}{\text{mol Au}}$$

The atomic mass values expressed in amu/atom and in g/mol *are exactly the same.* This equality holds for all elements: the mass in grams of 1 mole of an atom, a quantity called its **molar mass ($\mathcal{M}$)**, has exactly the same numerical value as the mass of one atom of the element expressed in amu. Thus, the molar mass of gold is 196.97 g/mol (Figure 2.15a).

The concept of molar mass applies to compounds as well as elements. Just as the average atomic mass in amu of an element translates exactly into its molar mass in g/mol, the molar mass of a molecular compound is equivalent to its molecular mass in amu, and the molar mass of an ionic compound is equivalent to its formula mass in amu. Thus, the molecular mass of SO_2 that we calculated in the previous section, 64.063 amu, is equivalent to a molar mass of 64.063 g/mol. Molecules of sulfur trioxide, SO_3, have one more atom of oxygen (mass = 15.999 amu) per molecule, giving them a mass of (64.063 + 15.999) = 80.062 amu and giving sulfur trioxide a molar mass of 80.062 g/mol (Figure 2.15b). Similarly, the formula masses of ionic compounds, expressed in amu, have the same values as the molar masses of the compounds expressed in g/mol. As we saw in the previous section, the formula mass of $CaCl_2$ is 110.984 amu, which means its molar mass is 110.984 g/mol. If we didn't already know the molecular mass or formula mass of a compound, we could calculate its molar mass directly from the molar masses of the elements in it. For example, the molar mass of carbon dioxide, CO_2, is the sum of the masses of 1 mole of carbon atoms and 2 moles of oxygen atoms:

$$\left(\frac{1\ \cancel{\text{mol C}}}{\text{mol CO}_2} \times \frac{12.011\ \text{g}}{1\ \cancel{\text{mol C}}}\right) + \left(\frac{2\ \cancel{\text{mol O}}}{\text{mol CO}_2} \times \frac{15.999\ \text{g}}{1\ \cancel{\text{mol O}}}\right) = \frac{44.009\ \text{g}}{\text{mol CO}_2}$$

The mole enables us to know the number of particles in any sample of a given substance simply by knowing the mass of the sample. We can do this because the mole represents both a fixed number of particles (Avogadro's number) and a specific mass (the molar mass) of the substance. It may be useful to think about moles and mass using the following analogy: a box that contains 1 dozen golf balls and a box that contains 1 dozen baseballs have very different masses, but each contains 12 balls (Figure 2.16). Indeed, the mole is sometimes referred to as the "chemist's dozen." Figure 2.17 summarizes how to use Avogadro's number, chemical formulas, and molar masses to convert between the mass, the number of moles, and the number of particles in a given quantity of a substance.

FIGURE 2.16 A dozen golf balls weigh less than a dozen baseballs, but both quantities contain the same number of balls.

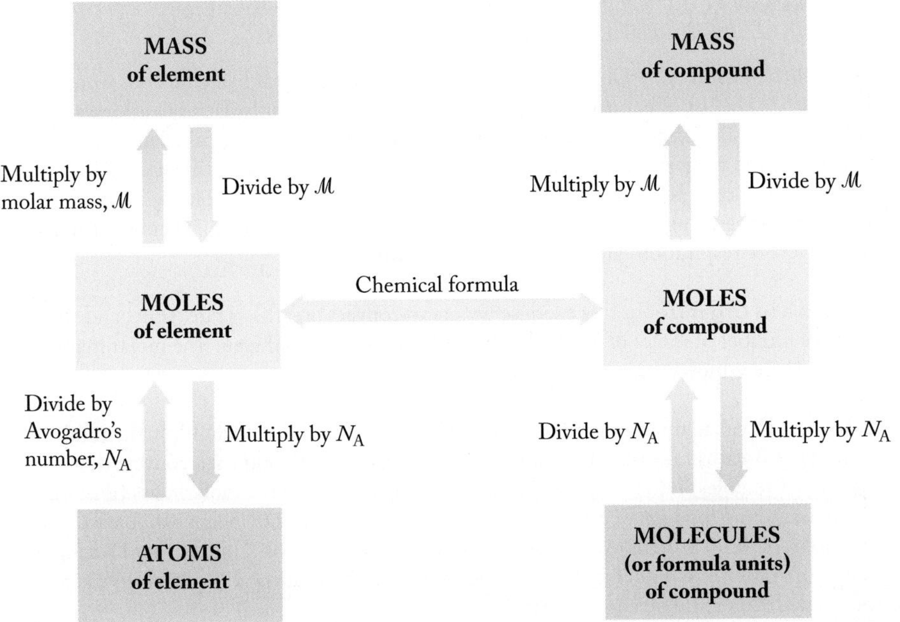

FIGURE 2.17 The mass of a pure substance can be converted into the equivalent number of moles or number of particles (atoms, ions, or molecules) and vice versa.

SAMPLE EXERCISE 2.7 Calculating the Molar Mass of a Compound **LO5**

Volcanic eruptions are often accompanied by the release of SO_2 and water vapor. When these two gases combine, they form sulfurous acid, H_2SO_3, which is a molecular compound. What is the molar mass of H_2SO_3?

Collect and Organize We are asked to calculate the molar mass of H_2SO_3. The molar mass of a molecular compound is the sum of the molar masses of the elements in its formula, each multiplied by the number of atoms of that element in a molecule of the compound. The molar masses of the relevant elements are 1.0079 g/mol H, 32.065 g/mol S, and 15.999 g/mol O.

Analyze One mole of H_2SO_3 contains 2 moles of H atoms, 1 mole of S atoms, and 3 moles of O atoms. To estimate the molar mass of H_2SO_3, we can multiply these numbers of moles by the molar masses of the three elements rounded to whole numbers. Working with whole numbers, we can estimate the results of the calculation: (2 mol H × 1 g/mol) + (1 mol S × 32 g/mol) + (3 mol O × 16 g/mol) = 82 g/mol H_2SO_3.

Solve

$$\left(\frac{2\ \cancel{\text{mol H}}}{\text{mol H}_2\text{SO}_3} \times \frac{1.0079\ \text{g}}{1\ \cancel{\text{mol H}}}\right) + \left(\frac{1\ \cancel{\text{mol S}}}{\text{mol H}_2\text{SO}_3} \times \frac{32.065\ \text{g}}{1\ \cancel{\text{mol S}}}\right) + \left(\frac{3\ \cancel{\text{mol O}}}{\text{mol H}_2\text{SO}_3} \times \frac{15.999\ \text{g}}{1\ \cancel{\text{mol O}}}\right)$$
$$= 82.078\ \text{g/mol H}_2\text{SO}_3$$

Think About It The calculated molar mass agrees quite well with our whole-number estimate because the average molar masses of H, S, and O are all close to whole-number values, which is not the case for many other elements.

Practice Exercise During photosynthesis, green plants convert water and carbon dioxide into glucose ($C_6H_{12}O_6$) and oxygen. What is the molar mass of glucose?

(Answers to Practice Exercises are in the back of the book.)

SAMPLE EXERCISE 2.8 Converting Number of Moles into Mass — LO5

In 2014 the World Anti-Doping Agency (WADA) announced that xenon gas would be added to the list of banned substances for athletes. Apparently, Russian athletes had been inhaling xenon since 2004 during training. The physiological effect of Xe is to boost the capacity of blood to carry oxygen, an important factor in aerobically demanding sports. If an athlete inhales an average of 1.68×10^{-4} moles of Xe per breath while sleeping, how many grams of Xe will be inhaled during 7.0 hours of sleep at an average respiration rate of 10.0 breaths per minute?

Collect and Organize In this exercise we are to convert breathing rate, time, and the average number of moles of Xe per breath into a mass of Xe inhaled. The molar mass of Xe is 131.29 g/mol.

Analyze To calculate the total quantity of Xe inhaled, we need to multiply the total number of breaths taken by the quantity of Xe in each breath and then convert that quantity (in moles) into an equivalent mass by multiplying by the molar mass (g/mol). To estimate the final answer, we note that 10 breaths per minute means 600 breaths per hour, or 4200 breaths over 7 hours. There are nearly 2×10^{-4} mol Xe per breath, or about 0.8 mol Xe in 4200 breaths. The molar mass of Xe is about 132 g/mol, so the mass of inhaled Xe should be about 100 g.

Solve

$$7.0\ \cancel{\text{h}} \times \frac{60\ \cancel{\text{min}}}{\cancel{\text{h}}} \times \frac{10.0\ \cancel{\text{breaths}}}{\cancel{\text{min}}} \times \frac{1.68 \times 10^{-4}\ \cancel{\text{mol Xe}}}{\cancel{\text{breath}}} \times \frac{131.29\ \text{g Xe}}{\cancel{\text{mol Xe}}} = 93\ \text{g Xe}$$

Think About It Our calculated answer is close to our estimate. In general, we multiply by molar mass to convert a number of moles into an equivalent mass in grams. The small number of moles of Xe per breath suggests that athletes did not need to breathe a high concentration of xenon. The mixture described in this Sample Exercise contained about 6 mL Xe/L of air.

Practice Exercise Some of the cylindrical silicon wafers used in the semiconductor industry are 10.00 cm in diameter and contain 3.42×10^{-3} mol Si. If the density of silicon is 2.33 g/cm^3, how thick is one of these wafers? Hint: The formula for the volume of a cylinder is $\pi r^2 h$.

(Answers to Practice Exercises are in the back of the book.)

SAMPLE EXERCISE 2.9 Converting Mass into Number of Moles LO5

If Geiger and Marsden used a piece of gold foil with dimensions 1.0 cm by 1.0 cm by 4.0×10^{-5} cm thick, how many moles of gold did the foil contain?

Collect and Organize We are given the dimensions of a piece of gold foil and its density, which we can use to calculate the mass of gold in the foil. The mass of gold and the equivalent number of moles of gold are related by its molar mass: 196.97 g/mol. The density of gold (Table A3.2) is 19.3g/cm^3.

Analyze To find the volume of the gold foil, we multiply the dimensions given to us in the problem. Density is mass per unit volume ($d = m/V$), so multiplying the volume of the gold foil by the density of gold gives us the mass of the foil. To convert grams of Au into moles of Au, we divide by the molar mass of Au. The volume of the gold is $(1.0 \times 1.0 \times 4.0 \times 10^{-5})$ cm^3, or 4.0×10^{-5} cm^3. Multiplying this value by a density of about 20 g/cm^3 and then dividing by a molar mass of about 200 g/mol should give us a value around 1/10 that of the volume of the foil, or 4×10^{-6} mol.

Solve Volume of gold foil:

$$1.0 \text{ cm} \times 1.0 \text{ cm} \times 4.0 \times 10^{-5} \text{ cm} = 4.0 \times 10^{-5} \text{ cm}^3$$

Moles of gold:

$$4.0 \times 10^{-5} \cancel{\text{cm}^3 \text{ Au}} \times \frac{19.3 \cancel{\text{g Au}}}{1 \cancel{\text{cm}^3 \text{ Au}}} \times \frac{1 \text{ mol Au}}{196.97 \cancel{\text{g Au}}} = 3.9 \times 10^{-6} \text{ mol Au}$$

Think About It The foil sample is very thin, so we would expect it to contain only a small number of moles of Au. The answer reflects that fact and agrees with our predicted value.

Practice Exercise The mass of the diamond in Figure 2.18 is 3.25 carats (1 carat = 0.200 g). Assuming diamonds are nearly pure carbon, how many moles and how many atoms of carbon are in the diamond?

(Answers to Practice Exercises are in the back of the book.)

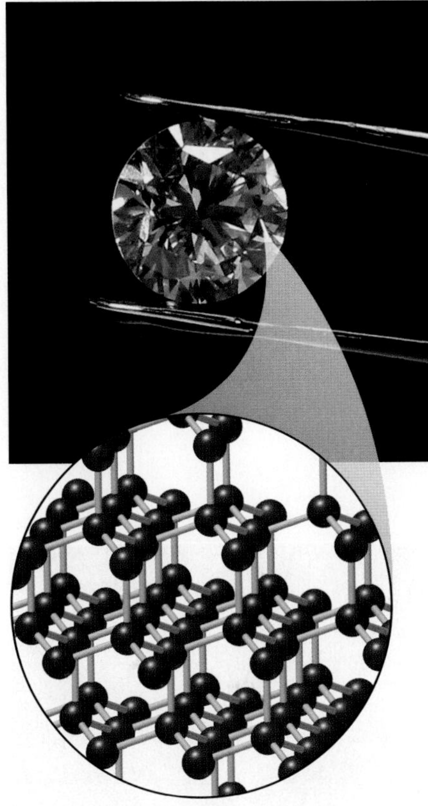

FIGURE 2.18 A 3.25-carat diamond.

CONCEPT TEST

Which contains more atoms: 1 gram of gold (Au) or 1 gram of silver (Ag)?

(Answers to Concept Tests are in the back of the book.)

SAMPLE EXERCISE 2.10 Interconverting Grams, Moles, Molecules, and Formula Units LO6

The ionic compound calcium carbonate, $CaCO_3$, is the active ingredient in a popular antacid tablet.

a. How many moles of $CaCO_3$ are in a tablet with a mass of 502 mg?
b. How many formula units of $CaCO_3$ are in that tablet?

mass spectrometer a device that separates, weighs, and counts ions based on their mass (m) to charge (z) ratio, m/z.

molecular ion (M^+) an ion formed in a mass spectrometer when a molecule loses an electron after being bombarded with high-energy electrons. The molecular ion has a charge of 1+ and has essentially the same molecular mass as the molecule from which it came.

mass spectrum a graph of the data from a mass spectrometer, where m/z ratios of the deflected particles are plotted against the number of particles with a particular mass. Because the charge on the ions typically is 1+, $m/z = m/1 = m$, and the mass of the particle may be read directly from the m/z axis.

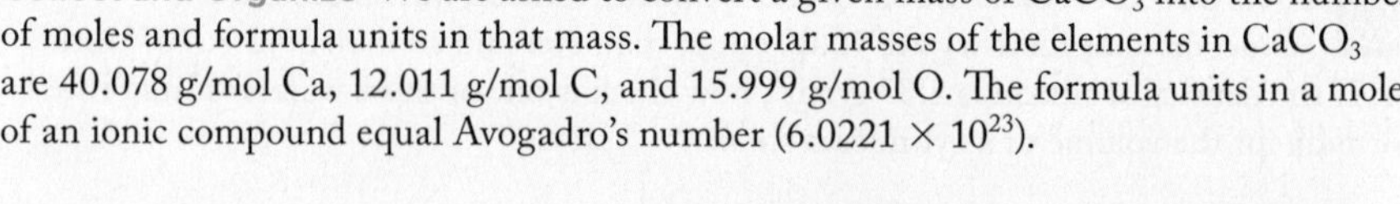

Collect and Organize We are asked to convert a given mass of $CaCO_3$ into the number of moles and formula units in that mass. The molar masses of the elements in $CaCO_3$ are 40.078 g/mol Ca, 12.011 g/mol C, and 15.999 g/mol O. The formula units in a mole of an ionic compound equal Avogadro's number (6.0221×10^{23}).

Analyze To convert a mass of a compound in milligrams into an equivalent number of moles, we need to convert the mass from milligrams into grams and then divide by the molar mass. The molar mass of $CaCO_3$ is the sum of the molar masses of Ca and C plus three times the molar mass of O. Multiplying the number of moles by Avogadro's number will give us the number of formula units. To estimate our answers, let's round the mass of the tablet to 500 mg, or 0.5 g, and approximate the molar mass of $CaCO_3$ using whole-number values for the molar masses of Ca, C, and O: (1 mol Ca × 40 g/mol) + (1 mol C × 12 g/mol) + (3 mol O × 16 g/mol) = 100 g/mol $CaCO_3$. Therefore, the answer to part a should be about 0.5/100 = 0.005 mole, and the answer to part b should be about $0.005 \times (6 \times 10^{23})$, or about 3×10^{21} formula units.

Solve

a. The molar mass of $CaCO_3$ is

$$\left(\frac{1\ \cancel{\text{mol Ca}}}{\text{mol CaCO}_3} \times \frac{40.078\ \text{g}}{1\ \cancel{\text{mol Ca}}}\right) + \left(\frac{1\ \cancel{\text{mol C}}}{\text{mol CaCO}_3} \times \frac{12.011\ \text{g}}{1\ \cancel{\text{mol C}}}\right) + \left(\frac{3\ \cancel{\text{mol O}}}{\text{mol CaCO}_3} \times \frac{15.999\ \text{g}}{1\ \cancel{\text{mol O}}}\right)$$
$$= 100.086\ \text{g/mol CaCO}_3$$

Converting from milligrams $CaCO_3$ to moles $CaCO_3$:

$$502\ \cancel{\text{mg CaCO}_3} \times \frac{1\ \text{g}}{1000\ \cancel{\text{mg}}} \times \frac{1\ \text{mol CaCO}_3}{100.086\ \cancel{\text{g CaCO}_3}} = 0.00502\ \text{mol CaCO}_3$$

b. Multiplying by Avogadro's number:

$$0.00502\ \cancel{\text{mol}}\ \text{CaCO}_3 \times \frac{6.0221 \times 10^{23}\ \text{formula units}}{1\ \cancel{\text{mol}}} = 3.02 \times 10^{21}\ \text{formula units CaCO}_3$$

Think About It The calculated values agree well with our estimates. As expected, the number of formula units in a relatively small mass is so enormous there is no practical way to "count them." However, combining the concept of molar mass with Avogadro's number allows us to calculate how many formula units there are in a sample of known mass.

Practice Exercise A standard aspirin tablet contains 325 mg of aspirin, which has the molecular formula $C_9H_8O_4$. How many moles and how many molecules of aspirin are in one tablet?

(Answers to Practice Exercises are in the back of the book.)

2.6 Mass Spectrometry: Isotope Abundances and Molar Mass

When you check in for a flight at many airports, a security worker may wipe the handle on your luggage or your hands with a small piece of cloth to check for the presence of explosives (Figure 2.19a). The instrument used to rapidly analyze the swab (Figure 2.19b) is a **mass spectrometer**, a technologically advanced version of Aston's positive-ray analyzer. In this section we explore how mass spectrometers can be used to determine the molar mass and even the identities of compounds, measure isotopic abundances, and detect explosives at airports or banned substances in athletes.

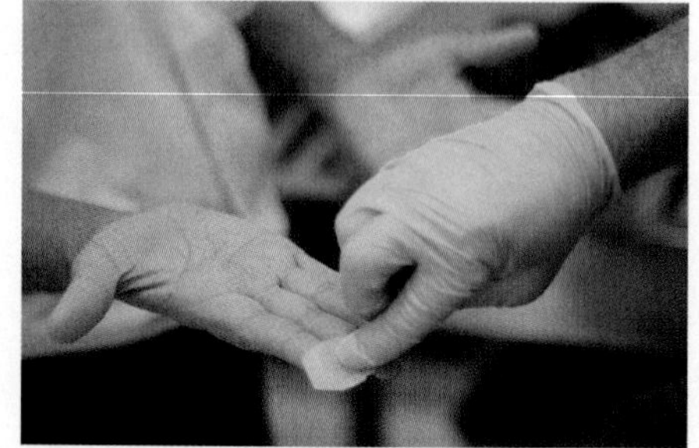

(a)

(b)

FIGURE 2.19 Airport security officials screen passengers for explosives by (a) swabbing a passenger's hands with a piece of cloth and (b) using a mass spectrometer to analyze the cloth.

Mass Spectrometry and Molecular Mass

Chemists often do not know the identities of compounds isolated from reaction mixtures or bioreactors. A key piece of information in identifying such an unknown is its molecular mass. In modern laboratories, this information is usually obtained with the aid of mass spectrometry.

Inside mass spectrometers, atoms and molecules are converted into ions that are then separated based on the ratio of their masses (m) to their electric charges (z). A common way to produce ions in a mass spectrometer is illustrated in Figure 2.20. This method involves vaporizing a sample and then bombarding the vapor with a beam of high-energy electrons. Collisions between these electrons and molecules can result in the molecules losing one of their own electrons, forming **molecular ions (M^+)**:

$$M + \text{high-energy } e^- \rightarrow M^+ + 2\,e^-$$

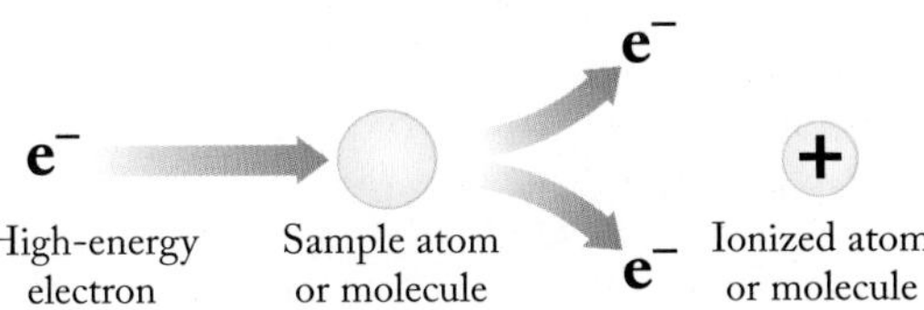

FIGURE 2.20 In some mass spectrometers, the atoms or molecules are bombarded with a beam of high-energy electrons to make atomic or molecular ions.

Other collisions may break apart molecules into fragments that also carry 1+ charges. When these ions and the molecular ion are separated based on their m/z values and then reach a detector, the resulting signals are used to create a graphical display called a **mass spectrum**, in which the m/z values of the ions are plotted on the horizontal axis and the intensity (the number of ions with a particular m/z value) on the vertical axis. Often the charge on every ion is 1+, so the m/z ratio is simply m. This means the mass of a molecular ion or fragment ion can be read directly from the position of its peak on the horizontal axis.

Figure 2.21 shows the mass spectra for acetylene (C_2H_2) and benzene (C_6H_6). The information in mass spectra such as these allow scientists to know with high precision and accuracy the molecular mass of compounds. For now, we concentrate on the molecular-ion peak, which is often the prominent peak in a mass spectrum with the largest mass.

In Figure 2.21a, the highest mass peak is at 26 amu, which corresponds to the molecular mass of C_2H_2:

$$\left(\frac{2\ \cancel{\text{atoms C}}}{\text{molecule } C_2H_2} \times \frac{12.011\ \text{amu}}{1\ \cancel{\text{atom C}}}\right) + \left(\frac{2\ \cancel{\text{atoms H}}}{\text{molecule } C_2H_2} \times \frac{1.0079\ \text{amu}}{1\ \cancel{\text{atom H}}}\right) = \frac{26.038\ \text{amu}}{\text{molecule } C_2H_2}$$

For benzene, the molecular-ion peak has a mass of 78 amu, consistent with the molecular mass of C_6H_6:

$$\left(\frac{6\ \cancel{\text{atoms C}}}{\text{molecule } C_6H_6} \times \frac{12.011\ \text{amu}}{1\ \cancel{\text{atom C}}}\right) + \left(\frac{6\ \cancel{\text{atoms H}}}{\text{molecule } C_6H_6} \times \frac{1.0079\ \text{amu}}{1\ \cancel{\text{atom H}}}\right) = \frac{78.113\ \text{amu}}{\text{molecule } C_6H_6}$$

The molecular-ion peak may not be the tallest peak in the mass spectrum, but it is usually the peak with the highest mass (not counting peaks from minor isotopes of the elements). For example, the small peaks at $m/z = 27$ and $m/z = 79$ (just to the right of the tall M^+ peaks at $m/z = 26$ and $m/z = 78$) in Figure 2.21a and b are due to fact that 99% of all carbon atoms are ^{12}C, but 1% are ^{13}C. Therefore, the chances that at least one of

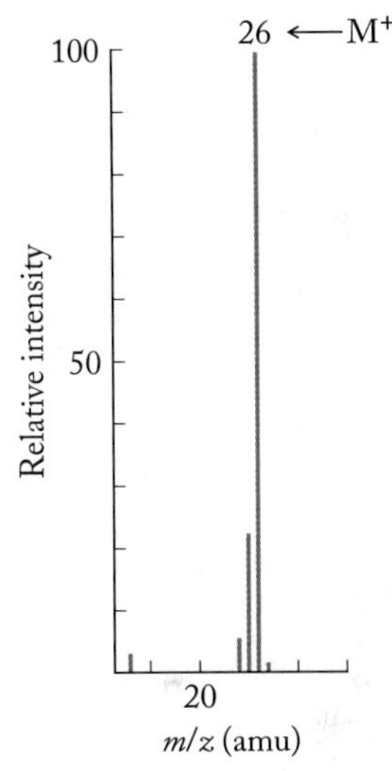

(a) Acetylene, C_2H_2

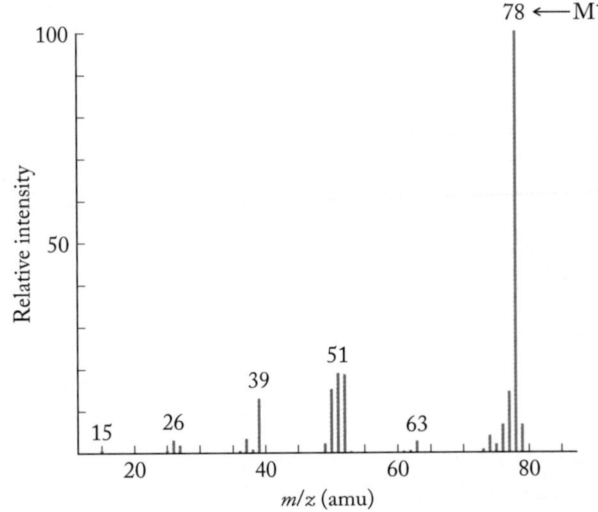

(b) Benzene, C_6H_6

FIGURE 2.21 Mass spectra of (a) acetylene and (b) benzene.

the carbon atoms in a molecule of C_2H_2 is ^{13}C are 2 × 1% or 2%. This means that 2% of the molecular ions in Figure 2.21b should be at m/z = 27. Similarly, the chances that one of the C atoms in a C_6H_6 molecular ion is ^{13}C are 6 × 1% = 6%, which is why there is a peak at m/z = 79 in Figure 2.21b that is about 6% the size of the one at m/z = 78.

The other peaks in mass spectra such as those in Figure 2.21 are also useful in confirming the identity of a compound. These fragment ion peaks represent sections of the molecule that survived electron bombardment intact, except for the loss of an electron. Distinctive fragmentation patterns are an effective way to confirm the presence of a target compound. For example, if an airport security swab produced a mass spectrum like the one shown in Figure 2.22, it could mean that the luggage contained the explosive trinitrotoluene, commonly known as TNT.

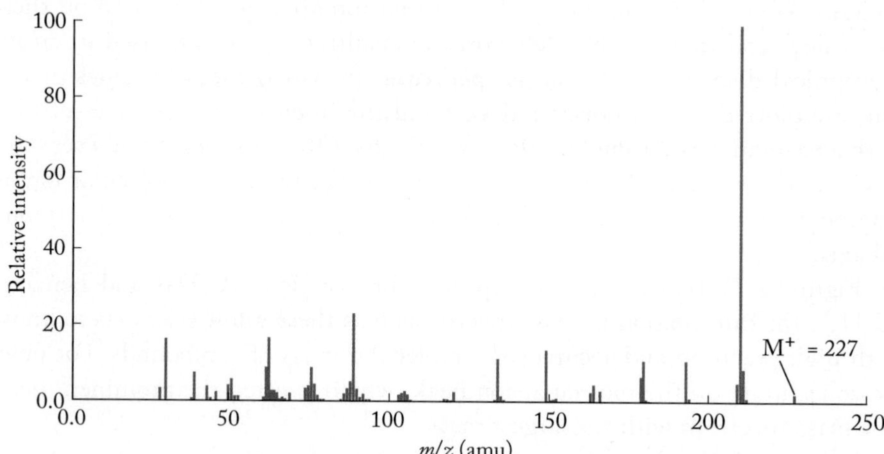

FIGURE 2.22 Mass spectrum of the common explosive compound trinitrotoluene (TNT, $C_7H_5N_3O_6$).

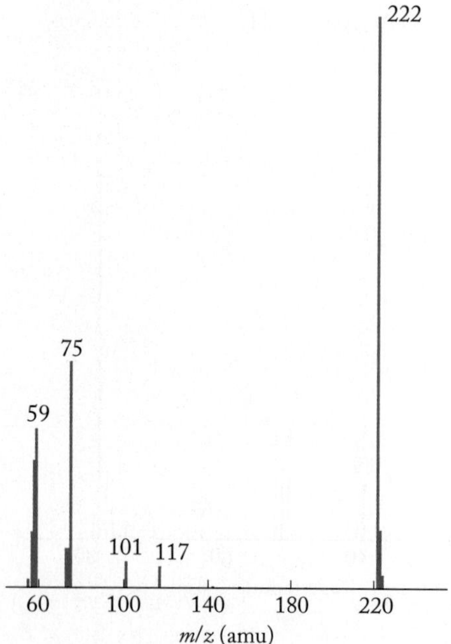

FIGURE 2.23 Mass spectrum of the explosive compound TATP.

SAMPLE EXERCISE 2.11 Determining Molecular Mass by Mass Spectrometry **LO7**

The explosive compound TATP is a major concern for law enforcement officials because it can be synthesized from readily available ingredients. Fortunately for airport security, TATP can be detected by its mass spectrum, shown in Figure 2.23.

a. What is the mass of the molecular-ion peak in Figure 2.23?
b. Show that this mass is consistent with the formula of TATP: $C_9H_{18}O_6$.

Collect, Organize, and Analyze The peak with the highest mass in a mass spectrum is often the molecular-ion peak. Its mass is the molecular mass of the compound. Once we determine the molecular mass, we can compare it to the molecular mass derived from the chemical formula. To obtain a compound's molecular mass, we need the average atomic masses of its elements: 1.0079 amu/atom H, 12.011 amu/atom C, and 15.999 amu/atom O.

Solve

a. The molecular ion for TATP is observed at m/z = 222 amu. Assuming z = 1+, the molecular mass of TATP is 222 amu/molecule.

b. To calculate the molecular mass of $C_9H_{18}O_6$, we simply add up the average atomic masses of the atoms in each of its molecules: 9 carbon atoms, 18 hydrogen atoms, and 6 oxygen atoms:

$$\frac{9 \text{ atoms C}}{\text{molecule } C_9H_{18}O_6} \times \frac{12.011 \text{ amu C}}{\text{atom C}} = \frac{108.099 \text{ amu}}{\text{molecule } C_9H_{18}O_6}$$

$$+ \frac{18 \text{ atoms H}}{\text{molecule } C_9H_{18}O_6} \times \frac{1.0079 \text{ amu H}}{\text{atom H}} = \frac{18.1422 \text{ amu}}{\text{molecule } C_9H_{18}O_6}$$

$$+ \frac{6 \text{ atoms O}}{\text{molecule } C_9H_{18}O_6} \times \frac{15.999 \text{ amu O}}{\text{atom O}} = \frac{95.994 \text{ amu}}{\text{molecule } C_9H_{18}O_6}$$

$$\frac{222.235 \text{ amu}}{\text{molecule } C_9H_{18}O_6}$$

Think About It Detection of the m/z ratio of a molecular ion in the mass spectrum that corresponds to the molecular mass of TATP is strong evidence of the presence of the compound. The fragmentation pattern of the sample should also match that of TATP when analyzed in a comparable mass spectrometer, confirming the presence of TATP in the sample.

Practice Exercise Mass spectrometry is also used to detect banned substances in athletes. The mass spectrum of testosterone is shown in Figure 2.24. Find the molecular ion in the mass spectrum. Does its mass match the molecular mass of testosterone, whose formula is $C_{19}H_{28}O_2$?

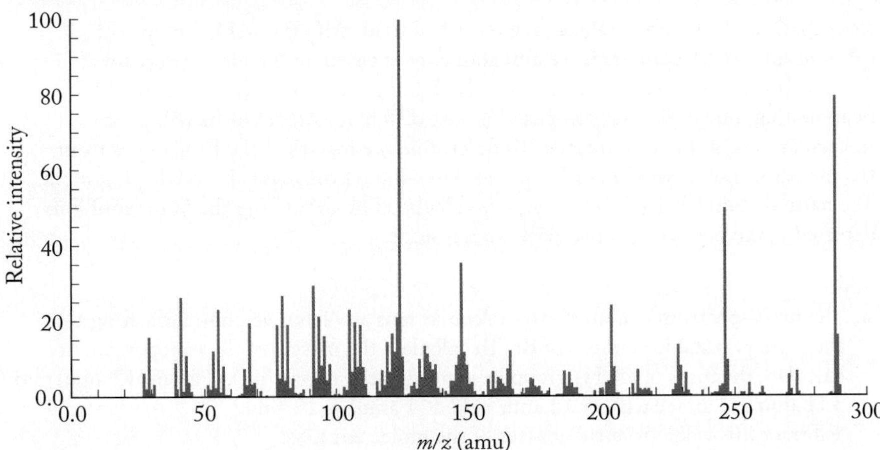

FIGURE 2.24 Mass spectrum of testosterone.

(Answers to Practice Exercises are in the back of the book.)

Mass Spectrometry and Isotopic Abundance

Mass spectrometry can also be used to determine the isotopic abundances of elements. For example, consider the portion of the mass spectrum of HCl shown in Figure 2.25. The two tallest peaks are at $m/z = 36$ and 38. These values are consistent with molecular ions containing ^{1}H atoms bonded to atoms of chlorine's two stable isotopes: ^{35}Cl and ^{37}Cl. The relative intensities of the two peaks are 100 and 31, which makes the natural abundance of the two Cl isotopes about 100/(131) = 76% ^{35}Cl and 24% ^{37}Cl. The two smaller peaks correspond to the loss of a hydrogen atom from these two molecular ions, yielding $^{35}Cl^+$ and $^{37}Cl^+$ ions with similar relative heights.

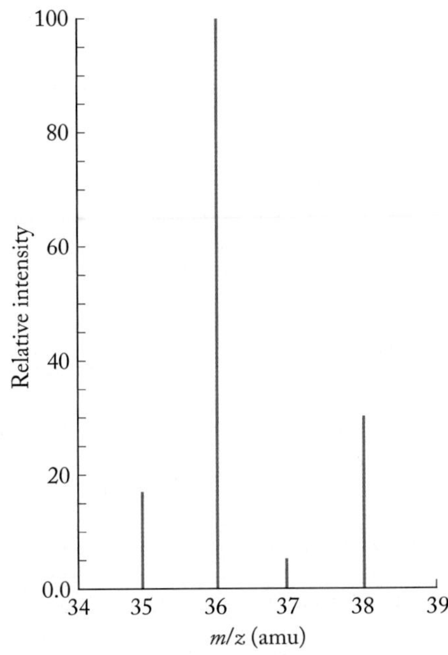

FIGURE 2.25 Mass spectrum of hydrogen chloride, HCl.

SAMPLE EXERCISE 2.12 Calculating Isotopic Abundances from Mass Spectra **LO8**

Bromomethane, CH_3Br, is used as an antifungal compound in agriculture. A mass spectrum of CH_3Br is shown in Figure 2.26. In this mass spectrum the two numbers beside the tallest two peaks indicate (1) the mass of the ions that produced the peak and (2) the number of ions that were detected. The second number is also represented by the height of the peak.

a. What are the masses of the stable isotopes of bromine (Br)?
b. What are the natural abundances of the stable isotopes of Br?

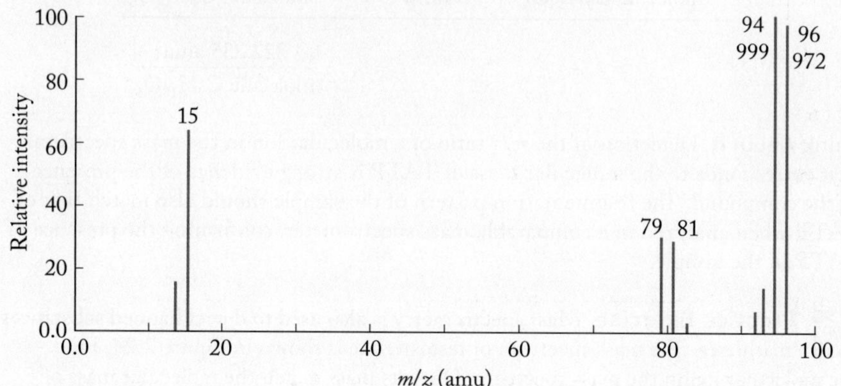

FIGURE 2.26 Mass spectrum of bromomethane, CH_3Br.

Collect, Organize, and Analyze We are asked to identify the stable isotopes of bromine and to calculate their natural abundances based on the mass spectrum of CH_3Br. If there is more than one relatively abundant isotope of Br, there should be more than one molecular-ion peak because different isotopes of Br will produce molecular ions of different masses. To determine the masses of the Br isotopes from the masses of the molecular ion(s), we need to subtract the mass of the CH_3 fragment. The natural abundance of each isotope is calculated by comparing the number of ions detected at the peaks associated with each isotope.

Solve

a. The mass spectrum contains two molecular ions at 94 and 96 amu, indicating that there are two stable isotopes of Br. To calculate the masses of the isotopes, we first calculate the mass of a CH_3 group by summing the masses of 1 C atom (12 amu) and 3 H atoms (1 amu each), or 12 amu + (3 × 1 amu) = 15 amu.
Subtract this value from the masses of the molecular ions:

$$m_{Br} = m_{CH_3Br} - m_{CH_3}$$
$$= 94 - 15 = 79$$
$$= 96 - 15 = 81$$

Therefore, the Br isotopes are ^{79}Br and ^{81}Br.

b. To determine the natural abundances of the two isotopes, we divide each of the ion counts of the molecular-ion peaks for a particular isotope by the sum of the ion counts, expressing the quotient as a percent.

$$\text{For the peak at 94 amu: } \frac{999}{(972 + 999)} \times 100\% = 50.7\%\ ^{79}Br$$

$$\text{For the peak at 96 amu: } \frac{972}{(972 + 999)} \times 100\% = 49.3\%\ ^{81}Br$$

Think About It The average atomic mass of bromine in the periodic table (79.904 amu) results from a nearly equal abundance of Br-79 and Br-81. The mass spectrum also contains fragment ions at 79 and 81 amu, confirming the presence of Br atoms within these masses. Because the masses in the mass spectrum were known only to the nearest amu, we rounded the atomic masses of C and H to whole amu values in the calculation. Normally rounding early in a calculation is not a good idea, but in this case their atomic masses (12.011 and 1.0079) are so close to whole mass numbers there is no danger of introducing an error. Moreover, the identities of isotopes are expressed using their mass *numbers*, so using whole-number atomic masses gives us the mass number superscripts in their isotopic symbols directly.

Practice Exercise The mass spectrum of chloromethane, CH_3Cl, is shown in Figure 2.27. (a) Use it to identify the stable isotopes of Cl and (b) estimate their natural abundances. Compare your answers with the relative abundances of chlorine isotopes determined from the mass spectrum of HCl.

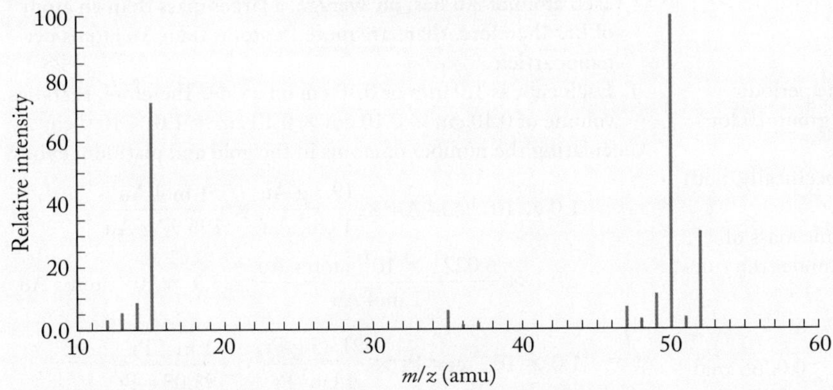

FIGURE 2.27 Mass spectrum of chloromethane, CH_3Cl.

(Answers to Practice Exercises are in the back of the book.)

CONCEPT TEST

Why is there no need to correct the ion counts of the molecular ions in Sample and Practice Exercises 2.12 for the presence of ^{13}C atoms in these ions?

(Answers to Concept Tests are in the back of the book.)

nanoparticle approximately spherical sample of matter with dimensions less than 100 nanometers (1×10^{-7} m).

SAMPLE EXERCISE 2.13 Integrating Concepts: Gold/Platinum Nanoparticles

For centuries, people believed that metallic gold had medicinal properties. In recent years, tiny gold **nanoparticles** with diameters less than 10^{-7} m have actually been used in medicine. The therapeutic properties of gold nanoparticles by themselves are limited, but mixing gold with platinum yields materials with antibiotic properties.

a. Do gold (Au) and platinum (Pt) belong to the same group in the periodic table?
b. Are gold and platinum best described as metals, nonmetals, or transition metals?
c. Platinum has six isotopes with the following natural abundances:

Symbol	Mass (amu)	Natural Abundance (%)
^{190}Pt	189.96	0.014
^{192}Pt	191.96	0.782
^{194}Pt	193.96	32.967
^{195}Pt	194.97	33.832
^{196}Pt	195.97	25.242
^{198}Pt	197.97	7.163

Use these data to calculate the average atomic mass of platinum.

d. Which has more neutrons, the most abundant isotope of platinum or ^{197}Au?
e. Will there be more atoms of gold or platinum in a nanoparticle containing 50.0% gold and 50.0% platinum by mass?
f. Suppose we have two cubes that are both 1.00 mm on a side. One is pure gold and the other pure platinum. How many atoms are in each cube?

Collect and Organize We need to locate two elements, Au and Pt, in the periodic table, classify them by their properties, and compare the composition of their nuclei. We are asked to determine the average atomic mass of platinum. Finally, we need to relate the number of atoms in samples of pure gold, pure platinum, and a mixture of the two metals. The atomic number of a nuclide is equal to the number of protons in its nucleus; its mass number is equal to the number of nucleons (protons plus neutrons). Equation 2.2 may be used to calculate the average atomic mass of an element, given the exact masses and natural abundances of its stable isotopes. The densities of the two metals (Table A3.2) are 19.3 g/cm^3 Au and 21.45 g/cm^3 Pt.

Analyze In part e we have a nanoparticle that contains equal masses of gold and platinum. There will be more atoms of the element that has the smaller atomic mass because it will take more of them to have the same mass as the other metal. In part f we first need to calculate the masses of two cubes that each have a volume of 1.00 mm^3. This involves converting each volume into cm^3 and then multiplying by density values expressed in g/cm^3. The number of atoms in each cube is calculated by dividing its mass by the appropriate molar mass to get the number of moles and then multiplying that value by Avogadro's number.

Solve

a. Platinum and gold are in different columns of the periodic table, which means they are in different groups: group 10 for platinum and group 11 for gold.
b. Both Pt and Au are classified as metals. More specifically, both are transition metals.
c. Using Equation 2.2 to calculate the average atomic mass of platinum (carrying one more digit than allowed under the rules concerning significant figures):

Average atomic mass =

$$\begin{array}{rcl} 189.96 \text{ amu} \times 0.00014 & = & 0.0266 \text{ amu} \\ +\ 191.96 \text{ amu} \times 0.00782 & = & 1.5011 \text{ amu} \\ +\ 193.96 \text{ amu} \times 0.32967 & = & 63.9428 \text{ amu} \\ +\ 194.97 \text{ amu} \times 0.33832 & = & 65.9622 \text{ amu} \\ +\ 195.97 \text{ amu} \times 0.25242 & = & 49.4667 \text{ amu} \\ +\ 197.97 \text{ amu} \times 0.07163 & = & 14.1806 \text{ amu} \\ \hline & & 195.0800 \text{ amu} \end{array}$$

Rounding the sum to the appropriate number of significant figures gives 195.080 amu.

d. The nuclei of atoms of Pt and Au contain 78 and 79 protons, respectively. The most abundant platinum isotope is ^{195}Pt. The numbers of neutrons in the nuclei of platinum-195 and gold-197 atom are

$$^{195}\text{Pt} = 195 - 78 = 117 \text{ neutrons}$$
$$^{197}\text{Au} = 197 - 79 = 118 \text{ neutrons}$$

Gold-197 has one more neutron than platinum-195.

e. There are equal masses of Au and Pt in each nanoparticle, but each atom of Au has, on average, a larger mass than an atom of Pt. Therefore, there are more Pt atoms than Au atoms per nanoparticle.
f. Each cube is 1.0 mm or 0.10 cm on a side. Therefore, each has a volume of $0.10 \text{ cm} \times 0.10 \text{ cm} \times 0.10 \text{ cm} = 1.0 \times 10^{-3} \text{ cm}^3$.

Calculating the number of atoms in the gold and platinum cubes:

$$1.0 \times 10^{-3} \cancel{\text{cm}^3 \text{ Au}} \times \frac{19.3 \cancel{\text{g Au}}}{1 \cancel{\text{cm}^3 \text{ Au}}} \times \frac{1 \cancel{\text{mol Au}}}{196.97 \cancel{\text{g Au}}} \times \frac{6.0221 \times 10^{23} \text{ atoms Au}}{1 \cancel{\text{mol Au}}} = 5.9 \times 10^{19} \text{ atoms Au}$$

$$1.0 \times 10^{-3} \cancel{\text{cm}^3 \text{ Pt}} \times \frac{21.45 \cancel{\text{g Pt}}}{1 \cancel{\text{cm}^3 \text{ Pt}}} \times \frac{1 \cancel{\text{mol Pt}}}{195.08 \cancel{\text{g Pt}}} \times \frac{6.0221 \times 10^{23} \text{ atoms Pt}}{1 \cancel{\text{mol Pt}}} = 6.6 \times 10^{19} \text{ atoms Pt}$$

The cube of platinum contains more atoms than the cube of gold.

Think About It In a nanoparticle containing equal masses of Au and Pt there are fewer gold atoms than platinum atoms because it takes more Pt atoms to have the same mass as a given mass of Au.

SUMMARY

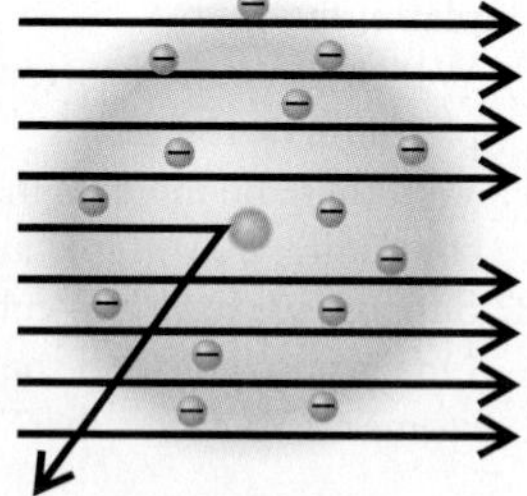

LO1 The values of the charge and mass of the electron were determined by J. J. Thomson's studies using cathode-ray tubes and by Robert Millikan's oil-drop experiments. Ernest Rutherford's group bombarded thin gold foil with **alpha (α) particles** and discovered that the positive charge and nearly all the mass of an atom are contained in its nucleus. (Section 2.1)

LO2 Atoms are composed of negatively charged **electrons** surrounding a **nucleus,** which contains positively charged **protons** and electrically neutral **neutrons.** The number of protons in the nucleus of an element defines its **atomic number (*Z*)**; the number of **nucleons** (protons + neutrons) in the nucleus defines the element's **mass number (*A*).** The different **isotopes** of an element consist of atoms with the same number of protons per nucleus but different numbers of neutrons. Symbols for subatomic particles and atoms list the symbol for the particle (X) with the value of A as a superscript and the value of Z as a subscript: ^A_ZX. (Section 2.2)

LO3 Elements are arranged in the **periodic table of the elements** in order of increasing atomic number and in a pattern based on their chemical properties, including the charges of the monatomic ions they form. Elements in the same column are in the same **group** and have similar properties. An atom or group of atoms having a net charge is called an ion. If the charge is positive, it is a **cation;** if the charge is negative, it is an **anion.** Elements in groups 1, 2, and 13–18 are **main group** (or **representative**) **elements.** The **transition metals** are in groups 3–12. **Metals** are mostly malleable, ductile solids; they form cations and are good conductors of heat and electricity. **Nonmetals** include elements in all three physical states; they form anions and are poor conductors of heat and electricity. **Metalloids,** or **semimetals,** have the physical properties of metals and chemical properties of nonmetals. (Section 2.3)

LO4 To calculate the **average atomic mass** of an element, multiply the mass of each of its stable isotopes by the **natural abundance** of that isotope as a percentage and then sum the products. (Section 2.4)

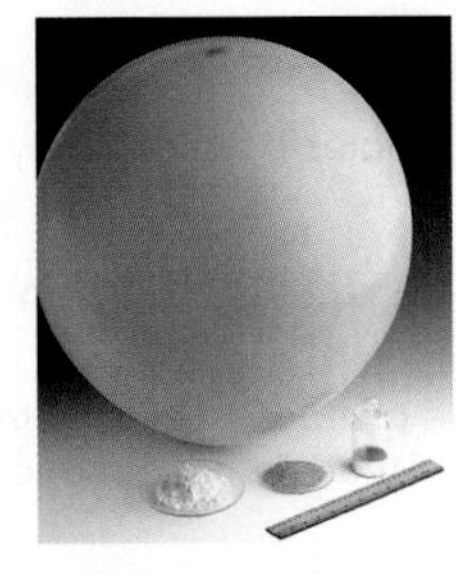

LO5 The **mole (mol)** is the SI base unit for quantity of substances. One mole of a substance consists of an **Avogadro's number** (N_A = 6.0221×10^{23}) of particles of the substance. The mass of 1 mole of a substance is its **molar mass** ($\mathcal{M}$). Avogadro's number and molar mass can be used to convert grams of a substance to moles and to the number of particles or to convert the number of particles to moles and to grams of the substance. (Section 2.5)

LO6 The **molecular mass** of a compound is the sum of the average atomic mass of each of the atoms in one of its molecules. The formula of an ionic compound defines the simplest combination of its ions that gives a neutral **formula unit** of the compound, which has a corresponding **formula mass**. (Sections 2.4 and 2.5)

LO7 The m/z values for the **molecular ion, M^+**, in the **mass spectrum** of atoms and molecules allow us to determine their molar masses. (Section 2.6)

LO8 The relative heights of the m/z peaks for the molecular ions, $M^{+\cdot}$, in the mass spectrum of an element or a molecule containing the element allow us to find the isotopic abundance of the elements. (Section 2.6)

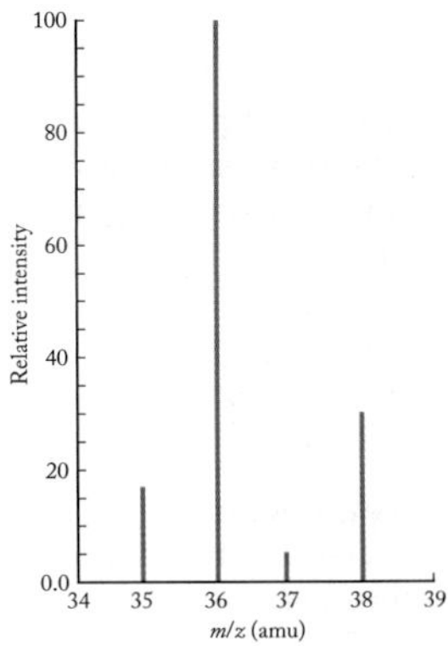

PARTICULATE PREVIEW WRAP-UP

Counting the numbers of protons and neutrons in the three nuclei in the figure yields the values in the following table where each mass number (A) is the sum of the numbers of protons and neutrons, and each atomic number (Z) is equal to the number of protons in that nucleus. Nuclei (b) and (c) have the same atomic number (6), which makes them isotopes of the same element (carbon).

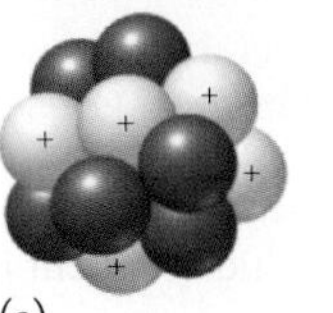

(a)

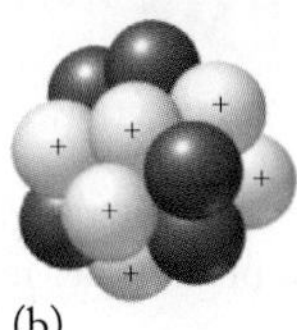

(b)

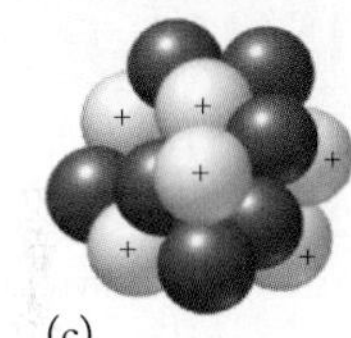

(c)

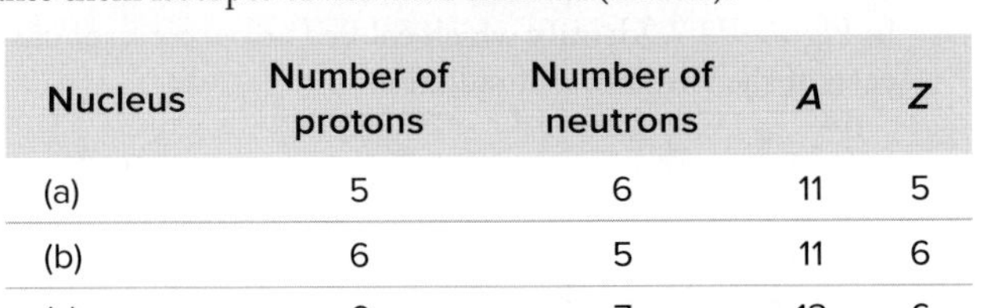

Nucleus	Number of protons	Number of neutrons	A	Z
(a)	5	6	11	5
(b)	6	5	11	6
(c)	6	7	13	6

PROBLEM-SOLVING SUMMARY

Type of Problem	Concepts and Equations	Sample Exercises
Writing symbols of nuclides and ions	Place a superscript for the mass number (A) and a subscript for the atomic number (Z) to the left of the element symbol. If the particle is a monatomic ion, add its charge as a superscript following the symbol.	**2.1, 2.2**
Navigating the periodic table	Use row numbers to identify periods in the periodic table, and use column numbers to identify groups. Groups with special names include the *alkali metals* (group 1), *alkaline earth metals* (group 2), *chalcogens* (group 16), *halogens* (group 17), and *noble gases* (group 18).	**2.3**
Calculating the average atomic mass of an element	Multiply the mass (m) of each stable isotope of the element by the natural abundance (a) of that isotope; then sum the products: $m_X = a_1m_1 + a_2m_2 + a_3m_3 + \cdots$ (2.2)	**2.4, 2.5**
Converting number of particles into number of moles (or vice versa)	Convert number of particles into number of moles by dividing by Avogadro's number. Convert number of moles into number of particles by multiplying by Avogadro's number.	**2.6, 2.10**

Type of Problem	Concepts and Equations	Sample Exercises
Converting mass of a substance into number of moles (or vice versa)	Convert mass of the substance to number of moles by dividing by the molar mass ($\mathcal{M}$) of the substance. Convert number of moles of the substance to mass by multiplying by the molar mass ($\mathcal{M}$) of the substance.	**2.8, 2.9, 2.10**
Calculating the molar mass of a compound	Sum the molar masses of the elements in the compound's formula, with each element multiplied by the number of atoms of that element in one molecule or formula unit of the compound.	**2.7, 2.10**
Determining molecular mass by mass spectrometry	Identify the molecular ion, $M^{+\cdot}$, the peak with the largest value of m/z.	**2.11**
Calculating isotopic abundances from mass spectra	Use the peak heights (ion counts) of molecular-ion peaks to calculate the natural abundances of the isotopes: $$\%\,^{A}X = \frac{\text{peak height of }^{A}X}{\text{sum of intensities of all }^{A}X} \times 100$$	**2.12**

VISUAL PROBLEMS

(Answers to boldface end-of-chapter questions and problems are in the back of the book.)

2.1. Atoms of which one of the highlighted elements in Figure P2.1 have the fewest protons per nucleus? Which element is this?

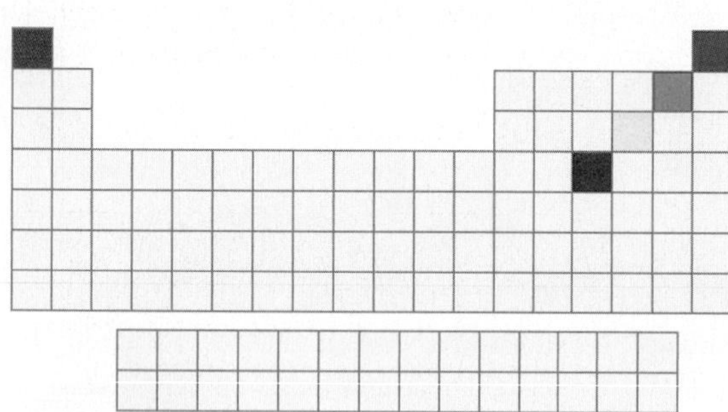

FIGURE P2.1

2.2. Atoms of which one of the highlighted elements in Figure P2.1 have, on average, the greatest number of neutrons?

2.3. Which one of the highlighted elements in Figure P2.1 has a stable isotope with no neutrons in its nucleus?

2.4. Which of the highlighted elements in Figure P2.4 has no stable isotopes?

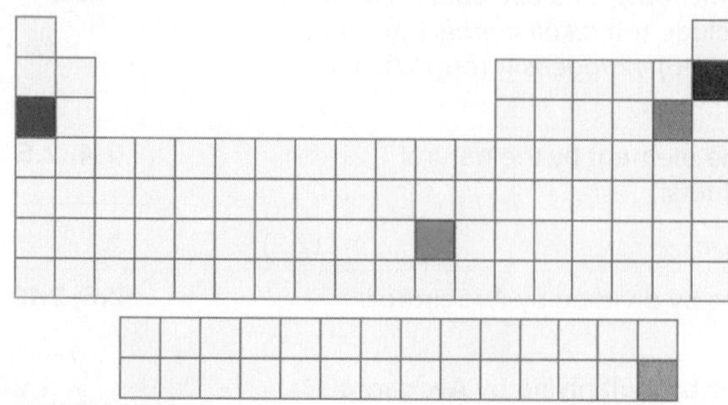

FIGURE P2.4

2.5. Which of the highlighted elements in Figure P2.4 is (a) a transition metal; (b) an alkali metal; (c) a halogen?

2.6. Which of the highlighted elements in Figure P2.4 is (a) a nonmetal; (b) a chemically inert gas; (c) a metal?

2.7. Alpha and beta particles emitted by a sample of pitchblende escape through a narrow channel in the shielding surrounding the sample and into an electric field as shown in Figure P2.7. Identify which colored arrow corresponds to each of the two forms of radiation.

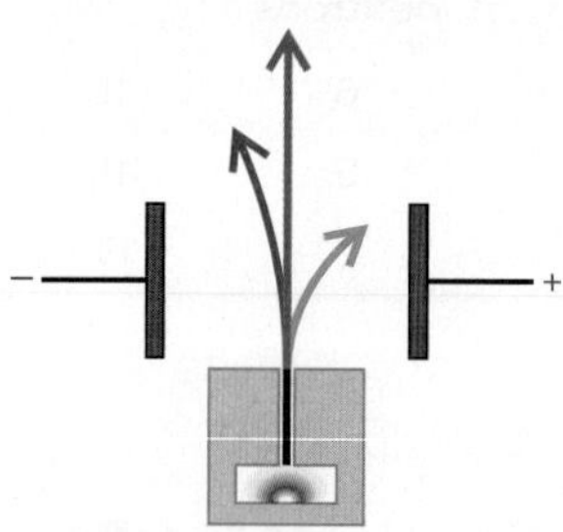

FIGURE P2.7

2.8. Which subatomic particle would curve in the same direction as the green arrow in Figure P2.7?

***2.9.** Dichloromethane (CH_2Cl_2, 84.93 g/mol) and cyclohexane (C_6H_{12}, 84.15 g/mol) have nearly the same molar masses. Which compound produced the mass spectrum shown in Figure P2.9? Explain your selection.

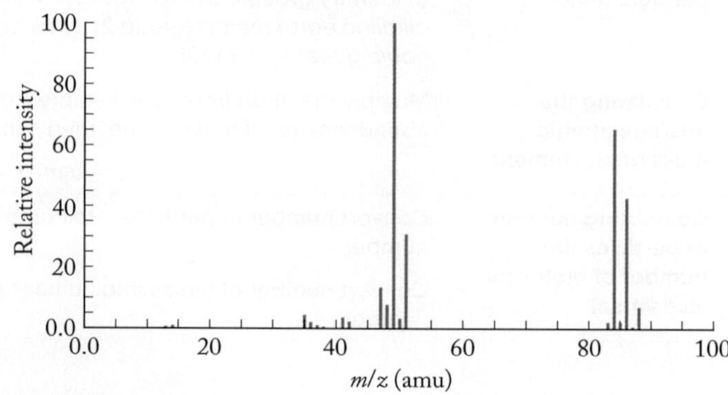

FIGURE P2.9

2.10. Krypton has six stable isotopes. How many neutrons are there in the most abundant isotope of krypton based on the mass spectrum in Figure P2.10?

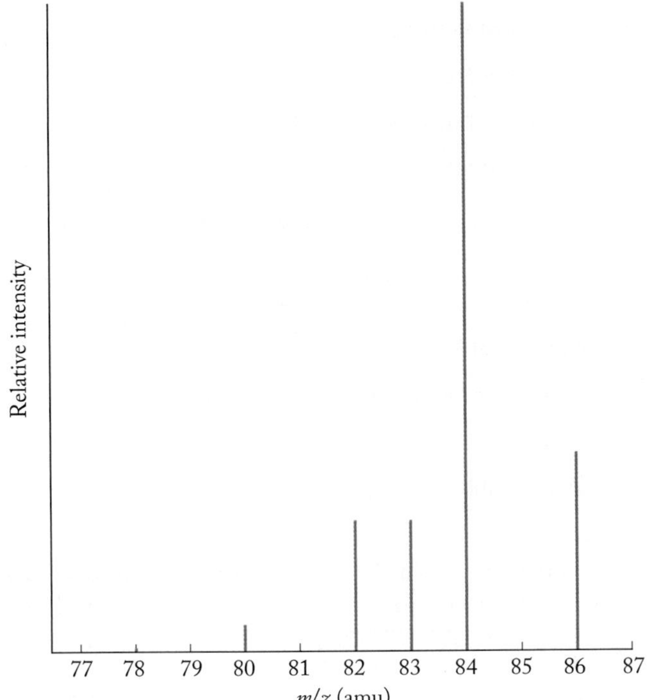

FIGURE P2.10

2.11. Which of the highlighted elements in Figure P2.11 forms monatomic ions with a charge of (a) 1+; (b) 2+; (c) 3+; (d) 1−; (e) 2−?

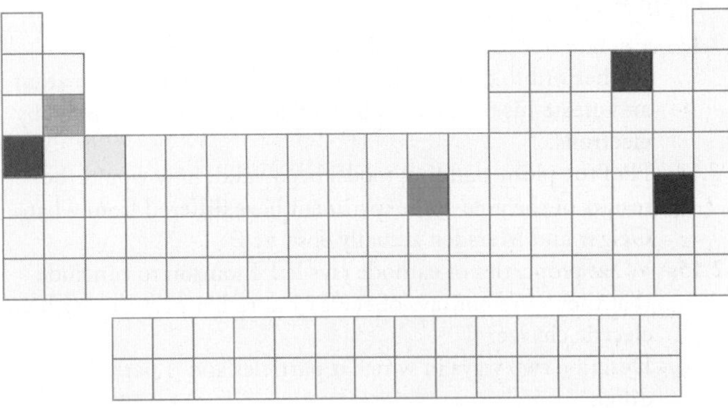

FIGURE P2.11

2.12. Use representations [A] through [I] in Figure P2.12 to answer questions a–f. (The atomic color palette is inside the back cover.)

a. Based on the ratio of cations to anions in representations [A] and [I], which compound is potassium iodide and which is potassium oxide?
b. Order the nuclei from the fewest neutrons to the most neutrons.
c. Which representations depict isotopes?
d. What is the mass of the molecule in representation [E]?
e. Which would contain more molecules, 100 g of [C] or 100 g of [G]?
f. Which would contain more sulfur atoms, 100 g of [C] or 100 g of [G]?

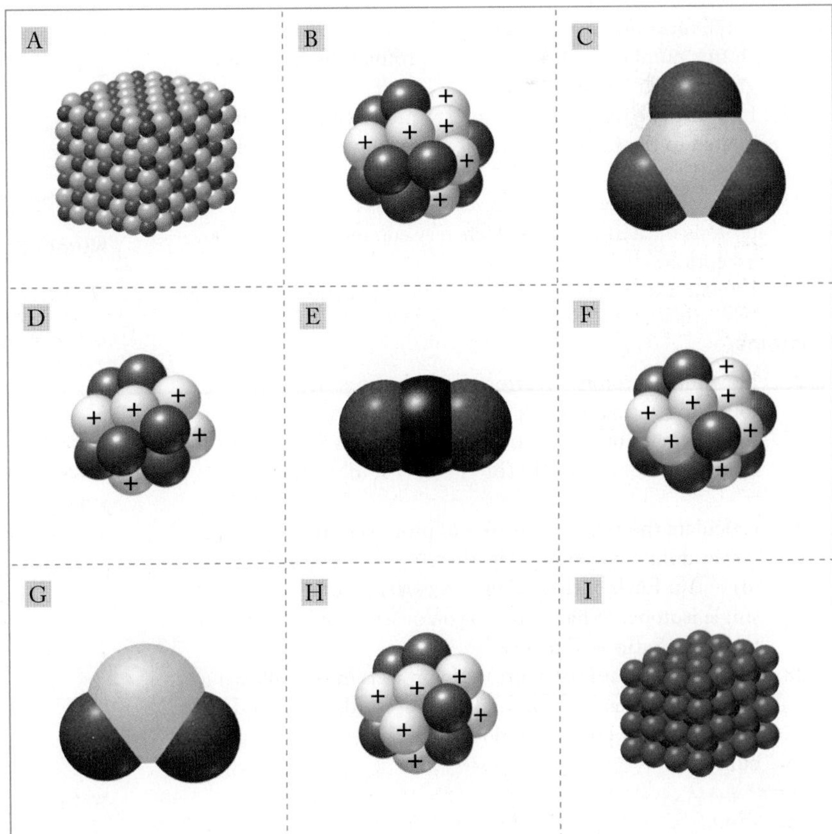

FIGURE P2.12

QUESTIONS AND PROBLEMS

The Rutherford Model of Atomic Structure

Concept Review

2.13. Explain how the results of the gold-foil experiment led Rutherford to dismiss the plum-pudding model of the atom and create his own model based on a nucleus surrounded by electrons.

2.14. Had the plum-pudding model been valid, how would the results of the gold-foil experiment have differed from what Geiger and Marsden actually observed?

2.15. What properties of cathode rays led Thomson to conclude that they were not rays of energy but rather particles with an electric charge?

2.16. Describe two ways in which α particles and β particles differ.

***2.17.** **Helium in Pitchblende** The element helium was first discovered on Earth in a sample of pitchblende, an ore of radioactive uranium oxide. How did helium get in the ore?

2.18. How might using a thicker piece of gold foil have affected the scattering pattern of α particles observed by Rutherford's students?

***2.19.** What would have happened to the gold atoms in Rutherford's experiment if their nuclei had absorbed α particles?

*2.20. In addition to gold foil, Geiger and Marsden tried silver and aluminum foils in their experiment. Why might foils of these metals have deflected fewer α particles than gold foil?

Nuclides and Their Symbols

Concept Review

2.21. If the mass number of a nuclide is more than twice the atomic number, is the neutron-to-proton ratio less than, greater than, or equal to 1?

2.22. How are the mass number and atomic number of a nuclide related to the number of neutrons and protons in each of its nuclei?

2.23. Nearly all stable nuclides have at least as many neutrons as protons in their nuclei. Which very common nuclide is an exception?

2.24. Explain the inherent redundancy in the nuclide symbol $^{A}_{Z}X$.

Problems

2.25. How many protons, neutrons, and electrons are in the following atoms? (a) ^{14}C; (b) ^{59}Fe; (c) ^{90}Sr; (d) ^{210}Pb

2.26. How many protons, neutrons, and electrons are there in the following atoms? (a) ^{11}B; (b) ^{19}F; (c) ^{131}I; (d) ^{222}Rn

---

2.27. Calculate the *ratio* of neutrons to protons in the following stable atomic nuclei: (a) ^{4}He; (b) ^{23}Na; (c) ^{59}Co; and (d) ^{197}Au. Each of these elements exists naturally as a single isotope. What trend do you observe for the neutron-to-proton ratio as Z increases?

2.28. Calculate the *ratio* of neutrons to protons in the following group 15 nuclei: (a) ^{14}N; (b) ^{31}P; (c) ^{75}As; (d) ^{121}Sb; and (e) ^{123}Sb. How does the ratio change with increasing atomic number?

2.29. Fill in the missing information about atoms of the four nuclides in the following table.

Symbol	^{23}Na	?	?	?
Number of Protons	?	39	?	79
Number of Neutrons	?	50	?	?
Number of Electrons	?	?	50	?
Mass Number	?	?	118	197

2.30. Fill in the missing information about atoms of the four nuclides in the following table.

Symbol	^{27}Al	?	?	?
Number of Protons	?	42	?	92
Number of Neutrons	?	56	?	?
Number of Electrons	?	?	60	?
Mass Number	?	?	143	238

2.31. Fill in the missing information about the monatomic ions in the following table.

Symbol	$^{37}Cl^{-}$	?	?	?
Number of Protons	?	11	?	88
Number of Neutrons	?	12	46	?
Number of Electrons	?	10	36	86
Mass Number	?	?	81	226

2.32. Fill in the missing information about the monatomic ions in the following table.

Symbol	$^{137}Ba^{2+}$	?	?	?
Number of Protons	?	30	?	40
Number of Neutrons	?	34	16	?
Number of Electrons	?	28	18	36
Mass Number	?	?	32	90

Navigating the Periodic Table

Concept Review

2.33. Mendeleev arranged the elements on the left side of his periodic table based on the formulas of the binary compounds they formed with oxygen, and he used the formulas as column labels. For example, group 1 in a modern periodic table was labeled "R_2O" in Mendeleev's table, where "R" represented one of the elements in the group. What labels did Mendeleev use for groups 2, 3, and 4 from the modern periodic table?

2.34. Mendeleev arranged the elements on the right side of his periodic table based on the formulas of the binary compounds they formed with hydrogen and used these formulas as column labels. Which groups in the modern periodic table were labeled "HR," "H_2R," and "H_3R," where "R" represented one of the elements in the group?

2.35. Mendeleev left empty spaces in his periodic table for elements he suspected existed but had yet to be discovered. However, he left no spaces for the noble gases (group 18 in the modern periodic table). Suggest a reason why he left no spaces for them.

2.36. Describe how the charges of the monatomic ions that elements form change as group number increases in a particular row of the periodic table and how ion charges change as the row number increases in a particular group.

Problems

2.37. **TNT** Molecules of the explosive TNT contain atoms of hydrogen and second-row elements in groups 14, 15, and 16. Which three elements are they?

2.38. **Phosgene** Phosgene was used as a chemical weapon during World War I. Despite the name, phosgene molecules contain no atoms of phosphorus. Instead, they contain atoms of carbon and the group 16 element in the second row of the periodic table and the group 17 element in the third row. What are the identities and atomic numbers of the two elements?

2.39. **Catalytic Converters** The catalytic converters used to remove pollutants from automobile exhaust contain compounds of several fairly expensive elements, including those described in the following list. Which elements are they?

a. The group 10 transition metal in the fifth row of the periodic table
b. The transition metal whose symbol is to the left of your answer to part a
c. The transition metal whose symbol is directly below your answer to part a

2.40. **Swimming Pool Chemistry** Compounds containing chlorine have long been used to disinfect the water in swimming pools, but in recent years a compound of a less corrosive halogen has become a popular alternative disinfectant. What is the name of this fourth-row element?

2.41. How many metallic elements are there in the third row of the periodic table?

2.42. Which third-row element in the periodic table has chemical properties of a nonmetal but physical properties of a metal?

The Masses of Atoms, Ions, and Molecules

Concept Review

2.43. What is meant by a *weighted average*?

2.44. Explain how percent natural abundances are used to calculate average atomic masses.

2.45. A hypothetical element consists of two isotopes (X and Y) with masses m_X and m_Y. If the natural abundance of the X isotope is exactly 50%, what is the average atomic mass of the element?

2.46. In calculating the formula masses of binary ionic compounds, we use the average masses of neutral atoms, not ions. Why?

***2.47.** The average mass of platinum is 195.08 amu, yet the natural abundance of ^{195}Pt is only 33.8%. Propose an explanation for this observation.

*2.48. The average atomic mass of europium (Eu, 151.96 amu), measured to five significant figures, is only 0.04 amu different from a whole number. Can we conclude that there is only one stable isotope of europium? Why or why not?

Problems

***2.49.** The argon in nature consists of three isotopes: ^{36}Ar, ^{38}Ar, and ^{40}Ar. Which one is the most abundant?

2.50. Manganese has only one stable isotope. How many neutrons are in each of its atoms?

2.51. Boron, lithium, and nitrogen each have two stable isotopes. Use the average atomic masses of the elements to determine which isotope in each of the following pairs of stable isotopes is the more abundant. (a) ^{10}B or ^{11}B; (b) ^{6}Li or ^{7}Li; (c) ^{14}N or ^{15}N

2.52. Rubidium, gallium, and vanadium each have two stable isotopes. Use the average atomic masses of the elements to determine which isotope in each of the following pairs of stable isotopes is the more abundant. (a) ^{85}Rb or ^{87}Rb; (b) ^{69}Ga or ^{71}Ga; (c) ^{50}V or ^{51}V

2.53. Copper in nature is a mixture of 69.17% copper-63 (62.9296 amu) and 30.83% copper-65 (64.9278 amu). Use this information to calculate the average atomic mass of copper.

2.54. Sulfur in nature is a mixture of four isotopes: ^{32}S (31.9721 amu, 95.04%); ^{33}S (32.9715 amu, 0.75%); ^{34}S (33.9679 amu, 4.20%); and ^{36}S (35.9671 amu, 0.01%). Use this information to calculate the average atomic mass of sulfur.

2.55. **Chemistry of Mars** Chemical analyses conducted by the first Mars rover robotic vehicle in its 1997 mission produced the magnesium isotope data shown in the table that follows. Is the average atomic mass of magnesium in this Martian sample the same as on Earth (24.31 amu)?

Isotope	Mass (amu)	Natural Abundance (%)
^{24}Mg	23.9850	78.70
^{25}Mg	24.9858	10.13
^{26}Mg	25.9826	11.17

2.56. The natural abundances of the four isotopes of strontium are 0.56% ^{84}Sr (83.9134 amu), 9.86% ^{86}Sr (85.9094 amu), 7.00% ^{87}Sr (86.9089 amu), and 82.58% ^{88}Sr (87.9056 amu). Calculate the average atomic mass of strontium and compare it to the value in the periodic table inside the front cover.

2.57. Use the data in the following table of abundances and masses of the five stable titanium isotopes to calculate the atomic mass of ^{48}Ti.

Isotope	Mass (amu)	Natural Abundance (%)
^{46}Ti	45.9526	8.25
^{47}Ti	46.9518	7.44
^{48}Ti	?	73.72
^{49}Ti	48.94787	5.41
^{50}Ti	49.94479	5.18
Average	47.867	

2.58. Use the following table of abundances and masses of the stable isotopes of zirconium to calculate the atomic mass of ^{92}Zr.

Symbol	Mass (amu)	Natural Abundance (%)
^{90}Zr	89.905	51.45
^{91}Zr	90.906	11.22
^{92}Zr	?	17.15
^{94}Zr	93.906	17.38
^{96}Zr	95.908	2.80
Average	91.224	

2.59. What are the masses of the formula units of each of the following ionic compounds? (a) CaF_2; (b) Na_2S; (c) Cr_2O_3

2.60. What are the masses of the formula units of each of the following ionic compounds? (a) KCl; (b) MgO; (c) Al_2O_3

2.61. How many carbon atoms are there in one molecule of each of the following compounds? (a) CH_4; (b) C_3H_8; (c) C_6H_6; (d) $C_6H_{12}O_6$

2.62. How many hydrogen atoms are there in each of the molecules in Problem 2.61?

2.63. Rank the following compounds based on increasing molecular mass. (a) CO; (b) Cl_2; (c) CO_2; (d) NH_3; (e) CH_4

2.64. Rank the following compounds based on decreasing molecular mass. (a) H_2; (b) Br_2; (c) NO_2; (d) C_2H_2; (e) BF_3

Moles and Molar Masses

Concept Review

2.65. In principle, we could use the more familiar unit *dozen* in place of mole when expressing the quantities of particles (atoms, ions, or molecules). What would be the disadvantage in doing so?

2.66. In what way is the molar mass of an ionic compound the same as its formula mass, and in what ways are they different?

2.67. Do equal masses of two isotopes of an element contain the same number of atoms?

2.68. The natural abundances of the isotopes of an element are given in % by mass. Does the same percentage apply to the percent natural abundance by moles?

Problems

2.69. Earth's atmosphere contains many volatile substances that are present in trace amounts. The following quantities of trace gases were found in a 1.0 mL sample of air. Calculate the number of moles of each gas in the sample.
a. 4.4×10^{14} atoms of Ne
b. 4.2×10^{13} molecules of CH_4
c. 2.5×10^{12} molecules of O_3
d. 4.9×10^{9} molecules of NO_2

2.70. The following quantities of trace gases were found in a 1.0 mL sample of air. Calculate the number of moles of each compound in the sample.
a. 1.4×10^{13} molecules of H_2
b. 1.5×10^{14} atoms of He
c. 7.7×10^{12} molecules of N_2O
d. 3.0×10^{12} molecules of CO

2.71. How many moles of iron are there in 1 mole of the following compounds? (a) FeO; (b) Fe_2O_3; (c) $Fe(OH)_3$; (d) Fe_3O_4

2.72. How many moles of Na^+ ions are there in 1 mole of the following compounds? (a) NaCl; (b) Na_2SO_4; (c) Na_3PO_4; (d) $NaNO_3$

2.73. What is the mass of 0.122 mol of $MgCO_3$?

2.74. What is the volume of 1.00 mol of benzene (C_6H_6) at 20°C? The density of benzene at 20°C is 0.879 g/mL.

2.75. How many moles of titanium and how many atoms of titanium are there in 0.125 mole of each of the following? (a) ilmenite, $FeTiO_3$; (b) $TiCl_4$; (c) Ti_2O_3; (d) Ti_3O_5

2.76. How many moles of iron and how many atoms of iron are there in 2.5 moles of each of the following? (a) wolframite, $FeWO_4$; (b) pyrite, FeS_2; (c) magnetite, Fe_3O_4; (d) hematite, Fe_2O_3

2.77. Which substance in each of the following pairs of quantities contains more moles of oxygen?
a. 1 mol Al_2O_3 or 1 mol Fe_2O_3
b. 1 mol SiO_2 or 1 mol N_2O_4
c. 3 mol CO or 2 mol CO_2

2.78. Which substance in each of the following pairs of quantities contains more moles of oxygen?
a. 2 mol N_2O or 1 mol N_2O_5
b. 1 mol NO or 1 mol $Ca(NO_3)_2$
c. 2 mol NO_2 or 1 mol $NaNO_2$

2.79. **Elemental Composition of Minerals** Aluminum, silicon, and oxygen form minerals known as aluminosilicates. How many moles of aluminum are in 1.50 moles of the following?
a. pyrophyllite, $Al_2Si_4O_{10}(OH)_2$
b. mica, $KAl_3Si_3O_{10}(OH)_2$
c. albite, $NaAlSi_3O_8$

2.80. **Radioactive Minerals** The uranium used for nuclear fuel exists in nature in several minerals. Calculate how many moles of uranium are in 1 mole of the following.
a. carnotite, $K_2(UO_2)_2(VO_4)_2$
b. uranophane, $CaU_2Si_2O_{11}$
c. autunite, $Ca(UO_2)_2(PO_4)_2$

2.81. Calculate the molar masses of the following gases. (a) SO_2; (b) O_3; (c) CO_2; (d) N_2O_5

2.82. Determine the molar masses of the following minerals.
a. rhodonite, $MnSiO_3$
b. scheelite, $CaWO_4$
c. ilmenite, $FeTiO_3$
d. magnesite, $MgCO_3$

2.83. **Flavors** Calculate the molar masses of the following common flavors in food.
a. vanillin, $C_8H_8O_3$
b. oil of cloves, $C_{10}H_{12}O_2$
c. anise oil, $C_{10}H_{12}O$
d. oil of cinnamon, C_9H_8O

2.84. **Sweeteners** Calculate the molar masses of the following common sweeteners.
a. sucrose, $C_{12}H_{22}O_{11}$
b. saccharin, $C_7H_5O_3NS$
c. aspartame, $C_{14}H_{18}N_2O_5$
d. fructose, $C_6H_{12}O_6$

2.85. How many moles of carbon are there in 500.0 grams of carbon?

2.86. How many moles of gold are there in 2.00 ounces of gold?

2.87. How many moles of Ca^{2+} ions are in 0.25 mol $CaTiO_3$? What is the mass in grams of the Ca^{2+} ions?

2.88. How many moles of O^{2-} ions are in 0.55 mol Al_2O_3? What is the mass in grams of the O^{2-} ions?

2.89. Suppose pairs of balloons are filled with 10.0 g of the following pairs of gases. Which balloon in each pair has the greater number of particles? (a) CO_2 or NO; (b) CO_2 or SO_2; (c) O_2 or Ar

2.90. If you had equal masses of the substances in the following pairs of compounds, which of the two would contain the greater number of ions? (a) NaBr or KCl; (b) NaCl or $MgCl_2$; (c) $CrCl_3$ or Na_2S

2.91. How many moles of SiO_2 are there in a quartz crystal (SiO_2) that has a mass of 45.2 g?

2.92. How many moles of NaCl are there in a crystal of halite that has a mass of 6.82 g?

2.93. The density of uranium (U; 19.05 g/cm^3) is more than five times as great as that of diamond (C; 3.514 g/cm^3). If you have a cube (1 cm on a side) of each element, which cube contains more atoms?

*2.94. Aluminum (d = 2.70 g/cm^3) and strontium (d = 2.64 g/cm^3) have nearly the same density. If we manufacture two cubes, each containing 1 mol of one element or the other, which cube will be smaller? What are the dimensions of this cube?

Mass Spectrometry

Concept Review

2.95. How does mass spectrometry provide information on the molecular mass of a compound?

2.96. How are isotopic abundances reflected in the mass spectrum of HBr?

2.97. Would you expect the mass spectra of CO_2 and C_3H_8 to have molecular ions with the same mass (to the nearest amu)?

*2.98. Would you expect the mass spectra of CO_2 and C_3H_8 to be the same?

Problems

2.99. **Screening for Explosives** Many of the explosive materials of concern to airport security contain nitrogen and oxygen. Calculate the masses of the molecular ions formed by (a) $C_3H_6N_6O_6$, (b) $C_4H_8N_8O_8$, (c) $C_5H_8N_4O_{12}$, and (d) $C_{14}H_6N_6O_{12}$.

*2.100. **Landfill Gas** Mass spectrometry has proven useful in analyzing the gases emitted from landfills. The principal component is methane (CH_4), but small amounts of dimethylsulfide (C_2H_6S) and dichloroethene ($C_2H_2Cl_2$) are often present, too. Calculate the masses of the molecular ions formed by these three compounds in a mass spectrometer.

2.101. The mass spectrum of chlorine, Cl_2, is shown in Figure P2.101. The natural abundances of its two stable isotopes are 75.78% ^{35}Cl and 24.22% ^{37}Cl.
a. Why are there peaks in the mass spectrum at 70, 72, and 74 amu?
b. Why is the peak at 70 amu so much taller than the peak at 74 amu?

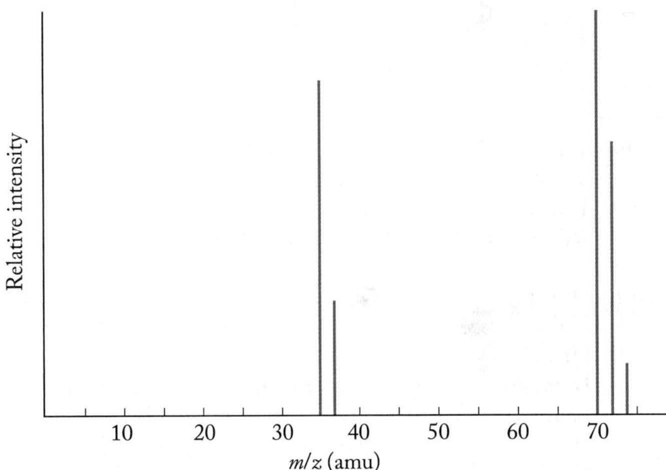

FIGURE P2.101

2.102. The mass spectrum of bromine, Br_2, is shown in Figure P2.102. The natural abundances of its two stable isotopes are 50.69% ^{79}Br and 49.31% ^{81}Br.

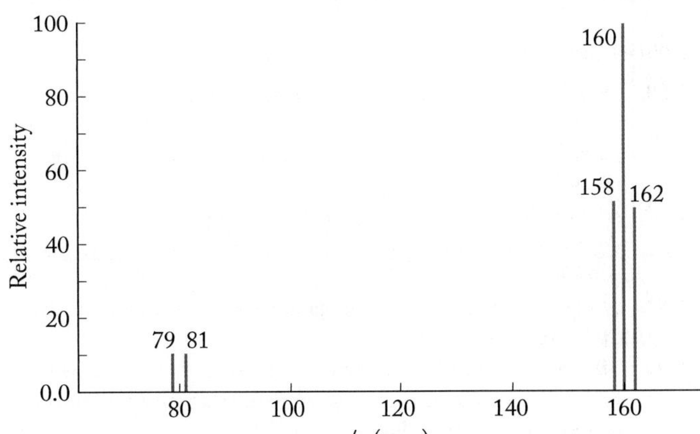

FIGURE P2.102

a. Why are there peaks in the mass spectrum at 158, 160, and 162 amu?

*b. How do we know that bromine doesn't have a third isotope, ^{80}Br?

***2.103. Sewer Gas** Hydrogen sulfide, H_2S, is a foul-smelling and toxic gas that may be present in wastewater sewers. Although the human nose detects H_2S in low concentrations, prolonged exposure to H_2S deadens our sense of smell, making it particularly dangerous to sewer workers who work in poorly ventilated areas. Sulfur in nature is a mixture of four isotopes: ^{32}S (94.93%), ^{33}S (0.76%), ^{34}S (4.29%), and ^{35}S (0.02%). Explain how the relative intensities of the peaks in the mass spectrum of H_2S (Figure P2.103) reflect the natural abundance of sulfur isotopes and the sequential loss of H atoms from molecules of H_2S.

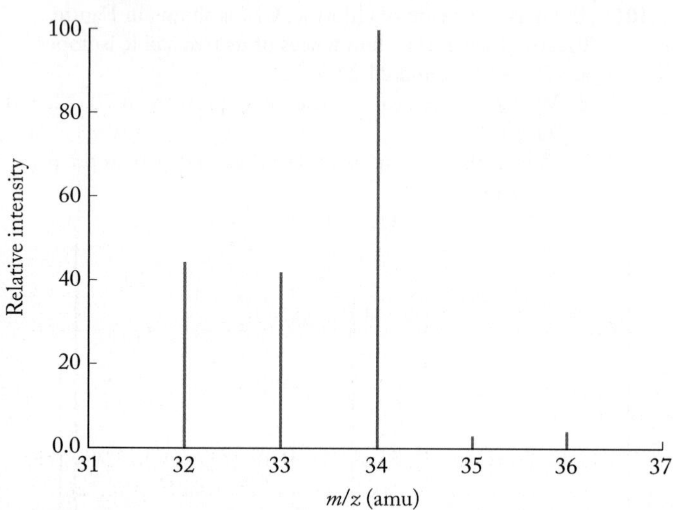

FIGURE P2.103

*2.104. **Bar Code Readers** Arsine, AsH_3, is a hazardous gas used in the manufacture of electronic devices, including the bar code readers used at the checkout counters of many stores. The mass spectrum of arsine is shown in Figure P2.104. Arsenic has only one stable isotope. What are the formulas of the ions responsible for the four peaks in the mass spectrum?

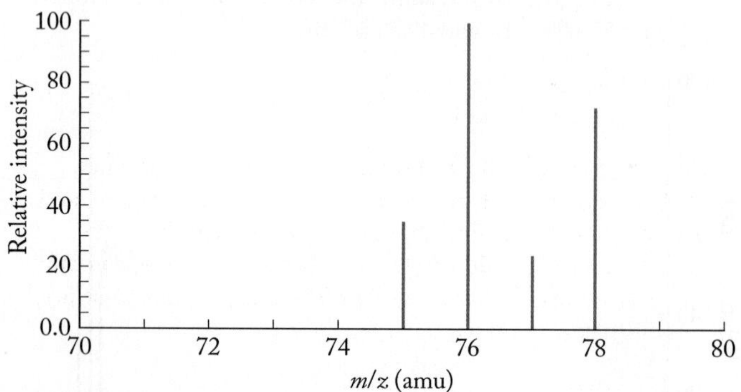

FIGURE P2.104

2.105. Detecting Illegal Drugs In July 2015, researchers in Britain reported on a new method for detecting cocaine on fingertips using mass spectrometry. The mass spectrum of cocaine is shown in Figure P2.105. What is the molar mass of cocaine?

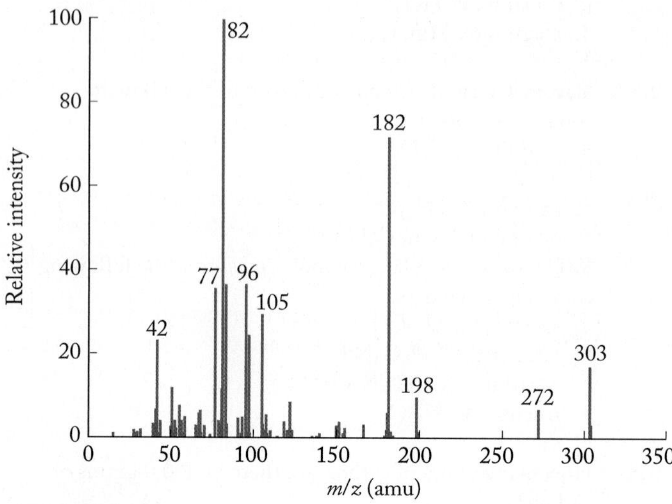

FIGURE P2.105

*2.106. Many main group elements form molecular compounds of general formula $(CH_3)_nM$, where n is a simple whole number. The mass spectrum of the compound where M = Sb is shown in Figure P2.106. The natural abundances of the two stable isotopes of antimony are 57.25% ^{121}Sb and 42.75% ^{123}Sb. What is the value of n in $(CH_3)_nSb$?

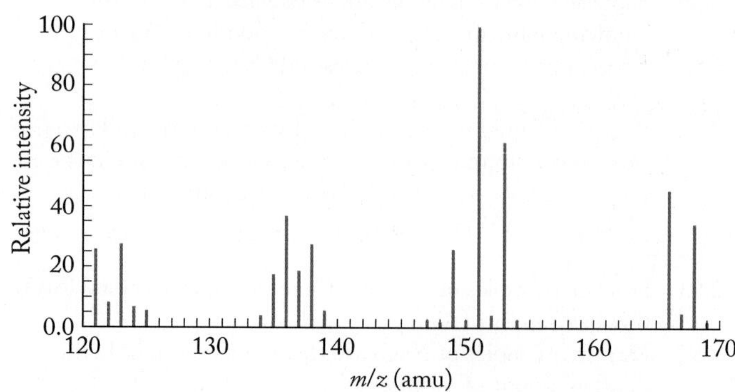

FIGURE P2.106

Additional Problems

2.107. In April 1897, J. J. Thomson presented the results of his experiment with cathode-ray tubes in which he proposed that the rays were actually beams of negatively charged particles, which he called "corpuscles."

a. What is the name we use for the particles today?

b. Why did the beam deflect when passed between electrically charged plates?

c. If the polarity of the plates were switched, how would the position of the light spot on the phosphorescent screen change?

2.108. Strontium has four isotopes: ^{84}Sr, ^{86}Sr, ^{87}Sr, and ^{88}Sr.

a. How many neutrons are there in an atom of each isotope?

b. Use the data in the following table to calculate the natural abundances of ^{87}Sr and ^{88}Sr.

Isotope	Mass (amu)	Natural Abundance (%)
^{84}Sr	83.9134	0.56
^{86}Sr	85.9094	9.86
^{87}Sr	86.9089	?
^{88}Sr	87.9056	?
Average	87.621	

2.109. There are three stable isotopes of magnesium. Their masses are 23.9850, 24.9858, and 25.9826 amu. If the average atomic mass of magnesium is 24.3050 amu and the natural abundance of the lightest isotope is 78.99%, what are the natural abundances of the other two isotopes?

2.110. Without consulting a periodic table, give the atomic number (Z) for each of the highlighted elements in Figure P2.110.

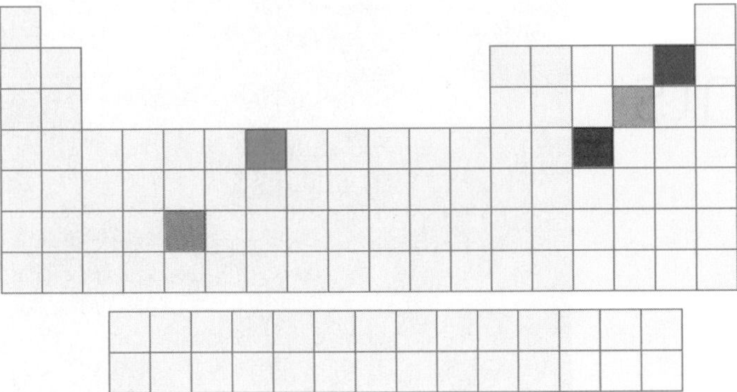

FIGURE P2.110

2.111. Silver Nanoparticles in Clothing The antimicrobial properties of silver metal have led to the use of silver nanoparticles in clothing (Figure P2.111) to reduce odors. If a silver nanoparticle with a diameter of 1×10^{-7} m contains 4.8×10^7 atoms of Ag, how many nanoparticles are in 1.00 g?

FIGURE P2.111

2.112. **HD Television** Some newer television sets utilize nanoparticles of cadmium sulfide (CdS) and cadmium selenide (CdSe), called "quantum dots," to produce the colors on the screen. Different-sized quantum dots lead to different colors.

a. Calculate the formula masses of CdS and CdSe.
b. If a nanoparticle of CdSe contains 2.7×10^7 atoms of Cd, how many atoms of Se are in the particle?
c. If a nanoparticle of CdS weighs 4.3×10^{-15} g, how many grams of Cd and how many grams of S does it contain?

***2.113. Greenhouse Gas Concentrations** Samples of air are collected daily at the Mauna Loa Observatory in Hawaii and analyzed for CO_2 content. During January 2016, the average result of these analyses was 402.5 μmoles (10^{-6} moles) of CO_2 per mole of air. If the average molar mass of the gases in air is 28.8 g/mol, how many μg of CO_2 per gram of air were in these samples?

2.114. **Performance-Enhancing Drugs** Mass spectrometry is used to detect performance-enhancing drugs in body fluids. Included on the list of banned substances for Olympic athletes is tetrahydrogestrinone, a compound that mimics the steroid testosterone and can be used to build muscle. The mass spectrum of tetrahydrogestrinone is shown in Figure P2.114. Identify the molecular ion and show that it has a mass consistent with the formula $C_{21}H_{28}O_2$.

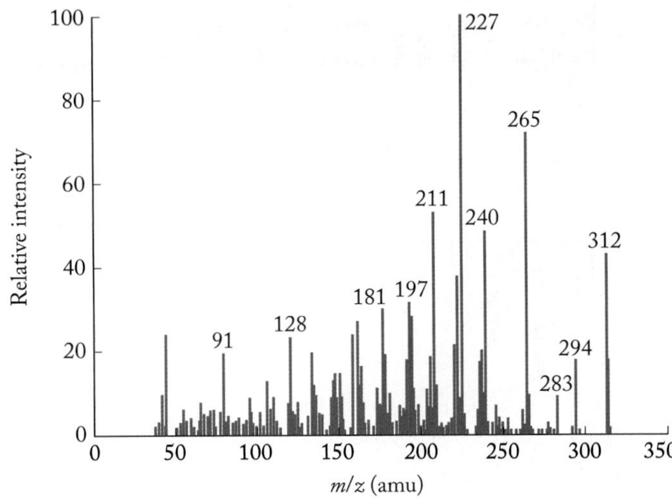

FIGURE P2.114

2.115. Hope Diamond The Hope Diamond (Figure P2.115) at the Smithsonian National Museum of Natural History has a mass of 45.52 carats.

a. How many moles of carbon are in the Hope Diamond (1 carat = 200 mg)?
b. How many carbon atoms are in the diamond?

FIGURE P2.115

2.116. Suppose we know the atomic mass of each of the three stable isotopes of an element to six significant figures, and we know the natural abundances of the isotopes to the nearest 0.01%. How well can we know the average atomic mass—that is, how many significant figures should be used to express its value?

smartwork5

If your instructor uses Smartwork5, log in at digital.wwnorton.com/atoms2.

3

Atomic Structure

Explaining the Properties of Elements

COLORS OF THE AURORA Some of the red and green colors of an aurora display are produced when atoms of oxygen in the upper atmosphere collide with high-speed charged particles emitted by the sun.

PARTICULATE **REVIEW**

Rutherford's Atom

In Chapter 2, we learned about Rutherford's gold-foil experiment, which revealed the structure of the atom.

- How many protons, neutrons, and electrons are in an atom of ^{197}Au?
- Which image depicts the model of the atom prevailing at the time when Rutherford began his experiments?
- What feature of the other image was inconsistent with the prevailing model and led Rutherford to propose a new one?

(Answers to Particulate Review questions are in the back of the book.)

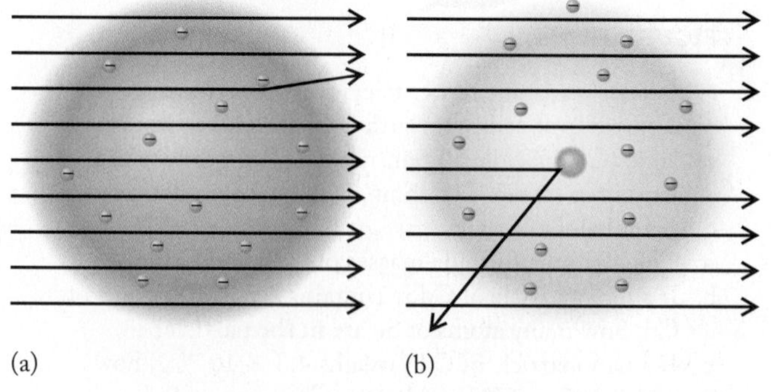

Oxygen atomic emission

Sunlight

400 500 600 700

Wavelength (nm)

PARTICULATE **PREVIEW**

Where Are the Electrons?

Additional models that further refined our view of atomic structure are described in Chapter 3. As you read Chapter 3, look for ideas that will help you answer these questions:

- What features of Rutherford's model are depicted in model (a)?
- What do the concentric circles in image a suggest about the arrangement of electrons in atoms?
- What is a key limitation of model (a) that is addressed by model (b)?

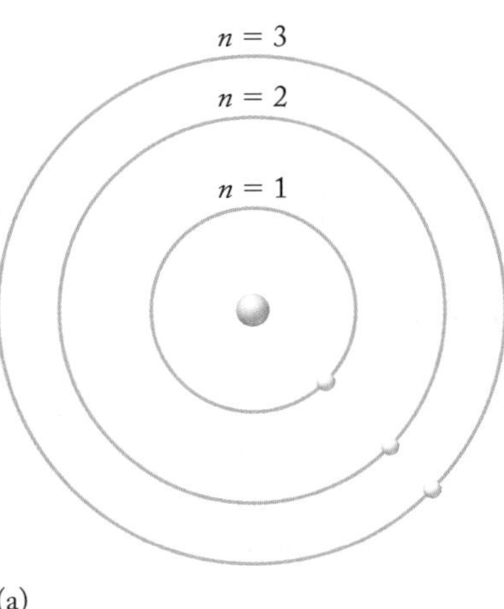

(a)

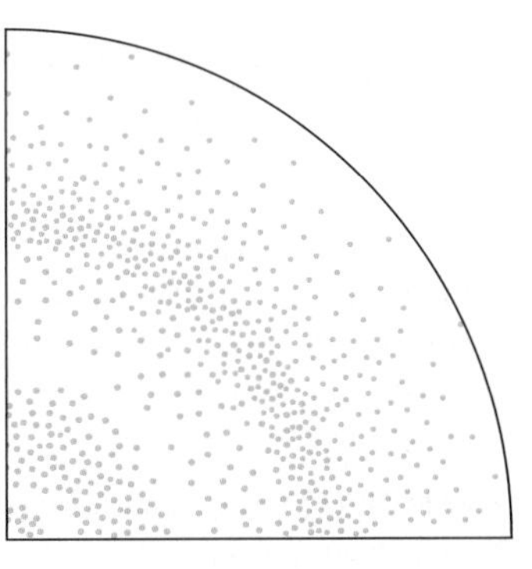

(b)

Learning Outcomes

LO1 Interconvert the energies, wavelengths, and frequencies of electromagnetic radiation and link their values to the appropriate regions of the electromagnetic spectrum
Sample Exercises 3.1, 3.2

LO2 Explain the photoelectric effect using quantum theory
Sample Exercise 3.3

LO3 Relate the energies and wavelengths of photons absorbed and emitted by atoms to electron transitions between atomic energy levels
Sample Exercises 3.4, 3.5

LO4 Apply the Heisenberg uncertainty principle to particles in motion and calculate their de Broglie wavelengths
Sample Exercises 3.6, 3.7

LO5 Assign quantum numbers to orbitals and use their values to describe the sizes, energies, and orientations of orbitals
Sample Exercises 3.8, 3.9

LO6 Use the aufbau principle and Hund's rule to write electron configurations and draw orbital diagrams of atoms and monatomic ions
Sample Exercises 3.10, 3.11, 3.12

LO7 Use the concept of effective nuclear charge to explain differences in the energies of atomic orbitals and to predict the relative sizes of atoms and monatomic ions
Sample Exercise 3.13

LO8 Relate the ionization energies and electron affinities of the elements to their positions in the periodic table
Sample Exercise 3.14

3.1 Nature's Fireworks and the Electromagnetic Spectrum

The opening photo of this chapter provides a view of the *Aurora borealis*, one of nature's most colorful light displays. Auroras are produced by electrons and positively charged particles flowing out from the sun at speeds approaching 1000 km/s. Earth's magnetic poles capture some of them, drawing them into the upper atmosphere where the particles collide with atoms and molecules of oxygen, nitrogen, and other atmospheric gases.

Some of the collisions between high-speed electrons and molecules produce molecular ions, in much the way beams of high-energy electrons ionize molecules in mass spectrometers. Sometimes there is not enough energy in these collisions to ionize atoms or molecules, but the energy may still be absorbed only to be re-emitted as visible light. Different particles emit different colors of light. For example, atoms of oxygen in the upper atmosphere may emit red light while O atoms at lower altitudes may emit green light, as shown in the chapter-opening photo. These two colors make up the visible emission spectrum of atomic oxygen. This *atomic* spectrum is not like the visible-light spectrum emitted by the sun or by artificial sources of "white" light, which contain all of the colors of the rainbow. Instead, O atoms, like those of other elements, produce relatively few characteristic colors of light. As we shall see in later sections of this chapter, different elements emit different colors. More importantly, we will also learn in this chapter how and why atoms produce their distinctive colors of light.

To understand the origins of the colors of auroras, we need to understand how atoms interact with the forms of energy that are collectively called *radiant* energy or **electromagnetic radiation**. Visible light is the most familiar form of electromagnetic radiation, but there are several others in the **electromagnetic spectrum** (Figure 3.1). Some forms, such as the gamma rays that accompany nuclear

electromagnetic radiation any form of radiant energy in the electromagnetic spectrum.

electromagnetic spectrum a continuous range of radiant energy that includes gamma rays, X-rays, ultraviolet radiation, visible light, infrared radiation, microwaves, and radio waves.

wavelength (λ) the distance from crest to crest or trough to trough on a wave.

frequency (ν) the number of crests of a wave that pass a stationary point of reference per second.

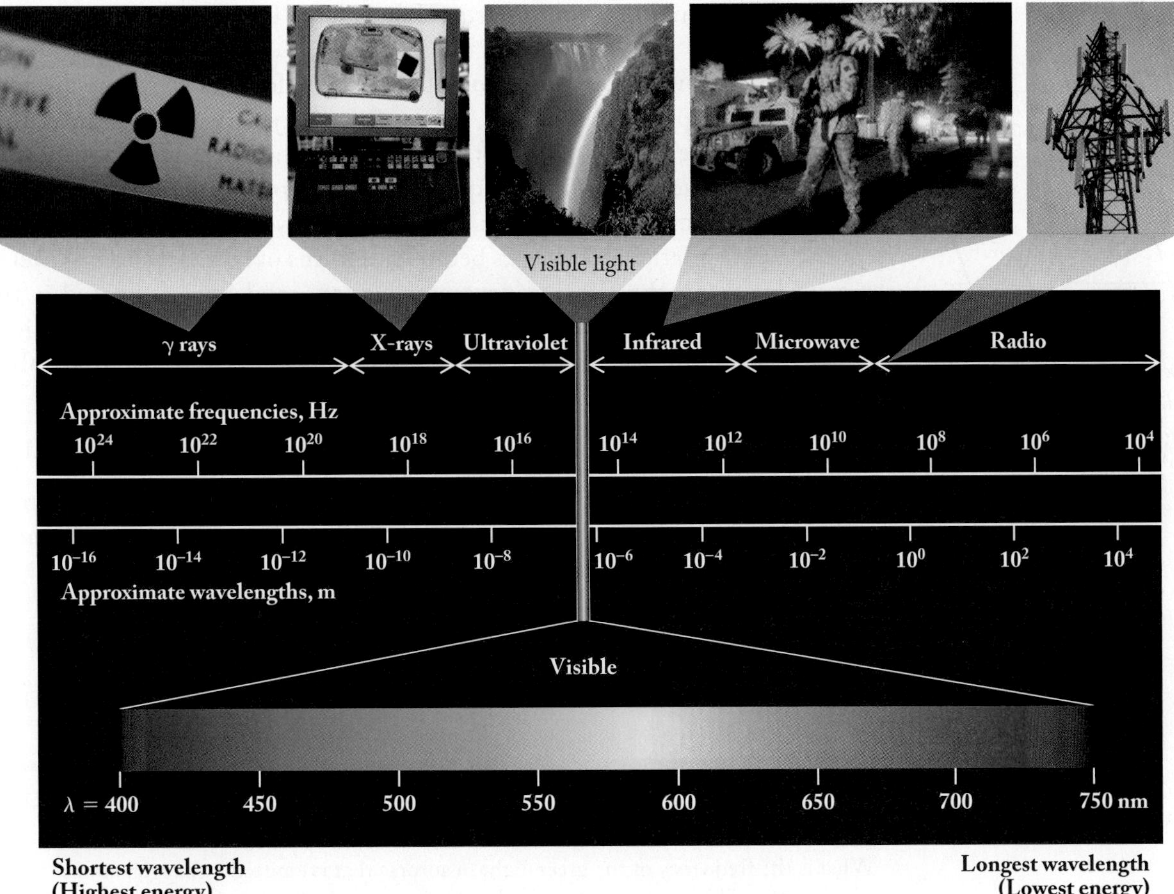

FIGURE 3.1 Visible light occupies a tiny fraction of the electromagnetic spectrum, which ranges from ultrashort-wavelength, high-frequency gamma (γ) rays to long-wavelength, low-frequency radio waves. Note that frequencies increase from right to left, whereas wavelengths increase from left to right.

reactions, the X-rays used in medical imaging, and the ultraviolet rays that can give you a sunburn, have more energy than visible light. Other forms have less energy, including the infrared radiation we feel as heat flowing from warm objects, the microwaves used in ovens, and the radio waves that connect devices in wireless networks.

All the forms of radiation in Figure 3.1 are considered *electromagnetic* because of a theory about their properties developed by Scottish scientist James Clerk Maxwell (1831–1879). According to Maxwell's theory, electromagnetic radiation moves through space (or through any transparent medium) as waves with two perpendicular components: an oscillating *electric* field and an oscillating *magnetic* field (Figure 3.2). Maxwell derived a set of equations based on his oscillating-wave model that accurately describes nearly all the observed properties of light and the other forms of radiant energy.

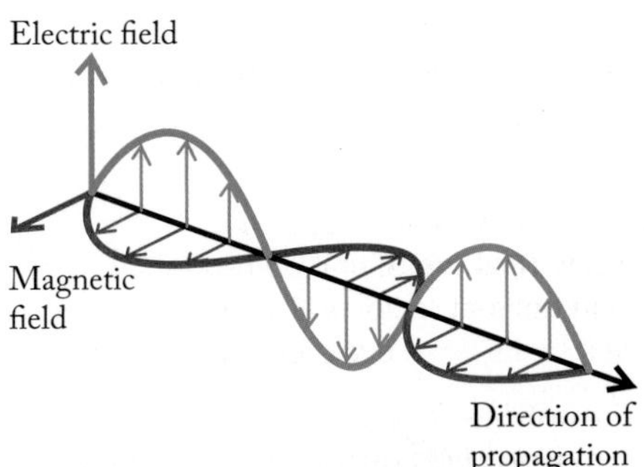

FIGURE 3.2 Electromagnetic waves consist of electric fields and magnetic fields that oscillate in planes oriented at right angles to each other.

A wave of electromagnetic radiation, like any wave traveling through any medium, has a characteristic **wavelength (λ)**, which is the distance from one wave crest to the next, as shown in Figure 3.3. Note that wave A in Figure 3.3 has twice the wavelength of wave B. Each wave also has a characteristic **frequency (ν)**, which is the number of crests that pass a stationary point in space per second. Frequency has units of waves per second, or simply *cycles per second* (s^{-1}). Scientists

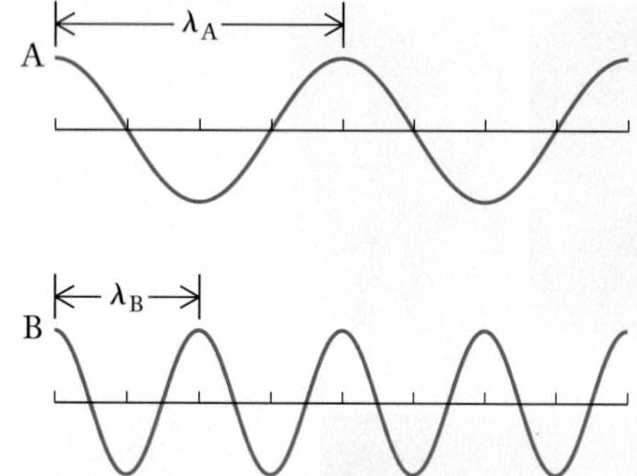

FIGURE 3.3 Every wave has a characteristic wavelength (λ) and frequency (ν). Wave A has a longer wavelength (and lower frequency) than wave B.

also use the frequency unit called **hertz (Hz)**, which is equal to 1 cycle per second ($1\ \text{Hz} = 1\ \text{s}^{-1}$).

Figure 3.3 shows that wave A completes two cycles over the same distance that wave B completes four cycles, so the frequency of wave B is *twice* that of wave A. Figure 3.3 also shows that wave B has a wavelength *half* that of wave A. The wavelengths and frequencies of electromagnetic radiation are inversely related in exactly this way: the shorter the wavelength, the higher the frequency. This relationship occurs because all forms of radiant energy travel at the same speed in the same transparent medium, and a wave's speed is the product of its wavelength and its frequency. Equation 3.1 shows this relationship for radiant energy traveling through a vacuum, where its speed is called the *speed of light*, *c*:

$$\lambda\nu = c \tag{3.1}$$

The value of c to four significant figures is 2.998×10^8 m/s. The reciprocal relationship between wavelength and frequency may be more evident if we rearrange the terms in Equation 3.1 as follows:

$$\nu = \frac{c}{\lambda} \tag{3.2}$$

SAMPLE EXERCISE 3.1 Calculating Frequency from Wavelength **LO1**

What is the frequency of the green light in auroras that is emitted by oxygen atoms? Its wavelength is 557.7 nm. (See the visible lines in oxygen's atomic emission spectrum in this chapter's opening photo.)

Collect and Organize We are given the wavelength of a color of visible light and asked to calculate its frequency. Frequency and wavelength are related by Equation 3.2:

$$\nu = \frac{c}{\lambda}$$

where c is 2.998×10^8 m/s.

Analyze The value c is in meters per second, but the wavelength is given in nanometers. Therefore, we need to convert nanometers to meters using the equality $1\ \text{nm} = 10^{-9}$ m as a conversion factor. The unit labels in Figure 3.1 indicate that frequencies of visible light are around 10^{14} Hz, or $10^{14}\ \text{s}^{-1}$, so we expect our answer to have an exponent in that range.

Solve Substituting the given values and appropriate unit conversion factor into Equation 3.2:

$$\nu = \frac{c}{\lambda} = \frac{2.998 \times 10^8\ \cancel{\text{m}}\ \text{s}^{-1}}{557.7\ \cancel{\text{nm}} \times \dfrac{10^{-9}\ \cancel{\text{m}}}{\cancel{\text{nm}}}} = 5.376 \times 10^{14}\ \text{s}^{-1}$$

Think About It The large value of ν is in the range that we expect for visible light. Remember that λ and ν are inversely proportional: as one increases, the other decreases.

 **Practice Exercise** If the radio waves transmitted by a radio station have a frequency of 90.9 MHz, what is the wavelength of the waves (in meters)?

(Answers to Practice Exercises are in the back of the book.)

hertz (Hz) the SI unit of frequency with units of reciprocal seconds: $1\ \text{Hz} = 1\ \text{s}^{-1} = 1$ cycle per second (cps).

Fraunhofer lines a set of dark lines in the otherwise continuous solar spectrum.

atomic emission spectra characteristic patterns of bright lines produced when atoms are vaporized in high-temperature flames or electrical discharges.

CONCEPT TEST

The ultraviolet (UV) region of the electromagnetic spectrum contains radiation with wavelengths from about 10^{-7} m to 10^{-9} m; the infrared (IR) region contains radiation with wavelengths from about 10^{-4} m to 10^{-6} m. Is UV radiation higher or lower in frequency than IR radiation?

(Answers to Concept Tests are in the back of the book.)

3.2 Atomic Spectra

In this section we explore how atoms and electromagnetic radiation interact. We start with observations that were made over 200 years ago by English scientist William Hyde Wollaston (1766–1828). In 1802, Wollaston was studying the spectrum of sunlight using carefully ground glass prisms when he discovered that the spectrum was not completely continuous. Instead, it contained a series of very narrow dark lines. Using even better prisms, German physicist Joseph von Fraunhofer (1787–1826) resolved and mapped the wavelengths of over 500 of these lines, now called **Fraunhofer lines** (Figure 3.4).

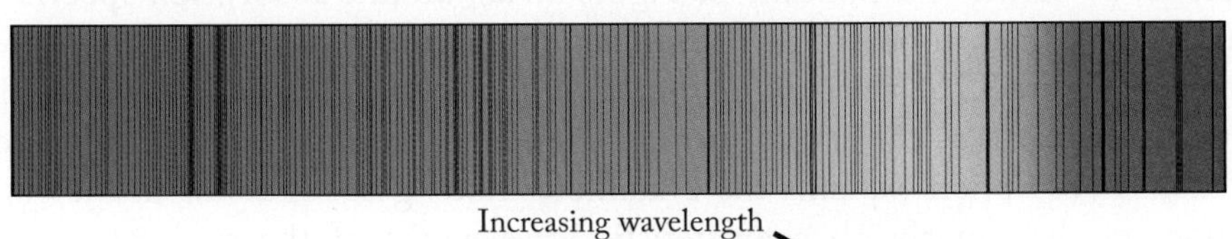

FIGURE 3.4 The spectrum of sunlight is not continuous but contains numerous narrow gaps, which appear as dark lines called Fraunhofer lines.

Fraunhofer did not know why those narrow lines were present in the sun's spectrum. That understanding came nearly a half century later from discoveries made by two other Germans, Robert Wilhelm Bunsen (1811–1899) and Gustav Robert Kirchhoff (1824–1887), using an instrument called a spectroscope (Figure 3.5a). They discovered that the light emitted by elements vaporized in the transparent flame of a gas burner developed by Bunsen produced spectra that were the opposite of Fraunhofer's. Whereas the sun's spectrum was nearly continuous except for a set of narrow dark lines, the spectra produced by the elements consisted of only a few narrow bright lines on a dark background. Bunsen and Kirchhoff discovered that the lines in the **atomic emission spectra** of certain elements exactly matched the wavelengths of some of the Fraunhofer lines in the spectrum of sunlight. For example, the Fraunhofer D line exactly matched the bright yellow-orange line in the emission spectrum produced by hot sodium vapor (Figure 3.5b).

Those experiments and others employing light sources called gas-discharge tubes showed that at very high temperatures the atoms of each element emit a characteristic spectrum (Figure 3.6). Conversely, atoms of elements in the gaseous state absorb electromagnetic radiation when illuminated by an external

(a)

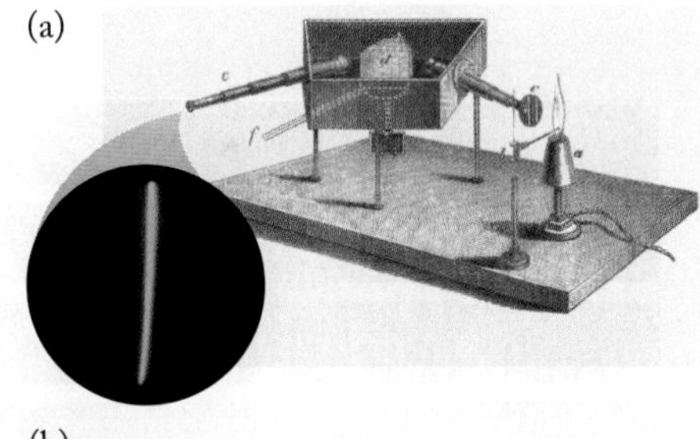

(b)

FIGURE 3.5 (a) Kirchhoff and Bunsen used spectroscopes such as this one to observe the atomic emission spectra of elements introduced to the flame of the burner on the right. (b) Emission from sodium atoms when viewed through a spectroscope.

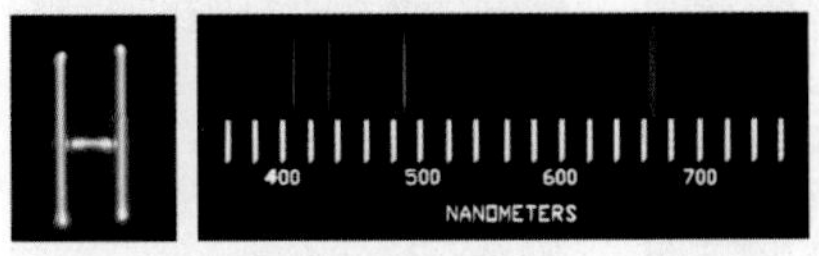

(a) Emission spectrum of hydrogen

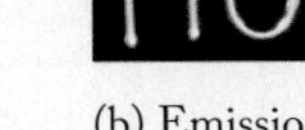

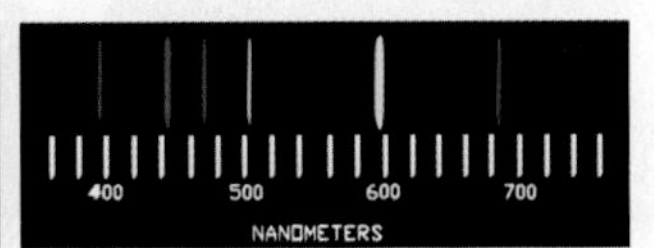

(b) Emission spectrum of helium

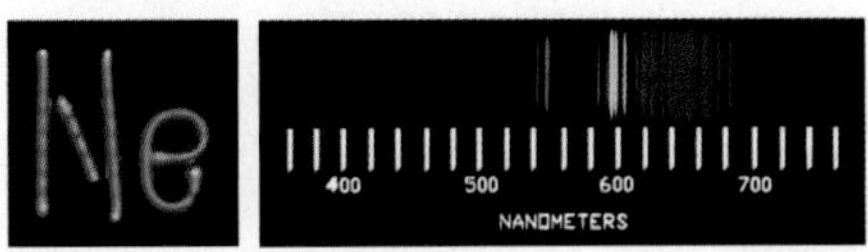

(c) Emission spectrum of neon

FIGURE 3.6 The colored light emitted from gas-discharge tubes filled with various gaseous elements produces atomic emission spectra that are characteristic of the element: (a) hydrogen, (b) helium, and (c) neon.

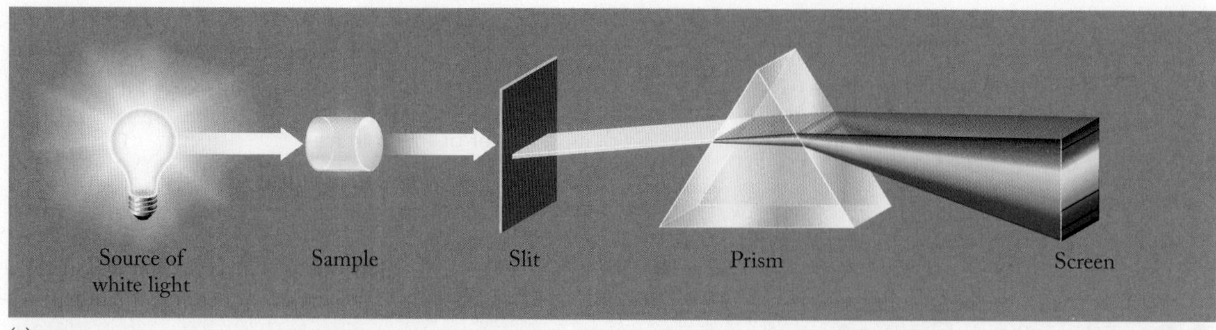

(a)

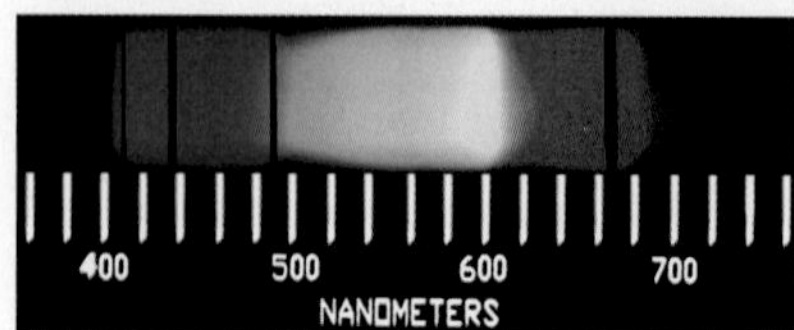

Absorption spectrum of hydrogen

Absorption spectrum of helium

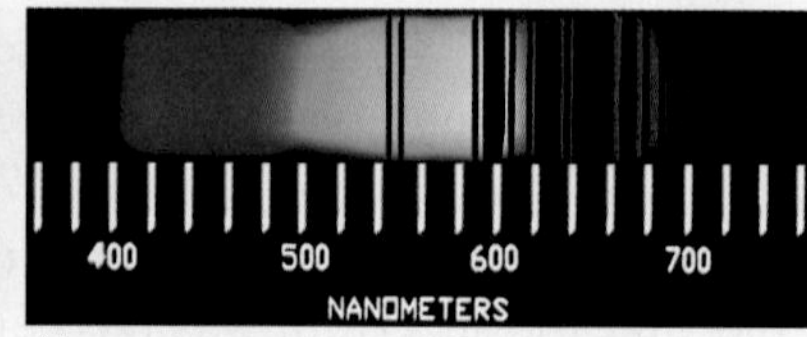

Absorption spectrum of neon

(b)

FIGURE 3.7 (a) When gaseous atoms of hydrogen, helium, and neon are illuminated by an external source of white light (containing all colors of the visible spectrum), the resultant atomic absorption spectra contain dark lines that are characteristic of the elements. (b) The dark lines have the same wavelengths as the bright lines in the elements' atomic emission spectra, shown in Figure 3.6.

source of radiation. When this radiation is passed through a narrow slit and then a prism, as shown schematically in Figure 3.7(a), **atomic absorption spectra** of narrow dark lines in otherwise continuous spectra are produced. For any given element (Figure 3.7b), the dark lines in its absorption spectrum are at exactly the same wavelengths as the bright lines in its emission spectrum. The phenomenon of atomic absorption explains the Fraunhofer lines: gaseous atoms in the outer regions of the sun absorb characteristic wavelengths of the sunlight passing through them on its way to Earth.

FIGURE 3.8 Atomic emission spectrum of mercury vapor (top) and four atomic absorption spectra (a–d).

CONCEPT TEST

Mercury lamps are used to illuminate large spaces such as sports arenas. The top spectrum in Figure 3.8 is the emission spectrum of mercury vapor. Select the absorption spectrum of mercury vapor from the four spectra labeled (a)–(d).

(Answers to Concept Tests are in the back of the book.)

3.3 Particles of Light: Quantum Theory

In addition to his studies of the narrow-line emission spectra of atoms, Kirchhoff also studied the continuous emission spectra produced by hot objects such as metals heated to incandescence (Figure 3.9).

atomic absorption spectra characteristic patterns of dark lines produced when an external source of radiation passes through free gaseous atoms.

quantum (plural *quanta*) the smallest discrete quantity of a particular form of energy.

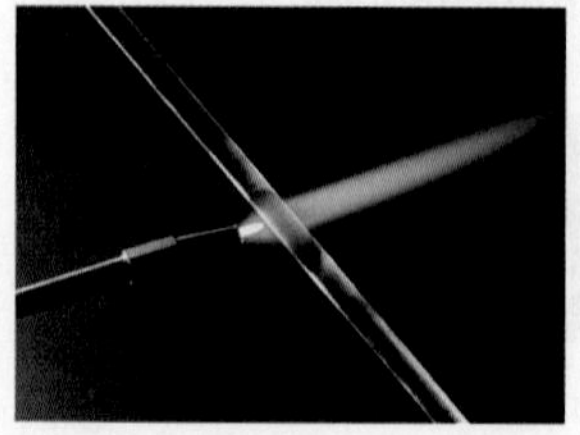
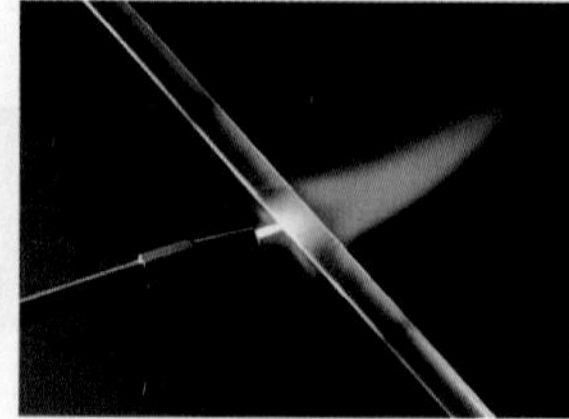

FIGURE 3.9 When a metal rod is heated, it glows red at first, then orange, and finally becomes white-hot as its temperature increases and light of shorter wavelengths is emitted.

Kirchoff and others recorded the spectra of sources of radiant energy that are called *blackbody* radiators because, when cold, their surfaces are jet black and they absorb all the light that strikes them. These perfect absorbers of light when cold become perfect radiators of light when hot. The spectra they emit depend only on their temperature: as they become hotter, they emit more radiation at shorter and shorter wavelengths as shown in the emission intensity profiles in Figure 3.10. These profiles presented a challenge for physicists in the late 19th century because none of the laws describing the behavior of electromagnetic radiation could explain them. A new model describing how radiant energy and matter interact was needed, and yet another German physicist developed it. His name was Max Planck (1858–1947, Figure 3.11).

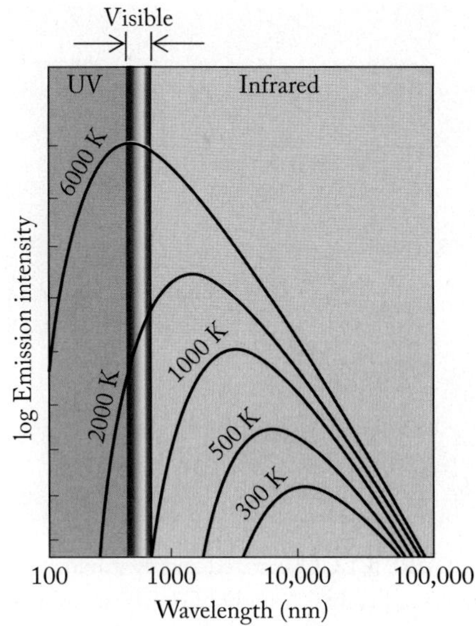

FIGURE 3.10 The solid curves show how the intensities and wavelengths of radiation emitted by blackbody radiators change with changing temperature. Logarithmic scales are used on both axes to include a wide range of intensities and wavelengths.

Photons of Energy

While a professor at the University of Berlin in 1900, Planck was commissioned by electric companies to maximize the output of the incandescent lightbulbs invented by Thomas Edison. His attempts at developing a theoretical model that accurately described their incandescence led him to discard classical physics and make a bold assumption about radiant energy: no matter what its source, it can never be truly continuous. Instead, Planck proposed that objects emit electromagnetic radiation only in integral multiples of an elementary unit, or **quantum**, of energy defined by the equation

$$E = h\nu \tag{3.3}$$

where ν is the frequency of the radiation and h is the **Planck constant**, 6.626×10^{-34} J · s. Combining Equations 3.2 and 3.3 yields an equation that relates the energy of a quantum of electromagnetic radiation to its wavelength:

$$E = \frac{hc}{\lambda} \tag{3.4}$$

Because Planck's model is characterized by these *quanta*, it has become known as **quantum theory**.

To visualize the meaning of Planck's quantum of energy, consider the difference between taking the steps or the ramp to get from the sidewalk to the entrance of a building (Figure 3.12). If you walk up the steps, you can stand at only discrete heights above the sidewalk. You cannot stand at a height between adjacent steps because there is nothing to stand on. If you walk up the ramp, however, you can stop at any height between the sidewalk and the entrance because height is continuous on the ramp. On the steps, height is **quantized**: the discrete changes in height model Planck's hypothesis that energy is released (analogous to walking down the steps) or absorbed (walking up the steps) in discrete packets, or quanta, of energy.

FIGURE 3.11 German scientist Max Karl Ernst Ludwig Planck is considered the father of quantum physics. He won the 1918 Nobel Prize in Physics for his pioneering work on the quantized nature of electromagnetic radiation.

CONCEPT TEST

Which of the following are quantized?

a. The volume of water in the Atlantic Ocean

b. The number of eggs remaining in a carton

c. The time it typically takes you to get ready for class in the morning

d. The number of red lights encountered when driving the length of Fifth Avenue in New York City

(Answers to Concept Tests are in the back of the book.)

Planck constant (h) the proportionality constant between the energy and frequency of electromagnetic radiation expressed in $E = h\nu$; $h = 6.626 \times 10^{-34}$ J · s.

quantum theory a model of matter and energy based on the principle that energy is absorbed or emitted in discrete packets, or quanta.

quantized having values restricted to whole-number multiples of a specific base value.

FIGURE 3.12 Quantized and continuously varying heights. A flight of stairs exemplifies quantization: each step rises by a discrete height to the next step. In contrast, the heights on a ramp are not quantized—they are continuous.

FIGURE 3.13 An ultra-high-definition television incorporating quantum dot technology.

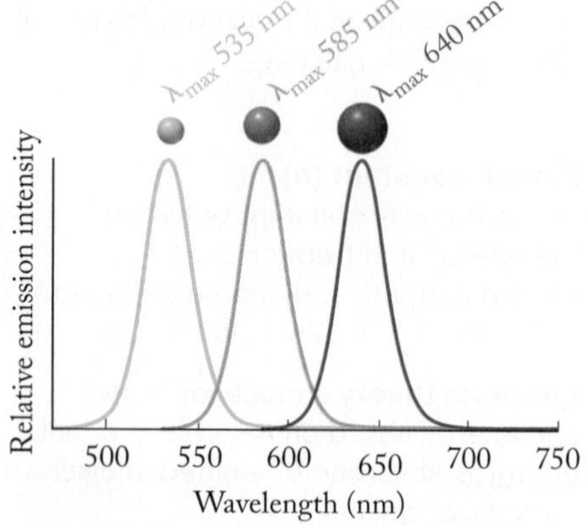

FIGURE 3.14 Emission intensity versus wavelength for three quantum dots of different size.

Given the extremely tiny value of the Planck constant (h), the energy values we obtain using Equation 3.4 tend to be very small because E represents the number of joules in a single quantum of energy. Today we call the tiny packets of radiant energy **photons**. They represent elementary building blocks of electromagnetic radiation in much the same way that atoms represent the building blocks of matter. The observed brightness of a source of radiant energy is the sum of the energies of the enormous number of photons it produces per unit of time.

SAMPLE EXERCISE 3.2 Calculating the Energies of Photons LO1

Some manufacturers of ultra-high-definition televisions (Figure 3.13) use tiny particles called *quantum dots* to more accurately reproduce the colors of transmitted images. The wavelengths of the light produced by quantum dots depend on their size. Figure 3.14 contains a plot of emission intensity as a function of wavelength for three quantum dots whose relative sizes are represented by the sizes of the spheres. What is the energy of a photon of light from the largest quantum dot that has a wavelength of 640 nm (6.40×10^2 nm)? Is this energy greater than or less than the energies of the photons produced by the smaller quantum dots?

Collect and Organize We need to calculate the energy of a photon of electromagnetic radiation starting with its wavelength. Equation 3.4

$$E = \frac{hc}{\lambda}$$

relates the energy of a photon to its wavelength. The value of the Planck constant (h) is 6.626×10^{-34} J · s, and the speed of light (c) is 2.998×10^8 m/s.

Analyze The wavelength is given in nanometers, but the value of the speed of light has units of meters per second, so we need to convert nanometers to meters. The value of h is extremely small, so the results of our calculation, even factoring in the speed of light, should be very small, too.

Solve

$$E = \frac{hc}{\lambda} = \frac{(6.626 \times 10^{-34}\ \text{J} \cdot \text{s})(2.998 \times 10^8\ \cancel{\text{m}}\ \text{s}^{-1})}{6.40 \times 10^2\ \cancel{\text{nm}} \times \dfrac{10^{-9}\ \cancel{\text{m}}}{\cancel{\text{nm}}}} = 3.10 \times 10^{-19}\ \text{J}$$

Figure 3.14 shows that smaller quantum dots produce photons with shorter wavelengths. The energy of a photon is inversely proportional to its wavelength, so these shorter-wavelength photons have more energy than those at 640 nm.

Think About It This calculated energy is indeed small, as it should be, because a photon is an atomic-level particle of radiant energy.

Practice Exercise Some instruments differentiate individual quanta of electromagnetic radiation based on their energies. Assume such an instrument has been adjusted to detect photons that have 5.00×10^{-19} J of energy. What is the wavelength of the detected radiation? Give your answer in nanometers and in meters.

(Answers to Practice Exercises are in the back of the book.)

The Photoelectric Effect

Although Planck's quantum model explained the emission spectra of hot objects, there was no experimental evidence in 1900 to support the existence of quanta of energy. In 1905, Albert Einstein (1879–1955) supplied that evidence. It came from his studies of a phenomenon called the **photoelectric effect**, in which electrons are emitted from metals and semiconductor materials when they are illuminated by and absorb electromagnetic radiation. Because light releases these electrons, they are called *photoelectrons*, derived from the Greek *photo*, meaning "light."

photon a quantum of electromagnetic radiation.

photoelectric effect the release of electrons from a material as a result of electromagnetic radiation striking it.

threshold frequency (ν_0) the minimum frequency of light required to produce the photoelectric effect.

Photoelectrons are emitted when the frequency of incident radiation exceeds some minimum **threshold frequency (ν_0)** (Figure 3.15). Radiation of frequencies less than the threshold value produces no photoelectrons, no matter how intense the radiation is. On the other hand, even a dim source of radiant energy produces at least a few photoelectrons when the frequencies it emits are equal to or greater than the threshold frequency. For example, some night vision goggles used by military and law enforcement personnel (Figure 3.16) can detect extremely low levels of light because they incorporate photoelectric sensors and amplifier circuits that produce over 10,000 electrons for each photoelectron initially emitted. When such large numbers of electrons strike a phosphorescent screen (like the ones used in cathode-ray tubes, as in Figure 2.2), they produce images that are tens of thousands of times brighter than the original.

Einstein used Planck's quantum model to explain the ability of the photoelectric materials to emit photoelectrons. He proposed that every photoelectric material has a characteristic threshold frequency ν_0 associated with the minimum quantum of absorbed energy needed to remove a single electron from the material's surface. The threshold frequency of a sensor used in night vision goggles, for example, is in the infrared region of the electromagnetic spectrum. When photons of radiation with this frequency or a higher one are focused on one of the sensors, photoelectrons are emitted.

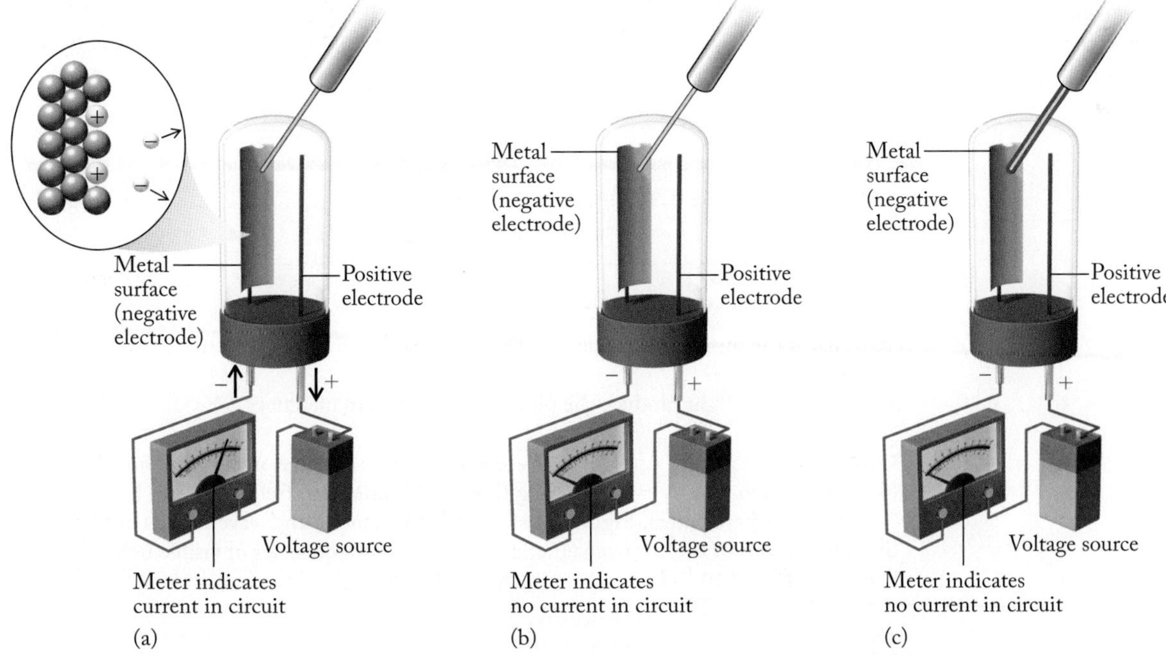

FIGURE 3.15 A phototube includes a positive electrode and a negative metal electrode. (a) If radiation of high enough frequency and energy (represented by the violet beam of light) illuminates the negative electrode, electrons are dislodged from the surface and flow toward the positive electrode. This flow of electrons completes the circuit and produces an electric current. The size of the current is proportional to the intensity of the radiation—to the number of photons per unit time striking the negative electrode. (b) Photons of lower frequency (red beam) and hence lower energy do not have sufficient energy to dislodge electrons and do not produce the photoelectric effect, no matter how many of them (c) bombard the surface of the metal. The circuit is not complete, and there is no current.

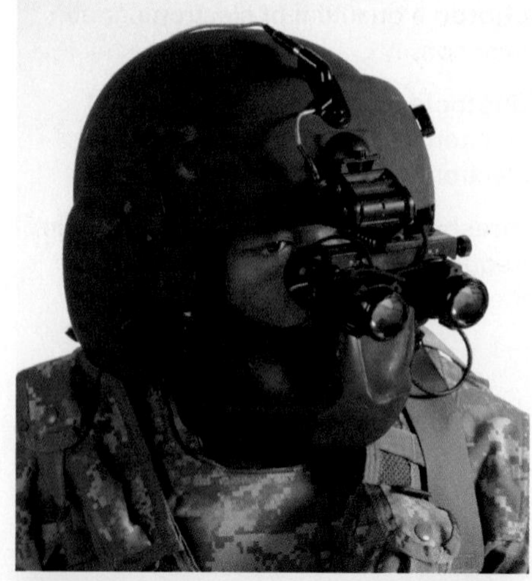

FIGURE 3.16 Night vision goggles incorporate photoelectric sensors and electronic amplifiers called image intensifiers to produce images that are tens of thousands of times brighter than those perceived by the human eye.

The minimum quantity of energy needed to emit photoelectrons from a photoelectric material is called the material's **work function (Φ)**:

$$\Phi = h\nu_0 \tag{3.5}$$

The value of Φ is related to the strength of the attraction between the nuclei of the metal's atoms and the electrons surrounding those nuclei. In Einstein's model, a photoelectron is emitted when a quantum of radiant energy (a photon) provides an electron with enough energy to break free of the surface of a photoelectric material. If the incoming beam includes photons with frequencies above the threshold frequency ($\nu > \nu_0$), then each photon has more than enough energy to dislodge an electron. This extra energy in excess of the work function is imparted to each ejected electron as kinetic energy: the higher the frequency above the threshold frequency, the greater the kinetic energy of the ejected electron. The kinetic energies of photoelectrons ($KE_{electron}$) can be determined using instruments that measure their speeds. If we know $KE_{electron}$ and the energy of the incident photons ($h\nu$), we can calculate the value of the work function of the target metal using Equation 3.6:

$$\Phi = h\nu - KE_{electron} \tag{3.6}$$

The photoelectric effect is used in many energy-saving devices, such as the *passive infrared devices* that turn on the lights when you enter a darkened room. These devices contain a material that detects the IR radiation ($\lambda \approx 10^{-5}$ nm) your body emits and triggers an electrical signal that turns on the lights. The same principle is used in remote control systems. A handheld remote control typically sends light with a wavelength of 940 nm to a detector in the device that translates the light into a flow of electrons. In the sophisticated remote control systems in use today, a sequence of pulses is used to distinguish the functions we wish the controller to accomplish (for example, change the channel or adjust the volume), but the underlying principle remains the photoelectric effect.

SAMPLE EXERCISE 3.3 Relating Work Functions and Threshold Frequencies — LO2

The work function of mercury is 7.22×10^{-19} J.

a. What is the minimum frequency of radiation required to eject photoelectrons from a mercury surface?
b. Could visible light produce the photoelectric effect in mercury?

Collect, Organize, and Analyze We are asked to use the value of the work function (Φ) of a metal to find its corresponding threshold (minimum) frequency (ν_0). Equation 3.5 relates the parameters. Figure 3.1 shows the frequencies and wavelengths of the different regions of the electromagnetic spectrum; the frequencies of visible light are between 10^{14} and 10^{15} Hz.

Solve

a. Rearrange Equation 3.5 to solve for the threshold frequency:

$$\nu_0 = \frac{\Phi}{h} = \frac{7.22 \times 10^{-19}\ \text{J}}{6.626 \times 10^{-34}\ \text{J}\cdot\text{s}} = 1.09 \times 10^{15}\ \text{s}^{-1}$$

b. The threshold frequency is greater than 10^{15} Hz and, according to Figure 3.1, is in the ultraviolet region of electromagnetic radiation. The frequencies of visible light are below the threshold frequency, so visible light cannot generate photoelectrons from mercury.

work function (Φ) the amount of energy needed to dislodge an electron from the surface of a material.

Think About It The calculated threshold frequency is close to, but still greater than, that of the highest-frequency (violet) visible light. Photons of visible light simply lack the energy needed to dislodge electrons from atoms on the surface of a drop of mercury.

Practice Exercise The work function of silver is 7.59×10^{-19} J. What is the longest wavelength (in nm) of electromagnetic radiation that can eject a photoelectron from the surface of a piece of silver?

(Answers to Practice Exercises are in the back of the book.)

3.4 The Hydrogen Spectrum and the Bohr Model

In formulating his quantum theory, Planck was influenced by the results of investigations of the emission spectra produced by free, gas-phase atoms (as described in Section 3.2). These line spectra led him to question whether any spectrum, even that of an incandescent lightbulb, was truly continuous. Among the early investigators of atomic emission spectra was a Swiss mathematician and schoolteacher named Johann Balmer (1825–1898). His work focused on the pattern of emission lines produced by high-temperature hydrogen atoms. In 1885 Balmer formulated an empirical equation that accurately predicted the wavelengths of the four brightest atomic emission lines in the visible spectrum of hydrogen. To express the wavelengths in nanometers, the units most often used for visible light, we can use the following form of the Balmer equation:

$$\lambda \text{ (nm)} = \left(\frac{364.56\, m^2}{m^2 - n^2}\right) \tag{3.7}$$

where m is an integer greater than 2 (the values 3, 4, 5, and 6 predict the wavelengths of the four visible hydrogen lines) and $n = 2$. If, for example, we let $m = 3$ and solve for λ, we get

$$\lambda \text{ (nm)} = \frac{(364.56) \times 3^2}{3^2 - 2^2} = 364.56 \times \frac{9}{5} = 656.21 \text{ nm}$$

which is the wavelength of the red line in the atomic emission spectrum of hydrogen (Figure 3.6a). By using m values of 4, 5, and 6 in Equation 3.7, we get the wavelengths of the blue-green (486.08 nm), blue (434.00 nm), and violet (410.13 nm) lines, respectively.

The Balmer equation is called an *empirical* equation because he derived it strictly from experimental data—namely, the wavelengths of light in the visible atomic emission spectrum of hydrogen. It had no theoretical foundation. No one in 1885 could explain why his equation fit the hydrogen spectrum, though other scientists of the time used it successfully to search for hydrogen lines corresponding to m values greater than 6, which no one could see because they are in the ultraviolet region of the electromagnetic spectrum.

In 1888 Swedish physicist Johannes Robert Rydberg (1854–1919) published a more general empirical equation for predicting the wavelengths of hydrogen's spectral lines:

$$\frac{1}{\lambda} = R_H\left(\frac{1}{{n_1}^2} - \frac{1}{{n_2}^2}\right) \tag{3.8}$$

where n_1 and n_2 are any positive integers (provided $n_2 > n_1$) and R_H is the Rydberg constant. The term on the left in Rydberg's equation is the reciprocal of wavelength ($1/\lambda$), not wavelength, and is called the *wavenumber* of a spectral line. To understand why Rydberg found wavenumbers useful in formulating his equation, remember that the energy of a photon is inversely proportional to its wavelength. Therefore, its energy is directly proportional to its wavenumber. Though Rydberg did not know it in 1888, the values of n_1 and n_2 on the right side of his equation correspond to energy levels inside hydrogen atoms, and the energy of a photon that a hydrogen atom absorbs or emits is exactly the same as the difference in energy between a pair of energy levels.

The value of Rydberg's constant, R_H, depends on the units used to express wavelength. Three common options are

$$\begin{aligned} R_H &= 1.097 \times 10^7 \text{ m}^{-1} \\ &= 1.097 \times 10^5 \text{ cm}^{-1} \\ &= 1.097 \times 10^{-2} \text{ nm}^{-1} \end{aligned}$$

When hydrogen lines are in the ultraviolet or visible regions of the electromagnetic spectrum (as happens when $n_1 = 1$ or 2), their wavelengths are usually expressed in nanometers, and the most convenient form of the Rydberg equation is

$$\frac{1}{\lambda} = (1.097 \times 10^{-2} \text{ nm}^{-1})\left(\frac{1}{n_1^2} - \frac{1}{n_2^2}\right) \tag{3.9}$$

When we let $n_1 = 2$ and $n_2 = 3$, 4, 5, or 6 in Equation 3.9 and then solve for λ, we get four wavelength values that match those we obtained with the Balmer equation when the value of his m parameter was 3, 4, 5, or 6. They are the wavelengths of the four lines in the visible emission spectrum of hydrogen. Actually, the Balmer equation represents a special case of the Rydberg equation for the hydrogen lines that correspond to $n_1 = 2$. An advantage of Rydberg's equation was that it allowed scientists to predict the wavelengths of other series of hydrogen emission lines for which $n_1 \neq 2$. None of these additional lines were in the visible region. In 1908, German physicist Friedrich Paschen (1865–1947) discovered a series of hydrogen emission lines in the infrared region with wavelengths corresponding to Rydberg equation values of $n_1 = 3$ and $n_2 = 4$, 5, 6, and so on. A few years later, Theodore Lyman (1874–1954) at Harvard University discovered another series of hydrogen emission lines in the ultraviolet region corresponding to $n_1 = 1$. By the 1920s the $n_1 = 4$ and $n_1 = 5$ series had been discovered. Like the $n_1 = 3$ series, they are also in the infrared region.

SAMPLE EXERCISE 3.4 Calculating the Wavelength of a Line in the Hydrogen Spectrum **LO3**

What is the wavelength in nanometers of the line in the hydrogen spectrum that corresponds to $m = 7$ in the Balmer equation (Equation 3.7)? Check your answer by also calculating this wavelength using the Rydberg equation.

Collect and Organize We need to calculate the wavelength of a line in hydrogen's atomic emission spectrum from its m value in the Balmer equation. The n value in the Balmer equation is always 2. We will then use the Rydberg equation to check the accuracy of our calculation. The requested units of wavelength are nanometers, so Equation 3.9 should be the most useful form of the Rydberg equation for this calculation.

Analyze To calculate a wavelength using the Balmer equation, we insert the appropriate values of m and n (7 and 2 in this case) and solve for λ. In the corresponding calculation based on the Rydberg equation, we let $n_2 = 7$ because it is the greater integer and $n_1 = 2$. Our answer should be near 400 nm because the wavelength corresponding to $m = 6$ in the Balmer equation is 410 nm, and the wavelengths generated by the Balmer equation become both smaller and closer to each other as m increases.

Solve Using the Balmer equation,

$$\lambda = 364.56\text{ nm}\left(\frac{m^2}{m^2 - n^2}\right) = 364.56\text{ nm}\left(\frac{7^2}{7^2 - 2^2}\right) = 396.97\text{ nm}$$

Using the Rydberg equation (Equation 3.9),

$$\frac{1}{\lambda} = (1.097 \times 10^{-2}\text{ nm}^{-1})\left(\frac{1}{n_1{}^2} - \frac{1}{n_2{}^2}\right)$$

$$= (1.097 \times 10^{-2}\text{ nm}^{-1})\left(\frac{1}{2^2} - \frac{1}{7^2}\right)$$

$$= (1.097 \times 10^{-2}\text{ nm}^{-1})(0.2296) = 2.519 \times 10^{-3}\text{ (nm}^{-1})$$

$$\lambda = 397.0\text{ nm}$$

Think About It The results of the two calculations are consistent, although the first one has one more significant figure. This occurs because there are five significant figures in the constant used in Equation 3.7 and only four in the constant used in Equation 3.9. The calculated values are also close to the wavelength that we estimated based on $m = 6$.

Practice Exercise What is the wavelength of the photon emitted by a hydrogen atom that corresponds to $m = 12$ in the Balmer equation? Would Balmer have been able to see this line?

(Answers to Practice Exercises are in the back of the book.)

CONNECTION We discussed Rutherford's gold-foil experiments and the model of the atom that evolved from them in Chapter 2.

The Bohr Model

When Balmer and Rydberg derived their empirical equations, they didn't know why their equations worked or what the integers in them physically represented. A few years later, Max Planck proposed that the discrete wavelengths of the lines meant that hydrogen atoms lost and gained only discrete quanta of energy, which indicated to him that there were discrete energy levels inside the atoms. However, classic (macroscopic-scale) physics could not explain the existence of quantized energy levels. A new model was needed to account for them.

In 1913 Danish scientist Niels Bohr (1885–1962) proposed such a model. With it he could explain (1) why hydrogen atoms lose and gain discrete quanta of energy and (2) why their electrons do not spiral into their nuclei. His explanation was based on Rutherford's planetary model, in which electrons orbit nuclei, combined with Planck's notion of quantized energy. In Bohr's model, the electron in a hydrogen atom revolves around the nucleus in one of an array of available orbits. Each orbit represents a discrete energy level inside the atom. Bohr assigned each orbit a number, n, starting with $n = 1$ for the orbit closest to the nucleus (Figure 3.17). In his model, orbits farther from the nucleus have larger values of n, and the electrons in them have higher (less negative) energies based on the following equation:

$$E = -2.178 \times 10^{-18}\text{ J}\left(\frac{1}{n^2}\right) \tag{3.10}$$

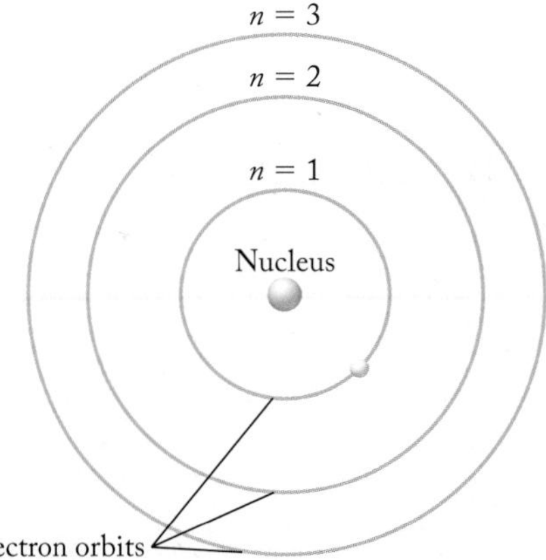

FIGURE 3.17 In the Bohr model of the hydrogen atom, its electron revolves around the nucleus in one of a series of concentric orbits. Each orbit represents an allowed energy level. An electron in the orbit closest to the nucleus ($n = 1$) has the lowest energy.

where $n = 1, 2, 3, \ldots, \infty$. According to Equation 3.10, an electron in the orbit closest to the nucleus ($n = 1$) has the lowest (most negative) energy:

$$E = -2.178 \times 10^{-18}\ \text{J}\left(\frac{1}{1^2}\right) = -2.178 \times 10^{-18}\ \text{J}$$

An electron in the next closest ($n = 2$) orbit has an energy of

$$E = -2.178 \times 10^{-18}\ \text{J}\left(\frac{1}{2^2}\right) = -5.445 \times 10^{-19}\ \text{J}$$

Note that this value is less negative than the value for the electron in the $n = 1$ orbit. The values of n for orbits farther and farther from the nucleus approach ∞, and as they do, E approaches zero:

$$E = -2.178 \times 10^{-18}\ \text{J}\left(\frac{1}{\infty^2}\right) = 0$$

Zero energy means that the electron at $n = \infty$ is no longer part of the hydrogen atom and that the atom no longer exists as a single entity. Rather, it has become two separate particles: a H^+ ion and a free electron.

Why is there a negative sign in Equation 3.10? What does it mean for an electron to have negative energy? To understand this, let's assume that the addition of energy is required to pry a negatively charged electron away from a positively charged nucleus. This is logical because oppositely charged particles are attracted to each other, and this attraction must be overcome to separate them. An electron that has been separated from an atom no longer interacts with the nucleus, and there is no energy of attraction between the particles ($E = 0$). So, if we had to add energy to get to zero energy, the initial energy of the electron in the atom must have been less than zero, as we would calculate using Equation 3.10.

An important feature of the Bohr model is that it provides a theoretical framework for explaining the experimental observations of Balmer, Rydberg, and others we have discussed. To see the connection, consider what happens when an electron moves between two allowed energy levels in Bohr's model. If we label the energy level where the electron starts n_{initial} and the level where the electron ends up n_{final}, then the change in energy of the electron is

$$\Delta E = -2.178 \times 10^{-18}\ \text{J}\left(\frac{1}{n_{\text{final}}^2} - \frac{1}{n_{\text{initial}}^2}\right) \tag{3.11}$$

Here we use the capital Greek letter delta (Δ) to represent change (and we will do so throughout this book).

If the electron moves to an orbit farther from the nucleus, then $n_{\text{final}} > n_{\text{initial}}$, and the overall value of the terms inside the parentheses in Equation 3.11 is negative because $1/n_{\text{final}}^2 < 1/n_{\text{initial}}^2$. This negative value multiplied by the negative coefficient gives us a positive ΔE and represents an increase in electron energy. On the other hand, if an electron moves from an outer orbit to one closer to the nucleus, then $n_{\text{final}} < n_{\text{initial}}$, and the sign of ΔE is negative. This means the electron loses energy.

When the electron in a hydrogen atom is in the lowest ($n = 1$) energy level, the atom is said to be in its **ground state**. According to the Bohr model, the electron cannot have any less energy than it has in the ground state, which means that it can't lose more energy and spiral into the nucleus.

ground state the most stable, lowest energy state of a particle.

excited state any energy state above the ground state.

electron transition movement of an electron between energy levels.

If the electron in a hydrogen atom is in an energy level above $n = 1$, then the atom is said to be in an **excited state**. An electron can move from the $n = 1$

(ground state) orbit to a higher level (for example, $n = 3$) by absorbing a quantum of energy (ΔE) that exactly matches the energy difference between the two states. Similarly, an electron in an excited state can move to an even higher energy level by absorbing a quantum of energy that exactly matches the energy difference between those two states. On the other hand, an electron in an excited state can move to a lower-energy excited state, or all the way down to the ground state, by emitting a quantum of energy that exactly matches the energy difference between those two states. The movement of an electron between any two energy levels is called an **electron transition**.

The energy-level diagram in Figure 3.18 shows some of the transitions that an electron in a hydrogen atom can make. The black arrow pointing upward represents the absorption of sufficient energy to completely remove the electron from a hydrogen atom (ionization). The downward-pointing colored arrows represent decreases in the internal energy of the hydrogen atom that occur when photons are emitted as the electron moves from a higher energy level to a lower energy level. If the colored arrows pointed up, they would represent the absorption of photons and would lead to increases in the internal energy of the atom. In every case, the energy of the photon absorbed or emitted matches the absolute value of ΔE.

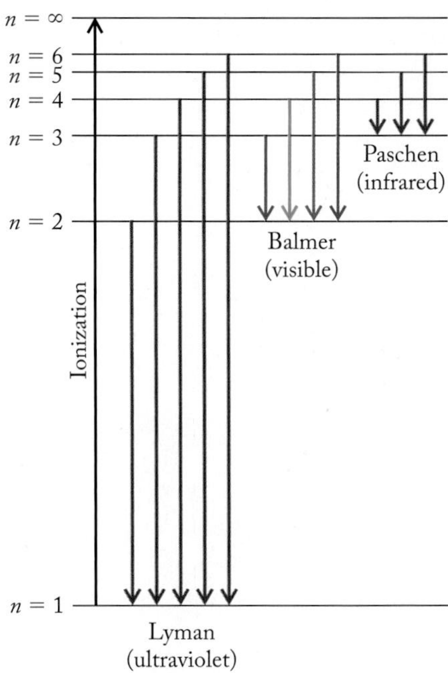

FIGURE 3.18 An energy-level diagram showing some of the possible electron transitions in a hydrogen atom. The arrow pointing up represents ionization. Arrows pointing down represent different electron transitions. During any transition from a higher to a lower energy level, the atom loses a particular quantity of energy that may be emitted as a photon of electromagnetic radiation.

CONCEPT TEST

Based on the lengths of the arrows in Figure 3.18, rank the following transitions in order from the greatest change in electron energy to the smallest change:

a. $n = 4 \rightarrow n = 2$

b. $n = 3 \rightarrow n = 2$

c. $n = 2 \rightarrow n = 1$

d. $n = 4 \rightarrow n = 3$

(Answers to Concept Tests are in the back of the book.)

SAMPLE EXERCISE 3.5 Calculating the Energy of a Transition in a Hydrogen Atom LO3

How much energy is required to ionize a ground-state hydrogen atom? That is, what is the *ionization energy* of hydrogen?

Collect and Organize We are asked to determine the energy required to remove the electron from a hydrogen atom in its ground state. Equation 3.11 enables us to calculate the energy change associated with any electron transition.

Analyze To use Equation 3.11, we need to identify the initial (n_{initial}) and final (n_{final}) energy levels of the electron. The electron begins in the ground state of a H atom, so $n_{\text{initial}} = 1$. If the atom is ionized, then $n_{\text{final}} = \infty$ and the electron is no longer associated with the nucleus.

Solve

$$\Delta E = -2.178 \times 10^{-18}\,\text{J}\left(\frac{1}{n_{\text{final}}^2} - \frac{1}{n_{\text{initial}}^2}\right)$$

$$= -2.178 \times 10^{-18}\,\text{J}\left(\frac{1}{\infty^2} - \frac{1}{1^2}\right)$$

Dividing by ∞^2 yields zero, so the value inside the parentheses simplifies to −1, which gives us

$$\Delta E = 2.178 \times 10^{-18}\ \text{J}$$

Think About It This energy is equal in magnitude but opposite in sign to the energy of an electron in the $n = 1$ orbit of a hydrogen atom. The sign of ΔE is positive because energy must be added to overcome the attraction between the negatively charged electron and the positively charged nucleus.

Practice Exercise Calculate the energy, in joules, required to ionize a hydrogen atom when its electron is initially in the $n = 3$ energy level. Before doing the calculation, predict whether this energy is greater than or less than the 2.178×10^{-18} J needed to ionize a ground-state hydrogen atom.

(Answers to Practice Exercises are in the back of the book.)

Before we end this section, let's compare Equation 3.11, which describes the differences in energy between pairs of energy levels in hydrogen atoms, and the Rydberg equation (Equation 3.9), which describes the wavenumbers for the lines in hydrogen's atomic emission and absorption spectra. Notice how much alike they are. The coefficients in them are different, because of the different units used to express wavenumber and energy, but the empirical Rydberg equation has the same form as the theoretical equation developed by Bohr to explain the internal structure of the hydrogen atom. Thus, atomic emission and absorption spectra reveal the energies of electrons inside hydrogen atoms and, as we will see, the atoms of other elements, too.

When we extend our exploration of atomic spectra and internal energies of atoms to other elements, we discover that the Bohr model of electrons revolving around nuclei in stable circular orbits works only for atoms or ions with only one electron. In multielectron atoms, the electrons interact with each other in ways that the Bohr model does not take into account. Thus, the picture of the atom provided by Bohr's model is limited. However, it was an important stepping-stone to the development of more complete models of atomic structure based on quantum theory.

3.5 Electrons as Waves

A decade after Bohr published his model of the hydrogen atom, a French graduate student named Louis de Broglie (1892–1987) developed another explanation for the stability of electrons orbiting the nuclei of hydrogen atoms. De Broglie based his hypothesis on the assumption that electrons could behave like *waves of matter* as well as particles of matter. This dual nature of electrons was modeled after the behavior of electromagnetic radiation, which could be explained not only by assigning it wavelike properties, as James Maxwell and many others had done, but also by treating it like it was made up of particles, or quanta, as Einstein had done to explain the photoelectric effect.

De Broglie Wavelengths

An electron traveling through space or around an atom's nucleus has a kinetic energy that, as we saw in Chapter 1, is related to its mass and the square of its speed. Assuming the electron can also behave as a wave, then its wavelength is inversely proportional to its energy. These two perspectives on the energy of a moving electron are contained in the **de Broglie equation** for calculating the wavelength of an electron (or any particle) in motion:

$$\lambda = \frac{h}{mu} \tag{3.12}$$

In this equation m is the mass of the particle, u is its speed, and h is Planck's constant. Note the inverse relationship between the particle's wavelength and the product of its mass times its speed. This product is the particle's *momentum*. The more momentum a particle has, the shorter its *de Broglie wavelength*.

De Broglie's equation tells us that any moving particle has wavelike properties; that is, the particle behaves as a **matter wave**. De Broglie predicted that moving particles much bigger than electrons, such as atomic nuclei, molecules, and even tennis balls and airplanes, have characteristic wavelengths that can be calculated using Equation 3.12. The wavelengths of large objects are extremely small, given the tiny size of the Planck constant and their considerable mass, so we never notice the wave nature of large objects in motion.

de Broglie equation relates the wavelength of any moving object to its mass and its speed.

matter wave the wave associated with any moving particle.

CHEMTOUR
De Broglie Wavelength

SAMPLE EXERCISE 3.6 Calculating the Wavelength of a Particle in Motion **LO4**

Calculate the de Broglie wavelength of a 142 g baseball thrown at 44 m/s (98 mi/hr) and compare this wavelength with the diameter of the ball, which is 7.35 cm.

Collect and Organize We know the mass and speed of a baseball and need to calculate its de Broglie wavelength so that we can compare its value to the size of the baseball. Equation 3.12 may be used to calculate this wavelength.

Analyze Given how small h is, the wavelength of a pitched baseball should be only a tiny fraction of the size of the baseball. The right side of Equation 3.12 has units of joule-seconds in the numerator and mass and speed in the denominator. To combine the units in a way that gives us a unit of length, we need to use the following conversion factor:

$$1\ \text{J} = 1\ \text{kg} \cdot \text{m}^2/\text{s}^2$$

To use this equality, we must convert the mass of the baseball from grams to kilograms: 142 g = 0.142 kg.

Solve The de Broglie wavelength of the baseball is

$$\lambda = \frac{h}{mu} = \frac{6.626 \times 10^{-34}\ \text{J} \cdot \text{s}}{(0.142\ \cancel{\text{kg}})(44\ \cancel{\text{m/s}})} \times \frac{1\ \cancel{\text{kg}} \cdot \text{m}^2/\text{s}^2}{1\ \cancel{\text{J}}} = 1.06 \times 10^{-34}\ \text{m}$$

This wavelength is

$$\frac{1.06 \times 10^{-34}\ \cancel{\text{m}}}{7.35\ \cancel{\text{cm}} \times (1\ \text{m}/100\ \cancel{\text{cm}})} \times 100\% = 1.4 \times 10^{-33}\%$$

of the ball's diameter.

Think About It The wavelength of the matter wave of the baseball is much too small to be observed, so its character contributes nothing to the behavior of the baseball. This is what we expected: large moving objects behave like large moving objects, not like waves.

 **Practice Exercise** The speed of the electron in the ground state of the hydrogen atom is 2.2×10^6 m/s. What is the wavelength of the electron in meters?

(Answers to Practice Exercises are in the back of the book.)

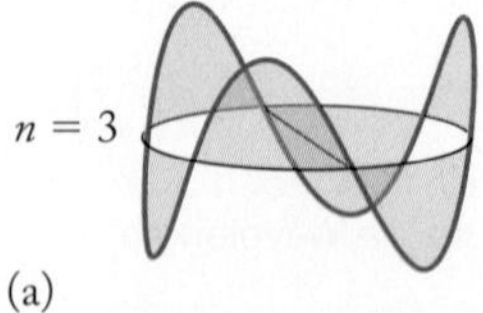

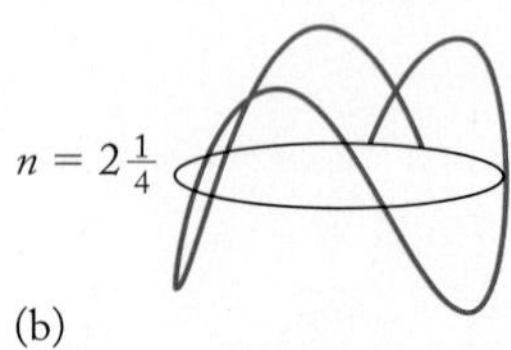

FIGURE 3.19 (a) The circular waves proposed by de Broglie account for the stability of the energy levels in Bohr's model of the hydrogen atom. Each stable wave must have a circumference equal to $n\lambda$, with n restricted to being an integer, such as the $n = 3$ circular standing wave shown here. (b) If the circumference is not an exact multiple of λ, as shown in the $n = 2\frac{1}{4}$ image, the wave is not continuous (or stable).

De Broglie used matter waves to explain the stability of the electron levels in Bohr's model of the hydrogen atom. He proposed that the orbiting electron behaves like a circular wave oscillating around the nucleus. However, a stable wave pattern is achieved only if, as shown in Figure 3.19(a), the circumference of the circle equals a whole-number multiple of the electron's wavelength:

$$\text{Circumference} = n\lambda \tag{3.13}$$

If the circumference is not a whole-number multiple, then the oscillating wave is discontinuous (Figure 3.19b) and unstable. Equation 3.13 gives a new meaning to Bohr's orbit label n: it represents the number of matter-wave wavelengths in that orbit's circumference.

De Broglie's matter-wave hypothesis created a quandary for the graduate faculty at the University of Paris, where he studied. Bohr's model of electrons moving between allowed energy levels had been widely criticized as an arbitrary suspension of well-tested physical laws. De Broglie's rationalization of Bohr's model seemed even more outrageous to many scientists. Before the faculty would accept his doctoral thesis based on matter waves, they wanted another opinion, so they sent it to Albert Einstein for review. Einstein wrote back that he found de Broglie's work "quite interesting." That endorsement was good enough for the faculty: de Broglie's thesis was accepted in 1924 and immediately submitted for publication.

The Heisenberg Uncertainty Principle

After de Broglie proposed that electrons exhibited both particle-like and wave-like behavior, questions arose about the impact of wave behavior on our ability to locate the electron, given that a wave by its very nature is spread out in space. The question, "Where is the electron?" was addressed by German physicist Werner Heisenberg (1901–1976), who proposed the following thought experiment: What if we tried to watch an electron orbiting the nucleus of an atom? We would need an extremely powerful microscope to see a particle as tiny as an electron. No such microscope exists, but if it did, it would need to use gamma rays for illumination because they are the only part of the electromagnetic spectrum with wavelengths short enough to match the diminutive size of electrons. Longer-wavelength radiation would pass right by an electron without being reflected by it. Unfortunately, the ultrashort wavelengths and high frequencies of gamma rays mean that they have enormous energies—so large that any gamma ray striking an electron would knock the electron off course. The only way not to affect the electron's motion would be to use a much lower-energy, longer-wavelength source of radiation to illuminate it, but then we would not be able to see the tiny electron clearly.

This situation presents an experimental dilemma. The only means for clearly observing an electron make it impossible to know the electron's motion or, more precisely, its momentum, which is defined as an object's velocity times its mass. Therefore, we can never know exactly both the position and the momentum of the electron simultaneously. This conclusion is known as the **Heisenberg uncertainty principle** and is mathematically expressed as

$$\Delta x \cdot m\Delta u \geq \frac{h}{4\pi} \quad (3.14)$$

Heisenberg uncertainty principle the principle that one cannot simultaneously know the exact position and the exact momentum of an electron.

where Δx is the uncertainty in the position of the electron, m is its mass, Δu is the uncertainty in its velocity, and h is the Planck constant. To Heisenberg, this uncertainty was the essence of quantum theory. Its message for us is that there are limits to what we can observe, measure, and therefore know about particles the size of electrons.

SAMPLE EXERCISE 3.7 Calculating Heisenberg Uncertainties **LO4**

Use the data in Sample Exercise 3.6 to compare the uncertainties in the speeds of a thrown baseball and an electron. Assume that the position of the baseball is known to within one wavelength of red light ($\Delta x_{\text{baseball}} = 6.80 \times 10^2 = 680$ nm) and that the position of the electron is known to within the radius of the hydrogen atom ($\Delta x_{\text{electron}} = 5.3 \times 10^{-11}$ m).

Collect and Organize We are asked to calculate the uncertainty in the speeds of two particles and are given the uncertainties in their positions. Equation 3.14 provides a mathematical connection between the variables. From Sample Exercise 3.6 we know that the mass of a baseball is 0.142 kg. The mass of an electron is 9.109×10^{-31} kg.

Analyze According to Equation 3.14, the uncertainty in the speed and position of a particle is inversely proportional to its mass. Therefore, we can expect little uncertainty in the speed of the baseball but much greater uncertainty in the speed of the electron. We need to rearrange the terms in the equation to solve for the uncertainty in speed (Δu):

$$\Delta u \geq \frac{h}{4\pi\Delta x \cdot m}$$

Solve For the baseball,

$$\Delta u \geq \frac{6.626 \times 10^{-34}\ \cancel{\text{J}} \cdot \text{s}}{4\pi(6.80 \times 10^{-7}\ \cancel{\text{m}})(0.142\ \cancel{\text{kg}})} \times \frac{1\ \cancel{\text{kg}} \cdot \text{m}^2/\text{s}^2}{1\ \cancel{\text{J}}} \geq 5.46 \times 10^{-28}\ \text{m/s}$$

For the electron,

$$\Delta u \geq \frac{6.626 \times 10^{-34}\ \cancel{\text{J}} \cdot \text{s}}{4\pi(5.3 \times 10^{-11}\ \cancel{\text{m}})(9.109 \times 10^{-31}\ \cancel{\text{kg}})} \times \frac{1\ \cancel{\text{kg}} \cdot \text{m}^2/\text{s}^2}{1\ \cancel{\text{J}}} \geq 1.1 \times 10^{6}\ \text{m/s}$$

Think About It The uncertainty in the speed of the baseball is insignificant compared, for example, to the speed of a major league fastball (> 40 m/s). That result is expected for objects in the macroscopic world. The uncertainty in the speed of the electron is many orders of magnitude larger because its mass (which is in the denominator of the expression we derived from Equation 3.14) is many orders of magnitude smaller.

Practice Exercise What is the uncertainty, in meters, in the position of an electron moving near a nucleus at a speed of 8×10^7 m/s? Assume the relative uncertainty in the speed of the electron is 1%—that is, 8×10^5 m/s.

(Answers to Practice Exercises are in the back of the book.)

Heisenberg was working with Bohr at the University of Copenhagen when he proposed his uncertainty principle. The two scientists had widely different views about the significance of the uncertainty principle and the idea that particles could behave like waves. To Heisenberg, uncertainty was a fundamental characteristic of nature. To Bohr, it was merely a mathematical consequence of the wave–particle duality of electrons; there was no physical meaning to an electron's position and path. The debate between the two gifted scientists was heated at times. Heisenberg later wrote about one particularly emotional debate:

> [A]t the end of the discussion I went alone for a walk in the neighboring park [and] repeated to myself again and again the question: "Can nature possibly be as absurd as it seems . . . ?"[1]

3.6 Quantum Numbers

Many of the leading scientists of the 1920s were unwilling to accept de Broglie's model of electron waves until it had a stronger theoretical foundation. They wanted a mathematical model that accurately described the behavior of matter waves and accounted for the atomic spectra of hydrogen. During a Christmas vacation in the Swiss Alps in 1925, Austrian physicist Erwin Schrödinger (1887–1961) created that mathematical foundation by developing what came to be called **wave mechanics** or **quantum mechanics**.

Schrödinger's mathematical description of electron waves is known as the **Schrödinger wave equation**. We do not examine it in detail in this book, but you should know that solutions to it are called **wave functions**: mathematical expressions represented by the Greek letter psi (ψ) that describe how the matter wave of an electron in an atom varies in both time and location inside the atom. Wave functions define the energy levels in the hydrogen atom. They can be simple trigonometric functions, such as sine or cosine waves, or they can be very complex.

What is the physical significance of a wave function? Schrödinger believed that a wave function depicted the "smearing" of an electron through three-dimensional space. However, this notion of subdividing a discrete particle was later rejected in favor of the model developed by German physicist Max Born (1882–1970), who proposed that the square of a wave function, ψ^2, defines an **orbital**: the space within an atom where the probability of finding an electron is high. Born later showed that his interpretation could be used to calculate the probability of electron transitions between orbitals, as happens when an atom absorbs or emits a quantum of energy.

To help visualize the probabilistic meaning of ψ^2, consider what happens when we spray ink onto a flat surface as in Figure 3.20. If we then draw a circle encompassing most of the ink spots, we are identifying the region of maximum probability for finding the spots.

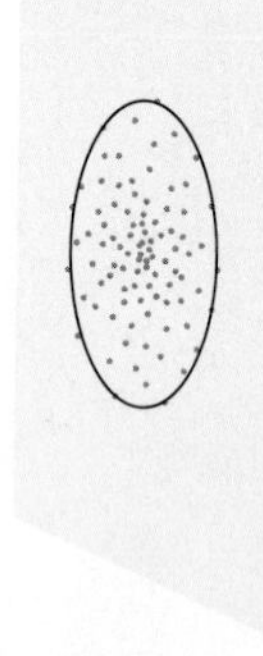

FIGURE 3.20 The probability of finding an ink spot in the pattern produced by a source of ink spray decreases with increasing distance from the center of the pattern, in much the way that electron density in the 1*s* orbital decreases with increasing distance from the nucleus.

Quantum mechanical orbitals are *not* two-dimensional concentric orbits, as in Bohr's model of the hydrogen atom, or even two-dimensional circles, as in the pattern of ink drops in Figure 3.20. Instead, they are three-dimensional regions of space with distinctive shapes, orientations, and average distances from the nucleus. Each orbital is a solution to Schrödinger's wave equation and is identified by a unique combination of three integers called **quantum numbers**, whose values flow directly from the mathematical solutions to the wave equation. The quantum numbers are as follows:

- The **principal quantum number** n is like Bohr's n value for the hydrogen atom in that it is a positive integer that indicates the relative size and energy of an orbital or of a group of orbitals in an atom. Orbitals (and the electrons in them) with the same value of n are in the same *shell*. Orbitals with larger values of n are in shells that are greater distances from the nucleus and have higher energies than those with lower values of n. In a hydrogen atom, an orbital's n value defines the energy of an electron in the orbital, consistent with Bohr's model. In

[1]Werner Heisenberg, *Physics and Philosophy: The Revolution in Modern Science* (Harper & Row, 1958), p. 42.

multielectron atoms, the relationship between energy levels and orbitals is more complex, but increasing values of n generally represent higher energy levels.

- The **angular momentum quantum number** ℓ is an integer with a value ranging from zero to $(n - 1)$ that defines the shape of an orbital. Orbitals with the same values of n and ℓ (and the electrons that reside in them) are in the same *subshell* and have the same energy. Orbitals with a given value of ℓ are identified with a letter according to the following scheme:

Value of ℓ	0	1	2	3	4
Letter Identifier	*s*	*p*	*d*	*f*	*g*

- The **magnetic quantum number** m_ℓ is an integer with a value from $-\ell$ to $+\ell$. It defines the orientation of an orbital in the space around the nucleus of an atom.

Each subshell has a two-part label that contains the appropriate value of n and a letter designation for ℓ. For example, orbitals with $n = 3$ and $\ell = 1$ are called 3*p* orbitals, and electrons in 3*p* orbitals are called 3*p* electrons. How many 3*p* orbitals are there? We can answer that question by finding all possible values of m_ℓ. Because *p* orbitals are those for which $\ell = 1$, they have m_ℓ values of -1, 0, and $+1$. The three values mean that there are three 3*p* orbitals, each with a unique combination of n, ℓ, and m_ℓ values. All the possible combinations of these three quantum numbers for the orbitals of the first four shells are listed in Table 3.1.

TABLE 3.1 Quantum Numbers of the Orbitals in the First Four Shells

Value of n	Allowed Values of ℓ	Subshell Letter	Allowed Values of m_ℓ	Number of Orbitals in: Subshell	Number of Orbitals in: Shell
1	0	*s*	0	1	1
2	0	*s*	0	1	
	1	*p*	−1, 0, +1	3	4
3	0	*s*	0	1	
	1	*p*	−1, 0, +1	3	
	2	*d*	−2, −1, 0, +1, +2	5	9
4	0	*s*	0	1	
	1	*p*	−1, 0, +1	3	
	2	*d*	−2, −1, 0, +1, +2	5	
	3	*f*	−3, −2, −1, 0, +1, +2, +3	7	16

SAMPLE EXERCISE 3.8 Identifying the Subshells and Orbitals in an Energy Level **LO5**

a. What are the names of all the subshells in the $n = 4$ shell?
b. How many orbitals are in all the subshells of the $n = 4$ shell?

Collect, Organize, and Analyze We are asked to describe the subshells in the fourth shell and to determine how many orbitals are in all the subshells. Subshell designations (Table 3.1) are based on the possible values of the quantum numbers n and ℓ. The allowed values of ℓ depend on the value of n, because ℓ is an integer between 0 and $(n - 1)$. The number of orbitals in a subshell depends on the number of possible values of m_ℓ, which range from $-\ell$ to $+\ell$.

wave mechanics or **quantum mechanics** a mathematical description of the wavelike behavior of electrons and other particles.

Schrödinger wave equation a description of how the electron matter wave varies with location and time around the nucleus of a hydrogen atom.

wave function (ψ) a solution to the Schrödinger wave equation describing how the matter wave of an electron varies in both time and location in an atom.

orbitals defined by the square of the wave function (ψ^2); regions in an atom where the probability of finding an electron is high.

quantum number one of four related numbers that specify the energy, shape, and orientation of orbitals in an atom and the spin orientation of electrons in the orbitals.

principal quantum number (*n*) a positive integer describing the relative size and energy of an atomic orbital or group of orbitals in an atom.

angular momentum quantum number (ℓ) an integer having any value from 0 to $(n - 1)$ that defines the shape of an orbital.

magnetic quantum number (m_ℓ) defines the orientation of an orbital in space; an integer that may have any value from $-\ell$ to $+\ell$, where ℓ is the angular momentum quantum number.

Solve

a. The allowed values of ℓ for $n = 4$ range from 0 to $(n - 1)$—that is, from 0 to 3—so they are 0, 1, 2, and 3. The ℓ values correspond to the subshell designations *s*, *p*, *d*, and *f*. The appropriate subshell names are thus 4*s*, 4*p*, 4*d*, and 4*f*.

b. The possible values of m_ℓ from $-\ell$ to $+\ell$ are

- $\ell = 0$; $m_\ell = 0$. This combination of ℓ and m_ℓ values for the $n = 4$ shell represents a single 4*s* orbital.
- $\ell = 1$; $m_\ell = -1, 0$, or $+1$. These three combinations of ℓ and m_ℓ values for the $n = 4$ shell represent the three 4*p* orbitals.
- $\ell = 2$; $m_\ell = -2, -1, 0, +1$, or $+2$. These five combinations of ℓ and m_ℓ values represent the five 4*d* orbitals.
- $\ell = 3$; $m_\ell = -3, -2, -1, 0, +1, +2$, or $+3$. These seven combinations of ℓ and m_ℓ values represent the seven 4*f* orbitals.

Thus, there are $1 + 3 + 5 + 7 = 16$ orbitals in the $n = 4$ shell.

Think About It We determined that there are 16 orbitals in the fourth shell. The number of orbitals in each shell is equal to n^2, the square of the principal quantum number of the shell.

Practice Exercise How many orbitals are there in the $n = 5$ shell? What are the names of all the subshells in the $n = 5$ shell?

(Answers to Practice Exercises are in the back of the book.)

The following relationships are worth noting in the quantum numbering system:

- There are n subshells in the nth shell: one subshell (1*s*) in the $n = 1$ shell, two subshells (2*s* and 2*p*) in the $n = 2$ shell, and so on.
- There are n^2 orbitals in the nth shell: $1^2 = 1$ in the $n = 1$ shell, $2^2 = 4$ in the $n = 2$ shell, and so on.
- There are $(2\ell + 1)$ orbitals in each subshell: one *s* orbital $(2 \times 0 + 1 = 1)$ in each *s* subshell, three *p* orbitals $(2 \times 1 + 1 = 3)$ in each *p* subshell, five *d* orbitals $(2 \times 2 + 1 = 5)$ in each *d* subshell, and so on.

The Schrödinger equation accounts for most, but not all, aspects of atomic spectra. The emission spectrum of hydrogen, for example, when viewed through a high-resolution spectrometer, contains a pair of red lines at 656 nm, where Balmer saw only one (Figure 3.7a). There are also pairs of lines in the spectra of multielectron atoms that have a single electron in their outermost shells.

In 1925, two students at the University of Leiden in the Netherlands, Samuel Goudsmit (1902–1978) and George Uhlenbeck (1900–1988), proposed that the pairs of lines, called *doublets*, were caused by a property they called *electron spin*. In their model, electrons spin in one of two directions, designated "spin up" and "spin down." A moving electron (or any charged particle) creates a magnetic field by virtue of its movement through space. The spinning motion produces a second magnetic field oriented up or down. To account for the two spin orientations, Goudsmit and Uhlenbeck proposed a fourth quantum number, the **spin quantum number**, m_s. There are two possible values of m_s: $+\frac{1}{2}$ for spin up and $-\frac{1}{2}$ for spin down.

spin quantum number (m_s) either $+\frac{1}{2}$ or $-\frac{1}{2}$, indicating the spin orientation of an electron.

Pauli exclusion principle principle that states no two electrons in an atom can have the same set of four quantum numbers.

Even before Goudsmit and Uhlenbeck proposed the electron-spin hypothesis, two other scientists, Otto Stern (1888–1969) and Walther Gerlach (1889–1979), observed the effect of electron spin when they shot a beam of silver ($Z = 47$) atoms through a magnetic field (Figure 3.21). Those atoms in which the net electron spin was "up" were deflected in one direction by the field, whereas those in which the net electron spin was "down" were deflected in the opposite direction.

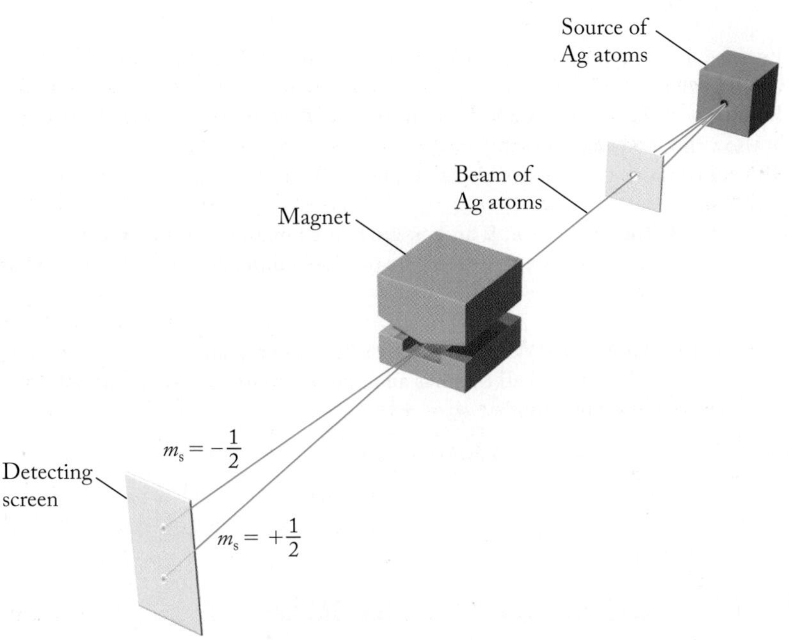

FIGURE 3.21 A narrow beam of silver atoms passed through a magnetic field is split into two beams because of the interactions between the field and the spinning electrons in the atoms. This observation led to the proposal of a fourth quantum number, m_s.

In 1925, Austrian physicist Wolfgang Pauli (1900–1958) proposed that no two electrons in a multielectron atom can have the same set of values for the four quantum numbers n, ℓ, m_ℓ, and m_s. This idea is known as the **Pauli exclusion principle.** We have seen how unique combinations of the three quantum numbers from Schrödinger's wave equation define each orbital in an atom. The fourth (spin) quantum number provides a unique address for each electron in each orbital.

SAMPLE EXERCISE 3.9 Identifying Valid Sets of Quantum Numbers **LO5**

Which of these five combinations of quantum numbers are valid?

	n	ℓ	m_ℓ	m_s
(a)	1	0	−1	$+\frac{1}{2}$
(b)	3	2	−2	$+\frac{1}{2}$
(c)	2	2	0	0
(d)	2	0	0	$-\frac{1}{2}$
(e)	−3	−2	−1	$-\frac{1}{2}$

Collect, Organize, and Analyze Table 3.1 contains valid combinations of the quantum numbers n, ℓ, and m_ℓ for the first four shells. The principal quantum number (n) can be any positive integer; the valid values of ℓ in a given shell are integers from 0 to $(n - 1)$, and the values of m_ℓ in a given subshell include all integers from $-\ell$ to $+\ell$, including 0. The only two options for m_s are $+\frac{1}{2}$ and $-\frac{1}{2}$.

Solve

a. Because $n = 1$, the maximum (and only) value of ℓ is $(n - 1) = 1 - 1 = 0$. Therefore, the values of n and ℓ are valid. However, if $\ell = 0$, then m_ℓ must be 0; it cannot be −1. Therefore, this set is not valid. The spin quantum number, $+\frac{1}{2}$, is a valid one, however.
b. Because $n = 3$, ℓ can be 2 and m_ℓ can be −2. Also, $m_s = +\frac{1}{2}$ is a valid choice for the spin quantum number. This set is valid.

c. Because $n = 2$, ℓ cannot be 2, making this set invalid. In addition, $m_s = 0$ is an invalid value.

d. Because $n = 2$, ℓ can be 0, and for that value of ℓ, m_ℓ must be 0, too. The value of m_s is also valid, and so is the set.

e. This set contains two impossible values, $n = -3$ and $\ell = -2$, so it is invalid.

Think About It The values of n, ℓ, and m_ℓ are related mathematically, and m_s can be either $+\frac{1}{2}$ or $-\frac{1}{2}$. Together the four numbers provide a unique address for every electron in an atom.

Practice Exercise Write all the possible sets of quantum numbers for an electron in the $n = 3$ shell that has an angular momentum quantum number $\ell = 1$ and a spin quantum number $m_s = +\frac{1}{2}$.

(Answers to Practice Exercises are in the back of the book.)

3.7 The Sizes and Shapes of Atomic Orbitals

We have learned that atomic orbitals have three-dimensional shapes that are graphical representations of ψ^2. In this section we examine how the shapes of orbitals impact the energies of the electrons in them.

s Orbitals

Figure 3.22 provides four different representations of the 1*s* orbital of hydrogen. In Figure 3.22(a), electron density is plotted against distance from the nucleus and seems to show that density decreases with increasing distance. However, Figure 3.22(b) provides a more useful profile of electron distribution. To understand why, think of the hydrogen atom as a tiny onion, made of many concentric spherical layers all of the same thickness. A cross section of that image of the atom is shown in Figure 3.22(c). What is the probability of finding the electron in one of the spherical layers? A layer very close to the nucleus has a very small radius, so it accounts for only a small fraction of the total volume of the atom. A layer with a larger radius makes up a much larger fraction of the volume of the atom because the volume of the layers increases as a function of r^2. Even though electron densities are higher closer to the nucleus (as Figure 3.22a shows), the volumes of the spherical shells closest to the nucleus are so small that the chances of the electron being near the center of an atom are extremely low. This low probability is shown in Figure 3.22(b), where the curve starts off at essentially zero for electron distribution values at distances very close to the nucleus. Farther from the nucleus, electron densities are lower but the volumes of the layers are much larger, so the probability of the electron being in one of the layers is relatively high, represented by the peak in the curve of Figure 3.22(b). At greater distances, volumes of the layers

FIGURE 3.22 (a) Probable electron density in the 1*s* orbital of the hydrogen atom represented by a plot of electron density (ψ^2) versus distance from the nucleus. (b) Electron distribution in the 1*s* orbital versus distance from the nucleus. The distribution is essentially zero, both for very short distances from the nucleus and for very long distances from the nucleus. (c) Cross section through the hydrogen atom, with the space surrounding the nucleus divided into an arbitrary number of thin, concentric hollow layers. Each layer has a unique value for radius *r*. The probability of finding an electron in a particular layer of radius *r* depends on the volume of the layer and the density of electrons in the layer. (d) Boundary–surface representation of a sphere within which the probability of finding a 1*s* electron is 90%.

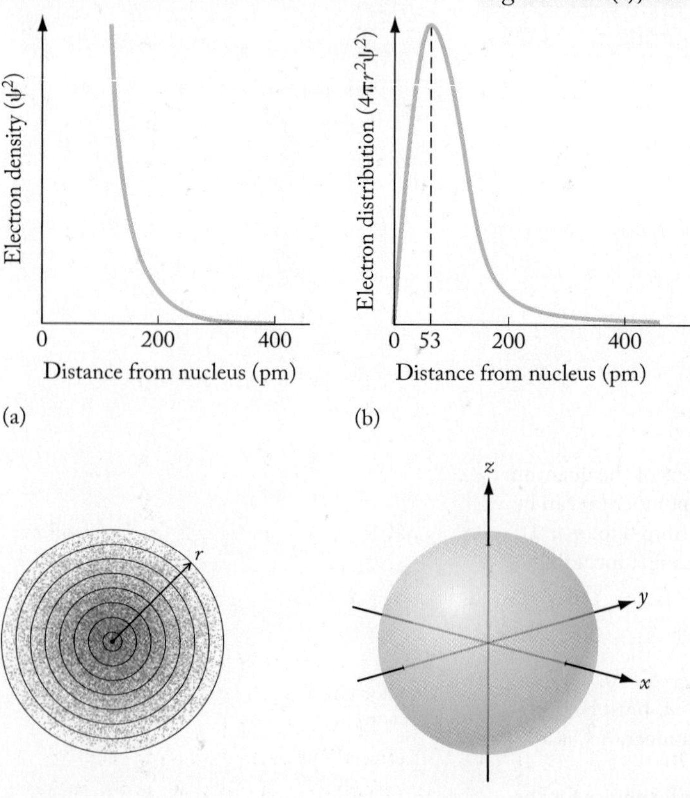

are very large but ψ^2 drops to nearly zero (see Figure 3.22a); therefore, the chances of finding the electron in layers far from the nucleus are very small.

Thus, Figure 3.22(b) represents a combination of two competing factors—increasing layer volume and decreasing probability of the electron being far from the nucleus. This combination produces a *radial distribution profile* for the electron. Figure 3.22(b) is a plot not of ψ^2 versus distance from the nucleus, as in Figure 3.22(a), but rather of $4\pi r^2\psi^2$ versus distance from the nucleus. In geometry, $4\pi r^2$ is the formula for the surface area of a sphere, but here it represents the volume of one of the very thin spherical layers in Figure 3.22(c). The maximum value of the curve in Figure 3.22(b), at 53 pm, corresponds to the most likely radial distance of a 1*s* electron from the nucleus.

Figure 3.22(d) provides a view of the spherical shape of this (or any other) *s* orbital. The surface of the sphere encloses the volume within which the probability of finding a 1*s* electron is 90%. This type of depiction, called a *boundary–surface representation*, is one of the most useful ways to view the relative sizes, shapes, and orientations of orbitals. All *s* orbitals are spheres, which means that they have no angular dependence on orientation.

Radial electron distribution profiles of hydrogen's 1*s*, 2*s*, and 3*s* orbitals are shown in Figure 3.23. Note that orbital size increases with increasing values of the principal quantum number *n*. Note also that in the quadrants above the profile curves there are bands in which the density of dots is high. The dots represent the probability of finding an electron at those locations, and each dark band represents a local maximum in electron distribution. In all three profiles, there is a local maximum close to the nucleus. This means that electrons in *s* orbitals, even *s* orbitals with high values of *n*, have some probability of being close to the nucleus.

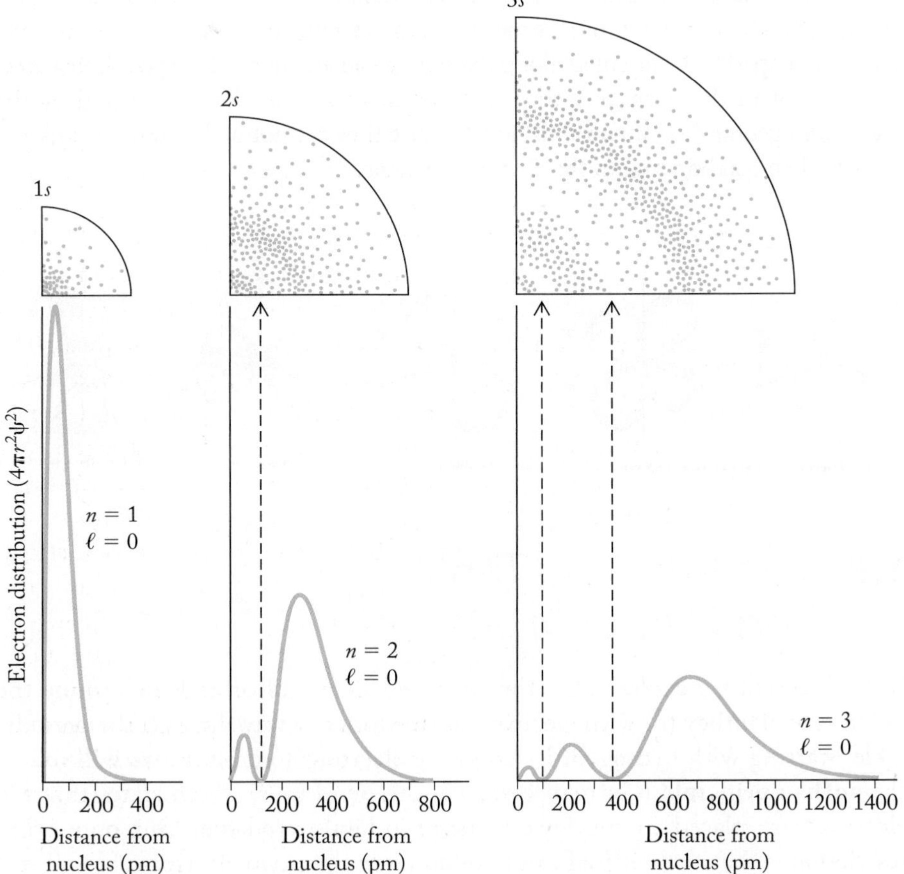

FIGURE 3.23 Radial distribution profiles of 1*s*, 2*s*, and 3*s* orbitals. Electrons in all these *s* orbitals have some probability of being close to the nucleus, but 3*s* electrons are more likely to be farther away from the nucleus than 2*s* electrons, which are more likely to be farther away than 1*s* electrons. The dashed lines connect two representations of the nodes in the radial distributions in the two diagrams.

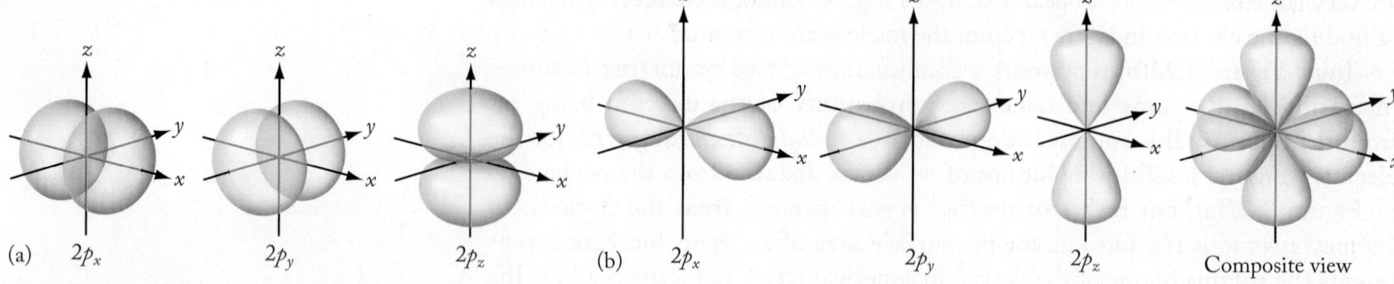

FIGURE 3.24 Boundary–surface views of the three $2p$ orbitals, showing their orientation along the x-, y-, and z-axes. (a) These views of $2p_x$, $2p_y$, and $2p_z$ are obtained from the wave functions of the orbitals. (b) We use an elongated version of the theoretical shapes of the orbitals throughout this book to make it easier to see the orientation of their lobes.

p and *d* Orbitals

All shells with $n \geq 2$ have a subshell containing three p orbitals ($\ell = 1$; $m_\ell = -1, 0, +1$). Each of the three p orbitals has two lobes oriented on either side of the nucleus along one of the three perpendicular Cartesian axes x, y, z. The true shape of the lobes is squashed and roundish, like mushroom caps (Figure 3.24a), but we draw them in an elongated teardrop shape to more easily visualize their orientation (Figure 3.24b). The orbitals are designated p_x, p_y, and p_z, depending on the axis along which the lobes are aligned. The two lobes of a p orbital are sometimes labeled with plus or minus signs to indicate the sign of the wave function that defines each lobe (but don't confuse the signs with electrical charges—all electrons have negative charges).

Shells with principal quantum numbers of 3 or higher have five d orbitals ($\ell = 2$; $m_\ell = -2, -1, 0, +1, +2$). Their shapes are shown in Figure 3.25. Four of them have teardrop-shaped lobes oriented like the leaves in a four-leaf clover. The lobes of three of the four, designated d_{xy}, d_{xz}, and d_{yz}, lie *between* the Cartesian axes. The lobes of the fourth orbital, $d_{x^2-y^2}$, are centered *on* the x- and y-axes. The fifth d orbital, d_{z^2}, is mathematically equivalent to the other four but has a much different shape with two teardrop-shaped lobes oriented along the z-axis and a doughnut shape called a *torus* in the x–y plane that surrounds the middle of the two lobes. We will not address the shapes and geometries of f orbitals here because they are not included in our discussions of chemical bonding in the chapters to come.

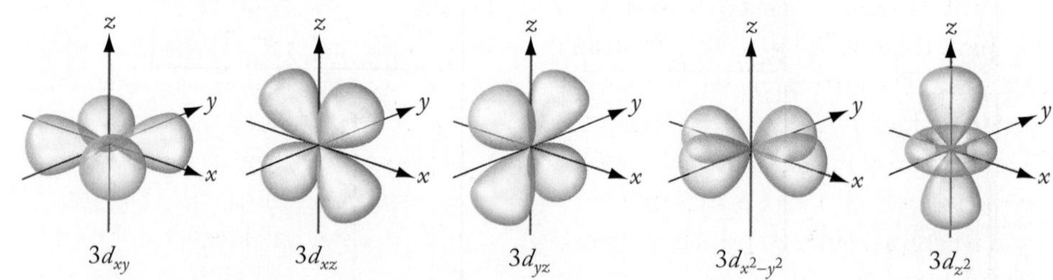

FIGURE 3.25 Boundary–surface views of the five $3d$ orbitals, showing their orientation relative to the x-, y-, and z-axes. As with the $2p$ orbitals in Figure 3.24(b), the theoretical boundary surfaces of the $3d$ orbitals are elongated to make it easier to see the orientation of their lobes. The d_{xy}, d_{xz}, and d_{yz} orbitals are not aligned along any axis; the $d_{x^2-y^2}$ orbital lies along the x- and y-axes; the d_{z^2} orbital consists of two teardrop-shaped lobes along the z-axis with a donut-shaped torus ringing the point where the two lobes meet.

3.8 The Periodic Table and Filling Orbitals

In this section we explore how the sizes and shapes of orbitals determine the order in which they fill with electrons as we work our way through the periodic table, starting with hydrogen. In assigning electrons to orbitals, we will follow the **aufbau principle** (German *aufbauen*, "to build up"), which states that the electrons are placed in the lowest energy orbitals available. Our only other restriction is that each orbital can have no more than two electrons.

aufbau principle the concept of building up ground state atoms so that their electrons occupy the lowest energy orbitals available.

electron configuration the distribution of electrons among the orbitals of an atom or ion.

Using these rules, let's begin by assigning the single electron in hydrogen ($Z = 1$) to the 1*s* orbital. We represent this arrangement with the **electron configuration** $1s^1$, where the first 1 indicates the principal quantum number (*n*) of the orbital, *s* indicates the subshell, and the superscripted 1 indicates that there is *one* electron in the 1*s* orbital. When we write the electron configurations of elements, we assume that their atoms are in their ground states.

The atomic number of helium is 2, which tells us there are two protons in the nucleus surrounded by two electrons in the neutral atom. Using the aufbau principle, we simply add another electron to the 1*s* orbital. The 1*s* orbital already contains one electron, so it has space for one more. The Pauli exclusion principle dictates that the spin quantum number for the second electron cannot be the same as the first: one must be $+\frac{1}{2}$ and the other must be $-\frac{1}{2}$. The two electrons with opposite spin quantum numbers are said to be *spin-paired*. Their presence gives helium a ground-state electron configuration of $1s^2$. With the two electrons, the 1*s* orbital is filled to capacity and so is the $n = 1$ shell.

The concept of filled shells and subshells is critical to understanding the chemical properties of the elements. Helium and the other group 18 elements are composed of atoms that have filled *s*, or *s* and *p*, orbitals in their outermost shells. These elements, the noble gases, are chemically stable and generally unreactive. Other main group elements have partially filled sets of *s* and *p* orbitals in their outermost shell. They engage in chemical reactions in which their atoms lose, gain, or share electrons in ways that produce filled outermost *s* and *p* subshells. In that way they acquire the stable electron configurations of group 18 elements.

CONNECTION In Chapter 2 we defined the main group elements as those in groups 1, 2, and 13 through 18 in the periodic table.

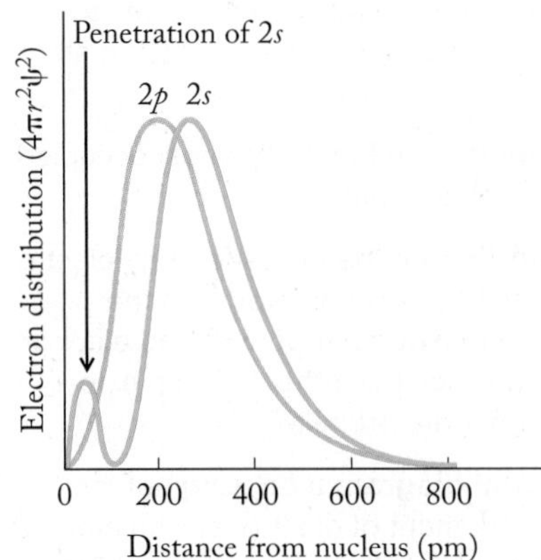

FIGURE 3.26 Radial distribution profiles of electrons in 2*s* and 2*p* orbitals. The 2*s* orbital is lower in energy because electrons in it penetrate more closely to the nucleus, as indicated by the local maximum in electron distribution about 50 pm from the nucleus. As a result, 2*s* electrons experience a greater effective nuclear charge than 2*p* electrons.

Effective Nuclear Charge

The location of lithium ($Z = 3$) in the periodic table—the first element in the second period—is a signal that lithium has one electron in its $n = 2$ shell. The row numbers in the periodic table correspond to the *n* values of the outermost shells of the elements in the rows. The second shell has four orbitals—one 2*s* and three 2*p*, which can hold up to eight electrons. Lithium's third electron occupies the lowest-energy orbital in the second shell, which is the 2*s* orbital, making the electron configuration of Li $1s^22s^1$.

Why is the 2*s* orbital lower in energy than any of the 2*p* orbitals? When we compare their radial distribution profiles in Figure 3.26, the small peak on the 2*s* curve near the nucleus means that an electron in the 2*s* orbital is closer to the nucleus more frequently than an electron in a 2*p* orbital, which has no such secondary peak. This proximity means that the 2*s* electron can get closer to the nucleus and experience more of its positive nuclear charge. An electron in a 2*s* orbital of a ground-state Li atom or in a 2*p* orbital of an excited-state Li atom is shielded from much of the nucleus's charge by the negative charges of the two electrons in the filled 1*s* orbital. This shielding means that a 2*s* or 2*p* electron experiences an **effective nuclear charge (Z_{eff})** that is only about (3+) + (2−) = 1+ (Figure 3.27). However, an electron in a 2*s* orbital can penetrate the cloud of 1*s* electrons better and experience a greater Z_{eff}. Therefore, a 2*s* electron has less energy than a 2*p* electron, which is why the 2*s* orbital fills first.

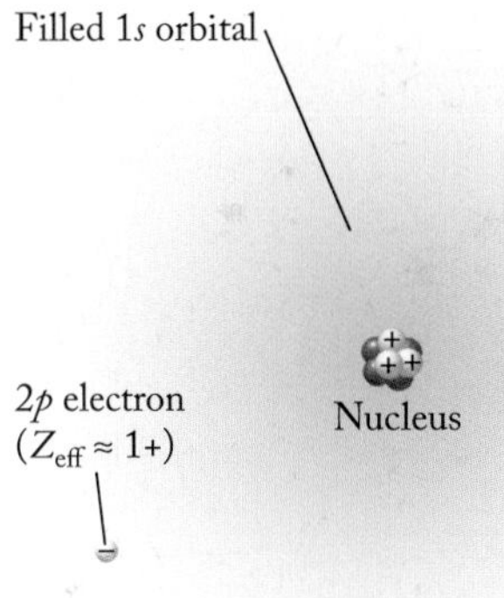

FIGURE 3.27 The effective nuclear charge (Z_{eff}) experienced by a 2*p* electron in an excited-state Li atom approximately equals the sum of the actual nuclear charge (3+) and the shielding effect of the negative charges of the two 1*s* electrons (2−). This shielding produces a net Z_{eff} of about 1+.

effective nuclear charge (Z_{eff}) the attraction toward the nucleus experienced by an electron in an atom; the positive charge on the nucleus is reduced by the extent to which other electrons in the atom shield the electron from the nucleus.

Condensed Electron Configurations

We can simplify the electron configurations of Li and all the elements that follow it in the periodic table by writing their *condensed* electron configurations. In this format the symbols representing all of the electrons in orbitals that were filled in

core electrons electrons in the filled, inner shells in an atom or ion that are not involved in chemical reactions.

valence electrons electrons in the outermost occupied shell of an atom having the most influence on the atom's chemical behavior.

valence shell the outermost occupied shell of an atom.

Hund's rule the lowest energy electron configuration of an atom has the maximum number of unpaired electrons, all of which have the same spin, in degenerate orbitals.

orbital diagram a depiction of the arrangement of electrons in an atom or ion using boxes to represent orbitals.

the rows above the element of interest are replaced by the symbol of the group 18 element at the end of the row above the element. For example, the condensed electron configuration of Li is $[He]2s^1$. Condensed electron configurations are useful because they eliminate the symbols of the **core electrons** in filled shells and subshells. Core electrons are not involved in the chemistries of the elements, so they are of less interest to us than those that reside in the outermost shells and subshells. The outermost electrons are the ones involved in bond formation and are called **valence electrons**. The outermost shell is called the **valence shell**. Notice that lithium has a single electron in its valence-shell *s* orbital, as does hydrogen, the element directly above it in the periodic table. Therefore, both elements have the same general valence-shell configuration, ns^1, where *n* represents both the number of the row in which the element is located and the principal quantum number of the valence shell.

Beryllium ($Z = 4$) is the fourth element in the periodic table and the first in group 2. Its electron configuration is $1s^22s^2$ or, in condensed form, $[He]2s^2$. The other elements in group 2 also have two spin-paired electrons in the *s* orbital of their valence shell. The second shell is not full at this point because it also has three *p* orbitals, which are all empty and fill as we move to the next elements in the periodic table.

Boron ($Z = 5$) is the first element in group 13. Its fifth electron is in one of its three $2p$ orbitals, resulting in the condensed electron configuration $[He]2s^22p^1$. It does not matter which of the three $2p$ orbitals contains the fifth electron because all three have the same energy. Chemists call orbitals with the same energy *degenerate* orbitals.

Hund's Rule and Orbital Diagrams

The next element is carbon ($Z = 6$). It has another electron in one of its $2p$ orbitals, giving it the condensed electron configuration $[He]2s^22p^2$. The second $2p$ electron resides in a different $2p$ orbital than the first. Why? Because electrons repel each other, and they will repel each other less if they are in separate orbitals rather than in the same orbital. This separation of the $2p$ electrons into different orbitals is an application of **Hund's rule**, named after German physicist Friedrich Hund (1896–1997), which states that the lowest-energy electron configuration for degenerate orbitals, like the three in the $2p$ subshell, is the one with the maximum number of unpaired valence electrons, all of which have the same spin.

We use **orbital diagrams** to show in detail how electrons, represented by single-headed arrows, are distributed among orbitals, which are represented by boxes. A single-headed arrow pointing upward represents an electron with spin up ($m_s = +\frac{1}{2}$), and a downward-pointing single-headed arrow represents an electron with spin down ($m_s = -\frac{1}{2}$). To obey Hund's rule, the orbital diagram for carbon must be

Carbon:	⇅	⇅	↑ ↑ _
	$1s$	$2s$	$2p$

The two $2p$ electrons are unpaired (in separate orbitals). By convention, the spin arrows of the single electrons in those orbitals are drawn pointed up.

The condensed electron configuration of the next element, nitrogen ($Z = 7$), is $[He]2s^22p^3$. According to Hund's rule, the third $2p$ electron resides alone in the third $2p$ orbital, so that the electron distribution is

Nitrogen:	⇅	⇅	↑ ↑ ↑
	$1s$	$2s$	$2p$

	Orbital diagram 1s	2s	2p	Electron configuration	Condensed electron configuration
H	[↑]	[ ]	[ ][ ][ ]	$1s^1$	
He	[↑↓]	[ ]	[ ][ ][ ]	$1s^2$	
Li	[↑↓]	[↑]	[ ][ ][ ]	$1s^22s^1$	$[He]2s^1$
Be	[↑↓]	[↑↓]	[ ][ ][ ]	$1s^22s^2$	$[He]2s^2$
B	[↑↓]	[↑↓]	[↑][ ][ ]	$1s^22s^22p^1$	$[He]2s^22p^1$
C	[↑↓]	[↑↓]	[↑][↑][ ]	$1s^22s^22p^2$	$[He]2s^22p^2$
N	[↑↓]	[↑↓]	[↑][↑][↑]	$1s^22s^22p^3$	$[He]2s^22p^3$
O	[↑↓]	[↑↓]	[↑↓][↑][↑]	$1s^22s^22p^4$	$[He]2s^22p^4$
F	[↑↓]	[↑↓]	[↑↓][↑↓][↑]	$1s^22s^22p^5$	$[He]2s^22p^5$
Ne	[↑↓]	[↑↓]	[↑↓][↑↓][↑↓]	$1s^22s^22p^6$	$[He]2s^22p^6 = [Ne]$

FIGURE 3.28 Orbital diagrams and condensed electron configurations for the first 10 elements. The boxes in the orbital diagrams represent orbitals. Each can hold up to two electrons of opposite spin. Condensed electron configurations for all elements are given in Appendix 3.

with all three spin arrows pointed up. As we proceed across the second row to neon ($Z = 10$), we fill the $2p$ orbitals as shown in Figure 3.28. The last three $2p$ electrons added (in oxygen, fluorine, and neon) pair up with the first three so that the spin orientations of each pair have opposite directions. All three $2p$ orbitals are filled to capacity in an atom of neon, and so is the $n = 2$ shell.

Sodium ($Z = 11$) follows neon. It is the third element in group 1 and the first in the third row. Its position means that its outermost electron is in the third shell. There are three types of orbitals in the third shell: $3s$, $3p$, and $3d$. Which type gets sodium's 11th electron? Based on the radial distribution profiles of electrons in these orbitals, shown in Figure 3.29, the $3s$ orbital gets the electron. Note the two peaks in the $3s$ profile and one of the peaks in the $3p$ profile that are all close to the nucleus. These peaks mean that electrons in $3s$ and $3p$ orbitals penetrate through the filled orbitals of the first two shells and experience greater effective nuclear charge than do $3d$ electrons. Therefore, their relative energies are lower, with $3s$ the lowest of all. So, in the third shell, the $3s$ orbital fills first and the $3d$ orbitals fill last. The same sequence applies to all the shells with $n > 3$. Shells for which $n \geq 4$ also contain f orbitals, which have higher energies than the d orbitals in the same shell, so the f orbitals fill last.

Because the s orbital in a shell always fills first, the condensed electron configuration of Na is $[Ne]3s^1$. Just as we write condensed electron configurations that focus on the valence shell, we can draw condensed orbital diagrams that do the same thing. The one for sodium is

Sodium: [Ne] [↑]
$3s$

This diagram reinforces the message that a sodium atom has a neon core of 10 electrons plus one more in its $3s$ orbital. Sodium has the same ns^1 valence-shell configuration as lithium and hydrogen, where n is the principal quantum number of the outermost shell. This pattern holds throughout the periodic table: all the elements in a given group have the same generic valence-shell electron configuration. For instance, the electron configuration of magnesium ($Z = 12$) is $[Ne]3s^2$,

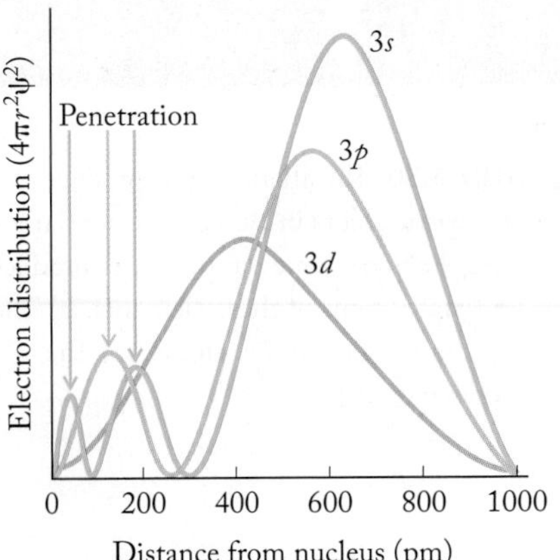

FIGURE 3.29 Radial distribution profiles of electrons in $3s$, $3p$, and $3d$ orbitals. The highlighted maxima in the $3s$ and $3p$ profiles mean that electrons in these orbitals penetrate closer to the nucleus, are less shielded by the electrons in filled inner shells, and have lower energy than $3d$ electrons.

(a)

(b)

FIGURE 3.30 The alkali metals produce characteristic colors in Bunsen burner flames because the high flame temperatures produce excited-state atoms of these elements. (a) The yellow-orange glow of Na atoms. (b) The lavender color of K atoms.

and the condensed electron configuration of every other element in group 2 consists of the immediately preceding noble gas core followed by ns^2.

CONCEPT TEST

What is the generic valence-shell configuration of the halogen (group 17) elements?

(Answers to Concept Tests are in the back of the book.)

The next six elements in the periodic table—aluminum, [Ne]$3s^23p^1$, through argon, [Ne]$3s^23p^6$—contain increasing numbers of $3p$ electrons until they achieve a filled $3p$ subshell in argon. As noted when we discussed helium, the filled $3s$ and $3p$ subshells of argon impart a chemical stability that is in keeping with the other noble gas elements.

Before leaving the third row, let's revisit the condensed electron configuration of sodium, [Ne]$3s^1$. This configuration represents a ground-state sodium atom because all of the electrons, and most importantly its valence electron, occupy the lowest energy orbitals available. Now think back to the discussion about atomic emission spectra in Section 3.2 and the distinctive yellow-orange glow that sodium makes in the flames of Bunsen burners, as illustrated previously in Figure 3.5 and here again in Figure 3.30(a). Each sodium atom absorbs a quantum of energy from the Bunsen burner that raises the valence electron from the ground state to an excited state (a transition represented by the yellow-orange arrows in Figure 3.31). The easiest excited state to populate is the one with the smallest energy above the ground state. The $3p$ orbitals fill after the $3s$ orbital, because they have the next lowest energy, so the lowest-energy (or *first*) excited state of sodium is the one in which its $3s$ electron has moved up to a $3p$ orbital. This excited state has the electron configuration [Ne]$3p^1$ (see Figure 3.31) and also has a very short lifetime. The electron typically takes less than a nanosecond to fall back to the ground state in a transition represented by the left-pointing yellow-orange arrow in Figure 3.31. This transition emits a quantum of energy ($h\nu$) equal to the difference in energy between the $3p$ and $3s$ orbitals in a Na atom—the energy of a photon of yellow-orange light.

CONCEPT TEST

Which of the following could describe the radiation emitted by sodium atoms initially in an excited state with the electron configuration [Ne]$4p^1$ as they transition to the ground state? (a) ultraviolet radiation; (b) the same yellow-orange light in Figure 3.5; (c) red light; (d) infrared radiation

(Answers to Concept Tests are in the back of the book.)

Potassium ($Z = 19$) is the first element of the fourth row. Like all group 1 elements, its generic valence-shell electron configuration is ns^1. For potassium, that translates into the condensed electron configuration [Ar]$4s^1$. Similarly, the condensed electron configuration of the next element, calcium ($Z = 20$), is [Ar]$4s^2$. At this point, the $4s$ orbital is filled. However, the $3d$ subshell *is still*

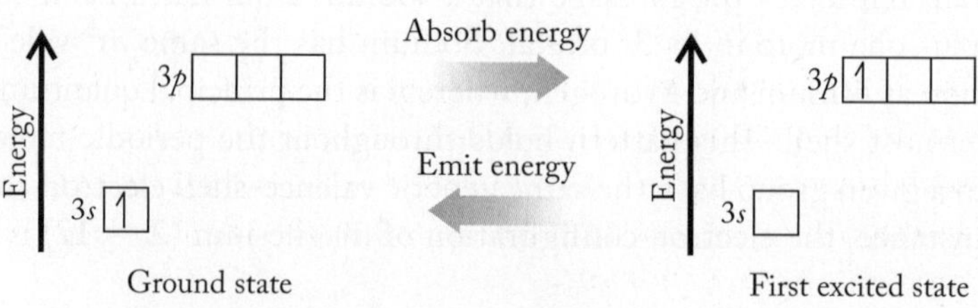

FIGURE 3.31 A ground-state Na atom absorbs a quantum of energy as its valence electron moves from the $3s$ orbital to a $3p$ orbital. This $3p$ electron in the excited-state atom spontaneously falls back to the empty $3s$ orbital, emitting a photon of yellow-orange light. The energy of the photon exactly matches the difference in energy between the $3p$ and $3s$ orbitals of Na atoms.

empty. Why does the 4*s* orbital fill before the 3*d* subshell? The answer to that question is linked to the relative energies of electrons in the orbitals. While it is true that energy levels increase with increasing *n* values, the increases get smaller as *n* values get larger, as shown in Figure 3.32. The differences in energy between orbitals in the third and fourth shells are so small that the 3*d* subshell is slightly higher in energy than the 4*s* orbital. Therefore, the 4*s* orbital fills first.

Only after the 4*s* orbital is full do the 3*d* orbitals begin to fill, starting with scandium ($Z = 21$), the first transition metal in the fourth row. Scandium has the condensed electron configuration $[Ar]3d^14s^2$. Note how the valence-shell orbitals are arranged in order of increasing principal quantum number in that electron configuration, not in the order in which they were filled. The reason we use such sequences will become clear in the next section, where we discuss which valence electrons are lost when transition metals form cations.

The 3*d* orbitals are filled in the fourth-row transition metals from scandium (group 3) to zinc (group 12). This pattern of filling the *d* orbitals of the shell whose principal quantum number is one less than the row number, abbreviated $(n - 1)d$, is followed throughout the periodic table. Thus, the 4*d* orbitals are filled in the transition metals of the fifth row, the 5*d* orbitals in the sixth row, and so on, as shown in Figure 3.33.

Titanium ($Z = 22$) has one more 3*d* electron than scandium, so it has the condensed electron configuration $[Ar]3d^24s^2$. At this point, you may feel that you can accurately predict the electron configurations of the remaining transition metals in the fourth period. However, because the energies of the 3*d* and 4*s* orbitals are so close together (see Figure 3.32), the sequence of *d*-orbital filling sometimes deviates from the pattern you might expect. The first deviation appears in the electron configuration of chromium ($Z = 24$). You might expect it to be $[Ar]3d^44s^2$, but it is $[Ar]3d^54s^1$ instead. The reason for this difference is that $[Ar]3d^54s^1$ puts one electron in each of the five *d* orbitals:

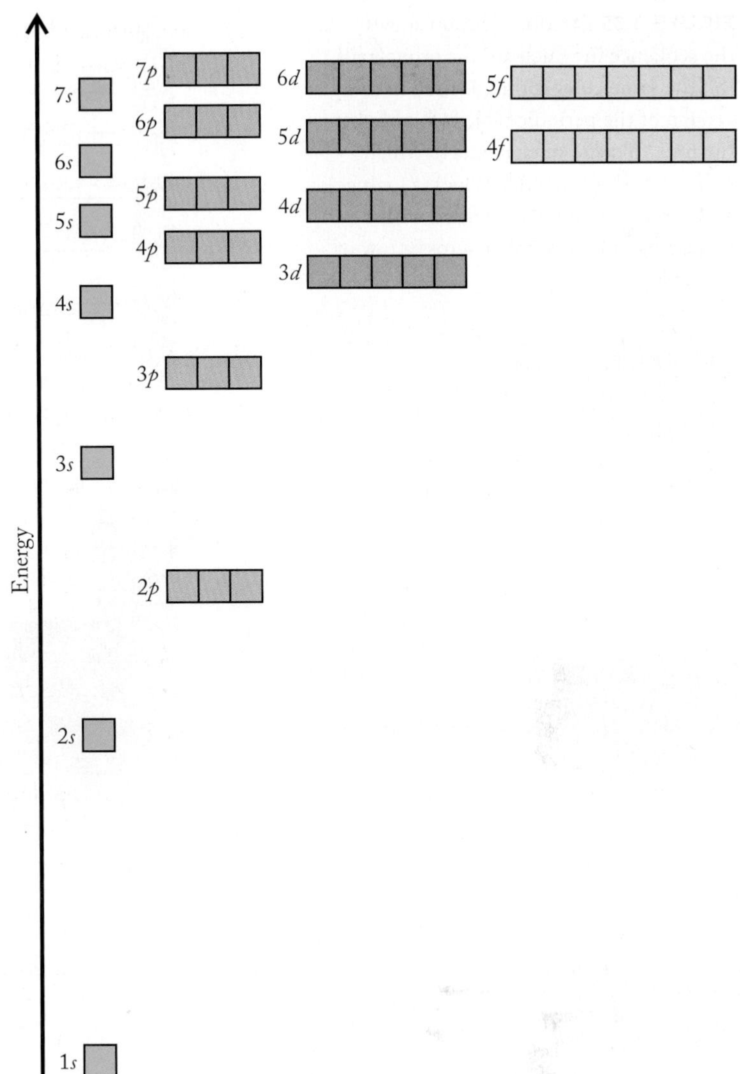

FIGURE 3.32 The energy levels in multielectron atoms increase with increasing values of *n* and with increasing values of ℓ within a shell. The difference in energy between adjacent shells decreases with increasing values of *n*, which may cause the energies of subshells in two adjacent shells to overlap. For example, electrons in 3*d* orbitals have slightly higher energy than those in the 4*s* orbital, resulting in the order of subshell filling $4s \rightarrow 3d \rightarrow 4p$.

Chromium: [Ar] [1|1|1|1|1] [1] ($3d^5$, $4s^1$)

This half-filled set of *d* orbitals represents a lower-energy electron configuration than $[Ar]3d^44s^2$. Its stability compensates for the energy needed to raise a 4*s* electron to a 3*d* orbital.

Another deviation from the expected filling pattern happens near the end of each row of transition metals. For example, the electron configuration of copper ($Z = 29$) is $[Ar]3d^{10}4s^1$ instead of $[Ar]3d^94s^2$ because the electron configuration with a completely filled *d* subshell represents a lower-energy configuration. By zinc ($Z = 30$), the 3*d* subshell is full and the next six electrons are added to 4*p* orbitals to reach the end of the fourth row, giving krypton ($Z = 36$) the condensed electron configuration $[Ar]3d^{10}4s^24p^6$. The pattern we have just described for the fourth row is repeated, though with additional deviations from the expected pattern, in the fifth row. In the fifth row, the deviations are due to the similar energies of 5*s* and 4*d* orbitals.

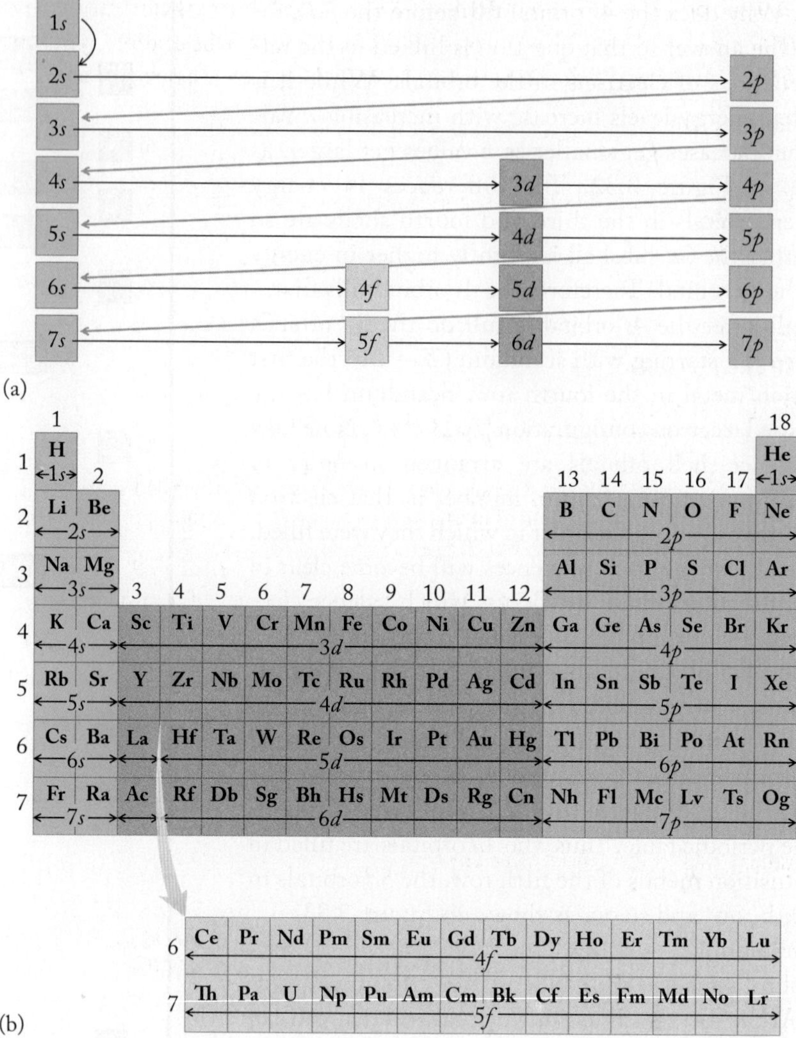

FIGURE 3.33 (a) This diagram shows the sequence in which atomic orbitals fill. (b) The same color coding is used in this version of the periodic table to highlight the four "blocks" of elements in which valence-shell *s* (green), *p* (blue), *d* (orange), and *f* (purple) orbitals are filled with increasing atomic number across a row of the table.

Figure 3.33 illustrates the overall orbital filling pattern just described. It also shows how the periodic table can be used to predict the electron configurations of the elements. The color patterns and labels in Figure 3.33 indicate which type of orbital is filled going across each row from left to right. For example, groups 1 and 2 are called *s* block elements because their outermost electrons are in *s* orbitals. Groups 13–18 (except for helium) are *p* block elements because their outermost electrons are in *p* orbitals. Note how the principal quantum numbers of the outermost orbitals in the *s* and *p* blocks match their row numbers: 2*s* and 2*p* in row 2, 3*s* and 3*p* in row 3, and so on.

The transition metals in groups 3–12 make up the *d* block. The principal quantum number of the outermost *d* orbitals in each row, starting with row 4, is always one less than the row number. The lanthanides (Z = 58 through 71) and actinides (Z = 90 through 103) at the bottom of the periodic table make up the *f* block. The principal quantum number of the outermost *f* orbitals in each of these rows is always *two* less than the row number of the element that precedes them. This means, for example, that the 4*f* orbitals, starting with cerium (Z = 58), do not fill until after the 6*s* orbital is full. The bottom two rows are each 14 elements long, because each *f* subshell contains seven orbitals ($\ell = 3$, $m_\ell = -3, -2, -1, 0, +1, +2$, and $+3$) that can hold up to 14 electrons.

With these patterns in mind, let's write the condensed electron configuration of lead (Z = 82), which is the group 14 element in the sixth row. The nearest noble gas above it is xenon ($Z = 54$). The difference in atomic numbers means that we need to account for $82 - 54 = 28$ electrons in the electron configuration symbols. The block labels in Figure 3.33 tell us that the 28 electrons are distributed as follows:

2 electrons in $6s$ 10 electrons in $5d$
14 electrons in $4f$ 2 electrons in $6p$

Therefore, the condensed electron configuration of lead is

$$\text{Pb:} \quad [\text{Xe}]4f^{14}5d^{10}6s^{2}6p^{2}$$

SAMPLE EXERCISE 3.10 Writing Electron Configurations **LO6**

Write the condensed electron configuration of silver ($Z = 47$).

Collect and Organize In a condensed electron configuration, the filled sets of inner-shell orbitals are represented by the atomic symbol of the noble gas immediately preceding the element of interest in the periodic table.

Analyze Silver ($Z = 47$) is the group 11 element in the fifth row of the periodic table. The noble gas at the end of the preceding row is krypton ($Z = 36$). The difference between the atomic numbers (47 – 36) means that we need to assign 11 electrons to orbitals.

Solve We would initially predict the first two of the 11 electrons would be in the $5s$ orbital and the next nine would be in $4d$ orbitals, resulting in a condensed electron configuration of $[\text{Kr}]4d^{9}5s^{2}$. However, a completely filled set of d orbitals is more stable than a partially filled set, as we saw for copper, the element just above Ag in the periodic table. Therefore, silver has 10 electrons in its $4d$ orbitals and only one in its $5s$ orbital, making its condensed electron configuration $[\text{Kr}]4d^{10}5s^{1}$.

Think About It We can generate a tentative electron configuration by simply moving across a row in Figure 3.33 until we come to the element of interest. However, in the transition metals, we have to remember the special stability of half-filled and full d orbitals and make the appropriate adjustments to our configuration.

Practice Exercise Write the condensed electron configuration of cobalt ($Z = 27$).

(Answers to Practice Exercises are in the back of the book.)

3.9 Electron Configurations of Ions

To write the electron configuration of monatomic ions, we begin with the electron configuration of the parent element. If the ion has a positive charge, we remove the appropriate number of electrons from the orbital(s) with the highest principal quantum number. If the ion has a negative charge, we add the appropriate number of electrons to one or more partially filled outer-shell orbitals.

Ions of the Main Group Elements

The s block elements (see Figure 3.33) form monatomic cations by losing all their outer-shell electrons, leaving their ions with the electron configuration of the noble gas immediately preceding them in the periodic table. For example, an atom of sodium forms a Na^+ ion by losing its single $3s$ electron:

$$\text{Na} \rightarrow \text{Na}^{+} + \text{e}^{-}$$
$$[\text{Ne}]3s^{1} \rightarrow [\text{Ne}] + \text{e}^{-}$$

isoelectronic describes atoms or ions that have identical electron configurations.

An element of the *p* block that forms a monatomic anion does so by gaining enough electrons to completely fill its valence-shell *p* orbitals. The ion it forms has the electron configuration of the noble gas at the end of its row in the periodic table. For example, an atom of fluorine ($[He]2s^22p^5$) forms a fluoride (F^-) ion by gaining one electron, which fills its set of three 2*p* orbitals and gives it the electron configuration of neon:

$$F + e^- \rightarrow F^-$$
$$[He]2s^22p^5 + e^- \rightarrow [He]2s^22p^6 = [Ne]$$

Thus, a Na^+ ion and a F^- ion have the same electron configuration as an atom of Ne. We say that the three species, Na^+, F^-, and Ne, are **isoelectronic**, meaning that they have the same electron configuration.

SAMPLE EXERCISE 3.11 Determining Isoelectronic Species in Main Group Ions — LO6

a. Determine the electron configuration of each of the following ions: Mg^{2+}, Cl^-, Ca^{2+}, and O^{2-}.
b. Which ions in part a are isoelectronic with neon?

Collect and Organize In part a, we are asked to determine the electron configurations of four ions. In part b, we are asked to identify which of the part a ions have the same electron configuration as neon, which has 10 electrons filling its $n = 1$ and $n = 2$ shells.

Analyze The ions include

- two from group 2, Mg and Ca, which form 2+ cations;
- one from group 16, O, which forms a 2− anion;
- one from group 17, Cl, which forms a 1− anion.

Let's arrange the atoms and ions in a table in order of increasing atomic number, remembering that atoms form cations by losing electrons and form anions by gaining them:

Element	Electron Configuration	Atomic Number (*Z*)	Charge on Ion	Electrons/Ion
O	$[He]2s^22p^4$	8	2−	10
Mg	$[Ne]3s^2$	12	2+	10
Cl	$[Ne]3s^23p^5$	17	1−	18
Ca	$[Ar]4s^2$	20	2+	18

Solve

a. The electron configurations for the ions are

O^{2-} $[He]2s^22p^6$ (same as Ne)
Mg^{2+} [Ne]
Cl^- $[Ne]3s^22p^6$ (same as Ar)
Ca^{2+} [Ar]

b. Two of the ions formed—O^{2-} and Mg^{2+}—are isoelectronic with Ne and with each other. The other two—Cl^- and Ca^{2+}—are isoelectronic with Ar and with each other.

Think About It Each of the ions has the stability we associate with the electron configuration of a noble gas.

Practice Exercise Write the electron configurations of K^+, I^-, Ba^{2+}, S^{2-}, and Al^{3+}. Which of these ions are isoelectronic with Ar?

(Answers to Practice Exercises are in the back of the book.)

Transition Metal Cations

As with the main group elements, writing the electron configurations of transition metal cations begins with the atoms from which the cations form. Nickel atoms, like those of many transition metals, form ions with 2+ charges by losing both electrons from their valence-shell *s* orbital:

$$Ni \rightarrow Ni^{2+} + 2e^-$$

$$[Ar]3d^84s^2 \rightarrow [Ar]3d^8 + 2e^-$$

We might have expected Ni atoms to form cations by losing 3*d* electrons, reasoning that the last orbitals to be filled are the highest in energy and so should be the first to be emptied. This does not happen for Ni and the other transition metals. Among the reasons are the following:

- The differences in energy between the valence-shell *s* orbitals (*ns*) and the $(n - 1)d$ orbitals of transition metals are very small.
- As the $(n - 1)d$ orbitals fill with increasing atomic number, the effective nuclear charge (Z_{eff}) felt by their electrons increases more than the Z_{eff} felt by the *ns* electrons. This happens because the $(n - 1)d$ electrons in a transition metal atom are shielded less than the *ns* electrons in the next higher shell. As a result, the *ns* electrons have higher energy and ionize first.

The rule that the electrons in orbitals with the highest *n* value ionize first applies to transition metals, too. Preferential loss of outer-shell *s* electrons explains why the most frequently encountered charge on transition metal ions is 2+.

Many transition metals form ions with charges greater than 2+ by losing *d* electrons in addition to their valence-shell *s* electrons. Atoms of scandium, $[Ar]3d^14s^2$, for example, lose both 4*s* electrons *and* their single 3*d* electron as they form Sc^{3+} ions. The chemistry of titanium, $[Ar]3d^24s^2$, is dominated by its tendency to lose two 4*s* and two 3*d* electrons, forming Ti^{4+} ions. In general, transition metals form 1+ and 2+ cations by losing all of their valence-shell *s* electrons (some, such as Ag, have only one to lose). They form cations with charges greater than 2+ by also losing $(n - 1)d$ electrons.

SAMPLE EXERCISE 3.12 Writing Electron Configurations of Transition Metal Ions **LO6**

What are the electron configurations of Fe^{2+} and Fe^{3+}?

Collect and Organize We are asked to write the electron configuration of two ions formed by iron ($Z = 26$). Iron is the group 8 element of the fourth row of the periodic table. Transition metals preferentially lose their outermost *s* electrons when they form ions.

Analyze The location of iron on the periodic table means that it has two 4*s* and six 3*d* electrons built on an argon core:

$$Fe: \quad [Ar]3d^64s^2$$

Solve We remove the two 4*s* electrons to form Fe^{2+}. We remove the two 4*s* electrons and one of the 3*d* electrons to form Fe^{3+}:

$$Fe^{2+}: \quad [Ar]3d^6 \qquad Fe^{3+}: \quad [Ar]3d^5$$

Think About It The electron configuration of Fe^{3+} means that each of its $3d$ orbitals contains one electron. This half-filled set of $3d$ orbitals should have enhanced stability. Actually, both Fe^{2+} and Fe^{3+} are common in nature. Most of the iron in your blood and tissues is Fe^{2+}.

Practice Exercise
Write the electron configurations for Mn^{3+} and Mn^{4+}.

(Answers to Practice Exercises are in the back of the book.)

CONCEPT TEST

The electron configuration of Eu ($Z = 63$) is [Xe]$4f^7 6s^2$, but the electron configuration of Gd ($Z = 64$) is [Xe]$4f^7 5d^1 6s^2$. Suggest a reason why the additional electron in Gd is not in a $4f$ orbital, which would result in the electron configuration [Xe]$4f^8 6s^2$.

(Answers to Concept Tests are in the back of the book.)

Bond length 198 pm

99 pm

(a) Radius of Cl

372 pm

186 pm

(b) Metallic radius of Na

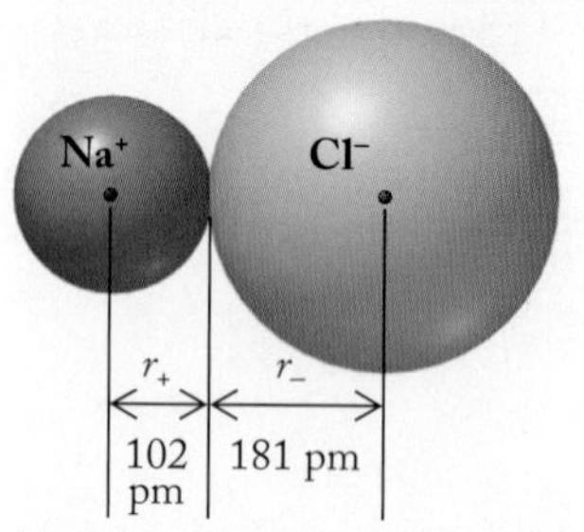

(c) Ionic radii of Na^+ and Cl^-

FIGURE 3.34 A comparison of atomic, metallic, and ionic radii. (a) An atomic radius is half the distance between identical nuclei in a molecule, such as the distance between chlorine nuclei in Cl_2. (b) A metallic radius is based on the distance of closest approach of adjacent atoms in a solid metal. (c) An ionic radius is determined by a series of comparisons among ionic compounds containing the ion of interest.

3.10 The Sizes of Atoms and Ions

In the previous sections we learned how electrons populate orbitals with higher values of n as we go down a column of elements in the periodic table. Earlier in the chapter we learned that electrons in orbitals with higher values of n are farther from the nucleus. For example, the radial distribution profiles in Figure 3.23 show how the distances from the nucleus to electrons in s orbitals increase with increasing principal quantum number n. In this section we explore how those ideas and the concept of effective nuclear charge can explain the relative sizes of the atoms and the common monatomic ions of the elements.

We usually express the sizes of atoms in terms of their radii. Given the wave-like behavior of electrons, the "edge" of a single atom cannot be exactly defined, so we must look at interactions between atoms to determine their radii. The atomic radius of an element that exists as a diatomic molecule, such as N_2, O_2, or Cl_2, is simply half the distance between the nuclear centers in the molecule (Figure 3.34a). The atomic radius of a metal, also called its *metallic* radius, is half the distance between the nuclear centers in the solid metal (Figure 3.34b). The values of ionic radii are derived from the distances between nuclear centers in solid ionic compounds (Figure 3.34c).

Trends in Atomic Size

Periodic trends in the relative sizes of the atoms of the main group elements are shown by the sizes of the spheres in Figure 3.35. Note how the sizes of atoms *increase* as we go down a group of the elements in the table (and their atomic numbers *increase*). Note, too, how the sizes of the atoms *decrease* as we move from left to right across a row, even though atomic number *increases* sequentially at the same time. Let's discuss both of these trends.

The value of the principal quantum number n of an orbital determines the most probable distance from the nucleus of the electrons in that orbital. As n increases, this distance increases. This factor explains the observed increase in atomic size with increasing atomic number (Z) in a group of elements. As Z increases, however, the positive charge of the nucleus increases, too. If electrons experience a stronger

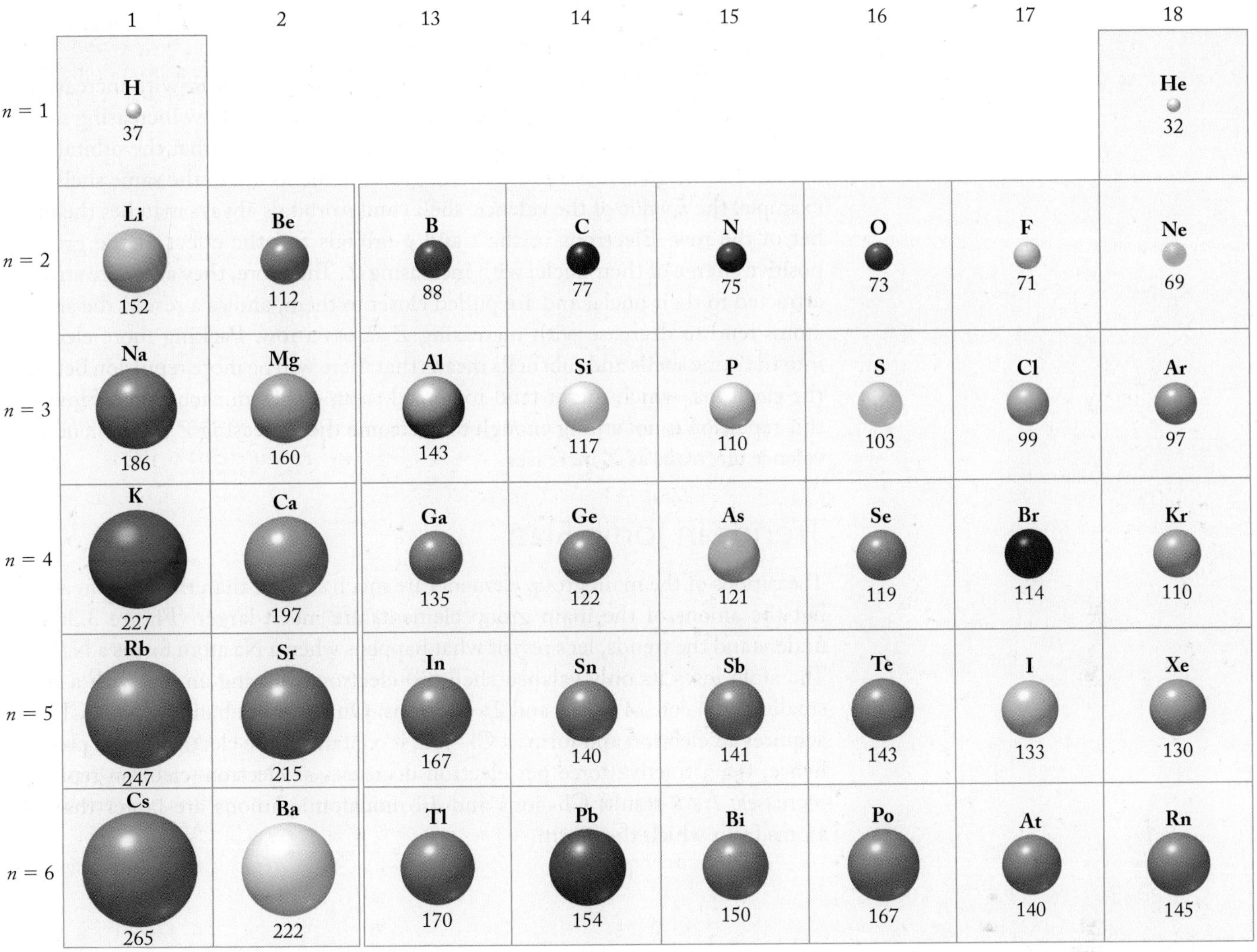

	1	2	13	14	15	16	17	18
$n = 1$	H 37							He 32
$n = 2$	Li 152	Be 112	B 88	C 77	N 75	O 73	F 71	Ne 69
$n = 3$	Na 186	Mg 160	Al 143	Si 117	P 110	S 103	Cl 99	Ar 97
$n = 4$	K 227	Ca 197	Ga 135	Ge 122	As 121	Se 119	Br 114	Kr 110
$n = 5$	Rb 247	Sr 215	In 167	Sn 140	Sb 141	Te 143	I 133	Xe 130
$n = 6$	Cs 265	Ba 222	Tl 170	Pb 154	Bi 150	Po 167	At 140	Rn 145

FIGURE 3.35 Atomic radii in picometers of the main group elements. Size generally increases from top to bottom in any group and generally decreases from left to right across any period.

nuclear charge, they should be pulled closer to the nucleus, shrinking the size of the atom. That does *not* happen as we go down a group of elements, but why not?

The answer has to do with the effective nuclear charge (Z_{eff}). As we discussed in Section 3.8, the negative charges of electrons in the filled inner shells of an atom shield electrons in the outermost shell from much of the positive nuclear charge. Shielding means that the Z_{eff} experienced by outer-shell (valence) electrons can be a tiny fraction of the total nuclear charge, Z. Therefore, the valence electrons experience a relatively weak attraction to the nucleus. Thus, the atomic radii of elements in the same group of the periodic table increase with increasing Z as we move from top to bottom. The atomic radii of the halogens in group 17 exemplify this trend:

Element	Atomic Number	*n* Value of Valence Shell	Atomic Radius (pm)
F	9	2	71
Cl	17	3	99
Br	35	4	114
I	53	5	133
At	85	6	140

As you can see in Figure 3.35, this pattern holds for all the other main group elements.

These same concepts explain why the sizes of atoms *decrease* with increasing Z across a row of the periodic table, especially if we focus on how increasing atomic number means more positive nuclear charge. We have seen that the orbitals that are filled as we go across a row of main group elements are in the same shells. For example, the n value of the valence-shell s and p orbitals always matches the number of the row. Electrons in the s and p orbitals feel the effect of the growing positive charge of their nuclei with increasing Z. Therefore, they are more strongly attracted to their nuclei and are pulled closer to them, and as a result, the sizes of atoms tend to decrease with increasing Z across a row. Packing more electrons into the same shells and subshells means that there will be more repulsion between the electrons, which might tend to spread them out from each other. However, this repulsion is not strong enough to overcome the increasing Z_{eff} experienced by valence electrons as Z increases.

Trends in Ionic Size

The cations of the main group elements are much smaller than their parent atoms, but the anions of the main group elements are much larger (Figure 3.36). To understand the trends, let's revisit what happens when a Na atom forms a Na^+ ion. The atom loses its only valence-shell (3s) electron, forming an ion with a much smaller neon core of 1s, 2s, and 2p electrons. On the other hand, when a Cl atom acquires an electron and forms a Cl^- ion, it contains more electrons than protons; hence, the attractive force per electron decreases as electron–electron repulsion increases. As a result, Cl^- ions and all monatomic anions are larger than the atoms from which they form.

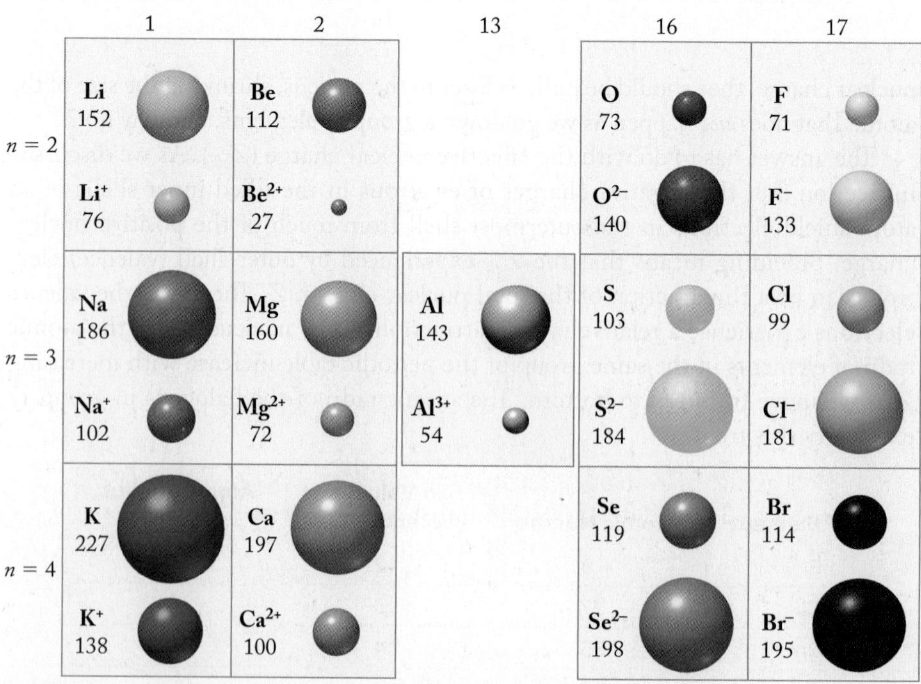

FIGURE 3.36 Comparison of atomic and ionic radii of selected main group elements, in picometers.

SAMPLE EXERCISE 3.13 Ranking Atoms and Ions by Size **LO7**

Using only the periodic table, arrange each set of particles by size, largest to smallest: (a) O, P, S; (b) Na^+, Na, K.

Collect and Organize We are asked to rank a set of three atoms based on their size and a set of two atoms and a cation of one of the atoms based on their size. The location of elements in the periodic table can be used to determine the relative sizes of their atoms.

Analyze All of the atoms and ions are of main group elements. Atomic radii of the elements decrease with increasing atomic number across rows because their valence electrons experience greater Z_{eff} as Z increases. Atomic radii increase with increasing atomic number down groups of elements because their valence electrons are in orbitals farther from the nucleus. Main group cations are always smaller than their parent atoms because they have lost their valence-shell electrons.

Solve

a. Sulfur is below oxygen in group 16, so S atoms are larger than O atoms. Phosphorus is to the left of sulfur in row 3, so P atoms are larger than S atoms. Therefore, the size order is $P > S > O$.

b. Cations are smaller than their parent atoms, so $Na > Na^+$. Size increases down a group of elements and K is below Na, so $K > Na$. Therefore, $K > Na > Na^+$.

Think About It The trend in the size of the atoms in set (a) reflects decreasing atomic size and increasing Z_{eff} across a row of elements in the periodic table, and increasing atomic size with increasing n values of their valence shells down a group. The relative sizes of the particles in set (b) are linked to increasing atomic size down a group of elements in the periodic table and to the smaller size of a cation relative to its parent atom.

Practice Exercise Arrange each set in order of increasing size (smallest to largest): (a) Cl^-, F^-, Na^+; (b) P^{3-}, Al^{3+}, Mg^{2+}.

(Answers to Practice Exercises are in the back of the book.)

3.11 Ionization Energies

We have seen how atomic emission spectra led to theories about the structure of atoms and the presence of quantized energy levels inside them. Another type of experimental evidence for energy levels in atoms and the electron configurations we have been exploring is obtained from the different energies required to remove electrons from atoms—that is, their ionization energies.

Ionization energy (IE) is the energy needed to remove 1 mole of electrons from 1 mole of gas-phase atoms or ions in their ground states. Removing the electrons always consumes energy because a negatively charged electron is always attracted to a positively charged nucleus, and energy is required to overcome that attraction. The amount of energy needed to remove 1 mole of electrons from 1 mole of atoms to make 1 mole of 1+ cations is called the *first ionization energy* (IE_1); the energy needed to remove 1 mole of electrons from 1 mole of 1+ cations to make 1 mole of 2+ cations is the *second ionization energy* (IE_2), and so forth. For example, for Mg atoms in the gas phase,

$$Mg \rightarrow Mg^+ + 1\,e^- \qquad IE_1 = 738 \text{ kJ/mol}$$
$$Mg^+ \rightarrow Mg^{2+} + 1\,e^- \qquad IE_2 = 1451 \text{ kJ/mol}$$

ionization energy (IE) the amount of energy needed to remove 1 mole of electrons from 1 mole of ground-state atoms or ions in the gas phase.

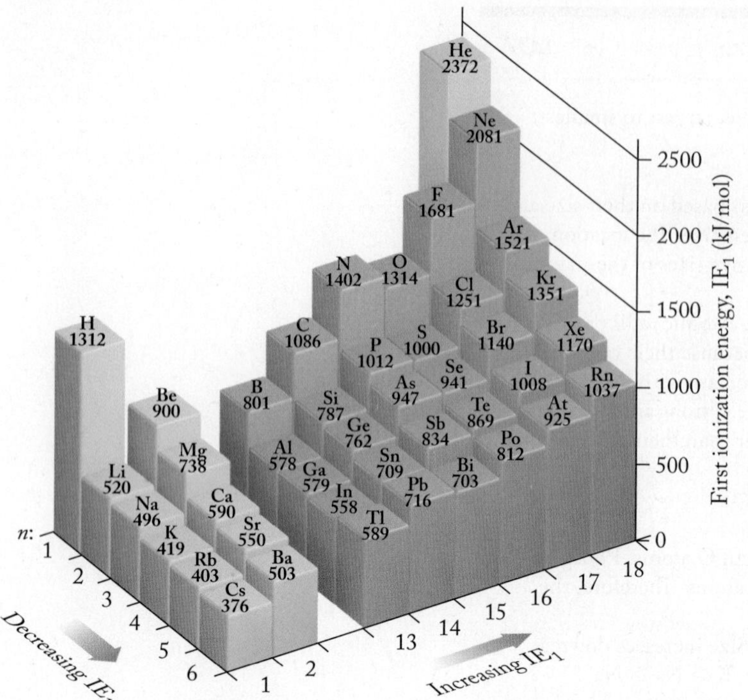

FIGURE 3.37 The first ionization energies of the main group elements generally increase from left to right in a period and decrease from top to bottom in a group.

The total energy required to make 1 mole of Mg^{2+} cations from 1 mole of Mg atoms in the gas phase is the sum of the two ionization energies:

$$\text{Mg} \rightarrow \text{Mg}^{2+} + 2\,\text{e}^-$$

$$\text{Total IE} = (738 + 1451)\ \text{kJ/mol} = 2189\ \text{kJ/mol}$$

Figure 3.37 shows how the first ionization energies vary in the main group elements. The IE_1 of hydrogen is 1312 kJ/mol, and the IE_1 of helium is nearly twice as big: 2372 kJ/mol. This difference seems reasonable because He atoms have two protons per nucleus, whereas H atoms have only one. In general, first ionization energies increase from left to right across a period; the easiest element to remove an electron from is in group 1, and the hardest is in group 18. This pattern arises because Z_{eff} increases with increasing atomic number across a row, and so do the strengths of the attractive forces between nuclei and valence electrons.

Two anomalies occur in the general trend of increasing IE_1 across each row. The first shows up between the group 2 and 13 elements that are next to each other in the second and third rows: there is a decrease in IE_1 between Be and B and between Mg and Al. The decrease occurs because B and Al lose a *p* electron when they ionize, whereas Be and Mg must lose an *s* electron. Recall from Figures 3.26 and 3.29 that electrons in an *s* orbital penetrate closer to the nucleus than electrons in the *p* orbitals in the same shell ($n > 2$). Therefore, a *p* electron experiences proportionally less Z_{eff} than an *s* electron in its shell and requires less energy to be removed from the atom.

The second anomaly occurs in the values of IE_1 between the group 15 and 16 elements. Recall that an electron lost by a group 15 element, such as nitrogen, was originally alone in a half-filled orbital:

N: [↑↓] [↑↓] [↑|↑|↑] → N^+: [↑↓] [↑↓] [↑|↑|] + e^-
1*s* 2*s* 2*p* 1*s* 2*s* 2*p*

However, the electron lost by a group 16 element, such as oxygen, was paired with another in a filled orbital:

O: [↑↓] [↑↓] [↑↓|↑|↑] → O^+: [↑↓] [↑↓] [↑|↑|↑] + e^-
1*s* 2*s* 2*p* 1*s* 2*s* 2*p*

Repulsion between the paired electrons raises their energy, which means less energy must be added to ionize one of them.

To see how ionization energy changes as we go down a group of elements in the periodic table, let's begin with group 1 and the IE_1 values of hydrogen and lithium. A lithium atom has three times the nuclear charge of a hydrogen atom, so we might expect its ionization energy to be about three times larger. However, its IE_1 value is only 520 kJ/mol, or less than half that of hydrogen (1312 kJ/mol), because the 2*s* electron in a lithium atom is shielded from the nucleus by the two electrons in the 1*s* orbital, so the Z_{eff} felt by the lithium valence electron is much less than 3+. In general, the combination of larger atomic size and more shielding by inner-shell electrons leads to decreasing IE_1 values from top to bottom in every group of the periodic table.

Another perspective on energy levels in atoms is provided by looking at the energies required to take more than one electron away from an atom. Consider

TABLE 3.2 Successive Ionization Energies[a] of the First 10 Elements

Element	Z	IE_1	IE_2	IE_3	IE_4	IE_5	IE_6	IE_7	IE_8	IE_9	IE_{10}
H	1	1312									
He	2	2372	5249								
Li	3	520	7296	12,040							
Be	4	900	1758	15,050	21,070						
B	5	801	2426	3660	24,682	32,508					
C	6	1086	2348	4617	6201	37,926	46,956				
N	7	1402	2860	4581	7465	9391	52,976	64,414			
O	8	1314	3383	5298	7465	10,956	13,304	71,036	84,280		
F	9	1681	3371	6020	8428	11,017	15,170	17,879	92,106	106,554	
Ne	10	2081	3949	6140	9391	12,160	15,231	19,986	23,057	115,584	131,236

[a]In kJ/mol.

the successive ionization energies for the first 10 elements, as shown in Table 3.2. For multielectron atoms, the energy required to remove an electron from an ion that has already lost one electron—that is, the second ionization energy, IE_2—is always greater than IE_1 because the second electron is being removed from an ion that already has a positive charge. Because electrons carry a negative charge, they are held more strongly in a cation than in the atom from which the cation is formed. The energy needed to remove a third electron is greater still because it is being removed from a 2+ ion.

Superimposed on this trend are much more dramatic increases in ionization energy (defined by the red line in Table 3.2) when all the valence electrons in ionizing atoms have been removed and the next electrons must come from inner shells. A core electron experiences much less shielding and a much greater Z_{eff}, so it requires much more energy to be removed from an atom than does an outer-shell electron. The impact of this energy difference between valence-shell and inner-shell electrons is illustrated in the first and second ionization energies of Na and Mg:

	Na	Mg
IE_1 (kJ/mol) (orbital of lost electron)	496 (3*s*)	738 (3*s*)
IE_2 (kJ/mol) (orbital of lost electron)	4562 (2*p*)	1451 (3*s*)

The IE_2 of Na is nearly 10 times its IE_1 because Na has only one 3*s* electron to lose. The second must come from an inner-shell 2*p* orbital. The IE_2 of Mg is only twice its IE_1 because the two electrons lost in forming a Mg^{2+} ion come from the same valence-shell 3*s* orbital.

SAMPLE EXERCISE 3.14 Recognizing Trends in Ionization Energies **LO8**

Without referring to Figure 3.37, arrange argon, magnesium, and phosphorus in order of increasing first ionization energy.

Collect, Organize, and Analyze We are asked to order three elements in the third row of the periodic table on the basis of their IE_1 values. First ionization energies generally increase from left to right across a row because the values of Z_{eff} increase.

electron affinity (EA) the energy change that occurs when 1 mole of electrons combines with 1 mole of atoms or ions in the gas phase.

Solve Assuming that increasing ionization energy with increasing atomic number is the dominant factor for these elements, their order of increasing first ionization energies should be

$$Mg < P < Ar$$

Think About It Magnesium forms stable 2+ cations, so we expect its ionization energy to be smaller than the values for phosphorus and argon, which do not form cations. Argon is a noble gas with a stable valence-shell electron configuration, so its first ionization energy should be the highest in the set. We can check our prediction against the IE_1 values in Figure 3.37: Mg, 738 kJ/mol; P, 1012 kJ/mol; Ar, 1521 kJ/mol.

Practice Exercise Without referring to Figure 3.37, arrange cesium, calcium, and neon in order from largest first ionization energy to smallest.

(Answers to Practice Exercises are in the back of the book.)

CONCEPT **TEST**

Why does ionization energy decrease with increasing atomic number down the group 18 elements?

(Answers to Concept Tests are in the back of the book.)

3.12 Electron Affinities

In the preceding section we examined the periodic nature of the energy required to remove electrons from atoms. Now we look at a complementary process and examine the change in energy when electrons are *added* to atoms to form monatomic anions. The energies involved are called **electron affinities (EA)**. They are the energy changes that occur when 1 mole of electrons is added to 1 mole of atoms or ions in the gas phase.

For example, the energy associated with adding 1 mole of electrons to 1 mole of chlorine atoms in the gas phase is

$$Cl + e^- \rightarrow Cl^- \quad EA = -349 \text{ kJ/mol}$$

The negative sign tells us that energy is lost, or released, when chlorine forms chloride ions. The same is true for most of the main group elements, as shown in Figure 3.38, because the association of a negatively charged electron with a positively charged nucleus should produce an ion that has less energy than the free atom and electron had before they came together.

Based on the EA values in Figure 3.38, the trends in EA are not as regular as the trends in atomic size and ionization energy. Electron affinities generally increase (become less negative) with increasing atomic number among the group 1 and group 17 elements (except for F and Cl), but other groups do not display a clear trend. In general, electron affinity becomes more negative with increasing atomic number across a row, but there are exceptions to that trend, too. The halogens of group 17 have the most negative EA values of the elements in each of their rows because of the relatively high Z_{eff} experienced by electrons in their valence-shell p orbitals (including

1	2	13	14	15	16	17	18
H −72.6							He (0.0)[a]
Li −59.6	Be >0	B −26.7	C −122	N +7	O −141	F −328	Ne (+29)[a]
Na −52.9	Mg >0	Al −42.5	Si −134	P −72.0	S −200	Cl −349	Ar (+35)[a]
K −48.4	Ca −2.4	Ga −28.9	Ge −119	As −78.2	Se −195	Br −325	Kr (+39)[a]
Rb −46.9	Sr −5.0	In −28.9	Sn −107	Sb −103	Te −190	I −295	Xe (+41)[a]
Cs −45.5	Ba −14	Tl −19.2	Pb −35.2	Bi −91.3	Po −183.3	At −270[a]	Rn (+41)[a]

[a]Calculated values.

FIGURE 3.38 Electron affinity (EA) values of main group elements are expressed in kilojoules per mole. The more negative the value, the more energy is released when 1 mole of atoms combines with 1 mole of electrons to form 1 mole of anions with a 1− charge. A greater release of energy reflects more attraction between the atoms of the elements and free electrons.

the electron each of their atoms acquires in becoming an anion) and because the anions they form have the stable electron configurations of noble gases.

On the other hand, adding an electron to a group 18 atom requires the addition of energy, because noble gas atoms have stable electron configurations already. Beryllium and magnesium have positive electron affinities because the added electrons have to occupy outer-shell *p* orbitals that are significantly higher in energy than the outer-shell *s* orbitals. We can also rationalize the positive electron affinity of nitrogen by noting that adding an electron to an N atom means that the atom loses the stability associated with a half-filled set of 2*p* orbitals:

$$\text{N: } \boxed{\uparrow\downarrow}_{1s}\ \boxed{\uparrow\downarrow}_{2s}\ \boxed{\uparrow\,|\,\uparrow\,|\,\uparrow}_{2p} + e^- \rightarrow \text{N}^-\text{: } \boxed{\uparrow\downarrow}_{1s}\ \boxed{\uparrow\downarrow}_{2s}\ \boxed{\uparrow\downarrow\,|\,\uparrow\,|\,\uparrow}_{2p}$$

CONCEPT TEST

Describe at least one similarity and one difference in the periodic trends in first ionization energies and electron affinities among the main group elements.

(Answers to Concept Tests are in the back of the book.)

To end this chapter, let's return to the first three decades of the 20th century, which saw remarkable advances in our understanding of the structure of atoms and how nature works at the atomic and subatomic levels. In this chapter we have tried to connect the advances, showing how one led to another (Figure 3.39), while also conveying a sense of how profoundly unsettling these new ideas about the laws of nature were to the leading scientists of the time.

Consensus in the scientific community on the ideas of quantum theory did not come easily. For example, we have seen how excited-state sodium atoms emit photons of yellow-orange light as they return to the ground state. Einstein puzzled over this phenomenon for several years before proposing that neither the exact moment when emission occurs nor the path of the emitted photon could be predicted exactly. He concluded that quantum theory allows us to calculate only the probability of a spontaneous electron transition; the details of the event are left to chance. In other words, no force of nature causes an excited-state sodium atom to emit a photon and return to a lower energy level at a particular instant.

This probabilistic view of the interaction of matter and energy was very different from the familiar cause-and-effect interactions involving energy exchange among large objects. For example, a marble rolling off a table drops immediately to the floor, yet an electron remains in an excited state for an indeterminate time before falling to a lower energy level. This lack of determinacy bothered Einstein and many of his colleagues. Had they discovered an underlying theme of nature—that some processes cannot be described or known with certainty? Are there fundamental limits to how well we can know and understand our world and the events that change it?

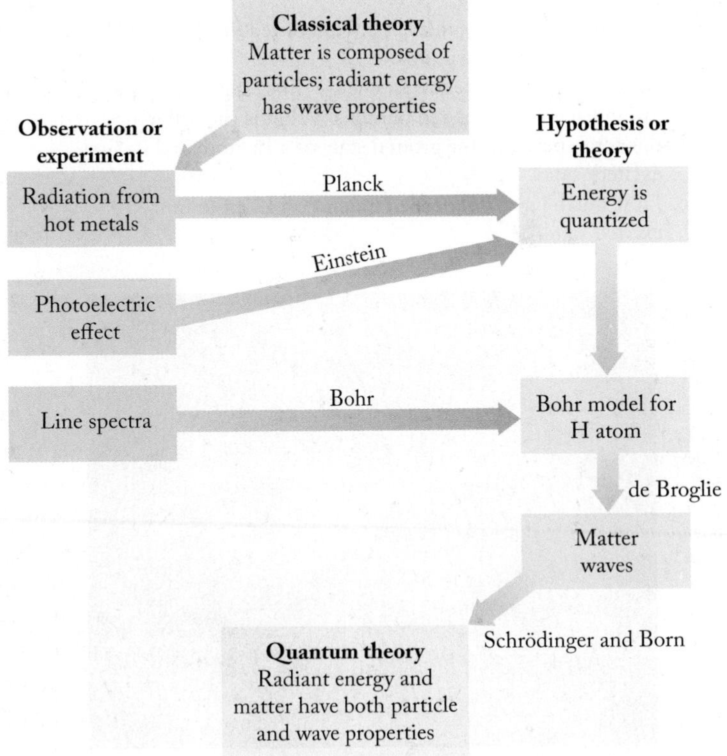

FIGURE 3.39 During the first three decades of the 20th century, quantum theory evolved from classical 19th-century theories of the nature of matter and energy. The arrows trace the development of modern quantum theory, which assumes that radiant energy has both wave properties and particle properties and that mass (matter) also has both wave properties and particle properties.

Many scientists in the early decades of the 20th century did not agree with this uncertain view of nature. They preferred the Newtonian view, where events occur for a reason and where there are clearly understood causes and effects. They believed that the more they studied nature with ever more sophisticated tools, the more they would understand why things happen the way they do. Soon after Max Born published his probabilistic interpretation of Schrödinger's wave functions in 1926, Einstein wrote Born a letter in which he contrasted the new theories with the certainties many people find in religious beliefs:

> *Quantum mechanics is very impressive. But an inner voice tells me that it is not yet the real thing. The theory produces a great deal but hardly brings us closer to the secret of the Old One. I am at all events convinced that He does not play dice.*[2]

SAMPLE EXERCISE 3.15 Integrating Concepts: Red Fireworks

The brilliant red color of some fireworks (Figure 3.40) is produced by the presence of lithium carbonate (Li_2CO_3) in shells that are launched into the night sky and explode at just the right moment. The explosions produce enough energy to vaporize the lithium compound and produce excited-state lithium atoms. The atoms quickly lose energy as they transition from their excited states to their ground states, emitting photons of red light that have a wavelength of 670.8 nm.

a. How much energy is in each photon of red light emitted by an excited-state lithium atom?
b. Lithium atoms in the lowest energy (first) state above the ground state emit the red photons. What is the difference in energy between the ground state of a Li atom and its first excited state?
c. How much energy is needed to ionize a Li atom in its first excited state?
d. What are the electron configurations of (1) a ground-state Li atom, (2) a first-excited-state Li atom, and (3) a ground-state Li^+ ion?

FIGURE 3.40 The red color of some fireworks is produced by the transition of electrons in lithium atoms from their first excited state to their ground state.

Collect and Organize We are given the wavelength of the red light produced by excited Li atoms and are asked to calculate the energy of the photons of this light, as well as the differences in energy between a ground-state and first-excited-state Li atom and between this excited-state Li atom and a ground-state Li^+ ion. We are also asked to write the electron configurations of the three states. The energy of a photon is related to its wavelength (λ) by Equation 3.4:

$$E = \frac{hc}{\lambda}$$

Lithium is the group 1 element in the second row of the periodic table. Its first ionization energy (Figure 3.37) is 520 kJ/mol (5.20×10^2 kJ/mol).

Analyze The energy of a photon (E) emitted by an excited-state atom is the same as the difference in energies (ΔE) between the excited state and the lower energy state (in this case, the ground state) of the transition that produced the photon. It takes 520 kJ to ionize one mole (6.0221×10^{23} atoms) of Li, so dividing 520 kJ by 6.0221×10^{23} gives us the energy required to ionize one Li atom. This should be greater than the energy needed to raise a Li atom to its first excited state, and the difference between the energies is that required to ionize the excited-state atom. We have seen in other calculations in this chapter that photons of visible light have energies in the 10^{-19} J range.

Solve

a. The energy of a photon with a wavelength of 670.8 nm is

$$E = \frac{hc}{\lambda} = \frac{(6.626 \times 10^{-34}\ \text{J} \cdot \text{s})(2.998 \times 10^{8}\ \cancel{\text{m}}/\text{s})}{670.8\ \cancel{\text{nm}}\left(\frac{10^{-9}\ \cancel{\text{m}}}{1\ \cancel{\text{nm}}}\right)}$$

$$= 2.961 \times 10^{-19}\ \text{J}$$

[2]Letter to Max Born, 12 December 1926; quoted in R. W. Clark, *Einstein: The Life and Times* (New York: HarperCollins, 1984), p. 880.

b. The difference in energies between the ground state of a Li atom and its lowest energy excited state must match exactly the energy of the photon emitted; therefore,

$$\Delta E = 2.961 \times 10^{-19}\ \text{J}$$

c. The energy required to ionize a ground-state Li atom (IE_1) is

$$IE_1 = \left(5.20 \times 10^2 \frac{\text{kJ}}{\text{mol}}\right)\left(\frac{1000\ \text{J}}{1\ \text{kJ}}\right)\left(\frac{1\ \text{mol}}{6.0221 \times 10^{23}\ \text{atoms}}\right)$$

$$= 8.635 \times 10^{-19}\ \text{J}$$

The energy required to ionize a Li atom that is in the first excited state (2.961×10^{-19} J above the ground state) is IE_1 less 2.961×10^{-19} J or

$$\begin{array}{r} 8.635 \times 10^{-19}\ \text{J} \\ -\ 2.961 \times 10^{-19}\ \text{J} \\ \hline 5.674 \times 10^{-19}\ \text{J or } 5.67 \times 10^{-19}\ \text{J} \end{array}$$

d. Li is the first atom in the second row of the periodic table, which means it has one electron in its 2*s* orbital. Therefore,

1. Ground-state Li atoms have the electron configuration $1s^22s^1$.
2. The first subshell (and electron energy level) above 2*s* is 2*p*, so the valence electron in a Li atom in the lowest excited state is in a 2*p* orbital, giving it the electron configuration $1s^22p^1$.
3. The valence electron of a Li atom is lost when it forms a Li^+ ion, which means that the electron configuration of a ground-state Li^+ ion is simply $1s^2$.

Think About It You might expect that the colors that lithium and other elements make in fireworks' explosions are the same ones they produce in Bunsen burner flames. You would be right. Many of these colors are produced by transitions from first excited states to ground states, because the lowest-energy excited state is the one most easily populated by the thermal energy available in a gas flame or an exploding fireworks shell.

SUMMARY

LO1 Visible light is one form of **electromagnetic radiation**. Like the other forms, it has wave properties described by characteristic **wavelengths (λ)** and **frequencies (ν)**, and it travels through a vacuum at the speed of light, $c = 2.998 \times 10^8$ m/s. (Section 3.1)

LO2 According to quantum theory there are discrete energy levels in atoms, which means they absorb or emit discrete amounts of energy called **photons**. Planck used these energy **quanta** to explain the radiation emitted by incandescent objects, and Einstein used them to explain the **photoelectric effect**. The radiant energy required to dislodge a photoelectron from a metal surface is called the **work function (Φ)** of the metal. (Section 3.3)

LO3 Free atoms in flames and in gas-discharge tubes produce **atomic emission spectra** consisting of narrow bright lines at characteristic wavelengths. When continuous radiation passes through atomic gases, absorption produces the dark lines of **atomic absorption spectra**. The bright lines of an element's emission spectrum and the dark lines of its absorption spectrum are at the same wavelengths. Balmer derived an equation that accounted for the bright lines in the visible emission spectrum of hydrogen and that predicted the existence of bright lines in the UV and IR regions of hydrogen's emission spectrum. Bohr proposed that the lines predicted by Balmer were related to energy levels occupied by electrons inside the hydrogen atom. A **ground-state** atom or ion has all its electrons in the lowest possible energy levels. Other arrangements are called **excited states**. **Electron transitions** from higher to lower energy levels in an atom cause the atom to emit particular frequencies of radiation; absorption of radiation of the same frequencies accompanies electron transitions from the same lower to higher energy levels. (Sections 3.2 and 3.4)

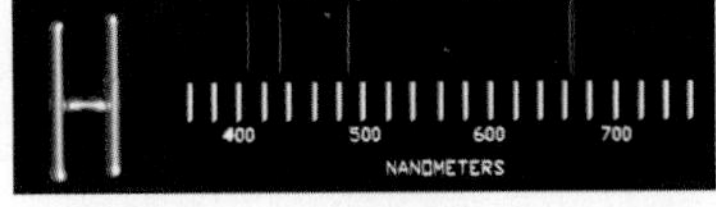

LO4 De Broglie proposed that electrons in atoms as well as all other moving particles have wave properties and can be treated as **matter waves**. He explained the stability of the electron orbits in the Bohr hydrogen atom in terms of **standing waves**: the circumferences of the allowed orbits had to be whole-number multiples of the hydrogen electron's characteristic wavelength. The **Heisenberg uncertainty principle** states that both the position and momentum of an electron cannot be precisely known at the same time. (Section 3.5)

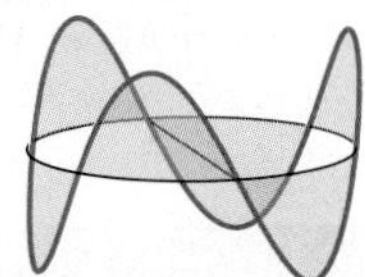

LO5 The solutions to **Schrödinger's wave equation** are mathematical expressions called **wave functions (ψ)** where ψ^2 defines the regions within an atom, called **orbitals**, where electron densities are high. Each orbital has a unique set of three **quantum numbers: principal quantum number *n***, which defines orbital size and energy level; **angular momentum quantum number ℓ**, which defines orbital shape; and **magnetic quantum number m_ℓ**, which defines orbital orientation in space. Two electrons in the same orbital have opposite **spin quantum numbers m_s**: $+\frac{1}{2}$ and $-\frac{1}{2}$. The **Pauli exclusion principle** states that no two electrons in an atom can have the same four values of n, ℓ, m_ℓ, and m_s. Orbitals have characteristic three-dimensional sizes, shapes, and orientations that are depicted by boundary–surface representations. All *s* orbitals are spheres that increase in size with increasing values of *n*. Each of the three *p* orbitals in any $n \geq 2$ shell has two lobes aligned along the *x*-, *y*-, or *z*-axis. The five *d* orbitals in $n \geq 3$ shells come in two forms: four are shaped like a four-leaf clover, and the fifth has two lobes oriented along the *z*-axis and a torus surrounding the middle of the two lobes. (Sections 3.6 and 3.7)

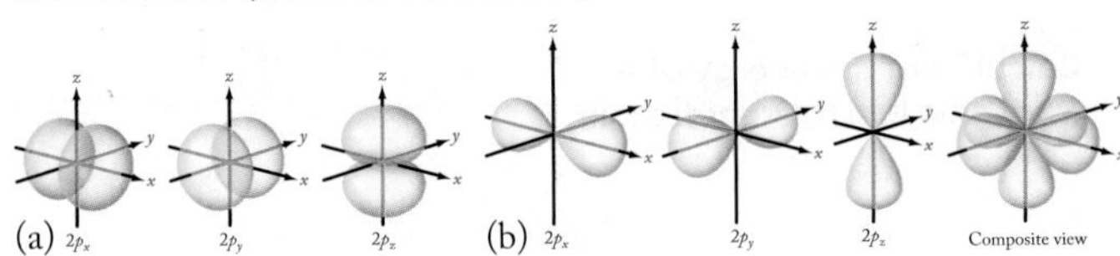

LO6 According to the **aufbau principle**, electrons fill the lowest-energy atomic orbitals of a ground-state atom first. **Effective nuclear charge (Z_{eff})** is the net nuclear charge felt by outer-shell electrons when they are shielded from the full nuclear charge by inner-shell electrons. Greater Z_{eff} means lower energy. An **electron configuration** is a set of numbers and letters expressing the number of electrons in each occupied orbital in an atom. The electrons in the outermost occupied shell of an atom are **valence electrons**. They are lost, gained, or shared in chemical reactions. All the orbitals of a given *p*, *d*, or *f* subshell are said to be degenerate, which means they all have the same energy. **Hund's rule** states that in any set of degenerate orbitals, each orbital must contain one electron before any orbital in the set can accept a second electron. We use **orbital diagrams** to show in detail how electrons are distributed among orbitals, which are represented by boxes. Atoms of group 1 and group 2 elements tend to lose electrons and form 1+ and 2+ cations, respectively. In so doing they become **isoelectronic** with the noble gas in the preceding period. Atoms in groups 16 and 17 tend to gain electrons to form 2− and 1− anions, respectively, thereby becoming isoelectronic with the noble gas at the end of their period.

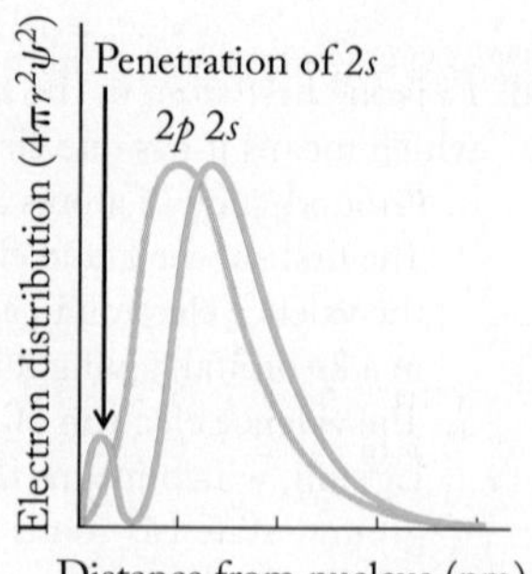

When transition metals form ions, the electrons are removed from the shell of highest *n* until the charge on the ion is achieved. (Sections 3.8 and 3.9)

LO7 The sizes of atoms increase with increasing atomic number in a group of elements because valence-shell electrons with higher *n* values are, on average, farther from the nucleus. However, the sizes of atoms decrease with increasing atomic number across a row of elements because the valence electrons experience higher effective nuclear charges. Anions are larger than their parent atoms due to additional electron–electron repulsion, but cations are smaller than their parent atoms—sometimes much smaller when all the electrons in the valence shell are lost. (Section 3.10)

LO8 **Ionization energy (IE)** is the amount of energy needed to remove 1 mole of electrons from 1 mole of atoms or ions in the gas phase. IE values generally increase with increasing effective nuclear charge across a row and decrease with increasing atomic number down a group in the periodic table. The energy differences between the different shells and subshells of atoms are reflected in the values of successive ionization energies (IE_1, IE_2, IE_3, . . .). **Electron affinity (EA)** is the energy change that occurs when 1 mole of electrons combines with 1 mole of atoms or ions in the gas phase. The EA values of many main group elements are negative, indicating that energy is released when they acquire electrons. (Sections 3.11 and 3.12)

PARTICULATE **PREVIEW WRAP-UP**

1. Model (a) shows electrons moving around the nucleus of atom, in keeping with the Rutherford model.
2. Concentric circles suggest electrons orbit the nucleus at different distances from it.
3. Model (a) suggests that electrons orbit the nuclei of atoms in concentric circular orbits; model (b) shows a more probabilistic picture of electron distribution, with regions of high electron densities at different distances from the nucleus.

PROBLEM-SOLVING SUMMARY

Type of Problem	Concepts and Equations		Sample Exercises
Calculating frequency from wavelength	$\nu = \frac{c}{\lambda}$	(3.2)	**3.1**
Calculating the energy of a photon	$E = \frac{hc}{\lambda}$	(3.4)	**3.2**
Using the work function	$\Phi = h\nu_0 = h\nu - KE_{electron}$	(3.5, 3.6)	**3.3**
Calculating the wavelength of a line in the hydrogen spectrum	$\lambda(\text{nm}) = \left(\frac{364.56\ m^2}{m^2 - n^2}\right)$	(3.7)	**3.4**
	$\frac{1}{\lambda} = R_H\left(\frac{1}{n_1^2} - \frac{1}{n_2^2}\right)$	(3.8)	
Calculating the energy of a transition in a hydrogen atom	$\Delta E = -2.178 \times 10^{-18}\ J\left(\frac{1}{n_{final}^2} - \frac{1}{n_{initial}^2}\right)$	(3.11)	**3.5**

Type of Problem	Concepts and Equations	Sample Exercises
Calculating the wavelength of a particle in motion	$\lambda = \frac{h}{mu}$ (3.12)	**3.6**
Calculating Heisenberg uncertainties	$\Delta x \cdot m\Delta u \geq \frac{h}{4\pi}$ (3.14)	**3.7**
Identifying the subshells and orbitals in an energy level and valid quantum number sets	n is the shell number; ℓ defines both the subshell and the type of orbital; an orbital has a unique combination of allowed n, ℓ, and m_ℓ values; ℓ is any integer from 0 to $(n - 1)$; m_ℓ is any integer from $-\ell$ to $+\ell$, including zero; m_s equals $+\frac{1}{2}$ or $-\frac{1}{2}$.	**3.8, 3.9**
Writing electron configurations of atoms and ions	Orbitals fill up in the following sequence: $1s^2$, $2s^2$, $2p^6$, $3s^2$, $3p^6$, $4s^2$, $3d^{10}$, $4p^6$, $5s^2$, $4d^{10}$, $5p^6$, $6s^2$, $4f^{14}$, $5d^{10}$, $6p^6$, $7s^2$, $5f^{14}$, $6d^{10}$, $7p^6$, where superscripts represent the maximum numbers of electrons. Orbitals in electron configurations are arranged based on increasing value of n. In forming transition metal ions, electrons in orbitals with the highest n value are removed first; there is enhanced stability in half-filled and filled d subshells.	**3.10–3.12**
Ordering atoms and ions by size	In general, sizes decrease left to right across a row and increase down a column of the periodic table; cations are smaller and anions are larger than their parent atoms.	**3.13**
Recognizing trends in ionization energies	First ionization energies increase across a row and decrease down a column.	**3.14**

VISUAL PROBLEMS

(Answers to boldface end-of-chapter questions and problems are in the back of the book.)

3.1. Which of the elements highlighted in Figure P3.1 consists of atoms with
a. a single s electron in their outermost shells? (More than one answer is possible.)
b. filled sets of s and p orbitals in their outermost shells?
c. filled sets of d orbitals?
d. half-filled sets of d orbitals?
e. two s electrons in their outermost shells?

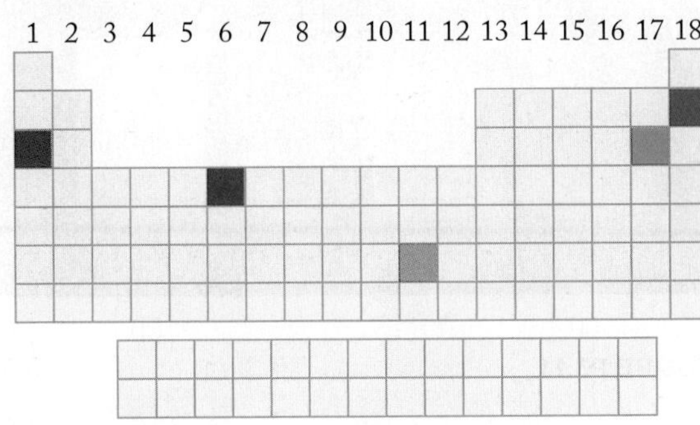

FIGURE P3.1

3.2. Which of the highlighted elements in Figure P3.1 has the greatest number of unpaired electrons per ground-state atom?

3.3. Which of the highlighted elements in Figure P3.1 form common monatomic ions that are (a) larger than their parent atoms and (b) smaller than their parent atoms?

3.4. Which of the highlighted elements in Figure P3.4 forms monatomic ions by
a. losing an s electron?
b. losing two s electrons?
c. losing two s electrons and a d electron?
d. adding an electron to a p orbital?
e. adding electrons to two p orbitals?

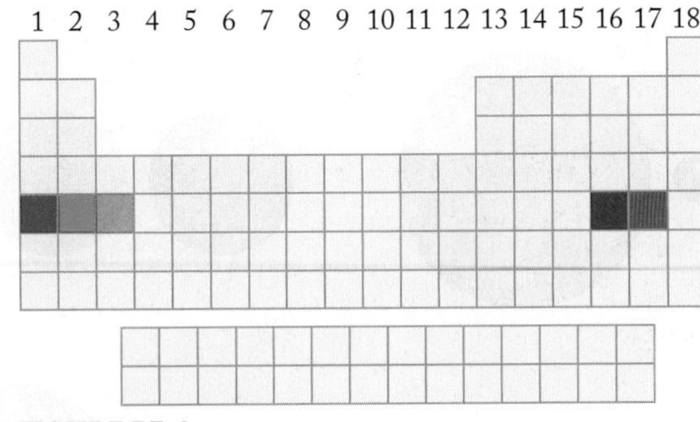

FIGURE P3.4

3.5. Which of the highlighted elements in Figure P3.4 form(s) common monatomic ions that are smaller than their parent atoms?

3.6. Rank the highlighted elements in Figure P3.4 based on (a) increasing size of their atoms and (b) increasing first ionization energy.

3.7. Suppose three beams of radiation are focused on a negatively charged metallic surface. The beam represented by the A waves in Figure P3.7 causes photoelectrons to be emitted from the surface. Which of the following statements accurately describes the abilities of the beams represented by waves B and C to also produce photoelectrons from this surface?

a. Neither the B nor the C wave can produce photoelectrons.
b. Both the B and C waves should produce photoelectrons if the sources of the waves are bright enough.
c. The B wave may or may not produce photoelectrons, but the C wave surely will.
d. The C wave may or may not produce photoelectrons, but the B wave surely will.

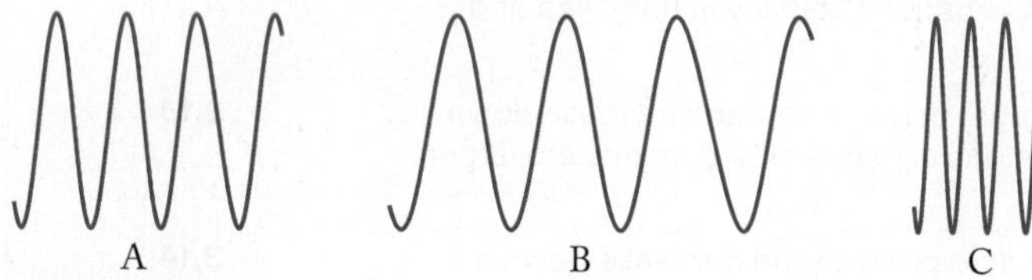

FIGURE P3.7

3.8. A group 2 element (M) and a group 17 element (X) have nearly the same atomic radius, as shown by the relative sizes of the blue and red spheres representing their atoms in Figure P3.8. Are the two elements in the same row of the periodic table? If not, which one is in the higher row?

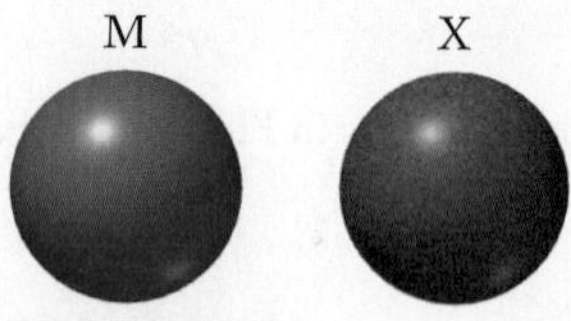

FIGURE P3.8

3.9. Which of the pairs of spheres in Figure P3.9 best depicts the relative sizes of the cation and anion formed by atoms of the elements M and X from Figure P3.8?

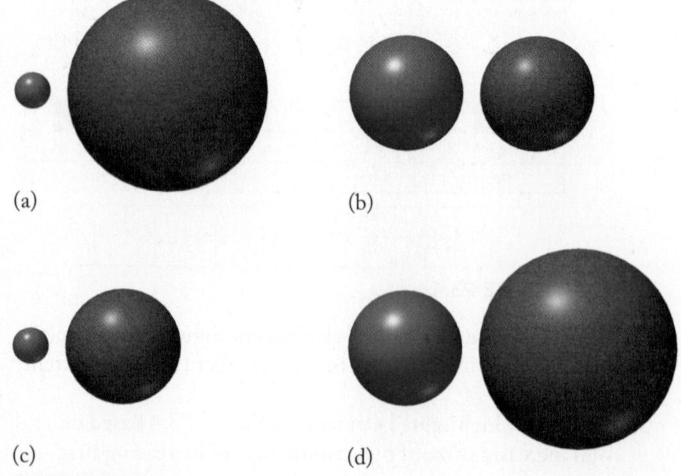

FIGURE P3.9

3.10. The arrows A, B, and C in Figure P3.10 show three transitions among the $n = 3$, 4, and 5 energy levels in a single-electron ion.

a. Assuming the transitions are accompanied by the loss or gain of electromagnetic energy, do the arrows depict absorption or emission of radiation?
b. The spectral lines of the transitions are, in order of increasing wavelength, 80, 117, and 253 nm. Which wavelength goes with which arrow?

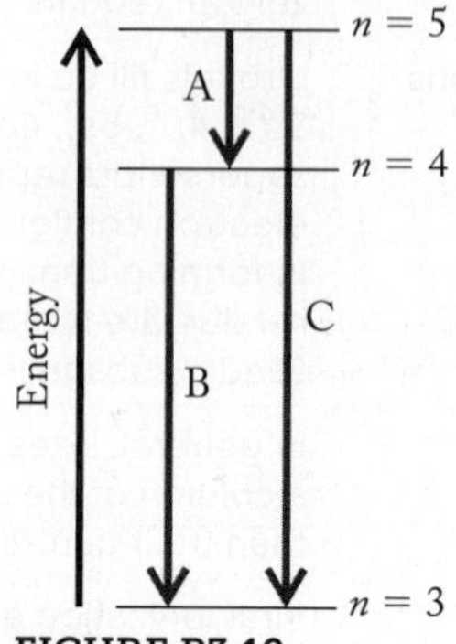

FIGURE P3.10

3.11. Quantum Dot Televisions Some flat panel displays and TV sets use quantum dots (nanocrystals of cadmium selenide and cadmium sulfide) to generate colors. The colors of the quantum dots in Figure P3.11 are determined by particle size and composition.

a. In which of the containers do the quantum dots emit photons with the greatest energies?
b. In which of the containers do the quantum dots emit photons with the longest wavelengths?
c. In which of the container are the quantum dots larger than in any of the others. (Hint: see Figure 3.14.)

FIGURE P3.11

3.12. Use representations [A] through [I] in Figure P3.12 to answer questions a–f.
   a. Between [A] and [I], which valence-shell *s* electron experiences the stronger attraction to its nucleus?
   b. Which representations depict the loss of an electron and which depict the gain of an electron?
   c. Which representation depicts quantized processes?
   d. Which representations depict the probabilistic nature of quantum mechanics?
   e. What processes are represented in [C], [D], [F], and [G]?
   f. Which representation is consistent with zero probability of an electron being in the nucleus of an atom?

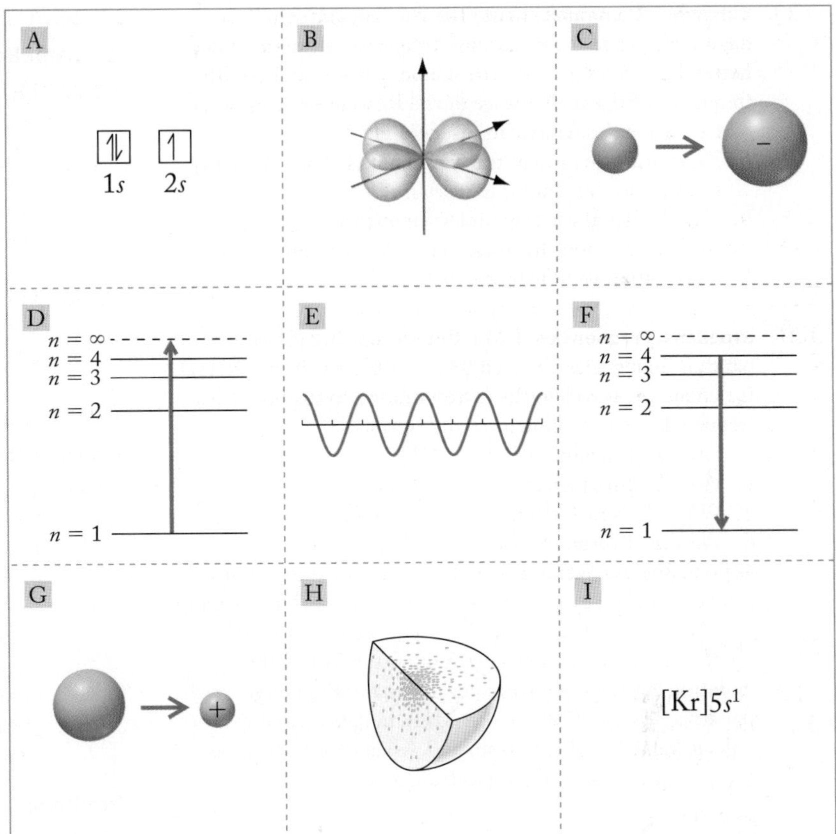

FIGURE P3.12

QUESTIONS AND PROBLEMS

Waves of Light

Concept Review

3.13. Why are the various forms of radiant energy called *electromagnetic* radiation?

3.14. Explain with a sketch why the frequencies of long-wavelength waves of electromagnetic radiation are lower than those of short-wavelength waves.

3.15. **Dental X-Rays** When X-ray images are taken of your teeth and gums in the dentist's office, your body is covered with a lead shield. Explain the need for this precaution.

3.16. **UV Radiation and Skin Cancer** Ultraviolet radiation causes skin damage that may lead to cancer, but exposure to infrared radiation does not seem to cause skin cancer. Why?

3.17. **Lava** As hot molten lava cools, it begins to solidify and no longer glows in the dark. Does this mean it no longer emits any kind of electromagnetic radiation? If not, what kind of radiation is it likely to emit once it is no longer "red" hot?

*3.18. If light consists of waves, why don't objects look "wavy" to us?

Problems

3.19. **Mercury Vapor Lamps** Lights containing mercury vapor (Figure P3.19) are used in sports arenas, factories, and streetlights. The wavelength of a bright line in the visible emission spectrum of mercury vapor is 546.1 nm. What is the frequency of this radiation? Does emission at this wavelength contribute to the greenish glow observed in a mercury vapor lamp?

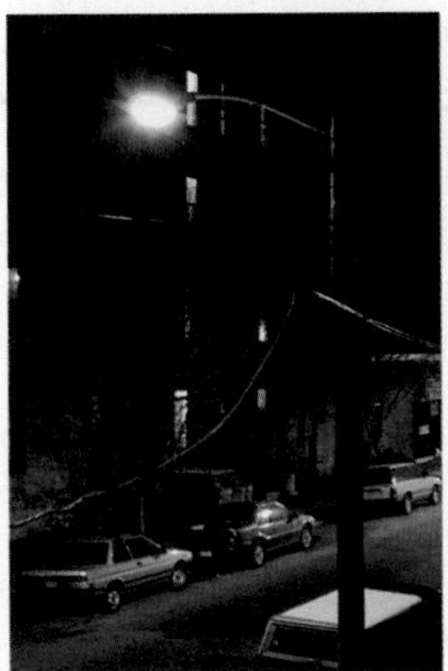

FIGURE P3.19

3.20. **Submarine Communications** The Russian and American navies developed extremely low-frequency communications networks to send messages to submerged submarines. The frequency of the carrier wave of the Russian network was 82 Hz, while the Americans used 76 Hz.
   a. What was the ratio of the wavelengths of the Russian network to the American network?
   b. To calculate the actual underwater wavelength of the transmissions in either network, what additional information would you need?

3.21. **Broadcast Frequencies** FM radio stations broadcast in a band of frequencies between 88 and 108 megahertz (MHz). Calculate the wavelengths corresponding to the broadcast frequencies of the following radio stations:
   a. KRNU (Lincoln, NE), 90.3 MHz
   b. WBRU (Providence, RI), 95.5 MHz
   c. WYLD (New Orleans, LA), 98.5 MHz
   d. WAAF (Boston, MA), 107.3 MHz

3.22. **Remote Keyless Systems** The doors of most automobiles can be unlocked by pushing a button on a key chain (Figure P3.22). These devices operate on frequencies in the MHz range. Calculate the wavelength of (a) the 315 MHz radiation emitted by remote keyless systems in North American cars and (b) the 434 MHz radiation used to unlock the doors of European and Asian cars. Where do these frequencies lie in the electromagnetic spectrum?

FIGURE P3.22

3.23. In 1895 German physicist Wilhelm Röntgen discovered X-rays. He also discovered that X-rays emitted by different metals have different wavelengths. Which X-rays have the higher frequency, those emitted by (a) Cu (λ = 0.154 nm) or (b) iron (λ = 194 pm)?

3.24. **Garage Door Openers** The remote control units for garage door openers transmit electromagnetic radiation. Before 2005 they operated on a frequency of 390 MHz, but since 2005, the operating wavelength has been 952 mm. Which radiation has the lower frequency, the pre-2005 or the post-2005 devices?

3.25. **Speed of Light** How long does it take light to reach Earth from the sun when the distance between them is 149.6 million kilometers?

3.26. **Exploration of the Solar System** The *Voyager 1* spacecraft was launched in 1977 to explore the outer solar system and interstellar space. By December 2012, it was 1.85×10^{10} km from Earth and still sending and receiving data. How long did it take a signal from Earth to reach *Voyager 1* over this distance?

Atomic Spectra

Concept Review

3.27. Describe the similarities and differences in the atomic emission and absorption spectra of an element.

3.28. Are the Fraunhofer lines the result of atomic emission or atomic absorption?

3.29. How did the study of the atomic emission spectra of the elements lead to the identification of the origins of the Fraunhofer lines in sunlight?

*3.30. How would the appearance of the Fraunhofer lines in the solar spectrum be changed if sunlight were passed through a flame containing high-temperature sodium atoms?

Particles of Light: Quantum Theory

Concept Review

3.31. What is a quantum?

3.32. What is a photon?

3.33. If a piece of tungsten metal were heated to 1000 K, would it emit light in the dark? If so, what color?

3.34. **Incandescent Lightbulbs** A variable power supply is connected to an incandescent lightbulb. At the lowest power setting, the bulb feels warm to the touch but produces no light. At medium power, the lightbulb filament emits a red glow. At the highest power, the lightbulb emits white light. Explain this emission pattern.

Problems

3.35. **Tanning Booths** Prolonged exposure to ultraviolet radiation in tanning booths significantly increases the risk of skin cancer. What is the energy of a photon of UV light with a wavelength of 3.00×10^{-7} m?

3.36. **Remote Control** Most remote controllers for AV equipment emit radiation with the wavelength profile shown in Figure P3.36. What type of radiation is shown and what is the energy of a photon at the peak wavelength?

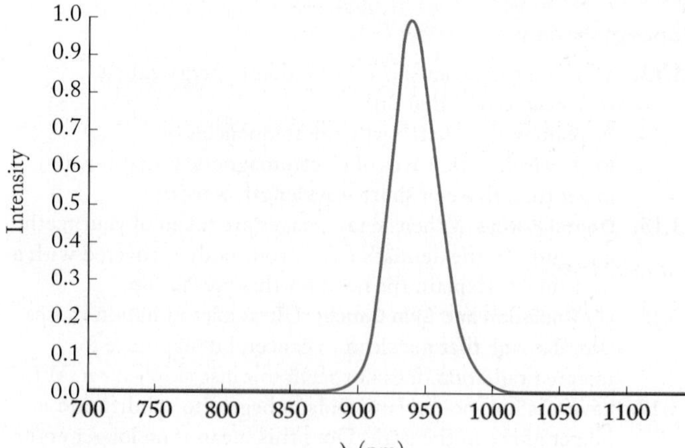

FIGURE P3.36

3.37. Which of the following have quantized values? Explain your selections.
   a. The elevation of the treads of a moving escalator
   b. The elevations at which the doors of an elevator open
   c. The speed of an automobile

3.38. Which of the following have quantized values? Explain your selections.
a. The pitch of a note played on a slide trombone
b. The pitch of a note played on a flute
c. The wavelengths of light produced by the heating elements in a toaster
d. The wind speed at the top of Mt. Everest

3.39. When a piece of gold foil is irradiated with UV radiation ($\lambda = 132$ nm), electrons are ejected with a kinetic energy of 7.34×10^{-19} J. What is the work function of gold?

3.40. The first ionization energy of a gas-phase atom of a particular element is 6.24×10^{-19} J. What is the maximum wavelength of electromagnetic radiation that could ionize this atom?

3.41. **Solar Power** Photovoltaic cells convert solar energy into electricity. Could germanium ($\Phi = 7.21 \times 10^{-19}$ J) be used to convert visible sunlight to electricity? Assume that most of the electromagnetic energy from the sun in the visible region is at wavelengths shorter than 600 nm.

3.42. With reference to Problem 3.41, could tin ($\Phi = 6.20 \times 10^{-19}$ J) be used to construct solar cells?

***3.43.** Pieces of potassium ($\Phi = 3.68 \times 10^{-19}$ J) and sodium ($\Phi = 4.41 \times 10^{-19}$ J) metal are exposed to radiation with a wavelength of 300 nm. Which metal emits electrons with the greater velocity? What is the velocity of the electrons?

3.44. Titanium ($\Phi = 6.94 \times 10^{-19}$ J) and silicon ($\Phi = 7.24 \times 10^{-19}$ J) surfaces are irradiated with UV radiation with a wavelength of 250 nm. Which surface emits electrons with the longer wavelength? What is the wavelength of the electrons emitted by the titanium surface?

3.45. **Red Lasers** The power of a red laser ($\lambda = 630$ nm) is 1.00 watt (abbreviated W, where 1 W = 1 J/s). How many photons per second does the laser emit?

*3.46. **Starlight** The energy density of starlight in interstellar space is 10^{-15} J/m^3. If the average wavelength of starlight is 500 nm, what is the corresponding density of photons per cubic meter of space?

The Hydrogen Spectrum and the Bohr Model

Concept Review

3.47. Why is the Balmer equation considered a special case of the Rydberg equation?

3.48. How does the value of n of an orbit in the Bohr model of hydrogen relate to the energy of an electron in that orbit?

3.49. Does the electromagnetic energy emitted by an excited-state H atom depend on the individual values of n_1 and n_2 or only on the difference between them $(n_1 - n_2)$?

3.50. Explain the difference between a ground-state H atom and an excited-state H atom.

3.51. Without calculating any wavelength values, predict which of the following four electron transitions in the hydrogen atom is associated with radiation having the shortest wavelength.
a. $n = 1 \rightarrow n = 2$
b. $n = 2 \rightarrow n = 3$
c. $n = 3 \rightarrow n = 4$
d. $n = 4 \rightarrow n = 5$

3.52. Without calculating any frequency values, rank the following transitions in the hydrogen atom in order of increasing frequency of the electromagnetic radiation that could produce them.
a. $n = 4 \rightarrow n = 6$
b. $n = 6 \rightarrow n = 8$
c. $n = 9 \rightarrow n = 11$
d. $n = 11 \rightarrow n = 13$

3.53. Electron transitions from $n = 2$ to $n = 3$, 4, 5, or 6 in hydrogen atoms are responsible for some of the Fraunhofer lines in the sun's spectrum. Are there any Fraunhofer lines due to transitions that start from the ground-state hydrogen atoms?

3.54. In the visible portion of the atomic emission spectrum of hydrogen, are there any bright lines due to electron transitions to the ground state?

3.55. Balmer observed a hydrogen emission line for the transition from $n = 6$ to $n = 2$, but not for the transition from $n = 7$ to $n = 2$. Why?

*3.56. In what ways should the emission spectra of H and He$^+$ be alike, and in what ways should they be different?

Problems

3.57. What is the wavelength of the photons emitted by hydrogen atoms when they undergo $n = 4 \rightarrow n = 3$ transitions? In which region of the electromagnetic spectrum does this radiation occur?

3.58. What is the frequency of the photons emitted by hydrogen atoms when they undergo $n = 5 \rightarrow n = 3$ transitions? In which region of the electromagnetic spectrum does this radiation occur?

***3.59.** The energies of photons emitted by one-electron atoms and ions fit the equation

$$E = (2.18 \times 10^{-18}\ \text{J})Z^2\left(\frac{1}{n_1^2} - \frac{1}{n_2^2}\right)$$

where Z is the atomic number, n_1 and n_2 are positive integers, and $n_2 > n_1$. Is the emission associated with the $n = 2 \rightarrow n = 1$ transition in a one-electron ion ever in the visible region? Why or why not?

*3.60. Can transitions from higher energy states to the $n = 2$ state in He$^+$ ever produce visible light? If so, for what values of n_2? Refer to the equation in Problem 3.59.

***3.61.** By absorbing different wavelengths of light, an electron in a hydrogen atom undergoes a transition from $n = 2$ to $n = 3$ and then from $n = 3$ to $n = 4$.
a. Are the wavelengths for the two transitions additive—that is, does $\lambda_{2\rightarrow 4} = \lambda_{2\rightarrow 3} + \lambda_{3\rightarrow 4}$?
b. Are the energies of the two transitions additive—that is, does $E_{2\rightarrow 4} = E_{2\rightarrow 3} + E_{3\rightarrow 4}$?

*3.62. The hydrogen atomic emission spectrum includes a UV line with a wavelength of 92.3 nm.
a. Is this line associated with a transition between different excited states or between an excited state and the ground state?
b. What is the energy of the longest-wavelength photon that a ground-state hydrogen atom can absorb?

Electrons as Waves

Concept Review

3.63. Identify the symbols in the de Broglie relation, $\lambda = h/mu$, and explain how the relation links the properties of a particle to those of a wave.

3.64. Why do matter waves *not* add significantly to the challenge of hitting a baseball thrown at 99 mph (44 m/s)?

3.65. Would the density or shape of an object have an effect on its de Broglie wavelength?

3.66. How does de Broglie's hypothesis that electrons behave like waves explain the stability of the electron orbits in a hydrogen atom?

3.67. Two objects are moving at the same speed. Which (if any) of the following statements about them are true?

a. The de Broglie wavelength of the heavier object is longer than that of the lighter one.
b. If one object has twice as much mass as the other, then its wavelength is one-half the wavelength of the other.
c. Doubling the speed of one of the objects will have the same effect on its wavelength as doubling its mass.

3.68. Which (if any) of the following statements about the frequency of a particle is true?

a. Heavy, fast-moving objects have lower frequencies than those of lighter, faster-moving objects.
b. Only very light particles can have high frequencies.
c. Doubling the mass of an object and halving its speed result in no change in its frequency.

Problems

3.69. Calculate the wavelengths of the following objects:

a. A muon (a subatomic particle with a mass of 1.884×10^{-28} kg) traveling at 325 m/s
b. Electrons ($m_e = 9.10938 \times 10^{-31}$ kg) moving at 4.05×10^6 m/s in an electron microscope
c. An 82 kg sprinter running at 9.9 m/s
d. Earth (mass $= 6.0 \times 10^{24}$ kg) moving through space at 3.0×10^4 m/s

3.70. How rapidly would each of the following particles be moving if they all had the same wavelength as a photon of red light ($\lambda = 750$ nm)?

a. An electron of mass 9.10938×10^{-28} g
b. A proton of mass 1.67262×10^{-24} g
c. A neutron of mass 1.67493×10^{-24} g
d. An α particle of mass 6.64×10^{-24} g

3.71. Particles in a Cyclotron The first cyclotron was built in 1930 at the University of California, Berkeley, and was used to accelerate molecular ions of hydrogen, H_2^+, to a velocity of 4×10^6 m/s. (Modern cyclotrons can accelerate particles to nearly the speed of light.) If the relative uncertainty in the velocity of the H_2^+ ion was 3%, what was the uncertainty of its position?

3.72. **Radiation Therapy** An effective treatment for some cancerous tumors involves irradiation with "fast" neutrons. The neutrons from one treatment source have an average velocity of 3.1×10^7 m/s. If the velocities of individual neutrons are known to within 2% of that value, what is the uncertainty in the position of one of them?

Quantum Numbers and the Sizes and Shapes of Atomic Orbitals

Concept Review

3.73. How does the concept of an orbit in the Bohr model of the hydrogen atom differ from the concept of an orbital in quantum theory?

3.74. What properties of an orbital are defined by each of the three quantum numbers n, ℓ, and m_ℓ?

3.75. How many quantum numbers are needed to identify an orbital?

3.76. How many quantum numbers are needed to identify an electron in an atom?

Problems

3.77. How many orbitals are there in an atom with each of the following principal quantum numbers? (a) 1; (b) 2; (c) 3; (d) 4; (e) 5

3.78. How many orbitals are there in an atom with the following combinations of quantum numbers?

a. $n = 3, \ell = 2$
b. $n = 3, \ell = 1$
c. $n = 4, \ell = 2, m_\ell = 2$

3.79. What are the possible values of quantum number ℓ when $n = 4$?

3.80. What are the possible values of m_ℓ when $\ell = 2$?

3.81. Which subshell corresponds to each of the following sets of quantum numbers?

a. $n = 6, \ell = 0$
b. $n = 3, \ell = 2$
c. $n = 2, \ell = 1$
d. $n = 5, \ell = 4$

3.82. Which subshell corresponds to each of the following sets of quantum numbers?

a. $n = 7, \ell = 1$
b. $n = 4, \ell = 2$
c. $n = 3, \ell = 0$
d. $n = 6, \ell = 5$

3.83. How many electrons could occupy orbitals with the following quantum numbers?

a. $n = 2, \ell = 0$
b. $n = 3, \ell = 1, m_\ell = 0$
c. $n = 4, \ell = 2$
d. $n = 1, \ell = 0, m_\ell = 0$

3.84. How many electrons could occupy orbitals with the following quantum numbers?

a. $n = 3, \ell = 2$
b. $n = 5, \ell = 4$
c. $n = 3, \ell = 0$
d. $n = 4, \ell = 1, m_\ell = 1$

3.85. Which of the following combinations of quantum numbers are allowed?

a. $n = 1, \ell = 1, m_\ell = 0, m_s = +\frac{1}{2}$

b. $n = 3, \ell = 0, m_\ell = 0, m_s = -\frac{1}{2}$

c. $n = 1, \ell = 0, m_\ell = 1, m_s = -\frac{1}{2}$

d. $n = 2, \ell = 1, m_\ell = 2, m_s = +\frac{1}{2}$

3.86. Which of the following combinations of quantum numbers are allowed?

a. $n = 3, \ell = 2, m_\ell = 0, m_s = -\frac{1}{2}$

b. $n = 5, \ell = 4, m_\ell = 4, m_s = +\frac{1}{2}$

c. $n = 3, \ell = 0, m_\ell = 1, m_s = +\frac{1}{2}$

d. $n = 4, \ell = 4, m_\ell = 1, m_s = -\frac{1}{2}$

The Periodic Table and Filling Orbitals; Electron Configurations of Ions

Concept Review

3.87. What is meant when two or more orbitals are said to be degenerate?

3.88. Explain how the electron configurations of the group 2 elements are linked to their location in the periodic table.

3.89. How do we know from examining the periodic table that the 4*s* orbital is filled before the 3*d* orbitals?

3.90. Why do so many transition metals form ions with a 2+ charge?

Problems

3.91. Identify the subshells with the following combinations of quantum numbers and arrange them in order of increasing energy in a multielectron atom:

a. $n = 3, \ell = 2$

b. $n = 7, \ell = 3$

c. $n = 3, \ell = 0$

d. $n = 4, \ell = 1$

3.92. Identify the subshells with the following combinations of quantum numbers and arrange them in order of increasing energy in an atom of gold:

a. $n = 2, \ell = 1$

b. $n = 5, \ell = 0$

c. $n = 3, \ell = 2$

d. $n = 4, \ell = 3$

3.93. What are the electron configurations of Li^+, Ca, F^-, Mg^{2+}, and Al^{3+}?

3.94. Which species in Problem 3.93 are isoelectronic with Ne?

3.95. What are the condensed electron configurations of K, K^+, Ba, Ti^{4+}, and Ni?

3.96. In what way are the electron configurations of C, Si, and Ge similar?

3.97. Which of the following orbital diagrams in Figure P3.97 best describes the ground-state electron configuration of Mn? Which one is the orbital diagram for Mn^{2+}?

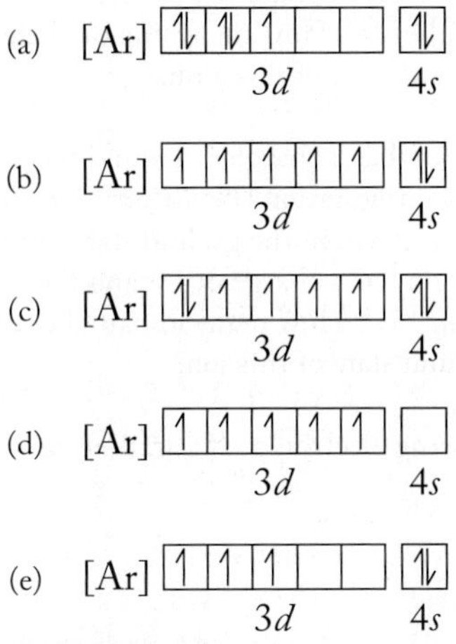

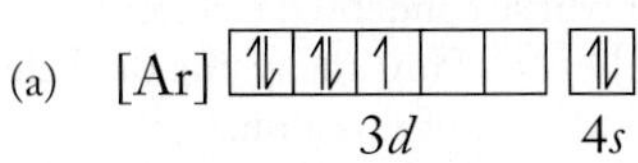

(a) [Ar] 3*d*: ↑↓ ↑↓ ↑ _ _ ; 4*s*: ↑↓

(b) [Ar] 3*d*: ↑ ↑ ↑ ↑ ↑ ; 4*s*: ↑↓

(c) [Ar] 3*d*: ↑↓ ↑ ↑ ↑ ↑ ; 4*s*: ↑↓

(d) [Ar] 3*d*: ↑ ↑ ↑ ↑ ↑ ; 4*s*: _

(e) [Ar] 3*d*: ↑ ↑ ↑ _ _ ; 4*s*: ↑↓

FIGURE P3.97

3.98. Which of the following orbital diagrams in Figure P3.98 best describes the ground-state electron configuration of Pb, Pb^{2+}, and Pb^{4+}?

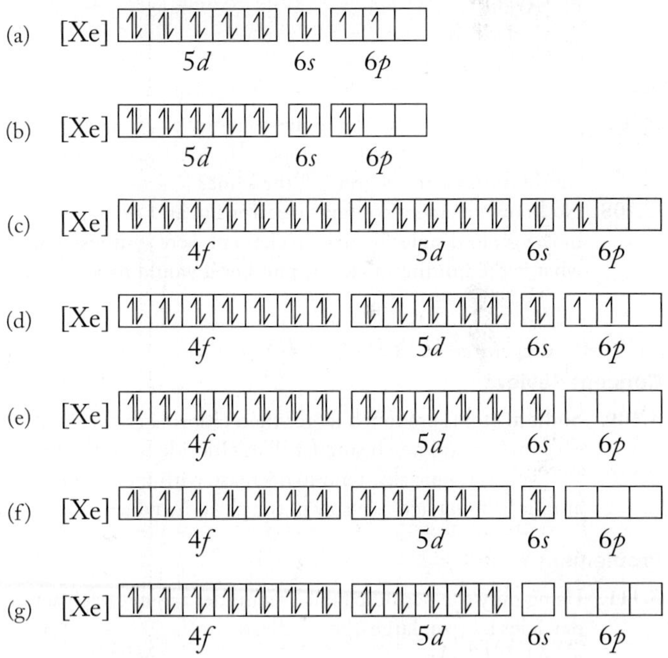

(a) [Xe] 5*d*: ↑↓ ↑↓ ↑↓ ↑↓ ↑↓ ; 6*s*: ↑↓ ; 6*p*: ↑ ↑ _

(b) [Xe] 5*d*: ↑↓ ↑↓ ↑↓ ↑↓ ↑↓ ; 6*s*: ↑↓ ; 6*p*: ↑↓ _ _

(c) [Xe] 4*f*: ↑↓ ↑↓ ↑↓ ↑↓ ↑↓ ↑↓ ↑↓ ; 5*d*: ↑↓ ↑↓ ↑↓ ↑↓ ↑↓ ; 6*s*: ↑↓ ; 6*p*: ↑↓ _ _

(d) [Xe] 4*f*: ↑↓ ↑↓ ↑↓ ↑↓ ↑↓ ↑↓ ↑↓ ; 5*d*: ↑↓ ↑↓ ↑↓ ↑↓ ↑↓ ; 6*s*: ↑↓ ; 6*p*: ↑ ↑ _

(e) [Xe] 4*f*: ↑↓ ↑↓ ↑↓ ↑↓ ↑↓ ↑↓ ↑↓ ; 5*d*: ↑↓ ↑↓ ↑↓ ↑↓ ↑↓ ; 6*s*: ↑↓ ; 6*p*: _ _ _

(f) [Xe] 4*f*: ↑↓ ↑↓ ↑↓ ↑↓ ↑↓ ↑↓ ↑↓ ; 5*d*: ↑↓ ↑↓ ↑↓ ↑↓ _ ; 6*s*: ↑↓ ; 6*p*: _ _ _

(g) [Xe] 4*f*: ↑↓ ↑↓ ↑↓ ↑↓ ↑↓ ↑↓ ↑↓ ; 5*d*: ↑↓ ↑↓ ↑↓ ↑↓ ↑↓ ; 6*s*: _ ; 6*p*: _ _ _

FIGURE P3.98

3.99. How many unpaired electrons are there in the following ground-state atoms and ions? (a) N; (b) O; (c) P^{3-}; (d) Na^+

3.100. How many unpaired electrons are there in the following ground-state atoms and ions? (a) Mn; (b) Ag^+; (c) Cu^{3+}; (d) Ti^{2+}

3.101. Identify the element whose condensed electron configuration is $[Ar]3d^24s^2$. How many unpaired electrons are there in the ground state of this atom?

3.102. Identify the element whose condensed electron configuration is $[Ne]3s^23p^3$. How many unpaired electrons are there in the ground state of this atom?

3.103. Which monatomic ion has a charge of 1− and the condensed electron configuration $[Ne]3s^23p^6$? How many unpaired electrons are there in the ground state of this ion?

3.104. Which monatomic ion has a charge of 1+ and the electron configuration $[Kr]4d^{10}5s^2$? How many unpaired electrons are there in the ground state of this ion?

3.105. Which of the following electron configurations represent an excited state?

a. $[He]2s^12p^5$
b. $[Kr]4d^{10}5s^25p^1$
c. $[Ar]3d^{10}4s^24p^5$
d. $[Ne]3s^23p^24s^1$

3.106. Which of the following condensed electron configurations represent an excited state? Could any represent *ground-state* electron configurations of 2+ ions?

a. $[Ar]4s^24p^1$
b. $[Ar]3d^{10}$
c. $[Kr]4d^{10}5s^1$
d. $[Ar]3s^23p^63d^1$

3.107. In which subshell are the highest-energy electrons in a ground-state atom of the isotope ^{131}I? Are the electron configurations of ^{131}I and ^{127}I the same?

*3.108. No known element contains electrons in an $\ell = 4$ subshell in the ground state. If such an element were synthesized, what is the minimum atomic number it would have to have?

The Sizes of Atoms and Ions

Concept Review

3.109. Sodium atoms are much larger than chlorine atoms, but sodium ions are much smaller than chloride ions. Why?

3.110. Why does atomic size tend to decrease with increasing atomic number across a row of the periodic table?

Problems

3.111. Using only the periodic table as a guide, arrange each set of particles by size, largest to smallest:

a. Al, P, Cl, Ar
b. C, Si, Ge, Sn
c. Li^+, Li, Na, K
d. F, Ne, Cl, Cl^-

3.112. Using only the periodic table, arrange each set of particles by size, largest to smallest:

a. Li, B, N, Ne
b. Mg, K, Ca, Sr
c. Rb^+, Sr^{2+}, Cs, Fr
d. S^{2-}, Cl^-, Ar, K^+

Ionization Energies

Concept Review

3.113. How do ionization energies change with increasing atomic number (a) down a group of elements in the periodic table and (b) from left to right across a row?

3.114. Explain the differences in ionization energy between (a) He and Li; (b) Li and Be; (c) Be and B; (d) N and O.

3.115. How does the wavelength of light required to ionize a gas-phase atom change with increasing atomic number down a group in the periodic table?

3.116. Why is the first ionization energy of Al less than that of Mg *and* less than that of Si?

Problems

3.117. Without referring to Figure 3.37, arrange the following groups of elements in order of increasing first ionization energy.

a. F, Cl, Br, I
b. Li, Be, Na, Mg
c. N, O, F, Ne

3.118. Without referring to Figure 3.37, arrange the following groups of elements in order of increasing first ionization energy.

a. Mg, Ca, Sr, Ba
b. He, Ne, Ar, Kr
c. P, S, Cl, Ar

Electron Affinities

Concept Review

3.119. An electron affinity (EA) value that is negative indicates that the free atoms of an element are higher in energy than the 1− anions they form by acquiring electrons. Does this mean that all of the elements with negative EA values exist in nature as anions? Give some examples to support your answer.

3.120. The electron affinities of the group 17 elements are all negative values, but the EA values of the group 18 noble gases are all positive. Explain this difference.

3.121. The electron affinities of the group 17 elements increase with increasing atomic number. Suggest a reason for this trend.

3.122. Ionization energies generally increase with increasing atomic number across the second row of the periodic table, but electron affinities generally decrease. Explain the opposing trends.

Additional Problems

***3.123. Colors of Fireworks** Barium compounds are a source of the green colors in many fireworks displays.

a. What is the ground-state electron configuration for Ba?
b. The lowest-energy excited state of Ba has the electron configuration $[Xe]5d^16s^1$. What are the possible quantum numbers n, ℓ, and m_ℓ of a $5d$ electron?

c. The excited state in part b is 1.79×10^{-19} J above the ground state. Could emission from this excited state account for the green color in fireworks?
d. Another Ba excited state has the electron configuration $[Xe]6s^16p^1$. Its energy is 3.59×10^{-19} J above the ground state. Could transitions from this state to the ground state account for the green color in fireworks?

*3.124. When an atom absorbs an X-ray of sufficient energy, one of its 2*s* electrons may be ejected, creating a hole that can be spontaneously filled when an electron in a higher-energy orbital—2*p*, for example—falls into it. A photon of electromagnetic radiation with an energy that matches the energy lost in the $2p \rightarrow 2s$ transition is emitted. Predict how the wavelengths of $2p \rightarrow 2s$ photons would differ between (a) different elements in the fourth row of the periodic table and (b) different elements in the same column (for example, between the noble gases from Ne to Rn).

***3.125.** Two helium ions (He^+) in the $n = 3$ excited state emit photons of radiation as they return to the ground state. One ion does so in a single transition from $n = 3$ to $n = 1$. The other does so in two steps: $n = 3$ to $n = 2$ and then $n = 2$ to $n = 1$. Which of the following statements about the two pathways is true?
a. The sum of the energies lost in the two-step process is the same as the energy lost in the single transition from $n = 3$ to $n = 1$.
b. The sum of the wavelengths of the two photons emitted in the two-step process is equal to the wavelength of the single photon emitted in the transition from $n = 3$ to $n = 1$.
c. The sum of the frequencies of the two photons emitted in the two-step process is equal to the frequency of the single photon emitted in the transition from $n = 3$ to $n = 1$.
d. The wavelength of the photon emitted by the He^+ ion in the $n = 3$ to $n = 1$ transition is shorter than the wavelength of a photon emitted by a H atom in an $n = 3$ to $n = 1$ transition.

3.126. Use your knowledge of electron configurations to explain the following observations:
a. Silver tends to form ions with a charge of 1+, but the elements to the left and right of silver in the periodic table tend to form ions with 2+ charges.
b. The heavier group 13 elements (Ga, In, and Tl) tend to form ions with charges of 1+ or 3+ but not 2+.
c. The heavier elements of group 14 (Sn and Pb) and group 4 (Ti, Zr, and Hf) tend to form ions with charges of 2+ or 4+.

3.127. Should the same trend in the first ionization energies for elements with atomic numbers $Z = 31$ through $Z = 36$ be observed for the second ionization energies of the same elements? Explain why or why not.

3.128. **Chemistry of Photo-Gray Glasses** "Photo-gray" lenses for eyeglasses darken in bright sunshine because the lenses contain tiny, transparent AgCl crystals. Exposure to light removes electrons from Cl^- ions, forming a chlorine atom in an excited state (indicated here by the asterisk):

$$Cl^- \xrightarrow{h\nu} Cl^* + e^-$$

The electrons are transferred to Ag^+ ions, forming silver metal:

$$Ag^+ + e^- \rightarrow Ag$$

Silver metal is reflective, producing the photo-gray color. How might substitution of AgBr for AgCl affect the light sensitivity of photo-gray lenses? In answering this question, consider whether more energy or less is needed to remove an electron from a Br^- ion than from a Cl^- ion.

3.129. Tin (in group 14) forms both Sn^{2+} and Sn^{4+} ions, but magnesium (in group 2) forms only Mg^{2+} ions.
a. Write condensed ground-state electron configurations for the ions Sn^{2+}, Sn^{4+}, and Mg^{2+}.
b. Which neutral atoms have ground-state electron configurations identical to Sn^{2+} and Mg^{2+}?
c. Which 2+ ion is isoelectronic with Sn^{4+}?

3.130. **Fog Lamp Technology** Sodium fog lamps and street lamps contain gas-phase Na atoms and Na^+ ions. Sodium atoms emit yellow-orange light at 589 nm. Do Na^+ ions emit the same yellow-orange light? Explain why or why not.

3.131. Effective nuclear charge (Z_{eff}) is related to atomic number (Z) by a factor called the shielding parameter (σ) according to the equation $Z_{eff} = Z - \sigma$.
a. Calculate Z_{eff} for the outermost *s* electrons of Ne and Ar, given that $\sigma = 4.24$ for Ne and $\sigma = 11.24$ for Ar.
b. Explain why the shielding parameter is much greater for Ar than for Ne.

3.132. **Millikan's Experiment** In his oil-drop experiment, Millikan used X-rays to ionize N_2 gas. Electrons lost by N_2 were absorbed by oil droplets. If the wavelength of the X-ray was 154 pm, could he also have filled his device with argon gas, which has an ionization energy of 1521 kJ/mol?

***3.133.** How can an electron get from one lobe of a *p* orbital to the other without going through the point of zero electron density between them?

3.134. Einstein did not fully accept the uncertainty principle, remarking that "He [God] does not play dice." What do you think Einstein meant? Niels Bohr allegedly responded by saying, "Albert, stop telling God what to do." What do you think Bohr meant?

smartwork5
If your instructor uses Smartwork5, log in at digital.wwnorton.com/atoms2.

4

Chemical Bonding

Understanding Climate Change

POWER PRODUCTION AND CLIMATE CHANGE
Coal-fired power plants, such as the one shown here, add to the increasing concentration of carbon dioxide in the atmosphere.

PARTICULATE **REVIEW**

Molecules and Valence Electrons

In Chapter 4, we learn how the electronic structure of atoms is related to the ways atoms bond together, forming molecules. Consider these models of three molecules that are abundant in Earth's atmosphere:

- What are the molecular formulas of these three atmospheric gases?
- What is the total number of electrons in each molecule?
- How many of these electrons come from the valence shells of the atoms in each molecule?

(a)

(b)

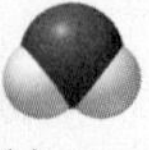

(c)

(Review Section 3.8 if you need help answering these review questions.)

(Answers to Particulate Review questions are in the back of the book.)

PARTICULATE **PREVIEW**

Bonds and Bonding Capacity

Molecules of two gases that contribute to climate change are shown in the figure. As you read Chapter 4, look for information that will help you answer these questions:

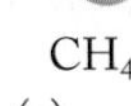

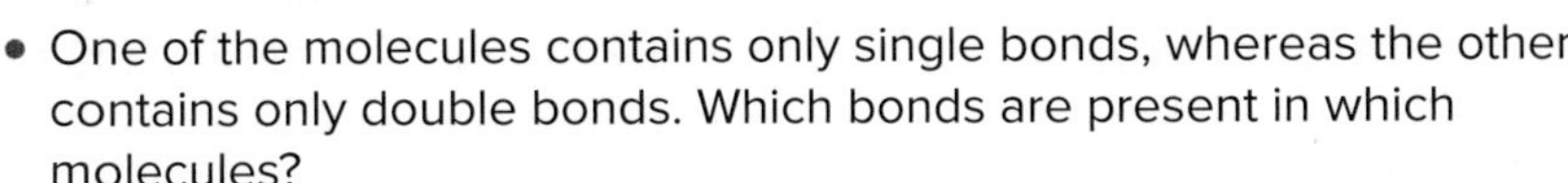

- One of the molecules contains only single bonds, whereas the other contains only double bonds. Which bonds are present in which molecules?

- How many bonds does the carbon atom in each molecule form?

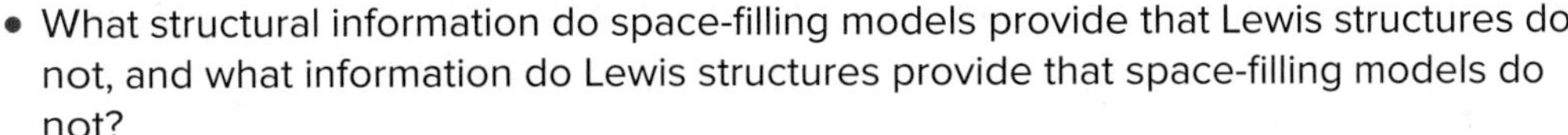

- What structural information do space-filling models provide that Lewis structures do not, and what information do Lewis structures provide that space-filling models do not?

Learning Outcomes

LO1 Describe ways in which covalent, ionic, and metallic bonds are alike and ways in which they differ

LO2 Calculate the relative strengths of ion–ion interactions
Sample Exercise 4.1

LO3 Name molecular and ionic compounds and write their formulas
Sample Exercises 4.2, 4.3, 4.4, 4.5, 4.6, 4.7

LO4 Draw Lewis structures of ionic compounds, molecular compounds, and polyatomic ions
Sample Exercises 4.8, 4.9, 4.10, 4.11, 4.12

LO5 Draw resonance structures and use formal charges to evaluate their relative importance
Sample Exercises 4.13, 4.16, 4.17, 4.18

LO6 Describe how bond order, bond energy, and bond length are related
Sample Exercise 4.14

LO7 Predict the polarity of covalent bonds based on differences in the electronegativity between the bonded elements
Sample Exercise 4.15

LO8 Explain how molecules of some atmospheric gases absorb infrared radiation and contribute to the greenhouse effect

4.1 Chemical Bonds and Greenhouse Gases

In Chapter 3 we learned how scientific advances in the early decades of the 20th century produced a radically different, probabilistic view of the structure of atoms and their interactions with electromagnetic radiation. In this chapter we continue our exploration of matter and energy, but we expand our focus to include the structure and behavior of the particles that make up ionic and molecular compounds and metallic solids, and we explore how molecules interact with electromagnetic radiation.

Let's consider one example of how radiant energy—in this case, infrared radiation—interacts with molecules in Earth's atmosphere, particularly molecules of carbon dioxide (CO_2). You are probably aware of the debate that is raging over global climate change and what should be done about it. Most scientists see a connection between recent increases in the average temperature of Earth's surface (about half a Celsius degree in the last half century) and increasing concentrations of CO_2 and other greenhouse gases in the atmosphere. These gases trap Earth's heat much like panes of glass trap heat in a greenhouse. Greenhouse gases are transparent to visible radiation, allowing sunlight to pass through the atmosphere and warm Earth's surface, but they trap heat that would otherwise flow from Earth's surface back into space. Were it not for the presence of greenhouse gases, Earth would be too cold to be habitable.

Increases in atmospheric concentrations of greenhouse gases have been linked to human activity, particularly to increasing rates of fossil fuel combustion and the destruction of forests that would otherwise consume CO_2 during photosynthesis. During the last 50 years, atmospheric concentrations of CO_2 have increased from about 315 parts per million (ppm) to about 400 ppm now. The atmosphere hasn't contained so much CO_2 in more than half a million years, long before our species evolved. In this chapter we do not address the social consequences of climate change, but we do explore why CO_2 in the atmosphere traps heat, whereas the two most abundant atmospheric gases, N_2 and O_2, do not.

Carbon dioxide is good at trapping heat because it absorbs infrared radiation. To understand why it does, we need to understand the properties of chemical bonds and the role of valence electrons in forming these bonds. We also need to understand why some pairs of atoms share bonding electrons equally while others do not, and how unequal sharing makes it possible for some compounds, such as CO_2, to absorb infrared radiation and contribute to climate change.

We begin this chapter by exploring how atoms combine, forming chemical bonds that link them together in molecular and ionic compounds. Fewer than 100 stable (nonradioactive) elements make up all the matter in the universe, yet these elements can combine to form more than 100 million compounds (with the number growing daily as chemists synthesize new compounds). How can such a relatively small number of elements combine to make such a large number of compounds? It turns out that most compounds are more stable than the elements from which they are made. Stable compounds exist because chemical bonds form when atoms or ions are attracted to one another.

We learned in Chapter 3 that an atom's outermost electrons are involved in forming chemical bonds and that these electrons are called *valence* electrons. Actually, chemists use the term **valence** all by itself to describe the capacity of the atoms of a particular element to form chemical bonds. Later in this chapter we will see how the atoms of some elements may form different numbers of chemical bonds and thus exhibit more than one *valency*. Before we explore how and why some elements display multiple valencies, let's first examine the three major categories of chemical bonds—namely, ionic bonds, covalent bonds, and metallic bonds.

valence the capacity of the atoms of an element to form chemical bonds.

electrostatic potential energy (E_{el}) the energy a charged particle has because of its position relative to another charged particle; it is directly proportional to the product of the charges of the particles and inversely proportional to the distance between them; also called *coulombic attraction*.

Ionic Bonds

Recall from Chapter 2 that binary (two-element) ionic compounds consist of cations formed from metallic elements and anions formed from nonmetals. These ionic compounds are held together by electrostatic attractions between ions of opposite charge. The strength of these attractions is a form of potential energy called **electrostatic potential energy (E_{el})**. It is also called *coulombic attraction*.

CONNECTION In Chapter 2 we learned that ions (charged particles) are further classified as cations (positively charged) and anions (negatively charged).

The value of E_{el} between a pair of ions is directly proportional to the product of their charges (Q_1 and Q_2) and inversely proportional to the distance (d) between their nuclei:

CONNECTION The coulomb is the SI unit of electrical charge (see Section 2.1).

$$E_{el} \propto \frac{Q_1 \times Q_2}{d}$$

We can turn this expression into an equation by replacing the "proportional to" symbol ($\propto$) with a constant of proportionality that gives us energy in joules when we express distance in nanometers and when the values of Q are the relative charges on the ions, such as 2+ for Ca^{2+} and 1− for Cl^-:

$$E_{el} = 2.31 \times 10^{-19}\ \text{J} \cdot \text{nm}\left(\frac{Q_1 \times Q_2}{d}\right) \tag{4.1}$$

Two ions with the same charge—either two positive ions or two negative ions—repel each other. According to Equation 4.1, this repulsion generates a positive E_{el} because multiplying together either two positive or two negative values yields a positive value. Conversely, the attraction between a positive ion and a negative ion produces a negative E_{el}. Moreover, the greater the attraction (that is, the larger the charges or the shorter the distance between them), the more negative the E_{el} value.

CONNECTION In Chapter 1 we defined potential energy as the energy of position or composition. The electrostatic potential energy between two ions depends on both the distance between them (their position) and their electrical charges (their composition).

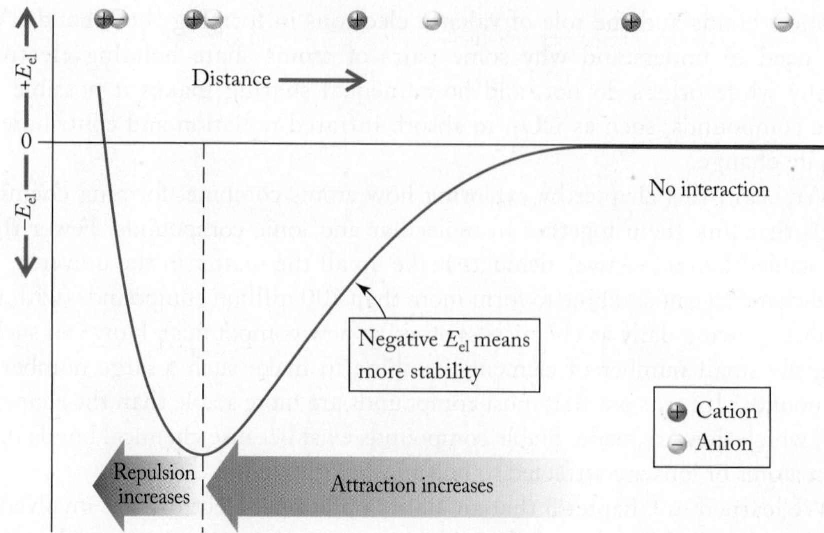

FIGURE 4.1 Changes in electrostatic potential energy (E_{el}) as an ionic bond forms. A pair of positive and negative ions that are far apart do not interact ($E_{el} = 0$). As the ions move closer together, the electrostatic potential energy between them becomes more negative, reaching a minimum corresponding to the energy of the ionic bond between the two. If the ions are forced even closer together, they repel each other and E_{el} rises to positive values.

The change in E_{el} that two oppositely charged particles experience as they approach each other is illustrated in Figure 4.1. When the ions are far apart (the right end of the curve) there is no attraction between them, and $E_{el} = 0$. As the ions approach each other, E_{el} decreases as electrostatic attractions increase, reaching an energy minimum that corresponds to maximum stability and the formation of an **ionic bond** between the ions. The absolute value of E_{el} at the energy minimum represents the amount of energy required to pull the ions apart and break the ionic bond. If the ions are pushed even closer together (to the left of the energy minimum position in Figure 4.1), E_{el} rises and the ion pair becomes less stable as the decreasing distance between their positively charged nuclei creates a growing electrostatic repulsion between the ions.

When the ions involved in Figure 4.1 are Na^+ and Cl^- ions, the distance between their nuclei at the energy minimum is 0.282 nm. This distance of maximum stability matches the distance between the nuclei of adjacent Na^+ and Cl^- ions in a crystal of table salt, sodium chloride.

CONNECTION We discussed how to predict the charges of monatomic ions from the positions of their parent elements in the periodic table in Chapter 2.

We can use Equation 4.1 to calculate the strength of the electrostatic attraction between a pair of oppositely charged ions if we know their charges and the distance between their nuclei. The latter value can be calculated for pairs of ions of the main group elements by summing their ionic radii (see Figure 3.36) as illustrated in Sample Exercise 4.1.

SAMPLE EXERCISE 4.1 Calculating the Electrostatic Potential Energy of an Ionic Bond — LO2

What is the electrostatic potential energy (E_{el}) of the ionic bond between a potassium ion and a chloride ion?

Collect and Organize We are asked to calculate the E_{el} of an ionic bond. Equation 4.1 relates E_{el} to the charges on a pair of ions and the distance between them. An ionic bond forms when E_{el} reaches a minimum, and that happens when the distance between the nuclei of the ions is close to the sum of their ionic radii. Ionic radii of main group elements are given in Figure 3.36.

Analyze Potassium is in group 1, so it forms 1+ ions. Chlorine is in group 17, so it forms 1− ions. According to Figure 3.36, the ionic radii of K^+ and Cl^- ions are 138 and 181 picometers, respectively.

ionic bond a bond resulting from the electrostatic attraction of a cation for an anion.

lattice energy (*U*) the energy released when one mole of an ionic compound forms from its free ions in the gas phase.

crystal lattice an ordered three-dimensional array of particles.

Solve The distance (d) between the nuclei of a pair of K^+ and Cl^- ions is

$$d = (K^+ \text{ ionic radius}) + (Cl^- \text{ ionic radius})$$
$$= (138 \text{ pm} + 181 \text{ pm}) = 319 \text{ pm}$$

The distance unit in Equation 4.1 is nm, so we need to convert this value for d to nanometers as part of the calculation of E_{el}:

$$E_{el} = 2.31 \times 10^{-19}\text{ J} \cdot \text{nm}\left(\frac{Q_1 \times Q_2}{d}\right)$$
$$= 2.31 \times 10^{-19}\text{ J} \cdot \cancel{\text{nm}}\left(\frac{(1+) \times (1-)}{319\ \cancel{\text{pm}} \times (1\ \cancel{\text{nm}}/1000\ \cancel{\text{pm}})}\right)$$
$$= -7.24 \times 10^{-19}\text{ J}$$

Think About It E_{el} has a negative value because K^+ and Cl^- are at a lower energy when they are connected by an ionic bond than when they are apart. It would take 7.24×10^{-19} J of energy to break the bond and separate the ions.

Practice Exercise What is the electrostatic potential energy (E_{el}) of the ionic bond between Ca^{2+} and S^{2-}? Before doing the calculation, predict whether it will be less than (more negative) or greater than the E_{el} value for KCl.

(Answers to Practice Exercises are in the back of the book.)

CONCEPT TEST

Without doing any calculations, rank the pairs of ions in each group in order of increasingly strong (more negative) electrostatic potential energy: (a) NaCl, KCl, RbCl; (b) KF, MgO, CaO.

(Answers to Concept Tests are in the back of the book.)

In Sample Exercise 4.1 we calculated the electrostatic potential energy in the ionic bond between one K^+ ion and one Cl^- ion. However, the energy associated with bonding is more conventionally expressed per *mole* of ions. If we multiply the result from Sample Exercise 4.1 by Avogadro's number, we obtain the following:

$$\frac{6.0221 \times 10^{23}\ \cancel{\text{ion pairs}}}{1\text{ mol}} \times \frac{-7.24 \times 10^{-19}\text{ J}}{\cancel{\text{ion pairs}}} = -4.36 \times 10^5\text{ J/mol, or } -436\text{ kJ/mol}$$

The answer is in the ballpark of the true value, but it's not exactly right. The actual change in energy that occurs when one mole of K^+ ions in the gas phase combines with one mole of gaseous Cl^- ions to form one mole of solid KCl is −720 kJ/mol, not −436 kJ/mol. What accounts for the difference? The answer is contained in the name we give the real value: the **lattice energy (*U*)** of KCl. Lattice energy is the energy released when free, gas-phase ions combine to form one mole of a crystalline solid. A crystalline solid consists of an ordered three-dimensional array of particles (atoms, ions, or molecules) called a **crystal lattice**. Figure 4.2 shows a photo of solid KCl and a drawing of its crystal lattice. Note the tightly packed array of K^+ ions (purple spheres) and Cl^- ions (green spheres).

In this array, every K^+ ion touches six Cl^- ions, not just one, and every Cl^- ion touches six K^+ ions. All of the additional interactions between ions of opposite charge make the ionic bonds stronger, which makes E_{el} in solid KCl more negative. The tight packing also produces some repulsions between ions of the same charge that are close to one another but do not touch. The net effect of the

FIGURE 4.2 Crystals of KCl contain ordered three-dimensional arrays of alternating K^+ ions (purple spheres) and Cl^- ions (green spheres).

TABLE 4.1 Lattice Energies (*U*) of Some Common Ionic Compounds

Compound	*U* (kJ/mol)
LiF	−1049
LiCl	−864
NaF	−930
NaCl	−786
NaBr	−754
KCl	−720
KBr	−691
MgO	−3791
$MgCl_2$	−2540

interactions is an overall decrease in E_{el}, corresponding to stronger ionic bonding than our calculation predicted based on one pair of ions.

Table 4.1 lists lattice energy values for several common ionic compounds. There are trends in these values that are consistent with the strengths of the ion-pair interactions predicted by Equation 4.1. For example, the only major difference between the crystal lattices of NaF, NaCl, and NaBr is the size of the anion ($F^- < Cl^- < Br^-$). Differences in ionic radii correspond to differences in the distance *d* between the nuclei of the pairs of ions. A large value of *d* in the denominator of Equation 4.1 means the attraction between the Na^+ ions and Br^- ions is weaker than the attraction between the Na^+ ions and F^- ions. Thus, NaF has the most negative lattice energy of the three compounds, and NaBr has the least negative *U* value.

The large difference between the lattice energies of NaF and MgO cannot be explained solely by differences in *d* because the sums of the radii of the ions involved are nearly the same. However, the crystal lattice of MgO consists of ions with charges of 2+ and 2−, whereas NaF is made of 1+ and 1− ions. Inserting these charges into the numerator of Equation 4.1 gives us Q_1Q_2 values of −4 for MgO and −1 for NaF. As a result, the lattice energy of MgO is about four times more negative than that of NaF.

Covalent Bonds

Figure 4.3 shows how the potential energy of two hydrogen atoms changes as they approach each other. The two are far enough apart on the right end of the curve that they do not interact with each other and have zero potential energy of interaction. As the two atoms approach each other, the proton in the nucleus of one is attracted to the electron of the other, and vice versa. This mutual attraction of both electrons for both nuclei is accompanied by a drop in potential energy that reaches a minimum at $d = 74$ pm. At this distance the two atoms share both of their electrons equally; they have formed a single **covalent bond**. They are no longer two separate H atoms but rather a more stable single molecule of H_2. If the atoms come any closer together, their potential energy rises as the repulsion between their two positive nuclei increases.

The distance of minimum energy at 74 pm represents the **bond length** of the H—H bond. Covalent bond lengths have values close to the sum of the atomic radii of the atoms involved in making the bonds. The minimum potential energy of −436 kJ/mol in Figure 4.3 is the energy that is released when two moles of H atoms form one mole of H_2 molecules. It is also the amount of energy required to break one mole of H_2 molecules into two moles of free H atoms. It is called the **bond energy** or *bond strength* of a H—H bond. The significance of bond lengths and bond energies is discussed in more detail in Section 4.5.

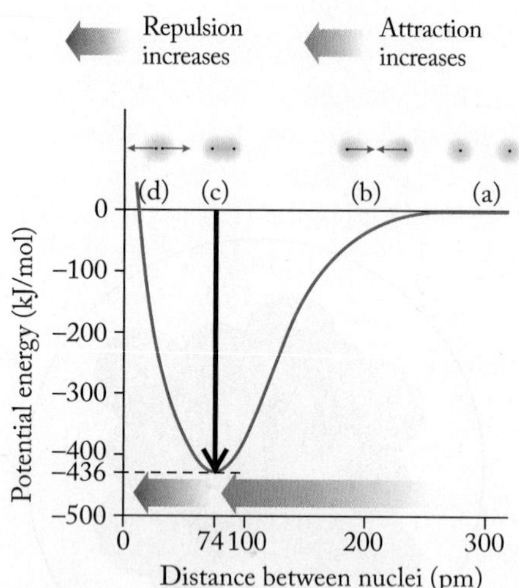

FIGURE 4.3 Changes in potential energy in the formation of an H—H covalent bond. Two hydrogen atoms that are far apart do not interact, but as the atoms move closer to one another, the positive nuclei and negative electrons on nearby atoms are attracted to each other, producing decreased electrostatic potential energy. At an internuclear distance of 74 pm, potential energy reaches a minimum value, corresponding to a covalent bond between the atoms. If the atoms are even closer together, repulsion between their nuclei causes potential energy to increase.

Metallic Bonds

A third type of chemical bond holds the atoms together in metallic solids. As with atoms in molecules, the positive nucleus of each atom in a metallic solid is attracted to the electrons of the atoms that surround it. These attractions, coupled with overlapping valence-shell orbitals, result in the formation of **metallic bonds**. These bonds are different from covalent bonds in that they are not pairs of electrons shared by pairs of atoms. Instead, the shared electrons in metallic bonds form a "sea" of mobile electrons that flows freely among all the atoms in a piece of metal.

To understand the mobility of metallic bonds, let's focus on the bonding in copper. The electron configuration of Cu ($Z = 29$) is $[Ar]3d^{10}4s^1$. If two Cu atoms are close enough to form a bond, we might expect them to share their 4*s* electrons and form a molecule of Cu_2—in much the way two hydrogen atoms form a covalent bond and a molecule of H_2 by sharing their 1*s* electrons. However, there is no experimental evidence for the existence of Cu_2 molecules in copper metal. Instead, there is a crystal lattice of copper atoms, as shown in Figure 4.4. Inside this lattice, each Cu atom is surrounded by 12 other Cu atoms. This means that each Cu atom shares its 4*s* electron with 12 other atoms, not just one, and those other 12 share their 4*s* electrons with 12 others, and so on. As a result, bonding electrons are not confined to particular pairs of Cu atoms but instead can easily move from one atom to another throughout a piece of copper. This electron mobility means that copper (and metals in general) are excellent conductors of electricity.

Table 4.2 summarizes some of the similarities and differences across the three types of chemical bonds. Keep in mind, though, that many bonds do not fall exclusively into just one category—they may, instead, have some covalent, ionic, and even metallic character simultaneously.

FIGURE 4.4 The atoms in copper metal are arranged in a crystal lattice in which each Cu atom forms metallic bonds with 12 other atoms. The electrons in metallic bonds are mobile, making copper and other metals good conductors of electricity.

TABLE 4.2 Types of Chemical Bonds

	Ionic	Covalent	Metallic
Elements Involved	Metals and nonmetals	Nonmetals and/or metalloids	Metals
Electron Distribution	Transferred	Shared	Delocalized
Example	KCl	H_2	Cu

4.2 Naming Compounds and Writing Formulas

Most of the matter on Earth consists of ionic and molecular compounds. As we explore the particulate nature of some of these compounds in this chapter, we need to be clear on how to write their names and their chemical formulas. We focus on simple compounds, starting with binary ionic compounds formed by combinations of main group elements.

Binary Ionic Compounds of Main Group Elements

Nearly all binary (two-element) ionic compounds formed by main group elements contain a cation formed by a group 1, 2, or 13 element and an anion formed by an element from groups 15–17. The compounds have two-word names, and there are only two rules for naming them:

1. The first word is the name of the cation, which is simply the name of its parent element.
2. The second word is the name of the anion, which is the name of its parent element with the ending changed to *-ide*.

When translating the name of a binary ionic compound into a formula, the positive and negative charges of the cations and anions must be in balance. Sample

covalent bond a bond created by two atoms sharing one or more pairs of electrons.

bond length the distance between the nuclei of two atoms joined together in a bond.

bond energy the energy needed to break one mole of a specific covalent bond in the gas phase; also called *bond strength*.

metallic bond a bond consisting of the nuclei of metal atoms surrounded by a "sea" of shared electrons.

Exercise 4.2 illustrates how to apply this concept and provides practice in naming ionic compounds.

SAMPLE EXERCISE 4.2 Naming and Writing Formulas of Binary Ionic Compounds **LO3**

What are the names and formulas of the compounds formed by ions of the following combinations of elements: (a) potassium and bromine, (b) calcium and oxygen, (c) sodium and sulfur, (d) magnesium and chlorine, and (e) aluminum and oxygen?

Collect, Organize, and Analyze The periodic table inside the front cover contains the names and symbols of the elements. Figure 2.10 shows how the charges of the monatomic ions of the elements are related to their locations in the periodic table. The name of a binary ionic compound starts with the name of the parent element of the cation followed by the name of the parent element of the anion in which the ending is changed to *-ide*. Writing the formulas of ionic compounds involves balancing the positive and negative charges on their ions by adding the appropriate subscripts after the symbols of the ions.

Solve The names of the compounds are listed in the accompanying table followed by the formulas of the ions in each compound from Figure 2.10. The ratios of the 1+ and 1− ions in (a) and the 2+ and 2− ions in (b) must both be 1 : 1 to balance their charges. However, twice as many Na^+ as S^{2-} ions must be present in compound (c) and there must be twice as many Cl^- ions as Mg^{2+} ions in compound (d). There must be two Al^{3+} ions for every three O^{2-} ions in compound (e). Thus, the names of the compounds and their formulas are the following:

	Name	Ions	Formula
(a)	Potassium bromide	K^+, Br^-	KBr
(b)	Calcium oxide	Ca^{2+}, O^{2-}	CaO
(c)	Sodium sulfide	Na^+, S^{2-}	Na_2S
(d)	Magnesium chloride	Mg^{2+}, Cl^-	$MgCl_2$
(e)	Aluminum oxide	Al^{3+}, O^{2-}	Al_2O_3

Think About It Subscripts indicate the ratio of the cations to anions in the neutral ionic compound. When no subscript is written for a particular ion, it is assumed to be 1. In this exercise we used Figure 2.10 to identify the charges of the ions. Try using only the periodic table inside the front cover as you complete the following Practice Exercise.

Practice Exercise Without referring to Figure 2.10, write the chemical formulas for (a) strontium chloride, (b) magnesium oxide, (c) sodium fluoride, and (d) calcium bromide.

(Answers to Practice Exercises are in the back of the book.)

Binary Ionic Compounds of Transition Metals

Some metallic elements, including many transition metals, form multiple cations with different charges. For example, most of the copper found in nature is present as Cu^{2+}; however, some copper compounds contain Cu^+ ions. Because the name *copper oxide* could apply to either CuO or Cu_2O (Figure 4.5), we need a naming system to distinguish between these two compounds. One system uses a Roman numeral after the name of the transition metal to indicate the positive charge on the

(a)

(b)

FIGURE 4.5 Two oxides of copper. (a) copper(II) oxide, CuO, and (b) copper(I) oxide, Cu_2O.

metal's ion. Thus, the formula of copper(II) oxide is CuO, and copper(I) oxide is Cu_2O. Roman numerals are used to indicate the charge of nearly all transition metal ions; Ag^+, Cd^{2+}, and Zn^{2+} are exceptions because those ions are the only ones that silver, cadmium, and zinc typically form. For example, the compounds they form with Cl^- are simply called silver chloride (AgCl) and zinc chloride ($ZnCl_2$).

For many years, chemists have also used historical names to identify different cations of the same element. In the older naming system, Cu^+ is called the *cuprous* ion and Cu^{2+} is called the *cupric* ion. Similarly, Fe^{2+} and Fe^{3+} are called *ferrous* and *ferric* ions, respectively. In both pairs of ions, the name of the ion with the lower charge ends in *-ous* and the name of the ion with the higher charge ends in *-ic*.

polyatomic ion a charged group of atoms joined together by covalent bonds.

oxoanion a polyatomic anion that contains at least one nonoxygen central atom bonded to one or more oxygen atoms.

SAMPLE EXERCISE 4.3 Writing Formulas of Transition Metal Compounds **LO3**

What are the chemical formulas of iron(II) sulfide and iron(III) oxide?

Collect, Organize, and Analyze We are asked to write the chemical formulas for two binary ionic compounds of iron. The Roman numerals in the name of the compounds indicate the charges in their iron ions: (II) means 2+ and (III) means 3+. Oxygen and sulfur are group 16 elements, which form 2− monatomic ions.

Solve A charge balance in iron(II) sulfide is achieved with a 1:1 ratio of Fe^{2+} to S^{2-} ions, so the chemical formula is FeS. To balance the charges on Fe^{3+} and O^{2-}, we need three O^{2-} ions for every two Fe^{3+} ions. Thus, the formula of iron(III) oxide is Fe_2O_3.

Think About It The Roman numeral system for conveying the charges on transition metal ions simplifies relating the names and formulas of the compounds they form. Writing the formulas of the two compounds based on their historical names would require that we memorize the charges of ferrous and ferric ions.

Practice Exercise Write the formulas of manganese(II) chloride and manganese(IV) oxide.

(Answers to Practice Exercises are in the back of the book.)

Polyatomic Ions

Table 4.3 lists the formulas and names of common **polyatomic ions**. Polyatomic ions consist of more than one atom joined by covalent bonds. The ammonium ion (NH_4^+) is the most common cation among the polyatomic ions; all the others in the table are anions.

Many of the polyatomic anions in Table 4.3 have the generic formula $XO_m^{\,n-}$. They are called **oxoanions**. An oxoanion's name is based on the name of the element (X) that appears before oxygen in its formula, with that element's ending changed to either *-ite* or *-ate*. The *-ate* oxoanion of an element has a larger number of oxygen atoms than its *-ite* oxoanion. For example, SO_4^{2-} is the sulfate ion and SO_3^{2-} is the sulfite ion.

If an element forms more than two oxoanions (for example, chlorine, bromine, and iodine form four), we add prefixes to distinguish among them. The oxoanion with the greatest number of oxygen atoms has the prefix *per-* and ends in *-ate*, and the oxoanion with the fewest oxygen atoms has the prefix *hypo-* and ends in *-ite*. The four oxoanions of chlorine listed in Table 4.3 that follow these naming rules are *per*chlor*ate* (ClO_4^-), chlor*ate* (ClO_3^-), chlor*ite* (ClO_2^-), and *hypo*chlor*ite* (ClO^-).

TABLE 4.3 Names, Formulas, and Charges of Some Common Polyatomic Ions

Name	Chemical Formula
Acetate	CH_3COO^-
Carbonate	CO_3^{2-}
Hydrogen carbonate or bicarbonate	HCO_3^-
Cyanide	CN^-
Hypochlorite	ClO^-
Chlorite	ClO_2^-
Chlorate	ClO_3^-
Perchlorate	ClO_4^-
Dichromate	$Cr_2O_7^{2-}$
Chromate	CrO_4^{2-}
Permanganate	MnO_4^-
Azide	N_3^-
Ammonium	NH_4^+
Nitrite	NO_2^-
Nitrate	NO_3^-
Hydroxide	OH^-
Phosphate	PO_4^{3-}
Hydrogen phosphate	HPO_4^{2-}
Dihydrogen phosphate	$H_2PO_4^-$
Disulfide	S_2^{2-}
Sulfate	SO_4^{2-}
Hydrogen sulfate or bisulfate	HSO_4^-
Sulfite	SO_3^{2-}
Hydrogen sulfite or bisulfite	HSO_3^-
Thiocyanate	SCN^-

Other groups of oxoanions differ by the presence of one or more H atoms in their formulas. For example, Table 4.3 includes phosphate (PO_4^{3-}), hydrogen phosphate (HPO_4^{2-}), and dihydrogen phosphate ($H_2PO_4^-$). Note how the negative charges of these ions decrease by 1 with each added H atom in the formula. Also note that those oxoanions in Table 4.3 that contain one H atom in their formula have historical names that begin with the prefix *bi-*. Sometimes these names are more commonly used than the modern ones, as is the case for HCO_3^-, which is more often called bicarbonate than hydrogen carbonate.

While there are several trends in the names and formulas of polyatomic ions, a name by itself does not tell you the chemical formula or the charge. For example, nitrate and sulfate both end in *-ate*, but nitrate is NO_3^- while sulfate is SO_4^{2-}. Therefore, you need to memorize the formulas, charges, and names of the common polyatomic ions in Table 4.3. Two important polyatomic ions that contain oxygen but are not considered oxoanions are the hydroxide (OH^-) and acetate (CH_3COO^-) ions. You will encounter them frequently in this book.

SAMPLE EXERCISE 4.4 Writing the Formulas of Compounds Containing Oxoanions LO3

What are the chemical formulas of (a) sodium sulfite; (b) magnesium phosphate; (c) ammonium nitrite?

Collect, Organize, and Analyze To write the formulas of the three named ionic compounds, we need to know the formulas and charges of the cations and anions that make up the compounds. Then we need to combine the ions in proportions that balance their positive and negative charges. Two of the cations in the compounds are Na^+ (a group 1 element) and Mg^{2+} (a group 2 element). The third cation, NH_4^+, is listed in Table 4.3, along with the formulas and charges of the other common polyatomic ions including sulfite (SO_3^{2-}), phosphate (PO_4^{3-}), and nitrite (NO_2^-).

Solve

a. We need a 2:1 ratio of Na^+ ions to SO_3^{2-} ions to balance their charges in sodium sulfite. Therefore, the formula is Na_2SO_3.
b. To balance the charges on Mg^{2+} and PO_4^{3-}, we need three Mg^{2+} ions for every two PO_4^{3-} ions, which means the formula of magnesium phosphate is $Mg_3(PO_4)_2$.
c. We need a 1:1 ratio of NH_4^+ to NO_2^- ions to balance their charges; so the formula of ammonium nitrite is NH_4NO_2.

Think About It In writing the formula of magnesium phosphate, we used parentheses around the phosphate ion to make it clear that the subscript 2 applies to all the atoms in PO_4^{3-}. We do not need parentheses in the formula of ammonium nitrite because there is only one of each ion in the formula.

 **Practice Exercise** What are the chemical formulas of (a) strontium nitrate; (b) potassium bicarbonate; (c) barium chromate?

(Answers to Practice Exercises are in the back of the book.)

SAMPLE EXERCISE 4.5 Naming Compounds Containing Oxoanions LO3

What are the names of the compounds with the following chemical formulas: (a) $CaCO_3$; (b) $LiNO_3$; (c) $Mg(ClO_4)_2$; (d) $(NH_4)_2SO_4$; (e) $KClO_3$; (f) $NaHCO_3$?

Collect, Organize, and Analyze We are given the formulas of six compounds, each containing at least one oxoanion, and are asked to name them. The names of ionic compounds start with the name of the cation followed by the name of the anion. The cations in five of the compounds are those formed by atoms of the following elements: (a) calcium, (b) lithium, (c) magnesium, (e) potassium, and (f) sodium. According to the list of polyatomic ion names in Table 4.3, the cation in formula (d) is the ammonium ion, and the oxoanions in the six compounds are (a) carbonate, (b) nitrate, (c) perchlorate, (d) sulfate, (e) chlorate, and (f) hydrogen carbonate.

Solve Combining the names of the cations and anions, we get (a) calcium carbonate, (b) lithium nitrate, (c) magnesium perchlorate, (d) ammonium sulfate, (e) potassium chlorate, and (f) sodium hydrogen carbonate.

Think About It As noted in the text, hydrogen carbonate is commonly called bicarbonate. Sodium bicarbonate is the principal ingredient in baking soda.

 **Practice Exercise** Name the following compounds: (a) $Ca_3(PO_4)_2$; (b) $MgSO_3$; (c) $LiNO_2$; (d) $NaClO$; (e) $KMnO_4$.

(Answers to Practice Exercises are in the back of the book.)

Binary Molecular Compounds

When two nonmetals combine, they form binary molecular compounds. To translate the molecular formula of such a compound into its two-word name, proceed as follows:

1. The first word is the name of the first element in the formula.
2. For the second word, change the ending of the name of the second element to *-ide.*
3. Use prefixes (Table 4.4) to indicate the number of atoms of each type in the molecule, but do *not* use the prefix *mono-* with the first element in a name.

For example, SO_2 is sulfur dioxide (not *mono*sulfur dioxide), SO_3 is sulfur trioxide, and N_2O is dinitrogen monoxide. When prefixes ending in *o-* or *a-* (like *mono-* and *tetra-*) precede a name that begins with a vowel (such as *oxide*), the *o* or *a* at the end of the prefix is deleted to make the combination of prefix and name easier to pronounce. Thus, the name of N_2O is dinitrogen monoxide, not dinitrogen mon*o*oxide.

The order in which the elements are named and written in formulas corresponds to their relative positions in the periodic table: the element with the lower group number appears first. When the elements are in the same group—for example, sulfur and oxygen—the name of the element with the higher atomic number goes first.

TABLE 4.4 Naming Prefixes for Molecular Compounds

one	*mono-*
two	*di-*
three	*tri-*
four	*tetra-*
five	*penta-*
six	*hexa-*
seven	*hepta-*
eight	*octa-*
nine	*nona-*
ten	*deca-*

SAMPLE EXERCISE 4.6 Naming Binary Molecular Compounds **LO3**

What are the names of the compounds with the following chemical formulas: (a) NO_2; (b) N_2O_4; (c) N_2O_5?

Collect and Organize All three compounds are binary nonmetal oxides, which means they are molecular compounds. Therefore, we use prefixes from Table 4.4 in their names to indicate the number of atoms of each element present in one molecule.

Analyze The first element in all three compounds is nitrogen, so the first word in each name is *nitrogen* with the appropriate prefix. There is only one nitrogen atom in the first compound, so no prefix is necessary. There are two nitrogen atoms in (b) and (c), so the prefix *di-* precedes nitrogen in their names. The second element in all three compounds is oxygen, so the second word in each name is *oxide* with the appropriate prefixes: *di-* to indicate two O atoms in (a); *tetra-* to indicate four O atoms in (b); *penta-* to indicate five O atoms in (c).

Solve

a. nitrogen dioxide
b. dinitrogen tetroxide
c. dinitrogen pentoxide

Think About It To avoid awkward *ao* vowel combinations in the middle of the second word in (b) and (c), we delete the last letter *a* of the prefixes *tetra-* and *penta-* before *oxide.*

 **Practice Exercise** Name the following compounds: (a) As_2O_5; (b) CO; (c) NCl_3.

(Answers to Practice Exercises are in the back of the book.)

Binary Acids

Some compounds have special names because of their particular chemical properties. Among these are acids. We will discuss the properties of acids in greater detail in later chapters, but for now we can consider an acid to be any compound that contains one or more *ionizable* hydrogen atoms that are released as H^+ ions when the compound dissolves in water.

For example, when the molecular compound hydrogen chloride (HCl) dissolves in water, it produces an aqueous solution we call *hydrochloric acid.* In this acidic solution, every molecule of HCl separates into a H^+ ion and a Cl^- ion. The simplest acids are called *binary* acids, and the simplest of these have the generic formula HX, where X is the symbol of a group 17 element. To name an acid such as HCl:

1. Add the prefix *hydro-* to the name of the second element in the formula.
2. Replace the last syllable in the second element's name with *-ic,* followed by the word *acid.*

To distinguish between a hydrogen halide (a molecular compound with covalent bonds between the hydrogen atom and the halogen atom) and the aqueous form (H^+ ions and X^- ions dissolved in water), we add the symbol (*aq*) for "aqueous" to the latter. For example, HBr is hydrogen bromide, but HBr(*aq*) is hydrobromic acid.

CONCEPT **TEST**

What is the name of the solution that is produced when HF dissolves in water?

(Answers to Concept Tests are in the back of the book.)

Oxoacids

oxoacid a compound composed of oxoanions bonded to H^+ ions.

Oxoanions that have bonded to H^+ ions form neutral **oxoacids**. If the oxoanion name ends in *-ate*, the name of the corresponding oxoacid ends in *-ic*. Thus, SO_4^{2-}

is the sulf*ate* ion and H_2SO_4 is sulfur*ic* acid. If the oxoanion name ends in *-ite*, the name of the oxoacid ends in *-ous*. Thus, NO_2^- is the nitr*ite* ion and HNO_2 is nitr*ous* acid.

For elements such as chlorine, bromine, and iodine that form more than two oxoanions, both a prefix (*hypo-* or *per-*) and a suffix (*-ate* or *-ic*) are necessary to distinguish the names of the oxoanions and their corresponding acids (Table 4.5).

TABLE 4.5 Oxoanions of Chlorine and Their Corresponding Acids

OXOANION		OXOACID	
Formula	Name	Formula	Name
ClO^-	hypochlorite	HClO	hypochlorous acid
ClO_2^-	chlorite	$HClO_2$	chlorous acid
ClO_3^-	chlorate	$HClO_3$	chloric acid
ClO_4^-	perchlorate	$HClO_4$	perchloric acid

SAMPLE EXERCISE 4.7 Naming Oxoacids **LO3**

What are the names and formulas of the oxoacids formed by the following oxoanions: (a) SO_3^{2-}; (b) ClO_4^-; (c) PO_4^{3-}?

Collect, Organize, and Analyze We are given the formulas of three oxoanions and are asked to name the oxoacids they form when they combine with H^+ ions. According to Table 4.3, the names of the oxoanions are (a) sulfite, (b) perchlorate, and (c) phosphoric acid. When the oxoanion name ends in *-ite*, the corresponding oxoacid name ends in *-ous*. When the anion name ends in *-ate*, the oxoacid name ends in *-ic*.

Solve Making the appropriate changes to the endings of the oxoanion names and adding the word *acid*, we get (a) sulfurous acid, H_2SO_3; (b) perchloric acid, $HClO_4$; and (c) phosphoric acid, H_3PO_4.

Think About It Once we know the names of the common oxoanions, naming the corresponding oxoacids is simply a matter of changing the ending of the oxoanion name from *-ate* to *-ic*, or from *-ite* to *-ous*, and then adding the word *acid.*

Practice Exercise Name the following acids: (a) HBrO; (b) $HBrO_2$; (c) H_2CO_3.

(Answers to Practice Exercises are in the back of the book.)

4.3 Lewis Symbols and Lewis Structures

In 1916, American chemist Gilbert N. Lewis (1875–1946) proposed that atoms form chemical bonds by sharing pairs of electrons. He further suggested that through this sharing each atom acquired enough valence electrons to mimic the outer-shell electron configuration of a noble gas. Today we associate these configurations with a filled 1*s* orbital (corresponding to a helium atom) or with a completely full set of *s* and *p* orbitals, which is the valence-shell configuration of the other noble gases. Lewis's view of chemical bonding predated quantum mechanics and the notion of atomic orbitals, but it was consistent with what he called the **octet rule**: all atoms except the very smallest (for example, hydrogen) tend to lose, gain, or share electrons so that each atom has eight valence electrons, or an *octet* of them. On the other hand, a hydrogen atom needs only one more electron to have a *duet* that mimics the electron configuration of a helium atom.

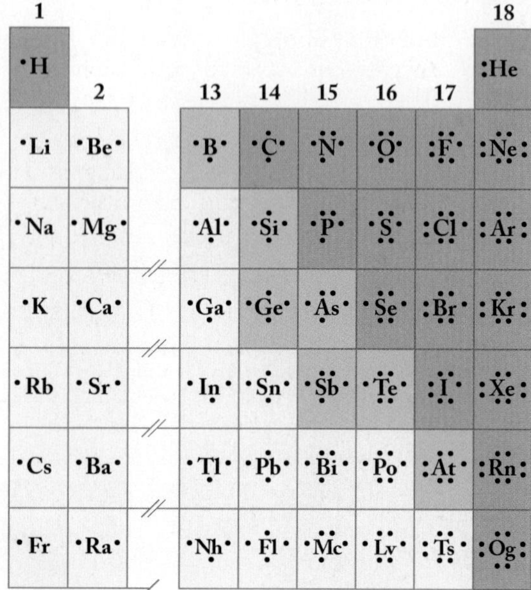

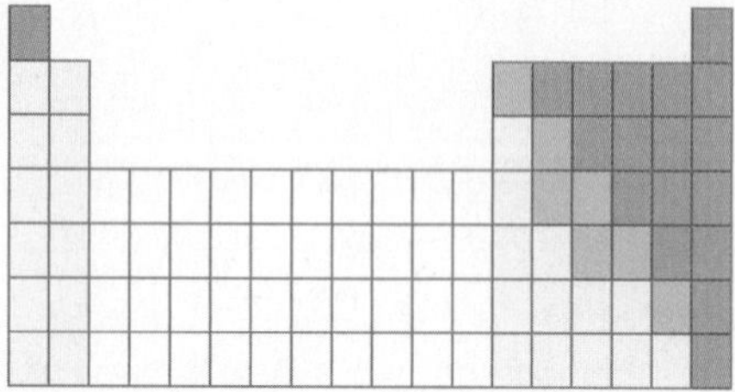

FIGURE 4.6 Lewis symbols of the main group elements. Elements in the same group have the same number of valence electrons and the same arrangement of dots representing the valence electrons in their Lewis symbols.

Lewis Symbols

Lewis developed a system of symbols, called **Lewis symbols** or *Lewis dot symbols*, to depict an atom's **bonding capacity**—that is, the number of chemical bonds that each atom of an element typically forms to complete its octet. A Lewis symbol consists of the symbol of an element surrounded by dots representing its valence electrons. The dots are placed on the four sides of the symbol (top, bottom, right, and left). The order in which they are placed does not matter as long as one dot is placed on each side before any dots are paired. The number of *unpaired* dots in a Lewis symbol indicates the typical bonding capacity for atoms of that element.

Figure 4.6 shows the Lewis symbols of the main group elements. Because all elements in a family have the same number of valence electrons, they all have the same arrangement of dots in their Lewis symbols. For example, the Lewis symbols of carbon and all other group 14 elements have four unpaired dots, representing four unpaired electrons. Four unpaired electrons means that carbon atoms each tend to form four chemical bonds. When they do, their four valence electrons plus four additional electrons from the atoms with which they bond yield an octet of eight valence electrons. Similarly, the Lewis symbols of nitrogen and all other group 15 elements have three unpaired dots, representing three unpaired electrons. A nitrogen atom has a bonding capacity of three and typically forms three bonds to complete its octet.

Before we start to use Lewis symbols to explore how elements form compounds, we need to keep in mind that the symbols indicate the number of bonds that the atoms of an element *typically* form. We will see in this chapter that atoms may exceed their bonding capacities in some molecules and polyatomic ions and fail to reach them in others. Bonding capacity is a useful concept, but we will treat it more like a guideline than a strict rule.

CONCEPT TEST

Devise a formula that relates the bonding capacity of the atoms of the elements in groups 14–17 to their group number.

(Answers to Concept Tests are in the back of the book.)

Lewis Structures of Ionic Compounds

A **Lewis structure** is a two-dimensional representation of a compound showing how its atoms are connected. Though Lewis structures are most often used to show the bonding in molecular compounds, they also provide insights into the composition of polyatomic ions and ionic compounds. The Lewis structures of binary ionic compounds such as NaCl are some of the simplest to draw, so we start with them.

Crystals of sodium chloride are held together by the attraction between oppositely charged Na^+ and Cl^- ions. We discussed in Chapter 3 how sodium and the other group 1 and 2 elements achieve noble gas electron configurations by losing their valence-shell *s* electrons and forming positively charged cations. Therefore, the ions of group 1 and 2 elements have no valence-shell electrons, and their Lewis structures have no dots around them. On the other hand, nonmetals such as chlorine, which acquire electrons to achieve noble gas electron configurations, have completely filled *s* and *p* orbitals in their valence shells. Therefore, their monatomic ions have Lewis structures with four pairs of dots. We place brackets around the dots to emphasize that all eight valence electrons are associated with the anion, and the charge of the ion is placed outside the bracket to indicate the overall charge of everything inside. Applying this notation to NaCl gives the following Lewis structure:

$$Na\cdot + \cdot\ddot{\underset{\cdot\cdot}{Cl}}: \rightarrow Na^+[:\ddot{\underset{\cdot\cdot}{Cl}}:]^-$$

octet rule atoms of main group elements make bonds by gaining, losing, or sharing electrons to achieve a valence shell containing eight electrons, or four electron pairs.

Lewis symbol the chemical symbol for an element surrounded by one or more dots representing valence electrons; also called a *Lewis dot symbol*.

bonding capacity the number of covalent bonds an atom forms to have an octet of electrons in its valence shell.

Lewis structure a two-dimensional representation of the bonds and lone pairs of valence electrons in an ionic or molecular compound.

bonding pair a pair of electrons shared between two atoms.

SAMPLE EXERCISE 4.8 Drawing the Lewis Structure of a Binary Ionic Compound **LO4**

Draw the Lewis structure of calcium fluoride.

Collect, Organize, and Analyze We are to draw the Lewis structure of the ionic compound formed by calcium and fluoride ions. Calcium is a group 2 element, which means its atoms form ions with 2+ charges by losing both of their valence (3*s*) electrons. Fluorine is a group 17 element, which means it forms 1− fluoride ions by gaining one valence-shell electron to give it a complete octet.

Solve A Ca^{2+} ion has no valence electrons left, so its Lewis structure is simply

$$Ca^{2+}$$

After acquiring an electron, a F ion has a complete octet and the following Lewis structure:

$$[:\ddot{\underset{\cdot\cdot}{F}}:]^-$$

There must be two chloride ions for every one calcium ion in $CaCl_2$ to have a neutral compound. A Lewis structure that reflects this 2:1 ratio is

$$Ca^{2+}\ 2[:\ddot{\underset{\cdot\cdot}{F}}:]^-$$

Think About It The lack of dots in the structure of Ca^{2+} reinforces the fact that Ca atoms, like those of the other main group metals, lose all their valence-shell electrons when they form monatomic cations. In contrast, atoms of the nonmetals form monatomic anions with complete octets.

Practice Exercise
Draw the Lewis structure of sodium oxide.

(Answers to Practice Exercises are in the back of the book.)

Lewis Structures of Molecular Compounds

Molecular compounds are held together by covalent bonds. Because covalent bonds are *pairs* of shared valence electrons, the Lewis structures of molecular compounds focus on how the shared pairs, also called **bonding pairs**, are distributed among the atoms in their molecules. A single bonding pair of electrons is

single bond a bond that results when two atoms share one pair of electrons.

lone pair a pair of electrons that is not shared.

called a **single bond**, often represented by a dash in Lewis structures, as in the structural formulas we first saw in Chapter 1. Electron pairs that are not involved in bonds appear as pairs of dots on one atom. The unshared electron pairs are called **lone pairs.**

A five-step process for drawing Lewis structures follows. It is particularly useful for drawing the structures of small molecules and polyatomic ions that have a single central atom bonded to atoms with lower bonding capacities.

Five Steps for Drawing Lewis Structures

1. *Determine the total number of valence electrons.* For a neutral molecule, count the valence electrons in all the atoms in the molecule. For an ion, sum the valence electrons for the atoms and then add (for an anion) or subtract (for a cation) the number of electrons needed to account for the charge on the ion.
2. *Arrange the symbols of the elements to show how their atoms are bonded together (a skeletal structure) and then connect them with single bonds (single pairs of bonding electrons).* Identify the element with the greatest bonding capacity and position it as the central atom. (If two elements have the same bonding capacity, choose the one that is less *electronegative*, an atomic property we will examine in Section 4.8.) Place the remaining atoms around the central atom, connecting each atom with a single bond. Some helpful hints: A hydrogen atom will never be a central atom because it has a bonding capacity of one (it can only form one bond). Sometimes the formulas of compounds and polyatomic ions suggest a skeletal structure. For example, the skeletal structures of HNCO and SCN^- are H—N—C—O and S—C—N, respectively.
3. *Complete the octets of all the atoms (except hydrogen) bonded to the central atom by adding lone pairs of electrons.* Place lone pairs on each of the outer atoms until each outer atom has an octet (including the two electrons used to connect it to the central atom in the skeletal structure).
4. *Compare the number of valence electrons in the Lewis structure to the number determined in step 1.* If valence electrons remain unused in the structure, place lone pairs of electrons on the central atom (even if doing so means giving it more than an octet of valence electrons), until all the valence electrons counted in step 1 have been included in the Lewis structure.
5. *Complete the octet on the central atom.* If there is an octet on the central atom, the structure is complete. If there is less than an octet on the central atom, create additional bonds to it by converting one or more lone pairs of electrons on the outer atoms into bonding pairs.

SAMPLE EXERCISE 4.9 Drawing the Lewis Structure of a Small Molecule I **LO4**

Chloroform is a volatile liquid once used as an anesthetic in surgery. It has the molecular formula $CHCl_3$. Draw its Lewis structure.

Collect, Organize, and Analyze The formula $CHCl_3$ tells us that a chloroform molecule contains one carbon atom, one hydrogen atom, and three chlorine atoms. Carbon is a group 14 element, so the carbon atom has four valence electrons and four more will complete its octet. Hydrogen, in group 1, has one valence electron, so one more will complete its duet. Chlorine, in group 17, has seven valence electrons, so one more will complete its octet. We can use the five-step process described previously to generate the Lewis structure.

Solve

Step 1. Summing the data in the last column of the following table gives us the total number of valence electrons in a molecule of $CHCl_3$: 26.

ELEMENT		VALENCE ELECTRONS	
Symbol	# of atoms	In one atom	Total*
C	1	4	4
H	1	1	1
Cl	3	7	21
	Valence electrons in $CHCl_3$		**26**

*Calculated by multiplying the values in the second and third columns together

Step 2. Carbon has the greatest bonding capacity of the three elements, so we put a carbon atom at the center of the skeletal structure of $CHCl_3$:

```
     H
     |
Cl—C—Cl
     |
     Cl
```

Step 3. The single bond to the hydrogen atom completes its duet. We need to add three lone pairs of electrons to each of the Cl atoms to complete their octets:

```
       H
       |
:C̤̈l—C—C̤̈l:
       |
      :C̤l:
```

Step 4. This structure contains four pairs of bonding electrons and nine lone pairs, for a total of $(4 \times 2) + (9 \times 2) = 26$ electrons, which matches the total number of valence electrons calculated in step 1.

Step 5. The carbon atom is surrounded by four bonds, which means it has eight valence electrons and a full octet. The structure is complete.

Think About It Carbon has a bonding capacity of four and it is bonded to four atoms in this structure, so there are no lone pairs of electrons on the carbon atom.

Practice Exercise Draw the Lewis structure of methane, CH_4.

(Answers to Practice Exercises are in the back of the book.)

SAMPLE EXERCISE 4.10 Drawing the Lewis Structure of a Small Molecule II **LO4**

Draw the Lewis structure of ammonia, NH_3.

Collect, Organize, and Analyze The chemical formula tells us that NH_3 contains one atom of nitrogen and three atoms of hydrogen per molecule. Nitrogen is a group 15 element with five valence electrons and a bonding capacity of three. Hydrogen atoms have one valence electron each and a bonding capacity of one. We can use the five-step process described previously to generate the Lewis structure.

Solve

Step 1. Summing the data in the last column of the following table gives us the total number of valence electrons in a molecule of NH_3: eight.

ELEMENT		VALENCE ELECTRONS	
Symbol	**# of atoms**	**In one atom**	**Total**
N	1	5	5
H	3	1	3
	Valence electrons in NH_3		**8**

Step 2. Nitrogen has the greatest bonding capacity of the three elements, so we put a nitrogen atom at the center of the skeletal structure:

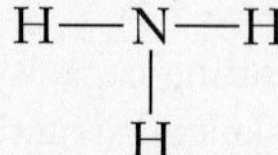

Step 3. The single bond to each hydrogen atom completes its duet.

Step 4. The three bonds in the structure represent $3 \times 2 = 6$ valence electrons, but we started with a total of eight valence electrons in step 1. Adding a lone pair of electrons to nitrogen gives us eight valence electrons and completes the octet on nitrogen:

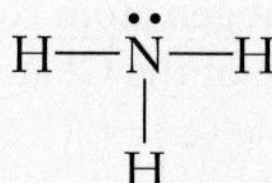

Step 5. The nitrogen has a complete octet and each hydrogen has a complete duet, so the Lewis structure is complete.

Think About It This structure makes sense because the nitrogen atom exhibits the bonding capacity predicted by its Lewis symbol. However, we will soon encounter a polyatomic ion, NH_4^+, in which the central nitrogen atom bonds to *four* hydrogen atoms. Keep in mind that Lewis bonding capacities are guidelines rather than strictly observed rules.

Practice Exercise

Draw the Lewis structure of phosphorus trichloride.

(Answers to Practice Exercises are in the back of the book.)

SAMPLE EXERCISE 4.11 Drawing the Lewis Structure of a Polyatomic Ion — LO4

Draw the Lewis structure of the hydroxide (OH^-) ion.

Collect, Organize, and Analyze Hydroxide ions each contain one atom of oxygen, a group 16 element with six valence electrons and a typical bonding capacity of two, and a hydrogen atom that has a bonding capacity of one. As in previous exercises, we use the five-step process to draw its Lewis structure.

Solve

Step 1. The charge of 1− means there is one more valence electron in the structure in addition to the six from the oxygen atom and one from the hydrogen atom for a total of $1 + 6 + 1 = 8$ valence electrons.

Step 2. Connecting the hydrogen atom to the oxygen atom with a covalent bond yields:

$$\mathrm{O—H}$$

Step 3. The bonded H atom has a complete duet of electrons.

Step 4. The one bond in the structure represents two valence electrons, but there are eight in the ion. So we add six more by placing three lone pairs of electrons around the O atom, which completes its octet:

$$\mathrm{:\ddot{\underset{\cdot\cdot}{O}}—H}$$

Step 5. The Lewis structure represents an ion, so square brackets are drawn around the structure and the ion's charge is placed outside the brackets. The Lewis structure is now complete.

$$\left[\mathrm{:\ddot{\underset{\cdot\cdot}{O}}—H}\right]^-$$

Think About It The bonding capacity of oxygen is two, yet there is only one bond in an OH^- ion. However, this ion also has the capacity to form an ionic bond with a positively charged cation, such as Na^+, producing a neutral ionic compound (NaOH) with equal numbers of ionic and covalent bonds.

Practice Exercise

Draw the Lewis structure of the ammonium ion, NH_4^+.

(Answers to Practice Exercises are in the back of the book.)

To draw the Lewis structure of a molecule with more than one "central" atom, it is often useful to draw pieces of the molecule, each containing one of the central atoms, as shown for hydrogen peroxide (H_2O_2):

$$\mathrm{H—\ddot{\underset{\cdot\cdot}{O}}—\ddot{\underset{\cdot\cdot}{O}}—H}$$

Here each oxygen atom is the "central atom" of a three-atom subunit and the bond between the two O atoms contributes two valence electrons toward completing the octets of both:

$$\mathrm{—\ddot{\underset{\cdot\cdot}{O}}—\ddot{\underset{\cdot\cdot}{O}}—H}$$

$$\mathrm{H—\ddot{\underset{\cdot\cdot}{O}}—\ddot{\underset{\cdot\cdot}{O}}—}$$

Lewis Structures of Molecules with Double and Triple Bonds

Lewis structures can also be used to show the bonding in molecules in which two atoms share more than one pair of bonding electrons. A bond in which two atoms share two pairs of electrons is called a **double bond**. For example, the two oxygen atoms in a molecule of O_2 share two pairs of electrons, forming an O═O double bond. When the two nitrogen atoms in a molecule of N_2 share three pairs of electrons, they form a N≡N **triple bond**. In Section 4.5 we discuss the characteristics of multiple bonds and compare them with those of single bonds.

How do we know when a Lewis structure has a double or triple bond? Typically we realize it when we apply steps 3 through 5 in the guidelines. Suppose

double bond a bond formed when two atoms share two pairs of electrons.

triple bond a bond formed when two atoms share three pairs of electrons.

we complete the octets of all the atoms attached to the central atom in step 3, and in doing so we use all the valence electrons available. If the central atom does not yet have an octet, we cannot just add valence electrons to the structure. But we can complete the central atom's octet in step 5 by converting one or more lone pairs of electrons from outer atom(s) into bonding pairs. The following example—drawing the Lewis structure of formaldehyde (CH_2O)—illustrates how this is done.

Step 1. The total number of valence electrons in a CH_2O molecule is

ELEMENT		VALENCE ELECTRONS	
Symbol	# of atoms	In one atom	Total
C	1	4	4
H	2	1	2
O	1	6	6
	Valence electrons in CH_2O		12

Step 2. Carbon has the greatest bonding capacity (four) and is the central atom. Connecting it with single bonds to the other three atoms produces this skeletal structure:

H—C—H
|
O

Step 3. Each H atom has a single covalent bond (two electrons), thus completing its valence shell. The addition of three lone pairs of electrons completes the octet for oxygen:

H—C—H
|
:Ö:

Step 4. There are 12 valence electrons in this structure, which matches the number in the molecule. No additional electrons (dots) can be added to the structure.

Step 5. The central C atom has only six electrons. To provide the carbon atom with the two additional electrons it needs to have an octet—without removing any electrons from oxygen, which already has an octet—we convert one of the lone pairs on the oxygen atom into a bonding pair between C and O:

H—C—H → H\C/H (C=O)

It does not matter which of the three lone pairs is converted into a bonding pair because all three are equivalent. The central carbon atom now has a complete octet, as does the oxygen atom. This structure is reasonable because the four covalent bonds around carbon—two single bonds and one double bond—match its bonding capacity. The double bond to oxygen is reasonable because oxygen is a group 16 element with a bonding capacity of two, and it has two bonds in this structure.

Notice that we have drawn the double bond and the two single bonds around the central carbon atom and the double bond and lone pairs of electrons on the oxygen atom in a way that maximizes the separation between the pairs. We will learn in Chapter 5 that such arrangements more closely represent the actual bonding and lone pair orientations in molecules. For now, be assured there is nothing wrong with drawing a Lewis structure for formaldehyde in which the bond angles are, for example, 90° instead of 120°. The purpose of Lewis structures is to show how atoms are bonded to each other in molecules, not necessarily how the bonds are oriented in space.

SAMPLE EXERCISE 4.12 Drawing Lewis Structures with Double and Triple Bonds **LO4**

Draw the Lewis structure of acetylene, C_2H_2, the fuel used in oxyacetylene torches for cutting steel and other metals.

Collect, Organize, and Analyze One molecule of acetylene contains two carbon atoms and two hydrogen atoms. Carbon is a group 14 element with a bonding capacity of four, and hydrogen has a bonding capacity of one. We follow the steps in the guidelines, using double or triple bonds as needed.

Solve

Step 1. The two carbon atoms contribute four valence electrons each, and the two hydrogen atoms each contribute one for a total of $(4 \times 2) + (2 \times 1) = 10$.

Step 2. The two C atoms serve as central atoms in 3-atom subunits, each bonded to the other and to one of the H atoms, which gives us the skeletal structure:

$$\mathrm{H{-}C{-}C{-}H}$$

Step 3. Each H atom has a single covalent bond (consisting of two electrons) and a complete valence shell.

Step 4. There are six valence electrons in our structure, but there are ten in the molecule, so we have to add four more. No additional electrons can be placed on the outer H atoms because they have complete duets.

Step 5. The four remaining valence electrons must be placed on the two central carbon atoms. One way to do that is to add two bonding pairs between the carbon atoms. This gives the structure the right number of valence electrons, and it completes the octets of both carbon atoms:

$$\mathrm{H{-}C{\equiv}C{-}H}$$

Think About It The carbon atoms each form four bonds, as carbon atoms usually do. In this case the four consist of one single bond and one triple bond.

Practice Exercise
Draw the Lewis structure of carbon dioxide.

(Answers to Practice Exercises are in the back of the book.)

4.4 Resonance

The atmosphere contains two kinds of oxygen molecules. Most of them are O_2, but a tiny fraction are O_3, which is a form of oxygen called ozone. Different molecular forms of the same element are called **allotropes**, and they have different chemical and physical properties. Ozone, for example, is an acrid, pale

allotropes different molecular forms of the same element, such as oxygen (O_2) and ozone (O_3).

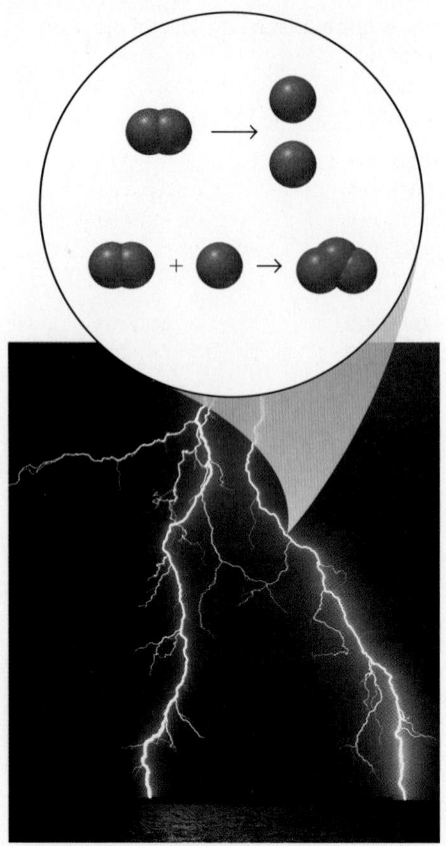

FIGURE 4.7 Lightning strikes contain sufficient energy to break oxygen–oxygen double bonds. The O atoms formed in this fashion collide with other O_2 molecules, forming ozone (O_3), an allotrope of oxygen.

blue gas that is toxic even at low concentrations, whereas O_2 is a colorless, odorless gas that is essential for most life-forms. Ozone is produced naturally by lightning (Figure 4.7) and is the source of the pungent odor you may have smelled after a severe thunderstorm. Ozone in the lower atmosphere is sometimes referred to as "bad ozone" because high levels in polluted air can damage crops, harm trees, and cause human health problems. Ozone in the upper atmosphere, on the other hand, is considered to be "good ozone" because it shields life on Earth from potentially harmful ultraviolet radiation from the sun.

Let's draw the Lewis structure of ozone by following our five-step process from Section 4.3. Oxygen is a group 16 element, so it has six valence electrons. The total number of valence electrons in an ozone molecule, then, is $3 \times 6 = 18$ (step 1). Connecting the three O atoms with single bonds (step 2) gives

O—O—O

Using 12 of the 14 remaining electrons to complete the octets of the atoms on the ends (step 3) gives

:Ö—O—Ö:

The last two electrons are added as a lone pair to the central oxygen atom (step 4):

:Ö—Ö—Ö:

We have used all 18 valence electrons, but this structure leaves the central atom two electrons shy of an octet, so we convert one of the lone pairs on the leftmost O atom into a bonding pair (step 5):

Ö=Ö—Ö:

We could also have formed the double bond with a lone pair from the rightmost O atom:

:Ö—Ö=Ö

CHEMTOUR Resonance

These two structures illustrate an important concept in Lewis theory called **resonance**: the existence of multiple Lewis structures, called **resonance structures**, which have the same arrangement of atoms but different arrangements of bonding electrons and lone pairs.

Having seen how it is possible to draw two equivalent Lewis structures for the same molecules, you may be wondering if either one accurately describes the bonding in ozone molecules. Experimental evidence indicates that, technically, neither structure is correct. Scientists have determined that the two bonds in ozone have exactly the same length, 128 pm. As we will see in Section 4.5, this value is about halfway between the length of an O—O single bond (148 pm) and an O=O double bond (121 pm). One way to explain this result is to assume that the actual bonding in an O_3 molecule is the average of the two resonance structures, which means identical bonds between the O atoms that are intermediate in length and strength between O—O single and O=O double bonds. It is as if two bonding pairs of electrons connect the center atom to the two others and that a third bonding pair is spread out across all three atoms.

This spreading out, or **delocalization**, of the third bonding pair can be explained by assuming bonding and lone pairs of electrons can be rearranged as shown by the red arrows in the following images.

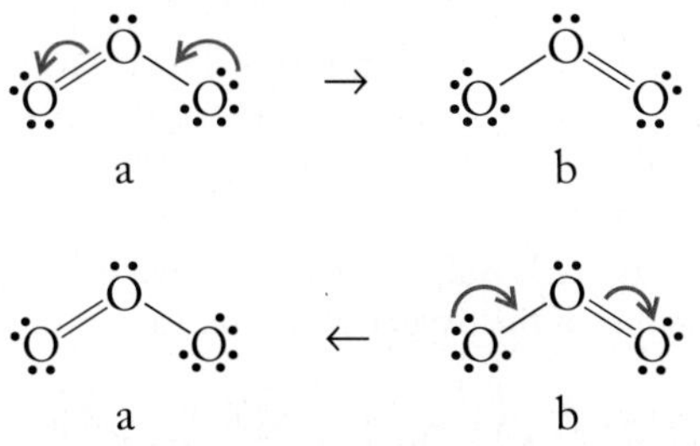

a → b

a ← b

This means that the bonding in a molecule of ozone is described by neither resonance structure (a) nor (b), but by an average of the two. A double-headed arrow is used between resonance forms to symbolize the averaging effect of bonding pair delocalization:

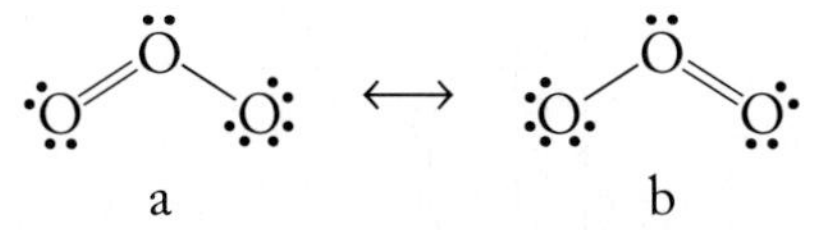

a ↔ b

A key point about delocalization is that it reduces the electrons' potential energy and lowers the energy of the molecule, a phenomenon called **resonance stabilization**. We will see in the chapters ahead that resonance can strongly influence the chemical properties of molecular substances. Resonance may also occur in polyatomic ions, as illustrated in Sample Exercise 4.13.

resonance a characteristic of electron distributions when two or more equivalent Lewis structures can be drawn for one compound.

resonance structure one of two or more Lewis structures with the same arrangement of atoms but different arrangements of bonding pairs of electrons.

electron-pair delocalization the spreading out of electron density over several atoms.

resonance stabilization the stability of a molecular structure due to the delocalization of its electrons.

SAMPLE EXERCISE 4.13 Drawing Resonance Structures **LO5**

Draw all the resonance structures of the nitrate ion, NO_3^-.

Collect, Organize, and Analyze We are asked to draw the resonance structures of the NO_3^- ion, which contains one atom of nitrogen (a group 15 element) and three atoms of oxygen (a group 16 element). The charge of the ion (1−) means it contains an additional valence electron.

Solve

Step 1. The number of valence electrons in a NO_3^- ion is

ELEMENT		VALENCE ELECTRONS	
Symbol	# of atoms	In one atom	Total
N	1	5	5
O	3	6	18
Plus one electron for the 1− charge			1
Valence electrons in ion			**24**

Step 2. Nitrogen has the higher bonding capacity, so N is the central atom. Connecting it with single bonds to the three O atoms gives us the following skeletal structure:

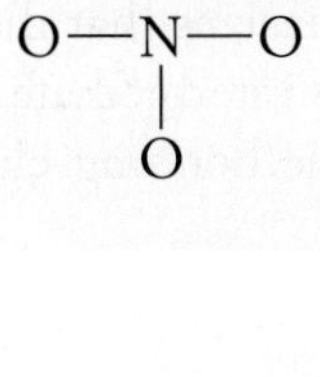

O—N—O
    |
    O

Step 3. Each O atom needs three lone pairs of electrons to complete its octet:

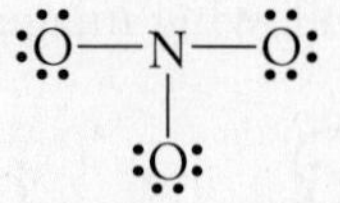

Step 4. There are 24 valence electrons in this structure, which matches the number determined in step 1.

Step 5. The central N atom has only six electrons, two short of an octet, so we convert a lone pair on one of the oxygen atoms into a bonding pair:

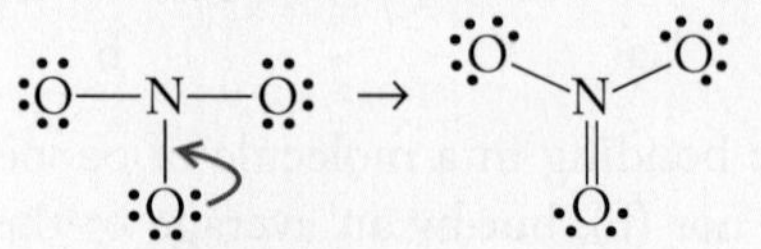

The nitrogen atom now has a complete octet. Adding brackets and the ionic charge, we have a complete Lewis structure:

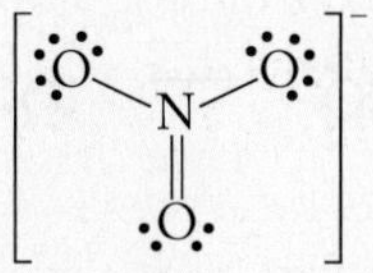

Using the O atom to the left or right of the central N atom to form the double bond creates two additional resonance forms, or three in all:

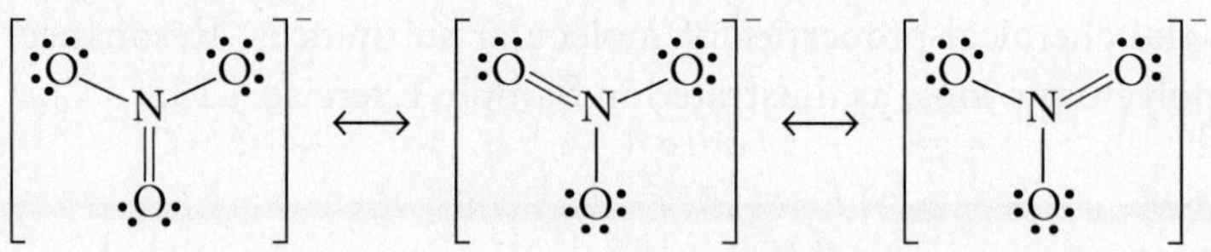

Think About It None of the three resonance structures accurately represents the nitrate ion; the true structure is an average of these three—a molecular ion in which there are three equivalent bonds rather than a double bond and two single bonds.

Practice Exercise Draw all the resonance forms of the azide ion, N_3^-, and the nitronium ion, NO_2^+.

(Answers to Practice Exercises are in the back of the book.)

FIGURE 4.8 The molecular structure of benzene is an average of the two equivalent structures at the top. It is frequently represented by a circle inside the hexagonal ring to indicate the completely uniform distribution of the electrons in the bonds around the ring.

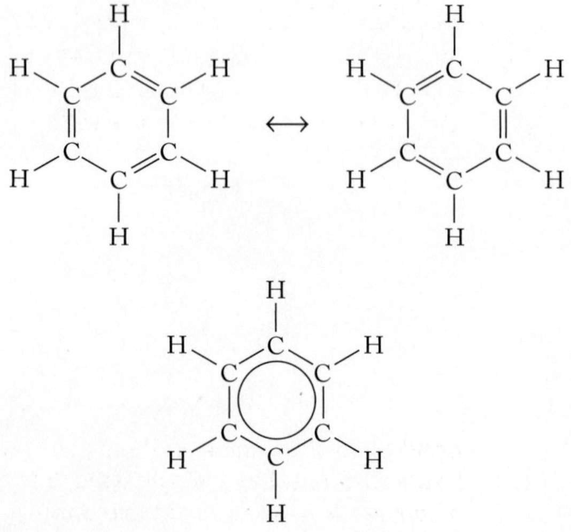

A test of whether a compound has multiple equivalent resonance forms—and therefore exhibits resonance stabilization—is the presence of one or more atoms having both single and double bonds to two or more atoms of another element, as in O_3 and NO_3^-. A molecule of benzene (C_6H_6) contains a ring of six carbon atoms with alternating single and double bonds (Figure 4.8), so it has that property, too. There are two equivalent ways to draw these single and double bonds, but chemists frequently draw benzene molecules with a circle in the center of the ring, as shown in Figure 4.8, to represent an averaging of the two resonance structures. The ring emphasizes that the six carbon–carbon bonds in the ring are all identical and intermediate in character between single and double bonds and that the bonding electrons are uniformly distributed around the ring.

4.5 The Lengths and Strengths of Covalent Bonds

The space-filling model of ozone in Figure 4.9(a) shows two equivalent oxygen–oxygen bonds. Both are 128 pm long, which is between the length of the typical O═O double bond (121 pm) and O—O single bond (148 pm) shown in Figure 4.9(b). In this section we use bond length and bond strength to rationalize and validate molecular structures.

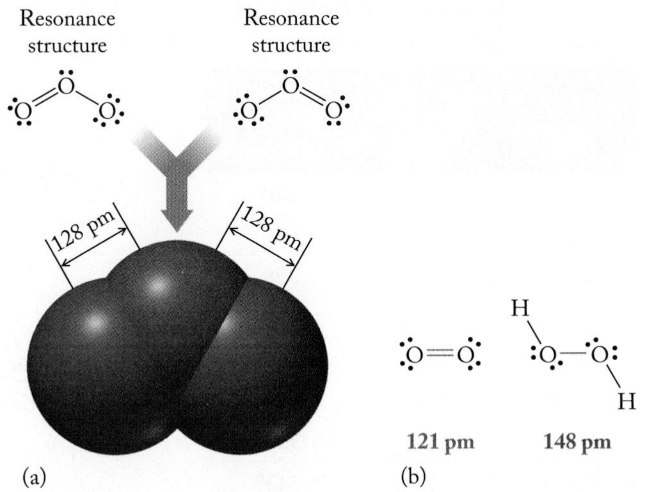

Resonance structure | Resonance structure

128 pm | 128 pm

121 pm | 148 pm

(a) (b)

FIGURE 4.9 (a) The molecular structure of ozone is an average of the two resonance structures shown at the top. Both bonds in ozone are 128 pm long. (b) The value falls between the average length of an O═O double bond (121 pm) and the average length of an O—O single bond (148 pm). The intermediate value for the ozone bond length indicates that the bonds in ozone molecules are neither single bonds nor double bonds but something in between, meaning that the true structure is neither resonance structure but an average of both.

Bond Length

The distance between atoms that are bonded to each other—that is, the length of the bond they form—depends on (1) the identities of the two atoms and (2) the number of bonds they form. The number of bonds is called the **bond order** and is equal to the number of pairs of electrons the atoms share. Experimental measurements of the lengths of the bonds in many molecular compounds indicate that the length of any given bond varies by only a few picometers in different compounds. For example, the C—H bond length in formaldehyde, CH_2O, is nearly the same as the C—H bond length in methane, CH_4. Similarly, the lengths of the C═O double bonds in formaldehyde and CO_2 are nearly identical (Figure 4.10).

As the bond order between two atoms increases, bond length decreases. We see this trend in the lengths of the O—O bond in hydrogen peroxide, H_2O_2, and the O═O bond in O_2 (Figure 4.9b); in the C═O bonds in CO_2 and the C≡O bond in CO (Figure 4.10); and in the average lengths of several series of bonds listed in Table 4.6. Bond length, then, can be used to determine bond order, and *vice versa*, even when bond order is not a simple whole number. For example, the actual bond order in O_3 is neither 1 nor 2, but rather 1.5, with each pair of O atoms sharing three electrons.

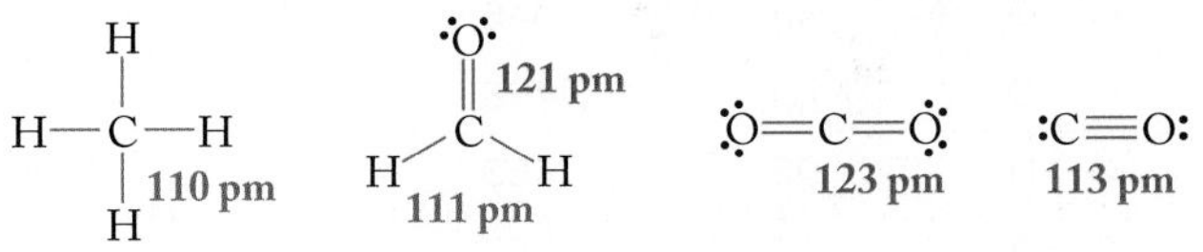

110 pm | 121 pm | 111 pm | 123 pm | 113 pm

FIGURE 4.10 Bond lengths depend on the identity of the two atoms forming the bond and decrease with increasing bond order. The lengths of the C—H single bonds in CH_2O and CH_4 are nearly the same, as are the lengths of the C═O double bonds in CH_2O and CO_2. However, the C≡O triple bond in CO is much shorter than the C═O double bonds in CH_2O and CO_2.

SAMPLE EXERCISE 4.14 Determining Bond Order and Bond Length from Resonance Structures **LO6**

Draw the resonance structures of the carbonate ion, CO_3^{2-}, and from these structures calculate the bond order of the carbon–oxygen bonds and estimate their length.

Collect and Organize We are asked to draw the resonance structures of a polyatomic ion and then to determine the order and length of its bonds. A five-step procedure for drawing Lewis structures is described in Section 4.3. Table 4.6 contains average bond lengths, which vary inversely with bond order.

Analyze We drew the resonance structures of the nitrate ion, NO_3^-, in Sample Exercise 4.13. Carbon atoms (group 14) have one fewer valence electron than nitrogen atoms (group 15), but carbonate ions have one more negative charge than nitrate ions. Therefore, CO_3^{2-} and NO_3^- are isoelectronic and should have similar Lewis structures.

bond order the number of bonds between atoms: 1 for a single bond, 2 for a double bond, and 3 for a triple bond.

TABLE 4.6 Average Lengths and Energies of Selected Covalent Bonds

Bond	Bond Length (pm)	Bond Energy (kJ/mol)
C—C	154	348
C=C	134	614
C≡C	120	839
C—N	147	293
C=N	127	615
C≡N	116	891
C—O	143	358
C=O	123	743[a]
C≡O	113	1072
C—H	110	413
C—F	133	485
C—Cl	177	328
N—H	104	391
N—N	147	163
N=N	124	418
N≡N	110	945
N—O	136	201
N=O	122	607
N≡O	106	678
O—O	148	146
O=O	121	498
O—H	96	463
S—O	151	265
S=O	143	523
S—S	204	266
S—H	134	347
H—H	74	436
H—F	92	567
H—Cl	127	431
H—Br	141	366
H—I	161	299
F—F	143	155
Cl—Cl	200	243
Br—Br	228	193
I—I	266	151

[a]The bond energy of the C—O bond in CO_2 is 799 kJ/mol.

Solve

Step 1. The number of valence electrons in a CO_3^{2-} ion is

ELEMENT		VALENCE ELECTRONS	
Symbol	# of atoms	In one atom	Total
C	1	4	4
O	3	6	18
Plus two electrons for the 2− charge			2
Valence electrons in ion			**24**

Step 2. Connecting the carbon atom with single bonds to the three O atoms gives the following skeletal structure:

O—C—O
|
O

Step 3. Adding lone pairs of electrons to complete the octets of the three O atoms:

:Ö—C—Ö:
|
:Ö:

Step 4. There are 24 valence electrons in this structure, which matches the number determined in step 1.

Step 5. The central C atom has only six electrons, so we convert a lone pair on one of the oxygen atoms into a bonding pair:

:Ö—C—Ö: → [structure with C=O double bond to bottom O]

The carbon atom now has a complete octet. Adding brackets and the ionic charge, we have a complete Lewis structure:

$[\ ]^{2-}$

Using the O atom to the left or right of the central C atom to form the double bond creates two additional resonance forms, or three in all:

$[\ ]^{2-} \longleftrightarrow [\ ]^{2-} \longleftrightarrow [\ ]^{2-}$

Each resonance structure contains two C—O bonds and one C=O bond, which means a total of four bonding pairs of electrons is distributed evenly among three pairs of bonded atoms. Therefore, each bond consists of 4/3 = 1.33 bonding pairs, giving it a bond order of 1.33.

According to Table 4.6, the average length of a C—O bond is 143 pm and the average length of a C=O bond is 123 pm. The length of the bonds in CO_3^{2-} ions should be in between these values.

Think About It In this exercise we used resonance structures to determine bond order and, in turn, to estimate bond length. The experimentally determined value of the bond lengths in CO_3^{2-} is 129 pm, which confirms our estimate. In the text we used an experimentally determined bond length in ozone to determine bond order and confirm the validity of the resonance structures we had drawn. The process seems to work well in both directions.

Practice Exercise Draw the resonance structures of HNO_3. Are all of the N—O bonds the same length?

(Answers to Practice Exercises are in the back of the book.)

polar covalent bond a bond resulting from unequal sharing of bonding pairs of electrons between atoms.

Bond Energies

The energy needed to break a H—H bond is represented by the depth of the potential energy minimum in Figure 4.3. That amount of energy (436 kJ/mol) is released when two H atoms come close enough together to form a bond. That same amount also must be used to break a single H—H bond—that is, to move the atoms so far apart that they are no longer electrostatically attracted to one another. Bond energy (or *bond strength*) for any bond is usually expressed in terms of the energy needed to break one mole of bonds in the gas phase. Average bond energies for some common covalent bonds, expressed in kilojoules per mole, are given in Table 4.6.

Bond energies (like bond lengths) are average values because they vary depending on the structure of the rest of the molecule. For example, the bond energy of a C═O bond in carbon dioxide is 799 kJ/mol, whereas the C═O bond energy in formaldehyde is only 743 kJ/mol. Bond energies are always positive quantities because breaking bonds requires the addition of energy.

Bond energies tend to increase as the bond order increases. For example, the bond energies of C—C, C═C, and C≡C bonds are 348, 614, and 839 kJ/mol, respectively. Table 4.6 shows similar trends in carbon–oxygen, nitrogen–nitrogen, and nitrogen–oxygen bonds.

4.6 Electronegativity, Unequal Sharing, and Polar Bonds

When Lewis proposed that atoms form chemical bonds by sharing electrons, he knew that electron sharing in covalent bonds did not necessarily mean *equal* sharing. For example, Lewis knew that molecules of HCl ionized to form H^+ ions and Cl^- ions when HCl dissolves in water. To explain this phenomenon, Lewis proposed that the bonding pair of electrons in a molecule of HCl is closer to the chlorine end of the bond than the hydrogen end. This unequal sharing makes the H—Cl bond a **polar covalent bond**. As a result, when the H—Cl bond breaks, the one shared pair of electrons remains with the Cl atom to form a Cl^- ion and simultaneously changes the H atom into an H^+ ion with no electrons.

The polarity of the bond in HCl means that the bond functions as a tiny electric *dipole*, in which case there is a slightly positive pole at the H end of the bond and a slightly negative pole at the Cl end, analogous to the positive and negative ends of the battery shown in Figure 4.11. This figure also shows two representations we use to depict unequal sharing of bonding pairs of electrons. One of them incorporates an arrow with a plus sign embedded in its tail. The

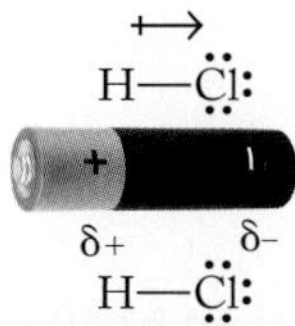

FIGURE 4.11 Just as a battery has positive and negative terminals, a polar bond such as the one in a molecule of HCl has positive and negative ends, represented here by the arrow with a positive tail above the H—Cl bond and by delta symbols (δ+ and δ−).

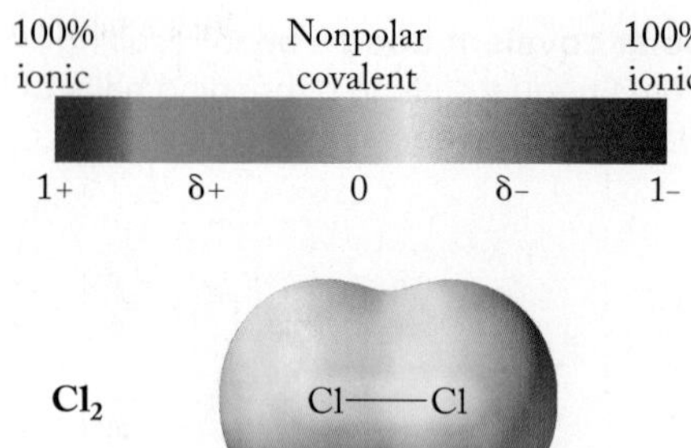

(a) Nonpolar covalent: even charge distribution

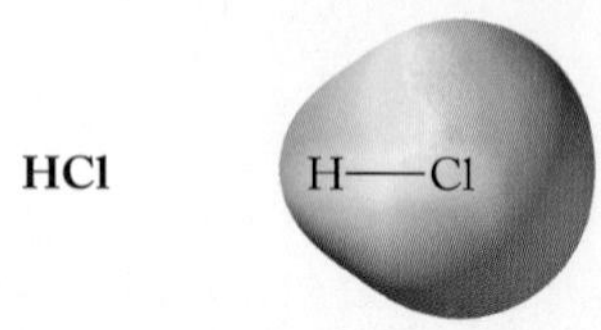

(b) Polar covalent: uneven charge distribution

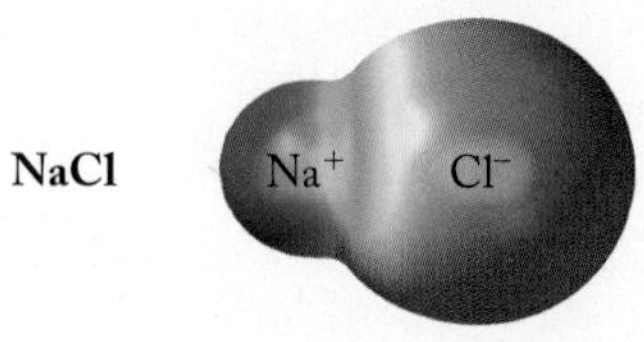

(c) Ionic: complete transfer of electron

FIGURE 4.12 Variations in valence electron distribution are represented using colored surfaces in these molecular models. (a) In the covalent bond in Cl_2, the same color pattern on both ends of the bond means that the two atoms share their bonding pair of electrons equally. (b) Unequal sharing of the bonding pair of electrons in HCl is shown by the partial negative change and orange-red color of the Cl atom and the partial positive charge and blue-green color around the H atom. (c) In ionic NaCl, the violet color on the surface of the sodium ion indicates that it has a 1+ charge, and the deep red of the chloride ion reflects its charge of 1−.

arrow points toward the more negative, electron-rich end of the bond, while the plus sign represents the more positive, electron-poor end. Another representation makes use of the lowercase Greek delta, δ, followed by a + or − sign. The deltas represent *partial* electrical charges, as opposed to the full electrical charges that designate the complete transfer of one or more electrons when atoms become cations or anions.

Figure 4.12 shows examples of the equal and unequal sharing of bonding pairs of electrons. In Figure 4.12(a), the Cl atoms of Cl_2 are connected by a **nonpolar covalent bond** because the bonding electrons are shared equally between two identical atoms. Figure 4.12(b), on the other hand, depicts the polar covalent bond in HCl, and Figure 4.12(c) depicts the ionic bond between a pair of Na^+ and Cl^- ions in NaCl. In NaCl, the valence electron of sodium has been completely transferred to the chlorine atom, creating a Na^+ ion and a Cl^- ion and resulting in *total*, rather than *partial*, separation of electrical charge: 1+ on Na and 1− on Cl. The degree of charge separation may be represented using color, as shown in the calibration bar at the top of Figure 4.12, where yellow-green represents equal sharing (nonpolar covalent bonding) and no partial charges. The red and violet ends represent full charge separation (ionic bonding), while the colors in between represent various degrees of partial charge separation (polar covalent bonding).

Another American chemist, Linus Pauling (1901–1994), explained why some covalent bonds are polar using the concept of **electronegativity**, which is represented by the Greek letter χ (chi) and defined as the tendency of an atom to attract electrons toward itself within a chemical bond. Central to this concept is the assumption that a bond between two atoms of the same element is 100% covalent, while the bonds between atoms of two different elements are not 100% covalent. Pauling developed electronegativity values for the most common elements (Figure 4.13). The degree of ionic character in a bond depends on the difference (Δχ) in the electronegativity values of the two elements the bond connects.

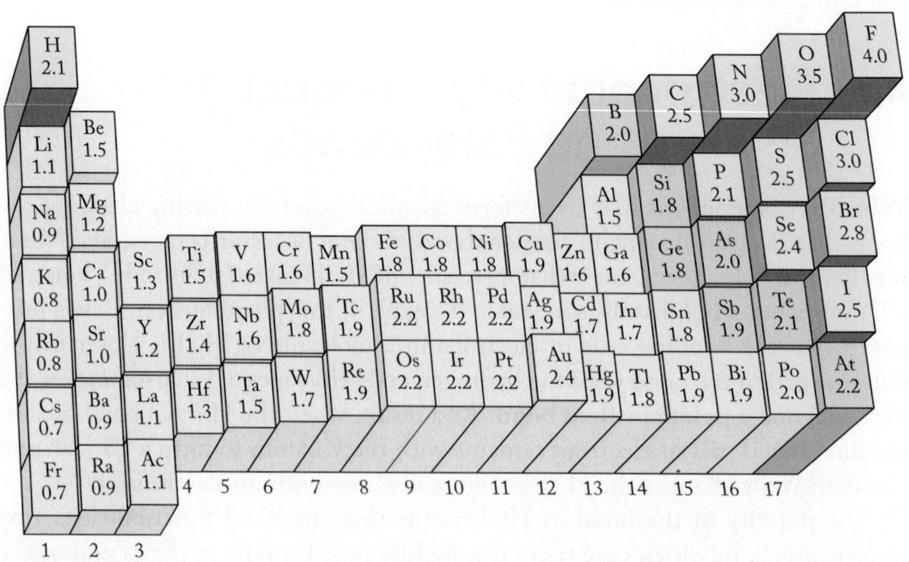

FIGURE 4.13 The electronegativity values of the elements increase from left to right across a period and decrease from top to bottom down a group. The greater an element's electronegativity, the greater the ability of an atom to attract electrons toward itself within a chemical bond.

nonpolar covalent bond a bond characterized by an even distribution of charge; electrons in the bond are shared equally by the two atoms.

electronegativity a relative measure of the ability of an atom to attract electrons to itself within a bond.

When $\Delta\chi$ is less than or equal to 0.4, a covalent bond is considered essentially nonpolar. The bond between Cl (χ = 3.0) and Br (χ = 2.8) is an example. When $\Delta\chi$ is between 0.4 and 2.0, bonds are considered polar covalent, and when $\Delta\chi$ is equal to or greater than 2.0, bonds are considered ionic. Thus, H (2.1), F (4.0), and Cl (3.0) form polar covalent bonds in HF and HCl, but the H—F bond ($\Delta\chi$ = 1.9) is *more* polar than the H—Cl bond ($\Delta\chi$ = 0.9). Calcium oxide is considered an ionic compound because the $\Delta\chi$ between Ca (1.0) and O (3.5) is 2.5. Keep in mind that these cutoff values are more like guidelines than strict limits. We revisit the characterization of bonds as either ionic or polar covalent in Chapter 5, where we discuss the polarity of entire molecules.

The data in Figure 4.13 show that electronegativity is a periodic property of the elements, with values generally increasing from left to right across a row in the periodic table and decreasing from top to bottom down a group. These trends arise for essentially the same reasons that produce the similar trends in first ionization energies (see Figure 3.37). Greater attraction between the nuclei of atoms and their outer-shell electrons produces both higher ionization energies and greater electronegativities across a row (Figure 4.14a). Down a group of elements, the weaker attraction between nuclei and valence-shell electrons as atomic number increases leads to lower ionization energies and smaller electronegativities (Figure 4.14b). For those two reasons, the most electronegative elements—fluorine, oxygen, nitrogen, and chlorine—are in the upper right corner of the periodic table, whereas the least electronegative elements—francium, cesium, rubidium, and potassium—are in the lower left corner.

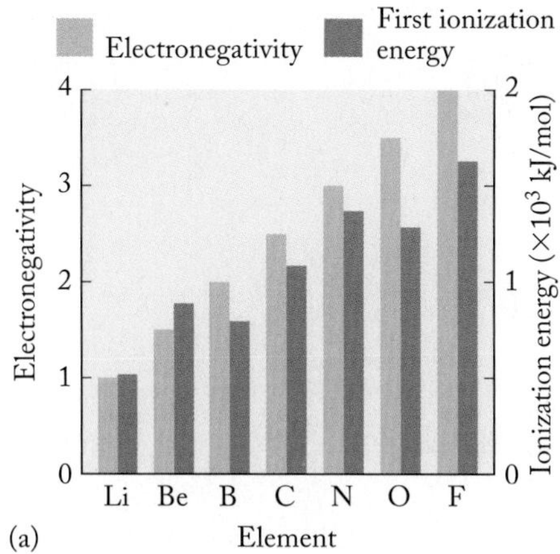

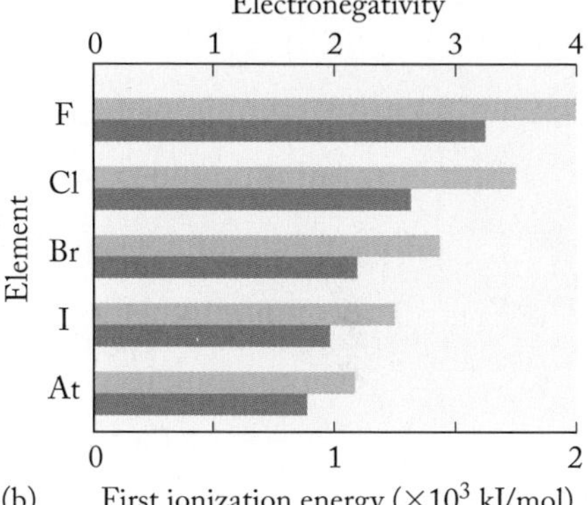

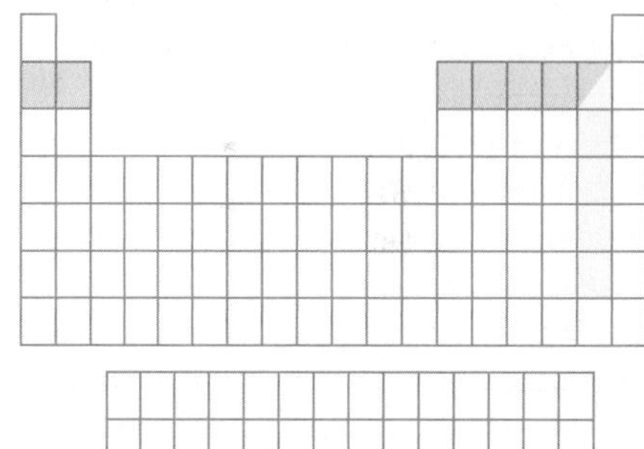

FIGURE 4.14 The trends in the electronegativities of the main group elements follow those of first ionization energies: both tend to (a) increase with increasing atomic number across a row and (b) decrease with increasing atomic number within a group.

CONCEPT TEST

Draw an arrow on the Lewis structure of CO to indicate the polarity of its carbon–oxygen bond.

(Answers to Concept Tests are in the back of the book.)

SAMPLE EXERCISE 4.15 Comparing Bond Polarities LO7

Rank the bonds formed between atoms of O and C, Cl and Ca, N and S, and O and Si in order of increasing polarity. Should any of the bonds be considered ionic?

Collect, Organize, and Analyze To rank the polarities of the bonds between different pairs of atoms, we need to calculate the differences in the parent elements' electronegativities ($\Delta\chi$, see Figure 4.13).

Solve Calculate the electronegativity differences:

O and C: $\Delta\chi = 3.5 - 2.5 = 1.0$

Cl and Ca: $\Delta\chi = 3.0 - 1.0 = 2.0$

N and S: $\Delta\chi = 3.0 - 2.5 = 0.5$

O and Si: $\Delta\chi = 3.5 - 1.8 = 1.7$

The electronegativity differences are proportional to the polarities of the bonds formed between the pairs of atoms. Therefore, ranking them in order of increasing polarity we have the following:

$$\text{N—S} < \text{O—C} < \text{O—Si} < \text{Cl—Ca}$$

The bond between Cl and Ca, with $\Delta\chi = 2.0$, is considered to be ionic based on the guidelines described in the text.

Think About It Ionic bonds tend to form between metals and nonmetals, so concluding that the bond between calcium (a metal) and chlorine (a nonmetal) is ionic is reasonable. We examined the Lewis structure for $CaCl_2$ in Section 4.3. Two of the other bonds, N—S and O—C, connect pairs of nonmetals, and the O—Si bond connects a nonmetal with a metalloid. We expect those three bonds to be covalent.

Practice Exercise Which pair forms the most polar bond: O and S, Be and Cl, N and H, or C and Br? Is the bond between that pair ionic?

(Answers to Practice Exercises are in the back of the book.)

4.7 Formal Charge: Choosing among Lewis Structures

Dinitrogen monoxide (N_2O), also known as nitrous oxide, is an atmospheric gas used as an anesthetic in dentistry and medicine. Its common name—laughing gas—is derived from the euphoria people feel when they inhale it.

To draw the Lewis structure of N_2O, we first count the number of valence electrons: five each from the two nitrogen atoms and six from the oxygen atom for a total of $(2 \times 5 + 6) = 16$. The central atom is a nitrogen atom because N has a higher bonding capacity than O. Connecting the atoms with single bonds we have

N—N—O

Completing the octets of the N and O atoms on the ends gives us a structure with 16 valence electrons, which is the number determined in step 1:

:N̤̈—N—Ö̤:

However, there are only two single bonds, and thus only four valence electrons, on the central N atom. To give it four more electrons and complete its octet, we need to convert lone pairs on the end atoms to bonding pairs. Which lone pairs should we choose? We have three choices. We could use two lone pairs from the N atom on the left to form a N≡N triple bond:

:N≡N—Ö̤:

Or we could use two lone pairs from the O atom to make a N≡O triple bond:

:N̤̈—N≡O:

Finally, we could use one lone pair from each end atom to make two double bonds:

:N̈=N=Ö:

Are these three resonance structures equivalent, as we saw with the carbonate ion CO_3^{2-} in Sample Exercise 4.14? So far, in the sets of resonance structures we have considered, such as those of O_3 and NO_3^-, all the structures have been equivalent, so no one of them has been more important than another in representing the actual bonding in the molecule. However, the three resonance structures for N_2O are not equivalent to one another, because one contains a N≡N triple bond, one contains a N≡O triple bond, and one contains both a N=N double bond and a N=O double bond. To help us decide which resonance form in a nonequivalent set is the most important in representing the actual bonding pattern in a molecule, we make use of the concept of formal charge.

A **formal charge (FC)** is not a real charge, but rather an accounting system for the number of electrons *formally assigned* to an atom in a structure of a molecule or polyatomic ion. To understand what formal charge means, let's go through the process of calculating it for each of the three atoms in N_2O.

formal charge (FC) the value calculated for an atom in a molecule or polyatomic ion by determining the difference between the number of valence electrons in the free atom and the sum of the lone pair electrons plus half of the electrons in the atom's bonding pairs.

Calculating Formal Charge

1. Determine the number of valence electrons in the free atom (the number of dots in its Lewis symbol).
2. Count the number of lone pair electrons on the atom in the structure.
3. Count the number of electrons in bonds to the atom and divide that number by 2.
4. Sum the results of steps 2 and 3, and subtract that sum from the number determined in step 1.

Summarizing these steps in the form of an equation, we have

$$\text{FC} = \begin{pmatrix}\text{number of} \\ \text{valence e}^-\end{pmatrix} - \begin{bmatrix}\text{number of e}^- \\ \text{in lone pairs} + \frac{1}{2}\begin{pmatrix}\text{number of} \\ \text{shared e}^-\end{pmatrix}\end{bmatrix} \quad (4.2)$$

The calculation of formal charge assumes that each atom is *formally assigned* all its lone pair electrons and half of the electrons shared in bonding pairs. If this number matches the number of valence electrons in a single atom of the element (as reflected in its Lewis symbol), then the formal charge on the atom is zero.

We can confirm that the three resonance forms of N_2O we have drawn are not equivalent by calculating the formal charges on the three atoms in each structure. The steps in this calculation are highlighted in the table that follows, where we have colored the lone pairs of electrons red and the shared pairs green to make it easier to track the quantities of the electrons in the formal charge calculations. The numbers of valence electrons in free atoms of N (5) and O (6) are shown in blue.

FORMAL CHARGE CALCULATIONS FOR THE RESONANCE STRUCTURES OF N_2O

STEP	:N≡N—Ö: A			N̈=N=Ö B			:N̈—N≡O: C		
1 Number of valence electrons	5	5	6	5	5	6	5	5	6
2 Number of lone pair electrons	2	0	6	4	0	4	6	0	2
3 Number of shared electrons	6	8	2	4	8	4	2	8	6
4 FC = valence – [lone pair + ½ (shared)]	0	+1	–1	–1	+1	0	–2	+1	+1

To illustrate one of the formal charge calculations in the table, consider the N atom at the end in structure A. It has two electrons in a lone pair and six electrons in three shared (bonding) pairs. Using Equation 4.2 to calculate the formal charge on this N atom,

$$\text{FC} = 5 - [2 + \tfrac{1}{2}(6)] = 0$$

The results of similar FC calculations for all the other atoms in the three resonance structures complete the bottom row in the table. Note that the formal charges of the three atoms in each of the three structures sum to zero, as they should for a neutral molecule. When we analyze the formal charges of the atoms in a polyatomic ion, the formal charges on its atoms must add up to the charge on the ion.

Having calculated the formal charges, we now need to use the results to decide which of the three N_2O structures is the best representation of the bonding in these molecules. To make this decision, we use three criteria:

1. The best structure is the one in which the formal charge on each atom is zero.
2. If no such structure can be drawn, or if the structure is that of a polyatomic ion, then the best structure is the one where most atoms have formal charges equal to zero or as close to zero as possible.
3. Any negative formal charges should be on the atom(s) of the more/most electronegative element.

Let's apply these criteria to the nonequivalent resonance structures of N_2O. First of all, none of the structures meet criterion (1) because each structure has at least two nonzero FC values. So we proceed to criterion (2) to find the structure with the most FC values that are closest to zero, such as -1 or $+1$. When we do, we have a tie between structure A (0, +1, -1) and structure B (-1, +1, 0).

To break the tie, we invoke criterion (3) by asking the question, "In which structure is the negative formal charge on the more electronegative atom?" Oxygen is more electronegative than nitrogen, so structure A, in which the formal charge on O is -1 and on N is 0, is the best representation of the actual bonding in N_2O.

How does calculating formal charges and choosing a best structure from among the nonequivalent possible structures compare to reality? The length of the bond between the two nitrogen atoms is shorter than the length of a $\text{N}{=}\text{N}$ bond but longer than the length of a $\text{N}{\equiv}\text{N}$ bond, so the true structure is in between the left structure and the middle structure. As a result, we say that the middle structure "contributes to" the bonding in N_2O. This reality check is important as we interpret the results of formal charge analyses. Just because a resonance structure scores the best in an FC analysis does not mean it precisely matches experimental data. The true structure is often intermediate between two contributing structures.

CONCEPT **TEST**

What is the formal charge on a sulfur atom that has three lone pairs of electrons and one bonding pair?

(Answers to Concept Tests are in the back of the book.)

Is there a link between the calculated formal charge on an atom in a resonance structure and the bonding capacity of that atom? The Lewis symbol of nitrogen has three unpaired electrons, so a nitrogen atom can complete its octet by forming three bonds. The Lewis symbol of oxygen has two unpaired electrons, so the bonding capacity of an oxygen atom is two. In the N_2O calculations, the nitrogen atom with three bonds has a formal charge of zero (structure A), and the oxygen atom with two bonds has a formal charge of zero (O in structure B). As a general rule—and assuming the octet rule is obeyed—atoms have formal charges of zero in resonance structures in which the numbers of bonds they form match their bonding capacities. If the number of bonds an atom forms is one more than its bonding capacity, such as an oxygen atom with three bonds (O in structure C), the formal charge is +1. If the number of bonds it forms is one fewer than the bonding capacity, such as an oxygen atom with one bond (O in structure A), the formal charge is −1.

SAMPLE EXERCISE 4.16 Selecting a Resonance Structure Based on Formal Charges **LO5**

Which of these resonance forms best describes the actual bonding in a molecule of CO_2?

$$:\ddot{\underset{..}{O}}-C\equiv O: \longleftrightarrow \ddot{\underset{}{O}}=C=\ddot{O} \longleftrightarrow :O\equiv C-\ddot{\underset{..}{O}}:$$

Collect, Organize, and Analyze We are given three nonequivalent resonance structures for CO_2. Formal charges can be used to select the structure most likely to match the experimentally determined bond lengths. The preferred structure is one in which the formal charges are closest to zero and any negative formal charges are on the more electronegative atom (O). In this case, oxygen is a group 16 element with six valence electrons and carbon is a group 14 element with four valence electrons.

Solve We use Equation 4.2 to calculate the formal charge on each atom. The results are tallied in the following table, where we have applied the same color-coding scheme used previously for the N_2O calculations:

FORMAL CHARGE CALCULATIONS FOR THE RESONANCE STRUCTURES OF CO_2

STEP	:Ö—C≡O:			Ö=C=Ö			:O≡C—Ö:		
1 Number of valence electrons	6	4	6	6	4	6	6	4	6
2 Number of lone pair electrons	6	0	2	4	0	4	2	0	6
3 Number of shared electrons	2	8	6	4	8	4	6	8	2
4 FC = valence − [lone pair + ½ (shared)]	−1	0	+1	0	0	0	+1	0	−1

The formal charges are all zero on the atoms in the structure with two double bonds. Therefore, this structure best represents the actual bonding in CO_2.

Think About It The sum of the formal charges is zero in all three resonance structures, as it should be for a neutral molecule.

Practice Exercise Which resonance structure(s) of the nitronium ion, NO_2^+ (see Practice Exercise 4.13), contribute(s) the most to the actual bonding in the ion?

(Answers to Practice Exercises are in the back of the book.)

4.8 Exceptions to the Octet Rule

Not all atoms achieve complete octets when forming covalent bonds. We have seen how a H atom can complete its duet by forming one covalent bond. The Lewis symbol of beryllium, •Be•, indicates that a Be atom has the capacity to form two bonds. However, adding two more electrons to the two it already has results in only four electrons in its valence shell, not eight. Similarly, the three dots in the Lewis symbols of boron and aluminum indicate that their atoms have the capacity to form three bonds, but doing so results in only six valence-shell electrons. Thus, according to Lewis theory, Be, B, and Al tend to form *electron-deficient* molecules. There are other examples, including those with odd numbers of electrons in them.

Odd-Electron Molecules

Traces of nitrogen monoxide (NO) enter the atmosphere from automobile exhaust and then react with O_2 to form nitrogen dioxide, NO_2, which plays a key role in the formation of air pollution known as photochemical smog. Molecules of NO and NO_2 have odd numbers of valence electrons: $5 + 6 = 11$ in NO and $5 + 2 \times 6 = 17$ in NO_2. To understand what that means, let's draw the Lewis structure of NO. There are only two atoms, so there is no *central* atom. Therefore, we start with a single bond between N and O and then complete the octet around O, which is the more electronegative element:

N—Ö:

Placing the remaining three electrons around the N atom

N—Ö:

leaves the N atom with only five valence electrons. To increase the number, we convert a lone pair on the O atom into a bonding pair:

N═O

This change has the added advantage of creating a double-bonded O atom, which gives it a formal charge of zero. The formal charge on nitrogen is also zero.

The only problem with this structure is that nitrogen has only seven electrons. Is there a way to give it eight electrons? Not really. We could use another of the O atom lone pairs to make a bonding pair:

N≡O:

but the N atom in this new structure now has *nine* valence-shell electrons, which is impossible for an atom with only four orbitals in its valence shell. Moving only one electron from the O atom to the N atom gives the N atom eight valence electrons, but now the O atom has only seven valence electrons:

:N≡O.

This makes matters worse, because now the more electronegative element (O), which attracts bonding electrons more strongly than the other (N), has an incomplete octet. As a result, it is reasonable to choose the structure where oxygen has an octet but nitrogen does not when there are insufficient electrons to complete the octets of both. Compounds that contain unpaired valence electrons are called **free radicals**. They are typically very reactive species because it is often energetically favorable for them to acquire or share an electron from another molecule or ion.

free radical an atom, ion, or molecule with unpaired electrons.

SAMPLE EXERCISE 4.17 Drawing the Lewis Structures of an Odd-Electron Molecule **LO5**

Draw the resonance structures of nitrogen dioxide (NO_2) that have formal charges closest to zero.

Collect, Organize, and Analyze We need to draw the resonance structures of NO_2 that best describe the actual bonding in its molecules. Each molecule of NO_2 contains an odd number (17) of electrons, which means one of its atoms has an incomplete octet.

Solve Nitrogen has the greater bonding capacity, so it is the central atom:

O—N—O

Completing the octets on the O atoms gives

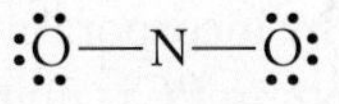

:Ö—N—Ö:

There are 16 valence electrons in the structure, so we add the one remaining valence electron to the N atom:

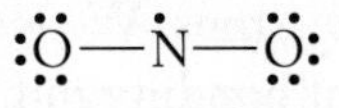

:Ö—Ṅ—Ö:

There are only five valence electrons around the N atom, leaving it three short of an octet. We increase the number by converting a lone pair on one of the O atoms to a bonding pair. Two equivalent resonance structures can be drawn with the formal charges shown in red:

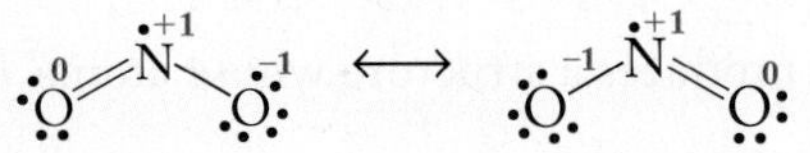

O(0)=N(+1)—O(−1) ⟷ O(−1)—N(+1)=O(0)

Think About It The two resonance structures are equivalent because each contains one N—O single bond and one N═O double bond. Neither satisfies the octet rule, but the formal charges of the atoms are close to zero, and the negative formal charge is on the more electronegative element (oxygen). Both O atoms have complete octets, leaving the less electronegative N atom one electron short in the odd-electron molecule. The odd electron on the N atom in NO_2 makes it chemically reactive. For example, two NO_2 molecules may combine by sharing the odd electrons on their N atoms, forming a N—N bond and a molecule of N_2O_4:

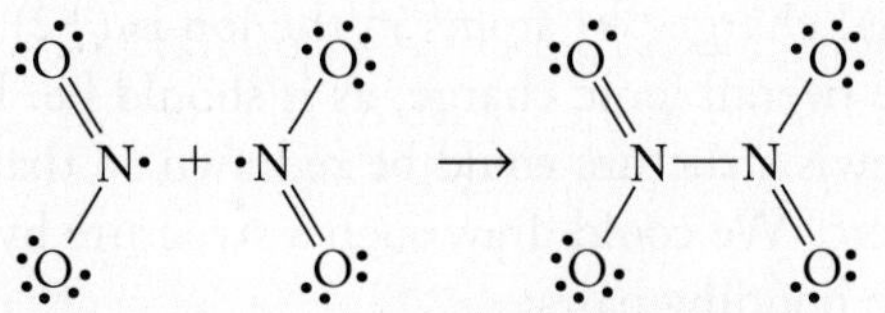

NO_2 (O═N•—O) + •N(O_2) → O_2N—NO_2

Practice Exercise Nitrogen trioxide (NO_3) may form in polluted air when NO_2 reacts with O_3. Draw its Lewis structure(s).

(Answers to Practice Exercises are in the back of the book.)

Expanded Octets

CHEMTOUR
Lewis Structures: Expanded Valence Shells

In some molecules, atoms of nonmetals in the third row and below in the periodic table ($Z > 12$) appear to have more than an octet of valence electrons. Consider, for example, the Lewis structures of PCl_5 and SF_6:

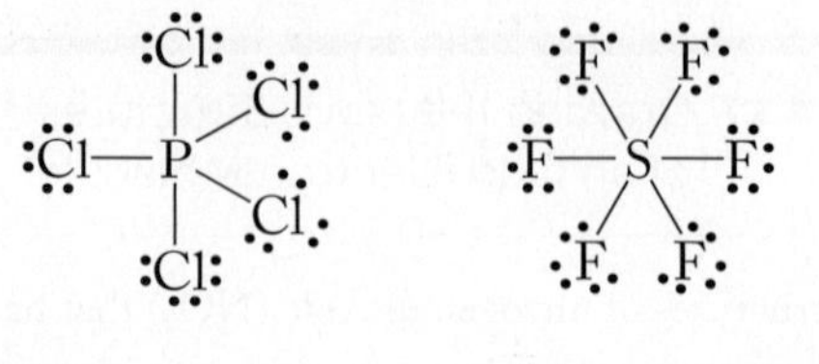

(a) (b)

The five covalent bonds in PCl_5 and the six in SF_6 indicate a total of 10 and 12 valence electrons around the central P and S atoms, respectively. All of the atoms in both structures have formal charges of zero, so the two structures seem to be perfectly acceptable representations of the bonding in these molecules. On the other hand, how can P and S atoms have more than eight valence electrons?

Over the years, chemists have devised several explanations for what many have called the *hypervalency* of phosphorus, sulfur, and other nonmetals in the third row and below in the periodic table. Some have proposed that these elements may have expanded octets by incorporating valence-shell *d* orbitals in bond formation. For example, the P atom in PCl_5 might use one of its empty 3*d* orbitals along with its 3*s* and its three 3*p* orbitals to form five covalent bonds. Others have proposed that hypervalency is really an illusion and that atoms can form five or six bonds using only eight valence electrons. We will revisit this possibility when we explore another theory of covalent bonding in Chapter 5.

The fact that some atoms may have expanded valence shells does not necessarily mean they always do. Rather, this tends to happen when

1. They bond with strongly electronegative elements—particularly F, O, and Cl.
2. An expanded shell produces a structure whose atoms' formal charges are closer to zero.

Consider, for example, the S atom in a $SO_4{}^{2-}$ ion. Following the usual steps to draw its Lewis structure and assign formal charges, we get

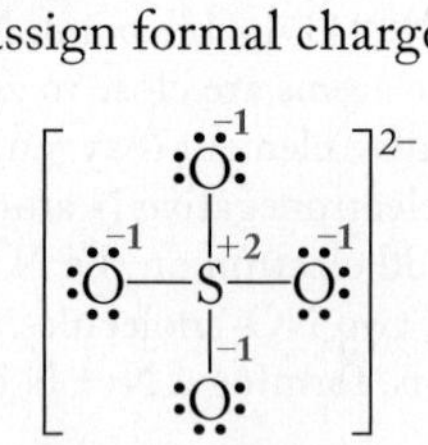

The sum of the formal charges on atoms in the ion is $(+2) + [4 \times (-1)] = -2$, which is equal to the overall ionic charge, as it should be. However, we need to consider how this Lewis structure could be redrawn so that at least some of the formal charges are zero. We could draw such a structure by converting two lone pairs of electrons into bonding pairs:

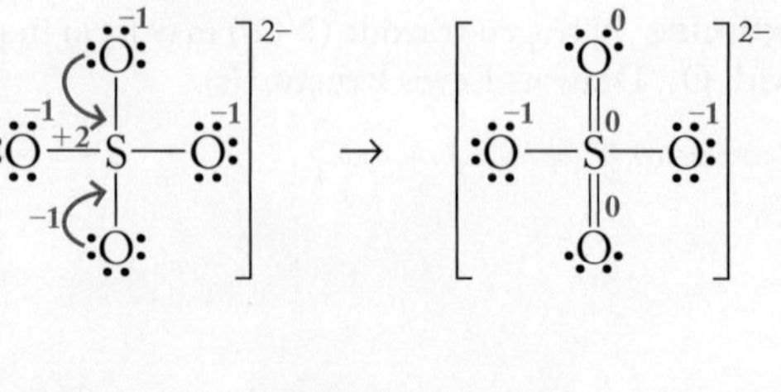

$\rightarrow$

Each O atom still has a complete octet, but the S atom has an expanded valence shell to accommodate 12 electrons. In this structure the formal charge on the S atom is 0, as are the formal charges on two of the four O atoms. There is still a formal charge of −1 on the other two oxygen atoms, which sums to an overall 2− charge of the ion. We could draw the two double bonds to any two of the O atoms, which means the structure is stabilized by resonance.

We can draw the Lewis structure of H_2SO_4 by combining two hydrogen ions (H^+) to the two oxygen atoms with 1− charges:

$$\begin{array}{c} \ddot{O} \\ \| \\ H-\ddot{O}-S-\ddot{O}-H \\ \| \\ \ddot{O} \end{array}$$

Each hydrogen atom has achieved its duet of electrons, each oxygen atom has an octet, and every atom has a formal charge of zero.

Based on the preceding formal charge analyses, we might conclude that the preferred bonding pattern in SO_4^{2-} ions and molecules of H_2SO_4 includes two S═O double bonds and zero formal charges all around. However, experimental evidence suggests that although the two structures with zero formal charges do contribute to the bonding of SO_4^{2-} and H_2SO_4, structures that obey the octet rule and have no S═O double bonds contribute as well. Thus, the actual bonding in these particles is an average of both the expanded octet and normal octet structures.

SAMPLE EXERCISE 4.18 Drawing a Lewis Structure Containing an Atom with an Expanded Valence Shell **LO5**

Draw a Lewis structure for the phosphate ion (PO_4^{3-}) that minimizes the formal charges on its atoms.

Collect, Organize, and Analyze Each ion contains one atom of phosphorus and four atoms of oxygen and has an overall charge of 3−. Phosphorus and oxygen are in groups 15 and 16 and have bonding capacities of three and two, respectively. Phosphorus is in row 3 ($Z = 15$), so it may exhibit hypervalency.

Solve The number of valence electrons is

ELEMENT		VALENCE ELECTRONS	
Symbol	# of atoms	In one atom	Total
P	1	5	5
O	4	6	24
Plus 3 electrons for the 3− charge			3
Valence electrons in ion			**32**

Phosphorus has the greater bonding capacity (3), so it is the central atom:

$$\begin{array}{c} O \\ | \\ O-P-O \\ | \\ O \end{array}$$

Each O atom needs three lone pairs of electrons to complete its octet:

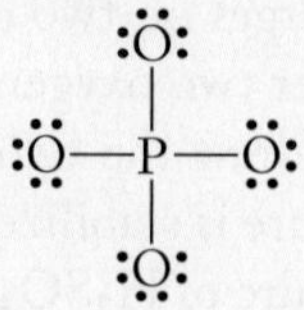

There are 32 valence electrons in the structure, which matches the number in the ion. To complete the Lewis structure we add brackets and its electrical charge:

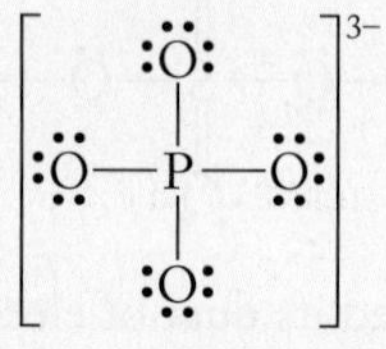

Each O has a single bond and a formal charge of −1; the four bonds around the P atom are one more than its bonding capacity, so its formal charge is +1. The sum of the formal charges, +1 + 4(−1) = −3, matches the charge on the ion, 3−.

We can reduce the formal charge on P by increasing the number of bonds to it, and we can do that by converting a lone pair on one of the O atoms into a bonding pair:

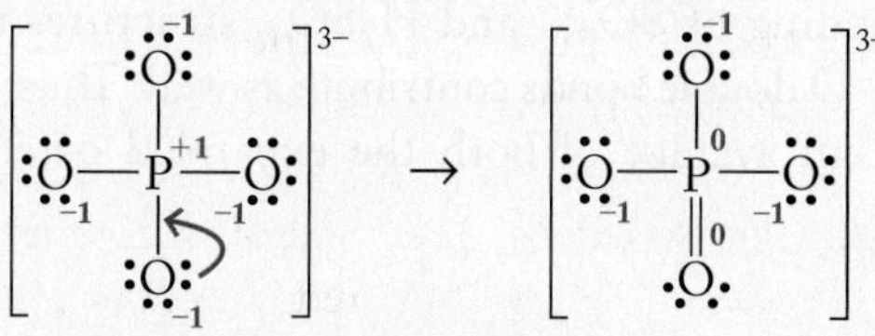

At the same time, we change a single-bonded O atom into a double-bonded O atom and thereby make its formal charge zero. Therefore, the structure on the right, in which the P atom has an expanded valence shell, is the best Lewis structure we can draw for the phosphate ion.

Think About It The phosphorus atom in the final structure has an expanded octet. It is also stabilized by resonance because we could draw the P═O double bond between any of the four O atoms and the central P atom.

Practice Exercise Draw the resonance structures of the selenite ion (SeO_3^{2-}) that minimize the formal charges on the atoms.

(Answers to Practice Exercises are in the back of the book.)

CONCEPT TEST

PF_5 exists, but NF_5 does not. Suggest a reason why.

(Answers to Concept Tests are in the back of the book.)

4.9 Vibrating Bonds and the Greenhouse Effect

Covalent bonds are not rigid. They all vibrate a little, stretching and bending like tiny atomic-sized springs (Figure 4.15). As polar bonds vibrate, the strengths of tiny electrical fields produced by the partial separations of charge in the bonds

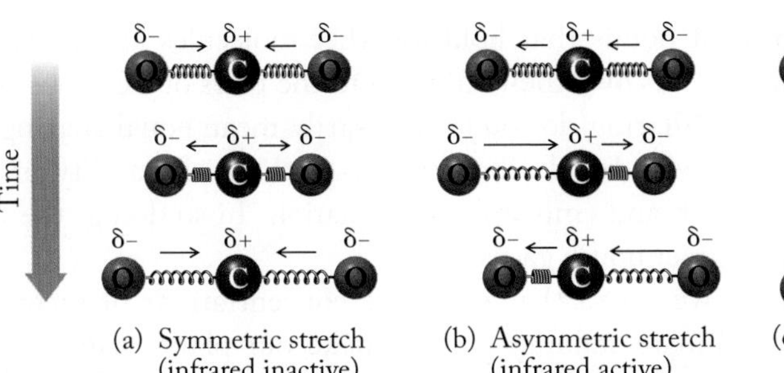

FIGURE 4.15 Three modes of bond vibration in a molecule of CO_2 include (a) symmetric stretching of the C═O bonds, which produces no overall change in the polarity of the molecule; (b) asymmetric stretching, which produces side-to-side fluctuations in polarity that may result in absorption of IR radiation; and (c) a bending mode, which produces up-and-down fluctuations that may also absorb IR radiation.

fluctuate at the same frequencies as their vibrations. The natural frequencies of the vibrations correspond to frequencies of infrared radiation. As we described in Chapter 3, all forms of radiant energy, including infrared rays, travel through space as oscillating electrical and magnetic fields. Now, suppose a photon of infrared radiation traveling through Earth's atmosphere strikes a molecule containing a polar bond that is vibrating at exactly the same frequency as the photon. The fluctuating fields of the photon and the vibrating bond may interact. The molecule might absorb that photon, temporarily increasing its internal energy, and later emit a photon of the same energy as it returns to its ground state. This molecule–photon interaction is at the heart of the greenhouse effect.

To understand the connection between vibrating bonds and a warming atmosphere, recall that infrared radiation is the part of the electromagnetic spectrum that we cannot see but that we feel as heat. Any warm object, including Earth's surface, emits infrared radiation. When infrared photons strike atmospheric molecules that contain polar bonds, such as CO_2, the photons may be absorbed. When they are reemitted, they are just as likely to move back toward Earth's surface as they are to go upward toward space. In this process of absorption and reemission, a significant fraction of the heat flowing from Earth's surface is trapped in the atmosphere, much in the way that the glass covering a greenhouse traps heat inside it.

CONNECTION The average temperature of Earth's surface is 287 K, which means that it emits its peak intensity of electromagnetic radiation in the infrared region (see Figure 3.11).

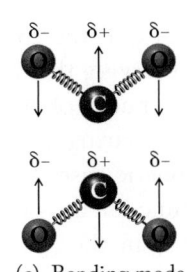

Not all polar bond vibrations result in the absorption and emission of infrared radiation. For example, two kinds of stretching vibrations can occur in a molecule of CO_2, which has two C═O bonds on opposite sides of the central C atom. One is a *symmetric* stretching vibration (Figure 4.15a) in which the two C═O bonds stretch and then compress at the same time. In that case the two fluctuating electrical fields produced by the two C═O bonds cancel each other out, and no infrared absorption or emission is possible. This vibration is said to be *infrared inactive.* However, when the bonds stretch such that one C═O bond gets shorter as the other gets longer (Figure 4.15b), the changes in charge separation do not cancel. This *asymmetric* stretch produces a fluctuating electrical field that enables CO_2 to absorb infrared radiation, so the vibration is *infrared active.* Molecules can also bend (Figure 4.15c) to produce fluctuating electrical fields that match the frequencies of other photons of infrared radiation.

CONCEPT TEST

Nitrogen and oxygen make up about 99% of the gases in the atmosphere. Is the stretching of the N≡N and O═O bonds in the molecules infrared active? Why or why not?

(Answers to Concept Tests are in the back of the book.)

In this chapter we have explored the electrostatic potential energy between oppositely charged ions that leads to ionic bond formation. We have also explored the nature of the covalent bonds that hold together molecules and polyatomic ions, observing that the bonds owe their strength to the pairs of electrons shared between nuclei of atoms. Sharing does not necessarily mean equal sharing, and unequal sharing coupled with bond vibration accounts for the ability of some atmospheric gases to absorb and emit infrared radiation. In so doing, the molecules function as potent greenhouse gases.

Early in the chapter we noted that moderate concentrations of greenhouse gases are required for climate stability and to make our planet habitable. The escalating concern of many is that Earth's climate is currently being destabilized by too much of a good thing. Policies being made by the world's governments today will have a significant impact on the problem of climate change, one way or the other. As an informed member of the world community, you have the opportunity to influence how those policy decisions are made. We hope that you will make the most of that opportunity.

SAMPLE EXERCISE 4.19 Integrating Concepts: Mothballs

A compound often referred to by the acronym PDB is the active ingredient in most mothballs. It is also used to control mold and mildew, as a deodorant, and as a disinfectant. Tablets containing it are often stuck under the lids of garbage cans or placed in the urinals in public restrooms, producing a distinctive aroma. Molecules of PDB have the following skeletal structure:

a. Draw the Lewis structure of PDB and note any nonzero formal charges.
b. Is the structure stabilized by resonance? If so, draw all resonance structures.
c. In the structure you drew, which of the bonds, if any, are polar?
d. Predict the average carbon–carbon bond length and bond strength in the structure you drew.

Collect and Organize We are given the skeletal structure of a molecule and are asked to draw its Lewis structure, including all resonance structures, and to perform a formal charge analysis. We are also asked to identify any polar bonds in the structure and to predict the length and strength of the carbon–carbon bonds. Bond polarity depends on the difference in electronegativities of the bonded atoms, which are given in Figure 4.13. Table 4.6 contains the average lengths and energies (strengths) of covalent bonds.

Analyze The five-step procedure used in Sample Exercises 4.9–4.13 to draw the Lewis structures of other small molecules should be useful in drawing the Lewis structure of PDB.

Solve

a and b. The number of valence electrons is

ELEMENT		VALENCE ELECTRONS	
Symbol	# of atoms	In one atom	Total
C	6	4	24
H	4	1	4
Cl	2	7	14
Valence electrons in molecule			**42**

Completing the octets on the Cl atoms:

gives us a structure with 36 valence electrons (12 bonding pairs and 6 lone pairs). We need six more. We add six by adding three more bonds between carbon atoms, turning three C—C single bonds into C═C double bonds. We have to distribute them evenly around the ring to avoid any C atoms with five bonds. Two equivalent resonance structures, analogous to those for benzene (Figure 4.8), can be drawn to show the bonding pattern:

Resonance stabilizes the structure of PDB. Each C atom has four bonds and each H and Cl atom has one bond, so every atom has the number of bonds that matches its bond capacity. This means that all formal charges are zero.

c. The differences in electronegativities for the bonded pairs of atoms are

C—C	$\Delta\chi = 0$
Cl—C	$\Delta\chi = 3.0 - 2.5 = 0.5$
C—H	$\Delta\chi = 2.5 - 2.1 = 0.4$

Of the three pairs, only the Cl—C bond meets our polar bond guidelines ($0.4 < \Delta\chi < 2.0$).

d. The even distribution of a total of nine bonding pairs of electrons among six C atoms means that, on average, each pair shares 1.5 pairs of bonding electrons. The corresponding bond length and bond strength should be about halfway between those of the C—C single and C═C double bonds given in Table 4.6:

Approximate bond length:

$$[(154 + 134)/2] \text{ pm} = 144 \text{ pm}$$

Approximate bond strength:

$$[(348 + 614)/2] \text{ kJ/mol} = 481 \text{ kJ/mol}$$

Think About It The resonance structures closely resemble those of benzene, which is reflected in the common name of PDB, *para*-dichlorobenzene. We will explore the rules for naming organic compounds like PDB in Chapter 19. For now, please note that the two polar C—Cl bonds in PDB are oriented *in opposite directions*. Thus, the unequal sharing of the bonding pair of electrons in the Cl—C bond on the left side of the molecule is offset by the unequal sharing of the bonding pair of electrons in the C—Cl bond on the right side. In Chapter 5 we will explain that offsetting bond polarities in symmetrical molecules like PDB explains why these substances are nonpolar overall.

SUMMARY

LO1 A chemical bond results from two ions being attracted to each other (an **ionic bond**) or from two atoms sharing electrons (a **covalent bond**). The atoms in metallic solids pool their electrons to form **metallic bonds**. (Section 4.1)

LO2 Electrostatic potential energy (E_{el}) is a measure of the strength of the attractions between cations and anions in an ionic compound. It is directly proportional to the product of the ion charges and inversely proportional to the distance between the nuclei of the ions. (Section 4.1)

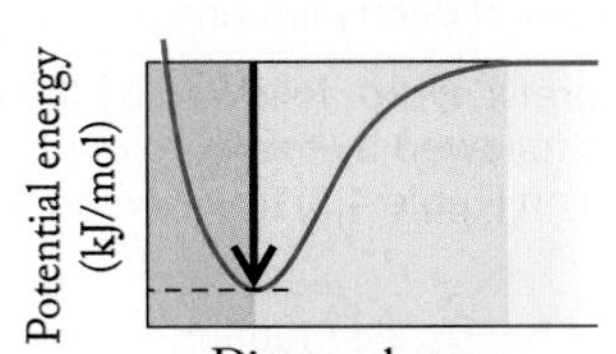

Distance between nuclei (pm)

LO3 To name binary ionic compounds, first write the name of the cation's parent element, and then write the name of the anion's parent element. Change the ending of the name of the second element to *-ide*. Roman numerals in parentheses indicate the charges on transition metal cations. The names of **oxoanions** (**polyatomic ions** containing oxygen atoms) end in *-ate* or *-ite* and may have a *per-* or *hypo-* prefix to indicate the relative number of oxygen atoms per ion. To name binary molecular compounds, first write the name of the element that is to the left of or, if the elements are in the same group, below the other one in the periodic table. Prefixes indicate the number of atoms of each element per molecule. The names of solutions of binary acids (general formula HX) begin with the prefix *hydro-* followed by the name of element X, but end in *-ic* followed by the word *acid*. The names of the **oxoacids** are similar to the names of their oxoanions, but their endings change from *-ate* to *-ic acid* and from *-ite* to *-ous acid*. (Section 4.2)

LO4 Lewis symbols use dots to represent paired and unpaired electrons in the ground states of atoms. The number of unpaired electrons indicates the number of bonds the element is likely to form—that is, its **bonding capacity**. Chemical stability is achieved when atoms have eight electrons in their valence *s* and *p* orbitals, following the **octet rule**. A **Lewis structure** shows the bonding pattern in molecules and polyatomic ions; pairs of dots represent **lone pairs** of electrons that do not contribute to bonding. A **single bond** consists of a single pair of electrons shared between two atoms; there are two shared pairs in a **double bond** and three shared pairs in a **triple bond**. (Section 4.3)

LO5 Two or more equivalent Lewis structures—called **resonance structures**—can sometimes be drawn for one molecule or polyatomic ion. The actual bonding pattern in a molecule is an average of equivalent resonance structures. The preferred resonance structure of a molecule is one in which the **formal charges (FC)** on its atoms are zero or as close to zero as possible, and any negative formal charges are on the more electronegative atoms. The formal charge on an atom in a Lewis structure is the difference between the number of valence electrons in the free atom and the sum of the number of electrons in lone pairs and half the number of electrons in bonding pairs on the bonded atom. **Free radicals** include reactive molecules that have an odd number of valence electrons and contain atoms with incomplete octets. Atoms of elements in the third row of the periodic table with $Z > 12$ and beyond may have expanded valence shells to accommodate more than an octet of electrons. (Sections 4.4, 4.7, and 4.8)

LO6 Bond order is the number of bonding pairs in a covalent bond. **Bond energy** is the energy change that accompanies the breaking of one mole of a particular covalent bond in the gas phase. As the bond order between two atoms increases, the bond length decreases and the bond energy increases. (Section 4.5)

LO7 Unequal electron sharing between atoms of different elements results in **polar covalent bonds**. Bond polarity is a measure of how unequally the electrons in covalent bonds are shared. More polarity results from greater differences between the **electronegativities** of the bonded atoms. Electronegativity generally increases with increasing ionization energy. (Section 4.6)

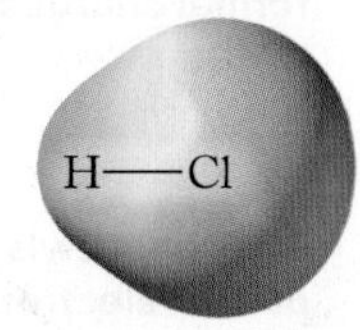

LO8 Covalent bonds behave more like flexible springs than rigid rods. They can undergo a variety of bond vibrations. The vibrations of polar bonds may create fluctuating electrical fields that allow molecules to absorb infrared (IR) electromagnetic radiation. When atmospheric gases absorb IR radiation, they contribute to the greenhouse effect. (Section 4.9)

PARTICULATE PREVIEW WRAP-UP

- There are four single (C—H) bonds in the molecule of CH_4 in image (a); there are two double (C=O) bonds in the molecule of CO_2 in image (b).
- Carbon atoms form four single bonds in CH_4 and two double bonds in CO_2.
- Space-filling models depict the sizes of atoms and the shapes of molecules, but they don't show the types of bonds (single, double, or triple) or lone pairs of electrons.

PROBLEM-SOLVING SUMMARY

Type of Problem	Concepts and Equations	Sample Exercises
Calculating the electrostatic potential energy of ionic bonds	$E_{el} = 2.31 \times 10^{-19}\ \text{J} \cdot \text{nm}\left(\frac{Q_1 \times Q_2}{d}\right)$ (4.1)	**4.1**
Naming binary ionic compounds and writing their formulas	First write the name of the cation's parent element; if it is a transition metal that forms ions with different charges, use a Roman numeral to represent the charge. Then write the name of the anion's parent element with its ending changed to *-ide*.	**4.2, 4.3**
Naming compounds of polyatomic ions and writing their formulas	As with a binary compound, write the name of the cation followed by the name of the anion. Use Table 4.3 to find the names of oxoanions (which end in *-ate* or *-ite*) and other polyatomic ions.	**4.4, 4.5**
Naming binary molecular compounds and writing their formulas	First write the name of the element that is to the left of or, if the elements are in the same group, below the other one in the periodic table. Then write the name of the other element, changing its ending to *-ide*. Use the prefixes in Table 4.4 to indicate the number of atoms of each element.	**4.6**
Naming acids and writing their formulas	For a binary acid (HX), begin with the prefix *hydro-* followed by the name of element X, but change its ending to *-ic* followed by the word *acid*. For an oxoacid, change the name of its oxoanion (Table 4.3) from *-ate* to *-ic acid*, or from *-ite* to *-ous acid*.	**4.7**
Drawing Lewis structures	Connect the atoms with single covalent bonds, distributing the valence electrons to give each outer atom eight valence electrons (except two for H); use multiple bonds where necessary to complete the central atom's octet.	**4.8–4.12**
Drawing resonance structures	Include all possible arrangements of covalent bonds in the molecule if more than one equivalent structure can be drawn.	**4.13**
Determining bond order and bond length from resonance structures	Draw resonance structures to determine the average bond order for the equivalent bonds. Relate bond order to bond length using Table 4.6.	**4.14**
Comparing bond polarities	Calculate the difference in electronegativity ($\Delta\chi$) between the two bonded atoms. If $\Delta\chi \geq 2.0$, the bond is considered ionic; if $0.4 < \Delta\chi < 2.0$, the bond is considered polar covalent; if $\Delta\chi \leq 0.4$, the bond is considered nonpolar covalent.	**4.15**
Selecting resonance structures based on formal charges	Calculate formal charge on each atom using $\text{FC} = \left(\begin{matrix}\text{number of}\\ \text{valence } e^-\end{matrix}\right) - \left[\begin{matrix}\text{number of } e^-\\ \text{in lone pairs}\end{matrix} + \frac{1}{2}\left(\begin{matrix}\text{number of}\\ \text{shared } e^-\end{matrix}\right)\right]$ (4.2) Select structures with formal charges closest to zero and with negative formal charges on the most electronegative atoms.	**4.16**
Drawing Lewis structures of odd-electron molecules	Distribute the valence electrons in the Lewis structure to leave the most electronegative atom(s) with eight valence electrons and the least electronegative atom with the odd number of electrons.	**4.17**
Drawing Lewis structures containing atoms with expanded valence shells	Distribute the valence electrons in the Lewis structure, allowing atoms of elements in period 3 and beyond to have more than eight valence electrons if more than four bonds are needed or if the structure with the expanded valence shell results in formal charges closer to zero.	**4.18**

VISUAL PROBLEMS

(Answers to boldface end-of-chapter questions and problems are in the back of the book.)

4.1. Which Lewis symbol in Figure P4.1 correctly portrays the most stable ion of aluminum?

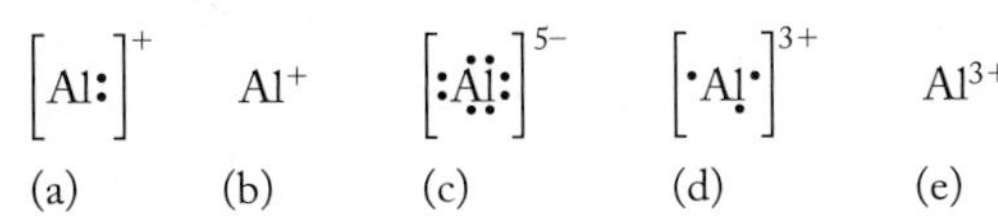

FIGURE P4.1

4.2. Which Lewis symbols in Figure P4.2 are correct?

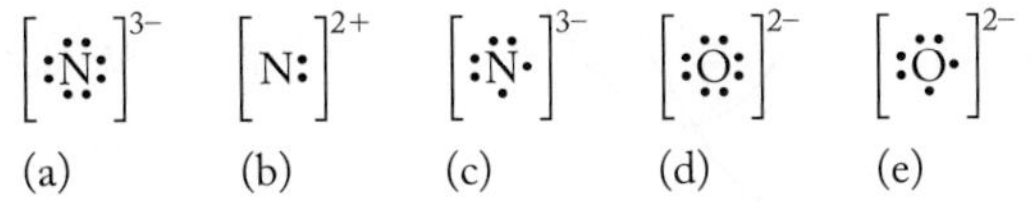

$[\text{:N:}]^{3-}$ $[\text{N:}]^{2+}$ $[\text{:N·}]^{3-}$ $[\text{:O:}]^{2-}$ $[\text{:O·}]^{2-}$

(a) (b) (c) (d) (e)

FIGURE P4.2

***4.3.** Which, if any, of the Lewis structures in Figure P4.3 are resonance structures for the thiocyanate ion (SCN^-)? Explain your selection(s).

$[\text{:S—N≡C:}]^{-}$ $[\text{:N≡C—S:}]^{-}$ $[\text{:N—S≡C:}]^{-}$

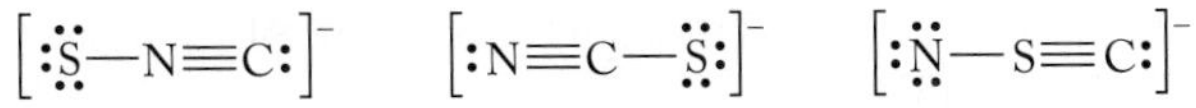

FIGURE P4.3

*4.4. Which of the Lewis structures in Figure P4.4 are resonance forms of the molecule S_2O? Explain your selections.

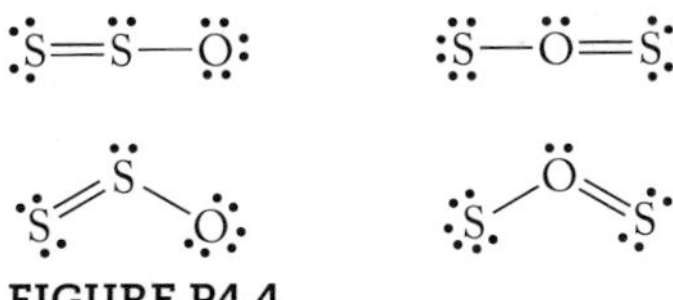

:S—O=S:

S O S

FIGURE P4.4

Note: The color scale used to indicate electron density in Problems 4.5–4.8 and 4.12 is the same as in Figure 4.12, where violet represents a charge of 1+, dark red is 1−, and yellow-green is 0.

4.5. Which image in Figure P4.5 is the best description of the distribution of electron density in BrCl?

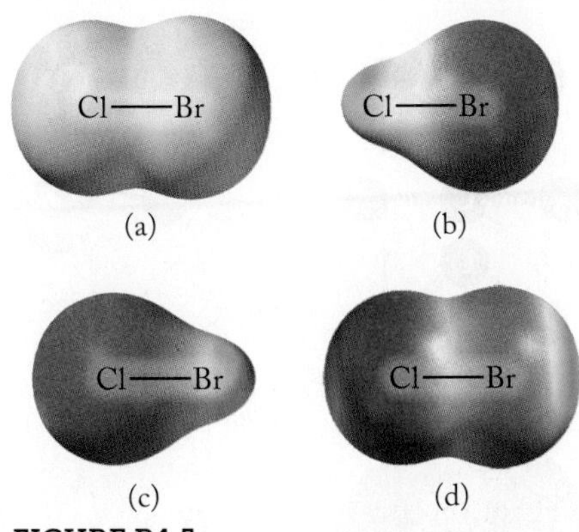

FIGURE P4.5

4.6. Which image in Figure P4.6 best describes the distribution of electron density in CsI?

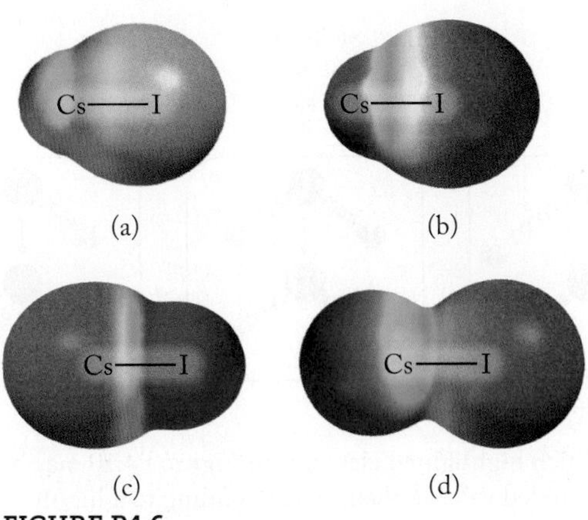

FIGURE P4.6

4.7. Which image in Figure P4.7 most accurately describes the distribution of electron density in SO_2? Explain your answer.

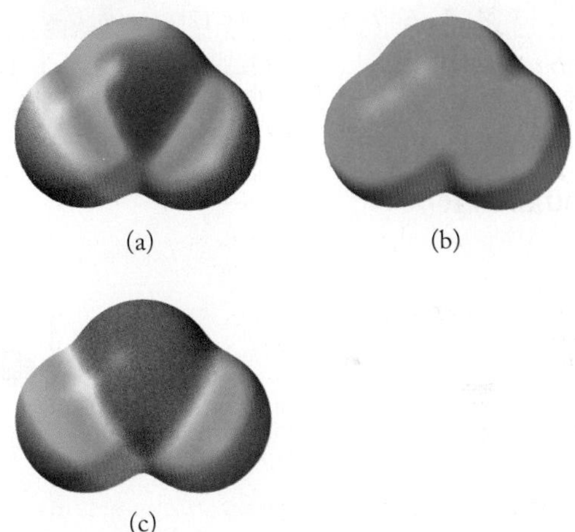

FIGURE P4.7

*4.8 The image in Figure P4.8 shows the electron density in a molecule of ozone, O_3. Note that electron density is higher at the ends of the molecule than in its center even though the bonding pairs of electrons in both O—O bonds are shared equally. Explain why electron density is not uniform across the whole molecule. (*Hint*: Calculate the formal charges on the three O atoms.)

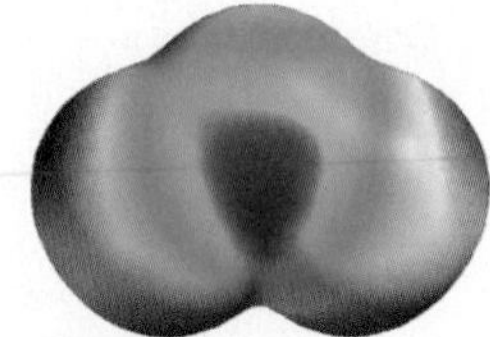

FIGURE P4.8

4.9. Water in the atmosphere is a greenhouse gas, which means its molecules are transparent to visible light but may absorb photons of infrared radiation. Which of the three modes of bond vibration shown in Figure P4.9 are infrared active? Note that molecules of H_2O are not linear.

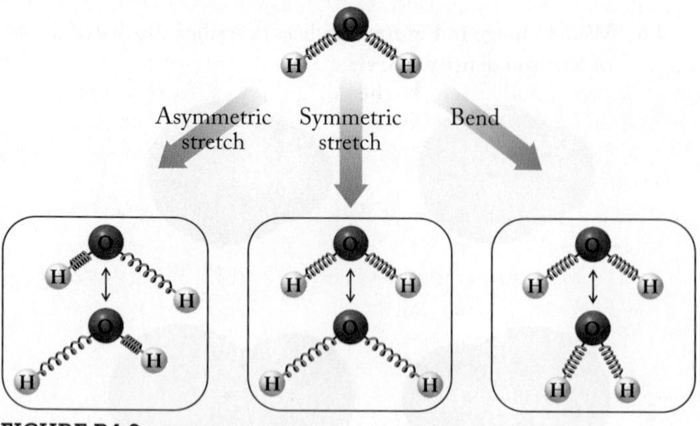

FIGURE P4.9

4.10. Which highlighted elements in Figure P4.10 may have expanded valence shells when bonding to a highly electronegative element?

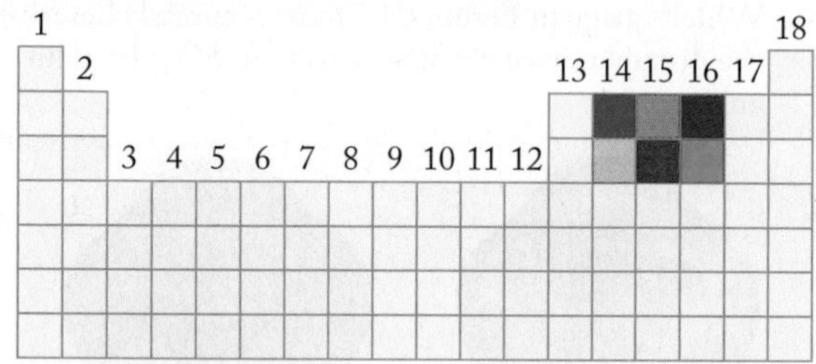

FIGURE P4.10

4.11. Figure P4.11 shows two graphs of electrostatic potential energy versus internuclear distance. One is for a pair of potassium and chloride ions, and the other is for a pair of potassium and fluoride ions. Which is which?

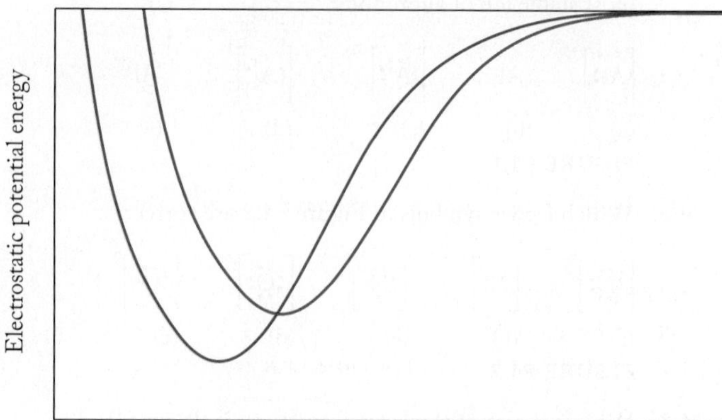

FIGURE P4.11

4.12. Use images [A] through [I] in Figure P4.12 to answer questions a–f.

a. Which processes require energy?
b. Which processes release energy?
c. Which process contributes to the greenhouse effect?
d. Which representations depict ionic bonding?
e. Which representations depict covalent bonding?
f. Which representation depicts both covalent bonding *and* ionic bonding?

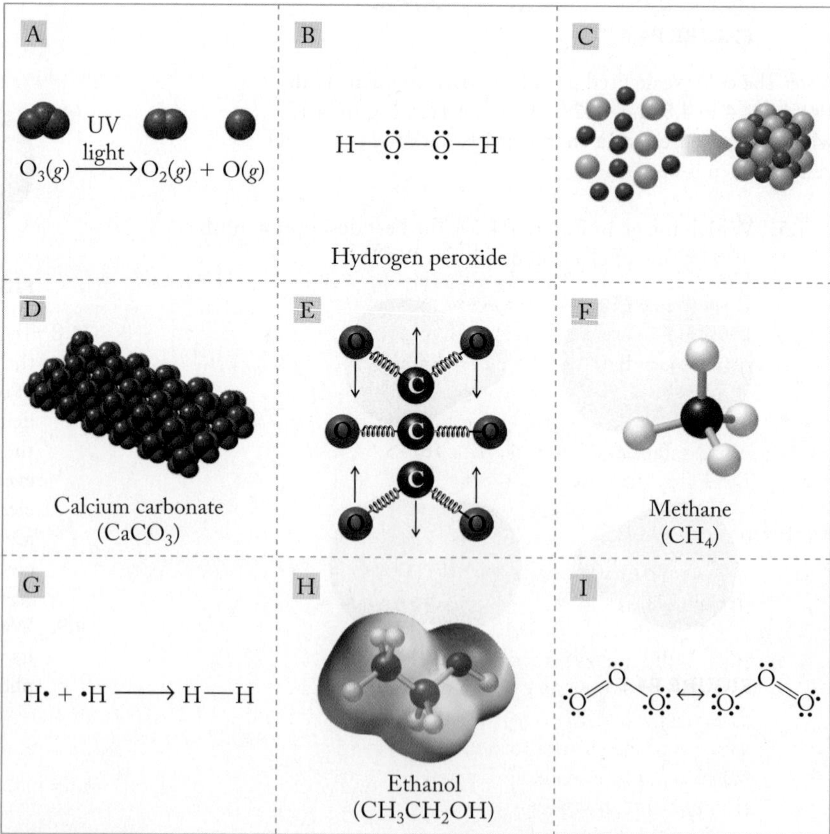

FIGURE P4.12

QUESTIONS AND PROBLEMS

Chemical Bonds

Concept Review

4.13. How does the number of valence electrons in the neutral atom of an element relate to the element's group number?

4.14. Which electrons in an atom are considered *valence* electrons?

4.15. Describe the differences in bonding between *covalent* and *ionic* compounds.

4.16. How is it possible for a compound to contain both covalent and ionic bonds?

Problems

4.17. What is the electrostatic potential energy between a pair of potassium and bromide ions in solid KBr? (*Hint*: See Figure 3.36.)

4.18. What is the electrostatic potential energy between a pair of aluminum and oxide ions in solid Al_2O_3?

4.19. Which of these substances has the most negative lattice energy? (a) KCl, (b) TiO_2, (c) $BaCl_2$, (d) KI

4.20. Rank the following ionic compounds, which have the same crystal structure, from least negative to most negative lattice energy: CsCl, CsBr, and CsI.

4.21. Rank the following ionic compounds in order of increasing coulombic attraction between their ions: KBr, $SrBr_2$, and CsBr.

4.22. Rank the following ionic compounds in order of increasing coulombic attraction between their ions: BaO, $BaCl_2$, and CaO.

Naming Compounds and Writing Formulas

Concept Review

4.23. What is the role of Roman numerals in the names of the compounds formed by transition metals?

4.24. Why does the name of a binary ionic compound in which the cation is from a group 1 or group 2 element not need a Roman numeral after the element's name?

4.25. Consider a mythical element X, which forms two oxoanions: XO_2^{2-} and XO_3^{2-}. Which of the two has a name that ends in *-ite*?

4.26. Concerning the oxoanions in Problem 4.25, would the name of either of them require a prefix such as *hypo-* or *per-*? Explain why or why not.

Problems

4.27. What are the names of these compounds of nitrogen and oxygen? (a) NO_3; (b) N_2O_5; (c) N_2O_4; (d) NO_2; (e) N_2O_3; (f) NO; (g) N_2O; (h) N_4O

4.28. More than a dozen neutral compounds containing only sulfur and oxygen have been identified. What are the chemical formulas of the following six? (a) sulfur monoxide; (b) sulfur dioxide; (c) sulfur trioxide; (d) disulfur monoxide; (e) hexasulfur monoxide; (f) heptasulfur dioxide

---

4.29. Predict the formula and give the name of the ionic compound formed by the following pairs of elements: (a) sodium and sulfur; (b) strontium and chlorine; (c) aluminum and oxygen; (d) lithium and hydrogen.

4.30. Predict the formula and give the name of the ionic compound formed by the following pairs of elements: (a) potassium and bromine; (b) calcium and hydrogen; (c) lithium and nitrogen; (d) aluminum and chlorine.

4.31. What are the names of the cobalt oxides that have the following formulas? (a) CoO; (b) Co_2O_3; (c) CoO_2

4.32. What are the formulas of the following copper minerals?

a. cuprite, copper(I) oxide
b. chalcocite, copper(I) sulfide
c. covellite, copper(II) sulfide

4.33. Give the formula and charge of the oxoanion in each of the following compounds: (a) sodium hypobromite; (b) potassium sulfate; (c) lithium iodate; (d) magnesium nitrite.

*4.34. Give the formula and charge of the oxoanion in each of the following compounds: (a) sodium tellurite; (b) potassium arsenate; (c) barium selenate; (d) potassium bromate.

4.35. What are the names of the following ionic compounds? (a) $NiCO_3$; (b) NaCN; (c) $LiHCO_3$; (d) $Ca(ClO)_2$

4.36. What are the names of the following ionic compounds? (a) $Mg(ClO_4)_2$; (b) NH_4NO_3; (c) $Cu(CH_3COO)_2$; (d) K_2SO_3

4.37. Give the name or chemical formula of each of the following acids: (a) HF; (b) H_2SO_3; (c) phosphoric acid; (d) nitrous acid.

*4.38. Give the name or chemical formula of each of the following acids: (a) HBr; (b) HIO_4; (c) selenous acid; (d) hydrocyanic acid.

4.39. Write the chemical formulas of the following compounds: (a) potassium sulfide; (b) potassium selenide; (c) rubidium sulfate; (d) rubidium nitrite; (e) magnesium sulfate.

4.40. Write the chemical formulas of the following compounds: (a) aluminum nitride; (b) ammonium sulfite; (c) rubidium chromate; (d) ammonium nitrate; (e) aluminum selenite.

4.41. What are the names of the following compounds? (a) MnS; (b) V_3N_2; (c) $Cr_2(SO_4)_3$; (d) $Co(NO_3)_2$; (e) Fe_2O_3

4.42. What are the names of the following compounds? (a) RuS; (b) $PdCl_2$; (c) Ag_2O; (d) WO_3; (e) PtO_2

4.43. Which is the formula of sodium sulfite? (a) Na_2S; (b) Na_2SO_3; (c) Na_2SO_4; (d) NaSH

4.44. Which is the formula of barium nitrate? (a) Ba_3N_2; (b) Ba_2NO_3; (c) $Ba_2(NO_3)_2$; (d) $Ba(NO_3)_2$

Lewis Symbols and Lewis Structures

Concept Review

4.45. Some critics described G. N. Lewis's approach to explaining covalent bonding as an exercise in double counting and therefore invalid. Explain the basis for the criticism.

4.46. Does the octet rule mean that a diatomic molecule must have 16 valence electrons?

4.47. Why is the bonding pattern in water H—O—H and not H—H—O?

4.48. Does each atom in a pair that is covalently bonded always contribute the same number of valence electrons to form the bonds between them?

Problems

4.49. Draw Lewis symbols of potassium, magnesium, and phosphorus.

4.50. Draw Lewis symbols of gallium, tellurium, and iodine.

4.51. Draw Lewis symbols for K^+, Al^{3+}, N^{3-}, and I^-.

4.52. Draw Lewis symbols for the most stable ions formed by lithium, magnesium, aluminum, and fluorine.

4.53. Which of the following ions have a complete valence-shell octet? B^{3+}, I^-, Ca^{2+}, or Pb^{2+}

4.54. How many valence electrons are in each of the following atoms or ions? Xe, Sr^{2+}, Cl, and Cl^-

4.55. How many valence electrons does each of the following species contain? (a) N_2; (b) HCl; (c) NH_4^+; (d) CN^-

4.56. How many valence electrons does each of the following species contain? (a) H^+; (b) H_3O^+; (c) CO_2; (d) CH_4

4.57. Draw Lewis structures for the following diatomic molecules and ions: (a) CO; (b) O_2; (c) ClO^-; (d) CN^-.

4.58. Draw Lewis structures for the following molecules and ions: (a) Br_2; (b) H_3O^+; (c) N_2; (d) HF.

4.59. Draw Lewis structures for the following molecular compounds and ions: (a) CCl_4; (b) BH_3; (c) SiF_4; (d) BH_4^-; (e) PH_4^+.

4.60. Draw Lewis structures for the following molecular compounds and ions: (a) $AlCl_3$; (b) PH_3; (c) H_2Se; (d) NO_2^-; (e) AlH_4^-.

4.61. **Greenhouse Gases** Chlorofluorocarbons (CFCs) are compounds linked to the depletion of stratospheric ozone. They are also greenhouse gases. Draw Lewis structures for the following CFCs:

a. CCl_3F (Freon 11)
b. CCl_2F_2 (Freon 12)
c. $CClF_3$ (Freon 13)
d. $Cl_2FC—CClF_2$ (Freon 113)
e. $ClF_2C—CClF_2$ (Freon 114)

4.62. The replacement of a halogen atom in a CFC molecule with a hydrogen atom makes the compound more environmentally "friendly." Draw Lewis structures for the following compounds:

a. $CHCl_2F$ (Freon 21)
b. CHF_2Cl (Freon 22)
c. CH_2ClF (Freon 31)
d. $F_3C—CHBrCl$ (Halon 2311)
e. $Cl_2FC—CH_3$ (HCFC 141b)

4.63. Draw Lewis structures for the following oxoanions: (a) ClO_2^-; (b) SO_3^{2-}; (c) HCO_3^-.

4.64. Draw Lewis structures for the following oxoanions: (a) BrO_4^-; (b) SeO_4^{2-}; (c) HPO_4^{2-}.

4.65. **Skunks and Rotten Eggs** Many sulfur-containing organic compounds have characteristically foul odors: butanethiol ($CH_3CH_2CH_2CH_2SH$) is responsible for the odor of skunks, and rotten eggs smell the way they do because they produce tiny amounts of pungent hydrogen sulfide, H_2S. Draw the Lewis structures for $CH_3CH_2CH_2CH_2SH$ and H_2S.

4.66. **Acid in Ants** Formic acid, HCOOH, is the smallest organic acid and was originally isolated by distilling red ants. Draw its Lewis structure if the atoms are connected as shown in Figure P4.66.

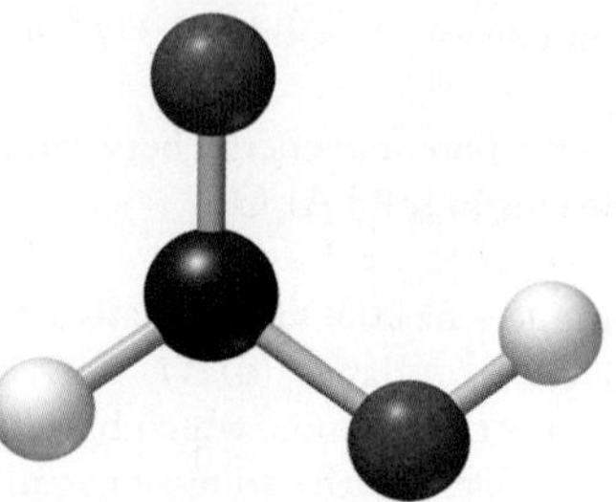

FIGURE P4.66

4.67. **Chlorine Bleach** Chlorine combines with oxygen in several proportions. Dichlorine monoxide (Cl_2O) is used in the manufacture of bleaching agents. Potassium chlorate ($KClO_3$) is used in oxygen generators aboard aircraft. Draw the Lewis structures for Cl_2O and ClO_3^-.

4.68. **Dangers of Mixing Cleansers** Labels on household cleansers caution against mixing bleach with ammonia (Figure P4.68) because they react with each other to produce monochloramine (NH_2Cl) and hydrazine (N_2H_4), both of which are toxic. Draw the Lewis structures for monochloramine and hydrazine.

FIGURE P4.68

***4.69.** Methanol is a toxic alcohol with the molecular formula CH_4O. Draw the Lewis structure for methanol.

4.70. Carbon disulfide, CS_2, is a flammable, low-boiling liquid. Draw the Lewis structure for CS_2.

Resonance

Concept Review

4.71. Explain the concept of resonance.

4.72. How does resonance influence the stability of a molecule or an ion?

4.73. What factors determine whether or not a molecule or ion exhibits resonance?

4.74. What structural features do all the resonance forms of a molecule or ion have in common?

4.75. Explain why NO_2 is more likely to exhibit resonance than CO_2.

*4.76. Three equivalent resonance structures can be drawn for a nitrate ion. How much of the time does the bonding in a nitrate ion match any one of them?

Problems

4.77. Draw two Lewis structures showing the resonance that occurs in cyclobutadiene (C_4H_4), a cyclic molecule with a structure that includes a ring of four carbon atoms.

*4.78. Pyridine (C_5H_5N) and pyrazine ($C_4H_4N_2$) have structures similar to benzene. Both compounds have structures with six atoms in a ring. Draw Lewis structures for pyridine and pyrazine showing all resonance forms. The N atoms in pyrazine are across the ring from each other.

***4.79.** Oxygen and nitrogen combine to form a variety of nitrogen oxides, including the following two unstable compounds that each have two nitrogen atoms per molecule: N_2O_2 and N_2O_3. Draw Lewis structures for the molecules and show all resonance forms.

*4.80. Oxygen and sulfur combine to form a variety of different sulfur oxides. Some are stable molecules and some, including S_2O_2 and S_2O_3, decompose when they are heated. Draw Lewis structures for these two compounds and show all resonance forms.

4.81. Draw Lewis structures for fulminic acid (HCNO) that show all resonance forms.

4.82. Draw Lewis structures for hydrazoic acid (HN_3) that show all resonance forms.

4.83. Draw Lewis structures that show the resonance that occurs in dinitrogen pentoxide. (*Hint*: N_2O_5 has an O atom at its center.)

4.84. **Bacteria Make Nitrites** Nitrogen-fixing bacteria convert urea $[H_2NC(O)NH_2]$ into nitrite ions. Draw Lewis structures for the two species. Include all resonance forms. (*Hint*: There is a C=O bond in urea.)

The Lengths and Strengths of Covalent Bonds

Concept Review

4.85. How does the nitrogen–oxygen bond length in the nitrate ion compare to the nitrogen–oxygen bond length in the nitrite ion?

4.86. Why is the oxygen–oxygen bond length in O_3 different than the one in O_2?

4.87. Explain why the nitrogen–oxygen bond lengths in N_2O_4 (which has a nitrogen–nitrogen bond) and N_2O are nearly identical (118 and 119 pm, respectively).

4.88. Do you expect the sulfur–oxygen bond lengths in sulfite (SO_3^{2-}) and sulfate (SO_4^{2-}) ions to be about the same? Why?

4.89. Rank the following ions in order of increasing nitrogen–oxygen bond lengths: NO_2^-, NO^+, and NO_3^-.

4.90. Rank the following substances in order of increasing carbon–oxygen bond lengths: CO, CO_2, and CO_3^{2-}.

4.91. Rank the following ions in order of increasing nitrogen–oxygen bond energy: NO_2^-, NO^+, and NO_3^-.

4.92. Rank the following substances in order of increasing carbon–oxygen bond energy: CO, CO_2, and CO_3^{2-}.

4.93. Which has the longer carbon–carbon bond: acetylene (C_2H_2) or ethane (C_2H_6)?

4.94. Which has the stronger carbon–carbon bond: acetylene (C_2H_2) or ethane (C_2H_6)?

Electronegativity, Unequal Sharing, and Polar Bonds

Concept Review

4.95. How can we use electronegativity to predict whether a bond between two atoms is likely to be covalent or ionic?

4.96. How do the electronegativities of the elements change across a row and down a group in the periodic table?

4.97. How are trends in electronegativity related to trends in atomic size?

4.98. Is the element with the most valence electrons in a row of the periodic table also the most electronegative?

4.99. What is meant by the term *polar covalent bond*?

4.100. Why are the electrons in bonds between different elements not shared equally?

Problems

4.101. Which of the following bonds are polar? C—Se, C—O, Cl—Cl, O=O, N—H, C—H. In the bond or bonds that you selected, which atom has the greater electronegativity?

4.102. Rank the following bonds from nonpolar to most polar: H—H, H—F, H—Cl, H—Br, H—I.

4.103. Which of the binary compounds formed by the following pairs of elements contain polar covalent bonds, and which are considered ionic compounds?

a. C and S
b. Al and Cl
c. C and O
d. Ca and O

4.104. Which of the beryllium halides, if any, are considered ionic compounds?

Formal Charge: Choosing among Lewis Structures

Concept Review

4.105. Describe how formal charges are used to choose between possible molecular structures.

4.106. How do the electronegativities of elements influence the selection of which Lewis structure is favored?

4.107. In a molecule containing S and O atoms, is a structure with a negative formal charge on sulfur more likely to contribute to bonding than an alternative structure with a negative formal charge on oxygen?

4.108. In a cation containing N and O, why do Lewis structures with a positive formal charge on nitrogen contribute more to the actual bonding in the molecule than those structures with a positive formal charge on oxygen?

Problems

4.109. Hydrogen isocyanide (HNC) has the same elemental composition as hydrogen cyanide (HCN), but the H in HNC is bonded to the nitrogen atom. Draw a Lewis structure for HNC, and assign formal charges to each atom. How do the formal charges on the atoms differ in the Lewis structures for HCN and HNC?

4.110. **Molecules in Interstellar Space** Hydrogen cyanide (HCN) and cyanoacetylene (HC_3N) have been detected in the interstellar regions of space. Draw Lewis structures for the molecules, and assign formal charges to each atom. The hydrogen atom is bonded to carbon in both cases.

4.111. Origins of Life The discovery of polyatomic organic molecules such as cyanamide (H_2NCN) in interstellar space has led some scientists to believe that the molecules from which life began on Earth may have come from space. Draw Lewis structures for cyanamide and select the preferred structure on the basis of formal charges.

4.112. Complete the Lewis structures for and assign formal charges to the atoms in five of the resonance forms of thionitrosyl azide (SN_4). Indicate which of your structures should be most stable. The molecule is linear with S at one end.

***4.113.** Nitrogen is the central atom in molecules of nitrous oxide (N_2O). Draw Lewis structures for another possible arrangement: N—O—N. Assign formal charges and suggest a reason why the structure is not likely to be stable.

4.114. **More Molecules in Space** Formamide ($HCONH_2$) and methyl formate (HCO_2CH_3) have been detected in space. Draw the Lewis structures of the compounds, based on the skeletal structures in Figure P4.114, and assign formal charges.

FIGURE P4.114

***4.115.** Nitromethane (CH_3NO_2) reacts with hydrogen cyanide (HCN) to produce $CNNO_2$ and CH_4.

a. Draw Lewis structures for CH_3NO_2 and show all resonance forms.

b. Draw Lewis structures for $CNNO_2$, showing all resonance forms, based on the two possible skeletal structures for it in Figure P4.115. Assign formal charges and predict which structure is more likely to exist.

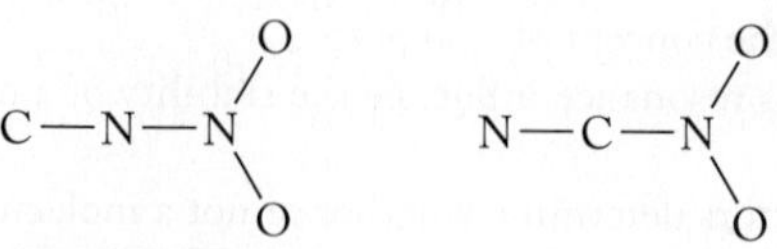

FIGURE P4.115

c. Are the two structures of $CNNO_2$ resonance forms of each other?

4.116. Use formal charges to determine which resonance form of each of the following ions is preferred: CNO^-, NCO^-, and CON^-.

Exceptions to the Octet Rule

Concept Review

4.117. Are all odd-electron molecules exceptions to the octet rule?

4.118. Describe the factors that contribute to the stability of structures in which the central atoms have more than eight valence electrons.

4.119. Why do C, N, O, and F atoms in covalently bonded molecules and ions have no more than eight valence electrons?

4.120. Do atoms in rows 3 and below always have expanded valence shells? Explain your answer.

Problems

4.121. In which of the following molecules does the sulfur atom have an expanded valence shell? (a) SF_6; (b) SF_5; (c) SF_4; (d) SF_2

4.122. In which of the following molecules does the phosphorus atom have an expanded valence shell? (a) $POCl_3$; (b) PF_5; (c) PF_3; (d) P_2F_4 (which has a P—P bond)

4.123. How many electrons are there in the covalent bonds surrounding the sulfur atom in the following species? (a) SF_4O; (b) SOF_2; (c) SO_3; (d) SF_5^-

4.124. How many electrons are there in the covalent bonds surrounding the phosphorus atom in the following species? (a) $POCl_3$; (b) H_3PO_4; (c) H_3PO_3; (d) PF_6^-

***4.125.** Draw the Lewis structures of NOF_3 and POF_3 in which the group 15 element is the central atom and the other atoms are bonded to it. What differences are there in the types of bonding in the molecules?

*4.126. The phosphate ion (PO_4^{3-}) is part of our DNA. The corresponding nitrogen-containing oxoanion, NO_4^{3-}, is not chemically stable. Draw Lewis structures that show any resonance forms of both oxoanions.

4.127. Dissolving NaF in selenium tetrafluoride (SeF_4) produces $NaSeF_5$. Draw the Lewis structures of SeF_4 and SeF_5^-. In which structure does Se have more than eight valence electrons?

4.128. The reaction between NF_3, F_2, and SbF_3 at 200°C and 100 atm pressure produces the ionic compound NF_4SbF_6. Draw the Lewis structures of the ions in the product.

4.129. **Ozone Depletion** The compound Cl_2O_2 may play a role in ozone depletion in the stratosphere. Draw the Lewis structure of Cl_2O_2 based on the arrangement of atoms in Figure P4.129. Does either of the chlorine atoms in the structure have an expanded valence shell?

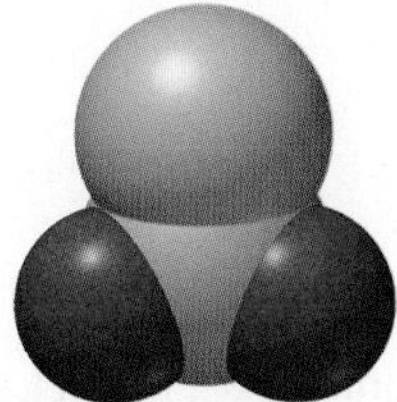

FIGURE P4.129

4.130. Cl_2O_2 decomposes to Cl_2 and ClO_2. Draw the Lewis structure of ClO_2.

4.131. Which of the following chlorine oxides are odd-electron molecules? (a) Cl_2O_7; (b) Cl_2O_6; (c) ClO_4; (d) ClO_3; (e) ClO_2

4.132. Which of the following nitrogen oxides are odd-electron molecules? (a) NO; (b) NO_2; (c) NO_3; (d) N_2O_4; (e) N_2O_5

4.133. In the following species, which atom is most likely to have an unpaired electron? (a) SO^+; (b) NO; (c) CN; (d) OH

4.134. In the following molecules, which atom is most likely to have an unpaired electron? (a) NO_2; (b) CNO; (c) ClO_2; (d) HO_2

4.135. Which of the Lewis structures in Figure P4.135 contributes most to the bonding in CNO?

a. ·Ċ̤—N≡O: c. :C≡N—Ö·

b. :C=N=Ö: d. ·C≡N—Ö:

FIGURE P4.135

4.136. Why is the Lewis structure in Figure P4.136 unlikely to contribute much to the bonding in NCO?

:N̤̈—C≡O·

FIGURE P4.136

Vibrating Bonds and the Greenhouse Effect

Concept Review

4.137. Describe how atmospheric greenhouse gases act like the panes of glass in a greenhouse.

*4.138. Water vapor in the atmosphere contributes more to the greenhouse effect than carbon dioxide, yet water vapor is not considered an important factor in climate change. Propose a reason why.

***4.139.** Increasing concentrations of nitrous oxide in the atmosphere may be contributing to climate change. Is the ability of N_2O to absorb IR radiation due to nitrogen–nitrogen bond stretching, nitrogen–oxygen bond stretching, or both? Explain your answer.

4.140. Is the ability of H_2O molecules to absorb photons of IR radiation due to symmetrical stretching or asymmetrical stretching of its O—H bonds, or both? Explain your answer. (*Hint*: The angle between the two O—H bonds in H_2O is 104.5°.)

4.141. Can molecules of carbon monoxide in the atmosphere absorb photons of IR radiation? Explain why or why not.

*4.142. How does the high-temperature conversion of limestone ($CaCO_3$) to lime (CaO) during the production of cement contribute to climate change?

4.143. Why does infrared radiation cause bonds to vibrate but not break (as UV radiation can)?

4.144. Argon is the third most abundant species in the atmosphere. Why isn't it a greenhouse gas?

***4.145.** Which C—O bond has a higher stretching frequency: the one in CO or the one in CH_2O? Explain your selection.

*4.146. Which compound, NO or NO_2, absorbs IR radiation of a longer wavelength?

Additional Problems

4.147. The unpaired dots in Lewis symbols of the elements represent valence electrons available for covalent bond formation. In Figure P4.147, which of the options for placing dots around the symbol for each element is preferred?

a. Be: or ·Be· b. :Al· or ·Ȧl·

c. ·Ċ̣· or :Ċ· d. He: or ·He·

FIGURE P4.147

4.148. Based on the Lewis symbols in Figure P4.148, predict to which group in the periodic table element X belongs.

a. ·Ẋ· b. ·Ẍ: c. :Ẍ: d. :Ẍ:

FIGURE P4.148

4.149. Use formal charges to predict whether the atoms in carbon disulfide are arranged CSS or SCS.

4.150. Use formal charges to predict whether the atoms in hypochlorous acid are arranged HOCl or HClO.

4.151. **Chemical Weapons** Draw the Lewis structure of phosgene, $COCl_2$, a poisonous gas used in chemical warfare during World War I.

4.152. The dinitramide anion $[N(NO_2)_2^-]$ was first isolated in 1996. The arrangement of atoms in $N(NO_2)_2^-$ is shown in Figure P4.152.

a. Complete the Lewis structure of $N(NO_2)_2^-$, including any resonance forms, and assign formal charges.

b. Explain why the nitrogen–oxygen bond lengths in $N(NO_2)_2^-$ and N_2O should (or should not) be similar.

c. $N(NO_2)_2^-$ was isolated as $[NH_4^+][N(NO_2)_2^-]$. Draw the Lewis structure of NH_4^+.

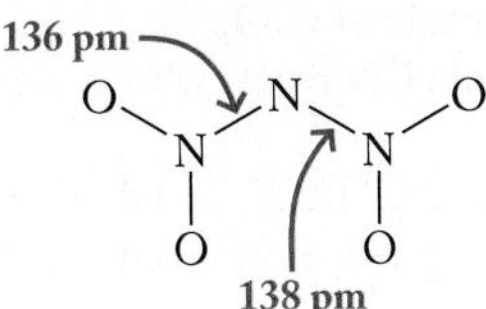

FIGURE P4.152

***4.153.** Silver cyanate (AgOCN) is a source of the cyanate ion (OCN^-). Under certain conditions, the species OCN is an anion with a charge of 1−; under others, it is a neutral, odd-electron molecule, OCN.

a. Two molecules of OCN combine to form OCNNCO. Draw the Lewis structure of the molecule, including all resonance forms.

b. The OCN^- ion reacts with BrNO, forming the unstable molecule OCNNO. Draw the Lewis structures of BrNO and OCNNO, including all resonance forms.

c. The OCN^- ion reacts with Br_2 and NO_2 to produce N_2O, CO_2, BrNCO, and OCN(CO)NCO. Draw the resonance structures of OCN(CO)NCO, which has the arrangement of atoms shown in Figure P4.153.

```
            O
            |
O—C—N—C—N—C—O
```

FIGURE P4.153

***4.154.** During the reaction of the cyanate ion (OCN^-) with Br_2 and NO_2, a very unstable substance called an *intermediate* forms and then quickly falls apart. Its formula is O_2NNCO.

a. Draw three of the resonance forms for O_2NNCO, assign formal charges, and predict which of the three contributes the most to the bonding in O_2NNCO. Its skeletal structure is shown in Figure P4.154(a).

```
O
 \
  N—N—C—O
 /
O
```

FIGURE P4.154(a)

b. Draw Lewis structures for the different arrangement of the N, C, and O atoms in O_2NNCO shown in Figure P4.154(b).

```
            O
           /
O—N—N—C
           \
            O
```

FIGURE P4.154(b)

4.155. A compound with the formula Cl_2O_6 decomposes to a mixture of ClO_2 and ClO_4. Draw two Lewis structures for Cl_2O_6: one with a chlorine–chlorine bond and one with a Cl—O—Cl arrangement of atoms.

*4.156. A compound consisting of chlorine and oxygen, Cl_2O_7, decomposes to ClO_4 and ClO_3.

a. Draw two Lewis structures of Cl_2O_7: one with a chlorine–chlorine bond and one with a Cl—O—Cl arrangement of atoms.

b. Draw the Lewis structure of ClO_3.

***4.157.** The odd-electron molecule CN reacts with itself to form cyanogen, C_2N_2.

a. Draw the Lewis structure of CN, and predict which arrangement for cyanogen is more likely: NCCN or CNNC.

b. Cyanogen reacts slowly with water to produce oxalic acid ($H_2C_2O_4$) and ammonia; the Lewis structure of oxalic acid is shown in Figure P4.157. Compare the structure to your answer in part (a). Do you still believe the structure you selected in part (a) is the better one?

```
   :Ö:        :Ö:
     \\       //
       C — C
      /       \
H—Ö:        :Ö—H
```

FIGURE P4.157

4.158. The odd-electron molecule SN forms S_2N_2, which has a cyclic structure (the atoms form a ring).

a. Draw a Lewis structure of SN and complete the possible Lewis structures for S_2N_2 in Figure P4.158.

b. Which of the two is the preferred structure for S_2N_2?

```
S—N        S—N
|  |        |  |
S—N        N—S
```

FIGURE P4.158

***4.159.** The molecular structure of sulfur cyanide trifluoride (SF_3CN) has been shown to have the arrangement of atoms with the bond lengths indicated in Figure P4.159. Using the observed bond lengths as a guide, complete the Lewis structure of SF_3CN and assign formal charges.

```
          N
116 pm →  |
          C
          | ← 174 pm
       F—S—F
          |  ↖ 160 pm
          F
```

FIGURE P4.159

4.160. **Strike-Anywhere Matches** Heating phosphorus with sulfur produces P_4S_3, a solid used in the heads of strike-anywhere matches. P_4S_3 has the skeletal structure shown in Figure P4.160. Complete its Lewis structure.

FIGURE P4.160

***4.161.** The $TeOF_6^{2-}$ anion was first synthesized in 1993. Draw its Lewis structure.

*4.162. **Sulfur in the Environment** Sulfur is cycled in the environment through compounds such as dimethyl sulfide (CH_3SCH_3), hydrogen sulfide (H_2S), and sulfite and sulfate ions. Draw Lewis structures for these four species. Are expanded valence shells needed to minimize the formal charges for any of these species?

4.163. **Antacid Tablets** Antacids commonly contain calcium carbonate and/or magnesium hydroxide. Draw the Lewis structures for calcium carbonate and magnesium hydroxide.

4.164. How many pairs of electrons does xenon share in the following molecules and ions? (a) XeF_2; (b) $XeOF_2$; (c) XeF^+; (d) XeF_5^+; (e) XeO_4

***4.165.** A short-lived allotrope of nitrogen, N_4, was reported in 2002.

a. Draw the Lewis structures of all the resonance forms of linear N_4 (N—N—N—N).
b. Assign formal charges and determine which resonance structure is the best description of N_4.
c. Draw a Lewis structure of a ring (cyclic) form of N_4 and assign formal charges.

*4.166. Scientists have predicted the existence of O_4, even though the molecule has never been observed. However, O_4^{2-} has been detected. Draw the Lewis structures for O_4 and O_4^{2-}.

4.167. Which of the following molecules and ions contains an atom with an expanded valence shell? (a) Cl_2; (b) ClF_3; (c) ClI_3; (d) ClO^-

4.168. Which of the following molecules contains an atom with an expanded valence shell? (a) XeF_2; (b) $GaCl_3$; (c) ONF_3; (d) SeO_2F_2

***4.169.** A linear nitrogen anion, N_5^-, was isolated for the first time in 1999.

a. Draw the Lewis structures for four resonance forms of linear N_5^-.
b. Assign formal charges to the atoms in the structures in part (a), and identify the structures that contribute the most to the bonding in N_5^-.
c. Compare the Lewis structures for N_5^- and N_3^-. In which ion do the nitrogen–nitrogen bonds have the higher average bond order?

*4.170. Carbon tetroxide (CO_4) was discovered in 2003.

a. Draw the Lewis structure of CO_4 based on the skeletal structure shown in Figure P4.170.
b. Are there any resonance forms of the structure you drew that have zero formal charges on all atoms?
c. Can you draw a structure in which all four oxygen atoms in CO_4 are bonded to carbon?

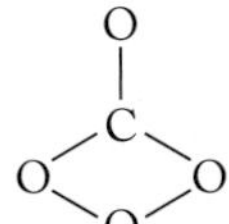

FIGURE P4.170

4.171. Plot the electronegativities of elements with $Z = 3$ to 9 (y-axis) versus their first ionization energy (x-axis). Is the plot linear? Use your graph to predict the electronegativity of neon, whose first ionization energy is 2081 kJ/mol.

*4.172. In the typical Lewis structure of BF_3 there are only six valence electrons on the boron atom and each B—F bond is a single bond. However, the length and strength of these bonds indicate that they have a small measure of double-bond character—that is, their bond order is slightly greater than 1.

a. Draw a Lewis structure of BF_3, including all resonance structures, in which there is one B=F double bond.
b. What is the formal charge on the B atom, and what is the average formal charge on each F atom?
c. Based on formal charges alone, what should be the bond order of each B—F bond in BF_3?
d. What factor might support a bond order slightly greater than 1?

4.173. The cation N_2F^+ is isoelectronic with N_2O.

a. What does it mean to be isoelectronic?
b. Draw the Lewis structure of N_2F^+. (*Hint*: The molecule contains a nitrogen–nitrogen bond.)
c. Which atom has the +1 formal charge in the structure you drew in part (b)?
d. Does N_2F^+ have resonance forms?
e. Could the middle atom in the N_2F^+ ion be a fluorine atom? Explain your answer.

4.174. **Ozone Depletion** Methyl bromide (CH_3Br) is produced naturally by fungi. Methyl bromide has also been used in agriculture as a fumigant, but its use is being phased out because the compound has been linked to ozone depletion in the upper atmosphere.

a. Draw the Lewis structure of CH_3Br.
b. Which bond in CH_3Br is more polar, carbon–hydrogen or carbon–bromine?

4.175. Draw the Lewis structure for dimethyl ether, C_2H_6O, given that the structure contains an oxygen atom bonded to two carbons: C—O—C.

*4.176. Draw another Lewis structure for C_2H_6O that has a different connectivity than that in Problem 4.175. (*Hint*: Remember that the bonding capacity of hydrogen is 1.)

4.177. Draw the Lewis structure for butane, C_4H_{10}, given the structure contains four carbon atoms bonded in a row: C—C—C—C.

*4.178. Draw another Lewis structure for C_4H_{10} that has a different connectivity than that in Problem 4.177. (*Hint*: Given that the bonding capacity of hydrogen is 1, how else might the carbon atoms be connected?)

5

Bonding Theories

Explaining Molecular Geometry

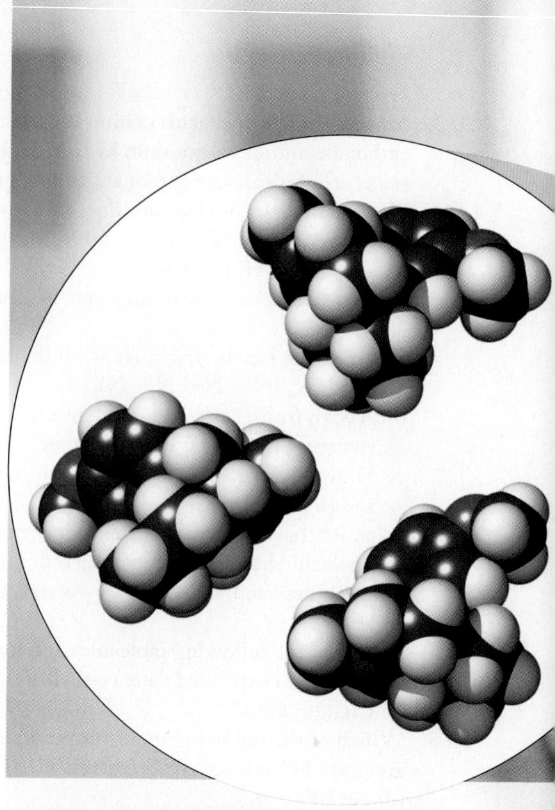

COUGH MEDICINE
Dextromethorphan has the molecular formula $C_{18}H_{25}NO$ and is a common ingredient in cough syrup. The bonds in dextromethorphan, and the direction those bonds point in space, are responsible for the shape of the molecule and its cough suppressant properties. Another molecule, levomethorphan, with the same atoms and bonds as dextromethorphan, has bonds that are oriented differently, giving levomethorphan a different shape and very different properties: the latter compound is an opioid narcotic.

PARTICULATE **REVIEW**

Lewis Structure and Polar Bonds

In Chapter 4, we learned to draw Lewis structures to represent bonds within a compound. Consider the ball-and-stick representation shown here of a molecule of acetaminophen, the active ingredient in Tylenol®:

- Draw the Lewis structure of acetaminophen.
- What does the circle inside the hexagon of carbon atoms represent?
- Which bond is the most polar in acetaminophen?

(Review Sections 4.3, 4.4, and 4.6 if you need help.)

(Answers to Particulate Review questions are in the back of the book.)

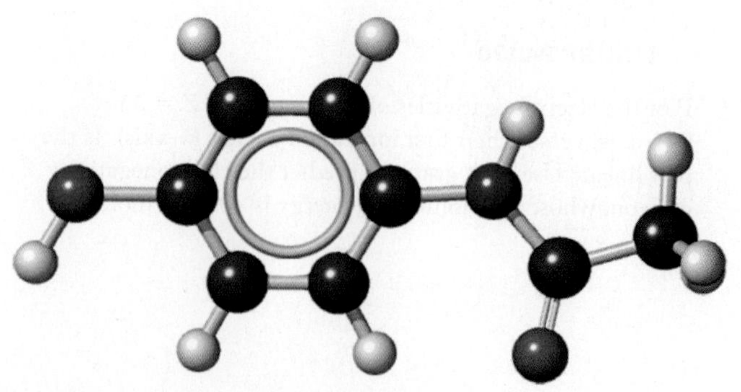

PARTICULATE **PREVIEW**

Bond Angles and Bonding Orbitals

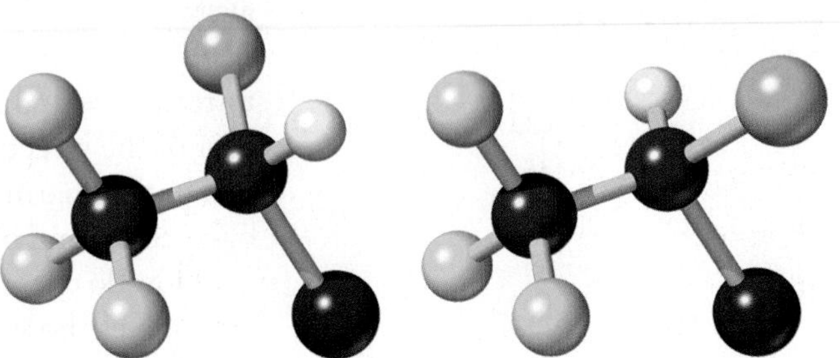

Shown here are two molecules of a compound commonly called halothane, $CF_3CHBrCl$. It is a powerful anesthetic used in surgery. As you read Chapter 5, look for ideas that will help you answer these questions:

- What are the bond angles between the F—C—F bonds?
- What orbitals overlap to form the C—H bonds?
- Do the two ball-and-stick representations of halothane depict the same molecule or two different molecules? If different, how are they related to one another?

Learning Outcomes

LO1 Use VSEPR and the concept of steric number to predict the bond angles in molecules and the shapes of molecules with one central atom
Sample Exercises 5.1, 5.2, 5.3

LO2 Predict whether a substance is polar or nonpolar based on its molecular structure
Sample Exercise 5.4

LO3 Use valence bond theory to account for bond angles and molecular shape
Sample Exercise 5.5

LO4 Use atomic orbital hybridization to explain molecular shape
Sample Exercise 5.6

LO5 Recognize chiral molecules
Sample Exercise 5.7

LO6 Draw molecular orbital (MO) diagrams of small molecules and use MO theory to predict bond order and explain magnetic properties and spectra
Sample Exercises 5.8, 5.9, 5.10

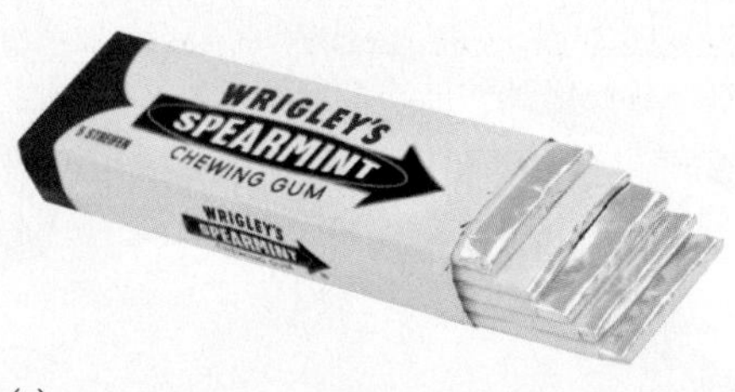

(a)

(b)

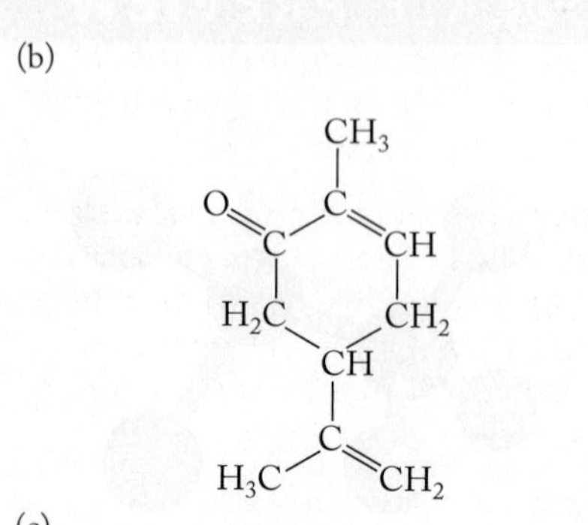

(c)

FIGURE 5.1 The smell and taste of (a) spearmint and (b) caraway are due to two different arrangements in space of the atoms in $C_{10}H_{14}O$, whose condensed molecular structure is drawn in (c).

5.1 Biological Activity and Molecular Shape

Hold your hands out in front of you, palms up, fingers extended. Now rotate your wrists inward so that your thumbs point straight up. Your right hand looks the same as the image your left hand makes in a mirror. Does that mean that your two hands have the exact same shape? If you have ever tried to put your right hand in a glove made for your left, then you know that they do *not* have the same shape—they are very similar but not the same. Many other objects in our world have a "handedness" about them, too, from scissors to golf clubs to the cough suppressants in many cold medicines.

This chapter focuses on the importance of shape at the molecular level. For example, the compound that produces the refreshing aroma of spearmint has the molecular formula $C_{10}H_{14}O$. The compound responsible for the musty aroma of caraway seeds in rye bread has the same molecular formula *and* the same Lewis structure (Figure 5.1). To understand how two compounds could be so much alike and still have different properties, we have to consider their structures in three dimensions.

We perceive a difference in their aromas in part because each molecule has a unique site where it attaches to our nasal membranes. Just as a left glove only fits a left hand, the spearmint molecule fits only the spearmint-shaped site, and the caraway molecule fits only the caraway-shaped site. This phenomenon is called *molecular recognition*, and it enables biomolecular structures, such as nasal membranes, to recognize and react when a particular molecule with a specific shape binds to a part of the structure known as an *active site*. Many substances in the foods we eat and in the pharmaceuticals we take exert physiological effects because they are recognized by, and bind to, specific sites in the molecules that make up our bodies.

How do chemists describe the three-dimensional shapes of molecules, and what determines those shapes? Can we predict shapes if we know how atoms are bonded together in molecules? In this chapter, we examine several theories of bonding that explain molecular shapes, and we begin to explore the impact of molecular shape on the physical, chemical, and biological properties of compounds.

The shape of a molecule can affect many properties of the substance, including its physical state at room temperature, its aroma, its biological activity, and its distribution in the environment. In Chapter 4 we drew Lewis structures to account for the bonding in molecules and polyatomic ions, but Lewis structures are only two-dimensional representations of how atoms and the electron pairs that surround them are arranged in molecules. Lewis structures show what atoms are *connected* to one another in molecules and polyatomic ions, but they don't show how the atoms are *oriented* in three dimensions, nor do they necessarily reveal the overall shape of the molecule.

To illustrate this point, look at the Lewis structures and ball-and-stick models of carbon dioxide and methane shown in Figure 5.2. The linear array of atoms and bonding electrons in the Lewis structure of CO_2 corresponds to the actual linear shape of the molecule as represented by the ball-and-stick model. The angle between the two C=O bonds is 180°, just as in the Lewis structure. On the other hand, the Lewis structure for methane shows 90° angles between the four C—H bonds, whereas the experimentally measured H—C—H **bond angles**, depicted in the ball-and-stick model, are 109.5°. This disconnect between the Lewis structure and experimental data suggests that the Lewis model of chemical bonding needs to be revised.

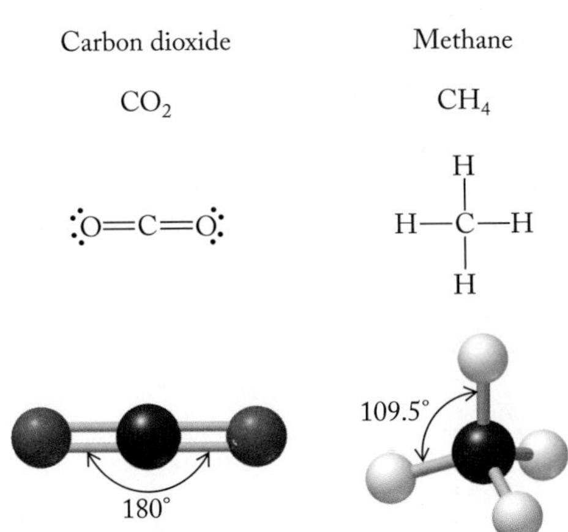

FIGURE 5.2 The Lewis structure of CO_2 matches its true molecular structure, but the Lewis structure of CH_4 does not because the C—H bonds in CH_4 actually extend in three dimensions.

In this chapter, we explore theories of covalent bonding that account for and predict the shapes of molecules. During this exploration you may wonder which bonding theory is the best one and ask which one you should use. Different models and different theories are well suited to answering some bonding questions but not others. For example, the Lewis structures we drew in Chapter 4 do a good job of explaining the bonding in molecules in which the atoms obey the octet rule, but they do not describe the three-dimensional shapes of those molecules or the angles between their bonds. As different theories of covalent bonding are described in this chapter, you will see which theory is best suited to answer a particular question about bonding and molecular structure.

5.2 Valence-Shell Electron-Pair Repulsion Theory (VSEPR)

Valence-shell electron-pair repulsion theory (VSEPR) is based on the fundamental chemical principle that electrons have negative charges and repel each other. VSEPR applies this principle by assuming that pairs of valence electrons are arranged about central atoms in ways that minimize repulsions between the pairs. To predict the shape of a molecule using VSEPR, we must consider its **electron-pair geometry**, which describes the relative positions in three-dimensional space of all the bonding pairs and lone pairs of valence electrons on the central atom, and its **molecular geometry**, which describes the relative positions of the atoms in a molecule. To accurately predict the molecular geometry, we first need to know the electron-pair geometry. If there are no lone pairs of electrons, then the process is simplified because the electron-pair geometry *is* the molecular geometry. Let's begin with this simpler case and consider the shapes of molecules that have different numbers of atoms bonded to a single central atom that has no lone pairs of electrons. To make our exploration even simpler, we will initially focus on molecules in which the same kind of atom is bonded to each central atom, so that all the bonds in each molecule are identical.

bond angle the angle (in degrees) defined by lines joining the centers of two atoms to a third atom to which they are chemically bonded.

valence-shell electron-pair repulsion theory (VSEPR) a model predicting the arrangement of valence electron pairs around a central atom that minimizes their mutual repulsion to produce the lowest energy orientations.

electron-pair geometry the three-dimensional arrangement of bonding pairs and lone pairs of electrons about a central atom.

molecular geometry the three-dimensional arrangement of the atoms in a molecule.

Central Atoms with No Lone Pairs

To determine the geometry of a molecule with a single central atom, we start by drawing its Lewis structure. From the Lewis structure, we determine a parameter called the **steric number (SN)** of the central atom, which is the sum of the number of atoms bonded to that atom and the number of lone pairs on it:

$$SN = \begin{pmatrix} \text{number of atoms} \\ \text{bonded to central atom} \end{pmatrix} + \begin{pmatrix} \text{number of lone pairs} \\ \text{on central atom} \end{pmatrix} \quad (5.1)$$

If the central atom has no lone pairs, then the *steric number equals the number of atoms bonded to the central atom.* In evaluating the shapes of the molecules, we will generate five common shapes—shown in Figure 5.3 and called linear, trigonal planar, tetrahedral, trigonal bipyramidal, and octahedral—which describe both electron-pair geometries and molecular geometries.

The simplest molecular structure with a central atom is one that has only two other atoms bonded to the central atom. As long as there are no lone pairs of valence electrons on the central atom, its steric number is 2. The electron pairs in the two bonds minimize their mutual repulsion by being on opposite sides of the central atom (Figure 5.3a). This gives a linear electron-pair geometry and a linear molecular geometry. The three atoms in the molecule are arranged in a straight line, and the angle between the two bonds is 180°.

If the central atom is bonded to three other atoms and has no lone pairs of valence electrons, then SN = 3. The three bonding atoms are as far apart as possible when they are located at the three corners of an equilateral triangle. The angle between each bond is 120° (Figure 5.3b). This electron-pair and molecular geometry is called **trigonal planar**.

If the central atom is bonded to four atoms and has no lone pairs, then SN = 4. The atoms bonded to the central atom occupy the four vertices of a *tetrahedron*, which is a four-sided pyramid (*tetra* is Greek, meaning "four"). The bonding pairs form bond angles of 109.5° with each other as shown in Figure 5.3(c). The electron-pair geometry and molecular geometry are both **tetrahedral**.

When a central atom is bonded to five atoms and has no lone pairs, SN = 5 and the five other atoms occupy the five corners of two triangular pyramids that share the same base. The central atom of the molecule is at the common center of the two bases as shown in Figure 5.3(d). One bonding pair points to the top of the upper pyramid, one points to the bottom of the lower pyramid, and the other three point to the three vertices of the shared triangular base. It may be helpful to think of the three vertices as three points along a circle at the equator of a sphere. In this model, the atoms that occupy the three sites and the bonds that connect them to the central atom are called *equatorial* atoms and bonds. The bond angles between the three equatorial bonds are 120° (just as in the trigonal planar geometry in Figure 5.3b).

FIGURE 5.3 Electron-pair geometries depend on the steric number (SN) of the central atom in a molecule. In these images there are no lone pairs of electrons on the central atoms (red dots), so SN is equal to the number of atoms bonded to the central atom, and molecular geometry is the same as electron-pair geometry. The black lines represent covalent bonds; the blue lines outline the geometric forms that give these shapes their names.

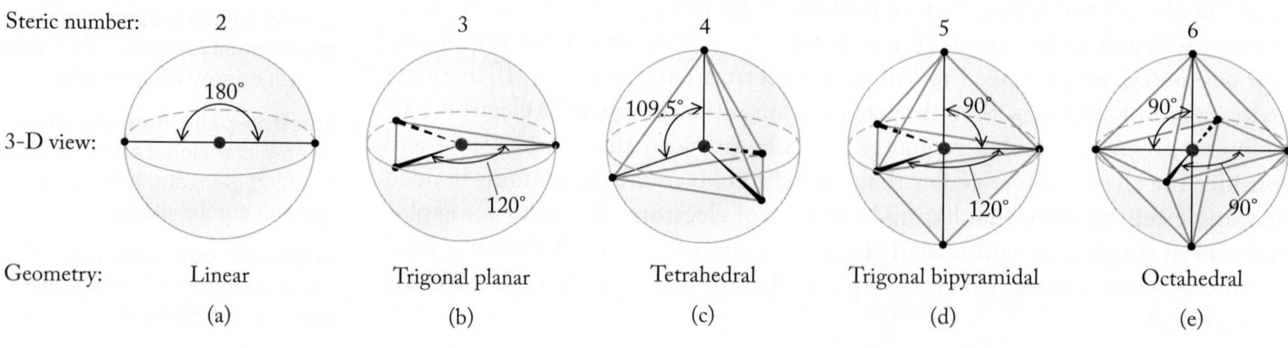

The bond angle between an equatorial bond and either vertical, or *axial*, bond is 90°, and the angle between the two axial bonds is 180°. A molecule in which the atoms are arranged this way is said to have a **trigonal bipyramidal** electron-pair and molecular geometry.

For SN = 6, picture two pyramids that have a square base (Figure 5.3e). Put them together, base to base, and you form a shape in which all six positions are equivalent. We can think of the six bonding pairs of electrons as three sets of two electron pairs each. The two pairs in each set are oriented at 180° to each other and at 90° to the other two pairs, just like the axes of an *xyz* coordinate system. In our spherical model, there are four equatorial atoms at the four vertices of the common square base. The bonds to them are 90° apart. A fifth atom lies at the top of the upper pyramid, and a sixth lies at the bottom of the lower one. This arrangement defines **octahedral** electron-pair and molecular geometries.

steric number (SN) the sum of the number of atoms bonded to a central atom plus the number of lone pairs of electrons on the central atom.

trigonal planar molecular geometry about a central atom with a steric number of 3 and no lone pairs of electrons.

tetrahedral molecular geometry about a central atom with a steric number of 4 and no lone pairs of electrons.

trigonal bipyramidal molecular geometry about a central atom with a steric number of 5 and no lone pairs of electrons, in which three atoms occupy equatorial sites and two other atoms occupy axial sites above and below the equatorial plane.

octahedral molecular geometry about a central atom with a steric number of 6 and no lone pairs of electrons, in which all six sites are equivalent.

CONCEPT TEST

Assume that all of the bonds in molecules with the five shapes described previously are single bonds and that none of the central atoms have lone pairs of electrons. Which SN values correspond to central atoms with less than an octet of valence electrons, which SN values correspond to central atoms with an expanded octet, and which SN value has a central atom with exactly eight valence electrons?

(Answers to Concept Tests are in the back of the book.)

Now let's consider a real-world analogy to these bond orientations. Beginning with fully inflated balloons, we tie together clusters of two, three, four, five, and six of them (Figure 5.4) so tightly that they push against each other. If the tie points of the clusters represent the central atom in our balloon model, then the opposite ends represent the atoms that are bonded to the central atom. Note how our clusters of balloons produce the same orientations that resulted from selecting points that were as far apart as possible. The long axes of the balloons in Figure 5.4 provide an accurate representation of the bond directions in Figure 5.3.

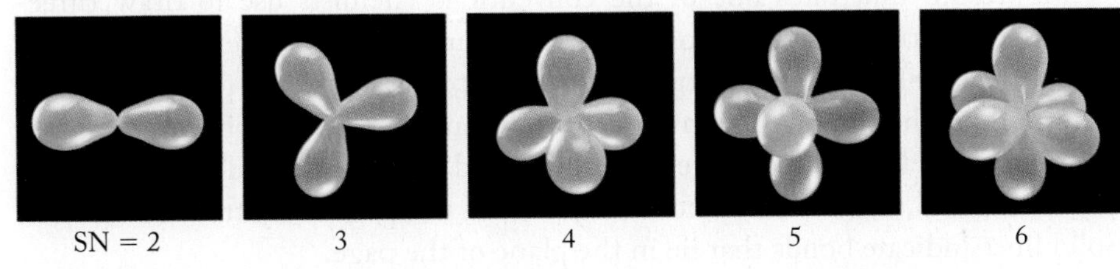

FIGURE 5.4 Balloons tied tightly together push against each other and orient themselves as far away from each other as possible. In doing so, they mimic the locations of different numbers of electron pairs about a central atom. Each balloon represents one electron pair. Note the similarities between the patterns of the balloons and the geometric shapes in Figure 5.3.

Now let's look at five simple molecules that have no lone pairs about the central atom and apply VSEPR and the concept of steric number to predict their electron-pair and molecular geometries. To do so, we follow three steps:

1. Draw the Lewis structure for the molecule.
2. Determine the steric number of the central atom.
3. Use the steric number and Figure 5.3 to predict the electron-pair and molecular geometries.

Example I: Carbon Dioxide, CO_2

1. Lewis structure:

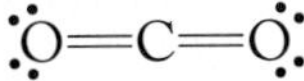

2. The central carbon atom has two atoms bonded to it and no lone pairs, so the steric number is 2.
3. Because SN = 2, the O═C═O bond angle is 180° (see Figure 5.3a) and the electron-pair and molecular geometries are both linear.

CONNECTION We learned in Chapter 4 that some molecules have central atoms with incomplete octets of electrons.

Example II: Boron Trifluoride, BF_3

1. The Lewis structure of BF_3 that results in zero formal charges on all atoms is

:F̤: — B — :F̤: with :F̤: bonded below B

2. There are three fluorine atoms bonded to the central boron atom, and there are no lone pairs on boron, so SN = 3.
3. If SN = 3, then all four atoms lie in the same plane. The three F atoms form an equilateral triangle, and the F—B—F bond angles are 120°. The electron-pair and molecular geometries are trigonal planar (Figures 5.3b and 5.5).

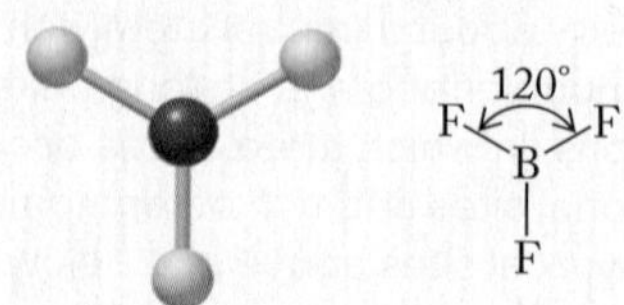

FIGURE 5.5 The ball-and-stick model shows the orientation of the atoms in boron trifluoride. All F—B—F bond angles are 120° in this trigonal planar molecular geometry.

Example III: Carbon Tetrachloride, CCl_4

1. Lewis structure:

:C̤l: — C — :C̤l: with :C̤l: above and below C

2. There are four chlorine atoms bonded to the central carbon atom, which gives carbon a full octet, so SN = 4.
3. If SN = 4, then the chlorine atoms are located at the vertices of a tetrahedron and all four Cl—C—Cl bond angles are 109.5°, producing tetrahedral electron-pair and molecular geometries (Figures 5.3c and 5.6).

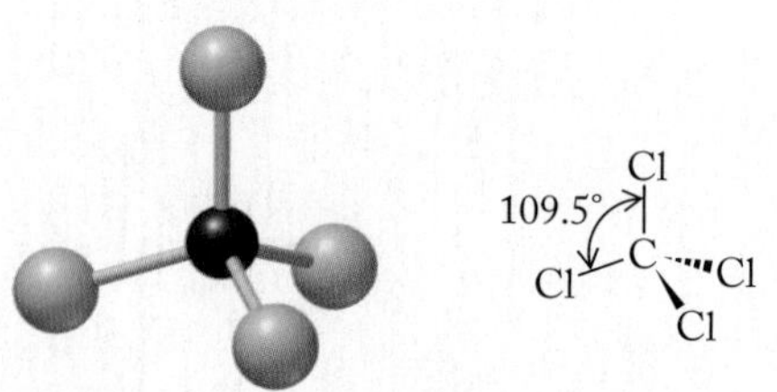

FIGURE 5.6 The ball-and-stick model shows the orientation of the atoms in carbon tetrachloride. All Cl—C—Cl bond angles are 109.5° in this tetrahedral molecular geometry.

Figure 5.6 illustrates one of the conventions chemists use to draw three-dimensional structures on a two-dimensional surface such as a textbook page. A solid wedge (▬) is used to indicate a bond that comes out of the page toward the viewer. Thus, the solid wedge in Figure 5.6 means that the chlorine atom in that position points toward the viewer at a downward angle. A dashed wedge (┈), on the other hand, indicates a bond that goes into the page away from the viewer. Solid lines indicate bonds that lie in the plane of the page.

Example IV: Phosphorus Pentafluoride, PF_5

1. Lewis structure:

P bonded to five :F̤: atoms

2. There are five fluorine atoms bonded to the central phosphorus atom, which has no lone pairs, so SN = 5.
3. If SN = 5, then the fluorine atoms are located at the vertices of a trigonal bipyramid and the electron-pair geometry and molecular geometry are trigonal bipyramidal (Figures 5.3d and 5.7).

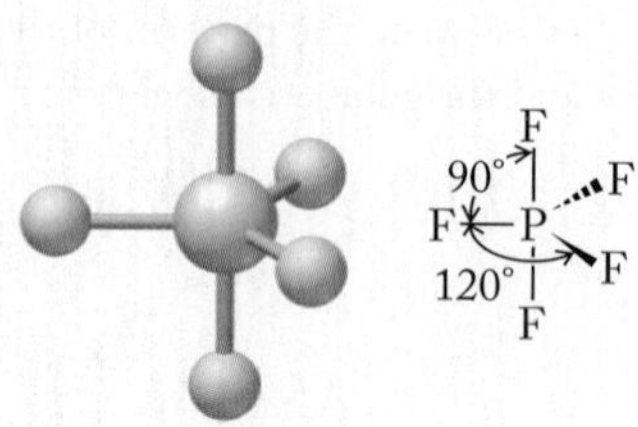

FIGURE 5.7 The ball-and-stick model shows the orientation of the atoms in phosphorus pentafluoride. The equatorial P—F bonds are 120° from each other. Each of them is 90° from the two bonds connecting the central phosphorus atom to the fluorine atoms in the axial positions. This molecular geometry is trigonal bipyramidal.

Example V: Sulfur Hexafluoride, SF_6

1. Lewis structure:

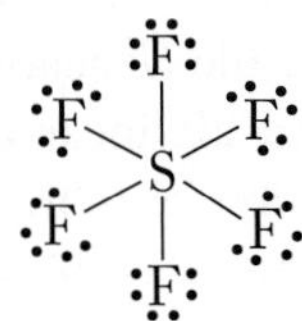

2. There are six fluorine atoms bonded to the central sulfur atom, which has no lone pairs, so SN = 6.
3. If SN = 6, then the fluorine atoms are located at the vertices of an octahedron and the electron-pair and molecular geometries are octahedral (Figures 5.3e and 5.8).

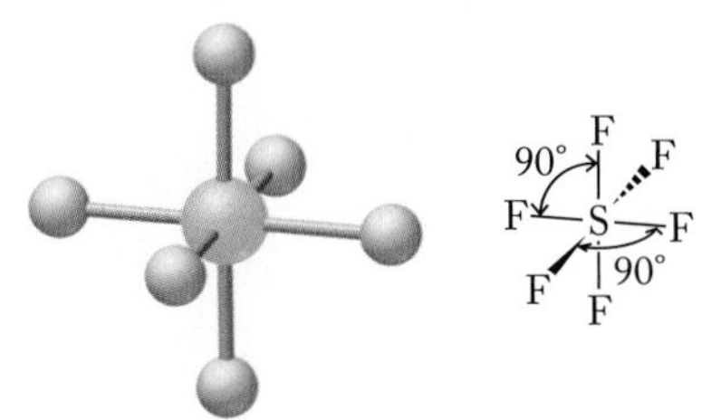

FIGURE 5.8 The ball-and-stick model shows how the six fluorine atoms are oriented in three dimensions about the central sulfur atom in this octahedral molecular geometry.

SAMPLE EXERCISE 5.1 Using VSEPR to Predict Molecular Geometry I **LO1**

Formaldehyde, CH_2O, is a gas at room temperature and an ingredient in solutions used to preserve biological samples. Use VSEPR to predict the molecular geometry of formaldehyde.

Collect, Organize, and Analyze To predict the geometry we need to (1) draw the Lewis structure, (2) determine the steric number of the central atom, and (3) identify the electron-pair and molecular geometries using Figure 5.3. Carbon has the greatest bonding capacity of the three component elements, so it is the central atom.

Solve Using the procedure developed in Chapter 4 for drawing Lewis structures, we obtain the following one for formaldehyde:

O
‖
C
H H

There are three atoms bonded to the central atom and no lone pairs on it, so SN = 3 and the electron-pair and molecular geometries are both trigonal planar.

Think About It The key to predicting the correct molecular geometry of a molecule with no lone pairs on its central atom is to determine the steric number, which is simply the number of atoms bonded to the central atom.

Practice Exercise Use VSEPR to determine the molecular geometry of the chloroform molecule, $CHCl_3$, and draw the molecule using the solid-wedge, dashed-wedge convention.

(Answers to Practice Exercises are in the back of the book.)

Measurements of the bond angles in formaldehyde show that the H—C—H bond angle is slightly smaller than the 120° predicted for a trigonal planar geometry (Figure 5.3b), and the H—C=O bond angles are about 1° larger. The C=O double bond consists of two pairs of bonding electrons that exert greater repulsion on adjacent bonds than a single bonding pair would. This greater repulsion increases the H—C=O bond angles and decreases the H—C—H bond angle. VSEPR does not enable us to predict the actual values of those bond angles, but it does allow us to correctly predict that bond angles deviate from the ideal values of an idealized trigonal planar molecule.

Central Atoms with Lone Pairs

If SN = 2 and one of the electron pairs on the central atom is a lone pair, then there is only one other atom in the molecule (Equation 5.1). Since two points define a straight line, two-atom molecules are all linear and they have no bond angles.

Moving on to molecules with three or more atoms, recall from Figure 4.9 that we drew ozone molecules with an angular shape:

To calculate the steric number of the central O atom in both resonance structures, notice that it is bonded to two other atoms and has one lone pair of electrons. Therefore, its SN value in both resonance structures is 2 + 1 = 3. When SN = 3, the arrangement of atoms and lone pairs about the central atom is trigonal planar (Figure 5.3b). This means that the electron-pair geometry, which takes into account both the atoms and the nonbonding electron pairs around the central atom, is trigonal planar (see Figure 5.9a). The molecular geometry, on the other hand, describes the relative positions of the *atoms* in the molecule only. With SN = 3 and two atoms attached to the central O atom, the ozone molecule has the **angular** (or **bent**) molecular geometry shown in Figure 5.9(b).

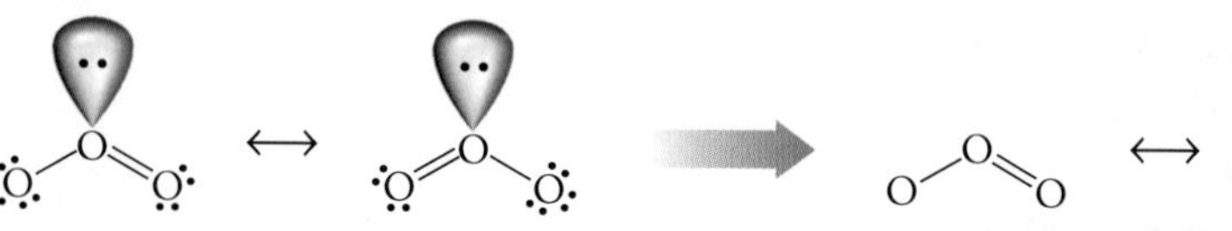

(a) Electron-pair geometry = trigonal planar

(b) Molecular geometry = bent

FIGURE 5.9 (a) The electron-pair geometry of O_3 is trigonal planar because the steric number of the central oxygen atom is 3 (it is bonded to two atoms and has one lone pair of electrons). (b) The molecular geometry is bent because there is no bonded atom, only a lone pair of electrons, on the central oxygen atom in the structure.

Experimental measurements confirm that O_3 is indeed a bent molecule with a bond angle of 117°. The angle is smaller than the 120° we would predict for a symmetrical trigonal planar electron-pair geometry. We can explain the smaller angle using VSEPR by comparing the amount of space near atoms occupied by bonding electrons with the amount of space occupied by electrons in a lone pair. Because the bonding electrons are attracted to two nuclei, they have a high probability of being located between the two atomic centers that share them. In contrast, the lone pair is not shared with a second atom and is spread out near the central O atom as shown in Figure 5.10. This puts the lone pair of electrons closer to the bonding pairs and produces greater repulsion. As a result, the lone pair pushes the bonding pairs closer together, thereby reducing the bond angle. In general,

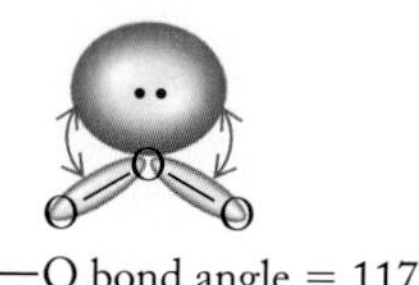

O—O—O bond angle = 117°

FIGURE 5.10 The lone pair of electrons on the central oxygen atom in O_3 occupies more space (larger purple region) than the electron pairs in the O—O bonds (smaller purple regions). The increased repulsion (represented by the double-headed arrows) resulting from the larger volume occupied by the lone pair forces the noncentral oxygen atoms closer together, making the bond angle 117° instead of 120°.

- Repulsion between lone pairs and bonding pairs is greater than repulsion between bonding pairs;
- Repulsion caused by a lone pair is greater than repulsion caused by a double bond;
- Repulsion caused by a double bond is greater than repulsion caused by a single bond;
- Two lone pairs of electrons on a central atom exert a greater repulsive force on the atom's bonding pairs than does one lone pair.

SAMPLE EXERCISE 5.2 Predicting Relative Sizes of Bond Angles **LO1**

Rank NH_3, CH_4, and H_2O in order of decreasing bond angles in their molecular structures.

Collect, Organize, and Analyze The bond angles surrounding a central atom are linked to the steric number of the central atom, as shown in Figure 5.3, and to the number of lone pairs of electrons on the central atom. To find the steric numbers we need to draw the Lewis structures of the three molecules.

Solve Using the method for drawing Lewis structures from Chapter 4, we obtain the following results:

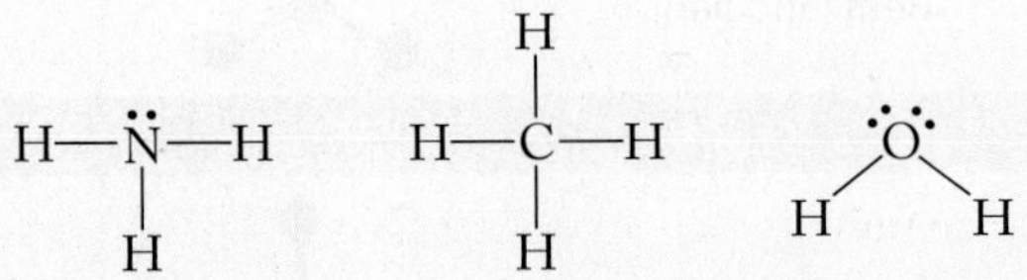

Because a total of four bonded atoms and lone pairs of electrons surrounds each of the central atoms in the three molecules, SN = 4 for all three. This means they all have tetrahedral electron-pair geometries, and nominal bond angles are 109.5°. In CH_4, all four tetrahedral electron pairs are equivalent bonding pairs, so all four bond angles are 109.5°. In NH_3 and H_2O, however, repulsion from the lone pair(s) of electrons pushes the bonds closer together, reducing the bond angles. The two lone pairs on the O atom in H_2O exert a greater repulsive force on its bonding pairs than the single lone pair on the N atom exerts on the bonding pairs in NH_3. Therefore, the bond angle in H_2O should be less than the bond angles in NH_3. Ranking the three molecules in order of decreasing bond angle, we have

$$CH_4 > NH_3 > H_2O$$

Think About It The logic used in answering this problem is supported by experimental evidence: the bond angles in molecules of CH_4, NH_3, and H_2O are 109.5°, 107.0°, and 104.5°, respectively.

Practice Exercise Determine which of the following species has the largest bond angle and which has the smallest: NO_2, N_2O, and NO_2^-.

(Answers to Practice Exercises are in the back of the book.)

angular or **bent** molecular geometry about a central atom with a steric number of 3 and one lone pair or a steric number of 4 and two lone pairs.

trigonal pyramidal molecular geometry about a central atom with a steric number of 4 and one lone pair of electrons.

As we saw in Sample Exercise 5.2, three different combinations of atoms and lone pairs are possible for a central atom with SN = 4: four atoms and no lone pairs, three atoms and one lone pair, or two atoms and two lone pairs (see Table 5.1). The first case is illustrated by the molecular structure of methane, CH_4, in which a tetrahedral electron-pair geometry translates into a tetrahedral molecular geometry. In ammonia, NH_3, which also has a tetrahedral electron-pair geometry by virtue of its three bonding pairs and one lone pair, one of the four vertices is the lone pair on the N atom (Figure 5.11). The resulting molecular geometry is called **trigonal pyramidal**. As we discussed in Sample Exercise 5.2, the strong repulsion produced by the diffuse lone pair of electrons on the N atom pushes the three N—H bonds closer together in NH_3 and reduces the angles between them from 109.5° (as in CH_4) to 107.0°.

FIGURE 5.11 The steric number of the N atom in NH_3 is 4 (it is bonded to three atoms and has one lone pair of electrons), so its electron-pair geometry is tetrahedral. However, one of the vertices of the tetrahedron is occupied by a lone pair of electrons, not an atom, so the molecular geometry is trigonal pyramidal.

The central O atom in a molecule of H_2O has SN = 4 and a tetrahedral electron-pair geometry because it is bonded to two H atoms and has two lone pairs of electrons (Figure 5.12). The molecular geometry of H_2O, on the other hand, is bent (or angular) because molecular geometries ignore lone pairs and take into account only the atoms bonded to the central atom. As we also discussed in Sample Exercise 5.2, the H—O—H bond angle in water is smaller than 109.5°

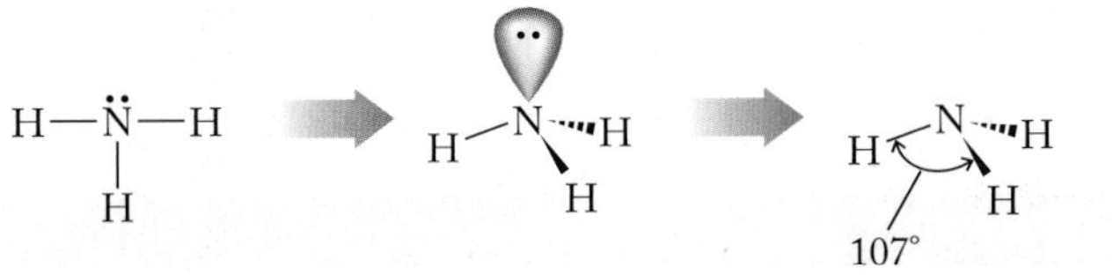

TABLE 5.1 Electron-Pair Geometries And Molecular Geometries

SN = 3	Electron-Pair Geometry	No. of Bonded Atoms	No. of Lone Pairs	Molecular Geometry	Structure	Theoretical Bond Angles
	Trigonal planar	3	0	Trigonal planar		120°
	Trigonal planar	2	1	Bent (angular)		<120°
SN = 4						
	Tetrahedral	4	0	Tetrahedral		109.5°
	Tetrahedral	3	1	Trigonal pyramidal		<109.5°
	Tetrahedral	2	2	Bent (angular)		<109.5°
SN = 5						
	Trigonal bipyramidal	5	0	Trigonal bipyramidal		90°, 120°
	Trigonal bipyramidal	4	1	Seesaw	=	<90°, <120°
	Trigonal bipyramidal	3	2	T-shaped	=	<90°
	Trigonal bipyramidal	2	3	Linear	=	180°
SN = 6						
	Octahedral	6	0	Octahedral		90°
	Octahedral	5	1	Square pyramidal		<90°
	Octahedral	4	2	Square planar		90°
	Octahedral	3	3	Although these geometries are possible, we will not encounter any molecules with them		
	Octahedral	2	4			

due to repulsion between the two lone pairs and each bonding pair. The actual bond angle is 104.5°.

Molecules with trigonal bipyramidal electron-pair geometry (SN = 5) have four possible molecular geometries depending on the number of lone pairs per molecule (see Table 5.1). These options arise because there are axial and equatorial vertices in a trigonal bipyramid (Figure 5.3d). VSEPR enables us to predict which vertices are occupied by bonded atoms and which are occupied by lone pairs. The key to these predictions is the fact that the repulsions between pairs of electrons increase as the angle between them decreases: two electron pairs at 90° experience a greater mutual repulsion than two at 120°, which in turn have a greater repulsion than two at 180°. To minimize repulsions involving lone pairs, VSEPR predicts that they preferentially occupy equatorial, rather than axial, vertices. Why? Because an equatorial lone pair has *two* 90° repulsions with the two axial electron pairs, as shown in Figure 5.13(a), but an axial lone pair has *three* 90° repulsions with three equatorial electron pairs, as shown in Figure 5.13(b).

When we assign one, two, or three lone pairs of valence electrons to equatorial vertices, we get three of the molecular geometries for SN = 5 in Table 5.1. When a single lone pair occupies an equatorial site, we get a molecular geometry called **seesaw** (Figure 5.14) because its shape, when rotated 90° clockwise, resembles a playground seesaw. (The formal name for this shape is *disphenoidal*.)

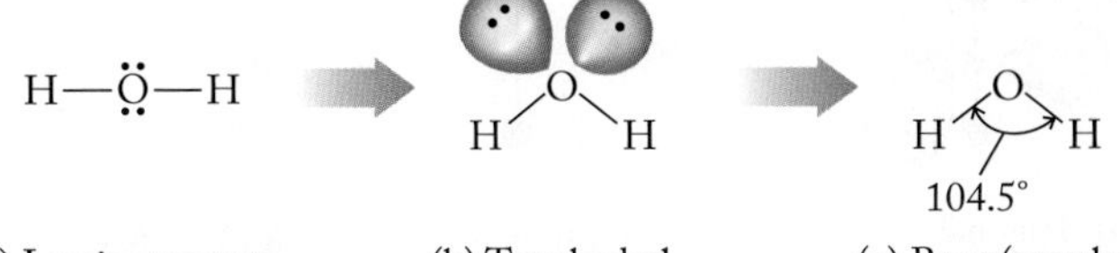

(a) Lewis structure (b) Tetrahedral electron-pair geometry (c) Bent (angular) molecular geometry

FIGURE 5.12 Two of the vertices of the tetrahedron are occupied by lone pairs of electrons, so only two atoms define the molecular geometry of H_2O, which is bent.

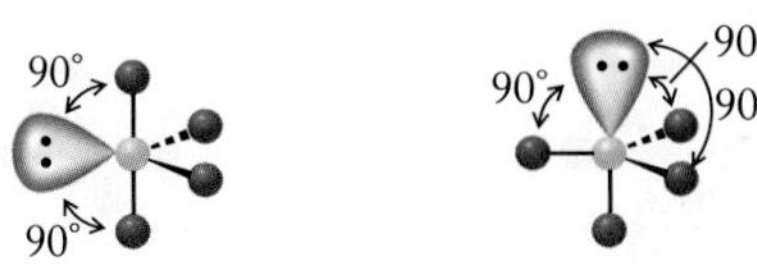

(a) Equatorial lone pair (b) Axial lone pair

FIGURE 5.13 A lone pair of electrons in an equatorial position of a molecule with trigonal bipyramidal electron-pair geometry interacts through 90° with two other electron pairs. A lone pair in an axial position interacts through 90° with *three* other electron pairs. Fewer 90° interactions reduce internal electron-pair repulsion and lead to greater stability.

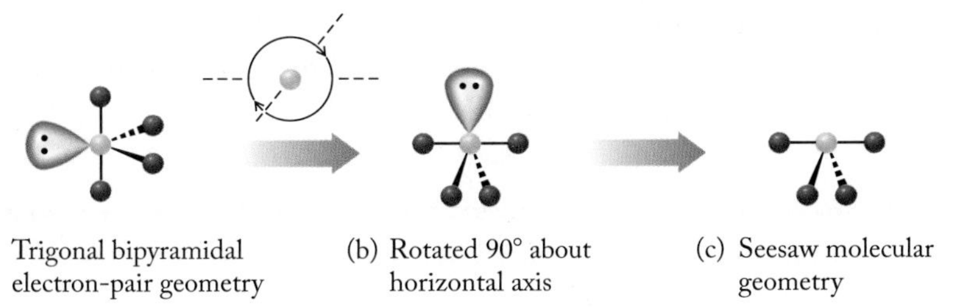

(a) Trigonal bipyramidal electron-pair geometry (b) Rotated 90° about horizontal axis (c) Seesaw molecular geometry

FIGURE 5.14 A single lone pair of electrons in an equatorial position of a trigonal bipyramidal electron-pair geometry produces a seesaw molecular geometry.

When two lone pairs occupy equatorial sites, the molecular geometry that results is called **T-shaped**. This designation becomes more apparent when the structure in Figure 5.15(a) is rotated 90° counterclockwise. Lone-pair repulsions result in bond angles that are slightly less than the 90° and 180° we would expect from a perfectly shaped "T" geometry.

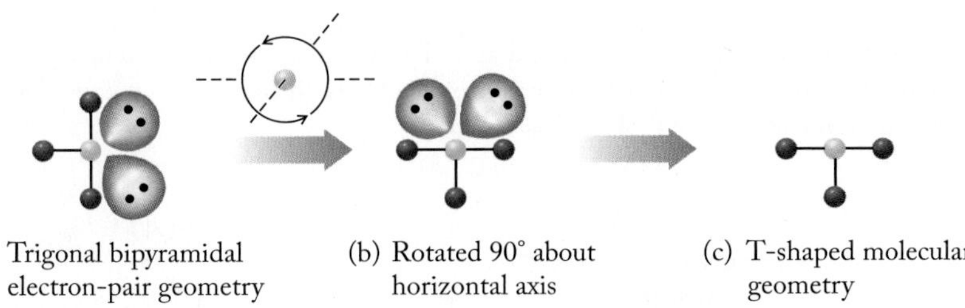

(a) Trigonal bipyramidal electron-pair geometry (b) Rotated 90° about horizontal axis (c) T-shaped molecular geometry

FIGURE 5.15 Two lone pairs of electrons in equatorial positions of a trigonal bipyramidal electron-pair geometry produce a T-shaped molecular geometry.

seesaw molecular geometry about a central atom with a steric number of 5 and one lone pair of electrons in an equatorial position; the atoms occupy two axial sites and two equatorial sites.

T-shaped molecular geometry about a central atom with a steric number of 5 and two lone pairs of electrons that occupy equatorial positions; the atoms occupy two axial sites and one equatorial site.

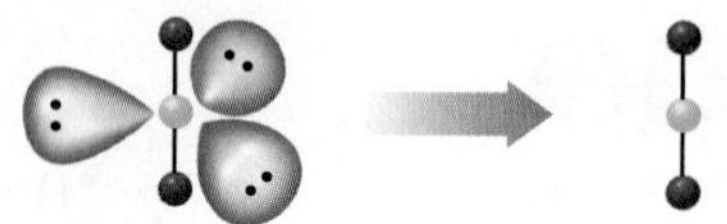

(a) Trigonal bipyramidal electron-pair geometry (b) Linear molecular geometry

FIGURE 5.16 Three lone pairs of electrons in equatorial positions of a trigonal bipyramidal electron-pair geometry produce a linear molecular geometry.

(a) Octahedral electron-pair geometry (b) Square pyramidal molecular geometry

FIGURE 5.17 A lone pair of electrons in an octahedral electron-pair geometry produces a square pyramidal molecular geometry.

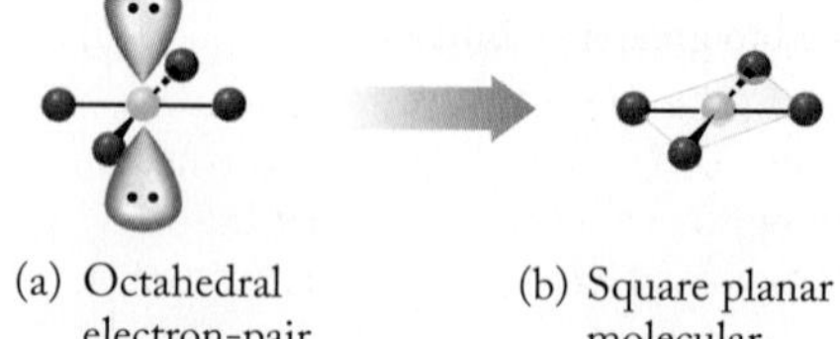

(a) Octahedral electron-pair geometry (b) Square planar molecular geometry

FIGURE 5.18 Two lone pairs of electrons on opposite sides of an octahedral electron-pair geometry produce a square planar molecular geometry.

Finally, a SN = 5 molecule with three lone pairs and two bonded atoms has a linear geometry because all three lone pairs occupy equatorial sites and the bonding pairs are in the two axial positions (Figure 5.16).

CONCEPT **TEST**

The bond angles in a trigonal bipyramidal molecular structure are either 90°, 120°, or 180°. Are the corresponding bond angles in a seesaw structure likely to be larger than, the same as, or smaller than those values?

(Answers to Concept Tests are in the back of the book.)

Molecules that have a central atom with SN = 6 have an octahedral electron-pair geometry (Figure 5.3e) and the molecular geometries listed in Table 5.1. With one lone pair of electrons, there is only one possible molecular geometry, because all the sites in an octahedron are equivalent (Figure 5.17). That one molecular geometry is called **square pyramidal** and it consists of a pyramid with a square base, four triangular sides, and the central atom "embedded" in the base.

Because of stronger repulsion between lone pairs and bonding pairs, the bond angles in a square pyramidal molecule should be slightly less than the ideal angle of 90°. In BrF_5, for example, the angles between its equatorial and axial bonds are 85°.

When two lone pairs are present at the central atom in a molecule with octahedral electron-pair geometry, they occupy vertices on opposite sides of the octahedron to minimize the repulsions between them. The resultant molecular geometry is called **square planar** because the molecule is shaped like a square and all five atoms reside in the same plane (Figure 5.18). Because the two lone pairs are on opposite sides of the bonding pairs, the presence of the lone pairs does not distort the bond angles: they are all 90°.

SAMPLE EXERCISE 5.3 Using VSEPR to Predict Geometry II **LO1**

The Lewis structure of sulfur tetrafluoride (SF_4) is

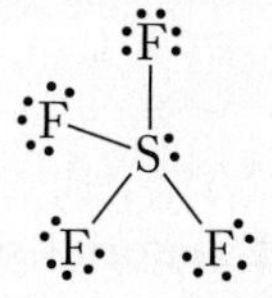

What is its molecular geometry and what are the angles between the S—F bonds?

Collect, Organize, and Analyze The Lewis structure of SF_4 shows that the central S atom is bonded to four F atoms and has a lone pair of electrons, which means the S atom has SN = 5. Table 5.1 contains molecular geometry options for molecules with different SN values.

Solve A steric number of 5 means the S atom has a trigonal bypramidal electron-pair geometry, but the presence of one lone pair of electrons on the S atom means that its molecular geometry (from Table 5.1) is seesaw:

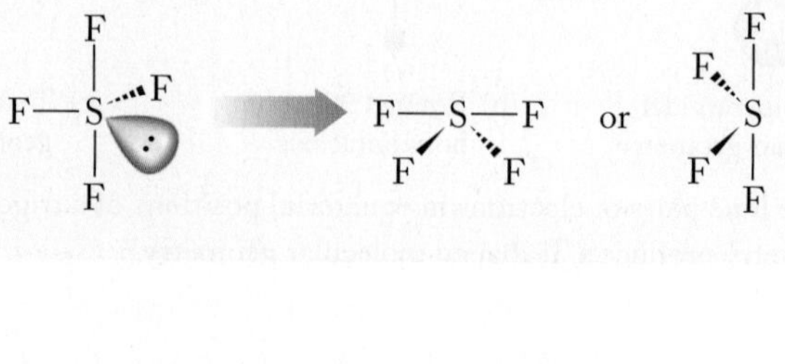

or

The proximity of the equatorial lone pair to the center of the S atom increases the electron-pair repulsion experienced by the bonding pairs, which decreases the bond angles between them from their normal 90° between the axial and equatorial bonds, 120° between the two equatorial bonds, and 180° between the two axial bonds.

Think About It Any molecule with SN = 5 and one lone pair should have a seesaw molecular geometry like SF_4.

Practice Exercise Determine the molecular geometry and the bond angles of SO_2Cl_2.

(Answers to Practice Exercises are in the back of the book.)

5.3 Polar Bonds and Polar Molecules

We learned in Chapter 4 that an unequal distribution of bonding electrons between two atoms produces a partial negative charge on one end of the bond and a partial positive charge on the other. This charge separation, caused by the difference in electronegativity between the two atoms, results in a **bond dipole**. Now that we know something about the three-dimensional geometry of molecules, we can understand how bond dipoles contribute to the *overall* polarity of a molecule.

The polarity of a molecule can be determined by summing the polarities of all the individual bond dipoles in the molecule. This summing must take into account both the strengths of the individual bond dipoles and their orientations with respect to one another. In CO_2, for example, the two dipoles are equivalent in strength because they involve the same atoms and the same kind of bond (C=O). The linear shape of the molecule means that the direction of one C=O bond dipole is opposite the direction of the other:

CONNECTION In Chapter 4 we introduced electronegativity and the unequal distribution of electrons in polar covalent bonds. See Figure 4.11 to review the use of arrows to indicate bond polarity and Figure 4.13 to review the periodic trends in electronegativity values.

$$:\ddot{O}\overset{\longleftrightarrow}{=}C\overset{\longleftrightarrow}{=}\ddot{O}:$$

Each dipole exactly offsets the other so that, even though the individual bonds are polar, CO_2 is a nonpolar molecule. Put another way, CO_2 has no overall **permanent dipole**. We use the adjective *permanent* here because, as we saw in Chapter 4, the vibration of the bonds in molecules like CO_2 can create fluctuating electric fields that allow CO_2 to absorb infrared electromagnetic radiation. The fluctuations create high-frequency *temporary* dipoles, even in nonpolar molecules.

Carbon tetrachloride, CCl_4, is also a nonpolar substance, even though its C—Cl bonds are polar. The four C—Cl bond dipoles all have the same strength because the same atoms are involved. CCl_4 has tetrahedral molecular geometry, however, so the bond dipoles offset one another and the CCl_4 molecule is nonpolar overall:

Cl / C / Cl Cl Cl

Water molecules are bent, not linear like molecules of CO_2. Therefore, the dipoles of the two O—H bonds in a molecule of H_2O do *not* offset each other.

square pyramidal molecular geometry about a central atom with a steric number of 6 and one lone pair of electrons; as typically drawn, the atoms occupy four equatorial sites and one axial site.

square planar molecular geometry about a central atom with a steric number of 6 and two lone pairs of electrons that occupy axial sites; the atoms occupy four equatorial positions.

bond dipole separation of electrical charge created when atoms with different electronegativities form a covalent bond.

permanent dipole permanent separation of electrical charge in a molecule due to unequal distributions of bonding and/or lone pairs of electrons.

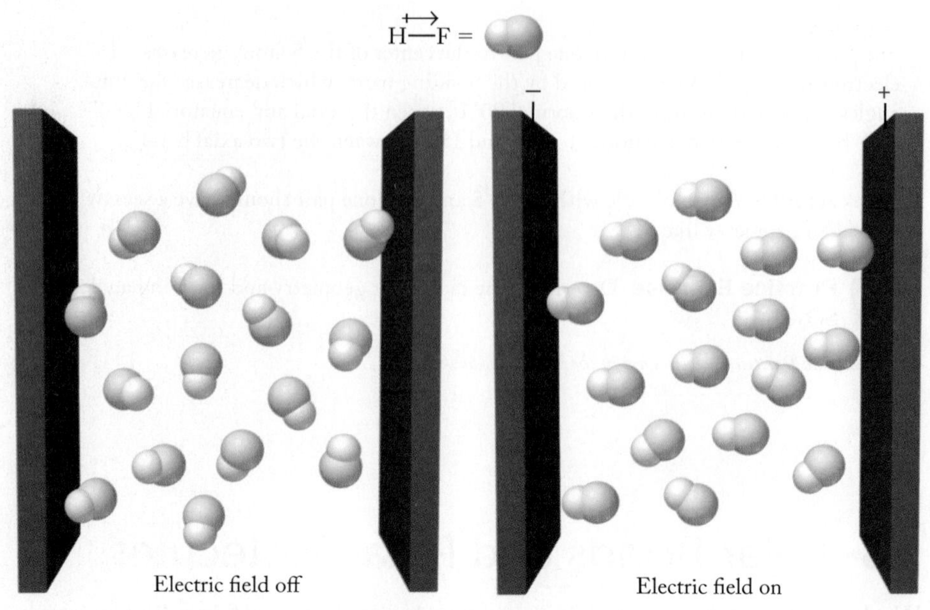

FIGURE 5.19 Gaseous HF molecules are oriented randomly in the absence of an electric field but align when an electric field is applied to two metal plates. The negative (fluorine) end of each molecule is directed toward the positively charged plate; the positive (hydrogen) end is directed toward the negative plate.

Instead, the molecule has an overall permanent dipole with the negative end directed toward the O atom:

H—Ö—H → H—Ö—H (Net dipole)

The presence of this overall dipole means that water is a polar molecule. Water's polarity leads to interactions between H and O atoms on adjacent H_2O molecules in liquid water or solid ice. Other polar molecules exhibit similar interactions. We discuss those interactions and their consequences in Chapter 6.

The polarity of a molecule can be determined experimentally by measuring its permanent **dipole moment (*μ*)**. The value of μ expresses the extent of the overall separation of positive and negative charge in the molecule, and it is determined by measuring the degree to which the molecule aligns with a strong electric field, as shown in Figure 5.19. Note how the negative (F) ends of the HF molecules in the figure are oriented toward the positive plate, and how the positive (H) ends are oriented toward the negative plate. The more polar a molecule is, the more strongly it aligns with an electric field.

Dipole moments are usually expressed in units of *debyes* (D), where 1 D = 3.34×10^{-30} coulomb-meter. The dipole moments of several polar substances are listed in Table 5.2. Chloroform ($CHCl_3$) has a relatively strong dipole moment (1.01 D) because chlorine is more electronegative than carbon, and the two electrons in each C—Cl bond are pulled away from C and toward Cl (Figure 5.20a). Because hydrogen is only slightly less electronegative than carbon ($\Delta EN = 0.4$), the H—C bond is considered to be nonpolar. Consequently, the electron distribution is away from the top part of the molecule as drawn and toward the bottom.

In trichlorofluoromethane (CCl_3F), all four bond dipoles point away from the central carbon atom because fluorine ($\chi = 4.0$) and chlorine ($\chi = 3.0$) are both more electronegative than carbon ($\chi = 2.5$; Figure 5.20b). However, a C—F bond dipole is stronger than a C—Cl bond dipole because fluorine is more electronegative than chlorine. As a result, bonding electrons are pulled more toward the C—F (top) part of the molecule than toward the C—Cl (bottom) part. The net direction of the summed bond dipoles is upward.

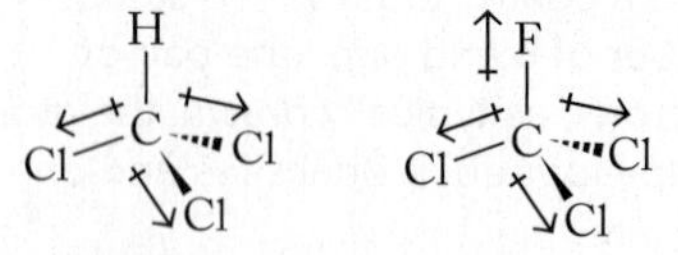

(a) $CHCl_3$ (b) CCl_3F

FIGURE 5.20 (a) The four bonds in chloroform ($CHCl_3$) are not equivalent, which causes an asymmetrical distribution of bonding electrons in the molecule, resulting in a permanent dipole directed toward the bottom of $CHCl_3$ as drawn. (b) The C—F bond dipole in CCl_3F is stronger than the C—Cl bond dipoles and is not completely offset by them, which makes CCl_3F a slightly polar substance with its dipole directed toward the fluorine atom.

dipole moment (μ) a measure of the degree to which a molecule aligns itself in an applied electric field; a quantitative expression of the polarity of a molecule.

TABLE 5.2 Permanent Dipole Moments Of Several Polar Molecules

Formula	Structure with Bond Dipole(s)	Direction of Overall Dipole	Dipole Moment (debyes)
HF	H—F	⟷	1.82
H_2O	H–O–H	↑	1.85
NH_3	H–N(H)–H	↑	1.47
$CHCl_3$	H–C(Cl)(Cl)–Cl	↓	1.01
CCl_3F	F–C(Cl)(Cl)–Cl	↑	0.45

SAMPLE EXERCISE 5.4 Predicting Whether or Not a Substance Is Polar **LO2**

Does formaldehyde (CH_2O) have a permanent dipole?

Collect, Organize, and Analyze To predict whether a molecular compound has a permanent dipole, we need to determine whether or not its molecules contain polar bonds and whether the bond dipoles offset each other. A bond's polarity depends on the difference in electronegativity between the bonded pair of atoms. The electronegativities of the elements in formaldehyde are H = 2.1, C = 2.5, and O = 3.5 (Figure 4.13).

Solve In Sample Exercise 5.1, we determined that formaldehyde has a trigonal planar structure:

O
‖
C
H H

The large difference in the electronegativities of C and O and the small difference in the electronegativities of C and H mean that the polar C=O bond gives formaldehyde a permanent dipole:

O
‖
C
H H

Think About It When more than one kind of atom is bonded to a central atom, it is highly likely that the molecule will have a permanent dipole because the orientation and/or magnitude of the individual bond dipoles do not cancel out.

Practice Exercise Does carbon disulfide (CS_2), a gas present in small amounts in crude petroleum, have a dipole moment?

(Answers to Practice Exercises are in the back of the book.)

valence bond theory a quantum mechanics–based theory of bonding that assumes covalent bonds form when half-filled orbitals on different atoms overlap or occupy the same region in space.

overlap a term in valence bond theory describing bonds arising from two orbitals on different atoms that occupy the same region of space.

sigma (σ) bond a covalent bond in which the highest electron density lies between the two atoms along the bond axis.

hybridization in valence bond theory, the mixing of atomic orbitals to generate new sets of orbitals that are then available to form covalent bonds with other atoms.

hybrid atomic orbital in valence bond theory, one of a set of equivalent orbitals about an atom created when specific atomic orbitals are mixed.

***sp*[3] hybrid orbitals** a set of four hybrid orbitals with a tetrahedral orientation produced by mixing one *s* and three *p* atomic orbitals.

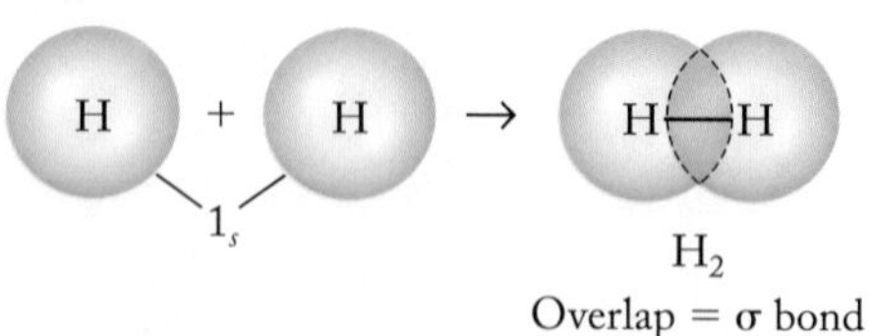

FIGURE 5.21 The overlap of the 1*s* orbitals on two hydrogen atoms results in a single σ bond between the two hydrogen atoms in a H_2 molecule.

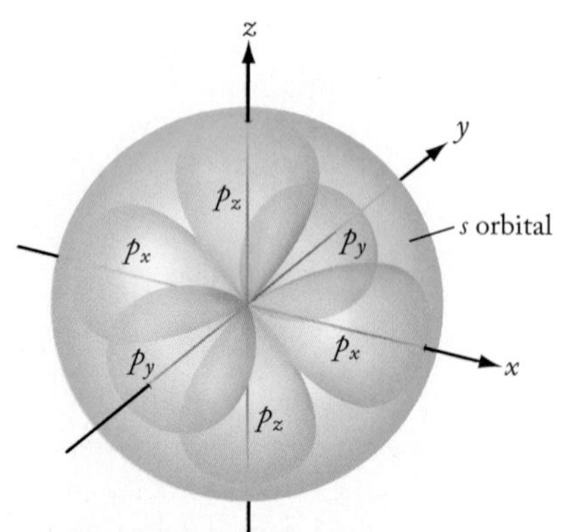

FIGURE 5.22 Relative orientation in space of one 2*s* orbital (spherical) and three 2*p* orbitals (teardrop-shaped) oriented at 90° to one another along the *x*-, *y*-, and *z*-axes of a coordinate system.

CONCEPT TEST

Water and hydrogen sulfide both have a bent molecular geometry with dipole moments of 1.85 D and 0.97 D, respectively. Why is the dipole moment of H_2S smaller than the dipole moment of H_2O?

(Answers to Concept Tests are in the back of the book.)

5.4 Valence Bond Theory and Hybrid Orbitals

Drawing Lewis structures and applying VSEPR theory enable us to reliably predict the geometry of many molecules. However, we have yet to make a connection between molecular geometry and the idea, first advanced in Chapter 3, that electrons exist in atomic orbitals. The absence of a connection between the electronic structure of atoms and the electronic structure of molecules led to the development of bonding theories that use atomic orbitals to account for the molecular geometries predicted by VSEPR. We explore one of those theories now.

Valence bond theory evolved in the late 1920s when Linus Pauling merged quantum mechanics with Lewis's model of shared electron pairs to develop a new theory to explain molecular bonding. According to valence bond theory, a chemical bond forms when the atomic orbitals of two atoms **overlap**. The electrons in overlapping orbitals are attracted to the nuclei of both bonded atoms. This attraction leads to lower potential energy and greater chemical stability than if the atoms were completely free (review Figure 4.3). Valence bond theory was an important advance over Lewis's model because it could be used to explain the molecular geometries we have been exploring in this chapter. Let's see how it works.

We begin by applying Pauling's valence bond theory to H_2, the simplest diatomic molecule. A free H atom has a single electron in a 1*s* atomic orbital. According to valence bond theory, overlap of two half-filled 1*s* orbitals leads to increased electron density between the nuclei of the bonding atoms as the two nuclei share the two 1*s* electrons. As a result, a H—H covalent bond forms (Figure 5.21). The region of highest electron density lies along the bond axis between the two atoms, which makes the H—H bond an example of a **sigma (σ) bond**.

sp[3] Hybrid Orbitals

Now let's consider the bonding in methane, CH_4. Recall from Chapter 3 that an individual carbon atom has the electron configuration $[He]2s^22p^2$, that *s* orbitals are spherical, and that *p* orbitals are usually drawn with teardrop shapes (actually, they're shaped more like mushroom caps—review Figure 3.24). The lobes of the three *p* orbitals in a subshell are oriented at 90° to one another, as shown in Figure 5.22. Further, we learned in Chapter 4 that carbon atoms form four covalent bonds to complete their octets.

We saw in Section 5.2 that the central carbon atom in methane has a steric number of 4 (four bonds, no lone pairs) and a tetrahedral molecular geometry. How can a carbon atom with a filled, spherical 2*s* orbital and two half-filled 2*p* orbitals that are oriented 90° from each other result in a tetrahedral molecule with bond angles of 109.5°? VSEPR cannot answer this question, but valence bond theory can. In valence bond theory, the 2*s* and 2*p* orbitals can be reconfigured in a process called **hybridization**.

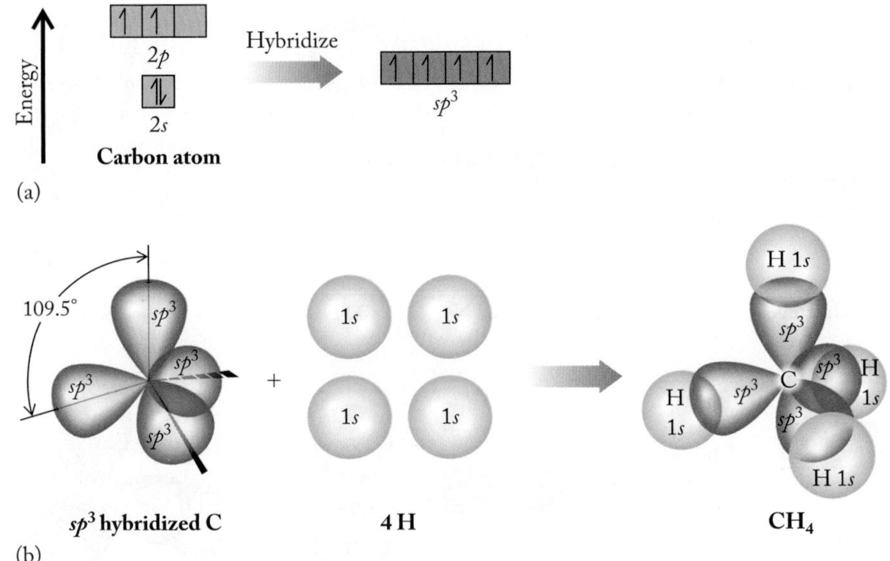

FIGURE 5.23 (a) When the $2s$ and $2p$ atomic orbitals of carbon are hybridized, the four orbitals are mixed to create four sp^3 hybrid orbitals, each containing one electron. (b) The hybrid orbitals are oriented 109.5° from one another, which is the orientation of the bonds in a molecule of methane and in other molecules with a tetrahedral molecular geometry.

Hybridization happens when, for example, four H atoms combine with a single C atom in the gas phase. As the valence-shell atomic orbitals of the C atom begin to overlap with the $1s$ orbitals of the four H atoms, the $2s$, $2p_x$, $2p_y$, and $2p_z$ orbitals of the C atom no longer represent the lowest energy orientations of the valence orbitals. What happens next can be viewed as a mathematical mixing of the wave functions defining the atom's $2s$ and $2p$ orbitals. The product of mixing is a new set of wave functions that defines four equivalent **hybrid atomic orbitals** with shapes and orientations unlike those of $2s$ and $2p$ orbitals (Figure 5.23). Because a single s orbital and three p orbitals are involved in this mixing process, we call the result a set of **sp^3 hybrid orbitals**, as shown in Figure 5.23(a). The lobes are all 109.5° from each other because the distance between them is *maximized*, which means the repulsion between the electrons in them is *minimized*. Overlap of the lobes with the $1s$ orbitals of four hydrogen atoms results in the formation of four σ bonds and one molecule of methane that has the tetrahedral molecular geometry shown in Figure 5.23(b).

CONNECTION Recall from Chapter 3 that the shapes of the p orbitals we use in this book are longer and narrower than their true shapes to make it easier to see how they are oriented (see Figure 3.24).

These sp^3 hybrid orbitals appear to have one lobe each, but in fact every sp^3 hybrid orbital consists of a major lobe and a minor lobe, as shown in Figure 5.24(a). Because our current focus is bonding and the minor lobe is not involved in bonding, we ignore it in this chapter and omit it from the illustrations. Also, the shapes of the hybrid orbitals that we use in this book are more elongated than the boundary surfaces that are calculated from quantum mechanics, as shown in Figure 5.24(b). As with our p orbital images, we stretch the lobes of hybrid orbitals a bit to better show their orientation.

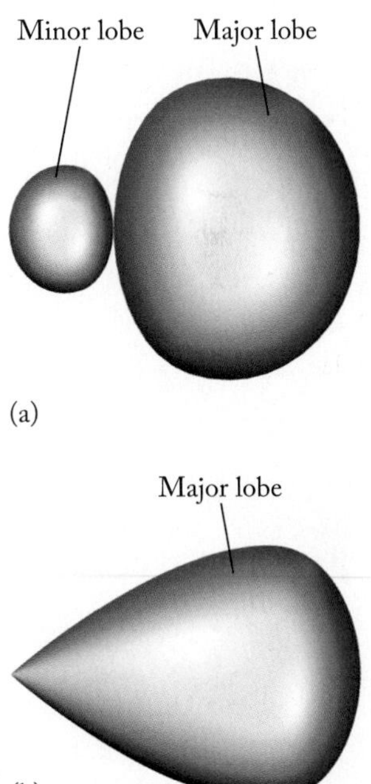

FIGURE 5.24 Two views of sp^3 hybrid orbitals. (a) Boundary surface of an sp^3 orbital derived from quantum mechanics, consisting of both major and minor lobes. (b) To better show the orientation of bonds formed by hybrid orbitals, we elongate the major lobe and ignore the minor one (which is not involved in bond formation).

According to valence bond theory, *any* atom with s and p orbitals in its valence shell can form a set of sp^3 hybrid orbitals when it is the center atom in a molecule with a tetrahedral electron-pair geometry. The hybrid orbitals may contain single electrons or pairs of electrons. For example, sp^3 hybridization of the nitrogen atom, which has five valence electrons, produces three hybrid orbitals that are half-filled and one hybrid orbital that is completely filled. In a molecule of ammonia (NH_3, Figure 5.25a), the three half-filled orbitals form three σ bonds by overlapping with the $1s$ orbitals of three hydrogen atoms, and the filled hybrid orbital contains the N atom's lone pair of electrons.

Similarly, the oxygen atom in water forms four sp^3 hybrid orbitals: two of them contain lone pairs, and two are half-filled and available for bond formation

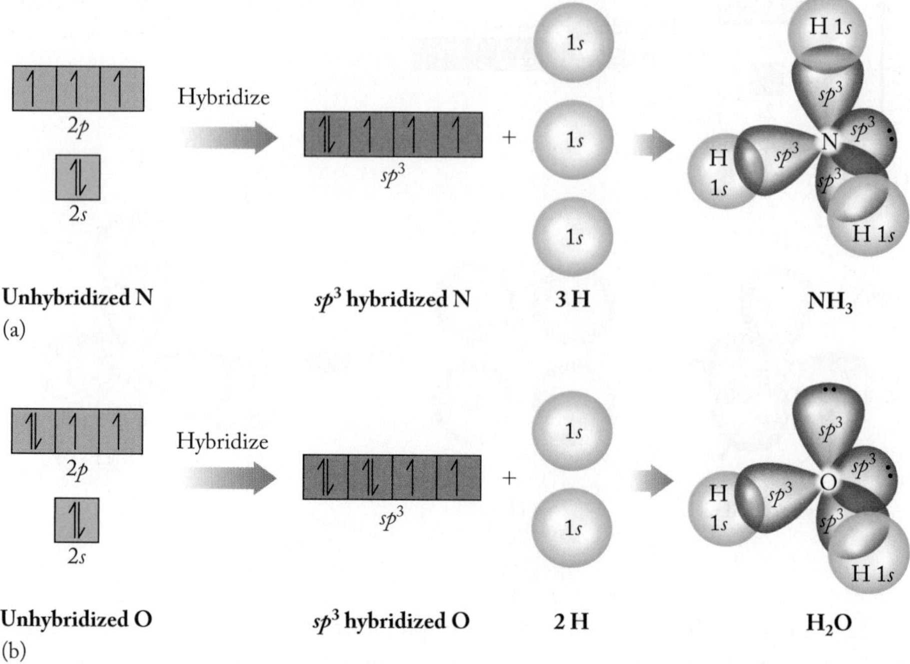

FIGURE 5.25 A set of four sp^3 hybrid orbitals points toward the vertices of a tetrahedron. (a) The nonbonding electron pair of the N in NH_3 occupies one of the four hybrid orbitals, giving the molecule a trigonal pyramidal molecular geometry. (b) The O atom in H_2O is also sp^3 hybridized. Only two of the four orbitals contain bonding pairs of electrons, giving H_2O a bent molecular geometry.

(Figure 5.25b). The half-filled orbitals overlap with 1*s* orbitals from hydrogen atoms and form two σ bonds, whereas the two filled orbitals represent the two lone pairs on O. Thus, a carbon atom that forms four σ bonds, a nitrogen atom that forms three σ bonds and has one lone pair of electrons, and an oxygen atom that forms two σ bonds and has two lone pairs of electrons all have the same steric number (4) and are all sp^3 hybridized atoms. This consistency is not accidental. As we will see, the SN of an atom with an octet of valence electrons is directly linked to both its hybridization and its molecular geometry.

sp^2 Hybrid Orbitals

Recall from Sample Exercise 5.1 that formaldehyde has a trigonal planar molecular geometry. According to the VSEPR model, the SN of the central carbon atom is 3, so three atomic orbitals must be mixed to make three equivalent hybrid orbitals. Instead of mixing all four atomic orbitals, as in sp^3 hybridization, we mix only three: the 2*s* orbital and two of the 2*p* orbitals. This leaves the third 2*p* orbital unhybridized (Figure 5.26a).

Mixing one *s* and two *p* orbitals generates three hybrid orbitals called ***sp*²** **hybrid orbitals.** The energy levels of sp^2 orbitals are slightly lower than those of sp^3 orbitals because only two *p* orbitals are mixed with the *s* orbital, forming a set of sp^2 orbitals. The orbitals in an sp^2 hybridized atom all lie in the same plane and are 120° apart. The two lobes of the unhybridized *p* orbital lie above and below the plane of the triangle defined by the sp^2 hybrid orbitals. An sp^2 hybridized atom can form up to three σ bonds with its three hybridized orbitals, and the carbon atom in formaldehyde does so—that is, it forms two σ bonds to the two H atoms and one σ bond to the O atom.

sp^2 hybrid orbitals three hybrid orbitals in a trigonal planar orientation formed by mixing one *s* and two *p* orbitals.

pi (π) bond a covalent bond in which electron density is greatest above and below the bonding axis.

For its part, the O atom also has a steric number of 3 (it's bonded to one atom and has two lone pairs of electrons), so it is also sp^2 hybridized (Figure 5.26b). Its unhybridized *p* orbital overlaps sideways with the one on the carbon atom, producing a **pi (π) bond** in which electron densities are highest above and below the internuclear axis, as shown in the last image in Figure 5.26(c). Note that the π

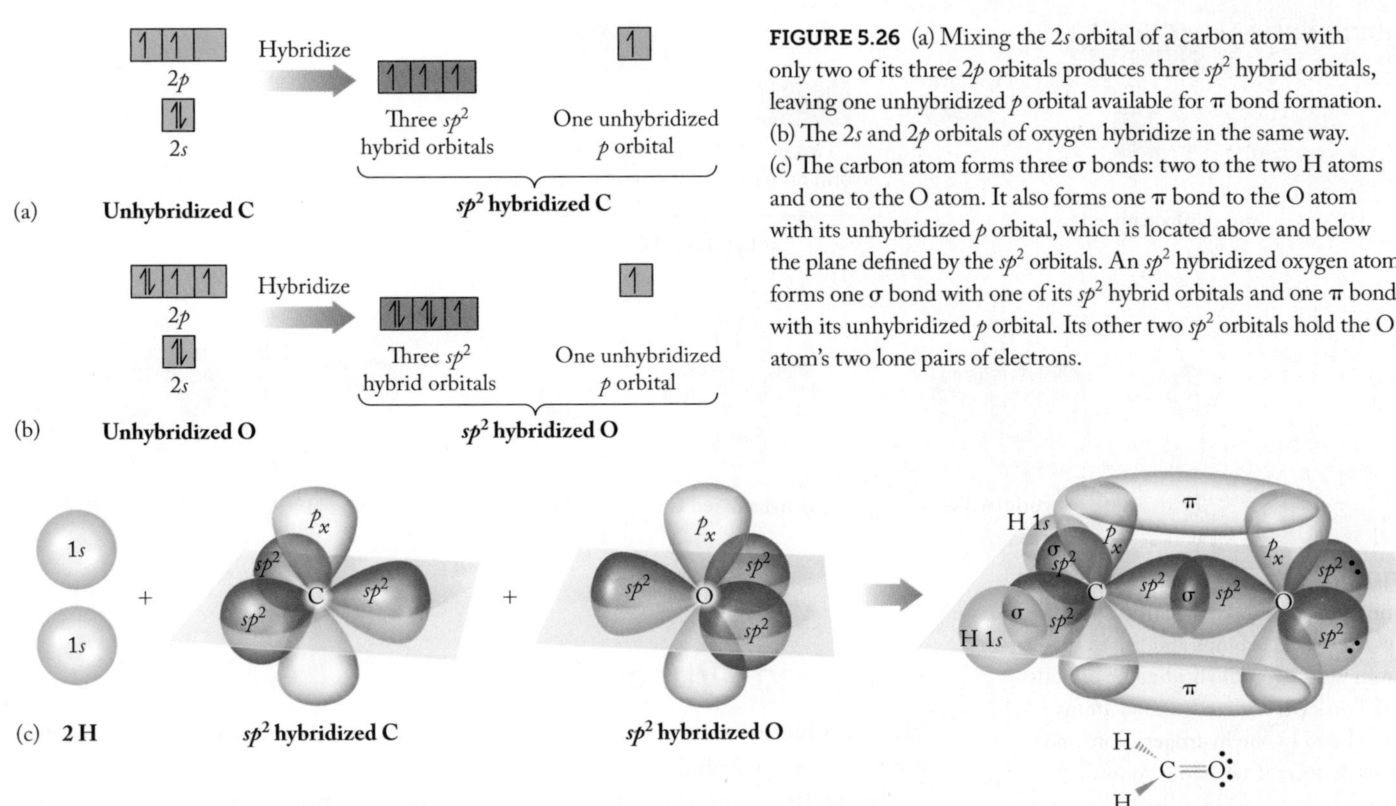

FIGURE 5.26 (a) Mixing the 2*s* orbital of a carbon atom with only two of its three 2*p* orbitals produces three sp^2 hybrid orbitals, leaving one unhybridized *p* orbital available for π bond formation. (b) The 2*s* and 2*p* orbitals of oxygen hybridize in the same way. (c) The carbon atom forms three σ bonds: two to the two H atoms and one to the O atom. It also forms one π bond to the O atom with its unhybridized *p* orbital, which is located above and below the plane defined by the sp^2 orbitals. An sp^2 hybridized oxygen atom forms one σ bond with one of its sp^2 hybrid orbitals and one π bond with its unhybridized *p* orbital. Its other two sp^2 orbitals hold the O atom's two lone pairs of electrons.

bond is *one* bond formed by the two atoms' *p* orbitals through *two* points of contact—it is *not* two bonds. This π bond and the σ bond between C and O together make up the C═O double bond in CH_2O (Figure 5.27). Note how the trigonal planar geometries around the C and O atoms align with each other. Also, the widths of the *p* orbitals' lobes and the distances between atoms are not drawn to scale in Figure 5.26(c). The lobes are actually much wider than drawn (see Figure 3.24), and therefore they overlap.

A nitrogen atom can also form sp^2 hybrid orbitals, as shown in Figure 5.28. When it does, its lone pair occupies one of the three hybrid orbitals; the other two sp^2 hybrid orbitals are half-filled and can form two σ bonds, while the unhybridized half-filled *p* orbital can form one π bond. This combination of one lone pair plus the capacity to form two σ bonds to two other atoms gives N a steric number of 3, the same SN value as sp^2 hybridized C and O atoms. Thus, central atoms with trigonal planar electron-pair geometries are often sp^2 hybridized.

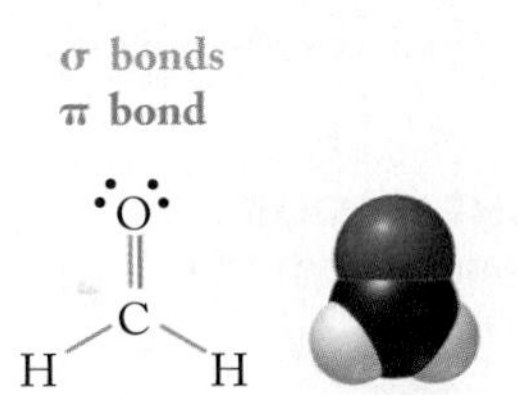

FIGURE 5.27 Formaldehyde has three σ bonds and one π bond.

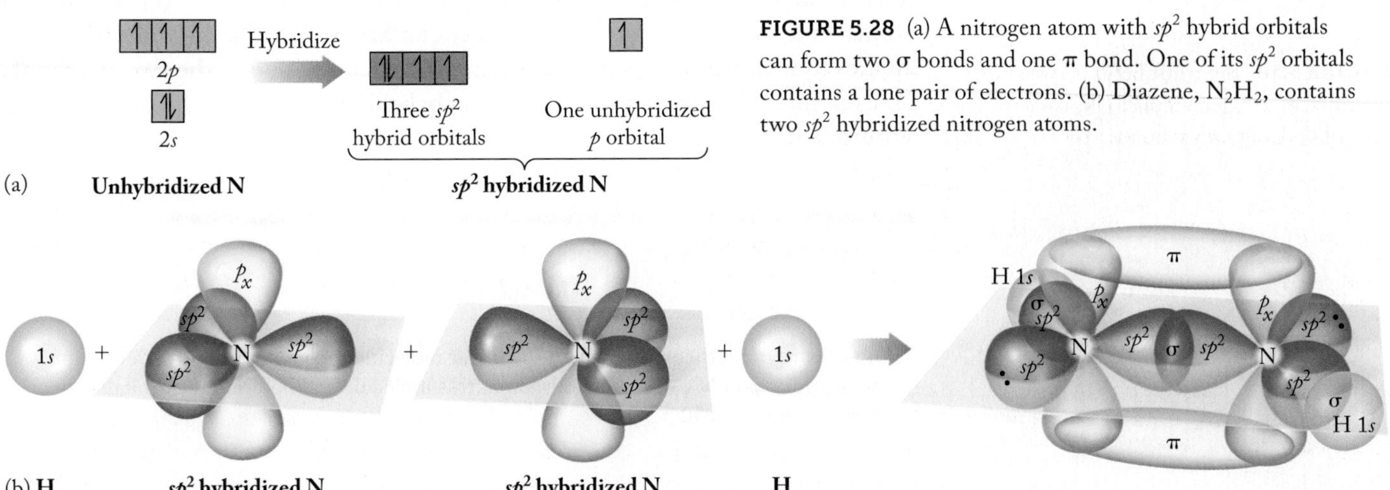

FIGURE 5.28 (a) A nitrogen atom with sp^2 hybrid orbitals can form two σ bonds and one π bond. One of its sp^2 orbitals contains a lone pair of electrons. (b) Diazene, N_2H_2, contains two sp^2 hybridized nitrogen atoms.

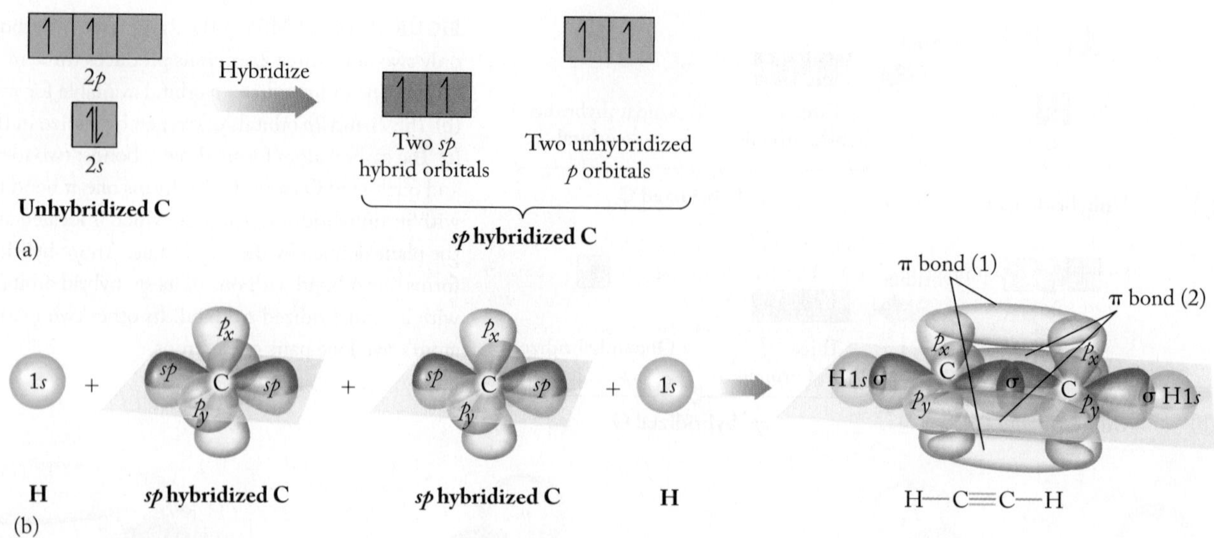

FIGURE 5.29 (a) Mixing one 2*s* orbital and one 2*p* orbital on a carbon atom creates two *sp* hybrid orbitals and leaves the carbon atom with two unhybridized *p* orbitals. (b) Two *sp* hybridized carbon atoms each bond to one hydrogen atom and to each other via σ bonds to form the linear molecule C_2H_2. The two pairs of unhybridized *p* orbitals on the two carbon atoms overlap sideways, one pair above and below and the other in front of and behind the σ bonding axis, forming two π bonds.

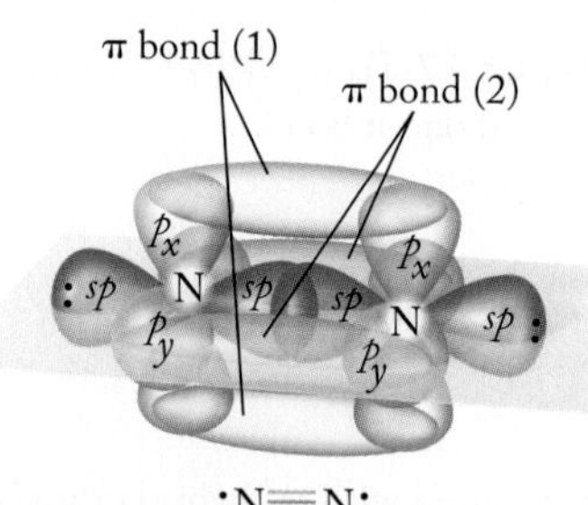

FIGURE 5.30 The triple bond between *sp* hybridized nitrogen atoms in N_2 consists of one σ bond and two π bonds.

sp Hybrid Orbitals

Hybrid orbitals can also be generated by mixing one *s* orbital with one *p* orbital, forming two *sp* orbitals.

In the linear molecule acetylene, C_2H_2, both carbon atoms have steric numbers equal to 2 according to VSEPR. Because steric number always indicates the number of hybrid orbitals that must be created, we must mix two atomic orbitals to make two hybrid orbitals on each carbon atom. Mixing one 2*s* and one 2*p* orbital on a carbon atom and leaving the other two 2*p* orbitals unhybridized produces two ***sp* hybrid orbitals** with major lobes that are on opposite sides of the carbon atom, as shown in Figure 5.29. The lobes of the two unhybridized *p* orbitals are above and below and in front of and behind the axis of the *sp* hybrid orbitals.

The valence bond view of C_2H_2 shows a σ-bonding framework in which a hydrogen 1*s* orbital overlaps with one *sp* hybrid orbital on each carbon atom. The remaining *sp* hybrid orbitals, one on each carbon atom, overlap with each other (Figure 5.29b). This arrangement brings the two sets of unhybridized *p* orbitals on the two carbon atoms into parallel alignment with each other. They then overlap to form two π bonds between the two carbon atoms, so that the one σ bond and the two π bonds form the triple bond of HC≡CH. The steric number of each *sp* hybridized carbon atom in acetylene is 2 (bonds to two atoms with no lone pairs). The link between SN = 2 and *sp* hybridization also applies to nitrogen atoms in N_2. A lone pair occupies one of the two *sp* orbitals on each nitrogen atom, and the other *sp* hybrid orbital forms the σ bond, as in Figure 5.30.

SAMPLE EXERCISE 5.5 Using Valence Bond Theory to Explain Molecular Shape **LO3**

Use valence bond theory to explain the linear molecular geometry of CO_2. Determine the hybridization of carbon and oxygen in this molecule and describe the orbitals that overlap to form the bonds.

Collect, Organize, and Analyze We know that CO_2 is linear and that the Lewis structure of the molecule is

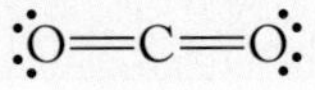

$$:\ddot{O}=C=\ddot{O}:$$

Solve The central carbon atom has no lone pairs, so SN = 2. Each of its two double bonds is composed of one σ bond and one π bond. Therefore, the carbon atom must be *sp* hybridized because its half-filled *sp* hybrid orbitals have the capacity to form two σ bonds and its two half-filled unhybridized *p* orbitals are available to form the two π bonds (Figure 5.31). The oxygen atoms must be sp^2 hybridized because that hybridization leaves the O atoms with one unhybridized *p* orbital to form a π bond. The σ bonds form when each *sp* orbital on the carbon atom overlaps with one sp^2 orbital on an oxygen atom. The lone pairs on oxygen occupy sp^2 orbitals.

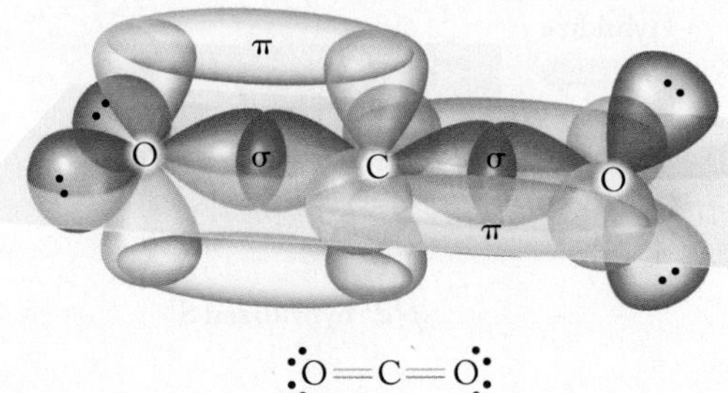

$$:\ddot{O}=C=\ddot{O}:$$

FIGURE 5.31 The bonding pattern in CO_2. The π bond on the left in the drawing is above and below the plane of the molecule; the π bond on the right is in front of and in back of the σ bonding axis.

Think About It Notice that the two unhybridized *p* orbitals of carbon are oriented 90° to each other, and that the plane containing the three sp^2 orbitals of one oxygen atom is rotated 90° with respect to the plane containing the three sp^2 orbitals of the other oxygen atom. The orientation of the lone pairs of electrons on each oxygen atom and its σ bond to carbon is trigonal planar, which minimizes their mutual repulsion.

Practice Exercise In which of these molecules does the central atom have sp^3 hybrid orbitals? (a) CCl_4; (b) HCN; (c) SO_2; (d) PH_3

(Answers to Practice Exercises are in the back of the book.)

CONCEPT TEST

Why can't a carbon atom with sp^3 hybrid orbitals form π bonds?

(Answers to Concept Tests are in the back of the book.)

Hybrid Schemes for Expanded Octets

Valence bond theory can also account for the molecular geometries of molecules with central atoms that may have more than eight valence electrons. We learned in Chapter 4 that one way to explain the hypervalency of central atoms is to assume that valence-shell *d* orbitals are involved in bond formation. For example, to expand the octet of the central phosphorus atom in PF_5 to form a fifth P—F bond, we might include one of the phosphorus atom's 3*d* orbitals in bond formation. To account for the trigonal bipyramidal molecular shape of PF_5, we could hybridize five of the P atom's atomic orbitals—its 3*s*, all three of its 3*p* orbitals, and one of its 3*d* orbitals—to produce a set of five sp^3d **hybrid orbitals** with lobes

***sp* hybrid orbitals** two hybrid orbitals on opposite sides of the hybridized atom formed by mixing one *s* and one *p* orbital.

sp^3d hybrid orbitals five equivalent hybrid orbitals with lobes pointing toward the vertices of a trigonal bipyramid that form by mixing one *s* orbital, three *p* orbitals, and one *d* orbital from the same shell.

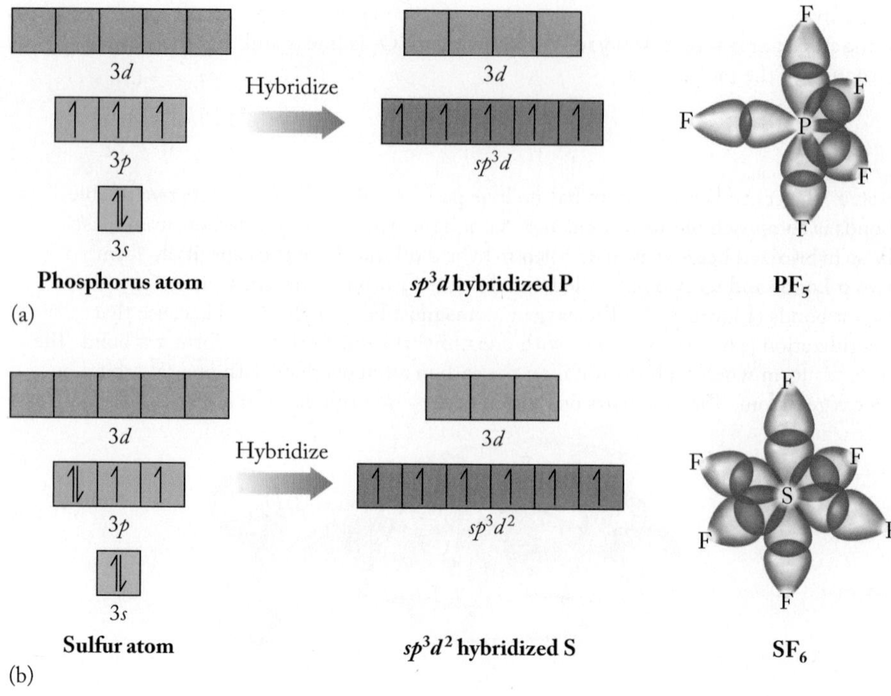

FIGURE 5.32 Hybrid orbitals can be generated by combining *d* orbitals with *s* and *p* orbitals in atoms that expand their octets. (a) One 3*s* orbital, three 3*p* orbitals, and one 3*d* orbital mix on the central phosphorus atom in PF_5 to form five sp^3d hybrid orbitals. (b) One 3*s*, three 3*p*, and two 3*d* orbitals mix on the central sulfur atom in SF_6 to form six sp^3d^2 hybrid orbitals. For simplicity, only the fluorine orbitals that overlap with the hybrid orbitals of P and S are shown.

TABLE 5.3 Summary Of Hybridization Schemes And Orbital Orientations

Hybridization	Orientation of Hybrid Orbitals	Numbers of σ Bonds	Molecular Geometries	Angles between Hybrid Orbitals
sp		2	Linear	180°
sp^2		3 2	Trigonal planar Bent	120°
sp^3		4 3 2	Tetrahedral Trigonal pyramidal Bent	109.5°
sp^3d		5 4 3 2	Trigonal bipyramidal Seesaw T-shaped Linear	90°, 120°, 180°
sp^3d^2		6 5 4	Octahedral Square pyramidal Square planar	90°, 180°

that point toward the vertices of a trigonal bipyramid (Figure 5.32a). To explain the presence of six S—F bonds and the octahedral shape of SF_6, we go one step further and incorporate *two* of the sulfur atom's *d* orbitals in a hybridization scheme that forms a set of six equivalent ***sp*^{3}*d*^{2} hybrid orbitals** (Figure 5.32b). The shapes associated with all of the hybridization schemes we have discussed so far are summarized in Table 5.3.

***sp*^{3}*d*^{2} hybrid orbitals** six equivalent hybrid orbitals pointing toward the vertices of an octahedron that form by mixing one *s* orbital, three *p* orbitals, and two *d* orbitals from the same shell.

Before ending our discussion of hybridization in hypervalent compounds, we need to acknowledge that many chemists don't support the use of *d* orbitals in the hybridization process. They note that the energies of *d* orbitals are significantly higher than the energies of the *s* and *p* orbitals in the same shell, which makes *s*–*p*–*d* orbital mixing problematic. Instead, other bonding models have been proposed to explain the bonding and geometries of hypervalent molecules—models that do not incorporate *d* orbitals. We will explore one of them, called molecular orbital (MO) theory, in Section 5.7.

SAMPLE EXERCISE 5.6 Predicting the Hybridization of Atoms in Molecular Structures LO4

Given the following Lewis structure of SeF_4,

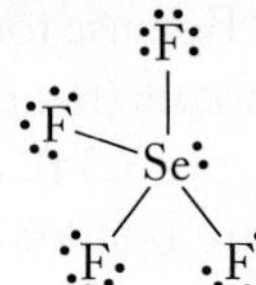

what is the shape of the molecule and what hybridization scheme could account for it?

Collect and Organize We are provided with the Lewis structure of SeF_4 and we need to determine the hybridization of the central atom. Hybridization schemes are used to explain molecular shapes, so we need to determine the shape of the molecule first.

Analyze The central atom has four single (σ) bonds and one lone pair, giving it SN = 5.

Solve The electron-pair geometry associated with SN = 5 is trigonal bipyramidal (Figure 5.3d). The presence of the lone pair means that the molecular geometry is not the same as the electron-pair geometry. Instead, according to Table 5.1, the molecular shape is seesaw.

A steric number of 5 also suggests that the Se atom has an expanded octet of 10 valence electrons. A hybridization scheme that incorporates 10 electrons (5 orbitals) uses one *d* orbital with one *s* and three *p* orbitals, or *sp*^{3}*d* hybridization.

Think About It The information in Table 5.3 confirms that a seesaw molecular geometry is consistent with *sp*^{3}*d* hybridization.

Practice Exercise What is the hybridization of the iodine atom in IF_5 that is consistent with the following Lewis structure?

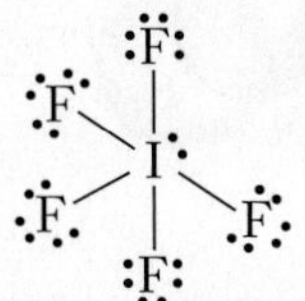

(Answers to Practice Exercises are in the back of the book.)

5.5 Molecules with Multiple "Central" Atoms

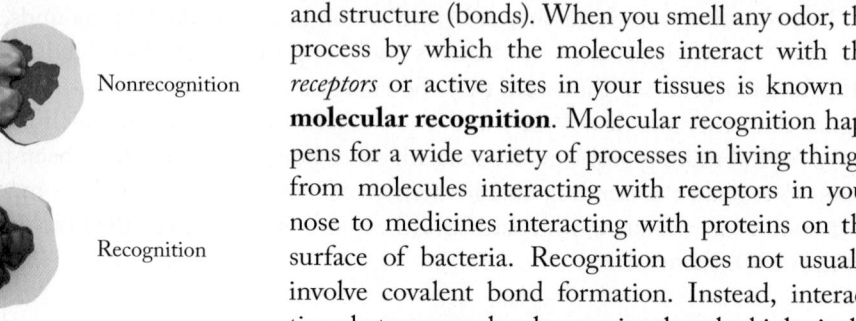

We began this chapter by pointing out that two very different smells—rye bread and spearmint—actually come from molecules with the same composition (atoms) and structure (bonds). When you smell any odor, the process by which the molecules interact with the *receptors* or active sites in your tissues is known as **molecular recognition**. Molecular recognition happens for a wide variety of processes in living things, from molecules interacting with receptors in your nose to medicines interacting with proteins on the surface of bacteria. Recognition does not usually involve covalent bond formation. Instead, interactions between molecules require that the biologically active molecules and the receptors that respond to them fit tightly together, which means that they must have complementary three-dimensional shapes (Figure 5.33). An example of a biological effect caused by molecular recognition is the process by which green tomatoes and other types of produce ripen. Tomatoes ripen faster when stored in a paper or plastic bag instead of sitting on a kitchen counter. Why? Because tomatoes give off ethylene gas (C_2H_4) as they ripen, and they also have receptors that respond to molecules of ethylene by accelerating the ripening process (Figure 5.34). A bag traps this gas.

FIGURE 5.33 Molecular recognition in biological systems. Only the blue molecule matches the shape of the active site and interacts with it.

FIGURE 5.34 Ripening tomatoes give off ethylene gas, which speeds up the ripening process. Ethylene molecules have a unique shape that interacts with the receptors that control ripening.

Both carbon atoms in ethylene are bonded to three other atoms and have no lone pairs of electrons (Figure 5.35a), so SN = 3. This means that the geometry around each carbon atom is trigonal planar and that the carbon atoms are both sp^2 hybridized (see Table 5.3). For each carbon atom, two of the sp^2 orbitals form σ bonds with hydrogen 1*s* orbitals; the third forms the C═C σ bond shown in Figure 5.35(b). The C═C π bond is formed by overlap of the unhybridized 2*p* orbitals on the carbon atoms. Taken together, the two trigonal planar carbon atoms produce an overall geometry for ethylene in which all six atoms lie in the same plane. The π bond in C_2H_4 molecules contributes to an overall distribution of electrons with maximum densities above and below the plane, as shown by the boundary surfaces in Figure 5.35(c).

Now let's consider a biologically active molecule with three "central" atoms. Acrolein is one of the components of barbeque smoke that contributes to the

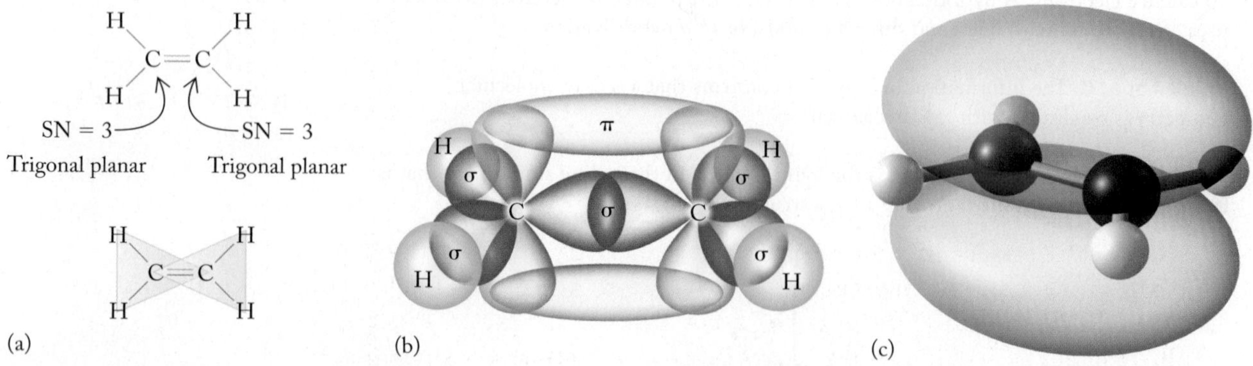

FIGURE 5.35 Ethylene molecules contain two carbon atoms that are (a) at the centers of two overlapping triangular planes that are also coplanar. This combination means that all of the atoms are in the same plane. (b) The carbon atoms are both sp^2 hybridized. (c) The π bond in ethylene results in electron density above and below the plane of the molecule.

distinctive odor of a cookout. It is also a possible cancer-causing compound. The Lewis structure of acrolein shows that the molecule contains three carbon atoms, each with a steric number of 3 and sp^2 hybridization:

This network of three carbon atoms, each representing a trigonal planar "center," means that *all* of the atoms in a molecule of acrolein could be in the same plane.

Another component of barbeque smoke, benzene, also consists of molecules in which all the atoms are in the same plane due to sp^2 hybridization and trigonal planar geometries on, in this case, six carbon atoms (Figure 5.36). As we saw in Chapter 4, a benzene molecule features a hexagon of carbon atoms, each bonded to two other carbon atoms and one hydrogen atom. This geometry is a perfect match for sp^2 hybridization because the 120° bond angles those orbitals produce are exactly the same as the 120° inside angles of a regular hexagon.

As drawn in Figure 5.36(a), bonding around the benzene ring consists of alternating single and double carbon–carbon bonds. This pattern is an example of *conjugation*—a term that applies to alternating single and multiple bonds in molecular compounds in which adjacent atoms have unhybridized *p* orbitals. The resonance structures in Figure 5.36(a) correspond to the two ways of arranging the π bonds formed by overlapping 2*p* orbitals on the six carbon atoms. All the carbon 2*p* orbitals are aligned perpendicular to the plane of the ring and are the same distance apart, so all are equally likely to overlap with the 2*p* orbitals on either of their neighbors around the ring. As a result, the electrons in the π bonds are delocalized over *all six* carbon atoms, forming two clouds of electrons above and below the hexagonal rings, as shown by the computer-generated images of π-electron density in Figure 5.36(b).

CONCEPT TEST

In which of these three molecules and polyatomic ions are the π electrons delocalized?

(a) (b) (c)

(Answers to Concept Tests are in the back of the book.)

The delocalization of bonding pairs is an important phenomenon because it spreads out their electrons, thereby reducing their mutual repulsion and lowering the potential energy of a molecule or polyatomic ion, which makes it more stable. In general, the greater the degree of delocalization, the greater the stability. This trend is evident in a class of compounds called **aromatic compounds** that includes benzene and other molecules with planar ring structures. An important subclass of aromatic compounds, known as *polycyclic aromatic hydrocarbons* (PAHs), consists of molecules containing multiple six-carbon rings

molecular recognition the process by which molecules interact with other molecules to produce a biological effect.

aromatic compound a cyclic, planar compound with delocalized π electrons above and below the plane of the molecule.

(a)

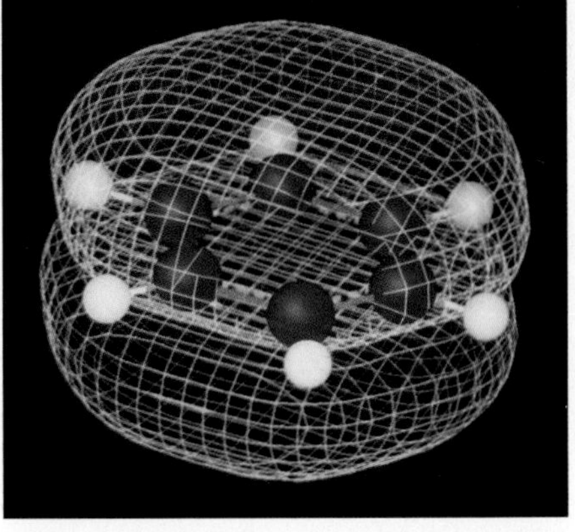

(b)

FIGURE 5.36 Two views of the resonance in benzene. (a) The Lewis resonance structures of benzene can be explained using valence bond theory by showing localized π bonds between adjacent carbon atoms. (b) Resonance leads to complete delocalization of the electrons in the π bonds around the benzene ring.

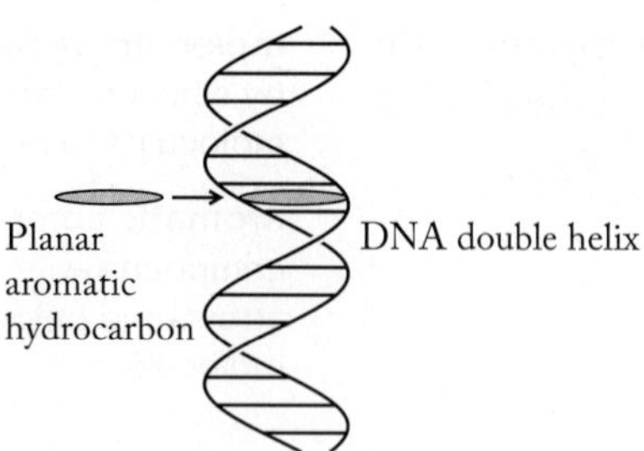

Intercalation of PAH in DNA

Naphthalene

Phenanthrene

Anthracene

Benzo[*a*]pyrene

FIGURE 5.37 The molecules of these polycyclic aromatic hydrocarbons, PAHs, consist of fused benzene rings whose π bonds are delocalized over all the rings in each molecule. Molecules with this planar shape can slip in between the strands of DNA and disrupt cell replication, which can lead to cell death or induce malignancy.

fused together, as shown in Figure 5.37. PAHs are formed in fires fueled by coal, oil, or natural gas, and they are found in cigarette smoke. In 2004 they were discovered in interstellar space.

The chemical stability of PAHs means that they persist in the environment. If we inhale or ingest the compounds, they may bind to the DNA in our cells in a process called *intercalation*. Because PAHs are flat, they can slide into the double helix that DNA forms, as shown in Figure 5.37. Once there, they may alter or prevent DNA replication and thereby damage or kill cells. Intercalation in DNA is one step in the process by which PAHs induce cancer.

CONNECTION Drawing molecular structures of benzene with a circle in the middle of the hexagon of carbon atoms (see Figure 4.7) is another way to represent the clouds of π electrons above and below the plane of the molecule.

5.6 Chirality and Molecular Recognition

Before ending our study of molecular shape, let's return to the subject of handedness introduced at the beginning of the chapter. There we described how two molecules with the same molecular formula and Lewis structure interact differently with receptors in our nasal membranes. As a result, one produces the smell of caraway seeds and the other spearmint leaves. You may find the different odors surprising when you consider the molecular structures of the two compounds:

(+)-Carvone (caraway)

(–)-Carvone (spearmint)

isomer one of a group of compounds having the same chemical formula but different molecular structures.

stereoisomers molecules with the same formula and bonding order, but with different spatial arrangements of their atoms.

enantiomer one of a pair of optical isomers of a compound.

chiral describes a molecule that is not superimposable on its mirror image.

At first glance they may seem identical. In fact, they have the same common chemical name, carvone. But look closely at the bonding pattern around the carbon atoms at the bottom of the rings highlighted with red circles. The hydrogen atom bonded to the bottom carbon atom is on the back side in the caraway compound, but it is on the front side of the ring in the spearmint compound. This minor difference in bond orientation creates a difference in molecular shape that triggers different receptors in our noses.

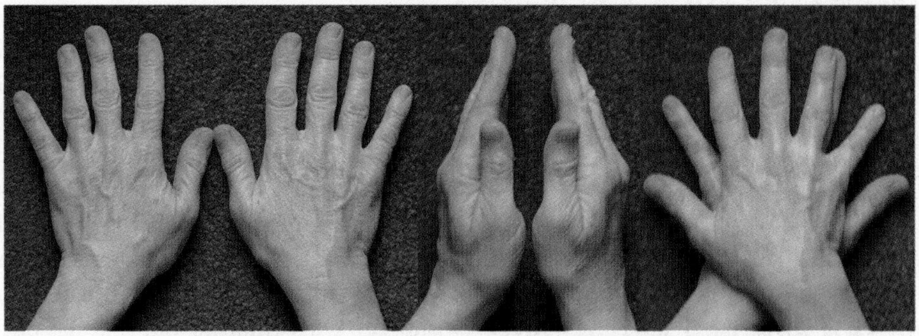

FIGURE 5.38 The left hand and right hand are mirror images of one another but cannot be superimposed upon each other. Molecules that have this relationship are known as enantiomers.

Compounds that have the same chemical formula but different molecular structures are called **isomers**. The term is derived from *iso* (Greek for "same") and *mer* (Greek for "unit" or "part"). In this chapter and those that follow we will encounter several kinds of isomerism. The carvone molecules above represent one kind: they have structures with the same atoms bonded together in the same ways—that is, they have identical Lewis structures, but some of the bonds are oriented in three-dimensional space in ways that are not the same. Isomers of this kind are called **stereoisomers**.

The two forms of carvone represent a particular kind of stereoisomerism—they are mirror images of one another (just as your right hand and left hand are mirror images of one another) but cannot be superimposed on one another (Figure 5.38). Compounds whose molecular structures are nonsuperimposable mirror images of each other—that is, they have the same composition and same bonds but different three-dimensional shapes—are called **enantiomers**.

Evidence for the existence of enantiomers comes from the fact that they interact differently with a kind of light called *plane-polarized* light. In a normal, unpolarized beam of light, the electric fields oscillate in all directions as the beam travels through space, as shown on the left in Figure 5.39. In plane-polarized light, however, the electric fields oscillate in only one plane. When a beam of plane-polarized light passes through a solution containing (+)-carvone, or any (+) stereoisomer, the plane rotates clockwise; if the beam passes through a solution of a (−) stereoisomer, such as (−)-carvone, it rotates counterclockwise. Stereoisomers that can rotate plane-polarized electromagnetic radiation in this way are said to be *optically active* and are called *optical* isomers.

Observing how a compound interacts with plane-polarized light is one way to determine if the compound is optically active. Another way is to examine its molecular structure and, in particular, to determine whether or not it has any sp^3 carbon atoms that are bonded to four different atoms or groups of atoms. Such a carbon atom is called a **chiral** carbon atom or *stereocenter*. A chiral compound contains one or more chiral atoms, which explains why the compound is optically active. Many compounds of biological importance, including the proteins and

CHEMTOUR
Chiral Centers

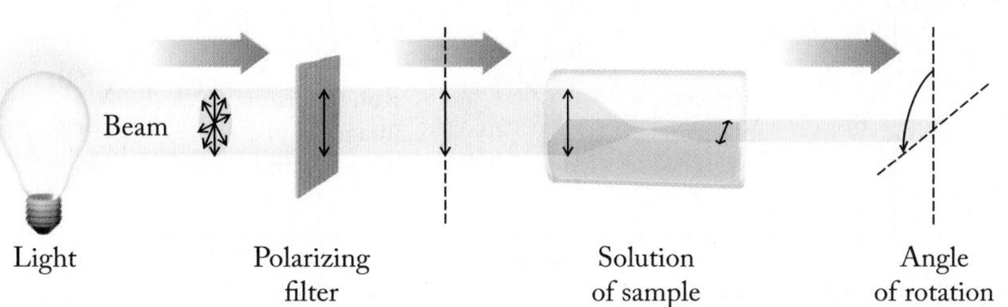

FIGURE 5.39 A beam of plane-polarized light consists of electric field vectors that oscillate in only one direction. The plane of oscillation rotates if the beam passes through a solution of one enantiomer of an optically active compound. The (+) enantiomer causes the beam to rotate clockwise; the (−) enantiomer causes the beam to rotate counterclockwise.

carbohydrates that we consume each day, are chiral compounds. "Chirality" comes from the Greek word *cheir*, or "hand," and is quite correctly called "handedness." Although several features within molecules can lead to chirality, the most common one is the presence of a chiral carbon atom.

To see how chirality works, let's look at a compound that contains a central carbon atom bonded to four different atoms. The compound is bromochlorofluoromethane (CHBrClF) (Figure 5.40a). It is used in fire extinguishers on airplanes. We will compare its molecular structure to that of dibromochloromethane ($CHBr_2Cl$), a compound that may form during the purification of municipal water supplies with chlorine and that has only *three* different atoms bonded to its central carbon atom (Figure 5.40b). In the first step of our comparison, we generate mirror images of both molecules. Then we rotate the mirror images 180° in an attempt to superimpose each mirror image on its original image. The reflected, rotated images of CHBrClF cannot be superimposed upon one another. For example, when we superimpose the F, C, and H atoms of the two structures in Figure 5.40(a), the Br and Cl atoms are not aligned. This means that CHBrClF is a chiral compound. However, the molecular images of $CHBr_2Cl$ *can* be superimposed in this way, so $CHBr_2Cl$ is not a chiral compound—it is an *achiral* (*not* chiral) compound.

Where are the chiral carbon atoms in (+)- and (−)-carvone? If you guessed the circled carbon atoms, you were right. One of the four different groups bonded to the chiral carbon is a hydrogen atom, and another is a carbon atom that has a single bond and a double bond to two other carbon atoms:

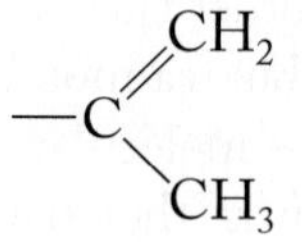

$$-C(=CH_2)-CH_3$$

The other two bonds from the chiral carbon are part of the six-carbon ring. They both connect to $-CH_2-$ groups, so how are they different? The answer lies in other groups to which the $-CH_2-$ groups are also bonded. Note that the left side of the ring contains a C=C double bond, whereas the right side contains a

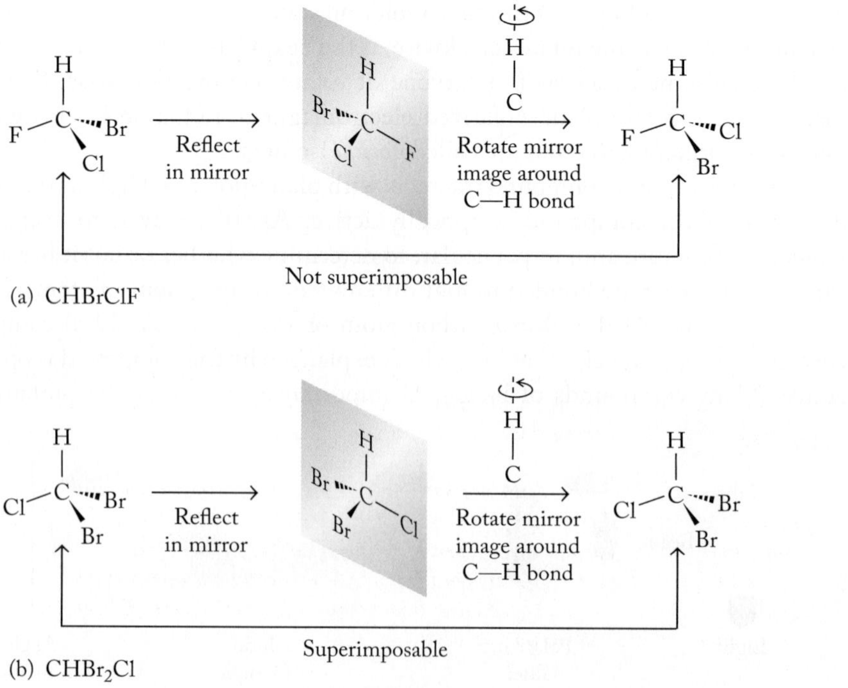

FIGURE 5.40 (a) A molecule of a chiral compound, such as CHBrClF, is not superimposable on its mirror image. (b) A molecule of a compound that is not chiral, such as $CHBr_2Cl$, is superimposable on its mirror image.

C═O double bond. This asymmetry in the ring is the reason that carvone is not superimposable on its mirror image.

SAMPLE EXERCISE 5.7 Recognizing Structures That Are Chiral **LO5**

Identify which of the following molecules are chiral and circle their stereocenters. Some molecules may have more than one chiral center.

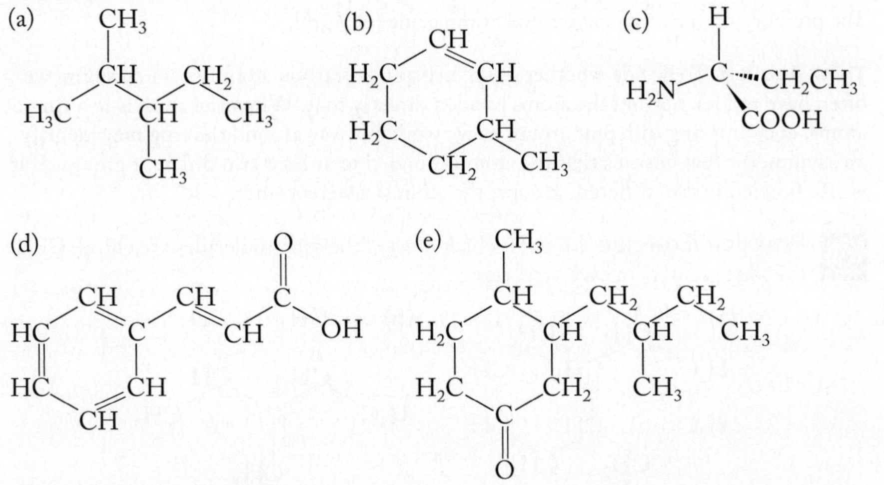

(a) CH_3 / CH CH_2 / H_3C CH CH_3 / CH_3

(b) CH / H_2C CH / H_2C CH / CH_2 CH_3

(c) H / H_2N—C····CH_2CH_3 / COOH

(d) O / CH CH C / HC C CH OH / HC CH / CH

(e) CH_3 / CH CH_2 CH_2 / H_2C CH CH CH_3 / H_2C CH_2 CH_3 / C / O

Collect, Organize, and Analyze A chiral compound is composed of molecules whose structure is not superimposable on its mirror images due to the presence of one or more stereocenters: carbon atoms that are bonded to four different atoms or groups of atoms.

Solve

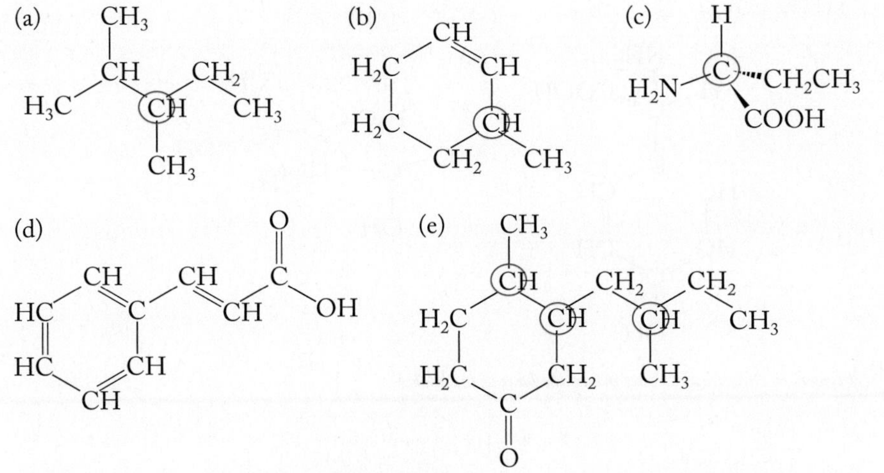

(a) CH_3 / CH CH_2 / H_3C (CH) CH_3 / CH_3

(b) CH / H_2C CH / H_2C (CH) / CH_2 CH_3

(c) H / H_2N—(C)····CH_2CH_3 / COOH

(d) O / CH CH C / HC C CH OH / HC CH / CH

(e) CH_3 / (CH) CH_2 CH_2 / H_2C (CH) (CH) CH_3 / H_2C CH_2 CH_3 / C / O

The circled carbon atom in compound (a) is bonded to four different groups: the CH_3 group below it, the H atom above it, a group to its right that contains two carbon and five hydrogen atoms, and a group to its left that contains three carbon and seven hydrogen atoms. Therefore, the circled carbon atom is a stereocenter and the compound is chiral.

In compound (b) the circled carbon atom is part of a six-carbon ring and bonded to a CH_3 group and a H atom. The ring is not symmetrical: one side has a C═C double bond and the other side does not. Therefore, the circled atom is chiral.

The circled carbon atom in (c) is bonded to four different groups of atoms: $-H$, $-NH_2$, $-COOH$, and $-CH_2CH_3$, so this compound is chiral. It is also an example of a class of compounds called amino acids.

racemic mixture a sample containing equal amounts of both enantiomers of a compound.

All of the carbon atoms in compound (d) are sp^2 hybridized, giving them planar molecular geometries. Planar molecules cannot be chiral because they have superimposable mirror images. Therefore, compound (d) is achiral.

Compound (e) has three chiral centers. Working from right to left, the first one is similar to the chiral center in compound (a). The other two are in the six-carbon ring. Each is bonded to different groups outside the ring, and the ring itself is not symmetrical: there is a CH_3 group bonded to the top carbon atom in the ring, and the bottom atom is double bonded to an oxygen atom. This asymmetry means that each of the two circled atoms is bonded to four different groups of atoms and is a chiral center. The presence of one stereocenter makes molecule (e) chiral.

Think About It To decide whether an sp^3 hybridized carbon atom is a stereocenter, we often have to look beyond the atoms bonded directly to it. When the atom is in a ring of atoms, encountering different groups as we work our way around the ring may identify an asymmetry that ensures that the atom is bonded to at least two different groups. If it is also bonded to two different groups, the atom is a stereocenter.

Practice Exercise Identify which of the following molecules are chiral. Circle the stereocenters in each structure.

(a)

(b)

(c)

(d)

(e)

(Answers to Practice Exercises are in the back of the book.)

Chirality in Nature

Many of the compounds formed by living systems are chiral. For example, the proteins in our tissues and in the foods we consume are made of amino acids that are nearly all chiral compounds. The two enantiomers of the amino acid alanine are shown in Figure 5.41. Our bodies can only make proteins from the (+) enantiomers of the 20 common amino acids. The origins of this biological preference for one enantiomer over another are unknown, though the amino acids in meteorites striking Earth have slightly more of the enantiomers found in living things in Earth, which may indicate that the chiral preferences observed in Earth's biosphere are not limited to our planet.

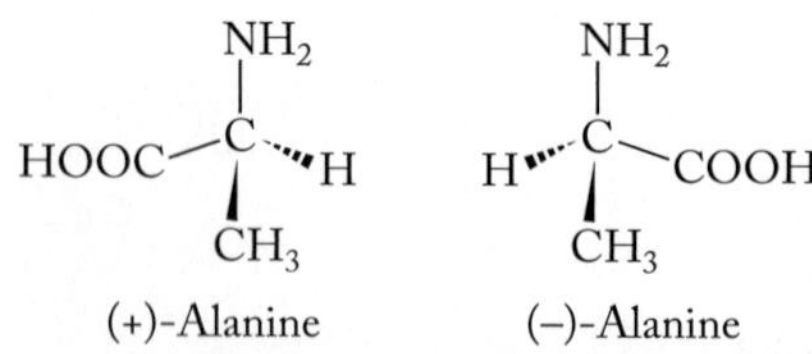

FIGURE 5.41 The two enantiomers of the amino acid alanine.

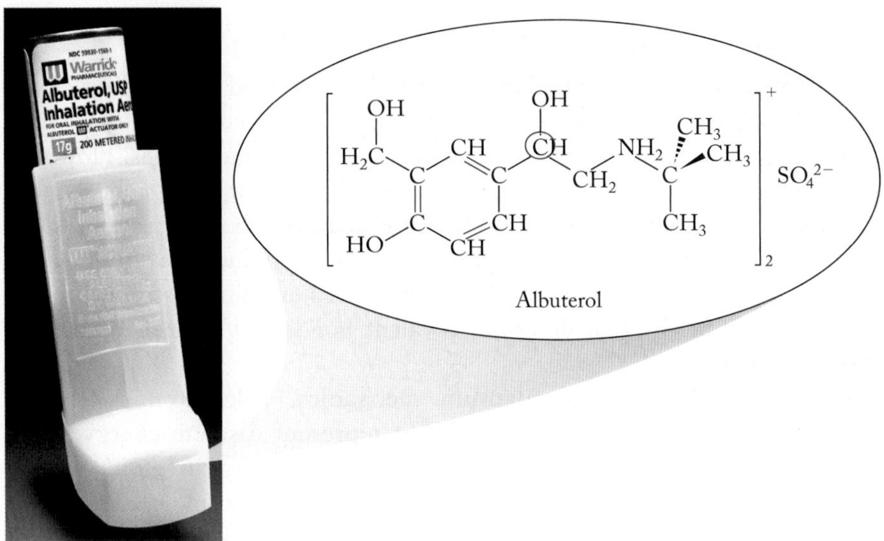

FIGURE 5.42 The antiasthma drug albuterol. The chiral center is circled.

Whatever the origin of the chirality, processes requiring molecular recognition in living systems often depend on the selectivity conveyed by chirality. The human body is a chiral environment, so the handedness of molecules matters, including those we take into our bodies for therapeutic purposes. As many as half of the prescription drugs on the market today are chiral and owe their function to recognition by a receptor that favors one enantiomer over the other. In 2014, nine of the ten top-selling drugs globally were chiral. Typically only one of the enantiomers of a chiral drug is active. A classic example is the drug albuterol (Figure 5.42), which is used to treat asthma and other respiratory disorders. One isomer causes bronchodilation (a widening of the air passages of the lungs, which eases breathing), while the other causes bronchial constriction and may actually be detrimental to the patient.

As noted previously, living organisms typically produce only one stereoisomer of a chiral compound. When a chiral compound such as albuterol is synthesized in a laboratory, however, both enantiomers are often produced in equal amounts. The resulting 50:50 mixture is called a **racemic mixture**. Because one isomer rotates the plane of polarized light in one direction, and the other to the same extent in the opposite direction, a racemic mixture does not rotate plane-polarized light at all. Because the two isomers interact differently with receptors, the pharmaceutical industry routinely faces two choices: devise a special synthetic procedure that yields only the isomer of interest or separate the two isomers during the manufacturing process. Both approaches are used in the production of chiral pharmaceuticals and both contribute to the cost of these drugs.

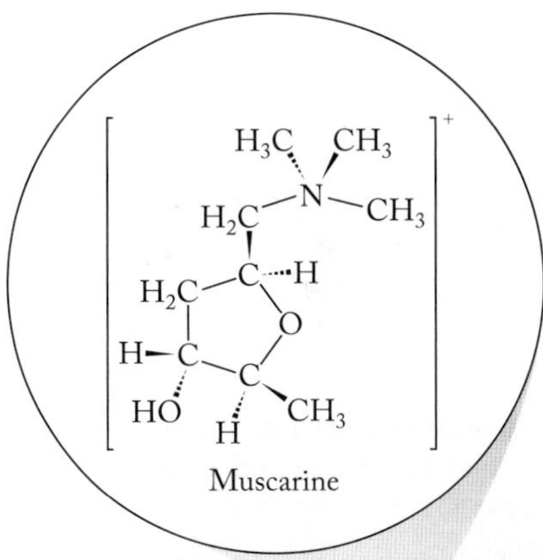

FIGURE 5.43 Several varieties of mushrooms, including *Amanita muscaria*, contain the toxic substance muscarine.

CONCEPT TEST

The Documents in the Case, a mystery written by Dorothy L. Sayers in 1930, involves the suspicious death of an authority on wild, edible mushrooms. Allegedly, the victim ate a stew made of poisonous mushrooms that contained a toxic natural product called muscarine (Figure 5.43). A forensic specialist evaluated the contents of the victim's stomach and found that they contained muscarine but that they did not rotate plane-polarized light. Based on the findings, the coroner concluded that the victim had been murdered. How did the coroner reach that conclusion?

(Answers to Concept Tests are in the back of the book.)

CONNECTION Molecular orbital theory also explains the sea of electrons model (Section 4.1) and the electrical conductivity of metals. This topic is revisited in Chapter 19.

FIGURE 5.44 Auroras are spectacular displays of color produced by excited-state atoms, ions, and molecules in Earth's upper atmosphere.

molecular orbital (MO) theory a bonding theory based on the mixing of atomic orbitals of similar shapes and energies to form molecular orbitals that belong to the molecule as a whole.

molecular orbital a region of characteristic shape and energy where electrons in a molecule are located.

bonding orbital a term in MO theory describing regions of increased electron density between nuclear centers that serve to hold atoms together in molecules.

antibonding orbital a term in MO theory describing regions of electron density in a molecule that destabilize the molecule because they do not increase the electron density between nuclear centers.

sigma (σ) molecular orbital in MO theory, the orbital that results in the highest electron density between the two bonded atoms.

molecular orbital diagram in MO theory, an energy-level diagram showing the relative energies and electron occupancy of the molecular orbitals for a molecule.

5.7 Molecular Orbital Theory

Lewis structures and valence bond theory help us understand the bonding capacities of individual atoms and the bonding patterns in molecules and ions, while VSEPR and valence bond theory account for their molecular shapes. However, there are phenomena that cannot be explained by any of these models. For example, none of the models explain why O_2 is attracted to a magnetic field, or how molecules and ions in Earth's upper atmosphere produce the shimmering, colorful display known as an *aurora* (Figure 5.44). To explain those phenomena, we need another model that describes covalent bonding. We need **molecular orbital (MO) theory**.

MO theory is based on quantum mechanics. Like atomic orbitals, **molecular orbitals** are wave functions that represent discrete energy states inside molecules. As with atomic orbitals, the lowest energy MOs always fill first. Electrons in them move to higher energy MOs when molecules absorb quanta of electromagnetic radiation. When the electrons return to lower energy MOs, distinctive wavelengths of UV and visible radiation are emitted, producing, for example, some of the colors in an aurora. Note how these processes of molecular absorption and emission of electromagnetic radiation parallel the absorption and emission of radiation by atoms, which we discussed in Chapter 3.

Unlike atomic orbitals, including the hybrid atomic orbitals of valence bond theory, MOs are not linked to single atoms, but rather belong to *all* the atoms in a molecule. Thus, any electron in a molecule could be anywhere in the molecule. This delocalized view of covalent bonding is particularly effective at describing the bonding in molecules such as benzene, in which conjugation allows π bonds to spread out over more than one pair of atoms. The availability of electrons from other atoms also means that we do not have to involve a particular atom's *d* orbitals to explain expanded octets, as we sometimes need to do in valence bond theory.

In MO theory, molecular orbitals are formed by combining atomic orbitals from each of the atoms in the molecule. In this book, we limit our discussion of MO theory to simple molecules in which MOs are formed from the atomic orbitals of only a few atoms. Some MOs have lobes of high electron density that lie *between* bonded pairs of atoms; they are called **bonding orbitals**. The energies of bonding MOs are lower than the energies of the atomic orbitals that combined to form them, so populating them with electrons (each MO can hold two electrons, just like an atomic orbital) stabilizes the molecule and contributes to the strength of the bonds holding its atoms together.

There are other MOs with lobes of high electron density that are not located between the bonded atoms. They are **antibonding orbitals** and have energies that are higher than the atomic orbitals that combined to form them. When electrons are in antibonding orbitals, they destabilize the molecule. If a molecule were to have the same number of electrons in its bonding and antibonding orbitals, then there would be no net energy holding the molecule together, and it would fall apart (actually, it would never have formed in the first place). Another key point in MO theory is that the *total number of molecular orbitals must match the number of atomic orbitals involved in forming them*. For example, combining two atomic orbitals, one from each of two atoms, produces one low-energy bonding orbital and one high-energy antibonding orbital. To better understand how atomic orbitals combine to form molecular orbitals, let's apply MO theory to the bond that holds together H_2.

Molecular Orbitals of H_2

According to MO theory, a hydrogen molecule is formed when the 1*s* atomic orbitals on two hydrogen atoms combine to form the two molecular orbitals shown in Figure 5.45(a). Molecular orbital theory stipulates that mixing two atomic orbitals creates two molecular orbitals. The lower energy, bonding molecular orbital is oval shaped and spans the two atomic centers. The high density of electrons between the two atoms makes this bonding MO a **sigma (σ) molecular orbital**. When two electrons occupy it, a single σ bond is formed. The σ bonding molecular orbital in H_2 is labeled σ_{1s} because it is formed by mixing two 1*s* atomic orbitals.

The high-energy, antibonding molecular orbital formed from two hydrogen atomic orbitals is designated σ^*_{1s} (pronounced "sigma star"). This antibonding orbital has two separate lobes of electron density oriented away from the internuclear axis and a region of zero electron density (a node) between the atoms, as shown in Figure 5.45(a).

Figure 5.45(b) is a **molecular orbital diagram**, analogous to the *atomic* orbital diagrams we discussed in Chapter 3. Note that the σ_{1s} MO is lower in energy and therefore more stable than the 1*s* atomic orbitals by about the same amount that the σ^*_{1s} MO is higher in energy than the 1*s* atomic orbitals. Therefore, the formation of the two MOs does not significantly change the total energy of the system. A hydrogen molecule has two valence electrons, one from each H atom, both residing in the lower energy σ_{1s} orbital. As in atomic orbitals, these two σ_{1s} electrons must have opposite spins. The electron configuration that corresponds to the molecular orbital diagram in Figure 5.45 is written $(\sigma_{1s})^2$, where the superscript indicates that there are two electrons in the σ_{1s} molecular orbital. Because the energy of the electrons in a $(\sigma_{1s})^2$ configuration is lower than the energy of the electrons in two isolated hydrogen atoms, H_2 molecules are lower in energy and therefore more stable than a pair of free H atoms.

Hydrogen is a diatomic gas, but helium exists as free atoms and not as molecular He_2. MO theory explains why. Each He atom has two valence electrons in its 1*s* atomic orbital. Mixing two He 1*s* orbitals yields the same set of molecular orbitals (Figure 5.46) as those of H_2. Adding four electrons (two from each He atom) fills both MOs. The presence of two electrons in the σ^*_{1s} orbital cancels the stability gained from having two electrons in the σ_{1s} orbital. Because there is no net gain in stability, He_2, $(\sigma_{1s})^2(\sigma^*_{1s})^2$, does not exist.

Another way to compare the bonding in H_2 and He_2 is to consider the *bond order* in each. We previously defined bond order as the number of bonds between two atoms: a bond order of 1 for Cl—Cl, 2 for O=O, and 3 for N≡N. In MO theory, we define bond order as follows:

$$\text{Bond order} = \frac{1}{2}\left[\begin{pmatrix}\text{number of} \\ \text{bonding electrons}\end{pmatrix} - \begin{pmatrix}\text{number of} \\ \text{antibonding electrons}\end{pmatrix}\right] \quad (5.2)$$

In a molecule of H_2, there are two electrons in the bonding MO and none in the antibonding MO, so

$$\text{Bond order in } H_2 = \tfrac{1}{2}(2 - 0) = 1$$

In nonexistent He_2, the bond order is 0 because the number of electrons in bonding and antibonding orbitals is equal:

$$\text{Bond order in } He_2 = \tfrac{1}{2}(2 - 2) = 0$$

In general, the strength of the bond and the stability of the molecule both increase as the bond order increases.

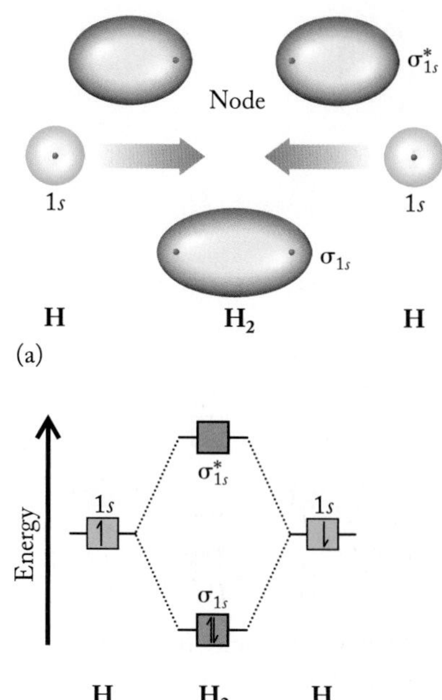

FIGURE 5.45 Mixing the 1*s* orbitals of two hydrogen atoms creates two molecular orbitals: a bonding σ_{1s} orbital containing two electrons and an empty antibonding σ^*_{1s} orbital. (a) The lower red oval is the bonding orbital. Dots show the locations of the two hydrogen nuclei. The two red ovals at the top together make up the antibonding orbital. *Note*: The two top ovals represent only *one* molecular orbital with a node of zero electron density in between. (b) A molecular orbital diagram shows the relative energies of the bonding and antibonding molecular orbitals and of the atomic orbitals that formed them.

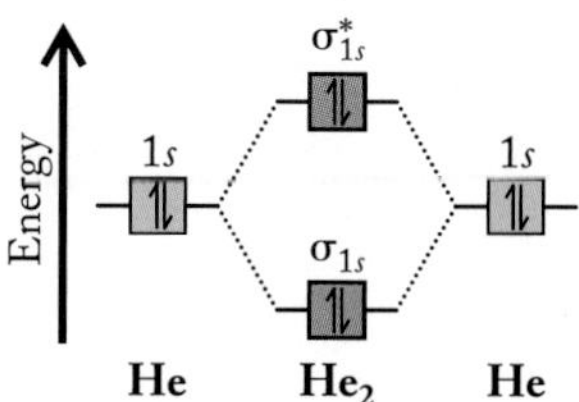

FIGURE 5.46 The molecular orbital diagram for nonexistent He_2 shows that the same number of electrons occupy the antibonding orbital and the bonding orbital. Therefore, the bond order is zero and the molecule is not stable.

CONNECTION We first defined bond order in Chapter 4 when we related the lengths and strengths of bonds to the number of electron pairs shared by two atoms.

SAMPLE EXERCISE 5.8 Using MO Theory to Predict Bond Order I **LO6**

Draw the MO diagram of H_2^-, determine the bond order of the ion, and predict whether or not it is stable.

Collect and Organize We apply MO theory to draw an orbital diagram for H_2^-, and then determine the bond order using Equation 5.2. If the bond order is greater than zero, the ion may be stable.

Analyze We should be able to base the MO diagram for H_2^- on the MO diagram for H_2 (Figure 5.45b) because the H_2^- ion has only one more electron than H_2 and the empty σ^*_{1s} orbital in H_2 can accommodate up to two electrons.

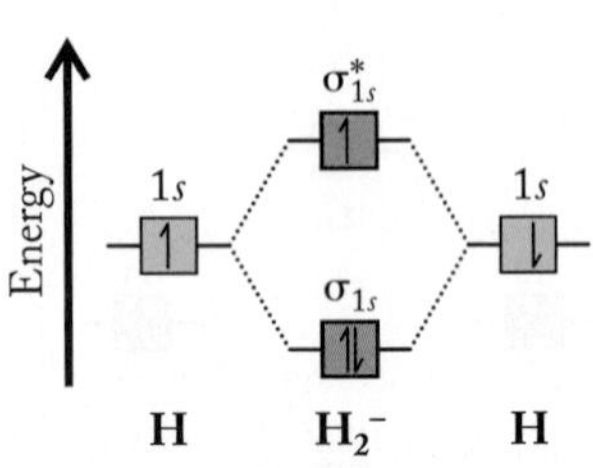

FIGURE 5.47 The molecular orbital diagram for H_2^-.

Solve The σ_{1s} orbital is filled in H_2. The third electron of H_2^- goes into the σ^*_{1s} orbital, so the MO diagram is as shown in Figure 5.47. The notation for this electron configuration is $(\sigma_{1s})^2\,(\sigma^*_{1s})^1$ (listing the molecular orbitals in order of increasing energy). The bond order is

$$\text{Bond order in } H_2^- = \tfrac{1}{2}(2 - 1) = \tfrac{1}{2}$$

The H_2^- ion exists, but it is not as stable as H_2.

Think About It We encountered fractional bonds in Chapter 4 in the discussion of resonance, and we encounter them again here in the MO treatment of H_2^-. A bond order of $\frac{1}{2}$ in MO theory means that the bond between the two atoms in H_2^- is weaker than the single bond in H_2, making H_2^- less stable than H_2.

Practice Exercise
Use MO theory to predict the bond order in a H_2^+ ion.

(Answers to Practice Exercises are in the back of the book.)

Molecular Orbitals of Other Homonuclear Diatomic Molecules

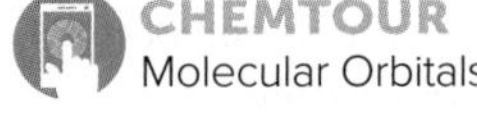

CHEMTOUR
Molecular Orbitals

Molecular orbital diagrams for homonuclear (same-atom) diatomic molecules like N_2 and O_2 are more complex than the diagram for H_2 because of the greater number and variety of atomic orbitals in N_2 and O_2. Not all combinations of atomic orbitals result in bonding, but there are some general guidelines for constructing the molecular orbital diagram for any molecule:

1. The number of molecular orbitals equals the number of atomic orbitals used to create them.
2. Atomic orbitals with similar energy and orientation mix more effectively than do those that have different energies and orientations.
3. Better mixing leads to a larger energy difference between bonding and antibonding orbitals and thus greater stabilization of the bonding MOs.
4. A molecular orbital can accommodate a maximum of two electrons; two electrons in the same MO have opposite spins.
5. Electrons in ground-state molecules occupy the lowest energy molecular orbitals available, following the aufbau principle and Hund's rule.

In mixing atomic orbitals to create molecular orbitals, we focus on the atom's valence electrons because core electrons do not participate in bonding. Focusing on N_2 and O_2 as examples, we first mix their 2*s* orbitals. The mixing process is analogous to the one we used for H_2, except that the resulting MOs are designated σ_{2s} and σ^*_{2s}.

Next we mix the three pairs of $2p$ orbitals, producing a total of six MOs. The different spatial orientations of the $2p_x$, $2p_y$, and $2p_z$ atomic orbitals result in different kinds of MOs (Figure 5.48). The $2p_z$ atomic orbitals point toward each other. When they mix, two molecular orbitals form: a σ_{2p} bonding orbital and a σ^*_{2p} antibonding orbital. The lobes of the $2p_x$ and $2p_y$ atomic orbitals are oriented at 90° to the bonding axis and also at 90° to each other. When the $2p_x$ orbitals mix together, and when the $2p_y$ orbitals mix together, they do so around the bonding axis instead of along it. This mixing produces **pi (π) molecular orbitals**. As with σ molecular orbitals, there are equal numbers of low-energy π and high-energy π^* molecular orbitals. Electrons that occupy π orbitals contribute to the formation of π bonds; those that populate π^* orbitals detract from π bond formation.

pi (π) molecular orbital in MO theory, an orbital formed by the mixing of atomic orbitals oriented above and below, or in front of and behind, the bonding axis; electrons in π orbitals form π bonds.

The relative energies of the σ and π molecular orbitals for N_2 and O_2 are shown in Figure 5.49. In each molecule, the energies of the σ_{2s} and σ^*_{2s} MOs are lower than the energy of the σ_{2p} MO for the same reason that a $2s$ atomic orbital is lower in energy than a $2p$ atomic orbital. However, the relative energies of the MOs formed by mixing the $2p$ orbitals in N_2 and O_2 are not the same. Let's begin with O_2 (Figure 5.49b) because its diagram is representative of most homonuclear

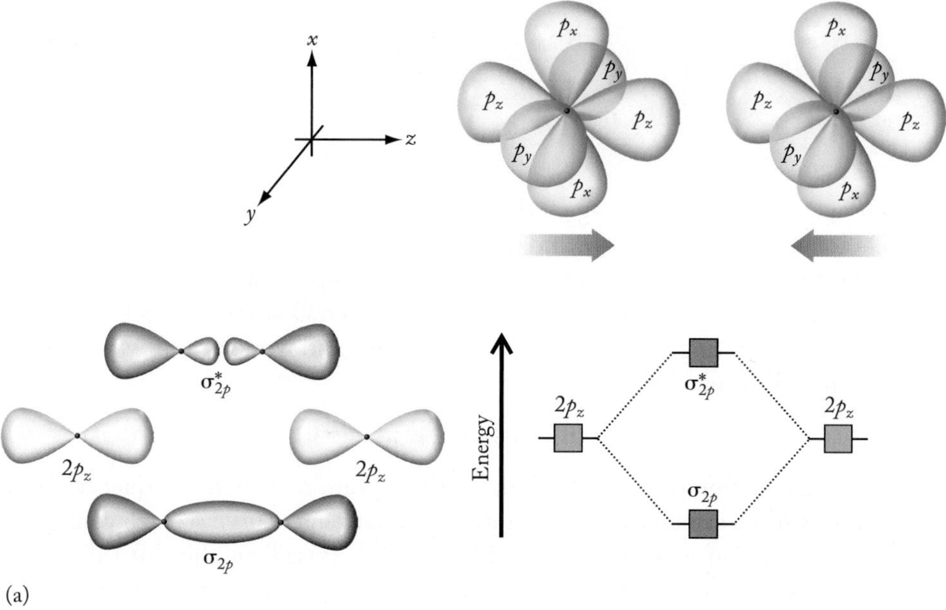

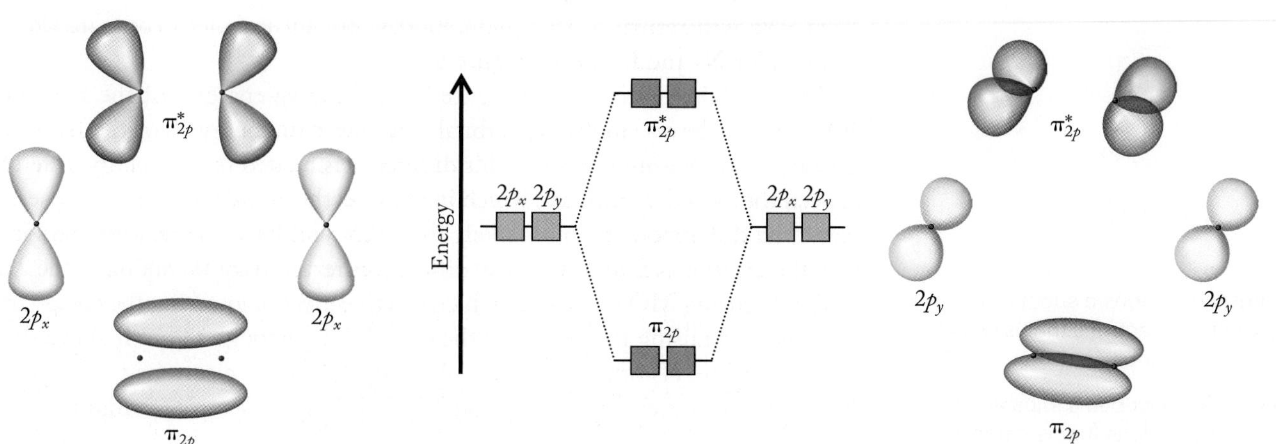

FIGURE 5.48 Two sets of three p atomic orbitals mix to form six molecular orbitals. (a) The $2p_z$ atomic orbitals create a σ_{2p} bonding orbital and a σ^*_{2p} antibonding orbital. (b) The $2p_x$ and $2p_y$ atomic orbitals mix to form two π_{2p} bonding molecular orbitals and two π^*_{2p} antibonding molecular orbitals.

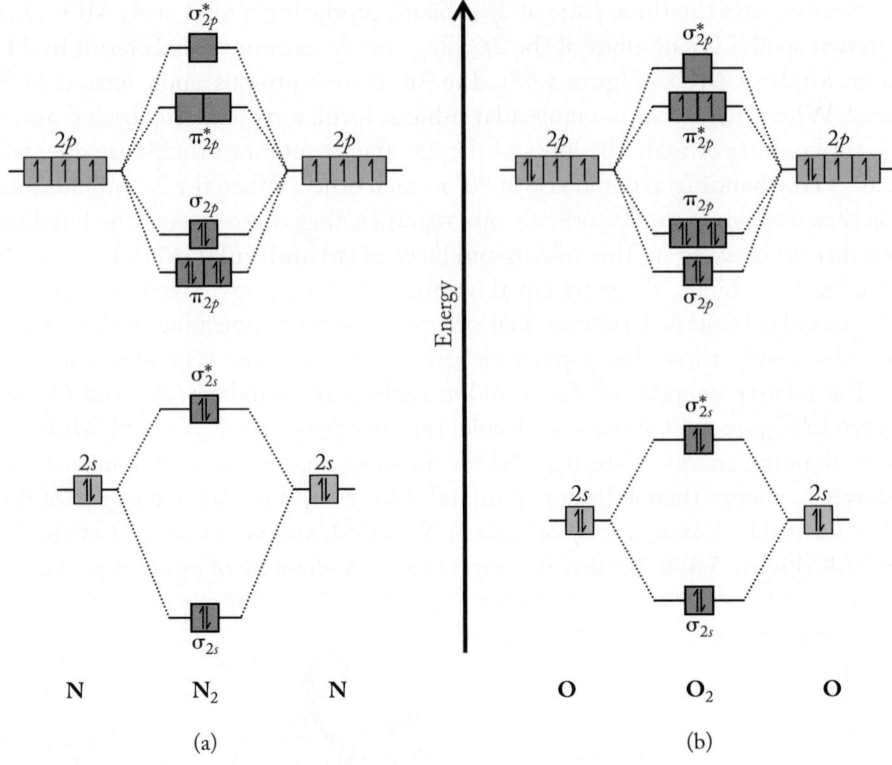

FIGURE 5.49 Molecular orbital diagrams of (a) N_2 and (b) O_2. The vertical sequence of orbitals in N_2 applies to homonuclear diatomic molecules of elements with $Z \leq 7$. The O_2 sequence applies to homonuclear diatomic molecules of elements beyond oxygen ($Z \geq 8$).

diatomic molecules, including all the halogens. In order of increasing energy, the MOs are σ_{2s}, σ^*_{2s}, σ_{2p}, π_{2p}, π^*_{2p}, and σ^*_{2p}. Keep in mind that there are groups of two π_{2p} and two π^*_{2p} orbitals. This means that each group of two can hold up to four electrons. Adding 12 valence electrons (six from each O atom in the O_2 molecule) to the MOs, starting with the lowest energy MO and working our way up, produces the following electron configuration:

$$O_2\text{: } (\sigma_{2s})^2(\sigma^*_{2s})^2(\sigma_{2p})^2(\pi_{2p})^4(\pi^*_{2p})^2$$

The distribution of electrons in the MO diagram reflects this sequence. Note that the two π^*_{2p} orbitals are degenerate (equivalent in energy), so each contains a single electron, in accordance with Hund's rule. Based on this MO diagram, there are two unpaired electrons in a molecule of O_2. This contradicts the representation of bonding obtained from the Lewis structure of O_2, from VSEPR, and from valence bond theory, all of which predict that the valence electrons are paired. We will return to this point shortly, but for now let's compare the MO diagrams for N_2 and for O_2 in Figure 5.49.

When we do, notice the difference in the relative energies of the π_{2p} and σ_{2p} MOs. In N_2, the π_{2p} molecular orbitals are lower in energy than the σ_{2p} orbital, whereas in O_2 they are reversed. This difference is related to the energy differences between the $2s$ and $2p$ orbitals, which increase as the atomic number increases. The greater the difference, the more likely the MOs will have the relative energies we see in the O_2 orbital diagram. Lesser differences result in some mixing of the $2s$ and $2p$ orbitals during MO formation, which has the effect of lowering the energy of the σ_{2s} orbital and raising the energy of the σ_{2p} orbital—enough to put it above the π_{2p} orbital, as in the N_2 orbital diagram. Adding 10 valence electrons to this stack of MOs in N_2, lowest energy first, produces the following electron configuration:

$$N_2\text{: } (\sigma_{2s})^2(\sigma^*_{2s})^2(\sigma_{2p})^4(\pi_{2p})^2$$

diamagnetic describes a substance with no unpaired electrons that is weakly repelled by a magnetic field.

paramagnetic describes a substance with unpaired electrons that is attracted to a magnetic field.

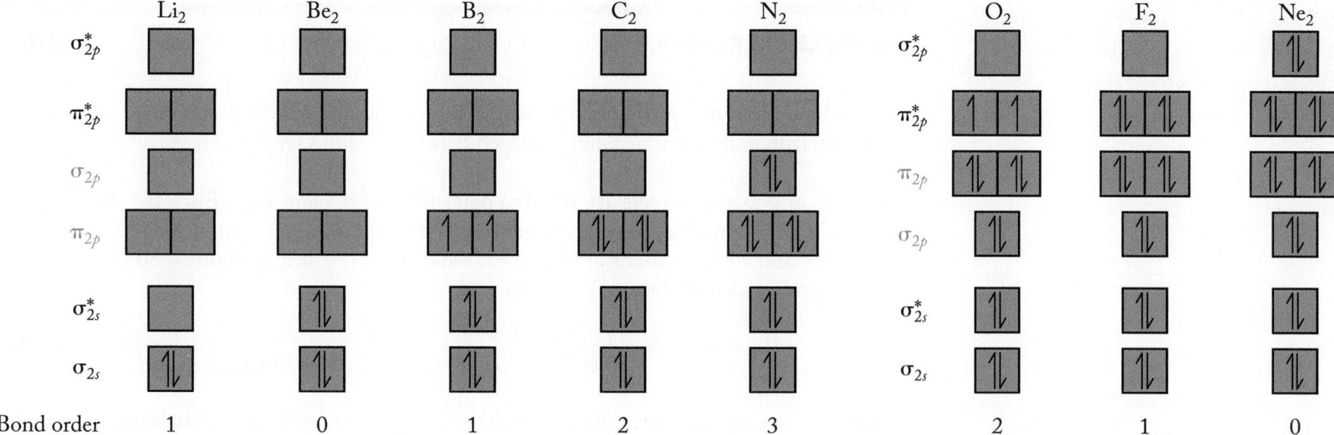

FIGURE 5.50 Valence-shell molecular orbital diagrams and bond orders of the homonuclear diatomic molecules of the second-row elements.

The distribution of electrons in the MO diagram in Figure 5.49(a) reflects that electron configuration. There are a total of eight electrons in bonding MOs and two in antibonding MOs. Using Equation 5.2 to calculate the bond order for N_2, we get

$$\text{Bond order in } N_2 = \tfrac{1}{2}(8 - 2) = 3$$

In O_2, eight electrons occupy bonding orbitals and four electrons occupy antibonding orbitals, so

$$\text{Bond order in } O_2 = \tfrac{1}{2}(8 - 4) = 2$$

On the basis of their Lewis structures (Section 4.3), we predicted a triple bond in N_2 and a double bond in O_2, and molecular orbital theory leads us to the same conclusions.

One of the strengths of MO theory is that it not only explains what we could with the early Lewis, VSEPR, and valence bond theories, but it also enables us to explain properties of molecular compounds that can't be explained by other bonding theories. The magnetic behavior of homonuclear diatomic molecules is a case in point. Electrons in atoms have two possible spin orientations depending on the value of their spin quantum number m_s. Figure 5.50 shows the MO-based electron configurations in diatomic molecules of the period 2 elements. In most of the molecules, all of the electrons are paired; the molecules that make up most substances contain only paired electrons. Complete electron pairing means that the substances are repelled slightly by a magnetic field. They are said to be **diamagnetic**. Most substances are diamagnetic, but if a substance's molecules (or ions) contain unpaired electrons, as do those of O_2, then the substance is attracted by a magnetic field (as shown in Figure 5.51) and is **paramagnetic**. The more unpaired electrons in a molecule, the greater the paramagnetism. Only MO theory accounts for the magnetic behavior of oxygen and other molecular substances.

FIGURE 5.51 Liquid O_2 poured from a cup is suspended in the space between the poles of this magnet because the unpaired electrons in its molecules make O_2 paramagnetic. Paramagnetic substances are attracted to magnetic fields.

CONCEPT TEST

a. Which of the diatomic molecules in Figure 5.50 does not exist?

b. Which one is paramagnetic (besides O_2)?

(Answers to Concept Tests are in the back of the book.)

CONNECTION In Chapter 3 we introduced the spin quantum number (m_s), which has a value of either $+\frac{1}{2}$ or $-\frac{1}{2}$ and defines the orientation of an electron in a magnetic field.

SAMPLE EXERCISE 5.9 Using MO Theory to Predict Bond Order II **LO6**

For which molecules in Figure 5.50 would there be an increase in bond order if they lost one electron each, forming singly charged diatomic cations (X_2^+)?

Collect and Organize We are asked to determine which ions Li_2^+, Be_2^+, B_2^+, C_2^+, N_2^+, O_2^+, F_2^+, and Ne_2^+ have a greater bond order than their parent molecules. The MO electron configurations of the parent molecules are given in Figure 5.50. Bond order can be calculated using Equation 5.2:

$$\text{Bond order} = \tfrac{1}{2}\left[\begin{pmatrix}\text{number of}\\\text{bonding electrons}\end{pmatrix} - \begin{pmatrix}\text{number of}\\\text{antibonding electrons}\end{pmatrix}\right] \quad (5.2)$$

Analyze The electron configurations of the ions of interest can be written by reducing by 1 the number of electrons in the highest energy occupied MOs (see the following table) in the parent molecules. There will be an increase in bond order if the electron is lost from an antibonding orbital and a decrease in bond order if an electron is lost from a bonding orbital.

Solve Removing one electron from each MO diagram in Figure 5.50 yields the results in the following table.

Ion	Electron Configuration	Bond Order	Bond Order in Parent Molecule
Li_2^+	$(\sigma_{2s})^1$	$\frac{1}{2}(1-0) = 0.5$	$\frac{1}{2}(2-0) = 1$
Be_2^+	$(\sigma_{2s})^2(\sigma^*_{2s})^1$	$\frac{1}{2}(2-1) = 0.5$	$\frac{1}{2}(2-2) = 0$
B_2^+	$(\sigma_{2s})^2(\sigma^*_{2s})^2(\pi_{2p})^1$	$\frac{1}{2}(3-2) = 0.5$	$\frac{1}{2}(4-2) = 1$
C_2^+	$(\sigma_{2s})^2(\sigma^*_{2s})^2(\pi_{2p})^3$	$\frac{1}{2}(5-2) = 1.5$	$\frac{1}{2}(6-2) = 2$
N_2^+	$(\sigma_{2s})^2(\sigma^*_{2s})^2(\pi_{2p})^4(\sigma_{2p})^1$	$\frac{1}{2}(7-2) = 2.5$	$\frac{1}{2}(8-2) = 3$
O_2^+	$(\sigma_{2s})^2(\sigma^*_{2s})^2(\sigma_{2p})^2(\pi_{2p})^4(\pi^*_{2p})^1$	$\frac{1}{2}(8-3) = 2.5$	$\frac{1}{2}(8-4) = 2$
F_2^+	$(\sigma_{2s})^2(\sigma^*_{2s})^2(\sigma_{2p})^2(\pi_{2p})^4(\pi^*_{2p})^3$	$\frac{1}{2}(8-5) = 1.5$	$\frac{1}{2}(8-6) = 1$
Ne_2^+	$(\sigma_{2s})^2(\sigma^*_{2s})^2(\sigma_{2p})^2(\pi_{2p})^4(\pi^*_{2p})^4(\sigma^*_{2p})^1$	$\frac{1}{2}(8-7) = 0.5$	$\frac{1}{2}(8-8) = 0$

When the values in the last two columns are compared, we find that the bonds in Be_2^+, O_2^+, F_2^+, and Ne_2^+ have a higher bond order than the bond in their parent molecules, although the bond order of zero in Be_2 and Ne_2 means that these molecules don't even exist.

Think About It Ionizing O_2, F_2, and (if they existed) Be_2 and Ne_2 reduces the number of electrons in antibonding molecular orbitals, which increases bond order. However, ionizing Li_2, B_2, C_2, and N_2, reduces the number of electrons in bonding molecular orbitals, which decreases bond order.

 **Practice Exercise** Which molecules in Figure 5.50 experience an increase in bond order when they form anions with 1− charges?

(Answers to Practice Exercises are in the back of the book.)

Molecular Orbitals of Heteronuclear Diatomic Molecules

Molecular orbital theory also enables us to account for the bonding in *heteronuclear* diatomic molecules, which are molecules made of two different atoms. The bonding in some of these molecules is difficult to explain using other bonding theories. For example, it is often difficult to draw a single Lewis structure for an odd-electron molecule like nitrogen monoxide, NO. In Chapter 4, we considered several arrangements of the valence electrons in NO, such as

$$:\dot{N}=\ddot{O}: \quad :\ddot{N}=\dot{O}:$$

We predicted that oxygen was more likely to have a complete octet of valence electrons because it is the more electronegative element. In addition, experimental evidence rules out structures with unpaired electrons on the oxygen atom. Therefore, the preferred structure is the one shown in red. However, the length of the NO bond (115 pm) is considerably shorter than the value in Table 4.6 for an average N═O double bond (122 pm).

Molecular orbital theory helps explain both the bonding in NO and the shorter-than-expected bond length. Let's look at the bonding first. The MO diagram for nitrogen monoxide is different from the diagrams of homonuclear diatomic gases. Nitrogen and oxygen atoms have different numbers of protons and electrons, and the difference in effective nuclear charge between N and O atoms means that their atomic orbitals have different energies, as Figure 5.52 shows for the 2*s* and 2*p* orbitals.

In constructing the MO diagram for NO, the guidelines described previously still apply. The number of MOs formed must equal the number of atomic orbitals combined, and the energy and orientation of the atomic orbitals being mixed must be considered. One additional factor influences the energies of the MOs in heteronuclear diatomic molecules: *bonding* MOs tend to be closer in energy to the atomic orbitals of the more electronegative atom (O, in this case), and *antibonding* MOs tend to be closer in energy to the atomic orbitals of the less electronegative atom (N). Note in Figure 5.52 how the energy of the bonding σ_{2s} orbital is closer to that of the 2*s* orbital of the oxygen atom, and the energy of the antibonding σ^*_{2s} orbital is closer to that of the 2*s* orbital of the nitrogen atom. Similarly, the π_{2p} MOs in NO are closer in energy to the 2*p* orbitals of oxygen, and the π^*_{2p} MOs are closer in energy to the 2*p* orbitals of nitrogen. The proximity of the nitrogen 2*p* atomic orbitals to the π^*_{2p} MOs is consistent with the single electron in the π^*_{2p} MOs being on nitrogen rather than on oxygen. This prediction is consistent with our Lewis structure in which the odd electron in NO is on the nitrogen atom.

Molecular orbital theory also enables us to rationalize the relatively short bond length in NO. Using Equation 5.2, the bond order is $\frac{1}{2}(8 - 3) = 2.5$, halfway between the bond orders for N═O and N≡O and consistent with a bond length of 115 pm, which is halfway between the lengths of the N═O bond (122 pm) and the N≡O bond (106 pm).

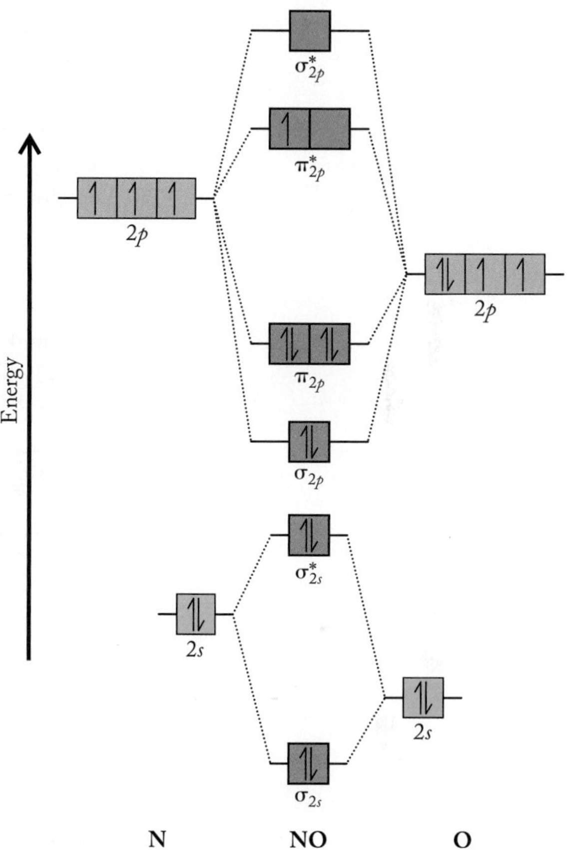

FIGURE 5.52 The molecular orbital diagram for NO shows that the unpaired electron occupies a π^*_{2p} antibonding orbital, which is closer in energy to the 2*p* atomic orbitals of nitrogen than to the 2*p* atomic orbitals of oxygen. As a result of this proximity, the electron has more nitrogen character and is more likely to be on the nitrogen atom than the oxygen atom.

SAMPLE EXERCISE 5.10 Using MO Theory to Determine the Bond Order of Heteronuclear Diatomic Molecules **LO6**

Nitrogen monoxide reacts with many transition metals, including the iron in our blood. In these compounds, NO may be NO^+ or NO^-. Use Figure 5.52 to predict the bond order of NO^+ and NO^-.

Collect and Organize We need to determine the bond order of the bonds in two diatomic ions based on the MO diagram of their parent molecule, NO, in Figure 5.52. Equation 5.2 relates bond order to the number of electrons in bonding and antibonding orbitals.

Analyze There is a single electron in the π^*_{2p} MO, which is the highest energy occupied MO in the molecule. This electron is lost when NO forms NO^+, and a second electron occupies this orbital in NO^-.

Solve The electron configuration of NO^+ is

$$NO^+: (\sigma_{2s})^2(\sigma^*_{2s})^2(\sigma_{2p})^2(\pi_{2p})^4$$

Adding a second electron to the π^*_{2p} orbital in NO yields the following electron configuration for NO^-:

$$NO^-: (\sigma_{2s})^2(\sigma^*_{2s})^2(\sigma_{2p})^2(\pi_{2p})^4(\pi^*_{2p})^2$$

The bond orders of the two ions are

$$NO^+: \text{Bond order} = \tfrac{1}{2}(8 - 2) = 3$$

$$NO^-: \text{Bond order} = \tfrac{1}{2}(8 - 4) = 2$$

Think About It The bond orders in N_2 and O_2 are 3 and 2, respectively. The cation NO^+ is isoelectronic with N_2, so our calculated bond order makes sense. The anion NO^- is isoelectronic with O_2, and so our calculated bond order for it is reasonable, too.

Practice Exercise Using Figure 5.52 as a guide, draw the MO diagram for carbon monoxide and determine the bond order for the carbon–oxygen bond.

(Answers to Practice Exercises are in the back of the book.)

Molecular Orbitals of N_2^+ and the Colors of Auroras

CONNECTION We used orbital diagrams in Chapter 3 to show electron transitions between energy levels as atoms absorb and emit electromagnetic radiation.

In addition to predicting the magnetic properties of molecules, MO theory is particularly useful for predicting their spectroscopic properties. Recall from Section 3.4 that the light emitted by excited free atoms is quantized because it is linked to the movement of electrons between atomic orbitals with discrete energies. Broadly speaking, the same is true in molecules: electrons can move from one molecular orbital to another when molecules absorb or emit light.

We can use this information to look again at how the colors of the auroras are produced. The principal chemical species involved are listed in Table 5.4. An asterisk indicates a molecule or molecular ion in an excited state. Excited N_2^+ ions produce blue-violet (391–470 nm) light, and excited N_2 molecules produce deep crimson red (650–680 nm) light. The MO diagrams of N_2^* and N_2, which are compared in Figure 5.53(a), show that one of the two electrons originally in the σ_{2p} orbital in N_2 has been raised to a π^*_{2p} orbital in N_2^*, leaving an unpaired σ_{2p} electron behind. Figure 5.53(b), which compares the MO diagrams of N_2^{+*} and N_2^+, shows us that N_2^{+*} also has one electron in a π^*_{2p} orbital, but its σ_{2p} orbital is empty because the other σ_{2p} electron originally in the N_2 molecule was lost when the molecule was ionized. As the π^*_{2p} electrons return from their antibonding orbital in the excited state to the bonding σ_{2p} orbital in the ground state, the distinctive celestial emissions of N_2^+ and N_2 are produced.

TABLE 5.4 Origins of Colors in the Aurora

Wavelength (nm)	Color	Chemical Species
650–680	Deep red	N_2^*
630	Red	O*
558	Green	O*
391–470	Blue-violet	N_2^{+*}

CONCEPT TEST

Are the bond orders of the excited-state species in Figure 5.53 the same as the ground-state species?

(Answers to Concept Tests are in the back of the book.)

Using MO Theory to Explain Fractional Bond Orders and Resonance

In Chapter 4 our attempts to draw the Lewis structure of ozone led to the conclusion that molecules of O_3 are held together by two O—O single bonds plus a shared pair of electrons distributed equally across both bonds. The concept of resonance was used to describe the delocalization of the second bonding pair and the equivalency of the bonds in O_3, which are shorter and stronger than O—O single bonds but not as short or strong as O═O double bonds. In this chapter we have learned how VSEPR explains the bent shape of O_3 molecules, and how valence bond theory and the formation of sp^2 hybrid orbitals help us visualize the trigonal planar orientation of bonding and lone pairs of electrons that account for ozone's molecular shape.

Now let's explore how MO theory provides another perspective on the delocalized bonding in O_3—one that does not require drawing resonance structures that, as we have admitted, do not accurately convey where all the bonding and lone pairs of electrons really are. MO theory is great at explaining bonding pair delocalization because MOs are not isolated between pairs of atoms but rather are spread across several atoms or even entire molecules. To simplify our exploration, we start with three O atoms that are connected by two single (σ) bonds and that have a total of five lone pairs of valence electrons (Figure 5.54). This trigonal planar pattern of electron pairs is what we would expect if each of the O atoms had sp^2 hybridized atomic orbitals. There are 14 valence electrons in Figure 5.54, leaving $(3 \times 6) - 14 = 4$ electrons unaccounted for. Let's distribute the four electrons among a set of molecular orbitals formed by mixing the three unhybridized p orbitals of the three O atoms.

In MO theory, mixing three p orbitals with lobes above and below the bonding plane produces three molecular orbitals. One of them is a low-energy π MO, and another is a high-energy π^* MO, but what about the third? It turns out that the third orbital is neither bonding nor antibonding. Instead, it is a *nonbonding* (n) molecular orbital. Electrons in nonbonding MOs neither lower the energy of a molecular structure, which would stabilize it, nor do they raise its energy and destabilize it. Rather, nonbonding MOs have the same energy as the atomic orbitals from which they formed, as shown in the partial MO diagram of ozone in Figure 5.55. Adding four electrons to these three molecular orbitals puts two electrons in the π orbital, two in the nonbonding orbital, and zero in π^*. Therefore, there is $\frac{1}{2}(2 - 0) = 1$ π bond distributed across the entire molecule. The overall bond order of 2 σ bonds + 1 π bond = 3 is divided equally between the two pairs of atoms, giving an average bond order for ozone of 1.5. The pair of nonbonding electrons is also delocalized. Each of the two noncentral O atoms gets half ownership of this nonbonding pair for a total of 2.5 lone pairs of electrons each, which is the average of two lone pairs on one noncentral O atom and three on the other in the resonance structures of O_3:

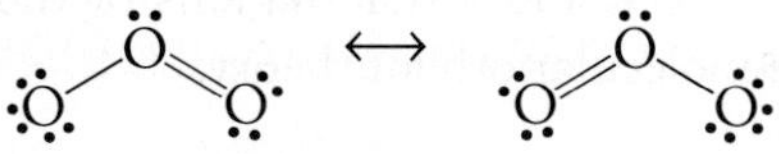

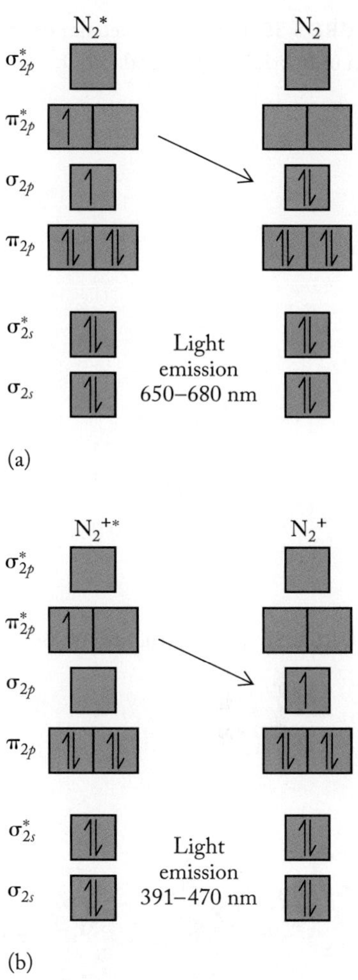

FIGURE 5.53 Molecular orbital diagrams for (a) N_2 and (b) N_2^+ show electronic transitions that result in the emission of visible light. High-energy collisions between particles in the upper atmosphere result in the promotion of electrons from the σ_{2p} orbital in N_2 molecules to the π^*_{2p} orbital and the formation of N_2^{+*} molecular ions that also have electrons in π^*_{2p} orbitals. When the electrons return to the ground state, red and blue-violet light is emitted.

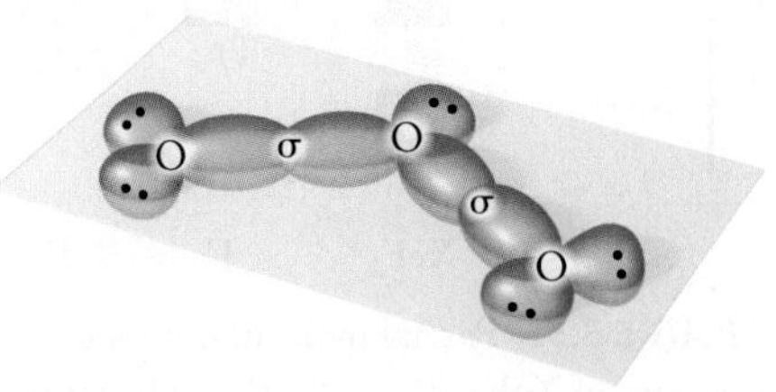

FIGURE 5.54 Two sigma bonds and five lone pairs of electrons in a molecule of O_3.

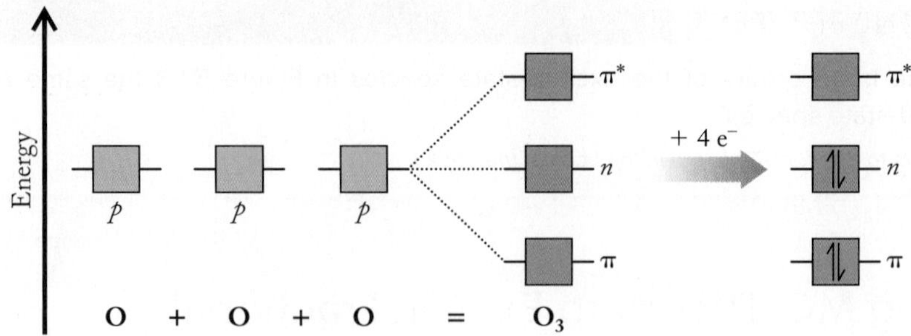

FIGURE 5.55 Forming molecular orbitals from unhybridized p orbitals in O_3.

MO Theory for SN > 4

In Chapter 4 we used expanded octets to explain the bonding in compounds such as SF_6 in which the central atom has a steric number greater than 4. In this chapter we have used hybrid schemes such as sp^3d and sp^3d^2 to help visualize the bonding in those compounds and to account for their molecular shapes. MO theory provides an alternate explanation of bonding and molecular shape that accommodates steric numbers greater than 4 without incorporating d orbitals.

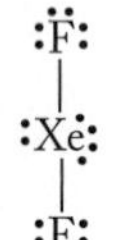

FIGURE 5.56 Lewis structure of XeF_2.

To see how the MO approach works, let's consider the bonding in XeF_2, which, in 1962, was one of the first noble gas compounds to be synthesized. Lewis theory predicts that there are two bonding pairs and three lone pairs of electrons around the central Xe atom. The molecule is linear, which makes sense if the lone pairs occupy the three equatorial positions of a trigonal bipyramidal electron-group array and the fluorine atoms are in the axial positions, as shown in Figure 5.56. This bonding pattern is consistent with a hybridization process involving all the 5s and 5p orbitals and one of the 5d orbitals of the central Xe atom—that is, sp^3d.

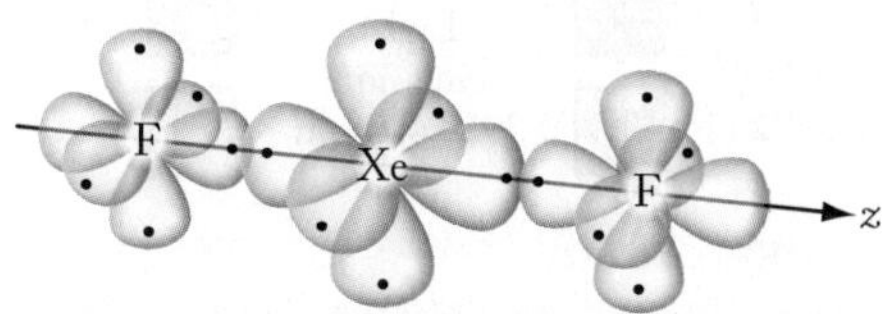

FIGURE 5.57 Alignment of valence-shell p_z orbitals of F and Xe prior to formation of bonding, nonbonding, and antibonding σ molecular orbitals in XeF_2.

MO theory can explain the linear shape of XeF_2 without using d orbitals. Let's start with a Xe atom and two F atoms aligned so that the lobes of their p_z orbitals are all in a single row (Figure 5.57). Aligning their lobes allows the orbitals to mix together and form the set of bonding, nonbonding, and antibonding σ molecular orbitals shown in Figure 5.58. There are a total of four electrons in these orbitals, and they fill the bonding and nonbonding MOs while leaving the antibonding orbital empty. This array produces a pair of bonding electrons and a bond order of 1 that is delocalized over the entire molecule. Therefore, the effective bond order of each of the two Xe—F bonds is only 0.5. You might think that a compound with bonds that are half the strength of single bonds might be unstable, and you would be correct. Xenon difluoride, though more stable than some other noble gas compounds, decomposes when it is exposed to sunlight and when it comes in contact with water.

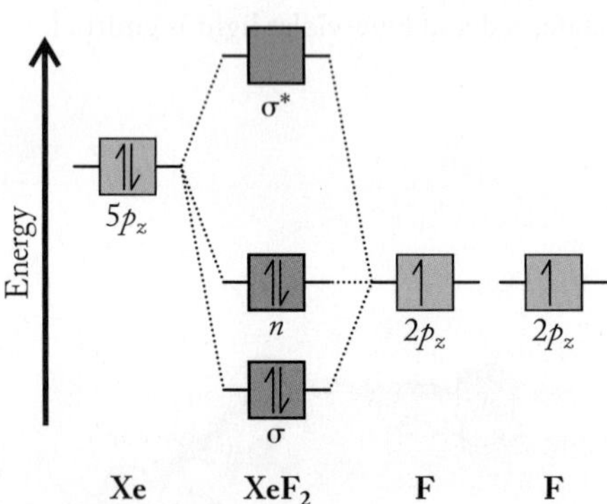

FIGURE 5.58 Partial molecular orbital diagram of XeF_2. These are the only MOs that contribute to bonding.

In Chapter 4 and in this chapter we have presented several theories of chemical bonding. Each theory has its particular strengths and capabilities. The best one to apply in a given situation depends on the question being asked. If our focus is on how the atoms in the second row of elements in the periodic table bond together in molecules, Lewis structures usually suffice. If we are also interested in visualizing the shapes of the molecules, then valence bond theory and hybridized atomic orbitals are useful tools. However, if we wish to explain the magnetic properties of molecular compounds and their ability to absorb and emit electromagnetic radiation in the UV and visible regions of the electromagnetic spectrum, then we must use molecular orbital theory.

SAMPLE EXERCISE 5.11 Integrating Concepts: Treatment for Alzheimer's Disease

In 2011 a compound with the molecular structure shown here was reported to be effective in treating neurodegenerative diseases, including Huntington's and Alzheimer's, in animals:

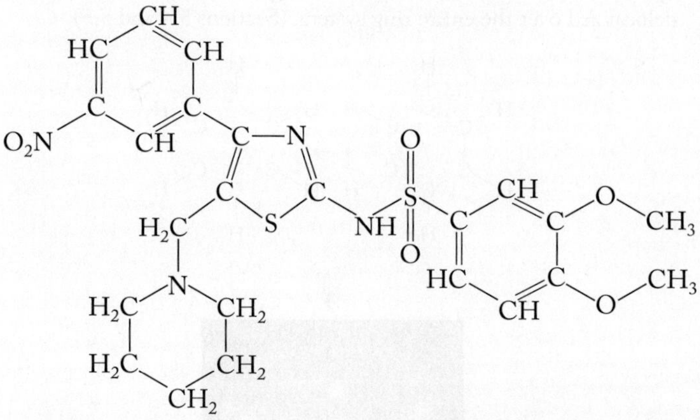

There are two S atoms in the structure. One has two bonds, which is consistent with its Lewis bonding capacity, but the other has six. Answer the following questions about this drug:

a. Many pharmaceuticals are chiral. Are any of the carbon atoms in this compound stereocenters?
b. What are the steric numbers and electron-pair geometries of the two sulfur atoms?
c. What is the bond angle between the two S═O double bonds?
d. There are two six-carbon rings in the structure. Are either of them aromatic rings? Which one(s)?
e. Does this compound have a permanent dipole?

Collect and Organize We are given the molecular structure of a biologically active compound and are asked questions about it. A chiral carbon atom is one that is bonded to four different atoms or groups of atoms. The steric number of an atom is the sum of its lone pairs of valence electrons and the number of atoms bonded to it. Steric numbers allow us to predict electron-pair geometries and bond angles and, indirectly, molecular geometries. Aromatic rings have six sp^2 hybridized C atoms and are stabilized by the delocalization of the electrons in three π bonds around the ring. Molecules in which bond dipoles do not cancel have permanent dipoles—that is, molecules with asymmetrical distributions of valence electrons.

Analyze

a. Only carbon atoms with four single (σ) bonds can be stereocenters, which eliminates all those in rings that are double bonded. Also, the C atoms in $-CH_2-$ groups can't be chiral because two of the four atoms or groups bonded to them are the same (H atoms).
b. and c. Sulfur atoms complete their octets by forming two covalent bonds, giving them two bonding pairs and two lone pairs. The sulfur atom in the five-atom ring fits that description. The other sulfur has two single bonds and two double bonds for a total of six pairs of bonding electrons—clearly more than an octet. This bonding pattern looks a lot like the one for sulfur in sulfuric acid (see Section 4.8), which has no lone pairs in its Lewis structure.
d. The atoms in both six-carbon rings are bonded to each other with alternating single and double bonds, which allows the π electrons in the three double bonds to be completely delocalized around each ring.
e. The molecule consists mostly of C and H atoms, which form bonds that are not very polar, but it also contains C—O, N═O, N—H, and S═O bonds, which are polar. There is also asymmetry in the overall structure of the molecule.

Solve

a. None of the sp^3 carbon atoms are bonded to four different atoms or groups of atoms, so none are chiral.
b. One of the two sulfur atoms has two atoms bonded to it and two lone pairs of valence electrons, which add up to SN = 4. The corresponding electron-pair geometry is tetrahedral. The other S atom has six bonds to four atoms and no lone pairs, which means its SN is also 4 and it also has a tetrahedral electron-pair geometry.
c. Given the tetrahedral electron-pair geometry around S and lack of lone pairs, we anticipate tetrahedral molecular geometry and an O═S═O bond angle of about 109°.
d. Due to the presence of delocalized π electrons, both six-carbon rings are aromatic.
e. Given its generally asymmetric structure and the presence of polar bonds, we can conclude that the molecule has a permanent dipole.

Think About It We established that the compound has no chiral carbon atoms, but does it have no stereocenters at all? Look carefully at the nitrogen atom in the middle of the structure. It is sp^3 hybridized and surrounded by (1) a H atom, (2) the right side of the molecule, which is different from (3) the left side, and (4) a lone pair of electrons. Do the four different groups (counting the lone pair as a "group") make this N atom a stereocenter? They would, except for a process known as *inversion*. If we label the right side of the molecule R and the left side L, then the structure surrounding the N atom could have two enantiomers:

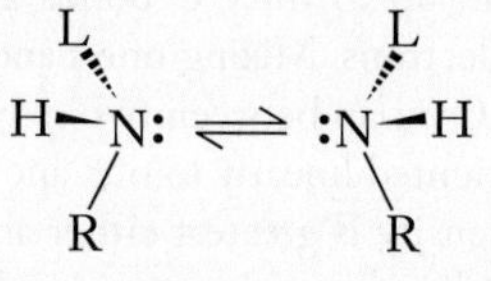

The double-arrow symbol between them means that the bonds of one enantiomer can reversibly flip around the N atom to form the other enantiomer. Because the bonds interconvert, we can't produce a sample of just one enantiomer.

SUMMARY

LO1 The shape of a molecule reflects the arrangement of the atoms in three-dimensional space and is determined largely by characteristic **bond angles.** Minimizing repulsion between pairs of valence electrons (the **VSEPR** model) results in the lowest energy orientations of bonding and nonbonding electron pairs and accounts for the observed **molecular geometries** of molecules. The shape of a molecule can be determined by its **steric number** (the sum of the number of bonded atoms and lone pairs around a central atom) and the **electron-pair geometry**, or arrangement of atoms and lone pairs. Molecules with SN = 2 and no lone pairs on the central atom have a **linear** electron-pair geometry and linear molecular geometry, while the electron-pair geometries of molecules with steric numbers 3 to 6 are **trigonal planar, tetrahedral, trigonal bipyramidal**, and **octahedral**, respectively. The presence of lone pairs of electrons in molecules with the preceding electron geometries produce these additional *molecular* geometries: **angular (bent), trigonal pyramidal, seesaw, T-shaped, square pyramidal**, and **square planar**. The observed bond angles in molecules deviate from the ideal values as a result of unequal repulsions between lone pairs and bonding pairs of electrons. (Sections 5.1 and 5.2)

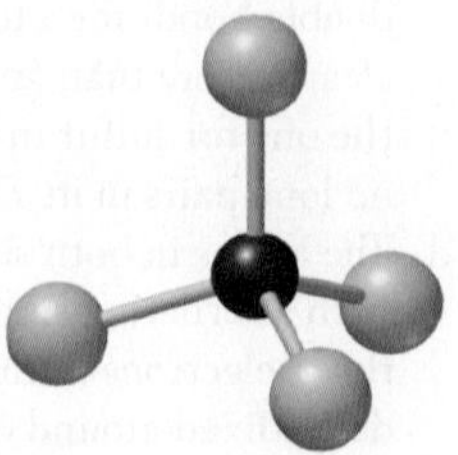

LO2 Two covalently bonded atoms with different electronegativities have partial electrical charges of opposite sign, creating a **bond dipole.** If the individual bond dipoles in a molecule do not offset each other, the molecule is polar. If they do offset each other, the molecule is nonpolar. A polar molecule has a permanent **dipole moment** (**μ**), which is a quantitative measure of the polarity of the molecule. (Section 5.3)

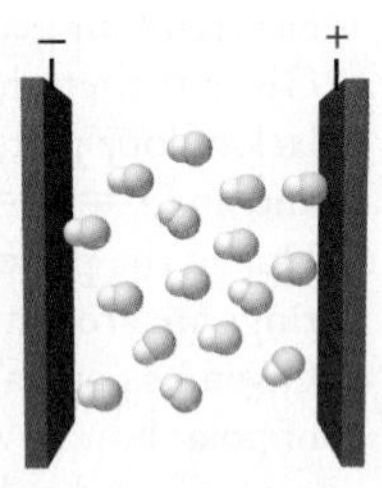

LO3 According to **valence bond theory**, the **overlap** of half-filled orbitals results in covalent bonds between pairs of atoms in molecules. Molecular geometry is explained by the mixing, or **hybridization**, of atomic orbitals to create **hybrid atomic orbitals**. Mixing one *s* and three *p* orbitals forms four ***sp*^{3} hybrid orbitals**. Overlap between *sp*^{3} orbitals and other atomic or hybrid orbitals results in up to four **sigma (σ) bonds** and a tetrahedral orientation of valence electrons. Mixing one *s* and two *p* orbitals forms three ***sp*^{2} hybrid orbitals**. Overlap between *sp*^{2} orbitals and other atomic or hybrid orbitals results in up to three σ bonds and a trigonal planar orientation of valence electrons. Mixing one *s* and one *p* orbital forms two ***sp* hybrid orbitals**. Overlap between two *sp* hybrid orbitals results in up to two σ bonds oriented linearly to one another. Covalent bonds in which the electron density is greatest either above and below or in front of and behind the bonding axis are **pi (π) bonds**. Mixing one *s* orbital, three *p* orbitals, and one *d* orbital yields five equivalent ***sp*^{3}*d* hybrid orbitals** with lobes that point toward the vertices of a trigonal bipyramid. Overlap between *sp*^{3}*d* orbitals and other atomic or hybrid orbitals results in up to five σ bonds. Mixing one *s* orbital, three *p* orbitals, and two *d* orbitals gives six equivalent ***sp*^{3}*d*^{2} hybrid orbitals** that point toward the vertices of an octahedron. Overlap between *sp*^{3}*d*^{2} orbitals and other atomic or hybrid orbitals results in up to six σ bonds. (Section 5.4)

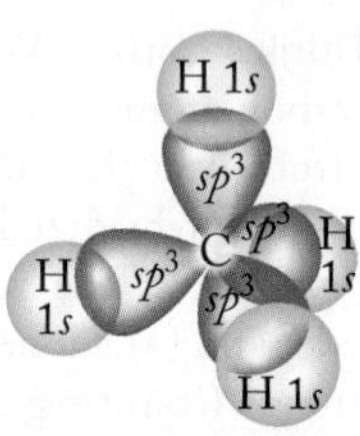

LO4 The shape of a molecule with more than one central atom is a result of overlapping geometries around the atoms. Molecules with only *sp*^{2} hybridized central atoms have extended planar geometries. The molecules of **aromatic compounds** contain planar rings of six *sp*^{2} hybridized carbon atoms with alternating π bonds whose electrons are delocalized over the entire ring system. (Sections 5.4 and 5.5)

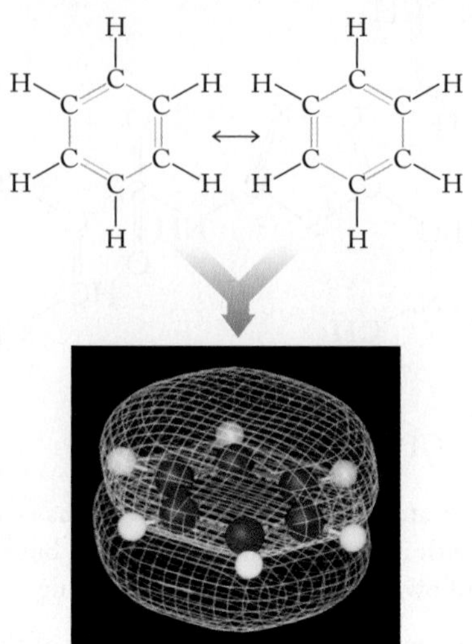

LO5 **Chiral** molecules have nonsuperimposable mirror images and have different properties. Many contain an *sp*^{3} hybridized carbon atom bonded to four different atoms or groups of atoms. (Section 5.6)

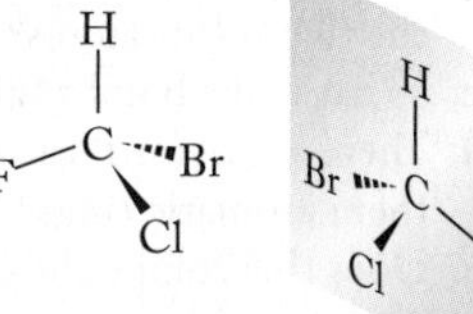

LO6 **Molecular orbital (MO) theory** is based on the formation of **molecular orbitals**, which are orbitals belonging to an entire molecule. MO theory describes the magnetic and spectroscopic properties of molecular compounds in ways that valence bond theory cannot. Mixing two atomic orbitals creates one **bonding orbital** and one **antibonding orbital**. The region of highest electron density lies along the bond axis in a **sigma (σ) molecular orbital**. Electrons in σ bonding orbitals form σ bonds. The regions of highest electron density of **pi (π) molecular orbitals** are above and below or behind and in front of the bonding axis. Electrons occupying π bonding orbitals form π bonds. A **molecular orbital diagram** shows the relative energies of the molecular orbitals in a molecule. MO electron configurations use the designations σ, σ^*, π, and π^* to describe the type of molecular orbitals occupied by electrons; subscripts to identify the atomic orbitals that combined to form the MOs; and superscripts to indicate the number of electrons in each MO. Atoms, ions, and molecules with no unpaired electrons are **diamagnetic** and are slightly repelled by an applied magnetic field. Atoms, ions, and molecules containing at least one unpaired electron are **paramagnetic** and are attracted by an external magnetic field. (Section 5.7)

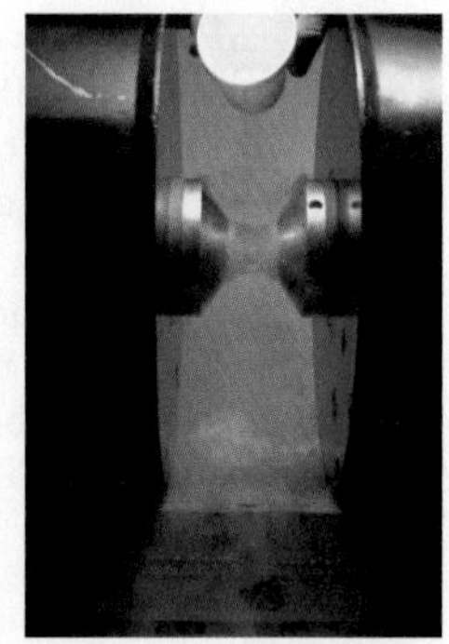

PARTICULATE PREVIEW WRAP-UP

- Both carbon atoms in halothane have a tetrahedral molecular geometry, so the F—C—F bond angles are 109.5°.
- The C—H bond is formed from the overlap of the C sp^3 hybrid orbital and the H 1*s* orbital.
- The two molecules shown here are not the same molecule; they are enantiomers of one another (note the orientation of the H, Cl, and Br atoms around carbon).

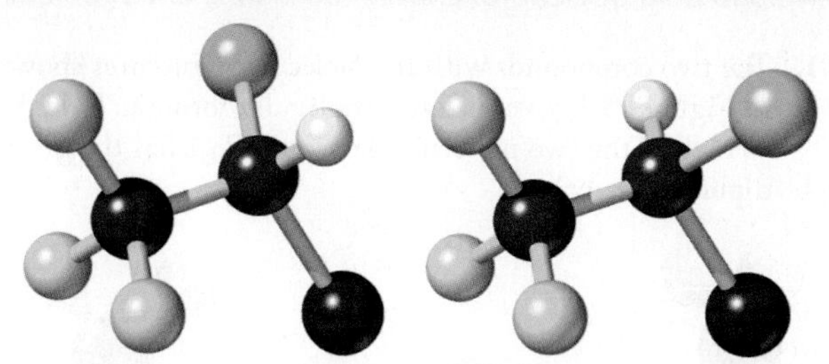

PROBLEM-SOLVING SUMMARY

Type of Problem	Concepts and Equations	Sample Exercises
Predicting molecular geometry	Draw the Lewis structure of the molecule. Determine the steric number (SN) of the central atom, where $$\text{SN} = \left(\begin{array}{c}\text{number of atoms}\\ \text{bonded to central atom}\end{array}\right) + \left(\begin{array}{c}\text{number of lone pairs}\\ \text{on central atom}\end{array}\right) \quad (5.1)$$ Choose the electron-pair geometry that corresponds to the SN. Choose a molecular geometry based on the electron-pair geometry that accounts for the number of lone pairs of valence electrons on the central atom.	**5.1, 5.3**
Predicting bond angles	Repulsion from lone pairs on a central atom pushes bonding pairs closer together and decreases bond angles.	**5.2**
Predicting whether or not a substance is polar	A substance is polar if a molecule of it has an overall permanent dipole—that is, if its bond dipoles and the locations of its lone pairs do not offset each other.	**5.4**
Using valence bond theory to explain molecular shape	Translate the observed electron-pair geometry into the appropriate central-atom hybridization following these guidelines: (see table below)	**5.5, 5.6**
Recognizing chiral molecules	Look for an sp^3 carbon atom that is bonded to four different atoms or groups of atoms.	**5.7**
Using MO theory to predict bond order	Draw the molecular orbital diagram and count the number of electrons in bonding and antibonding orbitals. $$\text{Bond order} = \tfrac{1}{2}\left[\left(\begin{array}{c}\text{number of}\\ \text{bonding electrons}\end{array}\right) - \left(\begin{array}{c}\text{number of}\\ \text{antibonding electrons}\end{array}\right)\right] \quad (5.2)$$	**5.8–5.10**

Steric Number	Electron-Pair Geometry	Hybridization
2	Linear	sp
3	Trigonal planar	sp^2
4	Tetrahedral	sp^3
5	Trigonal bipyramidal	sp^3d
6	Octahedral	sp^3d^2

VISUAL PROBLEMS

(Answers to boldface end-of-chapter questions and problems are in the back of the book.)

5.1. The two compounds with the molecular structures shown in Figure P5.1 have the same molecular formula: $C_2H_3F_3$. Which of the two molecules in Figure P5.1 has the greater dipole moment?

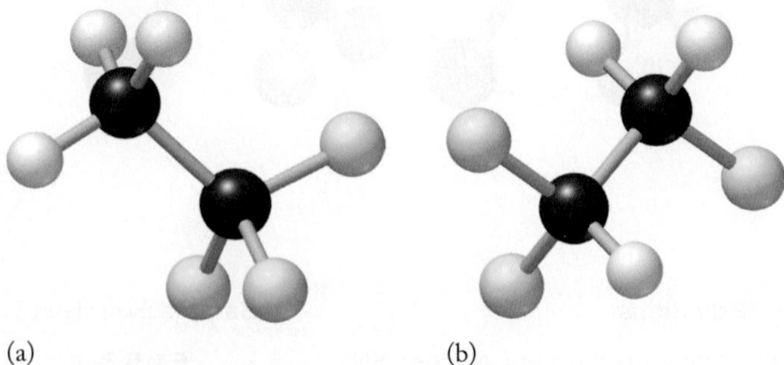

FIGURE P5.1

5.2. Could you distinguish between the two structures of N_2H_2 shown in Figure P5.2 by the magnitude of their dipole moments?

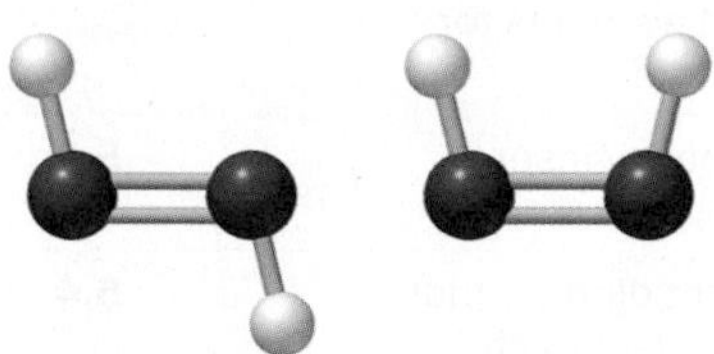

FIGURE P5.2

5.3. In which of the molecules shown in Figure P5.3 are all the atoms in a single plane? Are there delocalized π electrons in any of the molecules?

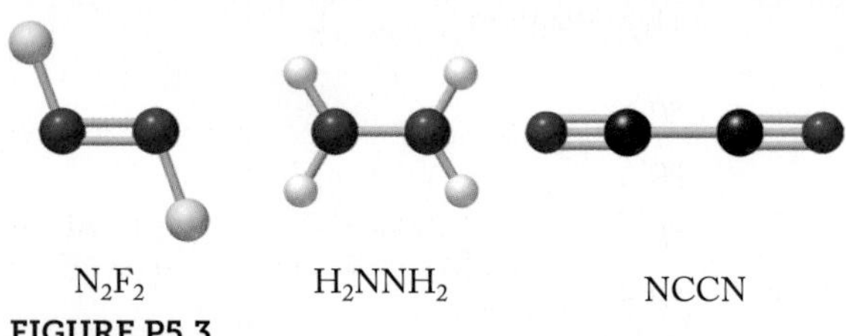

FIGURE P5.3

5.4. The ball-and-stick model in Figure P5.4 shows the skeletal structure of C_4H_4.

a. Draw the Lewis structure(s) of C_4H_4.
b. Are all the atoms in a single plane?
c. Are there delocalized π electrons in a molecule of C_4H_4?

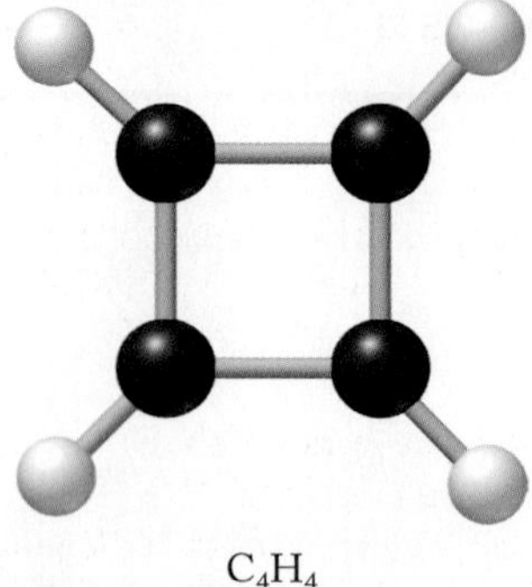

FIGURE P5.4

5.5. Use the MO diagram in Figure P5.5 to predict whether O_2^+ has more or fewer electrons in antibonding molecular orbitals than O_2^{2+}.

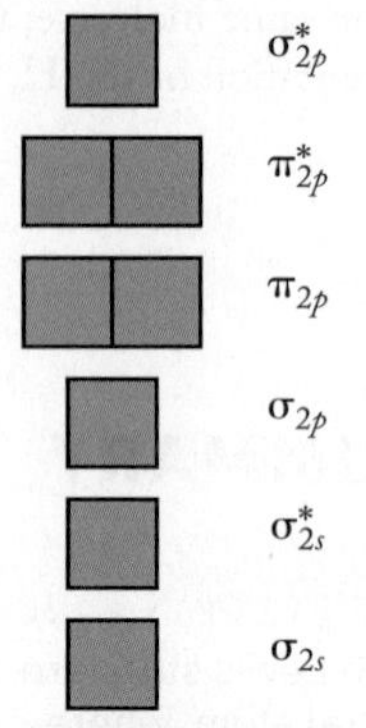

FIGURE P5.5

5.6. Under appropriate reaction conditions, diatomic molecules of iodine are ionized, forming I_2^+ cations. The corresponding anion, I_2^-, is unknown. Use the molecular orbital diagram in Figure P5.6 to explain why I_2^+ is more stable than I_2^-.

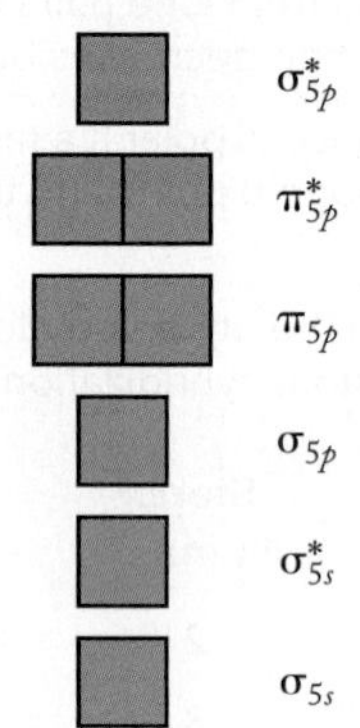

FIGURE P5.6

5.7. Figure P5.7 shows the molecular structure of a constituent of pine oil that contributes to its characteristic aroma. Is the compound chiral? Explain your answer.

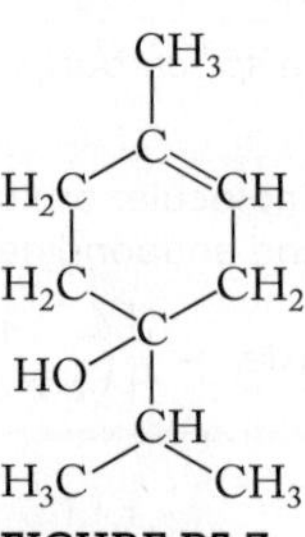

FIGURE P5.7

5.8. Figure P5.8 shows the molecular structure of menthol, a chiral compound that gives mint leaves their characteristic aroma. Locate the stereocenter(s) in the structure.

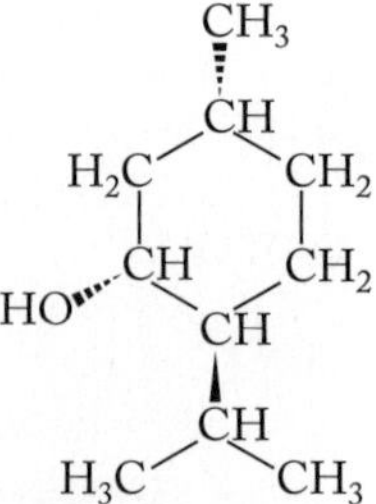

FIGURE P5.8

5.9. The molecular geometry of ReF_7 is an uncommon structure called a pentagonal bipyramid, which is shown in Figure P5.9. What are the bond angles in a pentagonal bipyramid?

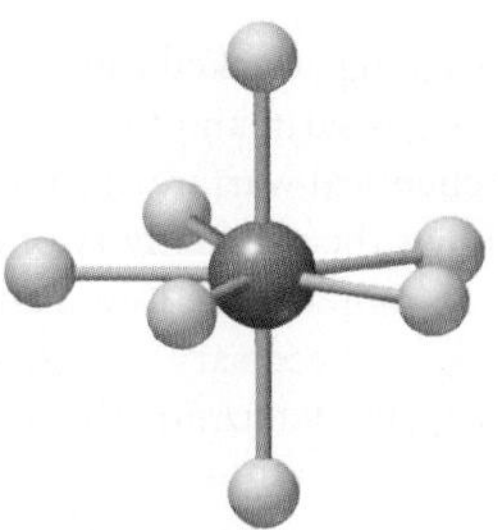

FIGURE P5.9

5.10. Use representations [A] through [I] in Figure P5.10 to answer questions a–f.
a. Which compounds contain delocalized π electrons?
b. Which compounds contain a π bond but not delocalized π electrons?
c. Of [B], [E], and [H], which is/are planar?
d. Locate the chiral carbon atom(s).
e. Compare the hybridization of the sulfur atoms in [D] and [G].
f. [C] is a CFC (chlorofluorocarbon) responsible for the destruction of the ozone layer. How is [F] related to [C]?

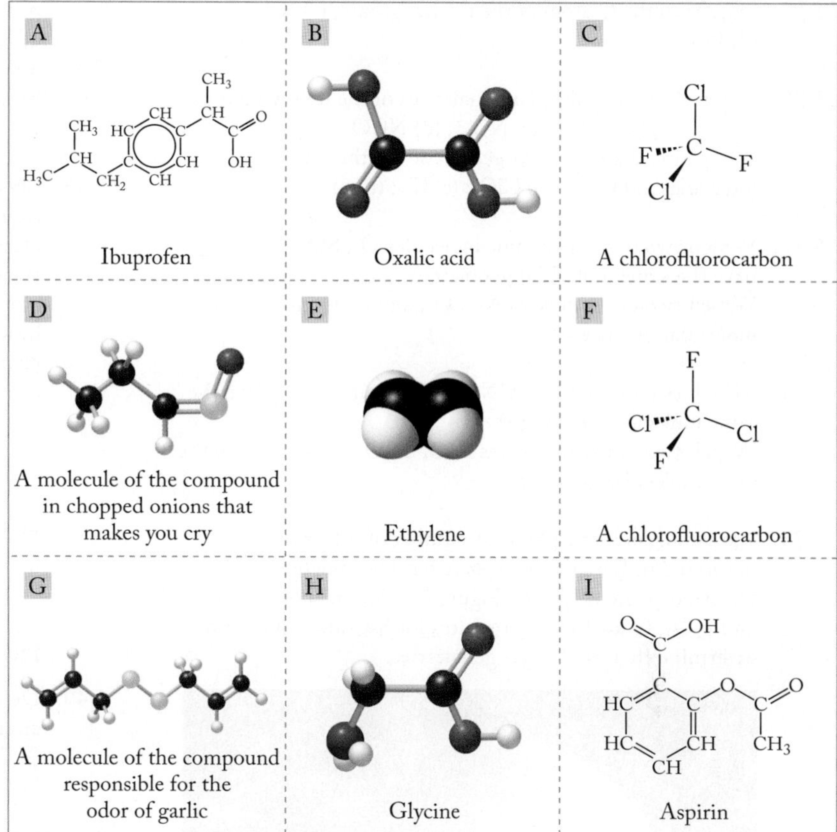

FIGURE P5.10

QUESTIONS AND PROBLEMS

Molecular Shape; Valence-Shell Electron-Pair Repulsion Theory (VSEPR)

Concept Review

5.11. Why is the shape of a molecule determined by repulsions between electron pairs and not by repulsions between nuclei?

5.12. In which molecular geometry do equatorial bonding pairs of electrons repel each other more: square pyramidal or trigonal bipyramidal?

5.13. Why do NO_3^- and NO_2^- ions have similar O—N—O bond angles, even though they have different numbers of N–O bonds?

5.14. Why do CF_4, SF_4, and XeF_4 have different molecular geometries, even though they all consist of a central atom bonded to four fluorine atoms?

5.15. Why are the bond angles in BH_3 and NH_3 different, even though they both consist of a central atom bonded to three hydrogen atoms?

5.16. The O–N–O bond angle in NO_2^+ is larger than it is in NO_2. Why?

5.17. Why are the H—C—H bond angles in molecules of CH_4 smaller than the H—C—H bond angles in molecules of CH_2O?

5.18. Why do we need to draw the Lewis structure of a molecule before predicting its geometry?

5.19. Why does the seesaw structure have lower energy than a trigonal pyramidal structure when SN = 5?

*5.20. Do all resonance forms of a molecule have the same molecular geometry? Explain your answer.

Problems

5.21. Rank the following molecules in order of increasing bond angles: (a) NH_2Cl; (b) CCl_4; (c) H_2S.

5.22. Rank the following molecular geometries in order of increasing bond angles: (a) trigonal pyramidal; (b) trigonal planar; (c) square planar.

5.23. Which of the following electron-pair geometries is not consistent with a linear molecular geometry, assuming three atoms per molecule? (a) tetrahedral; (b) octahedral; (c) trigonal planar

5.24. How many lone pairs of electrons would there have to be on a SN = 6 central atom for it to have a linear molecular geometry?

5.25. Determine the molecular geometries of the following molecules: (a) GeH_4; (b) PH_3; (c) H_2S; (d) $CHCl_3$.

5.26. Determine the molecular geometries of the following molecules and ions: (a) NO_3^-; (b) NO_4^{3-}; (c) NCN^{2-}; (d) NF_3.

5.27. Determine the molecular geometries of the following ions: (a) NH_4^+; (b) CO_3^{2-}; (c) NO_2^-; (d) XeF_5^+.

5.28. Determine the molecular geometries of the following ions: (a) SCN^-; (b) $CH_3PCl_3^+$ (P is the central atom and is bonded to the C atom of the methyl group); (c) ICl_2^-; (d) PO_3^{3-}.

5.29. Determine the molecular geometries of the following ions: (a) $S_2O_3^{2-}$; (b) PO_4^{3-}; (c) NO_3^-; (d) NCO^-.

5.30. Determine the molecular geometries of the following molecules: (a) ClO_2; (b) ClO_3; (c) IF_3; (d) SF_4.

5.31. Which two of the triatomic molecules O_3, SO_2, and CO_2 have the same molecular geometry?

5.32. Which two of the species N_3^-, O_3, and CO_2 have the same molecular geometry?

5.33. Which two of the ions SCN^-, CNO^-, and NO_2^- have the same molecular geometry?

5.34. Which two of the molecules N_2O, Se_2O, and CO_2 have the same molecular geometry?

5.35. The Venusian Atmosphere A number of sulfur oxides not found in Earth's atmosphere have been detected in the atmosphere of Venus (Figure P5.35), including S_2O and S_2O_2. Draw Lewis structures for S_2O and S_2O_2, and determine their molecular geometries.

FIGURE P5.35

5.36. The structures of NOCl, NO_2Cl, and NO_3Cl were determined in 1995. They have the skeletal structures shown in Figure P5.36. Draw Lewis structures for the three compounds and predict the electron-pair geometry at each nitrogen atom.

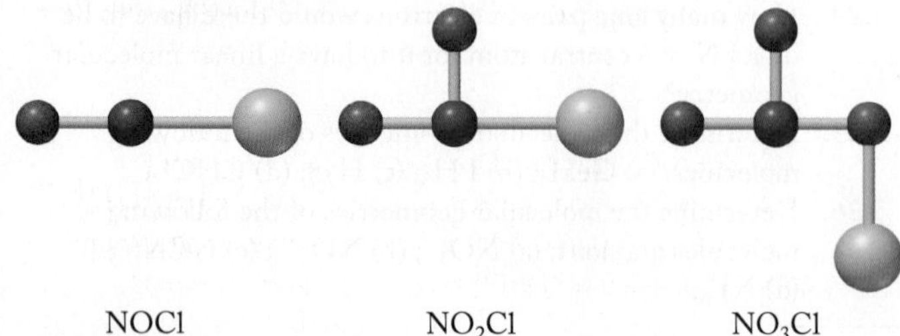

FIGURE P5.36

***5.37.** For many years, it was believed that the noble gases could not form covalently bonded compounds. However, xenon does react with fluorine. One of the products is the pentafluoroxenate anion, XeF_5^-. Draw the Lewis structure of XeF_5^- and predict its geometry.

*5.38. The first compound containing a xenon–sulfur bond was isolated in 1998. Draw a Lewis structure for HXeSH and determine its molecular geometry.

***5.39. Chemical Terrorism** In 1995 a gang attacked the Tokyo subway system with the nerve gas sarin and focused world attention on the dangers of chemical warfare agents. The structure in Figure P5.39 shows the connectivity of the atoms in the Sarin molecule. Complete the Lewis structure by adding bonds and lone pairs as necessary. Assign formal charges to the P and O atoms, and determine the molecular geometry around P.

```
                H
                |
 H    O    H—C—H
 |    |         |
H—C—P—O—C—H
 |    |         |
 H    F    H—C—H
                |
                H
```

FIGURE P5.39

5.40. Determine the electron-pair geometries around the nitrogen atoms in the following unstable nitrogen oxides: (a) N_2O_2; (b) N_2O_5; (c) N_2O_3. (N_2O_2 and N_2O_3 have N—N bonds; N_2O_5 does not.)

Polar Bonds and Polar Molecules

Concept Review

5.41. Explain the difference between a polar bond and a polar molecule.

5.42. Must a polar molecule contain polar covalent bonds? Why?

5.43. Can a nonpolar molecule contain polar covalent bonds?

5.44. What does a dipole moment measure?

Problems

5.45. Vinyl chloride (Figure P5.45) is used to make plastics. Which bond in vinyl chloride is most polar?

```
 H
  \
   C=CH2
  /
:Cl:
```

FIGURE P5.45

5.46. Which bond in Figure P5.45 has $\mu = 0$?

5.47. The following molecules contain polar covalent bonds. Which of them are polar molecules and which have no permanent dipoles? (a) CCl_4; (b) $CHCl_3$; (c) CO_2; (d) H_2S; (e) SO_2

5.48. Which of the following molecules has a permanent dipole? (a) C≡O; (b) N≡N; (c) H—C≡N; (d) H—C≡C—H

5.49. Compounds containing carbon, chlorine, and fluorine are known as chlorofluorocarbons (CFCs). Which of the following CFCs are polar and which are nonpolar? (a) $CFCl_3$; (b) CF_2Cl_2; (c) Cl_2FCCF_2Cl

5.50. Which of the following molecules has a permanent dipole? (a) C_4F_8 (cyclic structure); (b) $ClFCCF_2$; (c) $Cl_2HCCClF_2$

5.51. Predict which molecule in each of the following pairs is more polar: (a) $CClF_3$ or $CBrF_3$; (b) CF_2Cl_2 or CHF_2Cl.

5.52. Which molecule in each of the following pairs is more polar? (a) NH_3 or PH_3; (b) CCl_2F_2 or CBr_2F_2

5.53. **Chemical Warfare Gas** A compound with the formula $COCl_2$ has been used as a chemical warfare agent. It and two similar compounds, $COBr_2$ and COI_2, are all eye irritants and cause skin to blister. The severity of the skin reactions is influenced by the polarity of the compounds. Rank the compounds in order of increasing polarity of their C—X bonds (where X is a halogen atom).

5.54. **Interstellar Space** Among the diatomic molecules detected in interstellar space are CO, CS, SiO, SiS, SO, and NO. Arrange the molecules in order of increasing polarity.

Valence Bond Theory and Hybrid Orbitals

Concept Review

5.55. What must atomic orbitals have in common to mix together and form hybrid orbitals?

5.56. Why do atomic orbitals form hybrid orbitals?

Problems

5.57. What is the hybridization of the underlined carbon atom in each of these condensed structural formulas? (a) $CH_3—\underline{C}H_2—CH_3$; (b) $CH_2{=}\underline{C}H—CH_3$; (c) $CH_3—\underline{C}H(CH_3)—CH_3$; (d) $\underline{C}H{\equiv}C—CH_3$; (e) $CH{\equiv}\underline{C}—CH_3$.

5.58. What is the hybridization of nitrogen in each of the following ions and molecules? (a) NO_2^+; (b) NO_2^-; (c) N_2O; (d) N_2O_5; (e) N_2O_3

5.59. N_2F_2 has the two possible structures shown in Figure P5.59. Are the differences between the structures related to differences in the hybridization of nitrogen in N_2F_2? Identify the hybrid orbitals that account for the bonding in N_2F_2. Are they the same as those in acetylene, C_2H_2?

FIGURE P5.59

5.60. **Air Bag Chemistry** Azides such as sodium azide, NaN_3, are used in automobile air bags as a source of nitrogen gas. Another compound with three nitrogen atoms bonded together is N_3F. What differences are there between the arrangements of the electrons around the nitrogen atoms in the azide ion (N_3^-) and in N_3F? Is there a difference in the hybridization of the central nitrogen atom?

5.61. How does the hybridization of the central atom change in the series CO_2, NO_2, O_3, ClO_2?

5.62. How does the hybridization of the sulfur atom change in the series SF_2, SF_4, SF_6?

***5.63.** **Perchlorate Ion and Human Health** Perchlorate compounds adversely affect human health by interfering with the uptake of iodine in the thyroid gland. Because of that behavior, though, they are also used to treat hyperthyroidism, or overactive thyroid. Draw the Lewis structure(s) of the perchlorate ion, ClO_4^-, including all resonance forms, in which formal charges are closest to zero. What is the shape of the ion? Suggest a hybridization scheme for the central chlorine atom that accounts for this shape.

*5.64. **Bleaching Agents** Draw the Lewis structure of the chlorite ion, ClO_2^-, which is used as a bleaching agent. Include all resonance structures in which formal charges are closest to zero. What is the shape of the ion? Suggest a hybridization scheme for the central chlorine atom that accounts for the structures you have drawn.

5.65. Synthesis of the first compound of argon was reported in 2000. HArF was made by reacting Ar with HF. Draw the Lewis structure for HArF and determine the hybridization of Ar in this molecule.

5.66. Draw a Lewis structure for Cl_3^+. Determine its molecular geometry and the hybridization of the central Cl atom.

5.67. Do all resonance forms of N_2O have the same hybridization at the central N atom?

5.68. The Lewis structure for N_4O, with the skeletal structure O-N-N-N-N, contains one N—N single bond, one N=N double bond, and one N≡N triple bond. Is the hybridization of all the nitrogen atoms the same?

***5.69.** The trifluorosulfate anion was isolated in 1999 as the tetramethylammonium salt $(CH_3)_4NSOF_3$. Determine the geometry around sulfur in the anion and describe the bonding according to valence bond theory.

*5.70. Several resonance forms can be drawn for the anion $[C(CN)_3]^-$, including the two structures shown in Figure P5.70. Do they have the same geometry about the central carbon? What is the hybridization of each carbon atom?

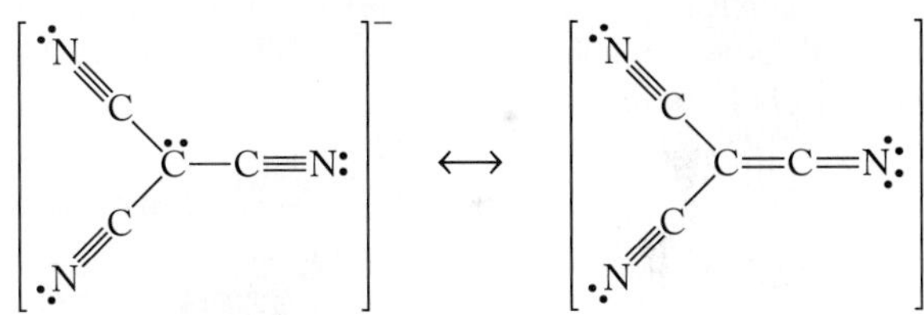

FIGURE P5.70

Molecules with Multiple "Central" Atoms

Concept Review

5.71. Can molecules with more than one central atom have resonance forms?

5.72. Why is it difficult to assign a single geometry to a molecule with more than one central atom?

***5.73.** Are resonance structures examples of electron delocalization? Explain your answer.

*5.74. Can hybrid orbitals be associated with more than one atom?

Problems

5.75. The two nitrogen atoms in nitramide are connected with two oxygen atoms on one terminal nitrogen and two hydrogen atoms on the other (Figure P5.75). What is the molecular geometry of each nitrogen atom in nitramide? Is the hybridization of both nitrogen atoms the same?

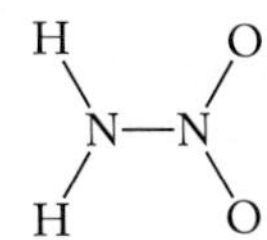

FIGURE P5.75

5.76. Cyclic structures exist for many compounds of carbon and hydrogen. Describe the molecular geometry and hybridization around each carbon atom in cyclopropene (C_3H_4), cyclobutane (C_4H_8), cyclobutene (C_4H_6), and cyclobutadiene (C_4H_4) (Figure P5.76).

Cyclopropene Cyclobutane Cyclobutene Cyclobutadiene

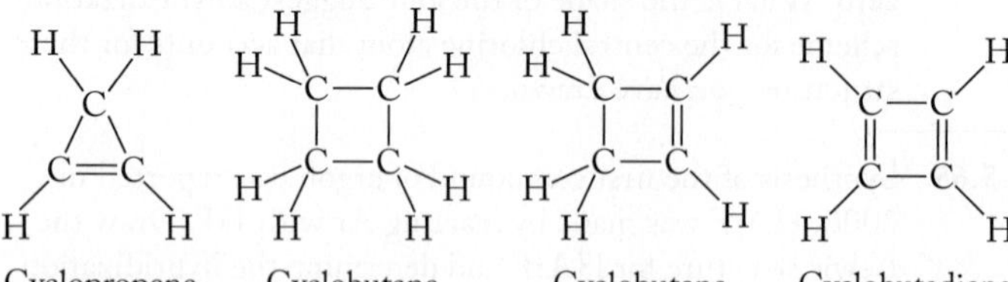

FIGURE P5.76

5.77. What is the molecular geometry around sulfur and nitrogen in the sulfamate anion shown in Figure P5.77? Which atomic or hybrid orbitals overlap to form the S—O and S—N bonds in the sulfamate anion?

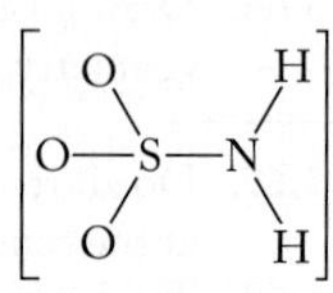

FIGURE P5.77

5.78. What is the geometry around each sulfur atom in the disulfate anion shown in Figure P5.78? What is the hybridization of the central oxygen atom?

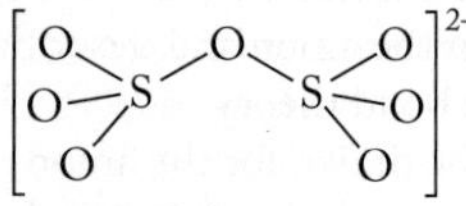

FIGURE P5.78

5.79. Butter The molecular structure of a compound that contributes to the flavor of butter is shown in Figure P5.79. What is the molecular geometry and bond angle at each carbon atom?

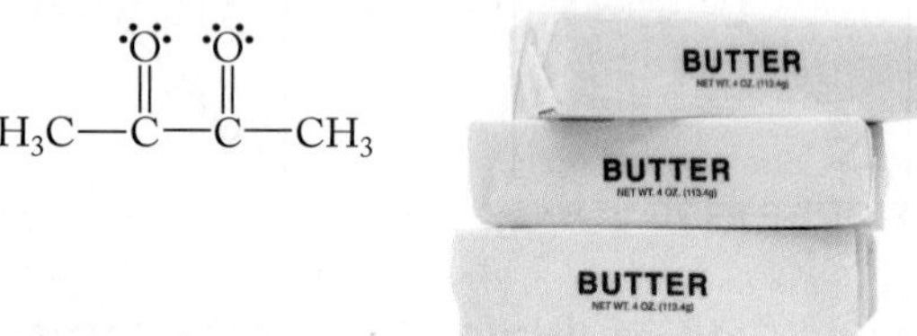

FIGURE P5.79

5.80. What is the hybridization of each carbon atom in Figure P5.79?

Chirality and Molecular Recognition

Concept Review

5.81. Which of the following objects are chiral? (a) a golf club; (b) a spoon; (c) a glove; (d) a shoe

5.82. Which of the following objects are chiral? (a) a key; (b) a screwdriver; (c) a fluorescent coil lightbulb; (d) a baseball

5.83. Could an *sp* hybridized carbon atom be a chiral center? Explain your answer.

5.84. Two compounds have the same Lewis structure and the same optical activity. Are they enantiomers or the same compound?

5.85. Are racemic mixtures homogeneous or heterogeneous?

5.86. Can a mixture of enantiomers rotate plane-polarized light? Explain your answer.

Problems

5.87. Which of the molecules in Figure P5.87 are chiral?

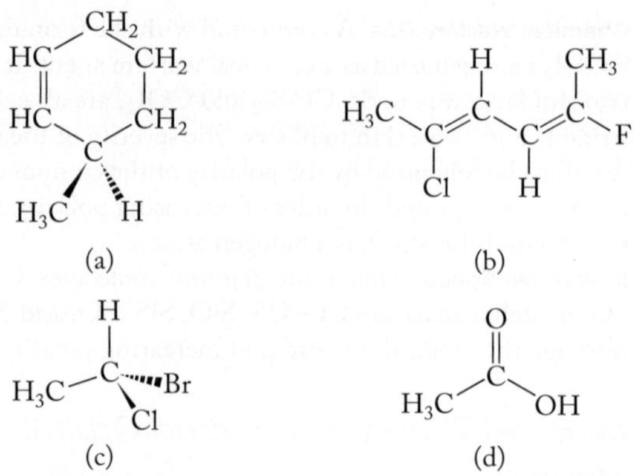

FIGURE P5.87

5.88. Which of the molecules in Figure P5.88 are chiral?

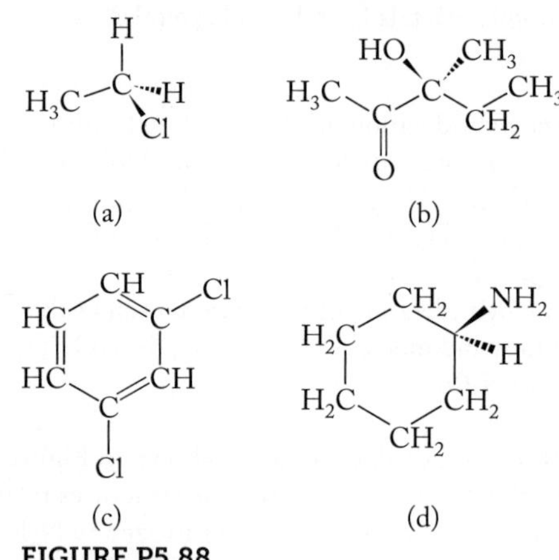

FIGURE P5.88

5.89. Which of the molecules in Figure P5.89 are chiral?

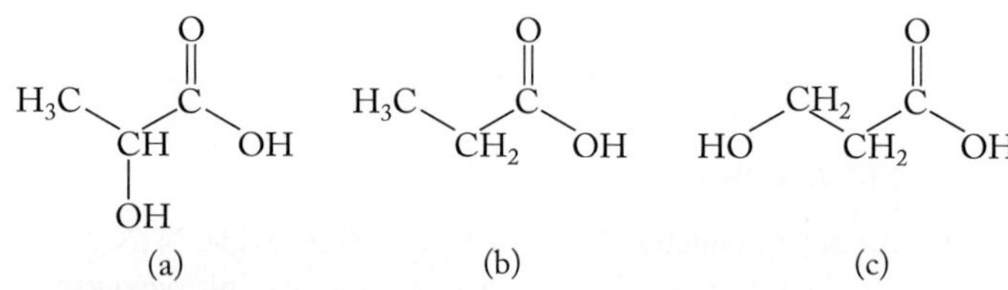

FIGURE P5.89

5.90. Which of the molecules in Figure P5.90 are chiral?

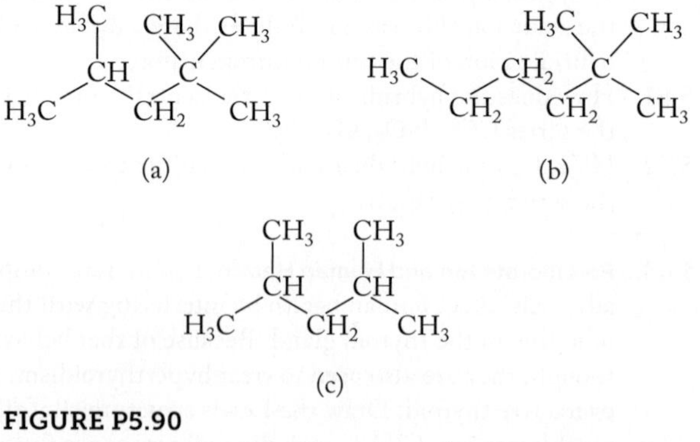

FIGURE P5.90

5.91. Artificial Sweeteners Figure P5.91 shows three artificial sweeteners that have been used in food and beverages. Saccharin is the oldest, dating to 1879. Cyclamates were banned in the United States in 1969 but are still available in over 50 other countries. Aspartame may be more familiar to you by the brand name NutraSweet. Each sweetener contains between zero and two chiral carbon atoms. Circle the chiral centers in each compound.

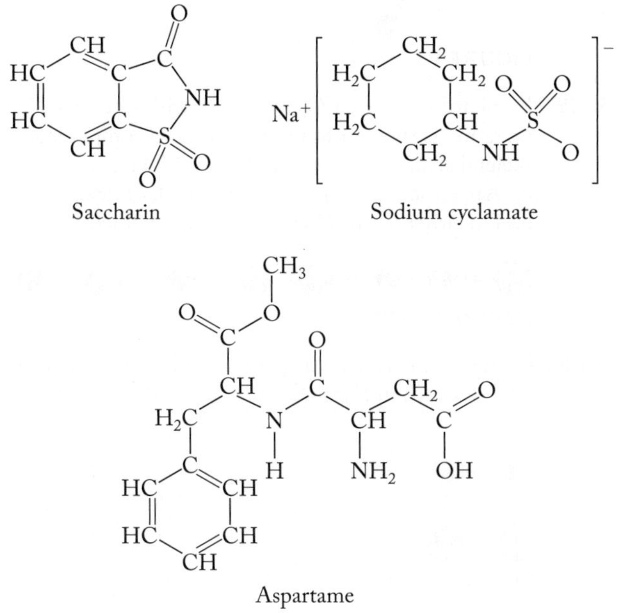

Saccharin

Sodium cyclamate

Aspartame

FIGURE P5.91

5.92. Identify the chiral centers in each of the molecules in Figure P5.92.

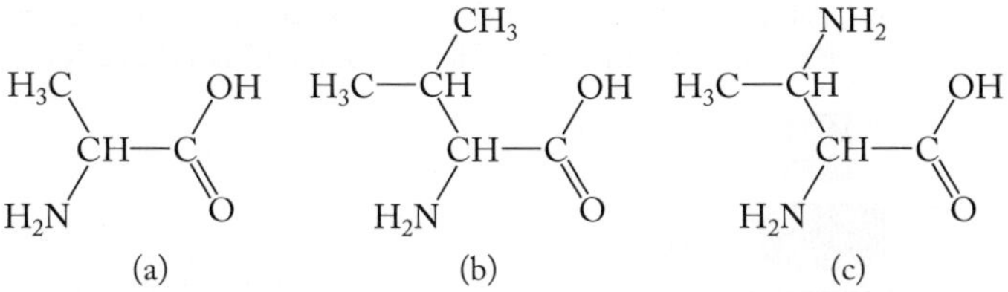

(a) (b) (c)

FIGURE P5.92

5.93. The Smell of Raspberries The compound with the structure in Figure P5.93 is a major contributor to the aroma of ripe raspberries. Identify any chiral center(s).

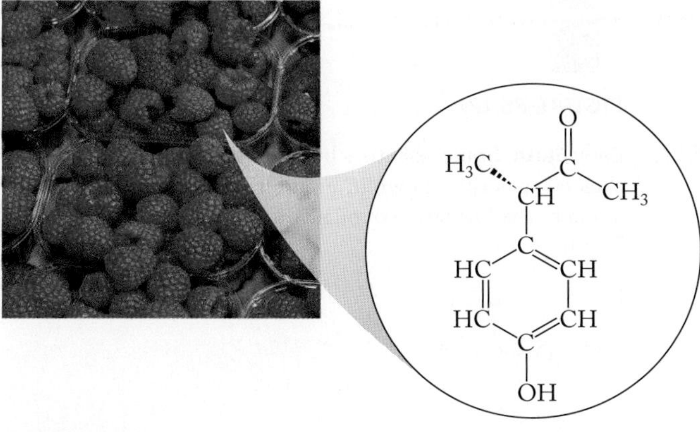

FIGURE P5.93

5.94. **Antidepressants** Figure P5.94 shows the structure of the drug bupropion, which is used to treat depression. Identify any chiral center(s).

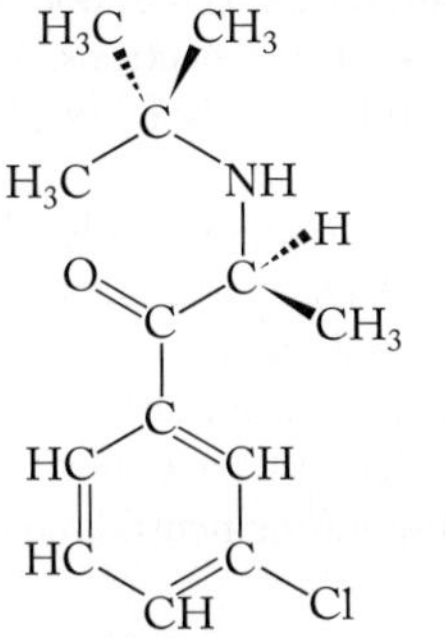

FIGURE P5.94

Molecular Orbital Theory

Concept Review

5.95. Which better explains the visible emission spectra of molecular substances: valence bond theory or molecular orbital theory?

5.96. Which better explains the magnetic properties of molecular substances: valence bond theory or molecular orbital theory?

5.97. Do all σ molecular orbitals result from the overlap of s atomic orbitals?

5.98. Do all π molecular orbitals result from the overlap of p atomic orbitals?

5.99. Are s atomic orbitals with different principal quantum numbers (n) as likely to overlap and form MOs as s atomic orbitals with the same value of n?

5.100. Are a $2p_x$ atomic orbital and a $2p_y$ atomic orbital on adjacent molecules as likely to form MOs as two $2p_z$ atomic orbitals on the same two atoms?

Problems

5.101. Make a sketch showing how two 1s orbitals overlap to form a σ_{1s} bonding molecular orbital and a σ^*_{1s} antibonding molecular orbital.

5.102. Make a sketch showing how two $2p_y$ orbitals overlap "sideways" to form a π_{2p} bonding molecular orbital and a π^*_{2p} antibonding molecular orbital.

5.103. Use MO theory to predict the bond orders of the following molecular ions: N_2^+, O_2^+, C_2^+, and Br_2^{2-}. Do you expect any of the species to exist?

5.104. Diatomic noble gas molecules, such as He_2 and Ne_2, do not exist. Would removing an electron create molecular ions, such as He_2^+ and Ne_2^+, that are more stable than He_2 and Ne_2?

5.105. Which of the following molecular ions are paramagnetic? (a) N_2^+; (b) O_2^+; (c) C_2^{2+}; (d) Br_2^{2-}

5.106. Which of the following molecular ions are diamagnetic? (a) O_2^-; (b) O_2^{2-}; (c) N_2^{2-}; (d) F_2^+

5.107. Which of the following molecular anions have electrons in π antibonding orbitals? (a) C_2^{2-}; (b) N_2^{2-}; (c) O_2^{2-}; (d) Br_2^{2-}

5.108. Which of the following molecular cations have electrons in π antibonding orbitals? (a) N_2^+; (b) O_2^+; (c) C_2^{2+}; (d) Br_2^{2+}

5.109. For which of the following diatomic molecules does the bond order increase with the gain of two electrons, forming the corresponding 2− anion? (a) B_2; (b) C_2; (c) N_2; (d) O_2

5.110. For which of the following diatomic molecules does the bond order increase with the loss of two electrons, forming the corresponding 2+ cation? (a) B_2; (b) C_2; (c) N_2; (d) O_2

5.111. Do the 1+ cations of homonuclear diatomic molecules of the second-row elements always have shorter bond lengths than the corresponding neutral molecules?

5.112. Do any of the anions of the homonuclear diatomic molecules formed by B, C, N, O, and F have shorter bond lengths than those of the corresponding neutral molecules? Consider only the 1− and 2− anions.

Additional Problems

5.113. Urea is the principal nitrogen-containing molecule in mammalian urine. It is also found in blood. Elevated concentrations of urea in blood may indicate kidney failure. Draw the Lewis structure of urea that corresponds to Figure P5.113. Is urea a planar molecule?

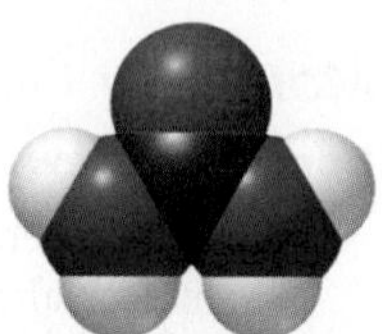

FIGURE P5.113

5.114. **Ant Venom** Formic acid occurs naturally in the venom of stinging ants. Draw the Lewis structure of formic acid that corresponds to Figure P5.114. Is formic acid a planar molecule?

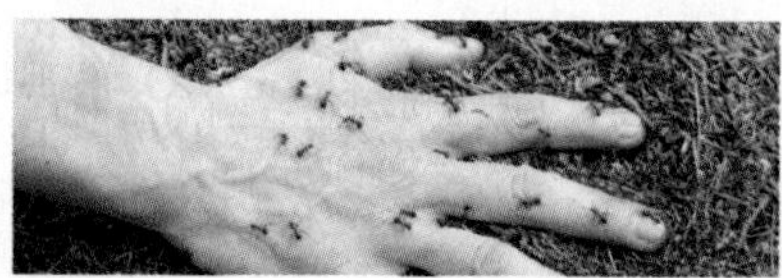

FIGURE P5.114

5.115. **Rocket Propellants** Draw the Lewis structure for the two ions in ammonium perchlorate (NH_4ClO_4), which is used as a propellant in solid fuel rockets, and determine the molecular geometries of the two polyatomic ions.

5.116. **Pressure-Treated Lumber** By December 31, 2003, concerns over arsenic contamination had prompted the manufacturers of pressure-treated lumber to voluntarily cease producing lumber treated with CCA (chromated copper arsenate) for residential use. CCA-treated lumber has a light greenish color and was widely used to build decks, sand boxes, and playground structures. Draw the Lewis structure for the arsenate ion (AsO_4^{3-}) that yields the most favorable formal charges. Predict the angles between the arsenic–oxygen bonds in the arsenate anion.

5.117. Consider the molecular structure in Figure P5.117. What is the angle formed by the N—C—C bonds in the structure? What are the O═C—O and C—O—H bond angles?

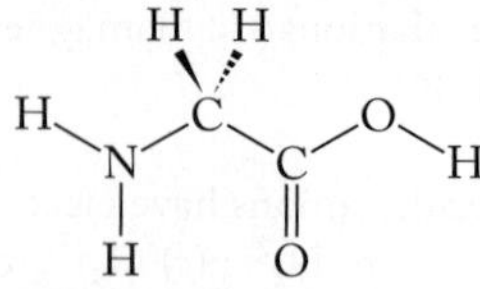

FIGURE P5.117

5.118. Cl_2O_2 may play a role in ozone depletion in the stratosphere. Draw the Lewis structure for Cl_2O_2 based on the skeletal structure in Figure P5.118. What is the geometry about the central chlorine atom?

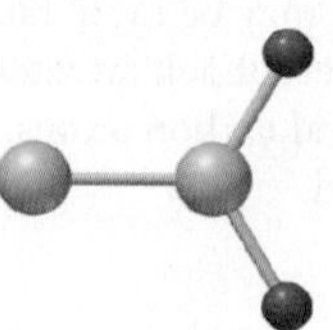

FIGURE P5.118

5.119. Bombardment of Cl_2O_2 molecules with intense UV radiation is thought to produce the two compounds with the skeletal structures shown in Figure P5.119.

a. Are either or both of these molecules linear?
b. Do either or both have a permanent dipole?

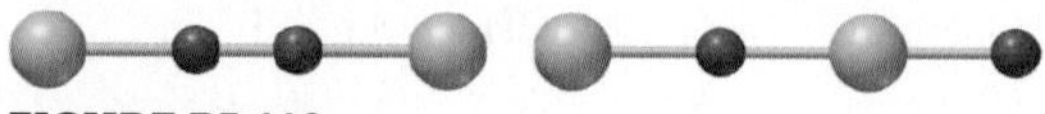

FIGURE P5.119

*5.120. Complete the Lewis structure for the cyclic structure of Cl_2O_2 shown in Figure P5.120. Is the cyclic Cl_2O_2 molecule planar?

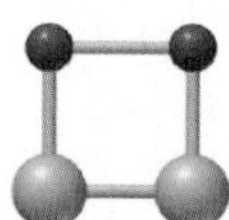

FIGURE P5.120

5.121. **Ozone Depletion** In 1999, the ClO^+ ion, a potential contributor to stratospheric ozone depletion, was isolated in the laboratory.

a. Draw the Lewis structure for ClO^+.
b. Using the molecular orbital diagram for ClO^+ in Figure P5.121, determine the order of the Cl–O bond in ClO^+.

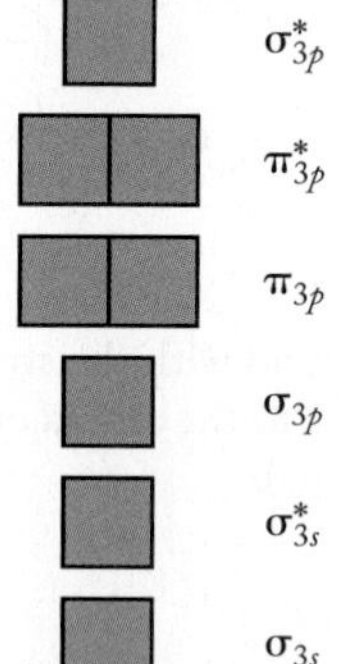

FIGURE P5.121

5.122. **Early Earth** Some scientists believe that an anion with the skeletal structure shown in Figure P5.122 may have played a role in the formation of nucleic acids before life existed on Earth.

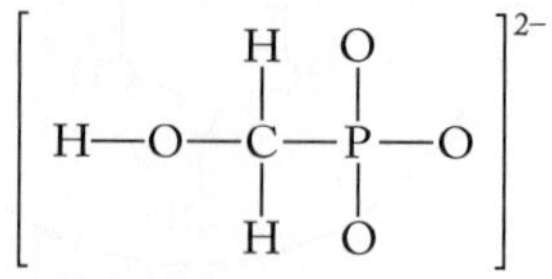

```
        H   O      2−
        |   |
[ H—O—C—P—O ]
        |   |
        H   O
```

FIGURE P5.122

a. Complete the Lewis structure of this anion.
b. Predict the C—P—O bond angle in the anion.

5.123. **Cola Beverages** Phosphoric acid imparts a tart flavor to cola beverages. The structure of phosphoric acid is shown in Figure P5.123. Complete the Lewis structure for phosphoric acid in which formal charges are closest to zero. What is the molecular geometry around the phosphorus atom in your structure?

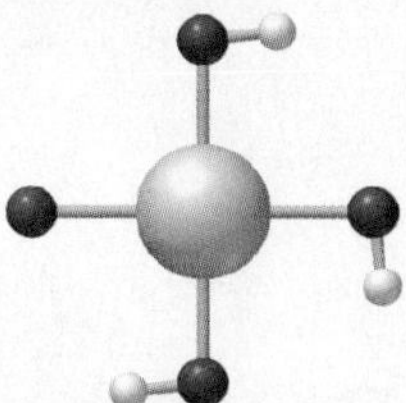

FIGURE P5.123

5.124. Fluoroaluminate anions AlF_4^- and AlF_6^{3-} have been known for over a century, but the structure of the pentafluoroaluminate ion, AlF_5^{2-}, was not determined until 2003. Draw the Lewis structures for AlF_3, AlF_4^-, AlF_5^{2-}, and AlF_6^{3-}. Determine the molecular geometry of each molecule or ion. Describe the bonding in AlF_3, AlF_4^-, AlF_5^{2-}, and AlF_6^{3-} using valence bond theory.

5.125. **Rocket Fuel** Hydrazine, NH_2NH_2, fueled the spacecraft that delivered the Mars rover *Curiosity* to the planet's surface in 2012.
a. Draw the Lewis structure of hydrazine.
b. What is the hybridization of its N atoms?
c. Is hydrazine a polar compound? Explain why or why not.

5.126. Two compounds formed by the reaction of boron with carbon monoxide have the following skeletal structures: B—B—C—O and O—C—B—B—C—O.
a. Draw the Lewis structures of both compounds that minimize formal charges.
b. What are the B–B–C bond angles in the molecules?

***5.127.** Boron reacts with NO, forming a compound with the formula BNO.
a. Draw the Lewis structure for BNO, including any resonance forms.
b. Assign formal charges and predict which structure provides the best description of the bonding in this molecule.
c. Predict the molecular geometry of BNO.

5.128. Borazine, $B_3N_3H_6$ (a cyclic compound with alternating B and N atoms in the ring), is isoelectronic with benzene (C_6H_6). Are there delocalized π electrons in borazine?

***5.129.** **Compounds That May Help Prevent Cancer** Broccoli, cabbage, and kale contain compounds that break down in the human body to form isothiocyanates, whose presence may reduce the risk of certain types of cancer. The simplest isothiocyanate is methyl isothiocyanate, CH_3NCS.
a. Draw the Lewis structure for methyl isothiocyanate, including all resonance forms. *Hint*: The nitrogen atom is bonded to the methyl (CH_3—) group.
b. Assign formal charges and determine which structure is likely to contribute the most to bonding.
c. Predict the molecular geometry of the molecule at both carbon atoms.

*5.130. **Toxic to Insects and People** Methyl thiocyanate (CH_3SCN) is used as an agricultural pesticide and fumigant. It is slightly water soluble and is readily absorbed through the skin, but it is highly toxic if ingested. Its toxicity stems in part from its metabolism to cyanide ion.
a. Draw the Lewis structure for methyl thiocyanate, including all resonance forms.
b. Assign formal charges and predict which structure contributes the most to bonding.
c. Predict the molecular geometry of the molecule at both carbon atoms.

5.131. Some chemists think HArF consists of H^+ ions and ArF^- ions. Using an appropriate MO diagram, determine the bond order of the Ar–F bond in ArF^-.

*5.132. To model the bonding in SF_6 gas, some chemists assume the existence of SF_4^{2+} cations surrounded by two F^- ions.
a. Draw the Lewis structure of a SF_4^{2+} ion.
b. What are the formal charges on S and F in the structure you drew?
c. What is the shape of the ion?
d. Does the S atom have an expanded octet?

***5.133.** Which of the unstable nitrogen oxides N_2O_2, N_2O_5, and N_2O_3 are polar molecules? (N_2O_2 and N_2O_3 have N–N bonds; N_2O_5 does not.)

5.134. Hydrogen atoms have one electron. Does this mean that hydrogen gas is paramagnetic? Why or why not?

5.135. Draw the molecular orbital diagram of the valence shell of a F_2^+ ion, and use it to determine the bond order in the ion.

5.136. Use molecular orbital diagrams to determine the bond order of the peroxide (O_2^{2-}) and superoxide (O_2^-) ions. Are the bond order values consistent with those predicted from Lewis structures?

***5.137.** Trimethylamine, $(CH_3)_3N$, has a trigonal *pyramidal* structure, while trisilylamine, $(SiH_3)_3N$, has a trigonal *planar* geometry. Draw Lewis structures for both compounds consistent with the observed geometries and explain your reasoning.

5.138. Elemental sulfur has several allotropic forms, including cyclic S_8 molecules. What is the orbital hybridization of sulfur atoms in the S_8 allotrope? The bond angles are about 108°.

5.139. Using an appropriate molecular orbital diagram, show that the bond order in the disulfide anion, S_2^{2-}, is equal to 1. Is S_2^{2-} diamagnetic or paramagnetic?

*5.140. Ozone (O_3) has a small dipole moment (0.54 D). How can a molecule with only one kind of atom have a dipole moment?

smartwork**5**

If your instructor uses Smartwork5, log in at digital.wwnorton.com/atoms2.

6

Intermolecular Forces

Attractions between Particles

FLARING NATURAL GAS The crude oil produced on offshore wells often contains natural gas that is not collected but instead is vented into the air and burned.

PARTICULATE **REVIEW**

Polar Bonds versus Polar Molecules

In Chapter 6, we explore the connections between the structure of molecules and the ways they interact with each other. Here are representations of four molecules: carbon dioxide, oxygen, water, and ozone.

- Which molecule is polar and contains polar bonds?
- Which molecule is nonpolar despite containing polar bonds?
- Which molecule is polar despite containing nonpolar bonds?

(Review Sections 5.2 and 5.3 if you need help.)

(Answers to Particulate Review questions are in the back of the book.)

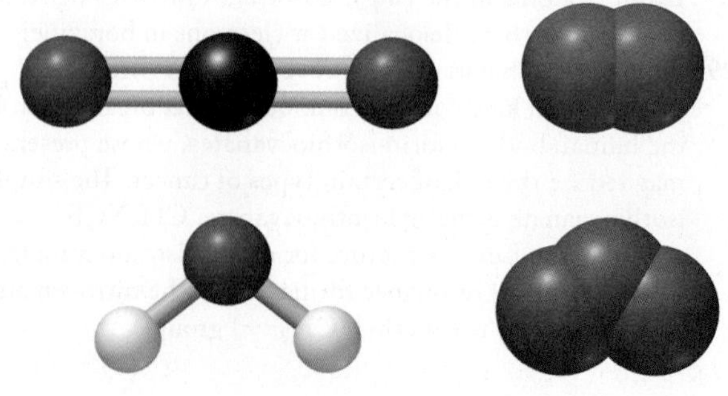

PARTICULATE **PREVIEW**

Bonds and Functional Groups

Each molecule shown here contains a carbon–oxygen bond. As you read Chapter 6, look for ideas that will help you answer these questions:

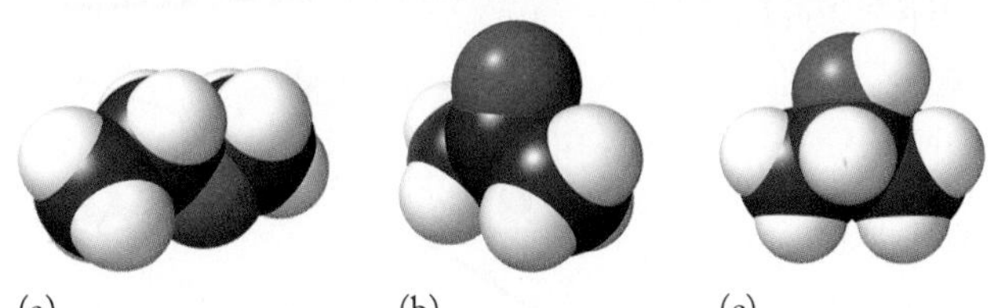

- Which molecule contains a carbon–oxygen double bond? What is the name of this functional group?
- What are the names of the functional groups in the other two molecules?
- Rank these three compounds from lowest boiling point to highest boiling point.

Learning Outcomes

LO1 Explain the origins of dispersion forces, ion–dipole forces, dipole–dipole forces, and hydrogen bonds and use these interactions to explain trends in boiling point, viscosity, and other physical properties of molecular compounds
Sample Exercise 6.1

LO2 Identify the solute(s) and solvent in a solution
Sample Exercise 6.2

LO3 Explain why ionic compounds and polar molecular compounds dissolve in polar solvents and nonpolar molecular compounds dissolve in nonpolar solvents
Sample Exercise 6.3

LO4 Identify the regions of a phase diagram and explain the effect of temperature and pressure on phase changes
Sample Exercise 6.4

LO5 Describe the unusual properties of water and how they relate to the hydrogen bonds formed by water molecules

hydrocarbon an organic compound whose molecules contain only carbon and hydrogen atoms.

alkane a hydrocarbon in which each carbon atom is bonded to four other carbon or hydrogen atoms.

temporary dipole the separation of charge produced in an atom or molecule by a momentary uneven distribution of electrons; also called *induced dipole*.

London dispersion force an intermolecular force between atoms or molecules caused by the presence of temporary dipoles in the molecules.

polarizability the relative ease with which the electron cloud in a molecule, ion, or atom can be distorted, inducing a temporary dipole.

6.1 London Dispersion Forces: They're Everywhere

The chapter-opening photo captures a common sight at production oil wells: a plume of burning natural gas at the end of a structure called a flare boom. Flaring (burning) natural gas is a common occurrence on offshore oil rigs because the crude oil pumped from deep beneath the ocean floor contains dissolved gases that bubble out of the oil as the pressure they are under is released—much as the dissolved carbon dioxide in a bottle of soda is released when the cap is removed.

The substances that make up natural gas and liquid petroleum belong to a class of compounds called **hydrocarbons**, so named because their molecules are composed of atoms of hydrogen and carbon. The crude oil that is pumped out of Earth's crust is a complex mixture of many hydrocarbons with a wide range of molar masses. Hydrocarbons composed of molecules that contain between one and four carbon atoms each are the principal components of natural gas. These are the compounds that fuel the flame in the chapter-opening photo. Hydrocarbons composed of larger molecules are liquids at room temperature and those with still larger molar masses are mostly solids that dissolve in the liquid components of crude oil.

FIGURE 6.1 Boiling points of C_1 to C_8 alkanes.

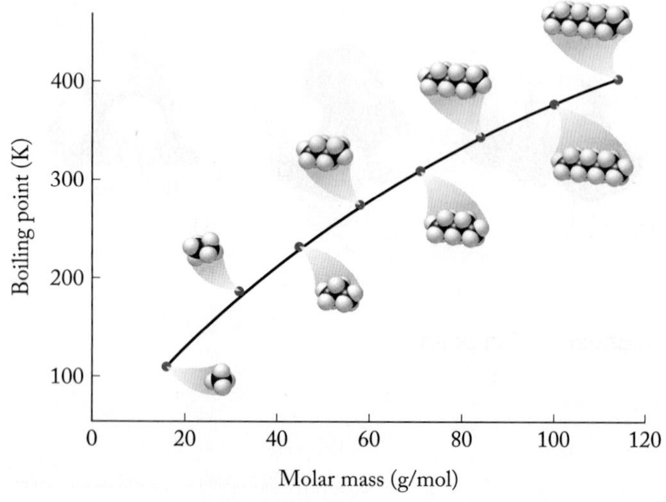

Figure 6.1 shows how the boiling points of eight hydrocarbons, including the principal components of natural gas, depend on the sizes of their molecules. These particular hydrocarbons are called **alkanes** because they are composed of molecules in which each carbon atom is bonded to four other carbon or hydrogen atoms. This means that all the bonds in their molecules are single bonds. Alkanes are the principal class of hydrocarbons in crude oil. The alkanes in Figure 6.1 contain carbon atoms that are bonded to no more than two other carbon atoms, which gives their molecules a chainlike appearance when there are four or more carbon atoms per molecule.

The data in Figure 6.1 show that the boiling points of alkanes with between one and eight carbon atoms per molecule increase with increasing molecular mass. Why do their boiling points increase? The reason is linked to how their molecules interact with each other and to the strengths of those interactions. Recall that the particles in liquids and solids are in direct contact with each other, whereas gas-phase particles are essentially free and independent. Vaporizing a liquid requires separating liquid-phase particles that are attracted to each other, which requires energy: the stronger the particles' attraction for each other, the greater the energy needed to separate them. Greater energy requirements mean higher boiling points.

How do molecules or single atoms interact with each other and why does the strength of these interactions increase as the mass of the particle increases? German-American physicist Fritz London (1900–1954) proposed one explanation for these interactions in 1930. His hypothesis was based on the notion that when atoms (Figure 6.2a) or molecules approach each other, they interact in ways that are similar to the electrostatic interactions involved in covalent bond formation: one atom's positive nucleus is attracted to the other atom's negative electrons, and vice versa, even as their electron clouds repel each other. These competing interactions can cause the electrons around each particle to be distributed unevenly, producing **temporary dipoles** of partial electrical charge (Figure 6.2b) that are attracted to regions of opposite partial charge on the adjacent atom. This attraction is represented by the arrows in Figure 6.2(b). In Chapter 4 we used different colors to represent the partial electrical charges created by the uneven sharing of bonding pairs of electrons. In this chapter we use the same colors (Figure 6.2c) to show partial electrical charges caused by the uneven distributions of electrons in neutral atoms and over entire molecules.

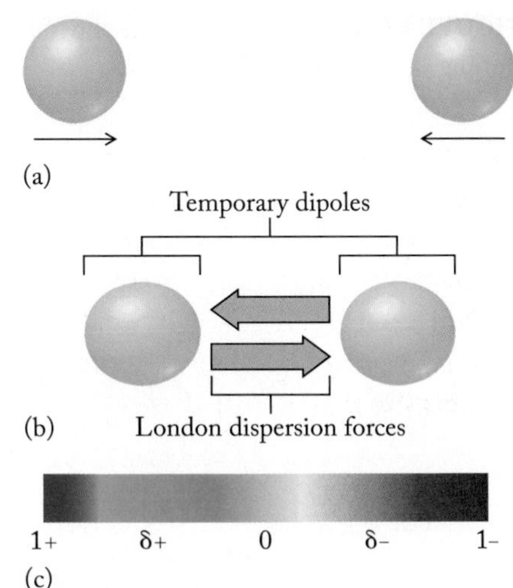

FIGURE 6.2 (a) Two atoms, each with a symmetrical distribution of electrons, approach each other and (b) create two temporary dipoles as their nuclei and electron clouds interact. (c) The strengths of temporary dipoles are shown with the same color scale used in Chapter 4 to represent the strengths of permanent dipoles in molecules.

Interactions based on the presence of temporary dipoles are called **London dispersion forces** in honor of Fritz London's pioneering work. The strengths of the interactions increase as the number of electrons in atoms and molecules increases. Because all atoms and molecules have electrons, all atoms and molecules experience London forces to some degree. The larger the cloud of electrons surrounding a nucleus in an atom or the multiple nuclei in a molecule, the more likely those electrons are to be distributed unevenly, or *polarized*. Greater **polarizability** leads to stronger temporary dipoles and stronger intermolecular interactions, so London dispersion forces become stronger as atoms and molecules become larger.

Increasing polarizability with increasing molecular mass explains the trend in the boiling points of the alkanes in Figure 6.1. It also explains the boiling points of the noble gases (Table 6.1), where the atomic number of each element provides a measure of the polarizability of electrons surrounding each of their atomic nuclei. A similar trend is also observed in the boiling points of the halogens (Table 6.2): their boiling points increase as their molar masses (and the number of electrons in each of their diatomic molecules) increase.

TABLE 6.1 Boiling Points of Noble Gases

Noble Gas	Atomic View	Molar Mass (g/mol)	Boiling Point (K)
He		2	4
Ne		10	27
Ar		18	87
Kr		36	120
Xe		54	165
Rn		86	211

CONCEPT TEST

Explain why CF_4 is a gas at room temperature but CCl_4 is a liquid.

(Answers to Concept Tests are in the back of the book.)

The Importance of Shape

Molecular shape, as well as molecular mass, helps determine the strength of London dispersion forces. The three hydrocarbons depicted in Figure 6.3 all have the same molecular formula, C_5H_{12}, and molar mass, 72 g/mol, but the bond locations and shapes of the molecules are different. Compounds such as these, with the same molecular formula but different connections between the atoms in their molecules,

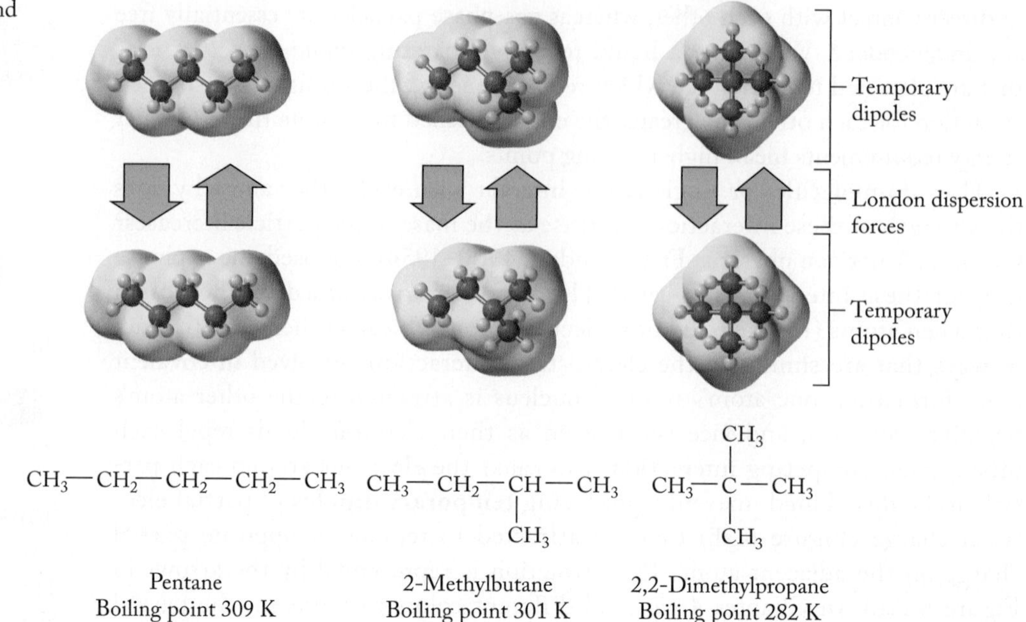

FIGURE 6.3 Molecular structures and boiling points of three C_5 isomers.

TABLE 6.2 Boiling Points of the Halogens

Halogen	Molecular View	Molar Mass (g/mol)	Boiling Point (K)
F_2		38	85
Cl_2		71	239
Br_2		160	332
I_2		254	457
At_2		420	610*

*estimated

are called **constitutional isomers** or *structural* isomers. Compounds that are constitutional isomers have different physical and chemical properties despite their shared molecular formulas. For example, all three of the C_5 (five carbon atoms per molecule) hydrocarbons in Figure 6.3 are nonpolar, which means none of them has a permanent dipole. Therefore, the only intermolecular forces they experience are London dispersion forces due to temporary dipoles. The carbon atoms in a molecule of pentane form a single chain, which gives the molecule the overall shape of a stubby piece of chalk. This shape gives the molecule a larger surface area and more opportunity to interact with other pentane molecules than molecules of 2,2-dimethylpropane, which are shaped more like lumpy spheres, have a smaller surface-area-to-volume ratio, and interact less strongly with adjacent molecules. Weaker London dispersion forces explain why 2,2-dimethylpropane has a lower boiling point than pentane. The remaining molecule, 2-methylbutane, is neither as straight as pentane nor as spherical as 2,2-dimethylpropane. As a result, 2-methylbutane boils at a temperature between the boiling points of the other two. In general, when molecules with similar molar masses and chemical features are compared, the ones with more branching in their structures have lower boiling points.

Viscosity

In addition to boiling point, another property of liquids that is influenced by London forces is viscosity. The **viscosity** of a fluid (a fluid is a liquid or a gas—anything that can flow) is a measure of its resistance to flow. Viscous or "thick" fluids such as honey, molasses, and the oil used to lubricate engines have high resistances to flow. Low-viscosity substances such as water and gasoline flow or pour easily. Gases have ultralow viscosities that are several orders of magnitude below those of the lowest-viscosity liquids.

TABLE 6.3 Viscosities of Some Liquid Alkanes

Compound	Molecular Structure	Molar Mass (g/mol)	Viscosity at 20°C (cP)
Hexane		86	0.29
Octane		114	0.54
Decane		142	0.92
Dodecane		170	1.34
Hexadecane		226	3.34

There are many ways to measure viscosity and to express the results of those measurements. One of the simplest measurements uses a device called a Zahn cup (Figure 6.4), which is a small metal cup with a hole in the bottom. The cup is dipped into a liquid sample and then lifted out. The time the liquid takes to drain through the hole is measured and converted into units of dynamic viscosity. (There are other types, but *dynamic* viscosity is the one most widely used in chemistry.) A common unit for expressing viscosity is the *centipoise* (cP), where 1.00 cP is the viscosity of pure water at 20°C.

Table 6.3 lists the viscosities of a group of liquid alkanes with between 6 (hexane) and 16 (hexadecane) carbon atoms per molecule. Notice how the viscosities increase with increasing molecular size. Molecules of hexadecane experience the greatest London dispersion forces in this group because they are the largest molecules. Stronger intermolecular attractions mean, in turn, that they do not slip past each other as easily as those experiencing weaker intermolecular attractions. More resistance to flow means higher viscosity.

FIGURE 6.4 Determination of viscosity using a Zahn cup.

CONCEPT **TEST**

Predict which compound has the higher boiling point in each of the following pairs: (a) CH_3Cl, CH_3Br; (b) CH_3CH_2OH, CH_3OH; (c) $(CH_3CH_2)_3N$, $(CH_3)_3N$.

6.2 Interactions Involving Polar Molecules

In Section 5.3 we saw how the unequal distributions of bonding pairs of electrons result in partial negative charges on some bonded atoms and partial positive charges on others. When bond dipoles are arranged asymmetrically within molecules, the molecules themselves have permanent dipoles. Molecules with permanent dipoles can interact with each other and with the ions in ionic compounds. The strengths of these interactions are much weaker than the strengths of ionic or covalent bonds, but they are strong enough that they need to be considered in

constitutional isomer one of a set of compounds with the same molecular formula but different connections between the atoms in their molecules; also called *structural isomer*.

viscosity a measure of the resistance to flow of a fluid.

dipole–dipole interaction an attraction between regions of polar molecules that have partial charges of opposite sign.

functional group a group of atoms in a molecule that is largely responsible for the physical and chemical properties of a molecular compound.

organic compounds compounds that contain carbon, hydrogen, and sometimes other elements, including oxygen, nitrogen, sulfur, and a halogen.

carbonyl group a functional group that consists of a carbon atom with a double bond to an oxygen atom.

ketone an organic compound that contains a carbonyl group bonded to two other carbon atoms.

hydrogen bond a strong dipole–dipole interaction, which occurs between a hydrogen atom bonded to an N, O, or F atom and another N, O, or F atom.

CONNECTION The bond energies of covalent bonds are discussed in Section 4.5.

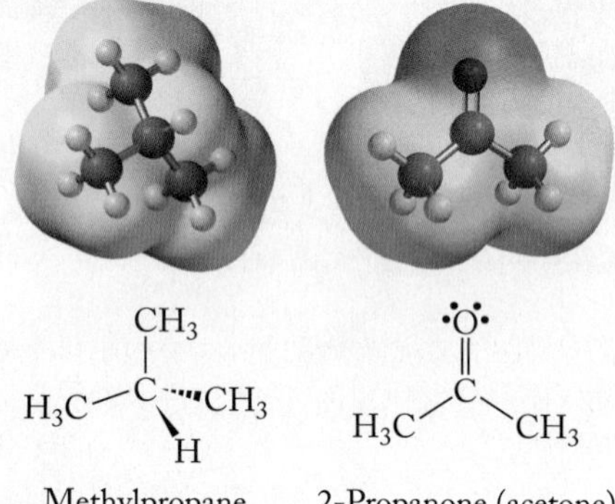

FIGURE 6.5 Molecular structures and boiling points of methylpropane and acetone.

addition to the London dispersion forces that all molecules experience. For example, interactions between polar liquids and ionic solids help salts dissolve in water, and interactions between the molecules of polar liquids are responsible for boiling points and viscosities that are higher than those of nonpolar substances with equivalent molar masses.

Dipole–Dipole Interactions

Consider the boiling points of two compounds: methylpropane and 2-propanone (commonly called acetone), which are composed of the molecules shown in Figure 6.5. The molar masses of these compounds are the same (58 g/mol), and the electron clouds surrounding their molecules have similar shapes. Therefore, they should experience similar London dispersion forces, yet the boiling point of acetone (329 K) is almost 70 K higher than the boiling point of methylpropane (261 K). Why?

The reason has to do with the differences in the polarities of these two molecules. Methylpropane is a nonpolar hydrocarbon, but the double bond between the highly electronegative ($\chi = 3.5$) oxygen atom and the less electronegative ($\chi = 2.5$) carbon atom in each molecule of acetone gives it an overall dipole moment of 2.9 D. Molecules with permanent dipole moments can interact with each other via a second kind of intermolecular force called **dipole–dipole interactions**. These interactions occur between acetone molecules when the partial negative charges on the oxygen atoms of some molecules are attracted to the partial positive charges on other acetone molecules. It is the strength of these interactions that must be overcome when molecules of liquid acetone vaporize, and these interactions are why the boiling point of acetone is over 25% higher than the boiling point of methylpropane.

The C═O bond constitutes one of the most common **functional groups** in carbon-based **organic compounds**. By functional group, we mean a particular group of atoms in a molecule that is largely responsible for the physical and chemical properties of a compound. The C═O functional group in organic compounds such as acetone is called a **carbonyl group**. When the carbon atom of a carbonyl group in a molecule is bonded to two other carbon atoms, the compound is called a **ketone**. Acetone is the ketone with the simplest molecular structure and smallest molar mass.

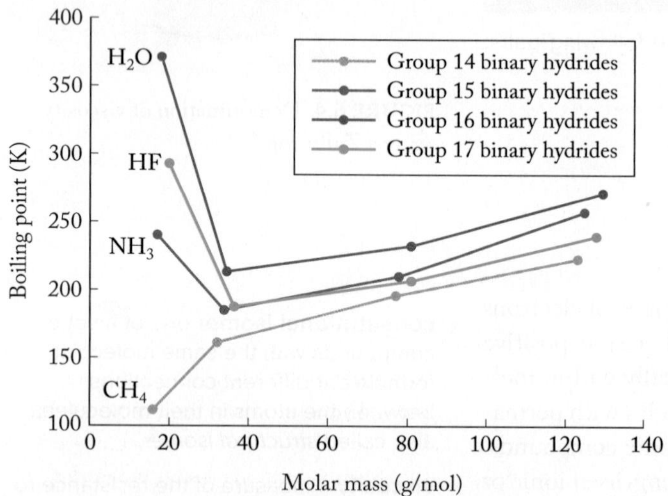

Hydrogen Bonds

The graph in Figure 6.6 compares the boiling points of compounds whose molecules each contain one atom of a group 14, 15, 16, or 17 element bonded to enough hydrogen atoms to have a complete valence-shell octet. Most of the boiling points increase with increasing molar masses, as we saw with nonpolar hydrocarbons in Figure 6.1. However, the boiling points of the lowest molar mass compounds of groups 15, 16, and 17—namely, NH_3, H_2O, and HF—are unusually high compared

FIGURE 6.6 The boiling points of most but not all of the group 14–17 binary hydrides increase with increasing molar mass. The boiling points of H_2O, HF, and NH_3 are much higher than expected. They are higher because of hydrogen bonding between their molecules.

to the others in their series. To understand why, we need to focus on the polar bonds formed between H and N, O, and F atoms. The H atoms share just two electrons, and when they are shared with a highly electronegative atom such as N, O, or F, the electron density on the surface of the H atom is reduced so much that the H atoms interact strongly with electronegative N, O, and F atoms on neighboring molecules. The resulting high-strength dipole–dipole interactions between molecules are called **hydrogen bonds** and are represented by the dotted lines in Figure 6.7.

Hydrogen bonds can also form between molecules of different substances, even when one of the substances has no H atoms bonded to N, O, or F atoms. For example, when acetone dissolves in water, hydrogen bonds form between the H atoms of water molecules and the O atoms of acetone molecules as shown in Figure 6.8. These interactions happen even though the H atoms in acetone are bonded to C atoms and cannot form hydrogen bonds. The key to hydrogen bonding between acetone and water molecules is the negative partial charge of the O atoms in molecules of acetone, as indicated by the red color of the electrostatic potential map of acetone in Figure 6.8, and the positive partial charge indicated by the blue-green color of the electron-deprived H atoms in molecules of H_2O.

CONNECTION Recall from Chapter 4 that the strength of attraction between particles of opposite electrical charge *increases* as the distance between them *decreases*.

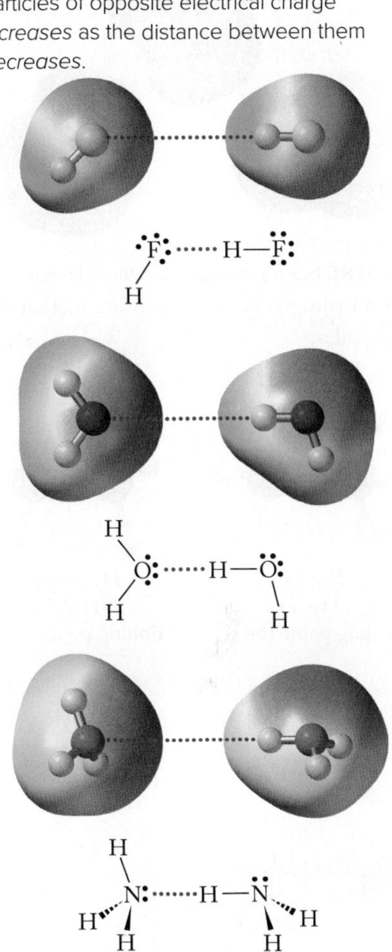

FIGURE 6.7 Hydrogen bonds (represented by the dotted blue lines) occur between hydrogen atoms bonded to F, O, or N atoms in one molecule and F, O, or N atoms in adjacent molecules.

CONCEPT TEST

Both CH_3F and HF contain one flourine atom and at least one hydrogen atom. Explain why only one of these compounds forms hydrogen bonds with molecules of itself.

To get another perspective on the importance of hydrogen bonding, let's consider the boiling points and intermolecular forces in three organic compounds: ethane, formaldehyde, and methanol. Their molar masses are nearly the same, so they should experience similar London dispersion forces. Table 6.4 shows electrostatic potential maps of their molecular structures and lists some of their properties. Note that the boiling point of formaldehyde is 70 K higher than the boiling point of ethane because the carbonyl functional group in formaldehyde gives it a dipole moment of 2.33 D. As a result, formaldehyde molecules experience dipole–dipole interactions that nonpolar ethane molecules do not. However, the difference between the boiling points of ethane and methanol is even

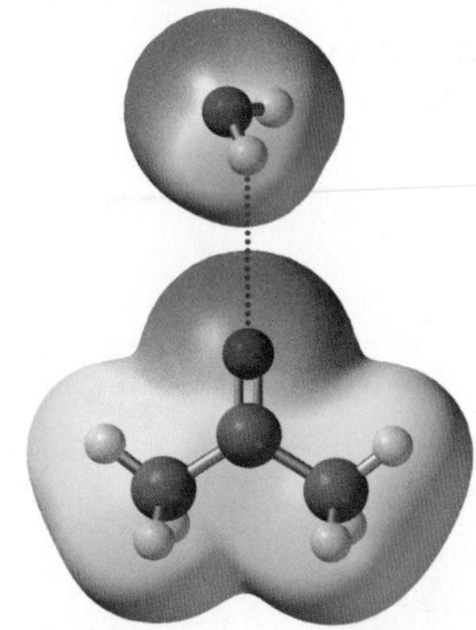

FIGURE 6.8 In solutions of acetone in water, hydrogen bonds form between the hydrogen atoms of water molecules and the oxygen atoms in acetone molecules.

TABLE 6.4 Some Properties of Ethane, Formaldehyde, and Methanol

	Ethane	Formaldehyde	Methanol
Formula	CH_3CH_3	CH_2O	CH_3OH
Structure			
$\mathcal{M}$ (g/mol)	30.0	30.0	32.0
Dipole Moment (D)	0.00	2.33	1.69
Boiling Point (K)	184	254	338

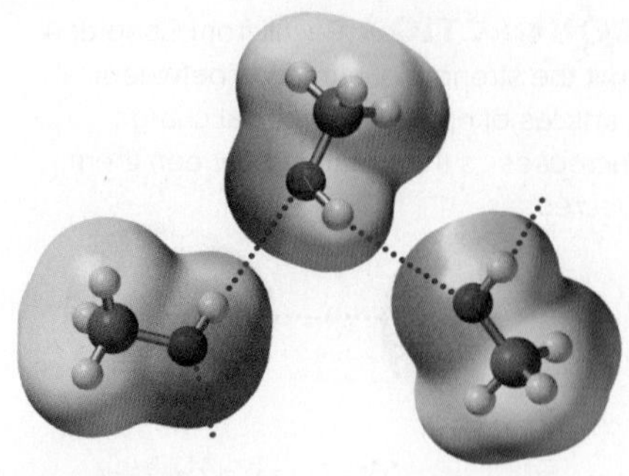

FIGURE 6.9 A methanol molecule can form hydrogen bonds with other methanol molecules.

greater, 154 K, despite the fact that the dipole moment of methanol is less than that of formaldehyde. The boiling point of methanol is so much higher because its OH group can form hydrogen bonds with neighboring methanol molecules, as represented by the dotted lines in Figure 6.9. The presence of the OH functional groups, called **hydroxyl groups**, makes methanol a member of a class of organic compounds called **alcohols**. The physical and chemical properties of alcohols are closely linked to the hydroxyl groups in their structures and their capacity to form hydrogen bonds. Sample Exercise 6.1 introduces another functional group called an **ether** group that consists of an oxygen atom with two single bonds to two carbon atoms.

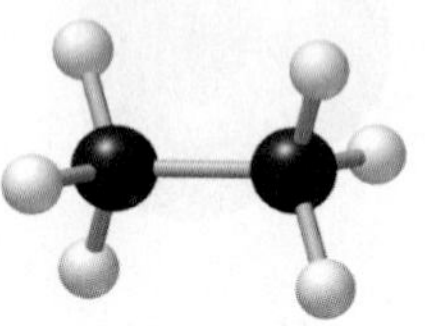

Ethane
CH_3CH_3
Boiling point 184 K

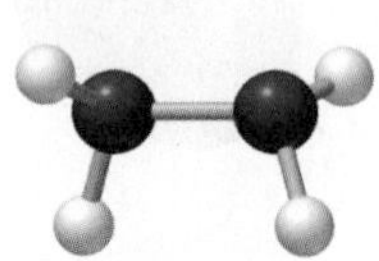

Hydrazine
NH_2NH_2
Boiling point 387 K

CONCEPT TEST

Ethane and hydrazine have similar molecular structures as shown on the left, yet the boiling point of hydrazine (387 K) is over 200 K greater than that of ethane (184 K). What intermolecular interactions account for this huge difference in boiling points?

CHEMTOUR Intermolecular Forces

SAMPLE EXERCISE 6.1 Explaining Differences in Boiling Point LO1

Dimethyl ether (C_2H_6O) has a molar mass of 46 g/mol and a boiling point of 248 K. Ethanol has the same chemical formula and molar mass but a boiling point of 351 K. Explain the difference in boiling point. Their structures are shown here.

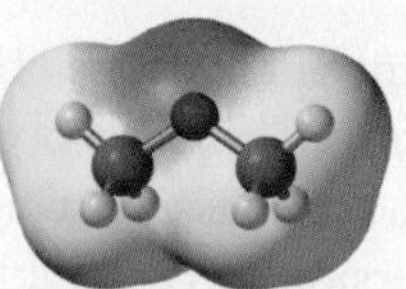

Dimethyl ether
CH_3OCH_3
Boiling point 248 K

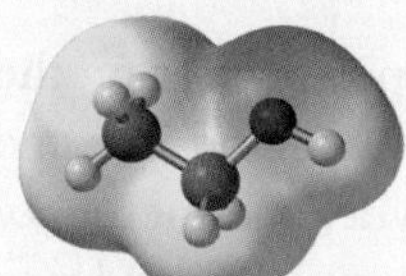

Ethanol
CH_3CH_2OH
Boiling point 351 K

Collect and Organize We need to explain the large difference between the boiling points of two compounds that have the same molar mass. Molecules of both compounds contain oxygen atoms bonded to atoms of less electronegative elements. The electrostatic potential maps of the compounds indicate that both have overall dipole moments.

Analyze Because their molar masses are the same, the molecules of both compounds should experience similar London dispersion forces. Both also contain polar groups: the OH group in the ethanol molecule is polar because of unequal sharing of the bonding pair of electrons between O and H, and the C—O—C group in dimethyl ether is polar because of the bond dipoles between C and O, and because the bond angle between the three atoms should be about 109°, given the sp^3 hybridization of the O atom. The resulting asymmetry in the molecule gives dimethyl ether and ethanol permanent dipoles. In addition, hydrogen bonds can form between the OH groups on adjacent ethanol molecules.

Solve Hydrogen bonds are particularly strong dipole–dipole interactions, so the hydrogen bonds between molecules of ethanol should be stronger than the dipole–dipole interactions between molecules of dimethyl ether. More energy is required to overcome those stronger interactions, which is why ethanol has a higher boiling point.

Think About It Dimethyl ether and ethanol are constitutional isomers, yet their boiling points differ by more than 100 K—a difference due to the hydrogen bonding between the OH groups present in molecules of ethanol but not present in molecules of dimethyl ether.

hydroxyl group a functional group that consists of an oxygen atom with a single bond to a hydrogen atom.

alcohol an organic compound whose molecular structure includes a hydroxyl group bonded to a carbon atom that is not bonded to any other functional group(s).

ether an organic compound that contains an oxygen atom with single bonds to two carbon atoms.

Practice Exercise Isopropanol (left molecule in the figure below) is the compound commonly known as rubbing alcohol. Its boiling point is 355 K. Ethylene glycol, which is the principal ingredient in automotive antifreeze, has about the same molar mass and boils at 469 K. Why do the substances have such different boiling points?

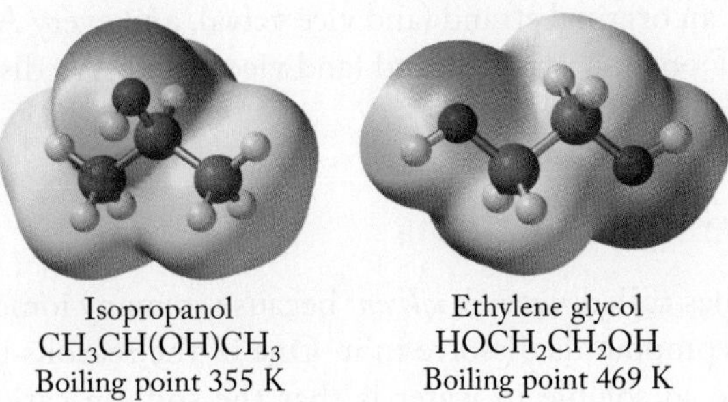

(Answers to Practice Exercises are in the back of the book.)

Hydrogen bonding also plays a role in defining the shapes of very large molecules, especially those of biological interest such as proteins and DNA. Proteins are so large that their long chains of atoms fold back and wrap around themselves, such that atoms hydrogen-bond with other atoms in the chain in much the way atoms can hydrogen-bond in adjacent molecules. In addition, hydrogen bonding between atoms on adjacent chains can induce them to interact in ways that maximize the number of hydrogen bonds they form. The double strands of DNA form a three-dimensional shape called a double helix in which pairs of the molecular building blocks of DNA, called nucleotides, form hydrogen bonds that keep the strands linked together. Pairs of two nucleotides named guanine (G) and cytosine (C) on adjacent DNA strands form three hydrogen bonds, and pairs of adenine (A) and thymine (T) form two hydrogen bonds (Figure 6.10).

FIGURE 6.10 Hydrogen bonds (dotted lines) that occur between hydrogen atoms and either nitrogen or oxygen atoms in adjacent strands of DNA stabilize the double-helix structure. The two detailed views show the four building blocks of DNA: guanine (G), cytosine (C), adenine (A), and thymine (T).

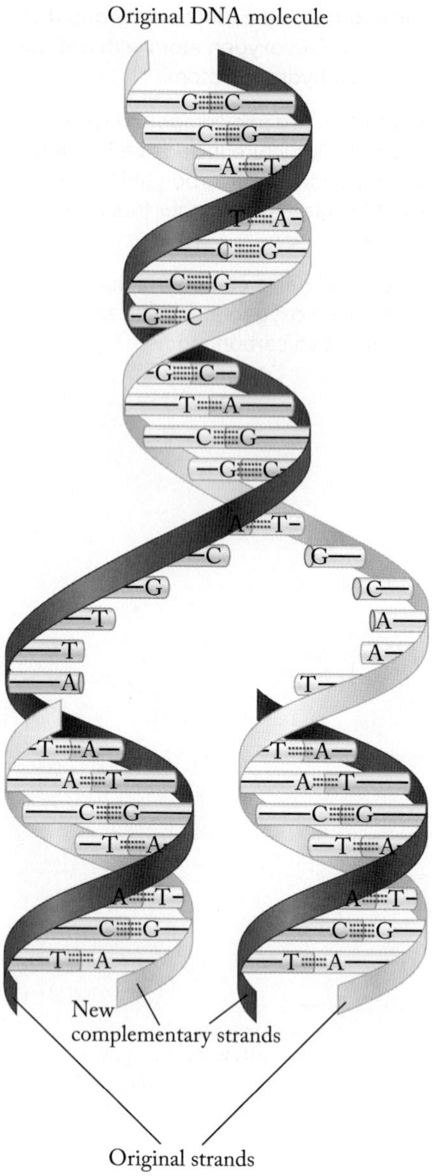

Original DNA molecule

New complementary strands

Original strands

FIGURE 6.11 Hydrogen bonding plays a key role in DNA replication. Nucleotides in the new strands pair up with those in the parent strands: A pairs with T and C pairs with G.

Hydrogen bond formation is a reversible process. When living cells prepare to divide, the hydrogen bonds in DNA break as the double helices pull apart. During cell division the DNA strands of the parent cell replicate, forming two new double helices as shown in Figure 6.11. The new DNA strands are synthesized in such a way that the G nucleotides of the new strands are hydrogen bonded to C nucleotides on an original strand (and vice versa), and every A on a new strand is paired with a T on an original strand (and vice versa). We discuss DNA replication in more detail in Chapter 20.

Ion–Dipole Interactions

Water is sometimes called *nature's solvent* because so many ionic solids as well as polar molecular compounds dissolve in it. One of the reasons ionic compounds such as NaCl are so soluble in water is that the sodium cations and chloride anions interact with the permanent dipoles of water molecules. These **ion–dipole interactions** compete with the ion–ion (coulombic) interactions holding the ions in a lattice (Figure 6.12). Ion–dipole interactions are stronger than dipole–dipole interactions because dipole–dipole interactions involve atoms with only partial charges. In contrast, an ion involved in an ion–dipole interaction has completely lost or gained one or more electrons and has a charge of at least 1+ or 1−.

As an ion is removed from its solid-state neighbors, it is surrounded by a cluster of water molecules that forms a **sphere of hydration** (Figure 6.13). The dissolved ions are said to be *hydrated*. The strengths of these multiple ion–dipole interactions (represented by the light green dashed lines in Figures 6.12 and 6.13) help overcome the lattice energy—that is, the energy of the ionic bonds that hold ions together in crystalline ionic solids. Within a sphere of hydration, the water molecules closest to the ion are oriented so that their oxygen atoms (negative poles) are directed toward cations and their hydrogen atoms (positive poles) are directed toward anions, as shown for Na^+ and Cl^- ions in Figure 6.13. The number of water molecules in the *inner sphere of hydration* depends on the size of the ion. Six water molecules hydrate most ions, but the number can range from four to nine. When a substance dissolves in a liquid other than water, the cluster of host molecules is called a sphere of *solvation* and the dissolved particles are said to be *solvated*.

The water molecules closest to the ions in Figure 6.13 are surrounded by, and form hydrogen bonds with, other water molecules that form an *outer sphere of hydration*. They in turn are surrounded by bulk water molecules that are not part of the sphere of hydration but that are hydrogen-bonded (the blue dotted lines in Figure 6.13) to each other and to molecules in the outer sphere. The molecules in the outer sphere are oriented more randomly than those in the inner sphere but not as randomly as those in bulk water.

CONNECTION Lattice energy is defined as the energy released when gas-phase cations and anions form one mole of an ionic solid (Section 4.1).

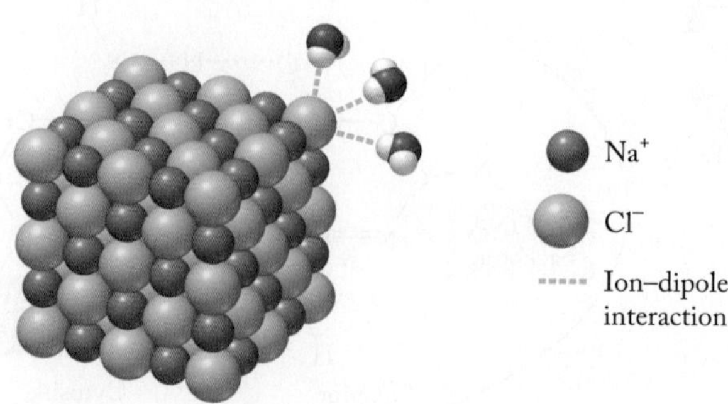

FIGURE 6.12 Ion–dipole interactions. The hydrogen atoms (positive poles) of H_2O molecules are attracted to the Cl^- ions of NaCl. In a similar fashion, the oxygen atoms (negative poles) of H_2O are attracted to the Na^+ cations.

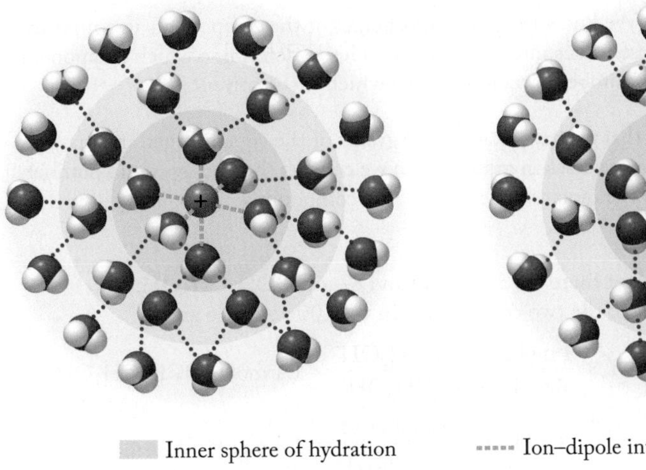

FIGURE 6.13 Each hydrated Na^+ and Cl^- ion in a solution of NaCl is surrounded by six water molecules oriented toward the center ion as a result of ion–dipole interactions (light green dashed lines). Those water molecules make up an inner sphere of hydration. Water molecules in an outer sphere of hydration are oriented as a result of hydrogen bonds (dark blue dotted lines) with molecules in the inner sphere. All water molecules experience particularly strong dipole–dipole interactions.

6.3 Trends in Solubility

Boiling point and viscosity are not the only properties influenced by the strengths of interactions between particles. Another is the extent to which different substances dissolve in each other. Before we explore the interactions involved in forming solutions, we need to define a few key terms. Recall from Chapter 1 that a solution is a homogeneous mixture of two or more substances. The substance that is present in the largest amount (based on numbers of moles) is called the **solvent**. All the other substances in the solution are dissolved in the solvent and are called **solutes**. For example, the water in seawater functions as the solvent, and sodium chloride and the other sea salts dissolved in it are solutes. **Solubility** is a measure of how much solute can dissolve in a given volume of solution.

The process by which an ionic salt dissolves in water, discussed in Section 6.2, consists of a competition between ion–dipole interactions and the lattice energy of the salt. Dissolving a gas in a liquid or one liquid in another are processes that are also influenced by the strengths of intermolecular forces. For example, low-molar-mass alcohols such as methanol, isopropanol, and ethylene glycol dissolve in water because of the hydrogen bonds that form between the OH groups in the alcohols and in water. The strengths and numbers of the hydrogen bonds, as shown for ethylene glycol in Figure 6.14, help these alcohols to dissolve in water and help water to dissolve them in all proportions. When two liquid compounds have unlimited solubility in each other, chemists say that they are **miscible** with each other. On the other hand, when two liquid compounds have only limited solubility in each other, they are said to be immiscible. Gasoline is immiscible with water.

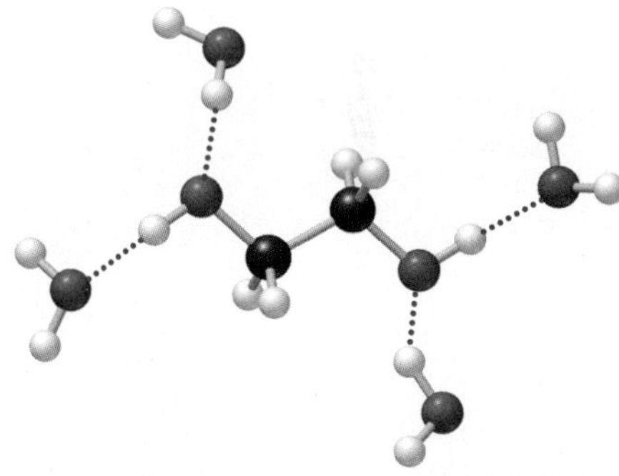

FIGURE 6.14 A single molecule of ethylene glycol ($HOCH_2CH_2OH$) can form multiple hydrogen bonds. The strength of these solute–solvent interactions makes ethylene glycol miscible with water.

ion–dipole interaction an attractive force between an ion and a molecule that has a permanent dipole.

sphere of hydration the cluster of water molecules surrounding an ion in an aqueous solution.

solvent the component of a solution for which the largest number of moles is present.

solute any component in a solution other than the solvent. A solution may contain one or more solutes.

solubility the maximum quantity of a substance that can dissolve in a given volume of solution.

miscible liquids that are mutually soluble in any proportion.

SAMPLE EXERCISE 6.2 Distinguishing Solute from Solvent **LO2**

The liquid in the cooling system of an automobile engine is a solution that protects the engine to temperatures as low as −51°C (−60°F). The mixture is 33% by mass water ($\mathcal{M}$ = 18 g/mol) and 67% by mass ethylene glycol, $HOCH_2CH_2OH$ ($\mathcal{M}$ = 62 g/mol). Which substance is the solvent?

Collect and Organize We know the mass percentages of the components in a mixture and need to convert these values into numbers of moles to determine which component is present in the greater number of moles, that is, which is the solvent.

Analyze If we assume that we have exactly 100 grams of the coolant, then the mass percentage values become masses in grams. Then we convert these masses into numbers of moles by dividing by the molar masses of the two compounds.

Solve

1. In 100 grams of coolant there are 33 g H_2O and 67 g $HOCH_2CH_2OH$.
2. The number of moles of the two components in the 100 g sample are

$$67\ \cancel{\text{g HOCH}_2\text{CH}_2\text{OH}} \times \frac{1\ \text{mol HOCH}_2\text{CH}_2\text{OH}}{62\ \cancel{\text{g HOCH}_2\text{CH}_2\text{OH}}} = 1.1\ \text{mol HOCH}_2\text{CH}_2\text{OH}$$

$$33\ \cancel{\text{g H}_2\text{O}} \times \frac{1\ \text{mol H}_2\text{O}}{18\ \cancel{\text{g H}_2\text{O}}} = 1.8\ \text{mol H}_2\text{O}$$

3. There are more moles of water than ethylene glycol, so water is the solvent.

Think About It Even though the coolant is 67% ethylene glycol and only 33% water by mass, the much larger molar mass of ethylene glycol means that there are fewer moles of it, so water is the solvent.

Practice Exercise Most of the rum sold in liquor stores is 30–40% ethanol, CH_3CH_2OH, by volume. However, a particular *overproof* variety used to prepare flaming drinks contains 75.5% ethanol by volume. Assuming the remaining 24.5% of the liquid is mostly water, which of the two compounds is the solute and which is the solvent in overproof rum? The densities of ethanol and water are 0.789 g/mL and 0.998 g/mL, respectively, at 20°C.

To predict whether a given solute is likely to be highly soluble in a given solvent or only slightly soluble, we need to consider the strengths and numbers of the interactions among solute particles and the strengths of the interactions among solvent particles. The stronger those interactions are, the harder it will be for the solute particles to separate from one another and dissolve in the solvent, and the harder it will be for the solvent particles to separate from each other to accommodate solute particles. On the other hand, the stronger the interactions are between solute and solvent molecules, the easier it will be for the solute to dissolve.

Let's discuss the solute–solvent interactions first because they help drive the dissolution process. Polar solutes tend to dissolve in polar solvents due to strong dipole–dipole interactions between molecules of solute and solvent—especially when hydrogen bonds form between the solute and solvent molecules (as when ethylene glycol or methanol dissolve in water). On the other hand, nonpolar solutes do not dissolve in polar solvents, or dissolve only a little, because the solute–solvent interactions that promote dissolution are much weaker than the interactions that keep solute molecules attracted to other solute molecules and the interactions that keep solvent molecules attracted to other solvent molecules. For example, octane does not dissolve in water because

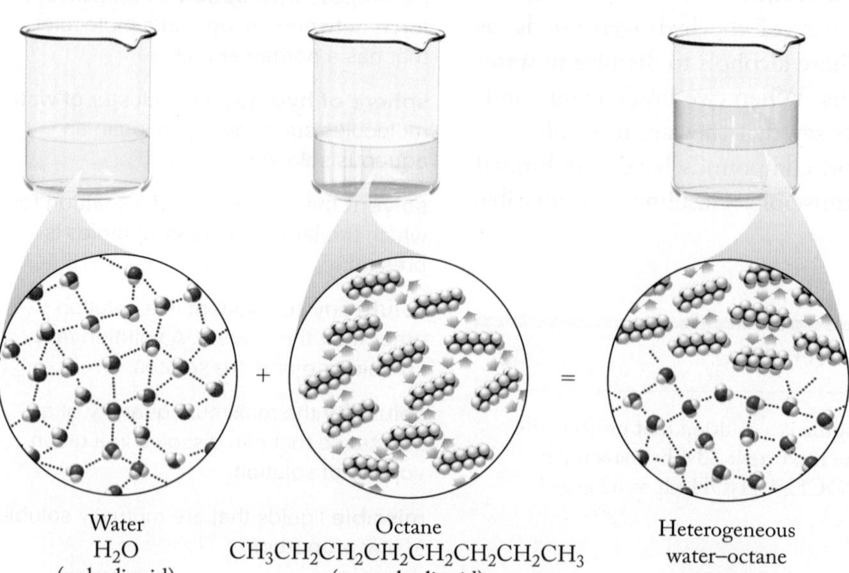

FIGURE 6.15 The strength of London dispersion forces, represented by the blue arrows, holds nonpolar molecules of octane together in the liquid phase. Water molecules are attracted to one another by hydrogen bonding. The lack of strong solute–solvent interactions needed to overcome these forces explains why octane and water are immiscible.

hydrogen bonding is the predominant attraction between H_2O molecules, while London dispersion forces are the predominant attractions between octane molecules. The result is two separate liquid phases that neither mix with nor dissolve in each other when octane and water are combined (Figure 6.15).

What little solubility octane or any nonpolar solute has in water (or any polar solvent) is promoted by **dipole–induced dipole interactions**. We saw in Section 6.1 how temporary dipoles occur in nonpolar molecules through transient perturbations in the normally symmetrical distribution of electrons in molecules. An even stronger perturbation may occur when a polar molecule such as H_2O collides with a nonpolar molecule such as O_2 (Figure 6.16). If the O atom in H_2O is near one end of an O_2 molecule, the partial negative charge on the O atom in H_2O repels the electrons in O_2, pushing them toward the other end of that molecule and creating, or *inducing*, a temporary dipole within the normally nonpolar O_2. The end of the O_2 molecule closest to the O atom in H_2O temporarily acquires a partial positive charge while the opposite end has a temporary partial negative charge. When the two molecules move apart, the partial charges in the O_2 molecule disappear. The transient induced dipole is not as strong as the permanent dipole that induced it, but it is strong enough that O_2 is slightly soluble in water—soluble enough that the amount of dissolved oxygen sustains a multitude of aquatic life-forms.

dipole–induced dipole interaction an attraction between a polar molecule and the oppositely charged pole it causes in another molecule.

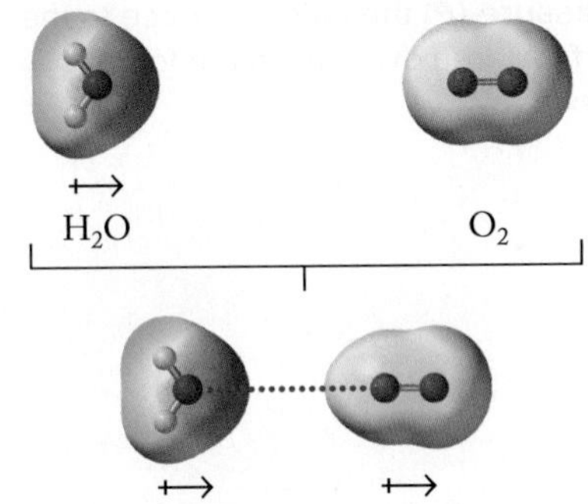

FIGURE 6.16 Dipole–induced dipole interactions between molecules of H_2O and O_2. The permanent dipole of H_2O induces a temporary dipole in a normally nonpolar molecule of O_2.

Competing Intermolecular Forces

If polar solutes readily dissolve in polar solvents, and nonpolar solutes do not, what is the solubility of solutes whose molecules contain both polar and nonpolar groups in a polar solvent such as water? To answer the question, let's consider the solubilities of several ketones (Table 6.5) in water. Note how their solubilities decrease as the number of nonpolar CH_2 groups per molecule increases. The

TABLE 6.5 Solubilities of Some Ketones in Water at 20°C

Compound	Condensed Molecular Structure	Solubility in Water (g/100 mL)
2-Propanone	$H_3C—C(=O)—CH_3$	Miscible
2-Butanone	$H_3C—C(=O)—CH_2CH_3$	25.6
2-Pentanone	$H_3C—C(=O)—CH_2CH_2CH_3$	4.3
2-Hexanone	$H_3C—C(=O)—(CH_2)_3CH_3$	1.4
2-Heptanone	$H_3C—C(=O)—(CH_2)_4CH_3$	0.4

hydrophobic describes a "water-fearing" or repulsive interaction between a solute and water that decreases water solubility.

hydrophilic describes a "water-loving" or attractive interaction between a solute and water that increases water solubility.

pressure (*P*) the ratio of a force to the surface area over which the force is applied.

additional nonpolar groups strengthen the London dispersion forces between the ketone molecules but contribute little to the interactions between the ketone molecules and water molecules, which are mostly through hydrogen bonding. As we have seen, strong interactions between solute molecules that are not offset by strong solute–solvent interactions tend to keep the solute molecules together in a single phase and inhibit their mixing with, and dissolving in, solvents.

Nonpolar interactions such as the London dispersion forces between the hydrocarbon chains of the ketones in Table 6.5 are called **hydrophobic** (literally, "water-fearing") interactions because they tend to keep the compounds from dissolving in water. On the other hand, dipole–dipole interactions, especially hydrogen bonding, promote solubility in water and are called **hydrophilic** ("water-loving") interactions. The solubility in water of compounds whose molecules contain both polar and nonpolar groups, as ketones do, is the net result of offsetting interactions. Hydrophilic interactions between molecules of solute and water promote solubility, whereas hydrophobic interactions among solute molecules inhibit it. As the sizes of the nonpolar regions of solute molecules increase, the strengths of solute–solute hydrophobic interactions increase and their solubilities in water decrease.

CONCEPT TEST

Rank the alcohols with the molecular structures shown here from most soluble to least soluble in water:

```
  H H H H                 H H H H H H                 H H H H H                 H H H
  | | | |                 | | | | | |                 | | | | |                 | | |
H-C-C-C-C-O-H           H-C-C-C-C-C-C-O-H           H-C-C-C-C-C-O-H           H-C-C-C-O-H
  | | | |                 | | | | | |                 | | | | |                 | | |
  H H H H                 H H H H H H                 H H H H H                 H H H
  Butanol                 Hexanol                     Pentanol                  Propanol
```

We have established that polar solutes tend to dissolve in polar solvents and that nonpolar solutes do not, at least not very much. Instead, nonpolar solutes tend to dissolve in nonpolar solvents, which leads to the following useful and general solubility guideline: *like dissolves like.* For example, the principal ingredient in some household cleaners (Figure 6.17) that remove grease, crayon wax, label adhesives, and other nonpolar materials from surfaces and fabrics is often labeled "petroleum distillate" or "petroleum naphtha." Those are common names for a mixture of hydrocarbons derived from crude oil, most of which have between 5 and 11 carbon atoms per molecule. They are effective in dissolving nonpolar substances because hydrocarbons are also nonpolar. Like dissolves like.

FIGURE 6.17 Household cleaners such as this one are effective at removing grease stains and other nonpolar materials from surfaces and fabrics because the principal ingredient is a mixture of nonpolar hydrocarbons.

SAMPLE EXERCISE 6.3 Predicting Solubility in Water **LO3**

Which of the following four compounds have only limited solubility in water: carbon tetrachloride (CCl_4), ammonia (NH_3), hydrogen bromide (HF), and nitrogen (N_2)?

Collect, Organize, and Analyze Water is polar, so compounds that are nonpolar should have limited solubility in it. From the Lewis structures of the compounds we should be able to determine which of the four compounds have the nonpolar bonds and/or molecular symmetry that means they have no permanent dipole moment and are nonpolar.

Solve The Lewis structures of the four compounds are shown below. Molecules of CCl_4 are perfectly tetrahedral and symmetrical, so this compound is nonpolar and should have limited solubility in water. Because molecules of N_2 have nonpolar bonds *and* they're symmetrical, the compound is nonpolar and should have limited solubility. However, molecules of both NH_3 and HF are asymmetrical and have polar bonds. Moreover, both have the capacity to form hydrogen bonds with water molecules, so these compounds should be very soluble in water.

CCl_4	NH_3	H—F	:N≡N:
Nonpolar	Polar	Polar	Nonpolar

Think About It Nonpolar solutes dissolve in nonpolar solvents but are not likely to dissolve in a polar solvent, such as water, in accordance with the "like dissolves like" principal.

Practice Exercise Deep-sea divers breathe a mixture of gases rich in helium because the solubility of helium gas in blood is lower than the solubility of nitrogen gas in blood. Assuming that blood behaves like water in terms of dissolving substances, why should helium be less soluble in water than nitrogen?

6.4 Phase Diagrams: Intermolecular Forces at Work

The strength of the attractive forces between particles determines whether a substance is a solid, liquid, or gas at a given temperature. Heating substances to higher temperatures provides the particles in them with more energy, until the kinetic energy of the particles eventually exceeds the energy of the intermolecular forces that hold them rigidly in place in a solid or keep them close to each other in a liquid. At this point, the solid should melt or the liquid should vaporize. It turns out, however, that pressure, in addition to temperature and the strength of intermolecular forces, influences the physical state of a substance.

Pressure

Why does **pressure (*P*)**, which is the ratio of force to surface area, influence the physical states of substances? The solid phases of most substances are the most dense and the vapor phases are the least dense, so the volume occupied by most substances (with the notable exception of H_2O) slightly expands when they melt, and the volume occupied by all substances expands significantly during vaporization (or sublimation). Increasing the pressure on substances inhibits them from expanding, thereby raising the temperatures at which most substances melt and raising the temperatures at which *all* substances vaporize and sublime.

The pressure experienced by most substances of interest to us is *atmospheric* pressure, which is the weight of Earth's atmosphere pressing down on its surface divided by its surface area (Figure 6.18). We can calculate how strong this pressure is by multiplying the mass of all the gases in the atmosphere by the acceleration due to gravity experienced by all objects on Earth's surface, which is 9.807 m/s^2. The resulting product is the force (*F*) that the mass of atmospheric gases exerts on Earth's surface due to gravity:

$$F = (5.270 \times 10^{18}\ \text{kg}) \times \left(9.807\ \frac{\text{m}}{\text{s}^2}\right) = 5.168 \times 10^{19}\ \frac{\text{kg m}}{\text{s}^2}$$

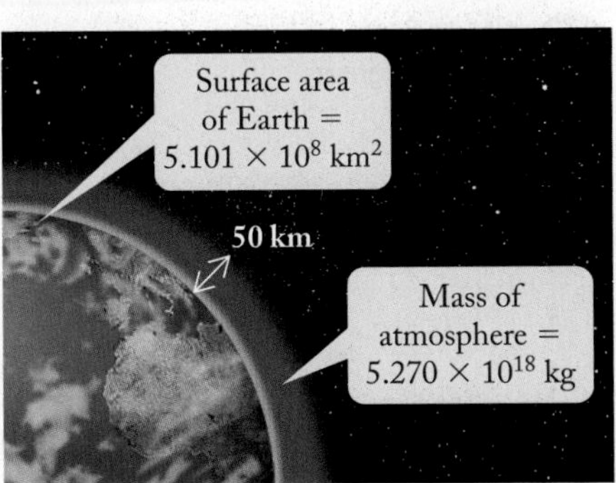

FIGURE 6.18 Atmospheric pressure results from the force exerted by the gases of Earth's atmosphere on Earth's surface.

standard atmosphere (atm) the average pressure at sea level on Earth.

phase diagram a graphical representation of the dependence of the stabilities of the physical states of a substance on temperature and pressure.

triple point the temperature and pressure where all three phases of a substance coexist. Freezing and melting, boiling and liquefaction, and sublimation and deposition all proceed at the same rate, so no net change takes place in the system.

This combination of units, kg · m/s², is defined as a newton (N) in honor of Sir Isaac Newton and is the SI unit of force. If we divide the force by Earth's surface area (A), we obtain the following average atmospheric pressure at sea level:

$$P = \frac{F}{A} \tag{6.1}$$

$$= \frac{(5.168 \times 10^{19}\ \text{N})}{5.101 \times 10^{8}\ \cancel{\text{km}}^2 \left(\frac{1000\ \text{m}}{\cancel{\text{km}}}\right)^2} = 1.013 \times 10^{5}\ \frac{\text{N}}{\text{m}^2}$$

One newton per square meter is defined as one pascal (Pa), the SI unit of pressure. The average pressure exerted by the atmosphere on Earth's surface (at sea level) is, therefore, 1.013×10^5 Pa. This average also defines a quantity of pressure called the **standard atmosphere (atm)**.

To avoid the need for exponents, atmospheric pressure is often expressed in units such as the kilopascal (kPa), bar, or millibar (mb):

$$1\ \text{atm} = 101.3\ \text{kPa} = 1.013\ \text{bar} = 1013\ \text{mb}$$

The bar is not an official SI unit, but it is based on one: 1 bar = 10^5 Pa, and it is widely used in science. Meteorologists tend to use millibars to express atmospheric pressures, as shown in Figure 6.19. Doing so simplifies pressure values on weather maps by avoiding the need for decimal points.

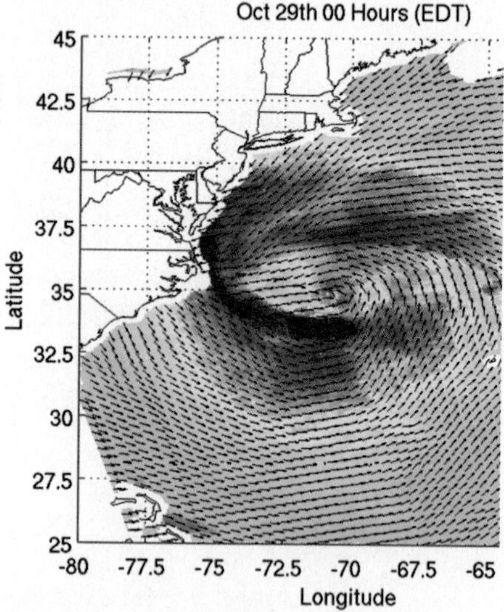

FIGURE 6.19 Superstorm Sandy caused significant damage to the eastern seaboard in 2012. The lines on this weather map mark different atmospheric pressures.

Phase Diagrams

Phase diagrams are graphs of pressure (y axis) versus temperature (x axis) that scientists use to represent which phases of a substance are the most stable at different combinations of temperature and pressure. The scale of the y axis is usually logarithmic rather than linear to depict a wide range of pressures.

Phase diagrams like that of water (Figure 6.20) have three regions corresponding to the three states of matter. The lines separating the regions represent

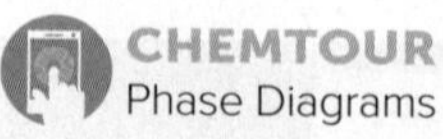

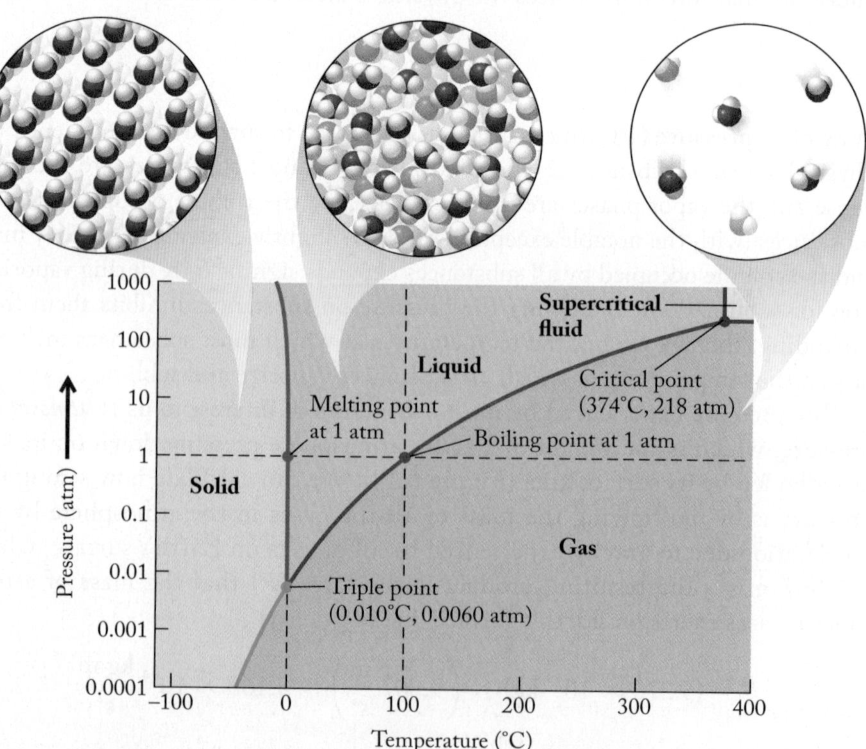

FIGURE 6.20 The phase diagram for water indicates the phase(s) of water that exist(s) at various combinations of pressure and temperature. The yellow region indicates the region where water exists as a supercritical liquid.

combinations of temperature and pressure at which the two phases on either side of each line coexist in equilibrium with each other. Thus, the blue line separating the solid and liquid regions in Figure 6.20 represents a collection of melting (and freezing) points, the red line separating the liquid and gas regions represents a series of boiling (and condensation) points, and the green line separating the solid and gas regions represents a series of sublimation (and deposition) points.

The red line curves from lower left to upper right, because when the pressure above a liquid is increased, as in a pressure cooker (Figure 6.21), more energy is required to overcome that pressure and convert liquid molecules into vapor, where they take up much more space. Heating the liquid to an even higher temperature gives it that additional energy. The solid–gas (green) line curves in a similar fashion because higher pressures make it more difficult for molecules of ice to sublime into water vapor. In both cases, a phase transition from a dense, condensed (liquid or solid) phase to a much less dense vapor phase occurs at higher temperatures under increasing pressure.

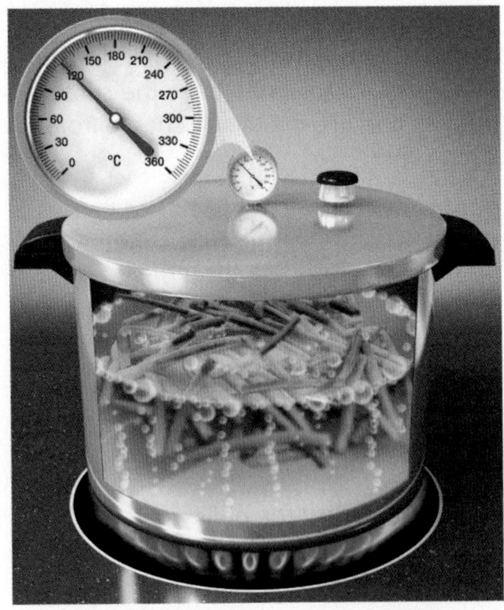

FIGURE 6.21 Steam is restricted from escaping a pressure cooker as the water inside it starts to boil. The higher resulting pressure causes the water to boil at a higher temperature, so the food inside cooks faster.

The blue melting/freezing equilibrium line for water is nearly vertical at low to moderate pressures and then bends to the left at high pressures. Thus, the temperature at which ice melts decreases as pressure increases. This trend is opposite that observed for almost all other substances (compare, for example, the curve of the blue line in the phase diagram of CO_2 in Figure 6.22). The reason for water's unusual melting/freezing behavior is that water expands as it freezes, whereas almost all other substances contract—that is, the solid phases of nearly all substances are more dense than their liquid states. Why does water expand and become less dense? As it freezes, more hydrogen bonds form between its molecules as they cease to flow past each other and instead take up positions in a solid structure where they are surrounded by, and hydrogen-bonded to, an extended array of other water molecules. These additional hydrogen bonds create a slightly more open structure in ice than in liquid water, which makes ice less dense. If we apply enough pressure to ice, we can force it to melt.

CONNECTION See Figure 1.11 to review the names of phase changes and the physical states involved.

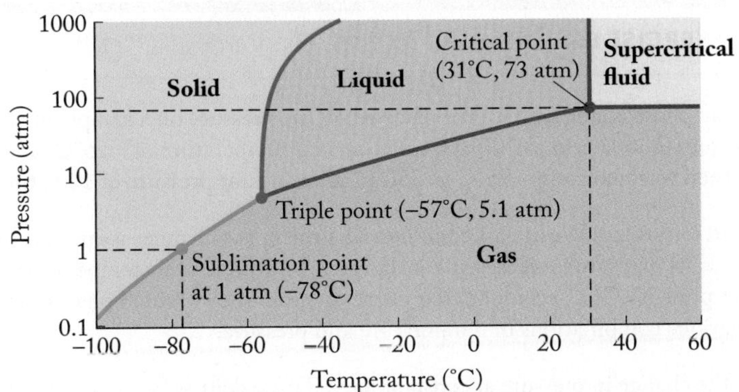

FIGURE 6.22 Phase diagram for carbon dioxide.

CONCEPT TEST

A length of wire with heavy weights on each end can pass downward through a block of ice without cutting it in two. The temperature of the ice stays below its melting point the whole time, and the ice is still a single block after the wire has passed all the way through it. Explain how the wire can pass through a frozen block of ice without cutting it in two.

The point on a phase diagram where all three phase transition lines meet is known as the **triple point** (see Figures 6.20 and 6.22). This is the combination of temperature and pressure at which the liquid, solid, and vapor states of a

critical point a specific temperature and pressure at which the liquid and gas phases of a substance have the same density and are indistinguishable from each other.

supercritical fluid a substance in a state that is above the temperature and pressure at the critical point, where the liquid and vapor phases are indistinguishable.

green chemistry laboratory practices that reduce or eliminate the use or generation of hazardous substances.

surface tension the ability of the surface of a liquid to resist an external force.

substance exist simultaneously. The triple point of water is at a temperature of 0.010°C—just above its normal (1 standard atmosphere) freezing point of 0°C—but at a pressure of only 0.0060 atm.

The **critical point**, on the other hand, is where the liquid/gas equilibrium line ends and the two states are indistinguishable from each other. The critical point is reached because thermal expansion at high temperature decreases the density of the liquid state while high pressure compresses the gas into a small volume, increasing its density. At the critical point the densities of the liquid and vapor states are equal. At temperature–pressure combinations above a substance's critical point, it exists as a **supercritical fluid** (see the yellow regions in Figures 6.20 and 6.22).

A supercritical fluid has the ultralow viscosity of a gas and can easily diffuse through many solid materials, yet it can dissolve substances in those materials as if it were a liquid. Supercritical CO_2 is used in the food processing industry to decaffeinate coffee beans and to remove fat from potato chips. It is also used to extract essential oils from plants to make perfumes and to dry-clean clothes. Because supercritical CO_2 can dissolve nonpolar substances, it provides an alternative to the organic solvents that were once widely used and is one example of the **green chemistry** principles that the use of hazardous substances derived from nonrenewable sources should be reduced or eliminated. In these applications, supercritical CO_2 has the advantage of being nontoxic and nonflammable and of forming at a relatively low temperature (31°C), which is an advantage in extracting pharmaceuticals and other compounds that may decompose at high temperature.

Notice in Figure 6.22 that the blue region representing liquid CO_2 does not exist below a pressure of 5.1 atm. This means that solid CO_2 does not melt into a liquid at normal (atmospheric) pressures. Rather, it sublimes directly to CO_2 gas. This behavior is why solid CO_2 is commonly called *dry ice*. It is a cold solid that does not melt at normal pressures; it just disappears as a colorless gas.

SAMPLE EXERCISE 6.4 Interpreting Phase Diagrams LO4

Describe the phase changes that take place when the pressure on a sample of water is increased from 0.0001 atm to 100 atm at a constant temperature of −25°C, and the sample is then warmed from −25°C to 350°C at a constant pressure of 100 atm.

Collect and Organize We are asked to describe the phase changes a sample of water undergoes as its pressure increases at constant temperature and its temperature increases at constant pressure. The accompanying phase diagram shows which phases of water are stable at various combinations of temperature and pressure.

Analyze The change in pressure at constant temperature defines two points on the phase diagram of water with the coordinates (−25°C, 0.0001 atm) and (−25°C, 100 atm). Connecting those points will give us a vertical straight line. If the line crosses a phase boundary, there will be a change in physical state. The change in temperature at constant pressure defines a third point (−25°C, 100 atm) on the phase diagram, which will connect to the second point via a horizontal line.

Solve The following figure shows a plot of our sample's changes in pressure and temperature. At the bottom end of vertical line 1, water is a gas (vapor). As pressure increases along line 1, it crosses the boundary between gas and solid, which means that vapor turns directly into solid ice, which is stable beyond 100 atm at −25°C. As the temperature of the ice increases along horizontal line 2, it intersects the solid/liquid boundary and the ice melts. At even higher temperatures, just below 350°C, the line intersects the liquid/gas boundary and the liquid water vaporizes.

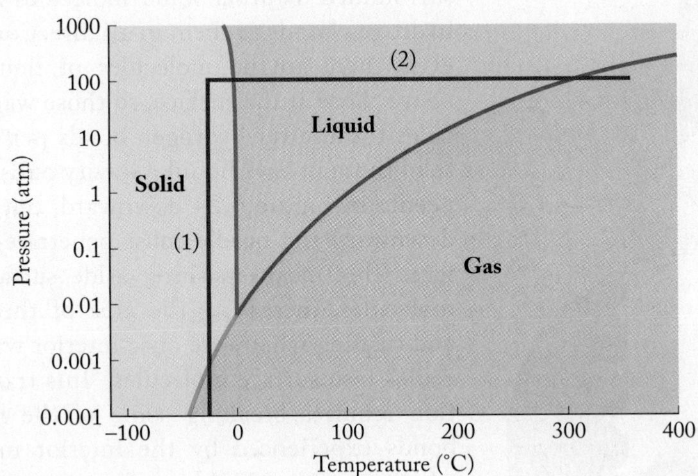

Think About It The solid-to-liquid and liquid-to-gas transitions with increasing temperature along line 2 are what we would expect when a solid substance is warmed to its melting point and then the liquid is heated to its boiling point at a given pressure. The transition along line 1 is less familiar because it is caused by increasing the pressure on a gas at a temperature that is below the triple point temperature, so the vapor never condenses. Instead, it is deposited as a solid.

Practice Exercise Describe the phase changes that occur when the temperature of CO_2 is increased from −100°C to 50°C at a pressure of 25 atm and the pressure is then increased to 100 atm.

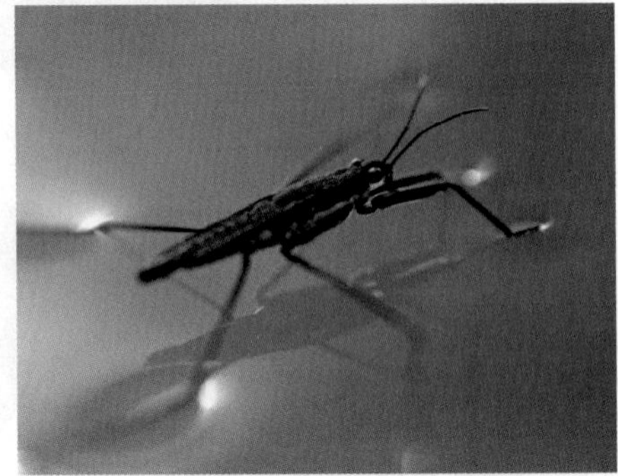

FIGURE 6.23 Surface tension allows a water strider to walk on water without sinking.

6.5 Some Remarkable Properties of Water

Water covers over 70% of Earth's surface, our muscle tissue is about 75% (by mass) water, and our blood is about 95% water. Water's omnipresence and familiarity may lead us to take its remarkable properties for granted. For example, nearly all substances are more dense as solids than as liquids, but water is not. Every other substance of comparable molar mass is a gas at room temperature and pressure, but water is a liquid. Objects that are more dense than water can still float on it because of a phenomenon called surface tension. Capillary action enables liquid water to overcome the force of gravity and climb to the top of the world's tallest trees. We have seen how hydrogen bonding between water molecules can explain its unusually high boiling point. In this section we explore how hydrogen bonding is linked to these other remarkable properties.

Surface tension is the ability of the surface of a liquid to resist an external force, such as the weight of an object. We usually think of surface tension as a property of liquids, which enables a water strider to literally walk on water (Figure 6.23), but it also applies to solids, where surface tension is called *surface energy* because it is expressed in units of energy per unit of surface area. It is also called *surface stress*, which is defined as the energy needed to stretch a surface so that its area increases by a unit amount.

To understand how surface tension depends on intermolecular forces, consider the microscopic view of a steel needle suspended on the surface of a dish of water (Figure 6.24). Water molecules in the interior of the water are

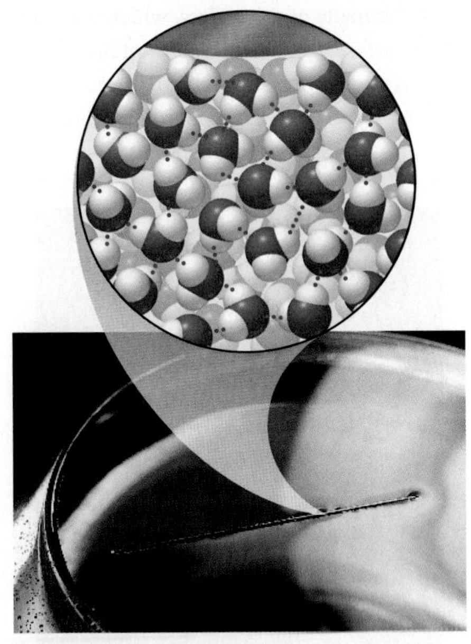

FIGURE 6.24 Intermolecular forces, including hydrogen bonding, are exerted equally in all directions in the interior of a liquid. However, there is no liquid water above a surface to exert attractive intermolecular forces. The resulting imbalance causes the surface water molecules to adhere tightly to one another, creating surface tension. This surface tension exceeds the downward force exerted on the surface water molecules by the needle, causing it to float even though it is much denser than water.

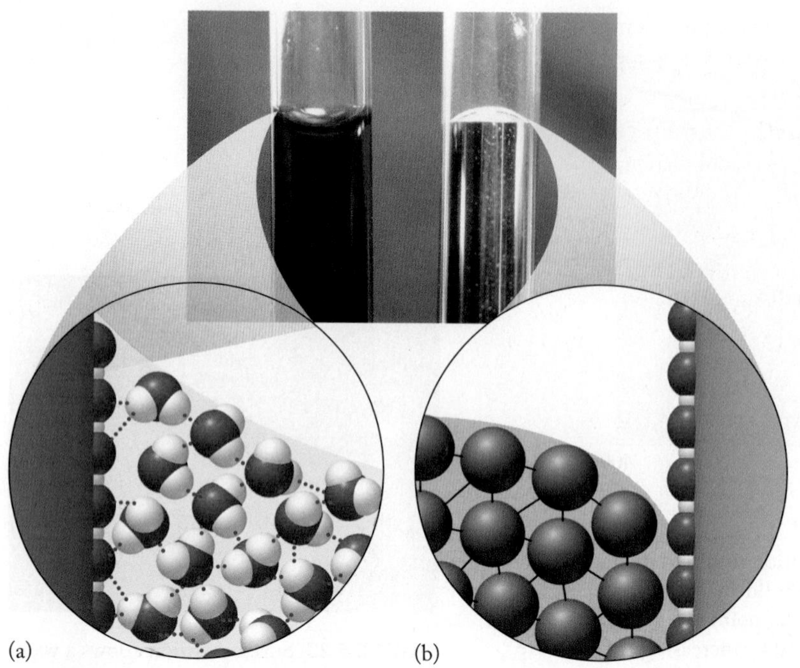

FIGURE 6.25 (a) A combination of adhesive and cohesive forces causes the water (dyed red here) to form a concave meniscus in a test tube made of silica glass, which contains polar surface groups. (b) Atoms of liquid mercury stick more strongly to other atoms of mercury than they adhere to glass, so mercury forms a convex meniscus.

FIGURE 6.26 Because of capillary action, water (containing blue dye) rises in a stalk of celery.

meniscus the concave or convex surface of a liquid.

capillary action the rise of a liquid in a narrow tube as a result of adhesive forces between the liquid and the tube and cohesive forces within the liquid.

surrounded by other water molecules and form hydrogen bonds to them in all directions. However, there are no molecules of liquid water above those at the surface, so those water molecules form fewer hydrogen bonds per molecule than those in bulk liquid. Gravity pulls the steel needle in Figure 6.24 downward, but to move downward the needle must penetrate the surface. This means pushing aside surface water molecules, increasing the area of the surface, and turning what were once interior water molecules into surface molecules. This transformation requires breaking some of the hydrogen bonds experienced by the interior molecules, which requires 23 kJ of energy per mole of hydrogen bonds. The downward pressure of the needle on the surface is insufficient to overcome this energy, so the needle floats on the surface.

Another phenomenon linked to the strength of intermolecular forces is the shape of a liquid's surface in a graduated cylinder or other small-diameter container made of glass, which is mostly SiO_2. As shown in Figure 6.25, the surface of water in a partially filled glass test tube is concave, but the surface of liquid mercury is convex. Either curved surface is called a **meniscus**. In both liquids, the meniscus is the result of two competing forces: *cohesive forces*, which are interactions between like particles (such as the hydrogen bonds between water molecules), and *adhesive forces*, which are interactions between materials made of different substances. In the water sample in Figure 6.25, the adhesive forces are dipole–dipole interactions between water molecules and polar Si—O—Si groups on the surface of the glass, and even hydrogen bonding between water molecules and Si—O—H groups on the surface. These adhesive forces are strong enough to cause the water to climb up the glass. Cohesive interactions with other water molecules pull them up to nearly the same height as the surface molecules. Water molecules farther from the inner wall of the tube are pulled up progressively less, creating a concave meniscus.

When mercury partially fills a glass test tube, it experiences strong cohesive forces in the form of metallic bonds between mercury atoms. The only adhesive forces are relatively weak interactions between induced dipoles in surface mercury atoms and the polar groups on the glass surface. In this case, mercury atoms are more strongly attracted to one another than to the glass. As a result, mercury pulls away from the surface, mounding up in a way that reduces its interaction with the test tube and increases cohesive interactions. The result is a convex meniscus.

The principle behind water's concave meniscus is taken to the extreme when water enters very narrow tubes called capillaries. The tiny inner diameter of capillaries means that all water molecules inside a capillary either experience adhesive intermolecular forces directly or are a small number of hydrogen bonds away from molecules that do. As a result, adhesion pulls the water molecules along the wall of the capillary and cohesion pulls all the others. The result of this combination of intermolecular forces is a phenomenon called **capillary action**, which is the ability of a liquid to flow against gravity, spontaneously rising in a narrow tube or in structures made up of narrow pores, such as celery stalks (Figure 6.26) or the trunks of trees.

Capillary action is also responsible for wicking, the movement of a fluid away from its source through a porous material. Wicking is the process by which paper towels absorb spills and by which some fabrics, such as those made of polyester microfibers, draw perspiration away from the skin. Undergarments made of such fabrics provide a base layer of clothing below other layers that provide thermal insulation. This combination keeps athletes warm while allowing them to practice and play in clothing that is not waterlogged with perspiration (Figure 6.27).

FIGURE 6.27 Athletes wear clothing made with fabrics that can wick away moisture.

CONCEPT TEST

Is mercury spontaneously drawn up into a glass capillary tube? Why or why not?

Another unusual property of water is the way its density changes as its temperature drops to near its freezing point. Like nearly all substances, water's density increases as its temperature decreases. However, the density of water reaches a maximum at 4°C (Figure 6.28) and actually *decreases* as it cools toward its freezing point. When water begins to freeze, its density drops even more, to about 0.92 g/mL, as the oxygen atom in each water molecule becomes the center of an array of two covalent bonds (to the H atoms within the molecule) and two hydrogen bonds (to two other H_2O molecules), as shown in Figure 6.29(a). The length of the hydrogen bonds is more than twice that of the covalent bonds in H_2O, but the larger distance is not the reason ice is less dense than water; rather, it is the *directionality* of the hydrogen bonds in ice. Recall from Chapter 5 that the electron-group geometry of the valence electrons in the O atoms (SN = 4) is tetrahedral. However, ice crystals feature *hexagonal* arrays of water molecules (Figure 6.29). This means that the angle between the H·····O—H hydrogen bond and covalent bond at each corner is about 120°, which is larger than the tetrahedral ideal of 109.5°. The larger corner angles

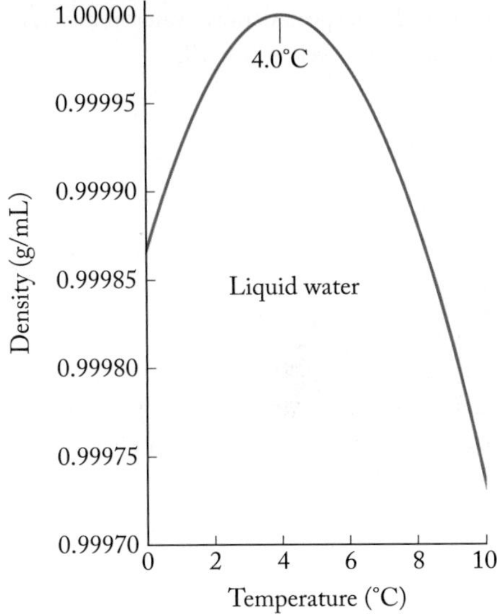

FIGURE 6.28 As water is cooled, its density increases until the temperature reaches 4°C. At that temperature the density has its maximum value, 1.000 g/mL. As the water cools from 4°C to its freezing point at 0°C, the density decreases.

FIGURE 6.29 (a) The oxygen atoms in ice form two covalent bonds to H atoms and two longer hydrogen bonds to the H atoms in nearby molecules. (b) Ice has a lower density than water because of its three-dimensional molecular structure created by the covalent and hydrogen bonds within and between its molecules.

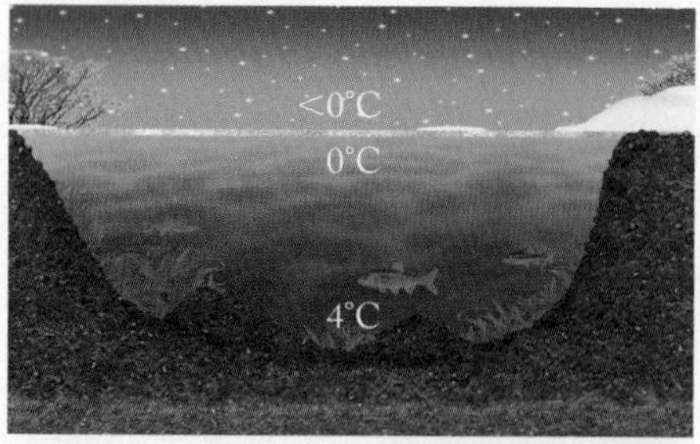

FIGURE 6.30 Wintering over. As the surface water of a lake in a temperate climate cools to 0°C and then freezes during the winter, the deepest, densest water in the lake remains a relatively warm 4°C.

CHEMTOUR
Capillary Action

mean that ice's hexagonal arrangement creates more space between the molecules than liquid water's largely tetrahedral orientation of hydrogen-bonded molecules, which allows liquid water molecules to get closer to their nearest neighbors.

Water and Aquatic Life

The expanded microscopic structure of ice plays a crucial role in aquatic ecosystems in temperate and polar climates. The lower density of ice means that lakes, rivers, and polar oceans freeze from the top down, allowing fish and other aquatic life to survive in the denser layer of liquid water below the frozen surface.

Let's follow how the temperature and density of the water in a deep lake in a temperate climate change over the course of a year. In early spring, rising air temperatures and more sunlight warm the surface waters of the lake. Continued heating creates an upper layer of warm, low-density water separated from the colder, denser water below by a *thermocline,* a sharp change in temperature between the two layers. Little mixing takes place between the two layers over the summer and into the fall. Then declining air temperatures cool the surface waters until they are colder and denser than the water below. The cold surface water sinks, forcing warmer water to the surface where it also cools and sinks. The process continues until all of the water in the lake reaches 4°C and is uniformly mixed.

As the surface water cools further with the approach of winter, it becomes less dense and ice eventually forms. The layer of ice floats above and insulates the water below, including the 4°C water that fills the deepest parts of the lake because it is the densest (Figure 6.30). This water provides a liquid haven for aquatic life-forms and allows them to survive the subfreezing air temperatures of winter.

SAMPLE EXERCISE 6.5 Integrating Concepts: Drug Efficacy and Partition Ratios

The effectiveness of most drugs, particularly those taken orally, depends on how well they can reach their target organs and tissues and on their therapeutic strength and persistence once they get to their targets. These factors are linked to how soluble a drug is in the aqueous environments of blood serum and cytosol (the liquid inside cells) and to how well the drug can penetrate nonaqueous, nonpolar parts of the body such as the epithelium surrounding the small intestine and the membranes surrounding individual cells. In other words, to be effective, a drug must have at least some hydrophilic *and* some hydrophobic character.

One way scientists can predict whether a molecular compound has a blend of hydrophilic and hydrophobic properties is to determine its relative solubility in water and in octanol, $CH_3(CH_2)_7OH$, which is immiscible in water. Typically, a quantity of the compound is added to a mixture of water and octanol. The mixture is shaken vigorously to allow the compound to dissolve in either liquid, the liquids are allowed to separate, and then the concentration of the compound in each liquid is

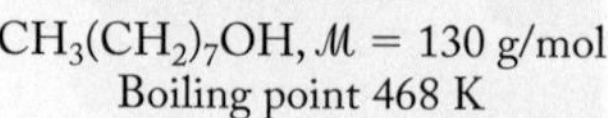
$CH_3(CH_2)_7OH$, $\mathcal{M}$ = 130 g/mol
Boiling point 468 K

(a) Octanol

$CH_3(CH_2)_7CH_3$, $\mathcal{M}$ = 128 g/mol
Boiling point 424 K

(b) Nonane

measured. The ratio of the concentrations defines the compound's partition ratio:

$$\text{Partition ratio} = \frac{\text{concentration in octanol}}{\text{concentration in water}} \quad (6.2)$$

a. What types of intermolecular forces are experienced by the molecules in a sample of octanol?
b. Which of the forces cited in part (a) accounts for most of the interaction between molecules of octanol? (The immiscibility

of octanol in water and the information in the accompanying figure may help you decide.)

c. Are compounds with large partition ratios more or less hydrophobic than compounds with small partition ratios? Why?

d. What if equal volumes of octanol and water were used in one determination of a compound's partition ratio, and then twice as much octanol was used in another determination? How would this difference in volumes influence the two experimental partition ratio values?

Collect and Organize We know the molecular structure of octanol and that it is immiscible in water. We are asked what types of intermolecular forces octanol molecules experience, which force is the strongest, and how the partitioning of a compound between octanol and water relates to its hydrophobicity. We are also asked to predict how changing the relative volumes of octanol and water affects the ratio of the concentrations of a solute in the two liquids.

Analyze Octanol is an alcohol with one OH group at the end of a chain of eight carbon atoms. The O and H atoms in OH groups can form hydrogen bonds to the H and O atoms on neighboring molecules. All molecules experience London dispersion forces that increase in strength with increasing molar mass. The immiscibility of octanol in water means that it is, on balance, hydrophobic and should be a good solvent for hydrophobic solutes. If the partition ratio of a compound is a constant, then the ratio of concentrations of the compound in octanol and water must be a constant that does not change with changes in the volumes of either or both of the solvents.

Solve

a. Molecules of octanol interact with each other via London dispersion forces, as all molecules do, and via hydrogen bonding between the OH groups on adjacent molecules.

b. The hydrophobicity of octanol is an indication that its molecules experience strong London dispersion forces due to their long, nonpolar chains of carbon and hydrogen atoms. The importance of these interactions is reinforced by how relatively small the difference is between the boiling points of octanol and nonpolar nonane, $CH_3(CH_2)_7CH_3$, a compound with nearly the same molar mass and molecular shape as octanol. Therefore, London dispersion forces are the principal intermolecular force among octanol molecules.

c. The long nonpolar chains in molecules of octanol make it an effective solvent for nonpolar, or hydrophobic, solutes. Those same solutes should have little solubility in water, which would give a relatively high concentration ratio in Equation 6.2 during a partition ratio determination. Thus, a compound with a large partition ratio is more hydrophobic than one with a smaller partition ratio.

d. Because the concentration ratio in Equation 6.2 is a constant, changing the relative volumes of octanol and water in a partition ratio determination should not alter the calculated value. On the other hand, doubling the volume of octanol would mean that twice as much of the total *quantity* of solute in the mixture would end up in the octanol phase.

Think About It Even though octanol contains an alcohol functional group, hydrogen bonding is not its predominant intermolecular force because the rest of the molecule (the larger proportion of the molecule—the hydrocarbon chain) experiences London dispersion forces. In addition to considering the presence (or absence) of functional groups, boiling points on the Kelvin scale also provide a measure of the strength of intermolecular forces. Theoretically, a substance that experiences no intermolecular interaction (admittedly, no such substance exists) would become a gas just as its temperature exceeded 0 K. Therefore, comparing the Kelvin-scale boiling points of compounds provides insights into the relative strengths of their intermolecular forces.

SUMMARY

LO1 All atoms and molecules experience **London dispersion forces** due to their **polarizability** and the existence of **temporary dipoles** even in particles that have no permanent dipole. Polarizability increases with increasing particle size, so larger atoms and molecules experience stronger London dispersion forces than smaller ones. Stronger London dispersion forces lead to higher boiling points and greater **viscosities. Hydrocarbons** are compounds whose molecules contain only hydrogen and carbon atoms. **Alkanes** are hydrocarbons in which each carbon atom is bonded to four other atoms. **Constitutional isomers** have the same molecular formulas but different connections between the atoms in their molecules. Ions in aqueous solution interact with water molecules through **ion–dipole interactions**, forming a **sphere of hydration** around the ion. Molecules with permanent dipoles interact through a combination of London dispersion forces and **dipole–dipole interactions**. The strongest dipole–dipole interactions are **hydrogen bonds**, which form between H atoms bonded to N, O, and F atoms and other N, O, and F atoms. Polar **organic compounds** contain polar **functional groups**. Among them are **carbonyl groups**, which contain C=O bonds. When the carbonyl carbon atom is bonded to two other carbon atoms in a molecular structure, the compound is called a **ketone**. Other polar functional groups include the **hydroxyl group** (OH) in **alcohols**. (Sections 6.1 and 6.2)

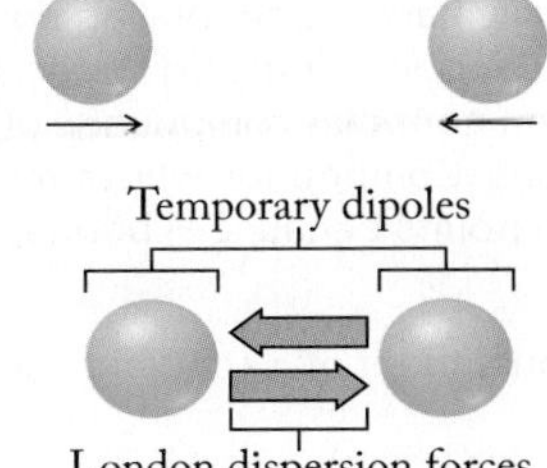

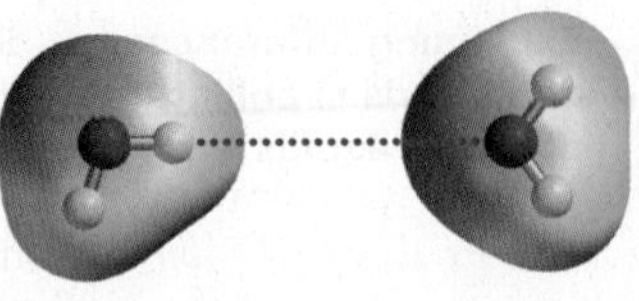

LO2 In a solution, the substance present in the largest amount (based on number of moles) is the **solvent**. All other substances are **solutes**.

LO3 Polar solutes dissolve in polar solvents when the dipole–dipole interactions between solute and solvent molecules offset the interactions that keep either solute molecules or solvent molecules together. The limited **solubility** of nonpolar solutes in polar solvents is a result of **dipole–induced dipole interactions. Hydrophilic** substances are more soluble in water than are **hydrophobic** substances, which are more soluble in nonpolar solvents. In general, polar solutes dissolve in polar solvents and nonpolar solutes dissolve in nonpolar solvents—that is, like dissolves like. (Section 6.3)

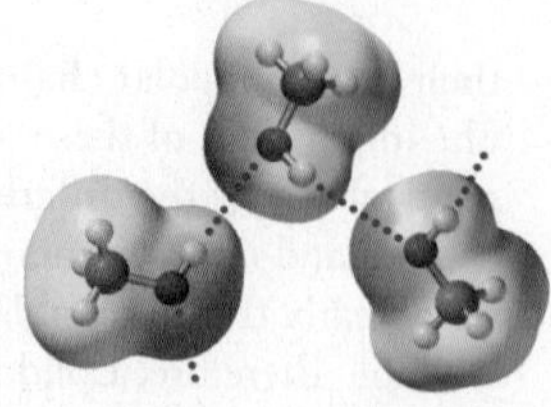

LO4 The ratio of force to surface area is defined as **pressure (*P*)**, which is often expressed in **standard atmospheres (atm)**, a unit equal to Earth's average atmospheric pressure at sea level. The **phase diagram** of a substance indicates whether it exists as a solid, liquid, gas, or **supercritical fluid** at a particular combination of pressure and temperature. Two adjoining regions in a phase diagram are separated by a line representing the pressures and temperatures at which the two phases can coexist. All three states (solid, liquid, and gas) exist in equilibrium at the **triple point**. Above the temperature and pressure of its **critical point**, a substance exists as a supercritical fluid with many of the physical properties of a gas but with the ability to dissolve other substances as if it were a liquid. (Section 6.4)

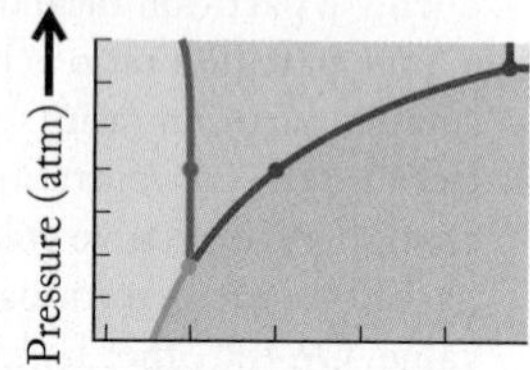

LO5 The remarkable behavior of water, including its unusually high **surface tension**, results from the strength of hydrogen bonding between its molecules. Those interactions also cause water to expand when it freezes and contribute to **capillary action**. (Section 6.5)

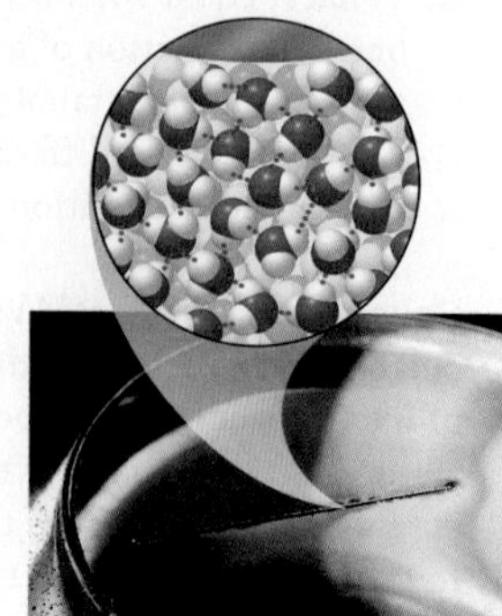

PARTICULATE **PREVIEW WRAP-UP**

The molecule in the middle (b) contains a C=O bond and the ketone functional group. The molecule on the left (a) contains an ether functional group, and the molecule on the right (c) contains an alcohol functional group. The boiling points of the compounds increase from left to right.

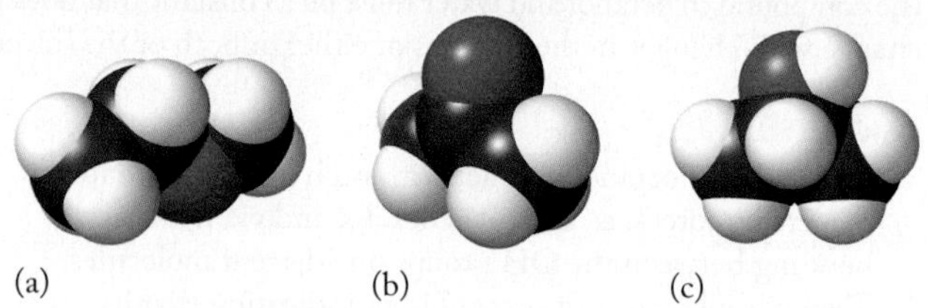

PROBLEM-SOLVING SUMMARY

Type of Problem	Concepts and Equations	Sample Exercises
Explaining differences and trends in boiling points of liquids	Substances made of large molecules usually have higher boiling points than those with smaller molecules because of London dispersion forces. Polar compounds have higher boiling points than nonpolar compounds of similar molar mass because of dipole–dipole interactions. Compounds whose molecules form hydrogen bonds have even higher boiling points because H bonds are especially strong dipole–dipole interactions.	**6.1**
Distinguishing solute from solvent	Calculate the number of moles of each constituent of a solution. The one with the greatest number of moles is the solvent.	**6.2**
Predicting solubility in water	Polar molecules are more soluble in water than nonpolar molecules. Molecules that form hydrogen bonds are more soluble in water than molecules that cannot form these bonds. Like dissolves like.	**6.3**
Interpreting phase diagrams	Locate the combination of temperature and pressure of interest on the phase diagram, and determine which physical state exists at that point. Changes in pressure at constant temperature are represented by vertical paths, and changes in temperature at constant pressure are represented by horizontal paths. If a path crosses a phase boundary line, there is a change in phase.	**6.4**

VISUAL PROBLEMS

(Answers to boldface end-of-chapter questions and problems are in the back of the book.)

6.1. Figure P6.1 shows molecular structures of four constitutional isomers of heptane (C_7H_{16}). Which one has the highest boiling point?

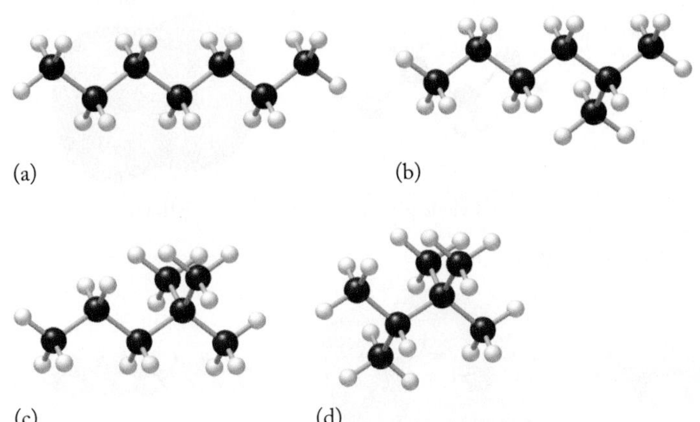

FIGURE P6.1

6.2. Which of the constitutional isomers of heptane in Figure P6.1 is the most viscous at 20°C?

6.3. Figure 6.3 contains the Lewis structures of ammonia (NH_3) and phosphine (PH_3). The boiling points of these compounds at $P = 1.00$ atm are 185 and 240 K. Which one boils at the higher temperature? Explain your selection.

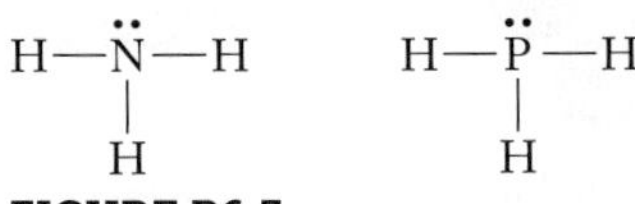

FIGURE P6.3

6.4. Figure P6.4 shows space-filling models of pentane (C_5H_{12}) and decane ($C_{10}H_{22}$). Which substance has the lower freezing point? Explain your answer.

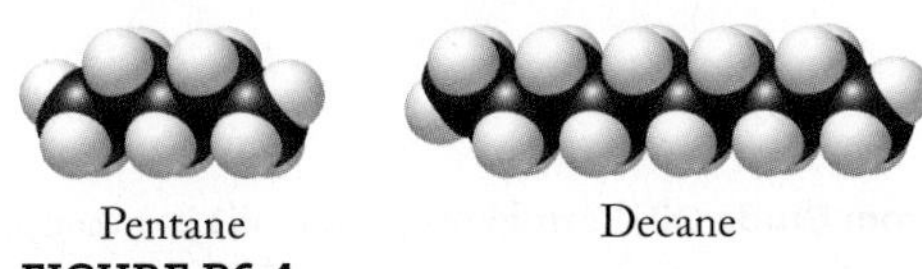

FIGURE P6.4

6.5. Identify the functional group in each molecule in Figure P6.5.

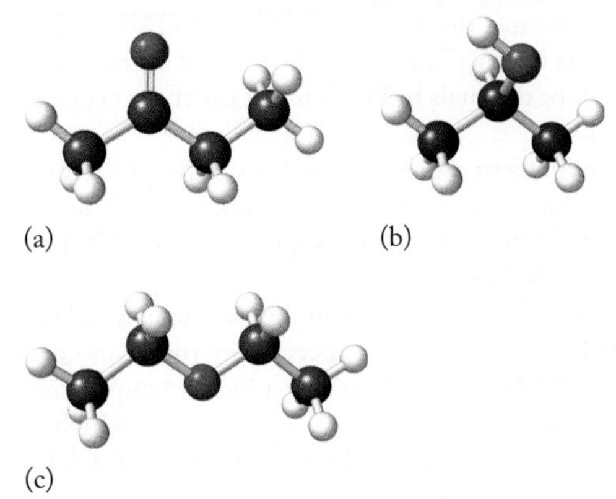

FIGURE P6.5

6.6. Which molecules in Figure P6.5 are constitutional isomers?

6.7. Figure P6.7 shows the phase diagram of imaginary molecular compound X. If a sample of X is left outside in a sealed container on a summer day, will the X in the container be a solid, liquid, or a gas at $P = 1.00$ atm?

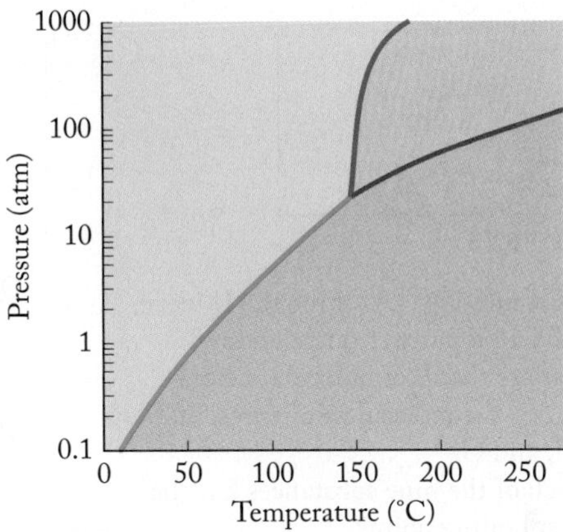

FIGURE P6.7

6.8. Suppose you bring the sample of X from Problem 6.7 inside and place it in an uncovered pot of boiling water on your kitchen stove. What phase changes, if any, will occur?

6.9. Another sample of X (Figure P6.7) is stored in a pressurized container ($P = 100$ atm) at 0°C. If the sample is then transferred to an oven and slowly warmed to 250°C, what phase changes, if any, will X undergo?

6.10. Does the green line in Figure P6.7 represent a series of (a) freezing points, (b) sublimation points, (c) boiling points, or (d) critical points?

6.11. Does the solid form of compound X (Figure P6.7) float on the liquid as the liquid begins to freeze at $P = 300$ atm?

6.12. The graph in Figure P6.12 is an expanded view of part of the phase diagram of water. In this version the pressure scale is linear instead of logarithmic. Which phases are represented by the blue and pink colors?

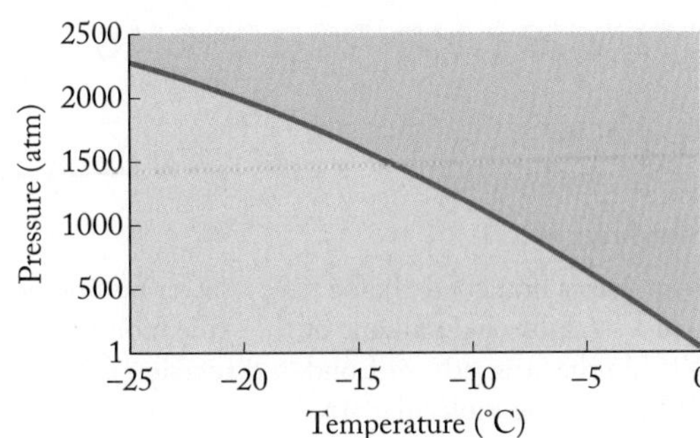

FIGURE P6.12

6.13. Referring to Figure P6.12, what phase change, if any, takes place if the temperature of a sample of water is increased from −25°C to −15°C while the pressure on it is decreased from 2500 atm to 1000 atm? Assume that the changes in temperature and pressure occur at constant rates over the same time interval.

*6.14. Suppose an ice skater weighing 50.0 kg and wearing newly sharpened ice skates stands on ice at a temperature of −10°C. If the surface area of the edges of the skates pressing down on the ice is 0.033 cm^2, will the ice under the skate edges melt? Figure P6.12 may help you make your prediction.

6.15. In Figure P6.15, identify the physical state (solid, liquid, or gas) of xenon and classify the attractive forces between the xenon atoms.

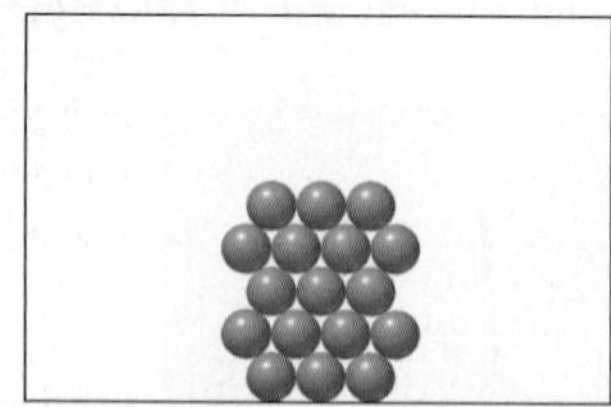

FIGURE P6.15

6.16. Use representations [A] through [I] in Figure P6.16 to answer questions a–f.

a. What are the intermolecular forces between the molecules in representations C, E, and G?
b. Which of the nine substances has the lowest boiling point?
c. Which representations are isomers of one another?
d. Which is more soluble in water: ethylene glycol or iodine? Which is more soluble in carbon tetrachloride? Explain your answers in terms of the intermolecular forces involved.
e. Which hydrocarbon has the higher boiling point?
f. Which substance does not form hydrogen bonds between its molecules but can form hydrogen bonds to another molecule?

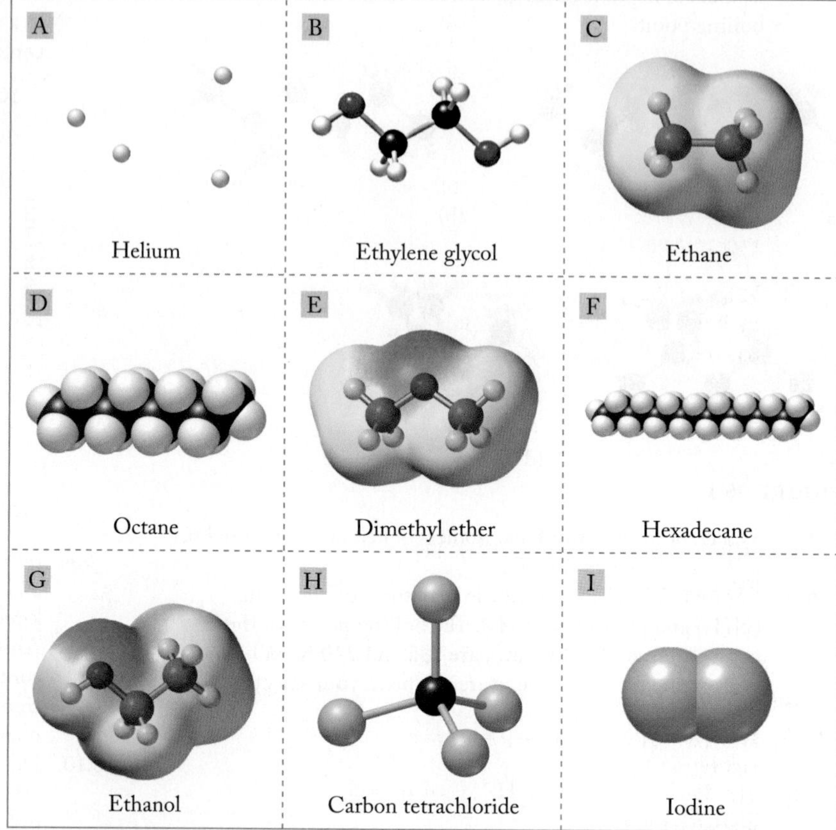

FIGURE P6.16

QUESTIONS AND PROBLEMS

London Dispersion Forces: They're Everywhere

Concept Review

6.17. Why does a branched alkane have a lower boiling point than a straight-chain alkane of the same molar mass?

6.18. Why do the strengths of London dispersion forces increase with increasing molecular size?

Problems

6.19. Select the compound in each of the following pairs whose molecules experience stronger London dispersion forces. (a) C_2Cl_6 or C_2F_6; (b) CH_4 or C_3H_8; (c) CS_2 or CO_2

6.20. The three most abundant gases in air are N_2, O_2, and Ar. Which of them has the highest boiling point, and which has the lowest boiling point?

6.21. **Fuels from Crude Oil** Petroleum (crude oil) is a complex mixture of mostly hydrocarbons that can be separated into useful fuels by distillation. Common petroleum-based fuels in order of increasing boiling point are gasoline, jet fuel, kerosene, diesel oil, and fuel oil. Which of the fuels contains hydrocarbons with the highest average molar mass?

6.22. Which of the fuels in Problem 6.21 is the most viscous at 20°C?

6.23. The permanent dipole moment of CH_2F_2 (1.93 D) is larger than that of CH_2Cl_2 (1.60 D), yet the boiling point of CH_2Cl_2 (40°C) is much higher than that of CH_2F_2 (−52°C). Why?

6.24. How is it that the permanent dipole moment of HCl (1.08 D) is larger than the permanent dipole moment of HBr (0.82 D), yet HBr boils at a higher temperature?

6.25. In each of the following pairs of molecules, which one experiences the stronger dispersion forces? (a) CCl_4 or CF_4; (b) CH_4 or C_3H_8

6.26. What kinds of intermolecular forces must be overcome as (a) solid CO_2 sublimes, (b) $CHCl_3$ boils, and (c) ice melts?

*6.27. The compound with the molecular structure in Figure 6.27 (b) melts at a higher temperature than the compound in (a). Explain why using the different types and strengths of the intermolecular forces experienced by the molecules.

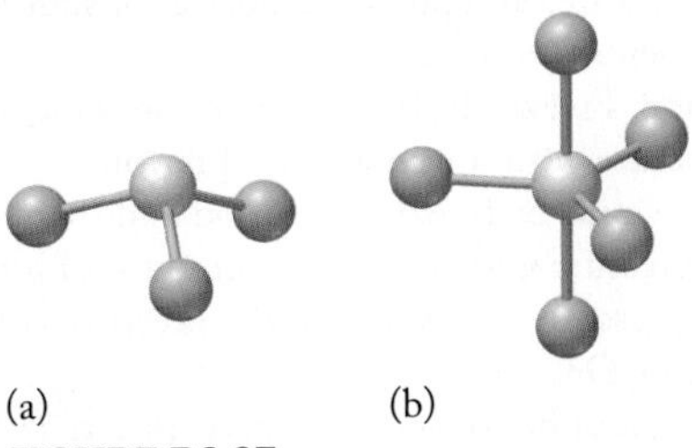

(a) (b)

FIGURE P6.27

*6.28. Which of the two compounds with the condensed molecular structures and formulas shown in Figure 6.28 has the higher boiling point?

$ClCH_2CH_2Cl$ $CHCl_2CH_3$

FIGURE P6.28

Interactions Involving Polar Molecules

Concept Review

6.29. How are water molecules oriented around anions in aqueous solutions?

6.30. How are water molecules oriented around cations in aqueous solutions?

6.31. Why are dipole–dipole interactions generally weaker than ion–dipole interactions?

6.32. Two liquids—one polar, one nonpolar—have the same molar mass. Which one is likely to have the higher boiling point?

6.33. Why are hydrogen bonds considered a special class of dipole–dipole interactions?

6.34. Can all polar hydrogen-containing molecules form hydrogen bonds?

Problems

6.35. Suggest two reasons why the boiling point of methyl fluoride, CH_3F, is higher than the boiling point of methane, CH_4.

6.36. Why is the boiling point of Br_2 lower than that of iodine monochloride, ICl, even though they have nearly the same molar mass?

6.37. Why do molecules of methanol (CH_3OH) form hydrogen bonds, but molecules of methane (CH_4) do not?

6.38. The boiling point of H_2S is lower than that of H_2O even though H_2S has twice the molar mass of H_2O. Why?

6.39. In which of the following compounds do the molecules experience dipole–dipole interactions? (a) CF_4; (b) CF_2Cl_2; (c) CCl_4; (d) $CFCl_3$

6.40. In which of the following compounds do molecules experience dipole–dipole interactions? (a) CO_2; (b) NO_2; (c) SO_2; (d) H_2S

6.41. Which of the following molecules can hydrogen-bond among themselves in pure samples of bulk material? (a) methanol (CH_3OH); (b) ethane (CH_3CH_3); (c) dimethyl ether (CH_3OCH_3); (d) acetic acid (CH_3COOH)

6.42. Which of the following molecules can hydrogen-bond with molecules of water? (a) methanol (CH_3OH); (b) ethane (CH_3CH_3); (c) dimethyl ether (CH_3OCH_3); (d) acetic acid (CH_3COOH)

(Before solving the following problems, you may find it useful to review Section 4.1 on the strengths of ionic bonds.)

6.43. In an aqueous solution containing chloride, bromide, and iodide salts, which anion would you expect to experience the strongest ion–dipole interactions with surrounding water molecules?

6.44. In an aqueous solution containing Na^+, Mg^{2+}, K^+, and Ca^{2+} salts, which cation would you expect to experience the strongest ion–dipole interactions?

Trends in Solubility

Concept Review

6.45. Which component in a solution is the solvent?

6.46. In a solution prepared by mixing equal masses of water and ethanol (CH_3CH_2OH), which ingredient is the solute and which is the solvent?

6.47. What is the difference, if there is any, between the terms *miscible* and *soluble*?

6.48. Which of the following substances have little solubility in water? (a) benzene, C_6H_6; (b) KBr; (c) Ar

6.49. In what context do the terms *hydrophobic* and *hydrophilic* relate to the solubilities of substances in water?

6.50. A series of alcohols has the generic molecular formula $C_nH_{(2n+1)}OH$. How does the value of n affect the solubility of these alcohols in water?

6.51. Which of the two compounds in Figure P6.27 is more soluble in water? Explain your reasoning.

*6.52. Are the two compounds in Figure P6.28 miscible with each other? Why do you think they are or are not?

Problems

6.53. In each of the following pairs of compounds, which compound is likely to be more soluble in water?

a. CCl_4 or $CHCl_3$
b. CH_3OH or $CH_3(CH_2)_4CH_2OH$
c. NaF or MgO
d. CaF_2 or BaF_2

6.54. In each of the following pairs of compounds, which compound is likely to be more soluble in CCl_4?

a. Br_2 or NaBr
b. CH_3CH_2OH or CH_3OCH_3
c. CS_2 or KOH
d. I_2 or CaF_2

6.55. Which of the following pairs of substances are likely to be miscible?
a. Br_2 and benzene (C_6H_6)
b. $CH_3CH_2OCH_2CH_3$ (diethyl ether) and CH_3COOH (acetic acid)
c. C_6H_{12} (cyclohexane) and hexane ($CH_3CH_2CH_2CH_2CH_2CH_3$)
d. CS_2 (carbon disulfide) and CCl_4 (carbon tetrachloride)

6.56. Which of the following pairs of substances is likely to be miscible?
a. CH_3CH_2OH (ethanol) and $CH_3CH_2OCH_2CH_3$ (diethyl ether)
b. CH_3OH (methanol) and methyl amine (CH_3NH_2)
c. CH_3CN (acetonitrile) and acetone (CH_3COCH_3)
d. CF_3CHF_2 (a Freon replacement) and $CH_3CH_2CH_2CH_2CH_3$ (pentane)

6.57. Which of the following compounds is likely to be the most soluble in water? (a) NaCl; (b) KI; (c) $Ca(OH)_2$; (d) CaO

6.58. Which sulfur compound would you predict to be more soluble in nonpolar solvents: SO_2 or CS_2?

6.59. Which of the following compounds is the most soluble in water?
a. $CH_3(CH_2)_2O(CH_2)_2CH_3$
b. $CH_3(CH_2)_3O(CH_2)_2CH_3$
c. CH_3OCH_3
d. $CH_3CH_2OCH_2CH_3$

6.60. Rank the ketones in Figure P6.60 from least soluble to most soluble in water.

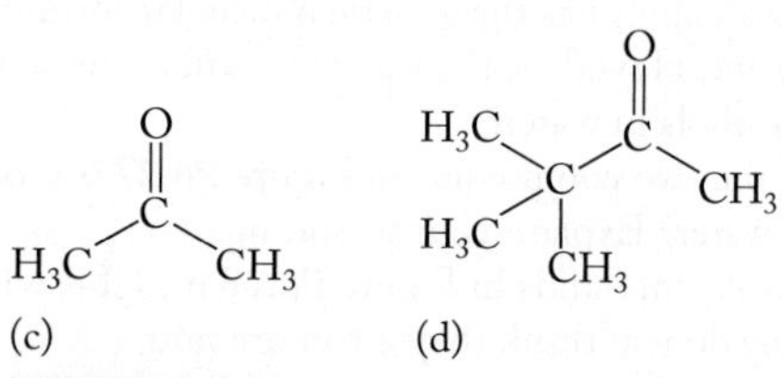

FIGURE P6.60

Phase Diagrams: Intermolecular Forces at Work

Concept Review

6.61. Explain the difference between sublimation and evaporation.

6.62. Can ice be melted merely by applying pressure? How about dry ice? Explain your answers.

6.63. Explain what is meant by the term *equilibrium line* as it is used in Section 6.4.

6.64. Explain how the solid–liquid line in the phase diagram of water (Figure 6.20) differs in character from the solid–liquid line in the phase diagrams of most other substances, such as CO_2 (Figure 6.22).

6.65. Which phase of a substance (gas, liquid, or solid) is most likely to be the stable phase: (a) at low temperatures and high pressures; (b) at high temperatures and low pressures?

6.66. At what temperatures and pressures does a substance behave as a supercritical fluid?

6.67. **Preserving Food** Freeze-drying is used to preserve food at low temperature with minimal loss of flavor. Freeze-drying works by freezing the food and then lowering the pressure with a vacuum pump to sublime the ice. Must the pressure be lower than the pressure at the triple point of H_2O?

6.68. Solid helium cannot be converted directly into the vapor phase. Does the phase diagram of He have a triple point?

Problems

(To solve Problems 6.69 through 6.78, you should consult Figures 6.20 and 6.22.)

6.69. What is the boiling point of water at a pressure of 200 atm?

6.70. What is the freezing point of water at $P = 1000$ atm?

6.71. What phase changes, if any, does liquid water at 100°C undergo if the initial pressure of 5.0 atm is reduced to 0.5 atm at constant temperature?

6.72. What phase changes, if any, occur if a sample of CO_2 initially at −80°C and 5.0 atm is warmed to −25°C and 5.0 atm?

6.73. Below what temperature can solid CO_2 (dry ice) be converted into CO_2 gas simply by lowering the pressure?

6.74. What is the maximum pressure at which solid CO_2 (dry ice) can be converted into CO_2 gas without melting?

6.75. Predict the phase of water that exists under the following conditions:
a. 2 atm of pressure and 110°C
b. 0.5 atm of pressure and 80°C
c. 7×10^{-3} atm of pressure and 3°C

6.76. Which phase or phases of water exist under the following conditions?
a. 0.32 atm and 70°C
b. 300 atm and 400°C
c. 1 atm and 0°C

6.77. List the steps you would take to convert a sample of water at 25°C and 1 atm pressure to water at its triple point.

6.78. List the steps you would take to convert a 10.0 g sample of water at 25°C and 2 atm pressure to ice at 1 atm pressure. At what temperature would the water freeze?

Some Remarkable Properties of Water

Concept Review

6.79. Explain why a needle floats on the surface of water but sinks in a container of methanol.

6.80. Explain why different liquids do not reach the same height in capillary tubes of the same diameter.

6.81. Explain why the plumbing in a house may burst if the temperature in the house drops below 0°C.

*6.82. A needle sinks when placed on the surface of hot (90°C) water but floats when placed on the surface of cold (5°C) water. Why?

6.83. The meniscus of mercury in a mercury thermometer (Figure P6.83) is convex. Why?

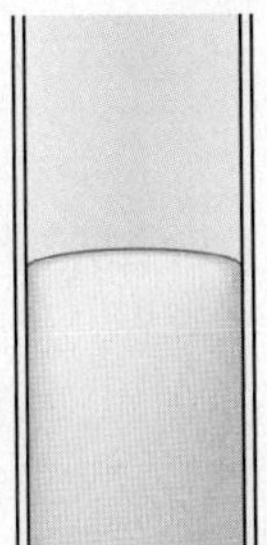

FIGURE P6.83

6.84. The mercury level in a capillary tube inserted into a dish of mercury is below the surface of the mercury in the dish. Why?

6.85. Describe the origin of surface tension in terms of intermolecular interactions.

6.86. Why does the density of water increase when it is cooled from 6°C to 4°C but then increase when it is cooled from 4°C to 2°C?

6.87. One of two glass capillary tubes of the same diameter is placed in a dish of water and the other is placed in a dish of ethanol (CH_3CH_2OH). Which liquid will rise higher in its tube?

6.88. Would you expect water to rise to the same height in a tube made of a polyethylene plastic as it does in a glass capillary tube of the same diameter? The molecular structure of polyethylene is shown in Figure P6.88.

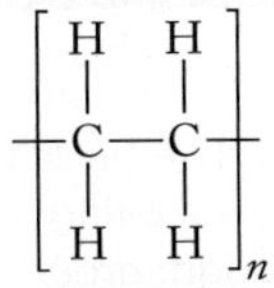

$$\left[-\mathrm{CH_2}-\mathrm{CH_2}- \right]_n$$

FIGURE P6.88

Additional Problems

6.89. Which type of intermolecular force exists in all substances?

6.90. Why does water climb higher in a capillary tube than in a test tube?

6.91. Methanol has a larger molar mass than water but boils at a lower temperature. Suggest a reason why.

6.92. What kinds of intermolecular forces must be overcome as (a) solid CO_2 sublimes; (b) $CHCl_3$ boils; (c) ice melts?

***6.93.** Which of these compounds has the lowest boiling point: (a) CH_4, (b) CH_3Cl, (c) CH_2Cl_2, (d) $CHCl_3$, (e) CCl_4.

*6.94. A partially filled bottle contains an equal number of moles of methanol (CH_3OH), ethanol (CH_3CH_2OH), and propanol ($CH_3CH_2CH_2OH$). Which of these alcohols is present in the highest mole percent in the vapor above the liquid mixture in the bottle?

6.95. Does the sublimation point of ice increase or decrease with increasing pressure?

6.96. Why is methanol miscible with water but methane is not?

***6.97.** The melting point of hydrogen is 15.0 K at 1.00 atm. The temperature of its triple point is 13.8 K. Does liquid H_2 expand or contract when it freezes?

6.98. Sketch a phase diagram for element Z, which has a triple point at 152 K and 0.371 atm, a boiling point of 166 K at a pressure of 1.00 bar, and a normal melting point of 161 K.

***6.99.** From among the compounds with the molecular structures shown in Figure P6.99, pick the ones that you think should be soluble in both water and octanol (see Sample Exercise 6.5).

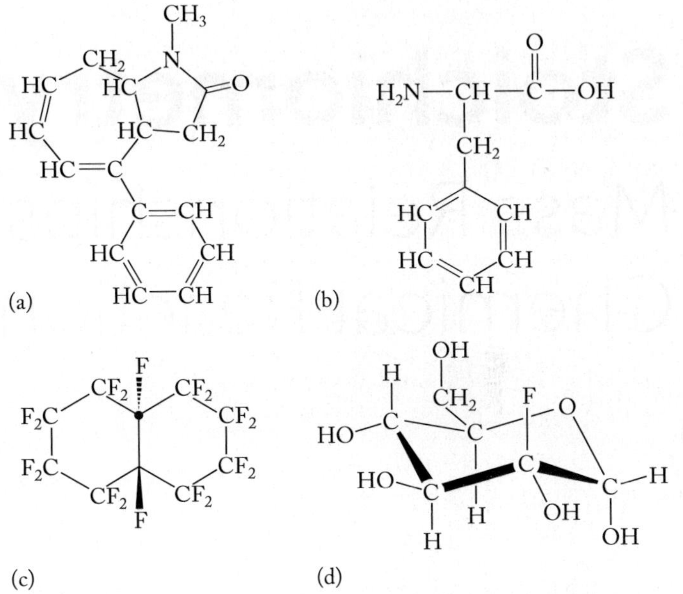

(a) (b) (c) (d)

FIGURE P6.99

*6.100. **First Aid for Bruises** Compounds with low boiling points may be sprayed on the skin as a topical anesthetic. They chill the skin as they evaporate and provide short-term relief from injuries. Predict which compound among those in Figure P6.100 has the lowest boiling point.

(a) (b) (c) (d)

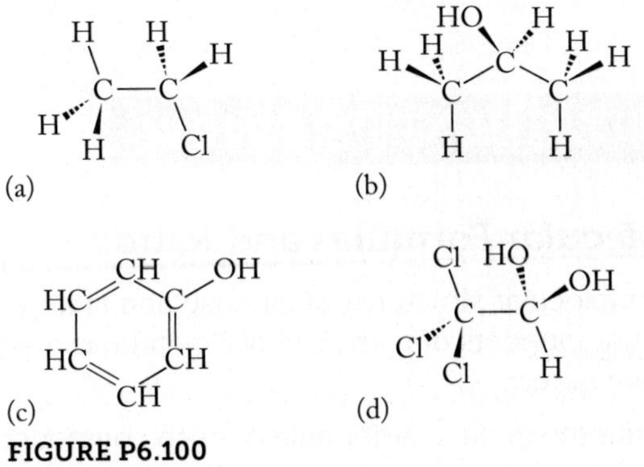

FIGURE P6.100

smartwork5

If your instructor uses Smartwork5, log in at digital.wwnorton.com/atoms2.

7

Stoichiometry

Mass Relationships and Chemical Reactions

PHOTOSYNTHESIS BY MICROORGANISMS Intense blooms of cyanobacteria turn a slow moving river in Puy-de-Dôme, France, bright green.

PARTICULATE **REVIEW**

Molecular Formulas and Ratios

The molecular structures of glucose and lactic acid—two components of carbohydrate metabolism—are shown here.

- Write the molecular formula of each compound.
- What is the ratio of carbon to hydrogen to oxygen in each compound?
- What functional groups are present in these compounds?

(Review Sections 1.6 and 6.2 if you need help answering these questions.)

(Answers to Particulate Review questions are in the back of the book.)

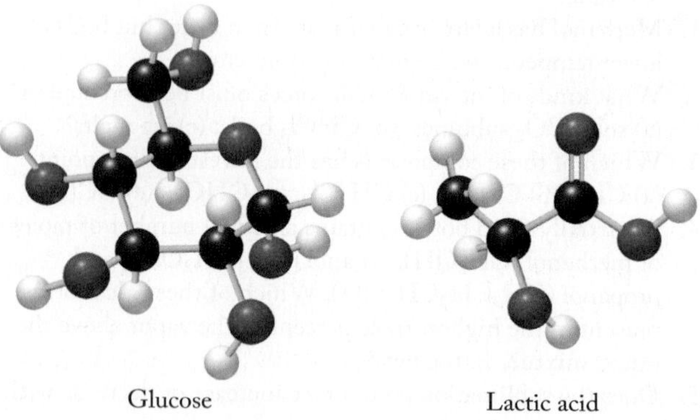

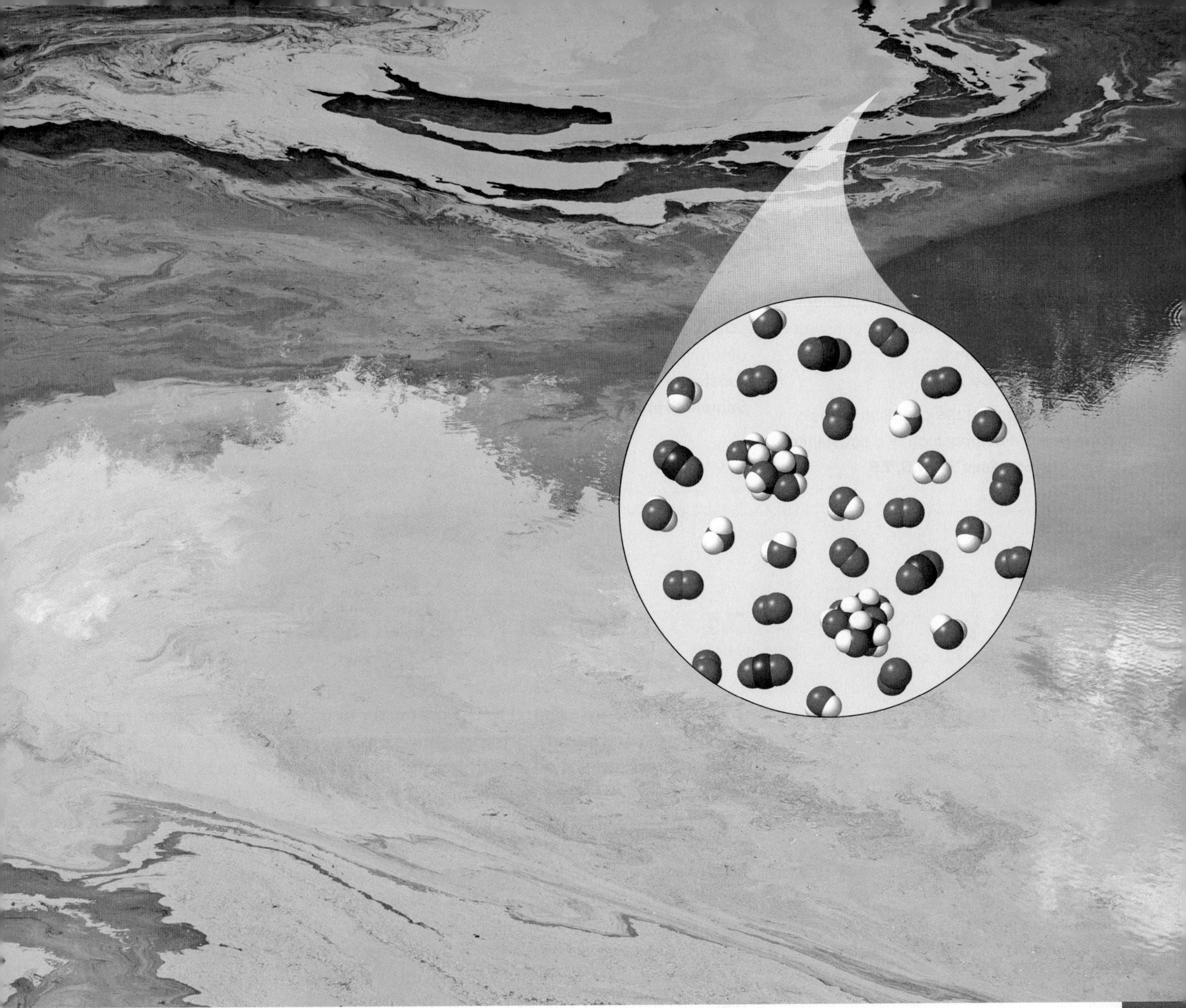

PARTICULATE **PREVIEW**

Unreacted Reactants

Nitromethane, CH_3NO_2, is a fuel used in drag racing. As you read Chapter 7, look for ideas that will help you to answer these questions about the combustion of nitromethane, as depicted here:

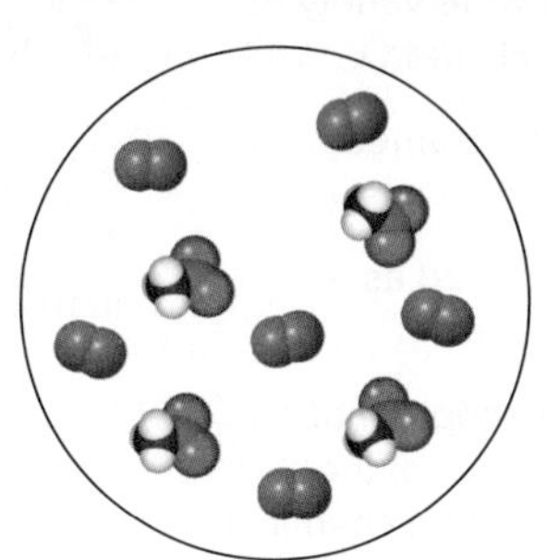

- What products are formed when a mixture of CH_3NO_2 and O_2 reacts? *Hint*: The nitrogen in the reactant is converted into $N_2(g)$ during the reaction.
- Write a balanced chemical equation describing the combustion of nitromethane.
- When the mixture shown in the circle reacts, how many CH_3NO_2 molecules remain unreacted? How many O_2 molecules remain unreacted?

Learning Outcomes

LO1 Write balanced chemical equations to describe chemical reactions
Sample Exercises 7.1, 7.2

LO2 Use balanced chemical equations to relate the masses of reactants consumed and products formed
Sample Exercise 7.3

LO3 Interconvert the chemical formula and percent composition of a substance
Sample Exercises 7.4, 7.5, 7.6

LO4 Determine the molecular formula of a substance from its percent composition and molecular mass
Sample Exercise 7.7

LO5 Use combustion analysis data to determine the empirical formula of a substance
Sample Exercise 7.8

LO6 Determine the limiting reactant and theoretical yield in a reaction mixture
Sample Exercises 7.9, 7.10

LO7 Calculate the percent yield in a chemical reaction
Sample Exercise 7.11

7.1 Chemical Reactions and the Carbon Cycle

CHEMTOUR Carbon cycle

Elements that are essential to the chemistry of life, including carbon, nitrogen, phosphorus, and sulfur, form intricate chemical webs, or cycles, that operate on both a global scale and locally within individual ecosystems. We begin this chapter by examining some of the chemical reactions in the carbon cycle. They are components of the six reaction pathways shown in Figure 7.1. Five of the six have been in operation for a very long time—about 2.5 billion years. They began following the evolution of a phylum of bacteria called *cyanobacteria* that is still an abundant component of Earth's biosphere as shown in this chapter's opening photo.

CONNECTION The energy of a photon can be calculated by multiplying the Planck constant (h) by the frequency (ν) of the photon (Section 3.3).

Cyanobacteria played a key role in launching the carbon cycle because they were among the first organisms capable of **photosynthesis**, a set of chemical reactions driven by energy from sunlight that, in its entirety, consumes carbon dioxide and water and turns these starting materials, or **reactants**, into different **products** depending on the organism involved. The photosynthesis that occurs in cyanobacteria and in green plants today produces O_2 gas and glucose ($C_6H_{12}O_6$), which serves as an energy source and as a building block for much of the biomass in all living things (see Pathways ① and ② in Figure 7.1).

Equation 7.1 provides a symbolic description of photosynthesis. It is an example of a **chemical equation**. Note that the formulas of the two reactants appear on the left, the formulas of the products are on the right, and a reaction arrow between them shows the direction of the reaction.

$$6\ CO_2(g) + 6\ H_2O(\ell) \xrightarrow{h\nu} C_6H_{12}O_6(aq) + 6\ O_2(g) \qquad (7.1)$$

The letters over the reaction arrow indicate that the reaction requires photons of electromagnetic energy (sunlight). The letters in parentheses indicate the physical states of the reactants and products: (ℓ) for liquids, (g) for gases, and (aq) for substances that are dissolved in water, which means they are in an *aqueous* solution. Had there been a solid reactant or product, its formula would have been followed by (s).

photosynthesis a set of chemical reactions driven by the energy of sunlight that convert carbon dioxide and water into oxygen and a wide variety of molecules containing C, H, and O.

reactants substance(s) consumed during a chemical reaction.

products substance(s) formed as a result of a chemical reaction.

chemical equation a description of the identities and proportions of the reactants and the products in a chemical reaction.

law of conservation of mass the principle that the sum of the masses of the reactants in a chemical reaction is equal to the sum of the masses of the products.

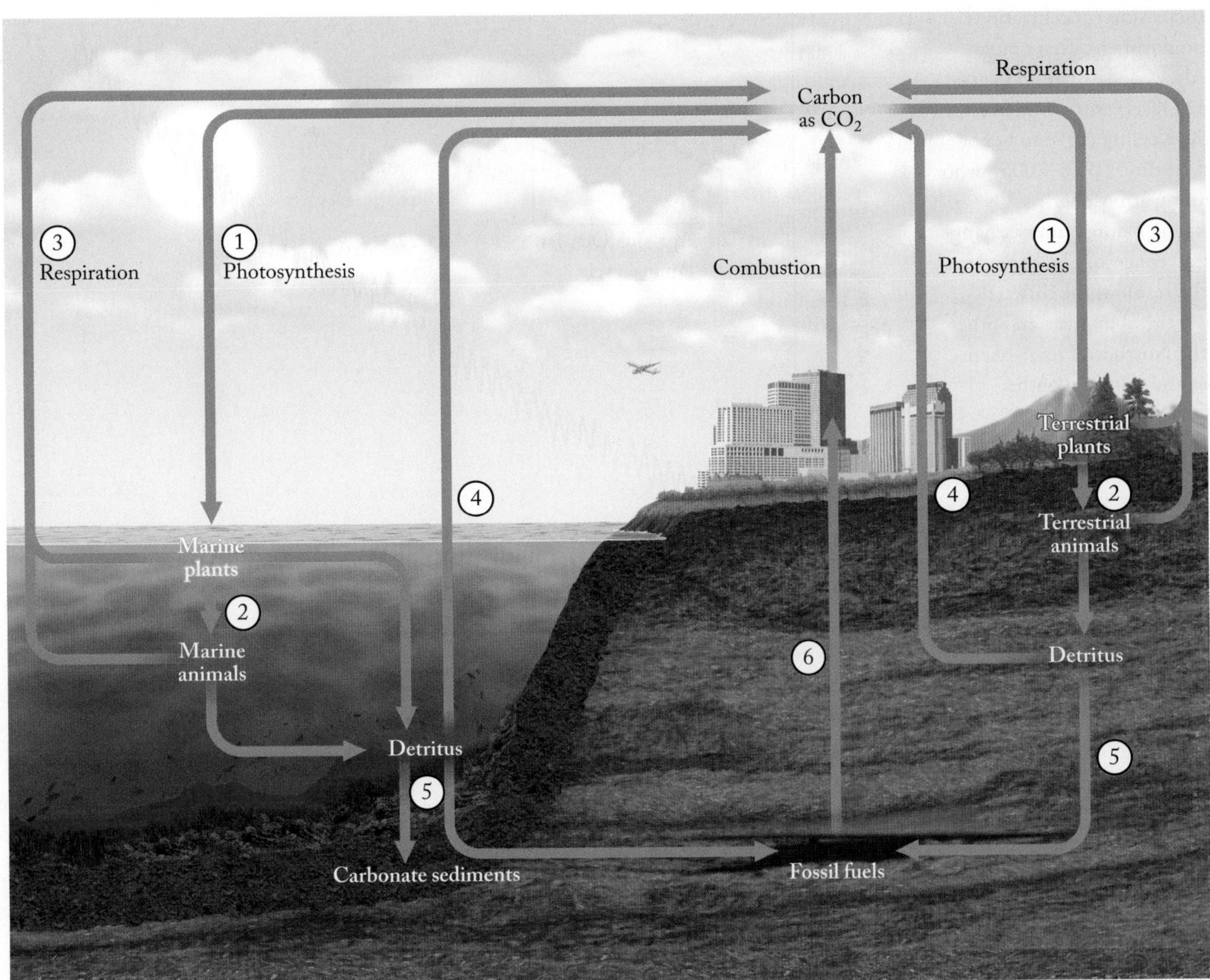

FIGURE 7.1 The carbon cycle. ①Terrestrial plants and marine plants incorporate CO_2 into their biomass. ②Some of the plant biomass becomes the biomass of animals. ③As plants and animals respire, they release CO_2 back into the environment. ④When they die, the decay of their tissues releases most of their carbon content as CO_2, but about 0.01% is incorporated into ⑤carbonate minerals and deposits of coal, petroleum, and natural gas (fossil fuels). ⑥Mining and the combustion of fossil fuels for human use are returning this carbon to the atmosphere.

Equation 7.1, like all chemical equations, is balanced, which means the total number of atoms of each element is the same on the left and the right sides of the reaction arrow. They have to be because atoms do not change their identities in chemical reactions. The need for balance is the reason the molecular formulas CO_2, H_2O, and O_2 are each preceded by the number 6, which means that six molecules of CO_2 and six molecules of H_2O are needed to produce one molecule of glucose and six molecules of O_2.

Equal numbers of atoms also means that the sum of the masses of the reactants always equals the sum of the masses of the products in a chemical equation. This equality is known as the **law of conservation of mass**, and it applies to all chemical reactions.

To harvest the chemical energy in glucose, plants and animals rely on a series of reactions called *respiration*, in which photosynthesis runs in reverse (Pathway ③ in Figure 7.1):

$$C_6H_{12}O_6(aq) + 6\ O_2(g) \rightarrow 6\ CO_2(g) + 6\ H_2O(\ell) \qquad (7.2)$$

On a global scale, photosynthesis and respiration proceed at nearly, but not exactly, the same rates. Each year about 0.01% of the decaying mass of dead plants and animals (called *detritus*) is incorporated into sediments and soils. Shielded from atmospheric O_2, this buried carbon is not converted back into CO_2, at least

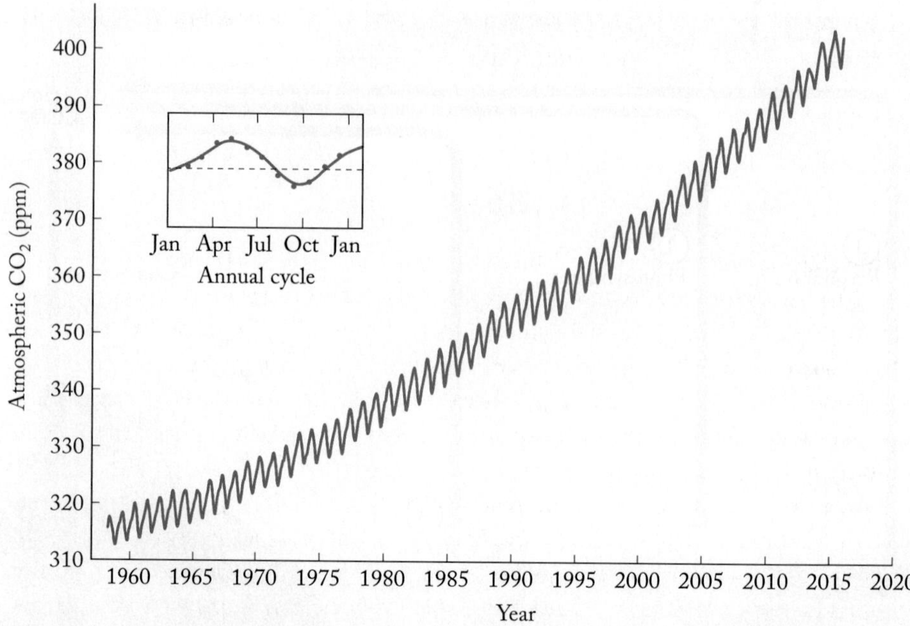

FIGURE 7.2 Increasing concentrations of CO_2 in the atmosphere. This graph is based on analyses done since 1958 at the Mauna Loa Observatory in Hawaii and is called the Keeling curve to honor Charles David Keeling (1928–2005), who began determining atmospheric CO_2 concentrations long before climate change was a widespread concern. The annual cycle in the data is the result of the different rates of photosynthesis by trees and other vegetation in the Northern Hemisphere during summer and winter months.

not right away. Although 0.01% may not seem like much, over more than 2 billion years it has resulted in the removal of about 10^{20} kilograms of carbon dioxide from the atmosphere. About 10^{15} kilograms of the buried carbon is in the form of fossil fuels: coal, petroleum, and natural gas.

CONNECTION The impact of increasing CO_2 concentration on global climate has been the subject of considerable debate, as we discussed in Chapter 4.

Fossil fuels are currently the primary source of energy in industrialized countries around the world. Each year approximately 8.2 trillion (8.2×10^{12}) kilograms of carbon are added to the atmosphere as CO_2. As a result, the concentration of CO_2 in the atmosphere has increased sharply, especially over the past 50 years (Figure 7.2). It is now over 400 ppm. To put this value in perspective, atmospheric CO_2 concentrations have oscillated between 190 and 280 ppm for the past 800,000 years and probably longer (Figure 7.3). The concentration of CO_2 in the atmosphere

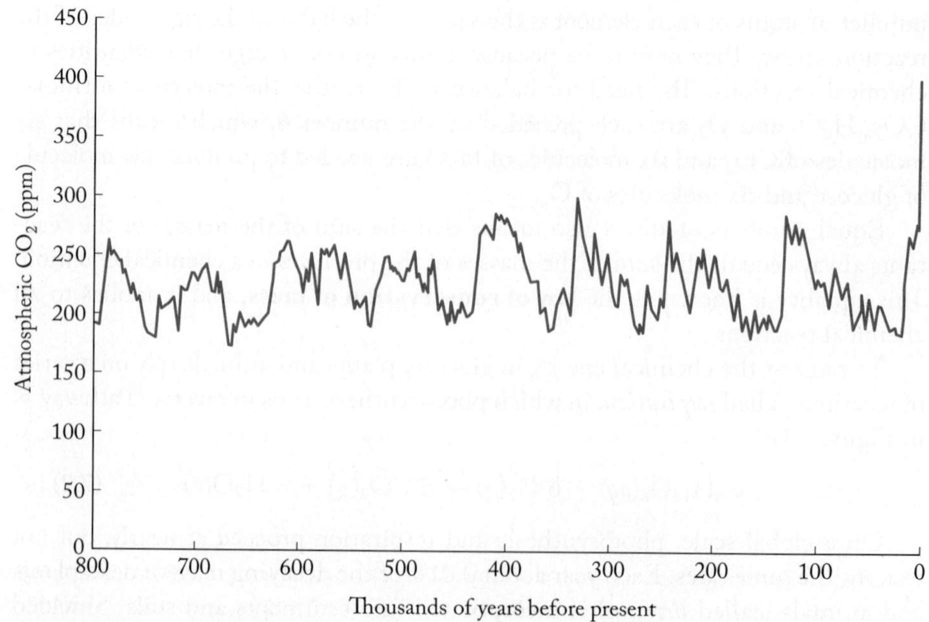

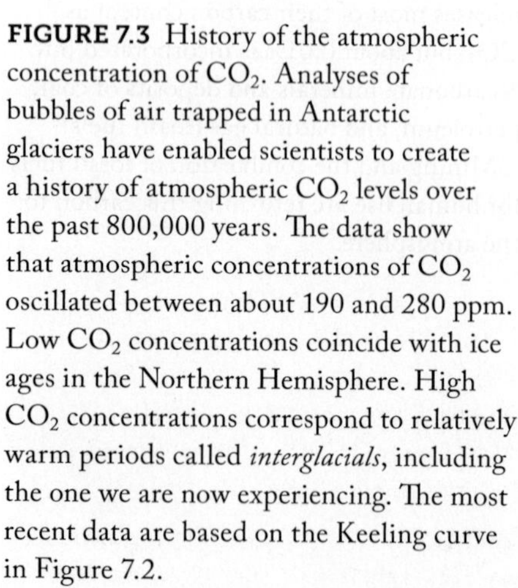

FIGURE 7.3 History of the atmospheric concentration of CO_2. Analyses of bubbles of air trapped in Antarctic glaciers have enabled scientists to create a history of atmospheric CO_2 levels over the past 800,000 years. The data show that atmospheric concentrations of CO_2 oscillated between about 190 and 280 ppm. Low CO_2 concentrations coincide with ice ages in the Northern Hemisphere. High CO_2 concentrations correspond to relatively warm periods called *interglacials*, including the one we are now experiencing. The most recent data are based on the Keeling curve in Figure 7.2.

today is more than 40% higher than at any time between our evolution as a species and the beginning of the Industrial Revolution (~1750–1850).

7.2 Writing Balanced Chemical Equations

CHEMTOUR Balancing Chemical Equations

In this chapter we use balanced chemical equations to describe several reactions linked to the carbon cycle and energy production and to relate the quantities of substances consumed and produced in these reactions. It's important that you be able to write complete, balanced chemical equations if given the identities of at least some of the reactants and products involved. To help you develop this ability, let's start with simple **combination reactions** in which two or more reactants combine to form a single product.

Some important combination reactions are also **combustion** reactions in which fuel rapidly reacts with O_2 and in the process releases energy. One of the simplest combustion reactions involves the complete combustion of carbon, the primary component of coal. The product of this reaction is carbon dioxide. At the atomic level, it takes just one atom of carbon and one molecule (two atoms) of oxygen to form one molecule of carbon dioxide:

Translating this atom-scale view of the reaction into a chemical equation, we get

$$C(s) + O_2(g) \rightarrow CO_2(g) \qquad (7.3)$$

Note that Equation 7.3 is already balanced when written with one atom or one molecule of each of the ingredients. Another combination reaction happens during the combustion of coal that contains sulfur impurities:

$$S(s) + O_2(g) \rightarrow SO_2(g) \qquad (7.4)$$

Like Equation 7.3, this one is balanced automatically with single particles of the reactants and product. When SO_2 enters the atmosphere, it may form sulfur trioxide in another combination reaction:

$$SO_2(g) + O_2(g) \rightarrow SO_3(g)$$

Taking an inventory of the atoms of S and O in this expression, we find that the S atoms are balanced, but not the O atoms:

$$SO_2(g) + O_2(g) \rightarrow SO_3(g)$$

$$\text{Atoms:} \qquad 1\,S + 4\,O \rightarrow 1\,S + 3\,O$$

The presence of an odd number of O atoms on the right is a problem because both sources of O atoms on the left contain 2 O atoms each. To solve this problem we add a coefficient of 2 in front of SO_3. Doing so gives us the following atom inventory:

$$SO_2(g) + O_2(g) \rightarrow \mathbf{2}\,SO_3(g)$$

$$\text{Atoms:} \qquad 1\,S + 4\,O \rightarrow 2\,S + 6\,O$$

Now both S and O are unbalanced. Since there is only one source of S atoms on the left and only one product, let's rebalance sulfur next by placing a coefficient of

combination reaction a reaction in which two or more substances form a single product.

combustion a rapid reaction between fuel and oxygen that produces energy.

2 in front of SO_2. Doing so gives us the following atom inventory and a balanced chemical equation:

$$\mathbf{2}\ SO_2(g) + O_2(g) \rightarrow \mathbf{2}\ SO_3(g) \tag{7.5}$$

$$\text{Atoms:}\quad 2\ S + 6\ O \rightarrow 2\ S + 6\ O \quad ✓$$

In this example balancing the number of S atoms also balanced the number of O atoms on both sides of the reaction arrow. Balancing two elements at once is not unusual in writing chemical equations. It is important to keep in mind that writing balanced chemical equations is an exercise in using the appropriate coefficients to make the number of atoms of each element the same on both sides of the reaction arrow. You may *not* change the subscripts in chemical formulas, because doing that changes the identity of a reactant or product and the nature of the reaction itself.

When SO_3 forms in the atmosphere, it may combine with water vapor, producing liquid sulfuric acid, H_2SO_4:

$$SO_3(g) + H_2O(g) \rightarrow H_2SO_4(\ell) \tag{7.6}$$

Atmospheric H_2SO_4 dissolves in rain, producing an aqueous solution of sulfuric acid known as acid rain. If this rain lands on objects made of limestone or marble, which are minerals composed of calcium carbonate ($CaCO_3$), the following chemical reaction occurs:

$$CaCO_3(s) + H_2SO_4(aq) \rightarrow CaSO_4(aq) + H_2O(\ell) + CO_2(g) \tag{7.7}$$

This is not a combination reaction because it has multiple products. One of them, calcium sulfate ($CaSO_4$), is more soluble in water than $CaCO_3$. Thus, the reaction in Equation 7.7 can result in the loss of distinctive features in statues and other structures made of limestone or marble (Figure 7.4). In these contexts the

1935

1994

FIGURE 7.4 Acidic precipitation reacts with limestone and marble statues such as this one of George Washington in New York City, converting calcium carbonate into more soluble calcium sulfate, which is washed away by rain and melting snow. The damage is particularly noticeable around Washington's mouth and chin.

reaction is an example of *chemical weathering*, a process that is continually changing minerals on Earth's surface and, sadly, objects sculpted from those minerals.

Combustion of Hydrocarbons

In recent years the use of coal as a fuel for heating and generating electricity has been declining, replaced in large part by natural gas, which is a cleaner burning fuel. Natural gas is mostly methane, CH_4. This simplest of all hydrocarbons produces more energy in the United States and in many other industrialized countries than any other single substance. The complete combustion of methane (or any hydrocarbon) rapidly consumes O_2 and produces CO_2 and H_2O. Let's write a balanced chemical equation describing the combustion of methane, starting with what we know about its reactants and products:

$$CH_4(g) + O_2(g) \rightarrow CO_2(g) + H_2O$$

$$\text{Atoms:} \quad 1\,C + 4\,H + 2\,O \rightarrow 1\,C + 2\,H + 3\,O$$

CONNECTION Hydrocarbons are compounds whose molecules contain only atoms of hydrogen and carbon (Section 6.1).

According to the inventory of atoms on each side of the reaction arrow, the number of C atoms is already balanced, but H and O atoms are not. There are H atoms in only one reactant and product, so let's balance it next. There are four H atoms in the molecule of methane on the left and only two in the molecule of water on the right, so we place a coefficient of 2 in front of H_2O:

$$CH_4(g) + O_2(g) \rightarrow CO_2(g) + \mathbf{2}\,H_2O$$

$$\text{Atoms:} \quad 1\,C + 4\,H + 2\,O \rightarrow 1\,C + 4\,H + 4\,O$$

Now C and H are balanced but O is not: there are four atoms of O on the right but only two on the left. In this chemical equation, balancing C and H did not automatically balance O because there is a source of O atoms, O_2, that does not contain C or H atoms. The good news is that we can increase the number of O atoms on the left in multiples of two by increasing the number of O_2 molecules in the equation, and doing so does not change the numbers of C and H atoms. So, we place a coefficient of 2 in front of O_2 to get four atoms of O on both sides of the reaction arrow and a balanced chemical equation:

$$CH_4(g) + \mathbf{2}\,O_2(g) \rightarrow CO_2(g) + \mathbf{2}\,H_2O$$

$$\text{Atoms:} \quad 1\,C + 4\,H + 4\,O \rightarrow 1\,C + 4\,H + 4\,O \quad ✓$$

We did not assign a physical state to H_2O in the balanced chemical equation, because we did not specify the temperature at which the reaction took place. It might have been reasonable to assume that, in the hot flame produced by the combustion of methane, the water would exist as water vapor. Under these conditions, the complete version of the chemical equation would be

$$CH_4(g) + 2\,O_2(g) \rightarrow CO_2(g) + 2\,H_2O(g) \qquad (7.8)$$

If the reaction mixture has cooled to room temperature, however, the complete reaction equation is

$$CH_4(g) + 2\,O_2(g) \rightarrow CO_2(g) + 2\,H_2O(\ell) \qquad (7.9)$$

FIGURE 7.5 Complete combustion of natural gas (mostly CH_4) produces a characteristic blue flame due to emission from excited-state molecular fragments, which are formed early in the course of the reaction.

This may seem like an unimportant detail, but the physical states of reactants and products influence how much energy is released or absorbed in a chemical reaction. We will study this phenomenon in detail in Chapter 9. For now, be assured that expressing the right physical states in chemical equations is important.

If we could peer inside the blue flame (Figure 7.5) produced by the complete combustion of methane and observe the molecules that are formed during the reaction, we would discover that the reactants are not transformed into final products all at once. Instead the reaction happens in stages. Both the speed at which a reaction takes place and the molecular-level view of the steps in the process by which reactants are converted into products are subjects in chemical kinetics that we deal with in Chapter 13. Here let us just note that, in an early stage of the reaction, molecules of CH_4 split apart into molecular fragments that combine with oxygen, forming a mixture of carbon monoxide, hydrogen gas, and water vapor:

$$CH_4(g) + O_2(g) \rightarrow CO(g) + H_2(g) + H_2O(g) \qquad (7.10)$$

Note that this equation is balanced as written. Next, the hydrogen formed in the reaction combines with more oxygen to form more water vapor. Writing single molecules of these reactants and products in a preliminary reaction expression gives us

$$H_2(g) + O_2(g) \rightarrow H_2O(g)$$

$$\text{Atoms:} \quad 2\,H + 2\,O \rightarrow 2\,H + 1\,O$$

According to our inventory, the H atoms are in balance but the O atoms are not. We need at least two O atoms on the right side because their source is diatomic molecules of O_2 on the left. Placing a coefficient of 2 in front of H_2O resolves the O imbalance but creates a new imbalance in H atoms:

$$H_2(g) + O_2(g) \rightarrow \mathbf{2}\,H_2O(g)$$

$$\text{Atoms:} \quad 2\,H + 2\,O \rightarrow 4\,H + 2\,O$$

We can balance the number of H atoms by placing a coefficient of 2 in front of H_2, which gives us a balanced chemical equation:

$$\mathbf{2}\,H_2(g) + O_2(g) \rightarrow \mathbf{2}\,H_2O(g) \qquad (7.11)$$

$$\text{Atoms:} \quad 4\,H + 2\,O \rightarrow 4\,H + 2\,O \quad ✓$$

Next, the carbon monoxide formed in the first step of the process (Equation 7.10) combines with more O_2, forming CO_2. Writing single molecules of these reactants and product in a preliminary reaction expression gives us

$$CO(g) + O_2(g) \rightarrow CO_2(g)$$

$$\text{Atoms:} \quad 1\,C + 3\,O \rightarrow 1\,C + 2\,O$$

In this case, the C atoms are in balance but the O atoms are not. Balancing the O atoms is complicated by the fact that the number on the right side must be an even number because there are two O atoms in each molecule of CO_2. Putting a coefficient of 2 in front of CO_2 in an attempt to increase the number of

O atoms on the right gives us one too many on the right, as well as one too many C atoms:

$$CO(g) + O_2(g) \rightarrow \mathbf{2}\ CO_2(g)$$

$$\text{Atoms:} \quad 1\ C + 3\ O \rightarrow 2\ C + 4\ O$$

Giving CO_2 a coefficient of 2 may not seem like a smart move, but it turns out that it is because we can now simultaneously bring both C and O into balance by adding a coefficient of 2 in front of CO:

$$\mathbf{2}\ CO(g) + O_2(g) \rightarrow \mathbf{2}\ CO_2(g) \tag{7.12}$$

$$\text{Atoms:} \quad 2\ C + 4\ O \rightarrow 2\ C + 4\ O \quad ✓$$

+ →

Writing Equation 7.12 demonstrates that balancing chemical equations is sometimes like playing a game of chess or solving a Rubik's Cube: you have to plan ahead, using one step to set up the next, even when it produces an expression that seems less balanced than the one that came before it.

We have just developed three balanced chemical equations that describe the stepwise combustion of methane:

Step 1	$CH_4(g) + O_2(g) \rightarrow CO(g) + H_2(g) + H_2O(g)$	(7.10)
Step 2	$2\ H_2(g) + O_2(g) \rightarrow 2\ H_2O(g)$	(7.11)
Step 3	$2\ CO(g) + O_2(g) \rightarrow 2\ CO_2(g)$	(7.12)

All three reactions occur rapidly in a flame, though the rate of the last one, conversion of CO to CO_2, is not quite as rapid as the other two. Inevitably, some CO escapes from the reaction mixture (the flame) before the third reaction is complete. This is a hazardous situation because CO is toxic, producing nausea and headaches at concentrations between 10 and 100 parts per million (ppm) in the air, and CO can be fatal at concentrations over 100 ppm. Formation of CO is a concern in the combustion of other hydrocarbons, including gasoline in automobile engines.

We can combine the three steps in the combustion of methane (Equations 7.10–7.12) to produce the overall reaction (Equation 7.8). Doing so is an exercise in balancing the numbers of molecules of substances that are products of early steps and reactants in later steps so that they cancel out when we add the reaction equations. For instance, H_2 is a product of step 1 and a reactant in step 2, and CO is a product of step 1 and a reactant in step 3. Before we combine the three steps, we need to balance the number of molecules of H_2 and CO in these steps. One way to do that is by multiplying all the coefficients in step 1 by two. This gives us two molecules of H_2 and CO on the product side, which balance the two molecules of H_2 and CO on the reactant sides of steps 2 and 3, respectively. So, multiplying step 1 by a factor of 2 and then adding it to steps 2 and 3 yields

2 × Step 1	$2\ CH_4(g) + 2\ O_2(g) \rightarrow 2\ CO(g) + 2\ H_2(g) + 2\ H_2O(g)$
+ Step 2	$2\ H_2(g) + O_2(g) \rightarrow 2\ H_2O(g)$
+ Step 3	$2\ CO(g) + O_2(g) \rightarrow 2\ CO_2(g)$

$$2\ CH_4(g) + 2\ H_2(g) + 2\ CO(g) + 4\ O_2(g) \rightarrow 2\ CO(g) + 2\ H_2(g) + 2\ CO_2(g) + 4\ H_2O(g)$$

We cancel out the terms that appear on both sides of the reaction arrow:

$$2\,CH_4(g) + \cancel{2\,H_2(g)} + \cancel{2\,CO(g)} + 4\,O_2(g) \rightarrow \cancel{2\,CO(g)} + \cancel{2\,H_2(g)} + 2\,CO_2(g) + 4\,H_2O(g)$$

or

$$2\,CH_4(g) + 4\,O_2(g) \rightarrow 2\,CO_2(g) + 4\,H_2O(g)$$

We can simplify this chemical equation by dividing all coefficients by 2:

$$CH_4(g) + 2\,O_2(g) \rightarrow CO_2(g) + 2\,H_2O(g)$$

This is Equation 7.8, which describes the overall reaction.

SAMPLE EXERCISE 7.1 Writing a Balanced Chemical Equation I **LO1**

Methane is the principal ingredient in natural gas, but ethane, C_2H_6, is also present. Write a balanced chemical equation for the complete combustion of C_2H_6.

Collect and Organize We need to write a balanced chemical equation describing the complete combustion of ethane. Combustion reactions involve the rapid reaction of fuels (C_2H_6 in this case) with O_2. The products of the complete combustion of hydrocarbons such as ethane are CO_2 and H_2O.

Analyze The first step involves writing a reaction expression with single molecules of the reactants (ethane + O_2) and the products (CO_2 + H_2O). At the high temperature of a combustion flame, all four species in the reaction are gases. Next we balance the number of C and H atoms because they are each present in only one reactant and one product, and we finish by balancing the total number of O atoms on each side of the reaction arrow.

Solve The preliminary reaction expression and its atom inventory are

$$C_2H_6(g) + O_2(g) \rightarrow CO_2(g) + H_2O(g)$$

$$\text{Atoms:}\quad 2\,C + 6\,H + 2\,O \rightarrow 1\,C + 2\,H + 3\,O$$

We balance C atoms first because, like the H atoms in this reaction, they are present in only one reactant and in only one product. Placing a coefficient of 2 in front of CO_2:

$$C_2H_6(g) + O_2(g) \rightarrow \mathbf{2}\,CO_2(g) + H_2O(g)$$

$$\text{Atoms:}\quad 2\,C + 6\,H + 2\,O \rightarrow 2\,C + 2\,H + 5\,O$$

followed by 3 in front of H_2O to balance the number of H atoms:

$$C_2H_6(g) + O_2(g) \rightarrow 2\,CO_2(g) + \mathbf{3}\,H_2O(g)$$

$$\text{Atoms:}\quad 2\,C + 6\,H + 2\,O \rightarrow 2\,C + 6\,H + 7\,O$$

This balances the number of C and H atoms but leaves an odd number of O atoms on the product side. The only source of O atoms is O_2, so we need an even number of O atoms on the right. One way to get an even number without disrupting the balance of C and H atoms is by multiplying all the coefficients in the expression by 2:

$$\mathbf{2}\,C_2H_6(g) + \mathbf{2}\,O_2(g) \rightarrow \mathbf{4}\,CO_2(g) + \mathbf{6}\,H_2O(g)$$

$$\text{Atoms:}\quad 4\,C + 12\,H + 4\,O \rightarrow 4\,C + 12\,H + 14\,O$$

We can now balance the number of O atoms by replacing the coefficient of 2 in front of O_2 with 7, which gives us five more O_2 molecules, or 10 more O atoms, and a balanced equation:

$$2\,C_2H_6(g) + \mathbf{7}\,O_2(g) \rightarrow 4\,CO_2(g) + 6\,H_2O(g)$$

$$\text{Atoms:}\quad 4\,C + 12\,H + 14\,O \rightarrow 4\,C + 12\,H + 14\,O \quad ✓$$

Think About It Writing a balanced chemical equation for the combustion of C_2H_6 required one more step (to get an even number of O atoms on both sides of the reaction arrow) than writing the chemical equation for the combustion of CH_4. This step was needed because we had an odd number of molecules of H_2O on the right side of the reaction arrow after balancing the number of H atoms. Can you see how this will happen every time the number of H atoms per molecule of a hydrocarbon fuel *is not divisible by 4*? For another illustration, try the following Practice Exercise.

Practice Exercise Write the chemical equation describing the complete combustion of butane (C_4H_{10}), the fuel in disposable lighters.

(Answers to Practice Exercises are in the back of the book.)

There are other ways to balance chemical equations, but the approach taken in Sample Exercise 7.1 works for many of them. The steps can be summarized as follows:

1. Write an expression using correct chemical formulas for all of the known reactants and products. Include symbols indicating physical states. Check whether the expression is already balanced. If so, your work is done.
2. If not, choose an element that appears in only one reactant and product to balance first. Insert the appropriate coefficient(s) to balance this element. Check whether the expression is now balanced. If so, your work is done.
3. If not, choose the element that appears in the next fewest reactants and products and balance it. Repeat the process for additional elements if necessary.

SAMPLE EXERCISE 7.2 Writing a Balanced Chemical Equation II — LO1

Write a balanced chemical equation for the chemical weathering reaction between acid rain that contains sulfuric acid and the iron-containing mineral hematite (Fe_2O_3), also known as iron(III) oxide. One of the products of the reaction is water-soluble $Fe_2(SO_4)_3$; the other is water itself.

Collect and Organize We are given the identities of the reactants and products. We also know something about their physical states: $Fe_2(SO_4)_3$ is soluble in water, so its symbol is (*aq*); H_2SO_4 is dissolved in rain, so its symbol is also (*aq*). A mineral is a solid substance of limited solubility in liquid water, so the symbol after Fe_2O_3 is (*s*).

Analyze To write a balanced chemical equation, we start with a preliminary expression that contains one molecule or formula unit of each reactant and product. Then we take inventories of the number of atoms of each of the elements in the reaction mixture and we balance them, starting with those that appear in only one reactant and product.

Solve

1. The preliminary expression relating reactants and products is

$$Fe_2O_3(s) + H_2SO_4(aq) \rightarrow Fe_2(SO_4)_3(aq) + H_2O(\ell)$$

When we take an inventory of the atoms on both sides, we get

$$\text{Atoms:}\quad 2\,\text{Fe} + 2\,\text{H} + 1\,\text{S} + 7\,\text{O} \rightarrow 2\,\text{Fe} + 2\,\text{H} + 3\,\text{S} + 13\,\text{O}$$

2. The iron atoms are already balanced, so let's focus on sulfur. There are three S atoms on the right but only one on the left. To balance sulfur, we place a coefficient of 3 in front of H_2SO_4 and recalculate the distribution of atoms on both sides:

$$Fe_2O_3(s) + \mathbf{3}\, H_2SO_4(aq) \rightarrow Fe_2(SO_4)_3(aq) + H_2O(\ell)$$

$$\text{Atoms:}\quad 2\,Fe + 6\,H + 3\,S + 15\,O \rightarrow 2\,Fe + 2\,H + 3\,S + 13\,O$$

3. Addressing the imbalance in hydrogen atoms next, there are six on the left and only two on the right. Placing a coefficient of 3 in front of H_2O balances not only hydrogen but also the entire equation:

$$Fe_2O_3(s) + 3\, H_2SO_4(aq) \rightarrow Fe_2(SO_4)_3(aq) + \mathbf{3}\, H_2O(\ell) \qquad (7.13)$$

$$\text{Atoms:}\quad 2\,Fe + 6\,H + 3\,S + 15\,O \rightarrow 2\,Fe + 6\,H + 3\,S + 15\,O \quad ✓$$

Think About It Balancing the equation required balancing the numbers of atoms of four elements. Balancing the three that were part of only one reactant and one product also balanced the fourth (oxygen), which was part of all four reactants and products.

 **Practice Exercise** Balance the chemical equation for the reaction between $P_4O_{10}(s)$ and liquid water that produces phosphoric acid, $H_3PO_4(\ell)$.

7.3 Stoichiometric Calculations

In Section 7.1 we noted that the combustion of fossil fuels adds 8.2×10^{12} kilograms of carbon to the atmosphere each year in the form of CO_2. What is the mass of this much CO_2? The answer must be more than 8.2×10^{12} kg, because this value represents only the mass due to carbon. The balanced chemical equation for the complete combustion of carbon is

$$C(s) + O_2(g) \rightarrow CO_2(g) \qquad (7.3)$$

Moles and Chemical Equations

CONNECTION In Chapter 2 we learned that the molecular mass of a compound is the sum of the average atomic masses of the atoms in one of its molecules and that its molar mass ($\mathcal{M}$) has the same numerical value as its molecular mass but is expressed in grams per mole.

Equation 7.3 tells us that the complete combustion of one atom of C produces one molecule of CO_2. Then it stands to reason that one *mole* of C atoms will produce one *mole* of CO_2 molecules. This mole ratio of reactant to product is a part of the **stoichiometry** of the reaction—that is, the mole ratios of all the reactants and products in a chemical reaction. Invoking the concept of molar mass, we can also state that one *molar mass* (12.01 g) of C will produce one *molar mass* (44.01 g) of CO_2. These equivalencies are illustrated in Figure 7.6.

The preceding interpretations of Equation 7.3 work for *any* number of moles (x mol) of C that produce x mol CO_2, because the ratio of the coefficients of C to CO_2 is 1:1. To see how, let's return to the conversion of 8.2×10^{12} kg C into a mass of CO_2. We first convert the mass of C into moles of C:

$$8.2 \times 10^{12}\ \cancel{\text{kg C}} \times \frac{10^3\ \cancel{\text{g C}}}{1\ \cancel{\text{kg C}}} \times \frac{1\ \text{mol C}}{12.01\ \cancel{\text{g C}}} = 6.83 \times 10^{14}\ \text{mol C}$$

stoichiometry the mole ratios among the reactants and products in a chemical reaction.

Note that the value of "moles of C" is expressed with three significant figures, even though the mass of C is known to only two significant figures. We carry

Chemical equation	$C(s)$ + $O_2(g)$ →	$CO_2(g)$
Particulate view	1 atom C 1 molecule O_2 →	1 molecule CO_2
Macroscopic view	1 mol C 1 mol O_2	1 mol CO_2
	$\times \mathcal{M}_C$ $\times \mathcal{M}_{O_2}$	$\times \mathcal{M}_{CO_2}$
	$12.01 \text{ g C} + 32.00 \text{ g } O_2$ =	$44.01 \text{ g } CO_2$

FIGURE 7.6 Particulate- and macroscopic-scale views of the combustion of carbon.

an additional digit because our result here is an intermediate value in this calculation. There is always a danger that rounding off an intermediate value in a multistep calculation can introduce a "rounding error." To avoid such errors, round off only the final calculated value to the appropriate number of significant figures. We know from the stoichiometry of the reaction that one mole of CO_2 is produced for every mole of C consumed; therefore,

$$6.83 \times 10^{14} \cancel{\text{mol C}} \times \frac{1 \text{ mol } CO_2}{1 \cancel{\text{mol C}}} = 6.83 \times 10^{14} \text{ mol } CO_2$$

To convert moles of CO_2 to mass, we first calculate the molar mass of CO_2:

$$\mathcal{M}_{CO_2} = 12.01 \text{ g/mol} + 2(16.00 \text{ g/mol}) = 44.01 \text{ g/mol } CO_2$$

Multiplying the moles of CO_2 by the molar mass of CO_2 and converting grams to kilograms gives us the mass of CO_2 added to the atmosphere:

$$6.83 \times 10^{14} \cancel{\text{mol } CO_2} \times \frac{44.01 \cancel{\text{g}} \text{ } CO_2}{1 \cancel{\text{mol } CO_2}} \times \frac{1 \text{ kg}}{10^3 \cancel{\text{g}}} = 3.0 \times 10^{13} \text{ kg } CO_2$$

Note that the answer in each of these steps is the starting point for the next step, so we could have combined the three separate calculations into one:

$$8.2 \times 10^{12} \cancel{\text{kg C}} \times \frac{10^3 \cancel{\text{g}}}{1 \cancel{\text{kg}}} \times \frac{1 \cancel{\text{mol C}}}{12.01 \cancel{\text{g C}}} \times \frac{1 \cancel{\text{mol } CO_2}}{1 \cancel{\text{mol C}}} \times \frac{44.01 \cancel{\text{g}} \text{ } CO_2}{1 \cancel{\text{mol } CO_2}} \times \frac{1 \text{ kg}}{10^3 \cancel{\text{g}}}$$
$$= 3.0 \times 10^{13} \text{ kg } CO_2$$

CONNECTION A summary of how to convert the number of moles of an element into the number of moles of a compound with that element in it and to relate the number of moles of a compound to the mass of the compound is shown in the accompanying figure, which is derived from Figure 2.17.

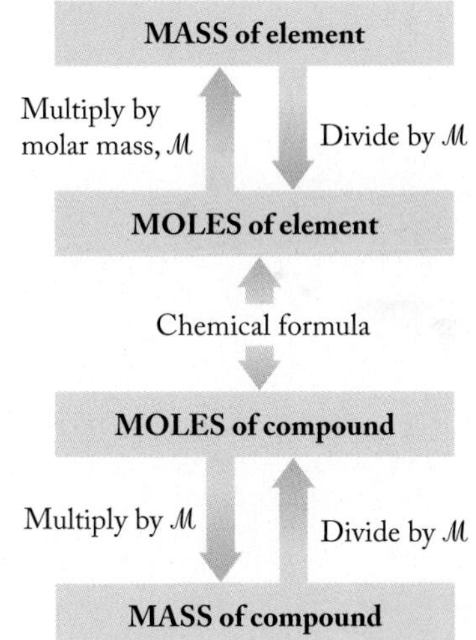

Note that in the second step of the calculation we converted kilograms into grams, and in the last step we converted grams into kilograms. These two steps had the effect of multiplying and then dividing by 1000, so they had no overall effect on the final answer. As you hone your skills in solving multistep calculations such as this one, check for offsetting conversion factors and simply cancel them out. Doing so will save you unnecessary calculation steps and reduce the chance of data entry errors.

The steps we followed in the preceding calculation can be used to determine the mass of any substance (reactant or product) involved in any chemical reaction, provided we know the quantity of another component of the reaction mixture and the stoichiometric relationship between the two substances—that is, their mole ratio in the balanced chemical equation describing the reaction.

CONCEPT TEST

The bright white patterns of light produced during fireworks displays are sometimes produced by the combustion of magnesium metal:

$$2\ \mathrm{Mg}(s) + \mathrm{O_2}(g) \rightarrow 2\ \mathrm{MgO}(s) \tag{7.14}$$

Is the total mass of Mg and O_2 consumed equal to the mass of MgO produced? Is the mass of Mg consumed equal to the mass of O_2 consumed? Explain both answers.

(Answers to Concept Tests are in the back of the book.)

SAMPLE EXERCISE 7.3 Calculating the Mass of a Product from the Mass of a Reactant — LO2

In 2014 power plants in the United States consumed 2.2×10^{11} kilograms of natural gas. How many kilograms of CO_2 ($\mathcal{M} = 44.01$ g/mol) were released into the atmosphere from these power plants? Natural gas is mostly methane (CH_4, $\mathcal{M} = 16.04$ g/mol), so base your solution on its combustion reaction (Equation 7.8):

$$\mathrm{CH_4}(g) + 2\ \mathrm{O_2}(g) \rightarrow \mathrm{CO_2}(g) + 2\ \mathrm{H_2O}(g)$$

Collect and Organize We are asked to calculate the mass of CO_2 released by the combustion of 2.2×10^{11} kg of CH_4 (our model compound for natural gas). We are given the molar masses of the two compounds and the balanced chemical equation relating the moles of CO_2 produced to the moles of CH_4 consumed.

Analyze To use Equation 7.8 we must convert the mass of CH_4 into moles of CH_4 by first converting kg CH_4 into g CH_4 and then dividing by the molar mass of CH_4. The number of moles of CH_4 consumed equals the number of moles of CO_2 produced. When we multiply this value by the molar mass of CO_2, we get the number of grams of CO_2, which we then convert to kilograms. The molar mass of CH_4 is about 16 g/mol, and the molar mass of CO_2 is about 44 g/mol, or about 2.5 times greater. Therefore, our final answer should be about 2.5 times the mass of natural gas given in the problem, or about 6×10^{11} kg CO_2.

Solve Converting kilograms of CH_4 into moles of CH_4 (and carrying an extra significant figure to avoid a rounding error):

$$2.2 \times 10^{11}\ \cancel{\mathrm{kg\ CH_4}} \times \frac{10^3\ \cancel{\mathrm{g}}}{1\ \cancel{\mathrm{kg}}} \times \frac{1\ \mathrm{mol\ CH_4}}{16.04\ \cancel{\mathrm{g\ CH_4}}} = 1.37 \times 10^{13}\ \mathrm{mol\ CH_4}$$

Converting moles of CH_4 into moles of CO_2:

$$1.37 \times 10^{13}\ \cancel{\mathrm{mol\ CH_4}} \times \frac{1\ \mathrm{mol\ CO_2}}{1\ \cancel{\mathrm{mol\ CH_4}}} = 1.37 \times 10^{13}\ \mathrm{mol\ CO_2}$$

Converting moles of CO_2 into kilograms of CO_2:

$$1.37 \times 10^{13}\ \cancel{\mathrm{mol\ CO_2}} \times \frac{44.01\ \cancel{\mathrm{g}}\ \mathrm{CO_2}}{1\ \cancel{\mathrm{mol\ CO_2}}} \times \frac{1\ \mathrm{kg}}{10^3\ \cancel{\mathrm{g}}} = 6.0 \times 10^{11}\ \mathrm{kg\ CO_2}$$

We can combine the three separate calculations into a single calculation, and in subsequent problems we will do this routinely and not show the individual steps:

$$2.2 \times 10^{11}\ \cancel{\mathrm{kg\ CH_4}} \times \frac{10^3\ \cancel{\mathrm{g}}}{1\ \cancel{\mathrm{kg}}} \times \frac{1\ \cancel{\mathrm{mol\ CH_4}}}{16.04\ \cancel{\mathrm{g\ CH_4}}} \times \frac{1\ \cancel{\mathrm{mol\ CO_2}}}{1\ \cancel{\mathrm{mol\ CH_4}}} \times \frac{44.01\ \cancel{\mathrm{g}}\ \mathrm{CO_2}}{1\ \cancel{\mathrm{mol\ CO_2}}} \times \frac{1\ \mathrm{kg}}{10^3\ \cancel{\mathrm{g}}}$$

$$= 6.0 \times 10^{11}\ \mathrm{kg\ CO_2}$$

Think About It The mass of CO_2 produced matches our estimated value. It represents 2% of the fossil carbon that is estimated to be entering the atmosphere as CO_2 each year.

 **Practice Exercise** How many grams of O_2 are consumed during the complete combustion of 1.00 g of butane, C_4H_{10}?

percent composition the composition of a compound expressed in terms of the percentage by mass of each element in the compound.

7.4 Percent Composition and Empirical Formulas

CHEMTOUR Percent Composition

During the Industrial Revolution, access to geological deposits of iron oxides was a key to economic development. The more accessible an iron-containing mineral was and the higher its iron content, the more valuable it was. There are several ways to express the elemental content of substances in mixtures. A popular one is **percent composition**: the percentage by mass of the constituent elements of a compound with respect to its total mass. Given the iron-containing minerals wustite (FeO) and hematite (Fe_2O_3), which has the higher iron content, and what is the iron content value expressed as a percent composition?

One way to answer the first question is to examine the mole ratios of Fe to O, which are 1:1 in FeO and 2:3 (or 1:1.5) in Fe_2O_3. There are 1.5 atoms of O for every atom of Fe in hematite, but only 1 atom of O per atom of Fe in wustite, which means there is a higher percentage of O and, hence, a lower percentage of Fe in Fe_2O_3. Therefore, FeO must have the higher iron content.

To find the percent composition of FeO, let's assume that we have 1 mole of it. If we do, then we have 1 mole of Fe and 1 mole of O. We can convert the number of moles to equivalent masses using the molar masses of Fe (55.85 g/mol) and O (16.00 g/mol):

$$\left(1\ \cancel{\text{mol}}\ \text{Fe} \times \frac{55.85\ \text{g}}{1\ \cancel{\text{mol}}}\right) + \left(1\ \cancel{\text{mol}}\ \text{O} \times \frac{16.00\ \text{g}}{1\ \cancel{\text{mol}}}\right) = 71.85\ \text{g FeO}$$

Of this 71.85 g, Fe accounts for 55.85 g, so the Fe content of FeO is

$$\frac{\text{mass of Fe}}{\text{total mass}} = \frac{55.85\ \text{g Fe}}{71.85\ \text{g FeO}} = 0.7773 \quad \text{or} \quad 77.73\%\ \text{Fe}$$

It follows that the oxygen content is

$$\frac{\text{mass of O}}{\text{total mass}} = \frac{16.00\ \text{g O}}{71.85\ \text{g FeO}} = 0.2227 \quad \text{or} \quad 22.27\%\ \text{O}$$

Because FeO contains only Fe and O, we could also have determined the percentage of O by subtracting the percentage of Fe from 100%:

$$100.00\% - 77.73\% = 22.27\%$$

SAMPLE EXERCISE 7.4 Calculating Percent Composition from a Chemical Formula **LO3**

What is the percent composition of the mineral forsterite, Mg_2SiO_4 (Figure 7.7)?

Collect, Organize, and Analyze The chemical formula of forsterite tells us that one mole of it contains 2 mol Mg, 1 mol Si, and 4 mol O. To convert these mole quantities into percent composition values based on mass, we must first convert the mole values to

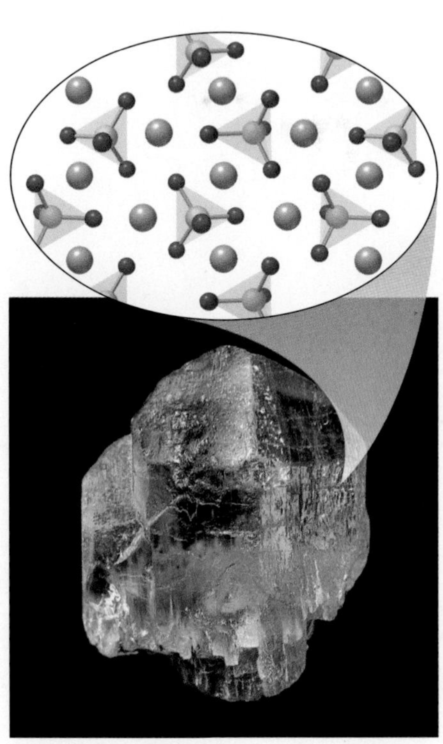

FIGURE 7.7 Crystals of pure Mg_2SiO_4 are colorless, but sometimes they contain impurities such as Fe^{2+} ions (in place of Mg^{2+} ions) that make the crystals green.

mass values by multiplying each by the appropriate molar mass. Summing these element masses, then dividing each one by the total mass, and then expressing the results as percentages, gives the percent composition of the mineral.

Solve Calculating the mass of each element in one mole of Mg_2SiO_4 and the molar mass of Mg_2SiO_4:

$$\begin{aligned} 2\ \cancel{\text{mol}}\ \text{Mg} \times 24.31\ \text{g}/\cancel{\text{mol}} &= 48.62\ \text{g Mg} \\ +\ 1\ \cancel{\text{mol}}\ \text{Si} \times 28.09\ \text{g}/\cancel{\text{mol}} &= 28.09\ \text{g Si} \\ +\ 4\ \cancel{\text{mol}}\ \text{O} \times 16.00\ \text{g}/\cancel{\text{mol}} &= 64.00\ \text{g O} \\ \hline \text{Molar mass of } Mg_2SiO_4 &= 140.71\ \text{g/mol} \end{aligned}$$

Calculating percent composition:

$$\%\text{Mg} = \frac{48.62\ \text{g Mg}}{140.71\ \text{g}} \times 100\% = 34.55\%\ \text{Mg}$$

$$\%\text{Si} = \frac{28.09\ \text{g Si}}{140.71\ \text{g}} \times 100\% = 19.96\%\ \text{Si}$$

$$\%\text{O} = \frac{64.00\ \text{g O}}{140.71\ \text{g}} \times 100\% = 45.48\%\ \text{O}$$

Think About It The molar masses of the three elements are similar, so it makes sense that the largest mass percent value belongs to O, followed by Mg and then Si because the mole ratios of these elements in forsterite are 4:2:1, respectively.

 **Practice Exercise** What is the percent composition of the antibiotic tetracycline, which has the molecular formula $C_{22}H_{24}N_2O_8$?

Note that the three percentages in Sample Exercise 7.4 sum to 99.99%. Percent composition values should always sum to 100%, or very close to it, if we have accounted for all the elements in the compound and done the math correctly.

We typically determine the percent composition of a substance in the laboratory by measuring the amount of each element in a given mass of the substance. We can use these data to determine the substance's empirical formula, which is a formula based on the lowest whole-number ratio of its component elements. The formulas of nearly all ionic compounds are empirical formulas because they are based on formula units, which contain the fewest numbers of cations and anions that together have no overall charge. The word *empirical* is synonymous with *experimental*, meaning "derived from experimental data." Its use here is appropriate because, as noted previously, empirical formulas are often derived from percent composition data that are obtained through chemical analyses.

CONNECTION We introduced the concept of formula units in Section 2.4. We used the term *empirical* to describe Balmer's equation (Equation 3.7) and Rydberg's equation (Equation 3.8) in Section 3.5.

To derive an empirical formula from percent composition data, we first assume that we have exactly 100 grams of the compound. This way, the percent composition values become the masses of each element expressed in grams. Suppose, for example, the results of an elemental analysis of a sample of an iron oxide show that it is 69.94% Fe and 30.06% O. In exactly 100 grams of the sample there would be 69.94 g Fe and 30.06 g O. Next, we convert these masses into equivalent numbers of moles by dividing by the appropriate molar masses:

$$69.94\ \cancel{\text{g Fe}} \times \frac{1\ \text{mol Fe}}{55.85\ \cancel{\text{g Fe}}} = 1.252\ \text{mol Fe}$$

and

$$30.06\ \cancel{g\ O} \times \frac{1\ mol\ O}{16.00\ \cancel{g\ O}} = 1.879\ mol\ O$$

To convert these values to a ratio of small whole numbers, we divide both by the smaller value:

$$\frac{1.879\ mol\ O}{1.252} = 1.501\ mol\ O \approx 3/2\ mol\ O$$

$$\frac{1.252\ mol\ Fe}{1.252} = 1.000\ mol\ Fe$$

To transform this mole ratio into whole numbers that can be the subscripts in an empirical formula, we need to recognize that 1.501 is really close to $1\frac{1}{2}$, or $\frac{3}{2}$. If we multiply each mole value by 2, we then have 2 moles of Fe and 3 moles of O, which translate into the empirical formula Fe_2O_3.

CONCEPT TEST

Which pairs of the following compounds have the same empirical formula?

a. ethylene (C_2H_4), a gas used to ripen green tomatoes
b. eicosene ($C_{20}H_{40}$), a compound used to trap Japanese beetles
c. acetylene (C_2H_2), a gas used in torches for cutting steel and other metals
d. benzene (C_6H_6), a carcinogenic compound found in petroleum

SAMPLE EXERCISE 7.5 Deriving an Empirical Formula from Percent Composition LO3

Structures containing the mineral magnetite (Figure 7.8) are found in birds' beaks and may help them navigate. Magnetite is an oxide of iron that is 72.36% iron by mass. What is its empirical formula?

FIGURE 7.8 The distinctive grey crystals in this mineral sample are composed of magnetite.

Collect and Organize We are given the iron content of magnetite and asked to determine its empirical formula—that is, the simplest ratio of iron ions to oxide ions in the mineral.

Analyze Because the only elements in the mineral are iron and oxygen, the percent composition by mass of the oxygen may be obtained by subtracting the percent Fe from 100%. We can use the steps described in the preceding discussion to

1. Convert percentage values into mass values in grams by assuming we have exactly 100 grams of magnetite.
2. Convert mass values into mole values by dividing by the molar masses of Fe and O.
3. Calculate a mole ratio by dividing both by the smaller of the two values.
4. Convert the mole ratio from step 3 into small whole numbers, if necessary.

Solve

1. In a 100 g sample there are 72.36 g Fe and

$$(100.00 - 72.36)\ g\ Fe = 27.64\ g\ O$$

2. Converting grams into moles:

$$72.36\ \cancel{g\ Fe} \times \frac{1\ mol\ Fe}{55.85\ \cancel{g\ Fe}} = 1.296\ mol\ Fe$$

and

$$27.64\ \cancel{\text{g O}} \times \frac{1\ \text{mol O}}{16.00\ \cancel{\text{g O}}} = 1.728\ \text{mol O}$$

3. Calculating a mole ratio of O to Fe:

$$\frac{1.728\ \text{mol O}}{1.296\ \text{mol Fe}} = 1.333\ \text{mol O/mol Fe}$$

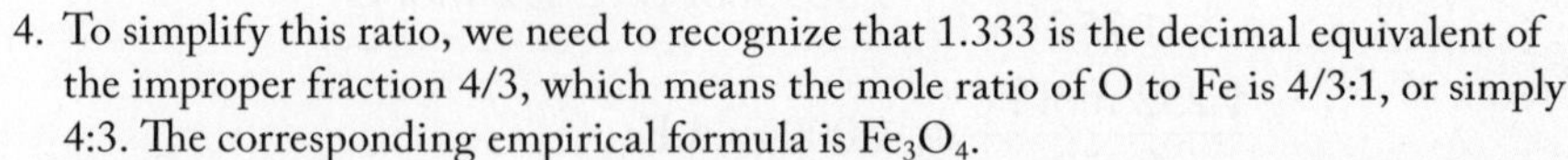

4. To simplify this ratio, we need to recognize that 1.333 is the decimal equivalent of the improper fraction 4/3, which means the mole ratio of O to Fe is 4/3:1, or simply 4:3. The corresponding empirical formula is Fe_3O_4.

FIGURE 7.9 Chalcocite is an important source of copper. The blue and blue-green colors in the sample are common for copper-containing compounds.

Think About It As we have seen again in this exercise, a key to deriving an empirical formula from percent composition data is translating mole ratios that are decimal values greater than 1 into equivalent improper fractions. Although we said in Chapter 2 that iron occurs in nature as Fe^{2+} and Fe^{3+} ions, the iron in magnetite *appears* to have a fractional charge of $\frac{8}{3}$, or 2.67. Actually, the formula unit of Fe_3O_4 contains two Fe^{3+} ions and one Fe^{2+} ion.

Practice Exercise For thousands of years, the mineral chalcocite (Figure 7.9) has been a valuable source of copper. Its chemical composition is 79.85% Cu and 20.15% S. What is its empirical formula?

SAMPLE EXERCISE 7.6 Deriving an Empirical Formula of a Compound Containing More than Two Elements **LO3**

A sample of dolomite, a carbonate mineral, is 21.73% Ca, 13.18% Mg, 13.03% C, and 52.06% O. What is its empirical formula?

Collect and Organize We are given the percent composition by mass of dolomite and asked to determine its empirical formula. The percent composition data should add up to 100% (which they do).

Analyze We follow the same steps in deriving the empirical formula of dolomite as we did in Sample Exercise 7.5, except that here we must calculate more than one mole ratio because there are four elements in the mineral.

Solve Converting percentage values into grams and then into moles:

$$21.73\ \cancel{\text{g Ca}} \times \frac{1\ \text{mol Ca}}{40.08\ \cancel{\text{g Ca}}} = 0.5422\ \text{mol Ca}$$

$$13.18\ \cancel{\text{g Mg}} \times \frac{1\ \text{mol Mg}}{24.31\ \cancel{\text{g Mg}}} = 0.5422\ \text{mol Mg}$$

$$13.03\ \cancel{\text{g C}} \times \frac{1\ \text{mol C}}{12.01\ \cancel{\text{g C}}} = 1.085\ \text{mol C}$$

$$52.06\ \cancel{\text{g O}} \times \frac{1\ \text{mol O}}{16.00\ \cancel{\text{g O}}} = 3.254\ \text{mol O}$$

Dividing each mole value by the smallest one (0.5422):

$$\frac{0.5422\ \text{mol Ca}}{0.5422} = 1.000\ \text{mol Ca} \qquad \frac{0.5422\ \text{mol Mg}}{0.5422} = 1.000\ \text{mol Mg}$$

$$\frac{1.085\ \text{mol C}}{0.5422} = 2.001\ \text{mol C} \qquad \frac{3.254\ \text{mol O}}{0.5422} = 6.001\ \text{mol O}$$

All results are either whole numbers or very close to whole numbers. Rounding them to single digits gives us a mole ratio of 1 Ca : 1 Mg : 2 C : 6 O, which corresponds to the empirical formula $CaMgC_2O_6$.

Think About It The problem states that dolomite is a carbonate mineral, which means that it contains CO_3^{2-} ions. Thus, we can rewrite the empirical formula in a way that indicates that the mineral is actually a blend of calcium carbonate and magnesium carbonate: $CaCO_3 \cdot MgCO_3$.

Practice Exercise Determine the empirical formula of a mineral that is 24.95% Fe, 46.46% Cr, and 28.59% O by mass.

7.5 Comparing Empirical and Molecular Formulas

An empirical formula provides the simplest mole ratios of the elements in a compound. It represents one formula unit of an ionic compound. It may—or may not—provide the number of atoms of each element in one molecule of a molecular compound. To explore these two possibilities, consider the molecular structures in Figure 7.10. The structure shown in Figure 7.10(a) is formaldehyde with the molecular formula CH_2O, which is also the simplest mole ratio of its component elements. Thus, formaldehyde's molecular and empirical formulas are the same. On the other hand, the structure shown in Figure 7.10(b) is glycolaldehyde, which has the molecular formula $C_2H_4O_2$. This molecular formula is not the simplest ratio of the moles of each element in the compound because it can be further simplified by dividing each subscript by 2. Doing so produces the same empirical formula, CH_2O, as formaldehyde and a whole family of compounds known as simple sugars, including the glucose ($C_6H_{12}O_6$) produced during photosynthesis in green plants. For simple sugars and many other molecular compounds, we need to know the molar masses to determine their molecular formulas from percent composition data.

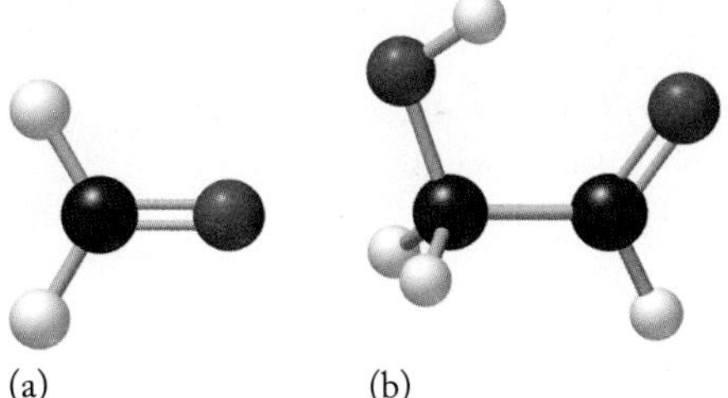

FIGURE 7.10 Molecular structures of (a) formaldehyde and (b) glycolaldehyde: two compounds with the same empirical formula, CH_2O.

CONCEPT TEST

Which of the following compounds have empirical formulas that match their molecular formulas?

a. ethylene glycol, $C_2H_6O_2$

b. isopropanol, C_3H_8O

c. glucose, $C_6H_{12}O_6$

To see how to use the empirical formula of a compound and its molecular mass to determine its molecular formula, let's reconsider glycolaldehyde and glucose. Their molecular formulas can be expressed in terms of their common empirical formula (CH_2O) as follows:

$$C_2H_4O_2 = (CH_2O)_2 \qquad C_6H_{12}O_6 = (CH_2O)_6$$

Extending these two equations to any molecular compound, we have the following general relation between molecular and empirical formulas:

$$(\text{Molecular formula}) = (\text{empirical formula})_n \tag{7.15}$$

where n is a positive integer equal to or greater than 1.

The key to deriving a molecular formula from an empirical formula is to determine the value of *n*. That's where molecular mass comes in. Using glucose as an example, let's first calculate the mass of one of its molecules:

$$\left(6\ \cancel{\text{atoms C}} \times \frac{12.01\ \text{amu}}{1\ \cancel{\text{atom C}}}\right) + \left(12\ \cancel{\text{atoms H}} \times \frac{1.008\ \text{amu}}{1\ \cancel{\text{atom H}}}\right)$$
$$+ \left(6\ \cancel{\text{atoms O}} \times \frac{16.00\ \text{amu}}{1\ \cancel{\text{atom O}}}\right) = 180.16\ \text{amu}$$

Now we calculate the mass of its empirical formula unit (CH_2O):

$$\left(1\ \cancel{\text{atoms C}} \times \frac{12.01\ \text{amu}}{1\ \cancel{\text{atom C}}}\right) + \left(2\ \cancel{\text{atoms H}} \times \frac{1.008\ \text{amu}}{1\ \cancel{\text{atom H}}}\right)$$
$$+ \left(1\ \cancel{\text{atoms O}} \times \frac{16.00\ \text{amu}}{1\ \cancel{\text{atom O}}}\right) = 30.03\ \text{amu}$$

Pretend that we did not already know the molecular formula of glucose, but we did know its empirical formula and could calculate the mass of one empirical formula unit (30.03 amu). If we also knew that its molecular mass was 180.16 amu, then we could divide the molecular mass by the mass of one empirical formula unit to obtain the value of *n* to use in Equation 7.15:

$$n = \frac{\text{molecular mass}}{\text{mass of 1 empirical formula unit}} \tag{7.16}$$
$$= \frac{180.16\ \cancel{\text{amu}}}{30.03\ \cancel{\text{amu}}} = 5.999 \approx 6$$

Inserting this value of *n* into Equation 7.15 yields the actual molecular formula of glucose:

$$(CH_2O)_6 = C_6H_{12}O_6$$

Molecular Mass and Mass Spectrometry Revisited

In Chapter 2, we introduced *mass spectrometry* as a technique to determine the molecular mass of compounds and measure isotopic abundances. The following provides a brief review of how a *mass spectrometer* works and how we interpret data in a graphical display called a *mass spectrum*.

Mass spectrometers ionize molecules and then separate the ions based on the ratio of their masses (m) to their electric charges (z). In many mass spectrometers, samples are vaporized and then bombarded with a beam of high-energy electrons (Figure 7.11). These electrons smash into gas-phase molecules with such force that they knock electrons off the molecules, forming positively charged *molecular ions* (M^+):

$$M + e^- \rightarrow M^+ + 2\ e^-$$

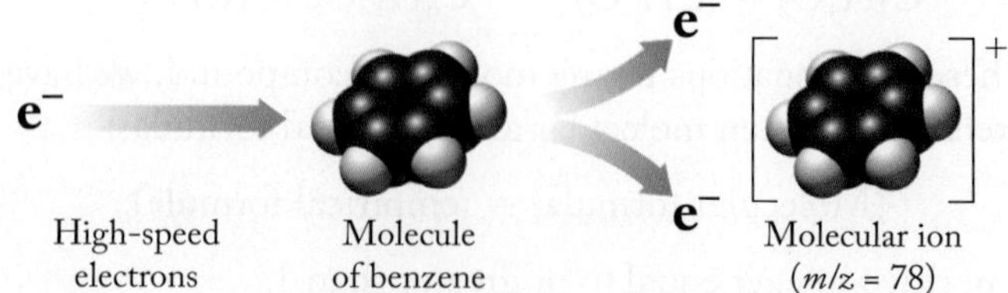

FIGURE 7.11 In many mass spectrometers, sample molecules are bombarded with beams of high-energy electrons, producing molecular ions with positive charges, as shown here for benzene.

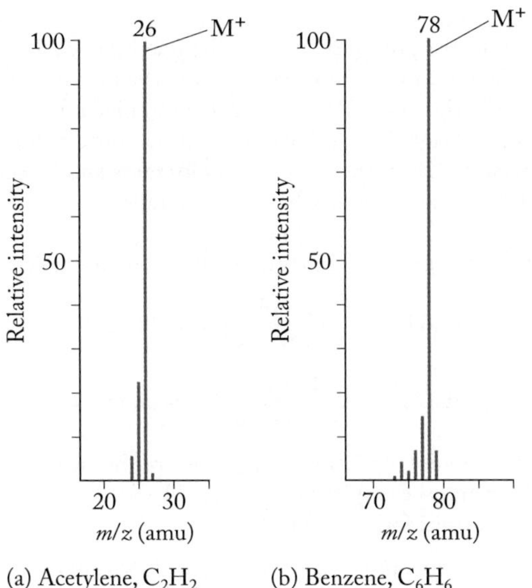

FIGURE 7.12 The molecular ions (M^+) formed by (a) acetylene and (b) benzene in a mass spectrometer.

Sometimes the collisions are so forceful that they break molecules into fragments and ionize the fragments. The charged fragments and molecular ions are carried into a second region of the mass spectrometer where they are separated based on their m/z ratios. Then they pass into a detector and are counted. The resulting data are displayed as a *mass spectrum* in which the horizontal axis is m/z and the spectrum itself is a series of vertical bars at various m/z values. The heights of the bars indicate the number of ions reaching the detector with that particular m/z value. The bar with the highest m/z value in the spectrum often represents the molecular ion, M^+. If they are singly charged ions ($z = 1+$), then the m/z value of that bar is simply m, the molecular mass of the compound.

CONNECTION The positive-ray analyzer that Francis Aston built and used to separate two of the isotopes of neon (see Section 2.2) was the forerunner of today's mass spectrometers.

Figure 7.12 shows the mass spectra of acetylene and benzene. In both spectra the tall bar with the greatest m/z value (at 26 amu in the acetylene spectrum and at 78 amu in benzene) is the molecular ion. The bars at lower m/z values represent fragments. Distinctive fragmentation patterns help scientists confirm molecular structures.

CONCEPT TEST

Does the molecular ion in a mass spectrum correspond to the mass associated with the empirical formula or the molecular formula of a compound?

SAMPLE EXERCISE 7.7 Using Percent Composition and Molecular Mass to Derive a Molecular Formula LO4

Pheromones are chemical substances secreted by members of a species to stimulate a response in other individuals of the same species. The percent composition of eicosene, a compound similar to the Japanese beetle mating pheromone, is 85.63% C and 14.37% H. Its molecular mass, as determined by mass spectrometry, is 280 amu. What is the molecular formula of eicosene?

Collect and Organize We are given the percent composition of eicosene and its molecular mass. The percent composition data can be used to derive the empirical formula. Equation 7.15 relates the empirical and molecular formulas of compounds. Equation 7.16 can be used to calculate the value of n from the mass of an empirical formula unit.

Analyze We can follow the procedure used in Sample Exercise 7.5 to determine the empirical formula of eicosene. Next we calculate the mass of one formula unit. Dividing that mass into the molecular mass (280 amu) given in the problem will give us the value of n for Equation 7.15 and will allow us to convert the empirical formula into a molecular formula. Given the large molecular mass and low atomic masses of C (12.01 amu) and H (1.008 amu), n may be a large number.

Solve Converting percentage values into grams and then into moles:

$$85.63 \text{ g C} \times \frac{1 \text{ mol C}}{12.01 \text{ g C}} = 7.130 \text{ mol C}$$

$$14.37 \text{ g H} \times \frac{1 \text{ mol H}}{1.008 \text{ g H}} = 14.26 \text{ mol H}$$

Dividing moles of H by moles of C (the smaller value of the two) gives us the simplified mole ratio of 1 C : 2 H and the empirical formula CH_2.

The mass of one empirical formula unit is

$$\left(1 \text{ atom C} \times \frac{12.01 \text{ amu}}{1 \text{ atom C}}\right) + \left(2 \text{ atoms H} \times \frac{1.008 \text{ amu}}{1 \text{ atom H}}\right) = 14.03 \text{ amu}$$

Using Equation 7.16 to calculate the formula multiplier n:

$$n = \frac{\text{molecular mass}}{\text{mass of 1 empirical formula unit}} = \frac{280 \text{ amu}}{14.03 \text{ amu}} = 19.96 \approx 20$$

Using Equation 7.15 to determine the molecular formula:

$$\text{Molecular formula} = (CH_2)_n = (CH_2)_{20} = C_{20}H_{40}$$

Think About It The molecular formula of eicosene represents the number of C and H atoms in one molecule of eicosene. The compound is one of many hydrocarbons with the empirical formula CH_2.

Practice Exercise Determine the empirical and molecular formulas of a compound that contains 43.64% P and 56.36% O and has a molar mass of 284 g/mol.

7.6 Combustion Analysis

Recall from Section 7.3 that combustion reactions involve burning substances in oxygen. In **combustion analysis**, the complete combustion of a compound followed by an analysis of the products enables chemists to determine the chemical composition of that compound. To ensure that combustion is complete, the process is carried out in an atmosphere of pure oxygen. This type of elemental analysis relies on complete combustion so that all the carbon in the sample is converted to CO_2 and all the hydrogen to H_2O. For a generic hydrocarbon C_aH_b, the overall reaction is

$$C_aH_b + \text{excess } O_2(g) \rightarrow a\ CO_2(g) + \frac{b}{2} H_2O(g) \qquad (7.17)$$

As shown in Figure 7.13, the products of combustion in an elemental analyzer flow through a tube packed with $Mg(ClO_4)_2(s)$, which selectively absorbs the $H_2O(g)$, and then through a tube containing $NaOH(s)$, which absorbs the $CO_2(g)$. The masses of the tubes are measured before and after combustion. The differences equal the masses of CO_2 and H_2O produced by the combustion of the sample. Suppose, for example, that complete combustion of a sample of a hydrocarbon results in a 1.320 g increase in the mass of the tube that traps $CO_2(g)$ and

combustion analysis a laboratory procedure in which a substance is burned completely in oxygen to produce known compounds whose masses are used to determine the composition of the original material.

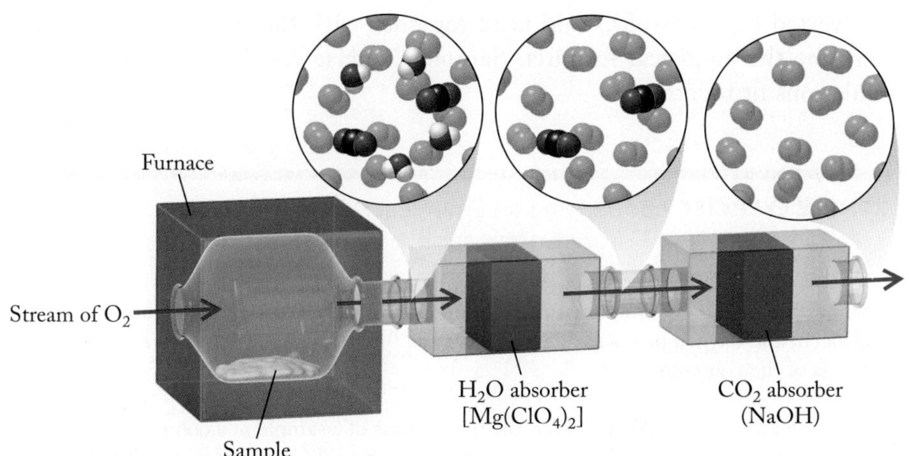

FIGURE 7.13 A carbon/hydrogen elemental analyzer relies on the complete combustion of organic compounds. The products are H_2O vapor and CO_2. Water vapor is absorbed by a cartridge packed with solid $Mg(ClO_4)_2$, while CO_2 is absorbed by solid NaOH. The empirical formula of the compound is calculated from the masses of H_2O and CO_2 absorbed.

a 0.541 g increase in the mass of the tube that traps $H_2O(g)$. How can we use these results to determine the empirical formula of the hydrocarbon?

Empirical formulas express the mole ratios of the elements in a compound, so we start by converting the masses of CO_2 and H_2O caught in the traps into moles of CO_2 and H_2O and then to equivalent numbers of moles of C and H in the original sample:

$$1.320\ \cancel{\text{g CO}_2} \times \frac{1\ \cancel{\text{mol CO}_2}}{44.01\ \cancel{\text{g CO}_2}} \times \frac{1\ \text{mol C}}{1\ \cancel{\text{mol CO}_2}} = 0.02999\ \text{mol C} \approx 0.0300\ \text{mol C}$$

$$0.541\ \cancel{\text{g H}_2\text{O}} \times \frac{1\ \cancel{\text{mol H}_2\text{O}}}{18.02\ \cancel{\text{g H}_2\text{O}}} \times \frac{2\ \text{mol H}}{1\ \cancel{\text{mol H}_2\text{O}}} = 0.0600\ \text{mol H}$$

The mole ratio of H to C is (0.0600 mol/0.0300 mol) = 2:1, so the empirical formula is CH_2.

If we want to extend this method to determining a molecular formula, we need to know the molecular mass. Suppose the mass spectrum of the hydrocarbon of interest has a molecular ion at 84 amu. The formula mass of CH_2 is 14 amu. Dividing the molecular mass by the formula mass, we get the multiplier, n:

$$n = \frac{84\ \cancel{\text{amu}}}{14\ \cancel{\text{amu}}} = 6$$

The molecular formula is therefore

$$(CH_2)_6 = C_6H_{12}$$

Note that in this problem we did not need to know the initial mass of the sample to determine its empirical formula. We only needed to know that the sample was a hydrocarbon and that it was completely converted into the stated amounts of CO_2 and H_2O.

What if we knew that our sample was not a hydrocarbon? What if we had isolated a pharmacologically promising compound from a tropical plant, and we knew that its molecules contained atoms of carbon, hydrogen, and oxygen? We would need to determine the percentage of oxygen in it, but there is no simple way to do that directly because the compound is burned in an atmosphere of O_2. However, the results of combustion analysis do allow us to calculate the number of moles of C and H in the original sample, which we can convert to grams of C and H. Subtracting their sum from the mass of the original sample provides a measure of the mass of O in the sample. That mass is converted to moles of O and used together with the moles of C and H to determine the empirical formula. Sample Exercise 7.8 illustrates how all the calculations fit together.

SAMPLE EXERCISE 7.8 Deriving an Empirical Formula from Combustion Analysis Data **LO5**

Combustion of 1.000 g of an organic compound known to contain only carbon, hydrogen, and oxygen produces 2.360 g CO_2 and 0.640 g H_2O. What is the empirical formula of the compound?

Collect and Organize We are given the initial mass of a sample of a compound that consists of C, H, and O, and we also know the masses of CO_2 and H_2O produced during its combustion. We are asked to determine the empirical formula of the compound.

Analyze We assume complete combustion of the sample, which means that its entire carbon content is converted into CO_2, and all its hydrogen content becomes H_2O. We need to determine the number of moles of C and H in 2.360 g CO_2 and 0.640 g H_2O, respectively. We also need to calculate the masses of C and H because the difference between the sum of the masses of C and H and the mass of the original sample is equal to the O content of the original sample. We then determine the C:H:O mole ratio and convert it into a ratio of small whole numbers.

Solve The numbers of moles of C and H in the CO_2 and H_2O collected during combustion are

$$2.360\ \cancel{\text{g CO}_2} \times \frac{1\ \cancel{\text{mol CO}_2}}{44.01\ \cancel{\text{g CO}_2}} \times \frac{1\ \text{mol C}}{1\ \cancel{\text{mol CO}_2}} = 0.05362\ \text{mol C}$$

$$0.640\ \cancel{\text{g H}_2\text{O}} \times \frac{1\ \cancel{\text{mol H}_2\text{O}}}{18.02\ \cancel{\text{g H}_2\text{O}}} \times \frac{2\ \text{mol H}}{1\ \cancel{\text{mol H}_2\text{O}}} = 0.07103\ \text{mol H}$$

The masses of C and H are

$$0.05362\ \cancel{\text{mol C}} \times \frac{12.01\ \text{g C}}{1\ \cancel{\text{mol C}}} = 0.6440\ \text{g C}$$

$$0.07103\ \cancel{\text{mol H}} \times \frac{1.008\ \text{g H}}{1\ \cancel{\text{mol H}}} = 0.07160\ \text{g H}$$

The difference between the original sample mass (1.000 g) and the sum of the masses of C and H (0.6440 g + 0.07160 g = 0.7156 g) is the mass of the oxygen in the sample:

$$\text{Mass of oxygen} = 1.000\ \text{g} - 0.7156\ \text{g} = 0.2844\ \text{g O}$$

The number of moles of O atoms in the sample is

$$0.2844\ \cancel{\text{g O}} \times \frac{1\ \text{mol O}}{16.00\ \cancel{\text{g O}}} = 0.01778\ \text{mol O}$$

Combining the above results, we get the mole ratio of the three elements in the sample:

$$0.05362\ \text{mol C} : 0.07103\ \text{mol H} : 0.01778\ \text{mol O}$$

To simplify the ratios we divide them by the smallest of their values (0.01778):

$$\frac{0.05362\ \text{mol C}}{0.01778} = 3.016 \approx 3\ \text{mol C}$$

$$\frac{0.07103\ \text{mol H}}{0.01778} = 3.995 \approx 4\ \text{mol H}$$

$$\frac{0.01778\ \text{mol O}}{0.01778} = 1\ \text{mol O}$$

The whole-number mole ratio that most closely matches the calculated values is 3:4:1, making the empirical formula of the sample C_3H_4O.

Think About It The results of our mole ratio calculations for C and H are very close to small whole numbers, which means that our calculations and the formula based on them are probably correct.

Practice Exercise Vanillin is the compound containing carbon, hydrogen, and oxygen that gives vanilla beans their distinctive flavor. The combustion of 30.4 mg of vanillin produces 70.4 mg of CO_2 and 14.4 mg of H_2O. The mass spectrum of vanillin shows a molecular-ion peak at 152 amu. Use this information to determine the molecular formula of vanillin.

limiting reactant a reactant that is consumed completely in a chemical reaction. The amount of product formed depends on the amount of the limiting reactant available.

7.7 Limiting Reactants and Percent Yield

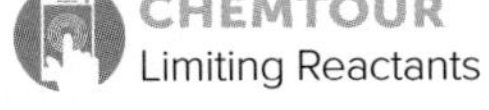

Let's return to the first reaction that we discussed in Section 7.3, the combustion of methane:

$$CH_4(g) + 2\ O_2(g) \rightarrow CO_2(g) + 2\ H_2O(\ell) \quad (7.9)$$

The stoichiometry of the reaction tells us that one mole of CH_4 requires two moles of O_2 for complete combustion. In reality, the furnaces and hot water heaters fueled by natural gas are surrounded by an abundant supply of O_2, as illustrated in the molecular view in Figure 7.14(a). Therefore, the quantities of reactants consumed, products formed, and energy generated are limited only by the fuel supply. In this reaction mixture, CH_4 is said to be the **limiting reactant**, which

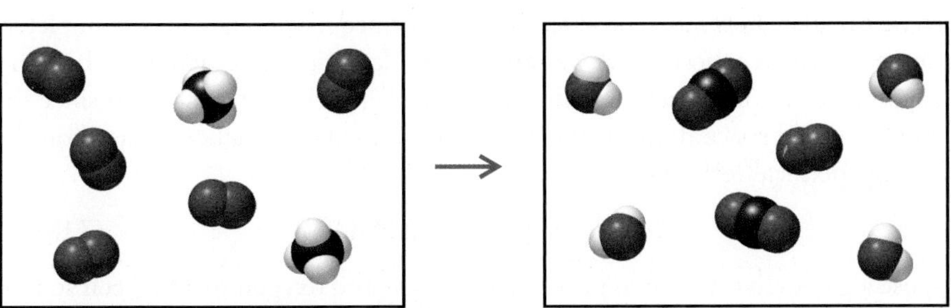

(a)

2 slices of bread + 1 slice of cheese + 1 slice of salami ⟶ 1 sandwich

(b)

FIGURE 7.14 (a) A combustion reaction mixture in which O_2 molecules outnumber CH_4 molecules 5 to 2 produces two molecules of CO_2 and four molecules of H_2O, leaving one molecule of O_2 left over. Methane is the limiting reactant. (b) A macroscopic illustration of the principle of the limiting reactant (or ingredient): how many modest salami and cheese sandwiches (with only one slice of salami and one slice of cheese per sandwich) can be prepared from eight slices of bread, four slices of cheese, and three slices of salami? Only three: salami is the limiting ingredient, and there are two slices of bread and one slice of cheese left over.

means that the extent to which the reaction proceeds is limited only by the quantity of CH_4 available. Figure 7.14(b) provides another—perhaps more familiar—macroscopic view of the concept of a limiting reactant.

CONCEPT TEST

Describe another chemical reaction in which at least one of the reactants is usually present in excess so that another is the limiting reactant.

Calculations Involving Limiting Reactants

Sometimes it is hard to say whether one reactant or another is the limiting reactant in a chemical reaction. We might know the masses of two reactants and be tempted to select the one with the smaller mass as the limiting reactant, but that is often not the case. Take the combustion of methane as an example. Suppose a sealed reaction vessel contains 16 g CH_4 and 48 g O_2. When the mixture is ignited, which reactant is the limiting reactant? You might select CH_4 because there is three times more O_2 (by mass) than CH_4 in the reaction mixture. However, the chemical equation for the combustion reaction tells us that we need two moles of O_2 for every one mole of CH_4, and the molar mass of O_2 (32.00 g/mol) is about twice that of CH_4 (16.04 g/mol). Twice as many moles of a reactant that has twice the molar mass means that we need four times as much of it by mass. However, the mass of O_2 in the reaction vessel is only three times the mass of CH_4, so O_2 is the limiting reactant.

Let's set up a procedure for identifying the limiting reactant for any reaction in which we know the quantities of two or more reactants. There are several ways to solve such a problem. We explore one of them here, starting with the generic chemical equation

$$A + B \rightarrow C$$

First, we calculate how much product would be formed if reactant A were completely consumed in the reaction. Then we repeat the calculation for reactant B. Let's try this approach with the reaction that has produced electricity in spacecraft since the dawn of manned space flight in the 1960s. It makes use of fuel cells based on the following chemical reaction:

$$2\,H_2(g) + O_2(g) \rightarrow 2\,H_2O(\ell) \tag{7.18}$$

Suppose a gas-delivery system provides hydrogen and oxygen to a fuel cell so that the mixture in the cell consists of 0.055 g H_2 and 0.48 g O_2. Which gas is the limiting reactant?

Equation 7.18 tells us that two moles of H_2 combine with one mole of O_2 to produce two moles of water, so let's calculate the amount of water each reagent would produce if it were completely consumed.

$$0.055\ \cancel{g\ H_2} \times \frac{1\ \cancel{mol\ H_2}}{2.016\ \cancel{g\ H_2}} \times \frac{2\ \cancel{mol\ H_2O}}{2\ \cancel{mol\ H_2}} \times \frac{18.02\ g\ H_2O}{1\ \cancel{mol\ H_2O}} = 0.49\ g\ H_2O$$

Now we calculate how much water would be produced if O_2 were completely consumed:

$$0.48\ \cancel{g\ O_2} \times \frac{1\ \cancel{mol\ O_2}}{32.00\ \cancel{g\ O_2}} \times \frac{2\ \cancel{mol\ H_2O}}{1\ \cancel{mol\ O_2}} \times \frac{18.02\ g\ H_2O}{1\ \cancel{mol\ H_2O}} = 0.54\ g\ H_2O$$

Less H_2O is produced from the H_2 supply than from O_2, so H_2 must be the limiting reactant. Some O_2 will be left unreacted. We can calculate how much by calculating the mass of O_2 that is consumed for the given amount of H_2. We use the 2:1 stoichiometric ratio between H_2 and O_2 and the molar mass of O_2:

$$0.055\ \cancel{g\ H_2} \times \frac{1\ \cancel{mol\ H_2}}{2.016\ \cancel{g\ H_2}} \times \frac{1\ \cancel{mol\ O_2}}{2\ \cancel{mol\ H_2}} \times \frac{32.00\ g\ O_2}{1\ \cancel{mol\ O_2}} = 0.44\ g\ O_2$$

The difference between the mass of O_2 supplied (0.48 g) and the mass consumed (0.44 g) means 0.04 g O_2 is left over.

This method not only identifies the limiting reagent in a process but also determines the **theoretical yield** of the product—the maximum amount of product that could be obtained from the given quantities of reactants. Theoretical yield is sometimes called *stoichiometric yield* or 100% yield. In this case, the theoretical yield of water from the given amounts of reactants is 0.49 g H_2O.

theoretical yield the maximum amount of product possible in a chemical reaction for the given quantities of reactants; also called the stoichiometric yield.

CONCEPT TEST

Suppose the reaction A + 2B + 3C yields a desired product (D). If a reaction mixture contains 1 mol A, 2.5 mol B, and 1.5 mol C, (a) which is the limiting reactant, and (b) how much of the other reactants remain after the limiting reactant is consumed?

SAMPLE EXERCISE 7.9 Identifying the Limiting Reactant in a Reaction Mixture LO6

The flame in an oxyacetylene torch reaches temperatures as high as 3500°C as a result of the combustion of a mixture of acetylene (C_2H_2) and pure O_2. If these two gases flow from high-pressure tanks at the rates of 52.0 g C_2H_2 and 188 g O_2 per minute, which reactant is the limiting reactant, or is the mixture stoichiometric?

Collect and Organize We are given the rates (g/min) at which two reactants are introduced to a reaction system (the torch) and asked to determine whether one of them is limiting and, if so, which one. Complete combustion of hydrocarbons produces CO_2 and H_2O.

Analyze We need to write a balanced chemical equation describing the combustion reaction to determine the required mole ratio of C_2H_2 to O_2. The rates at which the reactants flow to the torch can be converted directly into masses in grams by assuming the torch is operated for exactly one minute. These masses of the two reactants can then be converted into moles of product using the stoichiometry of the reaction. The lesser of the two quantities is the answer we seek.

Solve To write a balanced chemical equation, we start with one molecule of each reactant and product and take an elemental inventory on both sides of the reaction arrow:

$$C_2H_2(g) + O_2(g) \rightarrow CO_2(g) + H_2O(g)$$

$$\text{Atoms:}\quad 2\,C + 2\,H + 2\,O \rightarrow 1\,C + 3\,O + 2\,H$$

The H atoms are already balanced. To balance the C atoms, we place a coefficient of 2 in front of CO_2:

$$C_2H_2(g) + O_2(g) \rightarrow \mathbf{2}\,CO_2(g) + H_2O(g)$$

$$\text{Atoms:}\quad 2\,C + 2\,H + 2\,O \rightarrow 2\,C + 2\,H + 5\,O$$

Balancing the O atoms requires multiplying all coefficients by 2 to get an even number of O atoms on the right side (without disrupting the previously balanced numbers of C and H atoms):

$$\mathbf{2}\,C_2H_2(g) + \mathbf{2}\,O_2(g) \rightarrow \mathbf{4}\,CO_2(g) + \mathbf{2}\,H_2O(g)$$

$$\text{Atoms:}\quad 4\,C + 4\,H + 4\,O \rightarrow 4\,C + 4\,H + 10\,O$$

Adding six more atoms of O to the left side by changing the coefficient in front of O_2 from 2 to 5 yields a balanced chemical equation:

$$2\,C_2H_2(g) + \mathbf{5}\,O_2(g) \rightarrow 4\,CO_2(g) + 2\,H_2O(g)$$

$$\text{Atoms:}\quad 4\,C + 4\,H + 10\,O \rightarrow 4\,C + 4\,H + 10\,O \quad ✓$$

Calculating the quantity of CO_2 that could be produced by each reactant if it were the limiting reactant:

$$188\ \cancel{g\ O_2} \times \frac{1\ \cancel{mol\ O_2}}{32.00\ \cancel{g\ O_2}} \times \frac{4\ mol\ CO_2}{5\ \cancel{mol\ O_2}} = 4.70\ mol\ CO_2$$

$$52.0\ \cancel{g\ C_2H_2} \times \frac{1\ \cancel{mol\ C_2H_2}}{26.04\ \cancel{g\ C_2H_2}} \times \frac{4\ mol\ CO_2}{2\ \cancel{mol\ C_2H_2}} = 3.99\ mol\ CO_2$$

The smaller quantity of CO_2 that could be produced by acetylene tells us that (1) it is the limiting reactant and (2) the acetylene/oxygen ratio entering the torch was not a stoichiometric one.

Think About It We compared the amount of CO_2 produced in the reaction to determine the limiting reagent. We could also have compared the amount of water produced. We also could have calculated the amount of one of the reactants required for the complete consumption of the other.

Practice Exercise Any fuel–oxygen mixture that contains less fuel than oxygen needed to burn the fuel completely is called a *fuel-lean* (or simply *lean*) mixture, and a combustion mixture containing too little oxygen is called a *fuel-rich* mixture. A high-performance torch that burns propane, C_3H_8, is adjusted so that 88 g O_2 flows into its flame for every 23 g C_3H_8. Is the mixture fuel-rich, fuel-lean, or stoichiometric?

SAMPLE EXERCISE 7.10 Calculating the Theoretical Yield **LO6**

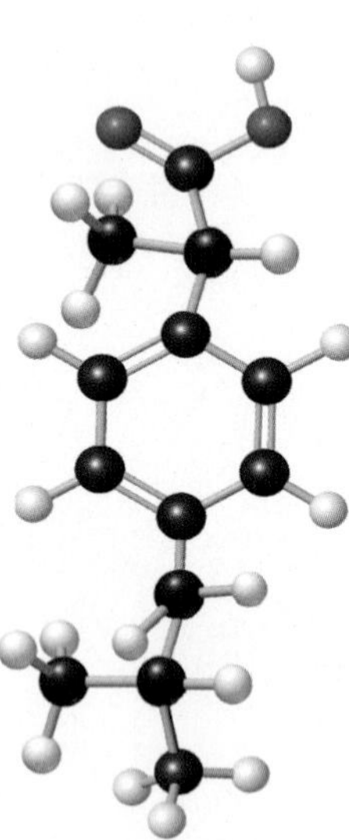

FIGURE 7.15 Ibuprofen is the active ingredient in the pain relievers Advil™ and Motrin™.

Ibuprofen (Figure 7.15) is an over-the-counter (OTC) pharmaceutical (for example, Advil™, Motrin™) for pain relief. Like aspirin, it is an NSAID (nonsteroidal anti-inflammatory drug). The original six-step synthesis patented in the 1960s resulted in significant waste, so a green chemistry alternative synthesis is now used that is cheaper and more environmentally friendly. The first step in the modern industrial synthesis involves the reaction of isobutylbenzene ($C_{10}H_{14}$; $\mathcal{M}$ = 134.21 g) with acetic anhydride ($C_4H_6O_3$; $\mathcal{M}$ = 102.09 g), which produces 4-isobutylacetophenone ($C_{12}H_{16}O$; $\mathcal{M}$ = 176.25 g). Acetic acid (CH_3COOH) is also produced:

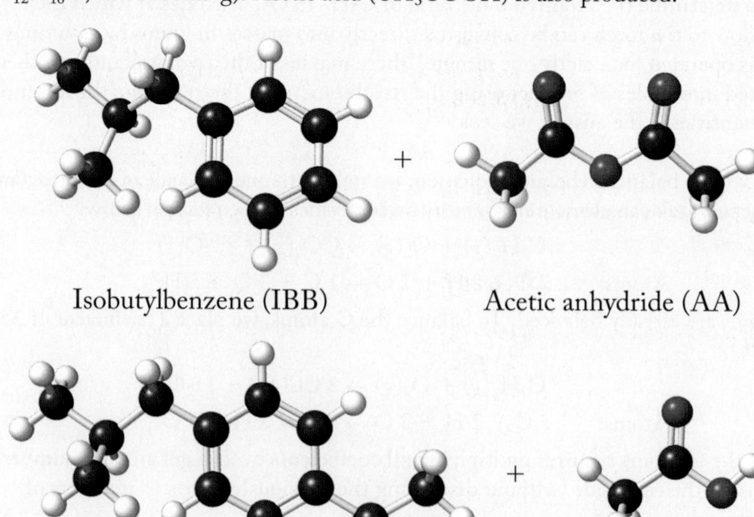

Suppose this reaction is studied on a laboratory scale in which 265 g of isobutylbenzene (IBB) is reacted with 198 g of acetic anhydride (AA).

a. Which is the limiting reactant?
b. How many grams of 4-isobutylacetophenone (IBA) would be produced if the limiting reactant was completely consumed?
c. How much of the other reactant would be left over?

Collect and Organize We are given the quantities of two reactants, the molar masses of the reactants and the desired product, and a balanced equation describing their reaction. We are asked to determine which is the limiting reactant, how much product could be made, and how much of the nonlimiting reactant will be left over.

Analyze Start with a balanced equation. To identify which compound is the limiting reactant, let's determine the quantity of product that could be made by first assuming that IBB is the limiting reactant and then by assuming AA is the limiting reactant. The limiting reactant will produce the smaller mass of IBA, which is also the answer to part (b). To calculate the quantity of the nonlimiting reactant that will be left over, we use the stoichiometry of the reaction to calculate how much of it is consumed during the reaction and then subtract that value from the initial amount present in the reaction mixture.

Solve

a. The balanced chemical equation for the reaction of IBB with AA is: $C_{10}H_{14} + C_4H_6O_3 \rightarrow C_{12}H_{16}O + C_2H_4O_2$. Calculating the mass of IBA that could be produced if IBB is the limiting reactant:

$$265\ \cancel{\text{g IBB}} \times \frac{1\ \cancel{\text{mol IBB}}}{134.21\ \cancel{\text{g IBB}}} \times \frac{1\ \cancel{\text{mol IBA}}}{1\ \cancel{\text{mol IBB}}} \times \frac{176.25\text{ g IBA}}{1\ \cancel{\text{mol IBA}}} = 348\text{ g IBA}$$

Repeating the preceding calculation with 198 g AA:

$$198\ \cancel{\text{g AA}} \times \frac{1\ \cancel{\text{mol AA}}}{102.09\ \cancel{\text{g AA}}} \times \frac{1\ \cancel{\text{mol IBA}}}{1\ \cancel{\text{mol AA}}} \times \frac{176.25\text{ g IBA}}{1\ \cancel{\text{mol IBA}}} = 342\text{ g IBA}$$

AA produces the smaller amount of product; therefore, it is the limiting reactant.

b. The maximum amount of IBA that can be made is 342 g.

c. The mass of IBB consumed in the reaction is

$$198\ \cancel{\text{g AA}} \times \frac{1\ \cancel{\text{mol AA}}}{102.09\ \cancel{\text{g AA}}} \times \frac{1\ \cancel{\text{mol IBB}}}{1\ \cancel{\text{mol AA}}} \times \frac{134.21\text{ g IBB}}{1\ \cancel{\text{mol IBB}}} = 260\text{ g IBB}$$

The mass of IBB left over is $(265 - 260)\text{ g} = 5\text{ g}$.

Think About It For a variety of scientific and economic reasons, industrial processes are often run with nonstoichiometric amounts of reagents. We will encounter some of these reasons as we explore chemical kinetics and equilibrium in subsequent chapters.

Practice Exercise Sodium borohydride ($NaBH_4$; $\mathcal{M} = 37.83$ g) is a solid material that can be used to store hydrogen gas. One way to prepare it is the reaction of trimethylborate [$B(OCH_3)_3$; $\mathcal{M} = 103.91$ g] with sodium hydride (NaH; $\mathcal{M} = 24.00$ g):

$$B(OCH_3)_3(\ell) + 4\ NaH(s) \rightarrow NaBH_4(s) + 3\ NaOCH_3(s)$$

If 71.5 g of trimethylborate is reacted with 56.2 g of sodium hydride, how many grams of sodium borohydride can be produced?

Percent Yield: Actual versus Theoretical

Recall that the *theoretical* yield is the maximum quantity of product possible from the given quantities of reactants. In nature, industry, or the laboratory, the *actual* yield is often less than the theoretical yield for several reasons. Sometimes reactants combine to produce products other than the ones desired. Other reactions

percent yield the ratio, expressed as a percentage, of the actual yield of a chemical reaction to the theoretical yield.

are so slow that some reactants remain unreacted even after long reaction times. Still other reactions do not go to completion no matter how long they are allowed to run. That can happen when a reaction also runs in reverse, turning products back into reactants. When this happens, the result can be a mixture of reactants and products in which the composition does not change with time. These reactions have reached a state of *chemical equilibrium*, a concept we explore in considerable detail in Chapter 14. So, for various reasons, it is useful to distinguish between the theoretical and actual yields of a chemical reaction and to calculate the **percent yield**:

$$\text{Percent yield} = \frac{\text{actual yield}}{\text{theoretical yield}} \times 100\% \quad (7.19)$$

SAMPLE EXERCISE 7.11 Calculating Percent Yield **LO7**

The industrial process for making the ammonia used in fertilizer, explosives, and many other products is based on the reaction between nitrogen and hydrogen at high temperature and pressure:

$$N_2(g) + 3\,H_2(g) \rightarrow 2\,NH_3(g)$$

If 18.2 kg NH_3 ($\mathcal{M}$ = 17.03 g/mol) is produced by a reaction mixture that initially contains 6.00 kg H_2 ($\mathcal{M}$ = 2.016 g/mol) and an excess of N_2, what is the percent yield of the reaction?

Collect and Organize We know that the actual yield of NH_3 is 18.2 kg. We also know that H_2 must be the limiting reactant because there is an excess of N_2 and that the reaction mixture initially contained 6.00 kg H_2. The mole ratio of NH_3 to H_2 in the chemical equation is 2:3.

Analyze One way to calculate percent yield involves the following steps:

1. Convert the mass of H_2 into moles of H_2.
2. Convert the moles of H_2 consumed into moles of NH_3 produced using the 3:2 stoichiometric ratio in the balanced chemical equation.
3. Convert moles of NH_3 into grams of NH_3 by multiplying by the molar mass of NH_3.
4. Convert grams of NH_3 into kilograms of NH_3 to calculate the theoretical yield of the reaction.
5. Divide the actual yield (18.2 kg NH_3) by the theoretical yield from step 4 to obtain the percent yield.

Solve Calculating the theoretical yield (combining steps 1–4 into a single calculation):

$$6.00\ \cancel{\text{kg H}_2} \times \frac{10^3\ \text{g}}{1\ \cancel{\text{kg}}} \times \frac{1\ \cancel{\text{mol H}_2}}{2.016\ \cancel{\text{g H}_2}} \times \frac{2\ \cancel{\text{mol NH}_3}}{3\ \cancel{\text{mol H}_2}} \times \frac{17.03\ \text{g NH}_3}{1\ \cancel{\text{mol NH}_3}} \times \frac{1\ \text{kg}}{10^3\ \text{g}}$$
$$= 33.8\ \text{kg NH}_3$$

Dividing the actual yield by this theoretical yield (step 5):

$$\frac{18.2\ \cancel{\text{kg NH}_3}}{33.8\ \cancel{\text{kg NH}_3}} \times 100\% = 53.8\%$$

Think About It A yield of about 54% may seem low, but it is not unusual for a reaction such as this one, which reaches chemical equilibrium before the limiting reactant has been completely consumed. In Chapter 14 we will discuss ways that chemists and chemical engineers manipulate reaction conditions to improve the yields of reactions that reach equilibrium.

 **Practice Exercise** The industrial synthesis of H_2 begins with the steam-reforming reaction, in which methane reacts with high-temperature steam:

$$CH_4(g) + H_2O(g) \rightarrow CO(g) + 3\,H_2(g)$$

What is the percent yield when a reaction vessel that initially contains 64 kg CH_4 and excess steam yields 18.1 kg H_2?

SAMPLE EXERCISE 7.12 Integrating Concepts: Synthesizing Hydrogen Gas

Practice Exercise 7.11 describes the first stage in the *steam-reforming reaction* for synthesizing H_2 gas. The carbon monoxide that is a by-product of the reaction can be reacted with more steam in a second stage to produce more hydrogen and carbon dioxide.

a. Write a balanced chemical equation for the reaction between CO and steam that produces H_2 and CO_2.
b. Combine your answer from part (a) with the chemical equation for the first stage of the steam-reforming reaction from Practice Exercise 7.11 to write an overall reaction in which methane and steam react to form hydrogen gas and carbon dioxide.
c. Suppose a reaction vessel for the two-stage process initially contains 48.0 kg CH_4 and 118 kg $H_2O(g)$ and that the process yields 16 kg H_2. Is the initial reaction mixture stoichiometric? If not, which is the limiting reactant?
d. What is the percent yield of the reaction mixture in part (c)?

Collect and Organize We need to write a balanced chemical equation for the reaction between CO and H_2O that produces H_2 and CO_2 and then combine it with the steam-reforming reaction to write an overall equation describing the formation of H_2 and CO_2 from CH_4 and H_2O. We must then determine whether or not a reaction mixture of CH_4 and H_2O is stoichiometric and calculate the percent yield of the reaction. We know the masses of CH_4 and H_2O in the reaction mixture and the mass of H_2 they produce. We also know the chemical equation of the steam-reforming reaction, which is the first step in the two-step process. From previous calculations in this chapter, we know the following molar masses: 18.02 g/mol H_2O, 16.04 g/mol CH_4, and 2.016 g/mol H_2.

Analyze To combine two chemical equations, we need to make sure that the coefficients of products in the first equation that are reactants in the second equation are the same, so that they cancel out when the equations are combined. One way to identify whether or not a reaction mixture is stoichiometric is to calculate the quantities of product each reactant could produce if it were completely consumed. Using this approach makes sense because we need to determine the percent yield of the reaction, which involves calculating the theoretical yield and comparing it to the actual yield.

Solve

a. First we write the chemical equation for the reaction between CO and steam that produces H_2 and CO_2. Starting with one molecule of each reactant and product and taking an inventory of all the atoms on both sides of the reaction arrow,

$$CO(g) + H_2O(g) \rightarrow H_2(g) + CO_2(g)$$

$$\text{Atoms:}\quad 1\,C + 2\,H + 2\,O \rightarrow 1\,C + 2\,H + 2\,O \quad ✓$$

we find that the equation is balanced already.

b. Comparing this equation to the one for the steam-reforming reaction,

$$CH_4(g) + H_2O(g) \rightarrow CO(g) + 3\,H_2(g)$$

we find that CO is a product of the steam-reforming reaction and a reactant in the second-stage reaction. Its coefficient is 1 in both reactions, so the two can be combined without modification:

$$\begin{array}{l} CH_4(g) + H_2O(g) \rightarrow \cancel{CO(g)} + 3\,H_2(g) \\ + \cancel{CO(g)} + H_2O(g) \rightarrow H_2(g) + CO_2(g) \\ \hline CH_4(g) + 2\,H_2O(g) \rightarrow 4\,H_2(g) + CO_2(g) \end{array}$$

c. To determine whether the reaction mixture is stoichiometric, we calculate the theoretical yields of H_2 produced by each of the reactants:

$$48.0\ \cancel{\text{kg CH}_4} \times \frac{10^3\ \cancel{\text{g}}}{1\ \cancel{\text{kg}}} \times \frac{1\ \cancel{\text{mol CH}_4}}{16.04\ \cancel{\text{g CH}_4}} \times \frac{4\ \cancel{\text{mol H}_2}}{1\ \cancel{\text{mol CH}_4}} \times \frac{2.016\ \text{g H}_2}{1\ \cancel{\text{mol H}_2}} \times \frac{1\ \text{kg}}{10^3\ \text{g}} = 24.13\ \text{kg H}_2$$

$$118\ \cancel{\text{kg H}_2\text{O}} \times \frac{10^3\ \cancel{\text{g}}}{1\ \cancel{\text{kg}}} \times \frac{1\ \cancel{\text{mol H}_2\text{O}}}{18.02\ \cancel{\text{g H}_2\text{O}}} \times \frac{4\ \cancel{\text{mol H}_2}}{2\ \cancel{\text{mol H}_2\text{O}}} \times \frac{2.016\ \text{g H}_2}{1\ \cancel{\text{mol H}_2}} \times \frac{1\ \text{kg}}{10^3\ \text{g}} = 26.40\ \text{kg H}_2$$

Methane produces less H_2, so the reaction mixture is not stoichiometric and CH_4 is the limiting reactant.

d. The percent yield calculation is based on the lesser theoretical yield (24.13 kg H_2):

$$\frac{16\ \cancel{\text{kg H}_2}}{24.13\ \cancel{\text{kg H}_2}} \times 100\% = 66\%\ \text{yield}$$

Think About It The second-stage reaction in the synthesis of H_2 is called the *water–gas shift reaction*. Many important industrial processes rely on chemical reactions with yields that are less than 100%, and the production of hydrogen gas is one of them. As in other multistep calculations, we retained extra significant figures for intermediate values and did not round off to two digits until the end. Another way to avoid rounding errors is to use values in calculator memory for each new step in a calculation rather than reentering the results of previous calculations.

SUMMARY

LO1 **Chemical equations**, which contain the chemical formulas of the **reactants** on the left and **products** on the right, provide concise, symbolic descriptions of chemical reactions. The proportions of the reactants and products are expressed by coefficients preceding their formulas. Writing balanced chemical equations is an exercise in adjusting these coefficients until the number of atoms of each element is the same on the reactant and product sides of the equation. The **law of conservation of mass** states that the sum of the masses of the reactants in a chemical reaction is equal to the sum of the masses of the products. (Sections 7.1 and 7.2)

LO2 The mole ratios of reactants and products in a chemical reaction are called the **stoichiometry** of the reaction. These ratios can be used to relate the masses of the reactants consumed and the products that are made in a chemical reaction. (Section 7.3)

LO3 The **percent composition** of a compound is the percentage by mass of each element in the compound. The **empirical formula** of a compound represents the lowest whole-number ratio of its elements' (Section 7.4)

LO4 A compound's empirical formula may or may not be the same as its molecular formula, which indicates the number of each type of atom in one molecule. Converting the empirical formula of a compound into a molecular formula requires knowing the molecular mass of the compound. (Section 7.5)

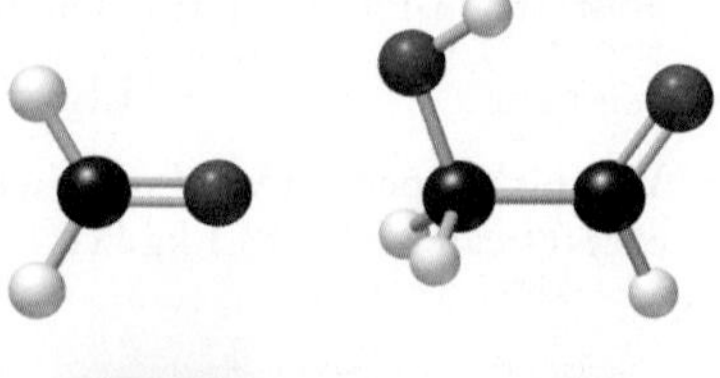

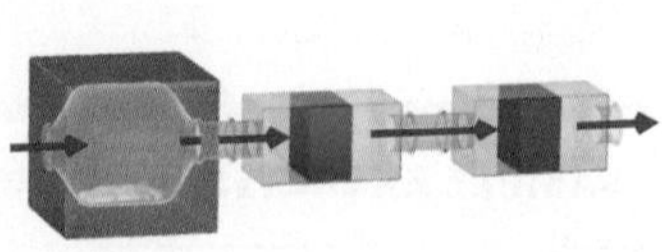

LO5 In **combustion analysis**, a known mass of an organic compound is burned in a stream of oxygen gas. The carbon in the sample is converted into CO_2 and the hydrogen is converted into H_2O. The masses of CO_2 and H_2O are measured and used to determine the masses of C and H in the organic compound and then the compound's empirical formula. (Section 7.6)

LO6 Chemical reactions are not always run with exact stoichiometric amounts of material. The **limiting reactant** in a reaction mixture is the reactant that limits how much product can be made. The maximum amount of product that can form in a chemical reaction is the **theoretical yield**. (Section 7.7)

LO7 The measured amount of product formed in a reaction is the actual yield, and the ratio of actual yield to theoretical yield, expressed as a percentage, is the **percent yield** for the reaction. (Section 7.7)

PARTICULATE **PREVIEW WRAP-UP**

Nitromethane burns in oxygen to produce carbon dioxide, water, and nitrogen gas:

$$4\ CH_3NO_2(\ell) + 3\ O_2(g) \rightarrow 4\ CO_2(g) + 2\ N_2(g) + 6\ H_2O(g).$$

If the mixture shown were to react, no molecules of CH_3NO_2, but 3 molecules of $O_2(g)$ would remain unreacted.

PROBLEM-SOLVING SUMMARY

Type of Problem	Concepts and Equations	Sample Exercises
Balancing a chemical reaction	Adjust coefficients to balance number of atoms of each element on both sides of the reaction arrow. Start with elements that are in only one reactant and product on each side.	**7.1, 7.2**
Writing and balancing a chemical equation for a combustion reaction	C and H in organic compounds react with O_2 to form CO_2 and H_2O; for example: $CH_4(g) + 2\ O_2(g) \rightarrow CO_2(g) + 2\ H_2O(g)$ Balance the atoms of C first, then H, then O.	**7.1**

Type of Problem	Concepts and Equations	Sample Exercises
Calculating the mass of a product from the mass of a reactant	Convert mass of reactant into moles of reactant; use reaction stoichiometry to calculate moles of product, and then convert moles of product into mass of product.	**7.3**
Calculating percent composition from a chemical formula	Calculate the mass of each element in one mole of the compound. Divide each element's mass by the molar mass; express the results as percentages.	**7.4**
Deriving an empirical formula from percent composition	Assume a 100 g sample so that percentage values become masses in grams. Convert the mass of each element into moles by dividing by its molar mass. Divide these numbers of moles by the smallest of their values. If necessary, multiply by an appropriate factor to remove any fractions and obtain whole-number mole values, which are the subscripts in the empirical formula of the compound.	**7.5, 7.6**
Using percent composition and molecular mass to derive a molecular formula	Derive an empirical formula from percent composition data. Calculate the multiplier n by dividing the molecular mass by the mass of one empirical formula unit. Multiply subscripts in the empirical formula by n to obtain the molecular formula.	**7.7**
Deriving an empirical formula from combustion analysis data	For hydrocarbons, convert given masses of CO_2 and H_2O into moles of CO_2 and H_2O, and then to moles of C and H. For compounds also containing O, convert moles of C and H into masses of C and H and subtract the sum of these values from the sample mass to calculate the mass of O. Convert mass of O to moles of O. Simplify mole ratios of C to H to O.	**7.8**
Identifying the limiting reactant in a reaction mixture and the theoretical yield of product	Calculate how much product each reactant could make; the reactant making the least amount of product is the limiting reactant.	**7.9, 7.10**
Calculating percent yield	Calculate the theoretical yield of product using the quantity of the limiting reactant. Divide actual yield (given or measured) by theoretical yield: $$\text{Percent yield} = \frac{\text{actual yield}}{\text{theoretical yield}} \times 100\% \quad (7.19)$$	**7.11**

VISUAL PROBLEMS

(Answers to boldface end-of-chapter questions and problems are in the back of the book.)

7.1. The molecular models in Figure P7.1 represent five oxides of nitrogen. Write the empirical and molecular formulas of the one(s) for which the two formulas are not the same.

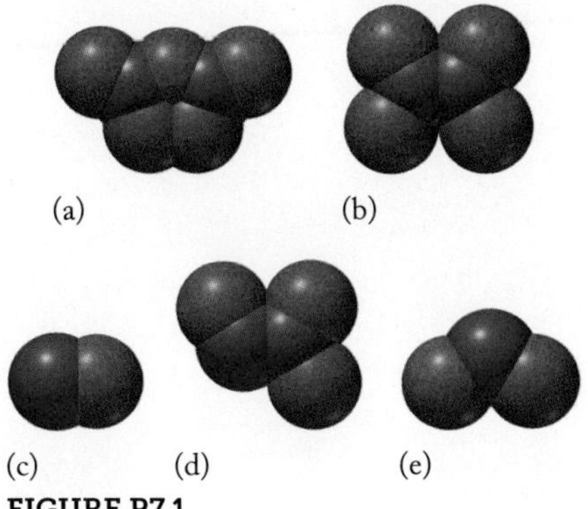

FIGURE P7.1

7.2. Which of the C_2 hydrocarbons in Figure P7.2 have different empirical and molecular formulas? What are those formulas?

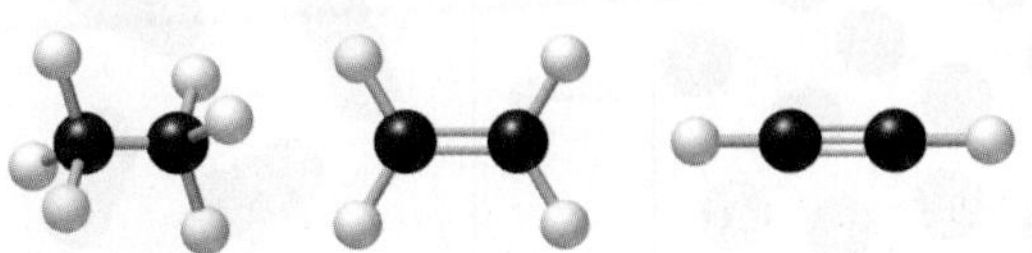

FIGURE P7.2

7.3. Each of the pairs of images in Figure P7.3 contains substances composed of two elements: X (red spheres) and Y (blue spheres). Write a balanced chemical equation for the reaction taking place in each pair of images. Be sure to indicate the physical states of the reactants and products using the appropriate symbols in parentheses.

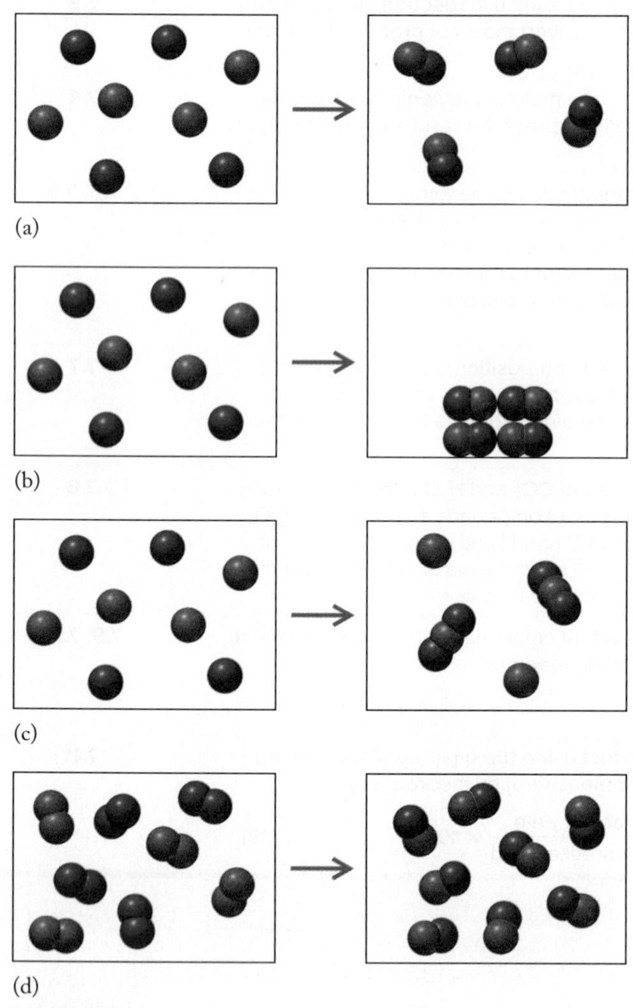

FIGURE P7.3

7.4. Identify the limiting reactant in each of the pairs of containers pictured in Figure P7.4. The red spheres represent atoms of element X, whereas the blue spheres are atoms of element Y. Each question mark means that there is unreacted reactant left over.

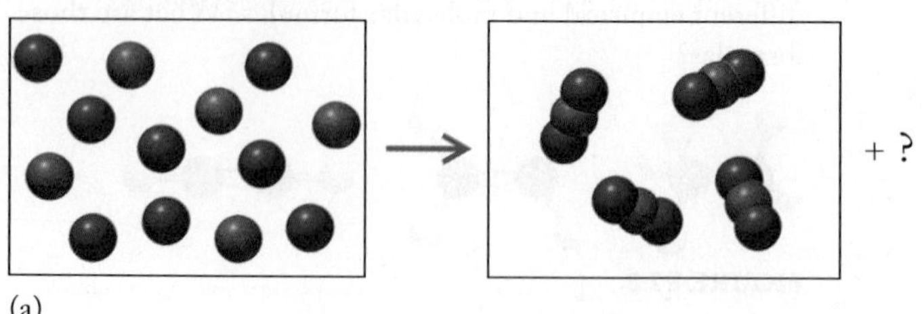

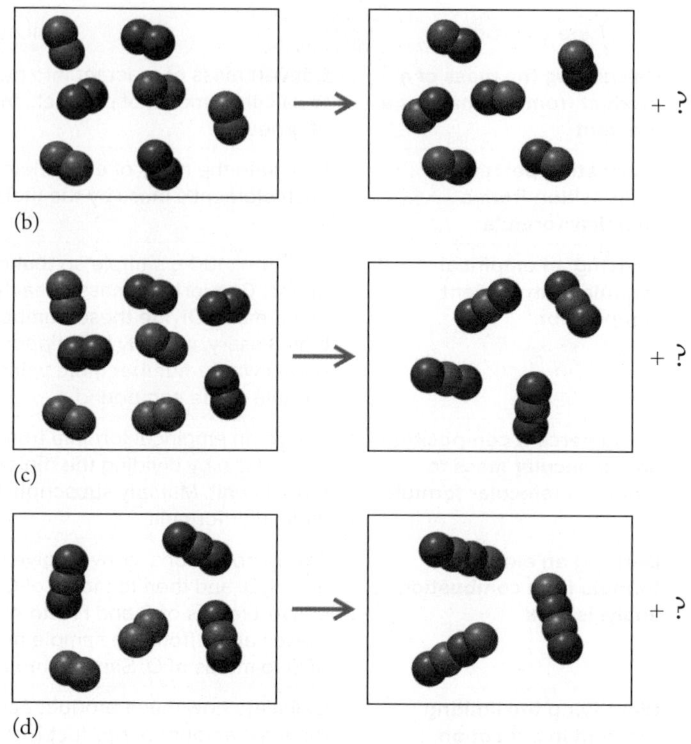

FIGURE P7.4

7.5. Which two hydrocarbons with the molecular structures shown in Figure P7.5 have the same percent composition?

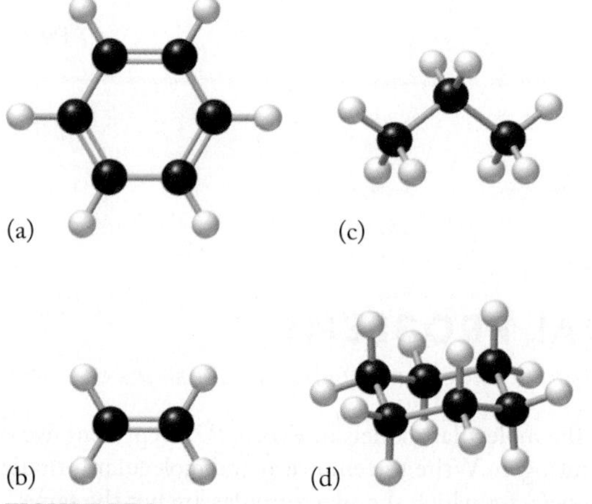

FIGURE P7.5

7.6. Use representations [A] through [I] in Figure P7.6 to answer questions a–f.
   a. Which represent compounds with the same empirical formula?
   b. Which represent compounds with the same percent composition?
   c. Combustion of which one(s) produce(s) the most molecules of CO_2?
   d. Which represents the compound with the larger percent sulfur by mass?
   e. Which reacts with water vapor, producing sulfuric acid in the atmosphere?
   f. Which represents the compound with the largest percent oxygen by mass?

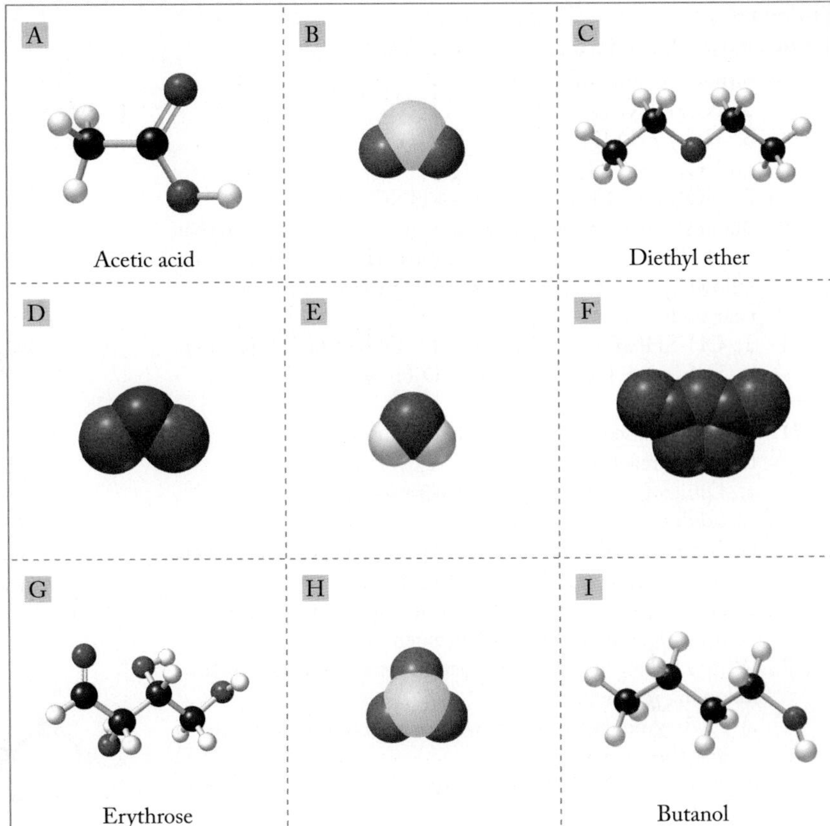

FIGURE P7.6

QUESTIONS AND PROBLEMS

Chemical Reactions and the Conservation of Mass

Concept Review

7.7. In the combination reaction A + 2 B → C, 1.00 gram of substance A and 4.00 grams of substance B are consumed. How many grams of substance C are formed?

7.8. In the reaction A + B → C + D, 3.00 grams of substance C and 3.00 grams of substance D are produced as 2.00 grams of substance A are consumed. How many grams of substance B are also consumed?

7.9. Among the four reactants and products in Question 7.8, which one has the largest molar mass?

*7.10. Among the three substances in Question 7.7 and the four substances in Question 7.8, which ones *could* be elements?

Problems

7.11. Depending on reaction conditions, O_2 may combine with N_2 to form NO or NO_2. If x grams of O_2 combine with y grams of N_2 to form NO, how many grams of O_2 combine with y grams of N_2 to form NO_2?

7.12. Combustion of sulfur may, depending on reaction conditions, produce SO_2 or SO_3. If x grams of O_2 combine with y grams of sulfur to form SO_2, how many grams of O_2 combine with y grams of sulfur to form SO_3?

7.13. Two of the more common oxides of iron have the formulas FeO and Fe_2O_3. How much more oxygen combines with a given mass of iron to form Fe_2O_3 than combines with the same mass of iron to form FeO?

7.14. Tin(IV) chloride is prepared by the following combination reaction:

$$Sn(s) + 2\,Cl_2(g) \rightarrow SnCl_4(\ell)$$

If x grams of chlorine combine with y grams of tin to form tin(IV) chloride, how many grams of chlorine are there in a sample of tin(II) chloride that contains y grams of tin?

Balancing Chemical Equations; Combustion Reactions

Concept Review

7.15. In a balanced chemical equation, does the number of atoms in the reactants always equal the number of atoms in the products?

7.16. In a balanced chemical equation for the complete combustion of a hydrocarbon, what is the ratio of atoms of C in the hydrocarbon to molecules of CO_2 produced?

7.17. How many moles of water vapor are produced for every mole of methane consumed in the combustion reaction $CH_4(g) + 2\,O_2(g) \rightarrow CO_2(g) + 2\,H_2O(g)$?

7.18. Why is CO produced during the combustion of hydrocarbons even when enough O_2 is present for complete combustion?

Problems

7.19. Balance the following reactions for the formation of nitrogen compounds:
a. $N_2(g) + O_2(g) \rightarrow NO(g)$
b. $N_2(g) + O_2(g) \rightarrow N_2O(g)$
c. $NO(g) + NO_3(g) \rightarrow NO_2(g)$
d. $NO(g) + O_2(g) + H_2O(\ell) \rightarrow HNO_2(\ell)$

7.20. **Chemistry of Geothermal Vents** Some scientists believe that life on Earth originated near geothermal vents. Balance the following reactions, which are among those taking place near such vents:
a. $CH_3SH(aq) + CO(aq) \rightarrow CH_3COSCH_3(aq) + H_2S(aq)$
b. $H_2S(aq) + CO(aq) \rightarrow CH_3CO_2H(aq) + S_8(s)$

***7.21.** Write a balanced chemical equation for each of the following reactions:
a. Dinitrogen pentoxide reacts with sodium metal to produce sodium nitrate and nitrogen dioxide.
b. A mixture of nitric acid and nitrous acid is formed when water reacts with dinitrogen tetroxide.
c. At high pressure, nitrogen monoxide decomposes to dinitrogen monoxide and nitrogen dioxide.

7.22. Write a balanced chemical equation for each of the following reactions:
a. Hydrazine, N_2H_4, reacts with oxygen to produce water and the element nitrogen.
b. Ammonia (NH_3) burns in oxygen to produce elemental nitrogen and water.
c. Silicon dioxide reacts with carbon to produce the element silicon and carbon monoxide.

7.23. Complete and balance the following chemical equations for the complete combustion of several hydrocarbons.
a. $C_3H_8(g) + O_2(g) \rightarrow$
b. $C_4H_{10}(g) + O_2(g) \rightarrow$
c. $C_6H_6(\ell) + O_2(g) \rightarrow$
d. $C_8H_{18}(\ell) + O_2(g) \rightarrow$

7.24. Complete and balance the following chemical equations for the complete combustion of several hydrocarbons.
a. $C_5H_{10}(\ell) + O_2(g) \rightarrow$
b. $C_6H_{14}(\ell) + O_2(g) \rightarrow$
c. $C_8H_{10}(\ell) + O_2(g) \rightarrow$
d. $C_9H_{12}(\ell) + O_2(g) \rightarrow$

7.25. Write balanced chemical equations for the complete combustion of the gaseous hydrocarbons with the molecular structures in Figure P7.25.

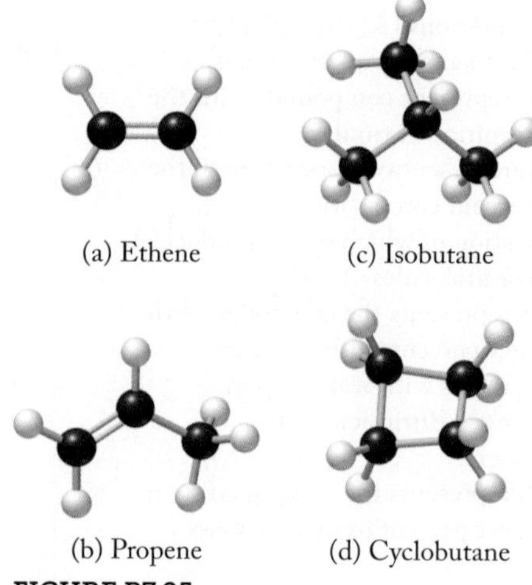

FIGURE P7.25

7.26. Write balanced chemical equations describing the complete combustion of the liquid hydrocarbons with the molecular structures in Figure P7.26.

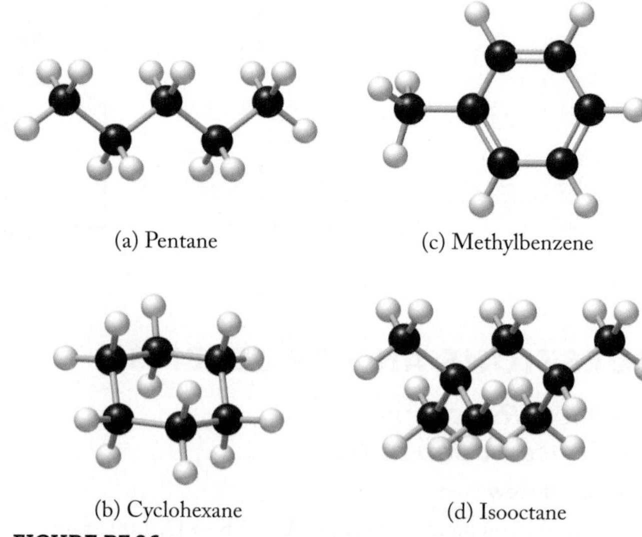

FIGURE P7.26

7.27. **Chemistry of Volcanic Gases** Balance the following reactions that occur during volcanic eruptions:
a. $SO_2(g) + O_2(g) \rightarrow SO_3(g)$
b. $H_2S(g) + O_2(g) \rightarrow SO_2(g) + H_2O(g)$
*c. $H_2S(g) + SO_2(g) \rightarrow S_8(s) + H_2O(g)$

*7.28. Copper was one of the first metals used by humans because it can be recovered from several copper minerals, including cuprite (Cu_2O), chalcocite (Cu_2S), and malachite [$Cu_2CO_3(OH)_2$]. Balance the following reactions for converting these minerals into copper metal:
a. $Cu_2O(s) + C(s) \rightarrow Cu(s) + CO_2(g)$
b. $Cu_2O(s) + Cu_2S(s) \rightarrow Cu(s) + SO_2(g)$
c. $Cu_2CO_3(OH)_2(s) + C(s) \rightarrow Cu(s) + CO_2(g) + H_2O(g)$

Stoichiometric Calculations and the Carbon Cycle

Concept Review

7.29. There are two ways to write the equation for the combustion of ethane:

$$C_2H_6(g) + \tfrac{7}{2}\,O_2(g) \rightarrow 3\,H_2O(g) + 2\,CO_2(g)$$

$$2\,C_2H_6(g) + 7\,O_2(g) \rightarrow 6\,H_2O(g) + 4\,CO_2(g)$$

Do the different ways of writing the equation affect the calculation of how much CO_2 is produced from a known quantity of C_2H_6?

7.30. Suppose the same mass of each of these components of natural gas was completely combusted: (a) CH_4, (b) C_2H_6, (c) C_3H_8, and (d) C_4H_{10}. Which one would produce the greatest mass of CO_2?

Problems

7.31. Land Management It has been estimated that better management of cropland, grazing land, and forests could reduce the amount of carbon dioxide in the atmosphere by 5.4×10^9 kilograms of carbon per year.

a. How many moles of carbon are present in 5.4×10^9 kilograms of carbon?
b. How many kilograms of carbon dioxide does this quantity of carbon represent?

7.32. Most of the CO_2 emitted by industrial sources comes from the combustion of fossil fuels, but some is also produced by cement manufacturing and the conversion of limestone ($CaCO_3$) to lime (CaO):

$$CaCO_3(s) \rightarrow CaO(s) + CO_2(g)$$

How many tons of CO_2 are produced per ton of lime?

7.33. When $NaHCO_3$ is heated above 270°C, it decomposes to $Na_2CO_3(s)$, $H_2O(g)$, and $CO_2(g)$.

a. Write a balanced chemical equation for the decomposition reaction.
b. Calculate the mass of CO_2 produced from the decomposition of 25.0 g $NaHCO_3$.

7.34. **Pigments for Stoplights** Cadmium yellow (cadmium sulfide) is a lemon-yellow pigment used in the lenses of stoplights. Its formula is CdS, and it is very insoluble in water. The recommended recipe for cadmium yellow is to mix cadmium nitrate with sodium sulfide in water. The cadmium yellow forms as a solid, while the other product, sodium nitrate, remains dissolved in the water.

a. Write a balanced chemical equation for the reaction.
b. Calculate the mass of cadmium nitrate you must start with to make 125 g of CdS.

7.35. One step in the conversion of aluminum ore into aluminum metal involves the synthesis of cryolite (Na_3AlF_6) in the following reaction:

$$6\,HF(g) + 3\,NaAlO_2(s) \rightarrow Na_3AlF_6(s) + 3\,H_2O(\ell) + Al_2O_3(s)$$

How much $NaAlO_2$ (sodium aluminate) is required to produce 1.00 kg of Na_3AlF_6?

7.36. Chromium metal can be produced from high-temperature reactions of chromium(III) oxide with silicon or aluminum:

$$Cr_2O_3(s) + 2\,Al(\ell) \rightarrow 2\,Cr(\ell) + Al_2O_3(s)$$

$$2\,Cr_2O_3(s) + 3\,Si(\ell) \rightarrow 4\,Cr(\ell) + 3\,SiO_2(s)$$

a. Calculate the mass of aluminum required to prepare 400.0 grams of chromium metal by the first reaction.
b. Calculate the mass of silicon required to prepare 400.0 grams of chromium metal by the second reaction.

7.37. Oxygen for First Responders In self-contained breathing devices used by first responders, potassium superoxide, KO_2, reacts with exhaled carbon dioxide to produce potassium carbonate and oxygen:

$$4\,KO_2(s) + 2\,CO_2(g) \rightarrow 2\,K_2CO_3(s) + 3\,O_2(g)$$

How much O_2 could be produced from 85 g KO_2?

7.38. **Capturing CO_2** Carbon dioxide can be removed from a gas stream by reacting it with potassium carbonate in the presence of water:

$$CO_2(g) + K_2CO_3(s) + H_2O(\ell) \rightarrow 2\,KHCO_3(s)$$

If a resting human exhales 36 mg CO_2 in one breath, then how much potassium carbonate would be required to capture it all?

***7.39.** The uranium minerals found in nature must be refined and enriched in ^{235}U before the uranium can be used as a fuel in nuclear reactors. One procedure for enriching uranium relies on the reaction of UO_2 with HF at high temperatures to form UF_4, which is then converted into UF_6 in another high-temperature reaction with fluorine gas:

$$(1)\ UO_2(g) + 4\,HF(aq) \rightarrow UF_4(g) + 2\,H_2O(\ell)$$

$$(2)\ UF_4(g) + F_2(g) \rightarrow UF_6(g)$$

a. How many kilograms of HF are needed to completely react with 5.00 kg UO_2?
b. How much UF_6 can be produced from 850.0 g UO_2?

7.40. The mineral bauxite, which is mostly Al_2O_3, is the principal industrial source of aluminum metal. How much aluminum can be produced from 1.00 metric ton (1.00×10^3 kg) of Al_2O_3?

***7.41.** Chalcopyrite ($CuFeS_2$) is an abundant copper mineral that can be converted into elemental copper. How much Cu is there in 1.00 kg $CuFeS_2$?

*7.42. **Mining for Gold** Unlike most metals, gold occurs in nature as the pure element. Miners in California in 1849 searched for gold nuggets and gold dust in streambeds, where the denser gold could be easily separated from sand and gravel. However, larger deposits of gold are found in veins of rock and can be separated chemically in the following two-step process:

$$(1)\ 4\,Au(s) + 8\,NaCN(aq) + O_2(g) + 2\,H_2O(\ell) \rightarrow 4\,NaAu(CN)_2(aq) + 4\,NaOH(aq)$$

$$(2)\ 2\,NaAu(CN)_2(aq) + Zn(s) \rightarrow 2\,Au(s) + Na_2[Zn(CN)_4](aq)$$

If 23 kilograms of ore is 0.19% gold by mass, how much Zn is needed to react with the gold in the ore? Assume that reactions 1 and 2 are 100% efficient.

Percent Composition and Empirical and Molecular Formulas

Concept Review

7.43. What is the difference between an empirical formula and a molecular formula?

7.44. Do the empirical and molecular formulas of a compound have the same percent composition values?

7.45. Is the element with the largest atomic mass always the element present in the highest percentage by mass in a compound?

*7.46. Sometimes the composition of a compound is expressed in units of mol % or atom %. Are the values of these parameters likely to be the same or different for a given compound?

Problems

7.47. Calculate the percent composition of (a) Na_2O, (b) NaOH, (c) $NaHCO_3$, and (d) Na_2CO_3.

7.48. Calculate the percent composition of (a) sodium sulfate, (b) dinitrogen tetroxide, (c) strontium nitrate, and (d) aluminum sulfide.

7.49. **Organic Compounds in Space** The following compounds have been detected in space. Which contains the greatest percentage of carbon by mass?
a. naphthalene, $C_{10}H_8$
b. chrysene, $C_{18}H_{12}$
c. pentacene, $C_{22}H_{14}$
d. pyrene, $C_{16}H_{10}$

7.50. **Lead Compounds as Pigments** Ancient Egyptians used lead compounds, including PbS, $PbCO_3$, and $Pb_2Cl_2CO_3$, as pigments in cosmetics, and many people suffered from chronic lead poisoning as a result. Calculate the percentage of lead in each of the compounds.

7.51. A 3.556 g sample of a pure aluminum oxide decomposes under high heat to produce 1.674 g of oxygen in addition to pure aluminum metal. What is the empirical formula of the aluminum oxide?

*7.52. Oxygen gas may be prepared in the laboratory by decomposing a compound of potassium, chlorine, and oxygen. The other product of the decomposition is KCl(*s*). Complete decomposition of 2.917 g of the compound produces 1.143 g of oxygen gas. What is the empirical formula of the compound?

7.53. Do any two of the following compounds, which have been detected in outer space, have the same empirical formula?
a. naphthalene, $C_{10}H_8$
b. chrysene, $C_{18}H_{12}$
c. anthracene, $C_{14}H_{10}$
d. pyrene, $C_{16}H_{10}$
e. benzoperylene, $C_{22}H_{12}$
f. coronene, $C_{24}H_{12}$

7.54. Which of the following nitrogen oxides have the same empirical formulas? (a) N_2O; (b) NO; (c) NO_2; (d) N_2O_2; (e) N_2O_4

7.55. **Surgical-Grade Titanium** Medical implants and high-quality jewelry items for body piercings are frequently made of a material known as G23Ti, or surgical-grade titanium. The percent composition of the material is 64.39% titanium, 24.19% aluminum, and 11.42% vanadium. What is the empirical formula of surgical-grade titanium?

7.56. A sample of an iron-containing compound is 22.0% iron, 50.2% oxygen, and 27.8% chlorine by mass. What is the empirical formula of this compound?

7.57. Phosphorus burns in pure oxygen with a brilliant white light. The product of combustion is 43.64% phosphorus and 56.36% oxygen.
a. What is the empirical formula of this compound?
b. The molar mass of the compound is 284 g/mol. What is its molecular formula?

7.58. Ferrophosphorus (Fe_2P) reacts with pyrite (FeS_2), producing iron(II) sulfide and a compound that is 27.87% P and 72.13% S by mass and has a molar mass of 444.56 g/mol.
a. Determine the empirical and molecular formulas of the compound.
b. Write a balanced chemical equation for the reaction.

7.59. **Asbestos and Lung Disease** Inhalation of asbestos fibers may lead to a lung disease known as asbestosis and to a form of lung cancer called mesothelioma. One form of asbestos, chrysotile, is 26.31% magnesium, 20.27% silicon, and 1.45% hydrogen by mass, with the remainder of the mass as oxygen. What is the empirical formula of chrysotile?

7.60. **Chemistry of Soot** A piece of glass held over a candle flame becomes coated with soot, which is the result of the incomplete combustion of candle wax. Elemental analysis of a compound extracted from a sample of soot gave the following results: 7.74% H and 92.26% C by mass. Calculate the empirical formula of the compound.

7.61. What is the empirical formula of the compound that is 24.2% Cu, 27.0% Cl, and 48.8% O by mass?

7.62. A chlorine oxide used to kill anthrax spores in contaminated buildings is 52.6% Cl by mass. What is its empirical formula?

Combustion Analysis

Concept Review

7.63. Explain why it is important for combustion analysis to be carried out in an excess of oxygen.

7.64. Why is the quantity of CO_2 obtained in a combustion analysis not a direct measure of the oxygen content of the starting compound?

7.65. Can the results of a combustion analysis ever give the true molecular formula of a compound?

7.66. A chemical reaction that is used to analyze a compound is expected to have a 100% yield. However, when a compound is synthesized, a 100% yield is almost never expected. Explain this difference.

Problems

7.67. A 0.100 g sample of a compound containing C, H, and O is burned in oxygen, producing 0.1783 g CO_2 and 0.0734 g H_2O. What is the empirical formula of the compound?

7.68. **GRAS List for Food Additives** The alcohol geraniol is on the U.S. Food and Drug Administration's GRAS (generally recognized as safe) list and can be used in foods and personal care products. By itself, geraniol has a roselike odor, but it is frequently blended with other scents to produce the fruity fragrances of some personal care products. Complete combustion of 175 mg of geraniol produces 499 mg CO_2 and 184 mg H_2O. What is the empirical formula of geraniol?

7.69. Combustion of 135.0 mg of a hydrocarbon produces 440.0 mg CO_2 and 135.0 mg H_2O. The molar mass of the hydrocarbon is 270 g/mol. What are the empirical and molecular formulas of this compound?

7.70. The combustion of 40.5 mg of a compound extracted from the bark of the sassafras tree and known to contain C, H, and O produces 110.0 mg CO_2 and 22.5 mg H_2O. The molar mass of the compound is 162 g/mol. What are its empirical and molecular formulas?

Limiting Reactants and Percent Yield

Concept Review

7.71. If a reaction vessel contains equal masses of Fe and S, a mass of FeS corresponding to which of the following could theoretically be produced?
a. the sum of the masses of Fe and S
b. more than the sum of the masses of Fe and S
c. less than the sum of the masses of Fe and S

7.72. A reaction vessel contains equal masses of magnesium metal and oxygen gas. The mixture is ignited, forming MgO. After the reaction has gone to completion, the mass of the MgO is less than the mass of the reactants. Is this result a violation of the law of conservation of mass? Explain your answer.

7.73. Explain how the parameters of theoretical yield and percent yield differ.

7.74. Can the percent yield of a chemical reaction ever exceed 100%?

7.75. If fewer moles of A are present in a reaction between A and B, then A must be the limiting reagent. What is wrong with this statement?

7.76. A chemical reaction produces less than the expected amount of product. Is this result a violation of the law of conservation of mass?

Problems

7.77. **Making Hollandaise Sauce** A recipe for 1 cup of hollandaise sauce calls for $\frac{1}{4}$ cup of butter, $\frac{1}{2}$ cup of hot water, 4 egg yolks, and the juice of a medium-sized lemon. How many cups of sauce can be made from a pound (2 cups) of butter, a dozen eggs, 4 medium-sized lemons, and an unlimited supply of hot water?

7.78. A factory making toy wagons has 13,466 wheels, 3360 handles, and 2400 wagon beds in stock. What is the maximum number of wagons with four wheels that the factory can make from these components?

***7.79.** **Sulfur in Coal** Suppose 75 metric tons of coal that is 3.0% sulfur by mass is burned at a power plant. During combustion, the sulfur is converted into SO_2. Antipollution scrubbers installed in the smokestacks of the power plant capture 3.9 metric tons of SO_2. How efficient are the scrubbers in capturing SO_2? How many metric tons of SO_2 escape?

7.80. One reaction in the production of sulfuric acid involves the conversion of sulfur dioxide to sulfur trioxide. In the presence of excess O_2, 88 kg SO_2 produces 106 kg SO_3. What is the percent yield?

***7.81.** An important industrial use of chloroform [$CHCl_3(\ell)$] is a reaction with $HF(g)$ to produce $CHClF_2(g)$:
i. $CHCl_3(\ell) + 2\,HF(g) \rightarrow CHClF_2(g) + 2\,HCl(g)$
The product $CHClF_2$ is then heated to a high temperature in the absence of oxygen to produce C_2F_4, which is used to make the polymer known as Teflon™.
ii. $2\,CHClF_2(g) \rightarrow C_2F_4(g) + 2\,HCl(g)$
a. In a reaction mixture of 775 g $CHCl_3$ and 775 g HCl, which is the limiting reactant?
b. Assume the yield in reaction (i) is stoichiometric and in reaction (ii) is 95%. How many grams of C_2F_4 could be prepared in reaction (ii) from the product formed in reaction (i)?
c. How much of which reactant is left over in the reaction mixture in part (i)?

7.82. A reaction vessel contains 10.0 g CO and 10.0 g O_2, which combine to form CO_2:

$$2\,CO(g) + O_2(g) \rightarrow 2\,CO_2(g)$$

a. Which reactant is limiting?
b. How many grams of CO_2 could be produced?
c. How many grams of the nonlimiting reactant are left over?

***7.83.** **Syngas** An industrial process for producing hydrogen gas is based on the reaction between carbon heated to incandescence and steam that produces a mixture of CO and H_2 called synthesis gas, or *syngas*.
a. Write a balanced chemical equation for the production of syngas.
b. If a reaction vessel that initially contains 66 kilograms of incandescent carbon and excess steam produces 6.8 kg H_2, what is the percent yield?

*7.84. Baking soda ($NaHCO_3$) is produced on an industrial scale by the Solvay process. A key reaction in the process is

$$NaCl(aq) + NH_3(aq) + CO_2(aq) + H_2O(\ell) \rightarrow NaHCO_3(s) + NH_4Cl(aq)$$

Suppose a reaction vessel initially contains 58.5 kg NaCl, 18.8 kg NH_3, and excess CO_2 and H_2O. If 66 kg $NaHCO_3$ is produced, what is the percent yield?

7.85. **Chemistry of Fermentation** Yeast converts glucose ($C_6H_{12}O_6$) in aqueous solution into ethanol (CH_3CH_2OH, $d = 0.789$ g/mL) in a process called fermentation. Carbon dioxide is also produced.
a. Write a balanced chemical equation for the fermentation reaction.
b. If 100.0 grams of glucose yields 50.0 mL of ethanol, what is the percent yield for the reaction?

*7.86. **Black Powder** Gunpowder, known also as black powder, is generally thought to have been invented by the Chinese in the 9th century. The first description in English was given by Roger Bacon in the 13th century. The basic formulation has not changed since then: 40% potassium nitrate,

30% carbon, and 30% sulfur by weight. The products of the reaction that provide the explosive force when the powder is ignited are three gases: carbon monoxide, carbon dioxide, and nitrogen. An additional product is potassium sulfite. Consider a sample of 100 g of black powder.

a. Which reagent is limiting?
b. How much of each of the reagents in excess is left over after the explosion takes place?

Additional Problems

***7.87. Artificial Bones for Medical Implants** The material often used to make artificial bones is the same material that gives natural bones their strength. Its common name is hydroxyapatite, and its formula is $Ca_5(PO_4)_3OH$.

a. Propose a systematic name for this compound.
b. What is the mass percentage of calcium in it?
c. When treated with hydrogen fluoride, hydroxyapatite becomes fluorapatite [$Ca_5(PO_4)_3F$], an even stronger substance. Does the percent mass of Ca increase or decrease as a result of this substitution?

***7.88.** As a solution of copper sulfate slowly evaporates, beautiful blue crystals form. Their chemical formula is $CuSO_4 \cdot 5H_2O$.

a. What is the percent water in this compound?
b. At high temperatures the water is driven off as steam. What fraction of the original sample's mass is lost as a result?

7.89. Aluminum is obtained from the mineral bauxite. Suppose that a sample of bauxite from Jamaica is 86% aluminum oxide. If 2.3 metric tons of aluminum metal is recovered from 5.1 metric tons of the bauxite ore, what is the percent yield of the recovery process?

7.90. Chemistry of Copper Production The mineral chalcopyrite ($CuFeS_2$) is an important source of copper metal, though recovering the metal requires several chemical reactions that transform $CuFeS_2$ into CuS, then Cu_2S, and finally Cu metal. The pennies minted in the United States between 1909 and 1982 weighed 3.11 g and were 95% by mass copper.

a. How much chalcopyrite had to be mined to produce one dollar's worth of these pennies?
b. How much chalcopyrite had to be mined to produce one dollar's worth of the pennies if the first reaction had a percent yield of 85% and the second and third reactions had percent yields of essentially 100%?
c. How much chalcopyrite had to be mined to produce one dollar's worth of the pennies if each of the reactions proceeded with 85% yield?

***7.91. Mining for Gold** Gold can be extracted from the surrounding rock using a solution of sodium cyanide. While effective for isolating gold, toxic cyanide finds its way into watersheds, causing environmental damage and harming human health.

$$4\,Au(s) + 8\,NaCN(aq) + O_2(g) + 2\,H_2O(\ell) \rightarrow 4\,NaAu(CN)_2(aq) + 4\,NaOH(aq)$$

$$2\,NaAu(CN)_2(aq) + Zn(s) \rightarrow 2\,Au(s) + Na_2[Zn(CN)_4](aq)$$

a. If a sample of rock contains 0.009% gold by mass, how much NaCN is needed to extract the gold as $NaAu(CN)_2$ from 1 metric ton (10^3 kg) of rock?
b. How much zinc is needed to convert the $NaAu(CN)_2$ from part (a) to metallic gold?
c. The gold recovered in part (b) is manufactured into a gold ingot in the shape of a cube. The density of gold is 19.3 g/cm^3. How big is the block of gold in cm^3?

***7.92.** Phosgenite, a lead compound with the formula $Pb_2Cl_2CO_3$, is found in Egyptian cosmetics. Phosgenite was prepared by the reaction of PbO, NaCl, and CO_2. An unbalanced expression of the reactant mixture is

$$PbO(s) + NaCl(aq) + H_2O(\ell) + CO_2(g) \rightarrow Pb_2Cl_2CO_3(s) + NaOH(aq)$$

a. Balance the equation.
b. How many grams of phosgenite can be obtained from 10.0 g PbO and 10.0 g NaCl in the presence of excess water and CO_2?
c. Phosgenite can be considered a mixture of two lead compounds. Which compounds appear to be combined to make phosgenite?

***7.93.** Uranium oxides used in the preparation of fuel for nuclear reactors are separated from other metals in minerals by converting the uranium to $UO_x(NO_3)_y(H_2O)_z$, where uranium has a positive charge ranging from 3+ to 6+.

a. Roasting $UO_x(NO_3)_y(H_2O)_z$ at 400°C leads to loss of water and decomposition of the nitrate ion to nitrogen oxides, leaving behind a product with the formula U_aO_b that is 83.22% U by mass. What are the values of a and b? What is the charge on U in U_aO_b?
b. Higher temperatures produce a different uranium oxide, U_cO_d, which is 84.8% U by mass. What are the values of c and d? What is the charge on U in U_cO_d?
c. The values of x, y, and z in $UO_x(NO_3)_y(H_2O)_z$ are found by gently heating the compound to remove all of the water. In a laboratory experiment, 1.328 g $UO_x(NO_3)_y(H_2O)_z$ produced 1.042 g $UO_x(NO_3)_y$. Continued heating generated 0.742 g U_nO_m. Using the information in parts (a) and (b), calculate x, y, and z.

***7.94.** Corn farmers in the American Midwest typically use 5.0×10^3 kilograms of ammonium nitrate fertilizer per square kilometer of cornfield per year. Some of the fertilizer washes into the Mississippi River and eventually flows into the Gulf of Mexico, promoting the growth of algae and endangering other aquatic life.

a. Ammonium nitrate can be prepared by the following reaction:

$$NH_3(g) + HNO_3(aq) \rightarrow NH_4NO_3(aq)$$

How much nitric acid would be required to make the fertilizer needed for 1 km^2 of cornfield per year?

b. Ammonium ions dissolved in groundwater may be converted into NO_3^- ions by bacterial action:

$$NH_4^+(aq) + 2\,O_2(g) \rightarrow NO_3^-(aq) + H_2O(\ell) + 2\,H^+(aq)$$

If 10% of the ammonium component of 5.0×10^3 kilograms of fertilizer ends up as nitrate ions, how much oxygen would be consumed?

7.95. Fiber in the Diet Dietary fiber is a mixture of many compounds, including xylose ($C_5H_{10}O_5$) and methyl galacturonate ($C_7H_{12}O_7$).

a. Do these compounds have the same empirical formula?
b. Write balanced chemical equations for the complete combustion of xylose and methyl galacturonate.

7.96. Some catalytic converters in automobiles contain two manganese oxides: Mn_2O_3 and MnO_2.
a. What are the names of these compounds?
b. What is the manganese content of each (expressed as a percent by mass)?
c. Explain how Mn_2O_3 and MnO_2 are consistent with the law of multiple proportions.

***7.97.** A number of chemical reactions have been proposed for the formation of organic compounds from inorganic precursors, including the following:

$$H_2S(g) + FeS(s) + CO_2(g) \rightarrow FeS_2(s) + HCO_2H(\ell)$$

a. Identify the ions in FeS and FeS_2.
b. What are the names of FeS and FeS_2?
c. How much HCO_2H is obtained by reacting 1.00 g FeS, 0.50 g H_2S, and 0.50 g CO_2 if the reaction results in a 50.0% yield?

*7.98. Organic compounds called *carbohydrates* may be formed in reactions between iron(II) sulfide and carbonic acid:

$$2\,FeS + H_2CO_3 \rightarrow 2\,FeO + \tfrac{1}{n}(CH_2O)_n + 2\,S$$

a. What is the empirical formula of these carbohydrates?
b. How much carbohydrate is produced from a reaction mixture that initially contains 211 g FeS and excess H_2CO_3 if the reaction results in a 78.5% yield?
c. If the carbohydrate product has a molecular mass of 300 amu, what is its molecular formula?

***7.99.** **Marine Chemistry of Iron** On the seafloor, solid iron(II) oxide may react with water to form solid Fe_3O_4 and hydrogen gas.
a. Write a balanced chemical equation for the reaction.
b. When CO_2 is also present, the product of the reaction is methane, not hydrogen. Write a balanced chemical equation for this reaction.

7.100. Titanium dioxide and zinc oxide are common names of two of the active ingredients approved by the U.S. FDA for use in sunscreens.
a. What are the chemical formulas of these compounds?
b. What are the proper names of the compounds based on the rules for naming described in Chapter 4?
c. Which of the two compounds contains the higher percentage of oxygen by mass?

***7.101.** **Ethanol in Fuel** E-85 is an alternative fuel for automobiles and light trucks that consists of 85% (by volume) ethanol (CH_3CH_2OH) and 15% gasoline. The density of ethanol is 0.79 g/mL.
a. How many moles of ethanol are in a gallon of E-85?
b. How many moles of carbon dioxide are produced by the complete combustion of the ethanol in a gallon of E-85 fuel?

7.102. **Military Balloons** The ferrosilicon process is used in the military to produce hydrogen quickly to inflate balloons. A small truck-mounted generator contains a steel vessel that is charged with a source of silicon (usually an iron–silicon mixture from which the process gets its name) and sodium hydroxide. When water is added, a reaction ensues between Si, NaOH, and water that produces sodium silicate (Na_2SiO_3) and hydrogen gas. How many grams of silicon and sodium hydroxide would you need to produce 1.55×10^4 g of hydrogen gas sufficient to inflate a reconnaissance balloon?

***7.103.** You are given a 0.6240 g sample of a substance with the generic formula $MCl_2 \cdot 2\,H_2O$. After completely drying the sample (which means removing the 2 mol of H_2O per mole of MCl_2), the sample has a mass of 0.5471 g. What is the identity of element M?

7.104. A compound found in crude oil is 93.71% C and 6.29% H by mass. The molar mass of the compound is 128 g/mol. What is its molecular formula?

7.105. A reaction vessel for synthesizing ammonia by reacting nitrogen and hydrogen is charged with 6.04 kg H_2 and excess N_2. If 28.0 kg NH_3 is produced, what is the percent yield of the reaction?

7.106. If a cube of table sugar, which is made of sucrose, $C_{12}H_{22}O_{11}$, is added to concentrated sulfuric acid, the acid "dehydrates" the sugar: it removes the hydrogen and oxygen from it, leaving behind a lump of carbon. What percentage of the initial mass of sugar is carbon?

7.107. **Reducing SO_2 Emissions** One way in which SO_2 is removed from the "stack" gases of coal-burning power plants is by spraying the gases with fine particles of solid calcium oxide suspended in O_2 gas. The product of the reaction of SO_2, CaO, and O_2 is calcium sulfate.
a. Write a balanced chemical equation for this reaction.
b. How many metric tons of calcium sulfate would be produced from each ton of SO_2 that is trapped?

7.108. **Gas Grill Reaction** The burner in a gas grill mixes 24 volumes of air for every one volume of propane (C_3H_8) fuel. Like all gases, the volume that propane occupies is directly proportional to the number of moles of it at a given temperature and pressure. Air is 21% (by volume) O_2. Is the flame produced by the burner fuel-rich (excess propane in the reaction mixture), fuel-lean (not enough propane), or stoichiometric (just right)?

7.109. A common mineral in Earth's crust has the chemical composition 34.55% Mg, 19.96% Si, and 45.49% O. What is its empirical formula?

7.110. **Ozone Generators** Some indoor air-purification systems work by converting a little of the oxygen in the air to ozone, which kills mold and mildew spores and other biological air pollutants. The chemical equation for the ozone generation reaction is

$$3\,O_2(g) \rightarrow 2\,O_3(g)$$

It is claimed that one such system generates 4.0 g O_3 per hour from dry air passing through the purifier at a flow of 5.0 L/min. If exactly 1 liter of indoor air contains 0.28 g O_2, what percentage of the O_2 is converted to O_3 by the air purifier?

smartwork5

If your instructor uses Smartwork5, log in at digital.wwnorton.com/atoms2.

8

Aqueous Solutions

Chemistry of the Hydrosphere

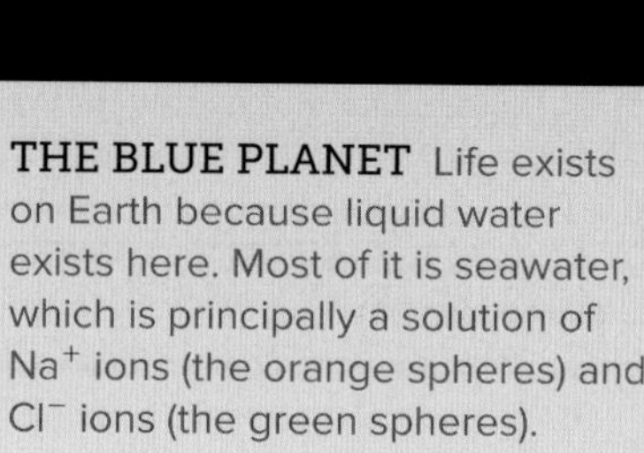

THE BLUE PLANET Life exists on Earth because liquid water exists here. Most of it is seawater, which is principally a solution of Na^+ ions (the orange spheres) and Cl^- ions (the green spheres).

PARTICULATE **REVIEW**

Empirical Formulas and Ionic Compounds

In Chapter 8, we explore the types of reactions that can occur when aqueous solutions of ionic compounds are mixed together.

- Which particle shown here could be one of the cations formed by iron? Which particle represents a nitrate ion?
- Name two compounds composed of iron cations and nitrate ions.
- Write the empirical formulas for these two compounds.

(Review Sections 4.1 and 4.2 if you need help answering these questions.)

(Answers to Particulate Review questions are in the back of the book.)

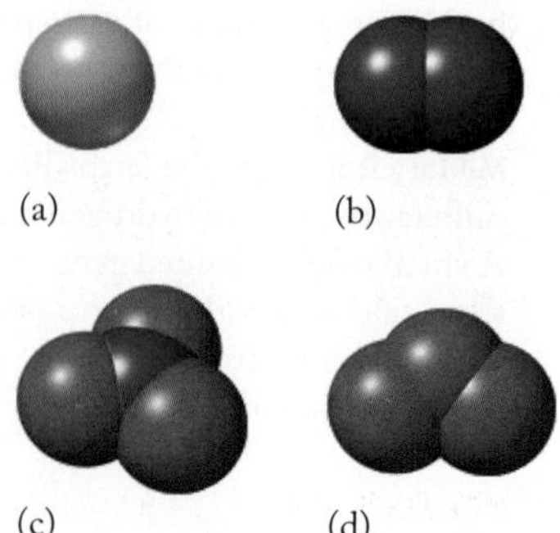

(a) (b) (c) (d)

PARTICULATE **PREVIEW**

H Atom versus Acidic Proton

Hydrogen nuclei can be part of molecules and ions. As you read Chapter 8, look for ideas that will help you determine when a hydrogen atom in a molecule is acidic and can ionize to become a proton and answer these questions:

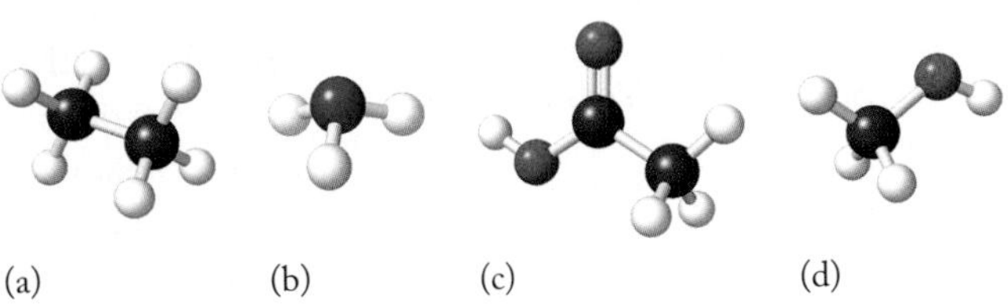

- Why is a hydrogen ion called a proton?
- Which of the molecules shown here ionize when they dissolve in water, donating H^+ ions to water molecules?
- Which of these molecules ionize completely in water—that is, each one donates at least one H^+ ion?

Learning Outcomes

LO1 Express the concentrations of solutions in different units, including molarity
Sample Exercises 8.1, 8.2

LO2 Calculate the mass of a solute or the volume of a highly concentrated solution required to make a solution of specified volume and concentration
Sample Exercises 8.3, 8.4

LO3 Distinguish between strong electrolytes, weak electrolytes, and nonelectrolytes

LO4 Identify Brønsted–Lowry acids and bases and write molecular, total ionic, and net ionic equations for reactions in solution
Sample Exercises 8.5, 8.6

LO5 Predict precipitation reactions using solubility rules

LO6 Calculate the concentration of a solute or the quantity of a precipitate from stoichiometry data
Sample Exercises 8.7, 8.8

LO7 Determine oxidation numbers and use them to identify redox reactions, oxidizing agents, and reducing agents
Sample Exercises 8.9, 8.10

LO8 Use changes in oxidation numbers to write balanced equations for redox reactions
Sample Exercises 8.11, 8.12

LO9 Determine the concentration of a solute from titration data
Sample Exercise 8.13

8.1 Solutions and Their Concentrations

Depressions in Earth's crust contain over 10^{21} L of freshwater and seawater—enough to cover 70% of the planet. All of this water contains dissolved ionic and molecular compounds. The concentrations of solutes vary considerably from one body of freshwater to the next, but the proportions of the major ions in seawater are essentially the same all over the world. Perhaps more remarkably, their concentrations appear to have changed little over the last billion years.

The constant composition of the oceans may come as a surprise, given that rivers deliver about 4×10^{12} kg of dissolved ionic compounds to the sea every year. On the other hand, the huge volume of freshwater flowing to the sea (4×10^{16} L per year) means that the total concentration of all ionic compounds in river water is typically less than 1/100 the salinity of seawater. In addition, the proportions of the ions in river water don't match those of seawater.

Given these differences in composition, how did the sea become so salty in the first place, and how has it stayed that way? Not all of the salt got there by physical erosion and chemical weathering of the land. Much of the Cl^- content, for example, probably came from HCl released by underwater volcanoes billions of years ago, and most of the Na^+ ions probably leached out of the ocean's floor as it first filled with water. Today, an elaborate system of physical and chemical processes operates within and above Earth's oceans to maintain their composition. About 90% of the water vapor in the atmosphere—the source of all that river water flowing into the sea—is evaporated seawater. Therefore, river water does not really dilute the saltiness of seawater because the evaporation process that created it actually made the sea a little saltier in the first place.

On the other hand, Ca^{2+} ions flowing into the sea don't remain in seawater very long. They are taken up by corals, shellfish, and other sea creatures to grow the hard parts of their bodies (made mostly of $CaCO_3$). Other ions, such as Fe^{3+} and Mn^{2+}, are soluble in freshwater—which is, on average, slightly acidic—but become insoluble when they reach the sea, which is slightly basic.

The composition of the sea and the survival of the creatures in it are linked to chemical and biochemical reactions that depend on acid–base balance, the formation of insoluble ionic compounds, and reactions in which some elements are oxidized as others are reduced. In this chapter we explore each of these types of reactions. When we discuss reactions that take place in solution, we need to be able to quantify the amounts of materials present so that we can carry out calculations based on stoichiometry. To do this, we have to understand the various modes of expressing the concentration of dissolved substances.

In Chapter 1 we defined a *solution* as a homogeneous mixture of two or more substances. Though we usually think of solutions as liquids, homogeneous mixtures of solids and gases also exist. For example, familiar metals such as brass (a mixture of copper and zinc), bronze (copper and tin), and stainless steel (chromium and other metals mixed with iron) are solutions; so is filtered air. The substance present in the greatest proportion in a solution is called the solvent, and all the substances dissolved in it are solutes. Solutions in which water is the solvent are called *aqueous* solutions, and they are the focus of this chapter.

CONNECTION Solvents and solutes were introduced in Chapter 6, where we used intermolecular forces to predict the solubility of molecular substances.

The concentration of an aqueous solution describes the quantity of solute dissolved in a specific quantity of either solvent or total solution. This ratio may be expressed in terms of the mass of the solute in a given mass or volume of solvent or in a given mass or volume of solution. Concentration units based on mass-to-mass ratios, such as milligrams of solute per kilogram of solvent (mg/kg), are commonly used in disciplines requiring exact measurements over a range of temperature and pressure, because volumes change with conditions but masses do not. Probably the most common concentration units you see in daily life are based on mass-to-volume ratios, such as milligrams of solute per liter of solution (mg/L). Actually, these two units (mg/kg and mg/L) are equivalent for dilute aqueous solutions near room temperature because the densities of these solutions are nearly the same as that of water, which is 1.00 kg/L at 20°C.

Another concentration unit that is synonymous with mg/kg is parts per million (ppm). These units are equivalent because 1 mg = 10^{-3} g and 1 kg = 10^3 g. Therefore,

$$1 \text{ mg/kg} = 10^{-3} \text{ g}/10^3 \text{ g} = 1 \text{ g}/10^6 \text{ g} = 1 \text{ ppm}$$

One-millionth of a gram per gram is the same as one gram per million grams, or simply one part per million. Even smaller concentrations may be expressed in micrograms per kilogram (μg/kg) or micrograms per liter (μg/L), which are also equivalent for dilute aqueous solutions and are the same as parts per billion (ppb). Environmental pollutants reported in the press are frequently expressed in mg/L or ppm. For example, the maximum contaminant level for arsenic in drinking water is 0.010 mg/L or 0.010 ppm. Results from clinical laboratories, such as those for blood tests, are also expressed as mass-to-volume ratios. For example, the acceptable range for fasting blood sugar is 70–99 mg/dL, where "dL" is deciliter (1/10 of a liter, or 100 mL).

Because we are interested in describing reactions in solutions, we will work with quantities of dissolved reactants and products that are related by how many *moles* of each of them there are in reaction mixtures. Consequently, we need to express their concentrations in units based on moles of solute per mass or volume of solution. One such unit is **molarity (M)**, which is the number of moles (n) of solute in a volume (V) of 1 liter of solution.[1] In equation form:

$$M = \frac{\text{moles of solute}}{\text{liter of solution}} = \frac{n}{V} \quad (8.1)$$

molarity (M) the number of moles of solute per liter of solution: $M = \frac{n}{V}$; also called *molar concentration*.

[1]Note that molarity is symbolized by M, whereas molar mass is $\mathcal{M}$ throughout this text.

Often we know the volume and concentration of a solution and need to calculate the number of moles of solute in it. Equation 8.1 can be rearranged to solve for n:

$$n = V \times M = \cancel{\text{L}} \times \frac{\text{mol}}{\cancel{\text{L}}} \tag{8.2}$$

To calculate the mass (m) in grams of the solute in a solution of a known volume and molar concentration, we can multiply the number of moles of solute obtained using Equation 8.2 by the molar mass ($\mathcal{M}$) of the solute:

$$m = (V \times M) \times \mathcal{M} = \left(\cancel{\text{L}} \times \frac{\cancel{\text{mol}}}{\cancel{\text{L}}}\right) \times \frac{\text{g}}{\cancel{\text{mol}}} \tag{8.3}$$

Equation 8.3 is particularly useful when we need to calculate the mass of solute required to prepare a solution of a desired volume and concentration.

In many environmental and biological systems solute concentrations are much less than 1.0 M. In Table 8.1, for instance, the average concentrations of most of the major ions in seawater and in human serum are more conveniently expressed in *milli*moles per liter (mmol/L), or millimolarity (mM), where 1 mM = 10^{-3} M. Even smaller concentrations of minor and trace substances in seawater and in biological samples such as blood and urine are expressed in micromolarity (μM, 1 $\mu M = 10^{-6}$ M), nanomolarity (nM, 1 nM = 10^{-9} M), and even picomolarity (pM, 1 pM = 10^{-12} M).

The concentrations in the first column of numbers in Table 8.1 are expressed in millimoles of solute per kilogram of seawater. Environmental scientists and especially oceanographers prefer to use concentration units based on the masses of water samples rather than their volumes because, as we mentioned previously, the volume of a given mass of water varies with temperature and pressure, whereas its mass and the quantities of solutes dissolved in that mass remain constant.

TABLE 8.1 Average Concentrations of 11 Major Constituents of Seawater and Human Serum

	SEAWATER		HUMAN SERUM
Constituent	mmol/kg	mM	mM
Na^+	468.96	480.57	135–145
K^+	10.21	10.46	3.5–5.0
Mg^{2+}	52.83	54.14	0.08–0.12
Ca^{2+}	10.28	10.53	0.2–0.3
Sr^{2+}	0.0906	0.0928	$<3 \times 10^{-4}$
Cl^-	545.88	559.40	98–108
SO_4^{2-}	28.23	28.93	0.3
HCO_3^-	2.06	2.11	22–30
Br^-	0.844	0.865	0.04–0.06
$B(OH)_3$	0.416	0.426	$<8 \times 10^{-4}$
F^-	0.068	0.070	5–6

CONCEPT TEST

Rank the following solutions from most concentrated to least concentrated: (a) 0.0053 M NaCl; (b) 54 mM NaCl; (c) 550 μM NaCl; (d) 56,000 nM NaCl.

(Answers to Concept Tests are in the back of the book.)

SAMPLE EXERCISE 8.1 Converting Mass-per-Volume Concentrations into Molarity LO1

Vinyl chloride (Figure 8.1) is one of the most widely used industrial chemicals (about 3×10^{10} kg is produced each year, mostly for making polyvinyl chloride plastic). It is also one of the most toxic chemicals and is a known carcinogen. The maximum concentration of vinyl chloride allowed in drinking water in the United States is 0.002 mg/L. What is that concentration in moles per liter?

Collect, Organize, and Analyze Because the solute mass is given in milligrams, we need to convert it to grams and then to moles. Given the number of C and Cl atoms in its molecular structure, the molar mass of vinyl chloride is likely to be between 50 and 100 g/mol, so 0.002 mg/L should translate into about 0.00002 mmol/L—or 2×10^{-8} *M*, or 0.02 μ*M*.

Solve The molar mass of vinyl chloride (C_2H_3Cl) is

$$\mathcal{M} = 2(12.01 \text{ g/mol}) + 3(1.008 \text{ g/mol}) + 35.45 \text{ g/mol} = 62.49 \text{ g/mol}$$

The molarity equivalent of 0.002 mg/L of vinyl chloride is

$$\frac{0.002 \cancel{\text{mg}}}{\text{L}} \times \frac{1 \cancel{\text{g}}}{10^3 \cancel{\text{mg}}} \times \frac{1 \text{ mol}}{62.49 \cancel{\text{g}}} = \frac{3 \times 10^{-8} \text{ mol}}{\text{L}} = 3 \times 10^{-8}\, M$$

Think About It This concentration is close to the value we predicted. Expressing it using micromolarity units to one significant figure is appropriate because the initial concentration value was known to only one significant figure. Given the hazards vinyl chloride poses to human health, it is reasonable that the allowed concentration of it in drinking water would be very small.

Practice Exercise When a 1.00 L sample of water from the surface of the Dead Sea is evaporated, 179 g of $MgCl_2$ is recovered. What was the molarity of $MgCl_2$ in the water sample?

(Answers to Practice Exercises are in the back of the book.)

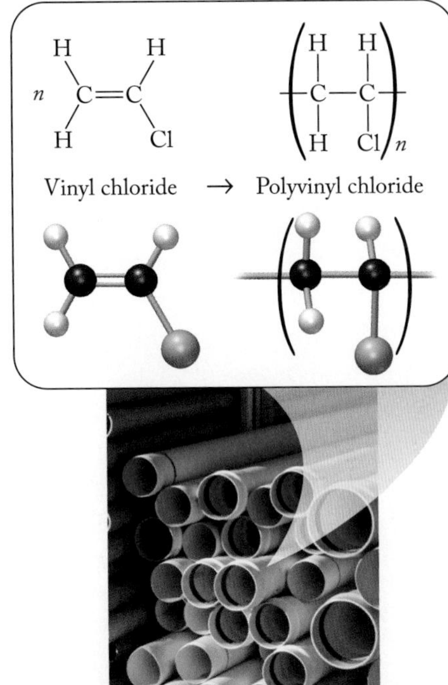

FIGURE 8.1 Vinyl chloride is the common name of a small molecule, called a monomer (Greek for "one unit"), from which very large molecules or polymers ("many units") called polyvinyl chloride (PVC) are made. PVC is used to make many products, including the pipes used for drainage and sewer systems.

SAMPLE EXERCISE 8.2 Converting Mass-per-Mass Concentrations into Molarity LO1

A water sample from the Great Salt Lake in Utah contains 83.6 mg Na^+ per gram of lake water. What is the molar concentration of Na^+ ions if the density of the lake water is 1.160 g/mL?

Collect and Organize Our task is to convert concentration units from mg Na^+ per gram of solution to mol Na^+ per liter of solution. We know the lake water's density, which we will use to convert a concentration based on mass of solution to molarity, which is based on volume of solution. The molar mass of Na is 22.99 g/mol.

Analyze The initial concentration value of 83.6 mg Na^+/g is the same as 83.6 g/kg. This mass of Na^+ corresponds to about 4 moles of Na^+, and the volume of a kilogram of lake water will be only a little less than a liter, given a density of about 1.2 kg/L, so the answer should be close to 4 *M*.

Solve Calculating the moles of Na^+ in 83.6 mg Na^+:

$$83.6 \cancel{\text{mg}}\ Na^+ \times \frac{1 \cancel{\text{g}}}{1000 \cancel{\text{mg}}} \times \frac{1 \text{ mol}}{22.99 \cancel{\text{g}}} = 36.36 \times 10^{-3} \text{ mol } Na^+$$

The volume of exactly one gram of lake water is

$$1\ \text{g} \times \frac{1\ \text{mL}}{1.160\ \text{g}} \times \frac{1\ \text{L}}{1000\ \text{mL}} = 8.621 \times 10^{-4}\ \text{L}$$

and the molarity of Na^+ ions in the lake is

$$M = \frac{36.36 \times 10^{-3}\ \text{mol Na}^+}{8.621 \times 10^{-4}\ \text{L}} = 4.22\ M$$

Think About It According to Table 8.1, the average concentration of Na^+ in seawater is 480.57 mM or 0.48057 M. The concentration of Na^+ in the Great Salt Lake is nearly 10 times greater. A clue to the lake's salinity is provided by its density: 1.160 g/mL is 16% greater than the density of pure water.

Practice Exercise If the density of seawater at a depth of 10,000 m is 1.071 g/mL, and a 25.0 g sample of water from that depth contains 99.7 mg K^+, what is the molarity of potassium ions in the sample?

SAMPLE EXERCISE 8.3 Calculating the Quantity of Solute Needed to Prepare a Solution LO2

An aqueous solution called phosphate-buffered saline (PBS) is used in biological research to wash and store living cells. Its ingredients include 10.0 mM Na_2HPO_4. How many grams of this solute would you need to prepare 10.0 L of PBS?

Collect and Organize We are asked to calculate the mass of solute needed to prepare a solution of specified volume and concentration. We know the formula of the solute. Equation 8.3 relates the mass of a solute in a solution to the volume and concentration of the solution and the molar mass of the solute.

Analyze We need to use the formula of the solute to calculate its molar mass and then use that value and the volume and concentration of the solution in Equation 8.3 to calculate the mass we need. We need enough solute to make 10.0 L of 10.0 mM (or 0.0100 M) solution, which translates into 10.0 L × 0.01 mol/L or 0.1 mole. Based on its formula, the molar mass of the solute should be around 200 g/mol, so 0.1 mol has a mass of about 20 g.

Solve Calculating the mass of 1 mol Na_2HPO_4, which contains a total of 2 mol Na, 1 mol H, 1 mol P, and 4 mol O:

$$\mathcal{M} = 2(22.99\ \text{g/mol}) + 1.008\ \text{g/mol} + 30.97\ \text{g/mol} + 4(16.00\ \text{g/mol})$$
$$= 141.96\ \text{g/mol}$$

Calculating the mass of Na_2HPO_4 using Equation 8.3:

$$m = (V \times M) \times \mathcal{M}$$
$$= \left(10.0\ \text{L} \times 10.0\ \frac{\text{mmol}}{\text{L}} \times \frac{1\ \text{mol}}{1000\ \text{mmol}}\right) \times \frac{141.96\ \text{g}}{1\ \text{mol}} = 14.2\ \text{g}$$

Think About It In solving the problem we converted "mmol/L" to "mol/L" so that the product ($V \times M$) yielded moles of solute, which we then converted into a mass in grams. The result of the calculation is reassuringly close to our estimate.

Practice Exercise An aqueous solution known as Ringer's lactate is administered intravenously to trauma victims suffering from blood loss or severe burns. The solution contains the chloride salts of sodium, potassium, and calcium and is 4.00 mM in sodium lactate, $CH_3CH(OH)COONa$. How many grams of sodium lactate are needed to prepare 10.0 L of Ringer's lactate?

standard solution a solution of accurately known concentration that is used in chemical analysis.

dilution the process of lowering the concentration of a solution by adding more solvent.

8.2 Dilutions

Scientists and technicians who routinely analyze environmental, biological, or clinical samples often use commercially available **standard solutions**. Standard solutions contain accurately known concentrations of one or more solutes. These concentrations are often higher than those required for laboratory work, so they must be diluted before use. **Dilution** is the process of reducing solute concentration by adding more solvent to a solution. Adding solvent does not change the number of particles in the solution, only the volume they occupy, as shown in Figure 8.2.

In chemistry laboratories dilution often involves transferring a precisely measured volume of a concentrated solution to a volumetric flask and then filling the flask to its calibration mark with solvent (Figure 8.3) and thoroughly mixing its contents. The number of particles of solute transferred from the concentrated solution to the flask is the same number of particles present in the entire volume in the flask after dilution. Figure 8.3 shows how to prepare 250.0 mL of 0.100 *M* $CuSO_4$ starting with a 1.00 *M* solution. Note that a 25 mL volumetric pipette is

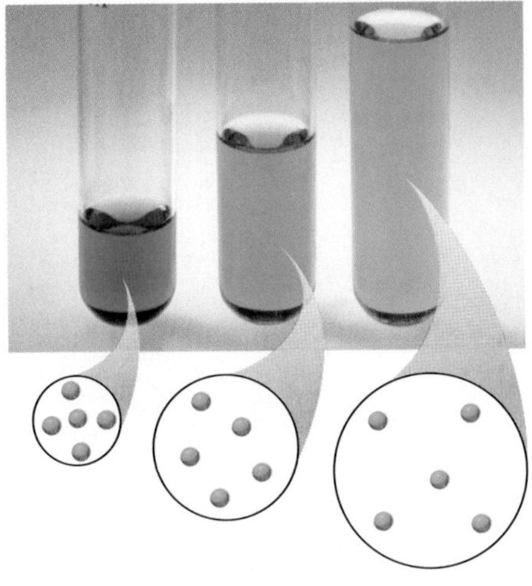

FIGURE 8.2 Adding more solvent to the same quantity of colored solute dilutes the concentration of the solutions and the intensity of their color.

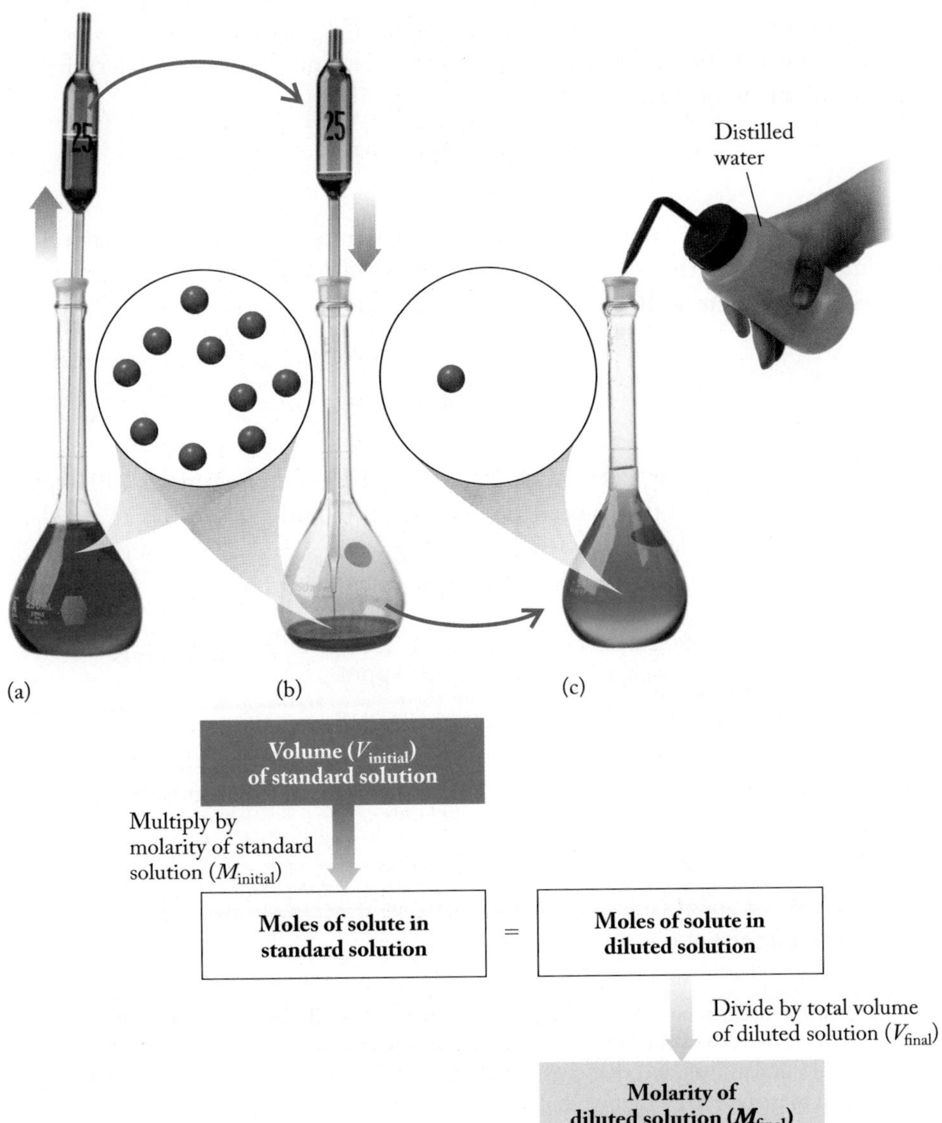

FIGURE 8.3 To prepare 250.0 mL of 0.100 *M* $CuSO_4$, (a) a pipette is used to withdraw 25.0 mL of 1.00 *M* $CuSO_4$. (b) This volume of standard solution is transferred to a 250.0 mL volumetric flask. (c) Distilled water is added to bring the volume of the dilute solution to 250.0 mL. The color of the dilute solution is lighter than the standard solution. The number of blue spheres in each molecular view represents the relative concentrations of Cu^{2+} and sulfate ions.

used to transfer that volume of the more concentrated solution to the flask in which it will be diluted to obtain the desired solution. To show how this volume was calculated, we start with a variation of Equation 8.2:

$$n_{\text{initial}} = V_{\text{initial}} \times M_{\text{initial}} \quad (8.4)$$

where the subscripts refer to volume, molarity, and number of moles of Cu^{2+} ions in the *initial* (concentrated) solution used in the dilution. The number of moles of Cu^{2+} ions in the *final* (dilute) solution:

$$n_{\text{final}} = V_{\text{final}} \times M_{\text{final}} \quad (8.5)$$

is the same as the number transferred. Because the two n values are equal, the right-hand sides of Equations 8.4 and 8.5 are also equal:

$$V_{\text{initial}} \times M_{\text{initial}} = V_{\text{final}} \times M_{\text{final}}$$

Rearranging the terms in this equation to solve for V_{initial}, we obtain

$$V_{\text{initial}} = \frac{V_{\text{final}} \times M_{\text{final}}}{M_{\text{initial}}} \quad (8.6)$$

To use Equation 8.6 to calculate the volume of 1.00 M $CuSO_4$ needed to prepare 250.0 mL of 0.100 M $CuSO_4$, you might think V_{final} would first have to be converted from mL to L. Actually, it doesn't because our goal is to determine an initial volume to be transferred that is also in milliliters:

$$V_{\text{initial}} = \frac{250.0 \text{ mL} \times 0.100 \cancel{M}}{1.00 \cancel{M}} = 25.0 \text{ mL}$$

Equation 8.6 can be modified for use with *any* units of concentration as well as volume—provided the units used to express the initial and final volumes are the same, and the units used to express the initial and final concentrations (C) are the same:

$$V_{\text{initial}} = \frac{V_{\text{final}} \times C_{\text{final}}}{C_{\text{initial}}} \quad (8.7)$$

Let's use Equation 8.7 to calculate the volume (in mL) of the commercial 1000 ppm standard in Figure 8.4 that is needed to prepare 100.0 mL of a 2.00 ppm standard solution for use in a laboratory where drinking water samples are being tested for lead contamination. Applying Equation 8.7 to calculate this volume:

$$V_{\text{initial}} = \frac{V_{\text{final}} \times C_{\text{final}}}{C_{\text{initial}}} = \frac{100.0 \text{ mL} \times 2.00 \cancel{\text{ppm}}}{1000 \cancel{\text{ppm}}} = 0.200 \text{ mL}$$

Commercial standards are usually sold with certificates of analysis that specify their concentrations to three significant figures. Therefore, a pipette should be used to transfer the calculated volume that is accurate to the nearest 0.001 mL.

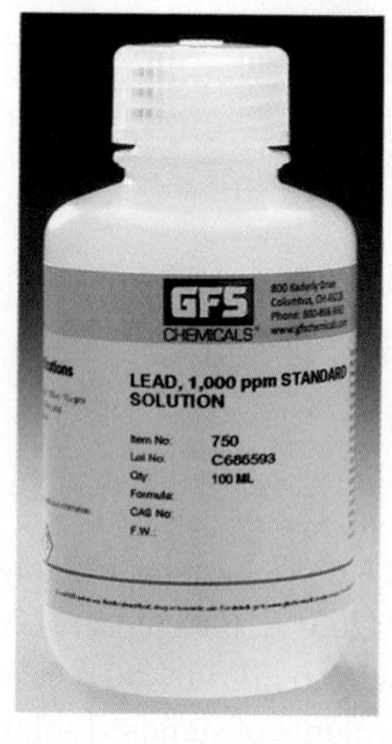

FIGURE 8.4 Commercial standard solutions such as this 1000 ppm lead standard are widely used in testing laboratories that analyze many environmental, clinical, and other samples. Their solute concentrations are often higher than those required for laboratory work, so they must be diluted before use.

SAMPLE EXERCISE 8.4 Diluting Solutions **LO2**

The solution used in hospitals for intravenous infusion—called *physiological saline* or *normal saline*—is 0.155 M NaCl. It may be prepared by diluting a commercially available standard solution that is 1.50 M NaCl. What volume of standard solution is required to prepare 10.0 L of physiological saline?

Collect, Organize, and Analyze We know the volume and concentration of the dilute solution (normal saline) to be prepared and the concentration of the standard solution to be diluted. Calculating the volume of the standard solution to be transferred is a matter of inserting the three quantities given into Equation 8.7, where $V_{\text{final}} = 10.0$ L, $C_{\text{final}} = 0.155\ M$, and $C_{\text{initial}} = 1.50\ M$. The final concentration is about 1/10 the initial concentration, so the volume to be transferred should be about 1/10 the final volume, or about 1 L.

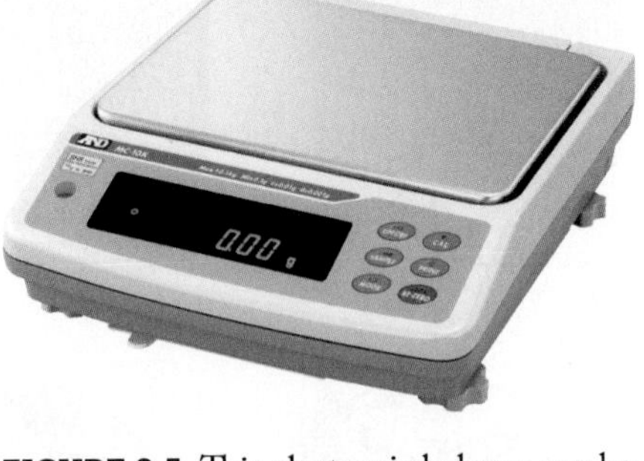

FIGURE 8.5 This electronic balance can be used to measure masses up to 10 kg to the nearest 0.01 g.

Solve

$$V_{\text{initial}} = \frac{V_{\text{final}} \times C_{\text{final}}}{C_{\text{initial}}} = \frac{10.0\text{ L} \times 0.155\ \cancel{M}}{1.50\ \cancel{M}} = 1.03\text{ L}$$

Think About It Our calculated volume is close to our estimate. How would we measure out 1.03 L of the standard solution with acceptable precision and accuracy? If we knew (or could accurately determine) the density of the initial solution, we might use an electronic balance such as the one shown in Figure 8.5 that could be used to dispense up to 10 kg of solution with a precision of 0.01 g.

Practice Exercise The concentration of Ca^{2+} in a standard solution is 10.00 mg/mL. What volume of the solution should be used to prepare 0.500 L of a solution in which the Ca^{2+} concentration is 50.0 mg/L?

(a) (b) (c) (d) (e)

FIGURE 8.6 Aqueous solutions of cough syrup.

CONCEPT **TEST**

Figure 8.6 shows several solutions of red cough syrup dissolved in water. Order these solutions from the most dilute to the most concentrated.

8.3 Electrolytes and Nonelectrolytes

The images in Figure 8.7 show an apparatus in which a lightbulb and two graphite rods are connected in series to a source of electricity (not shown) so that the bulb lights up only when enough electric current flows between the rods. As you can see, this happens in some but not all of the cases in Figure 8.7. It does *not* happen in Figure 8.7(a) when the two rods are immersed in distilled water. The lack of electric current means that distilled water is a poor conductor of electricity. The bulb does light when the rods are immersed in 0.50 *M* NaCl (Figure 8.7b), so this solution is a *good* conductor of electricity. How does adding NaCl make water a good conductor? Is there something inherently conductive about crystals of solid NaCl? As shown in Figure 8.7(c), the bulb does *not* light when the rods are immersed in a beaker of salt crystals, so solid NaCl is a poor conductor of electricity.

Thus, a solution of NaCl must be conductive because the salt is dissolved in water. When NaCl dissolves, its Na^+ and Cl^- ions are liberated from their crystal lattice as shown in the particulate view of Figure 8.7(b), and they are able to migrate independently through the solvent. As they migrate, they take their electric charges with them. If they migrate in opposite directions—as they do when the Na^+ ions are attracted to and migrate toward the graphite rod, or **electrode**, connected to the battery's *negative* terminal, while the Cl^- ions move toward the *positive* graphite electrode—then this ion migration allows electricity to flow through the solution.

electrode a solid electrical conductor that is used to make contact with a solution or other nonmetallic component of an electrical circuit.

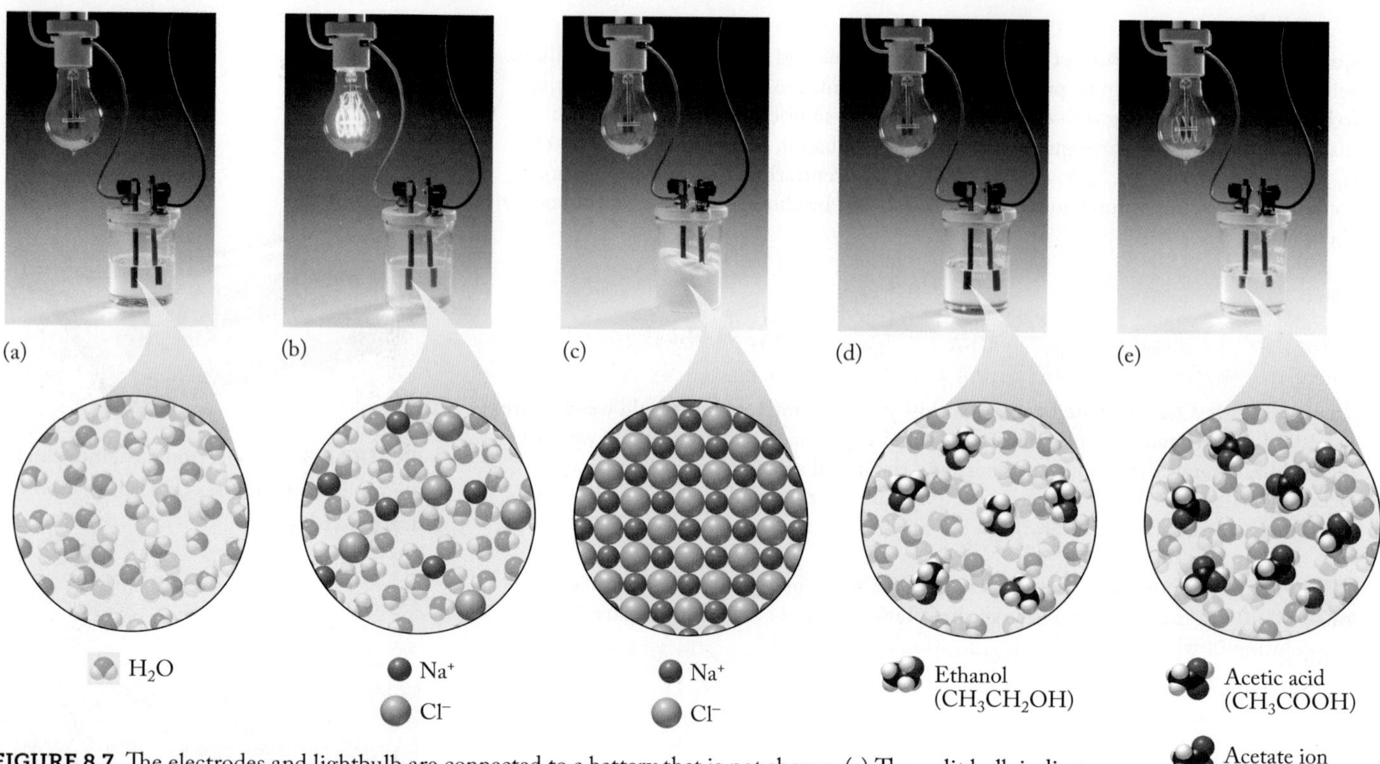

FIGURE 8.7 The electrodes and lightbulb are connected to a battery that is not shown. (a) The unlit bulb indicates that pure water, which consists almost entirely of neutral H_2O molecules, conducts electricity very poorly. (b) The brightly lit bulb indicates that a 0.50 *M solution* of NaCl conducts electricity very well, whereas (c) the unlit bulb indicates that *solid* sodium chloride (NaCl) does *not* conduct electricity. (d) The unlit bulb indicates that a 0.50 *M* solution of ethanol (CH_3CH_2OH) is no better a conductor than pure water. (e) The dimly lit bulb indicates that a 0.50 *M* solution of acetic acid (CH_3COOH) is a better conductor than pure water but not as good as 0.50 *M* NaCl.

Any ionic solute (such as NaCl) that produces a solution containing free ions is called an **electrolyte** because migration of the dissolved ions allows the solution to conduct electricity. NaCl dissociates completely into Na^+ and Cl^- ions when it dissolves in water, so it is called a **strong electrolyte**, even though solid NaCl alone is not a good conductor of electricity. On the other hand, if NaCl is heated to a temperature above its melting point, the molten NaCl that is produced *is* a good conductor of electricity because the ions in liquid NaCl are free to migrate.

In Figure 8.7(d), the graphite electrodes are immersed in a 0.50 *M* solution of ethanol (CH_3CH_2OH) in water. The unlit bulb indicates that ethanol conducts electricity no better than pure water. Thus, when molecules of CH_3CH_2OH dissolve in water, they remain intact as whole molecules—they do not dissociate into ions. Solutes that do not form ions when they dissolve are called **nonelectrolytes**.

Finally, the lightbulb in Figure 8.7(e) produces a faint glow, which means that some current is flowing through the 0.50 *M* solution of acetic acid in the beaker—just not as much as flows through 0.50 *M* NaCl. Acetic acid, like ethanol, is a molecular compound. Unlike ethanol, though, some of the acetic acid molecules ionize when they dissolve in water, donating H^+ ions to molecules of H_2O to form **hydronium ions, H_3O^+**, as shown in the molecular view of Equation 8.8. Less than 1% of the acetic acid molecules in this solution ionize, but that is enough to produce a faint glow in the bulb and enough to qualify acetic acid as a **weak electrolyte**: a class of molecular compounds that partially ionize when they dissolve in water.

electrolyte a solute that produces ions in solution, which enable its solutions to conduct electricity.

strong electrolyte a substance that dissociates completely when it dissolves in water.

nonelectrolyte a molecular substance that does not dissociate into ions when it dissolves in water.

hydronium ion (H_3O^+) an H^+ ion plus a water molecule, H_2O; the form in which the hydrogen ion is found in an aqueous solution.

weak electrolyte a substance that only partly dissociates into ions when it dissolves in water.

weak acid an acid that only partially dissociates in aqueous solutions.

strong acid an acid that completely dissociates into ions in aqueous solution.

The following chemical equation and molecular models describe the partial ionization of acetic acid:

$$CH_3COOH(aq) + H_2O(\ell) \rightleftharpoons CH_3COO^-(aq) + H_3O^+(aq) \quad (8.8)$$

The reactants and products in Equation 8.8 are not connected by the usual reaction arrow but instead by two *half-arrows* pointing in opposite directions. The upper half-arrow represents the reaction of acetic acid with water to form the acetate and hydronium ions, whereas the lower half-arrow represents the reverse reaction, in which hydronium and acetate ions recombine to form molecules of acetic acid and water. Recall from Chapter 7 that reversible reactions may not go to completion but rather may reach a state of chemical equilibrium, in which the reaction mixture contains quantities of both reactants and products that show no net change with time. An aqueous solution of acetic acid, such as the one in Figure 8.7(e), achieves chemical equilibrium after relatively few of its molecules have ionized. Chemical equilibria are discussed in much greater detail in Chapters 14, 15, and 16.

CONCEPT TEST

Identify the following compounds as electrolytes or nonelectrolytes: (a) $Ca(NO_3)_2$; (b) NH_4Cl; (c) CH_3OH (methanol); (d) $C_6H_{12}O_6$ (fructose, or fruit sugar); (e) $LiClO_4$.

8.4 Acids, Bases, and Neutralization Reactions

Compounds like acetic acid, which produce H_3O^+ ions when they dissolve in water, are called *acids*. More precisely, they are called *Arrhenius* acids in honor of Swedish chemist Svante August Arrhenius (1859–1927), whose research on the behavior of electrolytic solutions was recognized with the 1903 Nobel Prize in Chemistry. Most acids, like acetic acid, only partially ionize when they dissolve in water. Therefore, they are weak electrolytes and are also categorized as **weak acids**.

A few acids are categorized as **strong acids** because they ionize completely when they dissolve in water. They include the volatile hydrogen halides HCl, HBr, and HI, which are completely transformed from polar molecules in the gas phase into H_3O^+ and halide ions in aqueous solution. For example, when a molecule of hydrogen chloride dissolves in water it completely ionizes, forming one chloride and one hydronium ion:

$$HCl(g) + H_2O(\ell) \rightarrow Cl^-(aq) + H_3O^+(aq) \quad (8.9)$$

As a result, hydrogen chloride gas in water forms a solution we call hydrochloric acid, in which every molecule of HCl reacts with a molecule of H_2O and produces one H_3O^+ ion (Figure 8.8a). Hydrogen bromide and hydrogen iodide react the same way, forming hydrobromic and hydroiodic acids, which are also strong acids.

Hydrofluoric acid (formed from hydrogen fluoride, HF) is a weak acid, however, because it produces far fewer H_3O^+ ions than the number of HF molecules dissolved in solution (Figure 8.8b). Why? For one thing, the covalent bonds that

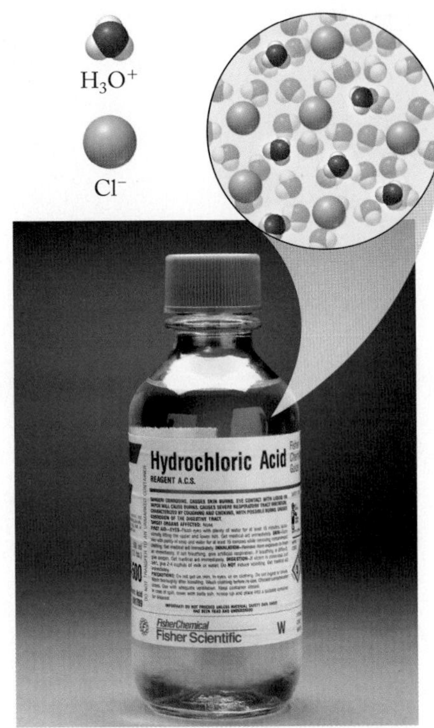

(a)

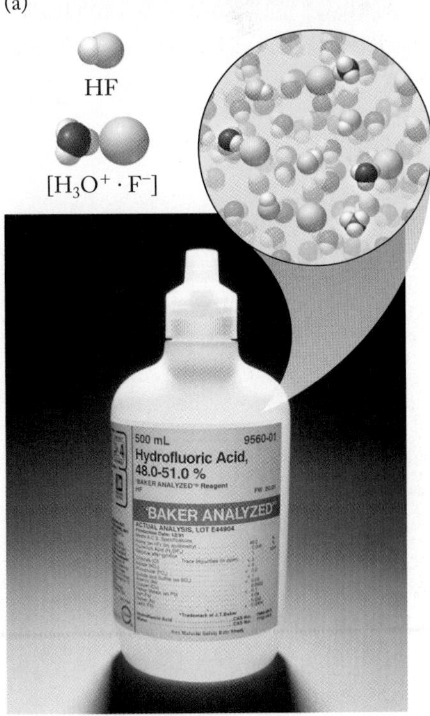

(b)

FIGURE 8.8 (a) Hydrochloric acid is a stronger acid than (b) hydrofluoric acid because molecules of HCl react with molecules of H_2O to form H_3O^+ and Cl^- ions. However, only some of the HF molecules react with molecules of H_2O, and even when they do, they mostly form $[H_3O^+ \cdot F^-]$ ion pairs rather than free H_3O^+ ions.

hold HF molecules together are much stronger than the bonds in the other hydrogen halides, and they don't break as easily. For another, even if the bonds do break and HF ionizes,

$$HF(g) + H_2O(\ell) \rightarrow F^-(aq) + H_3O^+(aq)$$

the ions that form may not move about freely in solution. Dissolved fluoride ions are much smaller than those of the other halogens, so they can get closer to and are more strongly attracted to hydronium ions. This electrostatic attraction is so strong that the ions form stable ion pairs:

$$F^-(aq) + H_3O^+(aq) \rightleftharpoons [H_3O^+ \cdot F^-](aq) \quad (8.10)$$

The reaction in Equation 8.10 is reversible and does not consume *all* the H_3O^+ ions in solution. The relatively few that are still free give hydrofluoric acid its weakly acidic properties.

Chemists often simplify the chemical equations describing the behavior of acids in aqueous solution by leaving H_2O out of the reactants. Instead, they use (*aq*) symbols to indicate the physical state of all solutes, and they leave out the H_2O part of the hydronium ion formula and instead write it $H^+(aq)$. Thus, the chemical equation for the ionization of hydrogen chloride when it dissolves in water (Equation 8.9) is sometimes written

$$HCl(g) \xrightarrow{H_2O(\ell)} H^+(aq) + Cl^-(aq) \quad (8.11)$$

We will encounter many other chemical equations in the paragraphs and chapters ahead that have been simplified by leaving out water molecules as individual molecules or parts of hydronium ions.

Table 8.2 lists the names and formulas of the most common strong acids and the gases that form them when those gases dissolve in water (or combine with water vapor). We have already discussed the acids formed by HCl, HBr, and HI, so let's focus on those formed by oxides of sulfur and nitrogen. The two strong acids they make, sulfuric and nitric, are among the most important industrial chemicals in the world. Sulfuric acid is made by combining sulfur trioxide and water vapor as discussed in Chapter 7:

$$SO_3(g) + H_2O(g) \rightarrow H_2SO_4(\ell) \quad (7.6)$$

Nitric acid can be made from nitrogen dioxide gas and water vapor:

$$3\ NO_2(g) + H_2O(g) \rightarrow NO(g) + 2\ HNO_3(\ell) \quad (8.12)$$

TABLE 8.2 Strong Acids and the Gases That Form Them

Gases	Acids
Hydrogen chloride, HCl	Hydrochloric acid, $HCl(aq)$
Hydrogen bromide, HBr	Hydrobromic acid, $HBr(aq)$
Hydrogen iodide, HI	Hydroiodic acid, $HI(aq)$
Sulfur trioxide, SO_3	Sulfuric acid, $H_2SO_4(\ell)$
Nitrogen dioxide, NO_2	Nitric acid, $HNO_3(\ell)$
[a]	Perchloric acid, $HClO_4(aq)$

[a] Perchloric acid is prepared by reacting concentrated solutions of sodium perchlorate and hydrochloric acid: $NaClO_4(aq) + HCl(aq) \rightarrow HClO_4(aq) + NaCl(s)$.

carboxylic acid a compound containing the —COOH functional group.

Aqueous solutions of HNO_3 and H_2SO_4 are completely ionized:

$$HNO_3(\ell) \xrightarrow{H_2O(\ell)} H^+(aq) + NO_3^-(aq) \quad (8.13)$$

$$H_2SO_4(\ell) \xrightarrow{H_2O(\ell)} H^+(aq) + HSO_4^-(aq) \quad (8.14)$$

The hydrogen sulfate (HSO_4^-) ion formed in the second reaction has another H atom that is partially ionized:

$$HSO_4^-(aq) \rightleftharpoons H^+(aq) + SO_4^{2-}(aq) \quad (8.15)$$

Thus, H_2SO_4 is a strong acid, but HSO_4^- is a weak acid. The capacity of sulfuric acid to donate up to two H^+ ions (protons) per molecule makes it a *diprotic* acid. Like most diprotic acids, the singly charged anion that forms when a molecule of the acid donates one H^+ ion is a weaker acid than the original one. The same is true for *triprotic* acids. For example, phosphoric acid (H_3PO_4, Table 8.3) is a stronger acid than the dihydrogen phosphate ion $H_2PO_4^-$, which is stronger acid than the hydrogen phosphate ion, HPO_4^{2-}.

The molecular structures of several common weak acids are shown in Table 8.3. They include acetic acid, CH_3COOH, which is the principal solute in vinegar and an example of a **carboxylic acid**, which means that its molecular structure contains a —COOH functional group as shown in Table 8.3. The H atom in this group is partially ionized when acetic acid and other carboxylic acids dissolve in water. Also in the table are carbonic acid, H_2CO_3, which is really a solution of carbon dioxide in water, and hydrosulfuric acid, H_2S, which forms when highly toxic and intensely foul-smelling hydrogen sulfide gas dissolves in water.

TABLE 8.3 Structures of some common weak acids

Name	Molecular Structure	Origin or Use
Acetic acid		Vinegar is a dilute aqueous solution of acetic acid
Carbonic acid		CO_2 gas dissolved in water
Hydrosulfuric acid		Hydrogen sulfide gas dissolved in water
Phosphoric acid		$P_4O_{10}(s)$ dissolved in water

CONCEPT TEST

A molecule with the common name EDTA has four ionizable hydrogen atoms. To highlight this property, it is sometimes written as H_4EDTA. Rank the following species in order of increasing acid strength: H_4EDTA, H_3EDTA^-, H_2EDTA^{2-}, $HEDTA^{3-}$, $EDTA^{4-}$.

In 1923 chemists Johannes Brønsted (1879–1947) and Thomas Lowry (1874–1936) independently proposed a new definition of what it means to be an acid. In the **Brønsted–Lowry model**, an acid is any compound that donates H^+ ions. Arrhenius acids are **Brønsted–Lowry acids** because producing H_3O^+ ions in water (as Arrhenius acids do) involves donating H^+ ions to H_2O molecules. However, the Brønsted–Lowry definition also includes compounds that donate H^+ ions to molecules other than H_2O, such as when hydrogen chloride and ammonia gas combine to form solid NH_4Cl:

$$HCl(g) + NH_3(g) \rightarrow NH_4Cl(s) \tag{8.16}$$

Molecules of HCl donate H^+ ions to NH_3 molecules, forming $NH_4{}^+$ and Cl^- ions. In this reaction, HCl is a Brønsted–Lowry acid but not an Arrhenius acid.

The substances that accept the H^+ ions that acids donate are called *bases*. In particular, they're called **Brønsted–Lowry bases.** Just as there are strong and weak acids, there are **strong bases** and **weak bases,** too. Among the strongest bases are hydroxide ions, OH^-, which combine with H^+ ions to form molecules of water:

$$H^+(aq) + OH^-(aq) \rightarrow H_2O(\ell) \tag{8.17}$$

Important sources of OH^- ions include the alkali metal hydroxides, which are solid ionic compounds that have the generic formula MOH (where M represents any alkali metal). They all dissociate completely into M^+ and OH^- ions when they dissolve in water. The hydroxides of Ca^{2+}, Sr^{2+}, and Ba^{2+} (group 2) also dissociate completely when they dissolve in water, but they are not as soluble as the group 1 hydroxides. The limited solubility of the group 2 hydroxides means that the concentrations of dissolved OH^- ions in their solutions cannot match those in concentrated solutions of the group 1 hydroxides.

Ammonia (NH_3) is a weak base. When it dissolves in water, it produces a solution that conducts electricity weakly: if the apparatus in Figure 8.7 were immersed in 0.50 *M* NH_3, the bulb would light about as brightly as it does in 0.50 *M* acetic acid (Figure 8.7e). The partial ionization of ammonia in aqueous solution is described in the following chemical equation:

$$NH_3(aq) + H_2O(\ell) \rightleftharpoons NH_4{}^+(aq) + OH^-(aq) \tag{8.18}$$

At equilibrium, this solution contains mostly aqueous ammonia molecules, with smaller concentrations of ammonium and hydroxide ions.

Some compounds can act as acids or bases because they have the capacity to accept H^+ ions in the presence of acids and donate them in the presence of bases. One compound that exhibits this behavior is water:

$$HCl(g) + H_2O(\ell) \rightarrow Cl^-(aq) + H_3O^+(aq) \tag{8.9}$$

$$NH_3(aq) + H_2O(\ell) \rightleftharpoons NH_4{}^+(aq) + OH^-(aq) \tag{8.18}$$

In a solution of hydrochloric acid, molecules of H_2O accept H^+ ions created by the ionization of molecules of HCl. In an aqueous solution of ammonia, molecules of H_2O ionize, donating H^+ ions to molecules of NH_3. Substances that can function as either a Brønsted–Lowry base or a Brønsted–Lowry acid (such as water) are said to be **amphiprotic**. Another common amphiprotic substance is sodium hydrogen carbonate, $NaHCO_3$, which is more commonly called sodium bicarbonate. It's the principal ingredient in baking soda. In a solution that also contains a strong acid, bicarbonate ions act like bases, accepting H^+ ions from the acid and forming the weak acid H_2CO_3, which decomposes to water and CO_2.

$$H^+(aq) + HCO_3^-(aq) \rightarrow H_2CO_3(aq) \rightarrow H_2O(\ell) + CO_2(g) \quad (8.19)$$

In a solution that also contains a strong base, bicarbonate acts like an acid, donating H^+ ions to the OH^- ions from the base:

$$OH^-(aq) + HCO_3^-(aq) \rightarrow H_2O(\ell) + CO_3^{2-}(aq) \quad (8.20)$$

Neutralization Reactions and Net Ionic Equations

The reaction in which H^+ and OH^- ions combine to form H_2O (Equation 8.17) is at the heart of an important class of chemical reactions between acids and bases called **neutralization reactions**. In any neutralization reaction, *there must be one H^+ ion for every OH^- ion* in aqueous solution. For example, in the neutralization reaction between hydrochloric acid and sodium hydroxide:

$$HCl(aq) + NaOH(aq) \rightarrow NaCl(aq) + H_2O(\ell) \quad (8.21)$$

we need one mole of NaOH for every mole of HCl, because each mole of HCl can donate one mole of H^+ ions and each mole of NaOH contains one mole of OH^- ions—that is, one mole of H^+ ion acceptors.

Equation 8.21 describes a classic neutralization reaction between a strong acid, HCl(*aq*), and a strong base, NaOH(*aq*). The two products, a salt and water, are common to most neutralization reactions. In fact, neutralization reactions provide one operational definition of a **salt**: the ionic compound produced in the neutralization reaction of an acid and a base.

Equation 8.21 is also an example of a **molecular equation**, which means each reactant and product are written as a neutral compound, which is the format we used for all the chemical equations in Chapter 7. To get a particle-level view of the reaction in Equation 8.21, let's rewrite it to show all of the ionic species in the aqueous reaction mixture:

$$\underbrace{H^+(aq) + Cl^-(aq)}_{HCl(aq)} + \underbrace{Na^+(aq) + OH^-(aq)}_{NaOH(aq)} \rightarrow \underbrace{Na^+(aq) + Cl^-(aq)}_{NaCl(aq)} + H_2O(\ell) \quad (8.22)$$

Equation 8.22 is an example of a **total ionic equation**. The following rules explain how to write a total ionic equation:

1. Write all strong electrolytes as separate ions, with each of their symbols followed by (*aq*).
2. Write all soluble nonelectrolytes—and all weak electrolytes, such as acetic acid—using their molecular formulas.

Brønsted–Lowry model defines acids as H^+ ion donors and bases as H^+ ion acceptors.

Brønsted–Lowry acid a proton (H^+) donor.

Brønsted–Lowry base a proton (H^+) acceptor.

strong base a base that completely dissociates into ions in aqueous solution.

weak base a base that only partially dissociates or ionizes in aqueous solutions.

amphiprotic describes a substance that can behave as either a proton acceptor or a proton donor.

neutralization reaction a reaction that takes place when an acid reacts with a base and produces a solution of a salt in water.

salt the product of a neutralization reaction; it is made up of the cation of the base in the reaction and the anion of the acid.

molecular equation a balanced equation that describes a reaction in solution in which all reactants and products are written as neutral compounds.

total ionic equation a balanced equation that shows all species, including spectator ions, present in a reaction, either as molecular materials or separate ions.

spectator ion an ion that is unchanged by a chemical reaction.

net ionic equation a balanced equation that describes the actual reaction taking place in solution; it is obtained by eliminating the spectator ions from the total ionic equation.

3. Use the molecular formula $H_2O(\ell)$ for water when it is a reactant or product (as in a neutralization reaction).
4. Write insoluble solids and gases using their normal chemical formulas followed by (*s*) or (*g*) as appropriate.

Na^+ and Cl^- ions appear on both sides of the reaction arrow in Equation 8.22 because they are not changed by the reaction: they were free ions in solution before the reaction, and they are free ions in solution afterward. Unchanged ions such as these are called **spectator ions** because they do not participate in the reaction. We can simplify the total ionic equation by canceling out the spectator ions:

$$H^+(aq) + \cancel{Cl^-(aq)} + \cancel{Na^+(aq)} + OH^-(aq) \rightarrow \cancel{Na^+(aq)} + \cancel{Cl^-(aq)} + H_2O(\ell)$$

which leaves us with

$$H^+(aq) + OH^-(aq) \rightarrow H_2O(\ell) \qquad (8.17)$$

Equation 8.17 is an example of a **net ionic equation**, which is a chemical equation of a reaction that contains only those species that are changed by the reaction.

SAMPLE EXERCISE 8.5 Writing the Net Ionic Equation for a Neutralization Reaction **LO4**

Write the net ionic equation for the reaction that takes place when acid rain containing sulfuric acid reacts with the surface of a marble statue, as depicted in Figure 7.4 and as described by Equation 7.8:

$$CaCO_3(s) + H_2SO_4(aq) \rightarrow CaSO_4(aq) + H_2O(\ell) + CO_2(g) \qquad (7.8)$$

Collect, Organize, and Analyze To write the net ionic equation for the reaction, we must first write the total ionic equation, which contains the formulas of all the individual particles in the reaction mixture. Sulfuric acid (H_2SO_4) is the H^+ ion donor in the reaction, making $CaCO_3$ the H^+ ion acceptor. Each mole of H_2SO_4 can donate two moles of H^+ ions. The products of the reaction include $CaSO_4$ and two others, H_2O and CO_2, which must have formed when one mole of CO_3^{2-} ions in $CaCO_3$ accepted the two moles of H^+ ions as the statue dissolved:

$$2\,H^+(aq) + CaCO_3(s) \rightarrow Ca^{2+}(aq) + H_2O(\ell) + CO_2(g)$$

If this equation is correct, then carbonate ions are the species acting as the base in this reaction.

Sulfuric acid is a strong acid and is completely ionized, forming H^+ and HSO_4^- ions (Equation 8.15). However, HSO_4^- is a weak acid and is only partially separated into H^+ and SO_4^{2-} ions (Equation 8.16), so it remains as HSO_4^- in the total ionic equation. Calcium sulfate is a soluble ionic compound and is completely separated into Ca^{2+} and SO_4^{2-} ions in solution.

Solve Building a total ionic equation from the preceding analysis of the reaction in Equation 7.8:

$$CaCO_3(s) + H^+(aq) + HSO_4^-(aq) \rightarrow Ca^{2+}(aq) + SO_4^{2-}(aq) + H_2O(\ell) + CO_2(g)$$

To write a net ionic equation, we eliminate spectator ions from the total ionic equation. However, there aren't any in this case: none of the ions on the reactant side of the reaction arrow appear on the product side. Therefore, the total ionic equation *is* the net ionic equation:

$$CaCO_3(s) + H^+(aq) + HSO_4^-(aq) \rightarrow Ca^{2+}(aq) + SO_4^{2-}(aq) + H_2O(\ell) + CO_2(g)$$

Think About It Because one of the reactants was a solid and the products included a gas and liquid water, the number of possible spectator ions in the total ionic equation was not large to begin with. That, coupled with the partial ionization of HSO_4^- ions, led to the complete absence of any spectator ions. We have seen in this section that adding enough acid to compounds that contain either HCO_3^- ions (Equation 8.20) or CO_3^{2-} ions (in this exercise) produces the same two products: liquid H_2O and CO_2 gas.

Practice Exercise Write balanced molecular, total ionic, and net ionic equations for the reaction between aqueous solutions of phosphoric acid (H_3PO_4) and sodium hydroxide. The products are sodium phosphate and water.

Neutralization reactions such as the one in Sample Exercise 8.5 highlight an important point about the behavior of acids and bases: even though weak acids and bases are only partially ionized in their aqueous solutions, they become completely ionized during neutralization reactions. For example, it takes two moles of a strong base such as NaOH to completely neutralize one mole of H_2SO_4 because there are two *ionizable* hydrogen atoms in one molecule of H_2SO_4. It does not matter that both H atoms may not be ionized in a solution of sulfuric acid; they both *do* ionize during a neutralization reaction. Similarly, it takes one mole of NaOH to completely neutralize one mole of acetic acid, even though less than 1% of the molecules of acetic acid in solution are ionized before the reaction begins.

The same principle applies to weak bases. For example, it takes one mole of hydrochloric acid (a strong acid) to neutralize one mole of ammonia in solution, even though most of the molecules of NH_3 have not reacted with H_2O to form NH_4^+ and OH^- ions before HCl is added. As the neutralization reaction proceeds, more and more NH_3 reacts with H_2O to form NH_4^+ and OH^- ions, and the OH^- ions are consumed by combining with H^+ ions from the acid to form H_2O.

8.5 Precipitation Reactions

Some of the most abundant elements in Earth's crust, including silicon (Si), aluminum (Al), and iron (Fe), are not abundant in seawater because the minerals that contain these elements have limited solubility in water. Many minerals contain ionic compounds, so why aren't they soluble? Recall from Chapter 6 that compounds are *not* soluble when solute–solvent interactions (ion–dipole in this case) are not strong enough to offset solute–solute interactions (ion–ion in this case) and solvent–solvent interactions such as hydrogen bonding between water molecules.

The strengths of ion–ion interactions increase with increasing charge and decrease with increasing ion size (see Equation 4.1). These trends are reflected in the solubility guidelines listed in Table 8.4 for some common ionic compounds. For example, all salts that contain alkali metal cations are soluble—as we would expect, given their 1+ charges. Additionally, all nitrate and acetate compounds are soluble because both ions are relatively large polyatomic ions and have charges of only 1−. On the other hand, nearly all ionic compounds in which the cations' charges are 3+ or 4+ and the anions' charges are 2− or 3− (many transition metal oxides and sulfides fall into this category) have limited solubility, because the interactions between ions with multiple charges are strong. However, other

TABLE 8.4 Solubility Guidelines for Ionic Compounds

(1)	All compounds containing the following ions are *soluble* in water: • Cations: group 1 ions (alkali metals) and NH_4^+ • Anions: NO_3^-, ClO_4^-, and CH_3COO^- (acetate)
(2)	Compounds containing the following anions are *soluble* except as noted: • Cl^-, Br^-, and I^-, except the halides of Ag^+, Cu^+, Hg_2^{2+}, and Pb^{2+} • SO_4^{2-}, except the sulfates of Ba^{2+}, Ca^{2+}, Hg_2^{2+}, Pb^{2+}, and Sr^{2+}
(3)	Compounds that contain the following anions are only *slightly soluble* (unless they also contain a group 1 cation or NH_4^+): • O^{2-}, *S^{2-}, *OH^-, CO_3^{2-}, PO_4^{3-}, AsO_4^{3-}, CrO_4^{2-}, $C_2O_4^{2-}$ (oxalate) *See rule 4.
(4)	The solubilities of the group 2 hydroxides and sulfides increase with increasing cation atomic number. MgS decomposes in water, forming H_2S and $Mg(OH)_2$.

CONNECTION The role of ion–dipole forces in the dissolution of ionic compounds was discussed in Chapter 6.

factors influence the solubilities of ionic compounds, too, thereby complicating solubility trends based only on ionic charges. As a result, some compounds composed of 1+ and 2+ cations and 1− anions have limited solubility in water. These include the halides of silver(I), mercury(I), lead(II), and many transition metal hydroxides.

CONCEPT TEST

Write the chemical formulas of the following compounds and then use the solubility guidelines in Table 8.4 to classify them as soluble, or only slightly soluble, in water: (a) calcium chloride; (b) potassium phosphate; (c) iron(III) hydroxide; (d) sodium sulfide; (e) zinc oxide.

The solubility guidelines in Table 8.4 allow us to predict whether or not a **precipitate** (solid product) forms when two aqueous solutions are mixed together. For example, does a precipitate form when a solution of sodium iodide (NaI) is mixed with a solution of lead(II) nitrate, $Pb(NO_3)_2$? Put another way, will either of the cations in the two solutions combine with the anion from the other solution to form a compound that has limited solubility? To answer this question we need to do the following:

1. Identify the ions that are dissolved in the two solutions after the two ionic compounds dissolve and separate into their component ions. In this example, one solution contains Na^+ ions and I^- ions, whereas the other contains Pb^{2+} ions and NO_3^- ions.
2. Determine whether either of the new anion/cation combinations produces a product of limited solubility. When the ions in the two solutions in this example swap partners, the new pairings are $NaNO_3$ and PbI_2. According to the rules in Table 8.4, sodium nitrate is soluble because it consists of a group 1 cation (all of which are soluble) and because it's a nitrate compound (all of which are soluble, too). However, PbI_2 has limited solubility in water because it is a lead(II) halide, so it precipitates when the two solutions mix (Figure 8.9).

Let's write balanced molecular and ionic equations for this precipitation reaction. We start with single formula units of reactants and products:

$$Pb(NO_3)_2(aq) + NaI(aq) \rightarrow PbI_2(s) + NaNO_3(aq)$$

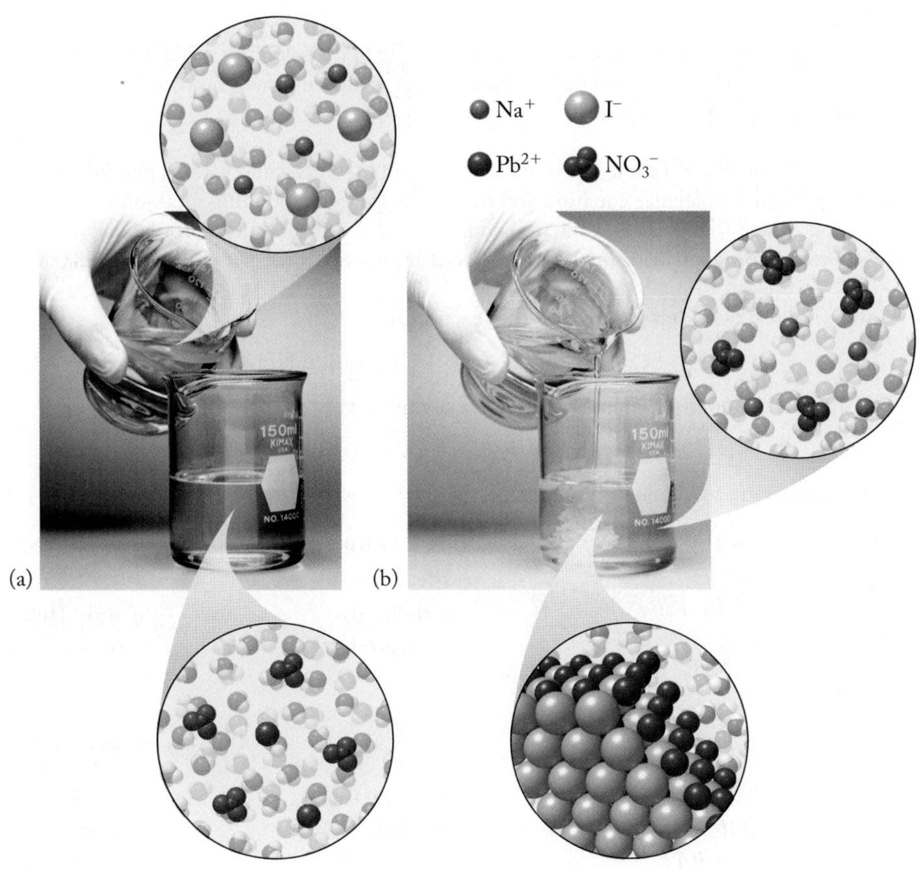

FIGURE 8.9 (a) One beaker contains 0.1 *M* $Pb(NO_3)_2$, and the other contains 0.2 *M* NaI. Both solutions are colorless. (b) As the NaI solution is poured into the $Pb(NO_3)_2$ solution, a yellow precipitate of PbI_2 forms. Sodium ions and nitrate ions remain in solution.

To produce a balanced molecular equation we need to balance the number of iodide and nitrate ions on each side of the reaction arrow. We do this by placing coefficients of 2 in front of NaI and $NaNO_3$:

$$Pb(NO_3)_2(aq) + 2\ NaI(aq) \rightarrow PbI_2(s) + 2\ NaNO_3(aq)$$

Next we write a total ionic equation in which we separate all of the soluble salts into their component ions:

$$Pb^{2+}(aq) + 2\ NO_3^-(aq) + 2\ Na^+(aq) + 2\ I^-(aq) \rightarrow PbI_2(s) + 2\ Na^+(aq) + 2\ NO_3^-(aq)$$

Note how the Na^+ and NO_3^- ions are in solution on both sides of the reaction arrow. They are spectator ions in the reaction, so we delete them to write the net ionic equation:

$$Pb^{2+}(aq) + 2\ I^-(aq) \rightarrow PbI_2(s) \tag{8.23}$$

SAMPLE EXERCISE 8.6 Writing the Net Ionic Equation for a Precipitation Reaction **LO4**

A precipitate forms when aqueous solutions of ammonium sulfate and calcium chloride are mixed. Write the net ionic equation for the reaction.

Collect and Organize We know that mixing solutions of ammonium sulfate, $(NH_4)_2SO_4$, and calcium chloride, $CaCl_2$, yields a product with limited solubility. We need to identify

precipitate a solid product formed from a reaction in solution.

it to be able to write the net ionic equation. According to the solubility guidelines in Table 8.4, ammonium salts and most sulfate salts are soluble in water; however, calcium sulfate is only slightly soluble.

Analyze To write a net ionic equation for a reaction in which solid $CaSO_4$ precipitates, we begin with a molecular equation and then write the corresponding total ionic equation, which includes all of the individual ions present in solutions of soluble reactants and products. Eliminating the spectator ions yields the net ionic equation.

Solve The reactants and products of the reaction are

$$(NH_4)_2SO_4(aq) + CaCl_2(aq) \rightarrow CaSO_4(s) + NH_4Cl(aq)$$

To balance the equation we just need a coefficient of 2 in front of NH_4Cl:

$$(NH_4)_2SO_4(aq) + CaCl_2(aq) \rightarrow CaSO_4(s) + 2\ NH_4Cl(aq)$$

The reactants are soluble ionic compounds, so they are written as separate ions in a total ionic equation, as is soluble NH_4Cl. However, solid calcium sulfate is written as $CaSO_4(s)$:

$$2\ NH_4^+(aq) + SO_4^{2-}(aq) + Ca^{2+}(aq) + 2\ Cl^-(aq) \rightarrow CaSO_4(s) + 2\ NH_4^+(aq) + 2\ Cl^-(aq)$$

Eliminating the NH_4^+ and Cl^- spectator ions

$$\cancel{2\ NH_4^+(aq)} + SO_4^{2-}(aq) + Ca^{2+}(aq) + \cancel{2\ Cl^-(aq)} \rightarrow CaSO_4(s) + \cancel{2\ NH_4^+(aq)} + \cancel{2\ Cl^-(aq)}$$

yields the net ionic equation

$$SO_4^{2-}(aq) + Ca^{2+}(aq) \rightarrow CaSO_4(s)$$

Think About It The solubility guidelines in Table 8.4 help us predict whether or not a precipitate will form when solutions of two ionic solutes are mixed together and their cations and anions have the opportunity to swap partners.

Practice Exercise Does a precipitate form when aqueous solutions of (a) sodium acetate and ammonium sulfate or (b) calcium chloride and mercury(I) nitrate are mixed together? Write the appropriate net ionic equation(s).

CONCEPT TEST

Lead(II) dichromate ($PbCr_2O_7$; the dichromate ion is $Cr_2O_7^{2-}$) is the solid pigment called school bus yellow that was once used to paint lines on highways. Propose a synthesis of school bus yellow that uses a precipitation reaction.

Precipitation reactions may be used to determine the concentrations of ions in solution. For example, NaCl (in the form of rock salt) is used to melt ice and snow on roads during the winter. Some of this NaCl may find its way into nearby drinking water supplies. To determine if sources of drinking water have been contaminated with road salt, the concentration of chloride ion can be determined by reacting a sample of the water with a solution containing more than enough silver nitrate, $AgNO_3$ to precipitate the Cl^- ions as AgCl:

$$NaCl(aq) + AgNO_3(aq) \rightarrow AgCl(s) + NaNO_3(aq) \quad (8.24)$$

The precipitate can be filtered, dried, and weighed. The mass of the AgCl precipitate is divided by its molar mass to calculate the number of moles of AgCl on the filter. This value is the same as the number of moles of Cl^- ions in the original water sample. Precipitation reactions used in analytical chemistry sometimes include indicators that provide a visual signal (like a color change) when there is an excess of the precipitation agent to ensure that all of the ions of interest have precipitated. That way the analyst can be sure Cl^- ions are the limiting reactant in the precipitation of AgCl.

CONNECTION The concept of a limiting reactant was introduced in Chapter 7.

SAMPLE EXERCISE 8.7 Calculating Solute Concentration from Mass of a Precipitate LO6

To determine the concentration of chloride ion in a 100.0 mL sample of groundwater, a chemist adds a large enough volume of a solution of $AgNO_3$ to precipitate all the Cl^- ions as AgCl. The mass of the resulting precipitate is 71.7 mg. What is the Cl^- concentration in the sample in milligrams per liter?

Collect and Organize We are given the sample volume and mass of AgCl formed. Our task is to determine the chloride concentration in milligrams of Cl^- per liter of groundwater.

Analyze We need to calculate the mass of the Cl^- ions in the weighed mass of AgCl and then divide that mass of Cl^- ions by the volume of the water sample to calculate a mg/L concentration. The molar mass of chlorine is about 1/3 that of silver and about 1/4 the formula mass of AgCl. So, 71.7 mg AgCl should contain a little less than 20 mg Cl^- ions, which were originally dissolved in a 100 mL sample. This mass-to-volume ratio is a little less than 200 mg/1000 mL, or 200 mg/L.

Solve The molar masses of Cl and Ag are 35.45 g/mol and 107.87 g/mol, respectively, so the molar mass of AgCl is 143.32 g/mol. The mass of Cl^- ions in 71.7 mg AgCl is

$$71.7\ \cancel{\text{mg AgCl}} \times \frac{1\ \text{g}}{1000\ \cancel{\text{mg}}} \times \frac{1\ \cancel{\text{mol AgCl}}}{143.32\ \cancel{\text{g AgCl}}} \times \frac{1\ \cancel{\text{mol Cl}^-}}{1\ \cancel{\text{mol AgCl}}} \times \frac{35.45\ \cancel{\text{g}}\ \text{Cl}^-}{1\ \cancel{\text{mol Cl}^-}} \times \frac{1000\ \text{mg}}{1\ \cancel{\text{g}}}$$

$$= 17.7\ \text{mg Cl}^-$$

This 17.7 mg mass of Cl^- ions was originally in a 100.0 mL sample. Converting this mass-to-volume ratio to mg/L yields

$$\frac{17.7\ \text{mg Cl}^-}{100.0\ \cancel{\text{mL}}} \times \frac{1000\ \cancel{\text{mL}}}{1\ \text{L}} = 177\ \text{mg Cl}^-/\text{L}$$

Think About It Did you notice that the first step in the first calculation converted milligrams to grams (of AgCl) and that the last step converted grams to milligrams (of Cl^- ions)? Numerically, the two steps involved dividing by 1000 and then multiplying by 1000. Had we recognized that the initial and final quantities were both expressed in milligrams, we could have eliminated those two steps and carried the *milli-* prefix through all of the steps from first to last, thereby simplifying the calculation without affecting the result.

Practice Exercise The concentration of SO_4^{2-} ions in a 50.0 mL sample of coastal seawater is determined by adding a solution of $BaCl_2$ to the sample and precipitating the SO_4^{2-} as $BaSO_4$. The precipitate is removed from the sample by filtration and then dried and weighed. If the mass of $BaSO_4$ recovered from the sample is 0.311 g, what is the sulfate concentration of the sample expressed in mmol/L?

SAMPLE EXERCISE 8.8 Predicting the Mass of a Precipitate LO6

Barium sulfate is used to enhance X-ray imaging of the upper and lower gastrointestinal (GI) tracts. In upper GI imaging, patients drink a suspension of solid $BaSO_4$ in water, which has the consistency of a dense, chalky milkshake and, even with flavoring added, tastes awful. The compound is not toxic because of its low solubility. To make pure $BaSO_4$, a precipitation reaction is employed: aqueous solutions of soluble barium nitrate and sodium sulfate are mixed together, and solid $BaSO_4$ is separated from the reaction mixture by filtration. If a 1 L volumetric flask (volume 1.00 L) is used to prepare a 1.55 *M* $Ba(NO_3)_2$ solution, how many grams of $BaSO_4$ ($\mathcal{M}$ = 233.40 g/mol) will be produced if this entire solution is reacted with excess Na_2SO_4?

Collect and Organize We know the soluble reactants [$Ba(NO_3)_2$ and Na_2SO_4] and the insoluble product ($BaSO_4$) of a precipitation reaction and the volume and molar concentration of a solution containing the limiting reactant. We are asked to calculate the mass of $BaSO_4$ that will be produced.

Analyze This stoichiometry problem is like the ones we solved in Chapter 7, except that this one is based on quantities of dissolved reactants [$Ba(NO_3)_2$ and Na_2SO_4] and a solid product ($BaSO_4$):

$$Ba(NO_3)_2(aq) + Na_2SO_4(aq) \rightarrow BaSO_4(s) + 2\ NaNO_3(aq)$$

There are about 1.5 moles of $Ba(NO_3)_2$ in 1 L of a 1.55 *M* solution, which translates into about 1.5 moles of Ba^{2+} ions and 1.5 moles of $BaSO_4$. Multiplying 1.5 by a molar mass of about 233 g/mol should give us an answer of about 350 g.

Solve We first multiply the volume and concentration of the $Ba(NO_3)_2$ solution to obtain an equivalent number of moles of $Ba(NO_3)_2$. Two more conversion factors are required to convert mol $Ba(NO_3)_2$ to mol $BaSO_4$, and finally into a mass of $BaSO_4$:

$$1.00\ \cancel{L} \times \frac{1.55\ \cancel{mol\ Ba(NO_3)_2}}{\cancel{L}} \times \frac{1\ \cancel{mol\ BaSO_4}}{1\ \cancel{mol\ Ba(NO_3)_2}} \times \frac{233.40\ g\ BaSO_4}{1\ \cancel{mol\ BaSO_4}} = 362\ g\ BaSO_4$$

Think About It As in many calculations based on reactions in solution, using "mol/L" instead of "*M*" helps us properly convert units and obtain a correct answer, which is close to our estimate.

Practice Exercise Vermilion, also known as Chinese red, is a very rare and expensive solid natural pigment used to print the artist's signature on works of art. It is mercury(II) sulfide, and it is insoluble in water. What mass of vermilion can be produced when 50.00 mL of 0.0150 *M* mercury(II) nitrate is mixed with a solution containing excess sodium sulfide?

Saturated Solutions and Supersaturation

The aqueous solubilities of many ionic and molecular solids increase with increasing temperature. For example, more table sugar (or sucrose, $C_{12}H_{22}O_{11}$) dissolves in hot water than in cold—a phenomenon familiar to anyone who has attempted to sweeten a glass of iced tea. When as much sugar as possible has dissolved in hot water, the resulting solution is said to be **saturated** with sugar because it can't hold any more. If we continue to heat the solution without adding more sugar, it becomes **unsaturated** because more sugar would dissolve in it if more were available. If the solution is then cooled to a temperature at which less sugar is soluble, but no sugar precipitates, then the solution is said to be **supersaturated**.

saturated solution a solution that contains the maximum concentration of a solute possible at a given temperature.

unsaturated solution a solution that contains less than the maximum quantity of solute predicted to be soluble in a given volume of solution at a given temperature.

supersaturated solution a solution that contains more than the maximum quantity of solute predicted to be soluble in a given volume of solution at a given temperature.

Supersaturated solutions can remain supersaturated for a long time if left undisturbed, but eventually mechanical shock or the addition of a *seed crystal* (a small crystal of the solute that provides a site for crystallization) causes the solute to rapidly precipitate (Figure 8.10). More dramatic examples of crystal growth from supersaturated solutions were discovered in 2000 by miners in Naica, Mexico (Figure 8.11). They happened upon a cave containing enormous crystals of the

(a)

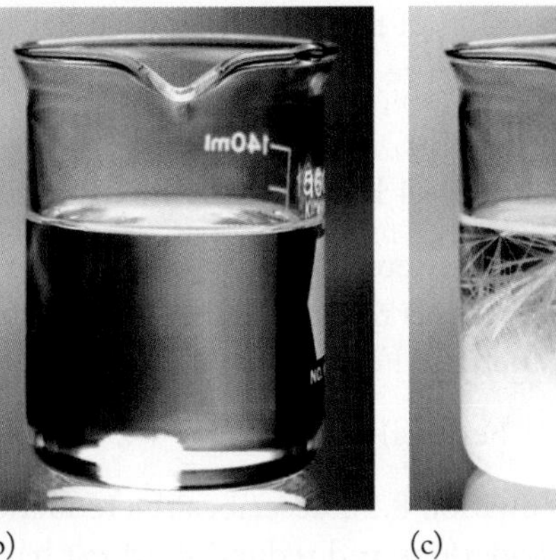

(b)

(c)

FIGURE 8.10 (a) A seed crystal is added to a supersaturated solution of sodium acetate. (b) The seed crystal becomes a site for rapid growth of sodium acetate crystals. (c) Crystal growth continues until the solution is no longer supersaturated. Instead, it is merely saturated with sodium acetate.

FIGURE 8.11 Gypsum ($CaSO_4 \cdot 2\ H_2O$) crystals in this cave in Naica, Mexico, grew to enormous size over half a million years as gypsum slowly precipitated.

mineral gypsum (calcium sulfate), some nearly 12 m long and 2 m across. Apparently the crystals formed from Ca^{2+} and SO_4^{2-} ions dissolved in groundwater that had been geothermally heated to nearly 60°C. As this water cooled in the cave to about 54°C, it became slightly supersaturated, and crystals of gypsum began to grow very slowly for over half a million years. When the supersaturated gypsum solution was pumped out of the cave as part of current mining operations, the crystals were revealed.

Solutions may also become supersaturated when solvent evaporates. For example, groundwater leaching through porous limestone becomes saturated with $CaCO_3$. If it then seeps into a cave where it simultaneously drips from the ceiling and evaporates, $CaCO_3$ precipitates, forming deposits (Figure 8.12) called stalactites when attached to the ceiling and called stalagmites when built up from the floor.

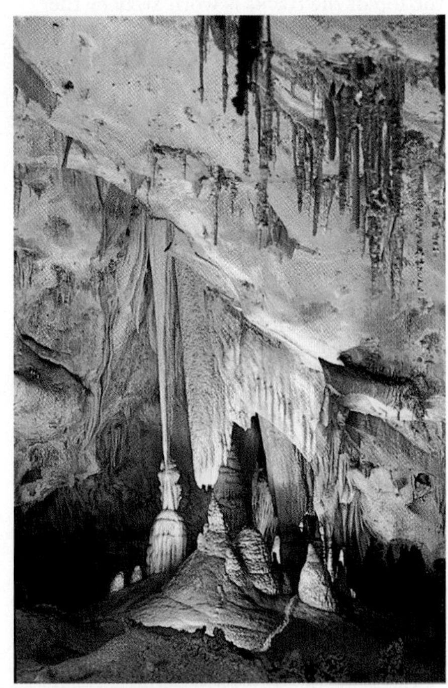

FIGURE 8.12 In Carlsbad Caverns in New Mexico, stalagmites of limestone grow up from the cavern floor and stalactites grow downward from the ceiling.

8.6 Oxidation–Reduction Reactions

Oxygen is one of the most abundant elements on Earth; it makes up 50% of the crust and 89% of the water covering Earth's surface by mass. Its molecules also make up 21% of all those in Earth's atmosphere. It is an essential element for nearly all creatures, even aquatic organisms, who must obtain their oxygen from the small concentrations of it that dissolve in water. Oxygen is at the core of many of the chemical reactions that release energy and sustain life, too: *oxidation–reduction* reactions, or *redox* reactions.

Although this chapter focuses on reactions in solution, we begin our coverage of redox chemistry with oxygen reactions that occur in the air because many of them are both familiar and simple to describe with balanced chemical equations. Oxygen combines with nonmetals such as carbon, sulfur, and nitrogen to produce volatile oxides. It also combines with metals and semimetals, producing solid oxides. All of these reactions are examples of redox reactions. Indeed, *oxidation* was once defined as a reaction that increased the oxygen content of a substance. Based on this definition, combustion of a hydrocarbon such as methane, which we discussed in Chapter 7, involves oxidation because the products, CO_2 and H_2O, contain more oxygen than CH_4:

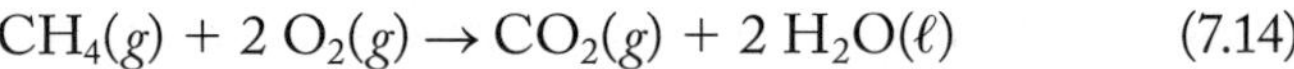

$$CH_4(g) + 2\ O_2(g) \rightarrow CO_2(g) + 2\ H_2O(\ell) \qquad (7.14)$$

FIGURE 8.13 Corrosion (oxidation) leaves steel railroad tracks covered in a thin layer of orange Fe_2O_3. Any rust that forms on the top of the tracks is worn away by the wheels of trains passing over them, leaving shiny metallic surfaces of mostly iron that is available to be oxidized further.

Similarly, the corrosion of metals is oxidation because the products of corrosion, such as rusted iron (Figure 8.13), have a higher oxygen content than the original metal (which had none):

$$4\ Fe(s) + 3\ O_2(g) \rightarrow 2\ Fe_2O_3(s) \quad (8.25)$$

Reduction reactions were originally defined as reactions in which the oxygen content of a substance was reduced. A classic example is the reduction of iron ores such as Fe_2O_3 to metallic Fe:

$$Fe_2O_3(s) + 3\ CO(g) \rightarrow 2\ Fe(s) + 3\ CO_2(g) \quad (8.26)$$

The reaction in Equation 8.26 describes the reduction of Fe_2O_3, but there is more to it than that. Oxidations and reductions don't happen in isolation; they happen together. This means that as Fe_2O_3 is reduced, an element in the other reactant must be oxidized. That other element is the carbon in CO, which is oxidized from CO to a compound with an even higher oxygen content, CO_2.

CONCEPT **TEST**

The following equation describes the conversion of one iron mineral called magnetite (Fe_3O_4) into another called hematite (Fe_2O_3):

$$4\ Fe_3O_4(s) + O_2(g) \rightarrow 6\ Fe_2O_3(s)$$

Is iron oxidized or reduced in the reaction?

Oxidation Numbers

Today we use *redox* to describe other reactions in addition to those that increase or decrease the oxygen content of a substance. An element is said to undergo **oxidation** when it *loses* electrons and **reduction** when it *gains* them.[2] These definitions are more general than the older ones, but they are consistent with them, too. To see how, let's reconsider the process of rust (Fe_2O_3) forming on an iron object.

$$4\ Fe(s) + 3\ O_2(g) \rightarrow 2\ Fe_2O_3(s) \quad (8.25)$$

Fe is oxidized according to the old definition—its oxygen content increases when it forms Fe_2O_3—but it does so by *losing* electrons. The Fe atoms in elemental iron have their normal free-atom complement of 26 electrons each, but the Fe^{3+} ions in Fe_2O_3 have only 23 electrons. This loss of three electrons means that iron is oxidized under the new definition, too. At the same time, oxygen is reduced from its elemental state (where the O atoms have their normal complement of eight electrons) to oxide (O^{2-}) ions with 10 electrons each. The simultaneous loss and gain of electrons are the key feature of all redox reactions: if one element loses electrons, another must gain them so that the losses and gains are in balance. In Equation 8.25 this balance is achieved because four Fe atoms each lose three electrons for a total loss of 12 electrons, while six O atoms gain two electrons each for a total gain of 12. Thus, Equation 8.25 is balanced because the numbers of Fe and O atoms on both sides of the reaction arrow are the same *and* because the number of electrons gained and lost is the same.

To help keep track of electron losses and gains in redox reactions, chemists follow changes in the oxidation states of the atoms in reactants and products. The

oxidation a chemical change in which an element loses electrons; the oxidation number of the element increases.

reduction a chemical change in which an element gains electrons; the oxidation number of the element decreases.

oxidation number (O.N.) or **oxidation state** a numerical value (which may be positive, negative, or zero) based on the number of electrons that an atom gains or loses when it forms an ion or that it shares when it forms a covalent bond with an atom of another element.

[2]The mnemonic OIL RIG may be helpful for remembering these expanded definitions: "Oxidation Is Loss; Reduction Is Gain."

TABLE 8.5 Rules for Assigning Oxidation Numbers (O.N.) to Atoms

1. O.N. = 0 for atoms in pure elements.
2. O.N. = the charge of single-atom ions.
3. O.N. of fluorine atoms = −1 in all compounds.
4. O.N. of oxygen atoms = −2 in *nearly* all compounds. Exceptions occur when oxygen atoms are bonded to fluorine atoms (for example, OF_2), where its O.N. = +2, or when O is bonded to another O atom in a peroxide (for example, H_2O_2, where its O.N. = −1), or in a superoxide (for example, KO_2, where its O.N. = $-\frac{1}{2}$).
5. O.N. of hydrogen atoms = +1 in *nearly* all compounds. The exception is hydrogen in metal hydrides (for example, LiH), where H is the more electronegative element and its O.N. = −1.

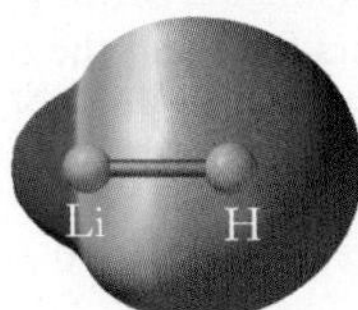

6. O.N. values of the atoms in a neutral compound sum to zero.
7. O.N. values of the atoms in a polyatomic ion sum to the charge on the ion.

oxidation state, or **oxidation number (O.N.),** of an atom in a molecule or ion is a measure of the number of electrons it has compared to the number it would have if it were a free atom. The O.N. values of monatomic ions are the same as their electrical charges. In molecules and polyatomic ions, atoms have O.N. values that are related to the number of covalent bonds they form, but their O.N. values also depend on the elements they are bonded to and the O.N. values of those elements. The rules for assigning O.N. values are summarized in Table 8.5.

There is a hierarchy in assigning O.N. values to the atoms in molecules that is based on electronegativity. Fluorine is always in a −1 oxidation state in molecular compounds because it is the most electronegative element and it forms one bond per atom. Other halogen atoms also have O.N. values of −1 in their compounds, but not when they are bonded to the more electronegative fluorine or oxygen atoms. As illustrated in Table 8.6, chlorine and the higher *Z* halogens can be in oxidation states up to +7 depending on the number of O atoms they are bonded to.

TABLE 8.6 Some Oxidation States of Chlorine

Compound	Chlorine O.N.
HCl	−1
HClO	+1
$HClO_2$	+3
ClO_2	+4
$HClO_3$	+5
$HClO_4$	+7

Oxygen nearly always has an O.N. of −2 in its compounds because it is the next most electronegative element after fluorine and tends to form two bonds per atom. However, its O.N. is +2 in OF_2 because of rule 3 in Table 8.5 (fluorine's O.N. is always −1 in its compounds) and rule 6 (the O.N. values of the elements in a compound must sum to zero). The oxidation state of O is also different when it bonds to itself in compounds called peroxides. These compounds have —O—O— functional groups in which oxygen's O.N. equals −1.

SAMPLE EXERCISE 8.9 Determining Oxidation Numbers **LO7**

What are the oxidation states of sulfur in (a) S_8, (b) SO_2, (c) Na_2S, and (d) $CaSO_4$?

Collect and Organize We are asked to assign oxidation numbers to sulfur in S_8 and in three sulfur compounds. The rules for assigning oxidation numbers are given in Table 8.5.

Analyze

a. S_8 is a molecular form of the element sulfur.
b. In SO_2, sulfur is bonded to oxygen, which is a more electronegative element and has an O.N. of −2 (rule 4).
c. Na_2S is a binary ionic compound; the charge of the sulfide ion is 2−.
d. Sulfur in $CaSO_4$ is part of the sulfate ion, which means its oxidation number added to those of the four O atoms must add up to the charge of the ion, which is 2− (rule 7).

Solve

a. The oxidation state of sulfur in elemental sulfur (S_8) is 0 (rule 1).
b. The sum of the oxidation numbers for S and the two O atoms in this neutral molecule must be zero (rule 6). Assigning an O.N. of −2 to each oxygen atom, and letting the oxidation state of sulfur be x, we can set up the following equation:

$$x + 2(-2) = 0$$

$$x = +4$$

So, the oxidation state of sulfur in SO_2 is +4.
c. The oxidation state of the atom in a monatomic ion is the same as the charge on the ion (rule 2). Therefore, the oxidation state of sulfur in the sulfide ion is −2.
d. The charge on the SO_4^{2-} ion is 2−. Each O atom has an O.N. of −2. Letting x be the O.N. of sulfur and applying rule 7:

$$x + 4(-2) = -2$$

$$x = +6$$

So, the oxidation state of S in $CaSO_4$ is +6.

Think About It We found that sulfur, like chlorine in the compounds in Table 8.6, can exist in oxidation states that range from negative to quite positive values. The most positive values occur when single sulfur atoms are bonded to multiple oxygen atoms: the more O atoms, the higher the oxidation state.

Practice Exercise Determine the oxidation states of nitrogen in (a) NO_2; (b) N_2O; (c) HNO_3.

Electron Transfer in Redox Reactions

Redox reactions involve the transfer of electrons between atoms, and these transfers result in changes in the oxidation states of the atoms. We know from rules 1 and 2 in Table 8.5 that when an atom acquires an electron, forming a 1− anion, its oxidation number is reduced from 0 to −1 (ΔO.N. = −1). On the other hand, when an atom loses an electron and forms a 1+ cation, its oxidation number increases from 0 to +1 (ΔO.N. = +1). We can generalize this pattern to describe the gain or loss of any number (n) of electrons:

Oxidation	Lose n electrons	ΔO.N. = $+n$
Reduction	Gain n electrons	ΔO.N. = $-n$

This general pattern can help us track the flow of electrons from the elements losing them (being oxidized) to the elements gaining them (being reduced) in a reaction. The tracking system involves calculating the changes in oxidation number of the elements that are oxidized and reduced.

To see how the system works, let's examine the redox properties of two neighbors in the periodic table: copper (Z = 29) and zinc (Z = 30). Elemental Zn has

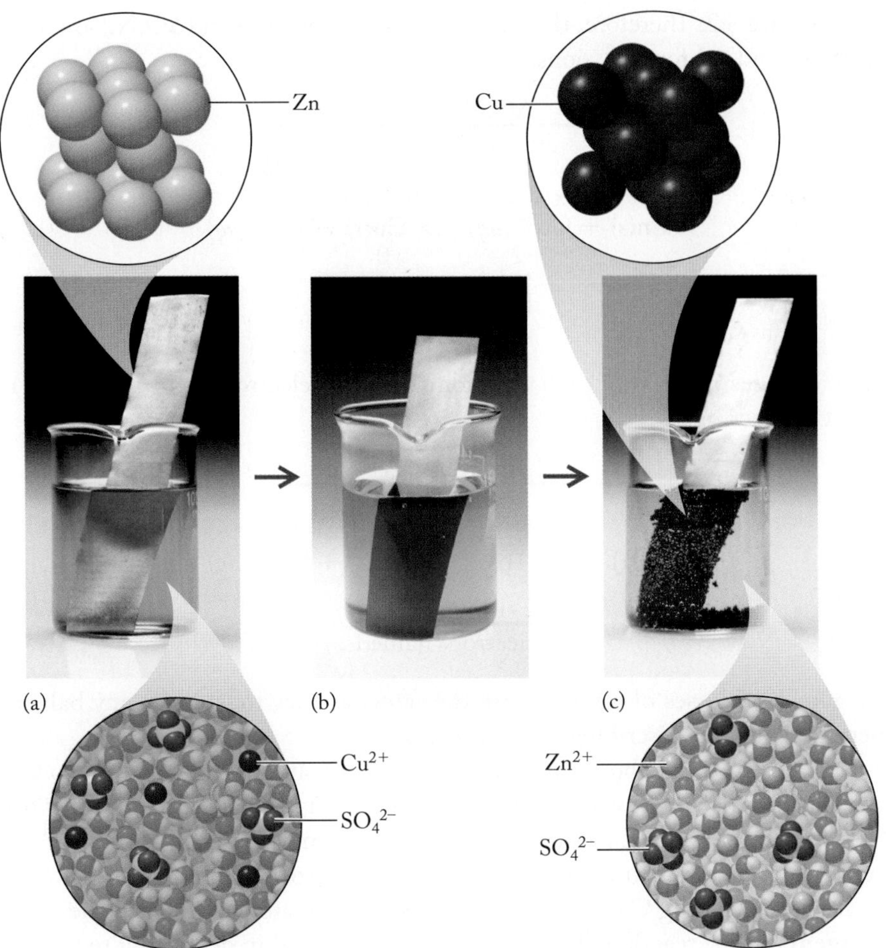

FIGURE 8.14 Redox reaction of Zn metal with Cu^{2+} ions. (a) A strip of Zn is placed in a solution of $CuSO_4$. (b) Over time, the surface of the Zn strip darkens as it is coated by a layer of Cu atoms. (c) Eventually the blue color of Cu^{2+} ions in solution fades as Cu atoms form. They accumulate on the Zn strip and then fall to the bottom of the beaker.

considerable electron-donating power, and Cu^{2+} ions tend to accept electrons. These complementary properties are illustrated in Figure 8.14.

When a strip of Zn metal is placed in a solution of $CuSO_4$—that is, a solution of Cu^{2+} ions and SO_4^{2-} ions—Zn atoms spontaneously donate electrons to Cu^{2+} ions, forming Zn^{2+} ions and Cu atoms. We can see this redox reaction happen: the shiny zinc surface turns dark brown as a textured layer of copper metal accumulates on it, and the distinctive blue color of $Cu^{2+}(aq)$ ions fades as they acquire electrons and become atoms of copper metal. The chemical equation describing this reaction is

$$Zn(s) + CuSO_4(aq) \rightarrow ZnSO_4(aq) + Cu(s) \qquad (8.27)$$

and the total ionic equation including all ions in solution is

$$Zn(s) + Cu^{2+}(aq) + SO_4^{2-}(aq) \rightarrow Cu(s) + Zn^{2+}(aq) + SO_4^{2-}(aq)$$

Eliminating SO_4^{2-} ions, which are unchanged (spectator ions) in the reaction, produces the following net ionic equation:

$$Zn(s) + Cu^{2+}(aq) \rightarrow Cu(s) + Zn^{2+}(aq) \qquad (8.28)$$

Now let's assign oxidation numbers to the Cu and Zn atoms and ions in Equation 8.28. According to rules 1 and 2 in Table 8.5, the oxidation states of the atoms in Zn and Cu metal are 0, and the oxidation states of the atoms in both of their ions are +2. Therefore, the changes in oxidation number (ΔO.N.) are +2 for Zn atoms and −2 for Cu^{2+} ions:

ΔO.N. = +2
(oxidation)

$$\overset{0}{Zn}(s) + \underset{+2}{Cu^{2+}}(aq) \rightarrow \underset{0}{Cu}(s) + \overset{+2}{Zn^{2+}}(aq)$$

(reduction)
ΔO.N. = −2

These changes mean that each Zn atom *loses* two electrons and each Cu^{2+} ion *gains* two electrons.

2 electrons lost
(oxidation)

$$Zn(s) + Cu^{2+}(aq) \rightarrow Cu(s) + Zn^{2+}(aq)$$

(reduction)
2 electrons gained

The gains and losses of electrons are the same, as they must be in any balanced chemical equation describing a redox reaction.

One way to think about this reaction is that Zn atoms give electrons to Cu^{2+} ions, causing the Cu^{2+} ions to be *reduced*, and Cu^{2+} ions take electrons away from Zn atoms, causing the Zn atoms to be *oxidized*. From this perspective, Cu^{2+} ions are forcing Zn atoms to be oxidized and are called **oxidizing agents**. Similarly, Zn atoms are causing Cu^{2+} ions to be reduced, so Zn atoms are called the **reducing agents** in the reaction. These terms are widely used in describing redox reactions, but they can take some getting used to because the *oxidizing* agent is actually *reduced* and the *reducing* agent is *oxidized*. Every redox reaction has an oxidizing agent and a reducing agent. The reducing agent is the electron donor, and the oxidizing agent is the electron acceptor.

CONCEPT TEST

Which of the following are redox reactions? For those that are, identify the oxidizing agent and the reducing agent:

a. $Sn^{2+}(aq) + Br_2(aq) \rightarrow Sn^{4+}(aq) + 2\ Br^-(aq)$

b. $2\ F_2(g) + 2\ H_2O(\ell) \rightarrow 4\ HF(aq) + O_2(g)$

c. $NaHCO_3(aq) + HCl(aq) \rightarrow NaCl(aq) + CO_2(g) + H_2O(\ell)$

oxidizing agent a reactant that accepts electrons from another in a redox reaction, thereby oxidizing the other reactant; the oxidizing agent is reduced in the reaction.

reducing agent a reactant that donates electrons to another in a redox reaction, thereby reducing the other reactant; the reducing agent is oxidized in the reaction.

SAMPLE EXERCISE 8.10 Identifying Oxidizing and Reducing Agents and Their Changes in Oxidation State LO7

Energy released by the reaction of hydrazine and dinitrogen tetroxide:

$$2\ N_2H_4(\ell) + N_2O_4(g) \rightarrow 3\ N_2(g) + 4\ H_2O(g)$$

is used to orient and maneuver spacecraft and to propel rockets into space (Figure 8.15). Identify the elements that are oxidized and reduced and their changes in oxidation state. Identify the oxidizing agent and the reducing agent.

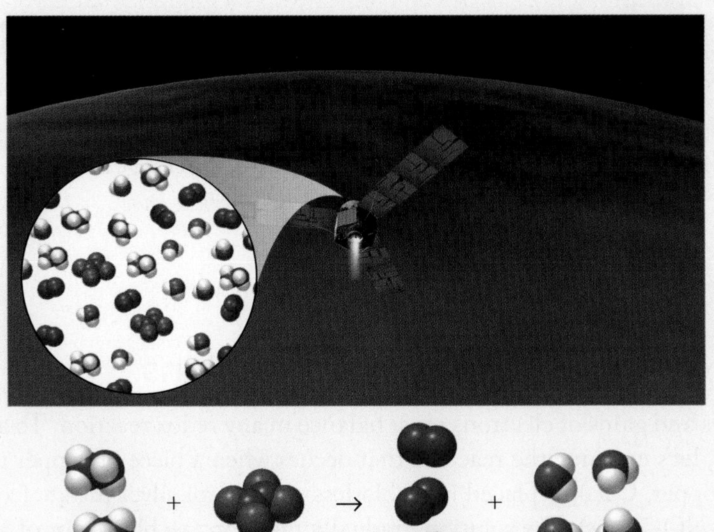

FIGURE 8.15 The NASA Juno spacecraft went into orbit around Jupiter in July, 2016, by firing its main engine, which is powered by the redox reaction between hydrazine (N_2H_4) and dinitrogen tetroxide (N_2O_4).

Collect, Organize, and Analyze The reactants are hydrazine, N_2H_4, and dinitrogen tetroxide, N_2O_4, and the products are N_2 and H_2O. According to the rules in Table 8.5, the oxidation state of nitrogen in molecules of N_2 (a pure element) is 0, and the O.N. values of O and H in the three reactant and product compounds are −2 and +1, respectively. Because the O.N. values of O and H do not change during the reaction, the only element that is oxidized and reduced in the reaction is nitrogen. Rule 6 in Table 8.5 can be used to calculate the oxidation states of nitrogen in N_2H_4 and N_2O_4.

Solve The oxidation numbers of N and H in N_2H_4 must sum to 0 (rule 6). The O.N. value of hydrogen is +1, so the O.N. of nitrogen is

$$2x + (+1)(4) = 0$$

$$x = -2$$

Applying rule 6 to N_2O_4, we find that the O.N. of nitrogen in this molecule is

$$2x + (-2)(4) = 0$$

$$x = +4$$

The O.N. of each N atom in N_2H_4 changes from −2 to 0 for a gain of +2. There are two of these N atoms per molecule, and there are two N_2H_4 molecules in the chemical equation, so the overall ΔO.N. is $2 \times 2 \times (+2) = +8$. The O.N. of each N atom in N_2O_4 changes from +4 to 0 for a ΔO.N. of −4. There are two N atoms in the chemical equation's single N_2O_4 molecule, so the overall ΔO.N. is $2 \times (-4) = -8$.

These overall changes in oxidation state are shown in the chemical equation below.

ΔO.N. = 2 × 2 × (+2) = +8

(oxidation)

−2 → 0

$$2\,N_2H_4(g) + N_2O_4(g) \rightarrow 3\,N_2(g) + 4\,H_2O(g)$$

+4 → 0

(reduction)

ΔO.N. = 2 × (−4) = −8

In this reaction N_2H_4 is the reducing agent because the N atoms in it are oxidized and N_2O_4 is the oxidizing agent because the N atoms in it are reduced.

Think About It We tend to identify the specific atoms that are oxidized or reduced, while whole molecules or ions are identified as the oxidizing agents or reducing agents. In the equation in this exercise, nitrogen is oxidized but N_2H_4 is the reducing agent.

Practice Exercise In the reaction between $O_2(g)$ and $SO_2(g)$ to make $SO_3(g)$, what is the element oxidized, the element reduced, the change in oxidation state of each element, the oxidizing agent, and the reducing agent?

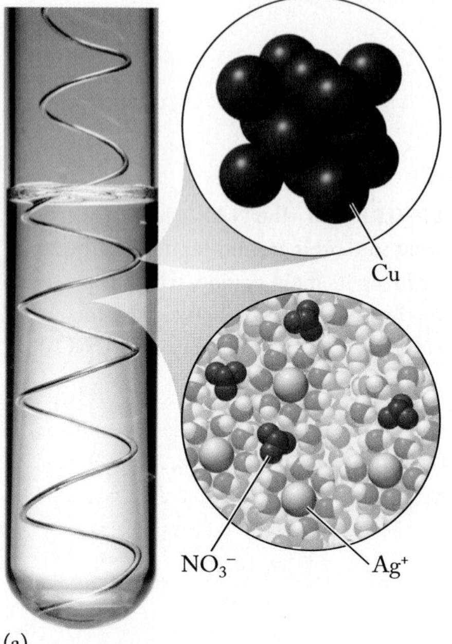

(a)

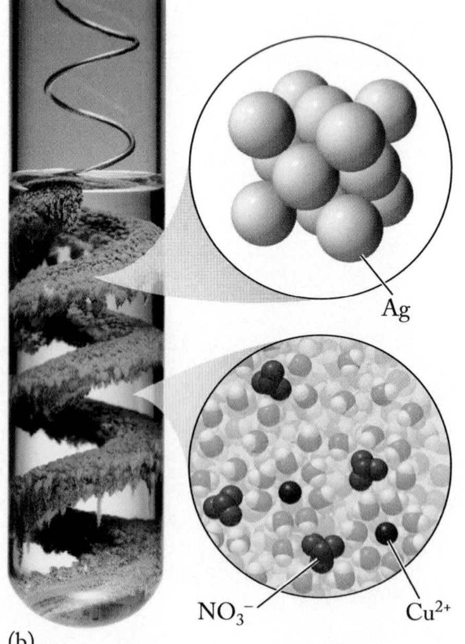

(b)

FIGURE 8.16 (a) When a Cu wire is immersed in a solution of $AgNO_3$, Cu metal oxidizes to Cu^{2+} ions as Ag^+ ions are reduced to Ag metal. (b) A day later, the solution has the blue color of Cu^{2+} ions in solution, and the wire is coated with Ag metal.

Balancing Redox Reaction Equations

The losses and gains of electrons must balance in any redox reaction. To apply this principle, let's examine the reaction that occurs when a piece of copper wire [elemental copper, Cu(*s*)] is placed in a colorless solution of silver nitrate (containing Ag^+ and NO_3^- ions). The solution gradually takes on the blue color of a solution of Cu^{2+} ions, and branchlike structures of solid Ag form on the copper wire (Figure 8.16). In this chemical reaction, Cu metal is oxidized to Cu^{2+} ions and Ag^+ ions are reduced to Ag metal. If we ignore the nitrate ions, which are only spectator ions in the reaction, the reactants and products that should appear in the net ionic equation are

$$Ag^+(aq) + Cu(s) \rightarrow Ag(s) + Cu^{2+}(aq)$$

However, this expression is *not* the net ionic equation because the electrical charges are not balanced: the sum of the charges is 1+ for the reactants but 2+ for the products. The charges are not balanced because the loss and gain of electrons by Cu metal and Ag^+ ions are not balanced, which is reflected in the changes in oxidation state: −1 for Ag^+ ions becoming Ag atoms, and +2 for Cu atoms becoming Cu^{2+} ions.

$$\overset{+1}{Ag^+}(aq) + \overset{0}{Cu}(s) \rightarrow \overset{0}{Ag}(s) + \overset{+2}{Cu^{2+}}(aq)$$

$\Delta O.N. = +2$ (Cu: 0 → +2); $\Delta O.N. = -1$ (Ag: +1 → 0)

To bring these changes in oxidation state and the loss and gain of electrons into balance, we need twice as many Ag^+ ions, which each consume one electron, as Cu atoms, each of which donates two electrons. So we place a coefficient of 2 in front of Ag^+ on the reactant side—and also in front of Ag on the product side to keep the number of Ag atoms in balance. These additions give us the following balanced net ionic equation:

$$2\,Ag^+(aq) + Cu(s) \rightarrow 2\,Ag(s) + Cu^{2+}(aq)$$

Now let's write a balanced net ionic equation for a slightly more complicated redox reaction. In this one, Mn^{2+} ions react with O_2 in a slightly acidic solution, forming solid MnO_2 (Figure 8.17). We start with a preliminary expression relating the known reactants and product:

$$Mn^{2+}(aq) + O_2(g) \rightarrow MnO_2(s)$$

At first glance this expression looks balanced, but closer inspection reveals that there is an ionic charge of 2+ on the reactant side and no charge on the product side. Oxygen is present as a free element (O.N. = 0) on the reactant side and in a compound (O.N. = −2) on the product side, so O_2 is reduced in the reaction. The reaction, then, is a redox reaction, which means that Mn^{2+} must be oxidized. The following O.N. analysis confirms that the oxidation state of Mn increases from +2 in Mn^{2+} to +4 in MnO_2:

$$\overset{+2}{Mn^{2+}}(aq) + \underset{0}{O_2}(g) \rightarrow \overset{+4}{Mn}\underset{-2}{O_2}(s)$$

ΔO.N. = +2 (Mn)

ΔO.N. = 2 × (−2) = −4 (O)

Based on ΔO.N. values, each Mn atom increases by +2 and each O atom decreases by −2. However, the O atoms come in pairs—as O_2 molecules—and the total change in oxidation number per molecule is 2 × (−2) = −4. Therefore, we need twice as many Mn^{2+} ions as O_2 molecules to balance the changes in O.N. and the losses and gains of electrons. So we place a coefficient of 2 in front of Mn^{2+}, and we do the same in front of MnO_2 to keep the number of Mn atoms in balance:

ΔO.N. = 2 × (+2) = +4

$$2\,\overset{+2}{Mn^{2+}}(aq) + \underset{0}{O_2}(g) \rightarrow 2\,\overset{+4}{Mn}\underset{-2}{O_2}(s)$$

ΔO.N. = 2 × (−2) = −4

Now the number of atoms of Mn is balanced, but not much else is. We have an ionic charge of 4+ on the reactant side but still no charge on the product side, and the number of O atoms is no longer balanced. Our first priority is to balance the charge; the question is how to do that. A clue is provided by the fact that the reaction takes place in a slightly acidic solution, which means that there are H^+ ions available. If we add four of them to the product side, we take care of the charge imbalance:

$$2\,Mn^{2+}(aq) + O_2(g) \rightarrow 2\,MnO_2(s) + 4\,H^+(aq)$$

Now we have an expression with four more H atoms and two more O atoms on the product side than the reactant side. At this point, the key is to recognize that this is the same number of H and O atoms as in two molecules of H_2O. There are plenty of them in an aqueous solution, so we can add two to the reactant side:

$$2\,Mn^{2+}(aq) + 2\,H_2O(\ell) + O_2(g) \rightarrow 2\,MnO_2(s) + 4\,H^+(aq)$$

A final check reveals that we have a balanced net ionic equation.

The procedure for writing a net ionic equation that we just employed can be used to balance other redox reaction equations that happen in acidic or neutral solutions. What about those in basic solutions? The difference comes in the charge-balancing step. Instead of adding H^+ ions, we add OH^- ions to either side of the reaction equation in basic solutions. Doing so will probably produce an imbalance in the number of H and O atoms that we can correct by adding molecules of H_2O to either side. Table 8.7 summarizes the steps to follow in writing balanced chemical equations for redox reactions in aqueous solution.

FIGURE 8.17 Oxidation of manganese(II) compounds on the surfaces of rocks in arid regions of the American Southwest builds up a layer of black MnO_2 that is commonly called desert varnish. Native Americans have etched petroglyphs such as these in desert varnish for thousands of years.

TABLE 8.7 Steps in Balancing Chemical Equations for Redox Reactions in Aqueous Solution

1. Write a preliminary reaction expression using one mole of each of the known reactants and products.
2. Calculate the ΔO.N. values of the elements that are reduced and oxidized.
3. Insert coefficients as appropriate to balance the ΔO.N. values (and the losses and gains of electrons). This step includes balancing the numbers of atoms of the elements that are oxidized and reduced, but skip balancing O atoms at this step.
4. Balance ionic charges by adding H^+ ions in acidic or neutral solutions or by adding OH^- ions in basic solutions.
5. Add H_2O molecules to balance the numbers of H and O atoms if necessary.

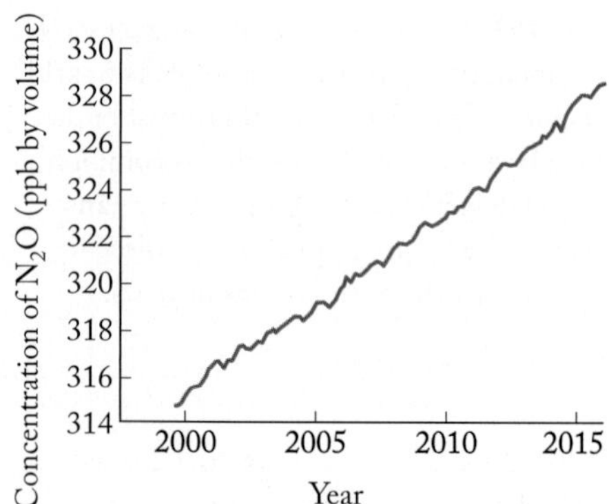

FIGURE 8.18 Atmospheric concentrations of N_2O varied between 200 and 300 ppb for at least 800,000 years before steadily increasing in recent years, due in part to increasing use of nitrogen-containing fertilizers in agriculture.

SAMPLE EXERCISE 8.11 Balancing Redox Reaction Equations I: Acidic Solutions — LO8

The concentration of N_2O in the atmosphere has been increasing in recent years (Figure 8.18). Environmental chemists believe that one reason for the increase is heavy agricultural use of fertilizers, which can result in elevated levels of nitrates in soils and groundwater. Microorganisms called denitrifying bacteria that grow in waterlogged soil convert NO_3^- ions into N_2O gas as they feed on dead plant tissue (empirical formula: CH_2O), converting it into CO_2 and H_2O. Write a balanced net ionic equation describing this conversion of dissolved nitrates to N_2O gas. Assume the reaction occurs in slightly acidic water.

Collect and Organize We need to write a balanced net ionic equation for a reaction in which the reactants include dissolved nitrate ions, $NO_3^-(aq)$, and dead plant tissue, $CH_2O(s)$; the products include N_2O gas, CO_2 gas, and liquid H_2O.

Analyze Table 8.7 outlines the steps to follow in writing a balanced net ionic equation for a redox reaction. The oxidation numbers of O and H have their usual values in these compounds: −2 and +1, respectively. That leaves N and C as the elements to be oxidized and reduced. The problem states that the water is slightly acidic, which means that we should add H^+ ions to balance ionic charges in step 4 of Table 8.7.

Solve

1. The following preliminary expression relates the known reactants and products:

$$NO_3^-(aq) + CH_2O(s) \rightarrow N_2O(g) + CO_2(g) + H_2O(\ell)$$

To begin balancing this equation, we need a coefficient of 2 in front of the NO_3^- ions on the left side to provide the two N atoms needed to form one molecule of N_2O:

$$2\,NO_3^-(aq) + CH_2O(s) \rightarrow N_2O(g) + CO_2(g) + H_2O(\ell)$$

2. We perform a ΔO.N. analysis for nitrogen and carbon:

$$\Delta O.N. = +4$$

$$\overset{0}{} \qquad \overset{+4}{}$$

$$2\,\underset{+5}{NO_3^-}(aq) + \overset{0}{C}H_2O(s) \rightarrow \underset{+1}{N_2}O(g) + \overset{+4}{C}O_2(g) + H_2O(\ell)$$

$$\Delta O.N. = 2 \times (-4) = -8$$

3. To balance these ΔO.N. values we need a coefficient of 2 in front of CH_2O. This also requires a 2 in front of CO_2 to balance the number of C atoms:

$$2\,NO_3^-(aq) + 2\,CH_2O(s) \rightarrow N_2O(g) + 2\,CO_2(g) + H_2O(\ell)$$

4. To balance the ionic charges (2− on the reactant side, 0 on the product side), we add two H^+ ions to the left side because the reaction is happening in slightly acidic water:

$$2\ NO_3^-(aq) + 2\ CH_2O(s) + 2\ H^+(aq) \rightarrow N_2O(g) + 2\ CO_2(g) + H_2O(\ell)$$

5. Taking an inventory of the atoms on both sides of the reaction arrow, we find that the left side has four more H atoms and two more O atoms than the right side. We correct this imbalance by adding two more molecules of H_2O to the right side for a total of three:

$$2\ NO_3^-(aq) + 2\ CH_2O(s) + 2\ H^+(aq) \rightarrow N_2O(g) + 2\ CO_2(g) + 3\ H_2O(\ell)$$

Now the equation is balanced. We do not have to rebalance the redox part of the reaction after adding H^+ or H_2O because the oxidation states of hydrogen and oxygen do not change.

Think About It Many organisms, including humans, get energy through respiration processes in which digested food (empirical formula: CH_2O) is oxidized to CO_2 and H_2O using molecular oxygen, O_2. Denitrifying bacteria are able to extract oxygen from nitrate ions to convert similar food (chemically) into CO_2 and H_2O and energy.

Practice Exercise Aqueous dichromate ion, $Cr_2O_7^{2-}$ (used to make school bus yellow pigments), reacts with iodide ion, $I^-(aq)$, to form $Cr^{3+}(aq)$ and $IO_3^-(aq)$ by the unbalanced chemical equation:

$$Cr_2O_7^{2-}(aq) + I^-(aq) \rightarrow Cr^{3+}(aq) + IO_3^-(aq)$$

Balance this equation in acidic solution.

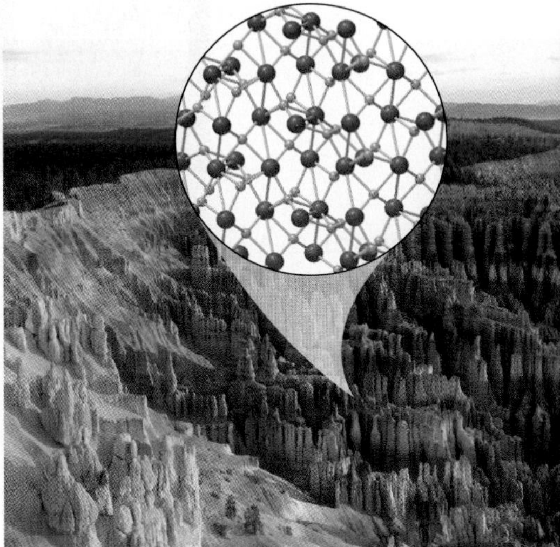

FIGURE 8.19 Red-orange rock formations called hoodoos in Bryce Canyon, Utah, owe their color to high concentrations of iron(III) oxide.

Redox processes play a major role in determining the character of Earth's rocks and soils. For example, the orange-red spires called hoodoos in Bryce Canyon National Park in Utah owe their distinctive color to an iron(III) oxide mineral called hematite (Figure 8.19). These rocks are also relatively fragile and susceptible to physical and chemical weathering, which has resulted in the fascinating shapes of the hoodoos for which Bryce is famous. Iron(III) oxide is also the form of iron known as rust, the crumbly, red-brown solid that forms on iron objects exposed to water and oxygen.

Soil color is influenced by mineral content and by the availability of molecular oxygen. Wetlands are areas that are either saturated with water year-round or flooded during much of the year. Even though wetland soils have considerable iron content, they do not have the distinctive red-orange color of iron(III) compounds, because the waterlogged soils have so little O_2 that the dominant oxidation state of Fe is +2 instead of +3. In fact, soil color is one of the ways environmental scientists can tell whether an area is a wetland (Figure 8.20).

We can observe the characteristic colors associated with soils rich in iron(II) or iron(III) compounds in the laboratory. Figure 8.21(a) shows a solution of iron(II) ammonium sulfate, $(NH_4)_2Fe(SO_4)_2$, as a solution of NaOH is added to it. Addition of base causes $Fe(OH)_2$ to precipitate as a blue-gray solid, similar in color to the wetland soil shown in Figure 8.20(a). When

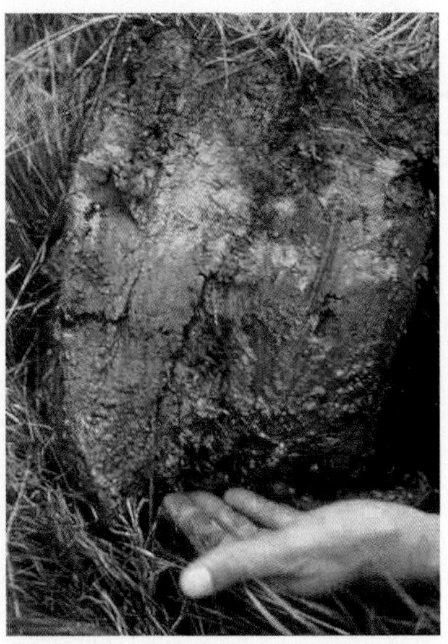

FIGURE 8.20 (a) Wetland soil has a blue-gray color because of the presence of iron(II) compounds. (b) Red-orange mottling indicates the presence of iron(III) oxides formed as a result of O_2 permeation through channels made by plant roots.

(a)

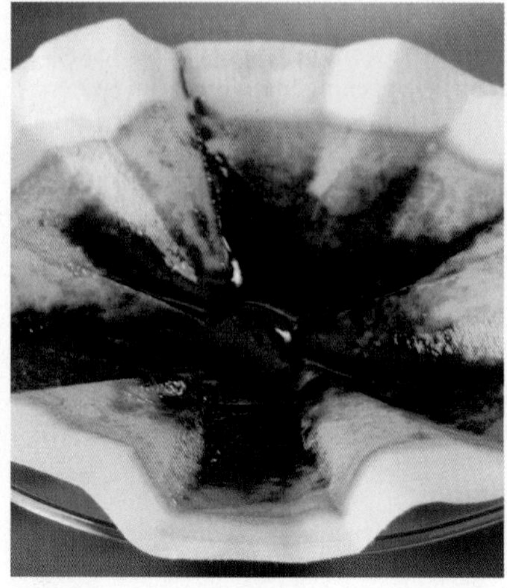

(b)

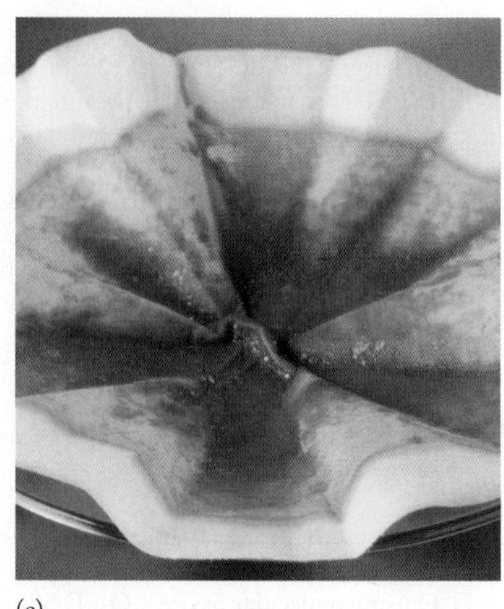

(c)

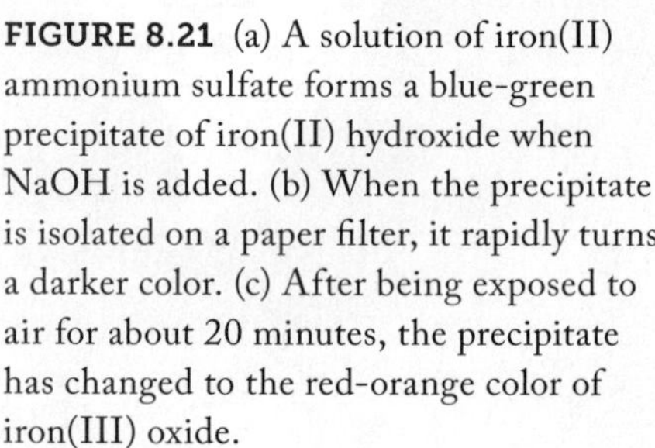

FIGURE 8.21 (a) A solution of iron(II) ammonium sulfate forms a blue-green precipitate of iron(II) hydroxide when NaOH is added. (b) When the precipitate is isolated on a paper filter, it rapidly turns a darker color. (c) After being exposed to air for about 20 minutes, the precipitate has changed to the red-orange color of iron(III) oxide.

this mixture is filtered, the iron(II) hydroxide residue immediately starts to darken (Figure 8.21b). After about 20 minutes of exposure to oxygen in the air, the precipitate has turned the orange-red color of iron(III) hydroxide (Figure 8.21c). The moist iron(II) hydroxide has been oxidized to iron(III) hydroxide—in much the way the iron(II) compounds in a wetland soil are oxidized when exposed to oxygen. Sample Exercise 8.12 illustrates how to write the net ionic equation for a redox reaction such as this one that is carried out in a basic solution.

SAMPLE EXERCISE 8.12 Balancing Redox Reaction Equations II: Basic Solutions **LO8**

The brown residues that collect along the inner walls and bottoms of toilet tanks are made mostly of solid $Fe(OH)_3$ that forms when Fe^{2+} ions in water are oxidized by dissolved O_2. Write a balanced net ionic equation for this reaction in a slightly basic solution.

Collect and Organize The reactants include dissolved O_2 and Fe^{2+} ions, and one of the products is solid $Fe(OH)_3$. We need to write a net ionic equation for the reaction—that is, an equation that represents all of the species involved in the reaction less those that are not changed by it.

Analyze Table 8.7 outlines the steps to follow in writing a balanced net ionic equation for a redox reaction. The reaction occurs in basic solution, so we balance charges by adding OH^- ions if necessary.

Solve

1. Write a preliminary expression relating the known reactants and products:

$$Fe^{2+}(aq) + O_2(aq) \rightarrow Fe(OH)_3(s)$$

2. Perform a ΔO.N. analysis for iron and oxygen:

ΔO.N. = +1

$$\overset{+2}{Fe^{2+}}(aq) + \underset{0}{O_2}(g) \rightarrow Fe(\underset{-2}{O}H)_3(s) \quad (Fe: +3)$$

ΔO.N. = 2 × (−2) = −4

3. To balance these ΔO.N. values (and the number of electrons gained and lost), we need a coefficient of 4 in front of Fe^{2+}. This also requires a 4 in front of $Fe(OH)_3$ to balance the number of Fe atoms:

$$4\ Fe^{2+}(aq) + O_2(g) \rightarrow 4\ Fe(OH)_3(s)$$

4. To balance the electrical charges (8+ on the left, 0 on the right) we add 8 OH^- ions to the reactant side:

$$4\ Fe^{2+}(aq) + O_2(g) + 8\ OH^-(aq) \rightarrow 4\ Fe(OH)_3(s)$$

5. Taking an inventory of the atoms on both sides of the reaction arrow, we find that the right side has four more H atoms and two more O atoms than the left side. We can correct this imbalance by adding two molecules of H_2O to the left side:

$$4\ Fe^{2+}(aq) + O_2(g) + 8\ OH^-(aq) + 2\ H_2O(\ell) \rightarrow 4\ Fe(OH)_3(s)$$

Now the equation is balanced. We do not have to rebalance the redox part of the reaction after adding OH^- or H_2O because the oxidation states of hydrogen and oxygen do not change.

Think About It We weren't just lucky to need twice as many H atoms as O atoms in step 5, an imbalance easily resolved by adding the necessary number of H_2O molecules. Actually, 2:1 imbalances in H and O atoms are frequently encountered in the final step of writing a net ionic equation describing a redox reaction and are often an indicator that the first four steps were done correctly.

Practice Exercise Hydroperoxide ions, HO_2^-, react with permanganate ions, MnO_4^-, producing MnO_2 and O_2 gas. Write a balanced net ionic equation for this reaction in a basic solution.

8.7 Titrations

Aqueous-phase reactions can be used to determine the concentrations of dissolved substances with excellent precision and accuracy. The analytical technique that is used is called **titration**. It is classified as a *volumetric* method of analysis because it is based on measuring the volumes of both a standard solution and a sample solution. The standard solution, called the **titrant**, contains a known concentration of a substance that reacts with a substance in the sample. The goal is to precisely determine the concentration of this second substance, called the **analyte**. The chemistries of titration reactions are often based on acid–base neutralization reactions, though precipitation and redox reactions are also widely used.

To get a feel for how titrations work, let's consider an application from environmental science: analysis of acidic water draining from abandoned coal mines. Sulfide-containing minerals called pyrites are often present in and around seams of coal. Mining the coal exposes the pyrite to chemical and biochemical weathering. Those processes oxidize the sulfur in the sulfides to sulfuric acid, which dissolves in groundwater seeping through abandoned mines. To assess and control the environmental impact of this acidic water, scientists and engineers need to know how much sulfuric acid is dissolved in it. That's where titrations come in.

Suppose we have a sample of mine drainage containing an unknown concentration of sulfuric acid. We can use an acid–base titration in which the sulfuric acid in the water sample is neutralized by reacting it with a standard solution of strong base, such as NaOH. The neutralization reaction is

$$H_2SO_4(aq) + 2\ NaOH(aq) \rightarrow Na_2SO_4(aq) + 2\ H_2O(\ell)$$

titration an analytical method for determining the concentration of a solute in a sample by reacting the solute with a solution of known concentration.

titrant the standard solution added to the sample in a titration.

analyte the substance whose concentration is to be determined in a chemical analysis.

(a)

(b)

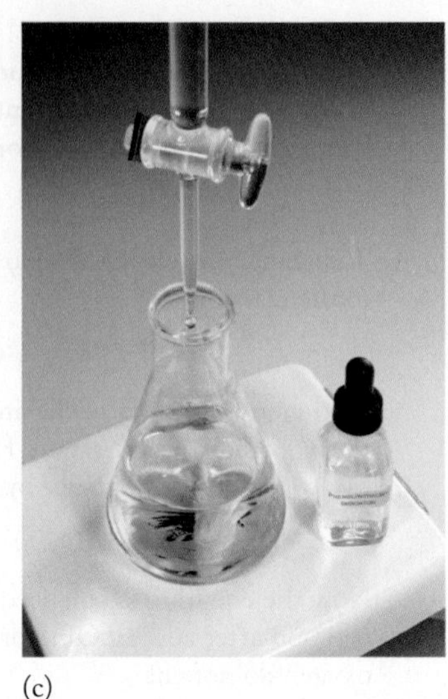
(c)

FIGURE 8.22 Determining a sulfuric acid concentration. (a) A known volume of a sample containing H_2SO_4 is placed in the flask. The buret is filled with the titrant, which is an aqueous NaOH solution of known concentration. A few drops of phenolphthalein indicator solution are added to the flask. (b) Titrant is carefully added to the flask until the indicator changes from colorless to faint pink, signaling that the acid has been neutralized and the end point has been reached. (c) If more titrant is added, the color becomes darker, signaling that more titrant than was needed to reach the end point has been added.

Note that both of the ionizable hydrogen atoms on each molecule of H_2SO_4 are neutralized in the reaction. Neutralizing both takes two moles of NaOH for every one mole of H_2SO_4.

To determine the concentration of analyte (sulfuric acid) in the sample, we need to measure the volume of titrant (standard solution of NaOH) needed to neutralize it. We start by carefully transferring a known volume of the sample into a reaction vessel such as an Erlenmeyer flask. A few drops of a color indicator are added to detect when just enough titrant has been added to neutralize the sulfuric acid (Figure 8.22). The titrant is added slowly using a buret, a narrow glass cylinder with volume markings and fitted with a stopcock to quickly stop the flow when just enough titrant has been added. This stage of a titration, when the reaction is just complete, is called the **equivalence point** because a stoichiometric quantity of titrant has been added that is *equivalent* to the quantity of the analyte.

The volume of titrant required to make the indicator change color may not be *exactly* the same as the volume of titrant needed to completely consume all of the analyte in the sample. The volume needed to make the indicator change color is actually the **end point** of the titration. It is the volume we record and use in calculating how much analyte was in the sample. With the right indicator and proper technique, the difference in volume between the equivalence point and the end point in acid–base titrations is insignificant.

Suppose it takes 22.40 mL of 0.00100 *M* NaOH to react completely with the H_2SO_4 in a 100.0 mL sample of acidic mine drainage. We know the volume and molarity of the NaOH standard solution, so we can adapt Equation 8.2 to calculate the number of moles of NaOH consumed after first converting the volume into liters:

$$n_{\text{NaOH}} = V_{\text{NaOH}} \times M_{\text{NaOH}}$$

$$= 0.02240\ \cancel{\text{L}} \times \frac{1.00 \times 10^{-3}\ \text{mol NaOH}}{\cancel{\text{L}}} = 2.240 \times 10^{-5}\ \text{mol NaOH}$$

equivalence point the point in a titration when just enough titrant has been added to react with all of the analyte in the sample.

end point the point in a titration when a color change or other signal indicates that enough titrant has been added to react with all of the analyte in the sample.

We also know from the stoichiometry of the reaction that two moles of NaOH are required to neutralize one mole of H_2SO_4, so the number of moles of H_2SO_4 in the 100.0 mL sample must have been

$$n_{\text{H}_2\text{SO}_4} = 2.240 \times 10^{-5}\ \cancel{\text{mol NaOH}} \times \frac{1\ \text{mol H}_2\text{SO}_4}{2\ \cancel{\text{mol NaOH}}} = 1.120 \times 10^{-5}\ \text{mol H}_2\text{SO}_4$$

Dividing this number of moles of H_2SO_4 by the volume of the sample (V_S) in liters gives us the molar concentration of H_2SO_4:

$$M_{H_2SO_4} = \frac{n_{H_2SO_4}}{V_S}$$

$$= \frac{1.120 \times 10^{-5} \text{ mol } H_2SO_4}{0.1000 \text{ L}} = \frac{1.120 \times 10^{-4} \text{ mol } H_2SO_4}{\text{L}}$$

$$= 1.12 \times 10^{-4}\, M\, H_2SO_4$$

To detect the end point in a sulfuric acid–sodium hydroxide titration, we might use the indicator *phenolphthalein*, which is colorless in acidic solutions but pink in basic solutions (Figure 8.22). The part of the titration that requires the most skill is adding just enough NaOH solution to reach the end point, the point at which a pink color first persists in the solution being titrated. To catch the end point precisely, the standard solution must be added no faster than one drop at a time, with thorough mixing between drops.

SAMPLE EXERCISE 8.13 Determining Solute Concentration from Titration Data **LO9**

Apple cider vinegar is an aqueous solution of acetic acid (CH_3COOH, $\mathcal{M} = 60.05$ g/mol) made from fermented apple juice. Commercial vinegar must contain at least 4 grams of acetic acid per 100 mL of vinegar. Suppose the titration of a 25.00 mL sample of vinegar requires 12.15 mL of 1.885 M NaOH. What is the molarity of acetic acid in the vinegar sample? Does it meet the 4 g/100 mL standard?

Collect and Organize We have the results of a titration of acetic acid in vinegar with a strongly basic titrant. We know the volume and concentration of the titrant and the volume of the sample and need to calculate the concentration of acetic acid in the sample.

Analyze We can use Equation 8.2 to calculate the number of moles of basic titrant consumed from its volume and concentration. We need a balanced chemical equation to relate the number of moles of titrant (base) to the number of moles of acid in the sample during the titration reaction. There is one ionizable H atom in a molecule of CH_3COOH (highlighted in red), and there is one OH^- ion per formula unit of NaOH. Therefore, the mole ratio of CH_3COOH to NaOH in the balanced chemical equation for the titration reaction is 1:1. The volume of titrant consumed is about half the sample volume, which means the concentration of acetic acid in the sample should be about half the concentration of NaOH in the titrant, or about 1 M.

Solve The balanced chemical equation for the neutralization reaction is

$$CH_3COOH(aq) + NaOH(aq) \rightarrow H_2O(\ell) + CH_3COONa(aq)$$

The number of moles of NaOH required to reach the end point is

$$n_{NaOH} = V_{NaOH} \times M_{NaOH}$$

$$= \left(12.15 \text{ }\cancel{\text{mL}} \times \frac{1 \text{ }\cancel{\text{L}}}{1000 \text{ }\cancel{\text{mL}}}\right) \times \frac{1.885 \text{ mol}}{\cancel{\text{L}}} = 0.02290 \text{ mol NaOH}$$

Converting moles of NaOH into moles of CH_3COOH:

$$n_{CH_3COOH} = 0.02290 \text{ }\cancel{\text{mol NaOH}} \times \frac{1 \text{ mol } CH_3COOH}{1 \text{ }\cancel{\text{mol NaOH}}} = 0.02290 \text{ mol } CH_3COOH$$

Dividing moles of CH_3COOH by the sample volume (in liters), we obtain the molar concentration of CH_3COOH in the sample:

$$M_{CH_3COOH} = \frac{n_{CH_3COOH}}{V_S}$$

$$= \frac{0.02290 \text{ mol } CH_3COOH}{25.00 \text{ mL}} \times \frac{1000 \text{ mL}}{1 \text{ L}} = 0.9160 \text{ mol/L} = 0.9160\ M$$

Converting this molar concentration into grams per 100 mL:

$$\frac{0.9160 \text{ mol}}{\text{L}} \times \frac{60.05 \text{ g}}{\text{mol}} \times \frac{1 \text{ L}}{1000 \text{ mL}} = 0.05501 \text{ g/mL} \quad \text{or} \quad 5.501 \text{ g/100 mL}$$

The sample meets the 4 g/100 mL standard for apple cider vinegar.

Think About It The calculated molar concentration of acetic acid in the vinegar sample was close to our 1 M estimate, and the equivalent concentration expressed in grams per 100 mL is just above the standard for apple cider vinegar, so our results are entirely reasonable.

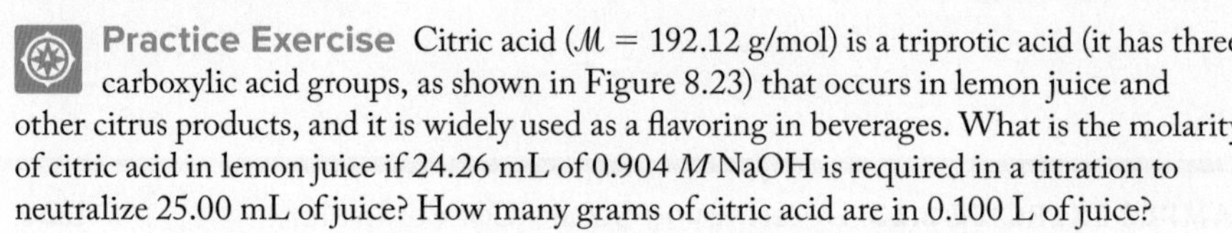

Practice Exercise Citric acid ($\mathcal{M}$ = 192.12 g/mol) is a triprotic acid (it has three carboxylic acid groups, as shown in Figure 8.23) that occurs in lemon juice and other citrus products, and it is widely used as a flavoring in beverages. What is the molarity of citric acid in lemon juice if 24.26 mL of 0.904 M NaOH is required in a titration to neutralize 25.00 mL of juice? How many grams of citric acid are in 0.100 L of juice?

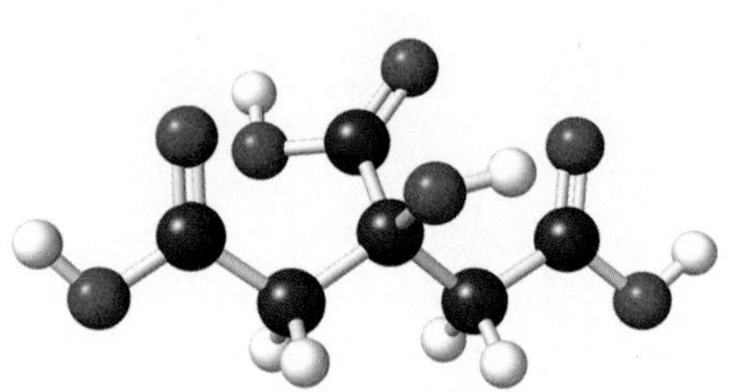

FIGURE 8.23 Citric acid, $C_6H_8O_7$.

8.8 Ion Exchange

In Section 8.5 we saw how dissolved ions can exchange partners, forming insoluble precipitates. Actually, ions exchanging partners in solution is at the heart of most aqueous-phase chemistry. It is also the basis for a method of water purification called **ion exchange**. Water containing certain metal ions—principally Ca^{2+} and Mg^{2+}—is called *hard water*, and it causes problems in industrial processes and in homes. Because hard water combines with soap to form a gray scum, clothes washed in hard water appear gray and dingy. Hard water forms scale (an incrustation) in boilers, pipes, and kettles, diminishing their ability to conduct heat and carry water, and hard water sometimes has an unpleasant taste.

To *soften* hard water (remove the ions responsible for its hardness), it is pumped through tanks filled with small plastic beads called resin, labeled "R" in Figure 8.24. In the tanks, Ca^{2+} and Mg^{2+} ions in the water are exchanged for Na^+ ions that are loosely bound to ion-exchange sites on the resin. In many resins these sites consist of carboxylate ($RCOO^-$) groups. As hard water flows through the resin, Ca^{2+} and Mg^{2+} ions exchange places with Na^+ ions on the resin because cations with 2+ and 3+ charges bind more strongly to the $RCOO^-$ groups than Na^+ ions. The following chemical equation illustrates this ion-exchange process:

$$2\,(RCOO^-)Na^+(s) + Ca^{2+}(aq) \rightarrow (RCOO^-)_2Ca^{2+}(s) + 2\,Na^+(aq)$$

Hard water that has been softened in this way contains increased concentrations of sodium ion. Although this may not be a problem for healthy children and adults, people suffering from high blood pressure often must limit their intake of Na^+ and so should not drink water softened by this kind of ion-exchange reaction.

ion exchange a process in which one ion is displaced by another.

zeolites natural crystalline minerals or synthetic materials consisting of three-dimensional networks of channels that contain sodium or other 1+ cations.

Naturally occurring minerals called **zeolites** are also capable of ion-exchange reactions. Zeolites are formed in nature as a result of chemical reactions between salt water and aluminum and silicon oxides in molten lava. They can also be

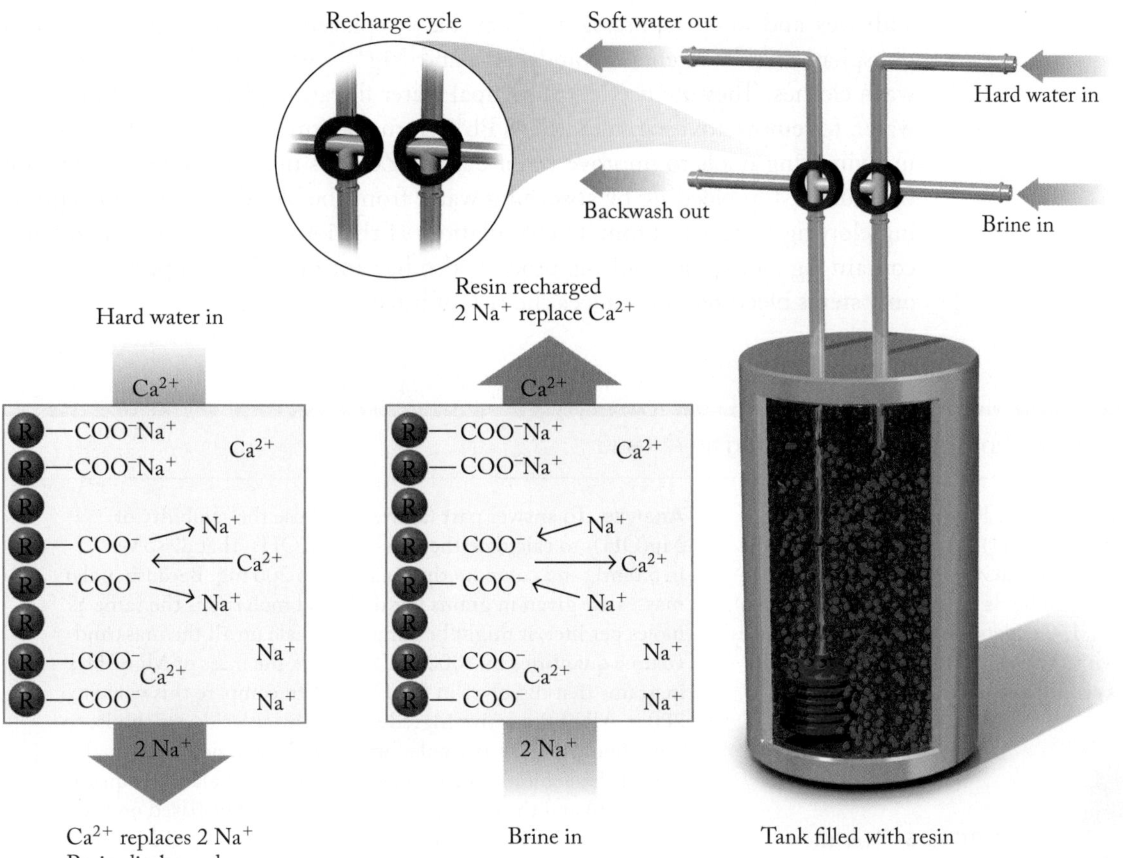

FIGURE 8.24 Residential water softeners use ion exchange to remove 2+ ions (such as Ca^{2+}) that make water hard. The ion-exchange resin contains cation-exchange sites that are initially occupied by Na^+ ions. These ions are replaced by 2+ "hardness" ions as water flows through the resin. Eventually most of the ion-exchange sites are occupied by 2+ ions, and the system loses its water-softening ability. The resin is then backwashed with a saturated solution of NaCl (*brine*), which displaces the hardness ions, restoring the resin to its Na^+ form.

synthesized by precipitating compounds called aluminosilicates from supersaturated solutions of aluminum and silicon oxides. On an atomic scale, natural zeolites consist of three-dimensional arrays of interconnecting tunnels and cages (Figure 8.25). The inner surfaces of the tunnels have negative charges that are balanced by cations electrostatically bound to them. They function as ion-exchange sites in much the same way that plastic resins do in water softeners: as water flows through tunnels (pores) lined with Na^+ ions, these ions exchange with Ca^{2+}, Mg^{2+}, and other cations dissolved in the water.

Zeolites are used in a variety of industrial processes and commercial products, from water softeners and purifiers to livestock feed

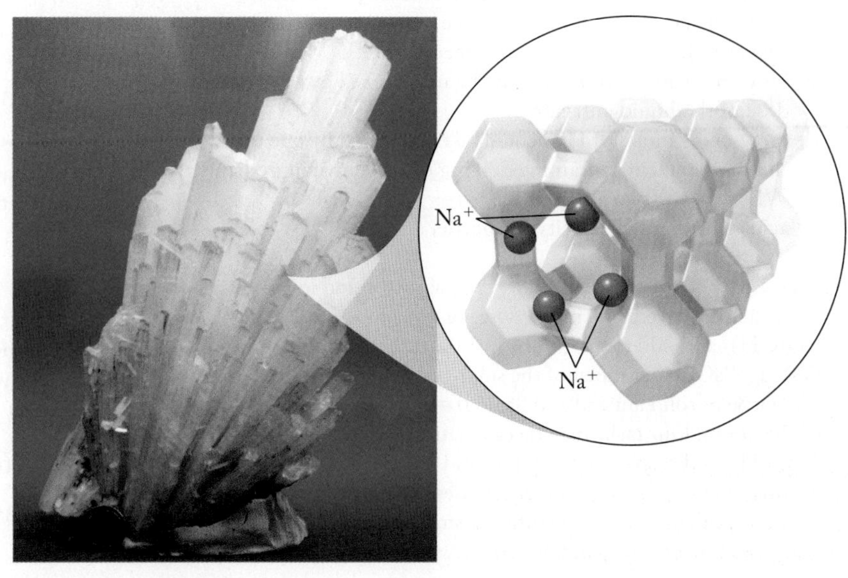

FIGURE 8.25 A sample of zeolite, which contains microscopic cages and tunnels that include cation-exchange sites, shown here in their Na^+ form.

additives and odor suppressants. They have replaced environmentally harmful phosphates in detergents to bind Ca^{2+} and Mg^{2+} and soften the water used to wash clothes. They are used in municipal water filtration plants to treat drinking water, to remove toxic metals such as Pb^{2+} from contaminated waste streams, and in swimming pools to improve water clarity. Zeolites may be poured directly on wounds to stop bleeding by absorbing water from the blood, thereby concentrating clotting factors to promote coagulation. If the ion-exchange sites in zeolites contain Ag^+ ions, the resulting product also has antimicrobial properties and not only stems bleeding but reduces the risk of infection.

SAMPLE EXERCISE 8.14 Integrating Concepts: Selecting an Antacid

A 5.0 mL dose of the liquid antacid in Figure 8.26(a) contains 200 mg $Mg(OH)_2$ and 200 mg $Al(OH)_3$. A tablet of the antacid in Figure 8.26(b) contains 500 mg $CaCO_3$. According to their labels, the bottle of liquid antacid holds 12 fluid ounces (355 mL), and the other bottle contains 150 tablets. Assume that the masses of the ingredients are known to two significant figures. In many drugstores, the two bottles sell for about the same price.

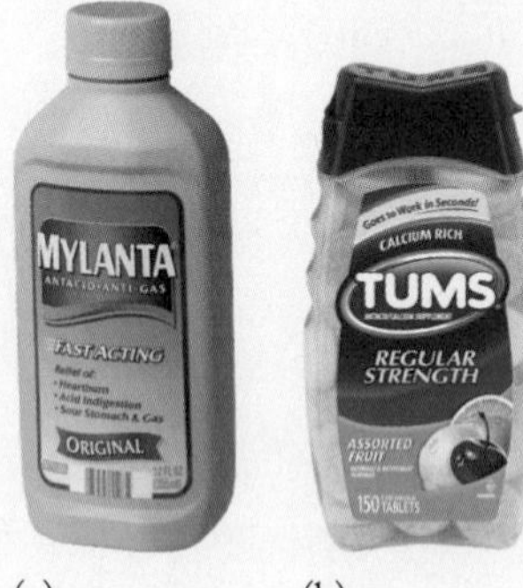

FIGURE 8.26 Over-the-counter antacids: (a) liquid Mylanta contains suspended $Mg(OH)_2$ and $Al(OH)_3$; (b) Tums tablets are composed mostly of $CaCO_3$.

a. If the maximum dissolved concentration (solubility) of $Mg(OH)_2$ is 1.2×10^{-4} *M*, what percentage of the $Mg(OH)_2$ in the liquid antacid is dissolved and not solid particles suspended in the viscous liquid?
b. Write net ionic equations for the neutralization of excess stomach acid, HCl(*aq*), by the three active ingredients in the two antacids. Assume that all three are solids.
c. How many moles of HCl could be neutralized by one 5.0 mL dose of the liquid antacid?
d. How many moles of HCl could be neutralized by one tablet of the solid antacid containing 500 mg of the active ingredient?
e. Which of the two antacids is a better buy in terms of acid-neutralizing capacity per bottle?

Collect and Organize We know the quantities and identities of the active ingredients in the doses of two antacids: 200 mg $Mg(OH)_2$ and 200 mg $Al(OH)_3$ in 5.0 mL of the liquid, and 500 mg $CaCO_3$ per tablet of the solid. There are 150 tablets in the bottle of solid antacid and 355/5.0 = 71 doses in one bottle of the liquid. Our tasks include calculating how much of the $Mg(OH)_2$ is dissolved in the liquid and not just suspended in it, writing balanced net ionic equations describing the neutralization reactions, calculating the quantity of stomach acid one dose of each antacid can neutralize, and determining which is the better buy.

Analyze To answer part (a), we could use the solubility of $Mg(OH)_2$ to calculate the mass of $Mg(OH)_2$ that dissolves in 5.0 mL and compare this quantity to 200 mg. Because molar masses are given in grams per mole and molarity is the same as moles per liter, it might be simpler to scale up all the mass and volume quantities by 1000 and calculate the mass of $Mg(OH)_2$ in grams that dissolves in 5.0 L and then compare this value to 200 g. All of the active ingredients in the antacids are solids, so they should appear as whole formula units and not individual ions in the net ionic equations in part (b). The balanced equations in part (b) and the calculations in part (c) will be based on the following stoichiometric considerations: (1) 1 mole of $Mg(OH)_2$ contains 2 moles of OH^- ions, so it reacts with 2 moles of HCl; (2) 1 mole of $Al(OH)_3$ contains 3 moles of OH^- ions, so it can neutralize 3 moles of HCl; (3) the 1 mole of CO_3^{2-} ions in a mole of $CaCO_3$ can accept 2 moles of H^+ ions, which means it can neutralize 2 moles of HCl. The molar masses of the three reactants are 58.32 g $Mg(OH)_2$/mol, 78.00 g $Al(OH)_3$/mol, and 100.09 g $CaCO_3$/mol.

Solve

a. If there are 200 mg of $Mg(OH)_2$ in 5.0 mL of the liquid antacid, then there are 200 g of $Mg(OH)_2$ in 5.0 L of it. Therefore, we convert the solubility of $Mg(OH)_2$ from molarity to the number of grams of it that will dissolve in 5.0 L:

$$5.0\ \cancel{L} \times \frac{1.2 \times 10^{-4}\ \cancel{mol\ Mg(OH)_2}}{\cancel{L}} \times \frac{58.32\ g\ Mg(OH)_2}{1\ \cancel{mol\ Mg(OH)_2}}$$
$$= 0.0350\ g\ Mg(OH)_2$$

We divide this result by 200 g, expressing the quotient as the percentage of $Mg(OH)_2$ that is dissolved (the rest is tiny particles in suspension):

$$\frac{0.0350\ g}{200\ g} \times 100\% = 0.017\%\ \text{dissolved}$$

b. Based on the stoichiometric ratios developed in the Analyze section, the molecular equations for the three neutralization reactions with the appropriate salts and moles of water are

$$Mg(OH)_2(s) + 2\ HCl(aq) \rightarrow MgCl_2(aq) + 2\ H_2O(\ell)$$
$$Al(OH)_3(s) + 3\ HCl(aq) \rightarrow AlCl_3(aq) + 3\ H_2O(\ell)$$
$$CaCO_3(s) + 2\ HCl(aq) \rightarrow CaCl_2(aq) + H_2O(\ell) + CO_2(g)$$

Separating the soluble ionic compounds into their free ions, we obtain total ionic equations:

$$Mg(OH)_2(s) + 2\,H^+(aq) + 2\,Cl^-(aq) \rightarrow Mg^{2+}(aq) + 2\,Cl^-(aq) + 2\,H_2O(\ell)$$

$$Al(OH)_3(s) + 3\,H^+(aq) + 3\,Cl^-(aq) \rightarrow Al^{3+}(aq) + 3\,Cl^-(aq) + 3\,H_2O(\ell)$$

$$CaCO_3(s) + 2\,H^+(aq) + 2\,Cl^-(aq) \rightarrow Ca^{2+}(aq) + 2\,Cl^-(aq) + H_2O(\ell) + CO_2(g)$$

In each of the preceding equations, Cl^- ions are the only spectator ions. Deleting them yields the following net ionic equations:

$$Mg(OH)_2(s) + 2\,H^+(aq) \rightarrow Mg^{2+}(aq) + 2\,H_2O(\ell)$$

$$Al(OH)_3(s) + 3\,H^+(aq) \rightarrow Al^{3+}(aq) + 3\,H_2O(\ell)$$

$$CaCO_3(s) + 2\,H^+(aq) \rightarrow Ca^{2+}(aq) + H_2O(\ell) + CO_2(g)$$

c. Calculating the number of moles of HCl neutralized by one dose (5.0 mL) of the liquid antacid (containing 200 mg of each active ingredient):

$$200\ \cancel{\text{mg Mg(OH)}_2} \times \frac{1\ \text{g}}{1000\ \cancel{\text{mg}}} \times \frac{1\ \cancel{\text{mol Mg(OH)}_2}}{58.32\ \cancel{\text{g Mg(OH)}_2}} \times \frac{2\ \text{mol HCl}}{1\ \cancel{\text{mol Mg(OH)}_2}} = 0.00686\ \text{mol HCl}$$

$$+\ 200\ \cancel{\text{mg Al(OH)}_3} \times \frac{1\ \text{g}}{1000\ \cancel{\text{mg}}} \times \frac{1\ \cancel{\text{mol Al(OH)}_3}}{78.00\ \cancel{\text{g Al(OH)}_3}} \times \frac{3\ \text{mol HCl}}{1\ \cancel{\text{mol Al(OH)}_3}} = 0.00769\ \text{mol HCl}$$

$$= 0.01455\ \text{mol HCl}$$

d. Calculating the number of moles of HCl neutralized by one tablet (500 mg) of the solid antacid:

$$500\ \cancel{\text{mg CaCO}_3} \times \frac{1\ \text{g}}{1000\ \cancel{\text{mg}}} \times \frac{1\ \cancel{\text{mol CaCO}_3}}{100.09\ \cancel{\text{g CaCO}_3}} \times \frac{2\ \text{mol HCl}}{1\ \cancel{\text{mol CaCO}_3}} = 0.00999\ \text{mol HCl}$$

The 5.0 mL dose of the liquid contains $0.01455/0.00999 \approx 1.5$ times more acid-neutralizing power than one tablet of the solid antacid.

e. There are 71 doses in one bottle of the liquid antacid, which have the capacity to neutralize (71 doses × 0.01455 mol HCl/dose) = 1.033 mol HCl. There are 150 tablets in a bottle of the solid antacid, which have the capacity to neutralize (150 tablets × 0.00999 mol HCl/tablet) = 1.50 mol HCl. Therefore, there is more neutralizing capacity in a bottle of the solid antacid, and it is the better buy.

Think About It All three active ingredients are solids with limited solubility in water. However, they react with—and dissolve in—acid solutions such as stomach acid. The solid antacid provides about half again as much acid-neutralizing capacity as the liquid for the same (or at least a comparable) price. In fairness to this particular liquid product, it also contains an ingredient that relieves gas.

SUMMARY

LO1 The concentration of solute in a solution can be expressed as mass of solute per mass of solution, such as milligrams of solute per kilogram of solution, which is the same as parts per million (ppm). Concentrations can also be expressed as mass of solute per volume of solution, or as moles of solute per liter of solution (mol/L), which is called **molarity (*M*)**. (Section 8.1)

LO2 During **dilution** the quantity of solute in a sample of a concentrated solution does not change, but adding solvent increases the volume of the sample and decreases the concentration of solute. (Sections 8.1 and 8.2)

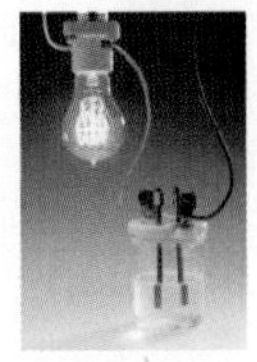

LO3 A solute that dissociates into ions in aqueous solution is called an **electrolyte**. Mobility of these ions makes the solution a better electrical conductor than pure water. **Strong electrolytes** dissociate completely in water, **weak electrolytes** dissociate partially, and **nonelectrolytes** do not dissociate at all. (Section 8.3)

LO4 **Brønsted–Lowry acids** are proton (H^+) donors and **Brønsted–Lowry bases** are proton acceptors. In a **neutralization reaction**, H^+ ions from the acid combine with OH^- ions from the base, forming H_2O and a **salt**. The **net ionic equation** of a reaction includes only the species that change during the reaction and omits the **spectator ions**. (Section 8.4)

LO5 In a precipitation reaction, mixing dissolved reactants produces an insoluble **precipitate** as cations and anions switch partners. (Section 8.5)

LO6 Balanced chemical equations and concentrations can be used to calculate the amount of precipitate formed when solutions are mixed. (Section 8.5)

LO7 In a redox reaction, substances either gain electrons and thereby undergo **reduction**, or they lose electrons and undergo **oxidation**. A reaction is a redox reaction if the **oxidation numbers (O.N.)**, or **oxidation states**, of elements in the reactants change during the reaction. (Section 8.6)

LO8 Oxidation and reduction are complementary processes; in balancing equations, the number of electrons lost by the substance being oxidized must match the number of electrons gained by the substance being reduced. (Section 8.6)

LO9 **Titrations** are volumetric methods of chemical analysis in which the concentrations of solutes called **analytes** are determined by reacting them with known volumes and concentrations of standard solutions called **titrants**. (Section 8.7)

PARTICULATE PREVIEW WRAP-UP

Most hydrogen atoms consist of 1 proton and 1 electron, so H^+ is literally a proton. CH_3COOH (molecule (c)) is a weak acid, meaning that a small fraction of these molecules ionize in water. Only the hydrogen atoms in COOH groups ionize; those in CH_3 groups do not ionize. None of the H atoms in molecules (a), (b), and (d) ionize in water.

PROBLEM-SOLVING SUMMARY

Type of Problem	Concepts and Equations	Sample Exercises
Converting concentration units	Use the solute/solvent ratios in concentration units as conversion factors. Use mol/L to represent molarity.	**8.1, 8.2**
Calculating mass of solute to prepare a solution	$m = (V \times M) \times \mathcal{M} = \left(\cancel{L} \times \frac{\cancel{mol}}{\cancel{L}}\right) \times \frac{g}{\cancel{mol}}$ (8.3)	**8.3**
Preparing dilute solutions	$V_{initial} = \frac{V_{final} \times C_{final}}{C_{initial}}$ (8.7)	**8.4**
Writing neutralization reaction equations	Balance the total ionic equation, using separate ions to represent strong electrolytes. Delete spectator ions to write the net ionic equation. Balance the number of H^+ ions donated by the acid and accepted by the base.	**8.5**
Writing a net ionic equation for a precipitation reaction	Balance the total ionic equation, writing the precipitates as solids. Delete spectator ions to write the net ionic equation.	**8.6**
Predicting the mass of a precipitate	Identify the limiting reactant if necessary. Use the stoichiometry of the net ionic equation to calculate moles of precipitate, and then convert moles to mass.	**8.7**
Calculating solute concentration from mass of a precipitate	Convert precipitate mass into moles by dividing by its molar mass. Convert moles of precipitate into moles of solute. Calculate molarity of the solute in the sample by dividing by the volume of the sample in liters.	**8.8**
Determining oxidation numbers (O.N.)	Follow the rules in Table 8.5. In general, the O.N. of a pure element is 0, the O.N. of a monatomic ion is the charge on the ion, and the O.N. values of O and H in most compounds are −2 and +1, respectively.	**8.9**
Identifying oxidizing and reducing agents and determining number of electrons transferred	The oxidizing agent contains an atom whose O.N. decreases during the reaction; the reducing agent contains an atom whose O.N. increases; the changes in O.N. determine the number of electrons transferred.	**8.10**
Balancing redox reaction equations	Follow the steps in Table 8.7. In general, use coefficients to balance ΔO.N. values (the gains and losses of electrons). In aqueous solutions, balance ionic charges by adding H^+ ions (in acidic solutions) or OH^- ions (in basic solutions). Balance H and O atoms by adding H_2O.	**8.11, 8.12**
Determining solute concentration from titration data	Calculate moles of titrant from its concentration and the volume needed to reach the end point. Use the stoichiometry of the titration reaction to calculate moles of analyte and then divide by sample volume to calculate analyte concentration.	**8.13**

VISUAL PROBLEMS

(Answers to boldface end-of-chapter questions and problems are in the back of the book.)

8.1. Figure P8.1 shows a solution containing three binary (HX) acids. One of them is a weak acid and the other two are strong acids. Which color sphere is formed by ionization of the weak acid?

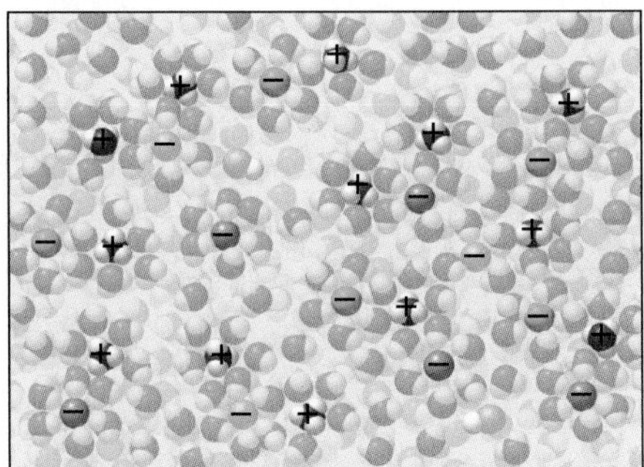

FIGURE P8.1

8.2. Figure P8.2 represents the products of the reaction that occurs when solutions of NaCl and $AgNO_3$ are mixed together. The silver spheres represent Ag^+ ions. Which spheres represent (a) Na^+; (b) Cl^-; (c) NO_3^- ions?

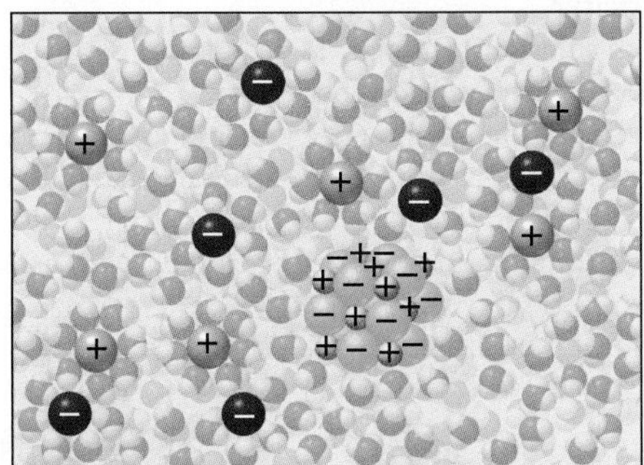

FIGURE P8.2

8.3. Figure P8.3 represents the reaction mixture at the equivalence point in a titration of a sample of battery acid (H_2SO_4) using a standard solution of NaOH as the titrant. Which ion is represented by the green spheres and which by the blue spheres?

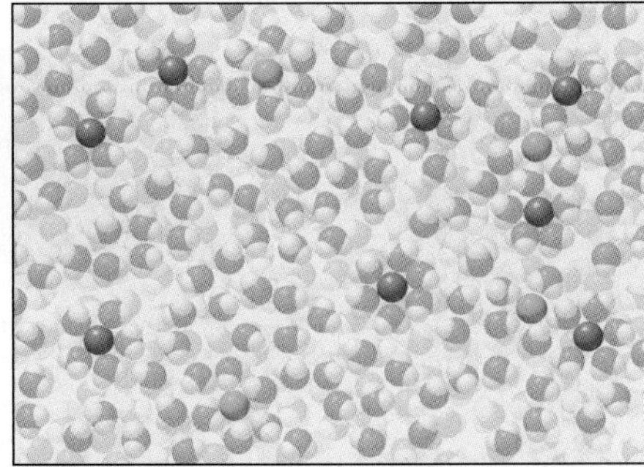

FIGURE P8.3

8.4. Suppose that two antacid tablets each containing 0.50 g $CaCO_3$ ($\mathcal{M}$ = 100 g/mol) react with 0.02 mol HCl. Which of the images in Figure P8.4, in which green spheres represent cations and blue spheres represent anions, accurately reflects the population of ions in solution after the reaction is over?

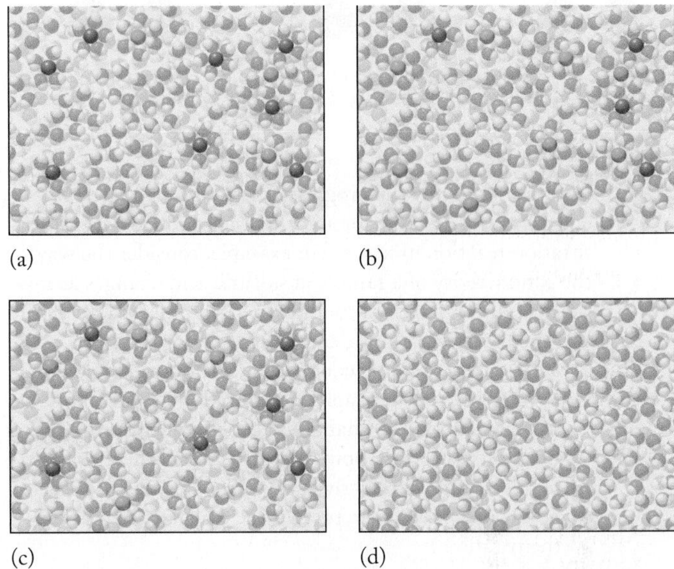

FIGURE P8.4

8.5. What is the change in oxidation state of nitrogen in the reaction described by the molecular models in Figure P8.5? (Blue spheres are nitrogen; red spheres are oxygen.)

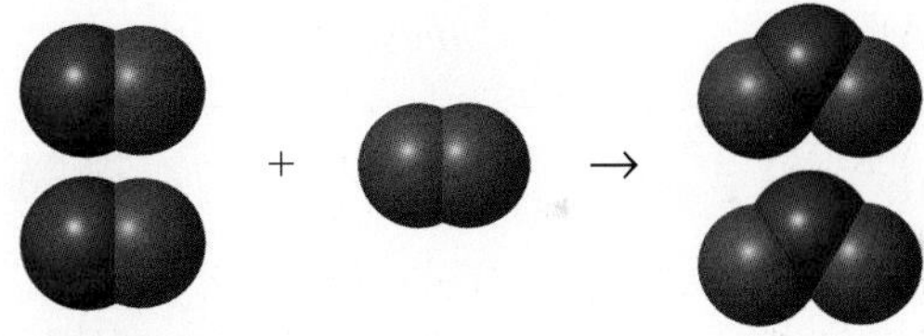

FIGURE P8.5

8.6. Identify the oxidizing and reducing agents in the reaction described by the molecular models in Figure P8.5.

8.7. Rank the nitrogen oxide molecules depicted by the models in Figure P8.7 in order of decreasing oxidation number of the nitrogen (blue spheres) in them. (The red spheres are oxygen atoms.)

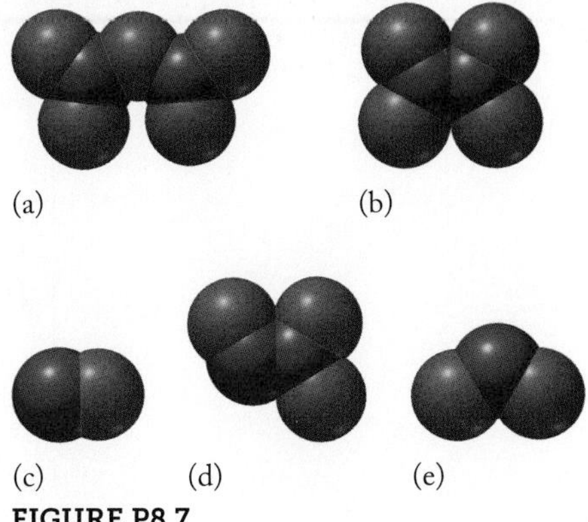

FIGURE P8.7

8.8. Rank the hydrocarbons depicted by the molecular models in Figure P8.8 based on decreasing oxidation state of the carbon (black spheres) in them.

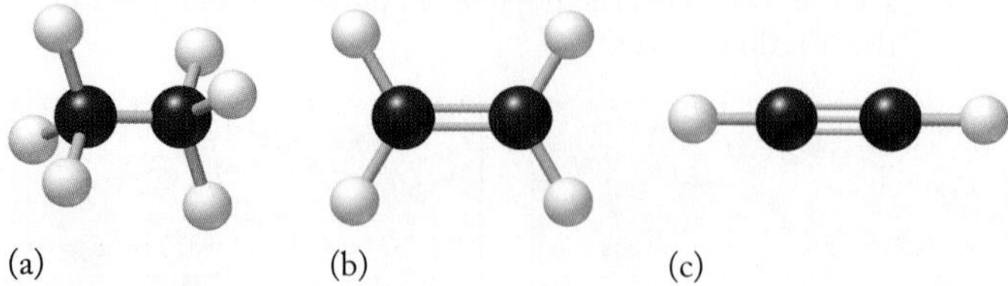

FIGURE P8.8

***8.9.** One way to follow the progress of a titration and detect its equivalence point is by monitoring the conductivity of the titration reaction mixture. For example, consider the way the conductivity of a sample of sulfuric acid changes as it is titrated with a standard solution of barium hydroxide before and then after the equivalence point.

a. Write the total ionic equation for the titration reaction.
b. Which of the four graphs in Figure P8.9 comes closest to representing the changes in conductivity during the titration? (The zero point on the y-axis of these graphs represents the conductivity of pure water; the break points on the x-axis represent the equivalence point in the titration.)

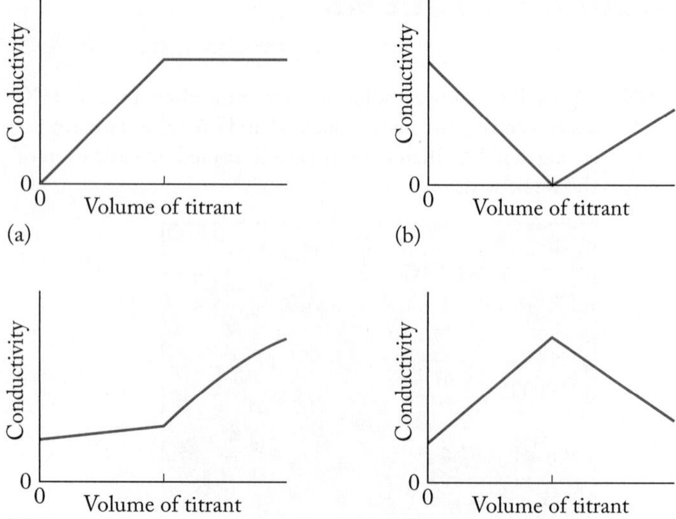

FIGURE P8.9

8.10. Use representations [A] through [I] in Figure P8.10 to answer questions a–f.

a. Which solids dissolve in water?
b. Which solids dissolve to create a solution that conducts electricity?
c. Which solutions can neutralize barium hydroxide? What would the spectator ions be?
d. Which solutions will form a precipitate upon mixing?
e. Name an aqueous solution that could be added to [A] to form the precipitate in [E].
f. Which combination of [A], [B], and [C] will cause a redox reaction?

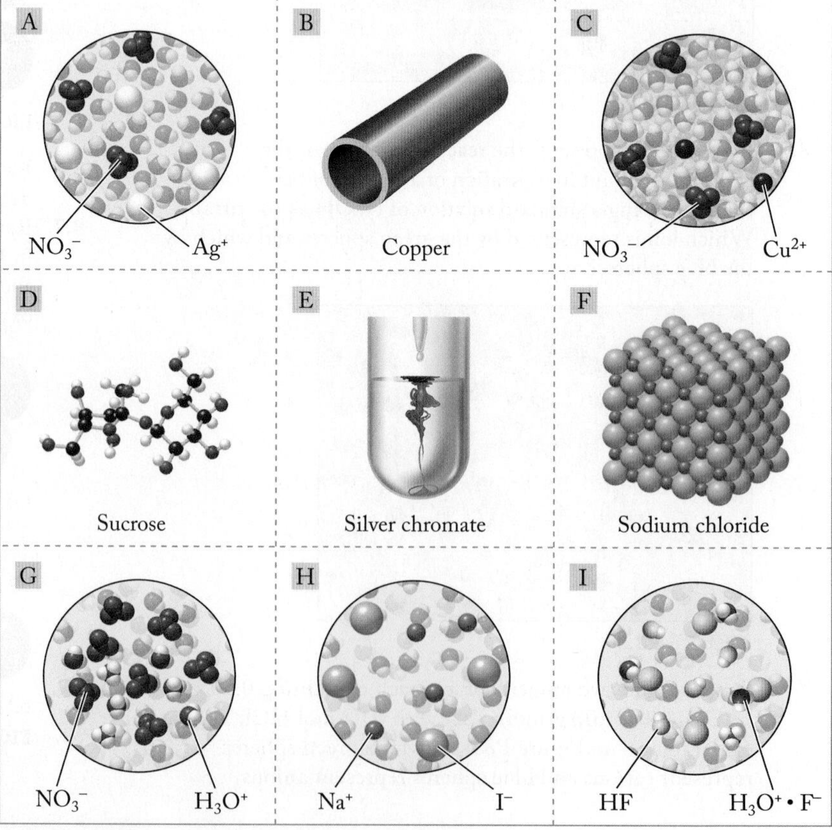

FIGURE P8.10

QUESTIONS AND PROBLEMS

Solutions and Their Concentrations

Problems

8.11. Calculate the molarity of each of the following solutions:
a. 0.56 mol $BaCl_2$ in 100.0 mL of solution
b. 0.200 mol Na_2CO_3 in 200.0 mL of solution
c. 0.325 mol $C_6H_{12}O_6$ in 250.0 mL of solution
d. 1.48 mol KNO_3 in 250.0 mL of solution

8.12. Calculate the molarity of each of the following solutions:
a. 0.150 mol urea $[(NH_2)_2CO]$ in 250.0 mL of solution
b. 1.46 mol $NH_4CH_3CO_2$ in 1.000 L of solution
c. 1.94 mol methanol (CH_3OH) in 5.000 L of solution
d. 0.045 mol sucrose ($C_{12}H_{22}O_{11}$) in 50.0 mL of solution

---

8.13. Calculate the molarity of Na^+ ions in each of the following solutions:
a. 0.29 M $NaNO_3$
b. 0.33 g NaCl in 25 mL of solution
c. 0.88 M Na_2SO_4
d. 0.46 g Na_3PO_4 in 100 mL of solution

8.14. Calculate the molarity of each of the following solutions:
a. 31.76 g $LiClO_4$ in 475.0 mL of solution
b. 6.37 g $(NH_4)_2SO_4$ in 250.0 mL of solution
c. 2.97 g KBr in 75.0 mL of solution
d. 0.773 g $Pb(NO_3)_2$ in 100.0 mL of solution

8.15. How many grams of solute are needed to prepare each of the following solutions?
a. 1.000 L of 0.200 M NaCl
b. 250.0 mL of 0.125 M $CuSO_4$
c. 500.0 mL of 0.400 M CH_3OH

8.16. How many grams of solute are needed to prepare each of the following solutions?
a. 500.0 mL of 0.250 M KBr
b. 25.0 mL of 0.200 M $NaNO_3$
c. 100.0 mL of 0.375 M CH_3OH

8.17. **River Water** The Mackenzie River in northern Canada contains, on average, 0.820 mM Ca^{2+}, 0.430 mM Mg^{2+}, 0.300 mM Na^+, 0.0200 M K^+, 0.250 mM Cl^-, 0.380 mM SO_4^{2-}, and 1.82 mM HCO_3^-. What, on average, is the total mass of these ions in 2.75 L of Mackenzie River water?

8.18. Zinc, copper, lead, and mercury ions are toxic to Atlantic salmon at concentrations of 6.42×10^2 mM, 7.16×10^{-3} mM, 0.965 mM, and 5.00×10^{-2} mM, respectively. What are the corresponding concentrations in milligrams per liter?

8.19. Calculate the number of moles of solute contained in the following volumes of aqueous solutions of four over-the-counter pharmaceuticals:
a. 0.250 L of 0.076 M acetaminophen (for pain relief)
b. 2.11 L of 0.193 mM chromalyn sodium (for hay fever)
c. 0.0475 L of 5.73 mM benzocaine (in cough syrup)
d. 14.6 L of 27.4 mM Benadryl (antihistamine)

8.20. A sample of crude oil contains 3.13 mM naphthalene, 12.0 mM methylnaphthalene, 23.8 mM dimethylnaphthalene, and 14.1 mM trimethylnaphthalene. What is the total number of moles of all the naphthalene compounds in 100.0 mL of the oil?

8.21. The pesticide DDT ($C_{14}H_9Cl_5$) was banned in the United States in 1972 because of dire environmental impacts. Just prior to its being banned, analyses of groundwater samples in Pennsylvania between 1969 and 1971 yielded the following results:

Location	Sample	Mass of DDT
Orchard	250.0 mL	0.030 mg
Residential	1.750 L	0.035 mg
Residential after a storm	50.0 mL	0.57 mg

Express these data in units of millimoles of DDT per liter.

8.22. Pesticide concentrations in the portion of the Rhine River that flows between Germany and France averaged 0.55 mg/L of hexachlorobenzene (C_6Cl_6), 0.06 mg/L of dieldrin ($C_{12}H_8Cl_6O$), and 1.02 mg/L of hexachlorocyclohexane ($C_6H_6Cl_6$) between 1969 and 1975. Express these concentrations in millimoles per liter.

8.23. Tap water in North America from groundwater sources contains an average of 48 mg/L Ca^{2+} ion. What is the molarity of calcium ion in this water?

*8.24. The concentration of manganese in one brand of soluble plant fertilizer is 0.05% by mass. If a 20 g sample of the fertilizer is dissolved in 2.0 L of solution, what is the molarity of dissolved Mn in the solution?

***8.25.** For which of the following compounds is it possible to make a 1.0 M solution at 0°C?
a. $CuSO_4 \cdot 5\,H_2O$, solubility = 23.1 g/100 mL
b. $AgNO_3$, solubility = 122 g/100 mL
c. $Fe(NO_3)_2 \cdot 6\,H_2O$, solubility = 113 g/100 mL
d. $Ca(OH)_2$, solubility = 0.185 g/100 mL

8.26. **Gold in the Ocean** About 6×10^9 g of gold is thought to be dissolved in the oceans of the world. If the total volume of the oceans is 1.5×10^{21} L, what is the average molarity of gold in seawater?

8.27. The concentration of Mg^{2+} in a sample of coastal seawater is 1.09 g/kg. What is the molarity of Mg^{2+} in this seawater if the seawater has a density of 1.02 g/mL?

8.28. **Hemoglobin in Blood** A typical adult body contains 6.0 L of blood. The hemoglobin content of blood is about 15.5 g/100.0 mL of blood. The approximate molar mass of hemoglobin is 64,500 g/mol. How many moles of hemoglobin are present in a typical adult?

Dilutions

Problems

8.29. Calculate the final concentrations of the following aqueous solutions after each has been diluted to a final volume of 25.0 mL:
a. 1.00 mL of 0.452 M Na^+
b. 2.00 mL of 3.4 mM LiCl
c. 5.00 mL of 6.42×10^{-2} mM Zn^{2+}

8.30. Chemists who analyze samples for dissolved trace elements may buy standard solutions that contain 1.000 g/L concentrations of the elements. If a chemist needs to prepare 0.500 L of a working standard that has a concentration of 5.00 mg/L, what volume of the 1.000 g/L standard is needed?

***8.31.** A puddle of coastal seawater, caught in a depression formed by some coastal rocks at high tide, begins to evaporate on a hot summer day as the tide goes out. If the volume of the puddle decreases to 23% of its initial volume, what is the concentration of Na^+ after evaporation if it was 0.449 *M* initially?

8.32. What volume of 2.5 *M* $SrCl_2$ is needed to prepare 500.0 mL of 5.0 m*M* solution?

***8.33.** **Dilution of Adult-Strength Cough Syrup** A standard dose of an over-the-counter cough suppressant for adults is 20.0 mL. A portion this size contains 35 mg of the active pharmaceutical ingredient (API). A pediatrician prescribes the drug for a 6-year-old child, but the child may take only 10.0 mL at a time and receive a maximum of 4.00 mg of the API. What is the concentration in mg/mL of the adult-strength medication, and how many milliliters of it should be diluted to make 100.0 mL of child-strength cough syrup?

*8.34. **Mixing Fertilizer** The label on a bottle of "organic" liquid fertilizer concentrate states that it contains 8 grams of phosphate per 100.0 mL and that 16 fluid ounces should be diluted with water to make 32 gallons of fertilizer to be applied to growing plants. What is the phosphate concentration in grams per liter in the diluted fertilizer? (1 gallon = 128 fluid ounces.)

Electrolytes and Nonelectrolytes

Concept Review

8.35. A solution of table salt is a good conductor of electricity, but a solution containing an equal molar concentration of table sugar is not. Why?

8.36. Why can scientists use conductivity to study the mixing of freshwater and seawater in estuaries?

8.37. What are present in solutions of electrolytes that are not present in solutions of nonelectrolytes?

8.38. **Fuel Cells** The electrolyte in an electricity-generating device called *a fuel cell* consists of a mixture of Li_2CO_3 and K_2CO_3 heated to 650°C. At this temperature these ionic solids melt. Explain how the mixture of molten carbonates can conduct electricity.

***8.39.** Rank the following solutions on the basis of their ability to conduct electricity, starting with the most conductive: (a) 1.0 *M* NaCl; (b) 1.2 *M* KCl; (c) 1.0 *M* Na_2SO_4; (d) 0.75 *M* LiCl.

8.40. Rank the conductivities of 1 *M* aqueous solutions of each of the following solutes, starting with the most conductive: (a) acetic acid; (b) methanol; (c) sucrose (table sugar); (d) hydrochloric acid.

Problems

8.41. What is the molarity of Na^+ ions in a 0.025 *M* aqueous solution of (a) NaBr; (b) Na_2SO_4; (c) Na_3PO_4?

8.42. What is the molarity of each ion in a 0.035 *M* aqueous solution of (a) NH_4Cl; (b) Li_2SO_4; (c) $MgBr_2$?

Acids, Bases, and Neutralization Reactions

Concept Review

8.43. What chemical property of an acid makes it an acid?

8.44. What is the difference between a strong acid and a weak acid?

8.45. Give the formulas of two strong acids and two weak acids.

8.46. Why is $HSO_4^-(aq)$ a weaker acid than H_2SO_4?

8.47. What chemical property of a base makes it a base?

8.48. What is the difference between a strong base and a weak base?

8.49. Give the formulas of two strong bases and two weak bases.

8.50. Write the net ionic equation for the neutralization of a strong acid by a strong base.

Problems

8.51. For each of the following acid–base reactions, identify the acid and the base and then write the net ionic equation:

a. $H_2SO_4(aq) + Ca(OH)_2(s) \rightarrow CaSO_4(aq) + 2\,H_2O(\ell)$
b. $PbCO_3(s) + H_2SO_4(aq) \rightarrow PbSO_4(s) + CO_2(g) + H_2O(\ell)$
c. $Ca(OH)_2(s) + 2\,CH_3COOH(aq) \rightarrow$ $Ca(CH_3COO)_2(aq) + 2\,H_2O(\ell)$

8.52. Complete and balance each of the following neutralization reactions, name the products, and write the net ionic equations.

a. $HI(aq) + LiOH(aq) \rightarrow$
b. $H_3PO_4(aq) + KOH(aq) \rightarrow$
c. $Al(OH)_3(s) + CH_3COOH(aq) \rightarrow$
d. $HNO_3(aq) + Ba(OH)_2(aq) \rightarrow$

8.53. Write a balanced molecular equation and a net ionic equation for the following reactions:

a. Solid magnesium hydroxide reacts with a solution of sulfuric acid.
b. Solid magnesium carbonate reacts with a solution of hydrochloric acid.
c. Ammonia gas reacts with hydrogen chloride gas.

8.54. Write a balanced molecular equation and a net ionic equation for the following reactions:

a. Solid aluminum hydroxide reacts with a solution of hydrobromic acid.
b. A solution of sulfuric acid reacts with solid sodium carbonate.
c. A solution of calcium hydroxide reacts with a solution of nitric acid.

8.55. **Toxicity of Lead Pigments** The use of lead(II) carbonate and lead(II) hydroxide as white pigments in paint was discontinued in the United States in 1978 because these compounds dissolved in the stomachs of young children who ingested paint chips. The Pb^{2+} ions released when the compounds dissolve interfere with neurotransmissions in the brain, causing neurological disorders. Using net ionic equations, show why lead(II) carbonate and lead(II) hydroxide dissolve in acidic solutions.

8.56. **Lawn Care** Many homeowners treat their lawns with $CaCO_3(s)$ to reduce the acidity of the soil. Write a net ionic equation for the reaction of $CaCO_3(s)$ with a strong acid.

Precipitation Reactions

Concept Review

8.57. What is the difference between a saturated solution and a supersaturated solution?

8.58. Calcium nitrate is soluble, but calcium phosphate is not. Explain this difference in solubility

8.59. Rain and snow are commonly called "precipitation." In what way are they like a solid that forms in an aqueous solution?

8.60. A precipitate may appear when two completely clear aqueous solutions are mixed. What circumstances are responsible for this event?

8.61. Is a saturated solution always a concentrated solution? Explain.

8.62. Honey is a concentrated solution of sugar molecules in water. Clear, viscous honey becomes cloudy after being stored for long periods. Explain how this transition illustrates supersaturation.

Problems

8.63. According to the solubility rules in Table 8.4, which of the following compounds have limited solubility in water? (a) barium sulfate; (b) barium hydroxide; (c) lanthanum nitrate; (d) sodium acetate; (e) lead hydroxide; (f) calcium phosphate

8.64. **Ocean Vents** The black "smoke" that flows out of deep ocean hydrothermal vents (Figure P8.64) is made of insoluble metal sulfides suspended in seawater. Of the following cations that are present in the water flowing up through these vents, which ones could contribute to the formation of the black smoke? Na^+, Li^+, Mn^{2+}, Fe^{2+}, Ca^{2+}, Mg^{2+}, Zn^{2+}, Pb^{2+}, Cu^{2+}

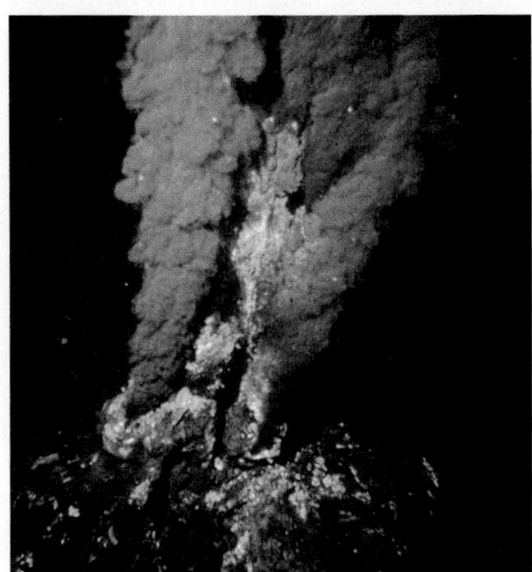

FIGURE P8.64

8.65. Complete and balance the chemical equations for the precipitation reactions, if any, between the following pairs of reactants, and write the net ionic equations:

a. $Pb(NO_3)_2(aq) + Na_2SO_4(aq) \rightarrow$
b. $NiCl_2(aq) + NH_4NO_3(aq) \rightarrow$
c. $FeCl_2(aq) + Na_2S(aq) \rightarrow$
d. $MgSO_4(aq) + BaCl_2(aq) \rightarrow$

*8.66. **Wastewater Treatment** Show, with appropriate net ionic reactions, how Cr^{3+} and Cd^{2+} can be removed from wastewater by treatment with solutions of sodium hydroxide.

***8.67.** An aqueous solution containing Ca^{2+}, Cl^-, CO_3^{2-}, and NO_3^- is allowed to evaporate. Which compound will precipitate first?

8.68. The solubility of $CaCl_2$ in water at 25°C is 81.1 g/100 mL; at 0°C its solubility decreases to 59.5 g/100 mL. Answer the following questions about an aqueous solution of 26.4 g of $CaCl_2$ in a volume of 37.5 mL:

a. At 25°C, is this a saturated solution?
b. If the solution is cooled to 0°C, do you expect a precipitate to form?
c. If the solution is slowly cooled to 0°C and no precipitate forms, then what kind of solution is it?

8.69. Calculate the mass of $MgCO_3$ precipitated by mixing 10.0 mL of a 0.200 *M* Na_2CO_3 solution with 5.00 mL of 0.0500 *M* $Mg(NO_3)_2$ solution.

*8.70. **Phosphates in Sewage** Eutrophication, the rapid growth of algae and the death of fish, may be caused by the presence of an excess of phosphates in water. Treatment plants that process sewage may add $Ca(OH)_2$ (slaked lime) to water to remove phosphates before returning the water to the environment. Although the phosphates may be present in several forms, we can use HPO_4^- as a representative phosphate in the net ionic equation:

$$5\ Ca^{2+}(aq) + 3\ HPO_4^-(aq) + 4\ OH^-(aq) \rightarrow Ca_5OH(PO_4)_3(s) + 3\ H_2O(\ell)$$

If phosphates (as HPO_4^-) are present at a level of 15.7 mg/L in wastewater, how much $Ca(OH)_2$ would need to be added to 1.00×10^5 L of water to precipitate 95% of the phosphate ion present?

8.71. Iron(II) can be precipitated from a slightly basic aqueous solution by bubbling oxygen through the solution, which converts Fe^{2+} to insoluble Fe^{3+}:

$$4\ Fe(OH)^+(aq) + 4\ OH^-(aq) + O_2(g) + 2\ H_2O(\ell) \rightarrow 4\ Fe(OH)_3(s)$$

How many grams of O_2 are consumed to precipitate all of the iron in 75 mL of 0.090 *M* Fe^{2+}?

8.72. Given the following equation, how many grams of $PbCO_3$ will dissolve when 1.00 L of 1.00 *M* H^+ is added to 5.00 g of $PbCO_3$?

$$PbCO_3(s) + 2\ H^+(aq) \rightarrow Pb^{2+}(aq) + H_2O(\ell) + CO_2(g)$$

***8.73.** **Analysis of Mineral Water** German mineral water typically contains 108–130 mg/L magnesium, whereas Italian varieties contain 50–60 mg/L. A 125.0 mL portion of mineral water of unknown origin is reacted with exactly 3.76 mL of a 0.0753 *M* solution of ammonium hydrogen phosphate [$(NH_4)_2HPO_4$] to produce 6.875 mg of magnesium ammonium phosphate ($MgNH_4PO_4$) as a solid precipitate. Is the mineral water likely to come from Germany, from Italy, or from some other source?

8.74. **Rhubarb Leaves** Rhubarb leaves contain 0.520 g of oxalic acid ($H_2C_2O_4$) per 100.0 g of leaves. The oxalic acid can react with calcium ion to make insoluble calcium oxalate, a major constituent of kidney stones. To study this reaction in the laboratory, 375 mL of a 0.866 *M* solution of oxalic acid is treated with excess 0.133 *M* calcium hydroxide solution.

a. How much calcium oxalate is formed in this reaction?
*b. What volume of calcium hydroxide solution is required if it must be present in 20% excess to ensure complete reaction?

Oxidation–Reduction Reactions

Concept Review

8.75. How are the gains or losses of electrons related to changes in oxidation numbers?

8.76. What is the sum of the oxidation numbers of the atoms in a molecule?

8.77. What is the sum of the oxidation numbers of all the atoms in each of the following polyatomic ions? (a) OH^-; (b) NH_4^+; (c) SO_4^{2-}; (d) PO_4^{3-}

8.78. Gold does not dissolve in concentrated H_2SO_4 but readily dissolves in H_2SeO_4 (selenic acid). Which acid is the stronger oxidizing agent?

8.79. Silver dissolves in sulfuric acid to form silver sulfate and H_2, but gold does not dissolve in sulfuric acid to form gold sulfate. Which of the two metals is the better reducing agent?

8.80. Rank the following oxoanions in order of decreasing oxidation number of chlorine: (a) ClO^-; (b) ClO_2^-; (c) ClO_3^-; (d) ClO_4^-.

***8.81.** The generic formula of normal alkanes is C_nH_{2n+2}. How does the oxidation state of carbon in these compounds change with increasing n?

8.82. Why is the oxidizing agent in a redox reaction reduced and the reducing agent oxidized?

Problems

8.83. What is the oxidation number of chlorine in each of the following oxoacids? (a) hypochlorous acid (HClO); (b) chloric acid ($HClO_3$); (c) perchloric acid ($HClO_4$)

8.84. What is the oxidation number of nitrogen in each of the following species? (a) elemental nitrogen (N_2); (b) hydrazine (N_2H_4); (c) ammonium ion (NH_4^+)

8.85. What is the change (if any) in the oxidation state of carbon in this reaction?

$$C_{12}H_{22}O_{11}(s) \rightarrow 12\ C(s) + 11\ H_2O(\ell)$$

8.86. **Nitrogen Cycle** In one stage of the nitrogen cycle, *Nitrosomonas* bacteria convert ammonia and oxygen into nitrite ions.

a. What is the change in oxidation state of nitrogen during the reaction?
b. Write a balanced net ionic equation for the reaction in acidic groundwater.

8.87. **Natural Weathering of Ores** Iron is oxidized in a number of chemical weathering processes. How many moles of O_2 are consumed when one mole of magnetite (Fe_3O_4) is converted into hematite (Fe_2O_3)?

*8.88. The mineral rhodochrosite [manganese(II) carbonate, $MnCO_3$] is a commercially important source of manganese. How many moles of O_2 are consumed when one mole $MnCO_3$ is converted into MnO_2 and CO_2?

8.89. **Earth's Crust** The following chemical reactions have helped to shape Earth's crust. Determine the oxidation numbers of all the elements in the reactants and products, and identify which elements are oxidized and which are reduced:

a. $3\ SiO_2(s) + 2\ Fe_3O_4(s) \rightarrow 3\ Fe_2SiO_4(s) + O_2(g)$
b. $SiO_2(s) + 2\ Fe(s) + O_2(g) \rightarrow Fe_2SiO_4(s)$
c. $4\ FeO(s) + O_2(g) + 6\ H_2O(\ell) \rightarrow 4\ Fe(OH)_3(s)$

8.90. Determine the oxidation numbers of each of the elements in the following reactions, and identify which of them, if any, are oxidized or reduced:

a. $SiO_2(s) + 2\ H_2O(\ell) \rightarrow H_4SiO_4(aq)$
b. $2\ MnCO_3(s) + O_2(g) \rightarrow 2\ MnO_2(s) + 2\ CO_2(g)$
c. $3\ NO_2(g) + H_2O(\ell) \rightarrow 2\ NO_3^-(aq) + NO(g) + 2\ H^+(aq)$

8.91. How many moles of O_2 are consumed in the conversion of one mole of $FeCO_3$ to each of the following compounds? Assume CO_2 is also produced. (a) Fe_2O_3; (b) Fe_3O_4

8.92. Uranium is found in Earth's crust as UO_2 and in an assortment of compounds containing UO_2^{n+} cations. How many moles of electrons are transferred in the conversion of one mole of UO_2 to each of the following species? In which of the conversions is uranium oxidized? (a) $UO_2(CO_3)_3^{4-}(aq)$; (b) $UO_2(HPO_4)_2^{2-}(aq)$

8.93. Nitrogen in the hydrosphere is found primarily as ammonium ions and nitrate ions. Complete and balance the following chemical equation for the oxidation of ammonium ions to nitrate ions in acid solution:

$$NH_4^+(aq) + O_2(g) \rightarrow NO_3^-(aq)$$

8.94. In sediments and waterlogged soil, dissolved O_2 concentrations are so low that the microorganisms living there must rely on other sources of oxygen for respiration. Some bacteria can extract the oxygen from sulfate ions, reducing the sulfur in them to hydrogen sulfide gas and giving the sediments or soil a distinctive rotten-egg odor.

a. What is the change in oxidation state of sulfur as a result of this reaction?
b. Write the balanced net ionic equation for the reaction under acidic conditions, which releases O_2 from sulfate and forms hydrogen sulfide gas.

8.95. The solubilities of Fe and Mn compounds in freshwater streams are affected by changes in the oxidation states of these metals. Complete and balance the following redox reaction for slightly acidic freshwater:

$$Fe(OH)_2^+(aq) + Mn^{2+}(aq) \rightarrow MnO_2(s) + Fe^{2+}(aq)$$

8.96. A method for determining the quantity of dissolved oxygen in natural waters requires a series of redox reactions. Balance the following chemical equations in that series under the conditions indicated:

a. $Mn^{2+}(aq) + O_2(g) \rightarrow MnO_2(s)$ (basic solution)
b. $MnO_2(s) + I^-(aq) \rightarrow Mn^{2+}(aq) + I_2(s)$ (acidic solution)
c. $I_2(s) + S_2O_3^{2-}(aq) \rightarrow I^-(aq) + S_4O_6^{2-}(aq)$ (neutral solution)

8.97. Silver can be extracted from rocks using cyanide ion. Complete and balance the following reaction for this process:

$$Ag(s) + CN^-(aq) + O_2(g) \rightarrow Ag(CN)_2^-(aq) \quad \text{(basic solution)}$$

8.98. Permanganate ion (MnO_4^-) is used in water purification to remove oxidizable substances. Complete and balance the following reactions for the removal of sulfide, cyanide, and sulfite. Assume that reaction conditions are basic:

a. $MnO_4^-(aq) + S^{2-}(aq) \rightarrow MnS(s) + S_8(s)$
b. $MnO_4^-(aq) + CN^-(aq) \rightarrow MnO_2(s) + CNO^-(aq)$
c. $MnO_4^-(aq) + SO_3^{2-}(aq) \rightarrow MnO_2(s) + SO_4^{2-}(aq)$

8.99. **Biocide Chemistry** The water-soluble gas ClO_2 is known as an oxidative biocide. It destroys bacteria by oxidizing their cell walls and viruses by attacking their viral envelopes. ClO_2 may be prepared for use as a decontaminating agent from several different starting materials in slightly acidic solutions. Complete and balance the following chemical reactions for the synthesis of ClO_2.

a. $ClO_3^-(aq) + SO_2(g) \rightarrow ClO_2(g) + SO_4^{2-}(aq)$
b. $ClO_3^-(aq) + Cl^-(aq) \rightarrow ClO_2(g) + Cl_2(g)$
c. $ClO_3^-(aq) + Cl_2(g) \rightarrow ClO_2(g) + O_2(g)$

*8.100. Toxic cyanide ions can be removed from wastewater by adding hypochlorite:

$$2\ CN^-(aq) + 5\ OCl^-(aq) + H_2O(\ell) \rightarrow N_2(g) + 2\ HCO_3^-(aq) + 5\ Cl^-(aq)$$

a. Identify the oxidizing agent in this reaction.
b. How many liters of 0.125 M OCl^- are required to remove the CN^- in 3.4×10^6 L of wastewater in which the CN^- concentration is 0.58 mg/L?

Titrations

Problems

8.101. How many milliliters of 0.100 M NaOH are required to neutralize the following solutions?
a. 10.0 mL of 0.0500 M HCl
b. 25.0 mL of 0.126 M HNO_3
c. 50.0 mL of 0.215 M H_2SO_4

8.102. How many milliliters of 0.100 M HNO_3 are needed to neutralize the following solutions?
a. 45.0 mL of 0.667 M KOH
b. 58.5 mL of 0.0100 M $Al(OH)_3$
c. 34.7 mL of 0.775 M NaOH

***8.103.** The solubility of slaked lime, $Ca(OH)_2$, in water at 20°C is 0.185 g/100.0 mL. What volume of 0.00100 M HCl is needed to neutralize 10.0 mL of a saturated $Ca(OH)_2$ solution?

8.104. The solubility of magnesium hydroxide, $Mg(OH)_2$, in water is 9.0×10^{-4} g/100.0 mL. What volume of 0.00100 M HNO_3 is required to neutralize 1.00 L of a saturated $Mg(OH)_2$ solution?

8.105. **Chlorinity of Seawater** Scientists can precisely determine the chloride ion concentration, or *chlorinity*, of seawater samples using a titration called the Mohr method. The titrant is a solution of $AgNO_3$. The indicator is a few drops of K_2CrO_4 solution, which imparts a light yellow color to the titration mixture before the equivalence point (Figure P8.105a). However, after all the Cl^- ions have been precipitated as AgCl, the first excess of Ag^+ ions combines with CrO_4^{2-} ions to form Ag_2CrO_4, which makes the milky suspension of AgCl look pink (Figure P8.105b). If it takes 27.80 mL of 0.5000 M $AgNO_3$ to titrate a 25.00 mL sample of seawater, what is the concentration of Cl^- in the sample? Express your answer in mM and in g/kg (the density of the sample is 1.025 g/mL).

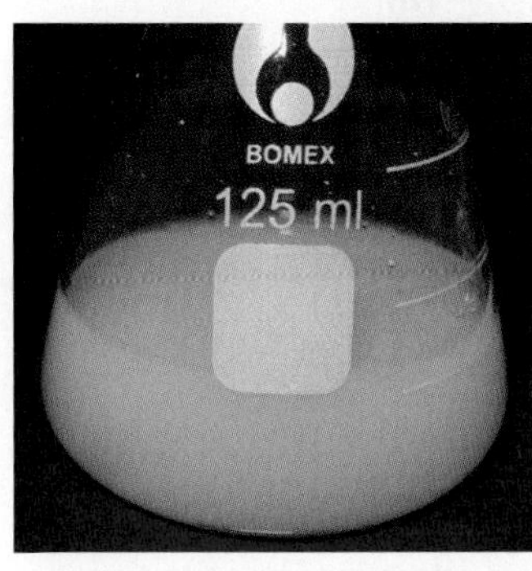

(a)

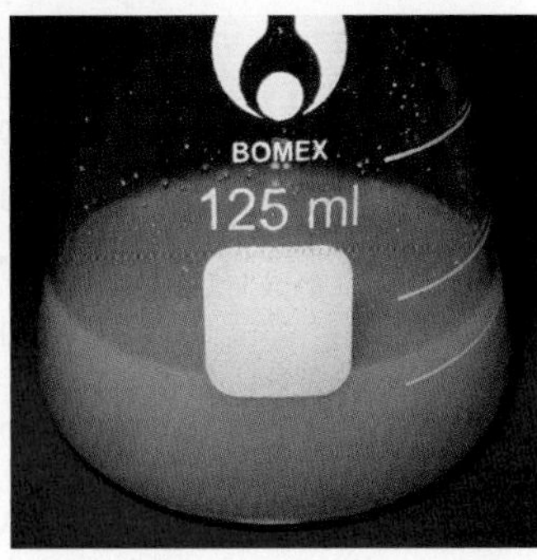

(b)

FIGURE P8.105

8.106. **Exercise Physiology** Lactic acid accumulates in muscle when glucose is metabolized under conditions where oxygen is the limiting reagent. Lactic acid has one carboxylic acid functional group per molecule (Figure P8.106). To determine the concentration of a solution of lactic acid, a chemist titrates a 20.00 mL sample of it with 0.1010 M NaOH and finds that 12.77 mL of titrant is required to reach the equivalence point. What is the molarity of the lactic acid solution?

FIGURE P8.106

Ion Exchange

Concept Review

8.107. Explain how a mixture of anion and cation exchangers can be used to deionize water.

8.108. Describe the process by which the ion exchanger in a home water softener is regenerated for further use.

8.109. What would be the advantage in using a K^+ resin rather than a Na^+ resin to soften water?

8.110. A piece of Zn metal is placed in a solution containing Cu^{2+} ions. At the surface of the Zn metal, Cu^{2+} ions react with Zn atoms, forming Cu atoms and Zn^{2+} ions. Is this reaction an example of ion exchange? Explain why or why not.

Additional Problems

8.111. To determine the concentration of SO_4^{2-} ion in a sample of groundwater, 100.0 mL of the sample is titrated with 0.0250 M $Ba(NO_3)_2$, forming insoluble $BaSO_4$. If 3.19 mL of the $Ba(NO_3)_2$ solution is required to reach the end point of the titration, what is the molarity of the SO_4^{2-}?

8.112. **Antifreeze** Ethylene glycol is the common name for the liquid used to keep the coolant in automobile cooling systems from freezing. It is 38.7% carbon, 9.7% hydrogen, and 51.6% oxygen by mass. Its molar mass is 62.07 g/mol, and its density is 1.106 g/mL at 20°C.
a. What is the empirical formula of ethylene glycol?
b. What is the molecular formula of ethylene glycol?
c. In a solution prepared by mixing equal volumes of water and ethylene glycol, which ingredient is the solute and which is the solvent?

***8.113.** According to the label on a bottle of concentrated hydrochloric acid, the contents are 36.0% HCl by mass and have a density of 1.18 g/mL.
a. What is the molarity of this concentrated HCl?
b. What volume of it would you need to prepare 0.250 L of 2.00 M HCl?
c. What mass of sodium hydrogen carbonate would be needed to neutralize the spill if a bottle containing 1.75 L of this concentrated HCl dropped on a lab floor and broke open?

8.114. **Synthesis and Toxicity of Chlorine** Chlorine was first prepared in 1774 by heating a mixture of NaCl and MnO_2 in sulfuric acid:

$$NaCl(aq) + H_2SO_4(aq) + MnO_2(s) \rightarrow Na_2SO_4(aq) + MnCl_2(aq) + H_2O(\ell) + Cl_2(g)$$

a. Assign oxidation numbers to the elements in each compound, and balance the redox reaction in acid solution.
b. Write a net ionic equation describing the reaction for the formation of chlorine.

c. If chlorine gas is inhaled, it causes pulmonary edema (fluid in the lungs) because it reacts with water in the alveolar sacs of the lungs to produce the strong acid HCl and the weaker acid HOCl. Balance the equation for the conversion of Cl_2 to HCl and HOCl.

***8.115.** When a solution of dithionite ions ($S_2O_4^{2-}$) is added to a solution of chromate ions (CrO_4^{2-}), the products of the ensuing chemical reaction that occurs under basic conditions include soluble sulfite ions and solid chromium(III) hydroxide. This reaction is used to remove Cr^{6+} from wastewater generated by factories that make chrome-plated metals.

a. Write the net ionic equation for this redox reaction.
b. Which element is oxidized and which is reduced?
c. Identify the oxidizing and reducing agents in the reaction.
d. How many grams of sodium dithionite would be needed to remove the Cr^{6+} in 100.0 L of wastewater that contains 0.00148 *M* chromate ion?

*8.116. A prototype battery based on iron compounds with large, positive oxidation numbers was developed in 1999. In the following reactions, assign oxidation numbers to the elements in each compound and balance the redox reactions in basic solution:

a. $FeO_4^{2-}(aq) + H_2O(\ell) \rightarrow FeOOH(s) + O_2(g) + OH^-(aq)$
b. $FeO_4^{2-}(aq) + H_2O(\ell) \rightarrow Fe_2O_3(s) + O_2(g) + OH^-(aq)$

***8.117.** **Polishing Silver** Silver tarnish is the result of silver metal reacting with sulfur compounds, such as H_2S, and O_2 in the air. The tarnish on silverware (Ag_2S) can be removed by soaking the silverware in a slightly basic solution of $NaHCO_3$ (baking soda) in a basin lined with aluminum foil.

a. Write a balanced chemical equation for the tarnish formation reaction.
b. Write a balanced net ionic equation for the tarnish removal process in which Ag_2S reacts with Al metal, forming $Al(OH)_3(s)$, Ag metal, and HS^- ions.

8.118. Give the formulas of the acids formed in the following chemical reactions of chlorine oxides.

a. $ClO + H_2O \rightarrow ? + ?$
b. $Cl_2O + H_2O \rightarrow HCl + ?$
c. $Cl_2O_6 + H_2O \rightarrow ? + ?$

8.119. Many nonmetal oxides react with water to form acidic solutions. Give the formulas of the acids produced in the following reactions:

a. $P_4O_{10} + 6\,H_2O \rightarrow ?$
b. $SeO_2 + H_2O \rightarrow ?$
c. $B_2O_3 + 3\,H_2O \rightarrow ?$

8.120. Write net ionic equations for the reactions that occur when

a. a sample of acetic acid is titrated with a solution of KOH.
b. a solution of sodium carbonate is mixed with a solution of calcium chloride.
c. calcium oxide dissolves in water.

8.121. One way to determine the concentration of hypochlorite ions (ClO^-) in solution is by first reacting them with I^- ions. Under acidic conditions the products of the reaction are I_2 and Cl^- ions. Then the I_2 produced in the first reaction is titrated with a solution of thiosulfate ions ($S_2O_3^{2-}$). The products of the titration reaction are $S_4O_6^{2-}$ and I^- ions. Write net ionic equations for the two reactions.

*8.122. **Fluoride Ion in Drinking Water** Sodium fluoride is added to drinking water in many municipalities to protect teeth against cavities. The target of the fluoridation is hydroxyapatite, $Ca_{10}(PO_4)_6(OH)_2$, a compound in tooth enamel. There is concern, however, that fluoride ions in water may contribute to skeletal fluorosis, an arthritis-like disease.

a. Write a net ionic equation for the reaction between hydroxyapatite and sodium fluoride that produces fluorapatite, $Ca_{10}(PO_4)_6F_2$.
b. The U.S. EPA currently restricts the concentration of F^- in drinking water to 4 mg/L. Express this concentration of F^- in molarity.
c. One study of skeletal fluorosis suggests that drinking water with a fluoride concentration of 4 mg/L for 20 years raises the fluoride content in bone to 6 mg/g, a level at which a patient may experience stiff joints and other symptoms. How much fluoride (in milligrams) is present in a 100 mg sample of bone with this fluoride concentration?

***8.123.** **Rocket Fuel in Drinking Water** Near Las Vegas, NV, improper disposal of perchlorates used to manufacture rocket fuel contaminated a stream flowing into Lake Mead, the largest artificial lake in the United States and a major supply of drinking and irrigation water for the American Southwest. The U.S. EPA has proposed an advisory range for perchlorate concentrations in drinking water of 4 to 18 μg/L. The perchlorate concentration in the stream averages 700.0 μg/L, and the stream flows at an average rate of 161 million gallons per day (1 gal = 3.785 L).

a. What are the formulas of sodium perchlorate and ammonium perchlorate?
b. How many kilograms of perchlorate flow from the Las Vegas stream into Lake Mead each day?
c. What volume of perchlorate-free lake water would have to mix with the stream water each day to dilute the stream's perchlorate concentration from 700.0 to 4 μg/L?
d. Since 2003, the states of Maryland, Massachusetts, and New Mexico have limited perchlorate concentrations in drinking water to 0.1 μg/L. Five replicate samples were analyzed for perchlorates by laboratories in each state, and the following data (μg/L) were collected:

MD	MA	NM
1.1	0.90	1.2
1.1	0.95	1.2
1.4	0.92	1.3
1.3	0.90	1.4
0.9	0.93	1.1

Which of the labs produced the most precise analytical results?

*8.124. **Analyzing Fruit Juice for Vitamin C** The amount of ascorbic acid (vitamin C) in fruit juice is determined by a titration using a redox reaction. Iodine is the titrant, but because iodine solutions in water are unstable, the iodine is generated by a reaction in which a titrant containing iodate (IO_3^-) ions is added to a solution containing the juice sample, iodide ions (I^-), and a few drops of a starch solution that turns dark blue in the presence of iodine. Iodate ions are reduced to iodine (I_2) while iodide ions are oxidized to iodine:

i. $IO_3^-(aq) + I^-(aq) \rightarrow I_2(aq)$ (not balanced)

However, any iodine formed by reaction (i) is immediately reduced back to iodide by ascorbic acid from the juice sample:

ii. $C_6H_8O_6(aq) + I_2(aq) \rightarrow C_6H_6O_6(aq) + 2\,I^-(aq) + 2H^+(aq)$

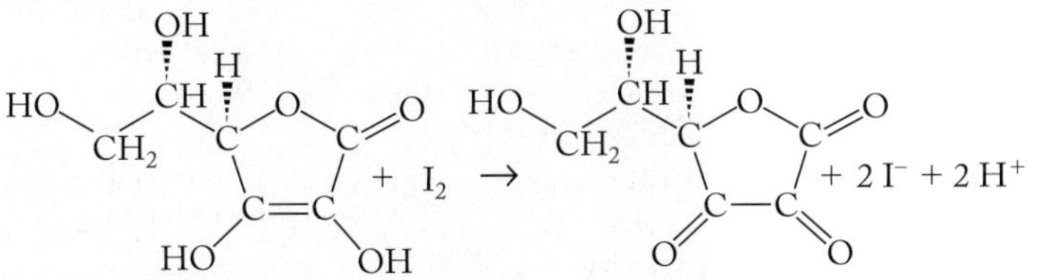

FIGURE P8.124

As long as there is ascorbic acid in the sample, no iodine can last in the solution and the starch indicator does not turn blue. As more iodate titrant is added, more iodine is generated by reaction (i) and consumed by reaction (ii), which also consumes the ascorbic acid in the sample. When enough iodate has been added and all of the ascorbic acid has been consumed, the next drop of iodate titrant generates iodine that is not consumed, and the starch indicator in the test solution turns blue. This is the end point of the titration.

a. Write a balanced chemical equation based on the reaction described previously in (i).
b. Combine equations (i) and (ii) to show the overall reaction in the titration.
c. A 25 mL portion of lemon juice is mixed with iodide solution, HCl, and water to a total volume of 100.0 mL. A 0.002455 M solution of KIO_3 is used as a titrant. A volume of 10.47 mL of titrant is required to reach the end point of the titration. What is the concentration in mg/L of ascorbic acid in the lemon juice?

***8.125. Making Apple Cider Vinegar** Some people who prefer natural foods make their own apple cider vinegar. They start with freshly squeezed apple juice that contains about 6% natural sugars. These sugars, which all have nearly the same empirical formula, CH_2O, are fermented with yeast in a chemical reaction that produces equal numbers of moles of ethanol (CH_3CH_2OH) and carbon dioxide. The product of fermentation, called hard cider, undergoes an acid fermentation step in which ethanol and dissolved oxygen gas react together to form acetic acid (CH_3COOH) and water. This acetic acid is the principal solute in vinegar.

a. Write a balanced chemical equation for the fermentation of natural sugars to ethanol and carbon dioxide. You may use in the equation the empirical formula given in the preceding paragraph.
b. Write a balanced chemical equation for the acid fermentation of ethanol to acetic acid.
c. What are the oxidation states of carbon in the reactants and products of the two fermentation reactions?
d. If a sample of apple juice contains 1.00×10^2 g of natural sugar, what is the maximum quantity of acetic acid that could be produced by the two fermentation reactions?

8.126. A food chemist determines the concentration of acetic acid in a sample of apple cider vinegar (see Problem 8.125) by acid–base titration. What is the concentration of acetic acid in the vinegar if the density of the sample is 1.01 g/mL, the titrant is 1.002 M NaOH, and the average volume of titrant required to titrate 25.00 mL subsamples of the vinegar is 20.78 mL? Express your answer the way a food chemist probably would: as percent by mass.

8.127. Cave Formations The stalactites and stalagmites in most caves are made of calcium carbonate (see Figure 8.10). In the Lower Kane Cave in Wyoming, however, they are made of gypsum (calcium sulfate). The presence of $CaSO_4$ is explained by the following sequence of reactions:

$$H_2S(aq) + 2\,O_2(g) \rightarrow H_2SO_4(aq)$$

$$H_2SO_4(aq) + CaCO_3(s) \rightarrow CaSO_4(s) + H_2O(\ell) + CO_2(g)$$

a. Which (if either) of these reactions is a redox reaction?
b. Write a net ionic equation for the reaction of H_2SO_4 with $CaCO_3$.
c. How would the net ionic equation be different if the reaction were written as follows?

$$H_2SO_4(aq) + CaCO_3(s) \rightarrow CaSO_4(s) + H_2CO_3(aq)$$

8.128. **Gardening Chemistry** Dolomite is a mixed carbonate mineral (Figure P8.128) with the formula $MgCa(CO_3)_2$. Gardeners add dolomite granules to soil and potting mixes to reduce acidity and provide a source of Mg^{2+} ions, which plants need to grow. Would a 50-pound bag of dolomite granules neutralize more acid than a 50-pound bag of limestone ($CaCO_3$) granules? How much more? Express your answer as a percentage.

FIGURE P8.128

8.129. Which of the following reactions of calcium compounds is or are redox reactions?

a. $CaCO_3(s) \rightarrow CaO(s) + CO_2(g)$
b. $CaO(s) + SO_2(g) \rightarrow CaSO_3(s)$
c. $CaCl_2(s) \rightarrow Ca(s) + Cl_2(g)$
d. $3\,Ca(s) + N_2(g) \rightarrow Ca_3N_2(s)$

8.130. HF is prepared by reacting CaF_2 with H_2SO_4:

$$CaF_2(s) + H_2SO_4(\ell) \rightarrow 2\,HF(g) + CaSO_4(s)$$

HF can be electrolyzed, in turn, when dissolved in molten KF to produce fluorine gas:

$$2\,HF(\ell) \rightarrow F_2(g) + H_2(g)$$

Fluorine is extremely reactive, so it is typically sold as a 5% mixture by volume in an inert gas such as helium. How much CaF_2 is required to produce 500.0 L of 5% F_2 in helium? Assume the density of F_2 gas is 1.70 g/L.

smartwork5

If your instructor uses Smartwork5, log in at digital.wwnorton.com/atoms2.

9

Thermochemistry

Energy Changes in Chemical Reactions

BEACH BONFIRE Burning wood for recreational and other purposes involves the combustion of cellulose and other large molecules rich in carbon and hydrogen. The products are CO_2, H_2O, and a considerable quantity of energy.

PARTICULATE **REVIEW**

Phase Changes and Energy

In Chapter 9, we explore the energy changes that accompany both physical and chemical changes. Particulate representations of the three phases of water are shown here.

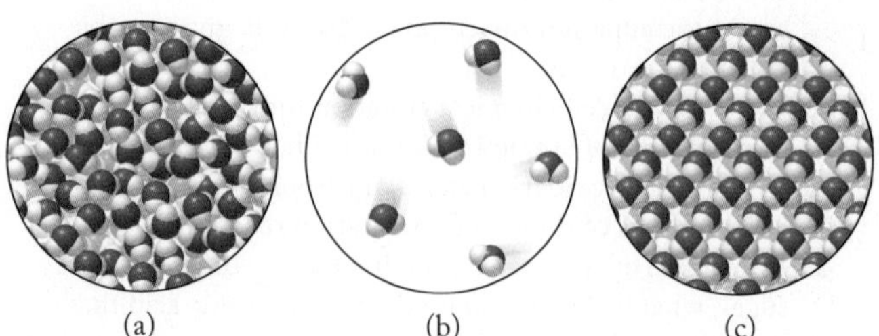

- Which representation depicts the solid phase of water? The liquid? The gaseous?
- Is energy added or released during the physical change from (a) to (b)? What intermolecular forces are involved?
- Describe the energy changes that accompany the physical changes from (a) to (c) and from (c) to (a).

(Review Section 1.4 and Section 6.2 if you need help answering these questions.)

(Answers to Particulate Review questions are in the back of the book.)

PARTICULATE **PREVIEW**

Breaking Bonds and Energy Changes

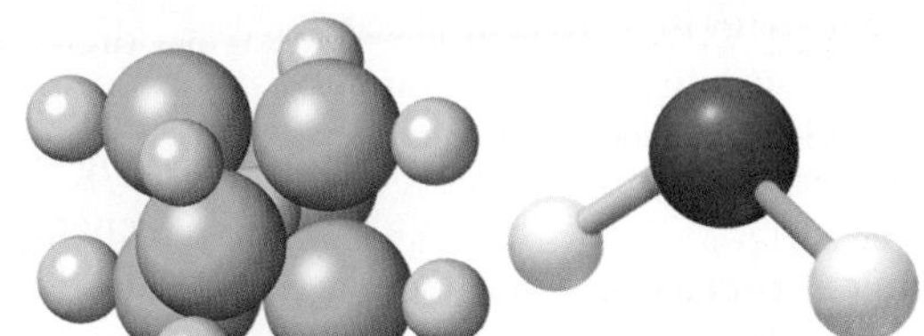

Calcium chloride, shown in the accompanying figure, is used to melt ice on sidewalks. As you read Chapter 9, look for ideas that will help you understand the energy changes that accompany the breaking and forming of bonds.

- What kind of bonds must be broken for calcium chloride to dissolve in water? Is energy absorbed or released in order to break these bonds?
- Which color spheres represent the chloride ions? Label the polar covalent bonds in water using $\delta+$ and $\delta-$.
- What intermolecular interactions form as the salt dissolves? Is energy absorbed or released as these attractions form?

Learning Outcomes

LO1 Distinguish between isolated, closed and open thermodynamic systems and between endothermic and exothermic processes
Sample Exercise 9.1

LO2 Relate changes in the internal energies of thermodynamic systems to heat flows and work done
Sample Exercises 9.2, 9.3

LO3 Calculate the heat gained or lost during changes in temperature and physical state
Sample Exercises 9.4, 9.5

LO4 Use calorimetry data to calculate enthalpies of reaction and heat capacities of calorimeters
Sample Exercises 9.6, 9.7

LO5 Calculate enthalpies of reaction using Hess's law and enthalpies of formation
Sample Exercises 9.8, 9.9, 9.10

LO6 Estimate enthalpies of reaction using average bond energies
Sample Exercise 9.11

LO7 Estimate enthalpies of solution and lattice energies using the Born–Haber cycle and Hess's law
Sample Exercise 9.12

LO8 Calculate and compare fuel values and fuel densities
Sample Exercises 9.13, 9.14

9.1 Energy as a Reactant or Product

FIGURE 9.1 Water about to go over a waterwheel has more potential energy than water at the bottom of the wheel. Waterwheels were developed by the ancient Greeks and widely used in medieval Europe and colonial America to convert this potential energy into mechanical energy to grind grain into flour, weave yarn into cloth, mill lumber, and shape metals.

Combustion reactions produce most of the energy we consume, but where does this energy come from? The combustion of coal (which is mostly carbon) generates much of the electricity in China, India, and the United States; the combustion of natural gas (mostly CH_4) heats water, warms homes, and is also used to generate electricity; and then there is the combustion of gasoline, which many of us rely on to get us where we need to go. All of the fuels in these combustion reactions are fossil fuels: the decomposed remains of plants and animals that lived millions of years ago. To trace the origin of the energy in fossil fuels, we need to analyze the food chains that supplied nutrition to those plants and animals. Nearly all of those food chains started with green plants that harvested the energy of the sun through photosynthesis.

The connection to the sun is perhaps more obvious when we consider renewable sources of energy. In many parts of the world, wood is the principal fuel for heating homes and for cooking food. Dry wood is mostly cellulose, the most common organic compound in the world, consisting of long chains of glucose molecules. As we discussed in Chapter 7, green plants harness the energy of the sun to synthesize glucose from carbon dioxide and water via photosynthesis. Trees require from hundreds to thousands of molecules of glucose to synthesize one molecule of cellulose.

Burning wood in a stove (or food inside us) turns this process around, converting cellulose (or glucose) and O_2 back into CO_2 and H_2O and liberating the energy of the sun that was originally harvested by green plants. R. Buckminster Fuller (1895–1983), an architect, inventor, and futurist, once described the release of energy that happens when wood burns as "all that sunlight unwinding."

chemical energy potential energy stored in chemical bonds.

internal energy (*E*) the sum of all the kinetic and potential energies of all of the components of a system.

thermochemistry the study of the changes in energy that accompany chemical reactions.

thermodynamics the study of energy and its transformations.

Forms of Energy

We learned in Chapter 1 that there are two broad categories of energy: *kinetic energy*, which is the energy of an object in motion, and *potential energy*, which can be the energy that an object has because of its position, as illustrated by the water in a mill pond in Figure 9.1. Another form of potential energy is the energy that a substance has because of its composition—that is, the way in which the atoms inside it are bonded together. This kind of potential energy is called **chemical energy**. The sum of all the kinetic and potential (including chemical) energies of an object, or a collection of many objects, is the **internal energy (E)** of the object or collection. When the collective internal energies of the reactants in a chemical reaction are different from those of the products—and they nearly always are—the difference constitutes the *change in energy* (ΔE) that accompanies the reaction.

As a practical matter, the *absolute* value of a system's internal energy is extremely difficult to determine, but *changes* in energy are fairly easy to measure because they correspond to changes in the system's physical state or temperature. The branch of chemistry that explores these changes in energy is called **thermochemistry**, our primary focus in this chapter. Thermochemistry is related to a more expansive scientific domain called **thermodynamics**, which is the study of energy and its transformations.

In our exploration of the energy changes that accompany chemical reactions, we will often encounter reactions in which the collective internal energy of the reactants is greater than that of the products, such as the combustion of hydrogen (Figure 9.2). The result is a release of energy, which has to go somewhere. It can't just disappear because that would violate the principle articulated by the law of conservation of energy and the **first law of thermodynamics**: energy cannot be created or destroyed, though it can change from one form to another. Expressed another way, the total energy of the universe is a constant. The impact of this principle may be clearer if we consider the universe to consist of only two parts: the part we are interested in, which we call the **system**, and everything else, which we call the **surroundings**. We can express these definitions in the form of an equation:

$$\text{Universe} = \text{system} + \text{surroundings} \tag{9.1}$$

Let's use Equation 9.1 to connect changes in the internal energy of a system and its surroundings to the principle of energy conservation. If energy cannot be created or destroyed, then the energy of the universe (E_{univ}) is a constant, which means that ΔE_{univ} during any process is always equal to zero. If so, then

$$\Delta E_{\text{univ}} = \Delta E_{\text{sys}} + \Delta E_{\text{surr}} = 0$$

and

$$\Delta E_{\text{sys}} = -\Delta E_{\text{surr}} \tag{9.2}$$

When a thermodynamic system releases energy, as when molecules of H_2 and O_2 combine to form molecules of H_2O, the energy flows into its surroundings. Whatever energy the system loses ($\Delta E_{\text{sys}} < 0$), its surroundings gain ($\Delta E_{\text{surr}} > 0$).

What effect does this increase in energy have on the surroundings? To get one answer to this question, let's consider the combustion of hydrogen that powers the engines of a Delta IV rocket (Figure 9.3). If the combustion gases (the reactants and products) are the system, then everything else in the universe—including the rocket, its launchpad, and the air through which the rocket travels on its way into space—makes up the surroundings. The most significant energy transfer from the system to the surroundings occurs when the rocket and its cargo are launched into space. This involves exerting a large[1] force (F) through a considerable distance (d), which meets our definition from Chapter 1 of doing work (w):

$$w = F \times d \tag{1.1}$$

[1]At full throttle, one of the Delta IV engines produces up to 3.2×10^6 newtons of thrust (force), which is about 15 times that of a Boeing 747 engine at takeoff.

first law of thermodynamics the principle that the energy gained or lost by a system must equal the energy lost or gained by the surroundings. Energy cannot be created or destroyed.

system the part of the universe that is the focus of a thermochemical study.

surroundings everything in a thermochemical study that is not part of the system.

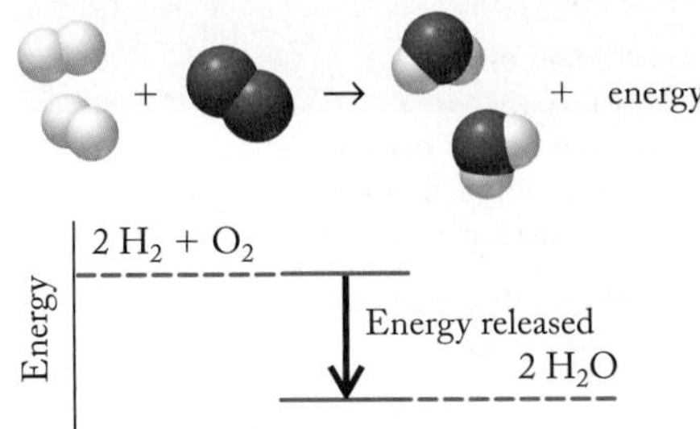

FIGURE 9.2 Energy is released during chemical reactions when higher-energy reactants, such as H_2 and O_2, form lower-energy products, such as H_2O.

CONNECTION In Chapter 1, we defined *energy* as the ability to do work. We also introduced the *law of conservation of energy* and the concept that energy cannot be created or destroyed but can be changed from one form of energy to another.

FIGURE 9.3 Up to three hydrogen-fueled engines make up the first stage of a Delta IV Heavy rocket such as this one, which can send a satellite weighing 23 metric tons into low Earth orbit.

TABLE 9.1 Flows of Heat and Work and Their Impact on E_{sys}

Processes That Increase E_{sys}	Result
Surroundings hotter than the system, so heat flows into the system	$q > 0$
Surroundings do work on the system	$w > 0$
Processes That Decrease E_{sys}	**Result**
System hotter than its surroundings, so heat flows into surroundings	$q < 0$
System does work on its surroundings	$w < 0$

In addition to doing work, the hot gases of the system heat their surroundings. These two processes, doing work and transferring heat, represent the two ways in which energy moves between a system and its surroundings. In general, if heat flows *out from* a system, or if work is done *by* the system on its surroundings, the internal energy E of the system decreases. On the other hand, if heat flows *into* a system or if work is done *on* a system, the internal energy of the system increases. We can express this last relationship in an equation, using the symbol q to represent heat:

$$\Delta E = q + w \tag{9.3}$$

Adding heat to a system means that q has a positive value. Doing work on a system means that w has a positive value. If both q and w are positive, then so is ΔE—that is, $\Delta E > 0$. When a system does work *on* its surroundings, like the reaction mixture in the rocket engines in Figure 9.3, then w is negative, and if heat flows out from the system (see again Figure 9.3), then q is also negative. Taken together, these changes produce a decrease in the internal energy of the system ($\Delta E < 0$). The various combinations of heat flow and work and their impact on ΔE are summarized in Table 9.1 and in Figure 9.4. Note that Equation 9.3 applies to all the changes of ΔE, q, and w in the figure and is essentially a statement of the first law of thermodynamics: any change in energy experienced by the system is balanced by a change of equal magnitude but opposite sign in the energy of its surroundings, as expressed by the quantity of heat that flowed between them and the quantity of work that one did on the other.

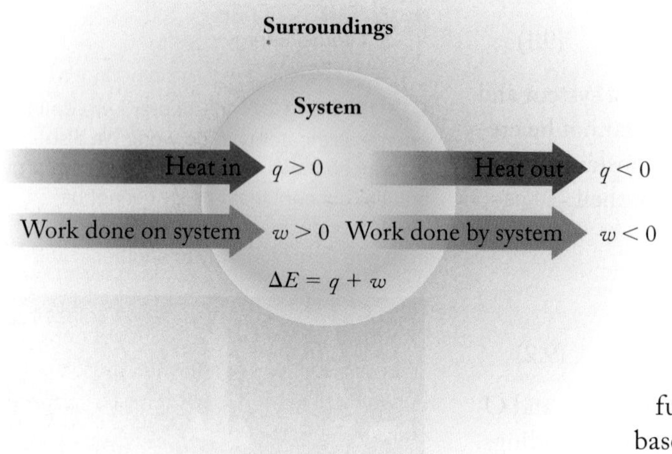

FIGURE 9.4 The change in internal energy (ΔE) of a system is positive when heat flows into it ($q > 0$) or work is done on it ($w > 0$). A system loses internal energy ($\Delta E < 0$) when heat flows out from it ($q < 0$) or the system does work on its surroundings ($w < 0$).

No matter how the internal energy of a system is changed, the final internal energy of the system is *independent* of how it was acquired. It makes no difference whether the system lost or gained heat, or whether it did work or had work done on it, because internal energy is a **state function**: it depends only on how much potential and kinetic energy the system has and not at all on the pathway or sequence of events that produced those levels of energy. Think of skiers ascending to the top of a mountain for their run: whether they ride straight up on a ski lift or hike up a winding path, they arrive at the same height in the end (Figure 9.5). Their altitude is a state function, as is the change in altitude they experienced from the base of the mountain to its summit: both are independent of the path the skiers take. Similarly, any change in the internal energy of a system is independent of the processes causing the change. Only the initial and final energy levels of the system define ΔE:

$$\Delta E = E_{\text{final}} - E_{\text{initial}}$$

We will encounter other examples of system properties that are state functions in later sections of this chapter and in later chapters of this book.

At this point we need to be clear on what we mean by *heat*. Heat is energy that is in the process of being transferred from a higher-temperature object to a lower-temperature one. An operating engine of a Delta IV rocket gets very hot. The quantity of **thermal energy** in the hot engine is determined not only by its temperature but also by the composition of the materials of which it is made and how massive it is (a concept called *heat capacity* that we explore in Section 9.4). By *thermal* energy, we mean that part of the total internal energy of a system that is linked to how hot it is.

CONNECTION In Chapter 1, we defined *heat* as the spontaneous transfer of energy from a warmer object to a cooler one.

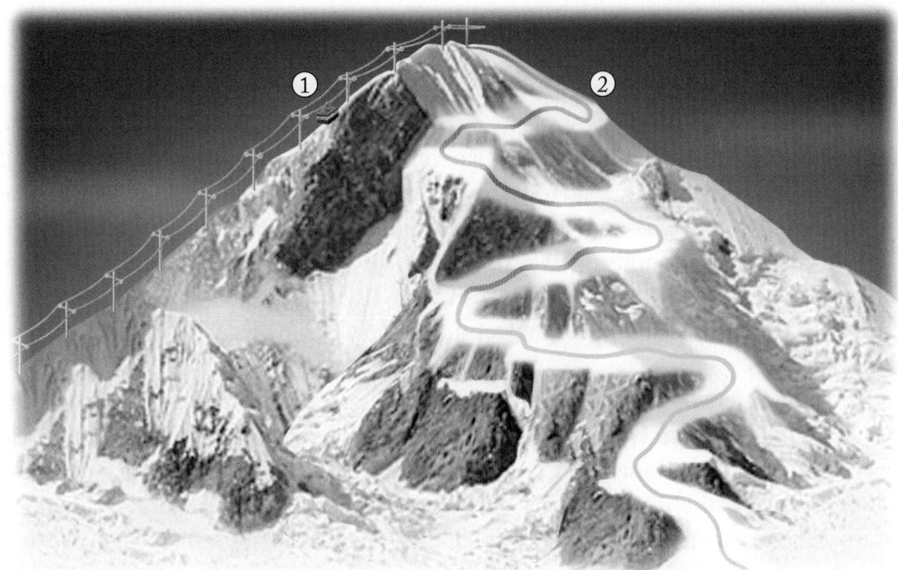

FIGURE 9.5 Whether a skier at the base of a mountain rides a ski lift or hikes up a mountain trail, the skier's increase in altitude and potential energy is the same.

At the atomic level, the thermal energy of a system is the sum of the kinetic energies of the particles that make up the system. We learned in Chapter 1 that the kinetic energy (KE) of any object in motion, whether it is a rocket or an atom, is a function of its mass (m) and its speed (u):

$$\text{KE} = \tfrac{1}{2}mu^2 \qquad (1.2)$$

The kinetic energies of atom-sized particles, such as molecules of H_2, O_2, or H_2O, depend only on their temperature. In fact, the average kinetic energy of a population of gaseous atoms and molecules *is directly proportional* to their *absolute* (kelvin) temperature: if T doubles, so does KE. Higher average kinetic energies mean higher average speeds, but because the speed term is squared in Equation 1.2, particle speed is not a linear function of temperature. For example, when the absolute temperature of a gas doubles, the average speed of the particles in it increases by only $\sqrt{2}$, or 1.414, times.

The motion of particles of gases as they bounce around inside the space they occupy is called *translational* kinetic energy (Figure 9.6a). Molecules also have nontranslational, or *internal,* forms of kinetic energy. These include *rotational* kinetic energy (Figure 9.6b) and *vibrational* kinetic energy (Figure 9.6c). Bond vibrations also provide additional forms of potential energy in molecules in much the way that stretching, compressing, or bending a spring increases its potential energy.

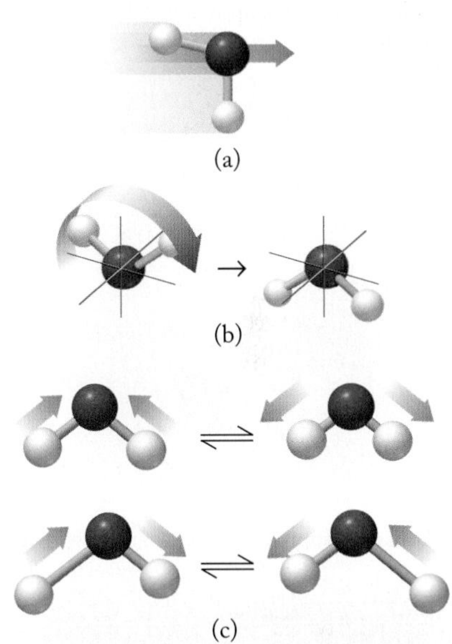

FIGURE 9.6 Some of the types of molecular motion that contribute to the internal energy of a system are (a) translational motion, the motion from place to place along a path; (b) rotational motion, the motion about a fixed axis; and (c) vibrational motion, the movement back and forth from some central position.

CONNECTION We noted in Chapter 4 that there are several kinds of bond vibrations in molecules with more than two atoms, including symmetric and asymmetric stretching and bending as the angles between bonds increase and decrease.

9.2 Transferring Heat and Doing Work

Systems come in different sizes. They can be as large as galaxies or as small as single living cells. Most of the systems we discuss in this chapter fit on a laboratory bench, and we usually limit our view of their surroundings to that part of the universe that can exchange energy and matter with the system. In evaluating how much energy is gained or lost in a chemical reaction, the system may be just the particles involved in the reaction, but it may also include the vessel that contains them.

state function a property of an entity based solely on its chemical or physical state or both but not on how it achieved that state.

thermal energy the portion of the total internal energy of a system that is proportional to its absolute temperature.

(a) **Isolated system:** A thermos bottle containing hot soup with the lid screwed on tightly

(b) **Closed system:** A cup of hot soup with a lid

(c) **Open system:** An open cup of hot soup

FIGURE 9.7 Illustrations of isolated, closed, and open systems: (a) Hot soup in a tightly sealed thermos bottle approximates an isolated system: no vapor escapes, no matter is added or removed, and no energy is transferred to the surroundings. (b) Hot soup in a cup with a lid is a closed system; the soup transfers heat to the surroundings as it cools, but no matter escapes and none is added. (c) Hot soup in a cup with no lid is an open system; it transfers both matter (steam) and energy (heat) to the surroundings as it cools. Matter in the form of pepper, grated cheese, or other flavor enhancers might also be added to the system.

Isolated, Closed, and Open Systems

As we explore a system's energy production and consumption, we need to be clear on how the system interacts with its surroundings. For example, can energy be transferred from the system to its surroundings? Can matter be exchanged between the system and its surroundings? The answers to these questions define the type of system we have. If both matter and energy can be transferred between a system and its surroundings, we have an *open* system. If only energy can flow between them, we have a *closed* system. And if a system is completely cut off from its surroundings so that neither matter nor energy can be transferred, we have an *isolated* system.

The containers of hot soup in Figure 9.7 illustrate the three types of systems. Figure 9.7(a) shows an isolated system: a serving of soup inside a thermos bottle that has a perfect seal and that is perfectly insulated, allowing neither heat nor matter to escape or enter. The mass of the system never changes, nor does its temperature or its energy content. No such thermos bottle actually exists, because even the best ones can't keep soup hot forever. For short times, however, substances in sealed, well-insulated containers approximate isolated systems.

Hot soup in a cup with a lid on it represents a *closed* thermodynamic system (Figure 9.7b). The lid keeps the soup and any vapor it produces from escaping, and it keeps any matter from getting into the soup. Only energy is exchanged between the soup and its surroundings as heat is transferred from the soup through the cup and into the air and the tabletop. Gradually the soup cools to room temperature. Many real thermodynamic systems are closed systems.

Soup in an open cup represents an *open* thermodynamic system because it can exchange both energy and matter with its surroundings (Figure 9.7c). Heat flowing from the soup into its surroundings and matter—in the form of water vapor, for example—is free to leave the system. Matter can also enter the system—perhaps a little seasoning to enhance the flavor.

CONCEPT TEST

Suppose two identical pots of water are heated on a stove until the water inside them begins to boil. Both pots are then removed from the stove. One of the two is covered with a tight lid; the other is not, and both are allowed to cool.

a. What type of thermodynamic system—open, closed, or isolated—describes each of the cooling pots?

b. Which pot cools faster? Why?

(Answers to Concept Tests are in the back of the book.)

Exothermic and Endothermic Processes

Chemists classify reactions based on whether they give off or absorb energy. A chemical reaction that results in the transfer of energy—usually in the form of heat—from the reaction mixture (the system) to its surroundings is an **exothermic** reaction. This energy flow can usually be detected by an increase in the temperature of the surroundings. Combustion reactions are good examples of exothermic reactions. Chemical reactions that absorb energy from their surroundings are **endothermic** reactions.

Nonchemical processes that give off or take in heat can also be described as exothermic or endothermic. For example, your body cools itself during strenuous exercise by sweating. Cooling happens as the sweat on your skin evaporates

FIGURE 9.8 A process that is exothermic in one direction, such as (a) the condensation of atmospheric water vapor on the outside of a cold water bottle, is endothermic in the reverse direction, such as (b) when perspiration evaporates from an athlete's skin and clothing.

because the conversion of a liquid to a gas is an endothermic process, requiring the flow of heat into the system (sweat) from its surroundings (your body), as shown in Figure 9.8. On the other hand, the reverse process, water vapor condensing, as on the outside of a bottle of cold water on a humid summer day, is exothermic. Heat is transferred from the system (condensing water vapor) to its surroundings, including the water in the bottle in Figure 9.8.

SAMPLE EXERCISE 9.1 Identifying Exothermic and Endothermic Processes **LO1**

Describe the flow of thermal energy (heat) during the desalination of seawater by distillation (Figure 9.9), identify the steps as either endothermic or exothermic, and give the sign of q associated with each step. Consider the water being purified to be the system.

Collect and Organize The water is the system, so we need to evaluate how the water gains or loses thermal energy during distillation. A positive value of q represents a gain in thermal energy.

Analyze During distillation, heat flows into the seawater, raising its temperature by increasing the kinetic energy of water molecules and particles of dissolved sea salts. Additional heating vaporizes the water by providing its molecules with enough kinetic energy to overcome the hydrogen bonds and other intermolecular forces that keep them together in the liquid phase. As water vapor passes through the cold condenser, heat flows from the vapor into the condenser. As a result, the kinetic energy of the vaporized molecules decreases, which allows their attractions for each other to pull them together. These molecules form a layer of liquid water on the inside wall of the condenser.

Solve Heat flows from the surroundings (hot plate) to heat the system (seawater) to its boiling point. That process is endothermic and the sign of q is positive. Additional heat

exothermic process one in which energy—usually in the form of heat—flows from a system into its surroundings.

endothermic process one in which energy—usually in the form of heat—flows from the surroundings into the system.

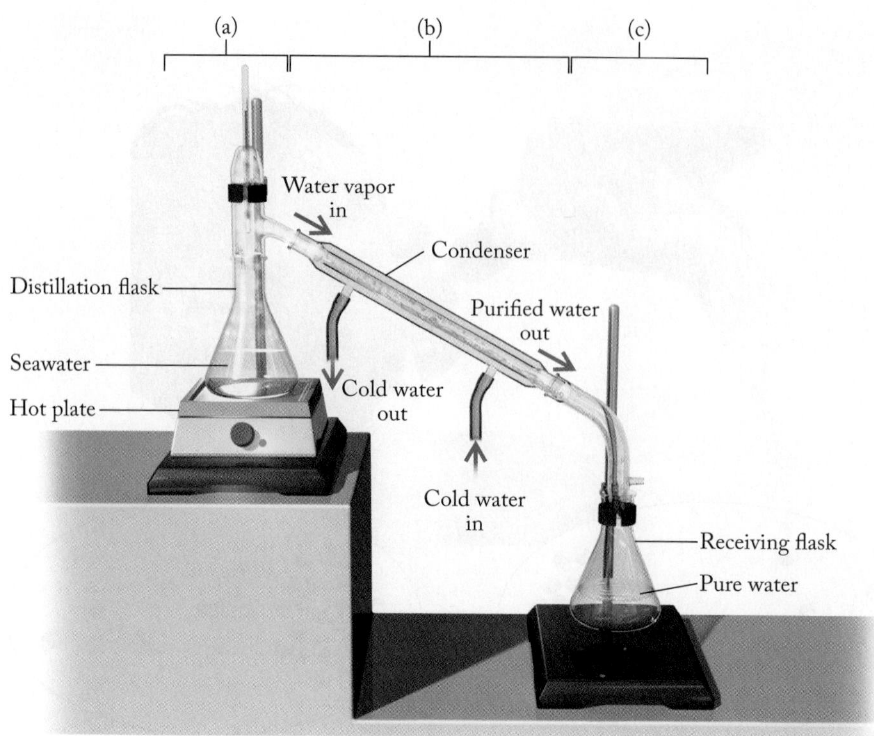

FIGURE 9.9 A laboratory setup for desalinating seawater. (a) Seawater is heated to the boiling point in the distillation flask. (b) Water vapor rises and enters the condenser, where it is liquefied. (c) The purified liquid is collected in the receiving flask.

flows from the surroundings into the system as the water vaporizes—this is another endothermic process for which q is positive. However, heat then flows from the system (in the form of water vapor) into its surroundings (cold condenser walls) as the water vapor condenses and further cools: these are both exothermic processes and the sign of q is negative for both of them.

Think About It As any liquid is heated to its boiling point and vaporized, the particles in it gain enough kinetic energy to overcome the intermolecular forces that kept them together in the liquid state. As any vapor cools and condenses, its particles lose the kinetic energy that allowed them to overcome their attraction for each other, and they come together in the liquid state.

Practice Exercise What is the sign of q as (a) a match burns, (b) drops of molten candle wax solidify, and (c) a piece of dry ice disappears at room temperature? In each case, define the system and indicate whether the process is endothermic or exothermic.

(Answers to Practice Exercises are in the back of the book.)

P–V Work

We have discussed how the internal energy E of a system increases ($\Delta E > 0$) when heat flows into it ($q > 0$) or work is done on it by its surroundings ($w > 0$), and how the internal energy of a system decreases ($\Delta E < 0$) when heat flows out from it ($q < 0$) or it does work on its surroundings ($w < 0$). Let's explore a system that rapidly cycles between having its surroundings do work on it and then it doing work on its surroundings.

The system is the gases in the combustion chamber of one of the cylinders in a diesel engine (Figure 9.10). First, the downward motion of the piston during the intake stroke (Figure 9.10a) allows air to enter the cylinder, where it mixes with

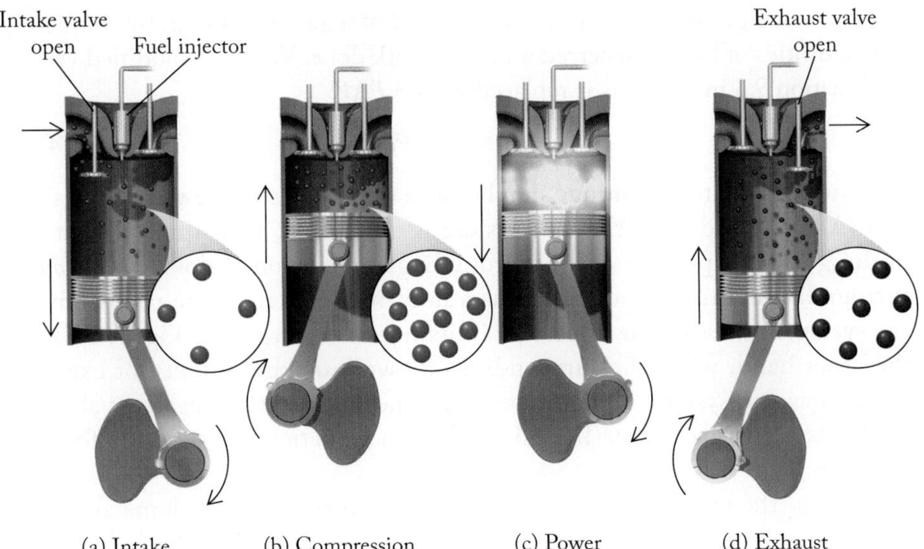

FIGURE 9.10 Diesel engines are based on a four-stroke cycle in which (a) fuel is injected and air is drawn into the combustion cylinder as the piston goes down, (b) the mixture is compressed as the piston moves up, (c) compression heating causes the mixture to ignite spontaneously and release energy that pushes the piston down, and (d) the products of combustion are pushed out as the piston moves up again.

an injection of a few microliters of diesel fuel. Then the intake valve closes, and the upward motion of the piston (Figure 9.10b) squeezes the air and fuel vapors (the system) in the cylinder into less than one-tenth of their initial volume. The compression heats up the reaction mixture (a clear sign its internal energy has increased) until, at the point of maximum compression, the mixture gets so hot it spontaneously ignites. The energy released by combustion rapidly raises the temperature of the reaction mixture even further, which increases the pressure it exerts on the top of the piston. This pressure drives the piston downward during the power stroke (Figure 9.10c) as the volume of the combustion mixture rapidly expands. Compressing the reaction mixture increased the mixture's internal energy because its surroundings (the piston) did work on the system. Then expansion of the hot products of combustion did work on their surroundings (the piston again), releasing energy they had gained through compression and combustion by pushing the piston downward (Figure 9.10d).

The work done *on* the gases in an engine cylinder as they are compressed and the work done *by* the gases as they expand are two examples of **pressure–volume (*P–V*) work**. The work done by the gases as they expand is the work that any gaseous system does when its volume increases by an amount ΔV (where $\Delta V = V_f - V_i$) against an opposing pressure P. The amount of work done is the product $P \times \Delta V$. This pressure–volume work is equivalent to force × distance ($F \times d$) work. To understand why, remember from Section 6.4 that pressure is force per unit of surface area:

$$P = \frac{F}{A} \tag{6.1}$$

CONNECTION In Chapter 6, we defined pressure and its units, including the pascal (Pa), the standard atmosphere (atm), and the bar.

Area has dimensions of distance squared (d^2), and volume is often expressed in units of distance cubed (d^3). If we substitute d^2 and d^3 for area and volume change, then

$$P \times \Delta V = \frac{F}{d^2} \times d^3 = F \times d$$

Now that we have established the equivalency of pressure–volume work and force–distance work, we need to know how to convert the units used to express $P–V$ work into units of energy (for example, joules). Doing so enables us to

pressure–volume (*P–V*) work the work associated with the expansion or compression of a gas.

calculate changes in the internal energy (ΔE) of a gaseous system by combining the quantities of heat transferred and P–V work done. We use a modified version of Equation 9.3 in which w is replaced with $-P\Delta V$:

$$\Delta E = q - P\Delta V \tag{9.4}$$

Equation 9.4 contains a minus sign because an increase in the volume of a system ($\Delta V > 0$) indicates that the system is expanding, which means that it is pushing aside a portion of its surroundings and involves doing work *on* its surroundings. This kind of work has a negative value from the perspective of the system and results in a decrease in its internal energy (see Figure 9.4). On the other hand, when the surroundings do work on the system (for example, when a gaseous system is compressed), volume change has a negative value, but work has a positive value. The minus sign in Equation 9.4 accounts for both processes.

Among the units commonly used to express pressure and volume are atmospheres (atm) and liters (L), respectively. When they are combined, as in Sample Exercise 9.2, the liter-atmosphere makes a handy conversion factor:

$$1 \text{ liter-atmosphere } (\text{L} \cdot \text{atm}) = 101.325 \text{ J}$$

SAMPLE EXERCISE 9.2 Calculating *P–V* Work **LO2**

A tank of compressed helium is used to inflate 100 balloons for sale at a carnival on a day when the atmospheric pressure is 1.01 atm. If each balloon is inflated with 4.8 L of He, how much P–V work is done by the compressed helium? Express your answer in joules.

Collect and Organize Our task is to determine how much P–V work is done by inflating the 100 balloons, each with 4.8 L of helium, and each expanding against an opposing pressure of 1.01 atm.

Analyze We focus on the work done on the atmosphere (the surroundings) by the helium (the system) as the volume of the balloons increases. Inflating each of 100 balloons with about 5 L of helium against a pressure of about 1 atm means that $P\Delta V$ will be about 500 liter-atmospheres and about 100 times that number of joules, or 50,000 J.

Solve The work (w) done by the compressed helium is

$$\begin{aligned} w &= -P\Delta V \\ &= -1.01 \cancel{\text{atm}} \times \frac{4.8 \cancel{\text{L}}}{1 \cancel{\text{balloon}}} \times 100 \cancel{\text{balloons}} \times \frac{101.325 \text{ J}}{1 \cancel{\text{L}} \cdot \cancel{\text{atm}}} \\ &= -4.9 \times 10^4 \text{ J or } -49 \text{ kJ} \end{aligned}$$

Think About It The internal energy of the compressed helium (the system) decreases when some of it expands to inflate the balloons because work is done *by* the system on its surroundings. Lower internal energy in this case corresponds to the lower kinetic energy (and lower temperature) of both the helium atoms in the balloons and those remaining in the tank.

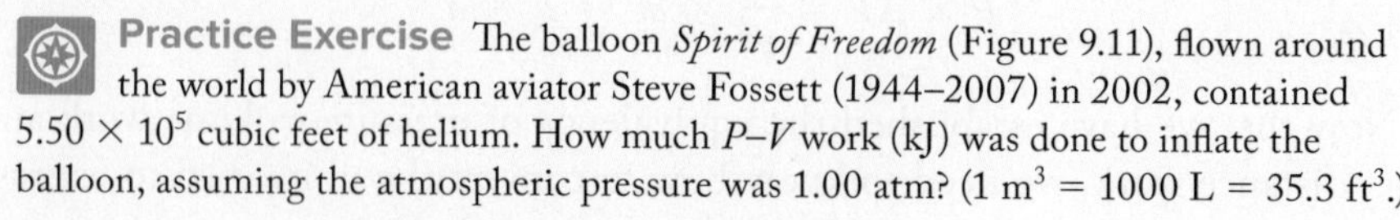

Practice Exercise The balloon *Spirit of Freedom* (Figure 9.11), flown around the world by American aviator Steve Fossett (1944–2007) in 2002, contained 5.50×10^5 cubic feet of helium. How much P–V work (kJ) was done to inflate the balloon, assuming the atmospheric pressure was 1.00 atm? (1 m^3 = 1000 L = 35.3 ft^3.)

FIGURE 9.11 The *Spirit of Freedom* was the first balloon to be flown nonstop around the world.

SAMPLE EXERCISE 9.3 Relating ΔE, q, and w LO2

The racing cars in Figure 9.12 are powered by V8 engines in which the motion of each piston in its cylinder displaces a volume of 0.733 L. If combustion of the mixture of gasoline vapor and air in one cylinder releases 1.68 kJ of energy, and if 33% of the energy does *P–V* work, how much pressure, on average, does the combustion reaction mixture exert on each piston? How much energy is transferred as heat from the reaction mixture to its surroundings?

Collect, Organize, and Analyze We know the quantity of energy (1.68 kJ) released by a combustion reaction and that 33% of this energy does *P–V* work on its surroundings. The rest (100 − 33 = 67%) must be lost as heat as described by Equation 9.4:

$$\Delta E = q - P\Delta V$$

The system loses energy; therefore, $\Delta E = -1.68$ kJ and the value of q will be about 2/3 of −1.68 kJ or about −1 kJ.

Solve Using the information given and Equation 9.4 to solve for q and P:

$$q = (0.67) \times (-1.68 \text{ kJ}) = -1.13 \text{ kJ}$$

$$P = -\frac{0.33 \times \Delta E}{\Delta V} = -\frac{0.33 \times (-1.68 \cancel{\text{kJ}})}{0.733 \cancel{\text{L}}} \times \frac{1 \cancel{\text{L}} \cdot \text{atm}}{101.325 \cancel{\text{J}}} \times \frac{1000 \cancel{\text{J}}}{1 \cancel{\text{kJ}}} = 7.5 \text{ atm}$$

Think About It We used a negative sign in the last equation because $P\Delta V$ contains positive values of pressure and volume change but represents work done *by* the system and is part of the energy lost by the system (−1.68 kJ). The heat transferred has a negative value because it is thermal energy lost by the system. Its value is close to the one we estimated.

Practice Exercise The air compressor used with a paint sprayer does 64 J of work pumping air into a tank. Warmed by the process, air in the tank gives off 32 J of energy to its surroundings. What is the change in the internal energy of the air in the tank as a result of the two processes?

FIGURE 9.12 These racing cars are powered by internal combustion engines in which hot gases do *P–V* work that is harnessed by the engine and drivetrain and propels the cars to speeds over 200 miles per hour (320 km/hr).

9.3 Enthalpy and Enthalpy Changes

Many physical and chemical changes take place at constant pressure. These include chemical and biochemical reactions that occur in organisms living on Earth's surface and the chemical reactions done in open beakers and flasks on laboratory benches. We use a thermodynamic parameter called *enthalpy* to track the changes in energy and the flow of heat into or out from these and other constant-pressure systems. **Enthalpy (*H*)** is a measure of the total energy of a system. By *total* we mean the system's internal energy (E) *and* the energy expended to push aside its surroundings to make space for the system at a given pressure. In equation form this definition is

$$H = E + PV \tag{9.5}$$

In this chapter our focus is not so much on the enthalpy of systems as on the *change* in enthalpy that accompanies, for example, a chemical reaction or a physical change. **Enthalpy change (ΔH)** is equal to the quantity of heat transferred into or out from a system during a chemical reaction or physical change that occurs at constant pressure—that is, q_P. To see how ΔH and q_P are equivalent, we rewrite Equation 9.5 to describe the changes in the three terms:

$$\Delta H = \Delta E + \Delta(PV)$$

enthalpy (*H*) a measure of the total energy of a system; the sum of the internal energy and the pressure–volume product of a system; $H = E + PV$.

enthalpy change (ΔH) the quantity of heat transferred into or out from a system during a chemical reaction or physical change at constant pressure.

During processes occurring at constant pressure, the $\Delta(PV)$ term becomes $P\Delta V$:

$$\Delta H = \Delta E + P\Delta V \tag{9.6}$$

Next, we apply Equation 9.4 ($\Delta E = q - P\Delta V$) to a process occurring at constant pressure so that $q = q_P$:

$$\Delta E = q_P - P\Delta V$$

If we rearrange the terms to isolate q_P on the left side, we get an equation with the same right side as Equation 9.6:

$$q_P = \Delta E + P\Delta V$$

In other words:

$$\Delta H = q_P \tag{9.7}$$

Like ΔE, ΔH is a state function, so the value of ΔH depends only on the difference in enthalpy between the initial and final states, not on the pathway followed in going from one to the other.

Recall that the terms *exothermic* and *endothermic* describe processes that release or absorb energy, usually as thermal energy or heat. These labels also describe processes in which the changes in enthalpy of the system are less than or greater than zero:

Description	Enthalpy Change
Exothermic	$\Delta H < 0$
Endothermic	$\Delta H > 0$

When an exothermic process runs in reverse, it becomes an endothermic process, and vice versa. For example, 6.01 kJ of thermal energy must be added to melt one mole of ice at 0°C, which is an endothermic process, but 6.01 kJ of thermal energy must be removed from a mole of liquid water at 0°C to freeze it, which is an exothermic process. This enthalpy change of 6.01 kJ/mol is called the **enthalpy of fusion (ΔH_{fus})** of water, where the subscript "fus" indicates melting (or *fusion*) of the solid. Sometimes ΔH_{fus} is referred to as the *heat of fusion*. The enthalpy changes associated with other paired phase changes—vaporization and condensation, and sublimation and deposition—also have complementary values. For example, it takes 40.7 kJ of thermal energy to vaporize a mole of liquid water at 100°C, and the same 40.7 kJ of thermal energy is released when one mole of water vapor condenses. As with ΔH_{fus}, the **enthalpy of vaporization (ΔH_{vap})** of water is often referred to as the *heat of vaporization*.

Table 9.2 lists ΔH_{fus} and ΔH_{vap} values for several common compounds. The values for the alkanes containing one to four carbons increase with increasing molar mass because these molecules interact via London dispersion forces that increase with increasing molecular size. Thus, the thermal energy required to convert them from solids into liquids and from liquids into gases also increases with increasing molecular size.

Most of the enthalpy of vaporization values in Table 9.2 are based on the energy required to vaporize substances at their normal boiling points. However, the ΔH_{vap} values for water include additional values at temperatures below the normal boiling point of 100°C. These ΔH_{vap} values decrease as temperature increases because the kinetic energy of the H_2O molecules in liquid water increases as temperature increases. More kinetic energy means greater molecular motion, which also means less interaction between molecules: the average number of hydrogen bonds per water molecule decreases from an estimated 3.6 at 25°C to only 3.2 at 100°C. As the

TABLE 9.2 Enthalpy of Fusion and Vaporization

Compound and Molecular Structure	$\mathcal{M}$ (g/mol)	ΔH_{fus} (kJ/mol)	ΔH_{vap} (kJ/mol)
Methane	16.04	0.94	8.2
Ethane	30.07	2.86	14.7
Propane	44.10	3.53	15.7
Butane	58.12	4.66	21.0
Methanol	32.04	3.18	35.3
Ethanol	46.07	5.02	38.6
Acetone	46.07	5.69	31.3
Ammonia	17.03	5.65	23.4
Water	18.02	6.01	44.0 at 25°C 43.5 at 40°C 42.5 at 60°C 41.6 at 80°C 40.7 at 100°C

interactions between molecules decrease, the energy required to separate them into independent, gas-phase particles decreases as well.

CONCEPT TEST

Using intermolecular forces, explain why water has the highest ΔH_{vap} value of all the compounds in Table 9.2 (and the highest boiling point, too).

9.4 Heating Curves and Heat Capacity

Hikers and cross-country skiers use portable stoves fueled by propane or butane to prepare hot meals. In the winter, the only source of water may be ice or snow. What changes in temperature and physical state does the water undergo as the energy from a portable stove turns a saucepan filled with snow initially at −18.0°C into boiling water? We can track the changes using a type of graph called a

enthalpy of fusion (ΔH_{fus}) the energy required to convert one mole of a solid substance at its melting point into the liquid state; also called the *heat of fusion*.

enthalpy of vaporization (ΔH_{vap}) the energy required to convert one mole of a liquid substance at its boiling point into the vapor state; also called the *heat of vaporization*.

(a)

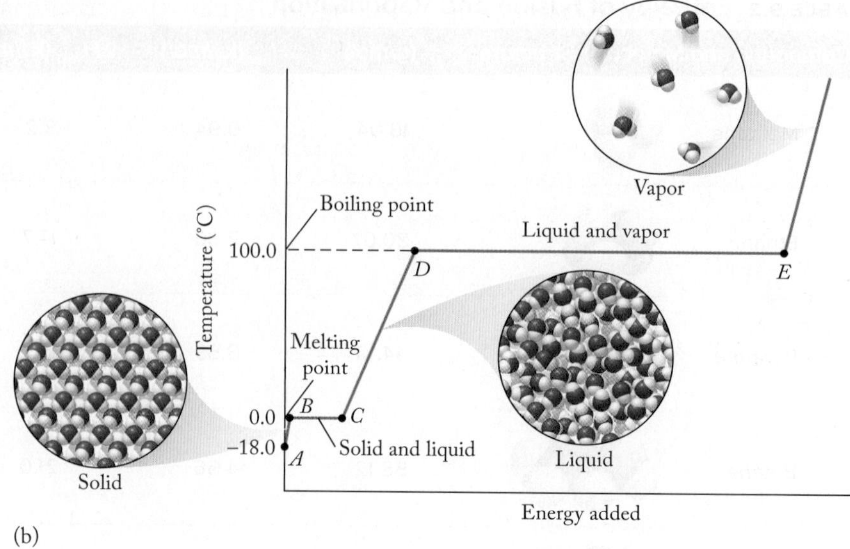

(b)

FIGURE 9.13 (a) The hiker in the photo is using a small cooking stove protected by a windscreen as a heat source. (b) The energy required to melt snow and boil the resultant water is illustrated by the four line segments on the heating curve of water: heating snow to its melting point ($\overline{AB}$); melting the snow to form liquid water ($\overline{BC}$); heating the water to its boiling point ($\overline{CD}$); and boiling the water to convert it to vapor ($\overline{DE}$).

heating curve (Figure 9.13), which plots the increasing temperature of a system while it is heated at a constant rate.

At first, the added energy raises the temperature of the snow as shown by the $\overline{AB}$ segment of the heating curve in Figure 9.13(b). The temperature stops rising when the snow reaches its melting point at 0.0°C, and it stays there (horizontal segment $\overline{BC}$) until all the snow has melted. Continued heating causes the temperature of the pot's contents, now liquid water, to rise again until the liquid reaches its boiling point at 100.0°C (segment $\overline{CD}$). At 100.0°C, the temperature of the water remains steady again as liquid water becomes water vapor (horizontal segment $\overline{DE}$). Only if all of the liquid water in the pan vaporizes (which would be too bad if the goal was to make a hot meal) can the temperature of the water vapor in the pan rise above 100.0°C, as indicated by the line segment beyond point E in Figure 9.13(b).

The difference between the x-axis coordinates at the beginning and end of each line segment in Figure 9.13(b) indicates how much energy is required during each step of the overall process. In the first step (line segment $\overline{AB}$), the energy required to raise the temperature of the snow from −18.0°C to 0.0°C is related to the snow's **heat capacity (C_P)**, a factor representing how much thermal energy the snow can store. Mathematically, C_P is the quantity of energy that must flow into an object at constant pressure (hence the "P" subscript) to raise its temperature by 1°C. Thus, the units of C_P are J/°C. If we know the value of C_P we can calculate how much energy (q) must be transferred to an object to raise its temperature by an amount ΔT:

$$q = C_P \Delta T \tag{9.8}$$

When an object is a pure substance (like the snow in the hiker's pot), its heat capacity can then be calculated using that substance's *specific heat capacity,* or **specific heat (c_P)**, which is the quantity of thermal energy required to raise the temperature of one gram of a substance by 1°C at constant pressure. Thus, the units of c_P are J/(g · °C). The product of the mass (m) in grams of an object composed of a single substance times the specific heat of the substance (c_P) is the same as the heat capacity of the object:

$$C_P(\text{J/°C}) = m(\text{g}) \times c_P[\text{J/(g · °C)}]$$

heat capacity (C_P) the energy required to raise the temperature of an object by 1°C at constant pressure.

specific heat (c_P) the energy required to raise the temperature of 1 g of a substance by 1°C at constant pressure.

molar heat capacity ($c_{P,n}$) the energy required to raise the temperature of one mole of a substance by 1°C at constant pressure.

Substituting (mc_P) for C_P in Equation 9.8 gives us an equation that relates energy flow to the change in temperature of a mass m of a pure substance:

$$q = mc_P\Delta T \tag{9.9}$$

Heat capacity is an extensive property that depends on how massive an object is as well as what it's made of, whereas specific heat is an intensive property that is characteristic of a particular substance in a particular physical state (see Table 9.3). For example, the c_P of liquid water, 4.18 J/(g · °C), is about twice that of ice. Why are they different? The answer has to do with the interactions between water molecules and their freedom of motion. Water molecules in solid ice have little freedom of motion, though they can vibrate in place. Some of these vibrations stretch and compress the O—H bonds in the molecules or change the angle between the bonds. Still others stretch and compress the hydrogen bonds between water molecules. These various modes of vibration provide ways for ice to disperse any energy flowing into it. The more ways there are, the greater the specific heat. The molecules in liquid water enjoy even more freedom of motion and even more ways to disperse absorbed thermal energy; therefore liquid water has a higher specific heat than ice.

Another intensive property related to specific heat is **molar heat capacity** (**$c_{P,n}$**), which is the quantity of energy required to raise the temperature of one mole of a substance by 1°C at constant pressure. If we know the number of moles (n) of

TABLE 9.3 Specific Heat and Molar Heat Capacity Values

Substance	Phase	c_P [J/(g · °C)]	$c_{P,n}$ [J/(mol · °C)]
Elements			
Aluminum	(s)	0.897	24.2
Carbon (graphite)	(s)	0.71	8.5
Carbon (diamond)	(s)	0.127	1.5
Chromium	(s)	0.449	23.3
Copper	(s)	0.385	24.5
Gold	(s)	0.129	25.4
Iron	(s)	0.45	25.1
Lead	(s)	0.129	26.7
Silicon	(s)	0.71	19.9
Silver	(s)	0.233	25.1
Tin	(s)	0.227	26.9
Titanium	(s)	0.523	25.0
Zinc	(s)	0.387	25.3
Compounds			
Silicon dioxide	(s)	0.703	42.2
Water (−10°C)	(s)	2.11	38.0
Water (25°C)	(ℓ)	4.18	75.3
Water (102°C)	(g)	1.89	34.1
Ammonia	(ℓ)	4.75	80.9
Mixture			
Air[a]	(g)	1.003	29.1

[a]Dry air at 0°C and 1 atmosphere of pressure.

a substance in a sample and also the temperature change in degrees Celsius (ΔT) that the substance experiences, we can calculate q for the process:

$$q = nc_{P,n}\,\Delta T \tag{9.10}$$

Hot Soup on a Cold Day

Suppose the hiker in Figure 9.13 decides to make soup starting with 275 g of snow. She puts the snow, which has an initial temperature (T_i) of −18.0°C, into the pan and starts to heat it up. At first, energy is consumed to warm the snow to its melting point, which makes the final temperature of the snow (T_f) 0.0°C. We can use Equation 9.9 to calculate how many joules of thermal energy must be added to do this. The value of m is 275 g, c_P is the specific heat of ice [2.11 J/(g · °C)], and the change in temperature is the difference between the final and initial temperatures, or

$$\Delta T = T_f - T_i = [0.0 - (-18.0)]°C = 18.0°C$$

We insert the values in Equation 9.9:

$$q = mc_P\Delta T$$

$$= 275\ \text{g} \times \frac{2.11\ \text{J}}{\text{g} \cdot \cancel{°\text{C}}} \times 18.0\ \cancel{°\text{C}} = 1.04 \times 10^4\ \text{J} = 10.4\ \text{kJ}$$

Notice that this value is positive, which means that the system (the snow) gains energy as it warms up. Also note that line segment $\overline{AB}$ on the heating curve has a positive slope because the temperature of the snow increases as the energy is added.

The thermal energy absorbed in the next step is that required to melt (or *fuse*) the snow, which is the process that occurs along line segment $\overline{BC}$ in Figure 9.13. The value of q for this step can be calculated using the heat of fusion (ΔH_{fus}) of water and the number of moles (n) of water in 275 g of snow:

$$q = n\Delta H_{fus} \tag{9.11}$$

$$= 275\ \text{g} \times \frac{1\ \cancel{\text{mol}}}{18.02\ \text{g}} \times \frac{6.01\ \text{kJ}}{\cancel{\text{mol}}} = 91.7\ \text{kJ}$$

This value is also positive because energy flows into the system even though its temperature doesn't change. Thermal energy breaks some of the hydrogen bonds that hold molecules of water in a rigid three-dimensional array in the solid phase. Disrupting these intermolecular forces increases the *potential* energy of the water molecules because they are no longer confined to a single location—they have gained the freedom to flow past their nearest molecular neighbors. However, this newly acquired ability does not result in actual motion of the H_2O molecules, and a corresponding increase in their average *kinetic* energies or temperature, until all of the ice has melted.

As energy continues to flow into the pot of 0.0°C water, its temperature rises. If enough energy is added, the temperature of the water reaches 100.0°C. The relation between this increase in temperature (ΔT) of 100.0°C and the quantity of energy that was absorbed by the water can again be calculated using Equation 9.9, but this time we use the specific heat of liquid water:

$$q = mc_P\Delta T$$

$$= 275\ \text{g} \times \frac{4.18\ \text{J}}{\text{g} \cdot \cancel{°\text{C}}} \times 100.0\ \cancel{°\text{C}} = 1.15 \times 10^5\ \text{J} = 115\ \text{kJ}$$

Note that line segment $\overline{CD}$ on the heating curve has a positive slope because the temperature of the water increases as energy is added.

At this point in our story we assume that our hiker–chef accidentally leaves the boiling water unattended and it vaporizes completely. As line segment $\overline{DE}$ in Figure 9.13(b) shows, the temperature of the water remains at 100.0°C until all of it is vaporized. The enthalpy change and the energy transferred are the product of the number of moles of liquid water and water's enthalpy of vaporization (ΔH_{vap}):

$$q = n\Delta H_{vap} \tag{9.12}$$

$$= 275 \text{ g} \times \frac{1 \text{ mol}}{18.02 \text{ g}} \times \frac{40.7 \text{ kJ}}{\text{mol}} = 621 \text{ kJ}$$

As happened when the snow sample melted, the energy absorbed by molecules of liquid H_2O during vaporization is converted into greater *potential* energy of the molecules, as essentially all of the hydrogen bonds between them are broken. However, their average kinetic energy and temperature do not increase. Only after all the liquid water has completely vaporized does the temperature of the water vapor increase above 100.0°C.

Notice in Figure 9.13(b) that line segment $\overline{DE}$, representing the phase change from liquid water to water vapor, is much longer than line segment $\overline{BC}$, which represents the change from solid snow to liquid water. The relative lengths of the lines reflect the fact that the enthalpy of vaporization of water (40.7 kJ/mol) is much larger than the enthalpy of fusion of ice (6.01 kJ/mol). It takes more energy to boil one mole of water than to melt one mole of ice, because only a fraction of the hydrogen bonds in ice are broken upon melting, whereas essentially all of the hydrogen bonds in liquid water are broken when it vaporizes.

Heating curves and the attendant calculations show the energy required as the temperature changes as a substance is heated. If we were to carry out the reverse of this process and cool a sample of a gas to form a liquid and then cool it further to form a solid, the resultant graph would be called a cooling curve. Cooling curves are the opposite of heating curves and show energy and temperature changes as a substance is cooled. Just like heating curves, cooling curves have horizontal segments where the state changes from gas to liquid or from liquid to solid.

FIGURE 9.14 An ice-resurfacing machine uses hot water to melt the ice, which then rapidly refreezes to create a smooth rink for skating.

SAMPLE EXERCISE 9.4 Calculating Energy Transfer **LO3**

Between periods of a hockey game, an ice-resurfacing machine (Figure 9.14) spreads 3.00×10^2 liters of hot (40.0°C) water across a skating rink. (a) Identify the segments on the following cooling curve in terms of the change occurring on the part of the water:

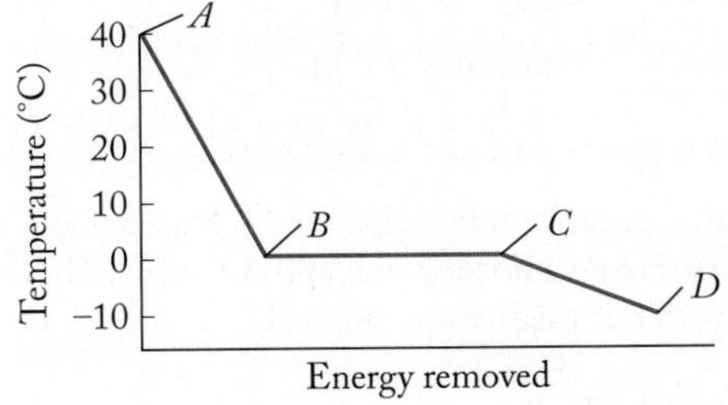

(b) How much energy must the water lose as it cools to its freezing point, freezes, and further cools to −10.0°C? Assume that the water is the system and that its density is 0.992 g/mL at 40.0°C.

Collect and Organize (a) We are given a cooling curve, and we know the temperature, volume, and density of water used to resurface a skating rink. We are asked to identify the changes that correlate to the segments on the curve and then to calculate how much energy the water loses as it cools from 40.0°C to −10.0°C. Table 9.3 lists specific heat and molar heat capacity values for liquid water and solid ice. The enthalpy of fusion (ΔH_{fus}) of water is 6.01 kJ/mol. (b) Dividing the volume of 40.0°C water applied to the ice by its density will give us the mass of water in grams. Dividing that number by the molar mass of H_2O (18.02 g/mol) will give us the number of moles, which is a value we can use with the molar heat capacities of liquid water and solid ice to calculate the energy losses during the cooling stages. When multiplied by the enthalpy of fusion, this value gives us how much energy must be removed to freeze the water. Given the large number of moles of H_2O in 300 L (~300 kg) of water and the high molar heat capacity of liquid water, we expect a large, negative value for the total energy removed.

Solve (a) First, we identify the line segments on the cooling curve with respect to the changes occurring: segment $\overline{AB}$, water cooled from 40°C to freezing point (0°C); segment $\overline{BC}$, liquid water converted to solid ice; segment $\overline{CD}$, ice at 0°C cooled to −10°C. (b) Next, we convert 3.00×10^2 L of water into moles:

$$3.00 \times 10^2 \text{ L} \times \frac{1000 \text{ mL}}{1 \text{ L}} \times \frac{0.992 \text{ g}}{1 \text{ mL}} \times \frac{1 \text{ mol}}{18.02 \text{ g}} = 1.651 \times 10^4 \text{ mol}$$

Then we calculate the energy lost in each stage of the cooling/freezing process.

1. Cooling 40.0°C water to its freezing point:

$$\begin{aligned} q_1 &= nc_{P,n}\Delta T \\ &= 1.651 \times 10^4 \text{ mol} \times \frac{75.3 \text{ J}}{\text{mol} \cdot {}^\circ\text{C}} \times (0.0 - 40.0)^\circ\text{C} \\ &= -4.973 \times 10^7 \text{ J} = -4.973 \times 10^4 \text{ kJ} \end{aligned}$$

2. Freezing the water:

$$\begin{aligned} q_2 &= n(-\Delta H_{fus}) \\ &= 1.651 \times 10^4 \text{ mol} \times \frac{-6.01 \text{ kJ}}{\text{mol}} = -9.923 \times 10^4 \text{ kJ} \end{aligned}$$

3. Cooling the ice:

$$\begin{aligned} q_3 &= nc_{P,n}\Delta T \\ &= 1.651 \times 10^4 \text{ mol} \times \frac{38.0 \text{ J}}{\text{mol} \cdot {}^\circ\text{C}} \times (-10.0 - 0.0)^\circ\text{C} \\ &= -6.274 \times 10^6 \text{ J} = -6.274 \times 10^3 \text{ kJ} \end{aligned}$$

Summing these three q values and rounding off the total to the correct number of significant figures:

$$\begin{array}{rl} q_1 & -4.973 \times 10^4 \text{ kJ} \\ +\, q_2 & -9.923 \times 10^4 \text{ kJ} \\ +\, q_3 & -0.6274 \times 10^4 \text{ kJ} \\ \hline & -15.5234 \times 10^4 \text{ kJ} = -1.55 \times 10^5 \text{ kJ} \end{array}$$

Think About It The large negative total value for q is reasonable given the large volume of water involved in resurfacing a hockey rink and the very high molar heat capacity of water. It is consistent with our prediction, too.

Practice Exercise The flame in a torch used to cut metal is produced by burning acetylene (C_2H_2) in pure oxygen. If the combustion of one mole of acetylene releases 1251 kJ of energy, what mass of acetylene is needed to cut through a piece of steel if the process requires 5.42×10^4 kJ of energy?

Water is an extraordinary substance for many reasons, and its high specific heat value (the second highest one in Table 9.3) is one of them. The ability of water to absorb large quantities of thermal energy is why it is used as a *heat sink* in both automobile radiators and our bodies. The term *heat sink* is often used to identify matter that can absorb energy without changing phase or significantly changing its temperature. Weather and climate changes are largely driven and regulated by cycles involving the retention of energy by Earth's oceans, which serve as enormous heat sinks.

Why does liquid water have such an unusually high specific heat? One of the principal reasons is hydrogen bonding. As we discussed in Chapter 6, molecules of H_2O can form up to four hydrogen bonds each. These strong intermolecular interactions contribute to a lower potential energy. As water absorbs thermal energy, its internal energy increases. Part of the increase goes toward increasing the kinetic energies of the water molecules and raising the temperature of the water. However, a significant part goes toward increasing the potential energy of the water by breaking some of the hydrogen bonds between its molecules via increased molecular motion. As we have seen in this section, increasing potential energy produces no increase in temperature. As a result, even more energy must be absorbed to produce a temperature change of, say, 1°C. In this way hydrogen bonding contributes to the rather high molar heat capacity of water (compared to the other compounds in Table 9.3). When we factor in its small molar mass compared to most other compounds, the extraordinarily high specific heat of liquid water seems reasonable.

CONNECTION As shown in Figure 6.29, a molecule of H_2O in ice or liquid water can form up to four hydrogen bonds: each of its H atoms can hydrogen-bond to the O atom of a nearby water molecule, and its O atom, with its two lone pairs of electrons, can form two hydrogen bonds to nearby H atoms.

Cold Drinks on a Hot Day

Let's consider another application of energy transfer. Suppose we throw a party and plan to chill three cases (72 aluminum cans, each containing 355 mL) of beverages by placing the cans, which are initially at a temperature of 25.0°C, in an insulated cooler and covering them with ice cubes that are initially at −18.0°C. If we assume that (1) the ice is sold in 10-pound bags; (2) the mass of aluminum in each can is 12.5 grams; (3) the cans contain mostly water, which has a density of 1.00 g/mL; and (4) the other ingredients are present in such small concentrations that they will not affect our energy transfer calculation, then how many bags of ice do we need to chill the cans and their contents to 0.0°C (that is, "ice cold")?

We can use the energy transfer relationships we have defined to predict how much ice is required. In doing so, we assume that whatever energy is absorbed by the ice is lost by the cans and the beverages in them. In equation form this assumption is

$$q_{\text{lost}} = -q_{\text{gained}} \tag{9.13}$$

and it applies to energy transfer processes that occur between warm and cold objects in any isolated system.

Let's first consider the energy lost by the cans and their contents. By substance, their masses are

$$72\ \cancel{\text{cans}} \times \frac{12.5\ \text{g Al}}{\cancel{\text{can}}} = 9.00 \times 10^2\ \text{g Al}$$

plus

$$72\ \cancel{\text{cans}} \times \frac{355\ \cancel{\text{mL}}\ \text{H}_2\text{O}}{\cancel{\text{can}}} \times \frac{1.00\ \text{g}}{\cancel{\text{mL}}} = 2.56 \times 10^4\ \text{g H}_2\text{O}$$

We can use Equation 9.9 and the specific heat values for aluminum and liquid water in Table 9.3 to calculate the quantity of energy that must be transferred to lower the temperatures of the beverages by −25.0°C:

$$q = mc_P\Delta T$$

$$= 9.00 \times 10^2 \cancel{\text{g Al}} \times \frac{0.897\ \text{J}}{\cancel{\text{g Al}} \cdot \cancel{°\text{C}}} \times (-25.0\cancel{°\text{C}})$$

$$+ 2.56 \times 10^4 \cancel{\text{g H}_2\text{O}} \times \frac{4.18\ \text{J}}{\cancel{\text{g H}_2\text{O}} \cdot \cancel{°\text{C}}} \times (-25.0\cancel{°\text{C}})$$

$$= -2.02 \times 10^4\ \text{J} + (-2.68 \times 10^6\ \text{J}) = -2.70 \times 10^6\ \text{J}) = -2.70 \times 10^3\ \text{kJ}$$

If the beverages transfer -2.70×10^3 kJ, then according to Equation 9.13 (and the law of conservation of energy), the ice gains $+2.70 \times 10^3$ kJ of energy. This energy serves two purposes: warming ice cubes to 0.0°C and then melting them. To calculate how many bags of ice melt, we'll need to solve for n, the number of moles of ice that we need. We focus on moles rather than grams because the enthalpy of fusion is expressed in kJ/mol, and we can calculate the energy required to warm the ice using its molar heat capacity [$c_{P,n}$ = 38.0 J/(mol · °C)] from Table 9.3 and Equation 9.10. To keep all energy units the same, let's convert $c_{P,n}$ to 0.0380 kJ/(mol · °C) before combining the energy absorbed by (1) warming and (2) melting the ice:

$$q_1 + q_2 = nc_{P,n}\Delta T + n\,\Delta H_{fus} = 2.70 \times 10^3\ \text{kJ}$$

$$\left(n \times \frac{0.0380\ \text{kJ}}{\text{mol} \cdot \cancel{°\text{C}}} \times 18.0\cancel{°\text{C}}\right) + \left(n \times \frac{6.01\ \text{kJ}}{\text{mol}}\right) = 2.70 \times 10^3\ \text{kJ}$$

$$n \times \left(\frac{0.684\ \cancel{\text{kJ}}}{\text{mol}} + \frac{6.01\ \cancel{\text{kJ}}}{\text{mol}}\right) = 2.70 \times 10^3\ \cancel{\text{kJ}}$$

$$n = 403\ \text{mol}$$

Converting moles into grams, then pounds, and then 10-pound bags:

$$403\ \cancel{\text{mol}} \times \frac{18.02\ \cancel{\text{g}}}{\cancel{\text{mol}}} \times \frac{1\ \cancel{\text{lb}}}{453.6\ \cancel{\text{g}}} \times \frac{1\ \text{bag}}{10\ \cancel{\text{lb}}} = 1.6\ \text{bags}$$

We can't buy a fraction of a bag, so we'll need to buy two bags of ice for our party.

SAMPLE EXERCISE 9.5 Calculating a Final Temperature from Energy Gain and Loss — LO3

Suppose you have exactly 1 cup (237 g) of hot (100.0°C) brewed tea in an insulated mug and that you add to it 2.50×10^2 g of ice initially at −18.0°C. If all of the ice melts, what is the final temperature of the tea? Assume that tea has the same thermal properties as water.

Collect and Organize We know the mass of tea, its initial temperature, and its specific heat (the c_P value for liquid water). We know the mass of ice and its initial temperature. From Tables 9.2 and 9.3 we know the specific heats and molar heat capacities of ice and liquid water and the enthalpy of fusion of ice. Our task is to find the final temperature of the tea.

Analyze The energy lost by the tea has the same absolute value as, but the opposite sign of, the energy gained by the ice cubes. In equation form:

$$q_{ice} = -q_{tea}$$

Based on the way the problem is presented, we have to assume that once all the ice has melted, the 250 grams of 0.0°C water produced will mix with and be warmed by the still warmer 237 grams of tea. Together they reach the same final temperature. Therefore, there are three parts to q_{ice}: the energy gained when the ice warms from −18.0°C to 0.0°C, the energy gained when the ice melts, and the energy gained when the temperature of the melted ice increases from 0.0°C to the final temperature (which we'll call x°C). We have to calculate the number of moles of H_2O in the ice to use ΔH_{fus}, so we'll use the molar heat capacities of ice and water to calculate q_{ice}. The tea undergoes no phase change, so its only loss of energy is due to its temperature change from 100.0°C to x°C.

Solve

1. Converting grams of ice to moles:

$$2.50 \times 10^2 \cancel{g} \times \frac{1 \text{ mol}}{18.02 \cancel{g}} = 13.87 \text{ mol}$$

2. The ice gains energy as it
   a. Warms to 0.0°C:

$$q_{ice} = nc_{P,n}\Delta T = 13.87 \cancel{\text{mol}} \times \frac{38.0 \text{ J}}{\cancel{\text{mol}} \cdot \cancel{°C}} \times (18.0\cancel{°C}) \times \frac{1 \text{ kJ}}{1000 \cancel{J}} = 9.49 \text{ kJ}$$

   b. Melts:

$$q_{ice} = n\Delta H_{fus} = 13.87 \cancel{\text{mol}} \times \frac{6.01 \text{ kJ}}{\cancel{\text{mol}}} = 83.4 \text{ kJ}$$

   c. Warms to final temperature x:

$$q_{ice} = nc_{P,n}\Delta T = 13.87 \cancel{\text{mol}} \times \frac{75.3 \cancel{J}}{\cancel{\text{mol}} \cdot \cancel{°C}} \times \frac{1 \text{ kJ}}{1000 \cancel{J}} \times (x - 0.0)\cancel{°C} = 1.044(x)\text{kJ}$$

3. The energy lost by the tea (q_{tea}) as it cools to x°C is

$$q_{tea} = mc_P\Delta T = 237 \cancel{g} \times \frac{4.18 \cancel{J}}{\cancel{g} \cdot \cancel{°C}} \times (x - 100.0)\cancel{°C} \times \frac{1 \text{ kJ}}{1000 \cancel{J}} = 0.991(x - 100.0)\text{kJ}$$

4. Balancing the loss and gain of energy:

$$q_{ice} = -q_{tea}$$

$$(9.49 + 83.4 + 1.044x)\cancel{\text{kJ}} = -(0.991x - 99.1)\cancel{\text{kJ}}$$

$$x = 3.1°C$$

Think About It The answer is consistent with the assumption that after all of the ice melted, the temperature of the liquid water thus produced would increase a little to reach the same temperature as the tea. There were no temperature units in the final values used to calculate x, but remember that x was assumed to have units of °C when we inserted it into the ΔT terms in the first steps of the calculation.

Practice Exercise Calculate the final temperature of a mixture of 0.350 kg of ice initially at −18°C and 237 g of water initially at 100.0°C.

Determining Specific Heat

Suppose your laboratory instructor gives you some metal pellets and asks you to determine whether the metal is aluminum, titanium, zinc, tin, or lead. You are not allowed to chemically alter the metal in order to determine its identity. In addition to the glassware in your lab drawer, you have access to a Bunsen burner, a balance, and Styrofoam coffee cups. How can you decide

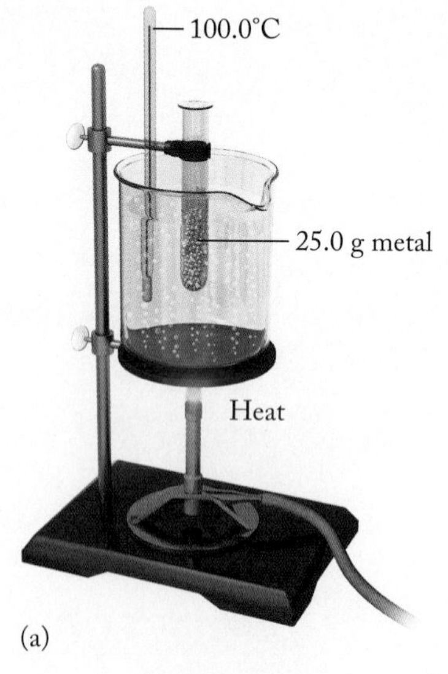

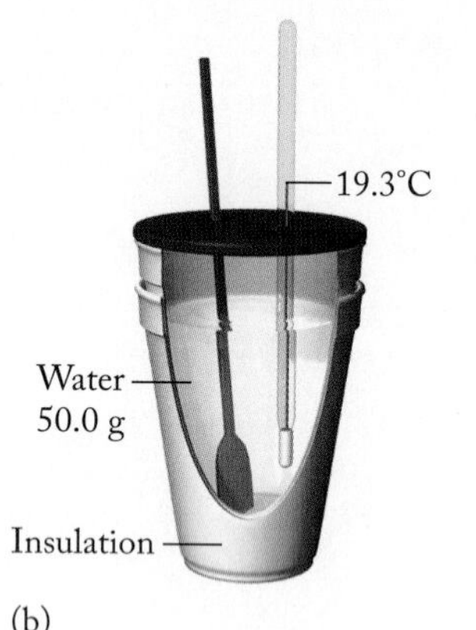

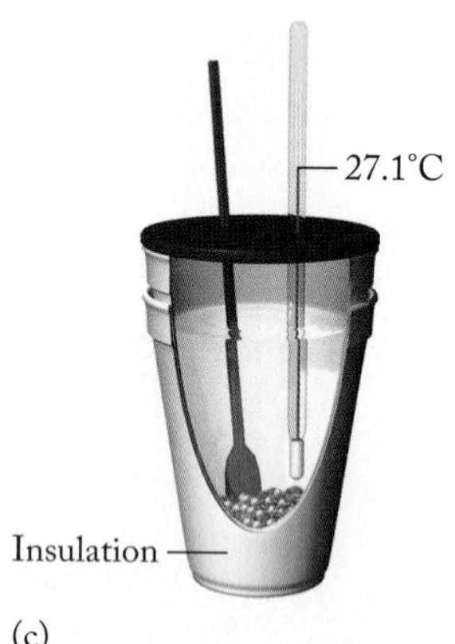

FIGURE 9.15 Experimental setup to determine the specific heat of a metal. (a) Heat 25.0 grams of the metal to 100.0°C in boiling water. (b) Add 50.0 grams of water at 19.3°C to nested Styrofoam cups. (c) Transfer the hot metal to the water, and record the peak temperature of the metal/water mixture, which reaches 27.1°C in this experiment.

which metal it is? (The title of this subsection and the data in Table 9.3 provide a hint.)

According to Table 9.3, the five possible metals have quite different specific heat values, ranging from 0.129 to 0.897 J/(g · °C). How can we determine the specific heat of the unknown metal? We need to link its mass (we have a balance) and a change in its temperature (easily measured with a thermometer) to the quantity of energy transferred during a temperature change. That is the challenge. One way to meet it is to couple the energy gained or lost by the metal to the energy lost or gained by a known substance. A good choice for this substance is water.

We can use an experimental setup like the one in Figure 9.15. It includes a Bunsen burner that we can use to heat a beaker of water to its boiling point: 100.0°C. Then we measure the mass of the pellets—let's say it is 25.0 g—and place them in a test tube to keep them dry. The test tube is immersed in the boiling water. After a while, the temperature of the metal should also be 100.0°C. Meanwhile we add 50.0 g of water to two nested Styrofoam coffee cups. We use the coffee cups because they have good insulating capability and negligible mass compared to the water.

The critical part of the experiment comes when we measure the temperature of the water in the nested cups—it's 19.3°C—and then quickly remove the test tube from the boiling water bath and pour the metal out of it and into the water in the cups. Then we measure the temperature of that water until it reaches a maximum value, which is 27.1°C. At this temperature we may assume that the energy gained by the water and the energy lost by the metal are in balance:

$$q_{\text{water}} = -q_{\text{metal}}$$

The left side of the equation is a quantity we can calculate using Equation 9.9:

$$q_{\text{water}} = mc_{\text{P}}\Delta T$$

$$= 50.0\ \cancel{\text{g}} \times \frac{4.18\ \text{J}}{\cancel{\text{g}} \cdot \cancel{°\text{C}}} \times (27.1 - 19.3)\cancel{°\text{C}} = 1.63 \times 10^3\ \text{J}$$

We apply this value to what we know about the right side of the equation:

$$1.63 \times 10^3\ \text{J} = -q_{\text{metal}} = -mc_{\text{P}}\Delta T$$

$$= -25.0\ \text{g} \times c_{\text{P}} \times (27.1 - 100.0)°\text{C} = (1823\ \text{g} \cdot °\text{C})\ c_{\text{P}}$$

$$c_{\text{P}} = 0.894\ \text{J/(g} \cdot °\text{C)}$$

According to the c_{P} values in Table 9.3, the mystery metal is aluminum.

CONCEPT TEST

In the preceding determination of the specific heat of a metal, how would the following factors have affected the experimental c_{P} value?

a. Slowly transferring the metal from the hot-water bath to the Styrofoam cups so that the metal's temperature was less than 100.0°C when it hit the water.

b. Drops of hot water adhering to the metal when it was transferred to the Styrofoam cups.

c. Ignoring the heat capacity of the thermometer.

d. Energy transfer through the Styrofoam cups or the open top. Assume the lab temperature was 22°C.

9.5 Enthalpies of Reaction and Calorimetry

In previous sections we have used specific heat, molar heat capacity, and enthalpies of fusion and vaporization to track flows of thermal energy as substances were warmed or cooled or changed physical state, but we have not explained how we know the values of those parameters. One of the ways we know them is through experiments based on calorimetry. **Calorimetry** is an analytical technique in which changes in the temperature of a device with a known heat capacity, called a **calorimeter**, are used to determine the energy released or absorbed by a process occurring inside the calorimeter. The process may be a physical exchange of thermal energy, or it can be a chemical reaction in which energy flows either from or into a reaction mixture.

The magnitude of the energy absorbed or released during a chemical reaction is proportional to the quantity of the reactants consumed and to a property of the reaction called its **enthalpy of reaction (ΔH_{rxn})**, often referred to as its *heat of reaction*. The subscript may be changed to reflect a specific type of reaction. For example, ΔH_{comb} is sometimes used to represent the enthalpy change that accompanies a combustion reaction.

The value of ΔH_{rxn} of a particular reaction depends on the difference in enthalpy between its products and reactants. It also depends on how we choose to represent the reaction in a balanced chemical equation. For example, suppose we write the balanced equation for the complete combustion of butane, the fuel in disposable lighters, in the following way:

$$2\ C_4H_{10}(g) + 13\ O_2(g) \rightarrow 8\ CO_2(g) + 10\ H_2O(\ell)$$

We can turn this chemical equation into a **thermochemical equation** by adding the enthalpy change that accompanies the combustion of 2 moles of C_4H_{10} by 13 moles of O_2 to form 8 moles of CO_2 and 10 moles of liquid H_2O:

$$2\ C_4H_{10}(g) + 13\ O_2(g) \rightarrow 8\ CO_2(g) + 10\ H_2O(\ell) \qquad \Delta H_{rxn} = -5754\ \text{kJ}$$

However, we might want to express the enthalpy change that accompanies the combustion of only 1 mole of butane. As you might expect, burning half the fuel releases half the energy and is accompanied by only half the change in enthalpy, as represented by the following thermochemical equation:

$$C_4H_{10}(g) + 13/2\ O_2(g) \rightarrow 4\ CO_2(g) + 5\ H_2O(\ell) \qquad \Delta H_{rxn} = -2877\ \text{kJ}$$

CONCEPT TEST

The enthalpy change accompanying the combustion of one mole of hydrogen gas, producing one mole of liquid water, is −286 kJ. What is the value of ΔH_{rxn} for the following reaction?

$$2\ H_2(g) + O_2(g) \rightarrow 2\ H_2O(\ell) \qquad \Delta H_{rxn} = ?$$

As noted in Section 9.3, enthalpy values are difficult to determine, but we can determine *changes* in enthalpy experimentally. The apparatus in Figure 9.15 is a kind of calorimeter, called a *coffee-cup calorimeter*, which is particularly useful for determining ΔH_{rxn} for reactions in aqueous solutions. Because the reactions happen at constant (ambient) pressure, the energy absorbed or released by the reaction (q_{rxn}) is equal to the enthalpy change that accompanies the reaction (see

calorimetry the experimental determination of the quantity of energy transferred during a physical change or chemical process.

calorimeter a device used to measure the absorption or release of energy by a physical change or chemical process.

enthalpy of reaction (ΔH_{rxn}) the enthalpy change that accompanies a chemical reaction; also called the *heat of reaction*.

thermochemical equation the chemical equation of a reaction that includes the change in enthalpy that accompanies the reaction.

Equation 9.7). If all of this energy is transferred to or from the contents of the calorimeter, then

$$q_{rxn} = -q_{calorimeter}$$

The value of $q_{calorimeter}$ can be calculated from the measured change in temperature (ΔT) and the heat capacity of the calorimeter ($C_{calorimeter}$):

$$q_{calorimeter} = C_{calorimeter}\,\Delta T$$

In reactions involving dilute aqueous solutions, nearly all of the mass (m) of the calorimeter comes from the mass of the water in it. This fact and the very high specific heat of water [$c_{P,H_2O} = 4.18$ J/(g · °C)] allow us to assume that the heat capacity of the calorimeter is essentially the same as the heat capacity of the water in it:

$$q_{calorimeter} = 4.18\ \text{J/(g}\cdot\text{°C)}\,m\Delta T \qquad (9.14)$$

SAMPLE EXERCISE 9.6 Calculating ΔH_{rxn} from Calorimetry Data **LO4**

When 0.200 L of 0.200 M HCl is mixed with 0.200 L of 0.200 M NaOH in a coffee-cup calorimeter, the temperature of the mixture increases from 22.15°C to 23.48°C. If the densities of the two solutions are 1.00 g/mL, what is the value of ΔH_{rxn} for the following reaction?

$$HCl(aq) + NaOH(aq) \rightarrow NaCl(aq) + H_2O(\ell) \qquad \Delta H_{rxn} = ?$$

Collect and Organize We know the (a) concentrations, (b) volumes, (c) densities, and (d) initial and final temperatures of two solutions: a strong acid and a strong base. We are asked to calculate the value of ΔH_{rxn} for the reaction in which they neutralize each other.

Analyze The densities of both aqueous solutions are nearly the same as that of water (1.00 g/mL), which confirms that they are dilute solutions with heat capacities that are essentially the same as that of water. Therefore, Equation 9.14 can be used to calculate $q_{calorimeter}$. Their common density also means that the total mass of the solutions in grams is the same as their total volume in milliliters: 400. Using this value in Equation 9.14 and a ΔT value of about 1.3 gives a $q_{calorimeter}$ of about $4 \times 400 \times 1.3 \approx 2000$ J, or 2 kJ. The energy gained by the solution represents about −2 kJ released by the exothermic neutralization reaction (q_{rxn}). To calculate ΔH_{rxn} for the reaction as written (one mole of each reactant), we will need to calculate the number of moles of each reactant in the reaction mixture. The volumes of each solution are 200 mL = 1/5 L and their concentrations are 1/5 molar, so there is only 1/5 × 1/5 = 1/25 of a mole of each reactant. Therefore, the value of ΔH_{rxn} for a reaction in which one mole of each reactant is consumed should be about −2 kJ × 25 = −50 kJ.

Solve Calculating the value of $q_{calorimeter}$ ($-q_{rxn}$):

$$\begin{aligned} q_{calorimeter} &= 4.18\ \text{J/(g}\cdot\text{°C)}\,m\Delta T \\ &= \frac{4.18\ \text{J}}{(\cancel{\text{g}}\cdot\cancel{\text{°C}})} \times (4.00 \times 10^2\ \cancel{\text{g}}) \times (23.48 - 22.15)\cancel{\text{°C}} = 2224\ \text{J} = 2.224\ \text{kJ} \end{aligned}$$

Therefore, $q_{rxn} = -2.224$ kJ. Calculating the value of ΔH_{rxn}:

$$\Delta H_{rxn} = \frac{-2.224\ \text{kJ}}{0.200\ \cancel{\text{L}} \times \dfrac{0.200\ \text{mol}}{\cancel{\text{L}}}} = -55.6\ \text{kJ/mol reactant, or } -55.6\ \text{kJ}$$

Think About It In the second step we converted the energy released by the reaction mixture (q_{rxn}) into the energy that would have been released had there been one mole of each reactant, which is the quantity of each of them in the balanced chemical equation.

Practice Exercise Addition of 1.31 g of zinc metal to 100.0 mL of 0.200 *M* HCl in a coffee-cup calorimeter causes the temperature to decrease from 15.21°C to 11.96°C. If the density of the HCl solution is 1.00 g/mL, what is the value of ΔH_{rxn} for the following reaction?

$$Zn(s) + 2\ HCl(aq) \rightarrow ZnCl_2(aq) + H_2(g)$$

Bomb Calorimetry

One of the most important categories of chemical reactions in terms of their transfer of thermal energy is combustion reactions. The quantities of energy they produce can be determined with devices called **bomb calorimeters** (Figure 9.16). To use these instruments, a combustible sample is placed in a sealed vessel (called a *bomb*) capable of withstanding high pressures and is submerged in a large volume of water in a heavily insulated container. Oxygen is introduced into the bomb, and the mixture is ignited with an electric spark. As combustion occurs, thermal energy generated by the reaction flows into the walls of the bomb and then into the water surrounding it.

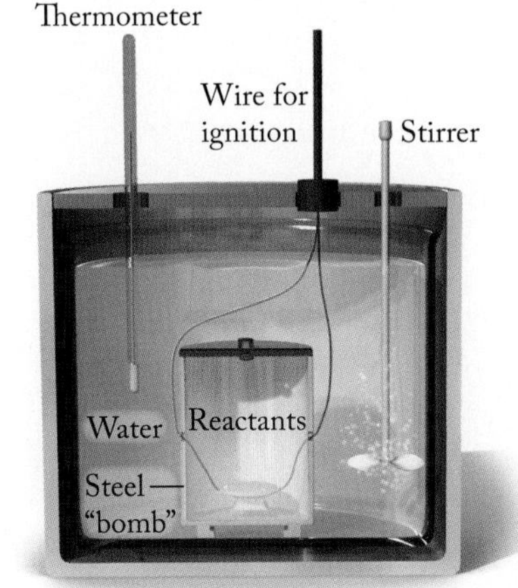

FIGURE 9.16 A bomb calorimeter.

A well-designed bomb calorimeter keeps the system (the chemical reaction) contained within the bomb and ensures that all energy released by the reaction stays inside the calorimeter. Specifically, the calorimeter consists of the bomb, the water, the insulated container, and minor components (stirrer, thermometer, and any other materials). The energy produced by the reaction is determined by measuring the temperature of the water before and after the reaction. The water is at the same temperature as all the other insulated parts, so the temperature change of the water tracks the temperature change of the entire calorimeter.

Measuring the change in temperature of the water is not the whole story. We also need to know the heat capacity of the water and all the other insulated components of the calorimeter—that is, its **calorimeter constant, $C_{calorimeter}$**. Why? Unlike in our coffee-cup calorimeter, where essentially all the energy warms a mass of water, the energy in a bomb calorimeter is absorbed by its many components, each of which has a heat capacity of its own. If we know the value of $C_{calorimeter}$ and if we can measure the change in water temperature, then we can calculate the quantity of energy that flows from a reaction mixture into a calorimeter.

How is $C_{calorimeter}$ determined? One way is to burn a known quantity of a material with a known enthalpy of combustion. The quantity of energy released by the reaction is then known, and the value of $C_{calorimeter}$ can be calculated from the change in temperature of the calorimeter produced by that quantity of energy. Benzoic acid (C_6H_5COOH) is often used for this purpose because very pure samples of it can be obtained. Complete combustion of exactly one gram of benzoic acid is known to release 26.38 kJ of thermal energy. Once $C_{calorimeter}$ has been determined, the calorimeter can be used to determine the quantities of energy produced by other combustion reactions on a per-gram or per-mole basis.

Because there is no change in the volume of the reaction mixture in a bomb calorimeter, this technique is referred to as *constant-volume calorimetry*. No *P–V* work is done, so Equation 9.4

$$\Delta E = q - P\Delta V$$

simplifies to

$$q = \Delta E$$

bomb calorimeter a constant-volume device used to measure the energy released during a combustion reaction.

calorimeter constant ($C_{calorimeter}$) the heat capacity of a calorimeter.

The energy lost by the reaction mixture in a bomb calorimeter is the energy gained by the calorimeter, so an increase in the temperature of a bomb calorimeter, which provides a measure of the energy flowing into it, also provides a measure of the energy released by the reaction happening inside it. Perhaps you are wondering how the ΔE measured in a bomb calorimeter is related to the ΔH of a reaction. For many reactions ΔE and ΔH are nearly the same, so we do not take into account the small differences between their ΔE and ΔH values.

SAMPLE EXERCISE 9.7 Determining the Heat Capacity of a Calorimeter — LO4

Before we can determine the energy change of a reaction run in a calorimeter, we must determine the heat capacity of the calorimeter, $C_{calorimeter}$. What is the value of $C_{calorimeter}$ if burning 1.000 g of benzoic acid increases the temperature of a calorimeter by 7.248°C? Combustion of benzoic acid releases 26.38 kJ of energy per gram of benzoic acid.

Collect, Organize, and Analyze We have been asked to find the calorimeter constant, which is the energy required to increase the temperature of a calorimeter by 1°C. We know that a reaction that releases 26.38 kJ of energy increases the temperature of the calorimeter by 7.248°C. Therefore, the calorimeter constant should be about 1/7 of 26.38, or a little less than 4 kJ/°C.

Solve Applying Equation 9.8 and solving for $C_{calorimeter}$:

$$q_{calorimeter} = C_{calorimeter}\Delta T$$

$$C_{calorimeter} = \frac{q_{calorimeter}}{\Delta T}$$

$$= \frac{26.38\ \text{kJ}}{7.248^\circ\text{C}} = 3.640\ \text{kJ/}^\circ\text{C}$$

Think About It The calorimeter constant is determined for a specific calorimeter. Once $C_{calorimeter}$ is known, the calorimeter can be used to determine the enthalpy of combustion of any combustible material. If any of the internal parts of the calorimeter change or are replaced, a new constant must be determined. Notice how large $C_{calorimeter}$ is compared to most of the substances in Table 9.3.

Practice Exercise When a 0.500 g sample of biodiesel fuel prepared from waste vegetable oil is burned in the bomb calorimeter from Sample Exercise 9.7, its temperature rises by 4.86°C. How much energy (in kilojoules) was released during combustion of the biodiesel sample?

9.6 Hess's Law and Standard Enthalpies of Reaction

As noted in the previous section, calorimetry can be used to determine the energy and enthalpy changes that accompany chemical reactions. However, there may be times when determining ΔH_{rxn} directly is not possible. For example, CO_2 is the principal product of the combustion of carbon in the form of charcoal:

$$(1) \qquad C(s) + O_2(g) \rightarrow CO_2(g) \qquad \Delta H_1$$

When the oxygen supply is limited, however, the products may include carbon monoxide:

$$(2) \qquad 2\ C(s) + O_2(g) \rightarrow 2\ CO(g) \qquad \Delta H_2$$

It is difficult to directly determine the enthalpy change that accompanies reaction (2) because as long as any oxygen is present, some of the CO formed may combine with O_2 to form CO_2, yielding a mixture of CO and CO_2. However, we can calculate ΔH for reaction (2) *indirectly* by starting with ΔH_{rxn} values that we *can* determine. For example, we can determine the enthalpy changes that accompany both the reaction in equation (1) and the combustion of a sample of pure CO gas:

$$(3) \qquad 2\ CO(g) + O_2(g) \rightarrow 2\ CO_2(g) \qquad \Delta H_3$$

Hess's law the principle that the enthalpy of reaction (ΔH_{rxn}) for a process that is the sum of two or more reactions is equal to the sum of the ΔH_{rxn} values of the constituent reactions; also called *Hess's law of constant heat of summation*.

Look closely at the reactants and products of chemical equations (1), (2), and (3). Equation (1) represents the complete combustion of carbon to CO_2, whereas equations (2) and (3) represent the stepwise combustion of carbon: first to CO and then to CO_2. As a result, the enthalpy change that accompanies the overall reaction (1) is related to the sum of the enthalpy changes associated with the stepwise reactions (2) and (3). Why? Because, as noted in Section 9.3, enthalpy change is a *state* function. This means that the ΔH_{rxn} value of an overall reaction that may occur in two or more steps, such as reaction (1), is the sum of the ΔH_{rxn} values of those steps. This holds true because the various steps consume the same reactants and eventually form the same products as the overall reaction.

Combining heats of reaction in this way is in accordance with **Hess's law**, also known as *Hess's law of constant heat of summation*, which states that the change in enthalpy that accompanies a process that occurs in more than one step is the sum of the enthalpy changes that occur in each of those steps. Let's put Hess's law to work by deriving an expression for enthalpy change that accompanies the incomplete combustion of C to CO as described in reaction (2) using the measurable ΔH_{rxn} values of reactions (1) and (3).

Combining both the equations that describe chemical reactions and their ΔH_{rxn} values is an exercise in pattern recognition. The key is to look for the reactants and products of the reaction whose ΔH_{rxn} we want to find in chemical equations describing the reactions whose ΔH_{rxn} values we know. In this example, we need to combine the chemical equations describing reactions (1) and (3) in a way that gives us the equation for reaction (2). Carbon and O_2 are reactants in equation (2), and CO is the only product. Inspecting the other two equations, we find that CO is a reactant in equation (3). Because CO is on the product side of equation (2), we flip equation (3) so that the reaction is written in reverse. Recall from Section 9.3 that reversing a reaction *changes the sign* of its ΔH value. Applying this principle to the reaction in equation (3), we get

$$(4) \qquad 2\ CO_2(g) \rightarrow 2\ CO(g) + O_2(g) \qquad \Delta H_4 = -\Delta H_3$$

Flipping equation (3) puts O_2 on the product side of equation (4), which is not where we want it. O_2 is on the reactant side of equation (1), so combining equations (1) and (4) may result in canceling out O_2 from the product side. However, we can't combine equations (1) and (4) just yet because there are two molecules of CO_2 on the reactant side of equation (4) but only one on the product side of equation (1). We'd prefer that these values be the same so that they cancel out when we combine equations (1) and (4), because there are no CO_2 terms at all in equation (2). To get two molecules of CO_2 on the product side of equation (1), we multiply all of the terms in the equation, *including* ΔH_1, by 2:

$$(5) \qquad 2\ C(s) + 2\ O_2(g) \rightarrow 2\ CO_2(g) \qquad \Delta H_5 = 2\Delta H_1$$

Now we combine equations (4) and (5), *which includes summing their ΔH_{rxn} values*:

$$
\begin{array}{llr}
(4) & 2\,CO_2(g) \rightarrow 2\,CO(g) + O_2(g) & \Delta H_4 = -\Delta H_3 \\
+\,(5) & 2\,C(s) + 2\,O_2(g) \rightarrow 2\,CO_2(g) & \Delta H_5 = 2\Delta H_1 \\
\hline
\multicolumn{3}{l}{\cancel{2\,CO_2(g)} + 2\,C(s) + 2\,O_2(g) \rightarrow 2\,CO(g) + \cancel{O_2(g)} + \cancel{2\,CO_2(g)}}
\end{array}
$$

or

$$(2) \quad 2\,C(s) + O_2(g) \rightarrow 2\,CO(g) \qquad \Delta H_2 = 2\Delta H_1 - \Delta H_3$$

Why did we multiply the enthalpy change that accompanies reaction (1) by 2 when we multiplied the coefficients of each of the reactants and product by 2? After all, why should changing the way we write the chemical equation impact the enthalpy change that accompanies the formation of the same product from the same reactants? Recall that the coefficients in chemical equations can represent the numbers of moles of reactants and products (as well as the numbers of atoms and molecules). Thus, reaction (1) describes the incomplete combustion of *one mole* of carbon, but reaction (5) describes the incomplete combustion of *two moles* of carbon. Burning twice as much fuel should generate twice as much heat—and be accompanied by twice the change in enthalpy.

Standard Enthalpy of Reaction (ΔH°_{rxn})

Now let's put some values on the enthalpy changes that accompany chemical reactions. Our focus will be the reactions that make up one industrial process for producing hydrogen gas (see Sample Exercise 7.12). In the first step of hydrogen production, methane reacts with a limited supply of high-temperature steam to produce carbon monoxide and hydrogen gas:

$$CH_4(g) + H_2O(g) \rightarrow CO(g) + 3\,H_2(g)$$

This reaction is called the steam-reforming reaction and is quite endothermic, requiring a flow of thermal energy into the reaction mixture to turn reactants into products. To describe *how* endothermic the reaction is, we use a thermodynamic property called the **standard enthalpy of reaction (ΔH°_{rxn})**, which is often called the *standard heat of reaction*. The adjective *standard*, indicated by the symbol °, describes the enthalpy change that accompanies a reaction under **standard conditions**, which means at a constant pressure of 1 atm. There is no universal standard temperature, though many tables of thermodynamic data, including those in the appendix of this book, apply to processes occurring at 25°C.

standard enthalpy of reaction (ΔH°_{rxn}) the enthalpy change associated with a reaction that takes place under standard conditions; also called the *standard heat of reaction*.

standard conditions in thermodynamics: a pressure of 1 atm (~1 bar) and some specified temperature, assumed to be 25°C unless otherwise stated; for solutions, a concentration of 1 *M* is specified.

standard state the most stable form of a substance under 1 atm pressure and some specified temperature (usually 25°C).

Implied in our notion of standard conditions is the assumption that parameters such as ΔH change with temperature and pressure. That assumption is correct, as we saw with enthalpies of vaporization in Section 9.3, but the changes are so small that we ignore them in the calculations in this chapter and those that follow. We also use the term **standard state** to describe the most stable physical state of a substance under standard conditions. For example, at $P = 1$ atm and $T = 25$°C, most metals and metalloids are solids, mercury and bromine are liquids, and H_2, N_2, O_2, F_2, Cl_2, and the group 18 elements are gases. These are the standard states of these elements.

Returning to the steam-reforming reaction and including its standard enthalpy of reaction value yields the following thermochemical equation:

$$(1) \quad CH_4(g) + H_2O(g) \rightarrow CO(g) + 3\ H_2(g) \qquad \Delta H_1^\circ = 206\ kJ$$

In the second step in hydrogen production, CO from the first step reacts with more steam to produce CO_2 and more H_2 gas in a reaction called the *water–gas shift reaction*:

$$(2) \quad CO(g) + H_2O(g) \rightarrow CO_2(g) + H_2(g) \qquad \Delta H_2^\circ = -41\ kJ$$

In accordance with Hess's law, we can write an overall thermochemical equation for the process by adding reactions (1) and (2) and their ΔH° values:

$$\begin{array}{lll} (1) & CH_4(g) + H_2O(g) \rightarrow \cancel{CO(g)} + 3\ H_2(g) & \Delta H_1^\circ = 206\ kJ \\ +\ (2) & \cancel{CO(g)} + H_2O(g) \rightarrow CO_2(g) + H_2(g) & \Delta H_2^\circ = -41\ kJ \\ \hline (3) & CH_4(g) + 2\ H_2O(g) \rightarrow CO_2(g) + 4\ H_2(g) & \Delta H_3^\circ = 165\ kJ \end{array}$$

This result is illustrated graphically in Figure 9.17.

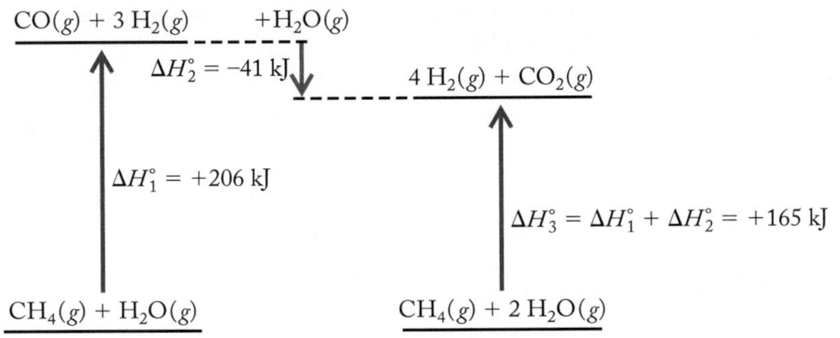

FIGURE 9.17 Hess's law predicts that the enthalpy change (ΔH_3°) for the reaction in which 1 mole of CH_4 and 2 moles of H_2O vapor produce 4 moles of H_2 and 1 mole of CO_2 is the sum of the enthalpy changes that accompany each of the two steps in the overall reaction: (1) 1 mole of CH_4 reacts with 1 mole of H_2O vapor to produce 1 mole of CO and 3 moles of H_2 (ΔH_1°), and (2) the reaction of the CO produced in the first step with another mole of H_2O vapor to form a fourth mole of H_2 and 1 mole of CO_2 (ΔH_2°).

SAMPLE EXERCISE 9.8 Calculating ΔH°_{rxn} Using Hess's Law — LO5

One reason furnaces and hot-water heaters fueled by natural gas need to be vented is that incomplete combustion can produce toxic carbon monoxide:

$$\text{Equation A:} \quad 2\ CH_4(g) + 3\ O_2(g) \rightarrow 2\ CO(g) + 4\ H_2O(g) \qquad \Delta H_A^\circ = ?$$

Use thermochemical equations B and C to calculate ΔH_A°:

$$\text{Equation B:} \quad CH_4(g) + 2\ O_2(g) \rightarrow CO_2(g) + 2\ H_2O(g) \qquad \Delta H_B^\circ = -802\ kJ$$

$$\text{Equation C:} \quad 2\ CO(g) + O_2(g) \rightarrow 2\ CO_2(g) \qquad \Delta H_C^\circ = -566\ kJ$$

Collect and Organize We are given two equations (B and C) with thermochemical data and a third (A) for which we are asked to find ΔH°. All the reactants and products in equation A are present in B and/or C.

Analyze We can manipulate equations B and C algebraically so that they sum to give the equation for which ΔH° is unknown. Then we can calculate the unknown value by applying Hess's law. Methane is a reactant in A and B, so we will use B in the direction written. CO is a product in A but a reactant in C, so we have to reverse C to get CO on the product side. Reversing C means that we must change the sign of ΔH_C°. If the coefficients in B and the reverse of C do not allow us to sum the two equations to obtain equation A, we will need to multiply one or both by appropriate factors.

Solve Comparing equation B as written and the reverse of C:

$$\text{(B)} \quad CH_4(g) + 2\,O_2(g) \rightarrow CO_2(g) + 2\,H_2O(g) \qquad \Delta H^\circ_B = -802 \text{ kJ}$$

$$\text{(C, reversed)} \quad 2\,CO_2(g) \rightarrow 2\,CO(g) + O_2(g) \qquad -\Delta H^\circ_C = +566 \text{ kJ}$$

with equation A, we find that the coefficient of CH_4 is 2 in A but only 1 in B, so we need to multiply all the terms in B by 2, including ΔH°_B:

$$\text{(2B)} \quad 2\,CH_4(g) + 4\,O_2(g) \rightarrow 2\,CO_2(g) + 4\,H_2O(g) \qquad 2\,\Delta H^\circ_B = -1604 \text{ kJ}$$

When we sum C (reversed) and 2B, the CO_2 terms cancel out and we obtain equation A:

$$\text{(C, reversed)} \quad \cancel{2\,CO_2(g)} \rightarrow 2\,CO(g) + \cancel{O_2(g)} \qquad -\Delta H^\circ_C = +566 \text{ kJ}$$

$$+\ \text{(2B)} \quad 2\,CH_4(g) + \overset{3}{\cancel{4}}\,O_2(g) \rightarrow \cancel{2\,CO_2(g)} + 4\,H_2O(g) \qquad 2\,\Delta H^\circ_B = -1604 \text{ kJ}$$

$$\text{(A)} \quad 2\,CH_4(g) + 3\,O_2(g) \rightarrow 2\,CO(g) + 4\,H_2O(g) \qquad \Delta H^\circ_A = -1038 \text{ kJ}$$

Think About It Our calculation shows that incomplete combustion of two moles of methane is less exothermic ($\Delta H^\circ_A = -1038$ kJ) than their complete combustion ($2\,\Delta H^\circ_B = -1604$ kJ), which makes sense because the CO produced in incomplete combustion reacts exothermically with more O_2 to form CO_2. In fact, the value of ΔH°_C for the reaction $2\,CO(g) + O_2(g) \rightarrow 2\,CO_2(g)$ is the difference between -1604 kJ and -1038 kJ.

Practice Exercise It does not matter how you assemble the equations in a Hess's law problem. Show that reactions A and C can be summed to give reaction B and result in the same value for ΔH°_B.

9.7 Enthalpies of Reaction from Enthalpies of Formation and Bond Energies

As noted in Sections 9.1 and 9.3, it is impossible to measure the *absolute* value of the internal energy of a substance, and the same is true for the enthalpy of a substance. However, we can establish *relative* enthalpy values that are referenced to a convenient standard: the **standard enthalpy of formation (ΔH°_f)** for a substance, which is defined as the enthalpy change that takes place at a constant pressure of 1 atm when one mole of a substance is formed from its constituent elements in their standard states. A reaction that fits this description is known as a **formation reaction**.

standard enthalpy of formation (ΔH°_f) the enthalpy change that takes place at a constant pressure of 1 atm when one mole of a substance is formed from its constituent elements in their standard states; also called the *standard heat of reaction*.

formation reaction a reaction in which one mole of a substance is formed from its component elements in their standard states.

The standard heat of formation of any pure element in its standard state is, by definition, zero. This is the zero point of all other enthalpy values (like using the freezing point of water as the zero point on the Celsius temperature scale). Standard heat of formation values for several compounds are given in Table 9.4, and a more complete list can be found in Appendix 4. Because the definition of a formation reaction specifies one mole of product, writing balanced equations for formation reactions may require the use of something we have tried to avoid until now: fractional coefficients in balanced equations. For example, the balanced thermochemical equation describing the formation of nitrogen monoxide is usually written

$$N_2(g) + O_2(g) \rightarrow 2\,NO(g) \qquad \Delta H^\circ_{rxn} = 180.6 \text{ kJ}$$

Although all reactants in the equation are in their standard states, it is not a formation reaction because two moles of product are formed. Therefore, ΔH°_{rxn} does

TABLE 9.4 Standard Enthalpies of Formation of Selected Substances at 25°C

Substance	ΔH_f° (kJ/mol)	Substance	ΔH_f° (kJ/mol)
$O_2(g)$	0	$CO_2(g)$	−393.5
$H_2(g)$	0	$CO(g)$	−110.5
$H_2O(g)$	−241.8	$N_2(g)$	0
$H_2O(\ell)$	−285.8	$NH_3(g)$, ammonia	−46.1
C(*s*, graphite)	0	$N_2H_4(g)$, hydrazine	95.35
$CH_4(g)$, methane	−74.8	$N_2H_4(\ell)$	50.63
$C_2H_2(g)$, acetylene	226.7	$NO(g)$	90.3
$C_2H_4(g)$, ethylene	52.4	$Br_2(\ell)$	0
$C_2H_6(g)$, ethane	−84.67	$CH_3OH(\ell)$, methanol	−238.7
$C_3H_8(g)$, propane	−103.8	$CH_3CH_2OH(\ell)$, ethanol	−277.7
$C_4H_{10}(g)$, butane	−125.6	$CH_3COOH(\ell)$, acetic acid	−484.5

not equal ΔH_f°. In fact, ΔH_{rxn}° is twice ΔH_f° because the equation as written describes the formation of *two* moles of NO. To write an equation describing the formation of one mole, we divide each coefficient (and ΔH_{rxn}°) in the preceding equation by 2:

$$\tfrac{1}{2}\,N_2(g) + \tfrac{1}{2}\,O_2(g) \rightarrow NO(g) \qquad \Delta H_f^\circ = 90.3 \text{ kJ}$$

SAMPLE EXERCISE 9.9 Recognizing Formation Reactions **LO5**

Which of the following ΔH_{rxn}° values are ΔH_f° values, assuming that each reaction takes place at 25°C and a constant pressure of 1 atm? For those that are not formation reactions, explain why not.

a. $H_2(g) + \tfrac{1}{2}\,O_2(g) \rightarrow H_2O(g)$ $\qquad \Delta H_{rxn}^\circ = -241.4$ kJ
b. $C_{graphite}(s) + 2\,H_2(g) + \tfrac{1}{2}\,O_2(g) \rightarrow CH_3OH(\ell)$ $\qquad \Delta H_{rxn}^\circ = -238.7$ kJ
c. $CH_4(g) + 2\,O_2(g) \rightarrow CO_2(g) + 2\,H_2O(g)$ $\qquad \Delta H_{rxn}^\circ = -802.3$ kJ
d. $P_4(s, \text{white}) + 6\,Cl_2(g) \rightarrow 4\,PCl_3(\ell)$ $\qquad \Delta H_{rxn}^\circ = -1278$ kJ

Collect, Organize, and Analyze We are given four balanced thermochemical equations and are asked to determine which of the ΔH_{rxn}° values are also ΔH_f° values. The standard heat of formation of a substance is the enthalpy of a reaction in which one mole of the substance is formed from its constituent elements, each in their standard state.

Solve

a. One mole of water vapor is formed from its constituent elements in their standard states. Therefore, this is the formation reaction for $H_2O(g)$, and its heat of reaction is ΔH_f° of water vapor.
b. One mole of liquid methanol is formed from its constituent elements in their standard states. Therefore, the equation describes a formation reaction, and the heat of reaction is ΔH_f°.
c. The reactants are not elements in their standard states and more than one mole of product is formed, so the heat of reaction is not ΔH_f°.
d. According to Table A4.3 in Appendix 4, the standard form of phosphorus is white phosphorus, which has the molecular formula P_4, so both reactants are elements in their standard states. However, the reaction produces *four* moles of PCl_3, so the heat of reaction is not the ΔH_f° of PCl_3. Actually, it is four times ΔH_f°.

Think About It Just because we can write formation reactions for substances like methanol does not mean that anyone would ever use that reaction to make methanol. Formation reactions are defined to provide a standard so that the flows of thermal energy in reactions can be evaluated.

 **Practice Exercise** Write formation reactions for (a) $CaCO_3(s)$; (b) $CH_3COOH(\ell)$ (acetic acid); (c) $KMnO_4(s)$.

Standard heats of formation can be used to predict standard heats of reaction. To do so, we break down each reaction into a series of formation reactions for the reactants and products. Then we apply Hess's law to combine the ΔH°_f values of the reactants and products into an overall ΔH°_{rxn} value. Let's use the combustion of methane to illustrate how this is done. As we saw in Chapter 7, the overall combustion reaction can be written

$$CH_4(g) + 2\ O_2(g) \rightarrow CO_2(g) + 2\ H_2O(g) \qquad (7.13)$$

The formation reactions for three of the four reactants and products (O_2 is a pure element in its standard state, so its $\Delta H^\circ_f = 0$) are the following:

$$\text{(A)} \quad C(s,\ \text{graphite}) + 2\ H_2(g) \rightarrow CH_4(g) \qquad \Delta H^\circ_f = -74.8\ \text{kJ}$$

$$\text{(B)} \quad C(s,\ \text{graphite}) + O_2(g) \rightarrow CO_2(g) \qquad \Delta H^\circ_f = -393.5\ \text{kJ}$$

$$\text{(C)} \quad H_2(g) + O_2(g) \rightarrow H_2O(g) \qquad \Delta H^\circ_f = -241.8\ \text{kJ}$$

To combine equations A, B, and C to end up with Equation 7.13, we need to first reverse equation A to put CH_4 on the reactant side. We also need to multiply equation C by 2 because we need a coefficient of 2 in front of H_2O in Equation 7.13, and also because we need a coefficient of 2 in front of H_2 to cancel out the H_2 term in the reverse of equation A. Making these changes and summing the three equations that result:

$$\text{(A, reversed)} \quad CH_4(g) \rightarrow C(s,\ \text{graphite}) + 2\ H_2(g) \qquad \Delta H^\circ_f = 74.8\ \text{kJ}$$

$$+\ \text{(B)} \quad C(s,\ \text{graphite}) + O_2(g) \rightarrow CO_2(g) \qquad \Delta H^\circ_f = -393.5\ \text{kJ}$$

$$+\ \text{(2C)} \quad 2\ H_2(g) + O_2(g) \rightarrow 2\ H_2O(g) \qquad \Delta H^\circ_f = -483.6\ \text{kJ}$$

$$CH_4(g) + \cancel{C(s,\ \text{graphite})} + 2\ O_2(g) + \cancel{2\ H_2(g)} \rightarrow \cancel{C(s,\ \text{graphite})} + \cancel{2\ H_2(g)} + CO_2(g) + 2\ H_2O(g)$$

or

$$CH_4(g) + 2\ O_2(g) \rightarrow CO_2(g) + 2\ H_2O(g) \qquad \Delta H^\circ_{rxn} = -802.3\ \text{kJ}$$

where $\Delta H^\circ_{rxn} = (74.8 - 393.5 - 483.6)\ \text{kJ} = -802.3\ \text{kJ}$.

Note how the reactant in the overall reaction is separated into its component elements in the first step in the preceding sequence. In the second and third steps the elements formed in the first step recombine to form the products. Thus, this sequence of steps describes the enthalpy changes that take place as the elemental building blocks in the reactant become the building blocks (with oxygen) in the products. We can create an analogous sequence for any chemical reaction by taking the following approach: sum the ΔH°_f values of the products, each multiplied by its coefficient in a balanced chemical equation describing the reaction, and then subtract the sum of the ΔH_f° values of the reactants, each multiplied by its coefficient. Expressing this sequence in equation form:

$$\Delta H^\circ_{rxn} = \sum n_{products}\, \Delta H^\circ_{f,products} - \sum n_{reactants}\, \Delta H^\circ_{f,reactants} \quad (9.15)$$

where $n_{products}$ is the number of moles of each product in the balanced equation and $n_{reactants}$ is the number of moles of each reactant. Applying Equation 9.15 to the combustion of methane:

$$\begin{aligned}\Delta H^\circ_{rxn} &= [(1\ \cancel{mol}\ CO_2)(-393.5\ kJ/\cancel{mol}) + (2\ \cancel{mol}\ H_2O)(-241.8\ kJ/\cancel{mol})]\\ &\quad - [(1\ \cancel{mol}\ CH_4)(-74.8\ kJ/\cancel{mol}) + (2\ \cancel{mol}\ O_2)(0.0\ kJ/\cancel{mol})]\\ &= [(-393.5\ kJ) + (-483.6\ kJ)] - [(-74.8\ kJ) + (0.0\ kJ)]\\ &= -802.3\ kJ\end{aligned}$$

SAMPLE EXERCISE 9.10 Calculating Standard Heats of Reaction from Standard Heats of Formation **LO5**

Use the appropriate standard heat of formation values to calculate ΔH°_{rxn} for the complete combustion of propane: $C_3H_8(g) + 5\ O_2(g) \rightarrow 3\ CO_2(g) + 4\ H_2O(g)$.

Collect and Organize The standard heats of formation of all the reactants and products are listed in Table 9.4. Our task is to use the balanced chemical equation and the ΔH_f° data from Table 9.4 to calculate the heat of combustion.

Analyze Equation 9.15 defines the relation between standard heats of formation of reactants and products and the standard heat of a reaction. Because O_2 gas is a pure element in its standard state, its ΔH_f° value is zero. We expect the reaction to be very exothermic ($\Delta H^\circ_{rxn} << 0$) because it represents combustion.

Solve Inserting ΔH_f° values for the products and reactants and their coefficients from the balanced chemical equation into Equation 9.15:

$$\begin{aligned}\Delta H^\circ_{rxn} &= \left[(3\ \cancel{mol}\ CO_2)\left(-393.5\ \frac{kJ}{\cancel{mol}}\right) + (4\ \cancel{mol}\ H_2O)\left(-241.8\ \frac{kJ}{\cancel{mol}}\right)\right]\\ &\quad - \left[(1\ \cancel{mol}\ C_3H_8)\left(-103.8\ \frac{kJ}{\cancel{mol}}\right) + (5\ \cancel{mol}\ O_2)\left(0.0\ \frac{kJ}{\cancel{mol}}\right)\right]\\ &= -2043.9\ kJ\end{aligned}$$

Think About It The large negative value is expected for the combustion of a hydrocarbon fuel. Propane is used in many backyard grills. Remember that the enthalpy value calculated in this problem only corresponds to the specific quantities of reactants and products described in the balanced chemical equation.

 **Practice Exercise** Use standard enthalpy of reaction values to calculate ΔH°_{rxn} for the water–gas shift reaction:

$$CO(g) + H_2O(g) \rightarrow CO_2(g) + H_2(g)$$

Enthalpies of Reaction and Bond Energies

The energy changes associated with chemical reactions depend on how much energy is required to break the bonds in the reactants and how much is released as the bonds in products form. For example, in the methane combustion reac-

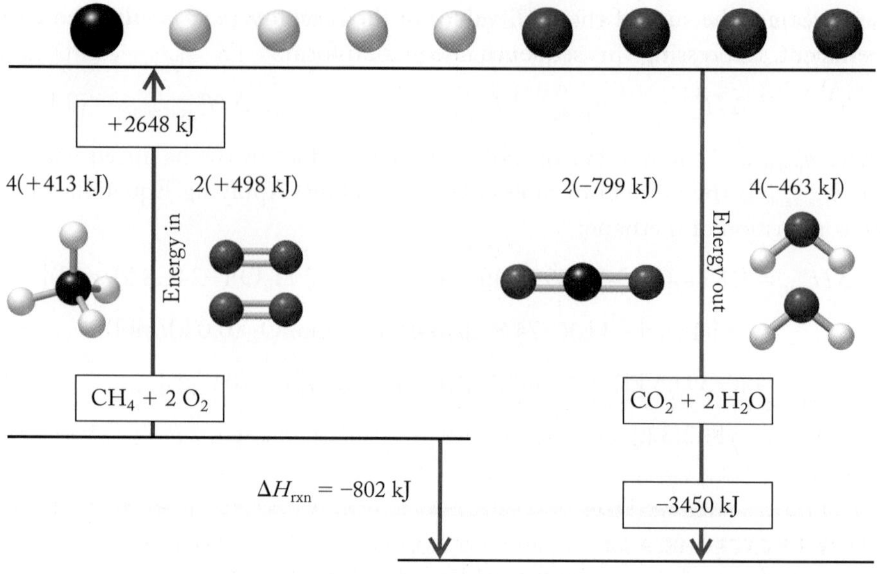

FIGURE 9.18 The complete combustion of 1 mole of methane requires that 4 moles of C—H bonds and 2 moles of O=O bonds be broken. Breaking bonds requires energy and is accompanied by an increase in enthalpy. Formation of 2 moles of C=O bonds and 4 moles of O—H bonds is accompanied by an even greater decrease in enthalpy, so the overall reaction is accompanied by a decrease in enthalpy and is exothermic.

tion [$CH_4(g) + 2\ O_2(g) \rightarrow CO_2(g) + 2\ H_2O(g)$], the C—H bonds in CH_4 and the O=O bonds in O_2 must be broken before the C=O bonds in CO_2 and the O—H bonds in H_2O can form. Breaking bonds is endothermic (the "Energy in" arrow in Figure 9.18), and forming bonds is exothermic (the "Energy out" arrows). If a chemical reaction is exothermic, as methane combustion is, more energy is released in forming the bonds in molecules of products than is consumed in breaking the bonds in molecules of reactants. If a chemical reaction is endothermic, on the other hand, more energy is consumed in breaking bonds in the reactants than is released in forming the bonds in the products.

Bond energy (or bond strength) is the enthalpy change (ΔH) that occurs when one mole of bonds in the gas phase is broken. The quantity of energy needed to break a particular bond is equal in magnitude but opposite in sign to the quantity of energy released when that same bond forms. In other words, breaking a bond requires an investment of energy and is a highly endothermic process ($\Delta H > 0$), whereas energy is released when atoms come together to make a covalent bond (and a more stable molecular structure). Thus, bond formation is a highly exothermic process ($\Delta H < 0$).

CONNECTION We first encountered bond energy in Chapter 4. Bond energies for some common covalent bonds are listed in Table 4.6, and a more complete list is found in Appendix 4.

During the combustion of one mole of CH_4 (Figure 9.18), we assume four moles of C—H bonds and two moles of O=O bonds are broken. The formation of one mole of CO_2 and two moles of H_2O requires the formation of two moles of C=O bonds and four moles of O—H bonds. The net change in energy resulting from breaking bonds in the reactants and forming bonds in the products can be estimated from the average bond energies. We start by taking an inventory of the bond energies involved:

	Bond	Number of Bonds (mol)	Bond Energy (kJ/mol)	ΔH
BONDS BROKEN	C—H	4	413	4 mol × 413 kJ/mol
	O=O	2	498	2 mol × 498 kJ/mol
BONDS FORMED	O—H	4	463	−(4 mol × 463 kJ/mol)
	C=O	2	799	−(2 mol × 799 kJ/mol)

Next we sum the positive ΔH values associated with breaking bonds and the negative ΔH values associated with forming them to obtain an estimate of the enthalpy change of the overall reaction:

$$\begin{aligned}\Delta H_{rxn} &= (4\ \cancel{mol} \times 413\ kJ/\cancel{mol}) \\ &\quad + (2\ \cancel{mol} \times 498\ kJ/\cancel{mol}) \\ &\quad - (4\ \cancel{mol} \times 463\ kJ/\cancel{mol}) \\ &\quad - (2\ \cancel{mol} \times 799\ kJ/\cancel{mol}) \\ &= -802\ kJ\end{aligned}$$

This estimate is essentially the same as the ΔH°_{rxn} value, −802.3 kJ, which we derived from the standard heats of formation of the reactants and products. We need to keep in mind that the energy of a particular type of bond, such as a C—H bond, can vary from one molecule to another or even within a molecule. However, bond energies are usually close enough to their average values that the averages can be used to calculate reliable estimates of the ΔH_{rxn} values of reactions that occur *in the gas phase*, where the impact of intermolecular forces on the energies of reactants and products can be ignored. The procedure we follow in estimating ΔH_{rxn} values for gas-phase reactions is expressed in equation form as follows:

CHEMTOUR Estimating Enthalpy Changes

$$\Delta H_{rxn} = \sum \Delta H_{\text{bond breaking}} - \sum \Delta H_{\text{bond forming}} \tag{9.16}$$

Here the negative sign is a reminder that we combine positive ΔH values (bond energies) for bonds that are broken with *negative* ΔH values (negative bond energies) for bonds that are formed.

SAMPLE EXERCISE 9.11 Estimating ΔH_{rxn} Values of Gas-Phase Reactions from Average Bond Energies **LO6**

Use average bond energies to estimate ΔH_{rxn} for the industrial synthesis of ammonia:

$$N_2(g) + 3\ H_2(g) \rightarrow 2\ NH_3(g)$$

Collect and Organize We need to estimate the value of ΔH_{rxn} of a gas-phase reaction from the bond energies of its reactants and products. Table A4.1 in Appendix 4 lists average values for the energies of covalent bonds. Equation 9.16 relates the average bond energies of the reactants and products to the heat of reaction.

Analyze First, we need to determine which types of bonds hold together molecules of N_2, H_2, and NH_3. This will require drawing their Lewis structures, as we learned to do in Chapter 4. Once we have the structures, we can take an inventory of the number of each type of bond that is broken or formed, look up the average bond energies for the bonds, and combine them using Equation 9.16. The reaction involves the formation of two moles of a compound from its component elements in their standard states. Therefore, the result of this calculation should be close to two times the standard heat of formation (ΔH°_f) of NH_3, which is −46.1 kJ/mol.

Solve Let's insert Lewis structures of the reactants and products into the balanced chemical equation to help get a count of all the bonds involved:

$$:\!N\equiv N\!: + 3\ H—H \rightarrow 2\ H—\ddot{N}(—H)—H$$

We then construct a table using the appropriate bond energies from Appendix 4:

	Bond	Number of Bonds (mol)	Bond Energy (kJ/mol)	ΔH
BONDS BROKEN	N≡N	1	945	1 mol × 945 kJ/mol
	H—H	3	436	3 mol × 436 kJ/mol
BONDS FORMED	N—H	6	391	−(6 mol × 391 kJ/mol)

Summing the values in the right-hand column to estimate ΔH_{rxn}:

$$\Delta H_{rxn} = (1 \text{ mol} \times 945 \text{ kJ/mol}) + (3 \text{ mol} \times 436 \text{ kJ/mol}) - (6 \text{ mol} \times 391 \text{ kJ/mol})$$
$$= -93 \text{ kJ}$$

Think About It An enthalpy change of −93 kJ means that the reaction is exothermic. The value for ΔH°_f of NH_3 (see Table 9.4 or Appendix 4) is −46.1 kJ/mol. Two moles of NH_3 are produced in the reaction, so multiplying by 2, we get −92.2 kJ, which is very close to the calculated value. Remember that the enthalpies of reaction calculated using bond energies are estimates because they use average bond energies. They generally vary from actual experimentally measured values.

Practice Exercise Use average bond energies to estimate ΔH_{rxn} for the water–gas shift reaction, which is part of the industrial process for making hydrogen gas:

$$CO(g) + H_2O(g) \rightarrow CO_2(g) + H_2(g)$$

9.8 Energy Changes When Substances Dissolve

FIGURE 9.19 Sand and salt (NaCl) help keep streets and highways clear of snow and ice during the winter.

In many cities and towns, winter brings subfreezing temperatures and snow—sometimes a lot of it. To keep walkways clear after plowing or shoveling, pellets of calcium chloride are often spread to melt packed snow and ice and to prevent new ice from forming. Meanwhile, road crews apply mixtures of sand and sodium chloride to keep streets and highways navigable (Figure 9.19).

In Chapter 11 we will explore why $CaCl_2$ and NaCl effectively melt ice, but for now, we explain why $CaCl_2$, which costs much more than NaCl, is so widely used to keep sidewalks free of ice. There are several reasons, but we focus here on a thermochemical one: when NaCl dissolves in melting ice, there is little change in the temperature of the ice or the meltwater, but when $CaCl_2$ dissolves, the water heats up, which helps melt more ice. Why is dissolving $CaCl_2$ a decidedly exothermic process, but dissolving NaCl is slightly endothermic (Figure 9.20)? To understand why, we need to analyze, step by step, how salts dissolve in water and to track the changes in enthalpy that accompany each of these steps.

In Section 9.7 we used covalent bond energies to estimate heats of reaction. In this section we consider energies associated with interactions between the ions in salts and molecules of water. As we have discussed in prior chapters, ionic compounds are held together by ionic bonds that have bond energies of many hundreds of kilojoules per mole. Molecules of H_2O interact with each other principally through hydrogen bonding. When an ionic solid dissolves in water, the ionic bonds that hold its ions together must be broken. Similarly, many hydrogen bonds must be broken as molecules of H_2O make space for and cluster around the dissolved ions.

enthalpy of solution ($\Delta H_{solution}$) the overall change in enthalpy that occurs when a solute is dissolved in a solvent; also called the *heat of solution*.

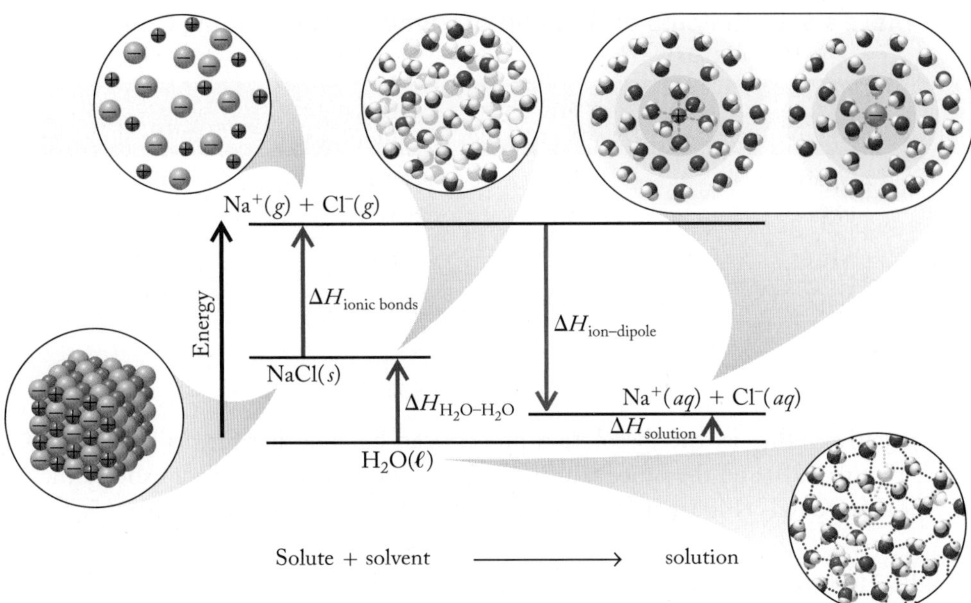

FIGURE 9.20 The enthalpy of solution ($\Delta H_{\text{solution}}$) of NaCl is the sum of the enthalpy changes that accompany breaking the hydrogen bonds between molecules of water (to make space for Na^+ and Cl^- ions), separating the Na^+ and Cl^- ions in solid NaCl, and forming ion–dipole interactions in solution.

These endothermic processes require investments of energy, which we can calculate by summing the enthalpy changes that accompany each process:

$$\text{Thermal energy invested} = \Delta H_{\text{ionic bonds}} + \Delta H_{H_2O-H_2O}$$

CONNECTION In Chapter 6 we discussed the intermolecular forces involved in determining solubility.

When ionic compounds dissolve in water, their ions interact with the permanent dipoles of the water molecules that form spheres of hydration around the ions. We use the symbol $\Delta H_{\text{ion-dipole}}$ to represent the change in enthalpy that accompanies these interactions. It is the exothermic part of the overall change in enthalpy, called the **enthalpy of solution ($\Delta H_{\text{solution}}$)** or the *heat of solution*, that accompanies the dissolution process (see Figure 9.20):

$$\Delta H_{\text{solution}} = \underbrace{[\Delta H_{\text{ionic bonds}} + \Delta H_{H_2O-H_2O}]}_{\text{Endothermic}} + \underbrace{\Delta H_{\text{ion-dipole}}}_{\text{Exothermic}} \quad (9.17)$$

Keep in mind that the energy required to break the ionic bonds in one mole of an ionic compound is equal in magnitude but opposite in sign to the lattice energy (U) of the compound:

$$\Delta H_{\text{ionic bonds}} = -U \quad (9.18)$$

We can determine the value of $\Delta H_{\text{solution}}$ experimentally using calorimetric methods such as the one shown in Figure 9.21. The heat of solution can be positive or

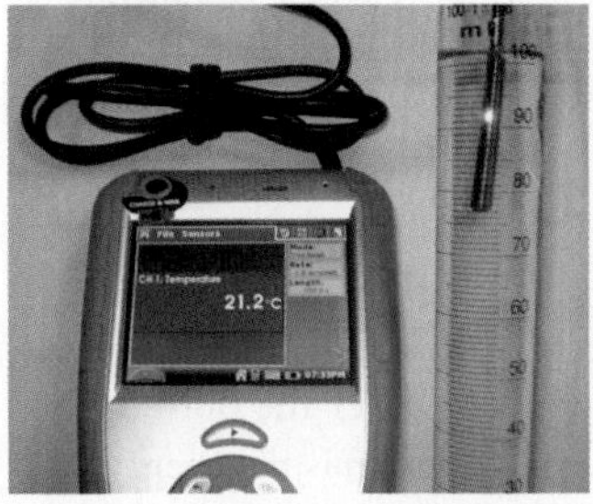

Initial water temperature = 21.2°C

Adding $CaCl_2$ to water

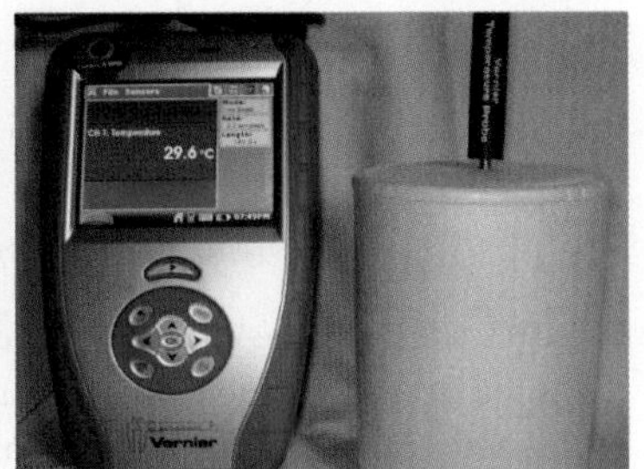

Final temperature of $CaCl_2$ solution = 29.6°C

FIGURE 9.21 Determining the enthalpy of solution. The temperature of 100 mL of water increases from 21.2°C to 29.6°C when 5.0 g of $CaCl_2$ dissolves in it. This temperature change and the heat capacity of water can be used to calculate the heat of solution of $CaCl_2$.

TABLE 9.5 Enthalpies of Solution of Some Common Ionic Compounds in Water

Endothermic Compound	$\Delta H_{solution}$ (kJ/mol)	Exothermic Compound	$\Delta H_{solution}$ (kJ/mol)
NH_4NO_3	25.7	$CaCl_2$	−82.2
KCl	17.2	KOH	−57.6
NH_4Cl	15.2	NaOH	−44.5
NaCl	4.0	LiCl	−37.1

CONNECTION Lattice energy is the energy released when gas-phase ions come together to form one mole of an ionic compound (see Chapter 4). This quantity of energy is consumed in separating a mole of the compound into individual gas-phase ions.

negative—that is, the dissolution process can be endothermic, as it is for NaCl (Figure 9.20) and the other salts in the left column of Table 9.5, or exothermic, as it is for $CaCl_2$ (Figure 9.21) and the salts in the right column of Table 9.5.

Calculating Lattice Energies Using the Born–Haber Cycle

Lattice energies are difficult to determine directly, but the lattice energy of a binary ionic compound can be calculated from its standard heat of formation and a judicious application of Hess's law (Section 9.7). To illustrate how this works, let's calculate the lattice energy of NaCl. We start with its standard heat of formation:

$$\text{Na}(s) + \tfrac{1}{2}\,\text{Cl}_2(g) \rightarrow \text{NaCl}(s) \qquad \Delta H^\circ_f = -411 \text{ kJ}$$

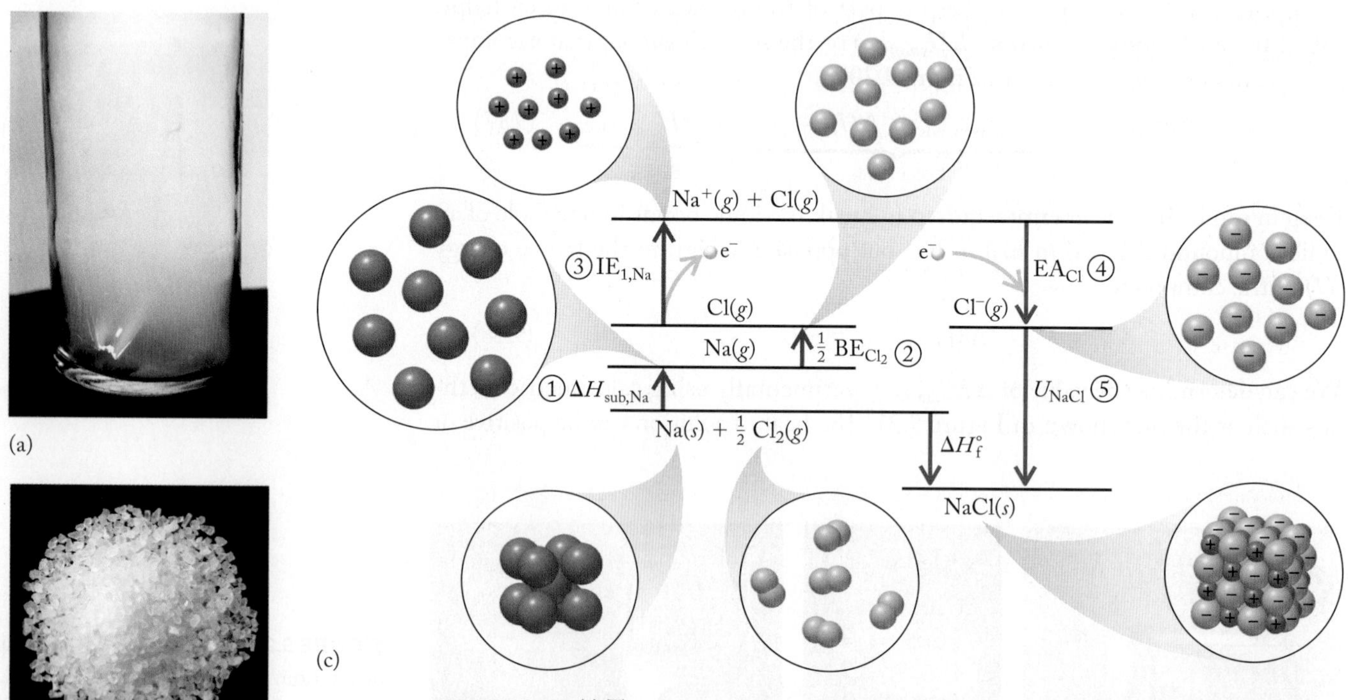

FIGURE 9.22 (a) The reaction between sodium metal and chlorine gas releases more than 400 kJ of energy per mole of NaCl produced. (b) Solid sodium chloride. (c) The Born–Haber cycle shows that the principal reason for this violently exothermic reaction is the energy released in step 5 when free sodium ions $Na^+(g)$ and free chloride ions $Cl^-(g)$ combine to form NaCl(*s*).

TABLE 9.6 Born–Haber Cycle for the Formation of NaCl(s)

Step	Process: Description	Process: Chemical Equation	Enthalpy Change (kJ)
1	Sublime 1 mol Na(s)	$Na(s) \rightarrow Na(g)$	$\Delta H_{sub,Na} = 108$
2	Break $\frac{1}{2}$ mol Cl—Cl bonds	$\frac{1}{2}\,Cl_2(g) \rightarrow Cl(g)$	$\frac{1}{2}\,BE_{Cl_2} = \frac{1}{2}\,(243) = 121.5$
3	Ionize 1 mol Na(*g*) atoms to form 1 mol Na^+ ions	$Na(g) \rightarrow Na^+(g) + e^-$	$IE_{1,Na} = 495$
4	1 mol Cl atoms acquires 1 mol electrons to form 1 mol Cl^- ions	$Cl(g) + e^- \rightarrow Cl^-(g)$	$EA_{Cl} = -349$
5	1 mol Na^+ ions combine with 1 mol Cl^- ions to form 1 mol NaCl(s)	$Na^+(g) + Cl^-(g) \rightarrow NaCl(s)$	$U_{NaCl} = ?$

The value of ΔH°_f can be determined experimentally with a calorimeter containing known quantities of elemental Na and Cl_2. Think of the formation reaction as consisting of five steps, as shown in Figure 9.22 and described as follows:

1. Subliming one mole of Na metal, producing one mole of gas-phase Na atoms. The enthalpy change, $\Delta H_{sub,Na}$, that accompanies this step is the heat of sublimation of sodium. Sublimation is an endothermic process (arrow points up in Figure 9.22).
2. Breaking the covalent bonds in 0.5 mole of Cl_2 molecules, producing one mole of gas-phase Cl atoms. The enthalpy change here is half the bond energy (BE) of one mole of Cl–Cl bonds, or $\frac{1}{2}\,BE_{Cl_2}$ (also endothermic).
3. Ionizing one mole of Na atoms to one mole of Na^+ ions in the gas phase, which requires adding a quantity of thermal energy equal to the first ionization energy of sodium, $IE_{1,Na}$ (also endothermic).
4. Combining one mole of Cl atoms with one mole of electrons to form one mole of Cl^- ions, which is accompanied by an enthalpy change equal to the electron affinity of chlorine, EA_{Cl} (an exothermic process for Cl and many other elements; arrow points down in Figure 9.22).
5. Forming one mole of solid NaCl from one mole each of gas-phase Na^+ and Cl^- ions. The enthalpy change that accompanies this step is the quantity we seek: the lattice energy (U_{NaCl}) of NaCl.

CONNECTION First ionization energy and electron affinity values are given in Figures 3.36 and 3.37, respectively.

This sequence of steps is diagrammed in Figure 9.22 and summarized in Table 9.6. It is called the **Born–Haber cycle**. The enthalpy changes that accompany the five steps add up to the standard heat of formation of NaCl, which is −411 kJ/mol: $\Delta H^\circ_{f,NaCl} = \Delta H_{sub,Na} + \frac{1}{2}\,BE_{Cl_2} + IE_{1,Na} + EA_{Cl} + U_{NaCl}$.

We insert the values from Table 9.6 and solve for U:

$$-411\ \text{kJ} = (108\ \text{kJ}) + \tfrac{1}{2}(243\ \text{kJ}) + (495\ \text{kJ}) + (-349\ \text{kJ}) + U_{NaCl}$$

$$U_{NaCl} = (-411\ \text{kJ}) - (108\ \text{kJ}) - (121.5\ \text{kJ}) - (495\ \text{kJ}) - (-349\ \text{kJ}) = -786\ \text{kJ}$$

The Born–Haber cycle can also be used to calculate enthalpy changes in other steps in the cycle that are difficult to determine experimentally, such as electron affinities. To do so, though, we need to know the enthalpy changes of all the other steps and the value of ΔH°_f.

SAMPLE EXERCISE 9.12 Calculating Lattice Energy **LO7**

Calcium fluoride occurs in nature as the mineral fluorite, which is the principal source of the world's supply of fluorine. (Highly reactive fluorine gas was first detected in

Born–Haber cycle a series of steps with corresponding enthalpy changes that describes the formation of an ionic solid from its constituent elements.

nature in a German fluorite mine in 2012.) Use the following data to calculate the lattice energy of CaF_2.

$$\Delta H_{\text{sub,Ca}} = 168 \text{ kJ/mol}$$
$$\text{BE}_{F_2} = 155 \text{ kJ/mol}$$
$$\text{EA}_F = -328 \text{ kJ/mol}$$
$$\text{IE}_{1,\text{Ca}} = 590 \text{ kJ/mol}$$
$$\text{IE}_{2,\text{Ca}} = 1145 \text{ kJ/mol}$$

Collect and Organize We are asked to calculate the lattice energy of CaF_2 using enthalpy changes that accompany steps in the Born–Haber cycle. Table A4.3 in Appendix 4 contains the standard heat of formation of solid CaF_2: −1228.0 kJ/mol.

Analyze The standard enthalpy of formation of CaF_2 is the enthalpy change for the overall process:

$$\text{Ca}(s) + \text{F}_2(g) \rightarrow \text{CaF}_2(s) \qquad \Delta H_f^\circ = -1228 \text{ kJ/mol}$$

When we break down this reaction into the steps of the Born–Haber cycle, we need to include both the first and second ionization energies of Ca because its atoms lose two electrons when forming Ca^{2+} ions. Two moles of fluorine atoms are needed to react with one mole of calcium atoms, so we need to break the F–F bonds in one mole of F_2, and we need to multiply the electron affinity of F by 2 to calculate the enthalpy change accompanying the formation of *two* moles of F^- ions. Figure 9.23 summarizes the Born–Haber cycle for calculating the lattice energy of CaF_2.

To estimate what a reasonable lattice energy value is, refer to Table 4.1, which lists the lattice energy of $MgCl_2$, another halide of an alkaline earth metal: $U_{MgCl_2} =$ −2540 kJ/mol. The charges on the ions in CaF_2 and $MgCl_2$ are the same. Even though Ca^{2+} ions are larger than Mg^{2+} ions, F^- ions are smaller than Cl^- ions, so the overall differences in ionic radii may offset, which would lead to comparable *d* values in Coulomb's law expressions (Equation 4.1) for the two compounds. Therefore, a U_{CaF_2} value near −2540 kJ/mol would be reasonable.

Solve The Born–Haber cycle for forming CaF_2 is

$$\Delta H_f^\circ = \Delta H_{\text{sub,Ca}} + \text{BE}_{F_2} + \text{IE}_{1,\text{Ca}} + \text{IE}_{2,\text{Ca}} + 2\,\text{EA}_F + U_{CaF_2}$$

Substituting in the values:

$$-1228 \text{ kJ} = 168 \text{ kJ} + 155 \text{ kJ} + 590 \text{ kJ} + 1145 \text{ kJ} + 2(-328 \text{ kJ}) + U_{CaF_2}$$

Solving for the lattice energy:

$$U_{CaF_2} = -2630 \text{ kJ/mol CaF}_2$$

Think About It The lattice energies of CaF_2 and $MgCl_2$ are similar in strength, just as we estimated.

 **Practice Exercise** Burning magnesium metal in air produces MgO and a very bright white light, making the reaction popular in fireworks and signaling devices:

$$\text{Mg}(s) + \tfrac{1}{2}\,\text{O}_2(g) \rightarrow \text{MgO}(s) \qquad \Delta H_f^\circ = -602 \text{ kJ}$$

Calculate the lattice energy of MgO from the following endothermic changes in enthalpy:

Process	Enthalpy Change (kJ/mol)	Process	Enthalpy Change (kJ/mol)
$Mg(s) \rightarrow Mg(g)$	148	$Mg(g) \rightarrow Mg^{2+}(g) + 2\,e^-$	2188
$O_2(g) \rightarrow 2\,O(g)$	498	$O(g) + 2\,e^- \rightarrow O^{2-}(g)$	605

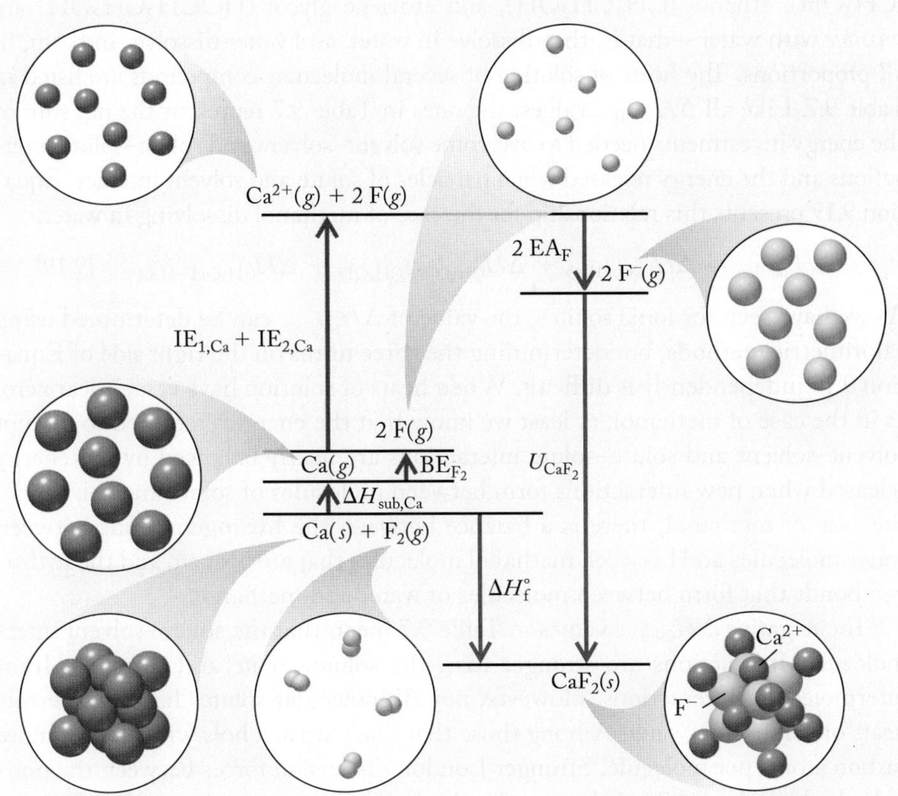

FIGURE 9.23 Born–Haber cycle for the formation of CaF_2.

Because the lattice energy of an ionic compound must be overcome if the compound is to dissolve, ionic compounds with very large lattice energies often have limited solubility in water. This result is consistent with the solubility rules described in Chapter 8—namely, ionic compounds formed by the group 1 cations (all 1+) and the ammonium ion (also 1+) are all soluble in water, as are all nitrates (1−), acetates (1−), and perchlorates (1−). We would expect compounds containing these ions to require less energy to separate into their component ions because of their small charges, which result in relatively small values in the numerator of Equation 4.1, which describes coulombic attraction:

$$E_{el} = 2.31 \times 10^{-19}\ J \cdot nm\left(\frac{Q_1 \times Q_2}{d}\right) \quad (4.1)$$

Moreover, NH_4^+, NO_3^-, and CH_3COO^- have relatively large ionic radii, which lead directly to relatively long distances between anion and cation centers and large d values in the denominator of Equation 4.1. Thus, NaF ($U = -910$ kJ/mol) is more soluble in water than MgO ($U = -3791$ kJ/mol) due to the greater attraction of the 2+ and 2− charges of the ions in MgO, as compared to the 1+ and 1− charges of the ions of comparable size in NaF. The solubility of NaF in water is 12 g/100 mL at 25°C, whereas the solubility of MgO is only 0.000062 g/100 mL.

Molecular Solutes

Until now our focus has been on enthalpy changes that accompany the dissolution of ionic solutes in water. However, many polar molecular compounds, especially those that can form hydrogen bonds with water molecules, are also soluble in water. In Chapter 6 we learned that low-molar-mass alcohols such as methanol

TABLE 9.7 Enthalpies of Solution of Some Common Molecular Compounds in Water

Compound	$\Delta H_{solution}$ (kJ/mol)
HCl	−74.8
NH_3	−30.5
CH_3CH_2OH	−10.6
CH_3OH	−3.0
CH_3COOH	−1.5

(CH_3OH), ethanol (CH_3CH_2OH), and ethylene glycol ($HOCH_2CH_2OH$) are *miscible* with water—that is, they dissolve in water, and water dissolves in them, in all proportions. The heats of solution of several molecular compounds are listed in Table 9.7. Like all $\Delta H_{solution}$ values, the ones in Table 9.7 represent the net sum of the energy investments needed to overcome solvent–solvent and solute–solute interactions and the energy released when particles of solute and solvent interact. Equation 9.19 presents this relationship for the case of methanol dissolving in water.

$$\Delta H_{solution} = \Delta H_{H_2O-H_2O} + \Delta H_{CH_3OH-CH_3OH} + \Delta H_{CH_3OH-H_2O} \quad (9.19)$$

As we have seen for ionic solutes, the value of $\Delta H_{solution}$ can be determined using calorimetric methods, but determining the three terms on the right side of Equation 9.19 independently is difficult. When heats of solution have values near zero, as in the case of methanol, at least we know that the energies required to disrupt solvent–solvent and solute–solute interactions are nearly balanced by the energy released when new interactions form between molecules of solute and solvent. In the case of methanol, there is a balance between the hydrogen bonds between water molecules and between methanol molecules that are broken, and the hydrogen bonds that form between molecules of water and methanol.

The negative $\Delta H_{solution}$ values in Table 9.7 mean that the solute–solvent intermolecular interactions are stronger than the solute–solute and solvent–solvent intermolecular interactions. However, not all molecular solutes have exothermic heats of solution in water. Among those that don't are alcohols with four or more carbon atoms per molecule. Stronger London dispersion forces between the nonpolar hydrocarbon tails of these molecules lead to stronger solute–solute interactions and contribute to endothermic heats of reaction. The increasing nonpolar (hydrophobic) nature of these alcohols also causes them to be less soluble in water.

CONNECTION The lower solubility in water of alcohols with greater molecular masses (and more nonpolar —CH_2— groups per molecule) was discussed in Chapter 6.

9.9 More Applications of Thermochemistry

In Section 9.7 we calculated the standard heats of reaction for the combustion of one mole of methane (−802.3 kJ) and one mole of propane (−2043.9 kJ). Does the much more negative (exothermic) value for propane make it an inherently better (higher-energy) fuel? Not necessarily. Expressing ΔH°_{rxn} values on a per-mole basis is the only way to ensure that we are talking about the same number of molecules. However, we do not purchase fuels, or anything else for that matter, in units of moles. Depending on the fuel, we are more likely to buy it by mass or by volume (gasoline and diesel fuel, for example).

To better compare methane and propane as fuels, let's calculate the enthalpy change that takes place when 1 g of each burns in air to produce CO_2 and water vapor. We divide the value of ΔH°_{rxn} (in kilojoules per mole) for each reaction by the molar mass of the hydrocarbon to determine the number of kilojoules of energy released per gram of substance:

$$CH_4: \quad \frac{-802.3\text{ kJ}}{\cancel{\text{mol}}} \times \frac{1\ \cancel{\text{mol}}}{16.04\text{ g}} = -50.02\text{ kJ/g}$$

$$C_3H_8: \quad \frac{-2043.9\text{ kJ}}{\cancel{\text{mol}}} \times \frac{1\ \cancel{\text{mol}}}{44.10\text{ g}} = -46.35\text{ kJ/g}$$

fuel value the quantity of energy released during the complete combustion of 1 g of a substance.

fuel density the quantity of energy released during the complete combustion of a particular volume of a liquid fuel.

From the perspective of the surroundings that absorb this energy, such as the air in a gas furnace, the water in a hot-water heater, or the food on a stove or backyard grill, equal masses of the two fuels provide comparable quantities of energy.

Energies like these, calculated on a per-gram basis, are called **fuel values**.

Fuel values of the C_1 to C_{10} alkanes are compared in Figure 9.24. Note how the fuel values *decrease* as the carbon numbers (and molar masses) *increase*. What is different about these compounds (besides the sizes of their molecules) that could explain this trend? One difference is the hydrogen-to-carbon ratios. As the number of carbon atoms per molecule increases, the hydrogen-to-carbon ratio decreases: from four atoms of H per atom of C in methane to about two atoms of H per atom of C in higher-molar-mass alkanes. To understand why this ratio is important, remember that it takes only 2.0 g of hydrogen atoms to make one mole of water vapor and release 241.8 kJ of thermal energy. However, it takes 12.0 g of carbon to form one mole of CO_2 and release 393.5 kJ. Thus, it takes six times as much carbon (by mass) to produce only about 60% more thermal energy than hydrogen. This is why pure hydrogen has the highest fuel value of all fuels.

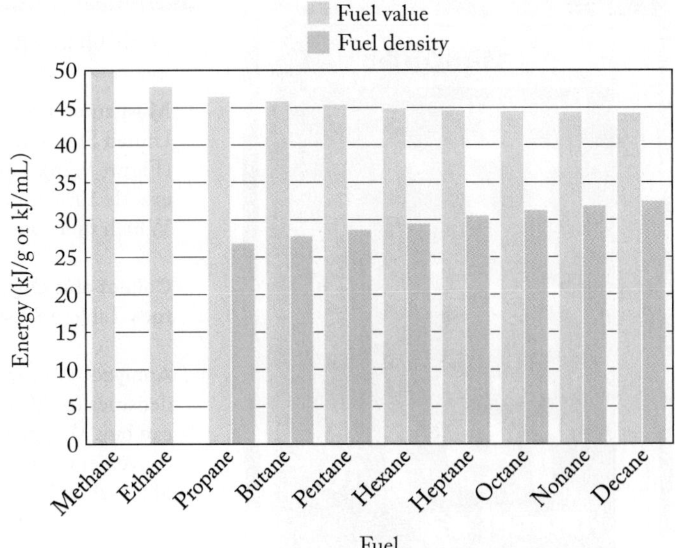

FIGURE 9.24 Fuel values and fuel densities of the C_1 to C_{10} alkanes based on their complete combustion to CO_2 and H_2O vapor.

Fuel values of the major fossil and renewable fuels are listed in Table 9.8. These data are based on combustion reactions that produce CO_2 and *liquid* H_2O, and they are somewhat higher than those plotted in Figure 9.24 for the same compounds (propane, for example) based on their combustion to CO_2 and H_2O *vapor*. Why are their fuel values different? Because it takes energy (water's enthalpy of vaporization) to vaporize a mole of liquid water. Therefore, liquid water is a lower-energy product than water vapor, and a combustion process that produces liquid water, such as the combustion of methane (Figure 9.25), is accompanied by a greater decrease in enthalpy than if water vapor had been produced.

Another term used to compare the energy content of liquid fuels is **fuel density**. Fuel density is the quantity of energy released per unit volume of a liquid fuel. Many liquid fuels are priced based on volume (gallons in the United States, liters in the rest of the world), so fuel density can be more useful than fuel value in comparing them as sources of thermal energy, as we do in Sample Exercise 9.13.

TABLE 9.8 Fuel Values of Some Common Fuels

Fuel	Fuel Value[a] (kJ/g)
Hydrogen	141.8
Natural gas	54.0
Propane	50.3
Butane	49.5
Gasoline	47.4
Diesel	44.8
Ethanol	29.8
Coal (anthracite)	27
Wood	17

[a]Based on complete combustion to CO_2 and liquid H_2O.

CONCEPT TEST

The data in Figure 9.24 show that the fuel densities of the C_3–C_{10} alkanes increase with increasing carbon number (and molar mass) even though their fuel values decrease. Why?

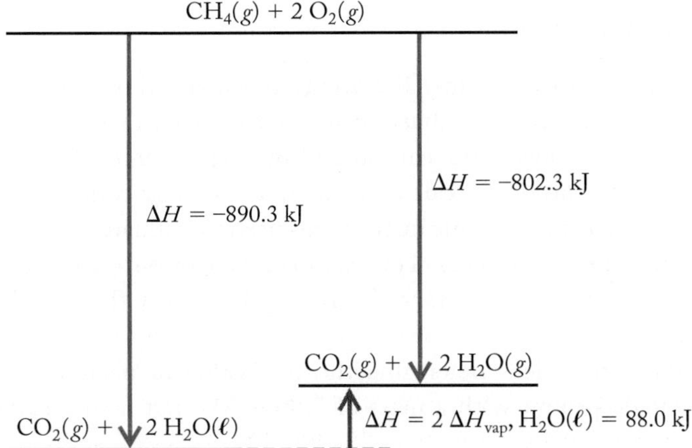

FIGURE 9.25 Methane's heats of combustion have different values depending on whether water vapor or liquid water is produced. The difference is due to the heat of vaporization of water.

FIGURE 9.26 Diesel fuel typically costs more than regular gasoline.

SAMPLE EXERCISE 9.13 Comparing Fuel Values and Fuel Densities **LO8**

Most automobiles run on either gasoline or diesel fuel. At most gas stations in the United States and Canada, the price of diesel fuel is higher than regular-grade gasoline (Figure 9.26). Suppose a particular station sells diesel fuel (density = 0.83 g/mL) for exactly 10% more per unit volume than regular-grade gasoline (density = 0.75 g/mL). Which fuel contains more thermal energy per dollar?

Collect and Organize We know the densities and relative prices of gasoline and diesel fuel. Table 9.8 lists their fuel values: 47.4 kJ/g for gasoline and 44.8 kJ/g for diesel fuel.

Analyze We can use the given density values to convert fuel values (in kJ/g) into fuel densities (in kJ/mL) and then compare the relative volumes the same amount of money can buy. We are not given the individual prices of the fuels, so to make the units convenient, let's make x the price of 1 mL of gasoline. That makes the price of 1 mL of diesel fuel $1.10x$. Gasoline has a slightly higher fuel value but significantly lower density, so it is likely diesel fuel will have the greater fuel density. However, diesel costs more per unit of volume, so the difference in energy content per x dollars (or cents) may be small.

Solve

$$\text{Gasoline:} \quad \frac{47.4\ \text{kJ}}{\text{g}} \times \frac{0.75\ \text{g}}{\text{mL}} \times \frac{1\ \text{mL}}{x} = 36\ \text{kJ}/x$$

$$\text{Diesel fuel:} \quad \frac{44.8\ \text{kJ}}{\text{g}} \times \frac{0.83\ \text{g}}{\text{mL}} \times \frac{1\ \text{mL}}{1.10\ x} = 34\ \text{kJ}/x$$

Therefore, gasoline provides slightly more energy content (about 6%) for the same amount of money x.

Think About It Comparable energy content for the money is only part of the energy story in selecting fuels (or buying cars). Another part is the efficiencies with which internal combustion (gasoline) engines and diesel engines convert the chemical energy in their fuels to mechanical work. On this score, diesel engines are considerably better, which translates into much greater distances traveled on the same fuel budget.

Practice Exercise In November 2011, an Atlas V rocket launched NASA's Mars Science Laboratory, including the Mars rover *Curiosity*, into space. The first-stage engine of the Atlas V (Figure 9.27) is fueled by kerosene, a hydrocarbon mixture with a density of 0.79 g/mL and a fuel value of 46 kJ/g. At full throttle the Atlas V engine generates 20 gigawatts (2.0×10^{10} J/s) of power. If the engine is 50% efficient at converting chemical energy into mechanical energy (thrust lifting the rocket into space), how rapidly (in liters per second) must kerosene be supplied to and completely burned in the engine?

FIGURE 9.27 Launch of the Mars Science Lab in 2011.

Energy from Food

Food is the fuel of living systems. The overall biochemical processes that convert foods into energy resemble combustion reactions, though each process involves many more steps to convert the carbon and hydrogen content of foods, such as carbohydrates, into carbon dioxide and water. The energy content of most foods can be determined using the same instruments used to evaluate the energy released by other combustible substances. For example, we can burn dry grain or dehydrated fruit in bomb calorimeters to determine the energy they would provide if we ate them.

As an illustration, let's determine the fuel value of peanuts. Suppose our sample is a single peanut with a mass of 1.89 g. We put it in a calorimeter for which $C_{calorimeter} = 19.31$ kJ/°C and burn it completely in excess oxygen. The resulting temperature increase of the calorimeter is 2.334°C. We determine the fuel value of the peanut much as we would determine the heat of combustion of any combustible material: by first calculating the quantity of heat required to raise the temperature of the calorimeter by 2.334°C:

$$\begin{aligned} q_{calorimeter} &= C_{calorimeter}\,\Delta T \\ &= (19.31\ \text{kJ}/\cancel{°\text{C}})(2.334\ \cancel{°\text{C}}) \\ &= 45.07\ \text{kJ} \end{aligned}$$

This quantity of thermal energy is generated by the combustion of 1.89 g, so the fuel value of the peanut is

$$\frac{45.07\ \text{kJ}}{1.89\ \text{g}} = 23.8\ \text{kJ/g}$$

The energy content of food is often expressed in Calories (Cal), where 1 Calorie $= 10^3$ cal $= 4.184$ kJ. The fuel value of the peanut in Calories is

$$23.8\ \cancel{\text{kJ}}/\text{g} \times \frac{1\ \text{Cal}}{4.184\ \cancel{\text{kJ}}} = 5.69\ \text{Cal/g}$$

SAMPLE EXERCISE 9.14 Calculating the Fuel Value of Food **LO8**

Glucose ($C_6H_{12}O_6$) is a simple sugar formed by photosynthesis in plants. The complete combustion of 0.5763 g of glucose in a calorimeter ($C_{calorimeter} = 6.20$ kJ/°C) raises the temperature of the calorimeter by 1.45°C. What is the fuel value of glucose in Calories per gram?

Collect and Organize We are asked to determine the fuel value of glucose, which means the energy given off when 1 g is completely converted to CO_2 and liquid H_2O. We know the mass of the glucose sample that is completely combusted and the temperature change produced by the heat released during combustion in a calorimeter with a known calorimeter constant.

Analyze The value of the heat released during combustion of the sample and gained by the calorimeter is the product of the calorimeter constant and the change in temperature. We can convert this quantity into Calories by using the conversion factor 1 Cal = 4.184 kJ. We anticipate a fuel value similar to that of the peanut.

Solve

$$q_{calorimeter} = C_{calorimeter}\,\Delta T = (6.20\ \text{kJ}/\cancel{°\text{C}})(1.45\cancel{°\text{C}}) = 8.99\ \text{kJ}$$

To convert this quantity of energy to a fuel value, we divide by the sample mass:

$$\frac{8.99\ \text{kJ}}{0.5763\ \text{g}} = 15.6\ \text{kJ/g}$$

and then convert into Calories:

$$(15.6\ \cancel{\text{kJ}}/\text{g})\left(\frac{1\ \text{Cal}}{4.184\ \cancel{\text{kJ}}}\right) = 3.73\ \text{Cal/g}$$

Think About It Food and nutrition scientists persist in using Calories to describe the energy content of foods even though the SI unit of energy, which is widely used in the physical sciences, is the joule. To add to the confusion, there are calories (lowercase "c") and there are Calories (uppercase "C"). They are not the same: 1 Cal = 1000 cal = 1 kcal. All three units are widely used in the health sciences and medicine.

Practice Exercise Sucrose (table sugar) has the formula $C_{12}H_{22}O_{11}$ ($\mathcal{M}$ = 342.30 g/mol) and a fuel value of 16.4 kJ/g. Determine the calorimeter constant of the calorimeter in which the combustion of 1.337 g of sucrose raises the temperature by 1.96°C.

FIGURE 9.28 Charles M. Hall (top) and Paul Louis-Toussaint Héroult (middle) independently developed the same electrolytic process for producing aluminum metal from aluminum ore. Hall's sister Julia (bottom), also a chemistry major at Oberlin College, assisted her brother in the lab, and her business skills made their aluminum production company a financial success. The company became the Aluminum Company of America, shortened to Alcoa.

CONCEPT TEST

Combustion of one gram of glucose ($C_6H_{12}O_6$) produces less energy than combustion of one gram of sucrose ($C_{12}H_{22}O_{11}$). Suggest a reason why.

Recycling Aluminum

Analyses of the energy required to carry out industrial procedures are frequently done to assess costs of operations and support new approaches to producing materials. The following evaluation of the energy requirements associated with the production and recycling of aluminum illustrates a practical application of thermochemical concepts.

Over the last century, aluminum, both alone and in combination with other metals, replaced steel for building structures in which high strength-to-weight ratios and corrosion resistance are paramount. These structures include major components of airplanes, motor vehicles, and the facades of buildings.

The industrial process for converting aluminum ore (Al_2O_3) into aluminum metal was developed by two 23-year-old chemists, Charles Hall and Paul Louis-Toussaint Héroult (both 1863–1914), working independently in the United States (Hall) and France (Héroult) (Figure 9.28). Their process is based on passing an electric current through a solution of aluminum oxide dissolved in molten cryolite (Na_3AlF_6). As electricity passes through the solution, Al^{3+} ions are reduced to aluminum metal while a positively charged carbon electrode is oxidized to carbon dioxide. The process is described by the following reaction:

$$2\,Al_2O_3(\text{in molten } Na_3AlF_6) + 3\,C(s) \rightarrow 4\,Al(\ell) + 3\,CO_2(g) \quad (9.20)$$

The principal energy cost of the Hall–Héroult process is the electricity needed to reduce Al^{3+} to Al. The major cost in recycling aluminum is the energy required to melt aluminum metal so that it can be reshaped into new products. We can use thermochemical principles to estimate the energy required for the Hall–Héroult process and compare it to the energy needed for recycling.

We can estimate the energy involved in producing aluminum metal from aluminum ore from the standard heats of formation of the reactants and products in Equation 9.20.

$$\begin{aligned}\Delta H^\circ_{\text{rxn}} &= [3(\Delta H^\circ_{\text{f},CO_2} + 4(\Delta H^\circ_{\text{f},Al(\ell)})] - [2(\Delta H^\circ_{\text{f},Al_2O_3}) + 3(\Delta H^\circ_{\text{f},C(s)})]\\ &= \left[(3\ \cancel{\text{mol CO}_2})\left(\frac{-393.5\text{ kJ}}{1\ \cancel{\text{mol CO}_2}}\right) + (4\ \cancel{\text{mol Al}})\left(\frac{10.79\text{ kJ}}{1\ \cancel{\text{mol Al}}}\right)\right]\\ &\quad - \left[(2\ \cancel{\text{mol Al}_2\text{O}_3})\left(\frac{-1675.7\text{ kJ}}{1\ \cancel{\text{mol Al}_2\text{O}_3}}\right) + (3\ \cancel{\text{mol C}})\left(\frac{0.0\text{ kJ}}{1\ \cancel{\text{mol C}}}\right)\right]\\ &= +2214.1\text{ kJ}\end{aligned}$$

Dividing this value by the four moles of aluminum produced in the reaction as written, we get

$$\frac{2214.1 \text{ kJ}}{4 \text{ mol Al}} = 553.52 \frac{\text{kJ}}{\text{mol Al}}$$

In this calculation we used a nonzero standard heat of formation value for aluminum metal. This may seem odd because ΔH_f° values of pure elements are normally zero. However, the zero value applies only to an element in its standard state, which for aluminum is solid, not the molten aluminum produced in the Hall–Héroult process. Therefore, we used the heat of fusion of aluminum (10.79 kJ/mol) as the standard heat of formation for liquid aluminum.

Our estimate of the energy requirements to reduce aluminum ore to aluminum metal did not include the energy needed to melt the cryolite in which the ore dissolves (its melting point is 1012°C), nor did we consider the inefficiency of using electricity to drive a chemical reaction. These and other factors make the energy requirements for producing one mole of Al from Al(III) ore in a modern refinery closer to 1000 kJ.

We can estimate the energy required to recycle one mole of aluminum by calculating the heat required to raise its temperature from 25°C to its melting point (660°C) and then melt it. The molar heat capacity of Al is 24.2 J/(mol · °C), so the heat needed to warm it to its melting point is

$$\begin{aligned} q &= nc_{\text{P,n}}\Delta T \\ &= 1.00 \cancel{\text{mol}} \times \frac{24.2 \cancel{\text{J}}}{\cancel{\text{mol}} \cdot \cancel{°\text{C}}} \times (660 - 25)\cancel{°\text{C}} \times \frac{1 \text{ kJ}}{1000 \cancel{\text{J}}} \\ &= 15.4 \text{ kJ} \end{aligned}$$

Once aluminum reaches its melting point, the energy required to melt it (ΔH_{fus} = 10.79 kJ/mol) is

$$\frac{10.79 \text{ kJ}}{\cancel{\text{mol}}} \times 1.00 \cancel{\text{mol}} = 10.8 \text{ kJ}$$

The estimated total energy to heat and melt 1.00 mole of aluminum is then

$$15.4 \text{ kJ} + 10.8 \text{ kJ} = 26.2 \text{ kJ}$$

This value represents only a small percentage of the energy needed to produce one mole of aluminum from its ore.

High energy costs make recycling aluminum economically attractive in addition to environmentally sound. The production and recycling of aluminum both incur energy costs, but detailed analyses of both processes have shown that recycling saves aluminum manufacturers about 95% of the energy required to produce the metal from ore. This energy saving has inspired the rapid growth of a global aluminum recycling industry. In the United States alone, aluminum recycling is a $1 billion per year business. In most industrialized countries nearly all of the aluminum in motor vehicles and building materials is recycled, as is about half the aluminum in food and beverage containers.

In this chapter we have explored how the internal energies of systems change as heat flows into, or out from, them and when work is done on, or by, them. Flows of heat or work produce (or accompany) changes in temperature, physical state, or chemical composition. The quantity of energy involved in physical changes can be calculated based on characteristic intensive properties such as the specific heat (or molar heat capacity), heat of fusion, and heat of vaporization. Energy is also a product or a reactant in virtually all chemical reactions and as such behaves stoichiometrically, which means that predictable quantities of energy are consumed or released during chemical reactions, depending on the quantities of the reactants consumed and the difference in chemical energy between them and the products. Combustion reactions supply most of the world's energy needs as chemical energy stored in different fuels is released when their carbon and hydrogen content is converted to CO_2 and H_2O. The fuel value of the foods that supply the energy needs of living organisms is similarly linked to a series of biochemical reactions that convert C and H atoms in the foods we eat to molecules of CO_2 and H_2O.

SAMPLE EXERCISE 9.15 Integrating Concepts: Selecting a Heating System

Suppose some friends who happen to be avid skiers decide to pool their resources and buy a vacation home near their favorite ski resort. They find one that meets their needs but then discover that it needs a new furnace, though it does have an operational wood stove. They collect information about replacement furnace options, focusing on two fuels: home heating oil, which is derived from crude oil and has a hydrocarbon distribution similar to that of diesel fuel, and propane. Their research on the two fuels yields the results summarized in the following table:

Fuel	Price of Fuel ($/U.S. gal)	Fuel Value (kJ/g)	Density (g/mL)	Furnace Efficiency
Heating oil	2.20	44.9	0.83	83%
Propane	2.88	50.3	0.62	95%

a. Based on these data, which of the two fuels is the better buy?
b. The brochure for the propane furnace they are considering describes it as a "condensing" furnace, noting that it is 10% more efficient than a noncondensing model. What does "condensing" mean in this context, and how much does it contribute to the higher efficiency of the furnace?
c. The most frugal of the friends suggests that they use the wood stove to save fuel costs. A local firewood dealer sells one cord of seasoned hardwood, which weighs about one metric ton, for $250. How much money would the friends save burning a cord of wood instead of the better fuel identified in part (a)? Assume the fuel value of the wood is 17 kJ/g and that the wood stove is 50% efficient.
d. Which of the fuels under consideration produces the most "fossil" CO_2 per kJ of heat delivered to the home's living space?

Collect and Organize We know the price per U.S. gallon of two fossil fuels and the price of wood, a renewable alternative. We also know the densities of the fossil fuels and the mass of a cord of wood, and we know the fuel values of all three fuels. We need to select which of the two fossil fuels is the better energy buy and then compare its cost-effectiveness with that of wood. The efficiencies with which stoves and furnaces convert the chemical energy of fuels into heat for the living space are also a factor, and we have to explain why a condensing furnace for burning propane is so efficient. Among the unit conversion factors in the back of the book is 1 U.S. gallon = 3.785 L.

Analyze One way to compare the costs of the three fuels is to start with the fuel values and convert them from kJ/g into kJ/$. The prices are based on volumes, so we will need to use the densities of the fuels as one of the conversion factors. Fuel values such as those given in Table 9.8 and in this problem are based on combustion reactions that convert the carbon and hydrogen content of the fuel into CO_2 and liquid H_2O. To compare fuel costs, we need to convert price-per-volume values into price per mass and then price per unit of energy produced. Then we use the efficiencies of the furnaces to calculate the price per unit of heat delivered to the vacation home's living space.

The CO_2 produced by the combustion of propane and heating oil is considered fossil CO_2 because the fuels are derived from natural gas and crude oil, making them fossil fuels. The fossil fuel with the higher carbon-to-hydrogen ratio will be the one that produces more fossil CO_2 per kilojoule of energy released.

Solve

a. Converting the fuel value of heating oil into kilojoules of heat for the home:

$$\frac{44.9\text{ kJ}}{\text{g}} \times \frac{0.83\text{ g}}{\cancel{\text{mL}}} \times \frac{1000\ \cancel{\text{mL}}}{\cancel{\text{L}}} \times \frac{3.785\ \cancel{\text{L}}}{\cancel{\text{gallon}}} \times \frac{1\ \cancel{\text{gallon}}}{\$2.20} \times \frac{83\%}{100\%} = 5.3 \times 10^4\text{ kJ/\$}$$

Doing the same conversion for propane:

$$\frac{50.3\text{ kJ}}{\text{g}} \times \frac{0.62\text{ g}}{\cancel{\text{mL}}} \times \frac{1000\ \cancel{\text{mL}}}{\cancel{\text{L}}} \times \frac{3.785\ \cancel{\text{L}}}{\cancel{\text{gallon}}} \times \frac{1\ \cancel{\text{gallon}}}{\$2.88} \times \frac{95\%}{100\%} = 3.9 \times 10^4\text{ kJ/\$}$$

Therefore, heating oil provides more energy per dollar.

b. The higher efficiency of the *condensing* propane furnace is the result of extracting additional heat from the reaction mixture as it cools and the water vapor in it *condenses*, giving the following overall combustion reaction:

$$C_3H_8(g) + 5\ O_2(g) \rightarrow 3\ CO_2(g) + 4\ H_2O(\ell)$$

instead of the one from Sample Exercise 9.10:

$$C_3H_8(g) + 5\ O_2(g) \rightarrow 3\ CO_2(g) + 4\ H_2O(g)$$
$$\Delta H^\circ_{\text{rxn}} = -2043.9\text{ kJ}$$

The difference between the ΔH°_{rxn} values of these two reactions is four times the heat of vaporization of one mole of water at 298 K:

$$H_2O(\ell) \rightarrow H_2O(g) \qquad \Delta H_{vap} = 44.0 \text{ kJ}$$

According to Hess's law, we can reverse this vaporization equation and multiply it by four to match the four moles of H_2O in the combustion reactions. We add it to the thermochemical equation from Sample Exercise 9.10 to get our desired equation:

$$C_3H_8(g) + 5\,O_2(g) \rightarrow 3\,CO_2(g) + \cancel{4\,H_2O(g)} \qquad \Delta H^\circ_{rxn} = -2043.9 \text{ kJ}$$

$$+\cancel{4\,H_2O(g)} \rightarrow 4\,H_2O(\ell) \qquad 4 \times (-\Delta H_{vap}) = -176.0 \text{ kJ}$$

$$C_3H_8(g) + 5\,O_2(g) \rightarrow 3\,CO_2(g) + 4\,H_2O(\ell) \qquad \Delta H^\circ_{rxn} = -2219.9 \text{ kJ}$$

The ratio of ΔH°_{rxn} values is

$$\frac{-2219.9 \text{ kJ}}{-2043.9 \text{ kJ}} = 1.0861$$

This nearly 9% increase in ΔH°_{rxn} accounts for most of the greater efficiency of the condensing propane furnace as compared to the noncondensing model.

c. Calculating the heat obtained by burning a cord of hardwood in the wood stove:

$$\frac{17 \text{ kJ}}{\text{g}} \times \frac{1000 \text{ g}}{\cancel{\text{kg}}} \times \frac{1000\,\cancel{\text{kg}}}{\cancel{\text{ton}}} \times \frac{1\,\cancel{\text{ton}}}{\cancel{\text{cord}}} \times 1\,\cancel{\text{cord}} \times \frac{50\%}{100\%}$$

$$= 8.5 \times 10^6 \text{ kJ}$$

The cost of the heating oil needed to produce this much heat is

$$8.5 \times 10^6\,\cancel{\text{kJ}} \times \frac{\$1}{5.32 \times 10^4\,\cancel{\text{kJ}}} = \$160$$

This cost is below that of a cord of hardwood, so burning wood would achieve no cost savings over heating oil.

d. We know from its formula that the ratio of C to H atoms in propane is 3:8, but we are not given this information for the hydrocarbons in heating oil. However, we do know that heating oil has a lower fuel value than propane, which means that it is composed of hydrocarbons with higher molar masses that have higher carbon-to-hydrogen ratios than propane. Therefore, more of heating oil's fuel value comes from the conversion of C to CO_2 than from the conversion of H to H_2O. This means it produces more CO_2 per kilojoule of heat released than propane. On top of that, the heating oil furnace is less efficient than the propane furnace, so heating oil produces even more fossil CO_2 to heat the home than propane. Wood is a renewable biofuel and produces no fossil CO_2 during combustion.

Think About It These calculations of energy costs considered the price and fuel values of three fuels and the efficiencies with which their chemical energies could be used to heat a living space. All three factors are important, though there are others we did not consider, such as the cost of installing a new furnace and connecting it to the home's existing heating system. Environmental concerns regarding climate change might also lead the skiing friends to use the wood stove to heat the home on ski weekends.

SUMMARY

Surroundings

System

$q > 0$

$w > 0$

LO1 Thermodynamics is the study of energy and its transformations, whereas **thermochemistry** is the study of the changes in energy that accompany chemical reactions. The **first law of thermodynamics** states that energy is neither created nor destroyed, so the energy gained or lost by a system equals the energy lost or gained by the surroundings. We use the symbol q to represent the *quantity* of thermal energy transferred between a system and its surroundings. In an **exothermic process** the system loses energy by heating its surroundings ($q < 0$); in an **endothermic process** the system absorbs energy from its warmer surroundings ($q > 0$). (Sections 9.1–9.3)

LO2 The sum of the kinetic and potential energies of a system is its **internal energy (E)**. The internal energy of a system is increased ($\Delta E = E_{final} - E_{initial}$ is positive) when it is heated ($q > 0$) or if work is done on it ($w > 0$). (Section 9.2)

LO3 The **enthalpy (H)** of a system is given by $H = E + PV$. The **enthalpy change (ΔH)** of a system is equal to the energy (q_P) added to or removed from the system at constant pressure: $\Delta H > 0$ for endothermic reactions and $\Delta H < 0$ for exothermic reactions. The **enthalpy of fusion (ΔH_{fus})** is the change in enthalpy when one mole of the solid substance melts at its melting point. The **enthalpy of vaporization (ΔH_{vap})** is the enthalpy change that occurs when one mole of the liquid substance is vaporized at its boiling point. (Section 9.4)

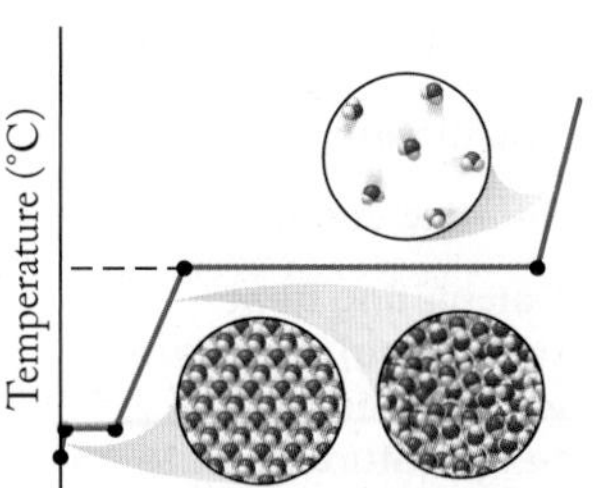

LO4 A **calorimeter** is a device for determining how much heat is transferred from a system (often a reaction mixture) to its surroundings. A coffee-cup calorimeter operates at constant (atmospheric) pressure and is useful for determining the enthalpy changes or **enthalpies of reaction (ΔH_{rxn})** of aqueous-phase chemical reactions. A **bomb calorimeter** is a constant-volume device used to determine the energy released during combustion reactions. A **thermochemical equation** includes the change in enthalpy that accompanies a chemical reaction. (Section 9.5)

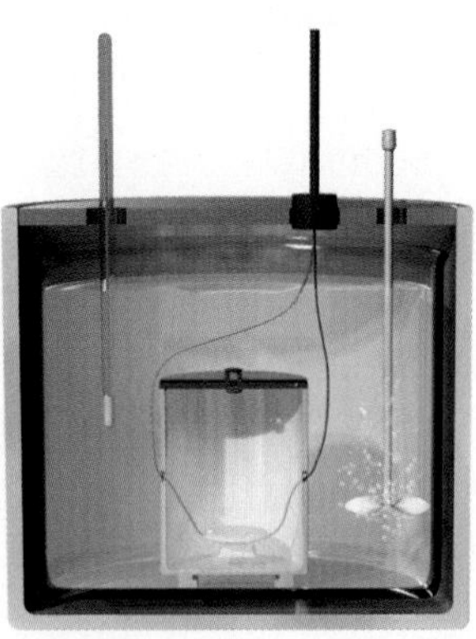

LO5 A **standard enthalpy of reaction (ΔH°_{rxn})** is the enthalpy change associated with a reaction occurring under **standard conditions**: a constant pressure of 1 atm and a specified temperature. **Hess's law** states that the heat of a reaction that occurs in more than one step is the sum of the heats of reaction of the steps. Hess's law can be used to calculate enthalpy changes in reactions that are hard or impossible to measure directly. The **standard enthalpy of formation (ΔH°_f)** of a substance is the enthalpy change that accompanies a **formation reaction,** in which one mole of the substance is made from its constituent elements in their standard states. Heats of reaction can be calculated from the heats of formation of reactants and products. (Sections 9.6 and 9.7)

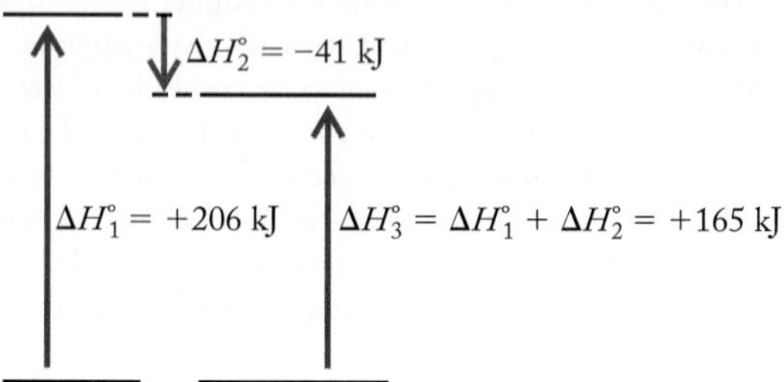

LO6 Enthalpies of reaction for gas-phase reactions can be estimated by considering the energy required to break bonds in reactants and energy released when bonds in products are formed. (Section 9.7)

LO7 The enthalpy of solution (**$\Delta H_{solution}$**) for an ionic compound dissolving in water is the sum of the energies consumed to overcome the compound's lattice energy and to separate water molecules to make space for the ions and the energy released due to the formation of ion–dipole interactions between solute ions and the water molecules surrounding them. Lattice energies can be calculated with a **Born–Haber cycle** in an application of Hess's law. (Section 9.8)

LO8 **Fuel value** is the amount of energy released during complete combustion of 1 g of a fuel, whereas **fuel density** is the amount of energy released per unit volume of a liquid fuel. Heats of combustion can be used to evaluate the fuel values of dried foods because the products (CO_2 + H_2O) are the same, so the enthalpy change that accompanies respiration in organisms is the same as that of combustion. (Section 9.9)

PARTICULATE **PREVIEW WRAP-UP**

The ionic bonds in calcium chloride must be broken by the addition of energy to overcome the electrostatic attractions between the calcium cations and the chloride anions (the green spheres). Ion–dipole forces form between the cations and the δ− oxygen atoms and between the anions and the δ+ hydrogen atom.

PROBLEM-SOLVING SUMMARY

Type of Problem	Concepts and Equations		Sample Exercises
Identifying endothermic and exothermic processes	During an endothermic process, heat flows into the system from its surroundings ($q > 0$). During an exothermic process, heat flows out from the system into its surroundings ($q < 0$).		**9.1**
Calculating *P–V* work	$w = -P\Delta V$		**9.2**
Relating ΔE, *q*, and *w*	$\Delta E = q + w = q - P\Delta V$	(9.3, 9.4)	**9.3**
Calculating heat transfer (*q*) associated with a change of temperature or state of a substance	Heating either an object: $q = C_P\Delta T$ or a mass (*m*) of a pure substance: $q = mc_P\Delta T$ or a quantity of a pure substance in moles (*n*): $q = nc_{P,n}\Delta T$ Melting a solid at its melting point: $q = n\Delta H_{fus}$ Vaporizing a liquid at its boiling point: $q = n\Delta H_{vap}$	 (9.8) (9.9) (9.10) (9.11) (9.12)	**9.4, 9.5**
Calculating $C_{calorimeter}$ and ΔH_{rxn} from calorimetry data	$q_{rxn} = -q_{calorimeter} = -C_{calorimeter}\Delta T$		**9.6, 9.7**
Calculating ΔH_{rxn} using Hess's law	Reorganize the information so that the reactions add together as desired. Reversing a reaction changes the sign of the reaction's ΔH_{rxn} value. Multiplying the coefficients in a reaction by a factor means the reaction's ΔH_{rxn} value has to be multiplied by the same factor.		**9.8**
Recognizing formation reactions	The reactants must be elements in their standard states and the product must be one mole of a single compound.		**9.9**

Type of Problem	Concepts and Equations	Sample Exercises
Calculating ΔH°_{rxn} from standard heats of formation	$\Delta H^\circ_{rxn} = \sum n_{products} \Delta H^\circ_{f,products} - \sum n_{reactants} \Delta H^\circ_{f,reactants}$ (9.15)	**9.10**
Estimating heats of gas-phase reactions from average bond energies	$\Delta H_{rxn} = \sum \Delta H_{bond\ breaking} - \sum \Delta H_{bond\ forming}$ (9.16)	**9.11**
Calculating lattice energy	The lattice energy of an ionic compound can be calculated from the enthalpy of formation of the compound by applying Hess's law to the steps involved in converting free elements into an ionic solid (the Born–Haber cycle).	**9.12**
Calculating fuel values and fuel densities	The fuel value of a substance is the energy released by the complete combustion of 1 g of the substance. Fuel density is the energy released during combustion of a unit volume (such as 1 mL) of a substance.	**9.13, 9.14**

VISUAL PROBLEMS

(Answers to boldface end-of-chapter questions and problems are in the back of the book.)

9.1. Figure P9.1 shows the compression stroke of a diesel engine. How does upward motion of the piston alter the internal energy of the gases trapped in the cylinder?

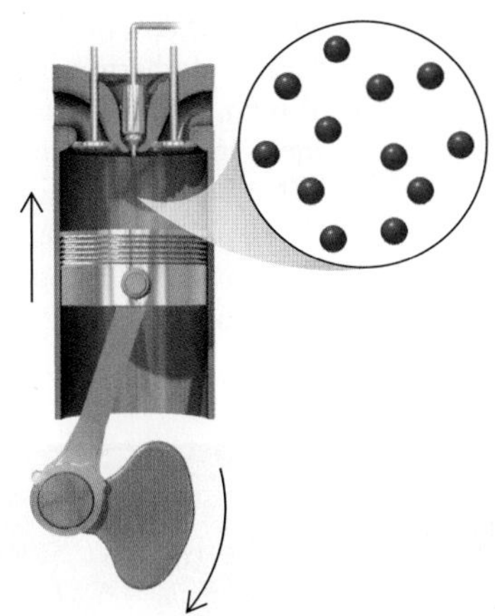

FIGURE P9.1

9.2. Figure P9.2 shows the power stroke of a diesel engine as energy released by the rapid combustion of the air and fuel vapor inside the cylinder pushes the cylinder downward. If the gases inside the cylinder are a thermodynamic system, how does the internal energy of the system change as a result of the combustion reaction and downward motion of the piston? In your description indicate the signs on ΔE, q, and w.

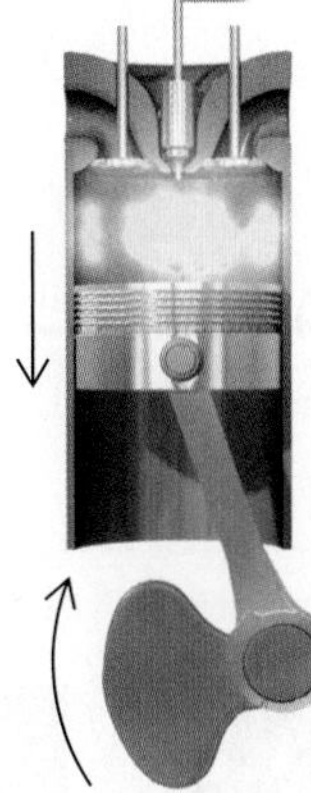

FIGURE P9.2

9.3. Based on their molecular structures, predict which of the four hydrocarbons in Figure P9.3 has the highest fuel value and which has the lowest.

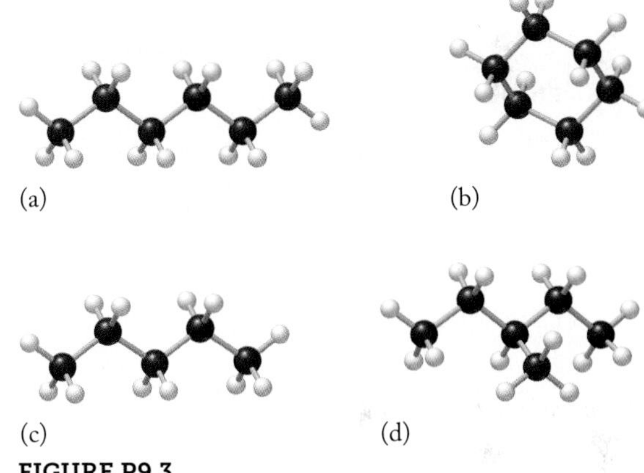

FIGURE P9.3

9.4. The diagram in Figure P9.4 shows how the volume of a reaction mixture at constant pressure and temperature changes as N_2 and H_2 combine, forming NH_3.

a. In this reaction, does the reaction mixture do work on the surroundings, or vice versa?
b. Use data from Appendix 4 to calculate ΔH°_{rxn} for the formation of one mole of product.
c. To achieve a final temperature that is the same as the initial one, does heat flow out from, or into, the reaction mixture?

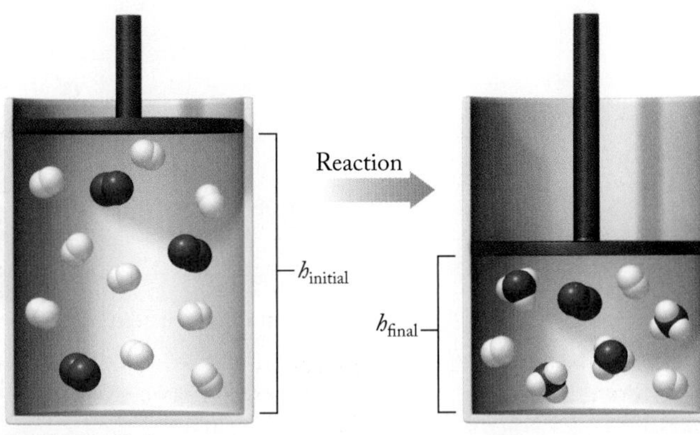

FIGURE P9.4

9.5. Assuming the reaction mixture in Figure P9.4 is a thermodynamic system, what are the signs on q, w, and ΔE? If each molecule in the figure represents one mole of reactant or product, what is the percent yield of the reaction?

9.6. The diagram in Figure P9.6 is based on the standard heats of formation (ΔH°_f) values for four compounds made from the elements listed on the "zero" line of the vertical axis.

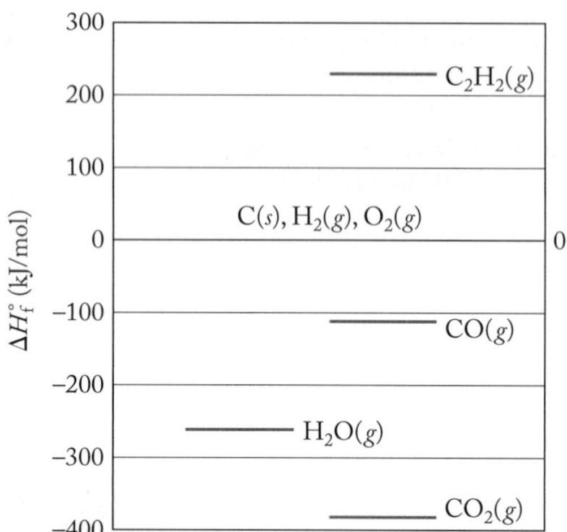

FIGURE P9.6

a. Why are the elements all on the same horizontal line?
b. Why is $C_2H_2(g)$ sometimes called an "endothermic" compound?
c. Based on these ΔH°_f values, predict which of these two reactions has the more negative enthalpy change:

$$2\ C_2H_2(g) + 3\ O_2(g) \rightarrow 4\ CO(g) + 2\ H_2O(g)$$

or

$$2\ C_2H_2(g) + 5\ O_2(g) \rightarrow 4\ CO_2(g) + 2\ H_2O(g)$$

9.7. Figure P9.7 represents a chemical reaction taking place at constant temperature and pressure.

a. Write a balanced chemical equation for the reaction.
b. Use data from Appendix 4 to calculate ΔH°_{rxn} for the formation of one mole of CO.
c. To achieve a final temperature that is the same as the initial one, does heat flow out from, or into, the reaction mixture?

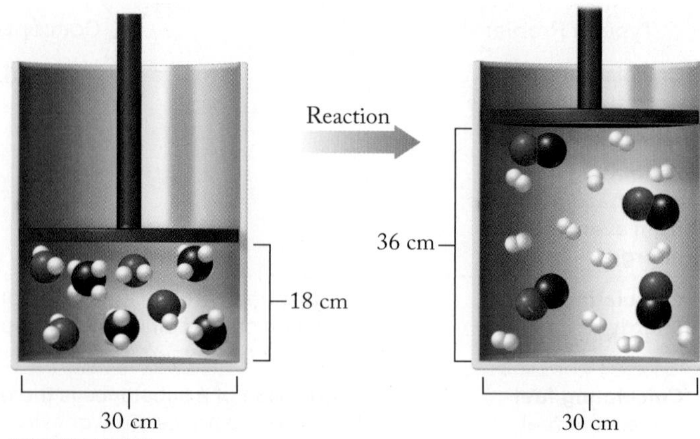

FIGURE P9.7

9.8. Use representations [A] through [I] in Figure P9.8 to answer questions a–f.

a. Match two of the particulate images to the phase change for liquid nitrogen in [B].
b. Match two of the particulate images to the phase change for dry ice (solid CO_2) in [H].
c. Which, if any, of the photos correspond to [D]? Are these endothermic or exothermic?
d. Which, if any, of the photos correspond to [F]? Are these endothermic or exothermic?
e. What bonds break when the solid ammonium nitrate in [E] dissolves in water to activate the cold pack?
f. Which particulate images show an element or compound in its standard state?

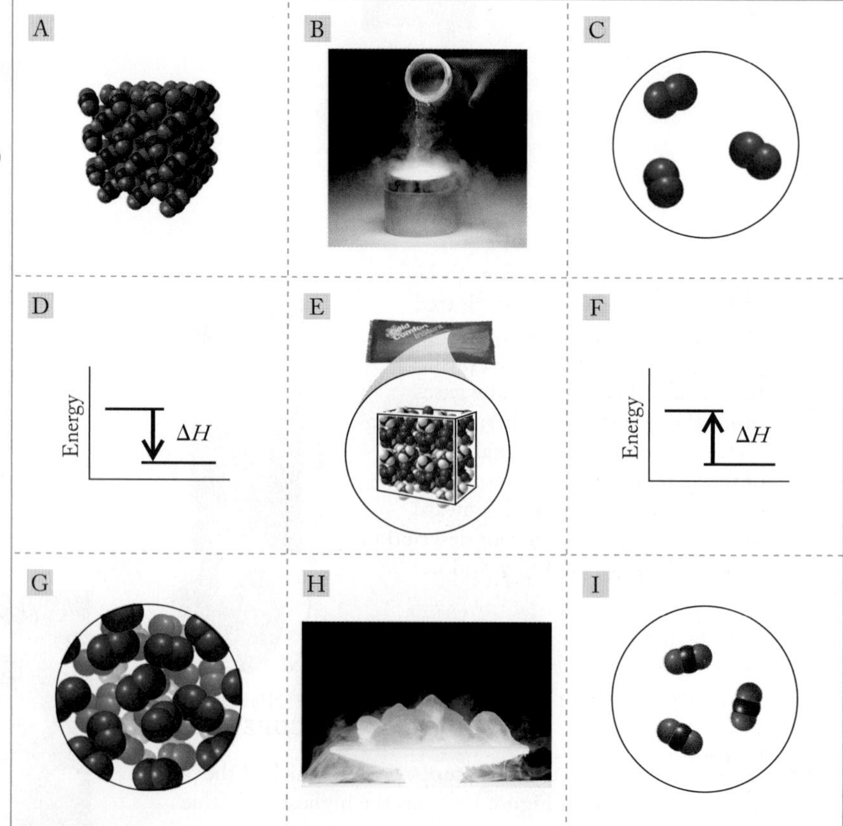

FIGURE P9.8

QUESTIONS AND PROBLEMS

Energy as a Reactant or Product

Concept Review

9.9. How are energy and work related?

9.10. Explain the difference between potential energy and kinetic energy in molecules.

9.11. Explain what is meant by a state function.

9.12. Are kinetic energy and potential energy both state functions?

9.13. If the potential energy of a particle increases as it is moved away from another particle, do the two particles attract or repel each other?

9.14. Explain how there can be kinetic energy in a stationary ice cube.

Transferring Heat and Doing Work

Concept Review

9.15. Describe two ways to increase the internal energy of a gas sample.

9.16. Assuming the kernels of unpopped popcorn in Figure P9.16 constitute a thermodynamic system, what are the signs of q, w, and ΔE during the popping process?

FIGURE P9.16

9.17. How can the product of pressure and volume ($P-V$ work) have energy units?

9.18. Why is there a negative sign in front of the $P\Delta V$ term in $\Delta E = q - P\Delta V$?

Problems

9.19. Which of the following processes are exothermic, and which are endothermic?

a. Molten aluminum solidifies.
b. Rubbing alcohol evaporates from the skin.
c. Fog forms over San Francisco Bay.

9.20. Which of the following processes are exothermic, and which are endothermic?

a. Ice cubes solidify in the freezer.
b. Ice cubes in a frost-free freezer slowly lose mass.
c. Dew forms on a lawn overnight.

9.21. What happens to the internal energy of a liquid at its boiling point when it vaporizes?

9.22. What happens to the internal energy of a gas when it expands (with no heat flow)?

9.23. How much P–V work does a gas system do on its surroundings at a constant pressure of 1.00 atm if the volume of gas triples from 250.0 mL to 750.0 mL? Express your answer in L · atm and joules (J).

9.24. An expanding gas does 150.0 J of work on its surroundings at a constant pressure of 1.01 atm. If the gas initially occupied 68 mL, what is the final volume of the gas?

9.25. Calculate ΔE when

a. $q = 100.0$ J; $w = -50.0$ J
b. $q = 6.2$ kJ; $w = 0.70$ L · atm
c. $q = -615$ kJ; $w = -3.25$ kilowatt-hours (1 kWh = 3600 kJ)

9.26. Calculate ΔE for a system that absorbs 726 kJ of heat from its surroundings and does 526 kJ of work on its surroundings.

9.27. Calculate ΔE for the combustion of a gas that releases 210.0 kJ of heat to its surroundings and does 65.5 kJ of work on its surroundings.

9.28. Calculate ΔE for a chemical reaction that releases 90.7 kJ of heat to its surroundings but does no work on them.

***9.29.** The following reactions take place in a cylinder equipped with a movable piston at atmospheric pressure (Figure P9.29). Which reactions will result in work being done on the surroundings? What is the sign of w? Assume the system returns to an initial temperature of 110°C. *Hint*: The volume of a gas is proportional to the number of moles (n) at constant temperature and pressure.

a. $CH_4(g) + 2\,O_2(g) \rightarrow CO_2(g) + 2\,H_2O(g)$
b. $C_3H_8(g) + 5\,O_2(g) \rightarrow 3\,CO_2(g) + 4\,H_2O(g)$
c. $N_2(g) + 2\,O_2(g) \rightarrow 2\,NO_2(g)$

FIGURE P9.29

*9.30. In which direction will the piston described in Problem 9.29 move when the following reactions are carried out at atmospheric pressure inside the cylinder and after the system has returned to its initial temperature of 110°C? What is the sign on w? *Hint*: The volume of a gas is proportional to the number of moles (n) at constant temperature and pressure.

a. $N_2(g) + 3\,H_2(g) \rightarrow 2\,NH_3(g)$
b. $C(s) + O_2(g) \rightarrow CO_2(g)$
c. $CH_3CH_2OH(g) + 3\,O_2(g) \rightarrow 2\,CO_2(g) + 3\,H_2O(g)$

Enthalpy and Enthalpy Changes

Concept Review

9.31. What is meant by an *enthalpy change*?

9.32. Describe the difference between an internal energy change (ΔE) and an enthalpy change (ΔH).

9.33. Why is the sign of ΔH negative for an exothermic process?

9.34. What happens to the magnitude and sign of the enthalpy change when a process is reversed?

Problems

9.35. Adding Drano to a clogged sink causes the drainpipe to get warm. What is the sign of ΔH when Drano dissolves in water?

9.36. **Instant Cold Pack Chemistry** Breaking the small pouch of water inside a chemical cold pack containing ammonium nitrate activates the pack, which is used by sports trainers for injured athletes. What is the sign of ΔH for the process taking place in the cold pack?

9.37. The stable form of oxygen at room temperature and pressure is the diatomic molecule O_2. What is the sign of ΔH for the following process?

$$O_2(g) \rightarrow 2\ O(g)$$

9.38. **Plaster of Paris** Gypsum is the common name of calcium sulfate dihydrate which has the formula $CaSO_4 \cdot 2\ H_2O$. When gypsum is heated to 150°C, it loses most of the water in its formula and forms plaster of Paris ($CaSO_4 \cdot 0.5\ H_2O$):

$$2\ CaSO_4 \cdot 2\ H_2O(s) \rightarrow 2\ CaSO_4 \cdot 0.5\ H_2O(s) + 3\ H_2O(g)$$

What is the sign of ΔH for making plaster of Paris from gypsum?

9.39. A solid with metallic properties is formed when hydrogen gas is compressed under extremely high pressures. What is the sign of ΔH for the deposition process: $H_2(g) \rightarrow H_2(s)$?

9.40. In a simple "kitchen chemistry" experiment, some vinegar is poured into an empty soda bottle. A deflated balloon containing baking soda is stretched over the mouth of the bottle. Holding up the balloon and shaking it allows the baking soda to fall into the vinegar, which starts the following reaction and inflates the balloon:

$$NaHCO_3(aq) + CH_3COOH(aq) \rightarrow CH_3COONa(aq) + CO_2(g) + H_2O(\ell)$$

If the contents of the bottle are the system, is work being done on the surroundings or on the system?

Heating Curves and Heat Capacity

Concept Review

9.41. What is the difference between *specific heat* and *heat capacity*?

9.42. Which has more heat capacity: one liter of water or one cubic meter of water? Which has more molar heat capacity?

9.43. Why is the heat of vaporization of water so much greater than its heat of fusion?

9.44. Figure P9.44 shows portions of the heating curves for two moles each of three liquids: chloroform (bp 142°C), water (bp 100°C), and ethanol (bp 78°C). Rank the liquids in order of their molar enthalpies of vaporization from lowest to highest.

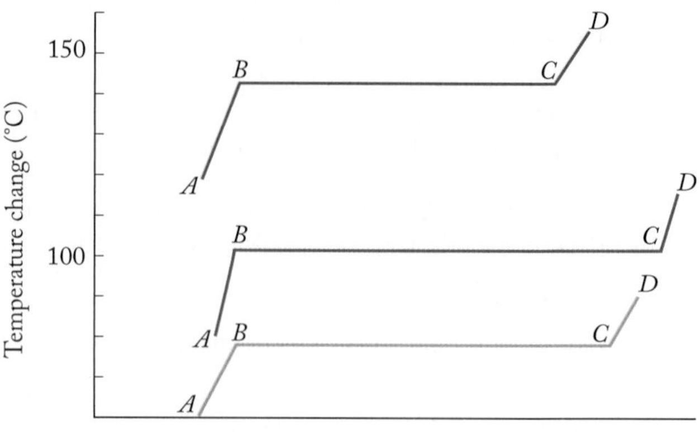

FIGURE P9.44

9.45. In Figure P9.44, why do line segments $\overline{AB}$ have slightly different slopes in each curve?

9.46. The same quantity of heat warms equal masses of metals A and B. Does the metal with the larger specific heat reach the higher temperature?

***9.47.** **Cooling an Automobile Engine** Most automobile engines are cooled by water circulating through them and a radiator. However, the original Volkswagen Beetle had an air-cooled engine. Why might car designers choose water cooling over air cooling?

*9.48. **Nuclear Reactor Coolants** The reactor-core cooling systems in some nuclear power plants use liquid sodium as the coolant. Sodium has a thermal conductivity of 1.42 J/(cm · s · K), which is quite high compared with that of water [6.1×10^{-3} J/(cm · s · K)]. The respective molar heat capacities are 28.3 J/(mol · K) and 75.3 kJ/(mol · K). What is the advantage of using liquid sodium over water in this application?

Problems

9.49. How much heat must be absorbed by 100.0 g of water to raise its temperature from 30.0°C to 100.0°C?

9.50. At an elevation where the boiling point of water is 93°C, 1.33 kg of water at 30°C absorbs 290.0 kJ from a mountain climber's stove. Is this amount of thermal energy sufficient to heat the water to its boiling point?

9.51. Use the following data to sketch a heating curve for one mole of methanol. Start the curve at −100°C and end it at 100°C.

Boiling point	65°C
Melting point	−94°C
Heat of vaporization	35.3 kJ/mol
Heat of fusion (ΔH_{fus})	3.18 kJ/mol
Molar heat capacity (ℓ)	81.1 J/(mol · °C)
Molar heat capacity (g)	43.9 J/(mol · °C)
Molar heat capacity (s)	48.7 J/(mol · °C)

9.52. Use the following data to sketch a heating curve for 1.5 moles of acetic acid. Start the curve at +16°C and end it at 130°C.

Boiling point	118°C
Melting point	16°C
Heat of vaporization	23.7 kJ/mol
Heat of fusion (ΔH_{fus})	11.7 kJ/mol
Molar heat capacity (ℓ)	123.1 J/(mol · °C)
Molar heat capacity (*g*)	63.4 J/(mol · °C)

9.53. Keeping an Athlete Cool During a strenuous workout, an athlete generates 233 kJ of thermal energy. What mass of water would have to evaporate from the athlete's skin to dissipate this energy?

9.54. The same quantity of thermal energy is added to equal masses of titanium, iron, silver, and lead. All of them are initially at the same temperature. Which metal has the highest final temperature?

9.55. Exactly 10.0 mL of water at 25.0°C is added to a hot iron skillet. All of the water is converted into steam at 100.0°C. The mass of the pan is 1.20 kg. What is the change in temperature of the skillet?

*9.56. A 20.0 g piece of iron and a 20.0 g piece of gold at 100.0°C are dropped into 1.00 L of water at 20.0°C. What is the final temperature of the water and the pieces of metal?

Heats of Reaction and Calorimetry

Concept Review

9.57. Why is it necessary to know the heat capacity of a calorimeter?

9.58. Could an endothermic reaction be used to determine the heat capacity of a calorimeter?

9.59. If we replace the water in a bomb calorimeter with another liquid, do we need to determine a new heat capacity of the calorimeter?

9.60. When measuring the heat of combustion of a very small amount of material, would you prefer to use a calorimeter having a heat capacity that is small or large?

Problems

9.61. Calculate the heat capacity of a calorimeter if the combustion of 5.000 g of benzoic acid produces a temperature increase of 16.397°C.

9.62. Calculate the heat capacity of a calorimeter if the combustion of 4.663 g of benzoic acid produces an increase in temperature of 7.149°C.

9.63. The complete combustion of 1.200 g of cinnamaldehyde (C_9H_8O, one of the compounds in cinnamon) in a bomb calorimeter ($C_{calorimeter} = 3.640$ kJ/°C) produced an increase in temperature of 12.79°C. How much thermal energy is produced during the complete combustion of one mole of cinnamaldehyde?

9.64. **Spices** The aromatic hydrocarbon cymene ($C_{10}H_{14}$) is found in nearly 100 spices and fragrances, including coriander, anise, and thyme. The complete combustion of 1.608 g of cymene in a bomb calorimeter ($C_{calorimeter} = 3.640$ kJ/°C) produced an increase in temperature of 19.35°C. How much thermal energy is produced during the complete combustion of one mole of cymene?

9.65. Hormone Mimics Phthalates that are used to make plastics flexible are among the most abundant industrial contaminants in the environment. Several have been shown to act as hormone mimics in humans by activating the receptors for estrogen, a female sex hormone. Combustion of one mole of one of these compounds, dimethyl phthalate ($C_{10}H_{10}O_4$), produces 4685 kJ of thermal energy. If 1.00 g of dimethyl phthalate is combusted in a bomb calorimeter whose heat capacity ($C_{calorimeter}$) is 7.854 kJ/°C, what is the change in temperature of the calorimeter?

9.66. **Flavorings** The flavor of anise is due to anethole, a compound with the molecular formula $C_{10}H_{12}O$. Combustion of one mole of anethole produces 5541 kJ of thermal energy. If 0.950 g of anethole is combusted in a bomb calorimeter whose heat capacity ($C_{calorimeter}$) is 7.854 kJ/°C, what is the change in temperature of the calorimeter?

9.67. A coffee-cup calorimeter contains 100.0 mL of 1.00 *M* HCl at 22.4°C. When 0.243 g of Mg metal is added to the acid, the ensuing reaction:

$$Mg(s) + 2\,HCl(aq) \rightarrow MgCl_2(aq) + H_2(g) \qquad \Delta H_{rxn} = ?$$

causes the temperature of the solution to increase to 33.4°C. What is the value of ΔH_{rxn} of the reaction? Assume the density of the solution is 1.01 g/mL and that its specific heat is 4.18 J/(g · °C).

9.68. **Chemical Cold Pack** When 4.00 g NH_4NO_3 ($\mathcal{M}$ = 80.04 g/mol)—the active ingredient in some chemical cold packs—is dissolved in 96.0 g H_2O, the temperature of the resulting solution is 3.07°C colder than the water and ammonium nitrate were before they were mixed together. What is the value of ΔH for the following dissolution process?

$$NH_4NO_3(s) \rightarrow NH_4NO_3(aq) \qquad \Delta H = ?$$

Hess's Law and Standard Heats of Reaction

Concept Review

9.69. How is Hess's law consistent with the law of conservation of energy?

9.70. Would Hess's law be valid if enthalpy were not a state function? Why or why not?

Problems

9.71. Use the ΔH°_{rxn} values of the following reactions:

$$2\,SO_2(g) + O_2(g) \rightarrow 2\,SO_3(g) \qquad \Delta H^\circ_{rxn} = -196 \text{ kJ}$$

$$\tfrac{1}{4}\,S_8(s) + 3\,O_2(g) \rightarrow 2\,SO_3(g) \qquad \Delta H^\circ_{rxn} = -790 \text{ kJ}$$

to calculate the ΔH°_{rxn} value of this reaction:

$$\tfrac{1}{8}\,S_8(s) + O_2(g) \rightarrow SO_2(g) \qquad \Delta H^\circ_{rxn} = ?$$

9.72. **Ozone Layer** The destruction of the ozone layer by chlorofluorocarbons (CFCs) can be described by the following reactions:

$$ClO(g) + O_3(g) \rightarrow Cl(g) + 2\,O_2(g) \qquad \Delta H^\circ_{rxn} = -29.90\text{ kJ}$$

$$2\,O_3(g) \rightarrow 3\,O_2(g) \qquad \Delta H^\circ_{rxn} = +24.18\text{ kJ}$$

Use the preceding ΔH°_{rxn} values to determine the value of the standard heat of reaction for this reaction:

$$Cl(g) + O_3(g) \rightarrow ClO(g) + O_2(g) \qquad \Delta H^\circ_{rxn} = ?$$

9.73. The mineral spodumene ($LiAlSi_2O_6$) exists in two crystalline forms called α and β. Use Hess's law and the following information to calculate ΔH°_{rxn} for the conversion of α-spodumene into β-spodumene:

$$Li_2O(s) + 2\,Al(s) + 4\,SiO_2(s) + \tfrac{3}{2}\,O_2(g) \rightarrow 2\,\alpha\text{-}LiAlSi_2O_6(s) \qquad \Delta H^\circ_{rxn} = -1870.6\text{ kJ}$$

$$Li_2O(s) + 2\,Al(s) + 4\,SiO_2(s) + \tfrac{3}{2}\,O_2(g) \rightarrow 2\,\beta\text{-}LiAlSi_2O_6(s) \qquad \Delta H^\circ_{rxn} = -1814.6\text{ kJ}$$

9.74. Use the following data to determine whether the conversion of diamond into graphite is exothermic or endothermic:

$$C(s,\text{ diamond}) + O_2(g) \rightarrow CO_2(g) \qquad \Delta H^\circ_{rxn} = -395.4\text{ kJ}$$

$$2\,CO(g) + O_2(g) \rightarrow 2\,CO_2(g) \qquad \Delta H^\circ_{rxn} = -566.0\text{ kJ}$$

$$2\,CO(g) \rightarrow C(s,\text{ graphite}) + CO_2(g) \qquad \Delta H^\circ_{rxn} = -172.5\text{ kJ}$$

$$C(s,\text{ diamond}) \rightarrow C(s,\text{ graphite}) \qquad \Delta H^\circ_{rxn} = ?$$

9.75. Given the following thermochemical data:

$$\tfrac{1}{2}\,N_2(g) + \tfrac{1}{2}\,O_2(g) \rightarrow \tfrac{1}{2}\,NO(g) \qquad \Delta H^\circ_{rxn} = +90.3\text{ kJ}$$

$$NO(g) + \tfrac{1}{2}\,Cl_2(g) \rightarrow NOCl(g) \qquad \Delta H^\circ_{rxn} = -38.6\text{ kJ}$$

what is the value of ΔH°_{rxn} for the decomposition of NOCl?

$$2\,NOCl(g) \rightarrow N_2(g) + O_2(g) + Cl_2(g) \qquad \Delta H^\circ_{rxn} = ?$$

9.76. **Gas from Coal or Oil Shale** Synthetic natural gas (SNG), sometimes called substitute natural gas, is a methane-containing mixture produced from the gasification of coal or oil shale directly at the site of the mine or oil field. One reaction for the production of SNG is:

$$4\,CO(g) + 8\,H_2(g) \rightarrow 3\,CH_4(g) + CO_2(g) + 2\,H_2O(g)$$

Use the following thermochemical equations to determine ΔH° for the reaction as written.

$$C(\text{graphite}) + 2\,H_2(g) \rightarrow CH_4(g) \qquad \Delta H^\circ = -74.8\text{ kJ}$$

$$C(\text{graphite}) + \tfrac{1}{2}\,O_2(g) \rightarrow CO(g) \qquad \Delta H^\circ = -110.5\text{ kJ}$$

$$CO(g) + \tfrac{1}{2}\,O_2(g) \rightarrow CO_2(g) \qquad \Delta H^\circ = -283.0\text{ kJ}$$

$$H_2(g) + \tfrac{1}{2}\,O_2(g) \rightarrow H_2O(g) \qquad \Delta H^\circ = -285.8\text{ kJ}$$

Heats of Reaction from Heats of Formation and Bond Energies

Concept Review

9.77. Why is the standard heat of formation of CO(g) difficult to measure experimentally?

9.78. Explain how the use of ΔH°_f to calculate ΔH°_{rxn} is an example of Hess's law.

9.79. Oxygen and ozone are both forms of elemental oxygen. Are the standard heats of formation of oxygen and ozone the same? Why or why not?

9.80. Why are the standard heats of formation of elements in their standard states assigned a value of zero?

9.81. Why must the stoichiometry of a reaction be known in order to estimate the enthalpy change from bond energies?

9.82. Why must the structures of the reactants and products be known in order to estimate the enthalpy change of a reaction from bond energies?

***9.83.** When calculating the enthalpy change for a chemical reaction using bond energies, why is it important that the reactants and products all be gases?

*9.84. If the energy needed to break two moles of C=O bonds is greater than the sum of the energies needed to break the O=O bonds in one mole of O_2 and vaporize one mole of carbon, why does the combustion of pure carbon release heat?

Problems

9.85. For which of the following reactions does ΔH°_{rxn} represent a heat of formation?

a. $C(s) + \tfrac{1}{2}\,O_2(g) \rightarrow CO(g)$
b. $6\,MnO(s) + O_2(g) \rightarrow 2\,Mn_3O_4(s)$
c. $2\,Na(s) + C(s,\text{ graphite}) + 3/2\,O_2(g) \rightarrow Na_2CO_3(s)$
d. $4\,H_2(g) + 2\,C(s) \rightarrow 2\,CH_4(g)$

9.86. For which of the following reactions does ΔH°_{rxn} also represent a heat of formation?

a. $2\,N_2(g) + 3\,O_2(g) \rightarrow 2\,NO_2(g) + 2\,NO(g)$
b. $N_2(g) + O_2(g) \rightarrow 2\,NO(g)$
c. $2\,NO_2(g) \rightarrow N_2O_4(g)$
d. $N_2(g) + 2\,O_2(g) \rightarrow 2\,NO_2(g)$

9.87. **Methanogenesis** Use standard heats of formation from Appendix 4 to calculate the standard heat of reaction for the following methane-generating reaction of methanogenic bacteria:

$$4\,H_2(g) + CO_2(g) \rightarrow CH_4(g) + 2\,H_2O(\ell)$$

9.88. Use standard enthalpies of formation from Appendix 4 to calculate the standard enthalpy of reaction for the following methane-generating reaction of methanogenic bacteria, given ΔH°_f of $CH_3NH_2(g) = -22.97$ kJ/mol:

$$4\,CH_3NH_2(g) + 2\,H_2O(\ell) \rightarrow 3\,CH_4(g) + CO_2(g) + 4\,NH_3(g)$$

9.89. Ammonium nitrate decomposes to N_2O and water vapor at temperatures between 250°C and 300°C. Write a balanced chemical reaction describing the decomposition of ammonium nitrate, and calculate its ΔH°_{rxn} using the appropriate ΔH°_f values from Appendix 4.

*9.90. **Chemistry of TNT** Trinitrotoluene (TNT) is a highly explosive compound. The thermal decomposition of TNT is described by the following chemical equation:

$$2\,C_7H_5N_3O_6(s) \rightarrow 12\,CO(g) + 5\,H_2(g) + 3\,N_2(g) + 2\,C(s)$$

If ΔH°_{rxn} for the reaction is −10,153 kJ/mol, how much TNT is needed to equal the explosive power of one mole of ammonium nitrate (see Problem 9.89)?

9.91. Explosives Fertilizer (ammonium nitrate) and fuel oil (a mixture of long-chain hydrocarbons similar to decane, $C_{10}H_{22}$) can form an explosive mixture. Determine the enthalpy change of the following explosive reaction using the appropriate enthalpies of formation (the standard enthalpy of formation of liquid $C_{10}H_{22}$ is 249.7 kJ/mol).

$$3\ NH_4NO_3(s) + C_{10}H_{22}(\ell) + 14\ O_2(g) \rightarrow 3\ N_2(g) + 17\ H_2O(g) + 10\ CO_2(g)$$

9.92. **Military Explosives** Explosives called amatols are mixtures of ammonium nitrate and trinitrotoluene (TNT) introduced during World War I when TNT was in short supply. The mixtures can provide 30% more explosive power than TNT alone. Above 300°C, ammonium nitrate decomposes to N_2, O_2, and H_2O. Write a balanced chemical reaction describing the decomposition of ammonium nitrate, and calculate its ΔH°_{rxn} using the appropriate ΔH°_f values from Appendix 4.

Note: Use the average bond energy values in Table A4.1 of Appendix 4 to answer Problems 9.93–9.98.

9.93. Use average bond energies to estimate the enthalpy changes of the following reactions under standard conditions:
a. $N_2(g) + 3\ H_2(g) \rightarrow 2\ NH_3(g)$
b. $N_2(g) + 2\ H_2(g) \rightarrow H_2NNH_2(g)$
c. $2\ N_2(g) + O_2(g) \rightarrow 2\ N_2O(g)$

9.94. Use average bond energies to estimate the enthalpy changes of the following reactions under standard conditions:
a. $CO_2(g) + H_2(g) \rightarrow H_2O(g) + CO(g)$
b. $N_2(g) + O_2(g) \rightarrow 2\ NO(g)$
*c. $C(s) + CO_2(g) \rightarrow 2\ CO(g)$
Hint: The heat of sublimation of graphite, C(*s*), is 719 kJ/mol.

9.95. Use average bond energies to estimate the difference in ΔH°_{rxn} values between the incomplete combustion of one mole of ethane to carbon monoxide and water vapor and the complete combustion of ethane to carbon dioxide and water vapor.

9.96. Use average bond energies to estimate the difference in ΔH°_{rxn} values between the reaction

$$C(s) + O_2(g) \rightarrow CO_2(g)$$

and the reaction

$$C(s) + \tfrac{1}{2}\ O_2(g) \rightarrow CO(g)$$

***9.97.** Use average bond energies to estimate ΔH°_{rxn} for the following reaction:

$$4\ NH_3(g) + 7\ O_2(g) \rightarrow 4\ NO_2(g) + 6\ H_2O(g)$$

*9.98. The value of ΔH°_{rxn} for the reaction

$$2\ H_2S(g) + 3\ O_2(g) \rightarrow 2\ SO_2(g) + 2\ H_2O(g)$$

is −1036 kJ. Estimate the energy of the bonds in SO_2.

Energy Changes When Substances Dissolve; More Applications of Thermochemistry

Concept Review

9.99. How do ion–dipole interactions influence whether an ionic compound's heat of solution is exothermic or endothermic?

9.100. Sodium hydroxide is more soluble in hot water than in cold water, but dissolving sodium hydroxide in water is an exothermic process. How can this be the case?

9.101. What is meant by *fuel value*?

9.102. What are the units of fuel values?

9.103. Without doing any calculations, predict which compound in each pair releases more energy during combustion:
a. 1 mole of CH_4 or 1 mole of H_2
b. 1 g of CH_4 or 1 g of H_2

9.104. Is fuel value or fuel density a more useful measure of energy content of liquid fuels? Explain your answer.

Problems

9.105. Use a Born–Haber cycle to calculate the lattice energy of potassium chloride (KCl) from the following data:

Ionization energy of K(*g*) = 419 kJ/mol
Electron affinity of Cl(*g*) = −349 kJ/mol
Energy to sublime K(*s*) = 89 kJ/mol
Bond energy of $Cl_2(g)$ = 243 kJ/mol
Standard heat of formation of KCl = −436.5 kJ/mol

9.106. Calculate the lattice energy of sodium oxide (Na_2O) from the following data:

Ionization energy of Na(*g*) = 495 kJ/mol
Electron affinity of O(*g*) for 2 electrons = 603 kJ/mol
Energy to sublime Na(*s*) = 109 kJ/mol
Bond energy of $O_2(g)$ = 498 kJ/mol
ΔH_{rxn} for $2\ Na(s) + \tfrac{1}{2}\ O_2(g) \rightarrow Na_2O(s)$ = −416 kJ/mol

---

9.107. If all the energy obtained from burning 1.00 pound of propane is used to heat water, how many kilograms of water can be heated from 20.0°C to 45.0°C?

9.108. An article in *Discover* magazine on world-class sprinters contained the following statement: "In one race, a field of eight runners releases enough energy to boil a gallon jug of ice at 0.0°C in ten seconds!" How much "energy" do the runners release in 10 seconds? Assume that the ice has a mass of 128 ounces.

9.109. Lightweight camping stoves typically use *white gas*, a mixture of C_5 and C_6 hydrocarbons.
a. Calculate the fuel value of C_5H_{12}, given that ΔH°_{comb} = −3535 kJ/mol.
b. How much heat is released during the combustion of 1.00 kg of C_5H_{12}?
c. How many grams of C_5H_{12} must be burned to heat 1.00 kg of water from 20.0°C to 90.0°C? Assume that all the heat released during combustion is used to heat the water.

9.110. The heavier (more dense) hydrocarbons in camp stove fuel are hexanes (C_6H_{14}).
a. Calculate the fuel value of C_6H_{14}, given that ΔH°_{comb} = −4163 kJ/mol.
b. How much heat is released during the combustion of 1.00 kg of C_6H_{14}?

c. How many grams of C_6H_{14} are needed to heat 1.00 kg of water from 25.0°C to 85.0°C? Assume that all of the heat released during combustion is used to heat the water.

d. Assume white gas is 25% C_5 hydrocarbons (see Problem 9.109) and 75% C_6 hydrocarbons; how many grams of white gas are needed to heat 1.00 kg of water from 25.0°C to 85.0°C?

Additional Problems

***9.111. Production of HCl** The industrial production of hydrogen chloride gas is most frequently carried out by direct synthesis from hydrogen and chlorine:

$$H_2(g) + Cl_2(g) \rightarrow 2\ HCl(g)$$

Smaller quantities of HCl(*g*) may be produced on the laboratory scale by the reaction of sodium chloride and sulfuric acid:

$$2\ NaCl(s) + H_2SO_4(\ell) \rightarrow 2\ HCl(g) + Na_2SO_4(s)$$

Apply concepts discussed in this chapter and data from the appendix to determine if either heating or cooling is required when these reactions are carried out.

9.112. **Burning Off Calories** A typical double-patty hamburger from a fast-food establishment contains about 563 Calories. (Remember that the dietary "Calorie" is actually a kilocalorie.) Walking at a brisk pace burns about 4.70 Calories per minute. How many minutes would you need to walk to "burn off" the Calories in one double burger?

9.113. A 100.0 mL sample of 1.0 *M* NaOH is mixed with 50.0 mL of 1.0 *M* H_2SO_4 in a large Styrofoam coffee cup; a thermometer is mounted in the lid of the cup to measure the temperature of the contents. The temperature of each solution before mixing is 22.3°C. After mixing, their temperature reaches 31.4°C. Assume that (1) the density of the mixed solutions is 1.00 g/mL, (2) the specific heat of the mixed solutions is 4.18 J/(g · °C), and (3) no heat is lost to the surroundings.

a. Write a balanced chemical equation for the reaction that takes place in the cup.

b. Is any NaOH or H_2SO_4 left in the cup when the reaction is over?

c. Calculate the enthalpy change per mole of H_2O produced in the reaction.

9.114. With reference to Problem 9.113, what if 65.0 mL of 1.0 *M* H_2SO_4 is mixed with 100.0 mL of 1.0 *M* NaOH? Will the increase in temperature be less than, more than, or the same as that measured in Problem 9.113?

***9.115.** An insulated container holds 50.0 g of water at 25.0°C. A 7.25 g sample of copper that had been heated to 100.1°C is dropped into the water. What is the final shared temperature of the copper and the water?

9.116. Magnetite (Fe_3O_4) is magnetic, whereas iron(II) oxide is not.

a. Write and balance the chemical equation for the formation of magnetite from iron(II) oxide and oxygen.

b. Given that 318 kJ of heat is released for each mole of Fe_3O_4 formed, what is the enthalpy change of the balanced reaction of formation of Fe_3O_4 from iron(II) oxide and oxygen?

9.117. Endothermic compounds have positive standard heats of formation. An example is acetylene, C_2H_2(ΔH°_f = 226.7 kJ/mol). Combustion of acetylene in pure oxygen produces a flame hot enough to cut and weld steel.

a. What is the standard heat of combustion of acetylene?

b. What is the fuel value of acetylene, assuming the products are CO_2 and H_2O vapor?

9.118. Balance the following chemical equation, name the reactants and products, and calculate the enthalpy change under standard conditions.

$$FeO(s) + O_2(g) \rightarrow Fe_2O_3(s) \qquad \Delta H^\circ_{rxn} = ?$$

***9.119. Polymer Chemistry** Use appropriate bond energies from Table A4.1 of Appendix 4 to predict whether the reaction in which ethylene forms polyethylene plastic is exothermic, endothermic, or involves no change in enthalpy. The reaction can be written:

$$n\ CH_2{=}CH_2 \rightarrow [-CH_2-CH_2-]_n$$

where the structure in the brackets is the *repeating unit* of polyethylene and the value of *n* is typically in the thousands.

9.120. **Metabolism of Methanol** In December 2011, over 100 people died in West Bengal, India, after drinking bootleg (illegal) liquor spiked with methanol. Methanol is toxic because it is metabolized in a two-step process that produces formic acid (HCOOH):

$$O_2(g) + 2\ CH_3OH(aq) \rightarrow 2\ HCOOH(aq) + 2\ H_2O(\ell) \qquad \Delta H^\circ_{rxn} = -1019.5\ kJ$$

a. Is the reaction endothermic or exothermic?

b. What change in enthalpy accompanies the metabolism of 60.0 g of methanol?

c. In the first step of the process, methanol is converted into formaldehyde (CH_2O), which is then converted into formic acid. Would you expect ΔH°_{rxn} for the metabolism of one mole of methanol to formaldehyde to be larger or smaller than the conversion of one mole of methanol to formic acid?

9.121. In a high-temperature gas-phase reaction, methanol (CH_3OH) reacts with N_2 to produce HCN and NH_3. The reaction is endothermic, requiring 164 kJ of thermal energy per mole of methanol under standard conditions.

a. Write a balanced chemical equation for this reaction.

b. Is energy a reactant or a product?

c. What is the change in enthalpy under standard conditions if 60.0 g of $CH_3OH(g)$ reacts with excess $N_2(g)$, forming HCN(*g*), and $NH_3(g)$?

9.122. Calculate ΔH°_{rxn} for the reaction

$$2\ Ni(s) + \tfrac{1}{4}\ S_8(s) + 3\ O_2(g) \rightarrow 2\ NiSO_3(s) \qquad \Delta H^\circ_{rxn} = ?$$

from the following data:

$$(1)\ NiSO_3(s) \rightarrow NiO(s) + SO_2(g) \qquad \Delta H^\circ_{rxn} = 156\ kJ$$

$$(2)\ \tfrac{1}{8}\ S_8(s) + O_2(g) \rightarrow SO_2(g) \qquad \Delta H^\circ_{rxn} = -297\ kJ$$

$$(3)\ Ni(s) + \tfrac{1}{2}\ O_2(g) \rightarrow NiO_2(s) \qquad \Delta H^\circ_{rxn} = -241\ kJ$$

9.123. Use the information in thermochemical equations (1) through (3) to calculate the value of ΔH°_{rxn} for the reaction in equation (4).

(1) $Pb(s) + \frac{1}{2} O_2(g) \rightarrow PbO(s)$ $\Delta H^\circ_{rxn} = -219$ kJ

(2) $C(s) + O_2(g) \rightarrow CO_2(g)$ $\Delta H^\circ_{rxn} = -394$ kJ

(3) $PbCO_3(s) \rightarrow PbO(s) + CO_2(g)$ $\Delta H^\circ_{rxn} = 86$ kJ

(4) $2\ Pb(s) + 2\ C(s) + 3\ O_2(g) \rightarrow 2\ PbCO_3(s)$ $\Delta H^\circ_{rxn} = ?$

9.124. **Ethanol as Automobile Fuel** Most of the new cars sold in Brazil are *flex-fuel* vehicles, which means they can run on blends of ethanol and gasoline or on pure ethanol alone. Ethanol is a popular liquid fuel in Brazil because it can be efficiently produced by fermenting sugar extracted from sugarcane. Use appropriate standard heats of formation to calculate the enthalpy change that accompanies the combustion of one mole of liquid ethanol in which the products are CO_2 and H_2O vapor.

9.125. Baking soda ($NaHCO_3$) thermally decomposes to soda ash (Na_2CO_3), CO_2, and H_2O:

$$2\ NaHCO_3(s) \rightarrow Na_2CO_3(s) + CO_2(g) + H_2O(g)$$

Use appropriate standard heats of formation to calculate the enthalpy change that accompanies the thermal decomposition of one mole of $NaHCO_3$.

*9.126. **Specific Heats of Metals** In 1819, Pierre Dulong and Alexis Petit reported that the product of the atomic mass of a metal times its specific heat is approximately constant, an observation called the *law of Dulong and Petit.*

a. Explain how the specific heat and molar heat capacity data for the metals in Table 9.3 support this law.
b. Use these data to predict the specific heat values of nickel and platinum.

9.127. **Odor of Urine** Urine odor gets worse with time because it contains the metabolic product urea, $CO(NH_2)_2$, a compound that is slowly converted to carbon dioxide and ammonia, which has a sharp, unpleasant odor:

$$CO(NH_2)_2(aq) + H_2O(\ell) \rightarrow CO_2(aq) + 2\ NH_3(aq)$$

This reaction is much too slow for ΔH°_{rxn} to be determined experimentally by measuring a change in temperature, but it can be calculated from the appropriate ΔH°_f values. Calculate ΔH°_{rxn}. [The value of ΔH°_f of $CO(NH_2)_2(aq)$ is -319.2 kJ/mol.]

9.128. Propane is a gas at 1 bar of pressure above −42°C, though it is transported as a liquid under pressure. How does the physical state of propane affect its fuel value and its fuel density?

9.129. **Rocket Fuels** The payload of a rocket includes a fuel and oxygen for combustion of the fuel. Reactions (1) and (2) describe the combustion of dimethylhydrazine and hydrogen, respectively. Pound for pound, which is the better rocket fuel?

(1) $(CH_3)_2NNH_2(\ell) + 4\ O_2(g) \rightarrow N_2(g) + 4\ H_2O(g) + 2\ CO_2(g)$ $\Delta H^\circ_{rxn} = -1694$ kJ

(2) $H_2(g) + \frac{1}{2} O_2(g) \rightarrow H_2O(g)$ $\Delta H^\circ_{rxn} = -241.8$ kJ

9.130. At high temperatures, such as those in the combustion chambers of automobile engines, nitrogen and oxygen form nitrogen monoxide:

$$N_2(g) + O_2(g) \rightarrow 2\ NO(g) \quad \Delta H^\circ_{comb} = +180 \text{ kJ}$$

Any NO released into the environment may be oxidized to NO_2:

$$2\ NO(g) + O_2(g) \rightarrow 2\ NO_2(g) \quad \Delta H^\circ_{comb} = -112 \text{ kJ}$$

Is the overall reaction,

$$N_2(g) + 2\ O_2(g) \rightarrow 2\ NO_2(g)$$

exothermic or endothermic? What is ΔH°_{comb} for this reaction?

smartwork5

If your instructor uses Smartwork5, log in at digital.wwnorton.com/atoms2.

10

Properties of Gases

The Air We Breathe

CLEAN AIR FOR FIRST RESPONDERS The compressed air in the tanks on their backs provides about 30 minutes of breathable air.

PARTICULATE **REVIEW**

How Many Moles of Gas?

In Chapter 10, we explore the physical and chemical properties of gases. The reactants for the combustion of methane are depicted in the figure on the right.

- Write a balanced equation for the combustion reaction of CH_4.
- Identify the limiting reactant in the reaction mixture in the image.
- Assuming each molecule represents one mole of gas, how many moles of gas are present after the reaction is complete?

(Review Sections 7.3 and 7.4 if you need help.)

(Answers to Particulate Review questions are in the back of the book.)

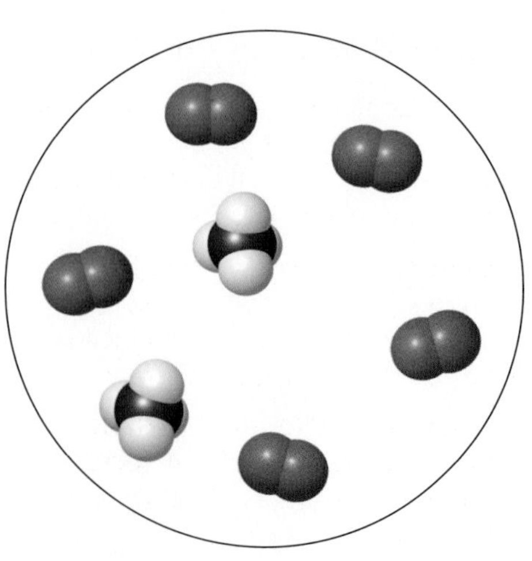

PARTICULATE **PREVIEW**

Expanding or Contracting Gases?

The balloons in the drawing are filled with an equal number of moles of helium. As you study Chapter 10, look for ideas that will help you understand the factors that can affect the volume of gases.

- Does the volume of the balloon expand from (a) to (b) or contract from (b) to (a) when the temperature of the helium increases? When atmospheric pressure increases?
- What happens to the frequency of collisions between helium atoms when the temperature of the helium increases? When atmospheric pressure increases?

(a)

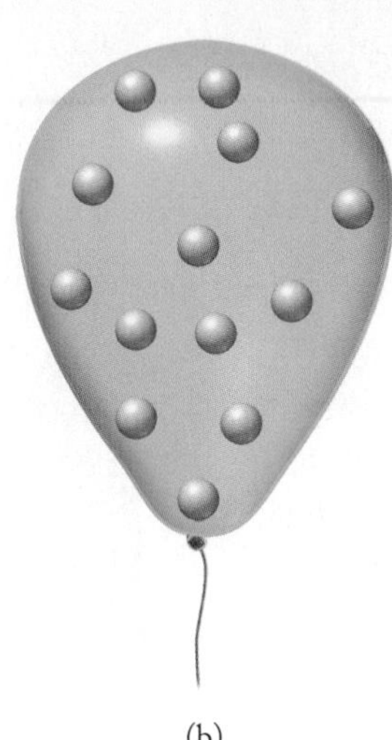

(b)

Learning Outcomes

LO1 Use kinetic molecular theory to explain the properties of gases

LO2 Use Graham's law to calculate the relative rates of effusion of gases
Sample Exercise 10.1

LO3 Calculate the root-mean-square speed of a gas from its mass and absolute temperature
Sample Exercise 10.2

LO4 Describe how a manometer works and interpret the results obtained with one
Sample Exercise 10.3

LO5 Calculate changes in the volume, temperature, pressure, and number of moles of an ideal gas using gas laws, including the ideal gas law
Sample Exercises 10.4, 10.5, 10.6, 10.7, 10.8

LO6 Use the ideal gas law to calculate the density and molar mass of an ideal gas
Sample Exercises 10.9, 10.10

LO7 Use balanced chemical equations to relate the volumes of gas-phase reactants and products using the stoichiometry of the reaction and the ideal gas law
Sample Exercise 10.11

LO8 Relate the mole fraction, partial pressure, and quantity of a gas in a mixture
Sample Exercises 10.12, 10.13

LO9 Use the van der Waals equation to correct for the nonideal behavior of gases
Sample Exercise 10.14

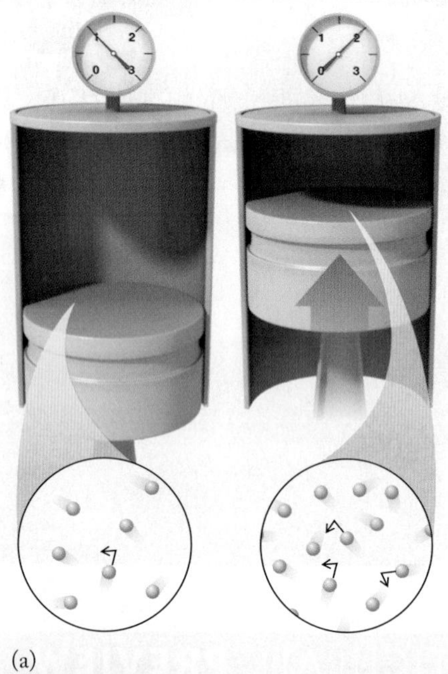

(a)

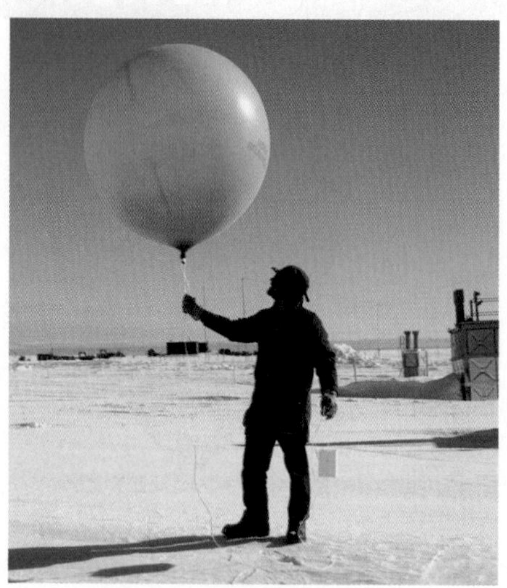

(b)

10.1 An Invisible Necessity: The Properties of Gases

How often do you think about the composition of the air you breathe? Chances are it is not often. Breathing is something you do unconsciously, assuming you are in good health and assuming your senses are not telling you there might be something unhealthy about the air around you. Air is supposed to be colorless, tasteless, and odorless. If it's not, you have good reason to quickly search for cleaner air to breathe. There is also the assumption that you are not engaged in an activity, such as fighting a fire or diving underwater, that limits your access to clean air and the precious oxygen in it.

Let's consider some situations where the preceding assumptions may not be true. For example, people suffering from impaired lung function may require supplemental oxygen. Often this O_2 flows from a tank of pure oxygen at a rate between one and two liters per minute. Even small, portable tanks contain enough oxygen to last several hours because the O_2 gas inside the tank has been compressed at high pressure. This means that a tank with an internal volume of less than 2.5 L can deliver hundreds of liters of O_2 at normal atmospheric pressure. Similarly, the air inside the 6 L tanks that firefighters often wear can last for half an hour or more because these tanks are filled with air at pressures as high as 300 atm.

Gases are compressible, but liquids and solids are not. This is why squeezing a liquid or a solid has little impact on its volume, but doubling the pressure on a gas, as happens by pushing up the piston in Figure 10.1(a), decreases the volume of the

FIGURE 10.1 The volume of a quantity of gas is inversely proportional to the pressure around it. (a) The volume of the helium gas in the cylinder decreases as the pressure on the piston increases. Decreasing the volume increases the concentration of the gas. (b) The volume of a weather balloon expands as it ascends into the sky and experiences decreasing atmospheric pressure with increasing altitude.

helium gas by half. On the other hand, the volume of the gas in a weather balloon (Figure 10.1b) *increases* as it ascends into the atmosphere because the pressure of the air surrounding it *decreases*. Both of these observations illustrate a well-known property of most gases:

- The volume a gas occupies is inversely proportional to its pressure ($V \propto 1/P$).

Following is a list of other properties of gases with some examples that you may have observed:

- The pressure of a quantity of gas in a rigid container is directly proportional to its absolute temperature ($P \propto T$). This property is one reason why a big drop in temperature in early winter may set off low-tire-pressure sensors in automobiles even though no air has leaked out of the tires.
- The gas pressure inside a rigid container at constant temperature is directly proportional to the quantity (number of moles) of gas in the container ($P \propto n$). For example, a pressure gauge attached to a scuba tank (Figure 10.2) not only measures the internal pressure of the tank but also provides a diver with a measure of how much air is left in the tank.
- All gases are miscible with all other gases—that is, gas mixtures are homogeneous. This means that when a diver takes a breath of air from a scuba tank, the proportion of O_2 in the air is always the same, whether it's the first breath taken at the beginning of a dive or the last one before reaching the surface.
- At a given temperature and pressure, the densities of gaseous substances are directly proportional to their molar masses and are much lower than their densities as solids or liquids. For example, the density of liquid butane (the fuel in disposable lighters) at its boiling point of 0°C is 0.60 g/mL, which is over 200 times greater than the density of butane vapor at 0°C (0.0026 g/mL).
- Gases expand to occupy the entire volume of their containers. For example, the shape of a party balloon is essentially the same whether it is partially or fully inflated: the gas in the balloon occupies all of the space available to it uniformly, pushing outward with the same pressure in all directions.
- Gases in inflated party balloons escape, or *effuse*, through the walls of the balloons at rates that are inversely proportional to the molar masses of the gases. Those with the smallest molar masses, such as helium, escape the most rapidly.

CONNECTION Average atmospheric pressure at sea level is, by definition, 1 atmosphere (atm), as described in Section 6.4.

CONNECTION As discussed in Section 9.1, the average kinetic energy of atoms and molecules in a gas sample is directly proportional to the absolute temperature of the sample.

FIGURE 10.2 The internal pressure of a scuba tank is a direct measure of the amount of air inside it.

In this chapter we investigate these and other properties of gases in more detail, and we explain why gases behave the way they do based on a theory of how their atoms and molecules move and interact with each other and with the walls of their containers. This theory provides particle-based explanations of all the macroscale properties listed previously.

CONCEPT TEST

Nitrogen dioxide, NO_2, is a red-brown gas that contributes to photochemical smog formation. Which of the images in Figure 10.3 best depicts a sealed container of NO_2?

(Answers to Concept Tests are in the back of the book.)

(a)

(b)

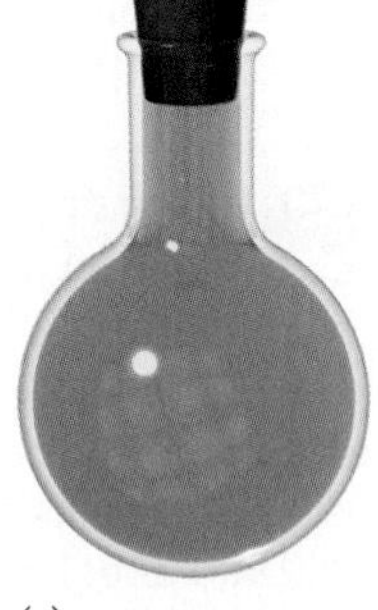

(c)

FIGURE 10.3

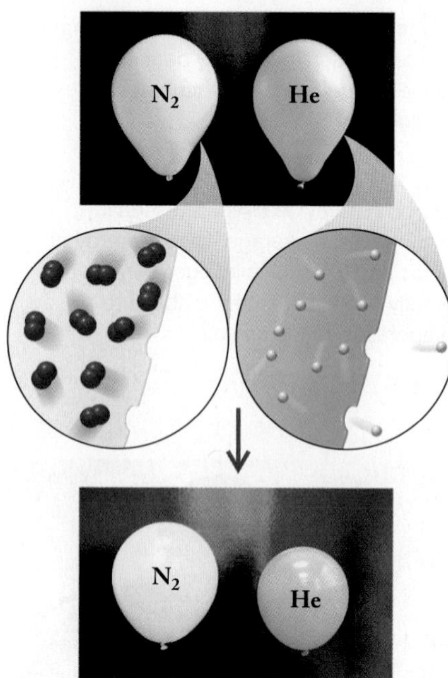

FIGURE 10.4 Two balloons at the same temperature and pressure, one filled with nitrogen gas and the other filled with an equal volume of helium gas. Over time, the volume of the helium balloon decreases much faster.

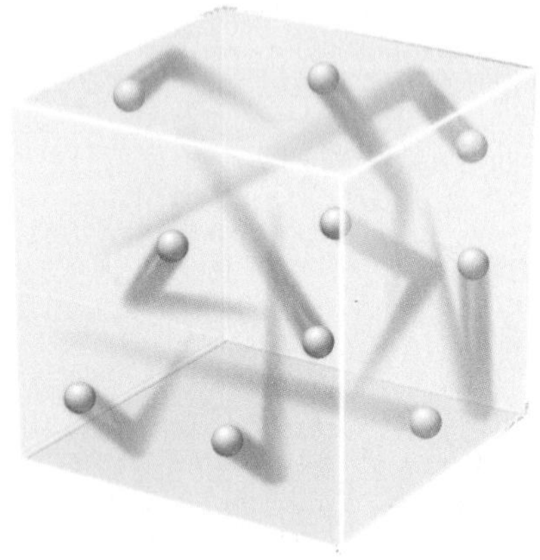

FIGURE 10.5 Helium molecules continually collide with one another and with their container walls.

effusion the process by which a gas escapes from its container through a tiny hole into a region of lower pressure.

Graham's law of effusion the rate of effusion of a gas is inversely proportional to the square root of its molar mass.

10.2 Effusion, Diffusion, and the Kinetic Molecular Theory of Gases

Let's begin our exploration of how and why gases behave the way they do by examining the last property of gases listed in Section 10.1. Why is it that helium-filled party balloons made of latex rubber deflate and are no longer buoyant after a day or two, whereas the same balloons filled with air remain inflated for many days? The answer to this question involves the structure of party balloons and the motion of the atoms and molecules inside them.

Tiny imperfections (Figure 10.4) in latex rubber balloons allow gas particles to slowly escape in a process called **effusion**. This phenomenon was investigated in the early 19th century by Scottish chemist Thomas Graham (1805–1869), who discovered that the rate of effusion of different gases was inversely proportional to the square root of their densities. As noted in Section 10.1, the densities of gases are proportional to their molar masses, so the modern interpretation of **Graham's law of effusion** states that the effusion rate of a gas is inversely proportional to the square root of its molar mass:

$$\text{Effusion rate} \propto \sqrt{\frac{1}{\text{density}}} \propto \sqrt{\frac{1}{\text{molar mass}}}$$

Why do gases with lower molar masses effuse faster? This behavior can be explained using the **kinetic molecular theory (KMT)** of gases. KMT is based on the following set of assumptions about the particulate nature of gases:

- Gas molecules have tiny volumes compared with the volume they occupy. Their individual volumes are so small as to be considered negligible, allowing particles in a gas to be treated as *point masses*—masses with essentially no volume. Gas molecules are separated by large distances; hence, a gas is mostly empty space. This assumption is consistent with the low density of gases compared to liquids and solids. For example, because liquid butane contains 223 times as many molecules as the same volume of butane vapor at 0°C, butane molecules must be spaced much farther apart in the vapor phase.
- Because gas particles are so spread out, they don't interact with each other as much as the particles in liquids and solids do. In fact, KMT assumes they don't interact *at all.*
- Gas molecules move constantly and randomly throughout the space they occupy, continually colliding with one another and with their container walls (Figure 10.5). These collisions are *elastic.* To get a mental picture of what an elastic collision is like, imagine a cue ball colliding with one or more stationary balls on a pool table (Figure 10.6). Most of the balls exit the collision in different directions at different speeds. Some of the kinetic energy of the cue ball has transferred to the others, but the *average* kinetic energy of all the balls is the same immediately after the collision as it was before.
- The average kinetic energy of the molecules in a gas is proportional to the absolute temperature of the gas. All populations of gas molecules at the same temperature have the same average kinetic energy.

Recall from Chapter 1 that the kinetic energy of any object in motion—including a gas-phase atom or molecule—is proportional to its mass (m) and the square of its speed (u):

$$\text{KE} = \tfrac{1}{2}mu^2 \tag{1.2}$$

We use the term *average* kinetic energy in describing the behavior of molecules of a gas because the speeds of particles with the same mass and at the same temperature are not all the same. Instead they are distributed over a range of values, as shown in the graph in Figure 10.7. The x-axis of the graph is particle speed; the y-axis represents the relative number of particles at each speed. The peak in the speed distribution curve represents the *most probable speed* (u_m). It is the speed of the largest number of particles. Because the distribution of speeds is not symmetrical, the *average speed* (u_{avg}), which is simply the arithmetic average of the individual speeds of all the particles, is a little higher than the most probable speed. A third kind of average speed, called the **root-mean-square speed (u_{rms})**, is larger still. Its name comes from its theoretical definition: the square root of the average of the squared speeds of all the particles of a gas. It is important to us because u_{rms} is the speed of a particle whose kinetic energy is exactly the same as the *average* kinetic energy of all the particles in a gas. The equation form of this definition is

$$KE_{avg} = \frac{1}{2}mu_{rms}^2 \quad (10.1)$$

FIGURE 10.6 The collisions between the balls on a pool table are a good model for elastic collisions, which means the average kinetic energy of all the balls is the same after a collision as it was before.

We can use Equation 10.1 to compare the u_{rms} speeds of different gases at the same temperature. For example, suppose we have a mixture of O_2 ($\mathcal{M}$ = 32.00 g/mol), N_2 ($\mathcal{M}$ = 28.01 g/mol), He ($\mathcal{M}$ = 4.003 g/mol), and H_2 ($\mathcal{M}$ = 2.016 g/mol). Because all four gases have the same temperature, their particles all have the same average kinetic energies. Therefore, the $\frac{1}{2}mu_{rms}^2$ term is also the same for each gas. However, the masses of their particles are not the same, which means their u_{rms} speeds cannot be the same: the least massive particles (the molecules of H_2) should have the highest u_{rms} speeds, whereas the most massive particles (the molecules of O_2) should have the slowest u_{rms} speeds. This prediction is supported by the speed distribution curves for these four elements in Figure 10.8. Note how particle speeds are inversely related to particle masses and that the least massive particles (the atoms of He and molecules of H_2) have not only the highest average speeds but also the most widely distributed speed profiles.

Now let's use Equation 10.1 to make a more quantitative connection between particle speed and particle mass. We start with a mixture of a heavier gas (N_2) and a lighter gas (He) from Figure 10.8. The mixture has a single temperature, so the two gases have the same average kinetic energies:

$$KE_{avg,He} = KE_{avg,N_2} \quad (10.2)$$

Substituting the right side of Equation 10.1 into both sides of Equation 10.2:

$$\frac{1}{2}m_{He}u_{rms,He}^2 = \frac{1}{2}m_{N_2}u_{rms,N_2}^2 \quad (10.3)$$

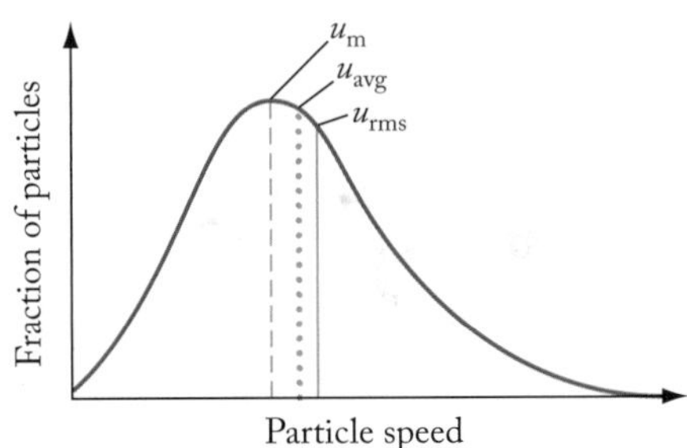

FIGURE 10.7 At any given temperature, the speeds of the particles that make up a gaseous substance cover a range of values. The dashed line represents the most probable particle speed (u_m); the dotted line is the arithmetic average speed (u_{avg}); and the solid line is the root-mean-square speed (u_{rms}), which is directly proportional to the average kinetic energy of the particles.

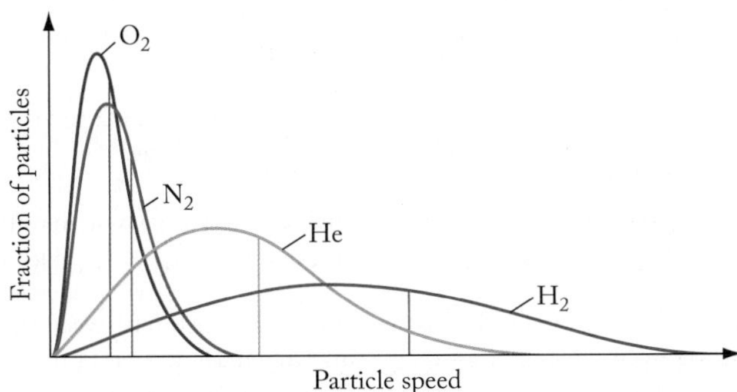

FIGURE 10.8 The speeds of gas particles at a given temperature are inversely related to their masses: the most massive (molecules of O_2 in this case) move, on average, the slowest, and the least massive (molecules of H_2) move the fastest and have the most widely distributed speeds.

kinetic molecular theory (KMT) a model that explains the behavior of gases based on the motion of the particles that make them up.

root-mean-square speed (u_{rms}) the square root of the average of the squared speeds of all the particles in a population of gas particles; a particle possessing the average kinetic energy moves at this speed.

Rearranging the terms in Equation 10.3 so that the masses are on one side of the equation and u_{rms} speeds are on the other:

$$\frac{u^2_{rms,He}}{u^2_{rms,N_2}} = \frac{m_{N_2}}{m_{He}}$$

Taking the square root of both sides:

$$\frac{u_{rms,He}}{u_{rms,N_2}} = \sqrt{\frac{m_{N_2}}{m_{He}}} \quad (10.4)$$

Now let's scale up the mass values on the right side of Equation 10.4 by multiplying both of them by Avogadro's number. Doing so converts molecular masses into molar masses without changing the value of their ratio:

$$\frac{u_{rms,He}}{u_{rms,N_2}} = \sqrt{\frac{\mathcal{M}_{N_2}}{\mathcal{M}_{He}}} = \sqrt{\frac{28.02 \cancel{g/mol}}{4.003 \cancel{g/mol}}} = 2.65 \quad (10.5)$$

CHEMTOUR Molecular Speed

Equation 10.5 tells us that the root-mean-square speeds of the two gases are inversely proportional to the square roots of their molar masses. Therefore, atoms of He collide with the walls of their container (such as the green party balloon in Figure 10.4) 2.65 times more frequently than molecules of N_2 at the same pressure and temperature. This means He atoms encounter and pass through atomic-scale escape routes in their balloon 2.65 times more frequently than N_2 molecules do. As a result, He effuses 2.65 times faster than N_2, and a helium-filled balloon deflates much faster than one filled with air (mostly N_2).

Equation 10.5 can be used with any pair of gases that behave as predicted by KMT. We can write a generic form of it for any two gases A and B:

$$\frac{u_{rms,A}}{u_{rms,B}} = \sqrt{\frac{\mathcal{M}_B}{\mathcal{M}_A}} \quad (10.6)$$

If particles of gas A have higher speeds than particles of gas B, then gas A particles collide more frequently with the walls of their container, which increases the frequency with which they encounter and pass through microscopic holes in the wall and undergo effusion. We express this in equation form as

$$\frac{\text{effusion rate}_A}{\text{effusion rate}_B} = \sqrt{\frac{\mathcal{M}_B}{\mathcal{M}_A}} \quad (10.7)$$

Equation 10.7 is a mathematical representation of Graham's law that can be used to calculate the relative rates of effusion of pairs of gases, as illustrated in Sample Exercise 10.1.

SAMPLE EXERCISE 10.1 Calculating Relative Rates of Effusion **LO2**

An odorous gas emitted by a hot spring was found to effuse at 0.342 times the rate at which helium effuses. What is the molar mass of the emitted gas?

Collect and Organize We are asked to determine the molar mass of an unknown gas based on its rate of effusion relative to the rate for helium ($\mathcal{M}_{He}$ = 4.003 g/mol).

Analyze Equation 10.7 provides the mathematical relationship between the effusion rates and molar masses of a pair of gases. Because the unknown gas effuses about 1/3 as fast as He does, the molar mass should be nearly 3^2 or 9 times the mass of He, or about $9 \times 4 = 36$ g/mol.

Solve Using the symbol X for the unknown gas in Equation 10.7:

$$\frac{\text{effusion rate}_X}{\text{effusion rate}_{He}} = \sqrt{\frac{\mathcal{M}_{He}}{\mathcal{M}_X}} = 0.342$$

Squaring both sides of the rightmost equality yields

$$\frac{\mathcal{M}_{He}}{\mathcal{M}_X} = (0.342)^2$$

$$\mathcal{M}_X = \frac{4.003 \text{ g/mol}}{(0.342)^2} = 34.2 \text{ g/mol}$$

Think About It The molar mass of the unidentified gas is 34.2 g/mol, which is consistent with our prediction. One possibility for the identity of this gas is $H_2S(g)$, $\mathcal{M}$ = 34.08 g/mol, a toxic gas with a foul smell associated with rotten eggs or low tide on a seaweed-covered shoreline.

Practice Exercise

Helium effuses 3.16 times as fast as which other noble gas?

(Answers to Practice Exercises are in the back of the book.)

Equation 10.6 allows us to compare u_{rms} values for any two gases, but it does not provide actual u_{rms} values. Instead, a different equation is needed:

$$u_{rms} = \sqrt{\frac{3RT}{\mathcal{M}}} \quad (10.8)$$

The value of R is a constant in Equation 10.8 and is called the **universal gas constant**. The origin of R is explored in Section 10.6, so for now we will simply treat it as a constant with the following value:

$$R = 8.314 \frac{\text{kg} \cdot \text{m}^2}{\text{s}^2 \cdot \text{mol} \cdot \text{K}}$$

In using Equation 10.8, we must express molar masses in kilograms (not grams) per mole to be consistent with the units on R, and the temperature must be expressed in kelvin. The units on u_{rms} are meters per second. Sample Exercise 10.2 shows how to calculate the speeds of gas-phase atoms and molecules.

SAMPLE EXERCISE 10.2 Calculating Root-Mean-Square Speeds **LO3**

Calculate the root-mean-square speed of nitrogen molecules at 25°C.

Collect and Organize We are asked to calculate the root-mean-square speed (u_{rms}) of N_2 molecules at 25°C (298 K). Equation 10.8 relates the u_{rms} speed of the particles in a gas to its absolute temperature and molar mass.

Analyze Nitrogen gas (N_2) has a molar mass of 28.01 g/mol. We can estimate the value of u_{rms} from the approximate values of R, T, and $\mathcal{M}$ on the right side of Equation 10.8: R is about 10 and T is about 300 K, so the value of the numerator is approximately $3 \times 10 \times 300$, or nearly 10,000. Expressing the molar mass of N_2 in kg/mol (to be compatible with the units on R) makes the denominator about 0.03, and the fraction inside the square root sign is about 10,000/0.03 or 333,000, the square root of which is between 500 and 600 m/s.

universal gas constant the constant R in the ideal gas equation; its value and units depend on the units used for the variables in the equation.

Solve

$$u_{rms} = \sqrt{\frac{3RT}{\mathcal{M}}}$$

$$= \sqrt{\frac{(3)\left(8.314\,\frac{\text{kg}\cdot\text{m}^2}{\text{s}^2\cdot\text{mol}\cdot\text{K}}\right)(273 + 25)\text{ K}}{\left(\frac{28.01\text{ g}}{1\text{ mol}}\right)\left(\frac{1\text{ kg}}{1000\text{ g}}\right)}} = 515\text{ m/s}$$

Think About It The calculated result is within our estimated range, and it suggests that the molecules in the air surrounding us are moving *very* rapidly. In fact, 515 m/s corresponds to 1150 miles per hour. An airplane flying that fast would be supersonic.

Practice Exercise Calculate the root-mean-square speed of helium at 25°C in meters per second, and compare your result with the root-mean-square speed of nitrogen calculated in Sample Exercise 10.2.

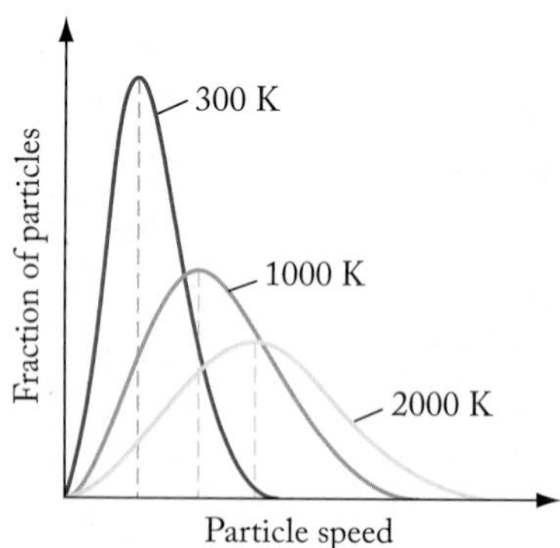

FIGURE 10.9 The most probable speeds (dashed lines) of the particles of a gas increase with increasing temperature. Notice that the distributions of particle speeds also broaden as the temperature increases.

Equation 10.8 and the foundations of kinetic molecular theory tell us that the root-mean-square speed of the particles of a gas is proportional to the square root of the absolute temperature of the gas. This increase in u_{rms} with increasing T is illustrated in Figure 10.9. Note that increasing u_{rms} values are accompanied by wider distributions in particle speeds with increasing temperature. The profiles in Figure 10.7 also show that no matter how hot a gas is, there is a finite possibility that at least some of the particles in it are hardly moving at all.

The effusion of gases is closely related to the **diffusion** of gases, which is defined as the spread of one substance through another. Gas-phase diffusion occurs when there are differences in composition of a mixture of gases within the space they occupy. Diffusion is one of the processes by which odors spread from their source through the air, such as the smell of perfume from the person wearing it or the smell of baking bread from the kitchen. Given the high speeds at which gas particles move, you might expect diffusion in air to be very rapid. It's not. If the air is still, it may take several minutes for the scent of an open bottle of perfume to spread throughout a space the size of a bedroom. It takes this long because the molecules of scent keep colliding with molecules of N_2, O_2, and other components of air (Figure 10.10). Even though ambient air is mostly empty space, these molecules still take up enough space to limit how far any molecule can travel before it collides with one of them.

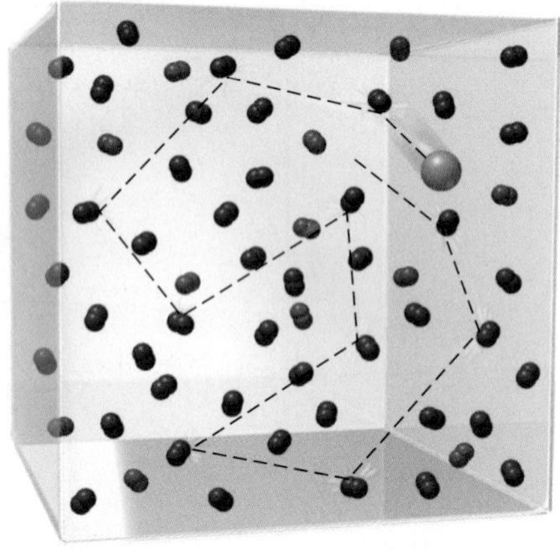

FIGURE 10.10 Diffusion rates of gases are slowed by frequent collisions between gas-phase particles. The dashed line shows the path followed by one particle as it collides with others. The average distance between collisions is called the mean free path.

The average distance that a particle can travel through air or any gas before colliding with another one is called the **mean free path** of the particle. Mean free paths depend on how densely the air particles are packed, and that depends on their pressure. At an atmospheric pressure of 1 atm, the mean free path is about 6.8×10^{-8} meter. For particles traveling at hundreds of meters per second, such a tiny distance translates into about 10^{10} collisions per second. Such a high collision frequency translates into a lot of bouncing around with little forward progress—and a slow rate of diffusion.

Thomas Graham also studied the diffusion of gases. One of his favorite experimental designs involved measuring the time it took different gases to diffuse through porous plugs made of plaster of Paris. The results of the experiments were essentially the same as those for gas effusion—namely, gases diffuse at rates that are inversely proportional to the square root of their densities

diffusion the spread of one substance (usually a gas or liquid) through another.

mean free path the average distance that a particle can travel through air or any gas before colliding with another particle.

barometer an instrument that measures atmospheric pressure.

or their molar masses. Thus, Graham's law of effusion and *Graham's law of diffusion* are interchangeable and Equation 10.7 applies to both the effusion and diffusion of gases.

CONCEPT TEST

Which graph in Figure 10.11 best describes the ratio of the rates of diffusion of two gases (X and Y) as a function of temperature?

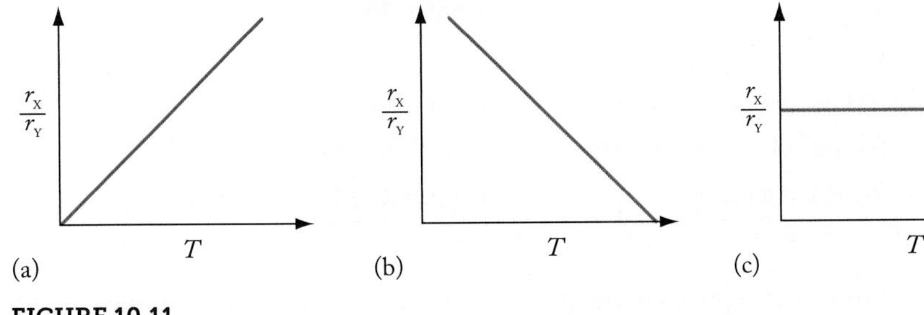

FIGURE 10.11

TABLE 10.1 Composition of Dry Air[a]

Compound	% (by volume)
Nitrogen	78.08
Oxygen	20.95
Argon	0.934
Carbon dioxide	0.0395[b]
Neon	0.0018
Helium	0.00052
Methane	0.00018
Krypton	0.00011

[a]Includes major and minor gases (with concentrations >1 ppm by volume).
[b]Value as of January 2013. Atmospheric CO_2 is increasing by about 2 ppm each year.

10.3 Atmospheric Pressure

In this section we continue our exploration of the behavior of gases by focusing on the ones we draw into our lungs with each breath. Earth is surrounded by an atmosphere that is about 50 km thick and is composed primarily of nitrogen (78% by volume), oxygen (21%), and lesser proportions of other gases (Table 10.1). To put the thickness of the atmosphere in perspective, if Earth were the size of an apple, the atmosphere would be about as thick as the apple's skin.

CONNECTION In Chapter 6 we discussed how the average value of Earth's atmospheric pressure is the ratio of the force exerted by the atmosphere to the surface area of the planet.

As we discussed in Chapter 6, Earth's atmosphere is pulled downward by gravity and exerts a force that is spread across the entire surface of the planet. The ratio of force, F, to surface area, A, defines atmospheric pressure, P:

$$P = \frac{F}{A} \tag{6.1}$$

Atmospheric pressure is measured with an instrument called a **barometer**. A simple but effective barometer design consists of a narrow tube about 1 meter long, sealed at one end and filled with mercury. The tube is inverted, and its open end is placed into a pool of mercury that is open to the atmosphere (Figure 10.12). Gravity pulls downward on the mercury in the tube, creating a vacuum at the top of the tube. At the same time, atmospheric pressure pushes downward on the mercury in the pool, which has the effect of pushing it up into the tube. The opposing forces on the mercury column in the tube create a stable column height that provides a measure of the atmospheric pressure.

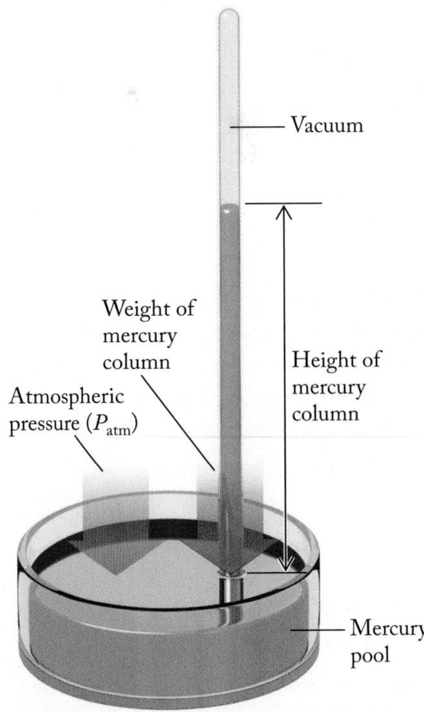

FIGURE 10.12 The height of the mercury column in this simple barometer designed by Evangelista Torricelli is proportional to atmospheric pressure.

Atmospheric pressure varies from place to place and with changing weather conditions. As we noted in Chapter 6, the average atmospheric pressure at sea level defines the standard atmosphere (1 atm, a non-SI unit) of pressure. This pressure is capable of supporting a column of mercury *exactly* 760 mm high, which is the basis for another non-SI unit of pressure: millimeters of mercury (mmHg). Pressure in millimeters of mercury is also expressed in a unit called the torr in honor of Evangelista Torricelli (1608–1647), the Italian mathematician and physicist who invented the barometer. Thus,

$$1 \text{ atm} = 760 \text{ mmHg} = 760 \text{ torr}$$

manometer an instrument for measuring the pressure exerted by a gas.

TABLE 10.2 Units for Expressing Pressure

Unit	Value
Standard atmosphere (atm)	1 atm
Pascal (Pa)	1 atm = 1.01325×10^5 Pa
Kilopascal (kPa)	1 atm = 101.325 kPa
Millimeter of mercury (mmHg)	1 atm = 760 mmHg
Torr	1 atm = 760 torr
Bar	1 atm = 1.01325 bar
Millibar (mbar or mb)	1 atm = 1013.25 mbar
Pounds per square inch (psi)	1 atm = 14.7 psi
Inches of mercury	1 atm = 29.92 inches of Hg

Other non-SI units for expressing the pressure of gases are derived from masses and areas in the U.S. Customary System, such as pounds per square inch (psi) and bars. Relationships between different units of pressure are summarized in Table 10.2.

The SI unit of pressure is the pascal (Pa), named in honor of French mathematician and physicist Blaise Pascal (1623–1662), who was the first to propose that atmospheric pressure decreases with increasing altitude. We can explain this phenomenon by noting that the atmospheric pressure at any given location on Earth's surface is related to the mass of the column of air *above* that location. As altitude increases, the mass of the column of air *above* that altitude decreases. Less mass means a smaller force exerted downward by the air at higher altitude, which means less pressure (Figure 10.13).

Scientists conducting experiments with gases usually need to know the pressures of gases in closed systems. A **manometer** is an instrument that was once

FIGURE 10.13 Atmospheric pressure decreases with increasing altitude because the mass of the column of air above a given area decreases with increasing altitude.

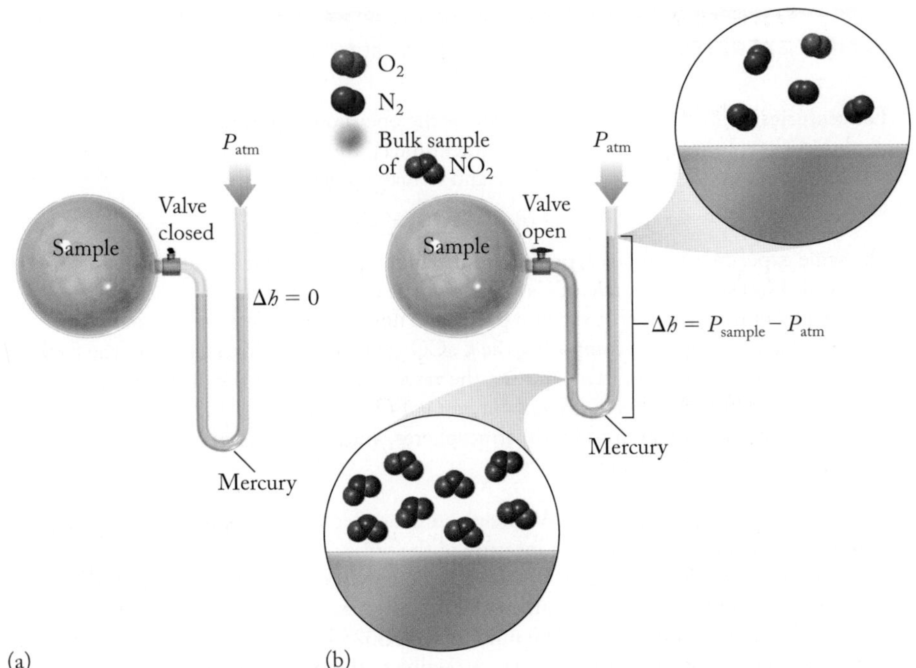

FIGURE 10.14 A manometer for measuring sample pressures that are somewhat greater than atmospheric pressure. (a) When the valve is closed there is no difference in the mercury levels. (b) When the valve is opened the pressure of the gas in the bulb pushes down on the mercury, making the two mercury levels different.

widely used to measure these pressures. One common type of manometer is illustrated in Figure 10.14. It is a U-shaped tube filled with mercury (or another dense liquid). One end of the tube is connected via a valve to a sample vessel containing the gas of interest; the other end is open to the atmosphere.

This type of manometer is particularly useful for determining gas pressures greater than atmospheric pressure. Initially there is no difference in their mercury levels ($\Delta h = 0$, Figure 10.14a). After the manometer is connected to the sample vessel and the valve is opened (Figure 10.14b), greater pressure in the sample flask pushes down on the mercury in the left side of the tube, creating a Δh that is a measure of the difference between the sample pressure and atmospheric pressure.

Today manometers have been largely replaced by electronic pressure sensors based on flexible metallic or ceramic diaphragms. As the pressure on one side of the diaphragm increases, it distorts away from that side. This is the same mechanism used to sense changes in atmospheric pressure in most barometers, including the recording barometer, or *barograph*, shown in Figure 10.15.

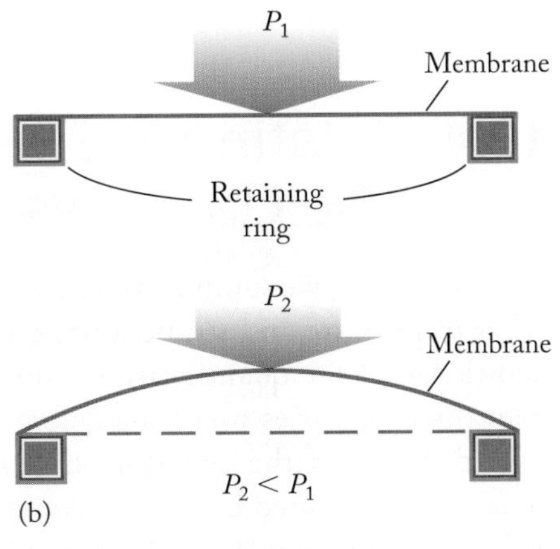

FIGURE 10.15 The pressure sensor in this barograph is a partially evacuated corrugated metal can. As atmospheric pressure decreases, the lid of the can distorts outward. This motion is amplified by a series of levers and transmitted via the horizontal arm to a pen tip that records the pressure on the graph paper as the drum slowly turns. A week's worth of barometric data can be recorded in this way.

SAMPLE EXERCISE 10.3 Measuring Gas Pressure with a Manometer 

FIGURE 10.16 A historic lime kiln.

For centuries, high-temperature kilns such as the one shown in Figure 10.16 have been used to convert limestone ($CaCO_3$) into quicklime (CaO), which is used to "sweeten" (reduce the acidity of) soil in agriculture and to make mortar and concrete:

$$CaCO_3(s) \rightarrow CaO(s) + CO_2(g)$$

The same experiment can be carried out in the apparatus shown in Figure 10.17. Figure 10.17(a) shows a sample of $CaCO_3$ before heating. All of the air has been removed from the flask with a vacuum pump and the flask connected to a manometer. Then the flask is heated, decomposing the $CaCO_3$ to CaO and CO_2 gas. After the flask has cooled back to its initial temperature, the valve is opened and the difference in levels of mercury (Δh) in the arms of the manometer is 144 mm (Figure 10.17b). Calculate the pressure of the CO_2 in (a) mmHg, (b) atmospheres, and (c) kilopascals on a day when atmospheric pressure is 756 mmHg.

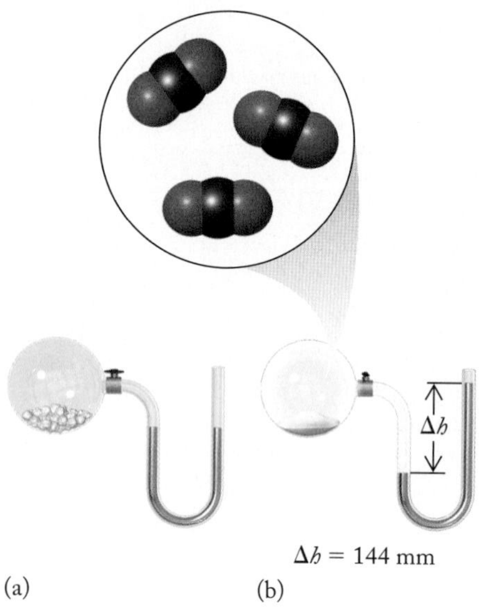

FIGURE 10.17 Apparatus for monitoring the thermal decomposition of $CaCO_3$ using a manometer. (a) A flask containing chunks of $CaCO_3$. (b) The same setup after a decomposition reaction that releases CO_2 into the flask and manometer.

Collect, Organize, and Analyze We are given the difference in height of the mercury column in a manometer, which is a measure of the pressure of CO_2 produced by the decomposition reaction above that of atmospheric pressure. We are asked to express this pressure in units of mmHg, atm, and kPa. The appropriate conversion factors, found in Table 10.2, are 1 atm = 760 mmHg = 101.325 kPa. The measured pressure of 144mm Hg is about 1/5 of 760, so we can estimate that the total pressure in the flask will be about 1.2 atm. There are about 100 kPa in 1 atmosphere, so 1.2 atm is equivalent to about 120 kPa.

Solve

a. The total pressure in mmHg is

$$756 + 144 = 900 \text{ mmHg} = 9.00 \times 10^2 \text{ mmHg}$$

b. Converting mmHg to atmospheres:

$$9.00 \times 10^2 \cancel{\text{mmHg}} \times \frac{1 \text{ atm}}{760 \cancel{\text{mmHg}}} = 1.18 \text{ atm}$$

c. Converting mmHg to kilopascals:

$$9.00 \times 10^2 \cancel{\text{mmHg}} \times \frac{101.325 \text{ kPa}}{760 \cancel{\text{mmHg}}} = 1.20 \times 10^2 \text{ kPa}$$

Think About It The calculated values are very close to our estimated values. All are expressed to three significant figures because there were three in the initial sum of the manometer and ambient pressures.

Practice Exercise Suppose the measurements in the preceding Sample Exercise were repeated, but the mass of the $CaCO_3$ sample was only 88.2% of the sample in the original measurement. What would be the value of Δh in mmHg in the final measurement?

10.4 Relating *P*, *T*, and *V*: The Gas Laws

In Section 10.1, we summarized some of the properties of gases and described the effect of pressure and temperature on volume, mostly in qualitative terms. Our knowledge of the quantitative relationships among *P*, *T*, and *V* goes back more than three centuries to a time before the field of chemistry as we know it even existed. Some of the experiments that led to our understanding of how gases behave were inspired by the development in the 18th century of the first hot-air balloons big enough (and safe enough) to carry passengers (Figure 10.18).

Boyle's Law: Relating Pressure and Volume

In Section 10.1 we explored the compressibility of gases and the inverse relation between the pressure and volume of a fixed quantity of gas at constant temperature. This relationship was discovered by British chemist Robert Boyle (1627–1691) and is known as **Boyle's law**:

$$P \propto \frac{1}{V} \quad \text{(at constant } n \text{ and } T\text{)} \tag{10.9}$$

In his experiments, Boyle used a J-shaped tube similar in construction to a manometer that is closed at one end and open at the other. With no mercury in the tube (Figure 10.19a), the pressure on the air inside the J-tube is simply atmospheric pressure (let's assume it to be 760 mmHg). When just enough mercury is added to fill up the bottom of the J-tube (Figure 10.19b), the pressure on the trapped air is still 760 mmHg. However, adding more mercury increases the pressure on the trapped air by an amount equal to the difference in height (Δh) of the mercury in the two sides of the tube (Figure 10.19c). The total pressure on the trapped air is 760 mmHg plus Δh. As the total pressure increases, the volume of the trapped air decreases as predicted by Equation 10.9.

We can turn Equation 10.9 from a proportionality into an equality by multiplying its right side by a constant:

$$P = (\text{constant}) \frac{1}{V}$$

or

$$PV = \text{constant} \tag{10.10}$$

where the value of the constant depends on how much air is trapped (n) and its temperature (T). If PV is a constant, then the product of the pressure and volume of a given quantity of gas under one set of conditions—that is, $P_1 \times V_1$—has the same value under any other set of conditions—say, $P_2 \times V_2$—as long as the temperature remains the same. Putting this equality in equation form gives us a handy mathematical expression of Boyle's law:

$$P_1V_1 = P_2V_2 \tag{10.11}$$

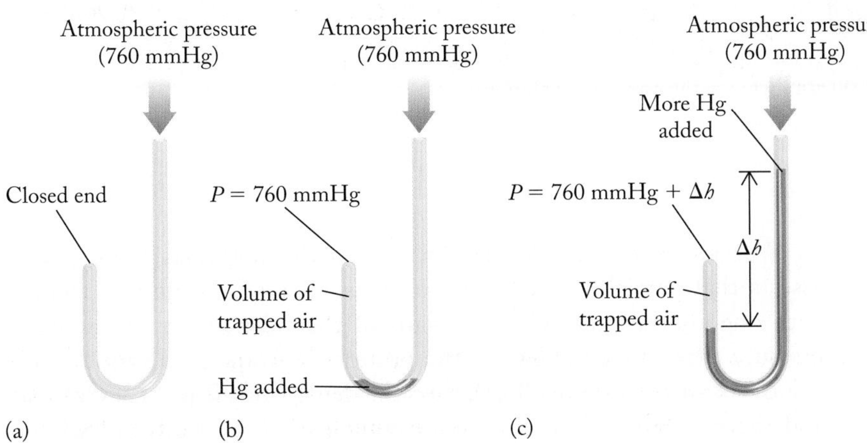

FIGURE 10.19 Boyle used a J-shaped tube for his experiments on the relationship between pressure and volume. The volume of the air trapped in the closed end of the tube decreased as more mercury was added and the difference in the height of mercury in the two sides of the tube—a measure of the pressure on the trapped gas—increased.

FIGURE 10.18 Hot-air balloons then and now. Fascination with hot-air ballooning in the late 18th and early 19th centuries led to important discoveries about the properties of gases.

Boyle's law the volume of a gas at constant temperature is inversely proportional to its pressure.

CONCEPT TEST

Which of the graphs in Figure 10.20 correctly describes the relationship between the product of pressure and volume (*PV*) as a function of pressure (*P*) for a given quantity of gas at constant temperature?

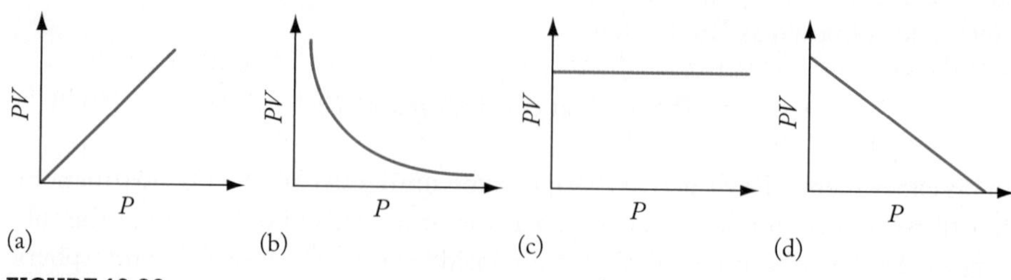

FIGURE 10.20

SAMPLE EXERCISE 10.4 Applying Boyle's Law — LO5

A balloon is partly inflated with 5.00 L of helium at sea level, where the atmospheric pressure is 1.00 atm. The balloon ascends to an altitude of 1600 m, where the pressure is 0.83 atm. What is the volume of the balloon at the higher altitude if the temperature of the helium does not change during the ascent?

Collect, Organize, and Analyze We are given the initial volume ($V_1 = 5.00$ L) of a gas and its initial pressure ($P_1 = 1.00$ atm) and we are asked to find the volume (V_2) of the gas when the pressure changes ($P_2 = 0.83$ atm). The balloon contains a fixed amount of gas and its temperature is constant, so pressure and volume are related by Boyle's law (Equation 10.11). Therefore, its volume should increase because pressure decreases.

Solve
Rearranging Equation 10.11 to solve for V_2 and inserting the given values of P_1, P_2, and V_1 gives

$$V_2 = \frac{P_1V_1}{P_2} = \frac{(1.00\ \cancel{\text{atm}})(5.00\ \text{L})}{0.83\ \cancel{\text{atm}}} = 6.0\ \text{L}$$

Think About It The prediction we made that the volume would increase as the pressure decreased is confirmed. Equation 10.11 can also be used to calculate the change in pressure that takes place at constant temperature when the volume of a quantity of gas changes.

Practice Exercise A scuba diver exhales 3.50 L of air while swimming at a depth of 20.0 m, where the sum of atmospheric pressure and water pressure is 3.00 atm. By the time the exhaled air rises to the surface, where the pressure is 1.00 atm, what is the change in volume?

FIGURE 10.21 Gas particles are in constant motion and exert pressure through collisions with the interior surface of their container. When a quantity of gas is squeezed into half its original volume, the particles collide more frequently with their container, causing the pressure to double.

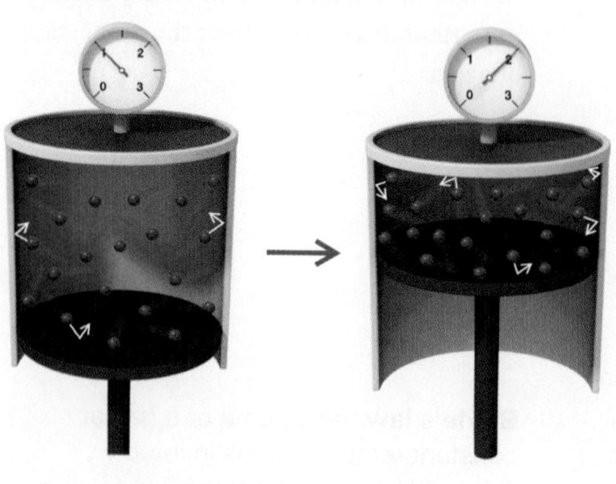

We can explain Boyle's law using kinetic molecular theory, which asserts that particles of a gas are in constant random motion, colliding with one another and with the walls of their container, producing the pressure exerted by the gas. If the particles are squeezed into a smaller volume of space (Figure 10.21), then the density of the particles is greater and the frequency with which they collide with the walls of their container is greater. The more frequent the collisions, the greater the force exerted by the particles per unit of interior surface area and the greater the pressure. The converse is also true: as the volume of a container of gas (such as the weather balloon in Figure 10.1b) increases, collision frequency decreases, and pressure drops.

Charles's law the volume of a gas at constant pressure is directly proportional to its absolute temperature.

Charles's Law: Relating Volume and Temperature

Nearly a century after Boyle's discovery of the inverse relation between the pressure exerted by a gas and the volume of the gas, French scientist Jacques Charles (1746–1823) documented the linear relation between the volume and temperature of a constant quantity of gas at constant pressure. Now known as **Charles's law**, the relation states that when the pressure exerted on a gas is held constant, the volume of a fixed quantity of gas is directly proportional to the *absolute* temperature (the temperature expressed in kelvin) of the gas:

$$V \propto T \quad \text{(at constant } P \text{ and } n\text{)} \tag{10.12}$$

The effects of Charles's law can be seen in Figure 10.22, in which a balloon has been attached to a flask, trapping a fixed amount of gas in the apparatus. Heating the flask causes the gas to expand, inflating the balloon. As with Boyle's law, an equals sign can replace the proportionality symbol in Equation 10.12 if we include a constant:

$$V = (\text{constant}) \times T$$

or

$$\frac{V}{T} = \text{constant} \tag{10.13}$$

The value of the constant depends on the number of moles of gas in the sample (n) and on the pressure of the gas (P). When these two parameters are held constant, the ratio V/T does not change, and any two combinations of volume and temperature are related as follows:

$$\frac{V_1}{T_1} = \frac{V_2}{T_2} \tag{10.14}$$

We can rearrange the terms in Equation 10.14 to obtain another expression of the proportionality between the volume of a gas and its absolute temperature:

$$\frac{V_1}{V_2} = \frac{T_1}{T_2} \tag{10.15}$$

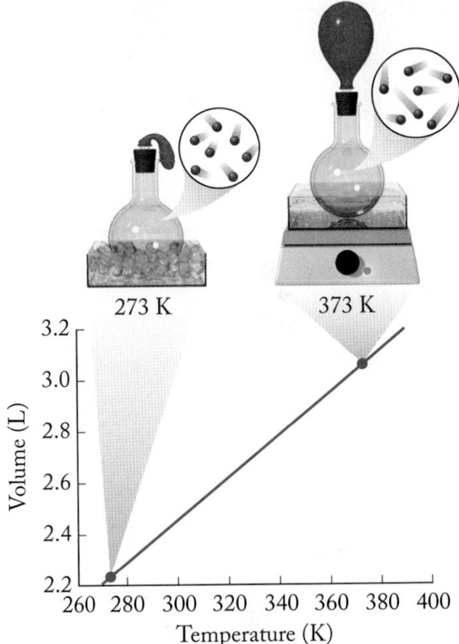

FIGURE 10.22 A balloon attached to a flask inflates as the temperature of the gas inside the flask increases from 273 K to 373 K at constant pressure. This behavior is described by Charles's law.

SAMPLE EXERCISE 10.5 Applying Charles's Law **LO5**

Several students at a northern New England college are hosting a party celebrating the mid-January start of "spring" semester classes. They decide to decorate the front door of their apartment building with party balloons. The air in the inflated balloons is initially 70°F (21°C). After an hour outside, the temperature of the balloons is −12°F (−24°C). Assuming that no air leaks from the balloons and that the pressure in them does not change significantly, how much does their volume change? Express your answer as a percentage of the initial volume.

Collect and Organize We are asked to calculate the change in volume that accompanies a temperature change from 21°C to −24°C. According to Charles's law, the volume of a fixed quantity of a gas at constant pressure is proportional to the *absolute* temperature of the gas.

Analyze There is more than one way to express Charles's law mathematically. Because we are given temperature values and need to calculate the corresponding ratio of volumes, Equation 10.15 is particularly well suited to our calculation:

$$\frac{V_1}{V_2} = \frac{T_1}{T_2}$$

The given temperatures are about 293 and 253 K. The ratio of the smaller to the larger temperatures (253/293) is a little more than 5/6, or about 85%. Therefore, the volume should *decrease* by about 15%.

Solve Letting T_1 be the lower temperature, the value of the right side of Equation 10.15 is

$$\frac{T_1}{T_2} = \frac{(-24 + 273)\text{K}}{(21 + 273)\text{K}} = \frac{249\text{ K}}{294\text{ K}} = 0.8469 = 84.7\%$$

This is the same value as the ratio of the volumes (V_1/V_2) on the left side, so the answer we seek is the difference between 100% and 85%, or a 15% decrease.

Think About It A substantial decrease in temperature (from a human's comfort perspective) translates into a relatively small decrease in volume, as we predicted. The small change makes sense because the difference in temperature on the Kelvin scale is the difference between two large values close to one another in value, which means their ratio is not very far from 1.

Practice Exercise Hot, expanding gases can be used to perform useful work in a cylinder fitted with a movable piston, such as the one in Figure 10.23. If the temperature of a gas confined to such a cylinder is raised from 245°C to 560°C, what is the ratio of the final volume to the initial volume if the pressure inside the cylinder remains constant?

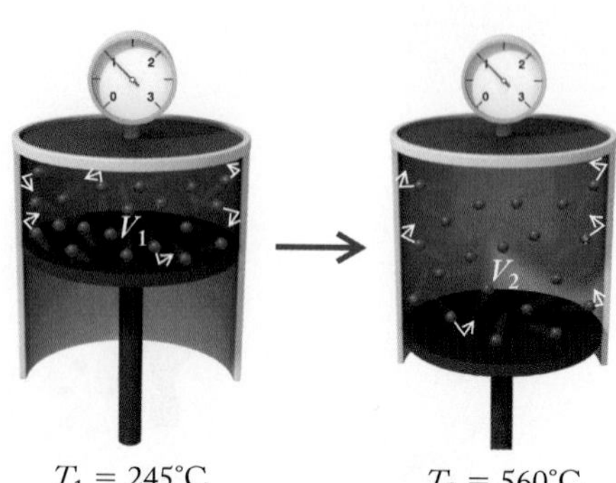

FIGURE 10.23 A molecular view of Charles's law. The volume of a gas increases when its temperature increases and pressure stays the same.

CONNECTION The internal energy of a system decreases as it does P–V work on its surroundings, as described in Section 9.2 and by Equation 9.4: $\Delta E = q - P\Delta V$.

How does kinetic molecular theory explain Charles's law? Temperature is directly related to molecular motion. The average speed at which gas-phase particles move increases with increasing temperature. For a given number of particles, increasing temperature increases their motion and therefore increases the frequency and average kinetic energy that they collide with the walls of their container. More frequent, higher-energy collisions would produce an increase in pressure if volume stayed the same. However, if pressure is to remain constant—as required by Charles's law—then the volume must expand so that the faster-moving particles have to travel farther to collide with the walls of their container. Thus, a higher temperature means more energetic particle–wall collisions, but a larger volume means that these collisions happen less frequently so that, overall, pressure does not change.

Charles's law can also be viewed from a thermodynamic perspective. Suppose a gaseous system is heated, which raises the average kinetic energy of the particles in the system and the internal energy (E) of the system itself. To maintain constant pressure, the system expands by an amount ΔV against a constant opposing pressure P, thereby doing P–V work (that is, $w = -P\Delta V$), which lowers the internal energy of the system.

Charles's law made possible the experimental determination that absolute zero (0 K) is equal to −273.15°C. The graph in Figure 10.24 shows what happens when we plot the volume of fixed quantities of helium at three different constant

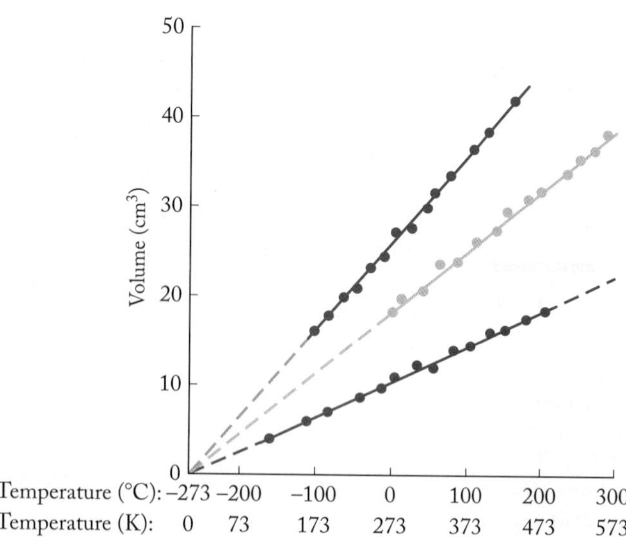

FIGURE 10.24 A plot of the volumes of a constant mass of helium gas at constant pressure versus temperature. Each line represents a different gas pressure. When the curves are extrapolated to $V = 0$, the corresponding temperatures are all −273.15°C on the Celsius scale, or 0 K (absolute zero).

pressures as a function of temperature. All three curves show a linear relationship between volume and temperature, but we are limited by how cold we can make the gases before we reach their boiling points—that is, the temperatures at which the gases condense. However, we can *extrapolate* from our experimental data points to the temperature at which the volume of a sample of gas would reach zero if condensation did not occur. The dashed lines in Figure 10.24 all cross the temperature axis ($V = 0$) at −273.15°C, which is 0.00 K, or absolute zero. Thus, the volume of any gas that obeys Charles's law is directly proportional to its absolute temperature at a given pressure.

Avogadro's law the volume of a gas at constant temperature and pressure is proportional to the quantity (number of moles) of the gas.

Avogadro's Law: Relating Volume and Quantity of Gas

Boyle's and Charles's laws apply to isolated systems that contain constant quantities of gases. However, many physical and chemical systems are open systems in which the quantities of gases *do* change. An example of an open system is the air in a scuba tank. As we noted in Section 10.1, the pressure of a scuba tank during a dive is a reliable indicator of how much air is left in the tank because when temperature and volume are constant, the pressure (P) of a gas, or of a mixture of gases, is proportional to the quantity (moles, n) of gas in its container:

$$P \propto n \text{ (at constant } V \text{ and } T\text{)}$$

or

$$\frac{P}{n} = \text{constant} \quad (10.16)$$

According to kinetic molecular theory, increasing the number of particles of a gas increases the frequency with which the particles collide with the walls of the container, assuming the volume and temperature are constant. If the number of particles doubles, for example, then the number of collisions doubles, and more collisions mean higher pressure.

Some containers (balloons, for example) are not rigid. When gas is added to them, their volumes increase. If temperature and pressure remain constant, then the increase in volume (V) is directly proportional to the quantity (moles, n) of gas added. Expressing this relationship mathematically:

$$V \propto n \text{ (at constant } P \text{ and } T\text{)}$$

or

$$\frac{V}{n} = \text{constant} \quad (10.17)$$

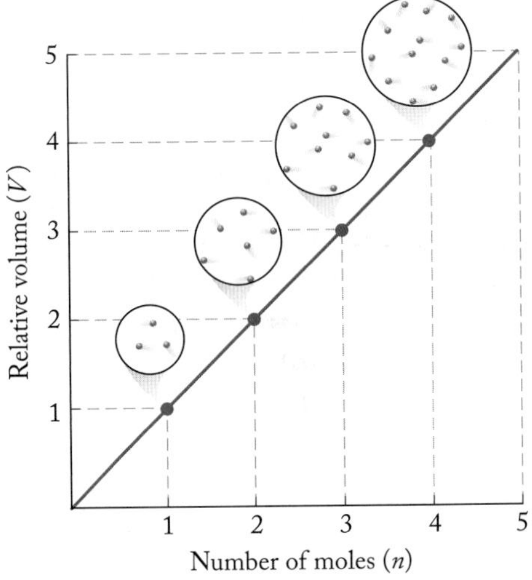

FIGURE 10.25 The volume of a gas at constant temperature and pressure is directly related to the quantity (number of moles) of the gas, a relationship known as Avogadro's law.

Amedeo Avogadro (1776–1856), whom we know from Avogadro's number (N_A), first recognized that the volume of a gas is proportional to the number of particles in it at constant temperature and pressure, and this relationship is called **Avogadro's law** in his honor. Avogadro's law can also be explained by kinetic molecular theory. As we just mentioned, adding particles of gas to a container increases their concentration, which increases the frequency that the particles collide with the walls of the container, thus increasing the pressure. However, if the volume of the container increases enough to maintain the same density of particles and the same particle collision frequency as before more particles were added, then the pressure will remain constant. Thus, pressure remains constant as long as the volume of the system is proportional to the number of particles of gas in it, as shown in Figure 10.25.

CONNECTION The number of particles in a mole is called Avogadro's number (or Avogadro's constant) in recognition of his early work with gases that led to the determination of atomic masses, as described in Chapter 2.

CONCEPT **TEST**

Which graph in Figure 10.26 correctly describes the relationship between *n* and the value of *V*/*n* (at constant *P* and *T*)?

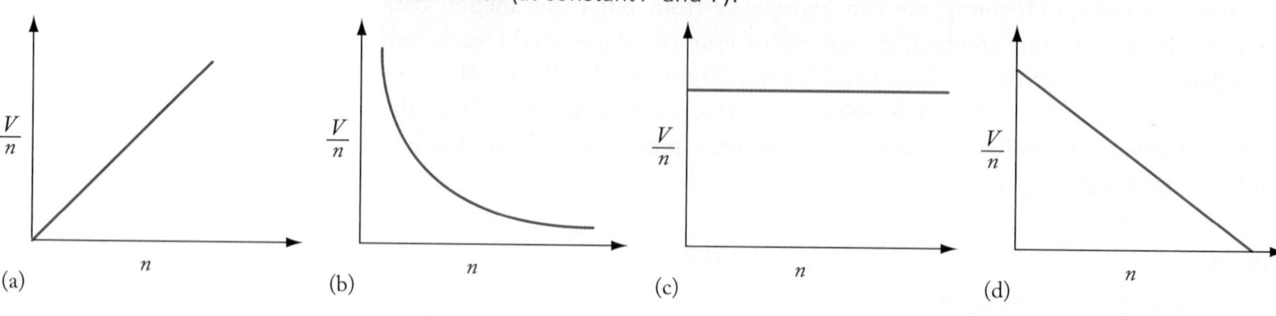

FIGURE 10.26

Amontons's Law: Relating Pressure and Temperature

According to Charles's law, the volume of a gas is proportional to its temperature at constant pressure. An increase in temperature increases the speed of gas particles, which in turn increases both the frequency and force of their collisions with the walls of their container. These increases produce an increase in pressure if the volume remains constant. However, an increase in volume relieves the increase in pressure, which is the essence of Charles's law. On the other hand, if the gas is confined to a rigid container, then the volume remains constant and the pressure increases. In general, the pressure of any quantity of gas is directly proportional to its absolute temperature if its volume does not change:

$$P \propto T \quad \text{or} \quad \frac{P}{T} = \text{constant} \quad (V \text{ and } n \text{ fixed}) \tag{10.18}$$

This relationship between *P* and *T* is called **Amontons's law** in honor of French physicist Guillaume Amontons (1663–1705), a contemporary of Robert Boyle who constructed a thermometer based on the observation that the pressure of a gas is directly proportional to its temperature.

Where do we see evidence of Amontons's law? Suppose we inflate a bicycle tire to a prescribed pressure of 100 psi (6.8 atm) on a warm afternoon in late autumn when the temperature is 25°C. The next morning, following an early freeze in which the temperature drops to 0°C, the pressure in the tire drops to about 91 psi (6.2 atm). Did the tire leak? Probably not. Instead, the decrease in pressure can be explained by Amontons's law. We can express the law in the form of an equation relating the two pressures and temperatures:

$$\frac{P_1}{T_1} = \frac{P_2}{T_2} \tag{10.19}$$

Solving for P_2 (the pressure at the lower temperature), we get the expected result:

$$P_2 = \frac{P_1 T_2}{T_1} = \frac{(6.8 \text{ atm})(273 \cancel{\text{K}})}{298 \cancel{\text{K}}} = 6.2 \text{ atm}$$

Amontons's law the pressure of a gas is proportional to its absolute temperature if its volume does not change.

SAMPLE EXERCISE 10.6 Applying Amontons's Law **LO5**

Labels on aerosol cans caution against incinerating them because the cans may explode when the pressure inside them exceeds 3.00 atm. At what temperature in degrees Celsius might an aerosol can burst if its internal pressure is 2.00 atm at 25°C?

Collect and Organize We are given the temperature (T_1 = 25°C) and pressure (P_1 = 2.00 atm) of a gas and asked to determine the temperature (T_2) at which the pressure (P_2) reaches 3.00 atm.

Analyze Because the gas is isolated in a rigid aerosol can, we know that the quantity of gas and its volume are constant. Amontons's law (Equation 10.19) relates the pressures of a confined quantity of gas at two different temperatures. To estimate our answer, we note that the pressure in the can must increase by 50% to reach 3.00 atm. Pressure is directly proportional to absolute temperature, so the temperature also must increase by 50%. The initial temperature of 25°C is nearly 300 K, so a 50% increase in absolute temperature corresponds to a final temperature near 450 K, or about 175°C.

Solve Rearranging Equation 10.19 to solve for T_2:

$$T_2 = \frac{T_1 P_2}{P_1}$$

and inserting the given T and P values:

$$T_2 = \frac{[(25 + 273)\text{K}](3.00 \cancel{\text{atm}})}{2.00 \cancel{\text{atm}}} = 447 \text{ K}$$

Converting T_2 to degrees Celsius:

$$T_2 = 447 \text{ K} - 273 = 174°\text{C}$$

Think About It This temperature is close to our estimated value and well below the design temperature of 850°C that solid-waste incinerators are supposed to achieve to decompose hazardous organic material. Such a high temperature makes the warning label on the can all the more important.

Practice Exercise Air pressure in each of the tires of an automobile is adjusted to 34 psi at a gas station in San Diego, California, where the air temperature is 68°F (20°C) and atmospheric pressure is 14.5 psi. After a three-hour drive along Interstate Highway 8, the car and driver are in Yuma, Arizona, where the temperature is 110°F (43°C). What is the pressure in the tires, assuming atmospheric pressure is still 14.5 psi? *Note*: The measured tire pressures are *gauge* pressures, meaning that they indicate the pressures in the tires *above atmospheric pressure*.

10.5 The Combined Gas Law

Boyle's law:

$$P_1V_1 = P_2V_2 \tag{10.11}$$

can be combined with Amontons's law:

$$\frac{P_1}{T_1} = \frac{P_2}{T_2} \tag{10.19}$$

to obtain a single equation that relates the pressure, volume, and temperature of a quantity of gas changing from one set of conditions (P_1, V_1, T_1) to another (P_2, V_2, T_2):

$$\frac{P_1V_1}{T_1} = \frac{P_2V_2}{T_2} \tag{10.20}$$

Equation 10.20 is known as the **combined gas law**. It is extremely useful for calculating the impact of changing temperature and pressure on the volume of a gaseous system, as shown in Sample Exercise 10.7.

SAMPLE EXERCISE 10.7 Applying the Combined Gas Law **LO5**

The pressure inside a weather balloon as it is released is 798 mmHg. If the volume and temperature of the balloon are 131 L and 20°C, what is the volume of the balloon when it reaches an altitude where its internal pressure is 235 mmHg and $T = -52°C$?

Collect and Organize We are given the initial temperature, pressure, and volume of a gas, and we are asked to determine the final volume after the pressure and temperature have both decreased.

Analyze The quantity of gas is a constant, so PV/T is also a constant according to the combined gas law (Equation 10.20) as long as we convert the Celsius temperatures to Kelvin temperatures. We expect the balloon's volume to decrease as its temperature decreases but to increase as its pressure decreases. The final pressure is only about 1/3 of the initial pressure, which would result in a tripling of the balloon's volume were it not for the simultaneous decrease in temperature. However, the relative change in absolute temperature is not as great as the decrease in pressure, so we can predict that the volume of the balloon should increase by almost, but not quite, a factor of 3.

Solve First we convert the given Celsius temperatures to Kelvin temperatures:

$$T_1 = 20°C + 273 = 293 \text{ K}$$

$$T_2 = -52°C + 273 = 221 \text{ K}$$

We then solve Equation 10.20 for V_2:

$$\frac{P_1V_1}{T_1} = \frac{P_2V_2}{T_2}$$

$$V_2 = V_1 \times \frac{P_1}{P_2} \times \frac{T_2}{T_1} = 131 \text{ L} \times \frac{798 \text{ mmHg}}{235 \text{ mmHg}} \times \frac{221 \text{ K}}{293 \text{ K}} = 336 \text{ L}$$

Think About It The volume increases by a factor of 336 L/131 L, or 2.56 times, which is in good agreement with our estimate of a nearly threefold increase.

Practice Exercise The balloon in Sample Exercise 10.7 is designed to continue its ascent to an altitude of 30 km, where it bursts, releasing a package of meteorological instruments that parachutes back to Earth. If the pressure inside the balloon at 30 km is 33 mmHg and the temperature is −45°C, what is the volume of the balloon when it bursts?

10.6 Ideal Gases and the Ideal Gas Law

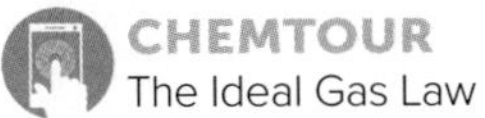

Gases that behave in accordance with the combined gas law are called **ideal gases**. Most gases exhibit ideal behavior at the pressures and temperatures we typically encounter in nature and while doing general chemistry lab experiments. Under these conditions, gases behave ideally because the assumptions about their composition that we described in Section 10.1 are valid. This means that the volumes of the individual gas particles are insignificant compared to the overall volume occupied by the gas, and the particles do not interact with one another. Instead they move independently with speeds that are related to their masses and to the temperature of the gas.

In Section 10.4 we explored how the pressure of an ideal gas is

1. Proportional to the moles of particles in it: $P \propto n$
2. Proportional to its absolute temperature: $P \propto T$
3. Inversely proportional to its volume: $P \propto \frac{1}{V}$

Combining these three expressions into one, we obtain

$$P \propto n \times T \times \frac{1}{V}$$

As in Section 10.4, we can convert the proportionality to an equality by introducing a constant:

$$P = (\text{constant}) \times n \times T \times \frac{1}{V}$$

The constant in this expression is the same *universal gas constant* R that we encountered in Section 10.2:

$$P = R \times n \times T \times \frac{1}{V}$$

Rearranging the terms and simplifying, we get

$$PV = nRT \tag{10.21}$$

Equation 10.21 is probably the most important one in this chapter. It is called the **ideal gas equation**—a mathematical expression of the **ideal gas law**. The value of R, however, is not the same as in Section 10.2; it depends on the units used for pressure and volume. (We always express the quantity of gas in moles, and we use the Kelvin scale for temperature.) If we use SI units for volume (cubic meters) and pressure (pascals), then

$$R = 8.314 \frac{\text{m}^3 \cdot \text{Pa}}{\text{mol} \cdot \text{K}}$$

As a practical matter, it is often more convenient to express volumes using units smaller than cubic meters, such as liters ($1 \text{ L} = 10^{-3} \text{ m}^3$), and to express pressures in larger units such as kilopascals ($1 \text{ kPa} = 10^3 \text{ Pa}$). Because these units are 1/1000 and 1000 times the size of the first set of units, the two factors cancel out and R has the same value:

$$R = 8.314 \frac{\text{L} \cdot \text{kPa}}{\text{mol} \cdot \text{K}}$$

combined gas law the ratio PV/T for a given quantity of gas is a constant.

ideal gas a gas whose behavior is predicted by the linear relations defined by the combined gas law.

ideal gas equation (also called the **ideal gas law**) the pressure, volume, number of moles, and temperature of an ideal gas are related by the equation $PV = nRT$, where R is the universal gas constant.

standard temperature and pressure (STP) 0°C and 1 bar as defined by IUPAC; 0°C and 1 atm are commonly used in the United States.

In many calculations in this book, we express pressure in standard atmospheres (atm), so we will also use an R value based on the equality 1 atm = 101.325 kPa. To avoid a rounding error we will also add one more significant figure to the value of R:

$$R = 8.3145 \frac{\text{L} \cdot \cancel{\text{kPa}}}{\text{mol} \cdot \text{K}} \times \frac{1 \text{ atm}}{101.325 \cancel{\text{kPa}}} = 0.08206 \frac{\text{L} \cdot \text{atm}}{\text{mol} \cdot \text{K}}$$

TABLE 10.3 Values of the Universal Gas Constant (R)

Value of R	Units
0.08206	L · atm/(mol · K)
8.314	kg · m^2/(s^2 · mol · K)
8.314	J/(mol · K)
8.314	m^3 · Pa/(mol · K)
62.37	L · torr/(mol · K)

Note how the numerators of all these values for R represent volume × pressure. In Chapter 9 we discussed how the product of volume and pressure is the kind of work, called P–V work, that a gas does when it expands against an opposing pressure. Energy is the ability to do work, so the numerator of R can also have energy units. The SI unit of energy is the joule, where 1 J is equivalent to 1 m^3 · Pa, which gives another set of units for R:

$$R = 8.314 \frac{\text{J}}{\text{mol} \cdot \text{K}}$$

The various values for R are summarized in Table 10.3.

Before closing this section, you need to know some useful reference conditions in the properties of gases. One of them is **standard temperature and pressure (STP)**. Since 1982 the International Union of Pure and Applied Chemistry (IUPAC) has defined STP as 0°C and 1 bar of pressure. Before that, STP was defined as 0°C and 1 standard atmosphere of pressure, and that definition has lived on, especially among some scientists and science teachers in the United States. One reason for the persistence of the old definition is that the standard atmosphere is equal to 1.01325 bar, so the difference between the two pressures is relatively small.

CONNECTION In Chapter 9 we defined standard conditions of temperature and pressure as they apply to thermochemistry. Note that *STP* and *standard conditions* are not the same. STP applies strictly to calculations involving the gas laws, whereas standard conditions apply to thermochemical data.

One value that depends on how we define STP is *molar volume*, which is the volume that one mole of an ideal gas occupies at STP (Figure 10.27). If we use the old definition of STP pressure (1 atm), the volume of one mole of an ideal gas at 0°C (273 K) is

$$V = \frac{nRT}{P} = \frac{(1 \cancel{\text{mol}})\left(0.08206 \frac{\text{L} \cdot \cancel{\text{atm}}}{\cancel{\text{mol}} \cdot \cancel{\text{K}}}\right)(273 \cancel{\text{K}})}{1 \cancel{\text{atm}}} = 22.4 \text{ L}$$

Molar volume based on the modern definition of STP (P = 1 bar or 100 kPa) is slightly larger:

$$V = \frac{(1 \cancel{\text{mol}})\left(8.314 \frac{\text{L} \cdot \cancel{\text{kPa}}}{\cancel{\text{mol}} \cdot \cancel{\text{K}}}\right)(273 \cancel{\text{K}})}{100 \cancel{\text{kPa}}} = 22.7 \text{ L}$$

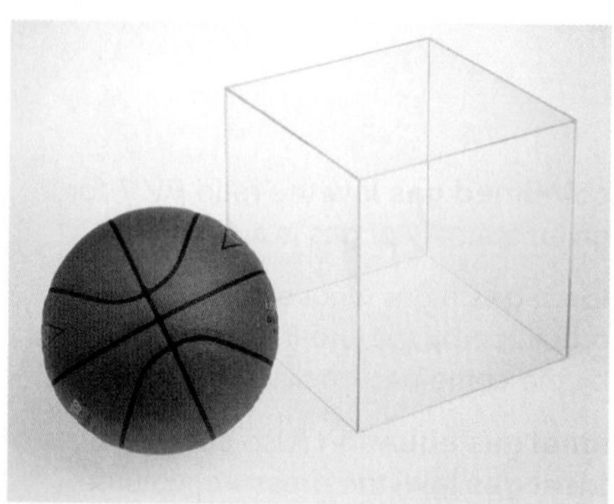

FIGURE 10.27 This box represents the volume of one mole of gas at STP. A basketball would fit loosely inside the box.

Many chemical and biochemical processes (including those that sustain us) take place at pressures near 1 atm (or 1 bar) and at temperatures between 0°C and 40°C. Under these conditions, the volume of one mole of a gas is no more than about 15% greater than the molar volume. Therefore, the volume of a gas can be estimated if the number of moles of the gas is known. An important feature of the molar volume of a gas at ambient temperatures and pressures is that it applies to essentially all gases, independently of their chemical composition, because they all behave like ideal gases.

SAMPLE EXERCISE 10.8 Applying the Ideal Gas Law **LO5**

Bottles of compressed O_2 carried by climbers ascending Mt. Everest contain 1.00 kg of the gas. What volume of O_2 can one bottle deliver to a climber at an altitude where the temperature is −38°C and the atmospheric pressure is 0.35 atm?

Collect and Organize We are given the pressure, mass, and temperature of a gas and asked to determine its volume.

Analyze The ideal gas equation enables us to relate P, V, T, and n, the number of moles of O_2. We are not given n, but we do know the mass of O_2, and we can use its molar mass ($\mathcal{M}$ = 32.00 g/mol) to calculate n and then use the ideal gas equation to calculate V. To estimate our answer, we note that a kilogram of O_2 contains 1000/32, or about 30 moles of the gas, which would occupy 30 × 22.4, or about 660 L at STP. However, conditions on the top of Mt. Everest are far from STP. The temperature on an absolute scale is somewhat lower, which would tend to reduce the volume of gas, but the pressure is only about 1/3 of what it is at STP, which means the volume should be nearly three times 660 L, or nearly 2000 L.

Solve Let's start by converting the mass of O_2 into moles:

$$1.00 \cancel{\text{kg O}_2} \times \frac{1000 \text{ g}}{1 \cancel{\text{kg}}} \times \frac{1 \text{ mol O}_2}{32.00 \cancel{\text{g O}_2}} = 31.3 \text{ mol O}_2$$

Solving the ideal gas equation for V and inserting the values known:

$$PV = nRT$$

$$V = \frac{nRT}{P} = \frac{(31.3 \cancel{\text{mol}})\left(0.08206 \dfrac{\text{L} \cdot \cancel{\text{atm}}}{\cancel{\text{mol}} \cdot \cancel{\text{K}}}\right)(273 - 38) \cancel{\text{K}}}{0.35 \cancel{\text{atm}}} = 1.7 \times 10^3 \text{ L}$$

Think About It Our answer is certainly reasonable based on our estimate. Most climbers require several of these bottles to climb Mt. Everest and return. We used the ideal gas law to solve this problem because all four gas law variables—P, V, T, and n—factored into the calculation.

 **Practice Exercise** Buoyancy for the Goodyear blimp *Spirit of Innovation* comes from 2.03 × 10^5 cubic feet of helium (Figure 10.28).

a. What is the mass of this much helium at 25°C and 1.00 atm of pressure?
b. What is the buoyancy of the balloon—that is, the difference between the mass of the He and the mass of the same volume of dry air at the same temperature and pressure?

FIGURE 10.28 The Goodyear blimp *Spirit of Innovation*.

10.7 Densities of Gases

Carbon dioxide is a relatively minor component of our atmosphere, but it has received considerable attention in recent years due to its increasing concentration in the atmosphere and the resulting impact on global climate. While this increase has been linked to human activity, many natural sources such as volcanoes also add CO_2 to the atmosphere. The quantities of CO_2 involved in volcanic eruptions are not normally a threat to human health, but when this gas collects in the waters of a deep crater lake and is then released all at once,

(a)

(b)

FIGURE 10.29 As many as 1800 people may have been asphyxiated by a dense cloud of CO_2 that was released by Lake Nyos in northwestern Cameroon on August 21, 1986. (a) The lake normally has a deep blue color. (b) The CO_2 event stirred up sediment from the lake floor, turning the water brown.

the results can be catastrophic. On August 21, 1986, volcanic Lake Nyos in Cameroon, Africa, suddenly released about a cubic kilometer of CO_2 into the air. The gas emerged from the lake in a frothy mist an estimated 100 meters high, which flowed out of the crater and into surrounding valleys, killing over 1800 people and their farm animals (Figure 10.29). Because CO_2 is denser than air, the deadly mist formed a ground-hugging layer that displaced the air and the oxygen necessary for life.

The density of any gas at STP can be calculated by dividing its molar mass by the molar volume. Carbon dioxide, for example, has a molar mass of 44.0 g/mol, so the density of CO_2 at 0°C and 1 atm is

$$\frac{44.0\ \text{g}/\cancel{\text{mol}}}{22.4\ \text{L}/\cancel{\text{mol}}} = 1.96\ \text{g/L}$$

The density of air, which is mostly N_2 ($\mathcal{M}$ = 28.01 g/mol) and O_2 ($\mathcal{M}$ = 32.00 g/mol), is only 1.29 g/L at STP, so 1 km^3 of CO_2, a colorless, odorless gas, could easily displace enough air to become a silent killer. The greater density of CO_2 and the fact that it does not support combustion also makes it effective in fighting small fires. The gas from a CO_2 fire extinguisher effectively blankets burning fuel, depriving it of O_2 and extinguishing the fire.

We can use the ideal gas equation to calculate the density of a gas at any temperature and pressure. Density is expressed in units of mass per unit volume, and the number of moles of a pure substance is related to mass by its molar mass. We can rearrange the ideal gas law to bring the n and V terms onto one side of the equation:

$$\frac{n}{V} = \frac{P}{RT}$$

If we multiply both sides of the equation by the molar mass in g/mol, $\mathcal{M}$, the left-hand side of the equation has units of g/L or density. Thus, Equation 10.22 relates the density of any gas to its molar mass and pressure:

$$\frac{\mathcal{M}n}{V} = \frac{\left(\frac{\text{g}}{\cancel{\text{mol}}}\right)(\cancel{\text{mol}})}{\text{L}} = d = \frac{\mathcal{M}P}{RT} \qquad (10.22)$$

We can determine the density of a gas experimentally using an apparatus such as the one shown in Figure 10.30. In this apparatus, a glass bulb of known volume is attached to a vacuum pump and the air is removed. The mass of the bulb is determined twice: first while it is evacuated, and again after the bulb is filled with the test gas. The difference in the masses divided by the volume of the bulb is the density of the gas.

If we also knew the temperature and pressure of the gas in the bulb in Figure 10.30, we could use the values of V, T, and P to calculate the number of moles of gas in the bulb:

$$n = \frac{PV}{RT}$$

Dividing the mass (m) of the gas in the bulb (in grams) by the number of moles of gas (n) gives the molar mass of the gas:

$$\mathcal{M} = \frac{m}{n}$$

CONCEPT TEST

Which graph in Figure 10.31 best approximates the following for an ideal gas?

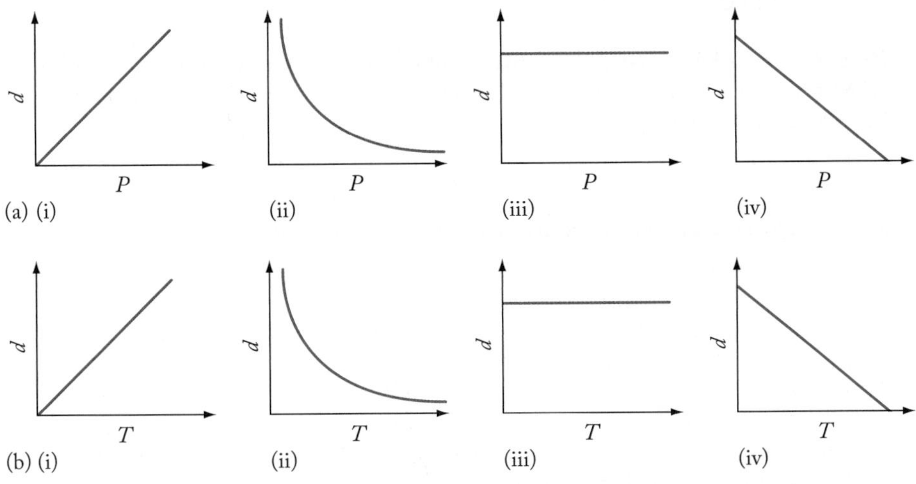

FIGURE 10.31

a. The relationship between density and pressure (n and T constant)

b. The relationship between density and temperature (n and P constant)

(a)

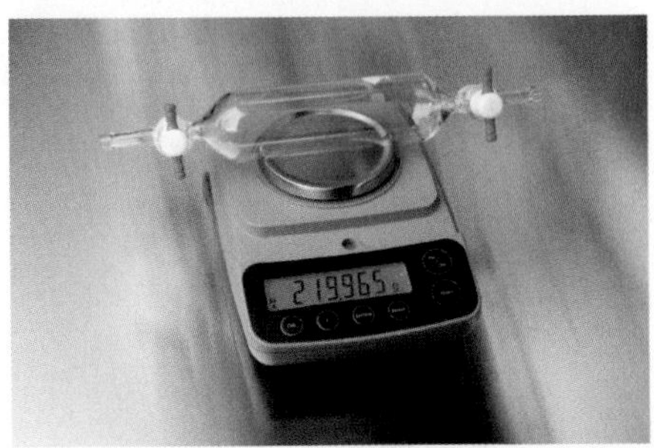

(b)

(c)

FIGURE 10.30 Apparatus for determining the density of a gas. (a) A gas collection tube with an internal volume of 235 mL is evacuated by connecting it to a vacuum pump. (b) The mass of the empty tube is measured. (c) The tube is then opened to the atmosphere to be refilled with air. The difference in mass between the filled and evacuated tube (0.273 g, or 273 mg) is the mass of the air inside it. The density of the air sample is 273 mg/235 mL = 1.16 mg/mL = 1.16 g/L.

SAMPLE EXERCISE 10.9 Calculating the Density of a Gas — LO6

According to the U.S. National Weather Service, the air temperature in Phoenix, Arizona, reached 78°F (26°C) on January 1, 2012, when atmospheric pressure was 1024 millibars. What was the density of the air? Assume the average molar mass of air is 28.8 g/mol, which is the weighted average of the molar masses of the various gases in dry air (see Table 10.1).

Collect and Organize We are provided with the average molar mass, temperature, and atmospheric pressure of an air mass and asked to calculate its density, where $d = m/V$.

Analyze We may assume that the air behaves like an ideal gas and obeys the ideal gas law. One way to calculate the density of air (or any ideal gas) is to divide the mass (m) of one mole of it (28.8 g in this case) by the volume (V) that one mole occupies. This volume can be calculated using the ideal gas equation and the temperature and pressure values provided. We will need to convert T and P to units that are compatible with one of our R values. Because the density of air at STP is about 1.3 g/L, we can estimate that the air density at Phoenix's warm temperature is about 10% lower than at STP, or about 1.2 g/L.

Solve There are several options for converting millibars to another pressure unit. Let's use the equality 1 atm = 1013.25 millibar (mb) and calculate the volume of one mole of air, starting with the ideal gas equation: $PV = nRT$. Solving for V:

$$V = \frac{nRT}{P} = \frac{(1\ \cancel{\text{mol}})\left(0.08206\ \frac{\text{L} \cdot \text{atm}}{\cancel{\text{mol}} \cdot \cancel{\text{K}}}\right)(299\ \cancel{\text{K}})}{(1024\ \cancel{\text{mb}})\left(\frac{1\ \text{atm}}{1013.25\ \cancel{\text{mb}}}\right)} = 24.28\ \text{L}$$

Calculating density:

$$d = \frac{m}{V} = \frac{28.8\ \text{g}}{24.28\ \text{L}} = 1.19\ \text{g/L}$$

Think About It The calculated result is consistent with our estimate and with the general trend that gases are about 10% less dense at comfortable ambient temperatures than at STP (0°C).

Practice Exercise Air is a mixture of mostly nitrogen and oxygen. A balloon is filled with pure oxygen and released in a room full of air. Will it sink to the floor or float to the ceiling?

SAMPLE EXERCISE 10.10 Calculating Molar Mass from Density **LO6**

Vent pipes at solid-waste landfills emit gases released by decomposing material. A sample of such a gas has a density of 0.65 g/L at 25.0°C and 757 mmHg. What is its molar mass?

Collect and Organize We are given the density (in g/L), temperature, and pressure (in mmHg) of a gaseous sample and are asked to calculate its molar mass.

Analyze Equation 10.22 relates the density, pressure, and molar mass of a gas. Before we can use this equation, we need to convert the Celsius temperature to a Kelvin temperature, and we need to convert mmHg to a more convenient pressure unit, such as atm (1 atm = 760 mmHg). We can use the concept of molar volume to estimate the molar mass of the gas. We saw in Sample Exercise 10.9 that, at 25°C and a pressure near 1 atm, the volume of one mole of an ideal gas is about 24 L. If the mass of one liter is 0.65 g, or about 2/3 of a gram, then the mass of 24 L is about 2/3 of 24, or 16 g/mol.

Solve Solving Equation 10.22 for $\mathcal{M}$ and substituting the values given for d, R, T, and P:

$$\mathcal{M} = \frac{dRT}{P} = \frac{\left(\frac{0.65\ \text{g}}{\cancel{\text{L}}}\right)\left(0.08206\ \frac{\cancel{\text{L}} \cdot \cancel{\text{atm}}}{\text{mol} \cdot \cancel{\text{K}}}\right)(25 + 273)\ \cancel{\text{K}}}{(757\ \cancel{\text{mmHg}})\left(\frac{1\ \cancel{\text{atm}}}{760\ \cancel{\text{mmHg}}}\right)} = 16\ \text{g/mol}$$

Think About It Equation 10.22 provides a clear route to solving problems like this one. However, if you hadn't memorized it or been provided it, you could still solve the problem using the central equation in this chapter: $PV = nRT$. The density of the gas tells us the mass of one liter of it. If we assumed V = exactly one liter, then we could use the given T and P values and $PV = nRT$ to calculate the number of moles (n) in one liter. Dividing 0.65 g/L by the calculated n/L yields g/mol of molar mass.

Practice Exercise When HCl(aq) and $NaHCO_3$(aq) are mixed together, a chemical reaction takes place in which a gas is one of the products. A sample of the gas has a density of 1.81 g/L at 1.00 atm and 23.0°C. What is the molar mass of the gas? Can you identify the gas?

10.8 Gases in Chemical Reactions

Many important chemical reactions, including those that supply us with most of our energy needs, involve gaseous reactants or products. Even when the fuel is a solid or a liquid, the actual combustion reaction takes place in the gas phase, such as when gasoline burns in the combustion chamber of a car engine, or the flames of a charcoal fire cook dinner at a backyard barbecue.

In a chemical reaction involving a gas as either reactant or product, the quantity of it (in moles) in the reaction mixture can be determined using the ideal gas law if we know the volume of the gas and the temperature and pressure at which the reaction proceeds. Once we know the value of n, we can calculate the quantities of the other reactants or products, including the quantity of energy released or consumed, from the stoichiometry of the reaction. Alternatively, we may know the quantity of another reactant and need to calculate the volume of a gaseous reactant that is needed for a complete reaction. Sample Exercise 10.11 provides an example of such a calculation from the world of backyard barbecues.

SAMPLE EXERCISE 10.11 Combining Stoichiometry and the Ideal Gas Law

LO7

The maximum flow rate of propane fuel through one of the burners in a gas grill is 5.5 g C_3H_8 per minute. What volume of air must mix with the propane ($\mathcal{M}$ = 44.1 g/mol) each minute to provide enough O_2 for complete combustion? Assume P = 1.00 atm and T = 25°C.

Collect and Organize We know that each minute 5.5 g C_3H_8 is burned, and we are asked to calculate the volume of air required for complete combustion. According to the data in Table 10.1, air is 20.95% O_2. The volume of an ideal gas is related to its amount, temperature, and pressure by the ideal gas law: $PV = nRT$.

Analyze The quantities of two reactants in any chemical reaction are related by the stoichiometry of the balanced chemical equation describing the reaction, so we need to write that equation. We also need to convert 5.5 g C_3H_8 to moles C_3H_8 and then to moles O_2 using the stoichiometry of the reaction. Moles of O_2 can be converted to a volume of O_2 using the ideal gas law and to a corresponding volume of air. To estimate our answer, we note that 5.5 g C_3H_8 / 44 g/mol is little more than 0.1 mol C_3H_8. We will need about five times that many mol O_2 to convert 3 mol C and 8 mol H into CO_2 and H_2O. Assuming we need about 0.5 mol O_2, the volume of O_2 will be a little more than half the molar volume, or about 12 liters. Air is about 1/5 O_2, so we will need about 12 × 5, or 60, liters of air.

Solve To write the balanced chemical equation for the reaction, we balance the number of C, then H, and finally the O atoms (as in Chapter 7), which yields

$$C_3H_8(g) + 5\ O_2(g) \rightarrow 3\ CO_2(g) + 4\ H_2O(g)$$

The number of moles of C_3H_8 is

$$5.5\ \cancel{\text{g } C_3H_8} \times \frac{1\ \text{mol } C_3H_8}{44.1\ \cancel{\text{g } C_3H_8}} = 0.125\ \text{mol } C_3H_8$$

Applying the 5:1 stoichiometric ratio of O_2 to C_3H_8, we obtain

$$0.125\ \cancel{\text{mol } C_3H_8} \times \frac{5\ \text{mol } O_2}{1\ \cancel{\text{mol } C_3H_8}} = 0.625\ \text{mol } O_2$$

The pressure is given as 1.00 atm, so we use the R value with atmospheres in the numerator in the ideal gas law to convert moles O_2 to liters of O_2:

$$V = \frac{(0.625\ \cancel{\text{mol}}\ O_2)\left(0.08206\ \frac{\text{L} \cdot \cancel{\text{atm}}}{\cancel{\text{mol}} \cdot \cancel{\text{K}}}\right)(25 + 273)\ \cancel{\text{K}}}{1.00\ \cancel{\text{atm}}} = 15.3\ \text{L } O_2$$

Air is 20.95% O_2, so the volume of air that contains 15.3 L of O_2 is

$$V_{\text{air}} = 15.3\ \text{L}\ \cancel{O_2} \times \frac{100.00\ \cancel{\%}\ \text{air}}{20.95\ \cancel{\%\ O_2}} = 73\ \text{L air}$$

Think About It Our calculated result, expressed with two significant digits, is close to the predicted value and reasonable given the size of a molar volume and the fact that air is only about 20% O_2 by volume.

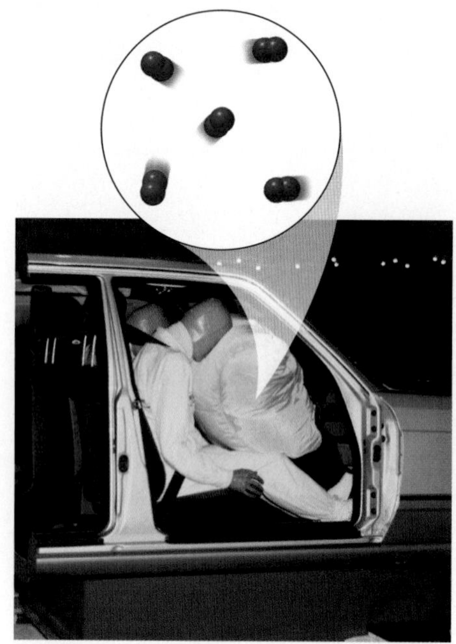

FIGURE 10.32 An automobile air bag inflates when solid NaN_3 rapidly decomposes, producing N_2 gas.

CONNECTION Dalton's law of partial pressures was articulated by the same English chemist who developed Dalton's law of multiple proportions and a theory that matter is composed of atoms. Both are described in Chapter 1.

Practice Exercise Automobile air bags (Figure 10.32) inflate during a crash or sudden stop by the rapid generation of nitrogen gas from sodium azide. The first step of the air bag reaction may be written as

$$2\ NaN_3(s) \rightarrow 2\ Na(s) + 3\ N_2(g)$$

How many grams of sodium azide are needed to provide enough nitrogen gas to fill a $45 \times 45 \times 25$ cm bag to a pressure of 1.20 atm at 15°C?

10.9 Mixtures of Gases

As noted in Table 10.1, dry air is mostly N_2 and O_2 plus much smaller concentrations of other gases, such as argon and carbon dioxide. Each gas in air, or in any mixture of gases, exerts its own pressure, called its **partial pressure**. Atmospheric pressure is the sum of the partial pressures of all of the gases in the air:

$$P_{atm} = P_{N_2} + P_{O_2} + P_{Ar} + P_{CO_2} + \cdots$$

A similar expression can be written for any mixture that contains "i" gases, where $i = 1, 2, 3, \ldots$:

$$P_{total} = P_1 + P_2 + P_3 + \cdots \quad (10.23)$$

Equation 10.23 is a mathematical expression of **Dalton's law of partial pressures**: the total pressure of any mixture of gases is the sum of the partial pressures of the gases in the mixture.

The most abundant gases in a mixture (such as N_2 and O_2 in air) have the greatest partial pressures and contribute the most to the total pressure of the mixture. The mathematical term used to express the abundance of component i in a mixture of gases is its **mole fraction (x_i)**,

$$x_i = \frac{n_i}{n_{total}} \quad (10.24)$$

where n_i is the number of moles of component i and n_{total} is the total number of moles of gas in the mixture. The partial pressure of each component is the product of its mole fraction times the total pressure of the mixture:

$$P_i = x_i P_{total} \quad (10.25)$$

Let's apply Equation 10.25 to the major gases in air. If the atmospheric pressure of a dry air mass is 998 mbar, the partial pressures of N_2 and O_2 are

$$P_{N_2} = x_{N_2}P_{total} = (0.7808)(998\ \text{mbar}) = 779\ \text{mbar}$$
$$P_{O_2} = x_{O_2}P_{total} = (0.2095)(998\ \text{mbar}) = 209\ \text{mbar}$$

Note that the partial pressure of each of these gases depends only on total pressure and its mole fraction; it does not depend on the identity of the gas or of the other gases in the mixture. Note, too, that mole fractions have no units. This means that they can be used when describing the behavior of any kind of homogeneous mixture (solid, liquid, or gas). Finally, the sum of the mole fractions of all the components in a mixture should always add up to 1 (or, because of rounding, very close to 1).

The kinetic molecular theory of gases explains the additive nature of partial pressures and the dependence of partial pressure on mole fraction. According to the theory, the particles of a pure gas, or of a mixture of gases, collide with the

CHEMTOUR Dalton's Law

partial pressure the contribution to the total pressure made by a component in a mixture of gases.

Dalton's law of partial pressures the total pressure of a mixture of gases is the sum of the partial pressures of all the gases in the mixture.

mole fraction (x_i) the ratio of the number of moles of a particular component i in a mixture to the total number of moles in the mixture.

walls of the mixture's container. The force of these collisions, when averaged over the surface area of the container, produces a total pressure. The more particles there are, the more frequent the collisions are and the greater the pressure is. The theory assumes that all particles have, on average, the same kinetic energy, which means that the frequency and force of their collisions, and the contributions these collisions make to overall pressure, depend only on how many of the particles are present. The theory further assumes that the particles do not interact with each other, so their identities are irrelevant—indeed, the particles of all of the components of a mixture are equal-opportunity colliders. The particles of each gas contribute to the total pressure of the mixture in exact proportion to how many of them are in the container as compared to the total number of particles.

SAMPLE EXERCISE 10.12 Calculating Mole Fractions and Partial Pressures LO8

Scuba divers who dive to depths below 50 meters may breathe a gas mixture called trimix during the deepest parts of their dives. One formulation of the mixture, called *trimix 10/70*, is 10% oxygen, 70% helium, and 20% nitrogen by volume. What is the mole fraction of each gas in this mixture, and what is the partial pressure of oxygen in the lungs of a diver at a depth of 60 meters (where the ambient pressure is 7.0 atm)?

Collect and Organize We are given the composition of a scuba diver's breathing mixture expressed in percent by volume of its three components: O_2, He, and N_2. We are asked to determine the mole fractions of the gases and the partial pressure of O_2 in the lungs of a diver breathing the mixture at a depth where the pressure is 7.0 atm.

Analyze According to Avogadro's law, equal volumes of ideal gases at the same temperature and pressure contain equal numbers of moles of gas. Therefore, the mole fraction of each gas in a mixture is equal to the volume of that gas divided by the sum of the volumes of all the gases that were mixed together. Thus, the mole fractions of the gases are the same as their percent by volume values. The partial pressure of a component of a gas mixture may be calculated using Equation 10.25:

$$P_i = x_i P_{\text{total}}$$

Solve Expressing the percent by volume values as mole fractions:

Gas	% by Volume	Mole Fraction
O_2	10	0.10
He	70	0.70
N_2	20	0.20

Calculating P_{O_2}:

$$P_{O_2} = x_{O_2} P_{\text{total}} = 0.10 \times 7.0 \text{ atm} = 0.70 \text{ atm}$$

Think About It This exercise introduces a useful equality: the concentrations of the gases in a mixture, such as those listed in Table 10.1 for dry air, are equivalent to the mole fractions of the gases, though decimal points may need to be moved—for example, two places to the left to convert percent values into mole fractions, or six places to the left to convert to parts per million values.

Practice Exercise What is the partial pressure, in atmospheres, of O_2 in the air outside an airliner cruising at an altitude of 10 km, where the atmospheric pressure is 190 mmHg? How much must the outside air be compressed to produce a cabin pressure in which $P_{O_2} = 0.200$ atm?

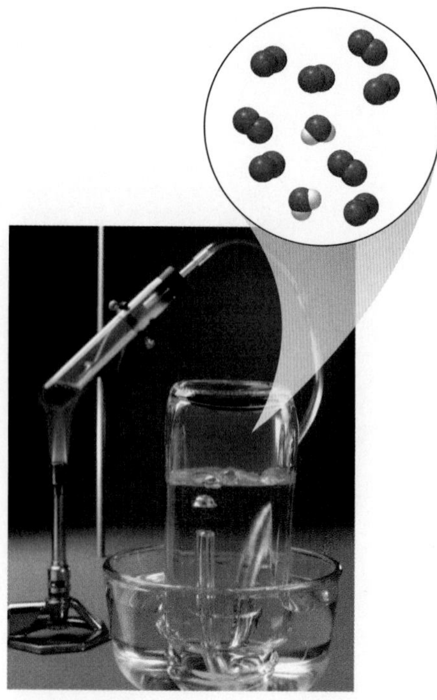

FIGURE 10.33 Collecting O_2 gas by water displacement. Oxygen is generated by the thermal decomposition of $KClO_3$. The delivery tube is removed when the height of the water in the collection jar matches the level of water around it, ensuring that the pressure in the jar is equal to atmospheric pressure.

CONCEPT TEST

Do each of the gases in an equimolar mixture of four gases have the same mole fraction?

Sample Exercise 10.12 and its Practice Exercise illustrate how our health and well-being rely on access to the appropriate breathing gases when we venture a few kilometers above sea level or a few meters below it. At more dweller-friendly altitudes, Dalton's law of partial pressures is used to determine the quantities of gaseous products in chemical reactions. For example, heating potassium chlorate ($KClO_3$) in the presence of MnO_2 causes it to decompose into $KCl(s)$ and $O_2(g)$. The oxygen gas produced by this reaction can be collected by bubbling the gas into an inverted bottle that is initially filled with water (Figure 10.33). As the reaction proceeds, $O_2(g)$ displaces the water in the bottle. When the reaction is complete, the volume of water displaced provides a measure of the volume of O_2 produced. If the temperature of the water and atmospheric pressure are known, the ideal gas law can be used to determine the number of moles of O_2 produced.

Water displacement can be used to collect and measure the volume of any gas that neither reacts with nor dissolves appreciably in water. However, one additional step is needed to apply the ideal gas law to calculating the number of moles of gas produced. At room temperature (nominally 20–25°C), any enclosed space above a pool of liquid water contains water vapor in addition to the gas produced by the reaction. Therefore, the gas collected in the $KClO_3$ reaction in Figure 10.33 is a mixture of O_2 and H_2O vapor. Each gas exerts its own partial pressure, and Dalton's law of partial pressures tells us the total pressure of the mixture when the water level inside the collection vessel matches the water level outside the vessel:

$$P_{\text{total}} = P_{\text{atm}} = P_{O_2} + P_{H_2O} \qquad (10.26)$$

To calculate the quantity of oxygen produced using the ideal gas law, we need to know P_{O_2}, which we get by subtracting P_{H_2O} (Table 10.4) from P_{atm}. If we know the values of P_{O_2}, T, and V, we can calculate the number of moles or number of grams of oxygen produced.

TABLE 10.4 Partial Pressure of Water Vapor at Selected Temperatures

Temperature (°C)	Pressure (mmHg)
5	6.5
10	9.2
15	12.8
20	17.5
25	23.8
30	31.8
35	42.2
40	55.3
45	71.9
50	92.5

SAMPLE EXERCISE 10.13 Calculating the Quantity of a Gas Collected by Water Displacement **LO8**

During the decomposition of $KClO_3$, 92.0 mL of gas is collected by the displacement of water at 25.0°C. If atmospheric pressure is 756 mmHg, what mass of O_2 is collected?

Collect and Organize We are given values for the total pressure (756 mmHg), volume (92.0 mL), and temperature (25.0°C) of the gas collected by water displacement during the decomposition of $KClO_3$, and we are asked to calculate the mass of O_2 in the gas. Table 10.4 contains values for the partial pressure of water vapor at several temperatures. The ideal gas law relates the number of moles of an ideal gas to its pressure, volume, and absolute temperature.

Analyze Oxygen is collected over water, so the collection vessel contains both O_2 and water vapor. To calculate P_{O_2}, we subtract the partial pressure of water vapor at 25°C (23.8 mmHg, according to Table 10.4) from atmospheric (total) pressure. To calculate the number of moles of O_2 using the ideal gas equation, we need to convert the pressure value to atm and the temperature to kelvin. Multiplying that number of moles by the molar mass of O_2 will give us the mass of O_2 collected. To estimate the result, let's round off the volume of the gas to 100 mL, or 0.1 L. At 25°C the molar volume of an ideal gas is nearly 25 L, so the volume collected corresponds to 0.1/25, or about 0.004 mole of gas. The partial pressure of water vapor is small compared to atmospheric pressure, so most of the gas collected is O_2. The mass of this quantity of O_2 is 0.004 mol × 32 g/mol, or about 0.13 g.

Solve The value of P_{O_2} is

$$P_{O_2} = P_{atm} - P_{H_2O} = (756 - 23.8)\ \text{mmHg} = 732\ \text{mmHg}$$

The number of moles of O_2 is

$$n = \frac{\left(732\ \cancel{\text{mmHg}} \times \dfrac{1\ \cancel{\text{atm}}}{760\ \cancel{\text{mmHg}}}\right)\left(92.0\ \cancel{\text{mL}} \times \dfrac{1\ \cancel{\text{L}}}{1000\ \cancel{\text{mL}}}\right)}{\left(0.08206\ \dfrac{\cancel{\text{L}} \cdot \cancel{\text{atm}}}{\text{mol} \cdot \cancel{\text{K}}}\right)(25 + 273)\ \cancel{\text{K}}} = 0.00362\ \text{mol}$$

and the mass of O_2 is

$$m = 0.00362\ \cancel{\text{mol}} \left(\frac{32.00\ \text{g}}{1\ \cancel{\text{mol}}}\right) = 0.116\ \text{g}$$

Think About It The calculated mass of O_2 is about what was expected. The value of P_{O_2} is (732/756) or 96.8% of P_{atm}—close enough to justify our ignoring the contribution of P_{H_2O} in our estimate, but not so close that it could be ignored in the actual calculation.

Practice Exercise Electrical energy can be used to separate water into O_2 and H_2. In one demonstration of this reaction, 27 mL of H_2 is collected over water at 25°C. Atmospheric pressure is 761 mmHg. How many milligrams of H_2 are collected?

10.10 Real Gases

Up to now all of our calculations of *P*, *V*, *T*, and *n* have assumed ideal gas behavior, and this is acceptable because, under typical atmospheric pressures and temperatures, most gases *do* behave ideally. We have assumed, according to the kinetic molecular theory, that the volume occupied by individual gas particles is negligible compared with the total volume occupied by the gas. In addition, we have assumed that no interactions other than random elastic collisions occur between particles. These assumptions are not valid, however, when gases are subjected to extremely high pressures or to temperatures so low that they approach the temperatures at which the gases condense.

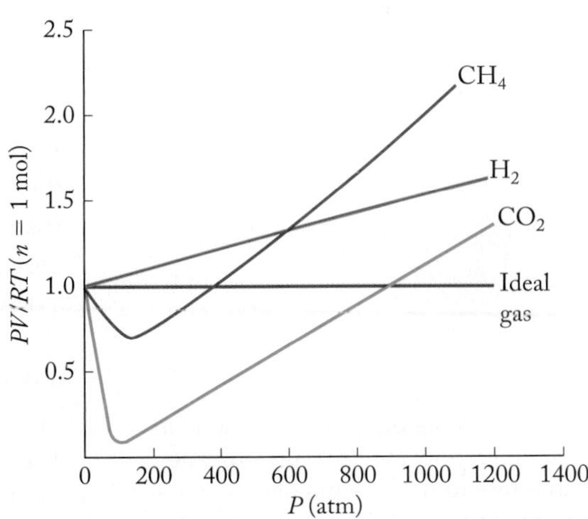

FIGURE 10.34 The effect of very high pressure on the behavior of one mole of real and ideal gases. The curves for real gases diverge from ideal behavior in a manner unique to each gas and due to two competing factors: interactions between particles (which lower the value of *PV*/*RT*) and the incompressibility of the particles themselves (which raises the value of *PV*/*RT*).

Deviations from Ideality

Let's start by considering the behavior of one mole of a gas as we increase the pressure on the gas. From the ideal gas law, we know that $PV/RT = n$, so for one mole of gas, *PV*/*RT* should remain equal to 1 regardless of how we change the pressure. This relationship between *PV*/*RT* and *P* for an ideal gas is shown by the horizontal purple line in Figure 10.34. However, the curves for CH_4, H_2, and CO_2 at pressures above 10 atm are not horizontal straight lines. Not only do their curves diverge from the ideal, but the shapes of the curves also differ for each gas, indicating that deviations from ideal behavior depend on the identity of the gas.

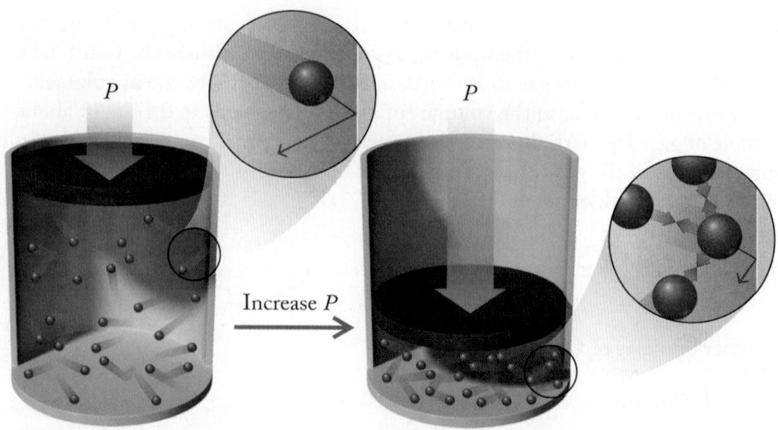

FIGURE 10.35 At very high pressures there are more interactions between particles (represented by the broad red arrows in the expanded view). These interactions reduce the frequency and force of particle collisions with the walls of the container (thin red arrow), thereby reducing the pressure.

Why don't real gases behave like ideal gases at high pressure? One reason is that particles of ideal gases do not interact with each other, but real atoms and molecules *do*, in the various ways discussed in Chapter 6. There are more opportunities for intermolecular interactions when extremely high pressures push gas-phase particles closer together, or when lower temperatures force the particles to move more slowly and they take longer to pass by one another. Both conditions—high pressure and low temperature—favor intermolecular interactions (represented by the broad red arrows in the right-side molecular view in Figure 10.35) that the ideal gas law does not take into account. The more often that particles interact with each other, the less frequently and forcefully they collide with the walls of their container, and the lower the pressure. Likewise, lowering the temperature of a gas causes molecules to move more slowly, leading to more attractions between molecules and more deviations from ideal gas behavior.

CONNECTION The nature and relative strengths of intermolecular forces were discussed in Chapter 6.

Another reason for nonideal behavior is that atoms and molecules of gases do have some volume. Under extremely high pressure, the volume of the particles themselves becomes a significant fraction of the total volume of the gas as the particles are squeezed more tightly together. However, it is only the empty space between particles that makes gases compressible. The particles themselves are not compressible: they have finite dimensions and do not shrink with increasing pressure. Therefore, when the pressure doubles, the volume of a real gas does not shrink by 50% (as it does in an ideal gas) because a significant part of that volume can't shrink at all. If the volume shrinks by less than 50% when pressure doubles, then the product $P \times V$ in the numerator of PV/RT is no longer a constant. Instead, $P \times V$ and the PV/RT ratio increase as P increases. Because all atoms and molecules have a finite volume, this positive deviation in PV/RT is observed for all gases, including the three in Figure 10.34, if P is high enough.

So two competing factors—namely, intermolecular interactions and the volumes of gas-phase particles—can cause PV/RT to decrease or increase with increasing external pressure. At extremely high pressures, the size of the particles and the space they occupy become the dominant factor and PV/RT increases with increasing P. At less extreme pressures, intermolecular interactions may offset the incompressibility of particles, and the value of PV/RT decreases with increasing P for some gases, including CH_4 and CO_2, as shown in Figure 10.34.

van der Waals equation an equation describing how the pressure, volume, and temperature of a quantity of a real gas are related; it includes terms that account for the incompressibility of gas particles and interactions between them.

van der Waals force any interaction between neutral atoms and molecules, including hydrogen bonds, other dipole–dipole interactions, and London dispersion forces; the term does not apply to interactions involving ions.

The van der Waals Equation for Real Gases

Because the ideal gas equation does not hold at extremely high pressures and low temperatures, we need to modify it if we are to relate the quantities, pressures, volumes, and temperatures of real gases under these nonideal conditions. An equation developed in 1873 by Dutch scientist Johannes Diderik van der Waals (1837–1923) incorporates the needed modifications:

$$\left[P + a\left(\frac{n}{V}\right)^2\right](V - nb) = nRT \qquad (10.27)$$

Equation 10.27 is called the **van der Waals equation** and includes two terms to correct for (1) intermolecular interactions that lower the number of independent particles and the pressure they create by colliding with container walls [the $a(n/V)^2$ term] and (2) the volume taken up by the particles of a gas, which are not compressible (the nb term). The values of a and b in these terms, called *van der Waals constants*, have been determined experimentally for many gases (Table 10.5). The values of both constants increase with increasing molar mass and with the number of atoms per molecule, which makes sense because larger molecules both take up more space and experience stronger London dispersion forces, which lead to stronger intermolecular interactions.

Note that the pressure correction term, $a(n/V)^2$, increases with the square of the concentration (n/V) of particles. This dependence makes sense if we consider that the chance that two particles will get close enough to interact with each other depends on the concentration of both particles—that is, $(n/V)^2$. In Chapter 6 we explored the many ways that molecules can interact: London dispersion, dipole–dipole, hydrogen bonding, and dipole–induced dipole interactions. All of these interactions may contribute to the value of the a constant of a gaseous substance, depending on its molecular structure, so they are collectively called **van der Waals forces**.

TABLE 10.5 Van der Waals Constants of Selected Gases

Substance	a ($L^2 \cdot atm/mol^2$)	b (L/mol)
He	0.0341	0.02370
Ar	1.34	0.0322
H_2	0.244	0.0266
N_2	1.39	0.0391
O_2	1.36	0.0318
CH_4	2.25	0.0428
CO_2	3.59	0.0427
CO	1.45	0.0395
H_2O	5.46	0.0305
NO	1.34	0.02789
NO_2	5.28	0.04424
HCl	3.67	0.04081
SO_2	6.71	0.05636

CONCEPT TEST

The van der Waals constant a for SO_2, 6.71 $L^2 \cdot atm/mol^2$, is nearly twice the value of a for CO_2, 3.59 $L^2 \cdot atm/mol^2$. Suggest a reason for this difference.

SAMPLE EXERCISE 10.14 Using the van der Waals Equation LO9

Patients suffering from chronic lung disease often rely on portable tanks of compressed O_2 when they are out and about (Figure 10.36). The tanks have an internal volume of 2.24 liters and typically contain 0.500 kg O_2 when filled to their recommended pressure. What is that pressure at 20°C? Calculate your answer using both the ideal gas equation and the van der Waals equation.

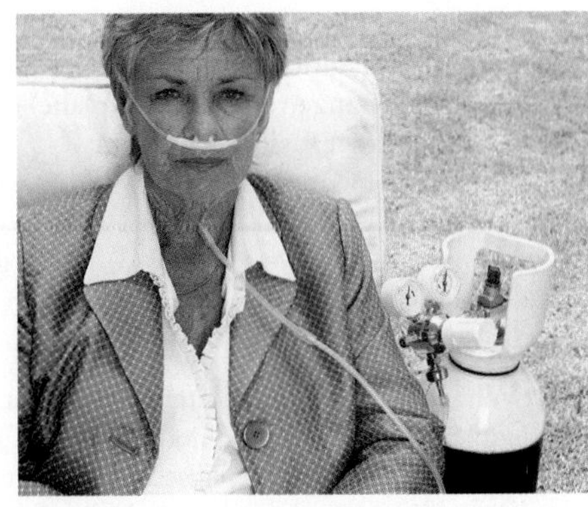

FIGURE 10.36 A patient breathing supplemental oxygen from a portable tank.

Collect, Organize, and Analyze We are given the mass, volume, and temperature of a quantity of O_2, and we are asked to calculate its pressure using both the van der Waals equation and the ideal gas equation.

The van der Waals constants for O_2, listed in Table 10.5, are $a = 1.36\ L^2 \cdot atm/mol^2$ and $b = 0.0318$ L/mol. These values are a little more than half the a and b values for CH_4, which exhibits the nonideal behavior shown in Figure 10.35—namely, intermolecular interactions produce a decrease in PV/RT below 400 atm. O_2 molecules should experience weaker, but still significant, intermolecular interactions that lead to pressures below those predicted by the ideal gas law.

Solve Let's start by converting the mass of O_2 into the corresponding number of moles because n appears several times in both the ideal gas and van der Waals equations.

$$0.500\ \cancel{kg\ O_2}\left(\frac{1000\ g}{1\ kg}\right)\left(\frac{1\ mol\ O_2}{32.00\ \cancel{g\ O_2}}\right) = 15.62\ mol\ O_2$$

FIGURE 10.37 A bus fueled by compressed natural gas.

Assuming ideal behavior and solving the ideal gas equation for P yields

$$P = \frac{nRT}{V} = \frac{(15.62\ \cancel{\text{mol}})\left(0.08206\ \frac{\cancel{\text{L}} \cdot \text{atm}}{\cancel{\text{mol}} \cdot \cancel{\text{K}}}\right)(273 + 20)\ \cancel{\text{K}}}{(2.24\ \cancel{\text{L}})} = 168\ \text{atm}$$

Repeating this calculation using the van der Waals equation, we obtain

$$P = \frac{nRT}{V - nb} - a\left(\frac{n}{V}\right)^2$$

$$= \frac{(15.62\ \cancel{\text{mol}})\left(0.08206\ \frac{\cancel{\text{L}} \cdot \text{atm}}{\cancel{\text{mol}} \cdot \cancel{\text{K}}}\right)(273 + 20)\ \cancel{\text{K}}}{2.24\ \cancel{\text{L}} - (15.62\ \cancel{\text{mol}})\left(0.0318\ \frac{\cancel{\text{L}}}{\cancel{\text{mol}}}\right)} - \left(1.36\ \frac{\cancel{\text{L}^2} \cdot \text{atm}}{\cancel{\text{mol}^2}}\right)\left(\frac{15.62\ \cancel{\text{mol}}}{2.24\ \cancel{\text{L}}}\right)^2$$

$$= 215.4\ \text{atm} - 66.1\ \text{atm}$$

$$= 149\ \text{atm}$$

Think About It As predicted, the "real" pressure is less than the "ideal" pressure due to intermolecular interactions between O_2 molecules. The last step in the calculation shows that the pressure in the tank would have been much higher (215 atm) than in an ideal gas (168 atm) because of the volume of the gas occupied by O_2 molecules—the nb correction term. However, this increase was more than offset by the decrease in pressure (−66 atm) due to interactions between molecules—the $a(n/V)^2$ correction term.

Practice Exercise Thousands of buses in the United States and other countries run on compressed natural gas (CNG) to reduce pollution and save fuel costs (Figure 10.37). CNG, which is mostly methane, is stored on the roofs of the buses in large tanks in which the density of CH_4 is as high as 165 g/L. Use the van der Waals equation and the ideal gas equation to calculate the pressure in one of these tanks at 25°C.

SAMPLE EXERCISE 10.15 Integrating Concepts: Air for a Jet Engine

The Boeing 767 (Figure 10.38) is one of the most popular wide-body commercial airliners (over 1000 have been built), and it's also one of the most fuel efficient. While cruising at 853 km/h (530 mph) at an altitude of 11,000 m (36,000 ft), a 767-200ER (the extended-range version of the plane) consumes about 1720 U.S. gallons of jet fuel per hour.

Use the following facts to solve this problem: (1) the density of jet fuel is 0.804 g/mL; (2) at an altitude of 11,000 m, P_{atm} = 210 mmHg and $T = -56$°C; (3) 1 U.S. gallon = 3.785 L; and (4) dodecane, $C_{12}H_{26}$, is considered an appropriate model hydrocarbon for jet fuel.

a. What volume of air, in liters, does a cruising 767 need so that it can completely burn an hour's worth of jet fuel?
b. Fuel efficiency is often based on the number of passengers times distance traveled per volume of fuel consumed. In the United States, this value is typically expressed in units of passenger-miles per gallon. However, in much of the rest of the world, efficiency units are inverted and are typically expressed in liters per 100 km per passenger. Express the fuel efficiency of a full 767-200ER (which holds 224 passengers) in both sets of units.

FIGURE 10.38 A Boeing 767-200ER.

Collect and Organize We know the quantity of dodecane to be combusted, and we are asked to calculate the volume of air needed for complete combustion—that is, to convert its C and H content into CO_2 and H_2O. We know the pressure and temperature of the air. According to Table 10.1, dry air is 20.95% O_2.

Analyze This exercise involves a chemical reaction, so writing a balanced chemical equation for it is a good place to start. Next we need to convert the volume of fuel into an equivalent number of moles of fuel, and then to use the stoichiometry of the reaction to convert that value to moles of O_2. We will then calculate the equivalent volume of O_2 using the ideal gas equation. (The pressure is below 1 atm, so there should be no need to correct for nonideal behavior.) Finally, we will convert the volume of O_2 to the corresponding volume of air. The volume of air needed each hour by the engines of a 767 should be enormous.

Solve

a. Calculating the volume of air needed in one hour:

1. Write the balanced chemical equation for the combustion reaction. The reactants and products are

$$C_{12}H_{26}(\ell) + O_2(g) \rightarrow CO_2(g) + H_2O(g)$$

We first balance the numbers of C and H atoms:

$$C_{12}H_{26}(\ell) + O_2(g) \rightarrow 12\ CO_2(g) + 13\ H_2O(g)$$

This leaves us with an odd number of O atoms on the right, requiring that we multiply all the terms by 2 and then balance the number of O atoms:

$$2\ C_{12}H_{26}(\ell) + 37\ O_2(g) \rightarrow 24\ CO_2(g) + 26\ H_2O(g)$$

2. Converting the volume of fuel consumed in 1 hour into an equivalent number of moles of $C_{12}H_{26}$ (and carrying an extra digit in intermediate values):

$$1720\ \cancel{\text{gal}}\left(\frac{3.785\ \cancel{\text{L}}}{\cancel{\text{gal}}}\right)\left(\frac{1000\ \cancel{\text{mL}}}{1\ \cancel{\text{L}}}\right)\left(\frac{0.804\ \cancel{\text{g}}}{\cancel{\text{mL}}}\right)\left(\frac{1\ \text{mol}}{170.33\ \cancel{\text{g}}}\right)$$
$$= 3.073 \times 10^4\ \text{mol}\ C_{12}H_{26}$$

3. Converting moles of $C_{12}H_{26}$ into moles of O_2:

$$3.073 \times 10^4\ \cancel{\text{mol}\ C_{12}H_{26}}\left(\frac{37\ \text{mol}\ O_2}{2\ \cancel{\text{mol}\ C_{12}H_{26}}}\right) = 5.685 \times 10^5\ \text{mol}\ O_2$$

4. In converting moles of O_2 into a volume of air using the ideal gas law, we need to convert pressure units from mmHg to atm and temperature to the Kelvin scale:

$$V = \frac{nRT}{P} = \frac{(5.685 \times 10^5\ \cancel{\text{mol}})\left(0.08206\ \frac{\text{L} \cdot \cancel{\text{atm}}}{\cancel{\text{mol}} \cdot \cancel{\text{K}}}\right)(273 - 56)\ \cancel{\text{K}}}{210\ \cancel{\text{mmHg}}\left(\frac{1\ \cancel{\text{atm}}}{760\ \cancel{\text{mmHg}}}\right)}$$
$$= 3.664 \times 10^7\ \text{L}\ O_2$$

5. Air is 20.95% O_2 by volume, so the volume of air the engines must take in each hour is

$$3.664 \times 10^7\ \cancel{\text{L}\ O_2}\left(\frac{100\ \text{L air}}{20.95\ \cancel{\text{L}\ O_2}}\right) = 1.75 \times 10^8\ \text{L air}$$

b. Calculating fuel efficiency in U.S. Customary units based on the distance traveled and fuel consumed in one hour:

$$\frac{224\ \text{passengers} \times 530\ \text{miles}}{1720\ \text{gal}} = 69.0\ \text{passenger-miles/gallon}$$

The corresponding efficiency value in liters per 100 km per passenger is

$$\frac{1720\ \cancel{\text{gal}} \times \frac{3.785\ \text{L}}{1\ \cancel{\text{gal}}}}{224\ \text{passengers} \times 853\ \text{km}} \times 100 = 3.41\ \text{L/100 km-passenger}$$

Think About It The calculated volume of air is 175 million liters. That is a lot of air, but then 1720 U.S. gallons of jet fuel is a lot of fuel. Perhaps a more interesting value is the 69.0 passenger-miles per gallon, which is much greater than that of most automobiles with two occupants, and you reach your destination a lot faster.

SUMMARY

LO1 The behavior of gases is explained by **kinetic molecular theory (KMT)**, which assumes that gases are composed of particles in constant random motion, that the volume of the particles is insignificant compared to the volume occupied by the gas, and the collisions of the particles are elastic. (Sections 10.1 and 10.2)

LO2 According to **Graham's law of effusion**, the rate of **effusion** (escape through a pinhole) of a gas at a fixed temperature is inversely proportional to the square root of its molar mass. Gas **diffusion** is the spread of one gas through another. Graham's law of effusion describes diffusion as well: at a fixed temperature, the rate at which a gas spreads is inversely proportional to the square root of its molar mass. (Section 10.2)

LO3 Gas particles move at **root-mean-square speeds** (u_{rms}) that are inversely proportional to the square root of their molar masses. (Section 10.2)

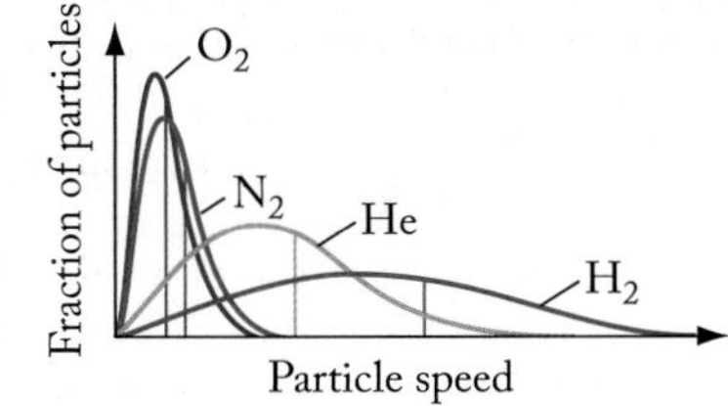

LO4 The mass of the gases in Earth's atmosphere combined with the force of gravity results in an atmospheric pressure on the surface of the planet. Pressure is defined as the ratio of force to surface area and is measured with **barometers** and **manometers**. (Section 10.3)

LO5 The laws describing how the pressure, volume, and temperature of a gas are related bear the names of the scientists who discovered them: **Boyle's law**, **Charles's law**, **Avogadro's law**, and **Amontons's law** (Figure 10.39). The **combined gas law** relates the pressure, volume, and temperature of a fixed quantity of gas. The **ideal gas law** and the **ideal gas equation** ($PV = nRT$, where R is the **universal gas constant**) describe the behavior of gases under normal conditions. At **standard temperature and pressure (STP)**, which in the United States is traditionally defined as 0°C and 1 atm, the *molar volume* of an **ideal gas** is 22.4 L. (Sections 10.4, 10.5, and 10.6)

LO6 The density of a gas is proportional to its molar mass and pressure and inversely proportional to its absolute temperature. (Section 10.7)

LO7 The ideal gas equation and the stoichiometry of a chemical reaction can be used to calculate the volumes of gases required or produced in the reaction. (Section 10.8)

(a) **Boyle's law:** volume inversely proportional to pressure; n and T fixed

PV = constant

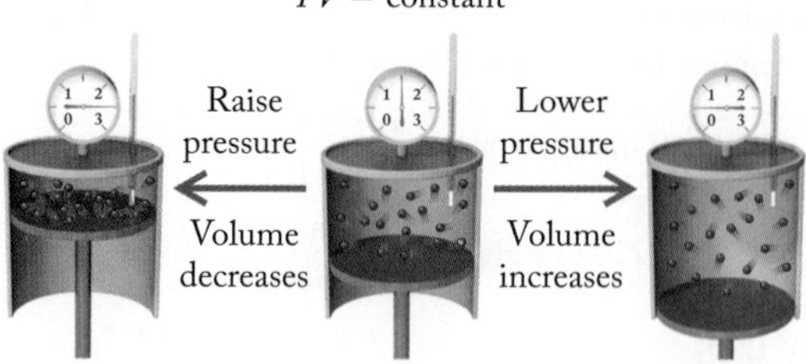

(b) **Charles's law:** volume directly proportional to temperature; n and P fixed

$\frac{V}{T}$ = constant

Lower temperature

Raise temperature

Volume decreases

Volume increases

(c) **Avogadro's law:** volume directly proportional to number of moles; T and P fixed

$\frac{V}{n}$ = constant

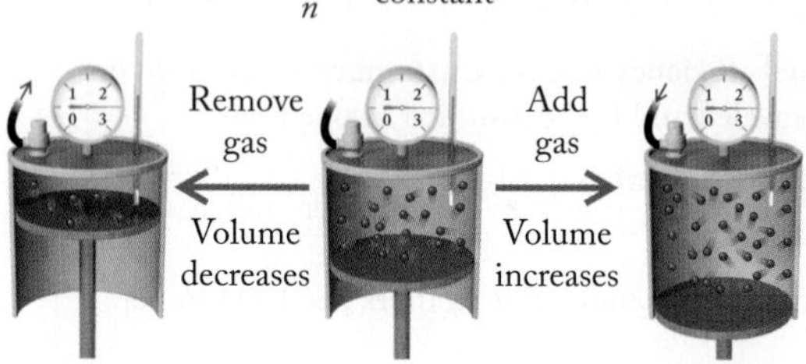

(d) **Amontons's law:** pressure directly proportional to temperature; n and V fixed

$\frac{P}{T}$ = constant

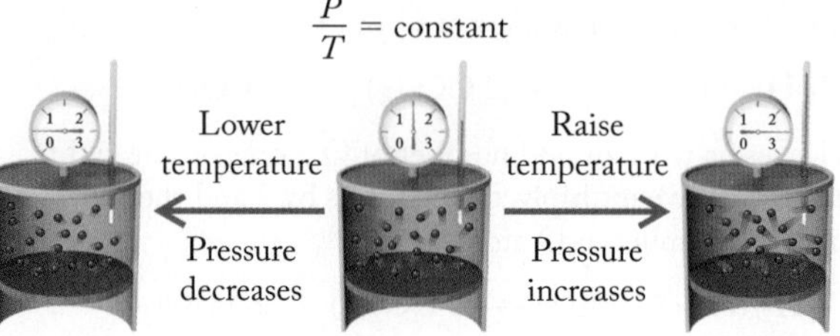

FIGURE 10.39 Summary of ideal gas behavior.

LO8 In a gas mixture, the contribution each component gas makes to the total gas pressure is called the **partial pressure** of that gas. **Dalton's law of partial pressures** states that the total pressure of a gas mixture is the sum of the partial pressures of its components. Dalton's law allows us to calculate the partial pressure (P_i) of any constituent gas i in a gas mixture if we know its **mole fraction** (x_i) and the total pressure. The total pressure of a gas sample collected over water is the sum of the partial pressure of water vapor and the partial pressure of the gas. (Section 10.9)

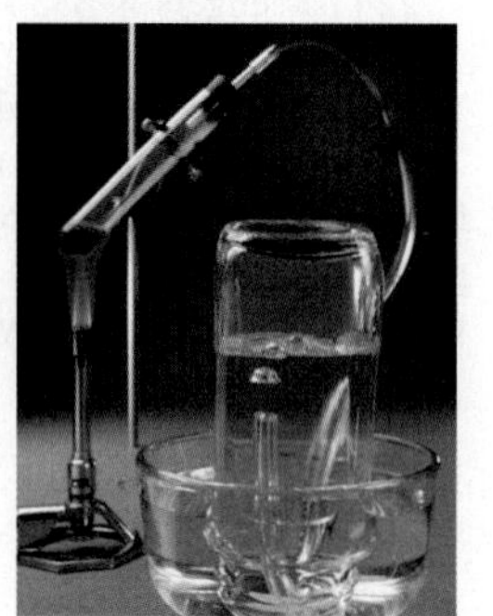

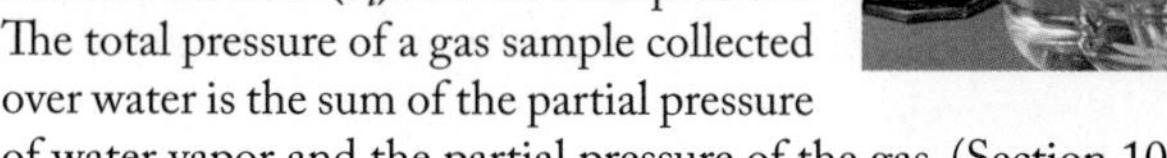

LO9 Ideal gas behavior is observed at moderate temperatures and low pressures, where $PV = nRT$ holds. Ideal gas behavior is characterized by elastic collisions and an absence of attractive forces between gas molecules. At high pressures, real gases deviate from the predictions of the ideal gas law. The **van der Waals equation**, a modified form of the ideal gas equation, accounts for the behavior of real gases. (Section 10.10)

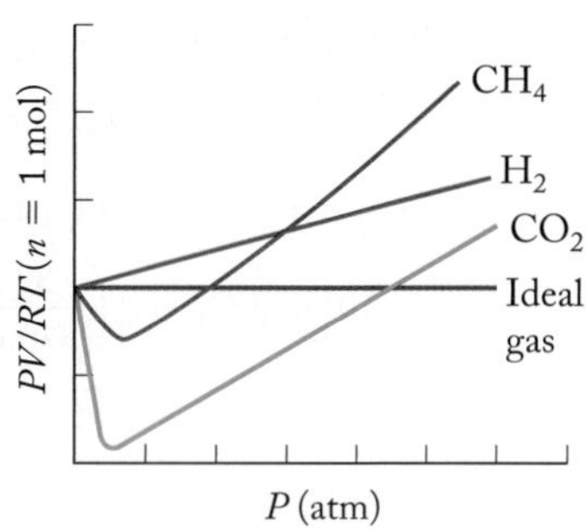

PARTICULATE PREVIEW WRAP-UP

As temperature increases, the balloon expands from (a) to (b) and the number of collisions between helium atoms decreases. As atmospheric pressure increases, the balloon contracts from (b) to (a), and the number of collisions between helium atoms increases.

PROBLEM-SOLVING SUMMARY

Type of Problem	Concepts and Equations		Sample Exercises
Calculating relative rates of effusion	$\frac{\text{effusion rate}_A}{\text{effusion rate}_B} = \sqrt{\frac{\mathcal{M}_B}{\mathcal{M}_A}}$	(10.7)	**10.1**
Calculating root-mean-square speeds	$u_{rms} = \sqrt{\frac{3RT}{\mathcal{M}}}$	(10.8)	**10.2**
Measuring gas pressure with a manometer; converting pressure units	See Table 10.2 and the conversion factors inside the back cover.		**10.3**
Applying Boyle's law	$P_1V_1 = P_2V_2$	(10.11)	**10.4**

Type of Problem	Concepts and Equations		Sample Exercises
Applying Charles's law	$\frac{V_1}{T_1} = \frac{V_2}{T_2}$	(10.14)	**10.5**
Applying Amontons's law	$\frac{P_1}{T_1} = \frac{P_2}{T_2}$	(10.19)	**10.6**
Applying the combined gas law	$\frac{P_1V_1}{T_1} = \frac{P_2V_2}{T_2}$	(10.20)	**10.7**
Applying the ideal gas law	$PV = nRT$	(10.21)	**10.8–10.11**
Calculating mole fractions and partial pressures	$x_i = \frac{n_i}{n_{total}}$	(10.24)	**10.12**
	$P_i = x_iP_{\text{total}}$	(10.25)	
Calculating the quantity of a gas collected by water displacement	Calculate the partial pressure (P_i) of the collected gas using the equation $P_i = P_{\text{atm}} - P_{\text{H}_2\text{O}}$		**10.13**
Using the van der Waals equation	$\left[P + a\left(\frac{n}{V}\right)^2\right](V - nb) = nRT$	(10.27)	**10.14**

VISUAL PROBLEMS

(Answers to boldface end-of-chapter questions and problems are in the back of the book.)

10.1. Which of the drawings in Figure P10.1 most accurately reflects the distribution of gas molecules in the balloon?

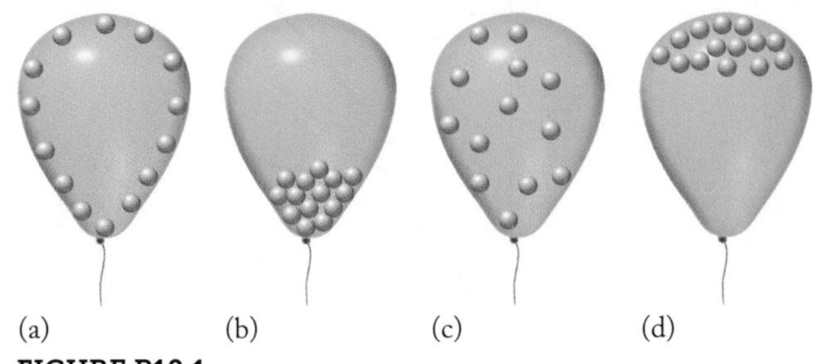

FIGURE P10.1

10.2. How does the downward motion of the piston in Figure P10.2(a) affect the pressure of the gas in the cylinder? Assume that the temperature of the gas does not change significantly, but the volume is reduced by half.

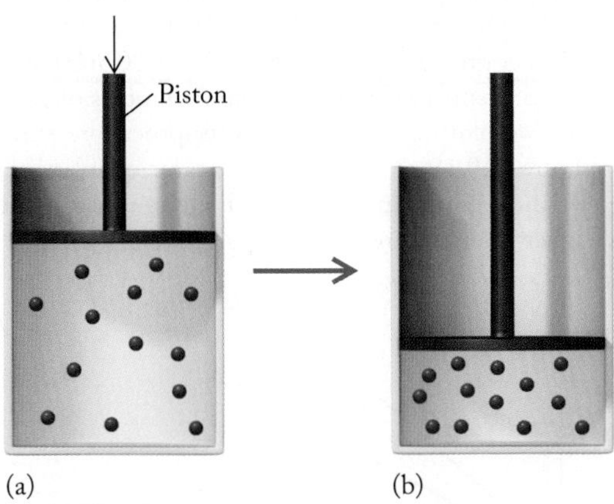

FIGURE P10.2

10.3. How does the motion of the piston in Problem 10.2 affect the root-mean-square speed of the gas particles in the cylinder?

10.4. Suppose the temperature of the gas in Figure P10.2(b) increases from 200 K to 400 K but that the pressure on the piston remains constant. Does the position of the piston change? If so, by how much does the volume of the gas change?

10.5. Enough gas is added isothermally to the cylinder in Figure P10.5(a) to double the number of particles inside the cylinder.

a. Assuming the position of the piston does not change, by how much does the pressure inside the cylinder change?
b. By how much does the frequency of the collisions between gas particles and the inner walls of the cylinder change?
c. By how much does the most probable speed of the gas particles change?

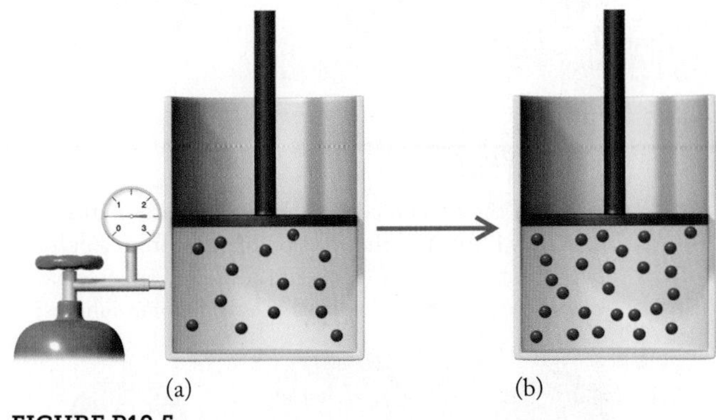

FIGURE P10.5

10.6. Suppose the same quantity of gas from Problem 10.5 was added to the cylinder in Figure P10.5 at constant temperature *and* constant pressure.

a. In which direction would the piston move?
b. By how much would the volume of the gas in the cylinder change?

10.7. Consider the two gas mixtures in Figure P10.7. Assume their volumes and temperatures are the same. Molecules of gas A are represented by red spheres.
a. In which mixture is the mole fraction of A greater?
b. In which mixture is the partial pressure of A greater?
c. In which mixture is the total pressure greater?

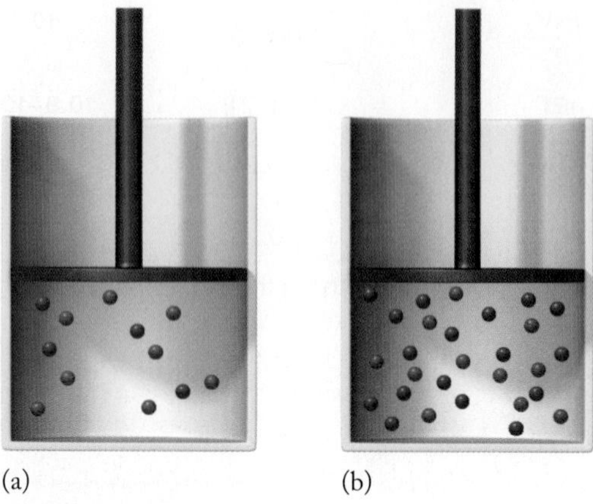

FIGURE P10.7

10.8. Suppose the piston in Figure P10.7(a) was pushed downward so that the volume of the gas mixture was about 2/3 its initial value. Which of the following values would increase, which would decrease, and which would remain the same, assuming no change in temperature or in the numbers of red and blue spheres occurs?
a. The mole fraction of A (whose molecules are represented by red spheres)
b. The partial pressure of A
c. The total pressure of the mixture
d. The most probable speed of the molecules of A

10.9. Which of the two outcomes diagrammed in Figure P10.9 more accurately illustrates the effusion of helium from a balloon at constant atmospheric pressure?

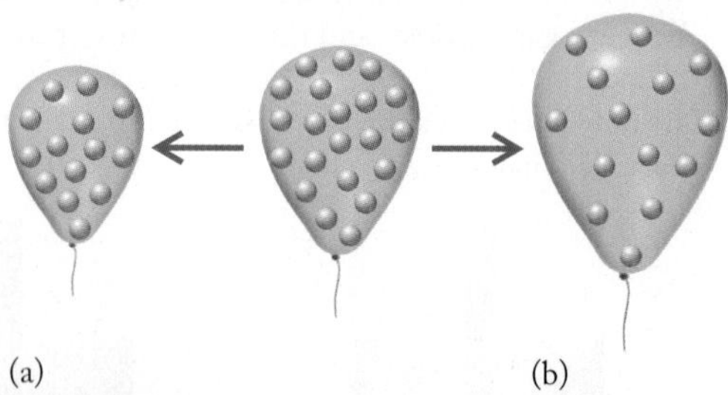

FIGURE P10.9

10.10. Which of the two outcomes shown in Figure P10.10 more accurately illustrates the effusion of gases from a balloon at constant atmospheric pressure if the red spheres have a greater root-mean-square speed than the blue spheres?

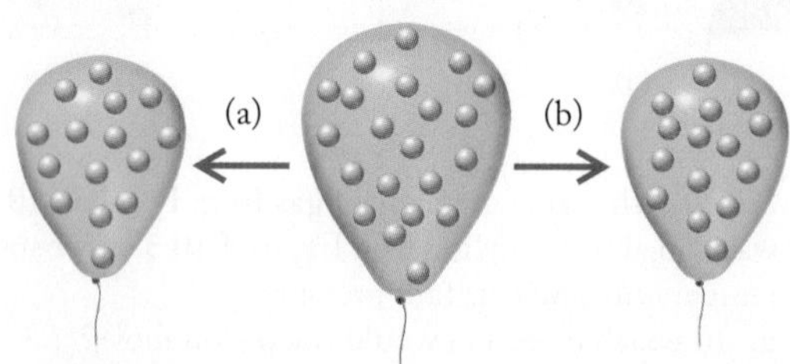

FIGURE P10.10

10.11. Figure P10.11 shows the distribution of molecular speeds of CO_2 and SO_2 molecules at 25°C. Which curve is the profile for SO_2? Which of the profiles should match that of propane (C_3H_8), a common fuel in portable grills?

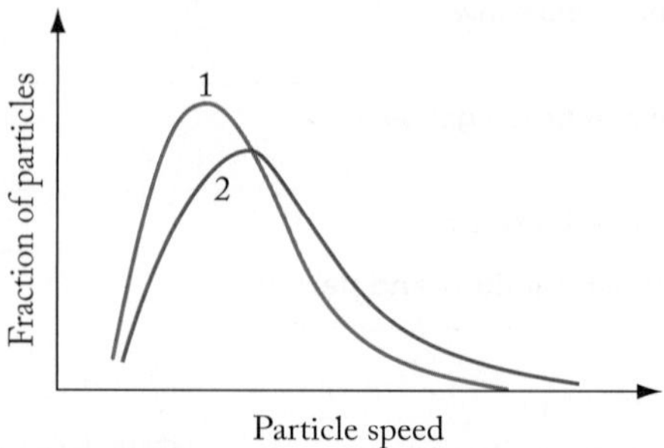

FIGURE P10.11

10.12. How would a graph showing the distribution of molecular speeds of CO_2 at −60°C differ from the curve for CO_2 shown in Figure P10.11?

10.13. Figure P10.13 describes the relationship between the volume of a gas and temperature at constant pressure (Charles's law).
a. If each of the lines represents 1.00 mole of gas, which line represents the highest pressure?
b. Estimate the slope of this line on the graph. What is the pressure of the gas?

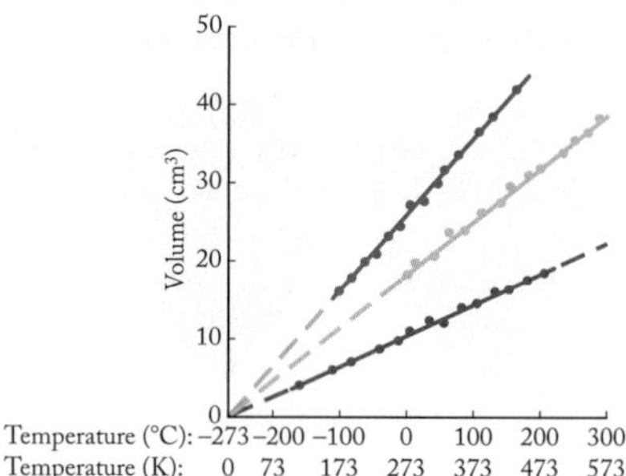

FIGURE P10.13

10.14. Use the graph in Figure P10.13 to answer the following questions.
a. Is the slope of the graph independent of the identity of the gas (for example, He versus N_2) if the number of moles of gas and the pressure are constant?
b. If each of the lines in Figure P10.13 is at the *same* exact pressure (for example, $P = 1.00$ atm), which line represents the greatest number of moles of gas?

10.15. The two plots of volume versus reciprocal pressure in Figure P10.15 correspond to the same quantity of gas at two different temperatures. Which line corresponds to the higher temperature?

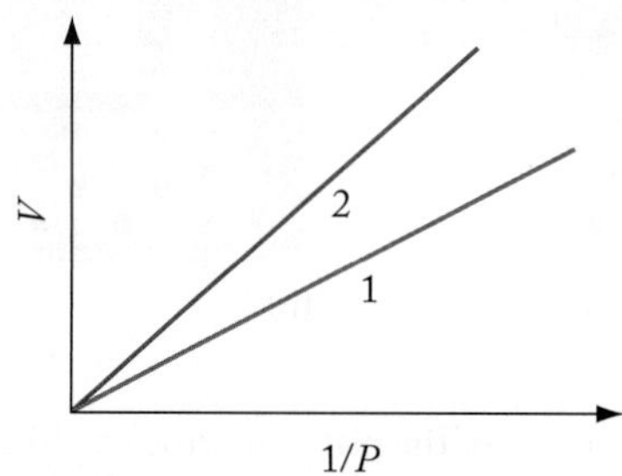

FIGURE P10.15

10.16. In Figure P10.16, which of the two plots of volume versus pressure at constant temperature is consistent with the ideal gas law?

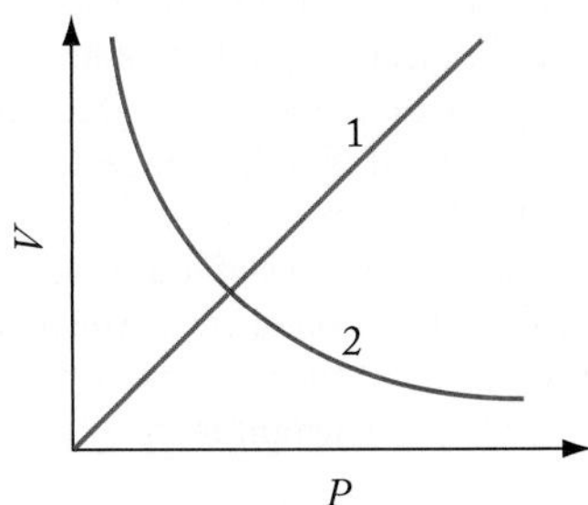

FIGURE P10.16

10.17. In Figure P10.17, which of the two plots of volume versus temperature at constant pressure is *not* consistent with the ideal gas law?

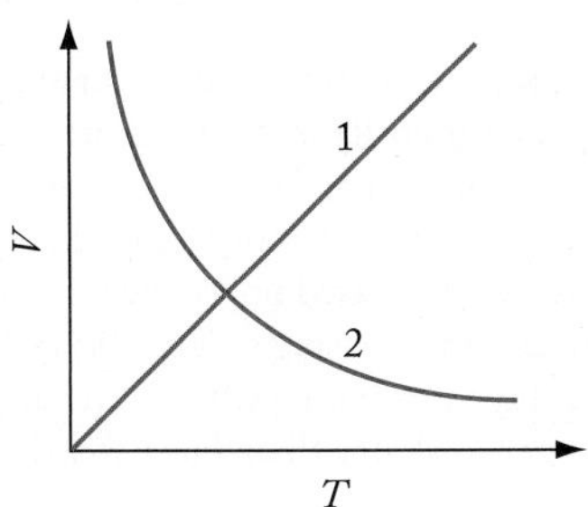

FIGURE P10.17

10.18. Figure P10.18 contains plots of the densities of propane (C_3H_8) and butane (C_4H_{10}) gas versus pressure at the same temperature. Which plot is based on propane densities?

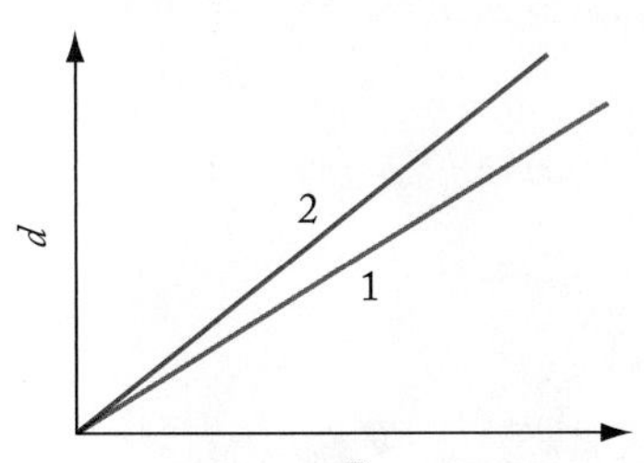

FIGURE P10.18

10.19. Consider the three Torricelli barometers in Figure P10.19. Which one is most likely sensing atmospheric pressure at each of the following locations?

1. San Diego, California (sea level)
2. The summit of Mount Everest (8.8 km above sea level)
3. The bottom of the Tau Tona gold mine in South Africa (3.9 km below sea level)

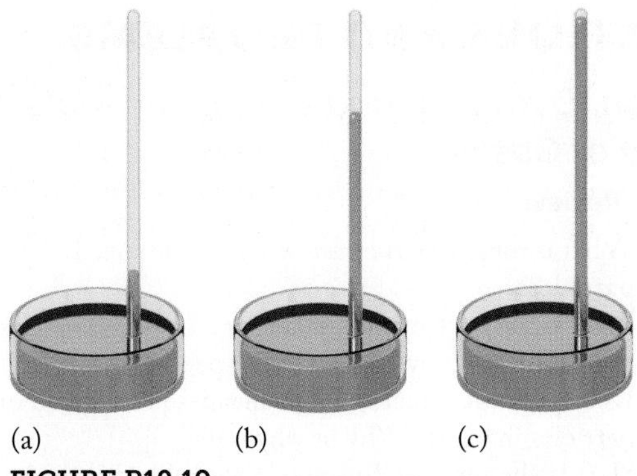

FIGURE P10.19

10.20. The blue arrows connecting the center [E] image to the others in Figure P10.20 represent changes in the temperature (T), volume (V), and (if there is a valve in the top of the other cylinder) the number of moles (n) of gas in the center [E] cylinder that produce the changes in pressure (P) indicated by the gauges.

a. Describe the changes in T, V, n, and/or P represented by each arrow.
b. Explain why pressure does not change during the [E] → [B], [E] → [F], and [E] → [I] transitions.
c. Explain why pressure decreased during the [E] → [A], [E] → [C], [E] → [G], and [E] → [H] transitions.
d. Explain why pressure increased during the [E] → [D] transition.
e. Does the root-mean-square speed of the molecules in the [E] cylinder increase, decrease, or remain the same in the [E] → [A] transition?
f. Are more molecules moving at the root-mean-square speed in [E] or in [I]?

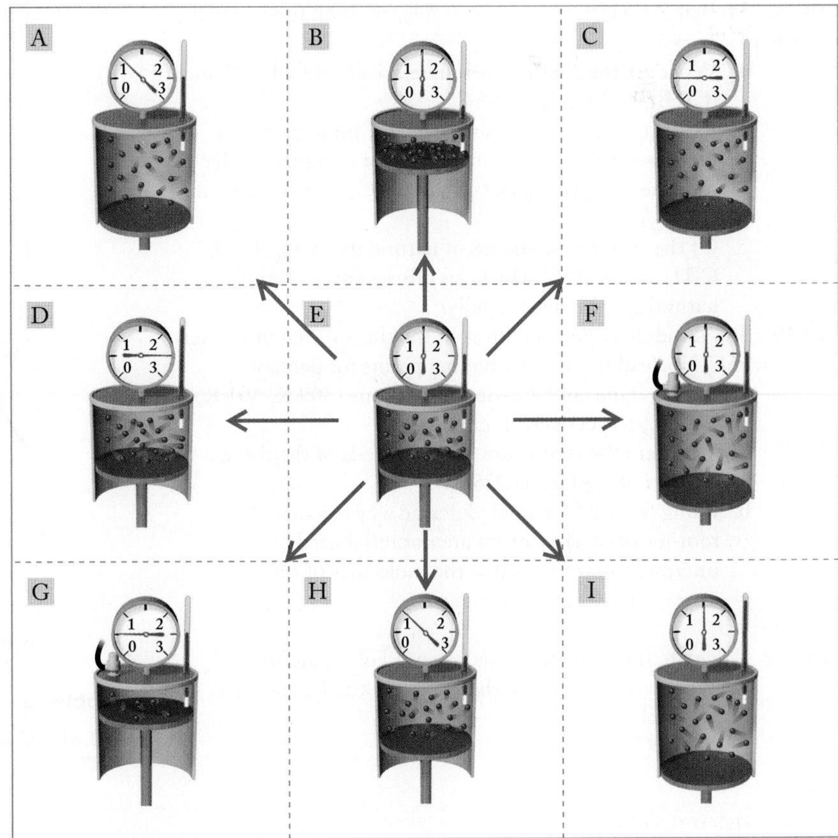

FIGURE P10.20

QUESTIONS AND PROBLEMS

Effusion, Diffusion, and the Kinetic Molecular Theory of Gases

Concept Review

10.21. What is meant by the *root-mean-square* speed of gas particles?

*10.22. Why is the root-mean-square speed of gas particles greater than the simple average of their speeds?

10.23. Does pressure affect the root-mean-square speed of the particles in a gas? Why or why not?

10.24. How is the rate of effusion of a gas related to each of the following?
a. Molar mass
b. Root-mean-square speed
c. Temperature
d. Density

Problems

10.25. Rank the gases NO, NO_2, N_2O_4, and N_2O_5 in order of increasing root-mean-square speed at 0°C.

10.26. **Chlorofluorocarbons** Chlorofluorocarbons with the general formula CF_nCl_{4-n} were used for decades in refrigerators and air conditioners until they were found to degrade Earth's ozone layer. For which value of n is the rate of effusion the slowest? Which compound will effuse at the fastest rate?

10.27. Molecular hydrogen effuses 4.0 times as fast as gas X at the same temperature. What is the molar mass of gas X?

10.28. Which noble gas effuses 3.2 times faster than argon at 0°C?

10.29. Methane, CH_4, and propane, C_3H_8, are both used as fuel for cooking.
a. What are the root-mean-square speeds of CH_4, ethane (C_2H_6), and C_3H_8 at 298 K?
b. Plot the root-mean-square speeds of these gases as a function of their molar mass and use it to predict the root-mean-square speed of butane (C_4H_{10}) and pentane (C_5H_{12}).
c. Of the major components of natural gas, CH_4, C_2H_6, C_3H_8, and C_4H_{10}, which gas effuses from a sample of natural gas the most rapidly?

10.30. The London dispersion forces in the three different isomers of C_5H_{12} lead to different boiling points for pentane, 2-methylbutane, and 2,2-dimethylbutane (309 K, 301 K, and 282 K, respectively, Figure 6.3).
a. What are the root-mean-square speeds of the three isomers of pentane at 325 K?
b. Using the KMT model, describe why calculated root-mean-square speeds are not influenced by the intermolecular forces that the molecules of these isomers apparently experience.

10.31. Calculate the root-mean-square speed of Ar atoms at the temperature at which their average kinetic energy is 5.18 kJ/mol.

10.32. Determine the root-mean-square speed of CO_2 molecules that have an average kinetic energy of 3.2×10^{-21} J per molecule.

10.33. A flask of ammonia is connected to a flask of an unknown acid HX by a 1.00 m glass tube. As the two gases diffuse down the tube, a white ring of NH_4X forms 68.5 cm from the ammonia flask. Identify element X.

10.34. **Enriching Uranium** The two isotopes of uranium, ^{238}U and ^{235}U, can be separated by diffusion of the corresponding UF_6 gases. What is the ratio of the root-mean-square speed of $^{238}UF_6$ to that of $^{235}UF_6$ at constant temperature?

Atmospheric Pressure

Concept Review

10.35. Describe the difference between force and pressure.

10.36. Why does atmospheric pressure decrease with increasing elevation?

10.37. Three barometers based on Torricelli's design are constructed using water (density d = 1.00 g/mL), ethanol (d = 0.789 g/mL), and mercury (d = 13.546 g/mL). Which barometer contains the tallest column of liquid?

10.38. In constructing a barometer based on Torricelli's design, what advantage is there in choosing a dense liquid?

10.39. Why does an ice skater exert more pressure on ice when wearing newly sharpened skates than when wearing skates with dull blades?

10.40. Why is it easier to travel over deep snow when wearing boots and snowshoes (Figure P10.40) rather than just boots?

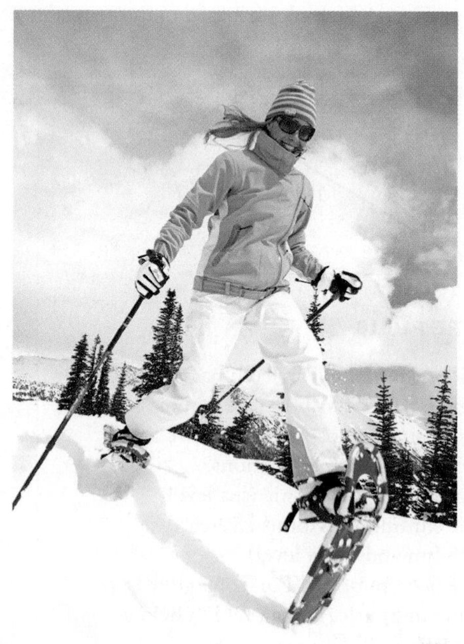

FIGURE P10.40

Problems

10.41. Convert the following pressures into atmospheres: (a) 2.0 kPa; (b) 562 mmHg.

10.42. Convert the following pressures into millimeters of mercury: (a) 0.541 atm; (b) 2.8 kPa.

10.43. Use the appropriate values from Appendix 3 to calculate the mass of a cube of tin that is 5.00 cm on a side and to calculate the downward pressure due to gravity exerted by the bottom face of the cube.

10.44. A gold brick is 4.0 cm wide, 12.0 cm long, and 3.0 cm high. Use the appropriate values from Appendix 3 to calculate the following:
a. The mass of the brick
b. The downward pressure due to gravity exerted by the bottom face of the brick
c. The downward pressure due to gravity exerted by the brick when it is stood up on its end

10.45. **Record High Atmospheric Pressure** The highest atmospheric pressure ever recorded on Earth was 108.6 kPa at Tosontsengel, Mongolia, on December 19, 2001. Express this pressure in (a) millimeters of mercury, (b) atmospheres, and (c) millibars.

10.46. **Record Low Atmospheric Pressure** Despite the destruction from Hurricane Katrina in August 2005, the lowest pressure for a hurricane in the Atlantic Ocean was measured several weeks after Katrina. Hurricane Wilma registered an atmospheric pressure of 88.2 kPa on October 19, 2005, about 2 kPa lower than Hurricane Katrina. What was the *difference* in pressure between the two hurricanes in (a) millimeters of mercury, (b) atmospheres, and (c) millibars?

Relating P, T, and V: The Gas Laws

Concept Review

10.47. How does kinetic molecular theory explain why pressure is directly proportional to temperature at fixed volume (Amontons's law)?

10.48. Explain Boyle's law using kinetic molecular theory.

10.49. A hot-air balloonist is rising too fast for her liking. Should she increase or decrease the temperature of the gas in the balloon?

10.50. Could the pilot of the balloon in Problem 10.49 reduce her rate of ascent by allowing some gas to leak out of the balloon? Explain your answer.

Problems

10.51. The volume of a quantity of gas at 1.00 atm is compressed from 3.25 L to 2.24 L. What is the final pressure of the gas if there is no change in temperature?

10.52. The pressure on a sample of an ideal gas is increased from 715 mmHg to 3.55 atm at constant temperature. If the initial volume of the gas is 485 mL, what is the final volume of the gas?

10.53. A 4.66 L sample of gas is warmed from 273 K to a final temperature of 398 K. Assuming no change in pressure, what is the final volume of the gas?

10.54. A 22.5 L sample of gas is cooled from 145°C to a temperature at which its volume is 18.3 L. What is the new temperature? Assume no change in pressure of the gas.

10.55. Balloons for a New Year's Eve party in Potsdam, New York, are filled to a volume of 5.0 L at a temperature of 20°C and then hung outside where the temperature is −25°C. What is the volume of the balloons after they have cooled to the outside temperature? Assume that atmospheric pressure inside and outside the house is the same.

10.56. The air inside a balloon is heated to 45°C and then cools to 25°C. By what percentage does the volume of the balloon change during cooling?

10.57. **Underwater Archaeology** A scuba diver releases a balloon containing 115 L of air attached to a tray of artifacts at an underwater archaeological site (Figure P10.57). When the balloon reaches the surface, it has expanded to a volume of 352 L. The pressure at the surface is 1.00 atm.
a. What is the pressure at the underwater site? Assume that the water temperature is constant.
b. If the pressure increases by 1.0 atm for every 10 m of depth, at what depth was the diver working?

FIGURE P10.57

*10.58. **Scuba Diving** A popular scuba tank for sport diving has an internal volume of 12.0 L and can be filled with air up to a pressure of 232 bar. Suppose a diver consumes air at the rate of 21 L/min while diving on a coral reef where the sum of atmospheric pressure and water pressure averages 2.2 bar. How long will it take the diver to use up a full tank of air if the temperatures of the air and water on the reef are both 28°C?

10.59. Which of the following actions would produce the greater increase in the volume of a gas sample?
a. Lowering the pressure from 760 mmHg to 700 mmHg at constant temperature
b. Raising the temperature from 10°C to 35°C at constant pressure

10.60. Which of the following actions would produce the greater increase in the volume of a gas sample?
a. Doubling the amount of gas in the sample at constant temperature and pressure
b. Raising the temperature from 244°C to 1100°C

10.61. What happens to the volume of gas in a cylinder with a movable piston under the following conditions?
a. Both the absolute temperature and the external pressure on the piston double.
b. The absolute temperature is halved, and the external pressure on the piston doubles.
c. The absolute temperature increases by 75%, and the external pressure on the piston increases by 50%.

10.62. What happens to the pressure of a gas under the following conditions?
a. The absolute temperature is halved and the volume doubles.
b. Both the absolute temperature and the volume double.
c. The absolute temperature increases by 75%, and the volume decreases by 50%.

10.63. A 150.0 L weather balloon contains 6.1 moles of helium but loses it at a rate of 10 mmol/h. What is the volume of the balloon after 24 h?

10.64. Which has the greater effect on the volume of a gas at constant temperature: doubling the number of moles of gas or reducing the pressure by half?

10.65. **Temperature Effects on Bicycle Tires** A bicycle racer inflates his tires to 7.1 atm on a warm autumn afternoon when temperatures reach 27°C. By morning the temperature has dropped to 5.0°C. What is the pressure in the tires if we assume that the volume of the tire does not change significantly?

10.66. The volume of a weather balloon is 200.0 L and its internal pressure is 1.17 atm when it is launched at 20°C. The balloon rises to an altitude in the stratosphere where its internal pressure is 63 mmHg and the temperature is 210 K. What is the volume of the balloon at this altitude?

Ideal Gases and the Ideal Gas Law

Concept Review

10.67. What is meant by standard temperature and pressure (STP)? What is the volume of one mole of an ideal gas at STP?

10.68. Which of the following are *not* characteristics of an ideal gas?
a. The molecules of gas have insignificant volume compared with the volume that they occupy.
b. Its volume is independent of temperature.
c. The density of all ideal gases is the same.
d. Gas atoms or molecules do not interact with one another.

Problems

10.69. How many moles of air must there be in a racing bicycle tire with a volume of 2.36 L if it has an internal pressure of 6.8 atm at 17.0°C?

10.70. At what temperature will 1.00 mol of an ideal gas in a 1.00 L container exert a pressure of 1.00 atm?

10.71. **Hyperbaric Oxygen Therapy** Hyperbaric oxygen chambers are used to treat divers suffering from decompression sickness (the "bends") with pure oxygen at greater than atmospheric pressure. Other clinical uses include treatment of patients with thermal burns and CO poisoning. What is the pressure in a chamber with a volume of 4.85×10^3 L that contains 15.00 kg of $O_2(g)$ at a temperature of 298 K?

10.72. **The *Hindenburg*** The German airship *Hindenburg* (Figure P10.72) made several transatlantic crossings in the early 1930s before a fiery crash in Lakehurst, New Jersey, in 1937 sealed the fate of "lighter-than-air" dirigibles. The *Hindenburg* held 1.6×10^4 kg of hydrogen gas. What was the volume of the hydrogen in the *Hindenburg* at 18°C and 0.98 atm?

FIGURE P10.72

10.73. In some cities public buses are fueled by compressed natural gas (mostly methane, CH_4). How many grams of CH_4 gas are in a 250 L fuel tank at a pressure of 255 bar at 20°C?

10.74. **Liquid Nitrogen–Powered Car** Students at the University of North Texas and the University of Washington built a car propelled by compressed nitrogen gas. The gas was obtained by boiling liquid nitrogen stored in a 182 L tank. What volume of N_2 is released at 0.927 atm of pressure and 25°C from a tank full of liquid N_2 (d = 0.808 g/mL)?

***10.75.** Suppose the atmospheric temperature and pressure at the top of a ski run are −5°C and 713 mmHg, respectively. At the bottom of the run, the temperature and pressure are 0°C and 734 mmHg, respectively. How many more moles of oxygen does a skier take in with a lungful of air at the bottom of the run than at the top? Express your answer as a percentage.

*10.76. A balloon vendor at a street fair is using a tank of helium to fill her balloons. The tank has an internal volume of 45.0 L and a pressure of 195 atm at 22°C. After a while she notices that the valve has not been closed properly and the pressure has dropped to 115 atm. How many moles of He have been lost?

10.77. The volume of a 4.0 g sample of a gaseous substance was measured at the temperatures listed in the table that follows. The pressure of the gas was 1.00 atm at all times.

a. How many moles of a gas were in the sample?
b. What was the probable identity of the gas?

V (L)	*T* (K)
7.88	96
3.94	48
1.97	24
0.79	9.6
0.39	4.8

10.78. The volume of 0.50 mol of a noble gas was measured at the temperatures listed in the following table. The pressure of the gas was 1.00 atm at all times. Use the data in the table to calculate the value of R, the ideal gas law constant.

V (L)	*T* (K)
3.94	96
1.97	48
0.79	24
0.39	9.6
0.20	4.8

Densities of Gases

Concept Review

10.79. Do all gases at the same pressure and temperature have the same density? Explain your answer.

10.80. Birds and sailplanes take advantage of thermals (rising columns of warm air) to gain altitude with less effort than usual. Why does warm air rise?

10.81. How does the density of a gas sample change when (a) its pressure is increased and (b) its temperature is decreased?

10.82. Rank the noble gases in order of decreasing density.

Problems

10.83. **Biological Effects of Radon Exposure** Radon is a naturally occurring radioactive gas found in the ground and in building materials. Cumulative radon exposure is a significant risk factor for lung cancer.
a. Calculate the density of radon at 298 K and 1 atm of pressure.
b. Are radon concentrations likely to be greater in the basement or on the top floor of a building?

*10.84. Four empty balloons, each with a mass of 10.0 g, are inflated to a volume of 20.0 L. The first balloon contains He, the second Ne, the third CO_2, and the fourth CO. If the density of air at 25°C and 1.00 atm is 1.17 g/L, how many of the balloons float in it?

10.85. A 30.0 mL flask contains 0.078 g of a volatile oxide of sulfur. The pressure in the flask is 750 mmHg, and the temperature is 22°C.
a. What is the density of the gas?
b. Is the gas SO_2 or SO_3?

10.86. A 100.0 mL flask contains 0.193 g of a volatile oxide of nitrogen. The pressure in the flask is 760 mmHg at 17°C.
a. Calculate the density of the gas.
b. Is the gas NO, NO_2, or N_2O_5?

10.87. The density of an unknown gas is 1.107 g/L at 300 K and 740 mmHg.
a. What is the molar mass of the gas?
b. Could this gas be CO or CO_2?

10.88. A gas containing chlorine and oxygen has a density of 2.875 g/L at 756 mmHg and 11°C.
a. Calculate the molar mass of the gas.
b. What is the most likely molecular formula of the gas?

Gases in Chemical Reactions

Problems

10.89. **Rescue Breathing Devices** Self-contained self-rescue breathing devices, like the one shown in Figure P10.89, convert CO_2 into O_2 according to the following reaction:

$$4\ KO_2(s) + 2\ CO_2(g) \rightarrow 2\ K_2CO_3(s) + 3\ O_2(g)$$

How many grams of KO_2 are needed to produce 100.0 L of O_2 at 20°C and 1.00 atm?

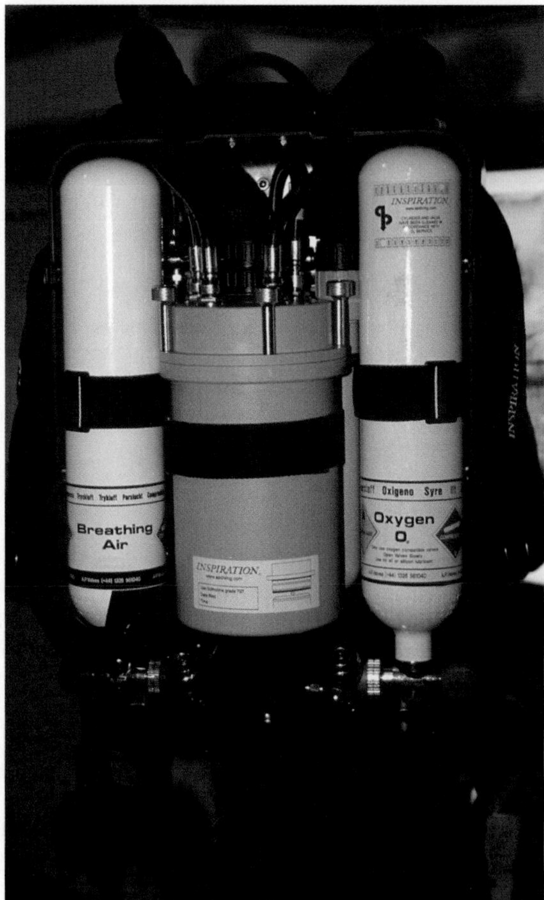

FIGURE P10.89

10.90. Acid precipitation dripping on limestone produces carbon dioxide by the following reaction:

$$CaCO_3(s) + 2\,H^+(aq) \rightarrow Ca^{2+}(aq) + CO_2(g) + H_2O(\ell)$$

Suppose that 8.6 mL of CO_2 were produced at 15°C and 760 mmHg.
a. How many moles of CO_2 were produced?
b. How many milligrams of $CaCO_3$ were consumed?

10.91. Oxygen is generated by the thermal decomposition of potassium chlorate:

$$2\,KClO_3(s) \rightarrow 2\,KCl(s) + 3\,O_2(g)$$

How many grams of $KClO_3$ are needed to generate 200.0 L of oxygen at 0.85 atm and 273 K?

10.92. Nitrogen gas is a product of the thermal decomposition of ammonium dichromate, $(NH_4)_2Cr_2O_7$:

$$(NH_4)_2Cr_2O_7(s) \rightarrow N_2(g) + Cr_2O_3(s) + 4\,H_2O(g)$$

How many liters of N_2 are produced during the decomposition of 100.0 grams of ammonium dichromate ($\mathcal{M}$ = 252.07 g/mol) at 22°C and a pressure of 757 torr?

10.93. Healthy Air for Submariners The CO_2 that builds up in the air of a submerged submarine can be removed by reacting it with an aqueous solution of 2-aminoethanol:

$$CO_2(g) + 2\,HOCH_2CH_2NH_2(aq) \rightarrow HOCH_2CH_2NH_3^+(aq) + HOCH_2CH_2NHCO_2^-(aq)$$

If a sailor exhales 125 mL of CO_2 per minute at 23°C and 1.02 atm, what volume of 4.0 *M* 2-aminoethanol is needed per sailor in a 24-hour period?

10.94. **Miners' Lamps** Before the development of reliable batteries, miners' lamps burned acetylene produced by the reaction of calcium carbide with water:

$$CaC_2(s) + H_2O(\ell) \rightarrow C_2H_2(g) + CaO(s)$$

Suppose a lamp uses 4.8 L of acetylene per hour at 1.02 atm pressure and 25°C.
a. How many moles of C_2H_2 are used per hour?
b. How many grams of calcium carbide are consumed for a four-hour shift?

Mixtures of Gases

Concept Review

10.95. What is meant by the *partial pressure* of a gas?

10.96. Can a barometer be used to measure just the partial pressure of oxygen in the atmosphere? Why or why not?

Problems

10.97. A gas mixture contains 0.70 mol N_2, 0.20 mol H_2, and 0.10 mol CH_4. What is the mole fraction of H_2 in the mixture?

10.98. A gas mixture contains 7.0 g N_2, 2.0 g H_2, and 16.0 g CH_4. What is the mole fraction of H_2 in the mixture?

10.99. Calculate the pressure of the gas mixture and the partial pressure of each constituent gas in Problem 10.97 if the mixture is in a 0.75 L vessel at 10°C.

10.100. Calculate the pressure of the gas mixture and the partial pressure of each constituent gas in Problem 10.98 if the mixture is in a 5.0 L vessel at 20°C.

10.101. A sample of oxygen is collected over water at 25°C and 1.00 atm. If the total sample volume is 0.480 L, how many moles of O_2 are collected?

10.102. Water vapor is removed from the O_2 sample in Problem 10.101. What is the volume of the dry O_2 at 25°C and 1.00 atm?

10.103. The following reactions are carried out in sealed containers. Will the total pressure after each reaction is complete be greater than, less than, or equal to the total pressure before the reaction? Assume all reactants and products are gases at the same temperature.
a. $N_2O_5(g) + NO_2(g) \rightarrow 3\,NO(g) + 2\,O_2(g)$
b. $2\,SO_2(g) + O_2(g) \rightarrow 2\,SO_3(g)$
c. $C_3H_8(g) + 5\,O_2(g) \rightarrow 3\,CO_2(g) + 4\,H_2O(g)$

10.104. In each of the following gas-phase reactions, determine whether the total pressure at the end of the reaction (carried out in a sealed, rigid vessel) will be greater than, less than, or equal to the total pressure at the beginning. Assume all reactants and products are gases at the same temperature.
a. $H_2(g) + Cl_2(g) \rightarrow 2\,HCl(g)$
b. $4\,NH_3(g) + 5\,O_2(g) \rightarrow 4\,NO(g) + 6\,H_2O(g)$
c. $2\,NO(g) + O_2(g) \rightarrow 2\,NO_2(g)$

***10.105. High-Altitude Mountaineering** Most alpine climbers breathe pure oxygen near the summits of the world's highest mountains. How much more O_2 is there in a lungful of pure O_2 at an elevation where atmospheric pressure is 266 mmHg than in a lungful of air at sea level? Express your answer as a percentage.

10.106. **Scuba Diving** A scuba diver is at a depth of 50 m, where the pressure is 5.0 atm. What should be the mole fraction of O_2 in the gas mixture the diver breathes to achieve the same P_{O_2} as at sea level?

10.107. A highly efficient hot-water heater fueled by natural gas can heat water at the rate of 33,500 BTU/hour (1 BTU or British thermal unit = 1.055 kJ). During exactly one hour of operation, what volume of natural gas would be consumed at STP? Assume natural gas is mostly methane (CH_4), which has a fuel value of 55.5 kJ/g.

10.108. The hot-water heater described in question 10.107 can be modified to burn propane (C_3H_8, fuel value = 50.4 kJ/g) instead of natural gas. What volume of propane gas would be consumed at STP by the water heater during exactly one hour of operation?

***10.109.** Ammonia is produced industrially from the reaction of hydrogen with nitrogen under pressure in a sealed reactor. What is the percent decrease in total pressure of a sealed reaction vessel during the reaction between H_2 at a partial pressure of 2.4 atm and N_2 at a partial pressure of 3.6 atm if half of the H_2 is consumed? No other gases are present initially in the reactor.

*10.110. A mixture of 0.156 mol C is reacted with 0.117 mol O_2 in a sealed, 10.0 L vessel at 500 K, producing a mixture of CO and CO_2. The total pressure is 0.640 atm. What is the partial pressure of CO?

Real Gases

Concept Review

10.111. Why do real gases behave nonideally at very low temperatures and very high pressures?

10.112. Under what conditions is the pressure exerted by a real gas *less* than that predicted for an ideal gas?

10.113. Why do the values of the van der Waals constant b of the noble gas elements increase with atomic number?

10.114. Why does the value of the constant a in the van der Waals equation generally increase with the molar mass of the gas?

Problems

10.115. Explain why the van der Waals constant a is greater for Ar than it is for He.

10.116. The van der Waals constant a for CO_2 is 3.59 $L^2 \cdot atm/mol^2$. Would you expect the value of a for CS_2 to be larger or smaller than that?

10.117. The graphs of PV/RT versus P (see Figure 10.34) for one mole of CH_4 and one mole of H_2 differ in how they deviate from ideal behavior. For which gas is the effect of the volume occupied by the gas molecules more important than the attractive forces between molecules at $P = 200$ atm?

10.118. Which noble gas is expected to deviate the most from ideal behavior in a graph of PV/RT versus P?

10.119. At high pressures, real gases do not behave ideally.

a. Use the van der Waals equation and data in the text to calculate the pressure exerted by 50.0 g of H_2 at 20°C in a 1.00 L container.

b. Repeat the calculation assuming that the gas behaves like an ideal gas.

10.120. Calculate the pressure exerted by 5.00 mol of CO_2 in a 1.00 L vessel at 300 K (a) assuming the gas behaves ideally and (b) using the van der Waals equation.

Additional Problems

10.121. Each of the cylinders in the engine of the sports car in Figure P10.121 contains 633 mL of air and gasoline vapor before the motion of a piston increases the pressure inside the cylinder by 9.0 times.

a. What is the volume of the air and gasoline vapor mixture after the compression, assuming the temperature of the mixture does not change significantly?

b. Is the assumption of no temperature change realistic? Why or why not?

FIGURE P10.121

10.122. **Planetary Atmospheres** Saturn's largest moon, Titan, has a surface atmospheric pressure of 1220 torr. The atmosphere consists of 82% N_2, 12% Ar, and 6% CH_4 by volume. Calculate the partial pressure of each gas in Titan's atmosphere. Chilly temperatures aside, could life as we know it exist on Titan?

10.123. Scientists have used laser light to slow atoms to speeds corresponding to temperatures below 0.00010 K. At this temperature, what is the root-mean-square speed of argon atoms?

10.124. **Blood Pressure** A typical blood pressure in a resting adult is "120 over 80," meaning 120 mmHg with each beat of the heart and 80 mmHg of pressure between heartbeats. Express these pressures in the following units: (a) torr; (b) atm; (c) bar; (d) kPa.

***10.125.** A popular style of scuba tank is called the "aluminum 80" because it can deliver 80 cubic feet of air at "normal" temperature (72°F) and pressure (1.00 atm) when filled with air at a pressure of 200 atm. A particular aluminum 80 tank has a mass of 15 kg empty. What is its mass when filled with air at 200 atm?

10.126. The flame produced by the burner of a gas (propane, C_3H_8) grill is a pale blue color when enough air mixes with the propane to burn it completely. For every gram of propane that flows through the burner, what volume of air is needed to burn it completely? Assume that the temperature of the burner is 200°C, the pressure is 1.00 atm, and the mole fraction of O_2 in air is 0.21.

***10.127.** **Anesthesia** Halothane (Figure P10.127) is a volatile liquid (its vapor pressure is 243 torr at 20°C) that is used as an anesthesia gas. If air at 1.00 atm of pressure is bubbled through liquid halothane at 20°C to produce anesthesia gas, what volume of air would be needed to completely vaporize 10.0 mL of liquid halothane (d = 1.87 g/mL)?

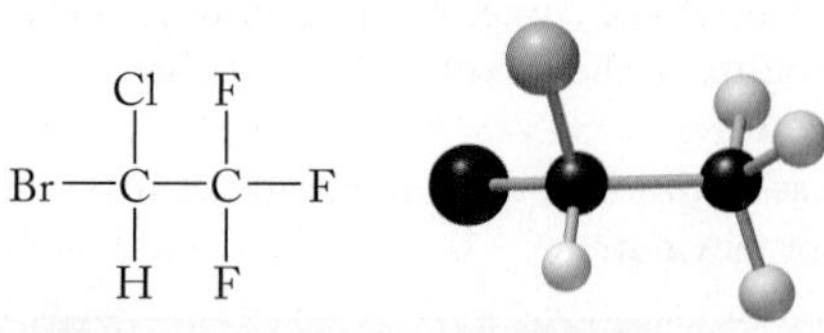

FIGURE P10.127

10.128. **Rocket Fuel** During the Second World War, German scientists and engineers developed the first operational rocket-propelled warplane (Figure P10.128). The engine was powered by the reaction of hydrogen peroxide and hydrazine, producing nitrogen and water vapor.

$$2\ H_2O_2(\ell) + N_2H_4(\ell) \rightarrow N_2(g) + 4\ H_2O(g)$$

What is the total pressure of the gases released by the reaction of 8.50 g H_2O_2 and 4.00 g N_2H_4 in a 125 L reaction vessel (rocket engine) at 1110°C?

FIGURE P10.128

***10.129.** The apparatus in Figure P10.129 is used in a series of experiments in which a cotton ball stuffed in the right end of a 1-meter-long glass tube is soaked in either hydrochloric acid (HCl) or acetic acid (CH_3COOH). A cotton ball at the left end of the tube is soaked in one of three amines (a class of basic organic compounds): CH_3NH_2, $(CH_3)_2NH$, or $(CH_3)_3N$. Vapors from the acids and the basic amines diffuse down the tube toward each other. When they meet, they neutralize each other, producing solid salts that form white rings in the tube.

a. In one combination of acid and amine, a white ring was observed halfway between the two ends. Which acid and which amine were used?
b. Which combination of acid and amine would produce a ring closest to the amine end of the tube?
c. Would two of the six possible combinations produce rings in the same position? Assume measurements can be made to the nearest centimeter.

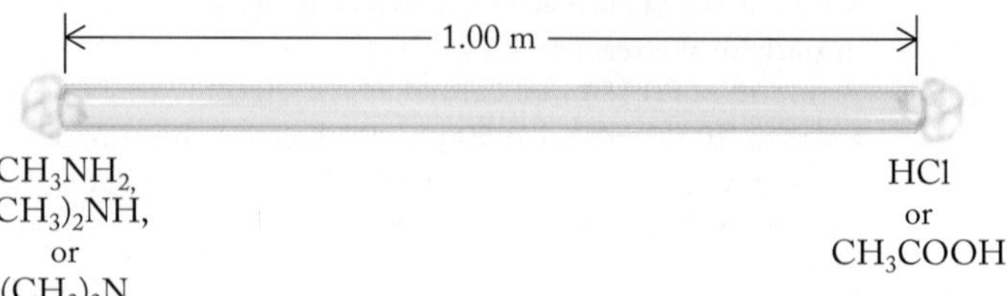

FIGURE P10.129

10.130. The pressure in an aerosol can is 1.2 atm at 27°C. The can will withstand a pressure of 3.0 atm. Will it burst if heated in a campfire to 450°C?

10.131. Uranus has a total atmospheric pressure of 130 kPa and consists of the following gases: 83% H_2, 15% He, and 2% CH_4 by volume. Calculate the partial pressure of each gas in Uranus's atmosphere.

10.132. Derive an equation that expresses the ratio of the densities (d_1 and d_2) of a gas under two different combinations of temperature and pressure: (T_1, P_1) and (T_2, P_2).

10.133. **Denitrification in the Environment** In some aquatic ecosystems, nitrate (NO_3^-) is converted to nitrite (NO_2^-), which then decomposes to nitrogen and water. As an example of this second reaction, consider the decomposition of ammonium nitrite:

$$NH_4NO_2(aq) \rightarrow N_2(g) + 2\ H_2O(\ell)$$

What is the change in pressure in a sealed 10.0 L vessel due to the formation of N_2 gas when the ammonium nitrite in 1.00 L of 1.0 M NH_4NO_2 decomposes at 25°C?

*10.134. When sulfur dioxide bubbles through a solution containing nitrite, chemical reactions that produce gaseous N_2O and NO may occur.

a. How much faster on average are NO molecules moving than N_2O molecules in such a reaction mixture?
b. If these two nitrogen oxides are separated based on differences in their rates of effusion, will unreacted SO_2 interfere with the separation? Explain your answer.

***10.135.** **Using Wetlands to Treat Agricultural Waste** Wetlands can play a significant role in removing fertilizer residues from rain runoff and groundwater. One way they do this is through denitrification, which converts nitrate ions to nitrogen gas:

$$2\ NO_3^-(aq) + 5\ CO(g) + 2\ H^+(aq) \rightarrow N_2(g) + H_2O(\ell) + 5\ CO_2(g)$$

Suppose 200.0 g of NO_3^- flows into a swamp each day.

a. What volume of N_2 would be produced at 17°C and 1.00 atm if the denitrification process were complete?
b. What volume of CO_2 would be produced?
c. Suppose the gas mixture produced by the decomposition reaction is trapped in a container at 17°C; what is the density of the mixture, assuming P_{total} = 1.00 atm?

10.136. Ammonium nitrate decomposes on heating. The products depend on the reaction temperature:

$$NH_4NO_3(s) \xrightarrow{300^\circ C} N_2(g) + \tfrac{1}{2}\ O_2(g) + 2\ H_2O(g)$$

$$NH_4NO_3(s) \xrightarrow{200\text{–}260^\circ C} N_2O(g) + 2\ H_2O(g)$$

A sample of NH_4NO_3 decomposes at an unspecified temperature, and the resulting gases are collected over water at 20°C.

a. Without completing a calculation, predict whether the volume of gases collected can be used to distinguish between the two reaction pathways. Explain your answer.
b. The gas produced during the thermal decomposition of 0.256 g of NH_4NO_3 displaces 79 mL of water at 20°C and 760 mmHg of atmospheric pressure. Is the gas N_2O or a mixture of N_2 and O_2?

10.137. **Air Bag Chemistry** Use the following chemical equation for the overall reaction in an automobile air bag:

$$20\ NaN_3(s) + 6\ SiO_2(s) + 4\ KNO_3(s) \rightarrow 32\ N_2(g) + 5\ Na_4SiO_4(s) + K_4SiO_4(s)$$

to calculate how many grams of sodium azide (NaN_3) are needed to inflate a 40 × 40 × 20 cm bag to a pressure of 1.25 atm at a temperature of 20°C. How much more sodium azide is needed if the air bag must produce the same pressure at 10°C?

10.138. **Decay Products of Uranium Minerals** Radon and helium are both by-products of the radioactive decay of uranium minerals. A fresh sample of carnotite, $K_2(UO_2)_2(VO_4)_2 \cdot 3\ H_2O$, is put on display in a museum. Calculate the relative rates of diffusion of helium and radon under fixed conditions of pressure and temperature. Which gas diffuses more rapidly through the display case?

10.139. The reaction between potassium superoxide and carbon dioxide is used to produce 0.200 L of O_2, which is collected over water at 25.0°C. The atmospheric pressure is 750.0 torr. The vapor pressure of water at 25.0°C is 24.0 torr. How many moles of O_2 have been collected?

$$4\ KO_2(s) + 2\ CO_2(g) \rightarrow 2\ K_2CO_3(s) + 3\ O_2(g)$$

10.140. On October 26, 2014, Alan Eustace set a record for the highest parachute jump when he dropped from a balloon at an altitude of 41,419 m, where the atmospheric pressure is only 5.6 mmHg.

a. What is the density of air at this height?
b. Is the mean free path of a gas molecule longer or shorter at this altitude than at sea level?

smartwork5

If your instructor uses Smartwork5, log in at digital.wwnorton.com/atoms2.

11

Properties of Solutions

Their Concentrations and Colligative Properties

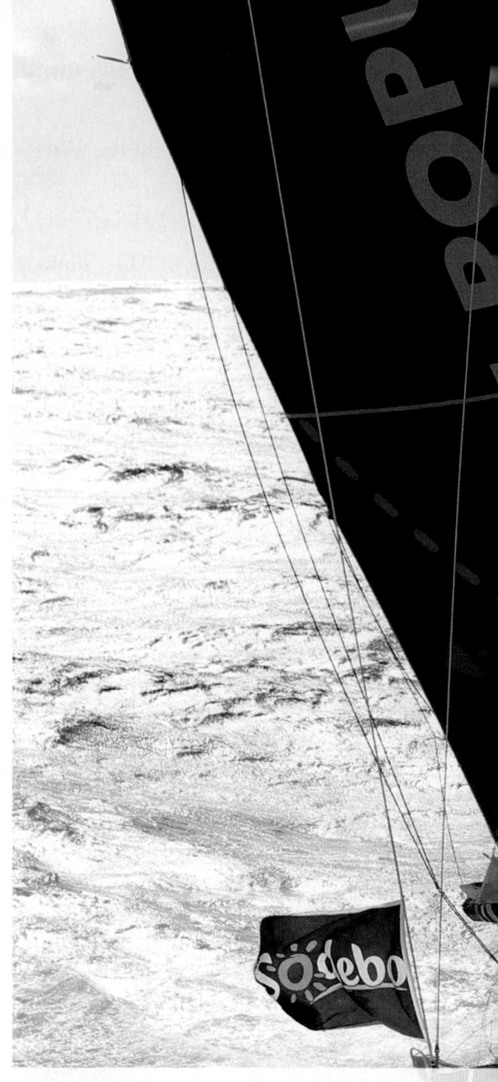

MAKING POTABLE WATER AT SEA Sailboats designed to race around the world nonstop have reverse osmosis systems onboard that turn seawater into potable drinking water.

PARTICULATE **REVIEW**

Same Composition, Different Properties

In Chapter 11, we explore the formation of solutions and compare their properties to those of the solvent and solute.

- What intermolecular forces exist between molecules of ethanol?
- What intermolecular forces exist between molecules of dimethyl ether?
- One of these compounds is a liquid at room temperature and the other is a gas. Which compound is more likely to be a liquid?

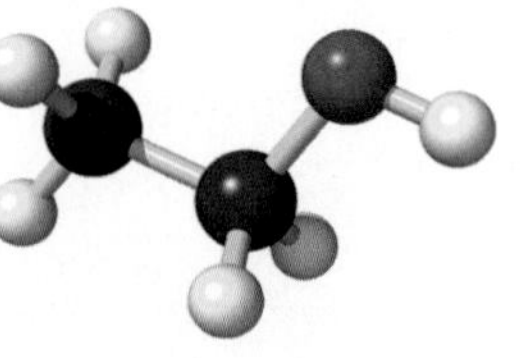

Ethanol

Dimethyl ether

(Answers to Particulate Review questions are in the back of the book.)

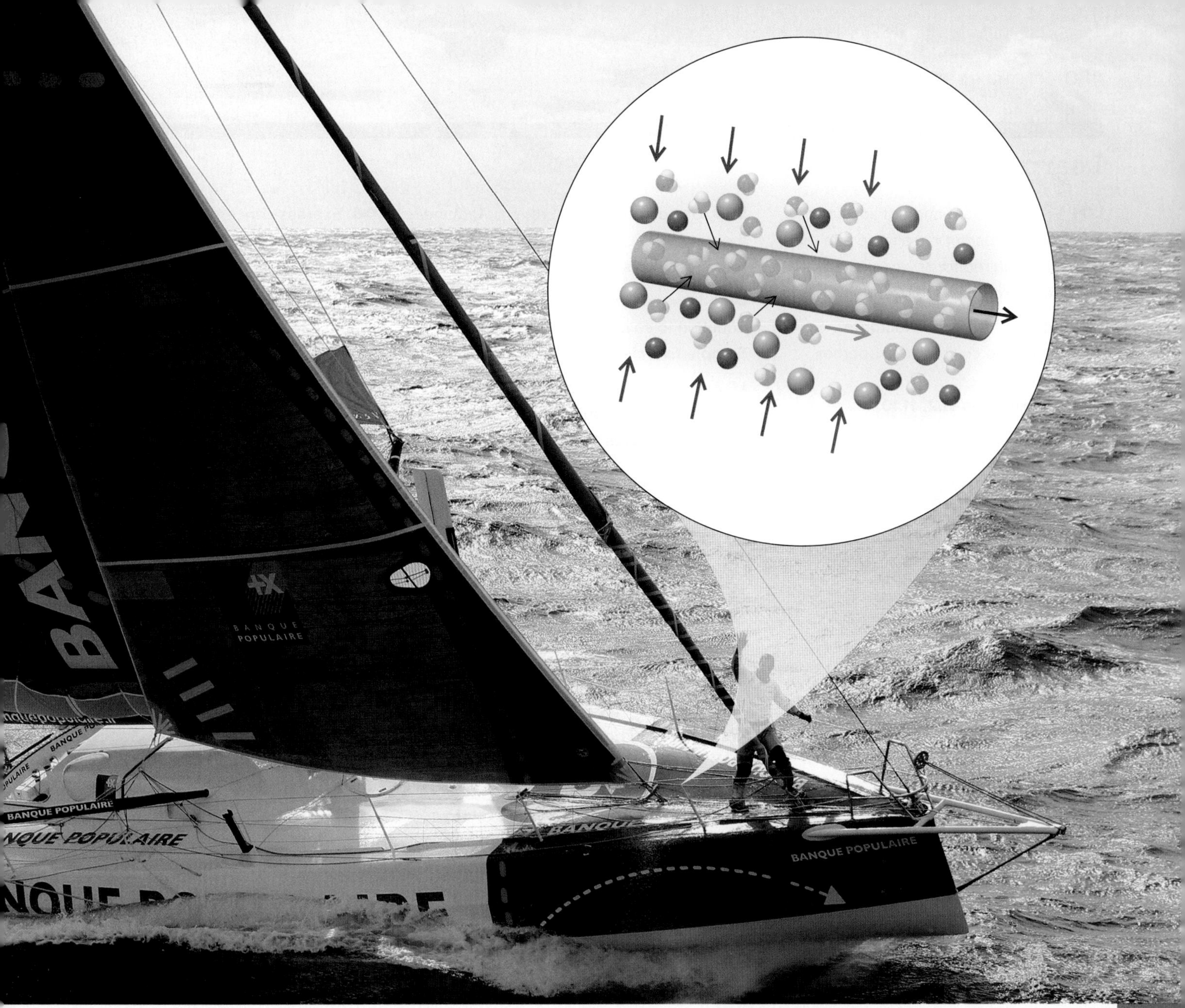

PARTICULATE **PREVIEW**

Solvent versus Solution

As you read Chapter 11, look for ideas that will help you answer these questions about the properties of pure water (on the left) and the aqueous solution of NaCl on the right.

- When NaCl is added to water, what intermolecular forces produce the solution on the right?
- What happens to the amount of water in the vapor phase when salt is added?
- How does the presence of a dissolved solute like salt affect the boiling point of water?

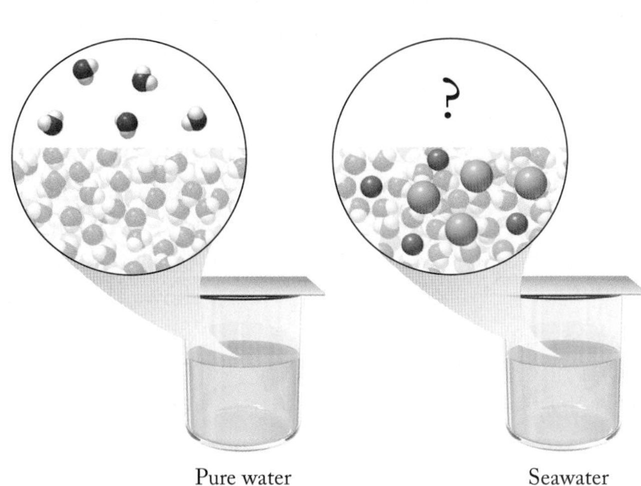

Learning Outcomes

LO1 Predict the direction of solvent flow and calculate osmotic pressure and the pressure required for reverse osmosis
Sample Exercises 11.1, 11.2

LO2 Use the van 't Hoff factor to account for the colligative properties of solutions
Sample Exercises 11.2, 11.10

LO3 Use osmotic pressure to determine molar mass
Sample Exercise 11.3

LO4 Use the Clausius–Clapeyron equation to calculate vapor pressure
Sample Exercise 11.4

LO5 Calculate vapor pressures of solutions using Raoult's law
Sample Exercises 11.5, 11.6

LO6 Express concentrations in molality
Sample Exercise 11.7

LO7 Calculate the freezing points and boiling points of solutions
Sample Exercises 11.8, 11.9

LO8 Calculate the solubility of a gas using Henry's law
Sample Exercise 11.11

11.1 Osmosis: "Water, Water, Everywhere"

In "Rime of the Ancient Mariner," British poet Samuel Taylor Coleridge (1772–1834) wrote: "Water, water, everywhere, and all the boards did shrink. Water, water, everywhere, nor any drop to drink." The water in Coleridge's verse is seawater, which is mostly an aqueous solution of NaCl and other salts. Even though all life-forms need water, sailors who drink seawater suffer from dehydration, and drinking enough of it can be fatal.

In this chapter we explore the science behind Coleridge's poetry—namely, how and why the properties of solutions depend on the concentration of the solutes in them. We also explore a process by which modern mariners, such as the sailor in this chapter's opening photo, are able to make seawater fit to drink.

Why is drinking seawater a really bad idea? The problem is the difference in salinity (salt concentration) between seawater and the fluid in most living cells (even those in the tissues of marine organisms). If human blood cells are immersed in seawater, they dehydrate, shrivel up, and die, because the concentration of the ions—mostly Na^+ and Cl^-—in seawater is much greater than in the aqueous medium inside the cell. Too high a concentration of ions means too low a concentration of H_2O: much lower than in most living cells. The difference in H_2O concentrations results in a net movement of water molecules from inside the cell, through the cell membrane, into the seawater surrounding the cell. This causes the cells to shrivel and collapse as shown in Figure 11.1(a).

This outward movement of water happens because the membranes surrounding red blood cells are *semipermeable*, which means they contain channels through which molecules of water can flow but not particles of hydrated solutes. Anytime a semipermeable membrane separates two aqueous solutions with different total concentrations of salts (or any solutes), water spontaneously flows through the membrane from the solution with the higher concentration of water molecules (lower solute concentration) into the solution with a lower concentration of water. This movement is the net result of water molecules flowing in both directions through the membrane, but the higher concentration of water molecules inside means that they pass through the cell membrane at a faster rate than those outside the cell flow into it. This spontaneous move-

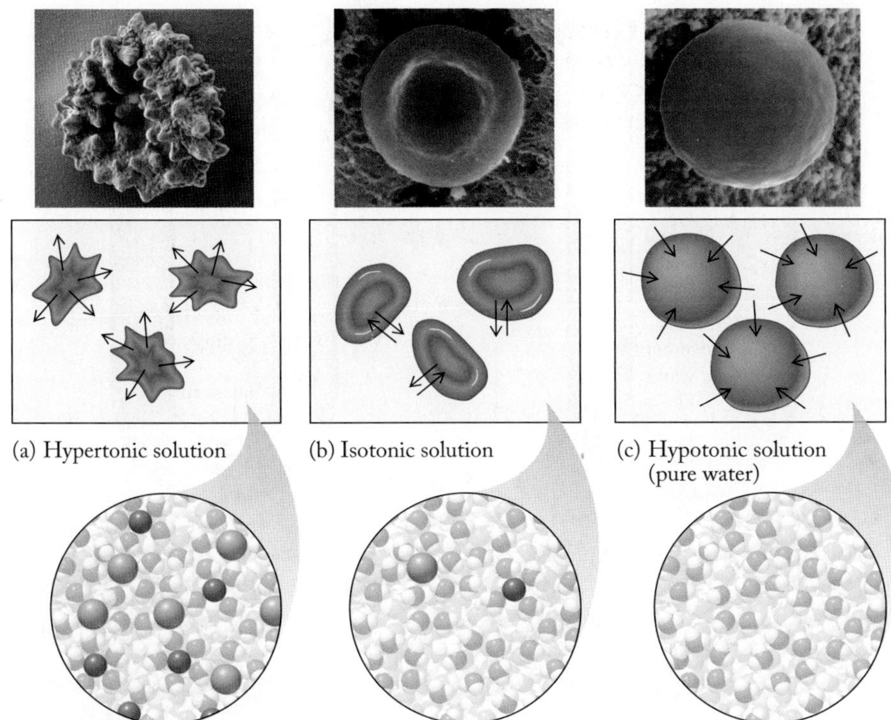

FIGURE 11.1 The membrane of a red blood cell is semipermeable, which means that water flows by osmosis into and out of the cell to equalize solute concentrations inside and out. (a) When a cell is bathed in a high-salinity *hypertonic* solution, there is a net osmotic flow out of the cell, and the cell shrinks. (b) When a cell is bathed in an *isotonic* solution in which the solute concentration matches that inside the cell, there is no net osmotic flow into or out of the cell, and its size does not change. (c) When the cell is bathed in pure water or a *hypotonic* solution, there is a net osmotic flow into the cell, and it expands.

ment of water through cell membranes (or any solvent through any semipermeable membrane) is called **osmosis**.

The outward flow of water from red blood cells by osmosis happens when they are immersed in seawater or any *hypertonic* solution, that is, a solution containing a higher concentration of solutes than inside the cell. Human cells do contain some dissolved electrolytes, which is why fluids administered by intravenous (IV) injection in hospitals to deliver medications or to rehydrate patients contain the equivalent of about 0.155 *M* NaCl (0.90% NaCl). This value matches the concentration of the electrolytes in most of our cells and is called *normal saline.* Such solutions are said to be *isotonic* solutions (Figure 11.1b).

Just as too high a total concentration of solute particles can be dangerous, too low a concentration can be, too. An IV injection without the needed electrolytes (a *hypotonic* solution, Figure 11.1c) would cause a patient to retain too much water, which could also have life-threatening consequences.

Although drinking salt water can be fatal to humans, saltwater fish thrive in the oceans because their kidneys have evolved to handle the higher salinity of the seawater they absorb. When placed in freshwater, though, saltwater fish find themselves immersed in a hypotonic solution and do not survive. Similarly, seawater is a lethal hypertonic solution for freshwater fish.

Figure 11.2 provides a molecular view of why the water in a red blood cell immersed in seawater flows out of the cell and into the seawater. The apparatus in the figure consists of two chambers: one containing seawater and the other normal saline. The two are separated by a semipermeable membrane that allows water molecules to move through it but not the hydrated ions present at a higher total concentration in seawater than in normal saline. Initially, the volumes and liquid levels in the two compartments are the same. Over time, however, the volume of the compartment containing normal saline decreases and the volume of the seawater increases due to a net flow of water molecules through the membrane from the normal saline side to the seawater side.

osmosis the flow of a fluid through a semipermeable membrane to balance the concentration of solutes in solutions on the two sides of the membrane. The flow of solvent molecules proceeds from the more dilute solution into the more concentrated one.

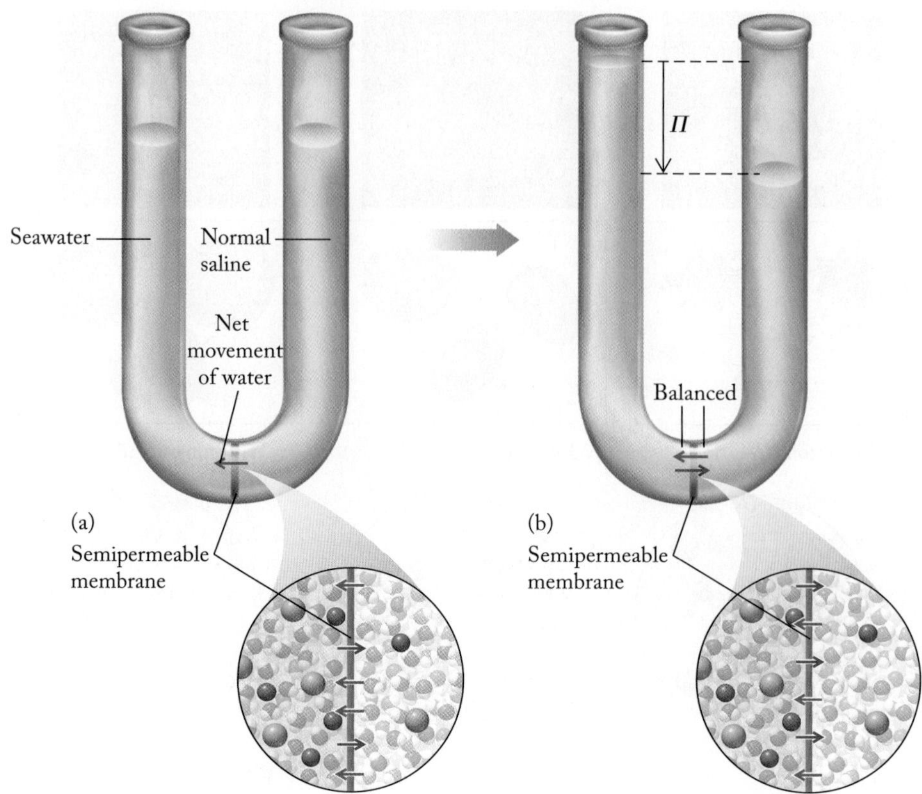

FIGURE 11.2 (a) When equal volumes of seawater and normal saline are separated by a semipermeable membrane, there is a net osmotic flow of water from the normal saline chamber to the seawater chamber. (b) Net osmotic flow stops when the difference between the heights of the two liquids creates a sufficiently large opposing osmotic pressure (Π) and additional flow from the seawater side to the normal saline side to balance osmotic flow.

Osmosis happens because the concentration of water molecules on the normal saline side of the membrane is higher than it is on the seawater side. This higher concentration means that the molecules pass through the semipermeable membrane at a faster rate than those on the seawater side pass through in the opposite direction, as symbolized by the greater number of blue arrows than red ones in Figure 11.2(a).

11.2 Osmotic Pressure and the van 't Hoff Factor

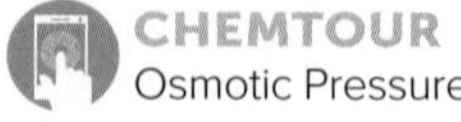

The osmotic flow of water shown in Figure 11.2 continues until the flow of water molecules passing through the membrane and into the seawater compartment is balanced by an opposing flow created by pressure due to the difference in the levels of the liquids in the two compartments. This opposing pressure is called the **osmotic pressure (Π)** of the solution. If we had filled the right-hand compartment with pure water instead of normal saline, then the difference in solute concentrations would have been even greater and the osmotic pressure would have been even higher. Equation 11.1 can be used to calculate just how high it would have been:

$$\Pi = MRT \tag{11.1}$$

Here the Greek capital letter pi (Π) is used as the symbol for osmotic pressure to distinguish it from the pressure (P) exerted by gases. The value of Π is equal to the total molar (M) concentration of all the solute particles in seawater (or any solution) times the gas law constant, R, and absolute temperature, T. When we use an R value of 0.08206 L · atm/(mol · K), we get osmotic pressure values in atmospheres.

osmotic pressure (Π) the pressure applied across a semipermeable membrane to stop the flow of water from the compartment containing pure solvent or a less concentrated solution to the compartment containing a more concentrated solution.

colligative properties characteristics of solutions that depend on the concentration and not the identity of particles dissolved in the solvent.

SAMPLE EXERCISE 11.1 Calculating Osmotic Pressure **LO1**

If the total concentration of solute particles in the fluid inside a red blood cell is 0.310 *M*, what is the osmotic pressure of this fluid at body temperature (37°C)? The membrane surrounding the cell is semipermeable.

Collect, Organize, and Analyze We are asked to calculate the osmotic pressure of a solution with a total particle concentration of 0.310 *M* at 37°C. Equation 11.1 relates the osmotic pressure of solutions to their solute particle concentrations and absolute temperatures.

Solve

$$\Pi = MRT = 0.310\,\frac{\cancel{\text{mol}}}{\cancel{\text{L}}} \times \frac{0.08206\ \cancel{\text{L}}\cdot\text{atm}}{\cancel{\text{mol}}\cdot\cancel{\text{K}}} \times 310\ \cancel{\text{K}} = 7.89\ \text{atm}$$

Think About It The calculated osmotic pressure is nearly 8 atm. That is a lot of pressure—comparable to that in high-performance bicycle tires. In the absence of this opposing pressure, water would migrate across the semipermeable membrane of the red blood cell if it were immersed in distilled water, causing the cell to swell up (Figure 11.1c) and eventually burst.

Practice Exercise Calculate the osmotic pressure across a semipermeable membrane separating pure water from a sample of seawater at 20°C. Assume the total concentration of all the ions in the sample is 1.148 *M*.

(Answers to Practice Exercises are in the back of the book.)

Is there a connection between osmotic pressure (Π) and the pressure exerted by ideal gases, as described in Chapter 10? Actually, the two kinds of pressure are analogous, as shown when we solve the ideal gas law $PV = nRT$ for pressure and write it this way:

$$P = \left(\frac{n}{V}\right)RT = MRT$$

The n/V part of the equation is the same as molarity (assuming V is expressed in liters).

Osmotic pressure is a **colligative property** of a solution, which means that it depends only on the total concentration of solute particles and not on their identities. This independence from solute identity arises because the key factor in defining the relative rates of migration of water molecules across a semipermeable membrane is the difference in the concentrations of water on both sides of the membrane. The more moles per liter of solute particles there are on one side, the smaller the concentration of water there is on that side and the greater the osmotic pressure.

When a semipermeable membrane separates two solutions that have different solute concentrations, such as seawater and normal saline, we can calculate the osmotic pressure between them by calculating the difference in solute concentrations on opposite sides of the membrane. In the case of seawater surrounding a red blood cell, this difference is 1.148 − 0.310 = 0.838 *M*. Inserting this value into Equation 11.1 and assuming $T = 20$°C:

$$\Pi = MRT = 0.838\,\frac{\cancel{\text{mol}}}{\cancel{\text{L}}} \times 0.08206\,\frac{\cancel{\text{L}}\cdot\text{atm}}{\cancel{\text{mol}}\cdot\cancel{\text{K}}} \times 293\ \cancel{\text{K}} = 20.1\ \text{atm}$$

In the absence of this opposing pressure, the water inside a red blood cell immersed in seawater spontaneously flows out through the cell membrane into the seawater, causing the cell to collapse as illustrated in Figure 11.1(a).

van 't Hoff Factors

If a solute is a weak or strong electrolyte, then one mole of it produces more than one mole of particles in solution. Therefore, the value of M in Equation 11.1 is greater than the molar concentration of the solute itself. Frequently we know the concentration of the solute in a solution (because we know the quantity of it used to prepare the solution) better than we know the concentration of particles present. This is the case for weak electrolytes that are only partially ionized in aqueous solutions, and it is even true for solutions of strong electrolytes, such as NaCl, especially at high concentrations.

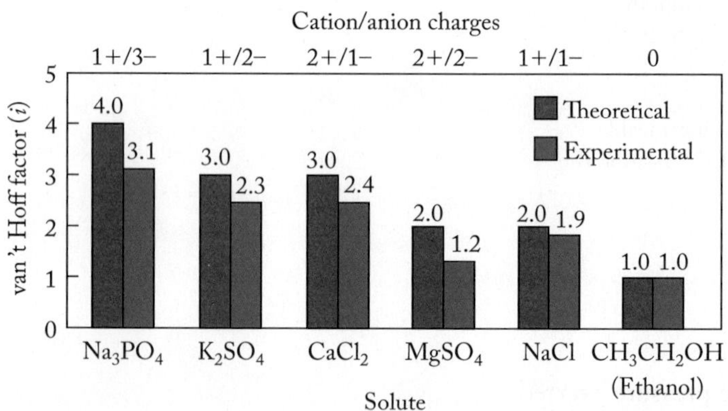

FIGURE 11.3 Theoretical and experimentally measured values for the van 't Hoff factors for 0.10 M solutions of several electrolytes and the nonelectrolyte ethanol. The higher the charge on the ions, the greater the difference between the theoretical and experimentally measured values.

Consider the data plotted in Figure 11.3. The height of each of the bars represents the factor that the concentration of each solute (they are all 0.10 M) should be multiplied by to calculate the total concentration of solute particles (ions and/or molecules) that are present in each solution. Each red bar represents a theoretical factor, such as 2 for NaCl because one mole of this salt should produce two moles of ions when it dissolves in water. Similarly the theoretical factor for $MgSO_4$ is also 2, but it is 3 for $CaCl_2$ and 4 for Na_3PO_4 because a mole of the latter two salts should produce three and four moles of ions, respectively, when they dissolve.

The blue bars in Figure 11.3 represent the actual ratios of particle concentrations to solute concentrations. For nonelectrolytes, such as ethanol, theoretical value and actual value are both 1.0 because one mole of dissolved solute produces one mole of dissolved molecules. For weak and strong electrolytes the actual ratios are determined experimentally by measuring the colligative properties of these solutions. These ratios are called **van 't Hoff factors (*i*)** to honor Dutch chemist Jacobus van 't Hoff (1852–1911) who studied the colligative properties of solutions and discovered that they are linked to the concentrations of solute particles in them.

In solutions of electrolytes below about 0.01 M the actual van 't Hoff factors are essentially the same as the theoretical values. However, at concentrations above 0.01 M, the experimental values decrease with increasing concentration, which is reflected in the lower heights of the blue bars in Figure 11.3 for solute concentrations of 0.10 M. Why does this happen? The answer is that cations and anions do not behave as completely independent particles in solution as their concentrations increase. Instead, they cluster together. The simplest and most common cluster is an **ion pair**: one cation and one anion acting as a single particle. This ion pairing reduces the total concentration of Na^+ and Cl^- ions in 0.10 M NaCl from 0.20 M to 0.19 M (Figure 11.3). Thus, the experimental van 't Hoff factor for 0.10 M NaCl is 1.9. Similar deviations from theoretical behavior are observed for other electrolytes, as shown by the data in Figure 11.3. Note how the difference between theoretical and experimental values, reflecting greater ion pairing, increases with increasing ionic charge. This behavior is consistent with Coulomb's law and the greater attraction between cations and anions of higher charge. Knowledge of the van 't Hoff factor (i) of a solution

van 't Hoff factor (*i*) the ratio of the concentration of solute particles in a solution to the concentration of particles that would be there if the solute did not dissociate.

ion pair a cluster formed when a cation and an anion associate with each other in solution.

reverse osmosis (RO) a purification process in which solvent is forced through semipermeable membranes, leaving dissolved impurities behind.

allows us to predict the osmotic pressure of the solution based on the molar concentration (M) of the solute:

$$\Pi = iMRT \tag{11.2}$$

For example, the osmotic pressure of 0.10 M K_2SO_4 at 25°C is

$$\Pi = iMRT = 2.3 \times 0.10 \frac{\cancel{\text{mol}}}{\cancel{\text{L}}} \times 0.08206 \frac{\cancel{\text{L}} \cdot \text{atm}}{\cancel{\text{mol}} \cdot \cancel{\text{K}}} \times 298 \cancel{\text{K}} = 5.6 \text{ atm}$$

CONCEPT TEST

If a 1.0 M glucose solution is inside a semipermeable membrane and 1.0 M KCl is on the outside, in which direction do water molecules migrate through the membrane?

(Answers to Concept Tests are in the back of the book.)

The preceding osmotic pressure calculations show that relatively small differences in solute concentration can produce large osmotic pressures that can distort and even destroy living cells. Fluids administered intravenously to hospital patients must have nearly the same concentration of solute particles as the patients' blood and blood cells.

Two common IV solutions are physiological saline, which contains 9.0 g/L NaCl, and D5W, which is 5.0×10^1 g/L dextrose ($\mathcal{M}$ = 180.16 g/mol). Let's convert these concentrations to molarity values:

$$\text{Physiological saline:} \quad \frac{9.0 \cancel{\text{ g NaCl}}}{1 \text{ L}} \times \frac{1 \text{ mol NaCl}}{58.44 \cancel{\text{ g NaCl}}} = 0.15 \, M \text{ NaCl}$$

$$\text{D5W:} \quad \frac{5.0 \times 10^1 \cancel{\text{ g dextrose}}}{1 \text{ L}} \times \frac{1 \text{ mol dextrose}}{180.16 \cancel{\text{ g dextrose}}} = 0.28 \, M \text{ dextrose}$$

While the *molarity* of D5W is nearly twice that of physiological saline, the van 't Hoff factor of a 0.15 M solution of NaCl is nearly 2. Dextrose, on the other hand, is a molecular compound that does not ionize in water and has a van 't Hoff factor of 1. Thus, the concentrations of dissolved particles in the two solutions are essentially the same.

Reverse Osmosis: Making Seawater Drinkable

We have developed a molecular-level explanation based on osmosis for why people shouldn't drink seawater. However, applying a technique based on osmosis can make seawater drinkable. The process known as *desalination* involves removing most of the ions from seawater. Distillation is a desalination method that is widely used in the desert countries surrounding the Persian Gulf; however, a less energy-intensive approach involves pumping seawater through semipermeable membranes. Molecules of water pass through the membranes, and the ions dissolved in the seawater are left behind. This process is called **reverse osmosis (RO)**. It works because very-high-pressure pumps overcome the osmotic pressure of seawater and force the normally spontaneous flow of H_2O molecules from pure water into seawater to reverse.

CONNECTION Simple distillation, such as the desalination of seawater as described in Chapter 1, is used to separate volatile solvents from nonvolatile solutes.

The RO system shown in Figure 11.4 consists of a cylindrical chamber containing many narrow tubes made of a semipermeable membrane. Seawater is forced through the chamber so that it washes over the exterior of the tubes, which

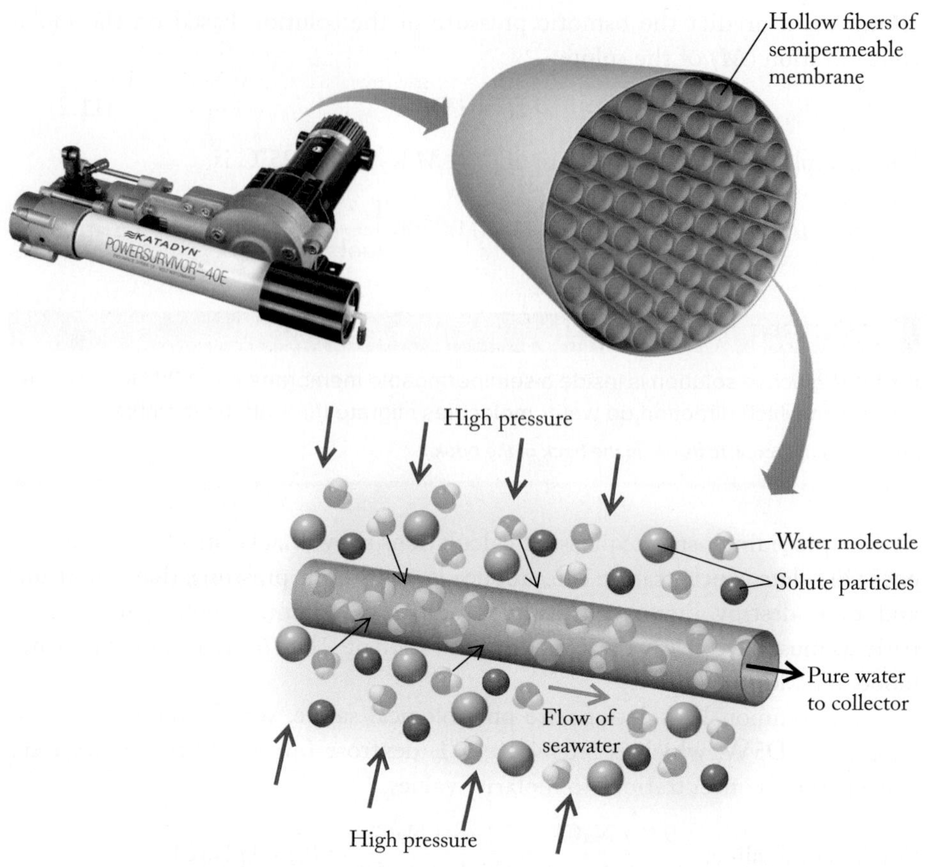

FIGURE 11.4 Seawater can be desalinated by reverse osmosis. In this apparatus, seawater flows at a pressure greater than its osmotic pressure around bundles of tubes with semipermeable walls. Water molecules pass from the seawater into the tubes and flow through the tubes to a collection vessel. Few hydrated ions can penetrate the walls of the tubes.

are filled with flowing pure water. Ordinarily, the direction of osmosis would be from the inner tubes into the surrounding seawater. However, when a pressure greater than the osmotic pressure of seawater is exerted on the seawater in the outer chamber, water flows in the other direction: from the chamber, through the membranes, into the tubes, and out of the apparatus to a pure-water collector.

Some municipal water-supply systems use reverse osmosis to make slightly saline (brackish) water fit to drink. Some industries use this method to purify conventional tap water, and it provides a renewable supply of drinking water for sailors on ocean-crossing yachts. Strong pumps and fittings and rugged membranes are required because reverse osmosis systems must operate at very high pressures (see Sample Exercise 11.2). Moreover, semipermeable membranes are not 100% efficient, so RO systems cannot turn seawater into completely deionized water. Some solvated Na^+ and Cl^- ions inevitably pass through the membrane, but their concentrations in the product water are reduced to levels that make the water drinkable.

SAMPLE EXERCISE 11.2 Calculating the Pressure Needed for Reverse Osmosis — LO1, 2

What is the reverse osmotic pressure required at 20°C to purify brackish well water that is equivalent to 0.177 *M* NaCl if the product water is to meet the drinking water standard that the concentration of Na^+ ions be no greater than 20 mg/L? Assume the *i* value of the well water electrolytes is 1.85.

Collect and Organize We are asked to calculate the pressure needed to purify water by reverse osmosis. We are given the temperature, the concentration of NaCl in the water to be purified, and the target concentration of Na^+ ions in the water to be produced.

Analyze We can adapt Equation 11.2 to calculate the difference in osmotic pressure between the source and the product water, but first we need to convert the concentration of Na^+ ions in the purified water to molarity. We may assume that this value also represents the concentration of Cl^- ions in the drinking water, which must be there to balance ionic charges. This concentration value will probably be small compared to the salt concentration of the well water, so we can estimate the osmotic pressure that must be overcome based on the fact that the NaCl concentration in the well water is close to that of physiological saline, which has an osmotic pressure of about 8 atm (see Sample Exercise 11.1).

Solve Convert 20 mg/L Na^+ to molarity:

$$\frac{20\ \cancel{\text{mg Na}^+}}{\text{L}} \times \frac{1\ \text{g}}{1000\ \cancel{\text{mg}}} \times \frac{1\ \text{mol Na}^+}{22.99\ \cancel{\text{g Na}^+}} = \frac{8.7 \times 10^{-4}\ \text{mol Na}^+}{\text{L}}$$

The difference in NaCl concentration between the well water and target drinking water is

$$(0.177 - 0.00087) = 0.176\ M\ \text{NaCl}$$

The corresponding osmotic pressure is

$$\begin{aligned} \Pi &= iMRT \\ &= 1.85 \times 0.176\ \frac{\cancel{\text{mol}}}{\cancel{\text{L}}} \times 0.08206\ \frac{\cancel{\text{L}} \cdot \text{atm}}{\cancel{\text{mol}} \cdot \cancel{\text{K}}} \times 293\ \cancel{\text{K}} \\ &= 7.8\ \text{atm} \end{aligned}$$

Think About It In the absence of any external pressure, solvent will flow from the product water to the brackish well water. However, an external pressure greater than 7.8 atm will force the spontaneous flow to reverse, forming drinkable product water from brackish well water.

Practice Exercise Calculate the minimum external pressure that must be applied in a reverse osmosis system at 20°C to reduce the salinity of a seawater sample from 3.5% by mass NaCl (d = 1.020 kg/L; i = 1.84) to 48 mg/L.

Using Osmotic Pressure to Determine Molar Mass

Measuring the osmotic pressure of the solution of a nonelectrolyte can also be used to determine its molar mass. Osmotic pressure measurements have the advantage of producing large, easily measured pressures, even from dilute solutions. This sensitivity is particularly important in determining the molar mass of small samples of biologically important substances that often have large molar masses.

CONNECTION In Chapter 10 we used measurements of density and the ideal gas law to calculate the molar masses of gases.

SAMPLE EXERCISE 11.3 Using Osmotic Pressure to Determine Molar Mass — LO3

A 47 mg sample of a water-soluble nonelectrolyte is isolated from a South African tree. The sample is dissolved in enough water to make 2.50 mL of solution. The osmotic pressure of the solution is 0.489 atm at 25°C. What is the molar mass of the compound?

Collect and Organize We know the concentration of a solution (in mg/mL) of an unknown substance and the osmotic pressure of the solution at 25°C. We need to find the molar mass of the substance.

Analyze Equation 11.2 can be used to calculate the molarity (M) of the solution from the known values of Π, i (1, nonelectrolyte), and T. We can then divide the calculated mol/L concentration value into the given mg/mL (which is the same as g/L) concentration value to obtain g/mol, or molar mass, of the solute.

Solve Rearranging Equation 11.2 to solve for M and substituting the relevant known values:

$$M = \frac{\Pi}{iRT} = \frac{0.489 \text{ atm}}{1 \times 0.08206 \frac{\text{L} \cdot \text{atm}}{\text{mol} \cdot \text{K}} \times 298 \text{ K}} = 2.00 \times 10^{-2} \text{ mol/L}$$

The sample concentration is also

$$\frac{47 \text{ mg}}{2.50 \text{ mL}} = 18.8 \text{ mg/mL} = 18.8 \text{ g/L}$$

To calculate molar mass we divide the g/L concentration value by the mol/L concentration value, which gives us g/mol:

$$\frac{18.8 \text{ g/L}}{2.00 \times 10^{-2} \text{ mol/L}} = 9.4 \times 10^{2} \text{ g/mol} = \mathcal{M}$$

Think About It One of the advantages of determining molar mass by osmotic pressure is that only a small quantity of sample is required to produce an osmotic pressure sufficiently large (0.489 atm) to allow for an accurate determination of a molar mass of nearly 1000. In most laboratories the most popular technique for determining molar mass is mass spectrometry, as we discussed in Chapter 2. It requires even less sample than does an osmotic pressure determination of molar mass.

CONNECTION In Chapter 7 we used molar masses determined by mass spectrometry to convert empirical formulas derived from elemental analyses into molecular formulas.

Practice Exercise A solution is made by dissolving 5.00 mg of a polysaccharide in water to give a final volume of 1.00 mL. The osmotic pressure of this nonelectrolyte solution is 1.91×10^{-3} atm at 25°C. What is the molar mass of the polysaccharide?

11.3 Vapor Pressure

Ask yourself what would happen if you left a glass of water on a kitchen counter for several days. If left undisturbed, the level of water would drop as the water slowly evaporated. At the particle level, some of the molecules of H_2O on the surface escape from the liquid and enter the air above it. They can escape because water molecules at the surface of the liquid have a distribution of kinetic energies (as we saw with gas-phase molecules in Chapter 10). Some of them have sufficient energy to overcome the intermolecular forces holding them on the surface. Once these high-energy molecules have left the liquid phase, taking their kinetic energies with them, the remaining water molecules redistribute their energies, drawing heat from their surroundings to do so, and the evaporation process continues. Eventually the glass will go dry without ever being above room temperature.

If an identical glass of water is left covered on the same counter (Figure 11.5), water molecules still evaporate at the same rate, but they are confined to the space above the water and below the cover. As the concentration of water molecules in this space increases, the rate at which they condense onto the cover and the walls of the glass and, more significantly, into the liquid surface also increases. In a short time the rates of evaporation and condensation equalize as water molecules

vapor pressure the pressure exerted by a gas in equilibrium with its liquid phase at a given temperature.

volatile having a significant vapor pressure at a given temperature.

FIGURE 11.5 A covered glass of water achieves a dynamic equilibrium where the rate at which liquid water is lost to evaporation equals the rate at which liquid water is gained by condensation.

in the liquid and gas phases achieve a state of dynamic equilibrium. Their concentration in the gas phase remains constant as long as the temperature does not change and the cover stays on.

The water vapor in the space above the water in the covered glass exerts a partial pressure that is equal to the **vapor pressure** of the water, which is defined as the partial pressure exerted by a gas in equilibrium with its liquid state at a given temperature. A liquid is said to be **volatile** when enough of its molecules vaporize to produce a significant vapor pressure at a given temperature: the higher its vapor pressure, the more volatile the liquid is.

The higher the temperature of a volatile liquid, the greater the fraction of its molecules with enough energy to break away from its surface and enter the gas phase, and the higher its vapor pressure. This trend is shown in Figure 11.6. The least volatile liquid in the figure, ethylene glycol, requires the highest temperatures

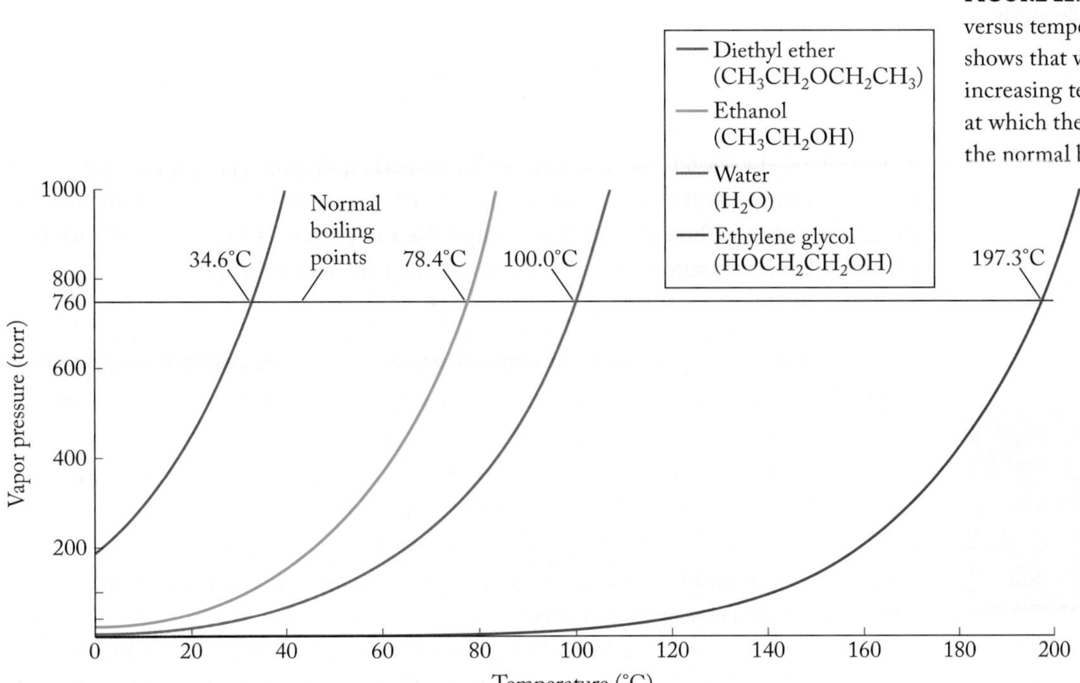

FIGURE 11.6 A graph of vapor pressure versus temperature for four liquids shows that vapor pressure increases with increasing temperature. The temperature at which the vapor pressure equals 1 atm is the normal boiling point of the liquid.

Clausius–Clapeyron equation relates the vapor pressure of a substance at different temperatures to its heat of vaporization.

to exert significant vapor pressures; the most volatile liquid, diethyl ether, vaporizes at much lower temperatures. If the temperature of a liquid is high enough, its vapor pressure reaches ambient atmospheric pressure (the horizontal line in Figure 11.6). The liquid vaporizes readily under these conditions because it has reached its boiling point. When this ambient pressure is a standard atmosphere (1 atm), the liquid is at its *normal* boiling point, such as 100.0°C for water.

CONCEPT **TEST**

Rank the compounds with these condensed molecular structures in order of increasing vapor pressure at 0°C:

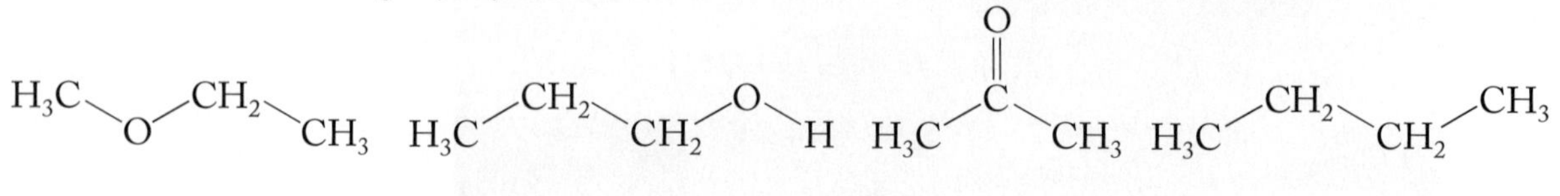

(a) $H_3C-O-CH_2-CH_3$ (b) $H_3C-CH_2-CH_2-O-H$ (c) $H_3C-C(=O)-CH_3$ (d) $H_3C-CH_2-CH_2-CH_3$

The Clausius–Clapeyron Equation

ln (P_{vap}) (P in torr)

7.00

6.00

5.00

3.25 3.50 3.75

$(1/T) \times 10^3$ (K^{-1})

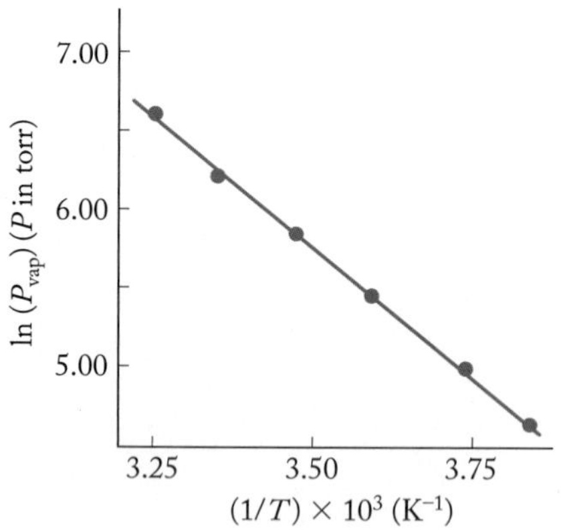

FIGURE 11.7 Plotting the natural logarithm of the vapor pressure versus the reciprocal of the absolute temperature gives a straight line described by the Clausius–Clapeyron equation. The graph shows the plot for pentane.

The vapor pressure of a volatile liquid increases with increasing temperature, and as Figure 11.6 shows, the increase is not linear. However, if we graph the natural logarithm of the vapor pressure versus $1/T$, where T is the absolute temperature, we do get a straight line (Figure 11.7). The slope of this line depends on the enthalpy of vaporization (ΔH_{vap}) of the liquid:

$$\ln(P_{vap}) = -\frac{\Delta H_{vap}}{R}\left(\frac{1}{T}\right) + C \quad (11.3)$$

where R is the universal gas constant [8.314 J/(mol · K)] and C is a constant that depends on the identity of the liquid. We can use Equation 11.3 to relate the vapor pressures at two temperatures (T_1 and T_2) to ΔH_{vap}:

$$\ln(P_{vap,T_1}) + \left(\frac{\Delta H_{vap}}{RT_1}\right) = C = \ln(P_{vap,T_2}) + \left(\frac{\Delta H_{vap}}{RT_2}\right)$$

or

$$\ln\left(\frac{P_{vap,T_1}}{P_{vap,T_2}}\right) = \frac{\Delta H_{vap}}{R}\left(\frac{1}{T_2} - \frac{1}{T_1}\right) \quad (11.4)$$

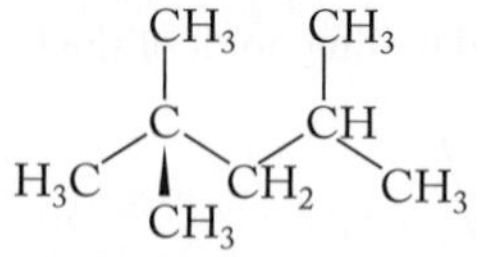

Equation 11.4 is called the **Clausius–Clapeyron equation**. We can use it to calculate the vapor pressure of a liquid at any temperature if we know its normal boiling point and ΔH_{vap} value, as shown in Sample Exercise 11.4, or to determine the value of ΔH_{vap} from measurements of vapor pressure at different temperatures.

$H_3C-C(CH_3)_2-CH_2-CH(CH_3)-CH_3$

FIGURE 11.8 Structural formula of isooctane.

Price per gallon All taxes included

MINIMUM OCTANE RATING (R+M)/2 METHOD

87 89 93

PRESS

FIGURE 11.9 Octane ratings for gasoline.

SAMPLE EXERCISE 11.4 Calculating Vapor Pressure Using the Clausius–Clapeyron Equation **LO4**

The compound with the common name *isooctane* (its official name is 2,2,4-trimethylpentane) has the structure shown in Figure 11.8. It defines the "100" value on the octane rating scale used to grade gasoline (Figure 11.9). Its normal boiling point is 99°C, and its heat of vaporization is 35.2 kJ/mol. What is the vapor pressure of isooctane at 25°C in torr?

Collect and Organize We are given the normal boiling point and heat of vaporization of isooctane. We are asked to calculate its vapor pressure at 25°C. The Clausius–Clapeyron equation relates the vapor pressure values of a liquid at two temperatures to its ΔH_{vap} value.

Analyze To use the Clausius–Clapeyron equation, we must express temperatures on the Kelvin scale and convert ΔH_{vap} to joules per mole to be compatible with the units on R, J/(mol · K). Vapor pressure decreases sharply (Figure 11.6) as temperatures decrease below the normal boiling point of a liquid, so the vapor pressure of isooctane at 25°C should be only a fraction of 760 torr.

Solve

ΔH_{vap} in joules per mole is

$$35.2\,\frac{\text{kJ}}{\text{mol}} \times \frac{10^3\ \text{J}}{\text{kJ}} = 3.52 \times 10^4\,\frac{\text{J}}{\text{mol}}$$

Entering these values in Equation 11.3 and solving for P_{vap,T_2} yields

$$\ln\left(\frac{P_{vap,T_1}}{P_{vap,T_2}}\right) = \frac{\Delta H_{vap}}{R}\left(\frac{1}{T_2} - \frac{1}{T_1}\right)$$

$$\ln\left(\frac{760\ \text{torr}}{P_{vap,T_2}}\right) = \frac{\left(3.52 \times 10^4\,\frac{\text{J}}{\text{mol}}\right)}{\left(8.314\,\frac{\text{J}}{\text{mol}\cdot\text{K}}\right)}\left(\frac{1}{298\ \text{K}} - \frac{1}{372\ \text{K}}\right) = 2.826$$

$$P_{vap,T_2} = 45.0\ \text{torr}$$

Think About It We expected isooctane to have a relatively low vapor pressure at a temperature well below its normal boiling point, so this number is reasonable. Its value and the value of other hydrocarbons in gasoline at summerlike temperatures are well above the vapor pressure of water (23.8 torr), which means that their evaporation rates are high enough to impact the quality of the air near gas stations and to make it economically feasible to use gas pump nozzles that trap some of these vapors.

Practice Exercise Pentane (C_5H_{12}) gas is used to blow the bubbles in molten polystyrene that turn it into Styrofoam, which is used in coffee cups and other products that have good thermal insulation properties. The normal boiling point of pentane is 36°C; its vapor pressure at 25°C is 505 torr. What is the heat of vaporization of pentane?

CONCEPT TEST

Diesel fuel is made of hydrocarbons with an average of 13 carbon atoms per molecule, and gasoline is made of hydrocarbons with an average of 7 carbon atoms per molecule. Which fuel has the higher vapor pressure at room temperature?

11.4 Solutions of Volatile Substances

Small differences in the vapor pressures of volatile liquids can be used to separate them from each other. Two familiar examples of such mixtures are gasoline and crude oil, the material from which gasoline is derived. Both are complex mixtures of hydrocarbons, molecular compounds composed entirely of atoms of carbon and hydrogen bonded together. Crude oil is the source of several important classes of fuels in addition to gasoline, including diesel oil, jet fuel, kerosene, and heating

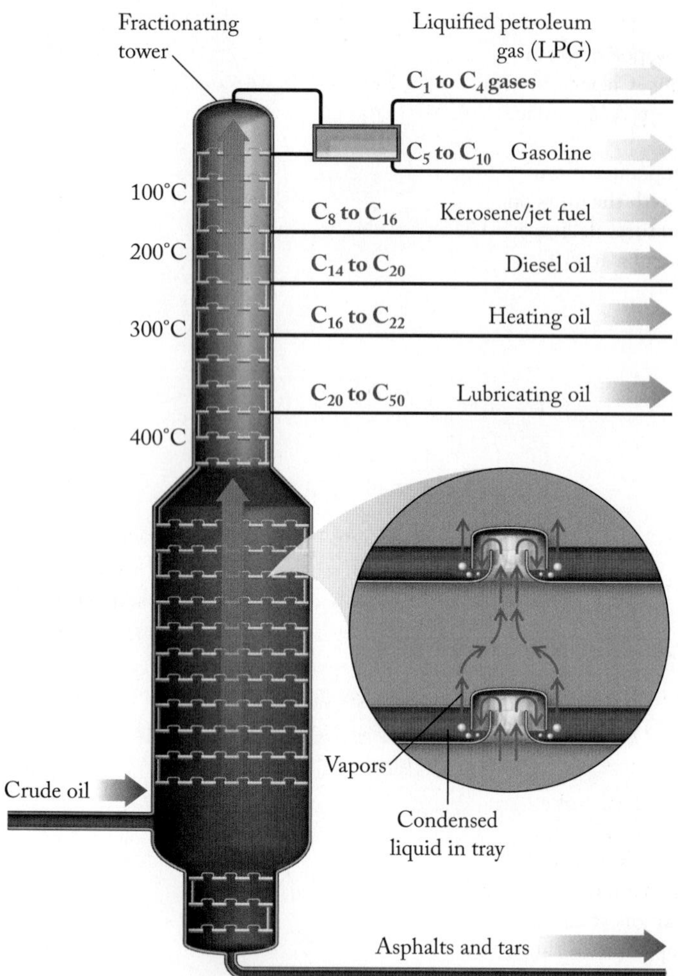

FIGURE 11.10 Fractional distillation separates crude oil into products that are used as fuels, lubricants, and building materials.

oil. The process by which crude oil is separated into these fuels is an example of **fractional distillation** (Figure 11.10). Like simple distillation, fractional distillation separates the components of mixtures based on differences in their volatilities through selective evaporation and condensation. Unlike simple distillation, fractional distillation employs a fractionating column (or fractionating tower in the industrial-scale system in Figure 11.10) in which many evaporation/condensation cycles separate substances that have only slightly different volatilities.

The separation depicted in Figure 11.10 begins with the injection of crude oil at the bottom of a fractionating tower, where the oil is heated to a temperature at which nearly all of its components vaporize. As the vapors percolate upward in the tower, they encounter cooler surfaces where the least volatile of them condense, collecting in trays. The more volatile vapors do not condense until they reach higher regions of the tower. As heat continues to flow upward in the tower, the substances collected in each tray reach temperatures at which they revaporize. However, the vapor produced by the liquid in each tray has a composition that is different from the liquid that condensed into the tray: the vapor-phase mixture is richer in the more volatile components of the liquid. When this enriched vapor condenses in another, cooler tray and then revaporizes, the vapor produced this time is even richer in the more volatile components. In a fractional distillation apparatus, these three steps of vaporization, condensation, and revaporization happen over and over again, which allows components with only slightly different vapor pressures to be separated from one another.

To explore how fractional distillation works in more detail, let's employ it to separate a 50:50 (by moles) mixture of two volatile hydrocarbons present in gasoline: heptane (C_7H_{16}, boiling point 98°C) and octane (C_8H_{18}, boiling point 126°C). Octane has the higher boiling point because its molecules are larger and experience stronger London dispersion forces. These forces keep octane molecules together in the liquid state and inhibit their vaporization until they acquire greater kinetic energies at higher temperatures.

CHEMTOUR Fractional Distillation

Suppose a sample of our 50:50 mixture is heated in a conventional distillation apparatus. The mixture starts to boil at a temperature at which the sum of their vapor pressures reaches ambient pressure (760 torr). This temperature turns out to be 108°C, which is between the normal boiling points of the two components of the mixture. At this temperature, heptane has the greater vapor pressure because it is more volatile, but octane has a significant vapor pressure, too, so the vapors that condense in the distillation apparatus are enriched in heptane but still have a significant octane component. This one-step distillation does not separate them completely.

To obtain a more complete separation, we use an apparatus that includes a fractionating column above the distillation flask (Figure 11.11). Like the towers used to refine crude oil, fractionating columns used in labs contain surfaces (sometimes they are packed with glass beads to increase surface area) on which hot vapors rising up from the boiling flask condense. As heptane-enriched vapors from a boiling 50:50 mixture reach the bottom of such a column, they condense,

fractional distillation a method of separating a mixture of compounds on the basis of their different boiling points.

forming a liquid that is also enriched in heptane. The process is shown graphically in Figure 11.12. The blue curve represents temperatures at which different liquid mixtures of the two components boil. The temperature at point 1 on the curve confirms that the initial boiling point of a 50:50 solution is about 108°C. The red line on the graph shows the composition of the vapor that is formed as these solutions boil. To find the composition of the vapor produced by a 50:50 mixture of the two liquids boiling at 108°C, we move horizontally to the left along the dashed line between points 1 and 2. The x-coordinate of point 2 tells us that the composition of the vapor produced by the liquid boiling at point 1 is about 65% heptane and only 35% octane.

This 65:35 vapor rises up in the distillation column, cools, and condenses as a 65:35 liquid in a process represented by the red arrow from point 2 to point 3. Continued heating of the column warms this liquid, and it vaporizes at about 104°C (the y-coordinate of point 3). To find the composition of the vapor above this boiling liquid, we again move left on the temperature axis until we intersect the red curve (point 4). Reading down from point 4 to the concentration axis, we see that the vapor concentration is now about 80% heptane and only 20% octane. This 80:20 vapor rises up, where it cools and condenses, and the distillation cycle is repeated.

If we continue this process of redistilling mixtures with increasing concentrations of heptane and then cooling and condensing the vapors, we eventually obtain a condensate that is pure heptane. If we monitor the temperature at which vapors condense at the very top of our distillation column, we will see a profile of temperature versus volume of distillate produced that looks like Figure 11.13. The first liquid to be produced is nearly pure heptane, which has a boiling point of 98°C. Ideally, all the heptane in the original sample is recovered before octane is collected as the temperature at the top rises to 126°C.

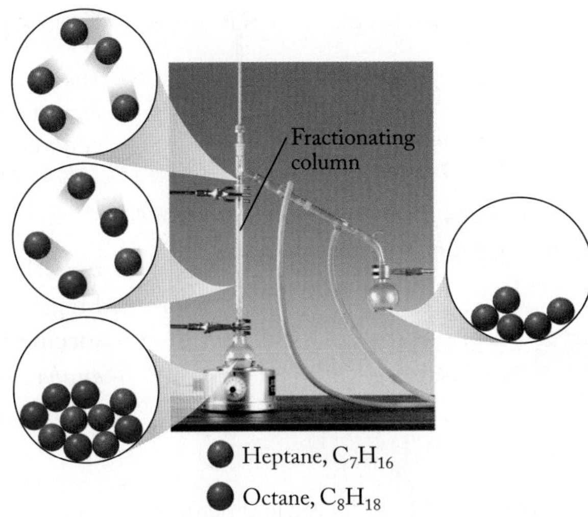

FIGURE 11.11 Fractional distillation apparatus. Vapors rise through a fractionating column, where they repeatedly condense and revaporize. The most volatile component distills first into the collecting flask. Increasingly less volatile, higher boiling components are distilled in turn. The progress of the distillation process is monitored using the thermometer at the top of the fractionating column.

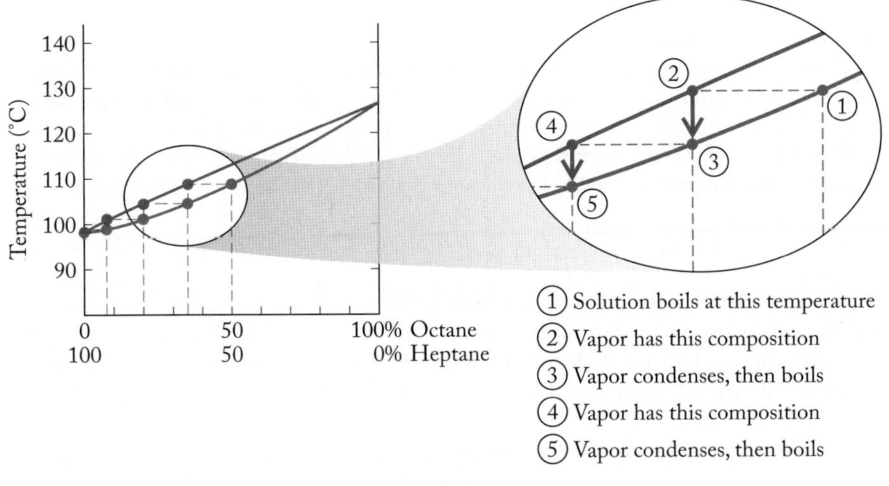

FIGURE 11.12 Stages in the fractional distillation of a 50:50 mixture of heptane and octane. The blue line tracks the boiling points of the original sample and the liquids that condense in the fraction column. Points on the red line show the composition of the vapor produced as the sample and the condensed liquids vaporize. With each vaporization and condensation cycle the composition of the liquid in the column comes closer to being pure heptane.

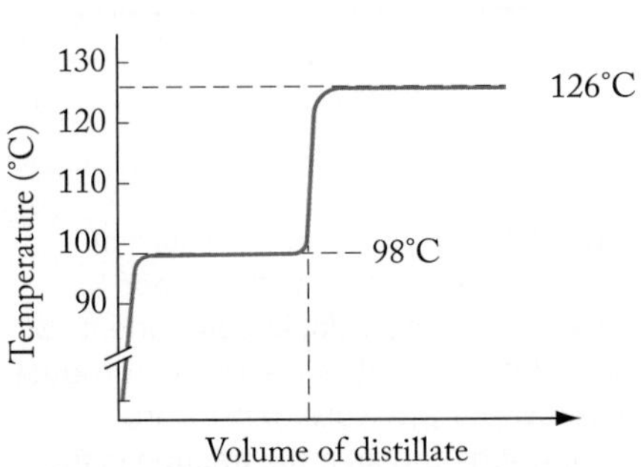

FIGURE 11.13 Temperature profile of the fractional distillation of a 50:50 mixture of heptane and octane.

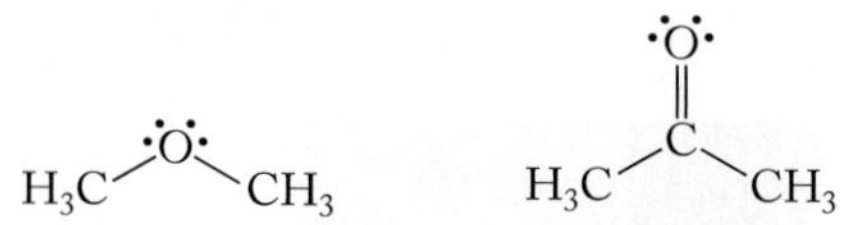

$H_3C-O-CH_3$

Dimethyl ether

$\mu = 1.30$ D

$H_3C-C(=O)-CH_3$

Acetone

$\mu = 2.88$ D

FIGURE 11.14 Lewis structures and dipole moments of dimethyl ether and acetone.

CHEMTOUR Raoult's Law

CONNECTION According to Dalton's law of partial pressures, each gas in a mixture contributes a partial pressure equal to the product of the mole fraction of each gas times the total pressure of the mixture (Chapter 10).

CONCEPT TEST

The Lewis structures and dipole moments of dimethyl ether and acetone are shown in Figure 11.14. Which of the two compounds would you expect to be present in higher concentration in the vapors of a boiling 50:50 (mol/mol) mixture? Explain your selection.

The fractional distillation process discussed above illustrates how most mixtures of volatile substances behave. This behavior was extensively studied by French chemist François Marie Raoult (1830–1901), who published the following succinct description of it in 1882: *The total vapor pressure of an ideal solution depends on the vapor pressure of each component in the solution and its mole fraction in the liquid mixture.* In equation form, this description is known as **Raoult's law**:

$$P_{\text{total}} = x_1P_1 + x_2P_2 + x_3P_3 + \cdots \quad (11.5)$$

where x_i is the mole fraction of each volatile component of a solution and P_i is the vapor pressure of the pure component. Each term in the series represents the contribution each volatile component of a solution makes to the total vapor pressure of the solution. Thus, Raoult's law is analogous to Dalton's law of partial pressures, but for volatile liquids.

SAMPLE EXERCISE 11.5 Calculating the Vapor Pressure of a Solution **LO5**

What is the vapor pressure of a solution prepared by dissolving 13 g of heptane (C_7H_{16}) in 87 g of octane (C_8H_{18}) at 25°C? How much higher is the mole ratio of heptane to octane in the vapor above the solution than in the solution itself? The vapor pressures of heptane and octane at 25°C are 31 torr and 11 torr, respectively.

Collect and Organize We are asked to calculate the vapor pressure of a solution of a volatile solute (heptane) dissolved in a volatile solvent (octane) and to predict the composition of the vapor produced by the solution at 25°C. We know the vapor pressures of the pure solute and pure solvent at this temperature. Raoult's law (Equation 11.5) relates the vapor pressure of a mixture of volatile substances to the individual vapor pressures of the components of the mixture.

Analyze To calculate the vapor pressures of the volatile components of a mixture (the x_iP_i terms in Equation 11.5), we need to first calculate their mole fractions in the mixture. The relative concentrations of the vapors trapped above a solution of volatile components should be proportional to the ratios of their vapor pressures—assuming the vapors behave like ideal gases. If they do, then they obey the ideal gas law, $PV = nRT$, and the partial pressure of each is proportional to its concentration (moles per liter) in the gas phase:

$$P \propto \frac{n}{V} \text{ at constant } T$$

The vapor pressure of the mixture should be between that of pure heptane and pure octane—and closer to the octane value because there is more of it. The vapor pressure of heptane is nearly three times that of octane, so the vapor should be enriched by about a factor of 3 in heptane compared to the composition of the liquid.

Solve The number of moles of each component is

$$87 \text{ g } C_8H_{18} \times \frac{1 \text{ mol } C_8H_{18}}{114.23 \text{ g } C_8H_{18}} = 0.762 \text{ mol } C_8H_{18}$$

$$13 \text{ g } C_7H_{16} \times \frac{1 \text{ mol } C_7H_{16}}{100.20 \text{ g } C_7H_{16}} = 0.130 \text{ mol } C_7H_{16}$$

Raoult's law the vapor pressure of a solution is the sum of the vapor pressures of the volatile components of the solution, which are each the product of the vapor pressure of the pure component and its mole fraction in the solution.

ideal solution one that obeys Raoult's law.

The mole fraction of each component in the mixture is

$$x_{\text{octane}} = \frac{0.762 \cancel{\text{mol}}}{(0.762 + 0.130) \cancel{\text{mol}}} = 0.854$$

$$x_{\text{heptane}} = 1 - x_{\text{octane}} = 0.146$$

Using these mole fraction values and the vapor pressures of the two hydrocarbons in Equation 11.5, we have

$$\begin{aligned} P_{\text{total}} &= x_{\text{heptane}}P_{\text{heptane}} + x_{\text{octane}}P_{\text{octane}} \\ &= 0.146(31 \text{ torr}) + 0.854(11 \text{ torr}) \\ &= 4.5 \text{ torr} + 9.4 \text{ torr} = 13.9 \text{ torr} \end{aligned}$$

As discussed above, the heptane/octane concentration ratio in the gas phase should be the same as the ratio of their vapor pressures:

$$\frac{4.5 \text{ torr}}{9.4 \text{ torr}} = 0.48$$

The mole ratio of heptane to octane in the liquid mixture is

$$\frac{0.13 \text{ mol}}{0.76 \text{ mol}} = 0.17$$

Therefore, the vapor phase is enriched in heptane by a factor of

$$\frac{0.48}{0.17} = 2.8$$

Think About It As expected, the vapor pressure of the mixture is between the vapor pressures of the separate components, and the vapors are enriched in the more volatile component, heptane, by about a factor of 3 compared to the liquid.

Practice Exercise Benzene (C_6H_6) is a trace component of gasoline. What is the mole ratio of benzene to octane in the vapor above a solution of 11% benzene and 89% octane by mass at 25°C? The vapor pressures of octane and benzene at 25°C are 11 torr and 95 torr, respectively.

Raoult's law applies to **ideal solutions**. What is an ideal solution? Solutions such as the hydrocarbons in gasoline and crude oil obey Raoult's law when the intermolecular interactions between all the molecules in the mixture have comparable strengths. In the case of a binary mixture in which the more abundant component (in terms of number of moles) is designated the *solvent* and the other component is the *solute*, an ideal solution is one in which the strengths of the solvent–solvent, solute–solute, and solute–solvent interactions are much the same.

Solute–solvent interactions, though, are sometimes stronger than solvent–solvent or solute–solute interactions. When this happens, the *adhesive* forces between solute and solvent molecules are greater than the *cohesive* forces between solute molecules and between solvent molecules. As a result, vaporization of both the solute and solvent is inhibited by their strong attraction for each other in the liquid phase, which produces *negative* deviations from the vapor pressures predicted by Raoult's law, as shown in Figure 11.15(a). Solutions of chloroform and acetone (Figure 11.16a) exhibit this behavior, probably due to the strong dipole–dipole interactions that form between acetone and chloroform molecules.

Positive deviations from Raoult's law (Figure 11.15b) are observed when the molecules of solute/solvent pairs experience *weaker* adhesive interactions than the cohesive forces between solute molecules and between solvent molecules. Solutions of chloroform and ethanol (Figure 11.16b) exhibit this behavior because ethanol

CONNECTION In Chapter 6 we learned that strong *adhesive* forces between the molecules of water and the molecular structure of a capillary coupled with strong *cohesive* forces between molecules of water produce capillary action.

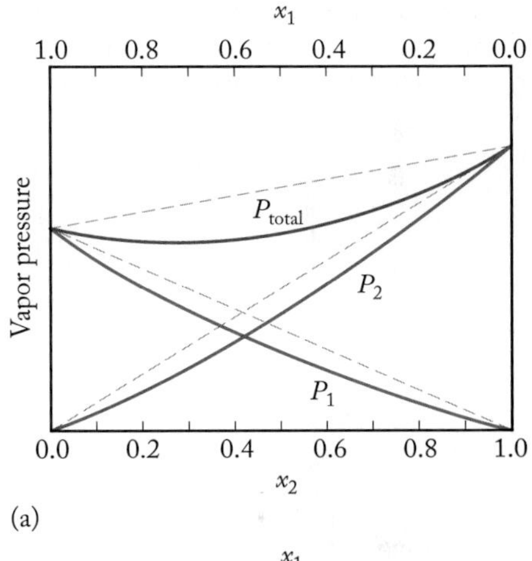

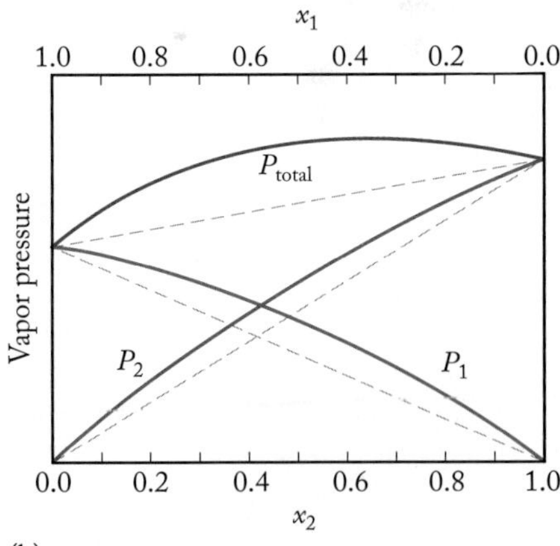

FIGURE 11.15 In a mixture of two volatile substances, the vapor pressures P_1 and P_2 may deviate from the ideal behavior predicted by Raoult's law and described by the dashed lines. (a) If solute–solvent interactions are *stronger* than solvent–solvent or solute–solute interactions, the deviations from Raoult's law are *negative*. (b) If solute–solvent interactions are *weaker* than solvent–solvent or solute–solute interactions, the deviations are *positive*.

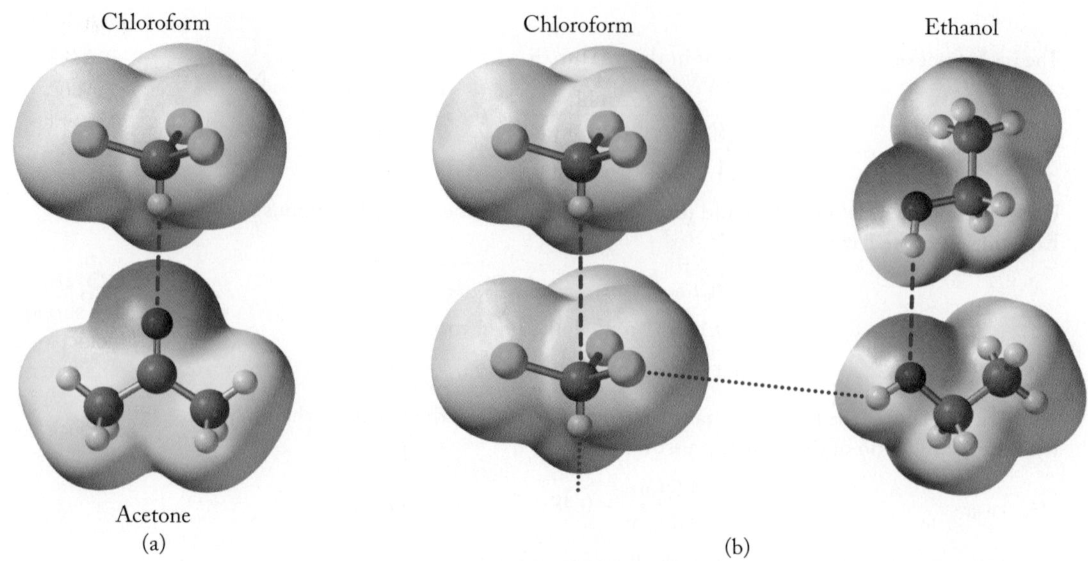

FIGURE 11.16 (a) In a solution of chloroform and acetone, solute–solvent interactions are stronger than solute–solute or solvent–solvent interactions. (b) In a solution of chloroform and ethanol, solute–solvent interactions are weaker (longer, thinner blue dotted line) than solute–solute or solvent–solvent interactions (shorter, thicker blue dashed lines).

molecules hydrogen-bond more strongly with each other than with molecules of $CHCl_3$, thereby reducing the likelihood of strong solute–solvent interactions.

CONCEPT TEST

Which of the following solutions is least likely to follow Raoult's law: (a) acetone and ethanol; (b) pentane and hexane; (c) methanol and water?

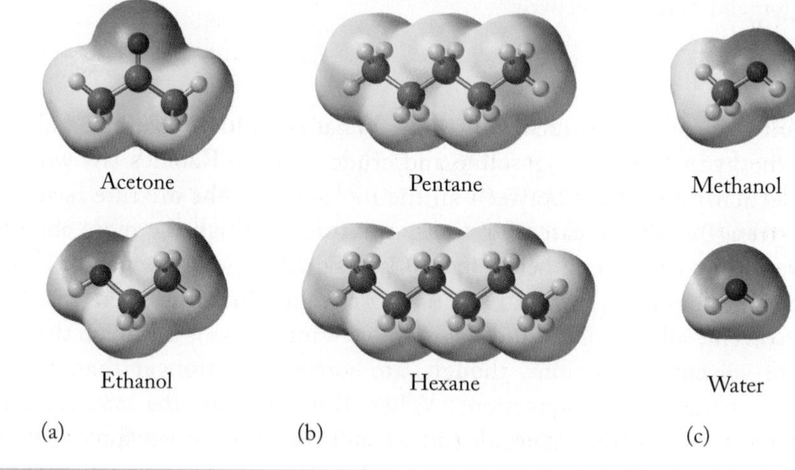

11.5 More Colligative Properties of Solutions

We have seen how Raoult's law lets us predict the vapor pressure of solutions of volatile compounds. In this section we examine the impact of *non*volatile solutes on the vapor pressure (and other properties) of volatile solvents. We start with the solution that covers most of Earth's surface: seawater. The salts in seawater have

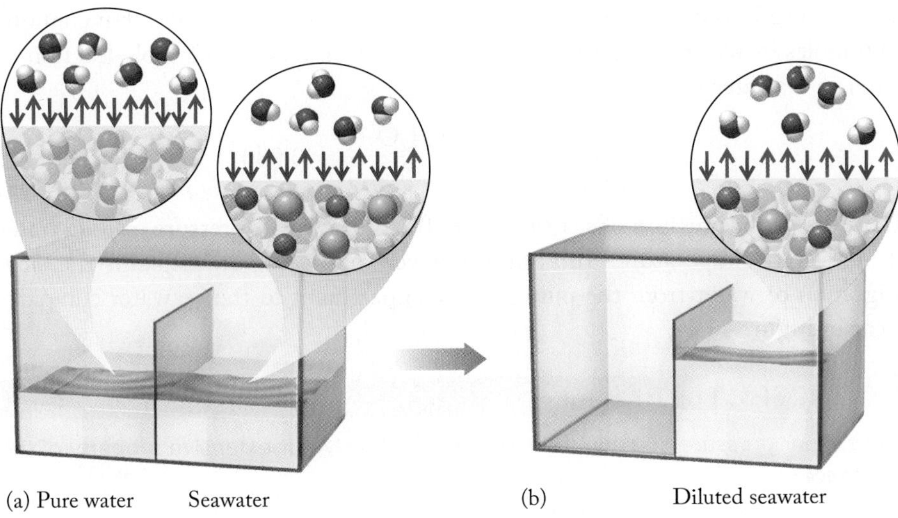

FIGURE 11.17 (a) Adjoining compartments are partially filled with pure water and seawater. (b) The slightly higher vapor pressure of the pure water leads to a net transfer of water from the pure-water compartment to the seawater compartment.

no vapor pressure of their own, but they can alter the volatility of the water they are dissolved in. Consider the images in Figure 11.17(a). The left-side compartment in the figure contains pure water, and the right side initially contains the same volume of seawater. Over time, the water level in the seawater compartment rises as the level in the pure-water compartment drops at a matching rate (assuming the system is well sealed). Eventually, nearly all the water ends up in the seawater compartment (Figure 11.17b).

How (and why) does water move from one compartment to the other? Water in both compartments evaporates, producing water vapor in the headspace above them. Eventually the concentration of water vapor stabilizes, exerting a partial pressure that matches the vapor pressure of water at the temperature inside the chamber. At this point, the overall rate of evaporation from the compartments matches the rate of condensation so that P_{H_2O} remains constant.

Because both compartments share a common headspace, the concentration of water vapor is the same throughout it. As a result, the rate of condensation into both compartments should be the same. If the rates of evaporation were also the same, the two liquid levels would not change over time. However, the levels clearly do change, so the rates of evaporation must not be the same. The rate of evaporation of H_2O from seawater must be lower because water is transported from the pure-water to the seawater compartment. Thus, the presence of solute particles apparently lowers the vapor pressure of the water in seawater. Let's explore why.

Raoult's Law Revisited

The lower vapor pressure of seawater can be understood in the context of Raoult's law. In this case the vapor pressure of a volatile solvent, $P_{solution}$, is related to its mole fraction times the vapor pressure of the pure solvent, $P_{solvent}$, according to the following variation of Raoult's law:

$$P_{solution} = x_{solvent}P_{solvent} \quad (11.6)$$

According to Equation 11.6, the more nonvolatile solute particles there are in a solution, the lower the mole fraction of volatile solvent particles, and the lower the vapor pressure of the solution. In addition, the value of $P_{solution}$ depends only on the number of solute particles and not on their nature. In other words, its lower vapor pressure is a colligative property of the solution.

Let's use Raoult's law to calculate the vapor pressure of seawater that contains 53.6 moles of water and 1.120 moles of ions per kilogram. The mole fraction of water in seawater is

$$x_{H_2O} = \frac{53.6 \cancel{\text{mol}}\ H_2O}{(53.6 + 1.120)\ \cancel{\text{mol}}} = 0.980$$

Thus, the vapor pressure of seawater is 98.0% of the vapor pressure of pure water at any given temperature. This slightly lower vapor pressure explains the slow migration of water from the pure-water compartment to the seawater compartment in Figure 11.17.

CONCEPT **TEST**

Is the vapor pressure of a liquid substance an intensive or extensive property of the substance?

SAMPLE EXERCISE 11.6 Calculating the Vapor Pressure of a Solution of One or More Nonvolatile Solutes **LO5**

Most of the world's supply of maple syrup is produced in Quebec, Canada, where the sap from maple trees is evaporated until the concentration of sugar (mostly sucrose) in the sap reaches at least 66% by weight. What is the vapor pressure in atm of a 66% aqueous solution of sucrose ($\mathcal{M}$ = 342.30 g/mol) at 100°C? Assume that the solution obeys Raoult's law.

Collect and Organize We are asked to calculate the vapor pressure of a 66% solution by weight of sucrose. The vapor pressure of the solution is a colligative property that depends on the mole fraction of solvent (x_{H_2O}):

$$P_{\text{solution}} = x_{H_2O} P_{H_2O}$$

The vapor pressure of pure water (P_{H_2O}) at 100°C (its normal boiling point) is 1.00 atm. Its molar mass is 18.02 g/mol.

Analyze A concentration of 66% means that a 100 g sample of the solution contains 66 g of sucrose and 34 g of water. To calculate the mole fraction of water in a 100 g sample, we need to convert these two masses to moles of sucrose and moles of water and then calculate the mole fraction of water. A mass of 66 g of sucrose ($\mathcal{M}$ = 342 g/mol) represents about 0.2 mole. There are about 2 moles of water in 34 g of water ($\mathcal{M} \approx$ 18 g/mol), so the mole fraction of water is about 2/2.2, or about 0.9, and the vapor pressure should be about 0.9 atm.

Solve In 100 g of syrup there are

$$66 \cancel{\text{g sucrose}} \times \frac{1 \text{ mol sucrose}}{342.30 \cancel{\text{g sucrose}}} = 0.193 \text{ mol sucrose}$$

$$34 \cancel{\text{g water}} \times \frac{1 \text{ mol water}}{18.02 \cancel{\text{g water}}} = 1.89 \text{ mol water}$$

The mole fraction of water is

$$x_{H_2O} = \frac{1.89 \cancel{\text{mol}}}{(1.89 + 0.193) \cancel{\text{mol}}} = 0.91$$

Vapor pressure of the solution is

$$P_{\text{solution}} = x_{H_2O} P_{H_2O} = 0.91 \times 1.00 \text{ atm} = 0.91 \text{ atm}$$

Think About It The calculated value is close to the estimated one and less than the vapor pressure of pure water, as we expect in a concentrated solution. Because its vapor pressure is less than atmospheric pressure, maple syrup does not boil at 100°C. Its boiling point at $P = 1.00$ atm, like those of other aqueous solutions containing nonvolatile solutes, is higher than 100°C.

Practice Exercise The liquid used in automobile cooling systems is prepared by dissolving ethylene glycol ($HOCH_2CH_2OH$, $\mathcal{M} = 62.07$ g/mol) in water. What is the vapor pressure at 50°C of a solution prepared by mixing equal volumes of ethylene glycol (density 1.114 g/mL) and water (density 1.000 g/mL)? Express your answer to the nearest torr. The vapor pressure of pure water at 50°C is 92 torr. The vapor pressure of ethylene glycol is less than 1 torr at 50°C.

boiling point elevation the decrease in the vapor pressure of the solution relative to that of pure solvent at all temperatures corresponds to an increase in the boiling point of the solution.

The impact of solute particles on the properties of a concentrated aqueous solution can be seen in the phase diagram in Figure 11.18. The orange line represents the boiling points of pure water at different pressures and temperatures. Keep in mind that the pressure values along this line, which represent a series of boiling points, are actually vapor pressure values at different temperatures. This is because water boils when its vapor pressure reaches ambient pressure. The more intense red line in Figure 11.18 represents vapor pressures of a concentrated aqueous solution, such as antifreeze, over the same range of temperatures. Notice that the vapor pressure of the solution is lower than that of pure water at all temperatures and that the solution reaches a vapor pressure of 1 atm at a temperature well above 100°C. This phenomenon is called **boiling point elevation.** The difference between the normal boiling point and the boiling point of the solution is labeled ΔT_b and is proportional to the total concentration of all the solute particles in the solution—it is, in fact, a colligative property.

At its lower end, the orange line intersects a faint blue line separating solid ice and pure liquid water (representing melting points) and a faint green line separating ice and steam (representing sublimation points). This point of intersection is the triple point of water. There is an analogous triple point for the solution, and

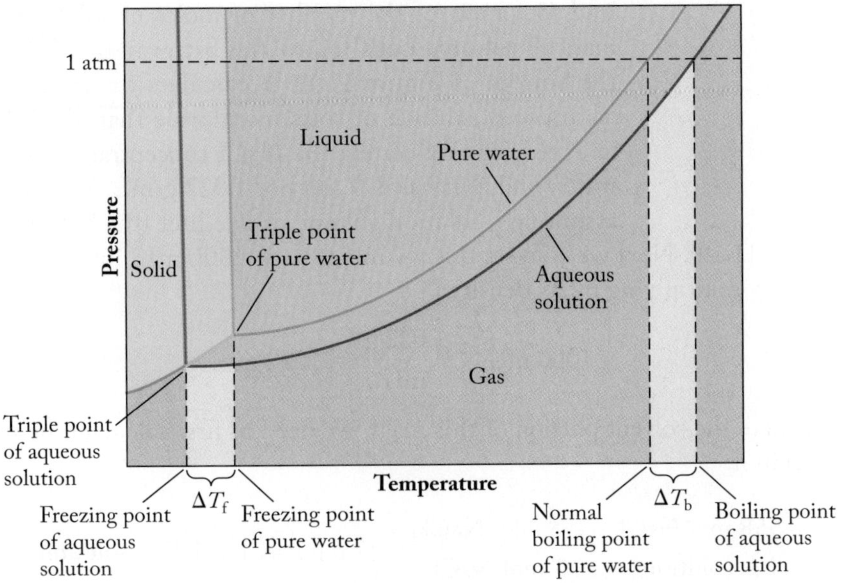

FIGURE 11.18 Combined phase diagram for pure water and a solution of a nonvolatile solute in water. The solution's boiling point is higher than the boiling point of the water, and the solution's freezing point is lower than the freezing point of the water.

freezing point depression solutions freeze at lower temperatures than their pure solvents.

molality (*m*) concentration expressed as the number of moles of solute per kilogram of solvent.

CONNECTION We introduced phase diagrams in Chapter 6 to show how the physical states of substances change with changing temperatures and pressures.

there are analogous lines for melting point (blue) and sublimation point (green). The blue lines are separated because solutions freeze at lower temperatures than their pure solvents—a phenomenon called **freezing point depression.** The symbol ΔT_f represents the difference in freezing points. As with boiling point elevation, ΔT_f is a colligative property; it is proportional to the total concentration of all the solute particles in solution. Before we use this proportionality to calculate the values of ΔT_b and ΔT_f, or to measure these temperature differences and use them to determine the concentrations of solutions, we need to become acquainted with the concentration unit known as *molality*.

Molality

Molarity, or moles of solute per liter of solution, is probably the most widely used unit of concentration in chemistry. Molality is closely related to molarity. It, too, is based on the number of moles of solute in solution, but instead of moles per liter of solution, **molality (*m*)** is based on moles of solute *per kilogram of solvent*:

$$m = \frac{n_{\text{solute}}}{\text{kg solvent}} \quad (11.7)$$

Molality, not molarity, is used to express solute concentrations when calculating boiling point elevation or freezing point depression because liquids, like all forms of matter, tend to expand as their temperatures increase and contract as their temperatures decrease. Exactly 1 kilogram of water has a volume of 1.0000 L at 4.0°C, but it expands to occupy 1.0434 L when heated to 100°C. This increase in volume means that the molarity of an aqueous solution is about 4% lower near its boiling point than it is at 4.0°C. However, the mole fractions of solute and solvent (which define shifts in vapor pressure, boiling point, and freezing point) remain unchanged. Molalities, on the other hand, do not change with temperature because the mass of solvent in which a given quantity of solute has dissolved does not change with temperature.

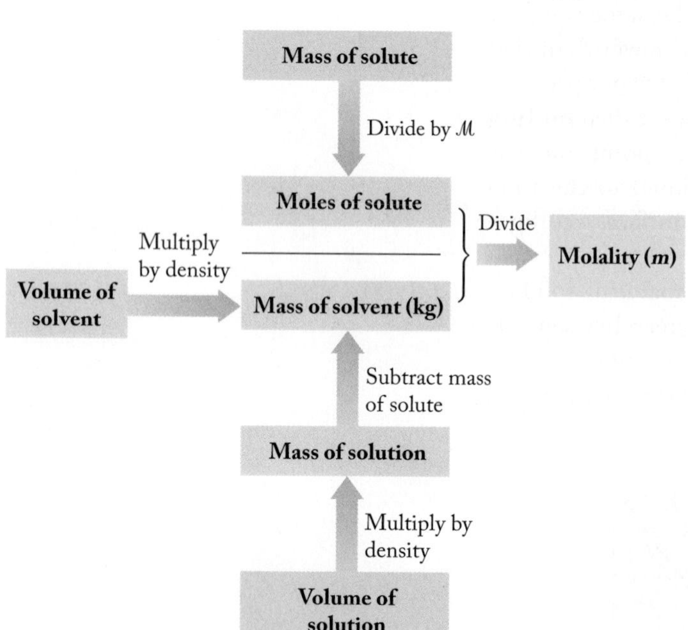

FIGURE 11.19 Flow diagram for calculating molality.

Figure 11.19 summarizes pathways that can be followed when calculating concentrations in molality. All lead to a ratio of the number of moles of solute to the mass of solvent. Let's follow the pathway that starts at the bottom of Figure 11.19 to calculate the molality of an aqueous solution of sodium chloride that is 0.558 *M* in NaCl (approximately the NaCl concentration in seawater) and that has a density of 1.022 g/mL. We start by assuming a solution volume of one liter (the bottom box in Figure 11.19). Next we convert this volume (1 L = 1000 mL) into a mass of solution by multiplying by its density:

$$1000 \text{ mL} \times \frac{1.022 \text{ g}}{\text{mL}} = 1022 \text{ g}$$

To calculate the solvent portion of this mass, we need to first calculate the mass of NaCl in it:

$$\frac{0.558 \text{ mol NaCl}}{1 \text{ L solution}} \times \frac{58.44 \text{ g NaCl}}{1 \text{ mol NaCl}} = 32.6 \text{ g NaCl in 1 L of solution}$$

We subtract this mass of solute from the total mass of solution (1022 g) to obtain the mass of solvent: (1022 g − 32.6 g) = 989 g = 0.989 kg. The molality of the solution (number of moles of solute per kilogram of solvent) is then

$$\frac{0.558 \text{ mol NaCl}}{0.989 \text{ kg solvent}} = 0.564\ m \text{ NaCl}$$

The molarity (0.558 *M*) and molality (0.564 *m*) are nearly the same, but the molality is greater than the molarity, which is true for all aqueous solutions.

CONCEPT TEST

The difference between the molar concentration and molal concentration of any dilute aqueous solution is small. Why?

SAMPLE EXERCISE 11.7 Calculating the Molality of a Solution LO6

A popular recipe[1] for preparing corned beef (a St. Patrick's Day favorite in many places) calls for preparing a seasoned brine (salt solution) that contains 0.50 pound of kosher salt (NaCl) dissolved in 3.0 quarts of water. Assuming the density of water is 1.00 g/mL, what is the molality of NaCl in this solution for "corning" beef?

Collect and Organize We are asked to calculate the molality of a salt solution prepared from a known mass of salt and volume of water. Molality is moles of solute per kilogram of solvent. We know the density of water is 1.00 g/mL, which is the same as 1.00 kg/L; the latter value should be more useful in this calculation given the volume of the brine. The table of conversion factors at the back of this book includes 1 pound = 453.59 g and 1 gallon = 3.785 L. There are four quarts in one gallon.

Analyze We need to convert 3.0 quarts of water into an equivalent mass in kilograms. We also need to calculate the number of moles of NaCl in 0.50 pound. Its molar mass is 58.44 g/mol. Half a pound of NaCl contains (453.59 g/2)/(58.44 g) moles, or about 4 mol. Three quarts of water is a little less than three liters, so its mass is a little less than 3 kg. Therefore, the results of our molality calculation should be about 4 mol/3 kg, or 1.3 *m*.

Solve The mass of water in the brine is

$$3.0 \cancel{\text{qt}} \times \frac{1 \cancel{\text{gal}}}{4 \cancel{\text{qt}}} \times \frac{3.785 \cancel{\text{L}}}{1 \cancel{\text{gal}}} \times \frac{1.00 \text{ kg}}{1 \cancel{\text{L}}} = 2.84 \text{ kg}$$

The number of moles of NaCl is

$$0.50 \cancel{\text{lb NaCl}} \times \frac{453.59 \text{ g}}{1 \cancel{\text{lb}}} \times \frac{1 \text{ mol NaCl}}{58.44 \cancel{\text{g NaCl}}} = 3.88 \text{ mol NaCl}$$

The molality of the brine is

$$\frac{3.88 \text{ mol NaCl}}{2.84 \text{ kg}} = 1.4\ m \text{ NaCl}$$

Think About It The calculated molality is consistent with our prediction. It is rounded off to only two significant figures, which may seem stingy, but it's appropriate given the low number of significant figures in the amounts of ingredients in the recipe. Actually, the quantities in this and most recipes are typically expressed with only one significant figure.

Practice Exercise What is the molality of a solution prepared by dissolving 78.2 g of ethylene glycol, $HOCH_2CH_2OH$, in 1.50 L of water? Assume the density of water is 1.00 g/mL.

[1]A. Brown, *Good Eats 3*, 2011, p. 32.

Boiling Point Elevation

Boiling point elevation is a colligative property, so it does not depend on the identity of the solute particles only on their number. However, each solvent has its own sensitivity to the presence of solute particles among its own molecules, so each solvent has its own *boiling-point-elevation constant*, K_b. This constant is used to relate boiling point elevation (ΔT_b) to the concentration of solute particles expressed in molality, or $i \cdot m$, where i is the van 't Hoff factor of the solute in a solution in which its molal concentration is m:

$$\Delta T_b = K_b\, i\, m \tag{11.8}$$

The K_b of water is 0.52°C/mol solute particles/kg H_2O, which means that dissolving one mole of solute particles in 1 kg of water ($i\,m = 1$) increases the boiling point of the water by 0.52°C.

SAMPLE EXERCISE 11.8 Calculating the Boiling Point Elevation of an Aqueous Solution **LO7**

What is the boiling point of seawater in which the concentration of sea salts is equivalent to 0.560 mol NaCl/kg seawater, and the van 't Hoff factor of NaCl is 1.85?

Collect and Organize We are asked to calculate the boiling point of 0.560 mol/kg NaCl. The K_b of water is 0.52°C/m, and its normal boiling point is 100.0°C. Boiling point elevation can be calculated using Equation 11.8: $\Delta T_b = K_b\, i\, m$.

Analyze To use Equation 11.8, we need to express the salt concentration in moles of dissolved particles per kilogram of *solvent*. The value we are given is in moles of NaCl per kilogram of *solution* (seawater). Therefore, we need to calculate the mass of H_2O in one kilogram of seawater by subtracting the mass of NaCl, and then use the mass of H_2O to calculate molality. Assuming the molal concentration we calculate is similar to the mol/kg concentration we started with, the value of ($i\,m$) will be about 2 × 0.56 m or 1.12 m. This concentration will elevate the boiling point of water by (1.12 × 0.52)°C, or about 0.6°C.

Solve Mass of H_2O in one kilogram of seawater:

$$\left(1.000\ \cancel{\text{kg}}\ \text{solution} \times \frac{1000\ \text{g}}{1\ \cancel{\text{kg}}}\right) - \left(0.560\ \cancel{\text{mol NaCl}} \times \frac{58.44\ \text{g NaCl}}{1\ \cancel{\text{mol NaCl}}}\right)$$
$$= 967\ \text{g} = 0.967\ \text{kg}$$

Concentration of NaCl:

$$\frac{0.560\ \text{mol NaCl}}{0.967\ \text{kg}\ H_2O} = 0.579\ m\ \text{NaCl}$$

Boiling point elevation:

$$\Delta T_b = K_b\, i\, m = 0.52\frac{°\text{C}}{\cancel{m}} \times 1.85 \times 0.579\ \cancel{m} = 0.56°\text{C}$$

To calculate the boiling point of seawater, we add ΔT_b to the boiling point of pure water

$$T_{b,\text{seawater}} = T_{b,\text{solvent}} + \Delta T_b = (100.0 + 0.56)°\text{C} = 100.6°\text{C}$$

Think About It As predicted, the boiling point of seawater is about 0.6°C higher than the boiling point of pure water. The change seems small given the saltiness of seawater, but consider that the value of K_b for water is itself small.

Practice Exercise Crude oil pumped out of the ground may be accompanied by saline *formation water*. If the boiling point of a sample of formation water is 2.3°C above the normal boiling point of pure water, what is the molality of dissolved particles in the sample?

Freezing Point Depression

Freezing point depression is another colligative property that is directly proportional to the molal concentration of dissolved solute particles:

$$\Delta T_f = K_f \, i \, m \qquad (11.9)$$

Here, ΔT_f is the change in the freezing temperature of the solvent, K_f is the *freezing-point-depression constant* of the solvent, and i is the van 't Hoff factor of the solute in a solution in which its molal concentration is m.

SAMPLE EXERCISE 11.9 Calculating the Freezing Point of a Solution **LO7**

What is the freezing point of automobile radiator fluid prepared by mixing equal volumes of ethylene glycol ($\mathcal{M}$ = 62.07 g/mol) and water at a temperature where the density of ethylene glycol is 1.114 g/mL and the density of water is 1.000 g/mL? The freezing-point-depression constant of water, K_f, is 1.86°C/m.

Collect and Organize We are asked to determine the freezing point of a 50:50 (by volume) solution of ethylene glycol in water. We are given the densities of the two liquids and the value of K_f, which relates the molality of ethylene glycol to freezing point depression (Equation 11.9). Ethylene glycol is a nonelectrolyte, so its i value is exactly 1.

Analyze To use Equation 11.9, we need to convert the 50:50 volume ratio into a solute concentration expressed in moles of ethylene glycol per kilogram of water. To simplify the calculation, let's assume we mix 1.000 liter of each of the two liquids. This is a handy volume because one liter of water has a mass of one kilogram (because a density of 1.000 g/mL is the same as 1.000 kg/L). The mass of one liter of ethylene glycol is 1114 g. It contains 1114/62.07 or about 1200/60 = 20 moles of ethylene glycol. The freezing point depression of a 20 m aqueous solution of ethylene glycol is (1.86 × 1 × 20)°C, or nearly 40°C. Therefore, the freezing point of the radiator fluid should be a bit above −40°C.

Solve The solvent mass is

$$1.000 \cancel{\text{L}}\ \text{H}_2\text{O} \times \frac{1000 \cancel{\text{mL}}}{1 \cancel{\text{L}}} \times \frac{1.000 \cancel{\text{g}}}{\cancel{\text{mL}}} \times \frac{0.001 \text{ kg}}{1 \cancel{\text{g}}} = 1.000 \text{ kg H}_2\text{O}$$

Moles of ethylene glycol in 1.000 L:

$$1.000 \cancel{\text{L ethylene glycol}} \times \frac{1000 \cancel{\text{mL}}}{1 \cancel{\text{L}}} \times \frac{1.114 \text{ g}}{1 \cancel{\text{mL}}} \times \frac{1 \text{ mol ethylene glycol}}{62.07 \cancel{\text{g ethylene glycol}}}$$

$$= 17.95 \text{ mol ethylene glycol}$$

Concentration of ethylene glycol:

$$m = \frac{17.95 \text{ mol ethylene glycol}}{1.000 \text{ kg H}_2\text{O}} = 17.95\ m$$

Using Equation 11.9 gives us

$$\Delta T_f = K_f \, i \, m = 1.86 \frac{°\text{C}}{\cancel{m}} \times 1 \times 17.95 \cancel{m} = 33.4°\text{C}$$

To calculate the freezing point of the solution we subtract ΔT_f from the freezing point of the pure solvent:

$$\text{Freezing point of radiator fluid} = T_{f,\text{solvent}} - \Delta T_f = 0.0°C - 33.4°C = -33.4°C$$

Think About It The answer is about what we expected. The freezing point of radiator fluid should be below the coldest expected temperatures, and −33°C (−27°F) is colder than most places in the United States ever get. In much of Alaska and Canada, however, 60–70% by volume solutions of ethylene glycol would be advisable to prevent engines from freezing up.

Practice Exercise Some Thanksgiving dinner chefs "brine" their turkeys prior to roasting them to help the birds retain moisture and to season the meat. Brining involves completely immersing a turkey for about 6 hours in a brine (salt solution) prepared by dissolving 2.0 pounds of salt (NaCl, $\mathcal{M}$ = 58.44 g/mol) in 2.0 gallons of water (a recipe from an American cookbook). Suppose a turkey soaking in such a solution is left for 6 hours on an unheated porch. At what temperature in °C is the brine in danger of freezing? The K_f of water is 1.86°C/m; assume that i = 1.82 for this NaCl solution.

SAMPLE EXERCISE 11.10 Assessing Particle Interactions in Solution **LO2**

The experimentally measured freezing point of a 1.90 m aqueous solution of NaCl is −6.57°C. What is the value of the van 't Hoff factor for this solution? Is the solution behaving ideally, or is there evidence that solute particles are interacting with one another? The freezing-point-depression constant of water is K_f = 1.86°C/m. Assume the freezing point of pure water is 0.00°C.

Collect and Organize We are asked to calculate the i value of a solution of known molality based on a measured freezing point and the K_f value. Equation 11.8 relates ΔT_f, K_f, the van 't Hoff factor (i) for this solution, and its molality (m). We are also asked whether the solution behaves ideally.

Analyze Ideally, i = 2 for solutions of NaCl. However, given the high concentration of NaCl (1.90 m), a value less than 2 is likely.

Solve Rearranging Equation 11.9 and solving for i:

$$i = \Delta T_f / K_f m = 6.57\cancel{°C}/(1.86 \frac{\cancel{°C}}{\cancel{m}} \times 1.90\ \cancel{m}) = 1.86$$

The value of i is less than 2 (the theoretical value for NaCl), so the solution is not behaving ideally. Ion pairs must be forming in solution.

Think About It As predicted, the value for i for this solution is less than the theoretical value of 2.

Practice Exercise The freezing point of a 1.12 M solution of $MgSO_4$ is 2.31°C below the freezing point of pure water. If the density of the solution is 1.126 g/ml what is the value of i?

11.6 Henry's Law and the Solubility of Gases

Henry's law the concentration of a sparingly soluble gas in a liquid is proportional to the partial pressure of the gas.

The last class of solutions we investigate in this chapter is the kind in which gases dissolve in liquids. Recall from Chapter 6 that relatively weak dipole–induced dipole forces between polar water molecules and nonpolar oxygen molecules lead to a concentration of dissolved oxygen sufficient to sustain aquatic plants and fish. The solubility of O_2 (or any sparingly soluble gas) in a liquid such as water is directly proportional to the partial pressure of the gas above the surface of the liquid. This relationship is known as **Henry's law**, in honor of William Henry (1775–1836), a British physician who first proposed the relationship. In equation form the relationship is:

$$C_{gas} = k_H P_{gas} \tag{11.10}$$

where C_{gas} represents the maximum concentration (solubility) of a gas in a particular solvent, k_H is the Henry's law constant for the gas in that solvent, and P_{gas} is the partial pressure of the gas in the environment surrounding the solvent. When C_{gas} is expressed in molarity (mol/L), the units of the Henry's law constant are moles per liter-atmosphere, mol/(L · atm). Table 11.1 lists k_H values for several common gases in water. The magnitude of k_H reflects both differences in the intermolecular forces in the dissolved gases and deviations from ideal gas behavior. For example, small, weakly polarizable noble gas atoms such as He have smaller Henry's law constants than larger, more polarizable molecules like O_2 or CH_4. Polar molecules like H_2S and CH_3Cl have even larger values of k_H.

According to Henry's law, the concentration of dissolved oxygen in blood is proportional to the partial pressure of oxygen in the air we inhale and thus proportional to atmospheric pressure. This is an accurate statement of Henry's law, but it does not tell the whole story. It does not mean, for example, that residents of Denver, Colorado (average P_{O_2} = 0.178 atm) live with less blood oxygen than residents of New York City or anywhere else near sea level (where P_{O_2} averages 0.209 atm).

The concentration of O_2 in the blood also depends on the concentration of hemoglobin, an oxygen-transporting protein that attaches to molecules of O_2 as blood passes through the lungs. The hemoglobin binding sites in most people can be saturated with O_2 even when the P_{O_2} in their lungs is as low as 0.110 atm. If P_{O_2} decreases to about 0.066 atm (as it does on high mountains), only about 80% of the hemoglobin molecules pick up molecules of O_2 passing through the lungs. Over several weeks, the body responds to low P_{O_2} by producing more hemoglobin so that the same amount of O_2 is delivered to tissues. Some endurance athletes choose to train at high altitudes so that, when they compete at lower altitudes, their bodies transport O_2 more efficiently.

The solubility of gases in water and other liquids decreases as the temperature of the liquid increases. Opening a can of warm soda leads to an explosive release of CO_2 (Figure 11.20) because the solubility of gas is lower in warm soda than in cold, but the increase in pressure caused by an increased amount in gas in the can forces more gas into solution than would normally dissolve at the higher temperature. When the can is opened, the pressure suddenly drops and the CO_2 gas escapes from the fluid, resulting in bubbles and fizz. The graph in Figure 11.21 shows that the solubility of O_2 in water decreases from 0°C to 50°C. The k_H valuc for O_2, which decreases as water temperature increases (Table 11.2), is consistent with this trend, too.

TABLE 11.1 Henry's Law Constants for Several Gases in Water at 20°C

Gas	k_H [mol/(L · atm)]
He	3.5×10^{-4}
O_2	1.3×10^{-3}
N_2	6.7×10^{-4}
CO_2	3.5×10^{-2}
H_2S	8.7×10^{-2}
CH_4	1.5×10^{-3}
CH_3Cl	1.3×10^{-1}

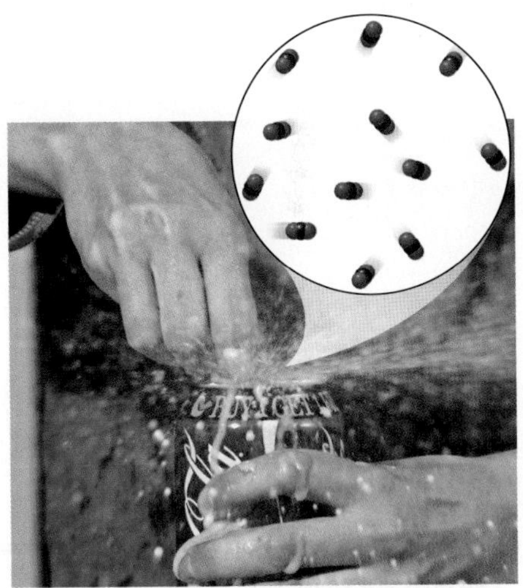

FIGURE 11.20 Opening a warm can of soda can be risky because dissolved CO_2 gas escapes from the fluid as the pressure is released.

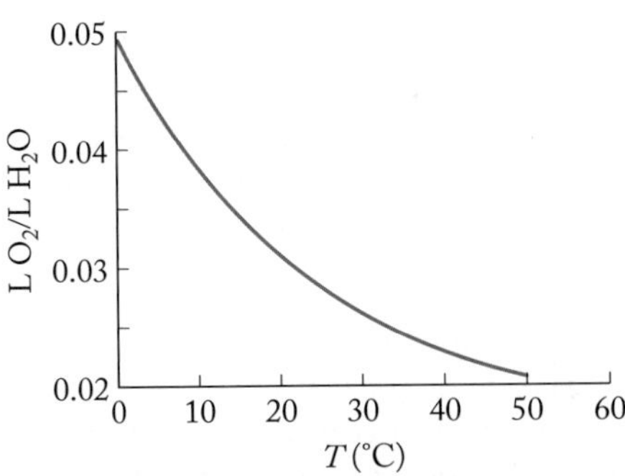

FIGURE 11.21 The solubility of O_2 in water decreases as the temperature increases (data are based on P_{O_2} = 1.00 atm).

SAMPLE EXERCISE 11.11 Calculating Gas Solubility Using Henry's Law **LO8**

Lake Titicaca is located high in the Andes Mountains between Peru and Bolivia. Its surface is 3811 m above sea level, where the average atmospheric pressure is 0.636 atm. During the summer, the average temperature of the water's surface rarely exceeds 15°C. What is the solubility of oxygen in Lake Titicaca at that temperature? Express your answer in molarity and in mg/L.

Collect and Organize We are asked to determine the solubility (the maximum concentration) of oxygen gas in water at 0.636 atm and 15°C. According to Table 10.1, dry air is 20.95% O_2. Henry's law (Equation 11.10) relates the solubility of gases to their

TABLE 11.2 Henry's Law Constants for O_2 Gas in Water

Temperature (°C)	k_H [mol/(L · atm)]
0	2.18×10^{-3}
5	1.90×10^{-3}
10	1.68×10^{-3}
15	1.50×10^{-3}
20	1.35×10^{-3}
25	1.23×10^{-3}
30	1.12×10^{-3}
35	1.04×10^{-3}
40	0.96×10^{-3}
45	0.90×10^{-3}
50	0.84×10^{-3}

partial pressures. The Henry's law constants for O_2 in water at different temperatures are listed in Table 11.2.

Analyze We need the partial pressure of oxygen to calculate its molar solubility using Equation 11.10. According to Equation 10.25, partial pressure is the product of the mole fraction of O_2 in air (20.95% expressed as a decimal value) and atmospheric pressure (0.636 atm). To express solubility in mg/L will require converting moles to grams by multiplying by the molar mass of O_2. To estimate our answer, we note that all the Henry's law constants for O_2 in water are about 10^{-3} mol/(L · atm), and the value of P_{O_2} is about 0.2×0.6, or 0.12 atm, so the product of $k_H \times P_{O_2}$ (the solubility according to Equation 11.10) will be around 10^{-4} mol/L. Multiplying this value by the molar mass of O_2 (32 g/mol) should give us a solubility near 3×10^{-3} g/L, or 3 mg/L.

Solve The partial pressure of oxygen is calculated using Equation 10.25:

$$P_{O_2} = x_{O_2}P_{total} = (0.2095)(0.636 \text{ atm}) = 0.133 \text{ atm}$$

We use this value for P_{O_2} in Equation 11.10 and the k_H for O_2 in water at 15°C from Table 11.2 to calculate the solubility (C_{O_2}):

$$C_{O_2} = k_H P_{O_2} = \left(1.50 \times 10^{-3} \frac{\text{mol}}{\text{L} \cdot \cancel{\text{atm}}}\right)(0.133 \cancel{\text{atm}}) = 2.00 \times 10^{-4} \text{ mol/L}$$

$$(2.00 \times 10^{-4} \cancel{\text{mol}}/\text{L})(32.00 \text{ g}/\cancel{\text{mol}}) = 6.40 \times 10^{-3} \text{ g/L} = 6.40 \text{ mg/L}$$

Think About It The calculated solubility values in mol/L and mg/L are in the same ballpark as those we predicted and illustrate the limited solubility of O_2 in water. Very little O_2 dissolves because it is a nonpolar solute and water is a polar solvent. As a result, the solute–solvent intermolecular forces that promote solubility are weak in this case.

Practice Exercise If the pressure of CO_2 inside a 1-liter bottle of seltzer is 44 psi at 20°C, how much CO_2 is dissolved in the seltzer? Express your answer in grams.

SAMPLE EXERCISE 11.12 Integrating Concepts: Antifreeze in Car Batteries

The aqueous electrolyte inside a fully charged lead–acid battery (the kind used to start automobile engines) is 35% (by volume) sulfuric acid, H_2SO_4. Pure sulfuric acid is a dense (1.84 g/mL), nonvolatile liquid (its vapor pressure at 25°C is <0.001 torr). It is also a strong acid that is completely ionized in concentrated aqueous solutions:

$$H_2SO_4(aq) \rightarrow H^+(aq) + HSO_4^-(aq)$$

The power to start the engine comes from the following chemical reaction as the battery is discharged:

$$Pb(s) + PbO_2(s) + 2\,H_2SO_4(aq) \rightarrow 2\,PbSO_4(s) + 2\,H_2O(\ell)$$

a. What is the vapor pressure of the solution at 25°C?
b. What is the freezing point of the solution?
c. How do vapor pressure and freezing point change as the battery is discharged?

Collect and Organize We know the concentration of a solution expressed as the volume ratio of a nonvolatile liquid solute (H_2SO_4, $\mathcal{M}$ = 98.08 g/mol) to solvent (H_2O). We are asked to calculate the freezing point of the solution and its vapor pressure at 25°C. We also need to describe how these freezing point and vapor pressure values change during the course of a reaction in which H_2SO_4 is a reactant and H_2O is a product. The vapor pressure of such a solution can be calculated using Equation 11.6: $P_{solution} = x_{solvent}P_{solvent}$. Its freezing point can be calculated using Equation 11.9: $\Delta T_f = K_f\, i\, m$. The vapor pressure of pure water at 25°C is 23.8 torr, and its K_f value is 1.86°C/m.

Analyze Solutions of nonvolatile solutes in volatile solvents have lower vapor pressures and freezing points than their pure solvents. To find out how much lower our solution's values are, we must first calculate the mole fraction of the solvent in solution (needed in Equation 11.6) and the molal concentration of solute (needed in Equation 11.9). As the battery is discharged, the concentration of H_2SO_4 decreases. Assume the H_2SO_4 completely dissociates into H^+ and HSO_4^- ions, so the van 't Hoff factor of sulfuric acid is 2.

Solve

a. If we assume a sample volume of 100 mL, then 35% H_2SO_4 consists of 35 mL H_2SO_4 and (100 − 35) = 65 mL H_2O. Converting these volumes into masses and then moles,

$$35 \cancel{\text{mL H}_2\text{SO}_4} \times 1.84 \frac{\cancel{\text{g}}}{\cancel{\text{mL}}} \times \frac{1 \cancel{\text{mol H}_2\text{SO}_4}}{98.08 \cancel{\text{g H}_2\text{SO}_4}} \times \frac{2 \text{ mol ions}}{1 \cancel{\text{mol H}_2\text{SO}_4}}$$
$$= 1.31 \text{ mol ions}$$

$$65\ \cancel{\text{mL H}_2\text{O}} \times \frac{1.00\ \text{g}}{\cancel{\text{mL}}} \times \frac{1\ \text{mol H}_2\text{O}}{18.02\ \cancel{\text{g H}_2\text{O}}}$$

$$= 3.61\ \text{mol H}_2\text{O}$$

The mole fraction of H_2O is then

$$3.61\ \cancel{\text{mol}}\ \text{H}_2\text{O}/(3.61 + 1.31)\ \cancel{\text{mol}} = 0.734$$

And the vapor pressure of the battery acid is

$$P_{\text{solution}} = x_{\text{solvent}}P_{\text{solvent}} = 0.734 \times 23.8\ \text{torr} = 17.5\ \text{torr}$$

b. To calculate freezing point depression, we first need to calculate the molality of solute particles:

$$\frac{1.31\ \text{mol ions}}{65\ \cancel{\text{mL}}\ \text{H}_2\text{O} \times 1.00\ \text{g}/\cancel{\text{mL}}} \times \frac{1000\ \text{g}}{\text{kg}} = 20.2\ \text{mol/kg} = 20.2\ m$$

Calculating freezing point depression (ΔT_f):

$$\Delta T_f = K_f m = 1.86°\text{C}/m \times 20.2\ m = 38°\text{C}$$

Therefore, the freezing point of battery acid is −38°C.

c. The chemical reaction that provides the battery's energy consumes solute, converting it into more solvent. The result is a less concentrated solution of sulfuric acid and a greater mole fraction of H_2O. Such a solution has a higher freezing point (closer to that of pure water) and a greater vapor pressure (again, closer to that of pure water).

Think About It We did not include the van 't Hoff factor of 2 when using Equation 11.9 in part b because it had already been used to calculate the number of moles of solute particles in part a. Thus, the m value we used in part b was actually the product of i and m.

SUMMARY

LO1 In **osmosis**, solvent flows through a semipermeable membrane from a solution of lower solute concentration into a solution of higher solute concentration. **Osmotic pressure (*Π*)** is defined as the pressure required to halt the net flow of solvent through a semipermeable membrane. It is a **colligative property** of a solution that depends only on the concentration of dissolved particles and not their identities. In **reverse osmosis (RO)** water is pumped through a semipermeable membrane with a pressure that reverses osmotic flow. The technique is used to purify water. (Sections 11.1 and 11.2)

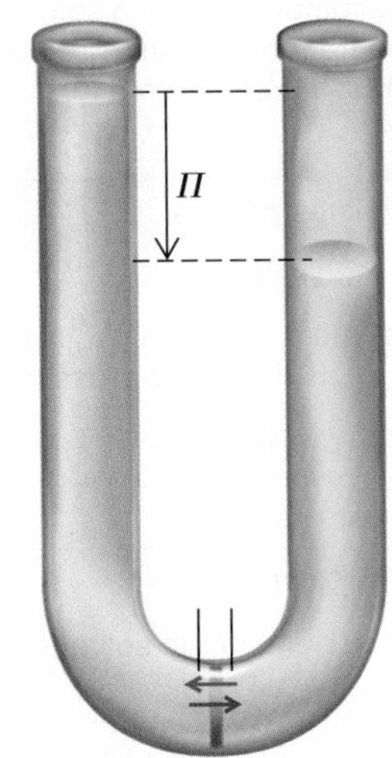

LO2 The **van 't Hoff factor (*i*)** accounts for the colligative properties of solutions of electrolytes, which may form solute clusters such as **ion pairs** that reduce the number of particles in solution. (Section 11.2)

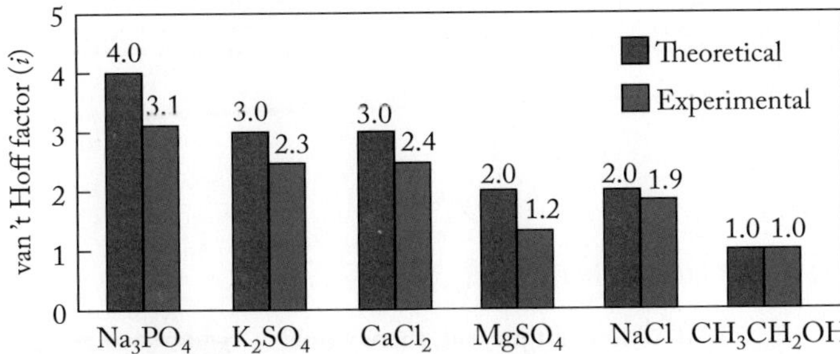

LO3 Colligative properties, such as osmotic pressure, of solutions of nonelectrolytes can be used to determine the molar mass of nonelectrolytes. (Section 11.2)

LO4 Molecules at the surface of a liquid evaporate by breaking intermolecular interactions with neighboring molecules and entering the gas phase. The **vapor pressure** of a **volatile** liquid is proportional to the fraction of its molecules that enter the gas phase and increases with increasing temperature as determined by the **Clausius–Clapeyron equation.** (Section 11.3)

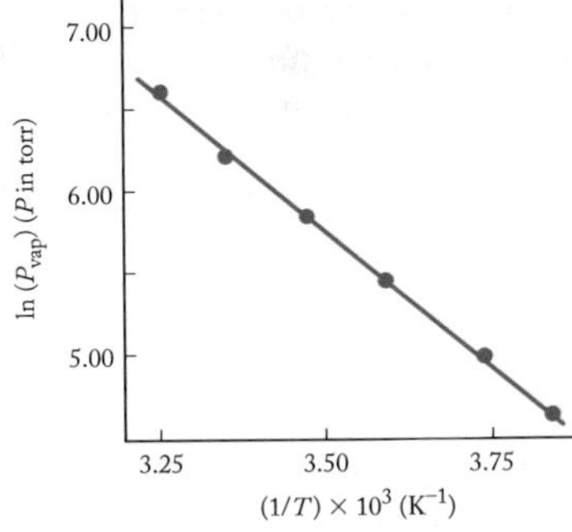

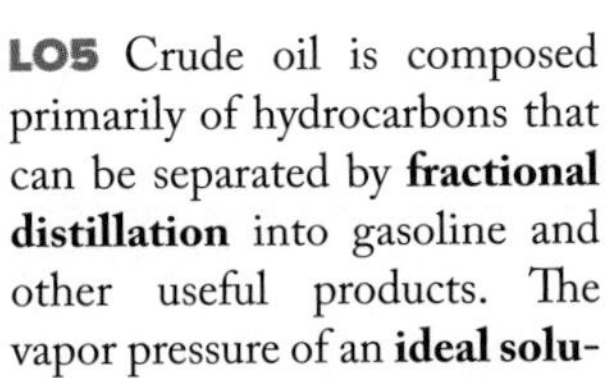

LO5 Crude oil is composed primarily of hydrocarbons that can be separated by **fractional distillation** into gasoline and other useful products. The vapor pressure of an **ideal solution** of volatile compounds follows **Raoult's law.** (Section 11.4)

LO6 The units used for describing concentration include molarity and **molality (*m*)**, which is the number of moles of solute per kilogram of solvent. (Section 11.5)

LO7 Nonvolatile solutes in solution decrease the solvent's vapor pressure, elevate the solvent's boiling point, and depress its freezing point. **Boiling point elevation** and **freezing point depression** are colligative properties. (Section 11.5)

LO8 According to **Henry's law**, the solubilities of gases increase with increasing partial pressure. Solubilities decrease with increasing temperature. (Section 11.6)

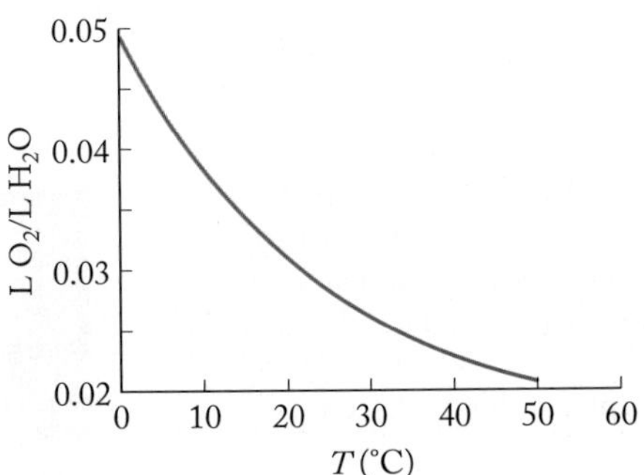

PARTICULATE PREVIEW WRAP-UP

Ion–dipole forces form between the sodium cations and the δ^- oxygen atoms and between the chloride anions and the δ^+ hydrogen atoms.

The amount of vapor is reduced above a solution when compared to pure solvent, and consequently the boiling point increases.

PROBLEM-SOLVING SUMMARY

Type of Problem	Concepts and Equations	Sample Exercises
Calculating osmotic pressure and reverse osmotic pressure	$\Pi = iMRT$ (11.2) where Π is the osmotic pressure, i is the van 't Hoff factor, M is the molar concentration of the solution, and T is the absolute temperature of the solution.	**11.1, 11.2**
Determining molar mass from osmotic pressure	Use $\Pi = iMRT$ to calculate the molarity of a solution containing a known mass of solute.	**11.3**
Calculating vapor pressure (ΔH_{vap}) using the Clausius–Clapeyron equation	$\ln\left(\frac{P_{vap,T_1}}{P_{vap,T_2}}\right) = \frac{\Delta H_{vap}}{R}\left(\frac{1}{T_2} - \frac{1}{T_1}\right)$ (11.4)	**11.4**
Calculating the vapor pressure of a solution of volatile compounds	Obtain the vapor pressures, P_i, of the components of a solution as pure compounds and determine their mole fractions (x_i) in the solution. Use Raoult's law to calculate the vapor pressure of the solution: $P_{total} = x_1P_1 + x_2P_2 + x_3P_3 + \ldots$ (11.5)	**11.5**
Calculating the vapor pressure of a solution of one or more nonvolatile solutes	Use the form of Raoult's law for nonvolatile solutes: $P_{solution} = x_{solvent}P°_{solvent}$ (11.6) where $P_{solution}$ is the vapor pressure of the solution, $x_{solvent}$ is the mole fraction of the solvent in the solution, and $P_{solvent}$ is the vapor pressure of the pure solvent.	**11.6**
Calculating the molality of a solution	$m = \frac{n_{solute}}{\text{kg solvent}}$ (11.7)	**11.7**
Calculating boiling points and freezing points of solutions	$\Delta T_b = K_b i\, m$ (11.8) where ΔT_b is the elevation in the boiling point of the solvent and $\Delta T_f = K_f i\, m$ (11.9) where ΔT_f is the depression in the freezing point of the solvent.	**11.8, 11.9**
Assessing ion clustering in electrolyte solutions	Use measured freezing point depression or boiling point elevation values and Equations 11.8 or 11.9 to determine the value of the van 't Hoff factor.	**11.10**
Calculating gas solubility using Henry's law	$C_{gas} = k_H P_{gas}$ (11.10)	**11.11**

VISUAL PROBLEMS

(Answers to boldface end-of-chapter questions and problems are in the back of the book.)

11.1. Figure P11.1 provides a particle-level view of a sealed container partially filled with a solution that has two components: X (blue spheres) and Y (red spheres). Which of the following statements about substances X and Y are true?

a. X is the solvent in this solution.
b. Pure Y is a volatile liquid.
c. If Y were not present, there would be fewer X particles in the gas above the liquid solution.
d. The presence of Y increases the vapor pressure of X.

FIGURE P11.1

11.2. Figure P11.2 provides a particle-level view of a sealed container partially filled with a solution of two miscible liquids: X (blue spheres) and Y

FIGURE P11.2

(red spheres). Which of the following statements about substances X and Y are true?

a. Y is the solvent in this solution.
b. Pure Y has a higher vapor pressure than pure X.
c. The presence of Y in the solution lowers the vapor pressure of X.
d. If Y were not present, there would be fewer total particles in the gas above the liquid solution.

11.3. Figure P11.3 provides particle-level views of 0.001 M aqueous solutions of the following four solutes: $C_6H_{12}O_6$, NaCl, $MgCl_2$, and K_3PO_4. The blue spheres represent particles of solute. Which compounds are represented in images (a)–(d)?

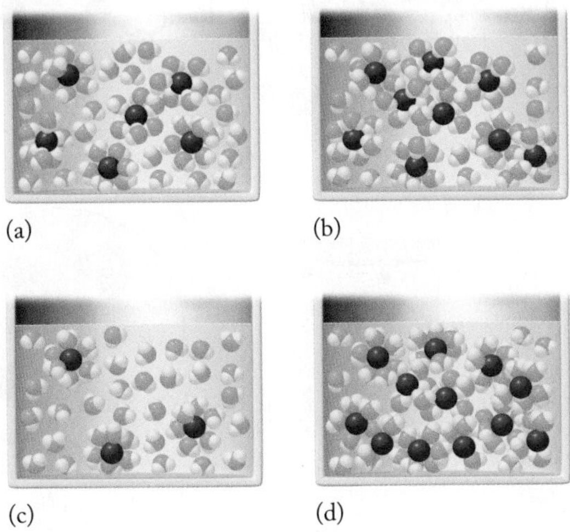

FIGURE P11.3

11.4. Which of the four solutions in Figure P11.3 has the highest (a) vapor pressure; (b) boiling point; (c) freezing point; (d) osmotic pressure compared to pure solvent?

11.5. Use the graph in Figure P11.5 to estimate the normal boiling points of substances X and Y. Molecules of which substance experience the stronger intermolecular forces?

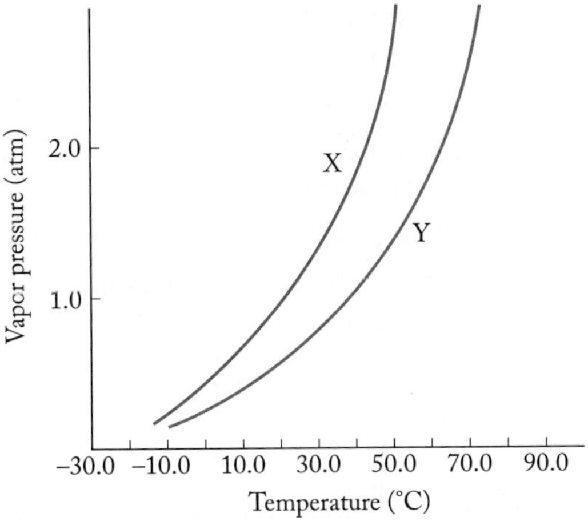

FIGURE P11.5

*11.6. **Kidney Dialysis** Semipermeable membranes of the sort used in kidney dialysis do not allow large molecules and cells to pass but do allow small ions and water to pass. Figure P11.6 shows such a membrane separating fluids of various compositions.

a. In which direction does the water flow in each solution?
b. In which direction do the sodium ions flow in each solution?
c. In which direction do the potassium ions flow in each solution?

(i)	(ii)
Pure H_2O	0.2 M KCl in H_2O
Membrane	Membrane
Proteins 0.5 M NaCl 0.1 M KCl in H_2O	Proteins 0.5 M NaCl 0.1 M KCl in H_2O

FIGURE P11.6

11.7. The arrow in Figure P11.7 indicates the direction of solvent flow through a semipermeable membrane in equipment designed to measure osmotic pressure. Which solution, A or B, is the more concentrated? Explain your answer.

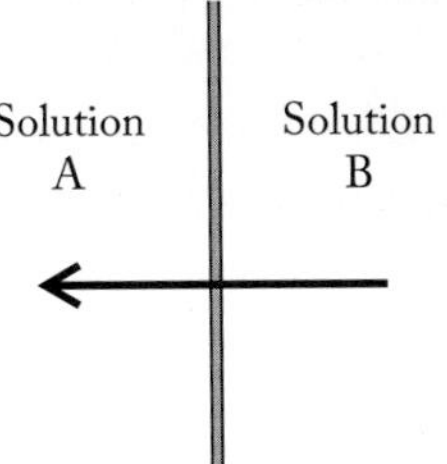

FIGURE P11.7

11.8. Which of the images in Figure P11.8 represents the gas with the greatest Henry's law constant, k_H?

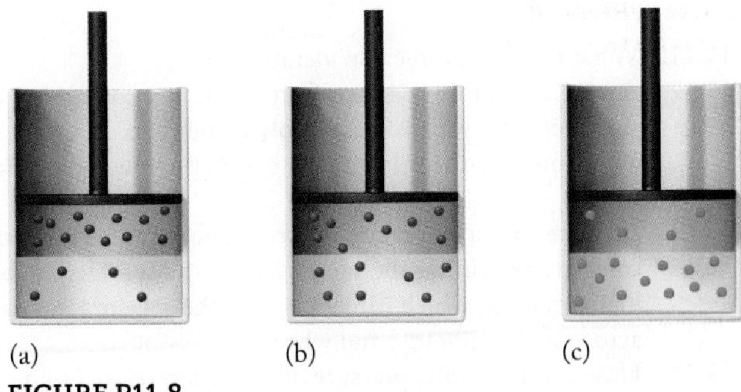

FIGURE P11.8

11.9. Which of the images in Figure P11.9 best describes the effect of pressure on the solubility of a gas?

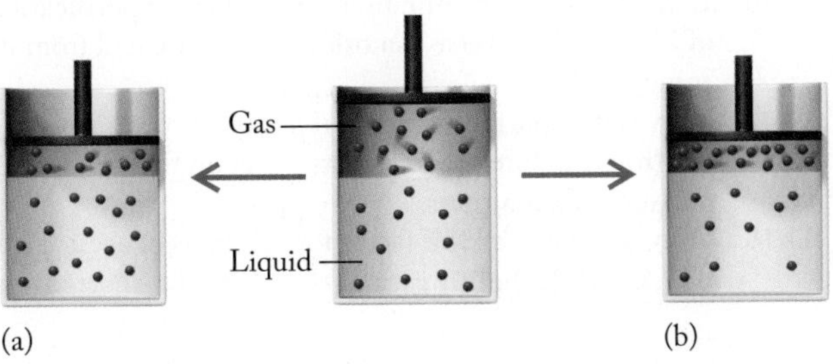

FIGURE P11.9

11.10. Use representations [A] through [I] in Figure P11.10 to answer questions a–f.
  a. Which compound, [B], [C], or [F] has the highest boiling point?
  b. Between [A] and [I], which has the higher vapor pressure?
  c. Between [A] and [I], which has the lower freezing point?
  d. If solutions [A] and [I] were separated by a semipermeable membrane, as shown in [E], what molecules or ions would flow across the membrane and in what direction?
  e. Name two variables that could be changed to increase the solubility of the gas in [D].
  f. Which beaker of water, [G] or [H], is at a lower temperature?

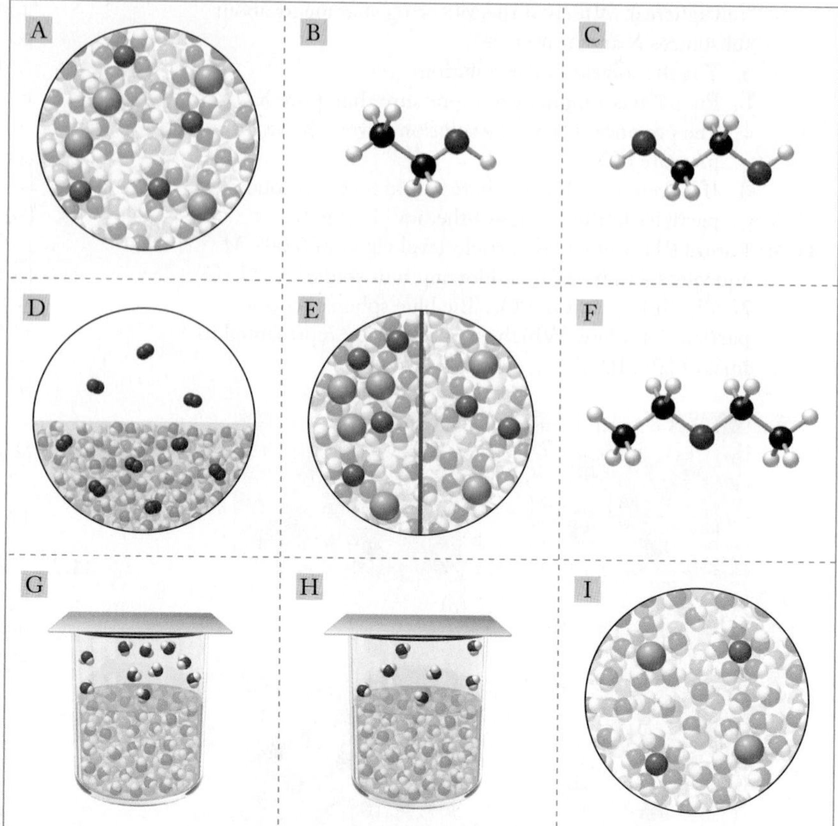

FIGURE P11.10

QUESTIONS AND PROBLEMS

Osmosis and Osmotic Pressure

Concept Review

11.11. What is a semipermeable membrane?

11.12. A pure solvent is separated from a solution containing the same solvent by a semipermeable membrane. In which direction does the solvent flow across the membrane, and why?

11.13. A dilute solution is separated from a more concentrated solution containing the same solvent by a semipermeable membrane. In which direction does the solvent tend to flow across the membrane, and why?

11.14. How is the osmotic pressure of a solution related to its molar concentration and its temperature?

11.15. What is reverse osmosis? List the basic components of equipment used to purify seawater by reverse osmosis.

11.16. Explain how the minimum pressure for the purification of seawater by reverse osmosis can be estimated from its composition.

11.17. Why is it important to know if a substance is a strong electrolyte before predicting its effect on the osmotic pressure of a solution?

11.18. Why is the van 't Hoff factor for solutions of acetic acid, CH_3COOH, slightly greater than 1 but far short of $i = 2$?

11.19. Why are theoretical and experimental van 't Hoff factors of electrolytes sometimes different?

11.20. Can an experimentally measured value of a van 't Hoff factor be greater than the theoretical value? Why or why not?

Problems

11.21. The following pairs of aqueous solutions are separated by a semipermeable membrane. In which direction will the solvent flow?
  a. A = 1.25 M NaCl; B = 1.50 M KCl
  b. A = 3.45 M $CaCl_2$; B = 3.45 M NaBr
  c. A = 4.68 M glucose; B = 3.00 M NaCl

11.22. The following pairs of aqueous solutions are separated by a semipermeable membrane. In which direction will the solvent flow?
  a. A = 0.48 M NaCl; B = 55.85 g of NaCl dissolved in 1.00 L of solution
  b. A = 100 mL of 0.982 M $CaCl_2$; B = 16 g of NaCl in 100 mL of solution
  c. A = 100 mL of 6.56 mM $MgSO_4$; B = 5.24 g of $MgCl_2$ in 250 mL of solution

11.23. Calculate the osmotic pressure of each of the following aqueous solutions at 20°C:
a. 2.39 *M* methanol (CH_3OH)
b. 9.45 m*M* $MgCl_2$
c. 40.0 mL of glycerol ($C_3H_8O_3$) in 250.0 mL of aqueous solution (density of glycerol = 1.265 g/mL)
d. 25 g of $CaCl_2$ in 350 mL of solution

11.24. Calculate the osmotic pressure of each of the following aqueous solutions at 27°C:
a. 10.0 g of NaCl in 1.50 L of solution
b. 10.0 mg/L of $LiNO_3$
c. 0.222 *M* glucose
d. 0.00764 *M* K_2SO_4

11.25. Determine the molarity of each of the following solutions from its osmotic pressure at 25°C. Include the van 't Hoff factor for the solution when the factor is given.
a. $\Pi = 0.674$ atm for a solution of ethanol (CH_3CH_2OH)
b. $\Pi = 0.0271$ atm for a solution of aspirin ($C_9H_8O_4$)
c. $\Pi = 0.605$ atm for a solution of $CaCl_2$, $i = 2.47$

11.26. Determine the molarity of each of the following solutions from its osmotic pressure at 25°C. Include the van 't Hoff factor for the solution when the factor is given.
a. $\Pi = 0.0259$ atm for a solution of urea [$H_2NC(O)NH_2$]
b. $\Pi = 1.56$ atm for a solution of sucrose ($C_{12}H_{22}O_{11}$)
c. $\Pi = 0.697$ atm for a solution of KI, $i = 1.90$

***11.27.** Which of the three curves of osmotic pressure versus temperature in Figure P11.27 represents a strong electrolyte if the concentrations of solutions A, B, and C are all 1.21×10^{-5} *M*?

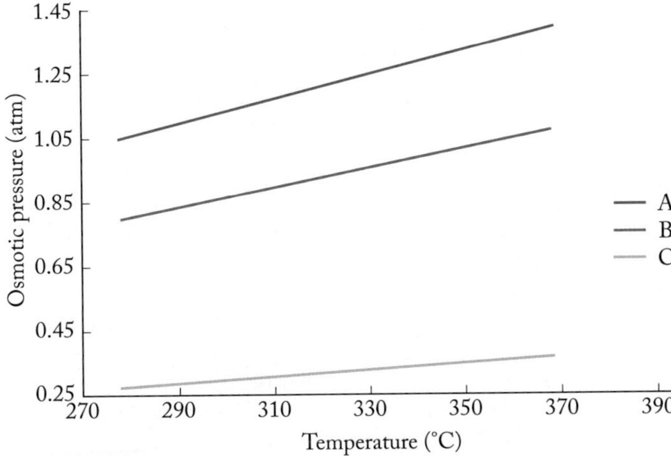

FIGURE P11.27

11.28. Suppose you have 1.00 *M* aqueous solutions of each of the following solutes: glucose ($C_6H_{12}O_6$), NaCl, and acetic acid (CH_3COOH). Which solution has the highest pressure requirement for reverse osmosis?

11.29. Throat Lozenges A 188 mg sample of a nonelectrolyte isolated from throat lozenges is dissolved in enough water to make 10.0 mL of solution at 25°C. The osmotic pressure of the resulting solution is 4.89 atm. Calculate the molar mass of the compound.

11.30. **Healing Herbs** Alpetin is a compound found in *Alpinia speciosa*, a tropical evergreen used historically for treating cold, flu, fever, flatulence, and indigestion. A 54.1 mg sample of alpetin is dissolved in 75.0 mL of water. The osmotic pressure of the solution is 0.0657 atm at 27°C. Assuming that alpetin is a nonelectrolyte, calculate the molar mass of alpetin.

11.31. Physiological Saline After 100.0 mL of a solution of physiological saline (0.90% NaCl by mass) is diluted by the addition of 250.0 mL of water, what is the osmotic pressure of the final solution at 37°C? Assume that NaCl dissociates completely into $Na^+(aq)$ and $Cl^+(aq)$.

*11.32. An unknown compound (152 mg) is dissolved in water to make 75.0 mL of solution. The solution does not conduct electricity and has an osmotic pressure of 0.328 atm at 27°C. Elemental analysis reveals the substance to be 78.90% C, 10.59% H, and 10.51% O. Determine the molecular formula of this compound.

Vapor Pressure

Concept Review

11.33. Use kinetic molecular theory to explain why the vapor pressure of a liquid increases with increasing temperature.

11.34. Why is the vapor pressure of neopentane (its condensed molecular structure is shown in Figure P11.34) higher than that of pentane at the same temperature?

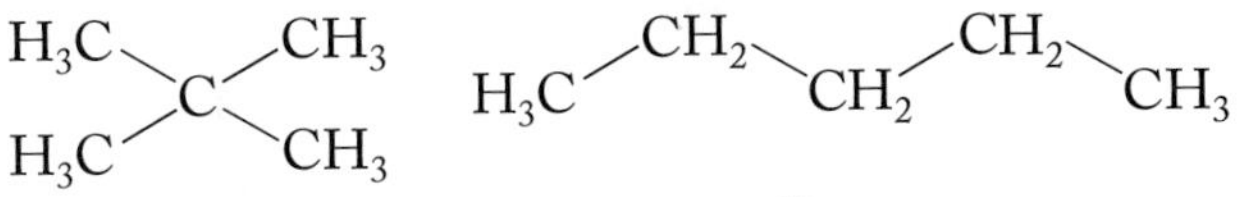

H_3C, CH_3, C, H_3C, CH_3 — Neopentane

H_3C–CH_2–CH_2–CH_2–CH_3 — Pentane

FIGURE P11.34

11.35. Generally speaking, how is the vapor pressure of a liquid affected by the strength of intermolecular forces?

11.36. Is vapor pressure an intensive or extensive property of a volatile substance?

Problems

11.37. Rank the following compounds in order of increasing vapor pressure at 298 K: (a) CH_3CH_2OH, (b) CH_3OCH_3, and (c) $CH_3CH_2CH_3$.

11.38. Rank the compounds in Figure P11.38 in order of increasing vapor pressure at 298 K.

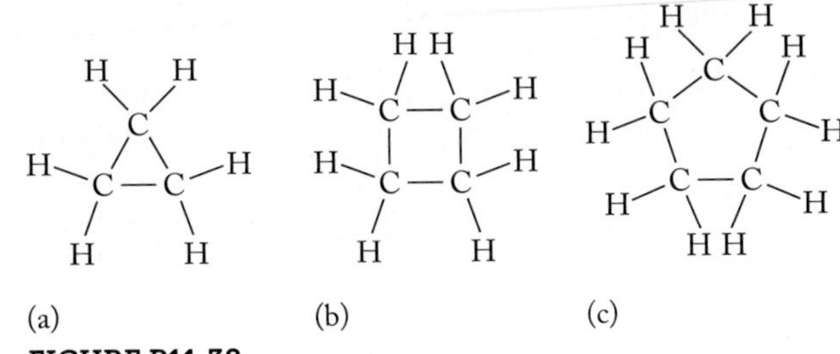

FIGURE P11.38

11.39. Pine Oil The smell of fresh-cut pine is due in part to the cyclic alkene pinene, whose carbon-skeleton structure is shown in Figure P11.39. Use the data in the table to calculate the heat of vaporization, ΔH_{vap}, of pinene.

CH_3, C, HC, CH_3, CH, C, CH_3, H_2C, CH_2, CH

Pinene

FIGURE P11.39

Vapor Pressure (torr)	Temperature (K)
760	429
515	415
340	401
218	387
135	373

11.40. **Almonds and Cherries** Almonds and almond extracts are common ingredients in baked goods. Almonds contain the compound benzaldehyde (shown in Figure P11.40), which accounts for the odor of the nut. Benzaldehyde is also partly responsible for the aroma of cherries. Use the data in the table to calculate the heat of vaporization, ΔH_{vap}, of benzaldehyde.

Vapor Pressure (torr)	Temperature (K)
50	373
111	393
230	413
442	433
805	453

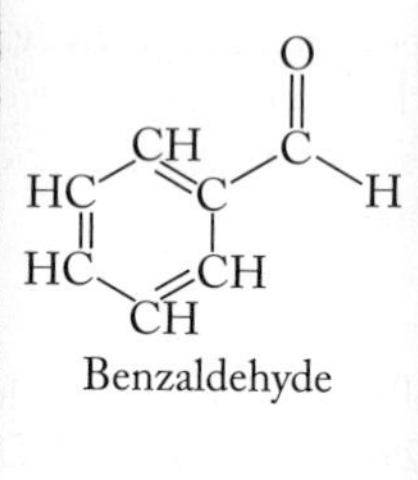

FIGURE P11.40

11.41. High-Octane Gasoline Gasoline is a complex mixture of hydrocarbons. It is sold with a variety of octane ratings that are based on the comparison of the gasoline with the combustion properties of isooctane, a compound with the molecular formula C_8H_{18}. The Lewis structures of isooctane and another compound with molecular formula C_8H_{18} are shown in Figure P11.41, along with their normal boiling points and heats of vaporization. Determine the vapor pressure of each isomer on a day when the temperature is 38°C.

Isooctane
bp = 98.2°C
ΔH_{vap} = 35.8 kJ/mol

Tetramethylbutane
bp = 106.5°C
ΔH_{vap} = 43.3 kJ/mol

FIGURE P11.41

11.42. **Camping Fuel** Portable lanterns and stoves used for camping and backpacking often use a mixture of C_5 and C_6 hydrocarbons known as "white gas." Figure P11.42 shows the carbon-skeleton structure of pentane, C_5H_{12}, along with its normal boiling point and heat of vaporization. Determine the vapor pressure of pentane on a morning when the temperature is 5.0°C.

H_3C, CH_2, CH_2, CH_2, CH_3

Pentane
bp = 36.0°C
ΔH_{vap} = 27.6 kJ/mol

FIGURE P11.42

Mixtures of Volatile Substances

Concept Review

11.43. What physical property of the components of crude oil is used to separate them?

11.44. What is the difference between simple distillation and fractional distillation?

11.45. In an equimolar mixture of C_5H_{12} and C_7H_{16}, which compound is present in higher concentration in the vapor above the solution?

11.46. Why does the boiling point of a mixture of volatile hydrocarbons increase over time during a distillation?

Problems

11.47. At 20°C, the vapor pressure of ethanol (CH_3CH_2OH) is 45 torr and the vapor pressure of methanol (CH_3OH) is 92 torr. What is the vapor pressure at 20°C of a solution prepared by mixing 25 g of methanol and 75 g of ethanol?

11.48. A bottle is half-filled with a 50:50 (by moles) mixture of heptane (C_7H_{16}) and octane (C_8H_{18}) at 25°C. What is the mole ratio of heptane vapor to octane vapor in the air space above the liquid in the bottle? The vapor pressures of heptane and octane at 25°C are 31 torr and 11 torr, respectively.

More Colligative Properties of Solutions

Concept Review

11.49. Explain how nearly all the water ends up in one compartment in Figure 11.17.

11.50. A dam separates the freshwater of the Charles River from the seawater of Boston Harbor. If the two bodies of water are at the same temperature, which one evaporates faster on a hot summer day?

11.51. What is the difference between molarity and molality?

11.52. As a solution of NaCl becomes more concentrated, does the difference between its molarity and its molality increase or decrease?

11.53. Why does seawater freeze at a lower temperature than normal saline?

*11.54. The thermostat in a refrigerator filled with cans of soft drinks malfunctions and the temperature of the refrigerator drops below 0°C. The contents of the cans of diet soft drinks freeze, rupturing many of the cans and causing an awful mess. However, none of the cans containing regular, nondiet soft drinks freeze and rupture. Why?

Problems

11.55. A solution contains 3.5 mol of water and 1.5 mol of glucose ($C_6H_{12}O_6$). What is the mole fraction of water in the solution? What is the vapor pressure of the solution at 25°C, given that the vapor pressure of pure water at 25°C is 23.8 torr?

11.56. A solution contains 4.5 mol of water, 0.3 mol of sucrose ($C_{12}H_{22}O_{11}$), and 0.2 mol of glucose. Sucrose and glucose are nonvolatile. What is the mole fraction of water in the solution? What is the vapor pressure of the solution at 35°C, given that the vapor pressure of pure water at 35°C is 42.2 torr?

11.57. Calculate the molality of each of the following solutions:
a. 0.875 mol of glucose ($C_6H_{12}O_6$) in 1.5 kg of water
b. 11.5 mmol of acetic acid (CH_3COOH) in 65 g of water
c. 0.325 mol of baking soda ($NaHCO_3$) in 290.0 g of water

11.58. Calculate the molality of each of the following solutions:
a. 11.7 g of NaCl in 1.0 kg of water
b. 4.99 g of $CuSO_4$ in 265 g of water
c. 6.41 g of CH_3OH in 375 g of water

11.59. Calculate the molality of each of the following aqueous solutions:
a. 1.30 *M* $CaCl_2$ (d = 1.113 g/mL)
b. 2.02 *M* fructose ($C_6H_{12}O_6$, d = 1.139 g/mL)
c. 8.94 *M* ethylene glycol (antifreeze, $HOCH_2CH_2OH$, d = 1.069 g/mL)
d. 1.97 *M* LiCl (d = 1.046 g/mL)

11.60. Table 8.1 lists molar concentrations of the major ions in seawater. Using a density of 1.025 g/mL for seawater, convert the concentrations into molalities.

11.61. What mass of the following solutions contains 0.100 mol of solute? (a) 0.334 *m* NH_4NO_3; (b) 1.24 *m* ethylene glycol, $HOCH_2CH_2OH$; (c) 5.65 *m* $CaCl_2$

11.62. How many moles of solute are there in the following solutions?
a. 0.150 *m* glucose solution made by dissolving the glucose in 100.0 kg of water
b. 0.028 *m* Na_2CrO_4 solution made by dissolving the Na_2CrO_4 in 1000.0 g of water
c. 0.100 *m* urea solution made by dissolving the urea in 500.0 g of water

---

11.63. Fish Kills High concentrations of ammonia (NH_3), nitrite ion, and nitrate ion in water can kill fish. Lethal concentrations of these species for rainbow trout are 1.1 mg/L, 0.40 mg/L, and 1361 mg/L, respectively. Express these concentrations in molality units, assuming a solution density of 1.00 g/mL.

11.64. The concentrations of six important elements in a sample of river water are 0.050 mg/kg of Al^{3+}, 0.040 mg/kg of Fe^{3+}, 13.4 mg/kg of Ca^{2+}, 5.2 mg/kg of Na^+, 1.3 mg/kg of K^+, and 3.4 mg/kg of Mg^{2+}. Express each of these concentrations in molality units.

11.65. Cinnamon Cinnamon owes its flavor and odor to cinnamaldehyde (C_9H_8O). Determine the freezing point of a solution of 75 mg of cinnamaldehyde dissolved in 1.00 g of benzene (K_f = 4.3°C/*m*; normal freezing point = 5.5°C).

11.66. **Spearmint** Determine the boiling point elevation of a solution of 125 mg of carvone ($C_{10}H_{14}O$, oil of spearmint) dissolved in 1.50 g of carbon disulfide (K_b = 2.34°C/*m*).

11.67. What molality of a nonvolatile, nonelectrolyte solute is needed to lower the melting point of camphor by 1.000°C (K_f = 39.7°C/*m*)?

11.68. What molality of a nonvolatile, nonelectrolyte solute is needed to raise the boiling point of water by 7.60°C (K_b = 0.52°C/*m*)?

11.69. Saccharin Determine the melting point of an aqueous solution made by adding 186 mg of saccharin ($C_7H_5O_3NS$) to 1.00 mL of water (density = 1.00 g/mL, K_f = 1.86°C/*m*).

11.70. Determine the boiling point of an aqueous solution that is 2.50 *m* ethylene glycol ($HOCH_2CH_2OH$); K_b for water is 0.52°C/*m*. Assume that the boiling point of pure water is 100.00°C.

11.71. Which aqueous solution has the lowest freezing point: 0.5 *m* glucose, 0.5 *m* NaCl, or 0.5 *m* $CaCl_2$?

11.72. Which aqueous solution has the highest boiling point: 0.5 *m* glucose, 0.5 *m* NaCl, or 0.5 *m* $CaCl_2$?

11.73. Which one of the following aqueous solutions should have the highest boiling point: 0.0400 *m* NH_4NO_3, 0.0165 *m* LiCl, or 0.0105 *m* $Cu(NO_3)_2$?

11.74. Which one of the following aqueous solutions should have the lowest freezing point: 0.0500 *m* $C_6H_{12}O_6$, 0.0450 *m* NaI, or 0.0125 *m* K_2SO_4?

11.75. Arrange the following aqueous solutions in order of increasing boiling point:
a. 0.06 *m* $FeCl_3$ (i = 3.4)
b. 0.10 *m* $MgCl_2$ (i = 2.7)
c. 0.20 *m* KCl (i = 1.9)

11.76. Arrange the following solutions in order of increasing freezing point depression:
a. 0.10 *m* $MgCl_2$ in water, i = 2.7, K_f = 1.86°C/*m*
b. 0.20 *m* toluene in diethyl ether, i = 1.00, K_f = 1.79°C/*m*
c. 0.20 *m* ethylene glycol in ethanol, i = 1.00, K_f = 1.99°C/*m*

***11.77. Caffeine** The freezing point of a solution prepared by dissolving 150 mg of caffeine in 10.0 g of camphor is lower by 3.07°C than that of pure camphor (K_f = 39.7°C/*m*). What is the molar mass of caffeine? Elemental analysis of caffeine yields the following results: 49.49% C, 5.15% H, 28.87% N, and the remainder is O. What is the molecular formula of caffeine?

*11.78. **Ethnobotany** One of the ingredients in the Native American stomachache remedy derived from common chokecherry is caffeic acid.

a. Combustion of 100 mg of caffeic acid yielded 220 mg CO_2 and 40.3 mg H_2O. Determine the empirical formula of caffeic acid.
b. A solution of 0.272 g caffeic acid in 10.0 g carbon tetrachloride causes the freezing point of the solution to decrease by 4.47°C. Given that K_f for CCl_4 is 29.8°C/m and that caffeic acid is a nonelectrolyte, calculate the molar mass and molecular formula of caffeic acid.

Solubilities of Gases and Henry's Law

Concept Review

11.79. Why is the Henry's law constant for CO_2 so much larger than those for N_2 and O_2 at the same temperature? *Hint*: Does CO_2 react with water?

11.80. The values of k_H for NO and CO gas in water at 20°C are 1.4×10^{-3} M/atm and 7.4×10^{-3} M/atm, respectively. Why is CO more soluble in water?

11.81. As water in a beaker is heated, bubbles form inside the beaker at temperatures well below the boiling point of water. What gas is in the bubbles?

*11.82. Air is primarily a mixture of nitrogen and oxygen. Is the Henry's law constant for the solubility of air in water the sum of k_H for N_2 and k_H for O_2? Explain why or why not.

Problems

***11.83.** **Arterial Blood** Arterial blood contains about 0.25 g of oxygen per liter at 37°C and standard atmospheric pressure. What is the Henry's law constant, in mol/(L · atm), for O_2 dissolution in blood at 37°C?

11.84. **Laughing Gas** Nitrous oxide, N_2O, is used in dental clinics as an anesthetic. The solubility of N_2O in water is 1.1 g/L at an atmospheric pressure of 1 atm and a temperature of 20°C.

a. Calculate the Henry's law constant of N_2O at 20°C.
b. Compare the value for k_H for N_2O with the value for O_2 in Table 11.1. Why are they different?

***11.85.** **Oxygen for Climbers and Divers** Use the Henry's law constant for O_2 dissolved in arterial blood from Problem 11.83 to calculate the solubility of O_2 in the blood of (a) a climber on Mt. Everest (P_{atm} = 0.35 atm) and (b) a scuba diver breathing air at a depth of 20 meters ($P \approx 3.0$ atm).

11.86. An air sample contains 1.89×10^{-4} % of methane gas (CH_4). The Henry's law constant for methane in water is approximately 1.3×10^{-3} M at 20°C and 1.0 atm.

a. Calculate the solubility of methane in water under a pure methane atmosphere of 1.0 atm.
b. Calculate the maximum solubility of the methane found in the air sample.

Additional Problems

***11.87.** **Melting Ice** $CaCl_2$ is used to melt ice on sidewalks. Could $CaCl_2$ melt ice at −20°C? Assume that the solubility of $CaCl_2$ at this temperature is 70.1 g $CaCl_2$/100.0 g of H_2O and that the van 't Hoff factor for a saturated solution of $CaCl_2$ is 2.5.

*11.88. **Making Ice Cream** A mixture of table salt and ice is used to chill the contents of hand-operated ice-cream makers. What is the melting point of a mixture of 2.00 lb of NaCl and 12.00 lb of ice if exactly half of the ice melts? Assume that all the NaCl dissolves in the melted ice and that the van 't Hoff factor for the resulting solution is 1.44.

11.89. The freezing points of 0.0935 m ammonium chloride and 0.0378 m ammonium sulfate in water were found to be −0.322°C and −0.173°C, respectively. What are the values of the van 't Hoff factors for these salts?

11.90. The following data were collected for three compounds in aqueous solution. Determine the value of the van 't Hoff factor for each salt (K_f for water = 1.86°C/m).

Compound	Concentration (g/kg)	Experimentally Measured ΔT_f(°C)
LiCl	5.0	0.410
HCl	5.0	0.486
NaCl	5.0	0.299

11.91. **Cloves** Eugenol is one of the compounds responsible for the flavor of cloves. A 111 mg sample of eugenol is dissolved in 1.00 g of chloroform (K_b = 3.63°C/m), increasing the boiling point of chloroform by 2.45°C. Calculate eugenol's molar mass. Eugenol is 73.17% C, 7.32% H, and 19.51% O by mass. What is the molecular formula of eugenol?

11.92. Suppose 100.0 mL of 2.50 mM NaCl is mixed with 80.0 mL of 3.60 mM $MgCl_2$ at 20°C. Calculate the osmotic pressure of each starting solution and that of the mixture, assuming that the volumes are additive and that both salts dissociate completely into their component ions.

11.93. A solution of 7.50 mg of a small protein in 5.00 mL aqueous solution has an osmotic pressure of 6.50 torr at 23.1°C. What is the molar mass of the protein?

11.94. **Kidney Dialysis** Hemodialysis, a method of removing waste products from the blood if the kidneys have failed, uses a tube made of a cellulose membrane that is immersed in a large volume of aqueous solution. Blood is pumped through the tube and is then returned to the patient's vein. The membrane allows small ions, urea, and water but not large protein molecules and cells to pass through it. Assume that a physician wants to decrease the concentration of sodium ion and urea in a patient's blood while maintaining the concentration of potassium ion and chloride ion in the blood. What materials must be dissolved in the aqueous solution in which the dialysis tube is immersed? How must the concentrations of ions in the immersion fluid compare to those in blood?

***11.95.** The Henry's law constant for the solubility of a gas and the van der Waal's constants a and b in Equation 10.27 depend on intermolecular forces.

a. Plot the values of a and b in Table 10.5 as a function of k_H (Table 11.1) for He, O_2, N_2, CH_4, and CO_2. Which gas seems to be an outlier?

b. Why might this gas be an outlier?

11.96. **Psoriasis Treatment** Xanthotoxin is a natural product extracted from the fruits of *Heracleum persicum* that seems to have some success in treating a skin disease known as psoriasis.

a. Combustion of 0.0100 g of xanthotoxin in excess oxygen yielded 0.0244 g CO_2 and 0.00333 g H_2O. Calculate the empirical formula of xanthotoxin.

b. A solution of 0.250 g xanthotoxin in 10.0 g carbon tetrachloride causes the freezing point of the solution to decrease by 3.46°C. Given that K_f for CCl_4 is 29.8°C/m and that xanthotoxin is a nonelectrolyte, calculate the molar mass and molecular formula of xanthotoxin.

smartwork5

If your instructor uses Smartwork5, log in at digital.wwnorton.com/atoms2.

12

Thermodynamics

Why Chemical Reactions Happen

CORROSION AT SEA Most metal objects in contact with seawater, including the remains of this ship, corrode as a result of spontaneous chemical reactions. Here, iron metal has turned into iron(III) oxide.

PARTICULATE **REVIEW**

Phase Changes and Energy Changes

In Chapter 12, we will investigate how energy changes affect the spontaneity of chemical reactions and physical processes.

- What two phase changes are represented in these particulate images?
- Which phase change occurs due to a flow of energy from the system into its surroundings on a cold day (−5°C)?
- Label each phase change as either endothermic or exothermic.

(Review Sections 9.1 and 9.2 if you need help.)

(Answers to Particulate Review questions are in the back of the book.)

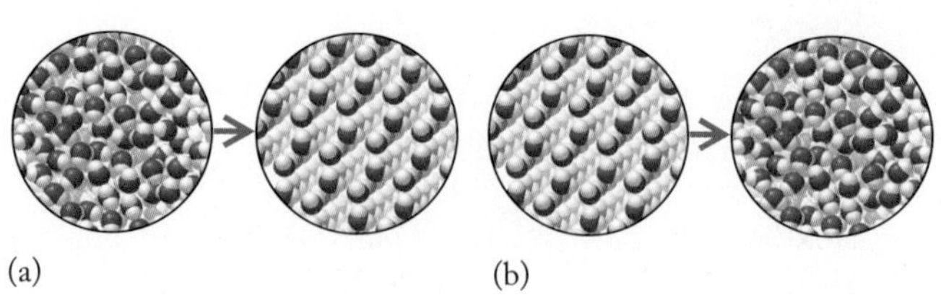

PARTICULATE **PREVIEW**

Spontaneous or Not?

When we exercise, we rely on the oxidation of glucose ($C_6H_{12}O_6$) to carbon dioxide and water for energy. When photosynthesis takes place in green plants, carbon dioxide and water react to produce glucose and oxygen. As you read Chapter 12, look for information that will help you answer these questions:

- Write balanced chemical equations describing these two reactions.
- Which reaction results in breaking larger molecules into smaller molecules? Does it occur spontaneously?
- Which reaction results in building larger molecules from smaller molecules? Does it occur spontaneously?

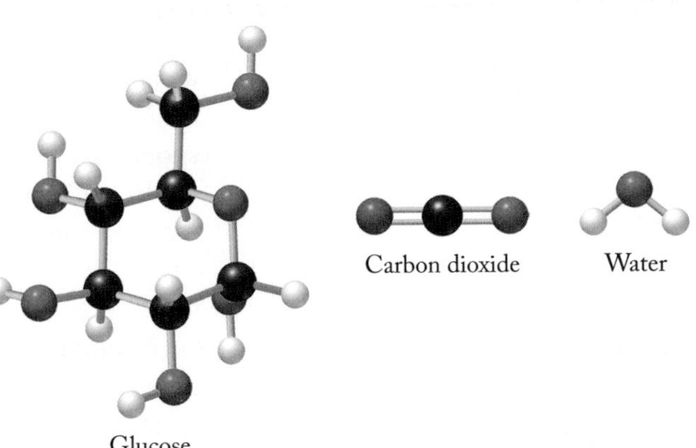

Learning Outcomes

LO1 Predict the signs of entropy changes for chemical reactions and physical processes
Sample Exercise 12.1

LO2 Predict the relative entropies of substances based on their molecular structures
Sample Exercise 12.2

LO3 Calculate entropy changes in chemical reactions using standard molar entropies
Sample Exercise 12.3

LO4 Calculate free-energy changes in chemical reactions
Sample Exercises 12.4, 12.5, and 12.6

LO5 Predict the spontaneity of a chemical reaction as a function of temperature
Sample Exercise 12.7

LO6 Calculate the net change in free energy of coupled spontaneous and nonspontaneous reactions
Sample Exercise 12.8

12.1 Spontaneous Processes

Some processes are so familiar that we don't question why they happen. If a car tire is punctured, the air inside rushes out and the tire goes flat; the air does *not* rush back into a punctured tire and re-inflate it. Discarded objects made of iron slowly turn into clumps of rust; they don't turn back into shiny metal all by themselves. A tray of ice cubes left on a kitchen counter turns into a tray of liquid water; as long as the kitchen stays warm, the water in the tray will not turn back into ice.

These processes are all **spontaneous** because they all happen without any ongoing intervention. In each case, the reverse process is nonspontaneous because it cannot happen on its own. So, why are some processes spontaneous and others nonspontaneous?

Let's answer this question by considering the spontaneous chemical reaction in 19th-century streetlights. Before Thomas Edison invented the lightbulb, city streets were often illuminated at night by gas-fueled lamps. Once lit, they burned through the night until their fuel was cut off as dawn approached. Chemical reactions like the combustion of gas-lamp fuel are examples of spontaneous reactions: once started, they proceed without outside assistance as long as there are sufficient quantities of reactants available. The reverse reaction—in this case, converting carbon dioxide and water back into fuel and oxygen—is **nonspontaneous**: it cannot happen on its own.

The word *spontaneous* can be misleading because many spontaneous reactions don't start all by themselves: they need a little energy boost, such as a spark or an external flame that ignites the flame in a gas lamp or the gasoline/air mixture in an automobile engine. However, once initiated, a spontaneous reaction continues on its own. It is the self-sustaining nature of these reactions that makes them spontaneous.

In addition, spontaneous does not necessarily mean fast. Though combustion reactions proceed rapidly, others do not, including iron rusting:

$$4\ Fe(s) + 3\ O_2(g) \rightarrow 2\ Fe_2O_3(s) \qquad \Delta H^\circ_{rxn} = -1648.4\ kJ$$

spontaneous process a process that proceeds without outside intervention.

nonspontaneous process a process that requires outside intervention to proceed.

Spontaneous is also *not* a synonym for exothermic, though many scientists once thought so. In the mid-19th century, many of them believed that exothermic reactions were always spontaneous, and it is true that most are. However, others are not spontaneous, and some endothermic reactions are spontaneous. For example, when baking soda (sodium bicarbonate) is added to room-temperature

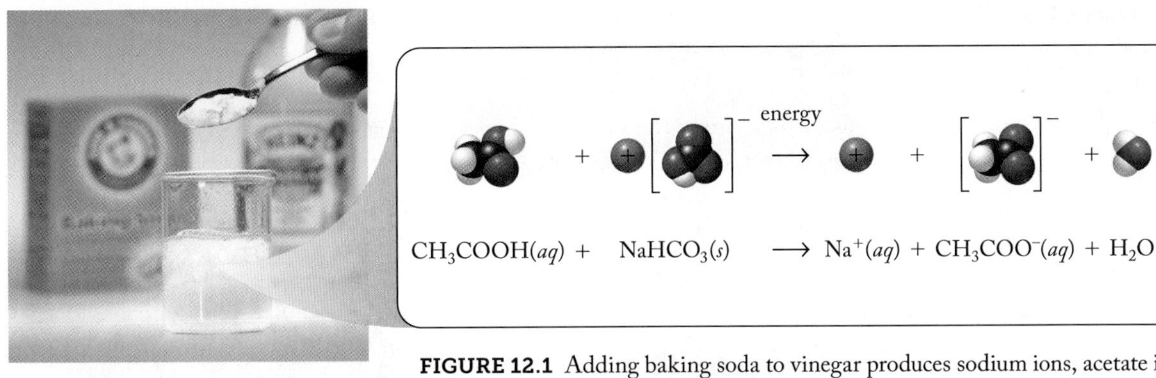

FIGURE 12.1 Adding baking soda to vinegar produces sodium ions, acetate ions, water, and carbon dioxide in an endothermic reaction ($\Delta H^\circ_{rxn} = 48.5$ kJ/mol) that is also spontaneous.

vinegar (dilute acetic acid) as shown in Figure 12.1, the foaming reaction mixture tells us that a chemical change is taking place, and a decrease in the temperature of the reaction mixture tells us that the reaction is endothermic ($\Delta H_{rxn} > 0$).

CONNECTION We learned in Chapter 9 that enthalpy change (ΔH) is a measure of how much energy flows into ($\Delta H > 0$) or out from ($\Delta H < 0$) a system during a process occurring at constant pressure.

Why is the vinegar and baking soda reaction spontaneous? The particles that make up the products are more spread out and/or have more freedom of motion than the particles in the reactants. For example, the reactants are liquid and solid, but the products include a gas (CO_2). The particles that make up these three states of matter have different freedoms of motion, as illustrated for the molecules of water in ice, liquid water, and steam in Figure 12.2. The particles that make up ice or any crystalline solid occupy fixed positions, and their motion is limited to vibrating in place without going anywhere (unless the whole crystal moves). The particles in liquids are more mobile: they are able to flow past and tumble over each other, which means they have some *translational* and *rotational* freedom of motion. Particles in the gas phase are much more spread out—which is why gases are compressible whereas liquids and solids are not—and they have much more translational and rotational freedom of motion. Therefore, when either solid or liquid reactants form one or more products that are gases, the particles are more spread out and have more freedom to move about.

An increase in particle motion also takes place in the process that makes instant cold packs cold (Figure 12.3). There are two compartments in an instant cold pack. One is filled with water, and the other contains a water-soluble compound such as ammonium nitrate that has a positive enthalpy of solution ($\Delta H_{solution} > 0$). When the membrane separating the two compartments is ruptured, the solid compound mixes with and dissolves in water. The resulting solution gets very cold, becoming an effective anti-inflammatory treatment for bruises and muscle sprains. The spontaneity of this process is due to the production of particles (dissolved NH_4^+ and NO_3^- ions) that are more spread out and have more freedom of motion than they had in the solid phase. This concept applies to the dissolution of all solids in liquid solvents. It also applies to melting solids because their particles are usually more spread out and always have more freedom of motion in the liquid state, as shown for ice and water in Figure 12.2.

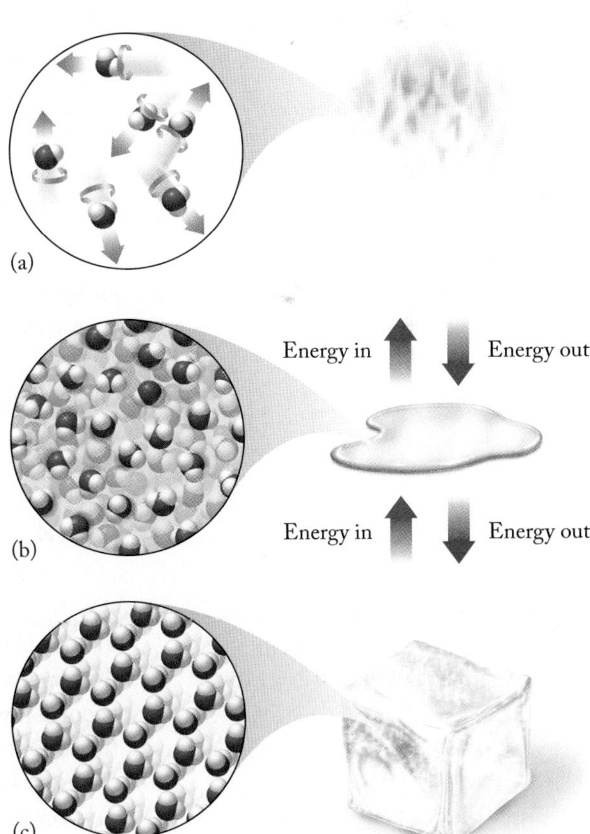

FIGURE 12.2 Molecular motion in the three phases of water. (a) Molecules of water vapor have much more translational and rotational motion than particles in either condensed phase. (b) Molecules in water can flow past and tumble over each other. (c) Molecules in ice can only vibrate in place.

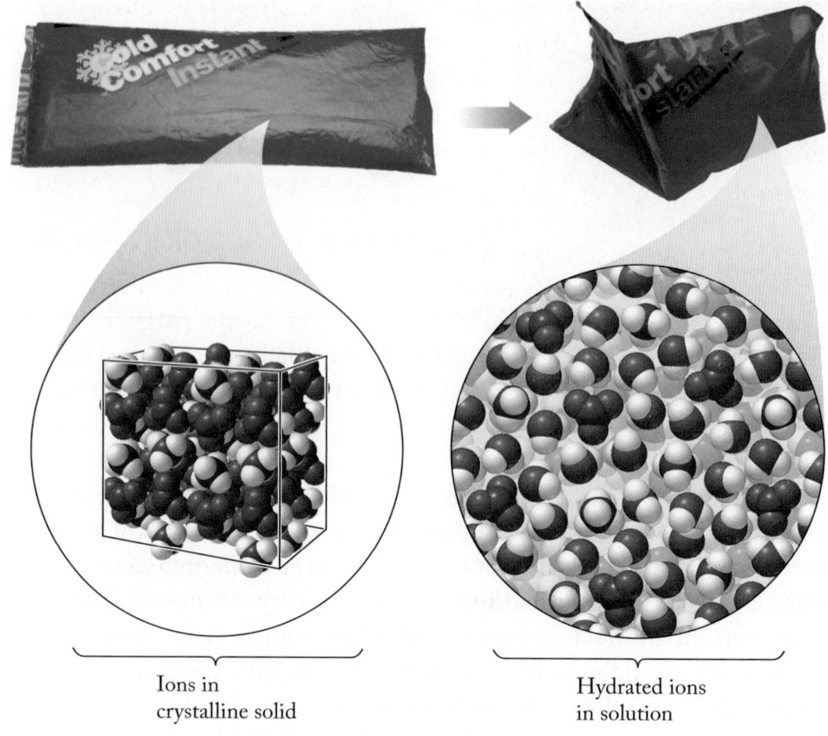

FIGURE 12.3 An instant cold pack gets cold when a membrane inside it is ruptured, allowing a water-soluble compound such as ammonium nitrate to dissolve. The NH_4^+ and NO_3^- ions in solid NH_4NO_3 experience increased freedom of motion as they form hydrated NH_4^+ and NO_3^- ions in solution. The temperature of the solution drops because the dissolution process is endothermic.

CONCEPT **TEST**

In which of the following processes do particles become more spread out and gain more freedom of motion?

a. Perspiration evaporates.

b. Dew forms overnight on blades of grass.

c. A spoonful of sugar dissolves in a cup of coffee.

d. A wooden log burns in a fireplace.

(Answers to Concept Tests are in the back of the book.)

12.2 Entropy and the Second Law of Thermodynamics

CONNECTION The first law of thermodynamics is described in Section 9.1.

The first law of thermodynamics tells us that energy cannot be created or destroyed. This means we cannot do more work than is allowed by the available energy. The **second law of thermodynamics** says, in effect, that not all of the energy released by a spontaneous reaction, such as burning gasoline in a car engine, is available to do useful work. These two principles of energy explain why so-called perpetual motion machines are impossible. In every process that releases energy, some of what is released—engineers call it "low-grade" energy—cannot be used to do anything useful, like propelling a car down the road.

second law of thermodynamics the principle stating that the total entropy of the universe increases in any spontaneous process.

entropy (*S*) a measure of how dispersed the energy in a system is at a specific temperature.

If energy cannot be destroyed, what happens to the energy that is unavailable to do useful work? Essentially, this energy simply spreads out, becoming less concentrated over time. In the case of both the automobile engine and solar energy, the low-grade energy flows uselessly into the surroundings in a much less concentrated form.

Whenever an isolated thermodynamic system undergoes a change in which the particles in the system disperse into a larger volume and/or gain freedom of motion, the process is *always* spontaneous. The system is said to experience an increase in **entropy (S)**. Entropy is a measure of how dispersed the energy of a system is. The formal description of the second law of thermodynamics states that the entropy of an isolated thermodynamic system always increases during a spontaneous process. What exactly is entropy?

The real meaning of entropy is linked to how energy is distributed in a thermodynamic system. Entropy is a state function, which means it depends only on the chemical and physical state of a system and not on the pathway followed to reach that state. To illustrate what we mean by energy dispersion of a system, let's revisit the experiment in Chapter 9 in which we determined the specific heat of a metal (see Figure 9.15). Recall how the metal pellets are first heated to 100°C in a bath of boiling water and then transferred to a small volume of cool water in an insulated container. The temperature of the water in the container increases, and the magnitude of that change can be used to calculate how much energy flows from the metal into the water as the metal and water (a closed thermodynamic system) come to thermal equilibrium. Energy initially confined to the metal flows spontaneously from the hot metal into the cool water and is dispersed throughout the metal/water system.

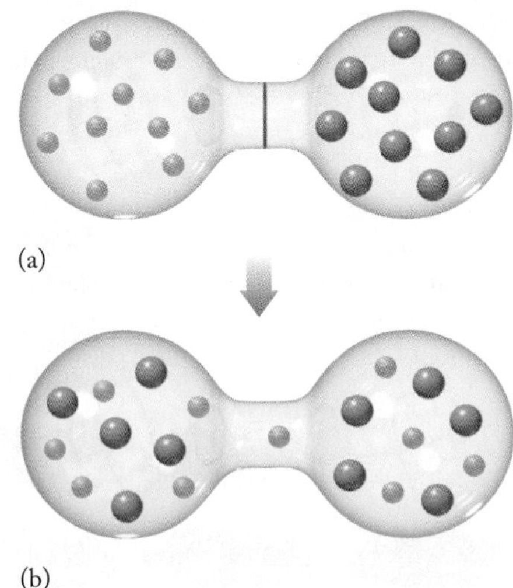

FIGURE 12.4 Entropy of mixing. (a) When the partition separating two containers with different noble gases is removed, (b) the two gases spontaneously diffuse into each other's compartment, resulting in a uniform mixture of randomly distributed atoms of the two gases.

CHEMTOUR Entropy

CONNECTION State functions were defined in Chapter 9.

CONNECTION In Chapter 10 we learned how kinetic molecular theory explains many of the physical properties of gases, including the rates at which they effuse and diffuse, as described by Graham's law.

Energy dispersion can happen even when all the components of a system, or the system and its surroundings, are at the same temperature. Let's revisit a spontaneous, isothermal process from Chapter 10: the diffusion of ideal gases. The apparatus in Figure 12.4 contains two equal-volume chambers. Suppose the one on the left contains 10 neon atoms, and the one on the right contains 10 argon atoms. We assume that the gases behave as ideal gases, which means their atoms are essentially tiny points of mass that do not interact with each other (except for frequent elastic collisions). We also assume the temperatures of both chambers are the same, which means that their pressures are also the same.

When the partition between the two chambers is removed, atoms of each gas spontaneously diffuse into the space originally occupied by the other gas. This behavior is explained by kinetic molecular theory and the random motion of the particles that make up ideal gases. According to Graham's law, the less massive neon atoms will diffuse faster into the space occupied by the more massive argon atoms, but, given enough time, the atoms of both gases will end up evenly distributed throughout the combined volume.

Though kinetic molecular theory is a powerful tool for predicting that gases mix spontaneously, there is another way to account for the spontaneity of mixing—one based on a statistical view of energy dispersion and entropy. To illustrate how it works, let's start with a vastly simplified model of two gases mixing (Figure 12.5). We start with two red marbles representing atoms of one

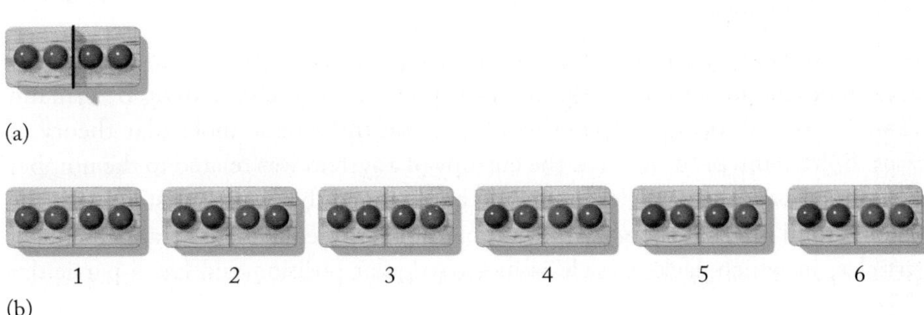

FIGURE 12.5 A simple model of the entropy of mixing. (a) Initially pairs of red and blue marbles are in separate compartments. (b) When mixed, the marbles can form six different patterns. In two of the arrangements (1 and 6), the marbles are still in separate compartments, but in four of them (2–5), they are mixed together: one of each color marble in both compartments. Thus, there are twice as many arrangements with red-blue pairs than there are single color pairs.

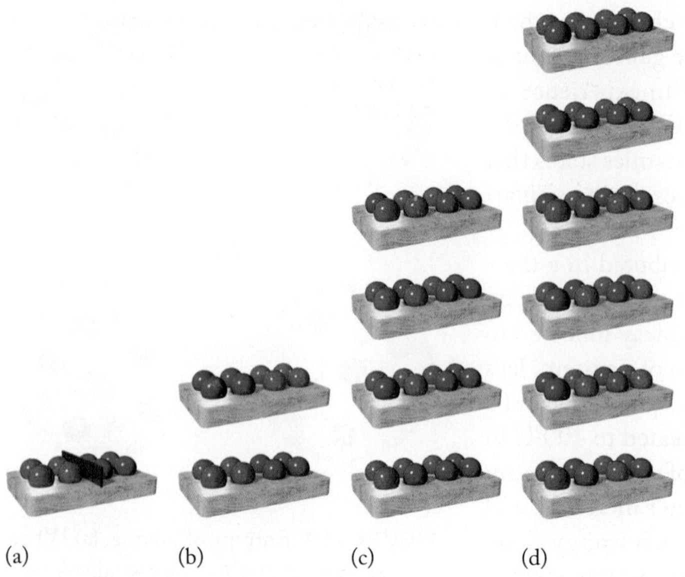

FIGURE 12.6 A larger model of the entropy of mixing. (a) Initially four red and four blue marbles are in separate compartments. (b) When mixed, there are two arrangements of the marbles in which the colors are still separate. (c) There are four possible arrangements in each compartment when it exchanges one marble with the other compartment. Therefore, there are $4 \times 4 = 16$ different arrangements for both compartments, and another 16 (not shown) when the compartments exchange three marbles. (d) The greatest number of arrangements: six for each compartment, or $6 \times 6 = 36$ for both, occurs when each compartment contains two red and two blue marbles. Therefore, the most probable arrangement is even mixing, the next most probable is uneven mixing, and by far the least probable is no mixing at all.

of the gases and two blue marbles representing atoms of the other. Initially they are separated into two compartments as shown in Figure 12.5(a). Then the partition separating them is removed and the marbles are able to occupy any of the sites in either compartment. There are six ways to arrange the two red and two blue marbles among the four sites as shown in Figure 12.5(b). In only two of these (numbered 1 and 6) are the same color marbles in the same compartment. In the other four they are mixed. If there is an equal chance any marble could be in any one of the positions, then it is twice as likely that they will be mixed together (arrangements 2–5) rather than separate (arrangements 1 and 6).

Now let's double the number of marbles and marble-holding sites as shown in Figure 12.6(a). If we randomly arranged all eight marbles among the eight available sites, there are again only two possible arrangements in which each compartment contains marbles of only one color (Figure 12.6b). There are several mixing options: either one or two or three of the red marbles could end up on the side that originally contained only blue marbles, and vice versa. The exchange of one marble produces four options for arranging the blue one and the remaining three red ones in the originally all-red compartment, as shown in Figure 12.6(c). For each of these four arrangements, there are four ways to arrange the one red marble and three blue ones in the other compartment. This means that there are $4 \times 4 = 16$ different ways to arrange the marbles if one red and one blue marble are exchanged between compartments. Can you see how there are also 16 different ways to arrange the marbles if three red and three blue ones are exchanged between compartments? The third option is equal mixing so that each compartment has two red and two blue marbles. There are six ways to arrange these marbles in each compartment (Figure 12.6d) for a total of $6 \times 6 = 36$ arrangements. If all arrangements are equally probable, then out of a total of $2 + 16 + 16 + 36 = 70$ arrangements, only two (or $2/70 = 0.029 = 2.9\%$) of them produce no mixing of the marbles. The remaining 68 arrangements (97.1%) produce at least some mixing, and the most probable mixing pattern (36 out of 70) produces a uniform distribution of marbles between the two compartments.

If we extend the message contained in mixing the small numbers of marbles in Figures 12.5 and 12.6 to extremely large numbers of particles (think Avogadro's number), we come to the following conclusions:

1. Substances (such as gases) that are miscible will mix with each other spontaneously because there is a higher probability of a mixed distribution.
2. The most probable mixing pattern produces a uniform distribution of particles throughout the volume they occupy.

This probabilistic view of particle behavior was first connected to the entropy of thermodynamic systems in the 1870s by German physicist Ludwig Boltzmann (1844–1906), following his pioneering work on the kinetic molecular theory of gases. Boltzmann proposed that the entropy of a system was related to the number of different ways the particles in it could be arranged that was consistent with the system's total internal energy. This number of energy-equivalent arrangements of particles, in which each particle is in a particular position and has a particular momentum (the product of its mass and speed), defined a variable Boltzmann called *Wahrscheinlichkeit* (W)—the German word for probability. He then linked the entropy (S) of the system to the number of probable arrangements (W) of its particles and particle energies, as shown in Equation 12.1:

$$S = k_B \ln W \tag{12.1}$$

where k_B is the Boltzmann constant, which is equal to the gas constant (R) divided by Avogadro's number (N_A):

$$k_B = \frac{R}{N_A} = \frac{8.314 \dfrac{\text{J}}{\cancel{\text{mol}} \cdot \text{K}}}{6.0221 \times 10^{23}/\cancel{\text{mol}}} = 1.381 \times 10^{-23}\ \text{J/K}$$

accessible microstate a unique arrangement of the positions and momenta of the particles in a thermodynamic system.

The probable number of arrangements of the particles in a system at a given temperature is called the number of **accessible microstates**. To get a feel for what this phrase means, let's return to Figure 12.5 and the arrangements numbered 2–5. For now, think of the marbles in these arrangements as stationary atoms in an isolated thermodynamic system. Each of the four arrangements represents an accessible microstate—an equivalent arrangement of the *macro*state in which two atoms of two elements are mixed so that each compartment has one atom of each element. Arrangements 1 and 6 are equivalent microstates of the macrostate in which the four atoms are not mixed. The fact that the mixed macrostate has more microstates means that it has greater entropy. This means that the mixing process is accompanied by an increase in the number of accessible microstates and an increase in entropy and, according to the second law, is spontaneous.

Boltzmann's probabilistic W also incorporates particle motion and position. In addressing particle motion, we need to think of all the ways that particles can move. Put another way, we need to consider the kinds of *motional* energy they have. As we discussed in Chapter 9 (see Figure 9.6), molecules (and atoms) have translational energy, rotational energy, and (in molecules) bond vibrational energy. Recall from Chapter 3 that energy on the atomic scale is quantized, not continuous. Thus, the translational and rotational energies of atoms and molecules and the vibrational energies of the bonds in molecules are all quantized. The differences between translational and rotational energy states of gas-phase atoms and molecules are so small that they have access to enormous numbers of energy states at room temperature. To illustrate how many, let's use Equation 12.1 to calculate the number of microstates that are accessible to a single neon atom at 25°C. To calculate the number of accessible microstates, we rearrange the terms in Equation 12.1 to solve for W and then insert the entropy value of one mole of Ne under standard conditions at 25°C (146.3 J/mol · K; see Section 12.3 and Appendix 4) and the quantity of Ne (one atom):

$$S = k_B \ln W$$

$$\ln W = \frac{S}{k_B}$$

$$\ln W = \frac{\left(146.3 \dfrac{\cancel{\text{J}}}{\cancel{\text{mol}} \cdot \cancel{\text{K}}}\right)\left(\dfrac{1\ \cancel{\text{atom}}}{6.0221 \times 10^{23}\ \frac{\cancel{\text{atoms}}}{\cancel{\text{mol}}}}\right)}{1.381 \times 10^{-23}\ \frac{\cancel{\text{J}}}{\cancel{\text{K}}}}$$

$$\ln W = 17.59$$

$$W = e^{17.59} = 4.358 \times 10^{7}$$

third law of thermodynamics the principle stating that the entropy of a perfect crystal is zero at absolute zero.

This value illustrates the enormous number of ways that the energy of a single gas-phase atom can be dispersed. The motion of an atom in a solid or a liquid is much more restricted, so it has access to many fewer microstates. An atom of carbon in the rigid crystalline structure of a diamond, for example, is restricted to vibrating within a very confined space. There is little uncertainty in the position of the atom, and it has little motional energy. These limitations are reflected in the small entropy value of diamond, 2.4 J/(mol · K), and in the number of microstates accessible to one of its carbon atoms:

$$S = k_B \ln W$$

$$\ln W = \frac{S}{k_B} = \frac{\left(2.4 \frac{\text{J}}{\cancel{\text{mol}} \cdot \text{K}}\right)\left(\frac{1 \cancel{\text{atom}}}{6.0221 \times 10^{23} \frac{\cancel{\text{atoms}}}{\cancel{\text{mol}}}}\right)}{1.381 \times 10^{-23} \frac{\text{J}}{\text{K}}} = 0.29$$

$$W = e^{0.29} = 1.3$$

Thus, there is on average little more than one microstate available to each atom in a diamond at 25°C and 1 bar, but that value may be misleading. When we attempt to calculate the number of accessible microstates in one *mole* of diamond at 25°C and 1 bar, the result is so enormous most calculators can't process it and will generate an error message:

$$\ln W = \frac{\left(2.4 \frac{\text{J}}{\cancel{\text{mol}} \cdot \text{K}}\right) \times 1 \cancel{\text{mol}}}{1.381 \times 10^{-23} \frac{\text{J}}{\text{K}}} = 1.7 \times 10^{23}$$

$$W = e^{1.7 \times 10^{23}} = [\text{Error:overflow}]$$

As a result, the number of microstates accessible to a mole of particles is enormous, even when they are locked in place in a crystalline solid.

The statistical definition of entropy based on accessible microstates is conceptually compatible with the thermodynamic, macroscopic view based on energy dispersion. After all, each accessible microstate represents one of the multitude of ways that energy can be dispersed in a thermodynamic system. Actually, some facets of entropy are easier to understand using the microstate model. One of them is the concept of absolute entropy, and the notion embedded in the **third law of thermodynamics**, which states that a perfect crystalline solid has zero entropy at absolute zero. If the particles of a crystalline solid are uniformly arranged in three-dimensional space (Figure 12.7) and in their lowest possible energy states, the crystal has only one microstate, and, according to Equation 12.1, its entropy is zero:

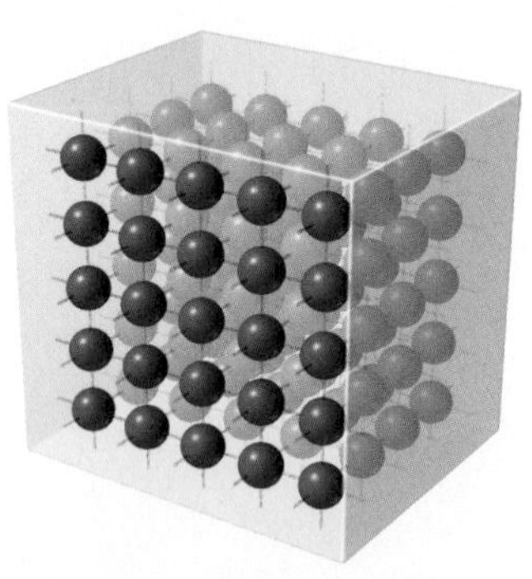

FIGURE 12.7 A perfect crystal at 0 K has zero entropy because all particles have no freedom of motion and no kinetic energy.

$$S = k_B \ln W = k_B \ln 1 = k_B \times 0 = 0$$

CONCEPT TEST

Predict which one of the following values comes closest to the number of microstates accessible to one molecule of liquid water at 25°C: (a) 1, (b) 10^6, (c) 10^{12}, (d) 10^{18}, (e) 10^{24}. Use the appropriate $S°$ value from Appendix 4 and Equation 12.1 to determine whether your selection was the best one.

12.3 Absolute Entropy and Molecular Structure

standard molar entropy ($S°$) the absolute entropy of one mole of a substance in its standard state.

The third law of thermodynamics provides a vital reference point for the absolute entropy scale: a perfect crystal at 0 K has zero entropy. From this starting point and by carefully measuring the rise in temperature of substances as known quantities of thermal energy are added to them, scientists have been able to establish absolute entropy values for many substances at many temperatures. The most common quantity for expressing these values is **standard molar entropy ($S°$)**, which is the entropy of one mole of a substance in its standard state (see Table 12.1) at a pressure of 1 bar (~1 atm). Some sample $S°$ values at 298 K are listed in Table 12.2.

TABLE 12.1 Standard States of Pure Substances and Solutions

Physical State	Standard State	Pressure[a]	Temperature[b]
Solid	Pure solid, most stable allotrope of an element	1 bar	25°C
Liquid	Pure liquid	1 bar	25°C
Gas	Pure gas	1 bar	25°C
Solution	1 *M*	1 bar	25°C

[a]Since 1982, 1 bar has been the standard pressure for tabulating all thermodynamic data. Prior to 1982, standard pressure was 1 atmosphere (atm) = 1.01325 bar.

[b]The thermodynamic data in Appendix 4 and used elsewhere in this book are based on a temperature of 25°C (298 K). This temperature is *not* the same as the STP value we use for ideal gases, which is 273 K.

The $S°$ values for liquid water and water vapor in Table 12.2 illustrate a difference between the entropies of liquids and gases that we have discussed before in this chapter: the molecules in a gas under standard conditions are much more widely dispersed than the molecules in a liquid, and the entropies of the different phases of a substance at a given temperature follow the order: $S_{solid} < S_{liquid} < S_{gas}$.

TABLE 12.2 Selected Standard Molar Entropy Values[a]

Formula	$S°$ [J/(mol · K)]	Formula	Name	$S°$ [J/(mol · K)]
$Br_2(g)$	245.5	$CH_4(g)$	Methane	186.2
$Br_2(\ell)$	152.2	$CH_3CH_3(g)$	Ethane	229.5
$C_{diamond}(s)$	2.4	$CH_3CH_2CH_3(g)$	Propane	269.9
$C_{graphite}(s)$	5.7	$CH_3(CH_2)_2CH_3(g)$	Butane	310.0
$CO(g)$	197.7	$CH_3(CH_2)_2CH_3(\ell)$		231.0
$CO_2(g)$	213.8	$CH_3OH(g)$	Methanol	239.9
$H_2(g)$	130.6	$CH_3OH(\ell)$		126.8
$N_2(g)$	191.5	$CH_3CH_2OH(g)$	Ethanol	282.6
$O_2(g)$	205.0	$CH_3CH_2OH(\ell)$		160.7
$H_2O(g)$	188.8	$C_6H_6(g)$	Benzene	269.2
$H_2O(\ell)$	69.9	$C_6H_6(\ell)$		172.9
$NH_3(g)$	192.5	$C_{12}H_{22}O_{11}(s)$	Sucrose	360.2

[a]Values for additional substances are given in Appendix 4.

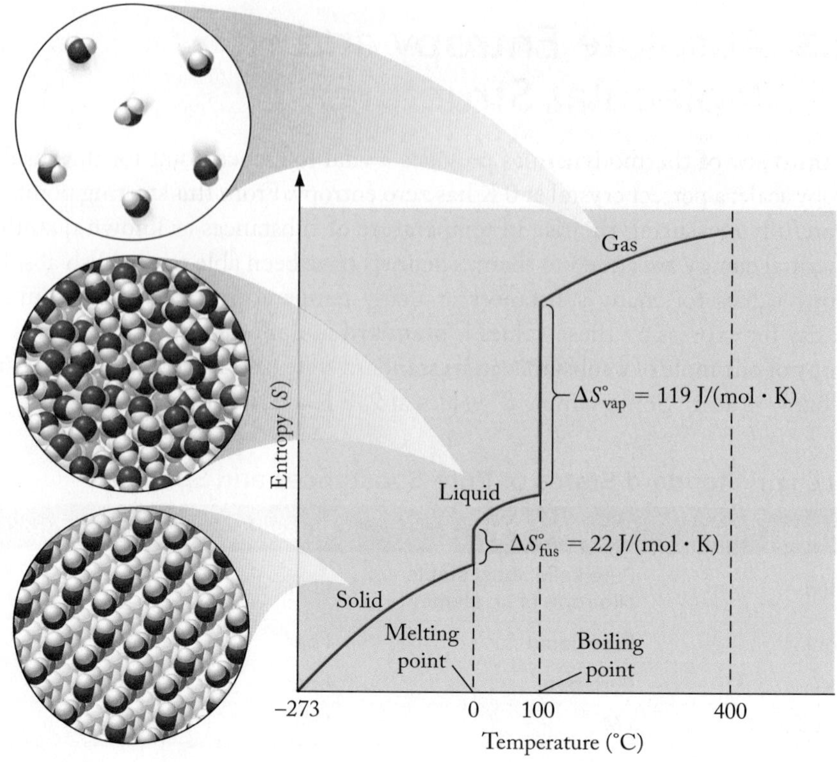

FIGURE 12.8 Abrupt increases in entropy accompany changes of state, as shown for water, with the greater increase occurring during the transition from liquid to gas. In both phase changes, entropy increases because each change increases the dispersion of the energy of a system's particles. In other words, the particles have access to more microstates. We can often make qualitative predictions about entropy changes that accompany chemical reactions based on these factors, even if we have no thermodynamic data about the reactants and products.

The entropy changes that occur as one mole of ice at 0 K is heated are shown in Figure 12.8. Note the jump in entropy as the ice melts at 0°C and the even bigger jump as the liquid water vaporizes at 100°C. Also note that the lines between the phase changes are curved. The change in entropy, ΔS, with temperature is not linear because heating a substance at a higher temperature produces a smaller entropy increase than adding the same quantity of heat to the same substance at a lower temperature.

The factors that affect entropy change can be summarized as follows:

1. Entropy increases when temperature increases.
2. Entropy increases when volume increases.
3. Entropy increases when the number of independent particles increases.

SAMPLE EXERCISE 12.1 Predicting the Sign of an Entropy Change **LO1**

Predict whether an increase or decrease in entropy accompanies each of the following processes when they occur at constant temperature:

a. $H_2O(\ell) \rightarrow H_2O(g)$
b. $NH_3(g) + HCl(g) \rightarrow NH_4Cl(s)$
c. $C_{12}H_{22}O_{11}(s) \rightarrow C_{12}H_{22}O_{11}(aq)$
d. $2\,H_2(g) + O_2(g) \rightarrow 2\,H_2O(g)$

Collect, Organize, and Analyze We are given chemical equations describing four processes and are asked to predict the signs of the accompanying entropy changes.

a. Molecules of liquid water become molecules of (gaseous) water vapor, which means the molecules and their kinetic energies are more dispersed.

b. Two gaseous compounds form a single solid compound, which means the particles of product occupy much less space and have much less freedom of motion.
c. A solid dissolves in water, forming molecules dispersed in an aqueous solution that have more freedom of motion.
d. A total of three moles of gaseous reactants combine to form two moles of a gaseous product. Having fewer gas-phase particles means less freedom of motion.

Solve

a. $\Delta S > 0$.
b. $\Delta S < 0$.
c. $\Delta S > 0$.
d. $\Delta S < 0$.

Think About It Entropy increases when solids melt and liquids vaporize because their particles experience increased freedom of motion and greater dispersion of their kinetic energies. When solids dissolve in liquids, they also gain freedom of motion and experience an increase in entropy. When gases combine to form liquids, solids, or fewer particles of gases, they lose freedom of motion and entropy decreases.

Practice Exercise Does each of the following chemical reactions result in an increase or decrease in the entropy of the system? Assume the reactants and products are at the same temperature.

a. $CaCO_3(s) + 2\ HCl(aq) \rightarrow CaCl_2(aq) + CO_2(g) + H_2O(\ell)$
b. $NH_3(g) + BF_3(g) \rightarrow NH_3BF_3(s)$
c. $C_3H_8(g) + 5\ O_2(g) \rightarrow 3\ CO_2(g) + 4\ H_2O(g)$

(Answers to Practice Exercises are in the back of the book.)

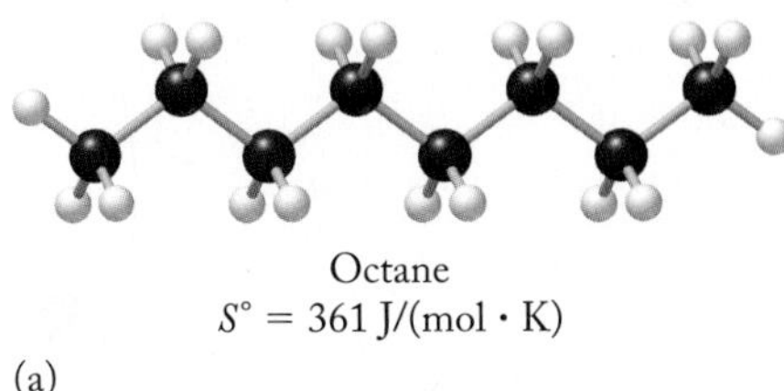

Octane
$S° = 361$ J/(mol · K)

(a)

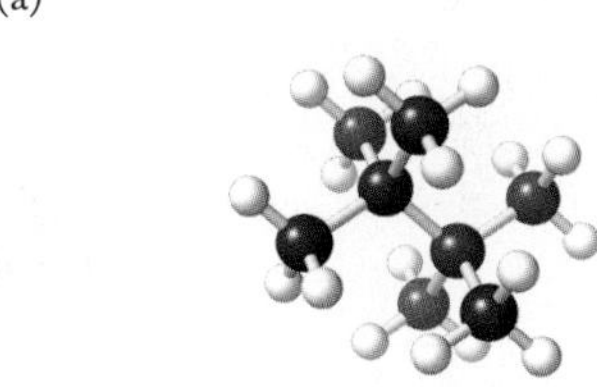

Tetramethylbutane
$S° = 274$ J/(mol · K)

(b)

FIGURE 12.9 Ball-and-stick models and $S°$ values of (a) octane and (b) tetramethylbutane.

The data in Table 12.2 contain an important message about the standard molar entropies of molecular substances: they are linked to the masses of their molecules and their molecular structures. To see the result of this influence, consider the standard molar entropies of the C_1 to C_4 alkanes in natural gas in Table 12.2. Note how $S°$ values increase with increasing molecular size. This trend is due to the different ways that these molecules move in addition to simple translational motion in three-dimensional space. For example, we have noted that molecules in the gas phase are free to rotate, which adds to the number of accessible microstates. Even liquid molecules tumble over each other as they flow past their nearest neighbors, adding to the ways that energy can be dispersed among them.

The quantities of energy involved in rotational motion depend on how massive the molecules are and on how their masses are distributed. Consider, for example, the $S°$ values of octane and tetramethylbutane in Figure 12.9. The isomers have the same molar masses but different molecular shapes. The longer, narrower shape of octane means that more of its mass is farther from the center of the molecule and farther from the axis of rotation when the molecule tumbles end over end. When the mass of a molecule is more spread out, the spacing between its rotational energy states is smaller and more are accessible. Therefore, molecules of octane have greater entropy than the same number of molecules of tetramethylbutane at the same temperature.

Another way to think about the distribution of mass and entropy is to consider how many different orientations could be drawn for the molecule due to free rotation about C—C single bonds. Octane is linear and has many more possible conformations than tetramethylbutane. Only a few of the possible orientations are shown in Figure 12.10.

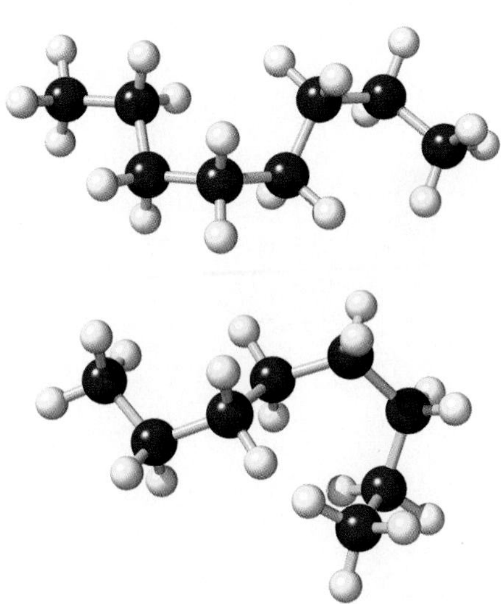

FIGURE 12.10 Linear alkanes don't have to be linear. These models represent a few of the many orientations of the atoms in octane as a result of rotations around its C—C bonds.

FIGURE 12.11 (a) Diamonds have a rigid three-dimensional structure which correlates with a small standard molar entropy. (b) The larger $S°$ for graphite reflects the ease with which layers of carbon atoms can slide past each other.

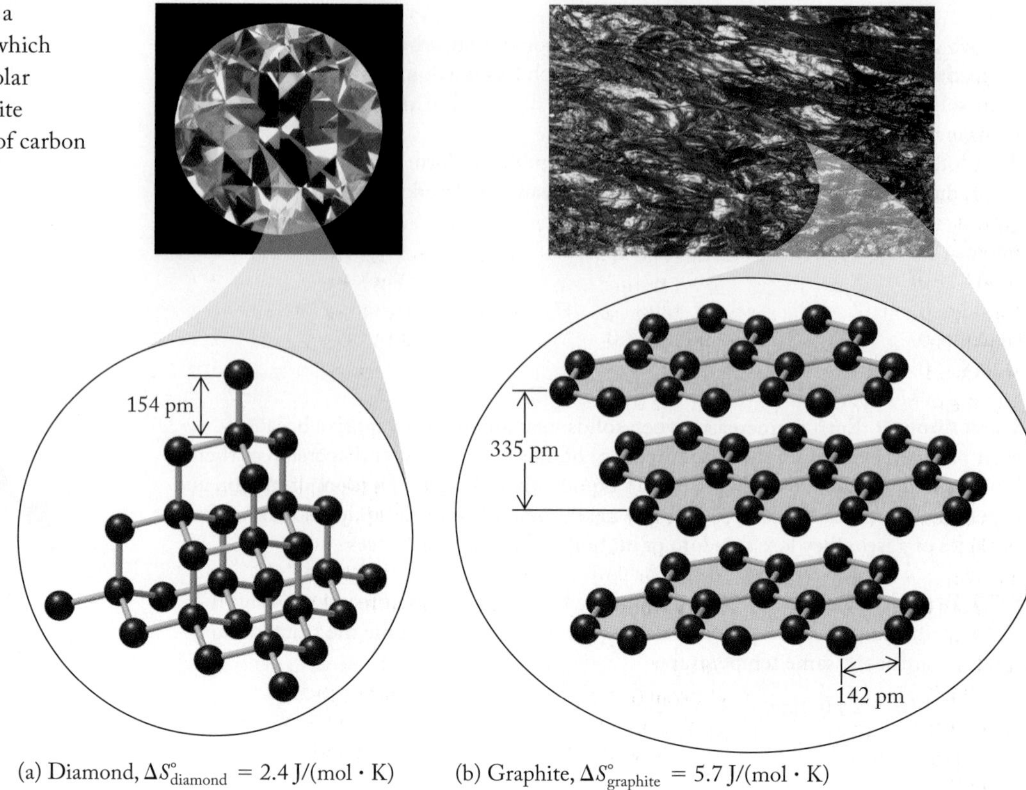

(a) Diamond, $\Delta S^\circ_{\text{diamond}} = 2.4\ \text{J/(mol} \cdot \text{K)}$ (b) Graphite, $\Delta S^\circ_{\text{graphite}} = 5.7\ \text{J/(mol} \cdot \text{K)}$

Another structural feature that influences entropy is rigidity. The two most common forms of carbon—namely, diamond and graphite—both consist of extended arrays of C—C bonds. However, diamond is an extremely hard substance because it has a rigid three-dimensional atomic structure (Figure 12.11a). Its atoms have access to fewer microstates and it has less entropy [$S° = 2.4$ J/(mol · K)] than soft, easily deformed graphite [$S° = 5.7$ J/(mol · K)], which consists of two-dimensional sheets (Figure 12.11b) that are separated by more than the length of a C—C bond. This means that they are attracted to each other by only London dispersion forces, which allows them to slide past each other when a modest shear force is applied.

SAMPLE EXERCISE 12.2 Comparing Standard Molar Entropy Values **LO2**

Without consulting any reference sources, select the component in each of the following pairs that has the greater standard molar entropy at 298 K. Assume there is one mole of each component in its standard state (i.e., the pressure of each gas is 1 bar and the concentration of each solution is 1 *M*).

a. $HCl(g)$, $HCl(aq)$
b. $CH_3OH(\ell)$, $CH_3CH_2OH(\ell)$
c.

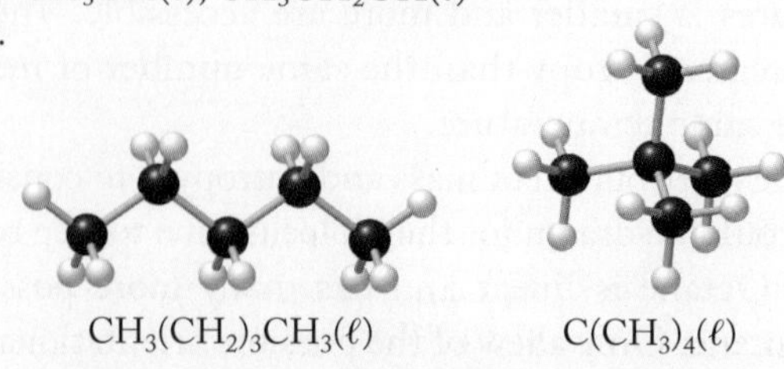

$CH_3(CH_2)_3CH_3(\ell)$ $C(CH_3)_4(\ell)$

Collect and Organize We are given chemical formulas or molecular models of pairs of substances, and we are asked to select which component of each pair has the greater entropy per mole under standard conditions—that is, the greater standard molar entropy ($S°$) at 298 K.

Analyze Particles in the vapor state of a substance are more dispersed and have more entropy than they do in the liquid state at the same temperature, which in turn have more entropy than particles in the solid state. Substances with larger molar masses tend to have greater $S°$ values than those in the same physical state that are composed of less massive particles. Substances composed of compact molecules have lower standard molar entropies than those composed of elongated molecules of the same mass. The volume of one mole of an ideal gas at $P = 1$ bar and $T = 298$ K is about 24.8 liters; the volume of one mole of solute in a 1 M solution is only 1 liter.

Solve

a. One mole of HCl dissolved in one liter of water occupies much less volume and its particles have much less freedom of motion than one mole of HCl gas. Therefore, HCl(g) has a greater $S°$ value than HCl(aq).

b. Both compounds are liquid alcohols, but molecules of ethanol (CH_3CH_2OH) are more massive than those of methanol (CH_3OH). Therefore, ethanol has a greater $S°$ value than methanol.

c. The two compounds are isomers with the same molar mass. However, molecules of $CH_3(CH_2)_3CH_3$ are more elongated and have more opportunities than $C(CH_3)_4$ to disperse their kinetic energy. Thus, there are more ways to draw molecules of $CH_3(CH_2)_3CH_3$ because of free rotation about the C—C single bonds. A few possibilities are shown here:

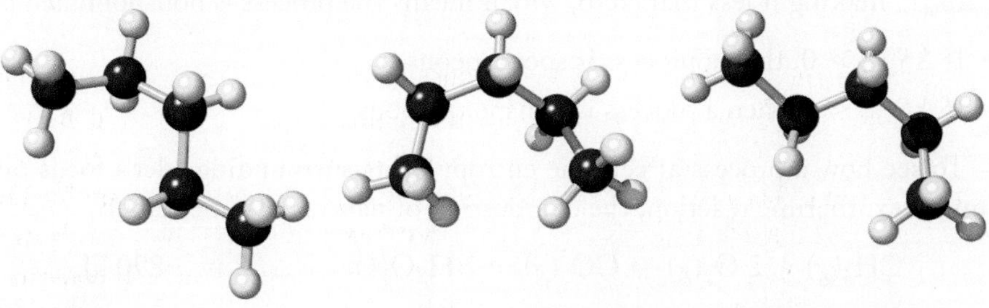

The greater number of possible orientations of C—C bonds in $CH_3(CH_2)_3CH_3$ means that linear $CH_3(CH_2)_3CH_3$ has more opportunities to disperse its kinetic energy and therefore it has a greater $S°$ value compared to $C(CH_3)_4$.

Think About It The comparison in part (a) is complicated by the fact that HCl is a strong acid, which means that each molecule produces two ions in solution. However, they are liquid-phase ions and have less total entropy than half as many gas-phase HCl molecules. [Compare the $S°$ values of HCl(g), $H^+(aq)$, and $Cl^-(aq)$ in Appendix 4 to see for yourself.]

Practice Exercise Figure 12.12 shows ball-and-stick models and structural formulas of four hydrocarbons that each contain six carbon atoms per molecule. Rank these compounds in order of decreasing $S°$ values.

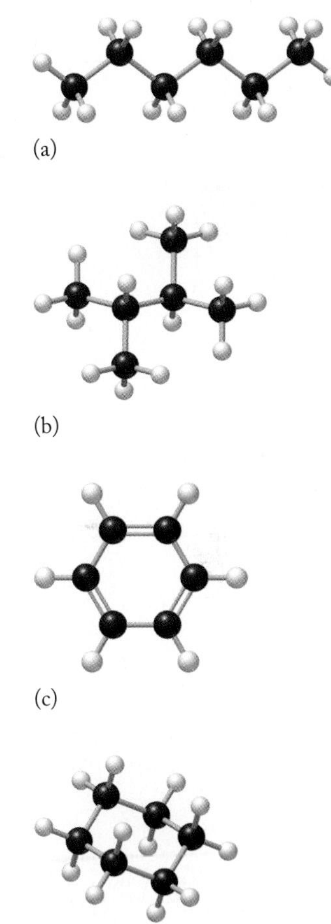

FIGURE 12.12

12.4 Applications of the Second Law

We saw in Section 12.2 how the second law of thermodynamics accounts for the spontaneity of processes occurring in isolated systems. It can also be applied to processes occurring in systems that are not isolated (which is good, because most systems are not—they are either open or closed). Recall from Chapter 9 that in

CONNECTION In Chapter 9 we defined an isolated thermodynamic system as one that exchanges neither energy nor matter with its surroundings. A closed system exchanges energy but not matter, and an open system exchanges both.

thermodynamics we divide up the universe into two parts: the part we are interested in (the system) and everything else (the surroundings). Mathematically,

$$\text{Universe} = \text{system} + \text{surroundings}$$

By analogy, the overall change in the entropy of the universe as a result of a physical or chemical change is the sum of the entropy changes experienced by the system and its surroundings:

$$\Delta S_{\text{univ}} = \Delta S_{\text{sys}} + \Delta S_{\text{surr}} \tag{12.2}$$

When a spontaneous process occurs in an isolated system, $\Delta S_{\text{sys}} > 0$. Because the system is isolated, the process has no impact on its surroundings, so $\Delta S_{\text{surr}} = 0$. Therefore, according to Equation 12.2, ΔS_{univ} must be greater than zero because $\Delta S_{\text{sys}} > 0$. The positive value of ΔS_{univ} is the basis for another way of expressing the second law of thermodynamics that applies to all systems: *a spontaneous process produces an increase in the entropy of the universe.*

The latter version of the second law assumes that a physical or chemical change in a closed or open thermodynamic system can alter the entropies of both the system and its surroundings. The second law says that a process is spontaneous when $\Delta S_{\text{univ}} > 0$. The second law also provides a thermodynamic requirement for nonspontaneity: a process that produces a decrease in the entropy of the universe will not occur spontaneously. Therefore, the reverse of any spontaneous process has to be nonspontaneous because it can be spontaneous in only one direction—the one for which $\Delta S_{\text{univ}} > 0$. Reversing a spontaneous process reverses the sign of ΔS_{univ}, making it less than zero, which means the process is nonspontaneous.

- If $\Delta S_{\text{univ}} > 0$, then a process is spontaneous.
- If $\Delta S_{\text{univ}} < 0$, then a process is nonspontaneous.

To see how a process affects the entropy of its surroundings, let's focus on a familiar exothermic reaction, the combustion of natural gas (methane):

$$CH_4(g) + 2\,O_2(g) \rightarrow CO_2(g) + 2\,H_2O(\ell) \qquad \Delta H^\circ = -890 \text{ kJ}$$

As written, the reaction consumes three moles of gases, and it produces two moles of a liquid product and one mole of CO_2 gas, so there are fewer moles of gases on the product side of the reaction equation. Given the much greater freedom of motion of particles in the gas phase, we predict that there will be a decrease in entropy of the reaction mixture, which is our thermodynamic system:

$$\Delta S_{\text{sys}} < 0$$

The reaction is spontaneous, however, so

$$\Delta S_{\text{univ}} > 0$$

How do we reconcile these opposing inequalities? Equation 12.2 supplies an explanation. If $\Delta S_{\text{univ}} > 0$, then the sum of ΔS_{sys} and ΔS_{surr} must also be greater than zero. The fact that $\Delta S_{\text{sys}} < 0$ simply means that ΔS_{surr} is not only greater than zero, but also it must have a large enough positive value to more than offset the negative value of ΔS_{sys}. Expressing this relationship in terms of the absolute values of ΔS_{surr} and ΔS_{sys}:

$$|\Delta S_{\text{surr}}| > |\Delta S_{\text{sys}}|$$

Is the combustion of methane likely to produce a large, positive ΔS_{surr}? Absolutely it will—because the reaction is highly exothermic: combustion of only one mole (16 grams) of methane releases 890 kJ of thermal energy. As energy flows from the system into its surroundings, dispersion of this energy produces a positive ΔS_{surr} that more than compensates for the unfavorable (negative) value of ΔS_{sys}.

The combinations of ΔS_{sys} and ΔS_{surr} that produce a positive value for ΔS_{univ} are shown in Figure 12.13. Note how processes in which ΔS_{sys} and ΔS_{surr} are both greater than zero (Figure 12.13a) inevitably result in an increase in ΔS_{univ}, which means they are always spontaneous. On the other hand, processes in which ΔS_{sys} and ΔS_{surr} are both less than zero (Figure 12.13d) always produce a decrease in ΔS_{univ}, which means they are always nonspontaneous. The other four cases in Figure 12.13 represent pairs of ΔS_{sys} and ΔS_{surr} values with opposite signs, which may or may not combine to produce positive ΔS_{univ} values. It all depends on the

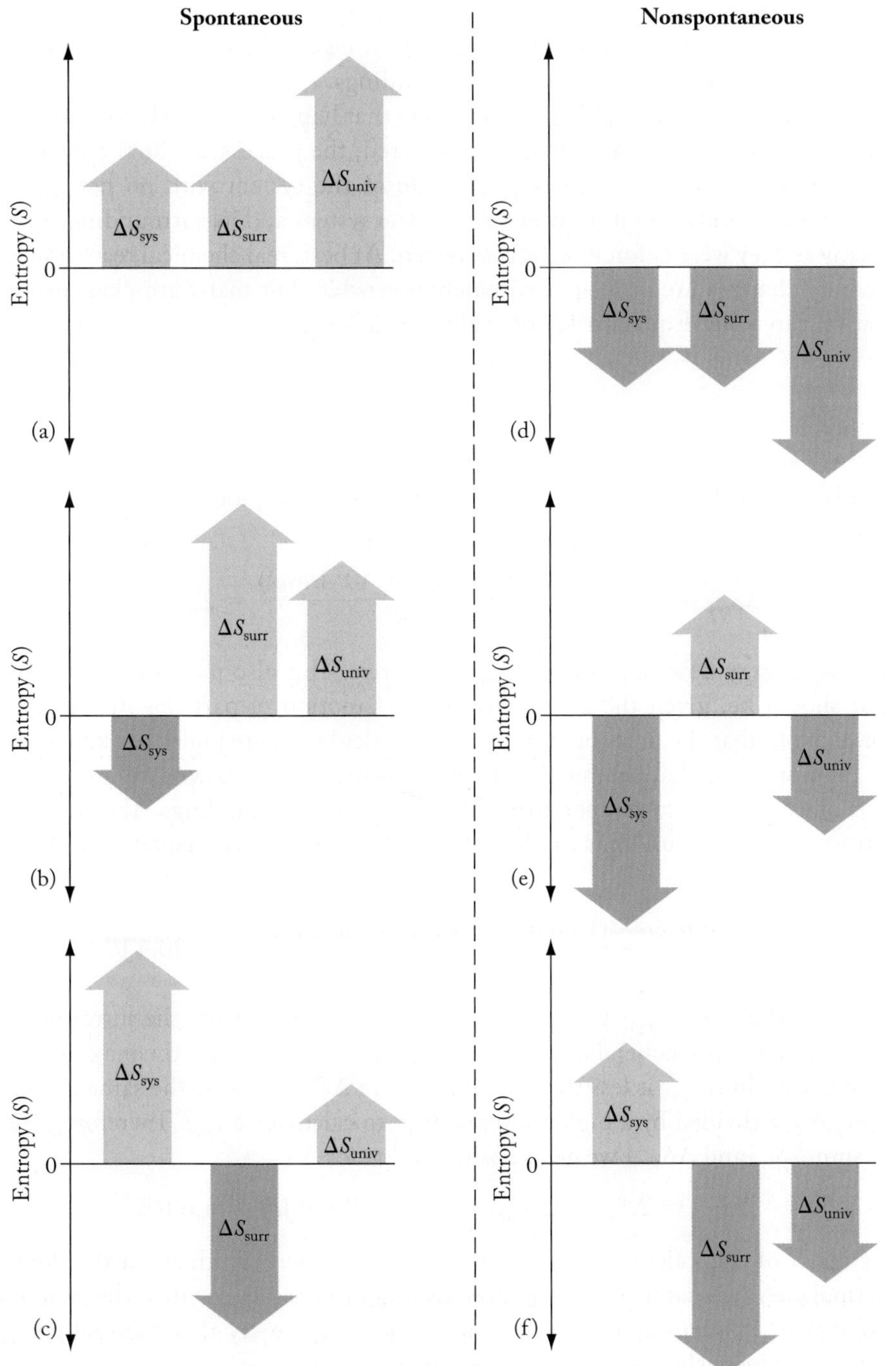

FIGURE 12.13 The relationships between ΔS_{sys}, ΔS_{surr}, and ΔS_{univ} in spontaneous and nonspontaneous processes. A process is spontaneous whenever $\Delta S_{univ} > 0$. A process is nonspontaneous whenever $\Delta S_{sys} + \Delta S_{surr} = \Delta S_{univ} < 0$.

reversible process a process that happens so slowly that an incremental change can be reversed by another tiny change, restoring the original state of the system with no net flow of energy between the system and its surroundings.

absolute values of ΔS_{sys} and ΔS_{surr}. If the larger of the two in magnitude has the positive value, then ΔS_{univ} increases, and the process is spontaneous.

All exothermic reactions have the capacity to raise the entropy of their surroundings. The more energy that flows into the surroundings, or into any collection of particles, the greater the dispersion of energy among the particles and the greater the increase in their entropy (ΔS). However, heating particles that are already hot produces a smaller gain in entropy than heating the same particles at a lower temperature. This inverse relationship between entropy gain and temperature was discussed with Figure 12.8 and is reflected in Equation 12.3:

$$\Delta S = \frac{q_{rev}}{T} \tag{12.3}$$

where q_{rev} represents the reversible energy flow caused by the difference in temperature between the system and its surroundings.

Theoretically, a **reversible process** is one that happens so slowly that, after an incremental change in the system has occurred, the process can be reversed by a tiny change that restores the original state of the system with no net flow of energy into or out from it. In other words, the system and its surroundings finish exactly as they were before the process began. At best, real chemical reactions and physical changes are only approximately reversible, but many are close enough that we can adapt Equation 12.3 to calculate ΔS_{sys}:

$$\Delta S_{sys} = \frac{q_{sys}}{T} \tag{12.4}$$

Let's apply Equation 12.4 to the melting of ice. It takes 6.01 kJ (or 6.01×10^3 J) of energy to melt 1.00 mol of ice at 0°C. Assuming the process occurs reversibly, then

$$\Delta S_{sys} = \frac{q_{sys}}{T} = \frac{(1.00\ \cancel{\text{mol}})(6.01 \times 10^3\ \text{J}/\cancel{\text{mol}})}{273\ \text{K}} = 22.0\ \text{J/K}$$

Because energy flows into the ice, q_{sys} is positive. This also makes ΔS_{sys} positive, as it should be, given the greater freedom of motion of particles in the liquid phase. Note that the units of entropy in this calculation are joules per kelvin.

Now suppose that the melting process occurs as heat flows into a one-mole cube of ice from room-temperature (22°C or 295 K) surroundings. The change in entropy of the surroundings can be calculated using another adaptation of Equation 12.3:

$$\Delta S_{surr} = \frac{q_{surr}}{T} = \frac{(1.00\ \cancel{\text{mol}})(-6.01 \times 10^3\ \text{J}/\cancel{\text{mol}})}{295\ \text{K}} = -20.4\ \text{J/K}$$

Note that the sign of q_{surr} is negative because energy flows from the surroundings into the system (ice cube). Note, too, that (1) $\Delta S_{surr} < 0$, and (2) the magnitude of the decrease in ΔS_{surr} is less than the increase in ΔS_{sys}, because the same absolute value of q is divided by a higher temperature to calculate ΔS_{surr}. Therefore, when we sum ΔS_{sys} and ΔS_{surr}, we get a positive value for ΔS_{univ}:

$$\Delta S_{univ} = \Delta S_{sys} + \Delta S_{surr} = (22.0 - 20.4)\ \text{J/K} = 1.6\ \text{J/K}$$

The result of this calculation is one example of a general truth about the flow of thermal energy—namely, it flows spontaneously into a system when the system is cooler than its surroundings. Even more generally, we say that thermal energy flows spontaneously from a warm object to an adjacent cooler object.

Why did we ignore the change in temperature of the surroundings as heat flowed from it into the melting ice? The answer lies in the sheer size of the surroundings (the universe minus a small cube of ice). The temperature of such a gigantic thermal mass does not change significantly.

Now let's consider how entropy changes when the temperature of the surroundings is *lower* than the temperature of the system. Suppose 1.00 mol of liquid water at 0°C is placed in a freezer at −10°C. Because the temperature of the surroundings (the freezer) is lower than the temperature of the liquid water, energy spontaneously flows from the water into its surroundings. The net entropy change for this process is

$$\begin{aligned}\Delta S_{\text{univ}} &= \Delta S_{\text{sys}} + \Delta S_{\text{surr}} \\ &= \frac{(1.00\ \cancel{\text{mol}})(-6.01 \times 10^3\ \text{J}/\cancel{\text{mol}})}{273\ \text{K}} + \frac{(1.00\ \cancel{\text{mol}})(+6.01 \times 10^3\ \text{J}/\cancel{\text{mol}})}{263\ \text{K}} \\ &= (-22.0\ \text{J/K}) + (+22.9\ \text{J/K}) \\ &= 0.9\ \text{J/K}\end{aligned}$$

Once again there is an increase in the entropy of the universe as energy flows spontaneously from the warmer object (liquid water at 0°C) into the colder surroundings (the freezer at −10°C).

CONCEPT **TEST**

Is ΔS_{univ} greater than, less than, or equal to zero when water vapor exhaled by the Inuit hunter in Figure 12.14 is deposited as crystals of ice on his beard?

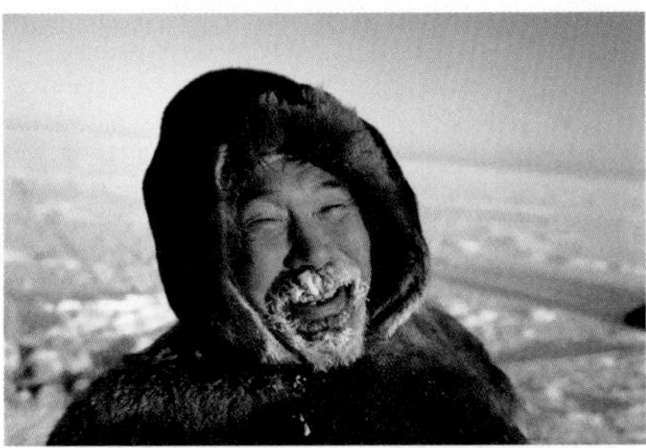

FIGURE 12.14 Water vapor exhaled by this Inuit hunter in the Northwest Territories of Canada was spontaneously deposited on his facial hair as frost—a testament to how cold it was when this photo was taken.

Entropy-change calculations based on Equation 12.3 assume process reversibility, which, as we have discussed, is an idealized, theoretical concept. In reality, the ΔS values calculated in this way are *minimum* ΔS values. When processes take place in the real world, the accompanying changes in entropy are inevitably greater than those calculated using Equation 12.3.

12.5 Calculating Entropy Changes

The entropy of a system (like its enthalpy and internal energy) is a state function, which means that the change in entropy that accompanies a process depends only on the initial and final states of the system, not on the pathway of the process. Therefore, the change in entropy experienced by the system is simply the difference between its initial and final absolute entropy levels:

CONNECTION State functions were defined in Chapter 9.

$$\Delta S_{\text{sys}} = S_{\text{final}} - S_{\text{initial}} \qquad (12.5)$$

We can adapt Equation 12.5 to calculate the change in entropy that accompanies a chemical reaction under standard conditions, $\Delta S^\circ_{\text{rxn}}$, from the difference between the standard molar entropies of $n_{\text{reactants}}$ moles of reactants (the equivalent of S_{initial} in Equation 12.5) and the standard molar entropies of n_{products} moles of products (i.e., S_{final}):

$$\Delta S^\circ_{\text{rxn}} = \sum n_{\text{products}} S^\circ_{\text{products}} - \sum n_{\text{reactants}} S^\circ_{\text{reactants}} \qquad (12.6)$$

Each individual S° value for a product or reactant is multiplied by the appropriate number of moles from the balanced chemical equation. In other words, $\Delta S^\circ_{\text{rxn}}$, like

ΔH°_{rxn}, is an extensive thermodynamic property that depends on the amounts of substances consumed or produced in a reaction. Standard molar entropies of selected substances are listed in Appendix 4.

SAMPLE EXERCISE 12.3 Calculating ΔS Values **LO3**

A process called methanol reforming is used to generate hydrogen for fuel cells from methanol and steam. Use the appropriate standard molar entropy values in Table 12.2 to calculate the value of ΔS°_{rxn} at 298 K for the methanol-reforming reaction:

$$CH_3OH(g) + H_2O(g) \rightarrow CO_2(g) + 3\ H_2(g)$$

Collect, Organize, and Analyze We are given a balanced chemical equation describing a reaction and are asked to use the appropriate standard molar entropy values to calculate the change in entropy that occurs during the reaction under standard conditions. Entropy changes associated with chemical reactions depend on the entropies of the reactants and products as described in Equation 12.6. Table 12.2 contains the following S° values:

Substance:	$CH_3OH(g)$	$H_2O(g)$	$CO_2(g)$	$H_2(g)$
S° (J/mol·K):	239.9	188.8	213.8	130.6

During the reaction two moles of gaseous reactants produce four moles of gaseous products, so it is likely the value of ΔS°_{rxn} will be greater than zero.

Solve

$$\begin{aligned}\Delta S^\circ_{rxn} &= \sum n_{products} S^\circ_{products} - \sum n_{reactants} S^\circ_{reactants}\\ &= \left[1\ \cancel{mol} \times 213.8 \frac{J}{\cancel{mol} \cdot K} + 3\ \cancel{mol} \times 130.6 \frac{J}{\cancel{mol} \cdot K}\right]\\ &\quad - \left[1\ \cancel{mol} \times 239.9 \frac{J}{\cancel{mol} \cdot K} + 1\ \cancel{mol} \times 188.8 \frac{J}{\cancel{mol} \cdot K}\right]\\ &= (605.6 - 428.7)\ J/K = 176.9\ J/K\end{aligned}$$

Think About It As predicted, $\Delta S^\circ > 0$ because of the greater freedom of motion and dispersion of energy with four moles of gas compared to two.

Practice Exercise Calculate the standard molar entropy change for the combustion of methane gas using S° values from Table 12.2. Before carrying out the calculation, predict whether the entropy of the system increases or decreases. Assume that liquid water is one of the products.

12.6 Free Energy

In Sample Exercise 12.3 we used standard molar entropy values to calculate ΔS°_{rxn}. The reactants and products together constituted a closed thermodynamic system, which meant that heat could flow into or out from it but matter was not exchanged with its surroundings. Therefore, the ΔS° value that was calculated applied only to the system. There was no evaluation of the change in the entropy of the system's surroundings accompanying the reaction. If the reaction were endothermic, heat flowing from the surroundings into the system would lower the temperature of the surroundings and produce a decrease in S_{surr}. If the reaction were exothermic, heat would flow from the system into its surroundings and ΔS_{surr} would be greater than zero.

Thus, the entropy change experienced by the surroundings of any chemical system depends on whether the process occurring in the system is exothermic or endothermic. When energy flows from an exothermic process occurring at constant pressure into a system's surroundings, the quantity of energy is equal in magnitude but opposite in sign to the enthalpy change of the system:

CONNECTION In Chapter 9 we defined a change in enthalpy (ΔH) as the energy gained or lost as heat during a process taking place at constant pressure.

$$q_{surr} = -\Delta H_{sys} \quad (12.7)$$

When a system undergoes an endothermic process, the direction of energy flow is reversed but Equation 12.7 still applies. Assuming these transfers of heat occur reversibly, we can calculate the value of ΔS_{surr} by substituting $-\Delta H_{sys}$ for q_{surr} in Equation 12.3:

$$\Delta S_{surr} = -\frac{\Delta H_{sys}}{T}$$

and combining the equation with Equation 12.2 ($\Delta S_{univ} = \Delta S_{sys} + S_{surr}$) gives

$$\Delta S_{univ} = \Delta S_{sys} - \frac{\Delta H_{sys}}{T} \quad (12.8)$$

The beauty of Equation 12.8 is that it allows us to predict whether or not a process is spontaneous at a particular temperature once we calculate the enthalpy and entropy changes accompanying the process. The downside of Equation 12.8 is that spontaneity relies on the value of a parameter (ΔS_{univ}) that is impossible to determine directly and that has little physical meaning. It would be useful if we could substitute a single thermodynamic parameter for ΔS_{univ} that is based only on the system and not the entire universe. Such a parameter exists, and it is called the change in the system's **free energy**. Note that *free* in this context does not mean "at no cost"; it means energy that is *freed* during a process to do useful work.

In chemistry we focus on a particular kind of free energy, called **Gibbs free energy (G)** in honor of American scientist J. Willard Gibbs (1839–1903). Gibbs free energy is the maximum amount of energy available to do useful work in processes happening at constant temperature and pressure or once the temperatures and pressures of reaction mixtures have returned to their initial values. These conditions are met in the chemical reactions and physical changes we discuss in the following sections. As a result, the changes in free energy that accompany these processes will be changes in Gibbs free energy, and we will use the symbol ΔG to represent these changes. Gibbs free energy, like enthalpy and entropy, is a state function.

CHEMTOUR Gibbs Free Energy

Like many of the thermodynamic properties we have examined, absolute free-energy values of substances are often of less interest than the *changes* in free energy that accompany chemical reactions and other processes. Gibbs proposed that the change in free energy (ΔG_{sys}) of a process occurring at constant temperature and pressure is linked directly to that temperature and ΔS_{univ}:

$$\Delta G_{sys} = -T\Delta S_{univ}$$

Because of the ($-T$) multiplier, *negative* values of ΔG_{sys} correspond to *positive* values of ΔS_{univ}. Therefore,

- If $\Delta G_{sys} < 0$, then $\Delta S_{univ} > 0$ and the reaction is spontaneous.
- If $\Delta G_{sys} > 0$, then $\Delta S_{univ} < 0$ and the reaction is nonspontaneous. Instead, the reaction running in reverse of the process is spontaneous.
- If $\Delta G_{sys} = 0$, then $\Delta S_{univ} = 0$, and the composition of the reaction mixture does not change with time. Instead, the reaction has reached equilibrium.

These points are summarized in Table 12.3.

free energy a measure of the maximum amount of work a thermodynamic system can perform.

Gibbs free energy (G) the maximum amount of energy released by a process occurring at constant temperature and pressure that is available to do useful work.

TABLE 12.3 Links between ΔG°_{sys}, ΔS°_{univ}, and Spontaneity

ΔG°_{sys}	ΔS°_{univ}	Spontaneity
< 0	> 0	Spontaneous
> 0	< 0	Nonspontaneous
0	0	No change with time; at equilibrium

The Gibbs equation ($\Delta G_{sys} = -T\Delta S_{univ}$) can be combined with Equation 12.8 once we multiply all of the terms in Equation 12.8 by ($-T$):

$$-T\Delta S_{univ} = -T\Delta S_{sys} + \Delta H_{sys} \quad (12.9)$$

The left side of Equation 12.9 is equal to ΔG_{sys}. Making that substitution and rearranging the terms on the right side gives

$$\Delta G_{sys} = \Delta H_{sys} - T\Delta S_{sys}$$

Since all of the parameters in this equation apply to the system, we typically simplify the equation by eliminating the subscripts:

$$\Delta G = \Delta H - T\Delta S \quad (12.10)$$

Equation 12.10 highlights the two thermodynamic factors that contribute to a decrease in free energy and to making a process spontaneous:

1. The system experiences an increase in entropy ($\Delta S > 0$).
2. The process is exothermic ($\Delta H < 0$).

Table 12.4 summarizes how the signs of ΔH and ΔS impact the sign of ΔG and how they define the conditions under which a process is spontaneous.

The change in free energy of a process can be calculated using Equation 12.10 if we first calculate the values of ΔH and ΔS. In the case of a chemical reaction occurring under standard conditions, the *standard* change in free energy ΔG°_{rxn} can be calculated using a modified version of Equation 12.10:

$$\Delta G^\circ_{rxn} = \Delta H^\circ_{rxn} - T\Delta S^\circ_{rxn} \quad (12.11)$$

In Chapter 9 we calculated ΔH°_{rxn} values from the difference in the standard heats of formation (ΔH°_f) values of products and reactants:

$$\Delta H^\circ_{rxn} = \sum n_{products}\Delta H^\circ_{f,products} - \sum n_{reactants}\Delta H^\circ_{f,reactants} \quad (9.15)$$

The value of ΔS°_{rxn} can be calculated using Equation 12.6:

$$\Delta S^\circ_{rxn} = \sum n_{products}S^\circ_{products} - \sum n_{reactants}S^\circ_{reactants} \quad (12.6)$$

The results of these two calculations are combined using Equation 12.11 to calculate ΔG°_{rxn}, as illustrated in Sample Exercises 12.4 and 12.5.

TABLE 12.4 Effects of ΔH, ΔS, and T on ΔG and Spontaneity

ΔH	ΔS	ΔG	Spontaneity
−	+	Always < 0	Always spontaneous
−	−	< 0 at lower temperature	Spontaneous at lower temperature
+	+	< 0 at higher temperature	Spontaneous at higher temperature
+	−	Always > 0	Never spontaneous

SAMPLE EXERCISE 12.4 Calculating ΔG° for a Chemical Reaction **LO4**

In automobile engines, nitrogen and oxygen react to form nitrogen monoxide (NO), which enters the atmosphere and combines with oxygen in the following reaction:

$$2\ NO(g) + O_2(g) \rightarrow 2\ NO_2(g)$$

What is the value of ΔG°_{rxn} for this reaction at 298 K if $\Delta H^\circ_{rxn} = -114.2$ kJ and $\Delta S^\circ_{rxn} = -146.5$ J/K?

Collect, Organize, and Analyze We are given the ΔH°_{rxn} and ΔS°_{rxn} values for a reaction and the temperature and asked to calculate the value of ΔG°_{rxn}. Equation 12.11 connects these three thermodynamic parameters. The change in entropy is given in J, so we must convert that value to kJ before combining it with ΔH°_{rxn}.

Solve

$$\begin{aligned}\Delta G^\circ_{rxn} &= \Delta H^\circ_{rxn} - T\Delta S^\circ_{rxn} \\ &= (-114.2 \text{ kJ}) - \left[298 \text{ K} \times (-146.5 \text{ J/K}) \times \frac{1 \text{ kJ}}{1000 \text{ J}}\right] \\ &= -70.5 \text{ kJ}\end{aligned}$$

Think About It The negative value of ΔG° means that the reaction is spontaneous under standard conditions and $T = 298$ K. Nitrogen dioxide is the substance responsible for the brown haze in the atmosphere in some areas with high vehicle traffic.

 **Practice Exercise** We opened the chapter discussing one of the reactions responsible for iron rusting:

$$4 \text{ Fe}(s) + 3 \text{ O}_2(g) \rightarrow 2 \text{ Fe}_2\text{O}_3(s)$$

Calculate ΔG°_{rxn} for this reaction at $T = 298$ K if $\Delta H^\circ_{rxn} = -1648.4$ kJ and $\Delta S^\circ_{rxn} = -549.3$ J/K.

SAMPLE EXERCISE 12.5 Predicting Reaction Spontaneity under Standard Conditions LO4

Consider the reaction of nitrogen gas and hydrogen gas (Figure 12.15) at 298 K to make ammonia at the same temperature:

$$\text{N}_2(g) + 3 \text{ H}_2(g) \rightarrow 2 \text{ NH}_3(g)$$

a. Before doing any calculations, predict the sign of ΔS°_{rxn}.
b. What is the actual value of ΔS°_{rxn}?
c. What is the value of ΔH°_{rxn}?
d. What is the value of ΔG°_{rxn} at 298 K?
e. Is the reaction spontaneous at 298 K and 1 bar of pressure?

Collect and Organize We are asked to predict the sign and calculate the value of the standard entropy change of a reaction. We are also asked to calculate the enthalpy change of the reaction and its change in free energy, and then to predict its spontaneity under standard conditions. Standard molar entropies and standard heats of formation of the reactants and product are listed in Table 12.2 and Appendix 4.

Analyze Figure 12.15 reinforces the point that there are more moles of gaseous reactants than products in the reaction. Equation 12.6 can be used to calculate entropy changes under standard conditions:

$$\Delta S^\circ_{rxn} = \sum n_{\text{products}} S^\circ_{\text{products}} - \sum n_{\text{reactants}} S^\circ_{\text{reactants}} \quad (12.6)$$

We learned in Chapter 9 how to calculate ΔH°_{rxn} from standard heats of formation values (in Appendix 4) using Equation 9.15:

$$\Delta H^\circ_{rxn} = \sum n_{\text{products}} \Delta H^\circ_{\text{f,products}} - \sum n_{\text{products}} \Delta H^\circ_{\text{f,reactants}} \quad (9.15)$$

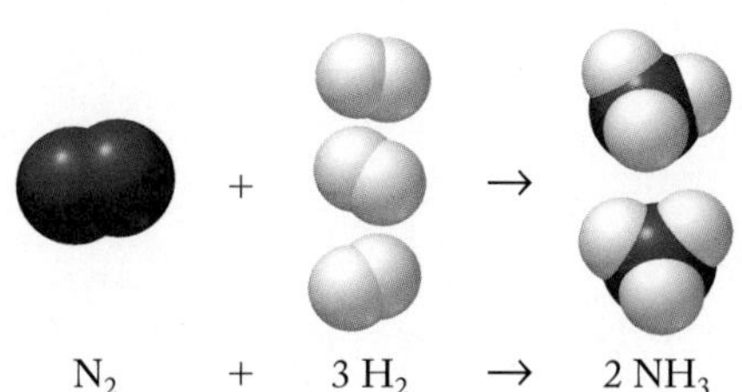

FIGURE 12.15 In the synthesis of ammonia, one molecule of nitrogen reacts with three molecules of hydrogen to yield two molecules of ammonia. All substances are gases.

The calculated values of ΔS°_{rxn} and ΔH°_{rxn} can be combined in Equation 12.11 to calculate ΔG°_{rxn}:

$$\Delta G^\circ_{rxn} = \Delta H^\circ_{rxn} - T\Delta S^\circ_{rxn} \qquad (12.11)$$

Whether the reaction is spontaneous ($\Delta G^\circ_{rxn} < 0$) or nonspontaneous ($\Delta G^\circ_{rxn} > 0$) under standard conditions (and $T = 298$ K) is difficult to predict without knowing the magnitudes of ΔH°_{rxn} and $T\Delta S^\circ_{rxn}$.

Solve

a. The number of gas-phase molecules decreases as the reaction proceeds, so the sign of ΔS°_{rxn} is probably negative.

b. Using data from Table 12.2 in Equation 12.6 to calculate ΔS°_{rxn},

$$\begin{aligned} S^\circ_{rxn} &= \sum n_{\text{products}} S^\circ_{\text{products}} - \sum n_{\text{reactants}} S^\circ_{\text{reactants}} \\ &= \left[2\ \cancel{\text{mol}} \times \left(192.5\ \frac{\text{J}}{\cancel{\text{mol}} \cdot \text{K}}\right)\right] \\ &\quad - \left[1\ \cancel{\text{mol}} \times \left(191.5\ \frac{\text{J}}{\cancel{\text{mol}} \cdot \text{K}}\right) + 3\ \cancel{\text{mol}} \times \left(130.6\ \frac{\text{J}}{\cancel{\text{mol}} \cdot \text{K}}\right)\right] \\ &= -198.3\ \text{J/K} \end{aligned}$$

The entropy change is negative, as predicted.

c. The change in enthalpy that accompanies the reaction under standard conditions is

$$\begin{aligned} \Delta H^\circ_{rxn} &= \sum n_{\text{products}} \Delta H^\circ_{\text{f,products}} - \sum n_{\text{reactants}} \Delta H^\circ_{\text{f,reactants}} \\ &= \left[2\ \cancel{\text{mol}} \times \left(-46.1\ \frac{\text{kJ}}{\cancel{\text{mol}}}\right)\right] - \left[1\ \cancel{\text{mol}} \times \left(0.0\ \frac{\text{kJ}}{\cancel{\text{mol}}}\right) + 3\ \cancel{\text{mol}} \times \left(0.0\ \frac{\text{kJ}}{\cancel{\text{mol}}}\right)\right] \\ &= -92.2\ \text{kJ} \end{aligned}$$

d. Using the values of S°_{rxn} and H°_{rxn} calculated in parts (b) and (c) in Equation 12.11,

$$\begin{aligned} \Delta G^\circ_{rxn} &= \Delta H^\circ_{rxn} - T\Delta S^\circ_{rxn} \\ &= -92.2\ \text{kJ} - (298\ \cancel{\text{K}})\left(-198.3\ \frac{\cancel{\text{J}}}{\cancel{\text{K}}}\right)\left(\frac{1\ \text{kJ}}{1000\ \cancel{\text{J}}}\right) \\ &= -33.1\ \text{kJ} \end{aligned}$$

e. $\Delta G^\circ_{rxn} < 0$, so the reaction is spontaneous under standard conditions and $T = 298$ K.

Think About It We were unable to predict the spontaneity of the reaction because we didn't have a feel for the magnitudes of ΔH°_{rxn} and ΔS°_{rxn}. In this case, an unfavorable entropy change ($\Delta S^\circ_{rxn} < 0$) is more than offset by a favorable enthalpy change ($\Delta H^\circ_{rxn} < 0$) so that, overall, ($\Delta G^\circ_{rxn} < 0$). Note that ΔH°_{rxn} is twice the ΔH°_f of ammonia.

Practice Exercise For the reaction $2\ H_2(g) + O_2(g) \rightarrow 2\ H_2O(\ell)$,

a. Predict the sign of the entropy change for the reaction.
b. What is the value of ΔS°_{rxn}?
c. What is the value of ΔH°_{rxn}?
d. Is the reaction spontaneous at 298 K and 1 bar pressure?

CONCEPT TEST

The preparation of ammonia from nitrogen and hydrogen in Sample Exercise 12.5 is predicted to be spontaneous under standard conditions, yet if we mix the two gases at 298 K, no reaction is observed. Suggest a reason why.

There is another way to calculate the change in free energy of a reaction under standard conditions. It is based on another thermodynamic property of substances listed in Appendix 4: the **standard free energy of formation (ΔG°_f)**. A compound's ΔG°_f value is the change in free energy associated with the formation of one mole of it in its standard state from its elements in their standard states.

standard free energy of formation (ΔG°_f) the change in free energy associated with the formation of one mole of a compound in its standard state from its elements in their standard states.

In much the way we calculated standard heats of reactions (ΔH°_{rxn}) from the difference in the standard heats of formation (ΔH°_f) of their products and reactants in Chapter 9, so, too, can we calculate the change in standard free energy of a reaction under standard conditions from the difference in the standard free energies of formation of its products and reactants. As with standard heats of formation, the standard free energy of formation of the most stable forms of elements in their standard states is zero. The similarities in the two calculations can be seen from the similar formats of the equations used to calculate ΔG°_{rxn}:

$$\Delta G^\circ_{rxn} = \sum n_{products} \Delta G^\circ_{f,products} - \sum n_{reactants} \Delta G^\circ_{f,reactants} \quad (12.12)$$

and ΔH°_{rxn}:

$$\Delta H^\circ_{rxn} = \sum n_{products} \Delta H^\circ_{f,products} - \sum n_{reactants} \Delta H^\circ_{f,reactants} \quad (9.15)$$

Sample Exercise 12.6 illustrates just how similar the two calculations are.

Octane
$\Delta G^\circ_f = 16.3$ kJ/mol
(a)

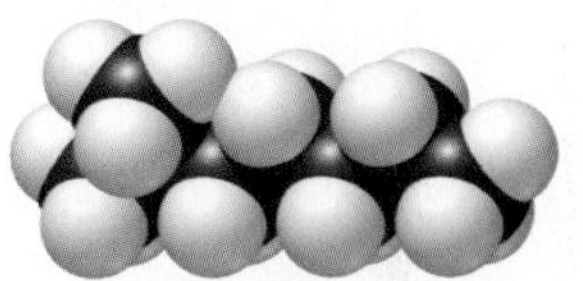

2-Methylheptane
$\Delta G^\circ_f = 11.7$ kJ/mol
(b)

3,3-Dimethylhexane
$\Delta G^\circ_f = 12.6$ kJ/mol
(c)

FIGURE 12.16 Molecular structures and ΔG°_f values of three C_8H_{18} isomers.

CONCEPT TEST

Molecular models and standard free energies of formation for three structural isomers with the molecular formula C_8H_{18} are shown in Figure 12.16. All three isomers burn in air, as described by the same chemical equation:

$$2\ C_8H_{18}(\ell) + 25\ O_2(g) \rightarrow 16\ CO_2(g) + 18\ H_2O(g)$$

Are the ΔG°_{rxn} values for the three combustion reactions also the same? Why or why not?

SAMPLE EXERCISE 12.6 Calculating ΔG°_{rxn} Using Appropriate ΔG°_f Values — LO4

Use the appropriate standard free energy of formation values in Appendix 4 to calculate the change in free energy as ethanol burns under standard conditions. Assume the reaction proceeds as described by the following chemical equation:

$$CH_3CH_2OH(\ell) + 3\ O_2(g) \rightarrow 2\ CO_2(g) + 3\ H_2O(\ell)$$

Collect and Organize We are asked to calculate the value of ΔG°_{rxn} for the combustion of ethanol using the "appropriate" ΔG°_f values, which are, according to Equation 12.12, the ΔG°_f values of the reactants and products in the combustion reaction:

Substance	$CH_3CH_2OH(\ell)$	$O_2(g)$	$CO_2(g)$	$H_2O(\ell)$
ΔG°_f (kJ/mol)	−174.9	0	−394.4	−237.2

Analyze The reaction consumes one mole of liquid ethanol and three moles of oxygen gas and produces two moles of CO_2 gas and three moles of liquid H_2O. Multiplying the ΔG°_f values by the appropriate numbers of moles and subtracting the sum of the reactant values from the sum of the product values, as described in Equation 12.12, will yield the value of ΔG°_{rxn}. The combustion of ethanol, a common additive in gasoline in the United States, is spontaneous, so ΔG°_{rxn} should be less than zero.

Solve Inserting the appropriate numbers of moles and ΔG_f° values in Equation 12.12 and doing the math,

$$\Delta G^\circ_{rxn} = \sum n_{products}\Delta G^\circ_{f,products} - \sum n_{reactants}\Delta G^\circ_{f,reactants}$$
$$= [2\ \cancel{mol\ CO_2} \times (-394.4\ kJ/\cancel{mol\ CO_2}) + 3\ \cancel{mol\ H_2O} \times (-237.2\ kJ/\cancel{mol\ H_2O})]$$
$$-[1\ \cancel{mol\ CH_3CH_2OH} \times (-174.9\ kJ/\cancel{mol\ CH_3CH_2OH}) + 3\ \cancel{mol\ O_2} \times (0\ kJ/mol\ O_2)]$$
$$= -1325.5\ kJ/mol$$

Think About It The negative value of ΔG°_{rxn} confirms our prediction that the reaction is spontaneous. The calculated value represents the maximum amount of the total energy released by the combustion of one mole of ethanol under standard conditions that is available to do useful work.

Practice Exercise Use the appropriate standard free energy of formation values in Appendix 4 to calculate the value of ΔG°_{rxn} for the steam–methane reforming reaction, which is used commercially to produce H_2 gas:

$$CH_4(g) + H_2O(g) \rightarrow CO(g) + 3\ H_2(g)$$

CONCEPT TEST

a. Given the thermodynamic data in Table A4.3, is the conversion of diamond to graphite spontaneous? Explain your answer.

b. If yes, does knowing that the conversion is spontaneous tell you how rapid the conversion is?

The Meaning of *Free* Energy

What exactly does "energy available to do useful work" mean? We answer this question by considering the internal combustion (gasoline) engines used to power most automobiles. A combustion reaction mixture is a thermodynamic system that experiences a decrease in internal energy (ΔE) as heat (q) flows out from it into its surroundings and as it does work (w) on its surroundings. These three variables are related by Equation 9.3:

$$\Delta E = q + w \tag{9.3}$$

and all three are less than zero from the perspective of the system. Internal combustion engines have cooling systems to manage the dissipation of heat, which is wasted energy that does nothing to power the car. The energy that does move the car is derived from the rapid expansion of the gaseous products of combustion in the cylinders of its engine. As Figure 12.17 shows, this expansion pushes down on the piston of a cylinder, increasing the volume of the reaction mixture. The product of the pressure exerted by the reacting gases and their products on the piston and the resulting volume change is P–V work (discussed in Section 9.2), which propels the car. At constant pressure:

FIGURE 12.17 Thermal expansion of the gases in a cylinder in a car engine pushes down on a piston with a pressure P represented by the blue arrow. The product of P and the change in volume of the gases ΔV is the work done by the expanding gases.

$$w = -P\Delta V$$

Free energy is a measure of the *maximum* amount of work that can be done by the energy released during combustion. To see how this theoretical quantity of work compares with the total energy released, we can rearrange Equation 12.10 to isolate the ΔH term:

$$\Delta H = \Delta G + T\Delta S \tag{12.13}$$

Equation 12.13 tells us that the enthalpy change that accompanies the making and breaking of chemical bonds during a chemical reaction may be divided into two parts, a ΔG part and a $T\Delta S$ part. The ΔG part is the energy that can theoretically be converted into motion and other useful work (like generating electricity for the car's electrical system). The $T\Delta S$ part, on the other hand, is unusable: it is the portion of energy that spreads out when, for example, hot gases flow out the end of an automobile exhaust pipe. This part of ΔH is wasted. Consequently, the conversion of chemical energy into useful mechanical energy (ΔG) is never 100% efficient. In addition, some portion of ΔG is also wasted because the combustion and energy conversion happen quickly and, therefore, irreversibly. Maximum efficiency comes with very slow, reversible reactions and energy conversion rates, but that is not how automobile engines operate. It turns out that gasoline engines convert only about 30% of the energy produced during combustion into useful work.

12.7 Temperature and Spontaneity

Let's revisit the process of ice melting (Figure 12.2), this time focusing on how the values of ΔH, $T\Delta S$, and ΔG change as the temperature of a mixture of ice and water increases from 263 to 283 K (Figure 12.18). As temperatures rise over this range, there is little impact on the enthalpy of fusion, ΔH, as shown by the nearly flat green line in Figure 12.18. Increasing the temperature, though, increases the value of $T\Delta S$ (as shown by the upward slope of the purple line). The ΔS of ice (or any solid) melting has a positive value, so $T\Delta S$ must increase as T increases. The green and purple lines intersect at 273 K or 0°C, which means ΔH is equal to $T\Delta S$ at that temperature. Put another way, the difference between ΔH and $T\Delta S$ is zero at 0°C—and so is ΔG (because $\Delta G = \Delta H - T\Delta S$). When $\Delta G = 0$ for a process, then that process is at equilibrium. Ice and water coexist and are in equilibrium with each other at 0°C.

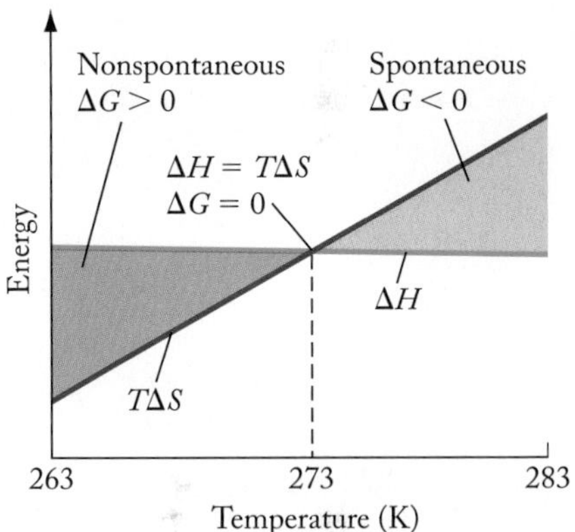

FIGURE 12.18 Changes in the values of ΔH, the quantity $T\Delta S$, and ΔG for ice melting as the temperature changes from −10°C (263 K) to +10°C (283 K).

Ice melts spontaneously above 0°C, so ΔG for the melting process must be less than zero. The graph in Figure 12.18 shows that indeed it is. Above 0°C the value of $T\Delta S$ is greater than ΔH. Therefore, the difference between them ($\Delta H - T\Delta S$) is less than zero and becomes more negative with increasing temperature, as illustrated by the distance from the purple line to the green line in Figure 12.18.

Below 0°C, ice does not melt spontaneously, which means $\Delta G > 0$. The graph shows why this is true: at $T < 273$ K (0°C), $T\Delta S$ values (the purple line) are less than the ΔH values (the green line), which means that $\Delta H - T\Delta S$ (and ΔG) is greater than zero. Positive ΔG values mean that the process is nonspontaneous below 273 K and ice does not melt below its freezing point. However, the opposite process—liquid water freezing—*is* spontaneous because reversing a process keeps the absolute values but switches the signs of ΔH, ΔS, and ΔG. Therefore, if the melting process is nonspontaneous, then the exothermic freezing process is spontaneous at low temperatures.

CONNECTION In Chapter 6 we discussed the equilibrium between phases of a substance and used phase diagrams to illustrate which physical states are stable at various combinations of temperature and pressure.

CONCEPT TEST

A process is spontaneous at lower temperatures and nonspontaneous at higher temperatures. Sketch a graph similar to Figure 12.18 for this process.

The temperature at the point in Figure 12.18 where $\Delta G = 0$ defines the melting (or freezing) point of water, which is the temperature at which solid ice and liquid water coexist in a state of thermodynamic equilibrium. Similarly, at 100°C and 1 atm, liquid water and water vapor coexist because the free-energy changes that accompany vaporization and condensation both equal zero.

The temperature at which $\Delta G = 0$ for a process can be calculated from Equation 12.10 if we know the values of ΔH and ΔS. These values for the fusion of ice are

$$H_2O(s) \rightarrow H_2O(\ell) \qquad \Delta H^\circ = 6.01 \times 10^3 \text{ J/mol}; \Delta S^\circ = 22.0 \text{ J/(mol} \cdot \text{K)}$$

Assuming ΔH° and ΔS° do not change significantly with small changes in temperature, we can use these values in calculations at temperatures other than 298 K. This turns out to be a reasonable assumption, so that

$$\Delta G = \Delta H - T\Delta S \approx \Delta H^\circ - T\Delta S^\circ$$

Inserting the values of ΔH° and ΔS° and using the fact that $\Delta G = 0$ for a process at equilibrium, we get

$$\Delta G = (6.01 \times 10^3 \text{ J/mol}) - T[22.0 \text{ J/(mol} \cdot \text{K)}] = 0$$

$$T = \frac{6.01 \times 10^3 \text{ J/}\cancel{\text{mol}}}{22.0 \text{ J/(}\cancel{\text{mol}} \cdot \text{K)}} = 273 \text{ K} = 0°\text{C}$$

which is the familiar value for the melting point of ice.

Like physical processes, chemical reactions can have zero change in free energy. As a spontaneous reaction ($\Delta G_{rxn} < 0$) proceeds, reactants become products and ΔG_{rxn} becomes less negative, as we will see in the next section. In fact, ΔG_{rxn} may reach zero before all the reactants are consumed. In this case, there is no further net increase in the quantities of products and no further net loss in the quantities of reactants in the reaction mixture. Rather, reactants and products coexist in equilibrium with each other. Reactants still react, and products are still formed, but the reverse reaction, in which products become reactants, proceeds at the same rate as the forward reaction. This is the state we know as chemical equilibrium. No net change in the amounts of reactants and products occurs once equilibrium is reached.

SAMPLE EXERCISE 12.7 Relating Reaction Spontaneity to ΔH_{rxn} and ΔS_{rxn} — LO6

The gas-phase reaction between methanol and water that produces carbon dioxide and three moles of hydrogen is accompanied by an increase in entropy, $\Delta S^\circ_{rxn} = 176.9$ J/K, under standard conditions (see Sample Exercise 12.3).

$$CH_3OH(g) + H_2O(g) \rightarrow CO_2(g) + 3\,H_2(g)$$

At what temperatures is this reaction spontaneous?

Collect and Organize We are asked to determine the temperatures at which a reaction with $\Delta S^\circ > 0$ is spontaneous. Equation 12.11 relates spontaneity ($\Delta G^\circ < 0$) to ΔH°_{rxn}, ΔS°_{rxn}, and T.

$$\Delta G^\circ_{rxn} \approx \Delta H^\circ_{rxn} - T\Delta S^\circ_{rxn} < 0$$

Analyze A reaction in which entropy increases ($\Delta S^\circ > 0$) will be spontaneous at all temperatures if it is also exothermic ($\Delta H^\circ_{rxn} < 0$). However, if the reaction is accompanied by a positive enthalpy change, $\Delta H^\circ_{rxn} > 0$ (endothermic), then the reaction will only be spontaneous above a certain temperature. We need to calculate ΔH°_{rxn} using standard enthalpies of formation and then solve the relationship:

$$T > \Delta H^\circ_{rxn}/\Delta S^\circ_{rxn}$$

Solve The enthalpy change for the reaction is calculated as follows:

$$\Delta H^\circ_{rxn} = [\Delta H^\circ_{CO_2,g} + 3\Delta H^\circ_{H_2,g}] - [\Delta H^\circ_{CH_3OH,g} + \Delta H^\circ_{H_2O,g}]$$
$$= [(1 \cancel{\text{ mol } CO_2})(-393.5 \text{ kJ}/\cancel{\text{mol}}) + (3 \cancel{\text{ mol } H_2})(0 \text{ kJ}/\cancel{\text{mol}})]$$
$$- [(1 \cancel{\text{ mol } CH_3OH})(-200.7 \text{ kJ}/\cancel{\text{mol}}) + (1 \cancel{\text{ mol } H_2O})(-241.8 \text{ kJ}/\cancel{\text{mol}})]$$
$$= 49.0 \text{ kJ}$$

Substituting into the relationship between T, ΔH°_{rxn}, and ΔS°_{rxn}:

$$T > \frac{\Delta H^\circ_{rxn}}{\Delta S^\circ_{rxn}} > \frac{49.0 \cancel{\text{ kJ}}}{0.1769 \cancel{\text{ kJ}}/\text{K}} > 277 \text{ K}$$

Thus, the conversion of methanol vapor and steam to hydrogen and carbon dioxide should be spontaneous above 277 K or 4°C.

Think About It The enthalpy change for the reaction is a relatively small value so we do not require an enormously high temperature for the reaction to be spontaneous.

Practice Exercise At high temperatures, ammonia decomposes into nitrogen and hydrogen gases.

$$2\ NH_3(g) \rightarrow N_2(g) + 3\ H_2(g)$$

ΔH°_{rxn} for the reaction is positive and ΔS°_{rxn} is positive. Predict whether the reaction is spontaneous at all temperatures or only at high temperatures.

12.8 Driving the Human Engine: Coupled Reactions

The laws of thermodynamics that govern chemical reactions in the laboratory also govern all the chemical reactions that take place in living systems (Figure 12.19). Organisms carry out reactions that release the energy contained in the chemical bonds of food molecules and then use that energy to do work and to sustain an array of other essential biological functions. Just like

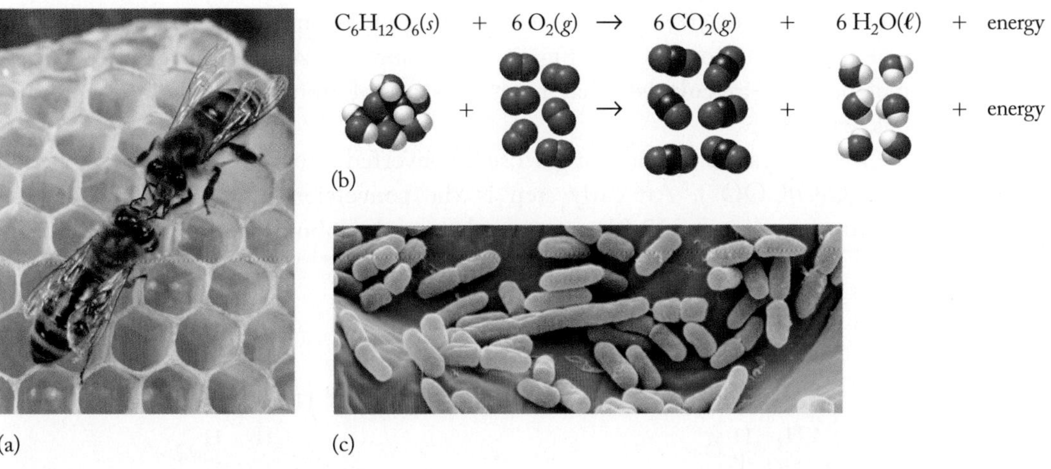

FIGURE 12.19 The rules of thermodynamics apply to all living systems. (a) Honeybees extract energy from nutrients to support life. As bees collect and process plant nectars, they store the energy as honey, which is then consumed by the bees, by humans, or by other animals. (b) When the sugar in honey is digested, carbon dioxide, water, and energy are released, increasing the entropy of the universe. (c) The spontaneous life processes of all organisms—including the *E. coli* bacteria and many others that live in our gastrointestinal tracts—increase the entropy of the universe.

glycolysis a series of reactions that converts glucose into pyruvate; a major anaerobic (no oxygen required) pathway for the metabolism of glucose in the cells of almost all living organisms.

phosphorylation a reaction resulting in the addition of a phosphate group to an organic molecule.

CONNECTION In Chapter 9 we defined 1 Calorie, the "calorie" used in discussing food, as equivalent to 1 kcal.

CONNECTION Photosynthesis and the carbon cycle were introduced in Chapter 7.

mechanical engines, however, humans and other life-forms are far from 100% efficient, which means that life requires a continuous input of energy. Thus, we humans must constantly absorb energy in the form of the caloric content of the food we eat and then release heat and waste products into our surroundings. Young women have an average daily nutritional need of 2100 Cal, whereas young men need 2900 Cal. This level of caloric intake provides the energy we need in order to function at all levels, from thinking to getting out of bed in the morning.

In living systems, spontaneous reactions ($\Delta G_{rxn} < 0$) typically involve breaking food down, while nonspontaneous reactions ($\Delta G_{rxn} > 0$) involve building molecules needed by the body. Living systems use the energy from spontaneous reactions to run nonspontaneous reactions; we say that the spontaneous reactions are *coupled* to the nonspontaneous reactions. Part of the study of biochemistry involves deciphering the molecular mechanisms that enable reaction coupling. This topic is covered in Chapter 20, but for now it is sufficient to know that elegant molecular processes have evolved to enable living systems to couple chemical reactions so that the energy obtained from spontaneous reactions can be used to drive the nonspontaneous reactions that maintain life. The metabolic chemical reactions we look at here are not presented for you to memorize. Rather, the intent is to aid your understanding of how changes in free energy allow spontaneous reactions to drive nonspontaneous reactions and to illustrate operationally what the phrase *coupled reactions* actually means.

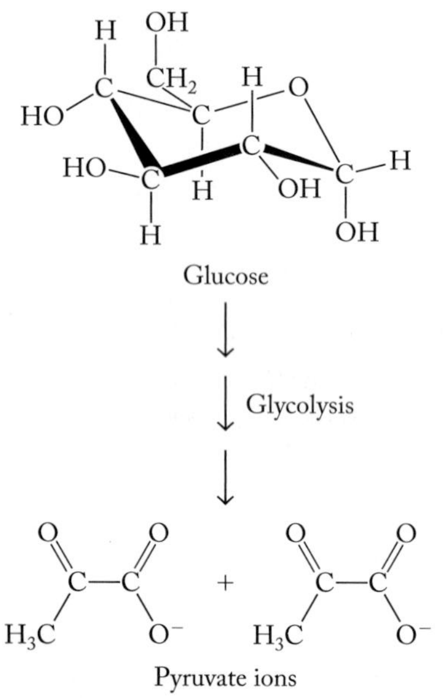

As noted in Chapter 9, the energy contained in the food we eat comes indirectly from solar energy. Green plants store energy from sunlight in their tissues in molecules such as glucose ($C_6H_{12}O_6$), which they produce from CO_2 and H_2O during photosynthesis.

The production of glucose by green plants is a nonspontaneous process, which is why the plants require the energy of sunlight to carry out this reaction. Animals that consume plants use the energy stored in glucose and other molecules, and they release CO_2 and H_2O back into the environment:

$$C_6H_{12}O_6(s) + 6\ O_2(g) \rightarrow 6\ CO_2(g) + 6\ H_2O(\ell)$$

This reaction increases the entropy of the universe, and is, therefore, spontaneous. However, the reaction takes place in living organisms in many steps, some of which are nonspontaneous. Figure 12.20 summarizes one portion of glucose metabolism—**glycolysis**—that involves a series of spontaneous and nonspontaneous reactions.

Glucose

Glycolysis

Pyruvate ions

FIGURE 12.20 In glycolysis, each molecule of glucose is converted into two pyruvate ions.

In glycolysis, each mole of glucose is converted into two moles of pyruvate ion ($CH_3C(O)COO^-$). An early step is the conversion of glucose into glucose 6-phosphate (Figure 12.21), an example of a **phosphorylation** reaction. Glucose

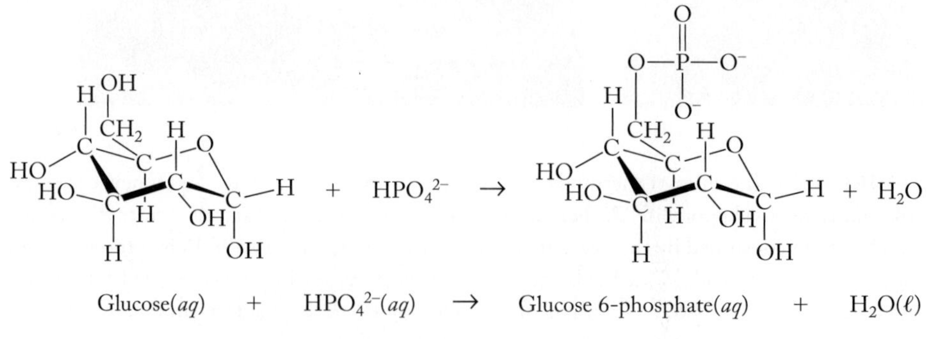

$$\text{Glucose}(aq) + HPO_4^{2-}(aq) \rightarrow \text{Glucose 6-phosphate}(aq) + H_2O(\ell)$$

FIGURE 12.21 The conversion of glucose into glucose 6-phosphate is an early step in glycolysis. This reaction is nonspontaneous, which means energy must be added to make the reaction go: $\Delta G^\circ_{rxn} = 13.8$ kJ/mol.

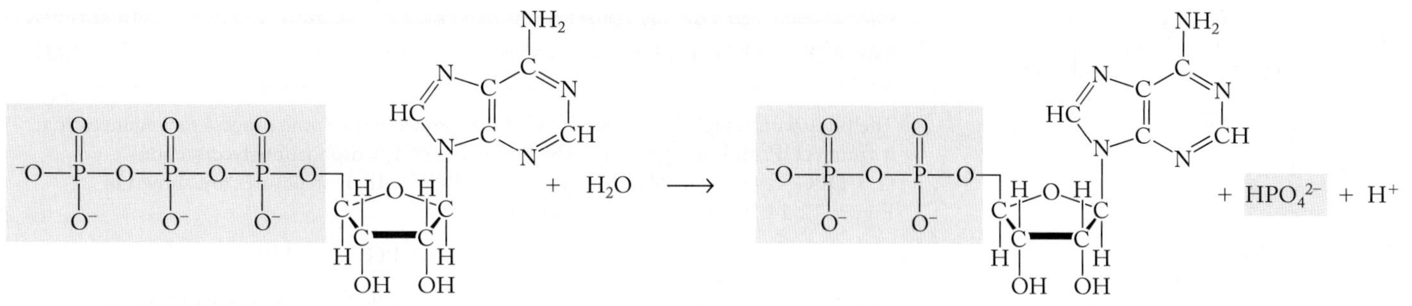

$+ H_2O \longrightarrow$ $+ HPO_4^{2-} + H^+$

Adenosine triphosphate (ATP^{4-}) Adenosine diphosphate (ADP^{3-})

FIGURE 12.22 The hydrolysis of ATP to ADP is a spontaneous reaction: $\Delta G° = -30.5$ kJ/mol. The body couples this reaction to nonspontaneous reactions so that the energy released can drive them (Figure 12.21).

reacts with hydrogen phosphate ion (HPO_4^{2-}), producing glucose 6-phosphate and water. This reaction is nonspontaneous ($\Delta G°_{rxn} = +13.8$ kJ/mol). The energy needed to make it happen comes from an ion called adenosine triphosphate (ATP^{4-}), which functions in our cells both as a storehouse of energy and as an energy-transfer agent: it hydrolyzes to the adenosine diphosphate ion (ADP^{3-}) in a reaction (Figure 12.22) that produces a hydrogen phosphate ion and energy ($\Delta G°_{rxn} = -30.5$ kJ/mol).

In a living system, the spontaneous hydrolysis of ATP^{4-} consumes water and produces HPO_4^{2-} and H^+, whereas the nonspontaneous phosphorylation of glucose consumes HPO_4^{2-} and produces water. Their complementary reactants and products allow the two reactions to couple: the spontaneous $ATP^{4-} \rightarrow ADP^{3-}$ reaction supplies the energy that drives the nonspontaneous formation of glucose 6-phosphate (Figure 12.23).

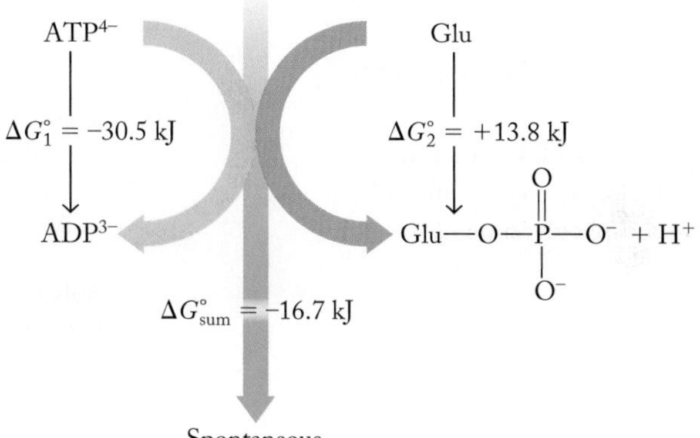

FIGURE 12.23 The spontaneous hydrolysis of ATP is coupled to the nonspontaneous phosphorylation of glucose (Glu). The overall reaction—the sum of the two individual reactions—is spontaneous.

This example illustrates another important general point about reactions and $\Delta G°_{rxn}$ values: those for coupled reactions (and also for sequential reactions) are additive. This is true for any set of reactions, not just those occurring in living systems.

The first two steps in glycolysis are as follows:

(1) $ATP^{4-}(aq) + H_2O(\ell) \rightarrow ADP^{3-}(aq) + HPO_4^{2-}(aq) + H^+(aq)$

$\Delta G°_{rxn} = -30.5$ kJ

(2) $C_6H_{12}O_6(aq) + HPO_4^{2-}(aq) \rightarrow C_6H_{11}O_6PO_3^{2-}(aq) + H_2O(\ell)$

Glucose — Hydrogen phosphate — Glucose 6-phosphate — $\Delta G°_{rxn} = 13.8$ kJ

If we add these reactions and their free energies, we get

$$ATP^{4-}(aq) + \cancel{H_2O(\ell)} + C_6H_{12}O_6(aq) + \cancel{HPO_4^{2-}(aq)} \rightarrow$$
$$ADP^{3-}(aq) + \cancel{HPO_4^{2-}(aq)} + H^+(aq) + C_6H_{11}O_6PO_3^{2-}(aq) + \cancel{H_2O(\ell)}$$
$$\Delta G°_{rxn} = (-30.5 + 13.8)\text{ kJ} = -16.7\text{ kJ}$$

Because equal quantities of H_2O and HPO_4^{2-} appear on both sides of the combined equation, they cancel out, leaving the following net ionic equation:

$$C_6H_{12}O_6(aq) + ATP^{4-}(aq) \rightarrow ADP^{3-}(aq) + C_6H_{11}O_6PO_3^{2-}(aq) + H^+(aq)$$
$$\Delta G°_{rxn} = -16.7\text{ kJ}$$

Since $\Delta G°_{rxn}$ for the net reaction is negative, the reaction is spontaneous.

CONNECTION In Chapter 9 we used Hess's law (the enthalpy change of a reaction that is the sum of two or more reactions equals the sum of the enthalpy changes of the constituent reactions) to calculate $\Delta H°_{rxn}$. We apply a similar principle here when adding $\Delta G°_{rxn}$ values of coupled reactions.

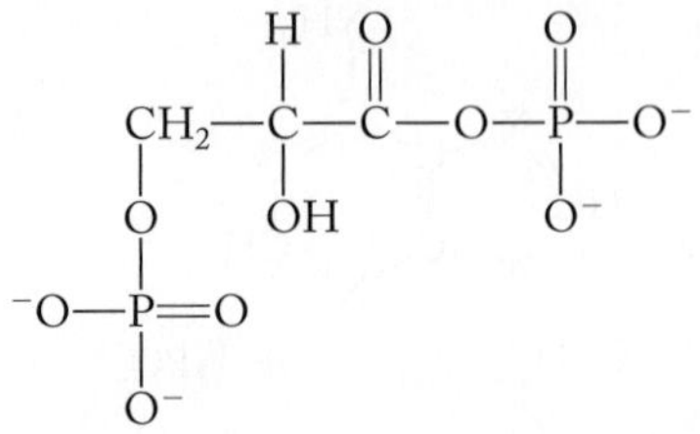

1,3-Diphosphoglycerate ($1,3\text{-}DPG^{4-}$)

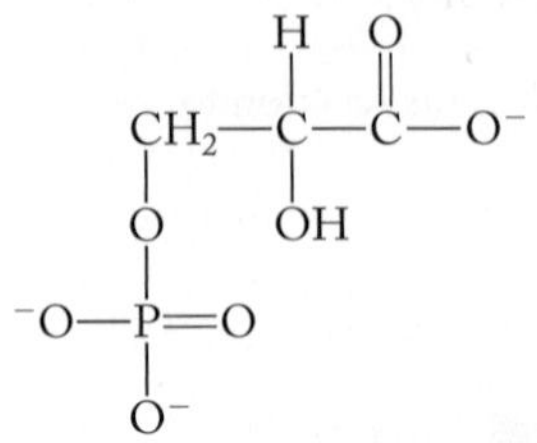

3-Phosphoglycerate ($3\text{-}PG^{3-}$)

FIGURE 12.24 Structures of 1,3-diphosphoglycerate and 3-phosphoglycerate.

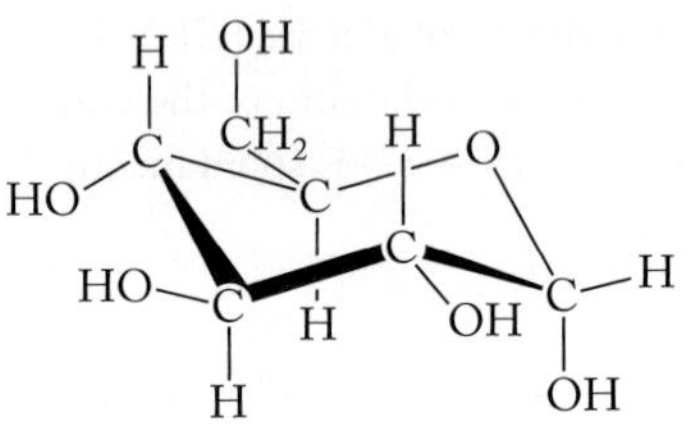

Glucose

Lactic acid

FIGURE 12.25 Structures of glucose and lactic acid.

SAMPLE EXERCISE 12.8 Calculating ΔG°_{rxn} of Coupled Reactions LO7

The body would rapidly run out of ATP if there were not some process for regenerating it from ADP, and that process is the hydrolysis of 1,3-diphosphoglycerate ions ($1,3\text{-}DPG^{4-}$) to 3-phosphoglycerate ions ($3\text{-}PG^{3-}$). Their structures are shown in Figure 12.24. The reaction can be written as

$$ADP^{3-} + 1,3\text{-}DPG^{4-} \rightarrow 3\text{-}PG^{3-} + ATP^{4-}$$

This hydrolysis is spontaneous. Calculate its ΔG° value from the following ΔG° values:

$$(1)\quad 1,3\text{-}DPG^{4-}(aq) + H_2O(\ell) \rightarrow 3\text{-}PG^{3-}(aq) + HPO_4^{2-}(aq) + H^+(aq) \qquad \Delta G^\circ_{rxn} = -49.0\ kJ$$

$$(2)\quad ADP^{3-}(aq) + HPO_4^{2-}(aq) + H^+(aq) \rightarrow ATP^{4-}(aq) + H_2O(\ell) \qquad \Delta G^\circ_{rxn} = 30.5\ kJ$$

Collect and Organize We need to calculate ΔG°_{rxn} for a reaction that is the sum of two reactions. If the reactions in equations 1 and 2 add up to the overall reaction, then the overall ΔG°_{rxn} is the sum of the ΔG°_{rxn} values for the individual reactions.

Analyze First, we add the reactions described by equations 1 and 2. Assuming that the overall reaction between ADP^{3-} and $1,3\text{-}DPG^{4-}$ is the sum of the reactions describing the hydrolysis of $1,3\text{-}DPG^{4-}$ and the phosphorylation of ADP^{3-}, we know that the sum of ΔG°_1 and ΔG°_2 will be less than zero because the overall reaction is spontaneous.

Solve Summing the reactions in equations 1 and 2 confirms that they equal the overall reaction:

$$\begin{array}{ll}(1) & 1,3\text{-}DPG^{4-}(aq) + H_2O(\ell) \rightarrow 3\text{-}PG^{3-}(aq) + HPO_4^{2-}(aq) + H^+(aq)\\ (2) & ADP^{3-}(aq) + HPO_4^{2-}(aq) + H^+(aq) \rightarrow ATP^{4-}(aq) + H_2O(\ell)\\ \hline \end{array}$$

$$1,3\text{-}DPG^{4-}(aq) + \cancel{H_2O(\ell)} + ADP^{3-}(aq) + \cancel{HPO_4^{2-}(aq)} + \cancel{H^+(aq)} \rightarrow 3\text{-}PG^{3-}(aq) + \cancel{HPO_4^{2-}(aq)} + \cancel{H^+(aq)} + ATP^{4-}(aq) + \cancel{H_2O(\ell)}$$

We sum the ΔG°_{rxn} values for steps 1 and 2 to determine ΔG°_{rxn} for the overall reaction:

$$\Delta G^\circ_{overall} = \Delta G^\circ_1 + \Delta G^\circ_2 = (-49.0) + (30.5)\ kJ = -18.5\ kJ$$

Think About It As predicted, the overall ΔG°_{rxn} value is less than zero for this spontaneous reaction. Hydrolysis of $1,3\text{-}DPG^{4-}$ provides more than enough free energy to convert ADP^{3-} into ATP^{4-}.

Practice Exercise Conversion of glucose into lactic acid (Figure 12.25) drives the phosphorylation of two moles of ADP to ATP:

$$\underset{\text{Glucose}}{C_6H_{12}O_6(aq)} + 2\ HPO_4^{2-}(aq) + 2\ ADP^{3-}(aq) + 2H^+(aq) \rightarrow \underset{\text{Lactic acid}}{2\ CH_3CH(OH)COOH(aq)} + 2\ ATP^{4-}(aq) + 2\ H_2O(\ell) \qquad \Delta G^\circ_{rxn} = -135\ kJ/mol$$

What is ΔG°_{rxn} for the conversion of glucose into lactic acid?

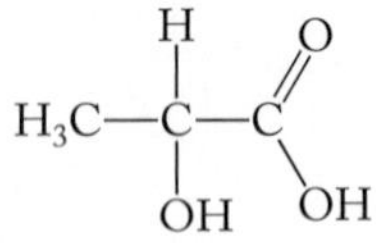

$$C_6H_{12}O_6(aq) \rightarrow 2\ CH_3CH(OH)COOH(aq)$$

The ATP produced from the breakdown of glucose (as, for example, via the first reaction in Practice Exercise 12.8) is used to drive nonspontaneous reactions in cells. The metabolism of fats and proteins also relies on a series of chemical cycles, all of which involve coupled reactions.

In this chapter we addressed the reasons why some reactions and processes are spontaneous while others are not. We have learned that the free-energy change (ΔG) determines the spontaneity of a chemical reaction or process. Together, enthalpy and entropy changes for chemical reactions allow us to determine ΔG, which also represents the maximum amount of work that can be done by the energy associated with a change. For changes carried out in the real world, the maximum amount of work is always less than ΔG, because—to paraphrase the second law of thermodynamics once more—in the game of energy, not only can you not win, you can't even break even.

SAMPLE EXERCISE 12.9 Integrating Concepts: Trouton's Rule

In 1883–1884, while still an undergraduate student at Trinity College, Dublin, Ireland, Frederick Trouton (1863–1922) published two short papers describing what we now call Trouton's rule. Trouton's rule states that the ratio of the enthalpy of vaporization (ΔH_{vap}) of a liquid to its normal boiling point (in K) is approximately constant:

$$\frac{\Delta H_{vap}}{T_b} \approx 88 \frac{\text{J}}{\text{mol} \cdot \text{K}}$$

This ratio is also known as the entropy of vaporization (ΔS_{vap}) and may be used to estimate ΔH_{vap} values of liquids whose boiling points are known.

a. Suggest why $\Delta H_{vap}/T$ should be approximately constant.
b. Calculate the values of ΔH_{vap} for the substances given in the table that follows (whose structures are shown in Figure 12.26) using Trouton's rule, and compare them to the experimentally determined values. Also calculate ΔS_{vap} for each substance by dividing the experimentally determined ΔH_{vap} values by the normal boiling points of the liquids. Identify those compounds whose actual ΔH_{vap} values deviate by more than 10% from those estimated using Trouton's rule.

Substance	Boiling Point T_b (°C)	Calculated ΔH_{vap} (J/mol)	Experimentally Determined ΔH_{vap} (J/mol)	Calculated Value of ΔS_{vap} [J/(mol · K)]
Ethanol	78.3		38,600	
Acetone	56.0		29,100	
Benzene	80.0		30,700	
Chloroform	61.1		29,200	
Formic acid	100.8		22,700	
Hexane	68.7		28,900	
Water	100.0		40,660	

c. Based on the results in part (b) and the molecular structures of the seven compounds, explain why some of them deviate from Trouton's rule.
d. Predict which of the four substances whose molecular structures are shown in Figure 12.27 obey Trouton's rule.
e. What is the sign of the change in standard free energy of vaporization (ΔG°_{vap}) of each of the seven liquids in the table at 298 K?

Collect and Organize We are given the boiling points of seven liquids. Trouton's rule relates the boiling points of liquids to their ΔH_{vap} values. In part (a) we are asked to explain why this relationship exists. In part (b) we use Trouton's rule to estimate the ΔH_{vap} values of seven liquids, calculate actual ΔS_{vap} values, and compare these results to their actual ΔH_{vap} values, explaining any deviations in part (c). In part (d) we predict how well Trouton's rule will apply to six other liquids, and in part (e) we are asked to predict the sign of ΔG°_{vap} of the seven liquids at 298 K.

Analyze The molecular structures of compounds influence how they interact with each other, and the strengths of their intermolecular interactions help define physical properties such as boiling points and heats of vaporization. Our task is to look for structural features that are associated with particularly strong intermolecular interactions. One of these is hydrogen bonds, which form in three of the liquids: ethanol, formic acid, and water. These three may be among those that do not obey Trouton's rule.

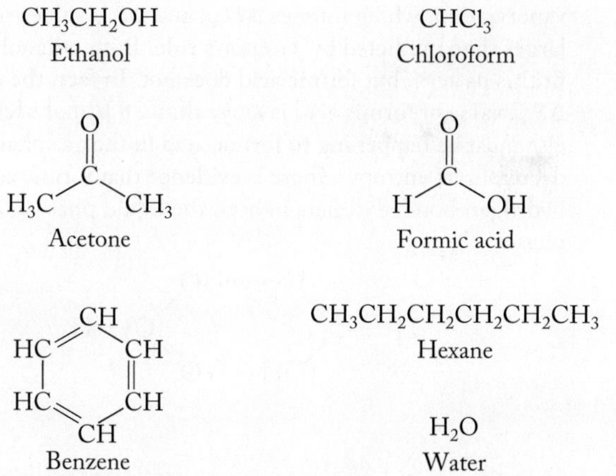

CH_3CH_2OH
Ethanol

$CHCl_3$
Chloroform

$H_3C{-}C({=}O){-}CH_3$
Acetone

$H{-}C({=}O){-}OH$
Formic acid

$CH_3CH_2CH_2CH_2CH_2CH_3$
Hexane

C_6H_6 (ring of six CH)
Benzene

H_2O
Water

FIGURE 12.26

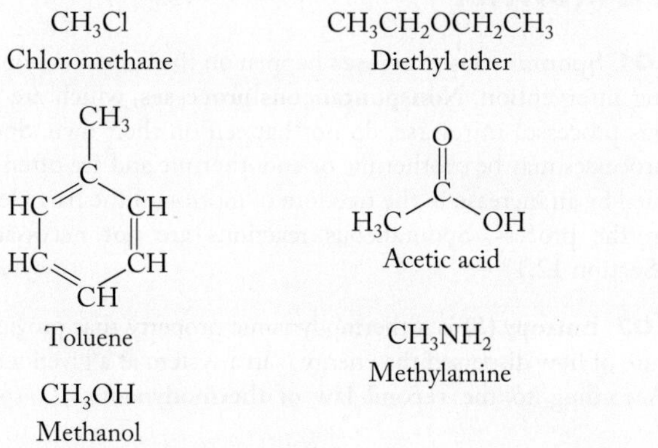

CH_3Cl
Chloromethane

$CH_3CH_2OCH_2CH_3$
Diethyl ether

$C_6H_5CH_3$
Toluene

$H_3C{-}C({=}O){-}OH$
Acetic acid

CH_3NH_2
Methylamine

CH_3OH
Methanol

FIGURE 12.27

Solve

a. The boiling points and ΔH_{vap} values of molecular compounds both depend on the strengths of interactions between molecules in the liquid phase. Given this mutual dependence, higher boiling points should be associated with higher ΔH_{vap} values so that the ratio between the two is (nearly) constant.

b. The calculated values of ΔH_{vap} for ethanol, formic acid, and water all vary by more than 10% from the measured values.

Substance	Boiling Point T_b (°C)	Calculated ΔH_{vap} (J/mol)	Experimentally Determined ΔH_{vap} (J/mol)	Calculated Value of ΔS_{vap} [J/(mol · K)]
Ethanol	78.3	30,900	38,600	110
Acetone	56.0	29,000	29,100	88
Benzene	80.0	31,100	30,700	87
Chloroform	61.1	29,400	29,200	87
Formic acid	100.8	32,900	22,700	61
Hexane	68.7	30,100	28,900	85
Water	100.0	32,800	40,660	109

c. The molecules of the three substances that deviate more than 10% from Trouton's rule all have O—H bonds, which means they can form hydrogen bonds that increase intermolecular interactions and decrease the energy and freedom of particle motions (entropy). These interactions must be overcome during vaporization, which implies ΔH_{vap} and ΔS_{vap} values that are larger than predicted by Trouton's rule. Both ethanol and water fit this pattern, but formic acid does not. In fact, the calculated ΔS_{vap} value of formic acid is *lower* than 88 J/(mol · K). What else must be happening to formic acid in the gas phase that decreases its entropy? There is evidence that formic acid forms hydrogen-bonded dimers in both the liquid phase *and* the gas phase:

$$\text{H—C}(=\text{O}\cdots\text{HO})(\text{—OH}\cdots\text{O}=)\text{C—H}$$

Their presence in both phases lowers the entropy of both, but the decrease in the gas phase is greater because the absolute entropy of the gas phase is much greater. Figure 12.28 illustrates the relative relationships between the entropies of the liquid and gas phases of these compounds.

d. The ΔH_{vap} values of (i) chloromethane, (ii) toluene, and (iv) diethyl ether that are calculated using Trouton's rule should be close to the experimental ΔH_{vap} values because these compounds do not form hydrogen bonds.

e. The boiling points of all seven liquids are above 25°C (298 K). Therefore, vaporization of their liquids under standard conditions and at $T = 298$ K is nonspontaneous, and the ΔG°_{vap} values of all seven should be greater than zero.

Think About It Trouton's rule was developed empirically, but our answer to part (a) provides an explanation of why it should be true. This exercise also shows that compounds experiencing intermolecular forces, such as hydrogen bonds that add significantly to London dispersion forces in the liquid phase, have ΔH_{vap} values that are significantly greater than those predicted by Trouton's rule.

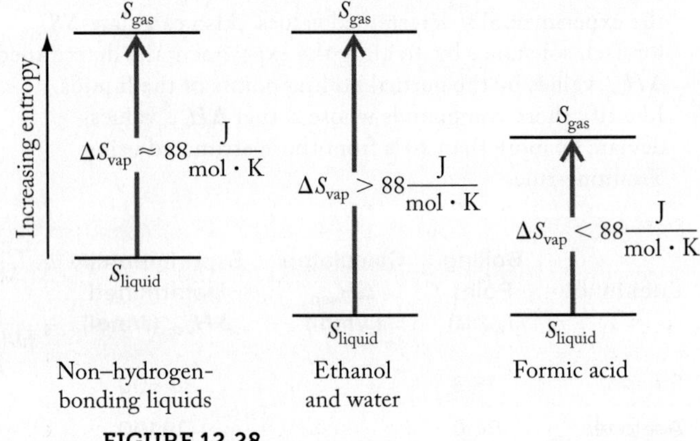

FIGURE 12.28

SUMMARY

LO1 Spontaneous processes happen on their own without continuing intervention. **Nonspontaneous processes**, which are spontaneous processes in reverse, do not happen on their own. Spontaneous processes may be exothermic or endothermic and are often accompanied by an increase in the freedom of motion of the particles involved in the process. Spontaneous reactions are not necessarily rapid. (Section 12.1)

LO2 Entropy (*S*) is a thermodynamic property that provides a measure of how dispersed the energy is in a system at a given temperature. According to the **second law of thermodynamics**, a spontaneous process is accompanied by an increase in the entropy of an isolated system or of the universe for a process occurring in any system. The entropy of a system increases as the number of probable arrangements of its particles, called **accessible microstates**, increases. According to the **third law of thermodynamics**, a perfect crystal of a pure substance has zero entropy at absolute zero. All substances have positive entropies at temperatures above absolute zero. (Section 12.2)

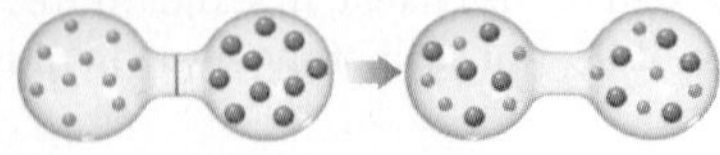

LO3 Standard molar entropies ($S°$) are entropy values for substances in their standard states. The entropy of a system increases with increasing molecular complexity and with increasing temperature. Any process is spontaneous if it produces an increase in the entropy of the universe. The entropy change in a reaction under standard conditions can be calculated from the standard entropies of the products and reactants and their coefficients in the balanced chemical equation. (Sections 12.4 and 12.5)

LO4 Gibbs free energy (G) is the maximum energy available to do useful work for a process occurring at constant temperature and pressure. When a process results in a decrease in **free energy** of a system ($\Delta G < 0$), the process is spontaneous; when $\Delta G > 0$, the process is nonspontaneous. The change in free energy of a reaction under standard conditions can be calculated either from the **standard free energies of formation (ΔG°_f)** of the products and reactants or from the values of ΔH°_{rxn} and ΔS°_{rxn}. (Section 12.6)

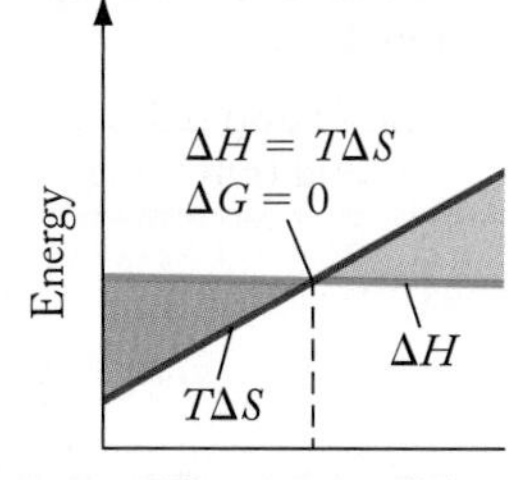

LO5 The temperature range over which a process is spontaneous depends on the signs and the relative magnitudes of ΔH and ΔS. (Section 12.7)

LO6 Many important biochemical processes, including **glycolysis** and **phosphorylation**, are made possible by the coupling of spontaneous and nonspontaneous reactions. The free energy released in the spontaneous processes going on in the body is used to drive nonspontaneous processes. (Section 12.8)

PARTICULATE PREVIEW WRAP-UP

The chemical equations describing (a) respiration and (b) photosynthesis are

$$\text{(a) } C_6H_{12}O_6(s) + 6\,O_2(g) \rightarrow 6\,CO_2(g) + 6\,H_2O(\ell)$$

$$\text{(b) } 6\,CO_2(g) + 6\,H_2O(\ell) \rightarrow C_6H_{12}O_6(s) + 6\,O_2(g)$$

During energy production relatively large $C_6H_{12}O_6$ molecules are converted into much smaller molecules of CO_2 and H_2O. This reaction is spontaneous. Photosynthesis is the reverse reaction: building larger $C_6H_{12}O_6$ molecules from much smaller molecules of CO_2 and H_2O in a reaction that is not spontaneous and relies on the energy in sunlight to happen.

PROBLEM-SOLVING SUMMARY

Type of Problem	Concepts and Equations	Sample Exercises
Predicting the sign of an entropy change	Look for fewer moles of gas as products than as reactants or for the precipitation of a solute from solution ($\Delta S < 0$ for both). For the reverse processes, $\Delta S > 0$.	**12.1**
Comparing standard molar entropy values	Substances composed of larger, less rigid molecules have more entropy. Among substances with similar molar masses, gases have more entropy than liquids, which have more than solids.	**12.2**
Calculating the entropy change of a chemical reaction	Use $\Delta S^\circ_{rxn} = \sum n_{products} S^\circ_{products} - \sum n_{reactants} S^\circ_{reactants}$ (12.6) where $S^\circ_{products}$ and $S^\circ_{reactants}$ are standard molar entropies of the products and reactants, and $n_{products}$ and $n_{reactants}$ are their stoichiometric coefficients in the balanced chemical equation describing the reaction.	**12.3**
Predicting reaction spontaneity under standard conditions	A reaction is spontaneous if $\Delta G^\circ_{rxn} < 0$, where $\Delta G^\circ_{rxn} = \Delta H^\circ_{rxn} - T\Delta S^\circ_{rxn}$ (12.11)	**12.4, 12.5**
Calculating ΔG°_{rxn} using appropriate ΔG°_f values	Use $\Delta G^\circ_{rxn} = \sum n_{products} \Delta G^\circ_{f,products} - \sum n_{reactants} \Delta G^\circ_{f,reactants}$ (12.12) where $\Delta G^\circ_{f,products}$ and $\Delta G^\circ_{f,reactants}$ are the standard molar free energies of formation of the products and reactants, and $n_{products}$ and $n_{reactants}$ are their stoichiometric coefficients in the balanced chemical equation describing the reaction.	**12.6**

Type of Problem	Concepts and Equations	Sample Exercises
Relating reaction spontaneity to ΔH_{rxn} and ΔS_{rxn}	Use $\Delta G_{rxn} = \Delta H_{rxn} - T\Delta S_{rxn}$	**12.7**
Calculating ΔG°_{rxn} of coupled reactions	Free-energy changes of coupled reactions are additive.	**12.8**

VISUAL PROBLEMS

(Answers to boldface end-of-chapter questions and problems are in the back of the book.)

12.1. The marbles in Figure P12.1 occupy two of the three depressions in each of the blocks. After the black divider is removed, the marbles in each block can occupy any of the sites in that block. How many arrangements of the marbles are possible before and after removal of the divider in the (a) block and in the (b) block?

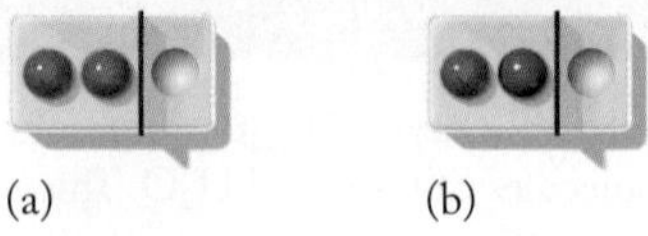

(a) (b)

FIGURE P12.1

12.2. The marbles in Figure P12.2 occupy two of the four depressions in each of the blocks. After the black divider is removed, the marbles in each block can occupy any of the sites in that block. How many arrangements of the marbles are possible before and after removal of the divider in the (a) block and in the (b) block?

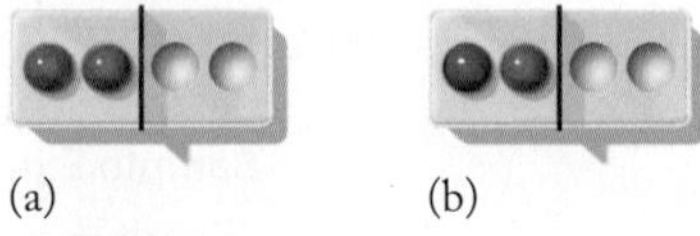

(a) (b)

FIGURE P12.2

12.3. Two tires shown in cross section in Figure P12.3 are inflated at the same temperature to the same volume, though more air is used to inflate the tire on the right. In which tire is the gas under greater internal pressure and in which does the gas have greater entropy?

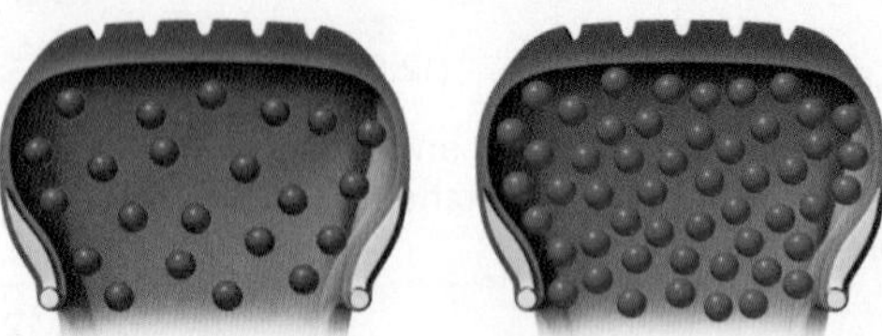

FIGURE P12.3

12.4. Two cubic containers (Figure P12.4) contain the same quantity of gas at the same temperature.

a. Which cube contains the gas with more entropy?

b. If the sample in cube (b) is left unchanged but the sample in cube (a) is cooled so that it condenses, which sample has the higher entropy?

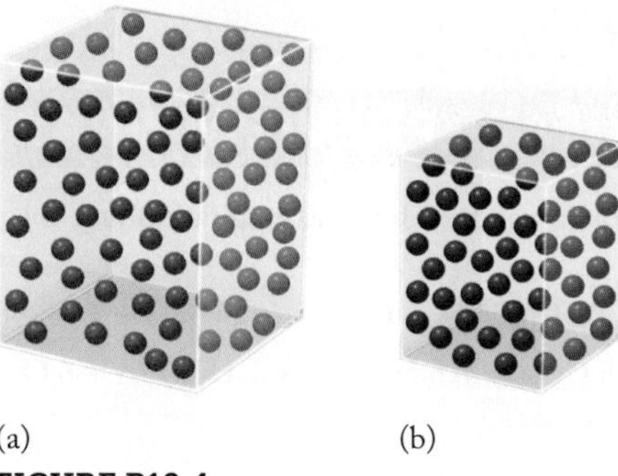

(a) (b)

FIGURE P12.4

12.5. Figure P12.5 shows two connected bulbs that have just been filled with a mixture of two ideal gases: A (red spheres) and B (blue spheres). If the molar mass of A is twice that of B, will the atoms of A eventually fill the bottom bulb and the atoms of B fill the top bulb? Why or why not?

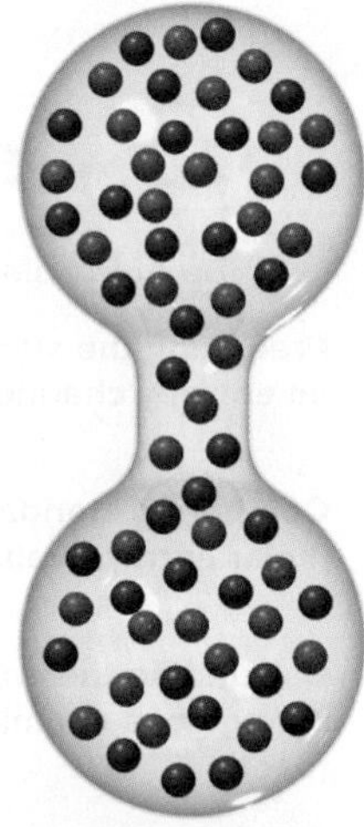

FIGURE P12.5

12.6. The box on the left in Figure P12.6 represents a mixture of two diatomic gases: A_2 (red spheres) and B_2 (blue spheres). How do the entropies of A_2 and B_2 change as a result of the process depicted by the arrow?

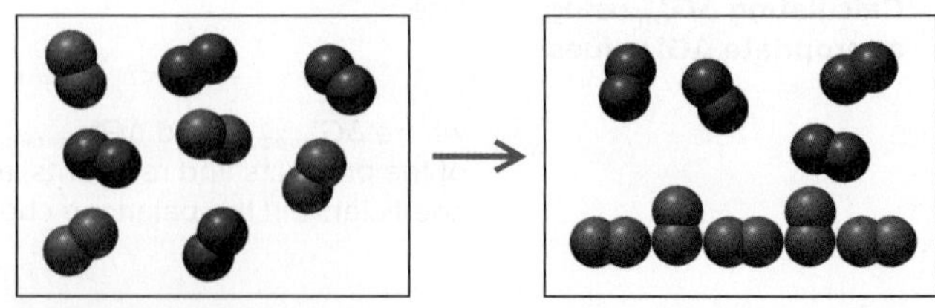

FIGURE P12.6

12.7. Is the process in Figure P12.6 more likely to be spontaneous at high temperature or low temperature, or is it unaffected by changing temperature?

12.8. Figure P12.8 shows the plots of ΔH and $T\Delta S$ for a phase change as a function of temperature.

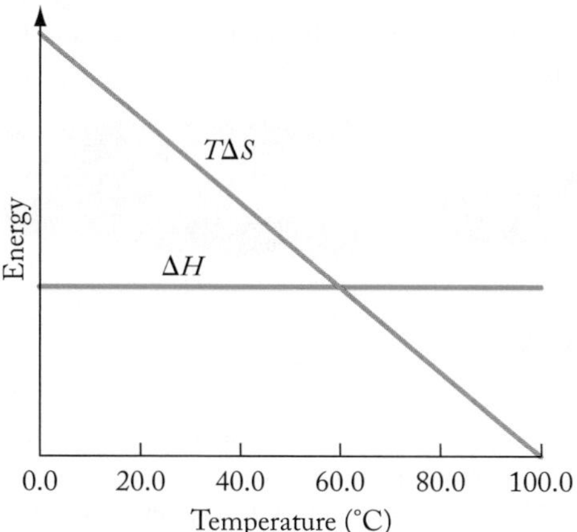

FIGURE P12.8

a. What is the status of the process at the point where the two lines intersect?
b. Over what temperature range is the process spontaneous?

12.9. Of the six phase changes—melting, vaporization, condensation, freezing, sublimation, and deposition—which ones have thermodynamic profiles that fit the pattern in Figure P12.8?

12.10. Use representations [A] through [I] in Figure P12.10 to answer questions a–f.

a. What is the sign of ΔS_{sys} for the formation of frost on a window pane in image [B]? Under what conditions would this process *not* be spontaneous?
b. Image [H] shows a neutralization reaction between antacid tablets (sodium bicarbonate) and stomach acid, HCl(*aq*). What is the sign of ΔS_{rxn}?
c. Image [D] depicts the mixing of two colorless solutions that results in image [F]. Is ΔS_{rxn} greater than, less than, or equal to zero?
d. If you mix the particulate substances in [A] and [C] to create the mixture in [E], does the entropy increase, decrease, or remain unchanged?
e. When you mix the particulate substances in [G] ad [I] to form mixture [E], how does the change in volume affect the entropy?
f. Does all mixing lead to an increase in entropy?

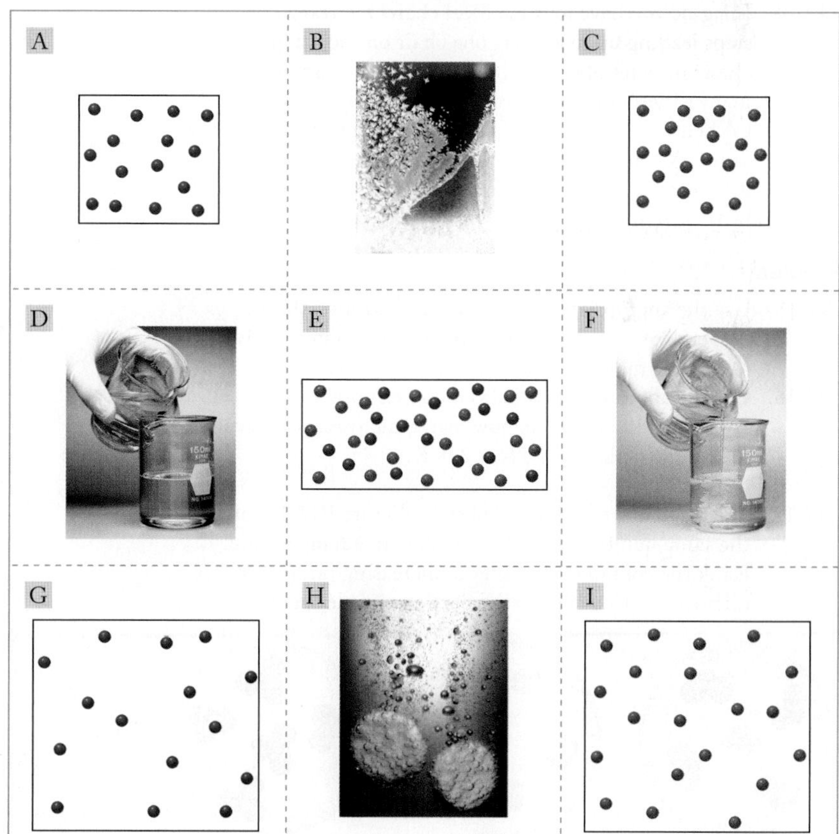

FIGURE P12.10

QUESTIONS AND PROBLEMS

Spontaneous Processes; Entropy

Concept Review

12.11. How is the entropy change that accompanies a reaction related to the entropy change that happens when the reaction runs in reverse?

12.12. Identify the following processes as spontaneous or nonspontaneous, and explain your choice.
   a. A photovoltaic cell in a portable device charges your cell phone.
   b. Dry ice (solid CO_2) sublimes at room temperature.
   c. A radiopharmaceutical imaging agent containing technetium emits gamma rays.

12.13. You flip three coins, assigning the values +1 for heads and −1 for tails. Each outcome of the three flips constitutes a microstate. How many different microstates are possible from flipping the three coins? Which value or values for the sums in the microstates are most likely? *Hint*: The sequence HHT (+1 +1 −1) is one possible outcome, or microstate. Note, however, that this outcome differs from THH (−1 +1 +1), even though the two sequences sum to the same value.

12.14. Imagine you have four identical chairs to arrange on four steps leading up to a stage, one chair on each step. The chairs have numbers on their backs: 1, 2, 3, and 4. How many different microstates for the chairs are possible? (When viewed from the front, all the microstates look the same. When viewed from the back, you can identify the different microstates because you can distinguish the chairs by their numbers.)

Problems

12.15. Use the appropriate standard molar entropy value in Appendix 4 to calculate how many microstates are available to a single molecule of liquid H_2O at 298 K.

12.16. Use the appropriate standard molar entropy value in Appendix 4 to calculate how many microstates are available to a single molecule of N_2 at 298 K.

12.17. The three identical glass spheres in Figure P12.17 contain the same number of particles at the same temperature. Rank the containers in order of increasing number of microstates accessible to the particles inside them.

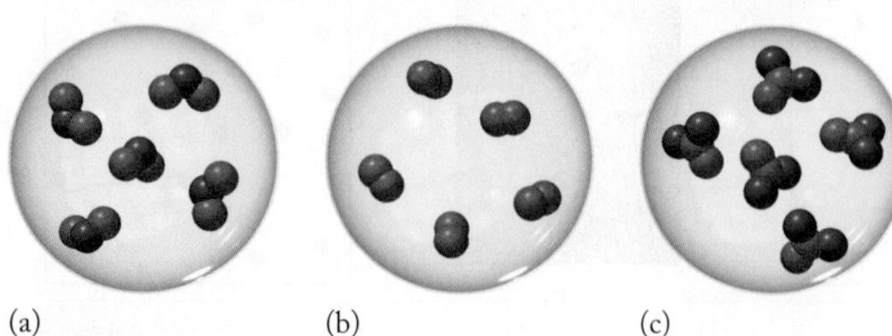

FIGURE P12.17

12.18. Figure P12.18(a) shows a cylinder within a cylinder that contains a population of gaseous molecules. The volume occupied by the molecules can be increased by pulling the inside cylinder out (Figure P12.18b), much like a telescope. The number of molecules within the cylinder remains the same during this operation. Compare the number of microstates available to the molecules in Figure P12.18(a) and (b).

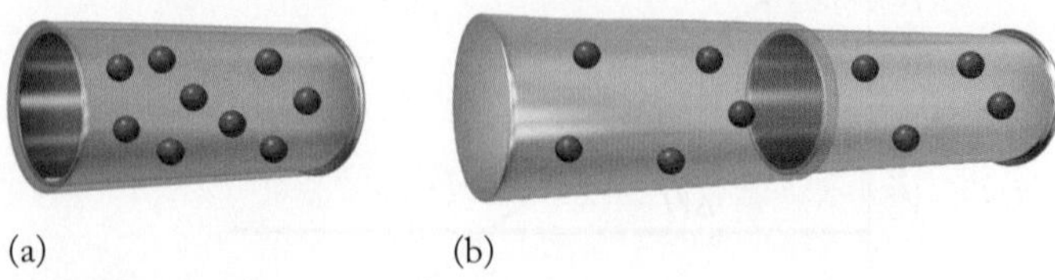

FIGURE P12.18

12.19. Which of the following ionic solutes experiences the greatest increase in entropy when 0.0100 mol of it dissolves in 1.00 liter of water? (a) $CaCl_2$, (b) NaBr, (c) KCl, (d) $Cr(NO_3)_3$, (e) LiOH

12.20. Which of the following molecular solutes experience an increase in entropy when dissolved in water? (a) $CO_2(g)$, (b) $HF(g)$, (c) $CH_3OH(\ell)$, (d) $CH_3COOH(\ell)$, (e) $C_{12}H_{22}O_{11}(s)$

Absolute Entropy and Molecular Structure

Concept Review

12.21. Which component in each of the following pairs has the greater entropy?
   a. 1 mole of $S_2(g)$ or 1 mole of $S_8(g)$
   b. 1 mole of $S_2(g)$ or 1 mole of $S_8(s)$
   c. 1 mole of $O_2(g)$ or 1 mole of $O_3(g)$

*12.22. **Metabolism of Pharmaceuticals** When some pharmaceutical agents like the pain killer morphine are metabolized (Fig. P12.22), our bodies add a molecule of sugar to the active compound to make it more polar and hence more readily excreted. Will the morphine molecule experience an increase or decrease in entropy as a result of this process?

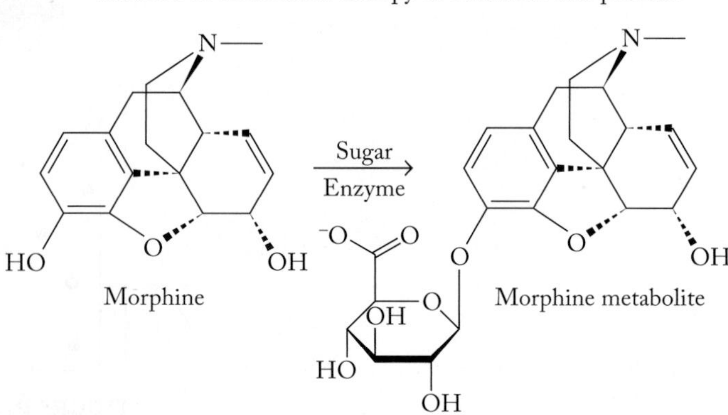

FIGURE P12.22

***12.23.** Diamond and the fullerenes are allotropes of carbon. On the basis of their different structures and properties, predict which has the higher standard molar entropy.

12.24. **Superfluids** The 1996 Nobel Prize in Physics was awarded to Douglas Osheroff, Robert Richardson, and David Lee for discovering *superfluidity* (apparently frictionless flow) in ^{3}He. When ^{3}He is cooled to 2.7 mK, the liquid settles into an *ordered* superfluid state. Predict the sign of the entropy change for the conversion of liquid ^{3}He into its superfluid state.

Problems

12.25. Rank the compounds in each of the following groups in order of increasing standard molar entropy ($S°$):
a. $CH_4(g)$, $CF_4(g)$, and $CCl_4(g)$
b. $CH_2O(g)$, $CH_3CHO(g)$, and $CH_3CH_2CHO(g)$
c. $HF(g)$, $H_2O(g)$, and $NH_3(g)$

12.26. Rank the compounds in each of the following groups in order of increasing standard molar entropy ($S°$):
a. $CH_4(g)$, $CH_3CH_3(g)$, and $CH_3CH_2CH_3(g)$
b. $CCl_4(\ell)$, $CHCl_3(\ell)$, and $CH_2Cl_2(\ell)$
c. $CO_2(\ell)$, $CO_2(g)$, and $CS_2(g)$

Applications of the Second Law

Concept Review

12.27. Ice cubes melt in a glass of lemonade, cooling the lemonade from 10.0°C to 0.0°C. If the ice cubes are the system, what are the signs of ΔS_{sys} and ΔS_{surr}?

12.28. Adding sidewalk deicer (calcium chloride) to water causes the temperature of the water to increase. If solid $CaCl_2$ is the system, what are the signs of ΔS_{sys} and ΔS_{surr}?

Problems

12.29. Which of the following combinations of entropy changes for a process are mathematically possible?
a. $\Delta S_{sys} > 0, \Delta S_{surr} > 0, \Delta S_{univ} > 0$
b. $\Delta S_{sys} > 0, \Delta S_{surr} < 0, \Delta S_{univ} > 0$
c. $\Delta S_{sys} > 0, \Delta S_{surr} > 0, \Delta S_{univ} < 0$

12.30. Which of the following combinations of entropy changes for a process are mathematically possible?
a. $\Delta S_{sys} < 0, \Delta S_{surr} > 0, \Delta S_{univ} > 0$
b. $\Delta S_{sys} < 0, \Delta S_{surr} < 0, \Delta S_{univ} > 0$
c. $\Delta S_{sys} < 0, \Delta S_{surr} > 0, \Delta S_{univ} < 0$

12.31. **Cleaning Natural Gas** Predict whether the entropy of the system increases or decreases for the following reaction, which describes the process used to remove hydrogen sulfide from natural gas:

$$8\,H_2S(g) + O_2(g) \rightarrow 3\,S_8(s) + 2\,H_2O(g)$$

12.32. **Making Sodium Hydroxide** Predict whether the entropy of the system increases or decreases for the following reaction, which describes the process used industrially to produce sodium hydroxide and chlorine by passing electric current through brine:

$$2\,NaCl(aq) + 2\,H_2O(\ell) \rightarrow 2\,NaOH(aq) + H_2(g) + Cl_2(g)$$

12.33. If the value of ΔS_{rxn} of the nonspontaneous reaction $A + B \rightarrow C$ is -66.0 J/K, what is the maximum entropy change in the reaction's surroundings?

12.34. The value of ΔS_{rxn} of the spontaneous reaction $D + E \rightarrow F$ is 72.0 J/K. What is the minimum value of the entropy change in the reaction's surroundings?

Calculating Entropy Changes

Concept Review

12.35. Under standard conditions, the products of a reaction have, overall, greater entropy than the reactants. What is the sign of $\Delta S°_{rxn}$?

12.36. Do decomposition reactions tend to have $\Delta S°_{rxn}$ values that are greater than zero or less than zero? Why?

12.37. Do precipitation reactions tend to have $\Delta S°_{rxn}$ values that are greater than zero or less than zero? Why?

12.38. For each of the reactions given, indicate whether ΔS should have a positive sign or a negative sign. If it is not possible to judge the sign of ΔS based on the information provided, indicate why that is the case.
a. $2\,Na(s) + Cl_2(g) \rightarrow 2\,NaCl(s)$
b. $4\,H_3PO_3(\ell) \rightarrow PH_3(g) + 3\,H_3PO_4(\ell)$
c. $CO(g) + H_2O(g) \rightarrow CO_2(g) + H_2(g)$
d. $Ca(OH)_2(s) + CO_2(g) \rightarrow CaCO_3(s) + H_2O(g)$

Problems

12.39. **Smog** Use the standard molar entropies in Appendix 4 to calculate $\Delta S°$ values for each of the following atmospheric reactions that contribute to the formation of photochemical smog.
a. $N_2(g) + O_2(g) \rightarrow 2\,NO(g)$
b. $2\,NO(g) + O_2(g) \rightarrow 2\,NO_2(g)$
c. $NO(g) + \frac{1}{2}\,O_2(g) \rightarrow NO_2(g)$
d. $2\,NO_2(g) \rightarrow N_2O_4(g)$

12.40. Use the standard molar entropies in Appendix 4 to calculate the $\Delta S°$ value for each of the following reactions of sulfur compounds.
a. $H_2S(g) + \frac{3}{2}\,O_2(g) \rightarrow H_2O(g) + SO_2(g)$
b. $2\,SO_2(g) + O_2(g) \rightarrow 2\,SO_3(g)$
c. $SO_3(g) + H_2O(\ell) \rightarrow H_2SO_4(aq)$
d. $S(g) + O_2(g) \rightarrow SO_2(g)$

12.41. **Ozone Layer** The following reaction plays a key role in the destruction of ozone in the atmosphere:

$$Cl(g) + O_3(g) \rightarrow ClO(g) + O_2(g)$$

The standard entropy change ($\Delta S°_{rxn}$) is 19.9 J/(mol · K). Use the standard molar entropies ($S°$) in Appendix 4 to calculate the $S°$ value of $ClO(g)$.

12.42. Calculate the $\Delta S°$ value for the conversion of ozone to oxygen,

$$2\,O_3(g) \rightarrow 3\,O_2(g)$$

in the absence of Cl atoms, and compare it with the $\Delta S°$ value in Problem 12.41.

Free Energy

Concept Review

12.43. What does the sign of ΔG tell you about the spontaneity of a process?

12.44. What does the sign of ΔG tell you about the rate of a reaction?

***12.45.** Many 19th-century scientists believed that all exothermic reactions were spontaneous. Why did so many of them share this belief?

12.46. In which direction does a reaction proceed when (a) $\Delta G_{rxn} < 0$; (b) $\Delta G_{rxn} = 0$; (c) $\Delta G_{rxn} > 0$?

12.47. What are the signs of ΔS, ΔH, and ΔG for the sublimation of dry ice (solid CO_2) at 25°C?

12.48. What are the signs of ΔS, ΔH, and ΔG for the formation of dew on a cool night?

12.49. Which of the following processes is/are spontaneous?
a. A tornado forms.
b. A broken cell phone fixes itself.
c. You get an A in this course.
d. Hot soup gets cold before it is served.

12.50. Which of the following processes is/are spontaneous?
a. Wood burns in air.
b. Water vapor condenses on the sides of a glass of iced tea.
c. Salt dissolves in water.
d. Photosynthesis occurs.

Problems

12.51. Calculate the free-energy change for the dissolution in water of one mole of NaBr and one mole of NaI at 298 K from the values in the following table.

	$\Delta H°_{solution}$ (kJ/mol)	$\Delta S°_{solution}$ [J/(mol · K)]
NaBr	−0.60	57
NaI	−7.5	74

12.52. The values of $\Delta H°_{rxn}$ and $\Delta S°_{rxn}$ for the reaction

$$2\,NO(g) + O_2(g) \rightarrow 2\,NO_2(g)$$

are −12 kJ and −146 J/K.
a. Use these values to calculate $\Delta G°_{rxn}$ at 298 K.
b. Explain why the value of $\Delta S°_{rxn}$ is negative.

12.53. A mixture of $CO(g)$ and $H_2(g)$ is produced by passing steam over hot charcoal:

$$H_2O(g) + C(s) \rightarrow H_2(g) + CO(g)$$

Calculate the $\Delta G°_{rxn}$ value for the reaction from the appropriate $\Delta G°_f$ data in Appendix 4.

12.54. Use the appropriate $\Delta G°_f$ data in Appendix 4 to calculate $\Delta G°_{rxn}$ for the complete combustion of methanol:

$$2\,CH_3OH(g) + 3\,O_2(g) \rightarrow 2\,CO_2(g) + 4\,H_2O(g)$$

12.55. **Photochemical Smog** Use the appropriate $\Delta G°_f$ data in Appendix 4 to calculate $\Delta G°_{rxn}$ for the oxidation of NO to NO_2—a key reaction in the formation of photochemical smog:

$$NO(g) + \tfrac{1}{2}\,O_2(g) \rightarrow NO_2(g)$$

12.56. Use the free energies of formation from Appendix 4 to calculate the standard free-energy change for the decomposition of ammonia in the following reaction:

$$2\,NH_3(g) \rightarrow N_2(g) + 3\,H_2(g)$$

Is the reaction spontaneous under standard conditions?

12.57. **Acid Precipitation** Aerosols (fine droplets) of sulfuric acid form in the atmosphere as a result of the following combination reaction:

$$SO_3(g) + H_2O(g) \rightarrow H_2SO_4(\ell)$$

Use the appropriate $\Delta G°_f$ data in Appendix 4 to calculate $\Delta G°_{rxn}$ for this reaction.

12.58. One source of sulfuric acid aerosols in the atmosphere is the combustion of high-sulfur fuels, which releases SO_2 gas that then is further oxidized to SO_3:

$$2\,SO_2(g) + O_2(g) \rightarrow 2\,SO_3(g)$$

Use the appropriate $\Delta G°_f$ data in Appendix 4 to calculate $\Delta G°_{rxn}$ for this combination reaction at 25°C. Is it spontaneous under standard conditions?

Temperature and Spontaneity

Concept Review

12.59. Are exothermic reactions spontaneous only at low temperature? Explain your answer.

12.60. Are endothermic reactions never spontaneous at low temperature? Explain your answer.

Problems

12.61. What is the lowest temperature at which the following reaction (see Problem 12.53) is spontaneous?

$$H_2O(g) + C(s) \rightarrow H_2(g) + CO(g)$$

12.62. Above what temperature does nitrogen monoxide form from nitrogen and oxygen?

$$N_2(g) + O_2(g) \rightarrow 2\ NO(g)$$

Assume that the values of ΔH°_{rxn} and ΔS°_{rxn} do not change appreciably with temperature.

12.63. Use the data in Appendix 4 to calculate ΔH° and ΔS° for the vaporization of hydrogen peroxide:

$$H_2O_2(\ell) \rightarrow H_2O_2(g)$$

Assuming that the calculated values are independent of temperature, what is the boiling point of hydrogen peroxide at $P = 1.00$ atm?

12.64. **Volcanoes** Deposits of elemental sulfur are often seen near active volcanoes. Their presence there may be due to the following reaction of SO_2 with H_2S:

$$SO_2(g) + 2\ H_2S(g) \rightarrow \tfrac{3}{8}\ S_8(s) + 2\ H_2O(g)$$

Assuming the values of ΔH°_{rxn} and ΔS°_{rxn} do not change appreciably with temperature, over what temperature range is the reaction spontaneous?

12.65. Which of the following reactions is spontaneous (i) only at low temperatures; (ii) only at high temperatures; (iii) at all temperatures?

a. $2\ NO(g) + O_2(g) \rightarrow 2\ NO_2(g)$
b. $2\ NH_3(g) + 2\ O_2(g) \rightarrow N_2O(g) + 3\ H_2O(g)$
c. $NH_4NO_3(s) \rightarrow 2\ H_2O(g) + N_2O(g)$

12.66. Which of the following reactions is spontaneous (i) only at low temperatures; (ii) only at high temperatures; (iii) at all temperatures?

a. $2\ Mg(s) + O_2(g) \rightarrow 2\ MgO(s)$
b. $2\ CH_3OH(\ell) + 3\ O_2(g) \rightarrow 2\ CO_2(g) + 4\ H_2O(\ell)$
c. $N_2(g) + O_2(g) \rightarrow 2\ NO(g)$

12.67. One method for the industrial production of methanol uses the following reaction:

$$CO(g) + 2\ H_2(g) \rightarrow CH_3OH(\ell)$$

a. Use the data in Appendix 4 to calculate ΔG° for this reaction at 298 K.
b. The reaction is normally run at a minimum temperature of 475 K. What is the value of ΔG at that temperature? Is the reaction spontaneous at that temperature?

12.68. Gas streams containing CO_2 are frequently passed through absorption tubes filled with $CaO(s)$, where the following reaction takes place to remove the CO_2 from the stream:

$$CaO(s) + CO_2(g) \rightarrow CaCO_3(s)$$

a. Use the data in Appendix 4 to calculate ΔG° at 298 K for this reaction.
b. Is the reaction spontaneous at 298 K?
c. Calculate ΔG for this reaction at 1500 K, a typical temperature for a lime kiln. (Assume ΔH and ΔS do not change with temperature.)
d. Is the reaction as written spontaneous at 1500 K?
e. In a lime kiln, calcium carbonate (in the form of oyster shells) is roasted to produce CaO and CO_2. Is this process spontaneous at the temperature of a kiln?

Driving the Human Engine: Coupled Reactions

Concept Review

12.69. Describe the ways in which two chemical reactions must complement each other so that the decrease in free energy of the spontaneous one can drive the nonspontaneous one.

12.70. How do you calculate the value of ΔG° for a reaction that is the result of coupling a spontaneous reaction ($\Delta G^\circ_{spon} < 0$) and a nonspontaneous reaction ($\Delta G^\circ_{nonspon} > 0$)?

12.71. Why is it important that at least some of the spontaneous steps in glycolysis convert ADP to ATP?

12.72. The second step in glycolysis converts glucose 6-phosphate into fructose 6-phosphate (Figure P12.72). Suggest a reason why ΔG° for this reaction is close to zero.

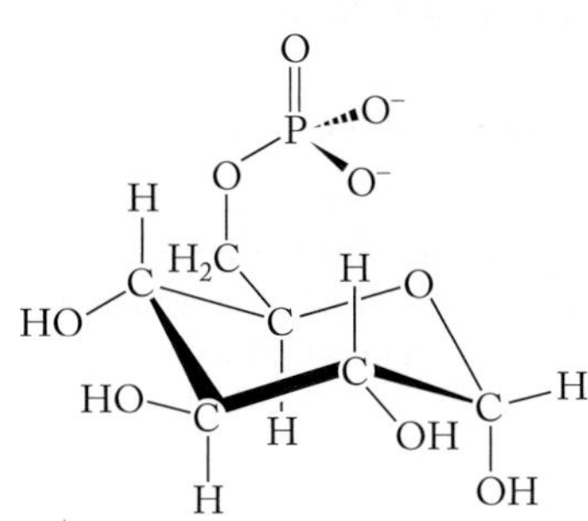

FIGURE P12.72

Problems

12.73. The methane in natural gas is an important starting material, or feedstock, for producing industrial chemicals, including H_2 gas.

a. Use the appropriate ΔG°_f value(s) from Appendix 4 to calculate ΔG°_{rxn} for the reaction known as *steam–methane reforming*:

$$CH_4(g) + H_2O(g) \rightarrow CO(g) + 3\ H_2(g)$$

b. To help drive this nonspontaneous reaction, the CO that is produced can be oxidized to CO_2 using more steam:

$$CO(g) + H_2O(g) \rightarrow CO_2(g) + H_2(g)$$

Use the appropriate ΔG°_f value(s) from Appendix 4 to calculate ΔG°_{rxn} for this reaction, which is known as the *water–gas shift reaction.*

c. Combine these two reactions and write the chemical equation of the overall reaction in which methane and steam combine to produce hydrogen gas and carbon dioxide.
d. Calculate the ΔG°_{rxn} value of the overall reaction. Is it spontaneous under standard conditions?

12.74. In addition to the reactions described in Problem 12.73, methane can, in theory, be used to produce hydrogen gas by a process in which it decomposes into elemental carbon and hydrogen:

$$(1) \quad CH_4(g) \rightarrow C(s) + 2\,H_2(g)$$

and the carbon produced in the first step is then oxidized to CO_2:

$$(2) \quad C(s) + O_2(g) \rightarrow CO_2(g)$$

a. Calculate the ΔG°_{rxn} values of reactions (1) and (2).
b. Write a balanced chemical equation describing the overall reaction obtained by coupling reactions (1) and (2), and calculate its ΔG°_{rxn} value. Is the coupled reaction spontaneous under standard conditions?

12.75. Making Steel Important industrial processes, such as converting iron ore to iron and then to steel, involve coupling a nonspontaneous reaction, such as reducing the iron in Fe_2O_3 to metallic iron, with a spontaneous one, such as the oxidation of the carbon in CO to CO_2. Use the appropriate thermodynamic data in Appendix 4 to calculate the ΔG°_{rxn} value of the following reaction at 1450°C:

$$Fe_2O_3(s) + 3\,CO(g) \rightarrow 2\,Fe(s) + 3\,CO_2(g)$$

12.76. One source of the carbon monoxide reactant in Problem 12.75 is pure hot carbon, called *coke*, which is derived from coal. What is the overall ΔG°_{rxn} value of the iron reduction reaction at 1450°C starting with carbon as the reducing agent instead of carbon monoxide? Assume coke has the thermodynamic properties of graphite.

Additional Problems

12.77. Chlorofluorocarbons (CFCs) are no longer used as refrigerants because they catalyze the decomposition of stratospheric ozone. Trichlorofluoromethane (CCl_3F) boils at 23.8°C and its molar heat of vaporization is 24.8 kJ/mol. What is the molar entropy of vaporization of $CCl_3F(\ell)$?

12.78. **Methane-Producing Bacteria** Methanogenic bacteria convert liquid acetic acid (CH_3COOH) into $CO_2(g)$ and $CH_4(g)$.
a. Is this process endothermic or exothermic under standard conditions?
b. Is the reaction spontaneous under standard conditions?

***12.79.** At what temperature is the free-energy change for the following reaction equal to zero?

$$NH_4Cl(s) \rightarrow NH_3(g) + HCl(g)$$

12.80. Consider the precipitation reactions described by the following net ionic equations:

$$Mg^{2+}(aq) + 2\,OH^-(aq) \rightarrow Mg(OH)_2(s)$$

$$Ag^+(aq) + Cl^-(aq) \rightarrow AgCl(s)$$

a. Predict the sign of ΔS°_{rxn} for the reactions.
b. Using the values for S° from Appendix 4, calculate ΔS° for these reactions.
c. Do your calculations support your prediction?

***12.81.** Calculate the standard free-energy change of the following reaction. Is it spontaneous?

$$2\,NO(g) + 2\,H_2(g) \rightarrow N_2(g) + 2\,H_2O(g)$$

12.82. Rudolf Clausius (1822–1888), considered one of the founders of thermodynamics, summed up the second law of thermodynamics once by reportedly saying, "The algebraic sum of all the transformations occurring in a cyclical process can only be positive, or, as an extreme case, equal to nothing." Use concepts and illustrations discussed in this chapter to explain his comment.

***12.83.** Show that hydrogen cyanide (HCN) is a gas at 25°C by estimating its normal boiling point from the following data:

	ΔH°_f (kJ/mol)	S° [J/(mol · K)]
HCN(ℓ)	108.9	113
HCN(g)	135.1	202

12.84. Write two equations for the complete combustion of one mole of acetylene, $C_2H_2(g)$, in oxygen at 298 K: in the first equation, the water produced as a product is a liquid; in the second equation, the water is in the gas phase.
a. Determine ΔG° for each reaction.
*b. Suggest a way you could determine the difference between the two ΔG° values without having to solve for ΔG° for both reactions.

12.85. Lightbulb Filaments Tungsten (W) is the favored metal for lightbulb filaments, in part because of its high melting point (3422°C). The enthalpy of fusion of tungsten is 35.4 kJ/mol. What is its entropy of fusion?

12.86. The absolute entropy (S) of a perfect, defect-free solid equals zero and has one accessible microstate. It is impossible to make such a material, but, silicon chip manufacturers strive for as few defects as possible in their products. Calculate the absolute molar entropy for a piece of silicon with a number of probable arrangements of (a) $W = 16$, (b) $W = 625$, and (c) $W = 2500$ per atom of Si.

12.87. Two allotropes (A and B) of sulfur interconvert at 369 K and 1 atm pressure:

$$S_8(s, A) \rightarrow S_8(s, B)$$

The enthalpy change in this transition is 297 J/mol. What is the entropy change?

*12.88. Over what temperature range is the reduction of tungsten(VI) oxide by hydrogen to give metallic tungsten and water spontaneous? The standard heat of formation of $WO_3(s)$ is −843 kJ/mol, and its standard molar entropy is 76 J/(mol · K).

*12.89. **Lime** Enormous amounts of lime (CaO) are used in steel industry blast furnaces to remove impurities from iron. Lime is made by heating limestone and other solid forms of $CaCO_3(s)$. Why is the standard molar entropy of $CaCO_3(s)$ higher than that of $CaO(s)$? At what temperature is the pressure of $CO_2(g)$ over $CaCO_3(s)$ equal to 1.0 atm?

	ΔH_f° (kJ/mol)	S° [J/(mol · K)]
$CaCO_3(s)$	−1207	93
$CaO(s)$	−636	40
$CO_2(g)$	−394	214

*12.90. Copper forms two oxides, Cu_2O and CuO.

a. Name these oxides.

b. Predict over what temperature range this reaction is spontaneous using the following thermodynamic data:

$$Cu_2O(s) \rightarrow CuO(s) + Cu(s)$$

	ΔH_f° (kJ/mol)	S° [J/(mol · K)]
$Cu_2O(s)$	−170.7	92.4
$CuO(s)$	−156.1	42.6

c. Why is the standard molar entropy of $Cu_2O(s)$ larger than that of $CuO(s)$?

*12.91. **Melting Organic Compounds** When dicarboxylic acids (compounds with two –COOH groups in their structures) melt, they frequently decompose to produce one mole of CO_2 gas for every mole of dicarboxylic acid melted (see Figure P12.91).

a. What are the signs of ΔH and ΔS for the process as written?

b. Do you think the dicarboxylic acid will re-form when the melted material cools? Why or why not?

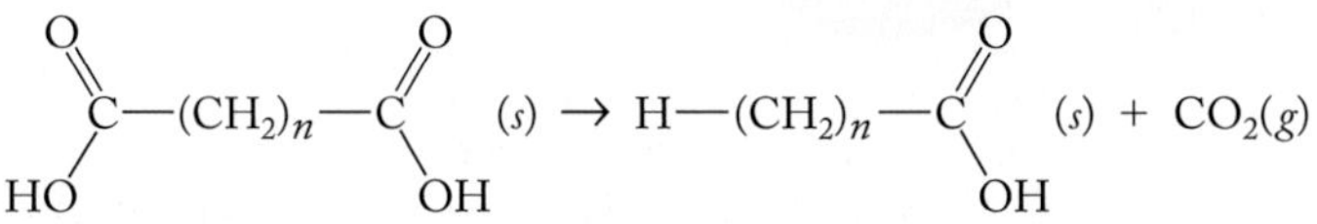

$$HO\text{—}C(=O)\text{—}(CH_2)_n\text{—}C(=O)\text{—}OH\,(s) \rightarrow H\text{—}(CH_2)_n\text{—}C(=O)\text{—}OH\,(s) + CO_2(g)$$

FIGURE P12.91

*12.92. **Melting DNA** When a solution of DNA in water is heated, the DNA double helix separates into two single strands.

a. What is the sign of ΔS for the separation process?

b. The DNA double helix re-forms as the system cools. What is the sign of ΔS for the process by which two single strands re-form the double helix?

c. The melting point of DNA is defined as the temperature at which $\Delta G = 0$. At that temperature, the melting reaction produces two single strands as fast as two single strands recombine to form the double helix. Write an equation that defines the melting temperature (T) of DNA in terms of ΔH and ΔS.

smartwork5

If your instructor uses Smartwork5, log in at digital.wwnorton.com/atoms2.

13

Chemical Kinetics

Clearing the Air

PHOTOCHEMICAL SMOG The brown color of the air in many large urban areas is caused by reactions that produce NO_2 gas.

PARTICULATE **REVIEW**

Temperature and Collisions

In Chapter 13, we examine the factors that influence the rates of reactions. One reaction that leads to the air pollution visible in the photo above is the reaction of nitrogen and oxygen to form nitrogen monoxide.

- Write a balanced equation for this reaction.
- Pictured are three samples of a mixture of nitrogen and oxygen gases. Which sample is at the highest temperature? The lowest temperature?

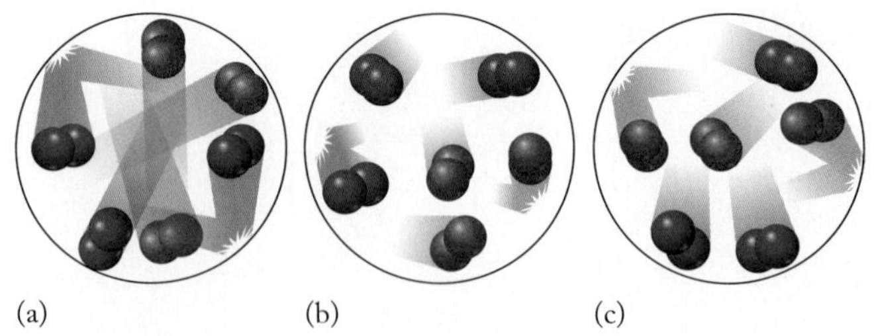

- Which mixture of gases contains the molecules colliding most frequently and with the most energy?

(Review Sections 10.3 and 10.5 if you need help.)

(Answers to Particulate Review questions are in the back of the book.)

PARTICULATE **PREVIEW**

Reaction Steps

Many reactions take place in multiple steps. These images provide a molecular view of how the decomposition of NO_2 takes place in two steps at high temperatures. As you read Chapter 13, look for ideas that will help you answer these questions:

- Write a balanced equation for each step in the reaction and add the two reactions together to obtain an overall reaction.
- What two processes do the dashed lines represent?
- Explain why one of the products in step 1 is *not* a product of the overall reaction.

Learning Outcomes

LO1 Use the stoichiometry of a reaction to relate the rates at which the concentrations of its reactants and products change as the reaction proceeds
Sample Exercises 13.1, 13.2

LO2 Determine average and instantaneous reaction rates
Sample Exercise 13.3

LO3 Determine the rate law and overall order for a chemical reaction using initial rate data
Sample Exercise 13.4

LO4 Use integrated rate laws to identify the orders of reactions and determine their rate constants
Sample Exercises 13.5, 13.7, 13.9

LO5 Calculate the half-life of a reactant
Sample Exercises 13.6, 13.8

LO6 Relate rates of reactions to temperature and their activation energies
Sample Exercises 13.10, 13.11

LO7 Use rate laws to assess the validity of a reaction mechanism
Sample Exercises 13.12, 13.13

LO8 Identify catalysts and describe their impact on reaction rates and mechanisms
Sample Exercise 13.14

13.1 Cars, Trucks, and Air Quality

In the photograph at the beginning of this chapter, a yellow-brown haze obscures the city skyline. This haze is called **photochemical smog**. Its photochemical origins are linked to chemical reactions involving nitrogen compounds emitted by vehicles and factories that are initiated by the energy in sunlight. The term *smog* was originally used to describe air pollution that occurred when *sm*oke mixed with *fog*.

How do the combustion reactions in car and truck engines produce these nitrogen compounds? A key first step takes place at the high temperatures inside these engines, which promote a highly endothermic combination reaction between N_2 and O_2:

$$N_2(g) + O_2(g) \rightarrow 2\ NO(g) \qquad \Delta H^\circ_{rxn} = 180.6\ kJ \qquad (13.1)$$

CONCEPT TEST

The reaction in Equation 13.1 requires extremely high temperatures to become spontaneous.

a. What is the sign of ΔS°_{rxn}?
b. Why is the sign of ΔS°_{rxn} difficult to predict based only on the information in Equation 13.1?

(Answers to Concept Tests are in the back of the book.)

When NO enters the atmosphere, it reacts with more oxygen, producing brown nitrogen dioxide gas:

$$2\ NO(g) + O_2(g) \rightarrow 2\ NO_2(g) \qquad \Delta H^\circ_{rxn} = -114.2\ kJ \qquad (13.2)$$

Sunlight provides sufficient energy to break the bonds in NO_2, forming NO and very reactive oxygen atoms:

$$NO_2(g) \xrightarrow{\text{sunlight}} NO(g) + O(g) \qquad \Delta H^\circ_{rxn} = 306.3\ kJ \qquad (13.3)$$

photochemical smog a mixture of gases formed in the lower atmosphere when sunlight interacts with compounds produced in internal combustion engines and with other pollutants.

This photochemically generated atomic oxygen combines with molecular oxygen, producing ozone,

$$O_2(g) + O(g) \rightarrow O_3(g) \qquad \Delta H^\circ_{rxn} = -106.5 \text{ kJ} \qquad (13.4)$$

CONNECTION In Chapter 4 we learned that odd-electron species such as OH belong to a highly reactive group of substances called free radicals.

It can also react with water vapor, producing hydroxyl radicals:

$$H_2O(g) + O(g) \rightarrow 2\,{}^{\bullet}OH(g) \qquad \Delta H^\circ_{rxn} = 70.6 \text{ kJ} \qquad (13.5)$$

In the atmosphere, O_3 and OH radicals react with volatile organic compounds (VOCs) to form a variety of noxious compounds. In one such reaction, acetaldehyde (a widely used industrial VOC) is transformed into peroxyacetyl nitrate, which is highly irritating to one's eyes, nose, and throat and can cause respiratory distress:

$$N_2(g) + 3\,O_2(g) + {}^{\bullet}OH(g) + CH_3CHO(g) \rightarrow CH_3C(O)O_2NO_2(g) + H_2O(g) + NO_2(g) \qquad (13.6)$$

Acetaldehyde Peroxyacetyl nitrate

Equations 13.1 through 13.6 show that many of the substances in photochemical smog are produced in some reactions and consumed in others. The rates of these and other atmospheric reactions influence when smog happens, how intense it is, and how long it persists. The graphs in Figure 13.1 show that the maximum NO concentration occurs during the morning rush hour. Later in the morning, the concentration of NO_2 reaches a maximum. This sequence makes sense because NO is a precursor of NO_2 (Equation 13.3). The highest ozone concentrations are reached in the middle of the afternoon, the result of the reactions shown in Equations 13.2 and 13.3 which produce a supply of free O atoms for the formation of ozone (Equation 13.4), hydroxyl radicals (Equation 13.5), and peroxyacetyl nitrate (Equation 13.6).

Ozone formed in the reaction in Equation 13.4 also reacts with NO to form NO_2 and O_2:

$$NO(g) + O_3(g) \rightarrow O_2(g) + NO_2(g) \qquad \Delta H^\circ_{rxn} = -199.8 \text{ kJ} \qquad (13.7)$$

However, NO_2 concentrations drop in the afternoon on smoggy days because NO_2, ozone, and hydrocarbons react to form an array of other (mostly unpleasant) compounds.

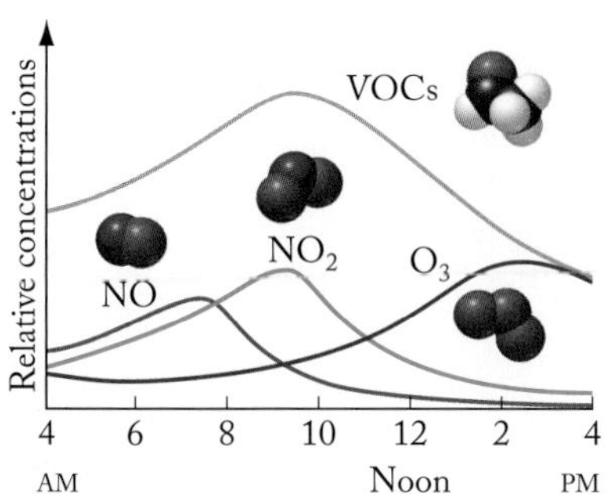

FIGURE 13.1 In photochemical smog, NO from engine exhaust builds up in the early morning and then decreases as it reacts with atmospheric O_2 to form NO_2, the concentration of which is highest in late morning. Photodecomposition of NO_2 leads to the formation of high levels of O_3 in the afternoon.

Photochemical smog became a significant environmental problem in many urban areas during the last half of the 20th century as the use of private automobiles expanded rapidly. By 1975 smog was so widespread, the U.S. Environmental Protection Agency required that pollutant emissions from cars and light trucks be dramatically reduced. Automotive engineers were able to meet these standards by designing engines that burned fuel more efficiently and cleanly and by developing devices called catalytic converters to further reduce pollutant concentrations in engine exhaust. In the years since 1975 these devices have removed billions of tons of pollution from urban air in the United States and around the world.

Designing catalytic converters required extensive study of the by-products of the combustion reactions in automobile engines and the rates of the reactions that convert these substances into less noxious gases. The target compounds

chemical kinetics the study of the rates of change of concentrations of substances involved in chemical reactions.

reaction rate a measure of how rapidly a reaction occurs; it is related to rates of change in the concentrations of reactants and products over time.

include unburned hydrocarbons and carbon monoxide in addition to various oxides of nitrogen. Catalytic converters speed up the oxidation of hydrocarbons and CO to CO_2 and H_2O and the conversion of nitrogen oxides into N_2 and O_2. These reactions happen during the fraction of a second that molecules spend in contact with the catalytic surfaces. These devices and cleaner-burning engines don't remove all the pollutants from engine exhaust, but they do remove most of them.

As we begin this chapter on rates of reactions, we will see how the rates of smog formation and other chemical reactions are measured. Knowledge of the rates of reactions allows us to better understand how they occur on molecular and atomic levels. The study of reaction rates, called **chemical kinetics**, enables chemists to understand how reactant particles interact with each other to form products. With that knowledge we can find ways to manipulate the rates of chemical reactions to improve the quality of our environment and our lives.

13.2 Reaction Rates

The concentration of NO in the exhaust from an automobile engine depends on how rapidly the reaction in Equation 13.1 proceeds. We express **reaction rate** as the change in the concentration of a product, such as NO, or a reactant, such as N_2 or O_2, that occurs over some interval of time. The rate of change in the concentration of a reactant (Δ[reactant]/Δt) has a negative value because reactant concentrations decrease as a reaction proceeds. Therefore, reactant concentration at the end of the interval ($[\text{reactant}]_{\text{final}}$) is always less than it was at the beginning ($[\text{reactant}]_{\text{initial}}$). Therefore, $\Delta[\text{reactant}] = [\text{reactant}]_{\text{final}} - [\text{reactant}]_{\text{initial}}$ must have a negative value, and so, too, must Δ[reactant]/Δt. However, the rate of any reaction is defined as a positive quantity because it describes the rate at which reactants form products, so a minus sign is used with Δ[reactant]/Δt values to obtain an overall positive value for reaction rate. Thus, the rate at which N_2 and O_2 combine to form NO can be expressed as follows:

$$\text{Reaction rate} = (-) \text{ the rate of change in } [N_2] = -\frac{\Delta[N_2]}{\Delta t} \tag{13.8}$$

We use brackets around N_2 to represent its concentration in moles per liter—a practice we will follow in the remaining chapters of this book. Reaction rates are usually expressed in units of concentration per time, as in the fraction on the right in Equation 13.8. A commonly encountered example is molarity per second (M/s).

Nitrogen and oxygen gas are consumed at the same rate in the balanced reaction in Equation 13.1 because their stoichiometric coefficients are both 1. However, two moles of NO form for every one mole of N_2 or O_2 consumed. Therefore, the rate of increase in the concentration of NO is twice the rate of decrease in the concentrations of N_2 and O_2. These relative rates of change can be expressed using the following equation:

$$\frac{\Delta[NO]}{\Delta t} = -2\frac{\Delta[N_2]}{\Delta t} = -2\frac{\Delta[O_2]}{\Delta t} \tag{13.9}$$

The negative signs are needed to convert the rates at which $[N_2]$ and $[O_2]$ are decreasing into positive values. When these values are multiplied by 2, they equal the rate at which [NO] is increasing. Now let's divide all of the terms in

Equation 13.9 by 2 and rearrange them so that they match the sequence of reactants and products in Equation 13.1:

$$-\frac{\Delta[N_2]}{\Delta t} = -\frac{\Delta[O_2]}{\Delta t} = \frac{1}{2}\frac{\Delta[NO]}{\Delta t} \quad (13.10)$$

Thus, the concentrations of N_2 and O_2 decrease at half the rate at which the concentration of NO increases, as shown by the curves in Figure 13.2, which depict the changing concentrations of these compounds as the reaction proceeds.

Note how the coefficient of 2 in front of NO in the chemical equation has become its reciprocal, $\frac{1}{2}$, in Equation 13.10. This pattern holds for all reactions: the coefficients in an equation expressing the relative rates of change of reactants and products are the reciprocals of the coefficients in a balanced chemical equation describing the reaction. To give all of the terms positive values, the rates representing decreasing reactant concentrations receive minus signs. Sample Exercise 13.1 provides practice in applying this approach to the changing concentrations of reactants and product in an important industrial process: the synthesis of ammonia.

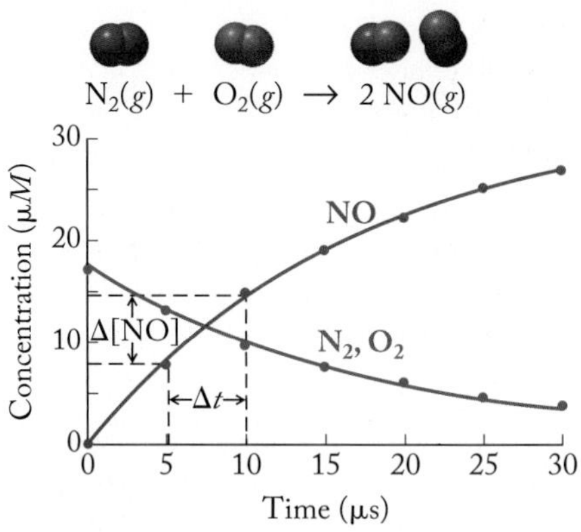

FIGURE 13.2 Concentrations of N_2, O_2, and NO over 30.0 μs for the reaction $N_2(g) + O_2(g) \rightarrow 2\,NO(g)$.

SAMPLE EXERCISE 13.1 Predicting Relative Rates of Concentration Change in a Chemical Reaction **LO1**

Each year about 130 million metric tons (1.3×10^{11} kg) of ammonia is produced worldwide, mostly for use as fertilizer in agriculture. The process for making ammonia is based on the following reaction:

$$N_2(g) + 3\,H_2(g) \rightarrow 2\,NH_3(g)$$

How is the rate of formation of NH_3 related to the rates of consumption of N_2 and H_2?

Collect, Organize, and Analyze The coefficients in an equation relating the rates of change in the concentrations of N_2, H_2, and NH_3 are the reciprocals of their coefficients (1, 3, and 2, respectively) in the balanced chemical equation. Nitrogen and hydrogen are consumed in the reaction, so their rate terms are preceded by minus signs.

Solve Inserting the reciprocals of 1, 3, and 2 in front of the rates of change in the concentrations of N_2, H_2, and NH_3, and placing minus signs before the two reactant terms give us the relative rates:

$$-\frac{\Delta[N_2]}{\Delta t} = -\frac{1}{3}\frac{\Delta[H_2]}{\Delta t} = \frac{1}{2}\frac{\Delta[NH_3]}{\Delta t}$$

Think About It The balanced equation indicates that two moles of NH_3 are formed for every one mole of N_2 consumed. This means that the rate of consumption of N_2 should be half the rate of formation of ammonia, which is what the first and last terms in the answer tell us. Similarly, we know that three moles of H_2 are consumed for every one mole of N_2 consumed. Therefore, the rate at which N_2 is consumed is one-third the rate at which H_2 is consumed, as described in the first and second terms of the answer.

Practice Exercise

In the reaction $2\,CO(g) + O_2(g) \rightarrow 2\,CO_2(g)$,

a. Which reactant (CO or O_2) is consumed at the higher rate during the oxidation of carbon monoxide?
b. How is the rate of change in the concentration of CO_2 related to the rate of change in the concentration of O_2?

(Answers to Practice Exercises are in the back of the book.)

Reaction Rate Values

Up to this point our focus has been on the relative rates at which reactants are consumed and products are formed in chemical reactions. We have seen how these rates differ depending on the values of the coefficients in balanced chemical equations—that is, depending on reaction stoichiometry. The questions to answer now are the following:

- How do we know the actual rates at which concentrations change in a chemical reaction?
- Which of these rates do we use to express the rate of the reaction?

The answer to the first question is that rates of change are determined experimentally. Suppose, for example, that analysis of the gas mixture in the ammonia synthesis in Sample Exercise 13.1 reveals that, over a 10.0-second time interval, the concentration of ammonia increases from 0.133 *M* to 0.605 *M*. These data mean that the average rate of change in ammonia concentration over the interval is

$$\frac{\Delta[NH_3]}{\Delta t} = \frac{(0.605 - 0.133)M}{10.0\ s} = 0.0472\ M/s$$

To calculate the rate of change in the concentration of N_2, we use the equation we derived in the exercise:

$$-\frac{\Delta[N_2]}{\Delta t} = \frac{1}{2}\frac{\Delta[NH_3]}{\Delta t} = \frac{1}{2}(0.0472\ M/s)$$

$$\frac{\Delta[N_2]}{\Delta t} = -0.0236\ M/s$$

Which of these two values should we use to express the rate of the reaction? Both values work as long as we have referenced the compound in the chemical equation whose concentration was monitored as a function of time. However, it is conventional to use the rate of change of the reactant or product with a coefficient of 1 in the chemical equation—in this case, N_2—as the basis for expressing the rate of the overall reaction:

$$\text{Rate} = -\frac{\Delta[N_2]}{\Delta t} = 0.0236\ M/s$$

If all of the coefficients in a chemical equation are greater than 1, then the rate of the reaction can be based on the rate of change in the concentration of any reactant or product as long as the rate is multiplied by the reciprocal of that substance's coefficient in the balanced chemical equation describing the reaction. For example, the rate of the reaction $2\ O_3(g) \rightarrow 3\ O_2(g)$ can be based on the rate of change of either $[O_3]$ or $[O_2]$:

$$\text{Rate} = \frac{1}{3}\frac{\Delta[O_2]}{\Delta t} = -\frac{1}{2}\frac{\Delta[O_3]}{\Delta t}$$

SAMPLE EXERCISE 13.2 Calculating Rates of Change in Reactant and Product Concentrations Using Stoichiometric Ratios **LO1**

In a high-temperature reaction between NO and H_2:

$$2\ NO(g) + 2\ H_2(g) \rightarrow N_2(g) + 2\ H_2O(g)$$

the initial rate of change in the concentration of NO is $-6.0 \times 10^{-5}\ M\,s^{-1}$. What are the initial rates of change in the concentrations of the other three substances in the reaction mixture?

Collect, Organize, and Analyze We have the balanced chemical equation for a reaction and an initial rate of change in concentration of one reactant. The stoichiometry enables us to relate the consumption of reactants (NO and H_2) to the formation of products (N_2 and H_2O).

Solve For every two moles of NO consumed in the reaction, two moles of H_2 are also consumed, and one mole of N_2 and two moles of H_2O are generated. Inserting the reciprocals of 2, 2, 1, and 2 in front of the rates of change in the concentrations of N_2, H_2, and NH_3 and placing minus signs before the two reactant terms give us the relative rates:

$$\text{Rate} = -\frac{1}{2}\frac{\Delta[\text{NO}]}{\Delta t} = -\frac{1}{2}\frac{\Delta[\text{H}_2]}{\Delta t} = \frac{\Delta[\text{N}_2]}{\Delta t} = \frac{1}{2}\frac{\Delta[\text{H}_2\text{O}]}{\Delta t}$$

The rates of change in the concentrations of all four species are

$$\frac{\Delta[\text{NO}]}{\Delta t} = -6.0 \times 10^{-5}\ M\,s^{-1}\ \text{(given)}$$

$$-\frac{1}{2}\frac{\Delta[\text{H}_2]}{\Delta t} = -\frac{1}{2}\frac{\Delta[\text{NO}]}{\Delta t} = -6.0 \times 10^{-5}\ M\,s^{-1}$$

$$\frac{\Delta[\text{N}_2]}{\Delta t} = -\frac{1}{2}\frac{\Delta[\text{NO}]}{\Delta t} = -\frac{1}{2} \times (-6.0 \times 10^{-5}\ M\,s^{-1}) = 3.0 \times 10^{-5}\ M\,s^{-1}$$

$$\frac{1}{2}\frac{\Delta[\text{H}_2\text{O}]}{\Delta t} = -\frac{1}{2}\frac{\Delta[\text{NO}]}{\Delta t} = -(-6.0 \times 10^{-5}\ M\,s^{-1}) = 6.0 \times 10^{-5}\ M\,s^{-1}$$

Think About It Reactants are consumed during the course of a reaction, so the rates of change in their concentrations have negative signs. However, product concentrations increase, so their rates of change are positive.

Practice Exercise Consider the following reaction:

$$8\ \text{A}(g) + 5\ \text{B}(g) \rightarrow 8\ \text{C}(g) + 6\ \text{D}(g)$$

If [C] is increasing at a rate of $4.0\ M\,s^{-1}$, at what rate are the concentrations of the other species changing?

Average and Instantaneous Reaction Rates

Suppose we ran an experiment to determine the rate of formation of NO in an automobile engine. In the laboratory, we use a reaction vessel as hot as the combustion chambers in the engine and we monitor the concentrations of reactants and products over time, obtaining the data in Table 13.1, which are plotted in Figure 13.2. Let's calculate the average rate of formation of NO between 5.0 and 10.0 μs:

$$\frac{\Delta[\text{NO}]}{\Delta t} = \frac{[\text{NO}]_{10.0\ \mu s} - [\text{NO}]_{5.0\ \mu s}}{t_{10.0\ \mu s} - t_{5.0\ \mu s}} = \frac{(14.8 - 7.8)\ \mu M}{(10.0 - 5.0)\ \mu s} = 1.4\ M/s$$

During the same interval, the average rate of change in the concentration of N_2 (and O_2) is

$$\frac{\Delta[\text{N}_2]}{\Delta t} = \frac{[\text{N}_2]_{10.0\ \mu s} - [\text{N}_2]_{5.0\ \mu s}}{t_{10.0\ \mu s} - t_{5.0\ \mu s}} = \frac{(9.6 - 13.1)\ \mu M}{(10.0 - 5.0)\ \mu s} = -0.70\ M/s$$

TABLE 13.1 Changing Concentrations of Reactants and Products for the Reaction $N_2(g) + O_2(g) \rightarrow 2\ NO(g)$

Time (μs)	$[N_2]$, $[O_2]$ (μM)	[NO] (μM)
0	17.0	0.0
5.0	13.1	7.8
10.0	9.6	14.8
15.0	7.6	18.6
20.0	5.8	22.2
25.0	4.5	24.8
30.0	3.6	26.7

These results give us two values on which to base the average rate of the reaction. In this example we select the rate of change in $[N_2]$ or $[O_2]$ because they each have a coefficient of 1 in the balanced chemical equation. Therefore, the average rate of this reaction between 5.0 and 10.0 μs is

$$\text{Rate} = -\frac{\Delta[N_2]}{\Delta t} = -(-0.70\ M/\text{s}) = 0.70\ M/\text{s}$$

The curvature of the lines in Figure 13.2 tells us that this average rate value applies only to that particular 5.0 μs time interval. Any other time interval has a different average rate. For instance, from $t = 25.0$ μs to $t = 30.0$ μs, the rate of the reaction is

$$-\frac{\Delta[N_2]}{\Delta t} = -\frac{[N_2]_{30.0\ \mu s} - [N_2]_{25.0\ \mu s}}{t_{30.0\ \mu s} - t_{25.0\ \mu s}} = -\frac{(3.6 - 4.5)\ \mu M}{(30.0 - 25.0)\ \mu s} = 0.18\ M/\text{s}$$

We can also determine the *instantaneous* rate of a reaction—that is, the rate of the reaction at a particular time after it began. The difference between average and instantaneous reaction rates is analogous to the difference between the average and instantaneous speeds of a runner. If a competitor in a marathon runs from mile 10 to mile 20 in 1 h, her average speed over that distance was 10 mi/h. At a given instant during the run, however, her instantaneous speed could have been 12 mi/h while going downhill or 8 mi/h while going uphill.

Let's again consider the exothermic conversion of NO into NO_2 in the atmosphere:

$$2\ NO(g) + O_2(g) \rightarrow 2\ NO_2(g) \qquad \Delta H^\circ_{\text{rxn}} = -114.2\ \text{kJ} \qquad (13.2)$$

The rate of this reaction based on the rate of consumption of O_2 is described by the data in Table 13.2, which are plotted in Figure 13.3. One way to determine the instantaneous rate of the reaction at a particular time involves drawing a tangent to the $[O_2]$ curve at that time and determining the slope of the tangent. Figure 13.3 illustrates how to do this at $t = 2000$ s. In this example, we select two convenient points along the tangent, at $t = 1000$ and 3000 s, determine the corresponding $[O_2]$ values at those times, and then use those time and concentration values to calculate the instantaneous reaction rate:

$$\text{Slope at } t = 2000\ \text{s} = \frac{\Delta[O_2]}{\Delta t} = \frac{(0.0072 - 0.0084)\ M}{(3000 - 1000)\ \text{s}}$$
$$= -6.0 \times 10^{-7}\ M\ \text{s}^{-1}$$

TABLE 13.2 Changing Concentrations of Reactants and Products for the Reaction $2\ NO(g) + O_2(g) \rightarrow 2\ NO_2(g)$ at 25°C

Time (s)	[NO] (M)	$[O_2]$ (M)	$[NO_2]$ (M)
0	0.0100	0.0100	0.0000
285	0.0090	0.0095	0.0010
660	0.0080	0.0090	0.0020
1175	0.0070	0.0085	0.0030
1895	0.0060	0.0080	0.0040
2975	0.0050	0.0075	0.0050
4700	0.0040	0.0070	0.0060
7800	0.0030	0.0065	0.0070

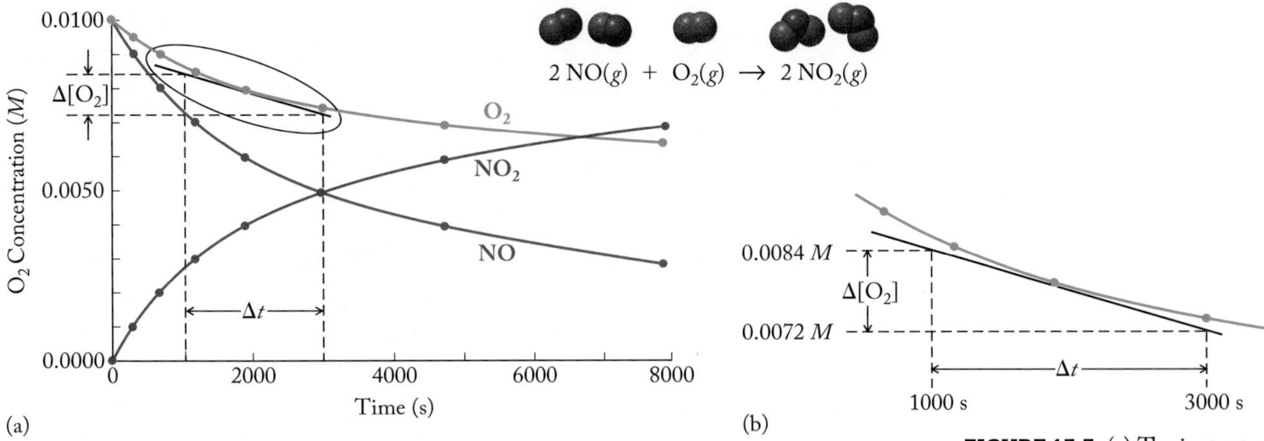

FIGURE 13.3 (a) The instantaneous rate of change in $[O_2]$ in the reaction $2\ NO(g) + O_2(g) \rightarrow 2\ NO_2(g)$ is equal to the slope of a tangent to the curve of $[O_2]$ versus time. (b) An expanded view of the instantaneous rate of change in $[O_2]$ at $t = 2000$ s.

$$\text{Instantaneous rate at } t = 2000 \text{ s} = -\frac{\Delta[O_2]}{\Delta t} = -(-6.0 \times 10^{-7}\ M\,s^{-1})$$
$$= 6.0 \times 10^{-7}\ M\,s^{-1}$$

Note that we use data points along the tangent line, not along the curve, to calculate instantaneous reaction rate.[1]

CONCEPT TEST

Which of the following statements is true about the instantaneous rate of the chemical reaction A → B as the reaction progresses? Assume the rate of the reaction decreases as [A] decreases.

a. $-\Delta[A]/\Delta t$ increases, $\Delta[B]/\Delta t$ decreases
b. $-\Delta[A]/\Delta t$ decreases, $\Delta[B]/\Delta t$ increases
c. $-\Delta[A]/\Delta t$ and $\Delta[B]/\Delta t$ both increase
d. $-\Delta[A]/\Delta t$ and $\Delta[B]/\Delta t$ both decrease

SAMPLE EXERCISE 13.3 Determining an Instantaneous Reaction Rate — LO2

a. What is the instantaneous rate of change of [NO] at $t = 2000$ s in the experiment that produced the data in Table 13.2?
b. What is the rate of the reaction based on your result in part (a)?

Collect, Organize, and Analyze We are asked to determine the instantaneous rate of change of [NO] and the corresponding reaction rate at $t = 2000$ s using the data in Table 13.2. The coefficient of NO is 2 in the balanced chemical equation: $2\ NO(g) + O_2(g) \rightarrow 2\ NO_2(g)$.

The corresponding reaction rate will be (−1/2) the value of the instantaneous rate of change of [NO].

Solve

a. First we plot [NO] versus time (Figure 13.4) and draw a tangent to the curve at $t = 2000$ s. Choosing two points along the tangent, $t = 1000$ and 3000 s, we

[1]If you have studied calculus, you may recall that the average rate of a reaction approaches the instantaneous rate as Δt approaches zero. Using calculus, we would say $\lim_{\Delta t \to 0} \Delta[O_2]/\Delta t = d[O_2]/dt$. The slope of a tangent to a curve is the derivative of the curve at a given point and can be calculated using many scientific calculators.

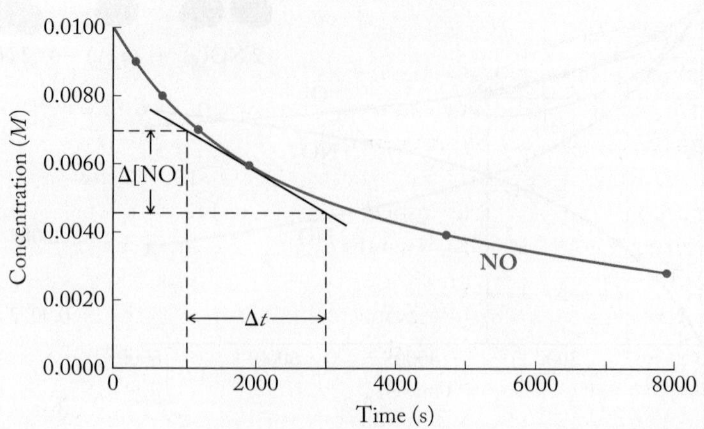

FIGURE 13.4 Plot of concentration versus time using data from Table 13.2.

determine the concentrations corresponding to those times along the vertical axis and use those values to calculate the slope of the line:

$$\frac{\Delta[\text{NO}]}{\Delta t} = \frac{(0.0046 - 0.0070)\ M}{(3000 - 1000)\ \text{s}} = -1.2 \times 10^{-6}\ M/\text{s}$$

b. The corresponding instantaneous reaction rate is

$$\text{Rate} = -\frac{1}{2}\frac{\Delta[\text{NO}]}{\Delta t} = -\frac{1}{2}(-1.2 \times 10^{-6}\ M/\text{s}) = 6.0 \times 10^{-7}\ M/\text{s}$$

Think About It Note that the instantaneous reaction rate calculated in this Sample Exercise is the same as the rate we calculated earlier based on the rate of change of $[O_2]$. This helps confirm the accuracy of the two determinations of reaction rate.

 **Practice Exercise** What is the instantaneous rate of change in $[NO_2]$ at $t = 2000$ s in the experiment that produced the data in Table 13.2?

13.3 Effect of Concentration on Reaction Rate

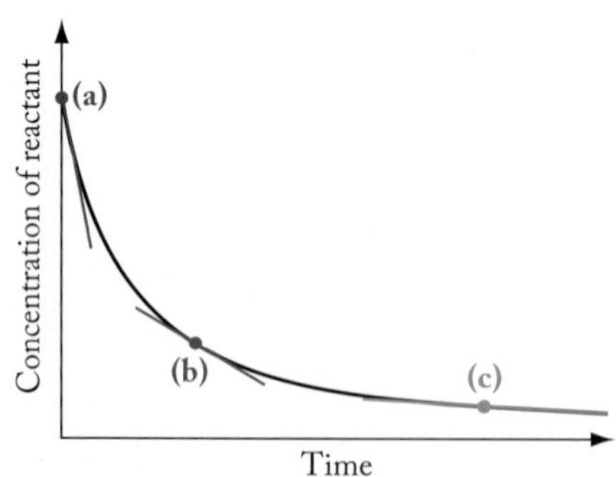

FIGURE 13.5 Typical plot of reactant concentration as a function of time: (a) tangent at $t = 0$; (b) tangent at the midpoint of the reaction; (c) tangent close to the end of the reaction.

CONNECTION We discussed kinetic molecular theory, a model that describes the behavior of gases, in Chapter 10.

Figure 13.5 shows a typical plot of reactant concentration versus time. Tangents are drawn to the line at three points: (a) at the instant the reaction begins, (b) at an instant when the reaction is about halfway to completion, and (c) when the reaction is nearly over. The slope of the tangent at point (a) defines the *initial* rate of the reaction, which is the rate observed at the instant the reactants are mixed at $t = 0$. The slopes of the other tangents become less negative as the reaction proceeds, which means the reaction is slowing down. Eventually the slopes of the tangents and the rate of the reaction approach zero. This behavior is typical of most reactions.

Kinetic molecular theory and our picture of molecules in the gas phase provide us with a way to explain this trend. If we assume that reactions take place as a result of collisions between reactant molecules, then the more reactant molecules there are in a given space, such as a flask, the more collisions per unit time and the more opportunities there are for reactants to turn into products. As reactant concentrations decrease, fewer reactant molecules occupy the space in the flask, so the frequency of collisions between reactant molecules decreases and the rate of conversion of reactants to products slows down.

Reaction Order and Rate Constants

Experimental observations and theoretical considerations tell us that reaction rates depend on reactant concentrations. However, they do not tell us to what extent rates depend on reactant concentrations. For example, if the concentration of a reactant decreases by half, does the reaction rate also decrease by half? The answer is linked to a parameter called **reaction order**, which tells us how reaction rates change with changing reactant concentrations. A key point here is that reaction order must be determined experimentally. There is no guarantee that the reaction order is linked to the coefficient of the reactant in the balanced chemical equation. This is because the equation that describes a reaction tells us which substances are consumed, which are produced, and in what proportions, but it does not tell us how the reaction takes place—which molecules collide with which other molecules as bonds break, new bonds form, and reactants are transformed into products.

reaction order an experimentally determined number defining the dependence of the reaction rate on the concentration of a reactant.

rate constant the proportionality constant that relates the rate of a reaction to the concentrations of reactants.

rate law an equation that defines the experimentally determined relation between the concentration of reactants in a chemical reaction and the rate of that reaction.

overall reaction order the sum of the exponents of the concentration terms in the rate law.

To understand what reaction order means quantitatively, let's revisit the chemical reaction that occurs between two components of smog, O_3 and NO:

$$O_3(g) + NO(g) \rightarrow O_2(g) + NO_2(g)$$

CHEMTOUR Reaction Order

Suppose different concentrations of O_3 and NO are injected into a reaction vessel at 298 K and the initial rates of reaction for each mixture are determined based on the instantaneous rate at $t = 0$. Table 13.3 contains the results of three determinations of initial reaction rate where we use a subscript 0 to denote the initial concentrations of the reactants.

In experiment 1 the initial concentrations of O_3 and NO are both 0.0010 M and the initial rate is 11 M/s. In experiment 2 the value $[O_3]_0$ is double that in experiment 1, but $[NO]_0$ is the same as in experiment 1. Comparing the reaction rates in experiments 2 and 1, we see that doubling $[O_3]_0$ doubles the initial reaction rate from 11 to 22 M/s. In experiment 3 the value of $[O_3]_0$ is the same as in experiment 1, but $[NO]_0$ has doubled and so, too, has the initial reaction rate.

These results tell us that the rate of the reaction is directly proportional to the concentration of O_3, and it's directly proportional to the concentration of NO. Expressed another way, reaction rate is a function of the reactants' concentrations each raised to the first power. In the language of chemical kinetics, this dependence means the reaction is first order in O_3 and first order in NO. These two concentration dependencies of reaction rate can be expressed mathematically using Equation 13.11:

$$\text{Rate} = k[O_3][NO] \tag{13.11}$$

where k is called the **rate constant** of the reaction and the entire equation is called the **rate law** of the reaction. Also, the first-order dependence on two reactants means that this reaction is second order overall. **Overall reaction order** is the sum of the powers in the rate law, which in this case is $1 + 1 = 2$.

TABLE 13.3 Effect of Reactant Concentrations on Initial Reaction Rates for $O_3(g) + NO(g) \rightarrow O_2(g) + NO_2(g)$

Experiment	$[O_3]_0$ (M)	$[NO]_0$ (M)	Initial Reaction Rate (M/s)
1	0.0010	0.0010	11
2	0.0020	0.0010	22
3	0.0010	0.0020	22

We can calculate the value k in Equation 13.11 by inserting the concentration and reaction rate data from any of the three experiments into the equation. Using the data from experiment 1:

$$\text{Rate} = k[O_3][NO] = 11\ M/s = k(0.0010\ M)^2$$

$$k = \frac{11\ \cancel{M}/s}{(0.0010\ \cancel{M})(0.0010\ M)} = 1.1 \times 10^7\ M^{-1}\,s^{-1}$$

Using this value of k in Equation 13.11 gives us a rate law for the reaction at 298 K:

$$\text{Rate} = 1.1 \times 10^7\ M^{-1}\,s^{-1}\,[O_3][NO] \quad (13.12)$$

CONCEPT TEST

If there had been data from a fourth experiment in Table 13.3 in which the initial concentrations of O_3 and NO had both been 0.0020 *M*, what would the initial rate of the reaction have been?

CHEMTOUR Collision Theory

Perhaps you are wondering why the two concentration terms in Equation 13.12 are multiplied together and not combined in some other way. A molecular view of the reason is provided in Figure 13.6, which shows how different numbers of NO and O_3 molecules in a reaction vessel might collide together and react with each other. Note how increasing the numbers of molecules in the containers in Figure 13.6(a)–(e) produces increasing numbers of collisions that are proportional to the product of the number of molecules of each reactant. Increasing the number of molecules of each type in the vessels is equivalent to increasing the concentrations of the two gases. Therefore, the rate of the reaction should depend on the product of the concentrations of NO and O_3.

The data in Table 13.3 made it relatively easy to see the dependence of reaction rate on reactant concentrations. This is not always the case. Consider, for example, the results in Table 13.4 from a kinetics study of another important reaction in atmospheric chemistry:

$$2\ NO(g) + O_2(g) \rightarrow 2\ NO_2(g)$$

The relationships between reaction rate and the reactants' concentrations in these data are not as evident as they were in the Table 13.3 data. So, to determine reaction order, we take a more mathematical approach, starting with a rate law expression for the reaction in which the [NO] and $[O_2]$ terms are given the generic exponents m and n:

$$\text{Rate} = k[NO]^m[O_2]^n \quad (13.13)$$

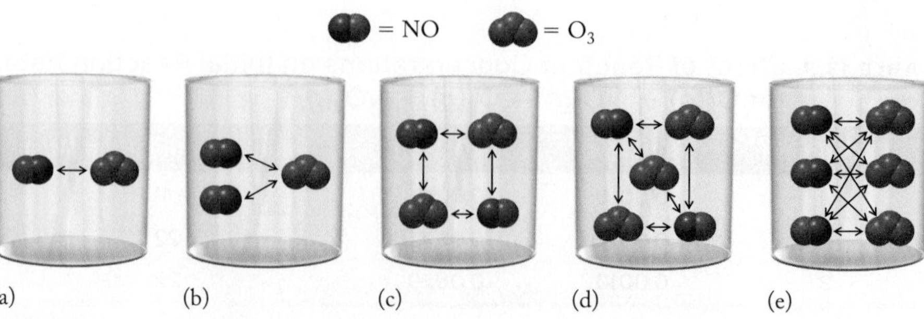

FIGURE 13.6 Increasing the concentration increases the number of possible effective collisions (double-headed arrows) and therefore the number of potential reaction events. Reaction rate depends on the number of collisions between molecules, which are shown for the reaction between NO and ozone (O_3) that produces NO_2 and O_2. (a) With only one NO molecule and one O_3 molecule, each molecule can collide only with the other, giving a relative reaction rate of $1 \times 1 = 1$. (e) Three molecules of NO can collide with three molecules of O_3 for a relative reaction rate of $3 \times 3 = 9$ times the relative rate in (a).

TABLE 13.4 Effect of Reactant Concentrations on Initial Reaction Rates for $2\ NO(g) + O_2(g) \rightarrow 2\ NO_2(g)$

Experiment	$[NO]_0$ (M)	$[O_2]_0$ (M)	Initial Reaction Rate (M/s)
1	0.025	0.025	1.6×10^{-5}
2	0.025	0.066	4.1×10^{-5}
3	0.050	0.025	6.2×10^{-5}

Our task is to determine the values of m and n from the data in Table 13.4. We start by comparing the results of experiments 1 and 2 in which [NO] is unchanged but in which $[O_2]$ and reaction rate are initially both higher in experiment 2. The ratio of the reaction rates in experiments 2 and 1, $\text{Rate}_2/\text{Rate}_1$, is related to the ratio of the concentrations of O_2 in experiments 2 and 1, $[O_2]_2/[O_2]_1$, and to the dependence of reaction rate on $[O_2]$, that is, on the value of n. Expressing this relationship in equation form and then making n a coefficient by taking the logarithm of both sides:

$$\frac{\text{Rate}_2}{\text{Rate}_1} = \left(\frac{[O_2]_2}{[O_2]_1}\right)^n \quad \text{or} \quad \log\left(\frac{\text{Rate}_2}{\text{Rate}_1}\right) = n \log\left(\frac{[O_2]_2}{[O_2]_1}\right)$$

Rearranging the terms and inserting the appropriate values from experiments 1 and 2:

$$n = \frac{\log\left(\frac{\text{Rate}_2}{\text{Rate}_1}\right)}{\log\left(\frac{[O_2]_2}{[O_2]_1}\right)} = \frac{\log \frac{4.1 \times 10^{-5}\ \cancel{M/\text{s}}}{1.6 \times 10^{-5}\ \cancel{M/\text{s}}}}{\log \frac{0.066\ \cancel{M}}{0.025\ \cancel{M}}} = 0.97 \text{ or } 1.0$$

The key to interpreting this value of n is to recognize that it is very close to 1, which means the reaction is first order in O_2.

Taking a similar approach with the data from experiments 1 and 3 in which the initial concentrations of O_2 are the same but those of NO differ, we calculate the value of m:

$$m = \frac{\log\left(\frac{\text{Rate}_3}{\text{Rate}_1}\right)}{\log\left(\frac{[NO]_3}{[NO]_1}\right)} = \frac{\log \frac{6.2 \times 10^{-5}\ \cancel{M/\text{s}}}{1.6 \times 10^{-5}\ \cancel{M/\text{s}}}}{\log \frac{0.050\ \cancel{M}}{0.025\ \cancel{M}}} = \frac{0.588}{0.301} = 1.95 \text{ or } 2.0$$

Therefore, the reaction is second order in NO, which gives us the following rate law expression:

$$\text{Rate} = k[NO]^2[O_2]$$

Using the data from experiment 3 to solve for k:

$$\text{Rate} = k[NO]^2[O_2] = 6.2 \times 10^{-5}\ M/\text{s} = k(0.050\ M)^2(0.025\ M)$$

$$k = \frac{6.2 \times 10^{-5}\ \cancel{M}/\text{s}}{(0.050\ M)^2(0.025\ \cancel{M})} = 0.99\ M^{-2}\ \text{s}^{-1}$$

Therefore, the rate law for the reaction at 298 K is

$$\text{Rate} = 0.99\ M^{-2}\ \text{s}^{-1}\ [NO]^2[O_2]$$

Thus, the reaction of NO with O_2 is second order in NO, first order in O_2, and third order overall.

In the preceding exercise, calculated exponent values that were very close to 1 and 2 were rounded to these values given the uncertainty in the data used to calculate them. This does not mean that the values of concentration exponents in rate law equations are always whole numbers, as we shall see in Sample Exercise 13.4. They may also be fractions, zero, or, in rare cases, negative. It is important to remember that rate laws and reaction orders are different from relative rates in that they must be determined experimentally. They cannot be predicted from the coefficients in a balanced chemical equation. The significance of reaction orders in describing how reactions take place is explored further in Section 13.5.

The values of the rate constants calculated using the data in Tables 13.3 and 13.4 are unique to these particular reactions and to the temperature at which the experiments were run. Keep in mind that reaction rates depend on the concentrations of reactants, but rate constants do not. However, rate constants do change with changing temperature. The units of reaction rates are usually expressed in *M*/s, but the units on a rate constant depend on the overall order of the reaction. The *k* values calculated previously had units of $M^{-2}\ s^{-1}$ for a reaction that was third order overall and $M^{-1}\ s^{-1}$ for a reaction that was second order overall. Do you see a pattern in these units and reaction order? If so, what are the units of *k* for a first-order reaction? If you said s^{-1}, you are right.

SAMPLE EXERCISE 13.4 Deriving a Rate Law from Initial Reaction Rate Data

LO3

Write the rate law for the reaction of N_2 with O_2

$$N_2(g) + O_2(g) \rightarrow 2\ NO(g)$$

using the data in Table 13.5. Determine the overall reaction order and the value of the rate constant.

TABLE 13.5 Initial Reaction Rates for $N_2(g) + O_2(g) \rightarrow 2\ NO(g)$

Experiment	$[N_2]_0$ (*M*)	$[O_2]_0$ (*M*)	Initial Reaction Rate (*M*/s)
1	0.040	0.020	707
2	0.040	0.010	500
3	0.010	0.010	125

Collect and Organize We are asked to determine the rate law and rate constant for a reaction from initial reaction rate data (note the subscript zero on the concentration terms in Table 13.5) for each of three sets of initial reactant concentrations. We could use equations similar to those used to derive a rate law expression from the [NO] and $[O_2]$ data in Table 13.4.

Analyze The general form of the rate law for this reaction is

$$\text{Rate} = k[N_2]^m[O_2]^n$$

We can use the experimental data given to find the values of *k*, *m*, and *n*. The overall order of the reaction is the sum of the orders of the individual reactants. Once we have established the rate law, the rate constant can be calculated using concentrations of reactants from any single row in Table 13.5. The rate constant for the reaction must have units that express the reaction rate in $M\ s^{-1}$.

Solve To determine the value of m we select a pair of experiments with the same O_2 and different N_2 concentrations, such as experiments 2 and 3, and insert rate and $[N_2]$ values into a version of Equation 13.14 tailored to this calculation:

$$m = \frac{\log\left(\frac{\text{Rate}_2}{\text{Rate}_3}\right)}{\log\left(\frac{[N_2]_2}{[N_2]_3}\right)} = \frac{\log\left(\frac{500\ M/\text{s}}{125\ M/\text{s}}\right)}{\log\left(\frac{0.040\ M}{0.010\ M}\right)} = \frac{0.602}{0.602} = 1.0$$

To calculate n, we use experiments 1 and 2 because they have different $[O_2]$ values, but $[N_2]$ is the same.

$$n = \frac{\log\left(\frac{\text{Rate}_1}{\text{Rate}_2}\right)}{\log\left(\frac{[O_2]_1}{[O_2]_2}\right)} = \frac{\log\left(\frac{707\ M/\text{s}}{500\ M/\text{s}}\right)}{\log\left(\frac{0.020\ M}{0.010\ M}\right)} = \frac{0.150}{0.301} = 0.50$$

Thus, the reaction is one-half order with respect to O_2, first order with respect to N_2, and $(\frac{1}{2} + 1) = \frac{3}{2}$ order overall:

$$\text{Rate} = k[N_2][O_2]^{1/2}$$

We can use the data from any experiment to obtain the value of k. Let's use experiment 1:

$$\text{Rate} = k[N_2][O_2]^{1/2} = 707\ M/\text{s} = k(0.040\ M)(0.020\ M)^{1/2}$$

$$k = \frac{707\ M/\text{s}}{(0.040\ M)(0.020\ M)^{1/2}} = 1.2 \times 10^5\ M^{-1/2}\ \text{s}^{-1}$$

Think About It The units on the rate constant, $M^{-1/2}\ \text{s}^{-1}$, are unusual but appropriate given the fractional order of the reaction with respect to O_2. A generic form of the equations used in this exercise can be written for any reactant "A":

$$m = \frac{\log\left(\frac{\text{Rate}_2}{\text{Rate}_1}\right)}{\log\left(\frac{[A]_2}{[A]_1}\right)} \qquad (13.14)$$

Practice Exercise Nitric oxide reacts rapidly with unstable nitrogen trioxide (NO_3), forming NO_2:

$$NO(g) + NO_3(g) \rightarrow 2\ NO_2(g)$$

Determine the rate law for the reaction and calculate the rate constant from the data in Table 13.6.

TABLE 13.6 Initial Reaction Rates for the Formation of NO_2 from the Reaction $NO(g) + NO_3(g) \rightarrow 2\ NO_2(g)$ at 25°C

Experiment	$[NO]_0$ (M)	$[NO_3]_0$ (M)	Initial Reaction Rate (M/s)
1	1.25×10^{-3}	1.25×10^{-3}	2.45×10^4
2	2.50×10^{-3}	1.25×10^{-3}	4.90×10^4
3	2.50×10^{-3}	2.50×10^{-3}	9.80×10^4

Integrated Rate Laws: First-Order Reactions

Determining a rate law using initial reaction rate data has two distinct disadvantages. The method requires several experiments with different concentrations of reactants, varied in a systematic fashion. We also must accurately determine the

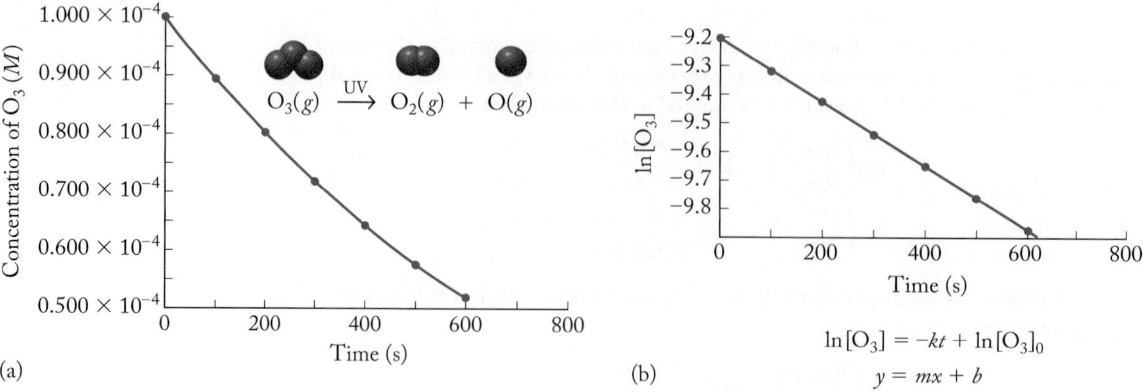

FIGURE 13.7 The decomposition of O_3 is plotted as (a) $[O_3]$ versus time and (b) $\ln[O_3]$ versus time. The line in (b) is straight, indicating that the decomposition reaction is first order in O_3.

reaction rate at the instant the reaction begins. It would be much easier if we could determine the rate law and calculate the rate constant for a reaction from the plot of concentration versus time alone. In fact, this can be accomplished for reactions in which the reaction rate depends on the concentration of only one substance. One such reaction is the photochemical decomposition of ozone:

$$O_3(g) \xrightarrow{\text{sunlight}} O_2(g) + O(g)$$

TABLE 13.7 Rate of Photochemical Decomposition of Ozone

Time (s)	$[O_3]$ (M)	$\ln[O_3]$
0.0	1.000×10^{-4}	−9.2103
100.0	0.896×10^{-4}	−9.320
200.0	0.803×10^{-4}	−9.430
300.0	0.719×10^{-4}	−9.540
400.0	0.644×10^{-4}	−9.650
500.0	0.577×10^{-4}	−9.760
600.0	0.517×10^{-4}	−9.870

This reaction can be studied in the laboratory using high-intensity ultraviolet light to simulate sunlight; one such study yielded the results listed in Table 13.7 and plotted in Figure 13.7(a). Because ozone is the only reactant, the rate law for the reaction should depend only on the ozone concentration. Let's start by assuming the reaction is first order in ozone and has the following rate law:

$$\text{Rate} = k[O_3]$$

Because the coefficient of O_3 in the balanced equation is 1, the O_3 consumption rate, $-\Delta[O_3]/\Delta t$, is equal to the reaction rate, and we can write

$$\text{Rate} = -\frac{\Delta[O_3]}{\Delta t} = k[O_3]$$

This rate law can be transformed using calculus into an expression that relates the concentration of ozone $[O_3]_t$ at any time (t) during the reaction to the initial concentration $[O_3]_0$:

$$\ln\frac{[O_3]_t}{[O_3]_0} = -kt \qquad (13.15)$$

This version of the rate law is called an **integrated rate law** because integral calculus is used to derive it. It describes the change in reactant concentration with time. The general integrated rate law for any reaction that is first order in reactant X is

$$\ln\frac{[X]_t}{[X]_0} = -kt \qquad (13.16)$$

Using the identity $\ln(a/b) = \ln a - \ln b$, we can rearrange this equation to

$$\ln[X]_t = -kt + \ln[X]_0 \qquad (13.17)$$

which is the equation of a straight line of the form

$$y = mx + b$$

integrated rate law a mathematical expression that describes the change in concentration of a reactant in a chemical reaction with time.

where $\ln[X]_t$ is the y variable and t is the x variable. The slope of the line (m) is $-k$, and the y-intercept (b) is $\ln[X]_0$. Rearranging Equation 13.15 to fit the format of Equation 13.17 gives

$$\ln[O_3] = -kt + \ln[O_3]_0 \quad (13.18)$$

Note in Figure 13.7(b) that a graph plotting the natural logarithm of $[O_3]$ versus time is indeed a straight line. This linearity means that our assumption was correct and the reaction is first order in O_3. If you have a graphing calculator, you can use its curve-fitting function to obtain the slope, $\Delta(\ln[O_3])/\Delta t$, of the straight line that best fits the data points. That value is $-1.10 \times 10^{-3}\ s^{-1}$, which means the value of k is $1.10 \times 10^{-3}\ s^{-1}$.

SAMPLE EXERCISE 13.5 Using Integrated Rate Laws to Determine Rate Law Expressions and Rate Constants **LO4**

Dinitrogen pentoxide is one of the least abundant oxides of nitrogen in the atmosphere because, if it forms, it rapidly decomposes:

$$2\ N_2O_5(g) \rightarrow 4\ NO_2(g) + O_2(g)$$

A kinetic study of the decomposition of N_2O_5 yields the data in Table 13.8(a).

a. Test the validity of the assumption that the decomposition of N_2O_5 is first order in N_2O_5.
b. Determine the value of the rate constant.

Collect and Organize We are given experimental data showing the concentration of a single reactant as a function of time and told to assume a first-order reaction. We can verify this assumption using the integrated rate law for a first-order reaction (Equation 13.17) and then calculate the rate constant.

Analyze Equation 13.17 has the form $y = mx + b$, which means that a plot of $\ln[N_2O_5]$ (y) versus time (x) should be linear if the decomposition of N_2O_5 is first order. The slope (m) of the graph corresponds to $-k$, the negative of the rate constant.

Solve Our first step is to determine $\ln[N_2O_5]$ values (Table 13.8b).

a. The plot of $\ln[N_2O_5]$ versus t is shown in Figure 13.8. It is a straight line, which means that the reaction is first order in N_2O_5.
b. The line that best fits the data points in Figure 13.8 has a slope of $-0.00693\ s^{-1}$. The slope of the line equals $-k$, so the rate constant $k = 0.00693\ s^{-1}$.

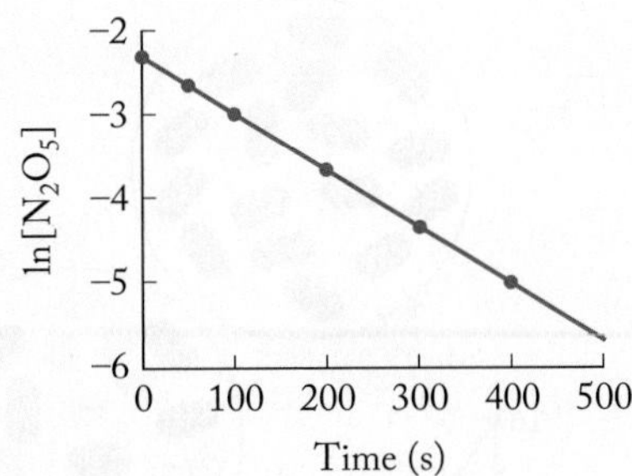

FIGURE 13.8 Plot of $\ln[N_2O_5]$ as a function of time.

Think About It We can test whether any reaction with a single reactant (X) is first order in X by determining whether a plot of $\ln[X]$ versus t is linear.

Practice Exercise Hydrogen peroxide (H_2O_2) decomposes into water and oxygen:

$$H_2O_2(\ell) \rightarrow H_2O(\ell) + \tfrac{1}{2}\,O_2(g)$$

Use the data in Table 13.9 to determine whether the decomposition of H_2O_2 is first order in H_2O_2, and calculate the value of the rate constant at the temperature of the experiment that produced the data.

TABLE 13.8(a) Concentration of N_2O_5 as a Function of Time

Time (s)	$[N_2O_5]$ (M)
0.0	0.1000
50.0	0.0707
100.0	0.0500
200.0	0.0250
300.0	0.0125
400.0	0.00625

TABLE 13.8(b) Concentration and Natural Logarithm of Concentration of N_2O_5 as a Function of Time

Time (s)	$[N_2O_5]$ (M)	$\ln[N_2O_5]$
0.0	0.1000	−2.303
50.0	0.0707	−2.649
100.0	0.0500	−2.996
200.0	0.0250	−3.689
300.0	0.0125	−4.382
400.0	0.00625	−5.075

TABLE 13.9 Concentration of Hydrogen Peroxide as a Function of Time

Time (s)	$[H_2O_2]$ (M)
0.0	0.500
100.0	0.460
200.0	0.424
500.0	0.330
1000.0	0.218
1500.0	0.144

half-life ($t_{1/2}$) the time interval during which the concentration of a reactant decreases by half in the course of a chemical reaction.

Half-Lives

The **half-life ($t_{1/2}$)** of a reactant is the interval during which the concentration of a reactant decreases by half, as shown for the decomposition of N_2O (laughing gas) in Figure 13.9:

$$2\,N_2O(g) \rightarrow 2\,N_2(g) + O_2(g)$$

Over each half-life interval, one-half of the quantity of reactant present at the beginning of the interval is consumed; the other half remains.

Half-life is inversely related to the rate constant of a reaction: the higher the reaction rate, the shorter the half-life. For first-order reactions, such as the decomposition of N_2O, we can derive a mathematical relation between the half-life $t_{1/2}$ and rate constant k by starting with Equation 13.16:

$$\ln \frac{[X]_t}{[X]_0} = -kt \tag{13.16}$$

After one half-life has passed ($t = t_{1/2}$), the concentration of X is half its original value: $[X]_t = \frac{1}{2}[X]_0$. Inserting these values for [X] and t into the equation yields

$$\ln \frac{\frac{1}{2}\cancel{[X]_0}}{\cancel{[X]_0}} = -kt_{1/2}$$

$$\ln \tfrac{1}{2} = -kt_{1/2}$$

The natural log of $\frac{1}{2}$ is -0.693, so

$$-0.693 = -kt_{1/2}$$

$$t_{1/2} = \frac{0.693}{k} \tag{13.19}$$

Thus, the half-life of a first-order reaction is inversely proportional to the rate constant, as noted at the beginning of this discussion. The absence of any concentration term in Equation 13.19 means that no matter what the initial concentration of the reactant in a first-order reaction is, half of it is consumed in one half-life.

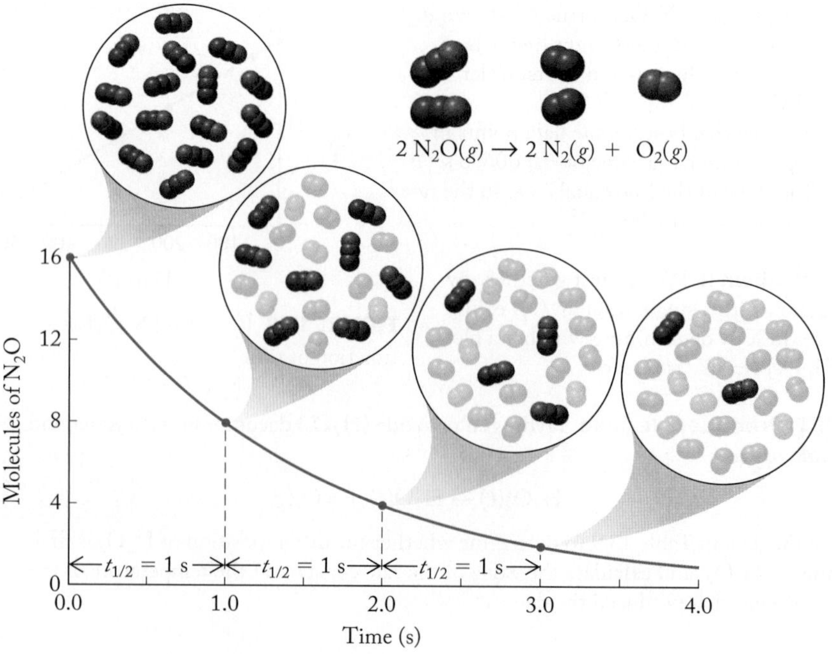

FIGURE 13.9 The decomposition of $N_2O(g)$ is first order in N_2O. At a particular temperature the half-life of the reaction is 1.0 s, which means that, on average, half of a population of 16 N_2O molecules decomposes in 1.0 s, half of the remaining 8 molecules decompose in the next 1.0 s, and so on.

Equations based on integrated rate laws can also be used to determine how much time it takes for a reactant concentration to drop to a particular value or how much of it remains after a particular reaction time. Calculations such as these are widely used in chemistry (see Sample Exercise 13.6) and especially in nuclear chemistry to determine, for example, the time it will take for the level of radioactivity in material removed from a nuclear reactor to reach acceptably low levels or to determine the age of an ancient wooden sculpture based on its concentration of a radioactive isotope that occurs naturally in all plant material. We explore these and other applications of radioactive decay rates, which always follow first-order kinetics, in Chapter 21.

SAMPLE EXERCISE 13.6 Relating Reactant Concentration, Reaction Time, and Half-Life in a First-Order Reaction **LO5**

The decomposition of dinitrogen pentoxide is first order in N_2O_5 and has a rate constant $k = 0.00693\ s^{-1}$.

$$2\ N_2O_5(g) \rightarrow 2\ N_2O_4(g) + O_2(g)$$

If this reaction is run under the same conditions as in Sample Exercise 13.5 and the initial concentration of N_2O_5 in the reaction vessel is 0.375 *M*,

a. What is the concentration of N_2O_5 after 3.00 minutes?
b. How long does it take (in minutes) for 95.0% of the N_2O_5 initially in the reaction vessel to decompose?
c. Use the value of k to calculate the half-life of N_2O_5 in minutes.

Collect and Organize We are given the rate constant for a first-order reaction and the initial concentration of reactant and are asked to calculate (a) the concentration of reactant remaining after 3.0 minutes, (b) the time it takes for 95% of the reactant to be consumed, and (c) the half-life of N_2O_5. Equations 13.16, 13.17, and 13.19 relate the variables of interest in a first-order reaction.

Analyze The rate constant k is given in seconds, but time is expressed in minutes in parts (a), (b), and (c), so we should convert the value of k into min^{-1}. We know all the terms in Equation 13.17: $\ln[X]_t = -kt + \ln[X]_0$ except $[X]_t$, which is the term we need to calculate in part (a). We know the $[X]_t/[X]_0$ ratio in the log term of Equation 13.16 so let's use that equation to estimate t in part (b). According to the data set for Sample Exercise 13.5, $[N_2O_5]$ dropped to half its initial value in 100 seconds, so the calculated $t_{1/2}$ value in part (c) should be $100\ s \times \frac{1\ min}{60\ s} = \frac{5}{3}$ or 1.67 minutes. It would take four half-lives to consume all but $(\frac{1}{2})^4 = 1/16 = 0.067$ or 6.7% of the initial concentration of N_2O_5 in the sample. Therefore, it should take a little more than four half-lives or $4 \times \frac{5}{3} = \frac{20}{3}$ or ~ 7 minutes to consume 95.0% of the initial quantity of N_2O_5 in the sample.

Solve

$$k = 0.00693\ s^{-1} \times (60\ s\ min^{-1}) = 0.416\ min^{-1}$$

a. Using the given concentration and time values in Equation 13.17:

$$\ln[X]_t = -kt + \ln[X]_0 = \ln[N_2O_5]_t = -(0.416\ min^{-1}) \times 3.00\ min + \ln(0.375\ M)$$

$$\ln[N_2O_5]_t = -2.229; \quad \text{therefore,} \quad [N_2O_5]_t = e^{-2.229} = 0.108\ M$$

b. If 95.0% of $[N_2O_5]_0$ is consumed, then $[N_2O_5]_t = (100.0 - 95.0) = 5.0\%$ or 0.050 $[N_2O_5]_0$. Using this value in Equation 13.16:

$$\ln\frac{[X]_t}{[X]_0} = -kt = \ln\frac{0.050\ \cancel{[N_2O_5]_0}}{\cancel{[N_2O_5]_0}} = -(0.416\ min^{-1})\ t = \ln(0.050) = -2.996$$

$$t = \frac{-2.996}{-0.416\ min^{-1}} = 7.2\ min$$

c. Using the calculated k value in Equation 13.19 to calculate $t_{1/2}$:

$$t_{1/2} = \frac{0.693}{k} = \frac{0.693}{0.416\ \text{min}^{-1}} = 1.67\ \text{min}$$

Think About It The calculated answers in parts (b) and (c) agree with the predicted results and the answer in (a) is reasonable given that the reaction time of 3.00 minutes represented nearly two half-lives. Therefore, a $[N_2O_5]_t$ value that is between $\frac{1}{3}$ and $\frac{1}{4}$ $[N_2O_5]_0$ makes sense.

Practice Exercise A wristwatch made in 1920 has a dial that glows in the dark because it contains a radioactive isotope of radium. The half-life of this isotope is 1600 years.

a. What is the rate constant of the radioactive decay process that makes the watch dial glow?
b. What fraction of the radioactive radium in the watch when it was made will still be there in 2020? (Express your answer as a percent.)
c. How long will it take for 99.0% of the radioactive radium in the watch to decay?

Integrated Rate Laws: Second-Order Reactions

In Section 13.1 we described how NO_2 exposed to sunlight decomposes to NO and O (Equation 13.3). Nitrogen dioxide may also undergo thermal decomposition, producing NO and O_2:

$$2\ NO_2(g) \rightarrow 2\ NO(g) + O_2(g) \qquad (13.20)$$

Because it has only one reactant, like the reactions of N_2O_5 and H_2O_2 in Sample Exercise 13.5 and its Practice Exercise, we might expect this reaction to be first order. However, when we use the data in Table 13.10 to evaluate the reaction order and rate constant, we find that the plot of $\ln[NO_2]$ versus time (Figure 13.10a) is curved, not linear, which means that the thermal decomposition of NO_2 is *not* first order.

TABLE 13.10 Rate of Decomposition of NO_2 to NO and O_2

Time (s)	$[NO_2]$ (M)	$\ln[NO_2]$	$1/[NO_2]$ (M^{-1})
0.0	1.00×10^{-2}	−4.605	100
100.0	6.48×10^{-3}	−5.039	154
200.0	4.79×10^{-3}	−5.341	209
300.0	3.80×10^{-3}	−5.573	263
400.0	3.15×10^{-3}	−5.760	317
500.0	2.69×10^{-3}	−5.918	372
600.0	2.35×10^{-3}	−6.057	426

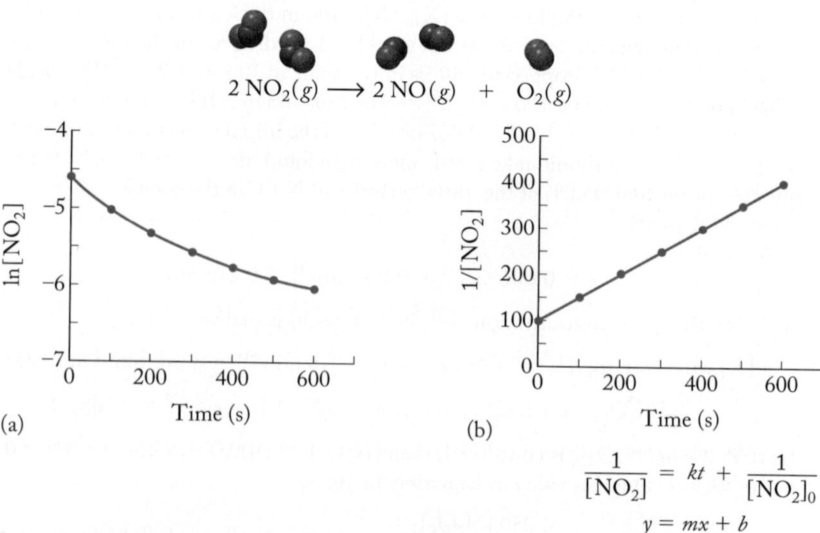

FIGURE 13.10 At high temperatures, NO_2 slowly decomposes into NO and O_2. (a) The plot of $\ln[NO_2]$ versus time is not linear, indicating that the reaction is not first order. (b) The plot of $1/[NO_2]$ versus time is linear, indicating that the reaction is second order in NO_2. The slope of the line in this graph equals the rate constant.

What is the reaction order? The answer to this question is hidden in how the reaction takes place. If each NO_2 molecule simply fell apart, the reaction would be first order, much like the decomposition of N_2O_5. However, if the reaction happens as a result of collisions between pairs of NO_2 molecules, it would be first order with respect to each NO_2 molecule and second order overall. In other words, if the reaction were second order, the decomposition described by Equation 13.20 would depend on collisions between pairs of molecules that just happen to be molecules of the same substance, NO_2. The rate law expression is

$$\text{Rate} = k[NO_2]^2 \qquad (13.21)$$

How can we determine whether this decomposition is really second order? One way is to assume that it is and then test that assumption. The test entails transforming the rate law in Equation 13.21 into the integrated rate law for a second-order reaction, again using calculus. The result of the transformation is

$$\frac{1}{[NO_2]} = kt + \frac{1}{[NO_2]_0} \qquad (13.22)$$

which, like Equation 13.18, has the form $y = mx + b$ and is the equation of a straight line, this time with $1/[NO_2]$ as the y variable and t as the x variable.

The graph obtained using data from columns 1 and 4 of Table 13.10 is shown in Figure 13.10(b). The plot is linear, so this NO_2 decomposition reaction is second order overall. The slope of the line provides a direct measure of k, which is $0.544\ M^{-1}\ s^{-1}$.

A general form of Equation 13.22 that applies to any reaction that is second order in a single reactant (X) is

$$\frac{1}{[X]} = kt + \frac{1}{[X_2]_0} \qquad (13.23)$$

SAMPLE EXERCISE 13.7 Distinguishing between First- and Second-Order Reactions **LO4**

Chlorine monoxide accumulates in the stratosphere above Antarctica each winter and plays a key role in the formation of the ozone hole above the South Pole each spring. Eventually, ClO decomposes according to the equation

$$2\ ClO(g) \rightarrow Cl_2(g) + O_2(g)$$

The kinetics of this reaction were studied in a laboratory experiment at 298 K; some of the results are listed in Table 13.11(a). Determine whether the reaction is first or second order in ClO and the value of k at 298 K.

Collect, Organize, and Analyze We are given experimental data describing the decomposition of ClO at 298 K and are asked to determine the order of the decomposition reaction of ClO, the rate law, and k. To distinguish between the two reaction orders for a reaction in which there is a single reactant, we plot ln[ClO] versus time and 1/[ClO] versus time. If the ln[ClO] plot is linear, the reaction is first order; if the 1/[ClO] plot is linear, the reaction is second order. We determine the rate constant from the slope of whichever plot is linear.

Solve To evaluate the two possibilities, we need to calculate ln[ClO] and 1/[ClO] values for each value of [ClO] in Table 13.11(a); sets of both of these values are listed in Table 13.11(b). The graphs for ln[ClO] and 1/[ClO] versus time are shown in Figure 13.11. The ln[ClO] plot is not linear, but the 1/[ClO] plot is, so the reaction is

TABLE 13.11(a) Concentration of Chlorine Monoxide as a Function of Time

Time (ms)	[ClO] (M)
0.0	1.50×10^{-8}
10.0	7.19×10^{-9}
20.0	4.74×10^{-9}
30.0	3.52×10^{-9}
40.0	2.81×10^{-9}
100.0	1.27×10^{-9}
200.0	6.60×10^{-10}

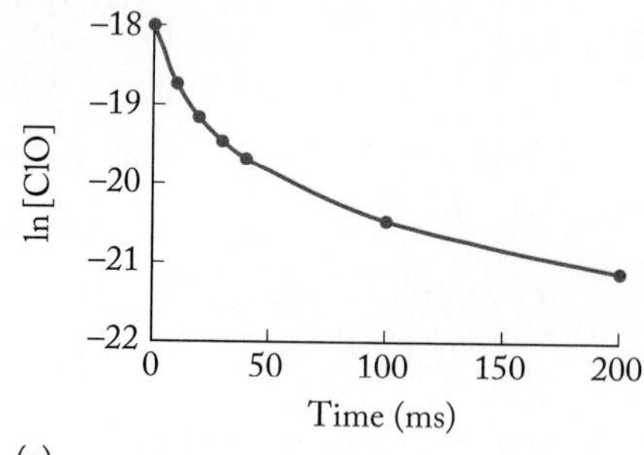

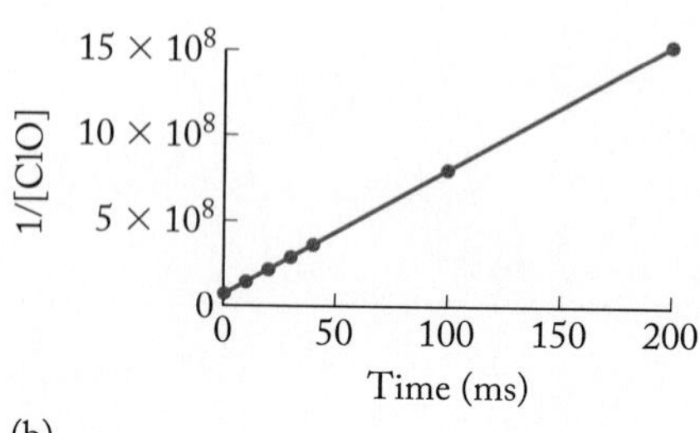

FIGURE 13.11 Plots of (a) ln[ClO] versus time and (b) 1/[ClO] versus time.

second order in ClO and second order overall. The line that best fits the data has a slope of $7.22 \times 10^9\ M^{-1}\ s^{-1}$, which is the value of k.

TABLE 13.11(b) Concentration of Chlorine Monoxide, ln[ClO], and 1/[ClO] as a Function of Time

Time (ms)	[ClO] (*M*)	ln[ClO]	1/[ClO] (M^{-1})
0.0	1.50×10^{-8}	−18.015	6.67×10^7
10.0	7.19×10^{-9}	−18.751	1.39×10^8
20.0	4.74×10^{-9}	−19.167	2.11×10^8
30.0	3.52×10^{-9}	−19.465	2.84×10^8
40.0	2.81×10^{-9}	−19.690	3.56×10^8
100.0	1.27×10^{-9}	−20.484	7.89×10^8
200.0	0.66×10^{-9}	−21.139	1.51×10^9

Think About It Sample Exercises 13.5 and 13.7 demonstrate that we can use integrated rate laws to distinguish between first- and second-order reactions in a single reactant.

Practice Exercise Experimental evidence shows that the rate of the following reaction depends only on the concentration of NO_2:

$$NO_2(g) + CO(g) \rightarrow NO(g) + CO_2(g)$$

Determine whether the reaction is first or second order in NO_2, and calculate the rate constant from the data in Table 13.12, which were obtained at 488 K.

TABLE 13.12 Concentration of NO_2 as a Function of Time

Time (h)	[NO_2] (*M*)
0.00	0.250
1.39	0.198
3.06	0.159
4.72	0.132
6.39	0.114
8.06	0.099
9.72	0.088
11.39	0.080

Second-order reactions have half-lives, too. The relation between k and $t_{1/2}$ for the decomposition of NO_2 can be derived from Equation 13.23 if we first rearrange the terms to solve for kt:

$$kt = \frac{1}{[X]} - \frac{1}{[X]_0}$$

After one half-life has elapsed ($t = t_{1/2}$), [X] has decreased to half its initial concentration. Substituting this information into the preceding equation, we have

$$kt_{1/2} = \frac{1}{\frac{1}{2}[X]_0} - \frac{1}{[X]_0}$$

$$= \frac{2}{[X]_0} - \frac{1}{[X]_0} = \frac{1}{[X]_0}$$

or

$$t_{1/2} = \frac{1}{k[X]_0} \tag{13.24}$$

Thus, the value of $t_{1/2}$ is inversely proportional to the initial concentration of X in a second-order reaction. This dependence on concentration contrasts with the $t_{1/2}$ values of first-order reactions, which are independent of concentration.

SAMPLE EXERCISE 13.8 Calculating the Half-Lives of Second-Order Reactions **LO5**

Calculate the half-life of the second-order decomposition of NO_2 (Equation 13.20) if $k = 0.544\ M^{-1}\ s^{-1}$ at a particular temperature and the initial concentration of NO_2 is 0.0100 *M*.

Collect and Organize We are asked to find the half-life of NO_2, the time required for the concentration of NO_2 to decrease by one-half. We are given the rate constant and initial concentration of NO_2 and told the reaction is second order. Equation 13.24 gives $t_{1/2}$ of a second-order reaction in which there is only one reactant.

Analyze The half-life is inversely proportional to the rate constant k and to the initial concentration of NO_2. The value of k is about 0.5 $M^{-1}\ s^{-1}$ at an initial concentration of $10^{-2}\ M$, so $t_{1/2}$ is inversely proportional to about $5 \times 10^{-3}\ s^{-1}$. Thus, we predict $t_{1/2} \approx 200$ s.

Solve

$$t_{1/2} = \frac{1}{k[NO_2]_0} = \frac{1}{(0.544\ M^{-1}\ s^{-1})(1.00 \times 10^{-2}\ M)} = 184\ s$$

Think About It The calculated value of $t_{1/2}$ is very close to the predicted value of 200 s. As in all second order reactions with one reactant, we need to know its initial concentration to calculate its half-life.

Practice Exercise The rate constant for the decomposition of ClO is $7.22 \times 10^9\ M^{-1}\ s^{-1}$ at 298 K. Determine $t_{1/2}$ of ClO when $[ClO]_0 = 1.50 \times 10^{-8}\ M$.

Pseudo-First-Order Reactions

The integrated rate law in Equation 13.23 applies only to reactions that are second order in a single reactant. It does not apply to reactions that are second order overall but are first order in two reactants, such as

$$NO(g) + O_3(g) \rightarrow NO_2(g) + O_2(g)$$

or

$$NO_2(g) + O_3(g) \rightarrow NO_3(g) + O_2(g)$$

Because the integrated rate law for a reaction that is first order in two reactants is complicated, chemists frequently adjust reaction conditions so that a simpler rate law can be used. One approach is to have one of the reactants present at a much higher concentration than the other. This condition is common for many components in the urban atmosphere where, for example, ozone concentrations are often hundreds to thousands of times greater than NO concentrations. With such a large excess, the ozone concentration remains virtually constant over the course of the reaction

$$NO(g) + O_3(g) \rightarrow NO_2(g) + O_2(g)$$

Thus, the rate law,

$$\text{Rate} = k[NO][O_3]$$

may be simplified to

$$\text{Rate} = k'[NO] \quad (13.25)$$

where

$$k' = k[O_3]_0 \quad (13.26)$$

and where $[O_3]_0$ is the initial concentration of ozone, which remains virtually constant throughout the reaction.

Equation 13.25 looks like the rate law for a first-order reaction (Rate = k[X]), but it is really a **pseudo-first-order** rate law because it only appears to obey

pseudo first order a reaction in which all the reactants but one are present at such high concentrations that they do not decrease significantly during the course of the reaction, so that the reaction rate is controlled by the concentration of the limiting reactant.

first-order kinetics. A pseudo-first-order reaction has the same integrated rate law as a first-order reaction, but the rate of the pseudo-first-order reaction depends on the concentration of more than one reactant.

TABLE 13.13(a) Concentration of NO as a Function of Time

Time (μs)	Concentration of NO (molecules/cm^3)
0.0	1.00×10^9
100.0	8.36×10^8
200.0	6.98×10^8
300.0	5.83×10^8
400.0	4.87×10^8
500.0	4.07×10^8
1000.0	1.65×10^8

SAMPLE EXERCISE 13.9 Deriving Pseudo-First-Order Rate Laws LO4

The data in Table 13.13(a) were obtained in a study of the oxidation of trace levels of NO in the presence of a large excess of ozone at 298 K:

$$NO(g) + O_3(g) \rightarrow NO_2(g) + O_2(g)$$

(Note the NO concentration units in Table 13.13a. When the molar concentrations of reactants are very low, as in the case of many atmospheric pollutants, it is easier to express them in units of molecules/cm^3.)

a. Verify that the reaction is pseudo first order and determine the pseudo-first-order rate constant k'.
b. If the ozone concentration is 100 times the NO concentration at $t = 0$, what is the second-order rate constant k? Express this k in $M^{-1}\,s^{-1}$.

Collect and Organize Under pseudo-first-order conditions, a large excess of one reactant is maintained so that we can use the integrated rate law for a first-order reaction. We are asked to verify that the reaction is pseudo first order and to determine the pseudo-first-order rate constant. We are given experimental data showing the change of concentration of NO with time. We are also given the initial concentration of O_3 and asked to calculate the second-order rate constant for the reaction.

TABLE 13.13(b) Concentration of NO and ln[NO] as a Function of Time

Time (μs)	[NO] (molecules/cm^3)	ln[NO]
0.0	1.00×10^9	20.723
100.0	8.36×10^8	20.544
200.0	6.98×10^8	20.364
300.0	5.83×10^8	20.184
400.0	4.87×10^8	20.004
500.0	4.07×10^8	19.824
1000.0	1.65×10^8	18.915

Analyze We verify our assumption by treating the data as we would for a first-order reaction and plotting ln[NO] versus time. The plot should be a straight line with a slope equal to $-k'$. We can solve for the second-order rate constant k using Equation 13.26. We expect the second-order rate constant to be rather large since the data (concentration changes) were collected on a microsecond scale.

Solve

a. The plot of ln[NO] values (Table 13.13b) versus time is linear as shown in Figure 13.12, which indicates that the reaction is indeed pseudo first order. The slope of the line that best fits the data is $-1.81 \times 10^{-3}\ \mu s^{-1}$. Changing the sign of the slope and converting the units from μs^{-1} to s^{-1} gives us $1.81 \times 10^3\ s^{-1}$ for the value of the pseudo-first-order rate constant k'.
b. We calculate the second-order rate constant k using Equation 13.26, $k' = k[O_3]_0$, where

$$[O_3]_0 = 100[NO]_0 = 100(1.00 \times 10^9 \text{ molecules/cm}^3)$$

Solving for k:

$$k = \frac{k'}{[O_3]_0} = \frac{1.81 \times 10^3\ s^{-1}}{1.00 \times 10^{11} \text{ molecules/cm}^3} = \frac{1.81 \times 10^{-8}\ \text{cm}^3}{\text{molecules} \cdot \text{s}}$$

To convert the units to $M^{-1}\,s^{-1}$, we need to convert the reciprocal concentration units of cm^3/molecule into M^{-1}:

$$k = \frac{1.81 \times 10^{-8}\ \cancel{\text{cm}^3}\ s^{-1}}{\cancel{\text{molecules}}} \times \frac{6.022 \times 10^{23}\ \cancel{\text{molecules}}}{1\ \text{mol}} \times \frac{1\ \text{L}}{1000\ \cancel{\text{cm}^3}}$$

$$= 1.09 \times 10^{13}\ \frac{\text{L} \cdot \text{s}^{-1}}{\text{mol}} = 1.09 \times 10^{13}\ M^{-1}\,s^{-1}$$

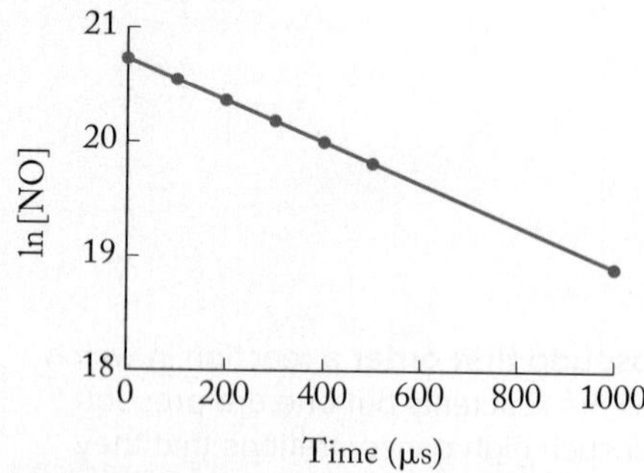

FIGURE 13.12 Plot of ln[NO] as a function of time.

Think About It As we predicted, the second-order rate constant for the reaction is large. The pseudo-first-order (k') and second-order (k) rate constants are very different because k' contains a term for the initial concentration of ozone, which is very small when expressed in moles per liter. The value of k' will be different for every initial concentration of O_3, but the value of k will be independent of $[O_3]_0$.

Practice Exercise The reaction

$$Cl(g) + O_3(g) \rightarrow ClO(g) + O_2(g)$$

is first order in both reactants. Determine the pseudo-first-order and second-order rate constants for the reaction from the data in Table 13.14 if $[O_3]_0 = 8.5 \times 10^{-11}$ *M*.

TABLE 13.14 Concentration of Chlorine Atoms as a Function of Time

Time (μs)	[Cl] (*M*)
0.0	5.60×10^{-14}
100.0	5.27×10^{-14}
600.0	3.89×10^{-14}
1200.0	2.69×10^{-14}
1850.0	1.81×10^{-14}

CONCEPT TEST

A student measured the pseudo-first-order rate constant for the reaction of NO with O_3 in Sample Exercise 13.9 at four initial concentrations of ozone. Assuming that all four $[O_3]_0$ values were much greater than [NO], how could the student determine the second-order rate constant graphically?

Zero-Order Reactions

We introduced the following reaction in the Practice Exercise accompanying Sample Exercise 13.7:

$$NO_2(g) + CO(g) \rightarrow NO(g) + CO_2(g)$$

The rate law for this reaction is

$$\text{Rate} = k[NO_2]^2 \qquad (13.27)$$

The rate of the reaction does not change when [CO] is changed, even when the concentrations of CO and NO_2 are comparable. This situation is not the same as in pseudo-first-order reactions, where the rate does not depend on the concentration of one reactant because that reactant is present in large excess.

One interpretation of Equation 13.27 is that it contains a [CO] term to the zeroth power, making the reaction *zero order* in that reactant. Because any value raised to the zeroth power is 1, we have

$$\text{Rate} = k[NO_2]^2[CO]^0 = k[NO_2]^2(1) = k[NO_2]^2$$

Reactions with a true zero-order rate law are rare, but let's consider a generic reaction involving a single reactant X, with the reaction zero order in reactant X and zero order overall. The rate law is

$$\text{Rate} = -\Delta[X]/\Delta t = k[X]^0 = k$$

and the integrated rate law is

$$[X] = -kt + [X]_0$$

The slope of a plot of reactant concentration versus time (Figure 13.13) equals the negative of the zero-order rate constant k.

We can calculate the half-life of a zero-order reaction by substituting $t = t_{1/2}$ and $[X] = [X]_0/2$ into the integrated rate law:

$$\frac{[X]_0}{2} = -kt_{1/2} + [X]_0$$

$$kt_{1/2} = [X]_0 - \frac{[X]_0}{2} = \frac{[X]_0}{2}$$

$$t_{1/2} = \frac{[X]_0}{2k}$$

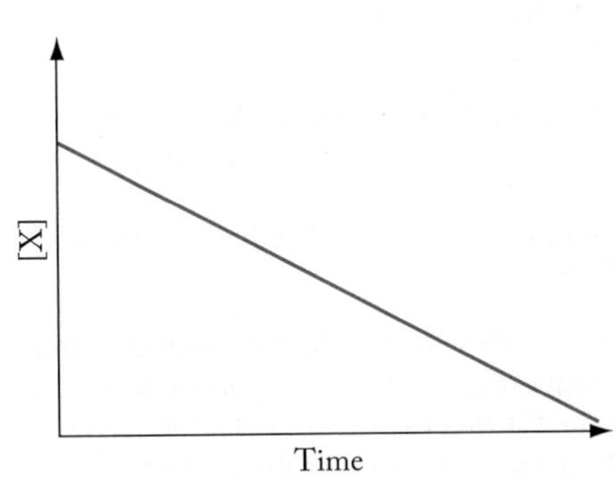

FIGURE 13.13 The change in concentration of the reactant X in the zero-order reaction $X \rightarrow Y$ is constant over time.

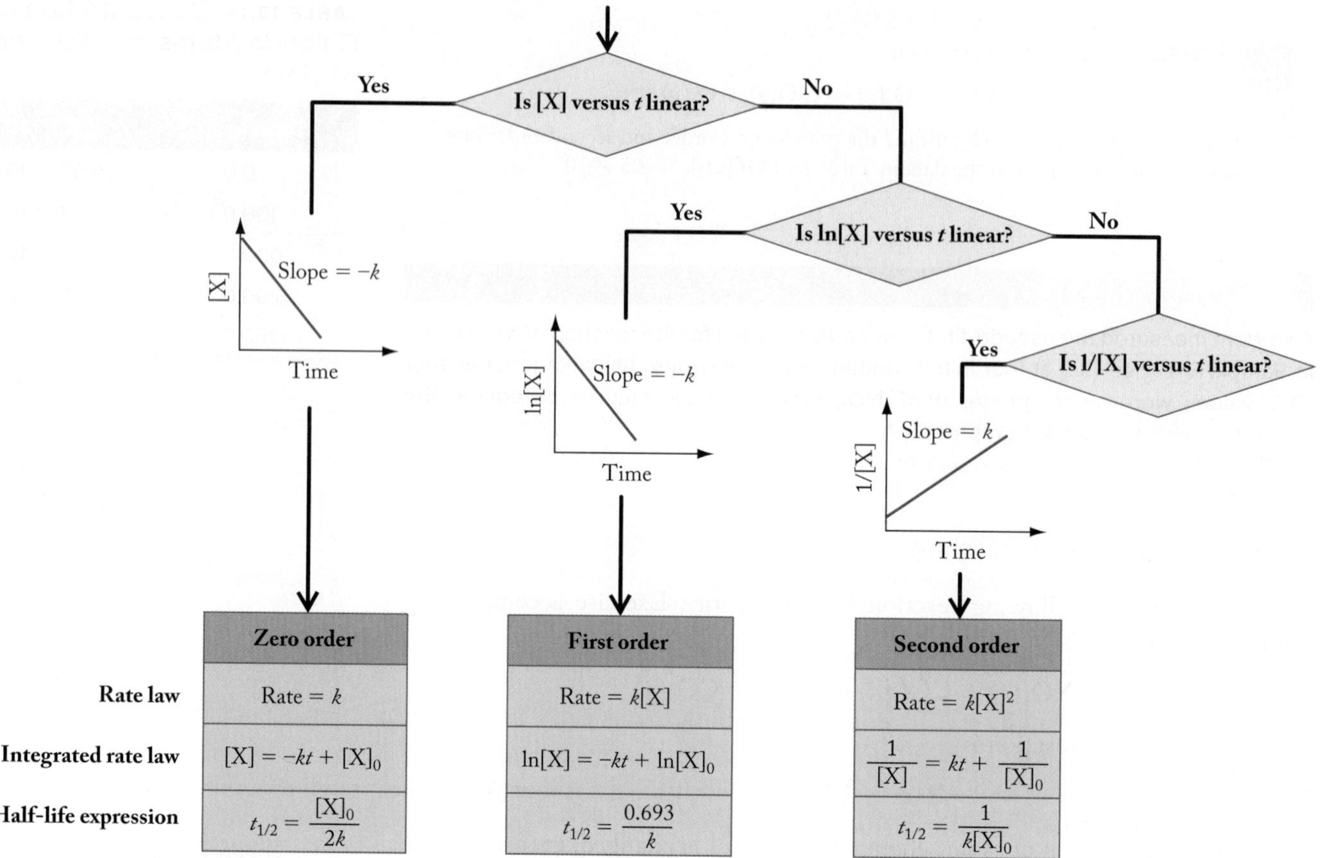

FIGURE 13.14 Summary of how to distinguish between zero-order, first-order, and second-order kinetics for reactions involving a single reactant (X).

For now we leave the discussion of zero-order reactions with this purely mathematical treatment. We return to them and examine their meaning at the molecular level in Sections 13.5 and 13.6.

Figure 13.14 summarizes how to determine whether a reaction in which a reactant X forms one or more products is zero, first, or second order. The figure also lists the rate laws and integrated rate laws for these reactions and the equations used to calculate $t_{1/2}$ of X. The decision-making process shown in Figure 13.14 can also be used to determine the kinetics of reactions involving two reactants (X and Y) if the reaction is zero order in Y. In such cases we focus on how the concentration of X changes with time. Finally, if a reaction is first order in both X and Y, which means second order overall, and Y is present in the reaction mixture at a much higher concentration than X, we can calculate the pseudo-first-order rate constant k' from a plot of ln[X] versus t, and then calculate the second-order rate constant by dividing k' by [Y].

activation energy (E_a) the minimum energy molecules need to react when they collide.

Arrhenius equation relates the rate constant of a reaction to absolute temperature (T), the activation energy of the reaction (E_a), and the frequency factor (A).

frequency factor (A) the product of the frequency of molecular collisions and a factor that expresses the probability that the orientation of the molecules is appropriate for a reaction to occur.

activated complex a short-lived species formed in a chemical reaction.

13.4 Reaction Rates, Temperature, and the Arrhenius Equation

Why do rate constants for different reactions have different values? Why do they have the particular numerical values that we measure? In Section 13.3, we introduced the idea that chemical reactions take place in the gas phase when molecules collide with sufficient energy to break bonds in reactants and allow bonds in products to form. The minimum amount of energy that enables this to happen is called the **activation energy (E_a)**, and every chemical reaction has a characteristic

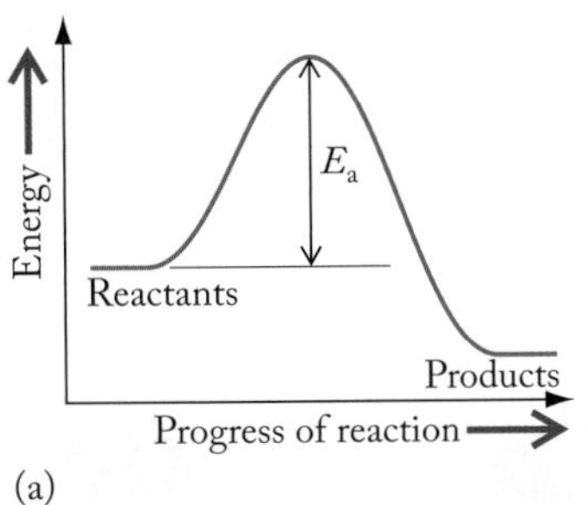

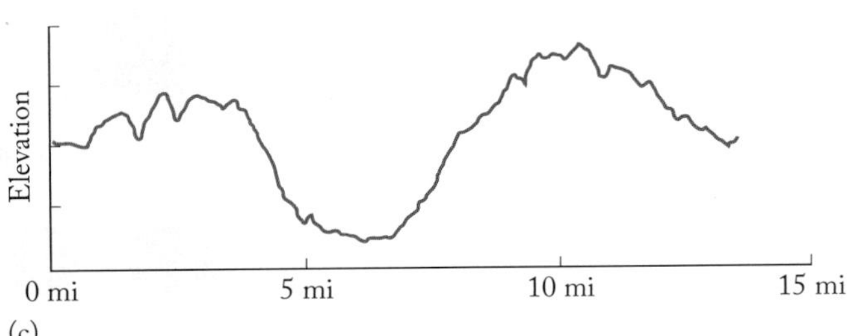

FIGURE 13.15 (a) The energy profile of a reaction includes an activation energy barrier E_a that must be overcome before the reaction can proceed. (b) A real-world analogy confronts a hiker climbing a series of hills. The Beartree Gap Trail forms part of the Iron Mountain Loop in the Mount Rogers National Recreation Area in southwestern Virginia. (c) From the starting point at mile zero, hikers must climb over two "energy barriers": Straight Mountain (at 3 mi) and Iron Mountain (at 10 mi) before returning to their starting point after a 13.5 mile (22 km) hike.

activation energy, usually expressed in kJ/mol. Activation energy is an energy barrier that must be overcome if a reaction is to proceed—like the hill that must be climbed to hike the trail shown in Figure 13.15. Generally, the greater the activation energy, the slower the reaction.

According to kinetic molecular theory, molecules move at higher speeds and have more kinetic energy at higher temperatures. This means that the fraction of molecules with kinetic energies greater than a given activation energy increases with increasing temperature, as shown in Figure 13.16(a). Therefore, the rates of chemical reactions should increase with increasing temperature, and indeed they do (Figure 13.16b).

In the late 19th century, experiments carried out in the laboratories of Jacobus van 't Hoff and Svante Arrhenius led to a fundamental advance in understanding how temperature affects rates of chemical reactions. The mathematical connection between temperature, the rate constant k for a reaction, and its activation energy is given by the **Arrhenius equation**:

$$k = Ae^{-E_a/RT} \tag{13.28}$$

where R is the gas constant in J/(mol · K) and T is the reaction temperature in kelvin. The factor A, called the **frequency factor**, is the product of collision frequency and a term that corrects for the fact that not every collision results in a chemical reaction. Because the exponential part of the equation is unitless, the frequency factor A must have the same units as the rate constant k.

Some collisions do not lead to products because the colliding molecules are not oriented relative to each other in the right way. To examine the importance of molecular orientation during collisions, let's revisit the reaction between O_3 and NO:

$$O_3(g) + NO(g) \rightarrow O_2(g) + NO_2(g) \tag{13.7}$$

Figure 13.17 shows two ways that ozone and nitric oxide molecules might approach each other. Only one of these orientations, the one in which an O_3 molecule approaches the nitrogen atom of NO, leads to a chemical reaction between the two molecules.

A collision between O_3 and NO molecules with the correct orientation and enough kinetic energy may result in the formation of the **activated complex** shown in Figure 13.17(a). In this species, one of the O—O bonds in the O_3

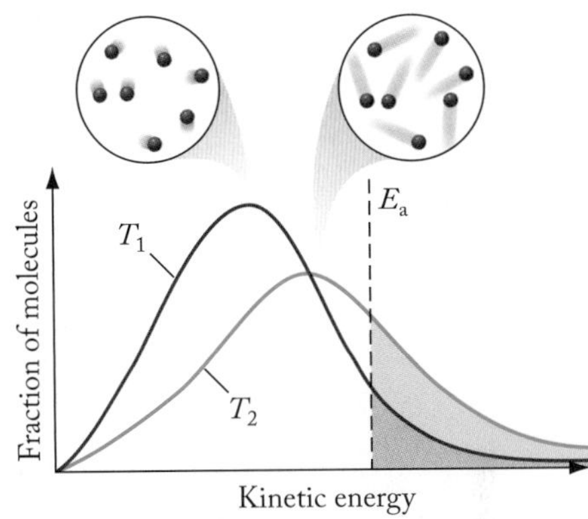

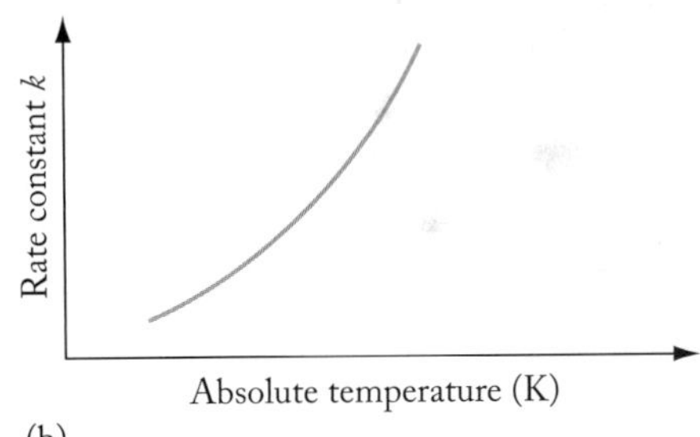

FIGURE 13.16 (a) According to kinetic molecular theory, a fraction of reactant molecules have kinetic energies equal to or greater than the activation energy (E_a) of the reaction. As temperature increases from T_1 to T_2, the number of molecules with energies exceeding E_a increases, leading to an increase in reaction rate. (b) The rate constant for any reaction increases with increasing temperature.

CONNECTION The distribution of speeds among particles in the gas phase was introduced in Chapter 10 (see Figure 10.8).

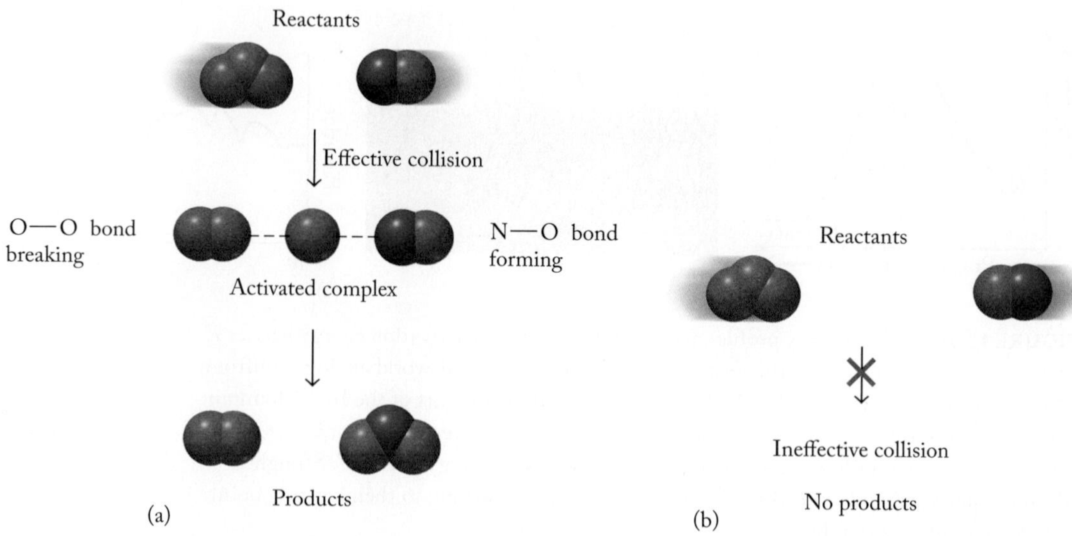

FIGURE 13.17 The effect of molecular orientation on reaction rate. (a) When an O_3 and a NO molecule are oriented such that the collision is between an O_3 oxygen and the NO nitrogen, the collision is effective and an activated complex forms, yielding the two product molecules O_2 and NO_2. (b) When the reactant molecules are oriented such that the collision is between an O_3 oxygen and the NO oxygen, no activated complex forms and no reaction occurs.

CONNECTION The van 't Hoff equation (Chapter 11) is named after the same Jacobus van 't Hoff who studied the temperature dependence of reaction rates. Arrhenius acids (Chapter 8) are named for the same Svante Arrhenius who proposed what we now call the Arrhenius equation.

molecule has started to break and the new N—O bond is beginning to form. Activated complexes represent midway points in chemical reactions. They have extremely brief lifetimes and fall apart rapidly, either forming products or re-forming reactants. Activated complexes are formed by reacting species that have acquired enough potential energy to react with each other. The internal energy of an activated complex represents a high-energy **transition state** of the reaction. In fact, the energy of an activated complex for a reaction defines the height of the activation energy barrier for the reaction. The magnitudes of activation energies can vary from a few kilojoules to hundreds of kilojoules per mole.

CHEMTOUR Arrhenius Equation

We can draw an **energy profile** for a chemical reaction that shows the changes in potential energy for the reaction as a function of the progress of the reaction from reactants to products. We have already seen one energy profile in Figure 13.15. Now consider the energy profile for the reaction between nitric oxide and ozone, shown in Figure 13.18(a). The *x*-axis represents the progress of the reaction and the *y*-axis represents chemical energy. The activation energy is equivalent to the difference in energy between the transition state and either the reactants or the products. The size of the activation energy barrier depends on the direction from which it is approached. E_a is smaller in the forward direction ($NO + O_3 \rightarrow NO_2 + O_2$, Figure 13.18a) than in the reverse direction ($NO_2 + O_2 \rightarrow NO + O_3$, Figure 13.18b): 10.5 kJ/mol versus 210 kJ/mol. A smaller activation energy barrier means that the forward reaction proceeds at a higher rate than the reverse reaction, assuming equal concentrations of reactants and products.

One of the many uses of the Arrhenius equation is to calculate the value of E_a for a chemical reaction. When we take the natural logarithm of both sides of Equation 13.28,

$$\ln k = -\frac{E_a}{R}\left(\frac{1}{T}\right) + \ln A \tag{13.29}$$

the result fits the general equation of a straight line ($y = mx + b$) if we make ($\ln k$) the *y* variable and ($1/T$) the *x* variable. We can calculate E_a by determining

transition state a high-energy state between reactants and products in a chemical reaction.

energy profile a graph showing the changes in potential energy for a reaction as a function of the progress of the reaction from reactants to products.

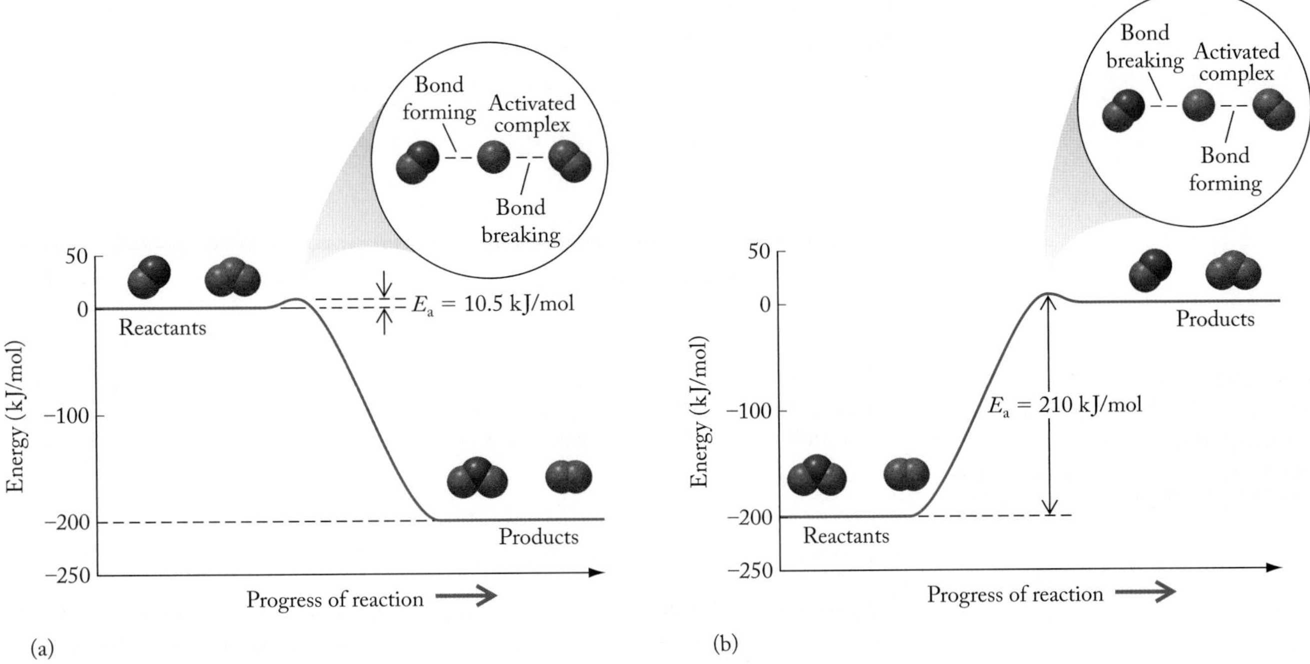

FIGURE 13.18 (a) The energy profile for the reaction $NO(g) + O_3(g) \rightarrow NO_2(g) + O_2(g)$ includes an activation energy barrier of 10.5 kJ/mol. (b) The reverse reaction has a much larger activation energy of 210 kJ/mol.

the rate constant k for the reaction at several temperatures. Plotting ln k versus $1/T$ should give a straight line, the slope of which is $-E_a/R$. Table 13.15 and Figure 13.19 show data for the reaction between NO and O_3 at six temperatures. The straight line that best fits the points in the figure has a slope of -1.26×10^3 K, and the activation energy of the reaction is

$$E_a = -\text{slope} \times R$$

$$= -(-1.26 \times 10^3\ \cancel{\text{K}}) \times \left(8.314 \frac{\text{J}}{\text{mol} \cdot \cancel{\text{K}}}\right) = 1.05 \times 10^4 \text{ J/mol}$$

$$= 10.5 \text{ kJ/mol}$$

TABLE 13.15 Temperature Dependence of the Rate of the Reaction $NO(g) + O_3(g) \rightarrow NO_2(g) + O_2(g)$

T (K)	k (M^{-1} s^{-1})	ln k	$1/T$ (K^{-1})
300	1.21×10^{10}	23.216	3.33×10^{-3}
325	1.67×10^{10}	23.539	3.08×10^{-3}
350	2.20×10^{10}	23.814	2.86×10^{-3}
375	2.79×10^{10}	24.052	2.67×10^{-3}
400	3.45×10^{10}	24.264	2.50×10^{-3}
425	4.15×10^{10}	24.449	2.35×10^{-3}

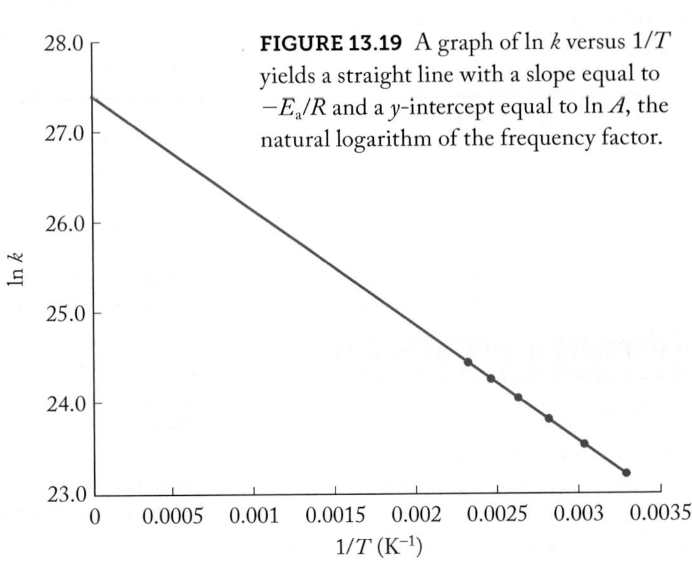

FIGURE 13.19 A graph of ln k versus $1/T$ yields a straight line with a slope equal to $-E_a/R$ and a y-intercept equal to ln A, the natural logarithm of the frequency factor.

The y-intercept ($1/T = 0$) in Figure 13.19 is 27.4. According to Equation 13.29, this value represents $\ln A$, which means that

$$A = e^{27.4} = 7.9 \times 10^{11}\ M^{-1}\ s^{-1}$$

Note that the units on A match those of the k values used to calculate it.

SAMPLE EXERCISE 13.10 Calculating Activation Energies and Frequency Factors **LO6**

As noted in Sample Exercise 13.7, ClO is a highly reactive gas that plays a key role in the destruction of ozone in the stratosphere. The data in Table 13.16(a) were collected in a study of the effect of temperature on the rate at which ClO decomposes into Cl_2 and O_2:

$$2\ ClO(g) \rightarrow Cl_2(g) + O_2(g)$$

Determine the activation energy (E_a) and frequency factor (A) for this reaction.

TABLE 13.16(a) Rate Constant as a Function of Temperature for the Decomposition of ClO

T (K)	k ($M^{-1}\ s^{-1}$)
238	1.9×10^9
258	3.1×10^9
278	4.9×10^9
298	7.2×10^9

Collect and Organize We are given rate constant values of a reaction at different temperatures, and we are asked to find the reaction's A and E_a values. According to the Arrhenius equation, the slope of a plot of $\ln k$ versus $1/T$ is equal to $-E_a/R$, and the y-intercept is equal to $\ln A$.

Analyze We need to convert the T and k values from Table 13.16(a) to $1/T$ and $\ln k$, respectively, to use them in the Arrhenius equation. The rate constants for the reaction are fairly large, about $10^9\ M^{-1}\ s^{-1}$, near room temperature, so the activation energy barrier should be relatively low.

TABLE 13.16(b) Summary of T, $1/T$, k, and $\ln k$ for the Decomposition of ClO

T (K)	$1/T$ (K^{-1})	k ($M^{-1}\ s^{-1}$)	$\ln k$
238	4.20×10^{-3}	1.9×10^9	21.365
258	3.88×10^{-3}	3.1×10^9	21.855
278	3.60×10^{-3}	4.9×10^9	22.313
298	3.36×10^{-3}	7.2×10^9	22.697

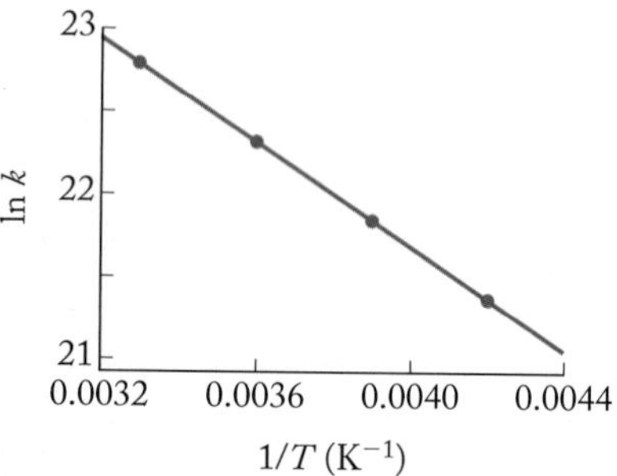

FIGURE 13.20 Plot of $\ln k$ versus $1/T$ for the decomposition of ClO(g). Note that the scale of the x axis does not start at zero.

Solve Expanding the data table to include columns of $\ln k$ and $1/T$ yields Table 13.16(b). A plot of $\ln k$ versus $1/T$ gives us a straight line, the slope of which is -1590 K (Figure 13.20) and the y-intercept is 28.0. Therefore, the values of A and E_a are

$$y\text{-intercept} = 28.0 = \ln A, \qquad A = e^{28.0} = 1.4 \times 10^{12}\ M^{-1}\ s^{-1}$$

$$E_a = -\text{slope} \times R$$

$$= -(-1590\ \text{K}) \times 8.314 \frac{\text{J}}{\text{mol} \cdot \text{K}} = 1.32 \times 10^4\ \text{J/mol} = 13.2\ \text{kJ/mol}$$

Think About It The activation energy is the height of a barrier that has to be overcome before a reaction can proceed. As predicted, the activation energy for ClO decomposition has a relatively small positive value.

 **Practice Exercise** The rate constant for the reaction

$$Br(g) + O_3(g) \rightarrow BrO(g) + O_2(g)$$

was determined at the four temperatures shown in Table 13.17. Calculate the values of E_a and A for this reaction.

TABLE 13.17 Rate Constants as a Function of Temperature for the Reaction of Br with O_3

T (K)	k [cm^3/(molecule · s)]
238	5.9×10^{-13}
258	7.7×10^{-13}
278	9.6×10^{-13}
298	1.2×10^{-12}

Calculating activation energies using the graphical method generally requires measuring the rate constant at a minimum of three different temperatures to verify that the plot of ln k versus $1/T$ is a straight line. Once we know the value of E_a, we can use it and the value of the rate constant (k_1) of a reaction at one temperature (T_1) to calculate the value of the rate constant (k_2) at another temperature (T_2). We start by substituting k_1, k_2, T_1, and T_2 into Equation 13.29:

$$\ln k_1 = -\frac{E_a}{R}\left(\frac{1}{T_1}\right) + \ln A \qquad \ln k_2 = -\frac{E_a}{R}\left(\frac{1}{T_2}\right) + \ln A$$

Subtracting these two expressions, $\ln k_1 - \ln k_2$, gives

$$\ln k_1 - \ln k_2 = \left[-\frac{E_a}{R}\left(\frac{1}{T_1}\right) + \ln A\right] - \left[-\frac{E_a}{R}\left(\frac{1}{T_2}\right) + \ln A\right]$$

Using the mathematical properties of logarithms, this equation can be rearranged as follows:

$$\ln \frac{k_1}{k_2} = \frac{E_a}{R}\left(\frac{1}{T_2} - \frac{1}{T_1}\right) \qquad (13.30)$$

SAMPLE EXERCISE 13.11 Calculating the Value of a Rate Constant at a Second Temperature **LO6**

The rate constant (k) for the reaction $2\ \mathrm{ClO}(g) \rightarrow \mathrm{Cl_2}(g) + \mathrm{O_2}(g)$ is $7.2 \times 10^9\ M^{-1}\ \mathrm{s}^{-1}$ at 298 K. In Sample Exercise 13.10 we determined that the activation energy of this reaction is 13.2 kJ/mol. What is the rate constant of the reaction in the lower stratosphere where $T = 217$ K?

Collect, Organize, and Analyze We are given a rate constant value at one temperature and asked to calculate its value at a lower temperature. We know the activation energy of the reaction, so we can use Equation 13.30 to solve for k at the lower temperature. Based on the rate constant data for this reaction in Table 13.16, the value of k at 217 K should be near $1 \times 10^9\ M^{-1}\ \mathrm{s}^{-1}$. As part of the calculation, we need to convert E_a from kilojoules to joules to be compatible with the value of R: 8.314 J/(mol · K).

Solve Letting 217 K be T_2 and 298 K be T_1 in Equation 13.30 and solving for k_2:

$$\ln \frac{k_1}{k_2} = \frac{E_a}{R}\left(\frac{1}{T_2} - \frac{1}{T_1}\right) = \left(\frac{13.2\ \frac{\cancel{\mathrm{kJ}}}{\cancel{\mathrm{mol}}} \times \frac{1000\ \cancel{\mathrm{J}}}{\cancel{\mathrm{kJ}}}}{8.314\ \frac{\cancel{\mathrm{J}}}{\cancel{\mathrm{mol}} \cdot \mathrm{K}}}\right)\left(\frac{1}{217\ \cancel{\mathrm{K}}} - \frac{1}{298\ \cancel{\mathrm{K}}}\right) = 1.989$$

$$\frac{k_1}{k_2} = \frac{7.2 \times 10^9\ M^{-1}\ \mathrm{s}^{-1}}{k_2} = e^{1.989} = 7.31$$

$$k_2 = 9.8 \times 10^8\ M^{-1}\ \mathrm{s}^{-1}$$

Think About It The calculated rate constant is close to the predicted value and about 1/5 the rate constant at 298 K. Equation 13.30 comes in handy when we know the rate constant of a reaction at a standard reference temperature, such as 298 K (25°C), but need to know it at a much higher or lower temperature.

Practice Exercise The activation energy for the reaction $\mathrm{O_3}(g) + \mathrm{NO}(g) \rightarrow \mathrm{O_2}(g) + \mathrm{NO_2}(g)$ is 10.2 kJ/mol. How many times larger is the rate constant for this reaction at 400 K than at 200 K?

reaction mechanism a set of steps that describes how a reaction occurs at the molecular level; the mechanism must be consistent with the rate law for the reaction.

intermediate a species produced in one step of a reaction and consumed in a subsequent step.

CONCEPT **TEST**

Which of the following statements is/are true about activation energies?

a. Fast reactions have large rate constants *and* large activation energies.

b. The forward reaction always has a lower activation energy than the reverse reaction.

c. Raising the temperature of a reaction mixture lowers the activation energy of the reaction.

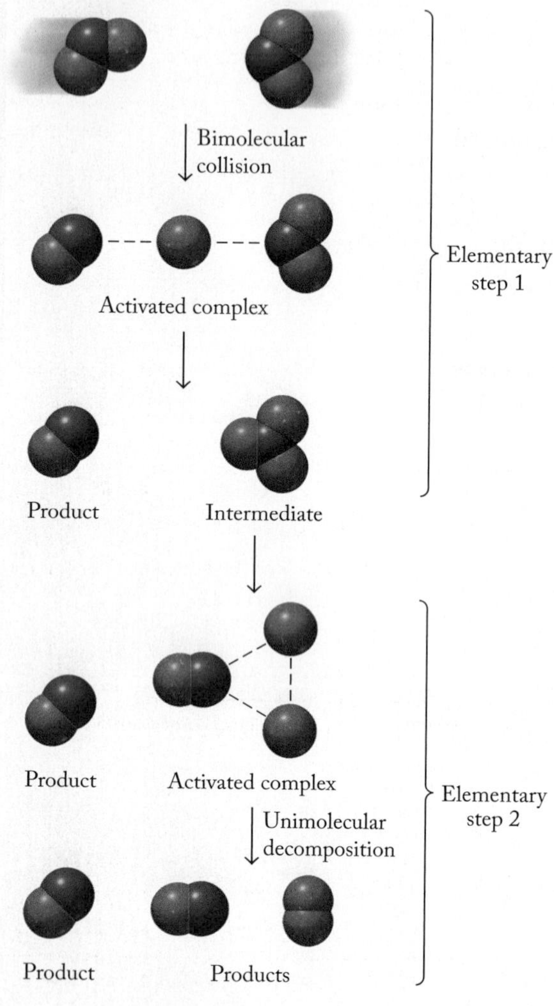

Overall reaction:

$2\ NO_2 \longrightarrow 2\ NO + O_2$

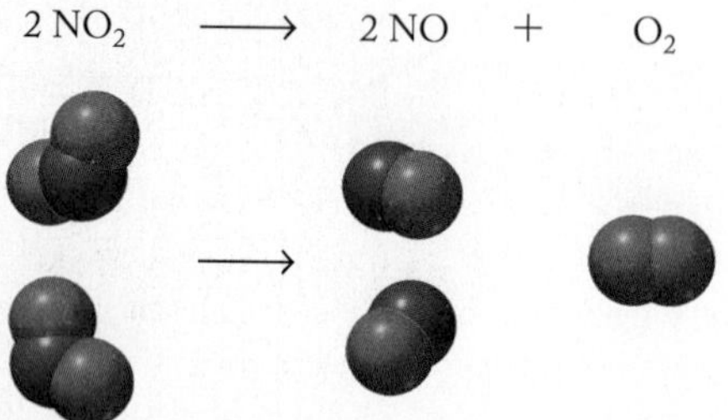

FIGURE 13.21 The decomposition of NO_2 begins when two NO_2 molecules collide, producing NO and NO_3 (elementary step 1). The NO_3 then rapidly decomposes into NO and O_2 (elementary step 2).

13.5 Reaction Mechanisms

We ended the previous section with an equation that enables us to calculate the rate constant of a reaction at any temperature from the value of k at a particular temperature and the value of E_a. This calculation assumes, however, that the reaction proceeds in the same way no matter what the temperature is.[2] In other words, Equation 13.30 holds as long as the reaction happens via the same set of collisions and involves the same sequence of bonds breaking in reactants and bonds forming in products at both T_1 and T_2. These molecular steps define its **reaction mechanism**. In this section we explore how kinetics data provides chemists with a molecular view of how chemical reactions happen.

Elementary Steps

Let's start by revisiting the thermal decomposition of NO_2:

$$2\ NO_2(g) \rightarrow 2\ NO(g) + O_2(g)$$

We noted in Section 13.3 that this reaction is second order in NO_2 because it takes place as a result of the collisions of pairs of NO_2 molecules. How do the atoms in two colliding NO_2 molecules rearrange themselves to form two NO molecules and one O_2 molecule? One answer to this question is contained in the two-step reaction pathway shown in Figure 13.21. In the first step, a collision between two NO_2 molecules produces a molecule of NO and a molecule of NO_3. In a second step, the NO_3 decomposes to NO and O_2. Both steps in the mechanism involve transient activated complexes. In the activated complex of the first step, two molecules share an oxygen atom. The bonds in the activated complex of the second step rearrange so that two oxygen atoms become bonded together, forming a molecule of O_2 and leaving behind a molecule of NO. The molecule NO_3 is an **intermediate** in this mechanism because it is produced in one step and consumed in a later step. Intermediates are not considered reactants or products and do not appear in the equation describing a reaction. In some cases, intermediates in chemical reactions are sufficiently long-lived to be isolated. In contrast, activated complexes have never been isolated, although they have been detected.

This reaction mechanism is a combination of two **elementary steps**. An elementary step that involves a single molecule is **unimolecular**, one that involves a collision between two molecules is **bimolecular**, and one that involves a collision between three

[2]This assumption is not always valid, especially when T_1 and T_2 are far apart. Under these circumstances, Equation 13.30 may not hold.

molecules is **termolecular**. Bimolecular elementary steps are much more common than termolecular elementary steps because the chance of three molecules colliding at exactly the same time is much smaller. The terms uni-, bi-, and termolecular are used by chemists to describe the **molecularity** of an elementary step, which refers to the number of reacting atoms, ions, or molecules involved in that step. The molecularity of an elementary step is the same as its reaction order.

A valid reaction mechanism must be consistent with the stoichiometry of the reaction. In other words, the elementary steps in Figure 13.21,

$$\text{Elementary step 1} \quad 2\ NO_2(g) \rightarrow NO(g) + NO_3(g)$$

$$\text{Elementary step 2} \quad NO_3(g) \rightarrow NO(g) + O_2(g)$$

must sum to yield the balanced chemical equation, which they do once the intermediate (NO_3) is canceled out:

$$2\ NO_2(g) + \cancel{NO_3(g)} \rightarrow 2\ NO(g) + \cancel{NO_3(g)} + O_2(g)$$

$$2\ NO_2(g) \rightarrow 2\ NO(g) + O_2(g)$$

elementary step a molecular-level view of a single process taking place in a chemical reaction.

unimolecular step a step in a reaction mechanism involving only one molecule on the reactant side.

bimolecular step a step in a reaction mechanism involving a collision between two molecules.

termolecular step a step in a reaction mechanism involving a collision among three molecules.

molecularity the number of ions, atoms, or molecules involved in an elementary step in a reaction.

How does the activation energy apply to a two-step reaction such as this one? The two elementary steps produce an energy profile with two maxima. In elementary step 1, collisions between pairs of NO_2 molecules result in the formation of an activated complex associated with the first transition state in Figure 13.22. As this activated complex transforms into NO and NO_3, the energy of the system drops to the bottom of the trough between the two maxima. In elementary step 2, NO_3 forms the activated complex associated with the second transition state. As this complex transforms into the final products, NO and O_2, the energy of the system drops to its final level.

Figure 13.22 shows that the energy barrier for elementary step 1 is much greater than that for elementary step 2. This difference is consistent with the relative rates of the two steps: step 1 is slower than step 2. If the reaction were to proceed in the reverse direction (as NO and O_2 react, forming NO_2), the first energy barrier would be the smaller of the two, and the first elementary step would be the more rapid one. Experimental evidence supports these expectations.

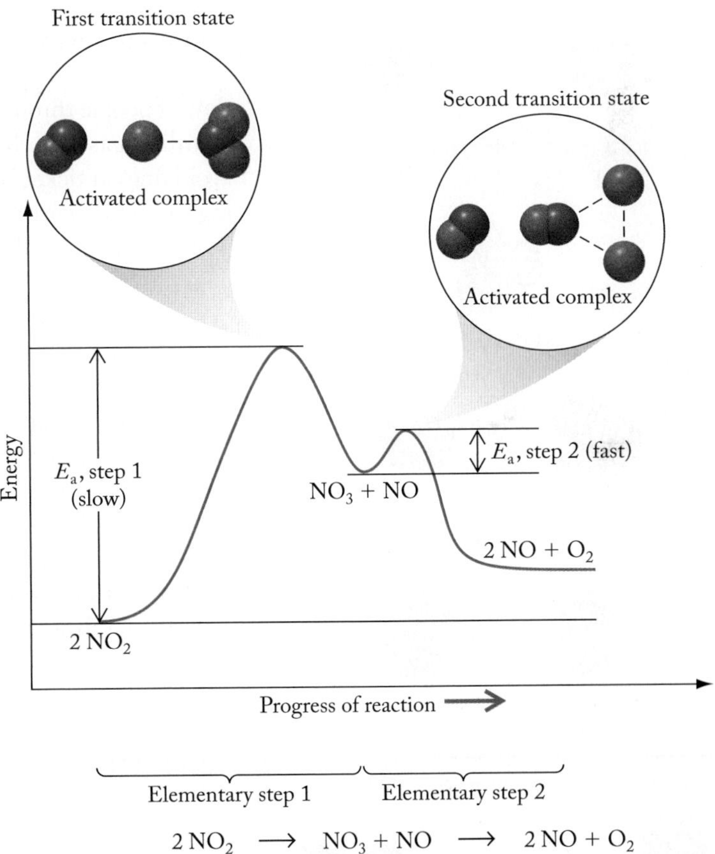

FIGURE 13.22 The energy profile for the decomposition of NO_2 to NO and O_2 shows activation energy barriers for both elementary steps. The activation energy of the first step is larger than that of the second step, so the first step is the slower of the two.

CONCEPT **TEST**

In a reaction mechanism with *n* elementary steps, how many activation energies does the overall reaction have?

Rate Laws and Reaction Mechanisms

Any mechanism proposed for a reaction must be consistent with the rate law derived from experimental data. As noted in Section 13.3, the rate law for the decomposition of NO_2 is second order in NO_2:

$$\text{Rate} = k[NO_2]^2 \qquad (13.21)$$

The balanced chemical equation for this (or any) reaction does not provide enough information to write the overall rate law for the reaction. However, we

rate-determining step the slowest step in a multistep chemical reaction.

can use the balanced chemical equation for any elementary step to write a rate law for that step. For the decomposition of NO_2, the rate law for the first elementary step, $2\ NO_2(g) \rightarrow NO_3(g) + NO(g)$, is

$$\text{Rate}_1 = k_1[NO_2]^2 \tag{13.31}$$

which we obtain by writing the concentration for each reactant raised to a power equal to the coefficient of that reactant in the balanced equation for that step. For the second elementary step in the NO_2 decomposition, $NO_3(g) \rightarrow NO(g) + O_2(g)$, the sole reactant, NO_3, has a coefficient of 1, so the rate law is

$$\text{Rate}_2 = k_2[NO_3] \tag{13.32}$$

How do we use the rate laws in Equations 13.31 and 13.32 to determine the validity of the mechanism—that is, how do we show that they conform to the observed rate law for the decomposition of NO_2 in Equation 13.21?

The two steps in the mechanism proceed at different rates, with different activation energies (Figure 13.22). Step 1 is slower (larger E_a) than step 2 (smaller E_a), so step 1 is the **rate-determining step** in the reaction. The rate-determining step is the slowest elementary step in a chemical reaction, and the rate of this step controls the overall reaction rate.

One way of visualizing the concept of a rate-determining step is to analyze the flow of people through a busy airport. Many travelers arrive at the airport with their boarding passes in hand or print them from convenient kiosks that allow them to avoid lines at ticket counters and move quickly. The next step in the process is passing through security. Typically, the number of people in the line outside the security point greatly exceeds the number of available security gates, so the time required to make it to a flight depends mostly on the time needed to pass through security. Security screening is the rate-determining step on the way through the airport.

If the first step in the mechanism in Figure 13.21 is the rate-determining step, then the value of k in Equation 13.21 is equal to k_1 from Equation 13.31. In addition, the value of k_1 must be smaller than the value of k_2. NO_3 is sufficiently stable that we can make small amounts of it and test whether $k_1 < k_2$. Experiments run at 300 K starting with NO_2 or NO_3 yield the vastly different values $k_1 \approx 1 \times 10^{-10}\ M^{-1}\ s^{-1}$ and $k_2 \approx 6.3 \times 10^4\ s^{-1}$. Therefore, as soon as any NO_3 forms in step 1, it rapidly falls apart to NO and O_2 in step 2.

Now let's consider the reaction that is the reverse of NO_2 decomposition—namely, the formation of NO_2 from NO and O_2:

$$\text{Overall reaction} \quad 2\ NO(g) + O_2(g) \rightarrow 2\ NO_2(g)$$

We determined in Section 13.3 that this reaction is second order in NO and first order in O_2:

$$\text{Rate} = k[NO]^2[O_2] \tag{13.33}$$

One proposed mechanism is shown in Figure 13.23. It has two elementary steps:

Step 1	$NO(g) + O_2(g) \rightleftharpoons NO_3(g)$	$\text{Rate} = k_1[NO][O_2]$	(13.34)
Step 2	$NO_3(g) + NO(g) \rightarrow 2\ NO_2(g)$	$\text{Rate} = k_2[NO_3][NO]$	(13.35)

The rate laws in Equations 13.34 and 13.35 were obtained by writing the concentration for each reactant raised to a power equal to the reactant's coefficient in the balanced equation. If step 1 were the rate-determining step, the reaction would be first order in NO and O_2, but that is not what the experimentally determined rate law indicates. If step 2 were the rate-determining step, the reaction would be first order in NO and NO_3, but that is inconsistent with the rate law, too. How can we account for the experimental rate law?

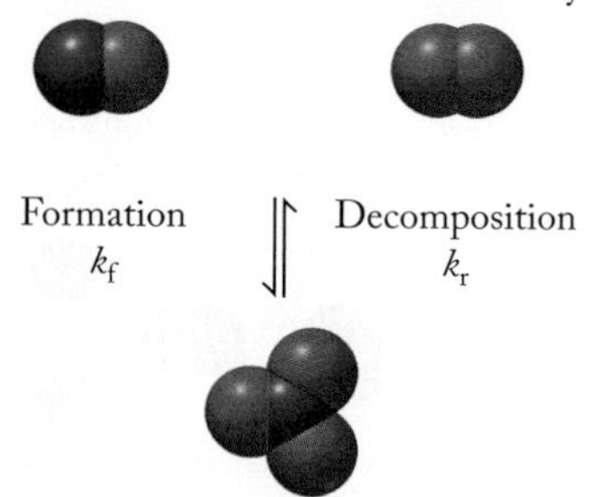

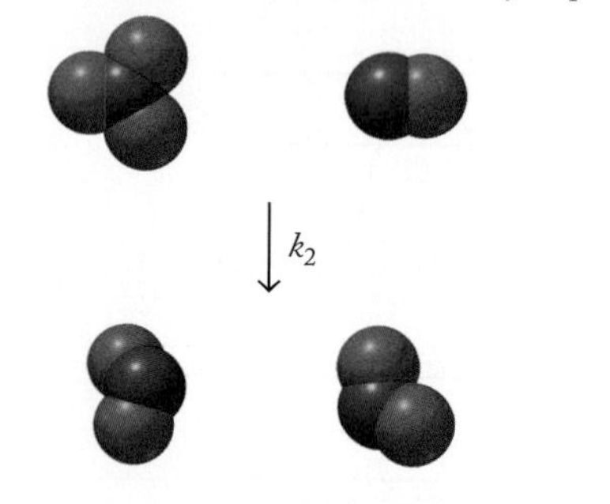

FIGURE 13.23 A mechanism for the formation of NO_2 from NO and O_2 has two elementary steps: (1) a fast, reversible bimolecular reaction in which NO and O_2 form NO_3 and (2) a slower, rate-determining bimolecular reaction in which NO_3 reacts with a molecule of NO to form two molecules of NO_2.

Consider what happens if step 2 is slow while step 1 is fast and reversible, which means NO_3 forms rapidly from NO and O_2 but decomposes just as rapidly back into NO and O_2. We indicate reversibility with a double arrow. Expressing this situation in equation form:

$$\text{Rate of forward reaction} = k_f[\text{NO}][\text{O}_2] = \text{fast}$$

$$\text{Rate of reverse reaction} = k_r[\text{NO}_3] = \text{equally fast}$$

The subscripts "f" and "r" refer to the "forward" and "reverse" reactions, respectively.

Combining these two expressions, we have

$$k_f[\text{NO}][\text{O}_2] = k_r[\text{NO}_3]$$

$$[\text{NO}_3] = \frac{k_f}{k_r}[\text{NO}][\text{O}_2] \tag{13.36}$$

Now, if we replace the $[NO_3]$ term in the rate law of Equation 13.35 (which we select because step 2 is the rate-determining step) with the right side of Equation 13.36, we get

$$\text{Rate} = k_2\frac{k_f}{k_r}[\text{NO}]^2[\text{O}_2]$$

The three rate constants can be combined,

$$k_{\text{overall}} = k_2\frac{k_f}{k_r}$$

and the rate law for the overall reaction becomes

$$\text{Rate} = k_{\text{overall}}[\text{NO}]^2[\text{O}_2]$$

This expression matches the overall rate law in Equation 13.33, so the proposed mechanism, a fast and reversible step 1 followed by a slow step 2, *may* be valid.

Even though the proposed reaction mechanism is consistent with the overall stoichiometry of the reaction and with the experimentally derived rate law, these consistencies do not prove that the proposed mechanism is correct. We need more experimental evidence, such as detecting the presence of an intermediate in the reaction mixture. On the other hand, not finding a reactive (and transient) intermediate would not necessarily disprove a reaction mechanism.

SAMPLE EXERCISE 13.12 Linking Reaction Mechanisms to Experimental Rate Laws **LO7**

The experimentally determined rate law for the reduction reaction,

$$2\,\text{H}_2(g) + 2\,\text{NO}(g) \rightarrow \text{N}_2(g) + 2\,\text{H}_2\text{O}(g)$$

which occurs at high temperatures is

$$\text{Rate} = k[\text{NO}]^2[\text{H}_2]$$

A proposed mechanism for the reaction, shown in Figure 13.24, consists of the following steps:

Elementary step 1	$H_2(g) + 2\,NO(g) \rightarrow N_2O(g) + H_2O(g)$
Elementary step 2	$H_2(g) + N_2O(g) \rightarrow N_2(g) + H_2O(g)$

Is this reaction mechanism consistent with the stoichiometry of the reaction and with the rate law? If so, which is the rate-determining step?

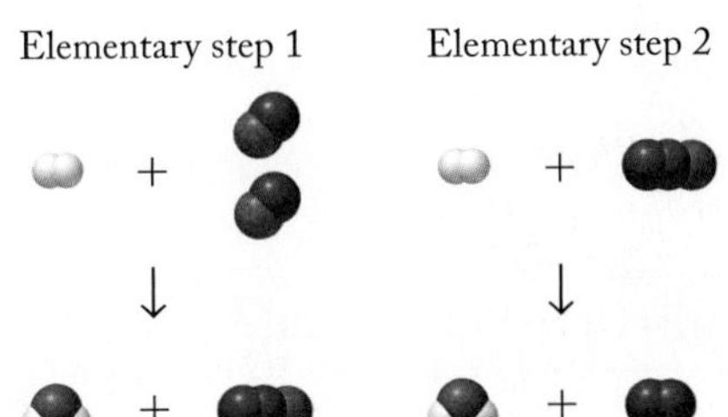

FIGURE 13.24 A proposed mechanism for the reaction between $NO(g)$ and $H_2(g)$.

Collect, Organize, and Analyze To answer the questions posed in this exercise we need to do the following:

1. Determine whether the chemical equations of the elementary steps add up to the overall reaction equation.
2. Write rate laws for each elementary step.
3. Compare the rate laws of the elementary steps to the experimental rate law of the overall reaction to assess the validity of the mechanism.
4. Determine which step is rate determining by matching its rate law to the observed rate law.

Solve Let's test whether the elementary steps add up to the overall reaction:

$$\begin{array}{ll} (1) & H_2(g) + 2\,NO(g) \rightarrow \cancel{N_2O}(g) + H_2O(g) \\ (2) & H_2(g) + \cancel{N_2O}(g) \rightarrow N_2(g) + H_2O(g) \\ \hline & 2\,H_2(g) + 2\,NO(g) \rightarrow N_2(g) + 2\,H_2O(g) \end{array}$$

This is indeed the equation of the overall reaction, so the elementary steps are consistent with the stoichiometry of the overall reaction.

Next we need to focus on the reaction mechanism. Elementary step 1 involves two molecules of NO colliding with one molecule of H_2 in a rare termolecular reaction. Elementary step 2 is a bimolecular reaction between the N_2O produced in step 1 and another molecule of H_2. The rate law for any elementary step can be written directly from the balanced equations, using the stoichiometric coefficients as exponents in the rate law:

Step 1	$H_2(g) + 2\,NO(g) \rightarrow N_2O(g) + H_2O(g)$	Rate $= k_1[H_2][NO]^2$
Step 2	$H_2(g) + N_2O(g) \rightarrow N_2(g) + H_2O(g)$	Rate $= k_2[H_2][N_2O]$

The rate law of step 1 matches the observed rate law of the overall reaction. Therefore, the proposed two-step mechanism is consistent with the experimental rate law, and step 1 is the rate-determining step.

Think About It We do not have direct proof that this is the correct mechanism, though it is consistent with the available data. A plausible mechanism for a reaction must yield a balanced equation whose stoichiometry matches the overall reaction and must be consistent with the rate law for the overall process. Both of these conditions are met here in Sample Exercise 13.12.

Practice Exercise The following mechanism is proposed for a reaction between compounds A and B:

Step 1	$2\,A(g) + B(g) \rightleftharpoons C(g)$	fast and reversible
Step 2	$B(g) + C(g) \rightarrow D(g)$	slow
Overall	$2\,A(g) + 2\,B(g) \rightarrow D(g)$	

What is the rate law for the overall reaction based on the proposed mechanism?

SAMPLE EXERCISE 13.13 Testing Proposed Reaction Mechanisms LO7

One proposed mechanism for the decomposition of N_2O_5 to NO_2 involves the following three elementary steps:

Step 1	$2\,N_2O_5(g) \rightleftharpoons N_4O_{10}(g)$	fast and reversible
Step 2	$N_4O_{10}(g) \rightarrow N_2O_3(g) + 2\,NO_2(g) + O_3(g)$	slow
Step 3	$N_2O_3(g) + O_3(g) \rightarrow 2\,NO_2(g) + O_2(g)$	fast
Overall:	$2\,N_2O_5(g) \rightarrow 4\,NO_2(g) + O_2(g)$	

What is the rate law for the overall reaction based on the proposed mechanism?

Collect and Organize We are asked to find the rate law for the overall reaction, which will reflect the rate laws for the elementary reactions preceding and including the rate-determining step. We are given three elementary steps and their relative rates.

Analyze The rate law for an elementary step can be written directly from the balanced chemical equation for that step. We are told that step 1 is fast *and reversible*, which means that as the reaction proceeds and $[N_4O_{10}]$ increases, the rate of step 1 in the forward direction is eventually matched by the rate of step 1 in the reverse direction. Only elementary steps 1 and 2 will contribute to the rate law for the overall reaction because step 2, the only slow step, must be the rate-determining step.

Solve The rate laws for step 1 in the forward and reverse directions are

$$\text{Rate of forward step 1} = k_f[N_2O_5]^2$$

$$\text{Rate of reverse step 1} = k_r[N_4O_{10}]$$

The rate law for step 2 is

$$\text{Rate of step 2} = k_2[N_4O_{10}]$$

N_4O_{10} is an intermediate in this reaction, so it should not appear in the overall rate law. To find the overall rate law, we need to express $[N_4O_{10}]$ in terms of $[N_2O_5]$ by setting the step 1 rates equal to each other and rearranging the terms:

$$[N_4O_{10}] = \frac{k_f}{k_r}[N_2O_5]^2$$

We then substitute this expression for $[N_4O_{10}]$ in the rate law for step 2, which is the rate-determining step, and then, combine all of the rate constants into one overall k:

$$\text{Rate} = k_2[N_4O_{10}] = k_2\frac{k_f}{k_r}[N_2O_5]^2 = k[N_2O_5]^2$$

Think About It The rate law for the overall reaction must depend on the concentration of N_2O_5. The order of the reaction in N_2O_5 depends on the proposed mechanism for the reaction. Notice that the rate of step 3 has no impact on the rate law for this mechanism because it follows the slowest step. In addition, note that the rate law for the mechanism here in Sample Exercise 13.13 does not match the experimentally determined rate law from Sample Exercise 13.4: Rate $= k[N_2O_5]$. Therefore, the mechanism that starts with the dimerization of N_2O_5 cannot be correct.

Practice Exercise Here is another proposed mechanism for the reaction of NO with H_2 (Sample Exercise 13.12):

Elementary step 1	$H_2(g) + NO(g) \rightarrow N(g) + H_2O(g)$
Elementary step 2	$N(g) + NO(g) \rightarrow N_2(g) + O(g)$
Elementary step 3	$H_2(g) + O(g) \rightarrow H_2O(g)$

Is this a valid mechanism?

Mechanisms and One Meaning of Zero Order

Before we leave reaction mechanisms, let's revisit the reaction between NO_2 and CO, which has an experimentally determined rate law that is zero order in CO, second order in NO_2, and second order overall:

$$NO_2(g) + CO(g) \rightarrow NO(g) + CO_2(g) \qquad \text{Rate} = k[NO_2]^2$$

What does this overall rate law tell us about the reaction? Remember that the overall rate law depends on the concentration of reactants in the rate-determining step, which means that CO is not a reactant in the rate-determining step. Carbon monoxide is clearly involved in the reaction—it is converted into CO_2—but whatever step involves CO must be more rapid than the rate-determining step. This leads us to conclude that the reaction must have at least two elementary steps.

It has been proposed that the reaction mechanism is the following:

$$(1)\quad 2\,NO_2(g) \rightarrow NO_3(g) + NO(g) \qquad \text{Rate} = k_1[NO_2]^2$$

$$(2)\quad NO_3(g) + CO(g) \rightarrow NO_2(g) + CO_2(g) \qquad \text{Rate} = k_2[NO_3][CO]$$

The experimentally determined overall rate law matches the rate law for the first step, which must be the slower, rate-determining step. The overall reaction is zero order in CO because CO is not a reactant in that step.

CONCEPT TEST

The rate law for the reaction $XO_2(g) + M(g) \rightarrow XO(g) + MO(g)$, where X and M represent metallic elements, is second order in $[XO_2]$ and independent of [M] for a wide variety of compounds and elements. Why can we conclude that these reactions likely proceed by the same mechanism?

13.6 Catalysts

We noted in Section 13.4 that reactions may be slow if they have a high activation energy. How can we increase the rate of such a reaction? One way is to increase the temperature of the reaction mixture. However, in some chemical reactions, elevated temperatures can lead to undesired products or to lower yields. Another way is to add a **catalyst**, a substance that increases the rate of a reaction but is not consumed in the process.

Catalysts and the Ozone Layer

Ozone in the stratosphere between 10 and 40 km above Earth's surface is necessary to protect us from UV radiation, but ozone at ground level is hazardous to our health. In this section, we discuss the role of catalysts in the breakdown of stratospheric ozone that has led to the annual formation of ozone holes over Antarctica during early spring in the Southern Hemisphere (September and October).

The natural photodecomposition of ozone in the stratosphere,

$$2\,O_3(g) \xrightarrow{\text{sunlight}} 3\,O_2(g) \qquad (13.37)$$

begins with the absorption of UV radiation from the sun and the generation of atomic oxygen:

$$(1)\quad O_3(g) \rightarrow O_2(g) + O(g)$$

The oxygen atom can react with an ozone molecule to form two more molecules of oxygen:

$$(2)\quad O_3(g) + O(g) \rightarrow 2\,O_2(g)$$

The second elementary step is slow because its activation energy is relatively high: 17.7 kJ/mol.

catalyst a substance added to a reaction that increases the rate of the reaction but is not consumed in the process.

In 1974, two American scientists, Sherwood Rowland (1927–2012) and Mario Molina (b. 1943), predicted significant depletion of stratospheric ozone because of the release of a class of volatile compounds called chlorofluorocarbons (CFCs) into the atmosphere at ground level that ultimately enter the stratosphere. This prediction was later supported by experimental evidence of a thinning of the ozone layer and formation of annual ozone holes over Antarctica. In September 2000, stratospheric ozone concentrations over Antarctica were the lowest ever observed—less than half of what they were in 1980—and the ozone hole covered nearly all of Antarctica and the tip of South America (Figure 13.25).

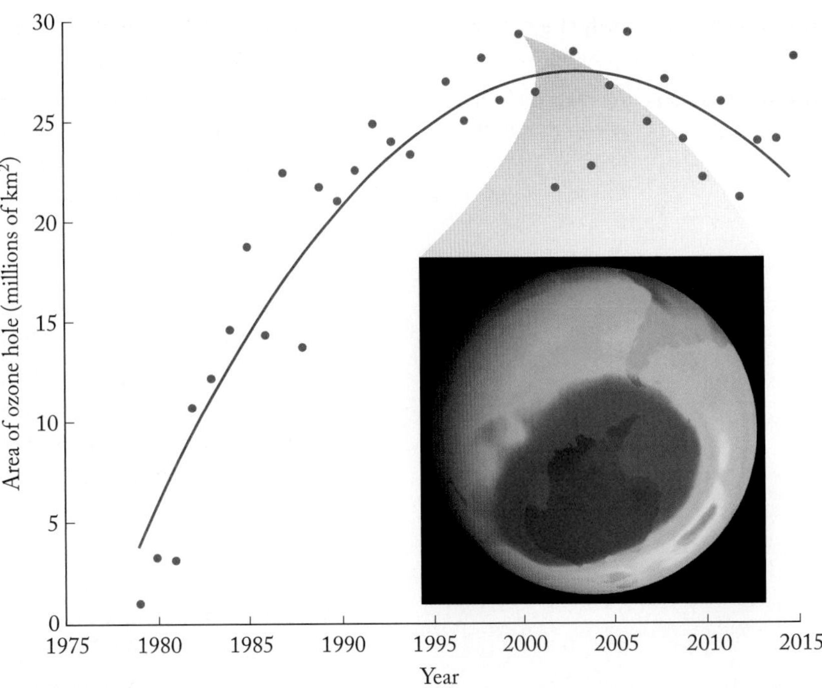

FIGURE 13.25 Trends in stratospheric ozone depletion over the South Pole. The data points represent the maximum area of the ozone hole observed during each spring in the Southern Hemisphere (usually in September or October). The satellite image shows the largest hole ever recorded, which occurred in 2000. The blue-violet color over Antarctica represents ozone concentrations that are less than half their normal values.

In 1987 an international agreement known as the Montreal Protocol called for an end to the production of ozone-depleting CFCs. The Montreal Protocol has had a dramatic effect on CFC production and emission into the atmosphere, and as the trend line in Figure 13.25 indicates, the ozone layer over Antarctica may be slowly recovering. In October 2012, the area of the ozone hole was 21.2 million km^2, down from the 2000 maximum of 29.4 million km^2.

How do CFCs contribute to the destruction of ozone? Three of the more widely used CFCs were CCl_2F_2, CCl_3F, and $CClF_3$. In the stratosphere, they encounter UV radiation with enough energy to break C—Cl bonds, releasing chlorine atoms:

$$CCl_3F(g) \rightarrow CCl_2F(g) + Cl(g) \tag{13.38}$$

Free chlorine atoms react with ozone, forming chlorine monoxide:

$$Cl(g) + O_3(g) \rightarrow ClO(g) + O_2(g) \tag{13.39}$$

Chlorine monoxide then reacts with ozone, producing oxygen and regenerating atomic chlorine:

$$ClO(g) + O_3(g) \rightarrow Cl(g) + 2\,O_2(g) \tag{13.40}$$

If we add Equations 13.39 and 13.40 and cancel species as needed, we get

$$2\,O_3(g) \rightarrow 3\,O_2(g)$$

The overall reaction is exactly the same as the natural photodecomposition of ozone in Equation 13.27. The difference is the presence of chlorine atoms in Equations 13.39 and 13.40. Chlorine atoms act as a catalyst for the destruction of ozone because they speed up the reaction but are not consumed by it. Rather, they are consumed in one elementary step but then regenerated in a later elementary step. Chlorine is a catalyst, not an intermediate, because a catalyst is consumed in an early step of a reaction mechanism and then regenerated in a later step, whereas an intermediate is produced before it is consumed. A single chlorine atom can catalyze the destruction of hundreds to thousands of stratospheric O_3 molecules before it combines with other atoms and forms a less reactive molecule.

The activation energy for the Cl-catalyzed reaction is only 2.2 kJ/mol, whereas the activation energy for the uncatalyzed reaction is 17.7 kJ/mol (Figure 13.26).

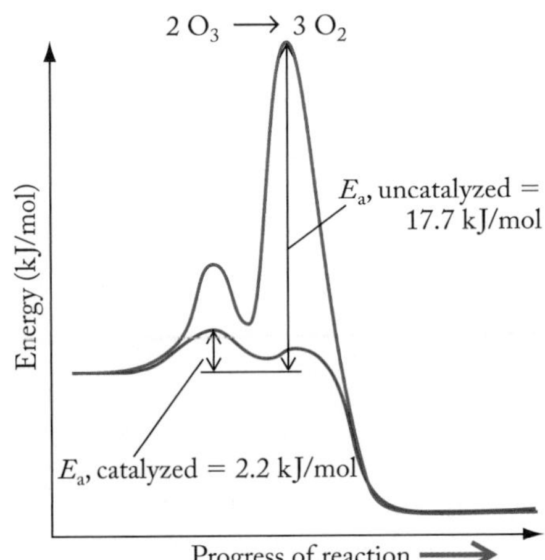

FIGURE 13.26 The decomposition of O_3 in the presence of chlorine atoms has a lower activation energy (2.2 kJ/mol) than the naturally occurring photodecomposition of O_3 to O_2 (17.7 kJ/mol). The catalytic effect of chlorine is a key factor in the depletion of stratospheric ozone and the formation of an ozone hole over the South Pole.

homogeneous catalyst a catalyst in the same phase as the reactants.

heterogeneous catalyst a catalyst in a different phase than the reactants.

Its smaller E_a value means that the catalyzed destruction of ozone is faster than the natural photodecomposition process. In the reaction describing the destruction of ozone, the catalyst, Cl(g), and reactant, $O_3(g)$, exist in the same physical phase. When a catalyst and the reacting species are in the same phase, the catalyst is a **homogeneous catalyst**.

Figure 13.26 illustrates another important feature of catalyzed reactions. Note that there are two activation energies for the catalyzed reaction (red line) while there is only a single E_a for the uncatalyzed reaction (blue line). The two energy barriers separated by an energy minimum for the catalyzed reaction represents a change in the mechanism for the Cl catalyzed decomposition of ozone. This is consistent with the formation of ClO as an intermediate in Equations 13.39 and 13.40.

The rate of ozone depletion accelerates over the South Pole toward the end of the Southern Hemisphere winter because of the clouds that cover much of Antarctica. These clouds form as crystals of ice and tiny drops of nitric acid in the winter when temperatures in the lower stratosphere dip to −80°C. The clouds become collection sites for chlorine compounds formed by the photodecomposition of CFCs, and they catalyze the transformation of these compounds into molecular chlorine, Cl_2. During late winter and early spring, the sun rises over Antarctica and its UV rays decompose Cl_2 into free Cl atoms. As we have seen, atomic chlorine is a powerful catalyst for ozone destruction, and high concentrations of newly formed atomic Cl mean rapid and widespread loss of stratospheric O_3 over Antarctica and the Southern Ocean (Figure 13.25).

SAMPLE EXERCISE 13.14 Identifying Catalysts in Reaction Mechanisms — LO8

A reaction mechanism proposed for the decomposition of ozone in the presence of NO at high temperatures consists of three elementary steps:

$$(1)\quad O_3(g) + NO(g) \rightarrow O_2(g) + NO_2(g)$$

$$(2)\quad NO_2(g) \rightarrow NO(g) + O(g)$$

$$(3)\quad O(g) + O_3(g) \rightarrow 2\,O_2(g)$$

If the rate of the overall reaction is higher than the rate of the uncatalyzed decomposition of ozone to oxygen (Equation 13.39), is NO a catalyst in the reaction?

Collect and Organize We are asked to determine whether NO is a catalyst in a reaction. A catalyst increases the rate of a reaction and is not consumed by the overall reaction. We are given the elementary steps of the reaction and are told that the reaction is more rapid in the presence of NO.

Analyze We can sum the reactions to determine the overall reaction. If NO is consumed in an early step before it is regenerated in a later one and it is not consumed in the overall process, it is a catalyst.

Solve Summing the three elementary steps gives the overall reaction:

$$O_3(g) + \cancel{NO(g)} + \cancel{NO_2(g)} + \cancel{O(g)} + O_3(g) \rightarrow$$
$$O_2(g) + \cancel{NO_2(g)} + \cancel{NO(g)} + \cancel{O(g)} + 2\,O_2(g)$$

$$2\,O_3(g) \rightarrow 3\,O_2(g)$$

This equation does not include NO, and the rate of the reaction is higher when NO is present. Thus, NO fulfills both requirements for being a catalyst.

Think About It NO behaves much like the Cl atoms in Equations 13.39 and 13.40. NO is not an intermediate because it is consumed *before* it is produced.

Practice Exercise The combustion of fossil fuels results in the release of SO_2 into the atmosphere, where it reacts with oxygen to form SO_3:

$$2\ SO_2(g) + O_2(g) \rightarrow 2\ SO_3(g)$$

In the atmosphere, SO_2 may react with NO_2, forming SO_3 and NO:

$$NO_2(g) + SO_2(g) \rightarrow NO(g) + SO_3(g)$$

The rate of reaction of SO_2 with NO_2 is faster than the rate of reaction of SO_2 with oxygen. If the NO produced in the reaction of NO_2 and SO_2 is then oxidized to NO_2,

$$2\ NO(g) + O_2(g) \rightarrow 2\ NO_2(g)$$

is NO_2 a catalyst in the reaction of SO_2 with O_2?

Catalytic Converters

We began this chapter with a discussion of air pollution caused by vehicles and of the technology that has been developed to clean the air. Figure 13.27 shows a catalytic converter in a car's exhaust system and how it provides a surface on which NO in engine exhaust decomposes to N_2 and O_2. A catalytic converter is an example of a **heterogeneous catalyst**, where reactants and catalyst have different phases. To maximize the surface area inside a catalytic converter, it is filled with a fine honeycomb mesh coated with transition metals such as palladium, platinum, and rhodium. The metals that make up catalytic coatings are first dissolved as metal salts and dispersed on the mesh, where they are reduced to metallic clusters 2 to 10 nm in diameter. The small size of these clusters and the large number of them add to the total surface area available to promote the removal of pollutants. Some of these clusters specifically speed up the decomposition of NO, while others promote the oxidation of carbon monoxide to CO_2 and unburned hydrocarbons to CO_2 and water vapor.

Many gas-phase reactions that occur on solid heterogeneous catalysts exhibit zero-order reaction kinetics. This happens when there are more reactant molecules in the system than there are catalytic surface sites available to promote their decomposition. Under these conditions, adding more reactant has no effect on the catalyzed reaction rate, and the reaction appears to be zero order.

In the reactions responsible for the production of photochemical smog, we saw that determinations of the rates of chemical reactions are crucial for us to comprehend processes on a molecular level. The mechanisms of the reactions of volatile

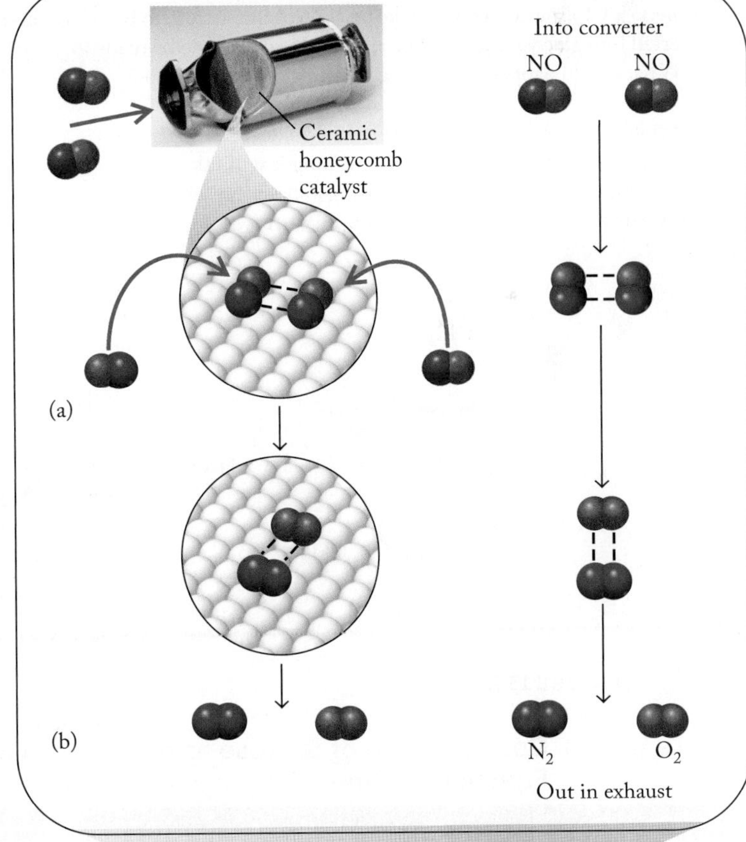

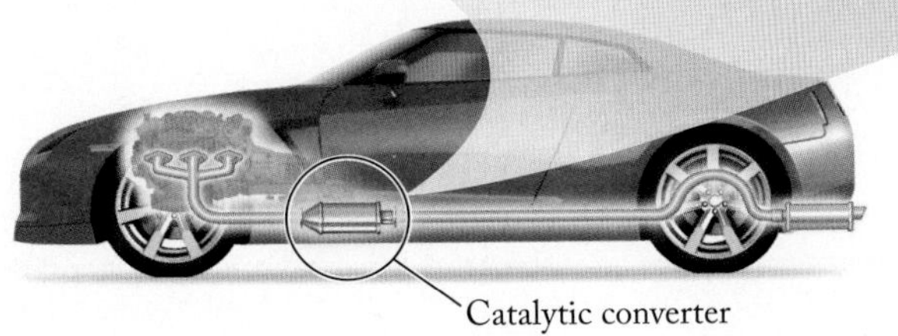

FIGURE 13.27 Catalytic converters in automobiles reduce emissions of NO by lowering the activation energy of its decomposition into N_2 and O_2. Metal catalysts are supported on a porous ceramic honeycomb. At the molecular level, (a) NO molecules are adsorbed onto the surface of metal clusters where their NO bonds are broken, and (b) pairs of O atoms and N atoms form O_2 and N_2. The O_2 and N_2 desorb from the surface and are released to the atmosphere.

oxides produced during combustion and by other natural events must be thoroughly understood so that problems arising because of the presence of these substances in the environment can be effectively managed. The continued development of catalytic converters for vehicles to diminish the problems caused by burning fossil fuels and to clean our air requires a thorough knowledge of the kinetics and mechanisms of many of the reactions discussed here.

SAMPLE EXERCISE 13.15 Integrating Concepts: Chocolate-Covered Cherries

Chocolate-covered cherries consist of a solid chocolate shell enclosing a sweet liquid that surrounds a cherry. To make chocolate-covered cherries, the fruits are initially covered with fondant, a dough-like mixture of powdered sugar (sucrose), butter, cherry juice, and the enzyme invertase. They are then dipped in chocolate and stored at 5°C.

During storage, sucrose (table sugar) hydrolyzes to produce two isomeric sugars, glucose and fructose, as shown in Figure 13.28. Invertase acts as a catalyst for this reaction. The mixture of glucose and fructose is called invert sugar, and it is a liquid. Table 13.18(a) lists data on the concentration of sucrose in the fondant as a function of time at 5°C. How long will it take for 95% of the sucrose to be converted into glucose and fructose and for the cherries to be ready to eat? What would happen if invertase were left out of the recipe?

Collect and Organize We are given data on decreasing reactant concentration with time and asked how long it will take for 95% of the reactant to be consumed. We must also describe the impact of leaving out a catalyst from the reaction mixture.

Analyze The sucrose data can be used to determine the order of the hydrolysis reaction and to calculate the value of its rate constant at 5°C. This k value can then be used in the integrated rate law with the available data to derive a rate law for the reaction and to estimate the time we have to wait for the cherries to be ready. The reaction takes place in an aqueous solution, so we may assume that the concentration of water is very large and does not change during the course of the reaction. Therefore, any change in the rate of the reaction is due only to the consumption of sucrose. Inspection of the data reveals that half of the initial concentration of sucrose is consumed between 48 and 96 hours or about 3 days of reaction time. For 95% of the sugar to be gone, a time equivalent to between 4 and 5 half-lives, or about 2 weeks, will be needed.

Solve Let's start by testing whether the reaction is first order in sucrose. If it is, a plot of ln[sucrose] versus time (Table 13.18b) will be linear.

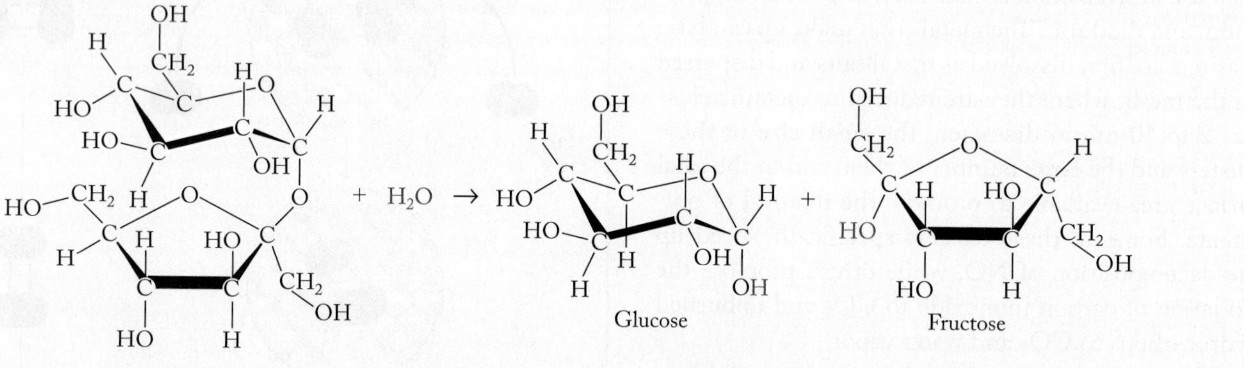

Sucrose + H_2O → Glucose + Fructose

FIGURE 13.28

TABLE 13.18(a) Concentration of Sucrose as a Function of Time

Time (h)	[Sucrose] in Fondant (g/cm³)
0	0.844
24	0.681
48	0.550
96	0.357
192	0.153

TABLE 13.18(b) Concentration of Sucrose and ln[Sucrose] as a Function of Time

Time (h)	[Sucrose] in Fondant (g/cm³)	ln[Sucrose]
0	0.844	−0.170
24	0.681	−0.384
48	0.550	−0.598
96	0.357	−1.030
192	0.153	−1.877

Plotting ln[sucrose] versus t does indeed yield a straight line (Figure 13.29), which means the reaction is first order in sucrose. The slope of the line that best fits the data points is $-0.00889\ h^{-1}$, which equals $-k$, so the rate law at 5°C is

$$\text{Rate} = (0.00889\ h^{-1})[\text{sucrose}]$$

Let's use this k value in the integrated rate law for a first-order reaction (Equation 13.16)

$$\ln\left(\frac{[\text{sucrose}]_t}{[\text{sucrose}]_0}\right) = -kt$$

to calculate the time t necessary for 95% of the sucrose to react (leaving 5%, or $0.05 \times [\text{sucrose}]_0$, unreacted):

$$t = \frac{1}{-k}\ln\left(\frac{[\text{sucrose}]_t}{[\text{sucrose}]_0}\right) = \frac{1}{-0.00889\ h^{-1}}\ln\left(\frac{0.05\ \cancel{[\text{sucrose}]_0}}{\cancel{[\text{sucrose}]_0}}\right)$$

$$= 337\ h$$

$$\left(337\ \cancel{h} \times \frac{1\ d}{24\ \cancel{h}}\right) = 14\ d\ (\text{or 2 weeks})$$

If the catalyst had not been included, the reaction rate would have been slower and the time to reach 95% conversion of sucrose to glucose and fructose would have been a lot longer.

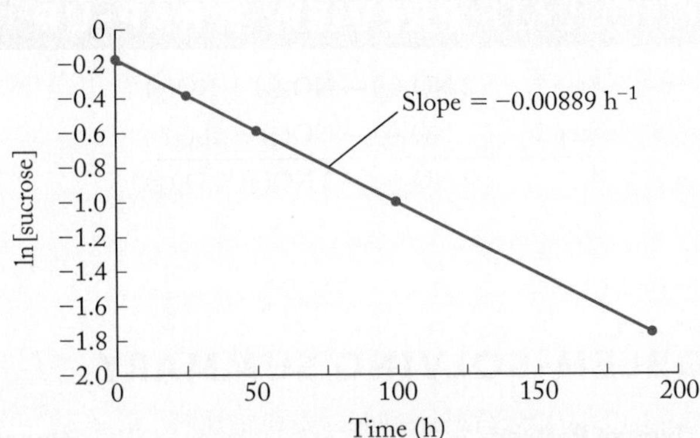

FIGURE 13.29 Plot of ln[sucrose] as a function of time.

Think About It The calculated reaction time agrees with our original estimate based on the approximate number of half-lives required to reach 95% conversion. The reaction was run at refrigerator temperatures, probably to preserve the cherries and keep the chocolate solid, but that likely slowed the rate of hydrolysis. Running the reaction at higher temperatures might have sped up the reaction but resulted in melted chocolate and spoiled cherries.

SUMMARY

LO1 The relative rates of disappearance of reactants and appearance of products are related by the stoichiometry of the reaction. (Sections 13.1 and 13.2)

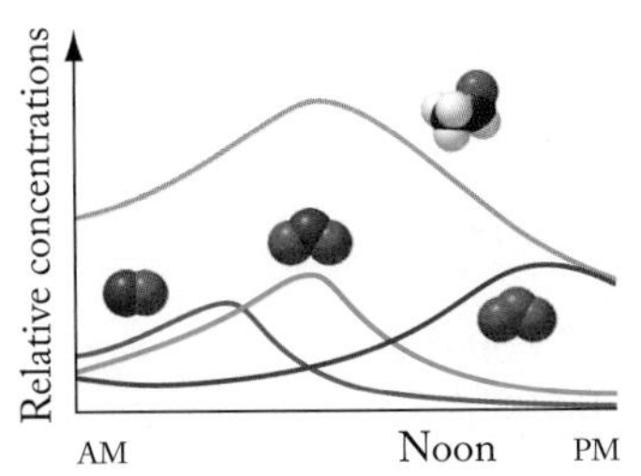

LO2 The overall rate of a reaction can be expressed as either an average rate or an instantaneous rate. **Reaction rates** are typically determined from experimental measurements. In most reactions, the reaction rate decreases as the reaction proceeds. (Section 13.2)

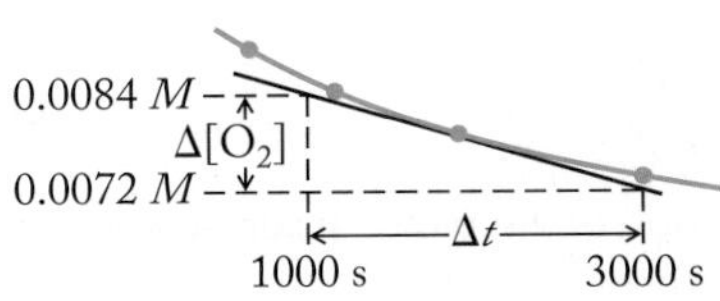

LO3 The dependence of the rate of a reaction X + Y → Z on reactant concentrations is expressed in the **rate law** for the reaction: Rate = $k[X]^m[Y]^n$, where m and n are the **reaction order** in reactants X and Y, respectively, and k is the **rate constant**. The units of a rate constant depend on the **overall reaction order**, which is the sum of the reaction orders with respect to individual reactants. (Section 13.3)

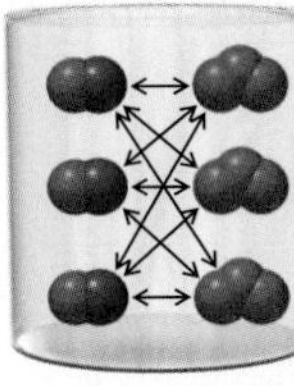

LO4 The order of a reaction and the rate law for the reaction can be determined from differences in the initial rates of reaction observed with different concentrations of reactants or from single kinetics experiments using **integrated rate laws**. (Section 13.3)

LO5 The **half-life ($t_{1/2}$)** of a reactant is the time required for the concentration of a reactant to decrease to one-half its starting concentration. (Section 13.3)

LO6 Increasing the temperature of a chemical reaction increases its rate. The **activation energy (E_a)** of a reaction is the minimum energy molecules need to react when they collide. Reactions with large activation energies are usually slow. Measuring the rate constant of a reaction at different temperatures allows the calculation of activation energies using the **Arrhenius equation**. (Section 13.4)

LO7 Rate studies give insight into **reaction mechanisms**, which describe what is happening at the molecular level. A reaction mechanism consists of one or more **elementary steps** that describe how the reaction takes place. The overall reaction is the sum of these elementary steps. The rate law for a reaction applies to the slowest elementary step, which is called the **rate-determining step**. The proposed mechanism for any reaction must be consistent with the observed rate law and with the stoichiometry of the overall reaction. (Section 13.5)

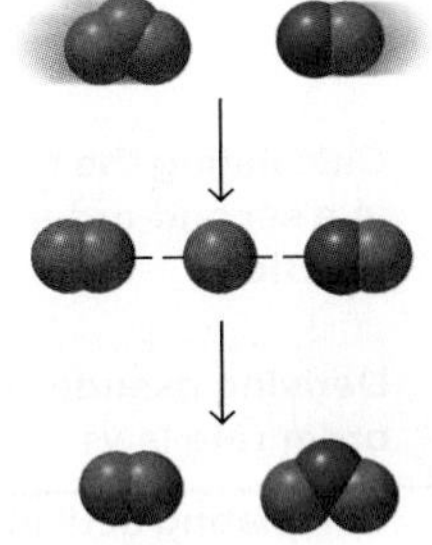

LO8 Catalysts increase reaction rates by changing the mechanism of a reaction and by decreasing the activation energy. A **homogeneous catalyst** is one that is in the same phase as the reactants in the reaction being catalyzed. A **heterogeneous catalyst** is one that is in a phase different from the phase of the reactant. (Section 13.6)

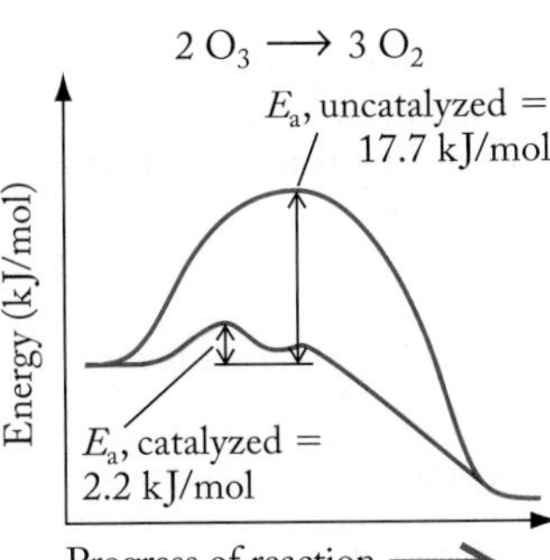

PARTICULATE PREVIEW WRAP-UP

Step 1:	$2\,NO_2(g) \rightarrow NO_3(g) + NO(g)$
Step 2:	$NO_3(g) \rightarrow NO(g) + O_2(g)$
	$2\,NO_2(g) \rightarrow 2\,NO(g) + O_2(g)$

The dashed lines represent the simultaneous breaking of bonds in the reactants and forming of bonds in the intermediate and/or products. The NO_3 molecule produced in step 1 is an intermediate that decomposes in step 2.

PROBLEM-SOLVING SUMMARY

Type of Problem	Concepts and Equations	Sample Exercises
Predicting relative reaction rates	Determine relative rates from the stoichiometry of the balanced chemical equation. For aA + bB → cC + dD $$\text{Rate} = -\frac{1}{a}\frac{\Delta[A]}{\Delta t} = -\frac{1}{b}\frac{\Delta[B]}{\Delta t} = \frac{1}{c}\frac{\Delta[C]}{\Delta t} = \frac{1}{d}\frac{\Delta[D]}{\Delta t}$$	**13.1, 13.2**
Determining an instantaneous reaction rate	Determine the slope of a line tangent to a point on the plot of concentration versus time.	**13.3**
Deriving a rate law from initial reaction rate data	Compare the change in rate as the concentration of each reactant is changed. $$m = \frac{\log\left(\frac{\text{Rate}_2}{\text{Rate}_1}\right)}{\log\left(\frac{[A]_2}{[A]_1}\right)} \quad (13.14)$$	**13.4**
Distinguishing between first- and second-order reactions	A linear plot of the natural logarithm of reactant concentration (ln[X]) versus time indicates a first-order reaction, whereas a linear plot of the reciprocal of reactant concentration (1/[X]) versus time indicates a second-order reaction.	**13.5, 13.7**
Relating reactant concentration, reaction time, and half-life	For first-order reactions: $$\ln\frac{[X]_t}{[X]_0} = -kt \quad (13.16)$$ $$\ln[X]_t = -kt + \ln[X]_0 \quad (13.17)$$ $$t_{1/2} = \frac{0.693}{k} \quad (13.19)$$	**13.6**
Calculating the half-life in a second-order reaction	In a second-order reaction, $$t_{1/2} = \frac{1}{k[X]_0} \quad (13.24)$$	**13.8**
Deriving pseudo-first-order rate laws	Plot ln [limiting reactant] versus t. The slope of the plot = [excess reactant] × k.	**13.9**
Calculating activation energies and frequency factors	Determine k values at different temperatures. Then plot ln k versus $1/T$ and solve for E_a and A: $$\ln k = -\frac{E_a}{R}\left(\frac{1}{T}\right) + \ln A \quad (13.29)$$	**13.10**
Calculating rate constants for a reaction at different temperatures	Solve for k_2 in $$\ln\frac{k_1}{k_2} = \frac{E_a}{R}\left(\frac{1}{T_2} - \frac{1}{T_1}\right) \quad (13.30)$$	**13.11**
Linking reaction mechanisms to experimental rate laws	The chemical equations describing the steps in the mechanism must sum to the equation that describes the overall reaction; the rate law of the rate-determining step must be consistent with the rate law of the overall reaction.	**13.12, 13.13**
Identifying catalysts in reaction mechanisms	A homogeneous catalyst is consumed in an early step and regenerated in a later one.	**13.14**

VISUAL PROBLEMS

(Answers to boldface end-of-chapter questions and problems are in the back of the book.)

13.1. Nitrous oxide decomposes to nitrogen and oxygen in the following reaction:

$$2\ N_2O(g) \rightarrow 2\ N_2(g) + O_2(g)$$

In Figure P13.1, which curve represents $[N_2O]$ and which curve represents $[O_2]$?

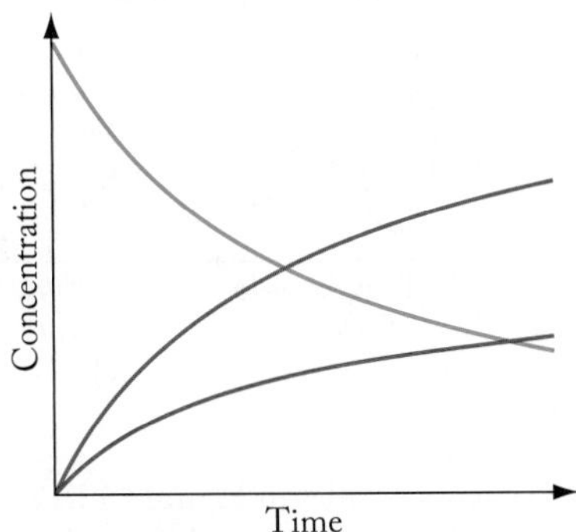

FIGURE P13.1

13.2. Sulfur trioxide is formed in the reaction

$$SO_2(g) + \tfrac{1}{2}\,O_2(g) \rightarrow SO_3(g)$$

In Figure P13.2, which curve represents $[SO_2]$ and which curve represents $[O_2]$? All three gases are present initially.

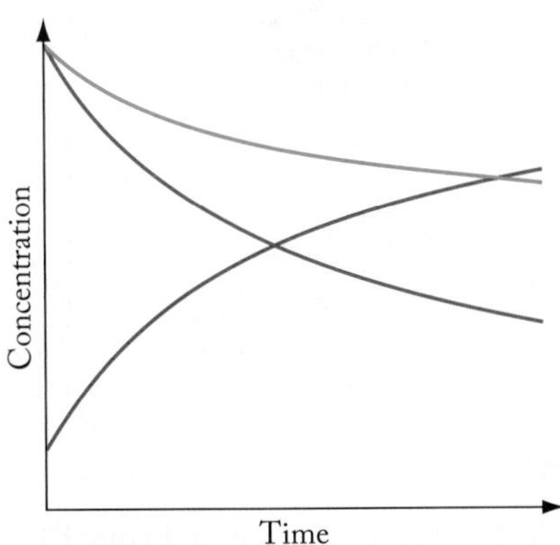

FIGURE P13.2

13.3. The rate law for the reaction $2\ A \rightarrow B$ is second order in A. Figure P13.3 represents samples with different concentrations of A, represented by red spheres. In which sample will the reaction $A \rightarrow B$ proceed most rapidly?

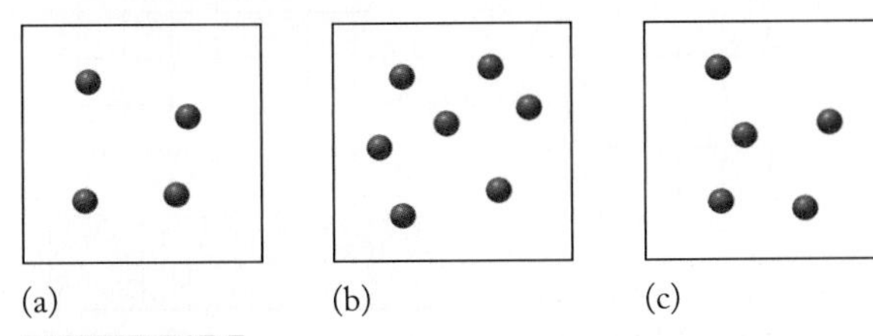

FIGURE P13.3

13.4. The rate law for the reaction $A + B \rightarrow C$ is first order in both A and B. Figure P13.4 represents samples with different concentrations of A (red spheres) and B (blue spheres). In which sample will the reaction $A + B \rightarrow C$ proceed most rapidly?

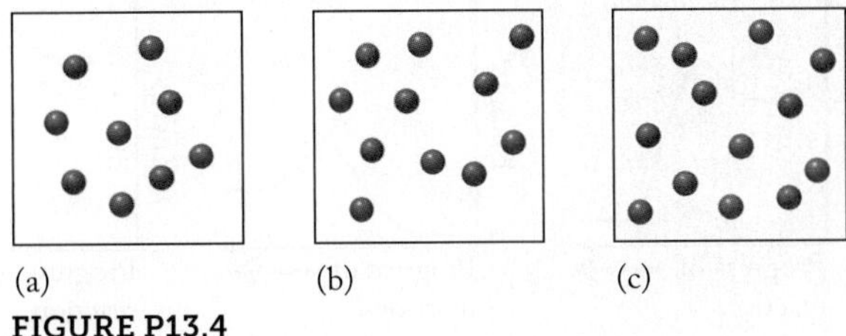

FIGURE P13.4

13.5. Using the numbers 1–5 on the reaction profile (Figure P13.5), identify the following:

a. The energy of the reactants
b. The energy of the products
c. The activation energy of the forward reaction
d. The activation energy of the reverse reaction
e. The energy change of the reaction

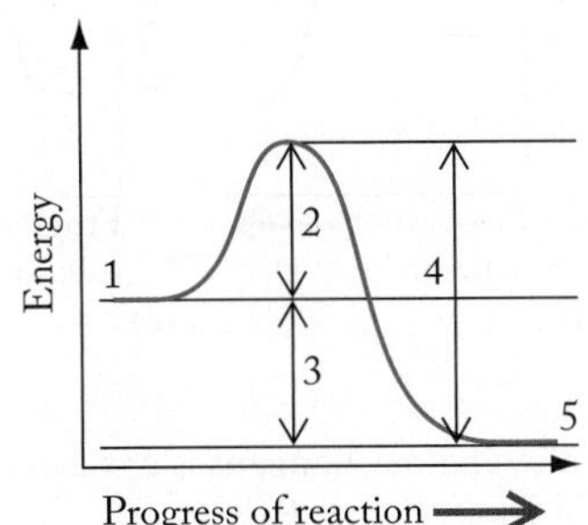

FIGURE P13.5

13.6. Select the energy profile from those given in Figure P13.6 that best corresponds to the following:

a. A highly exothermic reaction with a large activation energy
b. A highly endothermic reaction with a large activation energy
c. A reaction involving a stable intermediate

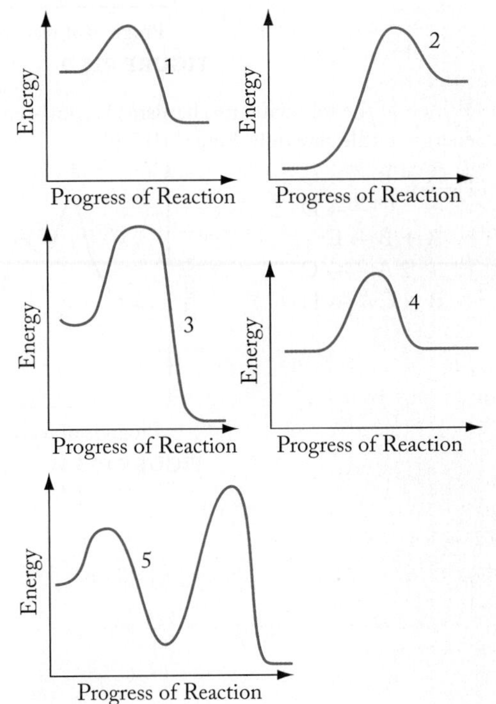

FIGURE P13.6

13.7. Which of the energy profiles in Figure P13.7 represents the reaction with the smallest rate constant at constant T?

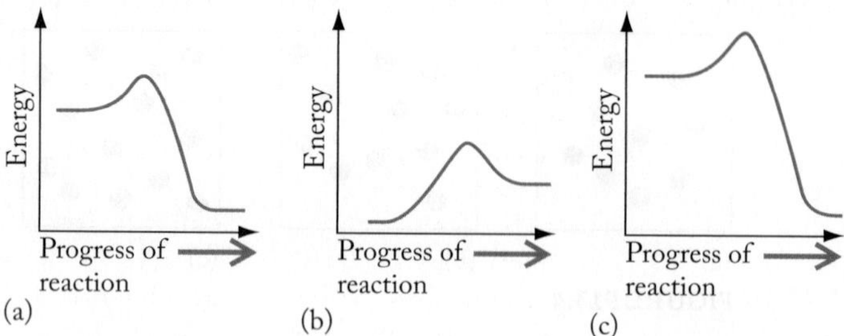

FIGURE P13.7

13.8. Which of the energy profiles in Figure P13.8 represents the reaction with the largest rate constant at constant T?

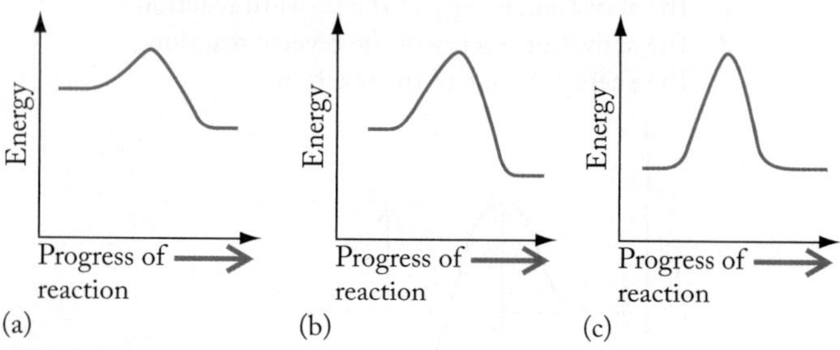

FIGURE P13.8

13.9. Which of the following mechanisms is consistent with the energy profile shown in Figure P13.9?

a. $2\,A \xrightarrow{\text{slow}} B$
   $B \xrightarrow{\text{fast}} C$

b. $A + B \rightarrow C$

c. $2\,A \underset{}{\overset{\text{fast}}{\rightleftharpoons}} B$
   $B \xrightarrow{\text{slow}} C$

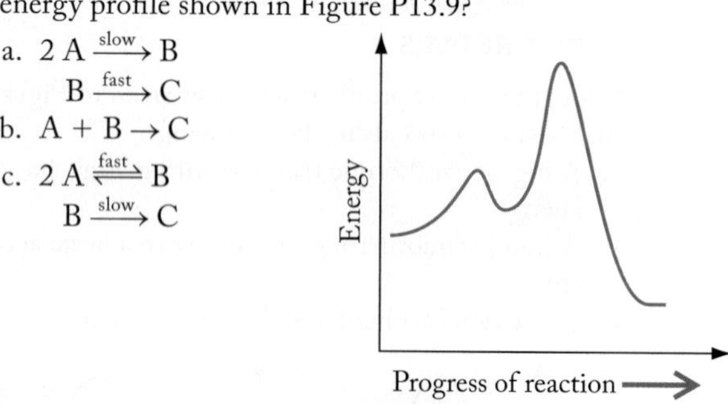

FIGURE P13.9

13.10. Which of the following mechanisms is consistent with the energy profile shown in Figure P13.10?

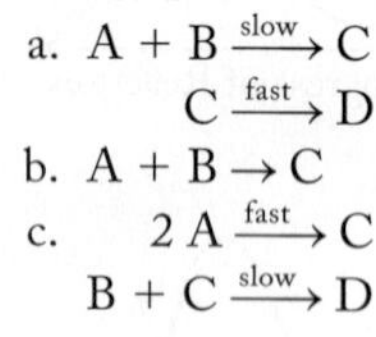

a. $A + B \xrightarrow{\text{slow}} C$
   $C \xrightarrow{\text{fast}} D$

b. $A + B \rightarrow C$

c. $2\,A \xrightarrow{\text{fast}} C$
   $B + C \xrightarrow{\text{slow}} D$

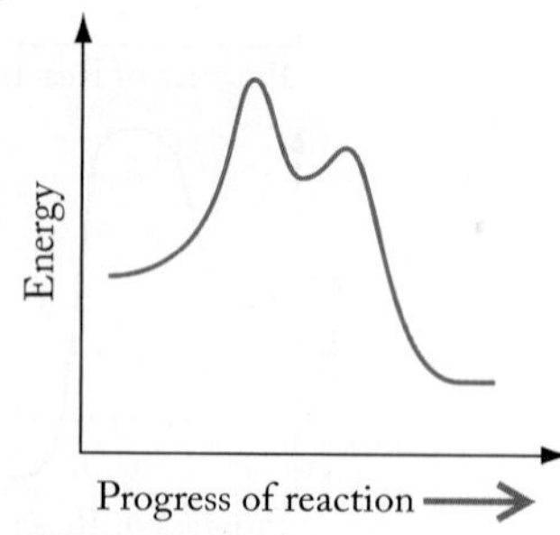

FIGURE P13.10

13.11. Which of the energy profiles in Figure P13.11 represents the effect of a catalyst on the rate of a reaction?

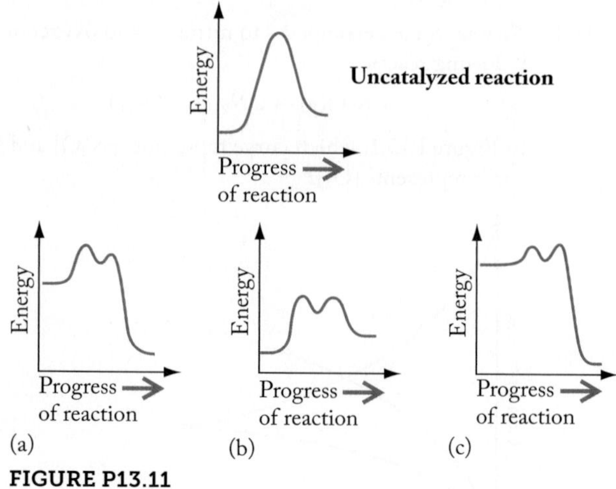

FIGURE P13.11

13.12. Explain why the energy profiles in Figure P13.12 (a) and (b) do not represent the effect of a catalyst on the rate of the reaction?

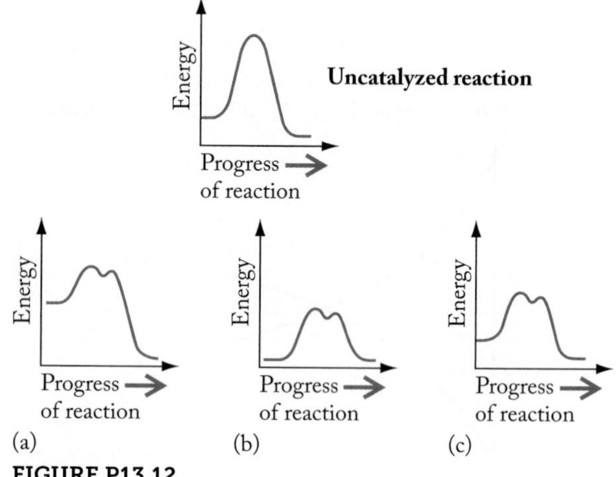

FIGURE P13.12

13.13. Which of the highlighted elements in Figure P13.13 forms volatile oxides associated with the formation of photochemical smog?

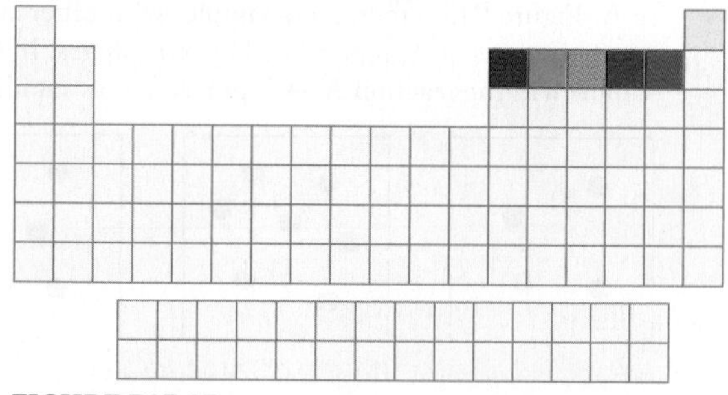

FIGURE P13.13

13.14. Which of the highlighted elements in Figure P13.13 forms noxious oxides that are removed from automobile exhaust as it passes through a catalytic converter?

13.15. Which of the highlighted elements in Figure P13.15 are widely used as heterogeneous catalysts?

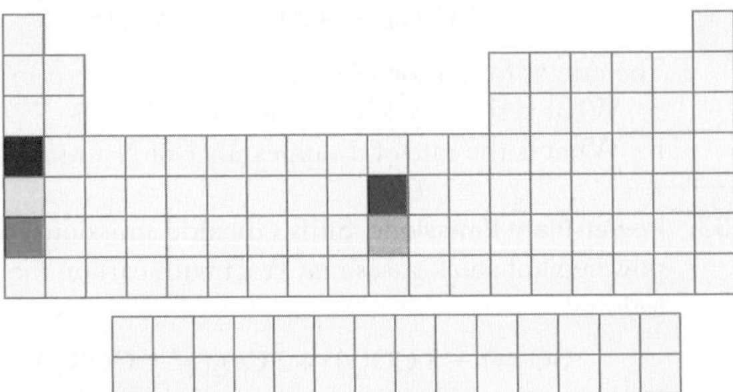

FIGURE P13.15

13.16. Use representations [A] through [I] in Figure P13.16 to answer questions a–f.

a. Which two elementary steps, when combined, would involve an intermediate of nitrogen trioxide?
b. What is the overall reaction for the two elementary steps in Problem 13.16(a)?
c. Which representation shows the photodecomposition of a chlorofluorocarbon?
d. Which two elementary steps, when combined, depict the chlorine-catalyzed destruction of ozone?
e. Which representation is the overall reaction for the chlorine-catalyzed destruction of ozone?
f. What happens to the number of collisions between chlorine radicals and ozone between representations [C], [F], and [I]?

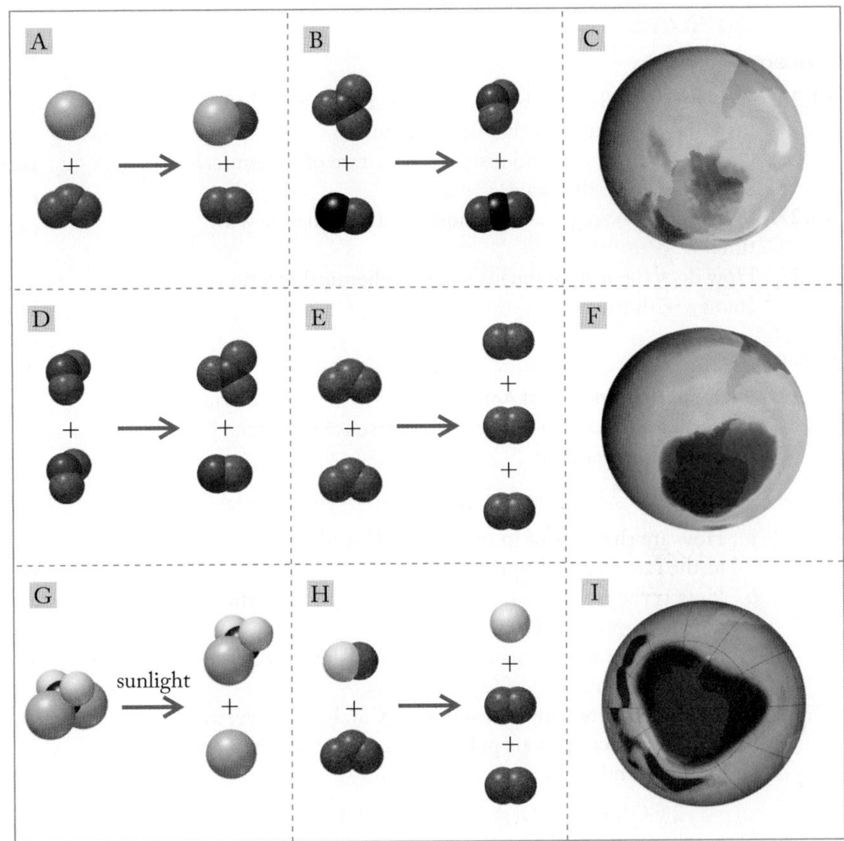

FIGURE P13.16

QUESTIONS AND PROBLEMS

Cars, Trucks, and Air Quality

Concept Review

13.17. Why does the maximum concentration of ozone in Figure 13.1 occur much later in the day than the maximum concentration of NO and NO_2?

13.18. If we plot the concentration of reactants and products as a function of time for any sequence of two spontaneous chemical reactions, such as

$$A \rightarrow B \rightarrow C$$

will the maximum concentration of final product C always appear after the maximum concentration of B?

13.19. Why isn't there an increase in NO concentration after the evening rush hour (see Figure 13.1)?

13.20. If ozone can react with NO to form NO_2, why does the ozone concentration reach a maximum in the early afternoon (see Figure 13.1)?

Problems

13.21. Using data in Appendix 4, calculate ΔH° for the reaction

$$2\ NO(g) + O_2(g) \rightarrow 2\ NO_2(g)$$

13.22. Using data in Appendix 4, calculate ΔH° for the reaction

$$O_3(g) + NO(g) \rightarrow O_2(g) + NO_2(g)$$

13.23. Nitrogen and oxygen can combine to form different nitrogen oxides that play a minor role in the chemistry of smog. Write balanced chemical equations for the reactions of N_2 and O_2 that produce (a) N_2O and (b) N_2O_5.

13.24. Nitrogen oxides such as N_2O and N_2O_5 are present in the air in low concentrations, in part because of their reactivity. Write balanced chemical equations for the following reactions:

a. The conversion of N_2O to NO_2 in the presence of oxygen
b. The decomposition of N_2O_5 to NO_2 and O_2

Reaction Rates

Concept Review

13.25. Explain the difference between the average rate and the instantaneous rate of a chemical reaction.

13.26. Can the average rate and instantaneous rate of a chemical reaction ever be the same?

13.27. Why do the average rates of most reactions change with time?

13.28. How does the instantaneous rate of a chemical reaction change with time?

Problems

13.29. **Bacterial Degradation of Ammonia** *Nitrosomonas* bacteria convert ammonia into nitrite in the presence of oxygen by the following reaction:

$$2\ NH_3(aq) + 3\ O_2(g) \rightarrow 2\ H^+(aq) + 2\ NO_2^-(aq) + 2\ H_2O(\ell)$$

a. How are the rates of formation of H^+ and NO_2^- related to the rate of consumption of NH_3?
b. How is the rate of formation of NO_2^- related to the rate of consumption of O_2?
c. How is the rate of consumption of NH_3 related to the rate of consumption of O_2?

13.30. **Catalytic Converters and Combustion** Catalytic converters in automobiles combat air pollution by converting NO and CO into N_2 and CO_2:

$$2\ CO(g) + 2\ NO(g) \rightarrow N_2(g) + 2\ CO_2(g)$$

a. How is the rate of formation of N_2 related to the rate of consumption of CO?
b. How is the rate of formation of CO_2 related to the rate of consumption of NO?
c. How is the rate of consumption of CO related to the rate of consumption of NO?

13.31. Nitryl chloride, NO_2Cl, is a reactive chlorine-containing species sometimes found in marine sediments in industrial areas. In the gas phase it decomposes to NO_2 and Cl_2:

$$2\ NO_2Cl(g) \rightarrow 2\ NO_2(g) + Cl_2(g)$$

Under laboratory conditions, the rate of formation of $NO_2(g)$ is $5.7 \times 10^{-6}\ M \cdot s^{-1}$.

a. What is the rate of formation of $Cl_2(g)$?
b. What is the rate of consumption of $NO_2Cl(g)$?

13.32. N_2O_5, when dissolved in water, decomposes to produce NO_2 and O_2:

$$2\ N_2O_5(aq) \rightarrow 4\ NO_2(aq) + O_2(aq)$$

The rate of formation of O_2 is $4.0 \times 10^{-3}\ M \cdot \text{min}^{-1}$.

a. What is the rate of formation of $NO_2(aq)$?
b. What is the rate of disappearance of $N_2O_5(aq)$?

13.33. **Power-Plant Emissions** Sulfur dioxide emissions in power-plant stack gases may react with carbon monoxide as follows:

$$SO_2(g) + 3\ CO(g) \rightarrow 2\ CO_2(g) + COS(g)$$

Write an equation relating each of the following pairs of rates:

a. The rate of formation of CO_2 to the rate of consumption of CO
b. The rate of formation of COS to the rate of consumption of SO_2
c. The rate of consumption of CO to the rate of consumption of SO_2

13.34. **Reducing Nitric Oxide Emissions from Power Plants** Nitric oxide (NO) can be removed from gas-fired power-plant emissions by reaction with methane as follows:

$$CH_4(g) + 4\ NO(g) \rightarrow 2\ N_2(g) + CO_2(g) + 2\ H_2O(g)$$

Write an equation relating each of the following pairs of rates:

a. The rate of formation of N_2 to the rate of formation of CO_2
b. The rate of formation of CO_2 to the rate of consumption of NO
c. The rate of consumption of CH_4 to the rate of formation of H_2O

13.35. **Stratospheric Ozone Depletion** Chlorine monoxide (ClO) plays a major role in the creation of the ozone holes in the stratosphere over Earth's polar regions.

a. If $\Delta[ClO]/\Delta t = -2.3 \times 10^7\ M/s$ at 298 K, what is the rate of change in $[Cl_2]$ and $[O_2]$ in the following reaction?

$$2\ ClO(g) \rightarrow Cl_2(g) + O_2(g)$$

b. If $\Delta[ClO]/\Delta t = -2.9 \times 10^4\ M/s$, what is the rate of formation of oxygen and ClO_2 in the following reaction?

$$ClO(g) + O_3(g) \rightarrow O_2(g) + ClO_2(g)$$

13.36. The chemistry of smog formation includes NO_3 as an intermediate in several reactions.

a. If $\Delta[NO_3]/\Delta t = -2.2 \times 10^5\ mM/\text{min}$ in the following reaction, what is the rate of formation of NO_2?

$$NO_3(g) + NO(g) \rightarrow 2\ NO_2(g)$$

b. What is the rate of change of $[NO_2]$ in the following reaction if $\Delta[NO_3]/\Delta t = -2.3\ mM/\text{min}$?

$$2\ NO_3(g) \rightarrow 2\ NO_2(g) + O_2(g)$$

13.37. Nitrite ion reacts with ozone in aqueous solution, producing nitrate ion and oxygen:

$$NO_2^-(aq) + O_3(g) \rightarrow NO_3^-(aq) + O_2(g)$$

The following data were collected for this reaction at 298 K. Calculate the average reaction rate between 0 and 100 μs (microseconds) and between 200 and 300 μs.

Time (μs)	$[O_3]$ (*M*)
0	1.13×10^{-2}
100	9.93×10^{-3}
200	8.70×10^{-3}
300	8.15×10^{-3}

13.38. At room temperature in the gas phase, dinitrogen pentoxide (N_2O_5) decomposes to dinitrogen tetroxide and oxygen:

$$2\,N_2O_5(g) \rightarrow 2\,N_2O_4(g) + O_2(g)$$

Calculate the average rate of this reaction between consecutive measurements listed in the following table.

Time (s)	$[N_2O_5]$ (*M*)
0	0.200
300	0.180
600	0.161
900	0.144
1200	0.130

13.39. The following data were collected for the dimerization of ClO to Cl_2O_2 at 298 K.

Time (s)	[ClO] (molecules/cm^3)
0	2.60×10^{11}
1	1.08×10^{11}
2	6.83×10^{10}
3	4.99×10^{10}
4	3.93×10^{10}
5	3.24×10^{10}
6	2.76×10^{10}

Plot [ClO] and $[Cl_2O_2]$ as a function of time and determine the instantaneous rates of change in both at 1 s.

13.40. **Tropospheric Ozone** Tropospheric (lower atmosphere) ozone is rapidly consumed in many reactions, including

$$O_3(g) + NO(g) \rightarrow NO_2(g) + O_2(g)$$

Use the following data to calculate the instantaneous rate of the reaction at $t = 0.000$ s and $t = 0.052$ s.

Time (s)	[NO] (*M*)
0.000	2.0×10^{-8}
0.011	1.8×10^{-8}
0.027	1.6×10^{-8}
0.052	1.4×10^{-8}
0.102	1.2×10^{-8}

Effect of Concentration on Reaction Rate

Concept Review

13.41. The decomposition of A is second order in A. What effect does doubling the concentration of A have on the rate constant?

13.42. Why are the units of the rate constants different for reactions of different order?

13.43. Can the half-life of a second-order reaction have the same units as the half-life of a first-order reaction?

13.44. How does the half-life in a first-order reaction depend on the concentration of the reactants?

13.45. What effect does doubling the initial concentration of a reactant have on the half-life in a reaction that is second order in the reactant?

13.46. Two first-order decomposition reactions of the form $A \rightarrow B + C$ have the same rate constant at a given temperature. Do the reactants in the two reactions have the same half-lives at this temperature?

Problems

13.47. For each of the following rate laws, determine the order with respect to each reactant and the overall reaction order.

a. Rate $= k[A][B]$
b. Rate $= k[A]^2[B]$
c. Rate $= k[A][B]^3$

13.48. Determine the overall order of the following rate laws and the order with respect to each reactant.

a. Rate $= k[A]^2[B]^{1/2}$
b. Rate $= k[A]^2[B][C]$
c. Rate $= k[A][B]^3[C]^{1/2}$

13.49. Write rate laws and determine the units of the rate constant (by using the units *M* for concentration and s for time) for the following reactions:

a. The reaction of oxygen atoms with NO_2 is first order in both reactants.
b. The reaction between NO and Cl_2 is second order in NO and first order in Cl_2.
c. The reaction between Cl_2 and chloroform ($CHCl_3$) is first order in $CHCl_3$ and one-half order in Cl_2.
*d. The decomposition of ozone (O_3) to O_2 is second order in O_3 and an order of -1 in O atoms.

13.50. Compounds A and B react to give a single product, C. Write the rate law for each of the following cases and determine the units of the rate constant by using the units *M* for concentration and s for time:

a. The reaction is first order in A and second order in B.
b. The reaction is first order in A and second order overall.
c. The reaction is independent of the concentration of A and second order overall.
d. The reaction is second order in both A and B.

13.51. Predict the rate law for the reaction $2\,BrO(g) \rightarrow Br_2(g) + O_2(g)$ under each of the following conditions:

a. The rate doubles when [BrO] doubles.
b. The rate quadruples when [BrO] doubles.
c. The rate is halved when [BrO] is halved.
d. The rate is unchanged when [BrO] is doubled.

13.52. Predict the order with respect to NO for the reaction $NO(g) + Br_2(g) \rightarrow NOBr_2(g)$ under each of the following conditions:

a. The rate doubles when [NO] is doubled and $[Br_2]$ remains constant.
b. The rate increases by 1.56 times when [NO] is increased 1.25 times and $[Br_2]$ remains constant.
c. The rate is halved when [NO] is doubled and $[Br_2]$ remains constant.

13.53. In the reaction of NO with ClO,

$$NO(g) + ClO(g) \rightarrow NO_2(g) + Cl(g)$$

the initial rate of reaction quadruples when the concentrations of both reactants are doubled. What additional information do we need to determine whether the reaction is first order in each reactant?

13.54. If the reaction in Problem 13.53 is not first order in each reactant, what other order(s) are possible?

13.55. Rate Laws for the Destruction of Tropospheric Ozone The reaction of NO_2 with ozone produces NO_3 in a second-order reaction overall:

$$NO_2(g) + O_3(g) \rightarrow NO_3(g) + O_2(g)$$

a. Write the rate law for the reaction if the reaction is first order in each reactant.
b. The rate constant for the reaction is $1.93 \times 10^4\ M^{-1}\ s^{-1}$ at 298 K. What is the rate of the reaction when $[NO_2] = 1.8 \times 10^{-8}\ M$ and $[O_3] = 1.4 \times 10^{-7}\ M$?
c. What is the rate of formation of NO_3 under these conditions?
d. What happens to the rate of the reaction if the concentration of $O_3(g)$ is doubled?

13.56. **Sources of Nitric Acid in the Atmosphere** The reaction between N_2O_5 and water is a source of nitric acid in the atmosphere:

$$N_2O_5(g) + H_2O(g) \rightarrow 2\ HNO_3(g)$$

a. The reaction is first order in each reactant. Write the rate law for the reaction.
b. When $[N_2O_5] = 0.132\ mM$ and $[H_2O] = 230\ mM$, the rate of the reaction is $4.55 \times 10^{-4}\ mM/min$. What is the rate constant for the reaction?

13.57. Each of the following reactions is first order in each reactant and second order overall. Which reaction is fastest if the initial concentrations of all the reactants are the same?

a. $ClO_2(g) + O_3(g) \rightarrow ClO_3(g) + O_2(g)$
$k = 3.0 \times 10^{-19}\ cm^3/(molecule \cdot s)$
b. $ClO_2(g) + NO(g) \rightarrow NO_2(g) + ClO(g)$
$k = 3.4 \times 10^{-13}\ cm^3/(molecule \cdot s)$
c. $ClO(g) + NO(g) \rightarrow Cl(g) + NO_2(g)$
$k = 1.7 \times 10^{-11}\ cm^3/(molecule \cdot s)$
d. $ClO(g) + O_3(g) \rightarrow ClO_2(g) + O_2(g)$
$k = 1.5 \times 10^{-17}\ cm^3/(molecule \cdot s)$

13.58. Two reactions in which there is a single reactant have nearly the same magnitude rate constant. One is first order; the other is second order.

a. If the initial concentrations of the reactants are both 1.0 mM, which reaction will proceed at the higher rate?
b. If the initial concentrations of the reactants are both 2.0 M, which reaction will proceed at the higher rate?

13.59. In the presence of water, NO and NO_2 react to form nitrous acid (HNO_2) by the following reaction:

$$NO(g) + NO_2(g) + H_2O(\ell) \rightarrow 2\ HNO_2(aq)$$

When the concentration of NO or NO_2 is doubled, the initial rate of reaction doubles. If the rate of the reaction does not depend on $[H_2O]$, what is the rate law for this reaction?

13.60. **Hydroperoxyl Radicals in the Atmosphere** During a smog event, trace amounts of many highly reactive substances are present in the atmosphere. One of these is the hydroperoxyl radical, HO_2, which reacts with sulfur trioxide, SO_3. The rate constant for the reaction

$$2\ HO_2(g) + SO_3(g) \rightarrow H_2SO_3(g) + 2\ O_2(g)$$

is $2.6 \times 10^{11}\ M^{-1}\ s^{-1}$ at 298 K. The initial rate of the reaction doubles when the concentration of SO_3 or HO_2 is doubled. What is the rate law for the reaction?

13.61. Disinfecting Municipal Water Supplies Chlorine dioxide (ClO_2) is a disinfectant used in municipal water-treatment plants (Figure P13.61). It dissolves in basic solution, producing ClO_3^- and ClO_2^-:

$$2\ ClO_2(g) + 2\ OH^-(aq) \rightarrow ClO_3^-(aq) + ClO_2^-(aq) + H_2O(\ell)$$

FIGURE P13.61

The following reaction rate data were obtained at 298 K:

Experiment	$[ClO_2]_0$	$[OH^-]_0$ (M)	Initial Rate (M/s)
1	0.060	0.030	0.0248
2	0.020	0.030	0.00827
3	0.020	0.090	0.0247

Determine the rate law and the rate constant for this reaction at 298 K.

13.62. The following data were collected for the reaction:

$$H_2(g) + NO(g) \rightarrow N_2O(g) + H_2O(g)$$

Experiment	Initial $[H_2]_0$	Initial $[NO]_0$	Initial Rate (M/s)
1	0.35	0.30	2.835×10^{-3}
2	0.35	0.60	1.134×10^{-2}
3	0.70	0.60	2.268×10^{-2}

Determine the rate law and the rate constant for the reaction at the temperature for which the data were collected.

13.63. Hydrogen gas reduces NO to N_2 in the following reaction:

$$2\,H_2(g) + 2\,NO(g) \rightarrow 2\,H_2O(g) + N_2(g)$$

The initial reaction rates of four mixtures of H_2 and NO were measured at 900°C with the following results:

Experiment	$[H_2]_0$ (M)	$[NO]_0$ (M)	Initial Rate (M/s)
1	0.212	0.136	0.0248
2	0.212	0.272	0.0991
3	0.424	0.544	0.793
4	0.848	0.544	1.59

Determine the rate law and the rate constant for the reaction at 900°C.

13.64. The rate of the reaction

$$NO_2(g) + CO(g) \rightarrow NO(g) + CO_2(g)$$

was determined in three experiments at 225°C. The results are given in the following table:

Experiment	$[NO_2]_0$ (M)	$[CO]_0$ (M)	Initial Rate (M/s)
1	0.263	0.826	1.44×10^{-5}
2	0.263	0.413	1.44×10^{-5}
3	0.526	0.413	5.76×10^{-5}

a. Determine the rate law for the reaction.
b. Calculate the value of the rate constant at 225°C.
c. Calculate the rate of appearance of CO_2 when $[NO_2] = [CO] = 0.500\ M$.

13.65. Nitrogen trioxide decomposes to NO_2 and O_2 in the following reaction:

$$2\,NO_3(g) \rightarrow 2\,NO_2(g) + O_2(g)$$

The following data were collected at 298 K:

Time (min)	$[NO_3]$ (μM)
0	1.470×10^{-3}
10	1.463×10^{-3}
100	1.404×10^{-3}
200	1.344×10^{-3}
300	1.288×10^{-3}
400	1.237×10^{-3}
500	1.190×10^{-3}

Calculate the value of the second-order rate constant at 298 K.

13.66. Two structural isomers of ClO_2 are shown in Figure P13.66.

O–Cl–O   Cl–O–O

FIGURE P13.66

The isomer with the Cl—O—O skeletal arrangement is unstable and rapidly decomposes according to the reaction $2\,ClOO(g) \rightarrow Cl_2(g) + 2\,O_2(g)$. The following data were collected for the decomposition of ClOO at 298 K:

Time (μs)	[ClOO] (M)
0.0	1.76×10^{-6}
0.7	2.36×10^{-7}
1.3	3.56×10^{-8}
2.1	3.23×10^{-9}
2.8	3.96×10^{-10}

Determine the rate law and the rate constant for the reaction at 298 K.

13.67. At high temperatures, ammonia spontaneously decomposes into N_2 and H_2. The following data were collected at one such temperature:

Time (s)	$[NH_3]$ (M)
0	2.56×10^{-2}
12	2.47×10^{-2}
56	2.16×10^{-2}
224	1.31×10^{-2}
532	5.19×10^{-3}
746	2.73×10^{-3}

Determine the rate law for the decomposition of ammonia and the value of the rate constant at the temperature of the experiment.

13.68. **Atmospheric Chemistry of Hydroperoxyl Radicals** Atmospheric chemistry involves highly reactive, odd-electron molecules such as the hydroperoxyl radical HO_2, which decomposes into H_2O_2 and O_2. Determine the rate law for the reaction and the value of the rate constant at 298 K by using the following data obtained at 298 K.

Time (μs)	$[HO_2]$ (μ*M*)
0.0	8.5
0.6	5.1
1.0	3.6
1.4	2.6
1.8	1.8
2.4	1.1

13.69. In addition to being studied in the gas phase, the decomposition of N_2O_5 has been evaluated in solution. In carbon tetrachloride (CCl_4) at 45°C,

$$2\,N_2O_5 \rightarrow 4\,NO_2 + O_2$$

is a first-order reaction and $k = 6.32 \times 10^{-4}\ s^{-1}$. How much N_2O_5 remains in solution after 1 h if the initial concentration of N_2O_5 was 0.50 mol/L? What percent of the N_2O_5 has reacted at that point?

13.70. Because the units of concentration in the term $\ln([X]/[X]_0)$ cancel out in the integrated rate law for first-order reactions (Equation 13.16), molar concentration can be replaced by any concentration term. With gases, for example, partial pressures may be used. The decomposition of phosphine gas (PH_3) at 600°C is first order in PH_3 with $k = 0.023\ s^{-1}$.

$$4\,PH_3(g) \rightarrow P_4(g) + 6\,H_2(g)$$

If the initial partial pressure of PH_3 is 375 torr, what percent of PH_3 reacts in 1 min?

13.71. **Laughing Gas** Nitrous oxide (N_2O) is used as an anesthetic (laughing gas) and in aerosol cans to produce whipped cream. It is a potent greenhouse gas and decomposes slowly to N_2 and O_2:

$$2\,N_2O(g) \rightarrow 2\,N_2(g) + O_2(g)$$

a. If the plot of $\ln[N_2O]$ as a function of time is linear, what is the rate law for the reaction?
b. How many half-lives will it take for the concentration of N_2O to reach 6.25% of its original concentration?

13.72. The unsaturated hydrocarbon butadiene (C_4H_6) in Figure P13.72 dimerizes to 4-vinylcyclohexene (C_8H_{12}). When data collected in studies of the kinetics of this reaction were plotted against reaction time, plots of $[C_4H_6]$ or $\ln[C_4H_6]$ produced curved lines, but the plot of $1/[C_4H_6]$ was linear.

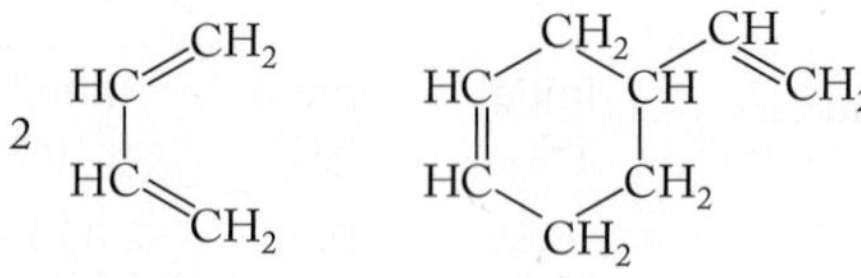

2 butadiene → vinylcyclohexene

FIGURE P13.72

a. What is the rate law for the reaction?
b. How many half-lives will it take for the $[C_4H_6]$ to decrease to 3.1% of its original concentration?

13.73. **Tracing Phosphorus in Organisms** Radioactive isotopes such as ^{32}P are used to follow biological processes. The following radioactivity data (in relative radioactivity values) were collected for a sample containing ^{32}P:

Time (days)	Relative Radioactivity
0	10.00
1	9.53
2	9.08
5	7.85
10	6.16
20	3.79

a. Write the rate law for the decay of ^{32}P.
b. Determine the value of the rate constant.
c. Determine the half-life of ^{32}P.
d. How many days does it take for 99% of a sample of ^{32}P to decay?

13.74. Nitrous acid slowly decomposes to NO, NO_2, and water in the following second-order reaction:

$$2\,HNO_2(aq) \rightarrow NO(g) + NO_2(g) + H_2O(\ell)$$

a. Use the following data to determine the rate constant for this reaction at 298 K:

Time (min)	$[HNO_2]$ (μ*M*)
0	0.1560
1000	0.1466
1500	0.1424
2000	0.1383
2500	0.1345
3000	0.1309

b. Determine the half-life for the decomposition of HNO_2.

c. If the experiment that yielded the results in the preceding table had been continued for 3000 minutes more, what would the concentration of HNO_2 have been?

13.75. The dimerization of ClO,

$$2\ ClO(g) \rightarrow Cl_2O_2(g)$$

is second order in ClO.

a. Use the following data to determine the value of k at 298 K.

Time (s)	[ClO] (molecules/cm^3)
0	2.60×10^{11}
1.00	1.08×10^{11}
2.00	6.83×10^{10}
3.00	4.99×10^{10}
4.00	3.93×10^{10}

b. Determine the half-life for the dimerization of ClO.

13.76. Kinetic data for the reaction $Cl_2O_2(g) \rightarrow 2\ ClO(g)$ are summarized in the following table.

Time (μs)	[Cl_2O_2] (M)
0	6.60×10^{-8}
172	5.68×10^{-8}
345	4.89×10^{-8}
517	4.21×10^{-8}
690	3.62×10^{-8}
862	3.12×10^{-8}

a. Determine the value of the rate constant.

b. Determine $t_{1/2}$ for the decomposition of Cl_2O_2.

13.77. **Kinetics of Sucrose Hydrolysis** The metabolism of table sugar (sucrose, $C_{12}H_{22}O_{11}$) begins with the hydrolysis of the disaccharide to glucose and fructose (both $C_6H_{12}O_6$):

$$C_{12}H_{22}O_{11}(aq) + H_2O(\ell) \rightarrow 2\ C_6H_{12}O_6(aq)$$

The kinetics of the reaction were studied at 24°C in a reaction system with a large excess of water, so the reaction was pseudo first order in sucrose. Determine the rate law and the pseudo–first-order rate constant for the reaction from the following data:

Time (s)	[$C_{12}H_{22}O_{11}$] (M)
0	0.562
612	0.541
1600	0.509
2420	0.484
3160	0.462
4800	0.442

13.78. Hydroperoxyl radicals react rapidly with ozone to produce oxygen and OH radicals:

$$HO_2(g) + O_3(g) \rightarrow OH(g) + 2\ O_2(g)$$

The rate of this reaction was studied in the presence of a large excess of ozone. Determine the pseudo-first-order rate constant and the second-order rate constant for the reaction from the following data:

Time (ms)	[HO_2] (M)	[O_3] (M)
0	3.2×10^{-6}	1.0×10^{-3}
10	2.9×10^{-6}	1.0×10^{-3}
20	2.6×10^{-6}	1.0×10^{-3}
30	2.4×10^{-6}	1.0×10^{-3}
80	1.4×10^{-6}	1.0×10^{-3}

Reaction Rates, Temperature, and the Arrhenius Equation

Concept Review

13.79. How does the magnitude of a reaction's activation energy influence the rate of a reaction?

13.80. Do all spontaneous reactions happen instantaneously at room temperature?

13.81. Under what circumstances is the activation energy of a reaction proceeding in the forward direction less than the activation energy of it happening in reverse?

13.82. Under what circumstances is the activation energy of a reaction proceeding in the forward direction greater than the activation energy of it happening in reverse?

***13.83.** The order of a reaction is independent of temperature, but the value of the rate constant varies with temperature. Why?

13.84. Does reducing the activation energy of a reaction by $\frac{1}{2}$ increase its rate constant by a factor of 2?

***13.85.** Two first-order reactions have activation energies of 15 and 150 kJ/mol. Which reaction will show the larger increase in rate as temperature is increased?

13.86. According to the Arrhenius equation, does the activation energy of a chemical reaction depend on temperature? Explain your answer.

Problems

13.87. The rate constant for the reaction of ozone with oxygen atoms was determined at four temperatures. Calculate the activation energy and frequency factor A for the reaction

$$O(g) + O_3(g) \rightarrow 2\ O_2(g)$$

given the following data:

T (K)	k [cm^3/(molecule · s)]
250	2.64×10^{-4}
275	5.58×10^{-4}
300	1.04×10^{-3}
325	1.77×10^{-3}

13.88. The rate constant for the decomposition of N_2O_5 in a solution of carbon tetrachloride

$$2\ N_2O_5 \rightarrow 2\ N_2O_4 + O_2$$

was determined over a temperature range of 20 K, with the following results:

T (K)	*k* × 10⁴ (*M*⁻¹ s⁻¹)
293	0.235
298	0.469
303	0.993
308	1.820
313	3.620

a. Determine the activation energy for the reaction.
b. Calculate the rate constant of the reaction at 273 K.

13.89. Activation Energy of Smog-Forming Reactions The initial step in the formation of smog is the reaction between nitrogen and oxygen. At the temperatures indicated, values of the rate constant of the reaction

$$N_2(g) + O_2(g) \rightarrow 2\ NO(g)$$

are as follows:

T (K)	*k* ($M^{-1/2}$ s^{-1})
2000	318
2100	782
2200	1770
2300	3733
2400	7396

a. Calculate the activation energy of the reaction.
b. Calculate the frequency factor for the reaction.
c. Calculate the value of the rate constant at ambient temperature, $T = 300$ K.

13.90. Values of the rate constant for the decomposition of N_2O_5 gas at four different temperatures are as follows:

T (K)	*k* (s^{-1})
658	2.14×10^5
673	3.23×10^5
688	4.81×10^5
703	7.03×10^5

a. Determine the activation energy of the decomposition reaction.
b. Calculate the value of the rate constant at 300 K.

13.91. Activation Energy of Stratospheric Ozone Destruction Reactions The kinetics of the reaction between chlorine dioxide and ozone are relevant to the study of atmospheric ozone destruction. The value of the rate constant for the reaction between chlorine dioxide and ozone was measured at four temperatures between 193 and 208 K. The results were as follows:

T (K)	*k* (M^{-1} s^{-1})
193	34.0
198	62.8
203	112.8
208	196.7

a. Calculate the values of the activation energy and the frequency factor for the reaction.
b. What is the value of the rate constant higher in the stratosphere where $T = 245$ K?

13.92. Chlorine atoms react with methane, forming HCl and CH_3. The rate constant for the reaction is $6.0 \times 10^7\ M^{-1}\ s^{-1}$ at 298 K. When the experiment was repeated at three other temperatures, the following data were collected:

T (K)	*k* (M^{-1} s^{-1})
303	6.5×10^7
308	7.0×10^7
313	7.5×10^7

a. Calculate the values of the activation energy and the frequency factor for the reaction.
b. What is the value of the rate constant in the lower stratosphere where $T = 218$ K?

13.93. A rule of thumb states that a reaction rate doubles for every 10°C rise in temperature. This is not always the case, but it is a convenient concept to apply when making estimations. Azomethane (CH_3NNCH_3) is a heat-sensitive explosive that decomposes to yield nitrogen gas and methyl radicals:

$$CH_3NNCH_3(g) \rightarrow N_2(g) + 2\ CH_3(g)$$

At 600°C, the reaction has an activation energy of 2.14×10^4 J/mol and $k = 2.00 \times 10^8\ s^{-1}$. Does the rule of thumb hold for this reaction?

13.94. The compound 1,1-difluoroethane decomposes at elevated temperatures to give fluoroethylene and hydrogen fluoride:

$$CH_3CHF_2(g) \rightarrow CH_2CHF(g) + HF(g)$$

At 460°C, $k = 5.8 \times 10^{-6}\ s^{-1}$ and $E_a = 265$ kJ/mol. To what temperature would you have to raise the reaction to make it go four times as fast?

Reaction Mechanisms

Concept Review

13.95. The rate law for the reaction between NO and H_2 is second order in NO and third order overall, whereas the reaction of NO with Cl_2 is first order in each reactant and second order overall. Do these reactions proceed by similar mechanisms?

13.96. The rate law for the reaction of NO with Cl_2 (Rate = k[NO][Cl_2]) is the same as that for the reaction of NO_2 with F_2 (Rate = k[NO_2][F_2]). Is it possible that these reactions have similar mechanisms?

***13.97.** Under what reaction conditions does a bimolecular reaction obey pseudo-first-order reaction kinetics?

*13.98. If a reaction is zero order in a reactant, does that mean the reactant is never involved in collisions with other reactants? Explain your answer.

Problems

13.99. The hypothetical reaction $A \rightarrow B$ has an activation energy of 50.0 kJ/mol. Draw a reaction profile for each of the following mechanisms:

a. A single elementary step
b. A two-step reaction in which the activation energy of the second step is 15 kJ/mol
c. A two-step reaction in which the activation energy of the second step is the rate-determining barrier

13.100. For the spontaneous reaction $A + B \rightarrow C \rightarrow D + E$, draw three reaction profiles, one for each of the following mechanisms:

a. C is an activated complex.
b. The reaction has two elementary steps; the first step is rate determining and C is an intermediate.
c. The reaction has two elementary steps; the second step is rate determining and C is an intermediate.

13.101. Write the rate laws for the following elementary steps and identify them as uni-, bi-, or termolecular steps:

a. $SO_2Cl_2(g) \rightarrow SO_2(g) + Cl_2(g)$
b. $NO_2(g) + CO(g) \rightarrow NO(g) + CO_2(g)$
c. $2\ NO_2(g) \rightarrow NO_3(g) + NO(g)$

13.102. Write the rate laws for the following elementary steps and identify them as uni-, bi-, or termolecular steps:

a. $Cl(g) + O_3(g) \rightarrow ClO(g) + O_2(g)$
b. $2\ NO_2(g) \rightarrow N_2O_4(g)$
*c. ${}^{14}_{6}C \rightarrow {}^{14}_{7}N + {}^{0}_{-1}\beta$

13.103. A common classroom demonstration of a reaction involves mixing 30% hydrogen peroxide with a solution of potassium iodide. The following mechanism has been proposed for the reaction:

Step 1	$H_2O_2(aq) + I^-(aq) \rightarrow H_2O(aq) + IO^-(aq)$	slow
Step 2	$H_2O_2(aq) + IO^-(aq) \rightarrow H_2O(aq) + O_2(g) + I^-(aq)$	fast

a. Write the equation for the overall reaction.
b. Write the rate law predicted by the mechanism for the overall reaction.
c. Which species is a catalyst?
d. Identify any intermediates in the reaction.

13.104. A proposed mechanism for the reaction of $NO_2(g)$ and $CO(g)$ is

Step 1	$2\ NO_2(g) \rightarrow NO(g) + NO_3(g)$	slow
Step 2	$NO_3(g) + CO(g) \rightarrow NO_2(g) + CO_2(g)$	fast

a. Write the equation for the overall reaction.
b. Write the rate law predicted by the mechanism for the overall reaction.
c. Identify the reactants and products of the reaction.
d. Identify any intermediates in the reaction.

***13.105.** In the following mechanism for NO formation, oxygen atoms are produced by breaking O=O bonds at high temperature in a fast reversible reaction. If $\Delta[NO]/\Delta t = k[N_2][O_2]^{1/2}$, which step in the mechanism is the rate-determining step?

(1)	$O_2(g) \rightleftharpoons 2\ O(g)$
(2)	$O(g) + N_2(g) \rightarrow NO(g) + N(g)$
(3)	$N(g) + O(g) \rightarrow NO(g)$
Overall:	$N_2(g) + O_2(g) \rightarrow 2\ NO(g)$

13.106. A proposed mechanism for the gas phase decomposition of hydrogen peroxide at an elevated temperature consists of three elementary steps:

$$H_2O_2(g) \rightarrow 2\ OH(g)$$
$$H_2O_2(g) + OH(g) \rightarrow H_2O(g) + HO_2(g)$$
$$HO_2(g) + OH(g) \rightarrow H_2O(g) + O_2(g)$$

If the rate law for the reaction is first order in H_2O_2, which step in the mechanism is the rate-determining step?

13.107. At a given temperature, the rate of the reaction between NO and Cl_2 is proportional to the product of the concentrations of the two gases: $[NO][Cl_2]$. The following two-step mechanism has been proposed for the reaction:

(1)	$NO(g) + Cl_2(g) \rightarrow NOCl_2(g)$
(2)	$NOCl_2(g) + NO(g) \rightarrow 2\ NOCl(g)$
Overall:	$2\ NO(g) + Cl_2(g) \rightarrow NOCl(g)$

Which step must be the rate-determining step if this mechanism is correct?

13.108. **Mechanism of Ozone Destruction** Ozone decomposes thermally to oxygen in the following reaction:

$$2\ O_3(g) \rightarrow 3\ O_2(g)$$

The following mechanism has been proposed:

$$O_3(g) \rightarrow O(g) + O_2(g)$$
$$O(g) + O_3(g) \rightarrow 2\ O_2(g)$$

The reaction is second order in ozone. What properties of the two elementary steps (specifically, relative rate and reversibility) are consistent with this mechanism?

13.109. **Mechanism of NO_2 Destruction** The rate laws for the thermal and photochemical decomposition of NO_2 are different. Which of the following mechanisms are possible for the thermal decomposition of NO_2, and which are possible for the photochemical decomposition of NO_2? For the thermal decomposition, Rate $= k[NO_2]^2$, and for the photochemical decomposition, Rate $= k[NO_2]$.

a. $NO_2(g) \xrightarrow{\text{slow}} NO(g) + O(g)$
$O(g) + NO_2(g) \xrightarrow{\text{fast}} NO(g) + O_2(g)$

b. $NO_2(g) + NO_2(g) \xrightarrow{\text{fast}} N_2O_4(g)$
$N_2O_4(g) \xrightarrow{\text{slow}} NO(g) + NO_3(g)$
$NO_3(g) \xrightarrow{\text{fast}} NO(g) + O_2(g)$

c. $NO_2(g) + NO_2(g) \xrightarrow{\text{slow}} NO(g) + NO_3(g)$
$NO_3(g) \xrightarrow{\text{fast}} NO(g) + O_2(g)$

13.110. The rate laws for the thermal and photochemical decomposition of NO_2 are different. Which of the following mechanisms are possible for the thermal decomposition of NO_2, and which are possible for the photochemical decomposition of NO_2? For the thermal decomposition, Rate = $k[NO_2]^2$, and for the photochemical decomposition, Rate = $k[NO_2]$.

a. $NO_2(g) + NO_2(g) \xrightarrow{\text{slow}} N_2O_4(g)$
$N_2O_4(g) \xrightarrow{\text{fast}} N_2O_3(g) + O(g)$
$N_2O_3(g) + O(g) \xrightarrow{\text{fast}} N_2O_2(g) + O_2(g)$
$N_2O_2(g) \xrightarrow{\text{fast}} 2\ NO(g)$

b. $NO_2(g) + NO_2(g) \xrightarrow{\text{slow}} NO(g) + NO_3(g)$
$NO_3(g) \xrightarrow{\text{fast}} NO(g) + O_2(g)$

c. $NO_2(g) \xrightarrow{\text{slow}} N(g) + O_2(g)$
$N(g) + NO_2(g) \xrightarrow{\text{fast}} N_2O_2(g)$
$N_2O_2(g) \xrightarrow{\text{fast}} 2\ NO(g)$

Catalysts

Concept Review

13.111. Does a catalyst affect both the rate and the rate constant of a reaction?

*13.112. Is the rate law for a catalyzed reaction the same as that for the uncatalyzed reaction?

13.113. Does a substance that increases the rate of a reaction also increase the rate of the reverse reaction?

13.114. The rate of the reaction between NO_2 and CO is independent of [CO]. Does this mean that CO is a catalyst for the reaction?

***13.115.** Can the concentration of a homogeneous catalyst appear in the rate law for the reaction it catalyzes?

*13.116. The rate of a chemical reaction is too slow to measure at room temperature. We could either raise the temperature or add a catalyst. Which would be a better solution for making an accurate determination of the rate constant?

Problems

13.117. Is NO a catalyst for the decomposition of N_2O in the following two-step reaction mechanism, or is N_2O a catalyst for the conversion of NO to NO_2?

(1) $NO(g) + N_2O(g) \rightarrow N_2(g) + NO_2(g)$

(2) $2\ NO_2(g) \rightarrow 2\ NO(g) + O_2(g)$

13.118. **NO as a Catalyst for Ozone Destruction** Explain why NO is a catalyst in the following two-step process that results in the depletion of ozone in the stratosphere:

(1) $NO(g) + O_3(g) \rightarrow NO_2(g) + O(g)$

(2) $O(g) + NO_2(g) \rightarrow NO(g) + O_2(g)$

Overall: $O(g) + O_3(g) \rightarrow 2\ O_2(g)$

13.119. On the basis of the frequency factors and activation energy values of the following two reactions, determine which one will have the larger rate constant at room temperature (298 K).

$$O_3(g) + O(g) \rightarrow O_2(g) + O_2(g)$$

$A = 8.0 \times 10^{-12}\ \text{cm}^3/(\text{molecules} \cdot \text{s})$ $\quad E_a = 17.1$ kJ/mol

$$O_3(g) + Cl(g) \rightarrow ClO(g) + O_2(g)$$

$A = 2.9 \times 10^{-11}\ \text{cm}^3/(\text{molecules} \cdot \text{s})$ $\quad E_a = 2.16$ kJ/mol

13.120. On the basis of the frequency factors and activation energy values of the following two reactions, determine which one will have the larger rate constant at room temperature (298 K).

$$O_3(g) + Cl(g) \rightarrow ClO(g) + O_2(g)$$

$A = 2.9 \times 10^{-11}\ \text{cm}^3/(\text{molecules} \cdot \text{s})$ $\quad E_a = 2.16$ kJ/mol

$$O_3(g) + NO(g) \rightarrow NO_2(g) + O_2(g)$$

$A = 2.0 \times 10^{-12}\ \text{cm}^3/(\text{molecules} \cdot \text{s})$ $\quad E_a = 11.6$ kJ/mol

Additional Problems

13.121. A student inserts a glowing wood splint into a test tube filled with O_2. The splint quickly catches on fire (Figure P13.121). Why does the splint burn so much faster in pure O_2 than in air?

FIGURE P13.121

*13.122. Methane gas leaking from the largest underground methane storage facility in the western United States caused thousands of people in southern California to be evacuated from their homes in October 2015. Methane is an explosion hazard, but a spark must be introduced into the mixture to cause it to react. Why is the spark needed?

13.123. On average, someone who falls through the ice covering a frozen lake is less likely to experience anoxia (lack of oxygen) than someone who falls into a warm pool and is underwater for the same length of time. Why?

*13.124. Why doesn't a quadrupling of the rate correspond to a reaction order of 4—for example, Rate $\propto [NO]^4$?

13.125. If the rate of the reverse reaction is much slower than the rate of the forward reaction, does the method used to determine a rate law from initial concentrations and initial rates also work at some other time t?

13.126. The rate at which drugs are metabolized depends upon age: children metabolize some drugs more rapidly than adults, while the elderly metabolize drugs more slowly. Diazepam is used to treat anxiety disorders and seizures in patients in all age groups. Its half-life in hours is estimated to be equal to the patient's age in years; in a 50-year-old, for example, diazepam would have a 50-hour half-life. How long will it take for 95% of a dose of diazepam to be metabolized in a

5-year-old child compared to a 50-year-old adult assuming a first order process?

13.127. How can a plot of $1/[X] - 1/[X]_0$ as a function of t be used to determine the value of a second order rate constant?

13.128. A teaching assistant is designing a synthesis experiment for use in a 3-hour laboratory. The literature preparation specifies that 125 mL of a 1.0 M solution of reactant A should be mixed with 125 mL of a 1.0 M solution of B at room temperature in a 500 mL flask. The rate of the reaction was such that, after 6 hours of sitting undisturbed on the lab bench, only 50% of the stoichiometric yield of the desired product C was produced. Suggest three reasonable changes in the procedure the assistant could try to improve the yield of product over a 3-hour period.

13.129. Why can't an elementary step in a mechanism have a rate law that is zero order in a reactant?

13.130. During the decomposition of dinitrogen pentoxide,

$$2\ N_2O_5(g) \rightarrow 4\ NO_2(g) + O_2(g)$$

how is the rate of consumption of N_2O_5 related to the rate of formation of NO_2 and O_2?

13.131. In the reaction between nitrogen dioxide and ozone,

$$2\ NO_2(g) + O_3(g) \rightarrow N_2O_5(g) + O_2(g)$$

how are the rates of change in the concentrations of the reactants and products related?

13.132. Use the initial rate data from the following table to determine the order of the decomposition reaction of N_2O_5:

Experiment	$[N_2O_5]_0$ (M)	Initial Rate (M/s)
1	0.050	1.8×10^{-5}
2	0.100	3.6×10^{-5}

13.133. At the temperature at which the experiments were carried out in Problem 13.132, what is the rate constant for the decomposition of N_2O_5? Write the complete rate law for the decomposition reaction.

13.134. The following table contains kinetics data for the reaction

$$2\ NO(g) + Cl_2(g) \rightarrow 2\ NOCl(g)$$

Experiment	$[NO]_0$ (M)	$[Cl_2]_0$ (M)	Initial Rate (M/s)
1	0.20	0.10	0.63
2	0.20	0.30	5.70
3	0.80	0.10	2.58
4	0.40	0.20	?

Predict the initial rate of reaction in experiment 4.

13.135. The following is an important reaction in the formation of photochemical smog:

$$NO(g) + O_3(g) \rightarrow NO_2(g) + O_2(g)$$

The reaction is first order in NO and O_3. The rate constant of the reaction is 80 $M^{-1}\ s^{-1}$ at 25°C and 3000 $M^{-1}\ s^{-1}$ at 75°C.

a. If this reaction were to occur in a single step, would the rate law be consistent with the observed order of the reaction for NO and O_3?

b. What is the value of the activation energy of the reaction?

c. What is the rate of the reaction at 25°C when $[NO] = 3 \times 10^{-6}\ M$ and $[O_3] = 5 \times 10^{-9}\ M$?

d. Predict the values of the rate constant at 10°C and 35°C.

13.136. Ammonia reacts with nitrous acid to form an intermediate, ammonium nitrite (NH_4NO_2), which decomposes to N_2 and H_2O:

$$NH_3(g) + HNO_2(aq) \rightarrow NH_4NO_2(aq) \rightarrow N_2(g) + 2\ H_2O(\ell)$$

a. The reaction is first order in ammonia and second order in nitrous acid. What is the rate law for the reaction? What are the units on the rate constant if concentrations are expressed in molarity and time in seconds?

b. The rate law for the reaction has also been written as

$$\text{Rate} = k[NH_4^+][NO_2^-][HNO_2]$$

Is this expression equivalent to the one you wrote in part (a)?

c. With the data in Appendix 4, calculate the value of ΔH°_{rxn} for the overall reaction $\Delta H^\circ_{f,HNO_2(aq)} = -128.9$ kJ/mol.

d. Draw a reaction-energy profile for the process with the assumption that E_a of the first step is lower than E_a of the second step.

***13.137.** When ionic compounds such as NaCl dissolve in water, the sodium ions are surrounded by six water molecules. The bound water molecules exchange with those in bulk solution as described by the reaction involving ^{18}O-enriched water:

$$Na(H_2O)_6^+(aq) + H_2{}^{18}O(\ell) \rightarrow Na(H_2O)_5(H_2{}^{18}O)^+(aq) + H_2O(\ell)$$

a. The following reaction mechanism has been proposed:

(1) $Na(H_2O)_6^+(aq) \rightarrow Na(H_2O)_5^+(aq) + H_2O(\ell)$

(2) $Na(H_2O)_5^+(aq) + H_2{}^{18}O(\ell) \rightarrow Na(H_2O)_5(H_2{}^{18}O)^+(aq)$

What is the rate law if the first step is the rate-determining step?

b. If you were to sketch a reaction-energy profile, which would you draw with the higher energy, the reactants or the products?

13.138. **Lachrymators in Smog** The combination of ozone, volatile hydrocarbons, nitrogen oxide, and sunlight in urban environments produces peroxyacetyl nitrate (PAN), a potent lachrymator (a substance that causes eyes to tear). PAN decomposes to acetyl radicals and nitrogen dioxide in a process that is second order in PAN, as shown in Figure P13.138:

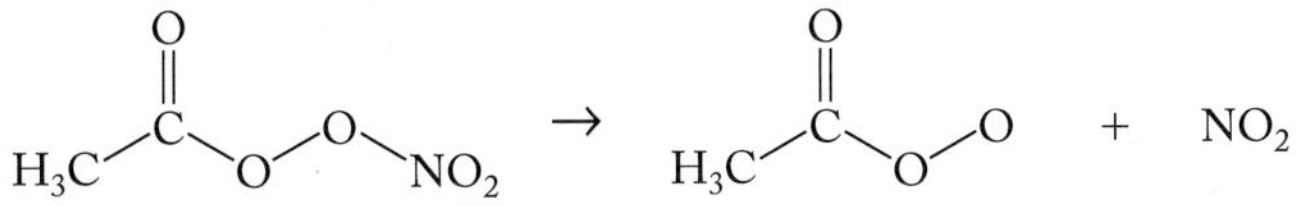

FIGURE P13.138

a. The half-life of the reaction, at 23°C and $P_{CH_3CO_3NO_2} = 10.5$ torr, is 100 h. Calculate the rate constant for the reaction.

b. Determine the rate of the reaction at 23°C and $P_{CH_3CO_3NO_2} = 10.5$ torr.

c. Draw a graph showing P_{PAN} as a function of time from 0 to 200 h starting with $P_{CH_3CO_3NO_2} = 10.5$ torr.

13.139. **Nitric Oxide in the Human Body** Nitric oxide (NO) is a gaseous free radical that plays many biological roles, including regulating neurotransmission and the human immune system. One of its many reactions involves the peroxynitrite ion ($ONOO^-$):

$$NO(g) + ONOO^-(aq) \rightarrow NO_2(g) + NO_2^-(aq)$$

a. Use the following data to determine the rate law and rate constant of the reaction at the experimental temperature at which these data were generated.

Experiment	$[NO]_0$ (*M*)	$[ONOO^-]_0$ (*M*)	Rate (*M*/s)
1	1.25×10^{-4}	1.25×10^{-4}	2.03×10^{-11}
2	1.25×10^{-4}	0.625×10^{-4}	1.02×10^{-11}
3	0.625×10^{-4}	2.50×10^{-4}	2.03×10^{-11}
4	0.625×10^{-4}	3.75×10^{-4}	3.05×10^{-11}

b. Draw the Lewis structure of peroxynitrite ion (including all resonance forms) and assign formal charges. Note which form is preferred.

c. Use the average bond energies in Appendix Table A4.1 to estimate the value of ΔH°_{rxn} using the preferred structure from part (b).

13.140. **Kinetics of Protein Chemistry** In the presence of O_2, NO reacts with sulfur-containing proteins to form *S*-nitrosothiols, such as $C_6H_{13}SNO$. This compound decomposes to form a disulfide and NO:

$$2\ C_6H_{13}SNO(aq) \rightarrow 2\ NO(g) + C_{12}H_{26}S_2(aq)$$

The following data were collected for the decomposition reaction at 69°C.

Time (min)	$[C_6H_{13}SNO]$ (*M*)
0	1.05×10^{-3}
10	9.84×10^{-4}
20	9.22×10^{-4}
30	8.64×10^{-4}
60	7.11×10^{-4}

Calculate the value of the first-order rate constant for the reaction.

13.141. Solutions of nitrous acid (HNO_2) in ^{18}O-labeled water undergo isotope exchange:

$$HNO_2(aq) + H_2{}^{18}O(\ell) \rightarrow HN^{18}O_2(aq) + H_2O(\ell)$$

a. Use the following data at 24°C to determine the dependence of the reaction rate on the concentration of HNO_2.

Time (min)	$[HNO_2]$
0	5.4×10^{-2}
20	1.5×10^{-3}
40	7.7×10^{-4}
60	5.2×10^{-4}

b. Does the reaction rate depend on the concentration of $H_2{}^{18}O$?

13.142. Ethylene (C_2H_4) reacts with ozone to form 2 mol of formaldehyde (a probable human carcinogen) per mole of ethylene as shown in Figure P13.142. The following kinetic data were collected at 298 K.

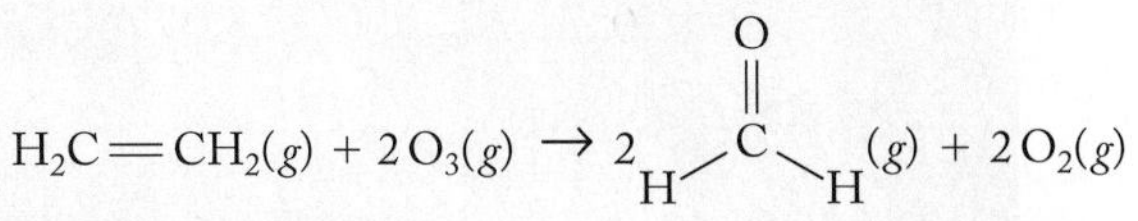

$$H_2C{=}CH_2(g) + 2\,O_3(g) \rightarrow 2\,H_2C{=}O(g) + 2\,O_2(g)$$

FIGURE P13.142

Experiment	$[O_3]_0$ (M)	$[C_2H_4]_0$ (M)	Rate (M/s)
1	0.86×10^{-2}	1.00×10^{-2}	0.0877
2	0.43×10^{-2}	1.00×10^{-2}	0.0439
3	0.22×10^{-2}	0.50×10^{-2}	0.0110

a. Determine the rate law and the value of the rate constant of the reaction at 298 K.
b. The rate constant was determined at several additional temperatures. Calculate the activation energy of the reaction from the following data.

T (K)	k (M^{-1} s^{-1})
263	3.28×10^2
273	4.73×10^2
283	6.65×10^2
293	9.13×10^2

13.143. **Reducing NO Emissions** Adding NH_3 to the stack gases at an electric power generating plant can reduce NO_x emissions. This selective noncatalytic reduction (SNR) process depends on the reaction between NH_2 (an odd-electron compound) and NO:

$$NH_2(g) + NO(g) \rightarrow N_2(g) + H_2O(g)$$

The following kinetic data were collected at 1200 K.

Experiment	$[NH_2]_0$ (M)	$[NO]_0$ (M)	Rate (M/s)
1	1.00×10^{-5}	1.00×10^{-5}	0.12
2	2.00×10^{-5}	1.00×10^{-5}	0.24
3	2.00×10^{-5}	1.50×10^{-5}	0.36
4	2.50×10^{-5}	1.50×10^{-5}	0.45

a. What is the rate law for the reaction?
b. What is the value of the rate constant at 1200 K?

smartwork5
If your instructor uses Smartwork5, log in at digital.wwnorton.com/atoms2.

14

Chemical Equilibrium

Equal but Opposite Reaction Rates

CHEMISTRY AND THE FOOD SUPPLY Most crops require nitrogen fertilizers, and the source of most of this nitrogen is ammonia, which is synthesized by the reaction $N_2(g) + 3\ H_2(g) \rightleftharpoons 2\ NH_3(g)$.

PARTICULATE **REVIEW**

Forward Rate versus Reverse Rate

In Chapter 14, we investigate reversible reactions that proceed until the rate of the reverse reaction matches the rate of the forward reaction.

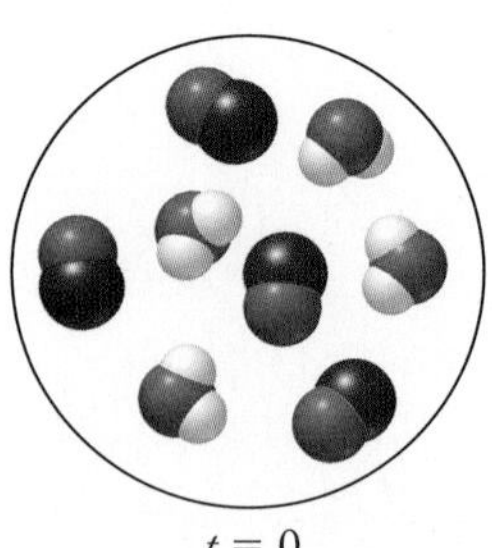

$t = 0$

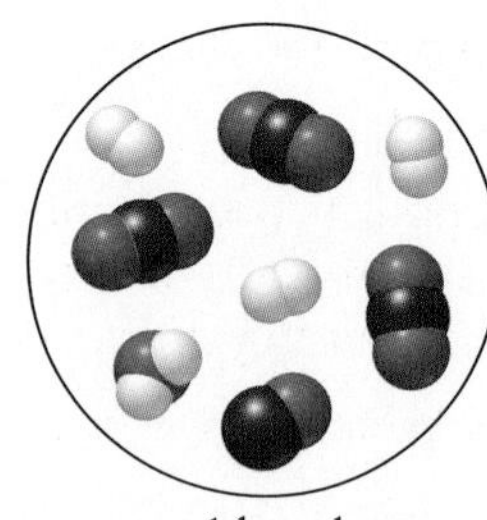

$t = 1$ hour later

- Write a balanced equation for this system that includes both a forward reaction and a reverse reaction.
- If more carbon monoxide were added to the system, how does this initially affect the frequency of collisions between the reactants? How would it affect the rate of the forward reaction?
- If carbon dioxide were removed form the system, how would it initially affect the frequency of collisions between the products? How would it affect the rate of the reverse reaction?

(Review Section 13.3 if you need help.)

(Answers to Particulate Review questions are in the back of the book.)

PARTICULATE PREVIEW

Spontaneous Shifts

The reaction $N_2O_4(g) \rightleftharpoons 2\ NO_2(g)$ is depicted under two different initial conditions. As you read Chapter 14, look for information that will help you answer these questions:

- Which condition spontaneously forms product?
- Which condition spontaneously forms reactant?
- What is the sign of ΔG for these two conditions? What would a condition of $\Delta G = 0$ mean for this reaction?

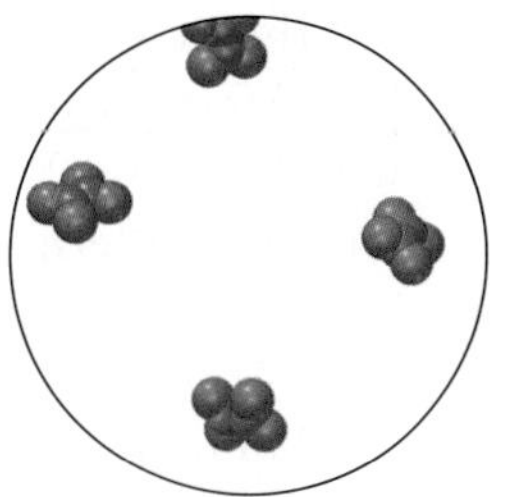

Condition A

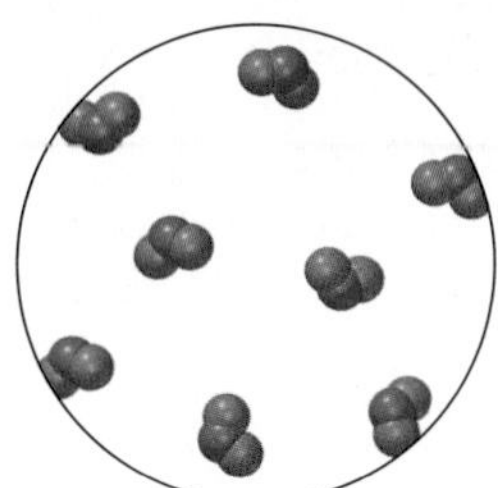
Condition B

Learning Outcomes

LO1 Describe the dynamic nature of chemical equilibria

LO2 Write mass action or equilibrium constant expressions for reversible reactions, including those involving heterogeneous equilibria
Sample Exercises 14.1, 14.8

LO3 Use equilibrium concentration or partial pressure data to calculate the value of K_c or K_p
Sample Exercises 14.2, 14.3

LO4 Interconvert K_c and K_p values of gas-phase reactions
Sample Exercises 14.4, 14.9

LO5 Calculate *K* values of related chemical reactions
Sample Exercises 14.5, 14.6

LO6 Calculate the value of a reaction quotient and use it to predict the direction of a reversible chemical reaction
Sample Exercise 14.7

LO7 Predict how a reaction at equilibrium responds to changes in reaction conditions
Sample Exercises 14.10, 14.11, 14.12

LO8 Calculate the concentrations or partial pressures of reactants and products in a reaction mixture at equilibrium from their starting values and the value of *K*
Sample Exercises 14.13, 14.14

LO9 Relate the standard free-energy change of a reversible reaction to its equilibrium constant
Sample Exercise 14.15

LO10 Predict the value of *K* at any temperature from thermodynamic data
Sample Exercise 14.16

14.1 The Dynamics of Chemical Equilibrium

For centuries farmers have applied fertilizer to their fields, providing their crops with nitrogen, potassium, and phosphorus compounds that are essential to plant growth. In the past farmers used mostly biological fertilizers: manure and pulverized bones from farm animals and the droppings of wild ones, including guano collected from bat caves. Today, large-scale production of corn, wheat, and other grains has created a demand for fertilizer that is too large to be met by traditional biological sources. Instead, most of the potassium and phosphorus come from mined minerals, and most of the nitrogen comes from a synthetic chemistry process in which atmospheric N_2 is combined with H_2 to form ammonia, NH_3. Each year about 200 million tons of ammonia and nitrogen compounds derived from it are used to enhance the world's food supply.

The industrial process for synthesizing ammonia was developed in Germany just before the outbreak of World War I, and much of the ammonia first produced with it was not used to grow food, but rather to manufacture explosives that sustained the German war effort. It is ironic that a process that sustains human life in many lands today contributed to the deaths of millions of people between 1914 and 1918.

Ammonia synthesis is based on the following chemical reaction:

$$N_2(g) + 3\,H_2(g) \rightleftharpoons 2\,NH_3(g)$$

The reaction is exothermic ($\Delta H^\circ_{rxn} < 0$) and spontaneous ($\Delta G^\circ_{rxn} < 0$) under standard conditions, but it's also very slow at 25°C. Catalysts and elevated temperatures increase the rate of the reaction but lower its yield. In other words, less ammonia is synthesized at higher temperatures, but what is produced is made faster. Why does yield decrease as temperatures rise? A clue is contained in the fact that it takes four moles of reactant gases to make two moles of ammonia gas.

This decrease in the number of gas-phase molecules indicates that the reaction mixture experiences a decrease in entropy ($\Delta S_{rxn} < 0$), which means that the value of ΔG_{rxn} in this key equation from Chapter 12,

$$\Delta G_{rxn} = \Delta H_{rxn} - T\Delta S_{rxn}$$

increases as T increases and eventually becomes greater than zero. In Chapter 12 we associated positive values of ΔG_{rxn} with nonspontaneous reactions. In this chapter we discover that reaction spontaneity is not a yes-or-no proposition, but rather a measure of how far a reaction proceeds until it reaches *chemical equilibrium* and there is no further increase in product concentration. The more positive the value of ΔG_{rxn}, the less product forms before equilibrium is achieved.

chemical equilibrium a dynamic process in which the concentrations of reactants and products remain constant over time and the rate of a reaction in the forward direction matches its rate in the reverse direction.

How, then, does the chemical industry produce millions of tons of ammonia each year? Scientists and engineers have developed ways to manipulate the equilibrium state in the reactors in which ammonia is synthesized. For example, they pump the equilibrium reaction mixture through condensers that liquefy and remove NH_3 but leave N_2 and H_2 in the gas phase, where these reactants combine to produce more ammonia. In this chapter we explore why this technique works, causing the reaction to make more ammonia and providing the fertilizer that helps feed billions of people.

Chemical equilibrium is a *dynamic* process because reactants form products and products form reactants at equal rates. Equilibrium is represented by a pair of single-headed arrows pointing in opposite directions:

$$\text{Reactants} \rightleftharpoons \text{Products}$$

and by the following equality in reaction rates at equilibrium:

$$\text{Rate}_{\text{forward}} = \text{Rate}_{\text{reverse}}$$

Some reactions reach equilibrium only after nearly all the reactants have formed products. We say that these equilibria *lie far to the right* (in the direction of the forward reaction arrow). In other reactions, little product is formed. These equilibria are said to favor reactants and *lie far to the left* (the direction of the reverse reaction arrow). Nothing can be inferred from the position of the equilibrium regarding how much time the reaction takes to reach equilibrium. Studies of equilibrium reveal only the extent to which a reaction proceeds, not how rapidly it proceeds.

To explore the dynamics of chemical equilibrium, let's look at a two-step industrial process for making H_2 gas. The reactants are methane and steam. In the first step, methane reacts with steam in the presence of a nickel or iron oxide catalyst at temperatures near 1000°C:

$$CH_4(g) + H_2O(g) \rightleftharpoons CO(g) + 3\,H_2(g) \qquad (14.1)$$

This reaction, called the steam-reforming reaction, is followed by another, called the water–gas shift reaction, in which the carbon monoxide formed in the first step is reacted with more steam in the presence of a Cu/ZnO catalyst at around 200°C:

$$H_2O(g) + CO(g) \rightleftharpoons H_2(g) + CO_2(g) \qquad (14.2)$$

Let's explore the dynamics of Equation 14.2. If we put an equal number of moles of water vapor and carbon monoxide in a closed chamber and allow them to react, the concentrations of CO and H_2O initially fall as the concentrations of H_2 and CO_2 increase, as shown in Figure 14.1. Because H_2O and CO react

FIGURE 14.1 Concentrations of reactants and products in the water–gas shift reaction change over time until equilibrium is reached. At equilibrium in the water–gas shift reaction, H_2 and CO_2 are being formed at the same rate at which they are reacting to re-form H_2O and CO.

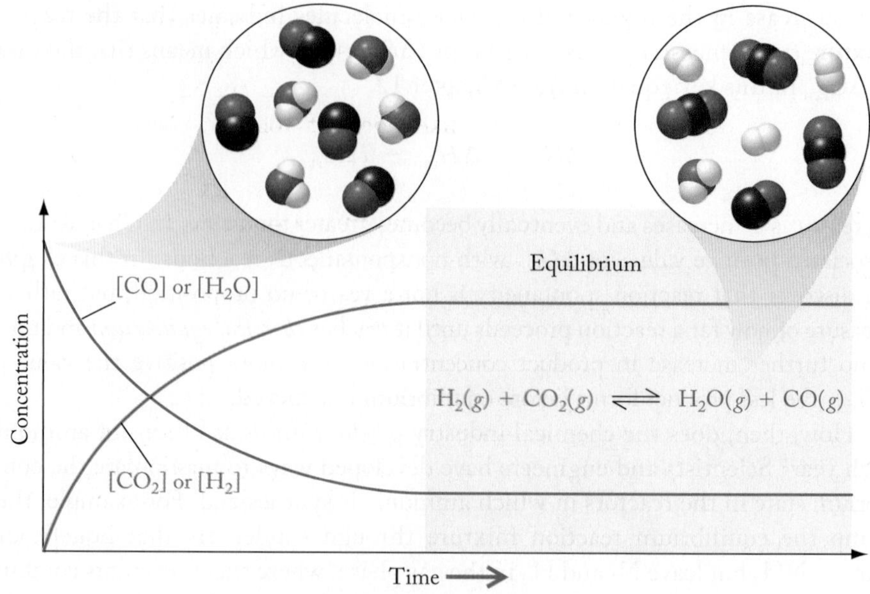

CONNECTION The method for calculating the percent yield of a reaction was described in Section 7.7.

CONNECTION In Chapter 13 we discussed how reaction rate depends on molecular collisions; the higher the concentrations of reactants, the more frequently molecules collide, and the faster a reaction proceeds.

in a 1:1 stoichiometric ratio, their concentrations decrease at the same rate and are always equivalent in the reaction chamber. Similarly, H_2 and CO_2 are formed in a 1:1 ratio, so their concentrations increase at the same rate and are always equal.

The water–gas shift reaction is reversible. Reversibility means that if we intervene in the course of the reaction, for example, by changing the reaction temperature or removing or adding a reactant or product, we might be able to force the reaction to run in reverse, re-forming reactants from products. One requirement for reversibility is that the products must remain in contact with each other. If one or more of the products is a gas, then we need to run the reaction in a sealed chamber.

The graph in Figure 14.1 shows that eventually the concentrations of reactants and products in the water–gas shift reaction no longer change with time, at which point the reaction has reached chemical equilibrium. Note, however, that the concentrations of the reactants (CO and H_2O) do not go to zero. The presence of these reactants in the reaction mixture when the reaction has reached equilibrium means that the yield of the reaction never reaches 100%.

How do the rates of the forward and reverse reactions change during the course of the water–gas shift reaction? When the CO and H_2O are initially mixed, their concentrations are

FIGURE 14.2 (a) The rates of forward and reverse reactions decrease and increase, respectively, until they are the same as equilibrium is achieved. (b) The composition of the reaction mixture remains constant after equilibrium is achieved.

Reactant
Product
Reaction rate
Forward rate
Reverse rate
Equilibrium is achieved: forward rate = reverse rate
Time
(a)

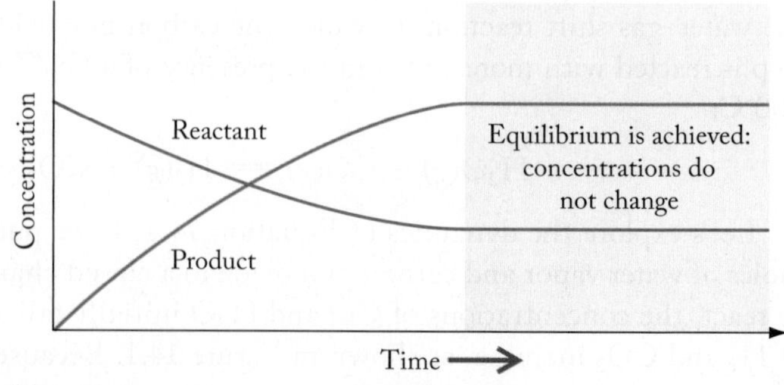

(b)

at their maximum and the rate of the forward reaction, as indicated by the slope of the [CO] and [H_2O] curve at $t = 0$ in Figure 14.1, is also at its maximum. As the reaction proceeds, reactant concentrations decrease, and the rate of the forward reaction also decreases, because the likelihood of collisions between reactant molecules decreases.

The concentrations of products are initially zero and so is the rate of the reverse reaction. As product molecules form, the likelihood that they will collide with one another to re-form reactant molecules increases and so does the rate of the reverse reaction. Ultimately, the rate of the reverse reaction equals the rate of the forward reaction, as shown in Figure 14.2. At this point equilibrium has been achieved.

Let's explore a chemical reaction of NO_2 where we can actually watch a system reach equilibrium. Nitrogen dioxide, NO_2, is the gas that gives photochemical smog its distinctive brown color (Figure 14.3). Pure nitrogen dioxide dimerizes, forming N_2O_4, which is a colorless gas with a boiling point of 21°C:

$$\underset{\text{(brown)}}{2\,NO_2(g)} \rightleftharpoons \underset{\text{(colorless)}}{N_2O_4(g)} \tag{14.3}$$

FIGURE 14.3 The brown layer in the atmosphere in this photo is caused by high concentrations of NO_2, a key ingredient in photochemical smog, as discussed in Chapter 13.

CONNECTION We discussed the rates of reactions of nitrogen oxides in the atmosphere in Chapter 13 in our discussion of the chemistry of photochemical smog.

The flask on the left in Figure 14.4 contains an equilibrium mixture of NO_2 and N_2O_4 at room temperature. The rate of the forward reaction (Rate_f) is second order in NO_2:

$$\text{Rate}_f = k_f[NO_2]^2 \tag{14.4}$$

At equilibrium, Rate_f equals Rate_r, the rate of the reverse reaction, which is first order in N_2O_4:

$$\text{Rate}_r = k_r[N_2O_4] \tag{14.5}$$

When an identical flask (on the right in Figure 14.4) is placed in an ice bath, most of the N_2O_4 in the mixture condenses, dramatically lowering its gas-phase concentration. A lower concentration of N_2O_4 gas means the rate of the reverse reaction in Equation 14.5 decreases. As a result, the forward reaction dominates, consuming NO_2 gas and causing the brown color of the reaction mixture to fade. As the concentration of NO_2 gas decreases, the rate of the forward reaction decreases until it again matches the rate of the reverse reaction:

$$\text{Rate}_f = \text{Rate}_r \tag{14.6}$$

and equilibrium is reestablished.

Replacing the terms in Equation 14.6 with the right sides of Equations 14.4 and 14.5 yields

$$k_f[NO_2]^2 = k_r[N_2O_4]$$

which we can rearrange to

$$\frac{k_f}{k_r} = \frac{[N_2O_4]}{[NO_2]^2} \tag{14.7}$$

FIGURE 14.4 Shifting equilibrium. The flask on the left contains an equilibrium reaction mixture of brown NO_2 gas and colorless N_2O_4 gas at room temperature. When an identical flask is placed in an ice bath, N_2O_4 condenses, which shifts the equilibrium to produce N_2O_4, lowering the concentration of NO_2 gas and causing the brown color of the gas mixture to fade.

equilibrium constant (*K*) the value of the ratio of concentration (or partial pressure) terms in the equilibrium constant expression at a specific temperature.

equilibrium constant expression the ratio of the equilibrium concentrations or partial pressures of products to reactants, each term raised to a power equal to the coefficient of that substance in the balanced chemical equation for the reaction.

law of mass action the ratio of the concentrations or partial pressures of products to reactants at equilibrium has a characteristic value at a given temperature when each term is raised to a power equal to the coefficient of that substance in the balanced chemical equation for the reaction.

mass action expression equivalent to the equilibrium constant expression but applied to reaction mixtures that may or may not be at equilibrium.

The ratio k_f/k_r is the ratio of two constants, which is simply another constant. This constant is called an **equilibrium constant (*K*)**. Equating it to the ratio of product concentrations to reactant concentrations on the right side of Equation 14.7 gives us the **equilibrium constant expression** for this reaction:

$$K = \frac{[N_2O_4]}{[NO_2]^2} \tag{14.8}$$

We just derived the *K* expression for the dimerization of NO_2 from the ratio of its forward and reverse rate laws. However, prior knowledge of the kinetics of reactions is not needed to write their equilibrium constant expressions. As described in the next section, all we need are balanced chemical equations. This is because the composition of an equilibrium mixture of reactants and products does not depend on the terms and exponents of the forward and reverse rate laws or on how rapidly the reaction proceeds. Some reactions achieve equilibrium quickly, whereas others take a very long time. The value of an equilibrium constant *K* says nothing about how rapidly a reaction proceeds toward equilibrium; it tells us only the extent to which the reaction proceeds before it reaches equilibrium.

CONCEPT **TEST**

In the reversible reaction $A \rightleftharpoons B$, the rate constant of the forward reaction at a particular temperature is 3.0 times the value of the rate constant of the reverse reaction. What is the value of the equilibrium constant at that temperature?

(Answers to Concept Tests are in the back of the book.)

14.2 Writing Equilibrium Constant Expressions

Let's revisit the water–gas shift reaction:

$$H_2O(g) + CO(g) \rightleftharpoons H_2(g) + CO_2(g) \tag{14.2}$$

The data in Table 14.1 describe the results of four experiments in which different quantities of H_2O vapor, CO, H_2, and CO_2 are injected into a sealed reaction vessel heated to 500 K. The four gases are allowed to react and the final concentrations of all four are determined when the reaction has reached equilibrium.

TABLE 14.1 Initial and Equilibrium Concentrations of the Reactants and Products in the Water–Gas Shift Reaction [$H_2O(g) + CO(g) \rightleftharpoons H_2(g) + CO_2(g)$] at 500 K

Experiment	INITIAL CONCENTRATION (*M*)				EQUILIBRIUM CONCENTRATION (*M*)			
	$[H_2O]$	[CO]	$[H_2]$	$[CO_2]$	$[H_2O]$	[CO]	$[H_2]$	$[CO_2]$
1	0.0200	0.0200	0	0	0.0034	0.0034	0.0166	0.0166
2	0	0	0.0200	0.0200	0.0034	0.0034	0.0166	0.0166
3	0.0100	0.0200	0.0300	0.0400	0.0046	0.0146	0.0354	0.0454
4	0.0200	0.0100	0.0200	0.0100	0.0118	0.0018	0.0282	0.0182

In experiment 1, the reaction vessel initially contains equimolar concentrations of H_2O and CO but no H_2 or CO_2. In experiment 2, the vessel initially contains equimolar concentrations of H_2 and CO_2 but no H_2O or CO. The data in Table 14.1 indicate that when the reaction mixtures in experiments 1 and 2 achieve chemical equilibrium, the concentrations of H_2O and CO are the same (0.0034 *M*) in both experiments and so too are the concentrations of H_2 and CO_2 (0.0166 *M*). (These statements about equilibrium concentrations are true in this case because the stoichiometric coefficients are all 1.) These results indicate that the composition of a reaction mixture at equilibrium is independent of the direction in which a particular reaction ran to achieve equilibrium.

Additional data in Table 14.1 show that the forward reaction takes place when the concentrations of the products are initially the same as the concentrations of the reactants (experiment 4) or even higher (experiment 3). The significance of these observations, taken with those from experiments 1 and 2, can be appreciated if we do the following math: multiply the equilibrium concentrations of the products (H_2 and CO_2) together and divide that value by the product of the equilibrium concentrations of the reactants (H_2O and CO):

$$\text{Experiments 1 and 2} \qquad \frac{[H_2][CO_2]}{[H_2O][CO]} = \frac{(0.0166)(0.0166)}{(0.0034)(0.0034)} = 24$$

$$\text{Experiment 3} \qquad \frac{[H_2][CO_2]}{[H_2O][CO]} = \frac{(0.0354)(0.0454)}{(0.0046)(0.0146)} = 24$$

$$\text{Experiment 4} \qquad \frac{[H_2][CO_2]}{[H_2O][CO]} = \frac{(0.0282)(0.0182)}{(0.0118)(0.0018)} = 24$$

In every experiment, this calculation yields the same result.

In fact, we would get the same ratio of product to reactant concentrations at equilibrium from any combination of initial concentrations of these four gases at 500 K. This constancy applies to other reaction mixtures and has been known since the mid-19th century, when Norwegian chemists Cato Guldberg (1836–1902) and Peter Waage (1833–1900) discovered that any reversible reaction eventually reaches a state in which the ratio of the concentrations of products to reactants, with each value raised to a power corresponding to the coefficient for that substance in the balanced chemical equation for the reaction, has a characteristic value at a given temperature. They called this phenomenon the **law of mass action**. This ratio of concentration terms is the equilibrium constant expression for the reaction. It is also called the **mass action expression**, but we will limit the use of that term to those reactions that are not at equilibrium. For the water–gas shift reaction at 500 K, the equilibrium constant expression is

$$K = \frac{[H_2][CO_2]}{[H_2O][CO]} = 24$$

The exponents in the equilibrium constant expression are the same as the coefficients in the balanced equation. This is in contrast to rate law expressions, where the exponents are frequently not the same as the coefficients in the balanced equation, because rate law exponents reflect the stoichiometry of the rate-determining step, not necessarily the overall reaction.

CONCEPT TEST

To run the water–gas shift reaction (Equation 14.2), equal numbers of moles of water vapor, carbon dioxide, carbon monoxide, and hydrogen gas are injected into a rigid, sealed reaction vessel and heated to 500 K. Which expression about the composition of the equilibrium reaction mixture is true?

a. $[CO] = [H_2] = [CO_2] = [H_2O]$

b. $[CO_2] = [H_2] > [CO] = [H_2O]$

c. $[CO_2] = [H_2] < [CO] = [H_2O]$

d. $[H_2O] = [H_2] > [CO_2] = [CO]$

e. $[H_2O] = [H_2] < [CO_2] = [CO]$

For the generic reaction in which *a* moles of reactant A react with *b* moles of B to form *c* moles of substance C and *d* moles of D,

$$a\,\mathrm{A} + b\,\mathrm{B} \rightleftharpoons c\,\mathrm{C} + d\,\mathrm{D}$$

the equilibrium constant expression is

$$K_c = \frac{[\mathrm{C}]^c[\mathrm{D}]^d}{[\mathrm{A}]^a[\mathrm{B}]^b} \tag{14.9}$$

CHEMTOUR Equilibrium in the Gas Phase

where the subscript "c" of the equilibrium constant represents concentration. If substances A, B, C, and D are gases, then the equilibrium constant may also be expressed in terms of their partial pressures:

$$K_p = \frac{(P_C)^c(P_D)^d}{(P_A)^a(P_B)^b} \tag{14.10}$$

As we'll see later in this chapter, the values of K_c and K_p for a given reaction and temperature may or may not be the same. It depends on whether the number of moles of gaseous reactants is the same as the number of moles of gaseous products.

Equilibrium constants never have units. This is true, even when there are different numbers of moles of reactants and products, because the concentrations and partial pressures in K_c or K_p expressions are actually ratios of concentration or pressure of substances to an ideal standard concentration (1.000 *M*) or partial pressure (1.000 atm). The units in such ratios cancel, so the terms in the equilibrium constant expression have no units. In practice, we use concentration and partial pressure values directly in equilibrium calculations and simply report the value of *K* as a number without units.

SAMPLE EXERCISE 14.1 Writing Equilibrium Constant Expressions **LO2**

A key reaction in the formation of acid rain involves the reversible combination of SO_2 and O_2 in the atmosphere, which produces SO_3:

$$2\,SO_2(g) + O_2(g) \rightleftharpoons 2\,SO_3(g)$$

Write the K_c and K_p expressions for this reaction.

Collect, Organize, and Analyze We are given the balanced chemical equation for a reaction and asked to write K_c and K_p expressions for the reaction. Equilibrium constant expressions are ratios of the concentrations (K_c) or partial pressures (K_p) of products to reactants in a reaction, with each term raised to the power equal to its coefficient in the balanced chemical equation. In the reaction of interest, the coefficients of SO_2 and SO_3 are both 2, so the SO_2 and SO_3 terms in the K_c and K_p expressions will be squared.

Solve

$$K_c = \frac{[SO_3]^2}{[SO_2]^2[O_2]}$$

$$K_p = \frac{(P_{SO_3})^2}{(P_{SO_2})^2(P_{O_2})}$$

Think About It The K_c and K_p expressions have the same form: their numerators and denominators contain terms for the same products and reactants, each raised to the same power. The difference between them is the nature of the terms: molar concentrations in the K_c expression and pressures in the K_p expression.

Practice Exercise Write the equilibrium constant expressions K_c and K_p for the following reaction:

$$CH_4(g) + H_2O(g) \rightleftharpoons CO(g) + 3\, H_2(g)$$

(Answers to Practice Exercises are in the back of the book.)

SAMPLE EXERCISE 14.2 Calculating the Value of K_c **LO3**

Table 14.2 contains data from four experiments on the dimerization of NO_2:

$$2\, NO_2(g) \rightleftharpoons N_2O_4(g)$$

The experiments were run at 100°C in a rigid, closed container. Use the data from each experiment to calculate a value of the equilibrium constant K_c for the dimerization reaction.

TABLE 14.2 Data for the Reaction $2\, NO_2(g) \rightleftharpoons N_2O_4(g)$ at 100°C

Experiment	INITIAL CONCENTRATION (*M*)		EQUILIBRIUM CONCENTRATION (*M*)	
	$[NO_2]$	$[N_2O_4]$	$[NO_2]$	$[N_2O_4]$
1	0.0200	0.0000	0.0172	0.00139
2	0.0300	0.0000	0.0244	0.00280
3	0.0400	0.0000	0.0310	0.00452
4	0.0000	0.0200	0.0310	0.00452

Collect and Organize We are given four sets of data that contain initial and equilibrium concentrations of a reactant and product. We are asked to determine the value of the equilibrium constant K_c in each experiment. The equilibrium constant expression (Equation 14.8) for this reaction is

$$K_c = \frac{[N_2O_4]}{[NO_2]^2}$$

Analyze In each of the four experiments, the concentration of NO_2 is nearly 10 times the concentration of N_2O_4 at equilibrium. However, both values in each experiment are much less than 1, and the $[NO_2]$ term is squared. These two factors taken together mean that the values of the numerators in the equilibrium constant expressions will probably be greater than the denominators, so the calculated values of K_c will be greater than 1.

Solve

$$\text{Experiment 1:} \quad K_c = \frac{[N_2O_4]}{[NO_2]^2} = \frac{0.00139}{(0.0172)^2} = 4.70$$

$$\text{Experiment 2:} \quad K_c = \frac{0.00280}{(0.0244)^2} = 4.70$$

$$\text{Experiment 3:} \quad K_c = \frac{0.00452}{(0.0310)^2} = 4.70$$

$$\text{Experiment 4:} \quad K_c = \frac{0.00452}{(0.0310)^2} = 4.70$$

Think About It The values calculated for K_c are the same, as they should be for the same reaction at the same temperature. They are greater than 1, too, as predicted.

Practice Exercise A mixture of gaseous CO and H_2, called *synthesis gas*, is used commercially to prepare methanol (CH_3OH), a compound considered to be an alternative fuel to gasoline. Under equilibrium conditions at 700 K, $[H_2] = 0.074\ M$, $[CO] = 0.025\ M$, and $[CH_3OH] = 0.040\ M$. What is the value of K_c for this reaction at 700 K?

SAMPLE EXERCISE 14.3 Calculating the Value of K_p — LO3

A sealed chamber contains an equilibrium mixture of NO_2 and N_2O_4 at 300°C. Their partial pressures are $P_{NO_2} = 0.101$ atm and $P_{N_2O_4} = 0.074$ atm. What is the value of K_p for the following reaction under these conditions?

$$2\ NO_2(g) \rightleftharpoons N_2O_4(g)$$

Collect and Organize We are asked to determine the value of the equilibrium constant K_p for the dimerization reaction of NO_2 to form N_2O_4. We are given the partial pressure values of both gases at equilibrium.

Analyze The form of the K_p expression for the reaction is the same as the K_c expression. The only difference is that the concentration terms in the K_p expression are replaced with partial pressure terms (Equation 14.9):

$$K_p = \frac{P_{N_2O_4}}{(P_{NO_2})^2}$$

The partial pressure of NO_2 is slightly greater than the partial pressure of N_2O_4 at equilibrium, but both values are less than 1. Moreover, the P_{NO_2} term is squared. These two factors taken together mean that the value of the numerator in the K_p expression will probably be greater than that of the denominator, so the calculated value of K_p will be greater than 1.

Solve

$$K_p = \frac{0.074}{(0.101)^2} = 7.3$$

Think About It As we predicted, $K_p > 1$, even though the equilibrium partial pressure of the product ($P_{N_2O_4} = 0.074$ atm) is actually less than that of the reactant ($P_{NO_2} =$ 0.101 atm). Relatively little N_2O_4 forms in this case because P_{NO_2} is so low, and two moles of NO_2 are required to make one mole of N_2O_4.

Practice Exercise A reaction vessel contains an equilibrium mixture of SO_2, O_2, and SO_3. If the partial pressures are $P_{SO_2} = 0.0018$ atm, $P_{O_2} = 0.0032$ atm, and $P_{SO_3} = 0.0166$ atm, what is the value of K_p for the following reaction?

$$2\ SO_2(g) + O_2(g) \rightleftharpoons 2\ SO_3(g)$$

The value of K indicates how far a reaction proceeds at a given temperature. The values we have seen thus far—$K_c = 24$ for the water–gas shift reaction at 500 K and $K_c = 4.7$ for the dimerization of NO_2 at 100°C, respectively—are considered intermediate values. Because the range of K values is so large ($0 < K < \infty$), all three of these values are considered close to 1, which means that comparable concentrations of reactants and products are likely to be present at equilibrium.

In contrast, the reaction between H_2 and O_2 to form water proceeds until one of the reactants is almost completely consumed. This observation is consistent with a large value of K at 25°C:

$$2\ H_2(g) + O_2(g) \rightleftharpoons 2\ H_2O(g) \qquad K_c = 3 \times 10^{81}$$

The decomposition of CO_2 to CO and O_2 at 25°C proceeds hardly at all and is consistent with a very small value of K and virtually no product being formed:

$$2\ CO_2(g) \rightleftharpoons 2\ CO(g) + O_2(g) \qquad K_c = 3 \times 10^{-92}$$

14.3 Relationships between K_c and K_p Values

As we noted in Section 14.2, the values of K_c and K_p for a given reaction and temperature may or may not be the same, depending on the numbers of moles of gaseous reactants and products. To better understand this relationship, we begin with the ideal gas law:

$$PV = nRT$$

If we solve for P and express volume in liters, then n/V has units of moles per liter, which is the same as molarity (M):

$$P = \frac{n}{V}RT = MRT \qquad (14.11)$$

Applying Equation 14.11 to the gases in the NO_2/N_2O_4 equilibrium from Sample Exercise 14.3 gives

$$P_{NO_2} = \frac{n_{NO_2}}{V}RT = [NO_2]RT$$

$$P_{N_2O_4} = \frac{n_{N_2O_4}}{V}RT = [N_2O_4]RT$$

Substituting these values into the expression for K_p from Sample Exercise 14.3, we get

$$K_p = \frac{P_{N_2O_4}}{(P_{NO_2})^2} = \frac{[N_2O_4]RT}{([NO_2]RT)^2} = \frac{[N_2O_4]\cancel{RT}}{[NO_2]^2(RT)^{\cancel{2}}}$$

The ratio of concentration terms in the expression on the right, $[N_2O_4]/[NO_2]^2$, is the same as the K_c expression for this reaction. Substituting K_c for those terms and simplifying the RT terms gives

$$K_p = K_c\frac{1}{RT}$$

This last equation defines the specific relationship between K_c and K_p for this reaction. A more general expression can be derived for the generic reaction of gases A and B to form gases C and D:

$$a\,A + b\,B \rightleftharpoons c\,C + d\,D$$

The K_p expression for this reaction is

$$K_p = \frac{(P_C)^c(P_D)^d}{(P_A)^a(P_B)^b} \qquad (14.10)$$

Replacing each partial pressure term in Equation 14.10 with the corresponding molar concentration term $\times$ RT (from Equation 14.11) gives us the general expression

$$K_p = \frac{([C]RT)^c([D]RT)^d}{([A]RT)^a([B]RT)^b} \qquad (14.12)$$

Combining the RT terms, we get

$$K_p = \frac{[C]^c[D]^d}{[A]^a[B]^b} \times (RT)^{[(c+d)-(a+b)]} \qquad (14.13)$$

The concentration ratio on the right side of this equation matches the K_c expression for this generic reaction (see Equation 14.9). Substituting this equality into Equation 14.13 gives us

$$K_p = K_c(RT)^{[(c+d)-(a+b)]} \qquad (14.14)$$

To simplify Equation 14.14, consider this: $(c + d)$ represents the sum of the coefficients of the gaseous products in the reaction—that is, the sum of the number of moles of gases produced. Similarly, $(a + b)$ represents the sum of the number of moles of gaseous reactants consumed. The difference between the two sums, $(c + d) - (a + b)$, represents the change in the number of moles of gases from the reactant side to the product side of the balanced chemical equation. We use the symbol Δn to represent this change. Substituting Δn for $(c + d) - (a + b)$ in Equation 14.14 gives us

$$K_p = K_c(RT)^{\Delta n} \qquad (14.15)$$

Equation 14.15 provides a quantitative interpretation of the opening statement of this section: the relationship between the K_p and K_c values of a chemical reaction involving gases depends on the number of moles of gaseous reactants and products. In reactions such as the water–gas shift reaction,

$$H_2O(g) + CO(g) \rightleftharpoons H_2(g) + CO_2(g) \quad (14.2)$$

in which the number of moles of gas on both sides of the reaction arrow is the same, $\Delta n = 0$ and $K_p = K_c$. However, in the steam-reforming reaction,

$$CH_4(g) + H_2O(g) \rightleftharpoons CO(g) + 3\ H_2(g) \quad (14.1)$$

two moles of gaseous reactants form four moles of gaseous products. Therefore,

$$\Delta n = 4 \text{ mol} - 2 \text{ mol} = 2 \text{ mol}$$

Inserting this value for Δn in Equation 14.15 gives us the following relationship between K_p and K_c for this reaction:

$$K_p = K_c(RT)^2$$

A final point about the relative sizes of K_p and K_c values: one mole of an ideal gas at STP (273 K and 1 atm) occupies a volume of 22.4 L. Therefore, its molar concentration is 1 mol/22.4 L = 0.0446 *M*. Thus, the pressure of a pure gas at STP is 22.4 times its molar concentration. The *RT* term in Equation 14.15 is essentially a conversion factor for changing molar concentrations into partial pressures. The value of *R* that we use depends on the units used to express concentration, pressure, and temperature. If we stick with molarity, atmospheres, and *T* in kelvin, the value of *R* is 0.08206 (L · atm)/(mol · K).

CONNECTION As stated in Section 10.6, the *molar volume* of an ideal gas is 22.4 L at T = 273 K and P = 1 atm. These values are widely considered *standard* temperature and pressure (STP).

SAMPLE EXERCISE 14.4 Calculating K_c from K_p — LO4

In Sample Exercise 14.3 we calculated the value of K_p (7.3) for the dimerization of NO_2 to N_2O_4 at 300°C. What is the value of K_c for this reaction at 300°C?

Collect and Organize We are given a K_p value and asked to calculate the corresponding K_c value for the same reaction at the same temperature. Equation 14.15 relates K_p and K_c values:

$$K_p = K_c(RT)^{\Delta n}$$

where Δn represents the change in the number of moles of gas when going from reactant to product.

Analyze We need a balanced chemical equation to determine the value of Δn. From Sample Exercise 14.3 we know that the equation is

$$2\ NO_2(g) \rightleftharpoons N_2O_4(g)$$

Because the number of moles of gas is not the same on both sides of the reaction arrow, the values of K_c and K_p will not be the same.

Solve According to the balanced equation, two moles of gaseous reactants yield one mole of gaseous product. Therefore:

$$\Delta n = 1 \text{ mol} - 2 \text{ mol} = -1 \text{ mol}$$

Inserting this value, the known value of K_p, and the temperature into Equation 14.15,

$$K_p = K_c(RT)^{\Delta n}$$

$$7.3 = K_c[0.08206 \times (573)]^{-1}$$

$$K_c = (7.3)(0.08206)(573) = 3.4 \times 10^2$$

Think About It The value of K_c differs from the value of K_p because two moles of gaseous reactants produce only one mole of gaseous product. Actually, the value of K_c is nearly 50 times larger than that of K_p.

Practice Exercise What is the value of K_p for the industrial synthesis of ammonia at 500°C?

$$N_2(g) + 3\,H_2(g) \rightleftharpoons 2\,NH_3(g) \qquad K_c = 5.8 \times 10^{-2} \text{ at } 500°C$$

14.4 Manipulating Equilibrium Constant Expressions

If $K_c = 100$ for the following hypothetical combination reaction,

$$A + B \rightleftharpoons 2\,C \qquad K_c = 100$$

then what is K_c for the reverse reaction?

$$2\,C \rightleftharpoons A + B \qquad K_{c,\text{reverse}} = ?$$

Moreover, what is K_c for the reaction in which only one mole of C is made—that is, in which all of the coefficients are half what they were in the original equation?

$$\tfrac{1}{2}\,A + \tfrac{1}{2}\,B \rightleftharpoons C \qquad K_{c,1/2} = ?$$

To answer questions like these we need to think about how the equilibrium constant expressions for these reactions are related. We explore these relationships in this section.

K for Reverse Reactions

The K_c expression for our combination reaction, $A + B \rightleftharpoons 2\,C$, is

$$K_c = \frac{[C]^2}{[A][B]} = 100$$

The K_c expression for the reverse reaction, $2\,C \rightleftharpoons A + B$, is

$$K_{c,\text{reverse}} = \frac{[A][B]}{[C]^2}$$

Comparing the K_c expressions for the forward and reverse reactions, we see that $K_{c,\text{reverse}}$ is the reciprocal of K_c, which means its value must be $1/100 = 0.0100$:

$$2\,C \rightleftharpoons A + B \qquad K_{c,\text{reverse}} = 0.0100$$

Generalizing this result for any chemical reaction, we have

$$K_{\text{forward}} = \frac{1}{K_{\text{reverse}}} \tag{14.16}$$

This inverse relationship between K_{forward} and K_{reverse} makes sense when we apply it to our hypothetical reaction. The relatively large value of K_{forward} (100) means that the concentration of product C in a reaction mixture at equilibrium is likely to be much higher than the concentrations of reactants A and B. The corresponding small value of K_{reverse} (0.0100) means the same thing in terms of the relative values of [A], [B], and [C]. The difference is that the equilibrium mixture in the

reverse reaction contains mostly reactant (C) and little of the products (A and B), as we would expect given the small value of $K_{reverse}$.

K for an Equation Multiplied by a Number

The K_c expression for our combination reaction, $\frac{1}{2}\,A + \frac{1}{2}\,B \rightleftharpoons C$, representing the preparation of only one mole of C, is

$$K_c = \frac{[C]}{[A]^{1/2}[B]^{1/2}}$$

Comparing this expression with the K_c expression for the forward reaction,

$$K_c = \frac{[C]^2}{[A][B]}$$

we see that the K_c expression for the preparation of only one mole of C is equal to the square root of the K_c expression for the forward reaction that produces two moles of C. The C term is squared while the A and B terms are raised to the first power in the latter, but the C term is only raised to the first power while the A and B terms are raised to the $\frac{1}{2}$ power in the former. Therefore, the value of $K_{c,1/2}$ is the square root of the value of K_c, or $\sqrt{100} = 10.0$. We can extend this pattern to all chemical equilibria. If the balanced chemical equation of a reaction is multiplied by some factor n, then the value of K is raised to the nth power (K^n).

SAMPLE EXERCISE 14.5 Calculating the K Values of Related Chemical Reactions **LO5**

Recall from Chapter 13 that a key reaction in the formation of photochemical smog is the reaction between NO and O_2:

$$(1) \qquad 2\,NO(g) + O_2(g) \rightleftharpoons 2\,NO_2(g)$$

The K_p value of this reaction is 2.4×10^{12} at 25°C.

a. What is the K_p value of the decomposition of $NO_2(g)$?

$$(2) \qquad 2\,NO_2(g) \rightleftharpoons 2\,NO(g) + O_2(g)$$

b. What is the K_p value of the reaction in which NO and O_2 combine to form only one mole of NO_2?

$$(3) \qquad NO(g) + \tfrac{1}{2}\,O_2(g) \rightleftharpoons NO_2(g)$$

Collect, Organize, and Analyze We know the K_p value of reaction 1, but not of reactions 2 and 3. Reaction 2 is the reverse of reaction 1. Therefore, the K_p value of reaction 2 is the reciprocal of the K_p value of reaction 1. Reaction 3 has the same reactants and product as reaction 1, but the stoichiometric coefficients in its chemical equation are one-half of the values used in the chemical equation for reaction 1. When the coefficients of a chemical equation are multiplied by a factor n ($\frac{1}{2}$ in this case), the value of the equilibrium constant is raised to the nth power, which in this case means that the K_p value of reaction 3 is the square root of the K_p value of reaction 1. The value of $K_{p,rxn1}$ is huge—a bit more than 10^{12}—which means the value of $K_{p,rxn2}$ for the reverse reaction should be very small—somewhat less than 10^{-12}. The value of $K_{p,rxn3}$ is equal to $\sqrt{K_{p,rxn1}}$ or about 10^6.

Solve

a. $K_{p,rxn2} = \dfrac{1}{K_{p,1}} = \dfrac{1}{2.4 \times 10^{12}} = 4.2 \times 10^{-13}$

b. $K_{p,rxn3} = \sqrt{K_{p,1}} = \sqrt{2.4 \times 10^{12}} = 1.5 \times 10^{6}$

Think About It As predicted, the value of $K_{p,rxn2}$ is very small and the value of $K_{p,rxn3}$ is quite large. Both values carry the same message: NO should (eventually) be oxidized to NO_2 in the atmosphere. How long the reaction takes, as we saw in Chapter 13, is another matter.

 **Practice Exercise** $K_c = 9.6$ for the synthesis of ammonia at 300°C.

$$N_2(g) + 3\ H_2(g) \rightleftharpoons 2\ NH_3(g)$$

What is the value of K_c for the following decomposition of ammonia at 300°C?

$$NH_3(g) \rightleftharpoons \tfrac{1}{2}\ N_2(g) + \tfrac{3}{2}\ H_2(g)$$

Given two equally legitimate equilibrium constant expressions for the same reaction, you may wonder how the same reaction with the same reactants and products can have two or more equilibrium constant values. Surely the same ingredients should be present in the same proportions at equilibrium no matter how we choose to write a balanced equation describing their reaction. In fact, they are. The difference in K values is not chemical; it's mathematical. It has to do with the two different ways we wrote balanced chemical equations describing the reaction, which resulted in two different sets of exponents on the concentration terms in the two K expressions. Thus, exactly the same concentrations of reactants and products, raised to different powers, produce different values of K.

Combining *K* Values

In Chapter 9, we applied Hess's law to calculate the enthalpies of combined reactions. We carry out a similar process here to determine the overall K for a reaction that is the sum of two or more other reactions.

Consider two reactions from Chapter 13 involved in the formation of photochemical smog, wherein the NO produced in a car's engine at high temperatures is oxidized to NO_2 in the atmosphere:

$$\begin{array}{ll} (1) & N_2(g) + O_2(g) \rightleftharpoons \cancel{2\ NO(g)} \\ (2) & \cancel{2\ NO(g)} + O_2(g) \rightleftharpoons 2\ NO_2(g) \\ \hline \text{Overall:} & N_2(g) + 2\ O_2(g) \rightleftharpoons 2\ NO_2(g) \end{array}$$

The equilibrium constant expression for the overall reaction is

$$K_c = \frac{[NO_2]^2}{[N_2][O_2]^2}$$

We can derive this expression from the equilibrium constant expressions for reactions 1 and 2,

$$K_1 = \frac{[NO]^2}{[N_2][O_2]} \quad \text{and} \quad K_2 = \frac{[NO_2]^2}{[NO]^2[O_2]}$$

if we multiply K_1 by K_2:

$$K_1 \times K_2 = \frac{\cancel{[NO]^2}}{[N_2][O_2]} \times \frac{[NO_2]^2}{\cancel{[NO]^2}[O_2]} = \frac{[NO_2]^2}{[N_2][O_2]} = K_{overall}$$

This approach works for all series of reactions, and as a general rule:

$$K_{overall} = K_1 \times K_2 \times K_3 \times \cdots \times K_n \tag{14.17}$$

The overall equilibrium constant for a sum of two or more reactions is the product of the equilibrium constants of the individual reactions. Thus, the value of K_c for the overall reaction for the formation of NO_2 from N_2 and O_2 at 1000 K is the product of the equilibrium constants for reactions 1 and 2:

$$K_1 = \frac{[NO]^2}{[N_2][O_2]} = 7.2 \times 10^{-9}$$

$$K_2 = \frac{[NO_2]^2}{[NO]^2[O_2]} = 0.020$$

$$K_{overall} = K_1 \times K_2 = (7.2 \times 10^{-9})(0.020) = 1.4 \times 10^{-10}$$

Remember that the equilibrium constant expression for the overall reaction must contain the appropriate terms for the products and reactants of that reaction. Just as with Hess's law in thermochemical calculations, we may need to reverse an equation or multiply an equation by a factor when we combine it with another to create the equation of interest. If we reverse a reaction, we must take the reciprocal of its K. If we multiply a reaction by a constant, we must raise its K to that power.

SAMPLE EXERCISE 14.6 Calculating Overall K Values of Combined Reactions **LO5**

At 1000 K, the K_c values of the following reactions are as given:

(1) $N_2O_4(g) \rightleftharpoons 2\ NO_2(g) \qquad K_c = 1.5 \times 10^6$

(2) $N_2(g) + 2\ O_2(g) \rightleftharpoons 2\ NO_2(g) \qquad K_c = 1.4 \times 10^{-10}$

What is the K_c value at 1000 K of the following reaction?

$$N_2(g) + 2\ O_2(g) \rightleftharpoons N_2O_4(g)$$

Collect and Organize We are given two reactions and their K_c values. We need to combine the two reactions in such a way that N_2 and O_2 are on the reactant side of the overall equation and N_2O_4 is on the product side and then calculate the K_c value of the overall reaction.

Analyze The overall reaction is the sum of the reverse of reaction 1 and reaction 2 as written:

Reaction 1 reversed	$\cancel{2\ NO_2(g)} \rightleftharpoons N_2O_4(g)$
Reaction 2	$N_2(g) + 2\ O_2(g) \rightleftharpoons \cancel{2\ NO_2(g)}$
Overall	$N_2(g) + 2\ O_2(g) \rightleftharpoons N_2O_4(g)$

Reversing a chemical reaction requires taking the reciprocal of its K_c value, and when two reactions are added, the value of K_c of the overall reaction is the product of the K_c values of the two reactions. The K_c value of reaction 1 is large ($>10^6$), which means its reciprocal is small ($<10^{-6}$). The product of the latter value and the even smaller K_c value of reaction 2 ($\sim 10^{-10}$) should be a very small value—about 10^{-16}.

Solve

$$K_{overall} = \frac{1}{K_1} \times K_2 = \left(\frac{1}{1.5 \times 10^6}\right)(1.4 \times 10^{-10}) = 9.3 \times 10^{-17}$$

Think About It The result of the calculation is close to our estimated value. The overall equilibrium of the combined reactions lies far to the left; very little N_2O_4 forms from N_2 and O_2 at 1000 K.

Practice Exercise Calculate the value of K_c for the hypothetical reaction,

$$Q(g) + X(g) \rightleftharpoons M(g)$$

given the following information:

$$2\,M(g) \rightleftharpoons Z(g) \qquad K_c = 6.2 \times 10^{-4}$$

$$Z(g) \rightleftharpoons 2\,Q(g) + 2\,X(g) \qquad K_c = 5.6 \times 10^{-2}$$

To summarize the key points for manipulating equilibrium constants:

- The K value of a reaction running in reverse is the reciprocal of K for the forward reaction.
- If the original chemical equation describing an equilibrium is multiplied by a factor n, the value of K of the new equilibrium constant expression is the value of the original K raised to the nth power.
- If an overall chemical reaction is the sum of two or more other reactions, the overall value of K is the product of the K values of the other reactions.

14.5 Equilibrium Constants and Reaction Quotients

In Sections 14.1 and 14.2, we introduced two key terms: *equilibrium constant expression* and *mass action expression*. Until now we have used the first term almost exclusively because we have been dealing with chemical reactions that have achieved equilibrium. Now it is time to reintroduce the concept of a mass action expression because we can apply it not only to the concentrations (or partial pressures) of products and reactants in reaction mixtures that have reached equilibrium, but also to reaction mixtures that are on their way to equilibrium but are not there yet.

Even if a reversible chemical reaction has not reached equilibrium, we can still insert reactant and product concentrations (or partial pressures) into its mass action expression. The mathematical result is not a K value because the reaction is not yet at equilibrium. Instead, it is a Q value, where $\boldsymbol{Q}$ stands for **reaction quotient**.

reaction quotient (Q) the numerical value of the mass action expression for any values of the concentrations (or partial pressures) of reactants and products; at equilibrium, $Q = K$.

The value of Q provides us with a kind of status report on how a reaction is proceeding. To see how, let's revisit the water–gas shift reaction,

$$H_2O(g) + CO(g) \rightleftharpoons H_2(g) + CO_2(g) \qquad K_c = 24 \text{ at } 500\text{ K}$$

and the data from experiment 3 in Table 14.1, where the initial concentrations of reactants and products are as follows:

$[H_2O]$ (M)	$[CO]$ (M)	$[H_2]$ (M)	$[CO_2]$ (M)
0.0100	0.0200	0.0300	0.0400

Inserting these values into the mass action expression for the reaction yields a value for Q based on concentration—that is, Q_c:

$$Q_c = \frac{[H_2][CO_2]}{[H_2O][CO]} = \frac{(0.0300)(0.0400)}{(0.0100)(0.0200)} = 6.00$$

$K_c = 24$ for this reaction, so $Q_c < K_c$, which means there are proportionally smaller concentrations of products and larger concentrations of reactants in the initial reaction mixture than there will be at equilibrium. To achieve equilibrium,

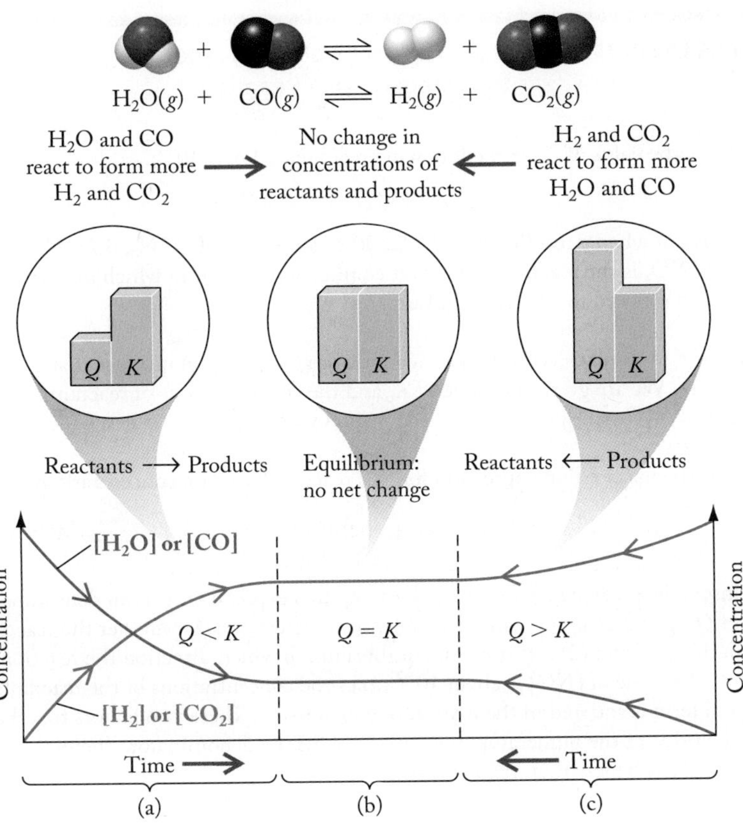

FIGURE 14.5 The value of the reaction quotient Q relative to the equilibrium constant K for the water–gas shift reaction. (a) Reactant concentrations (red) are higher than they are once equilibrium is reached, and product concentrations (blue) are lower than at equilibrium; $Q < K$, and the reactants form more products. (b) Equilibrium concentrations are achieved; $Q = K$, and no net change in concentrations takes place. (c) Product concentrations are higher than they are at equilibrium, and reactant concentrations are lower than at equilibrium; $Q > K$, and products form more reactants as the reaction runs in reverse.

some of the reactants must form products, increasing the value of Q_c until it matches the value of K_c and the following equilibrium concentrations of reactants and products are present in the reaction vessel:

[H_2O] (*M*)	[CO] (*M*)	[H_2] (*M*)	[CO_2] (*M*)
0.0046	0.0146	0.0354	0.0454

To put the results from the data in Table 14.1 in context, let's consider the curves in Figure 14.5. Starting at the left end of the graph (zone a), we have the initial conditions of experiment 1 from Table 14.1—namely, equal concentrations of reactants are present, but no products. Over time, reactant concentrations (the red curve) decrease as product concentrations (the blue curve) increase. At the right end of the graph (zone c), we have the initial conditions of experiment 2—namely, products are present, but no reactants. Over time, reactant concentrations increase as product concentrations decrease. In the middle of the graph (zone b), no net change occurs in the composition because the reaction is at equilibrium.

We can also characterize the three zones on the basis of the value of Q compared to K. In zone (a), $Q < K$ and there is a net conversion of reactants into products as the forward reaction dominates. In zone (c), $Q > K$ and a net conversion of products into reactants takes place as the reverse reaction dominates. In the middle, zone (b), $Q = K$ and no net change in the composition of the reaction mixture occurs over time. The relative values of Q and K and their consequences are summarized in Table 14.3.

TABLE 14.3 Comparison of Q and K Values

Value of Q	What It Means
$Q < K$	Reaction as written proceeds in forward direction (⟶).
$Q = K$	Reaction is at equilibrium (⇌).
$Q > K$	Reaction as written proceeds in reverse direction (⟵).

CONCEPT TEST

Do the initial reaction conditions in experiments 3 and 4 in Table 14.1 belong in zone (a), (b), or (c) in Figure 14.5?

SAMPLE EXERCISE 14.7 Using Q and K Values to Predict the Direction of a Reaction **LO6**

At 2300 K the value of K_c of the following reaction is 1.5×10^{-3}:

$$N_2(g) + O_2(g) \rightleftharpoons 2\,NO(g)$$

At the instant when a reaction vessel at 2300 K contains 0.50 M N_2, 0.25 M O_2, and 0.0042 M NO, is the reaction mixture at equilibrium? If not, in which direction will the reaction proceed to reach equilibrium?

Collect and Organize We are asked whether or not a reaction mixture is at equilibrium. We are given the value of K_c and the concentrations of reactants and product.

Analyze The mass action expression for this reaction based on concentrations is

$$Q_c = \frac{[NO]^2}{[N_2][O_2]}$$

If we insert the given concentration values into this expression, we can determine the value of Q_c. Comparing Q_c with K_c enables us to determine (1) whether the reaction is at equilibrium and (2) if it is not at equilibrium, in which direction the reaction will proceed. The value of [NO] is about 10^{-2} times the concentrations of the reactants, and the [NO] term is squared in the mass action expression. These two factors together make the value of the numerator about 10^{-4} that of the denominator. Therefore, the value of Q_c may be less than the value of K_c.

Solve

$$Q_c = \frac{(0.0042)^2}{(0.50)(0.25)} = 1.4 \times 10^{-4}$$

$Q_c < K_c$, so the reaction mixture is not at equilibrium. To achieve equilibrium, more reactants must form products so that the numerator of the mass action expression increases and the denominator decreases. This happens if the reaction proceeds in the forward direction.

Think About It Our prediction that Q_c would be less than K_c was correct. The value of K_c is small, but the value of Q_c is even smaller.

 **Practice Exercise** The value of K_c for the reaction

$$2\,NO_2(g) \rightleftharpoons N_2O_4(g)$$

is 4.7 at 373 K. Is a mixture of the two gases in which $[NO_2] = 0.025\ M$ and $[N_2O_4] = 0.0014\ M$ in chemical equilibrium? If not, in which direction does the reaction proceed to achieve equilibrium?

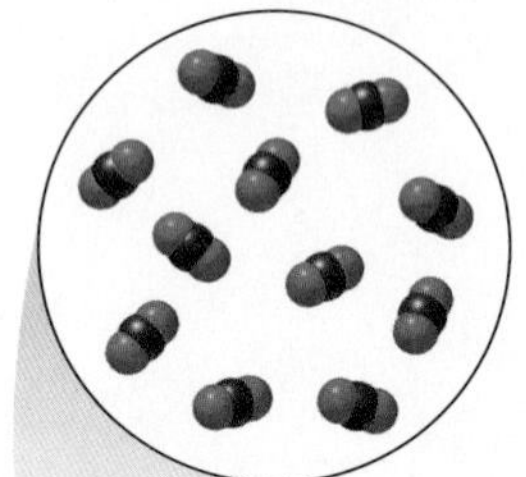

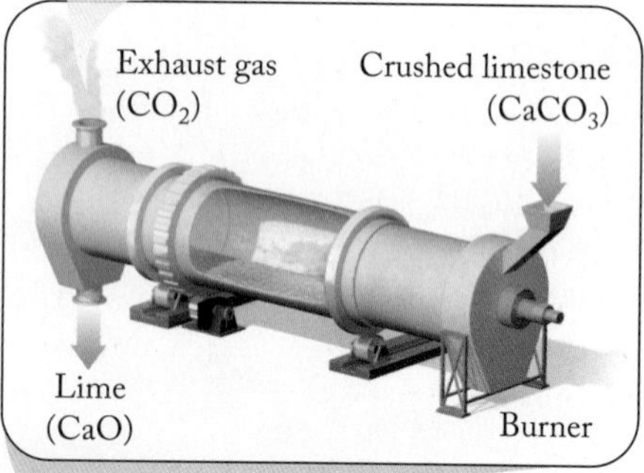

FIGURE 14.6 Rotary kilns operating at 1000°C are used to convert limestone into lime. Rotation of these large cylindrical kilns ensures that crushed limestone ($CaCO_3$) is uniformly heated and has thermally decomposed to lime (CaO) and CO_2 gas by the time it passes through the kiln.

14.6 Heterogeneous Equilibria

Thus far in this chapter we have focused on reactions in the gas phase. However, the principles of chemical equilibrium also apply to reactions in the liquid phase, particularly reactions in solution. Equilibria in which products and reactants are all in the same phase are **homogeneous equilibria**. Equilibria in which reactants and products are in different phases are **heterogeneous equilibria**.

In Sample Exercise 14.1 we considered the equilibrium associated with the oxidation of SO_2 to SO_3, a key step in forming aerosols of H_2SO_4 in the atmosphere. One way to prevent this reaction from happening is to "scrub" SO_2 from the exhaust gases emitted by factories where sulfur-containing fuels are burned.

Solid lime, CaO, is a widely used scrubbing agent. Sprayed into the exhaust gases, it combines with SO_2 to form calcium sulfite:

$$CaO(s) + SO_2(g) \rightleftharpoons CaSO_3(s)$$

The large quantities of lime needed for this reaction and for many other industrial and agricultural uses come from heating pulverized limestone, which is mostly $CaCO_3$, in kilns (Figure 14.6) operated at temperatures near 1000°C. At these temperatures, $CaCO_3$ decomposes into lime (CaO) and CO_2 gas:

$$CaCO_3(s) \rightleftharpoons CaO(s) + CO_2(g) \qquad \Delta H^\circ_{rxn} = 178.1 \text{ kJ}$$

Based on what we have learned so far, the concentration-based equilibrium constant expression for this reaction would be as follows:

$$K_c = \frac{[CaO][CO_2]}{[CaCO_3]}$$

This expression contains concentration terms for two solids: CaO and $CaCO_3$. But what do we mean by the concentration of a solid? Any pure solid has a constant concentration because its mass (and number of moles) per unit volume is always the same. As long as there is any CaO or $CaCO_3$ present, the effective "concentration" of both is 1, and there is no need for [CaO] and $[CaCO_3]$ terms in the equilibrium constant expression. This leaves us with

$$K_c = [CO_2]$$

This expression means that, as long as some CaO or $CaCO_3$ is present, the equilibrium concentration of CO_2 gas does not vary at a given temperature, as shown in Figure 14.7. Instead, the concentration of CO_2 is the same as the value of K_c at that temperature.

The same concept of constant concentration applies to pure liquids that are involved in reversible chemical reactions. As long as the liquid is present, its "concentration" is considered constant during the course of the reaction and does not appear in the equilibrium constant expression. Similarly, K_c expressions for most reactions in aqueous solutions do not include a term for $[H_2O]$, even when water

homogeneous equilibrium an equilibrium in which reactants and products are in the same phase.

heterogeneous equilibrium an equilibrium in which reactants and products are in more than one phase.

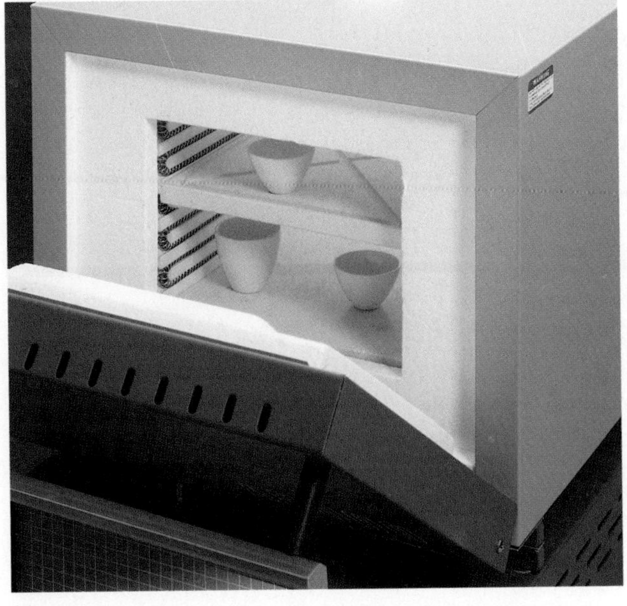

(a)

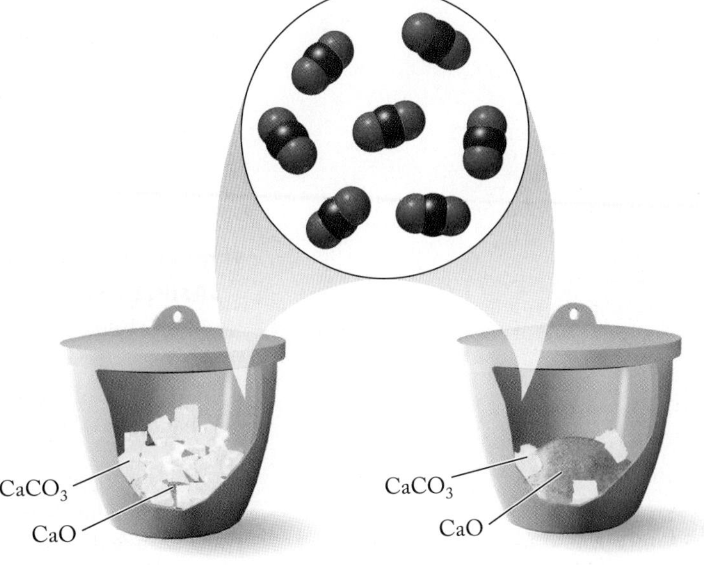

(b)

FIGURE 14.7 The position of a heterogeneous equilibrium between $CaCO_3$, CaO, and CO_2 at constant temperature depends only on the concentration of CO_2 gas present. As long as some solid is in the system, the equilibrium concentration of CO_2 remains the same. (a) Muffle furnace for heating crucibles containing $CaCO_3$. (b) Two crucibles containing different amounts of solid material at the same temperature have the same concentration of CO_2 gas.

is a reactant or product, because its concentration does not change significantly. In writing equilibrium constant expressions for heterogeneous equilibria, we follow the rules we learned earlier, with the additional rule that pure liquids and solids do not appear in the expression.

SAMPLE EXERCISE 14.8 Writing Equilibrium Constant Expressions for Heterogeneous Equilibria LO2

Write K_c expressions for the following reactions:

a. $CaO(s) + SO_2(g) \rightleftharpoons CaSO_3(s)$
b. $CO_2(g) + H_2O(\ell) \rightleftharpoons H_2CO_3(aq)$

Collect, Organize, and Analyze We are given reversible reactions involving reactants and products in more than one phase, and we are asked to write equilibrium constant expressions for their equilibria. We need to identify the pure liquids and solids involved in the equilibria and to omit terms for them from the equilibrium constant expressions, because they are considered to have constant concentrations. The first equilibrium involves two solids: CaO and $CaSO_3$. The second involves liquid H_2O.

Solve

a. There are no concentration terms for the two solids, CaO and $CaSO_3$, in the equilibrium constant expression for reaction (a), which leaves us with a term for only the gas-phase reactant, SO_2, in the denominator:

$$K_c = \frac{1}{[SO_2]}$$

b. In the equilibrium constant expression for reaction (b), $[H_2O]$ is a constant and is not included, leaving

$$K_c = \frac{[H_2CO_3]}{[CO_2]}$$

Think About It The equilibrium constant expressions we have written in this Sample Exercise and in previous exercises included terms for reactants and products whose concentrations or partial pressures were likely to change significantly during the course of the reaction. We omit terms for pure solids and liquids because their concentrations do not change.

Practice Exercise Write K_p expressions for the following reactions:

a. $C(s) + CO_2(g) \rightleftharpoons 2\,CO(g)$
b. $CO_2(g) + H_2(g) \rightleftharpoons CO(g) + H_2O(\ell)$

SAMPLE EXERCISE 14.9 Relating K_c and K_p Values of Heterogeneous Equilibria LO4

The K_p value of the reaction $CaO(s) + CO_2(g) \rightleftharpoons CaCO_3(s)$ is 9.6×10^{22} at 298 K. What is the value of K_c at the same temperature?

Collect, Organize, and Analyze We are given the K_p value of a heterogeneous equilibrium involving a gaseous reactant, a solid reactant, and a solid product at 298 K and are asked to calculate the value of K_c. Equation 14.15 relates the K_p and K_c values of a reversible reaction with one or more gaseous components: $K_p = K_c(RT)^{\Delta n}$.

Solve The chemical equation describes a reaction that consumes one mole of gas and produces none, so $\Delta n = -1$. Solving Equation 14.15 for K_c and inserting the values for K_p, temperature, and Δn:

$$K_p = K_c(RT)^{\Delta n}; \text{ so } K_c = K_p(RT)^{-\Delta n}$$
$$= 9.6 \times 10^{22}\,(0.08206 \times 298\text{ K})^{-(-1)} = 2.3 \times 10^{24}$$

Think About It The reaction has a very large value of K whether it is expressed as K_c or K_p, indicating that it goes to completion at 298 K. No wonder $CaCO_3$, not CaO, is one of the principle chemical forms of calcium in Earth's crust.

Practice Exercise What is the K_c value at 298 K of the decomposition of sodium bicarbonate if the value of K_p at 298 K is:

$$2\,NaHCO_3(s) \rightleftharpoons Na_2CO_3(s) + CO_2(g) + H_2O(g) \qquad K_p = 8.6 \times 10^{-7}$$

CONCEPT TEST

Explain why $K_c = [H_2O(g)]$ is the mass action expression for the equilibrium $H_2O(\ell) \rightleftharpoons H_2O(g)$.

14.7 Le Châtelier's Principle

CHEMTOUR Le Châtelier's Principle

We can perturb chemical reactions at equilibrium in several ways, including by changing the concentration or partial pressure of a reactant or product, or by changing the temperature of the system. Adding or removing an ingredient in the system alters the value of the reaction quotient Q so that it is no longer equal to the value of K. On the other hand, changing the temperature of a system changes the value of K. Either way, the perturbed system is not at equilibrium and the composition of the system must change to restore equilibrium.

One of the first scientists to study and then successfully predict how chemical equilibria respond to such perturbations was French chemist Henri Louis Le Châtelier (1850–1936). He articulated what is now known as **Le Châtelier's principle**, which states that, if a system at equilibrium is perturbed (or stressed), the position of the equilibrium shifts in the direction that relieves the stress. Through the years, chemists have used Le Châtelier's principle to increase the yields of chemical reactions that would otherwise have produced very little of a desired compound.

Effects of Adding or Removing Reactants or Products

When a reactant or product is added or removed, a system at chemical equilibrium is perturbed. Following Le Châtelier's principle, the system responds in such a way as to restore equilibrium.

To explore how industrial chemists exploit Le Châtelier's principle, we turn once again to the water–gas shift reaction for making hydrogen:

$$H_2O(g) + CO(g) \rightleftharpoons H_2(g) + CO_2(g) \qquad (14.2)$$

To shift the equilibrium toward the production of more H_2, chemists pass the reaction mixture through a scrubber containing a concentrated aqueous solution

Le Châtelier's principle a system at equilibrium responds to a stress in such a way that it relieves that stress.

of K_2CO_3. Doing this removes CO_2 from the gaseous mixture as a result of the following reaction:

$$CO_2(g) + H_2O(\ell) + K_2CO_3(aq) \rightleftharpoons 2\ KHCO_3(s)$$

Removing CO_2 means fewer molecules of it are available to collide with molecules of H_2 to drive the reverse reaction in Equation 14.2. As a result, the rate of the reverse reaction becomes slower than the rate of the forward reaction. This means the system is no longer in equilibrium. To return to equilibrium, the reaction proceeds in the forward direction (we say that the reaction *shifts to the right*), making more product to restore some of what was removed until a new equilibrium is achieved. The new equilibrium is like the old one in that the value of the mass action expression:

$$\frac{[H_2][CO_2]}{[H_2O][CO]}$$

is equal to the value of K. This is true even though the concentrations of the individual reactants and products have changed; the overall ratio in the K expression is restored.

Another way to shift a reaction mixture at equilibrium is to add more reactant or product. If the goal is to form more products, then adding more reactants increases their concentration in the reaction mixture, which increases the rate of the forward reaction. Some of the added reactants are converted into additional products. This approach works even when only one reactant is added in a multireactant reaction, as long as there is still some of the other reactant(s) available. To understand why, consider what happens when just one of the reactant concentration terms in the equilibrium constant expression is increased. The result is a reaction quotient Q that is less than K, which means the reaction will proceed in the forward direction, forming more products, until equilibrium is restored (i.e., until $Q = K$ again).

CONCEPT TEST

The reaction mixture in the water–gas shift reaction (Equation 14.2) is at equilibrium and some CO_2 is rapidly removed. When equilibrium is restored, which of the four compounds in the system is present at a higher concentration, which is present at a lower concentration, and which, if any, has the same concentration as in the original equilibrium mixture?

SAMPLE EXERCISE 14.10 Stressing an Equilibrium by Adding or Removing Reactants or Products — LO7

Suggest three ways the production of ammonia via the reaction

$$N_2(g) + 3\ H_2(g) \rightleftharpoons 2\ NH_3(g)$$

could be increased without changing the reaction temperature.

Collect, Organize, and Analyze We are given the balanced chemical equation of a reversible reaction and are asked to suggest three ways to increase its yield—that is, to shift its equilibrium to the right. We need to use the mass action expression for the reaction:

$$Q_P = \frac{(P_{NH_3})^2}{(P_{H_2})^3(P_{N_2})}$$

Effect of Temperature Changes

In Chapter 9 we explored the flow of energy that accompanies many chemical reactions. Now we explore the stresses produced on chemical equilibria when energy is added or removed by increasing or decreasing the temperature, respectively. We can write the synthesis of ammonia, an exothermic reaction, as follows,

$$N_2(g) + 3\ H_2(g) \rightleftharpoons 2\ NH_3(g) + \text{energy}$$

because energy is a product in the forward reaction. As a result, raising the temperature of the reaction mixture favors the reverse reaction, whereas lowering the temperature favors the forward reaction.

There is one key difference between applying Le Châtelier's principle to concentration or pressure changes and applying it to temperature changes: changing temperature does not perturb an equilibrium by changing the ratio of product to reactant concentrations or partial pressures—that is, by changing the value of Q. Instead, changing temperature changes the value of K. As a result, Q and K are no longer equal and the reaction proceeds in the direction that makes them equal again.

Increasing temperature reduces the yield of ammonia because increasing temperature reduces the value of K. In general, the value of K decreases as temperature increases for exothermic reactions. We will look more closely at the influence of temperature on K values in Section 14.10, but for now it is sufficient to note again that energy is essentially a reactant of an endothermic reaction and a product of an exothermic reaction; raising the temperature of an endothermic reaction mixture by heating it is the equivalent of adding more reactant, which promotes the reaction and raises its K value. On the other hand, heating an exothermic reaction mixture is the equivalent of adding more product, which promotes the reverse reaction and lowers the value of K.

SAMPLE EXERCISE 14.12 Predicting How Temperature Affects Chemical Equilibria — LO7

The color of an aqueous acidic solution of cobalt(II) chloride depends on the temperature (Figure 14.8). In aqueous HCl, the solution is pink at 0°C, magenta at 25°C, and dark blue at 75°C. Is the reaction producing the pink-to-blue color change exothermic or endothermic?

FIGURE 14.8 Two forms of cobalt(II) ion, one pink and one blue, are in equilibrium in aqueous hydrochloric acid solution. The position of equilibrium shifts to the right as temperature increases from 5°C to 75°C, causing the color of the solution to change from pink to blue as the concentration of $CoCl_4^{2-}$ ions increases.

Collect and Organize We are given a reversible reaction and asked to determine whether it is exothermic or endothermic. Asked another way, is energy a product or a reactant in this reaction?

Analyze If the reaction is exothermic, then increasing the temperature is the equivalent of adding a product, causing a shift toward the pink reactant side. If the reaction is endothermic, then increasing the temperature is the equivalent of adding a reactant, causing a shift toward the side of the blue product.

Solve When energy is added to the system, its color changes from pink to blue. This means energy is a reactant, and the reaction as written must be endothermic:

$$\text{Energy} + \text{pink} \rightleftharpoons \text{blue}$$

Think About It This exercise shows how determining the effect of changing the temperature of an equilibrium reaction mixture can tell us whether the reaction is exothermic or endothermic. The fact that the reaction mixture in this exercise is magenta at room temperature means that both the pink and blue forms are present, and that the value of K at room temperature is close to 1.

 **Practice Exercise** Predict how the value of the equilibrium constant of the reaction

$$N_2(g) + O_2(g) \rightleftharpoons 2\,NO(g) \qquad \Delta H^\circ_{rxn} = 181\text{ kJ}$$

changes with increasing temperature.

Table 14.4 summarizes how an exothermic system at equilibrium responds to various stresses.

CONCEPT TEST

If Table 14.4 had been based on an endothermic reaction, how would the equilibrium have shifted in the last two rows of the table?

TABLE 14.4 Responses of an Exothermic Reaction [$2\,A(g) \rightleftharpoons B(g)$] at Equilibrium to Different Kinds of Stress

Kind of Stress	How System Responds	Direction of Shift
Add A	Consume A	To the right
Remove A	Produce A	To the left
Add B	Consume B	To the left
Remove B	Produce B	To the right
Increase pressure by compressing the reaction mixture	Consume A to decrease pressure	To the right
Decrease pressure by expanding volume	Produce A to maintain equilibrium pressure	To the left
Increase temperature by adding energy	Consume some of the energy	To the left
Decrease temperature by removing energy	Produce energy	To the right

Catalysts and Equilibrium

The industrial production of ammonia (see Sample Exercise 14.10) was developed by the German chemists Fritz Haber (1868–1934) and Carl Bosch (1874–1940) in the early 20th century and is still widely referred to as the Haber–Bosch process. What makes the process commercially feasible is the use of catalysts. As discussed in Chapter 13, a catalyst increases the rate of a chemical reaction by lowering its activation energy. If a catalyst increases the rate of a reaction, does that catalyst affect the equilibrium constant of the reaction?

To answer this question, consider the energy profiles of the catalyzed and uncatalyzed reactions in Figure 14.9. The catalyst increases the rate of the reaction by decreasing the height of the energy barrier and changing the reaction mechanism. However, the barrier height is reduced by the same amount whether the reaction as written proceeds in the forward direction or in reverse. As a result, the increase in reaction rate produced by the catalyst is the same in both directions. Therefore, a catalyst has no effect on the equilibrium constant of a reaction or on the composition of an equilibrium reaction mixture. A catalyst does, however, decrease the amount of time needed for a reaction to reach equilibrium.

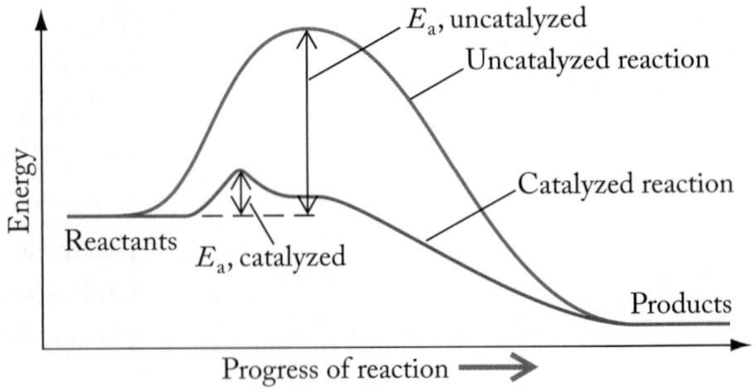

FIGURE 14.9 The effect of a catalyst on a reaction. A catalyst lowers the activation energy barrier, so the rate of the reaction increases. However, because both the forward reaction and the reverse reaction occur more rapidly, the position of equilibrium (that is, the value of K) does not change. The system comes to equilibrium more rapidly, but the relative amounts of product and reactant present at equilibrium do not change.

14.8 Calculations Based on K

CHEMTOUR Solving Equilibrium Problems

Reference books and the tables in Appendix 5 of this book contain lists of equilibrium constants for chemical reactions. These values are used in several kinds of calculations, including those in which

1. We want to determine whether a reaction mixture has reached equilibrium (Sample Exercise 14.7).
2. We know the value of K and the starting concentrations or partial pressures of reactants and/or products, and we want to calculate their equilibrium concentrations or pressures.

In this section we focus on the second type of calculation and introduce a useful way of handling such problems: a table of reactant and product concentration (or partial pressure) values called a *RICE table*. "RICE" is an acronym: the table starts with the balanced chemical equation describing the ***R***eaction followed by rows that contain ***I***nitial concentration values, ***C***hanges in those initial values as the reaction proceeds toward equilibrium, and ***E***quilibrium values.

In our first example, we calculate how much nitrogen monoxide forms in a sample of air heated to a temperature at which the K_p of the following reaction is 1.00×10^{-5}:

$$N_2(g) + O_2(g) \rightleftharpoons 2\,NO(g)$$

The initial partial pressures are $P_{N_2} = 0.79$ atm and $P_{O_2} = 0.21$ atm, and we assume no NO is present.

We start by writing the given (initial) information in our RICE table:

REACTION	$N_2(g)$	+	$O_2(g)$	$\rightleftharpoons$	$2\,NO(g)$
	P_{N_2} (atm)		P_{O_2} (atm)		P_{NO} (atm)
Initial	0.79		0.21		0
Change					
Equilibrium					

The reaction will proceed in the forward direction because there is no product initially present, so $Q_p = 0 < K_p$.

We need to use algebra to fill in rows C and E. We don't know how much N_2 or O_2 will be consumed or how much NO will be formed. We can define the change in partial pressure of N_2 as $-x$ because N_2 is consumed during the reaction. Because the mole ratio of N_2 to O_2 in the balanced chemical equation is 1:1, the change in O_2 is also $-x$. Two moles of NO are produced from each mole of N_2 and O_2, so the change in P_{NO} is $+2x$. Inserting these values in the C row, we have

REACTION	$N_2(g)$ +	$O_2(g)$ ⇌	$2\,NO(g)$
	P_{N_2} (atm)	P_{O_2} (atm)	P_{NO} (atm)
Initial	0.79	0.21	0
Change	$-x$	$-x$	$+2x$
Equilibrium			

Combining the I and C rows, we obtain expressions for the three partial pressures at equilibrium:

REACTION	$N_2(g)$ +	$O_2(g)$ ⇌	$2\,NO(g)$
	P_{N_2} (atm)	P_{O_2} (atm)	P_{NO} (atm)
Initial	0.79	0.21	0
Change	$-x$	$-x$	$+2x$
Equilibrium	$0.79 - x$	$0.21 - x$	$2x$

The next step is to substitute the terms from the E row into the K_p expression for the reaction:

$$K_p = \frac{(P_{NO})^2}{(P_{N_2})(P_{O_2})}$$

$$= \frac{(2x)^2}{(0.79 - x)(0.21 - x)} \quad (14.18)$$

Multiplying the terms in the denominator of Equation 14.18 gives

$$K_p = \frac{4x^2}{0.1659 - 1.00x + x^2} = 1.00 \times 10^{-5}$$

Cross-multiplying, we get

$$1.659 \times 10^{-6} - (1.00 \times 10^{-5})x + (1.00 \times 10^{-5})x^2 = 4x^2$$

Combining the x^2 terms and rearranging, we have

$$3.99999\,x^2 + (1.00 \times 10^{-5})x - 1.659 \times 10^{-6} = 0$$

This equation fits the general form of a quadratic equation:

$$ax^2 + bx + c = 0$$

which can be solved for x using a variety of tools, including the problem-solving functions in scientific calculators or in worksheet programs such as Excel, or by pencil and paper using the quadratic formula:

$$x = \frac{-b \pm \sqrt{b^2 - 4ac}}{2a}$$

Taking this low-tech approach:

$$x = \frac{-1.00 \times 10^{-5} \pm \sqrt{(1.00 \times 10^{-5})^2 - 4(3.99999)(-1.659 \times 10^{-6})}}{2(3.99999)}$$

Solving this equation yields two x values: 6.43×10^{-4} and -6.45×10^{-4}. We focus on the positive value because a gas cannot have a negative partial pressure. Calculating equilibrium partial pressures to two significant figures:

$$\begin{aligned} P_{O_2} &= 0.21 - x \\ &= 0.21 - (6.43 \times 10^{-4}) = 0.21 \text{ atm} \\ P_{N_2} &= 0.79 - x \\ &= 0.79 - (6.43 \times 10^{-4}) = 0.79 \text{ atm} \\ P_{NO} &= 2x \\ &= 2(6.43 \times 10^{-4}) = 1.286 \times 10^{-3} = 0.0013 \text{ atm} \end{aligned}$$

The small quantity of NO produced by the reaction means that there is no change in the partial pressures of N_2 or O_2 to two significant figures.

Because the x terms in the denominator of Equation 14.18 are very much smaller than the initial partial pressures, we can simplify the calculation of P_{NO} by ignoring the x terms in the denominator and using the initial values for P_{N_2} and P_{O_2} instead:

$$\begin{aligned} K_p = 1.0 \times 10^{-5} &= \frac{(P_{NO})^2}{(P_{N_2})(P_{O_2})} \\ &= \frac{4x^2}{(0.79 - x)(0.21 - x)} \approx \frac{4x^2}{(0.79)(0.21)} \\ 4x^2 &= (0.79)(0.21)(1.0 \times 10^{-5}) \\ &= 1.659 \times 10^{-6} \\ x^2 &= 4.148 \times 10^{-7} \\ x &= 6.44 \times 10^{-4} \text{ atm} \end{aligned}$$

This value of x is nearly the same as that obtained by solving the quadratic equation (6.43×10^{-4} atm) and actually is the same to two significant figures. Generally speaking, we can ignore the $-x$ component of a reactant concentration or partial pressure at equilibrium if the initial concentration or partial pressure is greater than 500 times the value of K. The reliability of this >500 test is illustrated in the preceding calculation: the ratio of the smaller initial partial pressure to the value of x is $0.21 \text{ atm} / (6.43 \times 10^{-4} \text{ atm}) = 330$. This ratio is less than 500, but even so, the values of x calculated with and without the $-x$ terms differed by only one unit in the third digit and were the same when expressed with only two significant figures. We will not need more than three significant figures in expressing the results of equilibrium calculations in this chapter and those that follow, so you may confidently rely on the >500 test.

Equilibrium calculations can also be simplified when initial reactant concentrations are the same. For example, suppose we have a vessel containing 0.100 M N_2 and 0.100 M O_2 at a temperature at which K_c for the NO formation

reaction is 0.100. What is the equilibrium concentration of NO? The RICE table in this case is

REACTION	$N_2(g)$	+	$O_2(g)$	$\rightleftharpoons$	$2\ NO(g)$
	$[N_2]$ (*M*)		$[O_2]$ (*M*)		$[NO]$ (*M*)
Initial	0.100		0.100		0
Change	$-x$		$-x$		$+2x$
Equilibrium	$0.100 - x$		$0.100 - x$		$2x$

Inserting the values from the E row into the expression for K_c gives

$$K_c = \frac{(2x)^2}{(0.100 - x)(0.100 - x)} = 0.100$$

Taking the square root of each side:

$$\frac{2x}{0.100 - x} = 0.316$$

To solve for x, we first cross-multiply and then combine the two x terms:

$$2x = (0.316)\,(0.100 - x) = 0.0316 - 0.316\,x$$

$$2x + 0.316\,x = 2.316\,x = 0.0316$$

$$x = 0.0136\ M$$

The equilibrium concentrations are $[N_2] = [O_2] = 0.100\ M - 0.0136\ M = 0.086\ M$ and $[NO] = 2x = 0.0272\ M$.

It's a good idea to check that these concentrations are consistent with the known value of K_c. Substituting into the K_c expression, we get

$$K_c = \frac{(0.0272)^2}{(0.086)^2} = 0.10$$

which is the value of K given for the reaction in question.

SAMPLE EXERCISE 14.13 Calculating an Equilibrium Partial Pressure I **LO8**

Much of the H_2 used in the Haber–Bosch process is produced by the water–gas shift reaction:

$$CO(g) + H_2O(g) \rightleftharpoons CO_2(g) + H_2(g) \qquad (14.2)$$

If a reaction vessel at 400°C is filled with an equimolar mixture of CO and steam such that $P_{CO} = P_{H_2O} = 2.00$ atm, what is the partial pressure of H_2 at equilibrium? The equilibrium constant $K_p = 10$ at 400°C.

Collect and Organize We are asked to find the partial pressure of a gaseous product in an equilibrium mixture given the initial partial pressures of reactants and the value of K_p. One way to approach this problem involves (1) setting up a RICE table, (2) using the partial pressures from the E row in the equilibrium constant expression, and (3) solving for P_{H_2}.

Analyze The system initially contains no product. This means that the reaction quotient Q_p is equal to zero and is thus less than K. Therefore, the reaction proceeds in

the forward direction, decreasing the partial pressures of the reactants while increasing those of the products. A K_p value of 10 means that most of the reactants should be converted into products, but there should be significant partial pressures of both reactants and products at equilibrium.

Solve Let x be the increase in partial pressure of H_2 as a result of the reaction. The stoichiometry of the reaction tells us that the change in P_{CO_2} is also x and that the changes in both P_{CO} and P_{H_2} are $-x$:

REACTION	$CO(g)$	+	$H_2O(g)$	$\rightleftharpoons$	$CO_2(g)$	+	$H_2(g)$
	P_{CO} (atm)		P_{H_2O} (atm)		P_{CO_2} (atm)		P_{H_2} (atm)
Initial	2.00		2.00		0.00		0.00
Change	$-x$		$-x$		$+x$		$+x$
Equilibrium	$2.00 - x$		$2.00 - x$		x		x

Inserting these equilibrium terms into the equilibrium constant expression for the reaction gives

$$K_p = \frac{(P_{CO_2})(P_{H_2})}{(P_{CO})(P_{H_2O})} = \frac{(x)(x)}{(2.00 - x)(2.00 - x)} = 10$$

This equation can be simplified by taking the square root of both sides:

$$\frac{x}{2.00 - x} = \sqrt{10} = 3.16$$

Solving for x gives $x = 1.52$ atm, which is the equilibrium partial pressure of H_2 (and CO_2).

Think About It Our prediction that most, but far from all, of the reactants would form products was correct. It is a good idea to substitute the results into the equilibrium constant expression as a check on the validity of the solution. The calculated partial pressures are $P_{H_2} = P_{CO_2} = 1.52$ atm and $P_{H_2O} = P_{CO} = 2.00 - 1.52 = 0.48$ atm. Inserting these values in the equilibrium constant expression gives $K_p = (1.52)^2/(0.48)^2 = 10.03$, which is not significantly different from the given value of 10 and which confirms that our calculation is correct.

 **Practice Exercise** The chemical equation for the formation of hydrogen iodide from H_2 and I_2 is

$$H_2(g) + I_2(g) \rightleftharpoons 2\,HI(g)$$

$K_p = 50$ for this reaction at 450°C. What is the partial pressure of HI in a sealed reaction vessel at 450°C if the initial partial pressures of H_2 and I_2 are both 0.100 atm and initially there is no HI present?

SAMPLE EXERCISE 14.14 Calculating an Equilibrium Partial Pressure II — LO8

Suppose that in a reaction vessel running the water–gas shift reaction at 400°C,

$$CO(g) + H_2O(g) \rightleftharpoons CO_2(g) + H_2(g)$$

the initial partial pressures are $P_{CO} = 2.00$ atm, $P_{H_2O} = 2.00$ atm, $P_{H_2} = 0.15$ atm, and $P_{CO_2} = 0.00$ atm. What is the partial pressure of H_2 at equilibrium, given $K_p = 10$ at 400°C?

Collect and Organize We are asked to calculate the partial pressure of a product at equilibrium. We know the initial partial pressures of all reactants and of both products as well as the value of K_p. The difference between this problem and Sample Exercise 14.13 is that here we have product present before the reaction starts.

Analyze Comparing the reaction quotient Q with the value of K lets us know in which direction the reaction proceeds to attain equilibrium. There is no CO_2 initially present, so $Q = 0$. Therefore, the reaction as written proceeds in the forward direction. Our strategy is to solve the problem by setting up a RICE table.

Solve Let x be the increase in P_{H_2}. The change in P_{CO_2} is also x, and the changes in P_{CO} and P_{H_2O} are $-x$:

REACTION	$CO(g)$	+	$H_2O(g)$	$\rightleftharpoons$	$CO_2(g)$	+	$H_2(g)$
	P_{CO} (atm)		P_{H_2O} (atm)		P_{CO_2} (atm)		P_{H_2} (atm)
Initial	2.00		2.00		0.00		0.15
Change	$-x$		$-x$		$+x$		$+x$
Equilibrium	$2.00 - x$		$2.00 - x$		x		$0.15 + x$

$$K_p = \frac{(P_{CO_2})(P_{H_2})}{(P_{CO})(P_{H_2O})} = \frac{(x)(0.15 + x)}{(2.00 - x)(2.00 - x)} = 10$$

Solving for x gives two values: 1.50 and 2.96. The 2.96 value is not useful because it produces negative partial pressures of CO and H_2O: $(2.00 - 2.96)$ atm $= -0.96$ atm, which is impossible. Therefore, $x = 1.50$, and at equilibrium $P_{H_2} = 0.15 + 1.50 =$ 1.65 atm.

Think About It Using the value of x in the terms in the E row of the RICE table, we find that the equilibrium partial pressures of CO, H_2O, and CO_2 are 0.50, 0.50, and 1.50, respectively. These values result in a K_p value of $(1.65)(1.50)/(0.50)(0.50) = 9.9$, which is acceptably close to the given value of 10. Comparing the results of this Sample Exercise to the previous one, we find that having an initial P_{H_2} of 0.15 atm results in a higher final P_{H_2} value (1.65 versus 1.52 atm); however, the presence of H_2 at the start of the reaction results in slightly less conversion of reactants to products than when no H_2 is present initially. This result makes sense because the presence of some product before the reaction starts means that less product has to form before $Q_p = K_p$.

Practice Exercise The value of K_c for the reaction

$$N_2O_4(g) \rightleftharpoons 2\ NO_2(g)$$

is 0.21 at 373 K. If a reaction vessel at that temperature initially contains 0.030 M NO_2 and 0.030 M N_2O_4, what are the concentrations of the two gases at equilibrium?

14.9 Equilibrium and Thermodynamics

In Section 12.6 we explored how the change in free energy, ΔG, of a chemical reaction provides us with an indication of whether or not it will proceed at a particular temperature and pressure. If ΔG is negative, a reaction is spontaneous as written and proceeds in the forward direction. If ΔG is positive, the reaction as written is nonspontaneous; the reverse reaction is spontaneous, and the reaction as written proceeds in the reverse direction. As a spontaneous reaction under constant temperature and pressure proceeds, the concentrations of reactants and products change, and the free energy of the system changes as well. Eventually ΔG reaches zero. When it does, no free energy is left to do useful work. The reaction has achieved chemical equilibrium.

The magnitude of ΔG—how far it is from zero in either a negative or positive direction—indicates how far a system is from its equilibrium position. Similarly, when Q is much larger or smaller than K, we know that a chemical system is far from equilibrium. It is reasonable, then, to think that the separation between the values of Q and K and the sign and magnitude of ΔG are somehow related. Indeed they are, and their mathematical relationship is one of the most important connections in chemistry because it enables us to relate the thermodynamics of a chemical reaction to the composition of a reaction mixture at equilibrium.

CONNECTION In Chapter 12 we defined the change in free energy of a reaction, ΔG_{rxn}, as the energy available to do useful work at a particular temperature and pressure.

The thermodynamic view of equilibrium and the relationship between ΔG and Q are described by the equation

$$\Delta G = \Delta G^\circ + RT \ln Q \tag{14.19}$$

where ΔG° is the change in free energy under standard conditions. To better understand what this means, let's look at a reaction we have examined several times in this chapter, the decomposition of N_2O_4:

$$N_2O_4(g) \rightleftharpoons 2\ NO_2(g)$$

We can calculate the change in standard free energy for the reaction (ΔG°_{rxn}) as we did in Chapter 12, using standard free energy of formation (ΔG°_f) values from Table A4.3 in Appendix 4 and the formula

$$\Delta G^\circ_{rxn} = \sum n_{products}\Delta G^\circ_{f,products} - \sum n_{reactants}\Delta G^\circ_{f,reactants} \tag{12.12}$$

$$= 2\ \cancel{mol}\ (51.3\ kJ/\cancel{mol}) - 1\ \cancel{mol}\ (97.8\ kJ/\cancel{mol}) = 4.8\ kJ$$

The positive value of ΔG°_{rxn} indicates that the reaction as written is not spontaneous in the forward direction at 298 K. However, this value of ΔG°_{rxn} applies only under standard conditions—namely, when $P_{NO_2} = P_{N_2O_4} = 1$ atm. That is not the case when the partial pressures vary from 1 atm.

What happens in a reaction vessel at 298 K if it initially contains just one mole of pure N_2O_4 at a partial pressure of 1 atm and no (or hardly any) NO_2? Under these conditions, $Q_p = 0$:

$$Q_p = \frac{(P_{NO_2})^2}{P_{N_2O_4}} = \frac{0}{1} = 0$$

Although we do not know the value of K, it must be greater than zero. Therefore, $Q_p < K$, and the system responds by spontaneously forming NO_2 from N_2O_4.

Because the reaction in the forward direction is spontaneous, the change in free energy of the reaction (ΔG) must be less than zero, even though the value of ΔG° is positive (4.8 kJ/mol). Furthermore, we know that the value of ΔG is negative because the ($RT \ln Q$) term in Equation 14.19 approaches $-\infty$ as Q approaches zero. Any reversible reaction, regardless of its ΔG° value, has a negative ΔG value and is spontaneous when there is only reactant and no product (or practically none) in the system.

The opposite situation occurs if we have two moles of NO_2 in our reaction vessel and essentially no N_2O_4. Under these conditions, the denominator of the reaction quotient is nearly zero, making the value of Q_p enormous. Likewise, the ($RT \ln Q_p$) term in Equation 14.19 has a large positive value, guaranteeing that $\Delta G > 0$. This means that the forward reaction is not spontaneous, but the reverse reaction is. We also know that NO_2 in the reaction vessel will combine to form

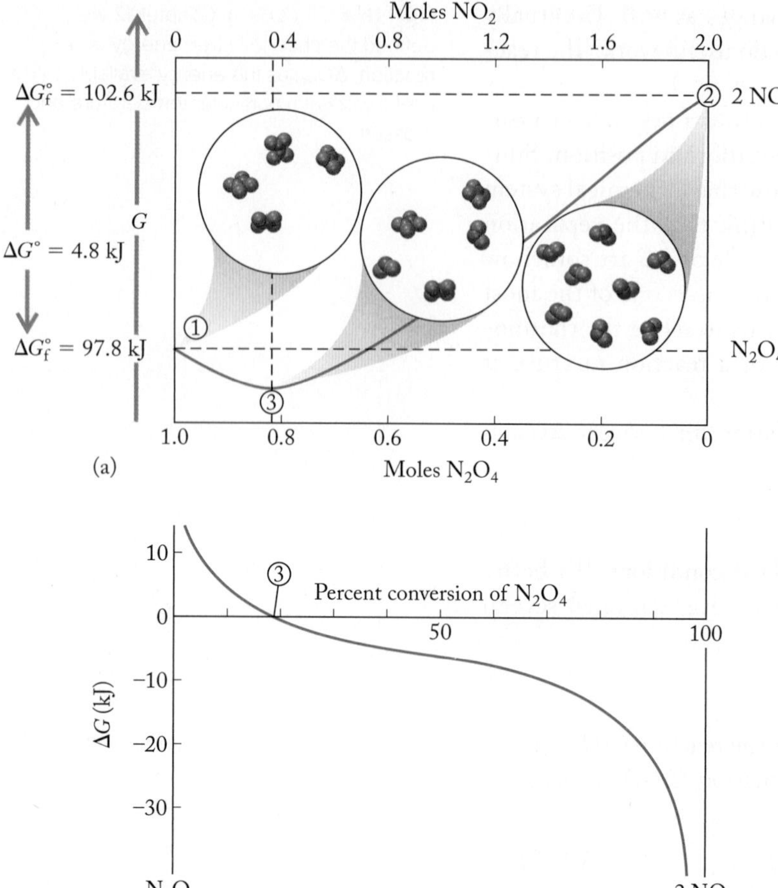

FIGURE 14.10 The change in free energy during the reaction $N_2O_4(g) \rightleftharpoons 2\ NO_2(g)$. At point ① the system consists of pure reactant N_2O_4; at point ②, the system is pure product, NO_2 gas. Neither system is thermodynamically stable. From point ① the reaction proceeds in the forward direction and the free energy of the system drops until point ③ is reached. From point ② the reaction proceeds in the reverse direction and the free energy of the system drops until, again, point ③ is reached. When the minimum in free energy at point ③ is reached, the system has come to equilibrium. (b) At equilibrium $\Delta G_{rxn} = 0$. The ΔG_{rxn} values of reaction mixtures not at equilibrium represent their "distance" from equilibrium in terms of free energy.

N_2O_4 because the enormous Q_p must be greater than K_p. Therefore, the reaction will run in reverse until $Q_p = K_p$.

Figure 14.10(a) shows how free energy changes as the quantities of N_2O_4 and NO_2 in the reaction mixture change. The minimum of the curve (point 3) corresponds to a free energy that is lower than that of either pure N_2O_4 (point 1) or pure NO_2 (point 2). The vertical dashed line, marking the minimum free energy, intersects the top and bottom axes at values that tell us the reaction mixture's composition at equilibrium: about 0.83 mol N_2O_4 and about 0.34 mol NO_2.

The curve in Figure 14.10(b) confirms what is shown in part (a). In part (b), the y-axis represents the distance from equilibrium of N_2O_4/NO_2 reaction mixtures of different composition. Mixtures that are either pure reactant or product are very far away from equilibrium, as we just discussed. The reaction curve crosses the x-axis ($\Delta G = 0$) at point 3, which defines the composition of the reaction mixture at equilibrium. This value, expressed as the percentage of N_2O_4 that has dissociated into NO_2, corresponds to the same N_2O_4/NO_2 ratio as the minimum in the curve in Figure 14.10(a).

CONCEPT TEST

Suppose $\Delta G^\circ_{rxn} = -3.0$ kJ/mol for the hypothetical chemical reaction A $\rightleftharpoons$ B. Which of the following statements about an equilibrium mixture of A and B at 298 K is true?

a. There is only A present.
b. There is only B present.
c. There is an equimolar mixture of A and B present.
d. There is more A than B present.
e. There is more B than A present.

Once a reaction has reached equilibrium, $Q = K$, $\Delta G = 0$, and Equation 14.19 becomes

$$\Delta G = \Delta G^\circ + RT \ln K = 0$$

or

$$\Delta G^\circ = -RT \ln K \tag{14.20}$$

Rearranging Equation 14.20 allows us to calculate the K value for a reaction from its change in standard free energy and absolute temperature. First, we rearrange the terms:

$$\ln K = \frac{-\Delta G^\circ}{RT} \tag{14.21}$$

and then take the antilogarithm of both sides:

$$K = e^{-\Delta G^\circ/RT} \tag{14.22}$$

Equation 14.22 provides the following interpretation of reaction spontaneity under standard conditions. Whenever $\Delta G°$ is negative, the exponent $-\Delta G°/RT$ in Equation 14.22 is positive, and $e^{-\Delta G°/RT} > 1$, making $K > 1$. Therefore, any reversible reaction with an equilibrium constant greater than 1 is spontaneous under standard conditions, as shown in Figure 14.11(a). This spontaneity has its limits. As reactants are consumed and products are formed, the value of the reaction quotient increases, making the value of ΔG less negative. When it reaches

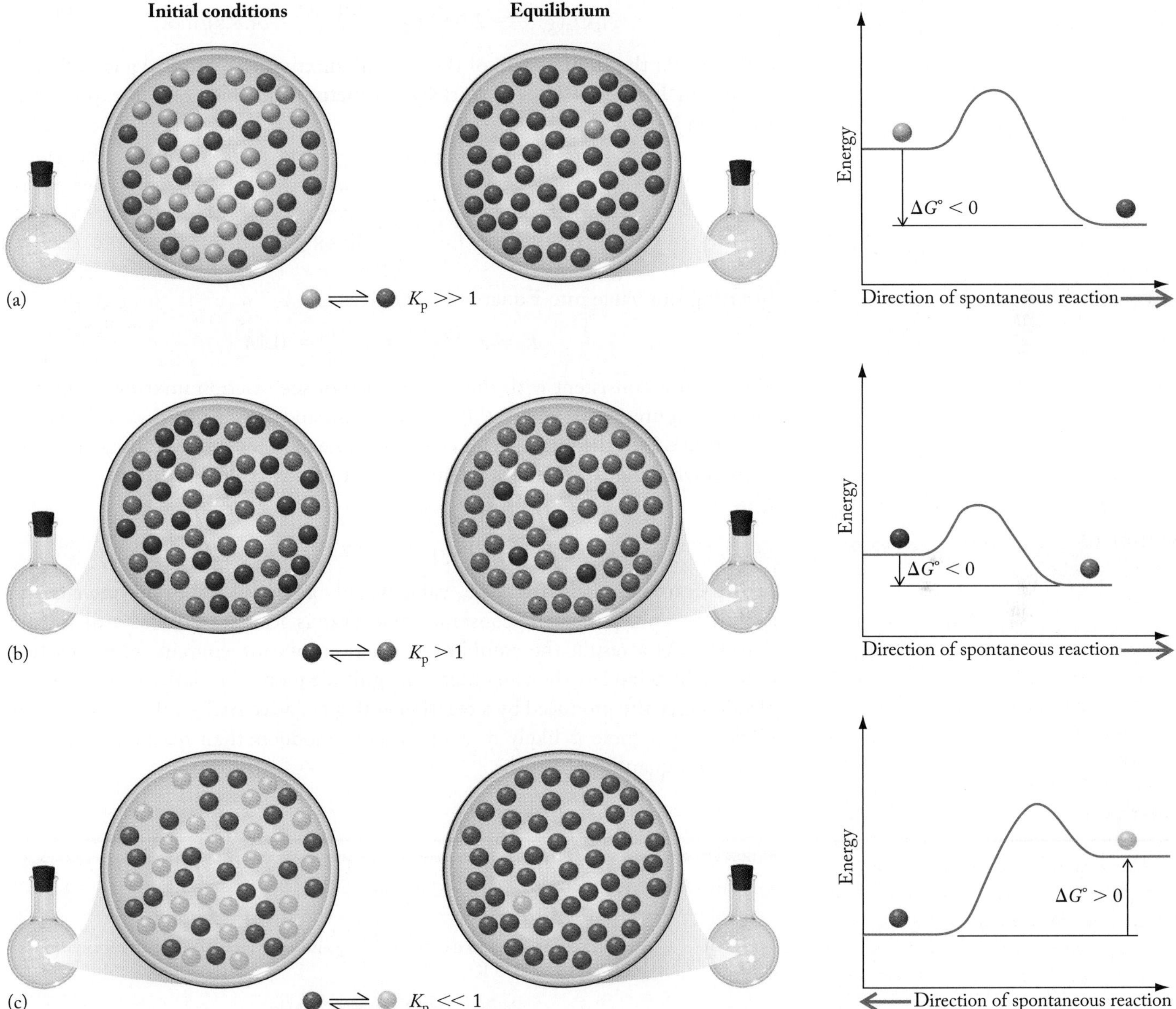

FIGURE 14.11 The equilibrium constant of a chemical reaction is linked to its $\Delta G°$ value. The initial conditions for the three flasks on the left each contain equimolar mixtures of different pairs of gases. The initial partial pressure of each gas is 1 atm. The three images on the right represent the mixtures in these same flasks at equilibrium. (a) The value of $\Delta G°$ for the formation of "red" gas from "green" gas has a large negative value, which makes K much greater than 1. The reaction proceeds in the forward direction, leaving little green gas unreacted. (b) If the value of $\Delta G°$ for the formation of "orange" gas from "blue" gas has a less negative value than in part (a), the value of K is smaller, though still greater than 1, and more orange gas is present at equilibrium than blue gas. (c) If the value of $\Delta G°$ for the formation of "yellow" gas from "purple" gas is positive, K is less than 1 and the reaction runs in the reverse direction, forming purple gas from yellow gas.

zero, there is no further change in the composition of the reaction mixture because chemical equilibrium has been achieved.

It follows that a reversible reaction with a less negative value of $\Delta G°$ (Figure 14.11b) is still spontaneous but has a smaller equilibrium constant, so less reactant is consumed and less product is formed before the value of ΔG reaches zero. Finally, a reaction that has a positive value of $\Delta G°$ (Figure 14.11c) has an equilibrium constant that is less than 1, and it is not spontaneous under standard conditions. Instead, the reverse of the reaction is spontaneous.

Let's apply this concept to the equilibrium between N_2O_4 and NO_2 at 298 K,

$$N_2O_4(g) \rightleftharpoons 2\ NO_2(g) \qquad \Delta G°_{rxn} = 4.8\text{ kJ/mol}$$

and focus on the composition of the reaction mixture of these two gases at equilibrium in Figure 14.11. We start by calculating the value of the exponent in Equation 14.22:

$$\frac{-\Delta G°_{rxn}}{RT} = -\frac{\left(\frac{4.8\ \cancel{\text{kJ}}}{\cancel{\text{mol}}}\right)\left(\frac{1000\ \cancel{\text{J}}}{1\ \cancel{\text{kJ}}}\right)}{\left(\frac{8.314\ \cancel{\text{J}}}{\cancel{\text{mol}} \cdot \cancel{\text{K}}}\right)(298\ \cancel{\text{K}})} = -1.94$$

Inserting this value into Equation 14.22 gives

$$K = e^{-\Delta G°_{rxn}/RT} = e^{-1.94} = 0.14$$

This result is consistent with the composition of the reaction mixture at equilibrium in Figure 14.10, where $P_{N_2O_4} \approx 0.83$ atm and $P_{NO_2} \approx 0.34$ atm. Inserting these values into the equilibrium constant expression for the reaction gives us an approximate value of K_p that is identical to the one we calculated from $\Delta G°_{rxn}$:

$$K_p = \frac{(P_{NO_2})^2}{P_{N_2O_4}} \approx \frac{(0.34)^2}{0.83} = 0.14$$

In this example, a positive $\Delta G°_{rxn}$ value of only a few kilojoules per mole corresponds to an equilibrium constant value that is less than 1 but still greater than 0.1. As a result, the equilibrium reaction mixture contains more reactant than product, but less than an order of magnitude more. Similarly, an equilibrium reaction mixture produced by a reaction with a negative $\Delta G°_{rxn}$ value of only a few kilojoules per mole is likely to contain more products than reactants, but with significant quantities of both.

SAMPLE EXERCISE 14.15 Relating K and $\Delta G°_{rxn}$ LO9

Use $\Delta G°_f$ values from Table A4.3 to calculate $\Delta G°_{rxn}$ and the value of K for the formation of NO_2 from NO and O_2 at 298 K:

$$NO(g) + \tfrac{1}{2}\ O_2(g) \rightleftharpoons NO_2(g)$$

Collect and Organize We are asked to calculate the values of $\Delta G°_{rxn}$ and K for a reaction starting with $\Delta G°_f$ values from Table A4.3: 51.3 kJ/mol for NO_2 and 86.6 kJ/mol for NO. Because O_2 gas is the most stable form of the element at STP, its $\Delta G°_f$ value is 0.0 kJ/mol.

Analyze We can use Equation 12.12 to calculate $\Delta G°_{rxn}$ and then Equation 14.22 to calculate the value of K. The $\Delta G°_f$ value of NO_2 is less than that of NO, because $\Delta G°_{rxn}$

is negative and the value of the exponent in Equation 14.22 is positive. That translates into a K value greater than 1.

Solve

$$\begin{aligned}\Delta G^\circ_{rxn} &= [\Delta G^\circ_f(NO_2)] - [\Delta G^\circ_f(NO) + \tfrac{1}{2}\Delta G^\circ_f(O_2)] \\ &= [1\ \text{mol}\ (51.3\ \text{kJ/mol})] - [1\ \text{mol}\ (86.6\ \text{kJ/mol}) + \tfrac{1}{2}\ \text{mol}\ (0.0\ \text{kJ/mol})] \\ &= -35.3\ \text{kJ or } -35{,}300\ \text{J per mole of } NO_2 \text{ produced}\end{aligned}$$

The exponent in Equation 14.22 is

$$\frac{-\Delta G^\circ_{rxn}}{RT} = -\frac{\left(\dfrac{-35{,}300\ \text{J}}{\text{mol}}\right)}{\left(\dfrac{8.314\ \text{J}}{\text{mol}\cdot\text{K}}\right)(298\ \text{K})} = 14.2$$

The corresponding value of K_p is

$$K_p = e^{-\Delta G^\circ_{rxn}/RT} = e^{14.2} = 1.5 \times 10^6$$

Think About It The exponential relationship between ΔG°_f and K_p means that a moderately negative free-energy change of −35.3 kJ/mol corresponds to a very large value of K_p: in this case, greater than 10^6.

 **Practice Exercise** The standard free energy of formation of ammonia at 298 K is −16.5 kJ/mol. What is the value of K_p for the reaction at 298 K?

$$N_2(g) + 3\,H_2(g) \rightleftharpoons 2\,NH_3(g)$$

Which equilibrium constant, K_c or K_p, is related to ΔG° by Equation 14.22? ΔG° represents a change in free energy under standard conditions, and the standard state for a gaseous reactant or product is one in which its partial pressure is 1 bar (≈1 atm). Thus, the ΔG° of a reaction in the gas phase is linked by Equation 14.22 to its K_p value. However, standard conditions for reactions in solution (the focus of Chapter 15) mean that all dissolved reactants and products are present at a concentration of 1.00 *M*. Thus, the ΔG° of a reaction in solution is related by Equation 14.22 to its K_c value.

The values of ΔG° and K are also linked by their dependence on how we choose to write the chemical equation describing a reaction. For example, multiplying the coefficients in an equation by n means multiplying its ΔG° value by n. However, in this situation, K is not multiplied by n but rather is raised to the nth power: K becomes K^n. This difference makes sense given the logarithmic relationship between ΔG° and K and the fact that $\ln (X^n) = n \ln X$. This logarithmic relationship also explains why reversing a reaction changes the sign of ΔG° but produces a K that is the reciprocal of the original: $\ln (1/X) = -\ln X$.

14.10 Changing *K* with Changing Temperature

We have noted often in this chapter that the value of K changes with changing temperature. In this section we look more closely at that relationship, developing several important equations that link K and T.

Temperature, *K*, and $\Delta G°$

Let's begin by combining a key equation from Chapter 12:

$$\Delta G° = \Delta H° - T\Delta S° \quad (12.11)$$

with one from this chapter:

$$\ln K = \frac{-\Delta G°}{RT} \quad (14.21)$$

to derive an equation that relates *K* to $\Delta H°$ and $\Delta S°$:

$$\ln K = \frac{-\Delta G°}{RT} = \frac{-\Delta H°}{RT} + \frac{T\Delta S°}{RT} = -\frac{-\Delta H°}{RT} + \frac{\Delta S°}{R} \quad (14.23)$$

Note how a negative value of $\Delta H°$ or a positive value of $\Delta S°$ contributes to a large value of *K*. These dependencies are predicted because negative values of $\Delta H°$ and positive values of $\Delta S°$ are the two factors that contribute to making reactions spontaneous.

Because we are discussing the influence of temperature on *K*, let's identify the factors affected by changes in *T* in Equation 14.23. First of all, the values of $\Delta H°$ and $\Delta S°$, such as those in Appendix 4 for $T = 298$ K, change very little with changing temperature, so we will treat $\Delta H°$ and $\Delta S°$ as constants in this discussion. However, the value of the first term on the right side of Equation 14.23 is inversely proportional to temperature, which means the influence of $\Delta H°$ on *K* decreases as temperature increases. Thus, a favorable (negative) $\Delta H°$ contributes less to increasing the value of *K* as temperature increases. This temperature dependence makes sense if we invoke Le Châtelier's principle and the notion that energy is a product of exothermic ($\Delta H° < 0$) reactions and a reactant in endothermic, $\Delta H° > 0$, reactions. Increasing temperature shifts a reaction toward the side opposite the energy term: it promotes endothermic reactions and inhibits exothermic reactions.

CONNECTION In Chapter 9 we defined exothermic reactions as those giving off energy as heat. They have a negative enthalpy change ($\Delta H < 0$), while endothermic reactions absorb energy ($\Delta H > 0$).

If $\Delta H°$ and $\Delta S°$ do not vary much with temperature, then Equation 14.23 predicts that ln *K* will be a linear function of $1/T$. Furthermore, we expect a graph of ln *K* versus $1/T$ to have a positive slope for an exothermic process and a negative slope for an endothermic process. We can determine $\Delta H°$ from the slope of the line and $\Delta S°$ from its *y*-intercept. Thus, we can calculate fundamental thermodynamic values of a reaction at equilibrium by determining its equilibrium constant at different temperatures. For example, let's determine the values of $\Delta H°$ and $\Delta S°$ for the reaction

$$2\ CO_2(g) \rightleftharpoons 2\ CO(g) + O_2(g)$$

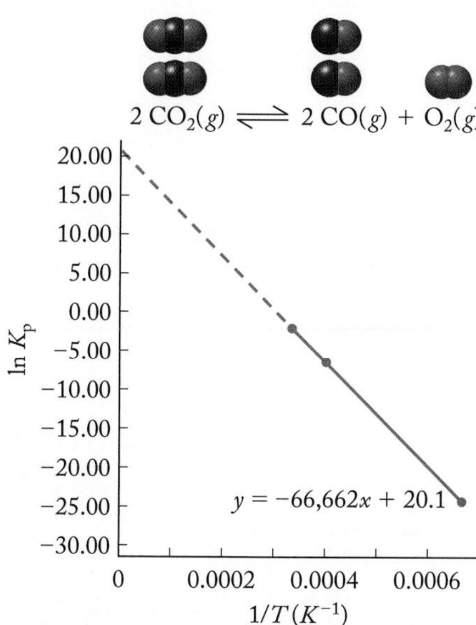

FIGURE 14.12 The graph of ln K_p versus $1/T$ is a straight line. The slope of the line is $\Delta H°_{rxn}/R$ and the *y*-intercept is $\Delta S°_{rxn}/R$. The negative slope indicates an endothermic process.

starting with its K_p values of 2.57×10^{-11} at 1500 K, 1.42×10^{-3} at 2500 K, and 0.112 at 3000 K. K_p increases as temperature increases, so energy must be a reactant and the reaction must be endothermic. We can use these data to determine the values of $\Delta H°$ and $\Delta S°$ by plotting ln K_p values versus $1/T$. To see how this graphical method works, let's rewrite Equation 14.23 so that it fits the form of the equation for a straight line ($y = mx + b$):

$$\ln K = \frac{-\Delta H°}{R}\left(\frac{1}{T}\right) + \frac{\Delta S°}{R} \quad (14.24)$$

The graph of ln K_p versus $1/T$ (Figure 14.12) is indeed a straight line. The slope of this line is −66,662 K, which is equal to $-\Delta H°_{rxn}/R$. The corresponding value of $\Delta H°_{rxn}$ is

$$\Delta H^\circ_{rxn} = -\text{slope} \times R$$
$$= -(-66{,}662\ \text{K})\left(\frac{8.314\ \text{J}}{\text{mol} \cdot \text{K}}\right)$$
$$= 554{,}228\ \text{J/mol} = 554.2\ \text{kJ/mol}$$

The y-intercept ($\Delta S^\circ_{rxn}/R$) of the graph is 20.1. The corresponding value of ΔS°_{rxn} is

$$\Delta S^\circ_{rxn} = (20.1)\left(\frac{8.314\ \text{J}}{\text{mol} \cdot \text{K}}\right) = 167\ \frac{\text{J}}{\text{mol} \cdot \text{K}}$$

The entropy change is positive because the forward reaction converts two moles of gaseous CO_2 into three moles of gaseous products.

We can use Equation 14.24 to compare the K values of a reaction at two different temperatures. In the two-temperature version of Equation 14.24, the $\Delta S^\circ/R$ terms drop out and we have

$$\ln\left(\frac{K_2}{K_1}\right) = -\frac{\Delta H^\circ}{R}\left(\frac{1}{T_2} - \frac{1}{T_1}\right) \quad (14.25)$$

This equation looks very much like the *Clausius–Clapeyron equation* we developed in Chapter 11, which describes the relationship between the vapor pressures of a liquid at two different temperatures:

$$\ln\left(\frac{P_{\text{vap},T_1}}{P_{\text{vap},T_2}}\right) = \frac{\Delta H_{\text{vap}}}{R}\left(\frac{1}{T_2} - \frac{1}{T_1}\right) \quad (11.4)$$

Indeed, the Clausius–Clapeyron equation is a special case of Equation 14.25 applied to a change in physical state (liquid $\rightarrow$ gas) instead of a change in chemical composition. The pressures in Equation 11.4 are actually equilibrium vapor pressures, and the enthalpy of reaction is the enthalpy of vaporization (ΔH_{vap}). Equation 14.25 is called the *van 't Hoff equation* because it was first derived by Jacobus van 't Hoff. It is particularly useful for calculating the value of K at a very high or very low temperature if we already know K at a standard reference temperature, such as 298 K. We use such a calculation in Sample Exercise 14.16.

SAMPLE EXERCISE 14.16 Calculating an Equilibrium Constant Value at a Specific Temperature — LO10

Use data from Appendix 4 to calculate the equilibrium constant K_p for the exothermic reaction

$$N_2(g) + 3\,H_2(g) \rightleftharpoons 2\,NH_3(g)$$

at 298 K and at 773 K, a typical temperature used in the Haber–Bosch process for synthesizing ammonia.

Collect and Organize We are asked to calculate K_p for a reaction at two temperatures. One of the temperatures is 298 K, which is the reference temperature for the standard thermodynamic data in Appendix 4. The ΔG°_f value of NH_3 is -16.5 kJ/mol. Note that ΔG°_f values are only valid for $T = 298$ K but that ΔH° and S° values may be used at other temperatures.

Analyze We can use Equation 14.22 to calculate the value of K_p at 298 K from the value of ΔG°_{rxn}. Then we can use Equation 14.25 to calculate the value of K_p at 773 K from the value of K_p at 298 K and the value of ΔH°_{rxn}. The ΔH°_{rxn} value can be calculated

from the enthalpy of formation of NH_3 gas (−46.1 kJ/mol). The negative value of ΔH_f° means that ΔH_{rxn}° is also negative, so the reaction is exothermic. Because energy is a product of the reaction, we predict that the value of K_p will be smaller at 773 K than at 298 K.

Solve The value of ΔG_{rxn}° for the reaction is

$$\Delta G_{rxn}^\circ = \frac{2\ \cancel{mol\ NH_3}}{1\ mol\ N_2} \times \frac{-16.5\ \cancel{kJ}}{1\ \cancel{mol\ NH_3}} \times \frac{1000\ J}{1\ \cancel{kJ}} = -33{,}000\ \frac{J}{mol\ N_2} = -3.30 \times 10^3\ J/mol$$

Using this value in Equation 14.22 to calculate the value of K_p,

$$K_p = e^{-\Delta G_{rxn}^\circ/RT} = \exp\left(\frac{-\left(-33{,}000\ \frac{J}{\cancel{mol}}\right)}{8.314\ \frac{J}{\cancel{mol \cdot K}} \times 298\ \cancel{K}}\right) = 6.09 \times 10^5$$

Once we know the value of K_p at 298 K, we can calculate K_p at 773 K using Equation 14.25. We must first calculate the value of ΔH_{rxn}°, which is twice the ΔH_f° of NH_3:

$$\Delta H_{rxn}^\circ = \frac{2\ \cancel{mol\ NH_3}}{1\ mol\ N_2} \times \frac{-46.1\ \cancel{kJ}}{1\ \cancel{mol\ NH_3}} \times \frac{1000\ J}{1\ \cancel{kJ}} = -92{,}200\ \frac{J}{mol\ N_2}$$

After substituting $K_1 = 6.1 \times 10^5$, $T_1 = 298$ K, and $T_2 = 773$ K into the equation, we solve for K_2:

$$\ln\left(\frac{K_2}{6.09 \times 10^5}\right) = -\frac{\left(-92{,}200\ \frac{J}{\cancel{mol}}\right)}{8.314\ \frac{J}{\cancel{mol \cdot K}}}\left(\frac{1}{773\ \cancel{K}} - \frac{1}{298\ \cancel{K}}\right)$$

$$= -22.87$$

$$\frac{K_2}{6.09 \times 10^5} = e^{-22.87} = 1.17 \times 10^{-10}$$

$$K_2 = 7.12 \times 10^{-5} \text{ at } 773\ K$$

Think About It The equilibrium constant decreases markedly when the temperature of the reaction is raised from 298 K to 773 K. This decrease agrees with our prediction for an exothermic reaction. Note that the standard thermodynamic data for the reaction (ΔH_{rxn}° and ΔG_{rxn}°) are expressed per mole of the reactant with a coefficient of 1 in the balanced chemical equation for N_2.

Practice Exercise Use data from Appendix 4 to calculate the value of K_p for the following reaction at 298 K and 2000 K.

$$2\ N_2(g) + O_2(g) \rightleftharpoons 2\ N_2O(g)$$

The equilibrium constant determined in Sample Exercise 14.16 for the ammonia synthesis reaction is greater than 10^5 at room temperature but about 10^{-5} at 773 K. That's ten orders of magnitude smaller. We now have a more complete picture of this reaction, which was the basis for Sample Exercise 14.9. Ammonia is usually synthesized from N_2 and H_2 at temperatures near 400°C. These high temperatures increase the rate of this exothermic reaction, even though they significantly decrease K_p and decrease the amount of ammonia produced at equilibrium. In this case, the practical benefit of a more favorable reaction rate outweighs the less favorable thermodynamics of running the reaction at high temperature. In practice, the ammonia produced in the reaction is continuously removed from

the reaction vessel by condensation, shifting the equilibrium toward the formation of more product as predicted by Le Châtelier's principle.

We have seen throughout this chapter that every reversible chemical reaction has a unique equilibrium constant at a specific temperature and that the value of that constant provides us with a measure of how much product can form from given quantities of reactants. We have also seen how the value of K is linked to the value of ΔG°_{rxn} and how this relationship provides us with a more sophisticated interpretation of what a negative value of ΔG°_{rxn} (and reaction spontaneity) means. Our new interpretation of $\Delta G^\circ_{rxn} < 0$ is that the reaction proceeds in the forward direction, producing reaction mixtures with an abundance of products and smaller quantities of reactants left over. The more negative the value of ΔG°_{rxn}, the greater the yield of products before the reaction reaches chemical equilibrium and there is no further increase in the ratio of products to reactants.

On the other hand, a positive value of ΔG°_{rxn} means more than that a reaction is simply nonspontaneous. It really means that a mixture of reactants may form some products (just not very much) before chemical equilibrium is reached and there is no further increase in the ratio of products to reactants. The more positive the value of ΔG°_{rxn}, the lower the yield of products.

SAMPLE EXERCISE 14.17 Integrating Concepts: Fire Extinguishers

Dry chemical fire extinguishers are the most common type purchased for use in the home. They are also used to fight fires at airport crash rescue sites. A dry chemical extinguisher sprays a very fine powder of sodium bicarbonate ($NaHCO_3$, baking soda) or potassium bicarbonate ($KHCO_3$). Both substances decompose at the elevated temperatures of a fire and release carbon dioxide to extinguish it. $K_p = 0.25$ for the baking soda decomposition reaction at 125°C:

$$2\,NaHCO_3(s) \rightleftharpoons Na_2CO_3(s) + CO_2(g) + H_2O(g)$$

a. Neither $NaHCO_3$ nor $KHCO_3$ decomposes at room temperature, but both decompose rapidly at high temperatures. Are their decomposition reactions exothermic or endothermic? What does this mean about the stability of these compounds in fire extinguishers?
b. $KHCO_3$ is marketed as being about two times as effective on fires involving flammable liquids or gases (class B fires) as $NaHCO_3$. Use thermochemical data from Appendix 4 to assess this claim by comparing the amounts of CO_2 generated when each material decomposes. Assume that the temperature of the fire is 1200°C.

Collect and Organize We are asked how long fire extinguishers containing $NaHCO_3$ or $KHCO_3$ can be kept at room temperature without losing their efficacy and whether manufacturers' claims about fighting fires with these compounds are valid. We are given an equilibrium constant for one reaction at 125°C, the fact that the compounds do not decompose at room temperature but do at elevated temperatures, and tables in Appendix 4 that contain thermochemical data on both compounds as well as the reaction products. To assess the claim of greater efficacy of $KHCO_3$, we need to determine K_p values for reactions of both substances at the fire temperature. We can then see if one produces more CO_2 than the other.

Analyze We can use ΔG°_f values for the reactants and products to calculate ΔG°_{rxn} at 298 K for the decomposition of $KHCO_3$ and then use that value to calculate K_p for the reaction at 298 K. Then we can use ΔH°_f values for the reactants and products to calculate ΔH°_{rxn} values for both reactions. These values and their K_p values at 298 K can be used with Equation 14.25 to calculate their K_p values at 1200°C. Comparing these two equilibrium constant values will help us determine whether one or the other bicarbonate compound releases more CO_2 at flame temperatures.

Solve

a. $NaHCO_3$ and $KHCO_3$ do not decompose at room temperature but do at flame temperatures, so we predict that their decomposition reactions are endothermic. Energy is essentially a reactant in endothermic reactions and adding heat means adding an essential reactant to the reaction mixture. These compounds should be stable in fire extinguishers stored at room temperature.
b. Let's first calculate the value of K_p for the decomposition of $NaHCO_3$ at 1200°C using Equation 14.25:

$$\ln\left(\frac{K_2}{K_1}\right) = -\frac{\Delta H^\circ}{R}\left(\frac{1}{T_2} - \frac{1}{T_1}\right)$$

We need ΔH° for the reaction of $NaHCO_3$, which we can calculate from data given in Appendix 4 and methods we learned in Chapter 9.

$$\Delta H^\circ = \sum n_{products}\,\Delta H^\circ_{f,products} - \sum n_{reactants}\,\Delta H^\circ_{f,reactions} \quad (9.15)$$

$$= (\Delta H^\circ_{f,Na_2CO_3} + \Delta H^\circ_{f,H_2O} + \Delta H^\circ_{f,CO_2}) - 2(\Delta H^\circ_{f,NaHCO_3})$$

$$= [(-1130.7) + (-241.8) + (-393.5)$$

$$-(2)(-950.8)]\text{ kJ/mol} = 135.6\text{ kJ/mol}$$

$$= 135{,}600\text{ J/mol}$$

Then

$$\ln \frac{K_2}{0.25} = -\frac{135{,}600\ \text{J/mol}}{8.314\ \dfrac{\text{J}}{\text{mol} \cdot \text{K}}}\left(\frac{1}{1473\ \text{K}} - \frac{1}{398\ \text{K}}\right)$$

$$\ln K_2 - \ln(0.25) = 29.91$$

$$\ln K_2 = 28.52$$

$$K_2 = 2.43 \times 10^{12} = K_{p,\text{NaHCO}_3} \text{ at } 1200°\text{C}$$

Now we need to calculate K_p for $KHCO_3$ at 1200°C using Equation 14.22:

$$K = e^{-\Delta G°/RT}$$

We can use $\Delta G°$ values from the Appendix to calculate $\Delta G°_{\text{rxn}}$, as we learned in Chapter 12, and from that we can determine K_p at 25°C. Then we can use Equation 14.25 and $\Delta H°$ values from the Appendix to estimate K_{p,KHCO_3} at 1200°C. The reaction of interest is

$$2\ \text{KHCO}_3(s) \rightleftharpoons \text{K}_2\text{CO}_3(s) + \text{CO}_2(g) + \text{H}_2\text{O}(g)$$

$$\Delta G° = \Sigma n_{\text{products}}\, \Delta G°_{\text{f,products}} - \Sigma n_{\text{reactants}}\, \Delta G°_{\text{f,reactants}} \quad (12.12)$$

$$= (\Delta G°_{\text{f,K}_2\text{CO}_3} + \Delta G°_{\text{f,H}_2\text{O}} + \Delta G°_{\text{f,CO}_2}) - 2(\Delta G°_{\text{f,KHCO}_3})$$

$$= [(-1063.5) + (-228.6) + (-394.4) - (2)(-863.6)]\ \text{kJ/mol}$$

$$= 40.7\ \text{kJ/mol} = 4.07 \times 10^4\ \text{J/mol}$$

Then

$$K = e^{4.07\times10^4(8.314\times298)} = e^{-16.43} = 7.34 \times 10^{-8}$$

We can now use $\Delta H°$ values from the Appendix to find $\Delta H°_{\text{rxn}}$ and apply Equation 14.25 again to calculate the value of K_{p,KHCO_3} at 1200°C:

$$\Delta H° = \sum n_{\text{products}}\, \Delta H°_{\text{f,products}} - \sum n_{\text{reactants}}\, \Delta H°_{\text{f,reactants}}$$

$$= (\Delta H°_{\text{f,K}_2\text{CO}_3} + \Delta H°_{\text{f,H}_2\text{O}} + \Delta H°_{\text{f,CO}_2}) - 2(\Delta H°_{\text{f,KHCO}_3})$$

$$= [(-1151.0) + (-241.8) + (-393.5) - (2)(-963.2)]\ \text{kJ/mol} = 140.1\ \text{kJ/mol}$$

$$= 1.401 \times 10^5\ \text{J/mol}$$

Then

$$\ln \frac{K_2}{7.43 \times 10^{-8}} = -\frac{1.401 \times 10^5\ \text{J/mol}}{8.314\ \dfrac{\text{J}}{\text{mol} \cdot \text{K}}}\left(\frac{1}{1473\ \text{K}} - \frac{1}{298\ \text{K}}\right)$$

$$\ln K_2 = 28.70 \qquad K_{p,\text{KHCO}_3} = 2.91 \times 10^{12}$$

The expressions for both equilibrium constants are the same:

$$K_{p,\text{NaHCO}_3} = [P_{\text{CO}_2}][P_{\text{H}_2\text{O}}]$$

$$K_{p,\text{KHCO}_3} = [P_{\text{CO}_2}][P_{\text{H}_2\text{O}}]$$

because the other ingredients are solids. We may assume the initial partial pressures of CO_2 and water vapor are insignificant compared to their equilibrium partial pressures. Their 1:1 mole ratio in the two reactions allows us to represent the equilibrium partial pressure of each with the same symbol. Let's use x in the $NaHCO_3$ reaction and y in the $KHCO_3$ reaction:

$$K_{p,\text{NaHCO}_3} = (P_{\text{CO}_2})(P_{\text{H}_2\text{O}}) = (x)(x) = 2.43 \times 10^{12}$$

$$K_{p,\text{KHCO}_3} = (P_{\text{CO}_2})(P_{\text{H}_2\text{O}}) = (y)(y) = 2.91 \times 10^{12}$$

Solving for x and y, we find that the partial pressure of CO_2 (P_{CO_2}) in a fire at 1200°C is 1.56×10^6 atm from the decomposition of $NaHCO_3$ and 1.71×10^6 atm from the decomposition of $KHCO_3$. These values are nearly the same, so the advantage of using $KHCO_3$ is apparently not based on its ability to produce more CO_2.

Think About It The $\Delta H°_{\text{rxn}}$ values for both compounds are positive, which agrees with our prediction that the reactions are endothermic. The similar P_{CO_2} values mean that $KHCO_3$ must do something else chemically that makes it a better fire-extinguishing agent than $NaHCO_3$. In fact, it does do something different: $KHCO_3$ better inhibits reactions involving free radical intermediates formed in a fire.

SUMMARY

LO1 **Chemical equilibrium** can be approached from either reaction direction and is achieved when the forward and reverse reaction rates are the same. At equilibrium, a reaction vessel may contain comparable amounts of reactants and products, may contain mostly reactants (the equilibrium *lies to the left*), or may contain mostly products (the equilibrium *lies to the right*). The value of the reaction's **equilibrium constant (*K*)** tells us whether products or reactants are favored at equilibrium. (Section 14.1)

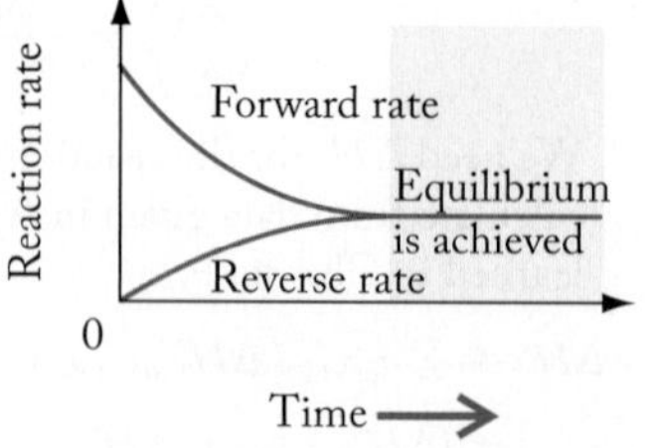

LO2 According to the **law of mass action**, the *equilibrium constant expression* (or **mass action expression**) for K_c is the ratio of the equilibrium molar concentrations of the products divided by the equilibrium molar concentrations of the reactants, each raised to the respective stoichiometric coefficient in the balanced equation. If the substances involved in an equilibrium are gases, then an equilibrium constant K_p may be written based on the partial pressures of the gases. **Heterogeneous equilibria** involve more than one phase. The concentrations of pure liquids and solids do not change during a reaction and are omitted from equilibrium constant expressions. Equilibrium constants have no units. (Sections 14.2, 14.6)

LO3 The value of K_c or K_p may be calculated using the equilibrium constant expression and concentrations or partial pressures of reactants and products at equilibrium. (Section 14.2)

LO4 The relationship between K_c and K_p depends on the number of moles of gaseous reactants and products in the balanced chemical equation. (Section 14.3)

LO5 The reverse of a reaction has an equilibrium constant that is the reciprocal of K for the forward reaction. If the balanced equation for a reaction is multiplied by some factor n, the value of K for that reaction is raised to the nth power. If reactions are summed to give an overall reaction, their equilibrium constants are multiplied together to obtain an overall equilibrium constant. (Section 14.4)

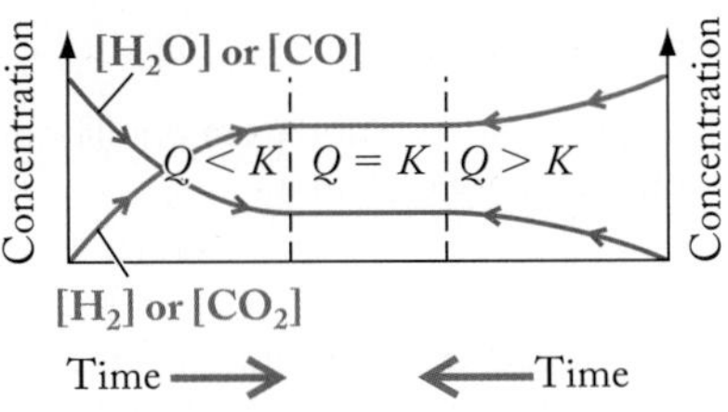

LO6 The **reaction quotient (Q)** is the value of the mass action expression at any instant during a reaction. At equilibrium, $Q = K$, but for nonequilibrium conditions, the value of Q indicates whether the concentrations of products ($Q < K$) or reactants ($Q > K$) will increase. (Section 14.5)

LO7 According to **Le Châtelier's principle**, chemical reactions at equilibrium respond to stress, such as removing or adding a reactant or product, by shifting position to relieve the stress. Changing its temperature also stresses an equilibrium mixture because it changes the value of K. A catalyst decreases the time it takes a system to achieve equilibrium but does not change the value of the equilibrium constant. (Section 14.7)

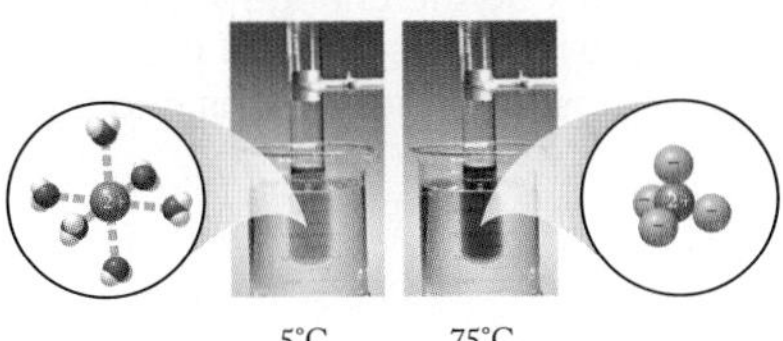

LO8 Equilibrium concentrations or partial pressures of reactants and products can be calculated from initial concentrations or pressures, the reaction stoichiometry, and the value of the equilibrium constant. (Section 14.8)

LO9 A negative value of $\Delta G°$ corresponds to $K > 1$ (products favored), and a positive value of $\Delta G°$ corresponds to $K < 1$ (reactants favored). Concentrations of reactants and products at equilibrium are similar if the value of $\Delta G°$ is near 0 (K is near 1). (Section 14.9)

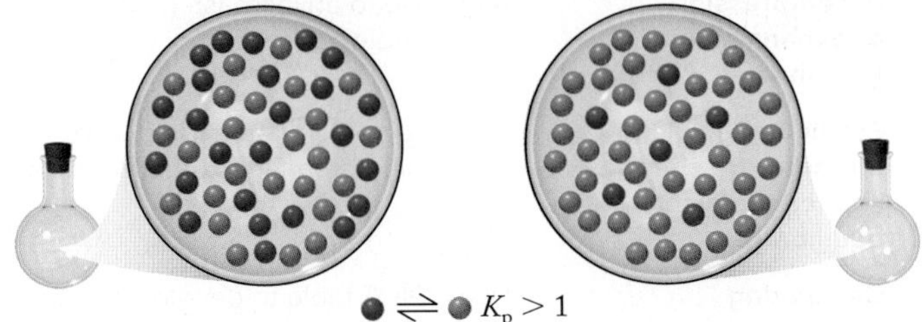

LO10 Higher reaction temperatures increase the equilibrium constant of an endothermic reaction, but decrease the equilibrium constant of an exothermic reaction. The slope of a plot of ln K versus $1/T$ for an equilibrium system is used to determine $\Delta H°_{rxn}$, and the y-intercept of the plot is used to determine the value of $\Delta S°_{rxn}$. (Section 14.10)

PARTICULATE PREVIEW WRAP-UP

Condition A spontaneously forms product (NO_2).
Condition B spontaneously forms reactant (N_2O_4).
Both conditions lead to spontaneous reactions, though in opposite directions. Therefore, there is a decrease in free energy ($\Delta G < 0$) starting with either condition. The forward and reverse reactions both proceed until $\Delta G = 0$, at which point the two reactions have produced the same equilibrium mixture of the two gases.

PROBLEM-SOLVING SUMMARY

Type of Problem	Concepts and Equations	Sample Exercises
Writing equilibrium constant expressions	For the reaction $$a\text{A} + b\text{B} \rightleftharpoons c\text{C} + d\text{D}$$ $$K_c = \frac{[\text{C}]^c[\text{D}]^d}{[\text{A}]^a[\text{B}]^b} \quad (14.9)$$ and $$K_p = \frac{(P_C)^c(P_D)^d}{(P_A)^a(P_B)^b} \quad (14.10)$$	**14.1**
Calculating K_c or K_p	Insert equilibrium molar concentrations or partial pressures into the equilibrium constant expression.	**14.2, 14.3**
Interconverting K_c and K_p	Use the relation $$K_p = K_c(RT)^{\Delta n} \quad (14.15)$$ where Δn is the number of moles of product gas minus the number of moles of reactant gas in the balanced chemical equation.	**14.4, 14.9**
Calculating K values of related reactions	To calculate K for the reverse of a reaction, take the reciprocal of K for the forward reaction. If all the coefficients in a chemical equation are multiplied by n, the value of K increases by the power of n. If reactions are summed to give an overall reaction, their equilibrium constants are multiplied together to obtain an overall K.	**14.5, 14.6**
Using Q and K values to predict the direction of a reaction	If $Q < K$, the reaction proceeds in the forward direction to make more products; if $Q = K$, the reaction is at equilibrium; if $Q > K$, the reaction proceeds in the reverse direction to make more reactants.	**14.7**

Type of Problem	Concepts and Equations	Sample Exercises
Writing equilibrium constant expressions for heterogeneous equilibria	Molar concentrations of pure liquids and pure solids are omitted from equilibrium constant expressions because these concentrations are constant.	**14.8**
Stressing an equilibrium by adding or removing reactants or products	Decreasing the concentration of a substance involved in an equilibrium shifts the equilibrium toward the production of more of that substance. Increasing the concentration of a substance shifts the equilibrium so that some of that substance is consumed in the reaction.	**14.10**
Predicting the effect of compression or expansion on gas-phase equilibria	Equilibria involving different numbers of gaseous reactants and products shift in response to an increase (or decrease) in volume toward the side with more (or fewer) moles of gases.	**14.11**
Predicting how temperature changes impact chemical equilibrium	The value of K for an endothermic reaction increases with increasing temperature; the value of K for an exothermic reaction decreases with increasing temperature.	**14.12**
Calculating concentrations or partial pressures of reactants and products at equilibrium	Use a RICE table to develop algebraic terms for each reactant's and product's partial pressure or concentration at equilibrium. Let x be the change in partial pressure or concentration of one component of the reaction. Express the changes in the other components in terms of x. Substitute these terms into the expression for K and solve for x.	**14.13, 14.14**
Relating K and ΔG°_{rxn}	Calculate ΔG°_{rxn} from $$\Delta G^\circ_{rxn} = \sum n_{products}\, \Delta G^\circ_{f,products} - \sum n_{reactants}\, \Delta G^\circ_{f,reactants} \quad (12.12)$$ Then use $$K = e^{-\Delta G^\circ/RT} \quad (14.22)$$ Be sure to convert ΔG° to joules per mole and use $R = 8.314$ J/(mol · K).	**14.15**
Calculating equilibrium constant values at specific temperatures	Use $$\ln\left(\frac{K_2}{K_1}\right) = -\frac{\Delta H^\circ}{R}\left(\frac{1}{T_2} - \frac{1}{T_1}\right) \quad (14.25)$$ Convert ΔH° from kilojoules per mole to joules per mole to match the units of R.	**14.16**

VISUAL PROBLEMS

(Answers to boldface end-of-chapter questions and problems are in the back of the book.)

14.1. Figure P14.1 shows the energy profiles of reactions $A \rightleftharpoons B$ and $C \rightleftharpoons D$, respectively. Which reaction has the larger forward rate constant? Which reaction has the smaller reverse rate constant? Which reaction has the larger equilibrium constant K_c?

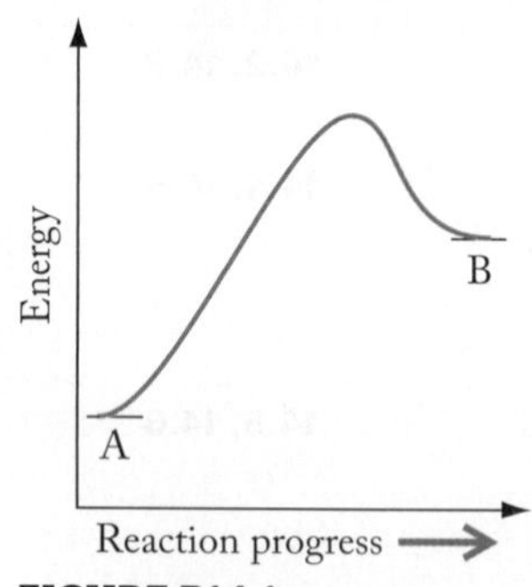

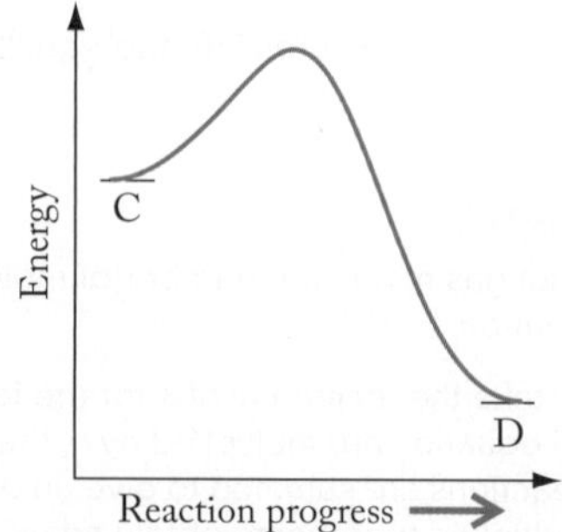

FIGURE P14.1

14.2. The progress with time of a reaction system is depicted in Figure P14.2. Red spheres represent the molar concentration of substance A and blue spheres represent the molar concentration of substance B.

a. Does the system reach equilibrium?
b. In which direction ($A \rightarrow B$ or $B \rightarrow A$) is equilibrium attained?

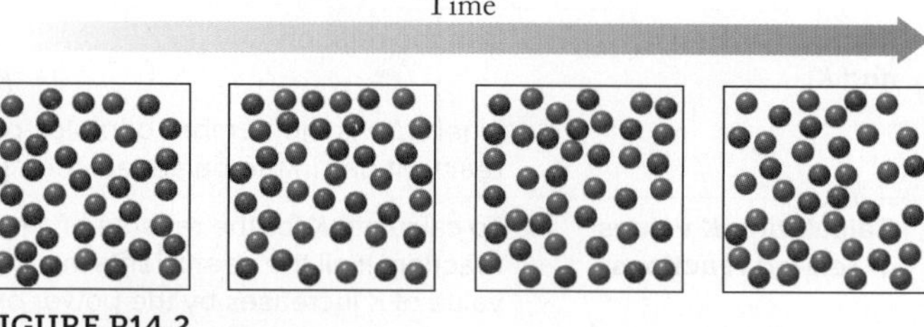

FIGURE P14.2

14.3. In Figure P14.3, the red spheres represent reactant A and the blue spheres represent product B in equilibrium with A.

a. Write a chemical equation that describes the equilibrium if the reaction is unimolecular.

b. What is the value of the equilibrium constant K_c?

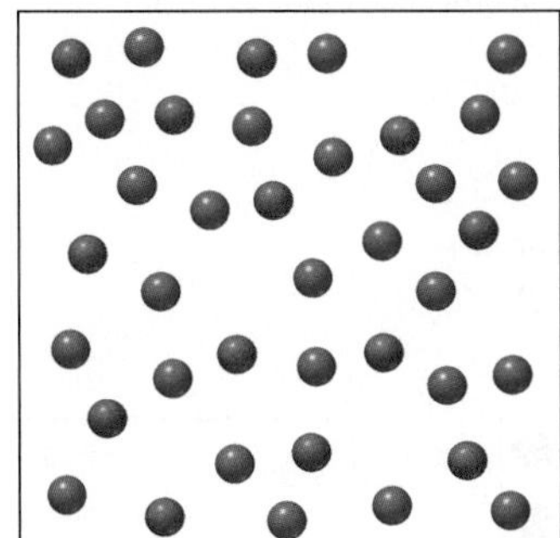

FIGURE P14.3

14.4. $K_c = 3.0$ for the following reaction at 300.0 K:

$$\text{A (red spheres)} + \text{B (blue spheres)} \rightleftharpoons \text{AB}$$

Does the situation depicted in Figure P14.4 correspond to equilibrium? If not, in what direction (to the left or to the right) will the system shift to attain equilibrium?

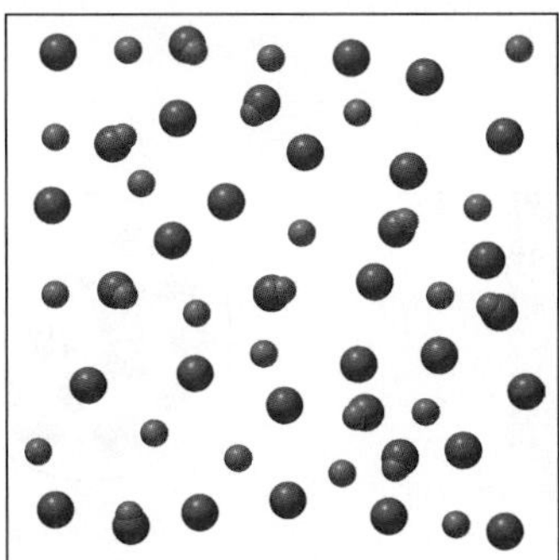

FIGURE P14.4

14.5. The left and right diagrams in Figure P14.5 represent equilibrium states of the reaction

$$\text{A (red spheres)} + \text{B (blue spheres)} \rightleftharpoons \text{AB}$$

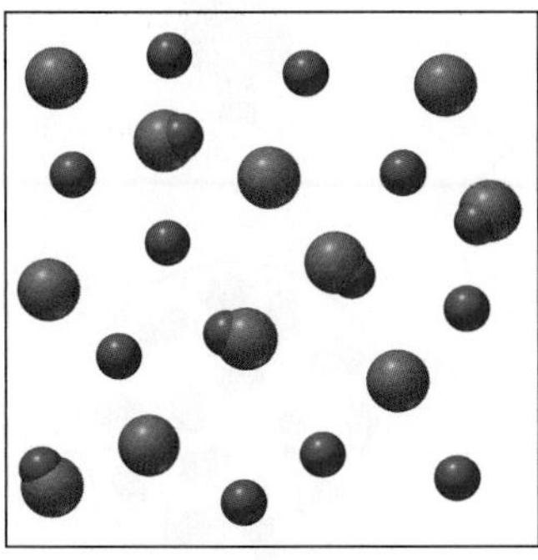

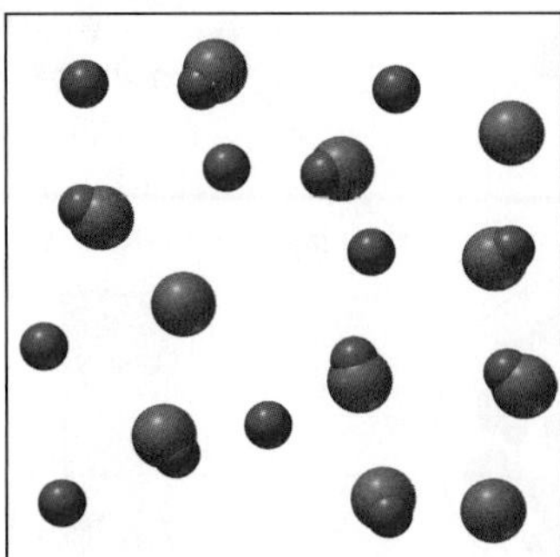

FIGURE P14.5

at 300 K and 400 K, respectively. Is this reaction endothermic or exothermic? Explain.

14.6. Figure P14.6 shows a plot of ln K_c versus $1/T$ for the reaction $\text{A} + 2\ \text{B} \rightleftharpoons \text{AB}_2$. Is the reaction endothermic or exothermic?

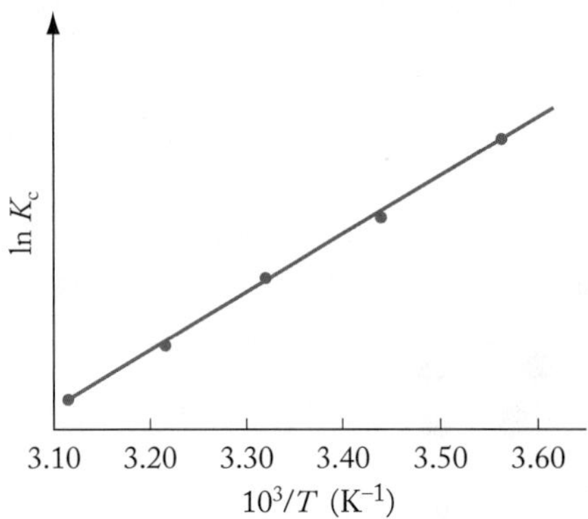

FIGURE P14.6

14.7. Does the reaction $\text{A} \rightarrow 2\ \text{B}$ represented in Figure P14.7 reach equilibrium in 20 μs? Explain your answer.

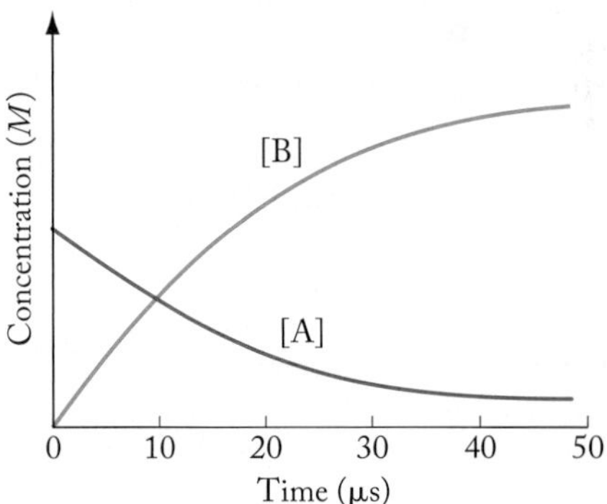

FIGURE P14.7

14.8. How are the concentrations of A and B in Figure P14.7 related at 10 μs? Is the system at equilibrium at that time? Why or why not?

14.9. The organic molecule on the left in Figure P14.9 is a β-diketone (A). It exists in equilibrium with its isomer (B), called an enol (alk**ene**-alcoh**ol**). Molecules of A are represented by red spheres and molecules of B by blue spheres. The boxes to the left of the reaction arrows in (a) and (b) contain 26 and 28 molecules of A, respectively.

a. Fill in the empty box with the appropriate number of molecules of each type at equilibrium if the equilibrium constant for the isomerization is $K = 0.3$.

b. Fill in the empty box with the appropriate number of molecules of each type at equilibrium if the equilibrium constant for the isomerization is $K = 3$.

$$H_3C-C(=O)-CH_2-C(=O)-CH_3 \rightleftharpoons H_3C-C(=O)-CH=C(OH)-CH_3$$

A B

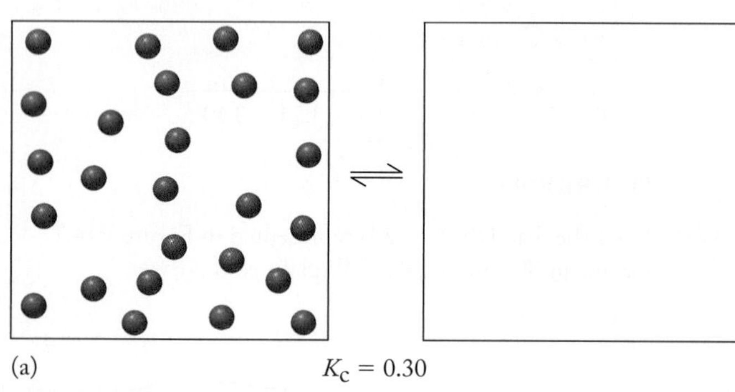

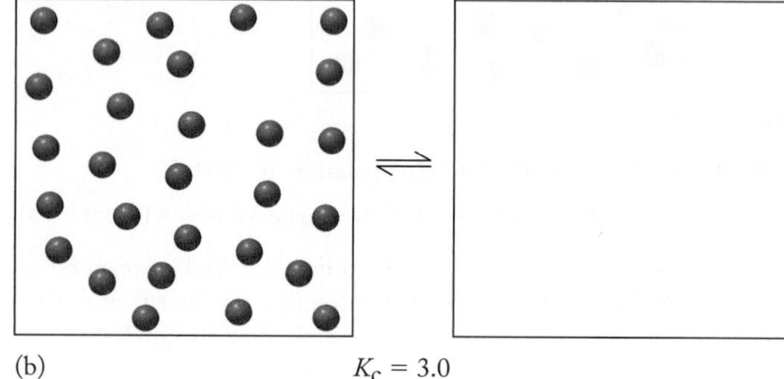

FIGURE P14.9

14.10. Use representations [A] through [I] in Figure P14.10 to answer questions a–f.

a. Write a balanced chemical equation describing the reversible reaction in [E], assuming NO_2 is the reactant.

b. Identify one stress to the system that would spontaneously transform [E] to [C].

c. While [E] is an equilibrium mixture, the reaction mixture in [C] is not yet at equilibrium. Which representation depicts the composition of [C] after equilibrium is re-established?

d. Are [G] and [I] at equilibrium or not? How can you tell?

e. Given that N_2O_4 is colorless and NO_2 is brown, match the pictures of the syringes to their respective particulate images.

f. Does the syringe in [H] contain any N_2O_4?

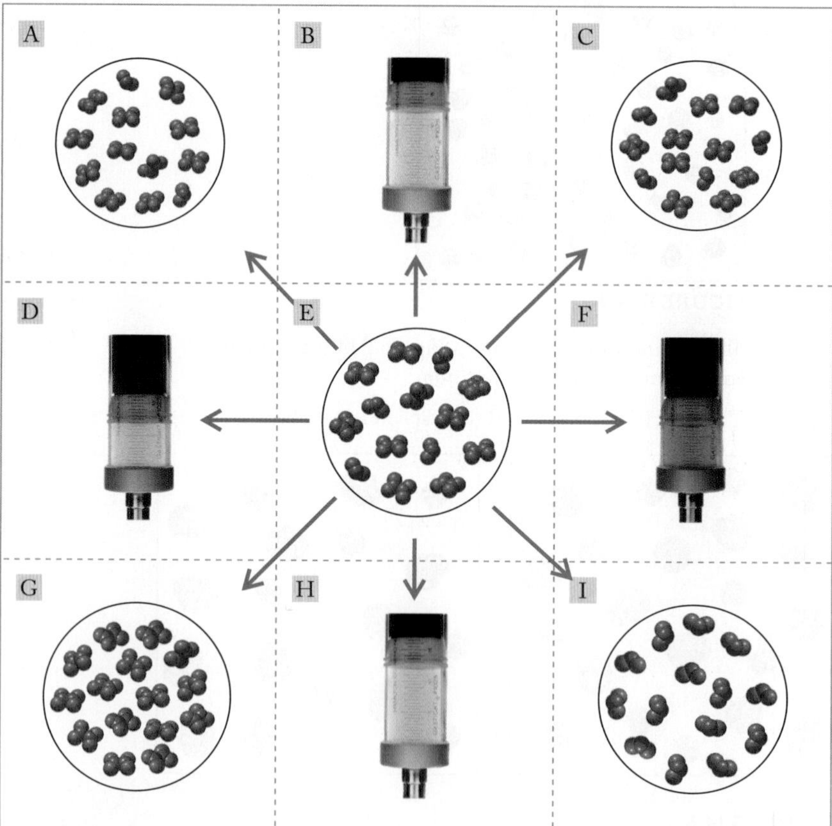

FIGURE P14.10

QUESTIONS AND PROBLEMS

The Dynamics of Chemical Equilibrium

Concept Review

14.11. Can a reaction that is not reversible achieve chemical equilibrium?

14.12. At equilibrium, is the sum of the concentrations of all the reactants always equal to the sum of the concentrations of the products? Explain.

14.13. Suppose the forward rate constant of the reaction $A \rightleftharpoons B$ is greater than the rate constant of the reverse reaction at a given temperature. Is the value of the equilibrium constant less than, greater than, or equal to 1?

14.14. Explain how it is possible for a reaction to have a large equilibrium constant but small forward and reverse rate constants.

Problems

14.15. In a study of the reaction,

$$2\ N_2O(g) \rightleftharpoons 2\ N_2(g) + O_2(g)$$

quantities of all three gases were injected into a reaction vessel. The N_2O consisted entirely of isotopically labeled $^{15}N_2O$. Analysis of the reaction mixture after 1 day revealed the presence of compounds with molar masses 28, 29, 30, 32, 44, 45, and 46 g/mol. Identify the compounds and account for their presence.

14.16. A mixture of ^{13}CO, $^{12}CO_2$, and O_2 in a sealed reaction vessel was used to follow the reaction

$$2\ CO(g) + O_2(g) \rightleftharpoons 2\ CO_2(g)$$

Analysis of the reaction mixture after 1 day revealed the presence of compounds with molar masses 28, 29, 32, 44, and 45 g/mol. Identify the compounds and account for their presence.

14.17. Suppose the reaction $A \rightleftharpoons B$ in the forward direction is first order in A with $k_f = 1.50 \times 10^{-2}\ s^{-1}$. The reverse reaction is first order in B with $k_r = 4.50 \times 10^{-2}\ s^{-1}$ at the same temperature. What is the value of the equilibrium constant for the reaction $A \rightleftharpoons B$ at this temperature?

14.18. At 700 K, $K_c = 8.7 \times 10^6$ for the gas-phase reaction between NO and O_2 forming NO_2. The rate constant for the reverse reaction at this temperature is $0.54\ M^{-1}\ s^{-1}$. What is the value of the rate constant for the forward reaction at 700 K?

Writing Equilibrium Constant Expressions; Relationships between K_c and K_p Values

Concept Review

14.19. Under what conditions are the numerical values of K_c and K_p equal?

14.20. At what temperature are the K_c and K_p values of the following reaction equal?

$$N_2(g) + O_2(g) \rightleftharpoons 2\ NO(g)$$

14.21. Nitrogen oxides play important roles in air pollution. Write expressions for K_c and K_p for the following reactions involving nitrogen oxides.

a. $N_2(g) + 2\ O_2(g) \rightleftharpoons N_2O_4(g)$
b. $3\ NO(g) \rightleftharpoons NO_2(g) + N_2O(g)$
c. $2\ N_2O(g) \rightleftharpoons 2\ N_2(g) + O_2(g)$

14.22. Write expressions for K_c and K_p for the following reactions, which contribute to the destruction of stratospheric ozone.

a. $Cl(g) + O_3(g) \rightleftharpoons ClO(g) + O_2(g)$
b. $2\ ClO(g) \rightleftharpoons 2\ Cl(g) + O_2(g)$
c. $2\ O_3(g) \rightleftharpoons 3\ O_2(g)$

Problems

14.23. Use the graph in Figure P14.23 to estimate the value of K_c for the following reaction:

$$N_2O(g) \rightleftharpoons N_2(g) + \tfrac{1}{2}\ O_2(g)$$

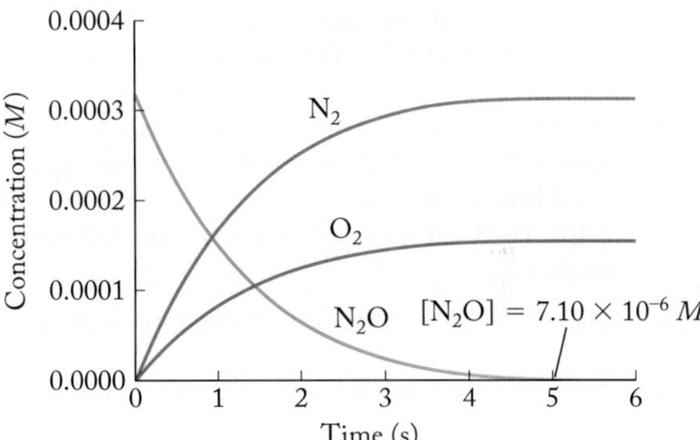

FIGURE P14.23

14.24. Use the data in Figure P14.24 to estimate the value of K_c for the following reaction:

$$2\ NO(g) + O_2(g) \rightleftharpoons 2\ NO_2(g)$$

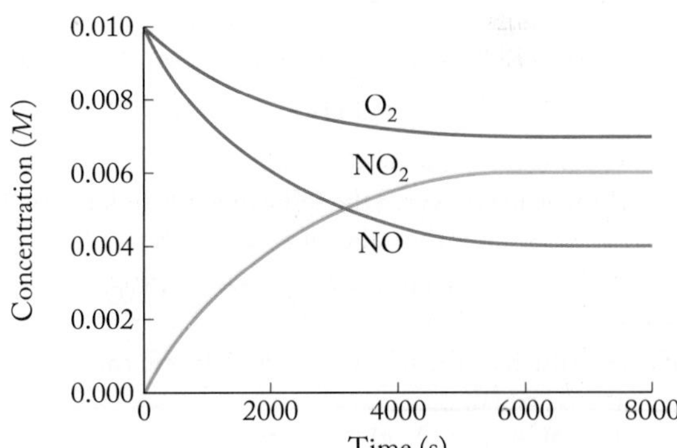

FIGURE P14.24

14.25. At 1200 K the partial pressures of an equilibrium mixture of H_2S, H_2, and S are 0.020, 0.045, and 0.030 atm, respectively. Calculate the value of K_p at 1200 K.

$$H_2S(g) \rightleftharpoons H_2(g) + S(g)$$

14.26. At 1045 K the partial pressures of an equilibrium mixture of H_2O, H_2, and O_2 are 0.040, 0.0045, and 0.0030 atm, respectively. Calculate the value of K_p at 1045 K.

$$2\ H_2O(g) \rightleftharpoons 2\ H_2(g) + O_2(g)$$

14.27. At equilibrium, the concentrations of gaseous N_2, O_2, and NO in a sealed reaction vessel are $[N_2] = 3.3 \times 10^{-3}\ M$, $[O_2] = 5.8 \times 10^{-3}\ M$, and $[NO] = 3.1 \times 10^{-3}\ M$. What is the value of K_c for the following reaction at the temperature of the mixture?

$$N_2(g) + O_2(g) \rightleftharpoons 2\ NO(g)$$

14.28. Analyses of an equilibrium mixture of gaseous N_2O_4 and NO_2 gave the following results: $[NO_2] = 4.2 \times 10^{-3}\ M$ and $[N_2O_4] = 2.9 \times 10^{-3}\ M$. What is the value of K_c for the following reaction at the temperature of the mixture?

$$2\ NO_2(g) \rightleftharpoons N_2O_4(g)$$

14.29. A sealed reaction vessel initially contains 1.50×10^{-2} mol water vapor and 1.50×10^{-2} mol CO. After the reaction

$$H_2O(g) + CO(g) \rightleftharpoons H_2(g) + CO_2(g)$$

has come to equilibrium, the vessel contains 8.3×10^{-3} mol CO_2. What is the value of K_c of the reaction at the temperature of the vessel?

*14.30. A 100 mL reaction vessel initially contains 2.60×10^{-2} mol NO and 1.30×10^{-2} mol H_2. At equilibrium, the concentration of NO in the vessel is 0.161 *M*. What is the value of K_c for the following reaction?

$$2\ H_2(g) + 2\ NO(g) \rightleftharpoons 2\ H_2O(g) + N_2(g)$$

14.31. $K_p = 32$ for the following equilibrium at 298 K.

$$A(g) + B(g) \rightleftharpoons AB(g)$$

What is the value of K_c for this same equilibrium at 298 K?

14.32. $K_p = 0.1764$ for the following equilibrium at 1773 K.

$$CO(g) + 3\ H_2(g) \rightleftharpoons CH_4(g) + H_2O(g)$$

What is K_c for this reaction at the same temperature?

14.33. At 500°C, $K_p = 1.45 \times 10^{-5}$ for the synthesis of ammonia:

$$N_2(g) + 3\ H_2(g) \rightleftharpoons 2\ NH_3(g)$$

What is the value of K_c at 500°C?

14.34. If the value of K_c for the following reaction is 5×10^5 at 298 K, what is the value of K_p at 298 K?

$$2\ CO(g) + O_2(g) \rightleftharpoons 2\ CO_2(g)$$

14.35. For which of the following reactions are the values of K_c and K_p equal?

a. $2\ SO_2(g) + O_2(g) \rightleftharpoons 2\ SO_3(g)$
b. $Fe(s) + CO_2(g) \rightleftharpoons FeO(s) + CO(g)$
c. $H_2O(g) + CO(g) \rightleftharpoons H_2(g) + CO_2(g)$

14.36. For which of the following reactions are the values of K_c and K_p different?

a. $SO_2Cl_2(g) \rightleftharpoons SO_2(g) + Cl_2(g)$
b. $2\ NO(g) + O_2(g) \rightleftharpoons 2\ NO_2(g)$
c. $2\ O_3(g) \rightleftharpoons 3\ O_2(g)$

14.37. **Bulletproof Glass** Phosgene ($COCl_2$) is used in the manufacture of foam rubber and bulletproof glass. It is formed from carbon monoxide and chlorine in the following reaction:

$$Cl_2(g) + CO(g) \rightleftharpoons COCl_2(g)$$

The value of K_c for the reaction is 5.0 at 327°C. What is the value of K_p at 327°C?

14.38. If the value of K_p for the following reaction

$$SO_2(g) + NO_2(g) \rightleftharpoons NO(g) + SO_3(g)$$

is 3.45 at 298 K, what is the value of K_c for the reaction?

Manipulating Equilibrium Constant Expressions

Concept Review

14.39. How is the value of the equilibrium constant affected by scaling up or down the coefficients of the reactants and products in the chemical equation describing the reaction?

14.40. Is the numerical value of K_p for the reaction

$$H_2(g) + I_2(g) \rightleftharpoons 2\ HI(g)$$

greater than, equal to, or less than the value of the equilibrium constant for the following reaction?

$$\tfrac{1}{2}\ H_2(g) + \tfrac{1}{2}\ I_2(g) \rightleftharpoons HI(g)$$

Problems

14.41. $K_c = 120$ for the following reaction at 425 K:

$$I_2(g) + Br_2(g) \rightleftharpoons 2\ IBr(g)$$

What is the value of K_c for the following reaction, also at 425 K?

$$\tfrac{1}{2}\ I_2(g) + \tfrac{1}{2}\ Br_2(g) \rightleftharpoons IBr(g)$$

14.42. The equilibrium constant K_p for the synthesis of ammonia,

$$N_2(g) + 3\ H_2(g) \rightleftharpoons 2\ NH_3(g)$$

is 4.3×10^{-3} at 300°C. What is the value of K_p for the following equilibrium, also at 300°C?

$$\tfrac{1}{2}\ N_2(g) + \tfrac{3}{2}\ H_2(g) \rightleftharpoons NH_3(g)$$

14.43. The following reaction is one of the elementary steps in the oxidation of NO:

$$NO(g) + NO_3(g) \rightleftharpoons 2\ NO_2(g)$$

Write an expression for the equilibrium constant K_c for this reaction and for the reverse reaction:

$$2\ NO_2(g) \rightleftharpoons NO(g) + NO_3(g)$$

How are the two K_c expressions related?

14.44. At a given temperature, the equilibrium constant K_c for the reaction

$$2\ NO(g) + 2\ H_2(g) \rightleftharpoons N_2(g) + 2\ H_2O(g)$$

is 0.11. What is the equilibrium constant K_p for the following reaction?

$$NO(g) + H_2(g) \rightleftharpoons \tfrac{1}{2}\ N_2(g) + H_2O(g)$$

14.45. **Air Pollutants** Sulfur oxides are major air pollutants. The reaction between sulfur dioxide and oxygen can be written in two ways:

$$SO_2(g) + \tfrac{1}{2}\ O_2(g) \rightleftharpoons SO_3(g)$$

and

$$2\ SO_2(g) + O_2(g) \rightleftharpoons 2\ SO_3(g)$$

Write expressions for the equilibrium constants for both reactions. How are they related?

14.46. At a given temperature, K_c for the reaction

$$2\ NO(g) + 2\ H_2(g) \rightleftharpoons N_2(g) + 2\ H_2O(g)$$

is 0.11. What is the equilibrium constant for the following reaction at the same temperature?

$$NO(g) + H_2(g) \rightleftharpoons \tfrac{1}{2}\ N_2(g) + H_2O(g)$$

14.47. At a given temperature, K_c for the reaction

$$2\ SO_2(g) + O_2(g) \rightleftharpoons 2\ SO_3(g)$$

is 2.4×10^{-3}. What is the value of the equilibrium constant for each of the following reactions at the same temperature?

a. $SO_2(g) + \frac{1}{2}\ O_2(g) \rightleftharpoons SO_3(g)$
b. $2\ SO_3(g) \rightleftharpoons 2\ SO_2(g) + O_2(g)$
c. $SO_3(g) \rightleftharpoons SO_2(g) + \frac{1}{2}\ O_2(g)$

14.48. If $K_c = 5 \times 10^{12}$ for the following reaction,

$$2\ NO(g) + O_2(g) \rightleftharpoons 2\ NO_2(g)$$

what is the value of the equilibrium constant of each of the following reactions at the same temperature?

a. $NO(g) + \frac{1}{2}\ O_2(g) \rightleftharpoons NO_2(g)$
b. $2\ NO_2(g) \rightleftharpoons 2\ NO(g) + O_2(g)$
c. $NO_2(g) \rightleftharpoons NO(g) + \frac{1}{2}\ O_2(g)$

14.49. Calculate the value of the equilibrium constant K_c for the reaction,

$$2\ D \rightleftharpoons A + 2\ B$$

given the following information:

$$A + 2\ B \rightleftharpoons C \qquad K_c = 3.3$$

$$C \rightleftharpoons 2\ D \qquad K_c = 0.041$$

14.50. **Reactions in Photochemical Smog** The brown gas responsible for the color in photochemical smog, NO_2, exists in equilibrium with its colorless dimer, $N_2O_4(g)$. In one of several possible decomposition reactions, N_2O_4 produces N_2O and oxygen. Calculate the value of the equilibrium constant K_c for the reaction

$$2\ N_2O_4(g) \rightleftharpoons 2\ N_2O(g) + 3\ O_2(g)$$

from the following information:

$$N_2(g) + \tfrac{1}{2}\ O_2(g) \rightleftharpoons N_2O(g) \qquad K_c = 2.7 \times 10^{-18}$$

$$N_2O_4(g) \rightleftharpoons 2\ NO_2(g) \qquad K_c = 4.6 \times 10^{-3}$$

$$N_2(g) + 2\ O_2(g) \rightleftharpoons 2\ NO_2(g) \qquad K_c = 1.7 \times 10^{-17}$$

Equilibrium Constants and Reaction Quotients

Concept Review

14.51. What is a reaction quotient?

14.52. How is an equilibrium constant different from a reaction quotient?

14.53. What does it mean when the reaction quotient Q is numerically equal to the equilibrium constant K?

14.54. Explain how knowing Q and K for an equilibrium system enables you to say whether it is at equilibrium or whether it will shift in one direction or another.

Problems

14.55. If $K_c = 22$ for the hypothetical reaction $A(g) \rightleftharpoons B(g)$ at a given temperature, and if $[A] = 0.10\ M$ and $[B] = 2.0\ M$ in a reaction mixture at that temperature, is the reaction at chemical equilibrium? If not, in which direction will the reaction proceed to reach equilibrium?

14.56. $K_c = 1.2 \times 10^3$ at 400°C for the reaction to produce the highly toxic nerve gas phosgene:

$$CO(g) + Cl_2(g) \rightleftharpoons COCl_2(g)$$

If equimolar amounts of CO, Cl_2, and $COCl_2$ are present at some point in the reaction in a 10.0 L reaction vessel, is the reaction at equilibrium? If not, in which direction will the reaction proceed to reach equilibrium?

14.57. Suppose the value of the equilibrium constant K_p of the following hypothetical reaction

$$A(g) + B(g) \rightleftharpoons C(g)$$

is 1.00 at 300 K. Are either of the following reaction mixtures at chemical equilibrium at 300 K?

a. $P_A = P_B = P_C = 1.0$ atm
b. $[A] = [B] = [C] = 1.0\ M$

14.58. In which direction will the following hypothetical reaction proceed to reach equilibrium under the conditions given?

$$A(g) + B(g) \rightleftharpoons C(g) \qquad K_p = 1.00 \text{ at } 300\text{ K}$$

a. $P_A = P_C = 1.0$ atm, $P_B = 0.50$ atm
b. $[A] = [B] = [C] = 1.0\ M$

14.59. If $K_c = 1.5 \times 10^{-3}$ for the reaction,

$$N_2(g) + O_2(g) \rightleftharpoons 2\ NO(g)$$

in which direction will the reaction proceed if the partial pressures of the three gases are all 1.00×10^{-3} atm?

14.60. At 650 K, the value of K_p for the ammonia synthesis reaction

$$N_2(g) + 3\ H_2(g) \rightleftharpoons 2\ NH_3(g)$$

is 4.3×10^{-4}. If a vessel contains a reaction mixture in which $[N_2] = 0.010\ M$, $[H_2] = 0.030\ M$, and $[NH_3] = 0.00020\ M$, will more ammonia form?

14.61. The hypothetical equilibrium $X + Y \rightleftharpoons Z$ has $K_c = 1.00$ at 350 K. If the initial molar concentrations of X, Y, and Z in a solution are all 0.2 M, in which direction will the reaction shift to reach equilibrium?

a. To the left, making more X and Y
b. To the right, making more Z
c. The system is at equilibrium and the concentrations will not change.

*14.62. Reactions between carboxylic acids and alcohols to produce esters typically do not go to completion. Ethyl acetate, a compound used industrially to decaffeinate coffee and tea, is made in the following reaction for which $K_c = 3.87$ at 75°C:

$$CH_3CH_2OH(\ell) + CH_3COOH(\ell) \rightleftharpoons CH_3COOCH_2CH_3(\ell) + H_2O(\ell)$$

a. Is a mixture of 125 g of ethanol (CH_3CH_2OH), 125 g of acetic acid (CH_3COOH), 125 g of ethyl acetate ($CH_3COOCH_2CH_3$), and 125 g of water in a 5 L reactor at equilibrium?
b. If not, in which direction will the reaction shift to reach equilibrium?

Heterogeneous Equilibria

Concept Review

14.63. Write the K_c expression for the following reaction:

$$CuS(s) \rightleftharpoons Cu^{2+}(aq) + S^{2-}(aq)$$

14.64. Write the K_c expression for the following reaction:

$$Al_2O_3(s) + 3\,H_2O(\ell) \rightleftharpoons 2\,Al^{3+}(aq) + 6\,OH^-(aq)$$

14.65. Why doesn't the K_c expression for the reaction

$$CaCO_3(s) \rightleftharpoons CaO(s) + CO_2(g)$$

contain terms for the concentrations of $CaCO_3$ and CaO?

14.66. How many partial pressure terms are there in the K_p expression for the thermal decomposition of sodium bicarbonate?

$$2\,NaHCO_3(s) \rightleftharpoons Na_2CO_3(s) + CO_2(g) + H_2O(g)$$

Le Châtelier's Principle

Concept Review

14.67. Does adding reactants to a system at equilibrium increase the value of the equilibrium constant?

14.68. Increasing the concentration of a reactant shifts the position of chemical equilibrium toward formation of more products. What effect does adding a reactant have on the rates of the forward and reverse reactions?

14.69. **Carbon Monoxide Poisoning** Patients suffering from carbon monoxide poisoning are treated with pure oxygen to remove CO from the hemoglobin (Hb) in their blood. The two relevant equilibria are

$$Hb + 4\,CO(g) \rightleftharpoons Hb(CO)_4$$

$$Hb + 4\,O_2(g) \rightleftharpoons Hb(O_2)_4$$

The value of the equilibrium constant for CO binding to Hb is greater than that for O_2. How, then, does this treatment work?

14.70. Is the equilibrium constant K_p for the reaction

$$2\,NO_2(g) \rightleftharpoons N_2O_4(g)$$

in air the same in Los Angeles as in Denver if the atmospheric pressure in Denver is lower but the temperature is the same?

14.71. Henry's law (Chapter 10) predicts that the solubility of a gas in a liquid increases with its partial pressure. Explain Henry's law in relation to Le Châtelier's principle.

*14.72. For the reaction

$$2\,CO(g) + O_2(g) \rightleftharpoons 2\,CO_2(g)$$

why does adding an inert gas such as argon to an equilibrium mixture of CO, O_2, and CO_2 in a sealed vessel increase the total pressure of the system but not affect the position of the equilibrium?

Problems

14.73. Which of the following equilibria will shift toward formation of more products if an equilibrium mixture is compressed into half its volume?

a. $2\,N_2O(g) \rightleftharpoons 2\,N_2(g) + O_2(g)$
b. $2\,CO(g) + O_2(g) \rightleftharpoons 2\,CO_2(g)$
c. $N_2(g) + O_2(g) \rightleftharpoons 2\,NO(g)$
d. $2\,NO(g) + O_2(g) \rightleftharpoons 2\,NO_2(g)$

14.74. Which of the following equilibria will shift toward formation of more products if the volume of a reaction mixture at equilibrium increases by a factor of 2?

a. $2\,SO_2(g) + O_2(g) \rightleftharpoons 2\,SO_3(g)$
b. $NO(g) + O_3(g) \rightleftharpoons NO_2(g) + O_2(g)$
c. $2\,N_2O_5(g) \rightleftharpoons 4\,NO_2(g) + O_2(g)$
d. $N_2O_4(g) \rightleftharpoons 2\,NO_2(g)$

14.75. What would be the effect of the changes listed on the equilibrium concentrations of reactants and products in the following reaction?

$$2\,O_3(g) \rightleftharpoons 3\,O_2(g)$$

a. O_3 is added to the system.
b. O_2 is added to the system.
c. The mixture is compressed to one-tenth its initial volume.

14.76. How will the changes listed affect the position of the following equilibrium?

$$2\,NO_2(g) \rightleftharpoons NO(g) + NO_3(g)$$

a. The concentration of NO is increased.
b. The concentration of NO_2 is increased.
c. The volume of the system is allowed to expand to 5 times its initial value.

14.77. How would reducing the partial pressure of $O_2(g)$ affect the position of the equilibrium in the following reaction?

$$2\,SO_2(g) + O_2(g) \rightleftharpoons 2\,SO_3(g)$$

*14.78. Ammonia is added to a gaseous reaction mixture containing H_2, Cl_2, and HCl that is at chemical equilibrium. How will the addition of ammonia affect the relative concentrations of H_2, Cl_2, and HCl if the equilibrium constant of reaction 2 is much greater than the equilibrium constant of reaction 1?

(1) $H_2(g) + Cl_2(g) \rightleftharpoons 2\,HCl(g)$

(2) $HCl(g) + NH_3(g) \rightleftharpoons NH_4Cl(s)$

14.79. In which of the following hypothetical equilibria does the product yield *increase* with increasing temperature?

a. $A + 2\,B \rightleftharpoons C \quad \Delta H > 0$
b. $A + 2\,B \rightleftharpoons C \quad \Delta H = 0$
c. $A + 2\,B \rightleftharpoons C \quad \Delta H < 0$

14.80. In which of the following hypothetical equilibria does the product yield *decrease* with increasing temperature?

a. $2\,X + Y \rightleftharpoons Z \quad \Delta H > 0$
b. $2\,X + Y \rightleftharpoons Z \quad \Delta H = 0$
c. $2\,X + Y \rightleftharpoons Z \quad \Delta H < 0$

Calculations Based on K

Concept Review

14.81. Why are calculations of how much product is formed in the reaction

$$X + Y \rightleftharpoons Z$$

often simpler when there is no Z initially present and the value of K is very small?

14.82. Could the quadratic equation be used to solve for the equilibrium concentration of NO_2 in the following reaction?

$$2\ NO(g) + O_2(g) \rightleftharpoons 2\ NO_2(g)$$

Problems

14.83. Consider the following reaction:

$$PCl_5(g) \rightleftharpoons PCl_3(g) + Cl_2(g) \qquad K_p = 23.6 \text{ at } 500\ K$$

a. Calculate the equilibrium partial pressures of the reactants and products if the initial pressures are $P_{PCl_5} = 0.560$ atm and $P_{PCl_3} = 0.500$ atm.
b. If more chlorine is added after equilibrium is reached, how will the concentrations of PCl_5 and PCl_3 change?

14.84. Enough NO_2 gas is injected into a cylindrical vessel to produce a partial pressure, P_{NO_2}, of 0.900 atm at 298 K. Calculate the equilibrium partial pressures of NO_2 and N_2O_4, given

$$2\ NO_2(g) \rightleftharpoons N_2O_4(g) \qquad K_p = 4.0 \text{ at } 298\ K$$

14.85. The value of K_c for the reaction between water vapor and dichlorine monoxide

$$H_2O(g) + Cl_2O(g) \rightleftharpoons 2\ HOCl(g)$$

is 0.0900 at 25°C. Determine the equilibrium concentrations of all three compounds if the starting concentrations of both reactants are 0.00432 *M* and no HOCl is present.

14.86. The value of K_p for the reaction

$$3\ H_2(g) + N_2(g) \rightleftharpoons 2\ NH_3(g)$$

is 4.3×10^{-4} at 648 K. Determine the equilibrium partial pressure of NH_3 in a reaction vessel that initially contained 0.900 atm N_2 and 0.500 atm H_2 at 648 K.

14.87. The value of K_p for the reaction

$$NO(g) + \tfrac{1}{2}\,O_2(g) \rightleftharpoons NO_2(g)$$

is 2.0×10^6 at 25°C. At equilibrium, what is the ratio of P_{NO_2} to P_{NO} in air at 25°C? Assume that $P_{O_2} = 0.21$ atm and does not change.

*14.88. **Making Hydrogen Gas** Passing steam over hot carbon produces a mixture of carbon monoxide and hydrogen:

$$H_2O(g) + C(s) \rightleftharpoons CO(g) + H_2(g)$$

The value of K_c for the reaction at 1000°C is 3.0×10^{-2}.
a. Calculate the equilibrium partial pressures of the products and reactants if $P_{H_2O} = 0.442$ atm and $P_{CO} = 5.0$ atm at the start of the reaction. Assume that the carbon is in excess.
b. Determine the equilibrium partial pressures of the reactants and products after sufficient CO and H_2 are added to the equilibrium mixture in part (a) to initially increase the partial pressures of both gases by 0.075 atm.

14.89. The value of K_p for the reaction

$$CO_2(g) + C(s) \rightleftharpoons 2\ CO(g)$$

is 1.5 at 700°C. Calculate the equilibrium partial pressures of CO and CO_2 if initially $P_{CO_2} = 5.0$ atm and $P_{CO} = 0.0$. Pure graphite is present in excess.

14.90. **Jupiter's Atmosphere** Ammonium hydrogen sulfide (NH_4SH) has been detected in the atmosphere of Jupiter, where it probably exists in equilibrium with ammonia and hydrogen sulfide:

$$NH_4SH(s) \rightleftharpoons NH_3(g) + H_2S(g)$$

The value of K_p for the reaction at 24°C is 0.126. Suppose a sealed flask contains an equilibrium mixture of NH_4SH, NH_3, and H_2S. At equilibrium, the partial pressure of H_2S is 0.355 atm. What is the partial pressure of NH_3?

***14.91.** A flask containing pure NO_2 is heated to 1000 K, a temperature at which $K_p = 158$ for the decomposition of NO_2:

$$2\ NO_2(g) \rightleftharpoons 2\ NO(g) + O_2(g)$$

The partial pressure of O_2 at equilibrium is 0.136 atm.
a. Calculate the partial pressures of NO and NO_2.
b. Calculate the total pressure in the flask at equilibrium.

14.92. The equilibrium constant K_p of the reaction

$$2\ SO_3(g) \rightleftharpoons 2\ SO_2(g) + O_2(g)$$

is 7.69 at 830°C. If a vessel at this temperature initially contains pure SO_3 and if the partial pressure of SO_3 at equilibrium is 0.100 atm, what is the partial pressure of O_2 in the flask at equilibrium?

***14.93.** **NO_x Pollution** In a study of the formation of NO_x air pollution, a chamber heated to 2200°C was filled with air (0.79 atm N_2, 0.21 atm O_2). What are the equilibrium partial pressures of N_2, O_2, and NO if $K_p = 0.050$ for the following reaction at 2200°C?

$$N_2(g) + O_2(g) \rightleftharpoons 2\ NO(g)$$

*14.94. The equilibrium constant K_p for the thermal decomposition of NO_2

$$2\ NO_2(g) \rightleftharpoons 2\ NO(g) + O_2(g)$$

is 6.5×10^{-6} at 450°C. If a reaction vessel at this temperature initially contains 0.500 atm NO_2, what will the partial pressures of NO_2, NO, and O_2 be in the vessel when equilibrium has been attained?

14.95. The value of K_c for the thermal decomposition of hydrogen sulfide

$$2\ H_2S(g) \rightleftharpoons 2\ H_2(g) + S_2(g)$$

is 2.2×10^{-4} at 1400 K. A sample of gas in which $[H_2S] = 6.00\ M$ is heated to 1400 K in a sealed high-pressure vessel. After chemical equilibrium has been achieved, what is the value of $[H_2S]$? Assume that no H_2 or S_2 was present in the original sample.

14.96. **Urban Air** On a very smoggy day, the equilibrium concentration of NO_2 in the air over an urban area reaches $2.2 \times 10^{-7}\ M$. If the temperature of the air is 25°C, what is the concentration of the dimer N_2O_4 in the air?

$$N_2O_4(g) \rightleftharpoons 2\ NO_2(g) \qquad K_c = 6.1 \times 10^{-3}$$

***14.97.** **Chemical Weapon** Phosgene, $COCl_2$, gained notoriety as a chemical weapon in World War I. Phosgene is produced by the reaction of carbon monoxide with chlorine:

$$CO(g) + Cl_2(g) \rightleftharpoons COCl_2(g)$$

$K_c = 5.0$ for this reaction at 600 K. What are the equilibrium partial pressures of the three gases if a reaction vessel initially contains a mixture of the reactants in which $P_{CO} = P_{Cl_2} = 0.265$ atm and $P_{COCl_2} = 0.000$ atm?

*14.98. At 2000°C, $K_c = 1.0$ for the following reaction:

$$2\ CO(g) + O_2(g) \rightleftharpoons 2\ CO_2(g)$$

What is the ratio of [CO] to [CO_2] in an atmosphere in which [O_2] = 0.0045 *M*?

***14.99.** The water–gas shift reaction is an important source of hydrogen. The value of K_c for the reaction

$$CO(g) + H_2O(g) \rightleftharpoons CO_2(g) + H_2(g)$$

at 700 K is 5.1. Calculate the equilibrium concentrations of the four gases if the initial concentration of each of them is 0.050 *M*.

*14.100. Sulfur dioxide reacts with NO_2, forming SO_3 and NO:

$$SO_2(g) + NO_2(g) \rightleftharpoons SO_3(g) + NO(g)$$

If $K_c = 2.50$ for the reaction, what are the equilibrium concentrations of the products if the reaction mixture was initially 0.50 *M* SO_2, 0.50 *M* NO_2, 0.0050 *M* SO_3, and 0.0050 *M* NO?

14.101. Nitrogen dioxide and nitric oxide react to form dinitrogen trioxide. $K_p = 0.535$ for this reaction at 25°C:

$$NO(g) + NO_2(g) \rightleftharpoons N_2O_3(g)$$

If a 4.0 L flask initially contains 15 g NO and 69 g NO_2, what are the concentrations of all species at equilibrium?

14.102. For the decomposition of HI(*g*) into $H_2(g)$ and $I_2(g)$ at 400°C, $K_c = 0.0183$:

$$2\ HI(g) \rightleftharpoons H_2(g) + I_2(g)$$

If 80.0 g of HI(*g*) is placed in a 2.5 L chamber at 400°C, what are the concentrations of all species when the system comes to equilibrium?

Equilibrium and Thermodynamics

Concept Review

14.103. Do all reactions with equilibrium constants < 1 have values of $\Delta G° > 0$?

*14.104. The equation $\Delta G° = -RT \ln K$ relates the value of K_p, not K_c, to the change in standard free energy for a reaction in the gas phase. Explain why.

14.105. Starting with pure reactants, in which direction will an equilibrium shift if $\Delta G° < 0$?

14.106. Starting with pure products, in which direction will an equilibrium shift if $\Delta G° < 0$?

Problems

14.107. Which of the following reactions has the largest equilibrium constant at 25°C?

a. $Cl_2(g) + F_2(g) \rightleftharpoons 2\ ClF(g)$ $\Delta G° = 115.4$ kJ
b. $Cl_2(g) + Br_2(g) \rightleftharpoons 2\ ClBr(g)$ $\Delta G° = -2.0$ kJ
c. $Cl_2(g) + I_2(g) \rightleftharpoons 2\ ICl(g)$ $\Delta G° = -27.9$ kJ

14.108. The value of $\Delta G°$ for the reaction

$$N_2O(g) + \tfrac{1}{2}\ O_2(g) \rightleftharpoons 2\ NO(g)$$

is 68.9 kJ. What is the value of the equilibrium constant for this reaction at 298 K?

***14.109.** At a temperature of 1000 K, $SO_2(g)$ combines with oxygen to make $SO_3(g)$:

$$2\ SO_2(g) + O_2(g) \rightleftharpoons 2\ SO_3(g)$$

Under these conditions, $K_p = 3.4$.

a. Use the appropriate thermodynamic data in Appendix 4 to calculate the value of $\Delta H°_{rxn}$ for this reaction.
b. What is the value of K_p for this reaction at 298 K?
c. Use the answer from part (b) to calculate the value of $\Delta G°_{rxn}$ at 298 K, and compare it to the value you obtained using the $\Delta G°_f$ values in Appendix 4.

*14.110. The value of the equilibrium constant K_p for the reaction

$$H_2(g) + CO_2(g) \rightleftharpoons H_2O(g) + CO(g)$$

is 0.534 at 700°C.

a. What is the value of K_p for this reaction at 298 K? (*Hint*: To perform this K_p calculation, you will need to calculate the value of $\Delta H°_{rxn}$ using the appropriate data from Appendix 4.)
b. Calculate the value of $\Delta G°_{rxn}$ at 298 K using your answer from part (a). Then compare it to the $\Delta G°_{rxn}$ value you obtained using the $\Delta G°_f$ values in Appendix 4.

14.111. Use the $\Delta G°$ data given here to calculate the value of K_p at 298 K for the following reaction:

$$N_2(g) + 2\ O_2(g) \rightleftharpoons 2\ NO_2(g)$$

$N_2(g) + O_2(g) \rightleftharpoons 2\ NO(g)$ $\Delta G° = 173.2$ kJ
$2\ NO(g) + O_2(g) \rightleftharpoons 2\ NO_2(g)$ $\Delta G° = -69.7$ kJ

14.112. Under the appropriate conditions, NO forms N_2O and NO_2:

$$3\ NO(g) \rightleftharpoons N_2O(g) + NO_2(g)$$

Use the values for $\Delta G°$ for the following reactions to calculate the value of K_p for the preceding reaction at 500°C.

$2\ NO(g) + O_2(g) \rightleftharpoons 2\ NO_2(g)$ $\Delta G° = -69.7$ kJ
$2\ N_2O(g) \rightleftharpoons 2\ NO(g) + N_2(g)$ $\Delta G° = -33.8$ kJ
$N_2(g) + O_2(g) \rightleftharpoons 2\ NO(g)$ $\Delta G° = 173.2$ kJ

Changing K with Changing Temperature

Concept Review

14.113. The value of the equilibrium constant of a reaction decreases with increasing temperature. Is this reaction endothermic or exothermic?

14.114. The reaction

$$2\ CO(g) + O_2(g) \rightleftharpoons 2\ CO_2(g)$$

is exothermic. Does the value of K_p increase or decrease with increasing temperature?

14.115. The value of K_p for the water–gas shift reaction

$$CO(g) + H_2O(g) \rightleftharpoons H_2(g) + CO_2(g)$$

increases as the temperature decreases. Is the reaction exothermic or endothermic?

14.116. Does the value of K_p for the reaction

$$CH_4(g) + H_2O(g) \rightleftharpoons 3\ H_2(g) + CO(g) \qquad \Delta H^\circ = 206\ kJ$$

increase, decrease, or remain unchanged as the temperature increases?

Problems

14.117. **Air Pollution** Automobiles and trucks pollute the air with NO. At 2000°C, K_c for the reaction

$$N_2(g) + O_2(g) \rightleftharpoons 2\ NO(g)$$

is 4.10×10^{-4}, and $\Delta H^\circ = 180.6$ kJ. What is the value of K_c at 25°C?

14.118. At 400 K the value of K_p for the reaction

$$N_2(g) + 3\ H_2(g) \rightleftharpoons 2\ NH_3(g)$$

is 41, and $\Delta H^\circ = -92.2$ kJ. What is the value of K_p at 700 K?

14.119. The equilibrium constant for the reaction

$$2\ NO(g) + O_2(g) \rightleftharpoons 2\ NO_2(g)$$

decreases from 1.5×10^5 at 430°C to 23 at 1000°C. From these data, calculate the value of ΔH° for the reaction.

14.120. The value of K_c for the reaction $A \rightleftharpoons B$ is 0.455 at 50°C and 0.655 at 100°C. Calculate ΔH° for the reaction.

Additional Problems

***14.121.** **CO as a Fuel** Is carbon dioxide a viable source of the fuel CO? Pure carbon dioxide ($P_{CO_2} = 1$ atm) decomposes at high temperatures. For the system

$$2\ CO_2(g) \rightleftharpoons 2\ CO(g) + O_2(g)$$

the percentage of decomposition of $CO_2(g)$ changes with temperature as follows:

Temperature (K)	Decomposition (%)
1500	0.048
2500	17.6
3000	54.8

Is the reaction endothermic? Calculate the value of K_p at each temperature and discuss the results. Is the decomposition of CO_2 an antidote for global warming?

14.122. Ammonia decomposes at high temperatures. In an experiment to explore this behavior, 2.00 mol of gaseous NH_3 is sealed in a rigid 1.00 L vessel. The vessel is heated to 785 K and some of the NH_3 decomposes in the following reaction:

$$2\ NH_3(g) \rightleftharpoons N_2(g) + 3\ H_2(g)$$

The system eventually reaches equilibrium and is found to contain 0.0040 mol of NH_3. What are the values of K_p and K_c for the decomposition reaction at 785 K?

***14.123.** Elements of group 16 form hydrides with the generic formula H_2X. When gaseous H_2X is bubbled through a solution containing 0.3 *M* hydrochloric acid, the solution becomes saturated and $[H_2X] = 0.1\ M$. The following equilibria exist in this solution:

$$H_2X(aq) + H_2O(\ell) \rightleftharpoons HX^-(aq) + H_3O^+(aq) \qquad K_1 = 8.3 \times 10^{-8}$$

$$HX^-(aq) + H_2O(\ell) \rightleftharpoons X^{2-}(aq) + H_3O^+(aq) \qquad K_2 = 1 \times 10^{-14}$$

Calculate the concentration of X^{2-} in the solution.

*14.124. Cobalt(II) oxide has been used for centuries as a glaze for pottery because of its deep blue color, which is known as cobalt blue. The oxide can be decomposed into cobalt metal by reduction with CO:

$$CoO(s) + CO(g) \rightleftharpoons Co(s) + CO_2(g)$$

For this equilibrium at 770 K, $K_c = 4.90 \times 10^2$.

a. What is the value of K_p at 770 K?
b. If the total pressure in the reactor in which the oxide is being reduced is 15.8 atm, what are the partial pressures of CO and CO_2?

***14.125.** Carbon disulfide is a foul-smelling solvent that dissolves sulfur and other nonpolar substances. It can be made by heating sulfur in an atmosphere of methane:

$$4\ CH_4(g) + S_8(s) \rightleftharpoons 4\ CS_2(g) + 8\ H_2(g)$$

Starting with the appropriate data in Appendix 4, calculate the values of K_p for the reaction at 25°C and 500°C.

*14.126. **Making Hydrogen** Debate continues on the practicality of using H_2 gas as a fuel for cars. The equilibrium constant K_c for the reaction

$$CO(g) + H_2O(g) \rightleftharpoons CO_2(g) + H_2(g)$$

is 1.0×10^5 at 25°C. Starting with this value, calculate the value of ΔG°_{rxn} at 25°C and, without doing any calculations, guess the sign of ΔH°_{rxn}.

***14.127.** **Air Pollution Control** Calcium oxide is used to remove the pollutant SO_2 from smokestack gases. The ΔG° of the overall reaction

$$CaO(s) + SO_2(g) + \tfrac{1}{2}\ O_2(g) \rightleftharpoons CaSO_4(s)$$

is -418.6 kJ. What is P_{SO_2} in equilibrium with air ($P_{O_2} = 0.21$ atm) and solid CaO?

14.128. **Hydrogen Production** The steam-methane reforming reaction plays a key role in producing hydrogen gas for use as a fuel and as a reactant in ammonia production. The equilibrium constant (K_p) of the reaction: $CH_4(g) + H_2O(g) \rightleftharpoons 3\ H_2(g) + CO(g)$, is 13.0 at 700°C.

a. Describe two advantages in running this endothermic ($\Delta H^\circ_{rxn} = 206$ kJ) reaction at 700°C instead of 100°C.
b. If the initial partial pressures of the two reactants are each 5.00 atm at 700°C and no products are present, what is the partial pressure of H_2 gas after equilibrium is achieved?

smartwork5

If your instructor uses Smartwork5, log in at digital.wwnorton.com/atoms2.

15

Acid–Base Equilibria

Proton Transfer in Biological Systems

SHADES OF PINK AND BLUE AND IN BETWEEN The colors of hydrangea blossoms depend on the acidity of the soil in which they grow.

PARTICULATE **REVIEW**

Strong Acids versus Weak Acids

In Chapter 15, we explore the properties of weak acids and bases. Compare dissolving $HCl(g)$ in water to dissolving $CH_3COOH(\ell)$ in water.

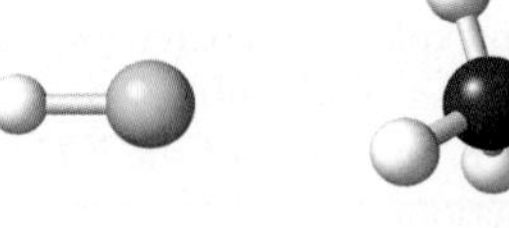

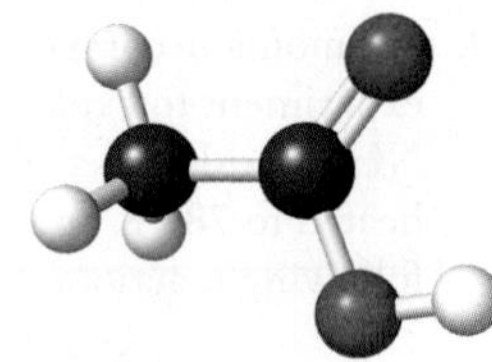

- What covalent bonds break? What covalent bonds form?
- What intermolecular forces are disrupted? What intermolecular interactions form? Between which species?
- What are the primary particles in the two aqueous solutions?

(Review Sections 8.3 and 8.4 if you need help.)

(Answers to Particulate Review questions are in the back of the book.)

PARTICULATE **PREVIEW**

Two Acids and Two Bases

The equilibrium present in an aqueous ammonia solution is depicted here. As you read Chapter 15, look for ideas that will help you understand the proton transfers in both the forward and reverse reactions.

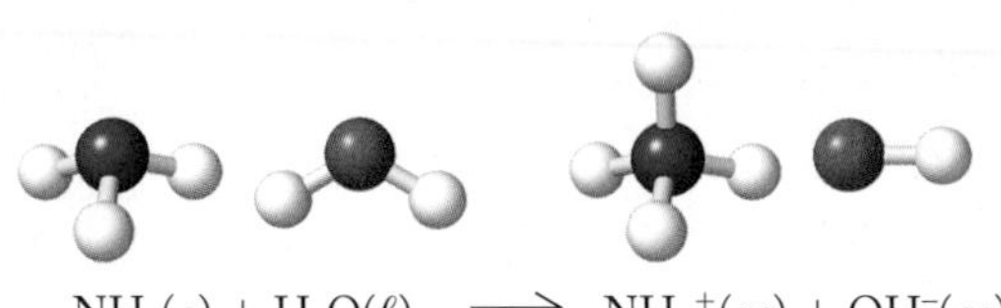

- Identify both the Brønsted–Lowry base and the Brønsted–Lowry acid in the forward reaction.
- What species functions as a Brønsted–Lowry base in the reverse reaction? Is there also a Brønsted–Lowry acid in the reverse reaction?
- Which acid and which base are the predominant species present in the solution if $K << 1$ for the reaction as written?

Learning Outcomes

LO1 Relate the strengths of acids and bases to their K_a and K_b values
Sample Exercise 15.1

LO2 Relate the strengths of acids to their molecular structures
Sample Exercises 15.2, 15.3

LO3 Recognize conjugate acid–base pairs and their complementary acid/base strengths
Sample Exercises 15.4, 15.5

LO4 Interconvert [H_3O^+], pH, pOH, and [OH^-]
Sample Exercises 15.6, 15.7, 15.8

LO5 Relate pH to percent ionization and the K_a values of weak acids or the K_b values of weak bases
Sample Exercises 15.9, 15.10

LO6 Calculate the pH of solutions of strong and weak acids and bases
Sample Exercises 15.11, 15.12, 15.13, 15.14

LO7 Calculate the pH of solutions of polyprotic acids
Sample Exercises 15.15, 15.16

LO8 Predict whether a salt is acidic, basic, or neutral, and calculate the pH of a salt solution
Sample Exercises 15.17, 15.18, 15.19

15.1 Acids and Bases: A Balancing Act

The pink and blue colors of the blossoms in the chapter-opening photograph are produced by the same species of plant (a hydrangea) grown in the same garden but in soils with different composition. The colors of hydrangea blossoms are controlled by the availability of aluminum ions (Al^{3+}) in the soil in which they grow. Aluminum ions are soluble in acidic soils, and hydrangeas grown in acidic soils form blue blossoms. In soils that are neutral or slightly basic, however, aluminum ions precipitate as aluminum hydroxide and are unavailable. As a result, hydrangea blossoms grown in those soils lack blue pigment and are pink.

For many biological systems, including the human body, acid–base balance is vital to normal function and good health. Our blood, for example, is normally slightly basic, regulated by the proportions of CO_2 gas and bicarbonate (HCO_3^-) ions dissolved in it. This regulation happens because of the following reversible chemical reactions:

$$CO_2(aq) + H_2O(\ell) \rightleftharpoons H_2CO_3(aq) \tag{15.1}$$

$$H_2CO_3(aq) + H_2O(\ell) \rightleftharpoons HCO_3^-(aq) + H_3O^+(aq) \tag{15.2}$$

Note how molecules of CO_2 and H_2O combine to form molecules of carbonic acid (H_2CO_3), which donate H^+ ions to more water molecules, thereby forming bicarbonate and hydronium (H_3O^+) ions.

As we learned in Chapter 8, compounds that produce H_3O^+ ions when dissolved in water are acids, whereas compounds that produce hydroxide ions (OH^-) in aqueous solution are bases. More precisely, these compounds are referred to as **Arrhenius acids** and **Arrhenius bases**, respectively, in honor of Svante Arrhenius, whose research on the behavior of electrolytic solutions was recognized with the 1903 Nobel Prize in Chemistry.

If we apply Le Châtelier's principle to Equation 15.2, the concentration of H_3O^+ ions will increase (as does the acidity of the solution) when the concentration of H_2CO_3 increases. In addition, the concentration of H_2CO_3 increases when the concentration of dissolved CO_2 increases because the equilibrium in Equation 15.1 shifts to the right. Therefore, an increase in the concentration of dissolved CO_2 makes a solution of carbonic acid more acidic. On the other hand,

Arrhenius acid a compound that produces H_3O^+ ions in aqueous solution.

Arrhenius base a compound that produces OH^- ions in aqueous solution.

Le Châtelier's principle also tells us that an increase in the concentration of HCO_3^- ions decreases the concentration of H_3O^+ ions and reduces acidity.

The equilibria in Equations 15.1 and 15.2 play a key role in human health. Normally, the concentration of HCO_3^- ions in our blood is about 20 times the dissolved concentration of CO_2, because the dissociation of carbonic acid is not our only source of HCO_3^- ions. The abundance of HCO_3^- ions in blood drives down the concentration of H_3O^+ ions to below 10^{-7} *M*. As we discuss later in this chapter, a $[H_3O^+]$ value slightly below 10^{-7} *M* means that our blood is actually slightly basic.

Unfortunately, some people suffer from medical conditions that alter the normal balance of CO_2 and HCO_3^- in their blood. For example, chronic lung disease impairs not only the ability to inhale O_2 but also the ability to exhale CO_2. As a result, CO_2 builds up in the blood and in the tissues where it is produced. The result is a condition called *respiratory acidosis*. Its symptoms include fatigue, lethargy, and shortness of breath. In severe cases, the disease can be fatal.

CONNECTION In Chapter 8 we learned that an aqueous solution's acidity is proportional to the concentration of H_3O^+ ions in it.

CONCEPT TEST

When people hyperventilate, they breathe too rapidly, resulting in carbon dioxide being removed from the bloodstream more quickly than the body produces it through metabolism. What effect does this have on the equilibria in Equations 15.1 and 15.2?

(Answers to Concept Tests are in the back of the book.)

In this chapter we examine acid–base balance in environmental waters and in our bodies. We study why changes in this balance occur, some of the consequences of these changes, and what we can do to control them.

15.2 Acid Strength and Molecular Structure

The Arrhenius model of acids and bases explains many but not all of the properties of acidic and basic compounds in aqueous solutions. For example, Arrhenius acids produce H_3O^+ ions in aqueous solution, but these compounds do not contain H_3O^+ ions (or even H^+ ions) in their molecular structures. Instead, they contain covalently bonded H atoms that ionize when the compounds disssolve in water. Similarly, very few basic compounds contain OH^- ions in their structures. The movement toward classifying substances based on the structures of their molecules and not just their macroscopic properties led to the development of the *Brønsted–Lowry* model of acids and bases, which defines an acid as a substance that *donates* H^+ ions and a base as a substance that *accepts* H^+ ions.

CONNECTION We introduced acids, bases, and neutralization reactions in Chapter 8, where we defined Brønsted–Lowry acids as substances that donate H^+ ions and bases as substances that accept H^+ ions.

We introduced the Brønsted–Lowry model in Chapter 8, where we also defined *strong acids* as those that ionize completely in aqueous solution and *weak acids* as those that only partially ionize. Later in this chapter we will discover that the degree of ionization of an acid depends on its concentration in solution: the more dilute it is, the higher its degree of ionization. Thus, even weak acids may ionize completely in very dilute solutions. This behavior requires us to refine our definitions of weak and strong acids. Throughout this chapter and those that follow, an acid (HA) will be considered a strong acid if the equilibrium constant for the reaction

$$HA(aq) + H_2O(\ell) \rightleftharpoons A^-(aq) + H_3O^+(aq) \tag{15.3}$$

is much greater than 1:

$$K_a = \frac{[A^-][H_3O^+]}{[HA]} >> 1 \tag{15.4}$$

Here the subscript "a" is used to identify the equilibrium constant K_a as an *acid* ionization constant. Note that this K_a expression has no $[H_2O]$ term. We do this with all aqueous equilibrium constants because the concentration of water in aqueous solutions is very large and rarely changes significantly due to reactions involving particles of solutes. Therefore, the value of $[H_2O]$ is a constant that is folded into equilibrium constant values for reactions in aqueous solutions. Table 15.1 contains the names of the common strong acids, and their molecular structures, including colored surfaces that show where valence electron densities are particularly high (red regions) or low (blue regions).

CONNECTION The use of molecular models with colored surfaces to show variations in the distribution of valence electrons was introduced in Chapter 4 (see Figure 4.12 on page 168.)

Weak acids have K_a values that are smaller, often much smaller, than 1. The smaller its K_a value, the weaker the acid. Table 15.2 contains the molecular structures and information about the acidic properties of several common weak acids, and ionizable H atoms are shown in red.

CHEMTOUR Acid–Base Ionization

TABLE 15.1 The Structures and Acidic Properties of the Common Strong Acids

Acid	Molecular Structure	Reaction in Water
Hydrochloric		$HCl(aq) + H_2O(\ell) \rightarrow Cl^-(aq) + H_3O^+(aq)$
Hydrobromic		$HBr(aq) + H_2O(\ell) \rightarrow Br^-(aq) + H_3O^+(aq)$
Hydroiodic		$HI(aq) + H_2O(\ell) \rightarrow I^-(aq) + H_3O^+(aq)$
Perchloric		$HClO_4(aq) + H_2O(\ell) \rightarrow ClO_4^-(aq) + H_3O^+(aq)$
Sulfuric		$H_2SO_4(aq) + H_2O(\ell) \rightarrow HSO_4^-(aq) + H_3O^+(aq)$
Nitric		$HNO_3(aq) + H_2O(\ell) \rightarrow NO_3^-(aq) + H_3O^+(aq)$

TABLE 15.2 The Molecular Structures and K_a Values of Some Common Weak Acids

Acid	Molecular Structure	K_a
Acetic	$H_3C{-}C(=O){-}O{-}H$	1.76×10^{-5}
Formic	$H{-}C(=O){-}O{-}H$	1.77×10^{-4}
Hydrofluoric	H—F	6.8×10^{-4}
Hypochlorous	H—O—Cl	2.9×10^{-8}
Nitrous	H—O—N=O	4.0×10^{-4}

SAMPLE EXERCISE 15.1 Relating K_a and Acid Strength **LO1**

Suppose 0.100 *M* aqueous solutions of each of the weak acids in Table 15.2 are prepared. Rank them from most acidic to least acidic.

Collect, Organize, and Analyze The acidity of an aqueous solution is proportional to the concentration of H_3O^+ ions in it, and the higher the value of K_a, the higher the concentration of H_3O^+ ions produced by a given concentration of acid.

Solve Ranking the acids in order of decreasing K_a values, we have hydrofluoric acid > nitrous acid > formic acid > acetic acid > hypochlorous acid.

Think About It The list of weak acids includes four in which the ionizable H atom is bonded to an O atom and one in which it is bonded to an F atom. The fact that all these H atoms are bonded to atoms of two of the most electronegative elements in the periodic table is not a coincidence, as we will see later in this section.

Practice Exercise Rank the following acids in order of decreasing acid strength.

Weak Acid	K_a
HN_3	1.9×10^{-5}
$CH_2ClCOOH$	1.4×10^{-3}
HCN	6.2×10^{-10}
$HClO_2$	1.1×10^{-2}
C_6H_5COOH	6.5×10^{-5}

(Answers to Practice Exercises are in the back of the book.)

TABLE 15.3 Average H—X Bond Dissociation Energies and X^- Ionic Radii for Four Binary Acids

Acid	Bond Energy (kJ/mol)	Ionic Radius (pm)
HF	567	133
HCl	431	181
HBr	366	195
HI	299	216

Strengths of Binary Acids

Let's examine the molecular structures of the strong acids in Table 15.1 to explore what makes them strong. There are two types of acids in the table. The first group includes three binary acids with the generic formula HX, where X is a group 17 element: Cl, Br, or I. Hydrofluoric acid is not in the table because HF is a weak acid, which is reflected in the small value of its acid ionization constant ($K_a = 6.8 \times 10^{-4}$).

The binary acids in Table 15.1 are listed in order of increasing atomic number of the group 17 element as well as increasing acid strength: HCl < HBr < HI. Why is this trend observed? One reason is that the bond dissociation energies of the H—X bonds in these acids (Table 15.3) decrease because an I ion is larger than a Br ion which is larger than a Cl ion. Weaker H—X bonds are easier to break, producing H^+ and X^- ions.

A second factor is the polarity of the H—X bond. Unequal sharing of each bonding pair of electrons leaves the H atoms with partial positive charges and the X atoms with partial negative charges, as shown for a molecule of HCl in Figure 15.1. This polarity leads to strong dipole–dipole interactions between the H atoms and electronegative O atoms in H_2O molecules (represented by the dotted blue line in Figure 15.1). This interaction facilitates ionization of the H—X bond, producing an X^- anion and a H^+ cation that bonds with a molecule of H_2O, forming a hydronium (H_3O^+) ion.

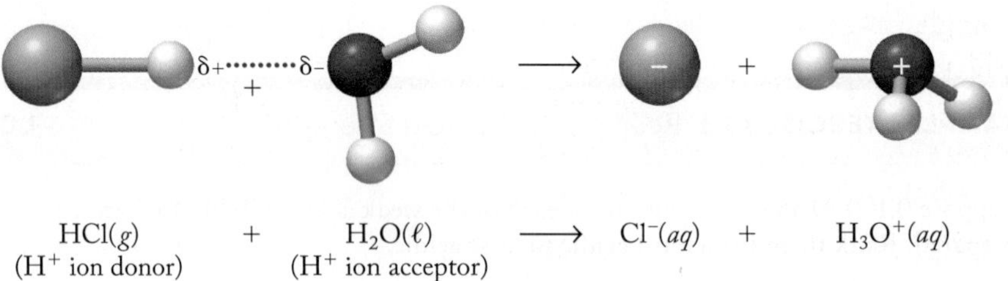

FIGURE 15.1 When hydrogen chloride gas dissolves in water, each molecule of HCl ionizes by reacting with a molecule of H_2O, forming Cl^- and H_3O^+ ions.

Based on bond polarity, we would expect HF to be the strongest binary acid. The fact that HF is the weakest tells us that HF's high bond dissociation energy offsets the strong dipole–dipole interactions (actually, they are hydrogen bonds) between molecules of HF and H_2O. However, we are about to see that bond polarity can play a key role in enhancing the strengths of other acids.

Oxoacids

CONNECTION The rules for naming oxoacids are described in Chapter 4.

The second class of strong acids in Table 15.1 are oxoacids. Not all oxoacids are strong acids. For example, $HClO_4$ is a strong acid, whereas $HClO_2$ in Table 15.2 is a weak acid. Note that all the ionizable H atoms in these acids are bonded to O atoms that are also bonded to central atoms of electronegative elements. Let's explore why some oxoacids are strong acids while others are not.

Consider two oxoacids (Figure 15.2) that contain the same X atom but different numbers of O atoms per molecule. We have seen that HNO_3 is a strong acid. An indication of its strength is given by the bar graph in Figure 15.2(a), which shows that essentially all (> 99.5%) the HNO_3 molecules in a 0.100 *M* aqueous solution are ionized:

$$HNO_3(\ell) + H_2O(\ell) \rightarrow NO_3^-(aq) + H_3O^+(aq) \qquad (15.5)$$

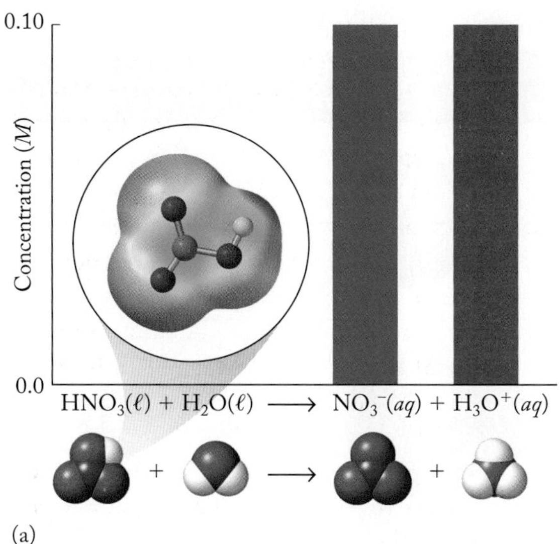

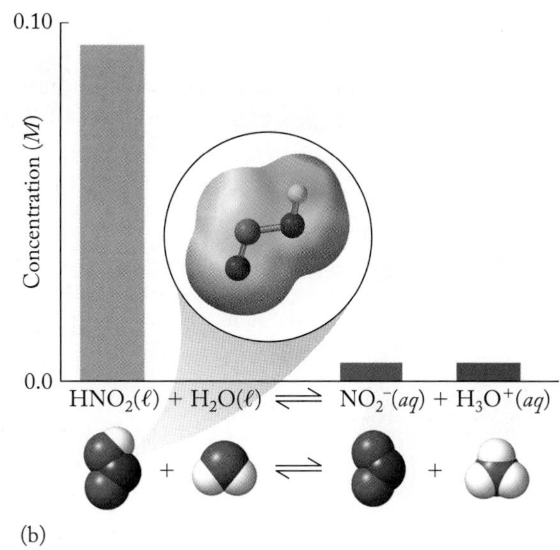

FIGURE 15.2 Degrees of ionization of a strong acid versus a weak acid. (a) Essentially all of the molecules of HNO_3 are ionized in a 0.100 *M* solution of the acid. (b) Only 6.1% of the nitrous acid molecules are ionized in 0.100 *M* HNO_2. The molecular structures show that the greater number of O atoms in molecules of HNO_3 compared to HNO_2 draws more electron density away from the N atoms, stabilizing the NO_3^- anion more than the NO_2^- anion in aqueous solutions. The concentration of water is not shown because it is much larger than the concentrations of the solutes.

The single-headed arrow tells us the equilibrium lies so far to the right, that we consider it to go nearly to completion.

In contrast, only 6.1% of the molecules of nitrous acid, HNO_2, in a 0.100 *M* solution of the acid are ionized (Figure 15.2b). This behavior is consistent with the much smaller K_a value of this weak acid:

$$HNO_2(\ell) + H_2O(\ell) \rightleftharpoons NO_2^-(aq) + H_3O^+(aq) \quad K_a = 4.0 \times 10^{-4} \qquad (15.6)$$

Why is HNO_3 so much stronger an acid than HNO_2? The key to answering this question is the electronegativity of oxygen and the presence of one more O atom in a molecule of HNO_3. The more O atoms there are bonded to the central atom of an oxoacid, the more electron density is pulled toward these atoms and away from the hydrogen end of the O—H bond. These variations in valence electron density are depicted by the different colors of the surfaces in the two molecular balloons in Figure 15.2. The deeper blue color near the H atom in the molecule of HNO_3 indicates less electron density than is near the H atom in HNO_2. The greater positive charges on the H atoms in HNO_3 mean stronger interactions with electronegative O atoms in water molecules contributing to the increased acidity of HNO_3.

Greater numbers of O atoms in oxoacids also increase the stability of the anions that form. For example, the four O atoms bonded to the central S atom in the HSO_4^- ion that forms when sulfuric acid ionizes:

$$H_2SO_4(\ell) + H_2O(\ell) \rightarrow HSO_4^-(aq) + H_3O^+(aq) \qquad (15.7)$$

allow its negative charge to be more delocalized (Figure 15.3) than the charge on the HSO_3^- ion that forms when a molecule of sulfurous acid ionizes:

$$H_2SO_3(\ell) + H_2O(\ell) \rightleftharpoons HSO_3^-(aq) + H_3O^+(aq) \quad K_a = 1.7 \times 10^{-2} \quad (15.8)$$

Greater delocalization of anionic charge helps make sulfuric acid a strong acid, whereas sulfurous acid is a weak acid. The images in Figure 15.3 also show that

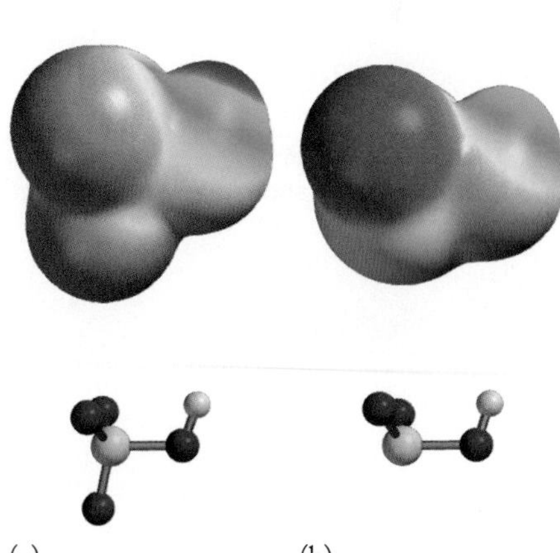

FIGURE 15.3 The greater number of O atoms in the HSO_4^- ion (a) delocalize its negative charge more than in the HSO_3^- ion (b). They also draw more electron density away from the remaining H atom, facilitating its ionization and helping make the HSO_4^- ion a stronger acid than the HSO_3^- ion.

TABLE 15.4 Structures and K_a Values of the Oxoacids of Chlorine

Acid	Structure	K_a
Hypochlorous $HClO$	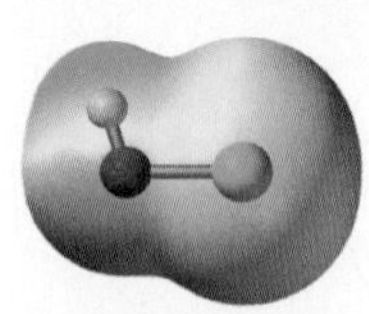	2.9×10^{-8}
Chlorous $HClO_2$	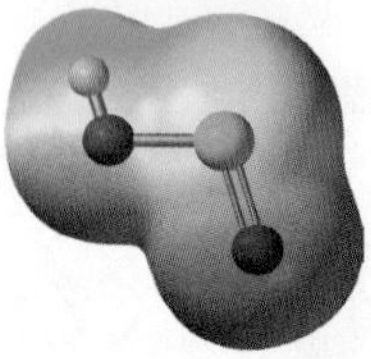	1.1×10^{-2}
Chloric $HClO_3$	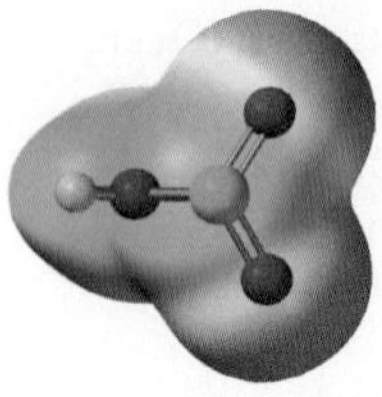	~1
Perchloric $HClO_4$	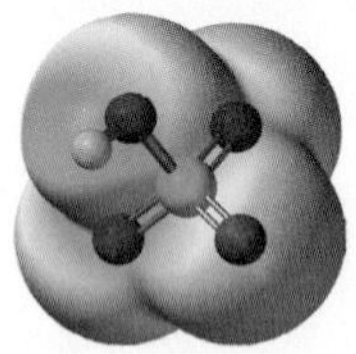	>>1

TABLE 15.5 Strengths of the Hypohalous Acids

Acid	Structure	Electronegativity of Halogen Atom	K_a
Hypochlorous $HClO$	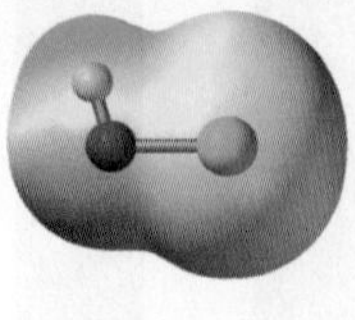	3.0	2.9×10^{-8}
Hypobromous $HBrO$	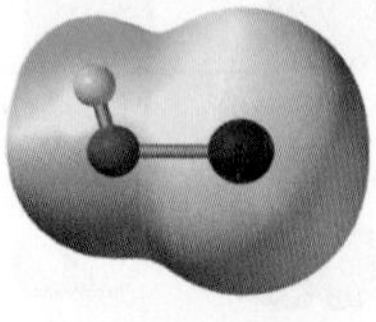	2.8	2.3×10^{-9}
Hypoiodous HIO	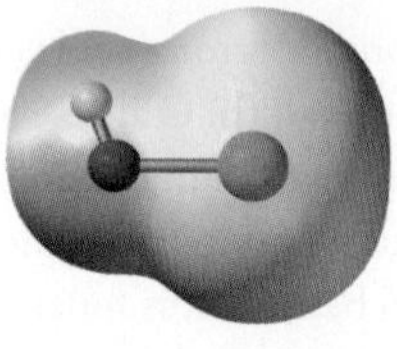	2.5	2.3×10^{-11}

the four O atoms in a HSO_4^- ion draw more electron density away from the remaining H atom, helping make the HSO_4^- ion a stronger acid ($K_{a_2} = 1.2 \times 10^{-2}$) than the HSO_3^- ion ($K_{a_2} = 6.2 \times 10^{-8}$).

There are even more dramatic differences in acid strength between members of other families of oxoacids. For example, the strengths of the four oxoacids of chlorine (Table 15.4) increase with increasing numbers of O atoms per molecule, as we saw with nitrous and nitric acids, but the K_a values in Table 15.4 span a range of over 10^{20}!

CHEMTOUR Acid Strength and Molecular Structure

Finally, let's consider the structural differences and relative strengths of three weak oxoacids, the *hypohalous* acids with the generic formula HXO in Table 15.5. The most electronegative of the three halogen atoms (Cl) in Table 15.5 has the greatest attraction for the pair of electrons it shares with oxygen. This attraction draws electron density away from hydrogen toward chlorine and toward the oxygen end of the already polar O—H bond. Thus, HClO(*aq*) is the strongest of the three acids, followed by hypobromous acid [HBrO(*aq*)] and hypoiodous acid [HIO(*aq*)].

Carboxylic Acids

The most abundant class of acidic compounds are carboxylic acids, which have the generic molecular structure shown in Figure 15.4. Their modest acid strength is linked to delocalization of the negative charge on the carboxylate ions they form when they undergo acid ionization. Note how resonance illustrates the

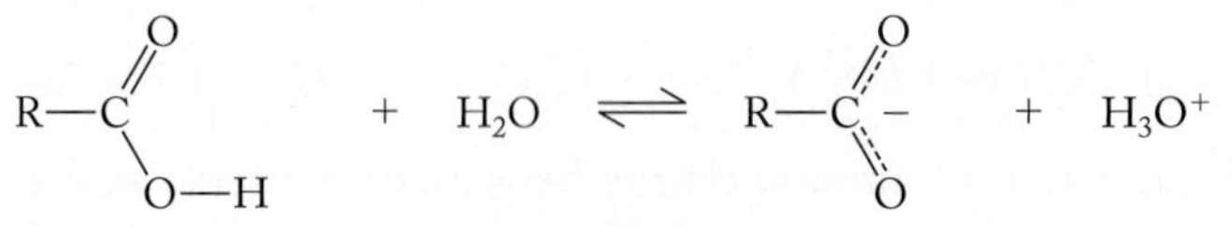

$$R{-}C({=}O){-}O{-}H + H_2O \rightleftharpoons R{-}CO_2^- + H_3O^+$$

Carboxylic acid — Carboxylate ion

FIGURE 15.4 Ionization of a carboxylic acid.

TABLE 15.6 Molecular Structures and K_a Values of Five Carboxylic Acids

Acid	Molecular Structure	K_a
Acetic, CH_3COOH		1.76×10^{-5}
Chloroacetic, $CH_2ClCOOH$		1.4×10^{-3}
Dichloroacetic, $CHCl_2COOH$		5.5×10^{-2}
Trichloroacetic, CCl_3COOH		2.3×10^{-1}
Trifluoroacetic, CF_3COOH		5.9×10^{-1}

delocalization of the 1− charge equally between two O atoms. Their acid strengths are also enhanced by the presence of atoms of electronegative elements in their R groups as shown in Table 15.6. Compare, for example, the K_a value of acetic acid (R = —CH_3) with those of the three acids beneath it in the table in which one, two, or all three of the H atoms in the acetic acid R group have been replaced with chlorine atoms. Note how the increasing number of atoms of electronegative chlorine pulls electron density away from the O—H bond. You can see this in the deepening blue color representing the partial positive charge surrounding the H atom and in the fading red color representing the partial negative charge on the O atom. In addition, drawing electron density away from the O—H bond further stabilizes the negative charge on the anions that form when these acids ionize. Finally, trifluoroacetic acid is even stronger than trichloroacetic acid because fluorine ($\chi = 4.0$) is more electronegative than chlorine ($\chi = 3.0$).

CONNECTION The electronegativities of the elements are given in Figure 4.8.

SAMPLE EXERCISE 15.2 Relating Acid Strength to Molecular Structure **LO2**

Rank the following compounds in order of decreasing acid strength: HBrO, $HBrO_2$, $HBrO_3$, $HBrO_4$, and $HClO_4$.

Collect, Organize, and Analyze All of the compounds are oxoacids of group 17 elements. The strengths of these acids increase with increasing electronegativity of the central atom and with increasing numbers of oxygen atoms bonded to it. Chlorine ($\chi = 3.0$) is slightly more electronegative than bromine ($\chi = 2.8$).

Solve The strengths of the four bromine oxoacids decrease in the order $HBrO_4 >$ $HBrO_3 > HBrO_2 >$ HBrO based on the number of O atoms bonded to the central Br atoms. Since Cl is more electronegative than Br, $HClO_4$ should be stronger than $HBrO_4$. Therefore, the overall rank-ordered list is $HClO_4 > HBrO_4 > HBrO_3 >$ $HBrO_2 >$ HBrO.

Think About It Had we compared the strength of HCl and HBr instead of $HClO_4$ and $HBrO_4$, the ranking would have been reversed: HBr > HCl, because the ionizable H atoms in these binary acids are bonded directly to the halogen atoms instead of O atoms. The bond dissociation energy of the H—Cl bond (Table 15.3) is $(431 - 366) =$ 65 kJ/mol stronger than the H—Br bond and much harder to break.

Practice Exercise Rank the following compounds in order of decreasing acid strength: fluoroacetic, difluoroacetic, trifluoroacetic, chloroacetic, and bromoacetic acid. Explain why you ranked the acids in the order that you did.

SAMPLE EXERCISE 15.3 Relating Acid Strength to Molecular Structure **LO2**

Rank the following compounds in order of decreasing acid strength: H_3PO_4, H_3AsO_4, H_3SbO_4, and H_3BiO_4.

Collect and Organize Each of these four oxoacids contains a different group 15 element. In each acid, the central atom is bonded to the same number of O atoms. The electronegativities of the elements decrease with increasing atomic number down a column in the periodic table.

Analyze In oxoacids, a more electronegative central atom draws electron density away from the H end of the O—H bond and helps disperse the negative charge that results when an oxoacid releases H^+ to form an oxoanion.

Solve Ranking the oxoacids in order of decreasing acid strength is a matter of ranking them in order of decreasing electronegativity of the central atom—that is, in order of increasing atomic number (and row number) in the periodic table: $H_3PO_4 > H_3AsO_4 >$ $H_3SbO_4 > H_3BiO_4$.

Think About It This trend of decreasing oxoacid strength with increasing atomic number holds for each of the other groups of nonmetals—as long as the central atoms are in the same oxidation state.

Practice Exercise Of H_2SeO_4, H_2SO_4, H_2SeO_3, and H_2SO_3, which is the strongest acid and which is the weakest?

15.3 Strong and Weak Bases

Like acids, bases are also classified as either weak or strong. Among the most common strong bases are the hydroxides of the elements of groups 1 and 2. Table 15.7 lists the ions present when these ionic compounds dissolve in water and dissociate completely. Note that one mole of a group 1 hydroxide produces one mole of OH^- ions when it dissolves in water and that one mole of a group 2 hydroxide produces two moles of OH^- ions.

Many bases don't contain hydroxide ions in their structures but produce them by reacting with water in solution. These compounds are strong bases if their reactions with water are complete. They include soluble ionic oxides such as calcium oxide, CaO, which has the common name quicklime:

$$CaO(s) + H_2O(\ell) \rightarrow Ca^{2+}(aq) + 2\ OH^-(aq) \qquad (15.9)$$

The water-soluble group 1 and group 2 oxides are all strong bases.

Weak bases also produce OH^- ions when they dissolve in and react with water, but their reactions reach equilibrium before all of the base is consumed. As we saw in Chapter 8, ammonia is a widely used weak base. Its basic properties are linked to the hydrogen bond that forms between the nitrogen atom in a molecule of NH_3 and a hydrogen atom in a molecule of H_2O. The strength of this interaction can lead to ionization of the water molecule and the formation of a fourth N—H bond, transforming NH_3 into an ammonium (NH_4^+) ion:

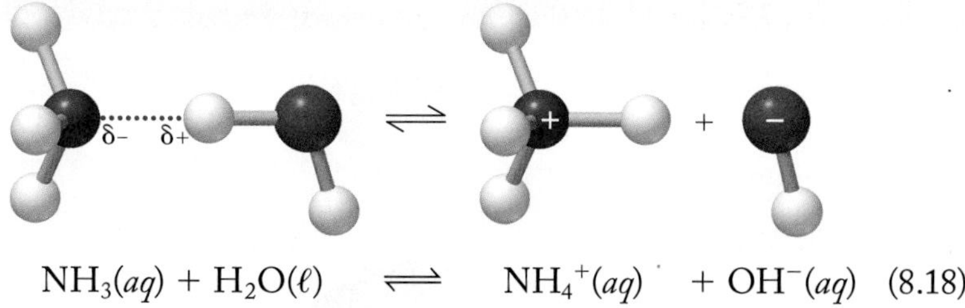

$$NH_3(aq) + H_2O(\ell) \rightleftharpoons NH_4^+(aq) + OH^-(aq) \qquad (8.18)$$

The degree to which a reaction such as this one proceeds before it reaches chemical equilibrium is reflected in the value of its equilibrium constant, K_b, where the "b" subscript tells us that it describes a *b*ase ionization reaction. The K_b expression and value for ammonia in water is

$$K_b = \frac{[NH_4^+][OH^-]}{[NH_3]} = 1.76 \times 10^{-5} \qquad (15.10)$$

As with K_a expressions, we simplify K_b expressions by leaving out the $[H_2O]$ term because of its high and nearly constant value in most reactions in aqueous solution.

TABLE 15.7 Some Strong Bases and Their Dissociation in Water

Strong Base	Solid	Solvated Ions
Lithium hydroxide	$LiOH(s)$	$Li^+(aq) + OH^-(aq)$
Sodium hydroxide	$NaOH(s)$	$Na^+(aq) + OH^-(aq)$
Potassium hydroxide	$KOH(s)$	$K^+(aq) + OH^-(aq)$
Calcium hydroxide	$Ca(OH)_2(s)$	$Ca^{2+}(aq) + 2\ OH^-(aq)$
Barium hydroxide	$Ba(OH)_2(s)$	$Ba^{2+}(aq) + 2\ OH^-(aq)$
Strontium hydroxide	$Sr(OH)_2(s)$	$Sr^{2+}(aq) + 2\ OH^-(aq)$

Amines

Amines are organic compounds with functional groups that resemble NH_3, except that N atoms in amines are bonded to one or more carbon atoms. Amines are labeled as primary, secondary, or tertiary amines depending on whether their nitrogen atoms are bonded to one, two, or three carbon atoms, respectively. Table 15.8 contains examples of different types of amines and their K_b values. The first amine in the table is also the simplest: methylamine, CH_3NH_2. Compare its K_b value to that of ammonia and you will see that it is over 20 times larger. One reason is a familiar one: differences in the electronegativities of the atoms in their molecules. The N atom in methylamine is more electronegative than the C and H atoms in the methyl group, which means the N atom draws electron density away from the methyl group and toward itself. (It also draws pairs of electrons in the three N—H bonds in ammonia toward itself, but there are many more electrons in a methyl group.) The greater electron density around the N atom in methylamine means that its lone pair of electrons is better able to bond with a hydrogen ion, which makes it a stronger Brønsted–Lowry base.

Ethylamine is a slightly stronger base than methylamine because of the greater number of valence electrons in the ethyl group that shift toward the N atom,

TABLE 15.8 Molecular Structures and K_b Values of Five Weak Bases

Name, Formula	Molecular Structure	K_a
Ammonia, NH_3		1.76×10^{-5}
Methylamine, CH_3NH_2		4.4×10^{-4}
Dimethylamine, $(CH_3)_2NH$		5.9×10^{-4}
Ethylamine, $CH_3CH_2NH_2$		5.6×10^{-4}
Aniline (Phenylamine), $C_6H_5NH_2$		4.0×10^{-10}

making it a slightly more effective H^+ ion acceptor. The same logic applies to dimethylamine, which is about as strong a base as ethylamine.

Other amines are weaker bases than ammonia because the atoms and bonds in their molecular structures draw electron density away from their nitrogen atoms. This happens in molecules of phenylamine, which is also called aniline. The lone pair of electrons on the nitrogen atom in these molecules is drawn into the cloud of delocalized π electrons around the benzene ring, which pulls the electrons away from the nitrogen atom and decreases its ability to accept hydrogen ions.

conjugate acid–base pair a Brønsted–Lowry acid and base that differ from each other only by a H^+ ion: acid $\rightleftharpoons$ conjugate base + H^+.

conjugate base the base formed when a Brønsted–Lowry acid donates a H^+ ion.

conjugate acid the acid formed when a Brønsted–Lowry base accepts a H^+ ion.

CONCEPT **TEST**

Is chloramine, NH_2Cl, a stronger or weaker base than ammonia? Explain your reasoning.

Conjugate Pairs

Let's revisit the acid ionization of nitrous acid in which HNO_2 functions as a Brønsted–Lowry acid by donating H^+ ions to molecules of H_2O (Figure 15.5a). In the reverse reaction, NO_2^- ions accept H^+ ions and function as a Brønsted–Lowry base. Structurally, the difference between HNO_2 and NO_2^- is the H^+ ion that a molecule of HNO_2 donates, resulting in the formation of a NO_2^- ion, and that a NO_2^- ion accepts in the reverse reaction to reform a molecule of HNO_2. An acid and a base that are related in this way are called a **conjugate acid–base pair**, or *conjugate pair*. An acid forms its **conjugate base** when it donates a H^+ ion, and a base forms its **conjugate acid** when it accepts a H^+ ion.

Figure 15.5(b) shows an ammonia molecule acting as a Brønsted–Lowry base, accepting a H^+ ion from a molecule of H_2O and producing an ammonium and a hydroxide ion. When NH_4^+ ions donate H^+ ions, they function as Brønsted–Lowry acids. The structures of NH_3 molecules and NH_4^+ ions differ only by the H^+ ion that a molecule of NH_3 accepts and that an NH_4^+ ion donates; so NH_4^+ and NH_3 are another conjugate acid–base pair.

Is water an acid or a base? In Figure 15.5(a), H_2O functions as a base because it accepts a H^+ ion from nitrous acid and forms H_3O^+, its conjugate acid. Thus, H_2O and H_3O^+ are a conjugate pair. In Figure 15.5(b), however, H_2O functions as an acid by donating a H^+ ion to ammonia to form the OH^- ion, its conjugate base. In this case, H_2O and OH^- are a conjugate pair. The two reactions in Figure 15.5 demonstrate that water can act as an acid or a base, depending on the acid–base properties of the substance dissolved in it.

The following generic chemical equations summarize the relationships between conjugate acid–base pairs in aqueous solutions:

$$\text{Acid}(aq) + H_2O(\ell) \rightleftharpoons \text{conjugate base}(aq) + H_3O^+(aq)$$

$$\text{Base}(aq) + H_2O(\ell) \rightleftharpoons \text{conjugate acid}(aq) + OH^-(aq)$$

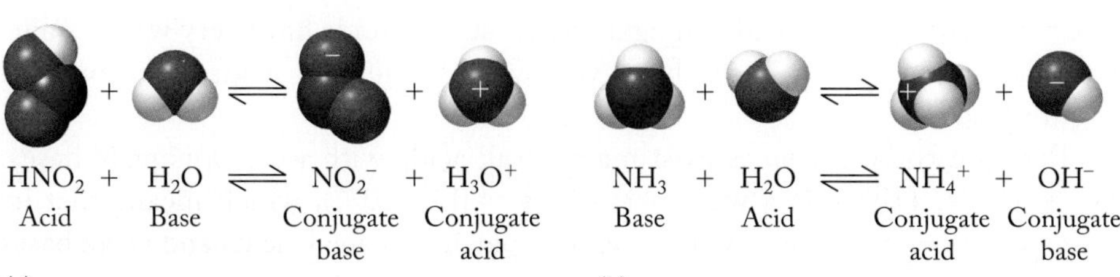

FIGURE 15.5 Acid–base conjugate pairs.

The difference within each conjugate acid–base pair is the H^+ ion that is donated by the acid to form its conjugate base and that a base accepts to form its conjugate acid.

SAMPLE EXERCISE 15.4 Identifying Conjugate Acid–Base Pairs **LO3**

Identify the conjugate acid–base pairs in the ionization reactions that occur when (a) perchloric acid ($HClO_4$) and (b) formic acid (HCOOH) dissolve in water.

Collect, Organize, and Analyze Brønsted–Lowry acids form their conjugate bases when they donate H^+ ions to molecules of H_2O. Therefore, the formulas of their conjugate bases are the formulas of the original acids minus a H^+ ion.

Solve

a. Perchloric acid, $HClO_4$, has only one H atom per molecule, so it must lose that one as a H^+ ion:

$$\underset{\text{Acid}}{HClO_4(aq)} + H_2O(\ell) \rightarrow \underset{\text{Conjugate base}}{ClO_4^-(aq)} + H_3O^+(aq)$$

b. Formic acid, HCOOH, has two H atoms per molecule, but only the one bonded to an O atom in the carboxylic acid group (see Table 15.2) is ionizable in water:

$$\underset{\text{Acid}}{HCOOH(aq)} + H_2O(\ell) \rightleftharpoons \underset{\text{Conjugate base}}{HCOO^-(aq)} + H_3O^+(aq)$$

In both reactions, H_3O^+ and H_2O are also a conjugate acid–base pair: H_2O is the base and H_3O^+ is its conjugate acid.

Think About It Perchloric acid is a strong acid and ionizes completely in aqueous solution. As a result, ClO_4^- is such a weak base that the reverse reaction, in which ClO_4^- adds a H^+ ion to reform $HClO_4$, essentially never occurs. This is why a single arrow is used to depict the ionization reaction of $HClO_4$.

 **Practice Exercise** Identify the conjugate acid–base pairs in the reaction that occurs when acetic acid (CH_3COOH) dissolves in water.

Relative Strengths of Conjugate Acids and Bases

HCl is a strong acid and ionizes completely:

$$HCl(g) + H_2O(\ell) \rightarrow Cl^-(aq) + H_3O^+(aq) \quad (15.11)$$

Essentially, the reverse reaction does not happen, which means the Cl^- ion (the conjugate base of HCl) is an extremely weak base whose presence in an aqueous solution has no significant impact on acidity. This contrast in relative strengths applies to all conjugate pairs: strong acids have very weak conjugate bases and strong bases have very weak conjugate acids, as shown in Figure 15.6.

Between these extremes exist many weak acids with weak conjugate bases. For example, HNO_2 is a weak acid ($K_a = 4.0 \times 10^{-4}$), which means that its conjugate base, NO_2^-, is a weak base. This pairing of weak acids and weak bases applies to the vast majority of the conjugate acids and bases.

leveling effect the observation that all strong acids completely ionize in aqueous solutions, forming H_3O^+ ions; strong bases are likewise leveled in water and are completely converted into their conjugate acids and OH^- ions.

All the strong acids in Figure 15.6 ionize completely in water: essentially every molecule of acid transfers a H^+ ion to a molecule of H_2O. In this context, water is said to *level* the strengths of these acids when they dissolve in water: they all are equally strong because they cannot be more than 100% ionized. This **leveling effect** means that H_3O^+, the conjugate acid of H_2O, is the strongest acid that can exist in water. An acid that is stronger than H_3O^+ simply donates all its ionizable H atoms to water molecules, forming H_3O^+ ions, when it dissolves in water. On the other hand, weak acids are differentiated by the fact that only a small fraction of the molecules donate their ionizable H atoms to water molecules. The weak acids higher on the list in Figure 15.6 form more acidic aqueous solutions than acids lower on the list because a greater fraction of the molecules ionize in water.

A similar leveling effect exists for strong bases. The strongest base that can exist in water is the OH^- ion, the conjugate base of H_2O. Any base that is stronger than OH^-, such as an oxide ion (O^{2-}), hydrolyzes in water, producing two OH^- ions:

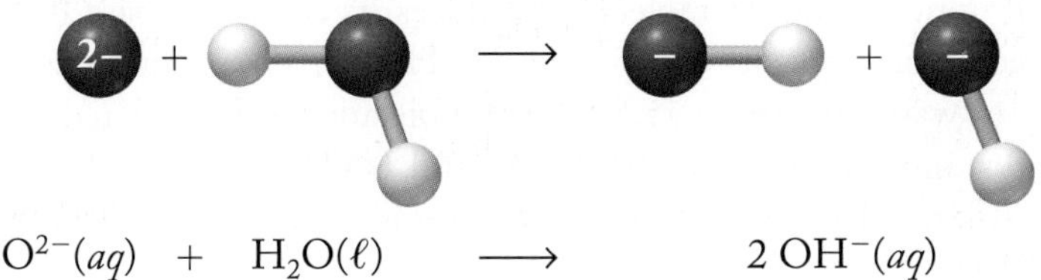

$$O^{2-}(aq) + H_2O(\ell) \longrightarrow 2\,OH^-(aq)$$

The strengths of bases weaker than OH^- ions can be differentiated by the fraction of their molecules that accept H^+ ions from water molecules in aqueous solutions. Weaker bases are higher on the list in Figure 15.6; stronger bases are lower on the list.

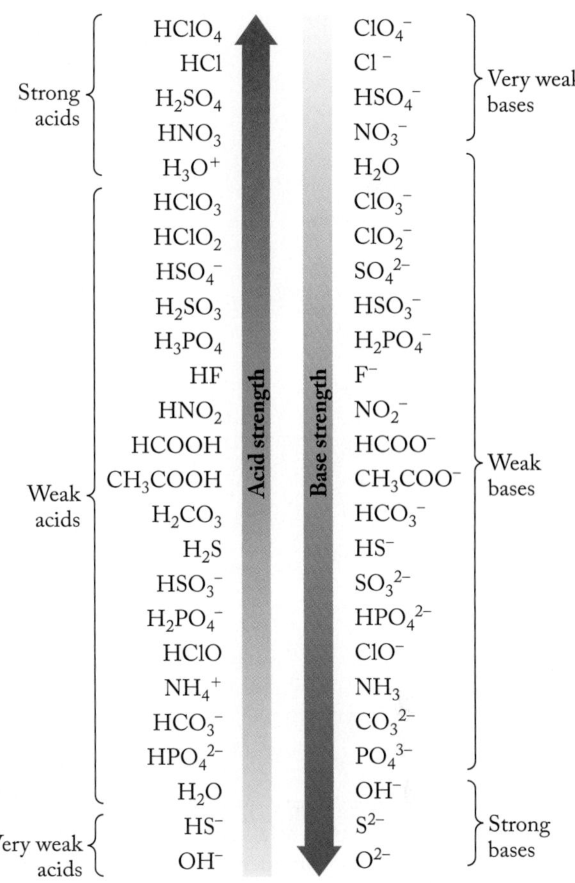

FIGURE 15.6 Opposing trends characterize the relative strengths of acids and their conjugate bases: the stronger the acid, the weaker its conjugate base. The same is true for bases: the stronger the base, the weaker its conjugate acid.

SAMPLE EXERCISE 15.5 Relating the Strengths of Conjugate Pairs **LO3**

List the following anions in order of decreasing strength as Brønsted–Lowry bases: F^-, Cl^-, OH^-, $HCOO^-$, and NO_2^-.

Collect, Organize, and Analyze All five anions are listed among the bases in Figure 15.6. Therefore, all we have to do is rank them according to their location in the figure: the strongest one will be the closest to the bottom and the weakest will be the closest to the top.

Solve The anions in decreasing order of strength as bases (H^+ ion acceptors) are OH^-, $HCOO^-$, NO_2^-, F^-, and Cl^-.

Think About It According to Figure 15.6, we have also sorted the anions in increasing order for the strength of their conjugate acids. The complementary sorting makes sense because the stronger the acid, the weaker its conjugate base, and vice versa.

Practice Exercise List the following anions in order of decreasing strength as Brønsted–Lowry bases: Br^-, S^{2-}, ClO^-, CH_3COO^-, and HSO_3^-.

15.4 pH and the Autoionization of Water

CHEMTOUR Autoionization of Water

We have seen that the acidity of a solution is directly related to its concentration of H_3O^+ ions. In this section we examine another way to quantify acidity. First, though, we need to understand what is meant by the **autoionization** of water. It is a process that occurs between water molecules and results in pairs of them essentially ionizing each other, forming equal but very small concentrations of H_3O^+ and OH^- ions in pure water:

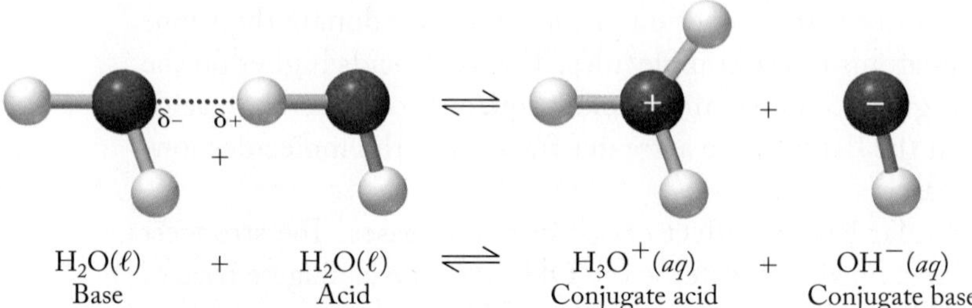

One water molecule, acting as an acid, donates a H^+ ion; the other, acting as a base, accepts a H^+ ion. The donor molecule produces its conjugate base (OH^-); the acceptor molecule produces its conjugate acid (H_3O^+). We encountered this acid–base duality of water in Figure 15.5, where molecules of H_2O acted as H^+ ion acceptors (bases) in solutions of acidic solutes and as H^+ ion donors (acids) in solutions of basic solutes. The autoionization of water is an example of its amphiprotic behavior.

CONNECTION *Amphiprotic* compounds, which have both acidic and basic properties, were first introduced in Chapter 8.

The equilibrium constant expression for the autoionization of water (K_w) is simply

$$K_w = [H_3O^+][OH^-] \quad (15.12)$$

because water concentration terms are not included in aqueous equilibria. In pure water at 25°C, $[H_3O^+] = [OH^-] = 1.0 \times 10^{-7}\ M$. Inserting these values into Equation 15.12 gives

$$K_w = [H_3O^+][OH^-] = (1.0 \times 10^{-7})(1.0 \times 10^{-7}) = 1.0 \times 10^{-14} \quad (15.13)$$

The value of K_w is tiny, thereby confirming that only a very small fraction of water molecules undergo autoionization at room temperature. The reverse of autoionization—the reaction between $[H_3O^+]$ and $[OH^-]$ to produce H_2O—has an equilibrium constant of $1/K_w = 1.0 \times 10^{14}$ at 25°C and essentially goes to completion:

$$H_3O^+(aq) + OH^-(aq) \rightleftharpoons 2\ H_2O(\ell) \qquad K = 1/K_w = 1.0 \times 10^{14}$$

Equation 15.13 means that an inverse relationship exists between $[H_3O^+]$ and $[OH^-]$ in any aqueous sample: as the concentration of one increases, the concentration of the other must decrease so that the product of the two is always 1.0×10^{-14}. A solution in which $[H_3O^+] > [OH^-]$ is acidic, a solution in which $[H_3O^+] < [OH^-]$ is basic, and a solution in which $[H_3O^+] = [OH^-] = 1.0 \times 10^{-7}\ M$ is neutral (neither acidic nor basic).

The tiny value of K_w means that the autoionization of water does not contribute significantly to $[H_3O^+]$ in solutions of most acids or to $[OH^-]$ in solutions of most bases, so we can ignore the contribution of autoionization in most calculations of acid or base strength. However, if acids or bases are extremely weak, or if their concentrations are extremely low, H_2O autoionization may need to be considered.

autoionization the process that produces equal and very small concentrations of H_3O^+ and OH^- ions in pure water.

pH the negative logarithm of the hydronium ion concentration in an aqueous solution.

The pH Scale

In the early 1900s, scientists developed a device called the *hydrogen electrode* to determine the concentrations of hydronium ions in solutions. The electrical voltage, or *potential*, produced by the hydrogen electrode is a linear function of the logarithm of $[H_3O^+]$. This relationship led Danish biochemist Søren Sørensen (1868–1939) to propose a scale for expressing acidity and basicity based on what he called "the potential of the hydrogen ion," abbreviated **pH**. Mathematically, we define pH as the negative logarithm of $[H_3O^+]$:

$$pH = -\log[H_3O^+] \quad (15.14)$$

For example, the pH of a solution in which $[H_3O^+] = 5.0 \times 10^{-3}\ M$ is

$$pH = -\log(5.0 \times 10^{-3}) = -(-2.30) = 2.30$$

Sørensen's pH scale has several attractive features. Because it is logarithmic, there are no exponents, as are commonly encountered in values of $[H_3O^+]$. The logarithmic scale also means that a change of one pH unit corresponds to a 10-fold change in $[H_3O^+]$, so that a solution with a pH of 5.0 has 10 times the concentration of H_3O^+ ions and is 10 times as acidic as a solution with a pH of 6.0. Similarly, the concentration of H_3O^+ ions in a solution with a pH of 12.0 is 1/10 that of a solution with a pH of 11.0. Conversely, the concentration of OH^- ions in a solution with a pH of 12.0 is 10 times that of a solution with a pH of 11.0.

The negative sign in front of the logarithmic term in Equation 15.14 means that pH values of most aqueous solutions are positive numbers between 0 and 14. It also means that *large* pH values correspond to *small* values of $[H_3O^+]$. Acidic solutions have pH values less than 7.00 ($[H_3O^+] > 1.0 \times 10^{-7}\ M$), and basic solutions have pH values greater than 7.00 ($[H_3O^+] < 1.0 \times 10^{-7}\ M$). A solution with a pH of exactly 7.00 is neutral. The pH values for some common aqueous solutions are shown in Figure 15.7.

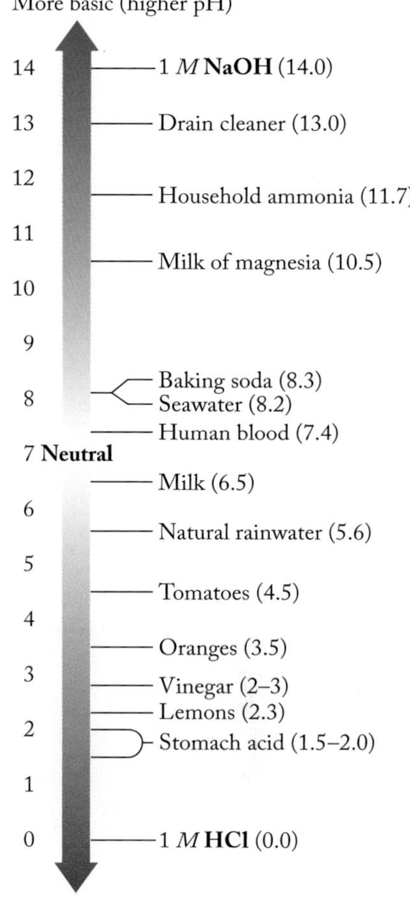

FIGURE 15.7 The pH scale is a convenient way to express the range of acidic or basic properties of some common materials.

CONCEPT TEST

Identify each of these five pH values as strongly acidic, weakly acidic, weakly basic, strongly basic or neutral: a. 0.22 b. 4.37 c. 7.00 d. 10.03 e. 13.77.

SAMPLE EXERCISE 15.6 Relating pH and $[H_3O^+]$ LO4

Suppose the pH of solution A is 8.0 and the pH of solution B is 4.0. Which of the following statements about the two solutions are true and which are false?

a. Solution A is 2 times as acidic as solution B.
b. Solution B is 10,000 times as acidic as solution A.
c. The concentration of OH^- ions in solution A is half the concentration in solution B.
d. The values of $[OH^-]$ and $[H_3O^+]$ are closer to each other in solution A than in solution B.

Collect, Organize, and Analyze The negative sign in the pH formula ($pH = -\log[H_3O^+]$) means that a higher pH value corresponds to a (much) lower $[H_3O^+]$. The pH scale is logarithmic, so a decrease of one pH unit corresponds to a 10-fold increase in $[H_3O^+]$.

Solve

a. False, because pH is a log scale, and because higher pH values mean lower $[H_3O^+]$.
b. True, a decrease of four pH units corresponds to a 10^4 increase in $[H_3O^+]$ and acidity.
c. False, because pH is a log scale. Actually, $[OH^-]$ in solution A is 10^4 times that of solution B.
d. True, because a pH of 8.0 is only one unit away from the pH (7.0) where $[H_3O^+] = [OH^-]$. A pH of 4.0 is three units below 7.0, so $[H_3O^+] = 1000 \times [OH^-]$ in solution B.

Think About It You may find it challenging at first to do log math in your head. It helps to understand that an increase of one pH unit corresponds to a decrease in $[H_3O^+]$ to only 1/10 or 10% of its initial value.

Practice Exercise Which of the following changes in pH corresponds to the greatest percent increase in $[H_3O^+]$? Which change corresponds to the greatest percent decrease in $[H_3O^+]$? a. pH 1 → pH 3; b. pH 7 → pH 4; c. pH 12 → pH 14; d. pH 5 → pH 9; e. pH 2 → pH 0.

CONNECTION The general rules for the use of significant figures in calculations involving measured quantities were discussed in Section 1.7.

When interconverting $[H_3O^+]$ and pH values, we need to know how to express pH values with the appropriate number of significant digits. Because a pH value is the negative logarithm of a hydronium ion concentration, the digit or digits before the decimal point in the pH value define the location of the decimal point in the $[H_3O^+]$ value. They do not define how precisely we know the concentration value, so they are not considered significant digits. For example, $[H_3O^+] = 2.7 \times 10^{-4}$ has two significant digits: the 2 and 7. The corresponding pH value, 3.57, should also have two significant digits after the decimal point and it does: the 5 and 7. The 3 in pH 3.57 is not a significant digit because its function is to tell us that the corresponding $[H_3O^+]$ value is between 10^{-3} and 10^{-4} *M*.

SAMPLE EXERCISE 15.7 Interconverting pH and $[H_3O^+]$ LO4

Cells located in the upper region of the human stomach called the fundus secrete gastric acid (about 0.16 *M* HCl) during digestion. After this acid mixes with the food being digested, the pH of the contents of the stomach is often between 3 and 4.

a. How much more acidic is 0.16 *M* HCl than stomach contents with a pH of 3.20?
b. Which of the two samples in part a has the lower $[H_3O^+]$?

Collect and Organize We are asked to compare the acidities (the ratio of the $[H_3O^+]$ values) of two solutions that both contain hydrochloric acid. We know the pH of one and the HCl concentration of the other, and we must also determine which sample has the lower $[H_3O^+]$.

Analyze Hydrochloric acid is a strong acid, so the $[H_3O^+]$ value of a solution of HCl is the same as the concentration of the acid. We can use Equation 15.14 to calculate the pH of 0.16 *M* HCl and to convert pH 3.20 into its corresponding $[H_3O^+]$ value. The result of the second calculation, when divided into 0.16 *M*, will give us the desired ratio of acidities.

Solve

a. To convert pH to $[H_3O^+]$, we need to solve Equation 15.14 for $[H_3O^+]$:

$$pH = -\log[H_3O^+]$$

or

$$\log [H_3O^+] = -pH$$

Taking the antilog (10^x) of both sides: $[H_3O^+] = 10^{-pH}$

Inserting pH = 3.20: $[H_3O^+] = 10^{-3.20} = 6.3 \times 10^{-4}\ M$

Calculating the ratio of acidities ($[H_3O^+]$ values):

$$\frac{[H_3O^+](\text{gastric acid})}{[H_3O^+](\text{stomach contents})} = \frac{0.16\ M}{6.3 \times 10^{-4}\ M} = 2.5 \times 10^2$$

Thus, the acidity of secreted gastric acid is 2.5×10^2 times the acidity of the digestion mixture in the stomach.

b. Using the values calculated in part a, the stomach contents have a lower $[H_3O^+]$.

Think About It The results of the two parts to this Sample Exercise are consistent: 0.16 *M* HCl is 2.5×10^2 times more acidic than the stomach contents sample, and its pH is 3.20 − 0.8 = 2.4 units lower. Note how this seemingly small change in pH corresponds to a more than 200-fold difference in acidity.

Practice Exercise The results of oceanographic studies indicate that the average pH of Earth's oceans is now 8.07. The hydrogen ion concentration of the ocean at the start of the industrial revolution (around 1750) was $6.61 \times 10^{-9}\ M$. How much more acidic is the ocean now? Does the ocean have a lower pH today than previously?

CONCEPT TEST

Is the pH of a 1.00 *M* solution of a weak acid higher or lower than the pH of a 1.00 *M* solution of a strong acid?

pOH, pK_a, and pK_b Values

The letter *p* as used in pH is also used with other symbols to mean *the negative logarithm* of the variable that follows it. For example, just as every aqueous solution has a pH value, it also has a **pOH** value, defined as

$$\text{pOH} = -\log[OH^-] \tag{15.15}$$

We can use the K_w expression, Equation 15.13, to relate pOH to pH. We start by taking the negative logarithm of both sides of the equation and then rearrange the terms:

$$K_w = [H_3O^+][OH^-] = 1.00 \times 10^{-14}$$

$$-\log K_w = -\log([H_3O^+][OH^-]) = -\log(1.0 \times 10^{-14})$$

$$pK_w = -\log[H_3O^+] - \log[OH^-] = -(-14.00)$$

$$pK_w = \text{pH} + \text{pOH} = 14.00 \tag{15.16}$$

Many tables of equilibrium constants list p*K* values rather than *K* values because doing so does not require the use of exponential notation, thus saving space. The tables in Appendix 5 of this book contain both sets of values. Use them whenever you need a *K* or p*K* value that is not provided in a problem.

SAMPLE EXERCISE 15.8 Relating $[H_3O^+]$, $[OH^-]$, pH, and pOH **LO4**

Digested food leaving your stomach and entering your small intestine undergoes an increase in pH from about 4.50 to 7.50. Convert these pH values to pOH, $[H_3O^+]$, and $[OH^-]$ values.

pOH the negative logarithm of the hydroxide ion concentration in an aqueous solution.

Collect, Organize, and Analyze We are given two pH values, and we want to determine the corresponding pOH, $[H_3O^+]$, and $[OH^-]$ values.

- Use Equation 15.14 to convert from pH to $[H_3O^+]$, as we did in Sample Exercise 15.6.
- Use Equation 15.16 to convert from pH to pOH.
- Use Equation 15.15 to convert from pOH to $[OH^-]$.

The $[H_3O^+]$ value we calculate for pH 4.50 should be 10^3, or 1000, times the $[H_3O^+]$ value we calculate for pH 7.50 because $7.50 - 4.50 = 3.00$. Correspondingly, the $[OH^-]$ value we calculate for pH 7.50 should be 10^3, or 1000, times the $[OH^-]$ value we calculate for pH 4.50. The inverse relationship between pH and pOH means that the pOH value we calculate for pH 4.50 should be three units larger than the one we get for pH 7.50.

Solve

1. Converting pH to $[H_3O^+]$: $\mathrm{pH} = -\log[H_3O^+]$ (15.14)

   or $[H_3O^+] = 10^{-\mathrm{pH}}$

   Stomach: $[H_3O^+] = 10^{-4.50} = 3.2 \times 10^{-5}\ M$

   Small intestine: $[H_3O^+] = 10^{-7.50} = 3.2 \times 10^{-8}\ M$

2. Converting pH to pOH: $\mathrm{pH} + \mathrm{pOH} = 14.00$ (15.16)

   $\mathrm{pOH} = 14.00 - \mathrm{pH}$

   Stomach: $\mathrm{pOH} = 14.00 - 4.50 = 9.50$

   Small intestine: $\mathrm{pOH} = 14.00 - 7.50 = 6.50$

3. Converting pOH to $[OH^-]$: $\mathrm{pOH} = -\log[OH^-]$ (15.15)

   or $[OH^-] = 10^{-\mathrm{pOH}}$

   Stomach: $[OH^-] = 10^{-9.50} = 3.2 \times 10^{-10}\ M$

   Small intestine: $[OH^-] = 10^{-6.50} = 3.2 \times 10^{-7}\ M$

Think About It As predicted, the $[H_3O^+]$ in the stomach is 1000 times that of the small intestine. The differences in pOH values are also as predicted, as are the differences in $[OH^-]$. All calculated concentration values were rounded off to two significant digits because each starting pH value had only two digits after the decimal point.

Practice Exercise What are the values of $[H_3O^+]$ and $[OH^-]$ in a sample of household ammonia, an aqueous solution of NH_3, that has a pH of 11.70?

Figure 15.8 summarizes the ways we can interconvert $[H^+]$, $[OH^-]$, pH, and pOH. The diagram may leave you wondering which path to follow to make conversions, such as $[OH^-]$ to pH. There are two equivalent sets of calculations: (i) from $[OH^-]$ to pOH to pH or (ii) from $[OH^-]$ to $[H_3O^+]$ to pH. Either results in the same answer.

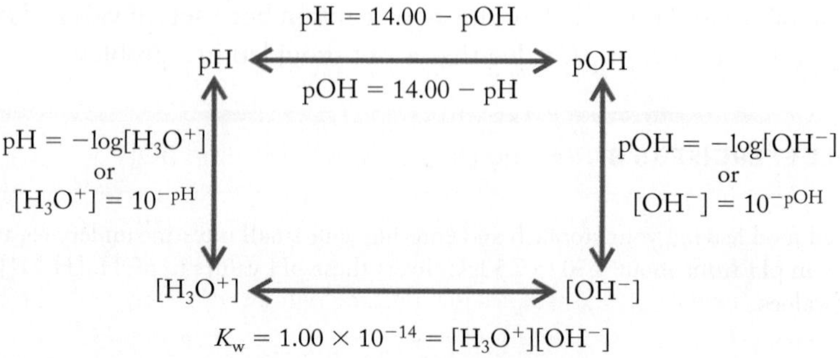

FIGURE 15.8 Pathways for interconverting $[H_3O^+]$, $[OH^-]$, pH, and pOH values.

15.5 K_a, K_b, and the Ionization of Weak Acids and Bases

Most of the acids and bases in nature are weak acids and bases with K_a and K_b values that are much less than 1. In this section and the one that follows, we explore how to quantify the degree to which these molecules react with molecules of water to form either H_3O^+ or OH^- ions.

Weak Acids

Let's begin with the ionization of a generic weak acid (HA), as expressed by the chemical equation

$$HA(aq) + H_2O(\ell) \rightleftharpoons A^-(aq) + H_3O^+(aq) \qquad (15.3)$$

The corresponding K_a expression is

$$K_a = \frac{[A^-][H_3O^+]}{[HA]} \qquad (15.17)$$

We can use Equation 15.17 to calculate the K_a value of an unknown weak acid if we know both the concentration of a solution of the acid and its pH. Suppose, for example, we determine that a 0.100 M solution of HA has a pH of 2.20. To calculate K_a, we first convert the pH value to $[H_3O^+]$:

$$[H_3O^+] = 10^{-2.20} = 6.3 \times 10^{-3}\ M$$

Next we assume that the only source of H_3O^+ ions is the ionization of HA. If so, then $[A^-]$ must also be $6.3 \times 10^{-3}\ M$, because one mole of HA ionizes to produce one mole of H_3O^+ and one mole of A^-. Also, if we assume no hydronium ions are initially present, then the $[H_3O^+]$ increases by $6.3 \times 10^{-3}\ M$, then [HA] must have decreased by $6.3 \times 10^{-3}\ M$. Therefore, the concentration of HA at equilibrium is

$$[HA] = (0.100 - 6.3 \times 10^{-3})\ M = 0.094\ M$$

Using the three calculated equilibrium concentrations in the K_a expression gives us

$$K_a = \frac{[A^-][H_3O^+]}{[HA]} = \frac{(6.3 \times 10^{-3})(6.3 \times 10^{-3})}{(0.094)} = 4.2 \times 10^{-4}$$

The small value of K_a confirms that HA is a weak acid.

The ratio of the equilibrium $[H_3O^+]$ to the initial [HA] represents the **degree of ionization** of HA, which is usually expressed as a percentage of the initial acid concentration. For this reason it is also called **percent ionization**. In equation form this relationship is

$$\text{Percent ionization} = \frac{[H_3O^+]}{[HA]} \times 100\% \qquad (15.18)$$

Inserting the data from the previous calculation into Equation 15.18:

$$\text{Percent ionization} = \frac{6.3 \times 10^{-3}\ M}{0.100\ M} \times 100\% = 6.3\%$$

degree of ionization the ratio of the quantity of a substance that is ionized to the concentration of the substance before ionization; when expressed as a percentage, called **percent ionization**.

CONCEPT **TEST**

Describe how the percent ionization of a weak acid is related to its K_a value.

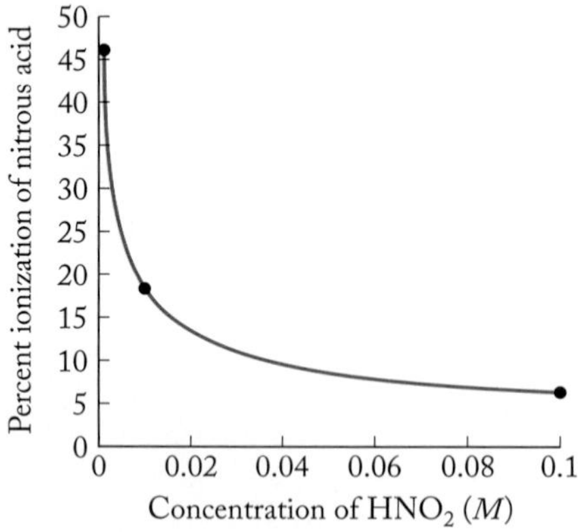

FIGURE 15.9 The degree of ionization of a weak acid increases with decreasing acid concentration. Here the degree of ionization of nitrous acid increases from about 6% in a 0.100 M solution to 18% in a 0.010 M solution to 46% in a 0.001 M solution.

As noted in Section 15.2, the degree of ionization of a weak acid increases with decreasing acid concentration. This trend is shown in the plot of percent ionization of nitrous acid as a function of the acid's initial concentration in Figure 15.9. This trend is observed for all weak acids.

We can explain this trend using Le Châtelier's principle. Suppose, for example, that we have a 1.0 M solution of a generic weak acid, HA, in which the concentrations of A^- and H_3O^+ ions at equilibrium are both 0.0010 M. Using these concentrations in Equation 15.17 to solve for K_a, we have

$$K_a = \frac{[A^-][H_3O^+]}{[HA]} = \frac{(0.0010)(0.0010)}{(1.0 - 0.0010)} = 1.0 \times 10^{-6}$$

Now suppose that more water is added to the HA solution, increasing its volume by a factor of 10. If no change in the degree of ionization occurred, then [HA] would be 0.10 M, and the concentrations of A^- and H_3O^+ ions would both be 1.0×10^{-4} M. However, if we insert these values into the equilibrium constant expression, we get a reaction quotient (Q) value of

$$Q = \frac{(1.0 \times 10^{-4})(1.0 \times 10^{-4})}{(0.10 - 1.0 \times 10^{-4})} = 1.0 \times 10^{-7}$$

This Q value is only 1/10 the value of K_a, which means the acid ionization reaction is no longer at equilibrium. Because $Q < K_a$, the reaction proceeds in the forward direction. This means the concentrations of A^- and H_3O^+ ions increase, as does the percentage of HA molecules that ionize. Further dilutions would produce even greater percent ionization values, following the trend we saw for nitrous acid in Figure 15.9.

SAMPLE EXERCISE 15.9 Relating pH, K_a, and Percent Ionization of a Weak Acid **LO5**

The pH of a 1.00 M solution of formic acid (HCOOH) is 1.88.

a. What is the percent ionization of 1.00 M HCOOH?
b. What is the K_a value of formic acid?

Collect and Organize We are asked to determine the K_a value of formic acid and its percent ionization in a solution of known concentration and pH. There are two H atoms in a molecule of formic acid, but only one is in a carboxylic acid group and can ionize in an aqueous solution. As a result, the acid ionization reaction is

$$HCOOH(aq) + H_2O(\ell) \rightleftharpoons HCOO^-(aq) + H_3O^+(aq)$$

Analyze The stoichiometry of the ionization reaction means that $[HCOO^-] = [H_3O^+]$ in a solution of HCOOH at equilibrium, so Equation 15.14, $pH = -\log[H_3O^+]$, will give us the values of $[H_3O^+]$ and $[HCOO^-]$ at equilibrium and Equation 15.18 will allow us to calculate the degree of ionization. At equilibrium, [HCOOH] is equal to its initial concentration minus the portion that ionized, or

$$[HCOOH]_{equilibrium} = [HCOOH]_{initial} - [H_3O^+]_{equilibrium}$$

We can calculate the value of K_a using the following equilibrium constant expression:

$$K_a = \frac{[HCOO^-][H_3O^+]}{[HCOOH]}$$

The pH of the 1.00 M solution is close to 2, which corresponds to $[H_3O^+] = 10^{-2}$ M. Therefore, the percent ionization of formic acid in this solution should be about 1%.

Solve

a. The pH of the 1.00 *M* solution is 1.88. The corresponding $[H_3O^+]$ is

$$[H_3O^+] = 10^{-1.88} = 1.32 \times 10^{-2}\ M$$

At equilibrium, $[H_3O^+] = [HCOO^-] = 1.32 \times 10^{-2}\ M$.
Inserting this value and the initial concentration of HCOOH into Equation 15.18:

$$\text{Percent ionization} = \frac{[HCOO^-]_{\text{equilibrium}}}{[HCOOH]_{\text{initial}}} \times 100\%$$

$$= \frac{1.32 \times 10^{-2}\ M}{1.00\ M} \times 100\% = 1.32\%$$

b. Since $[HCOO^-] = [H_3O^+]$, the equilibrium concentration of HCOOH is

$$(1.00 - 1.32 \times 10^{-2})\ M = 0.99\ M$$

Inserting these values in the expression for K_a:

$$K_a = \frac{[HCOO^-][H_3O^+]}{[HCOOH]} = \frac{(1.32 \times 10^{-2})(1.32 \times 10^{-2})}{(0.99)} = 1.8 \times 10^{-4}$$

Think About It The result of the percent ionization calculation is close to our estimate of about 1%, and the calculated K_a value is the same as the K_a value for formic acid listed in Table 15.2, 1.77×10^{-4}.

Practice Exercise The value of $[H_3O^+]$ in a 0.050 *M* solution of an organic acid is 5.9×10^{-3} *M*. What is the pH of the solution, the percent ionization of the acid, and its K_a value?

CONCEPT TEST

Three weak acids have the following K_a values:

Acid	A	B	C
K_a	3.6×10^{-5}	4.9×10^{-4}	9.2×10^{-4}

Which of the three acids is the most extensively ionized in a 0.10 *M* solution of the acid? Which of the three acids has the lowest percent ionization in a 1.00 *M* solution of the acid?

Weak Bases

The concept of percent ionization also applies to Brønsted–Lowry bases and their acceptance of protons in aqueous solutions. Consider the reaction of generic base B in an aqueous solution:

$$B(aq) + H_2O(\ell) \rightleftharpoons HB^+(aq) + OH^-(aq)$$

The corresponding equilibrium constant expression is

$$K_b = \frac{[HB^+][OH^-]}{[B]} \quad (15.19)$$

Analogous to weak acids, we can calculate the degree of ionization of a weak base and its K_b value if we know the pH and the initial concentration of the base in the solution:

$$\text{Percent ionization} = \frac{[OH^-]_{\text{equilibrium}}}{[B]_{\text{initial}}} \times 100\% \quad (15.20)$$

Doing so takes more steps than in the corresponding calculation for a weak acid as shown in Sample Exercise 15.10.

SAMPLE EXERCISE 15.10 Relating pH, K_b, and Percent Ionization of a Weak Base **LO5**

Trimethylamine, $(CH_3)_3N$, is a particularly foul-smelling volatile organic compound that forms during the decay of plant and animal matter. It is soluble in water, and it reacts with water molecules according to the following chemical equation:

$$(CH_3)_3N(aq) + H_2O(\ell) \rightleftharpoons (CH_3)_3NH^+(aq) + OH^-(aq)$$

The pH of a 0.50 *M* solution of trimethylamine is 11.75 at 25°C.

a. What is the percent ionization of trimethylamine in a 0.50 *M* solution?
b. What is the K_b value of trimethylamine?

Collect, Organize, and Analyze We are asked to determine the K_b value of trimethylamine and its percent ionization in a solution of known concentration and pH. According to the stoichiometry of the reaction, $[(CH_3)_3NH^+] = [OH^-]$ in an aqueous solution of $(CH_3)_3N$ at equilibrium. Additionally, $[(CH_3)_3N]$ is equal to the initial concentration of the base minus the portion of it that reacts with water to form OH^- ions:

$$[(CH_3)_3N]_{equilibrium} = [(CH_3)_3N]_{initial} - [OH^-]_{equilibrium}$$

Converting pH to pOH and then calculating $[OH^-]$ will give us $[OH^-]$ and $[(CH_3)_3NH^+]$ at equilibrium. We can then calculate the degree of ionization using Equation 15.16. Finally, to calculate the value of K_b, substitute the appropriate values into the equilibrium constant expression:

$$K_b = \frac{[(CH_3)_3NH^+][OH^-]}{[(CH_3)_3N]}$$

The pH of the 0.50 *M* solution is a little less than 12, which corresponds to a pOH that is a little more than 2 and a value for $[OH^-]$ that is a little less than 10^{-2} *M*. Therefore, the percent ionization of trimethylamine in this solution should be 1–2%.

Solve

a. The pH of the 0.50 *M* solution is 11.75. Calculating the corresponding pOH:

$$pH + pOH = 14.00$$

or $$pOH = 14.00 - pH = 14.00 - 11.75 = 2.25$$

Calculating $[OH^-]$: $$[OH^-] = 10^{-pOH} = 10^{-2.25} = 5.6 \times 10^{-3} M$$

This value is the same as $[(CH_3)_3NH^+]$ at equilibrium. Using it and the initial concentration of trimethylamine to calculate percent ionization:

$$\text{Percent ionization} = \frac{[HB^+]_{equilibrium}}{[B]_{initial}} \times 100\%$$

$$= \frac{5.6 \times 10^{-3}\ M}{0.50\ M} \times 100\% = 1.1\%$$

b. At equilibrium, $[(CH_3)_3NH^+] = [OH^-] = 5.6 \times 10^{-3}$ *M*, and $[(CH_3)_3N]$ is

$$(0.50 - 5.6 \times 10^{-3})\ M = 0.49\ M$$

Inserting these values in the K_b expression:

$$K_b = \frac{[(CH_3)_3NH^+][OH^-]}{[(CH_3)_3N]} = \frac{(5.6 \times 10^{-3})(5.6 \times 10^{-3})}{0.49}$$

$$= 6.4 \times 10^{-5}$$

Think About It The result of the percent ionization calculation is in our estimated range of 1–2%, which reflects the fact that although trimethylamine may have a strong odor, it is only a weak base. The latter point is reinforced by the small K_b value.

Practice Exercise The pH of a 0.100 *M* aqueous solution of ethylamine, $C_2H_5NH_2$, is 11.86 at 25°C. What percentage of the $C_2H_5NH_2$ molecules in this solution ionize as described in the chemical equation that follows, and what is the K_b value of ethylamine?

$$C_2H_5NH_2(aq) + H_2O(\ell) \rightleftharpoons C_2H_5NH_3^+(aq) + OH^-(aq)$$

15.6 Calculating the pH of Acidic and Basic Solutions

In this section we calculate the pH values of solutions of acids and bases. We begin with solutions of strong acids and strong bases.

Strong Acids and Strong Bases

We have seen that strong acids ionize completely in aqueous solutions. This means that in nearly all of their solutions, the concentration of H_3O^+ ions is the same as the initial concentration of the strong acid. For example, the pH of muriatic acid (7.4 *M* HCl), which is sold in hardware and building supply stores to clean concrete surfaces, is

$$pH = -\log[H_3O^+] = -\log[HCl] = -\log(7.4) = -0.87$$

This pH value is negative because $[H_3O^+] > 1$, which means $\log[H_3O^+] > 0$. The negative sign in the pH equation gives us a negative pH value.

To calculate the pH of a solution of a strong base, we need to calculate the concentration of OH^- ions from the initial concentration of the base and then convert $[OH^-]$ to pOH, and convert pOH to pH. Sample Exercise 15.11 illustrates the steps involved.

SAMPLE EXERCISE 15.11 Calculating the pH of a Solution of a Strong Base **LO6**

Liquid cleaners such as the one in the bottle shown in Figure 15.10 are used to remove clogs from bathroom and kitchen drains. They contain sodium hydroxide at concentrations as high as 1.2 *M*. What is the pH of 1.2 *M* NaOH?

Collect, Organize, and Analyze NaOH is a strong base that produces one mole of hydroxide ions per mole of NaOH in solution, so $[OH^-]$ is the same as the initial concentration of the base. Therefore, $[OH^-] = 1.2\ M$. We can convert this value to pOH and then to pH following the steps shown in Figure 15.8.

Solve

Calculating pOH: $pOH = -\log[OH^-] = -\log(1.2) = -0.079$

Calculating pH: $pH = 14.00 - pOH = 14.00 - (-0.079) = 14.08$

FIGURE 15.10 Cleaners used to remove grease and hair clogs from water pipes typically contain high concentrations of strong base, such as sodium hydroxide.

Think About It The calculated pH is slightly above 14 because $[OH^-]$ is slightly greater than one molar.

Practice Exercise Kalkwasser is the German word for "lime water," which is the common name for saturated solutions of $Ca(OH)_2$. It has many uses, including being added to water in aquarium tanks to adjust the pH and to provide Ca^{2+} ions for the plants and animals living in the tank. What is the pH of kalkwasser if the concentration of the solution is 0.0225 *M* $Ca(OH)_2$?

Weak Acids and Weak Bases

CONNECTION RICE tables were introduced in Chapter 14 to help us calculate the changes (C) from initial (I) to equilibrium (E) concentrations for a given reaction (R).

In Sample Exercises 15.12 and 15.13 we calculate the pH values of aqueous solutions of a weak acid and a weak base. We use the initial concentrations of the solutions to set up RICE tables similar to those used in Chapter 14.

SAMPLE EXERCISE 15.12 Calculating the pH of a Solution of a Weak Acid — LO6

FIGURE 15.11 This carabid beetle secretes concentrated formic acid as a defense against predators.

Formic acid is a common carboxylic acid and the one with the simplest molecular structure. A diluted sample of the secretions of the carabid beetle shown in Figure 15.11 contains 0.040 *M* formic acid, HCOOH. What is the pH of 0.040 *M* formic acid? The K_a value of the acid is 1.77×10^{-4}.

Collect and Organize We are asked to determine the pH of a known concentration of a weak acid. We also know its K_a value. The molecular structure of formic acid in Table 15.2 indicates that only one H atom per molecule is ionizable in aqueous solution.

Analyze The acid ionization reaction is

$$HCOOH(aq) + H_2O(\ell) \rightleftharpoons HCOO^-(aq) + H_3O^+(aq)$$

and the corresponding equilibrium constant expression is

$$K_a = \frac{[HCOO^-][H_3O^+]}{[HCOOH]} = 1.77 \times 10^{-4}$$

Calculating pH involves first solving for $[H_3O^+]$, which we can do by setting up a RICE table based on the above reaction. The ratio of the initial acid concentration to K_a value is $0.040/1.77 \times 10^{-4} = 225$. This value is below the 500× guideline for ignoring the $-x$ term in the denominator of the K_a expression.

Solve First we set up a RICE table with columns for the molar concentrations of the reactant and two products. The objective of the table is to solve for $[H_3O^+]$ at equilibrium, so we give it the symbol x. Filling in the other cells in the table based on the 1:1:1 stoichiometry of the reaction gives us

Reaction	$HCOOH(aq) + H_2O(\ell) \rightleftharpoons$	$HCOO^-(aq)$	+	$H_3O^+(aq)$
	$[HCOOH]$ (*M*)	$[HCOO^-]$ (*M*)		$[H_3O^+]$ (*M*)
Initial (I)	0.040	0		0
Change (C)	$-x$	$+x$		$+x$
Equilibrium (E)	$0.040 - x$	x		x

Inserting the equilibrium terms into the K_a expression,

$$K_a = \frac{[HCOO^-][H_3O^+]}{[HCOOH]} = \frac{(x)(x)}{(0.040 - x)} = 1.77 \times 10^{-4}$$

Solving for x by first cross-multiplying,

$$x^2 = 7.08 \times 10^{-6} - (1.77 \times 10^{-4})x$$

and rearranging the terms,

$$x^2 + (1.77 \times 10^{-4})x - 7.08 \times 10^{-6} = 0$$

gives us a quadratic equation with two solutions:

$$x = -0.00275\ M \quad \text{and} \quad x = 0.00257\ M$$

The negative x value has no physical meaning because it results in negative concentration values, so $[H_3O^+] = 0.00257\ M$.

Solving for pH: $\quad pH = -\log[H_3O^+] = -\log(0.00257) = 2.59$

Think About It The decision not to simplify the calculation by eliminating "$-x$" from the denominator [HCOOH] term is supported by percent ionization of the acid: $0.00257/0.040 = 6.4\%$. This represents a significant decrease in [HCOOH] due to acid ionization.

Practice Exercise Acetic acid, the main ingredient in vinegar, is also a carboxylic acid. What is the pH of a 0.035 M solution of acetic acid given that its $K_a = 1.76 \times 10^{-5}$?

SAMPLE EXERCISE 15.13 Calculating the pH of a Solution of a Weak Base

LO6

The concentration of NH_3 in the household ammonia used to clean windows ranges between 50 and 100 g/L, or from about 3 M to almost 6 M. What is the pH of 3.0 M NH_3? The K_b value for ammonia is 1.76×10^{-5}.

Collect, Organize, and Analyze We are asked to determine the pH of 3.0 M NH_3, which reacts with water to form NH_4^+ and OH^- ions:

$$NH_3(aq) + H_2O(\ell) \rightleftharpoons NH_4^+(aq) + OH^-(aq) \qquad K_b = 1.76 \times 10^{-5}$$

To calculate pH we must first determine the equilibrium concentration of OH^- ions and then convert $[OH^-]$ to pOH and finally to pH. Given the 1:1:1 stoichiometry of NH_3, NH_4^+, and OH^-, we know that $[NH_4^+] = [OH^-]$ at equilibrium. Letting these concentrations be x, the change in $[NH_3]$ during the course of the reaction is $-x$. The pH value we obtain for a fairly concentrated solution of a weak base should be well above 7 but below 14.

Solve We begin by setting up a RICE table in which we let $[NH_4^+] = [OH^-] = x$ at equilibrium:

Reaction	$NH_3(aq) + H_2O(\ell)$	$\rightleftharpoons$	$NH_4^+(aq)$	+	$OH^-(aq)$
	$[NH_3]$ (M)		$[NH_4^+]$ (M)		$[OH^-]$ (M)
Initial (I)	3.0		0		0
Change (C)	$-x$		$+x$		$+x$
Equilibrium (E)	$3.0 - x$		x		x

The initial concentration of base is $3.0/(1.76 \times 10^{-5}) = 170{,}000$ times the value of K_b, or well above the 500× guideline for simplifying the calculation by ignoring the $-x$ term in the denominator:

$$K_b = \frac{[NH_4^+][OH^-]}{[NH_3]} = \frac{(x)(x)}{3.0} = \frac{x^2}{3.0} = 1.76 \times 10^{-5}$$

so

$$x^2 = 5.28 \times 10^{-5}$$

Solving for x gives us

$$x = [OH^-] = \sqrt{5.28 \times 10^{-5}} = 7.3 \times 10^{-3}\ M$$

Taking the negative logarithm of $[OH^-]$ to calculate pOH:

$$pOH = -\log[OH^-] = -\log(7.3 \times 10^{-3}\ M) = 2.14$$

Then we subtract this value from 14.00 to obtain the pH:

$$pH = 14.00 - pOH = 14.00 - 2.14 = 11.86$$

Think About It The calculated pH value falls in the range we predicted, given the small K_b value but the relatively high initial concentration of ammonia. To check our simplifying assumption, let's compare the value of x to the initial $[NH_3]$ value of 3.0 M.

$$\frac{7.3 \times 10^{-3}\ M}{3.0\ M} = 0.0024 \times 100\% = 0.24\%$$

This small percentage change in $[NH_3]$ means our simplifying assumption is justified.

Practice Exercise What is the pH of a 0.200 M solution of methylamine (CH_3NH_2, $K_b = 4.4 \times 10^{-4}$)?

pH of Very Dilute Solutions of Strong Acids

Recall that the autoionization of water contributes very little to the equilibrium concentrations of H_3O^+ or OH^- ions due to the small value of K_w. Now that we have developed the concept of pH and explored how it helps us quantify the acidity of a solution, let's consider a situation where the concentration of hydronium ions produced by the autoionization of water is actually greater than the concentration of hydronium ions produced by an acid.

SAMPLE EXERCISE 15.14 pH Calculations Involving the Autoionization of Water — LO6

What is the pH of $1.0 \times 10^{-8}\ M$ HCl?

Collect and Organize We are asked to calculate the pH of a very dilute solution of HCl, which is a strong acid and ionizes completely. The autoionization of pure water at 25°C produces an equilibrium at $[H_3O^+] = 1.0 \times 10^{-7}\ M$.

Analyze In a $1.0 \times 10^{-8}\ M$ HCl solution, $[H_3O^+] = 1.0 \times 10^{-8}\ M$, and the pH should be

$$pH = -\log[H_3O^+] = -\log(1.0 \times 10^{-8}) = 8.00$$

This answer is not reasonable because a solution of a strong acid, no matter how dilute it is, cannot be basic (pH > 7). To calculate pH, we must also take into account the autoionization of water. Let's set up a RICE table based on the autoionization equilibrium in which x represents the increase in $[H_3O^+]$ resulting from autoionization and in which the initial value of $[H_3O^+]$ is 1.0×10^{-8} *M*.

Solve Setting up the RICE table and filling in the rows as described above:

Reaction	$H_2O(\ell) + H_2O(\ell)$	$\rightleftharpoons$	$H_3O^+(aq)$	+	$OH^-(aq)$
			$[H_3O^+]$ (*M*)		$[OH^-]$ (*M*)
Initial (I)			1.0×10^{-8}		0
Change (C)			$+x$		$+x$
Equilibrium (E)			$(1.0 \times 10^{-8}) + x$		x

Substituting equilibrium values into Equation 15.13 and solving for x:

$$K_w = [H_3O^+][OH^-] = 1.0 \times 10^{-14}$$

or,

$$1.0 \times 10^{-14} = (1.0 \times 10^{-8} + x)(x)$$

Rearranging this equation to solve for x gives the following quadratic equation:

$$x^2 + (1.0 \times 10^{-8})x - 1.0 \times 10^{-14} = 0$$

which has two solutions:

$$x = 9.5 \times 10^{-8}\ M \text{ and } x = -1.1 \times 10^{-7}\ M$$

Only the positive value makes physical sense; a negative value for x would mean a negative concentration of hydroxide ions at equilibrium. Thus, the total concentration of hydrogen ions in the solution is

$$[H_3O^+] = (9.5 \times 10^{-8}\ M) + (1.0 \times 10^{-8}\ M) = 10.5 \times 10^{-8}\ M = 1.05 \times 10^{-7}\ M$$

and the pH is

$$\text{pH} = -\log(1.05 \times 10^{-7}\ M) = 6.98$$

Think About It This value agrees with our prediction that the solution should be (very) slightly acidic.

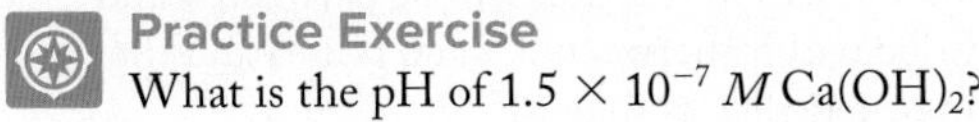

Practice Exercise
What is the pH of 1.5×10^{-7} *M* $Ca(OH)_2$?

15.7 Polyprotic Acids

Up to this point we have dealt with **monoprotic acids**, which have only one ionizable hydrogen atom per molecule. Acids that contain more than one ionizable hydrogen atom—such as sulfuric acid (H_2SO_4) and phosphoric acid (H_3PO_4)—are called **polyprotic acids**. For molecules with two and three ionizable hydrogen atoms, we use the more specific terms *diprotic acids* and *triprotic acids*, respectively. Let's consider the acidic properties of sulfuric acid, a strong, diprotic acid.

Acid Rain

Coal naturally contains sulfur impurities. When coal is burned to produce electricity, these impurities are released into the atmosphere as SO_2. Some of this SO_2 is oxidized to SO_3, which combines with water vapor to form particles of liquid sulfuric acid (H_2SO_4), a principal component of acid rain in many parts of the

monoprotic acid an acid that has one ionizable hydrogen atom per molecule.

polyprotic acid an acid that has two or more ionizable hydrogen atoms per molecule.

world. Sulfuric acid is a strong acid (see Table 15.1) that is essentially completely ionized in aqueous solutions:

$$H_2SO_4(\ell) + H_2O(\ell) \rightarrow HSO_4^-(aq) + H_3O^+(aq) \qquad K_{a_1} >> 1$$

However, the second ionization step has a much smaller K_a value and often does not result in complete ionization, depending on the concentration of the acid:

$$HSO_4^-(aq) + H_2O(\ell) \rightleftharpoons SO_4^{2-}(aq) + H_3O^+(aq) \qquad K_{a_2} = 0.012$$

Note that these K_a equilibrium constant symbols have an additional subscript number, with K_{a_1} and K_{a_2} corresponding to the donation of a first and then a second H^+ ion per molecule.

The combination of one complete and one incomplete ionization reaction means that many aqueous solutions of H_2SO_4 contain more than one mole but less than two moles of H_3O^+ ions for every mole of H_2SO_4 dissolved.

SAMPLE EXERCISE 15.15 Calculating the pH of a Solution of a Strong Diprotic Acid **LO7**

What is the pH of 0.100 M H_2SO_4?

Collect and Organize We are given the concentration of a solution of sulfuric acid and asked to calculate its pH. Sulfuric acid is a strong diprotic acid in that one hydrogen atom per molecule ionizes completely (K_{a_1} >>1), but the second hydrogen atom may not completely ionize for every molecule (K_{a_2} = 0.012).

Analyze We start with the first ionization reaction that goes to completion:

$$H_2SO_4(aq) + H_2O(\ell) \rightarrow HSO_4^-(aq) + H_3O^+(aq)$$

Therefore, as the second ionization step begins, $[HSO_4^-] = [H_3O^+] = 0.100\ M$. Ionization of HSO_4^- then produces additional H_3O^+ ions:

$$HSO_4^-(aq) + H_2O(\ell) \rightleftharpoons SO_4^{2-}(aq) + H_3O^+(aq)$$

The increase in $[H_3O^+]$ due to the second ionization reaction will be between 0 and 0.1 M, which means the total $[H_3O^+]$ value will be between 0.1 and 0.2 M and the pH of the solution at equilibrium should be a little less than 1.

Solve We begin by setting up a RICE table based on the second ionization reaction. Initially $[HSO_4^-] = [H_3O^+] = 0.100\ M$. We let the change in $[H_3O^+]$ during the second ionization step be $+x$. Filling in the other cells of the table based on the stoichiometry of the second ionization step:

Reaction	$HSO_4^-(aq) + H_2O(\ell)$	$\rightleftharpoons$	$SO_4^{2-}(aq)$	+	$H_3O^+(aq)$
	$[HSO_4^-]$ (M)		$[SO_4^{2-}]$ (M)		$[H_3O^+]$ (M)
Initial (I)	0.100		0		0.100
Change(C)	$-x$		$+x$		$+x$
Equilibrium (E)	$0.100 - x$		x		$0.100 + x$

Inserting the equilibrium concentrations in the equilibrium constant expression for K_{a_2}:

$$K_{a_2} = \frac{[H_3O^+][SO_4^{2-}]}{[HSO_4^-]} = \frac{(0.100 + x)(x)}{(0.100 - x)} = 1.2 \times 10^{-2}$$

Cross-multiplying and rearranging the terms:

$$x^2 + 0.112x - 1.2 \times 10^{-3} = 0$$

Solving this quadratic equation for x yields a positive value and a negative value:

$$x = 0.00985 \quad \text{and} \quad x = -0.122$$

The negative value for x has no physical meaning because it gives us a negative $[SO_4^{2-}]$ value. Therefore,

$$[H_3O^+] = (0.100 + x) = (0.100 + 0.00985) = 0.10985 = 0.110\ M$$

The corresponding pH is

$$pH = -\log[H_3O^+] = -\log(0.110) = 0.96$$

Think About It As predicted, the value of $[H_3O^+]$ is between 0.1 and 0.2 M and the pH of the solution is a little less than 1. The degree of ionization of HSO_4^- is

$$\frac{[SO_4^{2-}]_{\text{equilibrium}}}{[HSO_4^-]_{\text{initial}}} = \frac{0.00985\ M}{0.100\ M} = 0.0985 = 9.8\%$$

This means that ignoring the decrease in HSO_4^- to avoid solving a quadratic equation would have been a bad idea.

Practice Exercise What is the pH of a 0.200 M solution of H_2SO_4? How should doubling the concentration of H_2SO_4 affect the pH when compared to the 0.100 M solution of H_2SO_4 in Sample Exercise 15.15?

CONCEPT TEST

Identify the most abundant species present in an aqueous solution of phosphoric acid, H_3PO_4.

Normal Rain

Did you know that rainwater falling from the sky is naturally acidic? The fourth most abundant gas in the atmosphere is CO_2, which dissolves in water to form a small amount of carbonic acid, as we discussed in Section 15.1:

$$CO_2(g) + H_2O(\ell) \rightleftharpoons H_2CO_3(aq) \qquad (15.1)$$

The equilibrium constant for this reaction is only about 10^{-3}, so most of the dissolved carbon dioxide remains in the form of hydrated molecules of CO_2. However, to simplify equilibrium calculations involving solutions of carbon dioxide, we routinely represent the total concentration of $CO_2(aq)$ and $H_2CO_3(aq)$ as $[H_2CO_3]$, even though carbonic acid is not the principal species in solution. Thus, the acidic properties of dissolved CO_2 are represented by these chemical equations:

$$H_2CO_3(aq) + H_2O(\ell) \rightleftharpoons HCO_3^-(aq) + H_3O^+(aq) \qquad K_{a_1} = 4.3 \times 10^{-7}$$

$$HCO_3^-(aq) + H_2O(\ell) \rightleftharpoons CO_3^{2-}(aq) + H_3O^+(aq) \qquad K_{a_2} = 4.7 \times 10^{-11}$$

Calculating the pH of a solution of carbonic acid is simpler than the pH calculation for sulfuric acid. Because the K_{a_2} value of H_2CO_3 is so small, it does not contribute significantly to the concentration of H_3O^+ ions in a solution of the acid and can be ignored in calculating pH. We can explain why $K_{a_2} << K_{a_1}$ based on electrostatic attractions between oppositely charged ions. The first ionization step produces a negatively charged oxoanion, HCO_3^-. The second requires that a H^+ ion dissociate from HCO_3^- to produce an even more negative oxoanion, CO_3^{2-}. Separating oppositely charged ions that are naturally attracted to each other is not a favored process, which is confirmed by the much smaller value of K_{a_2}.

SAMPLE EXERCISE 15.16 Calculating the pH of a Solution of a Weak Diprotic Acid **LO7**

What is the pH of rainwater at 25°C in equilibrium with atmospheric CO_2, which gives the water a constant dissolved CO_2 concentration of 1.4×10^{-5} *M*?

Collect and Organize We are asked to determine the pH of a dilute solution of dissolved CO_2. There are two ionizable H atoms in H_2CO_3. The K_{a_1} and K_{a_2} values are given in the text. Any H_2CO_3 that dissociates to produce bicarbonate and hydronium ions will be replaced by more CO_2 dissolving from the atmosphere, so the $[H_2CO_3]$ value in the denominator of the K_{a_1} expression will be constant at 1.4×10^{-5} *M*.

Analyze The large difference between the K_{a_1} and K_{a_2} values means that the pH of the solution is controlled by the first ionization equilibrium:

$$H_2CO_3(aq) + H_2O(\ell) \rightleftharpoons HCO_3^-(aq) + H_3O^+(aq) \qquad K_{a_1} = 4.3 \times 10^{-7}$$

Because of the small value of K_{a_1} and the small concentration of dissolved CO_2, we should obtain a pH value that is less than 7, but a lot closer to 7 than 0.

Solve First we set up a RICE table in which $x = [H_3O^+] = [HCO_3^-]$ at equilibrium and the value of $[H_2CO_3]$ is a constant 1.4×10^{-5} *M*.

Reaction	$H_2CO_3(aq) + H_2O(\ell)$	$\rightleftharpoons$	$HCO_3^-(aq)$	+	$H_3O^+(aq)$
	$[H_2CO_3]$ (*M*)		$[HCO_3^-]$ (*M*)		$[H_3O^+]$ (*M*)
Initial (I)	1.4×10^{-5}		0		0
Change (C)	0		$+x$		$+x$
Equilibrium (E)	1.4×10^{-5}		x		x

$$K_{a_1} = \frac{[HCO_3^-][H_3O^+]}{[H_2CO_3]} = \frac{(x)(x)}{1.4 \times 10^{-5}} = 4.3 \times 10^{-7}$$

so

$$x^2 = 6.0 \times 10^{-12}$$

Therefore,

$$x = [H_3O^+] = 2.4 \times 10^{-6}\ M$$

Taking the negative logarithm of $[H_3O^+]$ to calculate pH:

$$pH = -\log[H_3O^+] = -\log(2.45 \times 10^{-6}\ M) = 5.61$$

Think About It Carbonic acid is a weak acid, and its concentration here is small, so obtaining a pH value that is only about 1.4 units below neutral pH is reasonable.

Practice Exercise The proximity of the calculated pH value to 7.00 raises the question of whether the autoionization of water contributes significantly to $[H_3O^+]$ in the rainwater sample. Recalculate the pH of the rainwater sample in Sample Exercise 15.16, assuming the initial concentration of $H_3O^+ = 1.00 \times 10^{-7}$ *M*.

Some acids have three ionizable H atoms per molecule. Two important triprotic acids are phosphoric acid (H_3PO_4) and citric acid, the acid responsible for the tart flavor of citrus fruits. Note in Table 15.9 that $K_{a_1} > K_{a_2} > K_{a_3}$ for both acids. This pattern is much like that for the $K_{a_1} > K_{a_2}$ values of diprotic acids and for the same reason: it is more difficult to remove a second H^+ ion from the negatively charged ion formed after the first H^+ ion is removed, and it is even more difficult to remove a third H^+ ion from an ion with a 2− charge.

TABLE 15.9 Ionization Equilibria for Two Triprotic Acids

	Reaction					Equilibrium constant
Phosphoric Acid						
(1)	$HO{-}P(=O)(OH){-}OH$	$+\ H_2O$	$\rightleftharpoons$	$HO{-}P(=O)(OH){-}O^-$	$+\ H_3O^+$	$K_{a_1} = 7.11 \times 10^{-3}$
(2)	$HO{-}P(=O)(OH){-}O^-$	$+\ H_2O$	$\rightleftharpoons$	$^-O{-}P(=O)(OH){-}O^-$	$+\ H_3O^+$	$K_{a_2} = 6.32 \times 10^{-8}$
(3)	$^-O{-}P(=O)(OH){-}O^-$	$+\ H_2O$	$\rightleftharpoons$	$^-O{-}P(=O)(O^-){-}O^-$	$+\ H_3O^+$	$K_{a_3} = 4.5 \times 10^{-13}$
Citric Acid						
(1)	$HO{-}C(CH_2COOH)(CH_2COOH){-}COOH$	$+\ H_2O$	$\rightleftharpoons$	$HO{-}C(CH_2COO^-)(CH_2COOH){-}COOH$	$+\ H_3O^+$	$K_{a_1} = 7.44 \times 10^{-4}$
(2)	$HO{-}C(CH_2COO^-)(CH_2COOH){-}COOH$	$+\ H_2O$	$\rightleftharpoons$	$HO{-}C(CH_2COO^-)(CH_2COOH){-}COO^-$	$+\ H_3O^+$	$K_{a_2} = 1.73 \times 10^{-5}$
(3)	$HO{-}C(CH_2COO^-)(CH_2COOH){-}COO^-$	$+\ H_2O$	$\rightleftharpoons$	$HO{-}C(CH_2COO^-)(CH_2COO^-){-}COO^-$	$+\ H_3O^+$	$K_{a_3} = 4.02 \times 10^{-7}$

CONCEPT TEST

Do you expect the second or third acid ionization step in phosphoric acid and citric acid to influence the pH of 0.100 *M* solutions of either acid? Why or why not?

15.8 Acidic and Basic Salts

Seawater and the freshwater in many rivers and lakes have pH values that range from weakly basic to weakly acidic. How can these waters be more basic than the acidic rainwater ($pH \leq 5.6$) that serves, directly or indirectly, as their water supply? When rain soaks into the ground, its pH changes as it flows through soils that contain basic components. To understand the chemical processes that produce neutral or slightly basic groundwater, we first need to examine the acid–base properties of some common ionic compounds present in these waters.

As discussed in Chapter 8, soluble ionic compounds separate into their component ions when they dissolve in water. For example, a 0.01 *M* solution of NaCl contains 0.01 M Na^+ ions and 0.01 M Cl^- ions. It is also a neutral solution. Neither Na^+ ions nor Cl^- ions hydrolyze (react with water) to form H_3O^+ or OH^- ions when they dissolve in water. The Cl^- ion is the conjugate base of a strong acid (HCl), so it is such a weak base that its base strength can be considered negligible.

When NaF dissolves in water, however, it produces F^- ions, which are the conjugate base of HF, a weak acid. Therefore, F^- ions are weakly basic, producing at least some OH^- ions when they dissolve in water:

$$F^-(aq) + H_2O(\ell) \rightleftharpoons HF(aq) + OH^-(aq)$$

Because Na^+ ions don't influence pH, solutions of NaF are weakly basic.

If salts that contain the conjugate bases of weak acids can be basic, then can salts that contain the conjugate acids of weak bases be acidic? One example of such a salt is NH_4Cl. The Cl^- ions in NH_4Cl do not affect pH, but the NH_4^+ ions are the conjugate acid of NH_3, a weak base. As a result, NH_4^+ ions are weakly acidic, producing at least some H_3O^+ ions:

$$NH_4^+(aq) + H_2O(\ell) \rightleftharpoons NH_3(aq) + H_3O^+(aq)$$

Consequently, aqueous solutions of NH_4Cl are weakly acidic.

Table 15.10 summarizes how salts can be acidic, basic, or neutral depending on whether they include cations that are the conjugate acids of weak bases, or anions that are the conjugate bases of weak acids, or both. Note that salts that contain both the conjugate base of a weak acid and the conjugate acid of a weak base may be acidic, basic, or neutral, depending on the relative strengths of the acid and base. Ammonium acetate represents the rare example of a salt in which the strengths of the conjugate acid (acetic acid) and the base (ammonia) happen to be exactly the same: the K_a of acetic acid = K_b of ammonia = 1.76×10^{-5}. As a result, ammonium acetate is a neutral salt.

TABLE 15.10 Acid–Base Properties of Salts

Anion of a	Cation of a	Aqueous Solutions Are	Example
Strong acid	Strong base	Neutral	NaCl
Strong acid	Weak base	Acidic	NH_4Cl
Weak acid	Strong base	Basic	NaF
Weak acid	Weak base	Neutral,[a] Acidic,[b] or Basic[c]	CH_3COONH_4 NH_4F NH_4HCO_3

[a]If K_a (of weak acid) = K_b (of weak base)
[b]If K_a (of weak acid) > K_b (of weak base)
[c]If K_a (of weak acid) < K_b (of weak base)

SAMPLE EXERCISE 15.17 Predicting Whether a Salt Is Acidic, Basic, or Neutral **LO8**

NaClO is the active ingredient in chlorine bleach. Is an aqueous solution of NaClO acidic, basic, or neutral?

Collect, Organize, and Analyze Sodium ions do not hydrolyze and do not affect the pH of aqueous solutions. However, ClO^- ions are the conjugate base of HClO, which is a weak acid (Table 15.2). Therefore, ClO^- ions are weak bases that partially hydrolyze in water, generating OH^- ions:

$$ClO^-(aq) + H_2O(\ell) \rightleftharpoons HClO(aq) + OH^-(aq)$$

Solve Because the hydrolysis of ClO^- ions produces OH^- ions, solutions of NaClO are weakly basic.

Think About It Any sodium salt that contains an anion that is the conjugate base of a weak acid produces weakly basic aqueous solutions.

 **Practice Exercise** Write a chemical equation for the hydrolysis reaction that explains why an aqueous solution of K_2SO_4 is basic.

Salts can be acidic, basic, or neutral, but how do we calculate the pH of their aqueous solutions? Our approach is much like the one we used to calculate the pH of solutions of weak acids and bases, but there is an extra step. For example, if we want to calculate the pH of a solution of ammonium chloride, we need to know the equilibrium constant of the reaction in which the NH_4^+ ion functions as a Brønsted–Lowry acid:

$$NH_4^+(aq) + H_2O(\ell) \rightleftharpoons NH_3(aq) + H_3O^+(aq)$$

K_a values are typically not listed for conjugate acids of weak bases. Instead, the K_b values for the weak bases are listed, as is the case for ammonia in Appendix 5:

$$NH_3(aq) + H_2O(\ell) \rightleftharpoons NH_4^+(aq) + OH^-(aq) \qquad K_b = 1.76 \times 10^{-5}$$

The strengths of conjugate acids and bases are complementary: the stronger one is, the weaker the other, so we can derive the K_a value for ammonium ions from the K_b value for ammonia. To see how, let's write the K_a and K_b equilibrium constant expressions for NH_4^+ and NH_3:

$$K_a = \frac{[NH_3][H_3O^+]}{[NH_4^+]} \qquad K_b = \frac{[NH_4^+][OH^-]}{[NH_3]}$$

These expressions are similar, although any shared terms that appear in the numerator of one expression are found in the denominator of the other. When we multiply the two expressions together,

$$K_a \times K_b = \frac{\cancel{[NH_3]}[H_3O^+]}{\cancel{[NH_4^+]}} \times \frac{\cancel{[NH_4^+]}[OH^-]}{\cancel{[NH_3]}} = [H_3O^+][OH^-]$$

we get the K_w expression for the autoionization of water:

$$K_a \times K_b = K_w \qquad (15.21)$$

Equation 15.21 is very handy because (1) it works for all conjugate acid–base pairs, and (2) it allows us to calculate either the K_b of the anion in a basic salt from the K_a of its conjugate acid or the K_a of the cation in an acidic salt from the K_b of its conjugate base.

FIGURE 15.12 The active ingredient in chlorine bleach is sodium hypochlorite.

SAMPLE EXERCISE 15.18 Calculating the pH of a Solution of a Basic Salt **LO8**

The bottle of chlorine bleach shown in Figure 15.12 holds an aqueous solution that contains 82.5 g/L of NaClO. What is the pH of this solution?

Collect and Organize We know the concentration of an aqueous solution of hypochlorite, NaClO, and we are asked to calculate its pH. We determined in Sample Exercise 15.17 that NaClO is a basic salt because the ClO^- ion is the conjugate base of HClO, a weak acid. The Na^+ ion plays no role in the acid–base properties of a salt. The K_a of HClO is 2.9×10^{-8}.

Analyze First we need to convert the K_a value for HClO into the K_b value for the ClO^- ion using Equation 15.21: $K_a \times K_b = K_w$. We also need to convert the concentration of NaClO from g/L to molarity. That value will be the initial value of the reactant in a RICE table based on its hydrolysis reaction: $ClO^-(aq) + H_2O(\ell) \rightleftharpoons HClO(aq) + OH^-(aq)$. We will solve for $[OH^-]$, then pOH, and then pH. The concentration of the solution should be about 1 *M*, and K_b should be a little more than 10^{-7}. Therefore, $[OH^-]$ of a ~1 *M* solution should be near the square root of that value, or ~10^{-3}, making pOH ≈ 3 and pH ≈ 11.

Solve

Calculate K_b:
$$K_b = \frac{K_w}{K_a} = \frac{1.0 \times 10^{-14}}{2.9 \times 10^{-8}} = 3.45 \times 10^{-7}$$

and then calculate the molar concentration of NaClO to use in the RICE table below:

$$\frac{82.5\ \text{g}}{\text{L}} \times \frac{1\ \text{mol}}{74.44\ \text{g}} = 1.108\ \text{mol/L}$$

Reaction	$ClO^-(aq)$ + $H_2O(\ell)$	⇌	$HClO(aq)$	+	$OH^-(aq)$
	$[ClO^-]$ (*M*)		$[HClO]$ (*M*)		$[OH^-]$ (*M*)
Initial (I)	1.108		0		0
Change (C)	$-x$		$+x$		$+x$
Equilibrium (E)	$1.108 - x$		x		x

To solve for *x*, we make the simplifying assumption that *x* will be small compared to 1.108 *M* because $[ClO^-]_{initial}$ is $1.108/3.45 \times 10^{-7} = 3.21 \times 10^6$ times the K_b value, which is much greater than our 500× guideline.

$$K_b = \frac{[HClO][OH^-]}{[ClO^-]} = \frac{(x)(x)}{1.108 - x} \approx \frac{x^2}{1.108} = 3.45 \times 10^{-7}$$

$$x = [OH^-] = 6.18 \times 10^{-4}$$

Calculating pOH: $pOH = -\log[OH^-] = -\log(6.18 \times 10^{-4}) = 3.21$

and pH: $pH = 14.00 - pOH = 14.00 - 3.21 = 10.79$

Think About It The calculated pH (10.79) is in the same ballpark as our estimate of 11. Also the calculated $[OH^-]$ value is much less than 5% of the initial $[ClO^-]$ value, so our assumption that we could ignore the $-x$ term in the denominator of the K_b expression is justified.

Practice Exercise The pH of swimming pools is made slightly basic by spreading solid Na_2CO_3 across the surface. Another approach involves preparing concentrated solutions of Na_2CO_3 and slowly adding them to the water circulating through the pool pump and filter. What is the pH of an aqueous solution of 0.100 *M* Na_2CO_3?

SAMPLE EXERCISE 15.19 Calculating the pH of a Solution of an Acidic Salt **LO8**

Aqueous solutions of ammonium chloride are used in organic chemistry when a weakly acidic solution is needed. What is the pH of 0.25 *M* NH_4Cl?

Collect and Organize We are asked to calculate the pH of a solution of NH_4Cl. When NH_4Cl dissolves in water, NH_4^+ and Cl^- ions are released into solution. The NH_4^+ ion is the conjugate acid of NH_3, a weak base. The Cl^- ion is the conjugate base of HCl, a strong acid.

Analyze The Cl^- ion has negligible strength as a Brønsted–Lowry base, so it does not contribute to the acid–base properties of NH_4Cl. Ammonia is a weak base ($K_b = 1.76 \times 10^{-5}$), which means that its conjugate acid, NH_4^+, is a weak acid, and the pH of the solution will be controlled by its hydrolysis:

$$NH_4^+(aq) + H_2O(\ell) \rightleftharpoons NH_3(aq) + H_3O^+(aq)$$

The K_a value of NH_4^+ can be calculated by dividing K_w by the K_b of ammonia. The K_b value of ammonia is close to 10^{-5}, so the K_a value of the ammonium ion will be close to 10^{-9}. Therefore, $[H_3O^+]$ of a 0.25 *M* solution should be the square root of $\sim 10^{-10}$, or $\sim 10^{-5}$, giving a pH $\approx$ 5.

Solve The K_a expression for the NH_4^+ ion is

$$K_a = \frac{[NH_3][H_3O^+]}{[NH_4^+]}$$

Rearranging Equation 15.21 to solve for K_a:

$$K_a = \frac{K_w}{K_b} = \frac{1.00 \times 10^{-14}}{1.76 \times 10^{-5}} = 5.68 \times 10^{-10} = \frac{[NH_3][H_3O^+]}{[NH_4^+]}$$

We set up a RICE table in which we make the usual assumptions that the reaction is the only significant source of H^+ and that $x = [H_3O^+] = [NH_3]$ at equilibrium:

Reaction	$NH_4^+(aq) + H_2O(\ell)$	$\rightleftharpoons$	$NH_3(aq)$	+	$H_3O^+(aq)$
	$[NH_4^+]$ (*M*)		$[NH_3]$ (*M*)		$[H_3O^+]$ (*M*)
Initial (I)	0.25		0		0
Change (C)	$-x$		$+x$		$+x$
Equilibrium (E)	$0.25 - x$		x		x

$$K_a = 5.68 \times 10^{-10} = \frac{[NH_3][H_3O^+]}{[NH_4^+]} = \frac{(x)(x)}{0.25 - x}$$

Given the very small value of K_a, we can make the simplifying assumption that $(0.25\ M - x) \approx 0.25\ M$, which gives us

$$\frac{x^2}{0.25} = 5.68 \times 10^{-10}$$

$$x^2 = 1.42 \times 10^{-10}$$

$$x = 1.19 \times 10^{-5} = [H_3O^+]$$

$$pH = -\log[H_3O^+] = -\log(1.19 \times 10^{-5}) = 4.92$$

Think About It This result matches our prediction quite closely. The calculated $[H_3O^+]$ is much less than 5% of the initial concentration of NH_4^+, so our simplifying assumption is valid.

 **Practice Exercise** What is the difference in pH between a 0.25 *M* solution of dimethyl ammonium chloride, $(CH_3)_2NH_2Cl$, and a 0.25 *M* NH_4Cl solution?

SAMPLE EXERCISE 15.20 Integrating Concepts: The pH of Human Blood

We began this chapter by noting how essential it is that our circulation and respiration systems efficiently remove from our bodies the CO_2 produced in our cells. An enzyme, carbonic anhydrase, plays a key role in this process by speeding up the rate at which CO_2 hydrolyzes:

$$CO_2(g) + H_2O(\ell) \rightleftharpoons H_2CO_3(aq)$$

and the rate at which H_2CO_3 ionizes:

$$H_2CO_3(aq) + H_2O(\ell) \rightleftharpoons HCO_3^-(aq) + H_3O^+(aq)$$

a. Does carbonic anhydrase increase the acid strength of dissolved CO_2—that is, the K_{a_1} of carbonic acid based on the total concentration of CO_2 in solution?
b. Suppose that, during strenuous exercise, the total concentration of dissolved CO_2 in the blood flowing through muscle tissues is 2.7×10^{-3} *M*. If the concentration of HCO_3^- ions in the blood is 0.028 *M*, what is the pH of the blood?

Collect and Organize We are asked whether the presence of an enzyme that increases the rates at which CO_2 hydrolyzes and undergoes acid ionization makes carbonic acid a stronger acid. We then need to calculate the pH of blood in which the initial $[HCO_3^-]$ and $[CO_2]$ values are known. Carbonic acid is a weak diprotic acid; its K_{a_1} is 4.3×10^{-7}.

Analyze
a. We learned in Chapter 14 that catalysts speed up reactions but do not alter their equilibrium constants.
b. Calculating the pH of a carbonic acid solution will require a RICE table based on K_{a_1} in which $[H_3O^+] = x$ at equilibrium, but in which the initial concentration of HCO_3^- is 0.028 *M*, not zero. The normal pH of human blood is close to 7.4, so the calculated pH should be close to 7.4.

Solve
a. Carbonic anhydrase should not affect the value of K_{a_1}, although it would produce increases in both $[H_2CO_3]$ and $[HCO_3^-]$. These increases in the numerator and denominator of the K_{a_1} expression offset each other and do not affect pH.
b. Set up a RICE table based on the K_{a_1} reaction, where the values in the $[H_2CO_3]$ column represent the total concentration of dissolved CO_2 ($[CO_2(aq)] + [H_2CO_3(aq)]$):

Reaction	$H_2CO_3(aq)$ + $H_2O(\ell) \rightleftharpoons$	$HCO_3^-(aq)$ +	$H_3O^+(aq)$
	$[H_2CO_3]$ (*M*)	$[HCO_3^-]$ (*M*)	$[H_3O^+]$ (*M*)
Initial (I)	2.7×10^{-3}	0.028	0
Change (C)	$-x$	$+x$	$+x$
Equilibrium (E)	$2.7 \times 10^{-3} - x$	$0.028 + x$	x

In solving for x, we should consider making the assumptions that we can ignore the $-x$ part of the $[H_2CO_3]$ term and the $+x$ part of the $[HCO_3^-]$ term. The initial $[H_2CO_3]$ value $2.7 \times 10^{-3}/4.3 \times 10^{-7} = 6.3 \times 10^3$ times the K_{a_1} value, which satisfies the 500× guideline for ignoring $-x$. We anticipate that the calculated pH value will be slightly basic (see Section 15.1), which means $[H_3O^+] = x$ should be less than 10^{-7}. Therefore, we may ignore the $+x$ component of the $[HCO_3^-]$ term.

$$K_{a_1} = \frac{[HCO_3^-][H_3O^+]}{[H_2CO_3]} = \frac{(0.028 + x)(x)}{2.7 \times 10^{-3}} \approx \frac{0.028x}{2.7 \times 10^{-3}}$$
$$= 4.3 \times 10^{-7}$$
$$x = 4.15 \times 10^{-8}$$

Therefore, the pH of the blood is

$$pH = -\log[H_3O^+] = -\log(4.15 \times 10^{-8}) = 7.38$$

Think About It As expected, the calculated pH is slightly basic. As discussed in Section 15.1, the high concentration of bicarbonate ions in the blood shifts the carbonic acid ionization equilibrium to the left, lowering $[H_3O^+]$ and raising pH. We explore the impact of having a second source of product ions on other equilibria in Chapter 16.

SUMMARY

LO1 The strengths of acids and bases are related to the values of their acid and base ionization constants, K_a and K_b. Most acids are weak, which means their K_a value is much less than 1 and they ionize only partially in water. (Section 15.2)

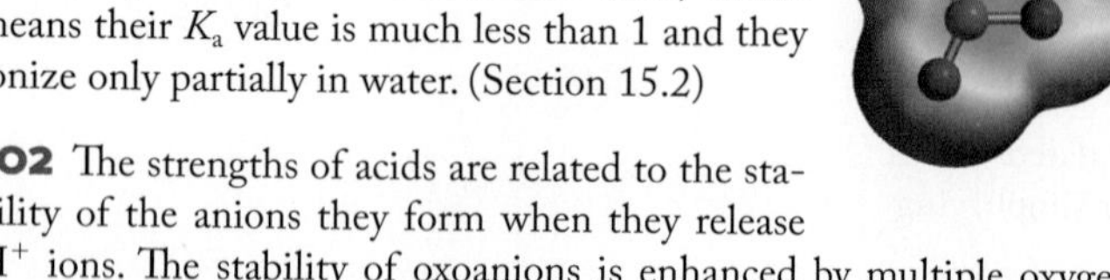

LO2 The strengths of acids are related to the stability of the anions they form when they release H^+ ions. The stability of oxoanions is enhanced by multiple oxygen atoms bonded to the central atom, which disperse the negative charge(s) over the anion. (Section 15.2)

LO3 When an acid (HA) ionizes, it forms its **conjugate base**, A^-. When a base (B) reacts with a H^+ ion, it forms its **conjugate acid**, HB^+. (Section 15.3)

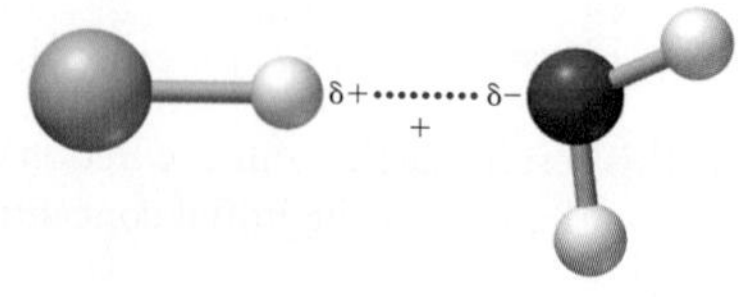

LO4 In a neutral solution, $[H_3O^+] = [OH^-] = 1.0 \times 10^{-7}$. The **pH** scale is a logarithmic scale for expressing the acidic or basic strength of solutions. Acidic solutions have pH values less than 7; basic solutions have pH values greater than 7. Because pH is the negative logarithm of the H_3O^+ concentration, the higher the pH, the lower the H_3O^+ concentration. An increase in one pH unit represents a decrease in $[H_3O^+]$ to 1/10 of its initial value. Likewise, pOH is the negative logarithm of the OH^- concentration. The sum of pH and pOH equals 14 in an aqueous solution at 25°C. (Section 15.4)

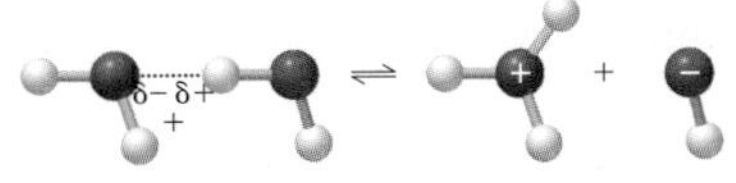

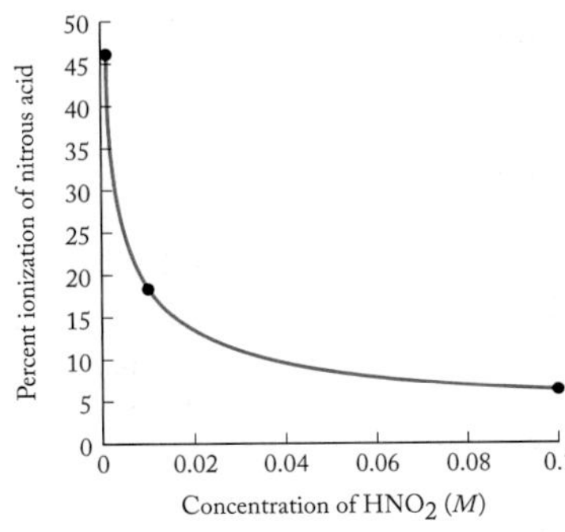

LO5 The K_a and K_b values of acids and bases can be used to calculate the extent to which their solutions are ionized—that is, their **degree of ionization** or **percent ionization**—and vice versa. (Section 15.5)

LO6 To calculate the pH of a weak acid or base, use a RICE table based on the acid or base ionization reaction to determine the equilibrium value of $[H_3O^+]$ or $[OH^-]$ in a solution of a weak acid or base. (Section 15.6)

LO7 Polyprotic acids can undergo more than one acid ionization reaction, but for most, the first ionization reaction is the one that controls pH. (Section 15.7)

LO8 A salt solution is acidic if the cation in the salt is the conjugate acid of a weak base and the anion is the conjugate base of a strong acid. A salt solution is basic if the anion in the salt is the conjugate base of a weak acid and the cation is the conjugate acid of a strong base. (Section 15.8)

PARTICULATE **PREVIEW WRAP-UP**

NH_3 is the Brønsted–Lowry base (proton acceptor) and H_2O is the Brønsted–Lowry acid (proton donor) in the forward reaction. In the reverse reaction, the ammonium ion donates a proton (Brønsted–Lowry acid), while the hydroxide ion accepts a proton (Brønsted–Lowry base). If $K << 1$, then the equilibrium lies far to the left and the predominant acid and base in solution are H_2O and NH_3, respectively.

PROBLEM-SOLVING SUMMARY

Type of Problem	Concepts and Equations	Sample Exercises
Ranking acids and bases from strongest to weakest	The strengths of acids and bases are proportional to their K_a and K_b values.	**15.1**
	The strengths of oxoacids increase with increasing electronegativity of the central atom and with increasing numbers of oxygen atoms bonded to it.	**15.2, 15.3**
Identifying acid–base conjugate pairs and predicting their relative strengths	The formula of the base in a conjugate pair is the formula of the acid with one less H^+ ion.	**15.4**
	The stronger an acid or base, the weaker its conjugate base or acid.	**15.5**
Interconverting $[H_3O^+]$, $[OH^-]$, pH, and pOH	Use the following: $$\mathrm{pH} = -\log[H_3O^+] \quad (15.14)$$ $$\mathrm{pOH} = -\log[OH^-] \quad (15.15)$$ $$pK_w = \mathrm{pH} + \mathrm{pOH} = 14.00 \quad (15.16)$$	**15.6, 15.7, 15.8**
Relating pH, K_a, and percent ionization of a weak acid	Use the following equations: $$[H_3O^+] = 10^{-\mathrm{pH}}$$ $$K_a = \frac{[A^-][H_3O^+]}{[HA]} \quad (15.17)$$ $$\text{Percent ionization} = \frac{[H_3O^+]_{\text{equilibrium}}}{[HA]_{\text{initial}}} \times 100\% \quad (15.18)$$	**15.9**

Type of Problem	Concepts and Equations	Sample Exercises
Determining percent ionization and K_b given the pH of a weak base	Use the following: $$[OH^-] = 10^{-pOH}$$ $$K_b = \frac{[HB^+][OH^-]}{[B]} \quad (15.19)$$ $$\text{Percent ionization} = \frac{[OH^-]_{equilibrium}}{[B]_{initial}} \times 100\% \quad (15.20)$$	**15.10**
Calculating the pH of a solution of strong acid or base	Assume 100% ionization so that $$[H_3O^+] = [HX] \quad \text{and}$$ $$[OH^-] = [MOH] \text{ or } [OH^-] = 2\,[M(OH)_2]$$ Calculate pH from $[H_3O^+]$ or $[OH^-]$ as described above.	**15.11**
Calculating the pH of a solution of weak acid HA	Set up a RICE table based on the K_a equilibrium $$HA(aq) + H_2O(\ell) \rightleftharpoons H_3O^+(aq) + A^-(aq)$$ Let $x = [H_3O^+] = [A^-]$ at equilibrium. Calculate x using $$K_a = \frac{[A^-][H_3O^+]}{[HA]}$$ Then calculate $pH = -\log[H_3O^+]$.	**15.12**
Calculating the pH of a solution of weak base B	Set up a RICE table based on the equilibrium $$B(aq) + H_2O(\ell) \rightleftharpoons HB^+(aq) + OH^-(aq)$$ Let $x = [OH^-] = [HB^+]$ at equilibrium. Calculate x using $$K_b = \frac{[BH^+][OH^-]}{[B]}$$ Then use $$pOH = -\log[OH^-] \quad \text{and} \quad pH = 14.00 - pOH$$	**15.13**
Calculating the pH of a solution of very dilute acidic (or very dilute basic) solution while considering the autoionization of water	Set up a RICE table based on the equilibrium $$H_2O(\ell) + H_2O(\ell) \rightleftharpoons H_3O^+(aq) + OH^-(aq)$$ Let $x = [H_3O^+] = [OH^-]$ due to autoionization. Add x to the $[H_3O^+]$ due to the dilute acid (or to the $[OH^-]$ due to the weak base). Solve for x using $$K_w = [H_3O^+][OH^-] \quad (15.12)$$ Then convert $[H_3O^+]$ to pH.	**15.14**
Calculating the pH of a solution of a strong diprotic acid	Assume $H_2A(aq) + H_2O(\ell) \rightarrow H_3O^+(aq) + HA^-(aq)$ is complete. Set up a RICE table based on the K_{a_2} equilibrium $$HA^-(aq) + H_2O(\ell) \rightleftharpoons H_3O^+(aq) + A^{2-}(aq)$$ Let x = additional $[H_3O^+]$ from the second ionization step; $[HA^-]_{initial\ (2nd\ step)} = [H_2A]_{initial}$. Calculate x using $$K_{a_2} = \frac{[H_3O^+][A^{2-}]}{[HA^-]}$$ Combine initial and additional $[H_3O^+]$ values and convert to pH.	**15.15**
Calculating the pH of a solution of a weak diprotic acid	Set up a RICE table based on the K_{a_1} equilibrium $$H_2A(aq) + H_2O(\ell) \rightleftharpoons H_3O^+(aq) + HA^-(aq)$$ Let $x = [H_3O^+] = [HA^-]$ at equilibrium. Calculate x using $$K_{a_1} = \frac{x^2}{[H_2A] - x}$$ Then convert $x = [H_3O^+]$ to pH.	**15.16**
Distinguishing acidic, basic, and neutral salts	The cations in acidic salts are the conjugate acids of weak bases. The anions in basic salts are the conjugate bases of weak acids.	**15.17**

Type of Problem	Concepts and Equations	Sample Exercises
Calculating the pH of a solution of a basic salt	Assume the salt (MA) completely dissociates into M^+ and A^-. Set up a RICE table for the equilibrium $$A^-(aq) + H_2O(\ell) \rightleftharpoons HA(aq) + OH^-(aq)$$ Let $x = [HA] = [OH^-] =$ at equilibrium. Calculate x using $K_b = K_w/K_{a\text{ (of conjugate acid, HA)}}$ $$K_b = \frac{K_w}{K_a} = \frac{x^2}{[A^-] - x}$$ Then convert $[OH^-]$ to pOH and then pOH to pH.	**15.18**
Calculating the pH of a solution of an acidic salt	Assume the salt (BHX) completely dissociates into BH^+ and X^-. Set up a RICE table for the equilibrium $$BH^+(aq) + H_2O(\ell) \rightleftharpoons B(aq) + H_3O^+(aq)$$ Let $x = [H_3O^+] = [B]$ at equilibrium. Calculate x using $K_a = K_w/K_{b\text{ (of conjugate base, B)}}$ $$K_a = \frac{K_w}{K_b} = \frac{x^2}{[BH^+] - x}$$ Then convert $x = [H_3O^+]$ to pH.	**15.19**

VISUAL PROBLEMS

(Answers to boldface end-of-chapter questions and problems are in the back of the book.)

15.1. Which of the lines in Figure P15.1 best represents the dependence of the degree of ionization of acetic acid on its concentration in aqueous solution?

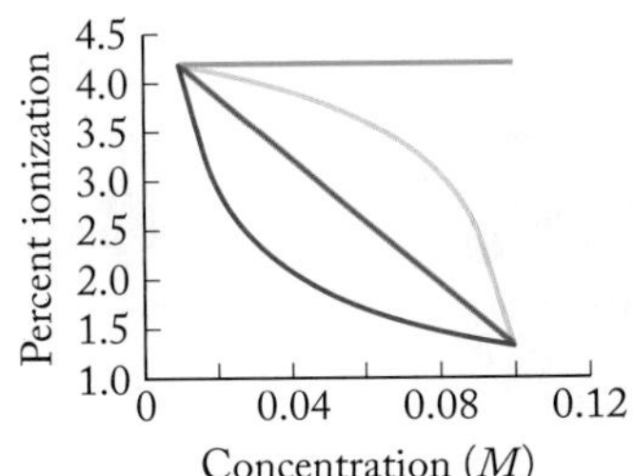

FIGURE P15.1

15.2. The graph in Figure P15.2 shows the percent ionization of two acids as a function of concentration in water. Which line describes the behavior of HNO_3, and which line describes the behavior of acetic acid (CH_3COOH)?

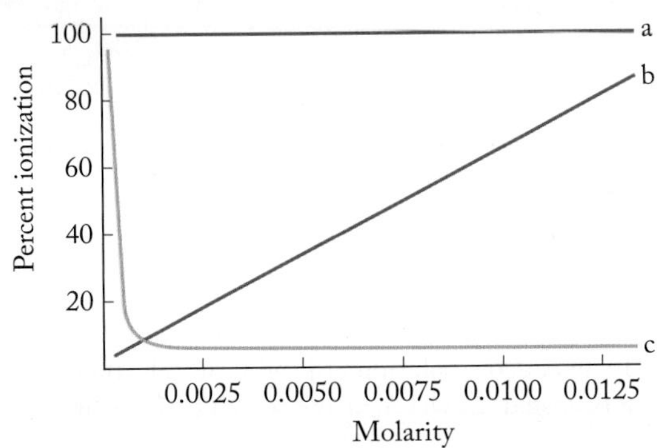

FIGURE P15.2

15.3. The bar graph in Figure P15.3 shows the degree of ionization of 1×10^{-3} *M* solutions of three hypohalous acids: HClO, HBrO, and HIO. Which bar corresponds to HIO?

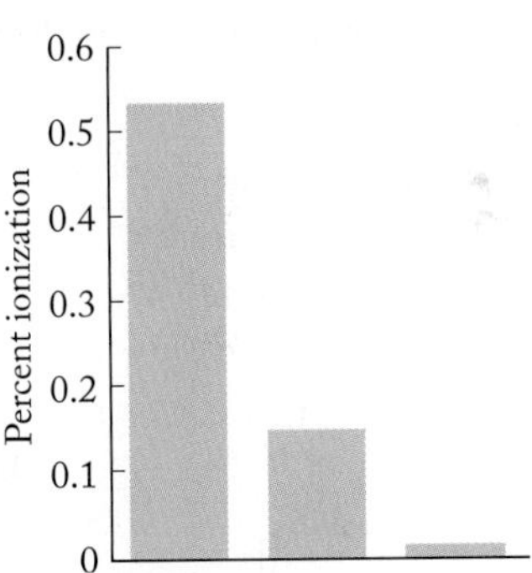

FIGURE P15.3

15.4. Figure P15.4 shows the molecular structure of alanine. Is an aqueous solution of alanine acidic, basic, or neutral? Explain your selection, or explain what additional information you need to make a more informed selection.

FIGURE P15.4

***15.5.** Figure P15.5 shows the condensed molecular structures of piperidine and morpholine. Which is the stronger base? Explain your selection.

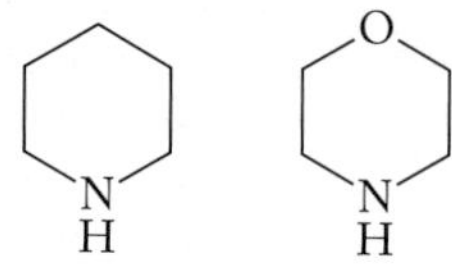

Piperidine Morpholine

FIGURE P15.5

15.6. Based on the skeletal structures shown in Figure P15.6, which is the stronger base, ethanolamine or ethylamine? Explain why you think so.

HO—NH_2 Ethanolamine

NH_2 Ethylamine

FIGURE P15.6

15.7. Figure P15.7 shows a condensed molecular structure of pseudoephedrine, a widely used decongestant and stimulant.

a. Is pseudoephedrine an acidic, basic, or neutral compound?
b. Which functional group in its structure gives it the property you selected in part a?

OH, H, N, CH_3, CH_3

FIGURE P15.7

*15.8. The pK_a values of the ions shown in Figure P15.8 increase as more methyl ($-CH_3$) groups are added. Do more methyl groups increase or decrease the strength of the neutral conjugate bases of these ions?

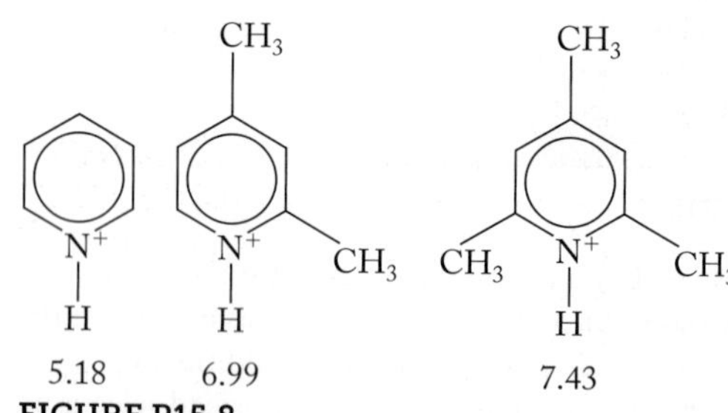

FIGURE P15.8

15.9. The value of K_{a_1} of phosphorous acid, H_3PO_3, is nearly the same as the K_{a_1} of phosphoric acid, H_3PO_4. Identify the ionizable hydrogen atoms in Figure P15.9.

O
‖
H—P—O—H
|
O
/
H

FIGURE P15.9

15.10. Use representations [A] through [I] in Figure P15.10 to answer questions a–f.

a. Identify the strong acid(s).
b. Identify the weak acid(s).
c. Identify the strong base(s).
d. Identify the weak base(s).
e. Which compound can function as both a weak acid and a weak base?
f. Which compound contains hydrogen atoms but is neither an acid nor a base?

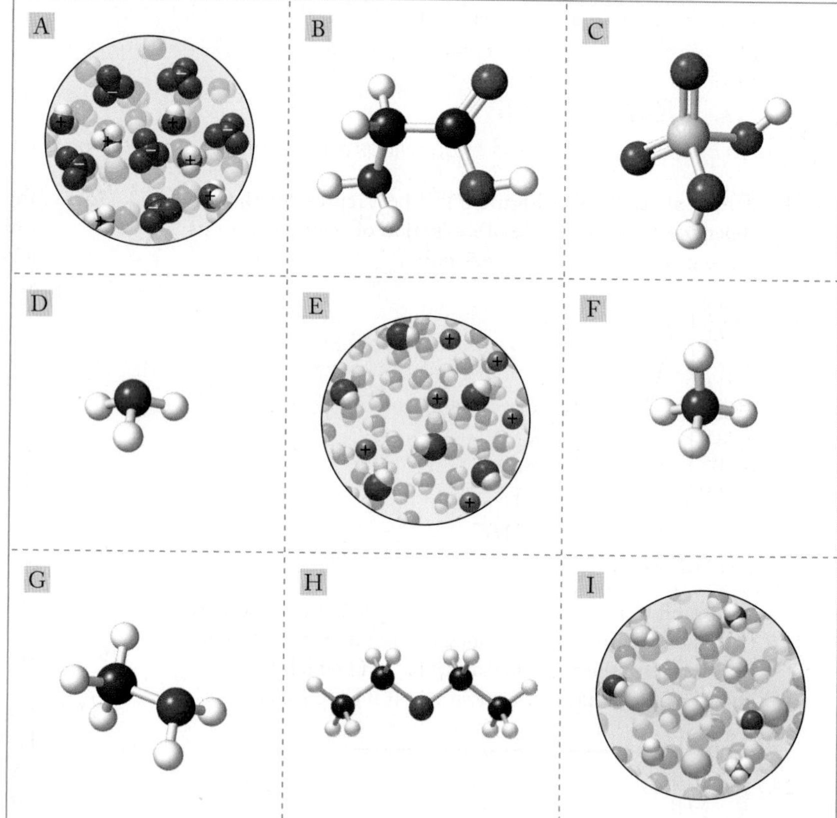

FIGURE P15.10

QUESTIONS AND PROBLEMS

Strong and Weak Acids and Bases

Concept Review

15.11. In an aqueous solution of HF, which compound acts as a Brønsted–Lowry acid and which is the Brønsted–Lowry base?

15.12. In an aqueous solution of HNO_3, which compound acts as a Brønsted–Lowry acid and which is the Brønsted–Lowry base?

15.13. In an aqueous solution of NH_3, which species acts as a Brønsted–Lowry acid and which is the Brønsted–Lowry base?

15.14. Both KOH and $Ba(OH)_2$ are strong bases. Does this mean that solutions of the two compounds with the same molarity have the same capacity to accept hydrogen ions? Why or why not?

15.15. Identify the acids and bases in the following reactions:
a. $HCl(aq) + NaOH(aq) \rightarrow NaCl(aq) + H_2O(\ell)$
b. $MgCO_3(s) + 2\,HCl(aq) \rightarrow MgCl_2(aq) + CO_2(g) + H_2O(\ell)$
c. $2\,NH_3(aq) + H_2SO_4(aq) \rightarrow (NH_4)_2SO_4(aq)$

15.16. Identify the acids and bases in the following reactions:
a. $(CH_3)_3N(aq) + H_2O(\ell) \rightleftharpoons (CH_3)_3NH^+(aq) + OH^-(aq)$
b. $CO_2(aq) + H_2O(\ell) \rightleftharpoons HCO_3^-(aq) + H_3O^+(aq)$
c. $(CH_3)_3COH(aq) + H_3O^+(aq) \rightleftharpoons (CH_3)_3COH_2^+(aq) + H_2O(\ell)$

15.17. Identify the conjugate base of each of the following compounds: HNO_2, $HClO$, H_3PO_4, and NH_3.

15.18. Identify the conjugate acid of each of the following species: $(CH_3)_3N$, CH_3COO^-, HSO_4^-, and OH^-.

15.19. What is the conjugate acid of the bisulfate ion, HSO_4^-, and what is its conjugate base?

*15.20. Compounds that do not ionize in water have been known to ionize in nonaqueous solvents. In such a solvent, what would be the conjugate acid and conjugate base of methanol, CH_3OH?

Problems

15.21. What is the concentration of H_3O^+ ions in 0.65 M HNO_3?

15.22. What is the concentration of H_3O^+ ions in a solution of hydrochloric acid that was prepared by diluting 7.5 mL of concentrated (11.6 M) HCl to a final volume of 100.0 L?

15.23. What is the value of $[OH^-]$ in 0.0205 M $Ba(OH)_2$?

15.24. Calcium hydroxide, also known as slaked lime, is the cheapest strong base available and is used in industrial processes in which low concentrations of base are required. Only 0.16 g of $Ca(OH)_2$ dissolves in 100 mL of water at 25°C. What is the concentration of hydroxide ions in 250 mL of a solution containing the maximum amount of dissolved calcium hydroxide?

Acid Strength and Molecular Structure

Concept Review

15.25. Explain why the K_{a_1} of H_2SO_4 is much greater than the K_{a_1} of H_2SeO_4.

15.26. Explain why the K_{a_1} of H_2SO_4 is much greater than the K_{a_1} of H_2SO_3.

15.27. Predict which acid in the following pairs of acids is the stronger acid: (a) H_2SO_3 or H_2SeO_3; (b) H_2SeO_4 or H_2SeO_3.

15.28. Trifluoroacetic acid, CF_3COOH, is over 10^4 times as strong as acetic acid, CH_3COOH. Explain why.

pH and the Autoionization of Water

Concept Review

15.29. Explain why pH values decrease as acidity increases.

15.30. Solution A is 100 times more acidic than solution B. What is the difference in the pH values of solution A and solution B?

15.31. Describe a solution (solute and concentration) that has a negative pH value.

15.32. Describe a solution (solute and concentration) that has a negative pOH value.

***15.33.** Draw the Lewis structures of the ions that would be produced if pure ethanol underwent autoionization.

*15.34. Liquid ammonia at a temperature of 223 K undergoes autoionization. The value of the equilibrium constant for the autoionization of ammonia is considerably less than that of water. Write an equation for the autoionization of ammonia and suggest a reason why the value of K for the process is less than that of water.

Problems

15.35. Calculate the pH and pOH of solutions with the following $[H_3O^+]$ or $[OH^-]$ values. Indicate which solutions are acidic, basic, or neutral.
a. $[H_3O^+] = 5.3 \times 10^{-3}\ M$
b. $[H_3O^+] = 3.8 \times 10^{-9}\ M$
c. $[H_3O^+] = 7.2 \times 10^{-6}\ M$
d. $[OH^-] = 1.0 \times 10^{-14}\ M$

15.36. Calculate the pH and pOH of the solutions with the following hydronium ion or hydroxide ion concentrations. Indicate which solutions are acidic, basic, or neutral.
a. $[OH^-] = 8.2 \times 10^{-11}\ M$
b. $[OH^-] = 7.7 \times 10^{-6}\ M$
c. $[H_3O^+] = 3.2 \times 10^{-4}\ M$
d. $[H_3O^+] = 1.0 \times 10^{-7}\ M$

15.37. Calculate the concentration of the following ions in the solution described:
a. $[H_3O^+]$ in $8.4 \times 10^{-4}\ M$ NaOH
b. $[H_3O^+]$ in $6.6 \times 10^{-5}\ M$ $Ca(OH)_2$
c. $[OH^-]$ in $4.5 \times 10^{-3}\ M$ HCl
d. $[OH^-]$ in $2.9 \times 10^{-5}\ M$ HCl

15.38. Determine the indicated pH or pOH values:
a. pH of a solution whose pOH = 5.5
b. pH of a solution whose pOH = 6.8
c. pOH of a solution whose pH = 9.7
d. pOH of a solution whose pH = 4.4

15.39. Calculate the pH and pOH of the following solutions:
a. stomach acid in which [HCl] = 0.155 M
b. 0.00500 M HNO_3
c. a 2:1 mixture of 0.0125 M HCl and 0.0125 M NaOH
d. a 3:1 mixture of 0.0125 M H_2SO_4 and 0.0125 M KOH

15.40. Calculate the pH and pOH of the following solutions:
a. 0.0450 M NaOH
b. 0.160 M $Ca(OH)_2$
c. a 1:1 mixture of 0.0125 M HCl and 0.0125 M $Ca(OH)_2$
d. a 2:3 mixture of 0.0125 M HNO_3 and 0.0125 M KOH

15.41. Calculate the pH of $1.33 \times 10^{-9}\ M$ LiOH.

15.42. Calculate the pH of $6.9 \times 10^{-8}\ M$ HBr.

Calculations Involving pH, K_a, and K_b

Concept Review

15.43. One-molar solutions of the following acids are prepared: CH_3COOH, HNO_2, $HClO$, and HCl.
a. Rank them in order of decreasing $[H_3O^+]$.
b. Rank them in order of increasing strength as acids (weakest to strongest).

15.44. On the basis of the following degree-of-ionization data for 0.100 M solutions, select which acid has the largest K_a.

Acid	C_6H_5COOH	HF	HN_3	CH_3COOH
Degree of Ionization (%)	2.5	8.5	1.4	1.3

15.45. A 1.0 M aqueous solution of HNO_3 is a much better conductor of electricity than is a 1.0 M solution of HNO_2. Explain why.

15.46. Hydrogen chloride and water are molecular compounds, yet a solution of HCl dissolved in H_2O is an excellent conductor of electricity. Explain why.

15.47. Hydrofluoric acid is a weak acid. Write the mass action expression for its acid ionization reaction.

15.48. **Early Antiseptic** The use of phenol, also known as carbolic acid, was pioneered in the 19th century by Sir Joseph Lister (after whom Listerine was named) as an antiseptic in surgery. Its formula is C_6H_5OH (the red hydrogen atom is ionizable). Write the mass action expression for the acid ionization equilibrium of phenol.

***15.49.** The K_a values of weak acids depend on the solvent in which they dissolve. For example, the K_a of alanine in aqueous ethanol is less than its K_a in water.

a. In which solvent does alanine ionize more?
b. Which is the stronger Brønsted–Lowry base: water or ethanol?

*15.50. The K_a of proline is 2.5×10^{-11} in water, 2.8×10^{-11} in an aqueous solution that is 28% ethanol, and 1.66×10^{-8} in aqueous formaldehyde at 25°C.

a. In which solvent is proline the strongest acid?
b. Rank these compounds on the basis of their strengths as Brønsted–Lowry bases: water, ethanol, and formaldehyde.

15.51. When methylamine, CH_3NH_2, dissolves in water, the resulting solution is slightly basic. Which compound is the Brønsted–Lowry acid and which is the base?

*15.52. When 1,2-diaminoethane, $H_2NCH_2CH_2NH_2$, dissolves in water, the resulting solution is basic. Write the formula of the ionic compound that is formed when hydrochloric acid is added to a solution of 1,2-diaminoethane.

Problems

15.53. **Muscle Physiology** During strenuous exercise, lactic acid builds up in muscle tissues. In a 1.00 M aqueous solution, 2.94% of lactic acid is ionized. What is the value of its K_a?

15.54. **Rancid Butter** The odor of spoiled butter is due in part to butanoic acid, which results from the chemical breakdown of butterfat. A 0.100 M solution of butanoic acid is 1.23% ionized. Calculate the value of K_a for butanoic acid.

15.55. At equilibrium, the value of $[H_3O^+]$ in 0.125 M of an unknown acid is 4.07×10^{-3} M. Determine the degree of ionization and the K_a of this acid.

15.56. Nitric acid (HNO_3) is a strong acid that is essentially completely ionized in aqueous solutions of concentrations ranging from 1% to 10% (1.5 M). However, in more concentrated solutions, part of the nitric acid is present as un-ionized molecules of HNO_3. For example, in a 50% solution (7.5 M) at 25°C, only 33% of the molecules of HNO_3 dissociate into H^+ and NO_3^-. What is the K_a value of HNO_3?

15.57. **Ant Bites** The venom of biting ants contains formic acid, HCOOH; $K_a = 1.77 \times 10^{-4}$ at 25°C. Calculate the pH of 0.055 M formic acid.

15.58. **Poisonous Plant** Gifblaar is a small South African shrub and one of the most poisonous plants known because it contains fluoroacetic acid. If a 0.480 M solution of fluoroacetic acid has a pH of 1.44, what is the K_a of the acid?

15.59. **Acid Rain I** A weather system moving through the American Midwest produced rain with an average pH of 5.02. By the time the system reached New England, the rain it produced had an average pH of 4.66. How much more acidic was the rain falling in New England?

15.60. **Acid Rain II** A newspaper reported that the "level of acidity" in a sample taken from an extensively studied watershed in New Hampshire in 1998 was "an astounding 200 times lower than the worst measurement" taken in the preceding 23 years. What is this difference expressed in units of pH?

15.61. The K_b of aminoethanol, $HOCH_2CH_2NH_2$, is 3.1×10^{-5}.

a. Is aminoethanol a stronger or weaker base than ethylamine, $pK_b = 3.36$?
b. Calculate the pH of 1.67×10^{-2} M aminoethanol.
c. Calculate the $[OH^-]$ concentration of 4.25×10^{-4} M aminoethanol.

15.62. **Food Dye** Quinoline is a weakly basic liquid used in the manufacture of quinolone yellow, a greenish-yellow dye for foods, and also in the production of niacin. Its pK_b is 9.15.

a. What is the pH of 0.0752 M quinoline?
b. What is the hydroxide ion concentration of the solution in part a?

15.63. **Painkillers** Morphine is an effective painkiller but is also highly addictive. Codeine is a popular prescription painkiller because it is much less addictive than morphine. Codeine contains a basic nitrogen atom that can be protonated to give the conjugate acid of codeine.

a. Calculate the pH of 1.8×10^{-3} M morphine if its $pK_b = 5.79$.
b. Calculate the pH of 2.7×10^{-4} M codeine if the pK_a of the conjugate acid is 8.21.

15.64. The awful odor of dead fish is due mostly to trimethylamine, $(CH_3)_3N$, one of three compounds related to ammonia in which methyl groups replace one, two, or all three of the H atoms in ammonia.

a. The K_b of trimethylamine $[(CH_3)_3N]$ is 6.5×10^{-5} at 25°C. Calculate the pH of 3.00×10^{-4} M trimethylamine.
b. The K_b of methylamine $[(CH_3)NH_2]$ is 4.4×10^{-4} at 25°C. Calculate the pH of 2.88×10^{-3} M methylamine.
*c. The K_b of dimethylamine $[(CH_3)_2NH]$ is 5.9×10^{-4} at 25°C. What concentration of dimethylamine is needed for the solution to have the same pH as the solution in part b?

Polyprotic Acids

Concept Review

15.65. Why is the K_{a_2} value of phosphoric acid less than its K_{a_1} value but greater than its K_{a_3} value?

15.66. In calculating the pH of 1.0 M H_2SO_3, we can ignore the H^+ ions produced by the ionization of the bisulfite (HSO_3^-) ion; however, in calculating the pH of 1.0 M sulfuric acid, we cannot ignore the H^+ ions produced by the ionization of the bisulfate ion. Why?

15.67. Carbonic acid, H_2CO_3, is a very weak diprotic acid ($K_{a_1} = 4.3 \times 10^{-7}$), but germanic acid, H_2GeO_3, is even weaker ($K_{a_1} = 9.8 \times 10^{-10}$). Suggest a reason why.

15.68. Figure P15.68 shows skeletal structures of the dicarboxylic acids malonic acid (left) and oxalic acid (right). The K_{a_1} of malonic acid is about 10^4 times as large as its K_{a_2}, whereas the K_{a_1} of oxalic acid is about 10^3 times as large as its K_{a_2}. Suggest a reason why the separation in K_a values is greater for malonic acid.

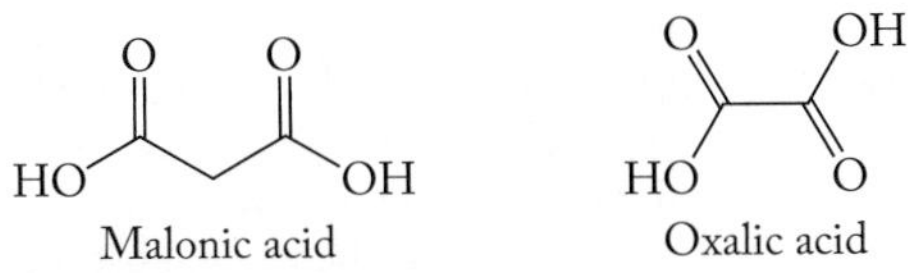

Malonic acid   Oxalic acid

FIGURE P15.68

Problems

15.69. What is the pH of 0.75 M H_2SO_4?

15.70. What is the pH of 5.00×10^{-4} M H_2SO_4?

15.71. Ascorbic acid (vitamin C) is a weak diprotic acid. What is the pH of 0.250 M ascorbic acid?

15.72. **Rhubarb Pie** The leaves of the rhubarb plant contain high concentrations of diprotic oxalic acid (HOOCCOOH) and must be removed before the stems are used to make rhubarb pie. What is the pH of 0.0288 M oxalic acid?

15.73. Nicotine Addiction Nicotine is responsible for the addictive properties of tobacco. What is the pH of 1.00×10^{-3} M nicotine?

15.74. Pseudoephedrine hydrochloride (Figure P15.74) is a common ingredient in cough syrups and decongestants. Its $pK_a = 9.22$. What is the pH of 0.0295 M pseudoephedrine hydrochloride?

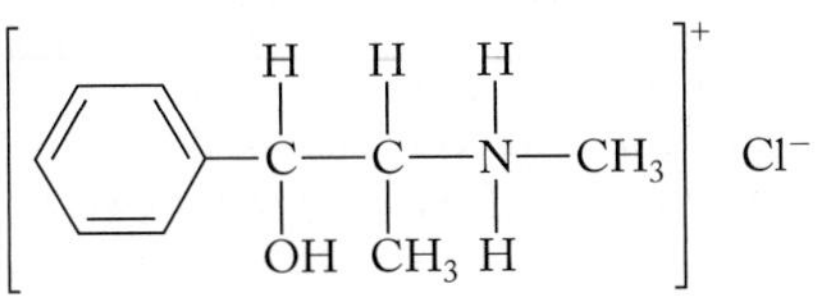

FIGURE P15.74

15.75. Malaria Treatment Quinine occurs naturally in the bark of the cinchona tree. For centuries it was the only treatment for malaria. Calculate the pH of 0.01050 M quinine.

15.76. Dozens of pharmaceuticals ranging from cyclizine for motion sickness to Viagra for impotence are derived from the organic compound piperazine, whose structure is shown in Figure P15.76.

a. Solutions of piperazine are basic ($K_{b_1} = 5.38 \times 10^{-5}$; $K_{b_2} = 2.15 \times 10^{-9}$). What is the pH of 0.0125 M piperazine?

*b. Draw the structure of the ionic form of piperazine that would be present in stomach acid (about 0.16 M HCl).

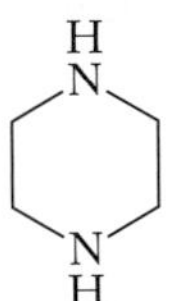

FIGURE P15.76

Acidic and Basic Salts

Concept Review

15.77. How is it that aqueous solutions of NaF are basic, but solutions of NaCl are neutral?

15.78. Why is it unnecessary to publish tables of K_b values of the conjugate bases of weak acids whose K_a values are known?

15.79. Which of the following salts produces an acidic solution in water: ammonium acetate, ammonium nitrate, or sodium formate?

15.80. Which of the following salts produces a basic solution in water: $NaNO_2$, KNO_3, or NH_4Cl?

15.81. Neutralizing the Smell of Fish Trimethylamine, $(CH_3)_3N$ ($K_b = 6.5 \times 10^{-5}$ at 25°C), contributes to the "fishy" odor of not-so-fresh seafood. Some people squeeze fresh lemon juice (which contains a high concentration of citric acid) on cooked fish to reduce the fishy odor. Why is this practice effective?

*15.82. **Nutritional Value of Beets** Beets contain high concentrations of the calcium salt of malonic acid (see Figure P15.68). Could the presence of the calcium salt of malonic acid affect the pH balance of beets? If so, in which direction? Explain.

Problems

15.83. The K_a of the conjugate acid of the artificial sweetener saccharin is 2.1×10^{-11}. What is the pK_b for saccharin?

15.84. The K_{a_1} value for oxalic acid (HOOCCOOH) is 5.9×10^{-2}, and the K_{a_2} value is 6.4×10^{-5}. What are the values of K_{b_1} and K_{b_2} of the oxalate anion ($^-OOCCOO^-$)?

15.85. Dental Health Sodium fluoride is added to many municipal water supplies to reduce tooth decay. Calculate the pH of 0.00339 M NaF at 25°C.

15.86. Calculate the pH of 1.25×10^{-2} M of the decongestant ephedrine hydrochloride if the pK_b of ephedrine (its conjugate base) is 3.86.

Additional Problems

15.87. Consider the following compounds: CH_3NH_2, CH_3COOH, $Ca(OH)_2$, and $HClO_4$.

a. Identify the Arrhenius acid(s).
b. Identify the Arrhenius base(s).
c. Identify the Brønsted–Lowry acid(s).
d. Identify the Brønsted–Lowry base(s).

15.88. Are all Arrhenius acids also Brønsted–Lowry acids? Are all Brønsted–Lowry acids also Arrhenius acids? If yes, explain why. If not, give a specific example to demonstrate the difference.

15.89. Are all Arrhenius bases also Brønsted–Lowry bases? Are all Brønsted–Lowry bases also Arrhenius bases? If yes, explain why. If not, give a specific example to demonstrate the difference.

15.90. Describe the intermolecular forces and changes in bonding that lead to the formation of a basic solution when methylamine (CH_3NH_2) dissolves in water.

***15.91.** Describe the chemical reactions of sulfur that begin with the burning of high-sulfur fossil fuel and that end with the reaction between acid rain and building exteriors made of marble ($CaCO_3$).

15.92. The K_{a_1} of phosphorous acid, H_3PO_3, is nearly the same as the K_{a_1} of phosphoric acid, H_3PO_4.

a. Draw the Lewis structure of phosphorous acid.
b. Identify the ionizable hydrogen atoms in the structure.
c. Explain why the K_{a_1} values of phosphoric acid and phosphorous acid are similar.

***15.93.** **pH of Natural Waters** In a 1985 study of Little Rock Lake in Wisconsin, 400 gallons of 18 *M* sulfuric acid were added to the lake over six years. The initial pH of the lake was 6.1 and the final pH was 4.7. If none of the acid was consumed in chemical reactions, estimate the volume of the lake.

15.94. **Acid–Base Properties of Pharmaceuticals I** Zoloft is a prescription drug for the treatment of depression. It is sold as its hydrochloride salt, which is produced as shown in Figure P15.94. When the hydrochloride dissolves in water, will the resulting solution be acidic or basic?

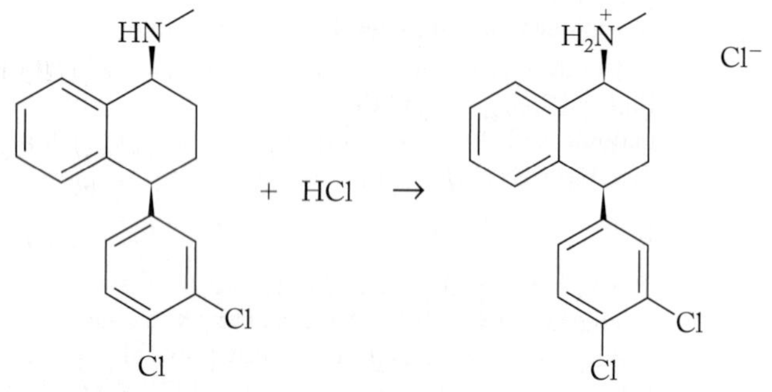

HN — Cl — Cl + HCl → $H_2\overset{+}{N}$ — Cl — Cl Cl^-

Zoloft

FIGURE P15.94

15.95. **Acid–Base Properties of Pharmaceuticals II** Prozac is a popular antidepressant drug. Its structure is given in Figure P15.95.

a. Is a solution of Prozac in water likely to be acidic or basic? Explain your answer.
b. Prozac is also sold as a hydrochloride salt. Which functional group is more likely to react with HCl?
c. Prozac is sold as its hydrochloride salt because the solubility of the salt in water is higher than unreacted Prozac. Why is the salt more soluble?

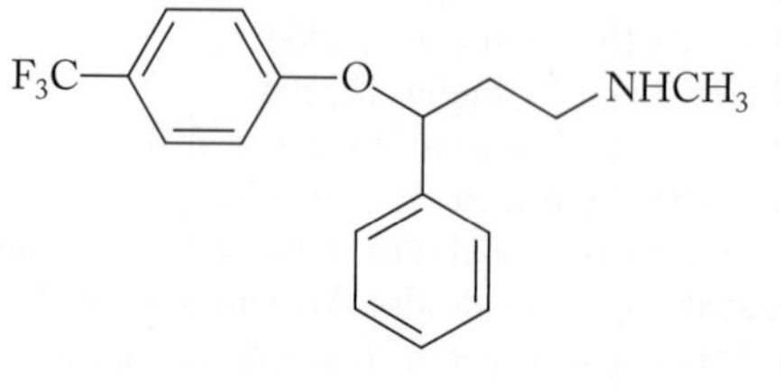

F_3C — O — $NHCH_3$

Prozac

FIGURE P15.95

15.96. Naproxen (sold as Aleve) is an anti-inflammatory drug used to reduce pain, fever, inflammation, and stiffness caused by conditions such as osteoarthritis and rheumatoid arthritis. Naproxen is an organic acid; its structure is shown in Figure P15.96. Naproxen has limited solubility in water, so it is sold as its sodium salt.

a. Draw the molecular structure of the sodium salt.
b. Is an aqueous solution of the salt acidic or basic? Explain why.
c. Explain why the salt is more soluble in water than naproxen itself.

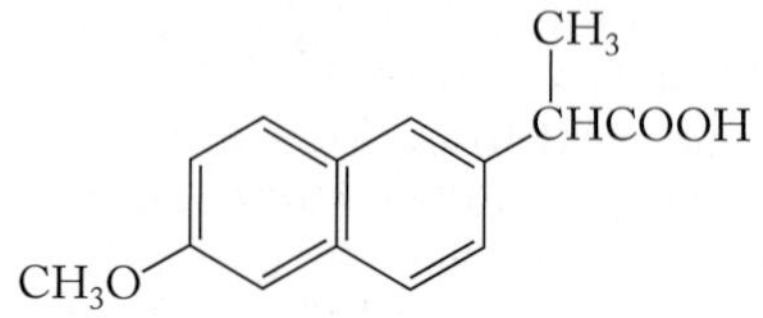

CH_3 | CHCOOH

CH_3O

Naproxen

FIGURE P15.96

***15.97.** Pentafluorocyclopentadiene, which has the structure shown in Figure P15.97, is a strong acid.

a. Draw the conjugate base of C_5F_5H.
b. Why is the compound so acidic when most organic acids are weak?

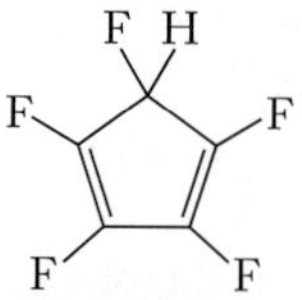

F H F F F F

FIGURE P15.97

15.98. **Ocean Acidification** Some climate models predict a decrease in the pH of the oceans of 0.3 to 0.5 pH unit by 2100 because of increases in atmospheric carbon dioxide.

a. Explain, by using the appropriate chemical reactions and equilibria, how an increase in atmospheric CO_2 could produce a decrease in oceanic pH.
b. How much more acidic would the oceans be if their pH dropped this much?
c. Oceanographers are concerned about the impact of a drop in oceanic pH on the survival of coral reefs. Why?

***15.99.** Sulfuric acid reacts with nitric acid as shown below:

$$HNO_3(aq) + 2\,H_2SO_4(aq) \rightarrow NO_2^+(aq) + H_3O^+(aq) + 2\,HSO_4^-(aq)$$

a. Is the reaction a redox process?
b. Identify the acid, base, conjugate acid, and conjugate base in the reaction. (*Hint*: Draw the Lewis structures for each.)

15.100. Thiosulfuric acid, $H_2S_2O_3$, can be prepared by the reaction of H_2S with HSO_3Cl:

$$HSO_3Cl(\ell) + H_2S(g) \rightarrow HCl(g) + H_2S_2O_3(\ell)$$

a. Draw a Lewis structure for $H_2S_2O_3$, given that it is isostructural with H_2SO_4.
b. Do you expect $H_2S_2O_3$ to be a stronger or weaker acid than H_2SO_4? Explain your answer.

15.101. Which of these solutions is the most acidic? Which is the most basic?
 i. 1.0 *M* H_2SO_3
 ii. 0.10 *M* H_2SO_4
 iii. 0.30 *M* $NaHSO_4$
 iv. 0.30 *M* Na_2SO_4
 v. 0.30 *M* Na_2SO_3

15.102. Predict which solution in each pair below will have the lower pH.
 a. 2.56×10^{-2} *M* HCl or 4.09×10^{-2} *M* HBr
 b. 1.00×10^{-5} *M* acetic acid ($K_a = 1.76 \times 10^{-5}$) or 1.00×10^{-5} *M* formic acid ($K_a = 1.77 \times 10^{-4}$)
 c. 22 m*M* CH_3NH_2 (pK_b = 3.36) or 22 m*M* $(CH_3)_2NH$ ($K_b = 5.9 \times 10^{-4}$)
 d. 158 m*M* NH_3 (pK_b = 4.75) or 158 m*M* acetic acid (pK_a = 4.75)
 e. 0.00395 *M* HNO_3 or 0.00145 *M* $HClO_4$
 f. 2.05×10^{-1} *M* propionic acid ($K_a = 1.4 \times 10^{-5}$) or 2.05×10^{-1} *M* fluoroacetic acid ($K_a = 2.6 \times 10^{-3}$)
 g. 375 m*M* pyridine (pK_b = 8.77) or 375 m*M* aniline (pK_b = 9.40)
 h. 0.555 *M* $Fe(H_2O)_6^{3+}$ ($K_a = 3 \times 10^{-3}$) or 0.355 *M* $Cr(H_2O)_6^{3+}$ ($K_a = 1 \times 10^{-4}$)

15.103. Predict which solution in each pair below will have the higher pH.
 a. 1.25×10^{-5} *M* HNO_3 or 1.00×10^{-3} *M* NaOH
 b. 0.345 m*M* HBrO (pK_a = 8.64) or 0.345 m*M* HClO ($K_a = 2.9 \times 10^{-8}$)
 c. 45 m*M* $Be(OH)_2$ ($K_b = 5 \times 10^{-11}$) or 45 m*M* $Ba(OH)_2$
 d. 1.6 m*M* [2-methylpyridine structure, N] (pK_b = 8.03) or
 0.60 m*M* [3-methylpyridine structure, N] (pK_b = 8.32)
 e. 1.67×10^{-3} *M* NaOH or 252 m*M* KOH
 f. 105 m*M* NH_3 (pK_b = 4.75) or 105 m*M* CH_3NH_2 (pK_b = 3.36)
 g. 1.50×10^{-5} *M* benzoic acid ($K_a = 6.25 \times 10^{-5}$) or 1.50×10^{-5} *M* pyridine ($K_b = 1.7 \times 10^{-9}$)
 h. 20 m*M* [structure: Cl, COOH] ($K_a = 1.20 \times 10^{-3}$), or
 5 m*M* [structure: Cl, COOH] (pK_a = 3.98)

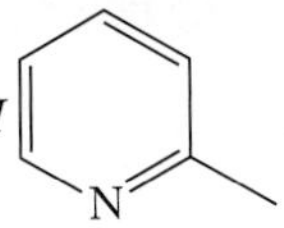

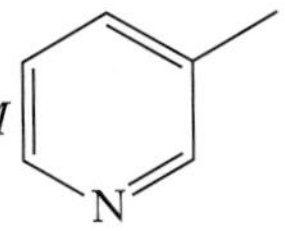

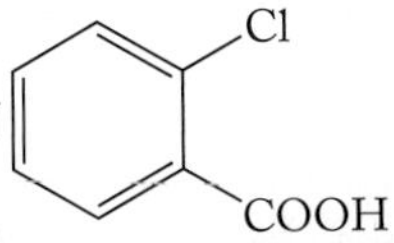

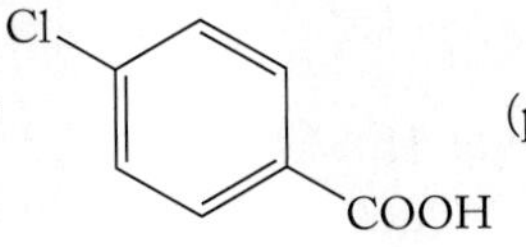

15.104. The value of K_w increases as temperature increases.
 a. If the pK_w = 13.017 at 60°C, what is the [H^+]?
 b. What is the pH of water at 60°C?

15.105. The pK_a of propanoic acid is 4.85. What is the pH of a 0.125 *M* aqueous solution of sodium propanoate?

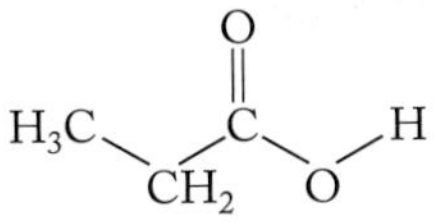

[structure: H_3C, CH_2, C, O, O, H]

15.106. Calculate the indicated value based on the information given:
 a. What is the K_b of the lactate ion? The K_a of lactic acid is 1.4×10^{-4}.
 b. What is the K_b of the conjugate base of pyruvic acid? The K_a of pyruvic acid is 2.8×10^{-3}.
 c. What is the K_a of the conjugate acid of aniline? Aniline has a K_b of 5.9×10^{-4}.
 d. Quinine has two basic nitrogen atoms in its structure. The K_b of the most basic nitrogen atom is 3.3×10^{-6}. What is the K_a of the HCl salt of that nitrogen atom?

15.107. Identify the conjugate base for the following weak acids and calculate K_b for each. Which conjugate base is the strongest?
 a. $ClCH_2COOH$
 b. NH_4^+
 c. HCN
 d. CH_3CH_2OH

15.108. Identify the conjugate acid for the following weak bases and calculate K_a for each. Which conjugate acid is the strongest?
 a. CH_3NH_2
 b. HPO_4^{2-}
 c. $H_2NC(O)NH_2$
 d. NO_2^-

15.109. For each of the molecular equations, write net ionic equations and identify the Brønsted-Lowry acids and bases:
 a. $2\,HNO_3(aq) + Ca(OH)_2(aq) \rightarrow 2\,H_2O(\ell) + Ca(NO_3)_2(aq)$
 b. $Na_2CO_3(aq) + H_2SO_4\,(aq) \rightarrow Na_2SO_4(aq) + CO_2(g) + H_2O(\ell)$
 c. $CH_3NH_2(aq) + HBr(aq) \rightarrow (CH_3NH_3)Br(aq)$
 d. $2\,CH_3COOH(aq) + Mg(OH)_2\,(s) \rightarrow (CH_3COO)_2Mg(aq) + 2\,H_2O(\ell)$
 e. $CaO(s) + H_2O(\ell) \rightarrow Ca(OH)_2(s)$
 f. $LiH(s) + H_2O(\ell) \rightarrow LiOH(aq) + H_2(g)$
 g. $Ba(OH)_2\,(aq) + H_2SO_4(aq) \rightarrow BaSO_4(s) + 2\,H_2O(\ell)$
 h. $NaSH(aq) + HNO_3(aq) \rightarrow NaNO_3(aq) + H_2S(g)$

15.110. Write the chemical and the net ionic equations describing the reactions that occur when aqueous solutions of these pairs of compounds are mixed together. For each reaction label the Brønsted-Lowry acids and bases.
 a. HCl and $Ca(OH)_2$
 b. H_3PO_4 and KOH
 c. HNO_3 and Na_2CO_3
 d. H_3PO_4 and $Ca(OH)_2$

smartwork5
If your instructor uses Smartwork5, log in at digital.wwnorton.com/atoms2.

16

Additional Aqueous Equilibria

Chemistry and the Oceans

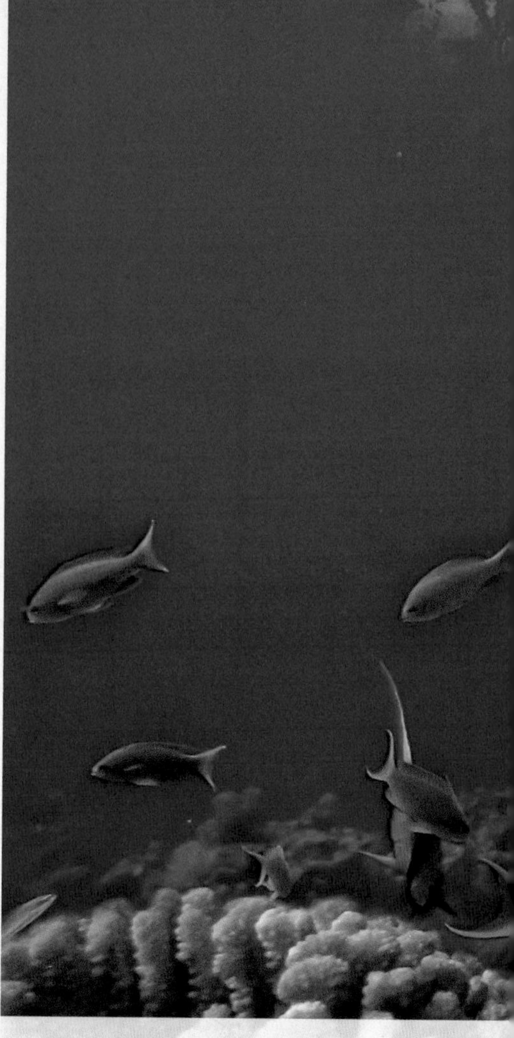

OCEAN ACIDIFICATION AFFECTS CORAL REEFS Increasing concentrations of CO_2 in the atmosphere are making seawater more acidic, which threatens corals and other marine life that form exoskeletons made of calcium carbonate.

PARTICULATE **REVIEW**

Ionic Compounds and Solubility

In Chapter 16, we revisit the solubility of ionic compounds. Ionic lattices for calcium carbonate, lithium oxide, and potassium chloride are shown here.

- Write the empirical formulas for these compounds.
- Match each of the ionic lattices to the name of the compound.
- If you added one mole of each compound to 1 L of water, which ion(s) would be present in the highest concentration?

(a)

(b)

(c)

(Review Section 8.5 if you need help.)

(Answers to Particulate Review questions are in the back of the book.)

PARTICULATE **PREVIEW**

Donating Protons versus Donating Electron Pairs

As you read Chapter 16, look for ideas that will help you compare the Brønsted–Lowry model of proton transfer to the Lewis model of acids and bases, which involves donating and accepting electron pairs. Answer these questions about the acids and bases shown here:

- Which of these particles can donate a proton? Which can accept a proton?
- Which of these particles has one or more lone pairs of electrons on the central atom?
- Which of these particles can donate a lone pair of electrons to another particle?

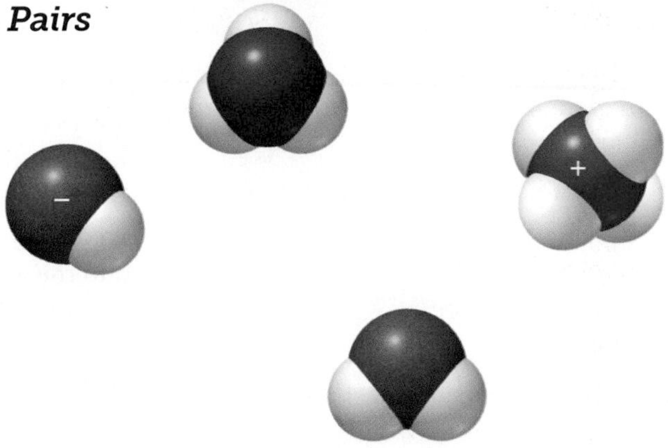

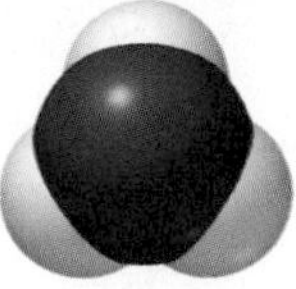

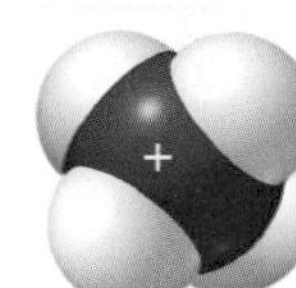

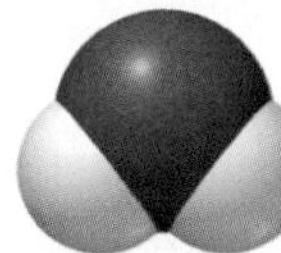

Learning Outcomes

LO1 Calculate the pH of a solution containing a weak acid or base and its conjugate base or acid.
Sample Exercises 16.1, 16.2

LO2 Prepare a buffer with a desired pH
Sample Exercises 16.3, 16.4, 16.5

LO3 Evaluate the capacity of a buffer to resist changes in its pH
Sample Exercises 16.6, 16.7, 16.8

LO4 Calculate and interpret the results of an acid–base titration
Sample Exercises 16.9, 16.10, 16.11, 16.12

LO5 Identify a compound as a Lewis acid or Lewis base in a reaction
Sample Exercise 16.13

LO6 Use formation constants to calculate the concentrations of free and complexed metal ions in solution
Sample Exercise 16.14

LO7 Relate the acid strength of hydrated metal ions to the charge on the ions

LO8 Relate the solubility of an ionic compound to its solubility product
Sample Exercises 16.15, 16.16, 16.17, 16.18

LO9 Separate mixtures of ionic compounds by selective precipitation reactions
Sample Exercise 16.19

16.1 Ocean Acidification: Equilibrium under Stress

In Chapter 7 we reviewed data that showed a dramatic increase in the concentration of carbon dioxide in the atmosphere since the Industrial Revolution began around 1750 (see Figures 7.2 and 7.3). The impact of an increasing concentration of atmospheric CO_2 on global temperature and climate change is a widely discussed topic these days, but another impact also worthy of our attention is ocean acidification. As we saw in Chapter 15, CO_2 gas dissolves in water to form carbonic acid:

$$CO_2(g) + H_2O(\ell) \rightleftharpoons H_2CO_3(aq) \qquad (16.1)$$

Although H_2CO_3 is present in only low concentrations in aqueous solutions, we use the symbol $[H_2CO_3]$ to represent the total amount of $CO_2(g)$ dissolved in water and to remind us that when $CO_2(g)$ dissolves in water, it forms an acidic solution:

$$H_2CO_3(aq) + H_2O(\ell) \rightleftharpoons HCO_3^-(aq) + H_3O^+(aq) \quad K_{a_1} = 4.3 \times 10^{-7} \qquad (16.2)$$

Bicarbonate ions may also donate H^+ ions in aqueous solutions:

$$HCO_3^-(aq) + H_2O(\ell) \rightleftharpoons CO_3^{2-}(aq) + H_3O^+(aq) \quad K_{a_2} = 4.7 \times 10^{-11} \qquad (16.3)$$

However, the much smaller value of K_{a_2} means that the concentration of CO_3^{2-} ions in slightly basic seawater is much smaller than the concentration of HCO_3^- ions.

The solubility of atmospheric CO_2 increases as its partial pressure increases, so higher concentrations of CO_2 in the atmosphere mean higher concentrations of dissolved CO_2 in the sea. Scientists estimate that about 30% of the "fossil" CO_2 added to the atmosphere by burning coal, natural gas, and petroleum-based fuels dissolves in the sea. Increasing concentrations of dissolved CO_2 have shifted the equilibrium in Equation 16.1 to the right—which, in turn, has shifted the equilibria in both Equations 16.2 and 16.3 to the right as well, resulting in an increase in the acidity of the sea. As a result, the average pH of seawater has dropped by about 0.11 pH unit since the 18th century.

CONNECTION Henry's law, discussed in Chapter 11, states that the solubility of a gas is proportional to its partial pressure.

This is an enormous problem—especially if oceanic pH continues to drop—because many marine organisms such as plankton, shellfish, and corals have exoskeletons made of $CaCO_3$. To make $CaCO_3$, these organisms need a supply of dissolved Ca^{2+} and CO_3^{2-} ions so that the following precipitation reaction can take place inside them:

$$Ca^{2+}(aq) + CO_3^{2-}(aq) \rightleftharpoons CaCO_3(s)$$

Unfortunately, increasing concentrations of H_3O^+ ions in the sea drive down the already low concentrations of CO_3^{2-} ions by shifting the equilibrium in Equation 16.3 to the left. There is a growing concern that the oceanic $[CO_3^{2-}]$ may become too low for marine organisms to form their $CaCO_3$ skeletons, and that they—and the aquatic systems that rely on them, such as tropical coral reefs—will not survive.

CONNECTION Le Châtelier's principle was introduced in Section 14.7 to describe how a system at equilibrium responds to stress.

In this chapter, we explore the equilibria that control the pH of the sea and of other environmental and biological systems. We also examine how changes in pH can impact other equilibria, including those that are key to the survival of marine and other species.

16.2 The Common-Ion Effect

The chemical equilibria that control acid–base balance in natural systems—from seawater to the fluids in all living cells—depend on the presence of both acidic and basic compounds. We encountered one example in Sample Exercise 15.20, where we calculated the pH of a sample of blood plasma (it was 7.38) from the concentrations of CO_2 gas (2.7×10^{-3} *M*) and HCO_3^- ions (0.028 *M*) dissolved in it. Now let's calculate the pH of a solution that initially contains 2.7×10^{-3} *M* dissolved CO_2 but no bicarbonate ions. We set up the appropriate RICE table with molar concentrations of the reactants and products. As in Chapter 15, we use $[H_2CO_3]$ to represent the total amount of dissolved $CO_2(g)$ in solution:

Reaction	$H_2CO_3(aq) + H_2O(\ell)$	$\rightleftharpoons$	$HCO_3^-(aq)$	+	$H_3O^+(aq)$
	$[H_2CO_3]$ (*M*)		$[HCO_3^-]$ (*M*)		$[H_3O^+]$ (*M*)
Initial	2.7×10^{-3}		0		0
Change	$-x$		$+x$		$+x$
Equilibrium	$(2.7 \times 10^{-3}) - x$		x		x

Taking the K_{a_1} value for carbonic acid (4.3×10^{-7}) from Equation 16.2 and solving for x:

$$K_{a_1} = \frac{[HCO_3^-][H_3O^+]}{[CO_2]} = \frac{(x)(x)}{(2.7 \times 10^{-3}) - x} \approx \frac{x^2}{2.7 \times 10^{-3}} = 4.3 \times 10^{-7}$$

$$x = [H_3O^+] = 3.4 \times 10^{-5}\ M$$

Calculating pH: $\mathrm{pH} = -\log[H_3O^+] = -\log(3.4 \times 10^{-5}) = 4.47$

Note how the pH value of this new solution (4.47) is nearly three units lower than the plasma sample (7.38), corresponding to a $[H_3O^+]$ value nearly 10^3 larger. This difference exists because the bicarbonate ions in blood plasma inhibit the forward reaction in Equation 16.2, as predicted by Le Châtelier's principle.

common-ion effect the shift in the position of an equilibrium caused by the addition of an ion taking part in the reaction.

Henderson–Hasselbalch equation an equation used to calculate the pH of a solution in which the concentrations of acid and conjugate base are known.

CONNECTION In Chapter 15, we introduced pH as the negative logarithm of hydronium ion concentration, pH = $-\log[H_3O^+]$, and pK_a as the negative logarithm of K_a.

The increase in pH caused by the presence of bicarbonate ions illustrates a principle known as the **common-ion effect**, which states that if the concentration of one of the ions produced in the ionization of a weak electrolyte, such as carbonic acid, is increased, the weak electrolyte ionizes less. Less ionization means that the initial concentrations of H_2CO_3 and HCO_3^- (or any conjugate acid–base pair) change very little before equilibrium is achieved. Therefore, we can use their initial concentrations to calculate the pH of their solution, as illustrated for the generic acid ionization equilibrium involving a solution of acid HA and its conjugate base, A^-:

$$K_a = \frac{[H_3O^+][A^-]_{\text{initial}}}{[HA]_{\text{initial}}}$$

If we take the negative logarithm of both sides of the K_a expression, we transform $[H_3O^+]$ into pH and K_a into pK_a:

$$pK_a = pH - \log\frac{[A^-]_{\text{initial}}}{[HA]_{\text{initial}}}$$

If we make the initial concentration symbols of the acid and its conjugate base simply [acid] and [base], the generic equation becomes

$$pK_a = pH - \log\frac{[\text{base}]}{[\text{acid}]}$$

or

$$pH = pK_a + \log\frac{[\text{base}]}{[\text{acid}]} \tag{16.4}$$

Equation 16.4 is called the **Henderson–Hasselbalch equation**. It can be used to calculate the pH of a solution in which there are separate sources of both a weak acid and its conjugate base (or both a weak base and its conjugate acid). It provides a mathematical shortcut for applying Le Châtelier's principle to calculating the pH of these mixtures.

What happens to the logarithm term in the Henderson–Hasselbalch equation when the concentrations of acid and base are equal to each other? The value of the fraction is 1, and the log of 1 is 0, so pH = pK_a. This equality serves as a handy reference point in working with solutions of conjugate acid–base pairs. If the concentration of the basic component is greater than that of the acid, the logarithmic term is greater than zero and pH > pK_a. If the concentration of the basic component is less than that of the acid, the logarithmic term is less than zero, and pH < pK_a.

Suppose, for example, the concentration of the base is 10 times the concentration of the acid—that is, [base] = 10[acid]. Substituting this equality into Equation 16.4, we have

$$pH = pK_a + \log\frac{10[\text{acid}]}{[\text{acid}]}$$

$$= pK_a + \log 10 = pK_a + 1$$

Thus, a 10-fold higher concentration of base produces a pH that is one unit above the pK_a value. Similarly, if the concentration of the acid component is 10 times that of the base, then pH = pK_a − 1. Sample Exercises 16.1 and 16.2 further illustrate how the Henderson–Hasselbalch equation simplifies pH calculations when we know the concentrations of both components of a conjugate pair.

SAMPLE EXERCISE 16.1 Calculating the pH of a Solution of a Weak Acid and Its Conjugate Base **LO1**

What is the pH of a sample of river water in which $[HCO_3^-] = 1.0 \times 10^{-4}$ *M* and the concentration of dissolved CO_2 in equilibrium with atmospheric CO_2 is 1.4×10^{-5} *M*?

Collect and Organize We are given the concentration of $CO_2(aq)$, which we will represent as $[H_2CO_3]$, as well as the concentration of HCO_3^-, which is the conjugate base of H_2CO_3. The pH of a solution of a weak acid and its conjugate base can be calculated using the Henderson–Hasselbalch equation:

$$\text{pH} = \text{p}K_a + \log \frac{[\text{base}]}{[\text{acid}]}$$

Analyze The conjugate acid–base pair relationship in this reaction is described by Equation 16.2:

$$H_2CO_3(aq) + H_2O(\ell) \rightleftharpoons HCO_3^-(aq) + H_3O^+(aq)$$

This reaction has a $\text{p}K_{a_1}$ value of 6.37 (see Table A5.1). The concentration of the base is nearly 10 times the concentration of the acid, so the log term in the Henderson–Hasselbalch equation should be almost 1, and the calculated pH should be a little higher than 7.

Solve Inserting the concentration and $\text{p}K_{a_1}$ values into the Henderson–Hasselbalch equation:

$$\text{pH} = \text{p}K_{a_1} + \log \frac{[\text{base}]}{[\text{acid}]} = 6.37 + \log \frac{1.0 \times 10^{-4}}{1.4 \times 10^{-5}}$$

$$= 6.37 + 0.85 = 7.22$$

Think About It This result is near the pH value we predicted. It contains two significant figures (the first 2 and the second 2) because there are two significant figures in all three of the values used to calculate it. The "7" in "7.22" is not significant because its purpose is to tell us that the corresponding $[H_3O^+]$ value is between 10^{-7} and 10^{-8} *M*.

Practice Exercise What is the pH of a solution in which $[HCOOH] = 2.5 \times 10^{-2}$ *M* and $[HCOO^-] = 7.8 \times 10^{-3}$ *M*?

(Answers to Practice Exercises are in the back of the book.)

We can also use the Henderson–Hasselbalch equation to calculate the pH of a solution of a weak base and its conjugate acid. An extra step is usually involved because we may know the $\text{p}K_b$ value of the base but not the $\text{p}K_a$ value of its conjugate acid, and only $\text{p}K_a$ values are used in the Henderson–Hasselbalch equation. However, we learned in Chapter 15 that the equilibrium constants for conjugate acid–base pairs are related by Equation 15.21 ($K_a \times K_b = K_w$). To recast Equation 15.21 into one with p*K* terms, we insert the numerical value of K_w at 25°C (1.0×10^{-14}):

$$K_a \times K_b = K_w = 1.0 \times 10^{-14}$$

and take the −log values of all three terms:

$$\text{p}K_a + \text{p}K_b = 14.00 \quad (16.5)$$

Thus, we can convert the $\text{p}K_b$ of a base into the $\text{p}K_a$ of its conjugate acid by subtracting the $\text{p}K_b$ value from 14.00.

SAMPLE EXERCISE 16.2 Calculating the pH of a Solution of a Weak Base and Its Conjugate Acid **LO1**

What is the pH of a solution that is 0.200 *M* in NH_3 and 0.300 *M* in NH_4Cl?

Collect, Organize, and Analyze We are asked to calculate the pH of a solution containing known concentrations of a weak base (NH_3) and a salt of its conjugate acid (NH_4^+). We can use the Henderson–Hasselbalch equation to calculate the pH of such a solution from the concentrations of the two components and the pK_a of the acid. Table A5.3 contains K_b and pK_b values of common bases. The pK_a and pK_b values of a conjugate acid–base pair are related by Equation 16.5:

$$pK_a + pK_b = 14.00$$

Our approach involves converting the pK_b value for NH_3 from Table A5.3 (4.75) into a pK_a value. Addition of ammonium ion to a solution of ammonia should produce a solution that is still basic, but not as basic as a solution that contains only ammonia.

Solve Inserting the pK_b value into Equation 16.5 and solving for pK_a:

$$pK_a = 14.00 - pK_b = 14.00 - 4.75 = 9.25$$

We can then use this value and the given concentrations of NH_3 and NH_4^+ in Equation 16.4:

$$\begin{aligned} pH &= pK_a + \log\frac{[\text{base}]}{[\text{acid}]} \\ &= 9.25 + \log\frac{0.200}{0.300} = 9.07 \end{aligned}$$

Think About It We predicted the solution would be basic, but not as basic as a solution of ammonia alone. To check whether or not this prediction is true, we can calculate the pH of 0.200 *M* NH_3 using the approach we followed in Sample Exercise 15.13. The result is a pH of 11.27—over two pH units higher (more basic) than the solution of ammonia and ammonium chloride in this exercise.

Practice Exercise Calculate the pH of a solution that is 0.25 *M* in methylamine, CH_3NH_2, and 0.75 *M* in methylamine hydrochloride, $[CH_3NH_3]Cl$.

16.3 pH Buffers

The common-ion effect plays a key role in controlling the pH of solutions that contain relatively high concentrations of both a weak acid and its conjugate base. These solutions have the capacity to withstand additions of acidic or basic substances with little or no measurable change in their pH. They are known as **pH buffers** and can control pH because the weak acid component of the buffer gives it the capacity to neutralize small additions of basic substances, while the conjugate base component gives it the capacity to neutralize small additions of acids. Ideally, a buffer has similar concentrations of both components of its conjugate pair so that it can neutralize small additions of other acids or bases equally well, as shown in Figure 16.1. When the ratio of the conjugate pair is close to one, the log term in the Henderson–Hasselbalch equation is close to zero, and the pH of the buffer is close to the pK_a of the weak acid in it. Actually, a buffer with different concentrations of its conjugate acid–base pair can still be effective at controlling pH over a range of pH values up to about one unit above or below the pK_a value of its acid.

pH buffer a solution that resists changes in pH when acids or bases are added to it; typically a solution of a weak acid and its conjugate base.

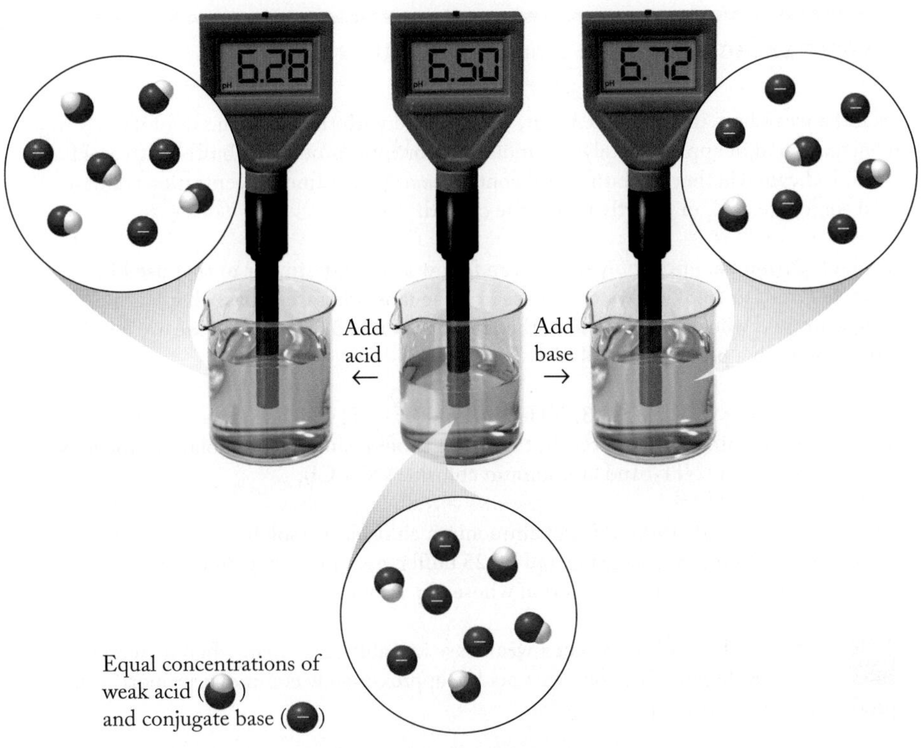

FIGURE 16.1 How buffers control pH. A pH 6.50 buffer (the middle beaker) contains an equal number of moles of a weak acid (HA) and its conjugate base (A^-). If enough strong acid contaminated the buffer to neutralize 25% of the buffer's conjugate base, converting it into weak acid, $H_3O^+(aq) + A^-(aq) \rightarrow HA(aq) + H_2O(\ell)$, the pH of the buffer would fall, but only by 0.22 pH unit. If a similar quantity of strong base contaminated the buffer, it would consume 25% of the buffer's weak acid: $OH^-(aq) + HA(aq) \rightarrow A^-(aq) + H_2O(\ell)$; pH would rise, but again only by 0.22 unit.

SAMPLE EXERCISE 16.3 Preparing an Acidic Buffer LO2

Select a weak acid in Table A5.1 of Appendix 5 that, when mixed with the sodium salt of its conjugate base in approximately equimolar proportions, produces a buffer with a pH of 2.80. Will the buffer contain *exactly* the same concentrations of acid and conjugate base, or slightly more acid or base?

Collect, Organize, and Analyze The weak acid we seek is one whose pK_a is close to 2.80, the target pH.

Solve Among the acids in Table A5.1 with pK_a values near 2.80 are bromoacetic acid ($pK_a = 2.70$) and chloroacetic acid ($pK_a = 2.85$). Either could be used to prepare a pH = 2.80 buffer, though neither would contain exactly the same concentration of the acid and its conjugate base: the bromoacetic acid buffer would require a slightly higher concentration of the conjugate base, whereas the chloroacetic acid buffer would require a little more of the acid.

Think About It Another criterion for selecting weak acids for aqueous buffers is that they be soluble in water. Both of the acids selected are carboxylic acids with relatively small molar masses, so they are quite soluble in water. However, organic acids with large hydrocarbon (hydrophobic) regions in their molecular structures, such as octanoic acid and benzoic acid, are only slightly soluble in water.

Octanoic acid   Benzoic acid

Practice Exercise Select a weak acid in Table A5.1 that, when mixed with the sodium salt of its conjugate base in approximately equimolar proportions, produces a buffer with a pH of 1.75.

SAMPLE EXERCISE 16.4 Preparing a Basic Buffer **LO2**

Select a weak base in Table A5.3 that, when mixed with the chloride salt of its conjugate acid in approximately equimolar proportions, produces a buffer with a pH of 9.25. Indicate whether the buffer will contain *exactly* the same concentrations of base and conjugate acid, or slightly more base or acid.

Collect, Organize, and Analyze We can follow a strategy similar to that used in Sample Exercise 16.3, but we need to search the table for a weak base that has a conjugate acid whose pK_a is close to the target pH of 9.25. This means we are looking for a base with a pK_b value of $14.00 - 9.25 = 4.75$.

Solve According to Table A5.3, NH_3 has a pK_b of 4.75, which exactly matches the value we are looking for. As a result, the buffer should contain equimolar proportions of aqueous ammonia (NH_3) and ammonium chloride (NH_4Cl).

Think About It Both ammonia and ammonium chloride are soluble in water, so they are good candidates for preparing a pH 9.25 buffer, assuming they do not chemically react with other solutes in the solution whose pH we wish to control.

Practice Exercise Select a weak base in Table A5.3 that, when mixed with the chloride salt of its conjugate acid in approximately equimolar proportions, produces a buffer with a pH of 10.75.

How do we determine the quantities of the components that are needed to prepare particular volumes of a buffer with a desired pH? Sample Exercise 16.5 illustrates this type of calculation.

SAMPLE EXERCISE 16.5 Preparing a Buffer Solution with a Desired pH **LO2**

A buffer system containing dihydrogen phosphate ($H_2PO_4^-$) and hydrogen phosphate (HPO_4^{2-}) helps regulate the pH of cytoplasm in living cells.

a. What is the mole ratio of HPO_4^{2-} ions to $H_2PO_4^-$ ions in a buffer with a pH of 6.75?
b. If the combined concentration of $H_2PO_4^-$ and HPO_4^{2-} ions in the buffer is to be 0.200 *M*, how many grams of NaH_2PO_4 ($\mathcal{M}$ = 120.0 g/mol) and how many grams of Na_2HPO_4 ($\mathcal{M}$ = 142.0 g/mol) are needed to prepare 20.0 liters of the buffer?

Collect and Organize We need to determine the mole ratio of HPO_4^{2-} ions to $H_2PO_4^-$ ions in a pH 6.75 buffer and to calculate the masses of the sodium salts of these ions that are needed to make 20.0 liters of 0.200 *M* buffer. According to Table A5.1, the pK_a of $H_2PO_4^-$ (actually the pK_{a_2} of H_3PO_4) is 7.19. The Henderson–Hasselbalch equation relates the pH of a buffer to the pK_a of its acid component and the concentrations of that acid and its conjugate base.

Analyze The pH of this buffer is controlled by the acid ionization equilibrium of dihydrogen phosphate ions:

$$H_2PO_4^-(aq) + H_2O(\ell) \rightleftharpoons HPO_4^{2-}(aq) + H_3O^+(aq)$$

The target pH (6.75) is less than the pK_a (7.19), so the buffer must contain a higher concentration of $H_2PO_4^-$ ions than HPO_4^{2-} ions. The sum of $[H_2PO_4^-]$ and $[HPO_4^{2-}]$ is 0.200 *M*, so if we let $x = [HPO_4^{2-}]$, then $[H_2PO_4^-] = (0.200 - x)$ *M*.

buffer capacity the quantity of acid or base that a pH buffer can neutralize while keeping its pH within a desired range.

Solve

a. Rearrange the Henderson–Hasselbalch equation to solve for the ratio of base to acid:

$$\log \frac{[\text{base}]}{[\text{acid}]} = \text{pH} - \text{p}K_a$$

Substituting the values for pH and pK_a gives

$$\log \frac{[HPO_4^{2-}]}{[H_2PO_4^-]} = \text{pH} - \text{p}K_a = 6.75 - 7.19 = -0.44$$

Then the ratio of $[HPO_4^{2-}]$ to $[H_2PO_4^-]$ is

$$\frac{[HPO_4^{2-}]}{[H_2PO_4^-]} = 10^{-0.44} = 0.36$$

b. Calculating the dissolved concentrations of HPO_4^{2-} (x) and $H_2PO_4^-$ ($0.200 - x$):

$$\frac{x}{0.200 - x} = 0.36$$

$$x = [HPO_4^{2-}] = 0.053\ M$$

$$(0.200 - x) = [H_2PO_4^-] = 0.147\ M$$

The masses of their sodium salts in 20.0 liters of buffer are

$$Na_2HPO_4\text{: } 20.0\ \cancel{\text{L}} \times 0.053\ \frac{\cancel{\text{mol}}}{\cancel{\text{L}}} \times 142.0\ \frac{\text{g}}{\cancel{\text{mol}}} = 150\ \text{g}$$

$$NaH_2PO_4\text{: } 20.0\ \cancel{\text{L}} \times 0.147\ \frac{\cancel{\text{mol}}}{\cancel{\text{L}}} \times 120.0\ \frac{\text{g}}{\cancel{\text{mol}}} = 353\ \text{g}$$

Think About It The buffer contains a higher concentration of $H_2PO_4^-$ ions than HPO_4^{2-} ions, which is consistent with our prediction and with the ratio of acid to conjugate base of any buffer that has a pH below the pK_a of the acid used to make it.

Practice Exercise How many kilograms of sodium ascorbate and ascorbic acid are needed to make 10.0 liters of pH 4.25 buffer if the total concentration of the two buffer components is 0.500 M?

Buffer Capacity

In addition to selecting the appropriate conjugate acid–base pair to prepare a buffer, chemists also need to decide how concentrated the buffer should be. The greater the concentrations of the conjugate pair components, the greater is its **buffer capacity**—that is, the greater is its ability to withstand additions of acid or base without a significant change in pH (Figure 16.2).

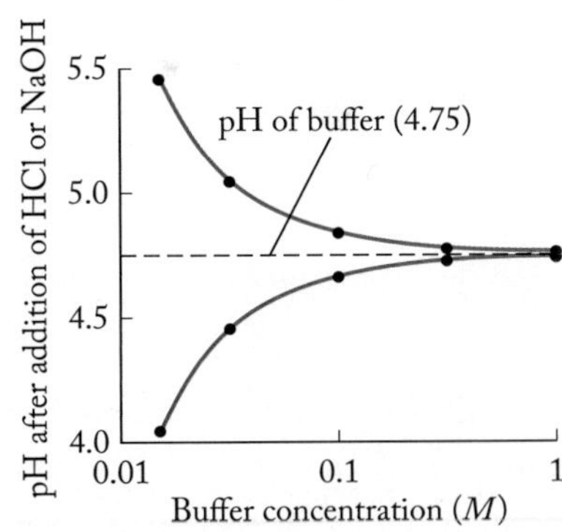

FIGURE 16.2 When strong acid (red line) or strong base (blue line) is added to a buffer solution, the extent to which the pH changes is inversely proportional to buffer concentration: the higher the concentrations of the buffer components, the smaller the change in pH. In this illustration, 100 mL samples of five solutions that are 0.015, 0.030, 0.100, 0.300, and 1.000 M acetic acid and sodium acetate all have an initial pH of 4.75 (dashed line). The graph shows the pH values of these solutions after 1.00 mL of 1.00 M HCl or 1.00 M NaOH has been added.

SAMPLE EXERCISE 16.6 Calculating Buffer Response to the Addition of Acid or Base **LO3**

a. What is the change in pH of a 1.00 L sample of the river water from Sample Exercise 16.1, in which the concentration of dissolved CO_2 (which we represent as $[H_2CO_3]$) = $1.4 \times 10^{-5}\ M$ and $[HCO_3^-] = 1.0 \times 10^{-4}\ M$, when 10.0 mL of $1.0 \times 10^{-3}\ M\ HNO_3$ is added to it?

b. Compare the pH change in part a to the pH change when the same quantity of strong acid is added to 1.00 L of pure water (pH = 7.00).

Collect and Organize We know the volume and composition of a river water sample and are asked to determine how much the pH of the sample changes as a result of adding a known volume and concentration of a nitric acid. We are also asked to compare that change in pH to the change that occurs when the same quantity of acid is added to pure water.

Analyze The river water contains a weak acid (dissolved CO_2) and its conjugate base (HCO_3^- ions), so the water has the capacity to function as a pH buffer. From Table A5.1, we know that the pK_{a_1} of carbonic acid is 6.37. Assuming nitric acid is the limiting reactant, adding x moles of this monoprotic, strong acid to the river water sample means adding x moles of H_3O^+ ions, which reacts with x moles of HCO_3^- ions, forming x moles of carbonic acid:

$$HCO_3^-(aq) + H_3O^+(aq) \rightarrow H_2CO_3(aq) + H_2O(\ell)$$

This sample decomposes to CO_2 gas:

$$H_2CO_3(aq) \rightarrow CO_2(g) + H_2O(\ell)$$

which is released into the atmosphere. Therefore, the concentration of the acid component of this particular buffer (H_2CO_3) is unchanged by the addition of acid.

If only a portion of the initial HCO_3^- concentration is consumed, the value of the [base] term in the Henderson–Hasselbalch equation will decrease somewhat from its original concentration, but because the function is logarithmic, the decrease in pH should be small. On the other hand, when strong acid is added to pure water, the change in pH depends only on dilution of the added acid. Using a rearranged form of Equation 8.6:

$$V_{\text{initial}} \times M_{\text{initial}} = V_{\text{final}} \times M_{\text{final}}$$

Solve

a. Calculating the number of moles of HCO_3^- initially present in the river water sample:

$$n_{HCO_3^-,\text{initial}} = 1.000\ \cancel{L} \times \left(1.0 \times 10^{-4}\ \frac{\text{mol } HCO_3^-}{\cancel{L}}\right) = 1.0 \times 10^{-4}\ \text{mol } HCO_3^-$$

Calculating x, the number of moles of H_3O^+ added to the river water sample in 10.0 mL of $1.0 \times 10^{-3}\ M\ HNO_3$:

$$x = 10.0\ \cancel{mL} \times \left(1.0 \times 10^{-3}\ \frac{\text{mol } H_3O^+}{\cancel{L}}\right) \times \left(10^{-3}\ \frac{\cancel{L}}{\cancel{mL}}\right) = 1.0 \times 10^{-5}\ \text{mol } H_3O^+$$

This quantity is equal to the number of moles of HCO_3^- in the river water sample consumed by the nitric acid. Therefore, the number of moles of bicarbonate present after the neutralization reaction is

$$n_{HCO_3^-,\text{final}} = [(1.0 \times 10^{-4}) - (1.0 \times 10^{-5})]\ \text{mol} = 9.0 \times 10^{-5}\ \text{mol } HCO_3^-$$

This final quantity of HCO_3^- is dissolved in a final volume of (1.00 L + 10.0 mL) = 1.01 L. Using these quantity and volume values, the constant [acid] value, and the pK_{a_1} value for carbonic acid from Table A5.1 in the Henderson–Hasselbalch equation to calculate pH:

$$\text{pH} = pK_a + \log\frac{[\text{base}]}{[\text{acid}]} = 6.37 + \log\frac{\left(\dfrac{9 \times 10^{-5}\ \cancel{mol}}{1.01\ \cancel{L}}\right)}{1.4 \times 10^{-5}\ \dfrac{\cancel{mol}}{\cancel{L}}} = 7.17$$

Thus, the addition of acid dropped the pH of the river water from 7.22 (see Sample Exercise 16.1) to 7.17 or by 0.05 pH unit.

b. Solving for M_{final}:

$$M_{\text{final}} = [H^+] = \frac{V_{\text{initial}} \times M_{\text{initial}}}{V_{\text{final}}}$$

$$= \frac{0.0100\ \text{L} \times (1.0 \times 10^{-3}\ M)}{1.01\ \text{L}} = 9.9 \times 10^{-6}\ M$$

$$\text{pH} = -\log[H_3O^+] = -\log(9.9 \times 10^{-6}\ M) = 5.00$$

Therefore, the change in pH is (5.00 − 7.00) = −2.00.

Think About It Addition of strong acid lowered the pH of the river water by only 0.04 pH unit because it consumed only 10% of the basic component of the buffer system controlling pH. This decrease in pH is a tiny fraction of the decrease (−2.00 pH units) that happens when the same quantity of acid is added to pure water.

Practice Exercise What is the pH of a buffer that contains 0.225 *M* acetic acid and 0.375 *M* sodium acetate? What is the pH of 100.0 mL of the buffer after 10.0 mL of 0.318 *M* NaOH is added to it?

SAMPLE EXERCISE 16.7 Effect of Concentration on Buffer Capacity LO3

Calculate the final pH after 0.0100 mole of H_3O^+ ions is added to (a) 0.100 L of buffer A, containing 1.00 *M* acetic acid and 1.00 *M* sodium acetate, and (b) 0.100 L of buffer B, containing 0.150 *M* acetic acid and 0.150 *M* sodium acetate.

Collect and Organize We are asked to calculate the final pH of equal volumes of two buffer solutions after the addition of the same quantity of acid (0.0100 mol H_3O^+). From Table A5.1, we know that the pK_a of CH_3COOH is 4.75. The pH of a solution with known concentrations of a conjugate acid–base pair can be calculated using the Henderson–Hasselbalch equation.

Analyze The pH of both buffers is controlled by the acid ionization equilibrium of acetic acid:

$$CH_3COOH(aq) + H_2O(\ell) \rightleftharpoons CH_3COO^-(aq) + H_3O^+(aq)$$

Addition of 0.0100 mol H_3O^+ to the two buffers will shift this equilibrium to the left, consuming 0.0100 mol CH_3COO^- and producing an additional 0.0100 mol CH_3COOH. To evaluate the impact of these changes, we need to calculate the initial numbers of moles of CH_3COOH and CH_3COO^- ions in both buffers. We can predict from Figure 16.2 that the more concentrated buffer will experience a smaller change in pH upon the addition of acid. We can modify the RICE tables used in Chapter 15 to track changes in the concentrations of the reactants and products.

Solve The initial quantities of CH_3COO^- and CH_3COOH in buffer A are both

$$0.100\ \cancel{L} \times \frac{1.00\ \text{mol}}{\cancel{L}} = 0.100\ \text{mol}$$

The initial quantities of CH_3COO^- and CH_3COOH in buffer B are both

$$0.100\ \cancel{L} \times \frac{0.150\ \text{mol}}{1\ \cancel{L}} = 0.0150\ \text{mol}$$

We use a modified RICE table to see how the amounts of CH_3COO^- and CH_3COOH in buffer A change as a result of the additional H_3O^+ ions. Note that the quantities in this RICE table are in moles, not molarity. After the addition of 0.0100 mol H_3O^+ ions to both buffers, the quantities in buffer A become:

Reaction	$CH_3COO^-(aq)$	+	$H_3O^+(aq)$	→	$CH_3COOH(aq) + H_2O(\ell)$
	CH_3COO^- (mol)		H_3O^+ (mol)		CH_3COOH (mol)
Initial	0.100		0.0100		0.100
Change	−0.0100		−0.0100		+0.0100
Final	0.0900				0.110

And the quantities in buffer B become:

Reaction	$CH_3COO^-(aq)$	+	$H_3O^+(aq)$	→	$CH_3COOH(aq) + H_2O(\ell)$
	CH_3COO^- (mol)		H_3O^+ (mol)		CH_3COOH (mol)
Initial	0.0150		0.0100		0.0150
Change	−0.0100		−0.0100		+0.0100
Final	0.0050				0.0250

The mole ratio of CH_3COO^- to CH_3COOH in each buffer can be substituted for the $[CH_3COO^-]/[CH_3COOH]$ ratio in the Henderson–Hasselbalch equation because both components are dissolved in the same volume of buffer. Using these substitutions and solving for pH:

$$\text{Buffer A: pH} = \text{p}K_a + \log\frac{[CH_3COO^-]}{[CH_3COOH]} = 4.75 + \log\frac{0.090}{0.110} = 4.66$$

$$\text{Buffer B: pH} = \text{p}K_a + \log\frac{[CH_3COO^-]}{[CH_3COOH]} = 4.75 + \log\frac{0.0050}{0.0250} = 4.05$$

Think About It Adding the same quantity of acid produced pH changes of 0.09 units in buffer A and 0.70 units in an equal volume of buffer B. As predicted, the buffer with a higher concentration—buffer A—is better able to resist a change in pH because the relative changes in the concentrations of its components are smaller.

Practice Exercise Calculate the change in pH when 0.24 mol OH^- is added to 1.00 L of two buffers: (a) a 1.16 *M* solution of sodium dihydrogen phosphate containing 1.16 *M* sodium hydrogen phosphate and (b) a 0.58 *M* solution of NaH_2PO_4 containing 0.58 *M* Na_2HPO_4.

Does the mole ratio of acid to conjugate base make a difference in how well a buffer controls pH? Suppose, for example, you wish to prepare an acidic buffer and you know that contamination by basic substances is more likely than contamination by acids. You might use a 1:1 mixture if you could find a weak acid with a pK_a that exactly matched the target pH—but what if there were another conjugate acid–base pair that you could use that required the acid component to be three times as concentrated as the base to achieve the desired pH? Would the second buffer, with the greater proportion of acid in it, do a better job of controlling pH against additions of base than the 1:1 buffer? Sample Exercise 16.8 helps answer this question.

SAMPLE EXERCISE 16.8 Effect of Base:Acid Ratio on Buffer Capacity LO3

Two pH 3.75 buffers are prepared. Buffer A contains equimolar concentrations of formic acid and sodium formate with a pK_a = 4.27 for the weak acid. Buffer B is prepared using the same overall concentration of conjugate pair components, but it contains a [base]/[acid] mole ratio of 1:3 to achieve the target pH of 3.75. What are the changes in pH of the two buffers if enough NaOH is added to equal volumes of both to neutralize 1/4, 1/2, and 3/4 of the formic acid in buffer A? Assume the NaOH additions do not significantly increase the volumes of the buffers.

Collect, Organize, and Analyze The composition of the buffer changes due to reaction with the added hydroxide ions:

$$HCOOH(aq) + OH^-(aq) \rightarrow HCOO^-(aq) + H_2O(\ell)$$

We are not given concentrations but rather mole ratios of the weak acid and its conjugate base in each buffer. If we let x be the initial concentrations of formic acid and sodium formate in buffer A, then the initial concentrations of the acid and base components in buffer B must be $1.5x$ and $0.5x$, respectively, to achieve the 1:3 ratio of [base]/[acid]. The three additions of NaOH will reduce the concentrations of the acid components values in both buffers by $-0.25x$, $-0.50x$, and $-0.75x$ and will increase the concentrations of the basic components by $+0.25x$, $+0.50x$, and $+0.75x$.

Solve Setting up Henderson–Hasselbalch equations with the initial [acid] and [base] values of the two buffers and the three changes in [acid] and [base] due to the three additions of NaOH:

$$\text{Buffer A, 25\% addition:}\quad \text{pH} = \text{p}K_\text{a} + \log\frac{[\text{base}]}{[\text{acid}]} = 3.75 + \log\frac{\cancel{x}(1.00 + 0.25)}{\cancel{x}(1.00 - 0.25)} = 3.97$$

$$\text{50\% addition:}\quad \text{pH} = \text{p}K_\text{a} + \log\frac{[\text{base}]}{[\text{acid}]} = 3.75 + \log\frac{\cancel{x}(1.00 + 0.50)}{\cancel{x}(1.00 - 0.50)} = 4.23$$

$$\text{75\% addition:}\quad \text{pH} = \text{p}K_\text{a} + \log\frac{[\text{base}]}{[\text{acid}]} = 3.75 + \log\frac{\cancel{x}(1.00 + 0.75)}{\cancel{x}(1.00 - 0.75)} = 4.60$$

$$\text{Buffer B, 25\% addition:}\quad \text{pH} = \text{p}K_\text{a} + \log\frac{[\text{base}]}{[\text{acid}]} = 4.27 + \log\frac{\cancel{x}(0.50 + 0.25)}{\cancel{x}(1.50 - 0.25)} = 4.05$$

$$\text{50\% addition:}\quad \text{pH} = \text{p}K_\text{a} + \log\frac{[\text{base}]}{[\text{acid}]} = 4.27 + \log\frac{\cancel{x}(0.50 + 0.50)}{\cancel{x}(1.50 - 0.50)} = 4.27$$

$$\text{75\% addition:}\quad \text{pH} = \text{p}K_\text{a} + \log\frac{[\text{base}]}{[\text{acid}]} = 4.27 + \log\frac{\cancel{x}(0.50 + 0.75)}{\cancel{x}(1.50 - 0.75)} = 4.49$$

The changes in pH compared to the target pH of 3.75 are depicted in Figure 16.3.

Think About It The greater capacity of buffer B to neutralize additions of base is evident when the $0.75x$ addition of NaOH consumes 75% of the acid in buffer A but only 50% of the acid in buffer B. While buffer B has more capacity to neutralize very large additions of base, Figure 16.3 shows that buffer A does a better job of controlling pH when small or moderate quantities of base are added.

FIGURE 16.3 Changes in pH produced by additions of NaOH to two buffers. Buffer A contains a 1:1 mole ratio of weak acid and its conjugate base. Buffer B contains the same total concentration of acid and conjugate base, but the mole ratio of base:acid is 1:3.

Practice Exercise Small volumes of strong acid, each containing 0.015 mole of $H_3O^+(aq)$ ions, are added to 1.00 liter samples of two basic buffers. Buffer A contains 0.150 mole of diethylamine and 0.150 mole of its conjugate acid. Buffer B contains 0.200 mole of methylamine and 0.100 mole of its conjugate acid. Predict which one will experience the greater change in pH before performing the calculation. Calculate the changes in pH in the two buffers as a result of adding the strong acid.

CONCEPT TEST

Which representation depicts the aqueous solution of formic acid and sodium formate that would experience the smallest changes of pH if small quantities of strong acid or strong base were added to the solution? The solvent has been omitted for clarity.

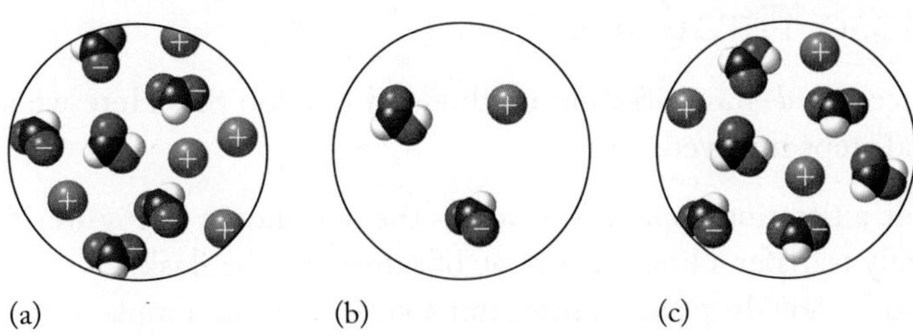

(Answers to Concept Tests are in the back of the book.)

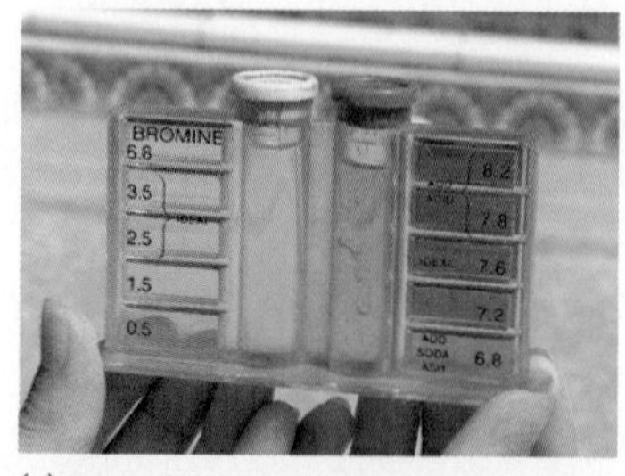
(a)

(b)

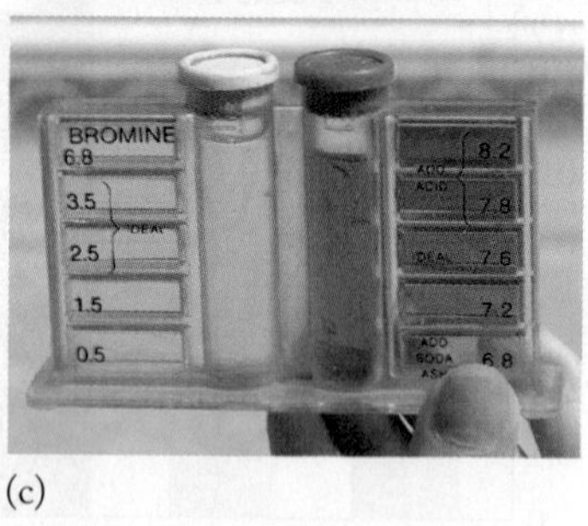
(c)

FIGURE 16.4 Many pool test kits include the pH indicator phenol red. A few drops are added to a sample of pool water collected in the tube with the red cap. (a) After a rainstorm, the pH of the pool water is 6.8 (or less), as indicated by the yellow color of the sample. (b) Sodium carbonate is added to the pool to raise the pH. (c) A follow-up test produces a red-orange color, indicating the pH of the pool has been properly adjusted.

16.4 Indicators and Acid–Base Titrations

Outdoor swimming pool operators routinely check the pH of pool water to make sure it is close to 7.3, which is the average pH of our eyes. They often add sodium carbonate to raise the water's pH after acidic summertime precipitation sends it below 7.3. To determine how much sodium carbonate to add, they test the pH of the water using a kit that includes a **pH indicator** (Figure 16.4), which is a substance that changes color as its pH changes.

One such substance is phenol red. It is a weak acid ($pK_a = 7.6$) that is yellow in its un-ionized form (which, for convenience, we assign the generic formula HIn) and violet in its ionized (In^-) form. At pH one unit above the pK_a—at pH 8.6—the ratio $[In^-]/[HIn]$ is 10:1 and a phenol red solution is violet. At a pH less than 6.6, where the ratio $[In^-]/[HIn]$ is 1:10, phenol red is largely un-ionized, and a solution of it is yellow. In the pH range from about 6.8 to 8.6, the color changes from yellow to orange to red to violet with increasing pH (Figure 16.5). These color changes allow pH to be determined to within about $\pm$ 0.1 unit.

As with buffers, a particular pH indicator is useful over a particular pH range that spans about two pH units. The midpoint of the range is defined by the pK_a of the indicator. In addition to their role in determining pH values, indicators are also used to detect the large changes in pH that occur at the equivalence points in acid–base titrations.

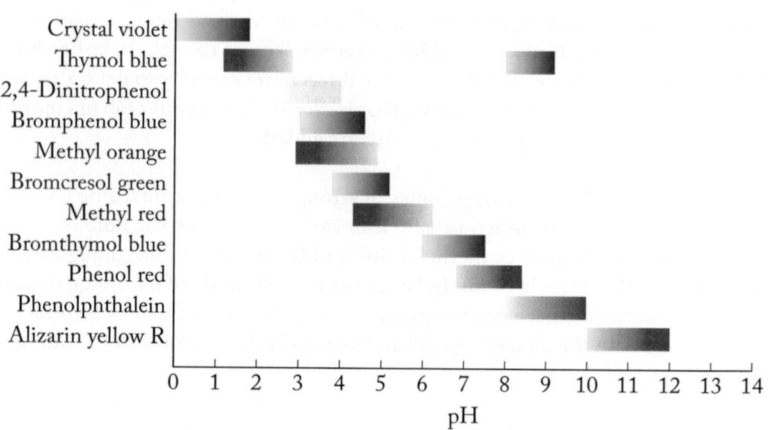

FIGURE 16.5 A pH indicator is useful within a range of one pH unit above and below the pK_a value of the indicator. This array of indicators could be used to determine pH values from 0 to 12.

CONNECTION In Section 8.7 we learned that when just enough titrant has been added to react with all of the analyte in a sample, the titration is at its equivalence point.

pH indicator a water-soluble weak organic acid that changes color as it ionizes.

Acid–Base Titrations

We introduced acid–base titration methods in Section 8.7. Here we summarize the principal steps involved:

1. Assemble a titration apparatus (such as the one shown in Figure 16.6).
2. Accurately transfer a known volume of sample to the flask.
3. Either add a few drops of an indicator solution to the sample or insert the probe of a pH meter.

4. Fill the buret with a solution (the *titrant*) of known concentration of a substance that reacts with a solute (the *analyte*) in the sample.
5. Slowly add titrant to the sample, and monitor the change in pH. When the volume of titrant needed to completely consume the analyte has been added, the equivalence point has been reached, as indicated by either a change in indicator color or a large change in pH as sensed by the pH electrode. This volume of titrant is a measure of the concentration of analyte in the sample.

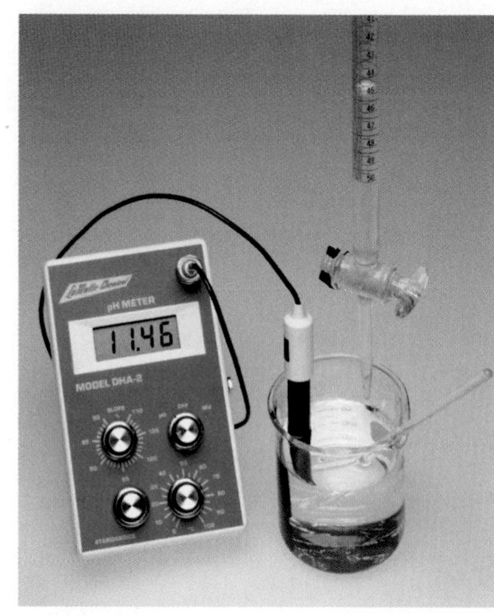

FIGURE 16.6 A digital pH meter is used to measure pH during a titration.

The neutralization titrations in Chapter 8 involved titrating strong acids with strong bases and vice versa. Here in Section 16.4, we begin with titrations of aqueous samples containing weak as well as strong monoprotic acids or *monobasic* bases. (A monobasic base accepts one hydrogen ion per molecule.) In the examples that follow, we monitor changes in pH during the titration using a pH electrode, and we plot the pH of the titration reaction mixture against the volume of titrant added.

First, let's compare the titration curves of two 20.0 mL samples: one contains 0.100 *M* HCl and the other contains 0.100 *M* acetic acid (CH_3COOH). Both are titrated with 0.100 *M* NaOH. The two graphs of pH versus titrant volume are shown in Figure 16.7. The initial pH of the HCl solution is lower than that of the acetic acid solution because HCl is a stronger acid, and the HCl titration curve stays below the acetic acid curve until they reach their equivalence points, where both acids have been completely neutralized.

As the base is added, the pH of the HCl sample (the red curve in Figure 16.7) does not change much until nearly all the acid has been consumed. Sample pH is determined only by the $[H_3O^+]$ remaining in it, which decreases in proportion to the volume of titrant added; however, the increase in pH is not a linear function of the volume of titrant added because of its log dependence on $[H_3O^+]$. As the equivalence point is approached, the principal ions present in the sample are Na^+, Cl^-, and H_3O^+.

When enough NaOH has been added to react with all the H_3O^+ ions in the sample, the equivalence point is reached. The solution now consists only of NaCl dissolved in water. Sodium and chloride ions do not hydrolyze and do not influence pH; therefore, the pH of the sample at the equivalence point is 7.00.

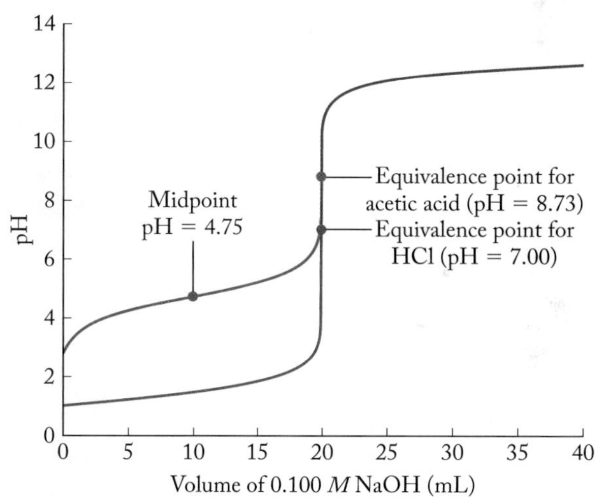

FIGURE 16.7 The titration curves of solutions of HCl and acetic acid with a standard solution of NaOH as the titrant.

SAMPLE EXERCISE 16.9 Calculating pH during Titration of a Strong Acid with a Strong Base — LO4

What is the pH of the solution being titrated in Figure 16.7 just before the equivalence point when 19.0 mL of 0.100 *M* NaOH has been added to 20.0 mL of 0.100 *M* HCl?

Collect, Organize, and Analyze We are asked to calculate the pH of a titration mixture: a sample containing HCl and a titrant containing NaOH, before the equivalence point, which means the mixture still contains some HCl and should have a pH less than 7. The titration reaction is

$$HCl(aq) + NaOH(aq) \rightarrow NaCl(aq) + H_2O(\ell)$$

The 1:1 mole ratio of HCl to NaOH means the concentration of H_3O^+ ions remaining in the mixture can be calculated by subtracting the number of moles of NaOH (OH^- ions)

CHEMTOUR Acid–Base Titrations

added from the number of moles of HCl (H_3O^+ ions) in the original sample and dividing by the sum of the volumes of original sample and the titrant added.

Solve The initial quantity of H_3O^+ ions is

$$20.0\ \cancel{\text{mL}} \times \frac{1\ \cancel{\text{L}}}{1000\ \cancel{\text{mL}}} \times \frac{0.100\ \text{mol HCl}}{\cancel{\text{L}}\ \text{solution}} = 2.00 \times 10^{-3}\ \text{mol HCl}$$

$$= 2.00 \times 10^{-3}\ \text{mol } H_3O^+$$

The quantity of OH^- ions added is

$$19.0\ \cancel{\text{mL}} \times \frac{1\ \cancel{\text{L}}}{1000\ \cancel{\text{mL}}} \times \frac{0.100\ \text{mol NaOH}}{\cancel{\text{L}}\ \text{solution}} = 1.90 \times 10^{-3}\ \text{mol NaOH}$$

$$= 1.90 \times 10^{-3}\ \text{mol } OH^-$$

The amount of H_3O^+ ions remaining is $(2.00 - 1.90) \times 10^{-3}$ mol $= 1.0 \times 10^{-4}$ mol. These H_3O^+ ions are dissolved in (20.0 mL + 19.0 mL) = 39.0 mL = 0.0390 L of solution. Therefore, the molarity of the unreacted H_3O^+ ions is calculated as follows:

$$\text{Concentration of unreacted ions} = \frac{1.0 \times 10^{-4}\ \text{moles } H_3O^+}{0.0390\ \text{L solution}} = 2.6 \times 10^{-3}\ M$$

The pH of the solution is $-\log(2.6 \times 10^{-3}) = 2.59$.

Think About It The pH of the solution before the equivalence point in a strong acid–strong base titration is determined by the amount of excess H_3O^+ ions. After the addition of another 1.00 mL NaOH, the equivalence point would be reached and the pH would jump from 2.59 to 7.00.

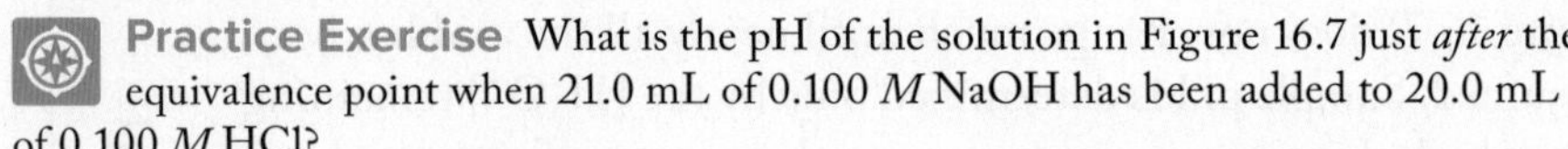

Practice Exercise What is the pH of the solution in Figure 16.7 just *after* the equivalence point when 21.0 mL of 0.100 *M* NaOH has been added to 20.0 mL of 0.100 *M* HCl?

CONCEPT **TEST**

The images shown here depict the solute particles present during the titration of a sample of LiOH(*aq*) with HBr(*aq*) as the titrant. Put the images in order from the beginning to the end of the titration. Which image depicts the equivalence point? The solvent has been omitted for clarity.

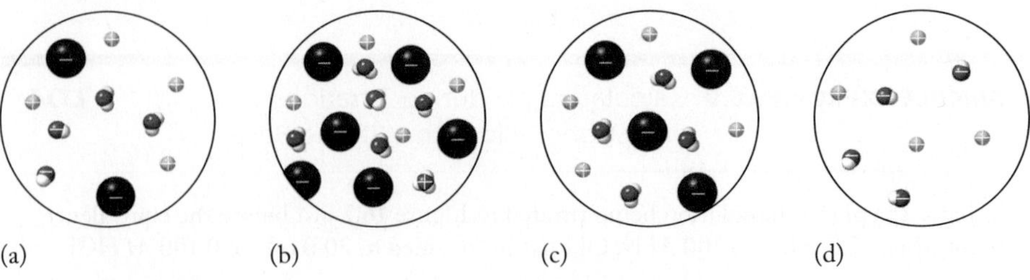

(a) (b) (c) (d)

The pH of the acetic acid sample (the blue curve in Figure 16.7) changes abruptly with the first few drops of added base, but then the changes become smaller and the titration curve levels out. In the nearly flat region before the equivalence point, the sample acts like a pH buffer as additions of NaOH titrant neutralize acetic acid, forming the sodium salt of its conjugate base:

$$CH_3COOH(aq) + NaOH(aq) \rightarrow CH_3COONa(aq) + H_2O(\ell)$$

The single reaction arrow tells us that the reaction goes to completion, or at least until all of the limiting reactant has been consumed. The titrant (NaOH) is the limiting reactant before the equivalence point, and the sample (CH_3COOH) is the limiting reactant after the equivalence point. The 1:1 mole ratio of the two reactants in this titration reaction means that one mole of CH_3COOH is consumed for every mole of NaOH added until the equivalence point is reached.

As long as there are significant concentrations of both acetic acid and acetate ions in the sample before the equivalence point, the changes in pH with added titrant are small. However, when enough titrant has been added to consume nearly all the acid in both samples depicted in Figure 16.7, pH rises sharply. At their equivalence points, the same volume of NaOH titrant has been added to both because the initial concentrations of the two acids and their sample volumes are the same. Therefore, the number of moles of acid in each sample is the same, and that is the quantity that defines the number of moles of base needed to reach the equivalence point. It does not matter whether or not the acids are completely ionized initially, because even a weak acid such as acetic acid is completely neutralized at its equivalence point.

Even though the volume of titrant needed to reach the equivalence points of the red and blue curves in Figure 16.7 is the same, the pH values at the two equivalence points are different—namely, 7.00 (neutral) in the HCl (strong acid) titration, but 8.73 in the acetic acid (weak acid) titration. The acetic acid solution is slightly basic because the product of the titration reaction is a solution of sodium acetate (CH_3COONa), which is a basic salt. Acetate ions in solution react with water, forming molecules of acetic acid and OH^- ions:

$$CH_3COO^-(aq) + H_2O(\ell) \rightleftharpoons CH_3COOH(aq) + OH^-(aq)$$

As a general rule, the pH at the equivalence point in the titration of any weak acid with a strong base is greater than 7 because the hydrolysis of the anion of the salt formed in the neutralization reaction produces OH^- ions. Beyond their equivalence points, the two curves overlap because the pH values of the solutions in this region are controlled only by the increasing concentration of NaOH.

Another important point in the acetic acid titration curve lies halfway to the equivalence point. At this *midpoint* in the titration, half of the acetic acid initially in the sample has been converted into acetate ions, its conjugate base. Therefore, the concentration of acetate ions produced by the neutralization reaction and the concentration of acetic acid remaining from the original sample are the same. If we insert this equality ([base] = [acid]) into the log term of the Henderson–Hasselbalch equation, the value of the term is zero. Therefore, the midpoint pH is equal to the pK_a of acetic acid, which is 4.75.

CONCEPT TEST

The images shown here depict the particles present during a titration of HClO with KOH. Which depicts the weak acid before the titration? Which represents the midpoint of the titration? Which represents the equivalence point? The solvent has been omitted for clarity.

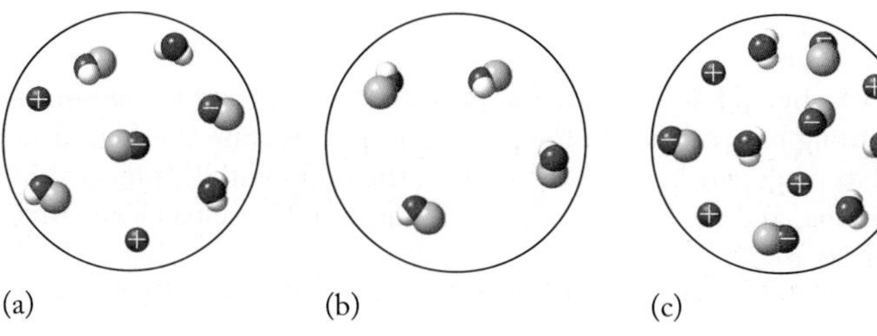

(a) (b) (c)

SAMPLE EXERCISE 16.10 Calculating pH during the Titration of a Weak Acid with a Strong Base **LO4**

What is the pH of the solution being titrated in Figure 16.7 after 10.0 mL of 0.100 *M* NaOH has been added to 20.0 mL of 0.100 *M* CH_3COOH?

Collect, Organize, and Analyze As in Sample Exercise 16.9, we can calculate the moles of weak acid present and the moles of OH^- ions added from their respective volumes and molarities. As the NaOH is added to the acetic acid solution, the following reaction occurs:

$$CH_3COOH(aq) + OH^-(aq) \rightarrow CH_3COO^-(aq) + H_2O(\ell)$$

We know how much of the weak acid remains and how much of its conjugate base is produced because one mole of acid reacts with one mole of OH^- ions to produce one mole of acetate ions. We can calculate $[CH_3COO^-]$ and then use the Henderson–Hasselbalch equation to determine the pH of the solution. The pK_a for acetic acid, which can be found in Appendix A5.1, equals 4.75.

Solve The initial quantity of CH_3COOH in the sample is

$$20.0\ \cancel{mL} \times \frac{1\ \cancel{L}}{1000\ \cancel{mL}} \times \frac{0.100\ mol\ CH_3COOH}{\cancel{L}} = 2.00 \times 10^{-3}\ mol\ CH_3COOH$$

The quantity of NaOH titrant added is

$$10.0\ \cancel{mL} \times \frac{1\ \cancel{L}}{1000\ \cancel{mL}} \times \frac{0.100\ mol\ NaOH}{\cancel{L}} = 1.00 \times 10^{-3}\ mol\ NaOH$$

We can use a modified RICE table to determine how many moles of CH_3COOH remain and how many moles of CH_3COO^- have been produced:

Reaction	$CH_3COOH(aq)$	+	$OH^-(aq)$	→	$CH_3COO^-(aq) + H_2O(\ell)$
	CH_3COOH (moles)		OH^- (moles)		CH_3COO^- (moles)
Initial	2.00×10^{-3}		1.00×10^{-3}		0
Change	-1.00×10^{-3}		-1.00×10^{-3}		$+1.00 \times 10^{-3}$
Final	1.00×10^{-3}		0		1.00×10^{-3}

The total sample volume is (20.0 mL + 10.0 mL) = 30.0 mL or 0.0300 L, and the concentrations of acetic acid and acetate ion are the same:

$$[CH_3COOH] = [CH_3COO^-] = \frac{1.00 \times 10^{-3}\ mol}{0.0300\ L} = 3.33 \times 10^{-2}\ M$$

Using these values in the Henderson–Hasselbalch equation gives us

$$pH = pK_a + \log\frac{[CH_3COO^-]}{[CH_3COOH]}$$

$$= 4.75 + \log\left(\frac{3.33 \times 10^{-2}}{3.33 \times 10^{-2}}\right) = 4.75 + \log(1) = 4.75 + 0 = 4.75$$

Think About It The pH at the midpoint in the titration of any monoprotic weak acid with a strong base is equal to the pK_a of the acid because the concentrations of the acid and its conjugate base are the same at the midpoint. This means the [base]/[acid] ratio is one, and the log term in the Henderson–Hasselbalch equation is zero. Therefore, $pH = pK_a$.

Practice Exercise What is the pH at the midpoint of the titration with a strong base of an aqueous sample that contains an unknown concentration of benzoic acid?

CONCEPT TEST

Why is it better to titrate a weak acid with a strong base than it is to titrate a weak acid with a weak base?

Titration curves of weakly or strongly basic analytes with strong acid titrants resemble the curves in Figure 16.7, but they're inverted: they start with high initial pH values and end with low ones. Figure 16.8 illustrates two such titrations. One sample contains 0.100 *M* NaOH (a strong base), whereas the other contains 0.100 *M* NH_3 (a weak base). Both are titrated with 0.100 *M* HCl (a strong acid). The initial pH of the NaOH titration curve is higher than that of the NH_3 sample because NaOH is a stronger base. The pH of the NaOH sample does not change much as acid is added until near the equivalence point. As the equivalence point is approached, the principal ions present in the reaction mixture are Na^+, Cl^-, and decreasing concentrations of OH^-. The sample pH, which is determined only by the $[OH^-]$ still present, starts to fall steeply. When enough acid has been added to completely consume all the OH^- ions in the sample, the equivalence point has been reached. The solution now consists of water and NaCl, which is neutral, so pH = 7.00 at the equivalence point.

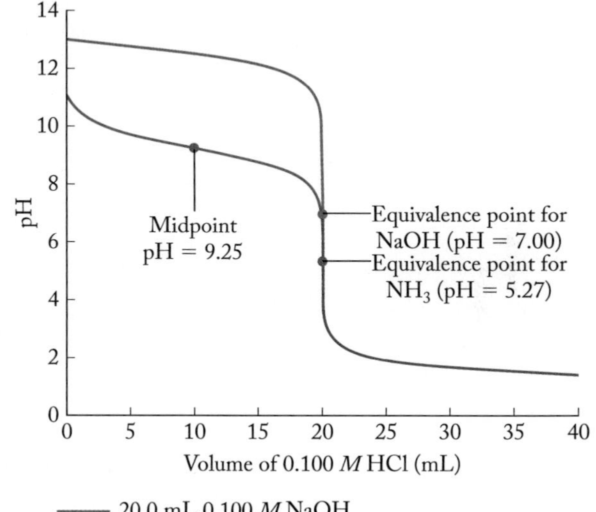

FIGURE 16.8 The titration curves of solutions of NaOH and NH_3 with a standard solution of HCl as the titrant.

The pH of the NH_3 solution changes abruptly with the first few drops of added acid, but then the changes become smaller and the titration curve levels out. In this nearly flat region, additions of acidic titrant are consumed as NH_3 reacts with H_3O^+ ions to form NH_4^+ ions. The sample acts like a pH buffer, just as the mixture of acetic acid and acetate ions did in the titration in Figure 16.7. The pH value at the equivalence point in the ammonia titration is 5.27 (slightly acidic) because the product of the titration reaction is an acidic salt, NH_4Cl, which reacts with water, forming ammonia molecules and H_3O^+ ions:

$$NH_4^+(aq) + H_2O(\ell) \rightleftharpoons NH_3(aq) + H_3O^+(aq)$$

Beyond their equivalence points, the NaOH and NH_3 curves are identical because the pH is determined only by the moles of HCl added after the equivalence point and the total volume of the titration reaction mixtures.

As in the acetic acid titration, the midpoint in the NH_3 titration curve is significant because it is the point at which half of the NH_3 initially in the sample has been converted into NH_4^+ ions. Therefore, $[NH_3] = [NH_4^+]$, and according to the Henderson–Hasselbalch equation,

$$pH = pK_a + \log\frac{[NH_3]}{[NH_4^+]} = pK_a + \log(1) = pK_a$$

That is, the pH at the midpoint of the titration is equal to the pK_a of the ammonium ion. This pK_a value can be calculated from the pK_b value for ammonia, which is 4.75 (see Table A5.3), using Equation 16.4: $pK_a = 14.00 - 4.75 = 9.25$. Thus, the pH at the midpoint of the titration is 9.25.

CONCEPT TEST

What is the pH at the midpoint of the titration of an aqueous sample of methylamine with a standard solution of hydrochloric acid?

CONCEPT TEST

The images shown here depict the particles present during a titration of $NH_3(aq)$ with $HCl(aq)$. Which represents the start of the titration? Which represents the region where the solution best functions as a buffer? Which represents the equivalence point? The solvent has been omitted for clarity.

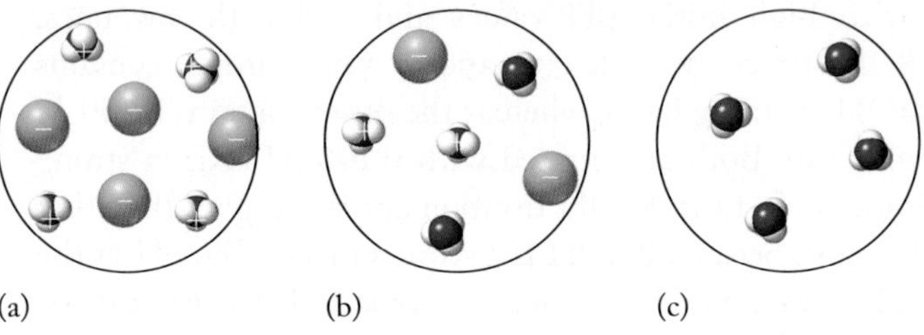

Titrations with Multiple Equivalence Points

So far, the titration curves in this chapter have each had only one equivalence point. An alkalinity titration (Figure 16.9), however, has two equivalence points. In environmental science, the alkalinity of a water sample is a measure of its capacity to neutralize additions of acid. Titrating the sample with strong acid provides a way to determine that capacity.

If carbonate ions are present in the sample, the first additions of titrant convert them into bicarbonate ions:

$$CO_3^{2-}(aq) + H_3O^+(aq) \rightarrow HCO_3^-(aq) + H_2O(\ell)$$

This reaction continues until all of the carbonate initially in the sample has been converted to bicarbonate. The titration has reached its first equivalence point, which is marked by a sharp drop in pH. In the second stage of the titration, the bicarbonate ions formed in the first stage plus any bicarbonate ions present in the original sample react with additional acidic titrant, forming carbonic acid, $H_2CO_3(aq)$:

$$HCO_3^-(aq) + H_3O^+(aq) \rightarrow H_2CO_3(aq) + H_2O(\ell)$$

As we have seen, carbonic acid produced in the second stage decomposes to carbon dioxide and water:

$$H_2CO_3(aq) \rightarrow CO_2(g) + H_2O(\ell)$$

In fact, bubbles of CO_2 gas may be seen forming in samples during the second stage of alkalinity titrations.

During the first stage, there is a buffer region where HCO_3^- and CO_3^{2-} ions function as a weak acid–conjugate base pair. At the first equivalence point, the dominant carbon-

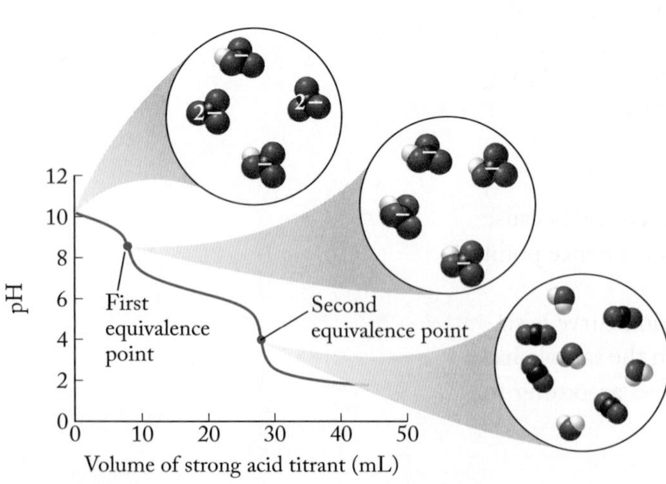

FIGURE 16.9 An alkalinity titration curve can have two equivalence points. The first marks the complete conversion of any carbonate in the sample into bicarbonate, whereas the second marks the conversion of bicarbonate into carbonic acid, which decomposes to CO_2 and H_2O.

containing species is HCO_3^- and the sample is still slightly basic (pH > 8). This basic pH indicates that the bicarbonate is more effective as a base (Equation 16.6) than as an acid (Equation 16.7):

$$HCO_3^-(aq) + H_2O(\ell) \rightleftharpoons H_2CO_3(aq) + OH^-(aq) \qquad (16.6)$$

$$HCO_3^-(aq) + H_2O(\ell) \rightleftharpoons CO_3^{2-}(aq) + H_3O^+(aq) \qquad (16.7)$$

We can confirm the dominance of the reaction in Equation 16.6 by considering the two-step acid ionization equilibria of carbonic acid. Equation 16.7 represents the second ionization step and its equilibrium constant is $K_{a_2} = 4.7 \times 10^{-11}$. The K_b of the base ionization described in Equation 16.6 is linked to the first ionization of carbonic acid:

$$H_2CO_3(aq) + H_2O(\ell) \rightleftharpoons HCO_3^-(aq) + H_3O^+(aq) \qquad K_{a_1} = 4.3 \times 10^{-7}$$

Converting K_{a_1} into K_b using Equation 15.21:

$$K_a \times K_b = K_w$$

$$K_b = \frac{K_w}{K_{a_1}} = \frac{1.0 \times 10^{-14}}{4.3 \times 10^{-7}} = 2.3 \times 10^{-8}$$

This K_b value is much greater than the value of K_{a_2}:

$$\frac{2.3 \times 10^{-8}}{4.7 \times 10^{-11}} = 4.9 \times 10^2$$

Therefore, solutions of compounds such as sodium bicarbonate (the active ingredient in baking soda), in which the cation has no influence on pH, are weakly basic.

During the second stage of the titration, the conversion of HCO_3^- ions to H_2CO_3 produces a second pH buffer and another plateau of slowly changing pH. When the HCO_3^- ions are completely consumed, the pH drops sharply for a second time at the second equivalence point. The pH of the second equivalence point is below 7, reflecting the acidic character and K_{a_1} value of carbonic acid.

Note that the initial pH of the sample in Figure 16.9 is slightly above 10, which is quite basic and above the pH range tolerated by many species of aquatic life. Such highly basic water may be found in arid regions such as the U.S. Southwest, where rocks containing $CaCO_3$ and other basic compounds are in contact with water. Seawater and most freshwater samples are not that basic, which means that the dominant carbonate species in them is bicarbonate. The alkalinity titration curves of these waters may have only one equivalence point, coinciding with the pH of the second equivalence point in Figure 16.9.

SAMPLE EXERCISE 16.11 Interpreting the Results of a Titration with Multiple Equivalence Points I **LO4**

If the volume of titrant needed to reach the first equivalence point in the titration shown in Figure 16.9 is 9.00 mL, and a total of 27.00 mL is required to reach the second equivalence point, what was the ratio of carbonate to bicarbonate in the original sample?

Collect, Organize, and Analyze Carbonate ions (CO_3^{2-}) in the sample react with H_3O^+ ions from the titrant and are converted to bicarbonate ions (HCO_3^-) in the first stage of the titration. Bicarbonate ions—including any in the original sample plus all those produced in the first-stage reaction—are converted to carbonic acid, H_2CO_3, in the second stage.

Solve The volume of titrant needed to titrate the CO_3^{2-} in the sample (9.00 mL) is exactly half the additional volume (27.00 − 9.00 = 18.00 mL) needed to titrate the HCO_3^- in the reaction mixture. However, 9.00 mL of the titrant consumed in the second stage was needed just to titrate the HCO_3^- produced in the first stage. This means the volume of titrant needed to titrate the HCO_3^- that was in the original sample was only (18.00 − 9.00) = 9.00 mL. Therefore, the original sample contained equal concentrations of CO_3^{2-} and HCO_3^- ions.

Think About It It would be tempting to interpret the volumes of titrant consumed in the two stages to mean that there was twice as much bicarbonate as carbonate in the original sample. It's important to remember that the HCO_3^- ions titrated in the second stage come from the original sample and from HCO_3^- ions produced from CO_3^{2-} ions during the first stage of the titration.

Practice Exercise Is the midpoint pH in the first stage of an alkalinity titration always the same as the pK_{a_2} of carbonic acid? Explain why or why not.

We can detect both equivalence points in an alkalinity titration using a pH electrode, or we can use appropriate indicators. Phenol red would not be a good choice for the titration in Figure 16.9 because it changes color between pH 6.8 and 8.4. This range is just below the pH of the first equivalence point and well above the pH of the second equivalence point. To detect the first equivalence point, we need an indicator with a pK_a near the pH of the solution at the first equivalence point, which is about 8.5. One candidate in Figure 16.5 is phenolphthalein (pK_a = 9.7), which is pink in its basic form and colorless at low pH.

To detect the second equivalence point, we could add bromcresol green (pK_a = 4.6) after the first equivalence point has been reached. We would not add it earlier because its blue-green color in basic solutions would obscure the pink-to-colorless transition of phenolphthalein. We do not need to be concerned about the phenolphthalein obscuring the bromcresol green color change because phenolphthalein is colorless in acidic solutions.

SAMPLE EXERCISE 16.12 Interpreting the Results of a Titration with Multiple Equivalence Points II — LO4

A 100.0 mL sample of water from the Sapphire Pool in Yellowstone National Park (Figure 16.10) is titrated with 0.0300 *M* HCl. A few drops of phenolphthalein are added at the beginning of the titration, and the solution turns pink. It takes 5.91 mL of titrant to reach the pink-to-clear equivalence point. Then a few drops of bromcresol green are added, and it takes an additional 26.02 mL of titrant before the blue-green color changes to yellow. What were the initial concentrations of carbonate and bicarbonate in the sample?

FIGURE 16.10 The water in Sapphire Pool in Yellowstone National Park is slightly alkaline, due mostly to carbonate and bicarbonate ions. It is also crystal clear and very hot.

Collect and Organize We are asked to determine the concentrations of CO_3^{2-} and HCO_3^- ions in a water sample from the results of a titration. These determinations are based on the volumes of titrant needed to reach two equivalence points. In the first stage, CO_3^{2-} ions in the sample are converted to HCO_3^- ions:

$$(1)\quad H_3O^+(aq) + CO_3^{2-}(aq) \rightarrow HCO_3^-(aq) + H_2O(\ell)$$

In the second stage, HCO_3^- ions are converted to H_2CO_3:

$$(2) \quad H_3O^+(aq) + HCO_3^-(aq) \rightarrow H_2CO_3(aq) + H_2O(\ell)$$

Analyze According to the stoichiometries of the reactions, it takes one mole of HCl to titrate one mole of carbonate to bicarbonate in the first stage, and it takes one mole of HCl to titrate one mole of bicarbonate to carbonic acid in the second stage. The HCO_3^- ions titrated in stage 2 include any in the original sample plus all the HCO_3^- ions produced by reaction 1. The difference between the titrant volumes needed to reach the two equivalence points, (26.02 − 5.91) = 20.11 mL, is the volume of acid required to react with the HCO_3^- that was present in the original sample. This difference is between three and four times the volume of titrant consumed in stage 1, so we can predict that there was three to four times as much HCO_3^- in the original sample as CO_3^{2-}.

Solve Calculating the concentrations of CO_3^{2-} and HCO_3^- in the original sample from the volumes and molarity of the HCl titrant consumed in stages 1 and 2:

$$[CO_3^{2-}] = \frac{\text{mol } CO_3^{2-}}{\text{L solution}}$$

$$= \frac{5.91 \text{ mL titrant} \times \frac{0.0300 \text{ mmol HCl}}{\text{mL titrant}} \times \frac{1 \text{ mol HCl}}{1000 \text{ mmol HCl}} \times \frac{1 \text{ mol } CO_3^{2-}}{1 \text{ mol HCl}}}{100.0 \text{ mL solution} \times \frac{1 \text{ L}}{1000 \text{ mL}}}$$

$$= 1.77 \times 10^{-3}\ M$$

$$[HCO_3^-] = \frac{\text{mol } HCO_3^-}{\text{L solution}}$$

$$= \frac{20.11 \text{ mL titrant} \times \frac{0.0300 \text{ mmol HCl}}{\text{mL titrant}} \times \frac{1 \text{ mol HCl}}{1000 \text{ mmol HCl}} \times \frac{1 \text{ mol } HCO_3^-}{1 \text{ mol HCl}}}{100.0 \text{ mL solution} \times \frac{1 \text{ L}}{1000 \text{ mL}}}$$

$$= 6.03 \times 10^{-3}\ M$$

Think About It The titration results confirm that the bicarbonate concentration in the original sample was just over three times the carbonate concentration.

Practice Exercise Suppose you titrate a 100.0 mL sample of water from another pool in Yellowstone National Park with a ratio of carbonate to bicarbonate of 3:1. If the total volume of titrant needed to reach the second equivalence point is 21.00 mL, what volume of titrant was used to reach the first equivalence point?

CONCEPT **TEST**

In a titration that initially contains both CO_3^{2-} and HCO_3^-, the volume of titrant required to reach the first equivalence point is less than that required to titrate from the first equivalence point to the second. Why?

16.5 Lewis Acids and Bases

Until now we have used the Brønsted–Lowry definition of acids (H^+ ion donors) and bases (H^+ ion acceptors). The time has come, however, to expand our concept of acids and bases to include acid–base interactions that may or may not involve

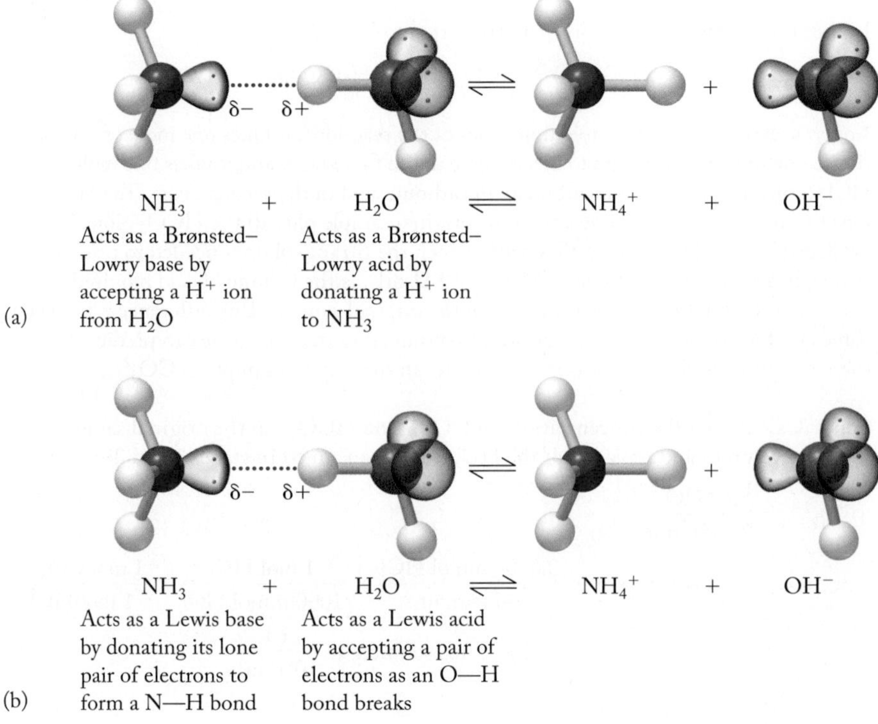
δ− δ+
NH_3 + H_2O ⇌ NH_4^+ + OH^-
Acts as a Brønsted–Lowry base by accepting a H^+ ion from H_2O
Acts as a Brønsted–Lowry acid by donating a H^+ ion to NH_3
(a)
δ− δ+
NH_3 + H_2O ⇌ NH_4^+ + OH^-
Acts as a Lewis base by donating its lone pair of electrons to form a N—H bond
Acts as a Lewis acid by accepting a pair of electrons as an O—H bond breaks
(b)

FIGURE 16.11 (a) The Brønsted–Lowry view of the reaction between H_2O (proton donor) and NH_3 (proton acceptor). (b) In the Lewis view of the reaction, H_2O acts as a Lewis acid (an electron-pair acceptor) and NH_3 acts as a Lewis base (an electron-pair donor).

the transfer of H^+ ions. Let's begin by revisiting what happens when ammonia gas dissolves in water:

$$NH_3(g) + H_2O(\ell) \rightleftharpoons NH_4^+(aq) + OH^-(aq)$$

Figure 16.11(a) shows a Brønsted–Lowry interpretation of this reaction: in donating H^+ ions to ammonia, H_2O acts as a Brønsted–Lowry acid. In accepting H^+ ions, NH_3 acts as a Brønsted–Lowry base.

Another way to view this reaction is illustrated in Figure 16.11(b). Instead of focusing on the transfer of a hydrogen ion, we consider the reaction as one in which one reactant donates and the other accepts a *pair of electrons.* In this view, the N atom in NH_3 donates its lone pair of electrons to one of the H atoms in H_2O. In the process, one of the H—O bonds in H_2O is broken in such a way that the bonding pair of electrons remains with the O atom. The donated lone pair from the N atom forms a fourth N—H covalent bond. The result is the same as in the Brønsted–Lowry model of acid–base behavior: a molecule of NH_3 bonds to a H^+ ion, forming an NH_4^+ ion, and a molecule of H_2O loses a H^+ ion, becoming a OH^- ion.

CONNECTION Lewis's theories on the nature of covalent bonding were described in Section 4.3.

When viewed from the perspective of the electron pair,

- A **Lewis base** is a substance that donates a lone pair of electrons in a chemical reaction.
- A **Lewis acid** is a substance that accepts a lone pair of electrons in a chemical reaction.

Lewis base a substance that donates a lone pair of electrons in a chemical reaction.

Lewis acid a substance that accepts a lone pair of electrons in a chemical reaction.

These definitions are named after their developer, Gilbert N. Lewis, who also pioneered research into the nature of chemical bonds. The Lewis definition of a base is consistent with the Brønsted–Lowry model we have used because a substance must be able to donate a pair of electrons if it is to bond with a H^+ ion. However, the same parallelism does not hold for acids. The Brønsted–Lowry

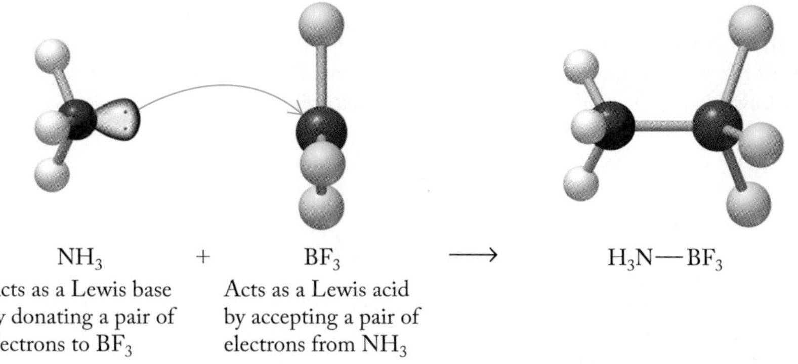

FIGURE 16.12 The reaction between NH_3 and BF_3 is an acid–base reaction in the Lewis sense, because NH_3 donates an electron pair (making it the Lewis base) and BF_3 accepts the electron pair (making it the Lewis acid). This is not an acid–base reaction in the Brønsted–Lowry sense, because no proton is transferred.

model defines an acid as a hydrogen-ion donor, but the Lewis definition includes species that have no hydrogen ions to donate but that can still accept electrons.

One such compound is boron trifluoride, BF_3. With only six valence electrons, the boron atom in BF_3 can accept another pair to complete its octet. NH_3 is a suitable electron-pair donor, as shown in Figure 16.12. No H^+ ions are transferred in this reaction, so it is not an acid–base reaction according to the Brønsted–Lowry model. However, NH_3 donates a lone pair of electrons and BF_3 accepts it, so it is an acid–base reaction according to the more general Lewis model.

Many important Lewis bases are anions, including the halide ions, OH^-, and O^{2-}. Oxide, O^{2-}, functions as a Lewis base in the reaction between SO_2 and CaO that is used to reduce SO_2 emissions from coal-burning power stations (Section 14.6):

$$CaO(s) + SO_2(g) \rightleftharpoons CaSO_3(s)$$

The oxide ion in CaO is the electron-pair donor and Lewis base. Sulfur dioxide is the electron-pair acceptor and is therefore a Lewis acid.

$$Ca^{2+}[:\ddot{O}:]^{2-} \quad :\ddot{O}{=}\ddot{S}{=}\ddot{O}: \rightarrow Ca^{2+}\left[:\ddot{O}-\ddot{S}(-\ddot{O}:)(=\ddot{O})\right]^{2-}$$

As a SO_2 molecule is absorbed onto the surface of a solid CaO particle, the oxide ion donates an electron pair to the sulfur atom resulting in an additional S—O covalent bond and formation of a sulfite anion, SO_3^{2-}. The formal charges on the S atom and the double-bonded O atom in SO_3^{2-} are both zero; the formal charges on the two single-bonded O atoms are −1, giving an overall charge of 2−.

CONNECTION The concept of formal charge and how to calculate it were described in Section 4.7.

SAMPLE EXERCISE 16.13 Identifying Lewis Acids and Bases **LO5**

In the following reaction, which species is a Lewis acid and which is a Lewis base?

$$AlCl_3 + Cl^- \rightarrow AlCl_4^-$$

Collect and Organize We are given a chemical reaction and asked to identify the Lewis acid (the reactant that accepts a pair of electrons) and the Lewis base (the reactant that donates that pair of electrons).

Analyze A Cl^- ion has four lone pairs of electrons in its valence shell, so it can *donate* one of them to form a covalent bond to aluminum. To determine whether $AlCl_3$ can act as a Lewis base, we need to draw its Lewis structure:

$$:\ddot{Cl}: \text{—} Al(\text{—}\ddot{Cl}:)_2$$

This structure accounts for all 24 of the valence electrons with three Al—Cl single bonds and *no lone pairs*. This leaves Al with only six valence electrons and thus the capacity to accept one more pair—that is, to act as a Lewis acid.

Solve In this reaction, $AlCl_3$ is a Lewis acid and the Cl^- ion is a Lewis base:

$$AlCl_3 + [:\ddot{Cl}:]^- \rightarrow [AlCl_4]^-$$

Think About It Drawing the Lewis structure of $AlCl_3$ and determining that the central Al atom has an incomplete octet is the key to identifying its capacity to act as a Lewis acid. Note that these Lewis structures assume that $AlCl_3$ is a molecular compound and not, like most other metal chlorides, an ionic compound. This assumption is supported by the physical properties of $AlCl_3$ and particularly by the fact that it sublimes at 178°C and 1 atm. Most metal halides do not even melt until heated to temperatures many hundreds of degrees higher than that.

Practice Exercise In the following reaction, which reactant is the Lewis acid and which is the Lewis base?

$$CO_2(g) + CaO(s) \rightarrow CaCO_3(s)$$

16.6 Formation of Complex Ions

In Chapter 6, we described how ions dissolved in water are *hydrated*—that is, they are surrounded by water molecules oriented with the positive ends of their dipoles directed toward anions and their negative ends directed toward cations (Figure 16.13). In some hydrated cations, ion–dipole interactions lead to the sharing of lone-pair electrons on the oxygen atoms of H_2O with empty valence-shell orbitals on the cations. This interaction between a lone pair and an empty orbital is another example of Lewis acid–base behavior. These shared electron pairs meet our definition of covalent bonds, but these particular bonds are called *coordinate* covalent bonds, or simply **coordinate bonds**. Coordinate bonds form when either a molecule or an anion donates a lone pair of electrons to an empty valence-shell orbital of an atom, cation, or molecule. Once formed, a coordinate bond is indistinguishable from any other kind of covalent bond. When a cation forms coordinate bonds to a cluster of molecular or ionic electron-pair donors, the resulting structure is called a **complex ion**. The electron-pair donors in complex ions are called **ligands**. For example, when six molecules of water form coordinate bonds to a Ni^{2+} cation in an aqueous solution, the resulting complex ion is written as $Ni(H_2O)_6{}^{2+}$ to show that it has six ligands (six water molecules bonded to the metal).

We can study the formation of complex ions using the same mathematical tools we have used to understand acid–base equilibria. These tools are appropriate

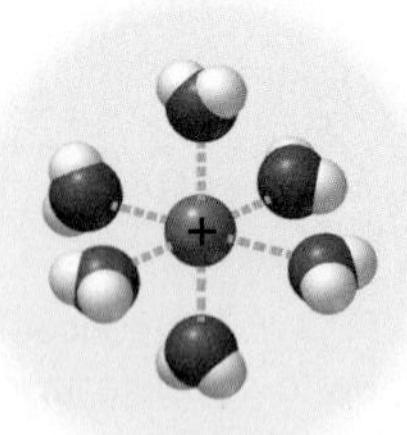

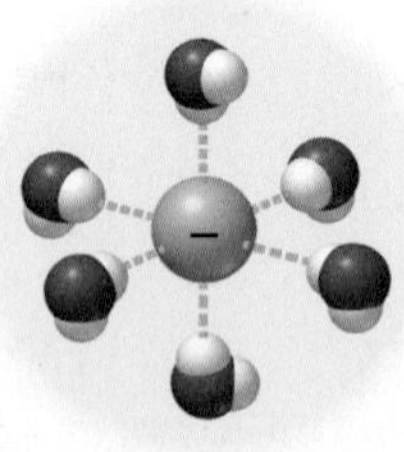

FIGURE 16.13 Ion–dipole interactions in a hydrated cation and a hydrated anion.

FIGURE 16.14 The beaker on the left contains a solution of $Cu^{2+}(aq)$, which is a characteristic robin's-egg blue. As a colorless solution of ammonia is added (from the bottle in the middle), the mixture of the two solutions turns dark blue (beaker on the right), which is the color of the $Cu(NH_3)_4{}^{2+}$ complex ion.

because complex formation processes are reversible, and many of them reach chemical equilibrium rapidly. We begin this investigation with two aqueous solutions, one containing copper(II) sulfate ($CuSO_4$) and the other containing NH_3 (Figure 16.14). The $CuSO_4$ solution is robin's-egg blue, the color characteristic of $Cu^{2+}(aq)$ ions dissolved in water, whereas the ammonia solution is colorless. When the solutions are mixed, the robin's-egg blue turns a dark navy blue, as shown on the right in Figure 16.14. This is the color of $Cu(NH_3)_4{}^{2+}$ complex ions. The change in color provides visual evidence that the following equilibrium lies far to the right, favoring complex ion formation:

CONNECTION Spheres of hydration around cations and anions were introduced in Section 6.2.

$$Cu^{2+}(aq) + 4\,NH_3(aq) \rightleftharpoons Cu(NH_3)_4{}^{2+}(aq)$$

This conclusion is supported by the large equilibrium constant for the reaction:

$$K_f = \frac{[Cu(NH_3)_4{}^{2+}]}{[Cu^{2+}][NH_3]^4} = 5.0 \times 10^{13}$$

The equilibrium constant K_f is called a **formation constant** because it describes the formation of a complex ion. For the general case in which one mole of metal ions (M^{m+}) combines with n moles of ligand (X^{x-}) to form the complex ion $MX_n{}^{(m-nx)+}$, the formation constant expression is

$$K_f = \frac{[MX_n{}^{(m-nx)+}]}{[M^{m+}][X^{x-}]^n}$$

Formation constants can be used to calculate the concentration of complex ions in solution or to calculate the concentration of free, uncomplexed metal ions, $M^{m+}(aq)$, in equilibrium with a given (usually larger) concentration of a ligand. Because K_f values are usually very large, equilibrium concentrations of uncomplexed metal ions are usually very small. One approach to calculating the concentration of an uncomplexed metal ion is to consider the reverse of the formation reaction and calculate how much $Cu(NH_3)_4{}^{2+}(aq)$ dissociates.

coordinate bond a covalent bond formed when one anion or molecule donates a pair of electrons to another ion or molecule.

complex ion an ionic species consisting of a metal ion bonded to one or more Lewis bases.

ligand a Lewis base bonded to the central metal ion of a complex ion.

formation constant (K_f) an equilibrium constant describing the formation of a metal complex from a free metal ion and its ligands.

The equilibrium constant for the dissociation reaction (K_d) is the reciprocal of the original equilibrium constant.

$$Cu(NH_3)_4^{2+}(aq) \rightleftharpoons Cu^{2+}(aq) + 4\,NH_3(aq) \quad K_d = \frac{1}{K_f} = \frac{[Cu^{2+}][NH_3]^4}{[Cu(NH_3)_4^{2+}]}$$

Using K_d to calculate the concentration of uncomplexed copper ion, $[Cu^{2+}]$, is illustrated in Sample Exercise 16.14.

SAMPLE EXERCISE 16.14 Calculating the Concentration of Free Metal in Equilibrium with a Complex Ion **LO6**

Ammonia gas is dissolved in a $1.00 \times 10^{-4}\,M$ solution of $CuSO_4$, to give an equilibrium concentration of $[NH_3] = 1.60 \times 10^{-3}\,M$. Calculate the concentration of $Cu^{2+}(aq)$ ions in the solution.

Collect and Organize The concentration of $CuSO_4$ means that $[Cu^{2+}]$ before complex formation is $1.00 \times 10^{-4}\,M$. The equilibrium concentration of the ligand (NH_3) is $1.60 \times 10^{-3}\,M$. The values of $[NH_3]$, $[Cu^{2+}]$, and $[Cu(NH_3)_4^{2+}]$ at equilibrium are related by the formation constant expression:

$$K_f = \frac{[Cu(NH_3)_4^{2+}]}{[Cu^{2+}][NH_3]^4} = 5.0 \times 10^{13}$$

Analyze Because K_f is large, we can assume that essentially all the Cu^{2+} ions are converted to complex ions and $[Cu(NH_3)_4^{2+}] = 1.00 \times 10^{-4}\,M$. Only a tiny concentration of free Cu^{2+} ions, x, remains at equilibrium. We can calculate the $[Cu^{2+}]$ by considering how much $Cu(NH_3)_4^{2+}$ dissociates:

$$Cu(NH_3)_4^{2+}(aq) \rightleftharpoons Cu^{2+}(aq) + 4\,NH_3(aq)$$

The equilibrium constant, K_d, for this reaction is the reciprocal of K_f:

$$K_d = \frac{1}{K_f} = \frac{[Cu^{2+}][NH_3]^4}{[Cu(NH_3)_4^{2+}]} = \frac{1}{5.0 \times 10^{13}} = 2.0 \times 10^{-14}$$

We can construct a RICE table incorporating the concentrations and concentration changes of the products and reactants. Given the small value of K_d, we should obtain a $[Cu^{2+}]$ value at equilibrium that is much less than $1.00 \times 10^{-4}\,M$.

Solve First, we complete the row of equilibrium concentrations of Cu^{2+}, NH_3, and $Cu(NH_3)_4^{2+}$ in the following RICE table:

Reaction (R)	$Cu(NH_3)_4^{2+}(aq)$	$\rightleftharpoons$	$Cu^{2+}(aq)$	+	$4\,NH_3(aq)$
	$[Cu(NH_3)_4^{2+}]$ (M)		$[Cu^{2+}]$ (M)		$[NH_3]$ (M)
Initial (I)	1.00×10^{-4}		0		1.60×10^{-3}
Change (C)	$-x$		$+x$		0
Equilibrium (E)	$1.00 \times 10^{-4} - x$		x		1.60×10^{-3}

Next, we make the simplifying assumption that x is much smaller than $1.00 \times 10^{-4}\,M$. Therefore, we can ignore the x term in the equilibrium value of $[Cu(NH_3)_4^{2+}]$ and use the simplified value in the K_d expression:

$$K_d = \frac{[Cu^{2+}][NH_3]^4}{[Cu(NH_3)_4^{2+}]}$$

$$= \frac{[x][1.60 \times 10^{-3}]^4}{[1.0 \times 10^{-4}]} = 2.0 \times 10^{-14}$$

$$x = 3.1 \times 10^{-7} = [Cu^{2+}]$$

Think About It This result validates our simplifying assumption and confirms our prediction that $[Cu^{2+}]$ at equilibrium is much less than $[Cu^{2+}]$ initially. In fact, more than 99% of the Cu(II) in the solution is present as $Cu(NH_3)_4^{2+}$.

Practice Exercise Calculate the equilibrium concentration of $Ag^+(aq)$ in a solution that is initially 0.100 *M* $AgNO_3$ and 0.800 *M* NH_3 after the following reaction takes place:

$$Ag^+(aq) + 2\,NH_3(aq) \rightleftharpoons Ag(NH_3)_2^+(aq) \qquad K_f = 1.7 \times 10^7$$

16.7 Hydrated Metal Ions as Acids

In Chapter 15 we saw that the strength of an oxoacid depends on the electronegativity of its central atom. For example, the relative strengths of the three hypohalous acids—HOCl > HOBr > HOI—correspond to the relative electronegativities of their halogen atoms: Cl > Br > I. The more electronegative the halogen, the more it draws electron density away from the oxygen in the polar O—H bond (see Table 15.5). These shifts make the negative ion formed by dissociation better able to bear a negative charge because of increased delocalization.

CONNECTION We discussed periodic trends in electronegativity in Section 4.6.

A similar shift in electron density occurs in hydrated metal ions having the generic formula $M(H_2O)_6^{n+}$ when $n \geq 2$. The electrons in the O—H bonds of the water molecules surrounding the metal ions are attracted to the positively charged ions. The resulting distortion in electron density increases the likelihood of one of these O—H bonds ionizing and donating a H^+ ion to a neighboring molecule of water:

$$M(H_2O)_6^{n+}(aq) + H_2O(\ell) \rightleftharpoons M(H_2O)_5(OH)^{(n-1)+}(aq) + H_3O^+(aq)$$

Figure 16.15 provides a molecular view of this reaction, using $Fe^{3+}(aq)$ as the central ion. Similar reactions allow other hydrated metal ions, particularly those with charges of 3+, to function as Brønsted–Lowry acids. The K_a values of several metal ions are listed in Table 16.1. Note how much stronger the 3+ ions are compared to the 2+ ions, but this is to be expected, given the greater electron-withdrawing power of the more highly charged central ions.

Figure 16.15 shows how one of the six water molecules of hydration surrounding a Fe^{3+} ion is converted into a hydroxide ion as a result of the acid ionization reaction. This reduces the charge of the complex ion (but *not* the iron ion) from 3+ to 2+. If the pH of a solution of $Fe(H_2O)_5(OH)^{2+}$ ions is raised by

TABLE 16.1 K_a **Values of Hydrated Metal Ions**

Ion	K_a
$Fe^{3+}(aq)$	3×10^{-3}
$Cr^{3+}(aq)$	1×10^{-4}
$Al^{3+}(aq)$	1×10^{-5}
$Cu^{2+}(aq)$	3×10^{-8}
$Pb^{2+}(aq)$	3×10^{-8}
$Zn^{2+}(aq)$	1×10^{-9}
$Co^{2+}(aq)$	2×10^{-10}
$Ni^{2+}(aq)$	1×10^{-10}

(Arrow: Acid strength)

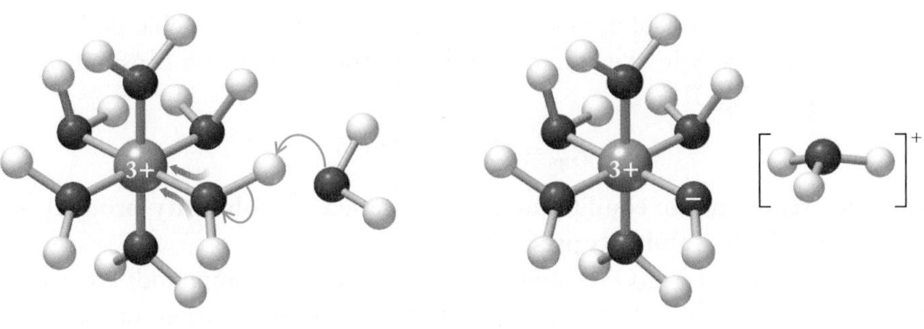

FIGURE 16.15 A hydrated Fe^{3+} cation draws electron density away from the water molecules of its inner coordination sphere, which makes it possible for one or more of these molecules to donate a H^+ ion to a water molecule outside the sphere. The hydrated Fe^{3+} ion is left with one fewer H_2O ligand and a bond between the complex ion and the OH^- ion.

adding a small quantity of a strong base such as NaOH, the ions undergo an additional acid ionization reaction, forming $Fe(H_2O)_4(OH)_2^+$ ions:

$$Fe(H_2O)_5(OH)^{2+}(aq) + OH^-(aq) \rightleftharpoons Fe(H_2O)_4(OH)_2^+(aq) + H_2O(\ell)$$

At still higher pH, solid iron(III) hydroxide forms:

$$Fe(H_2O)_4(OH)_2^+(aq) + OH^-(aq) \rightleftharpoons Fe(H_2O)_3(OH)_3(s) + H_2O(\ell)$$

For simplicity, we usually write the formula of iron(III) hydroxide as $Fe(OH)_3(s)$, even though each formula unit also contains three water molecules of hydration.

CONCEPT TEST

Is $Fe^{2+}(aq)$ a stronger or weaker Lewis acid than $Fe^{3+}(aq)$? Where would the K_a of $Fe^{2+}(aq)$ be listed in Table 16.1 relative to that of $Fe^{3+}(aq)$?

Cr^{3+} and Al^{3+} display similar behavior, but unlike Fe^{3+}, they are more soluble in strongly basic solutions than in weakly basic solutions. Why? Because solid $Cr(OH)_3$ and $Al(OH)_3$ may accept additional OH^- ions at high pH, forming soluble anionic complex ions:

$$Cr(OH)_3(s) + OH^-(aq) \rightleftharpoons Cr(OH)_4^-(aq) = Cr(H_2O)_2(OH)_4^-(aq)$$

$$Al(OH)_3(s) + OH^-(aq) \rightleftharpoons Al(OH)_4^-(aq) = Al(H_2O)_2(OH)_4^-(aq)$$

Zinc hydroxide, $Zn(OH)_2$, is the only other transition metal hydroxide that is soluble at high pH, because it forms $Zn(OH)_4^{2-}$ ions.

Nearly all transition metals exist as $M^{n+}(aq)$ ions only in strongly acidic solutions. They exist as complex ions, such as $M(H_2O)_5(OH)^{(n-1)+}$, in aqueous solutions that range from slightly acidic to slightly basic ($3 < pH < 9$). This pH range includes most environmental waters and biological fluids.

16.8 Solubility Equilibria

FIGURE 16.16 Milk of magnesia is milky because it is a *suspension* (not a solution) of $Mg(OH)_2$.

Recall from Section 15.3 that the common strong bases include the hydroxides of alkaline earth elements (group 2). One exception is $Mg(OH)_2$, which is a weak base because it has limited solubility in water. Magnesium hydroxide is the active ingredient in the antacid called *milk of magnesia*. This liquid appears "milky" (Figure 16.16) because it is an aqueous *suspension* (not solution) of solid, white $Mg(OH)_2$. We can express the limited solubility of solid $Mg(OH)_2$ by the following equation:

$$Mg(OH)_2(s) \rightleftharpoons Mg^{2+}(aq) + 2\,OH^-(aq)$$

Because $Mg(OH)_2$ is a solid, its effective concentration does not change as long as some of it is present in the system. Therefore, the equilibrium constant for the dissolution of $Mg(OH)_2$ is

$$K_{sp} = [Mg^{2+}][OH^-]^2$$

where K_{sp} represents an equilibrium constant called the **solubility-product constant** or simply the **solubility product**.

solubility-product constant (also called **solubility product, K_{sp}**) an equilibrium constant that describes the formation of a saturated solution of a slightly soluble salt.

The K_{sp} values of $Mg(OH)_2$ and other slightly soluble compounds are listed in Table A5.4. We can use these values to calculate the concentrations of these compounds in aqueous solutions. Two terms are widely used to describe how much of a solid dissolves in a solvent: *solubility*, which is often expressed in grams of solute per 100 mL of solution, and *molar solubility*, which is expressed in moles of solute per liter of solution. In Sample Exercise 16.15, we use the K_{sp} of $Mg(OH)_2$ to calculate its solubility and molar solubility.

SAMPLE EXERCISE 16.15 Calculating the Solubility of an Ionic Compound from K_{sp} **LO8**

What are the solubility (in grams per 100 mL of solution) and the molar solubility of $Mg(OH)_2$ at 25°C?

Collect, Organize, and Analyze The K_{sp} of $Mg(OH)_2$ is 5.6×10^{-12} (Table A5.4), and the K_{sp} expression is

$$K_{sp} = [Mg^{2+}][OH^-]^2 = 5.6 \times 10^{-12}$$

For every mole of $Mg(OH)_2$ that dissolves, one mole of Mg^{2+} ions and two moles of OH^- ions go into solution. If we let x be the molar solubility (mol $Mg(OH)_2$/L of solution) at 25°C, then $[Mg^{2+}] = x$ and $[OH^-] = 2x$. Converting molar solubility to g/100 mL is a matter of multiplying molar solubility by 58.32 g/mol, the molar mass of $Mg(OH)_2$, and factoring in the smaller volume.

Solve Inserting the x terms in the K_{sp} expression and solving for x:

$$K_{sp} = (x)(2x)^2 = (x)(4x^2) = 4x^3 = 5.6 \times 10^{-12}$$

$$x = 1.1 \times 10^{-4}\ M$$

Converting this molar solubility to g/100 mL:

$$\frac{1.1 \times 10^{-4}\ \cancel{\text{mol}}}{\cancel{\text{L}}} \times \frac{58.32\ \text{g}}{1\ \cancel{\text{mol}}} \times \frac{1\ \cancel{\text{L}}}{1000\ \cancel{\text{mL}}} \times 100\ \cancel{\text{mL}} = 6.4 \times 10^{-4}\ \text{g}$$

Think About It Note that the entire algebraic expression for $[OH^-]$, $2x$, is squared in this calculation: $(2x)^2 = 4x^2$. Forgetting to square the coefficient is a common mistake. Also note that the molar solubility of $Mg(OH)_2$ is a much larger value than its solubility product. This difference is true for all sparingly soluble ionic compounds, because K_{sp} values are the products of small concentration values multiplied together, producing even smaller K_{sp} values.

 **Practice Exercise** What are the molar solubility in water and the solubility in g/100 mL of solution of $Ca_3(PO_4)_2$ at 25°C?

CONCEPT TEST

In Sample Exercise 16.15, we calculated molar solubility from K_{sp}, but the reverse calculation is also possible. If the molar solubility of BaF_2 is $7.5 \times 10^{-3}\ M$, what is the K_{sp} for barium fluoride?

SAMPLE EXERCISE 16.16 Evaluating the Common-Ion Effect on Solubility **LO8**

The mineral barite (Figure 16.17) is mostly barium sulfate ($BaSO_4$) and is widely used in industry and in medical imaging. Calculate the molar solubility at 25°C of $BaSO_4$ in (a) pure water and (b) seawater in which the concentration of sulfate ions is 2.8 g/L.

FIGURE 16.17 Crystalline barite (barium sulfate).

Collect and Organize We want to calculate the molar solubility of $BaSO_4$ in both pure water and in seawater that already contains sulfate ions. The K_{sp} value of $BaSO_4$ given in Table A5.4 is 9.1×10^{-11}.

Analyze According to the dissolution reaction for barium sulfate,

$$BaSO_4(s) \rightleftharpoons Ba^{2+}(aq) + SO_4^{2-}(aq)$$

One mole of Ba^{2+} ions and one mole of SO_4^{2-} ions form from each mole of $BaSO_4$ that dissolves. If x mol/L of $BaSO_4$ dissolves in pure water, then $[Ba^{2+}] = [SO_4^{2-}] = x$. However, seawater has a background concentration of SO_4^{2-}. According to Le Châtelier's principle and the common-ion effect, the sulfate ion already in seawater should shift the dissolution equilibrium to the left, which means that less $BaSO_4$ should dissolve in seawater than in pure water.

Solve

a. In pure water,

$$K_{sp} = [Ba^{2+}][SO_4^{2-}] = (x)(x) = 1.08 \times 10^{-10}$$
$$x = 1.04 \times 10^{-5}\ M$$

b. In seawater, we first need to calculate the value of $[SO_4^{2-}]$ before any $BaSO_4$ dissolves:

$$[SO_4^{2-}]_{initial} = \frac{2.8\ g}{L} \times \frac{1\ mol}{96.06\ g} = \frac{0.0291\ mol}{L}$$

The value of $[SO_4^{2-}]$ at equilibrium is the sum of the background concentration (0.0291 mol/L) and the additional SO_4^{2-} ions from the dissolution of $BaSO_4$ (x). Incorporating this value into the $[SO_4^{2-}]$ term in the K_{sp} expression gives

$$K_{sp} = [Ba^{2+}][SO_4^{2-}] = (x)(0.0291 + x) = 1.08 \times 10^{-10}$$

Solving for x is simplified if we assume that the K_{sp} of $BaSO_4$ is so small that we can ignore its contribution to the total SO_4^{2-} concentration. Therefore,

$$(x)(0.0291 + x) \approx (x)(0.0291) = 1.08 \times 10^{-10}$$
$$x = 3.7 \times 10^{-9}\ M$$

Think About It The calculated molar solubility of $BaSO_4$ in seawater is much less than the initial $[SO_4^{2-}]$ value, so our simplifying assumption was justified. The lower solubility of $BaSO_4$ in seawater is another illustration of the common-ion effect: the dissolution of $BaSO_4$ is suppressed by the SO_4^{2-} ions already present in seawater.

Practice Exercise What is the molar solubility of $MgCO_3$ in alkaline spring water at 25°C in which $[CO_3^{2-}] = 0.0075\ M$?

The results of Sample Exercise 16.16 illustrate how the common-ion effect can suppress the solubility of an ionic compound. Other perturbations to solubility equilibria can actually promote solubility, such as when the anion of the compound is the conjugate base of a weak acid. The molar solubilities of such compounds increase in acidic solutions, as we see in Sample Exercise 16.17.

(a)

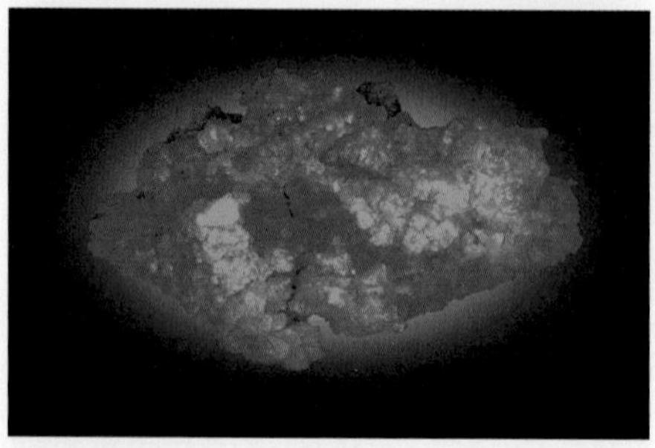

(b)

FIGURE 16.18 A sample of fluorescent fluorite (calcium fluoride) under (a) normal illumination and (b) under black light.

SAMPLE EXERCISE 16.17 Calculating the Effect of pH on Solubility **LO8**

Fluorite is a fluorescent mineral (Figure 16.18) composed mostly of calcium fluoride and the principal source fluoride ions for use in industry and medicine. What is the molar solubility of CaF_2 at 25°C in (a) pure water and (b) an acidic buffer in which $[H_3O^+]$ is a constant 0.050 M?

Collect and Organize We are asked to calculate the solubility of CaF_2 in both pure water and in an acidic buffer. The dissolution process is described by the following equilibrium:

$$(1) \quad CaF_2(s) \rightleftharpoons Ca^{2+}(aq) + 2\,F^-(aq) \qquad K_{sp} = 5.3 \times 10^{-9}$$

Analyze To account for the effect of acid on the solubility of a fluoride salt, we need to remember that the fluoride ion is the conjugate base of a weak acid (HF):

$$(2) \quad HF(aq) + H_2O(\ell) \rightleftharpoons H_3O^+(aq) + F^-(aq) \qquad K_a = 6.8 \times 10^{-4}$$

In a solution in which $[H_3O^+] = 0.050\ M$, the HF ionization reaction should proceed in the reverse direction:

$$(3) \quad H_3O^+(aq) + F^-(aq) \rightleftharpoons HF(aq) + H_2O(\ell)$$

The equilibrium constant for reaction 3 is the reciprocal of the K_a of HF:

$$K_3 = \frac{[HF]}{[H_3O^+][F^-]} = \frac{1}{6.8 \times 10^{-4}} = 1.47 \times 10^3$$

As reaction 3 proceeds, F^- ions are consumed, which shifts the equilibrium in reaction 1 to the right, increasing CaF_2 solubility. Therefore, we can anticipate an increase in the solubility of CaF_2 when acid is present.

Solve

a. Let x be the molar solubility of CaF_2 in pure water. According to the stoichiometry of reaction 1, $[Ca^{2+}] = x$ mol/L and $[F^-] = 2x$ mol/L. Inserting these variables in the K_{sp} expression:

$$K_{sp} = [Ca^{2+}][F^-]^2 = (x)(2x)^2 = 5.3 \times 10^{-9}$$

$$4x^3 = 5.3 \times 10^{-9}$$

$$x = 1.1 \times 10^{-3}\ M$$

b. In the acidic buffer, $[H_3O^+] = 0.050\ M$, and the F^- and H_3O^+ ions combine as shown in reaction 3. Assuming $[H_3O^+]$ is a constant 0.050 M, then

$$K_3 = \frac{[HF]}{[0.050][F^-]} = 1.47 \times 10^3$$

$$\frac{[HF]}{[F^-]} = 73.5$$

$$(4) \quad [HF] = 73.5[F^-]$$

According to this calculation, most of the F^- ions produced when CaF dissolves are converted into HF. The proportion that remain free F^- ions is defined by the following ratio:

$$(5) \quad \frac{[F^-]}{[F^-] + [HF]}$$

Here the numerator is the concentration of free F^- ions at equilibrium and the denominator is the concentration of all the F^- ions produced when CaF_2 dissolved. Combining equations 4 and 5 to calculate the fraction of the fluoride species that are free F^- ions and not in molecules of HF:

$$\frac{[F^-]}{[F^-] + [HF]} = \frac{[F^-]}{[F^-] + 73.5[F^-]} = \frac{[F^-]}{74.5[F^-]} = 0.0134$$

If x is the molar solubility of CaF_2 in the acid, x mol/L Ca^{2+} and $2x$ mol/L F^- are produced. However, most of the fluoride ions are converted into HF. The free F^- ion concentration is only $(0.0134)(2x) = 0.0268x$. Inserting this value in the K_{sp} expression:

$$K_{sp} = [Ca^{2+}][F^-]^2 = (x)(0.0268x)^2 = (7.18 \times 10^{-4})x^3$$

$$(7.18 \times 10^{-4})x^3 = 5.3 \times 10^{-9}$$

$$x = 1.9 \times 10^{-2}$$

Think About It Comparing the results from parts a and b shows that the molar solubility of CaF_2 is about 4 times higher in the acidic buffer, as we predicted, because most of the F^- ions produced when CaF_2 dissolves in the buffer are converted to molecules of HF. This conversion removes a product from the K_{sp} equilibrium mixture. According to Le Châtelier's principle, the result will be a shift in the position of the equilibrium to the right, in favor of forming product and increasing solubility.

 **Practice Exercise** What is the molar solubility of $Fe(OH)_2$ at 25°C in a buffer with pH 6.00?

CONCEPT TEST

Given aqueous solutions of sodium fluoride, hydrofluoric acid, and nitric acid, which would increase the solubility of PbF_2? Which would decrease the solubility of PbF_2? Which would have little effect?

The K_{sp} values listed in Table A5.4 shows that the K_{sp} values of the hydroxides of many metals, including all transition metals, are very small. However, these tiny K_{sp} values apply to equilibrium concentrations of free metal ions. As we have seen in this chapter, many metals form stable complex ions in aqueous solution, reducing their free metal ion concentration. Let's consider, for example, the solubility of Al^{3+} ions in pH 7.00 water. The K_{sp} of $Al(OH)_3$ is

$$K_{sp} = [Al^{3+}][OH^-]^3 = 1.9 \times 10^{-33}$$

Solving the K_{sp} expression for $[Al^{3+}]$ and inserting $[OH^-] = 1.0 \times 10^{-7}$, we get

$$[Al^{3+}] = \frac{K_{sp}}{[OH^-]^3} = \frac{1.9 \times 10^{-33}}{(1.0 \times 10^{-7})^3} = 1.9 \times 10^{-12}\ M \qquad (16.8)$$

This very small value seems to imply that no aluminum salt is soluble in water because the $Al^{3+}(aq)$ ions that it releases as it dissolves would immediately precipitate as $Al(OH)_3$. However, that conclusion is incorrect. $Al(NO_3)_3$, like all nitrate salts, is quite soluble in water. Its solubility can be explained by the acidic properties of hydrated Al^{3+} ions, $Al(H_2O)_6{}^{3+}$, which make $Al(NO_3)_3$ an acidic salt. The pH of 0.1 M $Al(NO_3)_3$ is about 3.0. At this pH, $[OH^-] = 1 \times 10^{-11}$. Inserting this value into Equation 16.8 and solving for $[Al^{3+}]$, we get

$$[Al^{3+}] = \frac{K_{sp}}{[OH^-]^3} = \frac{1.9 \times 10^{-33}}{(1.0 \times 10^{-11})^3} = 1.9\ M$$

This maximum value is more than 10 times the concentration of Al^{3+} ions in a 0.1 M solution of $Al(NO_3)_3$.

K_{sp} and Q

We can also use K_{sp} values to predict whether or not a particular concentration of an ionic compound is possible, or if a precipitate will form when the solutions of two salts are mixed together. In making these predictions, it is convenient to use the concept of the reaction quotient Q that we developed in Chapter 14. When applied to the equilibrium governing a slightly soluble salt, Q is sometimes called the *ion product*, because it is the product of the concentrations of the ions in solution after each is raised to a power equal to its subscript in the formula of the compound. If the calculated Q value is greater than the K_{sp} of the compound ($Q > K_{sp}$), the reaction will favor reactant formation, and the compound will precipitate (or never dissolve in the first place). If $Q < K_{sp}$, the reaction will favor product formation, and the compound will be soluble and will not precipitate.

SAMPLE EXERCISE 16.18 Predicting Whether or Not a Precipitate Forms When Two Solutions Are Mixed Together **LO8**

Lead(II) chloride is a white pigment used in 15th-century European sculpture. Will $PbCl_2$ precipitate when 275 mL of a 0.134 M solution of $Pb(NO_3)_2$ is added to 125 mL of a 0.0339 M solution of NaCl?

Collect and Organize We are asked if $PbCl_2$ will precipitate when two solutions containing Pb^{2+} ions and Cl^- ions are mixed. The solubility product of $PbCl_2$ is

$$K_{sp} = [Pb^{2+}][Cl^-]^2 = 1.70 \times 10^{-5}$$

Analyze To determine if a precipitate forms, we need to calculate the ion product Q and compare its value to K_{sp}. If $Q > K_{sp}$, $PbCl_2$ will precipitate; if $Q < K_{sp}$, it will not precipitate.

Solve First, we calculate the concentrations of the lead ions and chloride ions in the two solutions immediately after they are mixed together. Mixing the two solutions dilutes both, so the volumes of the solutions (275 mL and 125 mL) must be added together to get the final solution volume (400 mL = 0.400 L).

$$Pb^{2+}(aq): \quad 0.134\,\frac{\text{mol}}{\cancel{\text{L}}} \times 0.275\,\cancel{\text{L}} = 0.0369\text{ mol}$$

$$[Pb^{2+}] = \frac{0.0369\text{ mol}}{0.400\text{ L}} = 0.0921\,M$$

$$Cl^-(aq): \quad 0.0339\,\frac{\text{mol}}{\cancel{\text{L}}} \times 0.125\,\cancel{\text{L}} = 0.00424\text{ mol}$$

$$[Cl^-] = \frac{0.00424\text{ mol}}{0.400\text{ L}} = 0.0106\,M$$

The value of Q is

$$Q = [Pb^{2+}][Cl^-]^2 = (0.0921)(0.0106)^2 = 1.03 \times 10^{-5}$$

$Q < K_{sp}$, so *no* precipitate forms.

Think About It Lead(II) chloride was categorized as an *insoluble* compound in Table 8.4. In this scenario, $PbCl_2$ does not precipitate because the solutions of Pb^{2+} and Cl^- ions are too dilute to provide the concentrations required for the precipitate to form.

Practice Exercise Will calcium fluoride ($K_{sp} = 5.3 \times 10^{-9}$) precipitate when 175 mL of a 4.78×10^{-3} M solution of $Ca(NO_3)_2$ is added to 135 mL of a 7.35×10^{-3} M solution of KF?

We can use differences in the solubilities of ionic compounds to selectively separate ions, particularly cations, in solution. For example, suppose an aqueous solution contains 0.10 M Ca^{2+} ion and 0.020 M Mg^{2+} ion. Is it possible to selectively remove the Mg^{2+} ions from solution by precipitating them as $Mg(OH)_2$

while leaving the Ca^{2+} ions in solution? This approach might work because $Mg(OH)_2$ is much less soluble than $Ca(OH)_2$, as indicated by their K_{sp} values:

$$K_{sp} = [Mg^{2+}][OH^-]^2 = 5.6 \times 10^{-12}$$

$$K_{sp} = [Ca^{2+}][OH^-]^2 = 5.5 \times 10^{-6}$$

One way to address the question of ion separation involves calculating the maximum concentration of OH^- ions that will *not* cause the 0.10 M Ca^{2+} ion to precipitate and then determining whether that concentration is high enough to precipitate all of the Mg^{2+} ions. We can calculate the target $[OH^-]$ value from the K_{sp} of calcium hydroxide:

$$K_{sp} = 5.5 \times 10^{-6} = [Ca^{2+}][OH^-]^2 = (0.10)(x)^2$$

$$x = \sqrt{\frac{5.5 \times 10^{-6}}{0.10}} = 7.4 \times 10^{-3}\ M$$

Now we need to determine whether or not all of the Mg^{2+} ions in solution would precipitate as $Mg(OH)_2$ if $[OH^-]$ were 6.9×10^{-3} M. To do this, we calculate the $[Mg^{2+}]$ that would be in equilibrium with 6.9×10^{-3} M OH^- ions:

$$K_{sp} = 5.6 \times 10^{-12} = [Mg^{2+}][OH^-]^2 = (x)(6.9 \times 10^{-3})^2$$

$$x = \frac{5.6 \times 10^{-12}}{(6.9 \times 10^{-3})^2} = 1.2 \times 10^{-7}\ M$$

The original solution was 0.020 M Mg^{2+} ions, and 1.2×10^{-7} M Mg^{2+} represents $[(1.2 \times 10^{-7})/0.020](100\%)$, or only 0.00060%, of the original quantity of Mg^{2+} ions in the sample remaining in solution. This tiny percentage means the removal of Mg^{2+} is complete. (In general, reduction of a solute's concentration to 0.1% of its original value or less is considered complete removal.) This separation works as long as we keep $[OH^-]$ below about 10^{-3} M—that is, as long as we don't let the pH go much above 11. This approach of selectively precipitating $Mg(OH)_2$ has been used to separate these ions in seawater, where $[Ca^{2+}] = 0.0106$ M and $[Mg^{2+}] = 0.054$ M.

SAMPLE EXERCISE 16.19 Separating Anions in Solution — LO9

Both lead(II) chloride and lead(II) fluoride are slightly soluble salts. A solution of lead(II) nitrate is added to a solution that is 0.275 M in both $Cl^-(aq)$ and $F^-(aq)$. Can we use this method to separate the two halide ions? If complete precipitation is defined as there being less than 0.1% of a particular ion left in solution, is the precipitation of the first salt complete before the second salt begins to precipitate?

Collect and Organize We are given a solution that contains two ions that form slightly soluble lead(II) salts, and we are asked if one ion can be completely removed before the second one starts to precipitate when lead(II) ion is added to the solution. We have the K_{sp} values for both salts and the initial concentrations of both ions.

Analyze The equilibrium constant expressions for both ions are

$$K_{sp} = [Pb^{2+}][Cl^-]^2 = 1.7 \times 10^{-5}$$

$$K_{sp} = [Pb^{2+}][F^-]^2 = 3.2 \times 10^{-8}$$

The K_{sp} of $PbCl_2$ is $(1.7 \times 10^{-5}/3.2 \times 10^{-8}) = 500$ times the K_{sp} of PbF_2, so PbF_2 should precipitate first when Pb^{2+} ions are added to a solution containing the same concentrations of F^- and Cl^- ions. We need to determine the maximum $[Pb^{2+}]$ that could be added to precipitate PbF_2 but not cause $PbCl_2$ to precipitate. When we determine that value, we can calculate $[F^-]$ remaining in solution to determine if the precipitation of F^- as PbF_2 was complete.

Solve The maximum concentration of Pb^{2+} in the solution that will not cause the chloride ion to precipitate is

$$K_{sp} = 1.7 \times 10^{-5} = [Pb^{2+}][Cl^-]^2 = (x)(0.275)^2$$

$$x = \frac{1.7 \times 10^{-5}}{(0.275)^2} = 2.25 \times 10^{-4}\ M$$

The concentration of $F^-(aq)$ in the solution at this concentration of lead(II) ion is

$$K_{sp} = 3.5 \times 10^{-8} = (2.25 \times 10^{-4})(x)^2$$

$$x = \sqrt{\frac{3.5 \times 10^{-8}}{2.25 \times 10^{-4}}} = 0.0121\ M$$

The original solution was 0.275 M in F^- ions. A residual concentration of 0.0121 M F^- represents $(0.0123/0.275) \times 100\% = 4.5\%$ of the original $[F^-]$, which is greater than our 0.1% residual value that represents complete removal. Therefore, precipitation of $PbF_2(s)$ is incomplete and we cannot use this method to separate the two ions.

Think About It This attempt at selective precipitation did not work because the ratio of the K_{sp} values was only 500 *and* because the halide concentration terms in the K_{sp} expressions were both squared. If you look closely at the calculations, you can see that these squared concentration terms had the effect of producing a F^-/Cl^- ion ratio at equilibrium that was only the square root of their K_{sp} ratio: $(1/500)^{0.5} = 0.045$ or 4.5%.

Practice Exercise A water sample contains barium ions (0.0375 M) and calcium ions (0.0667 M). Can they be completely separated by selective precipitation of CaF_2? See Table A5.4 for the appropriate K_{sp} values.

We end this chapter with a Sample Exercise that integrates acid–base and solubility equilibria by revisiting a topic introduced in Section 16.1: ocean acidification. In this exercise we use the ionization equilibria of carbonic acid, as we have done in several other exercises, but this time we use different K_{a_1} and K_{a_2} values. Why? Because the K values we have used until now have all been *theoretical* values, which means they apply to ideal solutions in which each solute ion behaves as freely and independently as if it were the only ion present. Theoretical K values work best for very dilute solutions—they don't perform as well for solutions as concentrated (salty) as seawater. They don't account, for example, for ion pair formation. Therefore, we use *apparent* K'_{a_1} and K'_{a_2} values at 25°C in Sample Exercise 16.20 that apply to chemical equilibria in typical seawater, which contains 35 grams of dissolved sea salts per kilogram of seawater. Their symbols contain a prime (′) after the K to indicate that their values apply only to that particular sample.

CONNECTION Ion pair formation in 1.0 M solutions of ionic compounds significantly reduces the number of free ions in these solutions, as described in Chapter 11.

SAMPLE EXERCISE 16.20 Integrating Concepts: Evaluating the Impact of Ocean Acidification

As we discussed at the beginning of this chapter, increasing concentrations of atmospheric CO_2 have increased the concentration of CO_2 dissolved in the sea. As we saw in Equations 16.1–16.3, an increase in $[CO_2]$ shifts the following equilibria to the right:

(1) $H_2CO_3(aq) + H_2O(\ell) \rightleftharpoons HCO_3^-(aq) + H_3O^+(aq)$ $\quad pK'_{a_1} = 5.85$

(2) $HCO_3^-(aq) + H_2O(\ell) \rightleftharpoons CO_3^{2-}(aq) + H_3O^+(aq)$ $\quad pK'_{a_2} = 9.00$

These shifts raise oceanic $[H_3O^+]$ and lower pH.

a. The average pH of the Pacific Ocean near Hawaii dropped from 8.12 to 8.07 between 1989 and 2014. By how much did the average acidity ($[H_3O^+]$) of the ocean increase? Express your answer as a percentage of the 1989 acidity.

b. Ocean pH may drop by another 0.30 unit by the end of the 21st century. Assuming that today's concentrations of most other dissolved ions (e.g., $[Ca^{2+}] = 0.0106\ M$ and $[HCO_3^-] + [CO_3^{2-}] = 0.00211\ M$) remain the same, will the skeletal structures of marine organisms that are composed of $CaCO_3$ be soluble or insoluble in seawater in 2100? The K'_{sp} value for aragonite (the crystalline form of $CaCO_3$ in coral and seashells) is 6.46×10^{-7}.

Collect and Organize We are asked to convert a decrease in pH into an increase in acidity and to evaluate the impact of that increase on the solubility of $CaCO_3$. We know the concentration of Ca^{2+} ions and the total ($[CO_3^{2-}] + [HCO_3^{2-}]$) value as well as the appropriate K'_{sp} value for $CaCO_3$ in seawater. CO_3^{2-} and HCO_3^{2-} are a conjugate acid–base pair whose concentrations are linked by reaction 2 and the Henderson–Hasselbalch equation based on it:

$$pH = pK_a + \log\frac{[\text{base}]}{[\text{acid}]} = 9.00 + \log\frac{[CO_3^{2-}]}{[HCO_3^-]}$$

Analyze The ratio of the acidity of the seawater near Hawaii in 2014 compared to 1989 can be calculated using the equation $[H_3O^+] = 10^{-pH}$. To determine whether $CaCO_3$ will dissolve in seawater in 2100, we need to calculate the $[CO_3^{2-}]/[HCO_3^{2-}]$ ratio in equilibrium with pH $8.07 - 0.30 = 7.77$ seawater and then use that ratio and the total carbonate and bicarbonate value to calculate $[CO_3^{2-}]$. The product of that value times $[Ca^{2+}]$ will give us a Q value that can be compared with the K'_{sp} of $CaCO_3$ to determine whether seawater will be saturated with $CaCO_3$.

Solve

a. Converting the decrease in pH values to an increase in $[H_3O^+]$, expressed as a ratio of the 2014 value to the 1989 value:

$$\frac{[H_3O^+]_{2014}}{[H_3O^+]_{1989}} = \frac{10^{-8.07}\ M}{10^{-8.12}\ M} = 10^{0.05} = 1.12$$

Therefore, the 2014 sample contains 1.12 times as much acidity as the 1989 sample, which means acidity increased 12% during the 25 years before 2014.

b. Calculating the oceanic $[CO_3^{2-}]/[HCO_3^{2-}]$ ratio in 2100:

$$7.77 = 9.00 + \log\frac{[CO_3^{2-}]}{[HCO_3^-]}$$

$$\frac{[CO_3^{2-}]}{[HCO_3^-]} = 0.0589$$

If we let x be $[CO_3^{2-}]$, then

$$\frac{x}{(0.00211 - x)} = 0.0589$$

$$x = 1.17 \times 10^{-4}\ M$$

Calculating Q for $CaCO_3$:

$$Q = [Ca^{2+}][CO_3^{2-}] = (0.0106)(1.17 \times 10^{-4}) = 1.24 \times 10^{-6}$$

This value of Q ($1.24 \times 10^{-6}/6.46 \times 10^{-7}$) is 1.92 times the value of K'_{sp}, which means seawater will still be supersaturated with respect to $CaCO_3$.

Think About It The results of the calculations suggest that corals and seashells should still be able to form their skeletal structures in pH 7.77 seawater because it will still be supersaturated with respect to $CaCO_3$. However, the predicted degree of supersaturation (92%) in 2100 will have decreased from about 300% in 1989. (You are invited to repeat the part b calculation using pH 8.12 to confirm this value.) Moreover, some climate change models predict even greater decreases in oceanic pH by 2100. There appears to be cause for concern.

SUMMARY

LO1 As predicted by Le Châtelier's principle, adding conjugate base to a solution of a weak acid inhibits ionization of the acid, causing the pH of the solution to rise. Adding conjugate acid to a solution of a weak base lowers the pH of the solution. These shifts are examples of the **common-ion effect**. (Section 16.2)

LO2 A **pH buffer** is a solution that contains either a weak acid and a salt of its conjugate base or a weak base and a salt of its conjugate acid. The acidic component should have a pK_a close to the desired pH of the buffer. (Section 16.3)

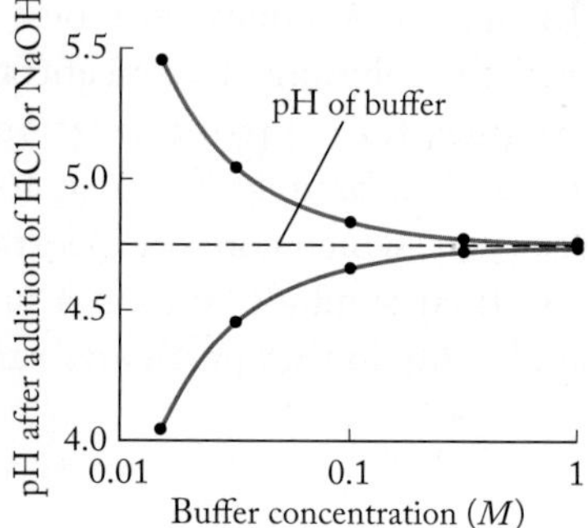

LO3 A pH buffer has the capacity to resist pH change because its acid component neutralizes additions of bases and its base component neutralizes additions of acids. (Section 16.3)

LO4 Color **pH indicators** or pH electrodes are used to detect the equivalence points in pH titrations, which are used to determine the concentrations of acids or bases in aqueous samples. (Section 16.4)

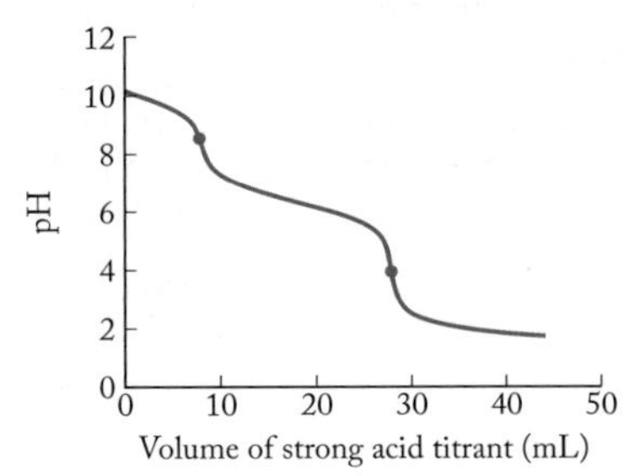

LO5 A **Lewis base** is a substance that donates pairs of electrons to a **Lewis acid**, defined as an electron-pair acceptor. The donated electron pair forms a **coordinate bond**. In some Lewis acid–Lewis base reactions, other bonds must break to accommodate the new one. (Sections 16.5 and 16.6)

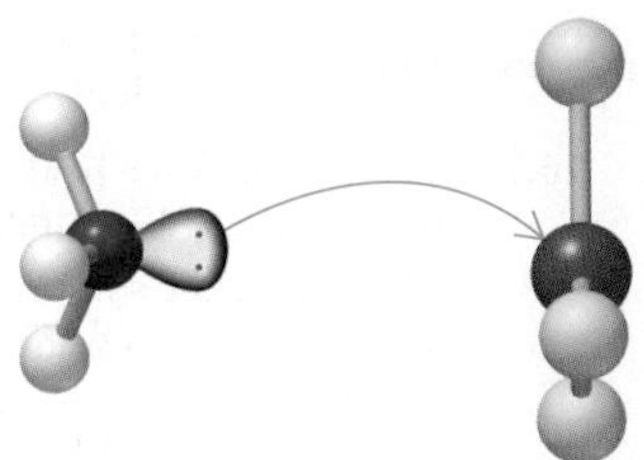

LO6 The stability of any complex ion is expressed mathematically by its **formation constant (K_f)**, which can be used to calculate the equilibrium concentration of free metal ions in a solution of complex ions. (Section 16.6)

LO7 Highly charged (e.g., 3+) hydrated metal ions are weak acids due to ionization of water molecules covalently bonded to them. (Section 16.7)

LO8 The solubility of slightly soluble ionic compounds is described by their **solubility product (K_{sp})**. Their solubility can be influenced by the common-ion effect, by complex ion formation, and by pH, especially if the anion is the conjugate base of a weak acid. (Section 16.8)

LO9 The relative solubilities of two slightly soluble ionic compounds can be used to selectively precipitate one from solution while the other remains soluble. (Section 16.8)

PARTICULATE **PREVIEW WRAP-UP**

NH_4^+ ions can donate protons (Brønsted–Lowry acid), while NH_3 molecules and OH^- ions can accept protons (Brønsted–Lowry bases). A molecule of H_2O can both donate and accept a proton. H_2O, NH_3, and OH^- each have at least one lone pair on the central atom that they can donate (Lewis base).

PROBLEM-SOLVING SUMMARY

Type of Problem	Concepts and Equations	Sample Exercises
Calculating pH of a solution of a weak base and its conjugate acid (or a weak acid and its conjugate base)	Insert the concentrations of the base and acid components and the pK_a of the acid in the Henderson–Hasselbalch equation: $$pH = pK_a + \log\frac{[\text{base}]}{[\text{acid}]} \quad (16.4)$$	**16.1, 16.2**
Preparing a buffer of given pH	Select a weak acid with a pK_a within one pH unit of the target pH. Add a Na^+ salt of its conjugate base in a proportion calculated using the Henderson–Hasselbalch equation.	**16.3, 16.4, 16.5**
Evaluating the effect of concentration and [base]:[acid] ratio on buffer capacity	Assume additions of strong acid or strong base react completely with the base or acid components of the buffer, respectively. Calculate pH from the concentrations of the components that remain using the Henderson–Hasselbalch equation.	**16.6, 16.7, 16.8**
Interpreting results of acid–base titrations	Use the volume and molarity of the titrant needed to reach the equivalence point to calculate the number of moles of it consumed by the analyte and, from the stoichiometry of the titration reaction, the number of moles of analyte in the sample.	**16.9–16.12**
Identifying Lewis acids and bases	Lewis bases donate pairs of electrons, whereas Lewis acids accept these pairs of electrons.	**16.13**
Using formation constants to calculate the concentration of free or complexed ion	Set up a RICE table based on the complex formation reaction. Let x be the concentration of free (not complexed) metal ion. Solve for x, which is usually much smaller than the concentration of the complex.	**16.14**
Calculating the solubility of an ionic compound from its K_{sp}	Express the concentrations of the cation and anion in the K_{sp} expression in terms of x moles of the compound that dissolve in one liter of solution.	**16.15, 16.16**

Type of Problem	Concepts and Equations	Sample Exercises
Calculating the effect of pH on solubility	If A^- is the conjugate base of a weak acid HA, calculate the fraction of A^- that remains as the free ion. Use this fraction as the coefficient for molar solubility in the K_{sp} expression.	**16.17**
Determining if a precipitate forms when solutions are mixed, and which precipitate forms first if more than one is possible	Compare the ion product Q to K_{sp} to determine if a precipitate will form; use K_{sp} expressions to calculate maximum concentrations of one ion in solution that will not cause another ion to precipitate.	**16.18, 16.19**

VISUAL PROBLEMS

(Answers to boldface end-of-chapter questions and problems are in the back of the book.)

16.1. The graph in Figure P16.1 shows the titration curves of a 1 *M* solution of a weak acid with a strong base and a 1 *M* solution of a strong acid with the same base. Which curve is which?

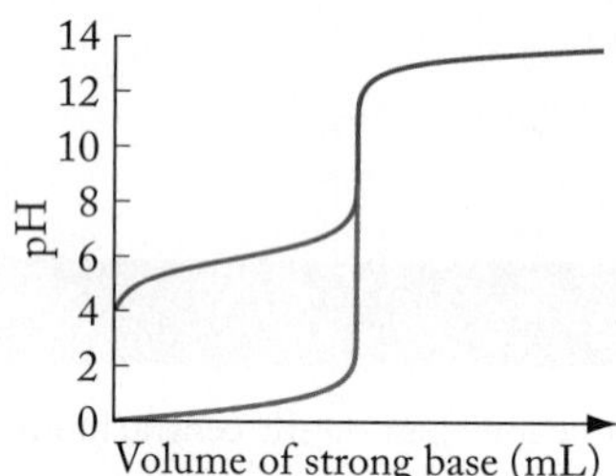

FIGURE P16.1

16.2. Estimate the pK_a of the weak acid in Problem 16.1.

16.3. Suppose you have four color indicators to choose from to detect the equivalence point of the titration reaction represented by the red curve in Figure P16.1. The pK_a values of the four indicators are 3.3, 5.0, 7.0, and 9.0. Which indicator would be the best one to choose?

16.4. Explain why the slope of the red titration curve in Figure P16.1 is nearly flat in the region extending about halfway from the start of the titration to its equivalence point.

16.5. One of the titration curves in Figure P16.5 represents the titration of an aqueous sample of Na_2CO_3 with strong acid; the other represents the titration of an aqueous sample of $NaHCO_3$ with the same acid. Which curve is which?

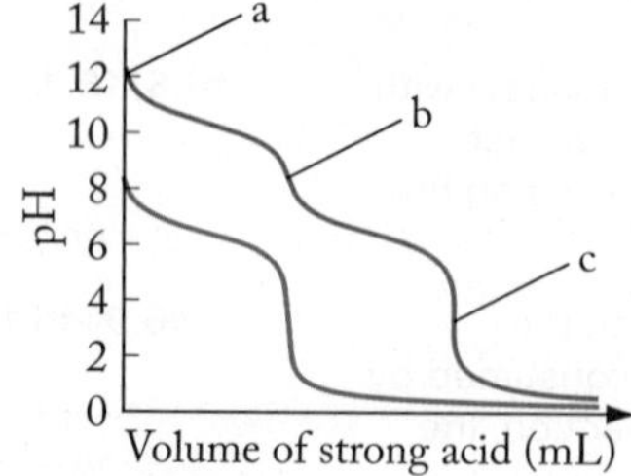

FIGURE P16.5

16.6. Identify the principal carbon-containing species in solution at points a, b, and c on the red titration curve in Figure P16.5.

16.7. Each of the three beakers in Figure P16.7 contains a few drops of the color indicator bromthymol blue, which is yellow in acidic solutions and blue in basic solutions. One beaker contains a solution of ammonium chloride, one contains ammonium acetate, and the third contains sodium acetate. Which beaker contains which salt?

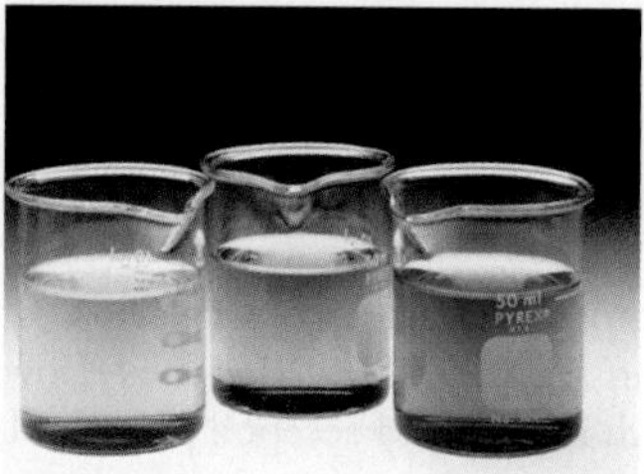

FIGURE P16.7

*16.8. The graphs in Figure P16.8 show the conductivity of a solution as a function of the volume of titrant added. Which of the graphs best represents the titration of (a) a strong acid with a strong base and (b) a weak acid with a strong base?

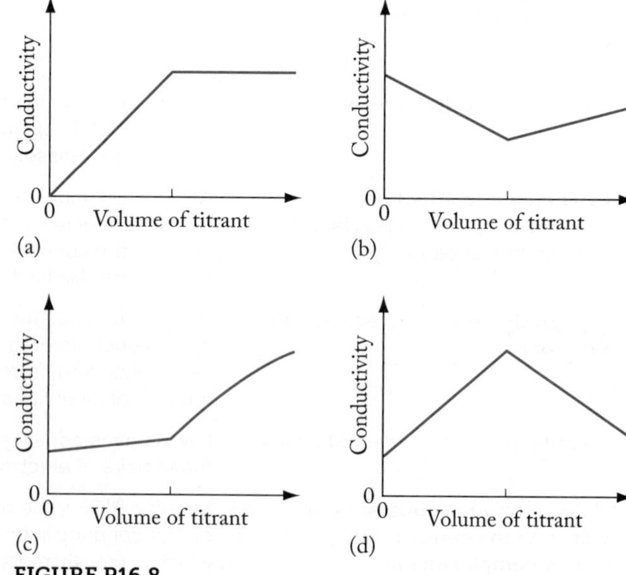

FIGURE P16.8

16.9. a. What kind of aqueous solution is represented in Figure P16.9: a weak acid, a weak base, or a buffer? The solvent molecules have been omitted for clarity.
b. What particles depicted in Figure P16.9 will change, and how will they change, upon addition of a strong base such as NaOH?

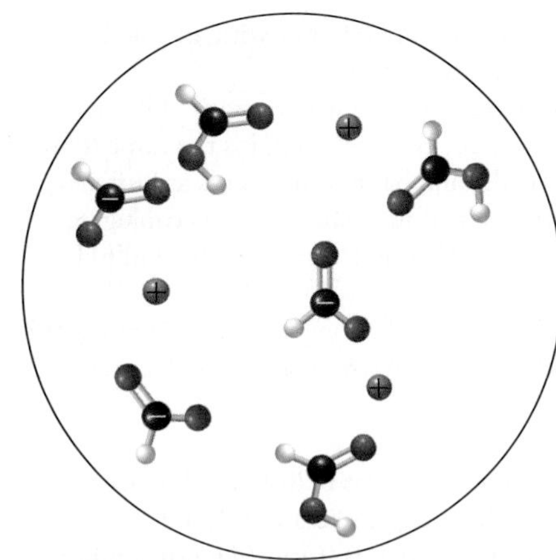

FIGURE P16.9

16.10. Use representations [A] through [I] in Figure P16.10 to answer questions a–f.
a. Which solution depicts a strong acid? Which depicts a strong acid after titration to its equivalence point with NaOH?
b. Which represents a buffer with equal concentrations of a weak acid and its conjugate base?
c. Which represents the buffer from part b in this question after the addition of strong acid?
d. Which represents the buffer from part b in this question after the addition of strong base?
e. Which structure shows the protonated form of the indicator methyl red? Which shows the deprotonated form of methyl red?
f. If methyl red is red at lower pH and yellow at higher pH, match the flasks in [C] and [G] to the structures in [A] and [I].

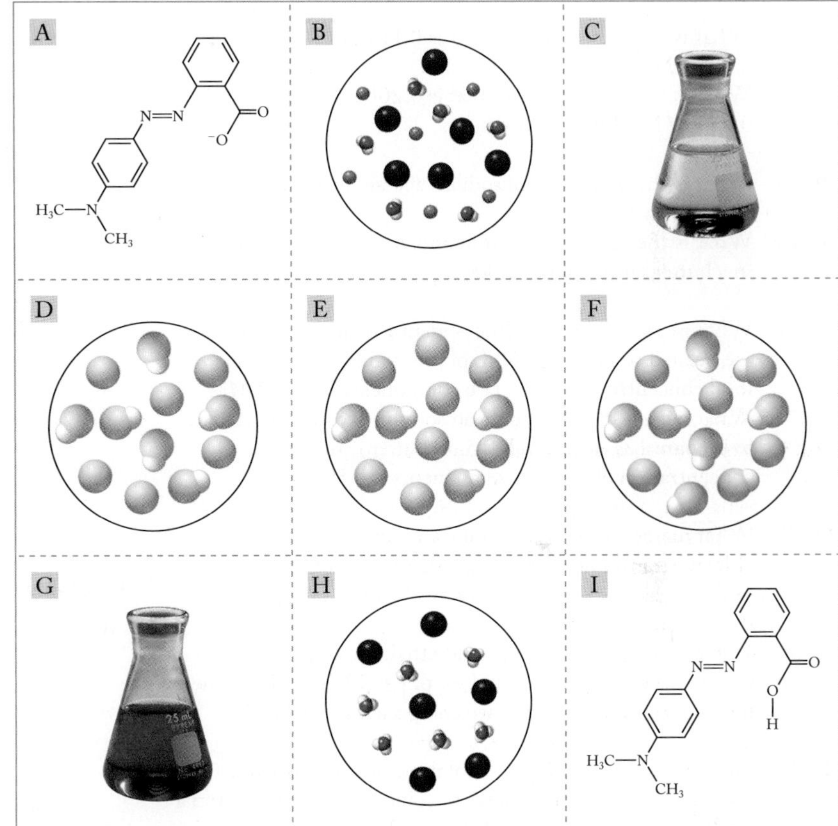

FIGURE P16.10

QUESTIONS AND PROBLEMS

Note: Tables A5.1 and A5.3 in Appendix 5 contain K_a, K_b, pK_a, pK_b, K_f, and K_{sp} values that may be useful in answering the questions and solving the problems in these sections.

The Common-Ion Effect and pH Buffers

Concept Review

16.11. Why does a solution of a weak acid and its conjugate base control pH better than a solution of the weak acid alone?

16.12. Why does a solution of a weak base and its conjugate acid control pH better than a solution of the weak base alone?

16.13. Identify a suitable buffer system to maintain a pH of 3.0 in an aqueous solution.

16.14. Identify a suitable buffer system to maintain a pH of 9.0 in an aqueous solution.

16.15. What does "buffer capacity" mean?

16.16. What effect does adding more NaF have on the pH and buffer capacity of an aqueous solution that is initially 1.0 *M* HF and 0.50 *M* NaF?

16.17. Three buffers are separately prepared using equal concentrations of formic acid and sodium formate, hydrofluoric acid and sodium fluoride, and acetic acid and sodium acetate. Rank the three buffers from highest to lowest pH.

16.18. Equal volumes of two buffers are prepared with equal concentrations of acid and conjugate base, but they use different weak acids with different pK_a values. Do the two buffers have the same buffer capacity?

16.19. How does diluting a pH = 4.00 buffer with an equal volume of pure water affect its pH?

*16.20. Buffer A contains nearly equal concentrations of its conjugate acid–base pair. Buffer B contains the same total concentration of acidic and basic components as buffer A, but B has twice as much of its weak acid as its conjugate base. Which buffer experiences a smaller change in pH when

a. the same small quantity of strong base is added to both?
b. the same small quantity of strong acid is added to both?

Problems

16.21. What is the pH of a buffer that is 0.200 *M* chloroacetic acid and 0.100 *M* sodium chloroacetate at 25°C?

16.22. What is the pH of a buffer that is 0.100 *M* methylamine and 0.175 *M* methylammonium chloride at 25°C?

16.23. What is the pH of a buffer that is 0.110 *M* HPO_4^{2-} and 0.220 *M* $H_2PO_4^-$ at 25°C?

16.24. What is the pH of a buffer that is 0.200 *M* H_2SO_3 and 0.250 *M* $NaHSO_3$ at 25°C?

16.25. What is the mole ratio of sodium acetate to acetic acid in a buffer with a pH of 5.75?

16.26. What is the mole ratio of ammonia to ammonium chloride in a buffer with a pH of 9.00?

16.27. What masses of bromoacetic acid and sodium bromoacetate are needed to prepare 1.00 L of pH = 3.00 buffer if the total concentration of the two components is 0.200 *M*?

16.28. What masses of acetic acid and sodium acetate are needed to prepare 125 mL of pH = 5.00 buffer if the total concentration of the two components is 0.500 *M*?

16.29. What masses of dimethylamine and dimethylammonium chloride do you need to prepare 0.500 L of pH = 11.00 buffer if the total concentration of the two components is 0.300 *M*?

16.30. What masses of ethylamine and ethylammonium chloride do you need to prepare 1.00 L of pH = 10.50 buffer if the total concentration of the two components is 0.250 *M*?

16.31. What is the pH at 25°C of a solution that results from mixing equal volumes of 0.05 *M* solution of ammonia and a 0.025 *M* solution of hydrochloric acid?

16.32. What is the pH at 25°C of a solution that results from mixing equal volumes of a 0.05 *M* solution of acetic acid and a 0.025 *M* solution of sodium hydroxide?

***16.33.** What volume of 0.422 *M* NaOH must be added to 0.500 L of 0.300 *M* acetic acid to raise its pH to 4.00 at 25°C?

*16.34. What volume of 1.16 *M* HCl must be added to 0.250 L of 0.350 *M* dimethylamine to produce a buffer with a pH of 10.75 at 25°C?

***16.35.** A buffer consists of 0.120 *M* HNO_2 and 0.150 *M* $NaNO_2$ at 25°C.

a. What is the pH of the buffer?
b. What is the pH after the addition of 1.00 mL of 11.6 *M* HCl to 1.00 L of the buffer solution?

*16.36. A buffer is prepared by mixing 50.0 mL of 0.200 *M* NaOH with 100.0 mL of 0.175 *M* acetic acid.

a. What is the pH of the buffer?
b. What is the pH of the buffer after 1.00 g NaOH is dissolved in it?

Indicators and Acid–Base Titrations

Concept Review

16.37. Do all titrations of samples of strong monoprotic acids with solutions of strong bases have the same pH at their equivalence points? Explain why or why not.

16.38. Do all titrations of samples of weak monoprotic acids with solutions of strong bases have the same pH at their equivalence points? Explain why or why not.

16.39. Describe two properties of phenolphthalein that make it a good choice as an indicator for detecting the first equivalence point in an alkalinity titration.

*16.40. Phenolphthalein can be used as a color indicator to detect the equivalence points of titrations of samples containing either weak acids or strong acids, even though the pH values of the equivalence points vary depending on the identity of the acid. Explain how this is possible.

16.41. In the titration of a solution of a weak monoprotic acid with a standard solution of NaOH, the pH halfway to the equivalence point was 4.44. In the titration of a second solution of the same acid, exactly twice as much of the standard solution of NaOH was needed to reach the equivalence point. What was the pH halfway to the equivalence point in this titration?

16.42. The pH of a solution of a strong monoprotic acid is lower than the pH of an equal concentration of a weak monoprotic acid, yet equal volumes of both require the same volume of basic titrant to reach the equivalence point. Explain why.

Problems

16.43. A 25.0 mL sample of 0.100 *M* acetic acid is titrated with 0.125 *M* NaOH at 25°C. What is the pH of the solution after 10.0, 20.0, and 30.0 mL of the base have been added?

16.44. A 25.0 mL sample of a 0.100 *M* solution of aqueous trimethylamine is titrated with a 0.125 *M* solution of HCl at 25°C. What is the pH of the solution after 10.0, 20.0, and 30.0 mL of acid have been added?

16.45. Sketch a titration curve for the titration of 50.0 mL of 0.200 *M* HNO_2 with 1.00 *M* NaOH. What is the pH of the sample after 2.50, 5.00, 7.50, and 10.00 mL of titrant have been added?

*16.46. Sketch a titration curve for the titration of 40.0 mL of a 0.100 *M* solution of oxalic acid with a 0.100 *M* solution of NaOH. What is the pH of the titration reaction mixture after 10.0, 20.0, 30.0, and 40.0 mL of titrant have been added?

16.47. What volume of 0.100 *M* HCl is required to titrate 250 mL of 0.0100 *M* Na_2CO_3 to the first equivalence point?

16.48. What volume of 0.0100 *M* HCl is required to titrate 250 mL of 0.0100 *M* Na_2CO_3 and 250 mL of 0.0100 *M* HCO_3^-?

***16.49.** **Window Cleaner** (a) What is the concentration of ammonia in a popular window cleaner if 25.34 mL of 1.162 *M* HCl is needed to titrate a 10.00 mL sample of the cleaner? (b) Suppose that the sample was diluted to about 50 mL with deionized water prior to the titration to make it easier to mount a pH electrode in it. What effect did this dilution have on the volume of titrant needed?

16.50. In an alkalinity titration of a 100.0 mL sample of water from a hot spring, 2.56 mL of a 0.0355 *M* solution of HCl is needed to reach the first equivalence point (pH = 8.3) and another 10.42 mL is needed to reach the second equivalence point (pH = 4.0). If the alkalinity of the spring water is due only to the presence of carbonate and bicarbonate, what are the concentrations of each?

16.51. For each titration, predict whether the pH of the equivalence point is less than, equal to, or greater than 7.
a. Quinine titrated with nitric acid
b. Pyruvic acid titrated with calcium hydroxide
c. Hydrobromic acid titrated with strontium hydroxide

16.52. For each titration, predict whether the pH of the equivalence point is less than, equal to, or greater than 7.
a. HCN titrated with $Ca(OH)_2$
b. LiOH titrated with HI
c. C_5H_5N titrated with KOH

16.53. When 100 mL of 0.0125 *M* ascorbic acid is titrated with 0.010 *M* NaOH, how many equivalence points will the titration curve have, and what pH indicator(s) could be used? Refer to Figure 16.5 for the colors of indicators.

16.54. Red cabbage juice is a sensitive acid–base indicator; its colors range from red at acidic pH to yellow in alkaline solutions. What color would red cabbage juice have at the equivalence point when 25 mL of a 0.10 *M* solution of acetic acid is titrated with 0.10 *M* NaOH?

Lewis Acids and Bases

Concept Review

16.55. Are all Lewis bases also Brønsted–Lowry bases? Explain why or why not.

16.56. Are all Brønsted–Lowry bases also Lewis bases? Explain why or why not.

16.57. Are all Brønsted–Lowry acids also Lewis acids? Explain why or why not.

16.58. Why is BF_3 a Lewis acid but not a Brønsted–Lowry acid?

Problems

16.59. Draw Lewis structures that show how electron pairs move and bonds form and break during the autoionization of water. Label the appropriate H_2O molecules as the Lewis acid and Lewis base.

16.60. Draw Lewis structures that show how electron pairs move and bonds form and break in the following reaction, and identify the Lewis acid and Lewis base.

$$MgO(s) + CO_2(g) \rightarrow MgCO_3(s)$$

16.61. Draw Lewis structures that show how electron pairs move and bonds form and break in the following reaction, and identify the Lewis acid and Lewis base.

$$SO_2(g) + H_2O(\ell) \rightarrow H_2SO_3(aq)$$

16.62. Draw Lewis structures that show how electron pairs move and bonds form and break in the following reaction, and identify the Lewis acid and Lewis base.

$$SeO_3(g) + H_2O(\ell) \rightarrow H_2SeO_4(aq)$$

16.63. Draw Lewis structures that show how electron pairs move and bonds form and break in the following reaction, and identify the Lewis acid and Lewis base.

$$B(OH)_3(aq) + 2\,H_2O(\ell) \rightleftharpoons B(OH)_4^-(aq) + H_3O^+(aq)$$

*16.64. Draw Lewis structures that show how electron pairs move and bonds form and break in the following reaction, and identify the Lewis acid and Lewis base.

$$SbF_5(s) + HF(g) \rightarrow HSbF_6(s)$$

(*Note*: $HSbF_6$ is an ionic compound and one of the strongest Brønsted–Lowry acids known.)

Formation of Complex Ions

Concept Review

16.65. When $CaCl_2$ dissolves in water, which molecules or ions occupy the inner coordination sphere around the Ca^{2+} ions?

16.66. When $AgNO_3$ dissolves in water, which molecules or ions occupy the inner coordination sphere around the Ag^+ ions?

***16.67.** A lab technician cleaning glassware that contains residues of AgCl washes the glassware with an aqueous solution of ammonia. The AgCl, which is insoluble in water, rapidly dissolves in the ammonia solution. Why?

*16.68. The procedure used in Problem 16.67 dissolves AgCl but not AgI. Why?

Problems

16.69. A solution is prepared in which 0.00100 mol $Ni(NO_3)_2$ and 0.500 mol NH_3 are dissolved in a total volume of 1.00 L. What is the concentration of $Ni(H_2O)_6^{2+}$ ions in the solution at equilibrium?

16.70. A 1.00 L solution contains 5.00×10^{-5} *M* $Cu(NO_3)_2$ and 1.00×10^{-3} *M* ethylenediamine. What is the concentration of $Cu(H_2O)_6^{2+}$ ions in the solution at equilibrium?

***16.71.** Suppose a solution contains 1.00 mmol $Co(NO_3)_2$, 0.100 mol NH_3, and 0.100 mol ethylenediamine in a total volume of 0.250 L. What is the concentration of $Co(H_2O)_6^{2+}$ ions in the solution?

*16.72. If 1.00 mL 0.0100 *M* $AgNO_3$, 1.00 mL 0.100 *M* NaBr, and 1.00 mL 0.100 *M* NaCN are diluted to 250 mL with deionized water in a volumetric flask and shaken vigorously, will the contents of the flask be cloudy or clear? Support your answer with the appropriate calculations.

Hydrated Metal Ions as Acids

Concept Review

16.73. Which, if any, aqueous solutions of the following chloride compounds are acidic? (a) $CaCl_2$; (b) $CrCl_3$; (c) NaCl; (d) $FeCl_3$

16.74. If 0.100 *M* aqueous solutions of each of these compounds were prepared, which one would have the lowest pH? (a) $BaCl_2$; (b) LiCl; (c) KCl; (d) $TiCl_4$

16.75. When ozone is bubbled through an aqueous solution of Fe^{2+} ions, the ions are oxidized to Fe^{3+} ions. How does the oxidation process affect the pH of the solution?

16.76. As an aqueous solution of KOH is slowly added to a stirred solution of $AlCl_3$, the mixture becomes cloudy but then clears when more KOH is added.
a. Explain the chemical changes responsible for the changes in the appearance of the mixture.
b. Would you expect to observe the same changes if KOH were added to a solution of $FeCl_3$? Explain why or why not.

16.77. Chromium(III) hydroxide is amphiprotic. Write chemical equations showing how an aqueous suspension of this compound reacts to the addition of a strong acid and the addition of a strong base.

16.78. Zinc hydroxide is amphiprotic. Write chemical equations showing how an aqueous suspension of this compound reacts to the addition of a strong acid and the addition of a strong base.

16.79. **Refining Aluminum** To remove impurities such as calcium and magnesium carbonates and Fe(III) oxides from aluminum ore (which is mostly Al_2O_3), the ore is treated with a strongly basic solution. In this treatment, Al^{3+} dissolves but the other metal ions do not. Why?

*16.80. Exactly 1.00 g of $FeCl_3$ is dissolved in each of four 0.500 L samples: 1 M HNO_3, 1 M HNO_2, 1 M CH_3COOH, and pure water. Is the concentration of $Fe(H_2O)_6{}^{3+}$ ions the same in all four solutions? Explain why or why not.

Problems

16.81. What is the pH of 0.25 M $Al(NO_3)_3$?

16.82. What is the pH of 0.50 M $CrCl_3$?

16.83. What is the pH of 0.100 M $Fe(NO_3)_3$?

16.84. What is the pH of 1.00 M $Cu(NO_3)_2$?

16.85. Sketch the titration curve (pH versus volume of 0.50 M NaOH added) for a 25 mL sample of 0.25 M $FeCl_3$.

16.86. Sketch the titration curve that results from the addition of 0.50 M NaOH to a sample containing 0.25 M $KFe(SO_4)_2$.

Solubility Equilibria

Concept Review

16.87. What is the difference between *molar solubility* and *solubility product*?

16.88. Give an example of how the common-ion effect limits the dissolution of a sparingly soluble ionic compound.

16.89. Which cation will precipitate first as a carbonate mineral from an equimolar solution of Mg^{2+}, Ca^{2+}, and Sr^{2+}?

16.90. If the solubility of a compound increases with increasing temperature, does K_{sp} increase or decrease?

16.91. The K_{sp} of strontium sulfate increases from 2.8×10^{-7} at 37°C to 3.8×10^{-7} at 77°C. Is the dissolution of strontium sulfate endothermic or exothermic?

16.92. Identify any of the following solids that are more soluble in acidic solution than in neutral water: $CaCl_2$, $Ba(HCO_3)_2$, $PbSO_4$, $Cu(OH)_2$. Explain your choices.

16.93. **Chemistry of Tooth Decay** Tooth enamel is composed of a mineral known as hydroxyapatite, which has the formula $Ca_5(PO_4)_3(OH)$. Explain why tooth enamel can be eroded by acidic substances released by bacteria growing in the mouth.

16.94. **Fluoride and Dental Hygiene** Fluoride ions in drinking water and toothpaste convert hydroxyapatite in tooth enamel into fluorapatite:

$$Ca_5(PO_4)_3(OH)(s) + F^-(aq) \rightleftharpoons Ca_5(PO_4)_3F(s) + OH^-(aq)$$

Why is fluorapatite less susceptible than hydroxyapatite to erosion by acids?

Problems

16.95. At a particular temperature, the $[Ba^{2+}]$ in a saturated solution of barium sulfate is 1.04×10^{-5} M. Starting with this information, calculate the K_{sp} value of barium sulfate at this temperature.

16.96. If only 0.160 g $Ca(OH)_2$ dissolves in 0.100 L of water, what is the K_{sp} value for calcium hydroxide at that temperature?

16.97. What are the equilibrium concentrations of Cu^+ and Cl^- in a saturated solution of copper(I) chloride at 25°C?

16.98. What are the equilibrium concentrations of Pb^{2+} and F^- in a saturated solution of lead(II) fluoride at 25°C?

16.99. What is the solubility of calcite ($CaCO_3$) in grams per milliliter at a temperature at which its $K_{sp} = 9.9 \times 10^{-9}$?

16.100. What is the solubility of silver iodide in grams per milliliter at 25°C?

16.101. What is the pH at 25°C of a saturated solution of silver hydroxide?

16.102. **pH of Milk of Magnesia** What is the pH at 25°C of a saturated solution of magnesium hydroxide (the active ingredient in the antacid milk of magnesia)?

16.103. Suppose you have 100 mL of each of the following solutions. In which will the most $CaCO_3$ dissolve?
(a) 0.1 M NaCl; (b) 0.1 M Na_2CO_3; (c) 0.1 M NaOH;
(d) 0.1 M HCl

16.104. In which of the following solutions will CaF_2 be most soluble? (a) 0.010 M $Ca(NO_3)_2$; (b) 0.01 M NaF;
(c) 0.001 M NaF; (d) 0.10 M $Ca(NO_3)_2$

16.105. **Composition of Seawater** The average concentration of sulfate in surface seawater is about 0.028 M. The average concentration of Sr^{2+} is 9×10^{-5} M. Is the concentration of strontium in the sea significantly controlled by the insolubility of its sulfate salt?

16.106. **Fertilizing the Sea to Combat Climate Change** Some scientists have proposed adding Fe(III) compounds to large expanses of the open ocean to promote the growth of phytoplankton that would in turn remove CO_2 from the atmosphere through photosynthesis. Assuming the average pH of open ocean water is 8.13, what is the maximum value of $[Fe^{3+}]$ in seawater if the K_{sp} value of $Fe(OH)_3$ is 1.1×10^{-36}?

16.107. Will calcium fluoride precipitate when 125 mL of 0.375 *M* $Ca(NO_3)_2$ is added to 245 mL of 0.255 *M* NaF at 25°C?

16.108. Will lead(II) chloride precipitate if 185 mL of 0.025 *M* sodium chloride is added to 235 mL of 0.165 *M* lead(II) perchlorate at 25°C?

16.109. A solution is 0.010 *M* in both Br^- and SO_4^{2-}. A 0.250 *M* solution of lead(II) nitrate is slowly added to it using a burette.

a. Which anion will precipitate first?
b. What is the concentration of the first anion when the second one starts to precipitate at 25°C?

*16.110. Solution A is 0.0200 *M* in Ag^+ ions and Pb^{2+} ions. You have access to two other solutions: (B) 0.250 *M* NaCl and (C) 0.250 *M* NaBr.

a. Which solution, B or C, would be the better one to add to solution A to separate Ag^+ ions from Pb^{2+} by selective precipitation?
b. Using the solution you selected in part a, is the separation of the two ions complete?

Additional Problems

***16.111. pH of Baking Soda** A cook dissolves a teaspoon of baking soda ($NaHCO_3$) in a cup of water and then discovers that the recipe calls for a tablespoon, not a teaspoon. If the cook adds two more teaspoons of baking soda to make up the difference, does the additional baking soda change the pH of the solution? Explain why or why not.

*16.112. **Antacid Tablets** Antacids contain a variety of bases such as $NaHCO_3$, $MgCO_3$, $CaCO_3$, and $Mg(OH)_2$. Only $NaHCO_3$ has appreciable solubility in water.

a. Write a net ionic equation for the reaction of each base with aqueous HCl.
b. Explain how substances sparingly soluble in water can act as effective antacids.

16.113. When silver oxide dissolves in water, the following reaction occurs:

$$Ag_2O(s) + H_2O(\ell) \rightarrow 2\,Ag^+(aq) + 2\,OH^-(aq)$$

If a saturated aqueous solution of silver oxide is 1.6×10^{-4} *M* in hydroxide ion, what is the K_{sp} of silver oxide?

*16.114. Why does adding $CaCl_2$ to a HPO_4^{2-}/PO_4^{3-} buffer increase the ratio of HPO_4^{2-} ions to PO_4^{3-} ions?

***16.115. Greenhouse Gases and Ocean pH** Some climate models predict the pH of the oceans will decrease by as much as 0.77 pH unit due to increases in atmospheric carbon dioxide.

a. Use the appropriate chemical reactions and equilibria to explain how increasing atmospheric CO_2 produces a decrease in oceanic pH.
b. How much more acidic (in terms of $[H_3O^+]$) would the oceans be if their pH dropped this much?
c. Oceanographers are concerned about the impact of a drop in oceanic pH on the survival of oysters. Why?

*16.116. A 125.0 mg sample of an unknown monoprotic acid was dissolved in 100.0 mL of distilled water and titrated with a 0.050 *M* solution of NaOH. The pH of the solution was monitored throughout the titration, and the following data were collected.

Volume of OH^- Added (mL)	pH	Volume of OH^- Added (mL)	pH
0	3.09	22	5.93
5	3.65	22.2	6.24
10	4.10	22.6	9.91
15	4.50	22.8	10.2
17	4.55	23	10.4
18	4.71	24	10.8
19	4.94	25	11.0
20	5.11	30	11.5
21	5.37	40	11.8

a. What is the K_a value for the acid?
b. What is the molar mass of the acid?

16.117. Based on the location of the equivalence points in the titration curve in the figure, which of the following statements is true about the alkalinity of the sample?

a. It is due only to the presence of carbonate ions
b. It is due mostly to the presence of carbonate ions
c. It is due only to the presence of bicarbonate ions
d. It is due mostly to the presence of bicarbonate ions
e. The ratio of bicarbonate to carbonate ions in the sample is exactly 50:50 (on a mole basis).

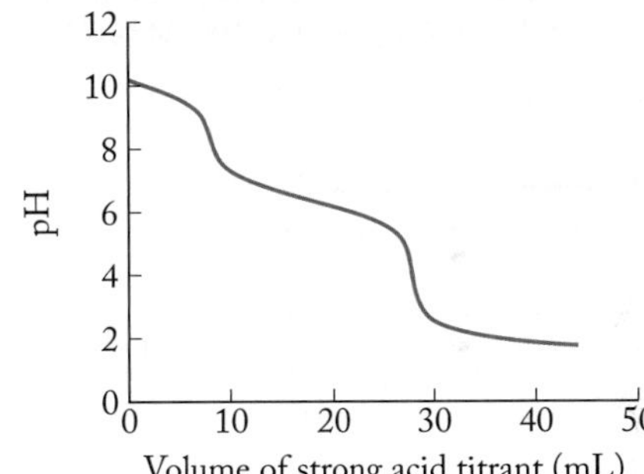

FIGURE P16.117

16.118. A pH 3.00 buffer is prepared by mixing solutions of nitrous acid and sodium nitrite. The total concentration of nitrous acid and sodium nitrate in the buffer is 0.100 *M*. Suppose 1.00 milliliter of 1.00 *M* HCl is added to a 100 mL sample of the buffer and 1.00 milliliter of 1.00 *M* NaOH is added to another 100 mL sample. In which sample would the addition of strong acid or base produce the greater change in pH? Explain your selection.

16.119. For each set of three acids, which one should you use to make a buffer with the given pH?

a. Select from acetic acid (pK_a 4.75), fluoroacetic acid (pK_a 2.59) and hypochlorous acid (pK_a 7.54) to make a buffer with a pH of 5.2.
b. Select from formic acid (pK_a = 3.75), hypobromous acid (pK_a = 7.54), and boric acid (pK_a 9.27) to make a buffer with a pH of 8.0.

16.120. Your task is to prepare a buffer that has a pH of exactly 8.00. The reagents you have available are H_3PO_4 ($\mathcal{M}$ = 98.0 g/mol), NaH_2PO_4 ($\mathcal{M}$ = 120.0 g/mol), Na_2HPO_4 ($\mathcal{M}$ = 142.0 g/mol), and Na_3PO_4 ($\mathcal{M}$ = 163.9 g/mol). How many grams of which of two of these reagents would you use to prepare exactly one liter of a buffer in which the total concentration of acid and basic components is 0.100 *M*?

16.121. Estimate the K_a values of the following indicators:
   a. Bromphenol blue, whose transition color occurs at a pH of about 3.8.
   b. Bromcresol green, whose transition color occurs at a pH of about 4.3.
   c. Alizarin yellow R, whose transition color occurs at a pH of about 10.9.

16.122. Figure 16.5 depicts an array of indicators. Use this figure to answer the following questions.
   a. The end-point in the titration of butanoic acid with sodium hydroxide occurs at a pH of about 8.52. Select a suitable indicator to use in this titration.
   b. Which of the color indicators could be used to detect the first equivalence point in the titration of a sample of phosphoric acid?

16.123. Calculate the pH at the equivalence point for the titration of 10.0 mL of 0.100 *M* formic acid ($K_a = 1.77 \times 10^{-4}$) with 0.100 *M* NaOH and for the titration of 10.0 mL of boric acid ($K_{a1} = 5.4 \times 10^{-10}$) with the same solution. Should the same indicator be used for both titrations?

16.124. The titration curve below shows the results of the titration of a 0.100 *M* aqueous solution of the propionic acid (CH_3CH_2COOH = HPr) with 0.100 *M* NaOH.

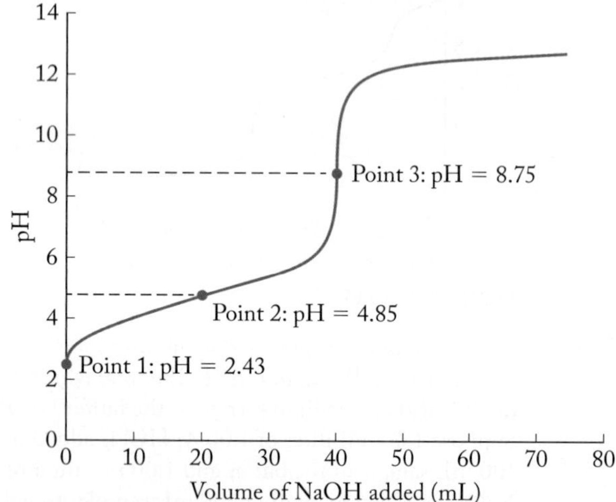

FIGURE P16.124

   a. At the points indicated on the titration curve, identify the relative concentrations of
      Point 1: $[Pr^-]$ and $[H_3O^+]$
      Point 2: [HPr] and $[Pr^-]$
      Point 3: [HPr] initially present at point 1 and $[OH^-]$ added.
   b. Estimate the pK_a of propionic acid.
   c. By referencing the number of mL of sodium hydroxide solution added (*x*-axis), define the buffer region of the titration curve.
   d. Refer to Figure 16.5 and select an appropriate indicator to determine the equivalence point of the titration.

16.125. Suppose it takes exactly 15.00 mL of NaOH titrant to reach the first equivalence point in the titration of a sample of sulfurous acid (H_2SO_3, pK_{a1} = 1.77, pK_{a2} = 7.21).
   a. How much more titrant (in milliliters) is required to reach the second equivalence point?
   b. What is the pH of the titration mixture after 7.50 mL of titrant have been added?
   c. What is the pH of the titration mixture after a total of 22.50 mL of titrant have been added?
   d. Identify the two most abundant ions in the titration reaction mixture at the first equivalence point.
   e. Identify the two most abundant ions in the titration reaction mixture at the second equivalence point.

16.126. Consider the titration of 21.5 mL of 0.120 *M* phenol (pK_a = 9.89) with 0.250 *M* NaOH.
   a. Calculate the pH at the equivalence point.
   b. Which indicator from Figure 16.5 would be the best choice for this titration?
   c. Calculate the pH for the titration in question after 0, 2.5, 5.0, 7.5, 9.8, 10, 10.2, 10.6, 10.8, 11, 12.5, 15, 17.5 and 20 mL of base have been added.
   d. Sketch the titration curve (pH vs mL of base added) for this titration and indicate which major species are present in solution when pH = pK_a, at the equivalence point and when the titration is complete.

16.127. Consider the titration of 15.8 mL of 367 m*M* pyridine (pK_b = 8.77) with 0.500 *M* HCl.
   a. Calculate the pH at the equivalence point.
   b. Which indicator from Figure 16.5 would be the best choice for this titration?
   c. Calculate the pH for the titration in question after 0.0, 1.0, 3.0, 7.0, 9.0, 10.0, 10.5, 11.0, 11.5, 11.8, 12.5, 15, and 20 mL of base have been added.
   d. Sketch the titration curve (pH vs mL of base added) for this titration and indicate which major species are present in solution when pH = pK_a, and at the equivalence point and when the titration is complete.

16.128. In each of the following pairs of compounds, which is less soluble in water?
   a. AgI ($K_{sp} = 8.3 \times 10^{-17}$) or AgCl ($K_{sp} = 1.8 \times 10^{-10}$)
   b. SrF_2 ($K_{sp} = 4.3 \times 10^{-9}$) or MgF_2 ($K_{sp} = 6.5 \times 10^{-9}$)
   c. $MnCO_3$ ($K_{sp} = 2.2 \times 10^{-1}$) or $PbCO_3$ ($K_{sp} = 1.5 \times 10^{-13}$)

16.129. Cobalt (II) and zinc (II) carbonate have nearly identical K_{sp} values. Do they have the same molar solubility? Do they have the same solubility in g/L?

16.130. The K_{sp} values for silver iodide and silver phosphate differ by less than 10%. Do they have the same molar solubility?

16.131. Which has a higher pH, a saturated solution of magnesium hydroxide ($K_{sp} = 5.6 \times 10^{-12}$) or a saturated solution of calcium hydroxide ($K_{sp} = 4.7 \times 10^{-6}$)?

16.132. Many monoprotic organic acids (HA's) have pK_a values between 4.0 and 5.0. This means that these acids exist in human serum (pH = 7.30) as:
i. Only molecules of HA
ii. About 100 times more HA molecules that A^- ions
iii. About 10 times more HA molecules that A^- ions
iv. About a 50:50 mix of HA molecules and A^- ions
v. Nearly all A^- ions

16.133. The activity of many enzymes is pH dependent. For example, the optimum activity of two common digestive enzymes trypsin and pepsin occur at pH 6.5 and pH 1.5, respectively. Using the data in Appendix 5 as a guide, which buffer would you choose for each enzyme to maximize their activity?

16.134. People with diabetes can have elevated levels of two acids in their blood: β-hydroxybutyric acid (pK_a = 4.72) and acetoacetic acid (pK_a = 3.58):

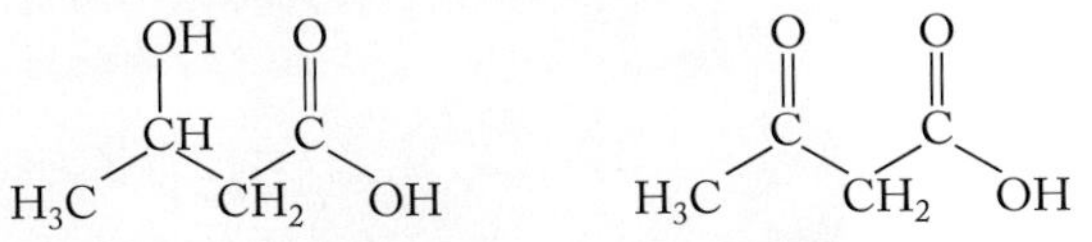

β-hydroxybutyric acid

acetoacetic acid

FIGURE P16.134

The presence of these acids lowers the pH of blood and can serve to diagnose diabetes.
a. Which acid yields a solution with the lower pH if 0.100 *M* solutions are prepared?
b. Which acid will dissociate to a greater degree?
c. What is the pH of a solution that contains 15.8 m*M* of acetoacetic acid and 10.8 m*M* sodium acetoacetate?
d. What is the pH of a solution of 90 mg/L β-hydroxybutyric acid and 90 mg/L β-hydroxybutyrate anion after 100 μL of 0.100 *M* HCl has been added?

16.135. Chickens do not have sweat glands, so when they become overheated, they cool themselves by panting (taking short, quick breaths). This lowers the concentration of CO_2 in their blood. If chickens pant too much, this causes the shells on chickens' eggs to become so thin that they break easily. Eggshells are about 95% calcium carbonate. Explain the effect of panting on shell thickness by considering the equilibrium between $CO_2(aq)$ and $CO_3^{2-}(aq)$ in blood.

16.136. Teeth in both humans and many animals are made of calcium hydroxyphosphate, $Ca_5(PO_4)_3OH$. Veterinarians have noticed that dogs' saliva has a pH that is more basic than the saliva of humans. This fact is suggested as one reason dogs are less subject to tooth decay than people. Explain this suggestion.

smartwork5

If your instructor uses Smartwork5, log in at digital.wwnorton.com/atoms2.

17

Electrochemistry

The Quest for Clean Energy

ZERO-EMISSION AUTOMOBILE Propulsion systems for all-electric cars such as this one are based on electric motors powered by lithium ion batteries.

PARTICULATE **REVIEW**

Redox: Metal versus Nonmetal

In Chapter 17, we investigate the transformation of chemical energy via redox reactions into electrical energy. Rust forms on abandoned cars such as the one in this photo through a series of reactions between iron in the car and oxygen in the atmosphere.

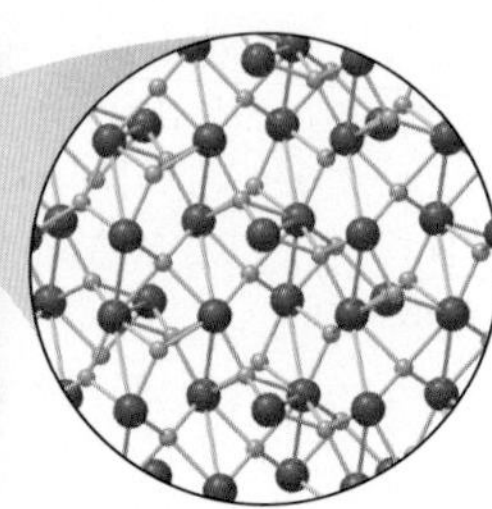

- Write a balanced chemical equation describing the formation of Fe_2O_3—a principal component of rust—from iron and oxygen.
- Which element is oxidized in this reaction? Which element is reduced?
- Describe the direction of electron transfer in the formation of rust.

(Review Section 8.6 if you need help.)

(Answers to Particulate Review questions are in the back of the book.)

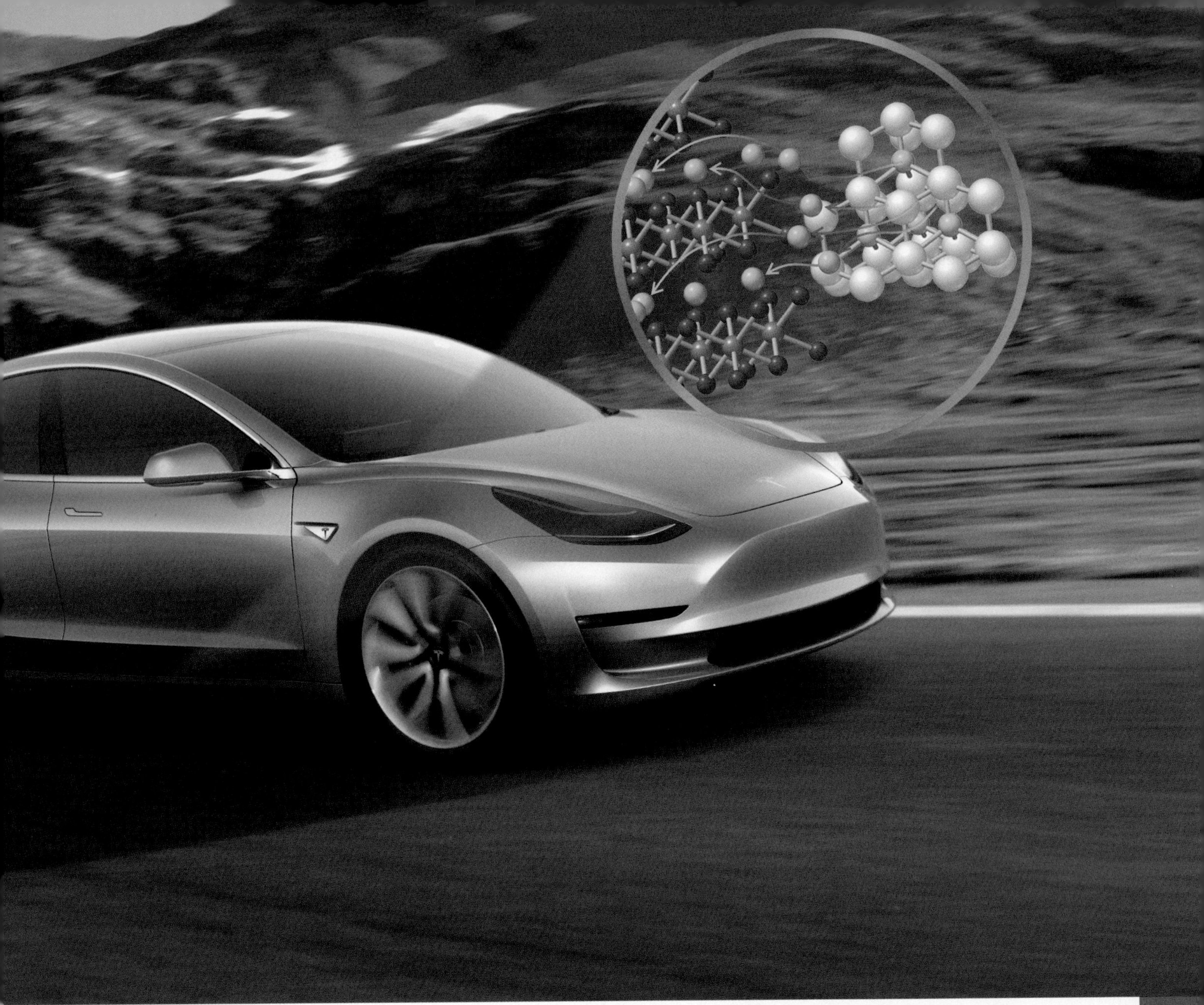

PARTICULATE **PREVIEW**

Redox: Electricity and Clean Fuel

Not all redox reactions involve the oxidation and reducton of metals. The redox chemistry used to create fuel cells involves hydrogen gas and oxygen gas as pictured here (the electrons are not drawn). As you read this chapter, look for ideas that will help you answer these questions:

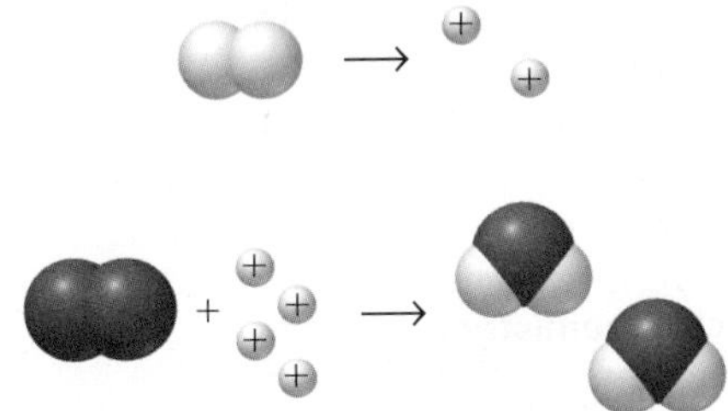

- Determine which half-reaction represents oxidation, which represents reduction, and write an equation for the overall chemical reaction in fuel cells that is balanced both in mass and charge.
- How do chemists manipulate electron transfer in this reaction to generate electricity to power cars?
- Why do chemists call hydrogen and oxygen a “clean” fuel mixture?

Learning Outcomes

LO1 Combine the appropriate half-reactions to write net ionic equations describing redox reactions
Sample Exercises 17.1, 17.2

LO2 Draw cell diagrams and describe the components of electrochemical cells and their roles in interconverting chemical and electrical energy
Sample Exercise 17.3

LO3 Calculate standard cell potentials from standard reduction potentials
Sample Exercise 17.4

LO4 Interconvert a cell's standard potential and the change in standard free energy of the cell reaction
Sample Exercise 17.5

LO5 Use the Nernst equation to calculate nonstandard cell potentials
Sample Exercise 17.6

LO6 Interconvert a cell's standard potential with the value of the equilibrium constant of the cell reaction under standard conditions
Sample Exercise 17.7

LO7 Interconvert masses of reactants in cell reactions with quantities of electrical charge
Sample Exercises 17.8, 17.9

17.1 Running on Electricity

The 21st century has seen wide swings in the price of crude oil and products derived from it, such as gasoline. These fluctuations, coupled with growing concerns over climate change and other pollution problems, invigorated the development of electric propulsion systems. Some vehicles, called hybrids, are propelled by combinations of electric motors powered by rechargeable batteries and small gasoline engines. Others, including the one in this chapter's opening photo, are powered only by electric motors and banks of high-performance batteries.

CONNECTION In Chapters 13 through 15, we examined some of the environmental problems associated with the combustion of fossil fuels and the operation of internal combustion engines.

Scientists and engineers continue to develop lighter, more reliable, and higher-capacity batteries to give all-electric cars the driving range and performance many motorists want. Batteries convert chemical energy into electrical energy; motors then convert that electrical energy into mechanical energy. These two processes together are more efficient than the conversion of chemical energy directly to mechanical energy in gasoline engines. Moreover, electric motors consume no energy when the vehicles they power are stopped in traffic. A vehicle's electric motor can even help recharge its batteries when the brakes are applied. As the car slows, the motor becomes an electric generator, turning the vehicle's kinetic energy into electrical energy.

In this chapter, we examine the chemistry of modern batteries and of another source of mobile electrical power, fuel cells. These devices are based on **electrochemistry**, the branch of chemistry that links chemical reactions to the production or consumption of electrical energy. At the heart of electrochemistry are chemical reactions in which electrons are transferred between substances at electrode surfaces. In other words, electrochemistry is based on *red*uction and *ox*idation, or *redox*, chemistry.

The principles of redox reactions, which were introduced in Section 8.6, can be summarized as follows:

- A redox reaction consists of two complementary processes: the *reduction* of a substance that gains electrons and the *oxidation* of a substance that loses electrons.
- Reduction and oxidation happen simultaneously so that the number of electrons gained during reduction exactly matches the number lost during oxidation.

electrochemistry the branch of chemistry that examines the transformations between chemical and electrical energy.

half-reaction one of the two halves of a redox reaction; one half-reaction is the oxidation component, and the other is the reduction component.

- A substance that is easily oxidized is one that readily gives up electrons. This electron-donating power makes the substance an effective reducing agent. In any redox reaction, the *reducing agent is always oxidized.*
- A substance that readily accepts electrons and is thereby reduced is an effective oxidizing agent. In any redox reaction, the *oxidizing agent is always reduced.*

CONNECTION Redox reactions were discussed in detail in Section 8.6.

In Chapter 8 we introduced redox chemistry using the reaction between zinc atoms and Cu^{2+} ions shown in Figure 17.1 and described in the following net ionic equation:

$$Zn(s) + Cu^{2+}(aq) \rightarrow Cu(s) + Zn^{2+}(aq) \quad (8.28)$$

Notice how the charge on Zn increases from 0 to +2, indicating the loss of two electrons, which means zinc metal is oxidized. At the same time, the charge on copper decreases from 2+ to 0 as copper ions gain two electrons and are, therefore, reduced. By the definitions given above, Zn is the reducing agent and Cu^{2+} is the oxidizing agent.

We can think of the Zn/Cu^{2+} reaction, or any redox reaction, as consisting of two halves, or **half-reactions**, one involving the oxidation of a substance (zinc in this case) and one involving the reduction of another (e.g., Cu^{2+} ion). In the zinc half-reaction each Zn atom that is oxidized *loses* two electrons:

$$Zn(s) \rightarrow Zn^{2+}(aq) + 2\ e^-$$

and each Cu^{2+} ion that is reduced *gains* two electrons:

$$Cu^{2+}(aq) + 2\ e^- \rightarrow Cu(s)$$

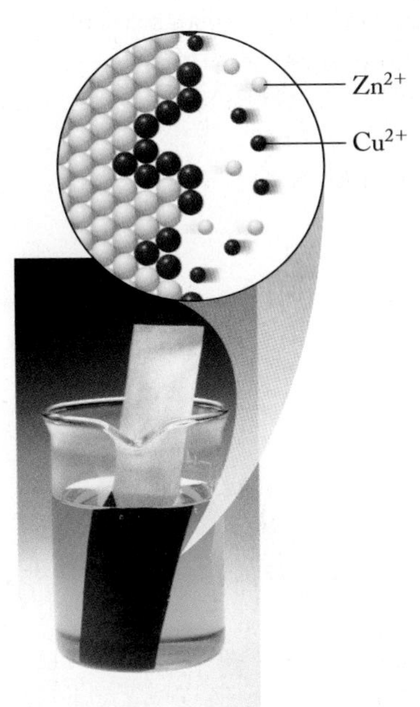

FIGURE 17.1 A strip of zinc is immersed in an aqueous solution of blue copper(II) sulfate. It becomes encrusted with a dark layer of copper as Cu^{2+} ions are reduced to Cu atoms and Zn atoms are oxidized to colorless Zn^{2+} ions. The solution becomes a paler blue as Cu^{2+} ions are reduced.

Because the number of electrons lost and gained is the same in the two half-reactions, writing a net ionic equation to describe the overall redox reaction is simply a matter of adding the two half-reactions together:

$$\begin{array}{r} Zn(s) \rightarrow Zn^{2+}(aq) + 2\ e^- \\ Cu^{2+}(aq) + 2\ e^- \rightarrow Cu(s) \\ \hline Zn(s) + Cu^{2+}(aq) + \cancel{2\ e^-} \rightarrow Cu(s) + Zn^{2+}(aq) + \cancel{2\ e^-} \end{array}$$

Canceling out the equal numbers of electrons gained and lost, we get Equation 8.28.

Combining half-reactions in this way can be a useful way to write the net ionic equation describing a redox reaction. It is especially useful when balanced equations for the two half-reactions are readily available. One handy source of such equations is in Appendix 6, which contains dozens of common half-reactions. Note that they are all written as reduction half-reactions. There is no need for a separate table of oxidation half-reactions because any reduction half-reaction can always be reversed to obtain the corresponding oxidation half-reaction. Therefore, you will find the half-reaction describing the reduction of Cu^{2+} ions to Cu metal in the table, but you will not find a half-reaction describing the oxidation of Zn metal to Zn^{2+} ions. However, you will find a half-reaction reduction of Zn^{2+} ions to Zn metal in the table. So, writing a net ionic equation describing the redox reaction depicted in Figure 17.1 involves (1) scanning the list of reactants in Table A6.1 for the product of the oxidation half-reaction (Zn^{2+} ions) and (2) making sure that the product of the reduction half-reaction in the table matches the reactant of our oxidation half-reaction (Zn metal).

Many of the half-reactions in Appendix 6 are written as if they occur in acidic solutions. In fact, H^+ ions are used to balance the number of H atoms in most of those in which water is a reactant or product. For example, the reduction of O_2 to H_2O is written

$$O_2(g) + 4\,H^+(aq) + 4\,e^- \rightarrow 2\,H_2O(\ell)$$

However, a few half-reactions in Table A6.1 contain OH^- ions, which means they apply to reactions in basic solutions. One of them involves the reduction of O_2:

$$O_2(g) + 2\,H_2O(\ell) + 4\,e^- \rightarrow 4\,OH^-(aq)$$

CONNECTION Keep in mind that hydrogen ions in aqueous solutions are really hydronium ions, $H_3O^+(aq)$, as described in Chapters 8 and 15.

In the following exercises we need to select the half-reaction that matches the pH conditions of the overall redox reaction.

SAMPLE EXERCISE 17.1 Writing Redox Reaction Equations by Combining Half-Reactions I — LO1

Write a balanced net ionic equation describing the oxidation of $Fe^{2+}(aq)$ by O_2 gas in an acidic solution. *Hint*: Water is a component of the O_2 reduction half-reaction.

Collect and Organize We find the following half-reactions for O_2 reduction in Table A6.1:

$$O_2(g) + 4\,H^+(aq) + 4\,e^- \rightarrow 2\,H_2O(\ell)$$
$$O_2(g) + 2\,H_2O(\ell) + 4\,e^- \rightarrow 4\,OH^-(aq)$$

All of the half-reactions in Table A6.1 are reductions, so there is no half-reaction for the oxidation of Fe^{2+} ions. However, Fe^{2+} ions appear as a product in the reduction of Fe^{3+}:

$$Fe^{3+}(aq) + e^- \rightarrow Fe^{2+}(aq)$$

We can reverse this reduction half-reaction to make Fe^{2+} the reactant in an oxidation half-reaction:

$$Fe^{2+}(aq) \rightarrow Fe^{3+}(aq) + e^-$$

Analyze The problem specifies acidic conditions, so we must use the O_2 half-reaction with H^+ rather than OH^- ions in it. In any redox reaction the number of electrons gained by the substance that is reduced must equal the number of electrons lost by the substance that is oxidized. There is a gain of four moles of electrons in the O_2 half-reaction and a loss of one mole of electrons in the Fe^{2+} half-reaction. Therefore, we multiply the Fe^{2+} half-reaction by 4 to obtain a balanced overall redox reaction equation.

Solve Multiplying the Fe^{2+} half-reaction by 4 and adding it to the O_2 half-reaction containing H^+ ions, we get

$$4\,Fe^{2+}(aq) \rightarrow 4\,Fe^{3+}(aq) + 4\,e^-$$
$$O_2(g) + 4\,H^+(aq) + 4\,e^- \rightarrow 2\,H_2O(\ell)$$
$$4\,Fe^{2+}(aq) + O_2(g) + 4\,H^+(aq) + \cancel{4\,e^-} \rightarrow 4\,Fe^{3+}(aq) + 2\,H_2O(\ell) + \cancel{4\,e^-}$$

Simplifying gives

$$4\,Fe^{2+}(aq) + O_2(g) + 4\,H^+(aq) \rightarrow 4\,Fe^{3+}(aq) + 2\,H_2O(\ell)$$

Think About It We can verify that this is a balanced net ionic equation by confirming that the numbers of Fe, O, and H atoms are the same on the two sides of the reaction arrow and that the total electrical charge is the same on the two sides (both are 12+). Note that combining these two half-reactions yields a net ionic equation, not a molecular equation. We can also check our work by comparing the changes in oxidation number to the number of electrons transferred. For example, the oxidation number of iron changes from +2 to +3. The loss of one electron by four Fe^{2+} ions equals the number of electrons on the right-hand side of the oxidation half-reaction. Similarly, the oxidation number of each oxygen changes from 0 to −2. Each oxygen atom in O_2 molecules gains two electrons on the left-hand side of the reduction half-reaction.

Practice Exercise Write a net ionic equation describing the oxidation of NO_2^- to NO_3^- by O_2 in a basic solution.

(Answers to Practice Exercises are in the back of the book.)

In Sample Exercise 17.1 we used half-reactions from Appendix 6 to write the net ionic equation for the oxidation of Fe^{2+} by O_2 under acidic conditions. What if we needed to write a balanced chemical equation describing the oxidation of iron under basic conditions, in which solid $Fe(OH)_2$ is oxidized to $Fe(OH)_3$? There is no half-reaction in Appendix 6 with these iron hydroxides, so we need to create our own. We begin with the formulas of the reduced and oxidized forms of the element of interest:

$$Fe(OH)_2(s) \rightarrow Fe(OH)_3(s)$$

In this expression there is one atom of Fe on both sides, but the right side has one more O and H atom than the left. The simplest solution to this imbalance is to add an OH^- ion (which would be present under basic reaction conditions) to the left side:

$$Fe(OH)_2(s) + OH^-(aq) \rightarrow Fe(OH)_3(s)$$

Because we are writing the chemical equation of a half-reaction, we use electrons to balance the imbalance in electrical charges. Adding one electron to the right side gives us a balanced chemical equation:

$$Fe(OH)_2(s) + OH^-(aq) \rightarrow Fe(OH)_3(s) + e^-$$

Note that adding one electron to the right side makes sense for another reason: the oxidation states of iron changed from +2 on the left to +3 on the right. This means that each Fe^{2+} ion must lose one electron, as reflected by the electron on the right side of the equation.

As we saw in Sample Exercise 17.1, Appendix 6 contains a half-reaction for the reduction of O_2 under basic conditions.

$$O_2(g) + 2\,H_2O(\ell) + 4\,e^- \rightarrow 4\,OH^-(aq)$$

This reduction half-reaction has four electrons on the left side, but the iron oxidation half-reaction has only one on the right, so we need to multiply the oxidation half-reaction by 4 so that the number of electrons lost and gained in the two half-reactions is the same:

$$4\,Fe(OH)_2(s) + 4\,OH^-(aq) \rightarrow 4\,Fe(OH)_3(s) + 4\,e^-$$

Adding the two half-reactions together and canceling the terms common to both sides,

$$\begin{array}{r} 4\,Fe(OH)_2(s) + 4\,OH^-(aq) \rightarrow 4\,Fe(OH)_3(s) + 4\,e^- \\ O_2(g) + 2\,H_2O(\ell) + 4\,e^- \rightarrow 4\,OH^-(aq) \\ \hline 4\,Fe(OH)_2(s) + \cancel{4\,OH^-(aq)} + O_2(g) + 2\,H_2O(\ell) + \cancel{4\,e^-} \rightarrow \\ 4\,Fe(OH)_3(s) + \cancel{4\,e^-} + \cancel{4\,OH^-(aq)} \end{array}$$

yields a balanced equation for the redox reaction:

$$4\,Fe(OH)_2(s) + O_2(g) + 2\,H_2O(\ell) \rightarrow 4\,Fe(OH)_3(s)$$

CONNECTION Balancing redox equations in acidic and basic solutions was described in Section 8.6.

SAMPLE EXERCISE 17.2 Writing Redox Reaction Equations by Combining Half-Reactions II **LO1**

One of the ways to determine the concentration of ethanol (CH_3CH_2OH) in blood involves a titration in which ethanol is reacted with a standard solution of dichromate ($Cr_2O_7^{2-}$) ions under acidic conditions. The products of the reaction are $Cr^{3+}(aq)$, $CO_2(g)$, and $H_2O(\ell)$. Write a balanced net ionic equation describing the titration reaction.

Collect and Organize We need to write the net ionic equation describing a reaction in which $CH_3CH_2OH(aq)$ and $Cr_2O_7^{2-}(aq)$ are reactants and $Cr^{3+}(aq)$, $CO_2(g)$, and $H_2O(\ell)$ are products. According to Appendix 6, the following reduction half-reaction contains both forms of chromium:

$$Cr_2O_7^{2-}(aq) + 14\ H^+(aq) + 6\ e^- \rightarrow 2\ Cr^{3+}(aq) + 7\ H_2O(\ell)$$

There is no half-reaction involving the oxidation or reduction of ethanol in Appendix 6.

Analyze Dichromate, which contains Cr(VI), is reduced to chromium(III) in the reduction half-reaction above, which means that ethanol is oxidized to carbon dioxide and water in the titration reaction. To write a net ionic equation describing the reaction, we need to first develop a half-reaction for the oxidation of ethanol and then balance the losses and gains of electrons before combining the chromium and ethanol half-reactions.

Solve Beginning with the formulas of the reactant and products in the ethanol half-reaction:

$$CH_3CH_2OH(aq) \rightarrow CO_2(g) + H_2O(\ell)$$

Let's first balance the number of carbon atoms on both sides by placing a coefficient of 2 in front of CO_2:

$$CH_3CH_2OH(aq) \rightarrow 2\ CO_2(g) + H_2O(\ell)$$

There is now 1 O atom on the left and 5 on the right. To balance them, we add 4 molecules of H_2O to the left side:

$$CH_3CH_2OH(aq) + 4\ H_2O(\ell) \rightarrow 2\ CO_2(g) + H_2O(\ell)$$

This equation simplifies to

$$CH_3CH_2OH(aq) + 3\ H_2O(\ell) \rightarrow 2\ CO_2(g)$$

Now we have 12 H atoms on the left and none on the right, so we add 12 H^+ ions to the right side:

$$CH_3CH_2OH(aq) + 3\ H_2O(\ell) \rightarrow 2\ CO_2(g) + 12\ H^+(aq)$$

Adding 12 electrons to the right side balances the charges of the 12 H^+ ions on that side and yields a balanced equation describing the oxidation half-reaction:

$$CH_3CH_2OH(aq) + 3\ H_2O(\ell) \rightarrow 2\ CO_2(g) + 12\ H^+(aq) + 12\ e^-$$

To balance the gain and loss of electrons, we multiply the reduction half-reaction, which consumes six electrons, by 2:

$$2\ Cr_2O_7^{2-}(aq) + 28\ H^+(aq) + 12\ e^- \rightarrow 4\ Cr^{3+}(aq) + 14\ H_2O(\ell)$$

Adding the two half-reactions and then simplifying the resulting equation,

$$2\ Cr_2O_7^{2-}(aq) + 28\ H^+(aq) + 12\ e^- \rightarrow 4\ Cr^{3+}(aq) + 14\ H_2O(\ell)$$

$$CH_3CH_2OH(aq) + 3\ H_2O(\ell) \rightarrow 2\ CO_2(g) + 12\ H^+(aq) + 12\ e^-$$

$$2\ Cr_2O_7^{2-}(aq) + \mathbf{16}\ \cancel{28}\ H^+(aq) + \cancel{12\ e^-} + CH_3CH_2OH(aq) + \cancel{3\ H_2O(\ell)} \rightarrow$$
$$4\ Cr^{3+}(aq) + \mathbf{11}\ \cancel{14}\ H_2O(\ell) + 2\ CO_2(g) + \cancel{12\ H^+(aq)} + \cancel{12\ e^-}$$

yields the following complete and balanced equation:

$$2\ Cr_2O_7^{2-}(aq) + 16\ H^+(aq) + CH_3CH_2OH(aq) \rightarrow$$
$$4\ Cr^{3+}(aq) + 11\ H_2O(\ell) + 2\ CO_2(g)$$

Think About It We could use the chromium half-reaction from Appendix 6 in this exercise because it contains the right chemical forms of the elements and because it is written for acidic reaction conditions. Acidic conditions are also the reason water molecules were used to balance the number of O atoms and H^+ ions were used to balance the number of H atoms in writing the ethanol half-reaction. The balanced net ionic equation contains a key piece of information for the analyst using this reaction to determine blood alcohol concentrations: dichromate and alcohol react in a mole ratio of 2:1.

Practice Exercise An acidic solution of permanganate (MnO_4^-) ions can oxidize nitrous acid to nitrate ions. The products include dissolved Mn^{2+} ions. Write chemical equations describing the oxidation of nitrous acid and the reduction of permanganate, and combine the two half-reactions to write a net ionic equation describing the overall redox reaction.

electrochemical cell an apparatus that converts chemical energy into electrical work or electrical work into chemical energy.

anode an electrode at which an oxidation half-reaction (loss of electrons) takes place.

cathode an electrode at which a reduction half-reaction (gain of electrons) takes place.

17.2 Electrochemical Cells

Having shown how the redox reaction between Zn metal and Cu^{2+} ions is the net result of two distinct half-reactions, one involving the oxidation of Zn metal and the other the reduction of Cu^{2+} ions, let's now physically separate the two half-reactions using a device called an **electrochemical cell**. Doing so forces the electrons lost by the Zn atoms to flow through an external circuit before they can be acquired by Cu^{2+} ions. Figure 17.2 depicts an electrochemical cell based on the Zn/Cu^{2+} reaction. One compartment in the Zn/Cu^{2+} cell contains a strip of zinc metal immersed in 1.00 *M* $ZnSO_4$; the other contains a strip of copper metal immersed in 1.00 *M* $CuSO_4$. Sulfate ions are spectator ions in this reaction and are not included in the half-reactions shown in the figure. The two metal strips serve as the *electrodes* in the cell, providing pathways along which the electrons produced and consumed in the two half-reactions flow through an external circuit.

As the cell reaction proceeds, the oxidation of Zn atoms produces electrons, which travel from the Zn electrode through the external circuit to the surface of the Cu electrode, where they combine with Cu^{2+} ions, forming Cu metal. In an electrochemical cell, the electrode at which the oxidation half-reaction takes place (the zinc electrode in this case) is always called the **anode**, and the electrode at which the reduction half-reaction takes place is always called the **cathode**.

You might think that the production of Zn^{2+} ions in the left compartment in Figure 17.2 would result in a buildup of positive charge on

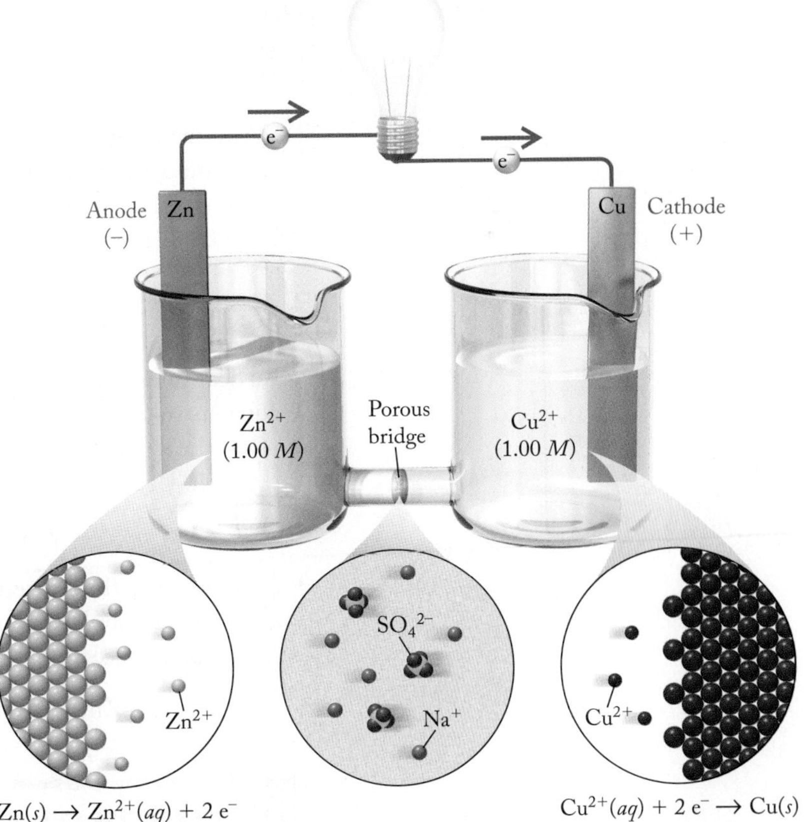

FIGURE 17.2 In this Zn/Cu^{2+} electrochemical cell, zinc metal is oxidized to Zn^{2+} ions at the anode in the left-hand compartment, while Cu^{2+} ions are reduced to Cu metal in the right-hand compartment. A porous bridge provides an electrical connection through which ions of a background electrolyte (Na_2SO_4) migrate from one compartment to the other.

cell diagram symbols that show how the components of an electrochemical cell are connected.

that side of the cell, and that the conversion of Cu^{2+} ions to Cu metal would create an excess of SO_4^{2-} ions and thus a negative charge in the Cu compartment. If that happened, the cell reaction would stop. However, no such charge buildup occurs because the two compartments are connected by a porous bridge that allows ions to migrate back and forth between the two compartments. Also the solutions surrounding both electrodes contain a *background electrolyte* made of ions that are not involved in either half-reaction. In the Zn/Cu^{2+} cell, Na_2SO_4 makes an excellent background electrolyte. Migration of Na^+ ions toward the Cu compartment and SO_4^{2-} ions toward the Zn compartment through the porous bridge, as shown in the middle molecular view in Figure 17.2, balances the flow of electrons in the external circuit and eliminates any accumulation of ionic charge in either compartment.

CONCEPT **TEST**

As the cell reaction in Figure 17.2 proceeds, is the increase in mass of the copper strip the same as the decrease in mass of the zinc strip?

(Answers to Concept Tests are in the back of the book.)

Figure 17.2 provides a view of the physical reality of a Zn/Cu^{2+} electrochemical cell, but we could use a more compact way to represent the components of the cell. A **cell diagram** uses a string of chemical formulas and symbols to show how the components of the cell are connected. A cell diagram does not convey stoichiometry, so any coefficients in the balanced equation for the cell reaction do not appear in the cell diagram.

We follow these steps when writing a cell diagram:

1. Write the chemical symbol of the anode at the far left of the diagram and the symbol of the cathode at the far right. Double vertical lines in the middle represent the connecting bridge between the anode and cathode compartments. For the Zn/Cu^{2+} cell:

$$Zn(s) \ldots\ldots\ldots \| \ldots\ldots\ldots Cu(s)$$

2. Working inward from the electrodes toward the connecting bridge, use vertical lines to indicate phase changes (e.g., between solid electrodes and aqueous solutions). Represent the electrolytes surrounding each electrode using the symbols of the ions or compounds that are changed by the cell reaction. Use commas to separate species in the same phase:

$$Zn(s) \mid Zn^{2+}(aq) \,\|\, Cu^{2+}(aq) \mid Cu(s)$$

3. If known, use the concentrations of the dissolved species in place of the (aq) phase symbols, and add the partial pressures of any gases within their (g) phase symbols:

$$Zn(s) \mid Zn^{2+}(1.00\ M) \,\|\, Cu^{2+}(1.00\ M) \mid Cu(s)$$

SAMPLE EXERCISE 17.3 Diagramming an Electrochemical Cell **LO2**

The photograph in Figure 17.3(a) shows what happens when an object made of copper (the Statue of Liberty) is exposed to air and ocean spray for over a century. The distinctive pale green color of the statue is due to the products of copper corrosion,

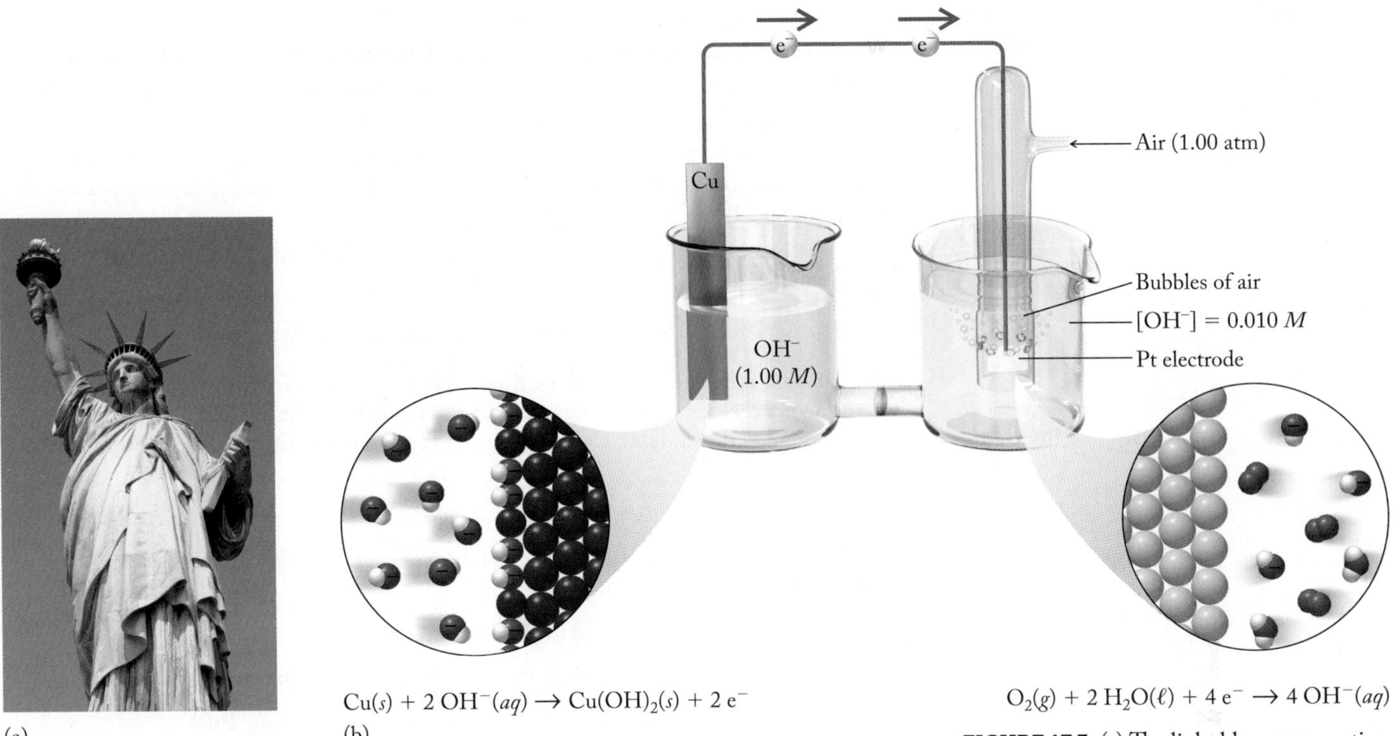

(a)

$Cu(s) + 2\,OH^-(aq) \rightarrow Cu(OH)_2(s) + 2\,e^-$

(b)

$O_2(g) + 2\,H_2O(\ell) + 4\,e^- \rightarrow 4\,OH^-(aq)$

FIGURE 17.3 (a) The light blue-green patina of the Statue of Liberty is caused by the accumulation of Cu(II) compounds on the surface of the copper sheets that make up its exterior. (b) A Cu/O_2 electrochemical cell.

including $Cu(OH)_2$, that form when copper is oxidized by atmospheric oxygen. The process can be modeled in the laboratory using an electrochemical cell (Figure 17.3b) in which a copper electrode is immersed in a solution containing OH^- ions. The second compartment contains an electrode made of chemically inert platinum, whose only role in the cell is to transport electrons away from the cell. (We revisit this use of Pt electrodes in Section 17.5.) The electrode is immersed in a solution that also contains OH^- ions and a stream of air bubbles as shown in Figure 17.3(b). Write a balanced chemical equation and a cell diagram for this cell reaction.

Collect and Organize We have a cell reaction in which electrons spontaneously flow from a copper electrode in contact with a solution containing OH^- ions through an external circuit to a platinum electrode in contact with a solution of OH^- ions and a stream of air that contains oxygen gas. The half-reactions in Table A6.1 involving oxygen in a basic solution, copper metal, and copper hydroxide are

$$O_2(g) + 2\,H_2O(\ell) + 4\,e^- \rightarrow 4\,OH^-(aq)$$
$$Cu(OH)_2(s) + 2\,e^- \rightarrow Cu(s) + 2\,OH^-(aq)$$

In a cell diagram, the anode and the species involved in the oxidation half-reaction are on the left and the cathode and the species involved in the reduction half-reaction are on the right. We use single lines to separate phases and a double line to represent the porous bridge separating the two compartments of the cell.

Analyze Copper is the anode and platinum is the cathode in Figure 17.3(b). This means that the copper half-reaction must run in reverse as an oxidation half-reaction:

$$Cu(s) + 2\,OH^-(aq) \rightarrow Cu(OH)_2(s) + 2\,e^-$$

Two moles of electrons are produced in the anode half-reaction, but four moles of electrons are consumed in the cathode half-reaction at which O_2 is reduced at the Pt electrode. Therefore, we need to multiply the copper half-reaction by 2 before combining the two equations.

Solve Multiplying the Cu^{2+} half-reaction by 2 and adding it to the O_2 half-reaction, we have

$$\begin{array}{r} O_2(g) + 2\,H_2O(\ell) + 4\,e^- \rightarrow 4\,OH^-(aq) \\ 2\,Cu(s) + 4\,OH^-(aq) \rightarrow 2\,Cu(OH)_2(s) + 4\,e^- \\ \hline O_2(g) + 2\,H_2O(\ell) + 2\,Cu(s) + \cancel{4\,OH^-(aq)} + \cancel{4\,e^-} \rightarrow \\ \cancel{4\,OH^-(aq)} + 2\,Cu(OH)_2(s) + \cancel{4\,e^-} \end{array}$$

or

$$O_2(g) + 2\,H_2O(\ell) + 2\,Cu(s) \rightarrow 2\,Cu(OH)_2(s)$$

The equation is balanced, and we are finished with the first part of the exercise.

Applying the rules for writing a cell diagram (anode on the left, cathode on the right, bridge in the middle), Cu is oxidized at the anode and O_2 is reduced at the Pt cathode:

$$Cu(s) \ldots\ldots\ldots \| \ldots\ldots\ldots Pt(s)$$

Adding electrode boundaries and the formulas of the substances produced and consumed in the cell reaction:

$$Cu(s) \mid Cu(OH)_2(s) \,\|\, O_2(g) \mid Pt(s)$$

and the hydroxide ions in the solutions surrounding the electrodes:

$$Cu(s) \mid Cu(OH)_2(s) \mid OH^-(aq) \,\|\, O_2(g) \mid OH^-(aq) \mid Pt(s)$$

yields a complete cell diagram.

Think About It The sequence of the phases in the cell diagram is linked to the progress of the reaction. Translating the cell diagram into words: Cu atoms at a copper anode are oxidized to copper(II) ions that form solid $Cu(OH)_2$ in an aqueous solution containing OH^- ions. Meanwhile oxygen gas is reduced to OH^- ions at a platinum cathode.

Practice Exercise Write a balanced chemical equation and the cell diagram for an electrochemical cell that has a copper cathode immersed in a solution of Cu^{2+} ions and an aluminum anode immersed in a solution of Al^{3+} ions.

17.3 Standard Potentials

Table A6.1 lists half-reactions in order of a parameter called the **standard reduction potential ($E°$)**, expressed in volts (V). The most positive $E°$ value in Table A6.1 is that for the reduction of fluorine:

$$F_2(g) + 2\,e^- \rightarrow 2\,F^-(aq) \qquad E° = 2.866\text{ V}$$

Fluorine's top position means that it is the most easily reduced reactant in Table A6.1. It also means that F_2 is the strongest oxidizing agent in the table. F_2 is capable of oxidizing any of the substances on the product side of the half-reactions lower in the table. We have seen fluorine's affinity for electrons before. Recall from Chapter 3 that the electron affinity of fluorine atoms is −328 kJ/mol (the second most negative EA value—meaning the second greatest affinity for electrons—of all the elements). And recall from Chapter 4 that fluorine is the most electronegative of all the elements.

The reactants with very negative $E°$ values at the bottom of the table include the major cations found in biological systems: Na^+, K^+, Mg^{2+}, and Ca^{2+}. Their negative $E°$ values mean that these ions are not easily reduced to their free metals.

In fact, none of these elements occurs in nature in the atomic state. Instead, they exist as the cations listed, which is consistent with the fact that they have the lowest ionization energies of all the elements.

Now let's consider what happens to E° when the half-reactions in the table run in reverse. This means that the products of reduction half-reactions become the reactants of oxidation half-reactions. It also means that the half-reaction at the very bottom of the table (the one with the most negative E° value),

$$Li^+(aq) + e^- \rightarrow Li(s) \qquad E^\circ = -3.05 \text{ V}$$

is likely to run in reverse as an oxidation half-reaction:

$$Li(s) \rightarrow Li^+(aq) + e^-$$

Thus, Li^+ ions in aqueous solution have little ability to oxidize anything, but Li metal is a very powerful reducing agent that can reduce any of the substances on the reactant side of the half-reactions above it in the table.

standard reduction potential (E°) the potential of a reduction half-reaction in which all reactants and products are in their standard states at 25°C.

standard cell potential (E°_{cell}) a measure of how forcefully an electrochemical cell, in which all reactants and products are in their standard states, can pump electrons through an external circuit.

electromotive force (emf) also called voltage, the force pushing electrons through an electrical circuit.

voltaic cell or **galvanic cell** an electrochemical cell in which chemical energy is transformed into electrical work by a spontaneous redox reaction.

cell potential (E_{cell}) the electromotive force with which an electrochemical cell can pump electrons through an external circuit.

CONCEPT TEST

Given the positions of the reactants and products in the table of standard reduction potentials, predict which of the following reactions are spontaneous under standard conditions:

a. $Cu(s) + 2\ Fe^{3+}(aq) \rightarrow Cu^{2+}(aq) + 2\ Fe^{2+}(aq)$

b. $2\ Ag(s) + Zn^{2+}(aq) \rightarrow 2\ Ag^+(aq) + Zn(s)$

c. $Hg(\ell) + 2\ H^+(aq) \rightarrow Hg^{2+}(aq) + H_2(g)$

CHEMTOUR
Cell Potential

We can use the standard reduction potentials to calculate the **standard cell potentials, E°_{cell}**, of electrochemical cells. Standard cell potentials are measures of the **electromotive force (emf)** generated by cell reactions, which reflects how forcefully cells pump electrons from their anodes through external circuits and into their cathodes. In the process of pumping electrons, these electrochemical cells convert the chemical energy of a spontaneous cell reaction into electrical work. They are called **voltaic cells** in honor of Italian physicist Alessandro Volta (1745–1827, Figure 17.4) who may have built the first operational battery. They can also be called **galvanic cells** after the physicist Luigi Galvani (1737–1798), a contemporary of Volta.

Consider the hypothetical case where two half-reactions both have the same standard reduction potential. That is, there would be no difference between their relative ability to function as oxidizing agents, and no electrons would spontaneously flow. For two half-reactions with different standard reduction potentials, the larger the difference between those potentials, the larger the E°_{cell} of the electrochemical cell that could be built from them. Therefore, E°_{cell} can be determined by calculating the difference between the standard reduction potentials of a voltaic cell's cathode and anode:

$$E^\circ_{cell} = E^\circ_{cathode} - E^\circ_{anode} \tag{17.1}$$

The superscript (°) in Equation 17.1 has its usual thermodynamic meaning—namely, all reactants and products are in their standard states: the concentrations of all dissolved substances are 1 M and the partial pressures of all gases are 1 atm. If we want to denote a generalized cell potential in which reactants and products are not necessarily in their standard states, we simply refer to the **cell potential (E_{cell})**.

FIGURE 17.4 Alessandro Volta is credited with building the first battery in 1798. It consisted of a stack of alternating layers of zinc, blotter paper soaked in salt water, and silver.

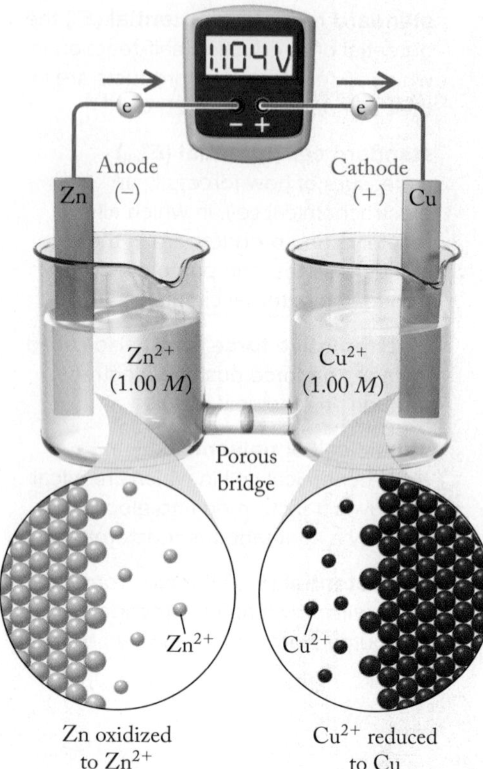

FIGURE 17.5 A voltmeter displays a cell potential of 1.104 V between a Zn electrode immersed in a 1.00 M solution of Zn^{2+} ions and a Cu electrode immersed in a 1.00 M solution of Cu^{2+} ions.

Let's use Equation 17.1 to calculate E°_{cell} for the Zn/Cu^{2+} voltaic cell in Figure 17.2. The standard reduction potential of the cathode half-reaction is

$$Cu^{2+}(aq) + 2\ e^- \rightarrow Cu(s) \qquad E^\circ = 0.342\ V$$

To obtain the standard potential for the oxidation half-reaction at the zinc anode, we start with the standard reduction potential of Zn^{2+} ions in Table A6.1:

$$Zn^{2+}(aq) + 2\ e^- \rightarrow Zn(s) \qquad E^\circ = -0.762\ V$$

Now we use Equation 17.1 to calculate E°_{cell}:

$$E^\circ_{cell} = E^\circ_{cathode} - E^\circ_{anode}$$

$$= 0.342\ V - (-0.762\ V) = 1.104\ V$$

This is the cell potential we measure if we connect a device called a voltmeter across the two electrodes, as shown in Figure 17.5.

To use Equation 17.1 we need to know which half-reaction occurs at the cathode (reduction) and which occurs at the anode (oxidation). This decision is made based on the data in Table A6.1. In the Zn/Cu^{2+} cell the value of E° for the reduction of Cu^{2+} ions to Cu metal is 0.342 V, which is greater than the value of E° for reducing Zn^{2+} ions to Zn metal (−0.762 V). Therefore, in the Zn/Cu^{2+} voltaic cell, Cu^{2+} ions are reduced and Zn metal is oxidized. We can generalize this observation to the cell reaction of any voltaic cell: the half-reaction with the more positive value of E° runs as a reduction and the other one runs in reverse as an oxidation.

CHEMTOUR Alkaline Battery

SAMPLE EXERCISE 17.4 Identifying Anode and Cathode Half-Reactions and Calculating the Value of (E°_{cell}) LO3

The standard reduction potentials of the half-reactions in single-use alkaline batteries are

$$ZnO(s) + H_2O(\ell) + 2\ e^- \rightarrow Zn(s) + 2\ OH^-(aq) \qquad E^\circ = -1.25\ V$$

$$2\ MnO_2(s) + H_2O(\ell) + 2\ e^- \rightarrow Mn_2O_3(s) + 2\ OH^-(aq) \qquad E^\circ = 0.15\ V$$

What is the net ionic equation for the cell reaction and the value of E°_{cell}?

Collect, Organize, and Analyze To calculate E°_{cell} using Equation 17.1:

$$E^\circ_{cell} = E^\circ_{cathode} - E^\circ_{anode}$$

we first need to decide which half-reaction occurs at the cathode and which one occurs at the anode. The MnO_2 half-reaction has the more positive E°, making it our reduction half-reaction. We must reverse the ZnO half-reaction, turning it into an oxidation half-reaction. Both half-reactions involve the transfer of two electrons, so they may be combined by simply adding them together.

Solve The oxidation half-reaction at the anode is

$$Zn(s) + 2\ OH^-(aq) \rightarrow ZnO(s) + H_2O(\ell) + 2\ e^-$$

and the reduction half-reaction at the cathode is

$$2\ MnO_2(s) + H_2O(\ell) + 2\ e^- \rightarrow Mn_2O_3(s) + 2\ OH^-(aq)$$

Combining these half-reactions to obtain the overall cell reaction, we get

$$2\ MnO_2(s) + \cancel{H_2O(\ell)} + Zn(s) + \cancel{2\ OH^-(aq)} + \cancel{2\ e^-} \rightarrow$$
$$Mn_2O_3(s) + \cancel{2\ OH^-(aq)} + ZnO(s) + \cancel{H_2O(\ell)} + \cancel{2\ e^-}$$

Simplifying gives us the net ionic equation for the cell reaction:

$$2\,MnO_2(s) + Zn(s) \rightarrow Mn_2O_3(s) + ZnO(s)$$

The overall E°_{cell} for this reaction is obtained using Equation 17.1:

$$E^\circ_{cell} = E^\circ_{cathode} - E^\circ_{anode} = 0.15\ V - (-1.25\ V) = 1.40\ V$$

Think About It The E°_{cell} value is reasonable because the potential of most alkaline batteries is nominally 1.5 V. In this particular cell reaction, the net ionic equation is also the complete molecular equation.

Practice Exercise The half-reactions in NiCad (nickel–cadmium) batteries are

$$Cd(OH)_2(s) + 2\,e^- \rightarrow Cd(s) + 2\,OH^-(aq) \qquad E^\circ = -0.81\ V$$

$$NiO(OH)(s) + H_2O(\ell) + e^- \rightarrow Ni(OH)_2(s) + OH^-(aq) \qquad E^\circ = 0.52\ V$$

Write the net ionic equation for the cell reaction and calculate the value of E°_{cell}.

Let's now examine what happens when an electrochemical cell is built using two half-reactions in which different numbers of electrons are gained and lost. This occurs in a type of battery that has a virtually limitless supply of one of its reactants. It is called the zinc–air battery (Figure 17.6), and it powers devices in which small battery size and mass are high priorities, such as hearing aids. Most of the internal volume of one of these batteries is occupied by an anode consisting of a paste of zinc particles packed in an aqueous solution of KOH. As in alkaline batteries (Sample Exercise 17.4), the anode half-reaction is

$$Zn(s) + 2\,OH^-(aq) \rightarrow ZnO(s) + H_2O(\ell) + 2\,e^-$$

which is the reverse of the standard reduction reaction:

$$ZnO(s) + H_2O(\ell) + 2\,e^- \rightarrow Zn(s) + 2\,OH^-(aq) \qquad E^\circ = -1.25\ V$$

The cathode consists of porous carbon supported by a metal screen. Air diffuses through small holes in the battery and across a layer of plastic film that lets gases pass through but keeps electrolyte from leaking out. As air passes through the cathode, oxygen is reduced to hydroxide ions:

$$O_2(g) + 2\,H_2O(\ell) + 4\,e^- \rightarrow 4\,OH^-(aq) \qquad E^\circ_{cathode} = 0.401\ V$$

To write the overall cell reaction, we need to multiply the oxidation (anode) half-reaction by 2 before combining it with the reduction (cathode) half-reaction:

$$2[Zn(s) + 2\,OH^-(aq) \rightarrow ZnO(s) + H_2O(\ell) + 2\,e^-]$$

$$O_2(g) + 2\,H_2O(\ell) + 4\,e^- \rightarrow 4\,OH^-(aq)$$

$$2\,Zn(s) + \cancel{4\,OH^-(aq)} + O_2(g) + \cancel{2\,H_2O(\ell)} + \cancel{4\,e^-} \rightarrow 2\,ZnO(s) + \cancel{2\,H_2O(\ell)} + \cancel{4\,OH^-(aq)} + \cancel{4\,e^-}$$

or

$$2\,Zn(s) + O_2(g) \rightarrow 2\,ZnO(s)$$

$$E^\circ_{cell} = E^\circ_{cathode} - E^\circ_{anode} = 0.401\ V - (-1.25\ V) = 1.65\ V$$

Note that when we multiplied the anode half-reaction by 2 and added it to the cathode half-reaction, we did not multiply the E° of the anode half-reaction by 2. E° is an *intensive* property of a half-reaction or a complete cell reaction, so it does not change when the quantities of reactants and products change. Thus, a zinc–air

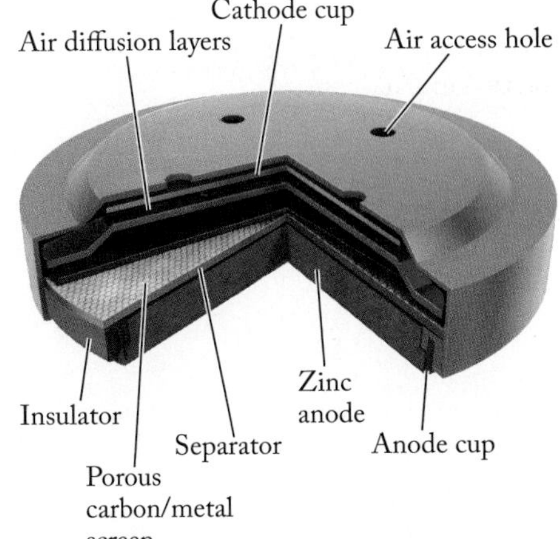

FIGURE 17.6 Most of the internal volume of a zinc–air battery is occupied by the anode: a paste of Zn particles in an aqueous solution of KOH surrounded by a metal cup that serves as the negative terminal of the battery. Oxygen from the air is the reactant at the cathode. Air enters through holes in an inverted metal cup that serves as the positive terminal of the battery. Once inside the battery, air diffuses through layers of gas-permeable plastic film that let air in but keep electrolyte from leaking out. Oxygen in the air is reduced at the porous carbon/metal cathode to OH^- ions that migrate toward the anode, where they are consumed in the Zn oxidation half-reaction.

battery the size of a pea has the same E°_{cell} as one the size of a book (like those being developed for electric vehicles). On the other hand, the amount of electrical work a zinc–air battery can do does depend on how much zinc is inside it because, as we are about to see, the electrical work that a voltaic cell can do depends on both cell potential and the quantity of charge it can deliver at that potential.

CONCEPT TEST

The balanced chemical equation for an aluminum–air battery is written as

$$4\ Al(s) + 3\ O_2(g) \rightarrow 2\ Al_2O_3(s)$$

The reaction between aluminum metal and oxygen gas that yields Al_2O_3 also represents the formation of Al_2O_3 (see Section 9.7) when written as

$$2\ Al(s) + \tfrac{3}{2}\ O_2(g) \rightarrow Al_2O_3(s)$$

Do these two reactions have the same value of E°_{cell}, or is E°_{cell} for the first reaction twice the value of E°_{cell} for the second?

17.4 Chemical Energy and Electrical Work

When the Zn and Cu electrodes in Figure 17.5 are connected to a digital voltmeter—the Zn electrode to the negative terminal of the meter and the Cu electrode to the positive terminal—the meter reads 1.104 V. These connections tell us that the battery is pumping electrons from the Zn electrode through the external circuit to the Cu electrode, and the reading tells us how much electromotive force (emf) is pushing the electrons through the circuit. Under standard conditions, this emf is the same as the standard cell potential of the cell (E°_{cell}); under any other conditions, it is simply E_{cell}.

When a voltaic cell pumps electrons through an external circuit, those moving electrons do electrical work, like lighting a lightbulb or turning an electric motor. The change in chemical free energy (ΔG_{cell}) that accompanies a spontaneous cell reaction is a measure of the electrical work (w_{elec}) that may result:

$$\Delta G_{cell} = w_{elec} \tag{17.2}$$

The sign of ΔG_{cell} is negative because the reaction in a voltaic cell *must be spontaneous*, which means there is a decrease in the free energy of the system (cell reaction mixture) as reactants become products. The sign of w_{elec} is also negative because it represents work done by the system on its surroundings, which has a negative value according to the sign convention summarized in Table 9.1 and Figure 9.4.

CONNECTION The sign conventions used for work done *on* a thermodynamic system (+) and the work done *by* the system (−) were explained in Section 9.1.

The work done by a voltaic cell on its surroundings is the product of the quantity of electric charge (Q) the cell pushes through an external circuit times the force (emf) pushing that charge. That force is the same as the cell potential, so the connection between E_{cell} and w_{elec} is

$$w_{elec} = -QE_{cell} \tag{17.3}$$

where the negative sign reflects the fact that work done by a voltaic cell on its surroundings (the external circuit) corresponds to energy lost by the cell.

The quantity of charge is proportional to the number of electrons flowing through the circuit. As noted in Chapter 2, the magnitude of the charge on a single electron is 1.602×10^{-19} coulomb (C). The magnitude of electric charge on one mole of electrons is

$$\frac{1.602 \times 10^{-19}\ \text{C}}{\cancel{e^-}} \times \frac{6.0221 \times 10^{23}\ \cancel{e^-}}{1\ \text{mol e}^-} = \frac{9.65 \times 10^4\ \text{C}}{\text{mol e}^-}$$

Faraday constant (*F*) the magnitude of electric charge in one mole of electrons; its value to three significant figures is 9.65×10^4 C/mol.

This quantity of charge, 9.65×10^4 C/mol, is called the **Faraday constant (*F*)** after Michael Faraday (1791–1867), the English chemist and physicist who discovered that redox reactions take place when electrons are transferred from one species to another. The quantity of charge Q flowing through an electrical circuit is the product of the number of moles of electrons, n, times the Faraday constant:

$$Q = nF \qquad (17.4)$$

Combining Equations 17.3 and 17.4 gives us an equation relating w_{elec} and E_{cell}:

$$w_{elec} = -nFE_{cell} \qquad (17.5)$$

Equations 17.2 and 17.5 relate the quantity of electrical work a voltaic cell does on its surroundings to the change in free energy in the cell, which under standard conditions is

$$\Delta G_{cell} = -nFE_{cell} \qquad (17.6)$$

The product on the right side of Equation 17.6 is the equivalent of energy because the units are

$$\cancel{\text{moles}} \times \frac{\text{coulombs}}{\cancel{\text{mole}}} \times \text{volts} = \text{coulombs-volts}$$

A coulomb-volt is the same quantity of energy as a joule:

$$1\ \text{coulomb-volt} = 1\ \text{joule}$$

$$1\ \text{C} \cdot \text{V} = 1\ \text{J}$$

The chemical reaction inside the cell will be spontaneous if the sign of ΔG_{cell} is negative, so the negative sign on the right side of Equation 17.6 indicates that E_{cell} of any voltaic cell must have a positive value. If E_{cell} is negative, then ΔG_{cell} will be positive and the cell reaction will be spontaneous in the opposite direction.

Let's use Equation 17.6 to calculate the change in standard free energy of the Zn/Cu^{2+} cell reaction. We begin with the standard cell potential calculated in Section 17.3:

$$E^\circ_{cell\ (Zn/Cu^{2+})} = 1.104\ \text{V}$$

We can convert this standard cell potential into a change in standard free energy (ΔG°_{cell}) using Equation 17.6 under standard conditions so that $\Delta G = \Delta G^\circ$ and $E = E^\circ$. We need to use $n = 2$ because two moles of electrons are transferred in the half-reactions for this cell:

$$\begin{aligned}\Delta G^\circ_{cell} &= -nFE^\circ_{cell} \\ &= -\left(2\ \cancel{\text{mol}} \times \frac{9.65 \times 10^4\ \text{C}}{\cancel{\text{mol}}} \times 1.104\ \text{V}\right) \\ &= -2.13 \times 10^5\ \text{C} \cdot \text{V} = -2.13 \times 10^5\ \text{J} = -213\ \text{kJ}\end{aligned}$$

To put this value in perspective, the Zn/Cu^{2+} reaction produces, on a mole-for-mole basis, nearly as much useful energy as the combustion of hydrogen gas:

$$H_2(g) + \tfrac{1}{2}O_2(g) \rightarrow H_2O(g) \qquad \Delta G^\circ = -228.6\ \text{kJ}$$

CONCEPT TEST

When a rechargeable battery, such as the one used to start a car's engine, is recharged, an external source of electrical power forces the voltaic cell reaction to run in reverse. What are the signs of ΔG_{cell} and E_{cell} during the recharging process?

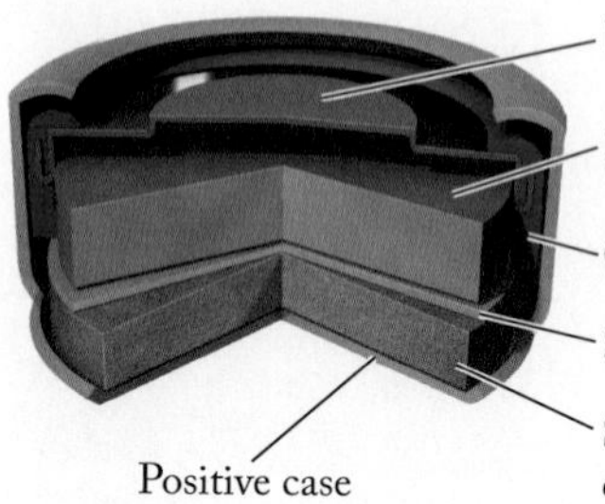

SAMPLE EXERCISE 17.5 Relating ΔG°_{cell} and E°_{cell} LO4

Many of the "button" batteries used in electric watches consist of a Zn anode and a Ag_2O cathode separated by a membrane soaked in a concentrated solution of KOH (Figure 17.7). Ag_2O is reduced to Ag metal at the cathode, and Zn is oxidized to solid $Zn(OH)_2$ at the anode. Write the net ionic equation for the reaction and, using the appropriate standard reduction potentials from Table A6.1, calculate the values of E°_{cell} and ΔG°_{cell}.

Negative cap
Zinc anode
Gasket
Separator
Silver oxide cathode
Positive case

FIGURE 17.7 Many of the button batteries that power small electronic devices incorporate a Zn anode and a Ag_2O cathode separated by a membrane containing KOH electrolyte.

Collect and Organize We know the reactants and products of the anode and cathode reactions and that the reactions occur in a basic solution. The following equations should be useful in calculating E°_{cell} and ΔG°_{cell} from the appropriate standard potentials:

$$E^\circ_{cell} = E^\circ_{cathode} - E^\circ_{anode}$$

$$\Delta G^\circ_{cell} = -nFE^\circ_{cell}$$

Analyze The half-reaction at the cathode is based on the reduction of Ag_2O to Ag. The appropriate half-reaction in Table A6.1 is

$$(1)\quad Ag_2O(s) + H_2O(\ell) + 2\,e^- \rightarrow 2\,Ag(s) + 2\,OH^-(aq) \qquad E^\circ_{cathode} = 0.342\text{ V}$$

We must reverse the anode's oxidation half-reaction to find an entry in Table A6.1 in which $Zn(OH)_2$ is the reactant and Zn is the product:

$$(2)\quad Zn(OH)_2(s) + 2\,e^- \rightarrow Zn(s) + 2\,OH^-(aq) \qquad E^\circ_{anode} = -1.249\text{ V}$$

However, in order to write the net ionic equation, we need to use the actual oxidation half-reaction at the anode—the reverse of this half-reaction—and combine it with the cathode's reduction half-reaction. The two half-reactions involve the transfer of the same number of electrons, so combining Equation (1) and the reverse of Equation (2) simply means adding them together. The value of E°_{cell} will be about [0.35 V − (−1.25 V)] or 1.60 V. This value is half again as large as the E°_{cell} of the Zn/Cu^{2+} cell. Therefore, the magnitude of its ΔG°_{cell} value should be 50% larger than −213 kJ/mol, or about −300 kJ/mol.

Solve Reversing the $Zn(OH)_2$ half-reaction and adding it to the Ag_2O half-reaction, we get

$$Zn(s) + 2\,OH^-(aq) \rightarrow Zn(OH)_2(s) + 2\,e^-$$
$$Ag_2O(s) + H_2O(\ell) + 2\,e^- \rightarrow 2\,Ag(s) + 2\,OH^-(aq)$$
$$Ag_2O(s) + H_2O(\ell) + Zn(s) + \cancel{2\,OH^-(aq)} + \cancel{2e^-} \rightarrow 2\,Ag(s) + \cancel{2\,OH^-(aq)} + Zn(OH)_2(s) + \cancel{2\,e^-}$$

or

$$Ag_2O(s) + H_2O(\ell) + Zn(s) \rightarrow 2\,Ag(s) + Zn(OH)_2(s)$$

Calculating E°_{cell},

$$E^\circ_{cell} = E^\circ_{cathode} - E^\circ_{anode} = 0.342\text{ V} - (-1.249\text{ V}) = 1.591\text{ V}$$

Calculating ΔG°_{cell},

$$\Delta G^\circ_{cell} = -nFE^\circ_{cell}$$
$$= -\left(2\,\cancel{\text{mol}} \times \frac{9.65 \times 10^4\text{ C}}{\cancel{\text{mol}}} \times 1.591\text{ V}\right)$$
$$= -3.07 \times 10^5\text{ C}\cdot\text{V} = -3.07 \times 10^5\text{ J} = -307\text{ kJ}$$

Think About It The positive value of E°_{cell} and the negative value of ΔG°_{cell} are expected because voltaic cell reactions are spontaneous. The calculated values are close to those we estimated.

Practice Exercise The alkaline batteries used in flashlights (Sample Exercise 17.4) produce a cell potential of 1.50 V. What is the value of ΔG_{cell}?

standard hydrogen electrode (SHE) a reference electrode based on the half-reaction $2\ H^+(aq) + 2\ e^- \rightarrow H_2(g)$ that produces a standard electrode potential defined to be 0.000 V.

Some final thoughts about the ΔG°_{cell} value calculated in Sample Exercise 17.5 are in order. First, it is based on the reaction of one mole of Ag_2O and one mole of Zn, which corresponds to 232 g of Ag_2O and 65 g of Zn. The energy stored in a button battery (Figure 17.7), which has a mass of only 1 or 2 g, would be a tiny fraction of the calculated value. Also, note that no ions appear in the net ionic equation. Because all the reactants and products in the silver oxide battery reaction are solids, its net ionic equation and its molecular equation are identical.

17.5 A Reference Point: The Standard Hydrogen Electrode

We can measure the value of E_{cell} using a voltmeter, but how do we measure the individual electrode potentials of the cathode and anode? The answer to this question is that we arbitrarily assign a value of zero volts to the standard potential for the reduction of hydrogen ions to hydrogen gas:

$$2\ H^+(aq) + 2\ e^- \rightarrow H_2(g) \qquad E^\circ = 0.000\ V \qquad (17.7)$$

An electrode that generates this reference potential, called the **standard hydrogen electrode (SHE)**, consists of a platinum electrode in contact with a solution of a strong acid ($[H^+] = 1.00\ M$) and hydrogen gas at a pressure of 1.00 atm (Figure 17.8). The platinum is unchanged—neither oxidized nor reduced—by the electrode reaction. Rather, it serves as a chemically inert conveyor of electrons. Electrons may be consumed at the electrode surface if H^+ ions are reduced to hydrogen gas, or they may flow away from the surface into the Pt electrode if hydrogen gas is oxidized to H^+ ions. The potential of the SHE, which is the same for both half-reactions, is defined to be 0.000 V.

To write the cell diagram for a cell in which the SHE serves as the anode, we represent the SHE half of the cell as follows:

$$Pt \mid H_2(g,\ 1.00\ atm) \mid H^+(1.00\ M) \parallel$$

This indicates that the anode half-reaction involves the oxidation of H_2 gas to H^+ ions. If the SHE is the cathode, then its half of the cell is diagrammed as follows:

$$\parallel H^+(1.00\ M) \mid H_2(g,\ 1.00\ atm) \mid Pt$$

This indicates that the cathode half-reaction involves the reduction of H^+ ions to H_2 gas.

By definition the standard reduction potential of the SHE is 0.000 V, so the measured E_{cell} of any voltaic cell in which a SHE is one of the two electrodes—either cathode or anode—is the potential produced by the other electrode. This means that if we attach a voltmeter to the cell, the meter reading is the electrode potential of the other electrode. Suppose, for example, that a voltaic cell consists of

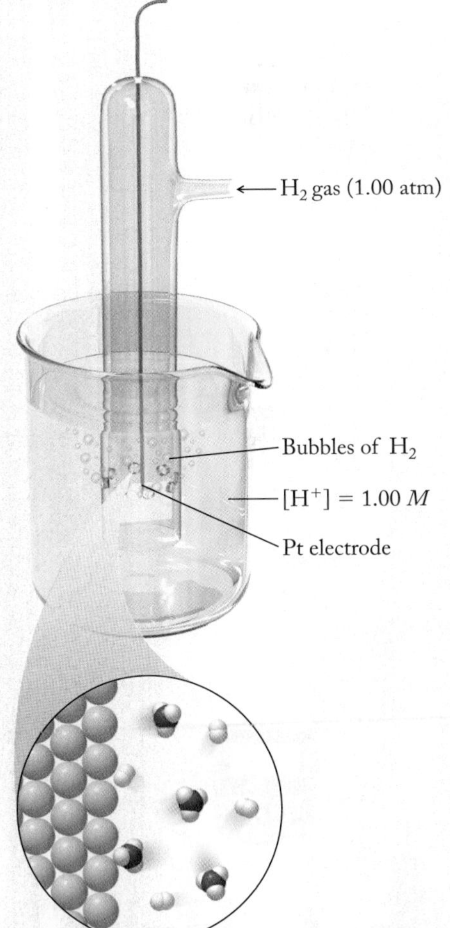

FIGURE 17.8 The standard hydrogen electrode consists of a platinum electrode immersed in a 1.00 M solution of $H^+(aq)$ and bathed in a stream of pure H_2 gas at a pressure of 1.00 atm. Its potential is the same (0.000 V) whether $H^+(aq)$ ions are reduced or H_2 gas is oxidized.

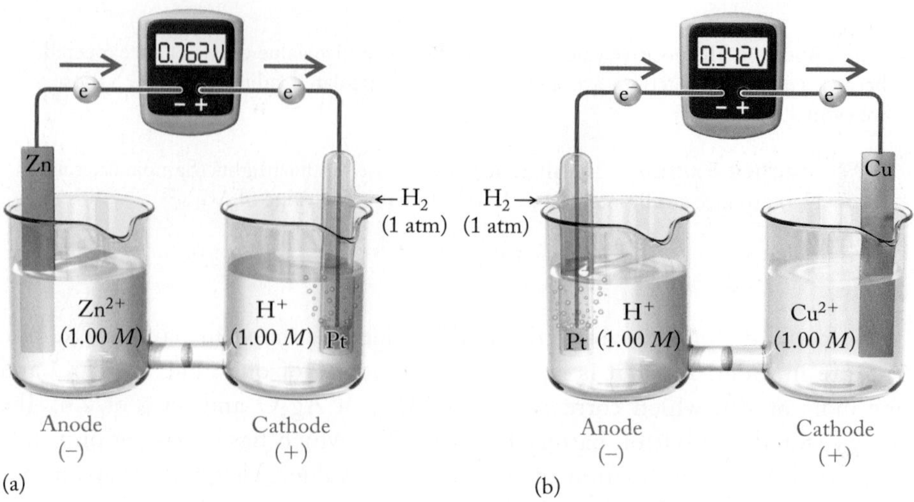

FIGURE 17.9 The standard hydrogen electrode allows us to determine the standard potential of any half-reaction. (a) When coupled to a Zn electrode under standard conditions, the SHE is the cathode and the Zn electrode is the anode. When the SHE is connected to the positive terminal of a voltmeter and the Zn electrode to the negative terminal, the meter measures a cell potential of 0.762 V. (b) When coupled to a Cu electrode under standard conditions, the SHE is the anode, the Cu electrode is the cathode, and the meter measures a cell potential of 0.342 V.

a strip of zinc metal immersed in a 1.00 *M* solution of Zn^{2+} ions in one compartment and a SHE in the other (Figure 17.9a). Also suppose that a voltmeter is connected to the cell so that it measures the potential at which the cell pumps electrons from the zinc electrode to the SHE. This direction of electron flow means that the zinc electrode is the cell's anode and the SHE is the cathode of the cell. At 25°C the meter reads 0.762 V. We know that the value of E°_{cathode} is that of the SHE (0.000 V) and that E°_{anode} is E°_{Zn}. Inserting these values and symbols into Equation 17.1:

$$\begin{aligned} E^\circ_{\text{cell}} &= E^\circ_{\text{cathode}} - E^\circ_{\text{anode}} \\ E^\circ_{\text{cell}} &= E^\circ_{\text{SHE}} - E^\circ_{\text{Zn}} \\ 0.762\text{ V} &= 0.000\text{ V} - E^\circ_{\text{Zn}} \\ E^\circ_{\text{Zn}} &= -0.762\text{ V} \end{aligned}$$

This value is equal to the standard reduction potential of Zn^{2+} in Table A6.1:

$$Zn^{2+}(aq) + 2\,e^- \rightarrow Zn(s) \qquad E^\circ = -0.762\text{ V}$$

In Figure 17.9(b), the SHE is coupled to a copper electrode immersed in a 1.00 *M* solution of Cu^{2+} ions. In this cell, electrons flow from the SHE through an external circuit to the copper electrode at a cell potential of 0.342 V at 25°C. The direction of current flow means that the electrons are consumed at the copper electrode, making it the cathode. The value of E° for the copper half-reaction is calculated as follows:

$$\begin{aligned} E^\circ_{\text{cell}} &= E^\circ_{\text{cathode}} - E^\circ_{\text{anode}} \\ &= E^\circ_{\text{Cu}} - E^\circ_{\text{SHE}} \\ 0.342\text{ V} &= E^\circ_{\text{Cu}} - 0.000\text{ V} \\ E^\circ_{\text{Cu}} &= 0.342\text{ V} \end{aligned}$$

This half-reaction potential matches the value of E° for the reduction of Cu^{2+} to Cu metal.

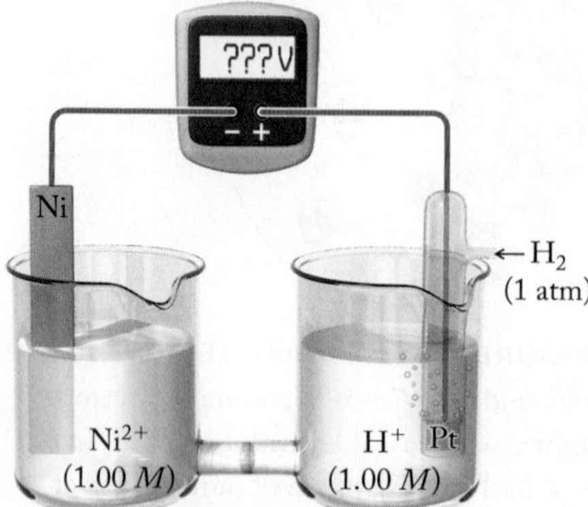

FIGURE 17.10 A SHE and a Ni electrode connected by a voltmeter.

CONCEPT TEST

A cell consists of a SHE in one compartment and a Ni electrode immersed in a 1.00 *M* solution of Ni^{2+} ions in the other. If a voltmeter is connected to the electrode as shown in Figure 17.10, what will be the value in the voltmeter's display?

17.6 The Effect of Concentration on E_{cell}

Reactions stop when one of the reactants is completely consumed. This concept was the basis for our discussion of limiting reactants in Chapter 7. However, a commercial battery usually stops operating at its rated cell potential—1.5 V for a flashlight battery—before its reactants are completely consumed. This happens because the cell potential of a voltaic cell is determined by the concentrations of the reactants and products.

The Nernst Equation

In 1889, German chemist Walther Nernst (1864–1941) derived an expression, now called the **Nernst equation**, which describes the dependence of cell potentials on reactant and product concentrations. We can reconstruct his derivation starting with Equation 14.19, which relates the change in free energy ΔG of any reaction to its change in free energy under standard conditions $\Delta G°$:

$$\Delta G = \Delta G° + RT \ln Q \qquad (14.19)$$

Note that Q in this equation is the *reaction quotient*, the mass action expression for a reaction that is not necessarily at equilibrium. (It has no relationship to the Q we use to represent electric charge.) Recall from Section 14.9 that as a reversible spontaneous reaction proceeds, the concentrations of products increase and the concentrations of reactants decrease until the positive value of $RT \ln Q$ offsets the negative value of $\Delta G°$. At that point, $\Delta G = 0$ and the reaction has reached chemical equilibrium.

CONNECTION In Chapter 14, we discussed the relationship between free-energy changes and the reaction quotient Q.

We can write an expression analogous to Equation 14.19 that relates E_{cell} to $E°_{cell}$ by substituting $-nFE_{cell}$ for ΔG_{cell} and $-nFE°_{cell}$ for $\Delta G°_{cell}$:

$$-nFE_{cell} = -nFE°_{cell} + RT \ln Q$$

Dividing all terms by $-nF$ gives

$$E_{cell} = E°_{cell} - \frac{RT \ln Q}{nF} \qquad (17.8)$$

This is the equation Walther Nernst developed in 1889. We can obtain a very useful form of Equation 17.8 if we insert values for R [8.314 J/(mol · K)] and F (9.65×10^4 C/mol), make $T = 298$ K, and convert the natural logarithm to a base-10 logarithm: $\ln Q = 2.303 \log Q$. Note that the units on RT/F are V · mol since 1 joule is equal to 1 coulomb-volt. With these changes the Nernst equation becomes

$$E_{cell} = E°_{cell} - \frac{0.0592 \text{ V}}{n} \log Q \qquad (17.9)$$

Equation 17.9 enables us to predict how the potential of a voltaic cell changes as the cell reaction proceeds and the concentrations of products inside the cell increase and the concentrations of reactants decrease. As they do, Q increases, so the value of $0.0592/n \times \log Q$ increases, too. The negative sign in front of this term in Equation 17.9 means that the value of E_{cell} decreases as reactants are converted into products. Eventually E_{cell} approaches zero. When it reaches zero, the cell reaction has achieved chemical equilibrium, and the cell can no longer pump electrons through an external circuit. In other words, the cell is dead.

Nernst equation an equation relating the potential of a cell (or half-cell) reaction to its standard potential ($E°$) and to the concentrations of its reactants and products.

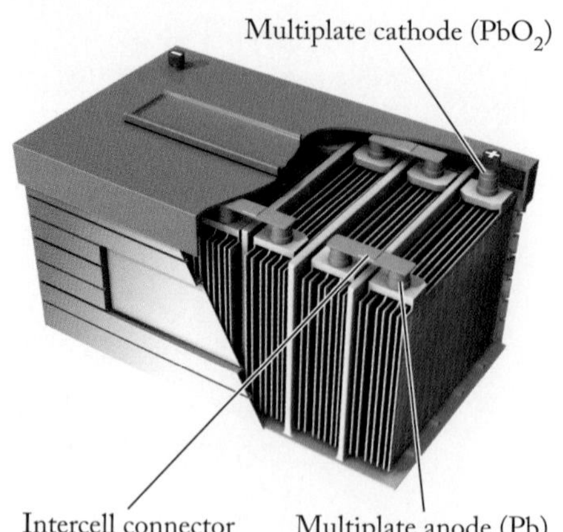

FIGURE 17.11 The lead–acid battery that provides power to start most motor vehicles contains six cells. Each has an anode made of lead and a cathode made of PbO_2 immersed in a background electrolyte of 4.5 M H_2SO_4. The electrodes are formed into plates and held in place by grids made of a lead alloy. The grids connect the cells together in series so that the operating potential of the battery (12.0 V) is the sum of the six cell potentials (each 2.0 V).

Batteries are voltaic cells, so their cell potentials should drop with use. Let's consider how much they drop by focusing on the lead–acid battery used to start most car engines (Figure 17.11). These batteries each contain six electrochemical cells. Their anodes are made of lead (Pb) and their cathodes are made of lead(IV) oxide (PbO_2). Both electrodes are immersed in 4.5 M H_2SO_4. The value of a fully charged cell is about 2.0 V. The six cells are connected in series so that the operating potential of the battery is the sum of the six cell potentials, or 12.0 V.

As the battery discharges, $PbO_2(s)$ is reduced to $PbSO_4(s)$ at the cathodes:

$$PbO_2(s) + 3\,H^+(aq) + HSO_4^-(aq) + 2\,e^- \rightarrow PbSO_4(s) + 2\,H_2O(\ell)$$

and Pb(s) is oxidized to $PbSO_4(s)$ at the anodes:

$$Pb(s) + HSO_4^-(aq) \rightarrow PbSO_4(s) + H^+(aq) + 2\,e^-$$

The reduction half-reaction consumes two moles of electrons, and the oxidation half-reaction involves the loss of two moles of electrons for each mole of Pb.

The net ionic equation for the overall cell reaction is the sum of the two half-reactions:

$$PbO_2(s) + Pb(s) + 2\,H^+(aq) + 2\,HSO_4^-(aq) \rightarrow 2\,PbSO_4(s) + 2\,H_2O(\ell)$$

Using the values of $E^\circ_{cathode}$ and E°_{anode} from Appendix A6.1, the value of E°_{cell} is

$$E^\circ_{cell} = E^\circ_{cathode} - E^\circ_{anode} = 1.685\text{ V} - (-0.356\text{ V}) = 2.041\text{ V}$$

Most of the reactants and products are pure solids or liquids, and the Q expression contains only terms for H^+ and HSO_4^-. As the battery discharges, $[H^+]$ and $[HSO_4^-]$ decrease, and so does the value of E_{cell} calculated from the Nernst equation:

$$E_{cell} = 2.041\text{ V} - \frac{0.0592\text{ V}}{2}\log\frac{1}{[H^+]^2\,[HSO_4^-]^2}$$

However, the decrease in E_{cell} is very gradual, not falling below 2.0 V until the battery is about 97% discharged, as shown in Figure 17.12. The gradual decrease is consistent with the logarithmic relationship between Q and E_{cell}. If $[H_2SO_4]$ decreased by an order of magnitude—for example, from 1.00 M to 0.100 M—the value of E_{cell} would decrease from 2.041 V by less than 6%:

$$E_{cell} = 2.041\text{ V} - \frac{0.0592\text{ V}}{2}\log\frac{1}{(0.100)^4} = 1.923\text{ V}$$

The logarithmic relationship between Q and E_{cell} means that most batteries can deliver current at a cell potential close to their fully charged potential until they are nearly completely discharged.

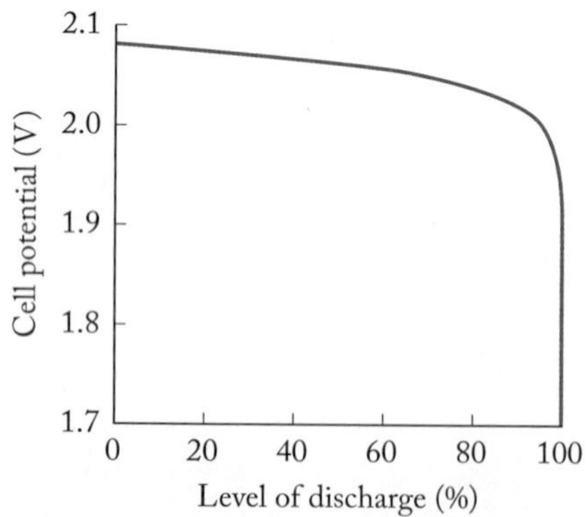

FIGURE 17.12 The potential of a cell in a lead–acid battery decreases as reactants are converted into products, but the change in potential is small until the battery is nearly completely discharged. While the value of the standard cell potential is 2.041 V, a fully charged commercial battery has a slightly higher potential (shown here as 2.08 V) due to the use of more concentrated sulfuric acid.

SAMPLE EXERCISE 17.6 Calculating E_{cell} from E°_{cell} and the Concentrations of Reactants and Products **LO5**

The standard potential (E°_{cell}) of a voltaic cell based on the Zn/Cu^{2+} ion reaction:

$$Zn(s) + Cu^{2+}(aq) \rightarrow Zn^{2+}(aq) + Cu(s)$$

is 1.104 V. What is the value of E_{cell} at 25°C when $[Cu^{2+}] = 0.100\ M$ and $[Zn^{2+}] = 1.90\ M$?

Collect, Organize, and Analyze We are given the standard cell potential and are asked to determine the value of E_{cell} when $[Cu^{2+}] = 0.100\ M$ and $[Zn^{2+}] = 1.90\ M$. The Nernst equation (Equation 17.9) enables us to calculate E_{cell} values for different concentrations of reactants and products. The only term in the numerator of the Q expression for this cell reaction is $[Zn^{2+}]$ and the only one in the denominator is $[Cu^{2+}]$, because no terms for pure solids, such as Zn(*s*) and Cu(*s*), appear in reaction quotients. Each Cu^{2+} ion acquires two electrons, and each Zn atom donates two, so $n = 2$ in the Nernst equation. The value of $[Zn^{2+}]$ is greater than $[Cu^{2+}]$, which makes $Q > 1$. The negative sign in front of the $0.0592/n \times \log Q$ term in Equation 17.9 means that the calculated value of E_{cell} should be less than the value of E°_{cell}.

Solve Substituting the values of $[Zn^{2+}]$ and $[Cu^{2+}]$ in the Nernst equation gives

$$E_{cell} = E^\circ_{cell} - \frac{0.0592\text{ V}}{n}\log Q = E^\circ_{cell} - \frac{0.0592\text{ V}}{n}\log\frac{[Zn^{2+}]}{[Cu^{2+}]} = 1.104\text{ V} - \frac{0.0592\text{ V}}{2}\log\frac{1.90}{0.100}$$

$$E_{cell} = 1.104\text{ V} - \frac{0.0592\text{ V}}{2}(1.28) = 1.066\text{ V}$$

Think About It The calculated E_{cell} value is only 0.038 V less than E°_{cell} because the logarithmic dependence of cell potential on reactant and product concentrations minimizes the impact of changing concentrations.

Practice Exercise The standard cell potential of the zinc–air battery (Figure 17.6) is 1.65 V. If at 25°C the partial pressure of oxygen in the air diffusing through its cathode is 0.21 atm, what is the cell potential? Assume the cell reaction is

$$2\text{ Zn}(s) + O_2(g) \rightarrow 2\text{ ZnO}(s)$$

E° and K

CHEMTOUR Cell Potential, Equilibrium, and Free Energy

When the cell reaction of a voltaic cell reaches chemical equilibrium, $\Delta G_{cell} = E_{cell} = 0$ and $Q = K$. Equation 17.9 then becomes

$$0 = E^\circ_{cell} - \frac{0.0592\text{ V}}{n}\log K$$

which we can rearrange to

$$\log K = \frac{nE^\circ_{cell}}{0.0592\text{ V}} \quad (17.10)$$

We can use Equation 17.10 to calculate the equilibrium constant for any redox reaction at 25°C, not just those in electrochemical cells. For the more general case, we substitute E°_{rxn} for E°_{cell}:

$$\log K = \frac{nE^\circ_{rxn}}{0.0592\text{ V}} \quad (17.11)$$

SAMPLE EXERCISE 17.7 Calculating K for a Redox Reaction from the Standard Potentials of Its Half-Reactions **LO6**

Many procedures for determining mercury levels in environmental samples begin by oxidizing the mercury to Hg^{2+} and then reducing the Hg^{2+} to elemental Hg with Sn^{2+}. Use the appropriate standard reduction potentials from Table A6.1 to calculate the equilibrium constant at 25°C for the following reaction:

$$Sn^{2+}(aq) + Hg^{2+}(aq) \rightarrow Sn^{4+}(aq) + Hg(\ell)$$

Collect and Organize We can calculate the equilibrium constant for a redox reaction from the standard potentials of its half-reactions. Equation 17.11 relates the equilibrium constant for any redox reaction to the standard potential E°_{cell}. To calculate E°_{cell}, we need to combine the appropriate half-reactions. Table A6.1 lists two reduction half-reactions involving our reactants and products:

$$Hg^{2+}(aq) + 2\,e^- \rightarrow Hg(\ell) \qquad E^\circ = 0.851\ V$$

$$Sn^{4+}(aq) + 2\,e^- \rightarrow Sn^{2+}(aq) \qquad E^\circ = 0.154\ V$$

Analyze The problem states that Hg^{2+} is reduced by Sn^{2+}, so Sn^{2+} is the reducing agent in the reaction, which means that it must be oxidized. Therefore, the second reaction runs in reverse, as an oxidation, and we must subtract its standard potential from that of the mercury half-reaction. The difference between the two half-reaction potentials is about +0.7 V. Therefore, the right side of Equation 17.11 will be about $(2 \times 0.7)/0.06 \approx 23$, and the value of K should be about 10^{23}.

Solve We obtain the standard potential for the reaction from a modified version of Equation 17.1, where $E^\circ_{rxn} = E^\circ_{cell}$ for any redox reaction:

$$E^\circ_{rxn} = E^\circ_{cathode} - E^\circ_{anode} = E^\circ_{Hg} - E^\circ_{Sn} = 0.851\ V - 0.154\ V = 0.697\ V$$

To obtain a value for K, we substitute this value for E°_{rxn} and $n = 2$ in Equation 17.11:

$$\log K = \frac{nE^\circ_{rxn}}{0.0592\ V} = \frac{2(0.697\ \cancel{V})}{0.0592\ \cancel{V}} = 23.5$$

$$K = 10^{23.5} = 3 \times 10^{23}$$

Think About It The calculated value is quite close to what we estimated. Note how a relatively small (<1 V) positive value of E°_{rxn} corresponds to a huge equilibrium constant, indicating that the reaction essentially goes to completion. The utility of this reaction for the reliable determination of mercury concentration in aqueous samples depends on the reaction going to completion.

Practice Exercise Use the appropriate standard reduction potentials from Table A6.1 to calculate the value of K at 25°C for the following reaction:

$$5\ Fe^{2+}(aq) + MnO_4^-(aq) + 8\ H^+(aq) \rightarrow 5\ Fe^{3+}(aq) + Mn^{2+}(aq) + 4\ H_2O(\ell)$$

Before ending our discussion of how to derive equilibrium constant values at 25°C from E°_{cell} values, note that measuring the potential of an electrochemical reaction allows us to calculate equilibrium constant values that may be too large or too small to determine from the values of equilibrium concentrations of reactants and products. A value of K as large as that calculated in Sample Exercise 17.7 could not be obtained by analyzing the composition of an equilibrium reaction mixture because the concentrations of the reactants would be too small to be determined accurately. Similarly, a cell potential of about −1 V would correspond to a tiny K value and concentrations of products that are too small to be determined quantitatively.

Table 17.1 summarizes how the values of K and E°_{cell} are related to each other and to the change in free energy (ΔG°_{cell}) of a cell reaction under standard conditions. Note that spontaneous electrochemical reactions have $E^\circ_{cell} > 0$ and $K > 1$. The connection between positive cell potential and reaction spontaneity applies even under nonstandard conditions. Also keep in mind that small positive values of E°_{cell} (only a fraction of a volt, for example) correspond to very large K values and to cell reactions that go nearly to completion.

TABLE 17.1 Relationships between K, E°_{cell}, and ΔG°_{cell} Values of Electrochemical Reactions

K	E°_{cell}	ΔG°_{cell}	Favors Formation of
<1	<0	>0	Reactants
>1	>0	<0	Products
1	0	0	Neither

17.7 Relating Battery Capacity to Quantities of Reactants

An important performance characteristic of a battery is its capacity to do electrical work—that is, to deliver electric charge at the labeled cell potential. This capacity—the amount of electrical work done—is defined by Equation 17.5,

$$w_{\text{elec}} = -nFE_{\text{cell}}$$

where nF is the quantity of electric charge delivered, in coulombs (C).

Another important unit in electricity is the *ampere* (A), which is the SI base unit of electric current. An ampere is defined as a current of 1 coulomb per second:

$$1 \text{ ampere} = 1 \text{ coulomb/second}$$

which we can rearrange to

$$1 \text{ coulomb} = 1 \text{ ampere} \cdot \text{second}$$

Multiplying both sides of the latter equation by volts, and recalling that 1 joule of electrical energy is equivalent to 1 coulomb · volt of electrical work, we get

$$1 \text{ (coulomb)(volt)} = 1 \text{ joule} = 1 \text{ (ampere} \cdot \text{second)(volt)} \quad (17.12)$$

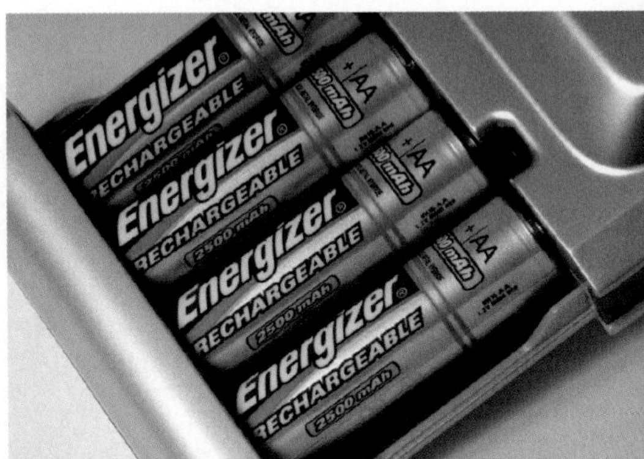

FIGURE 17.13 The electrical energy rating of these rechargeable nickel–metal hydride AA batteries is 2500 mA · h at 1.2 V.

Joules are small energy units, so battery capacities are usually expressed in energy units with time intervals longer than seconds. For example, the *energy ratings* of rechargeable AA batteries (Figure 17.13) are often expressed in ampere · hours or milliampere · hours at the rated cell potential.

We need even bigger units to express the capacities of the large battery packs used in hybrid vehicles. They are the *watt* (W), the SI unit of power, and the *kilowatt-hour*, a unit of energy equal to over 3 million joules, as shown in the following unit conversions:

$$\begin{aligned} 1 \text{ watt} &= 1 \text{ joule/second} = 1 \text{ ampere} \cdot \text{volt} \\ 1 \text{ kilowatt} &= 1000 \text{ W} = 1000 \text{ J/s} \\ 1 \text{ kilowatt} \cdot \text{hour} &= (1000 \text{ W})(1 \text{ h}) \\ &= (1000 \text{ J/s})(60 \text{ s/}\cancel{\text{min}})(60 \text{ }\cancel{\text{min}}\text{/}\cancel{\text{h}})(1 \text{ }\cancel{\text{h}}) \\ &= 3.6 \times 10^6 \text{ J} \end{aligned}$$

CONCEPT TEST

Which energy unit—watt-second (joule), watt-minute, watt-hour, or kilowatt-hour—would be most appropriate for expressing (a) the capacity of a cell phone battery, (b) the annual electrical energy consumed by an Energy Star dishwasher, and (c) the monthly electricity bill for a student apartment?

Nickel–Metal Hydride Batteries

Hybrid vehicles such as the Toyota Prius (Figure 17.14) are powered by combinations of small gasoline engines and electric motors. The motors are powered by battery packs made of dozens of nickel–metal hydride (NiMH) cells. At the cathodes in these cells, $NiO(OH)$ is reduced to $Ni(OH)_2$; at the anodes, which are made of one or more transition metals, hydrogen atoms are oxidized to

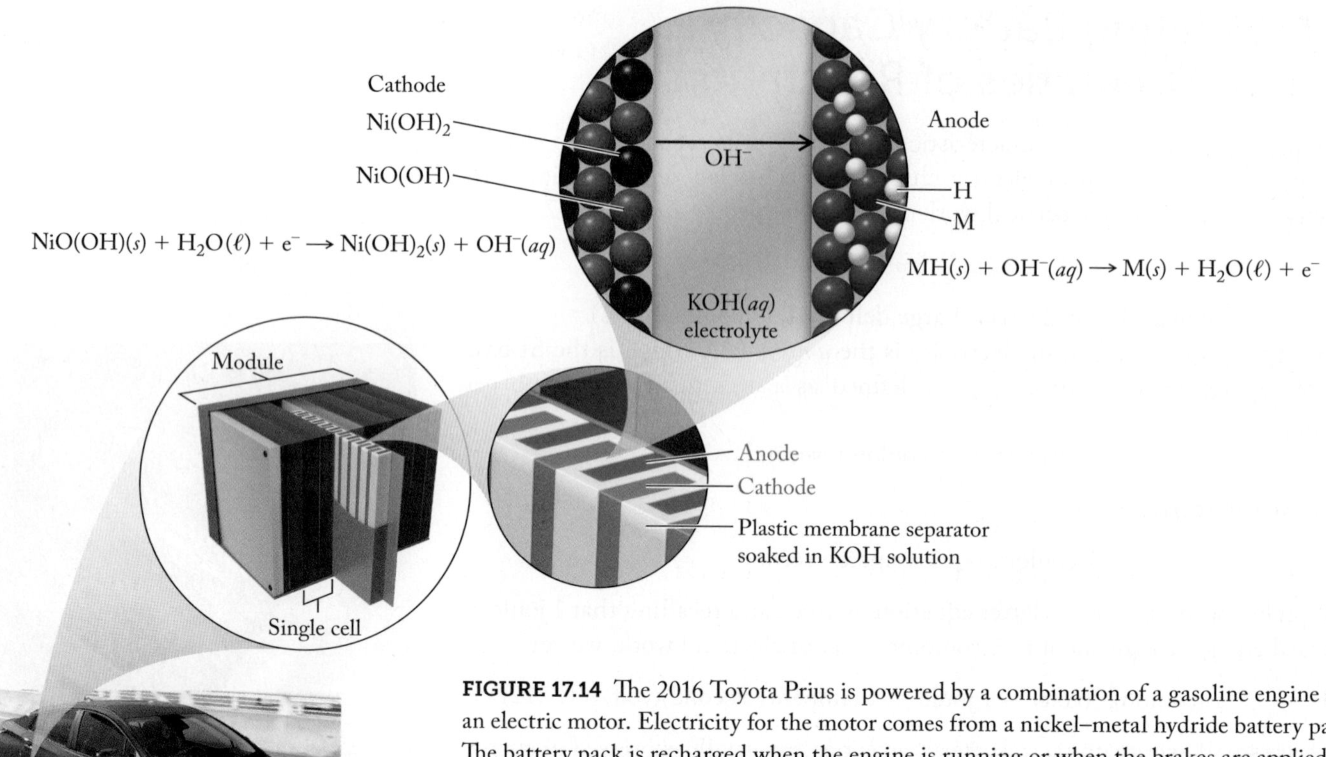

FIGURE 17.14 The 2016 Toyota Prius is powered by a combination of a gasoline engine and an electric motor. Electricity for the motor comes from a nickel–metal hydride battery pack. The battery pack is recharged when the engine is running or when the brakes are applied, as the car's electric motor acts like a generator to convert the car's kinetic energy into electrical energy. In the cells of a nickel–metal hydride battery pack, H atoms of the metal hydride are oxidized to H^+ ions at the anodes (blue plates), and NiO(OH) is reduced to $Ni(OH)_2$ at the cathodes (red plates). The OH^- ions produced by the cathode half-reaction migrate across a KOH-soaked porous membrane and are consumed in the anode half-reaction. The H^+ ions react with OH^- to form water. The anodes are connected to the case of the battery pack, which serves as the (−) terminal, and the cathodes are connected to the cap, which is the (+) terminal.

H^+ ions, which react with OH^- to form water. The electrodes are separated by aqueous KOH.

The cathode half-reaction is

$$NiO(OH)(s) + H_2O(\ell) + e^- \rightarrow Ni(OH)_2(s) + OH^-(aq) \qquad E^\circ = 0.52\text{ V}$$

At the anode, hydrogen is present as a metal hydride. To write the anode half-reaction, we use the generic formula MH, where M stands for a transition metal or metal alloy that forms a hydride. In a basic background electrolyte, the anode oxidation half-reaction is

$$MH(s) + OH^-(aq) \rightarrow M(s) + H_2O(\ell) + e^-$$

The standard potential of this half-reaction depends on the chemical properties of MH. The overall cell reaction from these two half-reactions is

$$MH(s) + NiO(OH)(s) \rightarrow M(s) + Ni(OH)_2(s)$$

The value of E°_{cell} for the NiMH battery cannot be calculated precisely because we have only an approximate value of E°_{anode}. Most NiMH cells are rated at about 1.2 V.

Now let's relate the electrical energy stored in a battery (in other words, its energy rating) to the quantities of reactants needed to produce that energy.

Consider a rechargeable AA NiMH battery rated to deliver 2.5 ampere-hours of electrical charge at 1.2 V. How much NiO(OH) has to be converted to $Ni(OH)_2$ to deliver this much charge? To answer the question, we need to relate the quantity of charge to a number of moles of electrons and then convert that to an equivalent number of moles of reactant and finally to a mass of reactant. Let's begin by recalling that an ampere is defined as a coulomb per second, which means the quantity of electrical charge delivered is

$$2.5\ \cancel{\text{A}} \cdot \cancel{\text{h}} \times \frac{1\ \text{C}}{1\ \cancel{\text{A}} \cdot \cancel{\text{s}}} \times \frac{60\ \cancel{\text{min}}}{1\ \cancel{\text{h}}} \times \frac{60\ \cancel{\text{s}}}{1\ \cancel{\text{min}}} = 9.0 \times 10^3\ \text{C}$$

The Faraday constant tells us that one mole of charge is equivalent to 9.65×10^4 C, so the number of moles of charge, which is equal to the number of moles of electrons that flow from the battery, is

$$9.0 \times 10^3\ \cancel{\text{C}}\left(\frac{1\ \text{mol e}^-}{9.65 \times 10^4\ \cancel{\text{C}}}\right) = 0.0933\ \text{mol e}^-$$

According to the stoichiometry of the cathode half-reaction, the mole ratio of NiO(OH) to electrons is 1:1, so the mass of NiO(OH) consumed is

$$0.0933\ \cancel{\text{mol e}^-}\left(\frac{1\ \cancel{\text{mol NiO(OH)}}}{1\ \cancel{\text{mol e}^-}}\right)\left(\frac{91.70\ \text{g NiO(OH)}}{1\ \cancel{\text{mol NiO(OH)}}}\right) = 8.6\ \text{g NiO(OH)}$$

The mass of a AA battery is about 30 g, so the mass of the NiO(OH) accounts for about 30% of the total.

Lithium–Ion Batteries

The NiMH batteries used in hybrid vehicles do not have the capacity to power the vehicles at highway speeds or for extended distances. Nor do these batteries have the energy capacity to power all-electric vehicles such as the Tesla sports sedan or plug-in hybrids such as the Chevrolet Volt (Figure 17.15). The electrical power demands of these vehicles require batteries with much greater ratios of energy capacity to battery size. The technology of choice in these applications is the lithium–ion battery, the same kind of battery that powers laptop computers, cell phones, and digital cameras.

In a lithium–ion battery, Li^+ ions are typically stored in a graphite anode, though other anode materials such as nanowires made of silicon have been recently developed. During discharge, these ions migrate through a nonaqueous electrolyte to a porous cathode. These cathodes are made of transition metal oxides or phosphates that can form stable complexes with Li^+ ions. One popular cathode material is cobalt(IV) oxide. Lithium–ion batteries with these cathodes and graphite anodes have cell potentials of about 3.6 V (three times that of a NiMH battery). The cell reaction is

$$Li_{1-x}CoO_2(s) + Li_xC_6(s) \rightarrow 6\ C(s) + LiCoO_2(s) \qquad (17.13)$$

In a fully charged cell, $x = 1$, which makes the cathode lithium-free CoO_2. As the cell discharges and Li^+ ions migrate from the carbon anode to the cobalt oxide cathode, the value of x falls toward zero. To balance this flow of positive charges inside the cell, electrons flow from the anode to the cathode through an external circuit. When fully discharged, the cathode is $LiCoO_2$, and the oxidation number of Co is reduced to +3. The electrodes in a lithium–ion battery may react with oxygen and water, so the electrolytes (for example, $LiPF_6$) are dissolved

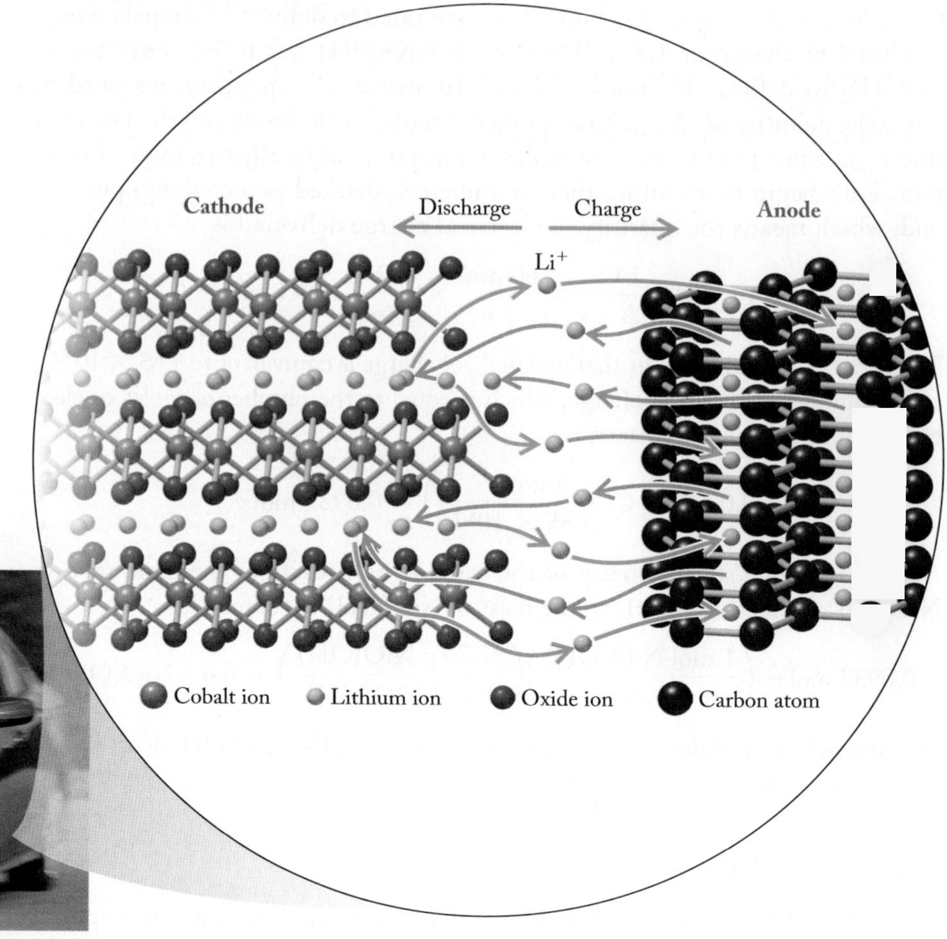

FIGURE 17.15 The 2016 Chevrolet Volt has a lithium–ion battery pack that can store 18.4 kW · h of electrical energy, giving the Volt a driving range of about 60 km (~37 mi). The battery pack is recharged either by plugging it into an electrical outlet or by running an onboard gasoline-powered generator. The generator gives the Volt an overall driving range of about 1000 km (~621 mi). In a discharging lithium–ion battery, Li^+ ions stored in graphite layers of the anode travel to the cathode, which is made of CoO_2 in this example. During recharging, the direction of ion migration reverses.

in polar organic solvents, such as tetrahydrofuran, ethylene carbonate, or propylene carbonate (Figure 17.16).

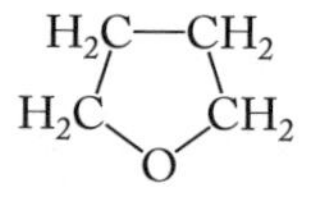

Tetrahydrofuran

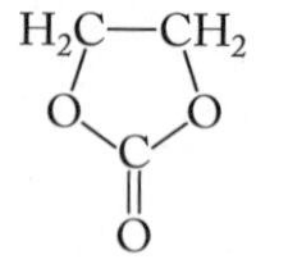

Ethylene carbonate

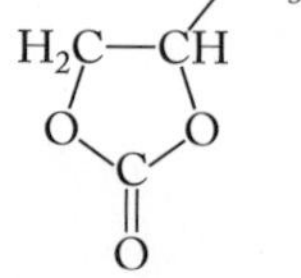

Propylene carbonate

FIGURE 17.16 Polar organic compounds such as these molecules are the solvents for the electrolytes in a lithium–ion battery.

SAMPLE EXERCISE 17.8 Relating the Mass of a Reactant in a Cell Reaction to a Quantity of Electrical Charge — LO7

The capacity of the lithium–ion battery in a digital camera is 3.4 W · h at 3.6 V. How many grams of Li^+ ions must migrate from anode to cathode to produce this much electrical energy?

Collect and Organize We are asked to relate the electrical energy generated by an electrochemical cell to the mass of the ions involved in generating that energy. We know the cell potential and its capacity in the energy unit of watt-hours. Given these starting points and our need to calculate moles and then grams of Li^+ ions, the following equivalencies may be useful:

$$1 \text{ watt (W)} = 1 \text{ ampere-volt (A} \cdot \text{V)}$$

$$1 \text{ coulomb (C)} = 1 \text{ ampere-second (A} \cdot \text{s)}$$

We may also need to use the Faraday constant, 9.65×10^4 C/mol.

Analyze The energy capacity of the battery is the product of the charge (electrons) it can deliver times the cell potential pumping that charge. This exercise focuses on the quantity of charge, so we need to separate the contribution that cell potential makes to the energy rating. To do that we need to divide the energy rating (3.4 W · h = 3.4 A · V · h) by the battery's cell potential (3.6 V), which translates into slightly less than 1 A · h. Converting this value into a mass of Li^+ ions involves unit conversions that take us from ampere-hours to ampere-seconds (equivalent to coulombs), to moles of electrons using the Faraday constant, to moles of Li^+ ions (equivalent to moles of electrons), and finally to grams of Li^+ ions.

Solve

$$\frac{3.4 \text{ A} \cdot \cancel{\text{V}} \cdot \cancel{\text{h}}}{3.6 \cancel{\text{V}}} \times \frac{3600 \text{ s}}{\cancel{\text{h}}} \times \frac{1 \cancel{\text{C}}}{\text{A} \cdot \text{s}} \times \frac{1 \cancel{\text{mol e}^-}}{96{,}500 \cancel{\text{C}}} \times \frac{1 \cancel{\text{mol Li}^+}}{1 \cancel{\text{mol e}^-}} \times \frac{6.941 \text{ g Li}^+}{1 \cancel{\text{mol Li}^+}}$$

$$= 0.24 \text{ g Li}^+ \text{ ions}$$

Think About It The battery that is the subject of this exercise has a mass of about 22 g, so Li^+ ions make up only about 1% of its mass. This small percentage is reasonable given the masses of the other required components of the cell, including an anode in which each Li^+ ion is surrounded by a hexagon of six carbon atoms and a cathode made of a transition metal compound with a molar mass many times that of Li metal.

 **Practice Exercise** Magnesium metal is produced by passing an electric current through molten $MgCl_2$. The reaction at the cathode is

$$Mg^{2+}(\ell) + 2\,e^- \rightarrow Mg(\ell)$$

How many grams of magnesium metal are produced if an average current of 63.7 A flows for 4 hours and 45 minutes? Assume all of the current is consumed by the half-reaction shown.

17.8 Corrosion: Unwanted Electrochemical Reactions

We began this chapter with a Particulate Review question based on the corrosion of the metal surface of a car. In this section we examine this process and some of the half-reactions that contribute to it in more detail. We have seen how half-reactions are physically separated in electrochemical cells. It turns out they are often separated in corrosion reactions as well; in this respect, the chemistry of corrosion is much like the electrochemical reactions in voltaic cells. In fact, we can define **corrosion** as the deterioration of metals due to spontaneous electrochemical reactions. This definition is reflected in several of the factors that promote corrosion:

1. *The presence of water.* Metals that are left out in the rain and snow rust or corrode more rapidly than those that are under cover. For example, objects made from iron spontaneously rust in moist air, as iron reacts with O_2 from the air in a series of reactions that includes

$$4\,Fe(s) + 3\,O_2(g) + 2\,H_2O(\ell) \rightarrow 4\,FeO(OH)(s)$$

2. *The presence of electrolytes.* Just as electrolytes carry electrical current between anodes and cathodes and facilitate cell reactions, corrosion is much more rapid in, for example, seawater than in freshwater.
3. *Contact between dissimilar metals.* Metals corrode more rapidly when in contact with other metals that are less likely to be oxidized—that is, other metals that have higher reduction potentials.

corrosion a process in which a metal deteriorates through spontaneous electrochemical reactions.

Let's explore the impact of these factors using the Statue of Liberty as a model. As the statue was built, its exterior copper sheets were attached to and supported by an interior network of iron beams (Figure 17.17). The French designers of the statue knew that these two metals in contact with each other might someday pose a corrosion problem because the two have very different electrochemical properties. The difference is reflected in the standard reduction potentials of these elements when they oxidize under neutral to slightly basic conditions, as occur in marine environments:

$$Cu(OH)_2(s) + 2\,e^- \rightarrow Cu(s) + 2\,OH^-(aq) \qquad E° = -0.230\ V$$

$$FeO(OH)(s) + H_2O(\ell) + 3\,e^- \rightarrow Fe(s) + 3\,OH^-(aq) \qquad E° = -0.87\ V$$

As we discussed in Section 17.3, the greater $E°$ of $Cu(OH)_2$ means that it is more easily reduced under standard conditions than FeO(OH), and Fe is more easily oxidized than Cu. We can confirm this by reversing the iron half-reaction and combining it with the copper half-reaction. We also need to multiply the iron half-reaction by 2 and the copper half-reaction by 3 to balance the loss and gain of electrons:

$$2[Fe(s) + 3\,OH^-(aq) \rightarrow FeO(OH)(s) + H_2O(\ell) + 3\,e^-]$$

$$3[Cu(OH)_2(s) + 2\,e^- \rightarrow Cu(s) + 2\,OH^-(aq)]$$

$$2\,Fe(s) + 3\,Cu(OH)_2(s) + \cancel{6\,OH^-(aq)} + \cancel{6\,e^-} \rightarrow 2\,FeO(OH)(s) + 3\,Cu(s) + 2\,H_2O(\ell) + \cancel{6\,OH^-(aq)} + \cancel{6\,e^-}$$

or

$$3\,Cu(OH)_2(s) + 2\,Fe(s) \rightarrow 3Cu(s) + 2\,FeO(OH)(s) + 2\,H_2O(\ell) \qquad (17.14)$$

The resulting equation could be that of an electrochemical cell that has a copper cathode coated in $Cu(OH)_2$ and an iron anode coated in FeO(OH). We obtain the standard potential for the reaction by substituting the reduction potentials for each half-reaction (Appendix A6.1) into Equation 17.1:

$$E°_{rxn} = E°_{cathode} - E°_{anode} = -0.230\ V - (-0.87\ V) = 0.64\ V$$

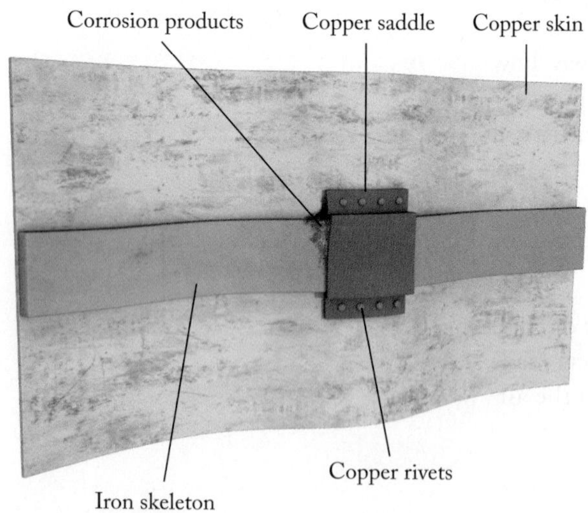

FIGURE 17.17 When the Statue of Liberty was built, its copper exterior was supported by an iron skeleton that corroded near the points of contact with the sheets and with the copper saddles that held the sheets and skeleton together.

The reaction has a positive standard cell potential, which means that under standard conditions, iron in contact with copper will spontaneously oxidize to FeO(OH) as $Cu(OH)_2$ is reduced.

To suppress the reaction in Equation 17.14, insulators made of asbestos mats soaked in shellac (Figure 17.17) were used to separate the statue's iron skeleton from its copper exterior. Unfortunately, these insulators did not stand up to the humid, marine environment of New York Harbor. They absorbed water vapor and seawater spray, eventually turning into electrolyte-soaked sponges that actually promoted rather than retarded iron oxidation.

What are the sources of the $Cu(OH)_2$ that are the oxidizing agents in Equation 17.14? Recall from Sample Exercise 17.3 that the light green patina of the Statue of Liberty is due to the presence of copper(II) compounds such as $Cu(OH)_2$, formed by the oxidation of Cu metal by atmospheric oxygen:

$$O_2(g) + 2\,H_2O(\ell) + 2\,Cu(s) \rightarrow 2\,Cu(OH)_2(s) \qquad (17.15)$$

This reaction occurs on an expansive surface of copper metal, which is also an excellent conductor of electricity. Unfortunately, this copper surface was in contact with the statue's iron skeleton through the water-logged, ion-rich asbestos mats. This connection meant the reactions in Equations 17.14 and 17.15 were linked together. Equation 17.16 describes the resulting interior and exterior reactions:

$$2[3\ Cu(OH)_2(s) + 2\ Fe(s) \rightarrow 3\ Cu(s) + 2\ FeO(OH)(s) + 2\ H_2O(\ell)] \qquad (17.14)$$

$$3[O_2(g) + 2\ H_2O(\ell) + 2\ Cu(s) \rightarrow 2\ Cu(OH)_2(s)] \qquad (17.15)$$

$$\cancel{6\ Cu(OH)_2(s)} + 4\ Fe(s) + 3\ O_2(g) + \cancel{6}^{2}\ H_2O(\ell) + \cancel{6\ Cu(s)} \rightarrow \cancel{6\ Cu(s)} + 6\ FeO(OH)(s) + \cancel{4\ H_2O(\ell)} + \cancel{6\ Cu(OH)_2(s)}$$

or

$$4\ Fe(s) + 3\ O_2(g) + 2\ H_2O(\ell) \rightarrow 4\ FeO(OH)(s) \qquad (17.16)$$

Note how the copper half-reaction has disappeared from the overall reaction. Thus, the statue's copper exterior functions as a giant electron delivery system, allowing electrons to flow from iron atoms—as they oxidize to FeO(OH) inside the statue—to molecules of atmospheric O_2 on the exterior surface. The overall effect was a dramatic acceleration of the rate of the air oxidation of the original iron skeleton.

The reaction in Equation 17.16 and others like it forming other components of rust resulted in severe deterioration of the skeletal network that held up the Statue of Liberty, so much so that in the 1980s the iron skeleton had to be replaced with one made of corrosion-resistant stainless steel.

Deterioration of the Statue of Liberty was not an isolated incident. According to one industrial estimate, the direct and indirect cost of corrosion to the U.S. economy in 2012 exceeded $1 trillion. Worldwide, the cost was over $2 trillion. These figures include the cost to repair or replace corroded equipment and structures and to protect them against corrosion. In the latter category is money spent on protective coatings, including paint and chemical or electrochemical modification of metal surfaces to make them less reactive.

Another widely used method for inhibiting corrosion involves chemically bonding metal oxide coatings to metal surfaces. For example, a method called bluing is widely used to protect steel tools, gun barrels, wood stoves, and other steel materials. The name comes from the distinctive very dark blue color of a surface layer of magnetite (Fe_3O_4), which can be reaction-bonded to steel. Formation of a protective oxide layer is also the mechanism that makes stainless steel resistant to corrosion. Though the main ingredient in all forms of stainless steel is iron, it also contains chromium. Oxidation of surface Cr atoms forms a durable protective layer of Cr_2O_3. Similarly, objects made of aluminum are protected by a surface layer of Al_2O_3 that strongly adheres to the underlying metal and inhibits further oxidation.

Another way to protect metal structures, especially those that come in contact with seawater (Figure 17.18), involves attaching to them objects made of even more reactive metals. These objects are called *sacrificial anodes.* As their name implies, their role is to form a voltaic cell

(a)

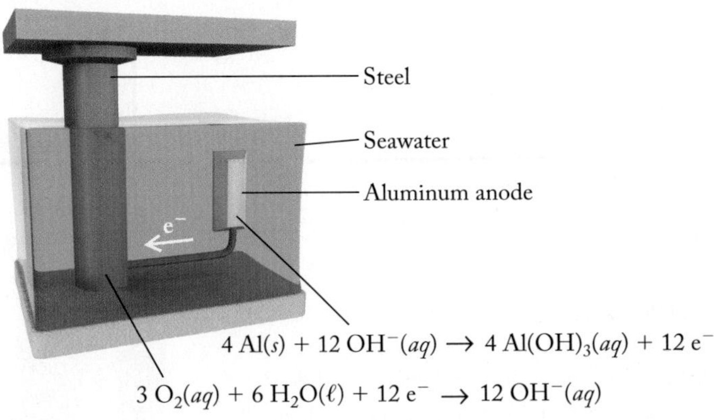

(b)

FIGURE 17.18 (a) Most metal structures immersed in seawater have serious corrosion problems unless (b) they have cathodic protection.

with the protected structure in which the object is the anode and the structure is the cathode. That way the sacrificial anodes oxidize and the structure does not. This preservation technique is called *cathodic protection*. Many of the sacrificial anodes used in the marine industry are made of zinc or aluminum/magnesium alloys.

$$Zn(s) + 2\ OH^-(aq) \rightarrow ZnO(s) + H_2O(\ell) + 2\ e^-$$

$$Mg(s) + 2\ OH^-(aq) \rightarrow MgO(s) + H_2O(\ell) + 2\ e^-$$

$$2\ Al(s) + 6\ OH^-(aq) \rightarrow Al_2O_3(s) + 3\ H_2O(\ell) + 6\ e^-$$

These materials have the benefit of forming oxide coatings as they oxidize, which partially protect them and slow the rate at which they oxidize further, thereby requiring less frequent replacement.

17.9 Electrolytic Cells and Rechargeable Batteries

Lead–acid, NiMH, and lithium–ion batteries are rechargeable, which means that their spontaneous ($\Delta G < 0$) cell reactions, which convert chemical energy into electrical energy, can be forced to run in reverse. Recharging happens when external sources of electrical energy are applied to the batteries. This electrical energy is converted into chemical energy as it drives nonspontaneous ($\Delta G > 0$) reverse cell reactions, re-forming reactants from products. To make this possible, the products of the original cell reactions must be substances that either adhere to or are embedded in the electrodes, and thus are available to react with the electrons supplied to the cathodes and drawn away from the anodes by the external power supply.

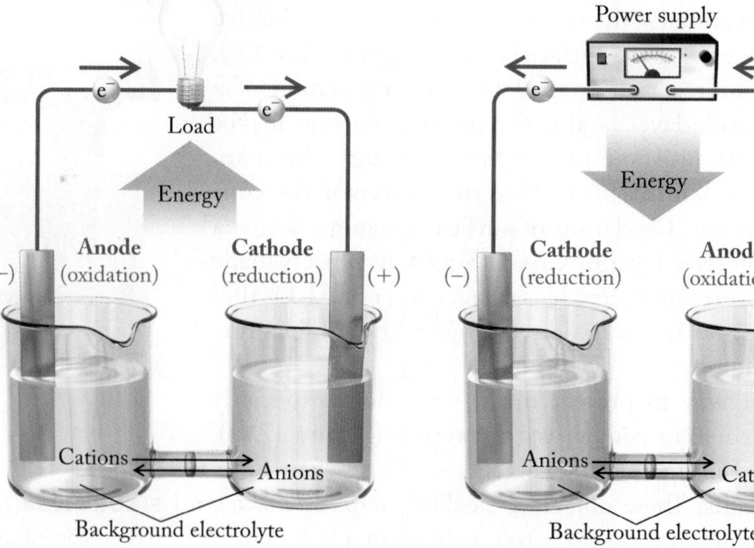

FIGURE 17.19 Voltaic versus electrolytic cells. (a) In a voltaic cell, a spontaneous reaction produces electrical energy and does electrical work on its surroundings, such as lighting a lightbulb. (b) In an electrolytic cell, an external supply of electrical energy does work on the chemical system in the cell, driving a nonspontaneous reaction.

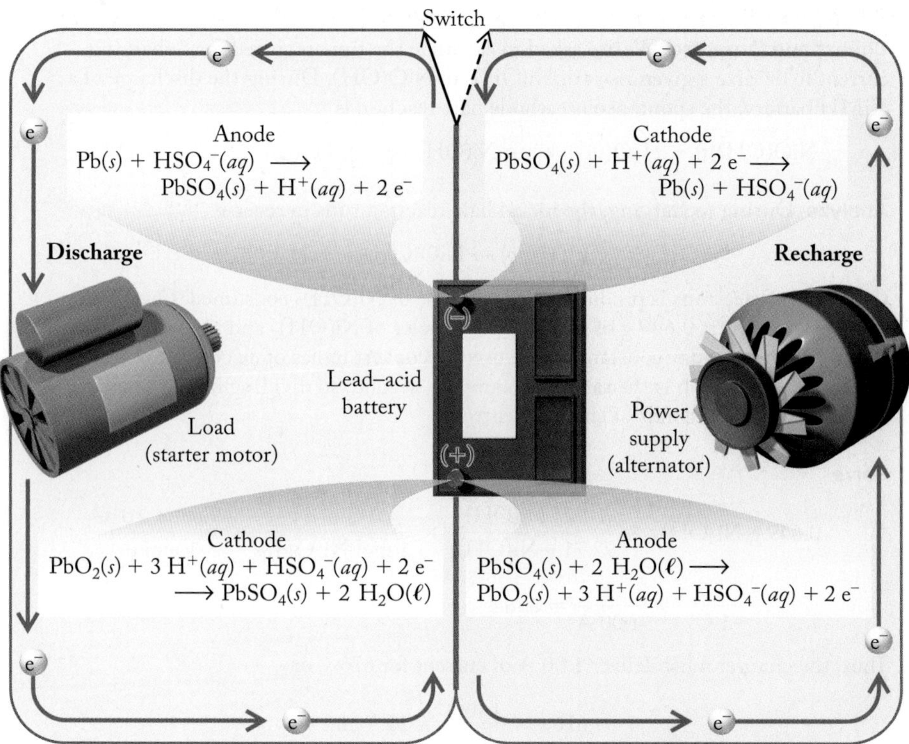

FIGURE 17.20 The lead–acid battery used in many vehicles is based on the oxidation of Pb and the reduction of PbO_2. As the battery discharges (left circuit), Pb is oxidized to $PbSO_4$ and PbO_2 is reduced to $PbSO_4$. When the engine is running (right circuit), a device called an alternator generates electrical energy that flows into the battery, recharging it as both electrode reactions are reversed: $PbSO_4$ is oxidized to PbO_2, and $PbSO_4$ is reduced to Pb.

Recharging a battery is an example of **electrolysis**, which is defined as any chemical reaction driven by electricity. During recharging, a battery is transformed from a voltaic cell into an **electrolytic cell** (Figure 17.19). To explore this transformation, let's revisit the lead–acid battery used to start car engines. Its discharge and recharge cycles are shown in Figure 17.20. Note that electrons flow in one direction when the battery discharges—out of the negative terminal and into the positive terminal—but in the opposite direction when the battery is recharging. Thus, the Pb electrodes connected to the negative terminal in Figure 17.20 serve as anodes during discharge but as cathodes during recharge. Any $PbSO_4$ that forms on these Pb electrodes during discharge is reduced back to Pb metal during recharge:

$$PbSO_4(s) + H^+(aq) + 2\ e^- \rightarrow Pb(s) + HSO_4^-(aq)$$

Similarly, the PbO_2 electrodes connected to the positive terminal serve as cathodes during discharge but as anodes during recharge. Any $PbSO_4$ that forms on these PbO_2 electrodes during discharge is oxidized back to PbO_2 during recharge:

$$PbSO_4(s) + 2\ H_2O(\ell) \rightarrow PbO_2(s) + 3\ H^+(aq) + HSO_4^-(aq) + 2\ e^-$$

electrolysis a process in which electrical energy is used to drive a nonspontaneous chemical reaction.

electrolytic cell a device in which an external source of electrical energy does work on a chemical system, turning reactant(s) into higher-energy product(s).

SAMPLE EXERCISE 17.9 Calculating the Time to Recharge a Battery **LO7**

If a charger for AA NiMH batteries supplies a current of 1.00 A, how many minutes does it take to oxidize 0.649 g of $Ni(OH)_2$ to $NiO(OH)$?

Collect and Organize We are asked to calculate the time required for a charging current to oxidize a given mass of $Ni(OH)_2$ to NiO(OH). During the discharge of a NiMH battery, the spontaneous cathode half-reaction is

$$NiO(OH)(s) + H_2O(\ell) + e^- \rightarrow Ni(OH)_2(s) + OH^-(aq) \qquad E° = 0.52 \text{ V}$$

Analyze During recharging, the nickel half-reaction runs in reverse:

$$Ni(OH)_2(s) + OH^-(aq) \rightarrow NiO(OH)(s) + H_2O(\ell) + e^-$$

One mole of electrons is produced for each mole of $Ni(OH)_2$ consumed. Our first steps are to convert 0.649 g of $Ni(OH)_2$ into moles of $Ni(OH)_2$ and then into moles of electrons. The Faraday constant can be used to convert moles of electrons to coulombs of charge. A coulomb is the same as an ampere-second, so dividing by the charging current will yield seconds of charging current.

Solve

$$0.649 \text{ g Ni(OH)}_2 \times \frac{1 \text{ mol Ni(OH)}_2}{92.71 \text{ g Ni(OH)}_2} \times \frac{1 \text{ mol e}^-}{1 \text{ mol Ni(OH)}_2} \times \frac{9.65 \times 10^4 \text{ C}}{1 \text{ mol e}^-}$$

$$\times \frac{1 \text{ A} \cdot \text{s}}{1 \text{ C}} \times \frac{1}{1.00 \text{ A}} = 676 \text{ s}$$

Thus, the charger must deliver 1.00 A of current for 676 s, or

$$676 \text{ s} \times \frac{1 \text{ min}}{60 \text{ s}} = 11.3 \text{ min}$$

Think About It No estimate was made. A charging time of 11.3 min may seem short, but the quantity of the $Ni(OH)_2$ to be oxidized (0.649 g) is much less than the quantity of NiO(OH) in a fully charged AA NiMH battery (8.6 g, as calculated in the previous section), so this battery was only slightly discharged.

Practice Exercise Suppose that a car's starter motor draws 225 amperes of current for 3.5 s to start the car's engine. What mass of Pb is oxidized in the battery to supply this much electricity?

FIGURE 17.21 The Oscar statuettes given out at the annual Academy Awards are made of cast bronze and immersed in a plating bath containing $Au(CN)_4^-$ ions. When connected to the negative terminal of an electric power supply, the statuette becomes the cathode in an electrolytic cell as the half-reaction $Au(CN)_4^-(aq) + 3\ e^- \rightarrow Au(s) + 4\ CN^-(aq)$ produces a layer of pure gold on the statuette.

Electrolysis is used in many processes other than recharging batteries. Electrolytic cells are used to electroplate thin layers of silver, gold, and other metals onto objects, giving these objects the appearance, resistance to corrosion, and other properties of the electroplated metal at a fraction of the cost of fabricating the entire object out of the metal (Figure 17.21).

In the chemical industry, the electrolysis of molten salts is used to produce highly reactive substances such as sodium, chlorine, and fluorine, as well as alkali and alkaline earth metals and aluminum. When NaCl, for instance, is heated to just above its melting point (above 800°C), it becomes an ionic liquid that can conduct electricity. If a sufficiently large potential is applied to carbon electrodes immersed in the molten NaCl, the sodium ions are attracted to the negative electrode and are reduced to sodium metal, while the chloride ions are attracted to the positive electrode and oxidized to Cl_2 gas:

$$2\ Na^+(\ell) + 2\ Cl^-(\ell) \rightarrow 2\ Na(\ell) + Cl_2(g)$$

A final note about anode and cathode polarity is in order. The reactions in voltaic cells are spontaneous. These cells pump electric current through external circuits and electrical devices with a force equal to their cell potentials. The anode in a voltaic cell is negative because an oxidation half-reaction supplies negatively charged electrons to the device powered by the cell. Electrons flow from the device into the positive battery terminal, which is connected to the cathode, where these electrons are consumed in a reduction half-reaction.

The reactions in electrolytic cells are nonspontaneous. They require electrical energy from an external power supply. When the negative terminal of such a power supply is connected to the cathode of the battery, the power supply pumps electrons into the cathode, where they are consumed in reduction half-reactions. Electrons are pumped away from the anode, where they must have been generated in an oxidation half-reaction, toward the positive terminal of the power supply. Thus, as we can see in Figure 17.19, the cathode of an electrolytic cell is the negative electrode, but the cathode of a voltaic cell is the positive electrode. Similarly, the anode of an electrolytic cell is the positive electrode, but the anode of a voltaic cell is the negative electrode. These "pole reversals" make sense if we keep in mind the fundamental definitions:

- Anodes are electrodes where oxidation takes place.
- Cathodes are electrodes where reduction takes place.

To verify your understanding of the pole reversal process, study the flow of electrons to and from the battery terminals in Figure 17.20 when the battery is being discharged (and working like a voltaic cell) and when it is being recharged (and operating as an electrolytic cell).

CONCEPT **TEST**

The electrolysis of molten NaCl produces liquid Na metal at the cathode and Cl_2 gas at the anode. However, the electrolysis of an aqueous solution of NaCl produces gases at both cathode and anode. Identify the gas produced at the cathode.

17.10 Fuel Cells

Fuel cells are promising energy sources for many applications, from powering office buildings to cruise ships to electric vehicles. Fuel cells are voltaic cells, but they are different from batteries in that their supplies of reactants are constantly renewed. Therefore, they do not "discharge"; they never run down, and they don't die unless their fuel supply is cut off. From a thermodynamic perspective, batteries are closed systems and fuel cells are open systems.

CONNECTION In Chapter 9 *open* thermodynamic systems are defined as those which exchange matter and energy with their surroundings; *closed* systems exchange energy but not matter.

In a typical fuel cell, electrons are supplied to an external circuit by the oxidation of H_2 at the anode, and electrons are consumed by the reduction of O_2 at the cathode. The fuel cell reaction, then, is

$$2\,H_2(g) + O_2(g) \rightarrow 2\,H_2O(\ell)$$

The fuel cells used to power electric vehicles consist of metallic or graphite electrodes separated by a hydrated polymeric material called a *proton-exchange membrane* (PEM). The PEM serves as both an electrolyte and a barrier that prevents crossover and mixing of the fuel and oxidant (Figure 17.22). The surfaces of the electrodes are coated with transition metal catalysts to speed up the electrode

fuel cell a voltaic cell based on the oxidation of a continuously supplied fuel; the reaction is the equivalent of combustion, but chemical energy is converted directly into electrical energy.

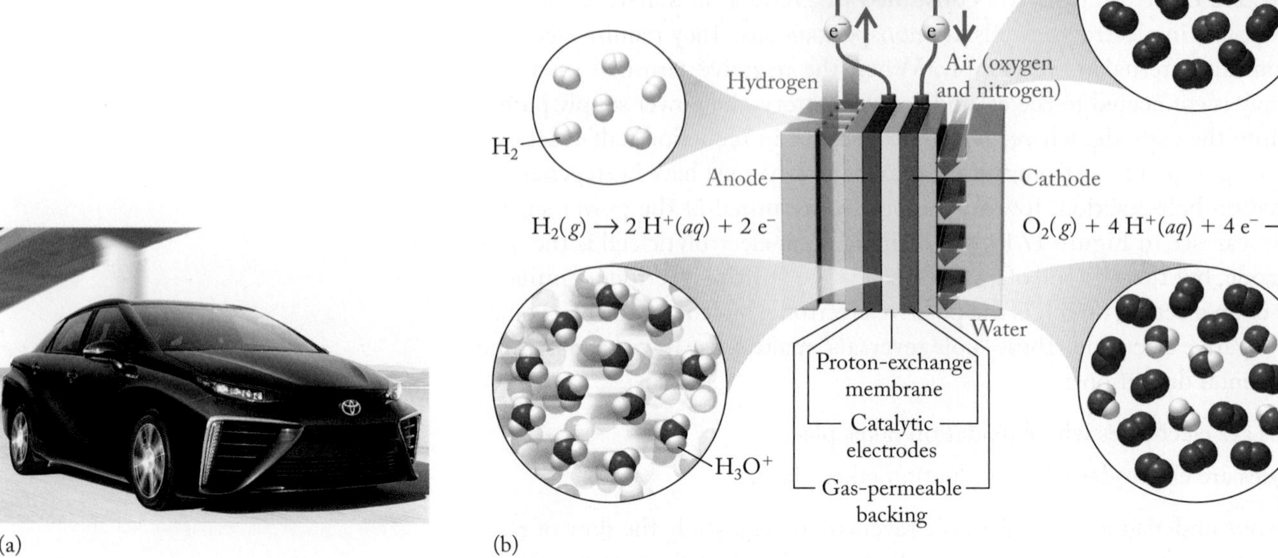

FIGURE 17.22 (a) The 2016 Toyota Mirai is powered by a 114 kW stack of 370 fuel cells and an electric motor that give the car a driving range of about 500 km (300 mi). (b) Most fuel cells used in vehicles have a proton-exchange membrane between the two halves of the cell. Hydrogen gas diffuses to the anode, and oxygen gas diffuses to the cathode. These electrodes are made of a porous material, such as graphite, that has a relatively high surface area for a given mass of material. Catalysts on the electrode surfaces also increase the rate of the half-reactions at the anode ($H_2(g) \rightarrow 2\,H^+(aq) + 2\,e^-$) and the cathode ($O_2(g) + 4\,H^+(aq) + 4\,e^- \rightarrow 2\,H_2O(\ell)$).

half-reactions. Platinum catalysts promote H—H bond breaking during the oxidation of H_2 gas to H^+ ions at the anode:

$$H_2(g) \rightarrow 2\,H^+(aq) + 2\,e^- \qquad E° = 0.000\text{ V}$$

and a platinum–nickel alloy with the formula Pt_3Ni is particularly effective in catalyzing the formation of free O atoms from O_2 molecules, which is part of the reduction half-reaction at the cathode:

$$O_2(g) + 4\,H^+(aq) + 4\,e^- \rightarrow 2\,H_2O(\ell) \qquad E° = 1.229\text{ V}$$

Hydrogen ions that form at the anode migrate through the PEM to the cathode, where they react with O_2. This migration of positive charges inside the fuel cell drives the flow of electrons in the electrical device attached to it.

A single PEM fuel cell typically has a cell potential of about 1.0 V. When hundreds of these cells are assembled into fuel cell *stacks*, they are capable of producing over 100 kW of electrical power. That is enough to give a midsize car such as the one in Figure 17.22 a top speed over 160 km/hr (100 mi/hr).

PEM fuel cells are well suited for use in vehicles because they are compact, lightweight, and operate at fairly low temperatures of 60 to 80°C. The performance of nearly all fuel cells is better at above-ambient temperatures because the higher rates of their half-reactions at these temperatures mean that they generate more electrical power.

Other fuel cells use basic electrolytes such as concentrated KOH. Pure O_2 is supplied to a cathode made of porous graphite containing a nickel catalyst, and H_2 gas is supplied to a graphite anode containing nickel(II) oxide. Hydroxide ions formed during O_2 reduction at the cathode,

$$O_2(g) + 2\,H_2O(\ell) + 4\,e^- \rightarrow 4\,OH^-(aq) \qquad E° = 0.401\text{ V}$$

migrate through the cell to the anode, where they react with H_2 as it is oxidized to H^+ and then react with hydroxide ion to form water:

$$H_2(g) + 2\,OH^-(aq) \rightarrow 2\,H_2O(\ell) + 2\,e^-$$

This oxidation half-reaction is the reverse of the following reduction half-reaction from Table A6.1:

$$2\,H_2O(\ell) + 2\,e^- \rightarrow H_2(g) + 2\,OH^-(aq) \qquad E° = -0.828\text{ V}$$

Note that this basic pair of standard electrode potentials yields the same $E°_{cell}$ value,

$$E°_{cell} = 0.401\text{ V} - (-0.828\text{ V}) = 1.229\text{ V}$$

as the acidic pair described earlier:

$$E°_{cell} = 1.229\text{ V} - 0.000\text{ V} = 1.229\text{ V}$$

This equality is logical because $\Delta G°$, the energy released under standard conditions by the oxidation of hydrogen gas to form liquid water,

$$2\,H_2(g) + O_2(g) \rightarrow 2\,H_2O(\ell)$$

should have only one value, which means that $E°_{cell}$ should have only one value independent of the pH of the electrolyte.

CONCEPT TEST

During the operation of molten alkali metal carbonate fuel cells, carbonate ions are generated at one electrode, migrate across the cell, and are consumed at the other electrode. Do the carbonate ions migrate toward the cathode or the anode?

The same chemical energy that is released in fuel cells could also be obtained by burning hydrogen gas in an internal combustion engine. However, only about 20–25% of the chemical energy in the fuel burned in such an engine is typically converted into mechanical energy; most is lost to the surroundings as heat. In contrast, fuel cell technologies can convert up to about 80% of the energy released in a fuel cell's redox reaction into electrical energy. The electric motors they power are also about 80% efficient at converting electrical energy into mechanical energy. Thus, the overall conversion efficiency of a fuel cell–powered car is theoretically as high as (80% × 80%) or 64%. Actually, the measured efficiency of the propulsion system of the car in Figure 17.22 is about 60%, which is still more than twice that of an internal combustion engine. In addition, H_2-fueled vehicles emit only water vapor; they produce no oxides of nitrogen, no carbon monoxide, and no CO_2.

As fuel cells become even more efficient and less expensive, the principal limit on their use in passenger cars will be the availability and cost of hydrogen fuel. Some people worry about the safety of storing hydrogen in a high-pressure tank in a car. In fact, H_2 has a higher ignition temperature than gasoline and spreads through the air more quickly, reducing the risk of fire. Still, hydrogen in air burns over a much wider range of concentrations than gasoline, and its flame is almost invisible.

CONNECTION Processes for producing hydrogen gas by reacting methane and other fuels with steam were discussed in Chapter 14.

SAMPLE EXERCISE 17.10 Integrating Concepts: The Electrolysis of Salt

Electrolysis of sodium chloride is used industrially to produce sodium metal and chlorine gas. One process uses a Downs cell (Figure 17.23), which has a carbon anode and an iron cathode and uses molten NaCl as the electrolyte and source of the reactants in the cell reaction. Calcium chloride is added to the sodium chloride (melting point 801°C; density 2.16 g/cm^3) to lower its melting point so that the process can be carried out at temperatures around 600°C. The products of the electrolysis are liquid sodium (melting point 97.8°C; density 0.93 g/cm^3) and chlorine gas. An iron screen in the cell separates the electrodes so that the two products do not come into contact with each other.

a. Write chemical equations for the half-reactions at the anode and cathode and for the overall cell reaction.
b. Use the appropriate standard reduction potentials in Appendix 6 to estimate the minimum potential difference between the anode and cathode that must be applied to electrolyze molten NaCl.
c. Migration of which ions carries electric charge between the anodes and cathodes in a Downs cell?
d. Calcium chloride is in the reaction mixture, but calcium metal is not produced in the cell. Why?
e. Why is it important to keep the two products separated?
f. In a Downs cell the electrodes are usually positioned at the bottom of the molten NaCl bath, but the products are collected at the top. Why?

Collect and Organize We are given a description of a Downs cell, the conditions under which it is run, and the physical properties of the reactants and products of the electrolytic process taking place in it. We are asked to write the cell half-reactions and overall reaction and to estimate the voltage needed to electrolyze molten NaCl. We are also asked how electricity flows through the cell, why Ca^{2+} ions are not also reduced at the cathode, why elemental sodium and chlorine need to be kept separated, and why they are harvested at the top of the cell even though the electrodes at which they are generated are near the bottom.

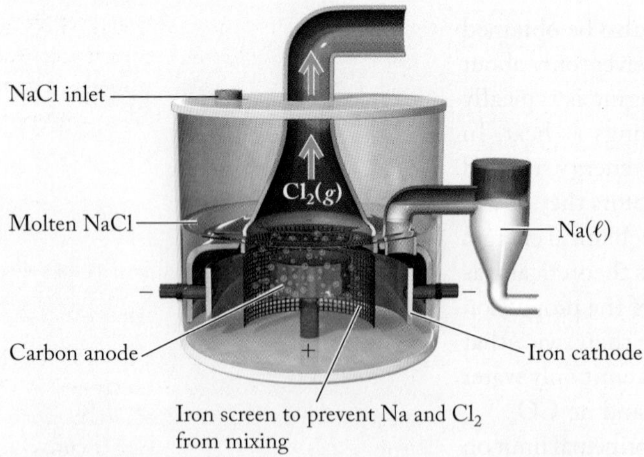

FIGURE 17.23 The Downs cell is the primary unit used to produce sodium commercially. It is also a minor source of industrially produced chlorine. An iron screen prevents contact between the two products.

Analyze Sodium ion is reduced at the cathode to elemental sodium, and chloride ion is oxidized at the anode to chlorine gas. Creating these reactive elements from a stable compound like NaCl requires a significant investment in electrical energy and, very likely, the application of a cell potential of several volts. Contact between the products would allow them to react and re-form NaCl. The spontaneity of this reaction is indicated by the ΔG°_f of NaCl in Appendix 4 (Table A4.3): −384.2 kJ/mol. Sodium ions will be preferentially reduced over Ca^{2+} ions if the standard reduction potential of Na^+ ions is less negative than that of Ca^{2+} ions. The manner of how the products are recovered from the electrolysis cell depends on their physical properties, especially their densities relative to the reactant, molten NaCl.

Solve

a. The anode half-reaction is

$$2\,Cl^-(\ell) \rightarrow Cl_2(g) + 2\,e^-$$

The cathode half-reaction is

$$Na^+(\ell) + e^- \rightarrow Na(\ell)$$

The overall reaction is the sum of the two reactions. Summing them requires first multiplying the Na reaction by 2 to balance the loss and gain of electrons:

$$2\,Na^+(\ell) + 2\,Cl^-(\ell) \rightarrow 2\,Na(\ell) + Cl_2(g)$$

b. Calculating E°_{cell} requires the appropriate $E^\circ_{cathode}$ and E°_{anode} values. Appendix 6 lists the following half-reactions and E° values:

$$Cl_2(g) + 2\,e^- \rightarrow 2\,Cl^-(aq) \qquad E^\circ = 1.3583\text{ V}$$
$$Na^+(aq) + e^- \rightarrow Na(s) \qquad E^\circ = -2.71\text{ V}$$

Calculating the difference between the standard reduction potentials,

$$E^\circ_{cell} = E^\circ_{cathode} - E^\circ_{anode} = -2.71\text{ V} - 1.3583\text{ V} = -4.07\text{ V}$$

The negative E°_{cell} value is a measure of how much voltage must be applied to drive the nonspontaneous cell reaction—that is, 4.07 V. However, the actual minimum voltage will not be exactly 4.07 V. For one thing, the sodium and chloride ions are in a nonaqueous environment of molten NaCl, and their concentrations are probably not 1 *M*. However, we have seen that cell potentials change relatively little as the concentrations of reactants and products change, so 4 volts is a reasonable estimate of the voltage needed to drive the reaction in a Downs cell.

c. Migration of Cl^- ions toward the positively charged anode, where they are oxidized to Cl_2 gas, carries part of the electrolysis current through the cell. The rest is carried by the migration of Na^+ ions toward the negatively charged cathode, where they are reduced to Na metal.

d. The standard reduction potential of Ca^{2+} ions (Appendix 6) is −2.868 V, which is more negative than the standard reduction potential of Na^+ ions (−2.71 V). Therefore, Na^+ ions are preferentially reduced in the presence of Ca^{2+} ions.

e. The ΔG°_f of solid NaCl is −384.2 kJ/mol, which indicates that elemental sodium and chlorine spontaneously react to form NaCl. Therefore, these two electrolysis products must be kept away from each other.

f. Liquid sodium is less dense than molten sodium chloride and, as it forms in the cell, it floats on top of the molten salt. Chlorine is a gas that bubbles out of molten NaCl and is collected in hoods above the liquid.

Think About It The Downs cell is an electrolytic cell that requires a large input of electrical energy to drive the nonspontaneous cell reaction. Sodium and chlorine are both so reactive that special care and extreme caution are required to collect and transport them.

SUMMARY

LO1 Electrochemistry is the branch of chemistry that links redox reactions to the production or consumption of electrical energy. Any redox reaction can be broken down into oxidation and reduction **half-reactions**. (Section 17.1)

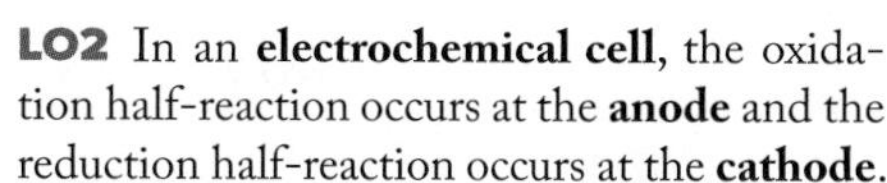

LO2 In an **electrochemical cell**, the oxidation half-reaction occurs at the **anode** and the reduction half-reaction occurs at the **cathode**. Migration of the ions in the cell's electrolyte allows electric charges to flow between the cathode and anode compartments as electrons flow through an external electrical circuit. A **cell diagram** shows how the components of the cathode and anode compartments of the cell are connected. (Section 17.2)

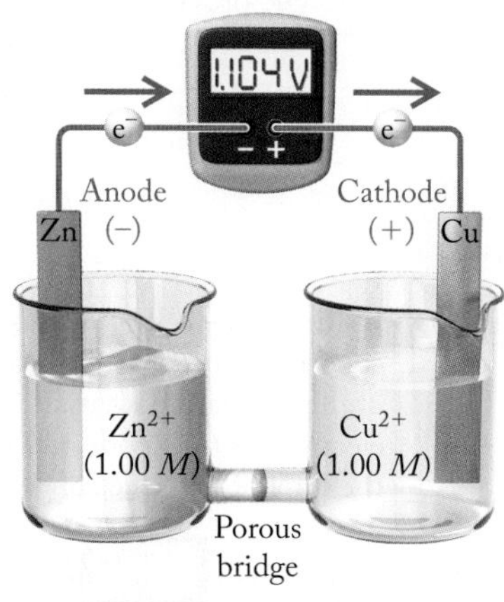

LO3 The difference between the **standard reduction potentials (E°)** of a cell's cathode and anode half-reactions is equal to the **standard cell potential (E°_{cell})**. The value of E°_{cell} is a measure of the **electromotive force (emf)** with which a **voltaic cell** can pump electrons through an external circuit under standard conditions. If reactants and products are not in their standard states, this value is simply called the **cell potential (E_{cell})**. All standard cell potentials are referenced to the cell potential of the **standard hydrogen electrode (SHE)**: $E^\circ_{SHE} = 0.000$ V. (Sections 17.3 and 17.5)

LO4 A voltaic cell has a positive cell potential ($E_{cell} > 0$) and its cell reaction has a negative change in free energy ($\Delta G^\circ_{cell} < 0$). This decrease in free energy in a voltaic cell is available to do work in an external electrical circuit. The **Faraday constant (F)** relates the quantity of electric charge to the number of moles of electrons and indirectly to the number of moles of reactants. (Section 17.4)

LO5 The potential of a voltaic cell decreases as reactants turn into products. The **Nernst equation** describes how cell potential changes with concentration changes. (Section 17.6)

LO6 The potential of a voltaic cell approaches zero as the cell reaction approaches chemical equilibrium, at which point $E_{cell} = 0$ and $Q = K$. (Section 17.6)

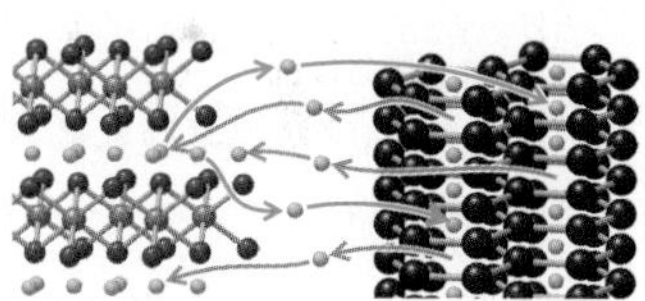

LO7 The quantities of reactants consumed in a voltaic cell reaction are directly proportional to the coulombs of electric charge delivered by the cell. Nickel–metal hydride batteries supply electricity when H atoms are oxidized to H^+ ions at the anodes and NiO(OH) is reduced to $Ni(OH)_2$ at the cathodes. In lithium–ion batteries, electricity is produced when Li^+ ions stored in graphite anodes migrate toward and are incorporated into transition metal oxide or phosphate cathodes. (Sections 17.7 and 17.9)

PARTICULATE PREVIEW WRAP-UP

The two half-reactions depict the oxidation of hydrogen and the reduction of oxygen:

$$H_2(g) \rightarrow 2\,H^+(aq) + 2e^-$$

$$O_2(g) + 4\,H^+(aq) + 4e^- \rightarrow 2H_2O(\ell)$$

which sum to a balanced, overall reaction of

$$2H_2(g) + O_2(g) \rightarrow 2\,H_2O(\ell)$$

When chemists physically separate two half-reactions, electrons can be forced to flow through an external circuit from one half-cell to the other. This external flow of electrons can be harnessed to do work as electricity. Chemists call hydrogen a "clean" fuel because it emits only water vapor, unlike traditional fuels, which when combusted, produce oxides of nitrogen and carbon that pollute the atmosphere.

PROBLEM-SOLVING SUMMARY

Type of Problem	Concepts and Equations	Sample Exercises
Writing redox reaction equations by combining half-reactions	Combine the reduction and oxidation half-reactions after balancing the gain and loss of electrons.	**17.1, 17.2**
Diagramming an electrochemical cell	Start with the symbol for the anode on the left; use double vertical lines to separate it from the symbol for the cathode. Using single lines to separate phases, list species in anode compartments and then those in the cathode compartment; separate species in the same phase with commas. Insert concentrations and/or partial pressures if known.	**17.3**
Identifying anode and cathode half-reactions and calculating the value of E°_{cell}	The half-reaction with the more positive standard reduction potential is the cathode half-reaction. $$E^\circ_{cell} = E^\circ_{cathode} - E^\circ_{anode} \quad (17.1)$$	**17.4**
Relating ΔG_{cell} and E_{cell}	$$\Delta G_{cell} = -nFE_{cell} \quad (17.6)$$ where n is the number of moles of electrons transferred in the cell reaction and F is the Faraday constant, 9.65×10^4 C/mol.	**17.5**
Calculating E_{cell} from E°_{cell} and the concentrations of reactants and products	E_{cell} at 25°C is related to E°_{cell} and the cell reaction quotient by the Nernst equation: $$E_{cell} = E^\circ_{cell} - \frac{0.0592\text{ V}}{n}\log Q \quad (17.9)$$	**17.6**
Calculating K for a redox reaction from the standard potentials of its half-reactions	E°_{cell} is related to K at 298 K by $$\log K = \frac{nE^\circ_{rxn}}{0.0592\text{ V}} \quad (17.11)$$	**17.7**
Relating the mass of a reactant in a cell reaction to a quantity of electrical charge	Determine the ratio of moles of reactants to moles of electrons transferred; use the Faraday constant to relate coulombs of charge to moles of electrons.	**17.8**
Calculating the time to oxidize a quantity of reactant	Use the Faraday constant and the relation between moles of electrons and moles of reactants to describe an electrolytic process.	**17.9**

VISUAL PROBLEMS

(Answers to boldface end-of-chapter questions and problems are in the back of the book.)

17.1. In the voltaic cell shown in Figure P17.1, the greater density of a concentrated solution of $CuSO_4$ allows a less concentrated solution of $ZnSO_4$ solution to be (carefully) layered on top of it. Why is a porous separator not needed in this cell?

FIGURE P17.1

17.2. In the voltaic cell shown in Figure P17.2, the concentrations of Cu^{2+} and Cd^{2+} are 1.00 *M*. On the basis of the standard potentials in Appendix 6, identify which electrode is the anode and which is the cathode. Indicate the direction of electron flow.

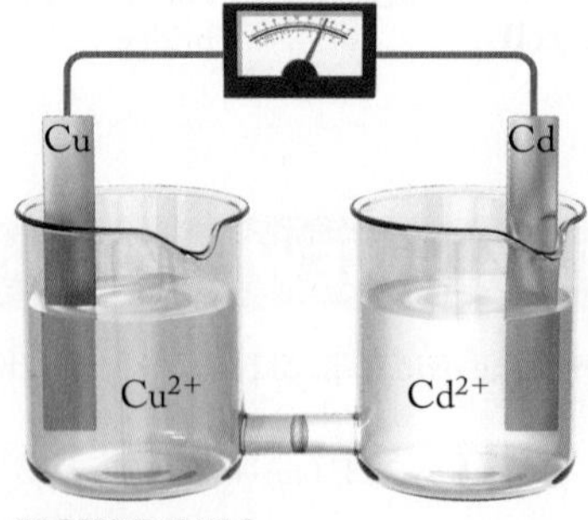

FIGURE P17.2

17.3. In the voltaic cell shown in Figure P17.3, $[Ag^+] = [H^+] = 1.00\ M$. On the basis of the standard potentials in Appendix 6, identify which electrode is the anode and which is the cathode. Indicate the direction of electron flow.

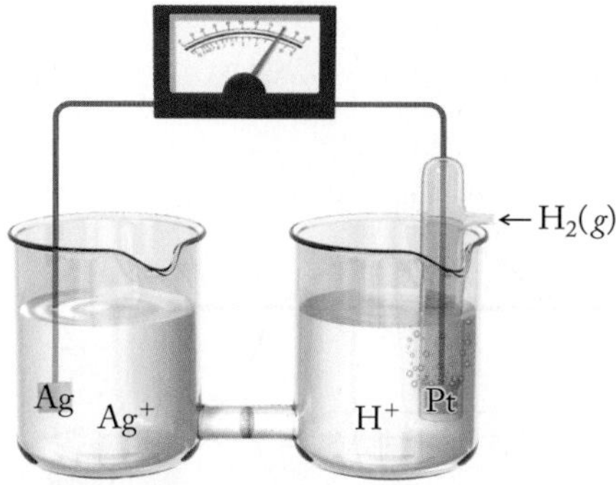

FIGURE P17.3

*17.4. In many electrochemical cells, the electrodes are metals that carry electrons to and from the cell but are not chemically changed by the cell reaction. Each of the highlighted clusters in the periodic table in Figure P17.4 consists of three metals. Which of the highlighted clusters is best suited to form inert electrodes?

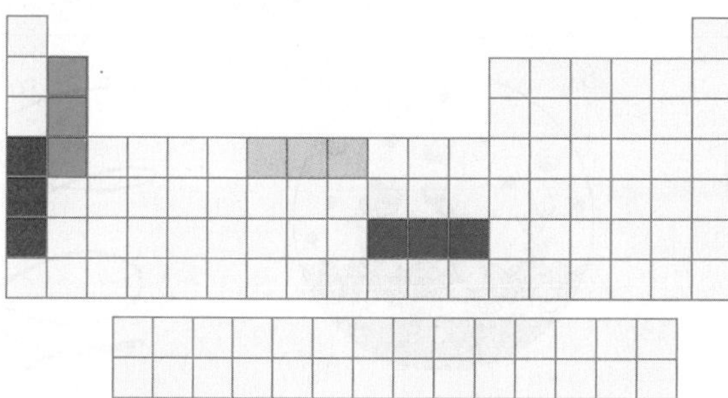

FIGURE P17.4

17.5. Which of the four curves in Figure P17.5 best represents the dependence of the potential of a lead–acid battery on the concentration of sulfuric acid? Note that the scale of the x-axis is logarithmic.

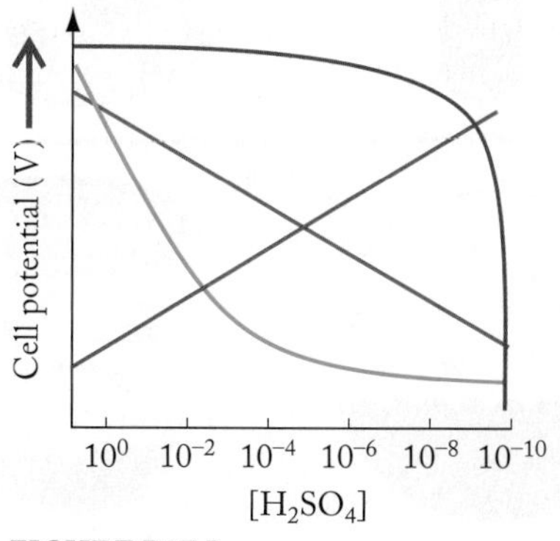

FIGURE P17.5

17.6. From top to bottom, the sizes of the batteries in Figure P17.6 are AAA, AA, C, and D, respectively. The performance of batteries like these is often expressed in units such as (a) volts, (b) watt-hours, or (c) milliampere-hours. Which of the values differ significantly between the four batteries?

FIGURE P17.6

17.7. The apparatus in Figure P17.7 is used for the electrolysis of water. Hydrogen and oxygen gas are collected in the two inverted burets. An inert electrode at the bottom of the left buret is connected to the negative terminal of a 6-volt battery; the electrode in the buret on the right is connected to the positive terminal. A small quantity of sulfuric acid is added to speed up the electrolytic reaction.

a. What are the half-reactions at the left and right electrodes and their standard potentials?
b. Why does sulfuric acid make the electrolysis reaction go more rapidly?

Overall cell reaction

$H_2O(\ell) \rightarrow H_2(g) + \frac{1}{2}\ O_2(g)$

FIGURE P17.7

17.8. An electrolytic apparatus identical to the one shown in Figure P17.7 is used to electrolyze water, but the reaction is speeded up by the addition of sodium carbonate instead of sulfuric acid.

a. What are the half-reactions and the standard potentials for the electrodes on the left and right?

b. Why does sodium carbonate make the electrolysis reaction go more rapidly?

17.9. **Redox Flow Batteries** Redox flow batteries are similar to fuel cells in that the reactants are continuously delivered to an electrochemical cell from large storage tanks. In the cell shown in Figure P17.9, an ion-exchange membrane separates the reactants from each other; the arrows indicate the direction of flow for the reactants. The reactions in one type of redox flow battery can be diagrammed as

$$C(s) \mid V^{2+}(aq), V^{3+}(aq) \parallel VO_2^{+}(aq), VO^{2+}(aq) \mid C(s)$$

a. Write balanced half-reactions for the processes that occur at the anode and cathode in acid solution.

b. Write a balanced chemical equation for the overall reaction and determine the number of electrons transferred.

c. Label the diagram in Figure P17.9 by putting the appropriate reaction in each compartment and show the direction of electron flow.

d. How does the value of E_{cell} (1.25 V) change with pH?

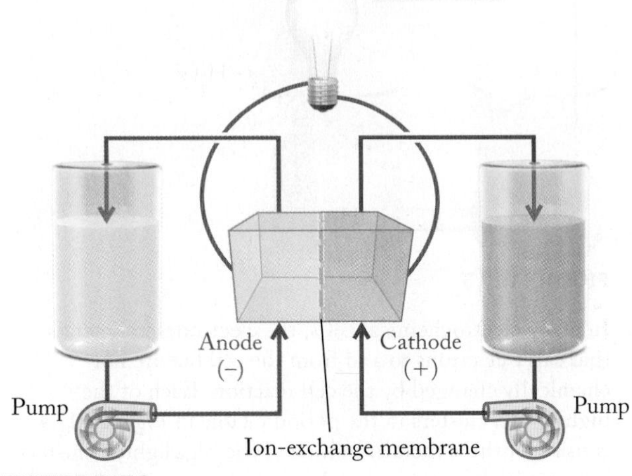

FIGURE P17.9

17.10. Use representations [A] through [I] in Figure P17.10 to answer questions a–f. [E] shows silver deposited onto copper. If the materials in [A], [C], [G], and [I] were combined along with a porous bridge and external circuit to generate an electrochemical cell that produced [E],

a. Which metal would be the cathode? Which solution would surround it?

b. Which metal would be the anode? Which solution would surround it?

c. When the reaction in [E] is finished, what will happen to the light blue color of the solution?

d. How could [E] be produced without a porous bridge or an external circuit?

e. Of the four particulate images [B], [D], [F], and [H] in Figure P17.10, which two correspond to the solutions [G] and [I]?

f. Which particulate image represents the copper wire with silver deposited onto it? What does the fourth particulate image depict?

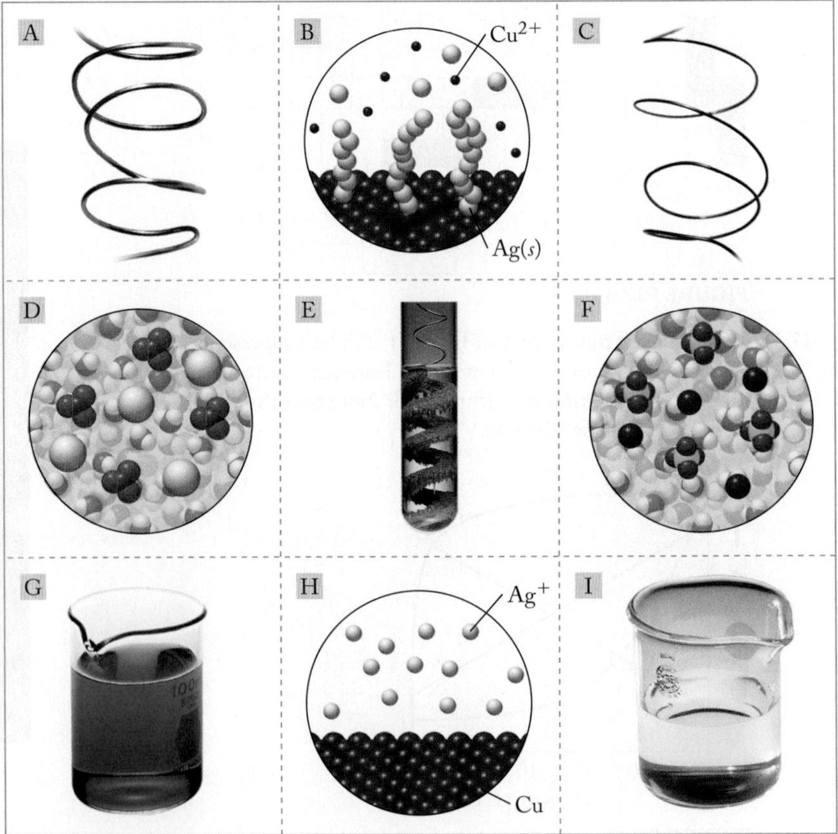

FIGURE P17.10

QUESTIONS AND PROBLEMS

Redox Chemistry Revisited and Electrochemical Cells

Concept Review

17.11. What is meant by a half-reaction?

17.12. The Zn/Cu^{2+} reactions in Figures 17.1 and 17.2 are the same. However, the reaction in the cell in Figure 17.2 generates electricity, whereas the reaction in the beaker in Figure 17.1 does not. Why?

17.13. Why can't a wire perform the same function as a porous separator in an electrochemical cell?

17.14. In a voltaic cell, why is the cathode labeled the positive terminal and the anode the negative terminal?

Problems

17.15. Balance the following half-reactions by adding the appropriate number of electrons. Identify the oxidation half-reactions and the reduction half-reactions.

a. $Br_2(\ell) \rightarrow 2\ Br^-(aq)$
b. $Pb(s) + 2\ Cl^-(aq) \rightarrow PbCl_2(s)$
c. $O_3(g) + 2\ H^+(aq) \rightarrow O_2(g) + H_2O(\ell)$
d. $H_2S(g) \rightarrow S(s) + 2\ H^+(aq)$

17.16. Balance the following half-reactions by adding the appropriate number of electrons. Which are oxidation half-reactions and which are reduction half-reactions?

a. $TiO^{2+}(aq) + 2\ H^+(aq) \rightarrow Ti^{3+}(aq) + H_2O(\ell)$
b. $S_4^{2+}\ (aq) \rightarrow 2\ S_2^{2-}\ (aq)$
c. $VO_2^+(aq) + 2\ H^+(aq) \rightarrow VO^{2+}(aq) + H_2O(\ell)$
d. $Fe(CN)_6^{3-} \rightarrow Fe(CN)_6^{4-}(aq)$

***17.17.** **Groundwater Chemistry** Write a half-reaction for the oxidation of magnetite (Fe_3O_4) to hematite (Fe_2O_3) in acidic groundwater.

*17.18. Write a half-reaction for the oxidation of the manganese in $MnCO_3$ to MnO_2 in neutral groundwater, where the principal carbonate species is HCO_3^-.

17.19. **Lithium–Iron Sulfide Batteries** Voltaic cells with nonaqueous electrolytes and based on the oxidation of Li(*s*) to $Li_2S(s)$ and the reduction of $FeS_2(s)$ to Fe(*s*) and S^{2-} ions provide the same voltage as traditional alkaline batteries but offer more storage capacity.

a. Write half-reactions for the cell's anode and cathode.
b. Write a balanced cell reaction.
c. Diagram the cell.

17.20. In a flow battery, two solutions are pumped into each voltaic cell, and oxidation and reduction half-reactions take place at two inert electrodes. One such cell is based on the reduction of $Sn^{4+}(aq)$ to $Sn^{2+}(aq)$ and the oxidation of $Fe^{2+}(aq)$ to $Fe^{3+}(aq)$.

a. Write half-reactions for the cell's anode and cathode.
b. Write a balanced cell reaction.
c. How many electrons are transferred in the cell reaction?

17.21. **Zinc–Nickel Batteries** Replacing cadmium with zinc in a NiCad battery avoids the use of the toxic element cadmium. A voltaic cell based on the reaction between NiO(OH)(*s*) and Zn(*s*) in alkaline electrolyte (OH^-) produces $Ni(OH)_2(s)$ and $Zn(OH)_2(s)$.

a. Write half-reactions for the anode and cathode.
b. Write a balanced cell reaction.
c. Diagram the cell.

17.22. **Super Iron Batteries** In 1999, scientists in Israel developed a battery based on the following cell reaction with iron(VI), nicknamed "super iron":

$$2\ K_2FeO_4(aq) + 3\ Zn(s) \rightarrow Fe_2O_3(s) + ZnO(s) + 2\ K_2ZnO_2(aq)$$

a. Determine the number of electrons transferred in the cell reaction.
b. What are the oxidation states of the transition metals in the reaction?
c. Draw the cell.

Standard Potentials

Concept Review

17.23. In some textbooks the formula used to calculate standard cell potentials from the standard reduction potentials of the half-reactions occurring at the cathode and anode is given as

$$E^\circ_{cell} = E^\circ_{reduction}(\text{cathode}) + E^\circ_{oxidation}(\text{anode})$$

Show how this equation is equivalent to Equation 17.1.

*17.24. Why is O_2 a stronger oxidizing agent in acid than in base? Use standard reduction potentials from Appendix 6 to support your answer.

Problems

17.25. Using standard reduction potentials from Appendix 6 and the following equations for half-reactions:

$$Cu^{2+}(aq) + 2\ e^- \rightarrow Cu(s)$$
$$Co^{2+}(aq) + 2\ e^- \rightarrow Co(s)$$
$$Hg^{2+}(aq) + 2\ e^- \rightarrow Hg(s)$$

a. Identify the combination of half-reactions that would lead to the largest value of E°_{cell}.
b. Identify the combination of half-reactions that would lead to the smallest positive value of E°_{cell}.

17.26. From the table of standard reduction potentials in Appendix 6,

a. Select an oxidizing agent that will oxidize Cr(*s*) to $Cr^{3+}(aq)$ but not Cd(*s*) to $Cd^{2+}(aq)$.
b. Select a reducing agent that will reduce $Br_2(\ell)$ to $Br^-(aq)$ but not $I_2(s)$ to $I^-(aq)$.

17.27. If a piece of silver is placed in a solution in which $[Ag^+] = [Cu^{2+}] = 1.00\ M$, will the following reaction proceed spontaneously?

$$2\ Ag(s) + Cu^{2+}(aq) \rightarrow 2\ Ag^+(aq) + Cu(s)$$

17.28. A piece of cadmium is placed in a solution in which $[Cd^{2+}] = [Sn^{2+}] = 1.00\ M$. Will the following reaction proceed spontaneously?

$$Cd(s) + Sn^{2+}(aq) \rightarrow Cd^{2+}(aq) + Sn(s)$$

17.29. In a voltaic cell similar to the Zn/Cu^{2+} cell in Figure 17.2, the Cu electrode is replaced with one made of Ni immersed in a 1.00 *M* solution of $NiSO_4$. Will the standard potential of this cell be greater than, the same as, or less than 1.10 V?

17.30. Suppose the copper half of the Zn/Cu^{2+} cell in Figure 17.2 was replaced with a silver wire in contact with 1 *M* $Ag^+(aq)$.
a. What would be the value of E°_{cell}?
b. Which electrode would be the anode?

17.31. Voltaic cells based on the following pairs of half-reactions are prepared so that all reactants and products are in their standard states. For each pair, write a balanced equation for the cell reaction, and identify which half-reaction takes place at the anode and which at the cathode.
a. $Hg^{2+}(aq) + 2\ e^- \rightarrow Hg(\ell)$
$Zn^{2+}(aq) + 2\ e^- \rightarrow Zn(s)$
b. $ZnO(s) + H_2O(\ell) + 2\ e^- \rightarrow Zn(s) + 2\ OH^-(aq)$
$Ag_2O(s) + H_2O(\ell) + 2\ e^- \rightarrow 2\ Ag(s) + 2\ OH^-(aq)$
c. $Ni(OH)_2(s) + 2\ e^- \rightarrow Ni(s) + 2\ OH^-(aq)$
$O_2(g) + 2\ H_2O(\ell) + 4\ e^- \rightarrow 4\ OH^-(aq)$

17.32. Voltaic cells based on the following pairs of half-reactions are constructed. For each pair, write a balanced equation for the cell reaction, and identify which half-reaction takes place at each anode and cathode.
a. $Cd^{2+}(aq) + 2\ e^- \rightarrow Cd(s)$
$Ag^+(aq) + e^- \rightarrow Ag(s)$
b. $AgBr(s) + e^- \rightarrow Ag(s) + Br^-(aq)$
$MnO_2(s) + 4\ H^+(aq) + 2\ e^- \rightarrow Mn^{2+}(aq) + 2\ H_2O(\ell)$
c. $PtCl_4^{2-}(aq) + 2\ e^- \rightarrow Pt(s) + 4\ Cl^-(aq)$
$AgCl(s) + e^- \rightarrow Ag(s) + Cl^-(aq)$

Chemical Energy and Electrical Work

Concept Review

17.33. The negative sign in Equation 17.3 ($w_{elec} = -QE_{cell}$) seems to indicate that a voltaic cell with a *positive* cell potential does *negative* electrical work. How is this possible?

*17.34. Mechanical work (w) is done by exerting a force (F) to move an object through a distance (d) according to the equation $w = F \times d$. Explain how this definition of work relates to electrical work.

Problems

17.35. Starting with the appropriate standard free energies of formation from Appendix 4, calculate the values of ΔG° and E°_{cell} of the following reactions:
a. $2\ Cu^+(aq) \rightarrow Cu^{2+}(aq) + Cu(s)$
b. $Ag(s) + Fe^{3+}(aq) \rightarrow Ag^+(aq) + Fe^{2+}(aq)$

17.36. Starting with the appropriate standard free energies of formation from Appendix 4, calculate the values of ΔG° and E°_{cell} of the following reactions:
a. $FeO(s) + H_2(g) \rightarrow Fe(s) + H_2O(\ell)$
b. $2\ Pb(s) + O_2(g) + 2\ H_2SO_4(aq) \rightarrow 2\ PbSO_4(s) + 2\ H_2O(\ell)$

17.37. **Flashlights** For many years the 1.50 V batteries used to power flashlights were based on the following cell reaction:

$$Zn(s) + 2\ NH_4Cl(s) + 2\ MnO_2(s) \rightarrow Zn(NH_3)_2Cl_2(s) + Mn_2O_3(s) + H_2O(\ell)$$

What is the value of ΔG_{cell}?

17.38. **Laptops** The first generation of laptop computers was powered by nickel–cadmium (NiCad) batteries, which generated 1.20 V based on the following cell reaction:

$$Cd(s) + 2\ NiO(OH)(s) + 2\ H_2O(\ell) \rightarrow Cd(OH)_2(s) + 2\ Ni(OH)_2(s)$$

What is the value of ΔG_{cell}?

17.39. **Nickel–Sodium Batteries** Researchers in England are developing a battery for electric vehicles based on the reaction between $NiCl_2(s)$ and $Na(s)$:

$$2\ Na(s) + NiCl_2(s) \rightarrow Ni(s) + 2\ NaCl(s)$$

The cells in the battery produce 2.58 V.
a. Assign oxidation numbers to each element in the nickel and sodium compounds.
b. How many electrons are transferred in the overall reaction?
c. What is the value of ΔG_{cell}?

17.40. Thomas Edison, the inventor of the incandescent lightbulb, developed a voltaic cell that delivers 1.4 V of cell potential in an alkaline electrolyte based on the following cell reaction:

$$3\ Fe(s) + 8\ NiO(OH)(s) + 4\ H_2O(\ell) \rightarrow 8\ Ni(OH)_2(s) + Fe_3O_4(s)$$

a. Assign oxidation numbers to each element in each of the nickel and iron compounds.
b. How many electrons are transferred in the overall reaction?
c. What is the value of ΔG_{cell}?

17.41. Starting with standard potentials listed in Appendix 6, calculate the values of E°_{cell} and ΔG° of the following reactions.
a. $Cu(s) + Sn^{2+}(aq) \rightarrow Cu^{2+}(aq) + Sn(s)$
b. $Zn(s) + Ni^{2+}(aq) \rightarrow Zn^{2+}(aq) + Ni(s)$

17.42. Starting with the standard potentials listed in Appendix 6, calculate the values of E°_{cell} and ΔG° of the following reactions.
a. $Fe(s) + Cu^{2+}(aq) \rightarrow Fe^{2+}(aq) + Cu(s)$
b. $Ag(s) + Fe^{3+}(aq) \rightarrow Ag^+(aq) + Fe^{2+}(aq)$

A Reference Point: The Standard Hydrogen Electrode

Concept Review

17.43. What is the function of platinum in the standard hydrogen electrode?

*17.44. What effect would replacing H_2 with O_2 in the standard hydrogen electrode (making it a standard *oxygen* electrode) have on the standard reduction potential values in Appendix 6?

The Effect of Concentration on E_{cell}

Concept Review

17.45. Why does the operating cell potential of most batteries change little until the battery is nearly discharged?

17.46. The standard potential of the Zn/Cu^{2+} cell reaction

$$Zn(s) + Cu^{2+}(aq) \rightarrow Zn^{2+}(aq) + Cu(s)$$

is 1.10 V. Would the potential of the cell differ from 1.10 V if the concentrations of both Cu^{2+} and Zn^{2+} were 0.25 *M*?

Problems

17.47. Calculate the E_{cell} value at 298 K for the cell based on the reaction

$$Fe^{3+}(aq) + Cr^{2+}(aq) \rightarrow Fe^{2+}(aq) + Cr^{3+}(aq)$$

when $[Fe^{3+}] = [Cr^{2+}] = 1.50 \times 10^{-3}$ *M* and $[Fe^{2+}] = [Cr^{3+}] = 2.5 \times 10^{-4}$ *M*.

17.48. Calculate the E_{cell} value at 298 K for the cell based on the reaction

$$Cu(s) + 2\ Ag^{+}(aq) \rightarrow Cu^{2+}(aq) + 2\ Ag(s)$$

when $[Ag^{+}] = 2.56 \times 10^{-3}$ *M* and $[Cu^{2+}] = 8.25 \times 10^{-4}$ *M*.

17.49. Using the appropriate standard potentials from Appendix 6, determine the equilibrium constant for the following reaction at 298 K:

$$Fe^{3+}(aq) + Cr^{2+}(aq) \rightarrow Fe^{2+}(aq) + Cr^{3+}(aq)$$

17.50. Using the appropriate standard potentials from Appendix 6, determine the equilibrium constant at 298 K for the following reaction between MnO_2 and Fe^{2+} in acid solution:

$$4\ H^{+}(aq) + MnO_2(s) + 2\ Fe^{2+}(aq) \rightarrow Mn^{2+}(aq) + 2\ Fe^{3+}(aq) + 2\ H_2O(\ell)$$

17.51. If the potential of a hydrogen electrode based on the half-reaction

$$2\ H^{+}(aq) + 2\ e^{-} \rightarrow H_2(g)$$

is 0.000 V at pH = 0.00, what is the potential of the same electrode at pH = 7.00?

17.52. **Glucose Metabolism** The standard potentials for the reduction of nicotinamide adenine dinucleotide (NAD^{+}) and oxaloacetate (reactants in the multistep metabolism of glucose) are as follows:

$$NAD^{+}(aq) + 2\ H^{+}(aq) + 2\ e^{-} \rightarrow NADH(aq) + H^{+}(aq) \quad E^{\circ} = -0.320\ V$$

$$\text{Oxaloacetate}(aq) + 2\ H^{+}(aq) + 2\ e^{-} \rightarrow \text{malate}(aq) \quad E^{\circ} = -0.166\ V$$

a. Calculate the standard potential for the following reaction:

$$\text{Oxaloacetate}(aq) + NADH(aq) + H^{+}(aq) \rightarrow \text{malate}(aq) + NAD^{+}(aq)$$

b. Calculate the equilibrium constant for the reaction at 298 K.

17.53. Permanganate ion can oxidize sulfite to sulfate in basic solution as follows:

$$2\ MnO_4^{-}(aq) + 3\ SO_3^{2-}(aq) + H_2O(\ell) \rightarrow 2\ MnO_2(s) + 3\ SO_4^{2-}(aq) + 2\ OH^{-}(aq)$$

Determine the potential for the reaction (E_{rxn}) at 298 K when the concentrations of the reactants and products are $[MnO_4^{-}] = 0.150$ *M*, $[SO_3^{2-}] = 0.256$ *M*, $[SO_4^{2-}] = 0.178$ *M*, and $[OH^{-}] = 0.0100$ *M*. Will the value of E_{rxn} increase or decrease as the reaction proceeds?

*17.54. Manganese dioxide is reduced by iodide ion in acid solution as follows:

$$MnO_2(s) + 2\ I^{-}(aq) + 4\ H^{+}(aq) \rightarrow Mn^{2+}(aq) + I_2(aq) + 2\ H_2O(\ell)$$

Determine the electrochemical potential of the reaction at 298 K when the initial concentrations of the components are $[I^{-}] = 0.225$ *M*, $[H^{+}] = 0.900$ *M*, $[Mn^{2+}] = 0.100$ *M*, and $[I_2] = 0.00114$ *M*. If the solubility of iodine in water is approximately 0.114 *M*, will the value of E_{rxn} increase or decrease as the reaction proceeds?

17.55. A copper penny dropped into a solution of nitric acid produces a mixture of nitrogen oxides. The following reaction describes the formation of NO, one of the products:

$$3\ Cu(s) + 8\ H^{+}(aq) + 2\ NO_3^{-}(aq) \rightarrow 2\ NO(g) + 3\ Cu^{2+}(aq) + 4\ H_2O(\ell)$$

a. Starting with the appropriate standard potentials from Appendix 6, calculate E°_{rxn} for this reaction.
b. Calculate E_{rxn} at 298 K when $[H^{+}] = 0.100$ *M*, $[NO_3^{-}] = 0.0250$ *M*, $[Cu^{2+}] = 0.0375$ *M*, and the partial pressure of NO = 0.00150 atm.

17.56. Chlorine dioxide (ClO_2) is produced by the following reaction of chlorate (ClO_3^{-}) with Cl^{-} in acid solution:

$$2\ ClO_3^{-}(aq) + 2\ Cl^{-}(aq) + 4\ H^{+}(aq) \rightarrow 2\ ClO_2(g) + Cl_2(g) + 2\ H_2O(\ell)$$

a. Determine E° for the reaction.
b. The reaction at 298 K produces a mixture of gases in the reaction vessel in which $P_{ClO_2} = 2.0$ atm and $P_{Cl_2} = 1.00$ atm. Calculate $[ClO_3^{-}]$ if, at equilibrium, $[H^{+}] = [Cl^{-}] = 10.0$ *M*.

***17.57.** The oxidation of NH_4^{+} to NO_3^{-} in acid solution is described by the following equation:

$$NH_4^{+}(aq) + 2\ O_2(g) \rightarrow NO_3^{-}(aq) + 2\ H^{+}(aq) + H_2O(\ell)$$

a. Calculate E° for the overall reaction.
b. If the reaction is in equilibrium with air ($P_{O_2} = 0.21$ atm) at pH 5.60, what is the ratio of $[NO_3^{-}]$ to $[NH_4^{+}]$ at 298 K?

17.58. What is the value of E° for the following reaction?

$$2\ AgCl(s) + H_2(g) \rightarrow 2\ Ag(s) + 2\ HCl(aq)$$

Relating Battery Capacity to Quantities of Reactants

Concept Review

17.59. One 12-volt lead–acid battery has a higher ampere · hour rating than another. Which of the following parameters are likely to be different for the two batteries?

a. Individual cell potentials
b. Anode half-reactions
c. Total masses of electrode materials
d. Number of cells
e. Electrolyte composition
f. Combined surface areas of their electrodes

17.60. In a voltaic cell based on the Zn/Cu^{2+} cell reaction,

$$Zn(s) + Cu^{2+}(aq) \rightarrow Cu(s) + Zn^{2+}(aq)$$

there is exactly one mole of each reactant and product. A second cell based on the following cell reaction:

$$Cd(s) + Cu^{2+}(aq) \rightarrow Cu(s) + Cd^{2+}(aq)$$

also has exactly one mole of each reactant and product. Which of the following statements about these two cells is true?

a. Their cell potentials are the same.
b. The masses of their electrodes are the same.
c. The quantities of electric charge that they can produce are the same.
d. The quantities of electric energy that they can produce are the same.

Problems

17.61. Which of the following voltaic cells, **A** or **B**, will produce the greater quantity of electric charge per gram of anode material?

Cell **A**: $Cd(s) + 2\ NiO(OH)(s) + 2\ H_2O(\ell) \rightarrow 2\ Ni(OH)_2(s) + Cd(OH)_2(s)$

Cell **B**: $4\ Al(s) + 3\ O_2(g) + 6\ H_2O(\ell) + 4\ OH^-(aq) \rightarrow 4\ Al(OH)_4^-(aq)$

17.62. Which of the following voltaic cells, **C** or **D**, will produce the greater quantity of electric charge per gram of anode material?

Cell **C**: $Zn(s) + MnO_2(s) + H_2O(\ell) \rightarrow ZnO(s) + Mn(OH)_2(s)$

Cell **D**: $Li(s) + MnO_2(s) \rightarrow LiMnO_2(s)$

***17.63.** Which of the following voltaic cell reactions, **E** or **F**, delivers more electrical energy per gram of anode material at 298 K?

Reaction **E**: $Zn(s) + 2\ NiO(OH)(s) + 2\ H_2O(\ell) \rightarrow 2\ Ni(OH)_2(s) + Zn(OH)_2(s)$ $\quad E^\circ_{cell} = 1.20\ V$

Reaction **F**: $Li(s) + MnO_2(s) \rightarrow LiMnO_2(s)$ $\quad E^\circ_{cell} = 3.15\ V$

*17.64. Which of the following voltaic cell reactions, **G** or **H**, delivers more electrical energy per gram of anode material at 298 K?

Reaction **G**: $Zn(s) + Ni(OH)_2(s) \rightarrow Zn(OH)_2(s) + Ni(s)$ $\quad E^\circ_{cell} = 1.50\ V$

Reaction **H**: $2\ Zn(s) + O_2(g) \rightarrow 2\ ZnO(s)$ $\quad E^\circ_{cell} = 2.08\ V$

Electrolytic Cells and Rechargeable Batteries

Concept Review

17.65. The positive terminal of a voltaic cell is the cathode. However, the cathode of an electrolytic cell is connected to the negative terminal of a power supply. Explain this difference in polarity.

17.66. The anode in an electrochemical cell is defined as the electrode where oxidation takes place. Why is the anode in an electrolytic cell connected to the positive (+) terminal of an external supply, whereas the anode in a voltaic cell battery is connected to the negative (−) terminal?

17.67. The salts obtained from the evaporation of seawater can act as a source of halogens, principally Cl_2 and Br_2, through the electrolysis of the molten alkali metal halides. As the potential of the anode in an electrolytic cell is increased, which of these two halogens forms first?

17.68. In the electrolysis described in Problem 17.67, why is it necessary to use molten salts rather than seawater itself?

17.69. Quantitative Analysis Electrolysis can be used to determine the concentration of Cu^{2+} in a given volume of solution by electrolyzing the solution in a cell equipped with a platinum cathode. If all of the Cu^{2+} is reduced to Cu metal at the cathode, the increase in mass of the electrode provides a measure of the concentration of Cu^{2+} in the original solution. To ensure the complete (99.99%) removal of the Cu^{2+} from a solution in which $[Cu^{2+}]$ is initially about 1.0 *M*, will the potential of the cathode (versus SHE) have to be more negative or less negative than 0.34 V (the standard potential for $Cu^{2+} + 2\ e^- \rightarrow Cu$)?

17.70. A high school chemistry student wishes to demonstrate how water can be separated into hydrogen and oxygen by electrolysis. She knows that the reaction will proceed more rapidly if an electrolyte is added to the water. She has access to 2.00 *M* solutions of the following compounds: H_2SO_4, HBr, NaI, NaCl, Na_2SO_4, and Na_2CO_3. Which one(s) should she use? Explain your selection(s).

Problems

17.71. Suppose the current from a battery is used to electroplate an object with silver. Calculate the mass of silver that would be deposited by a battery that delivers 1.7 A · h of charge.

17.72. A battery charger used to recharge the NiMH batteries in a digital camera can deliver as much as 0.50 A of current to each battery. If it takes 100 min to recharge one battery, how much $Ni(OH)_2$ (in grams) is oxidized to NiO(OH)?

17.73. A quantity of electric charge deposits 0.732 g of Ag(*s*) from an aqueous solution of silver nitrate. When that same quantity of charge is passed through a solution of a gold salt, 0.446 g of Au(*s*) is formed. What is the oxidation state of the gold ion in the salt?

17.74. What amount of current is required to deposit 0.750 g of Pt(*s*) from a solution containing the $[PtCl_4]^{2-}$ ion within a time of 2.5 hours?

17.75. A NiMH battery containing 4.10 g of NiO(OH) is 50% discharged when it is connected to a charger with an output of 2.00 A at 1.3 V. How long does it take to recharge the battery?

*17.76. How long does it take to deposit a coating of gold 1.00 μm thick on a disk-shaped medallion 4.0 cm in diameter and 2.0 mm thick at a constant current of 85 A? The density for the electroplating process is 19.3 g/cm^3. The electroplating solution contains gold(III).

***17.77.** **Oxygen Supply in Submarines** Nuclear submarines can stay under water nearly indefinitely because they can produce their own oxygen by the electrolysis of water.

a. How many liters of O_2 at 298 K and 1.00 bar are produced in 1 h in an electrolytic cell operating at a current of 0.025 A?
b. Could seawater be used as the source of oxygen in this electrolysis? Explain why or why not.

17.78. In the electrolysis of water, how long will it take to produce 125 L of H_2 at 20°C and a pressure of 750 torr using an electrolytic cell through which the current is 52 mA?

17.79. Calculate the minimum (least negative) cathode potential (versus SHE) needed to begin electroplating nickel from 0.35 M Ni^{2+} onto a piece of iron.

*17.80. What is the minimum (least negative) cathode potential (versus SHE) needed to electroplate silver onto cutlery in a solution of Ag^+ and NH_3 in which most of the silver ions are present as the complex $Ag(NH_3)_2{}^+$ and the concentration of $Ag^+(aq)$ is only 3.50×10^{-5} M?

Fuel Cells

Concept Review

17.81. Describe two advantages of hybrid (gasoline engine–electric motor) power systems over all-electric systems based on fuel cells. Describe two disadvantages.

17.82. Describe three factors limiting the more widespread use of cars and other vehicles powered by fuel cells.

17.83. Methane can serve as the fuel for electric cars powered by fuel cells. Carbon dioxide is a product of the fuel cell reaction. All cars powered by internal combustion engines burning natural gas (mostly methane) produce CO_2. Why are electric vehicles powered by fuel cells likely to produce less CO_2 per mile?

17.84. To make the refueling of fuel cells easier, several manufacturers are developing converters that turn readily available fuels—such as natural gas, propane, and methanol—into H_2 and CO_2. Although vehicles with such power systems are not truly "zero emission," they still offer significant environmental benefits over vehicles powered by internal combustion engines. Describe a few of these benefits.

Problems

17.85. Fuel cells with molten alkali metal carbonates as electrolytes can use methane as a fuel. The methane is first converted into hydrogen in a two-step process:

$$CH_4(g) + H_2O(g) \rightarrow CO(g) + 3\,H_2(g)$$

$$CO(g) + H_2O(g) \rightarrow H_2(g) + CO_2(g)$$

a. Assign oxidation numbers to carbon and hydrogen in the reactants and products.
b. Using the standard free energy of formation values from Appendix 4, calculate the standard free-energy changes in the two reactions and the overall $\Delta G°$ for the formation of $H_2 + CO_2$ from methane and steam.

*17.86. Molten carbonate fuel cells fueled with H_2 convert as much as 60% of the free energy released by the formation of water from H_2 and O_2 into electrical energy. Determine the quantity of electrical energy obtained from converting one mole of H_2 into $H_2O(\ell)$ in such a fuel cell.

Additional Problems

17.87. Suppose there were a scale for expressing electrode potentials in which the standard potential for the reduction of water in base:

$$2\,H_2O(\ell) + 2\,e^- \rightarrow H_2(g) + 2\,OH^-(aq)$$

was assigned an $E°$ value of 0.000 V. How would the standard potential values on this new scale differ from those in Appendix 6?

*17.88. To inhibit corrosion of steel structures in contact with seawater, pieces of other metals (often zinc) are attached to the structures to serve as "sacrificial anodes." Explain how these attached pieces of metal might protect the structures, and describe which properties of zinc make it a good selection.

17.89. **Lithium–Sulfur Dioxide Batteries** The U.S. military uses batteries based on the reduction of liquid sulfur dioxide at a carbon cathode in certain communications equipment. Lithium metal is used as the anode. The overall cell reaction is

$$2\,Li(s) + 2\,SO_2(\ell) \rightarrow Li_2S_2O_4(s)$$

a. Write half-reactions for the anode and cathode reactions.
b. How many electrons are transferred in the cell reaction?
c. Draw a Lewis structure for the $S_2O_4{}^{2-}$ anion.

*17.90. **Solar-Powered Lamps** Rechargeable nickel–cadmium (NiCad) batteries are used to store energy in solar-powered landscape lamps (Figure P17.90). The batteries contain Cd anodes and NiO(OH) cathodes. The products of the cell reaction include cadmium(II) and nickel(II) hydroxides. Write a net ionic equation describing the cell reaction.

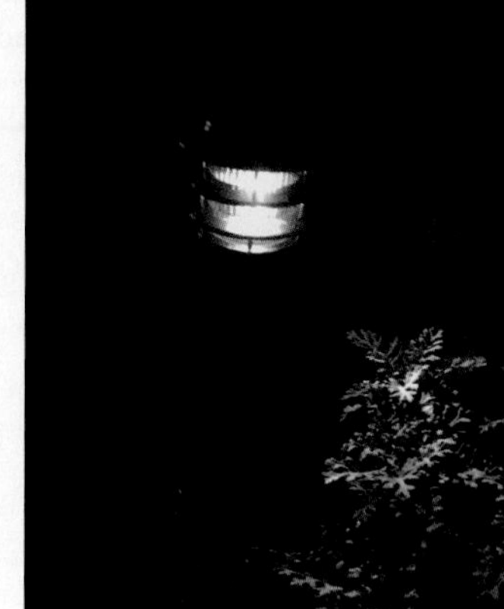

FIGURE P17.90

17.91. A *concentration cell* can be constructed by using the same half-reaction for both the cathode and anode. What is the value of E_{cell} of a concentration cell that combines copper electrodes in contact with 0.25 M copper(II) nitrate and 0.00075 M copper(II) nitrate solutions?

*17.92. **Lithium–Ion Batteries** Scientists at the University of Texas, Austin, and at MIT developed a cathode material for lithium–ion batteries based on $LiFePO_4$, which is the composition of the cathode when the battery is fully discharged. Batteries with this cathode are more powerful than those of the same mass with $LiCoO_2$ cathodes. They are also more stable at high temperatures.

a. What is the formula of the $LiFePO_4$ cathode when the battery is fully charged?
b. Is Fe oxidized or reduced as the battery discharges?
c. Is the cell potential of a lithium–ion battery with an iron phosphate cathode likely to differ from one with a cobalt oxide cathode? Explain your answer.

17.93. Starting with Equation 14.19 ($\Delta G = \Delta G^\circ + RT\ln Q$) and $\Delta G^\circ_{cell} = -n\,FE^\circ_{cell}$, derive an equation relating E°_{cell} and the equilibrium constant (K) of an electrochemical cell reaction. *Hint*: Recall that $Q = K$ and $\Delta G = 0$ in a reaction mixture at chemical equilibrium.

17.94. In a NiMH battery, what are the oxidation states of (a) Ni in NiO(OH), (b) H in MH, (c) M in MH, and (d) H in H_2O?

17.95. A magnesium battery can be constructed from an anode of magnesium metal and a cathode of molybdenum sulfide, Mo_3S_4. The half-reactions are

$$\text{Anode: } Mg(s) \rightarrow Mg^{2+}(aq) + 2\,e^- \qquad E^\circ_{anode} = 2.37\text{ V}$$

$$\text{Cathode: } Mg^{2+}(aq) + Mo_3S_4(s) + 2\,e^- \rightarrow MgMo_3S_4(s) \qquad E^\circ_{cathode} = ?$$

a. If the standard cell potential for the battery is 1.50 V, what is the value of E° for the reduction of Mo_3S_4?
b. What are the apparent oxidation states and electron configurations of Mo in Mo_3S_4 and in $MgMo_3S_4$?
*c. The electrolyte in the battery contains a complex magnesium salt, $Mg(AlCl_3CH_3)_2$. Why is it necessary to include Mg^{2+} ions in the electrolyte?

17.96. **Clinical Chemistry** The concentrations of Na^+ ions in red blood cells (11 mM) and in the surrounding plasma (140 mM) are quite different. Calculate the electrochemical potential (emf) across the cell membrane at 37°C as a result of this concentration gradient.

***17.97.** The element fluorine, F_2, was first produced in 1886 by the electrolysis of HF. Chemical syntheses of F_2 did not happen until 1986 when Karl O. Christe successfully prepared F_2 by the following reaction:

$$K_2MnF_6(s) + 2\,SbF_5(\ell) \rightarrow 2\,KSbF_6(s) + MnF_3(s) + \tfrac{1}{2}\,F_2(g)$$

a. Assign oxidation numbers to the elements in each compound and determine the number of electrons involved in the process.
b. Using the following ΔH°_f values, calculate ΔH° for the reaction.

$$\Delta H^\circ_{f,SbF_5(\ell)} = -1324\text{ kJ/mol} \qquad \Delta H^\circ_{f,K_2MnF_6(s)} = -2435\text{ kJ/mol}$$

$$\Delta H^\circ_{f,MnF_3(s)} = -1579\text{ kJ/mol} \qquad \Delta H^\circ_{f,KSbF_6(s)} = -2080\text{ kJ/mol}$$

c. If we assume that ΔS is relatively small, such that $\Delta G \approx \Delta H$, estimate E° for this reaction.
d. If ΔS for the reaction is greater than zero, is our value for E° in part (c) too high or too low?
e. The electrochemical synthesis of F_2 is described by the following electrolytic cell reaction:

$$2\,KHF_2(\ell) \rightarrow 2\,KF(\ell) + H_2(g) + F_2(g)$$

Assign oxidation numbers and determine the number of electrons involved in this process.

17.98. **Corrosion of Copper Pipes** The copper pipes frequently used in household plumbing may corrode and eventually leak. The corrosion reaction is believed to involve the formation of copper(I) chloride:

$$2\,Cu(s) + Cl_2(aq) \rightarrow 2\,CuCl(s)$$

a. Write balanced equations for the half-reactions in this redox reaction.
b. Calculate E°_{rxn} and ΔG°_{rxn} for the reaction.

17.99. Elemental uranium may be produced from uranium dioxide by the following two-step process:

$$UO_2(s) + 4\,HF(g) \rightarrow UF_4(s) + 2\,H_2O(\ell)$$

$$UF_4(s) + 2\,Mg(s) \rightarrow U(s) + 2\,MgF_2(aq)$$

a. Identify the reducing agent.
b. Identify the element that is reduced.
c. Using data from the table of standard reduction potentials in Appendix 6, find the maximum E° value for the reduction of UF_4 for the second reaction.
d. Will 1.00 g of Mg(s) be sufficient to produce 1.00 g of uranium?

*17.100. **Sodium–Sulfur Batteries** The low cost of sodium and sulfur relative to lithium makes voltaic cells based on sodium attractive to electric vehicle manufacturers, provided the technological hurdles of managing a battery that operates at 300°C can be overcome. The overall cell reaction is

$$2\ \text{Na}(s) + 3\ \text{S}(\ell) \rightarrow \text{Na}_2\text{S}_3(s) \qquad E^\circ_{\text{cell}} = 2.076\ \text{V}$$

a. Which element is oxidized and which is reduced?
b. How many electrons are transferred in the overall cell reaction?
c. What is the value of ΔG° for the reaction?
d. If a battery containing 5.25 kg Na is 50% discharged when it is connected to a charger with an output of 200 A, how long does it take to recharge the battery?
e. Draw a Lewis structure for the S_3^{2-} anion.

***17.101. Electrolysis of Seawater** Magnesium metal is obtained by the electrolysis of molten Mg^{2+} salts obtained from evaporated seawater.
a. Does elemental Mg form at the cathode or anode?
b. Do you think the principal ingredient in sea salt (NaCl) needs to be separated from the Mg^{2+} salts before electrolysis? Explain your answer.
c. Would electrolysis of an aqueous solution of $MgCl_2$ also produce elemental Mg?
d. If your answer to part (c) was no, what would the products of electrolysis be?

*17.102. **Silverware Tarnish** Low concentrations of hydrogen sulfide in air react with silver to form Ag_2S, more familiar to us as tarnish. Silver polish contains aluminum metal powder in a basic suspension.
a. Write a balanced net ionic equation for the redox reaction between Ag_2S and Al metal that produces Ag metal and $Al(OH)_3$.
b. Calculate E° for the reaction. *Hint*: Derive E° values for the half-reactions in which Ag_2S is reduced to Ag metal and $Al(OH)_3$ is reduced to Al metal. Then replace the $[Ag^+]$ and $[Al^{3+}]$ terms in the Nernst equations for these two half-reactions with terms based on the K_{sp} values of Ag_2S and $Al(OH)_3$ and the concentrations of sulfide and hydroxide ions (both of which are equal to one molar under standard conditions).

smartwork5

If your instructor uses Smartwork5, log in at digital.wwnorton.com/atoms2.

18

The Solid State

A Particulate View

STRONG ENOUGH TO REACH THE SKY One World Trade Center, the tallest building in the Western Hemisphere is clad in stainless steel, an alloy of iron in which about 1/4 of the atoms are chromium and nickel.

PARTICULATE **REVIEW**

Ionic Compounds versus Metals

In Chapter 18, we explore the structure and properties of solids. Most metals and ionic compounds are solids at room temperature, including platinum metal and potassium iodide whose structures are shown here. Answer the following questions about these substances:

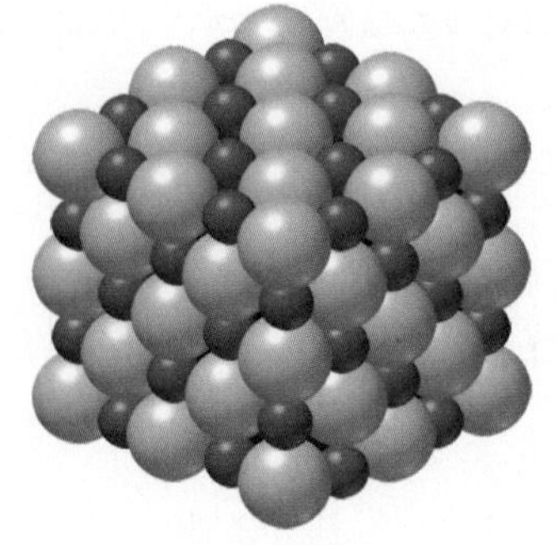

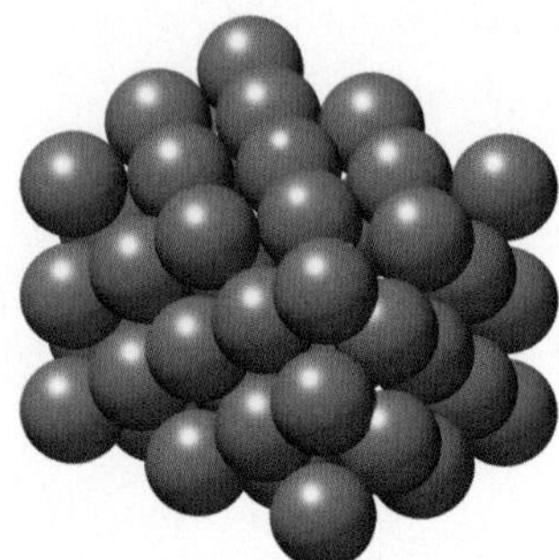

- Describe the nature of the bonds that hold the particles together in each solid.
- Which particles experience electrostatic attractions in potassium iodide? Which experience electrostatic repulsions?
- Which particles experience electrostatic interactions in platinum?

(Review Section 4.1 if you need help answering these questions.)

(Answers to Particulate Review questions are in the back of the book.)

PARTICULATE **PREVIEW**

Three-Dimensional Networks: Structure and Properties

Two allotropes of carbon, A and B, are depicted here. As you read Chapter 18, look for ideas that will help you answer these questions:

- Which allotrope has a planar repeating pattern? Which allotrope has a nonplanar repeating pattern?
- Which allotrope consists of carbon atoms covalently bonded to one another? Which allotrope has *both* covalent bonds *and* weak interactions between layers?
- One allotrope is extremely hard and is used as an abrasive; the other is soft and is used as a lubricant. Which is which?

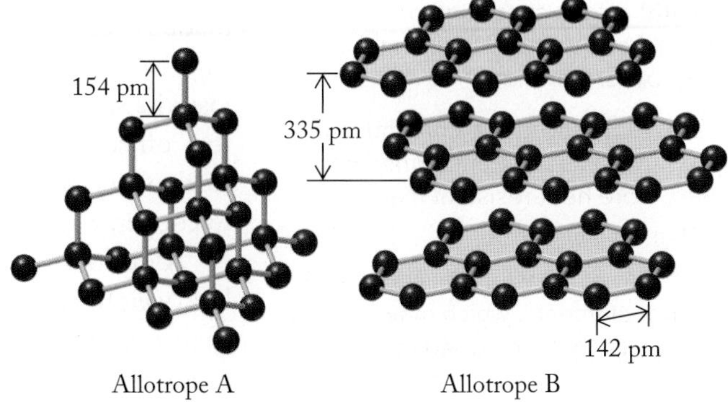

Learning Outcomes

LO1 Relate unit cell dimensions to the radii of the particles that make up the cell
Sample Exercises 18.1, 18.5

LO2 Relate the densities of solids to the masses of the particles in their unit cells and the dimensions of the cells
Sample Exercises 18.2, 18.3, 18.6

LO3 Distinguish between substitutional and interstitial alloys
Sample Exercise 18.4

LO4 Use band theory to explain the conductivity of metals and semiconductors

LO5 Distinguish between ionic solids, molecular solids, and covalent network solids

LO6 Calculate the interlayer spacing in crystalline solids using the Bragg equation
Sample Exercise 18.7

18.1 Stronger, Tougher, Harder

People have been using metals for tools, weapons, currency, and jewelry for thousands of years. For most of that time, only three elements—copper, silver, and gold—were used because these are the only ones found as free metals in Earth's crust. Even so, most silver and copper occur not as pure metals but in ores such as argentite (Ag_2S), chalcopyrite ($CuFeS_2$), and chalcocite (Cu_2S). **Ores** are naturally occurring compounds or mixtures of compounds from which elements can be extracted. Most ores are composed of **ionic solids** consisting of monatomic or polyatomic ions held together by ionic bonds. Earth's crust also yields materials used to make **ceramics**, solid inorganic compounds or mixtures of compounds that have been heated in such a way as to transform them into a harder and more heat-resistant material.

Gold rings and other gold jewelry may look like pure gold, but most are not. Gold is one of the softer metals, so it is very easy to bend jewelry made of pure gold. In contrast, jewelry made of gold blended with other metals such as palladium, silver, or copper is more resistant to physical damage and can last a lifetime. Any mixture of a host metal with one or more other elements is called an **alloy**, and there are many thousands of them. Some are solid solutions, which means they are homogeneous at the atomic level just as liquid solutions are, and others are heterogeneous mixtures. Varying the proportions of their constituent metals modifies the properties of alloys.

As we discussed at the beginning of Chapter 1, one of the first alloys was bronze, a mixture of copper and tin that was used by ancient Greeks and others to make tools and weapons that were harder, stronger, and more durable than those made of copper alone. Brass is another copper alloy, mixed with zinc instead of tin, that has long been used in doorknobs and hinges, even though it tarnishes and loses its shiny luster over time. Still, it is a popular material in hospitals and doctors' offices because it kills bacteria that come in contact with it, whereas stainless steel, which resists corrosion better, does not. The bactericidal action of brass is due to copper, a metal that has long been known for its antimicrobial properties.

Aluminum alloys are ideal for making aircraft because they are strong and lightweight. For example, the wing of a 747-400 airplane is about 1.8 m longer than the wing of a 747-300 but about 2 metric tons lighter because it is made of an alloy of 95% aluminum blended with copper, manganese, and iron. Even sports equipment benefits from the development of new alloys: a patented five-metal alloy of nickel, zirconium, titanium, copper, and beryllium is advertised as possibly changing the game of golf because the alloy produces a stronger,

ore naturally occurring compounds or mixtures of compounds from which elements can be extracted.

ionic solid a solid consisting of monatomic or polyatomic ions held together by ionic bonds.

ceramic a solid inorganic compound or mixture that has been transformed into a harder, more heat-resistant material by heating.

alloy a blend of a host metal and one or more other elements, which may or may not be metals, that are added to change the properties of the host metal.

crystalline solid a solid made of an ordered array of atoms, ions, or molecules.

lighter, more resilient club that enables a golfer to transfer more energy from the swing into the ball.

In this chapter we explore the links between the physical properties of solids at the macroscopic level and the structures of these solids at the atomic level. We start with metals and their alloys and conclude with ceramics, addressing questions such as the following: Why do metals bend? Why are they such good conductors of heat and electricity? Why are alloys so much stronger, tougher, and harder than the pure metals from which they are made?

18.2 Structures of Metals

Most elements are metals, which means they are typically hard, shiny, malleable (easily shaped), ductile (easily drawn out), and able to conduct electricity. In this section we take a detailed look at how the atoms in metals are arranged and explore how their atomic structure accounts for their physical properties.

Stacking Patterns

When a metallic element is heated above its melting point and then allowed to slowly cool, it usually solidifies into a **crystalline solid**—that is, a solid in which atoms are arranged in an ordered three-dimensional array called a *crystal lattice*. Think of a crystal lattice as stacked layers of atoms (designated *a*, *b*, *c*, . . .) that are packed tightly together. In the most tightly packed arrangements, each atom touches six others in its layer (denoted layer *a*) as shown in Figure 18.1(a). The blue and red dots between the atoms (seen as yellow spheres) represent the points above which the atoms in additional layers will be centered. The atoms represented by blue spheres in the second (*b*) layer nestle into the spaces created by the first layer that were marked by blue dots in Figure 18.1(a), in much the same way that oranges in a fruit-stand display or cannonballs at a 16th-century fort (Figure 18.2) might be stacked together.

The atoms in a third layer nestle among those in the second in one of two different alignments. They may sit directly above the atoms in the *a* layer as shown in Figure 18.1(c). This arrangement produces an *ababab* ... stacking pattern throughout the crystal. However, the atoms in the third layer may nestle between the atoms in the second layer in such a way that they are not aligned directly above the atoms in the *a* layer, but rather are above the red dots between them, creating a third (*c*) layer that is not aligned with either the *a* or *b* layers (Figure 18.1d). When this happens and a fourth layer of atoms is directly above those in the *a* layer, we have an *abcabc* ... stacking pattern. Other patterns are possible, but we will focus on these two.

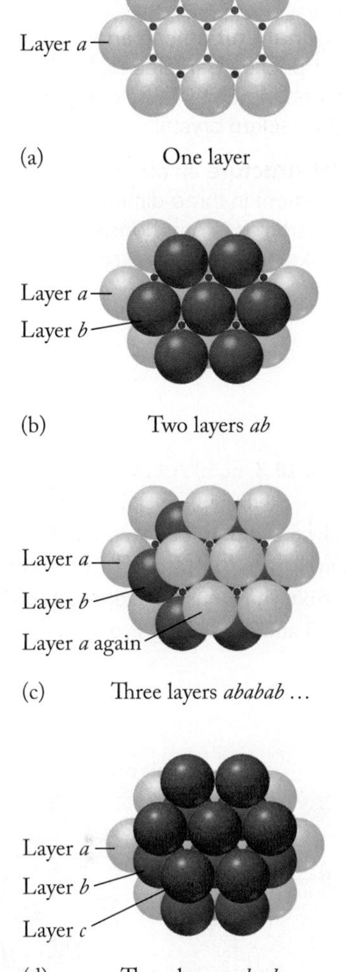

FIGURE 18.1 The *ababab* ... and *abcabc* ... stacking patterns represent two equally efficient ways to stack layers of atoms (or any particles of equal size).

(a)

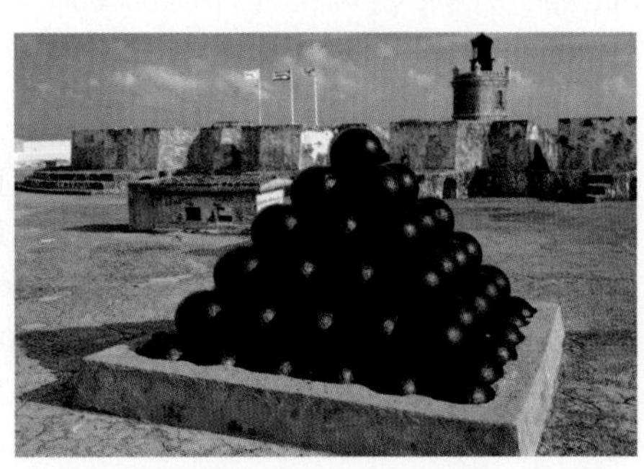
(b)

FIGURE 18.2 (a) Stacks of oranges in a grocery store and (b) cannonballs at El Morro in San Juan, Puerto Rico, illustrate closest-packed arrays of spherical objects.

hexagonal closest-packed (hcp) a crystal lattice in which the layers of atoms or ions have an *ababab...* stacking pattern.

unit cell the basic repeating unit of the arrangement of atoms, ions, or molecules in a crystalline solid.

hexagonal unit cell an array of nine closest-packed particles that are the repeating unit in a hexagonal closest-packed crystal.

crystal structure an ordered arrangement in three-dimensional space of the particles (atoms, ions, or molecules) that make up a crystalline solid.

Stacking Patterns and Unit Cells

Stacking patterns determine the shapes of the crystals that metals form. To explore why, let's take a closer look at a cluster of atoms in the *ababab...* stacking pattern (Figure 18.3). This cluster forms a *hexagonal* (six-sided) prism of closely packed atoms. In fact, they are as tightly packed as they can be, so the crystal structure is called **hexagonal closest-packed (hcp)**. In these lattices, the nine-atom cluster highlighted in Figure 18.3(c) serves as an atomic-scale building block—a pattern of atoms repeated over and over again in all three dimensions in the elements.

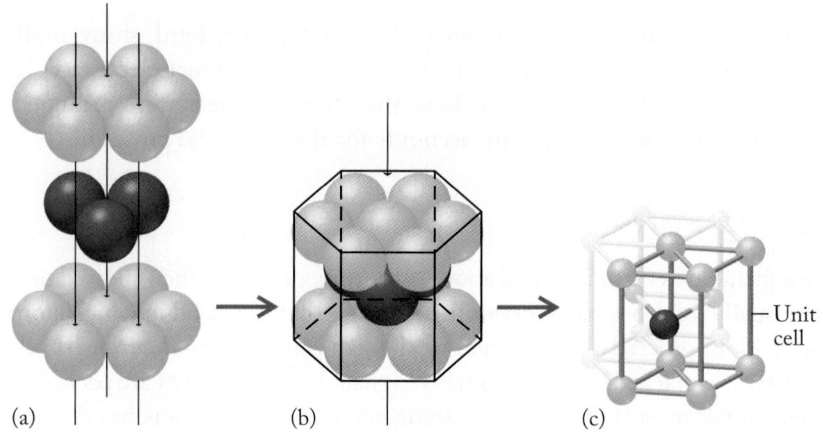

FIGURE 18.3 (a, b) A hexagonal closest-packed (hcp) crystal structure and (c) its hexagonal unit cell represent a highly efficient way to pack atoms in a crystalline solid. Sixteen metals, including all those in groups 3 and 4, have the hcp crystal structure.

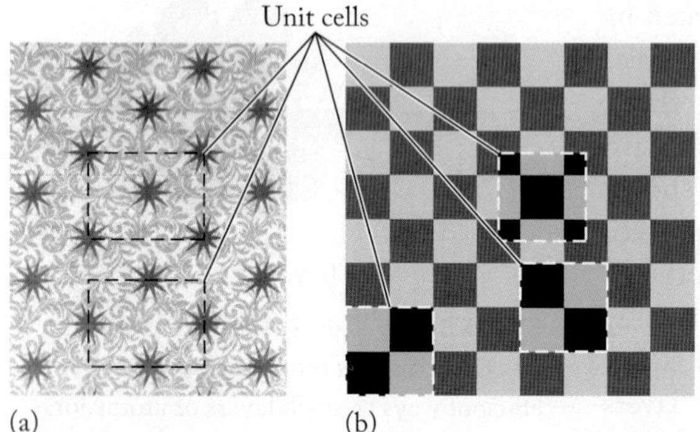

FIGURE 18.4 Two-dimensional repeating patterns. (a) The repeating patterns outlined with the dashed lines represent "unit cells" for this wrapping-paper design. (b) The portions of the squares inside the white-dashed lines show three ways to represent the repeating pattern in this checkerboard. Can you sketch an alternative way?

We call each of these building blocks a **unit cell**, and the structure in Figure 18.3 has a **hexagonal unit cell**. A unit cell represents the minimum repeating pattern that describes the array of atoms forming the crystal lattice of any crystalline solid, including metals. Think of unit cells as three-dimensional microscopic analogs of the two-dimensional repeating pattern in fabrics, wrapping paper, or even a checkerboard. Look closely at Figure 18.4 to confirm that the outlined portions represent the minimum repeating patterns in the paper and in the checkerboard. A unit cell has the same role in the **crystal structure** of a solid—that is, the arrangement in three-dimensional space of the atoms, ions, or molecules of which the solid is composed.

CONCEPT TEST

Try to find other portions of the images in Figure 18.4 that contain minimum repeating patterns.

(Answers to Concept Tests are in the back of the book.)

What is the unit cell in the *abcabc ...* stacking pattern? Look at what happens when we take the 14-atom cluster in Figure 18.5(a), rotate it, and tip it 45° to obtain the orientation shown in Figure 18.5(b). The black outline in Figure 18.5(b) shows that the atoms form a cube: one atom at each of the eight corners of the cube and one at the center of each of the six faces. Because the atoms are stacked together as closely as possible, this crystal lattice is called **cubic closest-packed (ccp)**, and the corresponding unit cell is called a **face-centered cubic (fcc) unit cell**. Because this unit cell is a cube, the edges are all of equal length, and the angle between any two edges is 90°. Note in Figure 18.5(b) how the atom in the center of each face touches each of the atoms at the four corners of that face. However, none of the corner atoms touch any other corner atoms.

So far we have introduced two *closest-packed* crystal lattices—hexagonal and cubic—and their associated unit cells (hexagonal and face-centered cubic). The hcp and ccp crystal lattices represent the most efficient ways of arranging solid spheres of equal radius. We can express the **packing efficiency** as the percentage of the total volume of the unit cell occupied by the spheres:

$$\text{Packing efficiency (\%)} = \frac{\text{volume occupied by spheres}}{\text{volume of unit cell}} \times 100\% \qquad (18.1)$$

$$= \frac{V_{\text{atoms}}}{V_{\text{unit cell}}} \times 100\%$$

For both hcp and ccp crystal lattices, the packing efficiency is approximately 74%. We will come back to this calculation at the end of the next subsection, but first, let's look at some other packing arrangements.

Stacking patterns also exist in which the atoms are arranged close together but not as efficiently as the hcp and ccp lattices. Two such patterns are shown in Figures 18.6 and 18.7. We can arrange the atoms in an *a* layer so that each atom touches four adjacent atoms, an arrangement called *square packing* (Figure 18.6a). If we add a second layer of spheres directly above the first, we create the *aaa* … stacking pattern shown in Figure 18.6(b) that is called *cubic packing*. The three-dimensional repeating pattern of this arrangement is called a **simple cubic (sc) unit cell**. It is the least efficiently packed of the cubic unit cells and is quite rare among metals: only radioactive polonium (Po) forms a simple cubic unit cell.

If each atom in a second layer is nestled in the space created by four atoms in a square-packed *a* layer (Figure 18.7a), we have two layers in an *ab* stacking pattern. If the atoms in the third layer are directly above those in the first, then we have an *ababab* … stacking pattern based on layers of square-packed atoms. The simplest three-dimensional repeating unit of this pattern is called a **body-centered cubic (bcc) unit cell**. It consists of nine atoms, one at each of the eight corners of a cube and one in the middle of the cube (Figure 18.7b). All the group 1 metals and many transition metals have bcc unit cells. Table 18.1 summarizes the different stacking patterns, packing efficiencies, and unit cells described in this section.

cubic closest-packed (ccp) a crystal structure composed of face-centered cubic unit cells and layers of particles having an *abcabc*... stacking pattern.

face-centered cubic (fcc) unit cell an array of closest-packed particles that has eight of the particles at the corners of a cube and six of them at the centers of each face of the cube.

packing efficiency the percentage of the total volume of a unit cell occupied by the spheres.

simple cubic (sc) unit cell a cell with atoms only at the eight corners of a cube.

body-centered cubic (bcc) unit cell a cell with atoms at the eight corners of a cube and at the center of the cell.

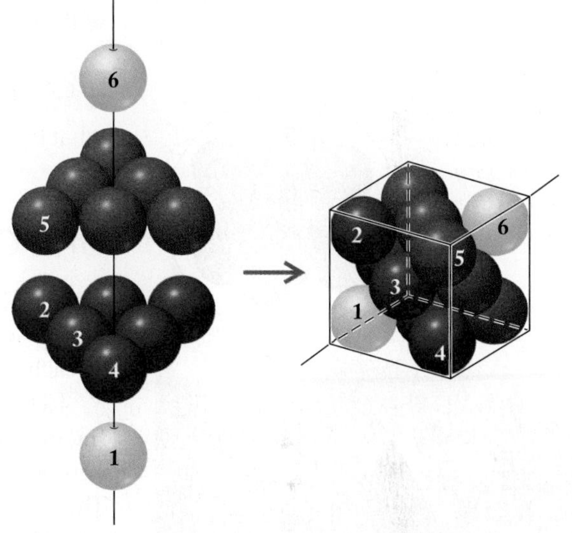

FIGURE 18.5 (a) The stacking pattern *abcabc*… has a face-centered cubic (fcc) unit cell. (b) The shape of the unit cell is more easily seen when the layers are tipped 45° and rotated. Note that atoms at adjacent corners do not touch each other, but the three atoms along the diagonal of any face of the cube—atoms 2, 3, and 4 here—do touch each other.

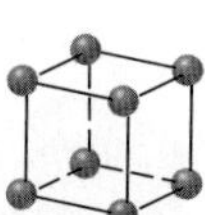

FIGURE 18.6 (a) In a square-packed *a* layer, each atom (like the one in the middle) touches four others. (b) If the atoms in all other layers are directly above those in the *a* layer, the stacking pattern is called *cubic packing*. (c) The repeating unit of this pattern is called a *simple cubic* (sc) unit cell with eight atoms at the eight corners of a cube.

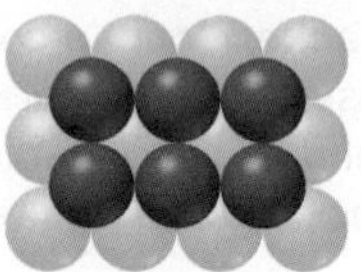

(a) Two square-packed layers in an *ab* pattern

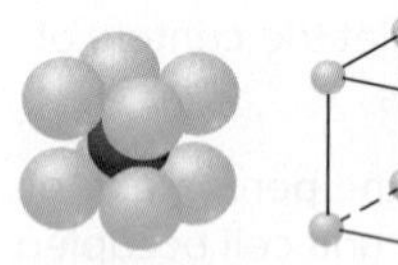

(b) Body-centered cubic unit cell

FIGURE 18.7 (a) The atoms represented by purple spheres in the *b* layer nestle into the spaces between the square-packed atoms (yellow spheres) in the *a* layer. (b) Atoms in the third layer are directly above those in the first, producing an *ababab*... stacking pattern and a *body-centered cubic* (bcc) unit cell.

TABLE 18.1 Summary of Unit Cell Types, Stacking Patterns, and Packing Efficiencies

Unit Cell	Stacking Pattern	Number of Nearest Neighbors	Packing Efficiency
Hexagonal	*ababab*...	12	74%
Face-centered cubic	*abcabc*...	12	74%
Body-centered cubic	*ababab*...	8	68%
Simple cubic	*aaaa*...	6	52%

Solids with cubic unit cells form crystals that often contain cubes or four-sided prisms (Figure 18.8), whereas crystalline solids with hexagonal unit cells tend to form hexagonal crystals (Figure 18.9).

As Figure 18.10 shows, the periodic table contains elements that exist in all of the crystal structures we have discussed in this section, as well as structures we will discuss in subsequent sections.

CONCEPT TEST

What is the difference between a crystal lattice and a unit cell?

FIGURE 18.8 Palladium (and other metals) with fcc unit cells may form crystals that contain four-sided prisms such as those in this photomicrograph.

Unit Cell Dimensions

Figure 18.11 shows whole-atom and cutaway views of sc, fcc, and bcc unit cells. These views provide us with a way to determine how many equivalent atoms are in each type of cubic unit cell.

Let's start with the simple cubic unit cell (Figure 18.11a). Note how only a fraction of each corner atom is inside the unit cell boundary. In a crystal lattice with this unit cell, each atom is a corner atom in eight unit cells (Figure 18.12a). Thus, each atom contributes the equivalent of one-eighth of an atom to the unit cell. There are eight corners in a cube, so there is a total of

$$\frac{\frac{1}{8}\text{ corner atom}}{\cancel{\text{corner}}} \times \frac{8\ \cancel{\text{corners}}}{\text{unit cell}} = 1\ \frac{\text{corner atom}}{\text{unit cell}}$$

This calculation applies to the corner atoms in any type of cubic unit cell. Note that the two corner atoms along each edge in Figure 18.11(a) touch each other. Therefore, the edge length ℓ in the simple cubic unit cell is equal to twice the atomic radius:

$$\ell = 2r$$

In an fcc unit cell (Figure 18.11b), there are eight corner atoms and one atom in the center of each of the six faces. Each face atom is shared by the two unit cells that

FIGURE 18.9 Materials with hexagonal unit cells tend to form hexagonal crystals. There are two such materials in this photo: quartz crystals (see also Figure 18.38) coated with a thin layer of hcp titanium.

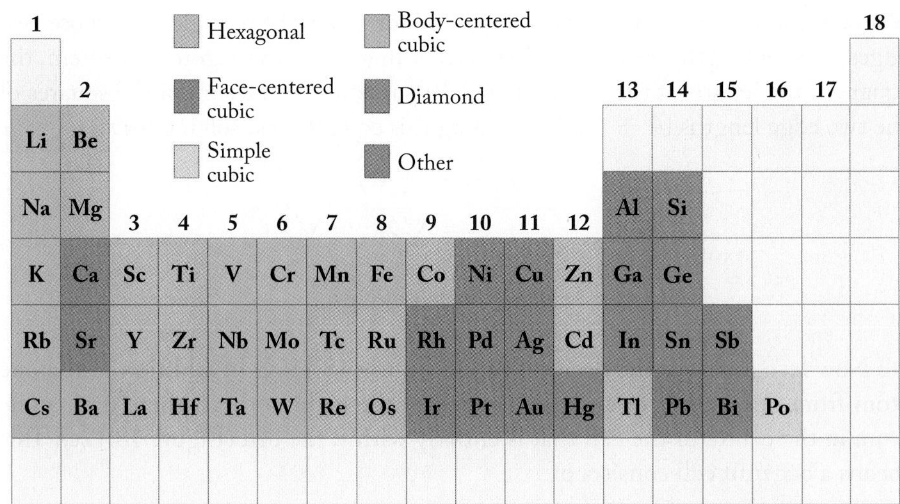

FIGURE 18.10 Crystal structures of metals and metalloids. The unit cells for the five elements designated "Other" are more complicated and beyond the scope of this book.

abut each other at that face (Figure 18.12b). Therefore, each unit cell "owns" half of each face atom, making a total of

$$\frac{\frac{1}{2}\text{ face atom}}{\cancel{\text{face}}} \times \frac{6\ \cancel{\text{faces}}}{\text{unit cell}} = 3\ \frac{\text{face atoms}}{\text{unit cell}}$$

Moreover, every cubic unit cell owns the equivalent of one corner atom, so an fcc unit cell consists of

$$1\text{ corner atom} + 3\text{ face atoms} = 4\text{ atoms per fcc unit cell}$$

To relate the dimensions of the fcc unit cell to the size of the atoms in it, note in the cutaway view of Figure 18.11(b) that the corner atoms do not touch one another but adjacent atoms along the face diagonal do touch each other. A face diagonal spans the radius r of two corner atoms and the diameter (2 radii = $2r$) of a face atom. Therefore, the length of a face diagonal is $1 + 2 + 1 = 4$ atomic radii $= 4r$. A face

(a) Simple cubic: Atoms touch along edge

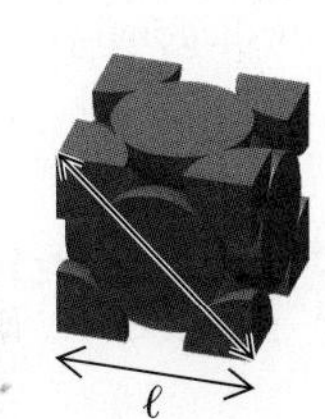

(b) Face-centered cubic: Atoms touch along face diagonal

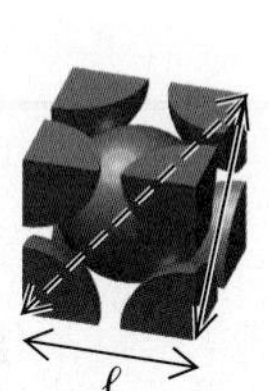

(c) Body-centered cubic: Atoms touch along body diagonal

FIGURE 18.11 Whole-atom and cutaway views of cubic unit cells. (a) In a simple cubic unit cell, each corner atom of the unit cell is part of eight unit cells. Atoms along each edge touch. (b) In a face-centered cubic unit cell, the face atoms are part of two unit cells. Atoms along the face diagonal touch. (c) In a body-centered cubic unit cell, one atom in the center lies entirely in one unit cell. The atoms along the body diagonal touch.

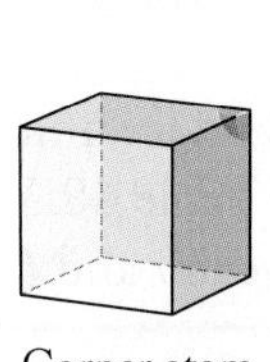

Corner atom in 1 unit cell

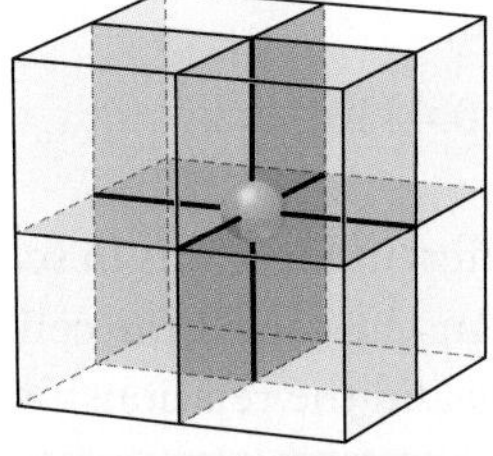

Corner atom shared by 8 unit cells

(a)

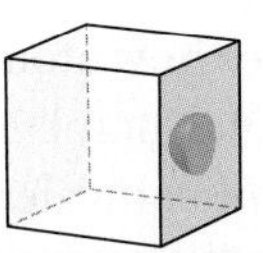

Face-centered atom in 1 unit cell

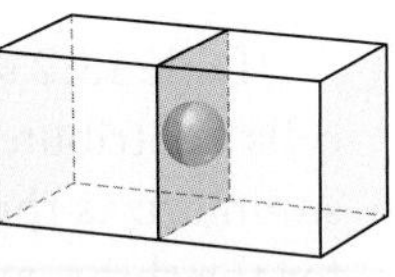

Face-centered atom shared by 2 unit cells

(b)

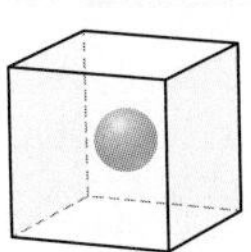

Body-centered atom in 1 unit cell

(c)

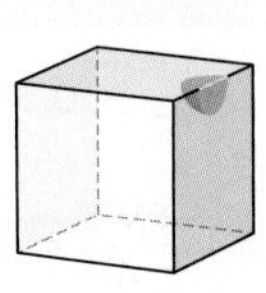

Edge atom in 1 unit cell

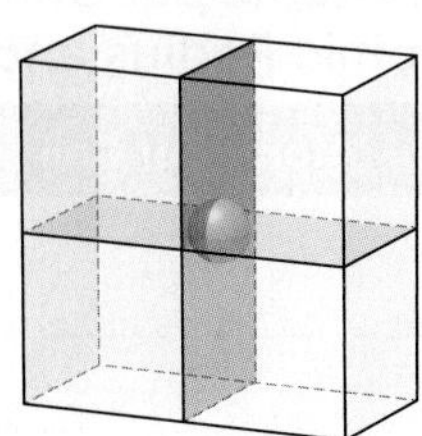

Edge atom shared by 4 unit cells

(d)

FIGURE 18.12 Crystal lattices illustrating (a) corner atoms shared by eight unit cells, (b) face atoms shared by two unit cells, (c) center atoms entirely in one unit cell, and (d) edge atoms shared by four unit cells.

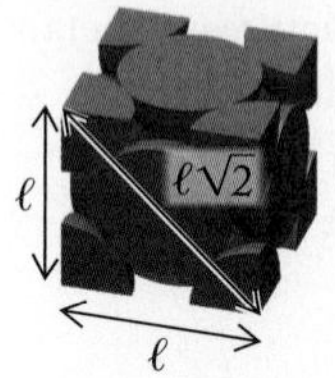

(a) fcc unit cell dimensions

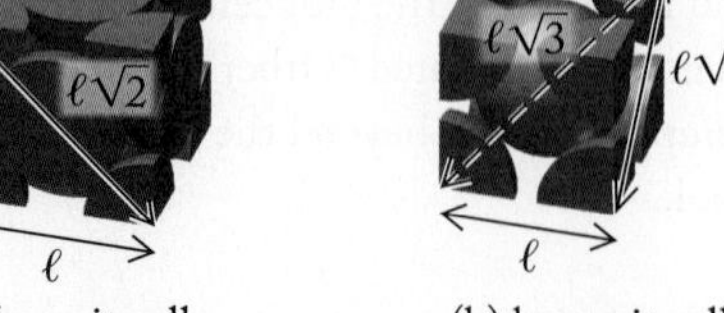

(b) bcc unit cell dimensions

FIGURE 18.13 (a) In a face-centered cubic unit cell, the face diagonal forms a right triangle with two adjoining edges each of length ℓ. Applying the Pythagorean theorem to this triangle yields $r = 0.3536\ell$. (b) In a body-centered cubic unit cell, the atoms touch along a body diagonal. Applying the Pythagorean theorem to the right triangle formed by an edge, a face diagonal, and a body diagonal yields $r = 0.4330\ell$.

CHEMTOUR Unit Cell

diagonal connects the ends of two edges and forms a right triangle with those two edges, each of length ℓ (Figure 18.13a). According to the Pythagorean theorem, the square of the length of the face diagonal $(4r)^2$ is equal to the sum of the squares of the two edge lengths $(\ell^2 + \ell^2)$. Combining this equality and solving for r:

$$(4r)^2 = (\ell^2 + \ell^2)$$

$$4r = \sqrt{\ell^2 + \ell^2} = \sqrt{2\ell^2} = \sqrt{2}\,\ell$$

$$r = \frac{\sqrt{2}}{4}\ell = 0.3536\ell \qquad (18.2)$$

Now let's focus on the bcc unit cell in Figure 18.11(c). In addition to the one atom from one-eighth of an atom at each of the eight corners, there is also one atom in the center of the cell that is entirely within the cell (Figure 18.12c). This means a bcc unit cell consists of

1 corner atom + 1 center atom = 2 atoms per bcc unit cell

Relating unit cell edge length to atomic radius in a bcc cell is complicated by the fact that, in addition to not touching along the edges, adjacent atoms do not touch along any face diagonal either. However, each corner atom does touch the atom in the center of the cell, which means that the atoms touch along a *body diagonal* that runs between opposite corners through the center of the cube. In the cutaway view in Figure 18.13(b), the diagonal runs from the bottom left corner of the front face to the top right corner of the rear face. It spans (1) the radius of the front-face bottom left corner atom, (2) the diameter (2 radii) of the central atom, and (3) the radius of the rear-face top right atom, making the length of the body diagonal equivalent to $4r$.

We can again apply the Pythagorean theorem, this time to a right triangle whose legs are an edge (ℓ) and a face diagonal ($\sqrt{2}\,\ell$), and whose hypotenuse is the body diagonal ($4r$). Squaring the hypotenuse and setting it equal to the sum of the squares of the other two sides:

$$(4r)^2 = \ell^2 + (\sqrt{2}\,\ell)^2$$

$$4r = \sqrt{\ell^2 + 2\ell^2} = \sqrt{3\ell^2} = \sqrt{3}\,\ell$$

$$r = \frac{\sqrt{3}}{4}\ell = 0.4330\ell \qquad (18.3)$$

Table 18.2 summarizes how atoms in different locations in sc, bcc, and fcc unit cells contribute to the total number of atoms in each unit cell, and Table 18.3 summarizes the number of equivalent atoms and the relationship between r and ℓ for the three cubic unit cells.

TABLE 18.2 Contributions of Atoms to Cubic Unit Cells

Atom Position	Contribution to Unit Cell	Unit Cell
Center	1 atom	bcc
Face	$\frac{1}{2}$ atom	fcc
Corner	$\frac{1}{8}$ atom	bcc, fcc, sc

TABLE 18.3 Summary of Atoms per Unit Cell and the Relation between Atomic Radius and Cell Edge Length

Unit Cell Type	Atoms per Cell	Relationship between r and ℓ
Simple cubic	1	$r = \frac{\ell}{2} = 0.5000\ell$
Body-centered cubic	2	$r = \frac{\ell\sqrt{3}}{4} = 0.4330\ell$
Face-centered cubic	4	$r = \frac{\ell\sqrt{2}}{4} = 0.3536\ell$

SAMPLE EXERCISE 18.1 Calculating Atomic Radii from Unit Cell Dimensions **LO1**

The structure of the most stable form of iron at room temperature, called *ferrite*, is shown in Figure 18.14. The unit cell in the figure has an edge length of 287 pm. What is the radius in picometers of the iron atoms in the cell? Check your answer against the data in Appendix 3.

Collect, Organize, and Analyze The unit cell in Figure 18.14 is body-centered cubic, which means the radii (r) of the atoms in the cell are related to the cell's edge length ($\ell = 287$ pm) by Equation 18.3.

Solve

$$r = 0.4330\ell$$
$$= 0.4330 \times 287 \text{ pm} = 124 \text{ pm} \quad (18.3)$$

Think About It The calculated value is quite close to average atomic radius of iron atoms (126 pm) in Table A3.1 in Appendix 3.

Practice Exercise At 1070°C, the most stable form of iron is *austenite* (Figure 18.15). The edge length of its unit cell is 361 pm. What is the atomic radius of iron in austenite?

(Answers to Practice Exercises are in the back of the book.)

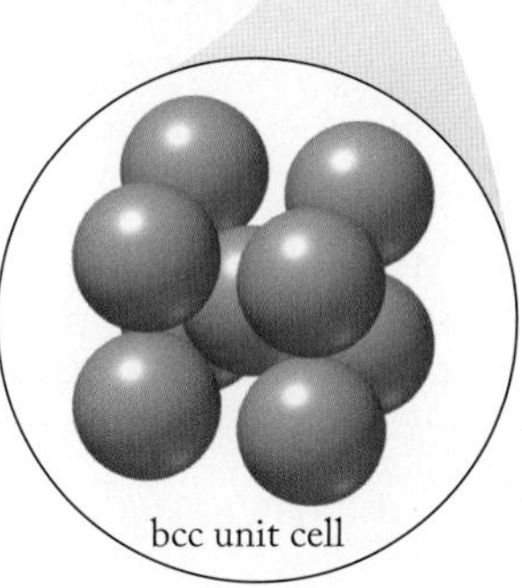

FIGURE 18.14 The pattern in this newly cut surface of the meteorite that caused the Odessa crater in Texas about 50,000 years ago is due to crystals of ferrite, which grew when the meteor formed from molten iron 4.5 billion years ago. The microview shows the bcc unit cell of ferrite.

SAMPLE EXERCISE 18.2 Using Unit Cell Dimensions to Calculate Density **LO2**

Calculate the density of iron (ferrite) in grams per cubic centimeter, given that its bcc unit cell has an edge length of 287 pm.

Collect and Organize We are to calculate the mass-to-volume ratio of ferrite iron, which has a bcc unit cell. There are two iron atoms per unit cell and 6.0221×10^{23} atoms/mol. The molar mass of Fe is 55.845 g/mol. The volume (V) of a cube of edge length ℓ is ℓ^3.

Analyze We have to assume the density of the unit cell matches the density of solid ferrite. To calculate the density of the unit cell, we need to calculate the mass of two Fe atoms starting with the molar mass of Fe and dividing by Avogadro's number to calculate the mass of each Fe atom in grams. The volume of the cell is the cube of its edge length $(287 \text{ pm})^3$. An object made of iron sinks when immersed in water, so the density of iron must be greater than 1 g/cm^3.

Solve Calculating the mass m of two Fe atoms:

$$m = \frac{55.845 \text{ g Fe}}{1 \cancel{\text{mol Fe}}} \times \frac{1 \cancel{\text{mol Fe}}}{6.0221 \times 10^{23} \cancel{\text{atoms Fe}}} \times 2 \cancel{\text{atoms Fe}} = 1.8547 \times 10^{-22} \text{ g Fe}$$

The volume V of the cell in cubic centimeters is

$$V = \ell^3 = (287 \cancel{\text{pm}})^3 \times \frac{(10^{-10} \text{ cm})^3}{(1 \cancel{\text{pm}})^3} = 2.364 \times 10^{-23} \text{ cm}^3$$

The density of the cell is

$$d = \frac{m}{V} = \frac{1.8547 \times 10^{-22} \text{ g}}{2.364 \times 10^{-23} \text{ cm}^3} = 7.85 \text{ g/cm}^3$$

FIGURE 18.15 The fcc unit cell of austenite.

Think About It The calculated density is larger than 1 g/cm^3, as expected, and quite close to the value (7.874 g/cm^3) in Appendix 3. Our assumption that density of the unit cell must equal the density of a bulk sample was reasonable because the latter is composed of a crystal lattice composed of bcc unit cells.

Practice Exercise Silver and gold both crystallize in face-centered cubic unit cells with edge lengths of 407.7 and 407.0 pm, respectively. Calculate the density of each metal and compare your answers with the densities listed in Appendix 3.

Earlier in this section we noted that ccp and hcp are the most efficient packing schemes for spheres. Now that we know more about the fcc unit cell, let's take another look at the calculation of packing efficiency using Equation 18.1. The unit cell for a ccp arrangement of solid spheres of radius r is an fcc unit cell that contains four equivalent spheres. The volume occupied by each of these spheres is

$$V_{\text{spheres}} = \frac{4}{3}\pi r^3$$

The radii of the atoms in an fcc unit cell are related to the edge length of the cell by Equation 18.2:

$$r = \frac{\sqrt{2}}{4}\ell = 0.3536\ell \qquad (18.2)$$

or

$$\ell = 2.828r$$

Expressing the volume of the unit cell in terms of r:

$$V_{\text{unit cell}} = \ell^3 = (2.828r)^3 = 22.62r^3$$

Substituting V_{spheres} and $V_{\text{unit cell}}$ into Equation 18.1:

$$\text{Packing efficiency (\%)} = \frac{V_{\text{spheres}}}{V_{\text{unit cell}}} \times 100\% = \frac{4(\frac{4}{3}\pi r^3)}{2.62r^3} \times 100\% = 74.1\%$$

This means that the most efficient packing of spheres results in about (100 − 74) = 26% of the total volume being empty.

SAMPLE EXERCISE 18.3 Using Atomic Radii and Unit Cell Structure to Calculate Density and Packing Efficiency **LO2**

Polonium is the only metal that crystallizes in a simple cubic structure. If the atomic radius of polonium is 167 pm, (a) what is the density of polonium in grams per cubic centimeter and (b) what is the packing efficiency of Po atoms?

Collect and Organize We are asked to calculate the density and packing efficiency of Po from its unit cell geometry and atomic radius. Po crystallizes in a simple cubic structure, so polonium atoms touch along the edge. A simple cubic unit cell contains the equivalent of a single atom. The molar mass of the most stable isotope of Po is 209 g/mol. The volume of a cube of edge length ℓ is $V = \ell^3$. The volume of a spherical atom of radius r is $\frac{4}{3}\pi r^3$.

Analyze As in Sample Exercise 18.2, we assume that the density of the unit cell is the same as the density of solid Po. The density of the Po unit cell is the mass of one Po atom divided by the cell volume, which is the edge length (2 atomic radii) cubed. We need to divide the molar mass of polonium by Avogadro's number to obtain the mass of a single Po atom in grams. The density of Po should be similar to that of Fe (~8 g/cm^3) because (1) Po atoms are more massive, but (2) they are not as efficiently packed. Calculating the packing efficiency with Equation 18.1 involves calculating the volume of a Po atom based on its atomic radius and dividing by the volume of the unit cell.

Solve

a. The mass m of a Po atom is

$$m = \frac{209\text{ g Po}}{1\text{ }\cancel{\text{mol Po}}} \times \frac{1\text{ }\cancel{\text{mol Po}}}{6.0221 \times 10^{23}\text{ }\cancel{\text{atoms Po}}} \times 1\text{ }\cancel{\text{atom Po}} = 3.47 \times 10^{-22}\text{ g Po}$$

and the volume of the unit cell in cubic centimeters is

$$\ell = 2r = 2(167\text{ }\cancel{\text{pm}}) \times \frac{10^{-10}\text{ cm}}{1\text{ }\cancel{\text{pm}}} = 3.34 \times 10^{-8}\text{ cm}$$

$$V_{\text{unit cell}} = \ell^3 = (3.34 \times 10^{-8}\text{ cm})^3 = 3.73 \times 10^{-23}\text{ cm}^3$$

The density of the unit cell (and bulk Po) is

$$d = \frac{m}{V_{\text{unit cell}}} = \frac{3.47 \times 10^{-22}\text{ g}}{3.37 \times 10^{-23}\text{ cm}^3} = 9.30\text{ g/cm}^3$$

b. The volume occupied by a Po atom is

$$V_{\text{atoms}} = \frac{4}{3}\pi r^3 = \frac{4}{3}\pi\left(167\text{ pm} \times \frac{10^{-10}\text{ cm}}{1\text{ pm}}\right)^3 = 1.95 \times 10^{-23}\text{ cm}^3$$

and the packing efficiency is

$$\text{Packing efficiency (\%)} = \frac{V_{\text{atoms}}}{V_{\text{unit cell}}} \times 100\% = \frac{1.95 \times 10^{-23}\text{ }\cancel{\text{cm}^3}}{3.73 \times 10^{-23}\text{ }\cancel{\text{cm}^3}} \times 100 = 52.3\%$$

Think About It As predicted, the calculated density of Po is similar to that of iron and it is close to the reference density value in Table A3.2 of Appendix 3: 9.32 g/mL. The calculated packing efficiency matches the reference value for simple cubic structures in Table 18.1.

Practice Exercise Tungsten filaments were widely used in incandescent lightbulbs. Tungsten has an atomic radius of 139 pm and a bcc unit cell. Calculate the density and packing efficiency of the atoms in tungsten. Compare your answer with the density listed in Appendix 3.

18.3 Alloys

Around 6000 years ago, metal technology took a giant leap forward when artisans in the Middle East and perhaps other parts of the world discovered how to convert copper ore, principally $CuFeS_2$, into copper metal. The process involved pulverizing the ore and then baking it in ovens. Baking initiated a chemical reaction with O_2 (from air), which converted the Cu in $CuFeS_2$ into CuO. In the second step in the process, CuO reacted with carbon monoxide, produced by burning wood or charcoal (mostly carbon) in a furnace with an insufficient supply of air:

$$CuO(s) + CO(g) \rightarrow Cu(s) + CO_2(g) \tag{18.4}$$

One disadvantage of primitive copper tools and weapons is that the metal is malleable, which means that copper objects are easily bent and damaged. The malleability of Cu (and other metals) is due to the relatively weak metallic bonds between their atoms, which makes it possible for the atoms in one layer, under stress, to slip past atoms in an adjacent layer (Figure 18.16). When the stress is

FIGURE 18.16 Copper and other metals are malleable because their atoms are stacked in layers that can slip past each other under stress. Slippage is possible because of the diffuse nature of metallic bonds and the relatively weak interactions between pairs of atoms in adjoining layers.

relieved and the atoms stop slipping, many have different atoms as their nearest neighbors, but the overall crystal structure is still the same. The ease with which copper atoms slip past each other made it easy for prehistoric metalworkers to hammer copper metal into spear points and shields, but it also meant that those objects could easily be damaged in battle.

Substitutional Alloys

About 5500 years ago, people living around the Aegean Sea discovered that mixing molten tin and copper produced bronze, an alloy that was much stronger than either tin or copper alone. Its discovery ushered in the Bronze Age.

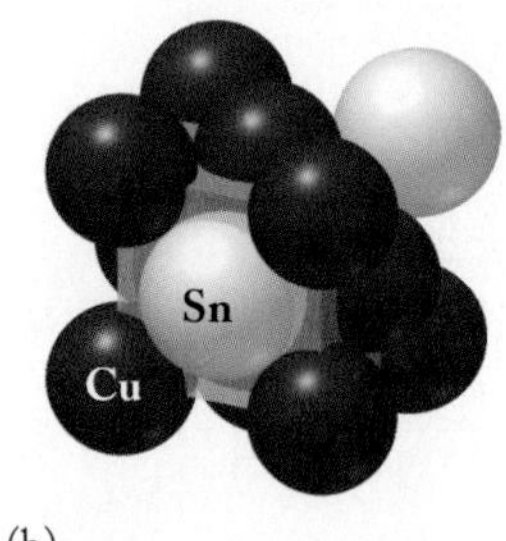

(a) (b)

FIGURE 18.17 Two atomic-scale views of one type of bronze, a substitutional alloy. (a) A layer of close-packed copper (Cu) atoms interspersed with a few atoms of tin (Sn). (b) One possible unit cell for bronze. In this case, tin atoms have replaced one corner Cu atom and one face Cu atom.

Figure 18.17 illustrates how the atoms in one layer of the crystal lattice of bronze might be arranged. The radii of copper and tin atoms are similar—128 pm and 140 pm, respectively. Inserting the slightly larger Sn atoms in the cubic closest-packed Cu crystal lattice disturbs the structure a little, making the planes of copper atoms "bumpy" instead of uniform (Figure 18.18). This atomic-scale roughness makes it more difficult for the copper atoms to slip past one another. Less slippage makes bronze less malleable than copper, but being less malleable also means that bronze is harder and stronger.

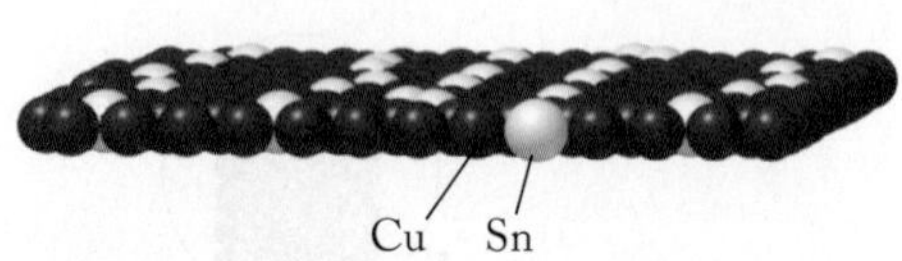

FIGURE 18.18 The larger Sn atoms in bronze disturb the Cu crystal lattice, producing atomic-scale bumps in the slip plane between layers of Cu atoms. These bumps make it more difficult for Cu atoms to slide by each other when an external force is applied.

Like the mixtures discussed in Chapter 1, alloys can be classified according to their composition as homogeneous or heterogeneous mixtures. Bronze is a *homogeneous alloy*, a solid solution in which the atoms of the added element(s) (in this case tin) are randomly but uniformly distributed among the atoms of the host metal (copper). *Heterogeneous alloys* consist of matrices of atoms of host metals interspersed with small "islands" made up of individual atoms of other elements. Alloy compositions may vary over limited ranges. In contrast, *intermetallic compounds* have a reproducible stoichiometry and constant composition (just like chemical compounds) but are still commonly referred to as alloys and are considered a subgroup within homogeneous alloys. An example of an intermetallic compound is Ag_3Sn, which is used in dental fillings. It is a homogeneous mixture of silver and tin atoms in exactly a 3:1 ratio.

CONNECTION The classification of homogeneous and heterogeneous mixtures and the difference between compounds and mixtures were described in Chapter 1.

Alloys are also classified by how the minor elements fit into the crystal structure of the host metals. A **substitutional alloy** is one in which atoms of the minor element(s) replace host atoms in the crystal lattice. Bronze is a *homogeneous, substitutional alloy* in which the tin concentration can be as high as 30% by mass. Substitutional alloys may form between metals that have atomic radii that are within about 15% of each other. It is helpful if both metals have the same crystal lattice as well, but many alloys, such as bronze, form despite having different crystal lattices for the pure metals. Other substitutional alloys include brass (zinc alloyed with copper) and pewter (tin alloyed with copper and antimony). These alloys are stronger than pure copper, and they are more resistant to corrosion.

substitutional alloy an alloy in which atoms of the nonhost element replace host atoms in the crystal lattice.

interstitial alloy an alloy in which the nonhost atoms occupy spaces between atoms of the host.

Most modern substitutional alloys are ferrous alloys, so called because the host metal is iron. The rust-resistant stainless steels, which contain about 10% nickel and up to 20% chromium, are an important class of ferrous alloys. When atoms of Cr on the surface of a piece of stainless steel combine with oxygen, they form a layer of Cr_2O_3 that bonds tightly to the surface and protects the metallic material beneath from further oxidation. This resistance to surface discoloration due to corrosion means that these alloys "stain less" than pure iron.

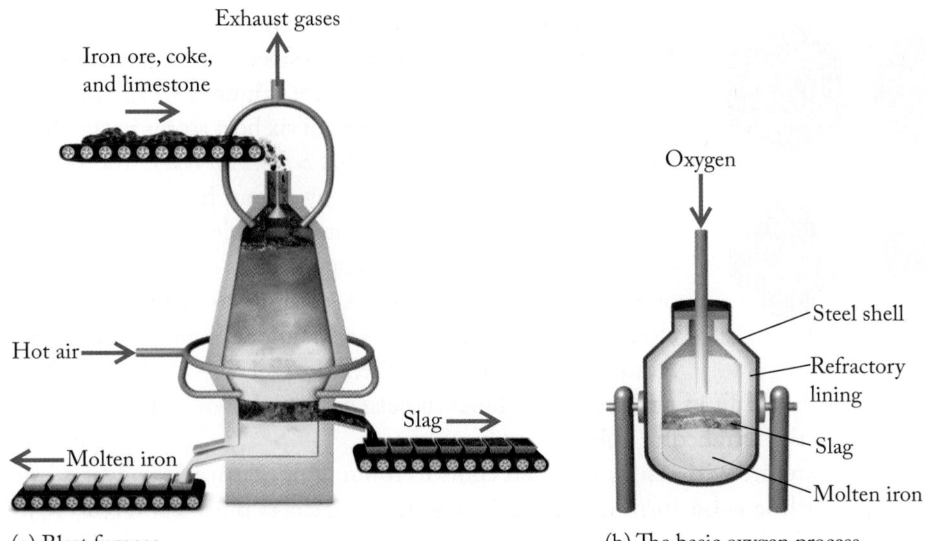

(a) Blast furnace

(b) The basic oxygen process

FIGURE 18.19 (a) Blast furnaces operate continuously at temperatures near 1600°C to convert iron ore into iron. Blasts of hot air inject $O_2(g)$ into the furnace, which converts C to CO. Limestone is added to react with Si and P impurities. The products of these reactions become part of the slag layer. (b) Molten iron from a blast furnace is further purified in a second furnace, where pure $O_2(g)$ is injected instead of air.

Interstitial Alloys

The Bronze Age began to wane about 3000 years ago when metalworkers were able to design furnaces that were hot enough to produce molten iron from the reduction of iron oxides. As in the conversion of CuO to Cu (Equation 18.4), these furnaces used carbon monoxide as the reducing agent:

$$Fe_2O_3(s) + 3\,CO(g) \rightarrow 2\,Fe(s) + 3\,CO_2(g) \qquad (18.5)$$

Iron replaced bronze as the metallic material of choice for fabricating tools and weapons because iron ore is much more abundant in Earth's crust and because tools and weapons made of iron and ferrous alloys are much stronger than those made of bronze.

Today, iron ore is reduced in enormous blast furnaces (Figure 18.19) that operate at about 1600°C. Iron ore, hot carbon (*coke*), and limestone are added to the top of the furnace. Solid impurities, called *slag*, float on top of the molten iron, which is harvested from the bottom. Blast furnaces get their name from blasts of hot air that are injected through nozzles near the bottom of the furnace and that suspend the reactants until iron reduction is complete. It may take as long as 8 hours for a batch of reactants to fall to the bottom of a blast furnace. On their way down, O_2 in the hot-air blasts partially oxidizes the coke to carbon monoxide, and the CO reduces the iron in iron ore as described in Equation 18.5.

Limestone ($CaCO_3$) in the reaction mixture decomposes to calcium oxide:

$$CaCO_3(s) \rightarrow CaO(s) + CO_2(g) \qquad (18.6)$$

which reacts with silica impurities in the ore, forming calcium silicate:

$$CaO(s) + SiO_2(s) \rightarrow CaSiO_3(s) \qquad (18.7)$$

Calcium silicate becomes part of the slag that floats on the denser molten iron at the bottom of the furnace.

When the molten iron cools to its melting point of 1538°C, it crystallizes in a body-centered cubic structure before undergoing a phase transition at around 1390°C to austenite, a form of solid iron made up of face-centered cubic unit cells. The spaces, or *holes*, between iron atoms in austenite can accommodate carbon atoms, forming an **interstitial alloy**, so named because the carbon atoms occupy spaces, or *interstices*, between the iron atoms (Figure 18.20).

FIGURE 18.20 Carbon steel is an interstitial alloy of carbon in iron.

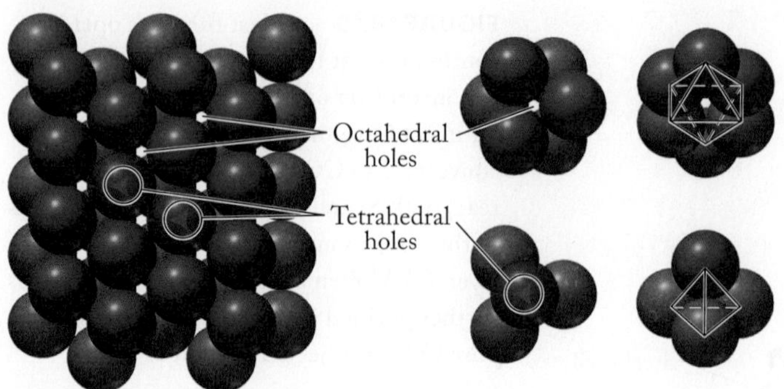

FIGURE 18.21 Close-packed atoms in adjacent layers of a crystal lattice produce octahedral holes surrounded by six host atoms and tetrahedral holes surrounded by four host atoms. Octahedral holes are larger than tetrahedral holes and can accommodate larger nonhost atoms in interstitial alloys. Note that all the atoms are identical here; the colors are only to distinguish one atom from another.

All interstices in a crystal lattice are not equivalent. Indeed, holes of two different sizes occur between the atoms in any closest-packed crystal lattice (Figure 18.21). The larger holes are surrounded by clusters of six host atoms in the shape of an octahedron and are called *octahedral holes.* The smaller holes are located between clusters of four host atoms and are called *tetrahedral holes.* The data in Table 18.4 show which holes are more likely to be occupied based on the relative sizes of minor and host atoms. According to Appendix 3, the atomic radii of C and Fe are 77 and 126 pm, respectively. According to Table 18.4, the ratio $77/126 = 0.61$ means that C atoms should fit in the octahedral holes of austenite, as shown in Figure 18.20, but not in the tetrahedral holes.

As austenite with its fcc unit cell cools to room temperature, it converts into the crystalline solid form of iron called ferrite, which is body-centered cubic. The holes in ferrite are smaller than those in austenite, so many fewer carbon atoms can be accommodated in a ferrite structure than in an austenite structure. The carbon that cannot be accommodated precipitates as clusters of carbon atoms, or it reacts with iron to form iron carbide, Fe_3C. The clusters of carbon and Fe_3C disrupt ferrite's crystalline lattice and inhibit the host iron atoms from slipping past each other when a stress is applied. This resistance to slippage, which is much like that experienced by the copper atoms in bronze (Figure 18.18), makes iron–carbon alloys, known as *carbon steel*, much harder and stronger than pure iron. In general, higher carbon concentrations correlate with stronger steel. However, there is a trade-off in this relationship. As Table 18.5 notes, increased strength and hardness come at the cost of increased brittleness.

CONNECTION In Chapter 5 we described how atoms surrounded by four or six bonding pairs of electrons and no lone pairs have tetrahedral or octahedral molecular geometries, respectively.

CONCEPT TEST

In Chapter 3 we learned that atomic radius increases down a group and decreases from left to right across each row of the periodic table of the elements. Which would you predict to have the larger radius ratio, a lithium–aluminum or a lithium–magnesium alloy?

TABLE 18.4 Atomic Radius Ratios and Locations of Nonhost Atoms in Unit Cells of Interstitial Alloys

Unit Cell Type	Hole Type	$r_{nonhost}/r_{host}$ [a]
Face-centered cubic or hexagonal	Tetrahedral	0.22–0.41
Face-centered cubic or hexagonal	Octahedral	0.41–0.73
Simple cubic	Cubic	0.73–1.00

[a]Radius ratios are uncertain because atoms are not truly solid spheres with constant radii.

TABLE 18.5 Effect of Carbon Content on the Properties of Steel

Carbon Content (%)	Designation	Properties	Used to Make
0.05–0.19	Low carbon	Malleable, ductile	Nails, cables
0.20–0.49	Medium carbon	High strength	Construction girders
0.5–3.0	High carbon	Hard but brittle	Cutting tools

SAMPLE EXERCISE 18.4 Predicting the Crystal Structure of a Two-Element Alloy **LO3**

Sterling silver, which is 93% Ag and 7% Cu by mass, is widely used in jewelry. The presence of Cu inhibits tarnishing and strengthens the alloy. Is this copper–silver alloy a substitutional or an interstitial alloy? Silver and copper metal form face-centered cubic unit cells.

Collect, Organize, and Analyze Atoms of two metals may form a substitutional alloy if their atoms are of similar size (within 15%) and if they form similar crystal structures as pure elements. A host metal with an fcc unit cell may form an interstitial alloy with another element if its atomic radius is 73% or less than that of the host element. According to the data in Appendix 3, the atomic radii of Ag and Cu are 144 and 128 pm, respectively.

Solve The ratio of the atomic radii of Cu to Ag is 128 pm/144 pm = 0.89 = 89%, which tells us that Cu atoms are too big to fit into either the tetrahedral or octahedral holes in the crystal lattice of metallic silver. Thus, an interstitial alloy is impossible, and sterling silver must be a substitutional alloy.

Think About It The atomic radii of Ag and Cu differ by only (144 – 128)/144 = 0.11 or 11%, and the unit cells of both metals are face-centered cubic. Therefore, we expect copper and silver to form a substitutional alloy.

Practice Exercise Would you expect gold (atomic radius 144 pm) to form a substitutional alloy with silver (atomic radius 144 pm)? With copper (atomic radius 128 pm)?

Table 18.4 lists a third type of hole, a *cubic hole*, which accommodates atoms as big as host atoms. What is a cubic hole? It is the space between the eight corner spheres of a simple cubic unit cell. The only metal in Figure 18.10 with a simple cubic unit cell is the radioactive metal polonium and little is known of its tendencies to form alloys.

Let's take another look at chromium-containing stainless steel alloys. The carbon atoms in steel fit in octahedral holes, but Cr atoms (r = 128 pm) substitute for Fe (r = 126 pm) in the bcc (ferrite) structure of iron. The fact that chromium also crystallizes with a bcc unit cell favors a substitutional alloy. While metallurgists do not think of chromium steel in this way, another way to view the crystal structures of these alloys is that they consist of simple cubic unit cells of Fe atoms with either an Fe or Cr atom in the cubic hole at the center of each cell.

Biomedical Alloys

Every year thousands of patients suffering from clogged arteries undergo a procedure called balloon angioplasty. It involves inserting a balloon and a small metal support, called a *stent*, into a patient's artery to increase its diameter and to keep it that way after the balloon is removed. Increasingly, stents are manufactured from an alloy—actually an intermetallic compound with the formula NiTi—called nitinol. Nitinol is particularly useful because objects made of NiTi change their crystal structure and macroscopic shape with changing temperature. As shown in Figure 18.22, a tube made of woven nitinol wire can be stretched, thereby shrinking the tube's diameter. If the temperature of the tube is less than 20°C, it retains this shape. However, if the tube is warmed to 37°C, or body temperature, the alloy undergoes a transition in its crystal structure that causes it to assume its original shape. Thus, a small-diameter, stretched-out stent can be inserted into an

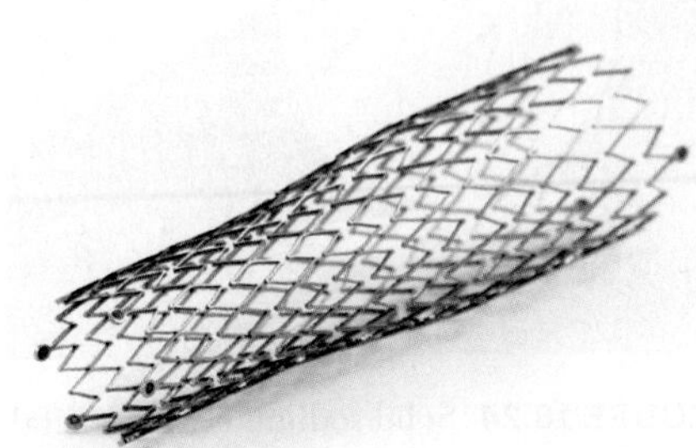

FIGURE 18.22 Shape memory alloys are used in stents for heart patients. This figure illustrates how a tube of woven NiTi wire can be stretched out at low temperatures so that it can be inserted into an artery. At body temperature the stent assumes its original, larger-diameter shape and keeps the artery open.

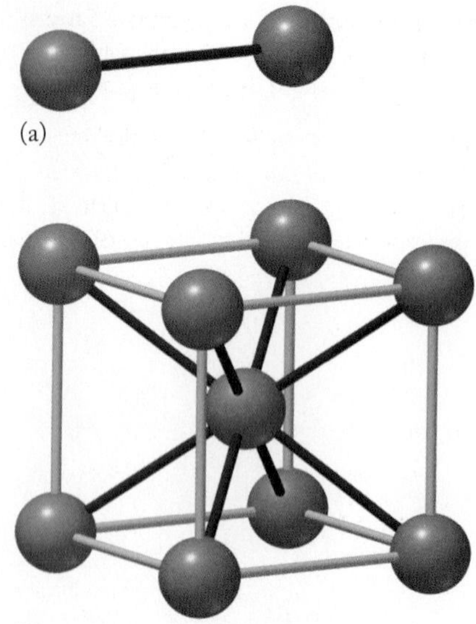

FIGURE 18.23 Covalent bonds differ from metallic bonds. (a) This ball-and-stick model of the molecule Na_2 is based on the two atoms sharing their 3*s* electrons and forming a covalent bond. (b) The atoms in solid sodium metal are actually arranged in a body-centered cubic structure in which each atom is bonded to eight others. As a result, the bonds in sodium and other metals are much more diffuse than the covalent bonds in small molecules.

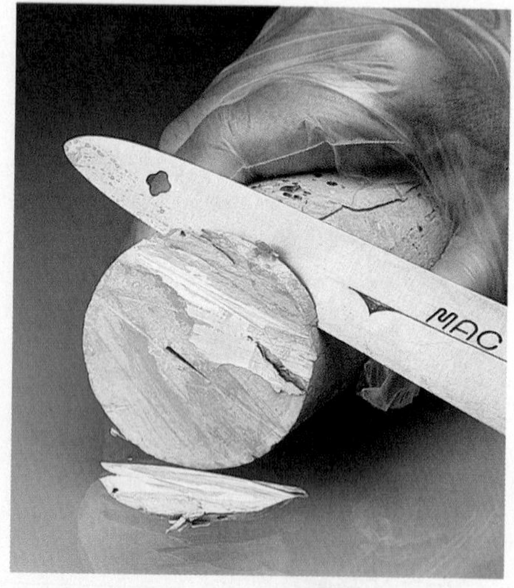

FIGURE 18.24 Solid sodium is a soft metal that can be cut with a knife because the metallic bonds between sodium atoms are weak.

band theory an extension of molecular orbital theory that describes bonding in solids.

valence band a band of orbitals that are filled or partially filled by valence electrons.

artery where the temperature of the body (aided by an inflating balloon) allows the stent to reacquire and then maintain its original larger diameter.

Alloys with good strength and corrosion resistance are essential for the manufacture of surgical tools and biomedical implants. Steel containing 13–16.5% chromium and 0.4–0.6% carbon by mass is appropriate for applications requiring "surgical steel." Dentists use chromium- and nickel-containing alloys in orthodontia.

18.4 Metallic Bonds and Conduction Bands

In the previous section we explored the malleability (the ability to be shaped) of metals and how this property can be modified by the formation of alloys. One of the reasons metals are malleable and ductile (the ability to be drawn out) is that the bonds between atoms in a solid are weak. Most metals are also able to conduct electricity. In this section we explore why they have these properties by examining metals at the atomic level and by exploring models describing the bonds that hold metal atoms together.

According to valence bond theory, a covalent bond forms between two atoms when partially filled atomic orbitals—one from each atom—overlap. Our focus in Chapter 5 was on covalent bonding in gas-phase molecules. In this section we explore the bonds that form between the densely packed atoms in metallic solids. Dense packing means that the valence orbitals of atoms overlap with orbitals of many nearby atoms. This large number of interactions makes metals strong. At the same time, sharing a limited number of valence electrons with many bonding partners makes the bond linking any two metal atoms relatively weak.

To understand this point, consider the bond that would form if two sodium atoms bonded together in a molecule of Na_2. As we learned in Chapter 3, sodium has the electron configuration [Ne]$3s^1$. When the partially filled 3*s* orbitals of the two Na atoms overlap, the result should be a diatomic molecule held together by a single covalent bond (Figure 18.23a). However, the Na atoms in solid sodium adopt a body-centered cubic structure (Figure 18.23b), where each Na atom is surrounded by and bonded to eight other Na atoms, each in turn bonded to eight of its neighbors. This means that each Na atom must share its 3*s* electron with eight other atoms, not just one. Inevitably, this dispersion of bonding electrons weakens the Na—Na bond between each pair of atoms. The weakness of these bonds contributes to the unusual softness of sodium metal: you can, literally, cut it with a knife (Figure 18.24).

In Chapter 4 we described metal atoms "floating" in seas of mobile bonding electrons, where the electrons are shared by all of the nuclei in the sample. The diffuse nature of metallic bonding described in the preceding paragraph certainly fits the sea-of-electrons model, but a more sophisticated approach, called **band theory**, better explains the bonding in metals and other solids. Let's apply band theory, which is an extension of molecular orbital theory, to explain the bonding between atoms in sodium metal. When the 3*s* atomic orbitals on two Na(*g*) atoms overlap to form Na_2, the atomic orbitals combine to form two molecular orbitals with different energies above and below the initial value (Figure 18.25). This is analogous to the formation of lower-energy bonding and higher-energy antibonding molecular orbitals (Section 5.7). If another two Na atoms join the first two to form a molecule of Na_4, the 3*s* atomic orbitals of four Na atoms combine to form four molecular orbitals. If we add another four atoms to make Na_8, a total of eight sodium 3*s* atomic orbitals combine to form eight molecular orbitals. In all these molecules the lower energy orbitals are filled with the available 3*s* electrons and the upper orbitals are empty, as shown in Figure 18.25. If we apply this model to the enormous number of atoms in a piece of solid Na, an equally enormous number of molecular orbitals are created. The lower energy half of them are occupied by electrons, whereas the higher energy half are empty. There are so many of these orbitals that they form a continuous *band* of energies with no gap between the occupied lower half and the empty upper half. Because this band of molecular orbitals was formed by combining valence-shell orbitals, it is called a **valence band**. This proximity between the energy of the occupied lower portion of the valence band and the empty upper portion means valence electrons can move easily from the filled lower portion to the empty upper portion, where they are free to move from one empty orbital to the next and flow throughout the solid.

conduction band an unoccupied band higher in energy than a valence band in which electrons are free to migrate.

conductor a material with partially filled valence bands or filled valence bands that overlap with empty conduction bands, leading to highly mobile electrons.

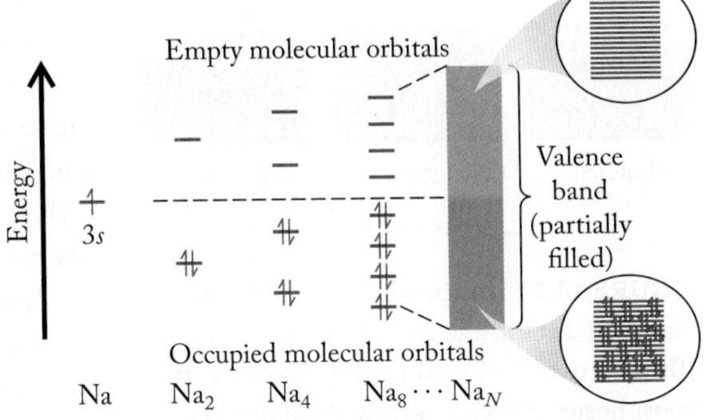

FIGURE 18.25 The half-filled 3*s* atomic orbitals of increasing numbers of sodium atoms combine to form molecular orbitals. As more molecular orbitals form, their energies get closer together until a continuous energy band is formed—a valence band that is half-filled with electrons. The electrons can move from the filled half (purple) to the slightly higher energy upper half (orange), where they are free to migrate throughout the entire solid.

The mobility of the electrons in partially filled valence bands explains the conductivity of many metals, though not all. Consider, for example, the excellent conductivity of zinc. Its electron configuration, $[Ar]3d^{10}4s^2$, tells us that all of its valence-shell electrons reside in filled orbitals, which means the valence band in solid zinc should be filled (Figure 18.26). With no empty space in the valence band to accommodate additional electrons, it might seem that the valence electrons in Zn would be immobile. They are actually quite mobile, however, and band theory explains why. The theory assumes that *all* atomic orbitals of comparable shape and energy, including the empty 4*p* orbitals in zinc, combine to form additional energy bands. The energy band produced by combining empty 4*p* orbitals, called a **conduction band**, is also empty and is broad enough to overlap the valence band. This overlap means that electrons from the valence band can move to the conduction band, where they are free to migrate from atom to atom in solid zinc, thereby conducting electricity.

Before ending this discussion, you need to understand that the two views of valence bands provided in Figures 18.25 and 18.26 are not mutually exclusive. The overlap of filled valence and empty conduction bands in Zn that is shown in Figure 18.26 can also be viewed as a partially filled valence band (as in Figure 18.25) if we assume that molecular orbitals can form from the mixing of the filled *s* and empty *p* orbitals of many Zn atoms. The take-home message is that either model supports the same conclusion: the valence electrons in zinc metal are quite mobile, making it a good conductor. We can also define any material with a partially filled valence band or a filled valence band that overlaps with an empty conduction band as an electrical **conductor**.

CONNECTION In Chapter 5 we discussed the valence bond theory of chemical bond formation as well as the combination of atomic orbitals into molecular orbitals.

CONCEPT TEST

Use band theory to explain the electrical conductivity of magnesium metal.

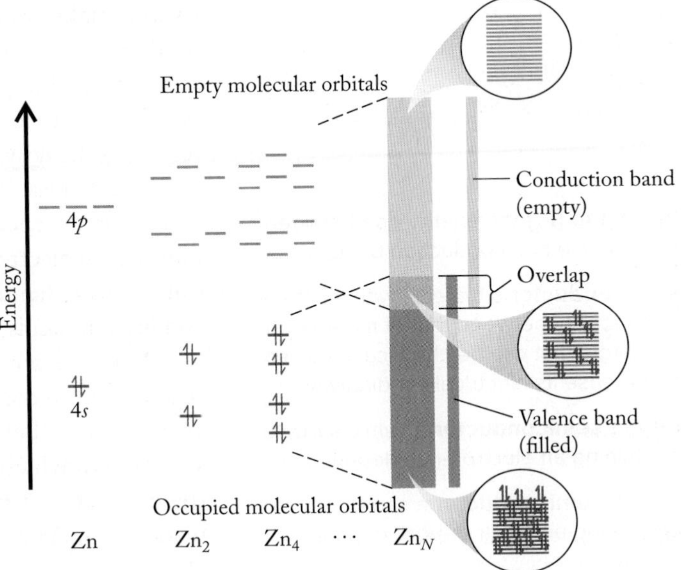

FIGURE 18.26 As the filled 4*s* atomic orbitals of an increasing number of Zn atoms overlap, they form a filled valence band (purple). An empty conduction band (gray) is produced by combining the empty 4*p* orbitals. The valence and conduction bands overlap each other, and electrons move easily from the filled valence band to the empty conduction band.

18.5 Semiconductors

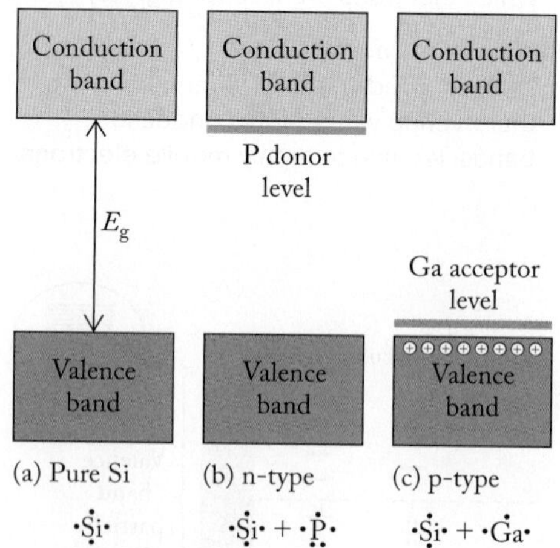

·S̈i· ·S̈i· + ·P̈· ·S̈i· + ·Ġa·

FIGURE 18.27 The band gaps of different materials. (a) Pure Si has a band gap E_g of 106 kJ/mol at 25°C. (b) Doping silicon with phosphorus creates a narrow P donor level just below the Si conduction band and an n-type semiconductor. (c) Doping silicon with gallium creates a narrow Ga acceptor level just above the Si valence electrons and a p-type semiconductor. Valence electrons move into the acceptor level, leaving behind positively charged holes (⊕) in the valence band that enhance conductivity.

To the right of the metals in the periodic table there is a "staircase" of elements—the metalloids—that tend to have the physical properties of metals and the chemical properties of nonmetals. The metalloids are not as good at conducting electricity as metals, but they are better at it than nonmetals. We can use band theory to explain this intermediate behavior. In metalloids, conduction and valence bands do not overlap but instead are separated by an energy gap. In silicon, the most abundant metalloid, band theory predicts an energy gap, or **band gap (E_g)**, of 106 kJ/mol at 25°C (Figure 18.27a).

Generally, only a few valence-band electrons in Si have sufficient energy to move to the conduction band, which limits silicon's ability to conduct electricity and makes it a **semiconductor**. However, we can enhance the conductivity of solid Si, or of any other elemental metalloid, by replacing some of the Si atoms with atoms of an element of similar atomic radius but with a different number of valence electrons. The replacement process is called *doping*, and the added element is called a *dopant*. In effect, doped semiconductors represent substitutional alloys.

Suppose the dopant is a group 15 element such as phosphorus. Each P atom has one more valence electron than the atom of silicon it replaces (the Lewis symbols of Si and P are compared in Figure 18.27). The energy of these additional electrons is different from the energy of the silicon electrons. They populate a narrow band labeled the P *donor level* (Figure 18.27b) that is only about 4 kJ/mol below the Si conduction band. This small energy difference means that the donor electrons can easily reach the Si conduction band, resulting in enhanced conductivity. Phosphorus-doped silicon is an example of an **n-type semiconductor**, so called because the dopant donates **n**egative charges (electrons) to the structure of the host element.

The conductivity of solid silicon can also be enhanced by replacing some Si atoms with atoms of a group 13 element, such as gallium (Figure 18.27c). Ga atoms have one fewer valence electrons than Si atoms, so substituting Ga in the Si structure means fewer valence electrons in the solid. The result is the creation of a narrow Ga *acceptor level* located about 7 kJ/mol above the Si valence band. Si valence electrons can move from the valence band into the acceptor level, leaving behind positively charged "holes" represented by the ⊕ symbols in Figure 18.27(c). The presence of the **p**ositive holes and the migration of electrons between them, filling one hole and creating another, enhances conductivity and makes Si doped with Ga a **p-type semiconductor**. The semiconductors used in solid-state electronics are combinations of n-type and p-type. The semiconducting properties of silicon and doped silicon make possible the myriad electronic devices in our lives, including the cell phones and computers we rely on so heavily.

CONNECTION We introduced metalloids in Section 2.3, when we described the structure of the periodic table.

CONNECTION Emission spectra obtained from gas-discharge tubes were discussed in Chapter 3.

To put these band gap values into perspective, an electrical insulator such as NaCl has an enormous band gap (6.8×10^5 kJ/mol) that prevents significant numbers of electrons from moving from its valence band to the conduction band. On the other hand, a metallic conductor such as Zn, with overlapping valence and conduction bands, has essentially no band gap and high electrical conductivity.

Doping is not the only way to change the conductivity of metalloids. Alloys prepared from combinations of group 13 and group 15 elements, such as gallium arsenide, also behave as semiconductors. The majority of these alloys adopt a structure in which the group 13 element occupies half of the tetrahedral holes in the cubic closest-packed arrangement of the group 15 element. Semiconductors made of GaAs have the same average number of valence electrons per atom as

band gap (E_g) the energy gap between the valence and conduction bands.

semiconductor a material with electrical conductivity between that of metals and insulators that can be chemically altered to increase its electrical conductivity.

n-type semiconductor a semiconductor containing an electron-rich dopant.

p-type semiconductor a semiconductor containing an electron-poor dopant.

(a)

(b)

FIGURE 18.28 Light from light-emitting diodes (LEDs) used in (a) indicator lights and (b) lightbulbs is generated by semiconductors made from group 13 and group 15 semiconducting materials.

silicon: (3 + 5)/2 = 4. Their band gaps are larger, however, and they function better at higher frequencies, which makes them better suited to applications like satellite communications and cell phones. Gallium arsenide is also capable of emitting infrared radiation (λ = 874 nm) when connected to an electrical circuit. This emission is used in many devices, including remote control units for televisions and DVD players. When electrical energy is applied to the material, electrons are raised to the conduction band. When they fall back to the valence band, they emit radiation. If aluminum is substituted for gallium in GaAs, the band gap increases, and the wavelength of emitted light decreases. For example, a material with the empirical formula $AlGaAs_2$ emits orange-red light (λ = 620 nm). Many of the multicolored indicator lights in electronic devices, called light-emitting diodes or LEDs (Figure 18.28a), use gallium arsenide, indium phosphide, and other semiconducting materials composed of group 13 and group 15 elements. LED lightbulbs (Figure 18.28b) based on gallium nitride offer higher efficiency and longer lifetimes than incandescent and fluorescent lightbulbs. One way of generating the familiar white light from LEDs is to mix the colors emitted by red, green, and blue LEDs.

Transition metal–containing alloys such as cadmium sulfide or cadmium selenide also behave as semiconductors. In Chapter 3 we described the use of CdS and CdSe quantum dots to produce the colors of ultra high-definition (HD) television sets. In these materials cadmium Cd^{2+} ions occupy tetrahedral holes in hexagonal closest-packed crystal structures of S^{2-} or Se^{2-} ions.

CONCEPT TEST

Gallium arsenide (GaAs) can be made an n-type or a p-type semiconductor by replacing some of the As atoms with another element. Which element—Se or Sn—would form an n-type semiconductor with GaAs?

18.6 Structures of Some Crystalline Nonmetals

In Figure 18.10, the color key identifies the unit cells of the group 14 elements silicon, germanium, and tin as "diamond." Carbon is a nonmetal (also from group 14), but the arrangement of carbon atoms in diamond (Figure 18.29a) is also found in some metallic materials. In addition, some diamonds may include metal impurities in the crystal lattice that give rise to distinctive colors.

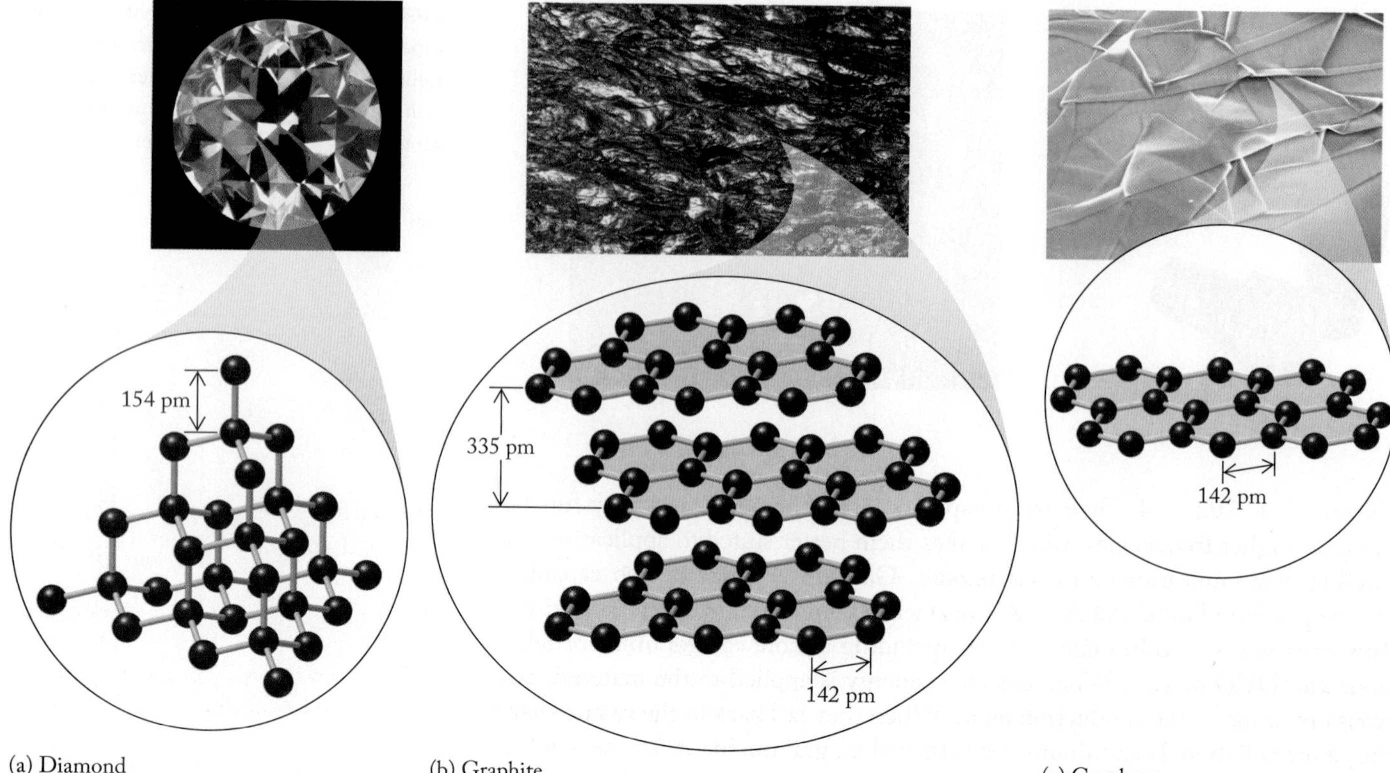

(a) Diamond (b) Graphite (c) Graphene

FIGURE 18.29 Two allotropes of carbon. (a) Diamond is a three-dimensional covalent network solid in which each carbon atom is connected by σ bonds to four adjacent carbon atoms. (b) Graphite is a collection of layers of carbon atoms connected by σ bonds and delocalized π bonds. (c) Graphene consists of single layers of graphitic carbon atoms.

CHEMTOUR Allotropes of Carbon

CONNECTION Allotropes are structurally different forms of the same physical state of an element, as explained in Chapter 4.

covalent network solid a solid consisting of atoms held together by extended arrays of covalent bonds.

Diamond is one of three allotropes of carbon, the other two being graphite (Figure 18.29a) and fullerenes (Figure 18.30a). Diamond is classified as a crystalline **covalent network solid** because it consists of atoms held together in an extended three-dimensional network of covalent bonds. Each carbon atom in diamond forms bonds by overlapping one of its sp^3 orbitals with an sp^3 orbital in each of four neighboring carbon atoms, creating a network of carbon tetrahedra. The atoms in these tetrahedra are connected by localized σ bonds, making diamond a poor electrical conductor. The sigma-bond network is extremely rigid, however, making diamond the hardest natural material known. The atoms of other group 14 elements also form covalent network solids based on the diamond crystal lattice.

Natural diamond forms from graphite under intense heat (>1700 K) and pressure (>50,000 atm) deep in Earth. Industrial diamonds synthesized at high temperatures and pressures from graphite or any other source rich in carbon are used as abrasives and for coating the tips and edges of cutting tools. Because diamond conducts heat better than any metal, cutting tools made from diamond do not overheat due to friction. However, industrial diamonds generally lack the size and optical clarity of gemstones, and laboratories also find it difficult to duplicate the naturally occurring impurities that produce rare and valuable colored diamonds.

By far the most abundant allotrope of carbon is graphite, another covalent network solid. Graphite is a principal ingredient in soot and smoke and is used to make pencils, lubricants, and gunpowder. Graphite contains sheets of carbon atoms connected by overlapping sp^2 orbitals to three neighboring carbon atoms in a two-dimensional covalent network of six-membered rings (Figure 18.29b). Each carbon–carbon σ bond is 142 pm, which is shorter than the C—C σ bond in diamond (154 pm). Overlapping unhybridized p orbitals on the carbon atoms form a network of π bonds that are delocalized across the plane defined by the

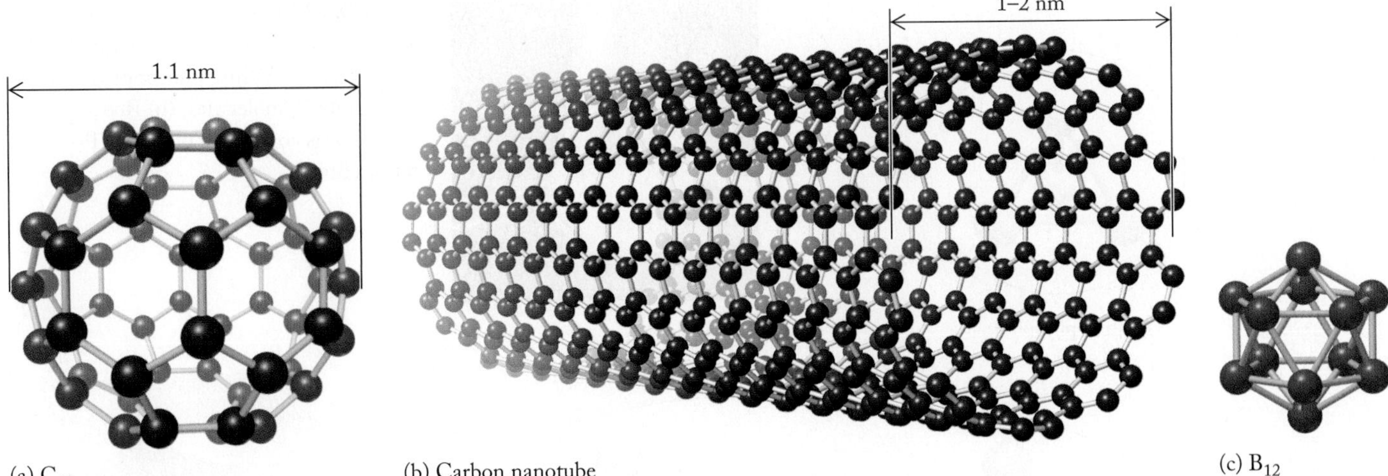

FIGURE 18.30 The atoms in some solids form clusters; structures between covalent network solids and molecular solids. In this category are fullerenes, an allotrope of carbon that includes structures known as (a) buckyballs and (b) nanotubes. (c) A crystalline form of boron consists of clusters of 12 boron atoms.

rings. The mobility of these delocalized electrons makes graphite a good conductor of electricity in this plane. However, graphite is a poor conductor in the direction perpendicular to this plane because the layers of fused rings are 335 pm apart (Figure 18.29b). This distance is much too long to be a covalent bonding distance, which means the sheets are held together only by London dispersion forces and not by shared electrons. On the other hand, the relatively weak interactions between adjacent sheets allow them to slide past each other, making graphite soft, flexible, and a good lubricant.

Another form of carbon is both the simplest and perhaps the most challenging to study: a single sheet of hexagonally bonded carbon atoms (Figure 18.29c). Material with this structure is called *graphene*. Isolating a single layer of graphitic carbon atoms is not difficult, but preserving it as a purely two-dimensional material is. It tends to curl and buckle, forming three-dimensional structures with less surface area. One of the first successful approaches to isolating a single sheet of graphene is elegant in its simplicity: single layers of C atoms can be pulled away from graphite using adhesive tape. Graphene is both the thinnest material known and the strongest—hundreds of times stronger than steel. It conducts heat better than any other known material and conducts electricity better than any metal. It is also nearly transparent, even though its carbon atoms and bonding electrons are so densely packed that atoms of helium gas cannot pass through it.

CONNECTION We introduced London dispersion forces between molecules in Chapter 6.

Yet another allotrope of carbon consists of networks of five- and six-atom carbon rings that form molecules of 60, 70, or more sp^2-hybridized carbon atoms. They look like miniature soccer balls (Figure 18.30a) and are called *fullerenes* because their shape resembles the geodesic domes designed by American architect R. Buckminster Fuller (1895–1983). Many chemists call them *buckyballs* for the same reason. When fullerenes were discovered in the 1980s, they were believed to be a form of carbon rarely found in nature. In recent years, however, analyses of soot and emission spectra from giant stars have shown that fullerenes are present in trace amounts throughout the universe.

These fullerenes are too small to be classified as covalent network solids but too large to be molecular solids (discussed below). They fall in an ambiguous zone between small molecules and large networks called *clusters*. However, some fullerenes stretch the meaning of the word *cluster*. They include structures known as nanotubes (Figure 18.30b), so named because they are 1–2 nanometers in diameter, or about the diameter of buckyballs, but nanotubes have been fabricated that are nearly 1 meter long. Like graphene, carbon nanotubes exhibit semiconductor properties. They also exhibit extraordinary strength and stiffness.

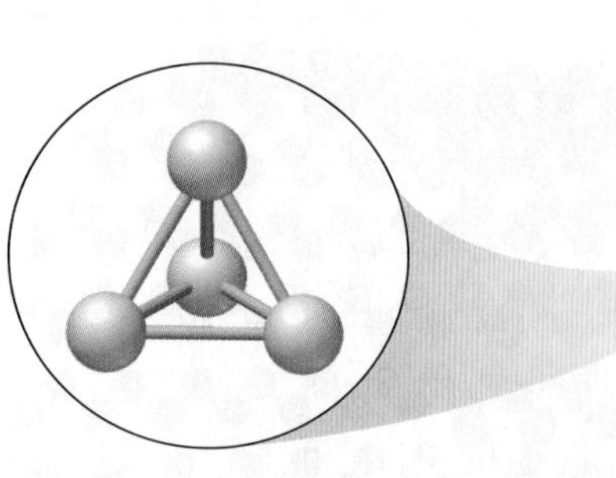
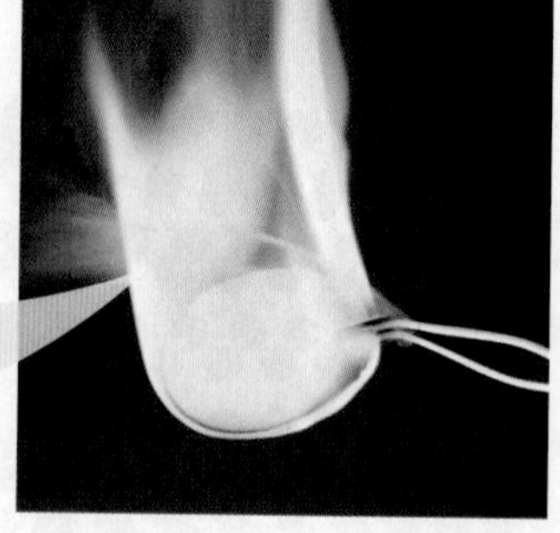

(a) White phosphorus

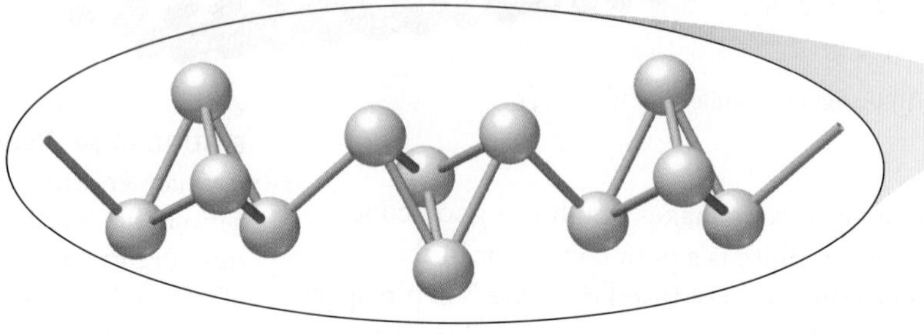

(b) Red phosphorus

FIGURE 18.31 The two most common allotropes of phosphorus are based on P_4 tetrahedra. (a) White phosphorus consists of discrete P_4 molecules. (b) Red phosphorus consists of chains of P_4 tetrahedra.

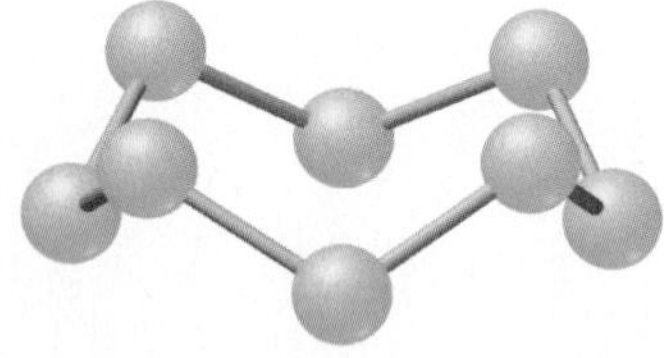

FIGURE 18.32 One form of sulfur is a molecular solid based on puckered S_8 rings.

molecular solid a solid formed by neutral, covalently bonded molecules held together by intermolecular attractive forces.

Some nonmetals, such as boron, also form clusters. One form of boron contains closest-packed arrays of 12-vertex, 20-sided *icosahedra* (singular *icosahedron*; Figure 18.30c) composed of 12 boron atoms.

Crystalline **molecular solids** consist of molecules held together by intermolecular forces. Ice, CO_2, glucose, and most organic molecules crystallize as molecular solids. One of the two most common allotropes of phosphorus, white phosphorus, is a molecular solid consisting of P_4 tetrahedra arranged in a cubic array (Figure 18.31a). White phosphorus is a waxy, soft material that can be cut with a knife. Because it burns in air, it is commonly stored in water. It gives off a yellow-green light in a phenomenon called *phosphorescence*.

The other common phosphorus allotrope, red phosphorus, is not a molecular solid but rather a covalent network solid made of chains of P_4 tetrahedra connected by covalent phosphorus–phosphorus bonds (Figure 18.31b). Both red phosphorus and white phosphorus melt to give the same liquid consisting of symmetrical P_4 tetrahedral molecules.

Sulfur has more allotropic forms than any other element. Most are molecular solids. This variety of forms exists because sulfur atoms form cyclic (ring) molecules of different sizes, which means that different crystalline arrangements of the molecules are possible. The most common allotropes of sulfur consist of puckered rings (Figure 18.32) containing eight covalently bonded sulfur atoms. London dispersion forces hold one ring to another in solid sulfur in staggered stacking patterns. The weakness of these interactions is the reason elemental sulfur is soft and melts at only 115.21°C.

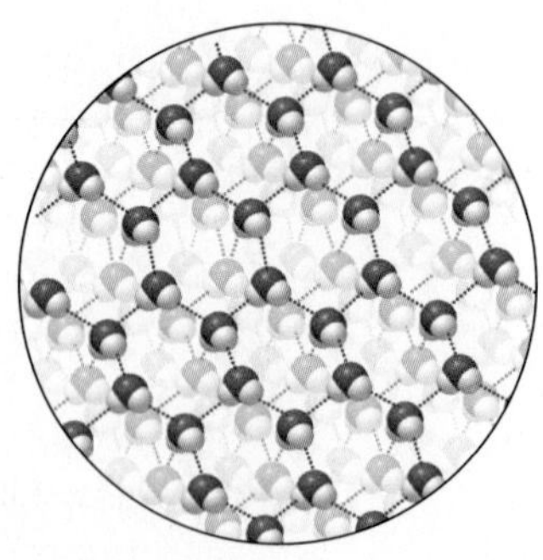

FIGURE 18.33 Hydrogen bonding in ice.

CONCEPT TEST

The crystal structure of ice is shown in Figure 18.33. Is ice better described as a molecular solid or a network solid?

18.7 Salt Crystals: Ionic Solids

Most of Earth's crust is composed of ionic solids consisting of monatomic or polyatomic ions held together by ionic bonds. Most of these solids are crystalline. The simplest crystal structures are those of binary salts, such as NaCl (Figure 18.34a). The cubic shape of large NaCl crystals is a reflection of the cubic shape of the NaCl unit cell.

The unit cell of NaCl (Figure 18.34b) is a face-centered cubic arrangement of Cl^- ions at the corners and in the center of each face with the smaller Na^+ ions occupying the 12 octahedral holes along the edges of the unit cell and the single octahedral hole in the middle of the cell. The Na^+ ions fit into the octahedral holes because the radius ratio of Na^+ to Cl^- is

$$\frac{r_+}{r_-} = \frac{102 \text{ pm}}{181 \text{ pm}} = 0.564$$

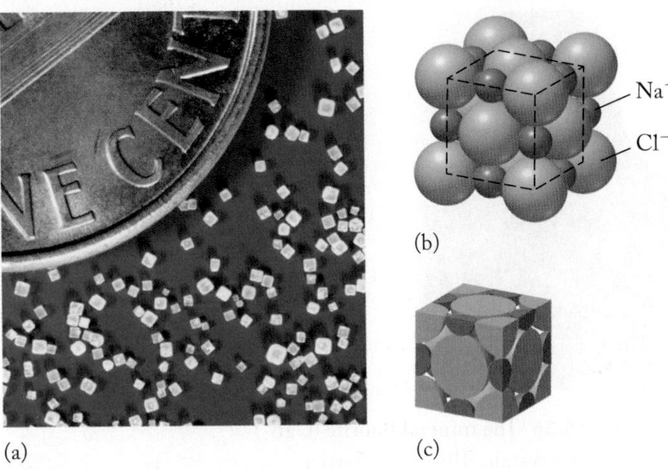

FIGURE 18.34 (a) Sodium chloride forms cubic crystals. (b) The NaCl crystal lattice is composed of fcc unit cells made of Cl^- ions with Na^+ ions in octahedral holes. (c) Cutaway view of a unit cell.

This ratio is too large for Na^+ to occupy a tetrahedral hole (see Table 18.4) but well within the range for occupying an octahedral hole.

Let's take an inventory of the ions in a unit cell of NaCl. Like the metal atoms in the fcc unit cell in Figure 18.11(b), an isolated unit cell contains portions of 14 Cl^- ions: one at each corner and one in each of the six faces of the cube. As in the analysis of Figure 18.11, accounting for partial contributions gives us a total of four Cl^- ions in the unit cell. To count the Na^+ ions, note that one Na^+ ion fits into each of the 12 octahedral holes along the edges of the cell. Because each Na^+ ion along an edge is shared by four unit cells, only one-fourth of each Na^+ ion on an edge is in each cell (see Figure 18.12d) for a total of three Na^+ ions. Where is the fourth Na^+ ion needed to match the empirical formula of NaCl? There is also an octahedral hole in the middle of the unit cell that is occupied by a Na^+ ion. The Na^+ ion in the center belongs completely to the unit cell. Therefore, the total number of Na^+ ions in the unit cell is

$$\left(12 \times \frac{1}{4}\right) + 1 = 4 \text{ Na}^+ \text{ ions}$$

The ratio of Na^+ to Cl^- ions in the unit cell is therefore 4:4, or 1:1, consistent with the chemical formula NaCl. Because the four Na^+ ions occupy all the octahedral holes in the unit cell, this calculation also tells us that each fcc unit cell contains the equivalent of four octahedral holes.

Note in Figure 18.34 that adjacent Cl^- ions along any face diagonal do not touch each other the way adjacent face-diagonal metal atoms do in Figure 18.11(b), because the Cl^- ions have to spread out a little to accommodate the Na^+ ions in the octahedral holes. Sodium and chloride ions touch along each edge of the unit cell, however, which means that each Na^+ ion touches six Cl^- ions and each Cl^- ion touches six Na^+ ions. This arrangement of positive and negative ions is common enough among binary ionic compounds to be assigned its own name: the *rock salt structure.*

In other binary ionic solids, the smaller ion is small enough to fit into the tetrahedral holes formed by the larger ions. For example, in the unit cell of the mineral sphalerite (zinc sulfide) the S^{2-} anions (ionic radius 184 pm) are arranged in an fcc unit cell (Figure 18.35), and half of the eight tetrahedral holes inside the cell are occupied by Zn^{2+} cations (74 pm). Therefore, the unit cell contains four Zn^{2+} ions that balance the charges on the four S^{2-} ions. This pattern of half-filled tetrahedral

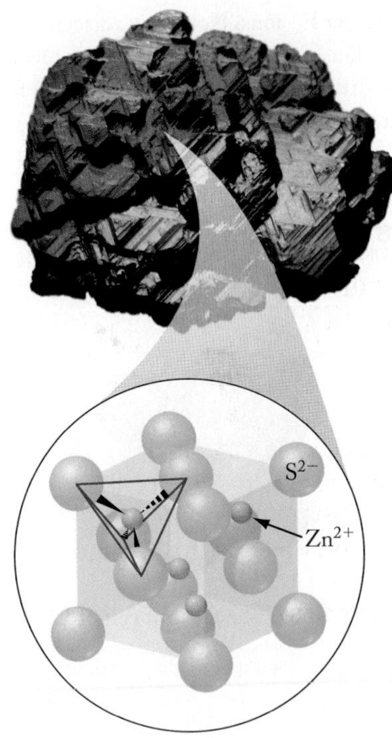

FIGURE 18.35 Many crystals of the mineral sphalerite (ZnS), like the largest one in this photograph, have a tetrahedral shape. The crystal lattice of sphalerite is based on an fcc unit cell of S^{2-} ions with Zn^{2+} ions in four of the eight tetrahedral holes. In the expanded view of the sphalerite unit cell, each Zn^{2+} ion is in a tetrahedral hole, such as the one outlined in red, which is formed by one corner S^{2-} ion and three face-centered S^{2-} ions.

FIGURE 18.36 The mineral fluorite (CaF_2) forms cubic crystals. The crystal lattice of CaF_2 is based on an fcc array of Ca^{2+} ions, with F^- ions occupying all eight tetrahedral holes. Because they are bigger than Ca^{2+} ions, the F^- ions do not fit in the tetrahedral holes of a cubic closest-packed array of Ca^{2+} ions. Instead, the Ca^{2+} ions, while maintaining an fcc unit cell arrangement, spread out to accommodate the larger F^- ions. Note how adjacent Ca^{2+} ions along any face diagonal do not touch each other the way they do in the ideal fcc unit cell in Figure 18.11(b).

holes in an fcc unit cell is called the *sphalerite structure*. This is also the unit cell for most of the group 13–group 15 semiconductors described in Section 18.5.

The crystal structure of the mineral fluorite (CaF_2) is based on an fcc unit cell of smaller Ca^{2+} ions at the cube's eight corners and six face centers, with all eight tetrahedral holes filled by larger F^- ions. Because there is a total of four Ca^{2+} ions in the unit cell and the eight F^- ions are all completely inside the cell, this arrangement satisfies the 1:2 mole ratio of Ca^{2+} ions to F^- ions. This structure is so common that it too has its own name: the *fluorite structure* (Figure 18.36). Other compounds having this structure are SrF_2, $BaCl_2$, and PbF_2.

Some compounds in which the cation-to-anion mole ratio is 2:1 have an *antifluorite structure.* In the crystal lattices of these compounds, which include Li_2O and K_2S, the smaller cations occupy the tetrahedral holes in an fcc unit cell formed by cubic closest-packing of the larger anions.

CONNECTION Periodic trends in atomic radii were discussed in Chapter 3.

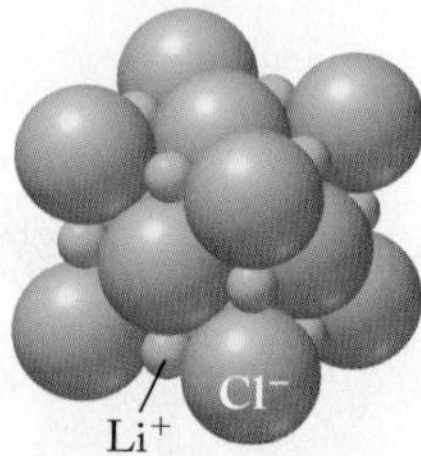

FIGURE 18.37 The fcc unit cell of LiCl.

SAMPLE EXERCISE 18.5 Calculating Ionic Radii from Unit Cell Dimensions LO1

The unit cell of lithium chloride (LiCl) contains an fcc arrangement of Cl^- ions (Figure 18.37). In LiCl, the Li^+ cations (radius 76 pm) are small enough to allow adjacent Cl^- ions to touch along any face diagonal.

a. If the edge length of the LiCl fcc cell is 513 pm, what is the radius of the Cl^- ion?
b. Based on the value calculated for the radius of the Cl^- ion, what type of hole does the Li^+ ion occupy?

Collect and Organize We are given the edge length of a LiCl unit cell, the radius of the Li^+ cation, and a picture of the unit cell. The unit cell is an fcc array of Cl^- ions that touch along the face diagonal.

Analyze The face diagonal of an fcc unit cell is equal to four times the radius of the spheres (chloride ions) that make up the cell, and it's also equal to $\sqrt{2}$ times the cell length (513 pm):

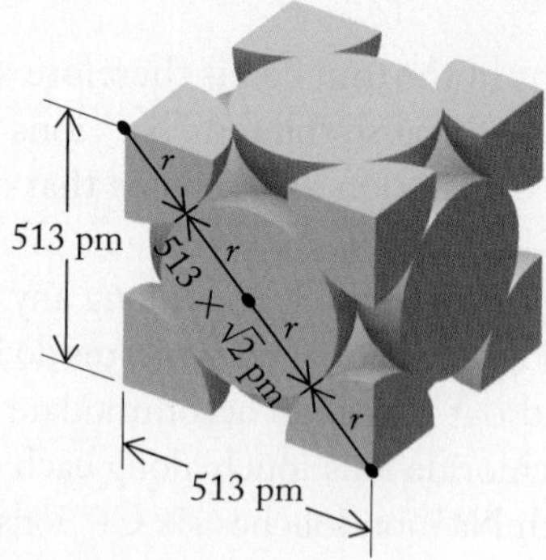

The radius ratio of Li^+ to Cl^- ions and Table 18.4 allow us to predict which type of hole Li^+ occupies. We expect Li^+ ions to be smaller than Cl^- ions, but they may not be small enough to fit in tetrahedral holes.

Solve

a. Substituting the edge length into Equation 18.2, the radius of Cl^- is

$$r = (513 \times \sqrt{2})/4 \text{ pm} = 181 \text{ pm}$$

b. The ratio of the radius of a Li^+ ion to the radius of a Cl^- ion is

$$76\ \cancel{\text{pm}}/181\ \cancel{\text{pm}} = 0.42$$

According to Table 18.4, the Li^+ cations occupy octahedral holes in the lattice formed by the larger Cl^- anions.

Think About It The average ionic radius value in Figure 3.35 for Cl^- ions is 181 pm, so our calculation is correct. Figure 18.37 shows the structure of LiCl to be similar to the rock salt structure of NaCl in Figure 18.34(b).

Practice Exercise Assuming the Cl^- radius in NaCl is also 181 pm, what is the radius of the Na^+ ion in NaCl if the edge length of the NaCl unit cell is 564 pm but the anions do *not* touch along any face diagonal?

SAMPLE EXERCISE 18.6 Calculating the Density of a Salt from Its Unit Cell Dimensions LO2

What is the density of LiCl if the edge length of its fcc unit cell is 513 pm?

Collect and Organize We are given the edge length of an fcc unit cell of LiCl and are asked to calculate its density. Density is the ratio of mass to volume, and the volume of a cubic cell is the cube of its edge length: $V = \ell^3$.

Analyze As in Sample Exercise 18.2, we assume that the density of the unit cell is the same as the density of the crystalline solid. The fact that LiCl has an fcc unit cell (like NaCl) means that there are four Cl^- ions and four Li^+ ions in the cell. The density of the unit cell is the sum of the masses of these eight ions divided by the volume of the cell, $(513\ \text{pm})^3$. As in Sample Exercise 18.2, we need to divide the molar masses of Li and Cl by Avogadro's number to obtain the masses of individual atoms (or monatomic ions) of Li and Cl in grams.

Solve Calculating the mass of four Cl^- ions,

$$m = \frac{35.45\ \text{g Cl}^-}{1\ \cancel{\text{mol Cl}^-}} \times \frac{1\ \cancel{\text{mol Cl}^-}}{6.0221 \times 10^{23}\ \cancel{\text{ions Cl}^-}} \times 4\ \cancel{\text{ions Cl}^-} = 2.355 \times 10^{-22}\ \text{g Cl}^-$$

and the mass of four Li^+ ions,

$$m = \frac{6.941\ \text{g Li}^+}{1\ \cancel{\text{mol Li}^+}} \times \frac{1\ \cancel{\text{mol Li}^+}}{6.0221 \times 10^{23}\ \cancel{\text{ions Li}^+}} \times 4\ \cancel{\text{ions Li}^+} = 0.4610 \times 10^{-22}\ \text{g Li}^+$$

Combining the masses of the two kinds of ions in the unit cell,

$$2.355 \times 10^{-22}\ \text{g} + 0.4610 \times 10^{-22}\ \text{g} = 2.816 \times 10^{-22}\ \text{g}$$

The volume of the cell in cubic centimeters is

$$V = \ell^3 = (513\ \cancel{\text{pm}})^3 \times \frac{(10^{-10}\ \text{cm})^3}{(1\ \cancel{\text{pm}})^3} = 1.350 \times 10^{-22}\ \text{cm}^3$$

Taking the ratio of mass to volume, we have

$$d = \frac{m}{V} = \frac{2.816 \times 10^{-22}\ \text{g}}{1.350 \times 10^{-22}\ \text{cm}^3} = 2.09\ \text{g/cm}^3$$

Think About It The result is reasonable because most minerals are more dense than water ($1\ \text{g/cm}^3$) but less dense than common metals ($>4\ \text{g/cm}^3$).

Practice Exercise What is the density of NaCl if the edge length of its fcc unit cell is 564 pm?

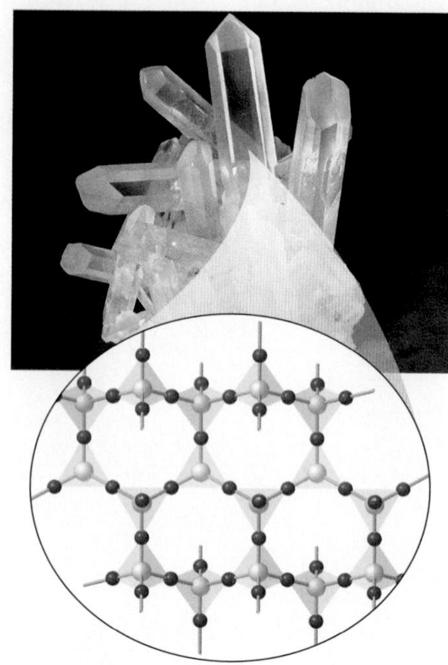

FIGURE 18.38 Quartz crystals are hexagonal. Their crystal structure features hexagonal arrays of silicon–oxygen tetrahedra that form an extended three-dimensional network. (All atoms are drawn undersized relative to the volume they actually occupy to make the structure easier to see.)

Ionic compounds such as NaCl are insulators, not conductors, even though aqueous solutions of NaCl conduct electricity quite nicely. Why are ionic solids insulators? The chloride ion has filled 3*s* and 3*p* atomic orbitals while the 3*s* orbital of Na^+ ions is empty. The crystal structure of NaCl in Figure 18.34 shows that the chloride ions touch along the edges and the face diagonal of the unit cell with the sodium ions also touching chloride ions. These interactions mean that the orbitals of all Cl^- ions form filled *s* and *p* valence bands while the Na^+ orbitals form an empty *s* band. The electronegativity difference between Cl (3.0) and Na (0.9) means that the filled bands from the chloride ion lie at much lower energy (6.8×10^5 kJ/mol lower) than the empty Na^+ band. This enormous band gap explains why NaCl, like most ionic solids, is an insulator.

18.8 Ceramics: Useful, Ancient Materials

The use of *ceramic* materials preceded metal technology by many thousands of years. The first ceramics were probably made of *clay*, the fine-grained soil produced by the physical and chemical weathering of igneous rocks (rocks of volcanic origin). Moist clay is easily molded into a desired shape and then hardened over fires or in wood-burning kilns, an ancient process still in use today.

In this section we examine some of the physical properties of both primitive earthenware (ceramics fired at low temperatures) and modern ceramic materials (typically fired at high temperatures) and relate those properties to the chemical composition of these materials and to the chemical changes that occur when they are heated. Most ceramics behave as electrical insulators.

Polymorphs of Silica

One of the most abundant families of minerals found in igneous rocks has the chemical composition SiO_2. The correct chemical name is silicon dioxide, but the more common name is *silica*. Silica is a covalent network solid in which each silicon atom is covalently bonded to four oxygen atoms, forming a tetrahedron with an oxygen atom at each corner and the silicon atom at the center (Figure 18.38). Each corner oxygen atom is covalently bonded to two silicon atoms, thereby linking the tetrahedra into an extended three-dimensional network. Because each oxygen atom is bonded to two silicon atoms, each silicon atom gets only half "ownership" of the four oxygen atoms to which it is bonded—hence, the formula SiO_2.

At least eight different minerals have the empirical formula SiO_2. The members of a family of substances with the same empirical formula but different crystal structures and properties are called *polymorphs*. The most abundant silica polymorph is quartz, a type of SiO_2 that can form impressively large, nearly transparent crystals (Figure 18.38). Note how the hexagonal ordering of the SiO_2 tetrahedra translates into hexagonal crystals.

Most, but not all, silica polymorphs are crystalline. When lava containing molten SiO_2 flows from a volcano into a sea or lake, it cools so quickly that the Si and O atoms may not have enough time to form an ordered crystal lattice as the lava solidifies. The solid formed in this way is an *amorphous* (disordered, noncrystalline) polymorph of silica known as either volcanic glass or obsidian (Figure 18.39). *Glass* is a term scientists and engineers use to describe any solid that has either no crystalline structure or only very tiny crystals surrounded by disordered arrays of atoms. This definition also applies to laboratory glassware and the drinking glasses we use at home.

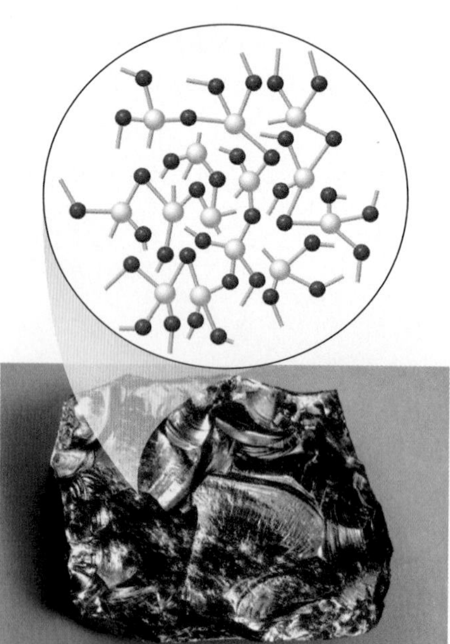

FIGURE 18.39 Obsidian (volcanic glass) is an unusual form of silica in that it is not crystalline. Obsidian contains mostly amorphous silica with random arrangements of silicon and oxygen atoms.

Ionic Silicates

In addition to covalent silica, igneous rocks also contain ionic minerals made of silicon and oxygen. These minerals have some of the tetrahedral crystal structure of silica, but not all the oxygen corner atoms are bonded to two Si atoms. Instead, some of the O atoms have an extra electron. The result is a *silicate* anion. One of the most common ionic silicates is chrysotile, a type of asbestos that consists of sheets of linked silicon–oxygen tetrahedra that form hexagonal clusters of six tetrahedra each (Figure 18.40). Each tetrahedron has three O atoms that it shares with other tetrahedra and one O atom—the one with the extra electron—that it does not share. Thus, the basic tetrahedral unit consists of one Si atom and $[3 + 1 \times (\frac{1}{2})] = 2.5$ oxygen atoms as well as a negative charge. This gives the sheet the empirical formula $SiO_{2.5}^{-}$. We generally use whole-number subscripts in chemical formulas. In this case, multiplying $SiO_{2.5}^{-}$ by 2 gives us the empirical formula $Si_2O_5^{2-}$. The subscript *n* in the formula in Figure 18.40 indicates that there are many empirical formula units in a single crystal.

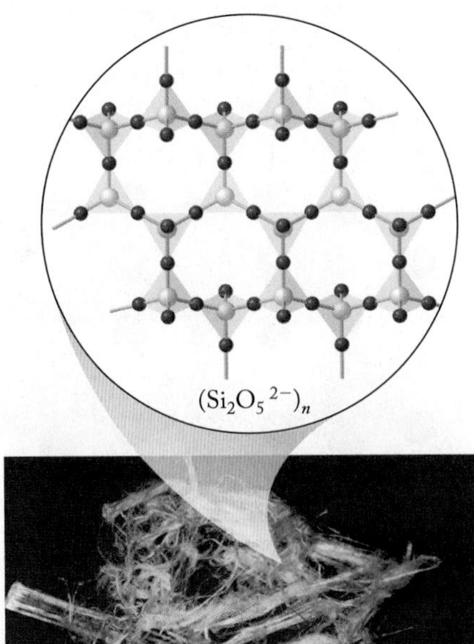

FIGURE 18.40 Chrysotile, one of the two principal forms of asbestos formerly used in building construction as thermal insulation, is an ionic silicate compound. The ease with which thin fibers of chrysotile can flake off is related to its layered crystal lattice and the relatively weak intermolecular interactions between layers. The O atoms that are bonded to only one Si atom have an extra electron and a negative charge.

Silicate minerals are neutral materials, so they must contain cations to balance the negative charges on the $Si_2O_5^{2-}$ layers. When this cation is Al^{3+}, the minerals are called *aluminosilicates*. One of the most common aluminosilicates is the clay mineral kaolinite (Figure 18.41). At least a little kaolinite is found in practically every soil, but rich deposits of nearly pure, brilliantly white kaolinite are found in highly weathered soils. For centuries these deposits have been mined for the clay used to make fine china and white porcelain. Today the greatest demand for kaolinite is in the production of the glossy white paper used in magazines and books (including this one).

The metal ions found in most igneous rocks: Na^+, K^+, Ca^{2+}, Mg^{2+}, and Fe^{3+}, are largely absent in kaolinite. Their absence indicates that kaolinite deposits form under acidic weathering conditions, during which H^+ ions displace other cations from ion-exchange sites. For example, $-O^-Na^+$ sites exchange H^+ for Na^+, leaving behind –OH groups such as those shown in Figure 18.41.

The strong ionic interactions and hydrogen bonds in kaolinite make it hard to separate its layers, and water molecules cannot penetrate between them. For thousands of years this property has made kaolinite pots handy vessels for carrying water. For the same reason, kaolinite does not expand when water is added, nor does it shrink as much as most other clays when dehydrated at high temperatures, which makes it a desirable starting material for ceramics. Finally, moist kaolinite is *plastic*, meaning that it can be molded into a shape, and it keeps that shape during heating and cooling.

CONNECTION Ion-exchange reactions were discussed in Chapter 8.

CONCEPT TEST

Magnesium ion, Mg^{2+}, can substitute for Al^{3+} in kaolinite to form $Mg_xAl_y(Si_2O_5)(OH)_4$. What are the values of *x* and *y* if *x* = *y*?

From Clay to Ceramic

Creating ceramic objects from kaolinite and other clays takes several steps. First, moist clay is formed into pots, bricks, and other objects on a potter's wheel or in molds or presses. Drying at just above 100°C removes much of the water that made the clay plastic. Further heating to about 450°C removes water that was adsorbed onto the surfaces of the clay particles or between the layers of nonkaolinite clays. Above 450°C, kaolinite begins to decompose, releasing molecules of H_2O as some of the aluminum ions that were bonded to octahedra of six oxide ions in kaolinite (Figure 18.41) end

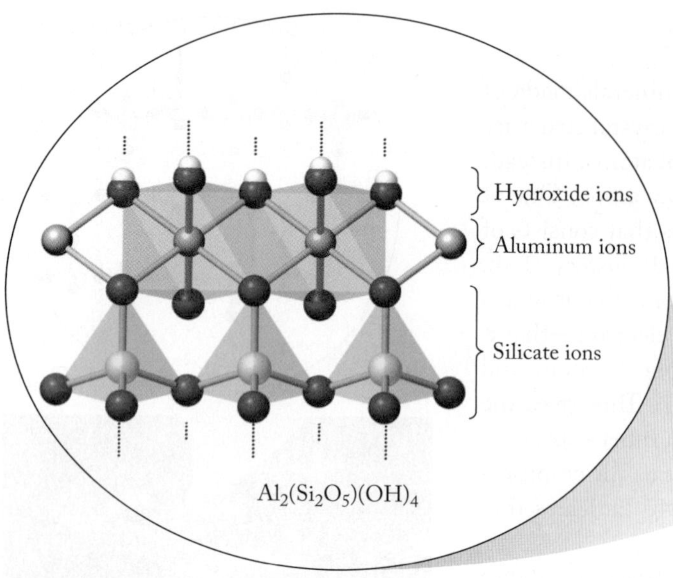

FIGURE 18.41 An edge-on view of the structure of kaolinite shows a top layer of OH^- ions bonded to a middle layer of Al^{3+} ions followed by a bottom layer of silicate ions ($Si_2O_5^{2-}$). The structure is repeated in subsequent layers, and the empirical formula of this crystal lattice is $Al_2(Si_2O_5)(OH)_4$. This enormous kaolinite mine in Bulgaria contributes to a worldwide production of about 40 trillion metric tons of the mineral per year.

up bonded to tetrahedra of only four oxide ions in a substance called metakaolin ($Al_2Si_2O_7$), which has an amorphous structure. The overall decomposition reaction, which is complete at temperatures near 900°C, is described by the equation

$$Al_2Si_2O_5(OH)_4(s) \xrightarrow{450\text{–}900^\circ C} Al_2Si_2O_7(s) + 2\ H_2O(g)$$

The next compositional change occurs just below 1000°C when $Al_2Si_2O_7$ decomposes into a mixture of yet another disordered aluminosilicate, $Al_4Si_3O_{12}$, and $SiO_2(s)$:

$$2\ Al_2Si_2O_7(s) \xrightarrow{\sim 950^\circ C} Al_4Si_3O_{12}(s) + SiO_2(s)$$

At even higher temperatures, $Al_4Si_3O_{12}$ continues forming mixtures of compounds with increasing aluminum content and SiO_2:

$$Al_4Si_3O_{12}(s) \rightarrow Al_4Si_2O_{10}(s) + SiO_2(s)$$

$Al_4Si_2O_{10}$ is a member of a family of ceramic materials called mullite. These materials are actually solid solutions containing different proportions of aluminum and silicon oxides. At the atomic level, their structures are a mixture of layers that contain octahedra of six-coordinate Al^{3+} ions each bonded to six oxide ions, or tetrahedra with either silicon atoms or four-coordinate Al^{3+} ions at their centers (Figure 18.42). Mullite is widely used in the manufacture of products that must tolerate temperatures as high as 1700°C: furnaces, boilers, ladles, and kilns. These products are used as containers of molten metals and in the glass, chemical, and cement industries. Mullite is also very hard and is widely used as an abrasive.

Other ceramics are used in high-temperature applications ranging from cookware to fireplace bricks. Ceramics are well suited to these uses because of their high melting points and because they are good thermal and electrical insulators. For example, the thermal conductivity of aluminum metal at 100°C is over eight times that of alumina (Al_2O_3). Pure alumina is found in nature as the mineral corundum, but it can also be prepared by heating aluminum(III) hydroxide to ~1200°C to form Al_2O_3 and water by the following equation:

$$2\ Al(OH)_3(s) \rightarrow Al_2O_3(s) + 3\ H_2O(g)$$

The crystal structure of alumina consists of a hexagonal closest-packed arrangement of oxide ions (O^{2-}) with Al^{3+} in octahedral holes. To account for the

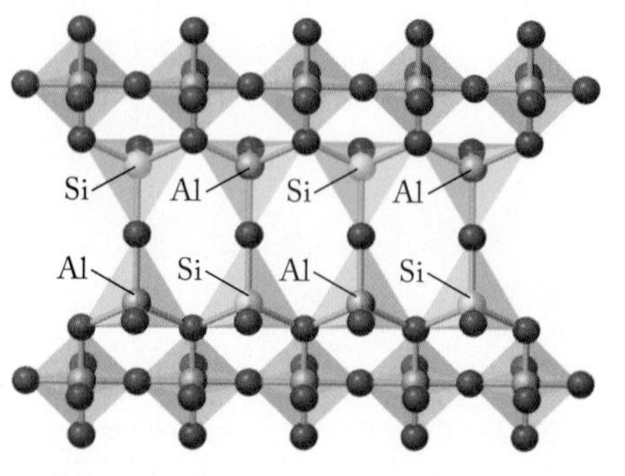

SiO_4 and AlO_4 tetrahedra

AlO_6 octahedra

FIGURE 18.42 One of many possible crystal structures of a family of ceramic materials called mullite.

observed stoichiometry in Al_2O_3, two-thirds of the octahedral holes are occupied. Alumina has myriad uses, including orthodontia, where colorless alumina brackets offer a more attractive option to traditional metal braces. Avoiding chromium- and nickel-containing dental alloys is becoming increasingly important as some patients develop allergic reactions to these metals. Two familiar gemstones: ruby and topaz, are made of alumina in which some Al^{3+} ions are replaced by Cr^{3+} and Fe^{3+} ions, respectively. In sapphires, trace quantities of Fe^{2+} and Ti^{4+} ions substitute for pairs of Al^{3+} ions in the alumina.

X-ray diffraction (XRD) a technique for determining the arrangement of atoms or ions in a crystal by analyzing the pattern that results when X-rays are scattered after bombarding the crystal.

18.9 X-ray Diffraction: How We Know Crystal Structures

Unit cell dimensions are determined using **X-ray diffraction (XRD)**. X-rays are well suited to the task of crystal structure determination because the wavelengths of X-rays (10^2 to 10^3 pm) are similar to the distances between the centers of neighboring atoms and ions in crystals. To see how XRD works, we will consider the atoms in a crystalline metal, but keep in mind that our description of XRD also applies to the particles in any crystalline solid.

Suppose a beam of monochromatic (same wavelength) X-rays is directed at the surface of a single metallic crystal as shown in Figure 18.43. The beam is directed toward the surface at an angle called the angle of incidence, which is represented by the Greek letter θ. Chances are, some of the X-rays will collide with the atoms in the crystal's surface layer and be scattered away from the atoms by these

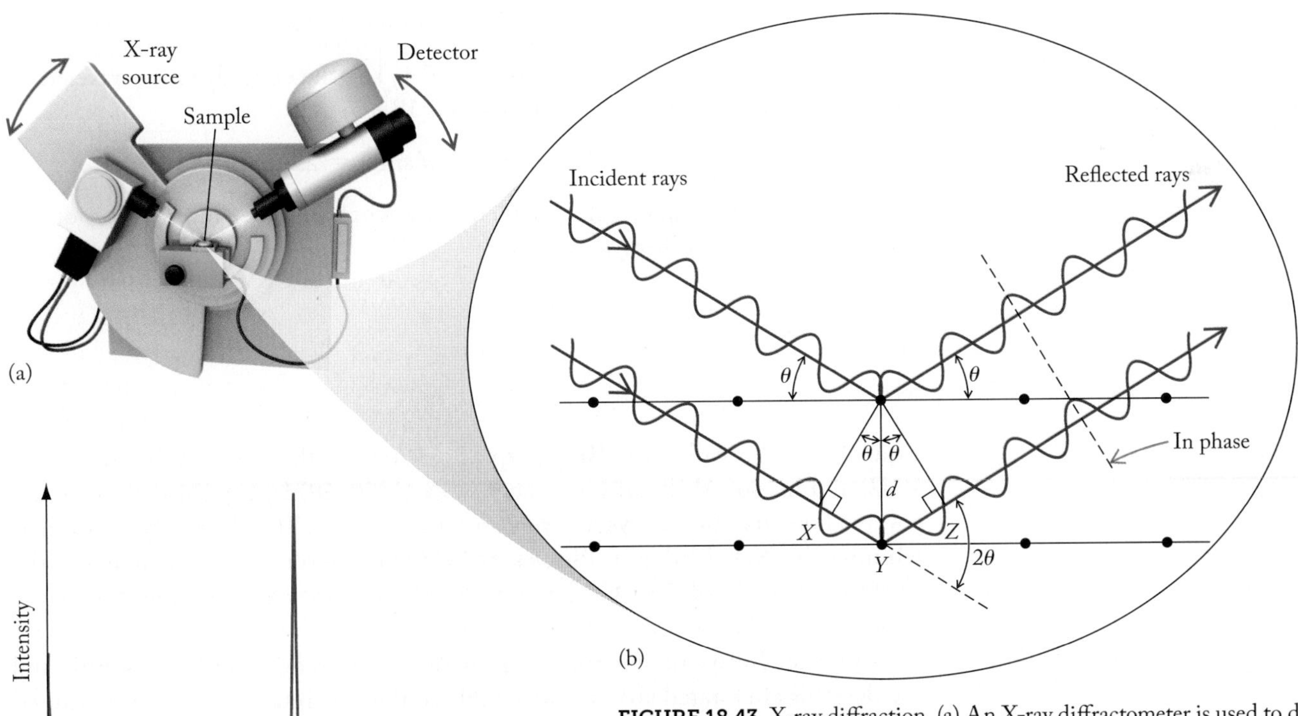

FIGURE 18.43 X-ray diffraction. (a) An X-ray diffractometer is used to determine the crystal lattices of solids. A source of X-rays and a detector are mounted so that they can rotate around the sample. (b) X-rays scattered from the surface layer of atoms and from the next layer interfere constructively when they are in phase, as they are at the distance *d* shown here. The angle between the incident and scattered X-rays is 2θ. (c) Moving the source and detector around the sample produces a scan such as this one for quartz. The peaks at different values of 2θ represent constructive interference among X-rays scattered from different layers of atoms in the solid.

collisions. Other X-rays may pass through the surface layer but then collide with and be scattered by atoms in the second layer as shown in Figure 18.43(b). Now suppose that a detector is mounted in such a way that it can detect those X-rays that have an angle of scatter that is equal to the angle of incidence, θ. These X-rays will have undergone a total change in direction, called their *angle of diffraction*, that is the sum of their angles of incidence and scatter, or 2θ.

Now let's consider the two X-rays in Figure 18.43(b). One is scattered by an atom in the surface layer; the other by an atom in the second layer. Both undergo an angle of diffraction of 2θ. However, they can be detected only if they and many others like them undergo *constructive interference*. This term applies to electromagnetic waves of the same wavelength that are in phase with each other; that is, the crests of the waves are exactly aligned, as are the troughs (bottoms) of the waves. Scattered X-rays that are out of phase with each other undergo *destructive interference* and are not detected.

To be in-phase with an X-ray scattering off a surface atom, the extra distance traveled by an X-ray scattering off a second-layer atom must be some whole-number multiple of the wavelength of the X-rays. This extra distance is the sum of the lengths of line segments $\overline{XY}$ and $\overline{YZ}$ in Figure 18.43(b). The two right triangles incorporating these line segments share a hypotenuse d. Geometry tells us that the angles opposite $\overline{XY}$ and $\overline{YZ}$ are both equal to θ. According to trigonometry, the ratio of either $\overline{XY}$ or $\overline{YZ}$ to d is

$$\frac{\overline{XY}}{d} = \sin\theta \qquad \frac{\overline{YZ}}{d} = \sin\theta$$

which means

$$d\sin\theta = \overline{XY} \qquad d\sin\theta = \overline{YZ} \tag{18.8}$$

As noted above, we are interested in $\overline{XY} + \overline{YZ}$, the extra distance traveled by the second ray. We therefore add Equations 18.8 to get

$$\overline{XY} + \overline{YZ} = d\sin\theta + d\sin\theta = 2d\sin\theta$$

When this extra distance $\overline{XY} + \overline{YZ}$ equals a whole-number multiple (n) of the wavelength (λ) of the X-rays, we have

$$\overline{XY} + \overline{YZ} = n\lambda$$

or

$$n\lambda = 2d\sin\theta \tag{18.9}$$

Equation 18.9 is called the **Bragg equation** after William Henry Bragg (1862–1942) and his son William Lawrence Bragg (1890–1971), Englishmen who discovered how to evaluate crystal structures using X-rays. This discovery led to their winning the Nobel Prize in Physics in 1915. Whenever $2d\sin\theta$ equals $n\lambda$, the crests and troughs of the two X-rays are in phase as they emerge from the metal, in which case they interfere constructively.

To detect X-rays undergoing constructive interference, the X-ray source and the detector are rotated around the sample so that the intensity of these scattered X-rays can be measured as a function of 2θ (the angle of diffraction). Peaks in the intensity of scattered X-rays (Figure 18.43c) occur at angles of diffraction that satisfy Equation 18.9. From these angles, and knowing the wavelength (λ) of the X-rays, scientists can calculate the distance (d) between the layers of particles in a crystalline sample and determine its type of lattice structure. In most X-ray diffraction patterns, the most intense peaks are those for which the value of n is 1.

Bragg equation an equation that relates the angle of diffraction (2θ) of X-rays to the spacing (d) between the layers of ions or atoms in a crystal: $n\lambda = 2d\sin\theta$.

SAMPLE EXERCISE 18.7 Determining Interlayer Distances by X-ray Diffraction **LO6**

An XRD analysis of a sample of copper has a major peak at $2\theta = 24.64°$ and much smaller peaks at $2\theta = 50.54°$ and $79.62°$. What distance d between layers of Cu atoms produces this diffraction pattern if the wavelength of the X-rays is 154 pm?

Collect and Organize We are given a series of diffraction angles 2θ and asked to find the distance between layers of atoms in a sample of copper metal. Equation 18.9 enables us to calculate this distance when we know the angles of diffraction associated with constructive interference in a beam of reflected X-rays.

Analyze We need to rearrange Equation 18.9 to solve for the parameter we are after: d.

$$d = \frac{n\lambda}{2 \sin \theta}$$

We are given the value of λ and can get values of θ from the given values of 2θ, but we need to determine the value(s) of n.

Solve The key to determining n is to look for a pattern in the θ values. Notice that the higher values of θ are approximately two and three times the lowest value:

$$24.64°/2 = 12.32 \qquad 50.54°/2 = 25.27 \qquad 79.62°/2 = 39.81$$

$$\frac{25.27°}{12.32°} = 2.051 \qquad \frac{39.81°}{12.32°} = 3.231$$

This pattern and the intensities of the peaks suggest that the values of n for this set of data are 1, 2, and 3, so let's use these combinations of n and θ to see whether they all give the same value of d:

$$d = \frac{(1)(154 \text{ pm})}{2 \sin 12.32°} = \frac{154 \text{ pm}}{(2)(0.2134)} = 361 \text{ pm}$$

$$d = \frac{(2)(154 \text{ pm})}{2 \sin 25.27°} = \frac{308 \text{ pm}}{(2)(0.4269)} = 361 \text{ pm}$$

$$d = \frac{(3)(154 \text{ pm})}{2 \sin 39.81°} = \frac{462 \text{ pm}}{(2)(0.6402)} = 361 \text{ pm}$$

We do indeed get the same value of d, so $d = 361$ pm is the distance between Cu atoms that produced the three peaks.

Think About It These consistent results mean that our assumption about the values of n for the three values of θ was correct. Moreover, the value of d is in the range of the edge lengths of unit cells we used in several Sample Exercises earlier in the chapter and is therefore also reasonable.

Practice Exercise An X-ray diffraction analysis of crystalline CsCl using X-rays of wavelength 142.4 pm has a prominent peak at $2\theta = 19.9°$. What is the spacing between the ion layers that produced the peak?

As we have seen throughout this text, the arrangement of atoms, ions, and molecules in a sample of matter is responsible for many of the properties of materials. XRD is a powerful tool for analyzing the atomic-level structure of a wide range of materials, including metals, minerals, polymers, plastics, ceramics, pharmaceuticals, and semiconductors. The technique is indispensable for scientific research and industrial production. The link between structure and properties is robust, and understanding the ordered arrangements of atoms and ions that characterize pure metals, alloys, semiconductors, ionic compounds, and ceramics enables us to design stronger, tougher, harder, and more heat-resistant solid materials.

SAMPLE EXERCISE 18.8 Integrating Concepts: Imperfect Crystals and Film Photography

Most of the crystals we have discussed and illustrated in this chapter have been perfect, in the sense that all sites in the crystal structure that should be occupied by an atom or ion *are* occupied. However, many crystals we find in the real world are imperfect, and several types of defects have been recognized as important. Scientists Walter Schottky (1886–1976) and Yakov Frenkel (1894–1952) studied defects known as *point defects* because they occur at specific points (locations) within a crystal structure. A Schottky defect forms when positively and negatively charged ions both leave their normal sites in the structure, creating vacancies at those sites. A Frenkel defect forms when an atom or ion leaves its location in the structure and occupies a site not usually occupied by a particle. Crystals of silver bromide (AgBr), having the rock salt structure like NaCl, may display both kinds of defects.

Silver bromide is used in black-and-white film photography. When light falls on a grain of AgBr embedded in film, a bromide ion loses an electron, which combines with a silver ion in a Frenkel defect and produces a silver atom. Later the film is bathed in a developer that reduces more silver ions to intensify the image. After the image is developed, the film is *fixed*, which involves dissolving any remaining silver bromide in an aqueous solution (about 0.2 *M*) of sodium thiosulfate, $Na_2S_2O_3$.

a. If a Schottky defect occurs in a unit cell of silver bromide (Figure 18.44a), does the overall stoichiometry of the unit cell change? If the defect occurs consistently throughout the structure, does the density of the silver bromide change?
b. If a Ag^+ ion that normally occupies an octahedral hole in the cubic closest-packed array of Br^- ions forms a Frenkel defect by moving to a tetrahedral hole, does the overall composition of the unit cell change? If the defect occurs consistently throughout a sample of AgBr, does its formula and its density change?
c. Why is AgBr, which is insoluble in water, soluble in a solution of sodium thiosulfate?
d. Silver ions are recovered from photographic processing solutions by passing them through columns containing steel wool (Fe). The reaction between silver(I) ions and Fe produces silver metal. Write a balanced net ionic equation describing this reaction and suggest why it is spontaneous.

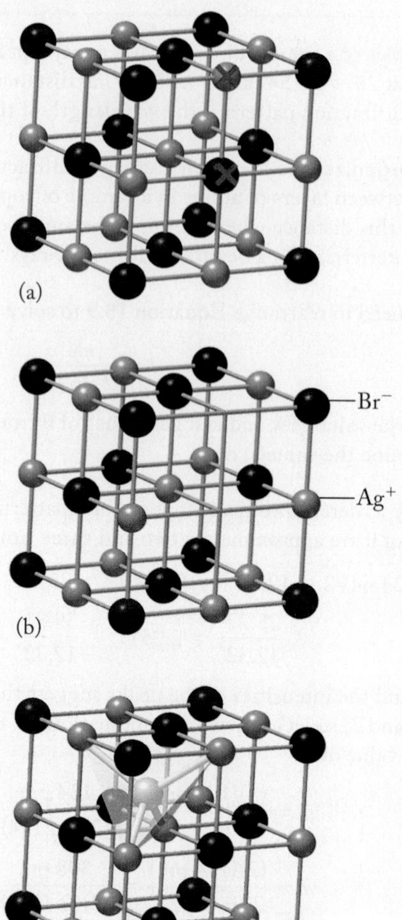

FIGURE 18.44 (a) A perfect crystal structure of AgBr, without defects. (b) One Ag^+ ion and one Br^- ion are missing from the front face, resulting in a Schottky defect. (c) One Ag^+ ion has moved from its normal location at the center to the highlighted tetrahedral hole, resulting in a Frenkel defect.

Collect and Organize We are asked if Schottky and Frenkel defects in AgBr cause changes in the stoichiometry of the unit cell and the density of the bulk material. We are also asked to explain the solubility of AgBr in a solution that contains thiosulfate ions and the chemistry behind the recovery of silver as the metal by reacting silver(I) compounds with iron metal.

Analyze To determine the impact of Schottky and Frenkel defects on AgBr, we start with a unit cell of AgBr (Figure 18.44a) and then incorporate the changes described. By determining the ratio of the ions in defective unit cells and the number of ions per unit cell, we can determine whether there will be changes in the overall formula and density of a sample. To explain the solubility of AgBr in thiosulfate solutions, we need to consider how Ag^+ and Br^- ions interact with thiosulfate ($S_2O_3^{2-}$) ions. According to Table A5.5 in Appendix 5, Ag^+ ions combine with $S_2O_3^{2-}$ ions to produce a complex ion, $Ag(S_2O_3)_2^{3-}$, which has a formation constant of 4.7×10^{13}. The K_{sp} of AgBr (Table A5.4) is 5.4×10^{-13}. Our solution to this problem will need to link these two equlibria. Finally, conversion of silver(I) to Ag metal with iron must be a redox process. Checking the standard reduction potentials in Table A6.1 in Appendix 6, we find the following relevant half-reactions:

$$Ag^+(aq) + e^- \rightarrow Ag(s) \qquad E^\circ = 0.7996 \text{ V}$$

$$Fe^{2+}(aq) + 2\,e^- \rightarrow Fe(s) \qquad E^\circ = -0.447 \text{ V}$$

The positive E° value of the silver half-reaction and the negative E° value of the iron half-reaction indicate that Fe may be a reducing agent strong enough to convert the silver(I) to free silver metal.

Solve

a. A Schottky defect in AgBr occurs when a pair of Ag^+ and Br^- ions is lost, perhaps producing a unit cell like the one shown in Figure 18.44(b). Though the composition of this particular unit cell has changed, because the missing ions occupied different positions that contributed differently to the total ion count in the cell, the overall ratio of silver ions to bromide ions in the solid should still be 1:1. However, if enough ion pairs are lost without changing the cell dimensions, the density of the solid will decrease.

b. If a silver ion moves from the center (octahedral) site to an unoccupied tetrahedral site (Figure 18.44c), the composition of the cell does not change, so neither the formula of the solid nor its density changes in the presence of Frenkel defects.

c. The K_{sp} of AgBr can be used to calculate the concentration of Ag^+ ions in equilibrium with solid AgBr as a result of the reaction $AgBr(s) \rightleftharpoons Ag^+(aq) + Br^-(aq)$:

$$K_{sp} = 5.4 \times 10^{-13} = [Ag^+][Br^-] = (x)(x) = x^2$$

$$x = [Ag^+] = 7.3 \times 10^{-7}\ M$$

Let's use this concentration to determine whether adding 0.2 *M* thiosulfate ions will result in the formation of $Ag(S_2O_3)_2^{\ 3-}$ complex ions. Inserting the known concentrations into the formation constant expression and solving for $[Ag(S_2O_3)_2^{\ 3-}]$,

$$K_f = \frac{[Ag(S_2O_3)_2^{\ 3-}]}{[Ag^+][S_2O_3^{\ 2-}]^2} = \frac{x}{(7.3 \times 10^{-7})(0.2)^2} = 4.7 \times 10^{13}$$

$$x = [Ag(S_2O_3)_2^{\ 3-}] = 1.4 \times 10^6\ M$$

This very large concentration, as compared to $[Ag^+]$, means that any Ag^+ ions that are produced when AgBr dissolves will form complex ions, effectively removing free Ag^+ ions from solution. According to Le Châtelier's principle, removing one of the product ions shifts the K_{sp} equilibrium described above toward making more dissolved ions, which keeps on happening until all of the AgBr has dissolved (assuming there is an adequate supply of thiosulfate ions).

d. The two half-reactions in the Analyze section can be combined to write a net ionic equation describing the reaction between silver(I) and iron if we (1) reverse the iron half-reaction and (2) multiply the silver half-reaction by 2 to balance the loss and gain of electrons:

$$2\,Ag^+(aq) + 2\,e^- \rightarrow 2\,Ag(s) \qquad E^\circ = 0.7996\ V$$

$$Fe(s) \rightarrow Fe^{2+}(aq) + 2\,e^- \qquad E^\circ = 0.447\ V$$

$$2\,Ag^+(aq) + Fe(s) \rightarrow 2\,Ag(s) + Fe^{2+}(aq) \qquad E^\circ_{rxn} = 1.247\ V$$

The positive reaction potential indicates that the reaction is spontaneous *under standard conditions*—that is, when $[Ag^+] = [Fe^{2+}] = 1\ M$. As noted above, the concentration of Ag^+ ions is much less than 1 *M*. However, we learned in Chapter 17 that the potentials of half-reactions change relatively little even as the concentrations of reactants and products change a lot. Therefore, we can confidently predict that the reaction will have a positive reaction potential and be spontaneous.

Think About It Physical and chemical properties of crystalline materials are closely related to their structure, and structural defects contribute to these properties as well. Other ionic substances with large size differences between the anion and the cation may also demonstrate Frenkel defects. AgCl, with a smaller anion (181 pm for Cl^- versus 195 pm for Br^-), is also photoreactive, but not to the same extent as AgBr because the Ag^+ ion (126 pm) is too big to fit into the tetrahedral holes in an array of Cl^- ions.

SUMMARY

LO1 Many metallic crystals are based on crystal lattices of the **cubic closest-packed (ccp)** and **hexagonal closest-packed (hcp)** types, which are the two most efficient ways of packing atoms in a solid. **Crystalline solids** contain repeating **unit cells**, which include **simple cubic (sc)**, **body-centered cubic (bcc)**, or **face-centered cubic (fcc)**. The dimensions of the unit cell in a crystalline solid can be used to determine the radius of the atoms or ions. (Sections 18.2 and 18.6)

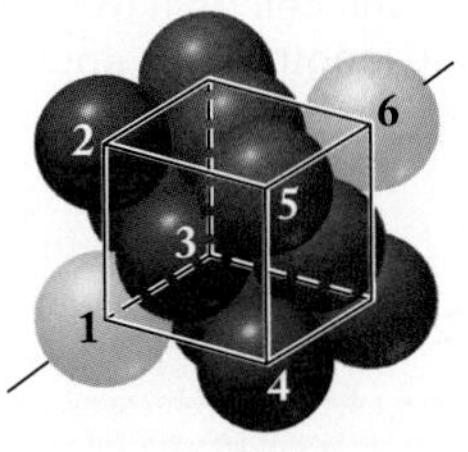

LO2 The dimensions a crystalline solid's unit cell can be used to calculate the density of the solid. (Sections 18.2 and 18.6)

LO3 **Alloys** are blends of a host metal and one or more other elements (which may or may not be metals) added to enhance the properties of the host, including strength, hardness, and corrosion resistance. In **substitutional alloys**, atoms of the added elements replace atoms of the host metal in its crystal lattice. In **interstitial alloys**, atoms of added elements are located in the tetrahedral and/or octahedral holes between atoms of the host metal. Aluminum and aluminum alloys are highly desirable for applications requiring corrosion resistance and low mass. (Section 18.3)

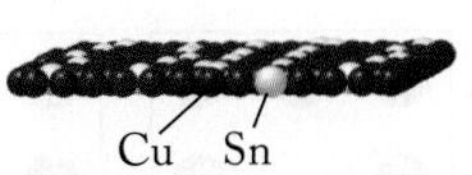

LO4 Most metals are malleable and ductile. The electrical conductivity of metals can be explained by **band theory** as the ease with which valence electrons can gain mobility by moving into empty energy levels in multiatom atructures. Metalloids are **semiconductors**, intermediate in electrical conducting ability between metals and nonmetals. In semiconductors, the filled valence band and empty conduction band are separated by a **band gap (E_g)**. Substituting electron-rich atoms into a semiconductor results in **n-type semiconductors**. Substituting electron-poor atoms results in **p-type semiconductors**. Both types of substitution increase the conductivity of the semiconductor by decreasing its band gap. (Sections 18.4 and 18.5)

Conduction band

E_g

Valence band

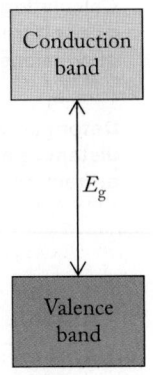

LO5 Many **ionic solids** consist of crystals with some number of either cations or anions forming the unit cell and the opposite ion occupying octahedral or tetrahedral holes in the unit cell. Two allotropes of carbon are the **covalent network solids** graphite and diamond. Many nonmetals form **molecular solids**, including sulfur, which forms puckered rings of eight sulfur

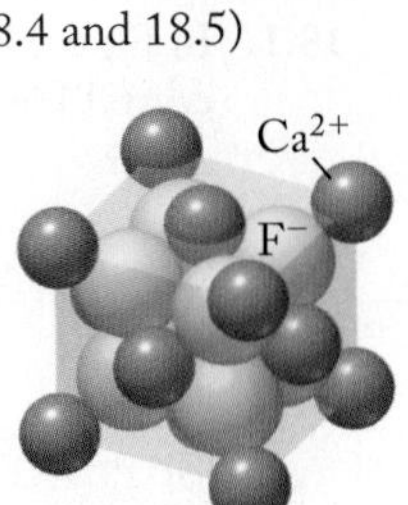

atoms, and phosphorus, which forms P_4 tetrahedra. Heating selected solid inorganic compounds (such as clays, which are aluminosilicate minerals) to high temperature alters their chemical composition and makes them harder, denser, and stronger. The resulting heat- and chemical-resistant materials (**ceramics**) are electrical insulators due to the large energy gap between their filled valence and empty conduction bands. The polymorphs of silica consist of tetrahedra made up of four O atoms at the corners surrounding a central Si atom. Each tetrahedron can share some or all of its oxygen atoms with other tetrahedra, forming Si—O—Si bridges and two- and three-dimensional covalent networks. (Sections 18.6, 18.7, and 18.8)

LO6 **X-ray diffraction (XRD)** is an analytical method that records constructive interference of X-rays reflecting off different layers of atoms or ions in a crystalline solid. The distances between layers are calculated using the **Bragg equation**. X-ray diffraction makes it possible to determine the crystal structure of crystalline solids. (Section 18.9)

PARTICULATE PREVIEW WRAP-UP

Allotrope B is planar, as each carbon atom is covalently bonded to three other carbon atoms in a trigonal planar geometry and is attracted to the planar layers above and below through dispersion forces. Allotrope A is nonplanar as each carbon atom is covalently bonded to four other carbon atoms in a tetrahedral geometry. Allotrope A is hard and abrasive because it has a rigid three-dimensional atomic structure. Allotrope B is soft and slippery because the layers of atoms are not covalently bonded together and can slip past each other.

PROBLEM-SOLVING SUMMARY

Type of Problem	Concepts and Equations	Sample Exercises
Calculating atomic or ionic radii from unit cell dimensions	Determine the length ℓ of an edge, face diagonal, or body diagonal along which adjacent atoms touch in a unit cell. Use the relationship between ℓ and the atomic/ionic radius r to calculate the value of r: $r = \ell/2$ for sc unit cells, $r = 0.3536\ell$ for fcc unit cells, and $r = 0.4330\ell$ for bcc unit cells.	**18.1, 18.5**
Using unit cells to calculate the density of a solid	Determine the mass of the atoms in the unit cell from their molar mass and the volume of the unit cell from the edge length, then calculate the density: $$d = \frac{m}{V}$$	**18.2, 18.3**
Predicting the crystal structure of two-element alloys	Compare the radii of the alloying elements. Similarly sized radii (within 15%) suggest a substitutional alloy. When the radius of the smaller atom is $<$73% of the radius of the larger atom, an interstitial alloy may form.	**18.4**
Calculating the density of a salt from its unit cell dimensions	Determine the masses of the ions in the unit cell from their molar masses and the volume of the unit cell from the edge length. Then apply $$d = \frac{m}{V}$$	**18.6**
Determining interlayer distances by X-ray diffraction	Apply the Bragg equation, $$n\lambda = 2d \sin \theta \quad (18.9)$$	**18.7**

VISUAL PROBLEMS

(Answers to boldface end-of-chapter questions and problems are in the back of the book.)

18.1. In Figure P18.1, which drawings are analogous to a crystalline solid, and which are analogous to an amorphous solid?

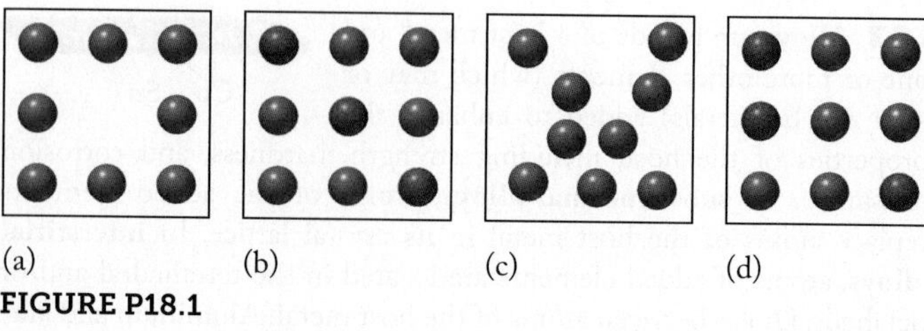

(a) (b) (c) (d)

FIGURE P18.1

18.2. The unit cells in Figure P18.2 continue infinitely in two dimensions. Draw a square around a unit cell in each pattern. How many light squares and how many dark squares are in each unit cell?

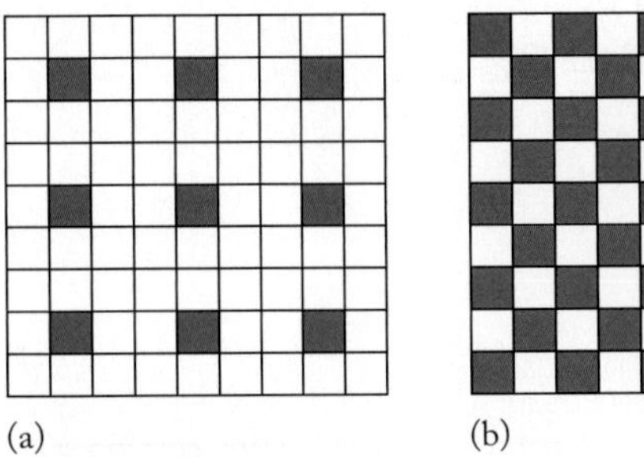

FIGURE P18.2

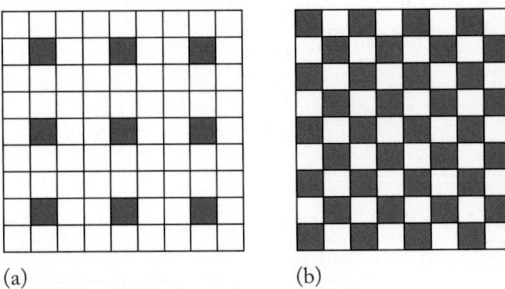

18.3. The pattern in Figure P18.3 continues indefinitely in three dimensions. Draw a box around the unit cell. If the red circles represent element A and the blue circles represent element B, what is the chemical formula of the compound?

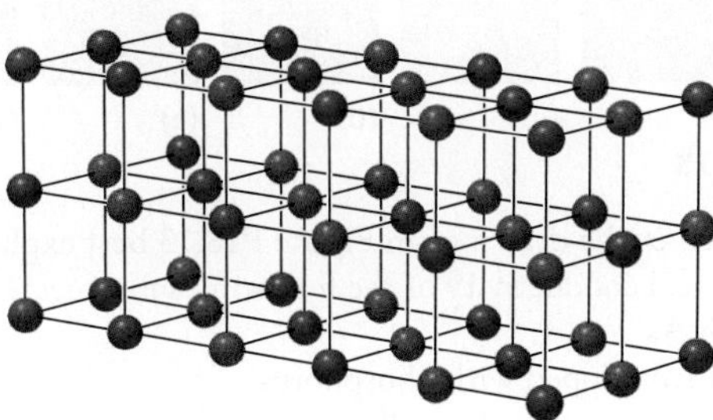

FIGURE P18.3

18.4. How many cations (A) and anions (B) are there in the unit cell in Figure P18.4?

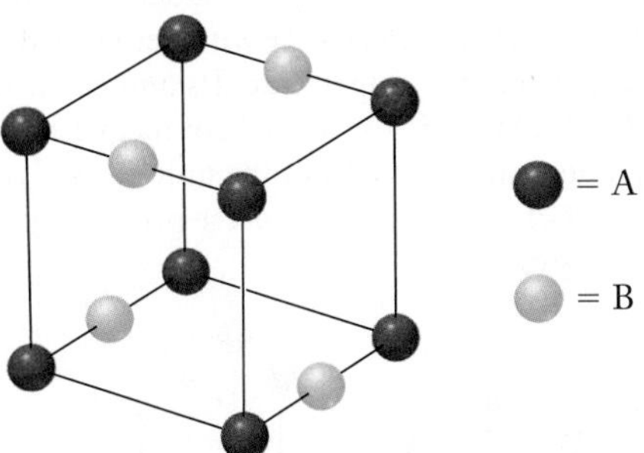

FIGURE P18.4

18.5. How many atoms of elements A and B are there in the unit cell in Figure P18.5? The blue spheres are in cell faces.

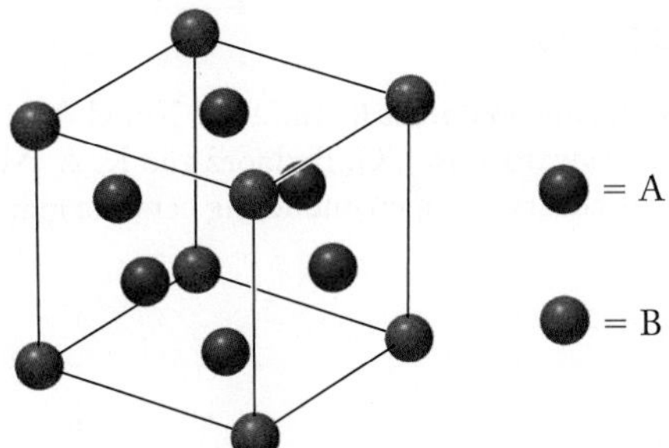

FIGURE P18.5

18.6. What is the chemical formula of the compound with the unit cell shown in Figure P18.6? The blue spheres are in the faces of the inner cube.

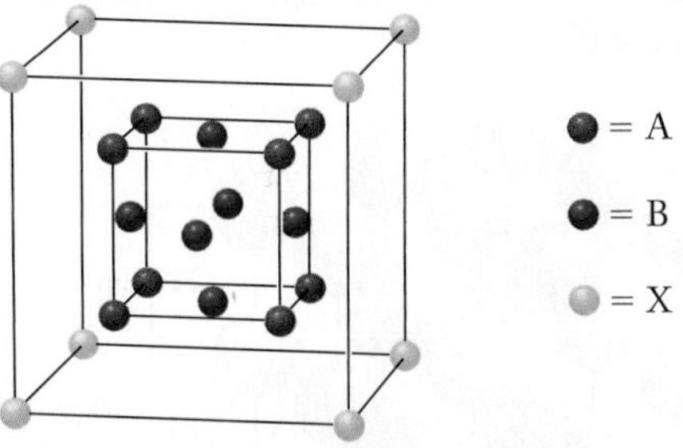

FIGURE P18.6

18.7. What is the chemical formula of the ionic compound with the unit cell shown in Figure P18.7 if A and B are cations, and X is an anion? The red spheres are in cell faces; the yellow sphere is in the center.

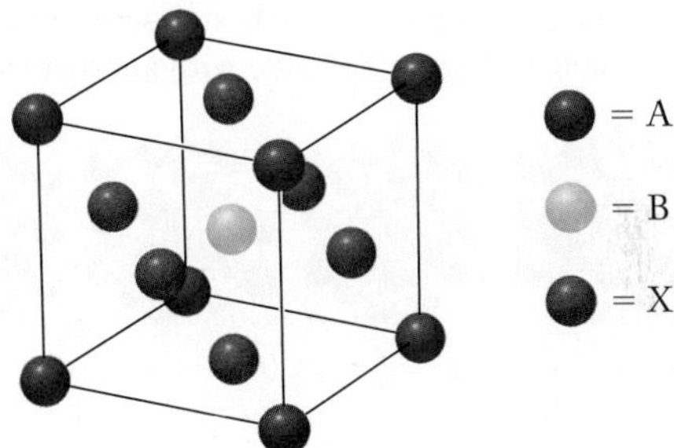

FIGURE P18.7

18.8. When amorphous red phosphorus is heated at high pressure, it is transformed into the allotrope black phosphorus, which can form fused six-membered rings of phosphorus atoms (Figure P18.8a). Why are the six-atom rings in black phosphorus puckered, whereas the six-atom rings in graphene (Figure P18.8b) are planar?

FIGURE P18.8

18.9. The distance between atoms in a cubic form of phosphorus is 238 pm (Figure P18.9).
a. Calculate the density of this form of phosphorus.
b. What type of hole is between the atoms in the cubic form of P?

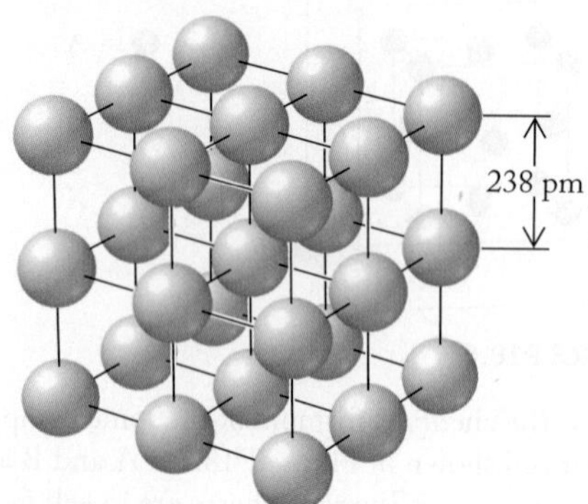

FIGURE P18.9

*18.10. Which drawing in Figure P18.10 best describes the result of doping cubic P (Figure P18.10a) with another element?

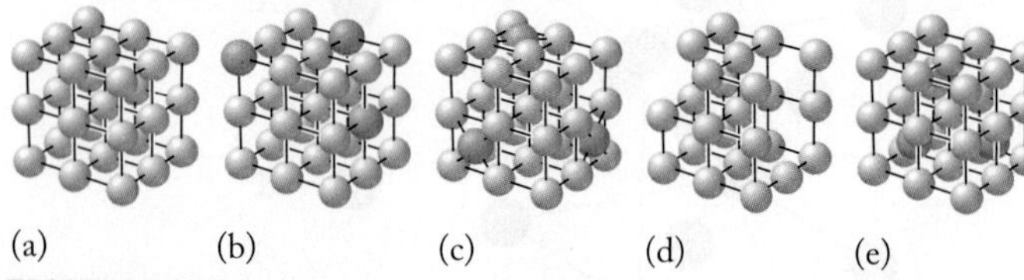

FIGURE P18.10

18.11. **Antibiotic Gold/Platinum Nanoparticles** In Chapter 2 we described how small particles containing gold and platinum have antibiotic properties. Which of the drawings in Figure P18.11 best describes the alloy formed by gold and platinum?

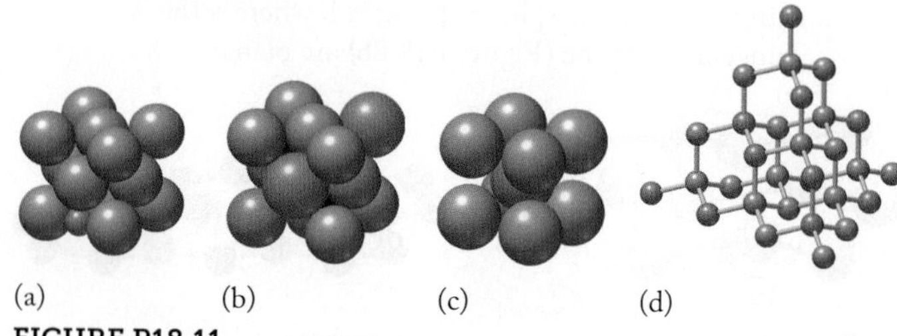

FIGURE P18.11

18.12. **Rose Gold** Jewelers take advantage of gold/copper alloys to bring a reddish hue to their products. Why is it unlikely that Au/Cu alloys will crystallize in structures (a) or (d) in Figure P18.11?

18.13. Which of the drawings in Figure P18.13 best explains the electrical conductivity of the following substances?
a. copper metal
b. magnesium metal
c. silicon
d. sodium chloride

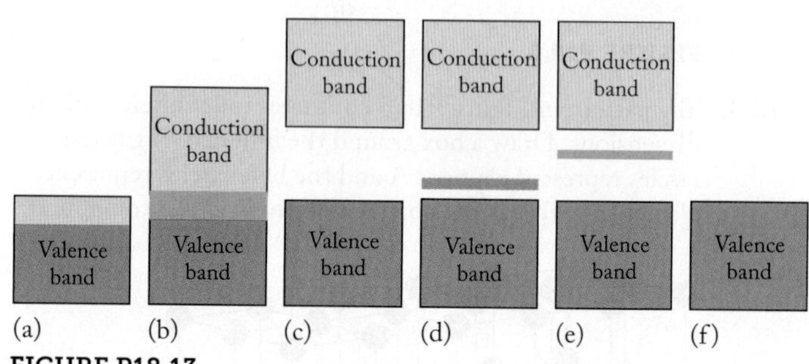

FIGURE P18.13

18.14. Which of the drawings in Figure P18.13 best explains the electrical conductivity of the following substances?
a. GaAs
b. Silicon doped with phosphorus
c. Silicon doped with gallium
*d. Zinc oxide, where some of the oxide ions are missing to give a formula ZnO_{1-x}

***18.15.** **Superconducting Materials I** In 2000, magnesium boride was observed to behave as a superconductor—that is, a material that conducts electricity with zero resistance below a characteristic temperature. Its unit cell is shown in Figure P18.15. What is the formula of magnesium boride? A boron atom is in the center of the unit cell (on the left), which is part of the hexagonal closest-packed crystal structure (right).

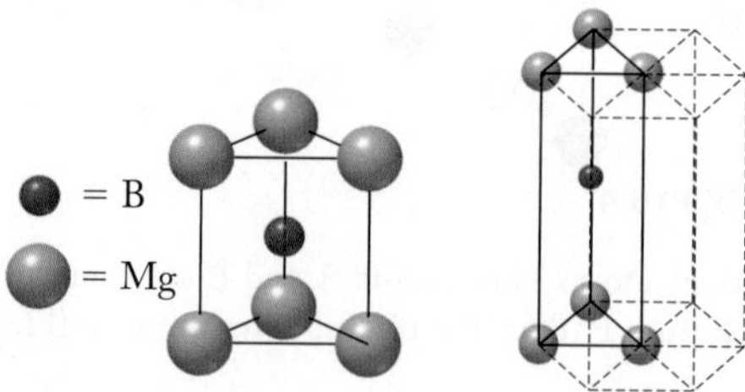

FIGURE P18.15

*18.16. **Superconducting Materials II** The 1987 Nobel Prize in Physics was awarded to J. G. Bednorz and K. A. Müller for their discovery of superconducting ceramic materials such as $YBa_2Cu_3O_7$. Figure P18.16 shows the unit cell of another yttrium–barium–copper oxide.

a. What is the chemical formula of this compound?
b. Removing four oxygen atoms from the unit cell in the figure produces the unit cell of $YBa_2Cu_3O_7$. Does it make a difference which oxygen atoms are removed?

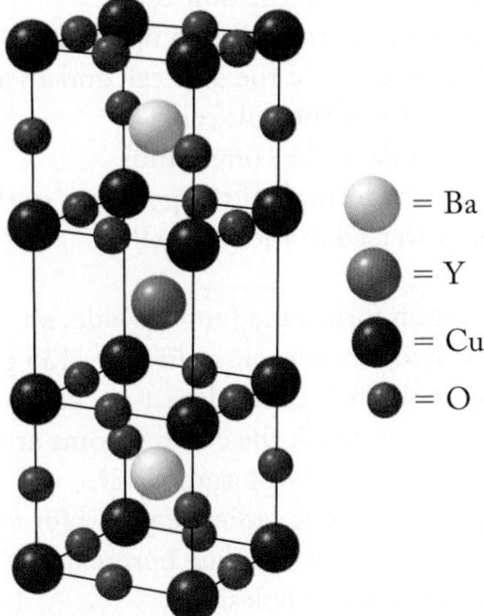

FIGURE P18.16

18.17. In the unit cell in Figure P18.17, the small spheres are in tetrahedral holes formed by the large spheres.

a. How many of the large spheres and how many of the small ones are assignable to the unit cell?
b. If the radius of the large spheres is 140 pm, what (theoretically) is the maximum size of the small spheres?

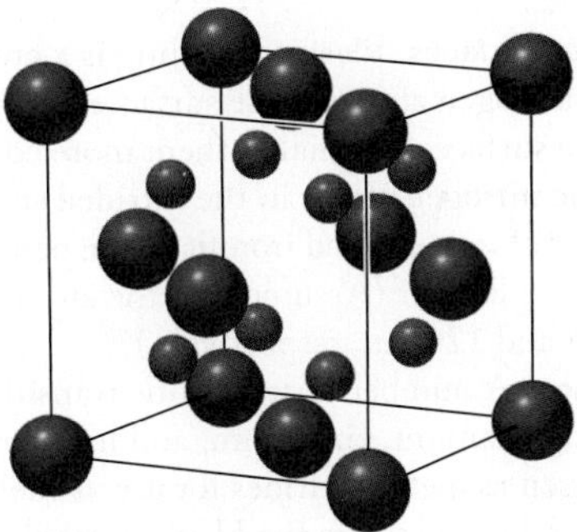

FIGURE P18.17

18.18. Use representations [A] through [I] in Figure P18.18 to answer questions a–f.

a. Two forms of iron are ferrite and austenite. One has a body-centered cubic structure, and the other takes a face-centered cubic structure. Which representations depict these forms of iron?
b. Two allotropes of phosphorus exist—white phosphorus is a molecular solid, whereas red phosphorus is a covalent network solid. Which representations depict white and red phosphorus?
c. Which representations depict allotropes of carbon?
d. Which representation depicts the unit cell of LiCl? Of CaF_2?
e. Which representations depict rock salt and fluorite structures?
f. Which process is depicted in representation [F]?

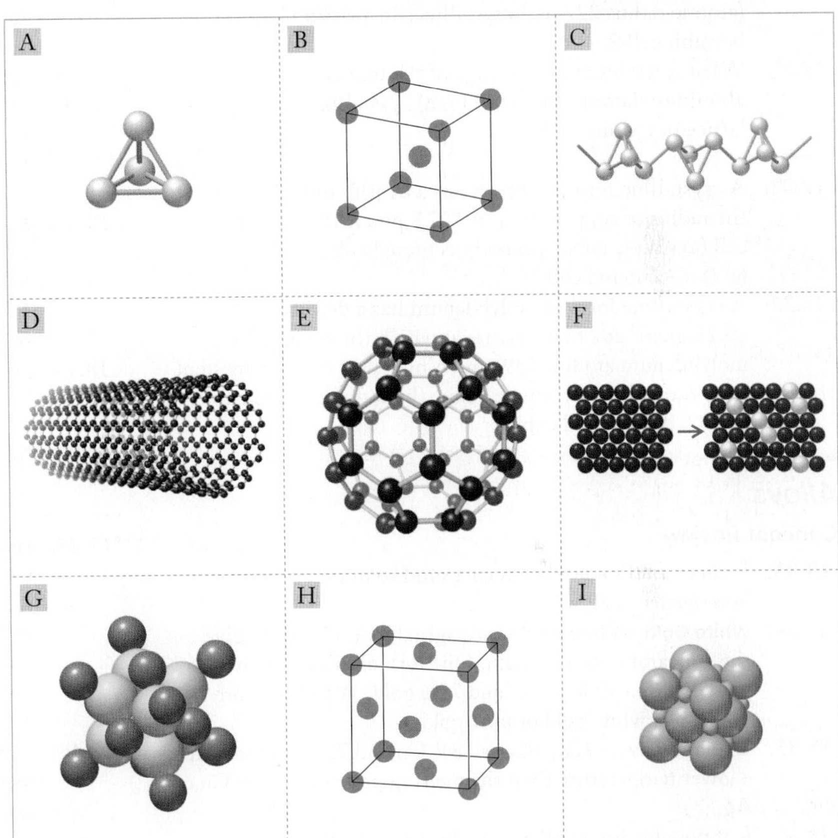

FIGURE P18.18

QUESTIONS AND PROBLEMS

Structures of Metals

Concept Review

18.19. Explain the difference between cubic closest-packed and hexagonal closest-packed arrangements of identical spheres.

*18.20. Is it possible to have a closest-packed crystal lattice with four different repeating layers, *abcdabcd*...?

18.21. Which unit cell has the greater packing efficiency, simple cubic or body-centered cubic?

18.22. Consult Figure 18.11 to predict which unit cell has the greater packing efficiency: body-centered cubic or face-centered cubic.

18.23. The unit cell in iron metal is either fcc or bcc, depending on temperature (see Sample Exercise 18.1). Are the fcc and the bcc forms of iron allotropes? Explain your answer.

*18.24. At low temperatures, the unit cell of calcium metal is found to be fcc. At higher temperatures, the unit cell of calcium metal is bcc. What might be a reason for this temperature dependence?

Problems

18.25. Derive the edge length in bcc and fcc unit cells in terms of the radius (r) of the atoms in the cells in Figure 18.11.

*18.26. Derive the length of the body diagonal in simple cubic and fcc unit cells in terms of the radius (r) of the atoms in the unit cells in Figure 18.11.

18.27. Fluorescent Lighting Europium is a lanthanide element that is used in the phosphors in fluorescent lamps. Metallic Eu forms bcc unit cells with an edge length of 240.6 pm. Use this information to calculate the radius of a europium atom.

18.28. Nickel has an fcc unit cell with an edge length of 350.7 pm. Use this information to calculate the radius of a nickel atom.

18.29. What is the length of an edge of the unit cell when barium (atomic radius 222 pm) crystallizes in a crystal lattice of bcc unit cells?

18.30. What is the length of an edge of the unit cell when aluminum (atomic radius 143 pm) crystallizes in a crystal lattice of fcc unit cells?

18.31. A crystalline form of copper has a density of 8.95 g/cm^3. If the radius of copper atoms is 127.8 pm, is the copper unit cell (a) simple cubic, (b) body-centered cubic, or (c) face-centered cubic?

18.32. A crystalline form of molybdenum has a density of 10.28 g/cm^3 at a temperature at which the radius of a molybdenum atom is 139 pm. Which unit cell is consistent with these data: (a) simple cubic, (b) body-centered cubic, or (c) face-centered cubic?

Alloys

Concept Review

18.33. Is there a difference between a solid solution and a homogeneous alloy?

18.34. **White Gold** White gold was originally developed to give the appearance of platinum. One formulation of white gold contains 25% nickel and 75% gold. Which is more malleable, white gold or pure gold?

***18.35.** Explain why an alloy that is 28% Cu and 72% Ag melts at a lower temperature than the melting points of either Cu or Ag.

18.36. Is it possible for an alloy to be both substitutional and interstitial?

18.37. A substitutional alloy is prepared using two metals: Mo and W, with the same unit cell (bcc), and atomic radius (r = 139 pm). Will the density of the alloy be less than, greater than, or the same as the density of pure molybdenum?

18.38. The nibs on some fountain pens were made from an alloy called Osmiroid that contains osmium and iridium. Using only the periodic table as a guide, do you expect osmium and iridium to form a substitutional or an interstitial alloy?

Problems

18.39. The unit cell of an intermetallic compound consists of a face-centered cube that has an atom of element X at each corner and an atom of element Y at the center of each face.

a. What is the formula of the compound?

b. What would be the formula if the positions of the two elements were reversed in the unit cell?

18.40. The bcc unit cell of an intermetallic compound has atoms of element A at the corners of the unit cell and an atom of element B at the center of the unit cell.

a. What is the formula of the compound?

b. What would be the formula if the positions of the two elements were reversed in the unit cell?

18.41. Vanadium and carbon form vanadium carbide, an interstitial alloy. Given the atomic radii of V (135 pm) and C (77 pm), which holes in a cubic closest-packed array of vanadium atoms do you think the carbon atoms are more likely to occupy—octahedral or tetrahedral?

18.42. What is the minimum atomic radius required for a cubic closest-packed metal to accommodate boron atoms (r = 88 pm) in its octahedral holes?

18.43. Dental Fillings Dental fillings are mixtures of several alloys, including one made of silver and tin. Silver (r = 144 pm) and tin (r = 140 pm) both crystallize in an fcc unit cell. Is this alloy likely to be a substitutional alloy or an interstitial alloy?

18.44. An alloy used in dental fillings has the formula Sn_3Hg. The radii of tin and mercury atoms are 140 pm and 151 pm, respectively. Which alloy has a smaller mismatch (percent difference in atomic radii), Sn_3Hg or bronze (Cu/Sn alloys)?

***18.45. Hardening Metal Surfaces** Plasma nitriding is a process for embedding nitrogen atoms in the surfaces of metals that hardens the surfaces and makes them more corrosion resistant. Do the nitrogen atoms in the nitrided surface of a sample of cubic closest-packed iron fit in the octahedral holes of the crystal lattice? (Assume that the atomic radii of N and Fe are 75 and 126 pm, respectively.)

*18.46. **Hydrogen Storage** A number of crystalline transition metals (including titanium, zirconium, and hafnium) can store hydrogen as metal hydrides for use as fuel in a hydrogen-powered vehicle. Are the H atoms (radius 37 pm) more likely to be in the tetrahedral or octahedral holes of these three metals whose atomic radii are 147, 160, and 159 pm, respectively?

18.47. An interstitial alloy is prepared from metals A and B, where B has the smaller atomic radius. The unit cell of metal A is fcc. What is the formula of the alloy if B occupies (a) all of the octahedral holes; (b) half of the octahedral holes; (c) half of the tetrahedral holes?

18.48. An interstitial alloy was prepared from two metals. Metal A, with the larger atomic radius, has a hexagonal closest-packed crystal lattice. What is the formula of the alloy if atoms of metal B occupy (a) all of the tetrahedral holes; (b) half of the tetrahedral holes; (c) half of the octahedral holes?

18.49. An interstitial alloy with a fcc unit cell contains one atom of B for every five atoms of host element A. What fraction of the octahedral holes is occupied in this alloy?

18.50. If the B atoms in the alloy described in Problem 18.49 occupy tetrahedral holes in A, what percentage of the holes would they occupy?

18.51. If the unit cell of a substitutional alloy of copper and tin has the same edge length as the unit cell of copper, will the alloy have a greater density than copper?

18.52. If the unit cell of an interstitial alloy of vanadium and carbon has the same edge length as the unit cell of vanadium, will the alloy have a greater density than vanadium?

Metallic Bonds and Conduction Bands

Concept Review

18.53. How does the sea-of-electrons model (Chapter 4) explain the high electrical conductivity of gold?

*18.54. How does band theory explain the high electrical conductivity of mercury?

18.55. The melting and boiling points of sodium metal are much lower than those of sodium chloride. What does this difference reveal about the relative strengths of metallic bonds and ionic bonds of the alkali metals?

*18.56. Which metal do you expect to have the higher melting point—Al or Na? Explain your answer.

18.57. Some scientists believe that the solid hydrogen that forms at very low temperatures and high pressures may conduct electricity. Is this hypothesis supported by band theory?

18.58. Would you expect solid helium to conduct electricity?

Semiconductors

Concept Review

18.59. Which groups in the periodic table contain metals with filled valence bands?

18.60. Insulators are materials that do not conduct electricity; conductors are substances that allow electricity to flow through them easily. Rank the following in order of increasing band gap: semiconductor, insulator, and conductor.

18.61. Why is it important to keep phosphorus out of silicon chips during their manufacture?

18.62. How might doping of silicon with germanium affect the conductivity of silicon?

***18.63.** Antimony (Sb) combines with sulfur to form the semiconductor compound Sb_2S_3. In which group of the periodic table might you find elements for doping Sb_2S_3 to form a p-type semiconductor?

*18.64. In which group of the periodic table might you find elements for doping Sb_2S_3 to form an n-type semiconductor?

Problems

18.65. Thin films of doped diamond hold promise as semiconductor materials. Trace amounts of nitrogen impart a yellow color to otherwise colorless pure diamonds.

a. Are nitrogen-doped diamonds examples of semiconductors that are p-type or n-type?

b. Draw a picture of the band structure of diamond to indicate the difference between pure diamond and N-doped (nitrogen-doped) diamond.

*c. N-doped diamonds absorb violet light at about 425 nm. What is the magnitude of E_g that corresponds to this wavelength?

18.66. **Hope Diamond** Trace amounts of boron give diamonds (including the Smithsonian's Hope Diamond) a blue color (Figure P18.66).

a. Are boron-doped diamonds examples of semiconductors that are p-type or n-type?

b. Draw a picture of the band structure of diamond to indicate the difference between pure diamond and B-doped diamond.

*c. What is the band gap in energy if blue diamonds absorb red-orange light with a wavelength of 675 nm?

FIGURE P18.66

***18.67.** The nitride ceramics AlN, GaN, and InN are all semiconductors used in the microelectronics industry. Their band gaps are 580.6, 322.1, and 192.9 kJ/mol, respectively. Which, if any, of these energies correspond to radiation in the visible region of the spectrum?

*18.68. Calculate the wavelengths of light emitted by the semiconducting phosphides AlP, GaP, and InP, which have band gaps of 241.1, 216.0, and 122.5 kJ/mol, respectively, and are used in the type of light source shown in Figure P18.68.

FIGURE P18.68

Structures of Some Crystalline Nonmetals

Concept Review

18.69. Molecules of S_8 are not flat octagons—why?

*18.70. Selenium exists either as Se_8 rings or in a structure with helical chains of Se atoms. Are these two structures of selenium allotropes? Explain your answer.

***18.71.** If the carbon atoms in graphite are replaced by alternating B and N atoms, would the resulting structure contain puckered rings like black phosphorus or flat ones like graphite? (See Figure P18.8.)

*18.72. Cyclic allotropes of sulfur containing up to 20 sulfur atoms have been isolated and characterized. Propose a reason why the bond angles in S_n (where $n = 10, 12, 18,$ and 20) are all close to 106°.

Problems

18.73. Ice is a molecular solid. However, theory predicts that, under high pressure, ice (solid H_2O) becomes an ionic compound composed of H^+ and O^{2-} ions. The proposed unit cell for ice under these conditions is a bcc unit cell of oxygen ions with hydrogen ions in holes.
a. How many H^+ and O^{2-} ions are in each unit cell?
b. Draw a Lewis structure for "ionic" ice.

18.74. **Ice under Pressure** Kurt Vonnegut's novel *Cat's Cradle* describes an imaginary, high-pressure form of ice called "ice nine." Assuming ice nine has a cubic closest-packed arrangement of oxygen atoms with hydrogen atoms in the appropriate holes, what type of hole will accommodate the H atoms?

18.75. A chemical reaction between H_2S_4 and S_2Cl_2 produces cyclic S_6. What are the bond angles in S_6?

18.76. Reaction between S_8 and six equivalents of AsF_5 yields $[S_4^{2+}][AsF_6^-]_2$ by the reaction

$$S_8 + 6\,AsF_5 \rightarrow 2\,[S_4^{2+}][AsF_6^-]_2 + 2\,AsF_3$$

The S_4^{2+} ion has a cyclic structure. Are all four sulfur atoms in one plane?

Salt Crystals: Ionic Solids

Concept Review

18.77. Crystals of both LiCl and KCl have the rock salt structure. In the unit cell of LiCl, adjacent Cl^- ions touch each other. In KCl they don't. Why?

18.78. Can $CaCl_2$ have the rock salt structure?

***18.79.** In some books the unit cell of CsCl is described as body-centered cubic (Figure P18.79); in others, as simple cubic (see Figure 18.11a). Explain how CsCl crystals might be described by either unit cell type.

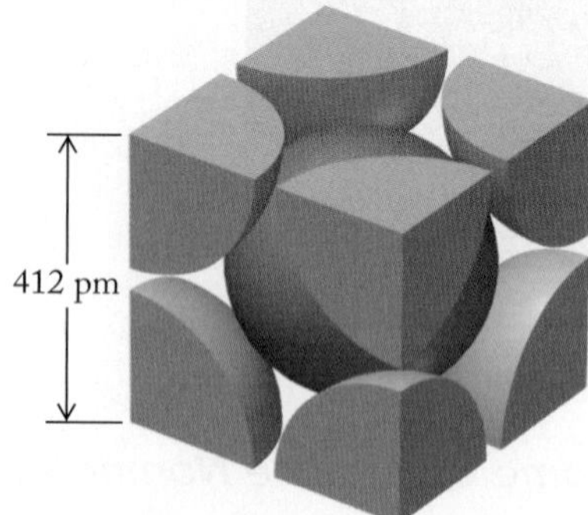

FIGURE P18.79

18.80. In the crystals of ionic compounds, how do the relative sizes of the ions influence the location of the smaller ions?

***18.81.** Instead of describing the unit cell of NaCl as a fcc array of Cl^- ions with Na^+ ions in octahedral holes, might we describe it as an fcc array of Na^+ ions with Cl^- ions in octahedral holes? Explain why or why not.

18.82. Why isn't crystalline sodium chloride considered a network solid? Why isn't sodium chloride considered an alloy of sodium and chlorine?

***18.83.** As the cation–anion radius ratio increases for an ionic compound with the rock salt crystal structure, is the calculated density more likely to be greater than or less than the measured value?

18.84. As the cation–anion radius ratio increases for an ionic compound with the rock salt crystal structure, is the length of the edge of the unit cell calculated from ionic radii likely to be greater than or less than the observed edge length?

Problems

18.85. What is the formula of the oxide that crystallizes with Fe^{3+} ions in one-fourth of the octahedral holes, Fe^{3+} ions in one-eighth of the tetrahedral holes, and Mg^{2+} in one-fourth of the octahedral holes of a cubic closest-packed arrangement of oxide ions (O^{2-})?

18.86. What is the chemical formula of the compound that crystallizes in a simple cubic arrangement of fluoride ions with Ba^{2+} ions occupying half of the cubic holes?

18.87. A compound of uranium and oxygen consists of a cubic close-packed arrangement of uranium ions with oxide ions in all the tetrahedral holes. What is the formula of this compound?

18.88. A mixture of gallium and arsenic is a widely used semiconductor. The arsenide ions are in a cubic close-packed arrangement and half the tetrahedral holes are occupied by the gallium ions. What is the formula of this compound?

18.89. **The Vinland Map** At Yale University there is a map, believed to date from the 1400s, of a landmass labeled "Vinland" (Figure P18.89). The map is thought to be evidence of early Viking exploration of North America. Debate over the map's authenticity centers on yellow stains on the map paralleling the black ink lines. One analysis suggests the yellow color is from the mineral anatase, a form of TiO_2 that was not used in 15th-century inks.
a. The crystal structure of anatase is approximated by a ccp arrangement of oxide ions with titanium(IV) ions in holes. Which type of hole are Ti^{4+} ions likely to occupy? (The radius of Ti^{4+} is 60.5 pm.)
b. What fraction of these holes is likely to be occupied?

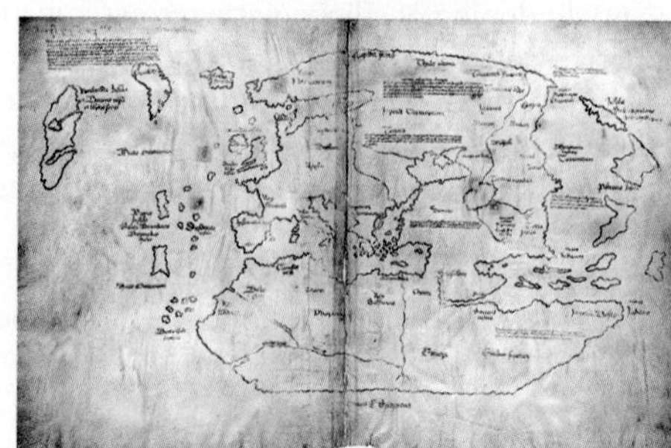

FIGURE P18.89

*18.90. The crystal structure of olivine—M_2SiO_4 (M = Mg, Fe)—can be viewed as a ccp arrangement of oxide ions with silicon(IV) ions in tetrahedral holes and metal ions in octahedral holes.
  a. What fraction of each type of hole is occupied?
  b. The unit cells of Mg_2SiO_4 and Fe_2SiO_4 have volumes of 2.91×10^{-26} cm^3 and 3.08×10^{-26} cm^3. Why is the volume of Fe_2SiO_4 larger?

***18.91.** The cadmium(II) sulfide (CdS) mineral hawleyite, has a sphalerite structure, and its density at 25°C is 4.83 g/cm^3. A hypothetical form of CdS with the rock salt structure would have a density of 5.72 g/cm^3. Why should the rock salt structure of CdS be denser? The ionic radii of Cd^{2+} and S^{2-} are 95 pm and 184 pm, respectively.

18.92. There are two crystalline forms of manganese(II) sulfide (MnS): the α form has a rock salt structure, whereas the β form has a sphalerite structure.
  a. Describe the differences between the two structures of MnS.
  b. The ionic radii of Mn^{2+} and S^{2-} are 67 and 184 pm, respectively. Which type of hole in a ccp lattice of sulfide ions could theoretically accommodate a Mn^{2+} ion?

18.93. The unit cell of rhenium trioxide (ReO_3) consists of a cube with rhenium atoms at the corners and an oxygen atom on each of the 12 edges. The atoms touch along the edge of the unit cell. The radii of Re and O atoms in ReO_3 are 137 and 73 pm, respectively.
  a. Sketch the unit cell of ReO_3.
  b. Calculate the density of ReO_3.
  c. Calculate the percent empty space in a unit cell of ReO_3.

18.94. Figure P18.79 shows the unit cell of CsCl. Use the information given and the radius of the chloride (corner) ions (181 pm).
  a. Calculate the radius of the Cs^+ ions.
  b. Calculate the density of simple cubic CsCl.
  c. Calculate the percent empty space in a unit cell of CsCl.

18.95. Magnesium oxide crystallizes in the rock salt structure. Its density is 3.60 g/cm^3. What is the edge length of the fcc unit cell of MgO?

18.96. Crystalline potassium bromide (KBr) has a rock salt structure and a density of 2.75 g/cm^3. Calculate the edge length of its unit cell.

Ceramics: Useful, Ancient Materials

Concept Review

18.97. Which of the following properties are associated with ceramics and which are associated with metals? (a) ductile; (b) thermal insulator; (c) electrically conductive; (d) malleable

18.98. Many ceramics such as TiO_2 are electrical insulators. What differences are there in the band structure of TiO_2 compared with Ti metal that account for the different electrical properties?

18.99. Replacing Al^{3+} ions in kaolinite [$Al_2(Si_2O_5)(OH)_4$] with Mg^{2+} ions yields the mineral antigorite. What is its formula?

18.100. What is the formula of the silicate mineral talc, obtained by replacing Al^{3+} ions in pyrophyllite [$Al_4Si_8O_{20}(OH)_4$] with Mg^{2+} ions?

Problems

18.101. Kaolinite [$Al_2(Si_2O_5)(OH)_4$] is formed by weathering of the mineral $KAlSi_3O_8$ in the presence of carbon dioxide and water, as described by the following unbalanced reaction:

$$KAlSi_3O_8(s) + H_2O(\ell) + CO_2(g) \rightarrow Al_2(Si_2O_5)(OH)_4(s) + SiO_2(s) + K_2CO_3(aq)$$

Balance the reaction and determine whether or not it is a redox reaction.

18.102. Albite, a feldspar mineral with an ideal composition of $NaAlSi_3O_8$, can be converted to jadeite ($NaAlSi_2O_6$) and quartz. Write a balanced chemical equation describing this transformation.

18.103. Under the high pressures in Earth's crust, the mineral anorthite ($CaAl_2Si_2O_8$) is converted to a mixture of three minerals: grossular [$Ca_3Al_2(SiO_4)_3$], kyanite (Al_2SiO_5), and quartz (SiO_2). (a) Write a balanced chemical equation describing this transformation. (b) Determine the charges and formulas of the silicate anions in anorthite, grossular, and kyanite.

*18.104. The calcium silicate mineral grossular is also formed under pressure in a reaction between anorthite ($CaAl_2Si_2O_8$), gehlenite ($Ca_2Al_2SiO_7$), and wollastonite ($CaSiO_3$):

$$\underset{\text{Anorthite}}{CaAl_2Si_2O_8} + \underset{\text{Gehlenite}}{Ca_2Al_2SiO_7} + \underset{\text{Wollastonite}}{CaSiO_3} \rightarrow \underset{\text{Grossular}}{Ca_3Al_2(SiO_4)_3}$$

  a. Balance this chemical equation.
  b. Express the composition of gehlenite the way mineralogists often do: as the percentage of the metal and metalloid oxides in it—that is, %CaO, %Al_2O_3, and %SiO_2.

18.105. The ceramic material barium titanate ($BaTiO_3$) is used in devices that measure pressure. The radii of Ba^{2+}, Ti^{4+}, and O^{2-} are 135, 60.5, and 140 pm, respectively. If the O^{2-} ions are in a closest-packed structure, which hole(s) can accommodate the metal cations?

18.106. The mixed metal oxide $LiMnTiO_4$ has a structure with cubic closest-packed oxide ions and metal ions in both octahedral and tetrahedral holes. Which metal ion is most likely to be found in the tetrahedral holes? The ionic radii of Li^+, Mn^{3+}, Ti^{4+}, and O^{2-} are 76, 67, 60.5, and 140 pm, respectively.

X-ray Diffraction: How We Know Crystal Structures

Concept Review

18.107. Why does an amorphous solid not produce an XRD scan with sharp peaks?

18.108. Why can we not use X-ray diffraction to determine the structures of compounds in solution?

18.109. Why are X-rays rather than microwaves chosen for diffraction studies of crystalline solids?

18.110. The radiation sources used in X-ray diffraction can be changed. Figure P18.110 shows a diffraction pattern made by a short-wavelength source. How would changing to a longer-wavelength source affect the pattern?

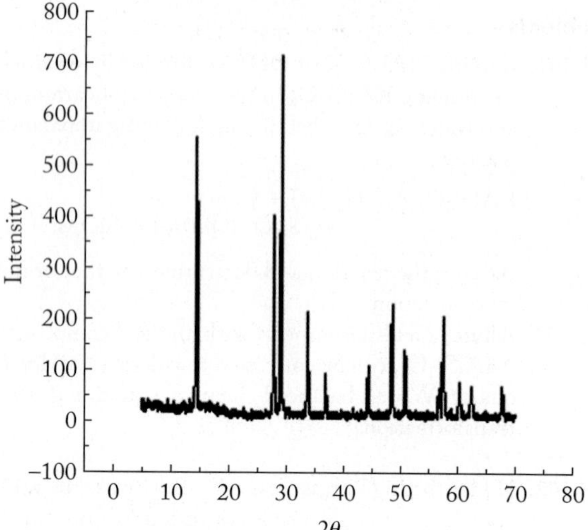

FIGURE P18.110

***18.111.** Why might a crystallographer (a scientist who studies crystal structures) use different X-ray wavelengths to determine a crystal structure? *Hint*: Consider the mechanical limits inherent in the design of the instrument depicted in Figure 18.43 and how those limits impact the 2θ scanning range.

*18.112. Where in earlier chapters did we use the interaction between matter and electromagnetic radiation to acquire insights into atomic or molecular structure?

Problems

18.113. The spacing between the layers of ions in sylvite (the mineral form of KCl) is larger than in halite (NaCl). Which crystal will diffract X-rays of a given wavelength through larger 2θ values?

18.114. Silver halides are used in black-and-white photography. In which compound would you expect to see a larger distance between ion layers, AgCl or AgBr? Which compound would you expect to diffract X-rays through larger values of 2θ if the same wavelength of X-ray were used?

18.115. Galena, Illinois, is named for the rich deposits of lead(II) sulfide (PbS) found nearby. When a crystal of PbS is exposed to X-rays with $\lambda = 71.2$ pm, the resulting XRD pattern contains strong peaks at $2\theta = 13.98°$ and 21.25°. Determine the values of n to which these peaks correspond, and calculate the spacing between the crystal layers.

18.116. **Pigments in Ceramics** Cobalt(II) oxide is used as a pigment in ceramics. It has the same type of crystal structure as NaCl. When cobalt(II) oxide is exposed to X-rays with $\lambda = 154$ pm, the XRD pattern contains strong peaks at $2\theta = 42.38°$, 65.68°, and 92.60°. Determine the values of n to which these peaks correspond, and calculate the spacing between the crystal layers.

18.117. Pyrophyllite [$Al_2Si_4O_{10}(OH)_2$] is a silicate mineral with a layered structure. The distance between the layers is 1855 pm. What is the smallest angle of diffraction of X-rays with $\lambda = 154$ pm from this solid?

18.118. Minnesotaite [$Fe_3Si_4O_{10}(OH)_2$] is a silicate mineral with a layered structure similar to that of kaolinite. The distance between the layers in minnesotaite is 1940 ± 10 pm. What is the smallest angle of diffraction of X-rays with $\lambda = 154$ pm from this solid?

Additional Problems

18.119. A unit cell consists of a cube that has an ion of element X at each corner, an ion of element Y at the center of the cube, and an ion of element Z at the center of each face. What is the formula of the compound?

18.120. An element crystallizes in the cubic closest-packed structure. The length of an edge of the unit cell is 408 pm. The density of the element is 19.27 g/cm^3. Identify the element.

***18.121.** What is the packing efficiency of the Si atoms in pure Si if the radius of one Si atom is 117 pm? The density of pure silicon is 2.33 g/cm^3.

*18.122. **The Composition of Light-Emitting Diodes** The colored lights on many electronic devices are light-emitting diodes (LEDs). One of the compounds used to make them is aluminum phosphide (AlP), which crystallizes in a sphalerite crystal structure.

a. If AlP were an ionic compound, would the ionic radii of Al^{3+} and P^{3-} be consistent with the size requirements of the ions in a sphalerite crystal structure?

b. If AlP were a covalent compound, would the atomic radii of Al and P be consistent with the size requirements of atoms in a sphalerite crystal structure?

18.123. Under the appropriate reaction conditions, small cubes of molybdenum, 4.8 nm on a side, can be deposited on carbon surfaces. These "nanocubes" are made of bcc arrays of Mo atoms.

a. If the edge of each nanocube corresponds to 15 unit cell lengths, what is the effective radius of a molybdenum atom in these structures?

b. What is the density of each molybdenum nanocube?

c. How many Mo atoms are in each nanocube?

18.124. In the fullerene known as buckminsterfullerene, molecules of C_{60} form a cubic closest-packed array of spheres with a unit cell edge length of 1410 pm.
a. What is the density of crystalline C_{60}?
b. If we treat each C_{60} molecule as a sphere of 60 carbon atoms, what is the radius of the C_{60} molecule?
c. C_{60} reacts with alkali metals to form M_3C_{60} (where M = Na or K). The crystal structure of M_3C_{60} contains cubic closest-packed spheres of C_{60} with metal ions in holes. If the radius of a K^+ ion is 138 pm, which type of hole is a K^+ ion likely to occupy? What fraction of the holes will be occupied?
d. Under certain conditions, a different substance, K_6C_{60}, can be formed in which the C_{60} molecules have a bcc unit cell. Calculate the density of a crystal of K_6C_{60}.

18.125. **Earth's Core** The center of Earth is composed of a solid iron core within a molten iron outer core. When molten iron cools, it crystallizes in different ways depending on pressure—in a bcc unit cell at low pressure and in a hexagonal unit cell at high pressures like those at Earth's center.
a. Calculate the density of bcc iron, assuming that the radius of an iron atom is 126 pm.
b. Calculate the density of hexagonal iron, given a unit cell volume of 5.414×10^{-23} cm^3.
*c. Seismic studies suggest that the density of Earth's solid core is less than that of hexagonal Fe. Laboratory studies have shown that up to 4% by mass of Si can be substituted for Fe without changing the hcp crystal structure built on hexagonal unit cells. Calculate the density of such a crystal.

18.126. The unit cell of an alloy with a 1:1 ratio of magnesium and strontium is identical to the unit cell of CsCl. The unit cell edge length of MgSr is 390 pm.
a. What is the density of MgSr?
b. Find the atomic radii of Mg and Sr in Appendix 3. Do atoms touch along the body diagonal of the unit cell?
*c. Why doesn't the formula of the alloy allow us to distinguish between a Mg atom in the cubic hole of simple cubic unit cell of Sr atoms and a Sr atom in the cubic hole of a simple cubic arrangement of Mg atoms?
*d. MgSr is a good electrical conductor. Do you expect a MgSr alloy to have a partially filled valence band or overlapping conduction and valence bands?

18.127. Substitutional alloys may form when the difference in atomic radii between the alloying elements is less than 15%.
a. Predict which of the following alloys has the greatest mismatch in atomic radii: AuZn, AgZn, or CuZn.
b. Find the atomic radii of Cu, Ag, Au, and Zn in Appendix 3, and calculate the percent difference in their atomic radii. Are all three alloys expected to form substitutional alloys?
c. If gold is alloyed with silver in a 1:1 ratio, do the atoms still touch along the face diagonal of a face-centered cubic unit cell?

*18.128. Removing two electrons from S_8 yields the dication S_8^{2+}. Will all of the sulfur atoms be in one plane in the S_8^{2+} cation?

18.129. Figure P18.129 shows the unit cell of a transition metal sulfide.
a. What is the formula of the compound?
b. How many M^{2+} and S^{2-} ions are there in the unit cell?
c. Which spheres represent the S^{2-} anions?
d. How many yellow spheres does each brown sphere touch?

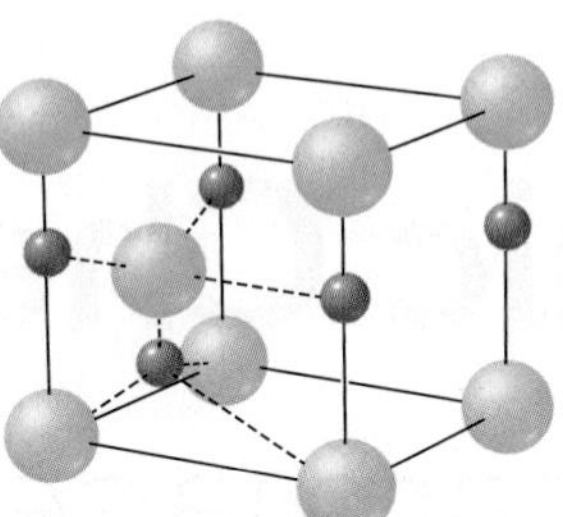

FIGURE P18.129

18.130. A transition metal sulfide crystallizes in the unit cell shown in Figure P18.130, where the metal (M^{n+}) ions occupy half of the tetrahedral holes in a cubic closest-packed arrangement of sulfide ions.
a. What is the formula of the compound?
b. How many equivalent M^{n+} and S^{2-} ions are there in the unit cell?
c. How does the unit cell in Figure P18.130 differ from the unit cell in Figure P18.129?
d. What percent of the unit cell is empty space?

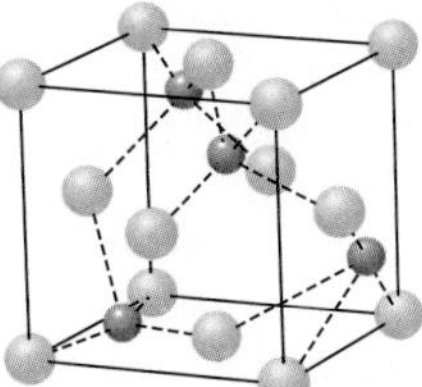

FIGURE P18.130

***18.131.** The alloy Cu_3Al crystallizes in a bcc unit cell. Propose a way that the Cu and Al atoms could be allocated between bcc unit cells that is consistent with the formula of the alloy.

18.132. Aluminum forms substitutional alloys with lithium (LiAl), gold ($AuAl_2$), and titanium (Al_3Ti).
a. Do these alloys fit the general size requirements for substitutional alloys? The atomic radii for Li, Al, Au, and Ti are 152, 143, 144, and 147 pm, respectively.
b. If the unit cell of LiAl is bcc, what is the density of LiAl?

19

Organic Chemistry

Fuels, Pharmaceuticals, and Modern Materials

TOUGH MATERIALS Physical properties of polymers, like the impact resistance of a hockey goalie's mask, are linked directly to the composition and structure of the molecules from which they are made.

PARTICULATE **REVIEW**

Carbon–Carbon Bonds

In Chapter 19, we explore the many ways carbon atoms bond to each other and to atoms of other elements in organic compounds, including the three hydrocarbons shown here.

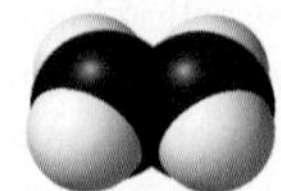

- Draw the Lewis structure of each compound.
- Which hydrocarbon has the strongest carbon–carbon bond? Which has the longest carbon–carbon bond?
- Which compound contains only sigma bonds?

(Review Section 6.1 if you need help.)

(Answers to Particulate Review questions are in the back of the book.)

PARTICULATE **PREVIEW**

Same Composition, Different Structures

Compound A and compound B have the same composition but different structures. As you read Chapter 19, look for ideas that will help you answer these questions:

- What word describes two compounds that have the same composition but different structures?
- What functional group is present in each molecule?
- The functional group in Compound A can be converted into the functional group in Compound B by reacting with a molecule that contains yet a third functional group. What is that functional group?

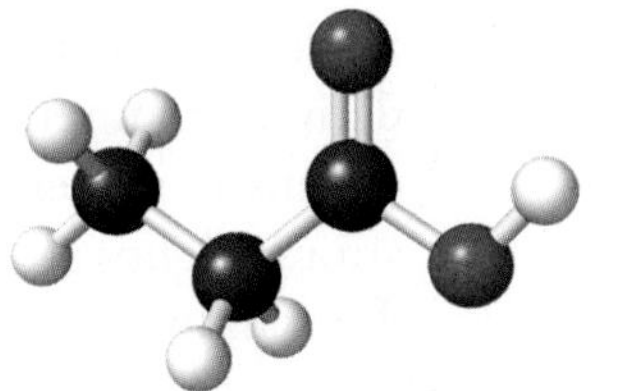

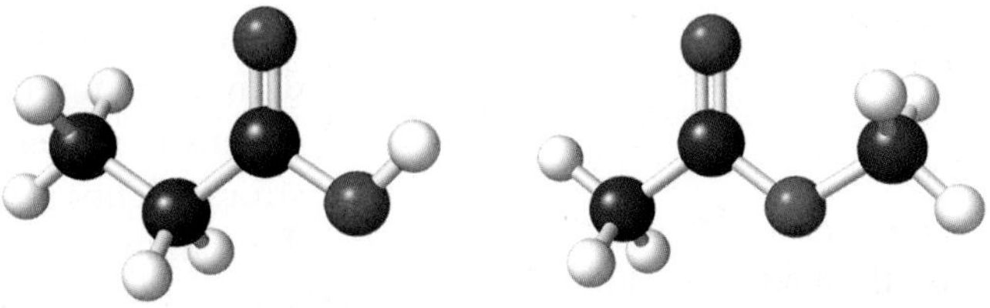

Learning Outcomes

LO1 Draw molecular structures of alkanes
Sample Exercises 19.1, 19.2

LO2 Recognize and name constitutional isomers of alkanes and draw their molecular structures
Sample Exercises 19.3, 19.4

LO3 Explain the differences in the chemical reactivities of alkanes, alkenes, and alkynes based on the bonding between carbon atoms in their molecules
Sample Exercise 19.5

LO4 Recognize and name constitutional isomers and stereoisomers of alkenes and draw their molecular structures
Sample Exercise 19.6

LO5 Identify the monomers that react and the polymers that are produced in addition and condensation polymerization reactions
Sample Exercises 19.7, 19.10, 19.12

LO6 Compare the fuel values of hydrocarbons and oxygenated fuels
Sample Exercise 19.8

LO7 Relate the properties of polymers to their molecular structures
Sample Exercises 19.9, 19.11

LO8 Explain the properties of organic compounds based on the size of their molecules and the reactivity of their functional groups

19.1 Carbon: The Stuff of Daily Life

We drive cars powered by gasoline or diesel fuel. Some of us cook food on grills that burn propane. Most of us live in homes heated by natural gas, fuel oil, or electricity, and about two-thirds of the electricity generated in the United States is produced by burning fossil fuels. All of these fuels are organic compounds—which means they are carbon-based compounds—and they are used as sources of energy because their combustion reactions are highly exothermic. The same is true of the food we eat. Most of the medicines we use are also organic compounds, whether they are natural products derived from living plants or animals or are synthesized by reacting other organic compounds together abiotically.

Your favorite T-shirt may be made of cotton, an organic natural fiber. Because of the composition, shape, and orientation of the large molecules that form cotton fibers, the material is soft and absorbent. In contrast, the soles of your running shoes are made of a synthetic rubber, a polymer manufactured from compounds of carbon and hydrogen that were originally derived directly from crude oil but now increasingly come from recycled materials. Because of the size, shape, and orientation of the polymer molecules in synthetic rubber, the material is springy, nonabsorbent, and resistant to wear.

Fans at a hockey game sit on molded plastic seats and munch on hot dogs wrapped in plastic wrap while watching players who wear helmets of impact-resistant polymeric materials. Plastics, cling wrap, and hockey helmets are all made from compounds derived from crude oil or, increasingly, from ethanol derived from renewable, sustainable resources like sugarcane as part of the green chemistry movement. Either way, they are organic compounds. The plastic seat is strong, lightweight, and capable of bearing considerable mass without breaking. The sandwich wrap must be flexible yet prevent oxygen from reaching the hot dog and bun before you open it to eat it, and the players' helmets must be lightweight but still able to withstand the impact of a hockey puck shot at 150 km/h. All these physical properties can be designed into the materials at the molecular level because they are directly linked to the structure, composition, and size of the molecules.

There are over 100 million carbon compounds, and they are so varied in size, shape, and properties that an entire field of chemistry, called **organic chemistry**,

organic chemistry the study of compounds containing C—C and/or C—H bonds.

R a symbol in a general formula standing for an organic group that has one available bond; it is used to indicate the part of a molecule that is not the functional group of interest.

is devoted to their study. This chapter is a brief introduction to organic chemistry. The designation *organic* for carbon-containing compounds was once limited to substances produced by living organisms, but that definition has been broadened for two reasons. First, since 1828, when Friedrich Wöhler (1800–1882) discovered how to prepare urea (Figure 19.1) in the laboratory "without the intervention of a kidney," scientists have learned to synthesize many materials previously thought to be the products only of living systems. Second, chemists have learned how to synthesize many carbon-based materials that have never been produced by living systems. Chemical Abstracts Service (CAS), an organization that tracks and collects all chemical information published worldwide, maintains a registry of all the known unique chemical substances. About 15,000 new ones are added to the registry each day. These statistics help explain why the study of organic chemistry occupies a special place within the discipline.

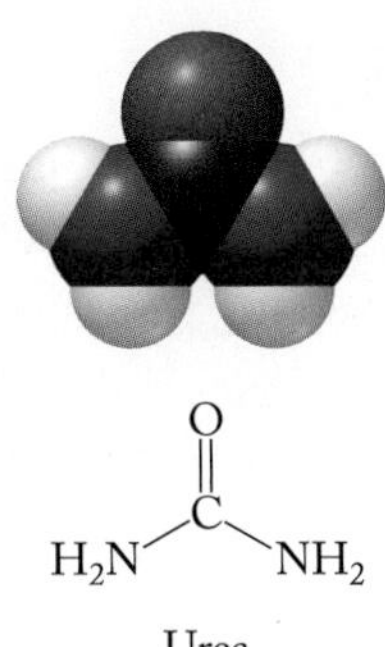

FIGURE 19.1 Urea was the first naturally occurring organic compound synthesized in the laboratory. It was prepared in 1828 by Friedrich Wöhler.

Much of the variety in the chemistry of carbon is derived from a carbon atom's ability to form four covalent bonds with a variety of other elements as well as other carbon atoms. The resulting compounds may contain a few carbon atoms or many thousands. Additional structural variety comes from the many patterns that carbon atoms can adopt, from long chains to heavily branched structures to planar and puckered rings.

CONNECTION In Chapter 5, we learned that compounds with the same molecular formula but different arrangements of atoms differ in physical properties such as melting point and boiling point.

With so many substances to deal with, some organizing principles are necessary. We use two of the most common ones in this chapter, and we have already introduced both of them elsewhere in this text. The first is based on *functional groups*: subunits of only a few atoms that confer particular chemical and physical properties to organic compounds. The second is based on molecular size. Molecules having masses less than 1000 amu are typically considered small molecules; molecules with masses up to and exceeding 1,000,000 amu are called *macromolecules*.

CONNECTION We first used the term *functional group* in Chapter 6 to describe the groups of atoms in molecules that are most responsible for the substance's physical and chemical properties.

The functional groups present in a molecule and the molecule's size and shape combine to determine the physical and chemical properties of materials at the microscopic and bulk levels. Indeed, the influence of the composition, structure, and size of molecules on bulk properties of materials is a theme of this chapter. In addition, the shape of many macromolecules in solution depends on their environment and their association with other molecules. The role of intermolecular interactions on properties of bulk materials is also an important theme here in Chapter 19 and in Chapter 20 on biochemistry.

Families Based on Functional Groups

We have already discussed several functional groups in prior chapters, including the –COOH group in carboxylic acids and the hydroxyl (–OH) group in alcohols. In this chapter we examine these and the other functional groups summarized in Table 19.1. When discussing functional groups, the convention is to use **R** to represent the entire molecule except the functional group. (Do not confuse it with the italic R representing the universal gas constant.)

Throughout this book we have seen numerous examples of organic compounds containing these functional groups. We focus on the functional groups in this chapter because much of organic chemistry deals with chemical reactions of organic substances and how those reactions can be used to modify substances, to find synthetic routes to known materials produced by nature, and to find ways to make totally new materials that have desirable properties. Understanding the characteristic reactivity of specific functional groups enables chemists to carry out these tasks in a logical fashion. Introducing the unique reactivities of the

TABLE 19.1 Functional Groups of Organic Compounds

Name	Structural Formula of Group	Example	Name
Alkane	R—H	$CH_3CH_2CH_3$	Propane
Alkene	$C{=}C$ (with four open bonds)	$H_2C{=}CH_2$	Ethylene (ethene)
Alkyne	$—C{\equiv}C—$	$H—C{\equiv}C—H$	Acetylene (ethyne)
Aromatic	six-membered ring of C atoms with alternating double bonds, one open bond on each C	C_6H_6	Benzene
Amine	R—NH_2 R—NHR R—NR_2	$H_3C—NH_2$	Methylamine
Alcohol	R—OH	CH_3CH_2OH	Ethanol
Ether	R—O—R	$H_3C—O—CH_3$	Dimethyl ether
Aldehyde	R—C(=O)—H	$H_3C—C(=O)—H$	Acetaldehyde
Ketone	R—C(=O)—R	$H_3C—C(=O)—CH_3$	Acetone
Carboxylic acid	R—C(=O)—OH	$H_3C—C(=O)—OH$	Acetic acid
Ester	R—C(=O)—OR	$H_3C—C(=O)—OCH_3$	Methyl acetate
Amide	R—C(=O)—NH_2	$H_3C—C(=O)—NH_2$	Acetamide

functional groups listed in Table 19.1 is beyond the scope of this text, but we will illustrate a small set of reactions of molecules bearing some of these functional groups that are used in the industrial production of polymers.

Monomers and Polymers

Many macromolecules consist of long chains made up of subunits called **monomers** that are chemically bonded together; the overall macromolecule is called a **polymer**. The mass boundaries that distinguish small molecules from large ones are somewhat arbitrary, and medium-sized organic molecules called *oligomers* inhabit the realm between small molecules and polymers. Organizing molecules by their size works because of the predictable ways properties vary as composition remains constant but size increases.

Organic compounds composed of small molecules have constant composition and well-defined properties such as melting points and boiling points. In contrast, many synthetic polymers do not have constant composition in terms of the number of monomers bonded together to make them. They may not have well-defined physical properties because they are mixtures of molecules that are similar but not identical: they may differ in the number and arrangement of monomers in their molecular structures. In this chapter we explore the structures of organic polymers and the intermolecular forces between them that define the properties of these polymers and the many familiar materials made from them.

monomer a small molecule that bonds with other monomers to form polymers.

polymer a very large molecule with high molar mass; the root word *meros* is Greek for "part" or "unit," so *polymer* literally means "many units."

saturated hydrocarbon an alkane.

Kekulé structure a structure using lines to show all of the bonds in a covalently bonded molecule, but not showing lone pairs on the atoms.

19.2 Alkanes

Recall from Chapter 6 that *hydrocarbons* are compounds whose molecules are composed of only carbon and hydrogen atoms, and *alkanes* are hydrocarbons whose molecules contain only single bonds. Alkanes are called **saturated hydrocarbons** because they contain the maximum ratio of hydrogen atoms to carbon atoms. The general molecular formula of alkanes is C_nH_{2n+2}, where n is the number of carbon atoms per molecule.

Drawing Organic Molecules

Figure 19.2 shows five different ways we represent the molecular structure of organic compounds—in this case, the five-carbon alkane pentane. In its Lewis structure, all of the bonds in the molecule are shown along with any lone pairs on the atoms. Alkanes do not have lone pairs on any of their atoms, but lone pairs are common in other organic molecules. When we draw structures of organic molecules by showing all the bonds with lines but leaving off any lone pairs, the structural formulas are called **Kekulé structures** after August Kekulé (1829–1896), the German chemist who first used this method of illustrating molecules.

Because writing Lewis or Kekulé structures for large organic molecules can be tedious, chemists use various shorter notations to convey structures in organic chemistry, such as the *condensed structures* shown in Figure 19.2(d). These structures do not show the individual bonds between atoms the way a Lewis structure does. Sometimes subscripts are used to indicate the number of times a particular subgroup, written in parentheses, is repeated. For example, the condensed structure of pentane can be written as $CH_3(CH_2)_3CH_3$ (Figure 19.2d). The numerical

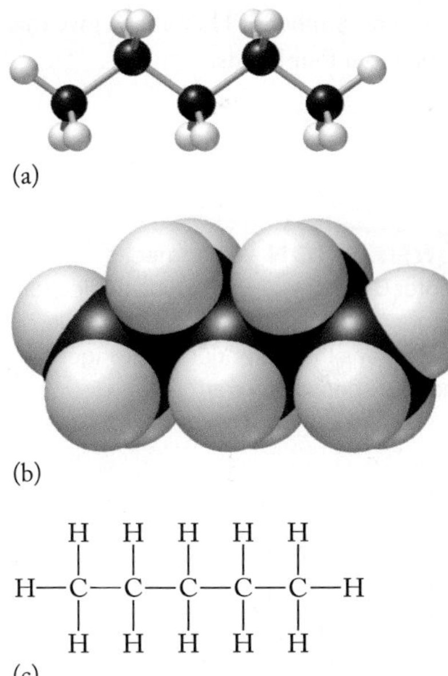

(c)

$CH_3CH_2CH_2CH_2CH_3$

$CH_3(CH_2)_3CH_3$

(d)

(e)

FIGURE 19.2 Different representations of the molecular structure of pentane: (a) ball-and-stick model; (b) space-filling model; (c) Lewis or Kekulé structure; (d) two kinds of condensed structures; (e) carbon-skeleton structure.

CONNECTION Kekulé structures are the same as the structural formulas we have been drawing throughout the text since Chapter 1.

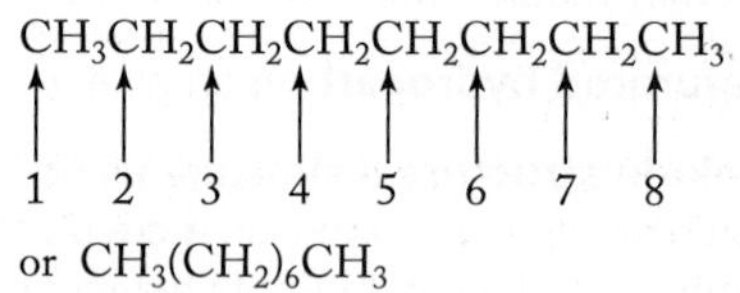

(a) Condensed structure

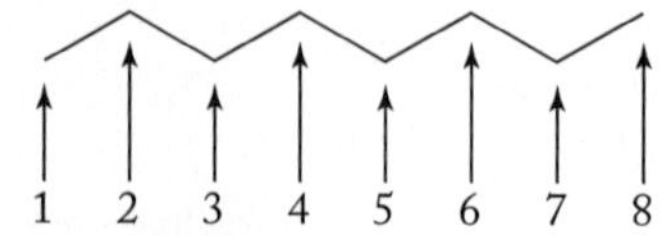

(b) Carbon-skeleton structure

FIGURE 19.3 When converting from (a) a condensed structure to (b) a carbon-skeleton structure, the carbon atoms are represented by junctions between lines, and each junction is assumed to have a sufficient number of H atoms to give that carbon atom four bonds.

subscript "3" after the parenthetical $-CH_2-$ group means that three of these groups connect the two terminal $-CH_3$ groups in this compound.

The most minimal notation, shown in Figure 19.2(e), is the *carbon-skeleton structure*, in which no alphabetic symbols are used for carbon and hydrogen atoms. (Atoms other than C and H are shown in carbon-skeleton structures, as we demonstrate in Section 19.3.) Figure 19.3 shows how a carbon-skeleton structure is created. Short line segments are drawn at angles to one another, and each line segment represents one carbon–carbon bond in the molecule. The angles represent the bond angle between the two carbon atoms (109.5° in the case of sp^3 hybridized carbon atoms in alkanes). Each end of the zigzag line is a $-CH_3$ group, and every intersection of two line segments is a $-CH_2-$ group. It is assumed that each carbon atom is bonded to enough hydrogen atoms to give it a steric number of 4.

Physical Properties and Structures of Alkanes

Alkanes are also known as *paraffins*, a name derived from Latin meaning "little affinity." This is a perfect description of the alkanes, which tend to be much less reactive than the other organic compounds. Despite their lack of reactivity, alkanes are important because they are widely used as fuels and lubricants. Like most fuels, alkanes often need some source of energy, such as a spark from an engine's spark plug, to initiate combustion. Figure 19.4 shows an energy profile for the progress of the combustion of methane. This reaction is highly exothermic, but a significant activation energy barrier must be overcome for the reaction to proceed.

CONNECTION We learned in Chapter 6 that the larger clouds of electrons in larger molecules are more polarizable, which means they experience stronger London dispersion forces.

In the language of organic chemistry, a **homologous series** is defined as a series of compounds in which members can be described by a general formula and similar chemical properties. The **linear (straight-chain) hydrocarbons** in the alkane family are a homologous series with the general formula C_nH_{2n+2}. Those whose molecules have three or more carbon atoms have the generic condensed structure $CH_3(CH_2)_nCH_3$. Each of these molecules differs from the next by one $-CH_2-$ unit, called a **methylene group**. The terminal CH_3- groups are **methyl groups**. As we noted in Chapter 6, the physical properties of linear alkanes, including their viscosities, melting points, and boiling points, increase with increasing molar mass (see Figure 6.2). This trend is typical of all homologous series: as molecules increase in size, they experience stronger London dispersion forces, which means, for example, that higher temperatures are required to melt them if they are solids or to vaporize them if they are liquids.

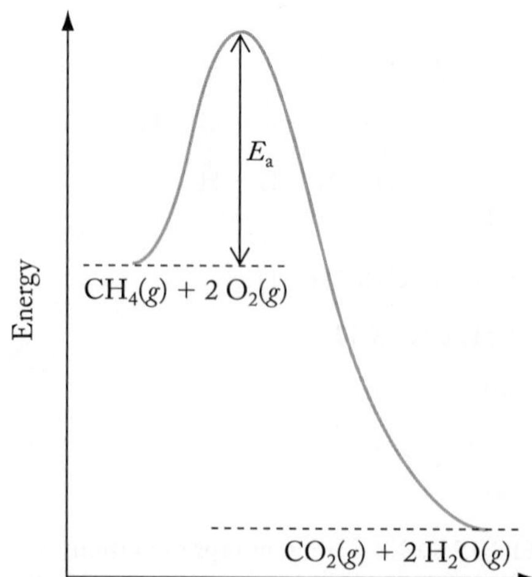

FIGURE 19.4 The energy profile for the combustion of methane illustrates that a large activation energy barrier E_a must be overcome before the reaction can proceed.

SAMPLE EXERCISE 19.1 Drawing Structures of Alkanes — LO1

Write the condensed structure and the carbon-skeleton structure for the following linear alkanes: (a) the three-carbon propane and (b) the 12-carbon dodecane.

Collect and Organize Figure 19.2 summarizes the differences between condensed and carbon-skeleton structures, which we need to apply to the hydrocarbons with three and 12 carbon atoms. Condensed structures show the symbol for each element in a molecule and subscripts indicating the numbers of atoms of each element, but they do not show C—H or single C—C bonds. Carbon-skeleton structures use short, zigzag line segments to represent the carbon–carbon bonds in molecules. The ends of these segments represent carbon atoms.

Analyze Each propane and dodecane chain has methyl groups (CH_3–) at both ends and one or more methylene groups (–CH_2–) in between. We can group methylene groups that are bonded together in a condensed structure into one term by using parentheses followed by a subscript showing the number of –CH_2– groups.

Solve

a. Propane has three carbon atoms, so the condensed structure is $CH_3CH_2CH_3$. The carbon-skeleton structure consists of two segments representing the two carbon–carbon bonds:

```
    H   H   H
    |   |   |
H — C — C — C — H
    |   |   |
    H   H   H
```

Lewis structure	Condensed structure	Carbon-skeleton structure
(above)	$CH_3CH_2CH_3$	(two-segment zigzag)

b. Dodecane is a continuous chain of 12 carbon atoms that has two methyl groups at the ends and 10 methylene groups in between. The condensed structure is therefore $CH_3CH_2CH_2CH_2CH_2CH_2CH_2CH_2CH_2CH_2CH_2CH_3$ or, gathering together the 10 –CH_2– groups, $CH_3(CH_2)_{10}CH_3$. The carbon-skeleton structure must contain as many ends and intersections (12) as there are carbon atoms in the molecule or 11 segments in all:

Think About It Condensed and carbon-skeleton structures must both contain the same number of atoms and the same number of bonds, even though the bonds are not shown in the former and not all of the atoms or bonds are shown in the latter.

 **Practice Exercise** Draw the carbon-skeleton structure of hexane, $CH_3(CH_2)_4CH_3$, and a condensed structure of heptane:

Heptane

(Answers to Practice Exercises are in the back of the book.)

homologous series a set of related organic compounds that differ from one another by the number of common subgroups, such as –CH_2–, in their molecular structures.

linear (straight-chain) hydrocarbon a hydrocarbon in which the carbon atoms are bonded together in one continuous line. Linear alkane chains have a methyl group at each end with methylene groups connecting them.

methylene group (–CH_2–), a structural unit that can make two bonds.

methyl group (CH_3–), a structural unit that can make only one bond.

branch a side chain bonded to the main (longest) chain in an organic molecule.

Structural Isomers Revisited

The alkane family would be huge even if it consisted of only straight-chain molecules, but another structural possibility makes the family even larger. We first saw this possibility in Chapter 6 when we compared the boiling points of alkanes having the same molecular formula (C_5H_{12}) but different shapes (see Figure 6.3). Similarly, the straight-chain molecule shown in the center of Table 19.2 is not the only shape possible for a four-carbon alkane, because with four or more carbon atoms per molecule *branched* structures also exist, as shown on the right side of the table. A **branch** is a side chain attached to the main (longest) chain.

The molecular formula of both structures in Table 19.2 is C_4H_{10}, but they represent different compounds with different properties and names (butane and 2-methylpropane). At this point it is not important to be able to name them, but you do need to be able to recognize that they are *constitutional isomers* (also called *structural isomers*). As we learned in Chapter 6, such compounds have the same

TABLE 19.2 Comparing Constitutional Isomers

	Butane	2-Methylpropane
Condensed Structure	$CH_3CH_2CH_2CH_3$	$CH_3CH(CH_3)CH_3$ or 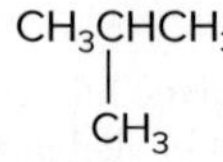 CH_3CHCH_3 with CH_3 bonded to the center carbon
Carbon-Skeleton Structure		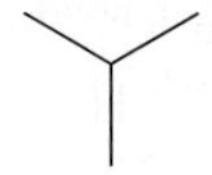
Melting Point (K)	135	114
Normal Boiling Point (K)	273	261

Temperatures measured at 1 atm.

molecular formula but have their atoms connected in different bonding patterns. This difference makes constitutional isomers chemically distinct from one another.

2-Methylpropane is an example of a branched hydrocarbon, which means that its molecular structure contains a main chain (the longest carbon chain in the molecule) and at least one side chain or branch. We can illustrate constitutional isomerism with either condensed or carbon-skeleton structures, as shown in Table 19.2. There are two ways to show the location of the methyl group. One way is to use condensed structures and to put the methyl group in parentheses next to the carbon to which it is bonded: $CH_3CH(CH_3)CH_3$. We can also show the carbon–carbon bond between CH_3 and the center carbon of the three-carbon main chain. In this instance, it may actually be easier to recognize the constitutional isomers of C_4H_{10} using the carbon-skeleton structures, which show that butane has no carbon atoms bonded to three other carbon atoms, but 2-methylpropane does.

SAMPLE EXERCISE 19.2 Identifying the Longest Chain in Organic Molecules **LO1**

Determine the number of carbon atoms in the longest chain in the following three branched, saturated hydrocarbons:

a. 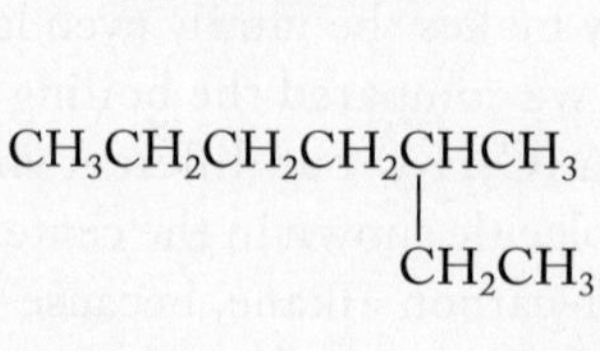 $CH_3CH_2CH_2CH_2CHCH_3$ (with CH_2CH_3 bonded to the CH carbon)

b. 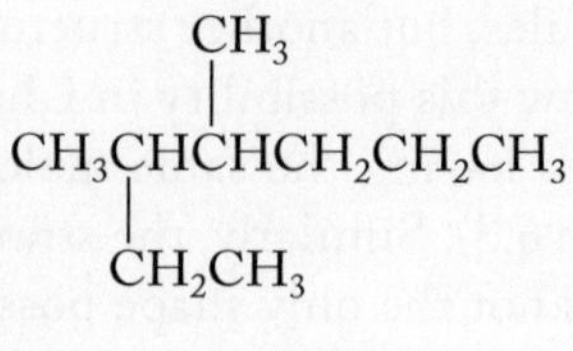 $CH_3CHCHCH_2CH_2CH_3$ (with CH_3 bonded to the third carbon and CH_2CH_3 bonded to the second carbon)

c. $(CH_3)_2CHCH(CH_3)C(CH_3)_2CH_2CH_3$

Collect and Organize We are asked to identify the longest carbon chain or *main chain* in three compounds that are represented by condensed structures.

Analyze We assume that there is a bond between each adjacent pair of carbon atoms in the condensed structures. Carbon bonds to side chains are shown using single lines in structures a and b. In structure c, groups in parentheses represent side-chain groups shown next to the carbon atoms to which they are bonded. We need to determine which branches belong to the main chain and which belong to the side chains.

Solve For a and b, we start at one end of any branch and assign numbers to each carbon atom. If the chain branches, we must choose one branch to follow. Later, we may wish to return to the compound and number the carbon atoms starting from a different end of the molecule or taking a different branch. For c, we first expand this form of condensed structure to more clearly show the branches.

a. Starting with the carbon atom furthest to the left and numbering consecutively from left to right, we reach a branch at carbon atom 5. If we continue straight across the page, we find that the chain contains six carbon atoms. If we take the branch at carbon atom 5, however, we find that the chain contains seven carbon atoms. This makes the longest chain in the compound seven carbon atoms long.

1 2 3 4 5 6
$CH_3CH_2CH_2CH_2CHCH_3$
CH_2CH_3

1 2 3 4 5
$CH_3CH_2CH_2CH_2CHCH_3$
CH_2CH_3
6 7

b. There are four places to start counting carbon atoms in structure b. Some possible ways to number the molecule are as follows:

CH_3
1 2 3 4 5 6
$CH_3CHCHCH_2CH_2CH_3$
CH_2CH_3

CH_3
5 4 3 2 1
$CH_3CHCHCH_2CH_2CH_3$
CH_2CH_3
6 7

5
CH_3
4 3 2 1
$CH_3CHCHCH_2CH_2CH_3$
CH_2CH_3

The longest chain in this molecule (the main chain) also has seven carbon atoms.

c. We draw out the condensed structure and then number the carbon atoms. In this case, there are two ways to number the molecule that results in the longest chain having six carbon atoms.

CH_3 CH_3
1 2 3 4 5 6
$CH_3CHCHCCH_2CH_3$
H_3C CH_3

1
CH_3 CH_3
2 3 4 5 6
$CH_3CHCHCCH_2CH_3$
H_3C CH_3

Think About It The main chain in an organic compound may not always be the one running horizontally across the page.

Practice Exercise Determine the number of carbon atoms in the longest chain in all three of the following branched, saturated hydrocarbons:

a.

b.

c. $CH_3CH_2C(CH_3)_2CH(CH_2CH_3)CH_2CH(CH_2CH_3)_2$

How do we determine whether two condensed structures or carbon-skeleton structures represent two different compounds, a single compound, or two constitutional isomers? Our first step is to translate each structure into a molecular formula.

If the molecular formulas are different, the structures represent two different compounds. If the molecular formulas are the same, we must compare the way the atoms are connected in the two structures to determine whether they are the same compound or constitutional isomers. Draw the structures with the longest chain horizontal, and then check to see if the same side chains are attached at the same positions along the longest chain. In doing so, we may have to reverse or rotate one of the structures. For example, structures 1 and 2 in the accompanying figure may seem to be different, but rotating structure 2 by 180° shows that it is the same as structure 1. The two represent the same compound: a four-carbon main chain with a one-carbon side chain connected to the carbon atom next to a terminal carbon.

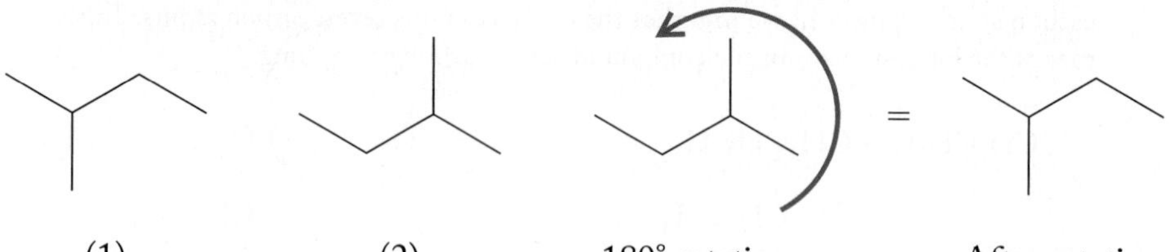

(1) (2) 180° rotation = After rotation

SAMPLE EXERCISE 19.3 Recognizing Constitutional Isomers LO2

Do the two structures in each set describe the same compound, constitutional isomers, or compounds with different molecular formulas?

a. $(CH_3)_2CHCH_2CH(CH_3)_2$

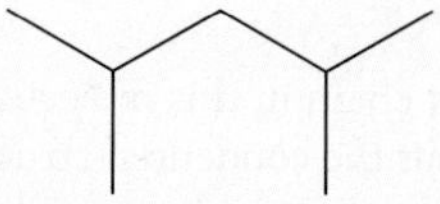

b.

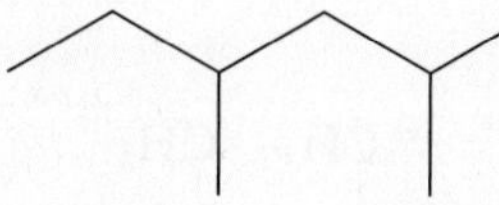

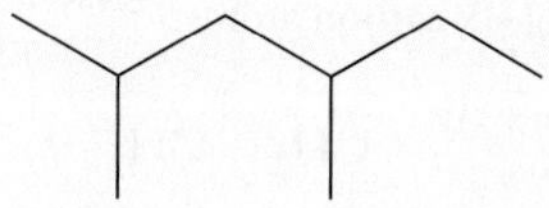

c.

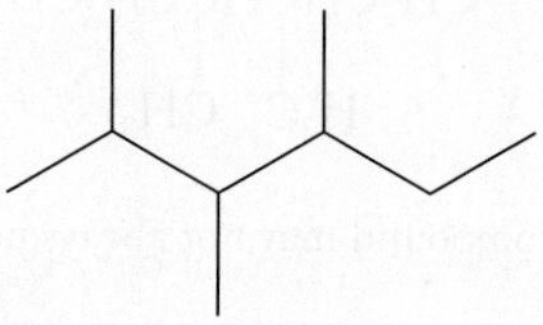

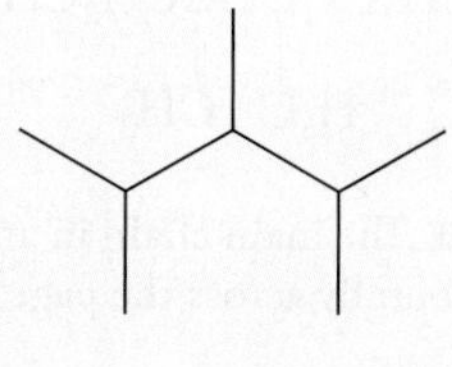

d. 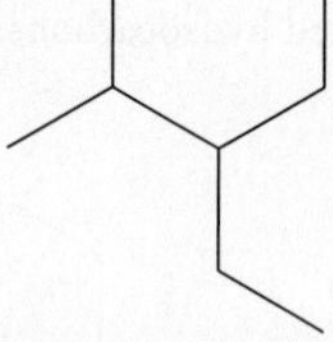 $CH_3CH_2C(CH_3)_2CH(CH_3)_2$

Collect and Organize Identical compounds have the same molecular formula and the same connectivity and spatial arrangement of the atoms. Constitutional isomers have the same molecular formula but different connectivity. Compounds with different molecular formulas are different compounds.

Analyze In each pair, we check first to see if the molecular formulas are the same. That means we count the number of carbon and hydrogen atoms. If the molecular formulas are the same, we may have one compound drawn two ways or a pair of constitutional isomers. If the carbon skeletons are the same in any pair of drawings, the drawings represent the same hydrocarbon.

Solve

a. The molecular formulas are the same: C_7H_{16}. Converting the condensed structure to a carbon-skeleton structure gives us

$(CH_3)_2CHCH_2CH(CH_3)_2 \longrightarrow$

This structure is identical to the carbon-skeleton structure in this set. Therefore, the condensed structure and carbon-skeleton structure represent the same molecule.

b. Both molecules in this set have the molecular formula C_8H_{18}. Both have six carbon atoms in the longest chain. Both have a methyl group attached to the second carbon atom from one end of the chain and another methyl group attached to the third carbon atom from the other end. Therefore, the two structures represent the same compound:

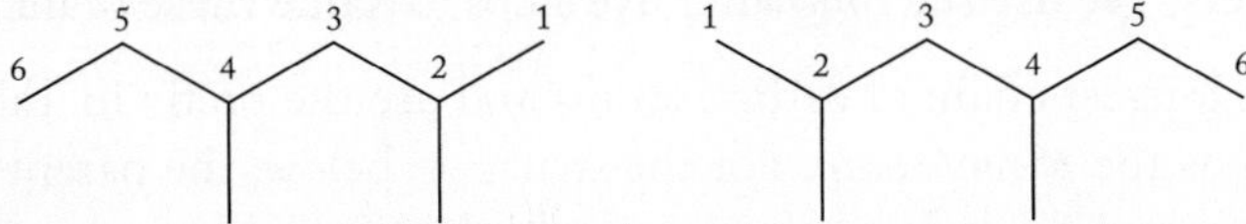

c. The left structure contains nine carbon atoms, but the right structure contains only eight. Therefore, the two structures represent different compounds.

d. The molecular formulas are the same: C_8H_{18}. Rotating the left hand structure 180° and converting the condensed structure to a carbon-skeleton structure gives us

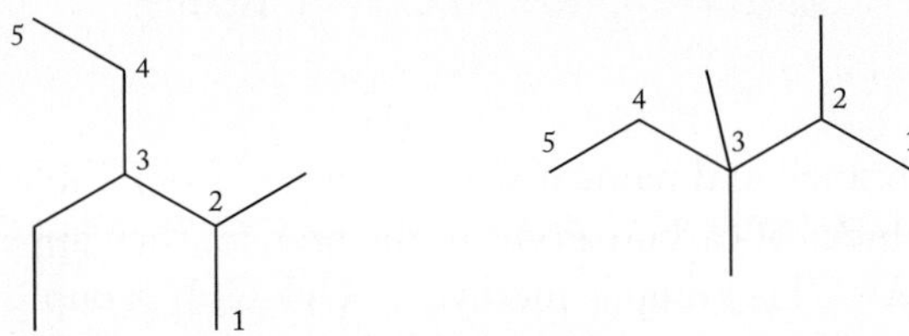

Both structures have a longest chain of five carbon atoms and a methyl group attached to C2 (the number 2 carbon atom). The structure on the left, however, has an ethyl group (CH_3CH_2–) on C3, whereas the structure on the right has two methyl groups (CH_3–). Thus, the two structures represent constitutional isomers.

Think About It Even if two compounds have the same molecular formula, they are not necessarily identical. We must determine whether they are constitutional isomers.

Practice Exercise Do the two structures in each set describe the same compound, two different compounds, or a pair of constitutional isomers?

a.

b.

c.

Naming Alkanes

Now that we know how to draw alkanes, let's look at a few simple rules for naming them. The nomenclature system follows the same pattern for all families of organic compounds, so we begin with rules for naming alkanes and will add a few more for other hydrocarbons. These rules are called IUPAC rules after the organization (International Union of Pure and Applied Chemistry) that defined them. Appendix 7 contains a more extensive treatment of nomenclature for organic compounds.

The names of the C_1 through C_{10} alkanes, in which all carbon atoms are bonded to no more than two other carbon atoms, start with the prefixes listed in Table 19.3 followed by *-ane*. However, we have seen that straight-chain alkanes have constitutional isomers in which at least one carbon atom is bonded to more than two others. We use the following five steps to name these branched alkanes:

TABLE 19.3 Prefixes for Naming Alkanes and Other Hydrocarbons

Prefix	Condensed Structure of Alkane	Name of Alkane
Meth-	CH_4	Methane
Eth-	CH_3CH_3	Ethane
Prop-	$CH_3CH_2CH_3$	Propane
But-	$CH_3(CH_2)_2CH_3$	Butane
Pent-	$CH_3(CH_2)_3CH_3$	Pentane
Hex-	$CH_3(CH_2)_4CH_3$	Hexane
Hept-	$CH_3(CH_2)_5CH_3$	Heptane
Oct-	$CH_3(CH_2)_6CH_3$	Octane
Non-	$CH_3(CH_2)_7CH_3$	Nonane
Dec-	$CH_3(CH_2)_8CH_3$	Decane

1. Select the longest chain of carbon atoms and use the prefix in Table 19.3 to name this as the *parent chain*. For the structure below, the parent chain is pentane because the longest chain is five carbon atoms long:

$$CH_3CH(CH_3)CH_2CH_2CH_3$$

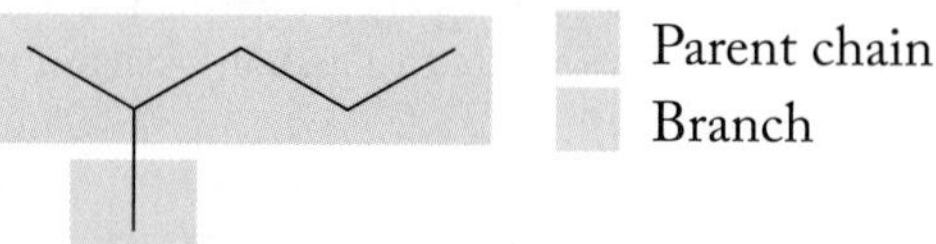

2. Identify each branch and name it with the prefix from Table 19.3 that defines the number of carbon atoms in the branch; then append the suffix *-yl* to the prefix. A $-CH_3$ group is methyl, a $-CH_2CH_3$ group is ethyl, and so on. The structure from step 1 has a methyl group branch. The name of the branch comes before the name of the parent chain, and the two are written together as one word: methylpentane.
3. To indicate where along the parent chain the branch is attached, we number the carbon atoms in the parent chain so that the branch (or branches, if there are more than one) has the lowest possible number. In the molecule from step 1, we start numbering from the left so that the methyl group is on carbon atom 2 in the parent chain (instead of on C4):

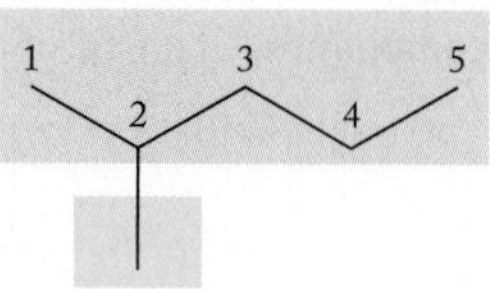

We indicate the position of the branch by a 2 followed by a hyphen in front of the name: 2-methylpentane.

4. If the same group is attached more than once to the parent chain, we use the prefixes *di-*, *tri-*, *tetra-*, and so on to indicate the number of groups present. The position of each group is indicated by the appropriate number before the group name, with the numbers separated by commas. Thus, the name of the following structure is 2,4-dimethylpentane:

$$\underset{1}{CH_3}\underset{2}{CH}(CH_3)\underset{3}{CH_2}\underset{4}{CH}(CH_3)\underset{5}{CH_3}$$

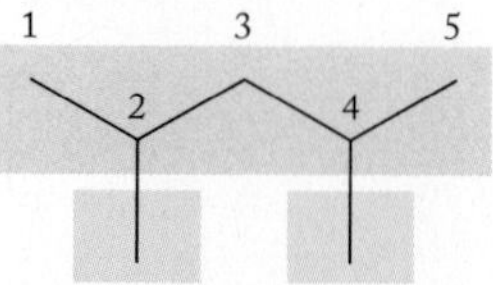

5. If different groups are attached to a parent chain, they are named in alphabetical order (for example, *e*thyl before *m*ethyl), as in 4-ethyl-3-methylheptane:

$$CH_3CH_2CH(CH_3)CH(CH_2CH_3)CH_2CH_2CH_3$$

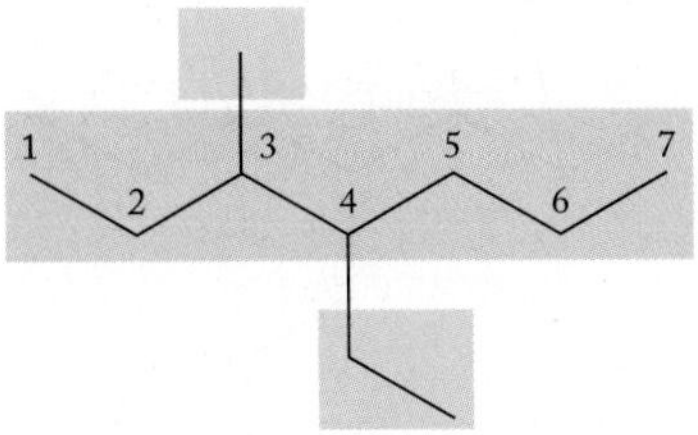

Numerical prefixes such as *di* and *tri* are not considered when determining alphabetical order. Thus, "ethyl" would precede "dimethyl" in the name of an alkane with one ethyl and two methyl groups. Note that we begin numbering the parent chain in 4-ethyl-3-methylheptane at the left end, as described in step 3, so that the first branch location has the lowest possible number. For example, the following compound is 5-ethyl-2-methyloctane, not 4-ethyl-7-methyloctane:

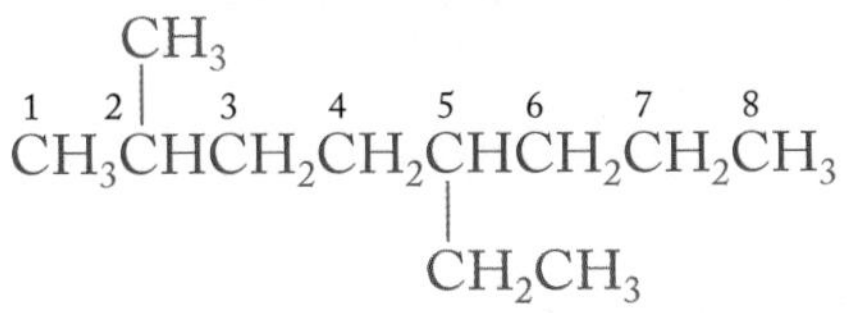

$$\underset{1}{CH_3}\underset{2}{C}H(CH_3)\underset{3}{C}H_2\underset{4}{C}H_2\underset{5}{C}H(CH_2CH_3)\underset{6}{C}H_2\underset{7}{C}H_2\underset{8}{C}H_3$$

Correct numbering

$$\underset{8}{CH_3}\underset{7}{C}H(CH_3)\underset{6}{C}H_2\underset{5}{C}H_2\underset{4}{C}H(CH_2CH_3)\underset{3}{C}H_2\underset{2}{C}H_2\underset{1}{C}H_3$$

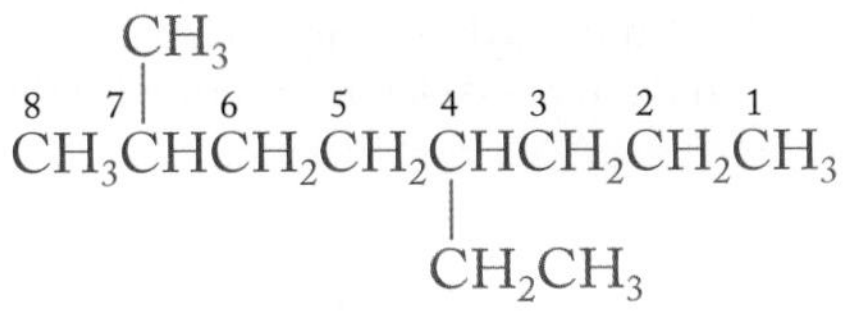

Incorrect numbering

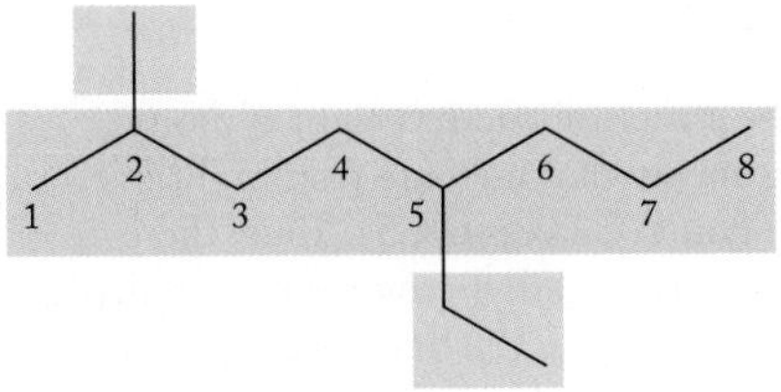

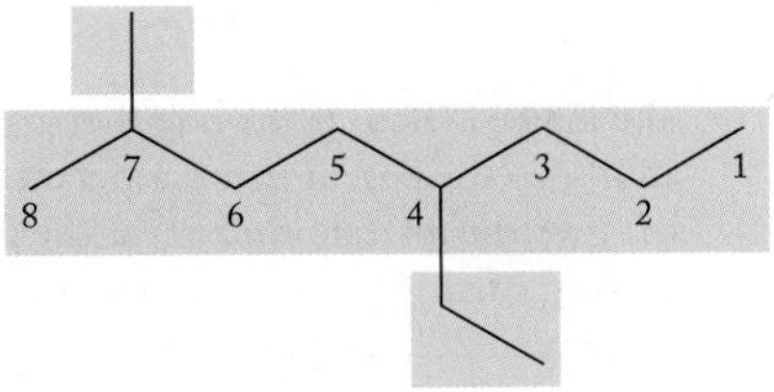

In practice, two or more paths of equal length may compete to be selected as the parent chain. We select the path in step 1 that (a) has the greatest number of side chains and (b) results in the lowest possible numbers for any branches as described in step 3. Therefore, we may need to do these two steps together and not in the order in which they are presented here.

SAMPLE EXERCISE 19.4 Naming Branched-Chain Alkanes **LO2**

Name the following branched-chain alkanes:

a. $CH_3CH_2C(CH_3)_2CH_2CH(CH_2CH_2CH_3)CH_2CH_2CH_2CH_3$

b. $CH_3CH_2CH_2CH_2C(CH_3)(CH_2CH_3)CH(CH_3)C(CH_3)_3$

Are they identical compounds, constitutional isomers, or different compounds with different molecular formulas?

Collect, Organize, and Analyze We are given two condensed structures. Identical compounds have identical names; constitutional isomers have identical formulas but different names; and different compounds have different molecular formulas and different names. To make the process of naming the compounds easier, we should redraw them as carbon-skeleton structures and then apply the rules for naming.

Solve

a. The carbon-skeleton structure for the molecule is

The longest chain has nine carbon atoms, so the parent name is nonane. If we begin numbering from the left, the chain with black numbers has three branches located at carbon atoms numbered 3, 3, and 5, whereas numbering from the right with red numbers results in branches at carbon atoms 5, 8, and 8. Because the lowest black number is lower than the lowest red number, we use the black numbering scheme. The branches are two methyl groups on C3 and one propyl group on C5, so they are named 3,3-dimethyl and 5-propyl. Alphabetizing the branches according to group name (that is, *m*ethyl before *p*ropyl), the compound is 3,3-dimethyl-5-propylnonane.

b. The carbon-skeleton structure for this molecule is

The longest chain is the one that goes across the structure and has eight carbon atoms, so the parent name is octane. The red numbers that increase from right to left give the lowest number for the first branch point (C2 versus C5). Thus, there are four methyl groups at positions 2, 2, 3, and 4, and an ethyl group at C4, so they are named 2,2,3,4-tetramethyl and 4-ethyl. Alphabetizing the branches according to group name (that is, *e*thyl before *m*ethyl), the compound is 4-ethyl-2,2,3,4-tetramethyloctane. The two structures represent different substances, because they have different names. However, they have the same molecular formula ($C_{14}H_{30}$), so they are constitutional isomers.

Think About It Being able to name molecules correctly is an important skill for chemists, but it is most useful for us in this course to simply understand what the names convey in terms of identifying and distinguishing between specific substances.

Practice Exercise Name both compounds and determine if they are unrelated compounds or constitutional isomers.

a. $CH_3CH_2CH_2CH(CH_2CH_3)CH(CH_2CH_3)CH(CH_3)CH_2CH_3$
b. $(CH_3)_2CHCH(CH_3)CH(CH_3)CH(CH_3)CH(CH_3)CH(CH_3)_2$

Cycloalkanes

Alkanes can form ring structures (Figure 19.5). These **cycloalkanes** have the general formula C_nH_{2n}, which is different from the general formula for straight-chain alkanes (C_nH_{2n+2}) because a cycloalkane has one more carbon–carbon bond and two fewer hydrogen atoms per molecule than a linear or branched alkane with the same number of carbon atoms. One mole of a cycloalkane can, at least in theory,

cycloalkane a ring-containing alkane with the general formula C_nH_{2n}.

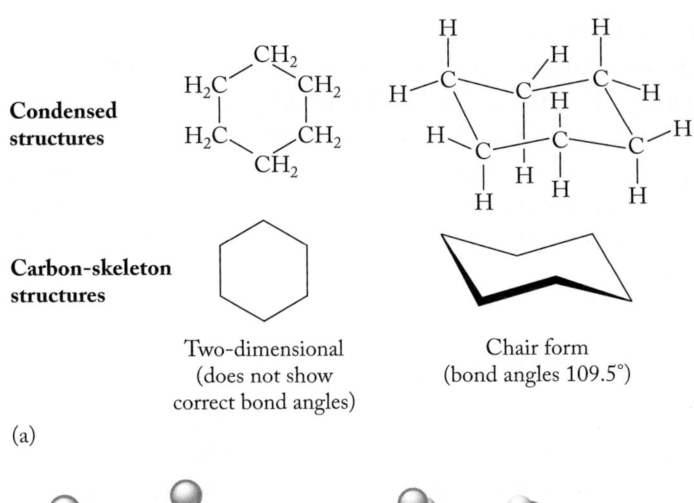

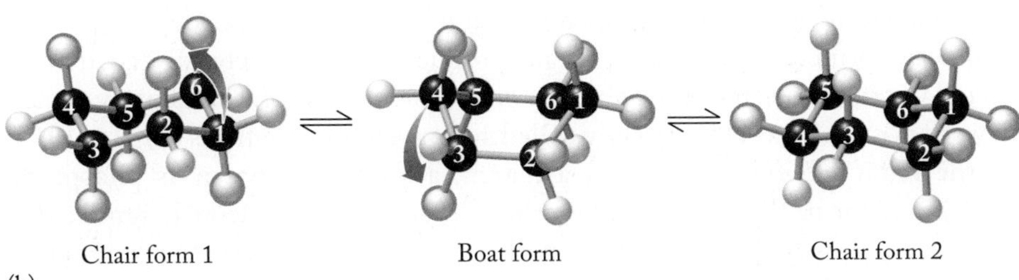

FIGURE 19.5 (a) The six-membered ring of cyclohexane may be drawn either flat or in styles that show its actual three-dimensional puckered conformations. (b) The most stable are two chair-shaped conformations shown in the left and right structures. A molecule can flip back and forth between them, but it has to temporarily adopt less stable shapes, such as the boat conformation shown in the middle structure, to do so.

CHEMTOUR Structure of Cyclohexane

react with one mole of hydrogen to make one mole of a straight-chain alkane. In practice, however, only cycloalkanes with three-carbon and four-carbon rings react with hydrogen.

The left column of Figure 19.5(a) shows the condensed structure and carbon-skeleton structure of cyclohexane drawn in two dimensions, where the C—C—C bond angles appear to be 120°. Because the carbon atoms have sp^3 hybrid orbitals, we expect the angles to be 109.5°, and indeed they are in this molecule. A more accurate representation of the ring is shown in the adjacent structures, in which the ring is puckered instead of flat and the bond angles are 109.5°.

There are two possible conformations of cyclohexane that have the greatest structural stability. They are called *chair* forms, and a molecule of cyclohexane spends over 99% of its time in one chair form or the other. It can also flip back and forth between the two chair conformations as shown in Figure 19.5(b). However, to get from one chair conformation to the other, the molecule must pass through higher-energy transition-state conformations. One of these high-energy states is called the *boat* conformation in which repulsion between the two hydrogen atoms across the ring from each other adds to the internal energy and reduces stability. Note how the H atoms in the left chair structure that are highlighted in light blue are above and below the ring (these are called *axial* positions), but these same H atoms are in positions more parallel to the ring (called *equatorial* positions) when the molecule flips to the chair structure on the right.

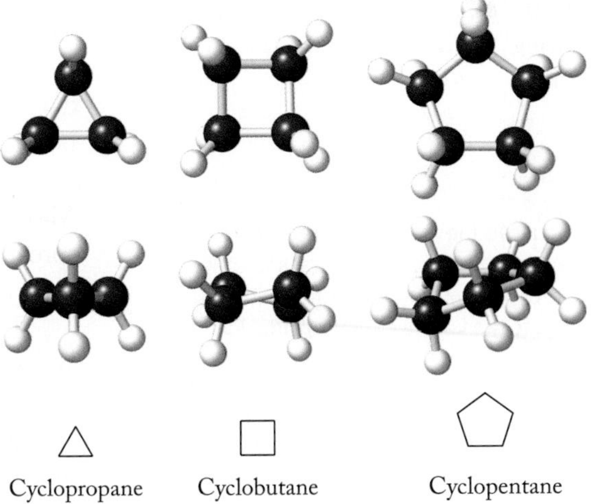

FIGURE 19.6 Cyclic alkanes have the general formula C_nH_{2n} and are considered to be unsaturated hydrocarbons. Shown here are cyclopropane (n = 3), cyclobutane (n = 4), and cyclopentane (n = 5). Cyclopropane is planar, but the side views of the ball-and-stick models show that cyclobutane and cyclopentane are not.

Cycloalkanes with fewer than six carbon atoms are possible, but they are not as stable as cyclohexane. One of the important features that determine stability is the C—C—C bond angle. For example, cyclopropane (a C_3 ring) exists but is a highly reactive species because the interior ring angles are 60°, far from the ideal bond angle of 109.5° for an sp^3 hybridized carbon atom (Figure 19.6). No puckering is possible in a three-membered ring to relieve the strain in this system, so cyclopropane tends to react in a fashion that opens up the ring and relieves the strain. The C—C—C bond angles in cyclobutane and cyclopentane are larger, so there is less ring strain. Rings of six sp^3 hybridized carbon atoms and beyond are essentially the same as straight-chain alkanes in terms of bond angle and have no ring strain. Six-membered rings are the most favored, because other thermodynamic factors have an impact on the formation of rings with seven or more carbon atoms.

Sources and Uses of Alkanes

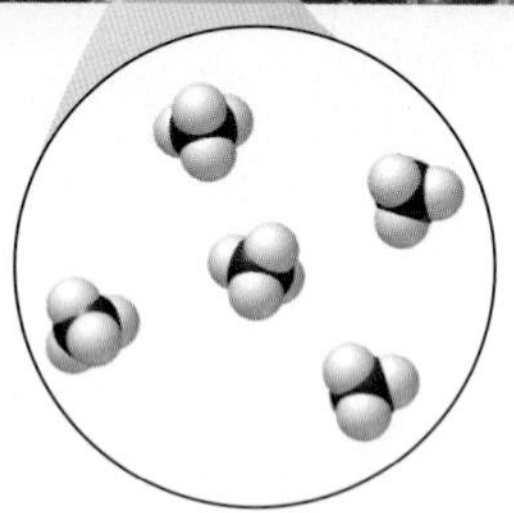

FIGURE 19.7 Methane bubbles trapped in a frozen pond. Methane is produced by rotting organic matter at the bottom of the pond.

Crude oil is the principal source of liquid alkanes. Natural gas is the major source for the simplest alkane, methane, as well as ethane, propane, and butanes. Methane is often associated with oil and natural gas deposits, but it is also the principal component of gas produced during bacterial decomposition of vegetable matter in the absence of air, a condition that frequently occurs in wetlands. This gas is commonly called swamp gas or marsh gas (Figure 19.7). In the reductive environment of a marsh, a molecule known as phosphine (PH_3) may form. Phosphine and methane together spontaneously ignite; this produces a ghostly flame known as will-o'-the-wisp that features prominently in some legends and gothic mysteries. Methane can also be produced in coal mines, where it can be especially dangerous because it creates explosive mixtures with humid air, giving rise to another common name for the gas: firedamp.

By far the most common use of alkanes is as fuels. Combustion reactions between alkanes and oxygen provide heat to power vehicles, generate electricity, warm our homes, and prepare our meals. Gasoline, kerosene, and diesel fuels are mostly mixtures of alkanes. Alkanes with higher boiling points are viscous liquids and used as lubricating oils. Low-melting solid alkanes (C_{20}–C_{40}) are used in candles and in manufacturing matches. Very heavy hydrocarbon gums and solid residues (C_{36} and up) are used as asphalt for paving roads and making roof shingles.

CONNECTION In Chapter 11 we discussed the distillation of crude oil to produce gasoline, kerosene, and other hydrocarbon mixtures useful as fuels and as feedstocks for the chemical industry.

CONNECTION In Chapter 4 we discussed methane's role as a greenhouse gas.

fracking the process of injecting liquids at high pressures into rock formations to force open existing fissures in the formations for the purpose of extracting oil or gas.

Reliance on fossil fuels to meet the energy needs of industrialized countries is currently a topic of considerable debate. Release of fossil carbon as CO_2, a greenhouse gas, is linked to a recent increase in the concentration of the gas in the atmosphere, as we discussed in Chapter 7 (see Figure 7.5), and to climate change. Moreover, Earth's supplies of fossil fuels are inevitably limited, although widespread use of novel recovery methods such as hydraulic fracturing, or **fracking,** has increased access to them.

The alkanes in natural gas and that are distilled from crude oil serve not only as fuels but also as the starting material for synthesizing more complex molecules in the chemical manufacturing industry. However, this reliance on oil and gas is changing as more companies develop green chemistry processes that begin with renewable, biological sources of starting materials.

Some alkanes have medicinal use. Mineral oil is a mixture of C_{15}–C_{24} alkanes used as a skin ointment ("baby oil") to treat diaper rash and to alleviate some forms of eczema. It is used in many cosmetics, creams, and ointments. Taken orally, mineral oil acts as a laxative. As we continue our study of organic compounds, we will return to the themes of fuels, pharmaceuticals, and materials at the bulk level and how composition, structure, and size at the microscopic level determine physical and chemical properties.

19.3 Alkenes and Alkynes

In Section 19.2 we focused on alkane hydrocarbons whose molecules contain only single bonds. We noted that alkanes are classified as saturated hydrocarbons because they contain the maximum ratio of hydrogen atoms to carbon atoms. There are two other major categories of hydrocarbons: **alkenes**, whose molecules have one or more carbon–carbon double bonds, and **alkynes**, which have one or more carbon–carbon triple bonds (Figure 19.8).

Alkenes and alkynes are **unsaturated hydrocarbons** that can combine with H_2 to form alkanes in a process called **hydrogenation**. A molecule with one C=C bond is described as having one *degree of unsaturation*. It combines with one molecule of H_2 to form one molecule of alkane (Figure 19.9), whereas a molecule with one C≡C bond has *two* degrees of unsaturation and combines with two molecules of H_2. A molecule with one double bond and one triple bond has three degrees of unsaturation, so it needs three molecules of H_2 to form one molecule of alkane. Determining the degree of unsaturation of an unknown compound using a hydrogenation reaction can help to identify its structure.

Alkene | Alkyne

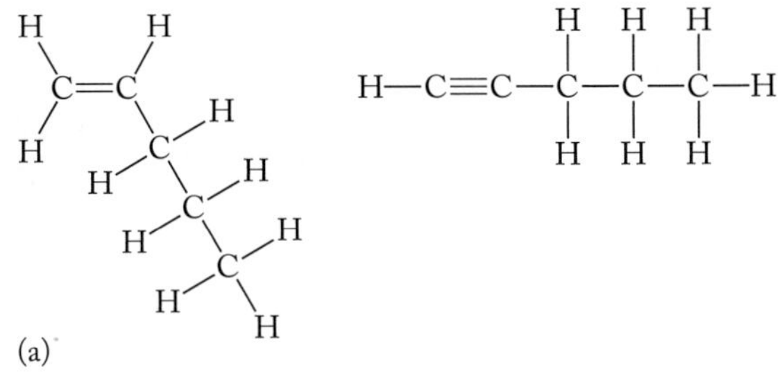

(a)

(b)

$H_2C{=}CHCH_2CH_2CH_3$ | $HC{\equiv}CCH_2CH_2CH_3$

$H_2C{=}CH(CH_2)_2CH_3$ | $HC{\equiv}C(CH_2)_2CH_3$

(c)

FIGURE 19.8 (a) Lewis structures, (b) condensed structures, and (c) carbon-skeleton structures of a C_5 alkene and a C_5 alkyne.

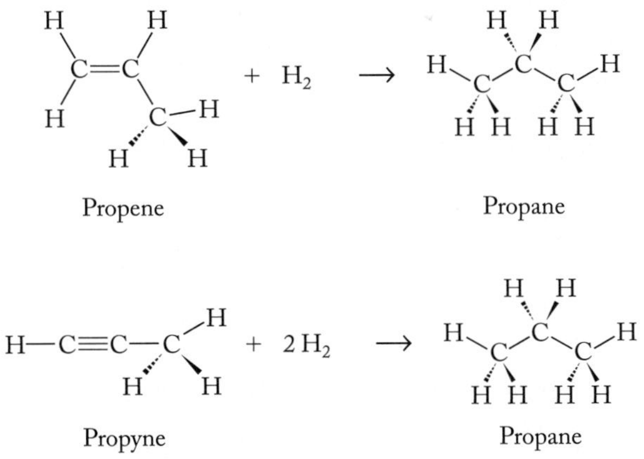

Propene + H_2 → Propane

Propyne + $2\,H_2$ → Propane

FIGURE 19.9 The hydrogenation of propene requires one mole of H_2 per mole of propene, whereas the hydrogenation of propyne requires two moles of H_2 per mole of propyne.

SAMPLE EXERCISE 19.5 Distinguishing among Alkanes, Alkenes, and Alkynes **LO3**

Three volatile hydrocarbons whose molecules each contain four carbon atoms are stored in separate, unlabeled gas tanks with the same internal pressure. One tank contains an alkane, another tank contains an alkene with one double bond per molecule, and the third tank contains an alkyne with one triple bond per molecule. Design an experiment based on the hydrocarbon's reactivity with H_2 gas to determine which tank contains the alkane, which contains the alkene, and which contains the alkyne.

Collect, Organize, and Analyze We are to evaluate the capacity of three hydrocarbons to combine with hydrogen to determine which of them is an alkane that does not react with hydrogen; which is an alkene whose molecules contain a C=C double bond that should each react with one molecule of H_2, and which is an alkyne whose molecules contain a C≡C triple bond that should each react with two molecules of H_2.

alkene a hydrocarbon containing one or more carbon–carbon double bonds.

alkyne a hydrocarbon containing one or more carbon–carbon triple bonds.

unsaturated hydrocarbon an alkene or alkyne.

hydrogenation a chemical reaction between molecular hydrogen (H_2) and another substance.

Solve Suppose we use a gas tight syringe to withdraw equal volumes containing exactly 1.00 millimole of each hydrocarbon from its tank and inject each sample into a separate reaction vessel that already contains 2.00 mmol H_2 gas. The pressure inside each reaction vessel is used to monitor the progress of each reaction. When all the vessel pressures no longer change, the following results will identify which tank contained which hydrocarbon:

$$1 \text{ mmol alkane} + 2 \text{ mmol } H_2 \xrightarrow{\text{no reaction}}$$
(no change in pressure)

$$1 \text{ mmol alkene} + 2 \text{ mmol } H_2 \rightarrow 1 \text{ mmol alkane} + 1 \text{ mmol } H_2 \text{ left over}$$
(pressure decreases by 1/3)

$$1 \text{ mmol alkyne} + 2 \text{ mmol } H_2 \rightarrow 1 \text{ mmol alkane}$$
(pressure decreases by 2/3)

Think About It Determining the quantity of hydrogen that reacts with a hydrocarbon is a way to distinguish saturated from unsaturated hydrocarbons and to determine their degree of unsaturation.

Practice Exercise The labels have fallen off two containers in the lab. One label has structure A printed on it, and the other has structure B printed on it as shown below:

$$CH_3{-}CH_2{-}CH{=}CH{-}CH{=}CH{-}CH_3$$
Compound A

$$CH_3{-}CH{=}CH{-}CH{=}CH{-}CH{=}CH_2$$
Compound B

Describe how to use hydrogenation reactions to determine which label belongs on which container. What is the structure of the product in each case?

CONCEPT TEST

Could we use information from hydrogenation reactions to distinguish a hydrocarbon with one triple bond from a hydrocarbon with two double bonds?

Simple alkenes occur at trace concentrations in crude oil, but alkenes with more complex molecular structures are abundant in plant products including pine oil, oil of celery, and oil of ginger (Figure 19.10). Alkynes are also rare ingredients

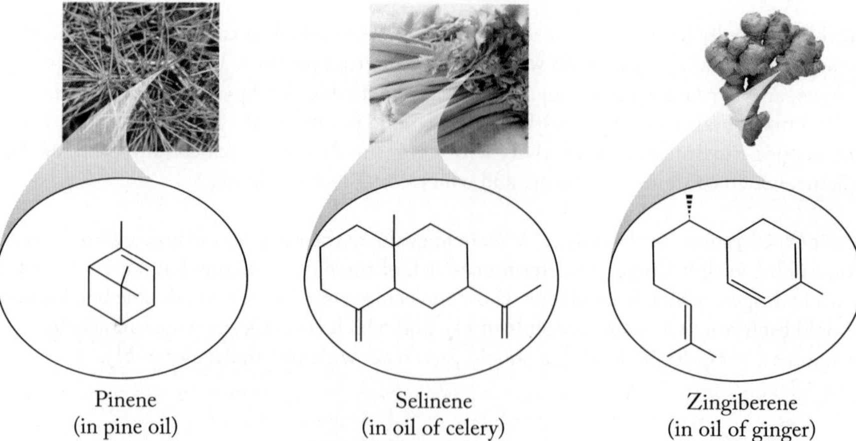

FIGURE 19.10 Some naturally occurring alkenes.

in crude oil, but some with value as pharmaceuticals or natural pesticides are produced by plants (Figure 19.11).

The simplest alkyne is ethyne, C_2H_2, which is better known by its common name, acetylene. It is the fuel used in the high-temperature torches that are used to cut through steel and other metals. Like other alkynes, acetylene may be synthesized by the partial oxidation of an alkane. For example, controlled oxidation of methane produces two fuels: acetylene and hydrogen gas.

$$6\ CH_4(g) + O_2(g) \rightarrow 2\ HC{\equiv}CH(g) + 2\ CO(g) + 10\ H_2(g) \quad (19.1)$$

This reaction illustrates the reduced state of the carbon atoms in methane and all alkanes. For another example, consider the progression of oxidation states of carbon from ethane to ethene to ethyne. The oxidation number of the carbon atoms in C_2H_6 (where each H is assigned an O.N. of +1) is −3. In C_2H_4 the O.N. of carbon is −2, and in C_2H_2 it's −1. Their capacity to be reduced makes the sp^2 carbon atoms in C=C double bonds and the *sp* carbon atoms in C≡C triple bonds more reactive than the sp^3 carbon atoms in alkanes. Actually, the carbon–carbon double bonds in alkenes are among the most versatile functional groups in all of organic chemistry.

Alkenes and alkynes, like alkanes, can be arranged in homologous series with predictable patterns in their physical properties. For example, the melting and boiling points of these compounds (Table 19.4) increase with increasing molar mass, just as with the alkanes.

Alkene and alkyne molecules may contain more than one C=C or C≡C bond, as we saw in Figures 19.10 and 19.11. In particular, many molecules produced by living systems have several double bonds. We discuss such molecules with multiple double bonds in Chapter 20,

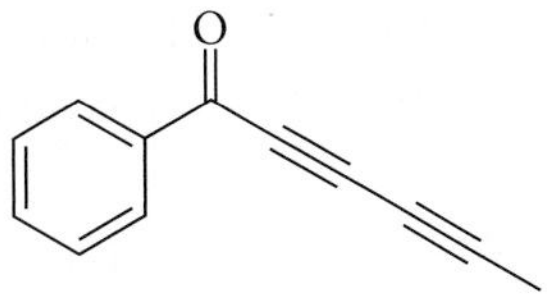

O

(a) Capillin

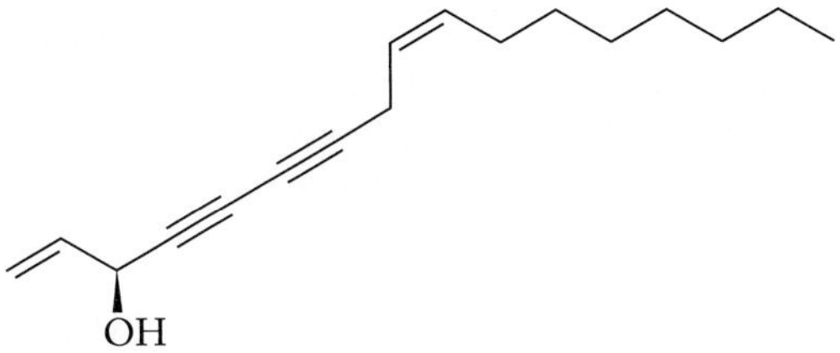

OH

(b) Falcarinol

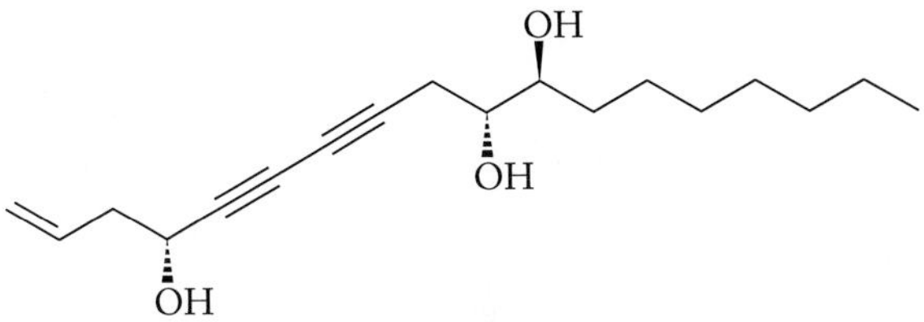

OH

OH

OH

(c) Panaxytriol

FIGURE 19.11 (a) Capillin is an antifungal produced by Oriental wormwood plants. (b) Falcarinol from carrots and English ivy has antifungal and antitumor activity. (c) Panaxytriol is a potent antitumor drug isolated from ginseng.

TABLE 19.4 Melting Points and Normal Boiling Points of Homologous Series of Alkenes and Alkynes

Condensed Structure	Melting Point (K)[a]	Normal Boiling Point (K)
$H_2C{=}CHCH_3$	88	226
$H_2C{=}CHCH_2CH_3$	88	267
$H_2C{=}CH(CH_2)_2CH_3$	135	303
$H_2C{=}CH(CH_2)_3CH_3$	133	336
$H_2C{=}CH(CH_2)_4CH_3$	154	367
$H_2C{=}CH(CH_2)_5CH_3$	169	396
$H_2C{=}CH(CH_2)_6CH_3$	192	419
$H_2C{=}CH(CH_2)_7CH_3$	207	444
$HC{\equiv}CCH_3$	171	250
$HC{\equiv}CCH_2CH_3$	147	281
$HC{\equiv}C(CH_2)_2CH_3$	168	313
$HC{\equiv}C(CH_2)_3CH_3$	141	344
$HC{\equiv}C(CH_2)_4CH_3$	192	373

[a]Melting points increase with molar mass but also depend on how molecules fit into crystal lattices. Melting points of alkenes with even numbers of carbon atoms form one series that follows this trend; alkenes with odd numbers of carbon atoms form another series.

addition reaction a reaction in which two molecules couple together and form one product.

cis isomer (also called ***Z* isomer**) a molecule with two like groups (such as two R groups or two hydrogen atoms) on the same side of the molecule.

trans isomer (also called ***E* isomer**) a molecule with two like groups (such as two R groups or two hydrogen atoms) on opposite sides of the molecule.

but for now we concentrate on the properties associated with small molecules containing only one or a small number of double or triple bonds.

Chemical Reactivities of Alkenes and Alkynes

Figure 19.12 shows the electron distributions in the π bonds in an alkene and an alkyne. Electron density in these bonds is greatest around the bonding axis, which makes π electrons more accessible to reactants than the electrons in the σ bonds between carbon atoms. For this reason, unsaturated hydrocarbons are more reactive than saturated ones.

The different reactivities of alkanes and alkenes with hydrogen halides (HCl, HBr, and HI) illustrate this difference. With an alkane, there is no reaction:

$$HX(g) + H_3C{-}CH_3(g) \rightarrow \text{no reaction} \qquad (19.2)$$

With an alkene, however, one molecule of hydrogen halide reacts with each double bond producing an alkyl halide (an alkane in which a halogen has been substituted for one of the hydrogen atoms):

$$\underset{\text{Hydrogen halide}}{HX(g)} + \underset{\text{Alkene}}{H_2C{=}CH_2(g)} \rightarrow \underset{\text{Alkyl halide}}{CH_3CH_2X} \qquad (19.3)$$

(At room temperature and 1 atm, CH_3CH_2X is a liquid when X = Br or I, and it is a gas when X = F or Cl.) This reaction is called an **addition reaction** because two reactants combine (that is, they add together) to form one product.

Alkynes react with *two* molecules of a hydrogen halide to produce an alkane bearing two halogen atoms:

$$2\,HBr(g) + HC{\equiv}CH(g) \rightarrow CH_3CHBr_2(\ell) \qquad (19.4)$$

Note how the reactivity patterns and the mole ratios of these addition reactions match the patterns and stoichiometry of the hydrogenation reactions we discussed at the beginning of this section. Recall that alkanes don't react with H_2, but each C=C double bond in an alkene can combine with one molecule of H_2 and each C≡C triple bond in an alkyne can combine with two molecules of H_2. In other words, the degree of unsaturation of an alkene or alkyne is directly linked to its capacity to engage in other addition reactions. Because both double and triple bonds react with many reagents in addition to hydrogen halides, alkenes and alkynes are useful substances in the industrial production of other compounds.

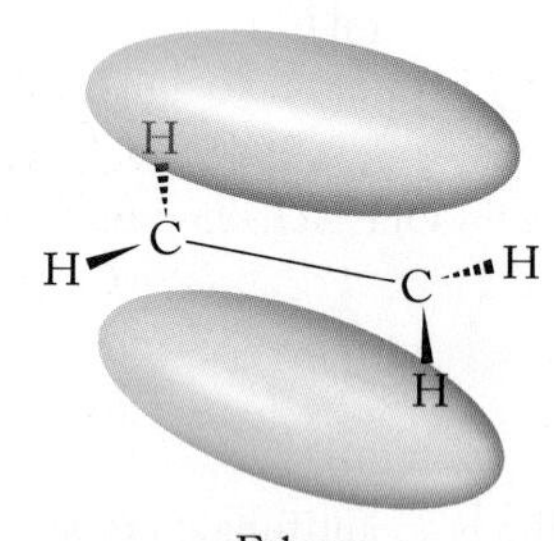

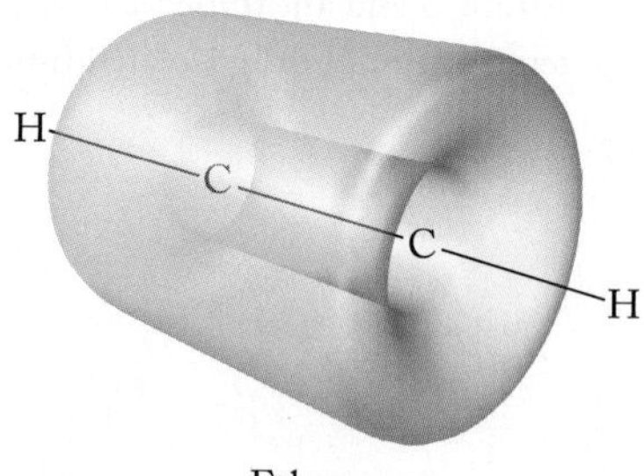

FIGURE 19.12 Shapes of the π bonds in ethene and ethyne.

Isomers of Alkenes and Alkynes

Molecules that contain alkene and alkyne functional groups can have straight or branched chains with the same types of constitutional isomers we saw with alkanes. Additionally, the location of the double or triple bond in a molecule can distinguish one constitutional isomer from another. For example, Figure 19.13 shows four possible structures for straight-chain isomers of pentene, the alkene that contains five carbon atoms and one double bond. If we apply the test used in Section 19.2 for alkanes, we find that structures (a) and (d) in Figure 19.13 both represent the same molecule. This is easier to see if we look at the carbon-skeleton structures in the figure. In both

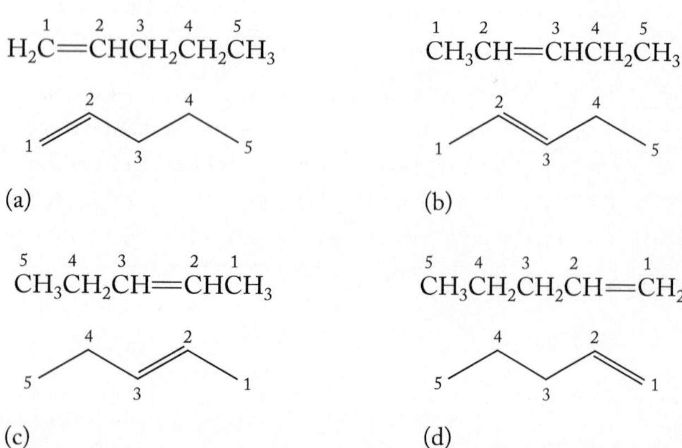

FIGURE 19.13 Four possible constitutional isomers of pentene, C_5H_{10}. Notice that structures (a) and (d) are identical, as are (b) and (c).

cases, the double bond is between C1 and C2. Recall with branched-chain alkanes that we number the carbon atoms from whichever end gives the carbon attached to the first branch the lowest number. The same rule applies to functional groups like the double bond shown here. Thus, the carbon atoms in structure (d) must be numbered from right to left to give the double bond the lowest possible numbers.

Structures (b) and (c) both represent the same molecule, which is a constitutional isomer of the molecule represented by (a) and (d) because the two different compounds have the same chemical formula but their double bonds are in a different location. Drawing the carbon-skeleton structures of (b) [or (c)], however, presents us with a new challenge. After we draw the first three atoms of structure (b), as in Figure 19.14(a), and draw a straight dashed line through the double bond, we see that we have two options for how to orient the rest of the molecule relative to the double bond. We can place the bond between C3 and C4 on the same side as the methyl group at C1 (Figure 19.14b) or on the opposite side (Figure 19.14c).

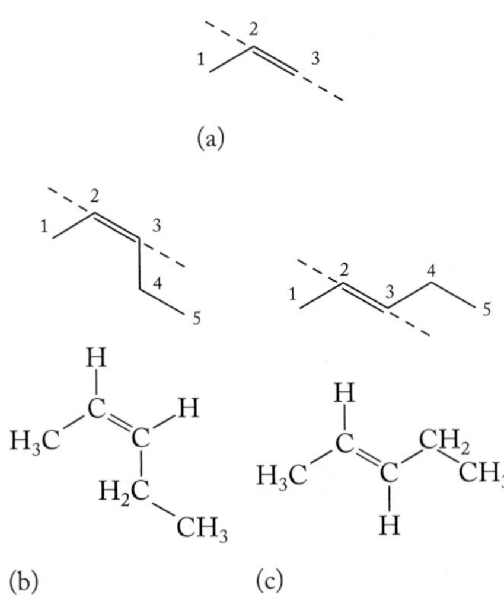

FIGURE 19.14 (a) The first three atoms of a carbon chain with a double bond between C2 and C3. (b) The chain continues on the same side of the double bond as C1 in the cis isomer. (c) The chain continues on the opposite side of the double bond as C1 in the trans isomer.

These two molecules are isomers of each other because they have the same molecular formula but different structures and therefore different properties. The isomer in Figure 19.14(b) is called either the ***Z* isomer** (*Z* stands for the German word *zusammen* or "together") or the **cis isomer** (*cis* is Latin for "on this side"), which in this case translates to "the methyl group and the chain after the double bond are both *together* or *on the same side* of the structure." The isomer in Figure 19.14(c) is called either the ***E* isomer** (*E* for *entgegen* or "opposite") or **trans isomer** (*trans* is Latin for "across"). Because these isomers are characterized by differences in the spatial arrangement of their atoms, and not by how those atoms are connected to each other, they are *stereoisomers*. The *cis* and *trans* prefixes are widely used in the names of alkenes with simple structures; the *E*/*Z* system is needed for more complex molecules in which more than two different substituents are attached to a double bond.

CONNECTION Stereoisomerism was introduced in Chapter 5.

Cis/trans isomers exist because there is no free rotation about the double bond. Recall from Chapter 5 that a double bond is formed from the overlap of two unhybridized *p* orbitals on adjacent carbon atoms. As Figure 19.15 shows, a carbon atom joined in a double bond cannot rotate freely about the bond axis without eliminating orbital overlap and breaking the bond. Breaking a π bond in one mole of an alkene costs about 290 kJ of energy, and that much energy is unavailable to molecules at room temperature. This situation gives rise to restricted rotation about a carbon–carbon double bond and to the existence of stereoisomers.

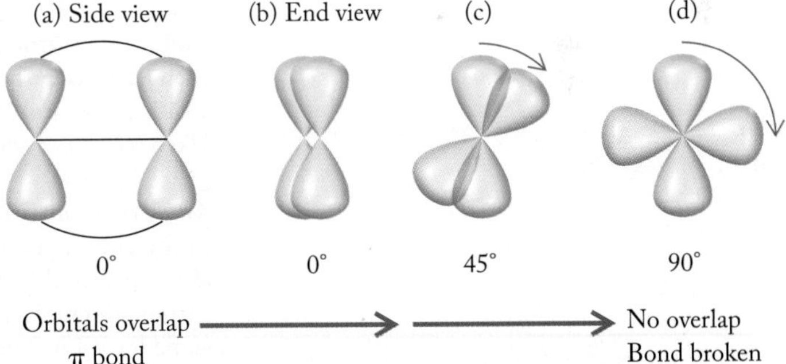

FIGURE 19.15 (a) To form a π bond, p_z orbitals overlap to establish a region of shared electron density above and below the plane of the carbon–carbon bond. (b) If you look down the bond axis, the orbitals line up. (c) If you rotate one carbon while keeping the other fixed, the orbitals are no longer parallel and do not overlap. (d) If you rotate one of the two bonded C atoms far enough, the π bond breaks.

Now let's examine the stereoisomers of structure (c) from Figure 19.14. In Figure 19.16(a), the two hydrogen atoms are on the same side of the double bond; this is the cis isomer. In Figure 19.16(b), the two hydrogen atoms are on opposite sides of the double bond; this is the trans isomer. The two molecules are stereoisomers.

A comparison of the stereoisomers in Figures 19.14 and 19.16 shows that the two cis structures are identical and the two trans structures are identical. Therefore, the straight-chain alkenes with five carbon atoms exist as three isomers: structure (a) from Figure 19.13, plus the cis and trans isomers of structure (b). All three isomers are chemically distinct molecules.

CONNECTION In Chapter 9 we introduced rotations about single bonds as one of the types of motion molecules experience as part of their overall kinetic energy.

CONCEPT TEST

Which of the following alkenes has cis and trans isomers?

a. $CH_2{=}CHCH_2CH_3$ b. $CH{=}CH$ bridged by CH_2 c. $(CH_3)_2C{=}C(CH_3)_2$

d. $(CH_3)_2C{=}CH_2$ e. $CH_3CH{=}CHCH_3$

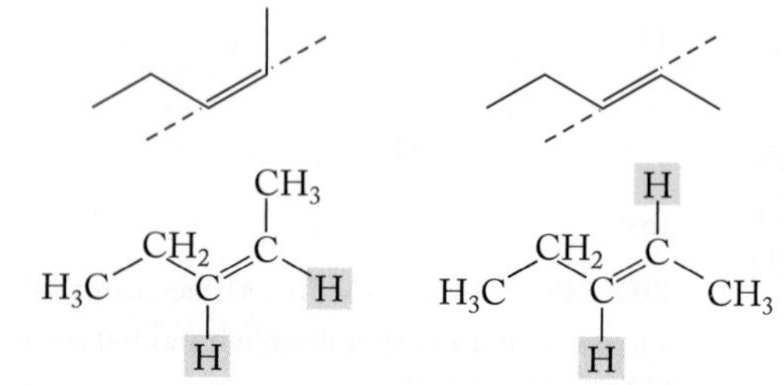

(a) The hydrogen atoms on the double bond are cis

(b) The hydrogen atoms on the double bond are trans

FIGURE 19.16 The positions of the H atoms in alkenes can be used to distinguish between (a) cis and (b) trans isomers.

Naming Alkenes and Alkynes

To name straight-chain alkenes and alkynes, the prefixes in Table 19.3 are used to identify the length of the chain. The suffix *-ene* is appended if the compound is an alkene and *-yne* if it is an alkyne. The carbon atoms in the chain are numbered so that the first carbon atom in the double or triple bond has the lowest number possible, and that number precedes the name, followed by a dash. Stereoisomers are identified by writing *cis-* or *trans-* before the number. Thus, the compounds in Figure 19.16(a) and (b) are *cis*-2-pentene and *trans*-2-pentene, respectively.

SAMPLE EXERCISE 19.6 Drawing the Molecular Structures and Naming the Stereoisomers and Constitutional Isomers of an Alkene **LO4**

Draw the condensed structures and carbon-skeleton structures of the five isomers of the six-carbon, straight-chain alkene containing one double bond. Name each isomer.

Collect and Organize We are asked to draw and name five isomers of molecules that each have six carbon atoms in a single chain and one C=C double bond. Figures 19.8, 19.13, 19.14, and 19.16 show examples of condensed and carbon-skeleton structures of alkenes.

Analyze Constitutional isomers, which depend on where the double bond is located in the chain, and stereoisomers, which depend on the orientation of the groups about the double bond (cis/trans isomers), are both possible. A straight-chain, six-carbon alkene has a maximum of five places where a C=C can be placed; however, some locations may result in identical molecules. Not all alkenes have cis and trans isomers.

Solve Let's start with the constitutional isomers, which have the double bond at different locations, and then draw the stereoisomers (cis/trans isomers) where possible. If all six carbon atoms are in one straight chain, then a double bond can be located between C1 and C2, C2 and C3, or C3 and C4. These isomers are named 1-hexene, 2-hexene, and 3-hexene, respectively, where the number indicates the location of the double bond along the chain. Chains with a double bond between C4 and C5 or C5 and C6 are identical to the isomers with a double bond between C2 and C3 or C1 and C2, respectively.

1-Hexene does not have stereoisomers because the carbon atoms that form the double bond have three H atoms and only one nonhydrogen atom attached. Only 2-hexene and 3-hexene can have stereoisomers. Both *cis*-2-hexene and *cis*-3-hexene have two R groups on the same side of the double bond, whereas *trans*-2-hexene and *trans*-3-hexene have R groups on opposite sides of the double bond.

1-Hexene

trans-2-Hexene

trans-3-Hexene

cis-2-Hexene

cis-3-Hexene

Think About It The number of constitutional isomers of a straight-chain alkene depends on the number of independent locations for its double bond. An alkene with only a terminal double bond does not have stereoisomers.

Practice Exercise Draw and name the five isomers of the molecule with the following carbon skeleton and one carbon–carbon double bond:

Polymers of Alkenes

Polyethylene (PE) is the most widely used plastic in the world with global production expected to reach 100 million metric tons in 2018. As its name suggests, PE is a polymer formed from ethylene ($H_2C{=}CH_2$) in a polymerization reaction that can be written this way:

$$n\,H_2C{=}CH_2 \rightarrow \left[CH_2{-}CH_2 \right]_n \quad (19.5)$$

The physical properties of ethylene and three of the common forms of PE made from it are listed in Table 19.5. Note that ethylene is a gas at room

TABLE 19.5 Some Physical Properties of the Monomer Ethylene and Three Polymers Made from It

	Ethene (Ethylene)	Polyethylene (PE) Polymers		
		LDPE (low density)	HDPE (high density)	UHMWPE (ultrahigh molecular weight)
Molecular mass (amu)	28	100,000–500,000	100,000–500,000	3,000,000–6,000,000
Physical state at 25°C, 1 atm	Gas	Flexible solid	Rigid solid	Very tough, abrasion-resistant solid
Density (g/cm^3) at 25°C, 1 atm	0.0012	0.910–0.940	≥ 0.941	0.930–0.935
Structure	$H_2C{=}CH_2$	Branched chains	Long straight chains	Very long straight chains
Melting point (°C)	−196	Softens from 85–110°C	Softens from 120–180°C	Softens from 130–136°C
Boiling point (°C)	−104	Decomposes at 430–470°C	Decomposes at 430–470°C	Decomposes at 440–500°C
Common use	Synthesis of PE; accelerates ripening of fruit	Grocery bags	Milk jugs and toys	Artificial joints and bulletproof vests

temperature, but the three polymers made from it are solids. All three have the same chemical composition as the monomer, but they have very different physical properties because of the different sizes and shapes of their enormous molecules.

CONCEPT TEST

Which type of intermolecular force between PE molecules is likely to be the dominant one?

PE is a **homopolymer**, which means it is composed of only one type of monomer. The condensed structure could be written $CH_3(CH_2)_nCH_3$, but there are so many more methylene groups than methyl groups that the repeating unit is frequently written as shown in Equation 19.5 to highlight the structure and composition of the monomer. Polyethylene is also an example of an **addition polymer**—that is, a polymer constructed by adding many molecules together to form the polymer chain.

In most products made of polyethylene, the n in Equation 19.5 ranges from 1000 to almost 1 million. As we showed in Table 19.5, polyethylene has a wide range of properties that depend on the value of n and on whether the polymer chains are straight or branched. In the particular type of low-density PE (LDPE) used in grocery bags, $n = 3500$–15,000 and the chains are branched (Figure 19.17). When the bagger at the grocery store asks, "Paper or plastic?" the plastic in question is LDPE.

homopolymer a polymer composed of only one kind of monomer unit.

addition polymer a macromolecule prepared by adding monomers to a growing polymer chain.

The molar mass of both the flexible LDPE polymer and the rigid, translucent high-density polyethylene (HDPE) used in milk bottles and toys ranges from 100,000 to 500,000 g/mol. The difference in properties between the two

FIGURE 19.17 Plastic shopping bags such as these are made of low-density polyethylene (LDPE), which consists of branched-chain polymers.

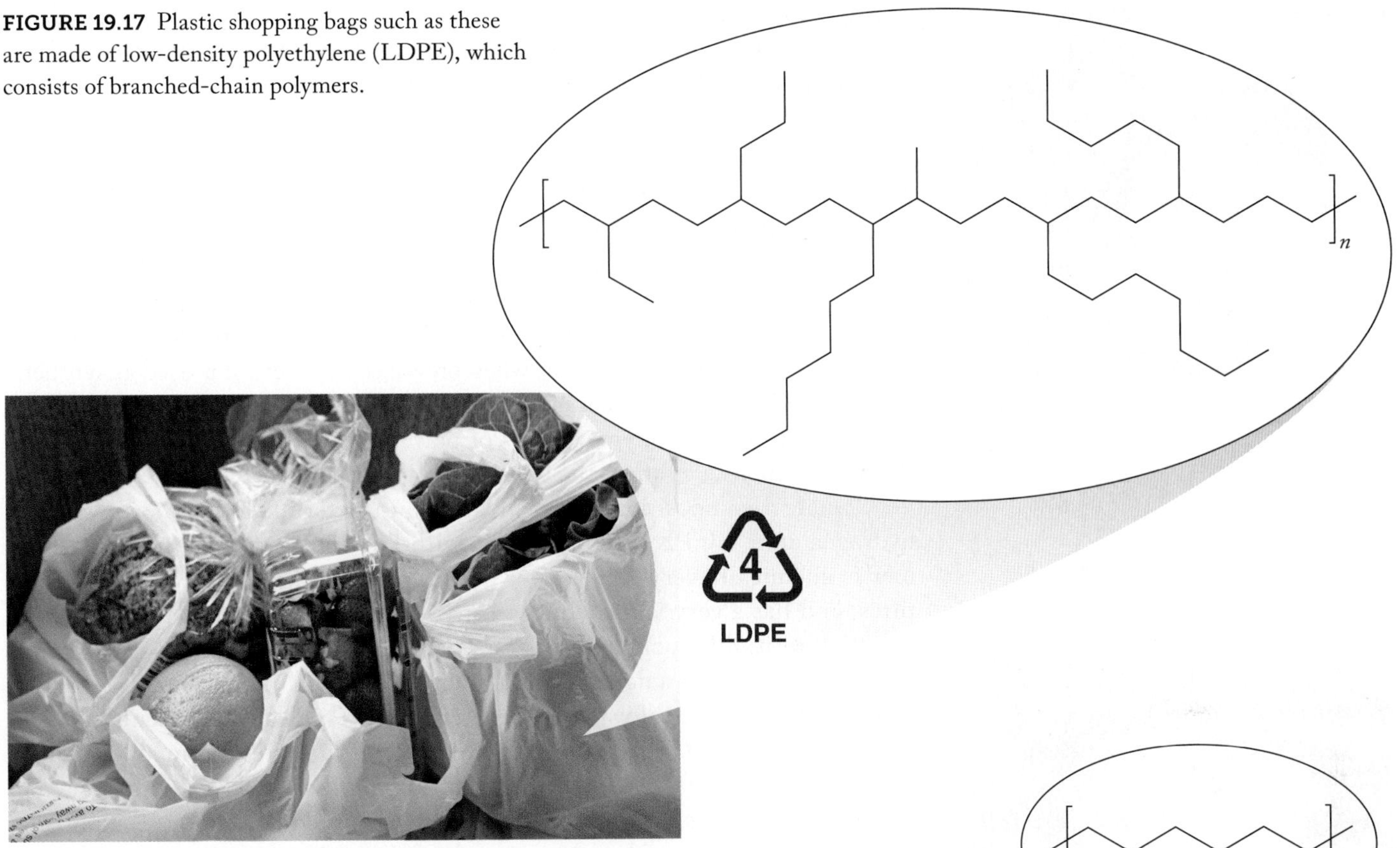

materials arises in large part from the structure of the individual chains. In HDPE the chains are straight with few branches (Figure 19.18).

What role does the presence or absence of branches play in determining properties? Think of the branched polymer as a tree branch with lots of smaller branches attached to it. The polymer can have branches that come out of the plane of the paper, so it has three dimensions. In contrast, the straight-chain polymer is like a long, straight pole. Suppose you had a pile of 100 tree branches and a pile of 100 poles, and your task was to stack each pile into the smallest possible volume to fit into a truck. You can certainly pile the branches on top of one another, but they will not fit together neatly, and probably the best you can do is to make the pile a bit more compact. In contrast, you can stack the poles into a very compact pile.

The same situation arises with the branched and linear molecules of polyethylene. The density of LDPE is low compared to HDPE, primarily because the branched molecules of LDPE stack less efficiently. Lower density, coupled with weaker intermolecular forces between polymer strands that are farther apart, makes LDPE more deformable and softer. HDPE is more rigid and even has some regions that are crystalline because the packing is so uniform and the polymers are more rigidly held in place. These different structures at the molecular level mean that objects made of HDPE and LDPE must be separated before they can be processed for recycling, as indicated by their different recycling symbols.

Ultrahigh-molecular-weight PE (UHMWPE; $n > 100{,}000$) is an even tougher material because its molecules are both straight and significantly larger

FIGURE 19.18 Many translucent plastic bottles are made of high-density polyethylene (HDPE), which consists of straight polymers that stack together efficiently, making the bottles rigid and tough.

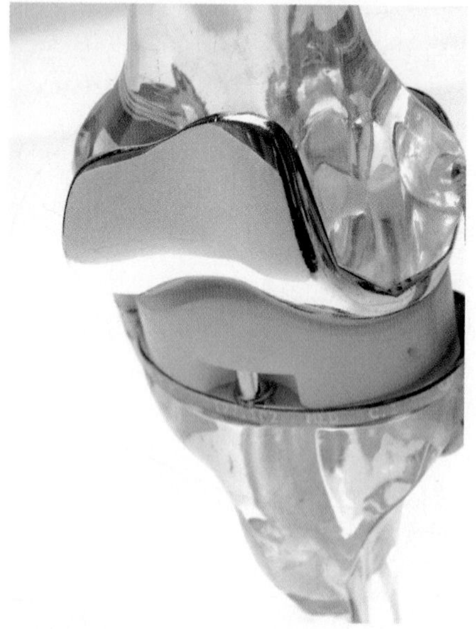
(a)

(b)

FIGURE 19.19 (a) Devices used in knee-replacement surgery often incorporate a pad made of durable ultrahigh-molecular-weight polyethylene (UHMWPE) to cushion the joint. (b) Warm places like Hobart, Australia, use the polymer for ice rinks.

than the molecules in HDPE. UHMWPE is used as a coating on some artificial ball-and-socket joints for hip replacements and to separate the prosthetic ends of bones in knee replacement surgery (Figure 19.19a). It is extremely resistant to abrasion and makes the joints last longer. When woven into threads, it is used in bulletproof vests and body armor; when produced in sheets, it is used as synthetic ice for skating rinks in climates not conducive to maintaining natural ice (Figure 19.19b). The different forms of polyethylene illustrate how the size and shape of its molecules affect the physical properties of a material.

Chemical composition also plays a role in determining properties. If the hydrogen atoms in PE are all replaced with fluorine atoms, the resultant polymer is chemically very unreactive, is capable of withstanding high temperatures, and has a very low coefficient of friction, which means other things do not stick to it. This polymer is Teflon, $(-CF_2-CF_2-)_n$, a linear polymer most familiar for its use as a nonstick surface in cookware. Teflon tubing is also used in the grafts inserted into small-diameter blood vessels during vascular surgery on limbs (Figure 19.20). The analogous material cannot be formed with chlorine, but a polymer does exist in which every other $-CH_2-$ in the polymer chain is $-CCl_2-$. Its repeating monomer unit is $(-CH_2-CCl_2-)_n$. The polymer was originally developed as a coating for fighter planes to protect them from the elements and also as mildew-resistant insoles for combat boots used in jungles. Later it was developed commercially as the thin, flexible plastic known as Saran wrap, but due to environmental concerns involving industrial processes using chlorine, as well as consumer concerns about the presence of halogenated materials in food wrapping, it has now been replaced in that product with a form of LDPE.

Hydrocarbons containing two or more double bonds are frequently used to manufacture polymers. For example, polymerization of butadiene yields a stretchy, synthetic rubber useful in rubber bands (Figure 19.21a). Polyisoprene, prepared from 2-methyl-1,3-butadiene, is used in surgical gloves (Figure 19.21b).

FIGURE 19.21 Monomer units and polymer structures for (a) butadiene and (b) a methylated butadiene known as isoprene.

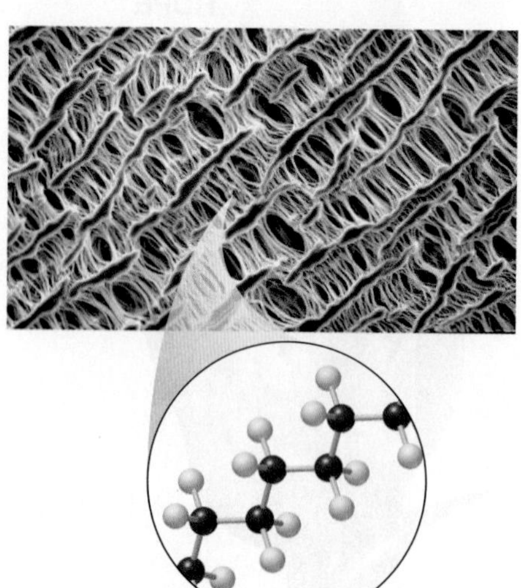
FIGURE 19.20 Fabrics made of woven strands of poly(tetrafluoroethylene) (Teflon) are used during surgical procedures to promote tissue growth. The strong, chemically inert mesh supports tissue that grows into its pores.

SAMPLE EXERCISE 19.7 Identifying Monomers — LO5

Polypropylene, $-\!\!\left[CH_2CH(CH_3)\right]\!\!-_n$, is an addition polymer used in the manufacture of fabrics, ropes, and other materials (Figure 19.22). Draw condensed and carbon-skeleton structures of the monomer used to prepare polypropylene, and name the monomer.

Collect and Organize We are given a condensed structure of a polymer, and we are asked to identify the monomer used to prepare it. We know that polypropylene is an addition polymer, so the monomer must be an alkene.

Analyze To understand the relationship between the polymer and the monomer from which it is made, let's look at Equation 19.5 in the reverse direction:

$$\text{+}\!\!\left(CH_2-CH_2 \right)\!\!\text{+}_n \rightarrow n\,CH_2{=}CH_2 \tag{19.6}$$

Breaking the blue bonds in Equation 19.6 and making the red bond a double bond illustrates the relationship between polyethylene and its monomer, ethylene. We need to apply a similar analysis to polypropylene.

Solve The relationship between polypropylene and its monomer is illustrated by a corresponding equation:

$$\text{+}\!\!\left(CH_2-CH(CH_3) \right)\!\!\text{+}_n \rightarrow n\,CH_2{=}CH(CH_3) \tag{19.7}$$

Breaking the two blue bonds in Equation 19.7 and making the red C–C bond a double bond yields the alkene shown in Figure 19.23. The official name of this three-carbon alkene is propene (though it is also known as propylene). There is no need to precede the name of the monomer with a number to indicate the position of the C═C bond, because it has to be between C1 and C2.

Think About It The constitutional difference between propylene and ethylene is the presence of a $-CH_3$ group bonded to one of the two sp^2 carbon atoms in propylene instead of the H atom found in ethylene.

Practice Exercise Draw the condensed and carbon-skeleton structures of the monomer used to make poly(methyl methacrylate), PMMA, a polymer used in shatterproof transparent plastic that can take the place of glass:

$$\left[\begin{array}{cc} H & CH_3 \\ | & | \\ -C- & -C- \\ | & | \\ H & C(=O)OCH_3 \end{array} \right]_n$$

FIGURE 19.22 Polypropylene is a common material for furniture, containers, clothing, lighting fixtures, and even objects of art. In addition to being moldable into many shapes, polypropylene is a good thermal insulator and does not absorb water easily.

$CH_3CH{=}CH_2$ 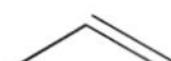

FIGURE 19.23 The monomer propene is polymerized to make polypropylene.

All addition polymers based on addition reactions of monosubstituted ethylene are called **vinyl polymers** because the $CH_2{=}CH-$ subunit is called the **vinyl group**, a name derived from *vinum* (Latin for "wine"). The name was given to the group by 18th-century chemists who prepared ethylene ($CH_2{=}CH_2$) from ethanol (CH_3CH_2OH), the alcohol in wine and other beverages. Polyvinyl chloride (PVC) is widely used in commercial articles ranging from plastic pipes for plumbing to computer cases. Classic vinyl phonograph records are also made from PVC. The condensed structures for the monomer and the polymer are

$$n\,CH_2{=}CHCl \rightarrow \text{+}\!\!\left(CH_2CHCl \right)\!\!\text{+}_n \tag{19.8}$$

Vinyl polymers are the world's second largest selling plastics, and polymers in this category are extraordinarily versatile. You probably encounter five to ten vinyl polymers before you leave your apartment or dorm room in the morning: vinyl shower curtains, vinyl drain pipes, vinyl flooring, and vinyl insulation around electrical conduits, among many others. As we explore more organic functional groups, the basic concepts developed for the vinyl polymers will apply to polymers in other categories: the features of functional group, size, and shape determine the chemical and physical properties of these extraordinarily useful materials.

vinyl polymer one of the family of polymers formed from monomers containing the subgroup $CH_2{=}CH-$.

vinyl group the subgroup $CH_2{=}CH-$.

19.4 Aromatic Compounds

Among the components of gasoline that play an important role in increasing its octane rating (a measure of the ignition temperature of the fuel and its ability to resist engine "knock") is the class of compounds called *aromatic* hydrocarbons. Benzene is one of them: a cyclic, planar molecule with delocalized π electrons above and below the plane containing its six carbon and six hydrogen atoms.

CHEMTOUR Structure of Benzene

Aromatic compounds are used as solvents, dyes, and starting materials for the production of pharmaceuticals and polymers. They are currently manufactured from petroleum, but renewable sources such as biomass are being developed. Many bacteria produce aromatic compounds and work is ongoing to enhance their production of compounds needed by industry. Technologies are also being developed to produce aromatic compounds directly from wood, agricultural waste, and cellulose.

As their class name implies, aromatic hydrocarbons have distinctive odors such as the smell of mothballs and recently cleaned public restrooms. To a chemist, *aromaticity* means more than odor; it means cyclic, planar molecules with sp^2 hybridized carbon atoms joined by a combination of σ and π bonds (Figure 19.24). Aromatic compounds may be considered relatives of alkenes because we frequently draw them using resonance structures having alternating single and double bonds. However, because their chemical and physical properties are unique and distinct from those of alkenes because of electron delocalization, they merit designation as a separate family.

The most common aromatic compound is benzene, C_6H_6. The different ways we view the bonding in benzene using Lewis theory and valence bond theory are summarized in Figure 19.24. As noted in Section 4.6, the three pairs of delocalized π electrons in benzene's planar ring lead to considerable resonance stability in this molecule, and this is true of all other aromatic molecules as well. Some aromatic molecules have other elements besides carbon in their rings, and some rings have more than, or fewer than, six atoms. However, to be aromatic, the number of π electrons in the ring must be either six or another number that satisfies the expression $4n + 2$ where n is 0 or any positive digit.

CONNECTION In Chapters 4 and 5 we described bonding in the benzene molecule using Lewis theory and valence bond theory.

The stability of aromatic systems has an impact on their chemical reactivity. For example, in Section 19.3 we saw that alkenes react rapidly with HBr, forming brominated alkanes. Benzene, on the other hand, does not react at all with HBr.

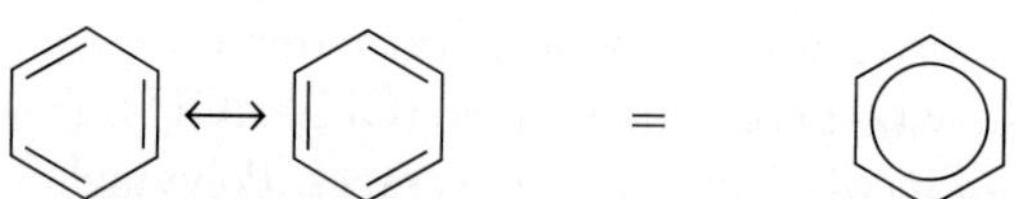

(a) Carbon-skeleton structures showing resonance forms of benzene

Skeletal symbol of benzene ring

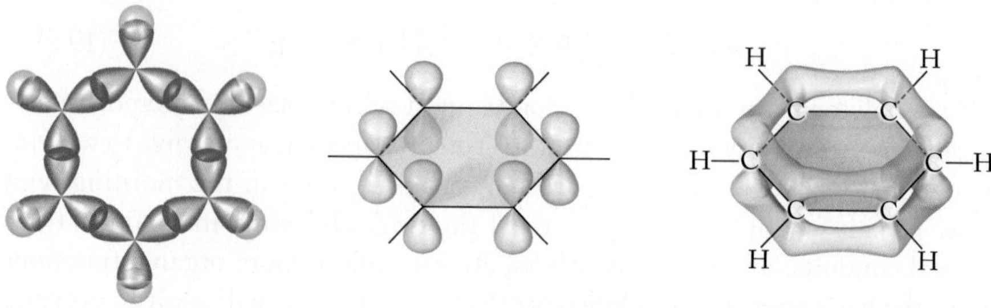

(b) Sigma bonds in benzene

(c) Unhybridized p orbitals of carbon atoms

(d) Delocalized π cloud of electrons above and below plane of ring

FIGURE 19.24 Different views of the bonding in benzene. (a) Carbon-skeleton structures showing resonance and double-bond delocalization. (b) Hexagonal array of σ bonds. (c) The unhybridized p_z orbitals on the sp^2 hybridized carbon atoms. (d) Delocalized π electrons above and below the ring.

Constitutional Isomers of Aromatic Compounds

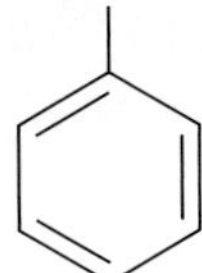

Toluene

Many compounds can be formed by replacing the hydrogen atoms in an aromatic ring with other substituents. For example, when one methyl group replaces a hydrogen atom in benzene, we get methylbenzene, also known by its common name, toluene. All of the positions around the benzene ring are equivalent, so it does not matter which carbon atom is bonded to the methyl group. This is why we do not have to include a position number in the name methylbenzene.

There are three different ways of attaching two methyl groups to a benzene ring, so there are three constitutional isomers of dimethylbenzene, also known as xylene.

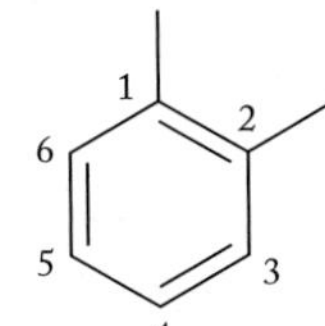

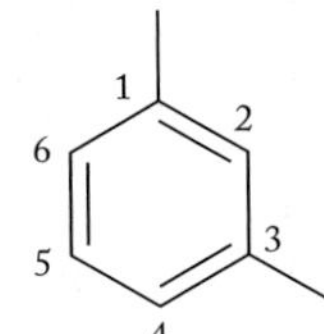

1 2 3 4 5 6

1,2-Dimethylbenzene 1,3-Dimethylbenzene 1,4-Dimethylbenzene

We distinguish between the three constitutional isomers by numbering the carbon atoms to give the substituents the lowest possible numbers. From left to right, the three dimethylbenzenes are called 1,2-dimethylbenzene, 1,3-dimethylbenzene, and 1,4-dimethylbenzene. Toluene and xylenes are used in inks, glues, and disinfectants.

The molecules of some aromatic compounds contain benzene rings that share one or more of their hexagonal sides. Three of these polycyclic aromatic hydrocarbons (PAHs) are naphthalene, anthracene, and phenanthrene:

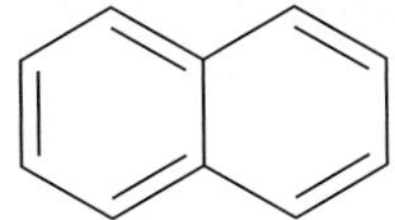

Naphthalene

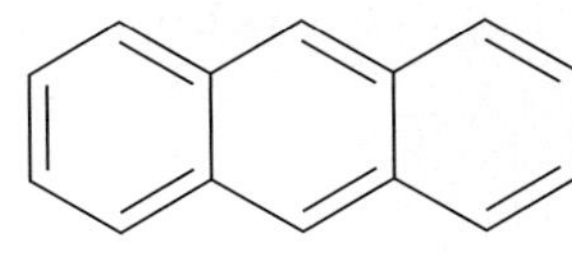

Anthracene

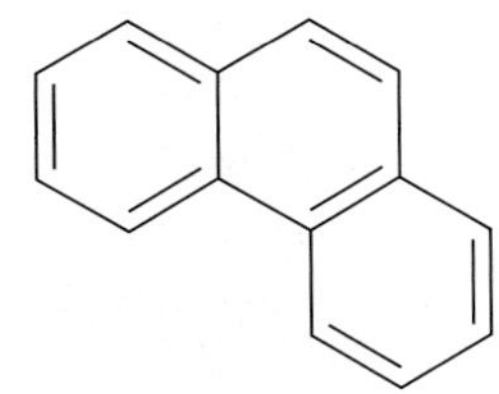

Phenanthrene

Extensive delocalization of the π electrons over all the rings makes PAHs particularly stable. In addition to being found in fossil fuels, they may be formed during the incomplete combustion of hydrocarbons and are present in particularly high concentrations in the soot from incinerators and diesel engines. They have also been identified in interstellar dust clouds and in blackened portions of grilled meat. When introduced into the environment, these compounds persist and are among the most long-lived of the hydrocarbons.

CONCEPT **TEST**

Why isn't 1,3,5-hexatriene considered an aromatic compound?

1 2 3 4 5 6

1,3,5-Hexatriene

Polymers Containing Aromatic Rings

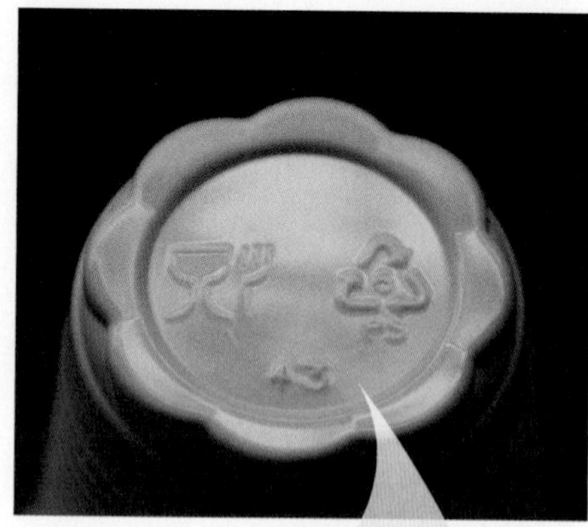

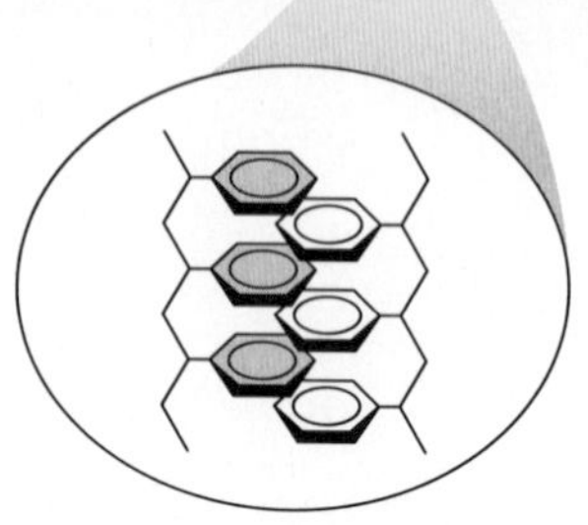

FIGURE 19.25 The aromatic rings on neighboring chains in polystyrene stack together and provide strength to the material.

Individual aromatic rings, as well as fused rings, are flat molecules. Intermolecular interactions between the rings causes them to stack (Figure 19.25), which gives rise to useful properties in materials that incorporate aromatic systems. Replacing one hydrogen atom in benzene with a vinyl group gives the monomer styrene, and the polymer made from this monomer is polystyrene (PS):

n Styrene $\longrightarrow$ Polystyrene

Solid PS is a transparent, colorless, hard, inflexible plastic. In this form it is used for compact disc cases and plastic cutlery. A more common form of PS, however, is the *expanded solid* made by blowing CO_2 or pentane gas into molten polystyrene, which then expands and retains voids in its structure when it solidifies. One form of this expanded PS is Styrofoam, the familiar material of coffee cups and take-out food containers (Figure 19.26).

FIGURE 19.26 In its nonexpanded form, polystyrene is rigid and strong, suitable for making plastic utensils. In its expanded form, it is Styrofoam, used in carry-out food containers, packing materials, and thermal insulation in buildings.

FIGURE 19.27 When the Styrofoam coffee cup on the left is placed under pressure, some of the air between the polystyrene chains is forced out. The cup shrinks to the size on the right but retains its overall shape.

The difference in properties between transparent, colorless, inflexible nonexpanded PS and opaque, white, pliable Styrofoam can be explained by considering the role of the aromatic ring in aligning the polymer chains. Branches in the chains have the same effect that we saw with polyethylene, but the aromatic rings and their tendency to stack provide additional interactions that make chain alignment more favorable energetically. The aromatic rings along two neighboring chains can stack, and this stacking makes nonexpanded PS rigid. When the chains are blown apart by a gas, the stacking is disrupted and the chains open to make cavities that fill with air, making expanded PS a good thermal insulator and packing material. The presence of air in Styrofoam is illustrated in Figure 19.27.

19.5 Amines

Nitrogen atoms are the defining components of functional groups in another important family of organic molecules called **amines.** In organic compounds, nitrogen atoms—and any atoms other than carbon, hydrogen, or metals—are called **heteroatoms** and are shown in Kekulé structures (without showing the lone pairs), condensed structures, and carbon-skeleton structures. For example, notice the heteroatom N at the junction of three lines in the carbon-skeleton structures of Benadryl and pargyline in Figure 19.28.

Amines containing an alkyl or aromatic group are thought of as being derived from ammonia, NH_3. If one hydrogen atom in ammonia is replaced by an R group, the compound is called a *primary amine.* If two hydrogen atoms are replaced by R groups, the compound is a *secondary amine*; if all three hydrogen atoms are replaced by R groups, it is a *tertiary amine*:

RNH_2	R_2NH	R_3N
Primary amine	Secondary amine	Tertiary amine

The R groups in a secondary or tertiary amine may be the same or different organic subunits. You may be familiar with the odor of trimethylamine, $(CH_3)_3N$, the compound responsible for the smell of decaying fish.

The amine functional group is found in many natural products and drugs. Some relatively simple examples include amphetamine (Figure 19.28), a stimulant that also contains an aromatic group, and Benadryl, an antihistamine used to treat the symptoms associated with allergies. Another amine, adrenaline, is produced by our bodies in glands near the kidneys and plays an important role in our nervous system.

Amines are organic bases, and their basicity is their defining chemical characteristic. They all react with water to some extent to produce hydroxide ions and protonated cations, just like ammonia:

$$NH_3(aq) + H_2O(\ell) \rightleftharpoons NH_4^+(aq) + OH^-(aq) \quad (19.9)$$

$$RNH_2(aq) + H_2O(\ell) \rightleftharpoons RNH_3^+(aq) + OH^-(aq) \quad (19.10)$$

Amines also react readily with acids like hydrochloric acid to form water-soluble salts:

$$RNH_2(\ell) + HCl(aq) \rightarrow RNH_3^+\,(aq) + Cl^-(aq) \quad (19.11)$$

Many pharmaceuticals (like Benadryl) that contain the amine functional group are sold as hydrochloride salts to improve their solubility in water.

Amine groups are polar and can form hydrogen bonds with water, which also promotes their solubility in water. On the other hand, the portions of amine molecules composed of carbon and hydrogen are nonpolar and diminish aqueous solubility because the larger the nonpolar portion, the stronger the London dispersion forces between amine (solute) molecules that inhibit their solubility in water.

Bacteria of the genus *Methanosarcina* convert primary, secondary, and tertiary methylamines to methane, carbon dioxide, and ammonia:

$$4\,CH_3NH_2(aq) + 2\,H_2O(\ell) \rightarrow 3\,CH_4(g) + CO_2(g) + 4\,NH_3(aq) \quad (19.12)$$

$$2\,(CH_3)_2NH(aq) + 2\,H_2O(\ell) \rightarrow 3\,CH_4(g) + CO_2(g) + 2\,NH_3(aq) \quad (19.13)$$

$$4\,(CH_3)_3N(aq) + 6\,H_2O(\ell) \rightarrow 9\,CH_4(g) + 3\,CO_2(g) + 4\,NH_3(aq) \quad (19.14)$$

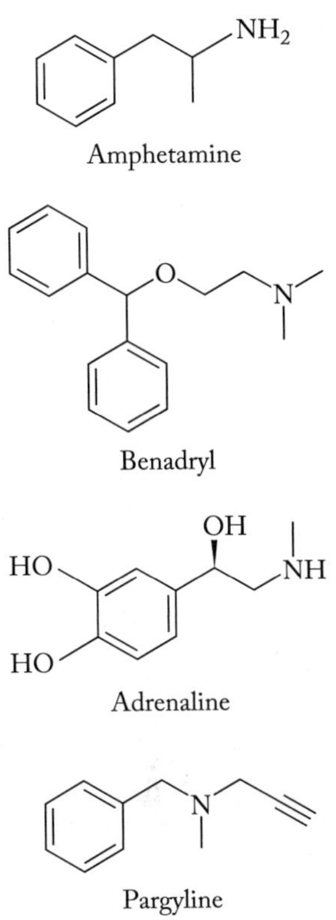

NH_2

Amphetamine

O N

Benadryl

HO OH NH HO

Adrenaline

N

Pargyline

FIGURE 19.28 Amphetamine, a drug known to produce increased wakefulness and focus; Benadryl, an antihistamine; adrenaline, a hormone produced in our bodies and involved in the fight or flight response; and pargyline, a drug used to treat hypertension. Each of these physiologically active compounds contains the amine functional group.

amine an organic compound that contains a group with the general formula RNH_2, R_2NH, or R_3N, where R is any organic subgroup.

heteroatom any atom other than carbon or hydrogen in an organic compound.

These reactions describe pathways by which methane can be produced from the decay of biomass, serving as potential sources of fuel for heating and transportation. Amines represent only a small fraction of the material in plants, however, and the industrial development of fuel production from amines has been slow, mainly because fossil fuels are still plentiful enough and cheap enough to make processing amines cost-ineffective. This economic imbalance may change as fossil fuels are depleted and become more expensive.

CONCEPT TEST

Are amphetamine, Benadryl, adrenaline, and pargyline (Figure 19.28) primary, secondary, or tertiary amines?

19.6 Alcohols, Ethers, and Reformulated Gasoline

In 1995, air-quality regulations went into effect in many U.S. cities mandating reductions in atmospheric pollutants from gasoline-fueled engines. The regulations led to the widespread use of "reformulated" gasoline containing additives to promote complete combustion and boost octane ratings. These additives are often organic compounds that contain oxygen in addition to hydrogen and carbon. Oxygen atoms in an organic compound are heteroatoms and are components of functional groups in two important families of organic molecules: alcohols and ethers.

Alcohols: Methanol and Ethanol

Alcohols have the general formula R—OH, where R is any alkyl group. The R group can be a straight chain, branched chain, or ring. Like the N atoms in amines, the O atoms in alcohols are shown in carbon-skeleton structures. The chemical and physical properties of alcohols can be understood if we recognize that an alcohol looks like a combination of an alkane and water:

R—H	H—OH	R—OH
Alkane	Water	Alcohol

TABLE 19.6 Solubilities of a Homologous Series of Alcohols in Water at 20°C

Condensed Structure	Water Solubility (g/100 mL)
CH_3OH	Miscible
CH_3CH_2OH	Miscible
$CH_3(CH_2)_2OH$	Miscible
$CH_3(CH_2)_3OH$	7.9
$CH_3(CH_2)_4OH$	2.3
$CH_3(CH_2)_5OH$	0.6
$CH_3(CH_2)_6OH$	0.2
$CH_3(CH_2)_7OH$	0.05

Recall from Section 6.3 that if the R group in the molecule is small, the alcohol behaves like water; as the R group gets larger, however, the alcohol behaves more like a hydrocarbon. See Table 19.6, which lists the water solubilities of a homologous series of alcohols. The polar –OH group makes one end of these compounds "water-like." As the number of carbon atoms increases, however, a greater proportion of the alcohol molecule is "oil-like." Therefore, the solubilities of these alcohols in water decrease until about C_8, beyond which their solubilities are comparable to those of the corresponding hydrocarbons.

As the names of the two simplest alcohols—methanol and ethanol—indicate, the chemical names of alcohols end in *-ol*; this suffix identifies the compound as an alcoh*ol*. Methanol (CH_3OH) is also known as methyl alcohol or wood alcohol. The latter name comes from one former source of this alcohol: it was made by collecting the vapors given off when wood was heated to the point of decomposition in the absence of oxygen. Methanol is a widely used industrial organic chemical. It is the starting material in the preparation of several organic compounds used to make polymers. Its industrial synthesis is based on reducing carbon monoxide with hydrogen:

$$CO(g) + 2\ H_2(g) \rightarrow CH_3OH(\ell)$$

The CO and H_2 used to make methanol come from the steam-reforming reaction of methane and water that we discussed in Chapter 9:

$$CH_4(g) + H_2O(g) \rightarrow CO(g) + 3\ H_2(g)$$

Methanol burns according to the thermochemical equation

$$2\ CH_3OH(\ell) + 3\ O_2(g) \rightarrow 2\ CO_2(g) + 4\ H_2O(\ell) \qquad \Delta H^\circ_{comb} = -1454\ kJ$$

If we divide the absolute value of ΔH°_{comb} by twice the molar mass of methanol (because the reaction consumes two moles of methanol), we get a fuel value for methanol of

$$\frac{1454\ kJ}{\left(\frac{32.04\ g}{\cancel{mol}}\right)(2\ \cancel{mol})} = 22.69\ kJ/g$$

The fuel value of octane is

$$2\ C_8H_{18}(\ell) + 25\ O_2(g) \rightarrow 16\ CO_2(g) + 18\ H_2O(\ell) \quad \Delta H^\circ_{comb} = -1.091 \times 10^4\ kJ$$

$$\frac{1.091 \times 10^4\ kJ}{\left(\frac{114.22\ g}{\cancel{mol}}\right)(2\ \cancel{mol})} = 47.76\ kJ/g$$

Thus, the fuel value of methanol is less than half that of octane (and most of the other hydrocarbons in gasoline). Why is this the case? The answer involves the molecular structure of methanol. The amount of energy released during combustion depends on the number of carbon atoms available for forming C═O bonds in CO_2 and the number of hydrogen atoms available for forming O—H bonds in H_2O. The presence of oxygen in CH_3OH adds significantly to its mass (methanol is 50% oxygen by mass) but adds nothing to its fuel value. The oxygen content of a combustible substance essentially dilutes its energy content: the more oxygen a fuel contains, the lower its fuel value.

CONNECTION We introduced fuel values in Chapter 9 as the amount of heat given off when one gram of fuel is burned.

As noted earlier, ethanol (CH_3CH_2OH), also known as ethyl alcohol, is the alcohol in alcoholic beverages. It is formed by the fermentation of sugar from an amazing variety of vegetable sources. Indeed, any plant matter containing sufficient sugar may be used to produce ethanol. Grains are commonly used, from which ethanol derives its trivial name *grain alcohol*. Ethanol may be the oldest organic chemical used by humans, and it is still one of the most important. For industrial purposes, ethanol is prepared by the reaction of ethylene and steam:

$$H_2C{=}CH_2(g) + H_2O(g) \rightarrow CH_3CH_2OH(g)$$

SAMPLE EXERCISE 19.8 Comparing Fuels for Camping Stoves **LO6**

Some of the small, single-burner stoves that campers and hikers use to prepare meals and boil water use ethanol as fuel; others use a petroleum distillate called white gas, which is a mixture of C_6 through C_{12} hydrocarbons. Assuming nonane ($\mathcal{M}$ = 128.26 g/mol; $\Delta H^\circ_{comb} = -6160$ kJ/mol) is a representative hydrocarbon in white gas, how much more thermal energy is available in 1 liter of white gas than in 1 liter of ethanol ($\mathcal{M}$ = 46.07 g/mol; $\Delta H^\circ_{comb} = -1367$ kJ/mol)? The densities of nonane and ethanol are 0.718 and 0.789 g/mL, respectively.

Collect and Organize We are given the densities, molar masses, and enthalpies of combustion of nonane and ethanol and are asked to determine how much more thermal energy is produced by the combustion of 1 liter of nonane than by burning the same volume of ethanol.

Analyze The heats of combustion are expressed in kilojoules per mole, so the comparison of the two fuels will require calculating how many moles of each are in 1 liter. The standard enthalpy of combustion of nonane is between 4 and 5 times that of ethanol, but the molar mass of nonane is about 3 times greater. Therefore, on balance, combustion of nonane should produce about 4.5/3 = 1.5 times more thermal energy than burning the same mass of ethanol. However, ethanol is about 10% more dense, so nonane should have about 1.4 times the energy content as the same volume of ethanol.

Solve Calculating the enthalpy change that accompanies the combustion of 1 liter of each fuel under standard conditions,

$$1\ \cancel{\text{L}}\ \text{nonane} \times \frac{1000\ \text{mL}}{1\ \cancel{\text{L}}} \times \frac{0.718\ \text{g}}{\cancel{\text{mL}}} \times \frac{1\ \cancel{\text{mol}}}{128.26\ \text{g}} \times \frac{-6160\ \text{kJ}}{\cancel{\text{mol}}} = -3.448 \times 10^4\ \text{kJ}$$

$$1\ \cancel{\text{L}}\ \text{ethanol} \times \frac{1000\ \text{mL}}{1\ \cancel{\text{L}}} \times \frac{0.789\ \text{g}}{\cancel{\text{mL}}} \times \frac{1\ \cancel{\text{mol}}}{46.07\ \text{g}} \times \frac{-1367\ \text{kJ}}{\cancel{\text{mol}}} = -2.341 \times 10^4\ \text{kJ}$$

Taking the ratio of the two values,

$$\frac{-3.448 \times 10^4\ \text{kJ}}{-2.341 \times 10^4\ \text{kJ}} = 1.47$$

Therefore, nonane (white gas) has 1.47 times the capacity to heat food and water as the same volume of ethanol.

Think About It The presence of oxygen accounts for 16 of ethanol's 46 grams of mass per mole, or 35%. It is reasonable, then, that an equal volume of a hydrocarbon of similar density would have significantly higher energy content. On the other hand, ethanol burns cleaner and, in principle, is a renewable fuel.

Practice Exercise Bottled propane and butane are also used as fuels in camping stoves. Which one has the greater fuel density? By how much (in kJ/mol)?

	Propane	Butane
$\Delta H^\circ_{\text{comb}}$ (kJ/mol)	−2200	−2877
Density (g/mL)	0.493	0.573

Most of the gasoline sold in the United States and Canada contains ethanol to promote complete combustion and reduce air pollution. Most of this ethanol is produced by the fermentation of sugar derived from corn, with annual production in the United States reaching 56 billion liters in 2015. There are, however, several challenges limiting the greater use of ethanol or other oxygenated compounds as automobile fuels. Their enthalpies of combustion are significantly less negative than that of the hydrocarbons in gasoline. The change in free energy that accompanies the combustion of ethanol is also less negative than the combustion of the same volume of the hydrocarbons. Therefore, there is less energy available to do useful work, such as propelling cars and trucks down the road. On the plus side, burning ethanol reduces CO_2 emissions by an average of 34% compared to gasoline and reduces CO emissions by as much as 30%.

To produce ethanol, considerable energy, fertilizer, irrigation water, and valuable farmland are needed to grow and harvest corn and to convert cornstarch into sugar and sugar into ethanol, and to separate the ethanol from the fermentation mixture. It is estimated that more than two-thirds of the energy released in the combustion of ethanol derived from corn is consumed in its production. Ethanol produced in this way is more expensive than gasoline. However, advances that consume less water and require less energy are being made in producing ethanol from grass, wood, and crop residues.

Alcohols are also components of many natural products and consumer goods derived from them. The distinctive odor of mint leaves comes from menthol, which is an alcohol, as is terpineol (oil of turpentine), an oil distilled from the resin of pine trees (Figure 19.29). Notice the similarity in the carbon-skeleton structures of these two compounds: both contain a six-carbon ring and an –OH group. Only the location of the OH group and the presence of a C═C double bond distinguish these two compounds.

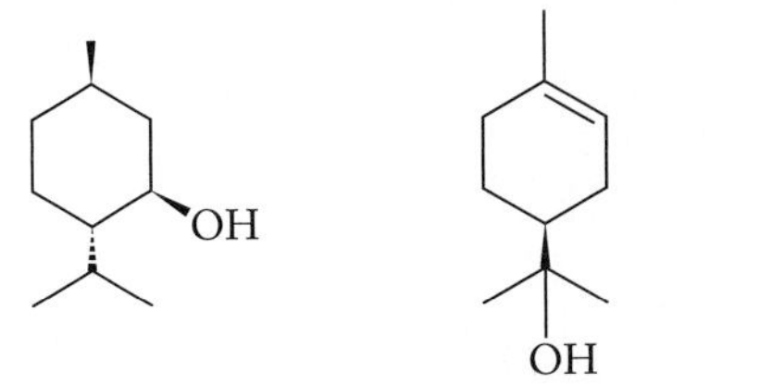

OH

OH

(a) Menthol (oil of mint)

(b) Terpineol (oil of turpentine)

FIGURE 19.29 (a) Menthol and (b) terpineol are two naturally occurring alcohols present in mint leaves and pine needles, respectively.

Ethers: Diethyl Ether

Ethers have the general formula R—O—R, where R is any alkyl group or an aromatic ring. Just as with alcohols, we can think of ethers as water molecules in which the two H atoms have been replaced by two organic groups (R and R′, which may be the same or different):

R—H	H—O—H	H—R′	R—O—R′
Alkane	Water	Alkane	Ether

Because the C—O—C bond angle is close to the tetrahedral bond angle of 109.5°, the bond dipoles of the two C—O bonds in an ether do not cancel, which means that ethers are polar molecules. This structural feature gives rise to the properties of typical ethers: their water solubilities are comparable to those of alcohols of similar molar mass, but their boiling points are about the same as alkanes of comparable molar mass (Table 19.7).

The most important ether industrially is diethyl ether, $CH_3CH_2OCH_2CH_3$. You may have heard of this as the substance simply called "ether" that has had wide use in medicine as an anesthetic since 1842. Although exactly how an anesthetic dulls nerves and puts patients to sleep is still unknown, certain properties of diethyl ether play a role in determining its behavior as a medicinal agent.

TABLE 19.7 Functional Groups Affect Physical Properties

	Molar Mass (g/mol)	Normal Boiling Point (K)	Solubility in Water (g/100 mL at 20°C)
CH_3CH_2—O—CH_2CH_3 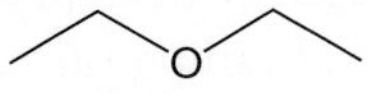 Diethyl ether	74	308	6.9
$CH_3CH_2CH_2CH_2CH_3$  Pentane	72	309	0.0038
$CH_3CH_2CH_2CH_2OH$  Butanol	74	390	7.9

Because diethyl ether has a low boiling point (35°C), it vaporizes easily, and a patient can inhale it. Because diethyl ether has a significant solubility in water, it is soluble in blood, which means that once inhaled, it can be easily transported throughout the body. Its low polarity and short saturated hydrocarbon chains combine to make it soluble in cell membranes, where it blocks stimuli coming into nerves. Ether has the unfortunate side effect of inducing nausea and headaches, and it has been replaced by newer anesthetics in modern hospitals. For many years, however, ether was the anesthetic of choice for surgical procedures.

Diethyl ether is extremely flammable, which was a liability in an operating room. This property becomes an advantage, however, when ether is sprayed into diesel engines to start them when it is too cold for diesel fuel to ignite. During the 1990s, another ether best known by the acronym MBTE (Figure 19.30), was added to gasoline to promote complete combustion. Unlike the nonpolar hydrocarbons in gasoline, MTBE is soluble in water. Consequently, gasoline spills, leaks from storage tanks, and releases from watercraft produced extensive MTBE contamination of groundwater and drinking water. After toxicity tests showed MTBE to be a possible carcinogen (cancer-causing agent), several states—including California, where more than 25% of the world's production of MTBE was used in gasoline—banned the use of MTBE as a gasoline additive. Most oil companies stopped adding MTBE to their gasolines in 2006. These changes raised the question of which oxygenated additive would replace MTBE. The leading candidate to date has been ethanol.

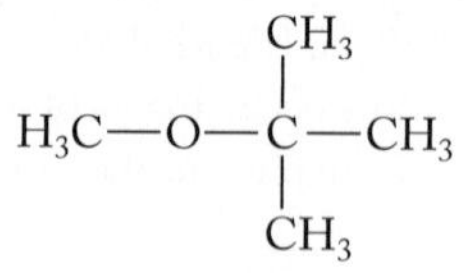

$$\begin{array}{c} CH_3 \\ | \\ H_3C\text{—}O\text{—}C\text{—}CH_3 \\ | \\ CH_3 \end{array}$$

MTBE

FIGURE 19.30 MTBE, methyl *tert*-butyl ether (*tert*- is an abbreviation for *tertiary*, referring to a carbon atom bonded to three other carbon atoms), was used in the 1990s as an oxygenated fuel additive.

CONCEPT **TEST**

Rank the following compounds in order of decreasing fuel value: diethyl ether, MTBE, methanol, and ethanol.

Polymers of Alcohols and Ethers

Over 1 million metric tons of the addition polymer poly(vinyl alcohol) (PVA) are used worldwide each year to make adhesives, emulsions, and materials known as sizing, which change the surface properties of textiles and paper to make them less porous, less able to absorb liquids, and smoother. PVA is the material of choice for laboratory gloves that are resistant to organic solvents. Because its chains are studded with –OH groups (Figure 19.31a), its surface is very polar and very water-like, and hydrocarbon solvents that are not soluble in water do not penetrate PVA barriers.

PVA is also impenetrable to carbon dioxide, and this property has led to its use in soda bottles, in which it is blended with the polymer poly(ethylene terephthalate) (PETE, Figure 19.31b). The two polymers do not mix but separate into layers (Figure 19.31c). The PETE makes the bottle strong enough to bear pressure changes due to temperature changes and survive the impact of falling off tables. CO_2, the dissolved gas that makes soda fizz, passes readily through PETE but not through the PVA layers, so the soda does not go flat. Polymers with different properties are frequently combined to create new materials with desired properties.

The monomer from which poly(vinyl alcohol) is made is not the one you might expect based on our discussion of how addition polymers are synthesized. The monomer "vinyl alcohol" does not exist:

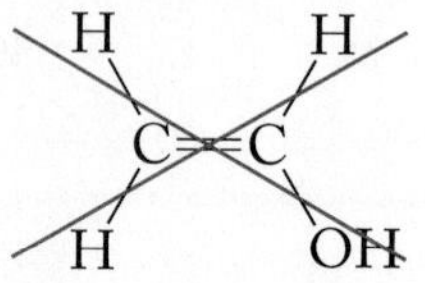

$$\begin{array}{ccc} H & & H \\ & C\!=\!C & \\ H & & OH \end{array}$$

copolymer a macromolecule formed from the chemical combination of two different monomers.

heteropolymer a polymer made of three or more different monomer units.

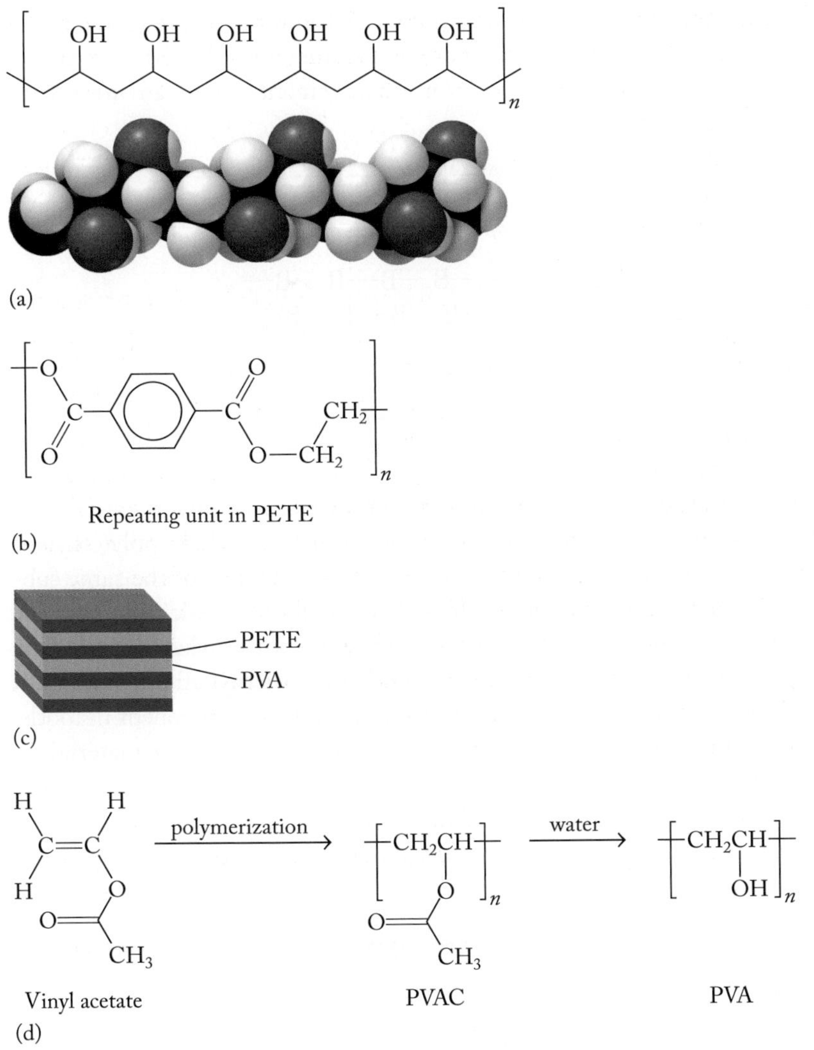

OH OH OH OH OH OH

(a)

Repeating unit in PETE

(b)

PETE

PVA

(c)

polymerization

water

Vinyl acetate PVAC PVA

(d)

FIGURE 19.31 (a) Representative section of a chain of PVA. (b) The repeating unit in PETE. (c) Layers of the polymers PVA and PETE are used to make soda bottles. PETE makes the bottle strong, whereas PVA keeps the carbon dioxide from escaping. (d) Poly(vinyl alcohol), or PVA, is synthesized from vinyl acetate. The intermediate polymer, poly(vinyl acetate), or PVAC, is reacted with water to produce PVA.

Instead, the monomer vinyl acetate is polymerized to make poly(vinyl acetate) (PVAC), which is then reacted with water to replace the acetate groups with –OH groups. This replacement reaction turns PVAC into PVA as shown in Figure 19.31(d).

The blend of PVA and PETE in early plastic soda bottles was a physical mixture of the two polymers. New materials can also be made by combining different monomer units in one polymer molecule. This type of molecule is called a **copolymer** when two different monomers are combined and a **heteropolymer** when three or more different monomers are combined. One example of an addition copolymer is a material called EVAL, made from ethylene and vinyl acetate (Figure 19.32). Food usually deteriorates in the presence of oxygen, but packages made of EVAL provide an excellent barrier to oxygen while retaining the flavor and fragrance of the packaged food.

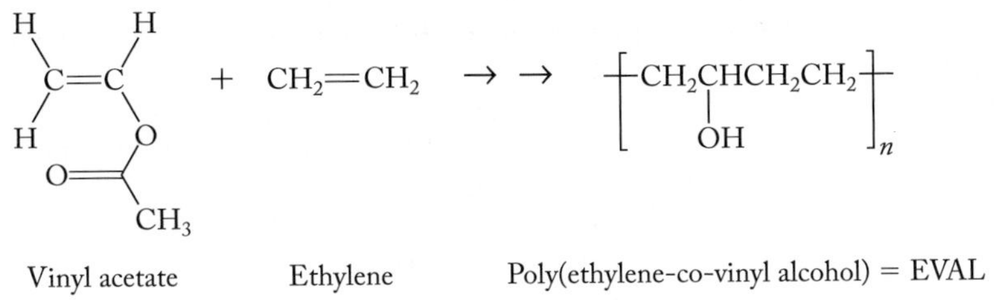

$$\text{Vinyl acetate} + CH_2{=}CH_2 \rightarrow \rightarrow -\!\!\left[CH_2CH(OH)CH_2CH_2\right]\!\!-_n$$

Vinyl acetate Ethylene Poly(ethylene-co-vinyl alcohol) = EVAL

FIGURE 19.32 EVAL is a copolymer of ethylene and vinyl acetate.

Monomers forming hetero- or copolymers can combine in different ways. If we represent the monomer units making up a copolymer with the letters A and B, one possible way they can combine is an arrangement called an *alternating copolymer*:

$$\text{+A—B—A—B—A—B—A—B—A—B+}$$

Another possibility is called a *block copolymer*:

$$\text{+A—A—A—A—B—B—B—B—}$$
$$\text{A—A—A—A—B—B—B—B+}$$

Finally, a *random copolymer* is also possible:

$$\text{+A—A—B—A—B—B—A—B—A—}$$
$$\text{A—A—A—B—B—A—B—B—B+}$$

EVAL is a random copolymer of the monomers ethylene and vinyl acetate.

Commercially important polymers made from ethers include poly(ethylene glycol) (PEG) and poly(ethylene oxide) (PEO), which are made of the same subunit (Figure 19.33). PEG is a low-molar-mass liquid oligomer made from ethylene glycol, and PEO is a higher-molar-mass solid made from ethylene oxide. As a polyether, PEG has properties closely related to those of diethyl ether—namely, it is soluble in both polar and nonpolar liquids. It is a common component in toothpaste because it interacts both with water and with the water-insoluble materials in the paste and keeps the toothpaste a uniform consistency both in the tube and during use. PEGs of many lengths are finding increasing use as attachments to pharmaceutical agents to improve their solubility and biodistribution in the body.

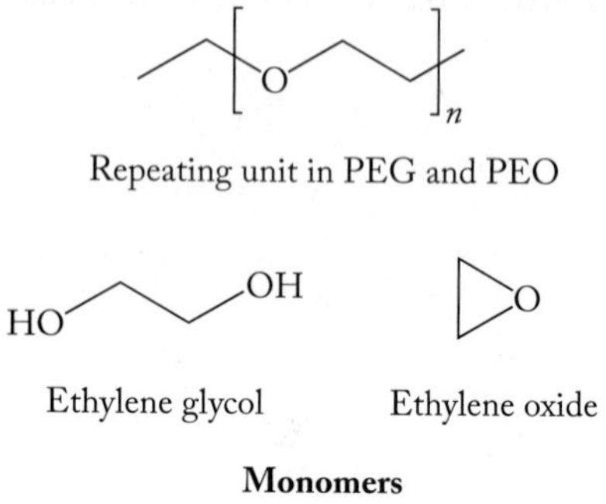

O
n

Repeating unit in PEG and PEO

HO OH
O

Ethylene glycol Ethylene oxide

Monomers

FIGURE 19.33 Poly(ethylene glycol) (PEG) and poly(ethylene oxide) (PEO) have the same repeating unit. The two polymers differ only in their molar masses. PEG is typically made from ethylene glycol, whereas ethylene oxide is the monomer of choice for making PEO.

SAMPLE EXERCISE 19.9 Assessing Properties of Polymers LO7

The polymer PEG (Figure 19.33) is used to blend materials that are not soluble in each other. It is soluble both in water and in benzene, a nonpolar solvent. Describe the structural features of PEG that make it soluble in these two liquids of very different polarities.

Collect, Organize, and Analyze The relationship between the structure and solubility of compounds was discussed in Chapter 6, where we learned that "like dissolves like." This general rule refers to the polarity of the molecules and to the attractive forces between polar groups and between nonpolar groups on solute and solvent molecules. Water is a polar molecule and benzene is a nonpolar molecule, so we predict that PEG contains both polar and nonpolar regions.

Solve The structure of PEG consists of $-CH_2CH_2-$ groups connected by oxygen atoms. The oxygen atoms are capable of hydrogen bonding with water molecules, so the attractive force between PEG and water is due to hydrogen bonding. Benzene is nonpolar and is attracted to the $-CH_2CH_2-$ groups in the polymer. Nonpolar groups interact via London dispersion forces, so those forces must be responsible for the solubility of PEG in nonpolar benzene.

Think About It As predicted, PEG contains both polar and nonpolar regions, allowing it to be solvated by both polar solvents like water and nonpolar solvents like benzene.

Practice Exercise When PEG is added to soft drinks, it keeps CO_2 in solution longer after the soda is poured. What intermolecular attractive forces between PEG and CO_2 might make this use possible?

19.7 Aldehydes, Ketones, Carboxylic Acids, Esters, and Amides

aldehyde an organic compound containing a carbonyl group bonded to one R group and one hydrogen; its general formula is RCHO.

Five functional groups—aldehydes, ketones, carboxylic acids, esters, and amides—all contain a carbon atom double-bonded to an oxygen atom (Figure 19.34). This configuration of carbon and oxygen atoms is known as a carbonyl group.

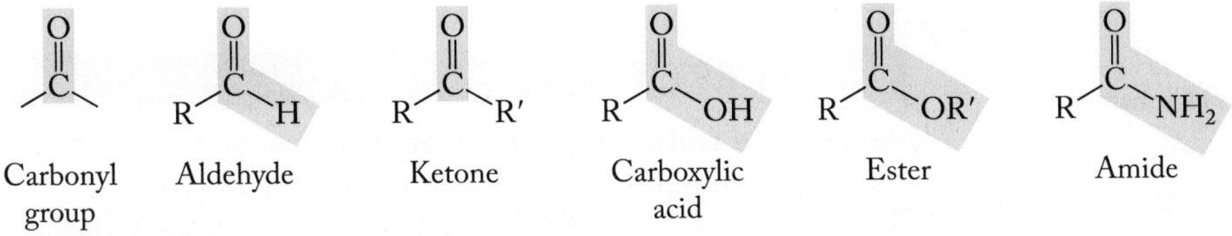

Carbonyl group | Aldehyde | Ketone | Carboxylic acid | Ester | Amide

FIGURE 19.34 The carbonyl group is found in five important functional groups: aldehydes, ketones, carboxylic acids, esters, and amides. The R and R′ groups may be any organic group.

The carbonyl groups in aldehydes and ketones are bonded only to R groups or H atoms. Carboxylic acids, as their name implies, are acidic, and their chemistry is determined by the –COOH subunit, referred to as a *carboxylic acid group*. Esters and amides can be made from carboxylic acids by reacting them with alcohols and amines.

CONNECTION We first introduced carboxylic acids in our discussion of acids in Chapter 4. We defined the carbonyl group and ketones in Chapter 6.

Aldehydes and Ketones

Aldehydes and ketones closely resemble each other in both chemical and physical properties. An **aldehyde** contains a carbonyl group bound to one R group and one hydrogen atom; its general formula is RCHO or R(C=O)H. A *ketone* contains a carbonyl group bound to two R groups; its general formula is RCOR or R(C=O)R. The R groups may be the same, as in acetone, $CH_3C(O)CH_3$, or different, as in 2-heptanone (Figure 19.35). 2-Heptanone is found in cloves, blue cheese, and many fruits and dairy products.

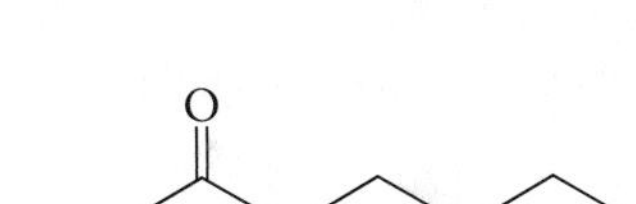

2-Heptanone

FIGURE 19.35 The ketone 2-heptanone is found in many plants and dairy products.

The double bond in the carbonyl group accounts for the reactivity of aldehydes and ketones. It is different from the double bond in an alkene, however, because it is polar (Figure 19.36). The electronegative oxygen pulls electron density toward itself, and the chemistry of aldehydes and ketones is linked to the polarity of the C=O bond. Other polar species tend to react with carbonyls when electron-rich (δ−) regions of their molecules approach the electron-poor (δ+) carbon atom of the carbonyl group.

Aldehydes and ketones are polar, and they tend to parallel the ethers with respect to water solubility. They cannot hydrogen-bond with other aldehyde and ketone molecules because they contain only carbon-bonded hydrogen atoms, so they have lower boiling points than alcohols of comparable molar mass.

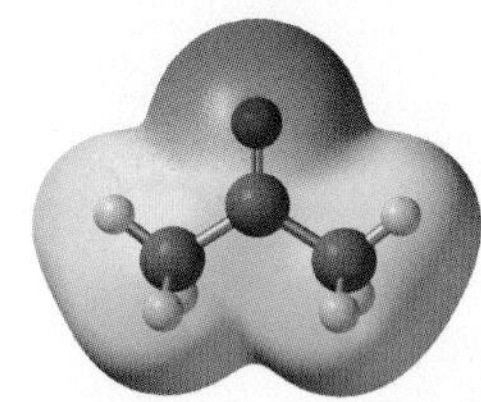

FIGURE 19.36 The electron distribution in a carbonyl group is skewed toward the oxygen end of the bond because oxygen is more electronegative than carbon.

Because of the C=O bond, aldehydes and ketones are in a higher oxidation state than alcohols, and indeed many of the smaller aldehydes and ketones are made by oxidizing alcohols of the same carbon number. Aldehydes and ketones do not polymerize through their carbonyl groups. Many polymers have carbonyl

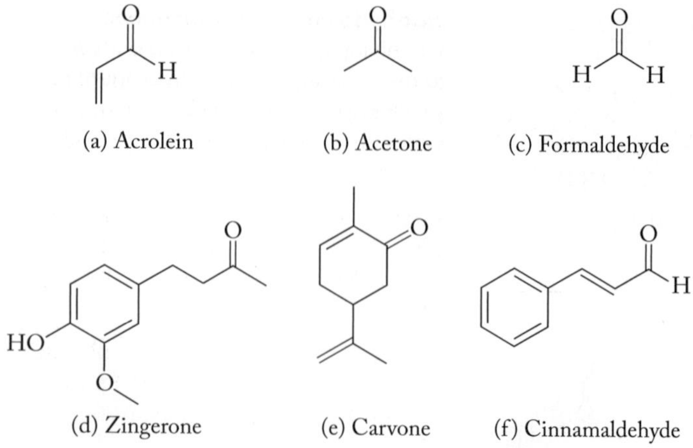

(a) Acrolein (b) Acetone (c) Formaldehyde

(d) Zingerone (e) Carvone (f) Cinnamaldehyde

FIGURE 19.37 The ketone and aldehyde functional groups are common among organic compounds: (a) acrolein is found in barbeque smoke, (b) acetone is used in nail polish remover, (c) aqueous solutions of formaldehyde are used to preserve biological specimens, (d) zingerone is found in the spice ginger, (e) carvone is found in the leaves of spearmint, and (f) cinnamon owes its flavor and odor to cinnamaldehyde.

functional groups as part of their structure, but these groups themselves do not react to form long chains.

We have already seen several examples of aldehydes and ketones in earlier chapters. In Chapter 5, we were introduced to formaldehyde and acrolein, two aldehydes shown in Figure 19.37. Acetone, $CH_3C(O)CH_3$, is a widely used solvent found in nail polish remover. The flavors and aromas of ginger (zingerone), spearmint (carvone), and cinnamon (cinnamaldehyde) all come from compounds that contain a carbonyl group.

CONCEPT TEST

Besides the carbonyl group, what other functional groups are present in zingerone, carvone, and cinnamaldehyde?

Carboxylic Acids

Carboxylic acids are organic compounds that are proton donors, which means they are Brønsted–Lowry acids (Sections 8.4 and 15.1). The R group in RCOOH may be any organic subunit. There is extensive hydrogen bonding between molecules of carboxylic acids. The hydrogen on the –COOH group of one molecule can hydrogen-bond to either O atom on a nearby carboxylic acid group (Figure 19.38). This interaction results in high boiling points relative to those of other organic compounds of comparable molar mass.

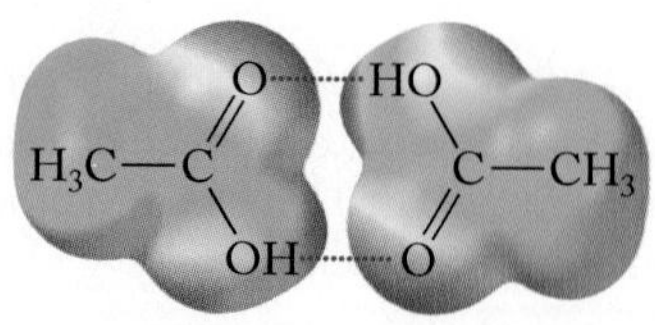

FIGURE 19.38 The high boiling points of carboxylic acids are the result of strong hydrogen bonds between neighboring molecules.

Donating a proton leaves the carboxylic acid with a negatively charged oxygen whose electron density is delocalized over the entire carboxylate group. This delocalization contributes to the stability of the carboxylate anion. The common carboxylic acids are weak acids, which means that they are present in aqueous solutions as mostly neutral molecules, a small fraction of which are ionized, donating H^+ ions to molecules of water (Figure 19.39).

CONNECTION We discussed quantitative aspects of equilibria involving weak acids and weak bases in Chapter 15.

Vinegar is a dilute aqueous solution of the carboxylic acid acetic acid. Large quantities of vinegar are produced commercially by the air oxidation of ethanol in the presence of enzymes from *Acetobacter* bacteria. Bacteria can also convert acetic acid and other constituents in biomass to methane. For example, the digestive systems of cows introduce about 100–200 liters of methane to the atmosphere per day per animal. Translating this process to an industrial scale is an attractive future source of hydrocarbons, provided the complexities of bacterial action can be adapted for large-scale production. If so, then converting organic matter into hydrocarbon fuel would be possible without waiting millennia for the anaerobic processes deep within Earth to do so.

CONNECTION We described the timescale for the formation of fossil fuels in Chapter 7.

The production of methane from plant residues that are mostly cellulose (a carbohydrate) requires the sequential action of several types of bacteria. In the

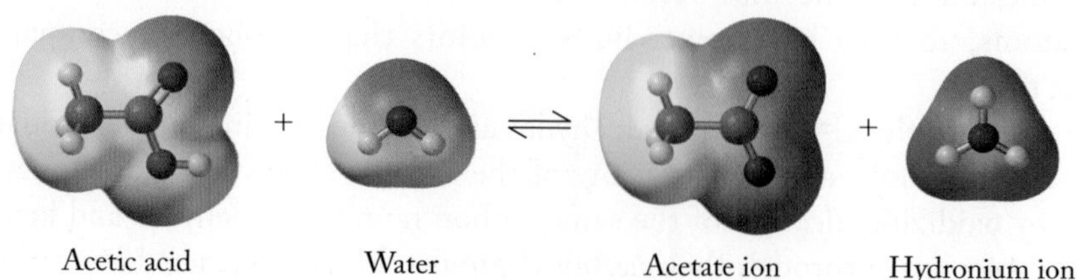

FIGURE 19.39 Carboxylic acids such as acetic acid (found in vinegar) are weak acids in water.

first stages, selected bacteria break up cellulose into mixtures of small molecules. Depending on the bacterial strain, these small-molecule products include H_2 and CO_2, acetic acid, formic acid, or methanol or some other small alcohol. All these products then undergo reactions promoted by the metabolism of **methanogenic** (methane-producing) **bacteria**, which consume hydrogen and simple organic compounds for energy and produce methane gas in the process:

$$4\,H_2(g) + CO_2(g) \rightarrow CH_4(g) + 2\,H_2O(\ell)$$

$$\underset{\text{Acetic acid}}{CH_3COOH(aq)} \rightarrow CH_4(g) + CO_2(g)$$

$$4\,\underset{\text{Formic acid}}{HCOOH(aq)} \rightarrow CH_4(g) + 3\,CO_2(g) + 2\,H_2O(\ell)$$

$$4\,\underset{\text{Methanol}}{CH_3OH(aq)} \rightarrow 3\,CH_4(g) + CO_2(g) + 2\,H_2O(\ell)$$

Methanogenic bacteria have a measurable effect on Earth's atmosphere and climate because methane is a potent greenhouse gas, trapping about 20 times more heat per molecule than carbon dioxide.

methanogenic bacteria bacteria using simple organic compounds and hydrogen for energy; their respiration produces methane, carbon dioxide, and water, depending on the compounds they consume.

ester an organic compound in which the –OH of a carboxylic acid group is replaced by –OR, where R can be any organic group.

condensation reaction two molecules combining to form a larger molecule and a small molecule (typically water).

Esters and Amides

A number of chemical families are closely related to the carboxylic acids. We consider only two of them here: esters and amides. In **esters**, the –COOH group of a carboxylic acid becomes a –COOR group, where R can be any organic group. The presence of these groups makes esters polar, and their boiling points are comparable to those of aldehydes and ketones of similar size. An ester is prepared by the *esterification* of an acid with an alcohol (Figure 19.40a). Esterification is a **condensation reaction**: two molecules combine ("condense") to create a larger molecule while a small molecule (typically water) is also formed.

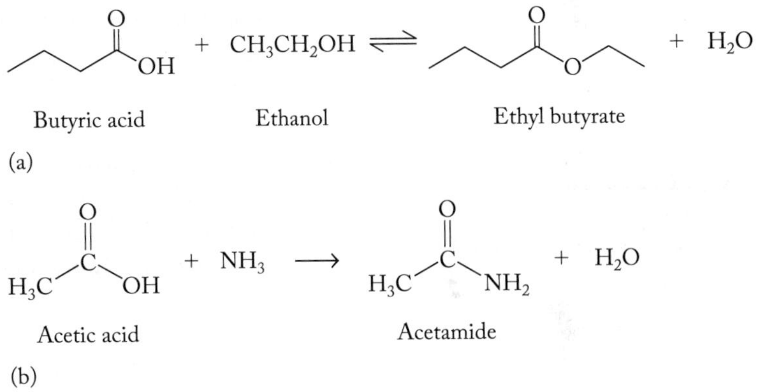

FIGURE 19.40 (a) Condensation reactions between carboxylic acids and alcohols produce esters. Here butyric acid reacts with ethanol, forming ethyl butyrate. (b) Condensation reactions between carboxylic acids and ammonia (or amines) produce amides. Here acetic acid reacts with ammonia, forming acetamide.

Esters frequently have very pleasant fragrances that are much different from the acids from which they are derived. For example, the carboxylic acid butyric acid, with a straight chain of four carbon atoms, is responsible for the odor of rancid butter. The ethyl ester of butyric acid (ethyl butyrate) is responsible for the aroma of ripe pineapple. Esters are widely used in the personal products industry to provide pleasant scents for products like shampoos and soaps.

Many medications consist of molecules that contain carboxylic acid and ester groups. Figure 19.41 illustrates the carbon-skeleton structures of three common pain relievers—aspirin, ibuprofen, and naproxen—each of which

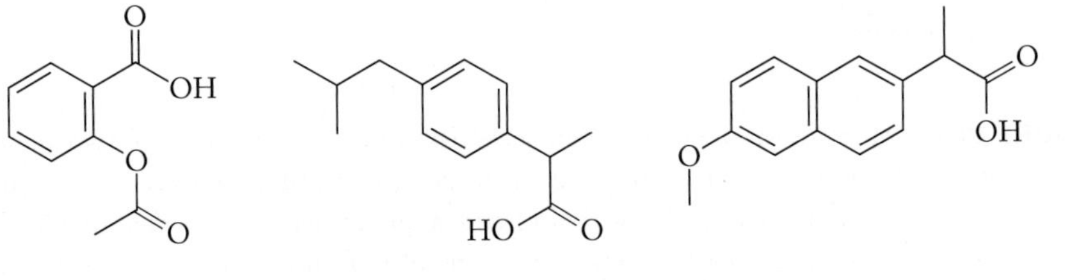

FIGURE 19.41 Three pain medications, all of which contain carboxylic acid functional groups. (a) Aspirin was the first medication to be available in tablet form. (b) Ibuprofen has fewer side effects than aspirin. (c) Naproxen is used for the management of mild to moderate pain. All three belong to a class of compounds called nonsteroidal anti-inflammatory drugs (NSAIDs).

amide an organic compound in which the –OH of a carboxylic acid group is replaced by $-NH_2$, –NHR, or $-NR_2$, where R can be any organic group.

condensation polymer a macromolecule formed by the reaction of monomers, yielding a polymer and water or another small molecule as products of the reaction.

CHEMTOUR Polymers

contains a carboxylic acid group. Aspirin also contains an ester group and naproxen, an ether.

Amides are made in condensation reactions between carboxylic acids and either ammonia or a primary or secondary amine (Figure 19.40b). Amides are polar, and hydrogen atoms bonded to nitrogen can hydrogen-bond with the oxygen atom of an adjacent amide group. These hydrogen bonds cause the boiling points of amides to be higher than those of esters of comparable molar mass.

Polyesters and Polyamides

Most of the compounds we have examined have been monofunctional, which means they have only one functional group that identifies their family. With the polymers of carboxylic acids and their derivatives, we enter the world of *difunctional* molecules, which are molecules with two functional groups. The key point to remember is that the functional groups for the most part still retain their individual chemical reactivity, even if they are in a molecule with another functional group. Also remember that the same features we enumerated for all other polymers still apply here: for polymers, function is determined by composition, structure, and size.

Look again at the esterification reaction in Figure 19.40(a). The –COOH group of the acid reacts with the –OH group of the alcohol to form a carbon–oxygen single bond and release a molecule of water. What would happen at the molecular level if we had a single compound that contained a carboxylic acid functional group at one end and an alcohol functional group at the other (Figure 19.42)? The carboxylic acid group of one molecule could react with the alcohol group of another molecule in a condensation reaction to generate a molecule that has a carboxylic acid group at one end, an alcohol at the other end, and an ester linkage in between. If this reaction happens repeatedly, a monomer containing one carboxylic acid and one hydroxyl group (a hydroxy acid) polymerizes, as shown in Figure 19.42, to form a polyester, a **condensation polymer**. PETE

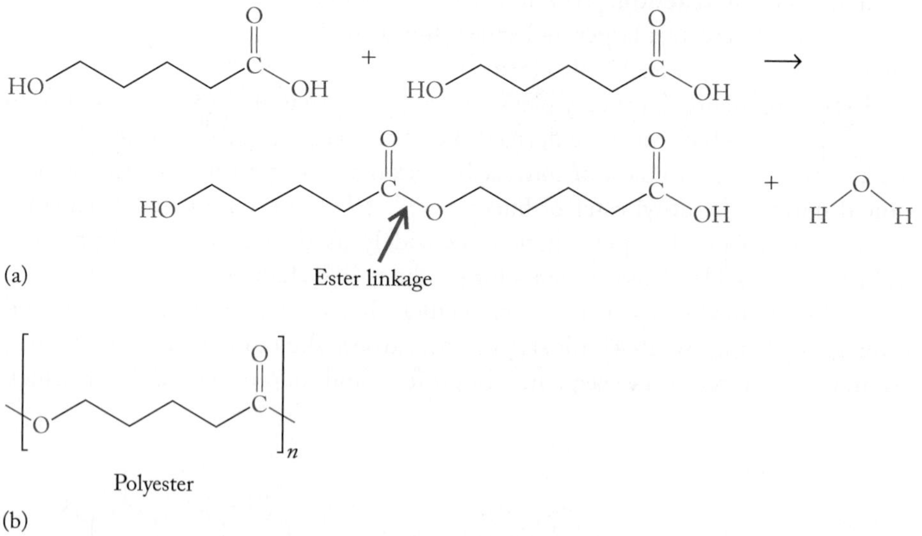

(a) Ester linkage

Polyester

(b)

FIGURE 19.42 (a) Synthesis of an ester from a condensation reaction between two identical difunctional molecules, each one containing an alcohol group and a carboxylic acid group. The diester can then react with additional difunctional molecules at its –OH and –COOH ends, forming a triester. The reaction repeats over and over, forming (b) the polyester made up of the repeating monomer unit shown.

$$x\,\text{HOCH}_2\text{COOH} + y\,\text{HOCH(CH}_3\text{)COOH} \longrightarrow \text{HO}[\text{CH}_2\text{C(=O)O}]_x[\text{CH(CH}_3\text{)C(=O)O}]_y\text{H} + \text{H}_2\text{O}$$

Glycolic acid Lactic acid A polyester

(a)

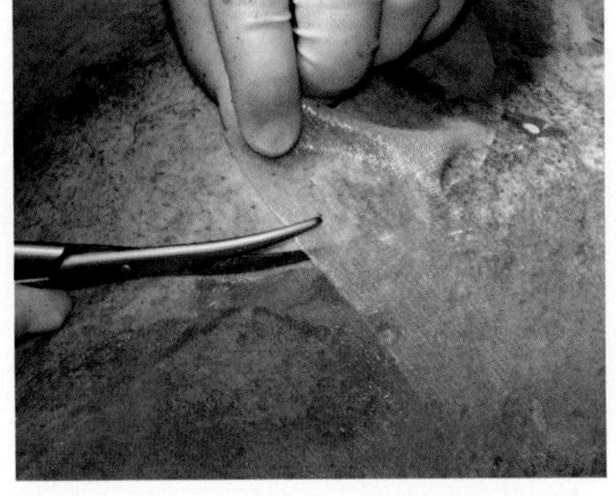

(b)

FIGURE 19.43 (a) The condensation polymer prepared from glycolic acid and lactic acid is used to make sutures that dissolve and artificial skins that protect against infection while promoting the regrowth of skin cells. (b) Synthetic skin being applied to a burn patient.

(Section 19.6 and Figure 19.31) is a condensation polymer, too. In general, condensation polymers are formed by the reaction of monomers that produce a polymer and water or another small molecule. In addition to its use in plastic soda bottles, PETE is used extensively in medicine to make artificial heart valves and grafts for arteries.

A copolymer formed by the reaction of glycolic acid and lactic acid (Figure 19.43) is used to support the growth of skin cells used in grafts for burn victims. This condensation polymer is also used to make dissolving sutures. Esterification reactions used to make polyesters can be reversed by the addition of water, breaking the ester linkage and forming alcohol and acid functional groups. We examine this process in greater detail in Chapter 20.

SAMPLE EXERCISE 19.10 Making a Polyester **LO5**

Show how a polyester can be synthesized from the difunctional alcohol $HO(CH_2)_3OH$ and the difunctional carboxylic acid $HOOC(CH_2)_3COOH$.

Collect, Organize, and Analyze An ester is the product of a reaction between a carboxylic acid and an alcohol. A polyester is a polymer with a repeating unit containing an ester functional group. We are given an alcohol and a carboxylic acid to react to make the ester monomer. Because the starting materials are difunctional, the alcohol can react with two molecules of carboxylic acid, and the acid can react with two molecules of alcohol.

Solve The reaction is

$$\text{HOCH}_2\text{CH}_2\text{CH}_2\text{OH} + \text{HO(O=)CCH}_2\text{CH}_2\text{CH}_2\text{C(=O)OH} \rightarrow$$

$$\text{HOCH}_2\text{CH}_2\text{CH}_2\text{O(O=)CCH}_2\text{CH}_2\text{CH}_2\text{C(=O)OH} + \text{H}_2\text{O}$$

The two OH groups shown in blue react to form an ester at one end of the carboxylic acid. The product molecule has an alcohol group on one end (shown in red) that can react with another molecule of carboxylic acid and a carboxylic acid group (green) on the other end that can react with another molecule of alcohol. Continuing these condensation reactions results in the formation of a polymer whose repeating unit is

$$\left[\!-CH_2CH_2CH_2O-\overset{O}{\overset{\|}{C}}CH_2CH_2CH_2\overset{O}{\overset{\|}{C}}-O-\!\right]_n$$

Think About It The repeating unit in the polyester contains one section that came from the alcohol and a second section that came from the carboxylic acid, which makes sense because esters are formed from alcohols and acids.

Practice Exercise A difunctional molecule may contain two different functional groups, such as this one with an alcohol group and a carboxylic acid group:

$$HO-CH_2CH_2CH_2\overset{O}{\overset{\|}{C}}-OH$$

Draw the repeating unit of the polyester made from this molecule.

Repeating unit in Dacron

(a)

SAMPLE EXERCISE 19.11 Comparing Properties of Polymers LO7

Clothes made from the polyester fabric known as Dacron may be less comfortable in hot weather than clothes made of cotton (also a polymer) because Dacron does not absorb perspiration as effectively as cotton. Based on the repeating units of these two polymers (Figure 19.44), suggest a structural reason why cotton absorbs perspiration (water) better than Dacron.

Collect, Organize, and Analyze Cotton and Dacron are polymers that differ in the functional groups in their repeating units. We need to identify the different functional groups and the interactions between these functional groups and water, a polar molecule.

Solve Each repeating unit in cotton has three –OH groups that are capable of forming hydrogen bonds to the molecules of water in perspiration and thereby draw them away from the body. The repeating unit in Dacron has polar ester groups, but no –OH groups, and it contains a nonpolar (hydrophobic) aromatic ring. Consequently, Dacron does not interact with water molecules as strongly, or absorb perspiration as well as cotton.

Think About It The principle of like interacting with like works for polymers just as it does for small molecules.

Practice Exercise Gloves made of a woven blend of cotton and polyester fibers protect the hands from exposure to oil and grease but are comfortable to wear because they "breathe"—they allow perspiration to evaporate and pass through them, thereby cooling the skin. Suggest how these gloves work at the molecular level.

Repeating unit in cotton: $n \approx 10{,}000$

(b)

FIGURE 19.44 Based on the molecular structures of (a) Dacron and (b) cotton, why is cotton the better material for making T-shirts worn during strenuous exercise?

Combining difunctional molecules in a condensation reaction can be used to make *polyamides*, another class of very useful synthetic polymers. The functional groups are a carboxylic acid and an amine. The difunctional monomer units can

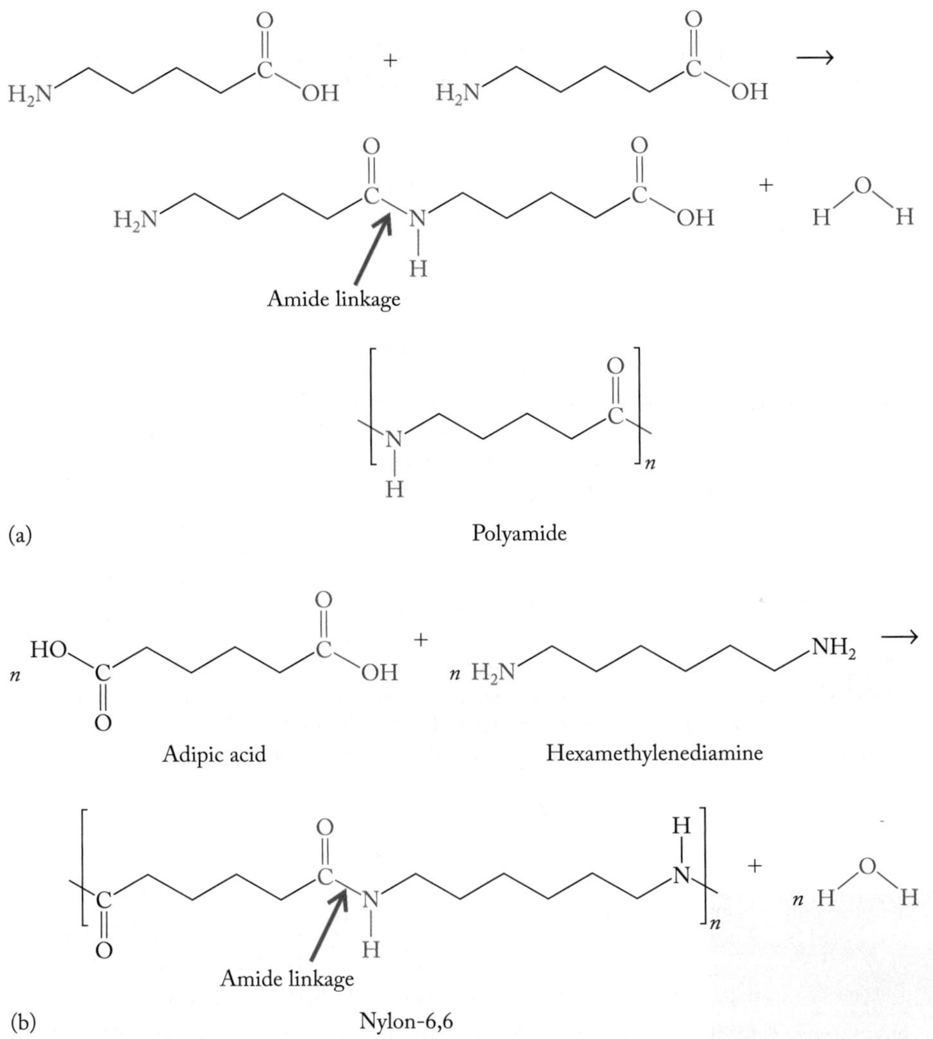

FIGURE 19.45 (a) Synthesis of a polyamide from two identical monomers, each containing a carboxylic acid functional group and an amine functional group. (b) Synthesis of the polyamide nylon-6,6 from nonidentical monomers: adipic acid (a dicarboxylic acid) and hexamethylenediamine (a diamine).

be identical (Figure 19.45a), each containing one carboxylic acid group and one amine group, or they can be different (Figure 19.45b), with one monomer containing two carboxylic acid groups (a dicarboxylic acid) and the other containing two amine groups (a diamine).

Nylon-6,6 is a polyamide made from the monomers shown in Figure 19.45(b). Each monomer contains six carbon atoms, which is what the digits in the name represent.

SAMPLE EXERCISE 19.12 Identifying Monomers **LO5**

Another form of nylon is nylon-6, with the single 6 indicating that the polymer is made from the reaction of a series of identical six-carbon monomers. By analogy with the polyester in Sample Exercise 19.10 and the accompanying Practice Exercise, draw the condensed structure of a six-carbon molecule that could polymerize to make nylon-6.

Collect, Organize, and Analyze Amides form from carboxylic acids and amines. By analogy to Sample Exercise 19.10 and its accompanying Practice Exercise, we can suggest a difunctional molecule that has a carboxylic acid on one end, an amine on the other, and enough carbon atoms in between to form a polyamide with a repeating unit six carbon atoms long.

HO(O=)C–C_6H_4–C(=O)OH

Benzene-1,4-dicarboxylic acid (terephthalic acid)

H_2N–C_6H_4–NH_2

1,4-Diaminobenzene

[–C(=O)–C_6H_4–C(=O)–N(H)–C_6H_4–N(H)–]$_n$

Repeating unit in Kevlar

FIGURE 19.46 The monomers used to make Kevlar are a dicarboxylic acid and a diamine. The amide bond in the repeating unit is highlighted.

Solve We can build the required monomer by starting with one of the functional groups. It doesn't matter which one, so let's begin with the amine:

$$\underbrace{H_2N}_{\text{Amine}}\text{—}CH_2CH_2CH_2CH_2CH_2\text{—}\underbrace{C(=O)OH}_{\text{Carboxylic acid}}$$

Five $-CH_2-$ groups plus one C from the $-COOH$ = six C

We then add a chain of five $-CH_2-$ units because the name nylon-6 indicates that six carbon atoms separate the ends of the repeating unit. Finally, we add the carboxylic acid functional group as the second functional group and the sixth carbon atom in the chain.

Think About It Many nylons with different properties can be made by varying the length of the carbon chain in a monomer like the one in this exercise or by varying the lengths of the chains in both the difunctional acid and difunctional amine in Figure 19.45(b).

Practice Exercise Draw the carbon-skeleton structures of two monomers that could react with each other to make nylon-5,4. Draw the carbon-skeleton structure of the repeating unit in the polymer. (NOTE: The first number refers to the carboxylic acid monomer; the second refers to the amine monomer.)

Polymers of nylon make long, straight fibers that are quite strong and excellent for weaving into fabrics. Nylon is flexible and stretchable because the hydrocarbon chains can bend and curl, much like a telephone cord or a Slinky spring toy. To produce an even stronger nylon, researchers had to find some way to reduce the ability of the chains to form coils. This was achieved by using monomer units containing functional groups that made it difficult for the chains to bend. One product of this work was a polyamide called Kevlar, invented by Stephanie Kwolek of DuPont in 1965. Kevlar is formed from a dicarboxylic acid of benzene and a diamine of benzene (Figure 19.46). When these two monomers polymerize, the flat, rigid aromatic rings keep the chains straight.

Two additional intermolecular interactions orient the chains and hold them tightly together (Figure 19.47). First, the –NH hydrogen atoms form hydrogen bonds with the oxygen atoms of carbonyl groups on adjacent chains. Second, the rings stack on top of one another (as in polystyrene) and provide additional interactions that

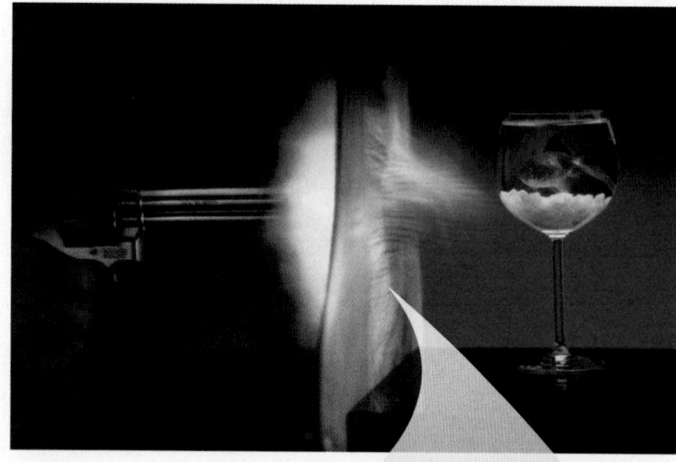

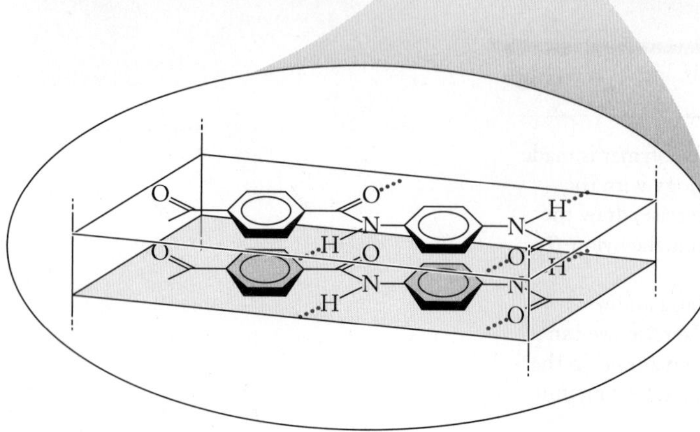

FIGURE 19.47 A bullet fired point-blank does not puncture fabric made of Kevlar. The strength is due in part to the strong interactions between the functional groups in the molecular structure of Kevlar.

TABLE 19.8 Summary of Common Polymers and Their Uses

Name	Abbreviation	Functional Group	Use
Addition Polymers			
Polyethylene	PE	Alkane	Plastic bags and films
Poly(tetrafluoroethylene)	Teflon	Fluoroalkane	Nonstick coatings
Polyvinyl chloride	PVC	Chloroalkane	Drain pipes
Poly(methyl methacrylate)	PMMA	Alkane and ester	Shatter-resistant glass, such as Plexiglas, Lucite
Polystyrene	PS	Aromatic hydrocarbon	Dishes, insulation
Poly(vinyl alcohol)	PVA	Alcohol	Gloves, bottles
Condensation Polymers			
Poly(ethylene glycol)	PEG	Ether	Pharmaceuticals, consumer products
Poly(ethylene terephthalate)	PETE	Ester	Plastic bottles
Nylon		Amide	Clothing
Kevlar		Amide	Protective equipment
Dacron		Ester	Clothing

hold the chains together in parallel arrays. The result is a fiber that is very strong but still flexible. Fabrics and helmets made of Kevlar resist puncture, even by bullets and hockey pucks fired at them, and they are also resistant to flames and reactive chemicals.

We have seen how the macroscopic properties of Kevlar and other polymers are influenced by their microscopic structure. Table 19.8 reinforces this message with a list of the functional groups in several addition and condensation polymers and examples of familiar materials made from them.

19.8 A Brief Survey of Isomers

The existence of isomers is partly responsible for the enormous number and wide variety of organic compounds in the world. We have already discussed isomers in Chapters 5 and 6, and in several sections of this chapter. This brief survey collects all the information we have presented so far, and Figure 19.48 summarizes the relationships between the types of isomers we have mentioned. Remember that isomers are molecules that have the same number and same kinds of atoms but differ in how those atoms are arranged. Because the molecules' structures are not the same, their physical and chemical properties differ as well.

As Figure 19.48 shows, isomers fall into two general categories: constitutional isomers and stereoisomers. Constitutional isomers, which we introduced in Chapter 6 and revisited with alkanes in Section 19.2, differ in the way in which the atoms comprising them are connected. Stereoisomers have their atoms connected in the same way, but the three-dimensional arrangement of their atoms differs. Let's review constitutional isomers first.

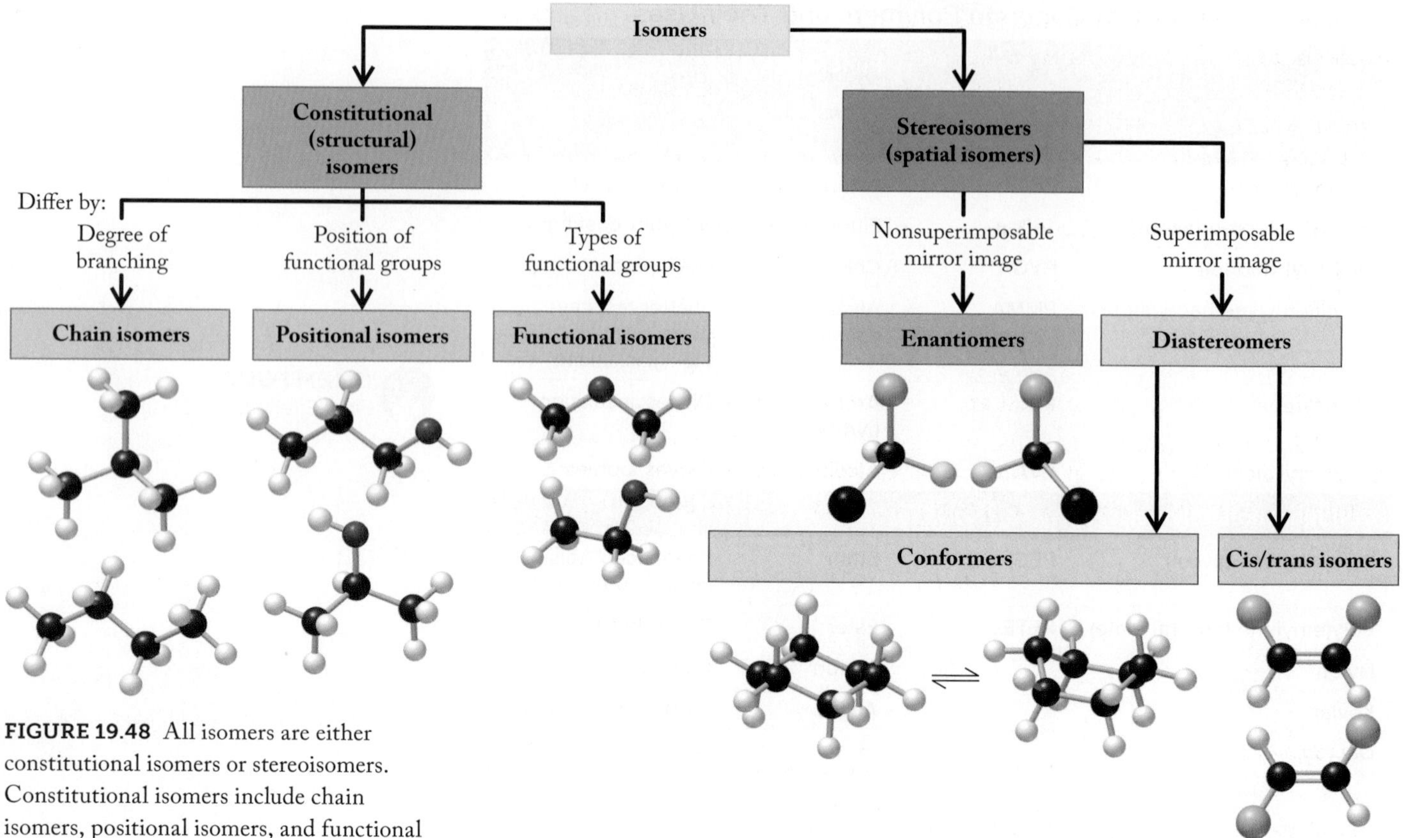

FIGURE 19.48 All isomers are either constitutional isomers or stereoisomers. Constitutional isomers include chain isomers, positional isomers, and functional isomers. Stereoisomers are further categorized as enantiomers (optically active compounds) or diastereomers.

In Chapter 6 we saw three kinds of constitutional isomers, although we did not name them at the time. Now that we have a better understanding of organic chemistry, we can categorize them more clearly as chain isomers, positional isomers, and functional isomers. *Chain isomers* are molecules having different arrangements of their carbon skeletons, like butane and 2-methylpropane. *Positional isomers* like propanol and isopropanol have the same functional group (in this case, the –OH group) bonded to different carbon atoms. *Functional isomers* like ethanol and dimethyl ether have different functional groups because of the arrangement of their atoms. The differing shapes and polarities of constitutional isomers give rise to differences in their physical properties, such as melting points and boiling points, as we saw in Figure 6.3.

We introduced stereoisomers in Chapter 5 in our discussion of chirality. We saw them again in Section 19.3 when discussing alkenes. Stereoisomers are different from constitutional isomers because their atoms have the same connectivity: the same atoms are connected to each other in the same way. What makes stereoisomers special is that the arrangement of the atoms in three-dimensional space differs. That's why the prefix *stereo-* is used in the name; it means we need to consider what the molecule looks like in three dimensions.

Stereoisomers can be *enantiomers* or *diastereomers*. Enantiomers are mirror-image isomers, whereas diastereomers are stereoisomers that are *not* enantiomers. Let's first review enantiomers.

Several structural features can give rise to enantiomers, but the most important one for our purposes is the presence of a carbon atom that has four different atoms or groups of atoms attached to it. This arrangement about a carbon atom makes a molecule nonsuperimposable on its mirror image, much like your right and left hands. Such a molecule is chiral. This molecule and its mirror image rotate beams of plane-polarized light in opposite directions, as we saw in Figure 5.39. This behavior gives rise to another name for enantiomers: *optically active compounds* or *optical isomers.*

Diastereomers include the cis and trans isomers that can occur because of the C=C double bonds in alkenes. *Conformers*, such as the chair and boat conformations of cyclohexane, are also diastereomers. Other structural features may give rise to enantiomers and diastereomers as well, but they are beyond the scope of this book.

In this chapter, we introduced the major functional groups in organic chemistry, and we discussed how the structure of a molecule and the functional groups it contains determine its chemical and physical properties. We examined a few reactions that produce several types of polymeric materials common in the modern world, ranging from carpets and insulation to bulletproof vests. Other organic compounds are used as fuels, as pharmaceuticals, and as building materials. Despite the range of materials we have discussed, this treatment can give only a brief overview of the importance of organic chemistry.

SAMPLE EXERCISE 19.13 Integrating Concepts: Taxol

Taxol, known generically as paclitaxel (Figure 19.49a; $C_{47}H_{51}NO_{14}$; molar mass 853.9 g/mol), is an important drug in the treatment of several types of cancer. The compound was originally harvested from the bark of the very slow-growing Pacific yew tree (*Taxus brevifolia*). Unfortunately, the concentration of taxol in the bark is only 0.01%. To conduct the clinical trials to test the effectiveness of taxol, *9000* mature trees were harvested. Using Pacific yew trees as the sole source of taxol would have led to their extinction, so other sources were needed. A number of research groups have reported the synthesis of taxol using either organic chemistry or biosynthetic processes using cultures of *Taxus* cells. The organic chemistry approaches start with other, simpler compounds found in nature. One of them uses a starting material called verbenone (Figure 19.49b), which is a commercially available derivative of pinene, a component of the resin exuded by pine trees. The synthesis of taxol from verbenone requires 36 steps and results in an overall yield of only 0.1% (moles of taxol/mole of verbenone).

a. How many aromatic rings are in taxol?
b. How many (i) ester, (ii) ether, (iii) alcohol, (iv) amide, (v) ketone, and (vi) alkene functional groups, and (vii) how many chiral centers are in the taxol molecule?
c. Suppose 10.0 kg of taxol is required for a clinical trial of the substance in cancer patients. How many kilograms of the starting material verbenone is needed to prepare that amount?

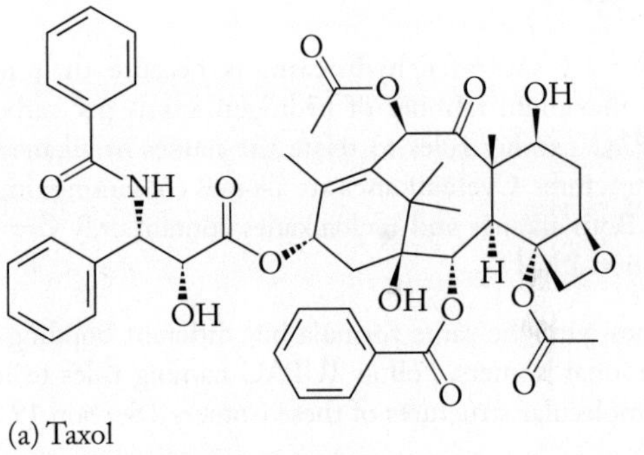

(a) Taxol

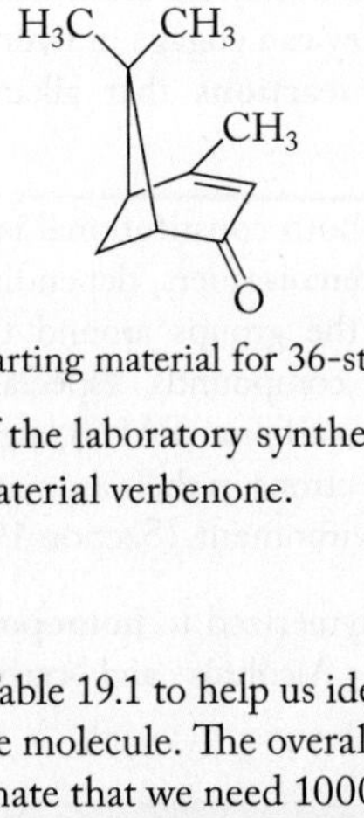

(b) Verbenone, the starting material for 36-step synthesis of taxol

FIGURE 19.49 One of the laboratory syntheses of (a) taxol begins with (b) the starting material verbenone.

Collect and Organize We have the structural formula of taxol, which consists of a variety of functional groups and chiral carbon centers. We also know the amount of taxol we need for a study and the overall yield of the 36-step synthesis used to prepare it. We are given the structure of the starting material used in step 1 of the process.

Analyze We can use Table 19.1 to help us identify the types of functional groups in the molecule. The overall yield of the synthesis is 0.1%, so we can estimate that we need 1000 times as much starting material to set up the reaction to prepare 10.0 kg of the final product. We can calculate the mass of verbenone we need for the first step of the reaction, but we need to know the number of moles of verbenone required. Verbenone is a much smaller molecule than taxol; its formula is $C_{10}H_{14}O$ and its molecular weight is 150.24 g, making its molar mass about 1/5 that of taxol. We estimate that we will need around 2000 kg of it for the reaction.

Solve The functional groups in taxol are highlighted in Figure 19.50.

a. There are 3 aromatic rings (gray).
b. (i) There are 4 esters (yellow); (ii) 1 ether (green); (iii) 3 alcohols (orange); (iv) 1 amide (blue); (v) 1 ketone (red); (vi) 1 alkene (pink); and (vii) 11 chiral centers (carbon atoms identified with asterisks).
c. The overall yield of the multistep process is 0.1%, or 0.001 mol taxol/mol verbenone, so the mass (kg) of verbenone needed to produce 10.0 kg (or 1.00×10^4 g) of taxol is

$$1.00 \times 10^4 \text{ g taxol} \times \frac{1 \text{ mol taxol}}{853.9 \text{ g taxol}} \times \frac{1 \text{ mol verbenone}}{0.001 \text{ mol taxol}}$$

$$\times \frac{150.24 \text{ g verbenone}}{1 \text{ mol verbenone}} \times \frac{1 \text{ kg}}{10^3 \text{ g}} = 2 \times 10^3 \text{ kg verbenone}$$

Thus, we need 2000 kilograms of verbenone to produce just 10 kilograms of taxol.

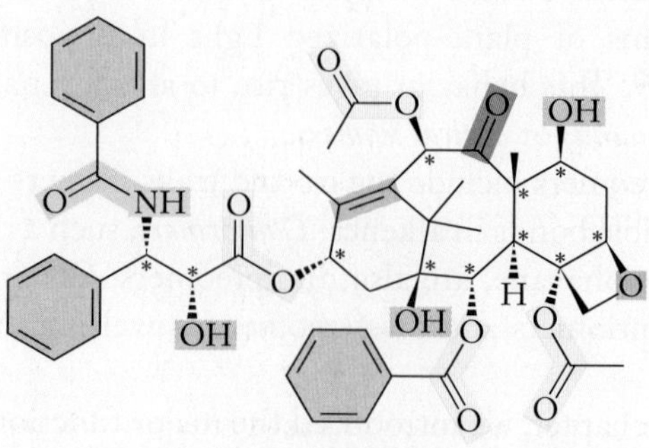

FIGURE 19.50

Think About It The answer matches our initial estimate (to one significant figure, which is all we are entitled to, given the 0.1% yield of the process). Like many natural substances with complex structures, taxol is a challenging substance to prepare synthetically. The overall yield of a series of reactions is equal to the product of the yields of 36 individual reactions. If the average yield of the reactions is n, then $n^{36} = 0.001$. Solving for n, we find that the average yield of each reaction is about 83%, which is quite good for most synthetic work.

SUMMARY

LO1 Alkanes are **saturated hydrocarbons** because their molecules contain the maximum number of hydrogen atoms per carbon atom. Follow IUPAC naming rules to relate the names of alkanes to their molecular structures. **Cycloalkanes** are alkanes containing rings of carbon atoms. Both alkanes and cycloalkanes contain only C—C single bonds. (Section 19.2)

LO2 Alkanes with the same formula but different bonding patterns are constitutional isomers. Follow IUPAC naming rules to relate the names and molecular structures of these isomers. (Section 19.2)

LO3 The -C═C- double bonds in alkenes and the -C≡C- triple bonds in **alkynes** mean that they are unsaturated hydrocarbons and that they can engage in **hydrogenation** and **addition reactions** that alkanes cannot. (Section 19.3)

LO4 Alkenes may have both constitutional isomers and **cis** or **trans** stereoisomers, depending on the arrangement of the groups around the double bond. Aromatic compounds, especially polycyclic aromatic hydrocarbons (PAHs), are stabilized by delocalization of the bonding electrons in their aromatic rings, which enables them to persist in the environment. (Section 19.3)

LO5 Alkenes can be polymerized to **homopolymers**, most of which are **addition polymers**. Alcohols and carboxylic acids react in **condensation reactions** to form **heteropolymers** called **polyesters**; alcohols and amines react similarly to produce polyamides. The properties of the polymers depend upon the identity and arrangement of the monomer units in the polymers. (Sections 19.3, 19.4, 19.6, 19.7)

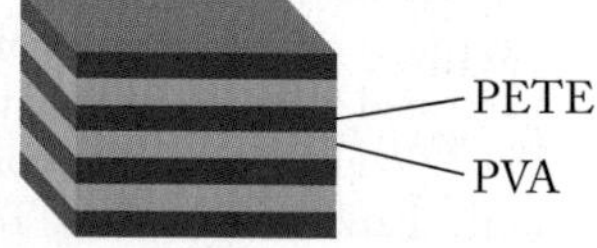

LO6 Most gasoline sold in the United States and Canada contains oxygenated additives such as ethanol to improve combustion and reduce pollution. However, the presence of these additives reduces fuel value and slightly decreases driving range. (Section 19.6)

LO7 Polymers made from ethers may be soluble in both polar and nonpolar liquids and are increasingly used to improve the biodistribution of pharmaceuticals in the body. The identity of monomer units in both natural and synthetic polymers determines their behavior as bulk materials. (Section 19.6)

LO8 The presence of heteroatoms in the functional groups of many organic compounds strongly influences their chemical properties and allows them to interact with polar inorganic compounds, such as water, in ways that nonpolar hydrocarbons cannot. (Sections 19.5, 19.7)

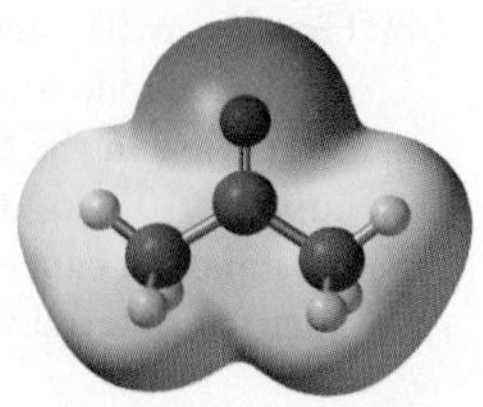

PARTICULATE PREVIEW WRAP-UP

Compounds A and B are isomers. Compound A contains a carboxylic acid functional group; compound B contains an ester functional group. A carboxylic acid can be converted into an ester by reacting with an alcohol.

PROBLEM-SOLVING SUMMARY

Type of Problem	Concepts and Equations	Sample Exercises
Naming alkanes and drawing their molecular structures	Start at one end of any branch and assign numbers to each carbon atom. If the chain branches, choose one branch to follow. Repeat the process to explore other side chains and identify the longest chain. The ends of alkane molecules are methyl groups; between those ends, gather all methylene groups inside parentheses to create the final condensed structure. To create a carbon-skeleton structure, use lines to depict single covalent bonds between carbon atoms. A sufficient number of H atoms to complete the valence of the carbon atoms is assumed.	**19.1, 19.2**
Recognizing constitutional isomers	Establish that the compounds have the same molecular formula, and if they do, look for different arrangements of C—C bonds.	**19.3, 19.4**
Distinguishing among alkanes, alkenes, and alkynes	Alkanes contain only C—C single bonds; they are saturated hydrocarbons and do not react with H_2. Alkenes contain at least one C═C double bond that can combine with a molecule of H_2. Alkynes contain at least one C≡C triple bond that can combine with two molecules of H_2.	**19.5**
Identifying and naming stereoisomers and constitutional isomers of alkenes	If molecules have the same formula, look for different arrangements of their bonds. Molecules with two groups on the same side of a C═C bond are cis isomers; those with groups on opposite sides are trans isomers. (cis: R, R / H, H; trans: R, H / H, R)	**19.6**
Identifying monomers in polymers and vice versa	Find the smallest portion of the polymer that is repeated. For condensation polymers, combine monomers with alcohol functional groups (R′OH) and carboxylic acid functional groups (RCOOH) to form water (H_2O) and ester groups (RCOOR′).	**19.7, 19.10, 19.12**
Comparing the energy content of fuel mixtures	Determine the number of moles of each component. Then multiply that number by the corresponding value of $\Delta H^\circ_{\text{comb}}$. Divide by molar masses to convert kJ/mol to kJ/g.	**19.8**
Assessing properties of polymers	Evaluate the polarity of the functional groups in the polymer and assess the relative importance of all types of intermolecular forces possible in the molecules.	**19.9, 19.11**

VISUAL PROBLEMS

(Answers to boldface end-of-chapter questions and problems are in the back of the book.)

19.1. How many degrees of unsaturation does each of the hydrocarbons shown in Figure P19.1 have?

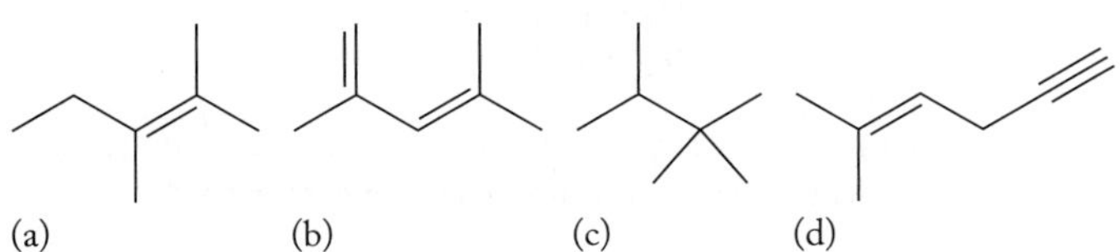

FIGURE P19.1

19.2. Which of the hydrocarbons in Figure P19.2 are constitutional isomers of each other?

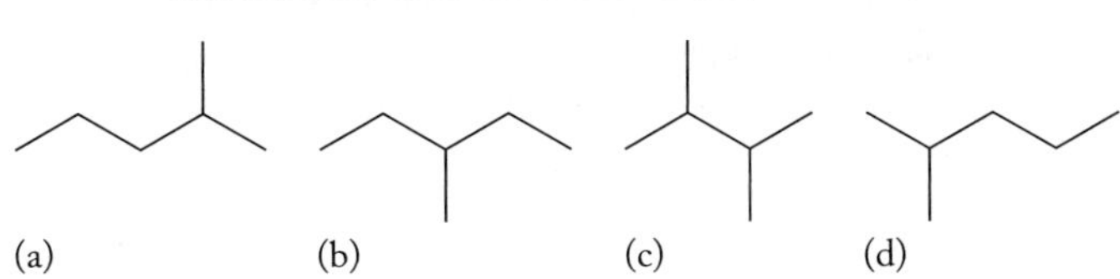

FIGURE P19.2

19.3. Figure P19.3 shows the carbon-skeleton structures of four organic compounds found in nature as fragrant oils. Which are alkenes?

OH

O

Pine oil Oil of peppermint Oil of celery Camphor

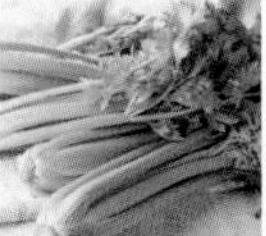

FIGURE P19.3

19.4. Which of the three molecules shown in Figure P19.4—acrylonitrile (found in barbeque smoke), capillin (an antifungal drug), and pargyline (an antihypertensive drug)—does *not* contain the alkyne functional group?

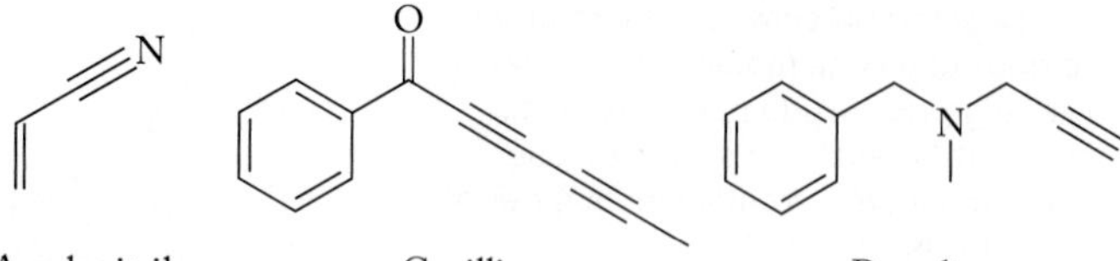

Acrylonitrile Capillin Pargyline

FIGURE P19.4

19.5. Which molecules in Figure P19.5 are considered aromatic compounds?

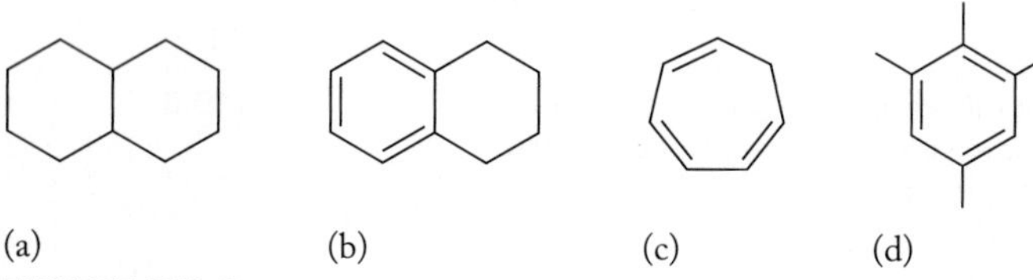

(a) (b) (c) (d)

FIGURE P19.5

19.6. Benzyl acetate, carvone, and cinnamaldehyde are all naturally occurring oils. Their carbon-skeleton structures are shown in Figure P19.6. Which ones contain an aromatic ring?

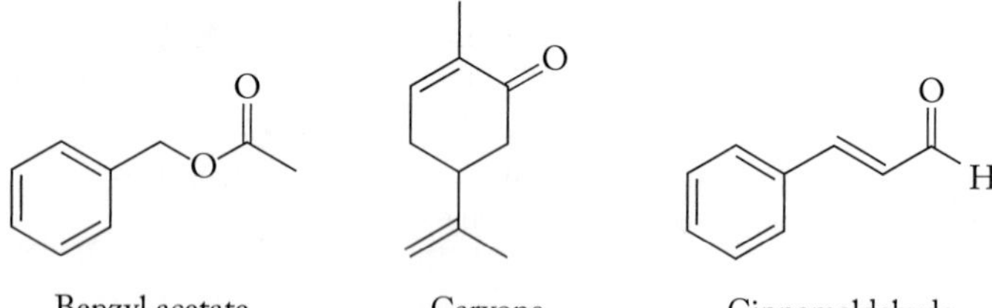

Benzyl acetate (oil of jasmine) Carvone (oil of spearmint) Cinnamaldehyde (oil of cinnamon)

FIGURE P19.6

19.7. In polypropylene, $\text{+}CH_2CH(CH_3)\text{+}_n$, the methyl groups can be in one of two positions with respect to each other (Figure P19.7): (a) on the same side of the carbon–carbon backbone or (b) on opposite sides. Many possible variations exist for (b). One form of the polymer in bulk is rigid and resists deformation; the other is soft and rubbery. Suggest which structure gives rise to which set of properties and explain your choice.

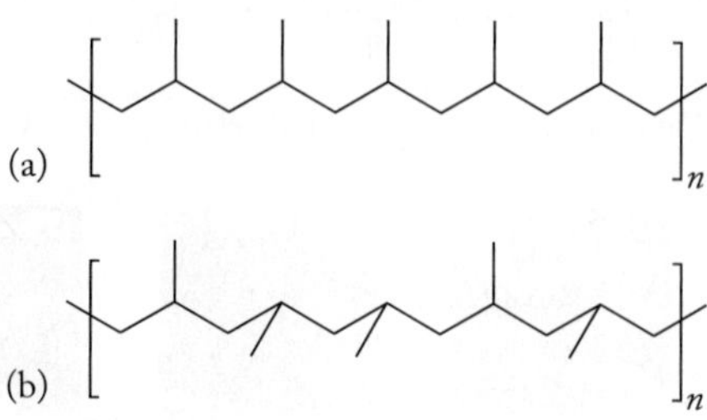

FIGURE P19.7

*19.8. The three polymers shown in Figure P19.8 are widely used in the plastics industry. In which of them are the intermolecular forces per mole of monomer the strongest?

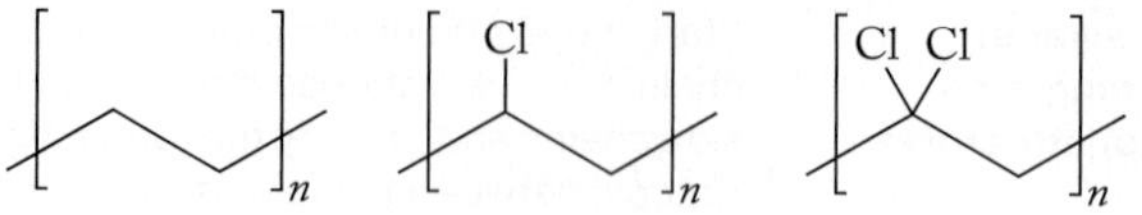

Cl Cl Cl

(a) Polyethylene (b) Poly(vinyl chloride) (c) Poly(1,1-dichloroethylene)

FIGURE P19.8

***19.9. Silly Putty** Silly Putty is a condensation polymer of dihydroxydimethylsilane (Figure P19.9). Draw the condensed structure of the repeating monomer unit in Silly Putty.

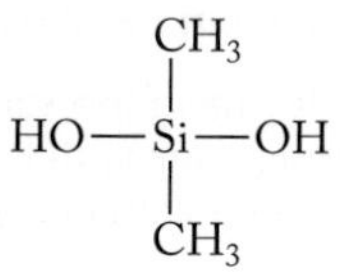

CH_3
HO—Si—OH
CH_3

Dihydroxydimethylsilane

FIGURE P19.9

19.10. **Orlon and Acrilon Fibers** Figure P19.10 shows the carbon-skeleton structure of polyacrylonitrile, which is marketed as Orlon and Acrilon. Draw the Kekulé structure of the monomeric reactant that produces this polymer.

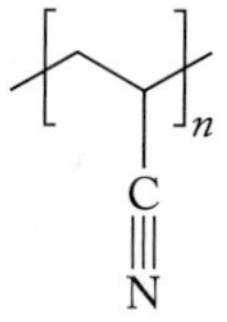

Polyacrylonitrile

FIGURE P19.10

19.11. Rubber is a polymer of isoprene. It is sometimes called polyisoprene. There are two forms of polyisoprene (Figure P19.11): *cis*-polyisoprene is the soft, flexible material we associate with the term *rubber*; gutta-percha, or *trans*-polyisoprene, is a much harder material.

a. Draw the monomeric units of *cis*- and *trans*-polyisoprene.
b. Consider the structures of the polymers and comment on why one form is rubbery and the other hard.

cis-Polyisoprene *trans*-Polyisoprene

FIGURE P19.11

19.12. Use representations [A] through [I] in Figure P19.12 to answer questions a–f.
  a. Identify the functional group present in each [B], [D], [E], [F] and [H].
  b. Which polymers were formed by addition?
  c. Which polymers were formed by condensation?
  d. Which functional groups reacted to form the condensation polymers?
  e. Which polymer structures likely result in rigid, hard materials?
  f. Which polymer structures likely result in flexible, stretchable, soft materials?

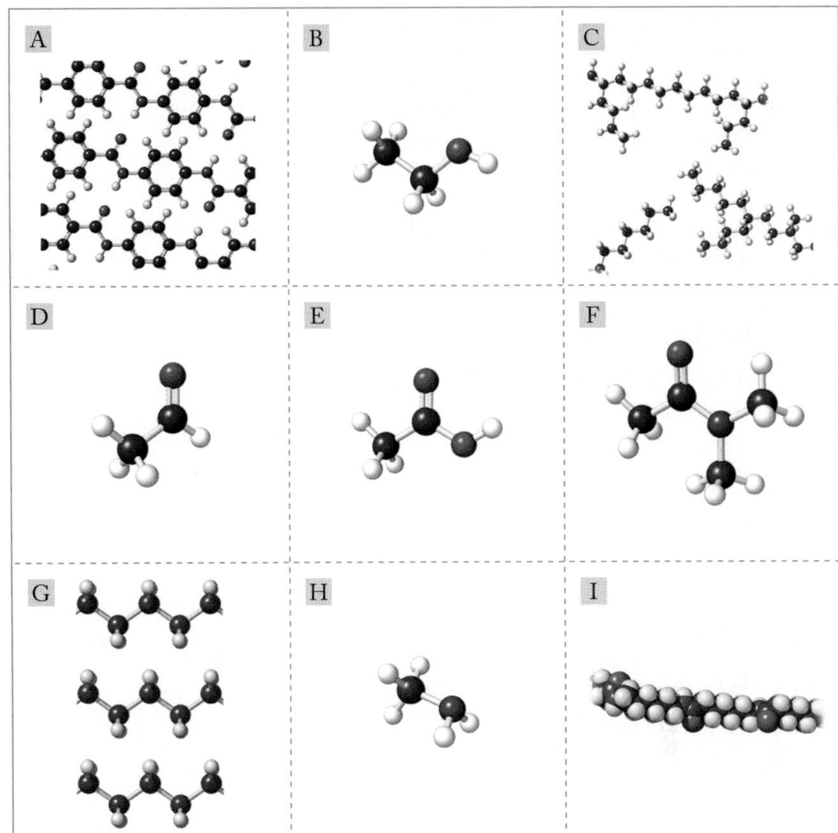

FIGURE P19.12

QUESTIONS AND PROBLEMS

Carbon: The Scope of Organic Chemistry

Concept Review

19.13. List the different sets of hybrid orbitals (valence bond theory) used to describe bonding in organic compounds. What combination of single and multiple bonds is possible with each hybridized set?

19.14. What is the principal difference between an oligomer and a polymer formed from the same monomer?

***19.15.** Is the interstitial alloy tungsten carbide (WC) considered to be an organic compound?

*19.16. Calcium carbide, CaC_2, was once used in miner's lamps. Reaction of CaC_2 with water yields acetylene which, when ignited, gives light. Is calcium carbide considered an organic compound?

19.17. Which of the functional groups in Table 19.1 are polar?

19.18. Which of the compounds in Table 19.1 should be water soluble?

19.19. If the average molar mass of a polyethylene sample A is twice that of sample B, which sample begins to soften at a higher temperature? Explain why.

19.20. Which of the following properties of polyethylene increases as the number of monomer units per molecule of the polymer increases? (a) melting point; (b) viscosity; (c) density; (d) C:H ratio; (e) fuel value

Alkanes

Concept Review

19.21. Do linear and branched alkanes with the same number of carbon atoms all have the same empirical formula?

19.22. If an alkane and a cycloalkane have equal numbers of carbon atoms per molecule, do they have the same number of hydrogen atoms?

19.23. What is the hybridization of carbon in alkanes?

19.24. Figure P19.24 shows the carbon-skeleton structures of hexane and cyclohexane. Are hexane and cyclohexane constitutional isomers?

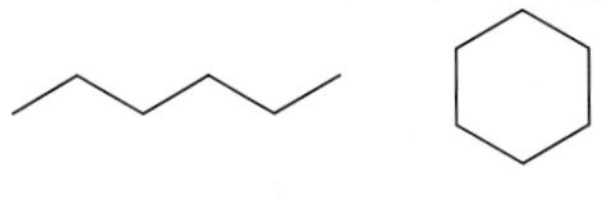

Hexane Cyclohexane

FIGURE P19.24

19.25. Why isn't cyclohexane a planar molecule?

19.26. Which of the simple cycloalkanes (C_nH_{2n}, n = 3–8) has a nearly planar geometry?

19.27. Are cycloalkanes saturated hydrocarbons?

19.28. Do constitutional isomers always have the same molecular formula?

19.29. Do constitutional isomers always have the same chemical properties?

19.30. Are constitutional isomers members of a homologous series?

Problems

19.31. Draw and name all the constitutional isomers of C_5H_{12}.
19.32. Draw and name all the constitutional isomers of C_6H_{14}.

19.33. Which of the molecules in Figure P19.33 are constitutional isomers of octane (C_8H_{18})? Name these molecules.

(a) (b) (c)

(d) (e)

FIGURE P19.33

19.34. Which of the molecules in Figure P19.34 are constitutional isomers of heptane (C_7H_{16})? Name these molecules.

(a) (b) (c) (d)

(e) (f)

FIGURE P19.34

19.35. Convert the carbon-skeleton structures in Problem 19.33 to molecular formulas.
19.36. Convert the carbon-skeleton structures in Problem 19.34 to molecular formulas.

19.37. **Crude Oil** Place the following molecules, all of which are products of the distillation of crude oil, in order in which they would appear in the distillate during simple distillation: C_6H_{14}, $C_{18}H_{38}$, $C_{12}H_{26}$, C_9H_{20}.
*19.38. Rank the molecules in Figure P19.38 in order of decreasing van der Waals forces.

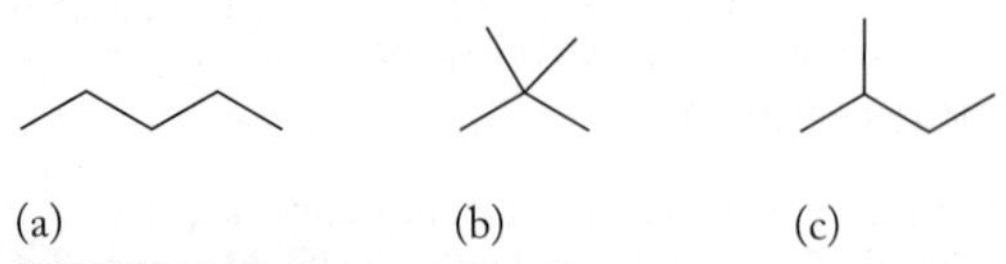

(a) (b) (c)

FIGURE P19.38

19.39. Which has a higher hydrogen-to-carbon ratio: hexane or cyclohexane?
*19.40. For linear alkanes, each methylene group (–CH_2–) in the structure contributes 658 kJ/mol to the heat of combustion. Heats of combustion for some cycloalkanes are listed below as a function of ring size. Explain the trend in the data.

Ring size	Heat of combustion per –CH_2– group (kJ/mol)
3	696
4	686
5	663
6	658

Alkenes and Alkynes

Concept Review

19.41. Can combustion analysis distinguish between an alkene and a cycloalkane containing the same number of carbon atoms?
19.42. Can their reaction with hydrogen distinguish between an alkene and a cycloalkane containing the same number of carbon atoms?
19.43. Why don't the alkenes in Figure P19.43 have cis and trans isomers?

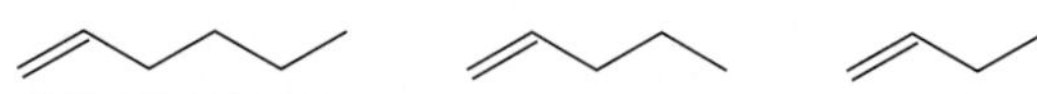

FIGURE P19.43

19.44. Why don't alkynes have cis and trans isomers?
***19.45.** Figure P19.45 shows the carbon-skeleton structure of carvone, which is found in oil of spearmint. Why doesn't the molecule carvone have cis and trans isomers?

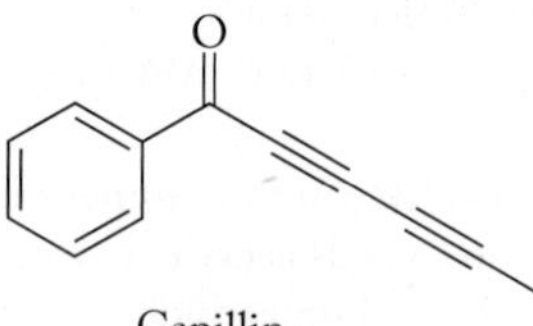

FIGURE P19.45

*19.46. Figure P19.46 shows the carbon-skeleton structure of the antifungal compound capillin. Are the π electrons in capillin delocalized?

O

Capillin

FIGURE P19.46

19.47. Ethylene reacts quickly with HBr at room temperature, but polyethylene is chemically unreactive toward HBr. Explain why these related substances have such different properties.
*19.48. Polymerization of butadiene ($CH_2{=}CHCH{=}CH_2$) does not yield the same polymer as polymerization of ethylene ($CH_2{=}CH_2$).
a. How could we convert poly(butadiene) into poly(ethylene)?
b. Predict the reactivity of poly(butadiene) with HBr.

Problems

19.49. Using the average bond strengths given in Appendix 4, estimate the molar heat of hydrogenation, $\Delta H_{\text{hydrogenation}}$, for the conversion of C_2H_4 to C_2H_6.

$$H_2C{=}CH_2(g) + H_2(g) \rightarrow CH_3CH_3(g)$$

19.50. Using the average bond strengths given in Appendix 4, estimate the molar heat of hydrogenation, $\Delta H_{\text{hydrogenation}}$, for the conversion of C_2H_2 to C_2H_6.

$$CH{\equiv}CH(g) + 2\ H_2(g) \rightarrow CH_3CH_3(g)$$

19.51. Cinnamon Label the isomers of cinnamaldehyde (oil of cinnamon) in Figure P19.51 as cis or trans and *E* or *Z*.

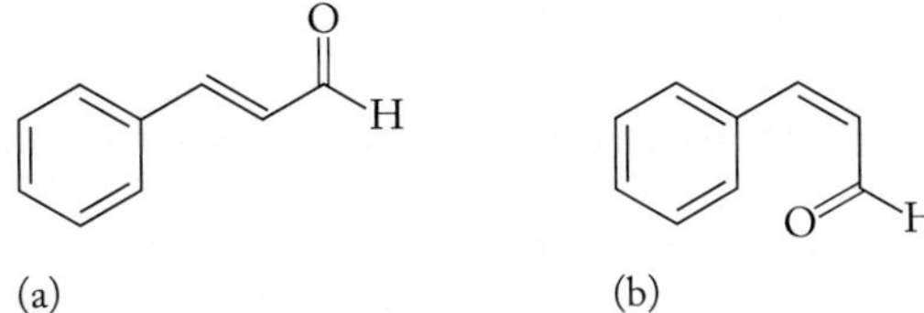

FIGURE P19.51

19.52. Prostaglandins, naturally occurring compounds in our bodies that cause inflammation and other physiological responses, are formed from arachidonic acid, an unsaturated hydrocarbon containing four C=C bonds and a carboxylic acid functional group. The stereoisomer containing all cis double bonds is shown in Figure P19.52. How many stereoisomers other than this one are possible? Draw the isomer containing all trans double bonds.

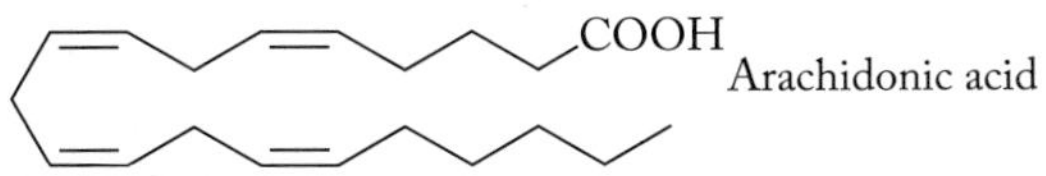

COOH
Arachidonic acid

FIGURE P19.52

19.53. Using data in Appendix 4, calculate ΔH_{rxn} for the production of acetylene from the controlled combustion of methane:

$$6\ CH_4(g) + O_2(g) \rightarrow 2\ C_2H_2(g) + 2\ CO(g) + 10\ H_2(g)$$

Is this an endothermic or an exothermic reaction?

19.54. Using data in Appendix 4, calculate ΔH_{rxn} for the production of acetylene from the reaction between calcium carbide and water, given that the ΔH_f° of CaC_2 is −59.8 kJ/mol:

$$CaC_2(s) + 2\ H_2O(\ell) \rightarrow C_2H_2(g) + Ca(OH)_2(s)$$

Is this an endothermic or an exothermic reaction?

***19.55.** Given the following two reactions and thermodynamic data from Appendix 4, estimate ΔH_{rxn} for the hydrogenation of acetylene (C_2H_2) with one mole of hydrogen gas to make ethylene (C_2H_4).

$$\underset{\text{Acetylene}}{HC{\equiv}CH(g)} + 2\ H_2(g) \rightarrow \underset{\text{Ethane}}{CH_3CH_3(g)}$$

$$\underset{\text{Ethylene}}{H_2C{=}CH_2(g)} + H_2(g) \rightarrow \underset{\text{Ethane}}{CH_3CH_3(g)}$$

*19.56. The heat of hydrogenation of *cis*-2-butene is −119.7 kJ/mol, whereas that of the trans isomer is −115.5 kJ/mol. Draw both isomers and the product of the hydrogenation reaction, and locate them on the graph of relative energy given in Figure P19.56. The condensed structural formula of 2-butene is $CH_3CH{=}CHCH_3$.

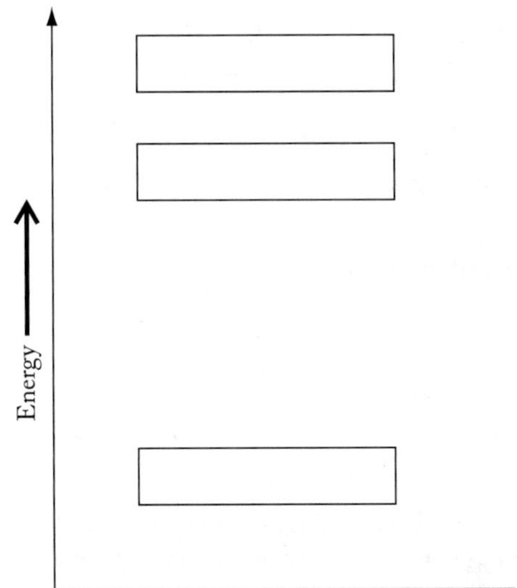

FIGURE P19.56

19.57. Making Glue Wood glue or "carpenter's glue" is made of poly(vinyl acetate). Draw the carbon-skeleton structure of this polymer. The monomer is shown in Figure P19.57.

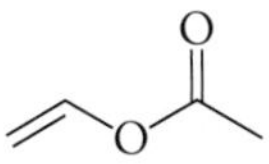

O
O
Vinyl acetate

FIGURE P19.57

19.58. The 2000 Nobel Prize in Chemistry was awarded for research on the electrically conductive polymer polyacetylene.

a. Draw the carbon-skeleton structure of three monomeric units of the addition polymer that results from the polymerization of acetylene, $HC{\equiv}CH$.

*b. There are two possible stereoisomers of polyacetylene. Describe the two isomeric forms.

***19.59.** Polyacetylene (see Problem 19.58) conducts electricity but has no commercial uses in large part because it is unstable in air, reacting with O_2 to produce polymers that contain carbonyl groups and ether linkages (Figure P19.59). Suggest why these changes in composition and structure decrease the conductivity of the polymer.

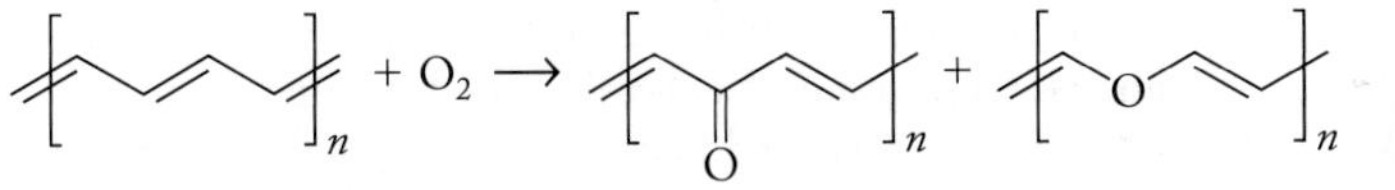

n + O_2 → n + O n
O

FIGURE P19.59

19.60. Substituted polyacetylene can be made by replacing hydrogen atoms with functional groups on the backbone of the polymer (Figure P19.60). The backbone may bend when this is done, decreasing the conductivity of the polymer. Why does bending the backbone of the chain decrease conductivity?

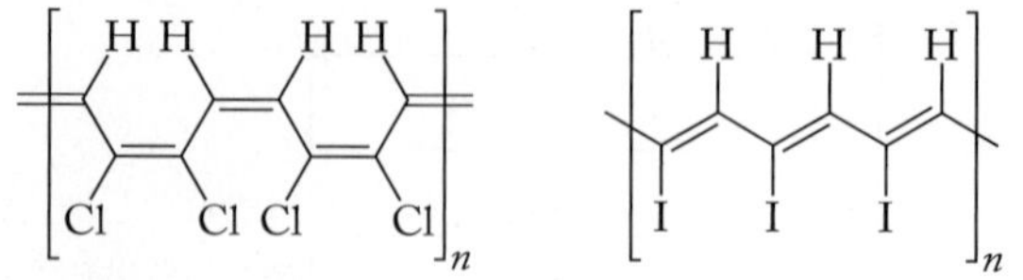

FIGURE P19.60

Aromatic Compounds

Concept Review

19.61. Why is benzene a planar molecule?

19.62. Why are aromatic molecules stable?

19.63. Do tetramethylbenzene and pentamethylbenzene have constitutional isomers?

19.64. Why aren't butadiene (C_4H_6) and 1,3-cyclohexadiene (C_6H_8) considered aromatic molecules? Their carbon-skeleton structures are shown in Figure P19.64.

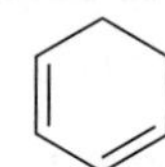

Butadiene 1,3-Cyclohexadiene

FIGURE P19.64

***19.65.** Pyridine (Figure P19.65) has the molecular formula C_6H_5N. Is pyridine an aromatic molecule?

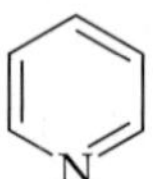

Pyridine

FIGURE P19.65

*19.66. Is cyclooctatetraene (Figure P19.66) an aromatic compound?

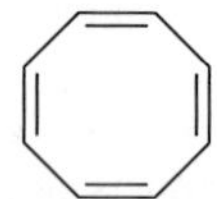

Cycloocatatetraene

FIGURE P19.66

Problems

19.67. Draw all the constitutional isomers of trimethylbenzene.

*19.68. The three isomeric trichlorobenzenes are labeled I, II, and III. As a result of a chemical reaction, one methyl group (CH_3–) is added to each isomer. Isomer I forms one monomethyl compound, isomer II forms two monomethyl compounds, and isomer III forms three monomethyl compounds. Draw the structures of the original trichlorobenzene isomers.

19.69. Calculate the fuel values of gaseous benzene (C_6H_6) and ethylene gas (C_2H_4). Does one mole of benzene have a higher or lower fuel value than three moles of ethylene?

19.70. Does one mole of gaseous benzene (C_6H_6) have a higher or lower fuel value than three moles of acetylene gas (C_2H_2)?

Amines

Concept Review

19.71. Explain why methylamine (CH_3NH_2) is more soluble in water than butylamine [$CH_3(CH_2)_3NH_2$].

19.72. Combustion of hydrocarbons in air yields carbon dioxide and water. What other product is expected in the complete combustion of amines?

Problems

19.73. **Are You Hungry?** Serotonin and amphetamine both contain the amine functional group (Figure P19.73). Serotonin is responsible, in part, for signaling that we have had enough to eat. Amphetamine, an addictive drug, can be used as an appetite suppressant. Identify the primary and secondary amine functional groups in these molecules.

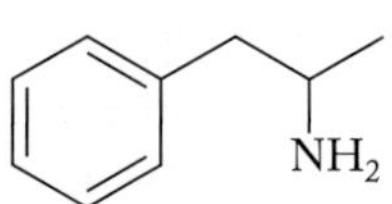

Serotonin Amphetamine

FIGURE P19.73

19.74. **Coffee** Caffeine, an active ingredient in coffee, contains four nitrogen atoms per molecule. Aspartame is an artificial sweetener containing two nitrogen atoms. Structures of caffeine and aspartame are shown in Figure P19.74. Which nitrogen atoms represent secondary amines and which ones represent tertiary amines?

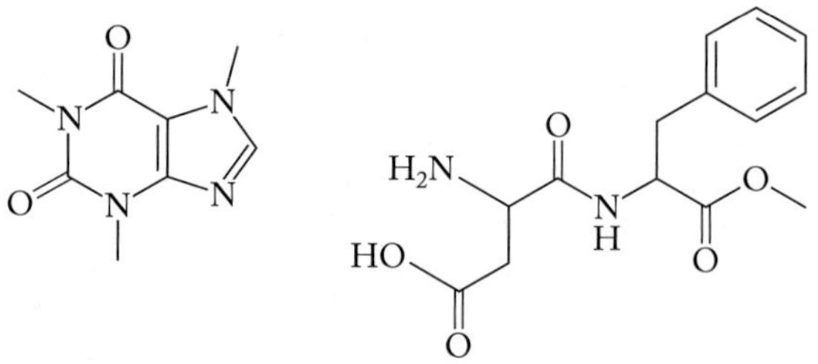

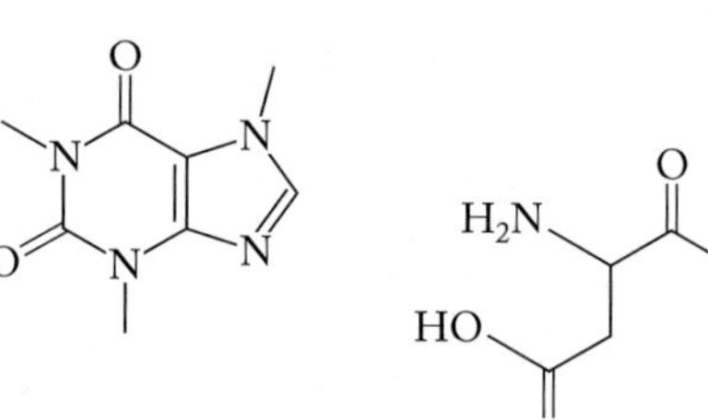

Caffeine Aspartame

FIGURE P19.74

19.75. **Renewable Energy** Bacteria of the genus *Methanosarcina* convert amines to methane. Their action helps make methane a renewable energy source. Determine the standard enthalpy of the following reaction from the appropriate standard enthalpies of formation ($\Delta H^\circ_{f,CH_3NH_2} = -23.0$ kJ/mol):

$$4\,CH_3NH_2(g) + 2\,H_2O(\ell) \rightarrow 3\,CH_4(g) + CO_2(g) + 4\,NH_3(g)$$

19.76. Determine the $\Delta H^\circ_{\text{rxn}}$ values of the following combustion reactions of methylamine ($\Delta H^\circ_{\text{f,CH}_3\text{NH}_2} = -23.0$ kJ/mol):

$$4\,CH_3NH_2(g) + 13\,O_2(g) \rightarrow 4\,CO_2(g) + 4\,NO_2(g) + 10\,H_2O(\ell)$$

$$4\,CH_3NH_2(g) + 6\,O_2(g) \rightarrow 4\,CO_2(g) + 4\,NH_3(g) + 4\,H_2O(\ell)$$

***19.77.** Methylamine is a weak base.

a. Use information in Appendix 5 to sketch the titration curve for the titration of 125 mL of a 0.015 *M* solution of methylamine with 0.100 *M* HCl.

b. Label the curve with the pH of the analyte solution, the pK_a of the analyte, and with the pH and titrant volumes halfway to the equivalence point and at the equivalence point.

c. Draw the structures of the species present in the solution at the equivalence point.

*19.78. The pK_b values in Appendix 5 tell us that methylamine is a stronger base than ammonia and that dimethylamine is even stronger. Use the differences in their molecular structures to explain this trend in the strengths of these three bases.

Alcohols, Ethers, and Reformulated Gasoline

Concept Review

19.79. Why are the fuel values of ethanol and dimethyl ether (Figure P19.79) lower than that of ethane?

Dimethyl ether Ethanol

FIGURE P19.79

19.80. Would you expect the fuel value of alcohols to increase or decrease as the number of carbon atoms in the alcohol increases?

19.81. Why do ethers typically boil at lower temperatures than alcohols with the same molecular formula?

19.82. Which do you expect to be more soluble in water, MTBE or 2,2-dimethylbutane (Figure P19.82)? Explain your answer.

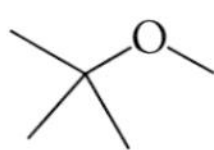

MTBE 2,2-Dimethylbutane

FIGURE P19.82

***19.83. Clean Skin** Disposable wipes used to clean the skin prior to getting an immunization shot contain ethanol. After wiping your arm, your skin feels cold. Why?

*19.84. **Dry Gas** In the winter months in cold climates, water condensing in a vehicle's gas tank reduces engine performance. An auto mechanic recommends adding "dry gas" to the tank during your next fill-up. Dry gas is typically an alcohol that dissolves in gasoline and absorbs water. Based on the structures shown in Figure P19.84, which product would you predict would do a better job—methanol or 2-propanol?

CH_3OH

Methanol 2-Propanol

FIGURE P19.84

Problems

19.85. Which of the compounds in Figure P19.85 are alcohols and which ones are ethers? Place them in order of increasing boiling point.

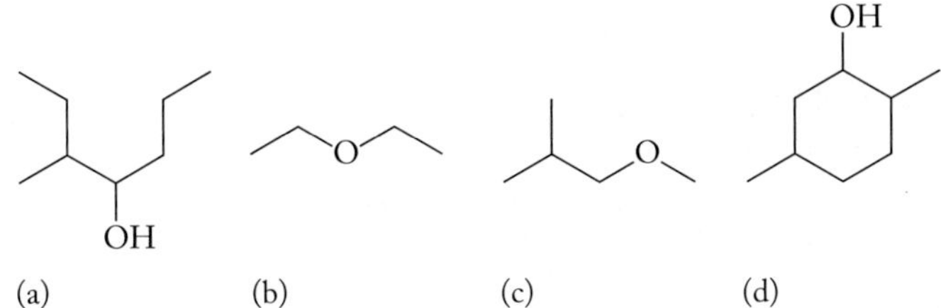

(a) (b) (c) (d)

FIGURE P19.85

19.86. Which of the compounds in Figure P19.86 are alcohols and which ones are ethers? Place them in order of increasing vapor pressure at 25°C.

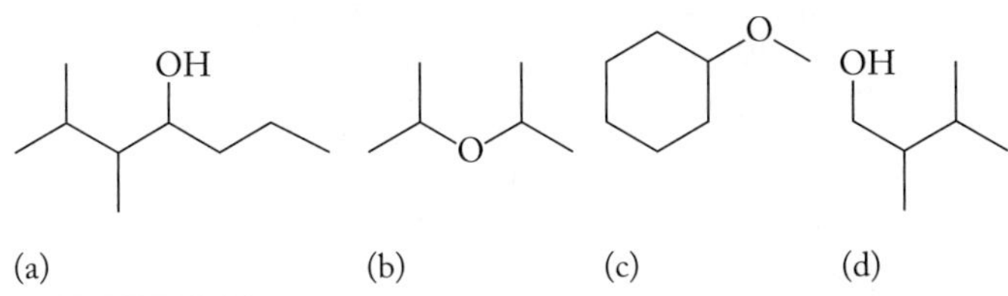

(a) (b) (c) (d)

FIGURE P19.86

Consult tables of thermochemical data in Appendix 4 for any values you may need to solve Problems 19.87 through 19.90.

19.87. Calculate the fuel values of diethyl ether and butanol (Figure P19.87). Which has the higher fuel value?

Diethyl ether Butanol

FIGURE P19.87

19.88. Calculate the fuel value of liquid diethyl ether and methyl propyl ether (Figure P19.88). Which has the higher fuel value? ($\Delta H^\circ_{\text{f,methyl propyl ether}} = -266.0$ kJ/mol.)

Diethyl ether Methyl propyl ether

FIGURE P19.88

19.89. Problem 19.80 asked you to predict whether the fuel value of alcohols increases or decreases with the number of carbon atoms in the alcohol. Calculate the fuel values of liquid methanol and ethanol (Figure P19.89). Does your answer support the prediction you made in Problem 19.80?

CH_3OH OH

Methanol Ethanol

FIGURE P19.89

19.90. Which has the greater fuel value: propanol or isopropanol (Figure P19.90)? ($\Delta H^\circ_{f,propanol,\ell} = -302.6$ kJ/mol and $\Delta H^\circ_{f,isopropanol,\ell} = -318.1$ kJ/mol.)

OH OH

Propanol Isopropanol

FIGURE P19.90

Aldehydes, Ketones, Carboxylic Acids, Esters, and Amides

Concept Review

19.91. Explain why carboxylic acids tend to be more soluble in water than aldehydes with the same number of carbon atoms.

19.92. In reference books, diethyl ether (Figure P19.92) is usually listed as "slightly soluble" in water, but 2-butanone is listed as "very soluble." Why is 2-butanone more soluble?

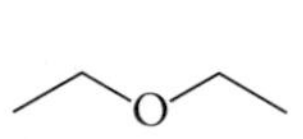

O

Diethyl ether 2-Butanone

FIGURE P19.92

19.93. Are butanal and 2-butanone (Figure P19.93) constitutional isomers?

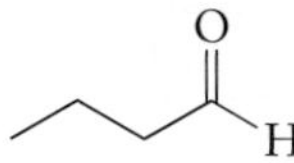

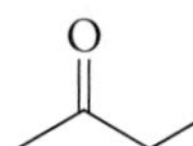

Butanal 2-Butanone

FIGURE P19.93

19.94. **Apples** Both of the esters shown in Figure P19.94 are found in apples and contribute to the flavor and aroma of the fruit. Are the two compounds identical, constitutional isomers, or stereoisomers, or do they have different molecular formulas?

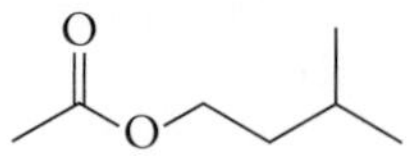

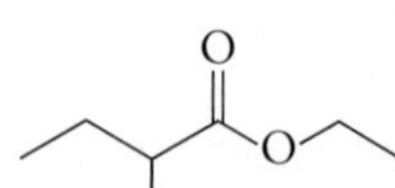

FIGURE P19.94

19.95. Can we use combustion analysis to distinguish between ketones and aldehydes with the same number of carbon atoms?

19.96. Can we use combustion analysis to distinguish between ethers and ketones with the same number of carbon atoms?

19.97. Resonance forms for acetic acid are shown in Figure P19.97. Which one contributes more to bonding? Explain your choice.

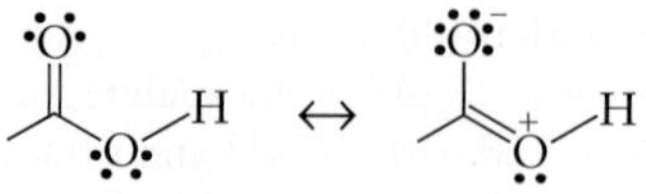

O H ↔ O H

(a) (b)

FIGURE P19.97

19.98. Figure P19.98 shows resonance forms for acetamide and acetic acid. Does resonance form (a) contribute more to the bonding in acetamide than resonance form (b) contributes to the bonding in acetic acid? Explain your answer.

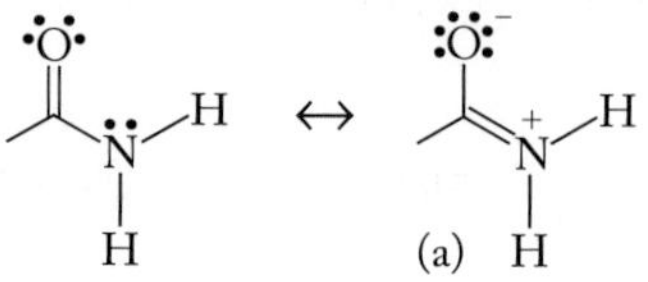

O N H H ↔ O N H (a) H

Acetamide

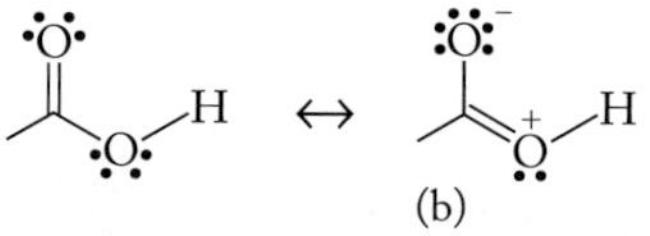

O O H ↔ O O H (b)

Acetic acid

FIGURE P19.98

19.99. What distinguishes an amine from an amide?

19.100. Explain why nylon reacts with acid, whereas polyethylene is inert toward acid.

Problems

19.101. Which of the compounds in Figure P19.101 are constitutional isomers of the aldehyde $C_5H_{10}O$?

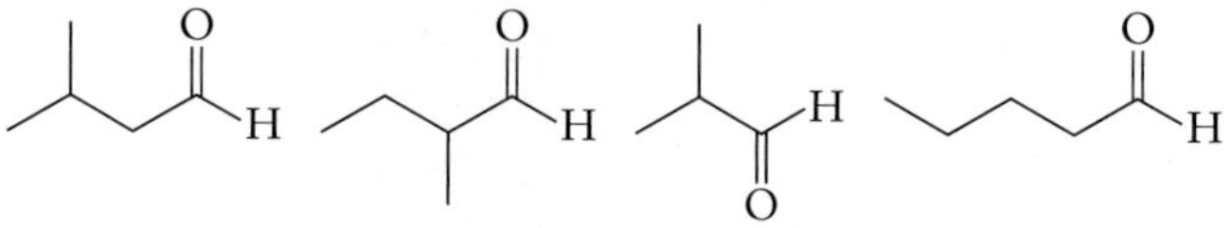

O H O H H O O H

(a) (b) (c) (d)

FIGURE P19.101

19.102. Each of the natural products in Figure P19.102 contains more than one functional group. Which of the compounds is an aldehyde?

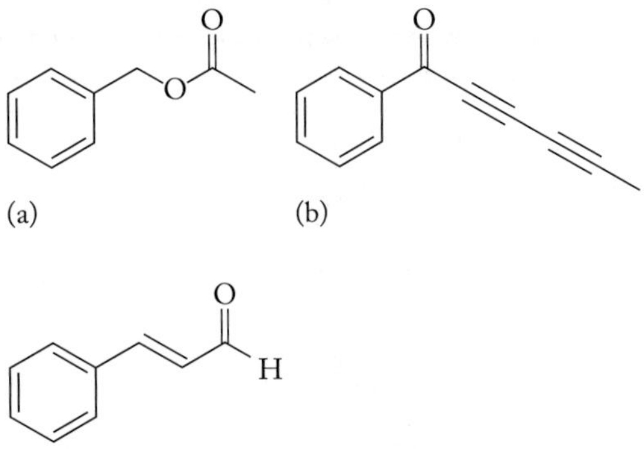

(a) (b) (c)

FIGURE P19.102

19.103. Which of the compounds in Figure P19.103 is a ketone?

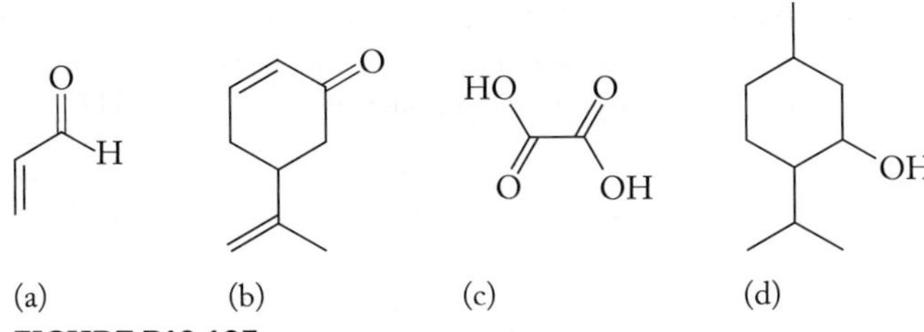

FIGURE P19.103

19.104. Propanal and acetone (2-propanone) have the same molecular formula, C_3H_6O, but different structures (Figure P19.104). Which compound is a ketone?

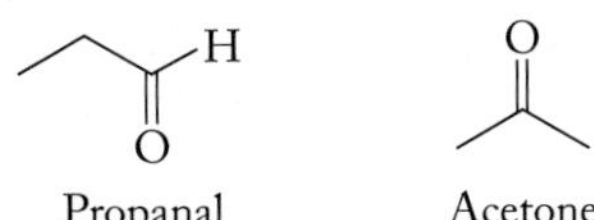

FIGURE P19.104

19.105. Plot the carbon-to-hydrogen ratio in aldehydes with one to six carbon atoms as a function of the number of carbon atoms. Does this graph correlate better with the plot of C:H ratios for alkanes or for alkenes?

19.106. Plot the carbon-to-hydrogen ratio in carboxylic acids with one to six carbon atoms as a function of the number of carbon atoms. Does this graph correlate better with the plot of C:H ratios for alkanes or for alkenes?

19.107. Esters are responsible for the odors of fruits, including apples, bananas, and pineapples. Figure P19.107 shows three esters from these fruits. Identify the alcohol and carboxylic acid that react to form these compounds.

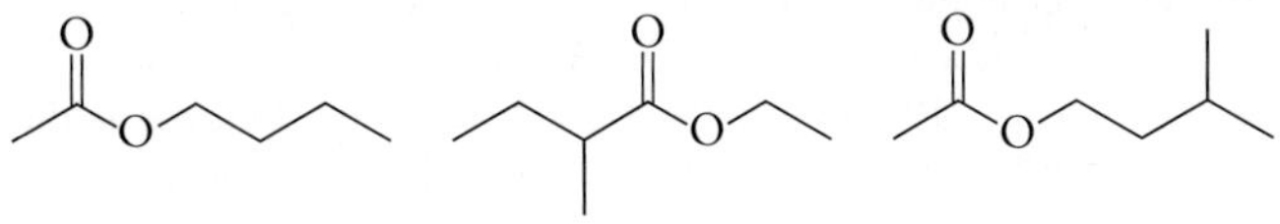

(a) In pineapples (b) In bananas (c) In apples

FIGURE P19.107

19.108. **Beeswax** The waxy material found in beehives (Figure P19.108) is an ester formed from an alcohol and a carboxylic acid with a long hydrocarbon chain. Identify the alcohol and acid in beeswax.

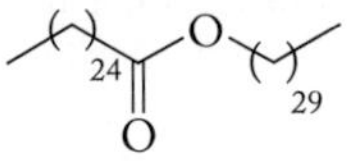

FIGURE P19.108

19.109. **Stimulants and Tranquilizers** Nicotine is a stimulant found in tobacco, whereas Valium is a tranquilizer. Both molecules contain two nitrogen atoms in addition to other functional groups. Do the nitrogen atoms shown in blue in Figure P19.109 belong to an amine or an amide?

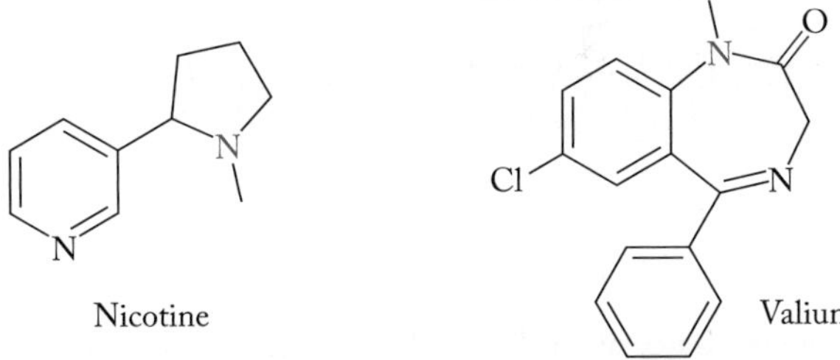

FIGURE P19.109

19.110. **Hot Peppers** Piperine and capsaicin are ingredients of peppers that give "heat" to spicy foods. Do the nitrogen atoms shown in blue in Figure P19.110 belong to an amine or an amide?

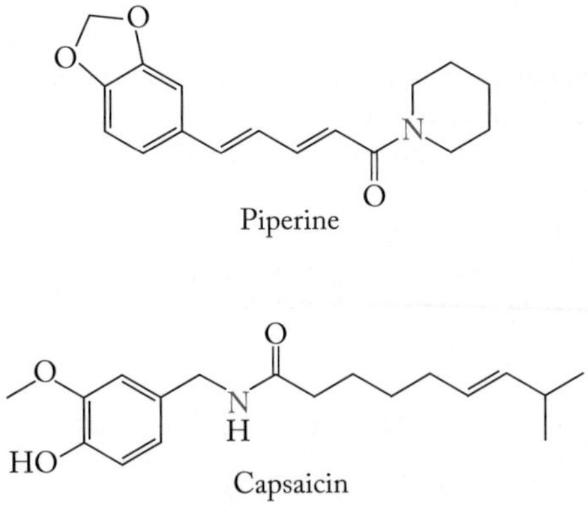

Piperine

Capsaicin

FIGURE P19.110

Consult tables of thermochemical data in Appendix 4 for any values you may need to solve Problems 19.111 through 19.114.

19.111. Calculate the fuel values of formaldehyde gas and liquid formic acid (Figure P19.111). The ΔH_f° of formaldehyde gas is −108.6 kJ/mol and the ΔH_f° of liquid formic acid is −425.0 kJ/mol. Which has the higher fuel value?

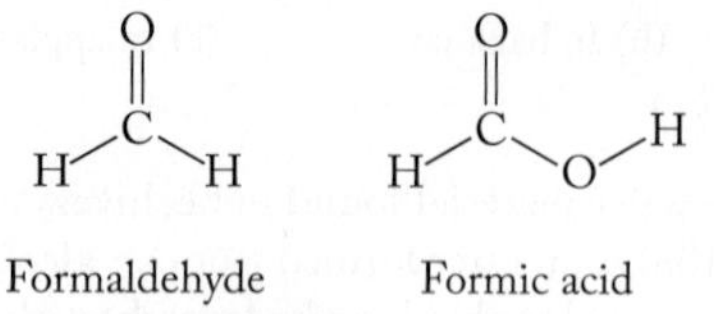

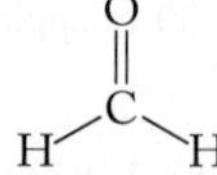

Formaldehyde Formic acid

FIGURE P19.111

19.112. Calculate the fuel values of liquid formamide and methyl formate (Figure P19.112), which have ΔH_f° values of −251 and −391 kJ/mol, respectively. Assume $NO_2(g)$ is a product of formamide combustion.

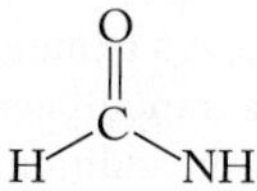

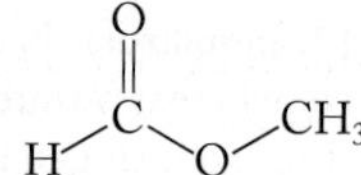

Formamide Methyl formate

FIGURE P19.112

19.113. Calculate ΔH°_{rxn} for the following reactions of methanogenic bacteria, given $\Delta H^\circ_{f,HCOOH} = -425.0$ kJ/mol:

(1) $CH_3COOH(\ell) \rightarrow CH_4(g) + CO_2(g)$

(2) $4\ HCOOH(\ell) \rightarrow CH_4(g) + 3\ CO_2(g) + 2\ H_2O(\ell)$

19.114. Calculate ΔH°_{rxn} for the following reactions of methanogenic bacteria:

(1) $4\ H_2(g) + CO_2(g) \rightarrow CH_4(g) + 2\ H_2O(\ell)$

(2) $4\ CH_3OH(\ell) \rightarrow 3\ CH_4(g) + CO_2(g) + 2\ H_2O(\ell)$

19.115. Reactions between 1,6-diaminohexane, $H_2N(CH_2)_6NH_2$, and different dicarboxylic acids, $HOOC(CH_2)_nCOOH$, are used to prepare polymers that have a structure similar to that of nylon. How many carbon atoms were in the dicarboxylic acids used to prepare the polymers with the repeating units shown in Figure P19.115?

(a)

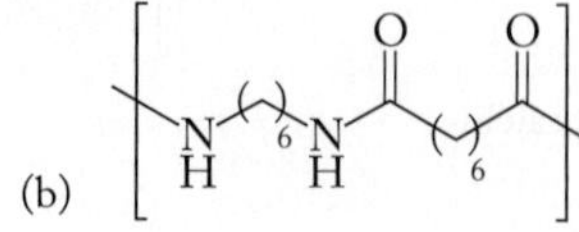

(b)

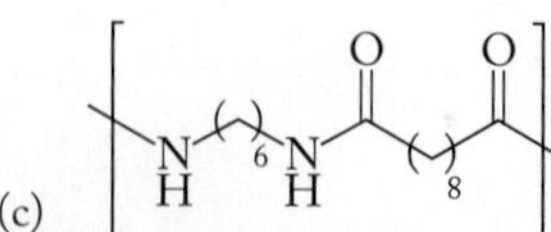

(c)

FIGURE P19.115

19.116. The two polymers in Figure P19.116 have the same empirical formula.

a. What pairs of monomers could be used to make each of them?
b. How might the physical properties of these two polymers differ?

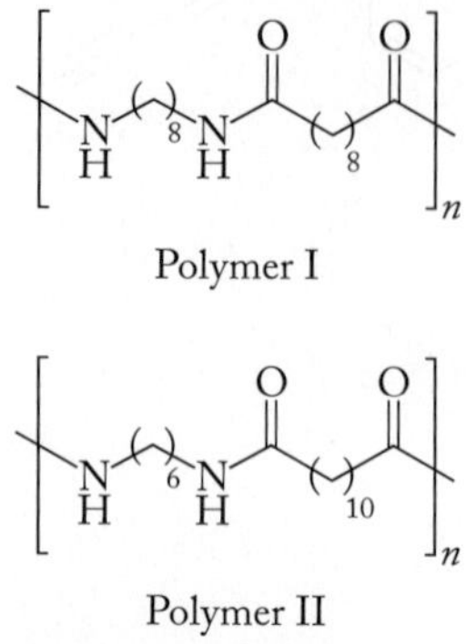

Polymer I

Polymer II

FIGURE P19.116

19.117. The polyester called Kodel is made with polymeric strands prepared by the reaction of dimethyl terephthalate with 1,4-di(hydroxymethyl)cyclohexane (Figure P19.117).

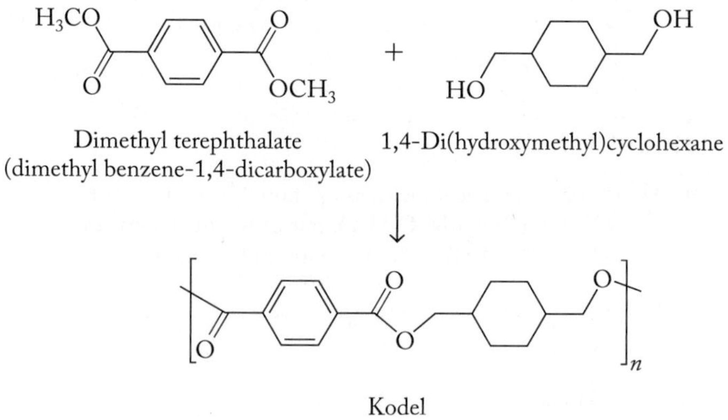

Dimethyl terephthalate (dimethyl benzene-1,4-dicarboxylate) 1,4-Di(hydroxymethyl)cyclohexane

Kodel

FIGURE P19.117

a. Is Kodel a condensation polymer or an addition polymer? What is the other product of the reaction?
*b. Dacron (Figure 19.44) is made from dimethyl terephthalate and ethylene glycol. What properties of Kodel fibers might make them better than Dacron as a clothing material?

19.118. Lexan is a polymer belonging to the class of materials called polycarbonates. Figure P19.118 shows the polymerization reaction for Lexan.

a. What other compound is formed in the polymerization reaction?

*b. Why is Lexan called a "polycarbonate"?

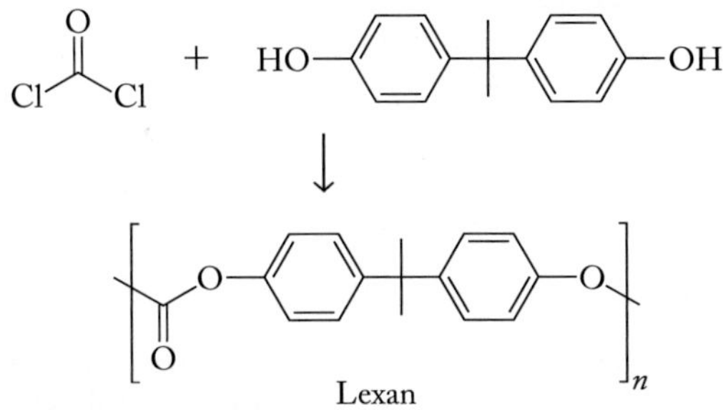

Lexan

FIGURE P19.118

A Brief Survey of Isomers

Concept Review

19.119. Can all of the terms *enantiomer*, *achiral*, and *optically active* be used to describe a single compound? Explain.

*19.120. Could a racemic mixture be distinguished from an achiral compound based on optical activity? Explain your answer.

19.121. Can stereoisomers of molecules, such as cis and trans RCH═CHR, also have optical isomers? (R may be any of the functional groups we have encountered in this textbook.) Explain your answer.

19.122. The compound $C_3H_6Br_2$ exists as a racemic mixture. Draw its structure.

Problems

19.123. Nutmeg The compound myristicin (Figure P19.123) is in part responsible for the flavor of nutmeg. Limited animal studies on myristicin have shown it may have a small influence on sexual behavior, but large doses cause very unpleasant side effects.

a. The structure shown for myristicin has four double bonds. Are cis and trans isomers of myristicin possible?

*b. How many structural isomers are possible based solely on the arrangement of the groups on the aromatic ring in myristicin?

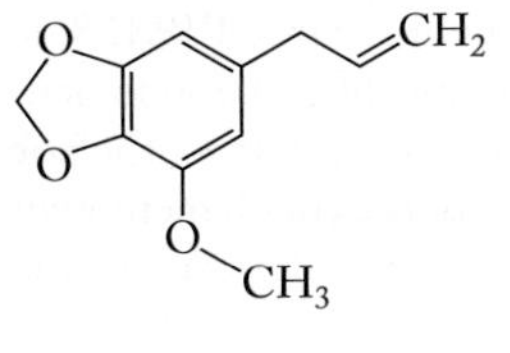

Myristicin

FIGURE P19.123

*19.124. The condensed structure of a molecule is $CH_3CH{=}C(CH_3)CH(OH)CH_2CH_3$. Draw all the possible stereoisomers of this compound.

19.125. Which of the molecules in Figure P19.125 do not have any possible constitutional isomers?

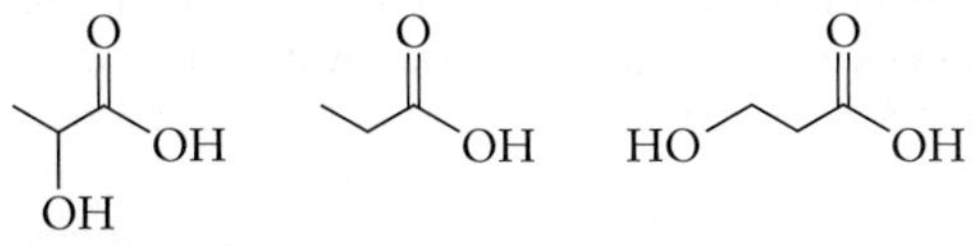

(a) (b) (c)

FIGURE P19.125

19.126. Which of the molecules in Figure P19.126 do not have any possible stereoisomers?

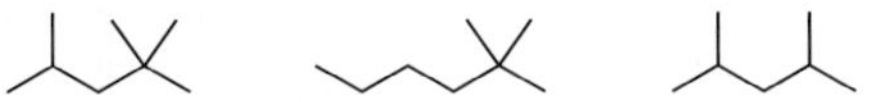

(a) (b) (c)

FIGURE P19.126

Additional Problems

19.127. How many grams of liquid methanol must be combusted to raise the temperature of 454 g of water from 20.0°C to 50.0°C? Assume that the transfer of heat to the water is 100% efficient. How many grams of carbon dioxide are produced in this combustion reaction?

19.128. Two compounds, both with molar masses of 74.12 g/mol, were combusted in a bomb calorimeter with $C_{\text{calorimeter}} = 3.640$ kJ/°C. Combustion of 0.9842 g of compound A led to an increase in temperature of 10.33°C, while combustion of 1.110 g of compound B caused the temperature to rise 11.03°C. Which compound is butanol and which is diethyl ether?

19.129. Benzene rings with two methyl groups are called xylenes. 1,4-Dimethylbenzene (*para*-xylene or *p*-xylene) and 1,3-dimethylbenzene (*ortho*-xylene or *o*-xylene) have very similar boiling points (138°C and 139°C, respectively) but very different melting points (13°C and −48°C, respectively). Explain this observation.

19.130. **Perfume from Whales** Ambergris comes from the intestines of sperm whales and was once used to make perfumes. Ambrein (Figure P19.130) is a constituent of ambergris that, when injected in rats, is known to cause an increase in their sexual activity.

a. What functional groups are present in ambrein?

b. How many chiral centers does ambrein have?

c. Are any cis/trans isomers possible for ambrein?

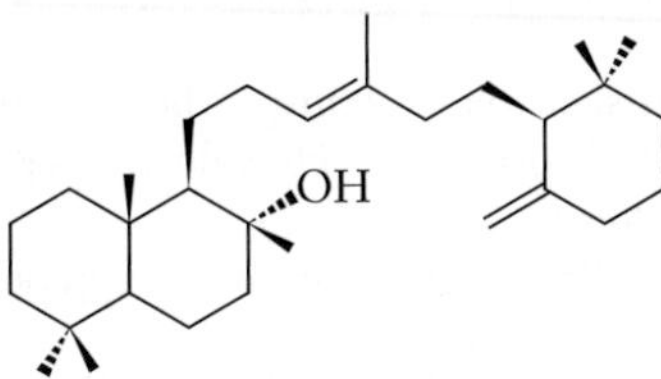

Ambrein

FIGURE P19.130

19.131. **Turmeric** Turmeric is commonly used as a spice in Indian and Southeast Asian dishes. Turmeric contains a high concentration of curcumin (Figure P19.131), a potential anticancer drug and a possible treatment for cystic fibrosis.

a. Are the substituents on the C=C bonds (non-aromatic) in cis or trans configurations?
b. Draw two other stereoisomers of this compound.
c. List all the types of hybridization of the carbon atoms in curcumin.

O O

HO OH

O O

Curcumin

FIGURE P19.131

19.132. Polycyclic aromatic hydrocarbons are potent carcinogens. They are produced during the combustion of fossil fuels and have also been found in meteorites. Can we use combustion analysis to distinguish between naphthalene and anthracene (Figure P19.132)?

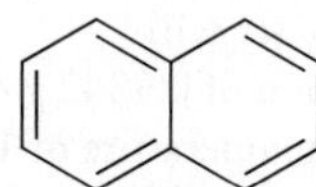

Naphthalene

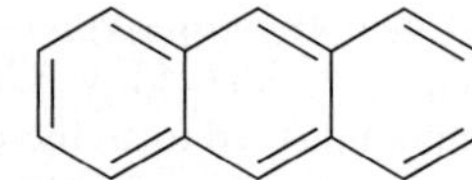

Anthracene

FIGURE P19.132

19.133. Identify the reactants in the polymerization reactions that produce the polymers shown in Figure P19.133.

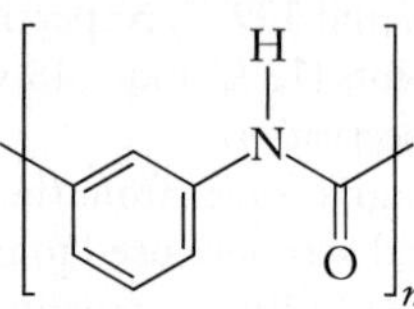

(a)

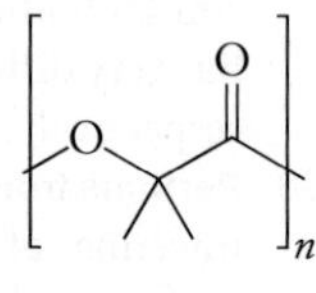

(b)

FIGURE P19.133

*19.134. **Raincoats** "Waterproof" nylon garments have a coating to prevent water from penetrating the hydrophilic fibers. Which functional groups in the nylon molecule make it hydrophilic?

19.135. Draw the carbon-skeleton structure of the condensation polymer of $H_2N(CH_2)_6COOH$. How does this polymer compare with nylon-6?

*19.136. Putrescine, $H_2N(CH_2)_4NH_2$, is one of the compounds that form in rotting meat.

a. Draw the carbon-skeleton structures of all the trimers (a molecule formed from three monomers) that can be formed from putrescine, adipic acid, and terephthalic acid (Figure P19.136). The three monomers forming the trimer do not have to be different from one another.
b. A chemist wishes to make a polymer containing putrescine and a 1:1 ratio of adipic acid to terephthalic acid. What should be the mole ratio of the three reactants?

$HOOCCH_2CH_2CH_2CH_2COOH$

Adipic acid

COOH

COOH

Terephthalic acid (benzene-1,4-dicarboxylic acid)

FIGURE P19.136

***19.137.** Polymer chemists can modify the physical properties of polystyrene by copolymerizing divinylbenzene with styrene (Figure P19.137). The resulting polymer has strands of polystyrene cross-linked with divinylbenzene. Predict how the physical properties of the copolymer might differ from those of 100% polystyrene.

Divinylbenzene (DVB)

Styrene (S)

$—CH—CH_2—CH—CH_2—$

$—CH—CH_2—CH—CH_2—$

S cross-linked with DVB

FIGURE P19.137

19.138. Maleic anhydride and styrene (Figure P19.138) form a polymer with alternating units of each monomer.

a. Draw two repeating monomer units of the polymer.
b. Based on the structure of the copolymer, predict how its physical properties might differ from those of polystyrene.

O O O

Maleic anhydride

Styrene

FIGURE P19.138

19.139. Superglue The active ingredient in Superglue is methyl 2-cyanoacrylate (Figure P19.139). The liquid glue hardens rapidly when methyl 2-cyanoacrylate polymerizes. This happens when it contacts a surface containing traces of water or other compounds containing –OH or –NH– groups. Draw the carbon-skeleton structure of two repeating monomer units of poly(methyl 2-cyanoacrylate).

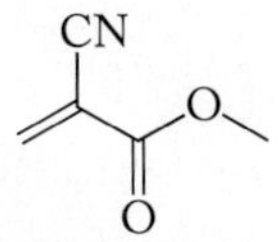

CN

O

O

Methyl 2-cyanoacrylate

FIGURE P19.139

*19.140. Silicones are polymeric materials with the formula $[R_2SiO]_n$ (Figure P19.140). They are prepared by the reaction of R_2SiCl_2 with water to yield the polymer and aqueous HCl. Imagine this reaction as taking place in two steps: (1) water reacts with one mole of R_2SiCl_2 to produce a new monomer and one mole of HCl(*aq*); (2) one new monomer molecule reacts with another new monomer molecule to eliminate one molecule of HCl and make a dimer with a Si—O—Si bond.

a. Suggest two balanced equations describing these reactions that occur over and over again to produce a silicone polymer.
b. Why are silicones water repellent?

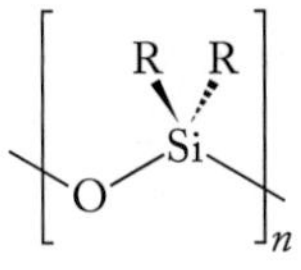

R R

Si

O

n

Silicone

FIGURE P19.140

19.141. Molecules of piperine and capsaicin (see Figure P19.110) contain amide functional groups.

a. Draw the amine and the carboxylic acid that could react to form these two compounds.
b. Are the non-aromatic double bonds in these molecules cis or trans?
c. Name the functional groups that contain the oxygen atoms in these compounds.

19.142. The heats of combustion of two constitutional isomers are the same when estimated from average bond energies, but they are different when determined experimentally. Why?

smartwork5

If your instructor uses Smartwork5, log in at digital.wwnorton.com/atoms2.

20

Biochemistry

The Compounds of Life

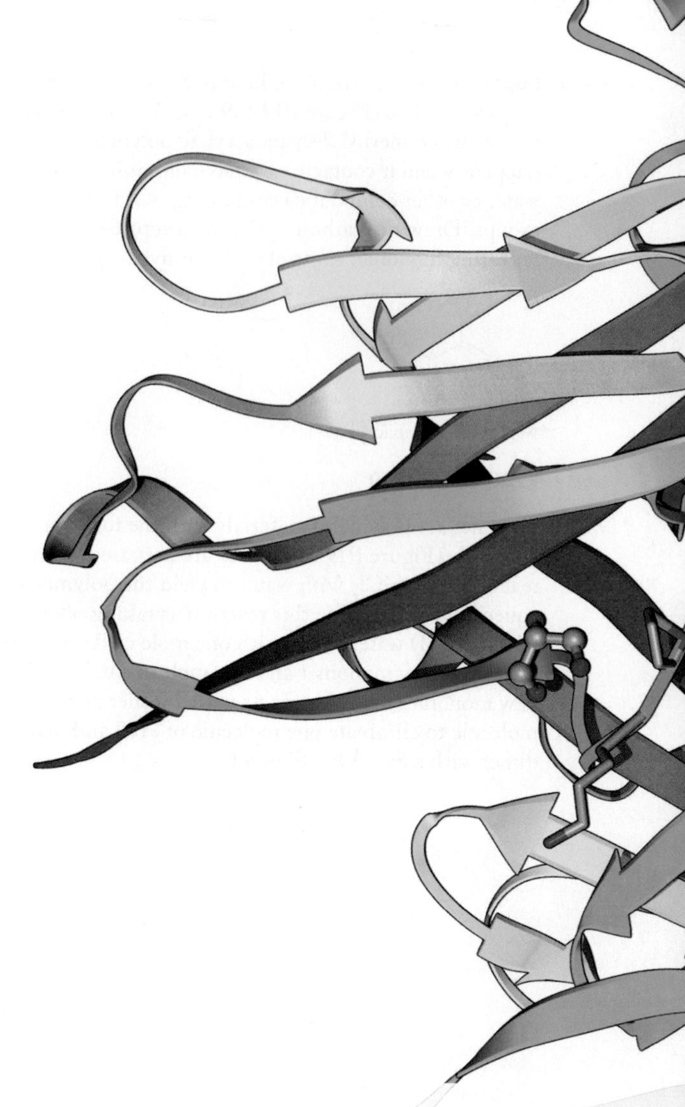

BIOLOGICAL MACROMOLECULES Titin is the largest known protein, consisting of over 34,000 amino acids. Its elasticity contributes to the function of skeletal muscles.

PARTICULATE **REVIEW**

Protons versus Hydrogen Bonds

In Chapter 20, we investigate the properties of proteins and the role of hydrogen bonding in defining protein structure and function.

- Hydrogen bonding is an important intermolecular force among which of the three molecules shown here?
- Identify any acidic hydrogen atoms in the three molecules.
- Which molecule(s) could function as a Lewis base?

(Review Sections 6.2 and 8.4 if you need help.)

(Answers to Particulate Review questions are in the back of the book.)

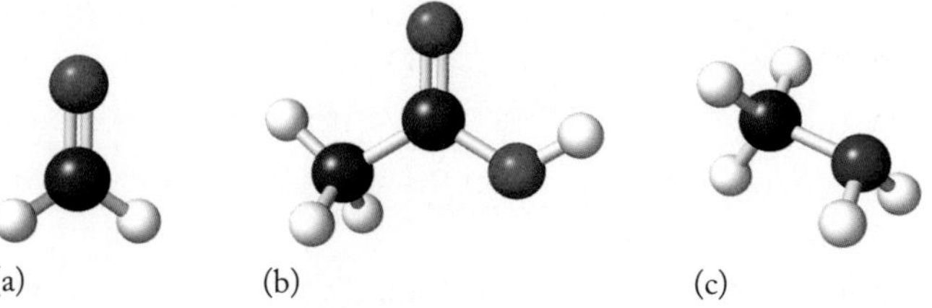

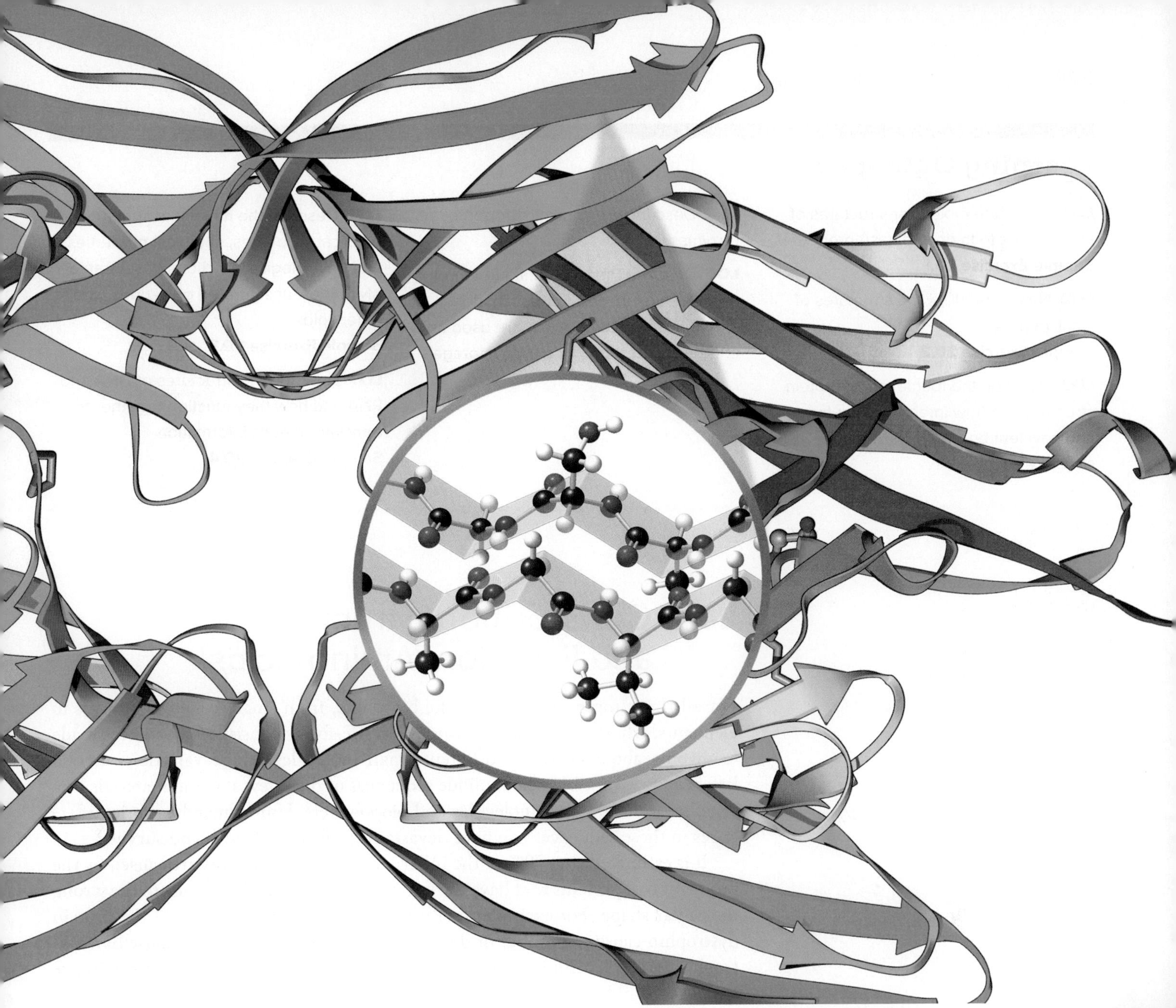

PARTICULATE **PREVIEW**

Functional Groups and pH

The molecule shown here contains two different functional groups. As you read Chapter 20, look for ideas that will help you answer these questions:

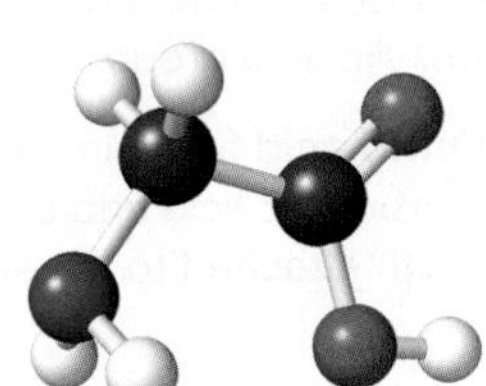

- What are the two functional groups in this molecule?
- What happens to each of these functional groups in moderately acidic (pH = 3) aqueous solutions?
- What happens to each of these functional groups in moderately basic (pH = 11) aqueous solutions?

Learning Outcomes

LO1 Relate the molecular structures of amino acids to their acid-base properties
Sample Exercise 20.1

LO2 Name and draw the structures of small peptides
Sample Exercise 20.2

LO3 Describe the four levels of protein structure and how intermolecular forces and covalent bonds stabilize these structures

LO4 Describe the catalytic properties of enzymes

LO5 Describe the molecular structures of simple sugars and polysaccharides and how these compounds are used as energy sources and for energy storage

LO6 Describe the molecular structure, physical properties, chemical properties, and physiological functions of saturated glycerides, unsaturated glycerides, and other lipids
Sample Exercise 20.3

LO7 Describe the structures of DNA and RNA and how they function together to translate genetic information
Sample Exercise 20.4

20.1 Composition, Structure, and Function: Amino Acids

For all the stunning diversity of the biosphere, from single-celled organisms to elephants, whales, and giant redwoods, all life-forms consist of substances made from only about 40 or 50 different small molecules. Huge variations among life-forms are possible when unique sequences of these few starting materials link together to form larger molecules and biopolymers. Unfortunately, subtle differences in these sequences can have devastating effects on health and survival.

It is estimated that 70% of the inherited diseases in humans are caused by the production of proteins that have the wrong composition, which leads to the wrong molecular shape. For example, malformation or the total absence of the protein dystrophin causes a number of debilitating diseases referred to as muscular dystrophy (MD). In mild forms of MD, misshapen molecules of dystrophin are unable to build muscle fibers of sufficient strength to function normally. As a result, muscle tissue wastes away and the patient becomes physically disabled. In a severe form of the disease, called Duchenne MD (DMD), functional dystrophin is absent, and the disability often results in early death. Understanding the mechanisms that produce dystrophin has led to promising therapies, some of which involve the grafting of healthy muscle cells that can synthesize normal dystrophin into the tissues of MD patients. For DMD patients, treatments involve the injection of stem cells that fuse with and genetically complement dystrophic muscle. Similar approaches have resulted in the development of new drugs and cell therapies for other inherited and contagious diseases.

In this chapter we explore the composition, structure, and function of proteins and three other major classes of biomolecular compounds: carbohydrates, lipids, and nucleic acids. The functions of these compounds are similar in all of the life-forms in which they occur. Many of these molecules are large and complex, but the knowledge we have developed about the chemical behavior of small molecules will serve as a useful framework on which to build an understanding of the behavior of the larger assemblies of molecules that form cells. The first small molecules we explore are the amino acids, the monomer units that combine to form proteins.

protein a biological polymer made of amino acids.

biomolecule an organic molecule present naturally in a living system.

amino acid a molecule that contains at least one amine group and one carboxylic acid group.

α-amino acid an amino acid in which the carboxylic acid and amine groups are both attached to the same (α) carbon atom.

α-carbon the carbon atom attached directly to a functional group.

essential amino acid any of the eight amino acids that make up peptides and proteins but are not synthesized in the human body and must be obtained through the food we eat.

Amino Acids: The Building Blocks of Proteins

Proteins are the most abundant class of **biomolecules** in all animals, including humans. Proteins account for about half of the mass of the human body that is not water. They are the major component in skin, muscles, cartilage, hair, and nails. Most of the enzymes that catalyze biochemical reactions are proteins, as are the molecules that transport oxygen to our cells and many of the hormones that regulate cell function and growth. Most proteins are biopolymers with molar masses of 10^5 g/mol or more.

The molecular structure and biological function of large biomolecules depend on the identities of their small-molecule building blocks and the sequence in which those small molecules occur. The molecular building blocks of proteins are **amino acids**, so named because each of them contains at least one amine ($-NH_2$) group and at least one carboxylic acid ($-COOH$) group. The amino acids in proteins are **α-amino acids** because each of their molecules contains an **α-carbon** atom (highlighted in blue in Figure 20.1) that is bonded to both an $-NH_2$ and a $-COOH$ group. The α-carbon is the carbon atom in a structure that is directly attached to a functional group. In amino acids, the α-carbon is directly attached to both the carbonyl group of the carboxylic acid group and to the nitrogen atom of the amine group.

α-carbon

(a) $H_2N-C(R)(H)-COOH$ (b) $H_2N-C(R)(H)\cdots COOH$

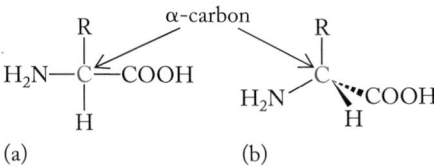

FIGURE 20.1 (a) General structure of an α-amino acid; (b) three dimensional structure.

CONNECTION As in Chapter 19, we use the generic symbol R to represent any group of covalently bonded atoms that is connected to the main structure of a molecule.

CONCEPT TEST

According to valence bond theory, what is the hybridization of the α-carbon atom in the amino acid in Figure 20.1, and what are the approximate angles between the four bonds surrounding it?

(Answers to Concept Tests are in the back of the book.)

CONNECTION The weakly basic properties of amines were described in Section 15.3.

In addition to its single bonds to the $-NH_2$ and $-COOH$ groups, the α-carbon atom in each of the amino acids that make up human proteins is bonded to a hydrogen atom and one of 20 common *R groups*. The structures of these R groups, often called *side-chain* groups, are highlighted in pink in the amino acid structures in Table 20.1. The amino acids are arranged in four categories based on the polarities or acid-base properties of their R groups. In the first category, which contains nine amino acids, the R groups contain mostly carbon and hydrogen atoms and are nonpolar. The R groups of the remaining 11 amino acids contain at least one heteroatom (O, N, or S) and are polar. Of these amino acids, two (aspartic acid and glutamic acid) have R groups that contain carboxylic acid functional groups, and three (histidine, lysine, and arginine) contain groups with nitrogen atoms that are weakly basic.

Our bodies can synthesize 12 of the 20 amino acids in Table 20.1, but the other eight must be present in the food we eat. These eight are marked with a superscript *b* and are referred to as the **essential amino acids**. Most proteins from animal sources, including those in meats, eggs, and dairy products, contain all the essential amino acids needed by the human body and in close to the needed proportions. These foods are sometimes referred to as sources of *complete proteins*. In contrast, most plant proteins from foods such as legumes and vegetables are incomplete proteins, so vegetarians must be careful to eat a combination of foods that provides all the essential amino acids. A good example is beans and rice (Figure 20.2), a traditional dish in Louisiana Creole and Latin American cuisine that provides a balance of these amino acids: rice has all of them but lysine, and beans lack only methionine.

FIGURE 20.2 A meal of beans and rice provides all the essential amino acids.

TABLE 20.1 Structures and Abbreviations of the 20 Common Amino Acids[a]

Nonpolar R Groups

Glycine (Gly; G)	Alanine (Ala; A)	Valine[b] (Val; V)	Leucine[b] (Leu; L)	Isoleucine[b] (Ile; I)
R = H	R = CH_3	R = $CH(CH_3)_2$	R = $CH_2CH(CH_3)_2$	R = $CH(CH_3)CH_2CH_3$

Proline (Pro; P)	Phenylalanine[b] (Phe; F)	Tryptophan (Trp; W)	Methionine[b] (Met; M)
H_2C, CH_2, CH_2 ring; $H{-}\overset{+}{N}(H){-}C(H){-}COO^-$	R = $CH_2C_6H_5$	R = CH_2–indole (HN)	R = $CH_2CH_2SCH_3$

Polar R Groups

Serine (Ser; S)	Threonine[b] (Thr; T)	Cysteine (Cys; C)	Tyrosine[b] (Tyr; Y)	Asparagine (Asn; N)	Glutamine (Gln; Q)
R = CH_2OH	R = $CH(CH_3)OH$	R = CH_2SH	R = $CH_2C_6H_4OH$	R = $CH_2C(=O)NH_2$	R = $CH_2CH_2C(=O)NH_2$

Acidic R Groups

Aspartic acid (Asp; D)	Glutamic acid (Glu; E)
R = $CH_2C(=O)O^-$	R = $CH_2CH_2C(=O)O^-$

Basic R Groups

Histidine (His; H)	Lysine[b] (Lys; K)	Arginine (Arg; R)
R = CH_2–imidazole ($\overset{+}{H}N$, NH)	R = $CH_2CH_2CH_2CH_2\overset{+}{N}H_3$	R = $CH_2CH_2CH_2NHC(=\overset{+}{N}H_2)NH_2$

General structure shown for each: $H_3\overset{+}{N}{-}C(R)(H){-}COO^-$

[a]R groups in pink, ionized forms at pH near 7 shown.

[b]The eight essential amino acids for adults (histidine is essential for children).

Chirality

The carbon atom in the generic structure of an amino acid (Figure 20.1) is bonded to four different groups. As a result, these α-carbon atoms are *chiral centers* (see Section 5.6). It also means that all the amino acids in Table 20.1, except glycine (R = H), are chiral compounds: their molecular structures are not superimposable on their mirror images, as illustrated for alanine in Figure 20.3. Instead, each amino acid and its mirror image constitute a pair of enantiomers.

Figure 20.3(b) shows another way to visualize the nonsuperimposable nature of alanine's mirror images. Think about standing up the two molecular structures as if they were micro-scale tripods with their $-COOH$ groups at the top and their $-CH_3$ and $-NH_2$ groups and H atoms forming their bases. Suppose the base constituents leave footprints as shown in Figure 20.3(b). Note that the two sets of footprints are not the same, and that it's impossible to rotate them so that they are superimposable. This inability to superimpose footprints is true no matter which constituent is at the top of the tripod, which reinforces the conclusion that the two structures are not the same.

For historical reasons, amino acid enantiomers are designated by the prefixes D- (for *dextro-*, right) and L- (*levo-*, left). These labels refer to how the four groups bonded to each chiral carbon atom are oriented in three-dimensional space. They do not refer to the rotation of plane-polarized light (see Figure 5.39) as it passes through a solution of the amino acid. In other words, there is no connection between the D- and L- prefixes in the names of amino acids and the *dextrorotary* and *levorotary* enantiomers that are designated with (+) and (−) signs based on their optical properties. All the chiral amino acids in the proteins in our bodies are L-enantiomers, even though nine of the 19 are actually dextrorotary.

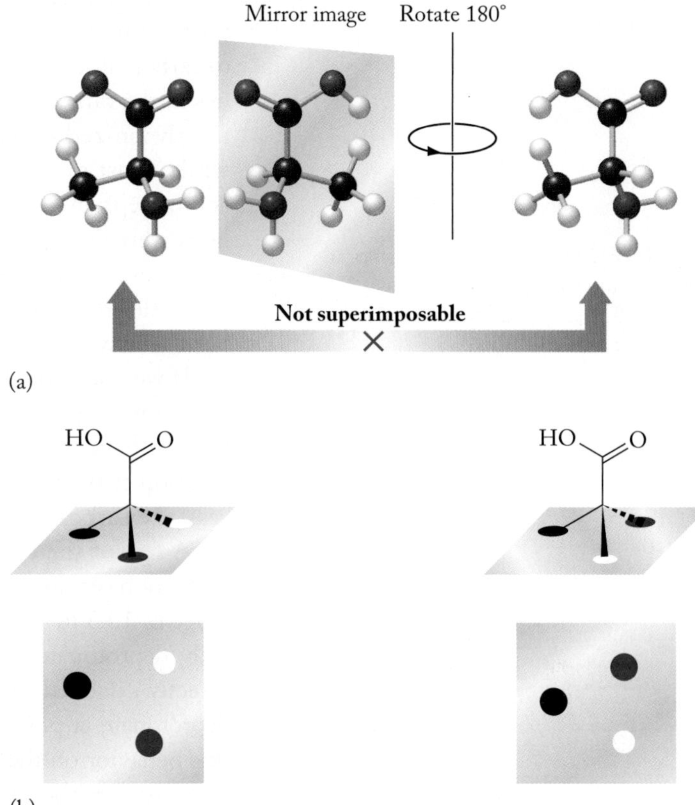

FIGURE 20.3 Like most α-amino acids, alanine is chiral, which means that its molecular structure is not superimposable on its mirror image. (a) Rotating the mirror image 180° so that the $-CH_3$ and $-COOH$ groups overlap does not produce a structure in which $-NH_2$ and $-H$ overlap. (b) To visualize alanine's chirality another way, keep the carboxylic acid groups on the two molecules aligned and make a footprint of the other three groups on a piece of paper. No matter how you turn the footprints, you cannot superimpose them.

CONNECTION Enantiomers and chirality were discussed in Section 5.6.

Zwitterions

If we dissolve an amino acid in a solution that already contains a strong acid (and therefore has a low pH), the carboxylic acid group on each molecule does not ionize. Each amine group, however, being a weak base, accepts a H^+ ion to form a protonated $-NH_3^+$ group. The result is a molecular ion with an overall positive charge, as illustrated for alanine in Figure 20.4(a).

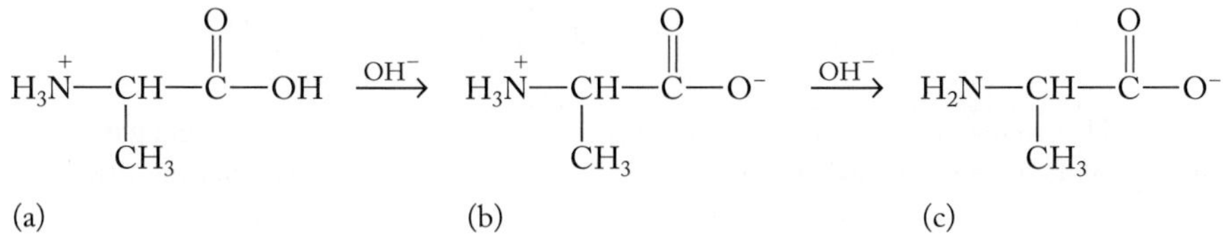

$$\underset{\text{(a)}}{H_3\overset{+}{N}-\underset{\underset{CH_3}{|}}{CH}-\overset{\overset{O}{\|}}{C}-OH} \xrightarrow{OH^-} \underset{\text{(b)}}{H_3\overset{+}{N}-\underset{\underset{CH_3}{|}}{CH}-\overset{\overset{O}{\|}}{C}-O^-} \xrightarrow{OH^-} \underset{\text{(c)}}{H_2N-\underset{\underset{CH_3}{|}}{CH}-\overset{\overset{O}{\|}}{C}-O^-}$$

FIGURE 20.4 (a) At low pH, alanine (like many other amino acids) exists as a 1+ ion. (b) At physiological pH (7.4), the $-COOH$ group is ionized and the molecule becomes a zwitterion with an overall charge of zero. (c) In basic solutions, the charge decreases to 1− as the $-NH_3^+$ group deprotonates.

zwitterion a molecule that has both positively and negatively charged groups in its structure.

CONNECTION The relationship between the strengths of acid–base conjugate pairs was discussed in Chapter 15 and illustrated in Figure 15.6.

Now suppose we add a strong base to this solution to neutralize the strong acid and to raise the pH to about 7.4, close to the pH of human blood and of the fluids in most of our tissues. At this *physiological pH* the –COOH groups on the α-carbon atoms in amino acids are mostly ionized, because they all have pK_a values in the range from 1.8 to 2.8. These values mean that the –COOH groups are mostly ionized above pH 3.0 and essentially completely ionized at pH 7.4. Meanwhile, most of the basic amine groups bonded to the α-carbon atoms of amino acid accept H^+ ions and form $-NH_3^+$ groups at pH 7.4. This combination of negatively charged $-COO^-$ ions and positively charged $-NH_3^+$ groups means that these amino acid molecules are **zwitterions**. This term is used to describe molecules that contain both positively and negatively charged functional groups, but that have no net electrical charge.

If we add more base to raise the pH well above 7.4, alanine loses a hydrogen ion from its $-NH_3^+$ group (we say that it *deprotonates*) to form an $-NH_2$ group, and we have a molecular ion with an overall charge of 1– (Figure 20.4c). The titration curve for alanine in Figure 20.5 shows this two-step neutralization process: the carboxylic acid is neutralized first, the protonated amine second.

A few of the common amino acids in Table 20.1 have R groups that contain acidic or basic units. These groups have pK_a values around 4 for the acids and in the range of 10.5 to 12.5 for the protonated amines, so the acids in the R groups typically deprotonate at pH values closer to physiological pH than do the groups directly attached to the α-carbon atom. The protonated amines in the R groups deprotonate at pH values lower than those of the α-amino groups. The impact of this behavior on the charges of the amino acids at physiological pH and on titration curves is the subject of Sample Exercise 20.1.

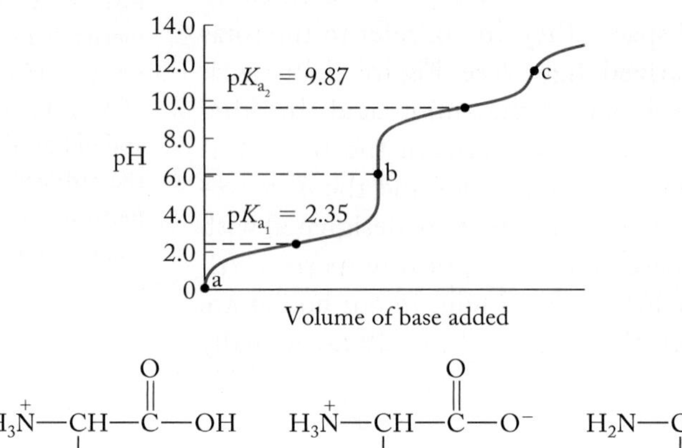

FIGURE 20.5 The titration curve of alanine resembles that of a weak diprotic acid. The carboxylic acid group (pK_a = 2.35) is neutralized first. The protonated amine group (pK_a = 9.87) is neutralized in the second step of the titration. The dominant forms of alanine present in solution are shown at (a), the starting point, and at (b) and (c), the two equivalence points.

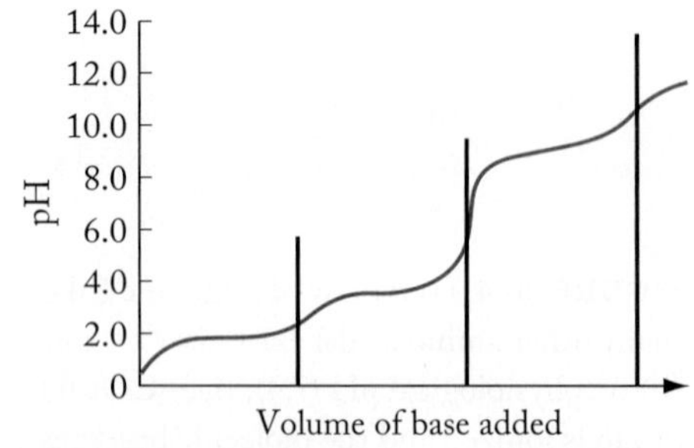

FIGURE 20.6 Titration curve of aspartic acid. The vertical lines identify equivalence points in the titration.

SAMPLE EXERCISE 20.1 Interpreting Acid–Base Titration Curves of Amino Acids LO1

The $-NH_2$ groups in α-amino acids are weak bases (pK_b ≈ 4–5), so their conjugate acids ($-NH_3^+$) are even weaker acids (pK_a ≈ 9–10). Figure 20.6 shows that the titration curve for aspartic acid has three equivalence points. Draw the molecular structures of the principal form of aspartic acid that is in solution at the start of the titration and at each equivalence point.

Collect, Organize, and Analyze The molecular structure of aspartic acid at physiological pH is shown in Table 20.1. There are two carboxylic acid groups and one amine group in a molecule of aspartic acid. At low pH the carboxylic acid groups are not ionized and the amine group is protonated. As the pH is raised, the –COOH groups ionize and then, under basic conditions, the protonated amine group releases its proton. The stronger of the two acid groups will ionize first. The –COOH group bonded to the α-carbon atom is only one carbon atom away from a protonated NH_3^+ group, which pulls electron density away from the –COOH group. The side-chain –COOH group is two carbon atoms away, so the carboxylate ion is further away and less affected by the NH_3^+ group. Therefore, the –COOH group bonded to the α-carbon atom should ionize first.

Solve At the beginning of the titration, the fully protonated 1+ ion is present:

At the first equivalence point, the –COOH group bonded to the α-carbon atom (highlighted in yellow) is ionized:

At the second equivalence point, the side-chain –COOH group is ionized:

And at the third equivalence point in the titration, the amine group is deprotonated:

peptide a compound of two or more amino acids joined by peptide bonds. Small peptides containing up to 20 amino acids are *oligopeptides*; the term *polypeptide* is used for chains longer than 20 amino acids but shorter than proteins.

peptide bond the result of a condensation reaction between the carboxylic acid group of one amino acid and the amine group of another.

Think About It We can estimate the pK_a values of the two carboxylic acid groups from the midpoint pH values in the first two steps of the titration: about 1.9 and 3.7. They differ by nearly 2 pH units, which means that the α-COOH group is almost 10^2 times stronger an acid than the side-chain –COOH group, due to the presence of the $\alpha\text{-NH}_3^+$ group.

Practice Exercise Sketch the titration curve for lysine starting at pH = 1.0, and draw the molecular structures of the principal form of lysine that is in solution at each equivalence point.

(Answers to Practice Exercises are in the back of the book.)

Peptides

Amino acids bond together to form chainlike molecules with a wide range of sizes. Amino acid *residues* (the name we give to amino acids that are part of a larger molecule) make up each link in the chain. The longest chains, some with molar masses as high as 3 million g/mol, are called proteins, but the shortest chains, called **peptides**, are only a few amino acid residues long. The smallest peptides contain only two or three amino acid residues and are called *dipeptides* and *tripeptides*, respectively. Peptides up to 20 residues long are called *oligopeptides*. Those made of more than 20 residues are called *polypeptides*. The size at which a polypeptide becomes a protein is arbitrary but is typically set at around 50–75 amino acid residues.

The bond linking the amino acids in peptides and proteins is called a **peptide bond** or *peptide linkage*. For example, the peptide bond that links molecules of valine (Val) and serine (Ser) is highlighted in blue in Figure 20.7. It forms when the α-carboxylic acid group of one amino acid (Val in this case) reacts with the α-amine group of another. The products are a peptide bond linking the two amino acids together and a molecule of water (which makes the reaction a condensation reaction). Note that peptide bonds have the same structure as the amide bonds that hold together the monomeric units of synthetic polyamides such as nylon.

The convention for drawing the structures of peptides begins by placing the amino acid that has a free α-amine group at the left end of the peptide chain and the amino acid with a free α-carboxylic acid at the right end. The left end is called the *amine* (or *N-*) *terminus* of the peptide and the right end is called the *carboxylic acid* (or *C-*) *terminus*. The name of a peptide is based on the names of its amino acids starting with the one at the N-terminus, except that the ending of all but the C-terminus amino acid is changed to *-yl*. Thus, the dipeptide formed by linking valine and serine in Figure 20.7 is called valylserine (abbreviated ValSer).

CONNECTION The formation of amide bonds in reactions between carboxylic acids and amines was described in Chapter 19 and illustrated in Figure 19.45.

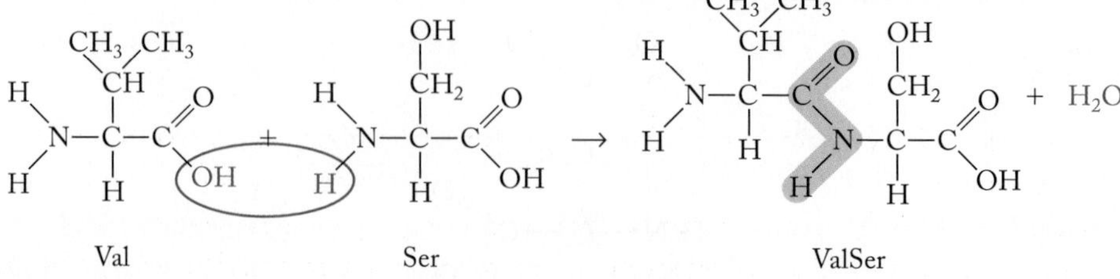

FIGURE 20.7 When the carboxylic acid group of valine forms a peptide bond with the amine group of serine, the products are the dipeptide valylserine and water.

The artificial sweetener aspartame is the methyl ester of the dipeptide aspartylphenylalanine (Figure 20.8). At pH 7.4, aspartame exists as a zwitterion because the aspartic acid amine group is protonated and the carboxylic acid in its R group is ionized. In contrast, its parent dipeptide has a net charge of 1− because both –COOH groups are ionized at that pH. In an amino acid that has an amine group in its R group, such as lysine or arginine, that amine group is probably protonated at physiological pH, giving the amino acid a net charge of 1+. To sum up, the overall charge on a peptide at physiological pH is the sum of the positive charges on protonated amine groups and the negative charges of ionized carboxylic acid groups.

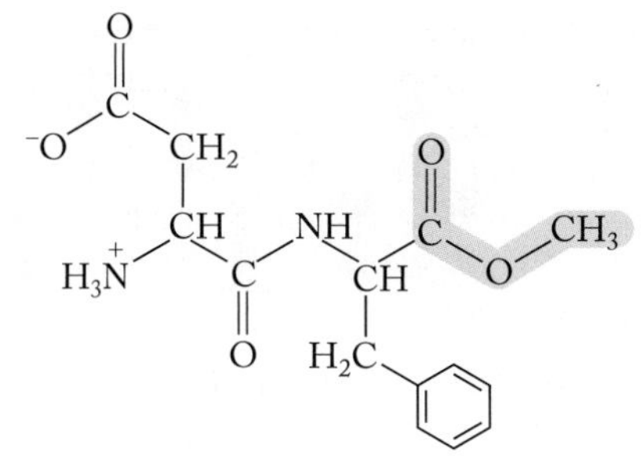

Aspartame

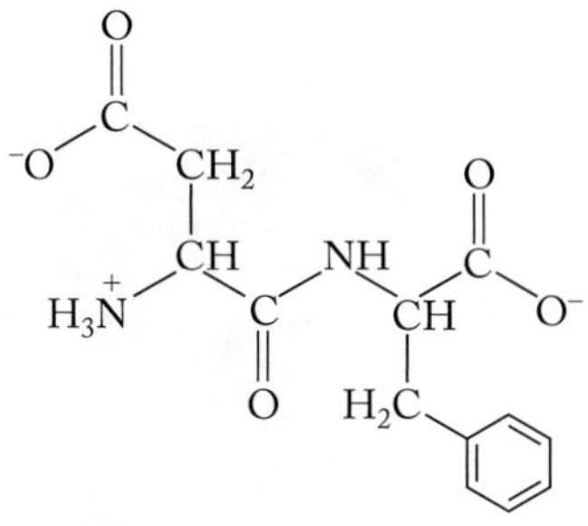

Aspartylphenylalanine

FIGURE 20.8 Aspartame is the methyl ester (highlighted in blue) of the dipeptide aspartylphenylalanine and has one fewer ionized –COOH groups than its parent dipeptide. Therefore, the overall charge of a molecule of aspartame at pH 7.4 is zero, whereas the charge on aspartylphenylalanine is 1−.

SAMPLE EXERCISE 20.2 Drawing and Naming Peptides **LO2**

a. Name all the dipeptides that can be made by reacting alanine with glycine.
b. Draw their molecular structures in solution at pH 7.4.

Collect, Organize, and Analyze Amino acid structures are in Table 20.1. Two different peptides can be made from alanine and glycine by changing their sequence from the N-terminus to the C-terminus: AlaGly and GlyAla. At pH 7.4, the carboxylic acid terminus –COOH groups should be ionized and the amine terminus $-NH_2$ groups should be protonated.

Solve

a. The names of the two peptides with the amino acid sequences GlyAla and AlaGly are glycylalanine and alanylglycine, respectively.
b. At pH 7.4, the $-NH_2$ groups of amino acids are protonated and the –COOH groups are ionized, so the principal forms of these dipeptides are

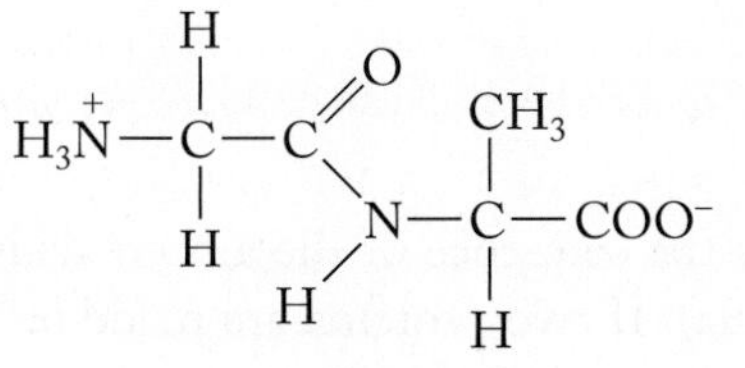

Glycylalanine (GlyAla)

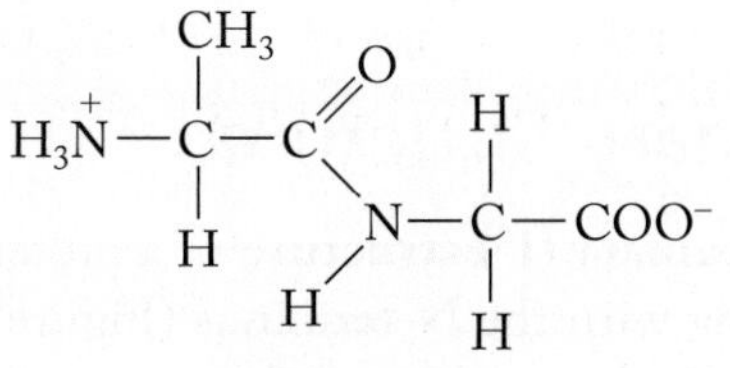

Alanylglycine (AlaGly)

Think About It Only the N-terminal $-NH_2$ and C-terminal –COOH groups can be protonated or ionized in these dipeptides. Therefore, most AlaGly and GlyAla molecules are zwitterionic with net charges of (1+) + (1−) = 0 at pH 7.4.

Practice Exercise How many different tripeptides can be synthesized from one molecule of each of three different amino acids A, B, and C?

20.2 Protein Structure and Function

The structure of a protein is crucial to its function. At the beginning of this chapter, we mentioned the catastrophic impact of a malformed muscle protein. The functions of nonstructural proteins, such as the ability of enzymes to catalyze biochemical reactions, are also closely linked to their structures. These large biomolecules must assume particular three-dimensional conformations to interact with other molecules and to fulfill their biological mission.

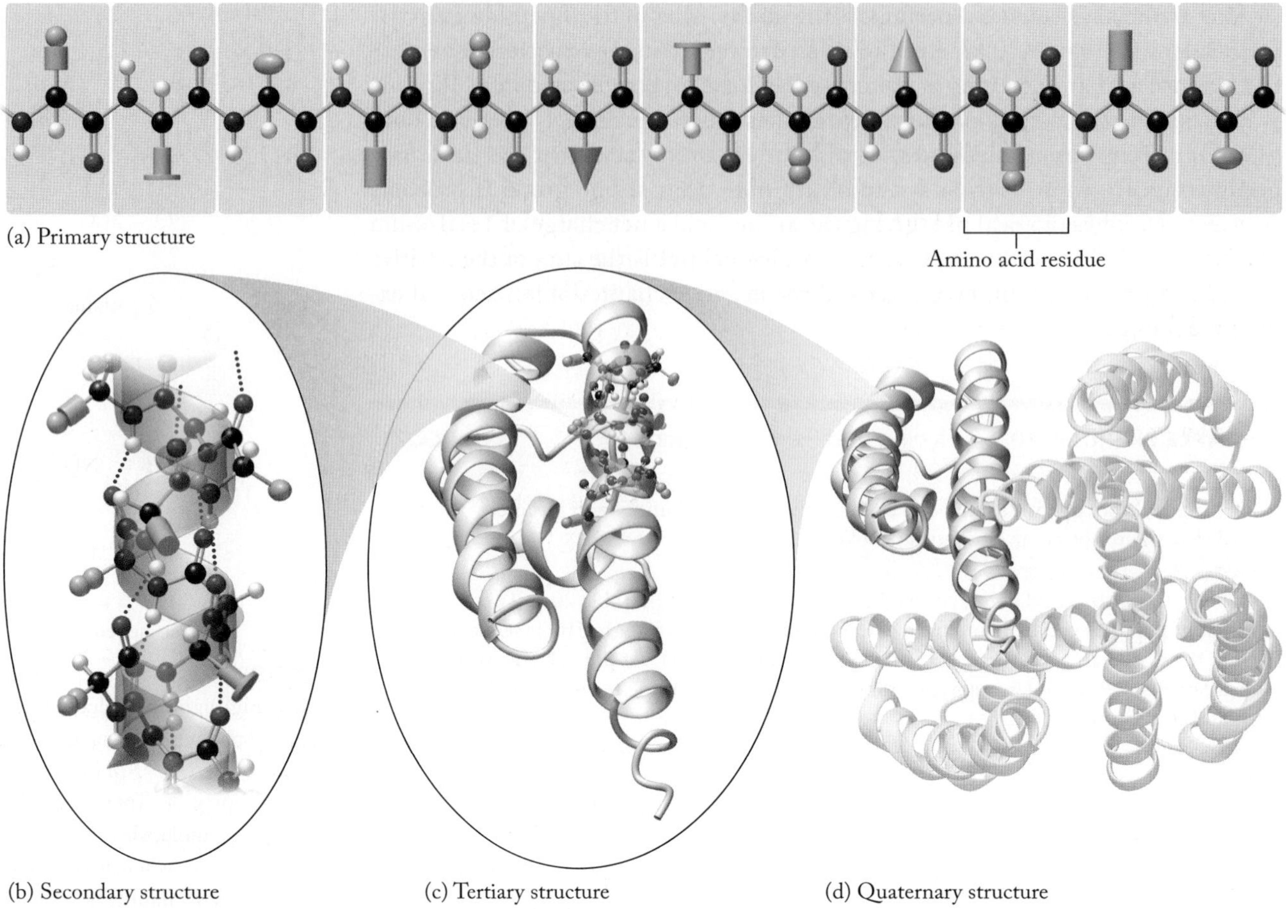

(a) Primary structure

Amino acid residue

(b) Secondary structure

(c) Tertiary structure

(d) Quaternary structure

FIGURE 20.9 The four levels of protein structure. (a) A protein's primary structure is its amino acid sequence. The green shapes represent different R groups. (b) Secondary structure (here, an α helix) describes the three-dimensional pattern adopted by segments of the protein strand. (c) Tertiary structure is the overall shape of the molecule as segments of it bend and fold. (d) Quaternary structure refers to the overall shape adopted by multiple protein strands that assemble into a single unit.

Primary Structure

The **primary (1°) structure** of a protein is the sequence of the amino acids in it, starting with the N-terminus (Figure 20.9a). If two proteins are made up of the same number and type of amino acids but have different amino acid sequences, they are different proteins.

Changing only one amino acid can dramatically alter a protein's function. In hemoglobin, for example, the sixth amino acid from the N-terminus of a protein strand that is 146 amino acids long is normally glutamic acid. In some people, valine substitutes for glutamic acid at this position. This one substitution alters the solubility of the protein and causes the red blood cell to take on a sickle shape instead of the normal, plump disc shape (Figure 20.10). These sickled cells do not pass through

CONNECTION We discussed in Chapter 6 how ion–dipole and dipole–dipole interactions are the key to understanding the solubility of solutes in water.

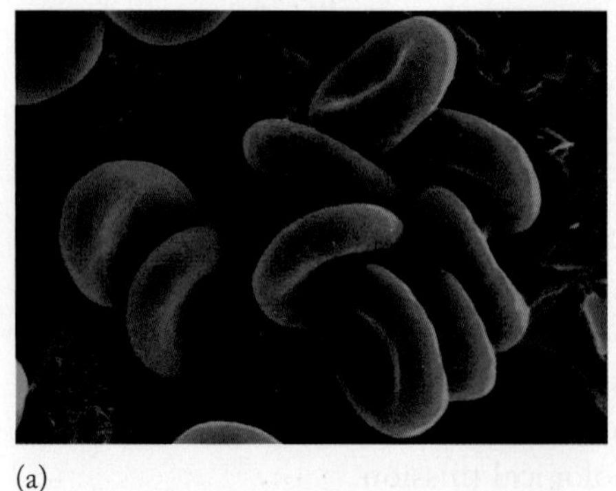

(a)

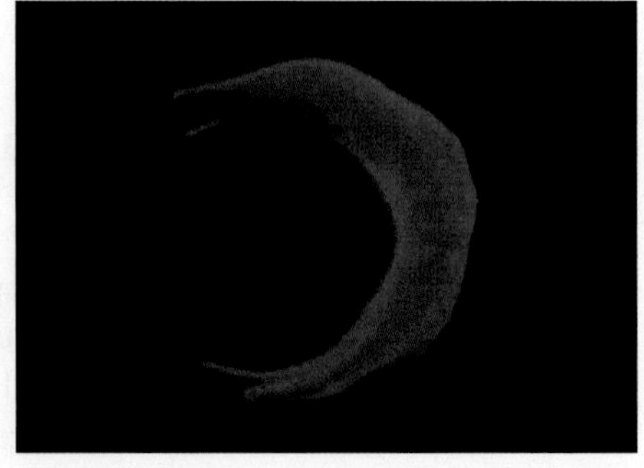

(b)

FIGURE 20.10 (a) Normal red blood cells are plump discs, whereas (b) those in patients suffering from sickle-cell anemia are distorted.

capillaries easily and may impede blood circulation. They do not last as long as normal blood cells, either. These factors diminish the capacity of the blood to carry oxygen, which is a characteristic of the disease called sickle-cell anemia.

Why does switching valine for glutamic acid affect the solubility of hemoglobin? The answer lies in the R groups of these two amino acids (Figure 20.11). The side-chain –COOH group of glutamic acid is ionized at physiological pH, and strong ion–dipole interactions with water molecules enhance the protein's solubility. However, if valine, with its nonpolar isopropyl R group, replaces glutamic acid, that strong intermolecular interaction with water is lost. The valine creates a hydrophobic region on the surface of the protein, which causes deoxygenated hemoglobins to stick to each other, deforming the cells.

Sickle-cell anemia is a debilitating disease, but it provides a survival advantage in regions where malaria is endemic. Having sickle-cell anemia does not protect people from contracting malaria or make them invulnerable to the parasite that causes it. However, children infected with malaria are more likely to survive the illness if they have sickle-cell anemia. Why sickle-cell anemia has this effect is not completely understood.

Primary structure

Normal protein:

Val - His - Leu - Thr - Pro - Glu - Lys - . . .

Abnormal protein:

Val - His - Leu - Thr - Pro - Val - Lys - . . .

(a)

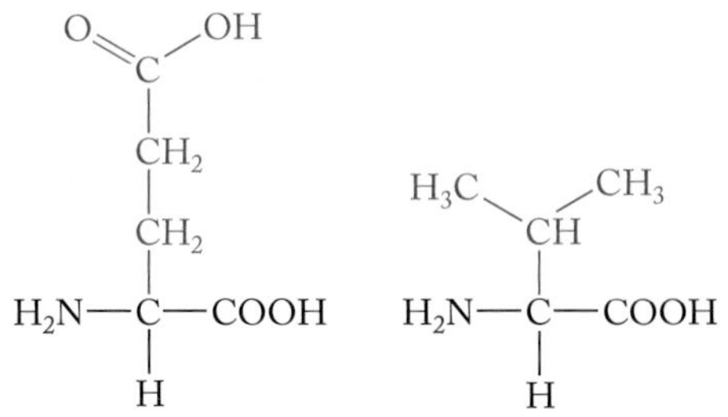

(b)

FIGURE 20.11 (a) The primary structure of the amine end of a protein in normal hemoglobin and in the abnormal hemoglobin responsible for sickle-cell anemia. (b) The replacement of glutamic acid (which has a hydrophilic R group) by valine (which has a hydrophobic R group) is responsible for the disease.

CONCEPT TEST

Which other amino acids besides valine might diminish hemoglobin solubility when substituted for glutamic acid?

Secondary Structure

In order for proteins to do what they are supposed to do, they must fold into a particular three-dimensional conformation. The first stage in the folding process is called a protein's **secondary (2°) structure**. Proteins acquire particular 2° structures as a result of a combination of intermolecular forces dominated by hydrogen bonds between the peptide linkages, dipole–dipole interactions, and hydrogen bonds between polar side groups. Among the most common geometric patterns that segments of amino acid chains make (Figure 20.9b) is the **α helix**, a coiled arrangement with the R groups pointing outward. The helical structure is maintained by hydrogen bonds between –NH groups on one part of the chain and C═O groups on nearby amino acids. The α helix looks very much like a spring. Muscle tissue that stretches and contracts is made largely of α-helical proteins, as are keratins, the proteins in hair and fingernails.

Another common 2° structure is the **β-pleated sheet**. These sheets are assemblies of multiple amino acid chains aligned side by side. The pleats are caused by the tetrahedral molecular geometries of the atoms along the chains (Figure 20.12). Adjacent chains are linked together by hydrogen bonds, and the collection of side-by-side zigzag chains forms a continuous sheet. R groups extend above and below the pleats. β-Pleated sheets may stack on top of one another like two pieces of corrugated roofing. Stacked sheets are held together by the same interactions that hold all proteins together, including ion–ion interactions, hydrogen bonding, and London dispersion forces, depending on the pairs of R groups involved.

The proteins that make up strands of silk form thin, planar crystals of β-pleated sheets that are only a few nanometers on a side. Enormous numbers of these crystals form long arrays of sheets stacked together like nanoscale pancakes. Hydrogen bonds hold the stacks together and reinforce adjacent sheets. As a result of these interactions, strands of silk are stronger than strands of steel with the same mass; indeed, silk is one of the toughest known materials—natural or synthetic.

primary (1°) structure the sequence in which amino acid monomers occur in a protein chain.

secondary (2°) structure the pattern of arrangement of segments of a protein chain.

α helix a coil in a protein chain's secondary structure.

β-pleated sheet a puckered two-dimensional array of protein strands held together by hydrogen bonds.

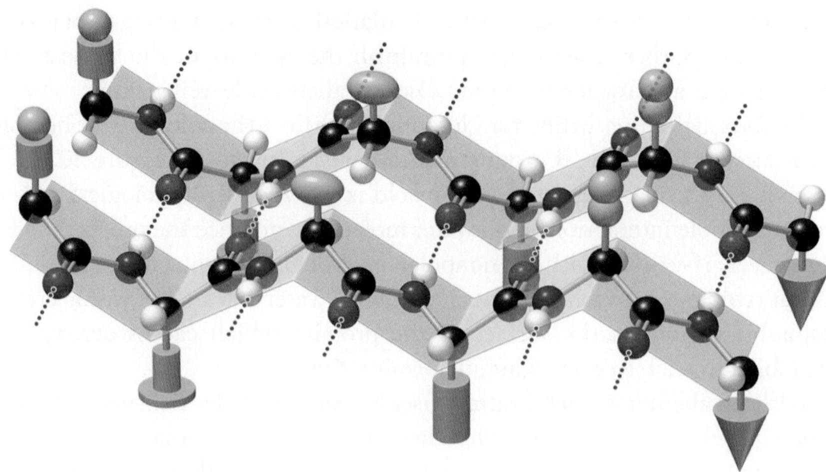

FIGURE 20.12 Each amino acid chain in a β-pleated sheet is folded in a zigzag pattern. Adjacent chains in a sheet are held together by hydrogen bonds (blue dotted lines). The R groups (green shapes) extend above and below the sheet, linking it to adjacent sheets via intermolecular forces.

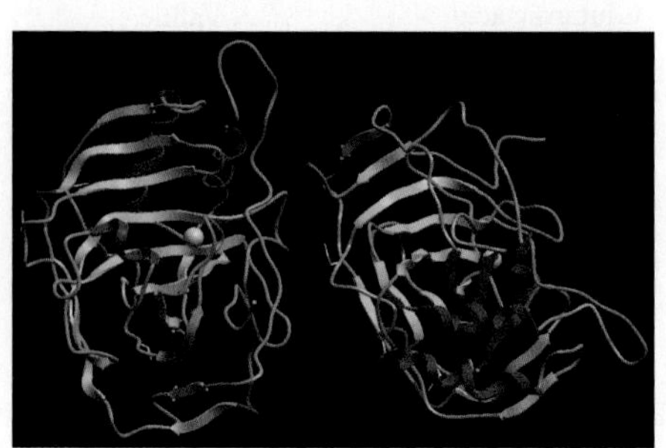

FIGURE 20.13 Two views of carbonic anhydrase. The structure of the protein has α-helical regions (red), β-pleated sheet regions (green), and random coil (blue). The light blue sphere in the center is a zinc ion.

Some single-strand proteins exist as α helices on their own but form β-pleated sheets when they clump together in multistrand aggregates. One consequence of this clumping is the formation of insoluble protein deposits called plaque. Abnormal accumulation of plaque can be a serious health risk: plaque formed by a protein called amyloid β has been linked to the onset of Alzheimer's disease.

If part of a protein is characterized by an irregular or rapidly changing structure, it is said to have a **random coil** 2° structure. The amino acid chain may fold back on itself and around itself, but it has no regular features the way an α helix or a β-pleated sheet does. When proteins *denature* due to heat or a change in pH, for example, they lose their secondary structure and may become random coils.

Large protein molecules may contain all three types of 2° structure. In describing a protein, scientists may indicate the percentage of amino acids involved in each type—for example, 50% α helix, 30% β-pleated sheet, and 20% random coil. Figure 20.13 shows a model of a protein called carbonic anhydrase, which illustrates all three types.

CONCEPT TEST

Why do proteins denature (lose their secondary structure) when they are heated?

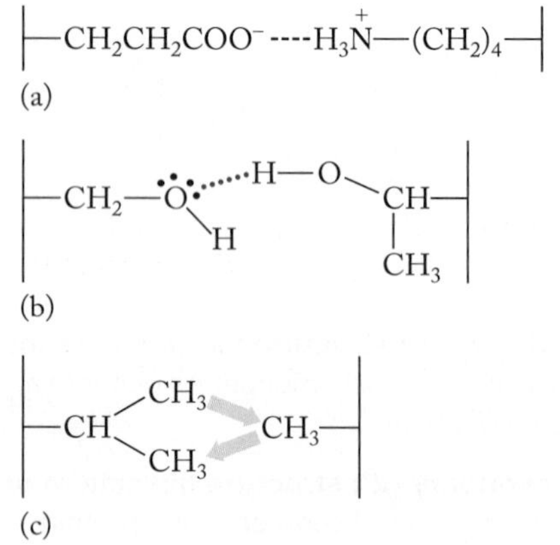

$-CH_2CH_2COO^- \cdots\cdots H_3\overset{+}{N}-(CH_2)_4-$

(a)

$-CH_2-\ddot{O}(H)\cdots\cdots H-O-CH(CH_3)-$

(b)

$-CH(CH_3)_2 \rightleftarrows CH_3-$

(c)

FIGURE 20.14 Interactions that influence the tertiary and quaternary structures of proteins include (a) ion–ion interactions between acidic and basic R groups, (b) hydrogen bonding, and (c) hydrophobic interactions between nonpolar side chains.

Tertiary and Quaternary Structure

Large proteins have structure beyond the 1° and 2° levels, folding back on themselves as a result of interactions between R groups on amino acids that can be considerable distances apart along the protein chain. These interactions may be ion–ion, ion–dipole, or van der Waals forces (Figure 20.14). They may even involve the formation of covalent bonds—specifically, disulfide bonds. The –SH groups on two cysteine residues may undergo an oxidation half-reaction that results in the formation of a disulfide linkage that holds two parts of the protein strand together via an intrastrand –S—S– covalent bond (Figure 20.15). Interactions and reactions such as these determine **tertiary (3°) structure**, the overall three-dimensional shape of the protein that is key to its biological activity (Figure 20.9c). Because the proteins in living systems exist in an aqueous environment, hydrophobic R groups tend to reside in the interiors of their 3° structures, whereas hydrophilic groups (as we saw in normal hemoglobin) are oriented toward the outside, where they interact with nearby molecules of water. Hydrophobic interactions are the primary cause of protein folding and compaction, but all the other modes of interaction help a large protein find its unique 3° structure.

Hemoglobin and some other proteins exhibit an even higher order of structure. One hemoglobin unit (Figure 20.16a) contains four protein strands, each of which enfolds a porphyrin ring containing one Fe^{2+} ion. The combination of four protein strands to make one hemoglobin assembly is an example of **quaternary (4°) structure** (Figure 20.9d). In the 4° structures of keratins (Figure 20.16b), protein strands with α-helical 2° structures coil around each other to make even larger coils. When the protein strands are held together mostly by van der Waals forces, as they are in the keratin in skin tissue, the structures are flexible and elastic. If they are also restrained by many covalent bonds, as in Figure 20.16(b), they produce tissues that are hard and less flexible, like fingernails and the beaks of birds.

$$\text{—}CH_2\text{—}SH \quad HS\text{—}CH_2\text{—}$$

$$\downarrow$$

$$\text{—}CH_2\text{—}S\text{—}S\text{—}CH_2\text{—}$$

$$+ 2\,H^+(aq) + 2\,e^-$$

FIGURE 20.15 The tertiary structure of some proteins is stabilized by intrastrand –S—S– bonds that form between the –SH groups on the side chains of cysteine residues.

Enzymes: Proteins as Catalysts

The chemical reactions involved with metabolism—both *catabolism* (breaking down of molecules) and *anabolism* (synthesis of complex materials from simple feedstocks)—are mediated in large part by proteins called **enzymes**. Enzymes are biological catalysts. Both catabolism and anabolism are organized in sequences of reactions called *metabolic pathways*, and each step in a metabolic pathway is catalyzed by a specific enzyme. For example, carbonic anhydrase (Figure 20.13) is an enzyme that speeds up the reaction shown in Equation 20.1. Note that this reaction is reversible, and remember that a catalyst speeds up a reversible reaction in both directions. This means that the hydrolysis of CO_2 and its ionization to bicarbonate and hydrogen ions are faster in the presence of carbonic anhydrase, and the reaction of the hydrogen ion with the bicarbonate ion to form CO_2 and water is also faster:

$$CO_2(aq) + H_2O(\ell) \rightleftharpoons HCO_3^-(aq) + H^+(aq) \qquad (20.1)$$

In the presence of carbonic anhydrase, this reaction proceeds about 10 million times faster than in its absence. Without carbonic anhydrase, we would not be able to expel carbon dioxide fast enough to survive (as the reaction in Equation 20.1 runs in reverse). The interconversion of CO_2 and HCO_3^- is a reaction

CONNECTION All the types of intermolecular forces described in Chapter 6 are involved in the intramolecular interactions that give proteins their unique 2° and 3° structures.

CONNECTION Recall from Chapter 14 that catalysts lower the activation energy barrier for a reaction and thereby increase the rate of both the forward and reverse reactions in an equilibrium.

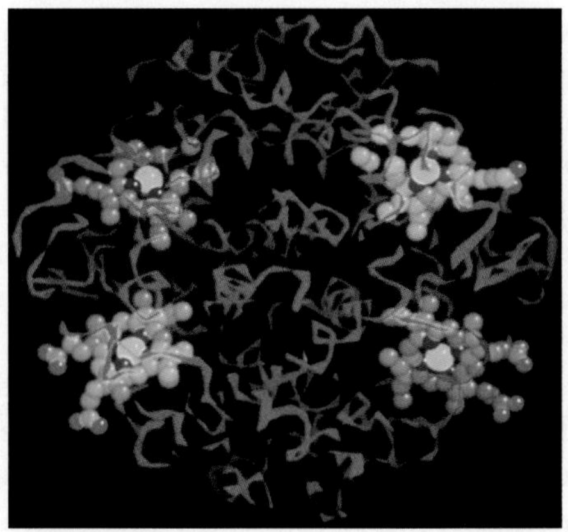

(a) Hemoglobin

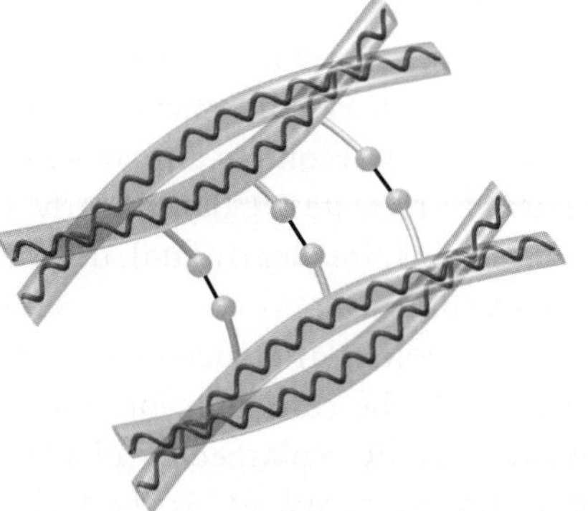

(b) Keratin

FIGURE 20.16 Quaternary structure of proteins. (a) Four protein chains form a single unit in the quaternary structure of hemoglobin. The iron-containing porphyrins are bright green. (b) Pairs of α-helical chains (blue) wound together and linked by interstrand –S—S– bonds (yellow) stabilize the quaternary structure of the keratin in hair and fingernails.

random coil an irregular or rapidly changing part of the secondary structure of a protein.

tertiary (3°) structure the three-dimensional, biologically active structure of the protein that arises because of interactions between R groups on amino acids.

quaternary (4°) structure the larger structure functioning as a single unit that results when two or more proteins associate.

enzyme a protein that catalyzes a reaction.

of extreme importance in the physiological system, because it is responsible for the maintenance of a relatively constant pH.

CONNECTION In Section 13.6 we discussed CFCs, inorganic homogeneous catalysts that contribute to the destruction of ozone. Enzymes are bio-organic homogeneous catalysts that selectively speed up biochemical reactions.

CONNECTION In Chapter 15 we examined the bicarbonate buffering system and the role of breathing in maintaining physiological pH.

One molecule of carbonic anhydrase can catalyze the hydrolysis and ionization of more than a million molecules of CO_2 in 1 s. This value is called the *turnover number* for the enzyme; in general, the higher the turnover number, the faster the enzyme-catalyzed reaction proceeds. Turnover numbers for enzymes typically range from 10^3 to 10^7. The higher the turnover number, the lower the activation energy of the catalyzed reaction. As we learned in Chapter 13, molecules must collide with the proper orientation for reactions to proceed. Carbonic anhydrase is essentially a perfect enzyme because it catalyzes the hydrolysis reaction nearly every time it collides with CO_2. Your saliva contains carbonic anhydrase, and this is why carbonated beverages fizz and release CO_2 quickly in your mouth, whereas it takes a much longer time for them to release dissolved CO_2 when they are simply open to the air.

Enzymes are highly selective: each catalyzes a particular reaction involving a particular reactant. For example, an enzyme called lactase catalyzes only the reaction by which lactose, the sugar in milk, is broken down during digestion. People who lack this enzyme cannot metabolize the sugar; they are said to be *lactose intolerant.* If they consume dairy products, unmetabolized lactose passes into their large intestines where bacteria ferment it, and unpleasant and painful abdominal disturbances result.

CONNECTION A racemic mixture contains equal proportions of the two enantiomers of a chiral compound, as we discussed in Section 5.6.

Enzyme selectivity sometimes extends to enantiomeric pairs, which is the main reason why natural products tend to be optically pure materials while the same products synthesized in the laboratory tend to be racemic mixtures. In the pharmaceutical industry, research is currently focused on using enzymes outside living systems to produce enantiomerically pure materials needed as drugs. This research area, called **biocatalysis**, uses enzymes to catalyze chemical reactions run in industrial-sized reactors. Because enzyme-catalyzed reactions frequently run at lower temperatures (between 4 and 40°C) and pressures than noncatalyzed processes and hence use less energy, their development and use are being encouraged in many industrial processes as part of the green chemistry initiative.

Synthetic reaction pathways catalyzed by enzymes can significantly reduce the production cost of, for example, biological pharmaceuticals. However, biocatalytic reactions often run best in very dilute solutions, which limits production. A key issue driving interest in such processes is that an enantiomerically pure pharmaceutical is likely to be more potent and produce fewer side effects than racemic mixtures of the same product. As an example, the drug thalidomide is a racemic mixture of two enantiomeric forms: one is very effective in treating morning sickness, but the second is a teratogen (an agent that disturbs the development of a fetus). In the late 1950s and early 1960s, more than 10,000 children were born with serious, frequently fatal, deformities as a result of their mothers having taken the racemic drug.

biocatalysis the strategy of using enzymes to catalyze reactions on a large scale.

active site the location on an enzyme where a reactive substance binds.

substrate the reactant that binds to the active site in an enzyme-catalyzed reaction.

The molecular structure of enzymes contains a region called an **active site** that binds the reactant molecule, called the **substrate**. The action of enzymes was originally explained by a lock-and-key analogy in which the substrate is the key and the active site is the lock (Figure 20.17). The substrate is held in the active site by the same kinds of intermolecular interactions that hold other biomolecules together. Some enzymes become covalently bound to intermediates in the catalytic process. Once in the active site, the substrate is converted into product via a reaction having a lower energy transition state than it would without the enzyme. The reaction of a substrate S with an enzyme E produces

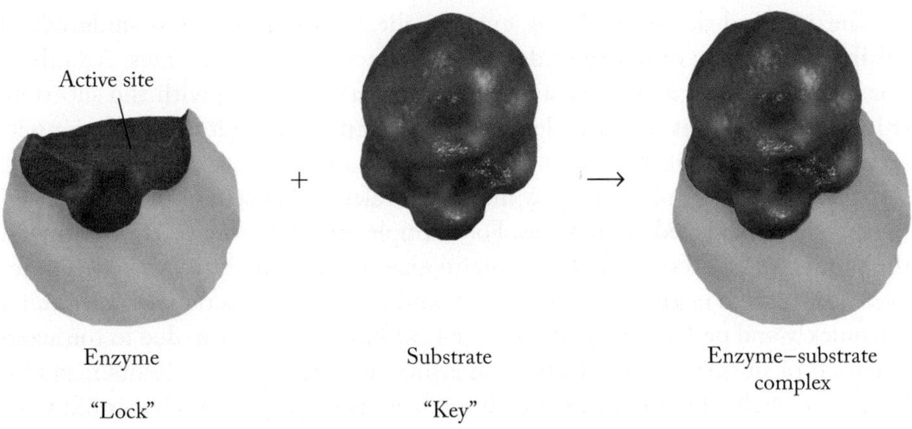

FIGURE 20.17 In the lock-and-key model of enzyme activity, the substrate fits exactly into the active site of the enzyme that catalyzes a chemical reaction involving the substrate.

an *enzyme–substrate* (E–S) *complex* that decomposes, forming a product P and regenerating the enzyme:

$$E + S \rightleftharpoons E\text{–}S \rightarrow E + P$$

CONNECTION The phenomenon of molecular recognition in biological systems based on complementary molecular shapes was introduced in Chapter 5.

The lock-and-key analogy does not fully account for enzyme behavior. A more accurate view is provided by the *induced-fit model*, which assumes that as the E–S complex forms, the binding site undergoes subtle changes in its shape to more precisely fit the three-dimensional structure of the transition state. A simplified illustration of such an interaction is shown in Figure 20.18(a).

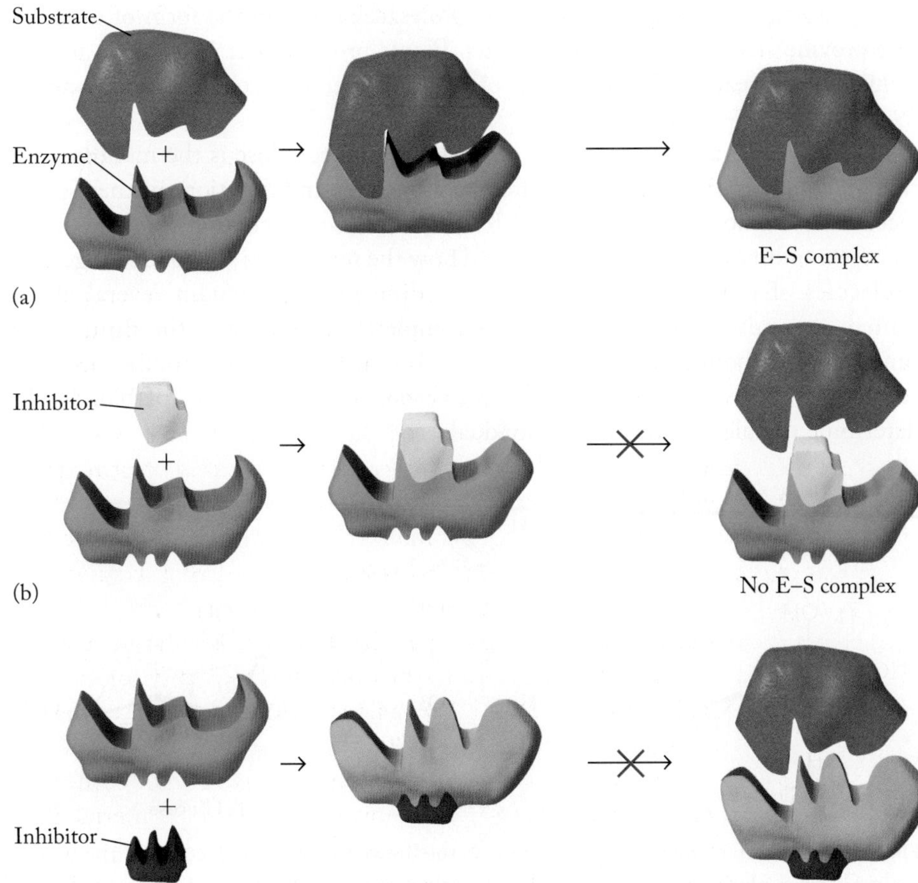

FIGURE 20.18 (a) The induced-fit model assumes that the shape of the enzyme changes to accommodate the substrate and form the enzyme–substrate (E–S) complex. Inhibitors (yellow and red molecules) may (b) block the enzyme's binding site or (c) cause a change elsewhere in the enzyme's structure that prevents the active site from attaining the shape it needs to form the E–S complex.

inhibitor a compound that diminishes or destroys the ability of an enzyme to catalyze a reaction.

carbohydrate an organic molecule with the generic formula $C_x(H_2O)_y$.

monosaccharide a single-sugar unit and the simplest carbohydrate.

polysaccharide a polymer of monosaccharides.

The induced-fit model helps explain the behavior of compounds called **inhibitors** that can diminish or destroy the effectiveness of enzymes. An inhibitor may bind to an active site and block it from interacting with the substrate (Figure 20.18b), or it may disable the enzyme by preventing it from assuming its active shape (Figure 20.18c).

Natural enzyme inhibitors play important roles in regulating the rates of reactions that are catalyzed by enzymes. For example, in a multistep reaction pathway, the product of a later step may inhibit an enzyme that catalyzes an earlier reaction. This kind of negative feedback keeps the sequence of reactions from running too quickly and perhaps jeopardizing the health of the organism due to the accumulation of undesirable products or intermediates. Enzyme inhibitors may also be used to fight disease. Powerful drugs have been developed that inhibit the enzymes that allow viruses to break down proteins. Several such drugs have been particularly effective in treating HIV.

20.3 Carbohydrates

Carbohydrates are a class of biological compounds with the generic formula $C_x(H_2O)_y$. Their name and formula suggest that these compounds are *hydrates of carbon*, but we shall see that this label does not fit their molecular structures. The smallest carbohydrates are **monosaccharides** (the name means "one sugar"), which bond together to form more complex carbohydrates called **polysaccharides.** Many organisms use monosaccharides as their main energy source but convert them to polysaccharides for the purpose of energy storage. Starch is the most abundant energy-storage polysaccharide in plants. Polysaccharides in the form of cellulose also provide structural support in plants. Plants produce over 100 billion tons of cellulose each year—the woody parts of trees are over 50% cellulose, and cotton is 99% cellulose.

The principal building block of both starch and cellulose is the monosaccharide glucose (Figure 20.19). The different properties and functions of these polysaccharides come from the subtle differences in the molecular geometries of the glucose monomers in their structures and how the monomers are bonded together. Molecules of most monosaccharides, including glucose, contain several chiral centers, so multiple isomers exist. This complexity gives rise to the third major function of carbohydrates: molecular recognition. For example, combinations of carbohydrates and proteins, called *glycoproteins*, on the surfaces of blood cells determine the blood type of an individual.

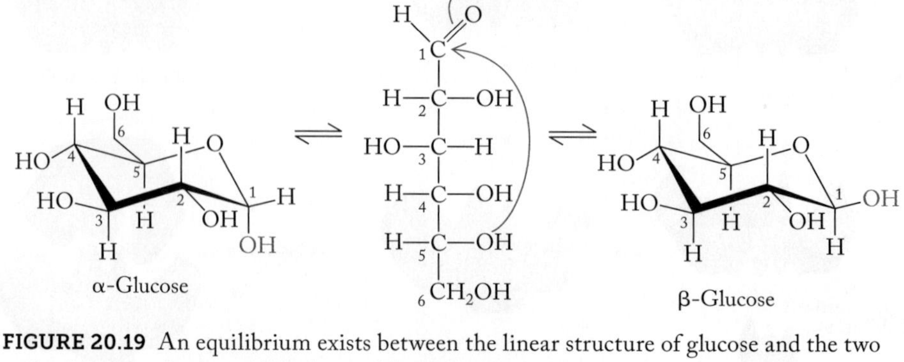

CONNECTION We introduced the chair conformation of six-membered rings in Chapter 19.

α-Glucose

β-Glucose

FIGURE 20.19 An equilibrium exists between the linear structure of glucose and the two cyclic forms, α-glucose and β-glucose. The difference between the two cyclic structures is the orientation of the –OH group on C1 (highlighted in blue).

Molecular Structures of Glucose and Fructose

Glucose is the most abundant monosaccharide in nature and in the human body. It is also called *dextrose*—a kind of abbreviation for *dextro*-glucose. Fructose is the principal monosaccharide in many fruits and root vegetables. Given the importance and abundance of these sugars, let's examine their structures and properties more closely.

Three molecular views of glucose are provided by the structures in Figure 20.19. All three have the same chemical formula ($C_6H_{12}O_6$), and they are all structural isomers of one another. Note that the middle structure contains an aldehyde (–CH═O) group and all three structures contain multiple alcohol (–C—OH) groups. The polarities of these groups and their capacities to form hydrogen bonds give glucose its high molar solubility in water (5.0 *M* at 25°C).

CONNECTION An aldehyde group (Chapter 19) includes the following atoms and bonds:

The cyclic structures of glucose form when the carbon backbone of the linear form curls around so that the –OH group bonded to the carbon atom labeled number 5 (C5) comes close to the aldehyde group on C1 (as shown by the red arrow in Figure 20.19). The aldehyde and alcohol groups react to form a six-membered ring made up of five carbon atoms (C1 to C5) and the oxygen atom of the C5 alcohol. There is a small but significant difference in the two cyclic structures. In the molecule on the left, called α-glucose, the –OH group on C1 points down. In the molecule on the right, β-glucose, the C1 –OH group points up. These two orientations are possible because there are two ways that the carbon chain in the middle structure can form a cyclic structure: the –OH group on C5 can approach C1 from either of the two sides of the plane defined by the C1 carbon atom and the O and H atoms bonded to it. When the C5 –OH group approaches C1 as shown in Figure 20.20, the α isomer is produced. Approaching C1 from the other side yields the β isomer. C1 is an aldehyde and is not a chiral center in the linear form, but the cyclization creates a new chiral center in the cyclic molecule.

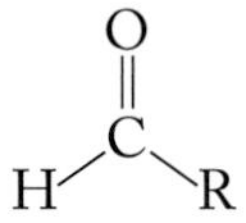

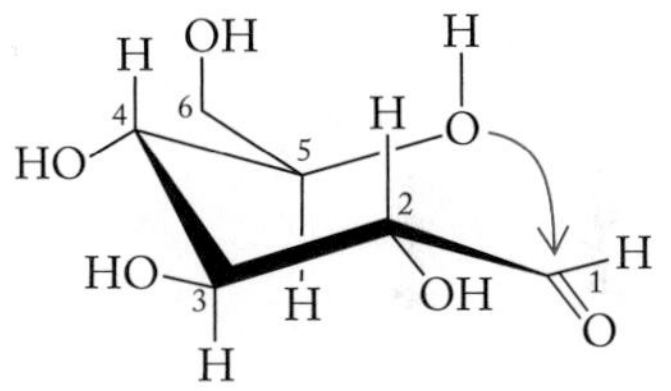

FIGURE 20.20 The –OH group on C5 may approach the aldehyde on C1 from either side of the plane defined by the C, H, and O atoms in the aldehyde group. For the approach shown here the product will be the α isomer.

The β form is slightly more stable than the α form and accounts for 64% of glucose molecules in aqueous solution; the α form accounts for the remaining 36%. Both cyclic forms are more stable than the straight-chain form, which exists only as an intermediate between the two cyclic forms. The energy barrier to flipping conformations is relatively small, so glucose molecules in solution are constantly opening and closing in a dynamic structural equilibrium.

Figure 20.21 shows the structures of the linear and cyclic forms of fructose, which also has the formula $C_6H_{12}O_6$. The C═O group in the linear form of fructose is not on the terminal carbon atom, so fructose is a ketone rather than an aldehyde. It forms a five-membered ring, not a six-membered ring, when the –OH group at C5 reacts with the carbonyl carbon atom at the C2 position. The product is a ring with two $-CH_2OH$ groups that are either on the same side of the ring (α-fructose) or on opposite sides (β-fructose).

CONNECTION A ketone group (Chapter 19) contains the following atoms and bonds:

α-Fructose ⇌ β-Fructose

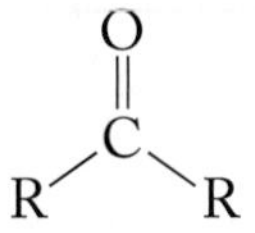

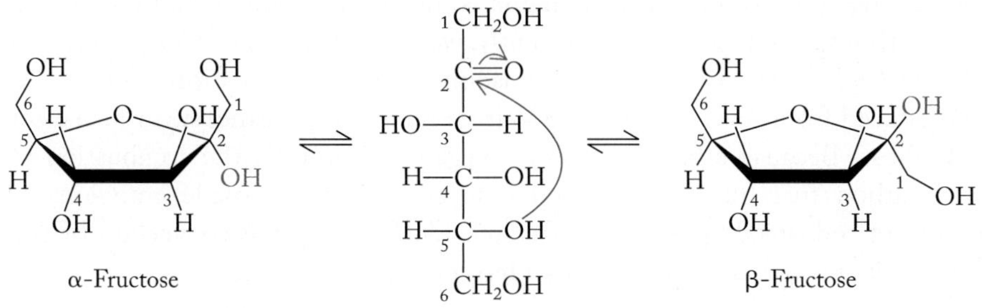

FIGURE 20.21 The cyclization of fructose proceeds via the –OH group on C5 and the ketone on C2.

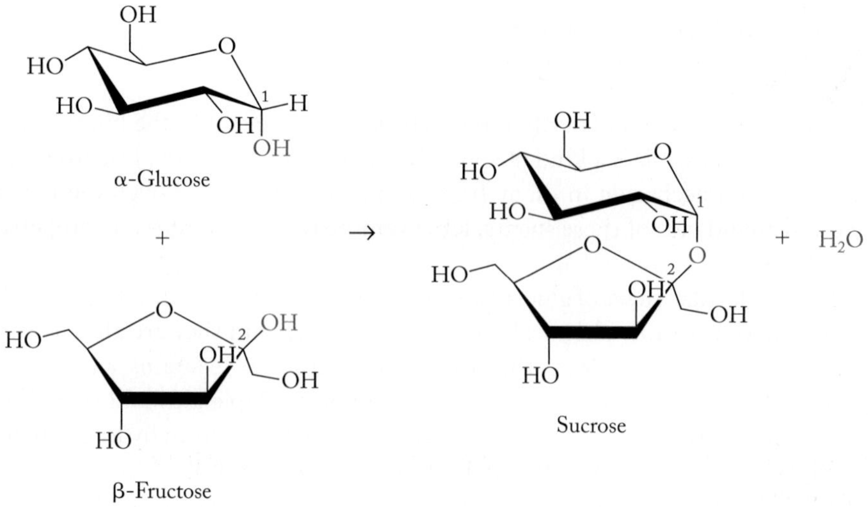

FIGURE 20.22 α-Glucose and β-fructose (both $C_6H_{12}O_6$) combine to form a molecule of sucrose ($C_{12}H_{22}O_{11}$)—ordinary table sugar—plus a molecule of water.

CHEMTOUR Formation of Sucrose

Disaccharides and Polysaccharides

Sucrose, or ordinary table sugar, is a *disaccharide* ("two sugars") that consists of one molecule of α-glucose bonded to one molecule of β-fructose (Figure 20.22). The bond between them forms when the –OH group on the C1 carbon atom of glucose reacts with the C2 carbon atom of fructose, producing a C—O—C **glycosidic bond** or *glycosidic linkage*. Water is also produced, making this reaction another example of a condensation reaction. The glycosidic bond in sucrose is called an α,β-1,2 linkage because of the orientations (α and β) of the two –OH groups involved and their positions (C1 and C2) in the cyclic structures of the two monosaccharides.

When glucose molecules form glycosidic linkages with the –OH groups on C1 and C4, they can form long-chain polysaccharides. If the starting monomer is α-glucose, the bonds between them are α-1,4 glycosidic linkages, and the product of their formation is starch (Figure 20.23a). The conversion of α-glucose into starch is an effective way for plants to store energy because formation of the α-1,4 linkage is reversible. With the aid of digestive enzymes, α-1,4 bonds can be hydrolyzed and starch can be converted back into glucose by plants that make the starch or by animals that eat the plants.

The cellulose that plants synthesize to build stems and other organs has a structure (Figure 20.23b) slightly different from that of starch because the building blocks of cellulose are β-glucose instead of α-glucose, so the monomers are linked by β-1,4 glycosidic bonds. This structural difference between α- and β-glycosidic bonds is important because it enables starch to coil and make granules for efficient storage while cellulose forms structural fibers.

Carbohydrates in plants are a major part of the total organic matter in any given ecological system—that is, they make up much of the system's **biomass**. People have used various forms of biomass, including wood and animal dung, as fuel for thousands of years. More recently, we have begun to convert biomass into a liquid *biofuel*, ethanol, for use as a gasoline additive or substitute. Ethanol can be produced from several renewable sources such as sugarcane, sugar beets, corn, and wheat. Because ethanol contains oxygen, it improves the combustion of a hydrocarbon fuel like gasoline and reduces carbon monoxide emissions. An important industrial application of starch hydrolysis is the conversion of cornstarch into glucose and then, through fermentation, into ethanol:

$$C_6H_{12}O_6(aq) \rightarrow 2\ CH_3CH_2OH(aq) + 2\ CO_2(g) \qquad (20.2)$$

glycosidic bond a C—O—C bond between sugar molecules.

biomass the sum total of the mass of organic matter in any given ecological system.

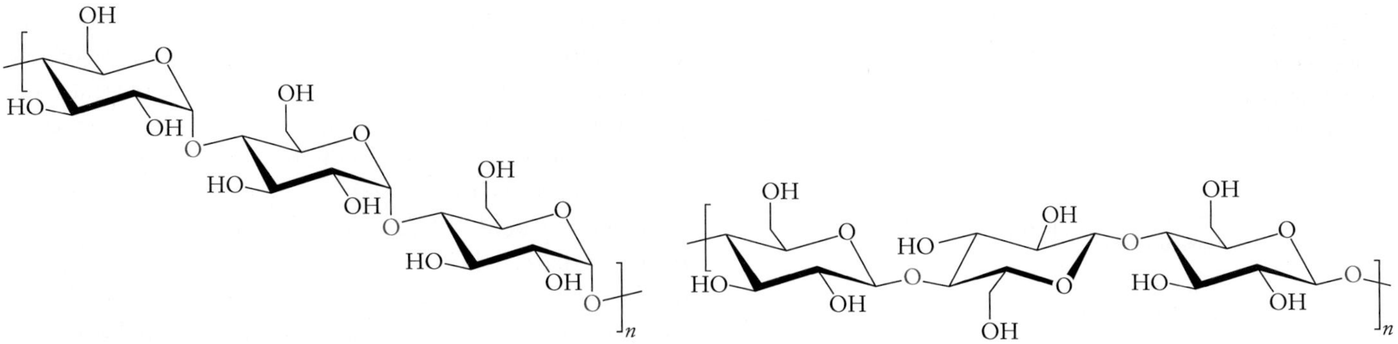

(a) Starch

(b) Cellulose

FIGURE 20.23 (a) Starch is a polysaccharide of α-glucose molecules joined by α-1,4 glycosidic bonds. (b) Cellulose is a polysaccharide of β-glucose molecules joined by β-1,4 glycosidic bonds.

This exothermic reaction provides energy to the yeast cells whose biological processes drive fermentation.

Unlike grazing animals, humans cannot digest cellulose because our digestive tracts do not contain microorganisms that have enzymes called cellulases, which catalyze hydrolysis of β-glycosidic bonds. The challenge of reproducing what cellulose-eating bacteria do in a laboratory or on an industrial scale is the focus of an enormous research effort as scientists try to develop efficient procedures for converting cellulose to glucose and then to ethanol. This research has focused on more efficient, less energy-intensive ways to break apart cellulose fibers. In addition, scientists are genetically engineering microorganisms like those in cattle stomachs to increase the supply of cellulases.

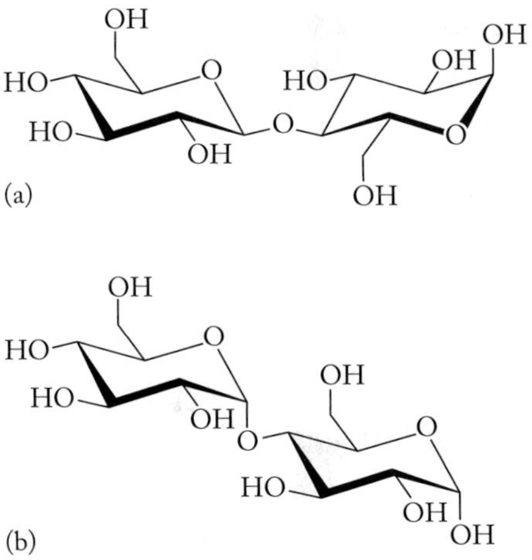

CONCEPT TEST

Cellobiose is a disaccharide made from the degradation of cellulose. We cannot digest cellobiose. Which of the two structures in Figure 20.24 represents a molecule of cellobiose?

(a)

(b)

FIGURE 20.24

If this research is successful, it will address several major problems associated with ethanol as a gasoline additive or alternative fuel. First of all, it will lower the cost of production. Today it costs more to produce ethanol from cornstarch than to produce gasoline from crude oil. One reason for this is that most of the mass of a corn plant, or any plant, is cellulose, not starch. Ethanol production from cornstarch is also energy intensive. More than 70% of the energy contained in ethanol is expended in producing it; therefore, the net energy value of ethanol from corn is less than 30%. If ethanol could be efficiently produced from cellulose instead of starch, its net energy value could be as high as 80%. Finally, the use of edible cornstarch in fuel production has driven up the price of foods derived from corn. The impacts of this inflation have been felt worldwide and have been particularly painful in developing countries. This competition with food has led to shifting research from the use of agricultural crops toward nonfood sources like cassava and sweet sorghum. However, the use of agricultural land and the consumption of increasingly scarce water resources for ethanol production raise further concerns.

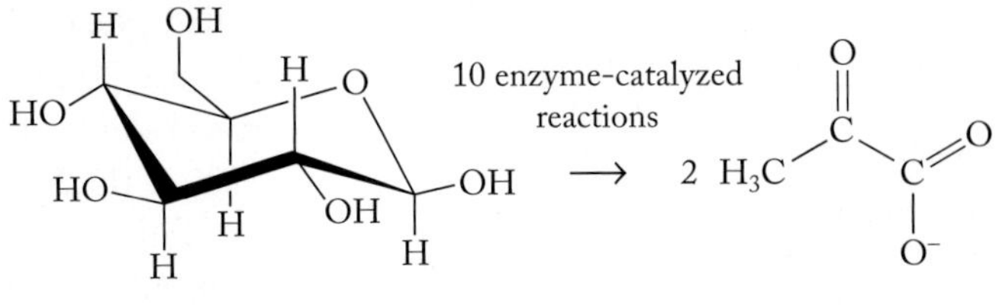

Glycolysis Revisited

Most cells, including those in human bodies, use glucose as a fuel in glycolysis (Figure 20.25), a series of reactions that oxidizes glucose to pyruvate ion, the conjugate base of pyruvic

FIGURE 20.25 During glycolysis, molecules of glucose are broken down to pyruvate ions.

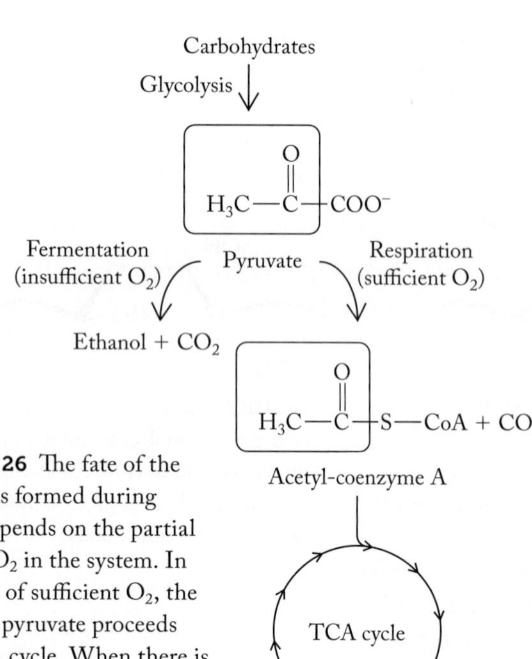

FIGURE 20.26 The fate of the pyruvate ions formed during glycolysis depends on the partial pressure of O_2 in the system. In the presence of sufficient O_2, the oxidation of pyruvate proceeds via the TCA cycle. When there is insufficient dissolved O_2 available, as in fermentation, the pyruvate may be converted to ethanol.

acid. Pyruvate sits at a metabolic crossroads and can be converted into different products depending on the type of cell in which it is generated, the enzymes that are present, and the availability of oxygen (Figure 20.26). In yeast cells growing under low-oxygen conditions, pyruvate is converted into ethanol and CO_2. Another series of reactions, called the **tricarboxylic acid (TCA) cycle**, occurs in the presence of sufficient dissolved O_2 and is fundamental to the conversion of glucose to energy in humans and other animals. A key step prior to the TCA cycle occurs when pyruvate loses CO_2 and forms an acetyl group, which then combines with coenzyme A. The resulting product, acetyl-coenzyme A, is a reactant in many biosynthetic pathways, including the production and breakdown of cholesterol.

Cholesterol is absolutely essential to maintain a healthy, functioning physiological system. It is a key component in the structure of cell membranes. It is also a precursor of the bile acids that aid in digestion and of steroid hormones, which regulate the development of the sex organs and secondary sexual traits, stimulate the biosynthesis of proteins, and regulate the balance of electrolytes in the kidneys. In a healthy person, the synthesis and use of cholesterol are tightly regulated to prevent overaccumulation and consequent deposition of cholesterol in coronary arteries. We need cholesterol, but persistent deposition in the arteries can lead to serious coronary disease (Figure 20.27). We will have more to say about this in the next section.

CONNECTION In Chapter 12 we introduced glycolysis in a discussion of the thermodynamics of coupled reactions.

Normal artery

Artery with cholesterol plaque

HO

Cholesterol

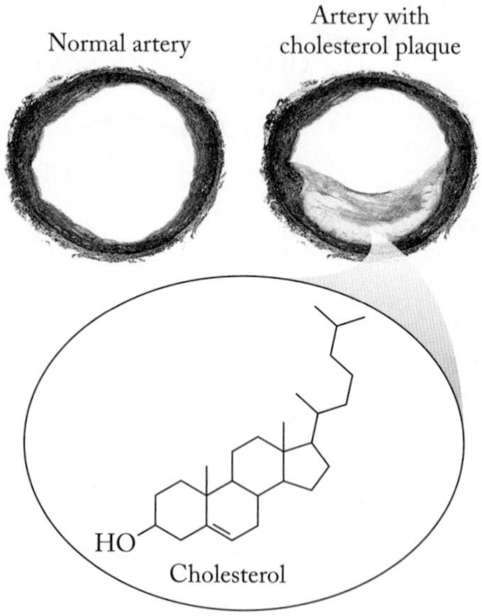

FIGURE 20.27 Cholesterol deposits are responsible for restricted blood flow, which results in a variety of sometimes catastrophic cardiovascular problems.

CONCEPT TEST

A newspaper article contained the wording, "Primarily made by the liver, cholesterol begins with tiny pieces of sugar..." What does this statement mean at the molecular level?

20.4 Lipids

Unlike carbohydrates and proteins, **lipids** are not biopolymers. Instead, lipids are best identified by a common physical property rather than a common structural subunit—they are all hydrophobic. As a result, lipids do not dissolve in water but are soluble in nonpolar solvents, and they are oily to the touch. Because they are insoluble in water, they are ideal components of cell membranes, which separate the aqueous solutions within cells from the aqueous environments outside them. This group of compounds includes fats, oils, waxes, compounds like cholesterol called sterols, and some other related compounds. An important class of lipids called **glycerides** are esters formed between glycerol and long-chain fatty acids (Figure 20.28). The three –OH groups on glycerol allow for mono-, di-, and triglycerides, with the latter being the most abundant. Glycerides account for over 98% of the lipids in the fatty tissues of mammals.

Table 20.2 lists some common fatty acids. The most abundant ones have an even number of carbon atoms because the fatty acids are built from two-carbon units. Their biosynthesis begins, as does the biosynthesis of cholesterol, with the conversion of pyruvate to acetyl-coenzyme A. Most fatty acids contain between 14 and 22 carbon atoms.

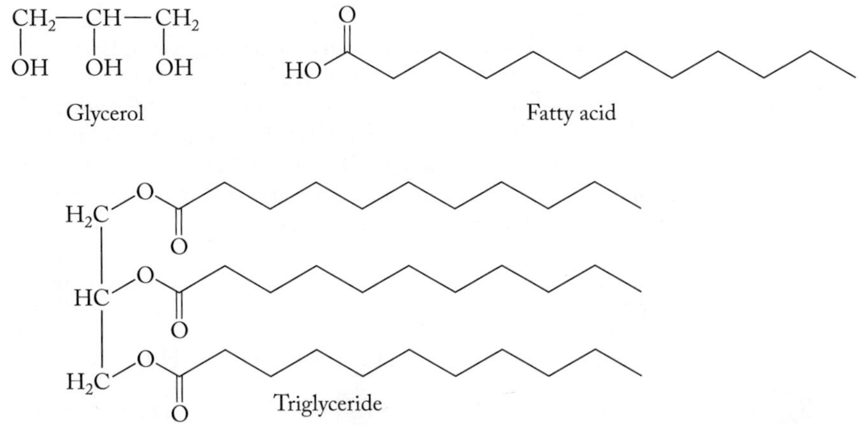

$$CH_2-CH-CH_2$$ with OH on each carbon — Glycerol

Fatty acid

Triglyceride

FIGURE 20.28 Glycerides are esters that form when glycerol combines with fatty acids. When all three –OH groups on glycerol react to form ester bonds, the product is a *tri*glyceride.

TABLE 20.2 Names, Formulas, and Structures of Common Fatty Acids

Common Name (chemical name) (source)	Formula
SATURATED FATTY ACIDS	
Lauric acid (dodecanoic acid) (coconut oil)	$CH_3(CH_2)_{10}COOH$
Myristic acid (tetradecanoic acid) (nutmeg butter)	$CH_3(CH_2)_{12}COOH$
Palmitic acid (hexadecanoic acid) (animal and vegetable fats)	$CH_3(CH_2)_{14}COOH$
Stearic acid (octadecanoic acid) (animal and vegetable fats)	$CH_3(CH_2)_{16}COOH$
UNSATURATED FATTY ACIDS	
Oleic acid (*cis*-9-octadecenoic acid) (animal and vegetable fats)	$CH_3(CH_2)_7CH{=}CH(CH_2)_7COOH$
Linoleic acid (*cis*,*cis*-9,12-octadecadienoic acid) (linseed oil, cottonseed oil)	$CH_3(CH_2)_4CH{=}CHCH_2CH{=}CH(CH_2)_7COOH$
Linolenic acid (*cis*,*cis*,*cis*-9,12,15-octadecatrienoic acid) (linseed oil)	$CH_3CH_2CH{=}CHCH_2CH{=}CHCH_2CH{=}CH(CH_2)_7COOH$

SAMPLE EXERCISE 20.3 Identifying Triglycerides **LO6**

How many different triglycerides (including structural isomers and stereoisomers) can be made from glycerol combining with two different fatty acids (symbolized by the letters X and Y to simplify their structures) if each molecule of triglyceride contains at least one molecule of each fatty acid?

Collect and Organize We are asked to determine how many different triglycerides can be made from glycerol and two different fatty acids. A triglyceride contains three fatty acid units.

tricarboxylic acid (TCA) cycle a series of reactions that continues the oxidation of the pyruvate formed in glycolysis.

lipid a class of hydrophobic, water-insoluble, oily organic compounds that are common structural materials in cells.

glyceride lipid consisting of esters formed between fatty acids and the alcohol glycerol.

Analyze Each fatty acid may bond to one of three –OH groups in glycerol. Each triglyceride has at least one X and one Y residue, which makes two combinations of X and Y possible: X_2Y and Y_2X. Each of these formulas has two structural isomers, depending on whether the single fatty acid in the formula is bonded to the middle carbon or to one of the end carbon atoms. Finally, if a structure has an X on one end carbon atom and a Y on the other, then the middle carbon atom is a chiral center, which means that there are two enantiomeric forms of that compound.

Solve We can generate four different molecular structures by attaching X and Y in four different sequences to the glycerol –OH groups:

(1)	(2)	(3)	(4)
$H_2C—O—X$	$H_2C—O—X$	$H_2C—O—Y$	$H_2C—O—Y$
$HC—O—X$	$HC—O—Y$	$HC—O—Y$	$HC—O—X$
$H_2C—O—Y$	$H_2C—O—X$	$H_2C—O—X$	$H_2C—O—Y$

Structures 1 and 3 have chiral centers (screened in yellow), so each of these isomers has two enantiomeric forms. Therefore, there is a total of six different triglycerides possible.

Think About It The central carbon atoms in structures 1 and 3 are chiral because they are each bonded to a H atom, an O atom, and to two C atoms that are themselves bonded to different fatty acids: X and Y.

Practice Exercise How many different triglycerides can be made from glycerol and one molecule each of three different fatty acids symbolized by the letters A, B, and C?

Function and Metabolism of Lipids

Lipids are an important energy source in our diets, providing more energy per gram than carbohydrates or proteins. As indicated in Table 20.2, fatty acids are either saturated, containing no carbon–carbon double bonds, or unsaturated, containing one or more carbon–carbon double bonds. **Fats** are glycerides composed primarily of saturated fatty acids. They are solids at room temperature. **Oils** are glycerides composed predominantly of unsaturated fatty acids and are liquids at room temperature. The consumption of too much saturated fat was once thought to be associated with coronary heart disease, but many recent studies indicate that there is no significant association between saturated fat intake and cardiovascular risk. Saturated fats were also thought to raise cholesterol levels in blood, but that link has also been questioned.

Olive oil, one of the main ingredients in the so-called Mediterranean diet, is a liquid composed of glycerides containing over 80% oleic acid, an unsaturated fatty acid. Oils can be converted into solid, saturated glycerides by hydrogenation. For example, in the hydrogenation of corn oil, which is a mixture of mostly two unsaturated fatty acids (oleic and linoleic acids), hydrogen is added to convert some or all of the –CH=CH– subunits into $–CH_2—CH_2–$ subunits. The resulting solid can be whipped with skim milk, coloring agents, and vitamins to produce the food spread we know as margarine.

A problem with hydrogenating vegetable oil arises when the oils are only partially hydrogenated. Partial hydrogenation alters the molecular structure around the remaining C=C bonds, changing them from their natural cis isomers into

fat a solid triglyceride containing primarily saturated fatty acids.

oil a liquid triglyceride containing primarily unsaturated fatty acids.

lipoprotein soluble proteins that combine with and transport fat or other lipids in the blood plasma.

trans isomers (Figure 20.29). Unsaturated trans fats like elaidic acid tend to be solids at room temperature because their molecules pack together more uniformly than do molecules of cis unsaturated fatty acids.

Consumption specifically of trans fatty acids (a.k.a. *trans fats*) is associated with increased levels of cholesterol in the blood and other health risks. Both saturated fats and *cis*-unsaturated fats occur naturally, and we have the necessary enzymes to metabolize these. A small number of trans fatty acids do appear in meat and dairy products, but the bulk of trans fats in our diets comes from hydrogenated fats in processed foods. There is no evidence that we produce the enzymes to metabolize trans fatty acids.

Cis and trans fats also play a role in the disposition of lipids, including cholesterol in our bodies. The discussion of the molecular-level relationships between proteins and cholesterol and fats are beyond the scope of this text, but you should be aware of the basic issues. Cholesterol is not soluble in blood, so it must be transported through the bloodstream by carriers called **lipoproteins**, which are made of fat (lipid) and soluble proteins. Two types of lipoproteins carry cholesterol to and from cells: low-density lipoprotein (LDL) and high-density lipoprotein (HDL). The main structural difference between LDL and HDL is their composition. Approximately 50% by weight of LDL particles is cholesterol and only 25% is protein. HDL particles consist of 20% cholesterol by weight and 50% protein. The remaining mass of both carriers is made up of other fats and water-insoluble molecules.

The main functional difference between the two is that they deliver cholesterol to different parts of the body. Low-density lipoproteins—the primary carriers of cholesterol—transport cholesterol to cells throughout the body where it is used to produce cell membranes and hormones. LDL particles can bind to artery walls. This sets off a cascade of events that may lead to the formation of atherosclerotic plaques (Figure 20.27). LDL is absolutely essential to health, but it is referred to as "bad" cholesterol because elevated levels of LDL are associated with a higher rate of atherosclerosis and increased risk for cardiovascular disease.

High-density lipoproteins are also absolutely essential to maintaining human health because they carry cholesterol in the direction opposite to LDL—that is, away from the heart and other organs, and deliver it back to the liver, where it is broken down and excreted. HDL cholesterol is considered "good" cholesterol because it helps remove LDL cholesterol from the arteries. Both HDL and LDL are essential to human health—it's the balance between the two that matters in determining "good" and "bad."

Cis and trans fats differ in their influence on the amounts of LDL and HDL in the blood. Cis fats promote HDL cholesterol, while trans fats increase LDL and decrease HDL. Consequently, trans fats are considered harmful to cardiovascular health. Trans fats that come from unnatural sources like hydrogenated oils in processed foods seem to be especially detrimental in terms of changing the HDL:LDL ratio. As a result, in 2013, the U.S. Food and Drug Administration (FDA) announced that it was requiring the food industry to completely phase out artificial trans fats.

The lipids in some prepared foods have been modified to reduce their caloric content but still provide the taste, aroma, and "mouth feel" we associate with lipid-rich foods. The active sites of enzymes that break down natural lipids accommodate triglycerides formed from glycerol and fatty acids. However, chemically modified esters made from the same fatty acids but attached to an alcohol other than glycerol cannot be metabolized by these enzymes. Such molecules, if they have the appropriate physical properties and are nontoxic, can be incorporated into foods without adding any calories because they are not metabolized.

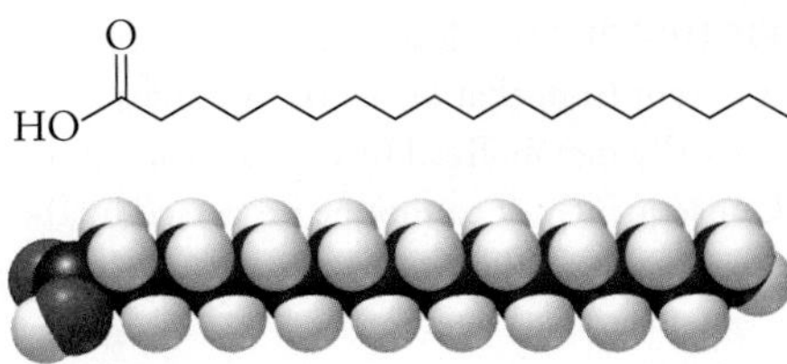

(a) Stearic acid: a saturated fatty acid

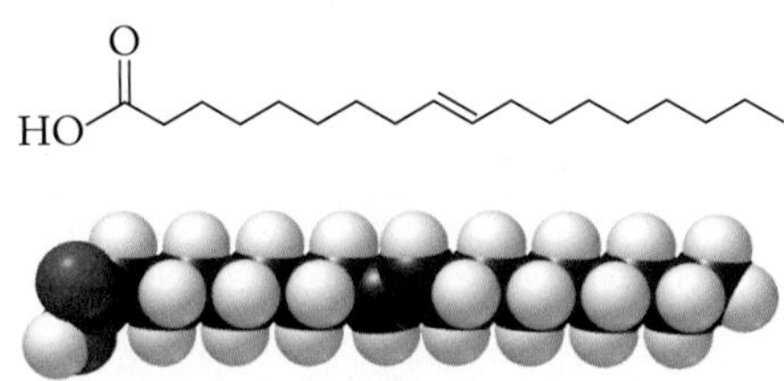

(b) Elaidic acid: a trans unsaturated fatty acid

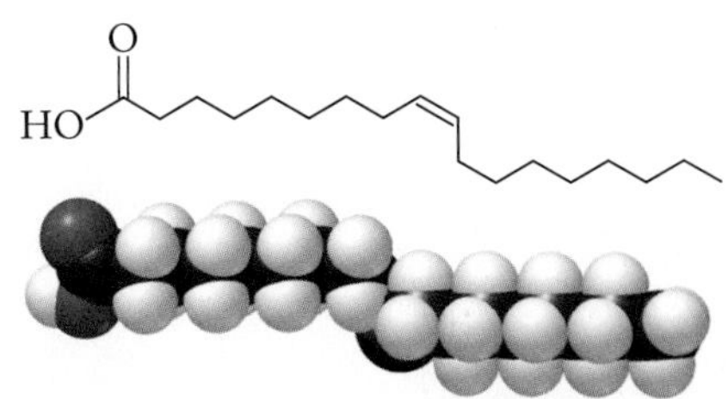

(c) Oleic acid: a cis unsaturated fatty acid

FIGURE 20.29 Types of fatty acids. (a) Stearic acid, a saturated C_{18} fatty acid. (b) Elaidic acid, an unsaturated C_{18} fatty acid (trans isomer). (c) Oleic acid, an unsaturated C_{18} fatty acid (cis isomer).

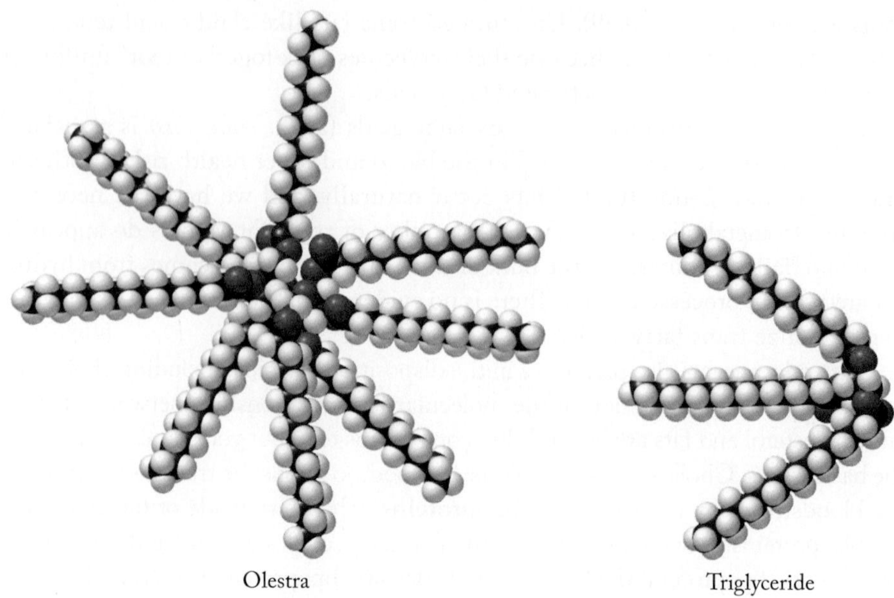

FIGURE 20.30 Olestra has a shape very different from that of the triglycerides typically metabolized by our bodies. Consequently, it cannot be processed by the enzymes that digest triglycerides.

Olestra is one such product (Figure 20.30). It is an ester made from long-chain fatty acids and the carbohydrate sucrose. (Remember, sugars have –OH groups and technically are alcohols.) Each of the eight –OH groups in a sucrose molecule reacts with a molecule of fatty acid to make the ester in olestra. The resultant material is used to deep-fry potato chips. Any olestra that remains on the chip does not add calories because it cannot be processed by enzymes that recognize only fatty acid esters on a glycerol scaffold. Olestra does, however, cause some side effects—namely abdominal cramping, nausea, diarrhea—in addition to decreasing the absorption of some vitamins and nutrients. These effects have decreased public acceptance of its use.

The enzymes that metabolize triglycerides hydrolyze the esters and release glycerol and the fatty acids bonded to it. Glycerol enters the metabolic pathway for glucose. The fatty acids are oxidized in a series of reactions known as β-oxidation, a process that removes two carbon atoms at a time. For example, stearic acid (the saturated C_{18} fatty acid) is transformed into a C_2 fragment and the C_{16} acid, palmitic acid. Palmitic acid yields another C_2 fragment and myristic acid, and so forth until the fatty acid is completely degraded. Electrons released from this oxidative process are eventually donated to O_2. The energy released by this process powers metabolism.

CONCEPT **TEST**

Olestra may be "calorie-free" as a food subject to metabolism in the human body, but how would it compare with a common triglyceride in terms of kilojoules of heat released per mole in a calorimeter experiment?

Other Types of Lipids

Cells contain other types of lipids in addition to triglycerides. One type, **phospholipids** (Figure 20.31a), plays a key role in cell structure. A phospholipid molecule consists of a glycerol molecule bonded to two fatty acid chains and to one phosphate group that is also bonded to a polar substituent. The presence of nonpolar fatty acid chains and a polar region in the same molecule

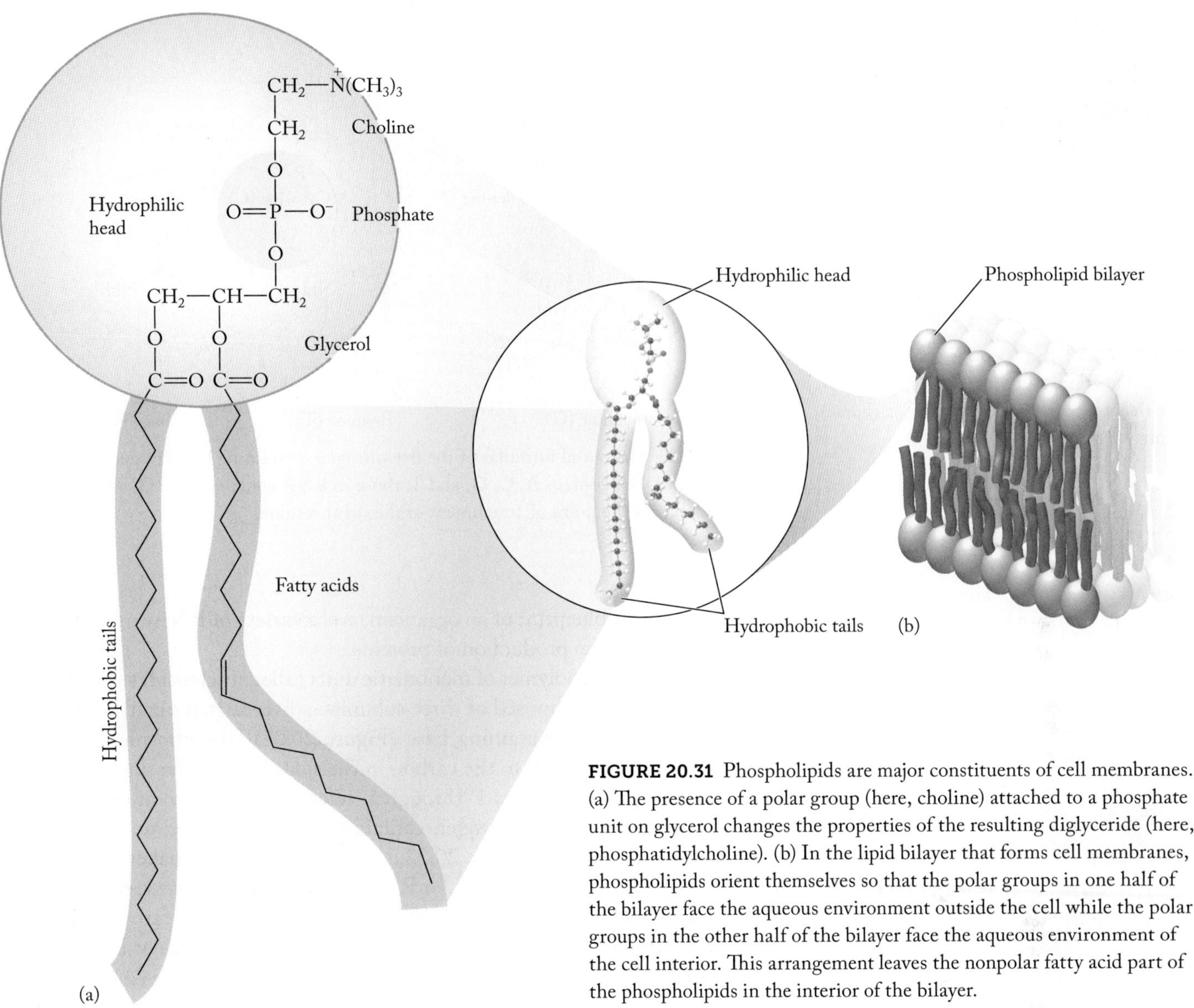

FIGURE 20.31 Phospholipids are major constituents of cell membranes. (a) The presence of a polar group (here, choline) attached to a phosphate unit on glycerol changes the properties of the resulting diglyceride (here, phosphatidylcholine). (b) In the lipid bilayer that forms cell membranes, phospholipids orient themselves so that the polar groups in one half of the bilayer face the aqueous environment outside the cell while the polar groups in the other half of the bilayer face the aqueous environment of the cell interior. This arrangement leaves the nonpolar fatty acid part of the phospholipids in the interior of the bilayer.

makes phospholipids ideal for forming cell membranes. In an aqueous medium, phospholipids form a **lipid bilayer**, a double layer enclosing each cell and isolating its interior from the outside environment. The phospholipid molecules of the bilayer align so that the nonpolar groups interact with each other within the membrane while the polar groups interact with water molecules outside the membrane (Figure 20.31b). Membranes exist both to isolate the contents of cells and to serve as the locus of communication between intracellular and extracellular processes.

phospholipid a molecule of glycerol with two fatty acid chains and one polar group containing a phosphate; phospholipids are major constituents of cell membranes.

lipid bilayer a double layer of molecules whose polar head groups interact with water molecules and whose nonpolar tails interact with each other.

20.5 Nucleotides and Nucleic Acids

nucleic acid one of a family of large molecules, which includes deoxyribonucleic acid (DNA) and ribonucleic acid (RNA), that contains the genetic blueprint of an organism and controls the production of proteins.

Nucleic acids are our fourth class of biomolecules and third class of biopolymers. We focus on two types: deoxyribonucleic acid (DNA) and ribonucleic acid (RNA). Even though nucleic acids make up only about 1% of an organism's mass, they code for the enzymes that control the metabolic activity of all its cells. DNA

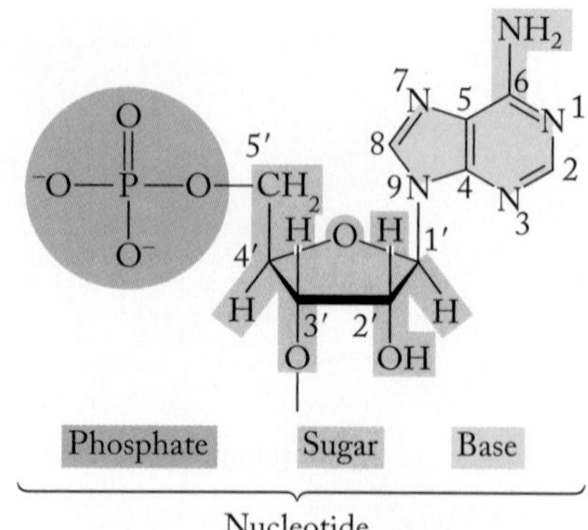

FIGURE 20.32 A nucleotide consists of a phosphate group and a nitrogen-containing base that are both bonded to a five-carbon sugar.

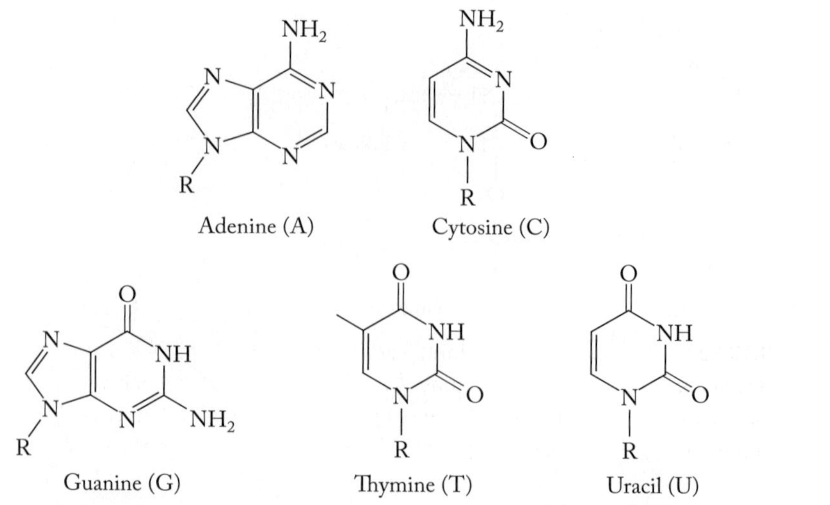

Adenine (A)
Cytosine (C)
Guanine (G)
Thymine (T)
Uracil (U)

FIGURE 20.33 Structural formulas of the five nitrogen-containing bases in nucleotides. The nucleotides in DNA contain A, C, G, and T; those in RNA contain A, C, G, and U. The R group identifies the point of attachment of the sugar residue.

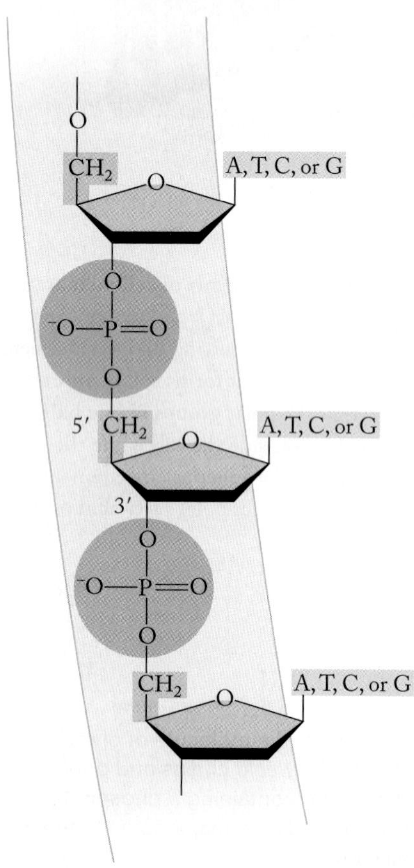

FIGURE 20.34 The backbone of the polymer chain in DNA consists of alternating sugar units (yellow) and phosphate units (pink). The bases (blue) are attached to the backbone through the 1′ carbon atom of the sugar unit.

nucleotide a monomer unit from which nucleic acids are made.

carries the genetic blueprint of an organism, and a variety of RNAs use that DNA blueprint to guide the production of proteins.

A nucleic acid is a polymer of monomeric units called **nucleotides**. Each nucleotide unit is in turn composed of three subunits: a five-carbon sugar, a phosphate group, and a nitrogen-containing base (Figure 20.32). The phosphate group in each nucleotide is attached to the carbon in the side chain of the sugar called the 5′ carbon atom. (The numbers 1′ through 5′ refer to the position of carbon atoms in the sugar molecule.) The nitrogen-containing base is attached to the sugar's 1′ carbon atom. The sugar in Figure 20.32 is called *ribose*, which makes this a nucleotide in a strand of *ribo*nucleic acid, or RNA. If the sugar were *deoxyribose* instead, there would be a H atom in place of the –OH group on the 2′ carbon atom, and the nucleotide would be a building block of *deoxyribo*nucleic acid, or DNA.

The nitrogen-containing base in Figure 20.32 is called adenine (A). Structural formulas of adenine and the other four bases in nucleic acids—cytosine (C), guanine (G), thymine (T), and uracil (U)—are shown in Figure 20.33. The point of attachment of the sugar–phosphate groups on each base is indicated by –R. In addition to the difference in their sugars, RNA and DNA also differ in one of the bases in their nucleotides: RNA contains A, C, G, and *U*, whereas DNA contains A, C, G, and *T*.

In a polymeric strand of nucleic acid, each phosphate is also linked to the 3′ carbon atom in the sugar of the monomer that precedes it in the chain, as shown for a strand of DNA in Figure 20.34. Both DNA and RNA strands are synthesized in the cell from the 5′ to the 3′ direction (downward in Figure 20.34). The structures of DNA and RNA are frequently written using only the single-letter labels of their bases, beginning with the free phosphate group on the 5′ end of the chain and reading toward the free 3′ hydroxyl group at the other terminus.

When scientists first isolated DNA and began to analyze its composition, they made a pivotal observation about the abundance of the nitrogen-containing bases. A typical molecule of DNA consists of thousands of nucleotides, and the percentages of the four bases in different samples of DNA can vary over a wide range. However, the percentage of A in a sample always matches the percentage of T. Likewise, the percentage of C always matches that of G. This result makes sense

(a) Adenine Thymine Guanine Cytosine

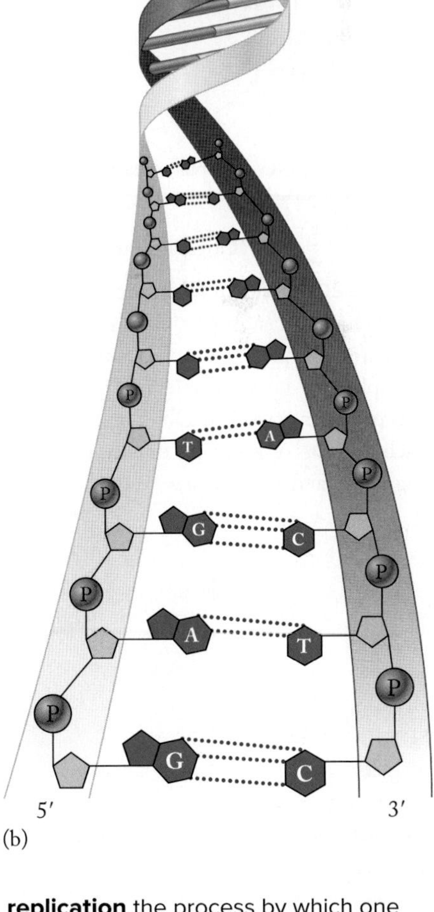

FIGURE 20.35 The nitrogen-containing bases on one strand of DNA pair with the bases on a second strand by hydrogen bonding. (a) Adenine and thymine pair via two hydrogen bonds; guanine and cytosine pair via three hydrogen bonds. (b) DNA as a double helix with the sugar–phosphate backbone on the outside and the base pairs on the inside.

if the bases are paired because a molecule of A can form two hydrogen bonds to a molecule of T, whereas a molecule of C can form three hydrogen bonds with a molecule of G. Therefore, A–T and G–C pairings maximize the number of hydrogen bonds possible (Figure 20.35a). More importantly, the base pairs assembled this way all have the same width and fit together in a regular structure.

The normal structure of DNA features two strands of nucleotides wrapped around each other in a form that is called a *double helix* (Figure 20.35b). The nucleotide backbone is on the outside of the spiraling strands, while hydrogen bonds between the complementary bases on the inside of the double helix keep the two strands together. The base pairs are parallel to each other and perpendicular to the helical axis. The fidelity of this base pairing—A always with T, and C always with G—gives DNA the ability to copy itself. If a pair of complementary strands is unzipped into two single strands, each strand provides a template on which a new complementary strand can be synthesized via the process called **replication** (Figure 20.36).

replication the process by which one double-stranded DNA forms two new DNA molecules, each one containing one strand from the original molecule and one new strand.

SAMPLE EXERCISE 20.4 Using Base Complementarity in DNA LO7

If 31.6% of the nucleotides in a sample of DNA are adenine, what are the percentages of cytosine, guanine, and thymine?

Collect and Organize We know how much adenine is in a DNA sample and need to calculate the remainder of the nucleotide composition. Nucleotides are paired: A always pairs with T, and C always pairs with G.

Analyze Because of base pairing, the percentage of T must equal the percentage of A, and the percentage of C must equal the percentage of G.

Solve If $A = 31.6\%$, then $T = 31.6\%$. This leaves $[100 - (2)(31.6)]\% = 36.8\%$ to be equally distributed between G and C. Therefore, $G = C = 36.8\%/2 = 18.4\%$.

Think About It The percentages should total 100%, and they do.

Practice Exercise Determine the sequence of the complementary strand on the double helix formed by each of the following sequences of nucleotides: (a) CGGTATCCGAT; (b) TTAAGCCGCTAG.

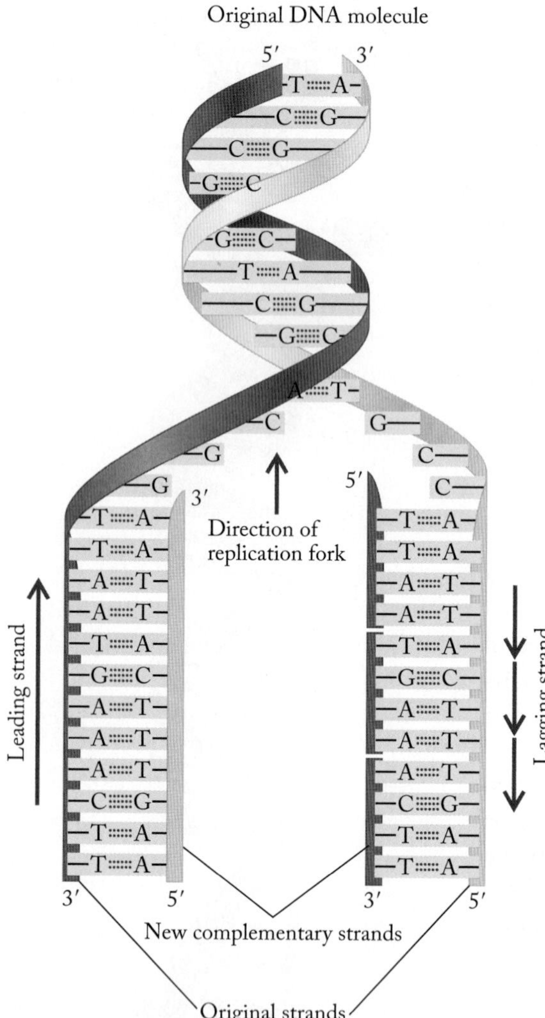

FIGURE 20.36 When DNA replicates, the two strands of a short portion of the double helix are unzipped, and each strand serves as a template for a new complementary strand, thus producing two new DNA molecules. Each new DNA molecule contains one original strand and one new strand.

During replication, the two strands are separated, forming a structure called the *replication fork*. The fork advances through the DNA as the process proceeds. One of the two resulting strands, the leading strand, is unzipped in the 3′ to 5′ direction, which allows the new complementary strand to be continuously synthesized in the 5′ to 3′ direction. The replication of the second strand, called the lagging strand, is more complicated. It proceeds in short fragments, which are then assembled into a continuous strand later in the process.

DNA's double-stranded structure is also the key to its ability to preserve genetic information. The two strands carry the same information, much like an old-fashioned photograph and its negative. Genetic information is duplicated every time a DNA molecule is replicated, a process that is essential whenever a cell divides into two new cells.

From DNA to New Proteins

Proteins are formed from amino acids in accordance with the *genetic code* contained in the base sequences of DNA strands. The bases A, T, G, and C are the alphabet in this code, and the "words" in the code are three-letter combinations of these four letters, with each word representing a particular amino acid or a signal to begin or end the protein synthesis process. Using four letters to write three-letter words means there are $4^3 = 64$ combinations possible, more than enough to encode the 20 amino acids found in cells. The genetic code specifies the protein's primary structure—the sequence of amino acids in proteins.

Protein synthesis begins with a process called **transcription** (Figure 20.37a), in which double-stranded DNA unwinds and its genetic information guides the synthesis of a single strand of a molecule called **messenger RNA (mRNA)**. This strand of mRNA has the complementary base sequence of the original DNA, although it does so using the bases A, C, G, and U, not A, C, G, and T:

DNA: . . . ACGTTGAGC . . .

mRNA: . . . UGCAACUCG . . .

The mRNA carries the three-letter words of the DNA, in the form of three-base sequences called **codons** (Table 20.3), from the nucleus of the cell into the cytoplasm, where the mRNA binds with a cellular structure called a ribosome.

At the ribosome the genetic information in the mRNA directs the synthesis of particular proteins in a process called **translation**. Another type of RNA, called **transfer RNA (tRNA)**, plays a key role in translation. There are 20 different forms of tRNA in the cell, one for each amino acid. Figure 20.37(b) shows how tRNA works. The first codon in this piece of an mRNA strand is AUG, which codes for the amino acid methionine (see Table 20.3). In the cytoplasm surrounding the ribosome, molecules of tRNA are reversibly bonded to molecules of every amino acid. The particular tRNA molecules that are bonded to methionine also contain the *anticodon* UAC, the complement of AUG, at a site that allows the tRNA to interact with mRNA. As Figure 20.37(b) shows, the segment of mRNA with the AUG codon links with the complementary anticodon on the tRNA molecule bonded to methionine. In doing so, the methionine is put into a position to unlink from the tRNA and to be the first amino acid residue in the protein being synthesized.

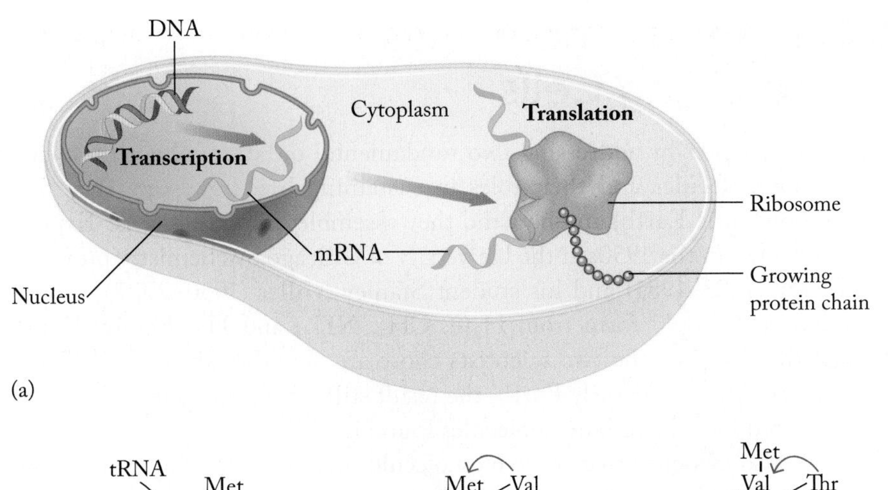

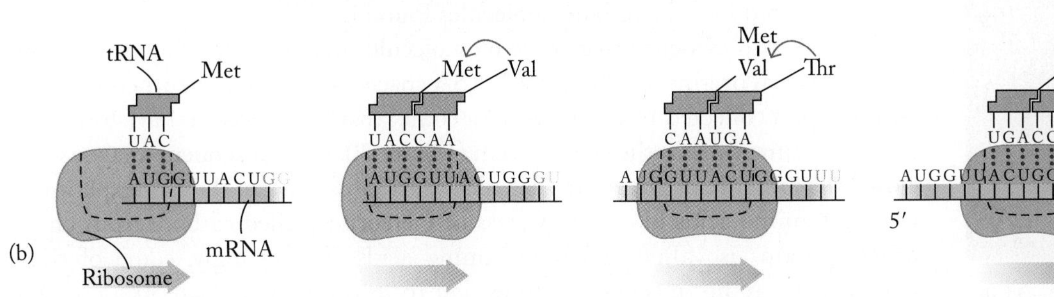

FIGURE 20.37 Transciption and translation. In protein synthesis, (a) DNA is *transcribed* into mRNA, and (b) the mRNA is then *translated* into proteins on the cell's ribosomes.

The sequence of events described in the preceding paragraph is repeated many times. In this illustration the next codon, GUU, links up with a molecule of tRNA that has a CAA binding site and a molecule of valine in tow. In this way valine moves into position to become the next amino acid residue in the protein and to form a peptide bond with the N-terminal methionine. Valine is followed by threonine, which is followed by glycine, and so on, until a "stop" codon signals the end of the translation process.

CONCEPT TEST

If a GUU codon attracts valine to the translation site, does a UUG codon do the same thing? Explain your answer.

TABLE 20.3 mRNA Codons

Amino Acid	Codon	Amino Acid	Codon
Ala	GCU, GCC, GCA, GCG	**Leu**	UUA, UUG, CUU, CUC, CUA, CUG
Arg	CGU, CGC, CGA, CGG, AGA, AGG	**Lys**	AAA, AAG
Asn	AAU, AAC	**Met**	AUG
Asp	GAU, GAC	**Phe**	UUU, UUC
Cys	UGU, UGC	**Pro**	CCU, CCC, CCA, CCG
Gln	CAA, CAG	**Ser**	UCU, UCC, UCA, UCG, AGU, AGC
Glu	GAA, GAG	**Thr**	ACU, ACC, ACA, ACG
Gly	GGU, GGC, GGA, GGG	**Trp**	UGG
His	CAU, CAC	**Tyr**	UAU, UAC
Ile	AUU, AUC, AUA	**Val**	GUU, GUC, GUA, GUG
Start	AUG	**Stop**	UAG, UGA, UAA

transcription the process of copying the information in DNA to mRNA.

messenger RNA (mRNA) the form of RNA that carries the code for synthesizing proteins from DNA to the site of protein synthesis in a cell.

codon a three-nucleotide sequence that codes for a specific amino acid.

translation the process of assembling proteins from the information encoded in mRNA.

transfer RNA (tRNA) the form of RNA that delivers amino acids, one at a time, to polypeptide chains being assembled by the ribosome–mRNA complex.

FIGURE 20.38 The apparatus used by Miller and Urey to simulate the synthesis of amino acids in the atmosphere of early (prebiotic) Earth. Discharges between the tungsten electrodes were meant to provide the sort of energy that might have come from lightning.

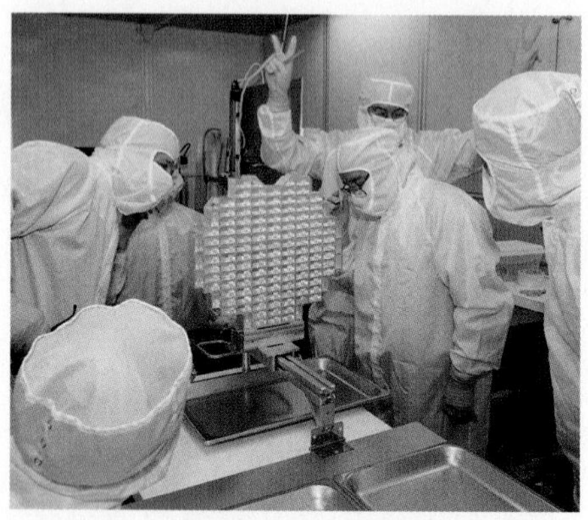

FIGURE 20.39 NASA scientists who analyzed the samples collected by the *Stardust* spacecraft were careful to avoid contaminating it with biological material from Earth. This meant isolating themselves from the sample as they prepared it for analysis.

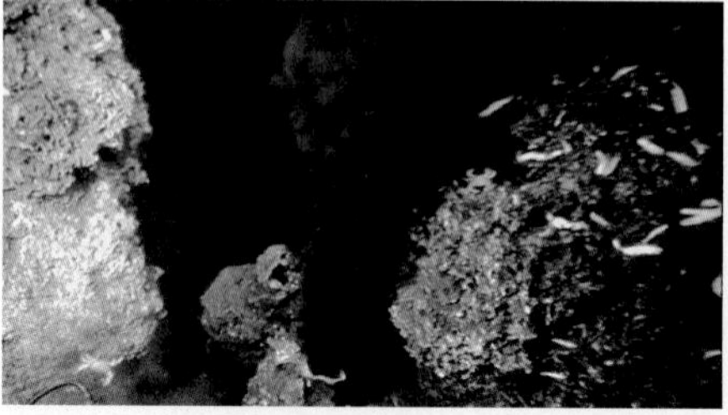

20.6 From Biomolecules to Living Cells

We end this chapter by addressing two fundamental questions about the major classes of biomolecules and their roles in sustaining life: how were they first formed on prebiotic Earth, and how did they assemble into living cells? Experiments conducted in the 1950s at the University of Chicago by chemistry professor Harold Urey (1893–1981) and his student Stanley Miller (1930–2007) showed that amino acids could form from H_2O, CH_4, NH_3, and H_2 (Figure 20.38). Although the reactants the two scientists chose are now thought to be different from those present on the early Earth, the result still stands: inorganic molecules can react to produce the organic molecules found in living systems.

There is also evidence that some biomolecules may have reached Earth from extraterrestrial origins. In 2006 the NASA spacecraft *Stardust* returned to Earth with samples collected from the tail of a comet that is believed to have formed at about the same time as the solar system (Figure 20.39). Subsequent analyses disclosed the presence of glycine in the comet dust. This was not the first experiment to detect amino acids in space. A class of meteorites called carbonaceous chondrites contains isovaline and other amino acids. Interestingly, some of these amino acids are not racemic mixtures, but instead are more than 50% L-amino acids, the enantiomeric form that dominates our biosphere. These observations have led some to suggest that L-amino acids are somehow "favored" and that life on Earth was "seeded" from elsewhere.

As we have seen in this chapter, RNA is needed to guide the assembly of amino acids into proteins. Since the 1990s, groups of scientists have explored the possibility that strands of RNA may have formed spontaneously from solutions of nucleotides in contact with mineral surfaces that guided the self-assembly of the nucleotides into long strands of RNA. Oligonucleotides made in this way might have been able to catalyze their own replication. This ability to self-replicate is crucial to life, and the observation that molecules are capable of speeding up their own replication on a clay surface suggests that processes essential to the formation of living cells could have happened spontaneously. Much controversy still exists about these ideas, but the capacity of RNA to both store information like DNA *and* act as a catalyst like an enzyme has motivated the *RNA world hypothesis*. This hypothesis proposes that a world filled with cellular or precellular life based on RNA alone predates the current world of life based on DNA and proteins.

Current research is also testing the theory that life on Earth may have evolved near deep-ocean hydrothermal vents. Entire ecosystems have been discovered at these locations since they were first explored in the 1970s. They are sustained by geothermal and chemical energy rather than energy from the sun. It may be that life actually began in such environments, with hydrothermal energy driving reactions in which inorganic compounds like carbon dioxide and hydrogen sulfide formed small organic compounds. As with the reactions on the surfaces of clay minerals, the synthesis reactions at hydrothermal vents may have been catalyzed and guided by reactants adsorbed on solid compounds such as FeS and MnO_2, which pour into the sea in dense black clouds near some vents (Figure 20.40). Among the known products of these reactions are acetate ions (CH_3COO^-).

FIGURE 20.40 Black clouds of transition metal oxides and sulfides flow into the sea through chimneys like this one at a deep-ocean hydrothermal vent. Some scientists believe that these particles may have guided and catalyzed the formation of the first self-replicating molecules on Earth.

Acetate is a key intermediate in many biosynthetic pathways in living organisms. In modern bacteria, the systems that make acetate depend on a catalyst made of iron, nickel, and sulfur that has a structure much like that of particles produced by "black smokers" on the ocean floor.

To take the next step toward forming living cells, large biomolecules must have assembled themselves into even larger structures, such as membranes, that allow cells and structures within them to collect materials and retain them at concentrations different from those in the surrounding medium. Molecules in these assemblies are not necessarily connected by covalent bonds, but rather are held together by the intermolecular interactions we have discussed in this chapter.

SAMPLE EXERCISE 20.5 Integrating Concepts: PKU Screening in Infants

Phenylketonuria (PKU) is a rare genetic disorder in which a baby is born without the ability to properly break down the amino acid phenylalanine (Phe). The gene for the liver enzyme phenylalanine hydroxylase, which is necessary for the metabolism of Phe to the amino acid tyrosine (Tyr) (Figure 20.41a), is mutated, and phenylalanine hydroxylase is absent or significantly decreased in activity. As a result, Phe accumulates, which hinders brain development, causing mental retardation. Eventually Phe is converted into phenylpyruvate (Figure 20.41b), which may be detected in the urine. A blood test may also be done to determine the amount of Phe present; normal levels are less than 2.0 mg/dL of blood. A different test using mass spectrometry (Figure 20.41c) can determine the ratio of Phe to Tyr; in a healthy infant the ratio should be less than 2.5 ([Phe]/[Tyr] < 2.5). Previous work has shown that IQ is impacted when Phe levels exceed 800 μmol/L.

a. Describe the molecular changes in Phe in terms of functional groups and charge when it is converted into Tyr and phenylpyruvate.
b. What should be the concentration of Tyr (in mg/dL) in a healthy infant if the concentration of Phe reaches 2 mg/dL?
c. Express the concentrations of Tyr and Phe from part b in mol/L.
d. Assuming the level of Tyr does not change from the value you calculated in part c, what is the [Phe]/[Tyr] ratio when [Phe] = 800 μmol/L?

Collect and Organize We are given the structural formulas for Phe, Tyr, and phenylpyruvate. We also know normal Phe levels in mg/dL, the normal [Phe]/[Tyr] ratio, and the level at which the amount of Phe damages mental abilities in μmol/L. We need to convert the numbers to one set of units to make the comparisons requested.

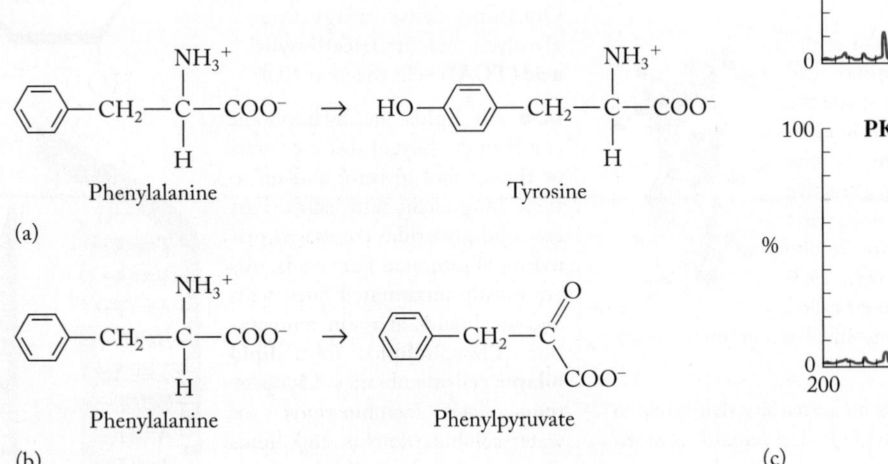

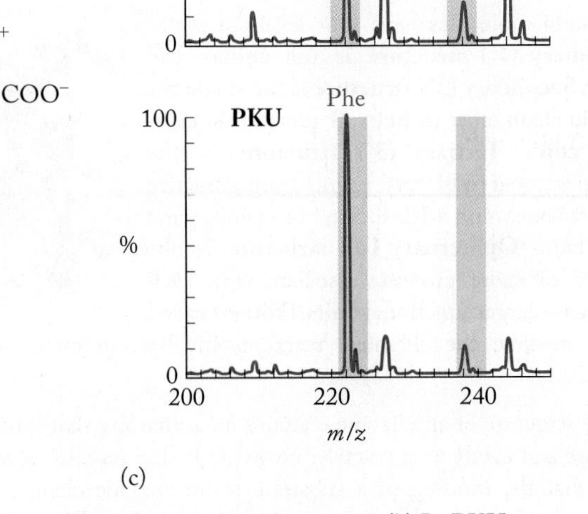

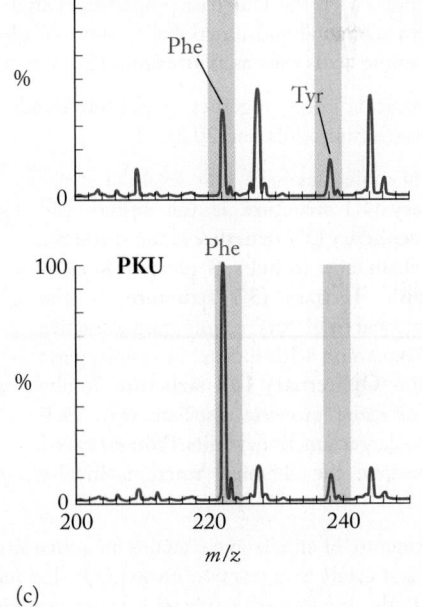

FIGURE 20.41 (a) The amino acid phenylalanine is normally converted into the amino acid tyrosine. (b) In PKU, phenylalanine is instead converted into phenylpyruvate. (c) Amino acid profiles from a mass spectrometer of normal blood (top) and blood from a newborn infant with PKU (bottom). Adapted from S. A. Banta and R. D. Steiner in *Journal of Perinatal and Neonatal Nursing*, 18(1), 41 (2004).

Analyze We can determine the molecular changes caused by metabolism from the structural formulas in Figure 20.41. We need to calculate the molecular masses of Phe and Tyr to convert between mg/dL and mol/L. The value of [Tyr] in part b will be $\frac{1}{2.5} \times 2.0$ mg/dL or a little less than 1 mg/dL. Based on their molecular structures, we can estimate that the molar masses of Tyr and Phe will be between 100 and 200 g/mol. Therefore, the molar concentrations in part c that are equivalent to 1 or 2 mg/dL (or 10–20 mg/L = 0.01–0.02 g/L) will be about (0.01 g/L)/(100 g/mol), or ~10^{-4} mol/L.

Solve

a. Phenylalanine is a neutral amino acid with a nonpolar R group. It contains an aromatic ring in addition to its amine and carboxylic acid functional groups. Phenylalanine hydroxylase converts Phe into Tyr, which has an –OH group on the aromatic ring and hence a polar R group. Tyr is also neutral. In the absence of phenylalanine hydroxylase, Phe is converted into phenylpyruvate, which is not an amino acid. Phenylpyruvate contains a ketone group and a carboxylic acid group that is ionized at physiological pH, so it is negatively charged.

b. If the concentration of Phe reaches 2 mg/dL, and if the Phe/Tyr ratio is to be <2.5, then the *minimum* concentration of Tyr (x) in a healthy infant would be

$$\frac{2.0 \text{ mg/dL}}{x} = 2.5, \qquad x = 0.80 \text{ mg/dL}$$

c. Using the structural formulas given, Phe is $C_9H_{11}NO_2$ with a molecular mass of 165.19 g/mol, and Tyr is $C_9H_{11}NO_3$ with a molecular mass of 181.19 g/mol. In units of mol/L, the concentrations of Phe and Tyr from part b are

$$[\text{Phe}] = \frac{2.0 \cancel{\text{mg}}}{\cancel{\text{dL}}} \times \frac{1 \cancel{\text{g}}}{1000 \cancel{\text{mg}}} \times \frac{1 \text{ mol}}{165.19 \cancel{\text{g}}} \times \frac{10 \cancel{\text{dL}}}{1 \text{ L}}$$
$$= 1.2 \times 10^{-4}\,M$$

$$[\text{Tyr}] = \frac{0.80 \cancel{\text{mg}}}{\cancel{\text{dL}}} \times \frac{1 \cancel{\text{g}}}{1000 \cancel{\text{mg}}} \times \frac{1 \text{ mol}}{181.19 \cancel{\text{g}}} \times \frac{10 \cancel{\text{dL}}}{1 \text{ L}}$$
$$= 4.4 \times 10^{-5}\,M$$

d. Converting [Phe] = 800 μmol/L into molarity:

$$800 \frac{\cancel{\mu\text{mol}}}{\text{L}} \times \frac{1 \text{ mol}}{10^6 \cancel{\mu\text{mol}}} = 8.0 \times 10^{-4}\,M$$

If we assume [Tyr] = $4.4 \times 10^{-5}\,M$, then the [Phe]/[Tyr] ratio is

$$\frac{8.0 \times 10^{-4} \cancel{M}}{4.4 \times 10^{-5} \cancel{M}} = 18$$

Think About It Mandatory screening programs identify infants with PKU at birth. Early diagnosis is essential for the successful management of this disorder. Infants diagnosed with PKU require a lifelong diet that is extremely low in Phe. This diet is challenging to maintain because Phe occurs in significant amounts in many common foods, including milk and eggs.

SUMMARY

LO1 In 19 of the 20 **α-amino** acids in human proteins, the α-carbon atom is a chiral center. The acid–base properties of amino acids are also important for structural and functional reasons. At physiological pH (7.4), most amino acids exist as **zwitterions**. (Section 20.1)

LO2 Amino acids linked together by **peptide bonds** form peptides and proteins. (Sections 20.1 and 20.2)

LO3 Protein molecules have four levels of structure. **Primary (1°) structure** is the amino acid sequence. **Secondary (2°) structure** is the shape the amino acid chain takes (**α helix, β-pleated sheet,** or **random coil**). **Tertiary (3°) structure** is the three-dimensional shape that results from attractive forces between amino acids located in various parts of the chain. **Quaternary (4°) structure** results when two or more proteins associate with each other to make larger functional units. Proteins called **enzymes** mediate the chemical reactions involved in metabolism. (Section 20.2)

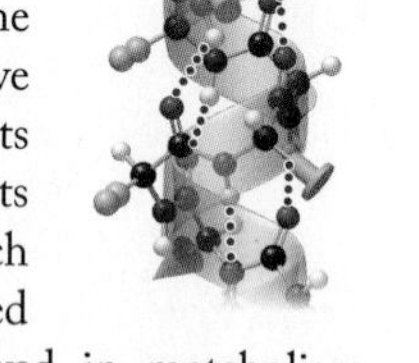

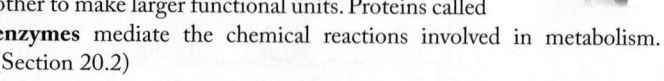

LO4 The structure of an enzyme contains an **active site** that binds to a **substrate** and catalyzes a reaction involving it. The *induced-fit model* suggests that the binding of a substrate to an enzyme changes the shape of the enzyme so that the reaction can take place. **Biocatalysis** seeks to replicate the selectivity of enzymes in industrial settings and is important in green chemistry processes. (Section 20.2)

LO5 Monosaccharides are joined into **polysaccharides** through **glycosidic bonds**. The isomers α-glucose and β-glucose form the polysaccharides starch and cellulose, respectively. Organisms derive energy from glycolysis and the **tricarboxylic acid (TCA) cycle**. (Section 20.3)

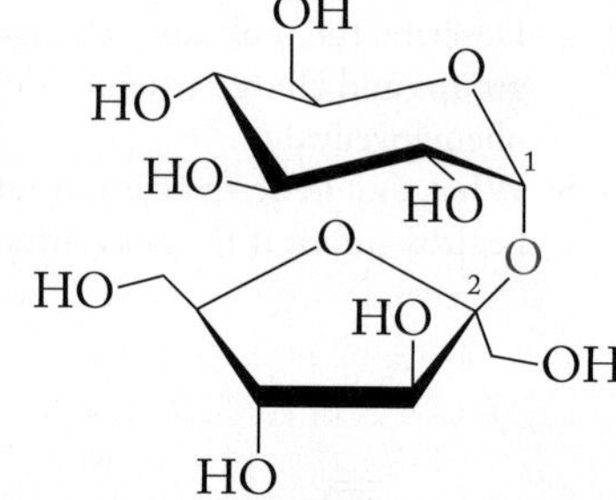

LO6 All **lipids** are hydrophobic compounds. **Glycerides** are esters of the alcohol glycerol and up to three long-chain fatty acids. **Fats** are solid glycerides composed primarily of saturated fatty acids; **oils** are mostly unsaturated fatty acids and are liquids at room temperature. **Phospholipids** form **lipid bilayer** cell membranes. **Lipoproteins** are combinations of water-soluble proteins and lipids that transport fats and other lipids in the bloodstream. (Section 20.4)

LO7 DNA and RNA are made of chains of **nucleotides**. DNA consists of two nucleotide chains

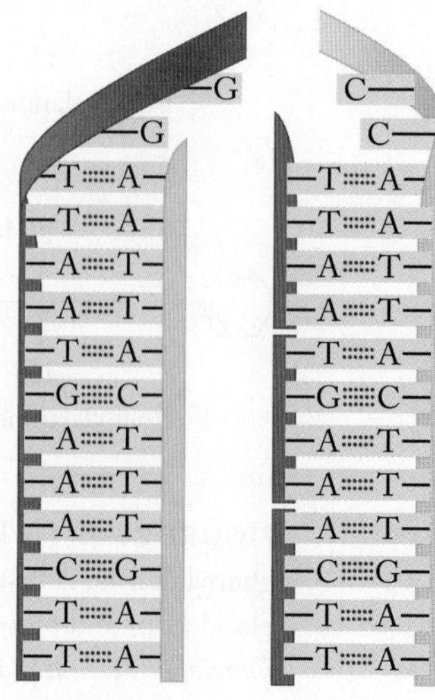

coiled into a double helix and connected via hydrogen bonds between complementary bases—A with T and C with G. RNA, in which the base pairings are A with U and C with G, is a single-stranded nucleic acid involved in protein synthesis through **messenger RNA (mRNA)** and **transfer RNA (tRNA)**. DNA and RNA contain an organism's genetic information and control protein synthesis through **transcription** and **translation**. Genetic information is transmitted to new cells during cell division in a process called **replication**. (Section 20.5)

PARTICULATE **PREVIEW WRAP-UP**

The two functional groups are a carboxylic acid and a primary amine. At pH < 3, both will be protonated: –COOH and $-NH_3^+$. At pH > 11 the carboxylic acid will be ionized: $-COO^-$, and the amine group will not be protonated: $-NH_2$.

PROBLEM-SOLVING SUMMARY

Type of Problem	Concepts and Equations	Sample Exercises
Interpreting acid–base titration curves of amino acids	At low pH all amino acids have at least two ionizable H atoms, one each from –COOH and $-NH_3^+$. Side-chain carboxylic acid and amine groups may also impart acidic and basic strength to amino acids.	**20.1**
Drawing and naming peptides	Connect the α-amine of one amino acid to the α-carboxylic acid of another with a peptide bond. Starting with the free amine (N-terminus) on the left, name each amino acid residue by changing the ending of the name of the parent amino acid to *-yl* in all but the last (C-terminal) amino acid.	**20.2**
Identifying triglycerides	The –COOH groups of fatty acids react with the –OH groups in glycerol to form triglycerides and water.	**20.3**
Using base complementarity in DNA	Identify the base pairs: A pairs with T; G pairs with C. The percentage of T should equal the percentage of A, and the percentage of C should equal the percentage of G.	**20.4**

VISUAL PROBLEMS

(Answers to boldface end-of-chapter questions and problems are in the back of the book.)

20.1. Figure P20.1 shows how a molecule of glucose changes from an α-glucose to a β-glucose conformation. Draw an energy diagram that shows how the energy of the molecule (or a mole of them) changes during the transformation from α-glucose to β-glucose.

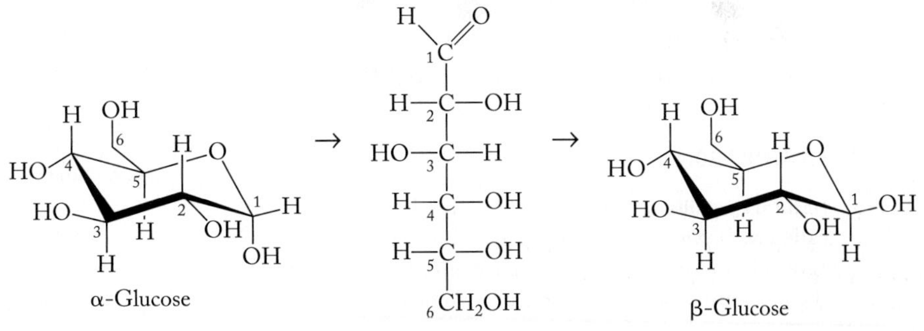

α-Glucose → → β-Glucose

FIGURE P20.1

20.2. The nucleotides in DNA contain the bases with the structures shown in Figure P20.2. How many primary and secondary amine groups are in each structure?

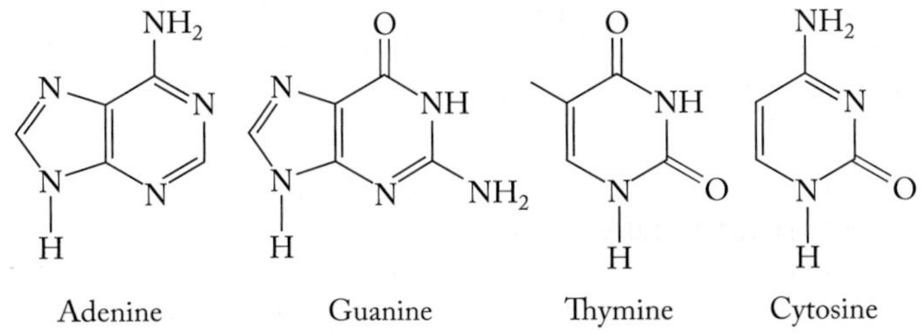

Adenine Guanine Thymine Cytosine

FIGURE P20.2

20.3. Olive Oil Olive oil contains triglycerides such as those shown in Figure P20.3.

a. Which of the fatty acids in these triglycerides is/are saturated?

b. Are the unsaturated fatty acids likely to be cis or trans isomers?

(a)
H_2C–O–C(=O)–$(CH_2)_{14}CH_3$
HC–O–C(=O)–$(CH_2)_7CH{=}CH(CH_2)_7CH_3$
H_2C–O–C(=O)–$(CH_2)_7CH{=}CHCH_2CH{=}CH(CH_2)_4CH_3$

(b)
H_2C–O–C(=O)–$(CH_2)_7CH{=}CH(CH_2)_7CH_3$
HC–O–C(=O)–$(CH_2)_{16}CH_3$
H_2C–O–C(=O)–$(CH_2)_7CH{=}CH(CH_2)_7CH_3$

FIGURE P20.3

20.4. The two compounds shown in Figure P20.4 are both amino acids. Which of them is an α-amino acid? Explain your choice.

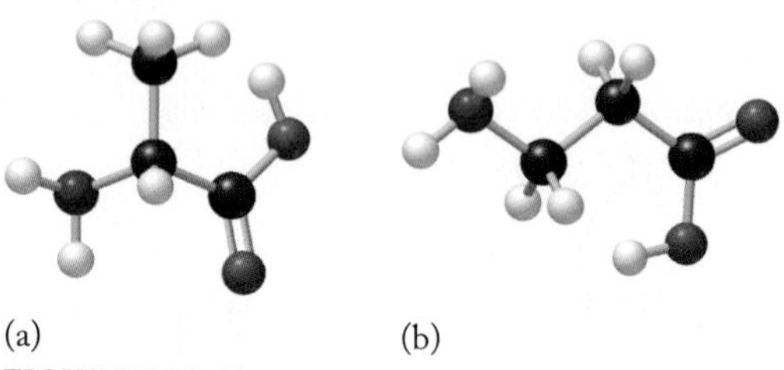

(a) (b)

FIGURE P20.4

20.5. Natural Painkillers The human brain produces polypeptides called *endorphins* that help in controlling pain. The endorphin in Figure P20.5 is called enkephalin. Identify the five amino acids that make up enkephalin.

Enkephalin

FIGURE P20.5

20.6. **Regulating Blood Pressure** Angiotensin II is a polypeptide that regulates blood pressure. Which amino acids make up the structure of angiotensin II shown in Figure P20.6?

Angiotensin II

FIGURE P20.6

20.7. Trans Fats The role of "trans fats" in human health has been extensively debated both in the scientific community and in the popular press. Which of the molecules in Figure P20.7 are trans fats?

(a) R =
(b) R =
(c) R =

FIGURE P20.7

20.8. **Cocoa Butter** Cocoa butter (Figure P20.8) is a key ingredient in chocolate. Cocoa butter is a triglyceride that results from the esterification of glycerol with three different fatty acids. Identify the fatty acids produced by the hydrolysis of cocoa butter.

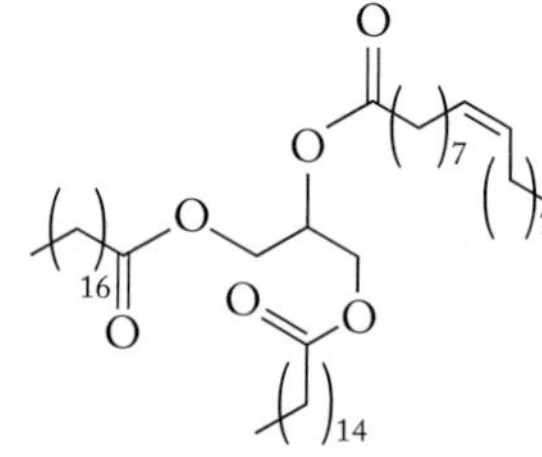

Cocoa butter

FIGURE P20.8

20.9. Sucralose The molecular structure of the artificial sweetener sucralose (trade name Splenda) is shown in Figure P20.9. Advertising for this product claims that it is made from sugar, implying that it is a natural product. What sugar might it be made from? Comment on the implication that it is a "natural" product.

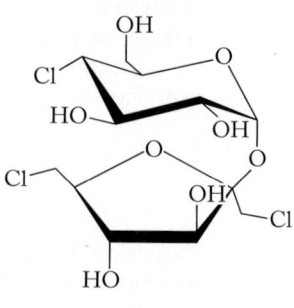

Sucralose

FIGURE P20.9

20.10. Use representations [A] through [I] in Figure P20.10 to answer questions a–f.

a. [A] depicts two strands of DNA in a double helix. Are the two strands held together by covalent bonds, intermolecular forces, or both?
[B] depicts two alpha helices linked together. Are the two helices held together by covalent bonds, intermolecular forces, or both?

b. [C] depicts a phospholipid bilayer. Is the bilayer held together by covalent bonds, intermolecular forces, or both?

c. [D] depicts a disaccharide. Are the two sugars held together by covalent bonds, intermolecular forces, or both?

d. Which of the single molecules depicted in the matrix are an important structural component of [C]?

e. Which of the single molecules depicted in the matrix react to form [E]?

f. When the amino acid depicted in [F] is incorporated into a protein, what intermolecular forces are involved?

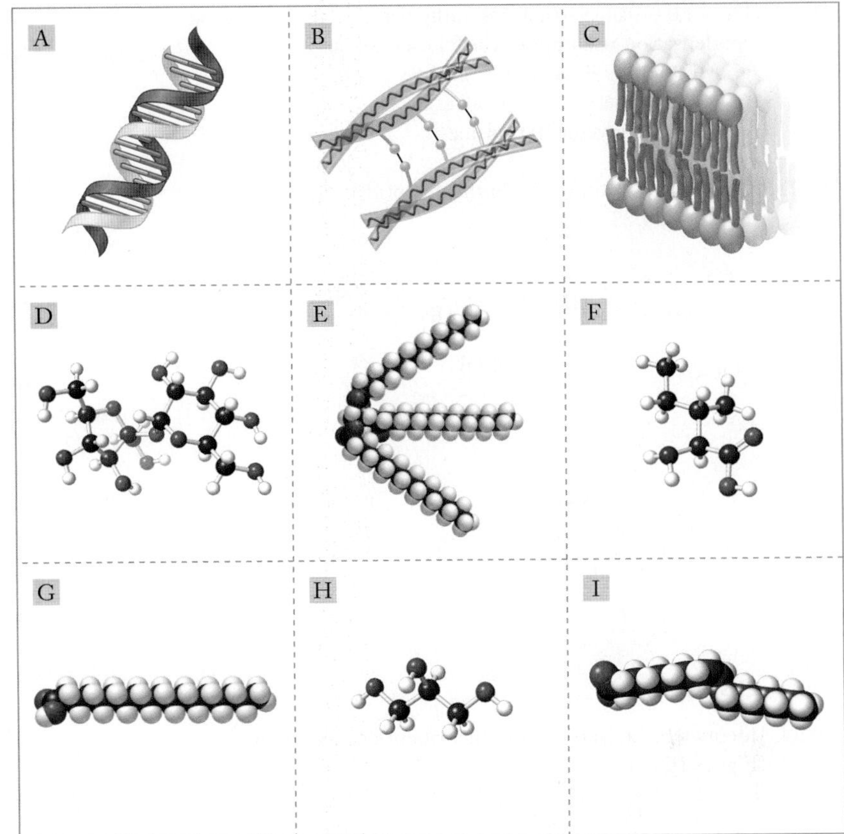

FIGURE P20.10

QUESTIONS AND PROBLEMS

The Composition of Proteins

Concept Review

20.11. In living cells, amino acids combine to make peptides and proteins. Are these processes accompanied by increases or decreases in the entropy of the reaction system?

20.12. What is the difference between a peptide bond and an amide bond?

20.13. What does the alpha mean in α-amino acid?

20.14. In 1806 French scientists were the first to isolate an amino acid. The source was asparagus shoots. The compound forms an anion in neutral aqueous solutions. Can you identify it?

20.15. Meteorites contain more L-amino acids, which are the forms that make up the proteins in our bodies, than D-amino acids. What do the prefixes L- and D- mean?

20.16. Do any of the amino acids in Table 20.1 have more than one chiral carbon atom per molecule?

20.17. Which of the compounds in Figure P20.17 is/are *not* an α-amino acid(s)?

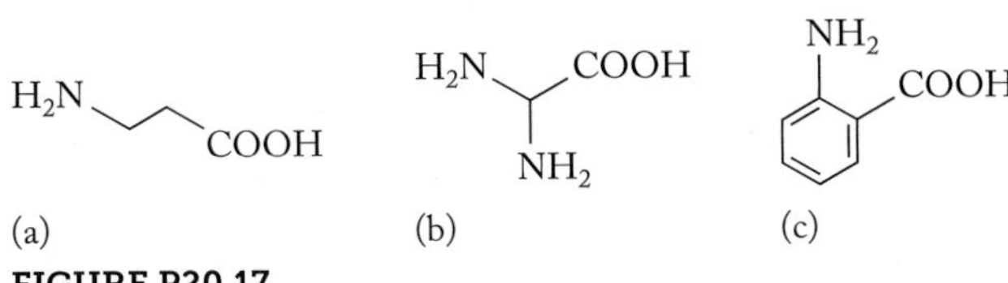

FIGURE P20.17

20.18. Which of the compounds in Figure P20.18 are α-amino acids?

O
NH_2
N–H
COOH
N
OH
O

(a) (b) (c)

FIGURE P20.18

20.19. Why do most amino acids exist in the zwitterionic form at physiological pH (pH ≈ 7.4)?

20.20. A simple organic acid with no functional groups other than an alkyl chain and a carboxylic acid typically has a pK_a between 4 and 5. The pK_a values of amino acids are all between 1.7 and 2.4. Why are amino acids more acidic than simple carboxylic acids?

Problems

20.21. Draw all possible structures of the peptides produced from condensation reactions of the following L-amino acids:

a. Alanine + serine
b. Alanine + phenylalanine
c. Alanine + valine

20.22. Draw all possible structures of the peptides produced from condensation reactions of the following L-amino acids:
a. Methionine + alanine + glycine
b. Methionine + valine + alanine
c. Serine + glycine + tyrosine

20.23. Identify the amino acids in the dipeptides shown in Figure P20.23.

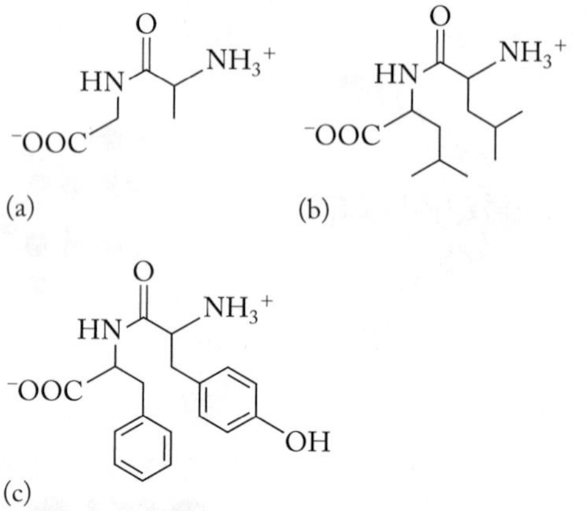

(a) (b)

(c)

FIGURE P20.23

20.24. Identify the amino acids in the tripeptides shown in Figure P20.24.

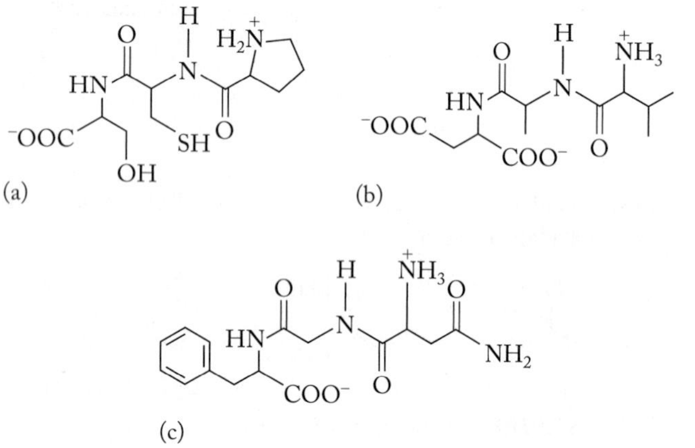

(a) (b)

(c)

FIGURE P20.24

20.25. Identify the missing product in the metabolic reaction shown in Figure P20.25.

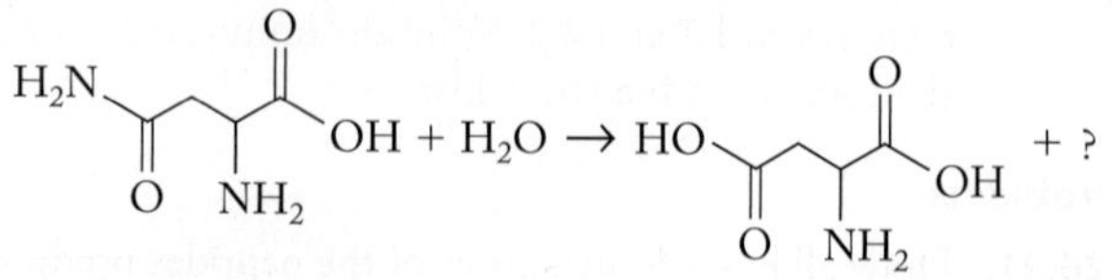

FIGURE P20.25

20.26. Identify the missing product in the metabolic reaction shown in Figure P20.26.

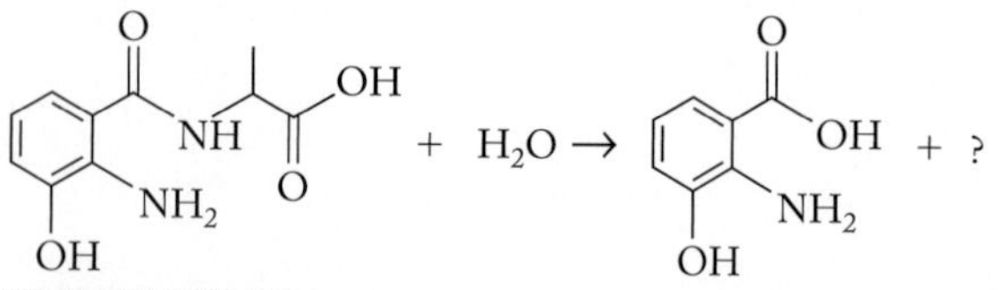

FIGURE P20.26

Protein Structure and Function

Concept Review

20.27. Which of the four levels of protein structure is most closely associated with the sequence of amino acids in a protein?

20.28. Which type of intermolecular interaction plays the dominant role in holding strands of proteins together in β-pleated sheets and stabilizing α helices?

20.29. Which level of protein structure is associated with ion–ion interactions and disulfide bond formation?

20.30. **Hard-Boiled Eggs** The protein in egg whites is ovalbumin. When an egg is hard-boiled, which is *least* affected in ovalbumin: its primary, secondary, tertiary, or quaternary structure? Explain your answer.

20.31. Describe the *induced-fit* theory of enzyme activity.

20.32. What will happen if an enzyme is added to a solution in which the substrate and product are in equilibrium?

20.33. Describe the role that molecular structure plays in the specificity of enzyme activity.

20.34. The rates of most reactions increase with increasing temperature, but above a critical temperature the rate of an enzyme-mediated reaction decreases with increasing temperature. Explain why.

20.35. When protein strands fold back onto themselves in forming stable tertiary structures, lysine residues are often paired up with glutamic acid residues. Why?

20.36. When the α-helical region in a protein unfolds, breaking the hydrogen bonds between the amino acids requires approximately 2 kJ/mol of energy. Does this mean that the hydrogen bonds in proteins are weaker or stronger than the hydrogen bonds in water? Explain your answer.

Carbohydrates

Concept Review

20.37. What are the structural differences between starch and cellulose?

20.38. Do cellulose fibers resemble proteins in α-helical, β-pleated sheet, or globular structure?

20.39. Is the fuel value of glucose in the linear form the same as that in the cyclic form?

*20.40. Without doing the actual calculation, estimate the fuel values of glucose and starch by considering average bond energies. Do you predict the fuel values of the two substances to be the same or different?

20.41. Describe in your own words the function of carbohydrates in the diet.

20.42. Do polysaccharides have a quaternary structure? Explain your answer.

20.43. How do we calculate the overall free-energy change of a process consisting of two steps?

20.44. During glycolysis, a monosaccharide is converted to pyruvate. Do you think this process produces an increase or decrease in the entropy of the system? Explain your answer.

Problems

20.45. Draw a diagram similar to Figure 20.19 that shows how the linear molecule in Figure P20.45 forms a six-atom ring.

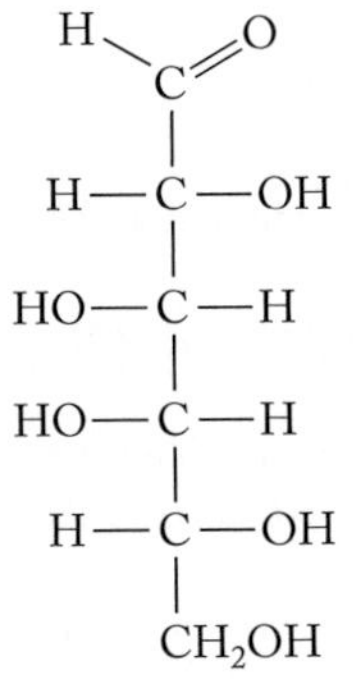

H—C=O
H—C—OH
HO—C—H
HO—C—H
H—C—OH
CH_2OH

Galactose

FIGURE P20.45

20.46. Draw a diagram similar to Figure 20.21 that shows how the linear molecule in Figure P20.46 forms a five-atom ring.

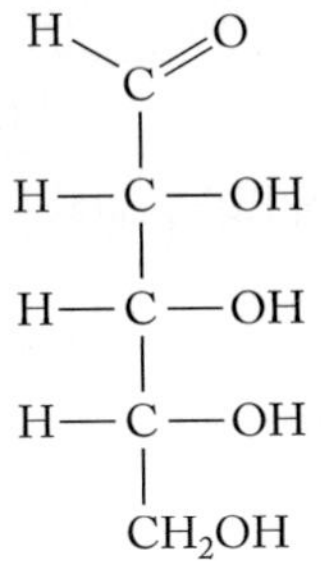

H—C=O
H—C—OH
H—C—OH
H—C—OH
CH_2OH

Ribose

FIGURE P20.46

20.47. Which of the structures in Figure P20.47, if any, are β isomers of a monosaccharide?

(a) (b) (c)

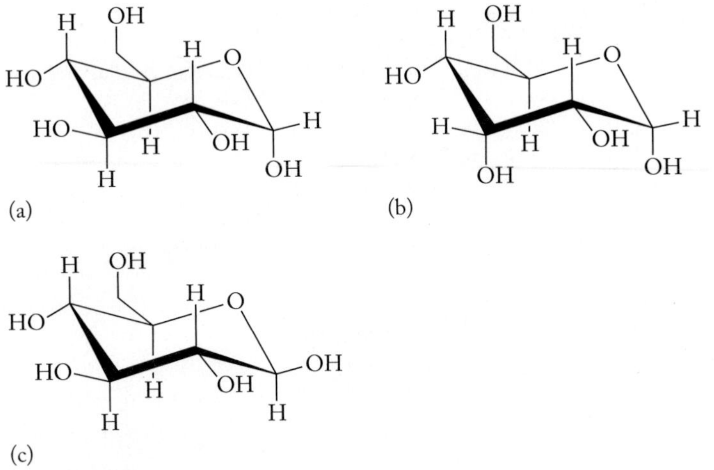

FIGURE P20.47

20.48. Which of the structures in Figure P20.48, if any, are β isomers of a monosaccharide?

(a) (b) (c)

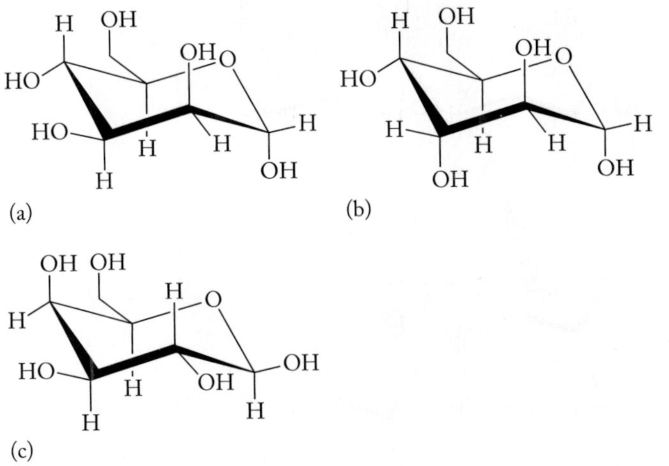

FIGURE P20.48

20.49. Which of the structures in Figure P20.49, if any, are α isomers of a monosaccharide?

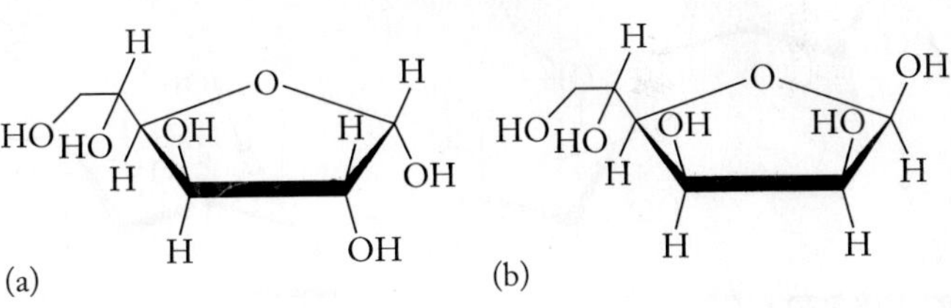

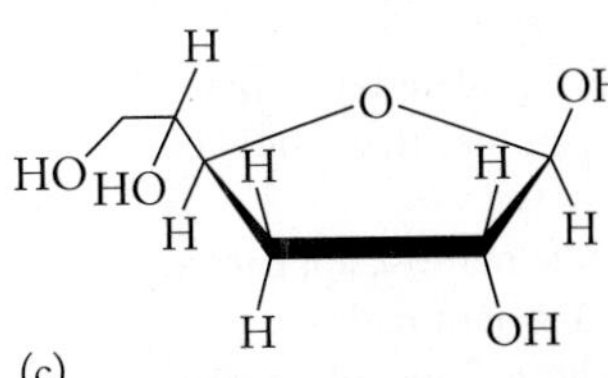

FIGURE P20.49

20.50. Which of the structures in Figure P20.50, if any, are α isomers of a monosaccharide?

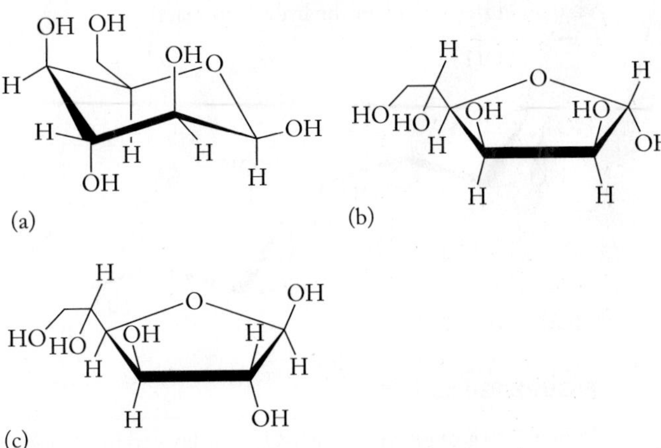

FIGURE P20.50

20.51. Which of the saccharides in Figure P20.51 is digestible by humans?

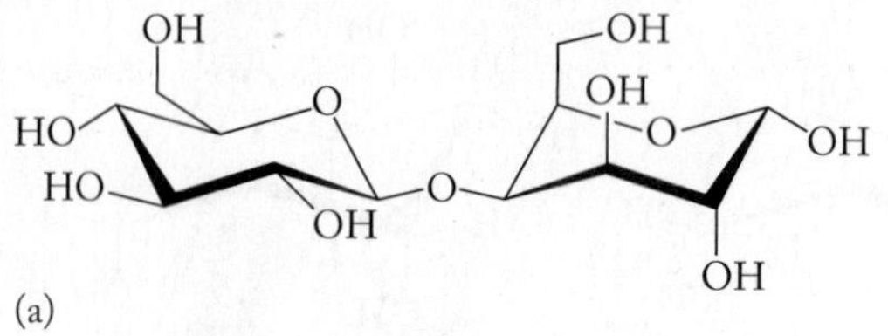

(a)

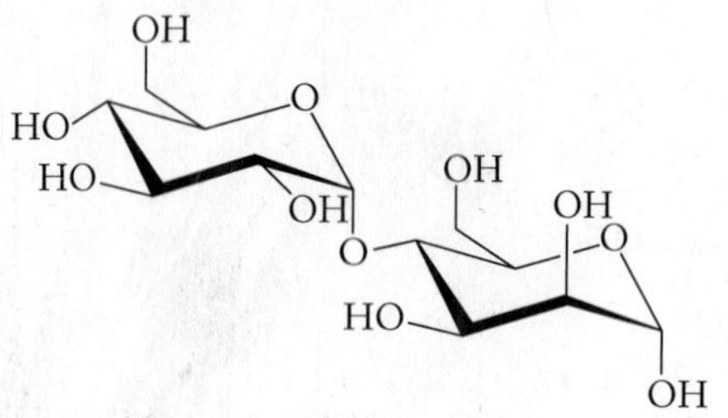

(b)

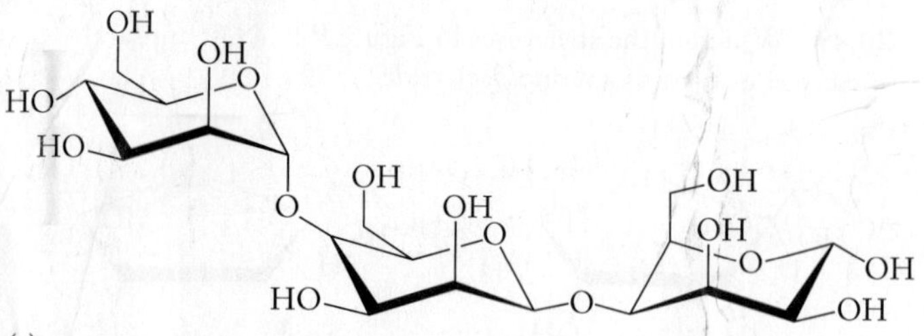

(c)

FIGURE P20.51

*20.52. For any of the disaccharides in Problem 20.51 that are *not* digestible by humans, draw an isomer that would be.

20.53. The structure of the disaccharide maltose appears in Figure P20.53. Hydrolysis of 1 mol of maltose ($\Delta G_f^\circ = -2246.6$ kJ/mol) produces 2 mol of glucose ($\Delta G_f^\circ = -1274.4$ kJ/mol):

$$\text{Maltose} + H_2O \rightarrow 2 \text{ glucose}$$

If the value of ΔG_f° of water is −285.8 kJ/mol, what is the change in free energy of the hydrolysis reaction?

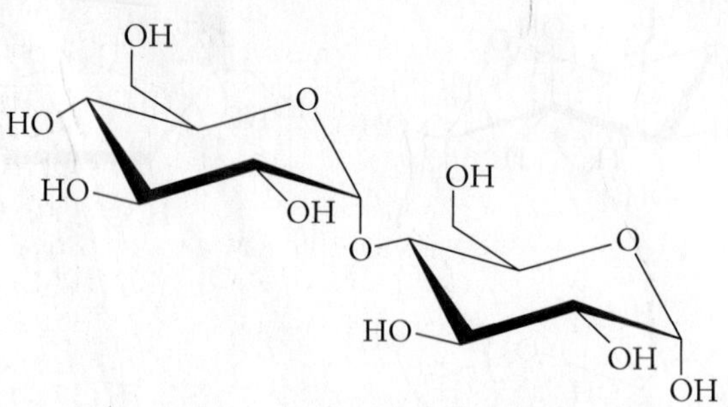

Maltose

FIGURE P20.53

20.54. When 5.00 g of glucose ($C_6H_{12}O_6$) was burned in a bomb calorimeter, the temperature of 1.750 kg of water was raised from 19°C to 39°C. The specific heat of water is 4.184 J/g · °C.

a. Calculate the fuel value (energy) of the glucose.
b. Fructose ($C_6H_{12}O_6$) has the same energy value as glucose when measured in a calorimeter, but dietarily, 1 oz of fructose provides 104 Calories of energy, while 1 oz of glucose provides 110 Calories. Explain why these values differ.

Lipids

Concept Review

20.55. What is the difference between a saturated and an unsaturated fatty acid?

20.56. Why are the average fuel values of fats higher than those of carbohydrates and proteins?

20.57. **Polar Exploration** Some Arctic explorers have eaten sticks of butter on their explorations. Give a nutritional reason for this unusual cuisine.

20.58. **Salad Dressing** Salad dressings containing oil and vinegar quickly separate on standing. Explain the observed separation of layers based on the structure and properties of aqueous vinegar and oil.

20.59. Do triglycerides have a chiral center? Explain your answer.

*20.60. Using your knowledge of molecular geometry and intermolecular forces, explain why polyunsaturated triglycerides are more likely to be liquids and not solids than saturated triglycerides?

Problems

20.61. Which of the triglycerides in Figure P20.61 are unsaturated fats?

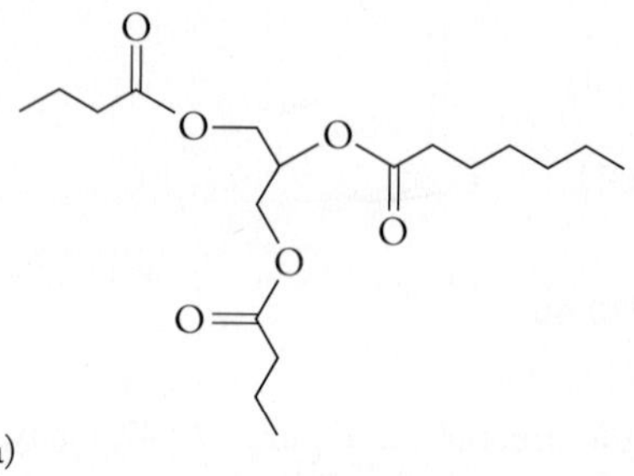

(a)

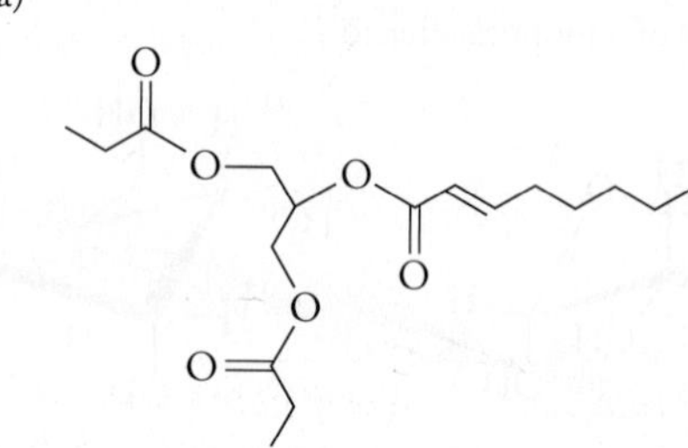

(b)

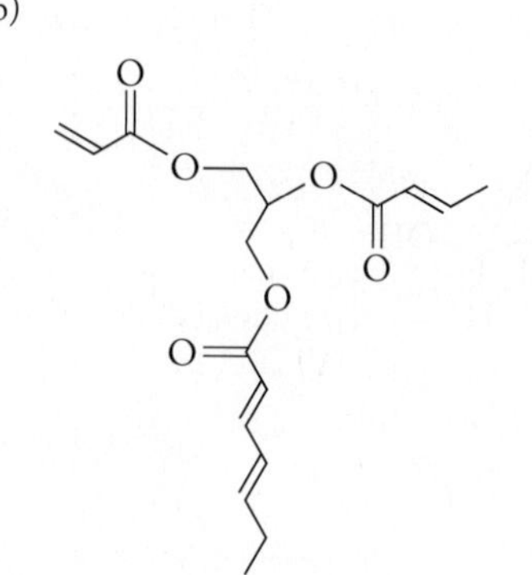

(c)

FIGURE P20.61

20.62. For each of the pairs of fatty acids in Figure P20.62, indicate whether they are structural isomers, stereoisomers, or unrelated compounds.

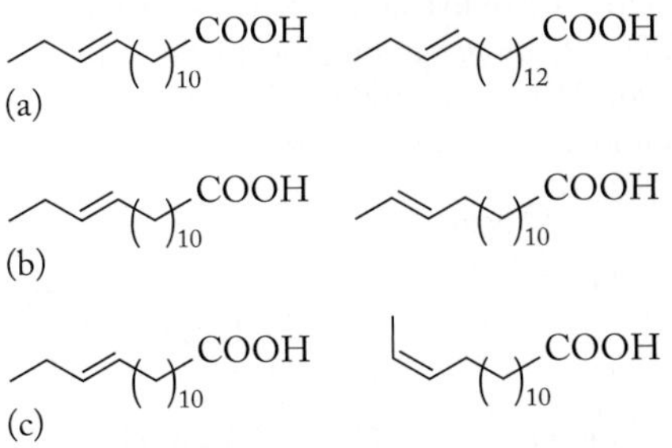

(a)

(b)

(c)

FIGURE P20.62

20.63. Draw the structures of the three fats formed by the reaction of glycerol with (a) octanoic acid ($C_7H_{15}COOH$), (b) decanoic acid ($C_9H_{19}COOH$), and (c) dodecanoic acid ($C_{11}H_{23}COOH$).

20.64. **Oil-Based Paints** Oil-based paints contain linseed oil, a triglyceride formed by the esterification of glycerol with linolenic acid (Figure P20.64).

a. Draw the carbon-skeleton structure of linolenic acid.

*b. Is the substitution around the double bonds in linolenic acid likely to be cis or trans? Why?

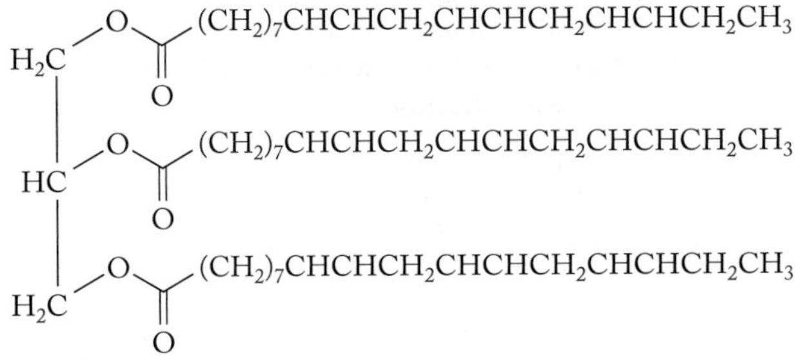

Linseed oil

FIGURE P20.64

Nucleotides and Nucleic Acids

Concept Review

20.65. What are the three kinds of molecular subunits in DNA? Which two form the "backbone" of DNA strands?

20.66. Why does a codon consist of a sequence of three, and not two, ribonucleotides?

20.67. What kind of intermolecular force holds together the strands of DNA in the double-helix configuration?

*20.68. Because of base pairing of the two strands in DNA, the percentage of T must equal the percentage of A, and the percentage of C must equal the percentage of G. In contrast, in RNA there is no relationship between the quantities of the four bases. What does this fact suggest about the structure of RNA?

Problems

20.69. Draw the structure of adenosine 5′-monophosphate, one of the four ribonucleotides in a strand of RNA.

20.70. Draw the structure of deoxythymidine 5′-monophosphate, one of the four nucleotides in a strand of DNA.

20.71. During the replication of DNA, a segment of an original strand has the sequence T-C-G-G-T-A. What is the sequence of the double-stranded helix formed in replication?

20.72. During transcription, a segment of the strand of DNA that is transcribed has the sequence T-C-G-G-T-A. What is the corresponding sequence of nucleotides on the messenger RNA that is produced in transcription?

Additional Problems

20.73. Olestra Olestra is a calorie-free fat substitute. The core of the olestra molecule (Figure P20.73) is a disaccharide that has reacted with a carboxylic acid, which results in the conversion of hydroxyl groups on the disaccharide into the depicted structure.

a. What is the name of the disaccharide core of the olestra molecule?

b. What functional group has replaced the hydroxyl groups on the disaccharide?

c. What is the formula of the carboxylic acid used to make olestra?

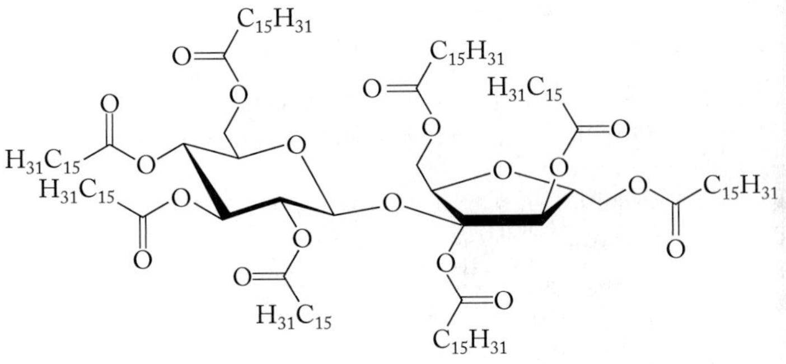

Olestra

FIGURE P20.73

20.74. **Amino Acids in Comets** When scientists at UC Santa Cruz directed UV radiation at an ice crystal containing methanol, ammonia, and hydrogen cyanide, three amino acids (glycine, alanine, and serine) were detected among the products of photochemical reactions. The formation of these amino acids suggests that they may also be synthesized in comets approaching the sun (and Earth). Determine the standard free-energy change of the hypothetical formation of glycine in comets, using standard free energies of formation of the reactants and products in this reaction [ΔG°_f for HCN(g) is +125 kJ/mol; ΔG°_f for solid glycine is −368.4 kJ/mol; other ΔG°_f values are in Appendix 4].

$$CH_3OH(\ell) + HCN(g) + H_2O(\ell) \rightarrow H_2NCH_2COOH(s) + H_2(g)$$

20.75. Homocysteine (Figure P20.75) is formed during the metabolism of amino acids. A mutation in some people's genes leads to high concentrations of homocysteine in the blood and a consequent increase in their risk of heart disease and their incidence of bone fractures in old age.

a. What is the structural difference between homocysteine and cysteine?

b. Cysteine is a chiral compound. Is homocysteine chiral?

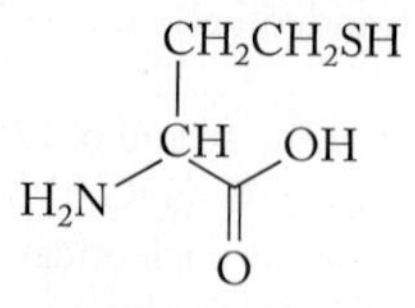

Homocysteine

FIGURE P20.75

20.76. **Molecules in Meteorites** Some scientists believe life on Earth can be traced to amino acids and other molecules brought to Earth by comets and meteorites. In 2004, a new class of amino acids called diamino acids (Figure P20.76) was found in the Murchison meteorite.

a. Which of these diamino acids is not an α-amino acid?

b. Which of these amino acids is chiral?

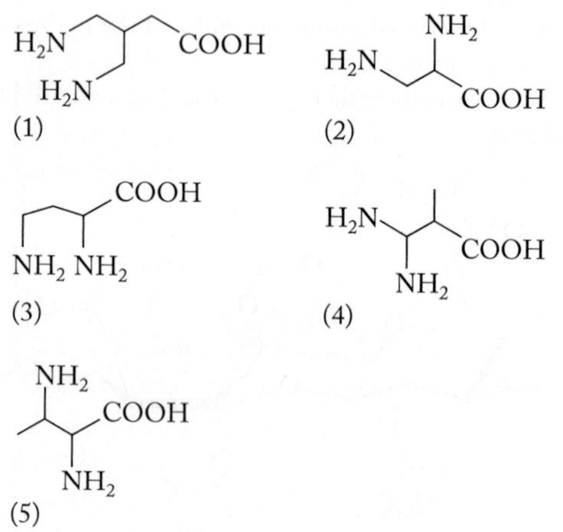

(1) (2) (3) (4) (5)

FIGURE P20.76

20.77. **Ackee** Ackee, the national fruit of Jamaica, is a staple in many Jamaican diets. Unfortunately, a potentially fatal sickness known as Jamaican vomiting disease is caused by the consumption of unripe ackee fruit, which contains the amino acid hypoglycin (Figure P20.77). Is hypoglycin an α-amino acid?

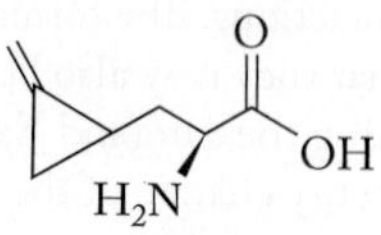

Hypoglycin

FIGURE P20.77

20.78. **E. coli** In late 2003, researchers at The Scripps Research Institute reported the development of genetically modified *E. coli* that could incorporate five new amino acids into proteins. These five amino acids, shown in Figure P20.78, are *not* among the 20 naturally occurring amino acids. Which naturally occurring amino acids are these compounds most similar to?

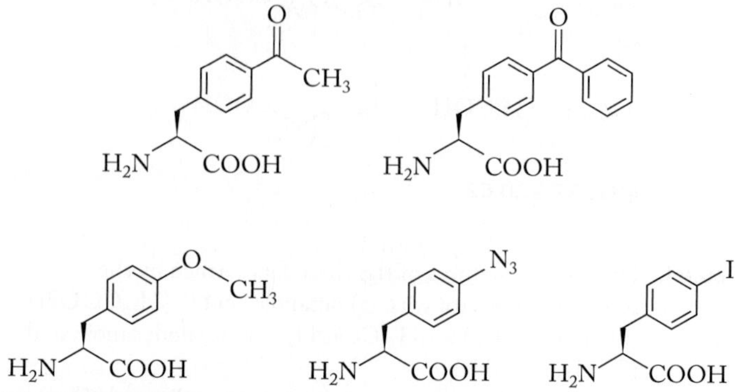

FIGURE P20.78

20.79. **Creatine** Creatine (Figure P20.79) is an amino acid produced by the human body. Body builders sometimes take creatine supplements to help gain muscle strength. A 2003 study reported that creatine may boost memory and cognitive thinking.

a. Is creatine an α-amino acid?

b. Draw the two dipeptides that can be formed from glycine and creatine.

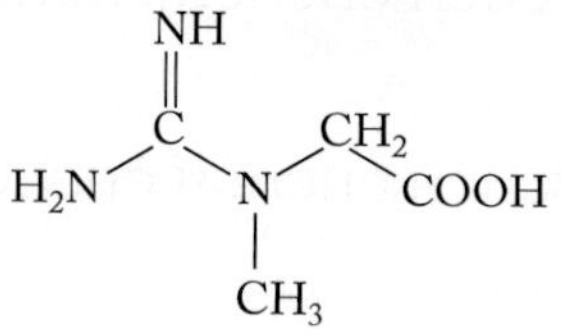

Creatine

FIGURE P20.79

20.80. Cytochrome c is an enzyme involved in oxidation–reduction reactions and is an intermediate in apoptosis, a controlled form of killing cells in response to infection or DNA damage. An elemental analysis of cytochrome c determined that it is 0.43% iron and 1.48% sulfur by mass. What is the minimum molecular weight of cytochrome c, and what is the minimum number of sulfur atoms per molecule?

20.81. Glutathione (Figure P20.81) is an essential molecule in the human body. It acts as an activator for enzymes and protects lipids from oxidation. Which three amino acids combine to make glutathione?

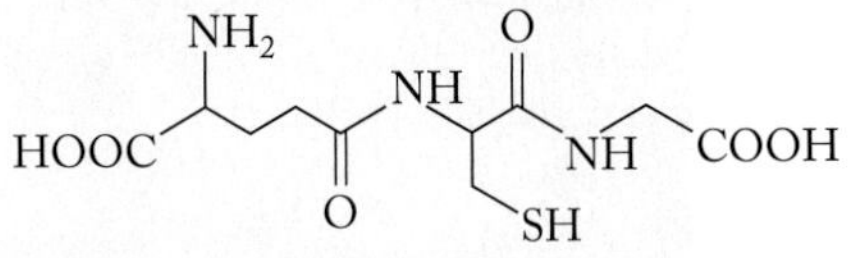

NH_2 O NH HOOC NH COOH O SH

Glutathione

FIGURE P20.81

20.82. The addition of ethanol to an aqueous solution of a globular protein causes the protein to denature. Ethanol also disrupts the structure of cell membranes. What interactions could be responsible for both of these effects?

***20.83.** Without doing the actual calculation, estimate the fuel values of leucine and isoleucine by considering average bond energies. Should the fuel values of the two amino acids be the same? Actual calorimetric measurements show that isoleucine has a lower fuel value than leucine. Explain why.

20.84. Sucralose (see Figure P20.9) is about 600 times sweeter than sucrose (see Figure 20.22). All substances that taste sweet have functional groups that form hydrogen bonds with "sweetness" receptor sites on the tongue. What does the difference in sweetness between sucralose and sucrose tell you about additional intermolecular interactions between sweet compounds and receptor sites that contribute to their sweet taste?

smartwork5

If your instructor uses Smartwork5, log in at digital.wwnorton.com/atoms2.

21

Nuclear Chemistry

The Risks and Benefits

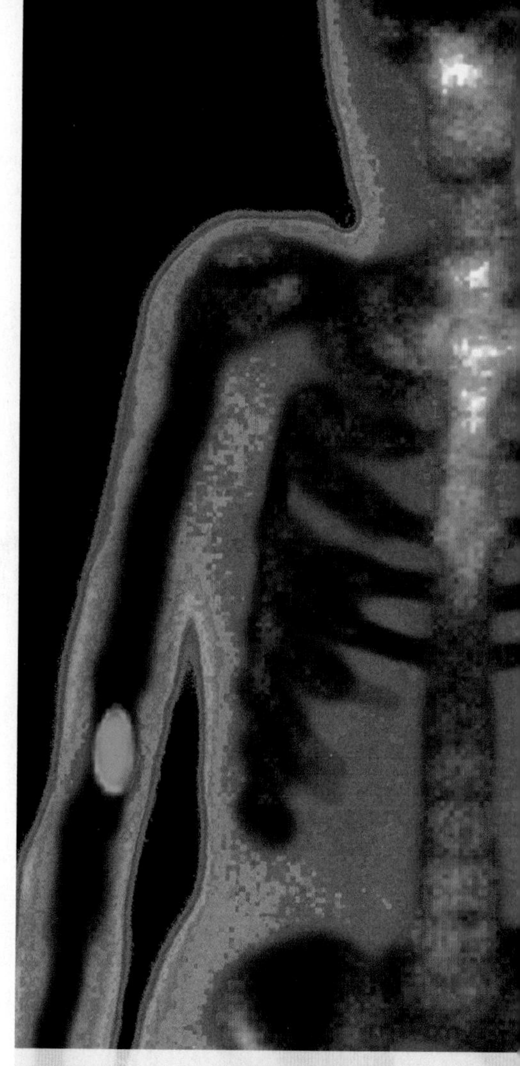

GAMMA SCANS OF THE HUMAN SKELETON Front and back views of a patient who was injected with a radionuclide that accumulated in bone tissues and emitted gamma rays.

PARTICULATE **REVIEW**

Isotopes Revisited

In Chapter 21, we investigate the stability and properties of radioactive nuclei. Three nuclides are depicted here:

- How many protons and how many neutrons does each nuclide contain?
- What is the mass of each nuclide in amu?
- Which two nuclides are isotopes of one another?

(Review Section 2.2 if you need help.)

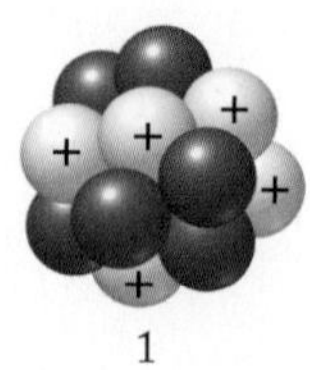

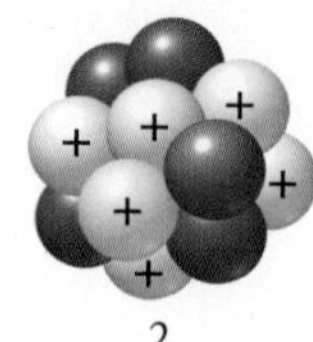

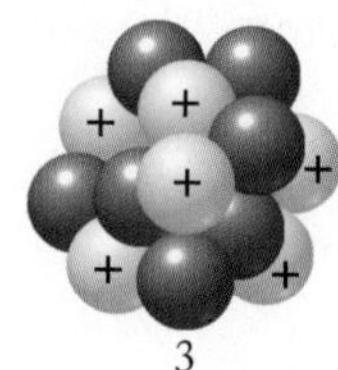

(Answers to Particulate Review questions are in the back of the book.)

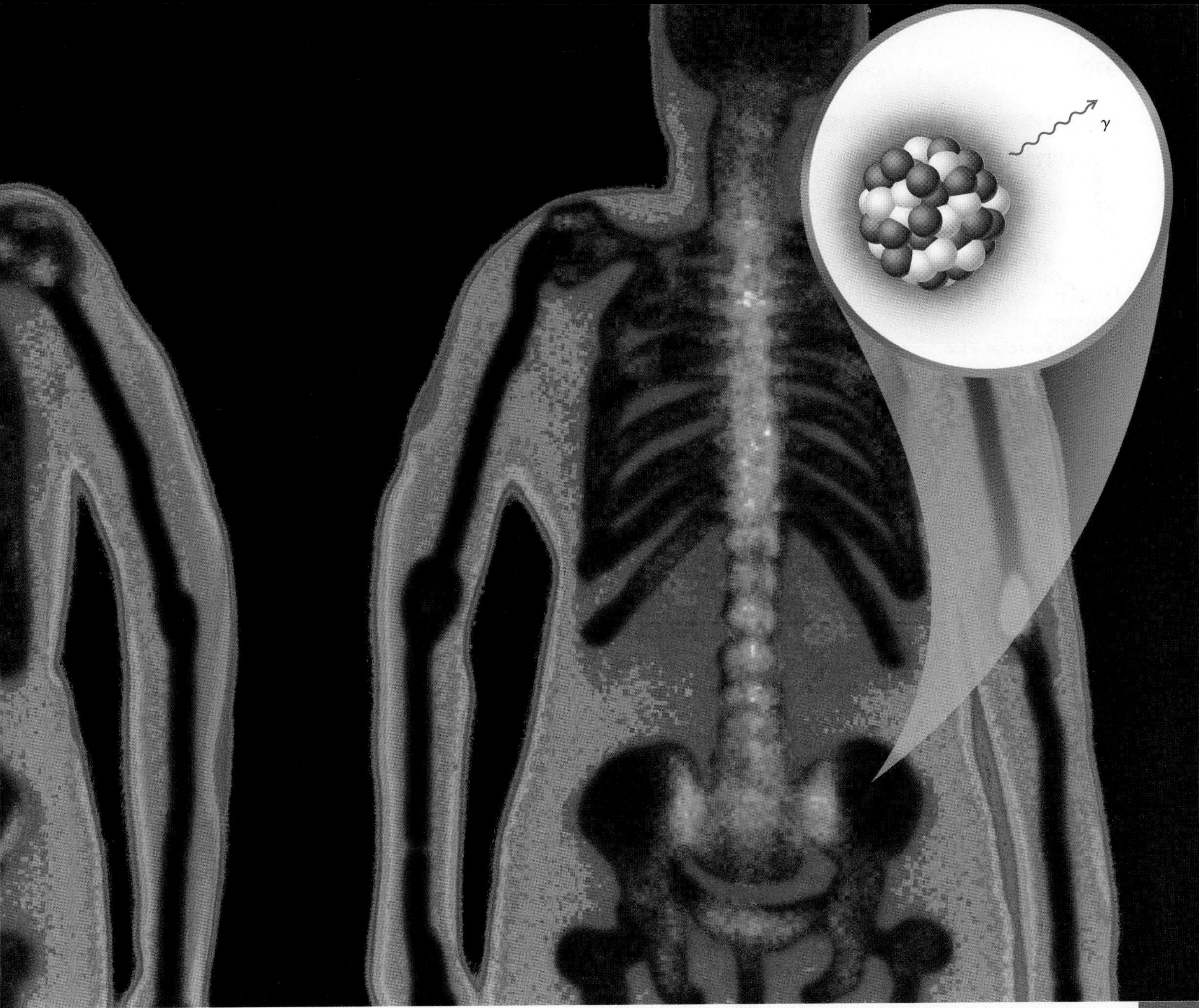

PARTICULATE **PREVIEW**

Unstable Nuclides versus Stable Nuclides

Radiocarbon dating uses the amount of carbon-14 in an artifact to determine its age. As you read Chapter 21, look for ideas that will help you answer these questions:

- Does the carbon-14 nucleus depicted here have more protons or more neutrons?
- How might the ratio of neutrons to protons affect the decay of an unstable nuclide?
- What nuclide is produced when carbon-14 undergoes decay? What is the neutron-to-proton ratio of this new nuclide?

Learning Outcomes

LO1 Write balanced equations to describe nuclear reactions
Sample Exercise 21.1

LO2 Predict the decay modes of radionuclides
Sample Exercise 21.2

LO3 Calculate the quantity of a radionuclide remaining after a defined decay time
Sample Exercise 21.3

LO4 Determine the age of a sample by radiometric dating
Sample Exercise 21.4

LO5 Calculate the binding energy of a nucleus and the energy released in a nuclear reaction from the masses of the products and reactants
Sample Exercises 21.5 and 21.6

LO6 Describe how elements are synthesized in the cores of giant stars

LO7 Compare and contrast nuclear fission and nuclear fusion

LO8 Predict the level of radioactivity in a sample of a radionuclide
Sample Exercise 21.7

LO9 Describe the dangers of exposure to nuclear radiation and calculate effective radiation doses
Sample Exercise 21.8

21.1 The Age of Radioactivity

Throughout this book, we have seen that the identities of atoms remain unchanged in chemical reactions in keeping with the law of conservation of mass. We now turn to *nuclear* reactions, in which the identities of atoms do change—because their nuclei change. The field of chemistry that studies these kinds of reactions is called **nuclear chemistry**.

Nuclear chemistry traces back to Henri Becquerel's 1896 discovery of the radioactivity emitted by pitchblende, a uranium-containing mineral. The α particles emitted by uranium were critical to Ernest Rutherford's efforts to unravel atomic structure. In 1898, Marie and Pierre Curie separated and purified two new radioactive elements from pitchblende: polonium and radium. These experiments and others launched a new scientific discipline called radiochemistry.

CONNECTION In Section 2.1 we defined α particles and described how the deflection of α particles by gold atoms led to Rutherford's model of the structure of the atom.

Interest in the applications of radium grew rapidly. Radium-containing paint was used to create luminous dials for watches and gauges. Ointments containing radium were prescribed as treatments for skin lesions. People were encouraged to visit health spas such as those in Saratoga Springs, New York, where the waters contained low concentrations of dissolved radium salts. Proliferation of radium-containing products also led to the discovery that nuclear radiation was dangerous. Young women hired to paint the dials of watches were instructed to lick the tips of their brushes, inadvertently introducing radium into their teeth and bones. Within a few years, many of these women developed bone cancer and died. Marie Curie herself succumbed to aplastic anemia caused by years of research with radioactive materials.

In recent years, a better understanding of the biological consequences of radiation has led to the development of methods for diagnosing and treating disease that minimize the hazards of radiation to patients and those who treat them. Nuclear reactors provide doctors and scientists with radionuclides that decay by predictable pathways within minutes or hours. Selective uptake of these nuclides by different organs in the body allows doctors to evaluate organ function and prescribe treatment when function is impaired. Radiation from other nuclides that concentrate in cancerous tissues can be used, often in conjunction with other therapies, to destroy malignant tumors.

In this chapter we examine the origins of nuclear reactions—that is, the interactions that take place in the nuclei of atoms. We address why some nuclei are stable and others are not, the kinds of spontaneous nuclear reactions that unstable nuclei undergo, and how the products of these reactions can be used to generate

nuclear chemistry the study of reactions that involve changes in the nuclei of atoms.

electrical power and to diagnose and treat disease. We also address some of the dangers that radioactive substances pose to human health and how we can shield ourselves from them.

21.2 Decay Modes of Radionuclides

Why does uranium undergo radioactive decay? The answer lies in the ratio of neutrons to protons in the nucleus of the uranium atom. The values of the atomic masses and mass numbers of the elements in the periodic table tell us about the ratios of neutrons to protons in the nuclei of their stable isotopes. The lighter elements have atomic masses that are about twice their atomic numbers and have neutron-to-proton ratios close to one. For example, ^{12}C has six neutrons and six protons, and most oxygen atoms have eight neutrons and eight protons.

As Z (atomic number) increases, however, so do the ratios of neutrons to protons. This trend is illustrated in Figure 21.1, where the green dots represent combinations of neutrons and protons that form stable nuclides. The band of green dots runs

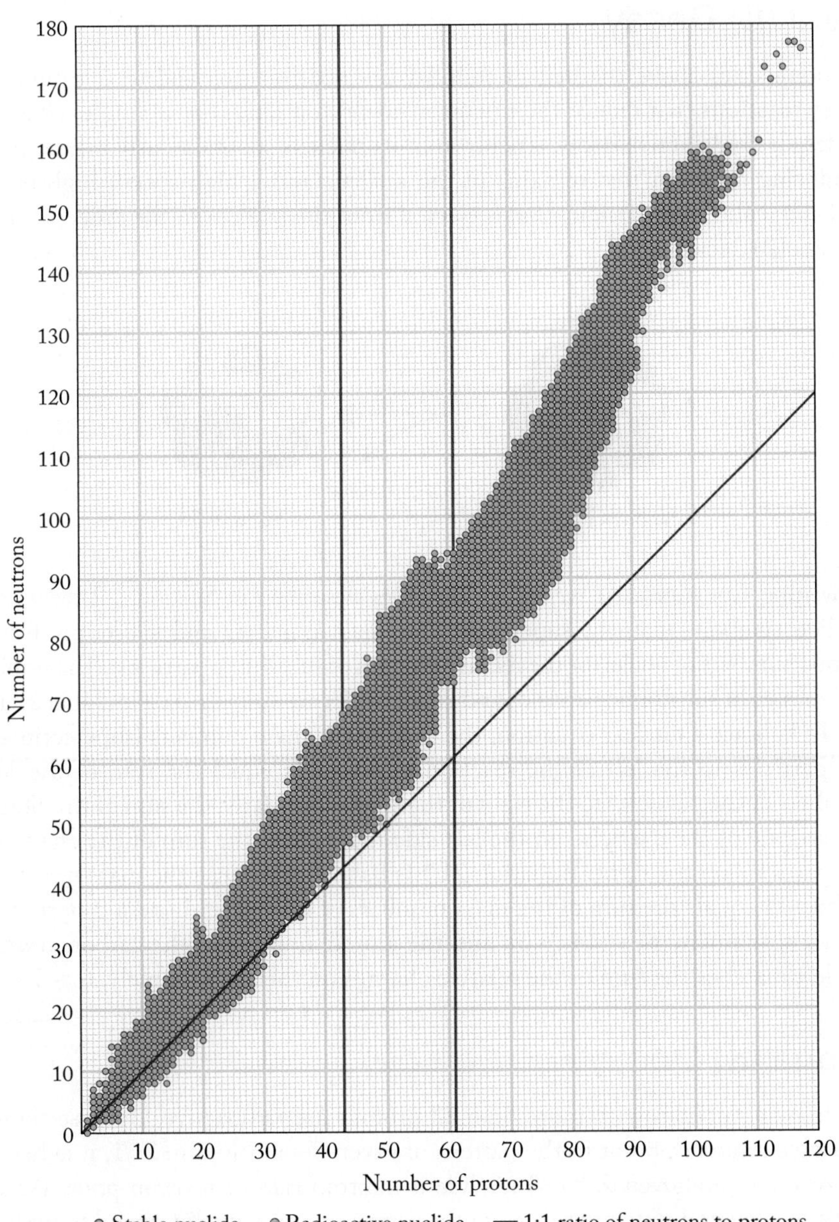

FIGURE 21.1 The belt of stability. Green dots represent stable combinations of protons and neutrons. Orange dots represent known radioactive (unstable) nuclides. Nuclides that fall along the purple line have equal numbers of neutrons and protons. Note that there are no stable nuclides (no green dots) for $Z = 43$ (technetium) and $Z = 61$ (promethium), as indicated by the two vertical red lines. These elements are the only two among the first 83 that are not found in nature.

belt of stability the region on a graph of number of neutrons versus number of protons that includes all stable nuclei.

radioactive decay the spontaneous disintegration of unstable particles accompanied by the release of radiation.

beta (β) decay the process by which a neutron in a neutron-rich nucleus decays into a proton and a β particle.

alpha (α) decay a nuclear reaction in which an unstable nuclide spontaneously emits an α particle.

diagonally through the graph, defining the **belt of stability**. Note how the belt curves upward away from the purple straight line, which represents a neutron-to-proton ratio of 1:1. This curvature shows that the neutron-to-proton ratio increases from about 1:1 for the lightest stable nuclides to about 1.5:1 for the most massive ones.

The nuclides represented by orange dots in Figure 21.1 are *radionuclides*. They can be classified as either neutron rich or neutron poor depending on their neutron-to-proton ratio relative to the belt of stability in Figure 21.1. Radionuclides are unstable and undergo **radioactive decay**, the spontaneous disintegration of radioactive nuclei accompanied by the release of nuclear radiation. Their particular mode of radioactive decay depends on whether they are above or below the belt of stability.

Four principal pathways—*alpha (α) decay, beta (β) decay, positron emission*, and *electron capture*—represent the most common decay mechanisms for unstable nuclei. Both α decay and β decay were introduced in Section 2.1, where the critical role they played in establishing the structure of the atom was described. We explore positron emission and electron capture in more detail here in Section 21.2.

Beta (β) Decay

Radionuclides above the belt of stability are neutron rich and tend to undergo decay reactions that reduce their neutron-to-proton ratio. For example, when ^{14}C undergoes radioactive decay, a neutron in its nucleus spontaneously disintegrates, producing a proton that remains in the nucleus and a high-speed, high-energy electron, called a β particle, that is ejected from the nucleus in the process known as **beta (β) decay**:

$$^{14}_{6}C \rightarrow ^{14}_{7}N + ^{0}_{-1}\beta \quad (21.1)$$

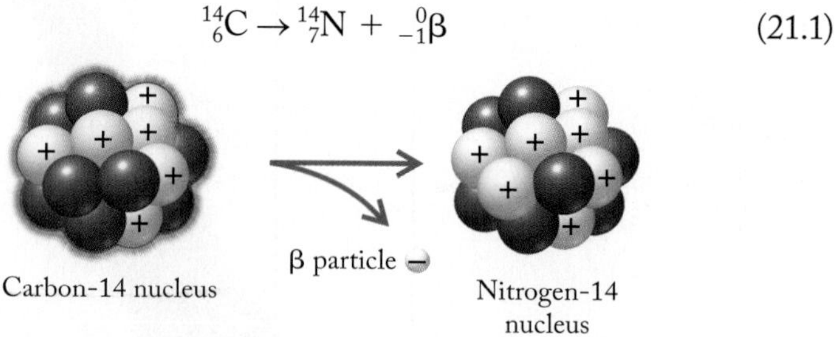
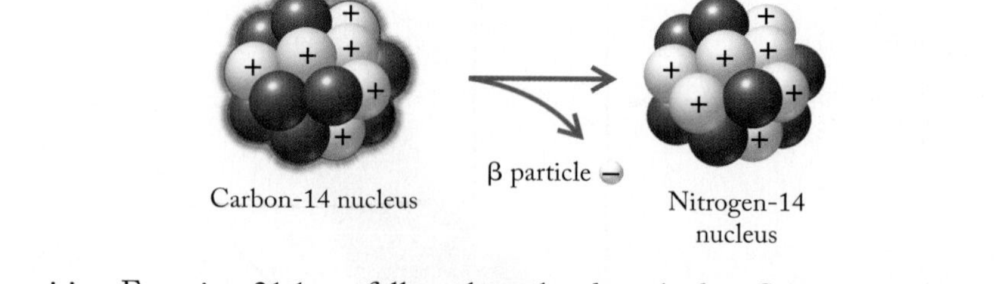

CHEMTOUR
Balancing Nuclear Equations

In writing Equation 21.1 we follow the rules described in Section 2.2 for writing nuclide symbols—namely, superscripts for mass numbers and subscripts for the atomic numbers of nuclei (or the relative charges of subatomic particles). The symbols and masses of subatomic particles and small nuclei are summarized in Table 21.1.

To balance a nuclear equation, the sum of the mass numbers (superscripts) of the particles to the left of the reaction arrow must equal the sum of the mass numbers of the particles to the right of the arrow. Similarly, the sum of the charges (subscripts) of the particles on the left side must equal the sum of the charges of the particles on the right. Equation 21.1 is balanced because the mass number of carbon-14 equals the sum of the mass numbers of nitrogen-14 and a β particle on the right (namely, 14 = 14 + 0), and the atomic number of the carbon atom on the left matches the sum of the relative charges on the right (namely, 6 = 7 − 1).

TABLE 21.1 Symbols and Masses of Subatomic Particles and Small Nuclei

Particle	Symbol	Mass (kg)
Neutron	$^{1}_{0}n$	1.67493×10^{-27}
Proton	$^{1}_{1}p$ or $^{1}_{1}H$	1.67262×10^{-27}
Electron (β particle)	$^{0}_{-1}\beta$ or $^{0}_{-1}e$	9.10939×10^{-31}
Deuteron	$^{2}_{1}D$ or $^{2}_{1}H$	3.34370×10^{-27}
α Particle	$^{4}_{2}\alpha$ or $^{4}_{2}He$	6.64465×10^{-27}
Positron	$^{0}_{1}\beta$	9.10939×10^{-31}

Alpha (α) Decay

All known nuclides with more than 83 protons are radioactive. Because there is no stable reference point in the pattern of green dots in Figure 21.1, it is hard to say whether any given $Z > 83$ nuclide is neutron rich or neutron poor. We can make one general statement though: these most massive nuclides tend to undergo either β decay or **alpha (α) decay**. Alpha decay produces a nuclide with two fewer protons and two fewer neutrons, as in the case of uranium-238 in Equation 21.2:

$$^{238}_{92}\text{U} \rightarrow {}^{234}_{90}\text{Th} + {}^{4}_{2}\alpha \qquad (21.2)$$

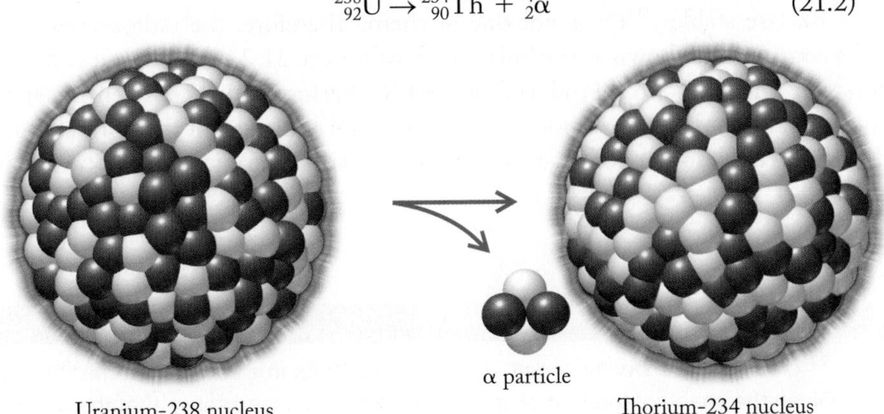

Uranium-238 provided the α particles critical to Rutherford's gold foil experiments to study the structure of the atom.

Among the most massive radioactive isotopes, one radioactive decay process often leads to another in what is referred to as a *radioactive decay series*. Consider, for example, the decay series that begins with the α decay of ^{238}U to ^{234}Th (Figure 21.2).

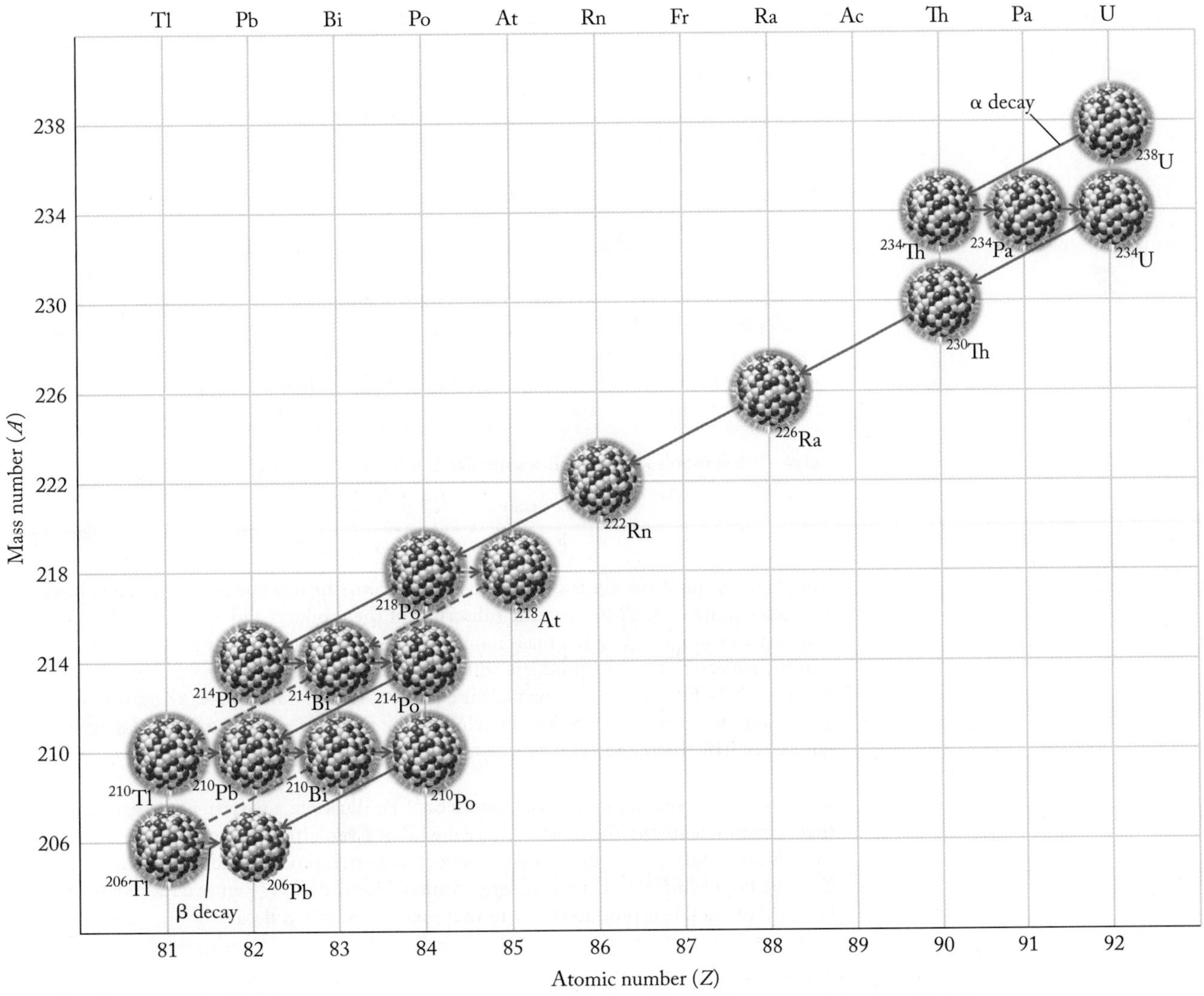

FIGURE 21.2 Uranium-238 radioactive decay series. The long diagonal arrows represent α decay events; the short horizontal ones represent β decay events. The dashed arrows are alternative pathways representing less than 1% of the decay events in this series. Whether decay proceeds by the solid-line or dashed-line pathway, the end product is always stable lead-206. (Note that the vertical axis in this figure is the mass number, not the number of neutrons, as in Figure 21.1.)

Thorium has no stable isotopes and undergoes two β decay steps to produce ^{234}U. In a series of subsequent α decay steps, ^{234}U turns into thorium-230, radium-226, radon-222, polonium-218, and finally lead-214. Although some isotopes of lead ($Z = 82$) are stable, ^{214}Pb is not one of them. Therefore, the radioactive decay series continues as shown at the bottom left of Figure 21.2 and does not end until the stable nuclide ^{206}Pb is produced. Ernest Rutherford and other scientists studied the uranium decay series for many years in the early part of the 20th century, identifying new isotopes of existing elements and discovering some new elements as well.

CONCEPT TEST

In the ^{234}U radioactive decay series, five α decay steps in a row transform ^{234}U into ^{214}Pb. Given the shape of the belt of stability, why does it make sense that the product of these α decay steps would be a neutron-rich nuclide that undergoes β decay?

(Answers to Concept Tests are in the back of the book.)

SAMPLE EXERCISE 21.1 Completing and Balancing Nuclear Equations

LO1

Starting with 7 tons of pitchblende, Pierre and Marie Curie were able to isolate a few milligrams of a new element, which they named polonium. Polonium-218 undergoes both α and β decay. Write balanced nuclear equations describing these nuclear reactions. Use symbols of the form $^{A}_{Z}X$ to represent the nuclides and any subatomic particles that may also have formed.

Collect, Organize, and Analyze Polonium (Po) has atomic number 84, α particles are helium-4 nuclei, and β particles are high-speed electrons. We can combine the element symbols and atomic numbers with the given mass numbers to write symbols of the nuclides involved in the reaction. To write a balanced nuclear equation, the sum of the subscripts (the atomic numbers) of the nuclides on the left side of the reaction arrow must equal the sum of the subscripts on the right. The sums of the superscripts (the mass numbers) must also match.

Solve The symbols of the nuclides involved in the reactions are

$$\alpha \text{ decay:} \quad ^{218}_{84}\text{Po} \rightarrow {}^{4}_{2}\text{He} + ?$$

$$\beta \text{ decay:} \quad ^{218}_{84}\text{Po} \rightarrow {}^{0}_{-1}\beta + ?$$

Let's first complete the alpha decay nuclear equation. The unknown product must have an atomic number of 82 (so that the subscripts of the products add up to 84), which makes it an isotope of Pb. The mass number of the unknown product must be 214 (so that the superscripts of the products add up to 218). Therefore, the product of the alpha decay is $^{214}_{82}Pb$. For the beta decay nuclear equation, the unknown product must have an atomic number of 85 (so that the subscripts of the products add up to 84) and a mass number of 218. Therefore, the product from the β decay of ^{218}Po is $^{218}_{85}At$.

Think About It The multiple decay modes of ^{218}Po illustrate a pattern of uncertainty that is common among the most massive nuclides. Even after undergoing an α decay, there is no certainty that the slightly more neutron-rich product nuclide will undergo β decay. For example, ^{234}U may undergo 5 successive α decays (Figure 21.2) before a radioactive nuclide is produced (^{214}Pb) that undergoes only β decay.

Practice Exercise Lead-214, ^{214}Pb, decays to ^{214}Bi and then to ^{210}Tl. Write balanced nuclear equations describing these reactions.

(Answers to Practice Exercises are in the back of the book.)

Positron Emission and Electron Capture

Nuclides below the belt of stability are neutron poor and undergo decay processes that increase their neutron-to-proton ratio. In one of these processes the radioactive nucleus emits a high-velocity particle that has the same mass as an electron but that has a positive charge. It is called a **positron ($^{0}_{1}\beta$)**, and its ejection from a neutron-poor nucleus is called **positron emission**. The net effect of positron emission is the production of a nucleus with one fewer proton and one more neutron, as illustrated in the decay of carbon-11 (Equation 21.3):

$$^{11}_{6}C \rightarrow ^{11}_{5}B + ^{0}_{1}\beta \qquad (21.3)$$

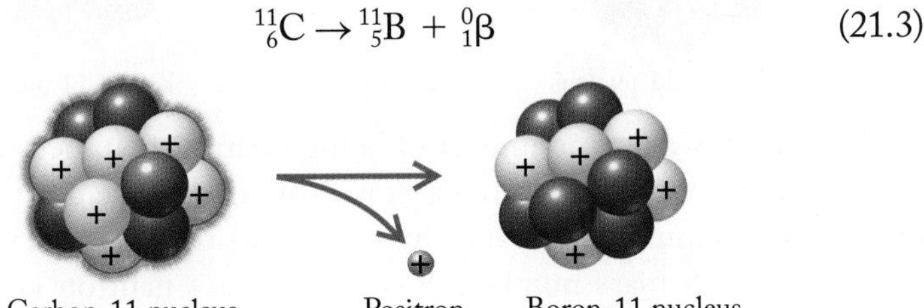

The boron-11 produced in this reaction is a stable isotope. In fact, 80.2% of all boron atoms in nature are boron-11.

Positrons belong to a group of subatomic particles that have the opposite charge but the same mass as particles typically found in atoms. In addition to these electrons with positive charges, there are protons with negative charges, called *antiprotons*, $^{1}_{-1}p$. These charge opposites are particles of **antimatter**.

Particles of matter and their antimatter opposites are like mortal enemies. If they collide, they instantly annihilate each other. In their mutual destruction, they cease to exist as matter, and all of their mass is converted to energy in the form of two or more gamma (γ) rays:

$$^{0}_{1}\beta + ^{0}_{-1}\beta \rightarrow 2\,\gamma \qquad (21.4)$$

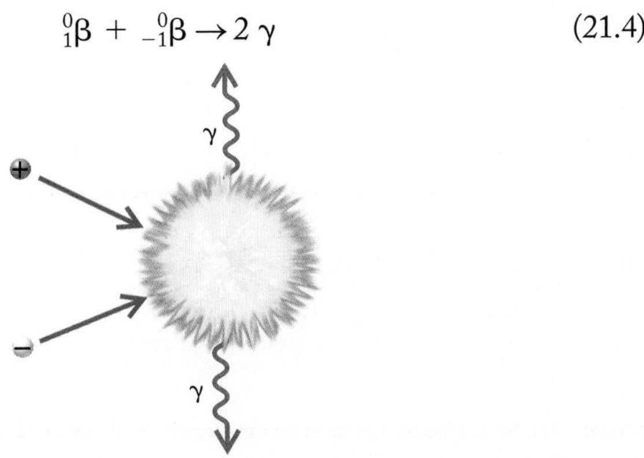

The yellow "sunburst" here symbolizes the energy released in the process depicted.

All nuclear reactions are accompanied by gamma (γ) ray emission. Gamma rays represent high-energy electromagnetic radiation that has essentially no mass. Gamma rays are also generated in the nuclear furnaces of stars and permeate outer space. Those that reach Earth are absorbed by the gases in our atmosphere. In the process, molecular gases are broken up into their component atoms, and

positron a particle with the mass of an electron but with a positive charge.

positron emission the spontaneous emission of a positron from a neutron-poor nucleus.

antimatter particles that are the charge opposites of normal subatomic particles.

electron capture a nuclear reaction in which a neutron-poor nucleus draws in one of its surrounding electrons, which transforms a proton in the nucleus into a neutron.

atomic nuclei may be broken up into their subatomic particles. Radioactive sources that emit gamma rays are also used by oncologists to disrupt the molecules in cancer cells and kill them.

CONNECTION Gamma rays are the highest-energy form of electromagnetic radiation (see Figure 3.1).

There is another way to increase the neutron-to-proton ratio of a neutron-poor nucleus: it can capture one of the inner-shell electrons of its atom. When it does, the negatively charged electron combines with a positively charged proton. The product of this combination reaction is a neutron. The effect of this **electron capture** process on the nucleus is the same as positron emission: the number of protons decreases by one and the number of neutrons increases by one. So, when carbon-11 undergoes electron capture, the product is identical to the product formed by positron emission, namely boron-11:

$$^{11}_{6}C + {}^{0}_{-1}e \rightarrow {}^{11}_{5}B \tag{21.5}$$

Carbon-11 nucleus   Electron   Boron-11 nucleus

Table 21.2 illustrates the impact of being neutron rich, neutron poor, or neither on the stability of isotopes of carbon. Note two are stable: ^{12}C and ^{13}C. The isotopes with mass numbers greater than 13 are neutron rich and undergo β decay, whereas those with mass numbers less than 12 are neutron poor and undergo either positron emission or electron capture. Figure 21.3 summarizes the changes in atomic number and mass number that result from the various modes of decay. Taken together, Figure 21.1 and Figure 21.3 allow us to make predictions about the stability of a nucleus with respect to radioactive decay and to identify likely decay pathways (see Sample Exercise 21.2).

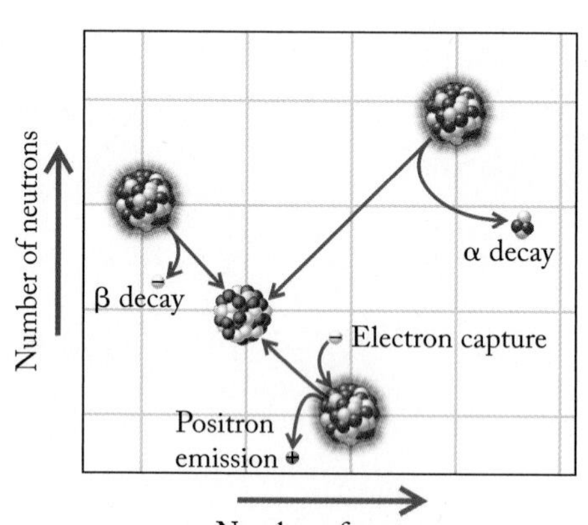

FIGURE 21.3 Radioactive decay results in predictable changes in the number of protons and neutrons in a nucleus. The red, shimmering shadow surrounding three of the nuclei indicates that they are radioactive. The gray nucleus represents a stable nucleus. The blue arrows describe three decay pathways that lead to a more favorable neutron–to–proton ratio. In alpha decay the nucleus loses two neutrons and two protons, so there is a decrease of 2 in atomic number and 4 in mass number. Beta decay leads to an increase of one proton at the expense of one neutron, so the atomic number increases by 1 but the mass number is unchanged. In positron emission and electron capture, the number of protons decreases by 1 and the number of neutrons increases by 1, so the atomic number decreases by 1, but the mass number remains the same.

TABLE 21.2 Isotopes of Carbon and Their Radioactive Decay Products

Name	Symbol	Mode(s) of Decay	Half-Life	Natural Abundance (%)
Carbon-10	$^{10}_{6}C$	Positron emission	19.45 s	—
Carbon-11	$^{11}_{6}C$	Positron emission, electron capture	20.3 min	—
Carbon-12	$^{12}_{6}C$	—	(Stable)	98.89
Carbon-13	$^{13}_{6}C$	—	(Stable)	1.11
Carbon-14	$^{14}_{6}C$	β decay	5730 yr	—
Carbon-15	$^{15}_{6}C$	β decay	2.4 s	—
Carbon-16	$^{16}_{6}C$	β decay	0.74 s	—

SAMPLE EXERCISE 21.2 Predicting the Modes and Products of Radioactive Decay **LO2**

Predict the mode of radioactive decay of ^{32}P and ^{60}Cu, which are two radionuclides used in biomedical research and treatment. Identify the nuclide that is produced in each decay process.

Collect, Organize, and Analyze The mode of decay of each radionuclide will depend on whether it is neutron rich or neutron poor. In Figure 21.4, ^{32}P (17 neutrons + 15 protons) is represented by the orange dot directly above the green dot for ^{31}P (the one and only stable phosphorus nuclide), which means that ^{32}P is radioactive and neutron rich. Neutron-rich radioisotopes of lighter elements undergo β decay. The orange dot for ^{60}Cu lies *below* the green dots for the two stable copper isotopes (^{63}Cu and ^{65}Cu), which means ^{60}Cu is a neutron-poor isotope and suggest either positron emission or electron capture are likely decay pathways. In the balanced nuclear equations, regardless of the nuclide or mode of decay, the sums of the superscripts on the left and right sides must be equal, as must the sums of the subscripts.

Solve A β particle must be one product of the decay reaction for ^{32}P, giving the incomplete nuclear equation:

$$^{32}_{15}P \rightarrow ? + ^{0}_{-1}\beta$$

The missing product must have an atomic number of 16 (so that the subscripts on the right side sum to 15), making it an isotope of S. Its mass number must be 32, so the product is sulfur-32:

$$^{32}_{15}P \rightarrow ^{32}_{16}S + ^{0}_{-1}\beta$$

The analogous equations for positron emission from ^{60}Cu are

$$^{60}_{29}Cu \rightarrow ? + ^{0}_{+1}\beta \text{ and}$$

$$^{60}_{29}Cu \rightarrow ^{60}_{28}Ni + ^{0}_{+1}\beta$$

The atomic number of copper-60 must equal the sum of the subscripts (29 = 28 + 1) on the right-hand side of the equation while the mass number of copper-60 must equal the sum of the superscripts (60 = 60 + 0).

Neutron-poor isotopes such as copper-60 can also decay by electron capture. Using Equation 21.5 as a model, we can write another nuclear equation for the decay of ^{60}Cu:

$$^{60}_{29}Cu + ^{0}_{-1}e \rightarrow ^{60}_{28}Ni$$

Note that the *total* number of nucleons (neutrons and protons) is equal for copper-60 and nickel-60 because they have the same mass number, consistent with the conversion of one proton and one electron in a Cu atom into one neutron.

Think About It By emitting a β particle, the ^{32}P nucleus has one additional proton and one fewer neutron, thereby reducing its neutron "richness" and becoming a stable isotope of sulfur. Because both positron emission and electron capture produce the same product in the decay of ^{60}Cu, it is difficult to predict which pathway will dominate. Because the number of protons increases by either pathway, the neutron-to-proton ratio increases and the product isotope, Ni-60, is stable as indicated by a green dot in Figure 21.4.

Practice Exercise Predict the mode of radioactive decay of ^{28}Al and ^{18}F. Identify the nuclide produced in each decay process.

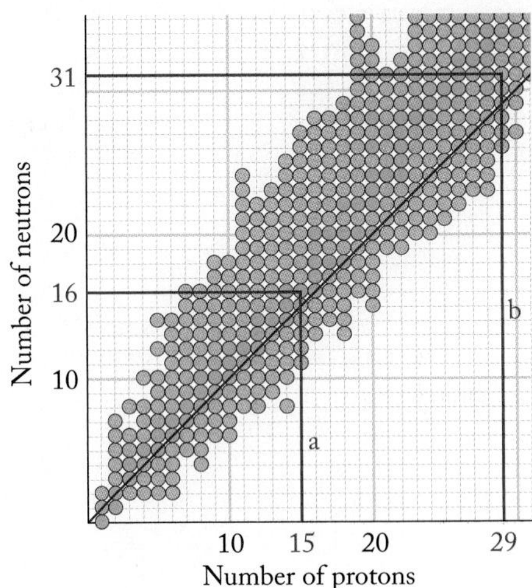

FIGURE 21.4 Region in the belt of stability for $Z \leq 30$.

21.3 Rates of Radioactive Decay

In Section 21.2 we examined how radionuclides undergo radioactive decay; in this section, we focus on how rapidly they decay. We have defined radioactive decay as the spontaneous disintegration of unstable nuclei, but, as with chemical reactions, spontaneous does not necessarily mean rapid.

First-Order Radioactive Decay

Because all radioactive decay processes follow first-order kinetics (Section 13.3), each nuclide has a characteristic *half-life* ($t_{1/2}$), the time interval during which the

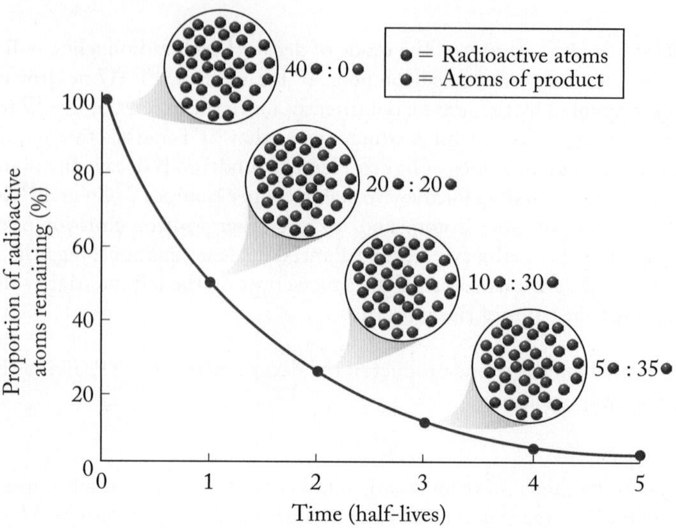

FIGURE 21.5 Radioactive decay follows first-order kinetics, which means, for example, that if a sample initially contains 40 radioactive atoms, it will contain only one-half that number after a time interval equal to one half-life. Half of the remaining half, or 10 radioactive atoms, remains after two half-lives, and so on.

quantity of radioactive particles decreases by one-half (Figure 21.5). The faster the decay process, the shorter the half-life. We can represent the number of half-lives, n, that have passed after time, t, using Equation 21.6:

$$n = \frac{t}{t_{1/2}} \tag{21.6}$$

To calculate what fraction of radionuclide remains after decay time t, we start with Equation 13.17 for a first-order process and substitute the number of nuclei at instant t (N_t) for $[X]_t$ and the number present at $t = 0$ (N_0) for $[X]_0$:

$$\ln \frac{[X]_t}{[X]_0} = \ln \frac{N_t}{N_0} = -kt \tag{13.17}$$

Next, we rearrange Equation 13.20,

$$t_{1/2} = \frac{0.693}{k} \tag{13.20}$$

so we are expressing k as a function of $t_{1/2}$ (Equation 21.7):

$$k = \frac{0.693}{t_{1/2}} \tag{21.7}$$

Substituting for k in Equation 13.17, we arrive at Equation 21.8:

$$\ln \frac{N_t}{N_0} = -0.693 \frac{t}{t_{1/2}} \tag{21.8}$$

Taking the antilog of both sides yields Equation 21.9:

$$\frac{N_t}{N_0} = 0.5^{t/t_{1/2}} = 0.5^n \tag{21.9}$$

CONNECTION We introduced the concept of *half-life*, $t_{1/2}$, and how its value is inversely proportional to the rate constant, k, of a first-order reaction in Chapter 13.

We can apply Equation 21.9 to various radioactive decay processes in Sample Exercise 21.3 and elsewhere in this chapter because all of these processes follow first-order reaction kinetics.

SAMPLE EXERCISE 21.3 Calculations Involving Half-Lives **LO3**

Free neutrons are radioactive with a half-life of 12 minutes, undergoing β decay to a proton and an electron. Starting with a population of 6.6×10^5 free neutrons, how many remain after 2.0 min?

Collect, Organize, and Analyze Equation 21.9 relates quantities of radioactive particles to decay times. The initial number of neutrons (N_0) is 6.6×10^5, and their half-life is $t_{1/2} = 12$ min. We are asked to solve for the number of neutrons that remain, N_t, after $t = 2.0$ min. The value of t is only a fraction of $t_{1/2}$. That is, far fewer than half of the initial number of neutrons will have decayed after 2.0 min.

Solve Solving Equation 21.9 for N_t and substituting the values,

$$N_t = 0.5^{t/t_{1/2}} \times N_0$$

$$= 0.5^{2.0\text{min}/12\text{min}} \times 6.6 \times 10^5 = 5.9 \times 10^5$$

Think About It The value of N_t is reasonable because, as we predicted, only a small fraction of the initial quantity of free neutrons decayed in 2.0 minutes.

Practice Exercise Cesium-131 is a short-lived radionuclide ($t_{1/2} = 9.7$ d) used to treat prostate cancer. If the therapeutic strength of the radionuclide is directly proportional to the number of nuclei present, how much therapeutic strength does a cesium-131 source lose over exactly 60 days? Express your answer as a percentage of the strength the source had at the beginning of the first day.

Radiometric Dating

The rate of radioactive decay has a practical application in **radiometric dating**, a term used to describe methods for determining the age of objects based on the tiny concentrations of radionuclides that occur naturally in them and the rates of radioactive decay of these nuclides. The concept was invented in the early 1900s by Ernest Rutherford, who had already recognized that radioactive decay processes have characteristic half-lives. Rutherford proposed to use this concept to determine the age of rocks and even the age of Earth itself. The basis for his initial attempt was the emission of α particles from uranium ore. He correctly suspected that alpha particles were part of helium atoms, and so he proposed to determine the age of uranium ore samples by determining the concentration of helium gas trapped inside them.

Rutherford's helium method did not yield very accurate results, but it did inspire a young American chemist, Bertram Boltwood (1870–1927), who had determined that the decay of radioactive uranium involves a series of decay events ending with the formation of stable lead (illustrated for ^{238}U in Figure 21.2). In 1907, Boltwood published the results of dating 43 samples of uranium-containing minerals based on the ratio of lead to uranium in them. The ages he reported spanned hundreds of millions to over a billion years and probably represent the first successful attempt at radiometric dating.

Much more recently, the development of the mass spectrometer for accurately determining the abundances of individual isotopes of elements, coupled with more accurate half-life values for decay events such as those in Figure 21.2, has allowed scientists to use the ratio of ^{206}Pb to ^{238}U in geological samples to determine their ages with a precision of about ±1%. Other methods, including one

radiometric dating a method for determining the age of an object based on the quantity of a radioactive nuclide and/or the products of its decay that the object contains.

CONNECTION In Section 2.4 we saw that mass spectrometry can be used to determine the abundances of the isotopes of elements in a sample.

based on the decay of ^{235}U to ^{207}Pb ($t_{1/2} = 7.0 \times 10^6$ yr), may be used to analyze the same samples, providing multiple independent determinations that mutually assure more accurate results. These analyses have shown that the oldest rocks on Earth are over 4.0 billion years old and that meteorites that formed as the solar system formed are 4.5 billion years old.

The radiometric methods described above yield reliable results only when the sample is a closed system, which means that the only loss of the radionuclide is via radioactive decay, and that all of the nuclides produced by the decay processes remain in the sample. In addition, those decay processes must be the only source of the product nuclides. For these reasons, scientists must exercise care in selecting the types of samples they subject to radiometric dating analysis. For example, the presence of the mineral zircon ($ZrSiO_4$) in a geological sample is good news for scientists interested in using radiometric dating because U^{4+} ions readily substitute for Zr^{4+} ions as crystals of $ZrSiO_4$ solidify from the molten state, but Pb^{2+} ions do not. Therefore, the only source of ^{206}Pb and ^{207}Pb in a zircon sample should be the decay of ^{238}U and ^{235}U, respectively.

In 1947, American chemist Willard Libby (1908–1980) developed a radiometric dating technique, called **radiocarbon dating**, for determining the age of artifacts from prehistory and early civilizations. The method is based on determining the carbon-14 content of samples derived from plants or the animals that consumed them. Carbon-14 originates in the upper atmosphere, where cosmic rays break apart the nuclei of atoms, forming free protons and neutrons. When one of these neutrons collides with a nitrogen-14 atom, they form an atom of radioactive carbon-14 and a proton:

$$^{14}_{7}N + ^{1}_{0}n \rightarrow ^{14}_{6}C + ^{1}_{1}p$$

Atmospheric carbon-14 combines with oxygen, forming $^{14}CO_2$. The atmospheric concentration of $^{14}CO_2$ amounts to only about 10^{-12} of all the molecules of CO_2 in the air. These traces of radioactive CO_2, along with the stable forms, $^{12}CO_2$ and $^{13}CO_2$, are incorporated into the structures of green plants during photosynthesis. The tiny fraction of the plant's mass that is ^{14}C gets even tinier after a plant dies, or after a part of it stops growing and photosynthesizing, because ^{14}C undergoes β decay (with a half-life of 5730 years) as we described in Section 21.2:

$$^{14}_{6}C \rightarrow ^{14}_{7}N + ^{0}_{-1}\beta$$

If we can determine the ^{14}C content (N_t) of an object of historical interest, such as a piece of wood from an ancient building, charcoal from a prehistoric campfire, or papyrus from an early Egyptian scroll, and if we know (or can predict) its ^{14}C content when the material in it was alive (N_0), then we can apply Equation 21.8 to determine its age:

$$\ln \frac{N_t}{N_0} = -0.693 \frac{t}{t_{1/2}}$$

Predicting the value of N_0 is usually done by analyzing samples from growing plants—that is, samples for which the ^{14}C decay time is zero. To facilitate radiocarbon dating calculations, let's rearrange the terms in Equation 21.8 to solve for the radiocarbon age t:

radiocarbon dating a method for establishing the age of a carbon-containing object by measuring the amount of radioactive carbon-14 remaining in the object.

$$t = -\frac{t_{1/2}}{0.693} \ln \frac{N_t}{N_0} \qquad (21.10)$$

Substitution for N_0, N_t, and $t_{1/2}$ in Equation 21.10 yields the time elapsed since the material was alive, as shown in Sample Exercise 21.4.

SAMPLE EXERCISE 21.4 Radiocarbon Dating LO4

The ^{14}C content of a wooden harpoon handle found in the remains of an Inuit encampment in western Alaska is 61.9% of the ^{14}C content of the same type of wood from a recently cut tree. How old is the harpoon?

Collect and Organize We are asked to calculate the age of a sample that contains 61.9% of the ^{14}C in a modern sample of the same material. The half-life of carbon-14 is 5730 years. Equation 21.10 provides the age t of the artifact if we know the ratio of the ^{14}C in it today to its initial ^{14}C content.

Analyze The ^{14}C content of the modern sample can be used as a surrogate for the initial ^{14}C content of the artifact. Therefore, 61.9% (or 0.619) represents the ratio N_t/N_0. This value is greater than 0.5, which means that the age of the sample is less than one half-life (5730 yr).

Solve

$$t = -\frac{t_{1/2}}{0.693}\ln\frac{N_t}{N_0}$$

$$= -\frac{5730\text{ yr}}{0.693}\ln(0.619)$$

$$= 3966\text{ yr} = 3.97 \times 10^3\text{ yr}$$

Think About It The resulting age is less than one half-life, which is reasonable because it contained more than half the original carbon-14 content. The result is expressed with three significant figures to match that of the starting composition (61.9%).

Practice Exercise The Old Testament describes the construction of the Siloam Tunnel, used to carry water into Jerusalem under the reign of King Hezekiah (727–698 BCE). An inscription on the tunnel has been interpreted as evidence that the tunnel was not built until 200–100 BCE. ^{14}C dating (in 2003) indicated a date close to 700 BCE. What is the ratio of ^{14}C in a wooden object made in 100 BCE to one made from the same kind of wood in 700 BCE?

The accuracy of radiocarbon dating can be checked by determining the ^{14}C content of the annual growth rings of very old trees, such as the bristlecone pines that grow in the American Southwest (Figure 21.6). When scientists plot the radiocarbon ages of these rings against their actual ages obtained by counting rings starting from the outer growth layer of the tree (representing $t = 0$), they find that the two sets of ages do not agree exactly, as shown in Figure 21.7. There are several reasons for this lack of agreement, including variability in the rates of ^{14}C production due to changing intensity of the cosmic rays striking Earth's upper atmosphere. To assure accurate ^{14}C results, scientists must correct for these and other variations, and they must be careful to avoid contaminating ancient samples with modern carbonaceous material. With proper analytical technique, radiocarbon dating results are generally accurate to within ±40 yr for samples that are 500–50,000 years old.

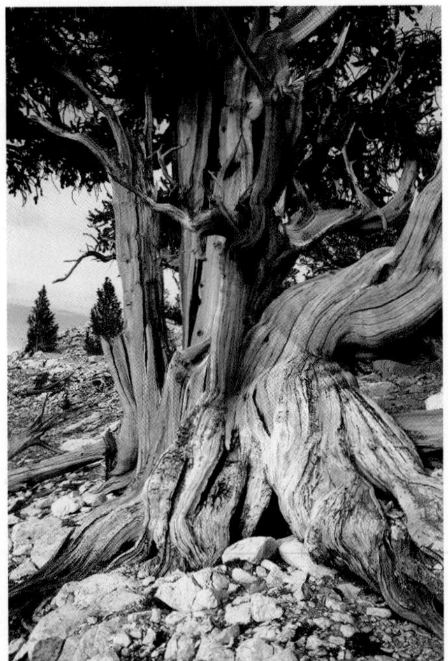

FIGURE 21.6 Radiocarbon dating relies on knowing the atmospheric concentration of carbon-14 over time. Ancient living trees, such as the bristlecone pines in the American Southwest, act as a check of the atmospheric carbon-14 levels over thousands of years. The ages of the rings can be determined by counting them, and their carbon-14 content can be determined by mass spectrometry.

CONCEPT TEST

How might the increased consumption of fossil fuels over the last century affect the ^{14}C content of growing plant tissues?

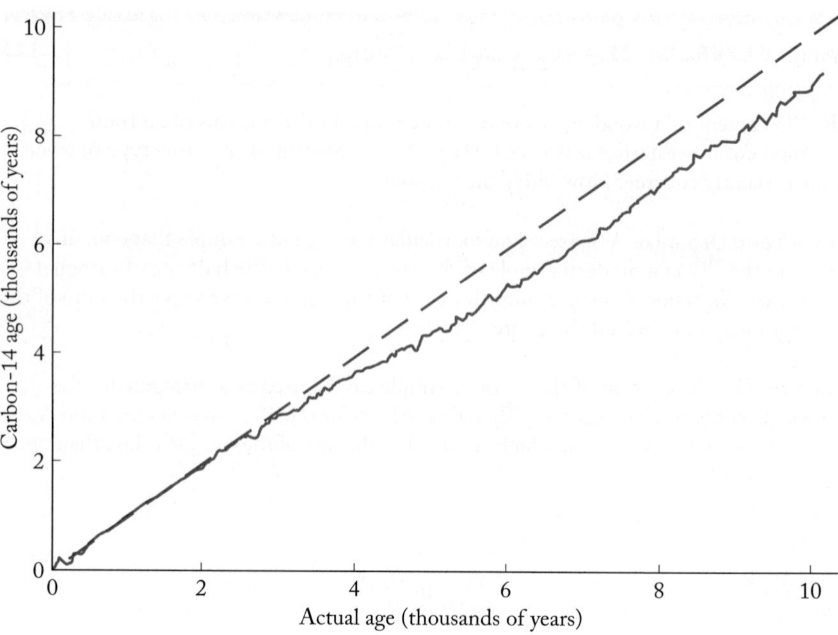

FIGURE 21.7 Calibration curves for radiocarbon dating allow scientists to accurately calculate the ages of archaeological objects. If the rate of ^{14}C production in the upper atmosphere were constant, then the age of objects based on their ^{14}C content would match their actual age—a condition represented by the red dashed line. However, analyses of tree rings, corals, and lake sediments indicate that the rate of ^{14}C production in the upper atmosphere is variable, so a correct plot of ^{14}C age versus actual age produces the jagged blue line. This plot allows scientists to convert ^{14}C ages into actual ages.

21.4 Energy Changes in Radioactive Decay

To understand why some nuclei are more stable than others, we need to explore the nature of the energy that keeps the nucleons in atomic nuclei together. By the 1930s, scientists had discovered that the mass of a stable nucleus is always less than the sum of the separate masses (Table 21.1) of its nucleons. For example, a helium-4 nucleus consists of two neutrons and two protons corresponding to a mass:

$$\begin{aligned} \text{Mass of 2 neutrons} &= 2(1.67493 \times 10^{-27}\ \text{kg}) \\ +\ \text{Mass of 2 protons} &= 2(1.67262 \times 10^{-27}\ \text{kg}) \\ \hline \text{Total mass of nucleons} &= 6.69510 \times 10^{-27}\ \text{kg} \end{aligned}$$

The difference between this value and the actual mass of the ^4He nucleus as listed in Table 21.1 at 6.64466×10^{-27} kg is

$$\begin{aligned} \text{Total mass of nucleons} &= 6.69510 \times 10^{-27}\ \text{kg} \\ -\ \text{Mass of } ^{4}_{2}\text{He nucleus} &= 6.64466 \times 10^{-27}\ \text{kg} \\ \hline &= 0.05044 \times 10^{-27}\ \text{kg} \end{aligned}$$

or 5.044×10^{-29} kg. This tiny difference in mass is called the **mass defect (Δm)** of the ^4He nucleus. It represents the energy that holds the four nucleons together in the nucleus—that is, the **binding energy (BE)** of the nucleus. The binding energy of a nucleus is the energy that would be released if free nucleons were to fuse together to form the nucleus. It is also the energy needed to split the nucleus apart into free nucleons. How big is this energy? We can calculate it using Albert Einstein's equation that relates mass (m) and energy (E):

$$E = mc^2 \tag{21.11}$$

mass defect (Δm) the difference between the mass of a stable nucleus and the masses of the individual nucleons that comprise it.

binding energy (BE) the energy that holds the nucleons together in a nucleus.

strong nuclear force the fundamental force of nature that keeps quarks together in subatomic particles and nucleons together in atomic nuclei.

where c is the speed of light in a vacuum, 2.998×10^8 m/s. Let's use Equation 21.11 to calculate the binding energy of an α particle (^{4}He nucleus) based on its mass defect of 5.044×10^{-29} kg.

$$\begin{aligned} \text{BE} &= (\Delta m)c^2 \\ &= 5.045 \times 10^{-29}\ \text{kg} \times (2.998 \times 10^8\ \text{m/s})^2 \\ &= 4.534 \times 10^{-12}\ \text{kg} \cdot (\text{m/s})^2 = 4.534 \times 10^{-12}\ \text{J} \end{aligned}$$

This amount of energy may seem very small, but it is the binding energy of just a single helium-4 nucleus. Its value per mole of ^{4}He is equivalent to billions of kilojoules:

$$\frac{4.534 \times 10^{-12}\ \cancel{\text{J}}}{\cancel{\text{atom}}} \times \frac{6.022 \times 10^{23}\ \cancel{\text{atoms}}}{\text{mol}} \times \frac{1\ \text{kJ}}{1000\ \cancel{\text{J}}} = 2.730 \times 10^9\ \text{kJ/mol}$$

By comparison, consider that the bond dissociation energy of H_2 at 436 kJ/mol is more than 6,000,000 times smaller than the binding energy of a mole of He atoms. This enormous amount of energy means that ^{4}He is a very stable nuclide.

Most of the stable nuclei with larger numbers of nucleons than ^{4}He have even larger binding energies. This makes sense because a nucleus with many protons in close proximity to each other must be held together by an enormous energy to overcome the electrostatic repulsion between all of the positively charged particles. In such close proximity, nucleons come under the influence of a fundamental force of nature known as the **strong nuclear force**. It operates only over very small distances, such as the diameters of atomic nuclei, but it is 100 times stronger than the repulsions the protons experience. The strong nuclear force binds nucleons together and stabilizes atomic nuclei.

To make comparisons of nuclear binding energies fair, they are usually divided by the number of nucleons in each nucleus. Expressing binding energy on a per nucleon basis allows us to compare the relative stabilities of different nuclides. When we plot binding energy per nucleon values against mass number, we get the curve shown in Figure 21.8.

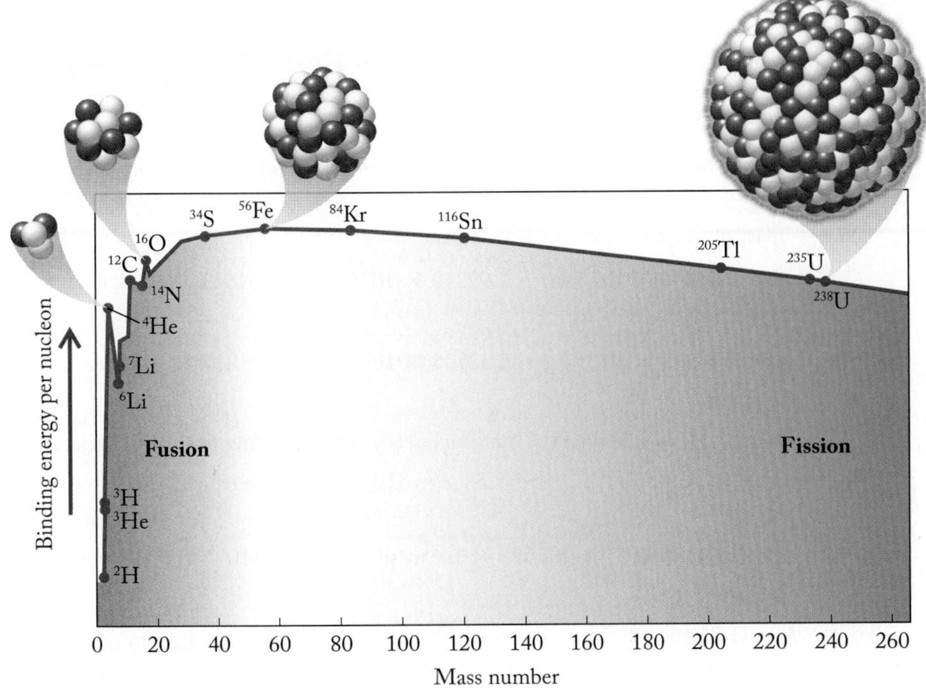

FIGURE 21.8 The stability of a nucleus is directly proportional to its binding energy per nucleon, which reaches a maximum at ^{56}Fe.

SAMPLE EXERCISE 21.5 Calculating the Binding Energy of an Isotope LO5

Recall from Section 21.2 that ^{12}C is a stable isotope of carbon but ^{14}C is radioactive. Calculate the binding energy per nucleon of these two nuclei, given their exact masses as 12.00000 amu and 14.00324 amu, respectively.

Collect and Organize We are asked to calculate the binding energy of two nuclei: ^{12}C and ^{14}C. We are given their exact masses, but we will also need the masses of a single neutron (1.67493×10^{-27} kg) and a single proton (1.67262×10^{-27} kg). We will also need a conversion factor to change the mass of ^{12}C and ^{14}C from amu to kg: 1.66054×10^{-27} kg/amu.

Analyze We must calculate the mass defect between the masses of the nuclei of the two isotopes and the masses of their subatomic particles. Using Equation 21.11, we convert the mass defect to energy. This energy represents the binding energy of each nucleus. We predict that the binding energy of the stable isotope of carbon, ^{12}C, will be greater than that of the radioactive (unstable) isotope, ^{14}C.

Solve First we convert the exact mass of one atom of each isotope from amu to kg:

$$^{12}_{6}\text{C: } 12.00000\ \cancel{\text{amu}} \times \frac{1.66054 \times 10^{-27}\text{ kg}}{\cancel{\text{amu}}} = 1.99265 \times 10^{-26}\text{ kg}$$

$$^{14}_{6}\text{C: } 14.00324\ \cancel{\text{amu}} \times \frac{1.66054 \times 10^{-27}\text{ kg}}{\cancel{\text{amu}}} = 2.32529 \times 10^{-26}\text{ kg}$$

Since we will be comparing the mass of the ^{12}C and ^{14}C nuclei, we need to subtract the mass of 6 electrons from the exact masses of the ^{12}C and ^{14}C atoms:

$$^{12}_{6}\text{C: } 1.99265 \times 10^{-26}\text{ kg} - 6(9.10939 \times 10^{-31}\text{ kg}) = 1.99210 \times 10^{-26}\text{ kg}$$

$$^{14}_{6}\text{C: } 2.3259 \times 10^{-26}\text{ kg} - 6(9.10939 \times 10^{-31}\text{ kg}) = 2.32535 \times 10^{-26}\text{ kg}$$

Next we calculate the mass of the protons and neutrons in the nucleus of one atom of each isotope:

$$^{12}_{6}\text{C: } \left(6\ \cancel{\text{protons}} \times \frac{1.67262 \times 10^{-27}\text{ kg}}{\cancel{\text{proton}}}\right) + \left(6\ \cancel{\text{neutrons}} \times \frac{1.67493 \times 10^{-27}\text{ kg}}{\cancel{\text{neutron}}}\right)$$
$$= 2.00853 \times 10^{-26}\text{ kg}$$

$$^{14}_{6}\text{C: } \left(6\ \cancel{\text{protons}} \times \frac{1.67262 \times 10^{-27}\text{ kg}}{\cancel{\text{proton}}}\right) + \left(8\ \cancel{\text{neutrons}} \times \frac{1.67493 \times 10^{-27}\text{ kg}}{\cancel{\text{neutron}}}\right)$$
$$= 2.34352 \times 10^{-26}\text{ kg}$$

The difference between the exact mass of the nucleus of the isotope and its subatomic particles (the mass defect, Δm) is

$$\Delta m = \text{mass of nucleus} - \text{mass of nucleons}$$

$$^{12}_{6}\text{C: } \Delta m = 2.00853 \times 10^{-26}\text{ kg} - 1.99210 \times 10^{-26}\text{ kg} = 0.01643 \times 10^{-26}\text{ kg}$$
$$= 1.643 \times 10^{-28}\text{ kg}$$

$$^{14}_{6}\text{C: } \Delta m = 2.34352 \times 10^{-26}\text{ kg} - 2.32535 \times 10^{-26}\text{ kg} = 0.01817 \times 10^{-26}\text{ kg}$$
$$= 1.817 \times 10^{-28}\text{ kg}$$

Using Equation 21.11, to calculate the binding energy corresponding to this difference in mass:

$$E = mc^2$$

$$^{12}_{6}\text{C: } E = (1.643 \times 10^{-28}\text{ kg})(2.998 \times 10^{8}\text{ m/s})^2 = 1.477 \times 10^{-11}\text{ kg}\cdot\text{m}^2/\text{s}^2$$
$$= 1.477 \times 10^{-11}\text{ J}$$

$$^{14}_{6}\text{C: } E = (1.817 \times 10^{-28}\text{ kg})(2.998 \times 10^{8}\text{ m/s})^2 = 1.633 \times 10^{-11}\text{ kg}\cdot\text{m}^2/\text{s}^2$$
$$= 1.633 \times 10^{-11}\text{ J}$$

To calculate the binding energy per nucleon, we divide each value of E by the number of nucleons

$$^{12}_{6}\text{C}: E = \frac{1.477 \times 10^{-11}\text{ J}}{12\text{ nucleons}} = \frac{1.231 \times 10^{-12}\text{ J}}{\text{nucleon}}$$

$$^{14}_{6}\text{C}: E = \frac{1.633 \times 10^{-11}\text{ J}}{14\text{ nucleons}} = \frac{1.166 \times 10^{-12}\text{ J}}{\text{nucleon}}$$

Think About It The difference in mass between the exact mass of a nucleus and its subatomic particles represents the binding energy. Stable nuclei should have larger binding energies per nucleon than unstable nuclei. The binding energy per nucleon of radioactive ^{14}C (1.166×10^{-12} J/nucleon) is indeed less than the binding energy per nucleon of stable ^{12}C (1.231×10^{-12} J/nucleon) as predicted.

Practice Exercise A "free" neutron undergoes β decay, forming a proton and an electron (a β particle). How much energy is released in this nuclear reaction? Express your answer in kJ/mol.

21.5 Making New Elements

Rutherford's experiments with gold foil led him to bombard other elements with α particles. In a 1919 experiment, Rutherford reported that passing α particles through nitrogen yielded two products: oxygen-17 and free protons. Bombardment of nuclei by α particles became a popular method for transmuting elements in the 1920s and 1930s. In 1933, French chemists Irène (1897–1956) and Frédéric (1900–1958) Joliot-Curie synthesized the first radionuclide not found in nature, phosphorus-30, by bombarding aluminum-27 with α particles:

$$^{27}_{13}\text{Al} + ^{4}_{2}\text{He} \rightarrow ^{30}_{15}\text{P} + ^{1}_{0}\text{n} \qquad (21.12)$$

In 1940 elements 93 (neptunium) and 94 (plutonium) were produced at the University of California, Berkeley, by bombarding uranium-238 with neutrons. Neither of these elements occur naturally and their synthesis opened the door for extending the periodic table beyond the known elements. Transmutation by neutron bombardment is easier than by α bombardment because both α particles and atomic nuclei have positive charges and so repel each other. The α particles must be traveling at high speeds to overcome this repulsion in a successful bombardment. Because neutrons have no charge, however, they are more readily captured by nuclei. For example, when nuclei of uranium-238 are bombarded with neutrons, a nuclear reaction may occur in which two β particles are ejected, leading to the formation of plutonium-239:

$$^{238}_{92}\text{U} + ^{1}_{0}\text{n} \rightarrow ^{239}_{94}\text{Pu} + 2\,^{0}_{-1}\beta \qquad (21.13)$$

Between 1944 and 1961, the Berkeley research team synthesized elements through $Z = 103$, lawrencium, by bombarding actinide nuclei with different combinations of neutrons and α particles. The research team was led by American chemist Glenn T. Seaborg (1912–1999, Figure 21.9), who won the Nobel Prize in Chemistry in 1951 for his team's ability to first synthesize and then characterize the chemical properties of these *transuranium* elements.

The methods used by Seaborg's team did not work for synthesizing elements with $Z > 103$ because nuclides more massive than californium-249 are highly unstable and rapidly lose α or β particles, so they are not useful target materials for making even more massive nuclides. However, by bombarding californium-249 targets with carbon, nitrogen, and oxygen nuclei instead of

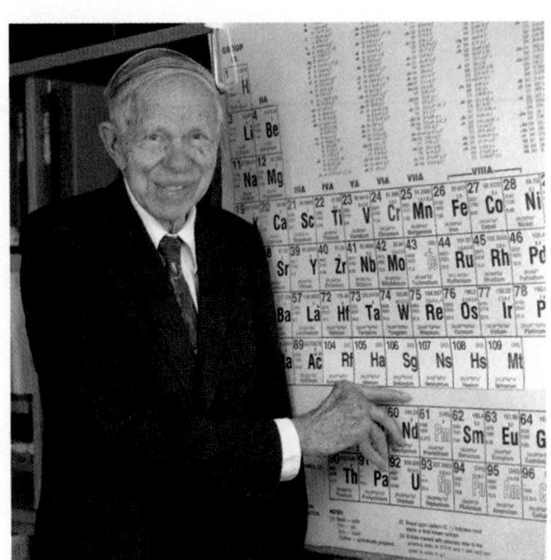

FIGURE 21.9 When element 106 was named seaborgium in 1994, Glenn T. Seaborg became the only living scientist to have an element named after him.

TABLE 21.3 Synthesis of Supermassive Nuclides

Bombarding Ion	Target	Nuclide Synthesized	Year First Synthesized
^{62}Ni	^{208}Pb	$^{269}_{110}Ds$	1994
^{64}Ni	^{209}Bi	$^{272}_{111}Rg$	1994
^{69}Zn	^{208}Pb	$^{277}_{112}Cn$	1996
^{70}Zn	^{209}Bi	$^{278}_{113}Nh$	2003
^{48}Ca	^{244}Pu	$^{289}_{114}Fl$	1999
^{48}Ca	^{243}Am	$^{288}_{115}Mc$	2003
^{48}Ca	^{248}Cm	$^{293}_{116}Lv$	2000
^{48}Ca	^{249}Bk	$^{293}_{117}Ts$	2009
^{48}Ca	^{249}Cf	$^{294}_{118}Og$	2002

α particles, scientists have been able to synthesize isotopes of rutherfordium ($Z = 104$), dubnium ($Z = 105$), and seaborgium ($Z = 106$):

$$^{249}_{98}\text{Cf} + ^{12}_{6}\text{C} \rightarrow ^{257}_{104}\text{Rf} + 4^{1}_{0}\text{n}$$

$$^{249}_{98}\text{Cf} + ^{15}_{7}\text{N} \rightarrow ^{260}_{105}\text{Dd} + 4^{1}_{0}\text{n}$$

$$^{249}_{98}\text{Cf} + ^{18}_{8}\text{O} \rightarrow ^{263}_{106}\text{Sg} + 4^{1}_{0}\text{n}$$

In recent years, scientists have reported synthesizing nuclides that have as many as 118 protons by bombarding targets as massive as californium-249 with medium-mass nuclei, such as calcium-48. In January 1999, for example, an atom with 114 protons and 175 neutrons was synthesized and lasted for 30 s (most supermassive nuclides have half-lives that are fractions of a second) before undergoing a series of α decay events that yielded isotopes of elements 112, 110, and 108. Some of these "supermassive" elements, and the bombarding ions and target nuclides used to make them, are listed in Table 21.3. Confirmation of new unstable elements can take decades. The syntheses of elements 113 (nihonium), 115 (moscovium), 117 (tennessine), and 118 (oganesson) were recognized in January 2016.

Why bother to make such short-lived elements? Their mere existence, no matter for how brief a time, can be a source of insight into the nature of nuclear structure. Their behavior illustrates the competition between the strong nuclear force that holds nucleons together and the electrostatic repulsion between protons. Supermassive elements are pieces of a puzzle that someday may tell us whether there is a limit to the size of atoms. Our ability to create new elements in the laboratory raises the following question: Where did the stable elements found on Earth come from?

21.6 Fusion and the Origin of the Elements

Theoretical physicists believe that our universe began about 13.7 billion years ago with an enormous release of energy that has come to be known as the Big Bang. Current theories suggest that within a few microseconds of the Big Bang, much of its energy had transformed into matter made of elementary particles such as electrons and **quarks** (Figure 21.10). Less than a millisecond later, the universe had expanded and "cooled" to a temperature of 10^{12} K, which allowed quarks to combine with one another to form neutrons and protons. Thus, in less than a second, the universe contained the three types of subatomic particles that would eventually make up atoms.

quarks elementary particles that combine to form neutrons and protons.

nucleosynthesis the natural formation of nuclei as a result of fusion and other nuclear processes.

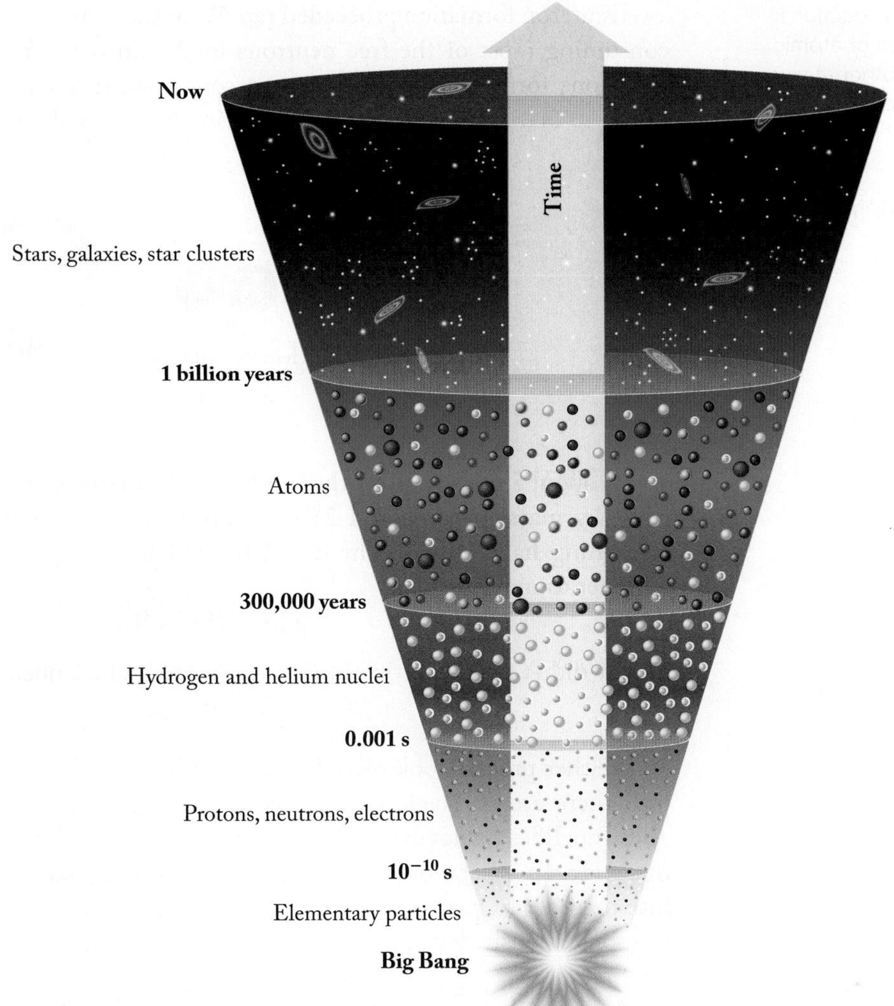

FIGURE 21.10 Time line for energy and matter transformations believed to have occurred since the universe began. In this model, protons, neutrons, and electrons were formed from quarks in the first millisecond after the Big Bang, followed by hydrogen and helium nuclei. Atoms of ^{1}H and ^{4}He did not form until after 300,000 years of expansion and cooling, and other elements did not form until the first galaxies appeared around 1 billion years after the Big Bang. According to this model, our solar system, our planet, and all life-forms on it are composed of elements synthesized in stars that were born, burned brightly, and then disappeared millions to billions of years after the Big Bang.

Primordial Nucleosynthesis

By about 4 minutes after the Big Bang, the universe had expanded and cooled to about 10^9 K. In this hot, dense subatomic "soup," neutrons and protons that collided with one another began to fuse together in a process called primordial **nucleosynthesis**. When a proton and a neutron collide and fuse together, they form a *deuteron* ($^{2}_{1}$D), which is the nucleus of the deuterium isotope of hydrogen:

$$^{1}_{1}\text{H} + {}^{1}_{0}\text{n} \rightarrow {}^{2}_{1}\text{D} \qquad (21.14)$$

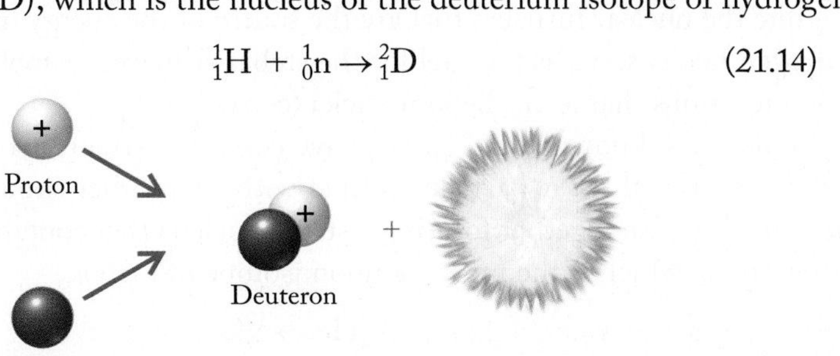

nuclear fusion a nuclear reaction in which subatomic particles or atomic nuclei collide with each other at very high speeds and fuse together, forming more massive nuclei and releasing energy.

neutron capture the absorption of a neutron by a nucleus.

Deuteron formation proceeded rapidly in the minutes following the Big Bang, consuming most of the free neutrons in the universe. However, no sooner did deuterons form than they, too, were consumed in collisions with one another, fusing to make $^{4}_{2}He$ nuclei (α particles) in a process called **nuclear fusion**:

$$2\,^{2}_{1}D \rightarrow \,^{4}_{2}He \quad (21.15)$$

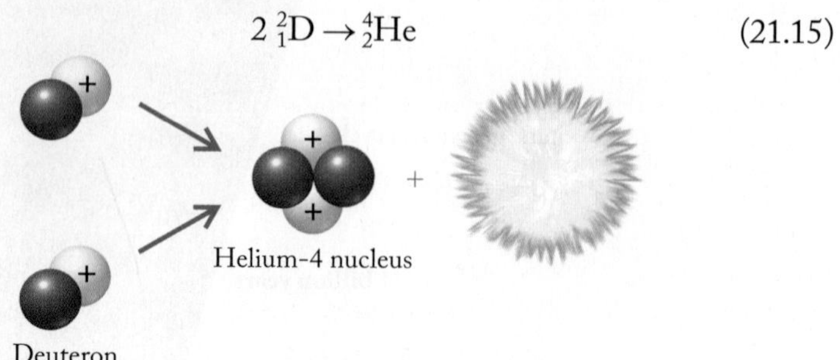

By about 5 minutes after the Big Bang, the matter of the universe had become 75% (by mass) protons and 25% α particles. Nucleosynthesis then came to a screeching halt. Why? Why didn't α particles and protons, for example, fuse together to make ^{5}Li:

$$^{4}_{2}He + \,^{1}_{1}H \xrightarrow{?} \,^{5}_{3}Li$$

or why didn't two α particles fuse together to make a nucleus of ^{8}Be?

$$2\,^{4}_{2}He \xrightarrow{?} \,^{8}_{4}Be$$

Neither process took place because neither ^{5}Li nor ^{8}Be is a stable nuclide. In fact, there are no stable nuclides with mass numbers of 5 or 8 in Figure 21.1. These nuclides do not exist because they have no more binding energy per nucleon than ^{4}He. In other words, no energy would be released if two ^{4}He nuclei were to fuse together to form ^{8}Be. Similarly, ^{8}Be nuclei would require no input of energy to spontaneously decompose into ^{4}He nuclei. Actually, ^{8}Be nuclei never form in the first place.

As the early universe continued to expand and cool, the kinetic energies of the particles of matter in it (hydrogen and helium nuclei and free electrons) decreased and their motions slowed, allowing nuclei to combine with free electrons to produce neutral atoms. The result was a universe that was 75% hydrogen and 25% helium. It remained that way for millions of years—until the first galaxies formed.

3 [α particle]

$^{12}_{6}C$ $^{4}_{2}α$

$^{16}_{8}O$ $^{4}_{2}α$

$^{20}_{10}Ne$ $^{4}_{2}α$

$^{24}_{12}Mg$

FIGURE 21.11 Fusion of three α particles forms carbon-12, followed by fusion of successively more massive nuclei to form oxygen-16, neon-20, magnesium-24, and so on. The fusion processes release the energy that fuels the nuclear furnaces of stars today.

Stellar Nucleosynthesis

Most of the matter detected and identified in the universe today is still hydrogen and helium, which is strong evidence supporting the Big Bang theory. But how did the other elements in the periodic table eventually form, including those that make up most of our planet? Scientists theorize that the synthesis of elements more massive than helium had to wait until nuclear fusion resumed in the first generation of stars. Inside the coalescing masses of hydrogen and helium that became the first stars, these gases underwent enormous compression heating, becoming hot enough to ignite the nuclear furnaces that are the source of the energy in all stars. Initially hydrogen nuclei serve as the fuel for the stellar furnaces, combining in a series of fusion reactions that forms helium nuclei (α particles).

Some stars, known as *red giants*, have cores so extraordinarily hot and dense that sometimes three α particles collide with each other simultaneously. When they do, they fuse together forming a stable nucleus that contains six protons and six neutrons, which is the most common isotope of carbon:

$$3\,^{4}_{2}He \rightarrow \,^{12}_{6}C \quad (21.16)$$

With the formation of ^{12}C, the barrier that had halted primordial nucleosynthesis was overcome. In the cores of giant stars, ^{12}C nuclei fuse with α particles to form ^{16}O (Figure 21.11). Then ^{16}O nuclei may fuse with more α particles to form ^{20}Ne, and so on. Additional fusion reactions involving nuclei with increasingly greater positive charges are possible in intensely hot (10^9 K) stars because the nuclei have enough kinetic energy, and are moving fast enough, to overcome the electrostatic repulsion experienced by particles with large positive charges.

For billions of years fusion reactions of this sort have simultaneously fueled the nuclear furnaces of stars and produced isotopes as heavy as ^{56}Fe (Figure 21.12). However, once the core of a star turns into iron, the star is in trouble because fusion reactions involving iron nuclei do not release energy; instead they consume it. This happens because the binding energy per nucleon (and, therefore, the nuclear stability) reaches a maximum with ^{56}Fe (Figure 21.8). Thus, a star with an iron core has essentially run out of fuel. Its nuclear furnace goes out, and the star begins to cool and collapse into itself.

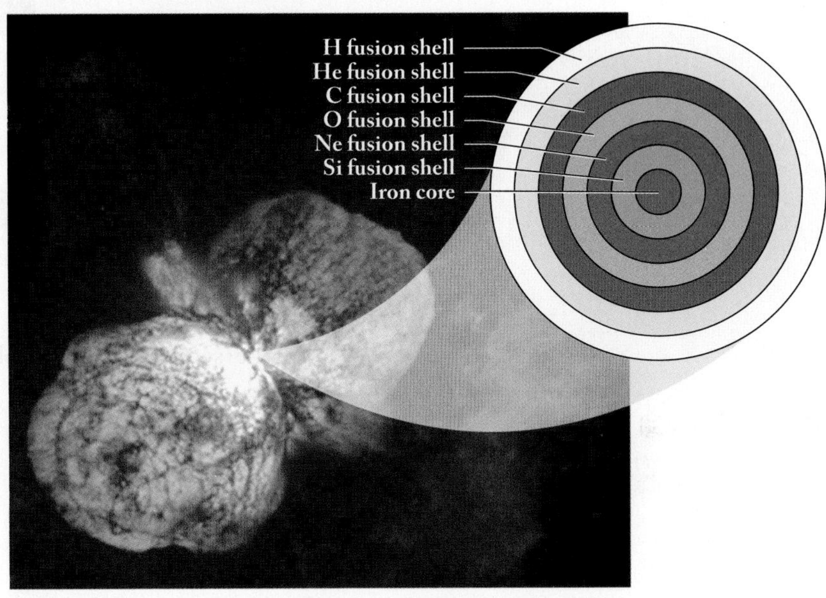

FIGURE 21.12 The star Eta Carinae is believed to be evolving toward an explosion. The outer regions are still fueled by energy released as hydrogen isotopes fuse, but the star is increasingly hotter and denser closer to the center. This central heating allows the fusion of larger nuclei and results in the production of ^{56}Fe in the core.

As the star collapses, compression reheats its core to above 10^9 K. At such temperatures, nuclei begin to disintegrate into free protons and neutrons. Free neutrons may collide and fuse with atomic nuclei in a process called **neutron capture**. If a stable nucleus captures enough neutrons, it becomes unstable. For example, if a nucleus of ^{56}Fe captures three neutrons, it forms the unstable, neutron-rich nuclide ^{59}Fe, which undergoes β decay to cobalt-59:

$$^{56}_{26}\text{Fe} + 3\,^{1}_{0}\text{n} \rightarrow {}^{59}_{26}\text{Fe} \rightarrow {}^{59}_{27}\text{Co} + {}^{0}_{-1}\beta \qquad (21.17)$$

The combination of repeated neutron capture and β decay events in the cores of collapsing stars produces the most massive stable nuclides in the periodic table.

Eventually, the enormous heating that occurs when a massive star collapses produces a gigantic explosion. Cosmologists call such an event a *supernova*. In addition to finishing the job of synthesizing the elemental building blocks found in the universe—up to and including the isotopes of uranium—a supernova serves as its own element-distribution system, blasting its inventory of elements throughout its galaxy (Figure 21.13). The legacies of supernovas are found in the elemental composition of later-generation stars like our sun and in the planets that orbit the stars. Indeed, all the matter in our solar system—including us—is essentially demolition debris from ancient exploding stars.

FIGURE 21.13 The Crab Nebula in the constellation Taurus is actually the debris field of a supernova that was observed on Earth in the year 1054.

Nucleosynthesis in Our Sun

The energy of our sun is derived from a nuclear fusion process that involves more steps than the process that took place during primordial nucleosynthesis when protons (hydrogen nuclei) and neutrons fused together to form deuterons (Equation 21.14).

This is because free neutron concentrations in the sun are far lower there than they were in the primordial universe. Instead, colliding protons may fuse together to form a deuteron and a positron:

$$2\,{}^{1}_{1}\mathrm{H} \rightarrow {}^{2}_{1}\mathrm{D} + {}^{0}_{1}\beta \quad (21.18)$$

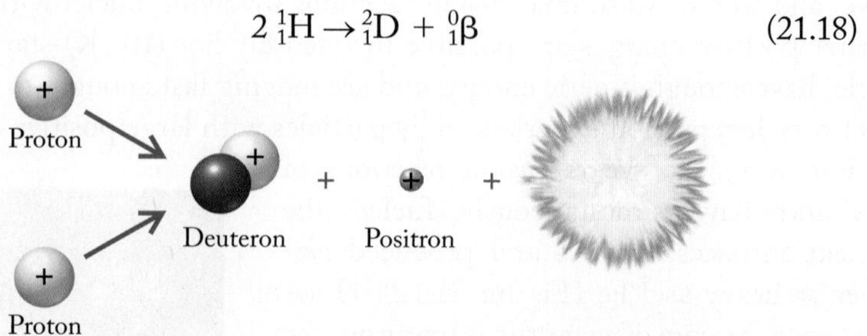

CONCEPT TEST

What particle is formed when a proton fuses with an electron?

In the second stage of solar fusion, protons fuse with deuterons to form helium-3 nuclei:

$${}^{1}_{1}\mathrm{H} + {}^{2}_{1}\mathrm{D} \rightarrow {}^{3}_{2}\mathrm{He} \quad (21.19)$$

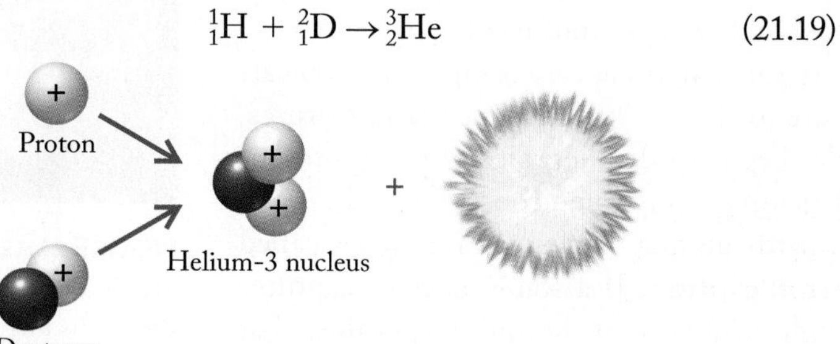

Finally, fusion of two helium-3 nuclei produces a helium-4 nucleus and two protons:

$$2\,{}^{3}_{2}\mathrm{He} \rightarrow {}^{4}_{2}\mathrm{He} + 2\,{}^{1}_{1}\mathrm{H} \quad (21.20)$$

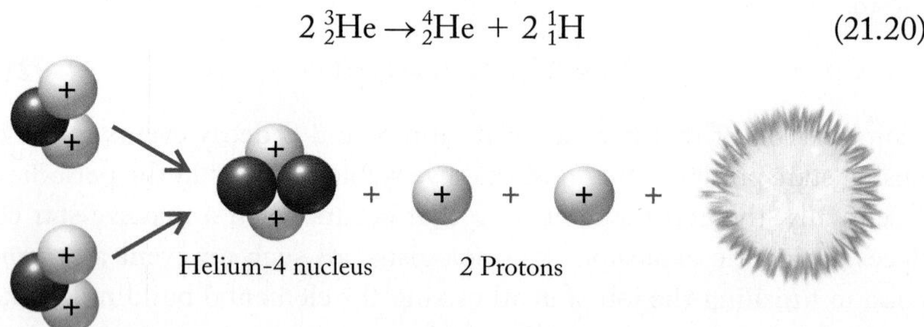

Deuterium and ^{3}He nuclei are intermediates in the hydrogen-fusion process because they are made in one step but then consumed in another. To write an overall equation for solar fusion, we combine Equations 21.18–21.20, multiplying Equations 21.18 and 21.19 by 2 to balance the production and consumption of the intermediate particles:

$$\begin{aligned} & 2\,[2\,{}^{1}_{1}\mathrm{H} \rightarrow {}^{2}_{1}\mathrm{D} + {}^{0}_{1}\beta] \\ +\; & 2\,[{}^{1}_{1}\mathrm{H} + {}^{2}_{1}\mathrm{D} \rightarrow {}^{3}_{2}\mathrm{He}] \\ +\; & 2\,{}^{3}_{2}\mathrm{He} \rightarrow {}^{4}_{2}\mathrm{He} + 2\,{}^{1}_{1}\mathrm{H} \\ \hline & 4\,\cancel{6}\,{}^{1}_{1}\mathrm{H} + \cancel{2\,{}^{2}_{1}\mathrm{D}} + \cancel{2\,{}^{3}_{2}\mathrm{He}} \rightarrow \cancel{2\,{}^{2}_{1}\mathrm{D}} + \cancel{2\,{}^{3}_{2}\mathrm{He}} + {}^{4}_{2}\mathrm{He} + \cancel{2\,{}^{1}_{1}\mathrm{H}} + 2\,{}^{0}_{1}\beta \end{aligned}$$

This reduces to

$$4\,{}^{1}_{1}\text{H} \rightarrow {}^{4}_{2}\text{He} + 2\,{}^{0}_{1}\beta \qquad (21.21)$$

4 Protons Helium-4 nucleus 2 Positrons

Annihilation reactions between the positrons produced in Equation 21.21 and electrons in the matter surrounding the reactants release considerable energy, but most of the energy from hydrogen fusion comes from the loss in mass as four protons are transformed into an alpha particle and two positrons. We calculate how much energy that is in Sample Exercise 21.6.

SAMPLE EXERCISE 21.6 Calculating the Energy Released in a Nuclear Reaction — LO5

How much energy in joules is released by the overall fusion process in which four protons undergo nuclear fusion, producing an α particle and two positrons (Equation 21.21)?

Collect and Organize We are asked to calculate the energy released in the following nuclear reaction:

$$4\,{}^{1}_{1}\text{H} \rightarrow {}^{4}_{2}\text{He} + 2\,{}^{0}_{1}\beta \qquad (21.21)$$

The energies associated with nuclear reactions are related to differences in the masses of the reactant and product particles by Einstein's equation, $E = mc^2$. The masses of the particles in Table 21.1 are given in kilograms, which is convenient because the units of energy and mass are related as follows:

$$1\text{ J} = 1\text{ kg (m/s)}^2$$

Analyze Given the value of the masses in Table 21.1, the difference in mass will probably be less than 10^{-27} kg. When multiplied by the square of the speed of light $(2.998 \times 10^8\text{ m/s})^2 \approx 10^{17}$, the calculated value of E should be less than 10^{-10} J.

Solve First we calculate the change in mass:

$$\begin{aligned}\Delta m &= (m_{\alpha\text{ particle}} + 2\,m_{\text{positron}}) - 4\,m_{\text{proton}}\\ &= [6.64465 \times 10^{-27} + (2 \times 9.10939 \times 10^{-31})]\text{ kg} - (4 \times 1.67262 \times 10^{-27})\text{ kg}\\ &= -4.40081 \times 10^{-29}\text{ kg}\end{aligned}$$

The energy corresponding to this loss in mass is calculated using Einstein's equation, where $m = -4.40081 \times 10^{-29}$ kg:

$$\begin{aligned}E &= mc^2\\ &= -4.40081 \times 10^{-29}\text{ kg} \times (2.998 \times 10^8\text{ m/s})^2\\ &= -3.955 \times 10^{-12}\text{ kg}\cdot(\text{m/s})^2 = -3.955 \times 10^{-12}\text{ J}\end{aligned}$$

Think About It The decrease in mass translates into energy lost by the reaction system to its surroundings. As we predicted, the absolute value of this energy is less (actually much less) than 10^{-10} J, which seems like an awfully small value compared to the world's energy needs. However, this value applies to the formation of a single α particle. If we multiply it by Avogadro's number and convert to a value in kilojoules per mole, a unit we typically use in thermochemistry, we get

$$\frac{-3.955 \times 10^{-12}\text{ J}}{\alpha\text{ particle}} \times \frac{6.022 \times 10^{23}\ \alpha\text{ particles}}{\text{mol}} \times \frac{1\text{ kJ}}{1000\text{ J}} = -2.382 \times 10^{9}\text{ kJ/mol}$$

nuclear fission a nuclear reaction in which the nucleus of an element splits into two lighter nuclei. The process is usually accompanied by the release of one or more neutrons and energy.

chain reaction a self-sustaining series of fission reactions in which the neutrons released when nuclei split apart initiate additional fission events and sustain the reaction.

critical mass the minimum quantity of fissionable material needed to sustain a chain reaction.

To put this value in perspective, it is about 10^7 times the change in free energy from the combustion of one mole of hydrogen gas.

Practice Exercise For decades scientists and engineers have sought to harness the enormous energy released during hydrogen fusion for peaceful purposes by using high-speed collisions between deuterium and tritium (^3H) nuclei to produce helium-4 nuclei in Equation 21.24. How much energy is released in the nuclear reaction described by Equation 21.24? Express your answer in kJ/mol. The mass of a tritium nucleus is 5.00827×10^{-27} kg.

$$^{2}_{1}H + ^{3}_{1}H \rightarrow ^{4}_{2}He + ^{1}_{0}n \quad (21.24)$$

21.7 Nuclear Fission

Unlike the nuclear fusion reactions described in Section 21.6, when an atom of uranium-235 captures a neutron, the nucleus of the unstable product, uranium-236, splits into two lighter nuclei in a process called **nuclear fission**. Several uranium-235 fission reactions can occur, including these three:

$$^{235}_{92}U + ^{1}_{0}n \rightarrow ^{141}_{56}Ba + ^{92}_{36}Kr + 3\,^{1}_{0}n$$

$$^{235}_{92}U + ^{1}_{0}n \rightarrow ^{137}_{52}Te + ^{97}_{40}Zr + 2\,^{1}_{0}n$$

$$^{235}_{92}U + ^{1}_{0}n \rightarrow ^{138}_{55}Cs + ^{96}_{37}Rb + 2\,^{1}_{0}n$$

In all these reactions, the sums of the masses of the products are slightly less than the sums of the masses of the reactants. As with hydrogen fusion, this difference in mass is converted to energy.

Fission reactions may also produce additional neutrons that smash into other uranium-235 nuclei and initiate more fission events in a **chain reaction** (Figure 21.14). The reaction proceeds as long as there are enough uranium-235 nuclei present to absorb the neutrons being produced. On average, at least one neutron from each fission event must cause another nucleus to split apart for the chain reaction to be self-sustaining. The quantity of fissionable material needed to assure that every fission event produces another is called the **critical mass**. For uranium-235, the critical mass is about 1 kg of the pure isotope.

Uranium-235 is the most abundant fissionable isotope, but it makes up only 0.72% of the uranium in the principal uranium ore, pitchblende (Figure 21.15a). The uranium in nuclear reactors must be at least 3% to 4% uranium-235, and enrichment to about 85% is needed for nuclear weapons. The most common method for enriching uranium ore involves extracting the uranium in a process that yields a material called yellowcake, which is mostly U_3O_8 (Figure 21.15b). This oxide is then converted to UF_6, which, despite a molar mass of over 300 g/mol, is a volatile solid that sublimes at 56°C. The volatility of this nonpolar molecular compound can be explained by the relatively weak London dispersion forces experienced by its compact, symmetrical molecules (Figure 21.16). Fissionable $^{235}UF_6$ is

FIGURE 21.14 Each fission event in the chain reaction of a uranium-235 nucleus begins when the nucleus captures a neutron, forming an unstable uranium-236 nucleus that then splits apart (fissions) in one of several ways. In the first process shown here, the uranium-236 nucleus splits into krypton-92, barium-141, and three neutrons. If, on average, at least one of the three neutrons from each fission event causes the fission of another uranium-235 nucleus, then the process is sustained in a chain reaction.

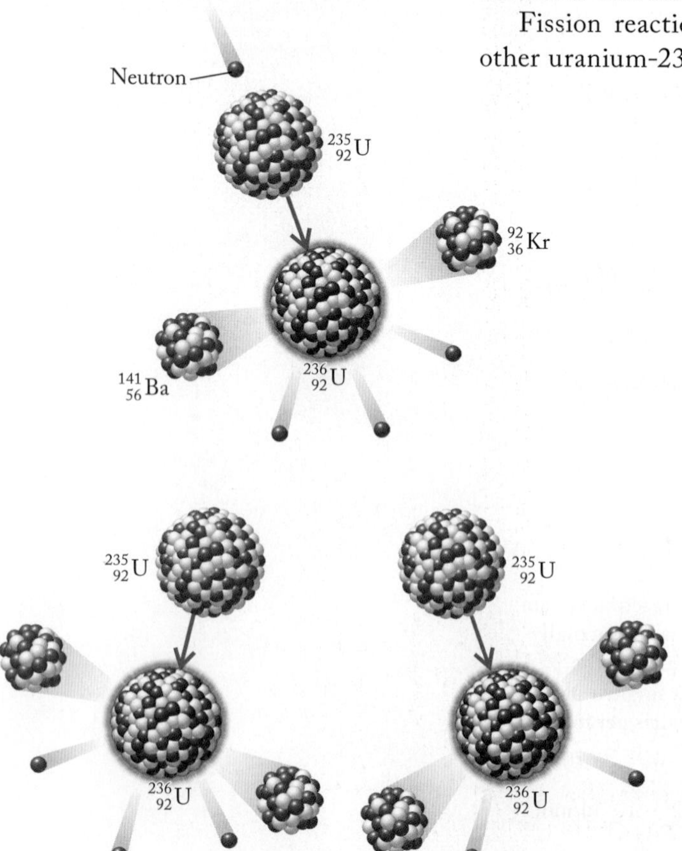

(a)

(b)

(c)

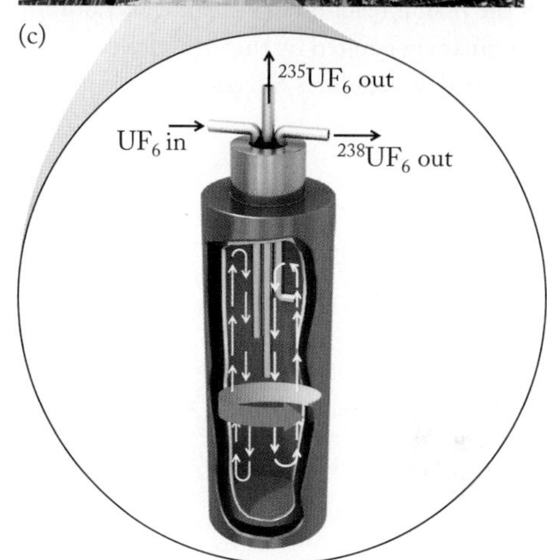

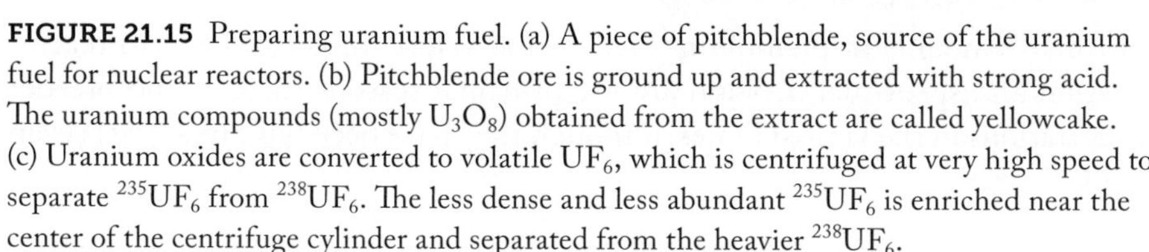

FIGURE 21.15 Preparing uranium fuel. (a) A piece of pitchblende, source of the uranium fuel for nuclear reactors. (b) Pitchblende ore is ground up and extracted with strong acid. The uranium compounds (mostly U_3O_8) obtained from the extract are called yellowcake. (c) Uranium oxides are converted to volatile UF_6, which is centrifuged at very high speed to separate $^{235}UF_6$ from $^{238}UF_6$. The less dense and less abundant $^{235}UF_6$ is enriched near the center of the centrifuge cylinder and separated from the heavier $^{238}UF_6$.

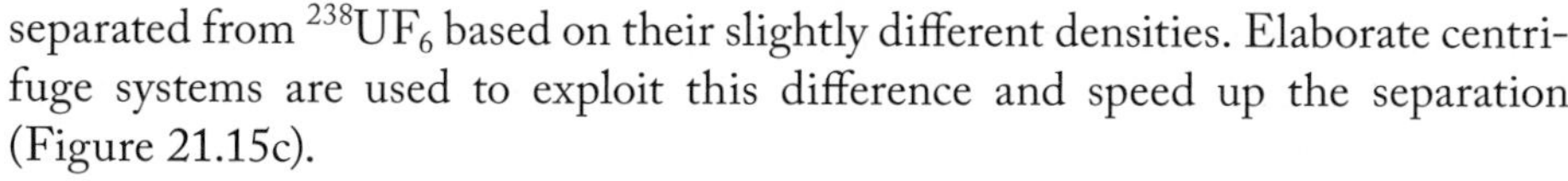

separated from $^{238}UF_6$ based on their slightly different densities. Elaborate centrifuge systems are used to exploit this difference and speed up the separation (Figure 21.15c).

Harnessing the energy released by nuclear fission to generate electricity began in the middle of the 20th century. In a typical nuclear power plant (Figure 21.17), fuel rods containing 3% to 4% uranium-235 are interspersed with rods of boron or cadmium that control the rate of the fission by absorbing some of the neutrons produced during fission. Pressurized water flows around the fuel and control rods, removing the heat created during fission and transferring it to a steam generator. The water also acts as a moderator, slowing down the neutrons and thereby allowing for their more efficient capture by ^{235}U atoms.

In 1952 the first **breeder reactor** was built, so called because in addition to producing energy to make electricity, the reactor makes ("breeds") its own fuel. The reactor starts out with a mixture of plutonium-239 and uranium-238. As the plutonium fissions and the energy from those reactions is collected to produce electricity, some of the neutrons that are produced sustain the fission chain reaction just as in the reactor depicted in Figure 21.17, while others convert the uranium into more plutonium fuel:

$$^{238}_{92}U + ^{1}_{0}n \rightarrow ^{239}_{92}U + \gamma \rightarrow ^{239}_{94}Pu + 2\ ^{0}_{-1}\beta$$

In less than 10 years of operation, a breeder reactor can make enough plutonium-239 to refuel itself and another reactor. Unfortunately, plutonium-239 is a carcinogen and one of the most toxic substances known. Only about half a kilogram is needed to make an atomic bomb, and it has a long half-life: 2.4×10^4 years. Understandably, extreme caution and tight security surround the handling of plutonium fuel and the transportation and storage of nuclear wastes containing

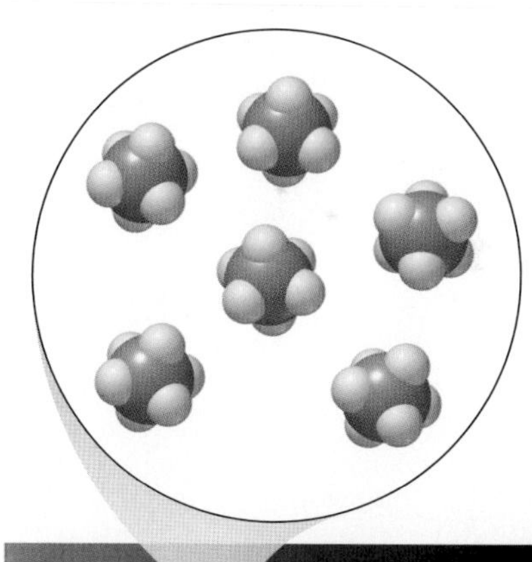

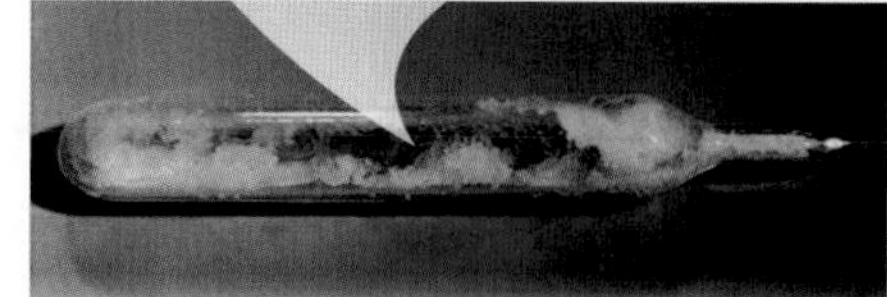

FIGURE 21.16 Uranium hexafluoride is a volatile solid that sublimes at a relatively low temperature of 56°C because it is composed of compact, symmetrical molecules that experience relatively weak London dispersion forces despite their considerable mass.

breeder reactor a nuclear reactor in which fissionable material is produced during normal reactor operation.

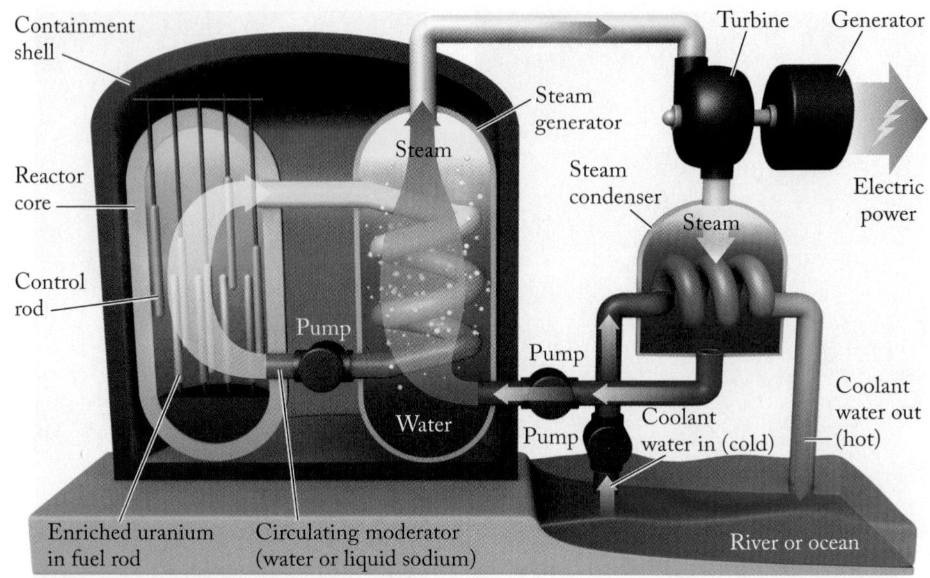

FIGURE 21.17 A pressurized, water-cooled nuclear power plant uses fuel rods containing uranium enriched to about 4% uranium-235. The fission chain reaction is regulated with control rods and a moderator that is either water or liquid sodium. The moderator slows down the neutrons released by fission so that they can be more efficiently captured by other uranium-235 nuclei. It also transfers the heat produced by the fission reaction to a steam generator. The steam generated by this heat drives a turbine that generates electricity.

even small amounts of plutonium. Health and safety matters related to reactor operation and spent-fuel disposal are the principal reasons there are no breeder power stations in the United States, although they have been built in at least seven other countries.

CONCEPT **TEST**

Nuclear reactors powered by the energy released by the fission of uranium-235 have been operating since the 1950s, but a reactor powered by the energy released by the fusion of hydrogen has yet to be built. Why is it taking so long to build a fusion reactor?

21.8 Measuring Radioactivity

French scientist Henri Becquerel discovered radioactivity in 1896 when he observed that uranium and other substances produce radiation that fogs photographic film. Photographic film is still used to detect radioactivity in the film dosimeter badges worn by people working with radioactive materials to record their exposure to radiation. Detectors called **scintillation counters** use materials called *phosphors* to absorb energy released during radioactive decay. The phosphors then release the absorbed energy as visible light, the intensity of which is a measure of the amount of radiation in the sample.

Radioactivity also can be measured with **Geiger counters**, which detect the common products of radioactivity—α particles, β particles, and γ rays—on the basis of their abilities to ionize atoms (Figure 21.18). A sealed metal cylinder, filled with gas (usually argon) and a positively charged electrode, has a window that allows nuclear radiation to enter. Once inside the cylinder, the particles and γ rays ionize argon atoms into Ar^+ ions and free electrons. If an electrical potential difference is applied between the cylinder shell and the central electrode, free electrons migrate toward the positive electrode and argon ions migrate toward the negatively charged shell. This ion migration produces a pulse of electrical current whenever radiation enters the cylinder. The current is amplified and read out to a meter and a speaker that makes a clicking sound.

scintillation counter an instrument that determines the level of radioactivity in samples by measuring the intensity of light emitted by phosphors in contact with the samples.

Geiger counter a portable device for determining nuclear radiation levels by measuring how much the radiation ionizes the gas in a sealed detector.

becquerel (Bq) the SI unit of radioactivity; one becquerel equals one decay event per second.

curie (Ci) non-SI unit of radioactivity; 1 Ci = 3.70×10^{10} decay events per second.

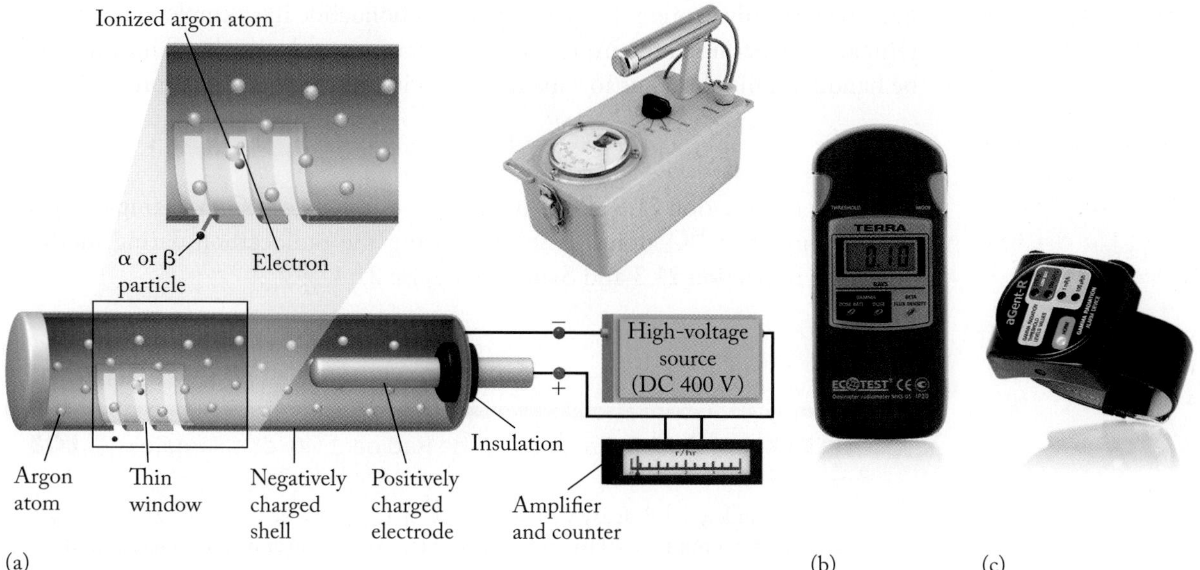

FIGURE 21.18 (a) In a Geiger counter, a particle produced by radioactive decay passes through a thin window usually made of beryllium or a plastic film. Inside the tube, the particle collides with atoms of argon gas and ionizes them. The resulting argon cations migrate toward the negatively charged tube housing, and the electrons migrate toward a positive electrode, creating a pulse of current through the tube. The current pulses are amplified and recorded via a meter and a speaker that produces an audible "click" for each pulse. (b) When radiation hits the silicon dioxide detector in this dosimeter, electrons are ejected, creating holes. The change in conductivity of the SiO_2 correlates with the amount of radiation absorbed. (c) This wristwatch is actually a γ-ray detector, illustrating advances in the miniaturization of radiation detection equipment.

One measure of radioactivity in a sample is the number of decay events per unit time. This parameter is called the radioactivity (A) of the sample. The SI unit of radioactivity is the **becquerel (Bq)**, named in honor of Henri Becquerel and equal to one decay event per second. An older radioactivity unit is the **curie (Ci)**, named in honor of Marie and Pierre Curie, where

$$1 \text{ Ci} = 3.70 \times 10^{10} \text{ Bq} = 3.70 \times 10^{10} \text{ decay events/s}$$

Both the becquerel and the curie quantify the rate at which a radioactive substance decays, which provides a measure of how much of it is in a sample. As noted in Section 21.3, all decay processes follow first-order kinetics. Recall from Section 13.3 that the rate constant k of a first-order reaction is related to the concentration of a reactant R according to the equation

$$\text{Rate} = k[\text{R}]$$

The same mathematical relationships apply to radioactive decay processes except that we refer to radioactivity (A) instead of rate of decay, and to the number of atoms (N) of a radionuclide in a sample instead of its concentration:

$$A = kN \tag{21.25}$$

Because radioactivity is the number of decay events per second, the units of the decay rate constant k are decay events per atom per second. As we saw earlier, the value of k is inversely proportional to the half-life of a radionuclide (Equation 21.7):

$$k = \frac{0.693}{t_{1/2}}$$

Scientists usually express the quantity of a radionuclide in a sample in terms of its radioactivity because it defines how the substance may be used and how it should be handled. This allows us to substitute activity, A, for N in Equation 21.8:

$$\ln \frac{N_t}{N_0} = \ln \frac{A_1}{A_0} = -0.693 \frac{t}{t_{1/2}}$$

This form of Equation 21.8 allows us to use the activity of ^{14}C in a sample in place of the number of ^{14}C atoms in it when using the radiocarbon dating methods described in Section 21.3 and Sample Exercise 21.4.

SAMPLE EXERCISE 21.7 Calculating the Radioactivity of a Sample **LO8**

Radium-223 undergoes β decay with a half-life of 11.4 days. What is the radioactivity of a sample that contains 1.00 μg of ^{223}Ra? Express your answer in becquerels and in curies.

Collect and Organize We are given the half-life and quantity of a radioactive substance and are asked to determine its radioactivity—that is, its rate of decay. All radioactive decay processes are first order, so the decay rate constant k is related to the half-life by Equation 21.7:

$$k = \frac{0.693}{t_{1/2}}$$

and the radioactivity is the product of the rate constant and the number of atoms of radionuclide in the sample (Equation 21.25, $A = kN$).

Analyze Before using Equation 21.7, we must convert the half-life into seconds because both of the radioactivity units we need to calculate are based on decay events per second. To calculate radioactivity, we determine the number of atoms in 1.00 μg of radium. This will likely be a very large number, which should translate into a large number of decay events per second given the relatively short half-life of ^{223}Ra.

Solve The half-life is

$$11.4\ \cancel{d} \times \frac{24\ \cancel{h}}{1\ \cancel{d}} \times \frac{60\ \cancel{min}}{1\ \cancel{h}} \times \frac{60\ s}{1\ \cancel{min}} = 9.85 \times 10^5\ s$$

Using this value in Equation 21.7 and solving for k,

$$k = \frac{0.693}{9.85 \times 10^5\ s}$$

$$= 7.04 \times 10^{-7}\ s^{-1} = 7.04 \times 10^{-7}\ \text{decay events/(atom} \cdot \text{s)}$$

The number of atoms (N) of ^{223}Ra is

$$N = 1.00\ \cancel{\mu g} \times \frac{1\ \cancel{g}}{10^6\ \cancel{\mu g}} \times \frac{1\ \cancel{mol\ Ra}}{223\ \cancel{g\ Ra}} \times \frac{6.022 \times 10^{23}\ Ra}{1\ \cancel{mol\ Ra}} = 2.70 \times 10^{15}\ \text{atoms Ra}$$

Inserting these values of k and N into Equation 21.25,

$$A = kN = \frac{7.04 \times 10^{-7}\ \text{decay events}}{\cancel{\text{atom}} \cdot s} \times 2.70 \times 10^{15}\ \cancel{\text{atoms Ra}} = 1.90 \times 10^9\ \text{decay events/s}$$

Because 1 Bq = 1 decay event/s, the radioactivity of the sample is 1.90×10^9 Bq. Expressing radioactivity in curies,

$$\frac{1.90 \times 10^9\ \cancel{\text{decay events}}}{\cancel{s}} \times \frac{1\ Ci}{3.70 \times 10^{10}\ \cancel{\text{decay events/s}}} = 0.0514\ Ci$$

Think About It We expected a large number of decay events per second because the sample contained a large number of radioactive atoms, which resulted in a high level of radioactivity.

Practice Exercise In March 2011, an earthquake and tsunami off the coast of northeastern Japan crippled nuclear reactors at a power station in Fukushima, Japan. The resulting explosions and fires released ^{133}Xe into the atmosphere. Determine the radioactivity in 1.00 μg of this radionuclide ($t_{1/2}$ = 5.25 days) in becquerels and in millicuries.

21.9 Biological Effects of Radioactivity

The γ rays and many of the α and β particles produced by nuclear reactions have more than enough energy to tear chemical bonds apart, producing odd-electron radicals or free electrons and cations. Consequently, these rays and particles are classified as **ionizing radiation**. Other examples are X-rays and short-wavelength ultraviolet rays. The ionization of atoms and molecules in living tissue can lead to radiation sickness, cancer, birth defects, and death. The scientists who first worked with radioactive materials were not aware of these hazards, and some of them suffered for it. Marie Curie died of aplastic anemia caused by her many years of radiation exposure, and leukemia claimed her daughter Irène Joliot-Curie, who continued the research program started by her parents.

In medicine, the term *ionizing radiation* is limited to photons and particles that have sufficient energy to remove an electron from water:

$$H_2O(\ell) \xrightarrow{1216\ \text{kJ/mol}} H_2O^+(aq) + e^-$$

The logic behind this definition is that the human body is composed largely of water. Therefore, water molecules are the most abundant ionizable targets when we are exposed to nuclear radiation. The cation H_2O^+ reacts with another water molecule in the body to form a hydronium ion and a hydroxyl free radical:

$$H_2O^+(aq) + H_2O(\ell) \rightarrow H_3O^+(aq) + OH(aq)$$

The rapid reactions of free radicals with biomolecules can threaten the integrity of cells and the health of the entire organism.

Radiation-induced alterations to the biochemical machinery that controls cell growth are most likely to occur in tissues in which cells grow and divide rapidly. One such tissue is bone marrow, where billions of white blood cells are produced each day to fortify the body's immune system. Molecular damage to bone marrow can lead to leukemia, an uncontrolled production of nonfunctioning white blood cells that spread throughout the body, crowding out healthy cells. Ionizing radiation can also cause molecular alterations in the genes and chromosomes of sperm and egg cells, increasing the chances of birth defects in offspring.

Radiation Dosage

The biological impact of ionizing radiation depends on how much of it is absorbed by an organism. If the radiation is coming from one radioactive source, then the amount absorbed depends on the radioactivity of the source and the energy of the radiation that is produced per decay event. Tables of radioactive isotopes often

ionizing radiation high-energy products of radioactive decay that can ionize molecules.

TABLE 21.4 RBE Values of Nuclear Radiation

Radiation	RBE
γ Rays	1.0
β Particles	1.0–1.5
Neutrons	3–5
Protons	10
α Particles	20

TABLE 21.5 Units for Expressing Quantities of Ionizing Radiation

Parameter	SI Unit	Description	Alternative Common Unit	Description
Radioactivity	Becquerel (Bq)	1 decay event/s	Curie (Ci)	3.70×10^{10} decay events/s
Ionizing energy absorbed	Gray (Gy)	1 J/kg of tissue	Rad	0.01 J/kg of tissue
Amount of tissue damage	Sievert (Sv)	1 Gy × RBE[a]	Rem	1 rad × RBE[a]

[a]RBE, relative biological effectiveness.

include information about their modes of decay and the energies of the particles and gamma rays they emit.

Absorbed dose is the quantity of ionizing radiation absorbed by a unit mass of living tissue. The SI unit of absorbed dose is the **gray (Gy)**. One gray is equal to the absorption of 1 J of radiation energy per kilogram of body mass:

$$1 \text{ Gy} = 1 \text{ J/kg}$$

Grays express dosage, but they do not indicate the amount of *tissue damage* caused by that dosage. Different products of nuclear reactions affect living tissue differently. Exposure to 1 Gy of γ rays produces about the same amount of tissue damage as exposure to 1 Gy of β particles. However, 1 Gy of α particles, which move about 10 times more slowly than β particles but have nearly 10^4 times the mass, causes 20 times as much damage as 1 Gy of γ rays. Neutrons cause 3 to 5 times as much damage as gamma rays. To account for these differences, values of **relative biological effectiveness (RBE)** have been established for the various forms of ionizing radiation (Table 21.4). When the absorbed dose in grays is multiplied by an RBE factor, the product is called the *effective* dose, a measure of tissue damage. The SI unit of effective dose is the **sievert (Sv)**.

Table 21.5 summarizes the various units used to express quantities of radiation and their biological impact. Two non-SI units are listed that predate their SI

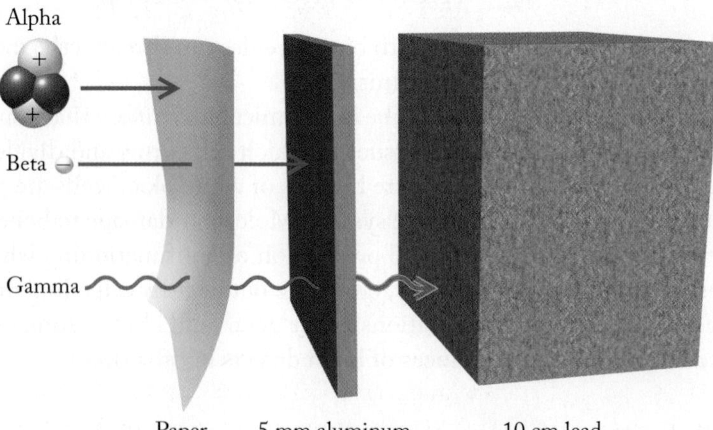

FIGURE 21.19 The tissue damage caused by α particles, β particles, and γ rays depends on their ability to penetrate materials shielding you from their source. Alpha particles are stopped by paper or clothing but are extremely dangerous if formed inside the body because they do not have to travel far to cause cell damage. Stopping gamma rays requires a thick layer of lead or several meters of concrete or soil.

gray (Gy) the SI unit of absorbed radiation; 1 Gy = 1 J/kg of tissue.

relative biological effectiveness (RBE) a factor that accounts for the differences in physical damage caused by different types of radiation.

sievert (Sv) SI unit used to express the amount of biological damage caused by ionizing radiation.

TABLE 21.6 Acute Effects of Single Whole-Body Effective Doses of Ionizing Radiation

Effective Dose (Sv)	Toxic Effect
0.05–0.25	No acute effect, possible carcinogenic or mutagenic damage to DNA
0.25–1.0	Temporary reduction in white blood cell count
1.0–2.0	Radiation sickness: fatigue, vomiting, diarrhea, impaired immune system
2.0–4.0	Severe radiation sickness: intestinal bleeding, bone marrow destruction
4.0–10.0	Death, usually through infection, within weeks
>10.0	Death within hours

FIGURE 21.20 The ruins of the nuclear reactor at Chernobyl, Ukraine, which exploded in 1986.

counterparts but are still often used. They are *radiation absorbed dose*, or *rad*, which is equivalent to 0.01 Gy, and the *rem* for tissue damage, which is an acronym for *roentgen equivalent man*. One rem is the product of one rad of ionization times the appropriate RBE factor. There are 100 rems in 1 Sv.

The RBE of 20 for α particles may lead you to believe that these particles pose the greatest health threat from radioactivity. Not exactly. Alpha particles are so big that they have little penetrating power; they are stopped by a sheet of paper, clothing, or even a layer of dead skin (Figure 21.19). However, if you ingest or inhale an α emitter, tissue damage can be severe because the relatively massive α particles do not have to penetrate far to cause cell damage. Gamma rays are considered the most dangerous form of radiation emanating from a source outside the body because they have the greatest penetrating power.

The effects of exposure to different single effective doses of radiation are summarized in Table 21.6. To put these data in perspective, the effective dose from a typical dental X-ray is about 25 μSv, or about 2000 times smaller than the lowest exposure level cited in the table.

Widespread exposure to very high levels of radiation occurred after the 1986 explosion at the Chernobyl nuclear reactor in what is now Ukraine (Figure 21.20). Many plant workers and first responders were exposed to more than 1.0 Sv of radiation. At least 30 of them died in the weeks after the accident. Many of the more than 300,000 workers who cleaned up the area around the reactor exhibited symptoms of radiation sickness, and at least 5 million people in Ukraine, Belarus, and Russia were exposed to fallout in the days following the accident. Studies conducted in the early 1990s uncovered high incidences of thyroid cancer in children in southern Belarus due to ^{131}I released in the Chernobyl accident, and children born in the region nearly a decade after the accident had unusually high rates of mutations in their DNA because of their parents' exposure to ionizing radiation. Genetic damage was also widespread among plants and animals living in the region (Figure 21.21).

(a)

(b)

FIGURE 21.21 Wildlife surrounding the destroyed nuclear reactor at Chernobyl, Ukraine, was exposed to intense ionizing radiation, which led to deaths and sublethal biological effects such as genetic mutations. One example of the latter is the (a) partially albino barn swallow. A normal swallow (b) has no white feathers directly beneath its beak.

Radiation exposure was not confined to Ukraine and Belarus. After the accident, a cloud of radioactive material spread rapidly across northern Europe, and within two weeks increased levels of radioactivity were detected throughout the Northern Hemisphere (Figure 21.22). The accident produced a global increase in human exposure to ionizing radiation estimated to be equivalent to 0.05 mSv per year.

As a result of an earthquake and tsunami in March 2011, nuclear reactors at a power station in Fukushima, Japan, released about 1/3 of the radiation released by

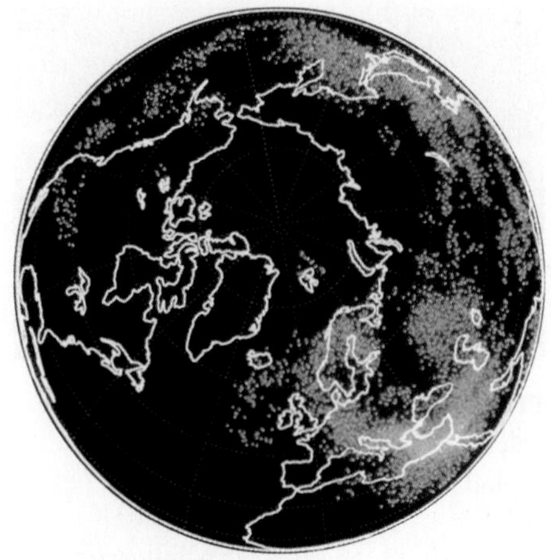

FIGURE 21.22 Radioactive fallout (shown in pink) from the Chernobyl accident in 1986 was detected throughout the Northern Hemisphere.

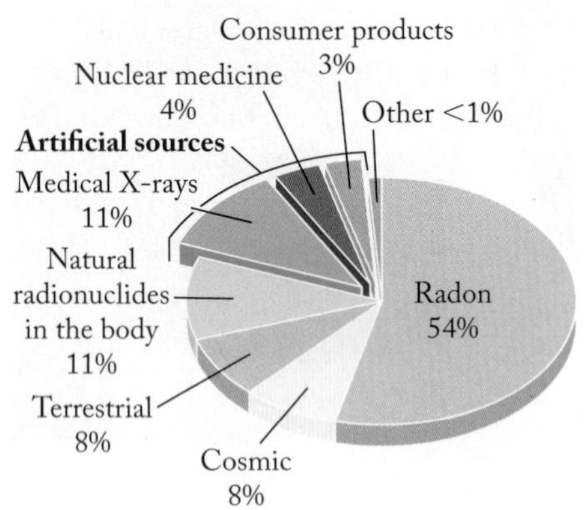

FIGURE 21.23 Sources of radiation exposure of the U.S. population. On average, a person living in the United States is exposed to 0.0036 Sv of radiation each year. More than 80% of this exposure comes from natural sources—mainly radon in the air and water. Artificial sources account for about 18% of the total exposure.

the Chernobyl disaster. There were no fatalities due to acute exposure, but as many as 500 deaths may someday be traceable to diseases caused by exposure to radiation released from the Fukushima reactors.

Evaluating the Risks of Radiation

To put global radiation exposure from Chernobyl and Fukushima in perspective, we need to consider typical annual exposure levels. For many people the principal source of radiation is radon gas in indoor air and in well water (Figure 21.23). Like all noble gases, radon is chemically inert. Unlike the others, all of its isotopes are radioactive. The most common isotope, radon-222, is produced when uranium-238 in rocks and soil decays to lead-206 (Figure 21.2). The radon gas formed in this decay series percolates upward and can enter a building through cracks and pores in its foundation.

If you breathe radon-contaminated air and then exhale before it decays, no harm is done. If radon-222 decays inside the lungs, however, it emits an α particle that can damage lung tissue. The nuclide produced by the α decay of ^{222}Rn is radioactive polonium-218, which may become attached to tissue in the respiratory system and undergo a second α decay, forming lead-214:

$$^{222}_{86}\text{Rn} \rightarrow ^{218}_{84}\text{Po} + ^{4}_{2}\alpha \qquad t_{1/2} = 3.8 \text{ d}$$

$$^{218}_{84}\text{Po} \rightarrow ^{214}_{82}\text{Pb} + ^{4}_{2}\alpha \qquad t_{1/2} = 3.1 \text{ min}$$

As we have seen, α particles are the most damaging product of nuclear decay when formed inside the body. How big a threat does radon pose to human health? Concentrations of indoor radon depend on local geology (Figure 21.24) and on how gastight building foundations are. The air in many buildings contains concentrations of radon in the range of 1 pCi per liter of air. How hazardous are such tiny concentrations? There appears to be no simple answer. The U.S. Environmental Protection Agency has established 4 pCi/L as an "action level," meaning that people occupying houses with higher concentrations should take measures to minimize their exposure.

This action level is based on studies of the incidence of lung cancer in workers in uranium mines. These workers are exposed to radon concentrations (and concentrations of other radionuclides) that are much higher than the concentrations in homes and other buildings. However, many scientists believe that people exposed to very low levels of radon for many years are as much at risk as miners exposed to high levels of radiation for shorter periods. Some researchers use a

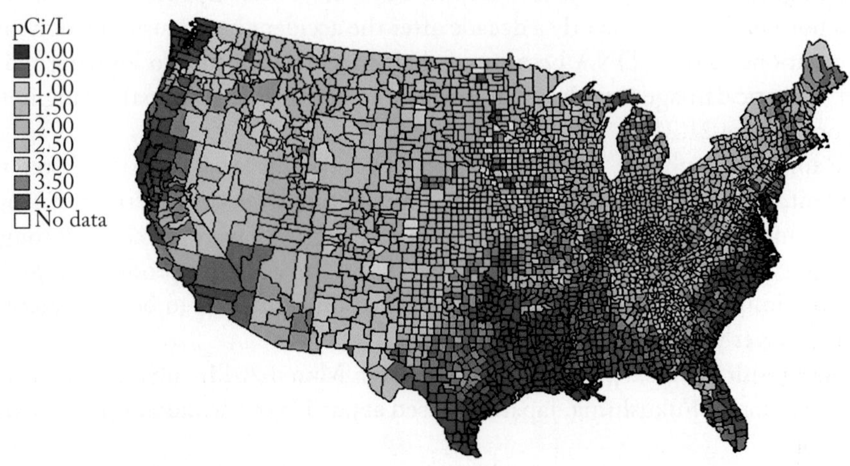

FIGURE 21.24 Levels of radon gas in soils and rocks across the United States.

model that assumes a linear relation between radon exposure and the risk of lung cancer. This model is represented by the red line in Figure 21.25. On the basis of this dose–response model, an estimated 15,000 Americans die of lung cancer each year because of exposure to indoor radon. This number comprises 10% of all lung-cancer fatalities and 30% of those among nonsmokers.

Is this linear model valid? Perhaps—but some scientists believe that there may be a threshold exposure below which radon poses no significant threat to public health. They advocate an S-shaped dose–response curve, shown by the blue line in Figure 21.25. Notice that the risk of death from cancer in the S-shaped curve is much lower than in the linear response model at low radiation exposure but rises rapidly above a critical value.

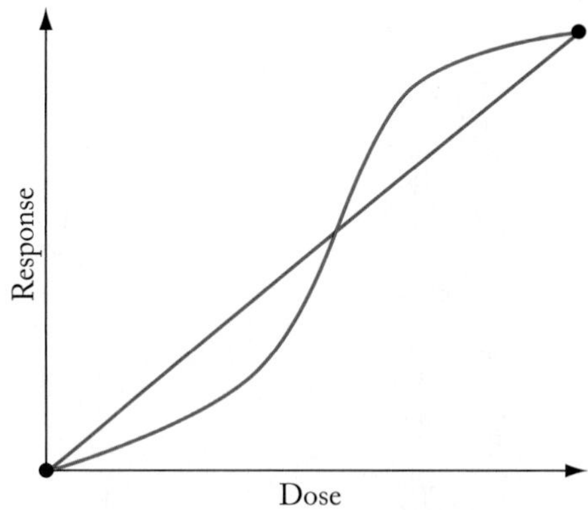

FIGURE 21.25 The risk of death from radiation-induced cancer may follow one of two models. In the linear response model (red line), risk is directly proportional to the radiation exposure. In the S-shaped model (blue line), risk remains low below a critical threshold and then increases rapidly as the exposure increases. In the S-shaped model, the risk is less than for the linear model at low doses but is higher at higher doses.

SAMPLE EXERCISE 21.8 Calculating Effective Dose **LO9**

It has been estimated that a person living in a home where the air radon concentration is 4.0 pCi/L receives an annual absorbed dose of ionizing radiation equivalent to 0.40 mGy. What is the person's annual effective dose, in millisieverts, from this radon? Use information from Figure 21.23 to compare this annual effective dose of radon with the average annual effective dose of radon estimated for persons living in the United States.

Collect, Organize, and Analyze We are given an absorbed radiation dose of 0.40 mGy. Radon isotopes emit α particles, which we know from Table 21.4 have a relative biological effectiveness of 20. The effective dose caused by an absorbed dose of ionizing radiation is the absorbed dose multiplied by the RBE of the radiation.

Solve

$$0.40 \text{ mGy} \times 20 = 8.0 \text{ mSv}$$

The caption for Figure 21.23 tell us that the average American is exposed to 3.6 mSv of radiation per year, with 54% of that amount, or 1.9 mSv, from radon. The person living in the home described in the problem has an effective dose slightly more than four times the average value.

Think About It The calculated value is more than twice the average annual effective dose of 3.6 mSv from all sources of radiation. The U.S. National Research Council has estimated that a nonsmoker living in air contaminated with 4.0 pCi/L of radon has a 1% chance of dying from lung cancer due to this exposure. The cancer risk for a smoker is close to 5%.

Practice Exercise A dental X-ray for imaging impacted wisdom teeth produces an effective dose of 15 μSv. If a dental X-ray machine emits X-rays with an energy of 6.0×10^{-17} J each, how many of these X-rays must be absorbed per kilogram of tissue to produce an effective dose of 15 μSv? Assume the RBE of these X-rays is 1.2.

21.10 Medical Applications of Radionuclides

Radionuclides are used in both the detection and the treatment of diseases, and they are key agents in the medical fields of diagnostic and therapeutic radiology. In diagnostic radiology, radionuclides are used alongside magnetic resonance imaging (MRI) and other imaging systems that involve only nonionizing radiation. Therapeutic radiology, however, is based almost entirely on the ionizing radiation that comes from nuclear processes.

Therapeutic Radiology

Because ionizing radiation causes the most damage to cells that grow and divide rapidly, it is a powerful tool in the fight against cancer. In the early days of radiochemistry, radium was prescribed in cancer treatment and also for ailments that included skin lesions, lethargy, and arthritis. Modern radiation therapy consists of exposing cancerous tissue to γ radiation. Often the radiation source is external to the patient, but sometimes it is encased in a platinum capsule and surgically implanted in a cancerous tumor. The platinum provides a chemically inert outer layer and acts as a filter, absorbing α and β particles emitted by the radionuclide but allowing γ rays to pass into the tumor.

A nuclide's chemical properties can be exploited to direct it to a tumor site. For example, most iodine in the body is concentrated in the thyroid gland, so an effective therapy against thyroid cancer starts with the ingestion of potassium iodide containing radioactive iodine-131. Some radionuclides used in cancer therapy are listed in Table 21.7.

Surgically inaccessible tumors can be treated with beams of γ rays from a radiation source outside the body. Unfortunately, γ radiation destroys both cancer cells and healthy ones. Thus, patients receiving radiation therapy frequently suffer symptoms of radiation sickness, including nausea and vomiting (the tissues that make up intestinal walls are especially susceptible to radiation-induced damage), fatigue, weakened immune response, and hair loss. To reduce the severity of these side effects, radiologists must carefully control the dosage a patient receives.

Diagnostic Radiology

The transport of radionuclides in the body and their accumulation in certain organs provide ways to assess organ function. A tiny quantity of a radioactive isotope is used, together with a much larger amount of a stable isotope of the same element. The radioactive isotope is called a *tracer*, and the stable isotope is the *carrier*. For example, the circulatory system can be imaged by injecting into the blood a solution of sodium chloride containing a trace amount of $^{24}NaCl$. Circulation is monitored by measuring the γ rays emitted by ^{24}Na as it decays.

The ideal isotope for medical imaging is one that has a half-life about equal to the length of time required to perform the imaging measurements. It should emit moderate-energy γ rays but no α particles or β particles that might cause tissue damage. Sodium-24 (a γ emitter with a half-life of 15 hours) meets both these criteria. Table 21.8 lists several other radionuclides that are used in medical imaging.

Positron emission tomography (PET) is a powerful tool for diagnosing organ and cell function. PET uses short-lived, neutron-poor, positron-emitting radionuclides.

TABLE 21.7 Some Radionuclides Used in Radiation Therapy

Nuclide	Radiation	Half-Life	Treatment
^{32}P	β	14.3 d	Leukemia therapy
^{60}Co	β, γ	5.3 yr	Cancer therapy
^{131}I	β	8.1 d	Thyroid therapy
^{131}Cs	γ	9.7 d	Prostate cancer therapy
^{192}Ir	β, γ	74 d	Coronary disease

TABLE 21.8 Selected Radionuclides Used for Medical Imaging

Nuclide	Radiation	Half-Life (h)	Use
^{99}Tc	γ	6.0	Bones, circulatory system, various organs
^{67}Ga	γ	78	Tumors in the brain and other organs
^{201}Tl	γ	73	Coronary arteries, heart muscle
^{123}I	γ	13.3	Thyroxine production in thyroid gland

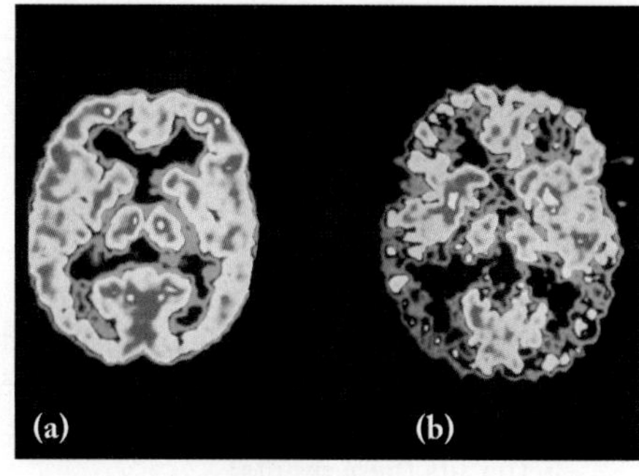

FIGURE 21.26 Positron emission tomography (PET) is used to monitor cell activity in organs such as the brain. (a) Brain function in a healthy person. The red and yellow regions indicate high brain activity; blue and black indicate low activity. (b) Brain function in a patient suffering from Alzheimer's disease.

For example, a patient might be administered a solution of a glucose derivative in which some of the sugar molecules' atoms have been replaced with atoms of ^{11}C, ^{15}O, or ^{18}F. The rate at which glucose is metabolized in various regions of the brain is monitored by detecting the γ rays produced by positron–electron annihilations (Equation 21.4). Unusual patterns in PET images of brains (Figure 21.26) can indicate damage from strokes, schizophrenia, manic depression, Alzheimer's disease, and even nicotine addiction in tobacco smokers.

SAMPLE EXERCISE 21.9 Integrating Concepts: Radium Girls and Safety in the Workplace

Radium was discovered by Pierre and Marie Curie in 1898. By 1902, the new element had its first practical use: radium compounds were mixed with zinc sulfide (ZnS) to make paint that glowed in the dark as alpha particles emitted by the decay of ^{226}Ra ($t_{1/2} = 1.60 \times 10^3$ years) caused ZnS crystals to emit a greenish fluorescence (Figure 21.27). The paint was used to make dials for watches, clocks, and instruments used on naval vessels and, a few years later, in military and civilian airplanes.

By 1914, U.S. companies were making radium-painted dials and employing young women in their late teens and early 20s as dial painters. Soon after their employment, many of the women became very sick, suffering from anemia and other symptoms we now associate with overexposure to nuclear radiation. Some developed bone cancer and over 100 of them, who became known around the world as the Radium Girls, died. Their deaths were linked to the practice of "pointing" the fine paint brushes they used, which meant using their lips to make fine points on the brushes to help them paint the tiny numerals and hands on watch faces (Figure 21.28). In the process they ingested some of

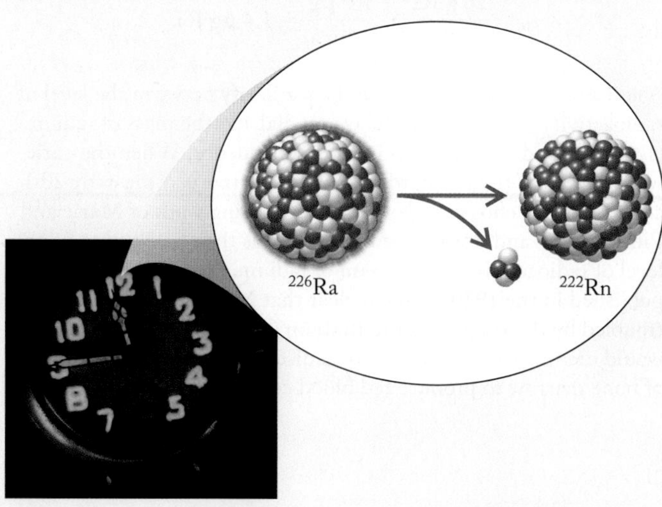

FIGURE 21.27 During the 20th century, many millions of watches and clocks had dials that glowed in the dark as high-energy α particles emitted by ^{226}Ra caused crystals of ZnS to fluoresce.

FIGURE 21.28 This editorial cartoon appeared in Sunday newspapers on February 28, 1926. It portrayed the deadly consequences of young women "pointing" their brushes with their lips as they painted the dials of watches with paint that contained radioactive radium.

the radioactive paint. Tests later determined that about 20% of the radium ingested was incorporated into their bones, where it attacked the bone marrow and caused malignancies known as osteosarcomas.

a. Suggest a reason why radium was concentrated in the victims' bones.
b. Studies of radiation levels and the incidence of cancer in over 1000 female dial painters yielded the results in the following table. Which of the two dose–response curves in Figure 21.25 better fits these results?

Radium Exposure (μg ingested)	Occurrence of Malignancy (% of workers exposed)
1	0
3	0
10	0
30	0
100	5
300	53
1000	85

c. The green luminescence of radium watch dials began to fade after a few years. Was this loss in luminosity due to decreased radioactivity in the paint? Explain why or why not.
d. In one study, the levels of radioactivity in pocket watches with radium-painted dials were found to be between 0.6 and 1.39 μCi per watch. How many micrograms of ^{226}Ra produce 1.39 μCi of radioactivity?

Collect and Organize We are asked (a) why ingested ^{226}Ra concentrates in bones; (b) whether malignancy in dial painters was proportional to their exposure to ^{226}Ra radiation or followed an S-shaped dose–response curve; (c) why the luminosity of radium-activated paint fades after a few years, and (d) how many micrograms of ^{226}Ra are needed to produce 1.39 μCi of radiation. One curie (Ci) is equal to 3.70×10^{10} decay events/s. The half-life ($t_{1/2}$) of ^{226}Ra is 1600 years and is related to the first-order rate constant (k) of the decay reaction by the equation $t_{1/2} = 0.693/k$. The level of radioactivity (A) in a sample of radium is equal to the product of the rate constant and the number (N) of ^{226}Ra atoms: $A = kN$.

Analyze Radium is a group 2 element and should have chemical and biochemical properties that are similar to those of the other elements in that group, including its association with biological tissues. High concentrations of another group 2 element, calcium, occur in teeth and bones. Worker exposure levels in the preceding table cover a wide range, but exposure of up to nearly 100 μg ^{226}Ra caused few malignancies, whereas concentrations above 100 μg caused many. The half-life of ^{226}Ra is so long that the radioactivity of a sample decreases little over a few years or even over many decades. Relating a half-life expressed in years to a level of radioactivity expressed in a multiple of decay events per second will require converting units of time and then quantities of radioactive atoms to moles and then micrograms.

Solve

a. Radium likely accumulates in bones because its chemistry is similar to that of calcium, which means that $^{226}Ra^{2+}$ ions are likely to take the place of Ca^{2+} ions in bone tissue.
b. Malignancies did not occur among the dial painters who ingested less than 100 μg of ^{226}Ra; however, the percentage of the women who suffered from them increased sharply with exposure between 100 and 1000 μg. This pattern is described by the S-shaped (blue) curve in Figure 21.25.
c. Given the 1600-year half-life of ^{226}Ra, the loss of watch dial luminescence was not the result of depleted radioactivity. Rather, it must have been due to less efficient conversion of the energy of radioactive decay into visible light by ZnS crystals.
d. Let's first convert the half-life of ^{226}Ra into a decay rate constant in units of s^{-1}:

$$k = \frac{0.693}{t_{1/2}} = \frac{0.693}{1.60 \times 10^3 \cancel{\text{yr}}} \times \frac{1 \cancel{\text{yr}}}{365.25 \cancel{\text{d}}} \times \frac{1 \cancel{\text{d}}}{24 \cancel{\text{h}}} \times \frac{1 \cancel{\text{h}}}{3600 \text{ s}}$$
$$= 1.372 \times 10^{-11} \text{ s}^{-1}$$

Next, we solve the equation $A = kN$ for N, and we use the above rate constant and the radioactivity of the watch to calculate the number of ^{226}Ra atoms in the dial:

$$N = \frac{A}{k} = \frac{1.39 \cancel{\mu\text{Ci}}}{1.372 \times 10^{-11} \text{ s}^{-1}} \times \frac{1 \cancel{\text{Ci}}}{10^6 \cancel{\mu\text{Ci}}}$$
$$\times \frac{3.70 \times 10^{10} \text{ atoms Ra s}^{-1}}{\cancel{\text{Ci}}} = 3.75 \times 10^{15} \text{ atoms of Ra}$$

The corresponding mass in micrograms is

$$3.75 \times 10^{15} \cancel{\text{atoms Ra}} \times \frac{1 \cancel{\text{mol Ra}}}{6.022 \times 10^{23} \cancel{\text{atoms Ra}}}$$
$$\times \frac{226 \text{ g Ra}}{1 \cancel{\text{mol Ra}}} \times \frac{10^6 \text{ μg}}{1 \text{ g}} = 1.4 \text{ μg Ra}$$

Think About It Did you notice the similarity between the level of radioactivity (1.39 μCi) in the watch dial and the mass of radium (1.41 μg) producing it? This is not a coincidence. When the curie was adopted as the standard unit of radioactivity in the early 20th century, it was chosen to honor the pioneering work of Marie and Pierre Curie, and it was based on what was then believed to be the level of radioactivity in one gram of radium. Newspaper articles published in the 1920s make it clear that Marie Curie was deeply troubled by the tragedy of the Radium Girls. Sadly, in 1934 she would die from aplastic anemia—a disease caused by the inability of bone marrow to produce red blood cells.

SUMMARY

LO1 **Nuclear chemistry** is the study and application of reactions that involve changes in atomic nuclei. The sum of the mass numbers of the reactants in a nuclear equation must be equal to the sum of the mass numbers of the products. The sum of the charges (subscripts) of the reactants must also equal the sum of the charges of the products. (Section 21.1)

LO2 Stable nuclei have neutron-to-proton ratios that fall within a range of values called the **belt of stability**. Unstable nuclides undergo **radioactive decay**. Neutron-rich nuclides (mass number greater than the average atomic mass) undergo **β decay**; neutron-poor nuclides undergo **positron emission** or **electron capture.** Very large nuclides ($Z > 83$) may undergo β decay or **α decay**. (Section 21.2)

LO3 Radioactive decay follows first-order kinetics, so the half-life ($t_{1/2}$) of a radionuclide is a characteristic value of the decay process. (Section 21.3)

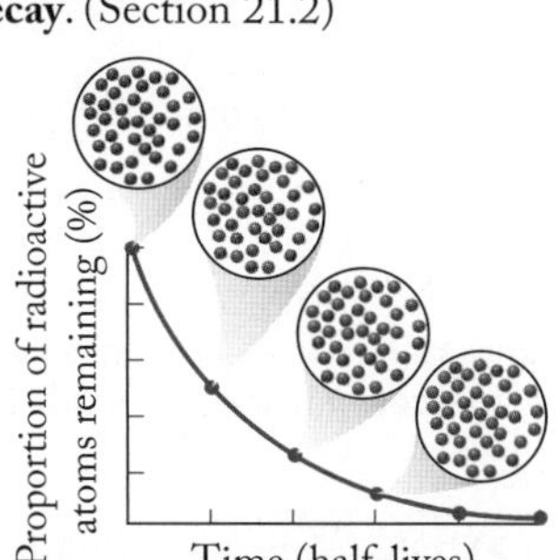

LO4 **Radiometric dating** is used to determine the age of an object based on its content of a radionuclide and/or its decay product. **Radiocarbon dating** involves measuring the amount of radioactive carbon-14 that remains in an object derived from plant or animal tissue to calculate the age of the object. To improve the accuracy of the technique, scientists calibrate the results of radiometric analysis of samples of known age, such as growth rings in ancient trees. (Section 21.3)

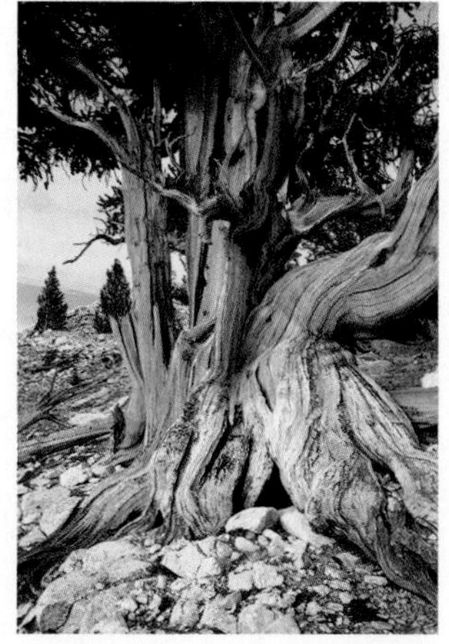

LO5 The **mass defect (Δ*m*)** of a nucleus is the difference between its mass and the sum of the masses of its nucleons. **Binding energy (BE)** is the energy released when the nucleons combine to form a nucleus. It is also the energy needed to split the nucleus into its nucleons. Binding energy per nucleon is a measure of the relative stability of a nucleus. The energy released in nuclear reactions is calculated using Einstein's equation: $E = mc^2$. (Section 21.4)

LO6 Neutrons, protons, and electrons formed within seconds of the Big Bang. During primordial **nucleosynthesis**, protons and neutrons fused to produce nuclei of helium. After galaxies formed, the nuclei of atoms with $Z \leq 26$ formed when the nuclei of lighter elements fused together in the cores of giant stars (stellar nucleosynthesis). The nuclei of elements with $Z > 26$ formed by a combination of neutron capture, beta (β) decay, and other nuclear reactions that occurred during supernovas (explosions of giant stars). As a result of these explosions, the elements produced were distributed throughout galaxies for possible inclusion in later-generation stars and in planets such as our own. Stellar nucleosynthesis continues today. Artificial nuclides are produced from high-speed collisions between atomic nuclei and subatomic particles and in collisions between two nuclei. The latter have produced supermassive elements with atomic numbers up to 118. (Sections 21.5 and 21.6)

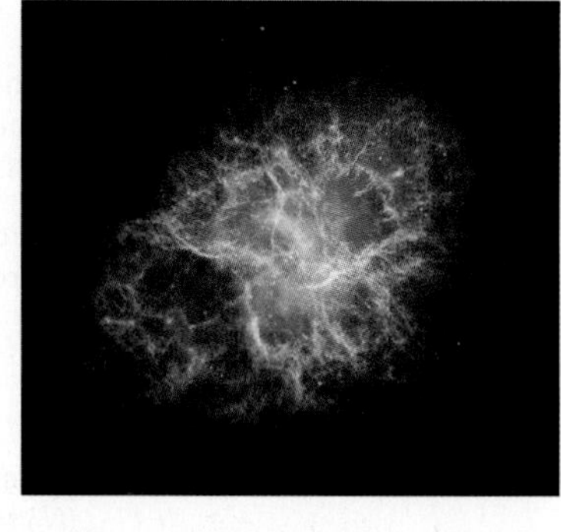

LO7 **Nuclear fusion** occurs when subatomic particles or atomic nuclei collide with each other and fuse together. When a particle of matter encounters a particle of **antimatter**, both are converted into energy (they annihilate one another), yielding γ rays. Neutron absorption by uranium-235 and a few other massive isotopes may lead to **nuclear fission** into lighter nuclei accompanied by the release of energy that can be harnessed to generate electricity. A **chain reaction** happens when the neutrons released during fission collide with other fissionable nuclei. They require a **critical mass** of a fissionable isotope. A **breeder reactor** is used to make plutonium-239 from uranium-238 while also producing energy to make electricity. (Sections 21.6 and 21.7)

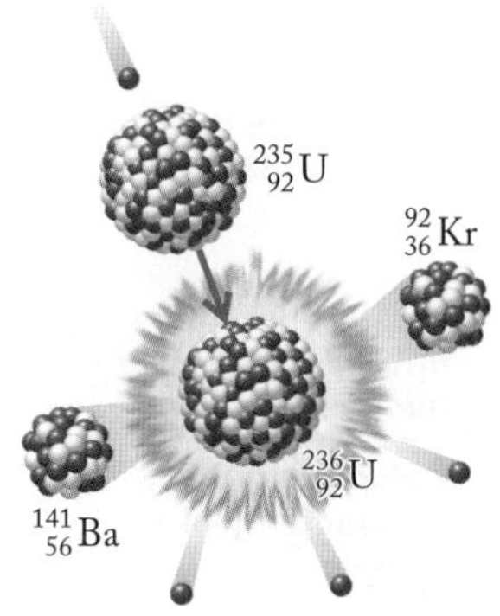

LO8 **Scintillation counters** and **Geiger counters** are used to measure levels of nuclear radiation. Radioactivity is the number of decay events per unit time. Common units are the **becquerel (Bq)** (1 decay event/s) and the **curie (Ci)** (1 Ci = 3.70×10^{10} Bq). (Section 21.8)

LO9 Alpha particles, β particles, and γ rays have enough energy to break up molecules into electrons and cations and are examples of **ionizing radiation** that can damage body tissue and DNA. The quantity of ionizing radiation energy absorbed per kilogram of body mass is called the *absorbed dose* and is expressed in **grays (Gy)**: 1 Gy = 1.00 J/kg. The effective dose of any type of ionizing radiation is the product of the absorbed dose in grays and the **relative biological effectiveness (RBE)** of the radiation; the unit of effective dose is the **sievert (Sv)**. Alpha particles have a larger RBE than β particles and γ rays but have the least penetrating power of these three types of ionizing radiation. Selected radioactive isotopes are useful as tracers in the human body to map biological activity and diagnose diseases. Other radioactive isotopes are used to treat cancers. (Sections 21.9 and 21.10)

PARTICULATE PREVIEW WRAP-UP

Carbon-14 has more neutrons than protons: six protons and eight neutrons. With a neutron-to-proton ratio greater than 1, it will decay to reduce that ratio. When ^{14}C radioactively decays by emitting a beta particle, it produces a nitrogen nuclide with a 1:1 neutron:proton ratio (seven protons and seven neutrons).

PROBLEM-SOLVING SUMMARY

Type of Problem	Concepts and Equations	Sample Exercises
Completing and balancing nuclear equations	Add nuclides or subatomic particles as needed to equalize the sum of the mass numbers (superscripts) and the sum of the atomic numbers or particle charges (subscripts) on the left and right sides of the equation.	**21.1**
Predicting the modes and products of radioactive decay	Neutron-rich nuclides tend to undergo β decay; neutron-poor nuclides undergo positron emission or electron capture.	**21.2**
Calculations involving half-lives	$\frac{N_t}{N_0} = 0.5^{t/t_{1/2}}$ (21.9) where N_t/N_0 is the ratio of the quantity of radionuclide present in a sample at time t (N_t) to the quantity at $t = 0$ (N_0).	**21.3**
Radiometric dating	Substitution for N_0, N_t, and $t_{1/2}$ in Equation 21.10 yields the time elapsed in a radioactive decay process. $t = -\frac{t_{1/2}}{0.693} \ln \frac{N_t}{N_0}$ (21.10)	**21.4**
Calculating the binding energy of a nucleus	Calculate the mass of the nucleons and compare it to the exact mass of the nucleus. Convert the difference in mass, m, to energy using Equation 21.11: $E = mc^2$ (21.11)	**21.5**
Calculating the energy released in a nuclear reaction	Substitute into a modified form of Equation 21.11, $E = \Delta mc^2$, where Δm is the loss in mass as reactants form products. When the units of Δm are kg and the units of c are m/s, E is in joules because $1\ J = 1\ kg \cdot (m/s)^2$.	**21.6**
Calculating the radioactivity of a sample	Insert the values of k and N into Equation 21.25: $A = kN$ (21.25) where $k = 0.693/t_{1/2}$. (21.7)	**21.7**
Calculating effective dose	Substitute for the absorbed dose and the relative body effectiveness (RBE) in equation: Effective dose = absorbed dose × RBE	**21.8**

VISUAL PROBLEMS

(Answers to boldface end-of-chapter questions and problems are in the back of the book.)

21.1. Which of the following nuclear processes do Figures P21.1(a) and P21.1(b) represent?

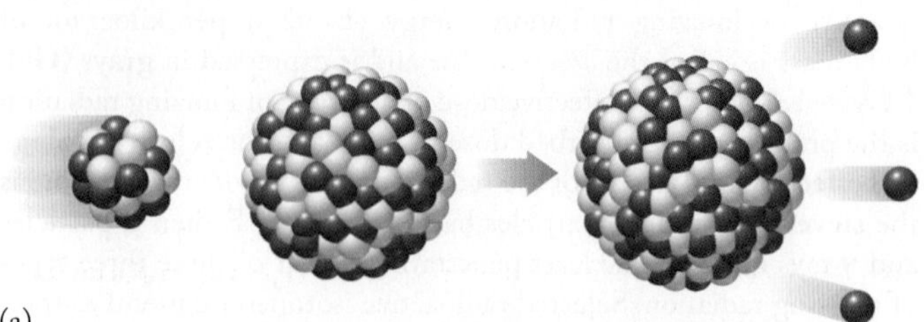

(a)

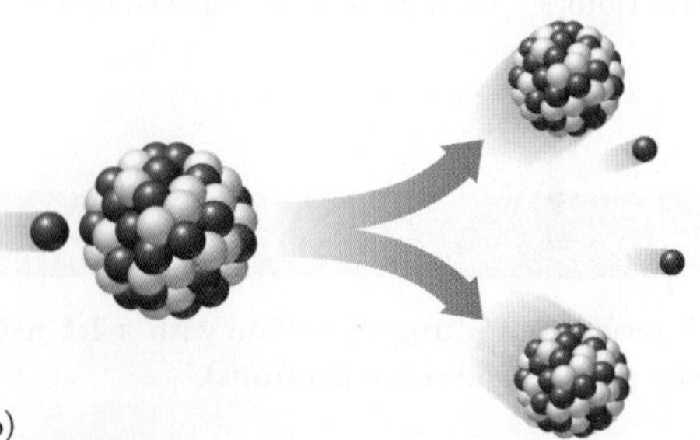

(b)

FIGURE P21.1

a. primordial nucleosynthesis
b. synthesis of a supermassive nuclide
c. solar fusion
d. fission
e. β decay

21.2. Exposure to which of the highlighted elements in Figure P21.2 could cause anemia and bone disease?

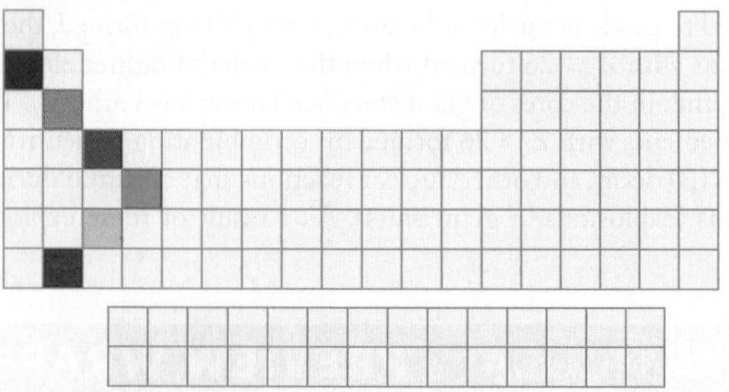

FIGURE P21.2

21.3. Which of the highlighted elements in Figure P21.2 are products of the decay of uranium-238?

21.4. Which of the graphs in Figure P21.4 illustrates α decay? Which decay pathway does the other graph illustrate?

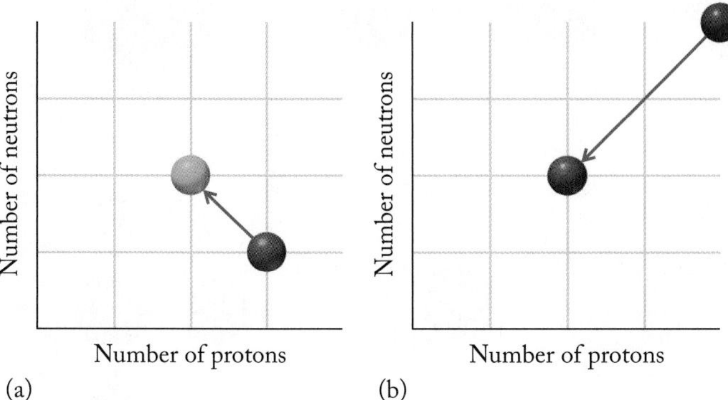

FIGURE P21.4

21.5. Which of the graphs in Figure P21.5 illustrates β decay?

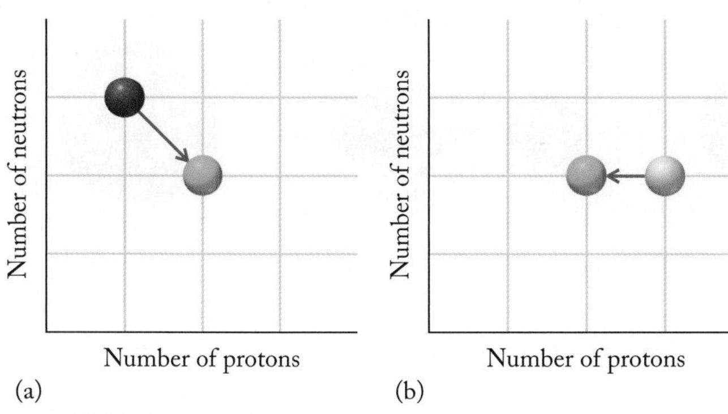

FIGURE P21.5

21.6. Which of the graphs in Figure P21.6 illustrates the overall effect of neutron capture followed by β decay?

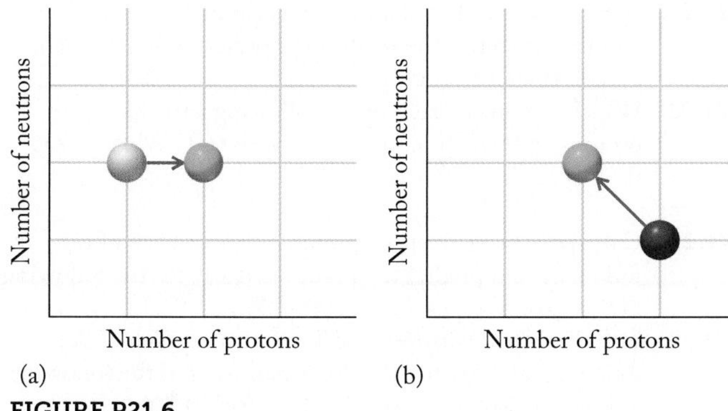

FIGURE P21.6

21.7. Which of the curves in Figure P21.7 represents the decay of an isotope that has a half-life of 2.0 days?

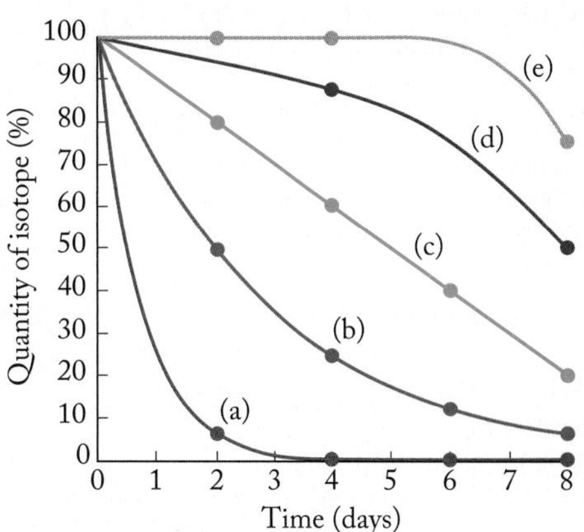

FIGURE P21.7

21.8. Which of the curves in Figure P21.7 do not represent a radioactive decay curve?

21.9. Which of the models in Figure P21.9 represents fission and which represents fusion?

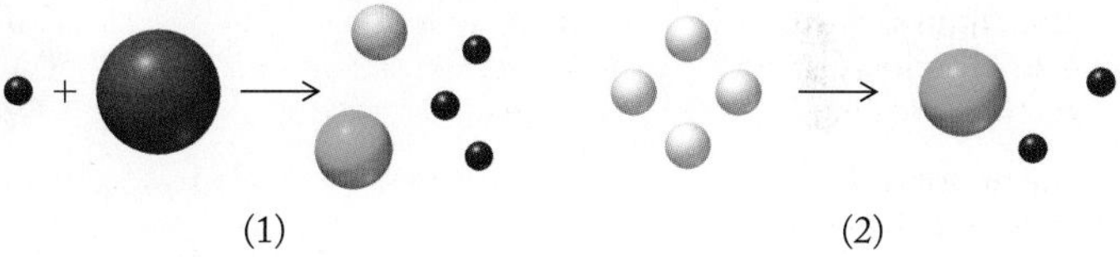

FIGURE P21.9

21.10. Isotopes in a nuclear decay series emit particles with a positive charge and particles with a negative charge. The two kinds of particles penetrate a column of water as shown in Figure P21.10. Is the "X" particle the positive or the negative one?

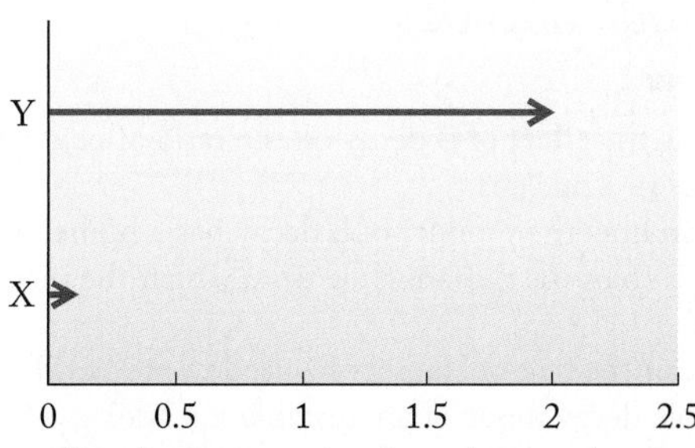

FIGURE P21.10

21.11. Does the radiation that follows the red path in Figure P21.11 penetrate solid objects better than the radiation following the green or blue path?

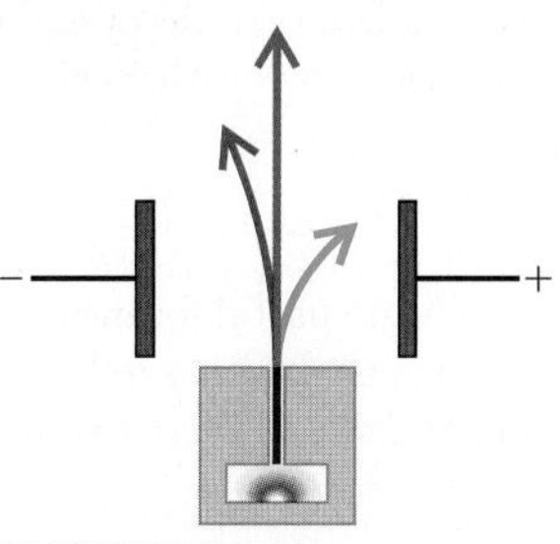

FIGURE P21.11

21.12. Use representations [A] through [I] in Figure P21.12 to answer questions (a)–(f).
a. Which image depicts beta decay? Write a balanced nuclear equation to represent the image.
b. Which image depicts fusion? Write a balanced nuclear equation to represent the image.
c. Which image represents positron emission? Write a balanced nuclear equation to represent the image.
d. Which image depicts nuclear fission? Write a balanced nuclear equation to represent the image.
e. Which nuclide will undergo alpha decay? Write a balanced nuclear equation to depict this process.
f. Which nuclide will undergo beta decay? Write a balanced nuclear equation to depict this process.

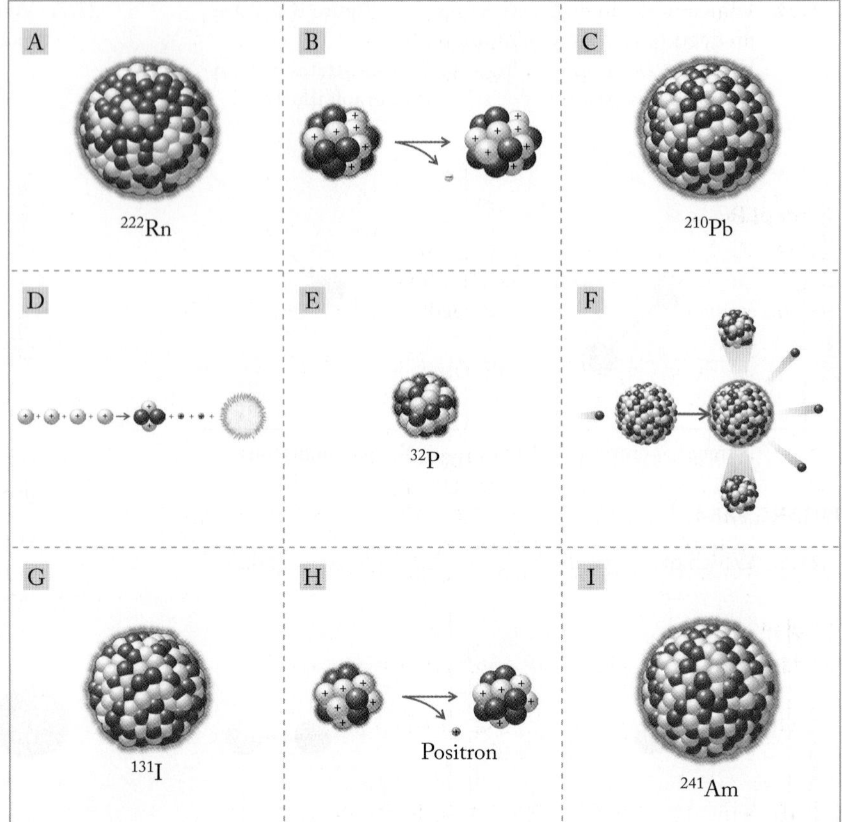

FIGURE P21.12

QUESTIONS AND PROBLEMS

Decay of Radionuclides

Concept Review

21.13. What is the effect of β decay on the ratio of neutrons to protons in a nucleus?

21.14. Explain how the product of β decay has a higher atomic number than the radionuclide from which the product forms.

21.15. How can the belt of stability be used to predict the probable decay mode of an unstable nuclide?

21.16. Compare and contrast positron-emission and electron-capture processes.

21.17. The ratio of neutrons to protons in stable nuclei increases with increasing atomic number. Use this trend to explain why multiple α decay steps in the ^{238}U decay series are often followed by β decay.

*21.18. Iridium-192 decays by two pathways: β decay and electron capture. Why does ^{192}Ir decay by two paths? (*Hint*: Find ^{192}Ir in Figure 21.1.)

Problems

21.19. Compounds containing ^{64}Cu are useful for measuring blood flow in the brain. Copper-64 decays to ^{64}Zn and to ^{64}Ni. What are the modes of decay in these two reactions?

21.20. Write a balanced nuclear equation for
a. Beta emission by ^{161}Tb
b. Alpha emission by ^{255}Lr
c. Electron capture by ^{67}Ga
d. Positron emission by ^{72}As

21.21. If the mass number of an isotope is more than twice the atomic number, is the neutron-to-proton ratio less than, greater than, or equal to 1?

21.22. Which isotope in each of the following pairs has more protons and which has more neutrons? (a) ^{92}Mo or ^{92}Zr; (b) ^{28}Si or ^{28}Ar; (c) ^{111}In or ^{114}In

21.23. Calculate the neutron-to-proton ratios for each of the following and predict the modes of decay for the following radioactive isotopes: (a) ^{47}Sc; (b) ^{89}Zr; (c) ^{230}Th.

21.24. Calculate the neutron-to-proton ratios for each of the following and predict the decay pathways of the following radioactive isotopes: (a) ^{238}U; (b) ^{186}Re; (c) ^{86}Y.

21.25. Iridium-192 can be inserted into tumors, where it undergoes both β decay and electron capture. Write balanced nuclear equations describing the decay of ^{192}Ir.

21.26. Arsenic-74 decays by β decay and by positron emission. Which nuclides are produced by each of these decay pathways?

21.27. **Elements in a Supernova** The isotopes ^{56}Co and ^{44}Ti were detected in supernova SN 1987A. Predict the decay pathway for these radioactive isotopes.

21.28. Nine isotopes of sulfur have mass numbers ranging from 30 to 38. Five of the nine are radioactive: ^{30}S, ^{31}S, ^{35}S, ^{37}S, and ^{38}S. Which of these isotopes do you expect to decay by β decay?

Rates of Radioactive Decay

Concept Review

21.29. Explain why radiocarbon dating is not reliable for artifacts and fossils older than about 50,000 years.

21.30. Which of the following statements about ^{14}C dating are true?

a. The amount of ^{14}C in all objects is the same.

b. Carbon-14 is unstable and is readily lost from the atmosphere.

c. The ratio of ^{14}C to ^{12}C in the atmosphere is a constant.

d. Living tissue will absorb ^{12}C but not ^{14}C.

21.31. Why is ^{40}K dating ($t_{1/2} = 1.28 \times 10^9$ years) useful only for rocks older than 300,000 years?

21.32. Where does the ^{14}C found in plants come from?

Problems

21.33. What percentage of a sample's original radioactivity remains after three half-lives?

21.34. What percentage of a sample's original radioactivity remains after six half-lives?

21.35. What is the half-life of a ^{199}Au if 16.5% of it decays in 168 hours?

21.36. A sample of ^{89}Zr is delivered to a hospital on Monday morning at 9 am for use in labeling antibodies. By Wednesday afternoon at 5 pm, 74.4% of it remains. What is the half-life of ^{89}Zr?

21.37. Fukushima Disaster Explosions at a disabled nuclear power station in Fukushima, Japan, in 2011 may have released more cesium-137 ($t_{1/2} = 30.2$ years) into the ocean than any other single event. How long will it take the radioactivity of this radionuclide to decay to 5.0% of the level released in 2011?

21.38. Spent fuel removed from nuclear power stations contains plutonium-239 ($t_{1/2} = 2.41 \times 10^4$ years). How long will it take a sample of this radionuclide to reach a level of radioactivity that is 2.5% of the level it had when it was removed from a reactor?

21.39. First Humans in South America Archaeologists continue to debate the origins and dates of arrival of the first humans in the Western Hemisphere. Radiocarbon dating of charcoal from a cave in Chile was used to establish the earliest date of human habitation in South America as 8700 years ago. What fraction of the ^{14}C present initially remained in the charcoal after 8700 years?

21.40. **Early Financial Records** For thousands of years native Americans living along the north coast of Peru used knotted cotton strands called *quipu* (Figure P21.40) to record financial transactions and governmental actions. A particular quipu sample is 4800 years old. Compared with the fibers of cotton plants growing today, what is the ratio of carbon-14 to carbon-12 in the sample?

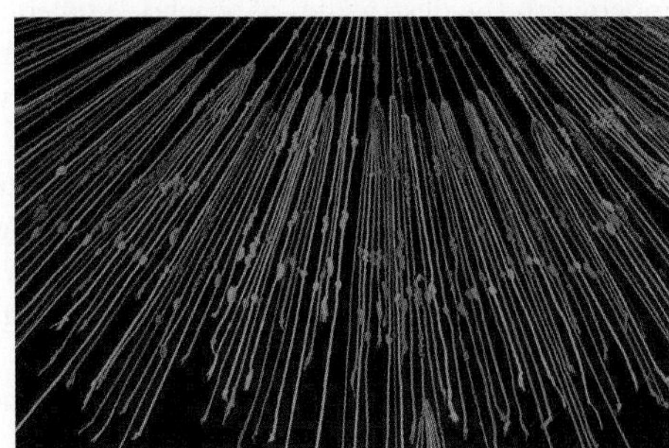

FIGURE P21.40

***21.41. The Ages of Sequoia Trees** Figure P21.41 is a close-up of the center of a giant sequoia tree cut down in 1891 in what is now Kings Canyon National Park. It contained 1342 annual growth rings. If samples of the tree were removed for radiocarbon dating today, what would be the difference in the $^{14}C/^{12}C$ ratio in the innermost (oldest) ring compared with that ratio in the youngest ring?

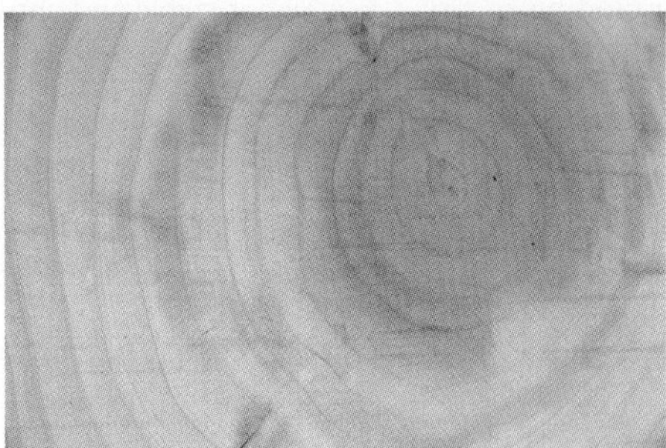

FIGURE P21.41

*21.42. **Dating Volcanic Eruptions** Geologists who study volcanoes can develop historical profiles of previous eruptions by determining the $^{14}C/^{12}C$ ratios of charred plant remains trapped in old magma and ash flows. If the uncertainty in determining these ratios is 0.1%, could radiocarbon dating distinguish between debris from the eruptions of Mt. Vesuvius that occurred in the years 472 and 512? (*Hint*: Calculate the $^{14}C/^{12}C$ ratios for samples from the two dates.)

21.43. Age of Mammoths Figure P21.43 shows a carved mammoth tusk that was uncovered at an ancient camp site in the Ural Mountains in 2001. The $^{14}C/^{12}C$ ratio in the tusk was only 1.19% of that in modern elephant tusks. How old is the mammoth tusk?

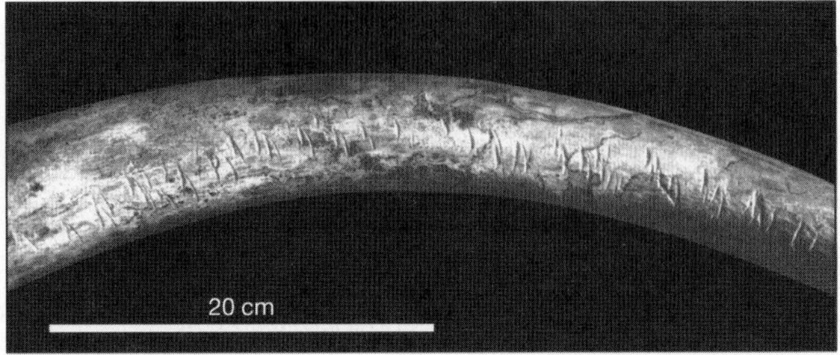

FIGURE P21.43

21.44. **The Destruction of Jericho** The Bible describes the Exodus as a period of 40 years that began with plagues in Egypt and ended with the destruction of Jericho. Archeologists seeking to establish the exact dates of these events have proposed that the plagues coincided with a huge eruption of the volcano Thera in the Aegean Sea.

a. Radiocarbon dating suggests that the eruption occurred around 1360 BCE, though other records place the eruption of Thera in the year 1628 BCE. What is the percent difference in the ^{14}C decay rate in biological samples from these two dates?

b. Radiocarbon dating of blackened grains from the site of ancient Jericho provides a date of 1315 BCE ±13 years for the fall of the city. What is the $^{14}C/^{12}C$ ratio in the blackened grains compared with that of grain harvested last year?

Energy Changes in Radioactive Decay

Concept Review

21.45. Why do all nuclear reactions produce heat?

21.46. Which isotope of phosphorus do you predict will have the larger binding energy: ^{31}P, which is stable and naturally occurring, or ^{32}P, which is radioactive with a half-life of 14.3 days?

Problems

21.47. Substitution of carbon-11 for some of the carbon-12 atoms in glucose yields a useful compound for imaging brain function.

a. Write a balanced nuclear equation for the decay of ^{11}C.

b. Calculate the binding energy for ^{11}C. The exact mass of ^{11}C is 1.82850×10^{-26} kg.

21.48. Fluorine-18 is often introduced into specific drug molecules for use as imaging agents.

a. Write a balanced nuclear equation for the decay of ^{18}F.

b. Calculate the binding energy for ^{18}F. The exact mass of ^{18}F is 2.98915×10^{-26} kg.

21.49. Calculate the binding energy of the two naturally occurring isotopes of chlorine: ^{35}Cl (34.9689 amu) and ^{37}Cl (36.9659 amu) in kJ/mol.

21.50. Which isotope has a larger binding energy: ^{10}B (10.0129 amu) or ^{11}B (11.0093 amu)?

Making New Elements

Problems

21.51. Write a balanced nuclear equation describing how a ^{209}Bi target might be bombarded with subatomic particles to form ^{211}At.

21.52. Bombardment of a ^{239}Pu target with α particles produces ^{242}Cm and another particle.

a. Use a balanced nuclear equation to determine the identity of the missing particle.

b. The synthesis of which other nuclide described in this chapter involves the same subatomic particles?

21.53. Complete the following nuclear equations describing the preparation of isotopes for nuclear medicine.

a. $^{197}Au + ? \rightarrow {}^{199}Hg + \beta$

b. $^{64}Ni + {}^{1}H \rightarrow {}^{64}Cu + ?$

c. $^{63}Cu + ? \rightarrow {}^{66}Ga + {}^{1}n$

d. $^{67}Zn + {}^{1}n \rightarrow {}^{67}Cu + ?$

21.54. Complete the following nuclear equations describing the preparation of isotopes for nuclear medicine.

a. $^{198}Pt + {}^{1}n \rightarrow ? \rightarrow {}^{199}Au + ?$

b. $^{203}Tl + {}^{2}H \rightarrow {}^{203}Pb + ?$

c. $^{60}Ni + ? \rightarrow {}^{60}Cu + {}^{1}n \rightarrow {}^{60}Ni$

d. $^{121}Sb + {}^{4}He \rightarrow ? + 2\,{}^{1}n$

21.55. Complete the following nuclear equations.

a. $^{131}_{52}Te \rightarrow {}^{131}_{53}I + ?$

b. $? + {}^{122}_{54}Xe + {}^{0}_{-1}\beta$

c. $? + {}^{4}_{2}He \rightarrow {}^{13}_{7}N + {}^{1}_{0}n$

d. $? + {}^{1}_{1}H \rightarrow {}^{67}_{31}Ga + 2\,{}^{1}_{0}n$

21.56. Complete the following nuclear equations.

a. $^{210}Po \rightarrow {}^{206}Pb + ?$

b. $^{3}_{1}H \rightarrow {}^{3}_{2}He + ?$

c. $^{14}_{6}C \rightarrow ? + {}^{0}_{-1}\beta$?

d. $? + {}^{1}_{0}n \rightarrow {}^{59}_{16}Fe$

Fusion and the Origin of the Elements

Concept Review

21.57. Arrange the following particles in order of increasing mass: electron, β particle, positron, proton, neutron, α particle, deuteron.

21.58. Electromagnetic radiation is emitted when a neutron and proton fuse to make a deuteron. In which region of the electromagnetic spectrum is the radiation?

21.59. Scientists at the Fermi National Accelerator Laboratory in Illinois announced in the fall of 1996 that they had created "antihydrogen." How does antihydrogen differ from hydrogen?

21.60. Describe an antiproton.

21.61. In what ways are the fusion reactions that formed alpha particles during primordial nucleosynthesis different from those that fuel our sun today?

21.62. Why is energy released in a nuclear fusion process when the product is an element preceding iron in the periodic table?

21.63. **Components of Solar Wind** Most of the ions that flow out from the sun in the solar wind are hydrogen ions. The ions of which element should be next most abundant?

*21.64. **Nucleosynthesis in Giant Stars** A star needs a core temperature of about 10^7 K for hydrogen fusion to occur. Core temperatures above 10^8 K are needed for helium fusion. Why does helium fusion require much higher temperatures?

21.65. **Origins of the Elements** Our sun contains carbon even though its core is not hot or dense enough to sustain carbon synthesis through the triple-alpha process. Where could the carbon have come from?

21.66. Early nucleosynthesis produced a universe that was more than 99% hydrogen and helium with less than 1% lithium. Why were the other elements not formed?

Problems

21.67. Calculate the energy and wavelength of the two gamma rays released by the annihilation of a proton and an antiproton.

21.68. Calculate the energy released and the wavelength of the two photons emitted in the annihilation of an electron and a positron.

21.69. Calculate the energy released in each of the following reactions from the masses of the isotopes: ^{2}H (2.0146 amu), ^{4}He (4.00260 amu), ^{10}B (10.0129 amu), ^{12}C (12.000 amu), ^{14}N (14.00307 amu), ^{16}O (15.99491 amu), ^{24}Mg (23.98504 amu), ^{28}Si (27.97693 amu).
a. $^{14}N + ^{14}N \rightarrow ^{28}Si$
b. $^{10}B + ^{16}O + ^{2}H \rightarrow ^{28}Si$
c. $^{16}O + ^{12}C \rightarrow ^{28}Si$
d. $^{24}Mg + ^{4}He \rightarrow ^{28}Si$

21.70. All of the following fusion reactions produce ^{32}S. Calculate the energy released in each reaction from the masses of the isotopes: ^{4}He (4.00260 amu), ^{6}Li (6.01512 amu), ^{12}C (12.000 amu), ^{14}N (14.00307 amu), ^{16}O (15.99491 amu), ^{24}Mg (23.98504 amu), ^{28}Si (27.97693 amu), ^{32}S (31.97207 amu).
a. $^{16}O + ^{16}O \rightarrow ^{32}S$
b. $^{28}Si + ^{4}He \rightarrow ^{32}S$
c. $^{14}N + ^{12}C + ^{6}Li \rightarrow ^{32}S$
d. $^{24}Mg + 2\ ^{4}He \rightarrow ^{32}S$

21.71. **Hydrogen Bombs** In January 2016, North Korea claimed to have tested a small hydrogen bomb based on the fusion of deuterium and tritium described in Equation 21.24. The large amount of tritium required to sustain the reaction in a hydrogen bomb is produced as shown in Figure P21.71. Calculate the energy change for the reaction between a neutron and lithium-6 nucleus that yields helium and tritium in Figure P21.71. NOTE: The exact masses of tritium and lithium-6 nuclei are 5.00827×10^{-27} kg and 9.98841×10^{-27} kg, respectively.

$$^{2}_{1}H + ^{3}_{1}H \rightarrow ^{4}_{2}He + ^{1}_{0}n$$

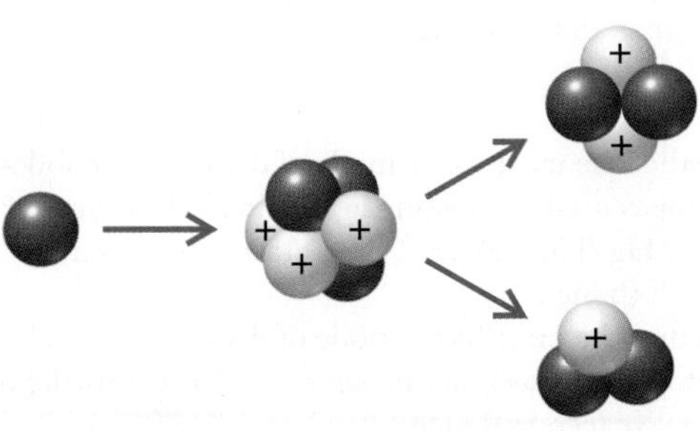

$$^{1}_{0}n + ^{6}_{3}Li \longrightarrow ^{4}_{2}He + ^{3}_{1}H$$

FIGURE P21.71

21.72. **Tokamak Radiochemistry** In 2009 construction began on ITER (originally an acronym for International Thermonuclear Experimental Reactor), a project to build the world's largest nuclear fusion reactor. When it is operational, ITER will use the fusion reactions described in Problem 21.71. Since the natural abundance of lithium-6 is only about 6.0%, ITER planned to use the more abundant ^{7}Li isotope. How much energy is released per nucleus of tritium produced during the reactions in Figure P21.72? NOTE: The exact masses of tritium and lithium-7 nuclei are 5.00827×10^{-27} kg and 1.16504×10^{-26} kg, respectively.

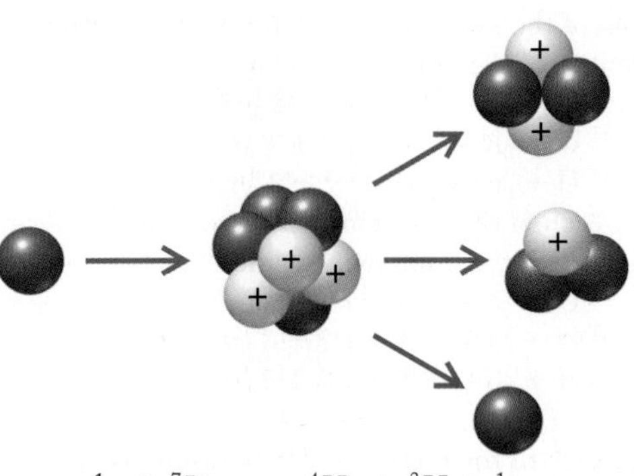

$$^{1}_{0}n + ^{7}_{3}Li \longrightarrow ^{4}_{2}He + ^{3}_{1}H + ^{1}_{0}n$$

FIGURE P21.72

21.73. What nuclide is produced in the core of a giant star by each of the following fusion reactions? Assume there is only one product in each reaction.
a. $^{12}C + ^{4}He \rightarrow$
b. $^{20}Ne + ^{4}He \rightarrow$
c. $^{32}S + ^{4}He \rightarrow$

21.74. What nuclide is produced in the core of a giant star by each of the following fusion reactions? Assume there is only one product in each reaction.
a. $^{28}Si + ^{4}He \rightarrow$
b. $^{40}Ca + ^{4}He \rightarrow$
c. $^{24}Mg + ^{4}He \rightarrow$

21.75. What nuclide is produced in the core of a collapsing giant star by each of the following reactions?
a. $^{96}_{42}Mo + 3\ ^{1}_{0}n \rightarrow ? + ^{0}_{-1}\beta$
b. $^{118}_{50}Sn + 3\ ^{1}_{0}n \rightarrow ? + ^{0}_{-1}\beta$
c. $^{108}_{47}Ag + ^{1}_{0}n \rightarrow ? + ^{0}_{-1}\beta$

21.76. What nuclide is produced in the core of a collapsing giant star by each of the following reactions?
a. $^{65}_{29}Cu + 3\ ^{1}_{0}n \rightarrow ? + ^{0}_{-1}\beta$
b. $^{68}_{30}Zn + 2\ ^{1}_{0}n \rightarrow ? + ^{0}_{-1}\beta$
c. $^{88}_{38}Sr + ^{1}_{0}n \rightarrow ? + ^{0}_{-1}\beta$

Nuclear Fission

Concept Review

21.77. How is the rate of energy release controlled in a nuclear reactor?

21.78. How does a breeder reactor create fuel and energy at the same time?

***21.79.** Why are neutrons always by-products of the fission of most massive nuclides? (*Hint*: Look closely at the neutron-to-proton ratios shown in Figure 21.1.)

21.80. Seaborgium (Sg, element 106) is prepared by the bombardment of curium-248 with neon-22, which produces two isotopes, ^{265}Sg and ^{266}Sg. Write balanced nuclear reactions for the formation of both isotopes. Are these reactions better described as fusion or fission processes?

Problems

21.81. The fission of uranium produces dozens of isotopes. For each of the following fission reactions, determine the identity of the unknown nuclide:
a. $^{235}U + {}^1_0n \rightarrow {}^{96}Zr + ? + 2\,{}^1_0n$
b. $^{235}U + {}^1_0n \rightarrow {}^{99}Nb + ? + 4\,{}^1_0n$
c. $^{235}U + {}^1_0n \rightarrow {}^{90}Rb + ? + 3\,{}^1_0n$

21.82. For each of the following fission reactions, determine the identity of the unknown nuclide:
a. $^{235}U + {}^1_0n \rightarrow {}^{137}I + ? + 2\,{}^1_0n$
b. $^{235}U + {}^1_0n \rightarrow {}^{137}Cs + ? + 3\,{}^1_0n$
c. $^{235}U + {}^1_0n \rightarrow {}^{141}Ce + ? + 2\,{}^1_0n$

21.83. For each of the following fission reactions, determine the identity of the unknown nuclide:
a. $^{235}U + {}^1_0n \rightarrow {}^{131}I + ? + 2\,{}^1_0n$
b. $^{235}U + {}^1_0n \rightarrow {}^{103}Ru + ? + 3\,{}^1_0n$
c. $^{235}U + {}^1_0n \rightarrow {}^{95}Zr + ? + 3\,{}^1_0n$

21.84. For each of the following fission reactions, determine the identity of the unknown nuclide:
a. $^{235}U + {}^1_0n \rightarrow {}^{147}Pm + ? + 2\,{}^1_0n$
b. $^{235}U + {}^1_0n \rightarrow {}^{94}Kr + ? + 2\,{}^1_0n$
c. $^{235}U + {}^1_0n \rightarrow {}^{95}Sr + ? + 3\,{}^1_0n$

Measuring Radioactivity; Biological Effects of Radioactivity

Concept Review

21.85. What is the difference between a *level* of radioactivity and a *dose* of radioactivity?

21.86. What are some of the molecular effects of exposure to radioactivity?

21.87. Describe the dangers of exposure to radon-222.

21.88. **Food Safety** Periodic outbreaks of food poisoning from *E. coli*–contaminated meat have renewed the debate about irradiation as an effective treatment of food. In one newspaper article on the subject, the following statement appeared: "Irradiating food destroys bacteria by breaking apart their molecular structure." How would you improve or expand on this explanation?

Problems

21.89. **Radiation Exposure from Dental X-rays** Dental X-rays expose patients to about 5 μSv of radiation. Given an RBE of 1 for X-rays, how many grays of radiation does 5 μSv represent? For a 50 kg person, how much energy does 5 μSv correspond to?

*21.90. **Radiation Exposure at Chernobyl** Some workers responding to the explosion at the Chernobyl nuclear power plant were exposed to 5 Sv of radiation, resulting in death for many of them. If the exposure was primarily in the form of γ rays with an energy of 3.3×10^{-14} J and an RBE of 1, how many γ rays did an 80 kg person absorb?

***21.91.** **Strontium-90 in Milk** In the years immediately following the explosion at the Chernobyl nuclear power plant, the concentration of ^{90}Sr in cow's milk in southern Europe was slightly elevated. Some samples contained as much as 1.25 Bq/L of ^{90}Sr radioactivity. The half-life of strontium-90 is 28.8 years.
a. Write a balanced nuclear equation describing the decay of ^{90}Sr.
b. How many atoms of ^{90}Sr are in a 200 mL glass of milk with 1.25 Bq/L of ^{90}Sr radioactivity?
c. Why would strontium-90 be more concentrated in milk than other foods, such as grains, fruits, or vegetables?

*21.92. **Radium Watch Dials** If exactly 1.00 μg of ^{226}Ra was applied to the glow-in-the-dark dial of a wristwatch made in 1914, how radioactive is the watch today? Express your answer in microcuries and becquerels. The half-life of ^{226}Ra is 1.60×10^3 years.

21.93. In 1999, the U.S. Environmental Protection Agency set a maximum radon level for drinking water at 4.0 pCi per milliliter.
a. How many decay events occur per second in a milliliter of water for this level of radon radioactivity?
b. If the above radioactivity were due to the decay of ^{222}Rn ($t_{1/2}$ = 3.8 days), how many ^{222}Rn atoms would there be in 1.0 mL of water?

21.94. **Death of a Spy** A former Russian spy died from radiation sickness in 2006 after dining at a London restaurant where he apparently ingested polonium-210. The other people at his table did not suffer from radiation sickness, even though they were very near the radioactive food the victim ate. Why were they not affected?

Medical Applications of Radionuclides

Concept Review

21.95. How does the selection of an isotope for radiotherapy relate to (a) its half-life, (b) its mode of decay, and (c) the properties of the products of decay?

21.96. Are the same radioactive isotopes likely to be used for both imaging and cancer treatment? Why or why not?

Problems

21.97. Predict the most likely mode of decay for the following isotopes used as imaging agents in nuclear medicine: (a) ^{197}Hg (kidney); (b) ^{75}Se (parathyroid gland); (c) ^{18}F (bone).

21.98. Predict the most likely mode of decay for the following isotopes used as imaging agents in nuclear medicine: (a) ^{133}Xe (cerebral blood flow); (b) ^{57}Co (tumor detection); (c) ^{51}Cr (red blood cell mass); (d) ^{67}Ga (tumor detection).

21.99. A 1.00 mg sample of ^{192}Ir was inserted into the artery of a heart patient. After 30 days, 0.756 mg remained. What is the half-life of ^{192}Ir?

21.100. A sample of dysprosium-165 with an radioactivity of 1100 counts per second was injected into the knee of a patient suffering from rheumatoid arthritis. After 24 hr, the radioactivity had dropped to 1.14 counts per second. Calculate the half-life of ^{165}Dy.

21.101. **Treatment of Tourette's Syndrome** Tourette's syndrome is a condition whose symptoms include sudden movements and vocalizations. Iodine isotopes are used in brain imaging of people suffering from Tourette's syndrome. To study the uptake and distribution of iodine in cells, mammalian brain cells in culture were treated with a solution containing ^{131}I with an initial radioactivity of 108 counts per minute. The cells were removed after 30 days, and the remaining solution was found to have a radioactivity of 4.1 counts per minute. Did the brain cells absorb any ^{131}I ($t_{1/2}$ = 8.1 days)?

21.102. **Mercury Test of Kidney Function** A patient is administered mercury-197 to evaluate kidney function. Mercury-197 has a half-life of 65 hours. What fraction of an initial dose of mercury-197 remains after 6 days?

21.103. Rhodium-105 is an isotope currently under investigation in diagnostic applications. The half-life of ^{105}Rh is 35.4 h, which is sufficiently long for transport from the supplier to a hospital. A supplier ships 250 mg of $^{105}RhCl_3$ overnight (12 hours).
a. What percentage of the ^{105}Rh remains upon arrival?
b. How long will it take for 95% of the ^{105}Rh to decay?

21.104. **Treating Prostate Cancer** Palladium-103 is used to treat prostate cancer by inserting a small (1 mm × 5 mm) cylindrical piece of ^{103}Pd directly into the tumor. How long will it take for 85% of the ^{103}Pd to decay, given that $t_{1/2}$ = 17 days?

***21.105.** **Boron Neutron-Capture Therapy** In boron neutron-capture therapy (BNCT), a patient is given a compound containing ^{10}B that accumulates inside cancer tumors. Then the tumors are irradiated with neutrons, which are absorbed by ^{10}B nuclei. The product of neutron capture is an unstable form of ^{11}B that undergoes α decay to ^{7}Li.
a. Write a balanced nuclear equation for the neutron absorption and α decay process.
b. Calculate the energy released by each nucleus of boron-10 that captures a neutron and undergoes α decay, given the following masses of the particles in the process: ^{10}B (10.0129 amu), ^{7}Li (7.01600 amu), ^{4}He (4.00260 amu), and ^{1}n (1.00866 amu).
c. Why is the formation of a nuclide that undergoes α decay a particularly effective cancer therapy?

21.106. **Balloon Angioplasty and Arteriosclerosis** Balloon angioplasty is a common procedure for unclogging arteries in patients suffering from arteriosclerosis. Iridium-192 therapy is being tested as a treatment to prevent reclogging of the arteries. In the procedure, a thin ribbon containing pellets of ^{192}Ir is threaded into the artery. The half-life of ^{192}Ir is 74 days. How long will it take for 99% of the radioactivity from 1.00 mg of ^{192}Ir to disappear?

Additional Problems

21.107. Thirty years before the creation of antihydrogen, television producer Gene Roddenberry (1921–1991) proposed to use this form of antimatter to fuel the powerful "warp" engines of the fictional starship *Enterprise.*
a. Why would antihydrogen have been a particularly suitable fuel?
b. Describe the challenges of storing such a fuel on a starship.

21.108. Tiny concentrations of radioactive tritium ($^{3}_{1}H$) occur naturally in rain and groundwater. The half-life of $^{3}_{1}H$ is 12 years. Assuming that tiny concentrations of tritium can be determined accurately, could the isotope be used to determine whether a bottle of wine with the year 1969 on its label actually contained wine made from grapes that were grown in 1969? Explain your answer.

21.109. In Section 21.6 we state that "no energy would be released if two ^{4}He nuclei were to fuse together to form ^{8}Be. Similarly, ^{8}Be nuclei require no energy to spontaneously decompose into ^{4}He nuclei, so they would immediately do so." Verify this statement by calculating the binding energy of ^{8}Be and comparing it to that of ^{4}He.

21.110. How much energy is required to remove a neutron from the nucleus of an atom of carbon-13 (mass = 13.00335 amu)? (*Hint*: The mass of an atom of carbon-12 is exactly 12.00000 amu.)

21.111. **Smoke Detectors** Americium-241 ($t_{1/2}$ = 433 yr) is used in smoke detectors. The α particles from this isotope ionize nitrogen and oxygen in the air, creating an electric current. When smoke is present, the current decreases, setting off the alarm.
a. Does a smoke detector bear a closer resemblance to a Geiger counter or to a scintillation counter?
b. How long will it take for the radioactivity of a sample of ^{241}Am to drop to 1% of its original radioactivity?
c. Why are smoke detectors containing ^{241}Am safe to handle without protective equipment?

*21.112. **Potassium/Argon Dating** Potassium-40 is a radioactive isotope of potassium, a very common element in terrestrial rocks, which decays to ^{40}Ar with a half-life of 1.28×10^{9} years. By measuring the ratio of ^{40}Ar to ^{40}K, geologists can to determine the age of ancient rocks.
a. Balance the nuclear equations for the decay of ^{40}K by identifying the missing isotope or particle.

$$^{40}K \rightarrow {}^{40}Ar + ?$$

$$^{40}K \rightarrow ? + {}^{0}_{-1}\beta$$

b. Why might ^{40}K decay by two different pathways?
c. Only about 11% of the ^{40}K decays to ^{40}Ar. If the ratio of ^{40}Ar to ^{40}K in a rock is found to be 0.435, how old is the rock?
d. Why don't geologists measure the ^{40}Ca:^{40}K ratio instead?

21.113. **Synthesis of a New Element** In 2006 an international team of scientists confirmed the synthesis of a total of three atoms of Og (oganesson) in experiments run in 2002 and 2005. They bombarded a ^{249}Cf target with ^{48}Ca nuclei.
a. Write a balanced nuclear equation describing the synthesis of $^{294}_{118}Og$.
b. The synthesized isotope of Og undergoes α decay ($t_{1/2}$ = 0.9 ms). What nuclide is produced by the decay process?
c. The nuclide produced in part b also undergoes α decay ($t_{1/2}$ = 10 ms). What nuclide is produced by this decay process?
d. The nuclide produced in part c also undergoes α decay ($t_{1/2}$ = 0.16 s). What nuclide is produced by this decay process?

e. If you had to select an element that occurs in nature and that has physical and chemical properties similar to Og, which element would it be?

21.114. Which element in the following series will be present in the greatest amount after 1 year?

$$\begin{array}{lccccccc} & {}^{214}_{83}\text{Bi} & \xrightarrow{\alpha} & {}^{210}_{81}\text{Ti} & \xrightarrow{\beta} & {}^{210}_{82}\text{Pb} & \xrightarrow{\beta} & {}^{210}_{83}\text{Bi} \rightarrow \\ t_{1/2} = & & 20\text{ min} & & 1.3\text{ min} & & 20\text{ yr} & 5\text{ d} \end{array}$$

21.115. Radiation exposure leads to the ionization of water to H_2O^+, which reacts to form H_3O^+ and OH. Draw Lewis structures for these three molecules or molecular ions.

*21.116. **Dating Cave Paintings** Cave paintings in Gua Saleh Cave in Borneo have been dated by measuring the amount of ^{14}C in calcium carbonate deposits that formed over the pigments used in the paint. The source of the carbonate ion was atmospheric CO_2.

a. What is the ratio of the ^{14}C radioactivity in calcium carbonate that formed 9900 years ago to that in calcium carbonate formed today?

b. The archaeologists also used a second method, uranium–thorium dating, to confirm the age of the paintings by measuring trace quantities of these elements present as contaminants in the calcium carbonate. Shown below are two candidates for the U–Th dating method. Which isotope of uranium do you suppose was chosen? Explain your answer.

$$\begin{array}{lccccc} & {}^{235}_{92}\text{U} & \rightarrow & {}^{231}_{90}\text{Th} \rightarrow & {}^{231}_{91}\text{Pa} & \rightarrow \\ t_{1/2} = & & 7.04 \times 10^8\text{ yr} & 25.6\text{ hr} & & 3.25 \times 10^4\text{ yr} \end{array}$$

$$\begin{array}{lccccc} & {}^{234}_{92}\text{U} & \rightarrow & {}^{230}_{90}\text{Th} \rightarrow & {}^{226}_{88}\text{Pa} & \rightarrow \\ t_{1/2} = & & 2.44 \times 10^5\text{ yr} & 7.7 \times 10^4\text{ hr} & & 1600\text{ yr} \end{array}$$

21.117. The synthesis of new elements and specific isotopes of known elements in linear accelerators involves the fusion of smaller nuclei.

a. An isotope of platinum can be prepared from nickel-64 and tin-124. Write a balanced equation for this nuclear reaction. (You may assume that no neutrons are ejected in the fusion reaction.)

b. Substitution of tin-132 for tin-124 increases the rate of the fusion reaction 10 times. Which isotope of Pt is formed in this reaction?

21.118. **Radon in Drinking Water** A sample of drinking water collected from a suburban Boston municipal water system in 2002 contained 0.5 pCi/L of radon. Assume that this level of radioactivity was due to the decay of ^{222}Rn ($t_{1/2}$ = 3.8 days).

a. What was the level of radioactivity (Bq/L) of this nuclide in the sample?

b. How many decay events per hour would occur in 2.5 L of the water?

21.119. Stone Age Skeletons The first attempt at radiocarbon dating six skeletons discovered in an Italian cave at the beginning of the 20th century indicated an age of 15,000 years. Redetermination of the age in 2004 indicated an older age for two bones of between 23,300 and 26,400 years. What is the ratio of ^{14}C in a sample 15,000 years old to one 25,000 years old?

21.120. There was once a plan to store radioactive waste that contained plutonium-239 in the reefs of the Marshall Islands. The planners claimed that the plutonium would be "reasonably safe" after 240,000 years. If the half-life is 24,400 years, what percentage of the ^{239}Pu would remain after 240,000 years?

21.121. Dating Prehistoric Bones In 1997 anthropologists uncovered three partial skulls of prehistoric humans in the Ethiopian village of Herto. Based on the amount of ^{40}Ar in the volcanic ash in which the remains were buried, their age was estimated at between 154,000 and 160,000 years old.

a. ^{40}Ar is produced by the decay of ^{40}K ($t_{1/2} = 1.28 \times 10^9$ yr). Propose a decay mechanism for ^{40}K to ^{40}Ar.

b. Why did the researchers choose ^{40}Ar rather than ^{14}C as the isotope for dating these remains?

*21.122. The carbon-14 radioactivity in papyrus growing along the Nile River today is 231 Bq per kilogram of carbon. If a papyrus scroll found near the Great Pyramid at Cairo has a carbon-14 radioactivity of 127 Bq per kilogram of carbon, how old is the scroll?

21.123. Thorium-232 slowly decays to bismuth-212 ($t_{1/2} = 1.4 \times 10^{10}$ yr). Bismuth-212 decays to lead-208 by two pathways: first β and then α decay, or α and then β decay. The intermediate nuclide in the second pathway is thallium-208. The thallium-208 can be separated from a sample of thorium nitrate by passing a solution of the sample through a filter pad containing ammonium phosphomolybdate. The radioactivity of ^{208}Tl trapped on the filter is measured as a function of time. In one such experiment, the following data were collected:

Time (s)	Counts/min
60	62
120	40
180	35
240	22
300	16
360	10

Use the data in the table to determine the half-life of ^{208}Tl.

21.124. The following nuclear equations are based on successful attempts to synthesize supermassive elements. Complete each equation by writing the symbol of the supermassive nuclide that was synthesized.

a. ${}^{58}_{26}\text{Fe} + {}^{209}_{83}\text{Bi} \rightarrow ? + {}^{1}_{0}\text{n}$

b. ${}^{64}_{28}\text{Ni} + {}^{209}_{83}\text{Bi} \rightarrow ? + {}^{1}_{0}\text{n}$

c. ${}^{62}_{28}\text{Ni} + {}^{208}_{83}\text{Pb} \rightarrow ? + {}^{1}_{0}\text{n}$

d. ${}^{22}_{10}\text{Ne} + {}^{249}_{97}\text{Bk} \rightarrow ? + 4\,{}^{1}_{0}\text{n}$

e. ${}^{58}_{26}\text{Fe} + {}^{208}_{82}\text{Pb} \rightarrow ? + {}^{1}_{0}\text{n}$

21.125. The absorption of a neutron by ^{11}B produces a radioactive nuclide that decays by either α decay or β decay. Write balanced nuclear equations describing the decay reactions.

21.126. An atom of darmstadtium-269 was synthesized in 2003 by the bombardment of a ^{208}Pb target with ^{62}Ni nuclei. Write a balanced nuclear equation describing the synthesis of ^{269}Ds.

21.127. The origins of the two naturally occurring isotopes of boron, ^{11}B and ^{10}B, are unknown. Both isotopes may be formed from collisions between protons and carbon, oxygen, or nitrogen in the aftermath of supernova explosions. Write balanced nuclear equations describing how ^{10}B might be formed from such collisions with ^{12}C and ^{14}N.

21.128. **Isotopes in Geochemistry** The relative abundances of the stable isotopes of the elements are not entirely constant. For example, in some geological samples (soils and rocks), the ratio of ^{87}Sr to ^{86}Sr is affected by the presence of a radioactive isotope of another element, which slowly undergoes β decay to produce more ^{87}Sr. What is this other isotope?

21.129. Rhodium-105 is made by neutron bombardment of ^{104}Ru, which decays to ^{105}Rh with a half-life of 4.4 h.

a. Write a balanced nuclear equation for the formation of ^{105}Rh.
b. Calculate how long it will take for 99% of the ^{105}Rh to decay.

21.130. **Biblical Archaeology** In 2012, Biblical scholar Karen King at Harvard University suggested that a piece of ancient papyrus believed to date from 400 CE (Figure P21.130) portrays Jesus as having a wife. Others have argued that the document is a forgery. Their argument relies on both linguistic interpretation and results of radiocarbon dating. Carbon-14 testing of the papyrus revealed a date in the 8th century CE.

a. By what percentage is the ratio of ^{14}C to ^{12}C in a sample from 750 CE smaller than the ratio in modern papyrus?
b. How much smaller is this ratio in papyrus from 400 CE, than in papyrus from 750 CE?

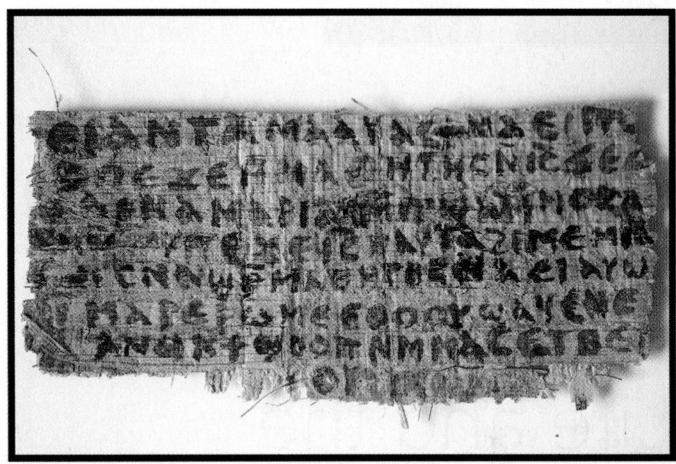

FIGURE P21.130

smartwork5

If your instructor uses Smartwork5, log in at digital.wwnorton.com/atoms2.

22

The Main Group Elements

Life and the Periodic Table

WHEELS OF PARMESAN CHEESE
Parmesan cheese contains calcium lactate, a natural source of the essential element calcium.

PARTICULATE **REVIEW**

Nitrogen, Sulfur, and Phosphorus: One, Two, or All Three?

In Chapter 22 we survey the roles of main group elements in living organisms. In addition to carbon, hydrogen, and oxygen, atoms of nitrogen, sulfur, and phosphorus are fundamental building blocks for important biological molecules in plants and animals.

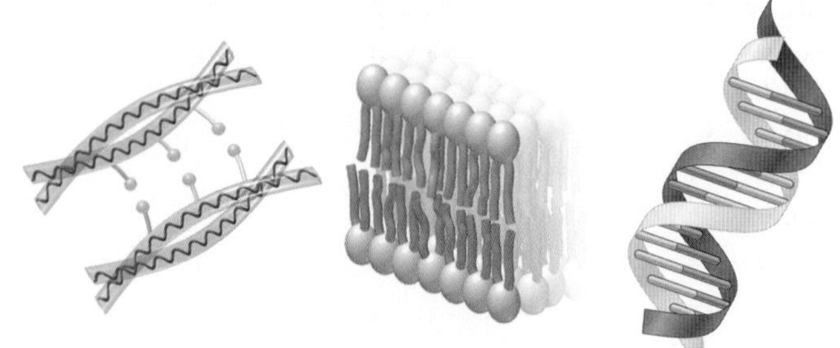

- Which of these structures contains sulfur atoms? Where are the sulfur atoms?
- Which two structures contain phosphorus atoms? Where are the phosphorus atoms?
- All three structures contain nitrogen atoms. List the functional groups that contain nitrogen atoms in each structure.

(Review Chapter 20 if you need help.)

(Answers to Particulate Review questions are in the back of the book.)

Charge versus Size: Selective Transport

Shown here are representations of four essential cations involved in selective ion transport through channels in cell membranes: Ca^{2+}, H_3O^+, K^+, and Na^+. The size of each cation is listed. As you read Chapter 22, look for ideas that will help you answer these questions:

100 pm

102 pm

113 pm

138 pm

- Identify each of the four ions pictured.
- The Na^+ channel is quite selective, allowing only Na^+ cations and one of the other three cations to pass through it. Which other ion— Ca^{2+}, H_3O^+, or K^+—is most likely to undergo selective transport through the sodium channel?
- Which of the other three ions, if any, is likely to pass through the K^+ channel?

Learning Outcomes

LO1 Distinguish between essential and nonessential elements and between major, trace, and ultratrace elements

LO2 Summarize the pathways and calculate the free energy for ion transport across cell membranes
Sample Exercise 22.1

LO3 Assign oxidation numbers, draw structures, and carry out calculations relevant to the behavior of major, trace, and ultratrace elements and their compounds *in vivo*
Sample Exercises 22.2, 22.3, 22.4

LO4 Describe the function of the essential and nonessential group 14–16 elements in the human body

LO5 Explain how radioactive isotopes of the main group elements are used in the diagnosis of disease
Sample Exercise 22.5

LO6 Describe how compounds containing main group elements are used in therapy and in other applications

22.1 Main Group Elements and Human Health

Have you ever wondered how many of the elements in the periodic table are in the human body? Or wondered which of them are important to human health?

Roughly one-third of the 90 naturally occurring elements have an identifiable role in human health and in organisms in general. **Essential elements** are defined as those that have a beneficial physiological effect, including those whose absence impairs functioning of the organism (Table 22.1). Some—like carbon, hydrogen, oxygen, nitrogen, sulfur, and phosphorus—are the principal constituents of all plants and animals. The alkali metal cations Na^+ and K^+ act as charge carriers, maintain osmotic pressure, and transmit nerve impulses. The alkaline earth cation Mg^{2+} is important in photosynthesis, and its family member Ca^{2+} forms structural materials such as bones and teeth. Chloride ions balance the charge of Na^+ and K^+ ions to maintain electrical neutrality in living cells. Other main group elements, such as iodine and selenium, are required in tiny amounts in our bodies to regulate metabolism and as components of the enzymes that catalyze biochemical reactions.

TABLE 22.1 Essential Elements in the Human Body

Major (>1 mg/g of body mass)	Trace (1–1000 μg/g of body mass)	Ultratrace (<1 μg/g of body mass)
Calcium	Fluorine	Chromium
Carbon	Iodine	Cobalt
Chlorine	Iron	Copper
Hydrogen	Silicon	Manganese
Magnesium	Zinc	Molybdenum
Nitrogen		Nickel
Oxygen		Selenium
Phosphorus		Vanadium
Potassium		
Sodium		
Sulfur		

Many **nonessential elements** are also present in the body but have no known function (Table 22.2). Some are useful in medicine as either diagnostic tools or therapeutic agents. Some radioactive isotopes may be used as imaging agents for organs and tumors; others are used to treat disease. Drugs containing lithium ions are used to treat depression. Compounds of bismuth act as mild antibacterial agents for treatment of diarrhea, and antimony compounds represent one of the few options for patients suffering from the tropical disease leishmaniasis, caused by protozoan parasites.

In some cases, the presence of a nonessential element has a **stimulatory effect**, which means that the consumption of small amounts of the element causes increased activity or growth in an organism. The effect may be beneficial or not, and often the mechanism of the effect is not understood. For example, small amounts of the nonessential element antimony promote growth in some mammals when added to their diets. Nonessential elements, and even toxic elements, are often incorporated into our bodies because their chemical properties are similar to those of an essential element. For example, Rb^+ ions are retained by the human body because they are similar to K^+ ions in size, charge, and chemistry, making rubidium the most abundant nonessential element in humans.

Oxygen, in the form of O_2 gas, occurs in the body in elemental form. Oxygen is also incorporated into myriad organic and inorganic compounds, the principal one being H_2O. When we speak of any element in the body other than oxygen, however, we usually refer to an ion or compound containing that element rather than the pure element. For example, when we describe calcium as an essential element, we are referring to calcium ions, Ca^{2+}, not calcium metal.

The essential elements are further classified as **major**, **trace**, or **ultratrace essential elements**. The six most abundant major elements—oxygen, carbon, hydrogen, nitrogen, calcium, and phosphorus—make up nearly 99% of the mass of the human body. Almost all of the foods we eat are rich in compounds containing these elements. Much of the remaining 1% of human body mass comes from the other five major elements: potassium, sulfur, sodium, chlorine, and magnesium. Salt is perhaps the most familiar dietary source of sodium and chlorine (in the form of sodium and chloride ions), although both are present in most food. Vegetables like broccoli and Brussels sprouts and fruits such as bananas are rich in potassium. Important sources of both calcium and magnesium include dairy products, green leafy vegetables, beans, and nuts.

Table 22.3 compares the elemental compositions of the human body, the universe, Earth's crust, and seawater. Note that the composition of our bodies most

TABLE 22.2 Nonessential Elements in the Human Body

Stimulatory	Unknown Role	No Role
Boron	Antimony	Barium
Titanium	Arsenic	Bromine
		Cesium
		Germanium
		Rubidium
		Strontium

TABLE 22.3 Comparative Composition[a] of the Universe, Earth's Crust, Seawater, and the Human Body

Element	Universe (%)	Crust (%)	Seawater (%)	Human Body (%)
Hydrogen	91	0.22	66	63
Oxygen	0.57	47	33	25.5
Carbon	0.021	0.019	0.0014	9.5
Nitrogen	0.042			1.4
Calcium		3.5	0.006	0.31
Phosphorus				0.22
Chlorine			0.33	0.03
Potassium		2.5	0.006	0.06
Sulfur	0.001	0.034	0.017	0.05
Sodium		2.5	0.28	0.01
Magnesium	0.002	2.2	0.033	0.01
Helium	9.1			
Silicon	0.003	28		
Aluminum		7.9		
Neon	0.003			
Iron	0.002	6.2		
Bromine			0.0005	
Titanium		0.46		
All other elements	< 0.1	< 0.1	< 0.1	< 0.1

[a]Values are expressed as the percentage of the total number of atoms. Because of rounding, the totals do not equal exactly 100%.

essential element element present in tissue, blood, or other body fluids that has a physiological function.

nonessential element element present in humans that has no known function.

stimulatory effect increased activity, growth, or other biological response to a stimulus.

major essential element essential element present in the body in average concentrations greater than 1 mg of element per gram of body mass.

trace essential element essential element present in the body in average concentrations between 1 and 1000 μg of element per gram of body mass.

ultratrace essential element essential element present in the body in average concentrations less than 1 μg of element per gram of body mass.

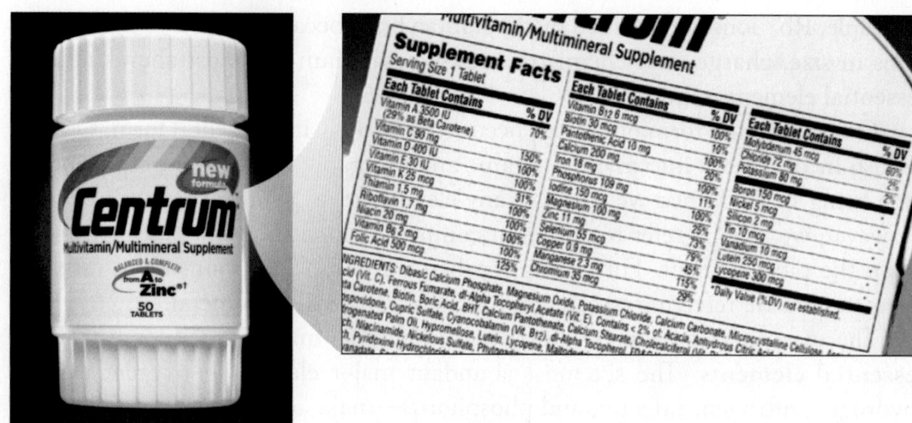

FIGURE 22.1 The labels on multivitamin supplements may not list DRI or RDA values but rather *% daily values* (DVs). Daily values are based on RDA or DRI values, but there can be inconsistencies, particularly among the ultratrace essential elements.

closely resembles the composition of seawater. The match would be even closer if it were not for the biological processes in the sea that remove essential elements, such as nitrogen and phosphorus, and that store others in solid structures, such as the $CaCO_3$ that makes up corals and mollusk shells.

Our diet should supply us with sufficient quantities of all essential elements. In the United States and Canada, these quantities are called *dietary reference intake* (DRI) values. They are based on the recommendations of the Food and Nutrition Board of the National Academy of Sciences and are frequently updated in response to research. For many essential elements, DRI values have replaced the *recommended dietary allowance* (RDA) values you may be familiar with from labels on food and vitamin packages (Figure 22.1). Table 22.4 lists DRI and RDA values for several major, trace, and ultratrace essential elements. The roles of transition metal ions such as Fe^{2+}, Fe^{3+}, and Zn^{2+} in biology are addressed in Chapter 23.

TABLE 22.4 Dietary Reference Intakes (DRI) and Recommended Dietary Allowances (RDA) for Selected Essential Elements[a]

Element	DRI	RDA
Calcium	1000 mg	1200 mg
Chlorine	2300 mg	2300 mg
Chromium	25–35 μg	35 μg
Copper	900 μg	900 μg
Fluorine	3–4 mg	4 mg
Iodine	150 μg	150 μg
Iron	8–18 mg	18 mg
Magnesium	420 mg	320–400 mg
Manganese	1.8–2.3 mg	2–5 mg
Molybdenum	45 μg	45 μg
Phosphorus	700 mg	700 mg
Potassium	4700 mg	4700 mg
Selenium	55 μg	55 μg
Sodium	1500 mg	1500 mg
Zinc	8–11 mg	11 mg

[a]DRI and RDA values in mg or μg per day from the U.S. Department of Agriculture (2009) and from the Council on Responsible Nutrition (CRN) for 19- to 30-year-olds.

Some main group elements, such as radon, beryllium, and lead, are toxic. As described in Chapter 21, inhaled radon gas poses serious health hazards from α decay taking place inside the body. Beryllium toxicity is a concern in industrial settings where beryllium-contaminated dust may be inhaled; the Be^{2+} ion replaces Mg^{2+} in the body, inhibiting Mg^{2+}-catalyzed RNA and DNA synthesis in cells. Lead ions, Pb^{2+}, are incorporated into teeth and bones because they are similar to Ca^{2+} ions in size and charge. Lead also interferes with the functioning of enzymes that require calcium ions, causing chronic neurological problems and blood-based disorders, especially in children. This issue became a major news story early in 2016 when it was discovered that a budget-cutting decision to switch drinking-water sources in Flint, Michigan, had exposed as many as 8000 children under age 6 to unsafe levels of lead. In June 2014, the city's source of water changed from Lake Huron to the Flint River. The river water proved more corrosive than the lake water and leached lead from old water pipes, causing lead levels to rise to more than 100 ppb; the EPA's standard for drinking water sets the action level at 15 ppb. Alarmingly, the crisis in Flint is not isolated, as unsafe levels of lead have been identified in other cities with aging water systems.

In this chapter we review the periodic properties of the main group elements and survey the roles of selected elements in the human body as well as their importance to good health. At the same time we call on the knowledge and skills you have acquired in your study of general chemistry to solve problems that link concepts from prior chapters to the central question of this chapter: What are the roles of the main group elements in the chemistry of life?

22.2 Periodic and Chemical Properties of Main Group Elements

The main group or representative elements, as defined in Chapter 2, appear in groups 1, 2, and 13–18 of the periodic table of the elements. These eight groups contain 44 naturally occurring elements, including five for which no stable isotopes are known: Fr, Ra, Po, At, and Rn. Relatively little is known about the chemistry of francium or astatine, both of which are exceedingly rare. Before we explore the biological roles of the others, it is useful to review what we have already learned about the physical and chemical properties of these elements and to look for periodic trends in these properties.

CONNECTION The organization of the periodic table was introduced in Chapter 2. Periodic properties were discussed in Chapter 3, and the stability of atomic nuclei was addressed in Chapter 21.

You may recall that the atomic radii of the main group elements increase as we descend a group and decrease across a row. Their first ionization energies and electronegativities decrease with increasing atomic number down a group and increase across a row. In general, electron affinities become more negative across a row. We have explored how all of these periodic trends are related to changes in the effective nuclear charge experienced by valence electrons as atomic numbers increase and inner shells fill with electrons.

A survey of the main group elements reveals a range of physical and chemical properties. The alkali metal and alkaline earth elements are relatively soft solids, while all of the group 18 elements are gases at standard temperature and pressure. All group 17 elements exist as diatomic molecules. Group 17 is the only group that contains elements in all three physical states at room temperature and pressure: bromine is one of only two liquid elements in the periodic table, iodine is a volatile solid, and the others are gases. Of the 20 naturally occurring elements in groups 13 to 16, six are nonmetals: C, N, O, P, S, and Se; six are semimetals: B, Si, Ge, As, Sb, and Te; and eight are metals: Al, Ga, In, Tl, Sn, Pb, Bi, and Po. Look at the periodic table and note how the nonmetals are clustered in the upper right hand corner of this block of elements; the metals are to the left and lower in the block, and the semimetals form a diagonal separating the metals and nonmetals. In every group metallic character increases with increasing atomic number.

The melting and boiling points of the metals in groups 1 and 2 decrease with increasing atomic number, but this trend is reversed for the nonmetals in groups 17 and 18 (Table 22.5). These opposing trends can be understood in terms of the different types of forces holding the atoms and molecules of these elements together. The increase in the size of the metallic elements when descending groups 1 and 2 leads to weaker metallic bonds and lower boiling points. In groups 17 and 18, however, increasing size leads to greater London dispersion forces as the polarizability of the atoms and molecules increases. The trends in melting points for groups 13–16 do not fit a clear pattern, in part because the properties of the elements change from nonmetallic to metallic down these groups.

Of the main group elements, only hydrogen (in trace amounts), carbon, nitrogen, oxygen, sulfur, and the six noble gases exist in nature in elemental form. The

TABLE 22.5 Summary of Periodic Trends for the Main Group Elements

Property	Group 1	Group 2	Group 13	Group 14	Group 15	Group 16	Group 17	Group 18
Melting point	Decreases down the group	Decreases down the group	No single trend	No single trend	No single trend	No single trend	Increases down the group	Increases down the group
Boiling point	Decreases down the group	Decreases down the group	No single trend	Decreases down the group	No single trend	No single trend	Increases down the group	Increases down the group

remaining elements are found exclusively in ionic and covalent compounds. The group 1 and 2 elements readily lose their valence electrons, forming 1+ and 2+ cations, respectively, which can combine with group 17 anions to give familiar ionic compounds such as NaCl and KI. Halide compounds of groups 13–17 generally contain covalent bonds; however, the heavier group 13 and 14 elements form insoluble ionic salts such as $PbCl_2$ and TlCl. The metallic elements in groups 1, 2, and 13 combine with oxygen forming oxides, including K_2O, CaO (quicklime, used in the manufacture of steel), and Al_2O_3 (alumina, used in orthodontics).

The atoms in compounds composed primarily of hydrogen and the lighter elements of groups 13–17 tend to share electrons and form covalent bonds rather than transfer electrons to form ionic bonds. As noted in Chapter 19, over 100 million such covalently bonded compounds have been discovered in nature or synthesized in chemistry laboratories.

Hydrogen, the least massive element, is difficult to classify. Most periodic tables include hydrogen in group 1 based on its electron configuration of a half-filled *s* orbital, and on the dissociation of acids to protons and anions, analogous to the dissolution of alkali metals salts to 1+ cations and anions. However, hydrogen can also complete its valence shell by gaining an electron to form a hydride ion, H^-, which is isoelectronic with He, a noble gas. Compounds called metal hydrides form between hydrogen and group 1, 2, or 13 metals, and they behave as salts containing a metal cation and a hydride anion. However, hydrogen most commonly forms covalently bonded compounds with oxygen (as water, H_2O) and other group 14–16 elements.

22.3 Major Essential Elements

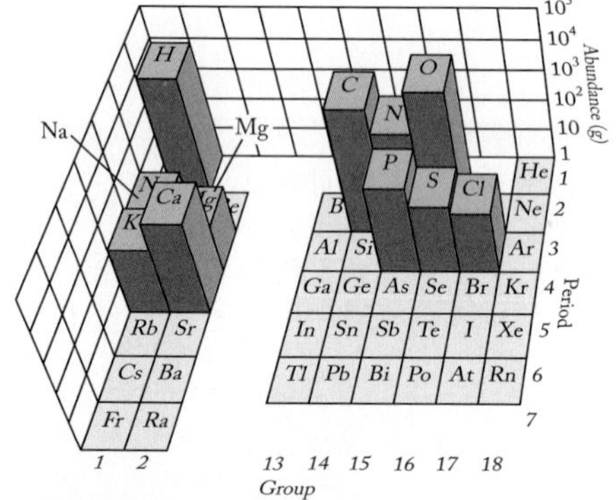

FIGURE 22.2 Abundance of the 11 major essential elements in a 70 kg adult human. Note that the *y*-axis has a log scale: abundances range from 35 g of magnesium to 46 kg of oxygen.

The 11 elements shown in red in Figure 22.2 and listed in the first column of Table 22.1 are the major essential elements. Together they account for 99.9% of the mass of the human body. Oxygen is the most abundant element by mass, followed by carbon and hydrogen. Although life depends on the presence of elemental oxygen in the form of O_2 gas, much of the oxygen in our bodies is combined with hydrogen in water molecules.

The 11 most abundant elements in the human body include seven nonmetals: C, H, O, Cl, S, P, and N. They are the building blocks for most of the body's molecular compounds and its principal polyatomic ions, HCO_3^-, SO_4^{2-}, and $H_2PO_4^-$, which are dissolved in body fluids. The average concentrations of the four major metals in the human body—Ca^{2+}, K^+, Na^+, and Mg^{2+}—are listed in Table 22.6. In this section we explore some of the roles that sodium, potassium, magnesium, calcium, chlorine, nitrogen, phosphorus, and sulfur play in the biochemistry of the human body. As we do, we revisit several of the chemical principles discussed in earlier chapters.

TABLE 22.6 Average Concentration of Four Metallic Elements in the Human Body

Element	mg/g of Body Mass
Calcium	15.0
Potassium	2.0
Sodium	1.5
Magnesium	0.5

Sodium and Potassium

Regulated concentrations of sodium and potassium ions are crucial to cell function. For example, too much Na^+ has been linked to hypertension (high blood pressure). To maintain a constant concentration of these two alkali metal ions in body fluids, the ions must be able to move into and out of cells. As noted in Section 20.4, the membrane surrounding a typical cell is a lipid bilayer, with polar groups containing phosphate groups on the two surfaces of the cell membrane and nonpolar fatty acids oriented toward the interior of the membrane. Direct diffusion of Na^+ and K^+ through the lipid bilayer is difficult because these polar cations do not dissolve in the nonpolar interior.

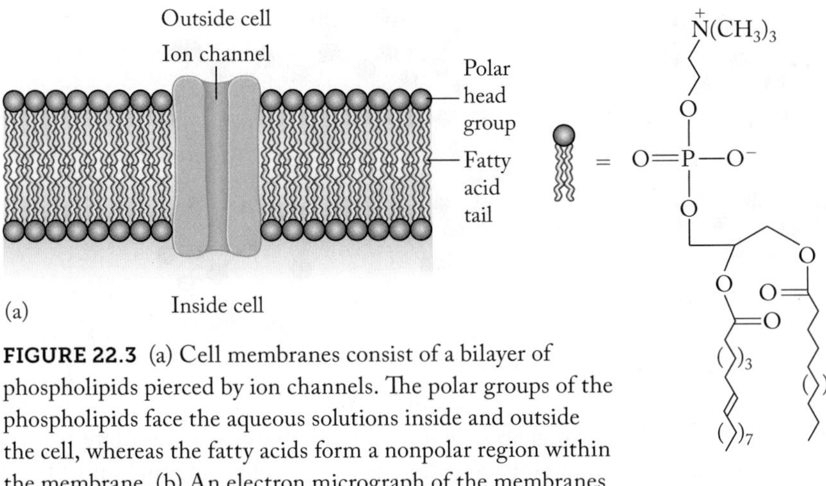

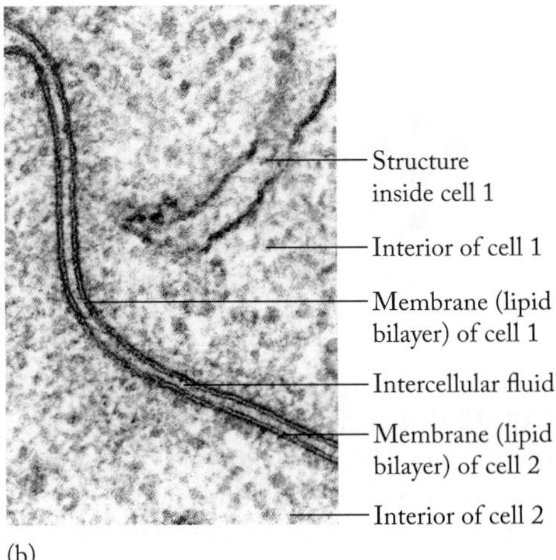

FIGURE 22.3 (a) Cell membranes consist of a bilayer of phospholipids pierced by ion channels. The polar groups of the phospholipids face the aqueous solutions inside and outside the cell, whereas the fatty acids form a nonpolar region within the membrane. (b) An electron micrograph of the membranes separating two adjacent cells.

As Figure 22.3 shows, cell membranes are pierced by **ion channels**, which are groups of protein complexes that allow selective transport of ions. Ion channels control which ions pass through the membrane, based on the size and charge of the ion as well as the shape of the protein. For example, the protein of the ion channel for potassium ions has its amino acids oriented in such a fashion that favorable ion–dipole interactions occur only for ions with the radius of a K^+ ion (138 pm) and not for Na^+ (102 pm) or any other cation. The sodium ion channel is also selective, excluding K^+ and Ca^{2+} even though the radii of Na^+ and Ca^{2+} (100 pm) differ by only 2 pm. Another difference between the Na^+ and K^+ channels is the ability of H_3O^+ (hydronium ion, radius 113 pm) to pass through sodium ion channels but not potassium ion channels.

Living organisms also contain oxygen-rich molecules such as nonactin (Figure 22.4) that interact with Ca^{2+}, K^+, Na^+, and Mg^{2+} ions through strong ion–dipole forces. The resulting complex ions consist of alkali metal ions encapsulated in nonpolar exteriors. Because each complex has a nonpolar exterior, it does not require a channel for passage through a cell membrane. Instead, the complex carries its alkali metal cation through the nonpolar regions of the bilayer, providing an alternative to ion channels for the transport of metal ions.

CONNECTION In Chapter 6 we described how ions dissolved in water are surrounded by water molecules that form spheres of hydration. As Na^+ and K^+ ions enter their respective ion channels, they lose some or all of their hydration spheres.

In addition to ion channels and diffusion of complex ions, alkali metal cations can be transported by a third mechanism, one involving Na^+– K^+ ion pumps. An **ion pump** is a system of membrane proteins that exchange ions inside the cell (for example, Na^+) with those in the intercellular fluid (for example, K^+). Unlike diffusion or transport through ion channels, transport via the Na^+– K^+ pump requires energy, which is provided by the hydrolysis of ATP to ADP. An example of how the Na^+– K^+ ion pump works is the response of a nerve cell to touch. Stimulation of the nerve cell causes Na^+ to flow into the cell and K^+ to flow out via ion channels; this two-way flow of ions produces the nerve impulse. The ion pump then "recharges" the system by pumping Na^+ out of the cell and K^+ into the cell so that another impulse can immediately be transmitted along the nerve.

The concentrations of several major ions, including Na^+ and Cl^-, are higher outside living cells than inside them, whereas others, including K^+ and Mg^{2+}, are more abundant inside. These unequal concentrations cause the ions to diffuse through a cell membrane from the high-concentration side to the low-concentration side, assuming that ion channels or other transport pathways are available. As ions pass through the cell membrane, they leave behind the ions of opposite charge, or *counter ions*, that made the aqueous medium on the high-concentration

ion channel a group of helical proteins that penetrates cell membranes and allows selective transport of ions.

ion pump a system of membrane proteins that exchanges ions inside the cell with those in the intercellular fluid.

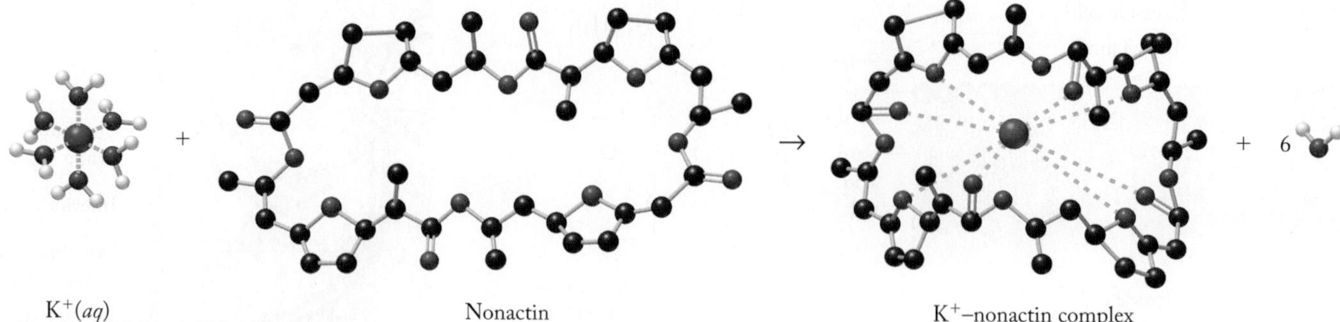

FIGURE 22.4 In living organisms, ligands such as nonactin form complex ions that have hydrophobic exteriors with any one of the four alkali metal and alkaline earth ions and that can transport these ions directly through cell membranes. Ion channels are not required in this transport pathway.

side neutral. Meanwhile, the ions and their electrical charges accumulate on the low-concentration side of the membrane. As a result, an electrical field builds up across the membrane until it reaches an **equilibrium potential** or **reversal potential, E_{ion}**. At this potential, the strength of the electrical field, which pushes ions back toward the high-concentration side of the membrane, balances the ions' concentration-driven diffusion toward the low-concentration side.

We can calculate the value of E_{ion} for a particular ion (X) using a modified version of the Nernst equation. The variables in the E_{ion} version include the charge of the ion, z, and the ratio of the concentrations of ion X outside and inside the cell:

$$E_{ion} = \frac{RT}{zF}\ln\frac{[X]_{outside}}{[X]_{inside}} \tag{22.1}$$

The brackets in Equation 22.1 represent molar concentration values, but other concentration units may be used for both values because they cancel each other during the calculation of E_{ion}. To see how they do, let's calculate E_{K^+} across the membrane of a human muscle cell for which the concentration of K^+ ions inside the cell is 145 mM and the concentration outside is 4.4 mM. To use Equation 22.1 for a human cell, we use $R = 8.314$ J/mol · K; $F = 96{,}500$ C/mol; $T = 37°C = (273 + 37 = 310)$ K, and 1 J = 1 C · V:

$$E_{K^+} = \frac{RT}{zF}\ln\frac{[K^+]_{outside}}{[K^+]_{inside}} = \frac{\left(8.314\frac{\text{J}}{\text{mol}\cdot\text{K}}\right)(310\ \text{K})}{(+1)\left(96{,}500\frac{\text{C}}{\text{mol}}\right)\left(\frac{1\ \text{J}}{\text{C}\cdot\text{V}}\right)}\ln\left(\frac{4.4\ \text{m}M}{145\ \text{m}M}\right)$$

$$= -0.093\ \text{V} = -0.093\ \text{V} = -93\ \text{mV}$$

The E_{ion} values of the major ions inside and outside living cells each contribute to an overall **membrane potential $E_{membrane}$** that is the weighted average of the E_{ion} values. Each weighting factor is the relative permeability of the membrane for a particular ion, which is strongly influenced by the number of open channels for that ion. The membrane potentials of most human cells are in the −50 to −70 mV range. In other words, the membrane surface facing the interior of the cell is 50 to 70 mV more negatively charged than the outside surface.

The difference between the membrane potential and the equilibrium potential for a given ion represents an electromotive force, $E_{transport}$, that pushes the ion across a cell membrane, from the inside toward the outside of the cell:

$$E_{transport} = E_{membrane} - E_{ion} \tag{22.2}$$

For example, the electromotive force that pushes potassium ions through a membrane having a membrane potential of −70 mV and an E_{K^+} value of −93 mV is

$$E_{transport} = E_{membrane} - E_{ion} = -70\ \text{mV} - (-93\ \text{mV}) = 23\ \text{mV}$$

equilibrium (reversal) potential, E_{ion} an electrochemical potential that results from a concentration gradient of a particular ion on opposite sides of a cell membrane.

membrane potential, $E_{membrane}$ a weighted average of the equilibrium (reversal) potentials of the major ions based on their concentrations and abilities to pass through the membrane.

This positive $E_{\text{transport}}$ value means that K^+ ions spontaneously flow through any available open channels from the inside of the cell into the aqueous medium (blood plasma) surrounding it. We can confirm that the outward flow of ions is spontaneous by calculating the change in free energy associated with it. To do this we modify Equation 17.6:

$$\Delta G_{\text{cell}} = -nFE_{\text{cell}}$$

to obtain an equation that enables us to calculate the free energy change that accompanies the transport of ions through a cell membrane:

$$\Delta G_{\text{transport}} = -zFE_{\text{transport}} \qquad (22.3)$$

The change in free energy for K^+ ions transported by an electromotive force of 23 mV is:

$$\Delta G_{\text{transport}} = -(+1)\left(96{,}500\frac{\cancel{\text{C}}}{\text{mol}}\right)(23\ \cancel{\text{mV}})\left(\frac{1\ \cancel{\text{V}}}{1000\ \cancel{\text{mV}}}\right)\left(\frac{1\ \text{J}}{\cancel{\text{CV}}}\right) = -2.2\ \text{kJ/mol}$$

The negative value of $\Delta G_{\text{transport}}$ confirms that the process is spontaneous. It also tells us that ion pumps must do at least 2.2 kJ/mol of work to pump K^+ ions in the opposite direction: from outside the cell, through the cell's membrane, into its interior.

CONNECTION The Nernst equation and the relationship between E and ΔG were introduced in Chapter 17.

SAMPLE EXERCISE 22.1 Calculating E_{ion} and Assessing Spontaneity of Ion Transport **LO2**

The concentrations of Na^+ are 0.050 *M* inside a squid axion cell and 0.440 *M* in the fluid surrounding the cell.

a. What is the equilibrium potential (E_{ion}) of the Na^+ ions at 280 K?
b. If the cell's membrane potential (E_{membrane}) is −50 mV, is the transport of Na^+ from inside to outside the cell spontaneous?

Collect, Organize, and Analyze We are given the concentrations of Na^+ on opposite sides of a cell membrane, and we are asked to calculate E_{Na^+} and to assess inside-out transport spontaneity. Equation 22.1 relates E_{ion} to ion concentrations surrounding a permeable cell membrane. Equation 22.2 allows us to use E_{ion} and E_{membrane} (−50 mV) to calculate $E_{\text{transport}}$, which, if positive, indicates that transport is spontaneous.

Solve

a. Using the given temperature, $[Na^+]$ values, and the equivalency 1 J = 1 C·V in Equation 22.1:

$$E_{Na^+} = \frac{RT}{zF}\ln\frac{[Na^+]_{\text{outside}}}{[Na^+]_{\text{inside}}} = \frac{\left(8.314\frac{\cancel{\text{J}}}{\cancel{\text{mol}\cdot\text{K}}}\right)(280\ \cancel{\text{K}})}{(+1)\left(96{,}500\frac{\cancel{\text{C}}}{\cancel{\text{mol}}}\right)\left(\frac{1\ \cancel{\text{J}}}{\cancel{\text{C}}\cdot\text{V}}\right)}\ln\left(\frac{0.440\ \cancel{M}}{0.050\ \cancel{M}}\right)$$

$$= 0.052\ \text{V} = 52\ \text{mV}$$

b. Using Equation 22.2 to calculate $E_{\text{transport}}$, which is the electromotive force available to push Na^+ ions across the membrane from the inside to the outside of the cell:

$$E_{\text{transport}} = E_{\text{membrane}} - E_{\text{ion}} = -50\ \text{mV} - 52\ \text{mV} = -102\ \text{mV}$$

The negative $E_{\text{transport}}$ value means Na^+ ions do not spontaneously pass through the membrane from inside to outside the cell.

Think About It The positive E_{Na^+} value is consistent with the much higher concentration of Na^+ ions outside the cell. The negative $E_{\text{transport}}$ value toward the outside of the cell also is reasonable because Na^+ ions would have to move up a steep concentration gradient and against a negative membrane potential that pulls them toward the inside of the cell. The fact that outward transport of Na^+ ions is not spontaneous means that their inward migration is spontaneous, or would be if Na^+ ion channels were open. In most cells few of them are.

Practice Exercise The concentrations of K^+ ions are 124 m*M* inside a frog muscle cell and 2.3 m*M* in the fluid surrounding the cell. What is the equilibrium potential for the K^+ ions at 293 K? If the cell membrane potential is −73 mV, how much work must be done to transport K^+ ions into the cell? Express you answer in kJ/mol.

(Answers to Practice Exercises are in the back of the book.)

CONCEPT **TEST**

Do ion pumps represent spontaneous or nonspontaneous processes?

(Answers to Concept Tests are in the back of the book.)

Magnesium and Calcium

CONNECTION The role of ATP and ADP in metabolism was described in Chapter 20.

CONNECTION The catabolism of glucose was described in Chapter 20.

The biological roles of Mg^{2+} and Ca^{2+} are more varied than those of Na^+ and K^+. We have mentioned that calcium is a major component of teeth and bones. A prolonged deficiency of calcium can lead to osteoporosis (a disease characterized by low bone density), whereas high concentrations of calcium in muscle cells contribute to cramps. Most kidney stones are made of calcium oxalate or calcium phosphate. Magnesium deficiencies can reduce physical and mental capacity because of the role of Mg^{2+} in the transfer of phosphate groups to and from ATP; slowing this transfer diminishes the amount of energy available to cells. The cellular concentrations of Mg^{2+} and Ca^{2+} are maintained by ion pumps.

Magnesium is a component of chlorophyll, which is one of several molecules used by plants to collect and capture light energy across the visible portion (400 to 700 nm) of the electromagnetic spectrum (Figure 22.5). Chlorophylls from different plants vary slightly in composition, but all of them contain magnesium coordinated to four nitrogen atoms. The presence of magnesium in chlorophyll does not account for the green color of the molecule, nor does it play a direct role in absorption of sunlight. The function of the Mg^{2+} ion is to orient the molecules in positions that allow energy to be transferred to the reaction centers where H_2O is consumed and O_2 is produced during photosynthesis. Carotene and related compounds are responsible for the orange colors of autumn leaves on deciduous trees when chlorophyll production ceases. Mg^{2+} ions play important roles in ATP hydrolysis and ADP phosphorylation. The many Mg^{2+}-mediated ATP $\rightarrow$ ADP processes include transferring phosphate to glucose in the conversion of glucose to pyruvate and driving Na^+– K^+ ion pumps.

To some extent, calcium ions are also capable of mediating ATP hydrolysis, but these ions play other roles in the cell. They are required to trigger muscle contractions, for example—the calcium ions used for this purpose are stored in proteins. Recall that the action of Na^+– K^+ pumps is responsible for the generation of nerve impulses. One effect of nerve impulses is to trigger the release of Ca^{2+} ions from their storage proteins into the intracellular fluid. In a multistep process, muscle cells contract and relax as calcium ions are released. Once the muscle action is complete, the ions are returned to their storage proteins in a process coupled to Mg^{2+}-mediated ATP hydrolysis.

Calcium ions also play a major role in the formation of teeth and bones. Mammalian bones are a *composite material*, defined as a material containing a mixture of different substances. About 30% of dry bone mass is elastic protein fibers. The

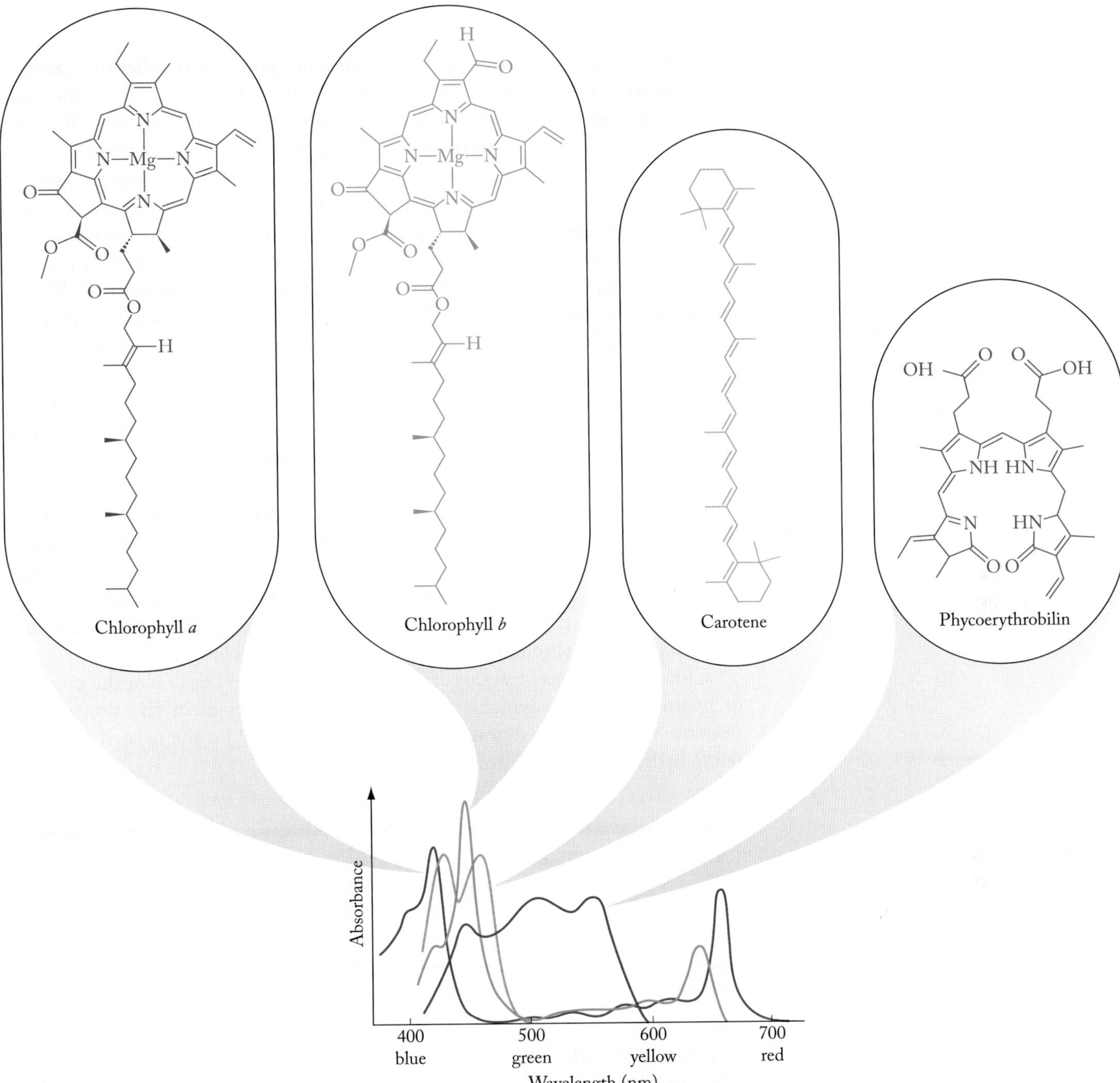

FIGURE 22.5 Photosynthetic bacteria, green plants, and algae use a variety of molecules to absorb all the visible wavelengths in sunlight. Among them, only chlorophylls contain magnesium and absorb blue-green and red-orange light. Carotene also absorbs in the blue-green region, whereas phycoerythrobilin absorbs a broad range of wavelengths from 400 to 600 nm.

remainder of the mass consists of calcium compounds, including the mineral hydroxyapatite, $Ca_5(PO_4)_3(OH)$, which is also a principal component of teeth. Hydroxyapatite crystals are bound to the protein fibers in bone through phosphate groups.

The shells of marine organisms are mostly calcium carbonate ($CaCO_3$) in a matrix of proteins and polysaccharides. Some magnesium is incorporated into the calcium carbonate outer shell of marine organisms that are capable of photosynthesis, such as algae and phytoplankton.

Chlorine

Of all the halogens, only chlorine (as chloride ions) is present in sufficient quantities to be considered a major essential element in humans. Chloride ions are the most abundant anions in the human body and are involved in many processes. The concentration of chloride ions in the human body (1.5 mg per gram of body mass) is slightly less than 10% of the concentration of Cl^- in seawater (19 mg per gram of water) but about 12 times greater than in Earth's crust (0.13 mg per gram of crust). Like the major essential cations, chloride ions are transported into and out of cells primarily via ion channels and ion pumps. To maintain electrical neutrality in a cell, the transport of alkali metal cations is accompanied by the transport of chloride anions. The *cotransport* of Na^+ and Cl^- is essential in kidney function, where the ions are reabsorbed by the body rather than eliminated with liquid waste products.

Malfunctioning chloride ion channels are the underlying cause of cystic fibrosis, a lethal genetic disease that causes patients to accumulate mucus in their airways such that breathing becomes difficult. The discovery of high concentrations of Na^+ and Cl^- in the sweat of cystic fibrosis patients led to an understanding of the role of chloride ion transport in patients with this disease.

Chloride ions also play a major role in the elimination of CO_2 from the body. Because it is nonpolar, carbon dioxide produced during glucose catabolism can pass from muscle cells (for example) into red blood cells, moving easily through the largely nonpolar cell membranes of these cells. Inside the red blood cells, CO_2 is converted to bicarbonate ion, HCO_3^-. When HCO_3^- is pumped out of the cell, Cl^- enters the cell through an ion channel to maintain charge balance.

Chloride ion concentrations are high in gastric juices because of the presence of hydrochloric acid, which catalyzes digestive processes in the stomach. In response to food in the digestive system, cells tap ATP for the needed energy to pump hydrochloric acid into the stomach.

SAMPLE EXERCISE 22.2 Calculating the Concentration of HCl in Stomach Acid — LO3

Acid reflux (sometimes called heartburn, though the heart is not involved) affects many people. It results from acid in the stomach leaking into the esophagus and causing discomfort. Stomach acid is primarily an aqueous solution of HCl. (a) Calculate the molarity of hydrochloric acid in gastric juice that has a pH of 0.80. (b) One treatment for the symptoms of acid reflux is to take an antacid tablet. What volume of gastric juice can be neutralized by a 750 mg tablet of calcium carbonate (a typical size for an over-the-counter antacid)?

Collect and Organize We are given the pH of a solution and are asked to calculate the concentration of HCl that corresponds to that pH. As we saw in Chapter 15, $pH = -\log[H_3O^+]$. Hydrochloric acid is a strong acid and ionizes completely to H_3O^+ and Cl^- in water:

$$HCl(aq) + H_2O(\ell) \rightarrow H_3O^+(aq) + Cl^-(aq)$$

We are also asked to calculate the volume of HCl solution that can be neutralized by a 750 mg tablet of calcium carbonate, $CaCO_3$. We need to write a balanced chemical equation for the neutralization reaction.

Analyze The equation describing the ionization of hydrochloric acid indicates that one mole of H_3O^+ ions is formed for every mole of HCl present. The pH of gastric juice falls between 1 and 0, so $[H_3O^+]$ will be between 10^{-1} (= 0.1) M and 10^0 (= 1) M.

The neutralization reaction is

$$CaCO_3(s) + 2\,H_3O^+(aq) \rightarrow Ca^{2+}(aq) + CO_2(g) + 3\,H_2O(\ell)$$

This equation indicates that two moles of H_3O^+ are consumed for every mole of $CaCO_3$. We are told that the tablet size is typical of an antacid tablet, so common sense leads us to predict that the volume of acid this tablet can neutralize will not be excessively large (greater than 1 L) or small (less than 10 mL): too large a tablet would be a waste of antacid, and too small a tablet would not relieve the symptoms.

Solve

a. Substitution into Equation 15.15 gives

$$pH = -\log[H_3O^+] = 0.80$$

We take the antilog of both sides to solve for $[H_3O^+]$:

$$[H_3O^+] = 10^{-0.80} = 0.16\,M\,H_3O^+$$

Therefore the concentration of HCl is

$$0.16\,M\,H_3O^+ \times \frac{1\text{ mol HCl}}{1\text{ mol }H_3O^+} = 0.16\,M\,\text{HCl}$$

b. First we calculate the number of moles of $CaCO_3$ present in 750 mg:

$$0.750\,\cancel{\text{g }CaCO_3} \times \frac{1\text{ mol }CaCO_3}{100.09\,\cancel{\text{g }CaCO_3}} = 7.49 \times 10^{-3}\text{ mol }CaCO_3$$

Next we use the stoichiometry of the neutralization reaction to calculate the volume of 0.16 M HCl this quantity of $CaCO_3$ can neutralize:

$$7.49 \times 10^{-3}\,\cancel{\text{mol }CaCO_3} \times \frac{2\,\cancel{\text{mol }H_3O^+}}{1\,\cancel{\text{mol }CaCO_3}} \times \frac{1\text{ L}}{0.16\,\cancel{\text{mol }H_3O^+}}$$

$$= 9.36 \times 10^{-2}\text{ L} = 94\text{ mL of }0.16\,M\,\text{HCl}$$

Think About It A concentration of 0.16 M seems reasonable because it is indeed within the range of values predicted for a solution with pH < 1. The volume of 0.16 M acid that a 750 mg tablet of $CaCO_3$ can neutralize is also reasonable; 94 mL represents about 3 ounces of gastric juice.

Practice Exercise Calculate the pH of a solution prepared by mixing 10.0 mL of 0.160 M HCl with 15.0 mL of water. How much antacid containing 4.00×10^2 mg of $Mg(OH)_2$ in 5.00 mL of water is needed to neutralize this volume of acid?

CONCEPT TEST

Taking an antacid tablet is often sufficient to treat an occasional case of mild acid reflux. Another remedy is a drug such as Prilosec, which inhibits a cell's proton pumps by binding to the site of the pump and disabling it for more than 24 hours. Is the equilibrium constant for the binding of a proton pump inhibitor likely to be less than or greater than 1?

Nitrogen

Nitrogen is a major essential element found primarily in proteins but also in DNA and RNA. Nitrogen is available in the atmosphere as N_2, and soil and water contain nitrate ions, but neither of these forms of nitrogen can be directly incorporated into amino acids, the building blocks of proteins. The biosynthesis

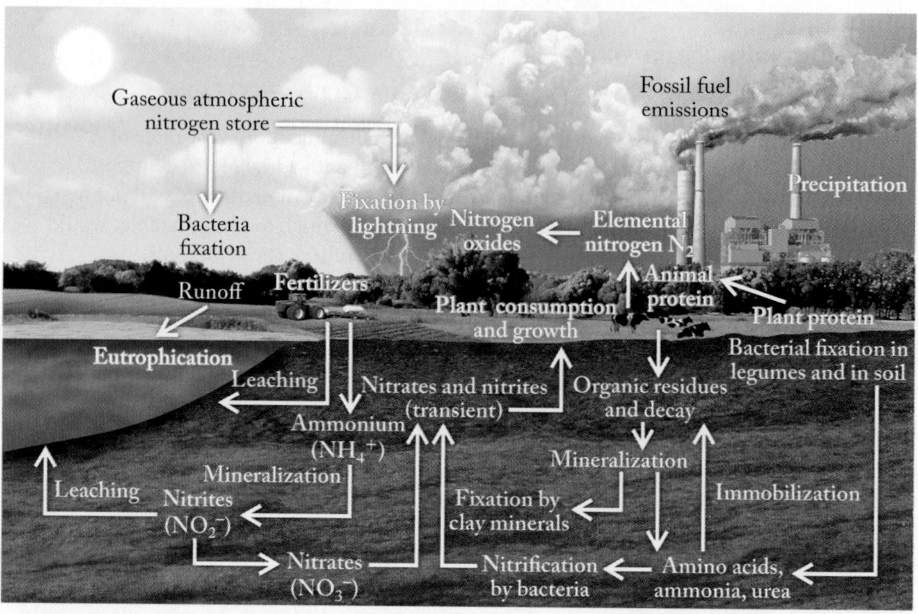

FIGURE 22.6 The nitrogen cycle: Enzymes interconvert the nitrogen-containing molecules and ions found in nature. Bacteria convert atmospheric nitrogen to ammonium ion, which is oxidized to nitrite (NO_2^-) and nitrate (NO_3^-) ions before being reduced back to N_2.

of amino acids requires ammonia or ammonium ions. For example, glycine (NH_2CH_2COOH), the simplest amino acid, is formed by reaction of CO_2 and ammonia in the presence of the appropriate enzyme. Interconversion of nitrogen-containing compounds in the environment is described by the nitrogen cycle illustrated in Figure 22.6. Certain bacteria use enzymes called *nitrogenases* to convert N_2 to ammonia. Plants convert NO_3^- ions to NO_2^- and then to NH_3 by using enzymes called *reductases*. Ultimately, the chemical reactions in these organisms begin a food chain that supplies the essential amino acids for human diets.

The reactions in Figure 22.7 are all redox reactions that illustrate the wide range of oxidation numbers found among nitrogen compounds in the nitrogen cycle. By definition, each atom in nitrogen gas, N_2, is assigned an oxidation

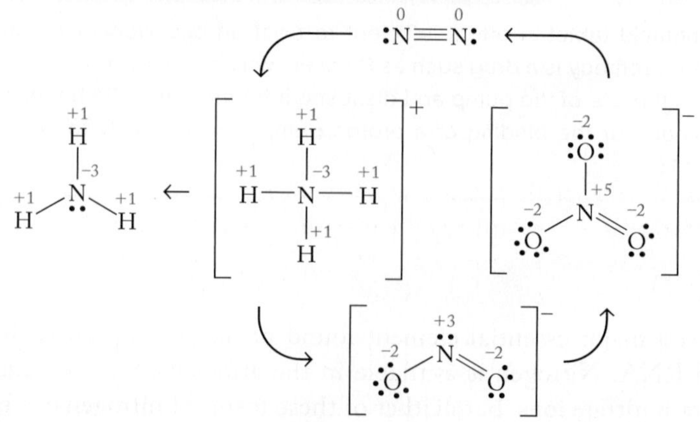

N_2 (0, 0) → $[NH_4]^+$ (N −3, H +1) ← NH_3 (N −3, H +1); $[NH_4]^+$ → $[NO_2]^-$ (N +3, O −2) → $[NO_3]^-$ (N +5, O −2) → N_2

FIGURE 22.7 Nitrogen-containing molecules and ions in the nitrogen cycle exhibit oxidation numbers ranging from −3 to +5.

number of 0. When N_2 is converted to NH_3, the oxidation number of N decreases to −3, a reduction. The oxidation numbers of N in nitrite (NO_2^-) and nitrate (NO_3^-) are +3 and +5, respectively, as a result of oxidation.

In Chapter 13 we encountered a different cycle for nitrogen in the environment: the conversion of N_2 and O_2 to NO and NO_2 in the engines of automobiles. Dinitrogen monoxide, N_2O, is a greenhouse gas. The nitrogen atoms in these volatile nitrogen oxides are assigned oxidation numbers of +1, +2, and +4 for N_2O, NO, and NO_2, respectively.

SAMPLE EXERCISE 22.3 Assigning Oxidation Numbers in the Reaction of Nitrate Reductases — LO3

Reduction of nitrate ions to ammonia is catalyzed by enzymes called reductases. The first step in this process is conversion of nitrate ions to nitrite ions. Assign oxidation numbers to nitrogen in these ions and in ammonia.

Collect, Organize, and Analyze We need to assign oxidation numbers based on the guidelines in Section 8.6. We know that the oxidation numbers of nitrogen and oxygen in each polyatomic ion must add up to the charge on the ion. Oxygen and hydrogen in compounds usually have oxidation numbers of −2 and +1, respectively.

Solve The oxidation number of nitrogen is unknown, so we call it x. The oxidation number of nitrogen in NO_3^- is

$$x + 3(-2) = -1$$
$$x = +5$$

The oxidation number of nitrogen in NO_2^- is

$$x + 2(-2) = -1$$
$$x = +3$$

The oxidation number of nitrogen in NH_3 is

$$x + 3(+1) = 0$$
$$x = -3$$

Think About It Assigning oxidation numbers confirms that nitrogen is reduced in the conversion of nitrate to nitrite to ammonia.

Practice Exercise Nitrogenases are enzymes that catalyze the conversion of nitrogen gas to ammonia. Is nitrogen oxidized or reduced in this process? Use oxidation numbers to support your answer.

In humans and other mammals, excess nitrogen is converted to urea in the liver and excreted via the kidneys. Plants use urea as a source of ammonia by the action of *ureases* via the reaction:

$$H_2N{-}C({=}O){-}NH_2 + H_2O \rightarrow 2NH_3 + CO_2$$

Unlike reactions catalyzed by nitrogenases and nitrate reductases, the conversion of urea to ammonia and carbon dioxide is not a redox reaction. It is a hydrolysis reaction, similar to the reaction of nonmetal oxides with water described in Chapter 8. Acidic solutions are observed when NH_4^+, NO_2, and N_2O_5 dissolve in water:

$$NH_4^+(aq) + H_2O(\ell) \rightleftharpoons NH_3(aq) + H_3O^+(aq)$$

$$2\ NO_2(g) + H_2O(\ell) \rightarrow HNO_2(aq) + HNO_3(aq)$$

$$N_2O_5(g) + H_2O(\ell) \rightarrow 2\ HNO_3(aq)$$

CONNECTION Hydrolysis of nitrite ion is described in Chapter 15.

Hydrolysis of nitrite ion, however, leads to weakly basic solutions:

$$NO_2^-(aq) + H_2O(\ell) \rightleftharpoons HNO_2(aq) + OH^-(aq)$$

Phosphorus and Sulfur

Phosphorus and sulfur are major essential elements found primarily in proteins and DNA, but they are also present in polyatomic anions prevalent in the environment. By comparison to the nitrogen cycle, the biological phosphorus cycle contains only a single major species: the phosphate ion, PO_4^{3-}, and its conjugate acids, HPO_4^{2-} and $H_2PO_4^-$, along with phosphate esters. Reduction of phosphorus(V) in PO_4^{3-} to phosphine (PH_3) does occur in swamps. An oxygen-free environment is needed because PH_3 spontaneously ignites in humid air, yielding phosphoric acid:

$$PH_3(g) + 2\ O_2(g) \rightarrow H_3PO_4(\ell)$$

Gases analogous to those seen in the nitrogen cycle, such as NO and NO_2, are absent from the phosphorus cycle. Slow weathering of insoluble phosphate minerals by weak acids in soil introduces phosphate ion into the environment, where it is eventually taken up by plants. Some of this phosphate is incorporated into biominerals such as hydroxyapatite [$Ca_5(PO_4)_3(OH)$], as seen previously in our discussion of calcium.

CONCEPT TEST

Why are phosphates more likely to precipitate with cations from aqueous solution than nitrates?

You may recall from Chapter 16 that phosphate ion is in equilibrium with its conjugate acid, HPO_4^{2-}, which in turn hydrolyzes to $H_2PO_4^-$ as shown in the following equations:

$$PO_4^{3-}(aq) + H_2O(\ell) \rightleftharpoons HPO_4^{2-}(aq) + OH^-(aq)$$

$$HPO_4^{2-}(aq) + H_2O(\ell) \rightleftharpoons H_2PO_4^-(aq) + OH^-(aq)$$

Aqueous phosphate ion can be transported into cells, where it is incorporated into familiar organic molecules such as ATP, glucose-6-phosphate, and nucleic acids (Figure 22.8). Note the presence of a carbon–oxygen bond between a monosaccharide and the phosphate group in all three of the molecules in Figure 22.8. In glucose-6-phosphate and nucleic acids, this new bond forms through a condensation reaction of an –OH group on the sugar molecule and HPO_4^- that produces water as a product. Adenosine diphosphate (ADP) is converted to ATP through a condensation reaction between HPO_4^{2-} and a phosphate group on ADP.

The processes in Figure 22.8 are all reversible; the P–O bond can be hydrolyzed, releasing HPO_4^{2-}. In the case of ATP, this reaction is exothermic and provides the energy for many cellular processes. Given the importance of phosphorus

$$ATP^{4-} + H_2O \rightarrow ADP^{3-} + HPO_4^{2-} + H^+$$

Adenosine triphosphate (ATP^{4-}) Adenosine diphosphate (ADP^{3-})

(a)

$$\text{Glucose}(aq) + HPO_4^{2-}(aq) \rightarrow \text{Glucose-6-phosphate}(aq) + H_2O(\ell)$$

(b)

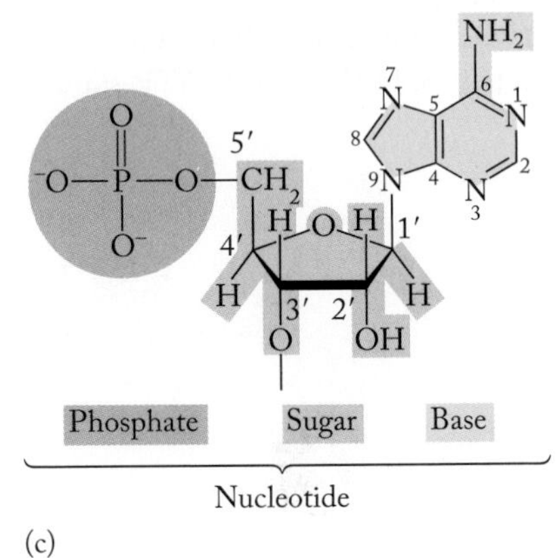

(c)

FIGURE 22.8 Condensation reactions between HPO_4^{2-} (*aq*) and –OH groups yield phosphate esters such as (a) ATP, (b) glucose-6-phosphate, and (c) the building blocks of nucleic acids.

to life, significant amounts of phosphates are used as fertilizer in agriculture. Agricultural runoff may stimulate rapid growth of algae in freshwater ponds and lakes. The explosive growth of algae can use up all the dissolved oxygen, killing other higher aquatic organisms.

The sulfur cycle in Figure 22.9 illustrates the array of sulfur compounds found in the environment. We have already encountered volatile sulfur oxides, SO_2 and

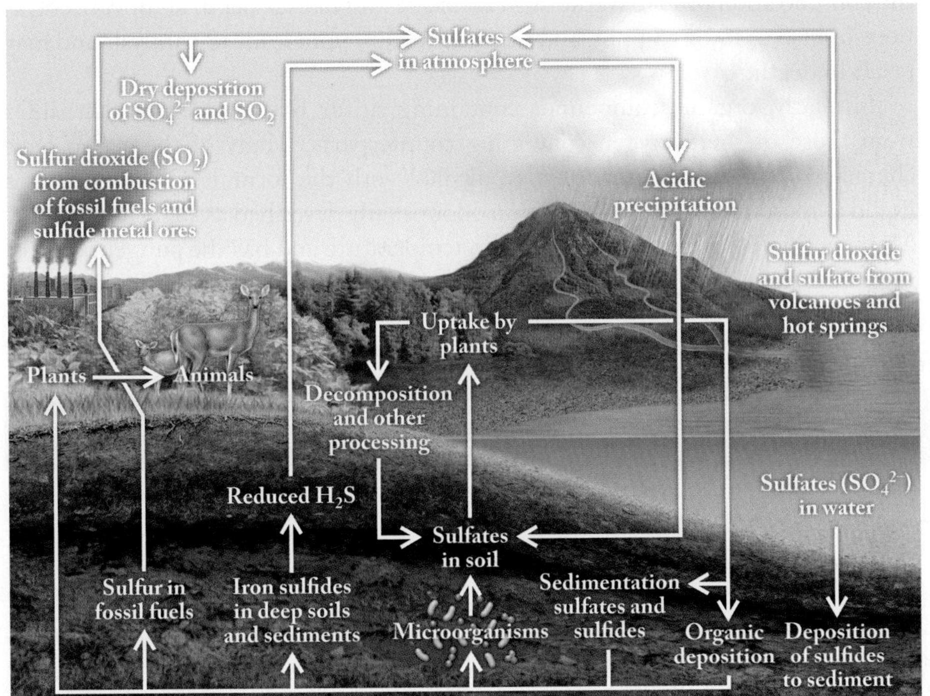

FIGURE 22.9 The key components in the sulfur cycle are sulfide ion (S^{2-}) in sediments and metal sulfides, sulfur oxides (SO_2 and SO_3), hydrogen sulfide (H_2S), and sulfate (SO_4^{2-}).

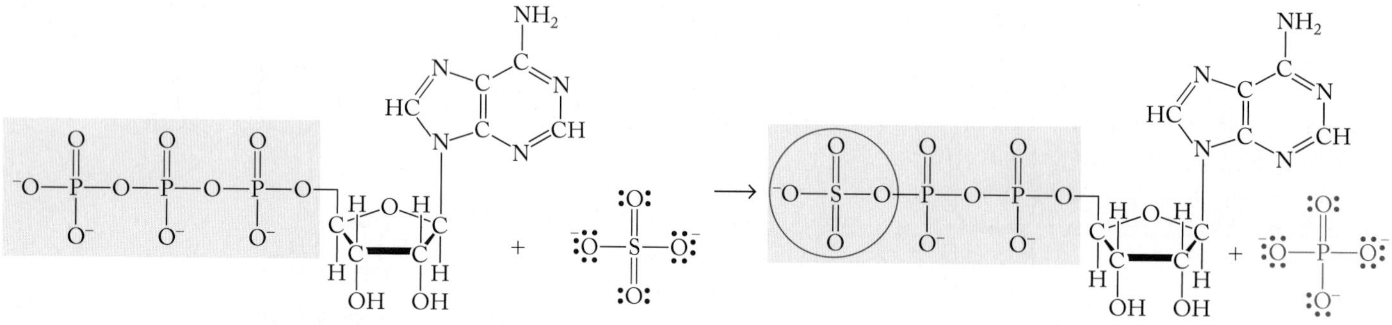

FIGURE 22.10 The reduction of sulfate ion begins by substitution of a phosphate (PO_4^{3-}) group on ATP by sulfate (SO_4^{2-}) catalyzed by the enzyme ATP-sulfurylase.

SO_3, in the context of acid rain on early Earth in Chapter 7. Like the nonmetal oxides of groups 14 and 15, SO_2 and SO_3 dissolve in water to produce the weak acid H_2SO_3 and the strong acid H_2SO_4, respectively. Sulfate ions produced by the dissociation of H_2SO_4, the sulfate ions derived from minerals in soil, and the sulfate ions produced by the oxidation of H_2S are all absorbed by plants.

The enzyme ATP sulfurylase promotes the introduction of SO_4^{2-} into ATP as adenosine-5′-phosphosulfate, APS^{2-} (Figure 22.10). The sulfur in APS^{2-} eventually finds its way into sulfur-containing amino acids (cysteine and methionine) and organic sulfur-containing compounds. It is also released to the environment as H_2S, the sulfur analogue of water.

Compounds of hydrogen with oxygen, sulfur, and the other group 16 elements provide interesting contrasts with respect to molecular shape and properties. The data in Table 22.7 show how different water is from the other compounds. Water is a liquid at ordinary temperatures and pressures, has a large negative heat of formation, and has a much larger bond angle than the other three hydrides. We also know that water is odorless and is absolutely essential for life. The hydrides of sulfur (S), selenium (Se), and tellurium (Te) are all gases under standard conditions, have bond angles close to 90°, and are foul-smelling and poisonous. Hydrogen sulfide is responsible for the smell of rotten eggs. It is especially dangerous because it tends to very quickly fatigue the nasal sensory sites responsible for detecting it. This means that the intensity of the odor is a very poor indicator of the concentration of H_2S in the air. Headache and nausea begin at air concentrations of H_2S as low as 5 ppm, and exposure to 100 ppm leads to paralysis and may result in death.

Similarly, organic compounds containing sulfur have properties that differ from those of their oxygen-containing counterparts. Many of them also have characteristic odors. Methanol is an alcohol with the formula CH_3OH. It is a liquid at room temperature and has an odor usually described as slightly alcoholic. Methanethiol, CH_3SH, is a gas at room temperature and has the pungent odor of rotten cabbage. It is produced in the intestinal tract of animals by the action of

TABLE 22.7 Comparison of the Properties of Group 16 Dihydrides

Hydride[a]	Melting Point (°C)	Boiling Point (°C)	Bond Length (pm)	Bond Angle (degrees)	Heat of Formation (kJ/mol)
H_2O	0	100	96	104.5	−285.8
H_2S	−86	−60	134	92	−20.17
H_2Se	−66	−41	146	91	73.0
H_2Te	−51	−4	169	90	99.6

[a] H_2Po is excluded; too little is known of its chemistry. Polonium has no stable isotopes and is present on Earth only in very small quantities.

bacteria on proteins and is one of the sulfur compounds responsible for the characteristic aroma of a feedlot or a barnyard.

Compounds with two carbon atoms include ethanol, CH_3CH_2OH, which is a liquid at room temperature. The corresponding sulfur compound is a volatile liquid at room temperature called ethanethiol, which has a penetrating and unpleasant odor, like very powerful green onions. The human nose can detect the presence of ethanethiol at levels as low as 1 ppb (part per billion) in the air, which is one reason why it is used as an odorant in natural gas. Natural gas has no odor, and natural gas leaks are such enormous fire hazards that ethanethiol is added to natural gas streams to make it easier to detect even small leaks.

If we rearrange the atoms in ethanol and ethanethiol, we produce two new compounds. In the case of ethanol we get dimethyl ether, CH_3OCH_3, a colorless gas used in refrigeration systems. Its counterpart, dimethyl sulfide, CH_3SCH_3, is one of the compounds responsible for the "low-tide" smell of ocean shorelines.

Three of the sulfur compounds described—hydrogen sulfide, methanethiol, and dimethyl sulfide—are referred to as volatile sulfur compounds (VSCs) by dentists. They are produced by bacteria in the mouth and are the principal compounds responsible for bad breath. One of the reasons the odors of these compounds differ from those of their oxygen counterparts is that their molecular sizes and shapes are slightly different. Also their polarities differ because of the electronegativity difference between oxygen and sulfur. Recall that we discussed the importance of molecular shape in determining the extent of interaction of a compound with receptors in nasal membranes in Chapter 5. In part, the vast differences in odor and sensory detectability of these compounds are due to their shapes and electron distributions.

Not all sulfur compounds have an odor, but many odiferous compounds do contain sulfur. The characteristic and unpleasant smell of urine produced by some people after eating asparagus results from the inability of their bodies to convert sulfur compounds (Figure 22.11) into odor-free sulfate ions. Not all people are able to convert the sulfur compounds in asparagus to sulfate, and not all people are able to smell the odiferous sulfur compounds. Apparently, genetic differences determine how we metabolize these compounds and how well we can sense their odors.

The odor of skunk is due mostly to butanethiol (Figure 22.12), and the odor of well-used athletic shoes is primarily due to the presence of sulfur compounds produced by bacteria. Not all sulfur compounds have aromas as unpleasant as these, however. A compound with the formula $C_{10}H_{18}S$ is responsible for the aroma of grapefruit. If the orientation of two atoms on one of the carbon atoms in its molecular structure is switched, the resulting compound has the same Lewis structure but no aroma at all. The carbon atom in question is a chiral center, and the two compounds are enantiomers.

H₃C—S—H Methanethiol   H₃C—S—S—CH₃ Dimethyl disulfide   H₃C—S—CH₂—S—CH₃ Bis(methylthio)methane

FIGURE 22.11 Structures of some of the volatile sulfur compounds responsible for the smell of "asparagus" urine. Compounds shown here have stronger (and more unpleasant) odors.

FIGURE 22.12 The pungent smell of skunk spray is due to butanethiol, $CH_3(CH_2)_3SH$.

SAMPLE EXERCISE 22.4 Drawing Lewis Structures for Molecules in the Sulfur Cycle LO3

Dimethyl sulfide, CH_3SCH_3, is one of the products of the sulfur cycle in Figure 22.9. In marine environments, dimethyl sulfide can be oxidized by bacteria to dimethyl sulfoxide, $(CH_3)_2SO$.

a. Draw Lewis structures for CH_3SCH_3 and $(CH_3)_2SO$, and determine the molecular geometry about the S atom in each.
b. Methanethiol, CH_3SH, is the simplest of the thiols and has a boiling point of 6°C. Dimethyl sulfide has a boiling point of 38°C. Explain the difference in boiling point between methanethiol and dimethyl sulfide.
c. Which hybrid orbitals does sulfur use in bonding to carbon in CH_3SCH_3 and CH_3SH?

Collect and Organize We need to draw Lewis structures for two sulfur compounds, determine their geometry, and compare their boiling points. Guidelines for drawing Lewis structures were discussed in Chapter 4. We are also asked to describe the hybrid orbitals of S in these compounds. Hybrid orbitals were described in Chapter 5. The effect of structure on boiling points was discussed in Chapter 6.

Analyze The Lewis structure for a molecule depends on the total number of valence electrons available, distributed over the atoms so that each atom has a complete octet (except H, which has a duet). Atoms with $Z \geq 13$ may have an expanded octet if such an arrangement leads to lower formal charges on the atoms. The arrangement of the bonding and lone pairs on the central atom allow us to predict intermolecular forces such as dipole–dipole interactions, hydrogen bonds, and dispersion forces. We can account for observed molecular geometries by combining s, p, and sometimes d orbitals to form hybrid atomic orbitals such as sp, sp^2, sp^3, sp^3d, and sp^3d^2.

Solve (a) Dimethyl sulfide has a total of 20 valence electrons: six from the S atom, six from the H atoms, and eight from the C atoms. These electrons can be distributed in six C—H single bonds and two S—C single bonds, with four electrons remaining as two nonbonding pairs on S. Dimethyl sulfoxide has an oxygen atom bonded to the sulfur in dimethyl sulfide. This oxygen brings an additional six valence electrons to the molecule, giving $(CH_3)_2SO$ a total of 26 valence electrons. Sharing one of the S lone pairs with O will complete the octets of both S and O but will leave O with a formal charge of −1 and sulfur with a formal charge of +1. Forming a S═O double bond makes the formal charges on both S and O equal to zero but requires that sulfur have an expanded octet. The two structures for $(CH_3)_2SO$ represent resonance forms:

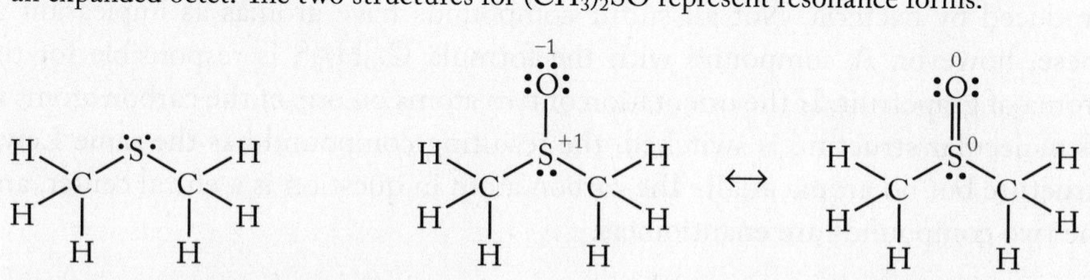

(b) Methanethiol has 14 valence electrons: six from the S atom, four from the H atoms, and four from the C atom. The electrons are distributed in three C—H single bonds, one C—S single bond, and one S—H single bond. As in dimethyl sulfide, there are two nonbonding pairs left on S. Both dimethyl sulfide and methanethiol contain a sulfur atom surrounded by two bonding pairs and two nonbonding pairs of electrons for a total of four electron pairs. The electron-pair geometry is tetrahedral, and the molecular geometry is bent:

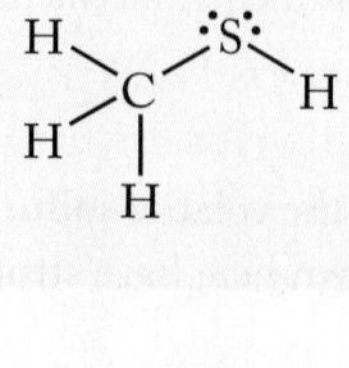

Both methanethiol and dimethyl sulfide are polar as a result of their molecular geometry and experience dipole–dipole interactions. Both molecules interact through dispersion forces as well. One might expect that the stronger dipole–dipole forces in CH_3SH would lead to a higher boiling point than for CH_3SCH_3, but we observe the opposite. We conclude that the dispersion forces in CH_3SCH_3 have a greater effect than the dipole–dipole interactions in CH_3SH, leading to dimethyl sulfide boiling at a temperature about 32°C higher than methanethiol.

(c) The similar geometry for CH_3SCH_3 and CH_3SH, with each molecule's central sulfur atom having two bonded atoms and two lone pairs, is consistent with sp^3 hybrid orbitals on the sulfur atom in both cases.

Think About It Methanethiol and dimethyl sulfide differ only in that the latter has a CH_3 group in place of a hydrogen atom. (This relationship between thiols and sulfides corresponds to the relationship between alcohols and ethers.) Sulfur does not expand its octet in these compounds because using the lone pairs on S to form multiple bonds is not needed. The boiling points of CH_3SCH_3 and CH_3SH reveal an important observation: many weaker bonds (dispersion forces) in CH_3SCH_3 can outweigh a few stronger forces (dipole–dipole forces) in CH_3SH.

Practice Exercise The nitrogen cycle in Figure 22.6 involves both neutral compounds such as NO_2 and polyatomic ions such as NO_2^-. Draw Lewis structures for both species and determine if they have the same molecular geometry about nitrogen. Identify which hybrid orbitals contain nitrogen lone pairs in NO_2 and NO_2^-.

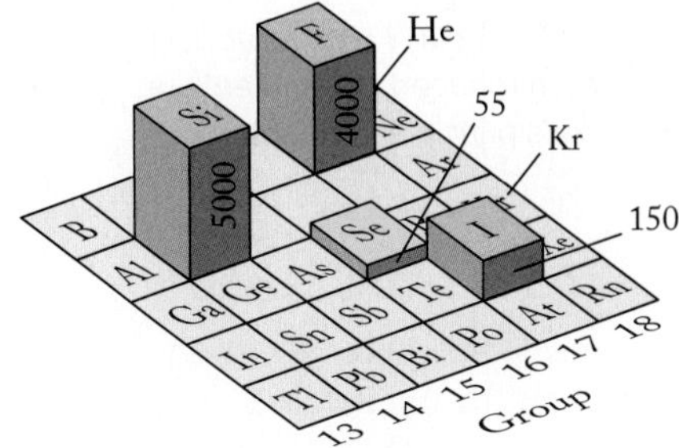

FIGURE 22.13 Silicon, fluorine, and iodine are trace essential elements, and selenium is an ultratrace essential element. The remaining labeled elements are nonessential. The vertical bars show DRI values for these elements in micrograms per day.

22.4 Trace and Ultratrace Essential Elements

Figure 22.13 shows the DRI values of four main group essential elements: silicon, selenium, fluorine, and iodine. Silicon, fluorine, and iodine are present in the body in average concentrations between 1 and 1000 μg of element per gram of body mass and are considered to be trace elements. Selenium is considered ultratrace, meaning that it is present in an average concentration of less than 1 μg of element per gram of body mass.

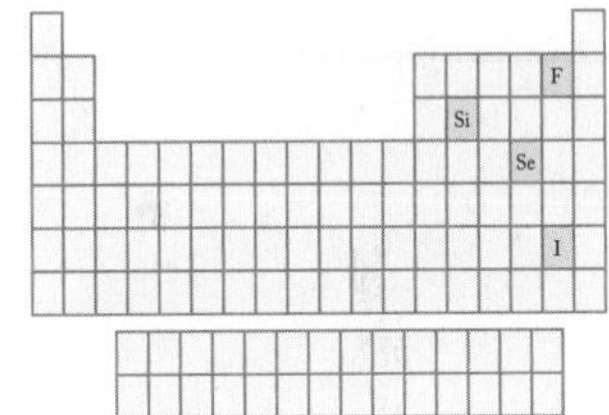

Selenium

The volatile selenium analogue to water, H_2Se, is toxic; however, selenium is considered an ultratrace essential element. The average concentration of Se in the human body is 0.3 μg per gram of body mass. Mounting scientific evidence points to a need for a minimum daily dose of approximately 55 μg of selenium. The effects of selenium toxicity, however, are apparent in people who ingest more than 500 μg per day. Most of the selenium we need is obtained from selenium-rich produce such as garlic, mushrooms, and asparagus, or from fish. Selenium occurs in the body as the amino acid selenocysteine (Figure 22.14) and is incorporated into enzymes.

Selenocysteine is an antioxidant. Our bodies need oxygen to survive, yet living in an oxygen-rich atmosphere can lead to the formation of potentially dangerous oxidizing agents in cells. For example, metabolism of fatty acids forms oxidizing agents called alkyl hydroperoxides, which can attack the lipid bilayer of cell membranes. It is believed that aging is related to the inability of the body to inhibit oxidative degradation of tissue. Selenocysteine participates in a series of reactions that results in the decomposition of these alkyl hydroperoxides.

FIGURE 22.14 Selenocysteine is the selenium-containing analogue of the amino acid cysteine. Much of the selenium in the human body is found in proteins containing selenocysteine.

Fluorine and Iodine

Fluorine is a trace essential element, and fluoride ions have significant benefits for dental health. Tooth enamel is composed of the mineral hydroxyapatite, $Ca_5(PO_4)_3(OH)$, which is essentially insoluble in water:

$$Ca_5(PO_4)_3(OH)(s) \rightleftharpoons Ca_5(PO_4)_3^{+}(aq) + OH^{-}(aq) \qquad K_{sp} \approx 2.4 \times 10^{-59}$$

When hydroxyapatite comes into contact with weak acids in your mouth, this equilibrium shifts to the right as the acid reacts with the hydroxide ions. This shift effectively increases the solubility of hydroxyapatite, so that your tooth enamel becomes pitted, and dental caries form. This is an example of Le Châtelier's principle. Fluoride ions reduce the likelihood of caries by displacing the OH^- ions in hydroxyapatite to form fluorapatite:

CONNECTION The solubility product, K_{sp}, was introduced in Chapter 16, as was Le Châtelier's principle in Chapter 14.

$$Ca_5(PO_4)_3(OH)(s) + F^{-}(aq) \rightleftharpoons Ca_5(PO_4)_3F(s) + OH^{-}(aq) \qquad K = 8.48$$

The solubility of fluorapatite is less dependent on pH than is the solubility of hydroxyapatite, so changing tooth enamel to fluorapatite makes your teeth more resistant to decay. This is why toothpaste contains fluoride compounds and why fluoride is added to drinking water in many communities in North America and Europe.

Of all the trace essential elements, iodine may have the best-defined role in human health. The body concentrates iodide ions in the thyroid gland, where they are incorporated into two hormones—thyroxine and 3,5,3′-triiodothyronine (Figure 22.15)—whose role is to regulate energy production and use. The conversion of thyroxine to 3,5,3′-triiodothyronine is catalyzed by selenocysteine-containing proteins. A deficiency of iodine or of either hormone can cause fatigue or feeling cold and can ultimately lead to an enlarged thyroid gland, a condition known as goiter. To help prevent iodine deficiency, table salt sold in the United States and many other countries is "iodized" with a small amount of sodium iodide. An excess of either hormone can cause a person to feel hot and is linked to Graves' disease, an autoimmune disease. The immune system in a patient with Graves' disease attacks the thyroid gland and causes it to overproduce the two hormones.

Silicon

The role of silicon in biological systems is less clear than for selenium and the halides. In mammals, a lack of the trace essential element silicon stunts growth. The presence of silicon as silicic acid, $Si(OH)_4$, is believed to reduce the toxicity of Al^{3+} ions in organisms by precipitating the aluminum as aluminosilicate minerals. Amorphous silica, SiO_2, is found in the exoskeletons of diatoms and in the cell membranes of some plants, such as the tips of stinging nettles.

Thyroxine

3,5,3′-Triiodothyronine

FIGURE 22.15 Thyroxine and 3,5,3′-triiodothyronine, two iodine-containing hormones found in the thyroid gland, regulate metabolism.

22.5 Nonessential Elements

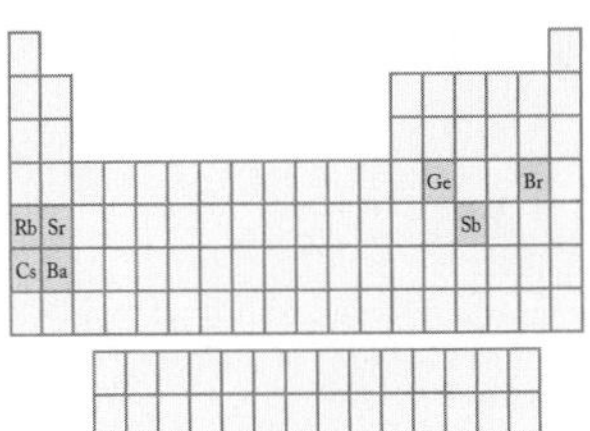

The ten elements listed in Table 22.2 are found in the human body but are classified as nonessential. In this section we discuss how some of these elements may end up in our bodies, working our way from left to right across the periodic table.

Rubidium and Cesium

Rubidium is generally regarded as nonessential in humans, yet it is the 15th most abundant element in the body. It is believed that Rb^+ is retained by the body because of the similarity of its size and chemistry to that of K^+. Like the other cations of group 1, cesium ions (Cs^+) are also readily absorbed by the body. Cesium cations have no known function, although they can substitute for K^+ and interfere with potassium-dependent functions. In most cases, the concentration of cesium in the environment is low, so exposure to Cs^+ is not a health concern. The nuclear accident at Chernobyl in 1986, however, released significant quantities of radioactive ^{137}Cs into the environment. The ability of Cs^+ to substitute for K^+ led to the incorporation of $^{137}Cs^+$ into plants, which rendered crops grown in the immediate area unfit for human consumption because of the radiation hazard posed by this long-lived ($t_{1/2} \approx 30$ yr) β emitter.

CONNECTION The biological effects of different types of nuclear radiation were described in Chapter 21.

Strontium and Barium

Some single-celled organisms build exoskeletons made with $SrSO_4$ and $BaSO_4$, but the human body appears to have no use for Sr^{2+} and Ba^{2+} ions. These ions do find their way into human bones, where they replace Ca^{2+} ions. At the low concentrations of Sr^{2+} and Ba^{2+} that are typically present in the human body, these elements appear to be benign. However, as in the case of radioactive ^{137}Cs, incorporation of ^{90}Sr ($t_{1/2} = 29$ yr) in bones can lead to leukemia. Atmospheric testing of nuclear weapons over the Pacific Ocean and in sparsely populated regions of the American West in the 1950s released ^{90}Sr into the environment. The full extent of the toxic effects of the fallout from these tests did not become apparent for several decades.

Germanium

It is generally agreed that germanium is a nonessential element and is barely detectable in the human body. Bis(carboxyethyl)germanium sesquioxide (Figure 22.16) has been touted as a nutritional supplement, but its efficacy remains controversial.

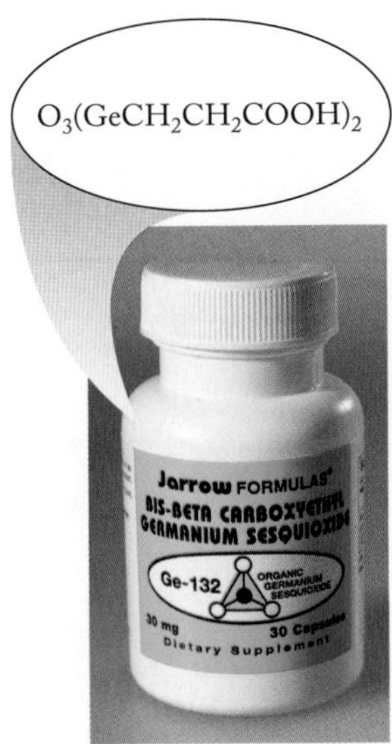

FIGURE 22.16 Bis(carboxyethyl) germanium sesquioxide has been sold as a nutrition supplement, but its benefits are not well established.

Antimony

The role of antimony is also poorly understood. Most antimony compounds are toxic because they cause liver damage. However, ultratrace amounts of antimony may have a stimulatory effect, and selected antimony compounds have been used medically as antiparasitic agents, as discussed in the next section.

Bromine

Bromine has no known function in the human body but is consumed in foods such as grains, nuts, and fish in amounts ranging from 2–8 mg per day, leading to average concentrations of Br^- in blood of about 6 mg/L. Br^- has sedative and anticonvulsive properties but becomes toxic at concentrations around 100 mg/L, limiting its use to veterinary medicine. Bromide ion concentrations in seawater typically range from 65–80 mg/L. A select group of aquatic species can metabolize Br^- into bromomethane, CH_3Br, and other brominated organic compounds.

CONCEPT TEST

Looking at groups 1, 2, 14, 15, and 17, what periodic trend do you see in the location of the nonessential elements relative to the essential elements in the same group?

22.6 Elements for Diagnosis and Therapy

So far we have talked about the biological roles of a number of essential and non-essential main group elements found in our bodies. Some of these elements are also useful in diagnosing or treating diseases, as are some of the other elements in groups 1–2 and 13–18 that we have not mentioned (Figure 22.17). In this section we describe some of the applications of radioactive isotopes in the diagnosis of diseases. We also explore how compounds of essential and nonessential elements have found application in the treatment of a wide range of illnesses.

Any diagnostic or therapeutic compound that is injected intravenously must be sufficiently soluble in blood to be delivered to the target. While in transit, the compound must be stable enough not to undergo chemical reactions that result in its precipitation or rapid elimination from the body. A medicinal chemist can also take advantage of substances that occur naturally in the body, such as antibodies, to carry a diagnostic or therapeutic metal ion to its target. Examples of elements and compounds containing elements from all 18 groups of the periodic table have been identified. In this section we focus on the main group elements (groups 1–2 and 13–18). Applications of the transition elements (groups 3–12) are described in Chapter 23.

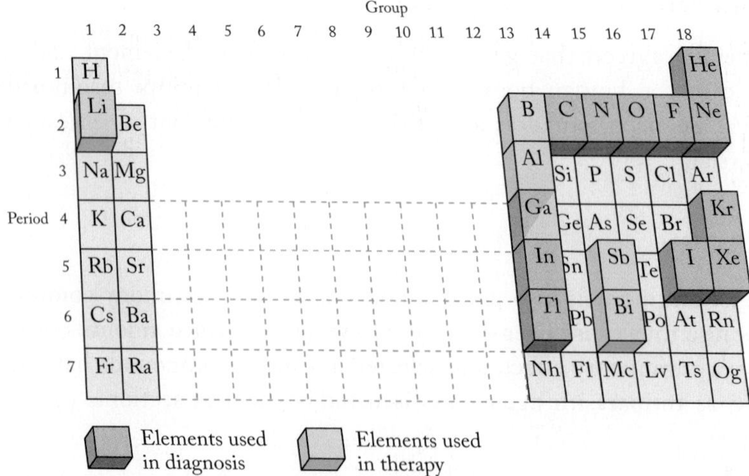

FIGURE 22.17 The elements shown in red are used in diagnostic imaging and those shown in green are used in therapy. Gallium is used in both diagnostic imaging and therapy.

Diagnostic Applications

Physicians in the 21st century have an array of imaging agents to help in diagnosing disease. Some methods use radionuclides with short half-lives that emit easily detectable gamma rays. Examples include the use of iodine-131 to image the thyroid gland (Figure 22.18) and of neutron-poor isotopes such as carbon-11 and fluorine-18 for positron emission tomography (PET). Not all imaging depends on radionuclides, however. In magnetic resonance imaging (MRI), for instance, which can diagnose soft-tissue injuries, stable isotopes of gadolinium are used as contrast agents to enhance images.

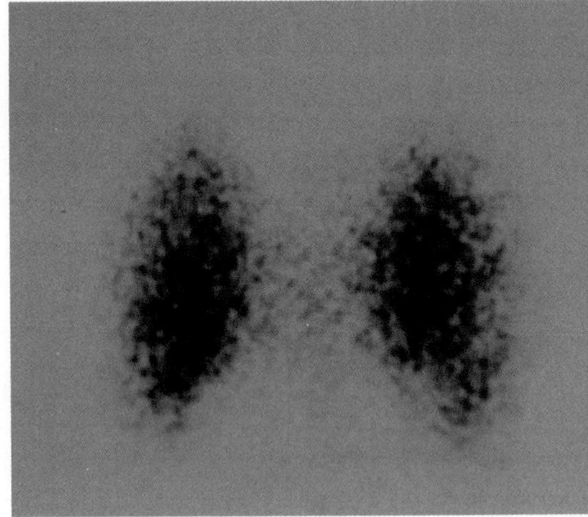

FIGURE 22.18 Gamma radiation that accompanies the decay of iodine-131 can be used to image the two butterfly-shaped lobes of the thyroid gland.

Imaging with Radionuclides The radionuclides used in medicine have short half-lives to limit the patient's exposure to ionizing radiation. If the half-life is too short, however, the nuclide may either decay before it can be administered or not reach the target organ rapidly enough to provide an image. Emission of relatively low-energy γ rays is essential to preventing collateral tissue damage.

Nuclide selection is also governed by the toxicity of both the parent element and the daughter nuclide. The speed at which the imaging agent is eliminated from the body can help mitigate toxic effects. Naturally, the cost and availability of a particular nuclide also factor into its usefulness in a clinical setting.

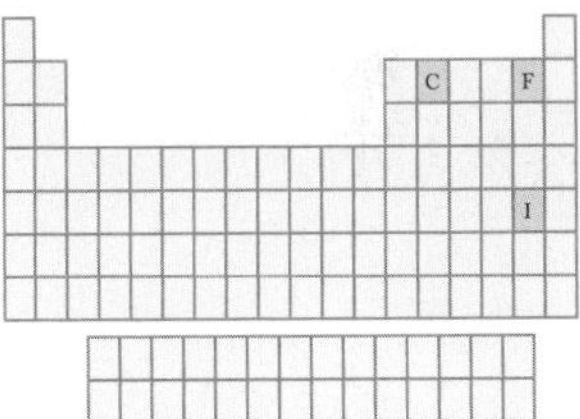

CONCEPT TEST

Why is it important to consider the nature of the decay products—α, β, or γ particles, or positrons—when choosing a radionuclide for medical imaging?

Gallium, Indium, and Thallium Gallium-66, gallium-67, gallium-68, and indium-111 are used as imaging agents for tumors and leukemia. All four nuclides decay by electron capture, and the γ radiation emitted in this nuclear reaction produces the images. All three gallium isotopes also decay by positron emission, which makes compounds containing these isotopes attractive for positron emission tomography. Their half-lives range from just over 1 hour for gallium-68 to 78 hours for gallium-67. The discovery that indium-111-containing compounds can image a range of cancers has led to the development of the drug Zevalin, currently used to treat some forms of non-Hodgkin's lymphoma.

CONNECTION Chapter 19 provides a more detailed discussion of nuclear chemistry and nuclear medicine, including an assessment of the effects of different types of radiation on living tissue.

The use of the gamma emitter thallium-201 ($t_{1/2}$ = 73 h) in diagnosing heart disease presents an interesting case for balancing the risks and benefits of using a particular isotope in medicine. Although thallium compounds are among the most toxic metal-containing compounds known, the nanogram quantities required for diagnosis pose few, if any, health hazards, meaning that the benefits outweigh the risks.

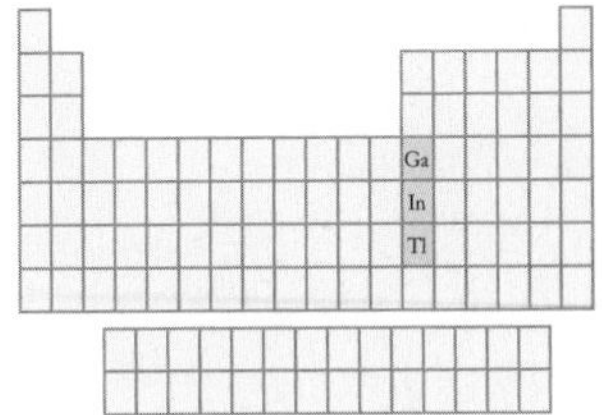

SAMPLE EXERCISE 22.5 Calculating Quantities of Radioactive Isotopes — LO5

Indium-111 ($t_{1/2}$ = 2.805 d) and gallium-67 ($t_{1/2}$ = 3.26 d) are both used in radioimaging to diagnose chronic infections. Which isotope decays faster? If we start with 10.0 mg of each isotope, how much of each remains after 24 hours?

Collect and Organize We are given the half-lives of two radionuclides and asked to predict which one will decay faster, and to calculate how much of each isotope remains after 24 hours of decay.

Analyze An isotope with a shorter half-life decays faster. Radioactive decay follows first-order kinetics. Quantitatively, the relationship between half-life and the amount of material remaining is described by the following equation from Chapter 21:

$$\ln \frac{N_t}{N_0} = -0.693 \frac{t}{t_{1/2}} \qquad (21.8)$$

Here N_0 and N_t refer to the amount of material present initially and the amount at time t, respectively. If our prediction for the relative decay rates of the two isotopes is correct, then more of the isotope with the longer half-life should remain after 24 hours. We need to complete two calculations to determine the amount of each sample present after 24 hours.

Solve Indium-111 has the shorter half-life, so it should decay faster.

For the amount of indium-111 remaining after 24 h, we have

$$\ln \frac{N_t}{10.0\ \text{mg}} = \frac{(-0.693)(24\ \cancel{\text{h}})}{(24\ \cancel{\text{h}}/\cancel{\text{d}})(2.805\ \cancel{\text{d}})} = -0.247$$

Taking the antilog of both sides, we get

$$\frac{N_t}{10.0\ \text{mg}} = 0.781$$

$$N_t = (0.781)(10.0\ \text{mg}) = 7.81\ \text{mg}$$

For gallium-67:

$$\ln \frac{N_t}{10.0\ \text{mg}} = \frac{(-0.693)(24\ \cancel{\text{h}})}{(24\ \cancel{\text{h}}/\cancel{\text{d}})(3.26\ \cancel{\text{d}})} = -0.213$$

$$\frac{N_t}{10.0\ \text{mg}} = 0.808$$

$$N_t = (0.808)(10.0\ \text{mg}) = 8.08\ \text{mg}$$

Think About It We predicted that indium-111 would decay faster, which means that after 24 h the quantity of this isotope should be less than the quantity of gallium-67, and it is.

Practice Exercise Two radioactive isotopes of bismuth are used to treat cancer. The half-lives are 61 min for bismuth-212 and 46 min for bismuth-213. If we start with 25.0 mg of each isotope, how much of each sample remains after 24 h?

Imaging with Noble Gases and MRI So far in this chapter we have had little opportunity to mention the noble gas elements. None of these elements are essential to the human body, although the World Anti-Doping Agency has added both Xe and Ar to the list of banned substances for athletes at Olympic and other sporting events since 2014. Apparently, in addition to behaving as an anesthetic, xenon increases the oxygen-carrying capacity of blood, providing an advantage in aerobically demanding sports. Similar chemistry is believed to occur with argon.

The lack of chemical reactivity of the group 18 elements and their ease of introduction into the body by inhalation, however, make selected isotopes of the noble gases—including helium-3, krypton-83, and xenon-129—attractive as agents for enhancing MRI images, particularly of the lungs. Krypton-83 provides greater sensitivity than xenon-129, allowing for better resolution in the images and, in principle, requiring the use of less gas.

Helium-3 has no known side effects and is preferable to xenon-129 for MRI, but it is present in only trace natural abundance. This isotope is obtained from β decay of tritium (^3H):

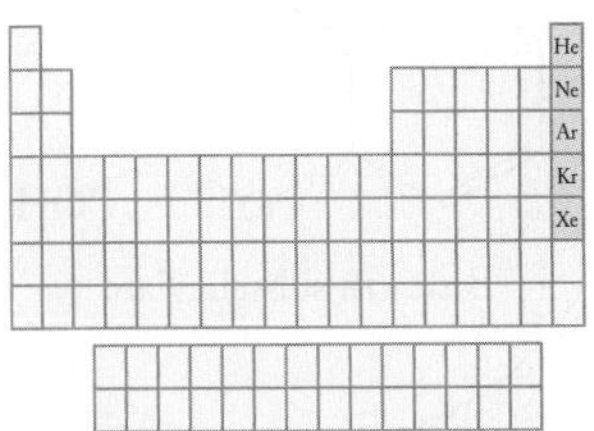

$$^{3}_{1}\text{H} \rightarrow ^{3}_{2}\text{He} + ^{0}_{-1}\beta$$

Neon-19 has been used in PET despite its short half-life (17.5 s). A patient positioned in a PET scanner breathes air containing a small amount of this isotope. Positron emission from the neon is recorded, and an image is created.

Therapeutic Applications

CONNECTION Tritium is produced in nuclear reactors by bombarding ^6Li and ^7Li with high-energy neutrons. The reactions involved were discussed in Chapter 19.

In this section we examine therapeutic agents that contain metallic and heavier main group elements in addition to carbon, hydrogen, nitrogen, oxygen, and sulfur.

Lithium, Boron, Aluminum, and Gallium The similar size of Li^+ (76 pm) and Mg^{2+} (72 pm) means that lithium ions can compete with magnesium ions in biological systems. The substitution of lithium for magnesium may account for its toxicity at high concentrations. Nevertheless, lithium carbonate is used to treat bipolar disorder, and other lithium compounds have been used to treat hyperactivity. In all cases, however, the use of lithium-containing drugs must be carefully monitored.

Of the elements of group 13, only boron and aluminum have been detected in humans. The role of boron in our bodies is not fully understood, but this element appears to play a role in nucleic acid synthesis and carbohydrate metabolism. Selected boron compounds appear to concentrate in human brain tumors. This property has opened the door to a treatment known as boron neutron-capture therapy (BNCT). Once a suitable boron compound has been injected and has made its way to a tumor, irradiation of the tumor with low-energy neutrons leads to the following nuclear reaction:

$$^{10}_{5}\text{B} + ^{1}_{0}\text{n} \rightarrow ^{7}_{3}\text{Li} + ^{4}_{2}\text{He}$$

The α particles generated in the reaction have a short penetration depth but high relative biological effectiveness (RBE), so they can kill the tumor cells without harming surrounding tissue. The identification of compounds suitable for BNCT remains an area of active research.

CONNECTION The relative biological effectiveness (RBE) of radioactive particles was introduced in Chapter 19.

Aluminum is found in some antacids as aluminum hydroxide, $Al(OH)_3$, or aluminum carbonate, $Al_2(CO_3)_3$. Sodium aluminum sulfate, $AlNa(SO_4)_2 \cdot 12\ H_2O$, is an ingredient in some brands of baking powder. Most of the aluminum in the human body can be traced to these sources. Aluminum is not considered essential to humans, but low-aluminum diets have been observed to harm goats and chickens. High concentrations of aluminum are clearly toxic; the effects are most noticeable in patients with impaired kidney function. The role of aluminum in Alzheimer's disease has been extensively debated but remains unresolved.

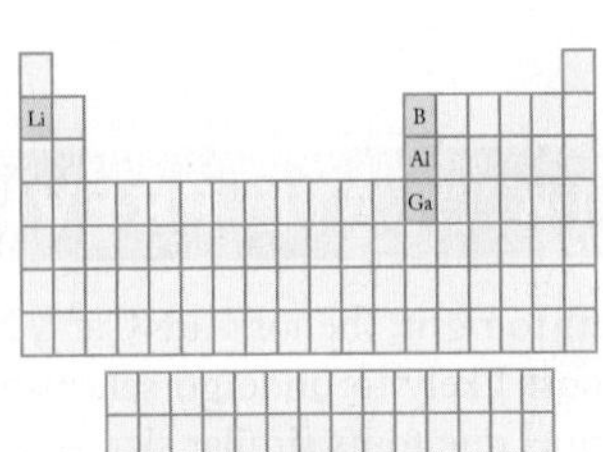

Soluble gallium(III) compounds, either alone or in combination with other drugs, have shown activity on bladder and ovarian cancers. The similar ionic radii of Ga^{3+} (62 pm) and Fe^{3+} (64.5 pm) allow gallium to block DNA synthesis by replacing iron in a protein called transferrin and in other enzymes. Because gallium compounds accumulate in tumors at a higher rate than in

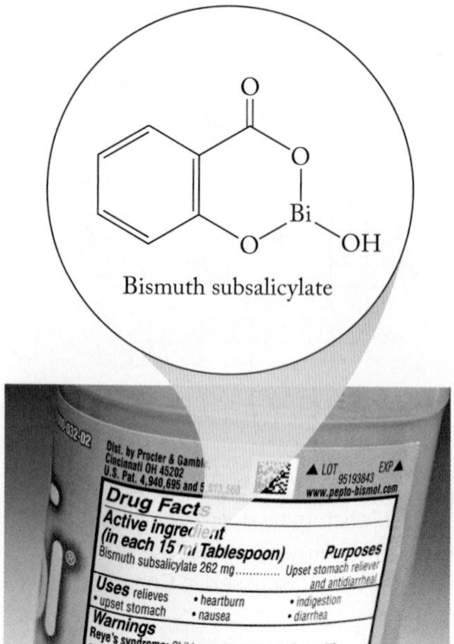

FIGURE 22.19 Bismuth subsalicylate is found in some antacids.

healthy tissue, the disruption of DNA synthesis in the tumor cells inhibits tumor growth.

Antimony and Bismuth Antimony compounds are generally considered to be toxic. It has been reported, for instance, that exposure of infants to antimony compounds used as fire retardants in mattresses may contribute to sudden infant death syndrome (SIDS). However, this element does have a medical use. Leishmaniasis, an insect-borne disease characterized by the formation of boils or skin lesions, is resistant to most treatments, but some patients have been successfully treated with sodium stibogluconate.

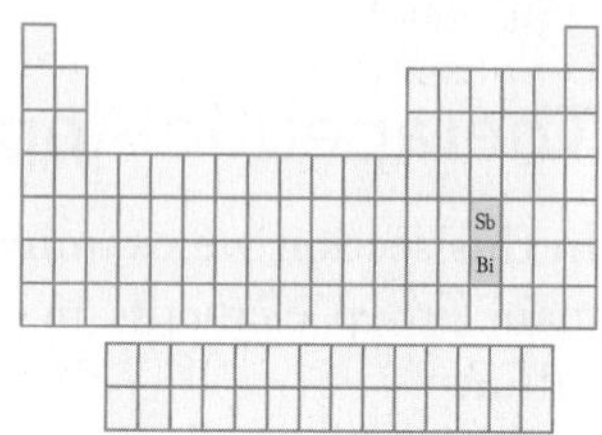

Popular over-the-counter remedies for indigestion, diarrhea, and other gastrointestinal disorders contain bismuth subsalicylate (Figure 22.19). The bismuth in these compounds acts as a mild antibacterial agent that reduces the number of diarrhea-causing bacteria.

The human body requires about 30 elements to function properly. To manage the problems of disease and injury, scientists, physicians, engineers, and scores of other people have turned to the properties of these and many other elements on the periodic table to develop treatments and to enhance quality of life. This chapter has provided the briefest of introductions into the roles of the main group elements in establishing and maintaining living systems.

SUMMARY

LO1 **Essential elements** have a physiological function in the body. **Nonessential elements** are present in the body but have no known functions. Some may have **stimulatory effects**. Essential elements are categorized as **major, trace,** or **ultratrace essential elements** depending on their concentrations in the body. (Sections 22.1 and 22.2)

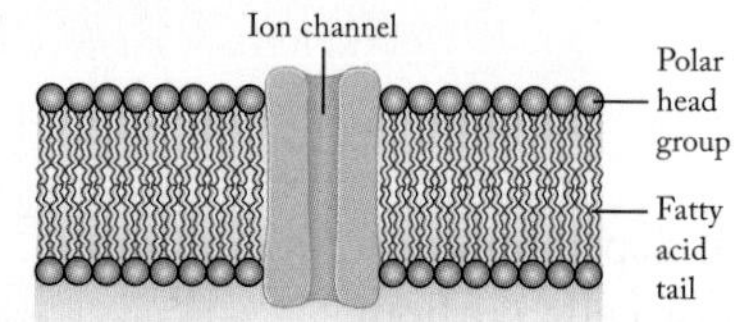

LO2 Transport of Na^+ and K^+ across cell membranes involves **ion pumps** or selective transport through **ion channels**. Chloride ion is the most abundant anion in the human body, facilitating transport of alkali metal cations and elimination of CO_2. Differences in concentrations of ions inside and outside cell membranes gives rise to a **membrane potential,** which determines the permeability of ions through the membrane. (Section 22.3)

LO3 Acid–base chemistry and redox reactions are of great significance in living systems. The principles of structure and electron distribution in molecules apply to interactions *in vivo* just as they do in the laboratory. (Sections 22.3, 22.4, 22.5, and 22.6)

LO4 The most abundant elements in the human body include seven nonmetals: C, H, O, Cl, S, P, and N, and the four metals: Ca, K, Na, and Mg. Nonessential elements from groups 1 and 2 may be absorbed into cells or substituted into bone or other tissues because they are similar size and charge to essential elements. Some nonessential elements are used in nutritional supplements, whereas others have therapeutic properties in low concentrations but are toxic at higher levels. (Sections 22.3, 22.4, and 22.5)

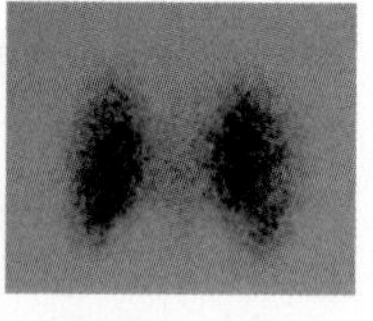

LO5 Radionuclides with short half-lives that emit low-energy γ rays are used in assessing function and diagnosing disease. The selection of a radionuclide for medical use is governed by the toxicities of the element and its daughter nuclides, its radioactive half-life, and the speed at which it is eliminated from the body. (Section 22.6)

LO6 Main group elements beyond the essential elements are useful in a wide variety of compounds with therapeutic value. Lithium salts are used in treating depression while aluminum compounds find use as antacids. Fluoride in toothpaste helps prevent cavities. (Section 22.6)

PARTICULATE PREVIEW WRAP-UP

From left to right, the ions are Ca^{2+}, Na^+, H_3O^+, and K^+. The H_3O^+ ion is most likely to undergo selective transport through the sodium ion channel due to its similar size and charge. None of the other three ions are likely to pass through the K^+ ion channel because of their smaller sizes.

PROBLEM-SOLVING SUMMARY

Type of Problem	Concepts and Equations	Sample Exercises
Calculating equilibrium potential, E_{ion} for ions and ΔG for ion transport	Calculate E_{ion} by using a modified form of the Nernst equation: $$E_{ion} = \frac{RT}{zF} \ln \frac{[X]_{outside}}{[X]_{inside}} \quad (22.1)$$ Calculate ΔG using the relationships between ΔG and E: $$E_{transport} = E_{membrane} - E_{ion} \quad (22.2)$$ $$\Delta G_{transport} = -nFE_{transport} \quad (22.3)$$	**22.1**
Calculating an acid concentration from its pH	Relate the pH of a solution to the $[H_3O^+]$ by the equation $$pH = -\log[H_3O^+] \quad (15.14)$$	**22.2**
Assigning oxidation numbers and identifying redox reactions	Assign oxidation numbers using the guidelines in Chapter 8.	**22.3**
Drawing Lewis structures for molecules	Use the guidelines in Sections 8.2 and 9.4.	**22.4**
Calculating quantities of radioactive isotopes	Use the equation $$\ln \frac{N_t}{N_0} = -0.693 \frac{t}{t_{1/2}}$$ where N_0 and N_t are the amounts of material present initially and at time t, respectively.	**22.5**

VISUAL PROBLEMS

(Answers to boldface end-of-chapter questions and problems are in the back of the book.)

22.1. Which part of Figure P22.1 best describes the periodic trend in monatomic cation radii moving up or down a group or across a period in the periodic table? (Arrows point in the direction of increasing radii.)

22.2. Which part of Figure P22.1 best describes the periodic trend in monatomic anion radii moving up or down a group or across a period in the periodic table? (Arrows point in the direction of increasing radii.)

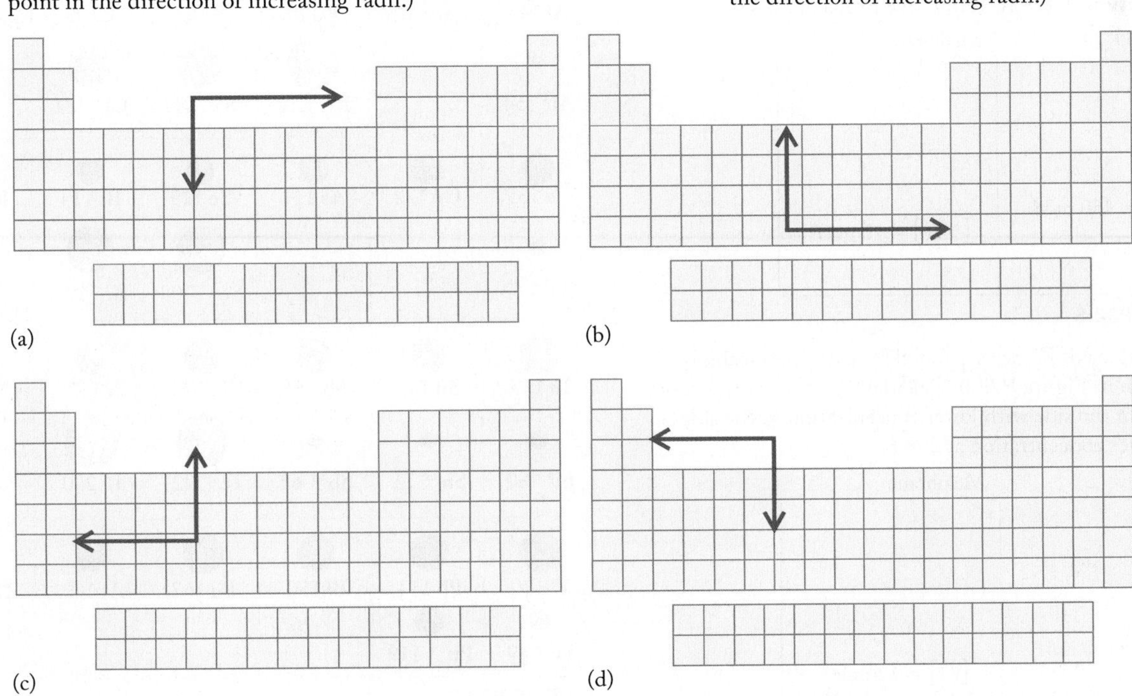

FIGURE P22.1

22.3. Which of the two groups highlighted in the periodic table in Figure P22.3 typically forms ions that have larger radii than the corresponding neutral atoms?

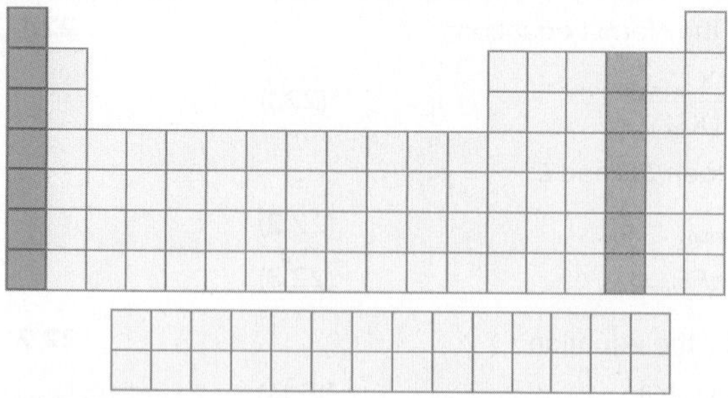

FIGURE P22.3

22.4. Which of the two groups highlighted in the periodic table in Figure P22.4 typically forms ions that have smaller radii than the corresponding neutral atoms?

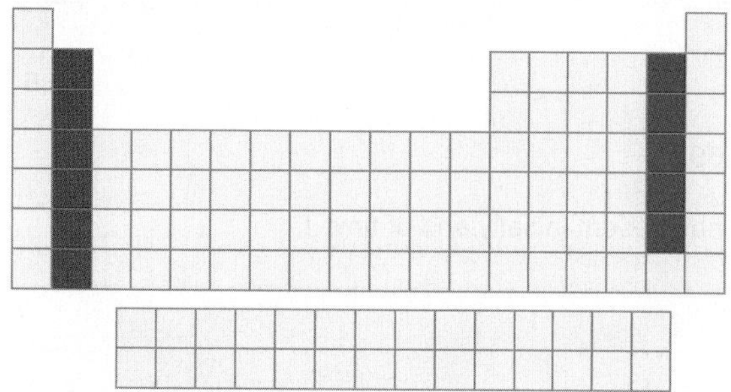

FIGURE P22.4

***22.5.** As we saw in Chapter 12, the free energy (ΔG) of a reaction is related to the cell potential by the equation $\Delta G = -nFE$. In Figure P22.5, two solutions of Na^+ of different concentrations are separated by a semipermeable membrane. Calculate ΔG for the transport of Na^+ from the side with higher concentration to the side with lower concentration at 298 K. (*Hint*: See Problem 22.45.)

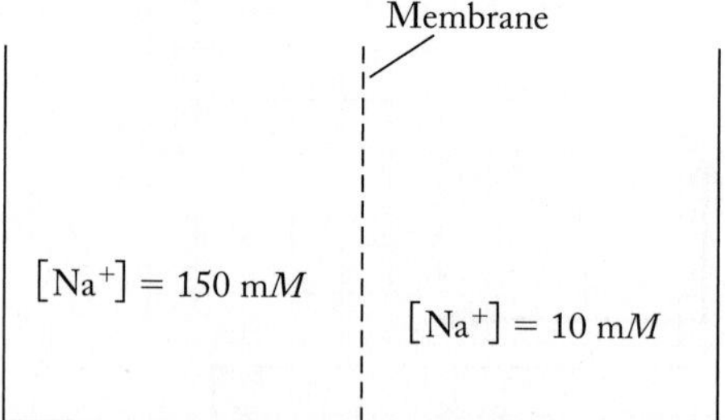

FIGURE P22.5

22.6. Two solutions of K^+ are separated by a semipermeable membrane in Figure P22.6. Calculate ΔG for the transport of K^+ from the side with lower concentration to the side with higher concentration at 298 K.

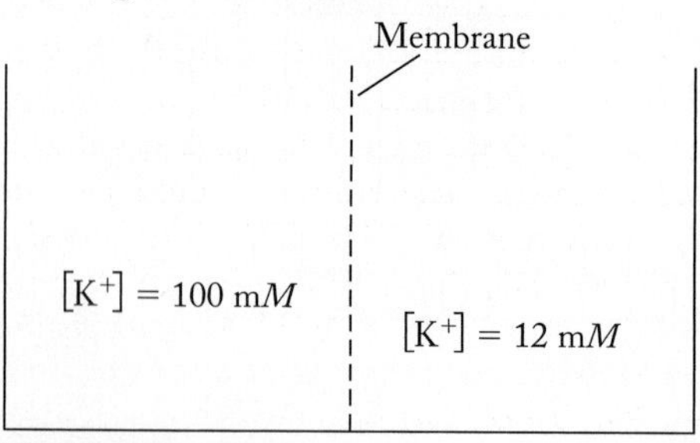

FIGURE P22.6

22.7. Describe the molecular geometry around each germanium atom in the germanium compound shown in Figure P22.7.

HOOC–CH_2–CH_2–Ge(=O)–O–Ge(=O)–CH_2–CH_2–COOH

FIGURE P22.7

22.8. Selenocysteine can exist as two enantiomers (stereoisomers). Identify the atom in Figure P22.8 responsible for the two enantiomers.

SeH–H_2C–CH(^+H_3N)–C(=O)–O^-

FIGURE P22.8

***22.9.** The relative sizes of the main group atoms and ions are shown in Figure P22.9 (all values are given in pm). Using this figure as a guide, which of the following polyatomic ions is likely to be the largest: sulfate, phosphate, or perchlorate?

					He 32
B 88	C 77	N 75	O 73	F 71	Ne 69
		N^{3-} 146	O^{2-} 140	F^- 133	
Al 143	Si 117	P 110	S 103	Cl 99	Ar 97
Al^{3+} 54		P^{3-} 212	S^{2-} 184	Cl^- 181	
Ga 135	Ge 122	As 121	Se 119	Br 114	Kr 110
Ga^{3+} 62			Se^{2-} 198	Br^- 195	
In 167	Sn 140	Sb 141	Te 143	I 133	Xe 130
In^{3+} 80	Sn^{4+} 71	Sb^{5+} 62	Te^{2-} 221	I^- 220	
Tl 170	Pb 154	Bi 150	Po 167	At 140	Rn 145
Tl^{3+} 89	Pb^{2+} 119				

FIGURE P22.9

22.10. Use representations [A] through [I] in Figure P22.10 to answer questions a–f.
  a. Which nuclide is the product of beta decay from tritium? Which will produce an alpha particle when bombarded with neutrons? Write nuclear equations to illustrate these two processes.
  b. The structure shown in [E] is also found in [D] and a similar structure is found in [F]. What metal binds to the nitrogen atoms in [D]? What metal binds to the nitrogen atoms in [F]?
  c. Identify the important structural features in [B] and label each as polar or nonpolar.
  d. What kinds of substances are likely to pass through the channel in [B]?
  e. Of the three substances shown in [G], [H], and [I], which must pass through a channel to enter a cell?
  f. Of the three substances shown in [G], [H], and [I], which can be transported directly through a cell membrane without the need for a channel?

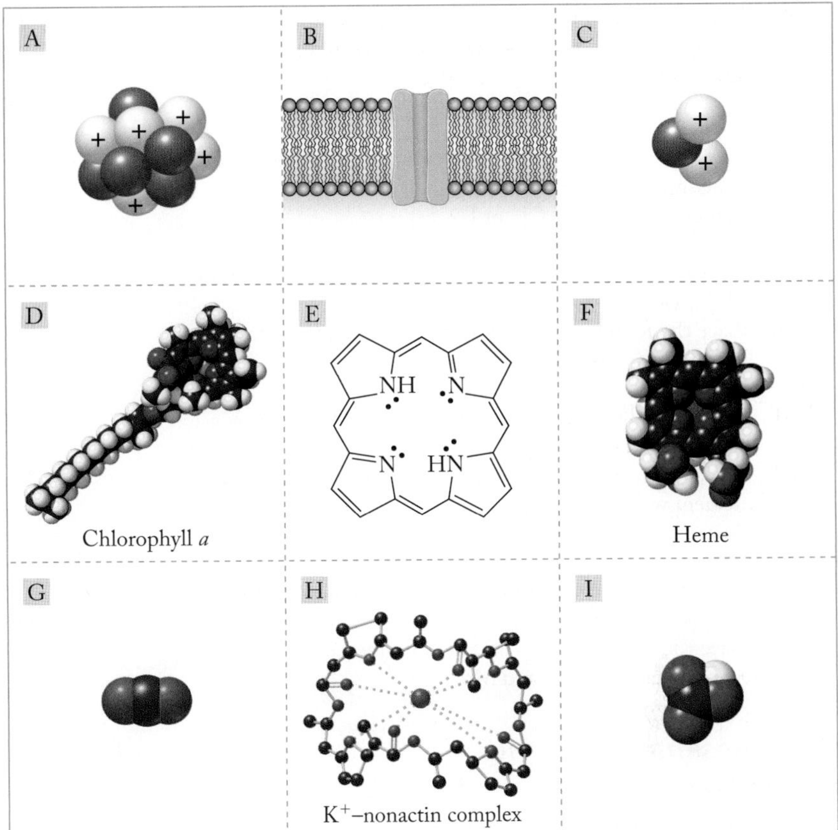

FIGURE P22.10

QUESTIONS AND PROBLEMS

Main Group Elements and Human Health

Concept Review

22.11. What is the difference between an essential element and a nonessential element?

22.12. Are all essential elements major essential elements?

22.13. What is the main criterion that distinguishes major, trace, and ultratrace essential elements from one another?

22.14. Should trace essential elements also be considered to be stimulatory?

Problems

22.15. The concentrations of very dilute solutions are sometimes expressed as parts per million. Express the concentration of each of the following trace and ultratrace essential elements in parts per million:
  a. Fluorine, 110 mg in 70 kg
  b. Silicon, 525 mg/kg
  c. Iodine, 0.043 g in 100 kg

22.16. In the human body, the concentrations of ultratrace essential elements are even lower than those of trace essential elements and therefore are sometimes expressed in parts per billion. Express the concentrations of each of the following elements in parts per billion:
  a. Bromine, 6 mg/L
  b. Boron, 0.014 g/100 kg
  c. Selenium, 5.0 mg/70 kg

22.17. In the following pairs, which element is more abundant in the human body? (a) silicon or oxygen; (b) iron or oxygen; (c) carbon or aluminum

22.18. In the following pairs, which element is more abundant in the human body? (a) H or Si; (b) Ca or Fe; (c) N or Cr

Periodic and Chemical Properties of Main Group Elements

Concept Review

22.19. In Chapter 2 we defined main group elements as those elements found in groups 1, 2, and 13–18 in the periodic table. Why do some chemists refer to these as the "*s*-block" and "*p*-block" elements?

22.20. Why do we classify the main group elements by group rather than period?

22.21. Lithium oxide (Li_2O) and carbon monoxide (CO) have nearly the same molar mass. Why is Li_2O a solid with a high melting point whereas CO is a gas?

22.22. The nonradioactive group 17 elements are found as diatomic molecules, X_2 (X = F, Cl, Br, I). Why is Br_2 a liquid at room temperature whereas Cl_2 is a gas?

22.23. Which of the following properties can be used to distinguish a metallic element from a semimetallic element: atomic radius, electrical conductivity, and/or molar mass?

22.24. Which of the following cannot be measured: ionization energy, electron affinity, ionic radius, atomic radius, or electronegativity?

22.25. Why is Be^{2+} more likely than Ca^{2+} to displace Mg^{2+} in biomolecules?

22.26. PbS, $PbCO_3$, and PbCl(OH) have limited solubility in water. Which of them is(are) more likely to dissolve in acidic solutions?

Problems

22.27. Which ion channel must accommodate the larger cation, a potassium or a sodium ion channel?

22.28. Which ion is larger: Cl^- or I^-?

22.29. Place the following ions in order of increasing ionic radius: Mg^{2+}, Li^+, Al^{3+}, and Cl^-.

22.30. Place the following ions in order of increasing ionic radius: Br^-, O^{2-}, K^+, and Ca^{2+}.

22.31. Place the following elements in order of increasing electronegativity: K, S, F, and Mg.

22.32. When comparing any two main group elements, is the element with the smaller atomic radius always more electronegative?

22.33. How is the electron affinity of Cl atoms related to the ionization energy of Cl^- ions?

22.34. Place the following ions in order of increasing ionization energy: Na^+, S^{2-}, F^+, and Mg^+.

Major Essential Elements

Concept Review

22.35. Ion Transport in Cells Describe three ways in which ions of major essential elements (such as Na^+ and K^+) enter and exit cells.

22.36. Which transport mechanism for ions requires ATP: diffusion, ion channels, or ion pumps?

22.37. Why is it difficult for ions to diffuse across cell membranes?

22.38. Why does Sr^{2+} substitute for Ca^{2+} in bones?

22.39. Which alkali metal ion is Rb^+ most likely to substitute for?

22.40. Why don't alkaline earth metal cations substitute for alkali metal cations in cases where the ionic radii are similar?

***22.41.** Why might nature have selected calcium carbonate over calcium sulfate as the major exoskeleton material in shells?

22.42. Bromide ion and fluoride ion are nonessential elements in the body. Do you expect their concentrations to be more similar to the concentrations of major essential elements or to the concentrations of ultratrace essential elements?

Problems

22.43. Osmotic Pressure of Red Blood Cells One of the functions of the alkali metal cations Na^+ and K^+ in cells is to maintain the cells' osmotic pressure. The concentration of NaCl in red blood cells is approximately 11 m*M*. Calculate the osmotic pressure of this solution at body temperature (37°C). (*Hint*: See Equation 11.2.)

22.44. Calculate the osmotic pressure exerted by a 92 m*M* solution of KCl in a red blood cell at body temperature (37°C). (*Hint*: See Equation 11.2.)

***22.45. Electrochemical Potentials across Cell Membranes** Very different concentrations of Na^+ ions exist in red blood cells (11 m*M*) and the blood plasma (160 m*M*) surrounding those cells. Solutions with two different concentrations separated by a membrane constitute a concentration cell. Calculate the electrochemical potential created by the unequal concentrations of Na^+ at 37°C.

22.46. The concentration of K^+ in red blood cells is 92 m*M*, and the concentration of K^+ in plasma is 10 m*M*. Calculate the electrochemical potential created by the two concentrations of K^+.

22.47. If the transport of K^+ across a cell membrane requires 5 kJ/mol, how many moles of ATP must be hydrolyzed to provide the necessary energy? The hydrolysis of ATP is described by the equation

$$ATP^{4-} + H_2O \rightarrow ADP^{3-} + HPO_4^{2-} + H^+ \qquad \Delta G° = -30.5 \text{ kJ}$$

*22.48. Removing excess Na^+ from a cell by an ion pump requires energy. How many moles of ATP must be hydrolyzed to overcome a cell potential of −0.07 V? The hydrolysis of one mole of ATP provides 34.5 kJ of energy.

22.49. Plankton Exoskeletons Exoskeletons of planktonic acantharia contain strontium sulfate. Calculate the solubility in moles per liter of $SrSO_4$ in water at 25°C given that $[SO_4^{2-}]$ in seawater = 0.028 *M*.

22.50. Algae in the genus *Closterium* contain structures built from barium sulfate (barite). Calculate the solubility in moles per liter of $BaSO_4$ in water at 25°C given that $K_{sp} = 1.08 \times 10^{-10}$.

Trace and Ultratrace Essential Elements; Nonessential Elements

Concept Review

22.51. What danger to human health is posed by ^{137}Cs ($t_{1/2} \approx 30$ yr)?

22.52. Why is ^{137}Cs ($t_{1/2} \approx 30$ yr) considered to be dangerous to human health when naturally occurring ^{40}K ($t_{1/2} = 1.28 \times 10^6$ yr) is benign?

22.53. What are the likely signs of ΔS and ΔG for the dissolution of tooth enamel?

22.54. Why does fluorapatite resist acid better than hydroxyapatite if both are insoluble in water?

***22.55.** Why do peroxide ions (O_2^{2-}) act as strong oxidizing agents?

*22.56. Why are thyroxine and 3,5,3′-triiodothyronine (Figure 22.15) considered to be amino acids? Why aren't they *essential* amino acids?

Problems

22.57. What are the products of radioactive decay of ^{137}Cs? Write a balanced equation for the nuclear decay reaction.

22.58. Potassium-40 decays by three pathways: β decay, positron emission, and electron capture. Write balanced equations for each of these processes.

22.59. Calculate the pH of a 1.00×10^{-3} *M* solution of selenocysteine ($pK_{a_1} = 2.21$, $pK_{a_2} = 5.43$).

22.60. Calculate the pH of a 1.00×10^{-3} *M* solution of cysteine ($pK_{a_1} = 1.7$, $pK_{a_2} = 8.3$). Is selenocysteine a stronger acid than cysteine?

22.61. Composition of Tooth Enamel Tooth enamel contains the mineral hydroxyapatite. Hydroxyapatite reacts with fluoride ion in toothpaste to form fluorapatite. The equilibrium constant for the reaction between hydroxyapatite and fluoride ion is $K = 8.48$. Write the equilibrium constant expression for the following reaction. In which direction does the equilibrium lie?

$$Ca_5(PO_4)_3(OH)(s) + F^-(aq) \rightleftharpoons Ca_5(PO_4)_3(F)(s) + OH^-(aq)$$

*22.62. **Effects of Excess Fluoridation on Teeth** Too much fluoride might lead to the formation of calcium fluoride according to the reaction

$$Ca_5(PO_4)_3(OH)(s) + 10\ F^-(aq) \rightleftharpoons 5\ CaF_2(s) + 3\ PO_4^{3-}(aq) + OH^-(aq)$$

Write the equilibrium constant expression for the reaction. Given the K_{sp} values for the following two reactions, calculate K for the reaction between $Ca_5(PO_4)_3(OH)$ and fluoride ion that forms CaF_2.

$$Ca_5(PO_4)_3(OH)(s) \rightleftharpoons 5\ Ca^{2+}(aq) + 3\ PO_4^{3-}(aq) + OH^-(aq) \quad K_{sp} = 2.3 \times 10^{-59}$$

$$CaF_2(s) \rightleftharpoons Ca^{2+}(aq) + 2\ F^-(aq) \quad K_{sp} = 3.9 \times 10^{-11}$$

22.63. Tooth enamel is actually a composite material containing both hydroxyapatite and a calcium phosphate, $Ca_8(HPO_4)_2(PO_4)_4 \cdot 6\ H_2O$ ($K_{sp} = 1.1 \times 10^{-47}$).
a. Is this calcium mineral more or less soluble than hydroxyapatite ($K_{sp} = 2.3 \times 10^{-59}$)?
b. Calculate the solubility in moles per liter of hydroxyapatite, $Ca_5(PO_4)_3(OH)$, $K_{sp} = 2.3 \times 10^{-59}$ in water at 25°C.
*c. Explain why the production of weak acids by bacteria on teeth and gums increases the solubility of hydroxyapatite.

22.64. The K_{sp} of actual tooth enamel is reported to be 1×10^{-58}.
a. Does this mean that tooth enamel is more soluble than pure hydroxyapatite ($K_{sp} = 2.3 \times 10^{-59}$)?
b. Does the measured value of K_{sp} for tooth enamel support the idea that tooth enamel is a mixture of hydroxyapatite, $Ca_5(PO_4)_3(OH)$, and a calcium phosphate, $Ca_8(HPO_4)_2(PO_4)_4 \cdot 6\ H_2O$ ($K_{sp} = 1.1 \times 10^{-47}$)?
c. Calculate the solubility in moles per liter of $Ca_8(HPO_4)_2(PO_4)_4 \cdot 6\ H_2O$ ($K_{sp} = 1.1 \times 10^{-47}$) in water at 25°C.

22.65. Some sources give the formula of hydroxyapatite as $Ca_{10}(PO_4)_6(OH)_2$. If the K_{sp} of $Ca_5(PO_4)_3(OH)$ is 2.3×10^{-59}, what is the K_{sp} of $Ca_{10}(PO_4)_6(OH)_2$?

22.66. The same sources mentioned in the previous problem cite the formula of fluorapatite as $Ca_{10}(PO_4)_6F_2$. If the K_{sp} of $Ca_5(PO_4)_3F$ is 3.2×10^{-60}, what is the K_{sp} of $Ca_{10}(PO_4)_6F_2$?

22.67. All of the group 16 elements form compounds with the generic formula H_2E (E = O, S, Se, or Te). Which compound is the most polar? Which compound is the least polar?

*22.68. All of the group 15 elements form compounds with the generic formula H_3E (E = N, P, As, Sb, and Bi). Which compound is the most polar? Which compound do you predict to have the smallest H–E–H bond angle?

Elements for Diagnosis and Therapy

Concept Review

22.69. When choosing an isotope for imaging, why is it important to consider the decay mode of the isotope as well as the half-life?

22.70. Why might an α emitter be a good choice for radiation therapy?

***22.71.** What advantage does a β emitter have over an α emitter for imaging?

22.72. Why do we sometimes use radioisotopes of toxic elements, such as thallium, for imaging?

22.73. Why does ^{213}Bi undergo β decay but ^{111}In decays by electron capture?

22.74. Several isotopes of arsenic are used in medical imaging. Which isotope, ^{72}As or ^{77}As, is more likely to be useful for PET imaging?

22.75. The World Anti-Doping Agency (WADA) added xenon and argon to the list of banned substances in 2014. Which intermolecular forces account for the solubility of Xe and Ar in blood?

22.76. Helium is used in scuba gear to prevent nitrogen narcosis. Do you expect the solubility of He in blood to be greater than or less than the solubility of Xe and Ar in blood?

Problems

22.77. **PET Imaging with Gallium** A patient is injected with a 5 μM solution of gallium citrate containing ^{68}Ga ($t_{1/2}$ = 9.4 h) for a PET study. How long is it before the activity of the ^{68}Ga drops to 5% of its initial value?

22.78. Iodine-123 ($t_{1/2}$ = 13.3 h) has replaced iodine-131 ($t_{1/2}$ = 8.1 d) for diagnosis of thyroid conditions. How long is it before the activity of ^{123}I drops to 5% of its initial value?

22.79. The bismuth in over-the-counter antacids is found as BiO^+. Draw the Lewis structure for the BiO^+ cation.

22.80. Some medicines used in treating depression contain lithium carbonate. Draw the Lewis structure for Li_2CO_3.

22.81. Aluminum hydroxide is used in some antacids. Write a balanced net ionic equation for the reaction of aluminum hydroxide with HCl.

22.82. Aluminum carbonate is used in some antacids. Write a balanced net ionic equation for the reaction of aluminum carbonate with hydrochloric acid.

22.83. The antacid known as Maalox contains a mixture of magnesium and aluminum hydroxides. Which substance will neutralize more acid on a per mole basis? Does the same substance also neutralize more acid on a per gram basis?

22.84. Sodium bicarbonate and calcium carbonate both act as antacids and are found in common stomach remedies. Which substance will neutralize more acid on a per mole basis? Does the same substance also neutralize more acid on a per gram basis?

22.85. How many grams of magnesium hydroxide are needed to neutralize 115 mL of 0.75 M stomach acid?

22.86. How many grams of aluminum hydroxide are needed to neutralize 115 mL of 0.75 M stomach acid?

smartwork5

If your instructor uses Smartwork5, log in at digital.wwnorton.com/atoms2.

23

Transition Metals

Biological and Medical Applications

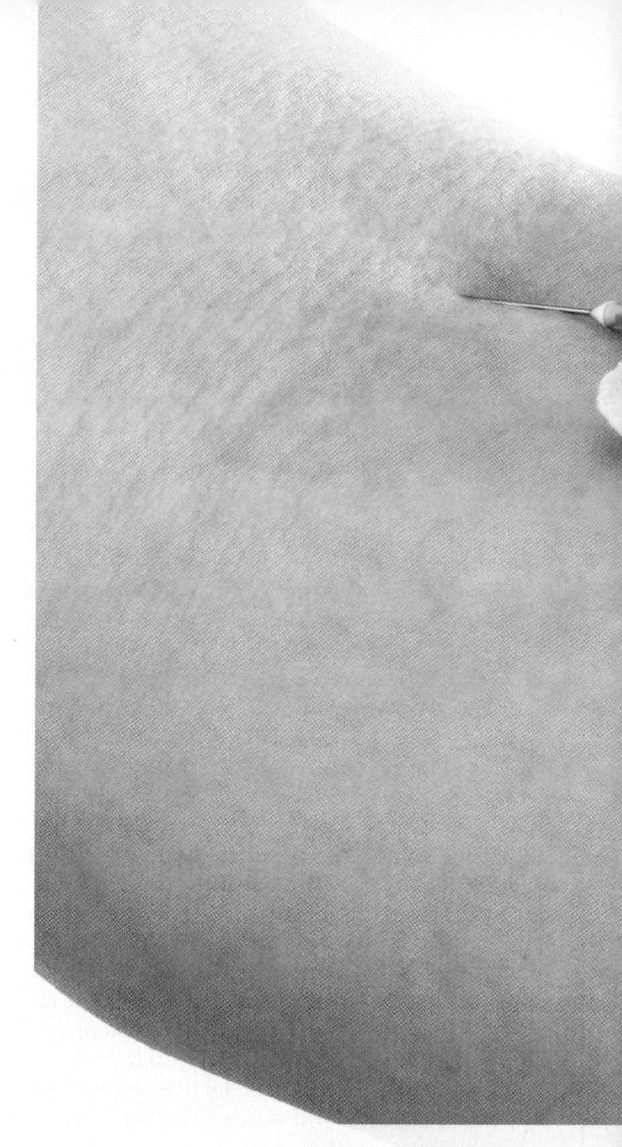

RED COLOR OF BLOOD The red color of blood comes from the heme group, a molecule containing a ring with four nitrogen atoms that bind to a central Fe^{2+} ion.

PARTICULATE **REVIEW**

Lewis Acid or Lewis Base?

In Chapter 23 we discuss the bonding and structure of compounds formed by transition metal ions and learn about their presence in biological systems and their applications to medicine. The compounds shown here are ammonia, borane, and water.

- Draw the Lewis structure for each compound.
- Which compound(s) has/have a lone pair of electrons on the central atom?
- Which compound(s) can function as a Lewis acid? Which can function as a Lewis base?

  (Review Chapters 4 and 16 if you need help.)

(Answers to Particulate Review questions are in the back of the book.)

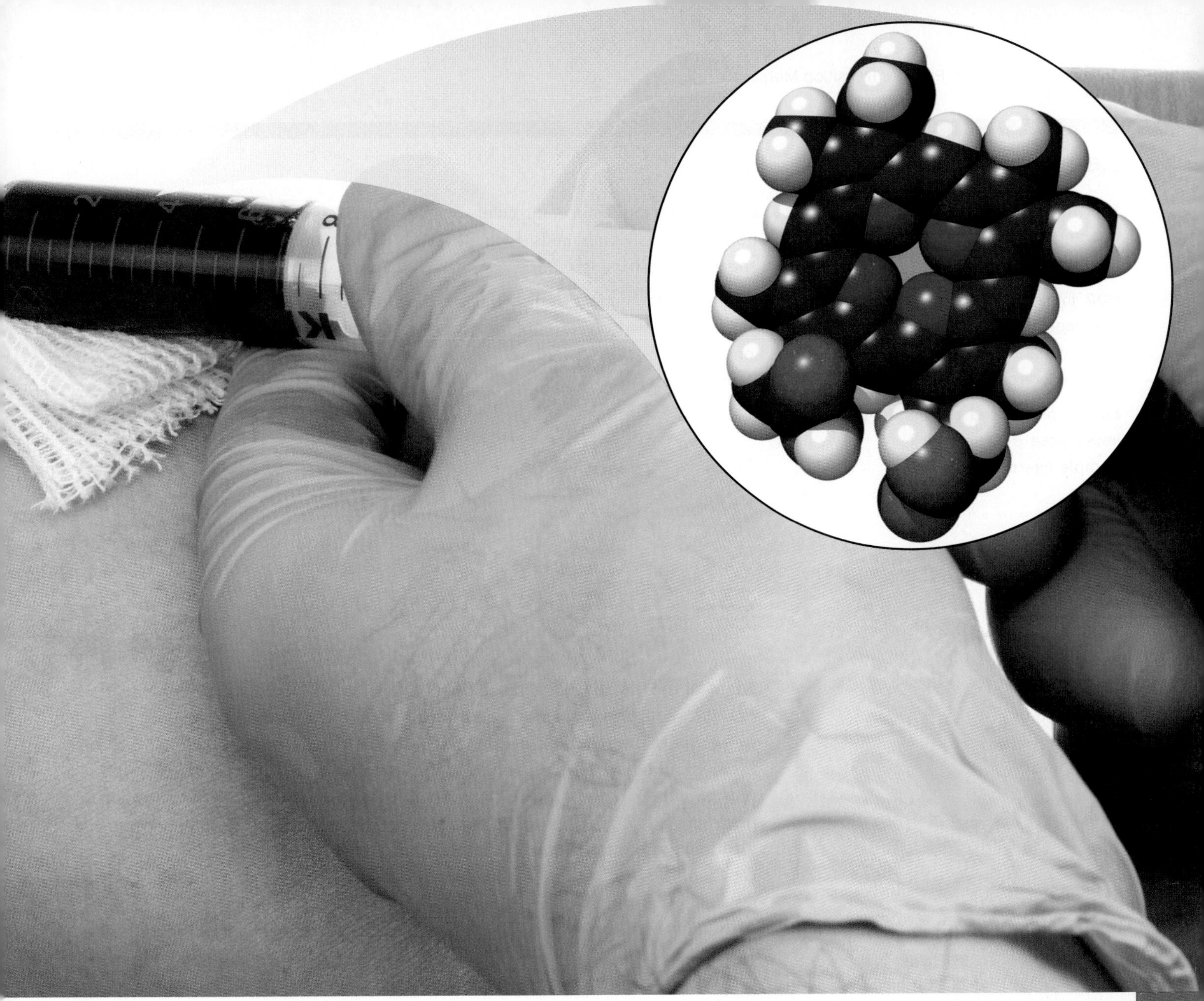

PARTICULATE **PREVIEW**

One Molecule, One Bond versus One Molecule, Two Bonds

Here are two complex ions, each of which contains a central metal cation bonded to several molecules. As you read Chapter 23, look for ideas that will help you answer these questions:

- Atoms of the same element form coordinate bonds to both of the central metals in the complex ions shown here. Which element is this?
- What is the name of the molecule bonded to the copper cation? How many of these molecules form coordinate bonds with copper?
- What molecule is bonded to the nickel cation? How many of these molecules form coordinate bonds with the nickel cation and how many coordinate bonds does each molecule form?

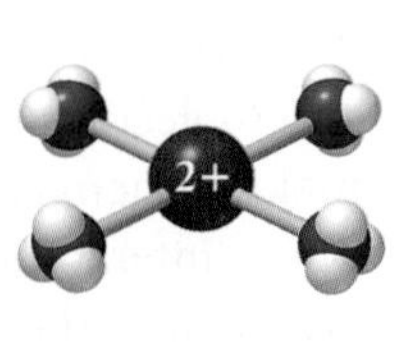

Cu^{2+} complex

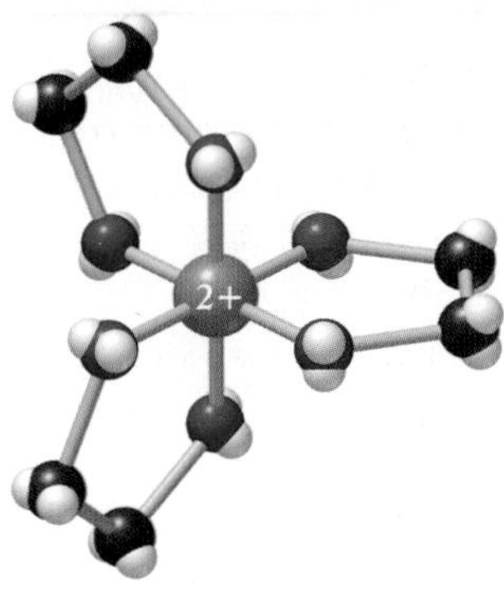

Ni^{2+} complex

Learning Outcomes

LO1 Recognize complex ions and their counter ions in chemical formulas
Sample Exercise 23.1

LO2 Interconvert the names and formulas of complex ions and coordination compounds
Sample Exercise 23.2

LO3 Explain the chelate effect and its importance
Sample Exercise 23.3

LO4 Explain the origin of the colors of transition metal compounds using the spectrochemical series

LO5 Describe the factors that lead to high-spin or low-spin states of complex ions
Sample Exercise 23.4

LO6 Identify stereoisomers of coordination compounds
Sample Exercise 23.5

LO7 Describe where metal complexes occur in biochemistry and how they are used as diagnostic or therapeutic compounds
Sample Exercise 23.6

23.1 Transition Metals in Biology: Complex Ions

Many of the metallic elements in the periodic table are essential to good health. For example, copper, zinc, and cobalt play key roles in protein function. Iron is needed to transport oxygen from our lungs to all the cells of our body. These and other essential metallic elements, several of which we mentioned in Chapter 22, should be present either in our diets or in the supplements many of us rely on for balanced nutrition. Table 23.1 lists the transition metals essential to our bodies in trace and ultratrace amounts. However, the mere presence of these elements is not sufficient—they must be in a form that our cells can use. Swallowing an 18-mg steel pellet as if it were an aspirin tablet would not be a good way for you to get your recommended daily allowance of iron. If we are to benefit from consuming essential metals in food and nutritional supplements, the metals need to be in compounds, not free elements, and the metals in the compounds must be biologically available.

TABLE 23.1 Essential Transition Elements Found in the Human Body

Trace (1–1000 μg/g Body Mass)	Ultratrace (<1 μg/g Body Mass)
Iron	Chromium
Zinc	Cobalt
	Copper
	Manganese
	Molybdenum
	Nickel
	Vanadium

All the metallic transition elements essential to human health occur in nature in ionic compounds, but not all ionic forms are absorbed equally well. For example, most of the iron in fish, poultry, and red meat is readily absorbed because it is present in a form called *heme iron*. However, the iron in plants is mostly nonheme and is not as readily absorbed. Eating a meal that includes both meat and vegetables improves the absorption of the nonheme iron in the vegetables, as does consuming foods high in vitamin C. All these dietary factors work together at the molecular level to provide us with the nutrients we need to survive.

Interactions between transition metal ions and accompanying nonmetal ions and molecules influence the solubility of the metals, which is a key factor toward making them chemically reactive and biologically available to plants and animals. These interactions also influence other properties, including the wavelengths of visible light the metal ions absorb and therefore the colors of their compounds and solutions. In this chapter we explore how the chemical environment of transition metal ions in solids and in solutions affects their physical, chemical, and biological properties. We answer questions such as why many, but not all, metal compounds have distinctive colors, and how, through the formation of complex ions with biomolecules, transition metals play key roles in many biological processes.

inner coordination sphere the ligands that are bound directly to a metal via coordinate bonds.

We start this chapter by examining the interactions between transition metal ions and the other ions and molecules that surround them in solids and solutions.

To understand these interactions, we need to review the definitions of Lewis acids and Lewis bases we used in Chapter 16:

- A *Lewis base* is a substance that donates a lone pair of electrons in a chemical reaction.
- A *Lewis acid* is a substance that accepts a lone pair of electrons in a chemical reaction.

CONNECTION Lewis's pioneering theories of the nature of covalent bonding were described in Chapter 4 and the formation of complex ions between metal ions (Lewis acids) and ligands (Lewis bases) was discussed in Chapter 16.

In Chapter 6, we described how ions dissolved in water are *hydrated*—that is, surrounded by water molecules oriented with their positive poles directed toward anions and their negative poles directed toward cations. When these ion–dipole interactions lead to the sharing of lone-pair electrons with empty valence-shell orbitals on the cations, they meet our definition of covalent bonds, and in this case are called *coordinate covalent bonds*, or simply *coordinate bonds*. Much of the chemistry of the transition metals is associated with their ability to form coordinate bonds with molecules or anions.

As we saw in Chapter 16, molecules or anions that function as Lewis bases and form coordinate bonds with metal cations are called *ligands*. The resulting species, which are composed of central metal ions and the surrounding ligands, are called *complex ions* or simply *complexes*. Direct bonding to a central cation means that the ligands in a complex occupy the **inner coordination sphere** of the cation. Take another look at Sample Exercise 14.12, in which we discussed the equilibrium between two forms of cobalt(II), one pink and one blue, in an aqueous solution of HCl (Figure 23.1). Both the pink and blue forms are complexes; the pink cobalt(II) species has six water ligands in its inner coordination sphere, and the blue species has four chloride ions. The charge on each complex ion is the sum of the charges of the metal ion and the ligands: 2+ for $Co(H_2O)_6^{2+}$ because the charge of the cobalt ion is 2+ and water molecules are neutral, and 2− for $CoCl_4^{2-}$ because the sum of the 2+ charge on the cobalt ion and four 1− charges on the chloride ions is 2−.

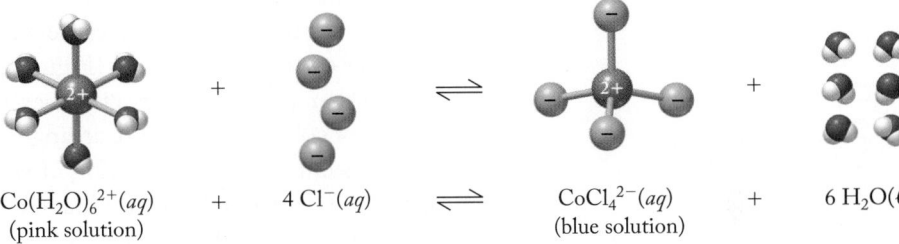

$$Co(H_2O)_6^{2+}(aq) + 4\,Cl^-(aq) \rightleftharpoons CoCl_4^{2-}(aq) + 6\,H_2O(\ell)$$

(pink solution) (blue solution)

FIGURE 23.1 Equilibrium between two forms of cobalt in aqueous HCl solution.

The foundation of our understanding of the bonding and structure of complex ions comes from the pioneering research of Swiss chemist Alfred Werner (1866–1919), for which he was awarded the Nobel Prize in Chemistry in 1913. Some of Werner's research addressed the unusual behavior of different compounds formed by oxidizing cobalt(II) chloride in aqueous ammonia to cobalt(III) by bubbling air through the solution. One of the redox reactions produces an orange compound (Figure 23.2a) that contains 3 moles of Cl^- ions for every 1 mole of Co^{3+} ions and 6 moles of ammonia. Another reaction produces a purple compound (Figure 23.2b) that has the same proportions of Cl^- and Co^{3+} ions, but with only 5 moles of ammonia per mole of cobalt(III). Both compounds are water-soluble solids that react with aqueous solutions of $AgNO_3$, forming solid AgCl. However, 1 mole of the orange compound produces 3 moles of solid AgCl, whereas 1 mole of the purple one produces only 2 moles of AgCl. Results like these inspired Werner to study the electrical conductivity of aqueous solutions of the two compounds. He found that the orange compound was the better conductor, indicating that it produced more ions in solution than the purple one.

FIGURE 23.2 Two compounds of cobalt(III) chloride and ammonia. (a) $[Co(NH_3)_6]Cl_3$, (b) $[Co(NH_3)_5Cl]Cl_2$.

Werner concluded that the differences in the two compounds' composition, in their capacities to react with Ag^+ ions, and in their electrolytic properties are all caused by the presence of two different types of Co–Cl bonds inside them. He

coordination number the number of sites occupied by ligands around a metal ion in a complex.

coordination compound a compound made up of at least one complex ion.

counter ion an ion whose charge balances the charge of a complex ion in a coordination compound.

CONNECTION We discussed in Chapter 8 how the electrical conductivity of aqueous solutions depends on the concentrations of dissolved ions in the solutions.

proposed that these bonds are all ionic in the orange compound, but that only two-thirds of them are ionic in the purple compound. The remaining one-third are coordinate covalent bonds.

This ability to form two different kinds of bonds means that Co^{3+} and other transition metal ions have two different kinds of bonding capacity, or *valence*. The first kind involves ionic bonds and is based on the number of electrons a metal atom loses when it forms an ion. This valence is equivalent to its oxidation number, which is +3 for cobalt in both the orange and purple compounds. The second kind of valence is based on the capacity of metal ions to form coordinate bonds. This property corresponds to an ion's **coordination number**, the number of sites around a central metal ion where bonds to ligands form.

The formulas of the reactants in the two chemical equations in Figure 23.3 fit the chemical composition and explain the properties of the orange and purple cobalt(III) compounds. Note how brackets in the formulas of these **coordination compounds** set off the complex ions from the ionically bonded chloride **counter ions**, which balance the charges on the complex ions: 3+ for the orange one but only 2+ for the purple one because it contains a negatively charged Cl^- ion within its inner coordination sphere (Figure 23.3).

Werner proposed correctly that the six ligands in the inner coordination sphere of cobalt(III) ions are arranged in an octahedral geometry. This geometry is consistent with the formation of six bonds to a central atom (or central ion in this case), as we learned in Chapter 5. Many other transition metal ions form six coordinate bonds and octahedral complex ions. Some are listed in Table 23.2, as are complex ions that have only two or four ligands. Those with two have linear structures, whereas those with four may be tetrahedral or square planar, with only four of the pairs of electrons involved in bond formation. Both geometries are seen in cobalt(II) ions: $CoCl_4^{2-}$ is tetrahedral whereas $Co(CN)_4^{2-}$ is square planar.

CONCEPT TEST

In the coordination compound $Na_3[Fe(CN)_6]$, which ions occupy the inner coordination sphere of the Fe^{3+} ion, and which ions are counter ions? An aqueous solution of which of the four cobalt complexes described in this section conducts electricity as well as an equal concentration of $Na_3[Fe(CN)_6]$?

(Answers to Concept Tests are in the back of the book.)

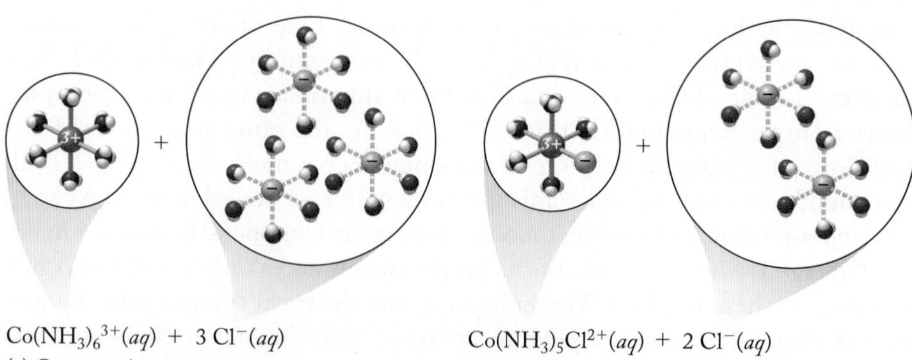

$Co(NH_3)_6^{3+}(aq) + 3\,Cl^-(aq)$
(a) Orange solution

$Co(NH_3)_5Cl^{2+}(aq) + 2\,Cl^-(aq)$
(b) Purple solution

FIGURE 23.3 The release of 3 moles of chloride ions, shown with their water molecules of hydration, from the orange compound, but only 2 from the purple compound, explains the compounds' chemical and electrolytic properties.

TABLE 23.2 Common Coordination Numbers and Shapes of Complex Ions

Coordination Number	Steric Number	Shape	Structure	Examples
6	6	Octahedral		$Fe(H_2O)_6^{3+}$ $Ni(H_2O)_6^{2+}$ $Co(H_2O)_6^{3+}$
4	6	Square planar		$Pt(NH_3)_4^{2+}$
4	4	Tetrahedral		$Zn(H_2O)_4^{2+}$
2	2	Linear		$Ag(NH_3)_2^{+}$

SAMPLE EXERCISE 23.1 Writing Formulas of Coordination Compounds **LO1**

The first modern synthetic color, Prussian blue, was made in 1704 in Berlin, Germany. It was heavily used in the 19th-century Japanese woodblock print *The Great Wave off Kanagawa* (Figure 23.4), and it continues to be popular with artists worldwide. The formula for the insoluble pigment is $Fe_4[Fe(CN)_6]_3$.

a. What is the formula of the complex ion in this compound and what is its charge if the counter ions are Fe^{3+}?
b. What is the oxidation state of iron in the complex ion?

Collect, Organize, and Analyze We have the formula of a coordination compound and are asked to identify the complex ion and the counter ion. The complex ion is contained in brackets, the counter ion is outside the brackets, and the compound must be electrically neutral.

Solve

a. The formula of the compound contains four Fe^{3+} counter ions, which have a total charge of $4 \times (3+) = 12+$. The compound is neutral, so the remainder of the formula, three $Fe(CN)_6^{n-}$ ions, must have an overall charge of 12−. This means each of the ions has a charge of $(12-)/3 = 4-$.

b. Each of the six cyanide ions in the complex ion has a charge of 1−, which means all six contribute a charge of 6−. We determined that the overall charge of the complex ion is 4−. Letting x be the charge (and oxidation state) of the central Fe ion:

$$(4-) = (6-) + x$$

$$x = 2+$$

Think About It Complex ions may have positive or negative charges. Correspondingly, counter ions will have charges opposite to that of the complex ion to ensure electrical neutrality.

FIGURE 23.4 The source of the blue color in *The Great Wave off Kanagawa*, a famous woodblock print by the Japanese artist Hokusai, is the pigment Prussian blue.

Practice Exercise A coordination compound of ruthenium, $[Ru(NH_3)_4Cl_2]Cl$, has shown some activity against leukemia in animal studies. Identify the complex ion, determine the oxidation state of the metal, and identify the counter ion.

(Answers to Practice Exercises are in the back of the book.)

23.2 Naming Complex Ions and Coordination Compounds

The names of complex ions and coordination compounds tell us the identity and oxidation state of the central ion, the names and numbers of ligands, the charge in the case of complex ions, and the identity of counter ions. To convey all this information, we need to follow some naming rules.

Complex Ions with a Positive Charge

1. Start with the identities of the ligand(s). Names of common ligands appear in Table 23.3. If there is more than one kind of ligand, list the names alphabetically.
2. Use the usual prefix(es) in front of the name(s) written in step 1 to indicate the number of each type of ligand (Table 23.4).
3. Write the name of the metal ion with a Roman numeral in parentheses indicating its oxidation state.

Examples are shown in the following table.

Formula	Name	Structure
$Ni(H_2O)_6^{2+}$	Hexaaquanickel(II)	2+
$Co(NH_3)_6^{3+}$	Hexaamminecobalt(III)	3+
$Cu(NH_3)_4(H_2O)_2^{2+}$	Tetraamminediaquacopper(II)	2+

TABLE 23.3 Names and Structures of Common Ligands

Ligand	Name within Complex Ion	Structure	Charge	Number of Donor Groups
Iodide	Iodo	$:\ddot{\underset{\cdot\cdot}{I}}:^{-}$	1–	1
Bromide	Bromo	$:\ddot{\underset{\cdot\cdot}{Br}}:^{-}$	1–	1
Chloride	Chloro	$:\ddot{\underset{\cdot\cdot}{Cl}}:^{-}$	1–	1
Fluoride	Fluoro	$:\ddot{\underset{\cdot\cdot}{F}}:^{-}$	1–	1
Nitrite	Nitro	$[O \cdots \ddot{N} \cdots O]^{-}$	1–	1
Hydroxide	Hydroxo	$[:\ddot{\underset{\cdot\cdot}{O}}—H]^{-}$	1–	1
Water	Aqua	H—$\ddot{\underset{\cdot\cdot}{O}}$—H	0	1
Pyridine (py)	Pyridyl	CH, HC, CH, HC, CH, $\ddot{N}$ (ring)	0	1
Ammonia	Ammine	$:NH_3$	0	1
Ethylenediamine (en)	(same)[a]	$H_2\ddot{N}—CH_2—CH_2—\ddot{N}H_2$	0	2
2,2'-Bipyridine (bipy)	Bipyridyl	HC—CH, HC, C—C, HC═N: / HC—CH, CH, :N═CH (two rings)	0	2
1,10-Phenanthroline (phen)	(same)[a]	HC═CH, HC—C, C—CH, HC, C—C, CH, HC═N:, :N═CH (fused rings)	0	2
Cyanide[b]	Cyano	$[:C≡N:]^{-}$	1–	1
Carbon monoxide[b]	Carbonyl	:C≡O:	0	1

[a]The names of some electrically neutral ligands in complexes are the same as the names of the molecules.
[b]Carbon atoms are the lone pair donors in these ligands.

TABLE 23.4 Common Prefixes Used in the Names of Complex Ions

Number of Ligands	Prefix
2	di-
3	tri-
4	tetra-
5	penta-
6	hexa-

It may seem strange having two *a*'s together in these names, but it is consistent with current naming rules. Prefixes are ignored in determining alphabetical order, which is why *ammine* comes before *aqua* rather than *di* before *tetra* in tetraamminediaquacopper(II).

In these three examples the ligands are all electrically neutral. This makes determining the oxidation state of the central metal ion a simple task because the charge on the complex ion is the same as the charge on the metal ion, which is the oxidation state of the metal. When the ligands are anions, determining the oxidation state of the central metal ion requires us to account for these charges.

Complex Ions with a Negative Charge

1. Follow the steps for naming positively charged complexes.
2. Add *-ate* to the name of the central metal ion to indicate that the complex ion carries a negative charge (just as we use *-ate* to end the names of oxoanions). Sometimes the name of the metal changes, too. The two most common examples are iron, which becomes *ferrate*, and copper, which becomes *cuprate*.

Examples are shown in the following table.

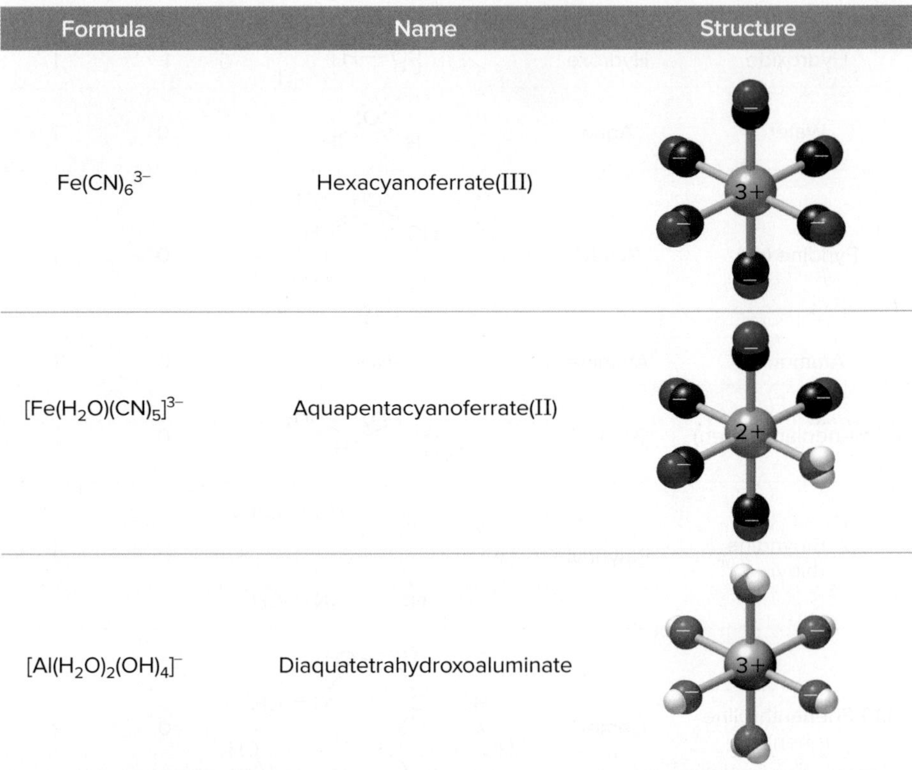

Formula	Name	Structure
$Fe(CN)_6^{3-}$	Hexacyanoferrate(III)	
$[Fe(H_2O)(CN)_5]^{3-}$	Aquapentacyanoferrate(II)	
$[Al(H_2O)_2(OH)_4]^-$	Diaquatetrahydroxoaluminate	

In the first two examples we must determine the oxidation state of Fe. We start with the charge on the complex ion and then take into account the charges on the ligand anions to calculate the charge on the metal ion. For example, the overall charge of the aquapentacyanoferrate(II) ion is 3−. It contains five CN^- ions. To reduce the combined charge of 5− from these cyanide ions to an overall charge of 3−, the charge on Fe must be 2+.

CONCEPT TEST

What is the name of the complex anion with the formula $PtCl_4^{2-}$?

Coordination Compounds

1. If the counter ion of the complex ion is a cation, the cation's name goes first, followed by the name of the anionic complex ion.

2. If the counter ion of the complex ion is an anion, the name of the cationic complex ion goes first, followed by the name of the anion.

Examples are shown in the following table.

Formula	Name	Structure
$[Ni(NH_3)]_6Cl_2$	Hexaamminenickel(II) chloride	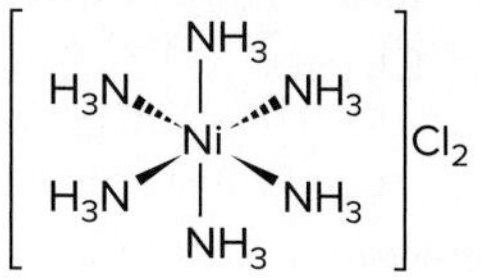 NH_3 H_3N NH_3 Ni H_3N NH_3 NH_3 Cl_2
$K_3[Fe(CN)_6]$	Potassium hexacyanoferrate(III)	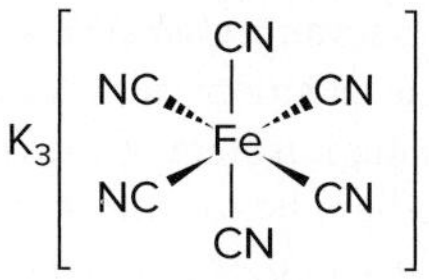 K_3 CN NC CN Fe NC CN CN
$[Co(NH_3)_5(H_2O)]Br_2$	Pentaammineaquacobalt(II) bromide	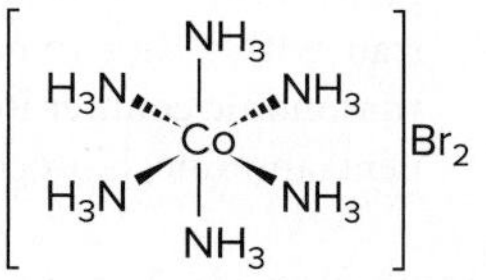 NH_3 H_3N NH_3 Co H_3N NH_3 NH_3 Br_2

A key to naming coordination compounds is to recognize from their formulas that they are coordination compounds. For help with this, look for formulas that have the atomic symbols of a metallic element and one or more ligands, all in brackets, either followed by the atomic symbol of an anion, as in $[Co(NH_3)_5(H_2O)]Br_2$, or preceded by the symbol of a cation, as in $K_3[Fe(CN)_6]$.

SAMPLE EXERCISE 23.2 Naming Coordination Compounds **LO2**

Name the coordination compounds (a) $Na_4[Co(CN)_6]$ and (b) $[Co(NH_3)_5Cl](NO_3)_2$.

Collect and Organize We are asked to write a name for each compound that unambiguously identifies its composition. The formulas of the complex ions appear in brackets in both compounds. Because cobalt, the central metal ion in both, is a transition metal, we express its oxidation state using Roman numerals. The names of common ligands are given in Table 23.3.

Analyze It is useful to take an inventory of the ligands and counter ions:

		LIGAND			
Compound	**Counter Ion**	**Formula**	**Name**	**Number**	**Prefix**
$Na_4[Co(CN)_6]$	Na^+	CN^-	Cyano	6	Hexa-
$[Co(NH_3)_5Cl](NO_3)_2$	NO_3^-	NH_3 Cl^-	Ammine Chloro	5 1	Penta- —

The oxidation state of each cobalt ion can be calculated by setting the sum of the charges on all the ions in both compounds equal to zero:

a. Ions:	($4\ Na^+$ ions)	+	(1 Co ion)	+	($6\ CN^-$ ions)		
Charges:	4+	+	x	+	6−		= 0
							x = 2+
b. Ions:	(1 Co ion)	+	($1\ Cl^-$ ion)	+	($2\ NO_3^-$ ions)		
Charges:	x	+	1−	+	2−		= 0
							x = 3+

Solve

a. Because the counter ion, sodium, is a cation, its name comes first. The complex ion is an anion. To name it, we begin with the ligand cyano, to which we add the prefix *hexa-* and write *hexacyano*. This is followed by the name of the transition metal ion: hexacyano*cobalt*. We add *-ate* to the ending of the name of the complex ion because it is an anion: hexacyanocobalt*ate*. We then add a Roman numeral to indicate the oxidation state of the cobalt: hexacyanocobaltate(*II*). Putting it all together, we get sodium hexacyanocobaltate(II).
b. The complex ion is the cation in this compound, and we begin by naming the ligands directly attached to the metal ion in alphabetical order: ammine and chloro. We indicate the number (5) of NH_3 ligands with the appropriate prefix: *penta*amminechloro. We name the metal next and indicate its oxidation state with a Roman numeral: pentaamminechloro*cobalt(III)*. Finally we name the anionic counter ion: *nitrate*. Putting it all together, we obtain the name: pentaamminechlorocobalt(III) nitrate.

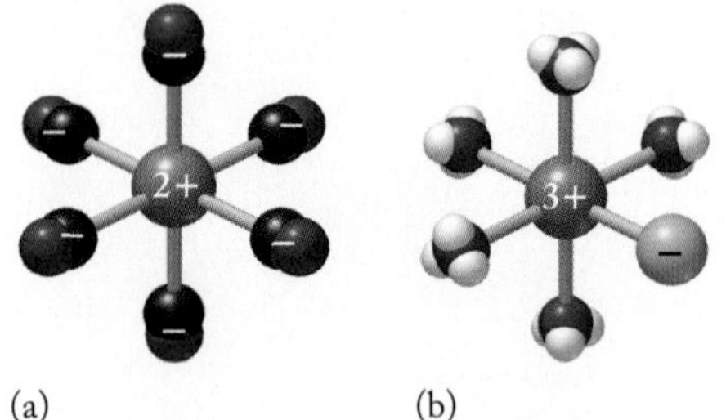

FIGURE 23.5 The structures of (a) sodium hexacyanocobaltate(II) and (b) pentaamminechlorocobalt(III) nitrate.

Think About It Naming coordination compounds requires us to (1) distinguish between ligands and counter ions and (2) recall which ligands are electrically neutral and which are anions. The structures of the complex ions in the named coordination compounds are shown in Figure 23.5.

Practice Exercise Identify the ligands and counter ions in (a) $[Zn(NH_3)_4]Cl_2$ and (b) $[Co(NH_3)_4(H_2O)_2](NO_2)_2$, and name each compound.

23.3 Polydentate Ligands and Chelation

We have seen that ligands are electron-pair donors—that is, Lewis bases. Let's explore the strengths of several ligands as Lewis bases by considering their affinity for $Ni^{2+}(aq)$ ions. Suppose we dissolve crystals of nickel(II) chloride hexahydrate, $NiCl_2 \cdot 6H_2O$, in water. The dot connecting the two halves of the formula and the prefix *hexa* indicate that there are six water molecules for every one Ni^{2+} ion in the solid hexahydrate. Four of these water molecules and two Cl^- ions surround each Ni^{2+} ion as show in Figure 23.6(a). Each complex is also hydrogen bonded to two other molecules of H_2O. When this compound dissolves in water, the two chloride ion are displaced by water molecules so that the formula of the complex ion in solution $Ni(H_2O)_6{}^{2+}$ (Figure 23.6b).

Now let's bubble colorless ammonia gas through a green solution of $Ni(H_2O)_6{}^{2+}$ ions. As shown in the middle test tube in Figure 23.6(b), the green solution turns blue. The color change means that different ligands are bonded to the Ni^{2+} ions. We may conclude that NH_3 molecules have displaced the H_2O molecules around

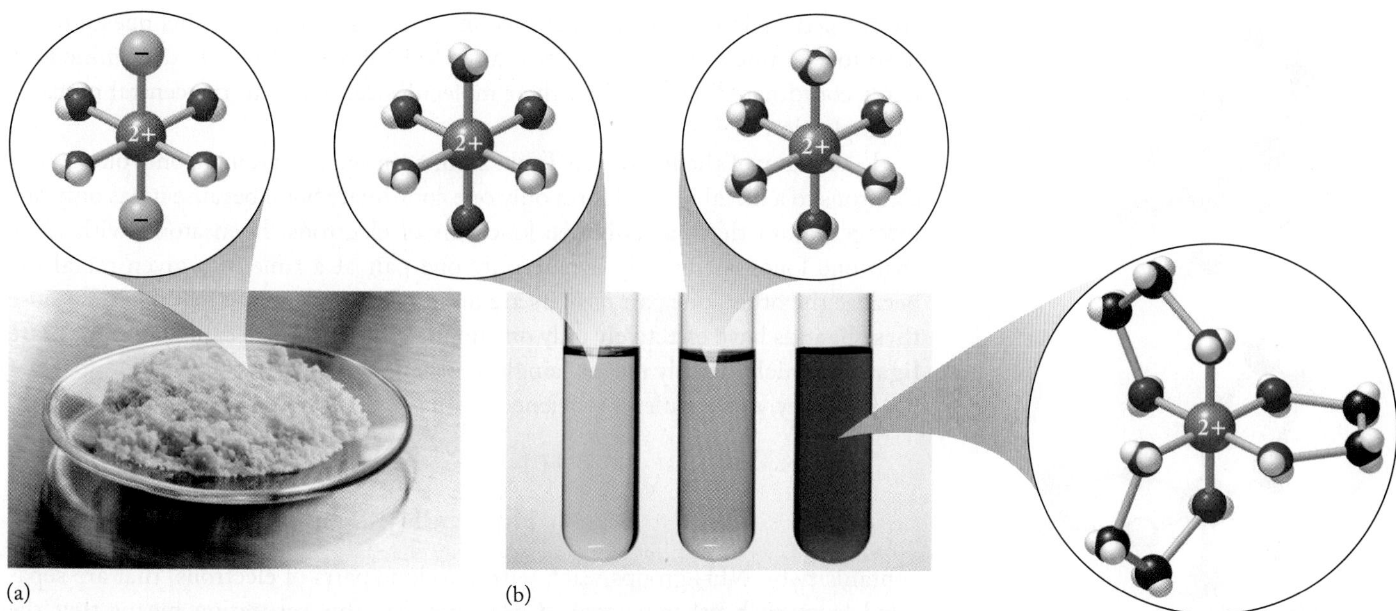

FIGURE 23.6 (a) Solid nickel(II) chloride hexahydrate is green. (b) When it dissolves in water, the resulting solution is also green, which suggests that each Ni^{2+} ion (gold sphere) is bonded to H_2O molecules both in the solid and in the solution. When ammonia gas is bubbled through a solution of $Ni(H_2O)_6{}^{2+}$, the color changes to blue as NH_3 replaces H_2O in the Ni^{2+} ion's inner coordination sphere. When ethylenediamine is added to a solution of $Ni(NH_3)_6{}^{2+}$, the color turns from blue to purple as the ethylenediamine displaces the ammonia ligands and the $Ni(en)_3{}^{2-}$ complex forms.

the Ni^{2+} ions. If all the molecules of H_2O are displaced, the complex $Ni(NH_3)_6{}^{2+}$ is formed. The following chemical equation describes this change:

$$Ni(H_2O)_6{}^{2+}(aq) + 6\ NH_3(g) \rightleftharpoons Ni(NH_3)_6{}^{2+}(aq) + 6\ H_2O(\ell) \qquad K_f = 5 \times 10^8$$

Keep in mind that the hydrated ion $Ni(H_2O)_6{}^{2+}$ is often expressed as $Ni^{2+}(aq)$.

This *ligand displacement* reaction illustrates that Ni^{2+} ions have a greater affinity for molecules of NH_3 than for molecules of H_2O. Many other transition metal ions also have a greater affinity for ammonia than for water. We may conclude that ammonia is inherently a better electron-pair donor and hence a stronger Lewis base than water. This conclusion is reasonable because we saw in Chapter 15 that ammonia was also a stronger Brønsted–Lowry base than H_2O.

Next we add the compound ethylenediamine (see Table 23.3) to the blue solution of $Ni(NH_3)_6{}^{2+}$ ions. The solution changes color again, from blue to purple (Figure 23.6b), indicating yet another change in the ligands surrounding the Ni^{2+} ions. Molecules of ethylenediamine displace ammonia molecules from the inner coordination sphere of Ni^{2+} ions. This affinity of Ni^{2+} ions for ethylenediamine molecules is reflected in the large formation constant for $Ni(en)_3{}^{2+}$ (where "en" represents ethylenediamine):

$$Ni(H_2O)_6{}^{2+}(aq) + 3\ en(aq) \rightleftharpoons Ni(en)_3{}^{2+}(aq) + 6\ H_2O(\ell) \qquad K_f = 1.1 \times 10^{18}$$

This value is more than 10^9 times the K_f value for $Ni(NH_3)_6{}^{2+}$.

Why should the affinity of Ni^{2+} ions for ethylenediamine be so much greater than their affinity for ammonia? After all, in both ligands the coordinate bonds

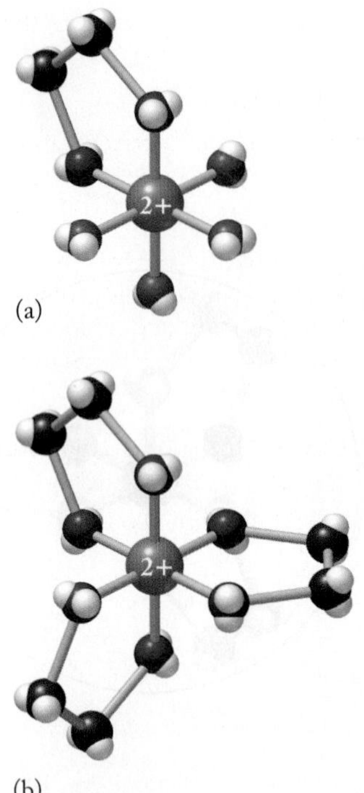

FIGURE 23.7 (a) The bidentate ligand ethylenediamine has two N atoms that can each donate a pair of electrons to empty orbitals of adjacent octahedral bonding sites on the same $Ni^{2+}(aq)$ ion (gold sphere), displacing two molecules of water. (b) Three ethylenediamine molecules occupy all six octahedral coordination sites of a Ni^{2+} ion.

are formed by lone pairs of electrons on N atoms. To answer this question, we need to examine the molecular structures of these two ligands to determine how many coordinate bonds each of their molecules can make to the central metal in a complex ion.

Like many of the ligands in Table 23.3, ammonia can donate only one pair of electrons to a metal ion and form only one coordinate bond because it has only one nitrogen atom that has only one lone pair of electrons. Even atoms with more than one lone pair usually donate only one pair at a time to a given metal ion because the other lone pair or pairs are oriented away from the metal ion. Because these ligands have effectively only one donor group, they are called **monodentate ligands**, which literally means "single-toothed."

However, a molecule of ethylenediamine

$$H_2\ddot{N}\text{—}CH_2\text{—}CH_2\text{—}\ddot{N}H_2$$

contains two $-NH_2$ groups, each with two lone pairs of electrons, that are separated from each other by two $-CH_2-$ groups. This separation means that the molecule can partially encircle a metal ion such as Ni^{2+}, as shown in Figure 23.7(a), and form two bonds on the same side of the metal ion.

Two more ethylenediamine molecules can bond to other bonding sites, forming a complex in which the Ni^{2+} ion is surrounded by three ethylenediamine molecules, as shown in Figure 23.7(b).

This capacity of an ethylenediamine molecule to donate more than one lone pair of electrons and therefore form more than one coordinate bond to a central metal ion puts it into a category of ligands called **polydentate ligands**, or more specifically *bidentate*, *tridentate*, and so on. One group of polydentate ligands is the polyamines, which include ethylenediamine.

Note that each ethylenediamine molecule forms a five-atom ring with the metal ion. If the ring were a regular pentagon (meaning all bond lengths and bond angles were exactly the same), each of its bond angles would be 108°. These pentagons are not perfect, but each ring's preferred octahedral bond angles of 90° for the N–Ni–N bond, and of 107° to 109° for all the other bonds, are accommodated with only a little strain on the ideal bond angles.

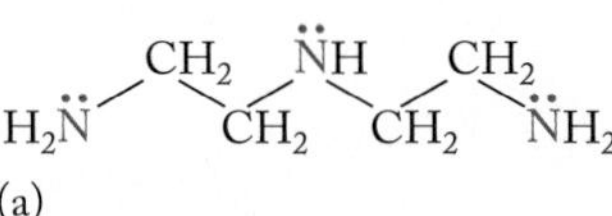

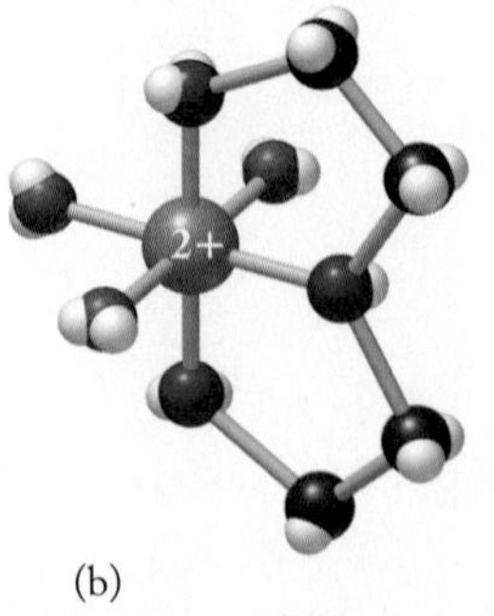

FIGURE 23.8 Tridentate chelation. (a) The three amine groups in the tridentate ligand diethylenetriamine are all potential electron-pair donor groups. (b) When these groups donate their lone pairs of electrons to a $Ni^{2+}(aq)$ ion (gold sphere), they occupy three of the six coordination sites on the ion.

An even larger ligand, diethylenetriamine ($H_2NCH_2CH_2NHCH_2CH_2NH_2$), is shown in Figure 23.8(a). The lone pairs of electrons on its three nitrogen atoms give diethylenetriamine the capacity to form three coordinate bonds to a metal ion, meaning this is a tridentate ligand (Figure 23.8b).

As you may imagine, larger molecules may have even more atoms per molecule that can bond to a single metal ion. The interaction of a metal ion with a ligand having multiple donor atoms is called **chelation** (pronounced *key-LAY-shun*). The word comes from the Greek *chele*, meaning "claw." The polydentate ligands that take part in these interactions are called *chelating agents*.

When we added ethylenediamine to the blue solution of $Ni(NH_3)_6^{2+}$ ions in Figure 23.6(b), molecules of ethylenediamine (en) displaced ammonia molecules from the inner coordination sphere of Ni^{2+} ions as described in the following chemical equation:

$$Ni(NH_3)_6^{2+}(aq) + 3\ en(aq) \rightleftharpoons Ni(en)_3^{2+}(aq) + 6\ NH_3(aq) \quad (23.1)$$

The color change tells us that this reaction as written is spontaneous. As we discussed in Chapter 17, spontaneous reactions are those in which free energy decreases ($\Delta G < 0$). Furthermore, under standard conditions the change in free energy ($\Delta G°$) is related to the changes in enthalpy and entropy that accompany the reaction:

$$\Delta G° = \Delta H° - T\Delta S°$$

The displacement of NH_3 by ethylenediamine is exothermic, but only slightly ($\Delta H° = -12$ kJ/mol). More important, $\Delta S° = +185$ J/(mol · K). This means that at 25°C,

$$T\Delta S° = 298\ \cancel{K} \times \frac{185\ \cancel{J}}{\text{mol} \cdot \cancel{K}} \times \frac{1\ \text{kJ}}{1000\ \cancel{J}} = 55.1\ \text{kJ/mol}$$

To understand why there is such a large increase in entropy, consider that there are 4 moles of reactants but 7 moles of products in Equation 23.1. Nearly doubling the number of moles of aqueous products over reactants translates into a large gain in entropy. It is this positive $\Delta S°$, more than the negative $\Delta H°$ value, that drives the reaction and makes it spontaneous. Entropy gains drive many complexation reactions that involve polydentate ligands. The entropy-driven affinity of metal ions for polydentate ligands is called the **chelate effect**.

Many chelating agents have more than one kind of electron-pair-donating group. Aminocarboxylic acids represent one family of such compounds. The most important of them is ethylenediaminetetraacetic acid, EDTA, the molecular structure of which is shown in Figure 23.9(a). Note that one molecule of EDTA contains two amine (nitrogen-containing) groups and four carboxylic acid (–COOH) groups. When the acid groups release their H^+ ions, they form four carboxylate anions, $–COO^-$, in which either of the O atoms can donate a pair of electrons to a central metal ion. When O atoms on all four groups do so and the two amine groups do as well, six octahedral bonding sites around the metal ion can be occupied, as shown in Figure 23.9(b).

EDTA forms very stable complex ions and is used as a metal ion *sequestering agent*, that is, a chelating agent that binds metal ions so tightly that they are "sequestered" and prevented from reacting with other substances. For example, EDTA is used as a preservative in many beverages and prepared foods because it sequesters iron, copper, zinc, manganese, and other transition metal ions often present in these foods that can catalyze the degradation of ingredients in the foods. Many foods are fortified with ascorbic acid (vitamin C), which is particularly vulnerable to metal-catalyzed degradation because it is also a polydentate ligand and is more likely to be oxidized when chelated to one of the above metal ions. EDTA effectively shields vitamin C from these ions.

monodentate ligand a species that forms only a single coordinate bond to a metal ion in a complex.

polydentate ligand a species that can form more than one coordinate bond per molecule.

chelation the interaction of a metal with a polydentate ligand (chelating agent); pairs of electrons on one molecule of the ligand occupy two or more coordination sites on the central metal.

chelate effect the greater affinity of metal ions for polydentate ligands than for monodentate ligands.

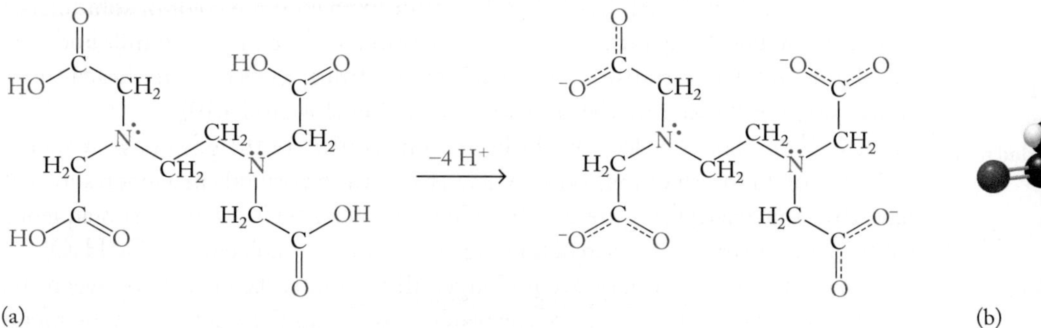

(b)

FIGURE 23.9 (a) In the hexadentate ligand EDTA, the six donor groups are the two amine groups and the four carboxylic acid groups. The acid groups ionize to form carboxylate anions. (b) All six Lewis base groups in ionized EDTA can form a coordinate bond with the same metal ion, such as Co^{3+} (the gold sphere) shown here. In the process they form four five-membered rings.

SAMPLE EXERCISE 23.3 Identifying the Potential Electron-Pair-Donor Groups in a Molecule **LO3**

How many donor groups does this polydentate ligand, nitrilotriacetic acid (NTA), have?

$$N(CH_2COOH)_3$$

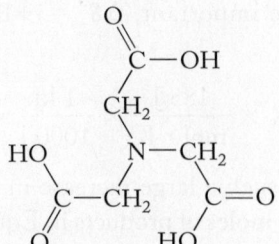

O
C—OH
CH_2
HO N—CH_2
C—CH_2 C=O
O HO

Collect, Organize, and Analyze We need to examine this molecular structure to find electron pairs that can be donated. Because there are three single bonds around the N atom, the atom's fourth sp^3 orbital must contain a lone pair of electrons. When all three carboxylic acid groups are ionized, there are three carboxylate groups in the molecule, and each carboxylate group can donate one nonbonding pair of electrons from one of its oxygen atoms to a metal atom.

Solve The central N atom and an O atom from each of the three carboxylate groups form a total of four coordinate bonds. Therefore, NTA is potentially a tetradentate ligand with four donor groups.

Think About It The tetradentate capacity of NTA is reasonable because, like EDTA, it is an aminocarboxylic acid. It has one fewer amino group and one fewer carboxylic acid group than the hexadentate EDTA.

Practice Exercise How many potential donor groups are there in citric acid, a component of citrus fruits and a widely used preservative in the food industry?

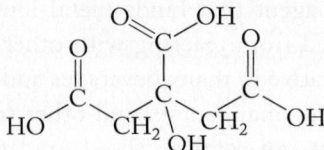

O OH
O C O
C C C
HO CH_2 OH CH_2 OH

CHEMTOUR Crystal Field Splitting

FIGURE 23.10 When chromium(III) nitrate dissolves in water, the resulting solution has a distinctive violet color due to the presence of $Cr(H_2O)_6{}^{3+}$ ions.

23.4 Crystal Field Theory

We have seen that formation of complex ions can change the color of solutions of transition metals. Why is this? The colors of transition metal compounds and ions in solution are due to transitions of *d*-orbital electrons. Let's explore these transitions using Cr^{3+} as our model transition metal ion (Figure 23.10).

A Cr^{3+} ion has the electron configuration $[Ar]3d^3$ in the gas phase. When a Cr^{3+} ion (or any atom or ion) is in the gas phase, all the orbitals in a given subshell have the same energy (Figure 23.11a). However, when a Cr^{3+} ion is in an aqueous solution and surrounded by an octahedral array of water molecules in $Cr(H_2O)_6{}^{3+}$, the energies of its 3*d* orbitals are no longer all the same. Repulsions between the lone pairs on the oxygen atoms of the water ligands and the electrons in the metal 3*d* orbitals raise the energies of all the *d* orbitals but to different extents. The $3d_{xy}$, $3d_{yz}$, and $3d_{xz}$ orbitals experience some increase in energy, but the energies of the $3d_{x^2-y^2}$ and $3d_{z^2}$ orbitals increase even more (Figure 23.11b) because the lobes of

CONNECTION Crystal field theory is an example of molecular orbital theory, which was introduced in Chapter 5.

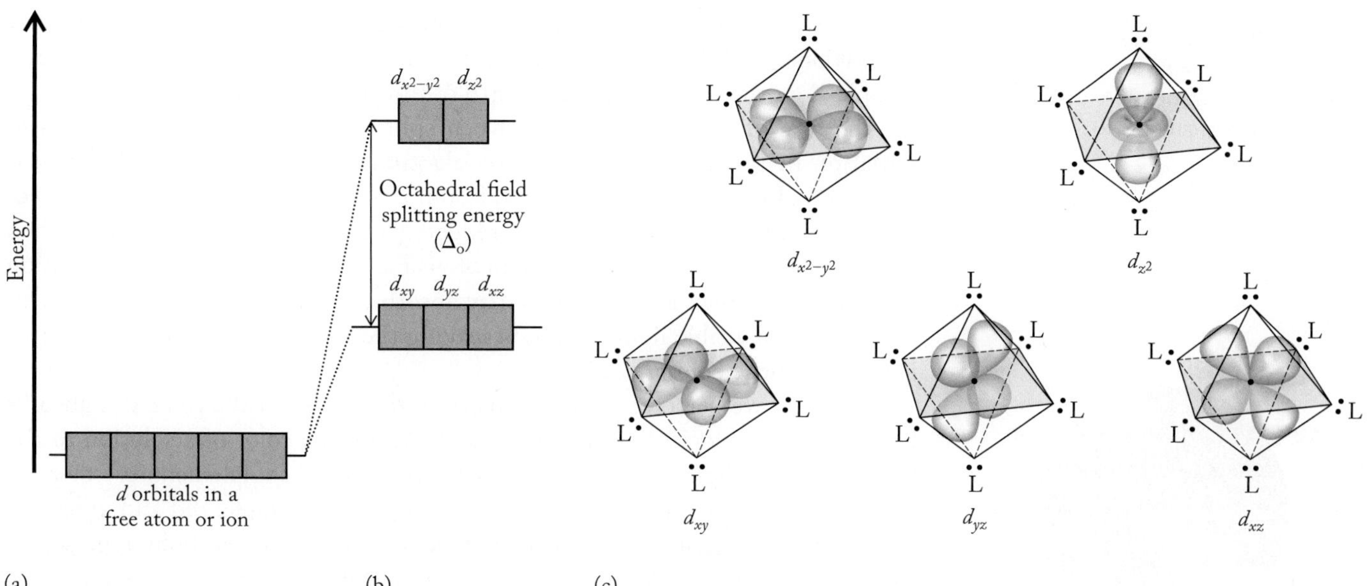

FIGURE 23.11 Octahedral crystal field splitting. (a) In an atom or ion in the gas phase, all orbitals in a subshell are degenerate, as shown here for the five 3*d* orbitals. (b) When an ion is part of a complex ion in a compound or solution, repulsions between electrons in the ion's *d* orbitals and ligand electrons raise the energy of the orbitals, as shown here for an octahedral field. (c) The greatest repulsion is experienced by electrons in the $d_{x^2-y^2}$ and d_{z^2} orbitals because the lobes of these orbitals are directed toward the corners of the octahedron and so are closest to the lone pairs on the ligands (L). The lobes of the lower-energy d_{xy}, d_{yz}, and d_{xz} orbitals are directed toward points that lie between the corners of the octahedron, so electrons in them experience less repulsion.

the $3d_{x^2-y^2}$ and $3d_{z^2}$ orbitals point directly toward the H_2O molecules' oxygen atoms at the corners of the octahedron formed by the ligands and are repelled more by the electrons on those O atoms (Figure 23.11c). The energies of the $3d_{xy}$, $3d_{yz}$, and $3d_{xz}$ orbitals are not raised as much because the lobes of these three orbitals do not point directly toward the corners of the octahedron, so they experience less electron repulsion.

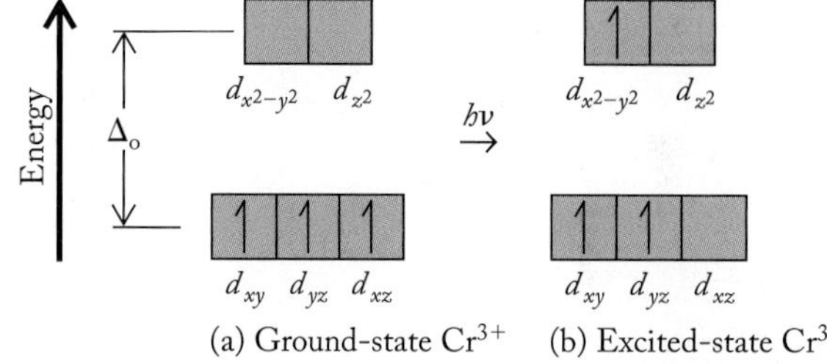

FIGURE 23.12 A Cr^{3+} ion, $[Ar]3d^3$, in an octahedral field can absorb a photon of light that has energy ($h\nu$) equal to Δ_o. This energy raises a 3*d* electron from (a) one of the lower-energy *d* orbitals to (b) one of the higher-energy *d* orbitals.

This change from degenerate (equal-energy) *d* orbitals to orbitals with different energies is known as **crystal field splitting**, and the difference in energy created by crystal field splitting is called **crystal field splitting energy (Δ)**. The name was originally used to describe splitting of *d*-orbital energies in minerals, but the theory also applies to species in aqueous solutions.

In a $Cr(H_2O)_6^{3+}$ ion, three electrons are distributed among five 3*d* orbitals. According to Hund's rule, each of the three electrons should occupy one of the three lower-energy orbitals, leaving the two higher-energy orbitals unoccupied, as shown in Figure 23.12(a). The energy difference between the two subsets of orbitals is symbolized by Δ_o, where the subscript "o" indicates that the energy split was caused by an *o*ctahedral array of electron repulsions.

What if an aqueous Cr^{3+} ion absorbs a photon whose energy is exactly equal to Δ_o? As the photon is absorbed, a 3*d* electron moves from a lower-energy orbital to a higher-energy orbital (Figure 23.12b). The wavelength λ of the absorbed photon is related to the energy difference between the two groups of orbitals—in other words, to the crystal field splitting energy—as follows:

$$E = \frac{hc}{\lambda} = \Delta_o \tag{23.2}$$

crystal field splitting the separation of a set of *d* orbitals into subsets with different energies as a result of interactions between electrons in those orbitals and lone pairs of electrons in ligands.

crystal field splitting energy (Δ) the difference in energy between subsets of *d* orbitals split by interactions in a crystal field.

CONNECTION In Chapter 3 we first used the equation $E = h\nu = hc/\lambda$ in discussing the energy of light in the electromagnetic spectrum.

As we discussed in Chapter 3, the energy and wavelength of a photon are inversely proportional to each other. Therefore the larger the crystal field splitting in a complex ion, the shorter the wavelength of the photons the ion absorbs.

The size of the energy gap between split *d* orbitals often corresponds to radiation in the visible region of the electromagnetic spectrum. This means that the colors of solutions of metal complexes depend on the strengths of metal–ligand interactions and are reflected in the magnitude of Δ_o. When white light (which contains all colors of visible light) passes through a solution containing complex ions, the ions may absorb energy corresponding to one or more colors of visible light. The light leaving the solution and reaching our eyes is missing those colors.

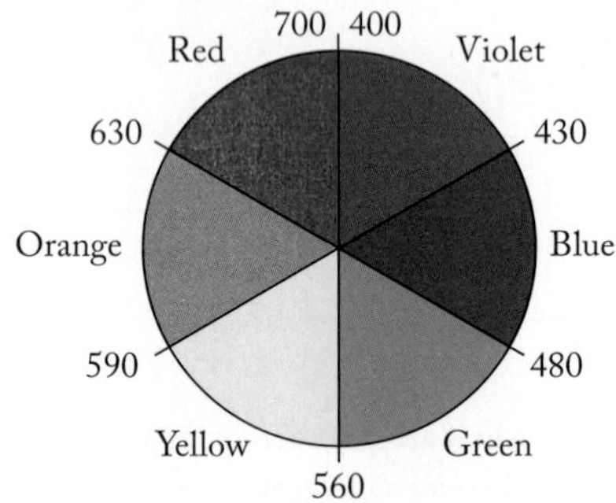

FIGURE 23.13 A color wheel. Colors on opposite sides of the wheel are complementary to each other. When we look at a solution or object that absorbs light corresponding to a given color, we see the complementary color. Wavelengths are in nanometers.

The color we perceive for any transparent object is not the color(s) it absorbs but rather the color(s) that it transmits. To relate the color of a solution to the wavelengths of light it absorbs, we need to consider complementary colors as defined by a simple color wheel (Figure 23.13). For example, red and green are complementary colors; therefore a solution that absorbs green light appears red to us.

The solution of $Cr^{3+}(aq)$ in Figure 23.10 is violet because the electron transition shown in Figure 23.12 absorbs yellow-orange light. Yellow-orange is the complement of violet, so $Cr^{3+}(aq)$ solutions appear to be violet.

A solution of $Cu(NH_3)_4^{2+}$ ions has a distinctive deep blue color (Figure 23.14). The light transmitted by such a solution features an absorption band that spans yellow, orange, and red wavelengths with a minimum transmission of light at 620 nm. Our eyes sense the range that is transmitted, which mostly spans violet to blue-green. Then our brain processes this band of transmitted colors and signals to us the average of these colors, a deep navy blue.

The nickel(II) and chromium(III) complexes we have examined up to this point were octahedral, and their central ions had a coordination number of 6. However, in the solution of $Cu(NH_3)_4^{2+}$, the copper(II) ion has a coordination number of 4. This means that the deep blue color of this complex ion is caused by a different crystal field.

Four ligands around a central metal may have either a tetrahedral arrangement or a square planar arrangement (Table 23.2). Square planar geometries tend to be limited to the transition metal ions with nearly filled valence-shell *d* orbitals, particularly those with d^8 or d^9 electron configurations. Cu^{2+} has the electron configuration $[Ar]3d^9$, and the $Cu(NH_3)_4^{2+}$ complex is square planar—which means that the strongest interactions occur between the 3*d*

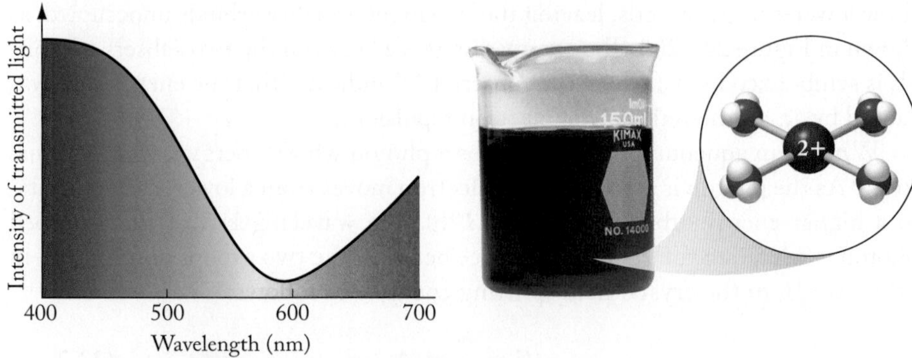

FIGURE 23.14 The visible light transmitted by a solution of $Cu(NH_3)_4^{2+}$ ions is missing much of the yellow, orange, and red portions of the visible spectrum because of a broad absorption band centered at 590 nm. Our eyes and brain perceive the transmitted colors as navy blue.

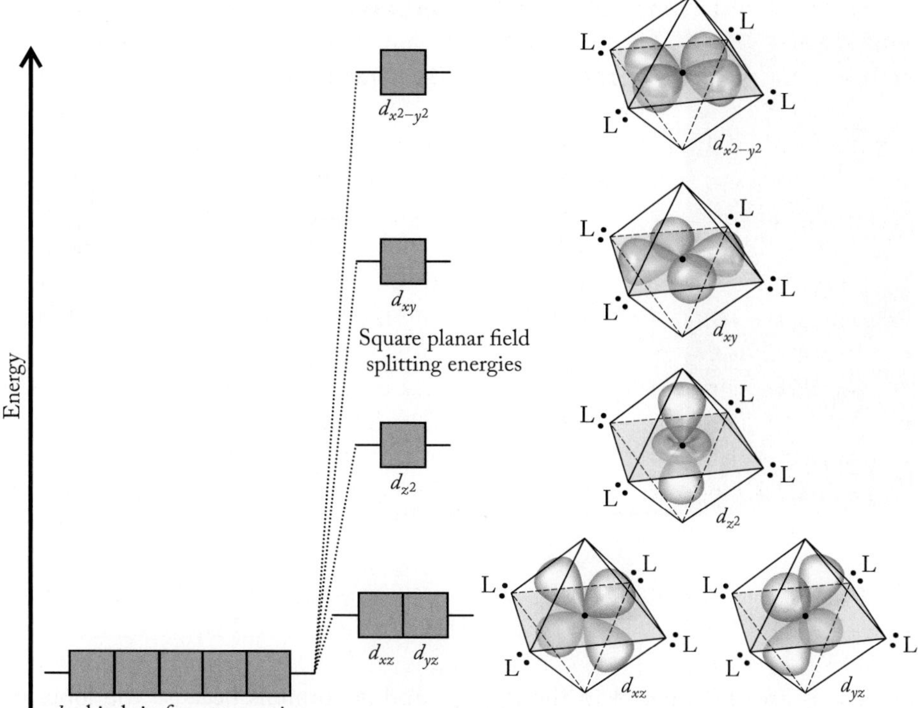

FIGURE 23.15 Square planar crystal field splitting. The d orbitals of a transition metal ion in a square planar field are split into several energy levels depending on the relative orientations of the metal orbitals and the ligand electrons at the four corners of the square. The $d_{x^2-y^2}$ orbital has the highest energy because its lobes are directed right at the four corners of the square plane.

orbitals on the central ion and the nitrogen atom lone pairs at the four corners of the equatorial plane of the octahedron, as shown in Figure 23.15. The $3d_{x^2-y^2}$ orbital has the strongest interactions and the highest energy because its lobes are oriented directly at the four corners of the plane. The d_{xy} orbital has slightly less energy because its lobes, although in the xy plane, are directed 45° away from the corners. Electrons in the three d orbitals with most of their electron density out of the xy plane interact even less with the lone pairs of the ligand and thus have even lower energies.

Finally, let's consider the d orbital crystal field splitting that occurs in a tetrahedral complex (Figure 23.16a). In this geometry, the greatest electron–electron

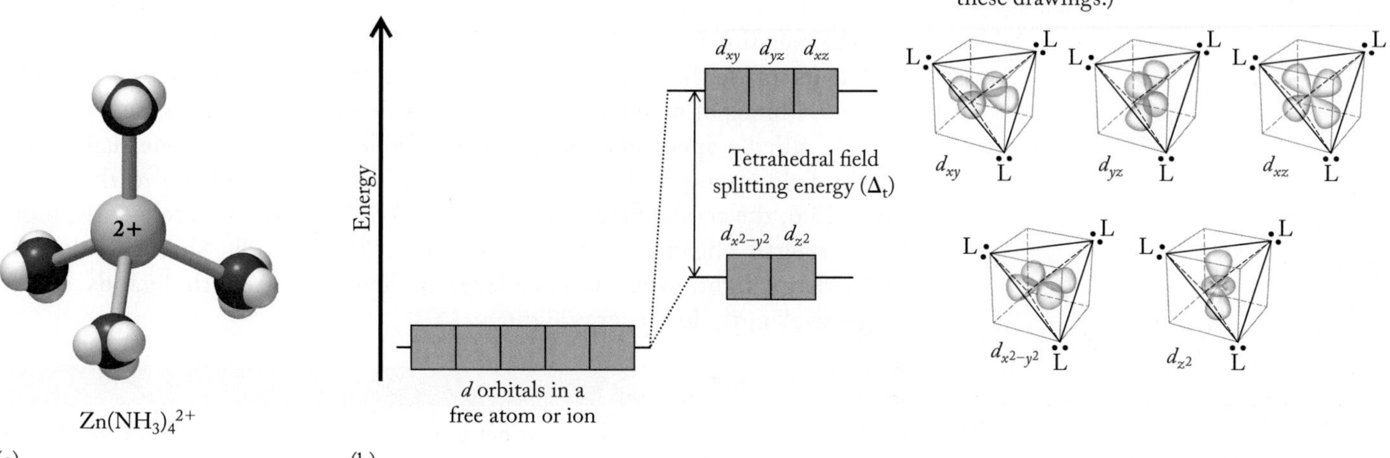

FIGURE 23.16 (a) In a tetrahedral complex ion, such as $Zn(NH_3)_4^{2+}$, the d orbitals of the metal ion are split by a tetrahedral crystal field. (b) The lobes of the higher-energy orbitals—d_{xy}, d_{yz}, and d_{xz}—are closer to the ligands at the four corners of the tetrahedron than the lobes of the lower-energy orbitals are. (One of the four corners of the tetrahedron is hidden in these drawings.)

spectrochemical series a list of ligands rank-ordered by their ability to split the energies of the *d* orbitals of transition metal ions.

TABLE 23.5 Light Transmitted and Absorbed by Three Ni^{2+} Complexes

Complex	$Ni(H_2O)_6^{2+}$	$\xrightarrow{NH_3}$	$Ni(NH_3)_6^{2+}$	$\xrightarrow{en}$	$Ni(en)_3^{2+}$
Appearance	Green		Blue		Violet
Absorbs	Red		Yellow		Green
Absorbed λ (nm)	725	>	570	>	545
$E = hc/\lambda$ ($J \times 10^{19}$)	2.7	<	3.3	<	3.6

TABLE 23.6 Spectrochemical Series of Some Common Ligands

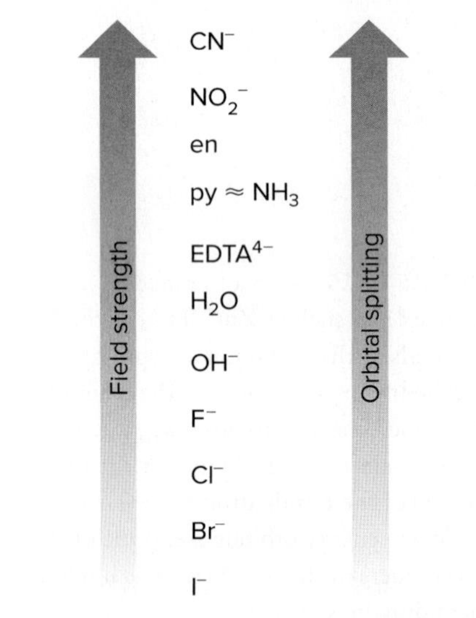

Ligand
CN^-
NO_2^-
en
py ≈ NH_3
$EDTA^{4-}$
H_2O
OH^-
F^-
Cl^-
Br^-
I^-

repulsions are experienced in the d_{xy}, d_{yz}, and d_{xz} orbitals because the lobes of these orbitals are oriented most directly to the corners of the tetrahedron, which are occupied by ligand electron pairs (Figure 23.16b). The two other *d* orbitals are less affected because their lobes do not point toward the corners. The difference in energy between the two subsets of *d* orbitals in a tetrahedral geometry is labeled Δ_t.

Before ending this discussion on the colors of transition metal ions, let's revisit the color changes we saw in Figure 23.6, when first ammonia and then ethylenediamine were added to a solution of Ni^{2+} ions. Let's think in terms of the colors these solutions absorb in the visible portion of the spectrum. A solution of $Ni^{2+}(aq)$ ions absorbs colors at the red end of the spectrum. Similarly, a solution of $Ni(NH_3)_6^{2+}$ absorbs yellow light and a solution of $Ni(en)_3^{2+}$ absorbs green light. Note how the colors these three solutions absorb are in the sequence red, yellow, and green. This sequence runs from longest wavelength to shortest, from about 725 nm for red to about 545 nm for green. Radiant energy is inversely proportional to wavelength (Equation 23.2); therefore the order of crystal field splitting energies of the *d* orbitals of Ni^{2+} ions is en > NH_3 > H_2O. Table 23.5 summarizes the observed colors of these Ni^{2+} ion complexes and the meaning we can infer from their colors.

Chemists use the parameter *field strength* to describe how much the energies of the *d* orbitals of metal ions are split by different ligands, ranking ligands in what is called a **spectrochemical series**. Table 23.6 contains one such series. As the field strength of the ligand increases from the bottom to the top of Table 23.6, the crystal field splitting energy (Δ) increases. Consequently, high-field-strength ligands form complexes that absorb short-wavelength, high-energy light, whereas complexes of low-field-strength ligands absorb long-wavelength, low-energy light.

Absorption

400 500 600 700

Wavelength (nm)

FIGURE 23.17

CONCEPT TEST

Figure 23.17 shows the absorption spectrum of a complex found in nature. What color is this complex ion?

23.5 Magnetism and Spin States

In addition to contributing to the color of transition metal ions, crystal field splitting influences their magnetic properties because these properties depend on the number of unpaired electrons in the valence shell *d* orbitals. The more unpaired electrons, the more paramagnetic the ion. For example, an Fe^{3+} ion has five 3*d* electrons (Figure 23.18a). In an octahedral field there are two ways to distribute the five 3*d* electrons among these orbitals. One arrangement conforms to Hund's rule and has a single electron in each orbital, leaving them all unpaired (Figure 23.18b). However, when Δ_o is large, as shown in Figure 23.18(c), all five electrons occupy the three lower-energy orbitals. This pattern of electron distribution occurs when the energy of repulsion between two electrons in the same orbital is less than the energy needed to promote an electron to a higher-energy orbital. In this configuration, only one electron is unpaired.

CONNECTION We introduced the magnetic behavior of matter in Chapter 5 in our discussion of molecular orbital theory.

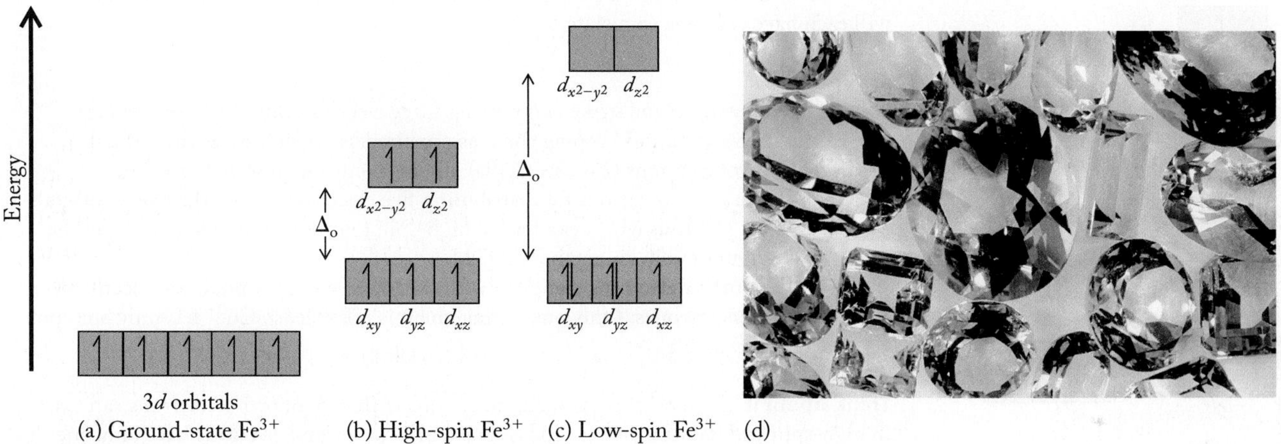

FIGURE 23.18 Low-spin and high-spin complexes. (a) The ground state of a free Fe^{3+} ion has a degenerate, half-filled set of 3*d* orbitals. (b) A weak octahedral field ($\Delta_o <$ electron-pairing energy) produces the high-spin state: five unpaired electrons, each in its own orbital. (c) In a strong octahedral field ($\Delta_o >$ electron-pairing energy), the energies of the 3*d* orbitals are split enough to produce the low-spin state: two sets of paired electrons, one unpaired electron, and two empty, higher-energy orbitals. (d) The Fe^{3+} ions in crystals of aquamarine are high spin.

The configuration with all five electrons unpaired is called the *high-spin state* because the spin on all five electrons is in the same direction, resulting in the maximum magnetic field produced by the spins. The configuration with only one electron unpaired is called the *low-spin state.* Both configurations are paramagnetic because both have at least one unpaired electron, but material containing high-spin Fe(III) ions would be more strongly attracted to an external magnet than a material containing low-spin Fe(III) ions.

Not all transition metal ions can have both high-spin and low-spin states. Consider, for example, Cr^{3+} ions in an octahedral field. Because each Cr^{3+} ion has only three 3*d* electrons (Figure 23.12), each electron is unpaired whether the orbital energies are split a lot or only a little. Therefore Cr^{3+} ions and any ion with less than four *d* electrons will have only one spin state. Metal ions with eight or more electrons in *d* orbitals will also have only one spin state because there cannot be more than one or two unpaired electrons in these orbitals no matter how the electrons are distributed.

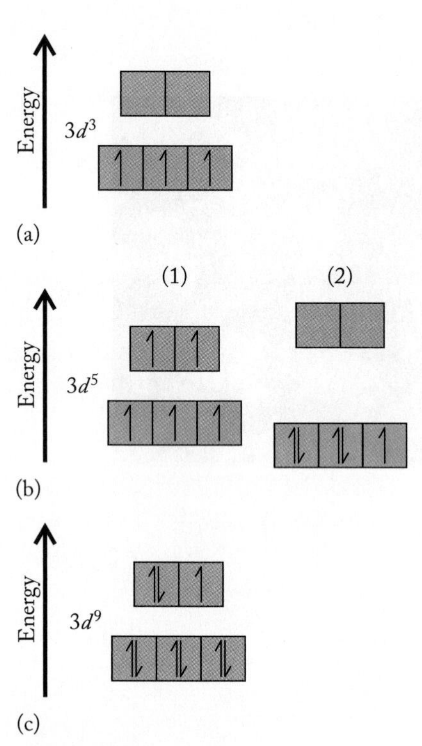

FIGURE 23.19

SAMPLE EXERCISE 23.4 Predicting Spin States **LO5**

Determine which of these ions can have high-spin and low-spin configurations when part of an octahedral complex: (a) Mn^{4+}; (b) Mn^{2+}; (c) Cu^{2+}.

Collect and Organize Mn and Cu are in groups 7 and 11 of the periodic table, so their atoms have 7 and 11 valence electrons, respectively. In an octahedral field, a set of five *d* orbitals splits into a low-energy subset of three orbitals and a high-energy subset of two orbitals.

Analyze To determine whether high-spin and low-spin states are possible, we need to determine the number of *d* electrons in each ion. Then we need to distribute them among sets of *d* orbitals split by an octahedral field to see if it is possible for the ions to have different spin states. The electron configurations of the atoms of the three elements are $[Ar]3d^54s^2$ for Mn and $[Ar]3d^{10}4s^1$ for Cu. When they form cations, the atoms of these transition metals lose their 4*s* electrons first and then their 3*d* electrons. Therefore the numbers of *d* electrons in the ions are 3 in Mn^{4+}, 5 in Mn^{2+}, and 9 in Cu^{2+}. It is likely that the ion with the fewest *d* electrons (Mn^{4+}) and the one with the most (Cu^{2+}) will each have only one spin state.

Solve

a. Mn^{4+}: Following Hund's rule and putting three electrons into the lowest-energy 3*d* orbitals available and keeping them as unpaired as possible gives this orbital distribution of electrons (Figure 23.19a). There is only one spin state for Mn^{4+}.
b. Mn^{2+}: There are two options for distributing five electrons among the five 3*d* orbitals (Figure 23.19b). Thus Mn^{2+} can have a high-spin (on the left) or a low-spin (on the right) configuration in an octahedral field.
c. Cu^{2+}: The nine 3*d* electrons completely fill the lower-energy orbitals and nearly fill the higher-energy ones. Only one arrangement is possible, so Cu^{2+} has only one spin state (Figure 23.19c).

Think About It In an octahedral field, metal ions with 4, 5, 6, or 7 *d* electrons can exist in high-spin and low-spin states. Those ions with 3 or fewer *d* electrons have only one spin state, in which all the electrons are unpaired and in the lower-energy set of orbitals. Ions with 8 or more *d* electrons have only one spin state because their lower-energy set of orbitals is completely filled. The magnitude of the crystal field splitting energy, Δ_o, determines which spin state an ion with 4, 5, 6, or 7 *d* electrons occupies.

Practice Exercise Which of these ions can have high-spin and low-spin configurations when part of an octahedral complex: (a) V^{4+}; (b) Cr^{3+}; (c) Ni^{3+}? Are any of the possible spin configurations diamagnetic?

As noted earlier, whether a transition metal ion is in a high-spin state or a low-spin state depends on whether less energy is needed to promote an electron to a higher-energy orbital or to overcome the repulsion experienced by two electrons sharing the same lower-energy orbital. Several factors affect the size of Δ_o. We have already discussed a major one in the context of the spectrochemical series shown in Table 23.6: the different field strengths of different ligands. Because molecules containing nitrogen atoms with lone pairs of electrons are stronger field-splitting ligands than H_2O molecules, hydrated metal ions (small Δ_o) are more likely to be in high-spin states, and metals surrounded by nitrogen-containing ligands (large Δ_o) are more likely to be in low-spin states. Another factor affecting spin state is the oxidation state of the metal ion. The higher the oxidation number (and ionic charge), the stronger the attraction of the electron pairs on the ligands for the ion. Greater attraction leads to more ligand–d orbital interaction and therefore to a larger Δ_o.

Complexes of transition metals in the fifth and sixth rows tend to be low spin because their 4d and 5d orbitals extend farther from the nucleus than do 3d orbitals. These larger d orbitals overlap more and interact more strongly with the lone pairs of electrons on the ligands, leading to greater crystal field splitting.

Our discussion of high-spin and low-spin states has focused entirely on d orbitals split by octahedral fields. What about spin states in tetrahedral fields? Almost all tetrahedral complexes are high spin because tetrahedral fields are weaker than octahedral fields. Weaker field strength means less d-orbital splitting—not enough to offset the energies associated with pairing two electrons in the same orbitals. Therefore Hund's rule is obeyed.

CONCEPT TEST

Explain the following:

a. $Mn(pyridine)_6^{2+}$ is a high-spin complex ion, but $Mn(CN)_6^{4-}$ is low spin.

b. $Fe(NH_3)_6^{2+}$ is high spin, but $Ru(NH_3)_6^{2+}$ is low spin.

23.6 Isomerism in Coordination Compounds

The coordination compound $Pt(NH_3)_2Cl_2$ was first described in 1849 and is now widely used as a chemotherapeutic agent for the treatment of several types of cancer. Both molecules of ammonia and the two chloride ions are coordinately bonded to the Pt^{2+} ion, so the name of the compound is diamminedichloroplatinum(II). No counter ions are present because the sum of the charges on the ligands and the platinum ion is zero and the complex is neutral. The compound is square planar, and when we draw it, there are two ways to arrange the ligands about the central metal ion: the two chloride ions and the two ammonia ligands can be at adjacent corners of the square (Figure 23.20a) or at opposite corners (Figure 23.17b). Coordination compounds like these two that have the same composition and the same connections between parts but differ in the three-dimensional arrangement of those parts are stereoisomers.

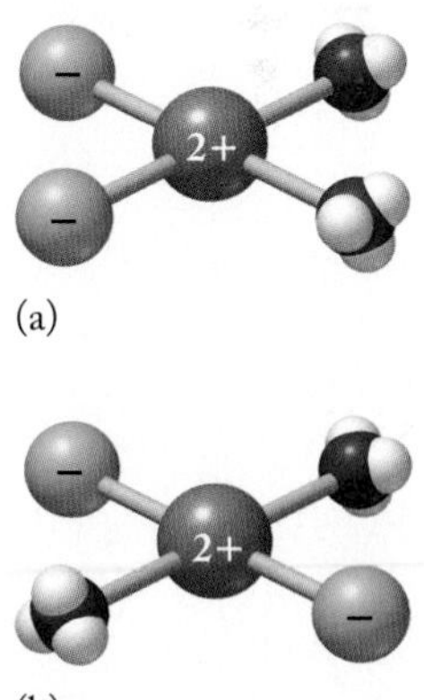

FIGURE 23.20 Two ways to orient the Cl^- ions and NH_3 molecules around a Pt^{2+} ion (gold sphere) in the square planar coordination compound $Pt(NH_3)_2Cl_2$. (a) The two members of each pair of ligands are on the same side of the square in *cis*-diamminedichloroplatinum(II). (b) The two members of each pair are at opposite corners in *trans*-diamminedichloroplatinum(II).

The two molecules in Figure 23.20 have different physical and chemical properties, and we must find a way to distinguish between them when we name them. Isomer (a), with two members of each pair of ligands at adjacent corners, is *cis*-diamminedichloroplatinum(II). Isomer (b), with pairs of ligands at opposite corners, is *trans*-diamminedichloroplatinum(II). By analogy to the stereoisomers of alkenes we saw in Chapter 19, these types of stereoisomers are called cis–trans isomers.

To illustrate the importance of stereoisomerism in coordination compounds, consider that *cis*-diamminedichloroplatinum(II) is a widely used anticancer drug with the common name *cisplatin*, but the trans isomer is much less effective in fighting cancer. The therapeutic power of cisplatin comes from its structurally specific reactions with DNA. During these reactions the two Cl atoms of $Pt(NH_3)_2Cl_2$ are replaced by two nitrogen-containing bases on a strand of DNA in the nucleus of a cell, as shown in Figure 23.21. The ability of cisplatin to

FIGURE 23.21 The attack of *cis*-diamminedichloroplatinum(II) on DNA. (a) After the drug is administered, one of its chloride ions is displaced by a molecule of water, turning each molecule into a 1+ ion. (b) The ion is attracted to a nitrogen-containing base on a strand of DNA, which displaces the water molecule and forms a coordinate bond to Pt. (c) A nearby base forms a bond to Pt by displacing the chloride ion. Forming bonds to the two DNA bases distorts the DNA molecule so much that it cannot function properly. (d) A magnified view of the bonds that form between platinum(II) and nitrogen atoms in DNA. We explored the structure and function of DNA in Chapter 20.

cross-link these bases distorts the molecular shape of the DNA (Figure 23.21d) and disrupts its normal function. Most importantly, this kind of cell DNA damage inhibits the ability of cancerous cells to grow and replicate. The trans isomer forms complexes with other intracellular compounds more readily than with DNA so it is much less effective as an anticancer agent.

Cis–trans isomers are also possible in octahedral complexes containing more than one type of ligand. For example, there are two possible stereoisomers of $[Co(NH_3)_4Cl_2]Cl$ (Figure 23.22). The two chloro ligands in $[Co(NH_3)_4Cl_2]^+$ are either on the same side of the complex with a 90° Cl–Co–Cl bond angle (the cis isomer) or across from each other so that the Cl–Co–Cl angle is 180° (the trans isomer). *cis*-Tetraamminedichlorocobalt(III) chloride is violet, and *trans*-tetraamminedichlorocobalt(III) chloride is green.

(a) *cis*-Tetraamminedichlorocobalt(III) chloride

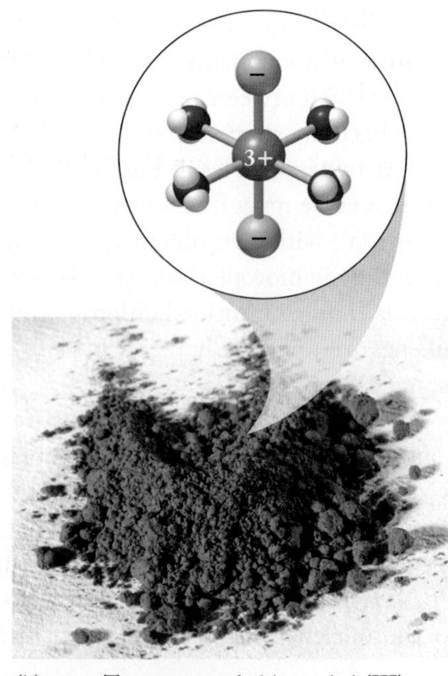

(b) *trans*-Tetraamminedichlorocobalt(III) chloride

FIGURE 23.22 The two stereoisomers of the coordination compound with the formula $[Co(NH_3)_4Cl_2]Cl$: (a) *cis*-tetraamminedichlorocobalt(III) chloride and (b) *trans*-tetraamminedichlorocobalt(III) chloride.

SAMPLE EXERCISE 23.5 Identifying Stereoisomers of Coordination Compounds **LO6**

Sketch the structures and name the stereoisomers of $Ni(NH_3)_4Cl_2$.

Collect and Organize We are given the formula and asked to name and draw the structures of the stereoisomers of a coordination compound. The Ni^{2+} complexes we have seen so far in the chapter have all been octahedral. Ammonia molecules and Cl^- ions are both monodentate ligands.

Analyze The formula contains no brackets, so the Cl^- ions are not counter ions; they must be covalently bonded to the Ni ion. There are two Cl^- ions, so the charge on Ni must be 2+. There is a total of six ligands, which confirms that the compound is octahedral.

Solve There are two ways to orient the chloride ions: opposite each other with a Cl–Ni–Cl bond angle of 180° or on the same side of the octahedron with a Cl–Ni–Cl bond angle of 90°:

Cl, NH_3, H_3N—Ni—NH_3, H_3N, Cl

H_3N, Cl, H_3N—Ni—Cl, H_3N, NH_3

The first isomer is *trans*-tetraamminedichloronickel(II); the second is *cis*-tetraamminedichloronickel(II).

Think About It We can draw other tetraamminedichloronickel(II) structures that do not look exactly like these two structures. If we flip or rotate them, however, they will match one of the two structures shown.

 **Practice Exercise** Sketch the stereoisomers of $[CoBr_2(en)(NH_3)_2]^+$ and name them.

Enantiomers and Linkage Isomers

Consider the octahedral Co(III) complex ion containing two ethylenediamine molecules and two chloride ions. There are two ways to arrange the chloride ions: on adjacent bonding sites in a cis isomer or on opposite sides of the octahedron in a trans isomer. These two molecules represent cis–trans isomers, but the cis isomer (Figure 23.23) illustrates another kind of stereoisomerism that is possible in complex ions and coordination compounds.

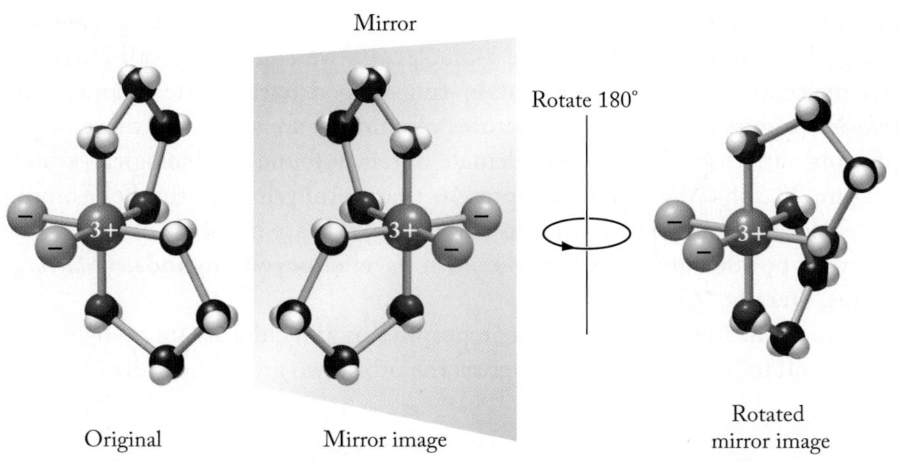

FIGURE 23.23 The complex ion *cis*-dichlorobis(ethylenediamine)cobalt(III) is chiral, which means that its mirror image is not superimposable on the original complex. To illustrate this point, we rotate the mirror image 180° about its vertical axis so that it looks as much like the original as possible. However, note that the top ethylenediamine ligand is located behind the plane of the page in the original but in front of the plane of the page in the rotated mirror image. Thus the mirror images are not superimposable.

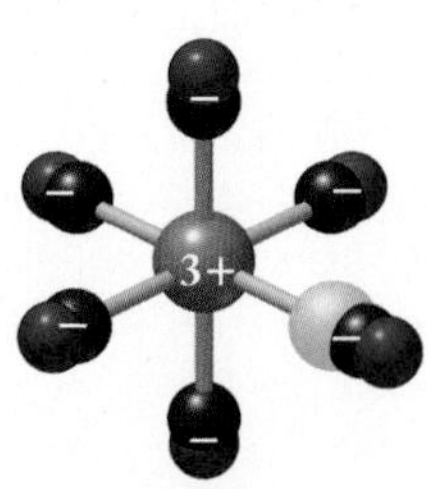

(a) $[Co(CN)_5SCN]^{3-}$

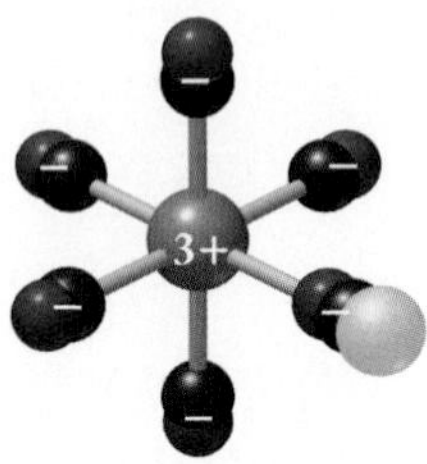

(b) $[Co(CN)_5NCS]^{3-}$

FIGURE 23.24 The SCN^- ligand is bound to the Co^{3+} ion through the S atom in (a) the pentacyanothiocyanatocobaltate(III) ion and through the N atom in (b) the pentacyanoisothiocyanatocobaltate(III) ion.

Figure 23.23 shows that *cis*-$Co(en)_2Cl_2^+$ is chiral: it has a mirror image that is not identical to the original. The difference is demonstrated by the fact that there is no way to rotate the mirror image so that its atoms align exactly with those in the original. In other words, the two structures are not superimposable. We encountered this phenomenon in Chapter 5 and noted that such nonsuperimposable stereoisomers are called *enantiomers*.

Naming this cis isomer also requires an additional rule. The complex ion is named *cis*-dichlorobis(ethylenediamine)cobalt(III). Note the prefix *bis*- just before (ethylenediamine). In naming complex ions containing a polydentate ligand, we use *bis*- instead of *di*- to indicate that two molecules of the ligand are present and to avoid the use of two *di*- prefixes in the same ligand name. Other prefixes used in this fashion are *tris*- for three polydentate ligands and *tetrakis*- for four.

CONCEPT TEST

Would four different ligands arranged in a square planar geometry produce a chiral complex ion? What about four different ligands in a tetrahedral geometry?

A third kind of isomerism called *linkage isomerism* occurs in coordination compounds when a ligand can bind to a central metal ion using either of two possible electron-pair-donating atoms. The two complexes of Co^{3+} shown in Figure 23.24 are a pair of linkage isomers. One complex contains the thiocyanate (SCN^-) ion as a ligand where the S atom covalently bonds to the metal (Figure 23.24a). The other contains the isothiocyanate (NCS^-) ion as a ligand where the N atom forms the bond (Figure 23.24b). Another ligand that forms linkage isomers with metals is NO_2^-, called nitro when the N atom donates an electron pair and nitrito when an O atom is the donor. Note that because these pairs of molecules have different connectivities, linkage isomers are *not* stereoisomers. Instead, linkage isomers are a type of constitutional isomer.

CONNECTION We introduced constitutional isomers, which have the same molecular formula but different connections between their atoms, in Chapter 6.

23.7 Coordination Compounds in Biochemistry

At the beginning of the chapter, we noted that metals essential to human health must be present in foods in forms the body can absorb. In this section we explore how transition metal cations form complexes by bonding to the nitrogen atoms of the amino acids in proteins and other Lewis bases in biological systems.

Manganese and Photosynthesis Let's begin with photosynthesis, a chemical process at the foundation of our food chain. Green plants can harness solar energy because they contain large biomolecules we collectively call *chlorophyll.* All molecules of chlorophyll contain ring-shaped tetradentate ligands called *chlorins* (Figure 23.25a). The structures of chlorins are similar to those of **porphyrins**, another class of tetradentate ligands found in biological systems (Figure 23.25b). We saw in Chapter 16 that porphyrins are the Fe^{2+}-binding molecules in hemoglobin. Chlorins and porphyrins are members of a larger category of polydentate compounds known as **macrocyclic ligands**. (*Macrocycle* means, literally, "big ring.")

porphyrin a type of tetradentate macrocyclic ligand.

macrocyclic ligand a ring containing multiple electron-pair donors that bind to a metal ion.

Two of the four nitrogen atoms in porphyrins and chlorins are sp^3 hybridized and bound to hydrogen atoms, whereas the other two are sp^2 hybridized with no

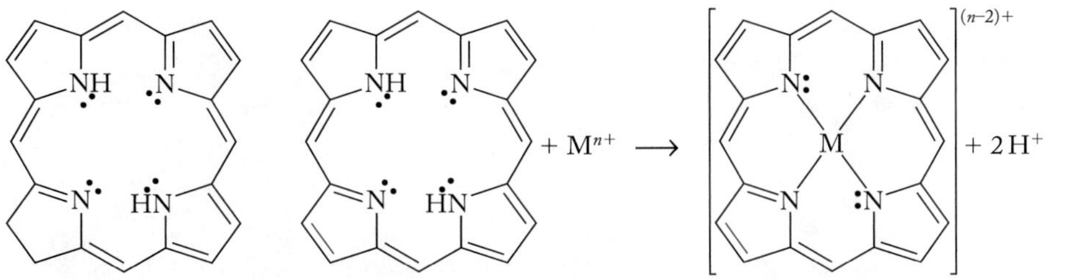

FIGURE 23.25 Skeletal structures of (a) chlorin and (b) porphyrin. (In these structures the bonds formed by carbon atoms are shown but not the atoms themselves or their bonds to hydrogen atoms.) The principal difference between the structures is a C=C in the porphyrin structure (shown in red) that is a single bond in chlorin rings. The innermost atoms in each ring are four nitrogen atoms with lone pairs of electrons. All four N atoms form coordinate bonds with a metal ion, as shown with the porphyrin ring in (c).

hydrogen atoms. When either compound forms coordinate bonds with a metal ion M^{n+} (Figure 23.25c), the two hydrogen atoms ionize, giving the ring a charge of 2− and the complex ion an overall charge of $(n - 2)$. The lone pairs of electrons on the N atoms in the ionized structure are oriented toward the ring center. These lone pairs can occupy either the four equatorial coordination sites in an octahedral complex ion or all four coordination sites in a square planar complex ion. In octahedral complex ions, each central metal ion still has two axial sites available for bonding to other ligands. Depending on the charge of the central ion, the coordination compound may be either ionic (a complex ion) or electrically neutral.

Porphyrin and chlorin rings are widespread in nature and play many biochemical roles. Their chemical and physical properties depend on the following:

1. the identity of the central metal ion.
2. the species that occupy the axial coordination sites of octahedral complexes.
3. the number and identity of organic groups attached to the outside of the ring.

The role of the major essential element Mg^{2+} in photosynthesis was described in Chapter 22. The ultratrace essential element manganese also plays a role in the production of oxygen during photosynthesis. To examine that role, let's write an equation for photosynthesis that is slightly different from the equation we are used to seeing. Instead of writing the formula $C_6H_{12}O_6$ for glucose, we use the generic carbohydrate formula $(CH_2O)_n$ so that the coefficient is 1 for all other species in the reaction:

$$H_2O(\ell) + CO_2(g) \rightarrow \tfrac{1}{n}(CH_2O)_n(aq) + O_2(g)$$

Writing the equation in this form makes it easier to see how the overall reaction between water and carbon dioxide involves electron transfer: photosynthesis is a redox reaction in which oxygen is oxidized and carbon is reduced.

The redox process in photosynthesis is catalyzed by manganese-containing enzymes in which Mn in the +3 and +4 oxidation states mediates the transfer of electrons from water molecules. Although the exact structure of these manganese compounds remains undetermined, it is believed that two Mn(III) ions and two Mn(IV) ions are present at the site of O_2 production.

CONCEPT **TEST**

Manganese ions involved in photosynthesis are often surrounded by six ligands in an octahedral geometry. Is reduction of a Mn(IV) ion to Mn(III) in an octahedral complex (see Section 23.5) accompanied by a change in spin state of the Mn ion?

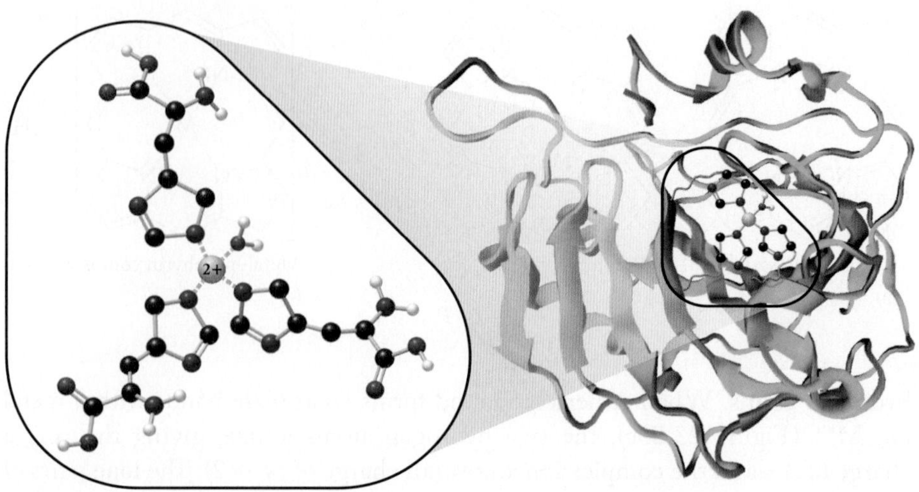

FIGURE 23.26 The active site of one form of carbonic anhydrase consists of a zinc ion (gray sphere) bonded to three histidine molecules and one H_2O molecule. The OH^- ion produced when this H_2O ionizes combines with a molecule of CO_2, forming a HCO_3^- ion.

Zinc and Carbonic Anhydrase Many of the enzymes in our bodies are *metalloenzymes* that contain transition metal ions. Among them is *carbonic anhydrase*, which catalyzes the reaction between water and carbon dioxide to form bicarbonate ions:

$$H_2O(\ell) + CO_2(aq) \rightleftharpoons HCO_3^-(aq) + H^+(aq)$$

The α-form of carbonic anhydrase contains 260 amino acid residues and a zinc ion at the active site. Note in Figure 23.26 that the zinc ion is coordinately bonded to three nitrogen atoms on histidine side chains and to one molecule of water. The presence of these ligands and a fourth histidine nearby facilitates ionization of the water molecule. Ionization leaves a OH^- ion attached to the Zn^{2+} ion and a H^+ ion bonded to the side-chain nitrogen atom of the fourth histidine. In addition, a space just the right size and shape to accommodate a CO_2 molecule is next to the active site. When a CO_2 molecule in this space bonds to a hydroxide ion, an HCO_3^- ion forms. As the bicarbonate ion pulls away, another water molecule occupies the fourth coordination site on the Zn^{2+} ion, another CO_2 molecule enters, the histidine is protonated, and the catalytic cycle repeats. This reaction is important because it helps eliminate CO_2 from cells during respiration and mediates the uptake of CO_2 during photosynthesis in some plants.

Iron and Cytochromes Iron-containing peroxidases and catalases are integral to the transfer of oxygen to biomolecules. For example, plants use a fatty acid peroxidase to catalyze the stepwise degradation of fatty acids. One $-CH_2-$ group at a time is removed from the fatty acid using hydrogen peroxide:

$$\underset{\text{Fatty acid}}{R{-}CH_2{-}COOH(aq)} + \underset{\text{Hydrogen peroxide}}{2\,H_2O_2(aq)} \xrightarrow{\text{fatty acid peroxidase}} 3\,H_2O(\ell) + \underset{\text{Aldehyde}}{R{-}CHO(aq)} + CO_2(aq)$$

Subsequent oxidation of the aldehyde back to a carboxylic acid yields a new fatty acid with one fewer $-CH_2-$ group:

$$\underset{\text{Aldehyde}}{2\,R{-}CHO(aq)} + O_2(aq) \rightarrow \underset{\text{Fatty acid}}{2\,RCOOH(aq)}$$

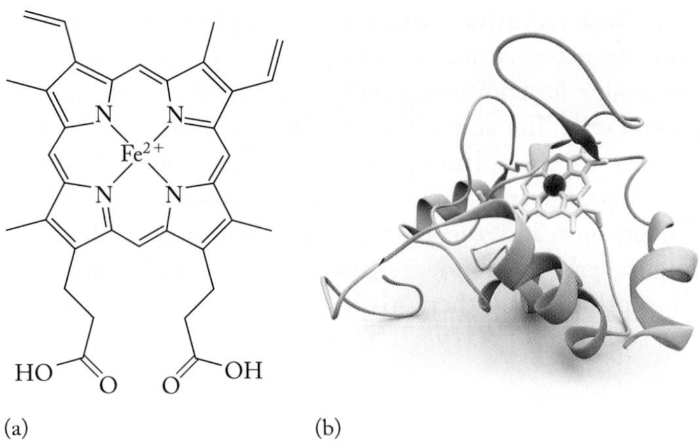

FIGURE 23.27 (a) Heme is the specific porphyrin that binds iron in the oxygen-transport protein hemoglobin and also in many enzymes of great significance in all life on Earth. (b) The structure of cytochrome proteins, such as cytochrome c shown here, includes one or more heme complexes (shown in gray) that mediate energy production and redox reactions in living cells. Different cytochromes have different ligands occupying the fifth and sixth octahedral coordination sites, and different groups in the protein may be attached to the porphyrin ring.

Iron-containing enzymes also catalyze the reduction of nitrite (NO_2^-) and sulfite (SO_3^{2-}) ions:

$$NO_2^-(aq) + 6\ e^- + 8\ H^+(aq) \xrightarrow{\text{nitrite reductase}} NH_4^+(aq) + 2\ H_2O(\ell)$$

$$SO_3^{2-}(aq) + 6\ e^- + 7\ H^+(aq) \xrightarrow{\text{sulfite reductase}} HS^-(aq) + 3\ H_2O(\ell)$$

In each case, the iron in the enzyme is oxidized, providing the electrons needed for reduction.

Proteins called *cytochromes* also contain one or more heme groups (Figure 23.27a). Cytochromes mediate oxidation and reduction processes connected with energy production in cells. The heme group conveys electrons as the half-reaction

$$Fe^{3+} + e^- \rightleftharpoons Fe^{2+}$$

rapidly and reversibly consumes or releases electrons needed in the biochemical reactions that sustain life. Cytochromes catalyze electron transport in photosynthesis and the metabolism of glucose to CO_2 and water. Their role in electron transport makes them key participants in *in vivo* reactions that produce ATP, which is required for intracellular energy production and transfer in living things. As catalysts for oxidation and reduction, cytochromes are essential for the removal of toxic substances and for the process of programmed cell death, both of which are required for the health of multicellular organisms. Cytochrome c (Figure 23.27b) is a component of the electron transport chain in mitochondria that is ubiquitous in life-forms ranging from microbes to human beings. Its amino acid sequence is very similar or the same across the spectrum of plants, animals, and many unicellular organisms. Consequently, studies of the cytochrome c molecule are seminal in evolutionary biology.

Many other kinds of cytochrome proteins have different substituents on the porphyrin rings and different axial ligands, each of which influences the function of the complex. This last point has been repeated several times in this chapter: the chemical properties and biological functions of transition metals that are essential to living organisms are linked to their molecular environments and to the formation of stable complex ions with ligands that are strong electron donors—that is, strong Lewis bases.

CONCEPT TEST

Which of the following small molecules could not function as a Lewis base and hence would not be expected to bond to iron in a heme protein? CO; H_2; NO_2; H_2S

Molybdenum and Vanadium in Metalloenzymes Many metalloenzymes are involved in transformations of nitrogen. Molybdenum-containing reductases are responsible for converting NO_3^- ions to NO_2^- ions and then to NH_3 in the nitrogen cycle. The active site of sulfite oxidase, which converts SO_3^{2-} ions to SO_4^{2-} ions in the sulfur cycle, also contains molybdenum. Sometimes more than one type of transition metal is found in a metalloenzyme. For example, both molybdenum and iron are required by xanthine oxidase, an important enzyme along the pathway for degradation of excess nucleic acids (adenine and guanine) to xanthine and then to uric acid for elimination through the kidneys.

$$\text{Xanthine} + H_2O + O_2 \xrightarrow[H^+]{\text{xanthine oxidase}} \text{Uric acid} + H_2O_2$$

Xanthine   Uric acid

The combination of iron and vanadium is essential to the function of haloperoxidases, a class of enzymes found in some algae, lichens, and fungi that replace C—H bonds with carbon–halogen bonds. The reaction products are thought to function in the defense systems of these organisms.

Copper and Superoxide Dismutase Copper-containing proteins perform several functions in both plants and animals, including oxygen transport in mollusks such as clams and oysters. Copper is also an essential element in the enzymes azurin and plastocyanin, which mediate electron transfer during photosynthesis.

Reactions catalyzed by xanthine oxidase may produce other reactive oxygen species besides H_2O_2. One of them is the superoxide ion, O_2^-. Superoxide ion is a strong oxidizing agent and must be eliminated to prevent cell damage. The removal of superoxide begins with the action of superoxide dismutases, which convert superoxide to hydrogen peroxide:

$$2\,O_2^-(aq) + 2\,H^+(aq) \xrightarrow{\text{superoxide dismutase}} H_2O_2(aq) + O_2(aq)$$

Researchers have isolated superoxide dismutases containing a variety of transition metals, including a copper–zinc enzyme. The hydrogen peroxide produced in this reaction is decomposed to water and oxygen by iron-containing catalase.

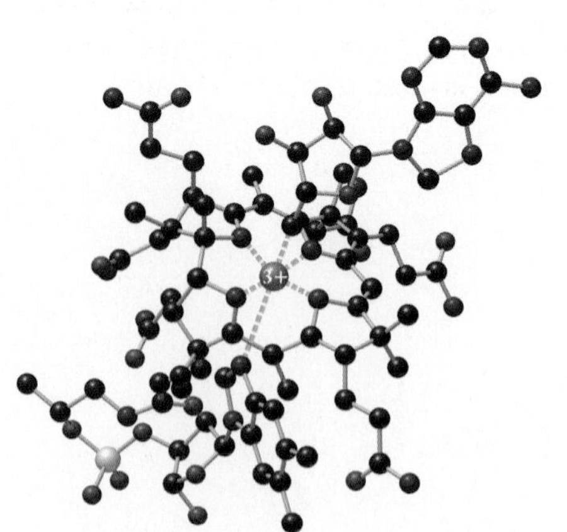

FIGURE 23.28 Coenzyme B_{12} contains cobalt(III), an ultratrace essential metal.

Nickel and Urease Ureases, the enzymes responsible for the conversion of urea to ammonia in plants, contain nickel(II). Nickel is also found with iron in enzymes called *hydrogenases*. Hydrogenases oxidize hydrogen gas to protons:

$$H_2(g) \xrightarrow{\text{hydrogenase}} 2\,H^+(aq) + 2\,e^-$$

Nickel and iron also combine in CO dehydrogenase, an enzyme that catalyzes the formation of acetyl-CoA, which is a key component in the tricarboxylic acid cycle discussed in Chapter 20. Finally, nickel-containing enzymes are found among the enzymes responsible for methane generation by bacteria.

Cobalt and Coenzymes Many enzymes require the presence of a **coenzyme**, which is an organic compound that cocatalyzes a biochemical reaction. The coenzyme B_{12} (Figure 23.28) contains the ultratrace essential element cobalt(III) and is a derivative of vitamin B_{12}. The cobalt(III) in coenzyme B_{12} is easily reduced to cobalt(II) and even cobalt(I) in the course of enzyme-catalyzed redox reactions.

coenzyme an organic molecule that, like an enzyme, accelerates the rate of biochemical reactions.

The change in oxidation state of Co allows for facile transfer of methyl groups, as in the conversion of methionine to homocysteine:

$$\text{Methionine} \longrightarrow \text{Homocysteine}$$

Coenzyme B_{12} is also critical to the function of mutases, which are enzymes that catalyze the rearrangement of the skeleton of a molecule, as in the interconversion of glutamate and methylaspartate:

$$\text{Glutamate} \xrightleftharpoons[\text{coenzyme } B_{12}]{\text{glutamate mutase}} \text{Methylaspartate}$$

Chromium and Diabetes Chromium in the +3 oxidation state is an ultratrace essential element in our diets. It is involved in regulating glucose levels in the blood through a molecule called chromodulin. Chromodulin is a polypeptide incorporating only 4 of the 20 naturally occurring amino acids: glycine, cysteine, glutamic acid, and aspartic acid. Four Cr^{3+} ions are bound to the peptide chain. Cereals and grains contain enough chromium for our daily needs, but certain plants (such as shepherd's purse) concentrate chromium and have been used as herbal remedies in diabetes treatment.

Chromium in the +6 oxidation state, as found in chromate ions (CrO_4^{2-}), is acutely toxic and also carcinogenic. Chromate ion enters cells through ion channels that transport SO_4^{2-} ions. Once inside, CrO_4^{2-} is reduced to Cr^{3+}, which binds to the phosphate backbone of DNA.

CONNECTION Magnetic resonance imaging (MRI) was mentioned in Chapter 3 in our discussion of the use of cryogenic helium for superconducting magnets.

23.8 Coordination Compounds in Medicine

Transition metal coordination compounds are becoming important in both the diagnosis and treatment of diseases (Figure 23.29). Any diagnostic or therapeutic compound that is injected intravenously must be sufficiently soluble in blood to be delivered to the target. While in transit, the compound must be stable enough not to undergo chemical reactions that result in its precipitation or rapid elimination from the body. Occasionally, the compound can be in the form of a simple salt, but more often a metal ion is introduced as a coordination complex or coordination compound. Ligands used in forming biologically active coordination complexes include amino acids and simple anions like the citrate ion. Chelating ligands like diethylenetriaminepentaacetate ($DTPA^{5-}$; Figure 23.30) are often used in biological applications. A medicinal chemist can also take advantage of substances that occur naturally in the body, such as antibodies, to transport a diagnostic or therapeutic metal ion to its target.

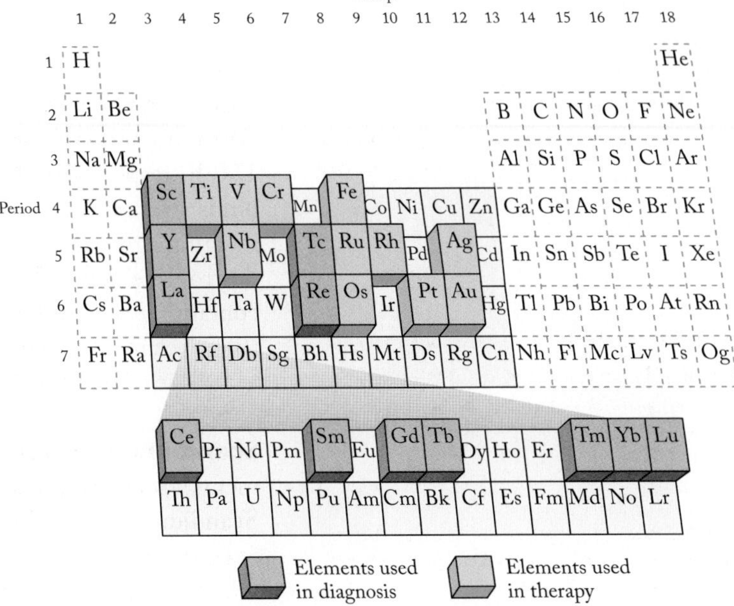

FIGURE 23.29 The elements shown in red are used in diagnostic imaging and those shown in green are used in therapy.

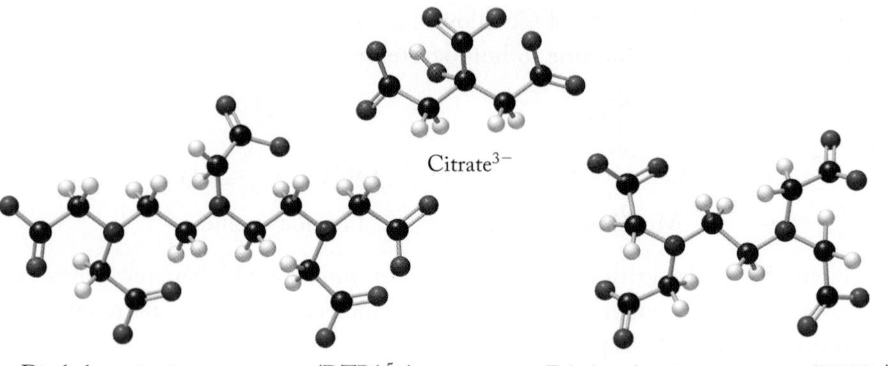

FIGURE 23.30 Diethylenetriaminepentaacetate ($DTPA^{5-}$), citrate^{3-}, and ethylenediaminetetraacetate ($EDTA^{4-}$) are often used as chelating ligands for diagnostic and therapeutic agents based on transition metals. These ions form stable complex ions with 2+ and 3+ metal cations. The solubilities of the complex ions are typically much greater than those of the hydrated ions at physiological pH (7.4).

Transition Metals in Medical Imaging and Diagnosis

CONNECTION Chapter 21 gave a more detailed discussion of nuclear chemistry and nuclear medicine, including an assessment of the effects of different types of radiation on living tissue.

Technetium, just below manganese in group 7 of the periodic table, is the most widely used radioactive isotope in medical imaging. Technetium is unusual in that it does not exist naturally on Earth in easily measurable amounts because it has no stable isotopes; in other words, all technetium isotopes are radioactive. These isotopes can be produced in nuclear reactors for use in medicine. The ^{99m}Tc used in hospitals for imaging is prepared in technetium generators in which a stable isotope of molybdenum, ^{98}Mo (23.78% natural abundance), is bombarded with neutrons. Technetium-99 has a half-life greater than 20,000 years; however, when it is produced in a nuclear reactor, its nucleus is in an excited state, called a *metastable* nucleus. The metastable state, designated by adding the letter "m" to the mass number, as in technetium-99m or ^{99m}Tc, has a half-life of six hours ($t_{1/2}$ of ^{99m}Tc = 6.0 h) and decays to the more stable technetium-99 nucleus.

Technetium-99m has been widely used as an imaging agent because it has a short half-life and emits low-energy γ rays. Patients can be injected with a variety of technetium compounds, depending on the target organ. For imaging the heart, the coordination compounds shown in Figure 23.31 are used.

The ability to target the delivery of radionuclides to a particular organ opens the possibility for selective irradiation of a tumor located in that organ. Therefore some radionuclides can be used to not only image but also to treat cancers. Certain tumors have highly selective receptor sites for particular molecules on their surfaces. Rhenium, for example, is being studied as both an imaging and a therapeutic agent for some tumors, including breast, liver, and skin cancer. Rhenium-186 and rhenium-188 undergo β decay with half-lives of 3.72 d and 17.0 h, respectively. By including a radioactive rhenium ion in a molecule that binds strongly and specifically to these receptors, physicians can deliver both an imaging agent and a therapeutic agent to the tumor. In principle, the β particles destroy the tumor. Several patents have been issued for the use of compounds containing rhenium isotopes, but therapies based on these compounds remain in the experimental stage. One example of a rhenium compound used in these applications is shown in Figure 23.32.

Scandium, Yttrium, and Lanthanide Elements Isotopes of two group 3 elements, scandium and yttrium, are used in medical imaging and therapy. Scandium-46 has been used to image the spleen, and scandium-47 shows promise

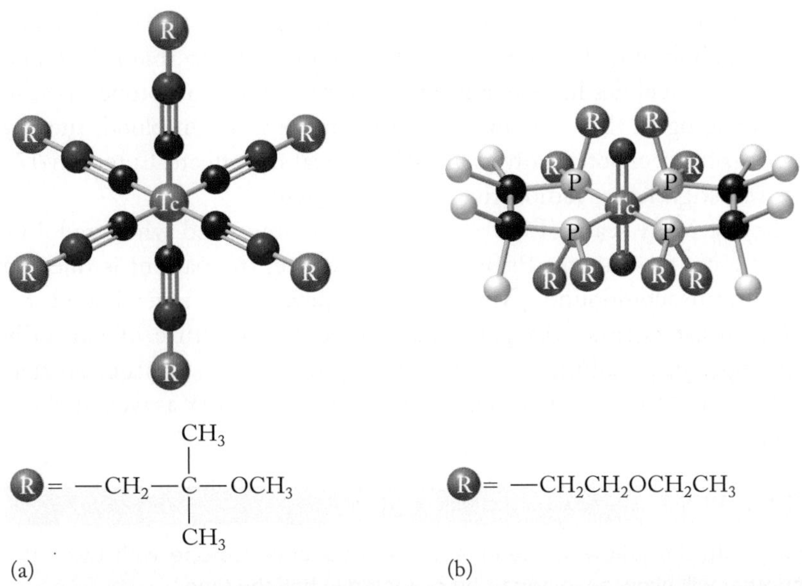

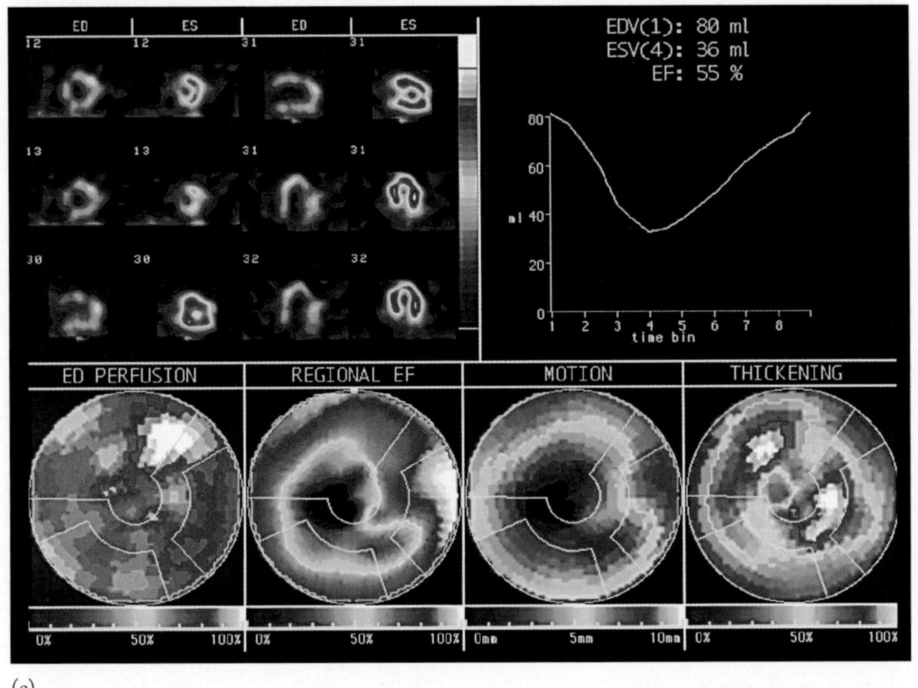

FIGURE 23.31 (a) Cardiolite [$Tc(CNR)_6$] and (b) Myoview [$TcO_2(RPCH_2CH_2PR)_2$] are used for imaging the heart. (c) Images of a patient's heart after intravenous injection with a technetium-containing drug. The images along the bottom show four measurements of heart function. Radioactive technetium compounds emit gamma rays that allow the blood to be tracked as it is pumped through the heart.

in diagnosing breast cancer. Yttrium-90 has been used to image a variety of tumors, including intestinal, breast, and thyroid cancers. Chelating ligands are used to complex the $^{90}Y^{3+}$ cation and deliver it to the tumor, where it binds strongly to receptors on the surface of the tumor. Yttrium-90 agents for treatment of non-Hodgkin's lymphoma are currently in clinical trials.

Radioactive isotopes of several lanthanides have been used in scintigraphic imaging, a procedure that uses a scintillation counter (Chapter 21) to measure the light emitted when radiation strikes a phosphor-coated screen in the instrument. The intensity of the emitted light is translated into a three-dimensional image of the organ of interest in a manner similar to the PET technique described in

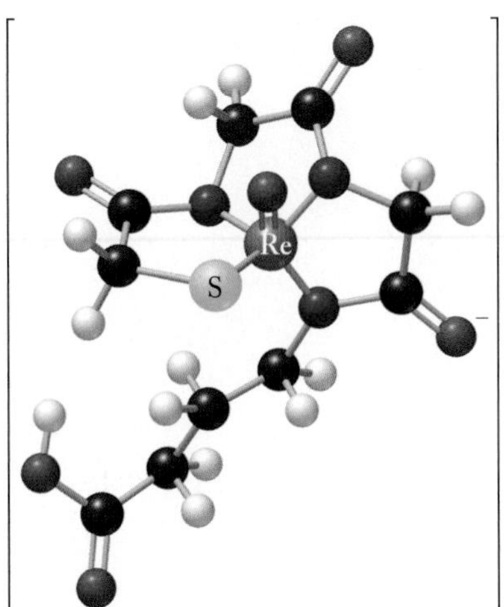

FIGURE 23.32 This rhenium–mercaptoacetylglycylglycyl-γ-amino acid complex is used to attach ^{186}Re or ^{188}Re to monoclonal antibodies or peptides.

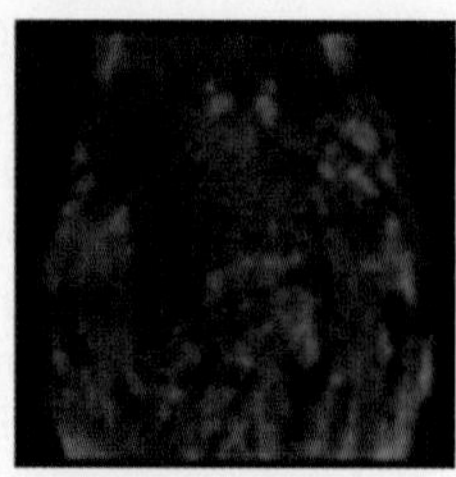

(a) Unenhanced

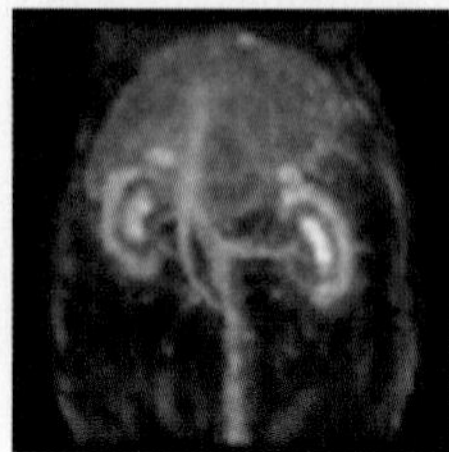

(b) Enhanced

FIGURE 23.33 MRI scans made (a) without a contrast agent and (b) enhanced by a gadolinium contrast agent.

Chapter 21. Among the radionuclides used for this purpose are cerium-141, samarium-153, gadolinium-153, terbium-160, thulium-170, ytterbium-169, and lutetium-177. These nuclides have an advantage over the gallium isotopes used in scintigraphic imaging in that they have lower retention times in blood, thereby reducing the possibility of potentially harmful biological side effects. Lutetium-177 is also under investigation as radioactive antitumor agent.

The quality of MRI scans (Figure 23.33) can be improved when gadolinium is used as a contrast agent. Prior to the procedure, the patient is injected with a gadolinium compound. Many ligands have been investigated for Gd^{3+} MRI contrast agents. The gadolinium used is a mixture of naturally occurring isotopes of gadolinium. Coordination compounds of other lanthanide ions (Dy^{3+} and Ho^{3+}) have been evaluated but do not work as well as those of gadolinium.

CONCEPT **TEST**

What is wrong with the following statement? "A radioactive isotope with twice the half-life of another will allow an image to be collected in half the time."

Transition Metals in Therapy

Like main group metal compounds described in Chapter 22, coordination compounds of the transition metals are finding their way into our expanding arsenal of useful therapeutic agents. We have already described the use of platinum compounds to treat cancer and the potential for chromium compounds to help mitigate diabetes. In this section we examine additional therapeutic agents that contain transition metals.

Iron The body must regulate the amount of iron in cells in order to produce enough hemoglobin to maintain good health. Deficiencies in iron lead to several diseases broadly classified as anemia. Mild forms of anemia are common among women of childbearing age and are treated with oral iron supplements. However, a more serious genetic form of anemia, known as thalassemia, must be treated with blood transfusions that leave patients with too much iron in their blood. To remove the excess iron, these patients are treated with chelating ligands that complex some of the iron and transport it out of the red blood cells and eventually out of the body in what is known as chelation therapy. Chelation therapy relies on a ligand exchange reaction:

$$\text{FeL}(aq) + \text{L}'(aq) \rightleftharpoons \text{FeL}'(aq) + \text{L}(aq)$$

in which iron bonded to ligand L (such as hemoglobin) in the blood is treated with ligand L′, which also binds to iron. If the formation constant for the FeL′ complex is greater than for FeL, then FeL dissociates and FeL′ forms. The ligand L′ is chosen so that the complex formed between iron and L′ is eliminated from the body, thereby reducing the concentration of iron and relieving the symptoms caused by thalassemia treatments. Often the chelating ligand forms chiral metal complexes in which one enantiomer of the iron complex often works better than the other in the treatment of diseases such as thalassemia. Chelation therapy is also the treatment of choice for heavy metal poisoning from toxic metals, including lead and mercury.

SAMPLE EXERCISE 23.6 Calculating the Equilibrium Constant of a Ligand Exchange Reaction **LO7**

The protein transferrin is involved in the transport of iron into cells. Iron accumulation in cells has been implicated in diseases such as Parkinson's, Alzheimer's, and thalassemia. The chelating ligand deferoxamine (DFO) is used to treat thalassemia; the complex between iron and DFO is eliminated from the body. Given the formation constants for the complexation of Fe(III) by transferrin (Equation A) and the reaction of deferoxamine with iron (Equation B), calculate the equilibrium constant for the ligand exchange reaction between deferoxamine and transferrin (Equation C).

$$Fe^{3+}(aq) + \text{transferrin}(aq) \rightleftharpoons \text{Fe(transferrin)}^{3+}(aq) \qquad K_{f,A} = 4.7 \times 10^{20} \quad (A)$$

$$Fe^{3+}(aq) + \text{DFO}(aq) \rightleftharpoons \text{Fe(DFO)}^{3+}(aq) \qquad K_{f,B} = 4.0 \times 10^{30} \quad (B)$$

$$\text{Fe(transferrin)}^{3+}(aq) + \text{DFO}(aq) \rightleftharpoons \text{Fe(DFO)}^{3+}(aq) + \text{transferrin}(aq) \qquad K_C = ? \quad (C)$$

Collect and Organize We are given the formation constants of two reactions involving iron and different ligands and are asked to calculate the equilibrium constant for the ligand exchange reaction. You may wish to refer to Chapter 16 for calculations involving formation constants of complexes, as well as to Chapters 14 and 15, where we manipulated equilibrium constants.

Analyze Transferrin is a reactant in Equation A and a product in the exchange reaction (C). Therefore, we need to reverse Equation A before adding it to Equation B to obtain Equation C. Combining the reverse of Equation A with Equation B to obtain Equation C means multiplying $(1/K_{f,A})$ and $K_{f,B}$ together to obtain K_C. The reciprocal of $K_{f,A}$ has a value of about 10^{-20}. Therefore, the value of K_C should be about $10^{-20} \times 10^{30}$, or about 10^{10}.

Solve Multiplying $(1/K_{f,A})$ by $K_{f,B}$ to obtain K_C:

$$K_C = \frac{1}{4.7 \times 10^{20}} \times (4.0 \times 10^{30}) = 8.5 \times 10^{9}$$

Think About It The equilibrium constant for the ligand exchange reaction is indeed $\sim 10^{10}$ and illustrates why DFO is an effective treatment for thalassemia—removing iron from the body.

Practice Exercise The equilibrium constants for the reactions of penicillamine and methionine with methyl mercury are

$$CH_3Hg^+ + \text{penicillamine} \rightleftharpoons CH_3Hg(\text{penicillamine})^+ \qquad K_f = 6.3 \times 10^{13}$$

$$CH_3Hg^+ + \text{methionine} \rightleftharpoons CH_3Hg(\text{methionine})^+ \qquad K_f = 2.5 \times 10^{7}$$

Calculate the equilibrium constant for the exchange reaction:

$$CH_3Hg(\text{methionine})^+ + \text{penicillamine} \rightleftharpoons CH_3Hg(\text{penicillamine})^+ + \text{methionine}$$

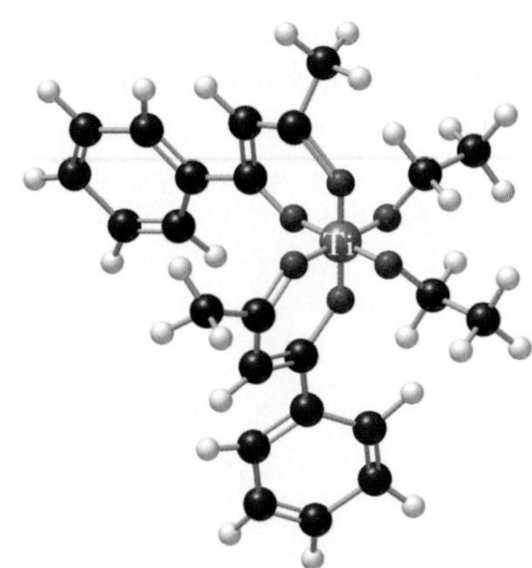

Budotitane

FIGURE 23.34 Budotitane is a titanium(IV) complex that shows activity against colon cancer.

Titanium, Vanadium, and Niobium Cancer Drugs Compounds of a large number of transition metals demonstrate antitumor activity. Titanium(IV) complexes such as budotitane (Figure 23.34) show promise against colon cancer. Encouraging results against breast, lung, and colon cancers were observed with a series of compounds with formulas $(C_5H_5)_2TiCl_2$, $(C_5H_5)_2VCl$, and $(C_5H_5)_2NbCl_2$. Unfortunately, at therapeutically useful doses the potential for liver

organometallic compound a molecule containing direct carbon–metal covalent bonds.

damage outweighs the benefits of these compounds, which contain carbon–metal bonds and belong to a class of substances called **organometallic compounds.**

Vanadium, Chromium, and Insulin Activity Insulin is a hormone needed for proper glucose metabolism and protein synthesis. People with diabetes either are unable to produce insulin, or they produce the hormone but are unable to use it effectively. The suggestion that vanadium plays a role in insulin production has prompted investigation of vanadium compounds as oral diabetes drugs. Encouraging results from animal studies using bis(allixinato)oxovanadium(IV) (Figure 23.35a) have been reported, but the compound is not currently approved for use in humans. Chromium(III) compounds have been shown to lower fasting blood sugar levels and reduce the amount of insulin required by some diabetics (Figure 23.35b).

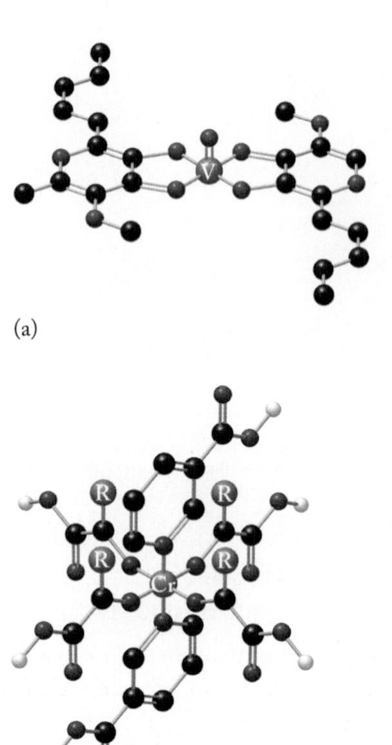

FIGURE 23.35 (a) Vanadium(IV) compounds like bis(allixinato) oxovanadium(IV) can act as insulin mimics. (b) The chromium(III) complex in a glucose tolerance factor contributes to our bodies' ability to regulate insulin levels. The light brown spheres are the side chain R groups of the amino acids in the structure. (To simplify the structures, the hydrogen atoms bonded to carbon atoms are not shown.)

Platinum Group and Coinage Metals The period 5 and period 6 elements in groups 8, 9, and 10 (Ru, Rh, Pd, Os, Ir, and Pt) are often referred to as the *platinum group metals.* The group 11 elements of these two periods (Ag and Au) along with Cu are called the *coinage metals.* Compounds of these metals are used in medications for arthritis, cancer, and other diseases.

The serendipitous discovery in the 1960s that *cis*-diamminedichloroplatinum(II) (cisplatin in Figure 23.36) is effective in treating testicular, ovarian, and other cancers spurred the development of a host of cancer drugs based on both platinum group metals and coinage metals. The results of this research include a compound known as carboplatin that shows the same activity as cisplatin but has fewer side effects. In addition to platinum compounds, the antitumor activities of complexes of gold, rhodium, ruthenium, and silver have been explored. The effectiveness of all of these drugs lies in their ability to bind to the nitrogen-containing bases in DNA and inhibit cell replication. If their cells cannot divide, tumors cannot grow. The greater toxicity of rhodium compounds relative to those of platinum and ruthenium has limited their clinical application.

Selected osmium compounds reduce inflammation in joints resulting from arthritis, although the use of such compounds has diminished with the development of other anti-inflammatory agents. The therapeutic effects of aqueous solutions of osmium tetroxide, OsO_4, were first investigated in the 1950s. The use of osmium tetroxide was superseded by the use of glucose polymers containing osmium, known as osmarins. The use of osmarins reduces the toxic effects of osmium and illustrates how even toxic metals can be adapted to therapy.

Although the historic use of gold for medicinal purposes dates back millennia, the effective use of gold-containing pharmaceuticals originated with the discovery in the 1920s and 1930s that a gold thiosulfate compound, sanochrysin, alleviates the symptoms of rheumatoid arthritis. The most commonly used gold drugs for the treatment of arthritis today are sold under the names myochrysine and auranofin (Figure 23.37). Myochrysine is injected; auranofin can be taken orally.

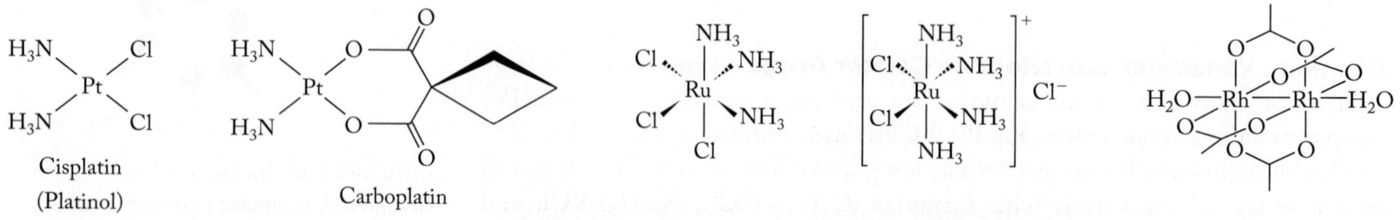

Cisplatin (Platinol)

Carboplatin

FIGURE 23.36 Cisplatin and carboplatin are effective antitumor agents. The ruthenium and rhodium compounds show activity against leukemia but have not yet seen widespread use.

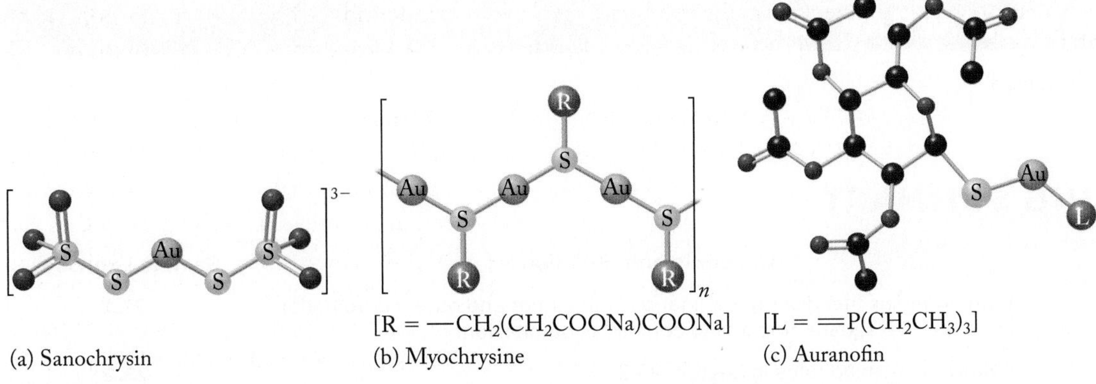

FIGURE 23.37 (a) Sanochrysin, (b) myochrysine, and (c) auranofin are three gold-containing drugs used to treat arthritis.

Eye drops containing silver salts are used to treat eye infections. Silver sulfadiazine is a broad-spectrum "sulfa" drug used to prevent and treat bacterial and fungal infections (Figure 23.38). It is also an active ingredient in creams used to treat thermal and chemical burns.

Although present in smaller amounts in the human body than the main group elements, the transition metals play crucial roles in our physiological system and in those of many organisms. Their rich and diverse chemistry has led to many applications of both essential and nonessential transition metal compounds to maintaining health and diagnosing and treating disease. Without a doubt, the smorgasbord of the periodic table will continue to provide us with useful substances and materials.

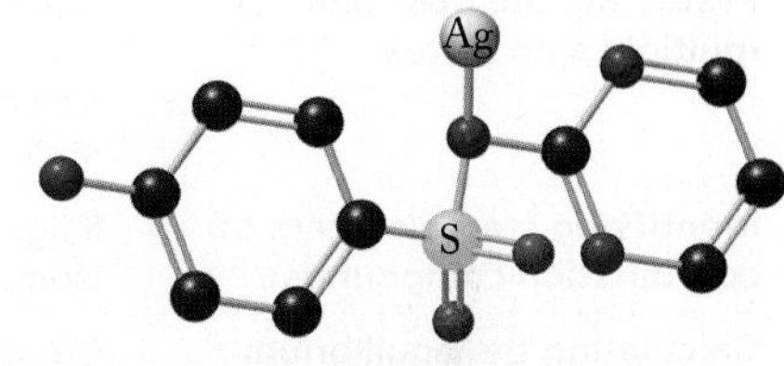

Silver sulfadiazine

FIGURE 23.38 Silver sulfadiazine is an effective antibiotic when applied to burns. (H atoms are not shown.)

SUMMARY

LO1 Molecules or anions that function as Lewis bases and form coordinate bonds with metal cations are called *ligands*. The resulting species, which are composed of central metal ions and the surrounding ligands, are called *complex ions* or simply complexes. Direct bonding to a central cation means that the ligands in a complex occupy the **inner coordination sphere** of the cation. (Section 23.1)

LO2 The names of complex ions and coordination compounds provide information about the identities and numbers of ligands, the identity and oxidation state of the central metal ion, and the identity and number of counter ions. (Section 23.2)

LO3 A **monodentate ligand** donates one pair of electrons in a complex ion; a **polydentate ligand** donates more than one pair in a process called **chelation.** Polydentate ligands are particularly effective at forming complex ions. This phenomenon is called the **chelate effect.** EDTA is a particularly effective sequestering agent, which is a chelating agent that prevents metal ions in solution from reacting with other substances.(Section 23.3)

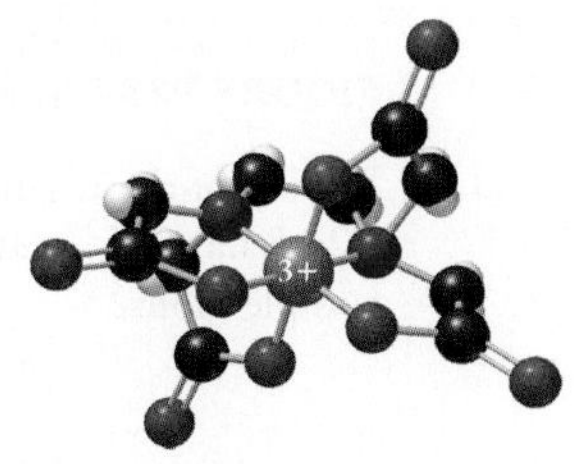

LO4 The colors of transition metal compounds can be explained by the interactions between electrons in different *d* orbitals and the lone pairs of electrons on surrounding ligands. These interactions create a **crystal field splitting** of the energies of the *d* orbitals. A **spectrochemical series** ranks ligands on the basis of their field strength and the wavelengths of electromagnetic radiation absorbed by their complex ions. (Section 23.4)

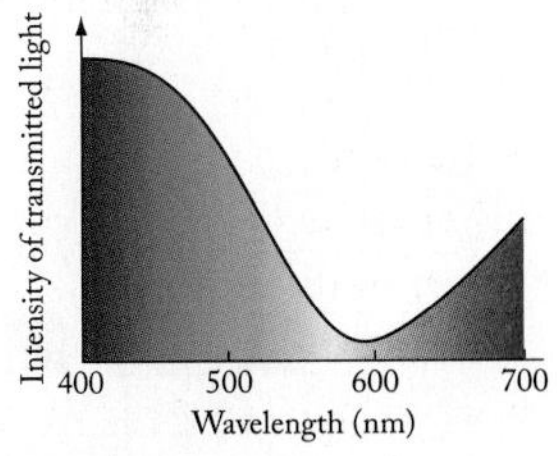

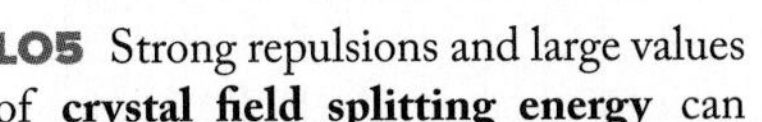

LO5 Strong repulsions and large values of **crystal field splitting energy** can lead to electron pairing in lower-energy orbitals and an electron configuration called a low-spin state. Metals and their ions with *d* electrons evenly distributed across all the *d* orbitals in the valence shell represent a high-spin state. (Section 23.5)

LO6 Complex metal ions containing more than one type of ligand may form stereoisomers. When one type occupies two adjacent corners of a square planar complex, the complex is a *cis* isomer; when the same ligand occupies opposite corners, it is a *trans* isomer. (Sections 23.6 and 23.7)

LO7 Many enzymes contain transition metal ions. Soluble coordination compounds and organometallic compounds of the transition metals appear in enzymes and are used for imaging and in medications for arthritis, cancer, and other diseases. (Section 22.8)

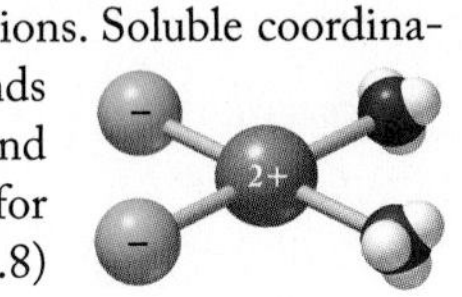

PARTICULATE PREVIEW WRAP-UP

Nitrogen; ammonia, 4; ethylenediamine, 3

PROBLEM-SOLVING SUMMARY

Type of Problem	Concepts and Equations	Sample Exercises
Writing formulas of coordination compounds	Write formulas that distinguish ligands in the inner and outer coordination spheres. Use square brackets to contain a complex ion.	**23.1**
Naming coordination compounds	Follow the naming rules in Section 23.2	**23.2**
Identifying potential electron-pair–donor groups in a molecule	Examine the molecular structure for nitrogen, oxygen, or other elements with lone pairs of valence electrons.	**23.3**
Predicting the possibility of multiple spin states	Sketch a *d*-orbital diagram based on crystal field splitting. Fill the lowest-energy orbitals with one valence electron each. If there are two ways of adding the additional valence electrons to the diagram, then two spin states are possible.	**23.4**
Identifying stereoisomers of coordination compounds	If ligands of one type are all on the same side of the complex ion, it is a *cis* isomer. If ligands of one type are on opposite sides, it is a *trans* isomer.	**23.5**
Calculating the equilibrium constant of a ligand exchange reaction	Combine the equilibrium constants for the overall reaction using the guidelines in Chapter 14.	**23.6**

VISUAL PROBLEMS

(Answers to boldface end-of-chapter questions and problems are in the back of the book.)

23.1. Two of the four highlighted elements in Figure P23.1 have cations which form colored compounds with Cl^-. Which ones?

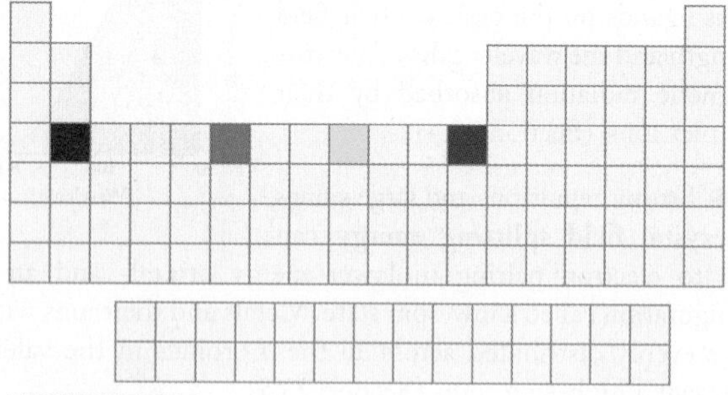

FIGURE P23.1

23.2. Which of the highlighted transition metals in Figure P23.2 form M^{2+} cations that cannot have high-spin and low-spin states?

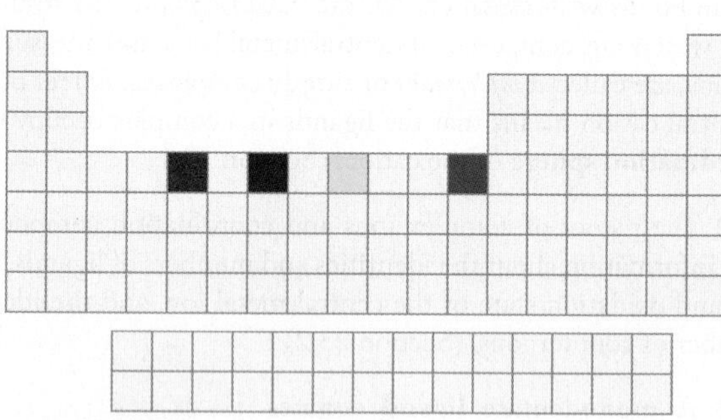

FIGURE P23.2

23.3. Which of the highlighted transition metals in Figure P23.2 have M^{2+} cations that form colorless tetrahedral complex ions?

23.4. Smoky quartz has distinctive lavender and purple colors due to the presence of manganese impurities in crystals of silicon dioxide. Which of the orbital diagrams in Figure P23.4 best describes the Mn^{2+} ion in a tetrahedral field?

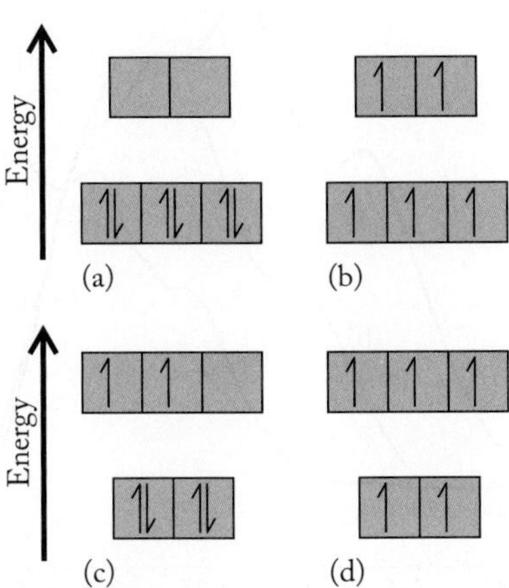

FIGURE P23.4

23.5. **Chelation Therapy I** The compound with the structure shown in Figure P23.5 is widely used in chelation therapy to remove excessive lead or mercury in patients exposed to these metals. How many electron-pair-donor groups ("teeth") does the sequestering agent have when the carboxylic acid groups are ionized?

HO–C(=O)–CH(SH)–CH(SH)–C(=O)–OH

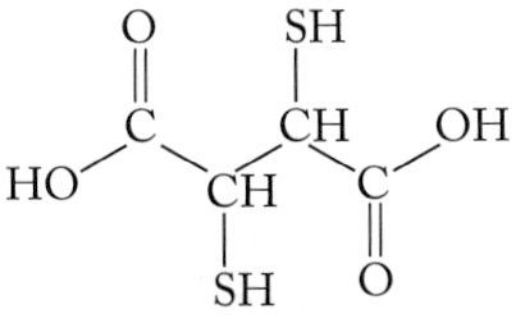

FIGURE P23.5

23.6. **Chelation Therapy II** The compound with the structure shown in Figure P23.6 has been used to treat people exposed to plutonium, americium, and other actinide metal ions. How many electron pair donor groups does the sequestering agent have when the carboxylic acid groups are ionized?

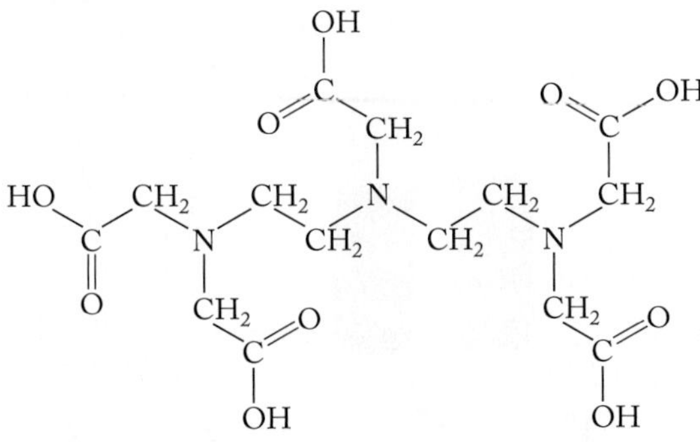

FIGURE P23.6

23.7. The three beakers shown in Figure P23.7 contain solutions of $[CoF_6]^{3-}$, $[Co(NH_3)_6]^{3+}$, and $[Co(CN)_6]^{3-}$. Based on the colors of the three solutions, which compound is present in each of the beakers?

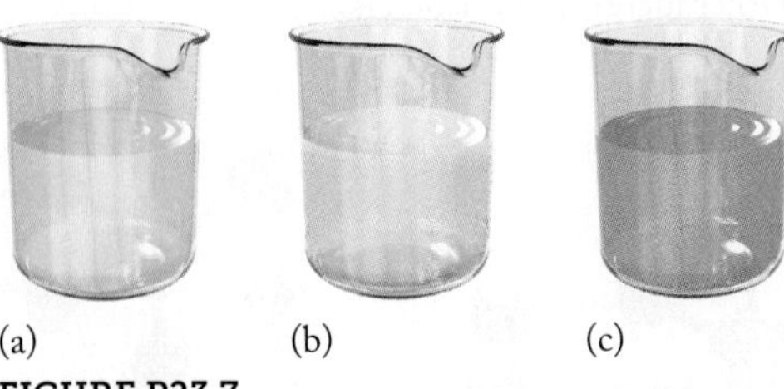

FIGURE P23.7

23.8. Figure P23.8 shows the absorption spectrum of a solution of $Ti(H_2O)_6{}^{3+}$. What color is the solution?

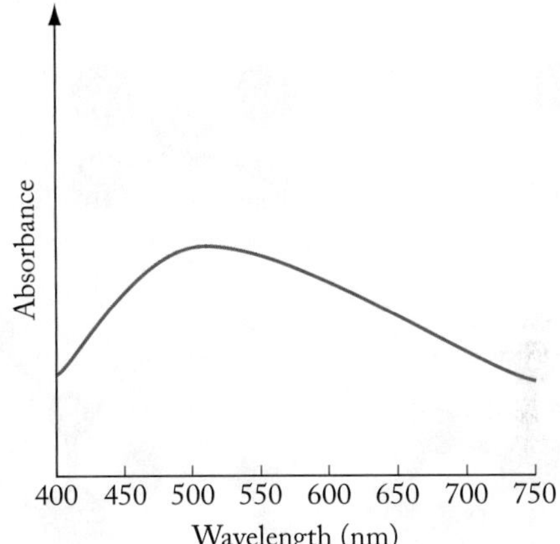

FIGURE P23.8

23.9. For each pair of complexes in Figure P23.9, indicate whether the complexes are (i) identical, (ii) isomers, or (iii) neither identical nor isomers.

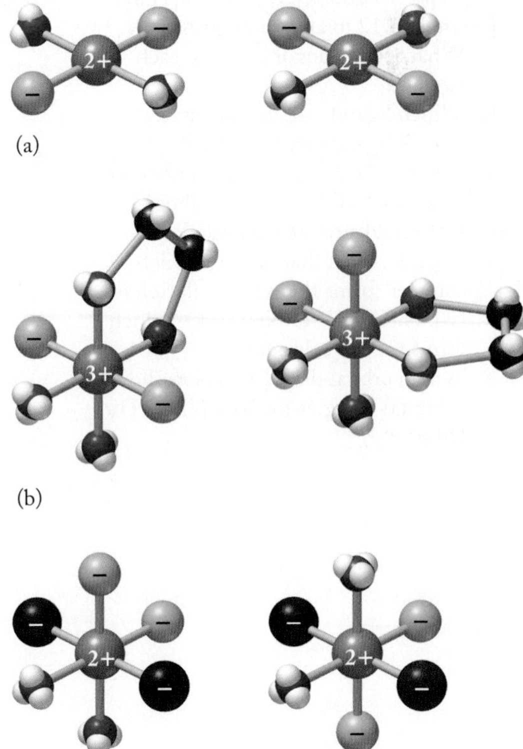

FIGURE P23.9

23.10. For each pair of complexes in Figure P23.10, indicate whether the complexes are (i) identical, (ii) isomers, or (iii) neither identical nor isomers.

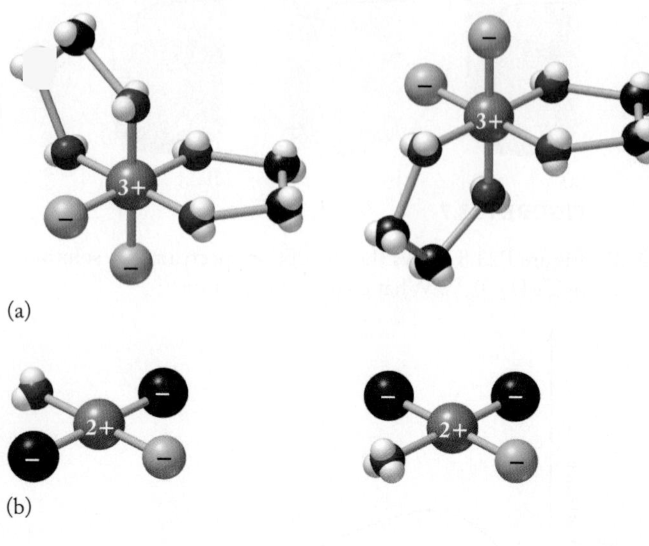

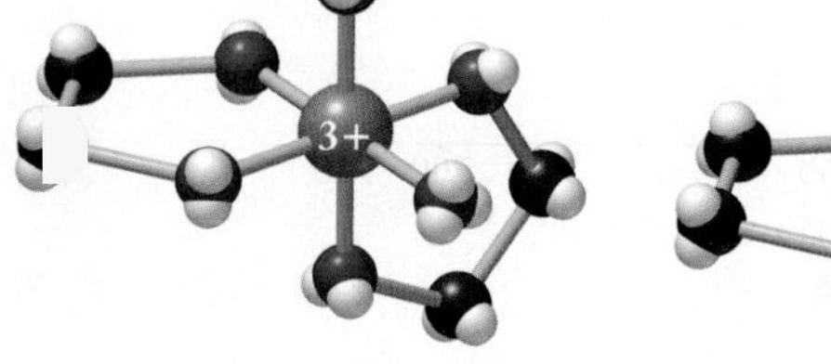

(c)

FIGURE P23.10

***23.11.** The periodic trend for iodide, bromide, and chloride in the spectrochemical series is $Cl^- > Br^- > I^-$. Which curve in Figure P23.11 represents the spectrum for $CoCl_4^{2-}$? Which curve represents CoI_4^{2-}?

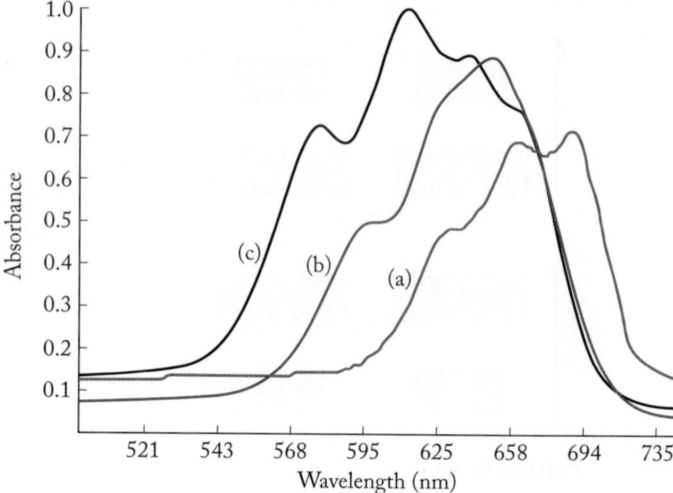

FIGURE P23.11

23.12. Use representations [A] through [I] in Figure P23.12 to answer questions a–f.

a. What are the donor atoms in each ligand?
b. Which ligands are monodentate?
c. Which ligands are bidentate?
d. What is the coordination number for nickel in $Ni(CN)_4^{2-}$? In $Ni(dmg)_2$?
e. If the field strength of cyanide ion is greater than that of dmg, which complex forms the yellow solution in [E]: $Ni(CN)_4^{2-}$ or $Ni(dmg)_2$? Which forms the red solution?
f. Which orbital diagram depicts the *d*-orbital energies for $Ni(CN)_4^{2-}$? For $Ni(dmg)_2$?

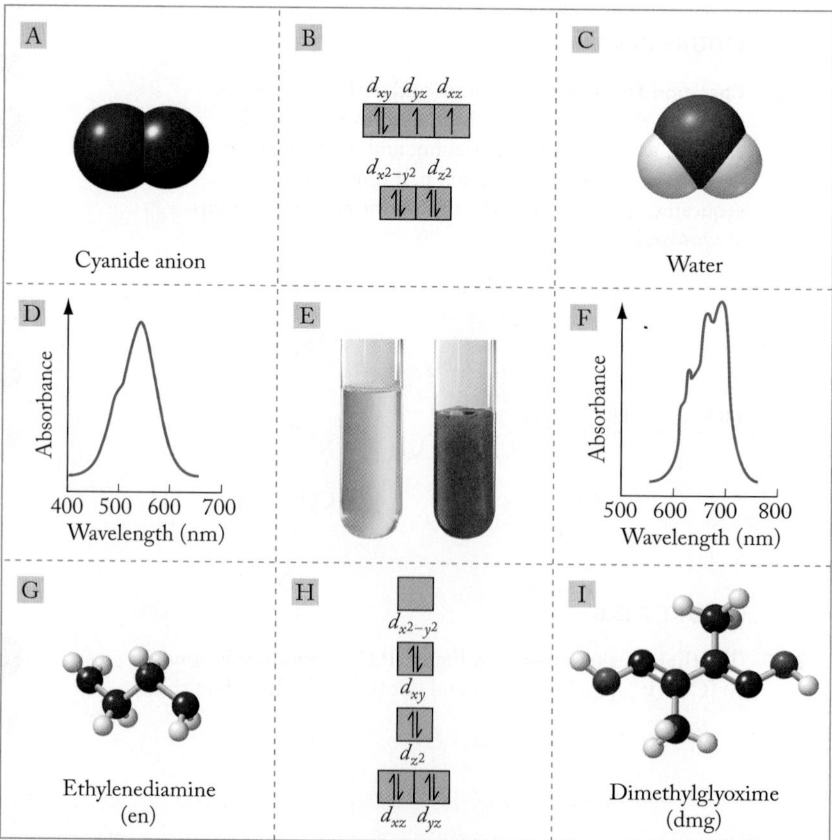

FIGURE P23.12

QUESTIONS AND PROBLEMS

Complex Ions

Concept Review

23.13. When NaCl dissolves in water, which molecules or ions occupy the inner coordination sphere around the Na^+ ions?

23.14. When $Cr(NO_3)_3$ dissolves in water, which of the following species are nearest the Cr^{3+} ions?
a. Other Cr^{3+} ions
b. NO_3^- ions
c. H_2O molecules with the O atoms closest to the Cr^{3+}
d. H_2O molecules with the H atoms closest to the Cr^{3+}

23.15. When $Ni(NO_3)_2$ dissolves in water, what molecules or ions occupy the inner coordination sphere around the Ni^{2+} ions?

23.16. When $[Ni(NH_3)_6]Cl_2$ dissolves in water, which molecules or ions occupy the inner coordination sphere around the Ni^{2+} ions?

23.17. Which ion is the counter ion in the coordination compound $Na_2[Zn(CN)_4]$?

23.18. Which ion is the counter ion in the coordination compound $[Co(NH_3)_4Cl_2]NO_3$?

23.19. The table below contains data from reactions of solutions of a series of octahedral platinum(IV) coordination compounds with $AgNO_3(aq)$. The compounds contain complex ions with the formulas shown in the table. Write the complete formulas of the five compounds.

Composition of Complex	Number of Moles of AgCl Produced per Mole of Complex
$Pt(NH_3)_6Cl_4$	4
$Pt(NH_3)_5Cl_4$	3
$Pt(NH_3)_4Cl_4$	2
$Pt(NH_3)_3Cl_4$	1
$Pt(NH_3)_2Cl_4$	0

23.20. The compositions of three compounds of chromium(III) and chloride ion are known: $Cr(H_2O)_6Cl_3$, $Cr(H_2O)_5Cl_3$, and $Cr(H_2O)_4Cl_3$. The table here summarizes their properties.

Compound	Color of Aqueous Solution	Number of Moles of AgCl Produced per Mole of Complex
A	Dark green	1
B	Gray-blue	3
C	Light green	2

Write the correct formulas for compounds A, B, and C.

Naming Complex Ions and Coordination Compounds

Problems

23.21. Name the coordination compounds of platinum(IV) in Problem 23.19.

23.22. Name the coordination compounds of chromium(III) in Problem 23.20.

23.23. What are the names of the following complex ions?
a. $Cr(NH_3)_6{}^{3+}$
b. $Co(H_2O)_6{}^{3+}$
c. $[Fe(NH_3)_5Cl]^{2+}$

23.24. What are the names of the following complex ions?
a. $Cu(NH_3)_2{}^+$
b. $Ti(H_2O)_4(OH)_2{}^{2+}$
c. $Ni(NH_3)_4(H_2O)_2{}^{2+}$

23.25. What are the names of the following complex ions?
a. $CoBr_4{}^{2-}$
b. $Zn(H_2O)(OH)_3{}^-$
c. $Ni(CN)_5{}^{3-}$

23.26. What are the names of the following complex ions?
a. $CoI_4{}^{2-}$
b. $CuCl_4{}^{2-}$
c. $[Cr(en)(OH)_4]^-$

23.27. What are the names of the following coordination compounds?
a. $[Zn(en)_2]SO_4$
b. $[Ni(NH_3)_5(H_2O)]Cl_2$
c. $K_4Fe(CN)_6$

23.28. What are the names of the following coordination compounds?
a. $(NH_4)_3[Co(CN)_6]$
b. $[Co(en)_2Cl](NO_3)_2$
c. $[Fe(H_2O)_4(OH)_2]Cl$

Polydentate Ligands and Chelation

Concept Review

23.29. What is meant by the term *sequestering agent*? What properties make a substance an effective sequestering agent?

23.30. The structures of two compounds that each contain two $-NH_2$ groups are shown in Figure P23.30. The one on the left is ethylenediamine, a bidentate ligand. Does the molecule on the right have the same ability to donate two pairs of electrons to a metal ion? Explain why you think it does or does not.

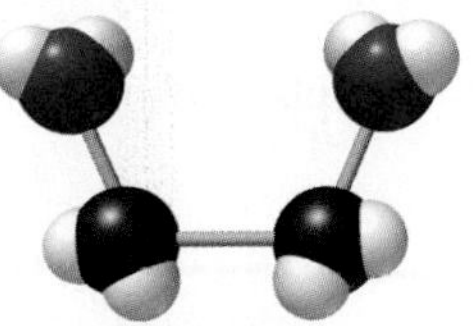
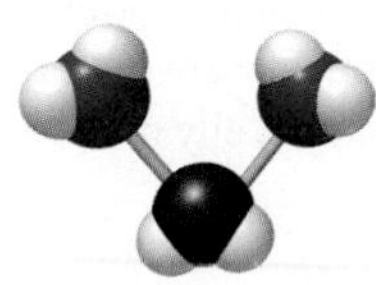

FIGURE P23.30

23.31. How does the chelating ability of an aminocarboxylic acid vary with changing pH?

*23.32. **Food Preservative** The EDTA that is widely used as a food preservative is added to food, not as the undissociated acid, but rather as the calcium disodium salt: $Na_2[CaEDTA]$. This salt is actually a coordination compound with a Ca^{2+} ion at the center of a complex ion. Draw the structure of this compound.

Crystal Field Theory

Concept Review

23.33. Explain why the compounds of most of the first-row transition metals are colored.

23.34. Unlike the compounds of most transition metal ions, those of Ti^{4+} are colorless. Why?

23.35. Why is the d_{xy} orbital higher in energy than the d_{xz} and d_{yz} orbitals in a square planar crystal field?

23.36. On average, the *d* orbitals of a transition metal ion in an octahedral field are higher in energy than they are when the ion is in the gas phase. Why?

Problems

23.37. Aqueous solutions of one of the following complex ions of Cr(III) are violet; solutions of the other are yellow. Which is which? (a) $Cr(H_2O)_6^{3+}$; (b) $Cr(NH_3)_6^{3+}$

23.38. Which of the following complex ions should absorb the shortest wavelengths of electromagnetic radiation? (a) $CuCl_4^{2-}$; (b) CuF_4^{2-}; (c) CuI_4^{2-}; (d) $CuBr_4^{2-}$

23.39. The octahedral crystal field splitting energy Δ_o of $Co(phen)_3^{3+}$ is 5.21×10^{-19} J/ion. What is the color of a solution of this complex ion?

23.40. The octahedral crystal field splitting energy Δ_o of $Co(CN)_6^{3-}$ is 6.74×10^{-19} J/ion. What is the color of a solution of this complex ion?

23.41. Solutions of $NiCl_4^{2-}$ and $NiBr_4^{2-}$ absorb light at 702 nm and 756 nm, respectively. In which ion is the split of *d*-orbital energies greater?

23.42. Chromium(III) chloride forms six-coordinate complexes with bipyridine, including *cis*-$[Cr(bipy)_2Cl_2]^+$, which reacts slowly with water to produce two products, *cis*-$[Cr(bipy)_2(H_2O)Cl]^{2+}$ and *cis*-$[Cr(bipy)_2(H_2O)_2]^{3+}$. In which of these complexes should Δ_o be the largest?

Magnetism and Spin States

Concept Review

23.43. What determines whether a transition metal ion is in a *high-spin* configuration or a *low-spin* configuration?

23.44. For which d^n electron configurations in a tetrahedral geometry are high spin and low spin configurations theoretically possible?

Problems

23.45. How many unpaired electrons are there in the following transition metal ions in an octahedral field? High-spin Fe^{2+}, Cu^{2+}, Co^{2+}, and Mn^{3+}

23.46. Which of the following cations can have either a high-spin or a low-spin electron configuration in an octahedral field? Fe^{2+}, Co^{3+}, Mn^{2+}, and Cr^{3+}

23.47. Which of the following cations can, have either a high-spin or a low-spin electron configuration in a tetrahedral field? Co^{2+}, Cr^{3+}, Ni^{2+}, and Zn^{2+}

23.48. How many unpaired electrons are in the following transition metal ions in an octahedral crystal field? High-spin Fe^{3+}, Rh^+, V^{3+}, and low-spin Mn^{3+}

23.49. The manganese minerals pyrolusite, MnO_2, and hausmannite, Mn_3O_4, contain manganese ions surrounded by oxide ions.

a. What are the charges of the Mn ions in each mineral?
b. In which of these compounds could there be high-spin and low-spin Mn ions?

23.50. **Dietary Supplement** Chromium picolinate is an over-the-counter diet aid sold in many pharmacies. The Cr^{3+} ions in this coordination compound are in an octahedral field. Is the compound paramagnetic or diamagnetic?

***23.51.** **Refining Cobalt** One method for refining cobalt involves the formation of the complex ion $CoCl_4^{2-}$. This anion is tetrahedral. Is this complex paramagnetic or diamagnetic?

*23.52. Why is it that $Ni(CN)_4^{2-}$ is diamagnetic, but $NiCl_4^{2-}$ is paramagnetic?

Isomerism in Coordination Compounds

Concept Review

23.53. What do the prefixes *cis*- and *trans*- mean in the context of an octahedral complex ion?

23.54. What do the prefixes *cis*- and *trans*- mean in the context of a square planar complex?

23.55. What minimum number of different types of donor groups are required in order to have stereoisomers of a square planar complex?

*23.56. A square planar complex contains four different ligands. Does this complex have stereoisomers?

Problems

23.57. Does the complex $Co(en)(H_2O)_2Cl_2$ have stereoisomers?

23.58. Does the complex ion $Fe(en)_3^{3+}$ have stereoisomers?

***23.59.** Sketch the stereoisomers of the square planar complex ion $CuCl_2Br_2^{2-}$. Are any of these isomers chiral?

*23.60. Sketch the stereoisomers of the complex ion $Ni(en)Cl_2(CN)_2^{2-}$. Are any of these isomers chiral?

Coordination Compounds in Biochemistry

Concept Review

23.61. Enzymes are large proteins.

a. What is the function of enzymes?
b. Are all proteins enzymes?

*23.62. Why is Cd^{2+} more likely than Cr^{2+} to replace Zn^{2+} in an enzyme like carbonic anhydrase?

23.63. What effect does an enzyme have on the activation energy of a biochemical reaction?

*23.64. Why might reductases also be described as reducing agents?

23.65. When a transition metal ion such as Cu^{2+} is incorporated into a metalloenzyme, is the formation constant likely to be much greater than one ($K >> 1$) or much less than one ($K << 1$)?

$$Cu^{2+} + \text{protein} \rightleftharpoons \text{metalloenzyme} \qquad K = \frac{[\text{metalloenzyme}]}{[Cu^{2+}][\text{protein}]}$$

23.66. Carbon monoxide binds more strongly to hemoglobin than O_2. Yet CO can be displaced from hemoglobin by breathing pure O_2. Explain why using Le Châtelier's principle.

23.67. What is the likely sign of ΔS for the reaction shown in Figure P23.67, which is catalyzed by carboxypeptidase?

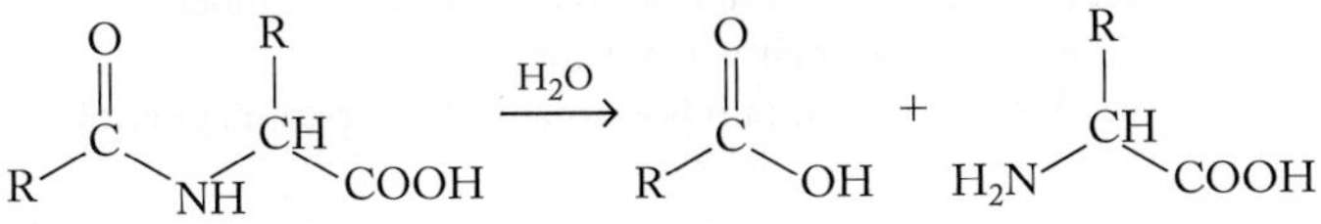

$$R{-}C({=}O){-}NH{-}CH(R){-}COOH \xrightarrow{H_2O} R{-}C({=}O){-}OH + H_2N{-}CH(R){-}COOH$$

FIGURE P23.67

23.68. What is the likely sign of ΔG for the reaction in Question 23.67?

Problems

***23.69.** The activation energy for the uncatalyzed decomposition of hydrogen peroxide at 20°C is 75.3 kJ/mol. In the presence of the enzyme catalase, the activation energy is reduced to 29.3 kJ/mol. Use the following form of the Arrhenius equation, $RT\ln(k_1/k_2) = E_{a_2} - E_{a_1}$, to calculate how much larger the rate constant of the catalyzed reaction is.

*23.70. **Enzymatic Activity of Urease** Urease catalyzes the decomposition of urea to ammonia and carbon dioxide (Figure P23.70). The rate constant for the uncatalyzed reaction at 20°C and pH = 8 is $k = 3 \times 10^{-10}\ s^{-1}$. A urease isolated from the jack bean increases the rate constant to $k = 3 \times 10^{4}\ s^{-1}$. By using the $RT\ln(k_1/k_2) = E_{a_2} - E_{a_1}$ form of the Arrhenius equation, calculate the difference between the activation energies.

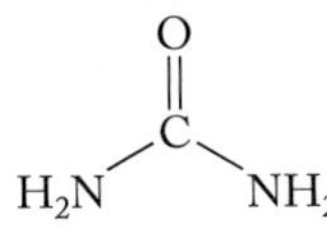

$$H_2N{-}C({=}O){-}NH_2(aq) + H_2O(\ell) \rightarrow 2\ NH_3(aq) + CO_2(g)$$

FIGURE P23.70

***23.71.** The initial rate of an enzyme-catalyzed reaction depends on the concentration of substrate as shown in Figure P23.71. What is the apparent reaction order at the far right side of the graph?

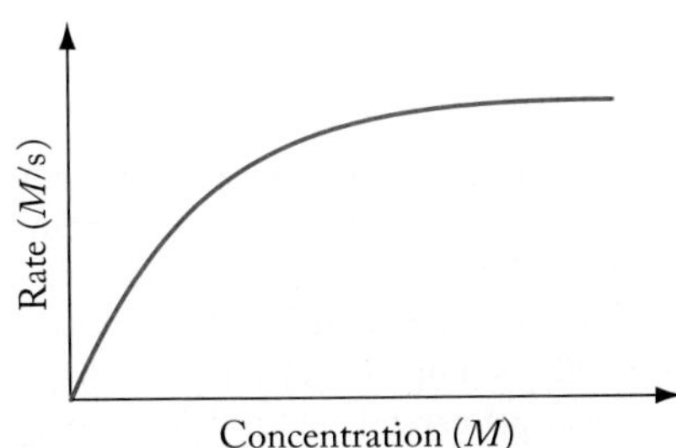

FIGURE P23.71

23.72. The rate of an enzyme-catalyzed reaction depends on the concentration of substrate. Which line in Figure P23.72 has the highest reaction rate?

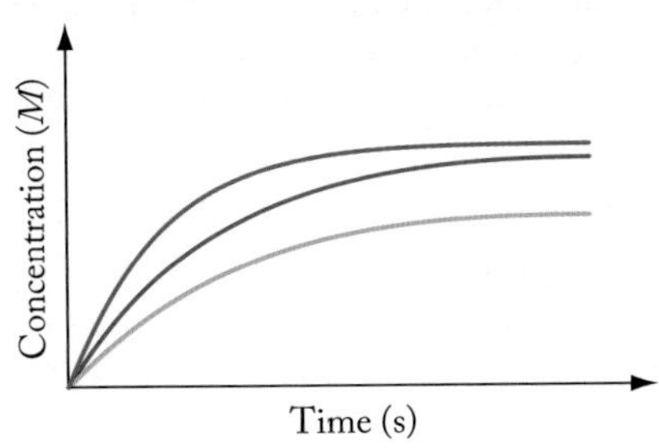

FIGURE P23.72

Coordination Compounds in Medicine

Concept Review

23.73. Under what circumstances might a coordination complex of a toxic metal be useful in medicine?

23.74. **Cancer Treatment** Brachiotherapy is a cancer treatment that involves surgical implantation of a small capsule of ^{192}Ir into a tumor. Iridium-192 lies between two stable isotopes of iridium, ^{193}Ir and ^{191}Ir. How does this account for the fact that ^{192}Ir decays by β decay, electron capture, and positron emission?

23.75. Gadolinium-153 decays by electron capture. What type of radiation does ^{153}Gd produce that makes it useful for imaging?

23.76. Gadolinium-153 and samarium-153 both have the same mass number. Why might ^{153}Gd decay by electron capture whereas ^{153}Sm decays by emitting β particles?

23.77. How do platinum- and ruthenium-containing drugs fight cancer?

23.78. Many transition metal complexes are brightly colored. Why is the titanium(IV) compound budotitane be colorless?

23.79. Is the glucose tolerance factor that contains chromium(III) paramagnetic or diamagnetic?

23.80. Coordination complexes containing paramagnetic transition metal ions make the best MRI contrast agents.

a. Which of the first-row transition metal cations with a 2+ cations have the most unpaired valence electrons in the gas phase?

b. Which of the first-row transition metal cations with a 2+ cations have the most unpaired valence electrons in octahedral coordination complexes?

Problems

23.81. The complexation of mercury(II) ion with methionine

$$Hg^{2+} + \text{methionine} \rightleftharpoons \text{Hg(methionine)}^{2+}$$

has a formation constant of $\log K_f = 14.2$, whereas the formation constant for the Hg^{2+} complex with penicillamine

$$Hg^{2+} + \text{penicillamine} \rightleftharpoons \text{Hg(penicillamine)}^{2+}$$

is $\log K_f = 16.3$. Calculate the equilibrium constant for the reaction

$$\text{Hg(methionine)}^{2+} + \text{penicillamine} \rightleftharpoons \text{Hg(penicillamine)}^{2+} + \text{methionine}$$

23.82. The complexation of mercury(II) ion with cysteine in aqueous solution

$$Hg^{2+} + \text{cysteine} \rightleftharpoons \text{Hg(cysteine)}^{2+}$$

has a formation constant of $\log K_f = 14.2$, whereas the formation constant for the Hg^{2+} complex with glycine

$$Hg^{2+} + \text{glycine} \rightleftharpoons \text{Hg(glycine)}^{2+}$$

is $\log K_f = 10.3$. Calculate the equilibrium constant for the reaction

$$\text{Hg(cysteine)}^{2+} + \text{glycine} \rightleftharpoons \text{Hg(glycine)}^{2+} + \text{cysteine}$$

23.83. The equilibrium constant of the reaction

$$CH_3Hg(\text{penicillamine})^+(aq) + \text{cysteine}(aq) \rightleftharpoons CH_3Hg(\text{cysteine})^+(aq) + \text{penicillamine}(aq)$$

is $K = 0.633$. Calculate the equilibrium concentrations of cysteine and penicillamine if we start with a 1.00 M solution of cysteine and a 1.00 mM solution of $CH_3Hg(\text{penicillamine})^+$.

23.84. The equilibrium constant of the reaction

$$CH_3Hg(\text{glutathione})^+(aq) + \text{cysteine}(aq) \rightleftharpoons CH_3Hg(\text{cysteine})^+(aq) + \text{glutathione}(aq)$$

is $K = 5.0$. Calculate the equilibrium concentrations of cysteine and glutathione if we start with a 1.20 mM solution of cysteine and a 1.20 mM solution of $CH_3Hg(\text{glutathione})^+$.

23.85. The two platinum compounds shown in Figure P23.85 have been studied in cancer therapy.

a. What is the difference between the two compounds?
b. Name the two complex ions.
c. Do both compounds have the same orbital diagram?
d. Sketch the orbital diagram for the platinum compound(s), including the appropriate number of electrons.

OH
H_3N Pt N_3
H_3N N_3
OH

OH
N_3 Pt NH_3
H_3N N_3
OH

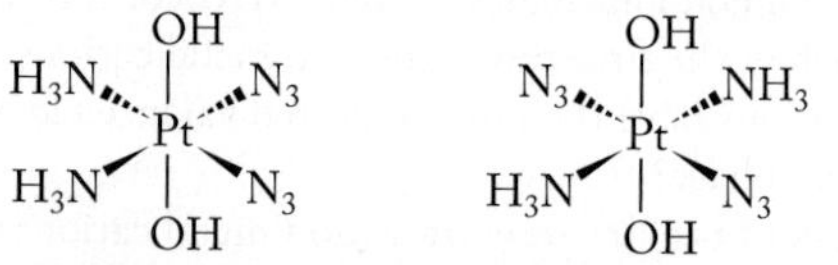

FIGURE P23.85

23.86. Anticancer treatments based on coordination complexes of ruthenium may have less severe side effects than cisplatin and other platinum-based medications. The complex shown in Figure P23.86 was tested against a colon cancer cell line with promising results.

a. Draw the other stereoisomer of $[Ru(py)_2Cl_4]^-$.
b. Name both isomers.
c. Sketch the orbital diagram for the ruthenium compound, including the appropriate number of electrons.

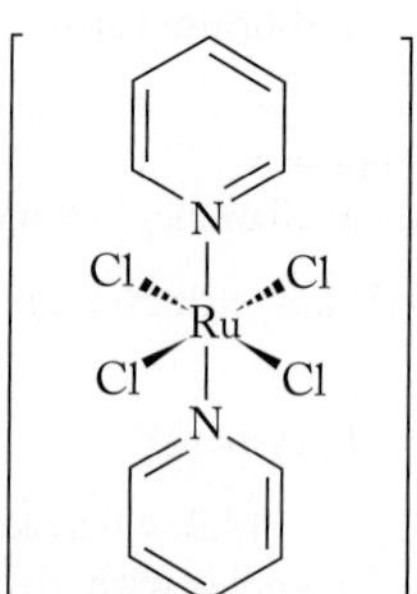

FIGURE P23.86

***23.87.** Two square planar gold and platinum complexes with medicinal properties are shown in Figure P23.87.

a. Do these two complexes have the same number of d electrons in their valence shells?
b. Are these compounds diamagnetic or paramagnetic?

NH_2 O O
Pt
NH_2 O O

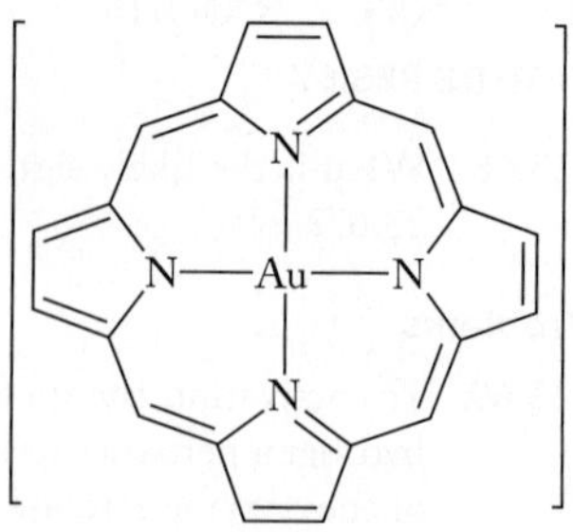

FIGURE P23.87

*23.88. In Chapter 9 we saw that polycyclic aromatic compounds can intercalate into DNA. Compounds like the rhodium and ruthenium complexes shown in Figure P23.88 behave in a similar fashion toward DNA.

a. What is the hydridization at the nitrogen and the carbon atoms in these ligands?
b. Why are these ligands planar?
c. Are these compounds diamagnetic or paramagnetic?

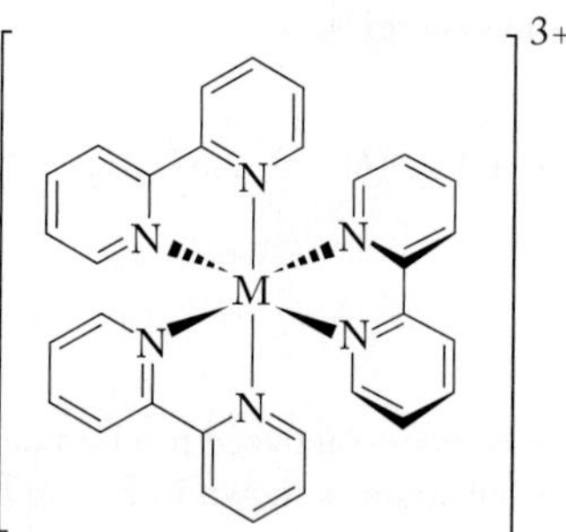

M = Ru, Rh

FIGURE P23.88

Additional Problems

23.89. Dissolving cobalt(II) nitrate in water gives a beautiful purple solution. There are three unpaired electrons in this cobalt(II) complex. When cobalt(II) nitrate is dissolved in aqueous ammonia and oxidized with air, the resulting yellow complex has no unpaired electrons. Which cobalt complex has the larger crystal field splitting energy Δ_o?

23.90. **Lead Poisoning** Children used to be treated for lead poisoning with intravenous injections of EDTA. If the concentration of EDTA in the blood of a patient is 2.5×10^{-8} M and the formation constant for the complex $Pb(EDTA)^{2-}$ is 2.0×10^{18}, what is the concentration ratio of the free (and potentially toxic) $Pb^{2+}(aq)$ in the blood to the much less toxic Pb^{2+}–EDTA complex?

***23.91.** A solid compound containing Fe(II) in an octahedral crystal field has four unpaired electrons at 298 K. When the compound is cooled to 80 K, the same sample appears to have no unpaired electrons. How do you explain this change in the compound's properties?

23.92. When Ag_2O reacts with peroxodisulfate ($S_2O_8^{2-}$) ion (a powerful oxidizing agent), AgO is produced. Crystallographic and magnetic analyses of AgO suggest that it is not simply Ag(II) oxide, but rather a blend of Ag(I) and Ag(III) in a square planar environment. The Ag^{2+} ion is paramagnetic but, like AgO, Ag^+ and Ag^{3+} are diamagnetic. Explain why.

23.93. The iron(II) compound $Fe(bipy)_2(SCN)_2$ is paramagnetic, but the corresponding cyanide compound $Fe(bipy)_2(CN)_2$ is diamagnetic. Why do these two compounds have different magnetic properties?

23.94. Aqueous solutions of copper(II)–ammonia complexes are dark blue. Will the color of the series of complexes $Cu(H_2O)_{(6-x)}(NH_3)_x^+$ shift toward shorter or longer wavelengths as the value of x increases from 0 to 6?

23.95. Manganese(II) chloride was one of the first compounds to be investigated as MRI contrast agents. How many unpaired electrons does the complex have when $MnCl_2$ dissolves in water to form the coordination compound $MnCl_2(H_2O)_4$?

23.96. Gadolinium(III) ions are used in contrast agents for MRI because they have unpaired electrons. What is the electron configuration for Gd^{3+} and how many unpaired electrons does it have?

smartwork5

If your instructor uses Smartwork5, log in at digital.wwnorton.com/atoms2.

Appendix 1

Mathematical Procedures

Working with Scientific Notation

Quantities that scientists work with are sometimes very large, such as Earth's mass, and other times very small, such as the mass of an electron. It is easier to work with these values when they are expressed in scientific notation.

The general form of standard scientific notation is a value between 1 and 10 multipled by 10 raised to an integral power. According to this definition, 598×10^{22} kg (Earth's mass) is not in standard scientific notation, but 5.98×10^{24} kg is. It is good practice to use and report values in standard scientific notation.

1. **To convert an "ordinary" number to standard scientific notation,** move the decimal point to the left for a large number, or to the right for a small one, so that the decimal point is located after the first nonzero digit.
   A. For example, to express Earth's average density (5517 kg/m^3) in scientific notation requires moving the decimal point three places to the left. Doing so is the same as dividing the number by 1000, or 10^3. To keep the value the same we multiply it by 10^3. So, Earth's density in standard scientific notation is 5.517×10^3 kg/m^3.
   B. If we move the decimal point of a value less than 1 to the right to express it in scientific notation, then the exponent is a negative integer equal to the number of places we moved the decimal point to the right. For example, the value of R used in solving ideal gas law problems is 0.08206 L · atm/(mol · K). Moving the decimal point two places to the right converts the value of R to scientific notation: 8.206×10^{-2} L · atm/(mol · K).
   C. Another value of R, 8.314 J/(mol · K), does not need an exponent, though it could be written 8.314×10^0 J/(mol · K) because any value raised to the zero power is equal to 1.

2. **For calculations with numbers in scientific notation,** most calculators have a function key for entering the exponents of values expressed in scientific notation. In many calculators it is labeled "E" or "EE" or "Exp." To enter, for example, the speed of light in meters per second, 2.998×10^8, we enter the value before the exponent, 2.998, followed by the exponent key and then the value of the exponent (8). To enter a value with a negative exponent, use the sign-change key. Sometimes it is labeled "(−)" or "+/−" or "±."

Working with Logarithms

A logarithm to the base 10 has the following form:

$$\log_{10} x = \log x = a, \text{ where } x = 10^a$$

We usually abbreviate the logarithm function "log" if the logarithm is to the base 10, which means the scale in which log 10 = 1.

A logarithm to the base e, called a *natural logarithm*, has the following form:

$$\log_e x = \ln x = b, \text{ where } x = e^b$$

Scientific calculators have (LOG) and (LN) keys, so it is easy to convert a number into its log or ln form. The directions below apply to most calculators.

Sample Exercise 1 Find the logarithm to the base 10 of 2.247 (log 2.247).

Solution In some calculators you enter 2.247 first and then press the (LOG) key. In others, such as the TI 84/89 series, you press the (LOG) key first followed by 2.247 and then the (ENTER) key. Either way, the answer should be 0.3516 (to four significant figures).

Sample Exercise 2 Find the natural logarithm of 2.247 (ln 2.247).

Solution Follow the same procedure as in Sample Exercise 1 except use the (LN) key. The answer should be 0.8096. This answer is (0.8096/0.3516) = 2.303 times larger than log value. That is,

$$\ln x = 2.303 \log x$$

Sample Exercise 3 Find the log of 6.0221×10^{23}.

Solution Following the procedures described above for entering a value with an exponent into your calculator and then taking its log to the base 10, you should obtain the value 23.77974796. We know the original value to five significant figures; the log value should have the same precision, but how do we express it? You might think the log value should be 23.780; however, the 23 to the left of the decimal point reflects the value of the exponent in the original value, and exponents don't count in determining the number of significant figures in a value. Therefore, the value of the logarithm to five significant figures is 23.77975. To understand better why this is so, calculate the log of 6.0221. You should get 0.77975 to five significant figures. Note how the difference between the two log values (23.77975 and 0.77975) is simply the value of the exponent in the first value.

Combining Logs The following equations summarize how logarithms of the products or quotients of two or more values are related to the individual logs of those values:

$$\text{logarithm } (ab) = \text{logarithm } a + \text{logarithm } b$$

and

$$\text{logarithm } (a/b) = \text{logarithm } a - \text{logarithm } b$$

Converting Logarithms into Numbers

If we know the value of log x, what is the value of x? This question is frequently asked when working with pH, which is the negative log of the concentration of hydrogen ions, $[H_3O^+]$, in solution:

$$\text{pH} = -\log[H_3O^+]$$

Suppose the pH of a solution of a weak acid is 2.50. The concentration of H_3O^+ is related to this pH value as follows:

$$2.50 = -\log[H_3O^+]$$

or

$$-2.50 = \log[H_3O^+]$$

To find the value of $[H_3O^+]$, we enter 2.5 into the calculator and press the appropriate key to change its sign to −2.5. The next step depends on the type of calculator. If yours has a (10^x) key (often accessible using a second function key), use it to find the value of $10^{-2.5}$, which is the value we are looking for. The corresponding keystrokes with many graphing calculators are (10^x), (−), 2.5, (ENTER). Some calculators, including the virtual one in many Windows operating systems, have an (x^y) key. We can use it for this problem by entering 10 and pushing the (x^y) key, then entering 2.5 and pushing the (+/−) key followed by the (=) key. All of these approaches do the same calculation, taking 10 to the −2.50 power, and give the same answer, $[H_3O^+] = 3.2 \times 10^{-3}$ to two significant figures. (Remember, pH is a log value, so the digit before the decimal point is not significant.)

Solving Quadratic Equations

If the terms in an equation can be rearranged so that they take the form

$$ax^2 + bx + c = 0$$

they have the form of a quadratic equation. The value(s) of x can be determined from the values of the coefficients a, b, and c by using the equation

$$x = \frac{-b \pm \sqrt{b^2 - 4ac}}{2a}$$

For example, if the solution to a problem yields the following expression where x is the concentration of a solute:

$$x^2 + 0.112x - 1.2 \times 10^{-3} = 0$$

Then the value of x can be determined as follows:

$$x = \frac{-b \pm \sqrt{b^2 - 4ac}}{2a}$$

$$= \frac{-0.112 \pm \sqrt{(0.112)^2 - 4(1)(-1.2 \times 10^{-3})}}{2(1)}$$

$$= \frac{-0.112 \pm \sqrt{0.01254 + 0.0048}}{2}$$

$$= \frac{-0.112 \pm 0.132}{2} = +0.010 \text{ or } -0.122$$

In this example, the negative value for x satisfies the equation, but it has no meaning because we cannot have negative concentration values; therefore we use only the +0.010 value.

Expressing Data in Graphical Form

Fitting curves to plots of experimental data is a powerful tool in determining the relationships between variables. Many natural phenomena obey exponential functions. For example, the rate constant (k) of a chemical reaction increases exponentially with increasing absolute temperature (T). This relationship is described by the Arrhenius equation:

$$k = A\, e^{-E_a/RT}$$

where A is a constant for a particular reaction (called the frequency factor), E_a is the activation energy of the reaction, and R is the ideal gas constant. Taking the natural logarithms of both sides of the Arrhenius equation gives

$$\ln k = \ln A - \left(\frac{E_a}{RT}\right)$$

This equation fits the general equation of a straight line ($y = mx + b$) if ($\ln k$) is the y-variable and ($1/T$) is the x-variable. Plotting ($\ln k$) versus ($1/T$) should give a straight line with a slope equal to $-E_a/R$. The slopes of these plots are negative because the activation energies, E_a, of chemical reactions are positive. The data for a reaction given in columns 2 and 4 of Table A1.1 are plotted in Figure A1.1. The program that generated the graph also gives us the equation of the straight line that best fits the data. The slope (−1281 K) of this line is used to calculate the value of E_a:

$$-1281 \text{ K} = -\frac{E_a}{R}$$

$$E_a = -(-1281 \text{ K})[8.314 \text{ J/(mol} \cdot \text{K)}]$$

$$= 10{,}650 \text{ J/mol} = 10.65 \text{ kJ/mol}$$

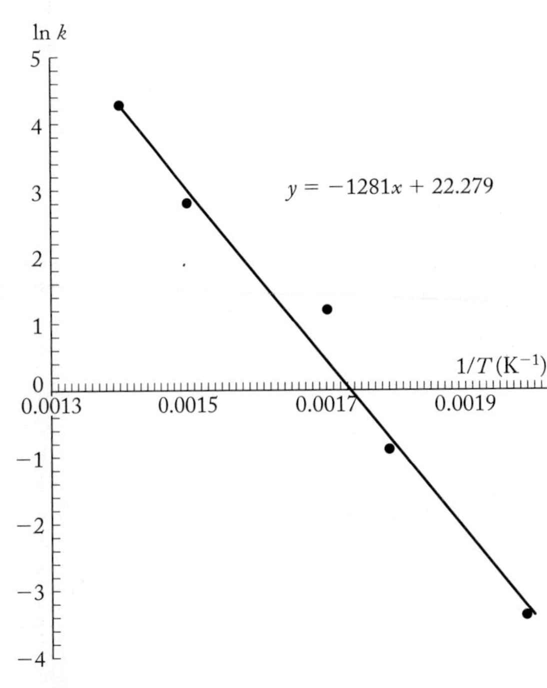

FIGURE A1.1

Expressing Precision and Accuracy

The precision in the results of a set of replicate measurements or analyses is determined by calculating the arithmetic mean ($\bar{x}$) of the data set and its standard deviation (s), which is a measure of the variation between the mean and the results of each measurement or analysis (x_i). The equation for calculating s is

$$s = \sqrt{\frac{\sum_i (x_i - \bar{x})^2}{n - 1}}$$

TABLE A1.1 Rate Constant k as a Function of Temperature T

Temperature T (K)	$1/T$ (K^{-1})	Rate Constant k	ln k
500	0.0020	0.030	−3.51
550	0.0018	0.38	−0.97
600	0.0017	2.9	1.06
650	0.0015	17	2.83
700	0.0014	75	4.32

where n is the number of data points. To calculate the confidence interval, which is the range that is predicted to contain the true mean value (μ) of the data set, we use this equation:

$$\mu = \bar{x} \pm \frac{t\,s}{\sqrt{n}}$$

where t is a statistic that depends on the value of n and the degree of certainty, or *confidence level*, in the prediction. Table A1.2 contains values of t for confidence levels of 90.0, 95.0, 99.0, and 99.9%.

TABLE A1.2 Values of t

	CONFIDENCE LEVEL			
$n - 1$	90.0%	95.0%	99.0%	99.9%
1	6.314	12.71	63.66	636.62
2	2.920	4.303	9.925	31.599
3	2.353	3.182	5.841	12.924
4	2.132	2.776	4.604	8.610
5	2.015	2.571	4.032	6.869
6	1.943	2.447	3.707	5.959
7	1.895	2.365	3.499	5.408
8	1.860	2.306	3.355	5.041
9	1.833	2.262	3.250	4.781
10	1.812	2.228	3.169	4.587
11	1.796	2.201	3.106	4.437
12	1.782	2.179	3.055	4.318
13	1.771	2.160	3.012	4.221
14	1.761	2.145	2.977	4.140
15	1.753	2.131	2.947	4.073
16	1.746	2.120	2.921	4.015
17	1.740	2.110	2.898	3.965
18	1.734	2.101	2.878	3.922
19	1.729	2.093	2.861	3.883
20	1.725	2.086	2.845	3.850
25	1.708	2.060	2.787	3.725
30	1.697	2.042	2.750	3.646
40	1.684	2.021	2.704	3.551
60	1.671	2.000	2.660	3.460
80	1.664	1.990	2.639	3.416
100	1.660	1.984	2.626	3.390
∞	1.645	1.960	2.576	3.291

Appendix 2

SI Units and Conversion Factors

TABLE A2.1 Six SI Base Units

SI Base Quantity	Unit	Symbol
length	meter	m
mass	kilogram	kg
time	second	s
amount of substance	mole	mol
temperature	kelvin	K
electric current	ampere	A

TABLE A2.2 Some SI-Derived Units

SI-Derived Quantity	Unit	Symbol	Dimensions
electric charge	coulomb	C	A · s
electric potential	volt	V	J/C
force	newton	N	$kg \cdot m/s^2$
frequency	hertz	Hz	s^{-1}
momentum	newton-second	N · s	kg · m/s
power	watt	W	J/s
pressure	pascal	Pa	N/m^2
radioactivity	becquerel	Bq	s^{-1}
speed or velocity	meter per second	m/s	m/s
energy	joule (newton-meter)	J (N · m)	$kg \cdot m^2/s^2$

TABLE A2.3 SI Prefixes

Prefix	Symbol	Multiplier	Prefix	Symbol	Multiplier
deci	d	10^{-1}	deka	da	10^{1}
centi	c	10^{-2}	hecto	h	10^{2}
milli	m	10^{-3}	kilo	k	10^{3}
micro	μ	10^{-6}	mega	M	10^{6}
nano	n	10^{-9}	giga	G	10^{9}
pico	p	10^{-12}	tera	T	10^{12}
femto	f	10^{-15}	peta	P	10^{15}
atto	a	10^{-18}	exa	E	10^{18}
zepto	z	10^{-21}	zetta	Z	10^{21}

TABLE A2.4 Special Units and Conversion Factors

Quantity	Unit	Symbol	Conversion[a]
energy[a]	electron-volt	eV	1 eV = 1.6022×10^{-19} J
energy[a]	kilowatt-hour	kWh	1 kWh = 3600 kJ
energy	calorie	cal	1 cal = 4.184 J
mass	pound	lb	1 lb = 453.59 g
mass[a]	atomic mass unit	amu	1 amu = 1.66054×10^{-27} kg
length	angstrom	Å	1 Å = 10^{-8} cm = 10^{-10} m
length	inch	in	1 in = 2.54 cm
length	mile	mi	1 mi = 5280 ft = 1.6093 km
pressure	atmosphere	atm	1 atm = 1.01325×10^{5} Pa
pressure	torr	torr	1 torr = 1/760 atm
temperature	Celsius scale	°C	$T(°C) = T(K) - 273.15$
temperature	Fahrenheit scale	°F	$T(°F) = \frac{9}{5}T(°C) + 32$
volume	liter	L	1 L = 1 dm^3 = 10^{-3} m^3
volume	cubic centimeter	cm^3, cc	1 cm^3 = 1 mL = 10^{-3} L
volume	cubic foot	ft^3	1 ft^3 = 7.4805 gal
volume	gallon (U.S.)	gal	1 gal = 3.785 L

[a]From http://physics.nist.gov/cuu/constants/.

TABLE A2.5 Physical Constants[a]

Quantity	Symbol	Value
acceleration due to gravity (Earth)	g	9.807 m/s^2
Avogadro's number	N_A	6.0221×10^{23} mol^{-1}
Bohr radius	a_0	5.29×10^{-11} m
Boltzmann constant	k_B	1.3806×10^{-23} J/K
electron charge-to-mass ratio	$-e/m_e$	1.7588×10^{11} C/kg
elementary charge	e	1.602×10^{-19} C
Faraday constant	F	9.65×10^{4} C/mol
mass of an electron	m_e	9.10938×10^{-31} kg
mass of a neutron	m_n	1.67493×10^{-27} kg
mass of a proton	m_p	1.67262×10^{-27} kg
molar volume of ideal gas at 0°C and 1 atm	V_m	22.4 L/mol
Planck constant	h	6.626×10^{-34} J · s
speed of light in vacuum	c	2.998×10^{8} m/s
universal gas constant	R	8.314 J/(mol · K) 0.08206 L · atm/(mol · K)

[a]From http://physics.nist.gov/cuu/constants/.

Appendix 3

The Elements and Their Properties

TABLE A3.1 Ground-State Electron Configurations, Atomic Radii, and First Ionization Energies of the Elements

Element	Symbol	Atomic Number Z	Ground-State Configuration	Atomic Radius (pm)	First Ionization Energy (kJ/mol)
hydrogen	H	1	$1s^1$	37	1312.0
helium	He	2	$1s^2$	32	2372.3
lithium	Li	3	$[He]2s^1$	152	520.2
beryllium	Be	4	$[He]2s^2$	112	899.5
boron	B	5	$[He]2s^22p^1$	88	800.6
carbon	C	6	$[He]2s^22p^2$	77	1086.5
nitrogen	N	7	$[He]2s^22p^3$	75	1402.3
oxygen	O	8	$[He]2s^22p^4$	73	1313.9
fluorine	F	9	$[He]2s^22p^5$	71	1681.0
neon	Ne	10	$[He]2s^22p^6$	69	2080.7
sodium	Na	11	$[Ne]3s^1$	186	495.3
magnesium	Mg	12	$[Ne]3s^2$	160	737.7
aluminum	Al	13	$[Ne]3s^23p^1$	143	577.5
silicon	Si	14	$[Ne]3s^23p^2$	117	786.5
phosphorus	P	15	$[Ne]3s^23p^3$	110	1011.8
sulfur	S	16	$[Ne]3s^23p^4$	103	999.6
chlorine	Cl	17	$[Ne]3s^23p^5$	99	1251.2
argon	Ar	18	$[Ne]3s^23p^6$	97	1520.6
potassium	K	19	$[Ar]4s^1$	227	418.8
calcium	Ca	20	$[Ar]4s^2$	197	589.8
scandium	Sc	21	$[Ar]4s^23d^1$	162	633.1
titanium	Ti	22	$[Ar]4s^23d^2$	147	658.8
vanadium	V	23	$[Ar]4s^23d^3$	135	650.9
chromium	Cr	24	$[Ar]4s^13d^5$	128	652.9
manganese	Mn	25	$[Ar]4s^23d^5$	127	717.3
iron	Fe	26	$[Ar]4s^23d^6$	126	762.5
cobalt	Co	27	$[Ar]4s^23d^7$	125	760.4
nickel	Ni	28	$[Ar]4s^23d^8$	124	737.1
copper	Cu	29	$[Ar]4s^13d^{10}$	128	745.5
zinc	Zn	30	$[Ar]4s^23d^{10}$	134	906.4
gallium	Ga	31	$[Ar]4s^23d^{10}4p^1$	135	578.8
germanium	Ge	32	$[Ar]4s^23d^{10}4p^2$	122	762.2
arsenic	As	33	$[Ar]4s^23d^{10}4p^3$	121	947.0
selenium	Se	34	$[Ar]4s^23d^{10}4p^4$	119	941.0

Continued on next page

TABLE A3.1 Ground-State Electron Configurations, Atomic Radii, and First Ionization Energies of the Elements *(Continued)*

Element	Symbol	Atomic Number Z	Ground-State Configuration	Atomic Radius (pm)	First Ionization Energy (kJ/mol)
bromine	Br	35	$[Ar]4s^2 3d^{10} 4p^5$	114	1139.9
krypton	Kr	36	$[Ar]4s^2 3d^{10} 4p^6$	110	1350.8
rubidium	Rb	37	$[Kr]5s^1$	247	403.0
strontium	Sr	38	$[Kr]5s^2$	215	549.5
yttrium	Y	39	$[Kr]5s^2 4d^1$	180	599.8
zirconium	Zr	40	$[Kr]5s^2 4d^2$	160	640.1
niobium	Nb	41	$[Kr]5s^1 4d^4$	146	652.1
molybdenum	Mo	42	$[Kr]5s^1 4d^5$	139	684.3
technetium	Tc	43	$[Kr]5s^2 4d^5$	136	702.4
ruthenium	Ru	44	$[Kr]5s^1 4d^7$	134	710.2
rhodium	Rh	45	$[Kr]5s^1 4d^8$	134	719.7
palladium	Pd	46	$[Kr]4d^{10}$	137	804.4
silver	Ag	47	$[Kr]5s^1 4d^{10}$	144	731.0
cadmium	Cd	48	$[Kr]5s^2 4d^{10}$	151	867.8
indium	In	49	$[Kr]5s^2 4d^{10} 5p^1$	167	558.3
tin	Sn	50	$[Kr]5s^2 4d^{10} 5p^2$	140	708.6
antimony	Sb	51	$[Kr]5s^2 4d^{10} 5p^3$	141	833.6
tellurium	Te	52	$[Kr]5s^2 4d^{10} 5p^4$	143	869.3
iodine	I	53	$[Kr]5s^2 4d^{10} 5p^5$	133	1008.4
xenon	Xe	54	$[Kr]5s^2 4d^{10} 5p^6$	130	1170.4
cesium	Cs	55	$[Xe]6s^1$	265	375.7
barium	Ba	56	$[Xe]6s^2$	222	502.9
lanthanum	La	57	$[Xe]6s^2 5d^1$	187	538.1
cerium	Ce	58	$[Xe]6s^2 4f^1 5d^1$	182	534.4
praseodymium	Pr	59	$[Xe]6s^2 4f^3$	182	527.2
neodymium	Nd	60	$[Xe]6s^2 4f^4$	181	533.1
promethium	Pm	61	$[Xe]6s^2 4f^5$	183	535.5
samarium	Sm	62	$[Xe]6s^2 4f^6$	180	544.5
europium	Eu	63	$[Xe]6s^2 4f^7$	208	547.1
gadolinium	Gd	64	$[Xe]6s^2 4f^7 5d^1$	180	593.4
terbium	Tb	65	$[Xe]6s^2 4f^9$	177	565.8
dysprosium	Dy	66	$[Xe]6s^2 4f^{10}$	178	573.0
holmium	Ho	67	$[Xe]6s^2 4f^{11}$	176	581.0
erbium	Er	68	$[Xe]6s^2 4f^{12}$	176	589.3
thulium	Tm	69	$[Xe]6s^2 4f^{13}$	176	596.7
ytterbium	Yb	70	$[Xe]6s^2 4f^{14}$	193	603.4
lutetium	Lu	71	$[Xe]6s^2 4f^{14} 5d^1$	174	523.5
hafnium	Hf	72	$[Xe]6s^2 4f^{14} 5d^2$	159	658.5
tantalum	Ta	73	$[Xe]6s^2 4f^{14} 5d^3$	146	761.3
tungsten	W	74	$[Xe]6s^2 4f^{14} 5d^4$	139	770.0
rhenium	Re	75	$[Xe]6s^2 4f^{14} 5d^5$	137	760.3
osmium	Os	76	$[Xe]6s^2 4f^{14} 5d^6$	135	839.4
iridium	Ir	77	$[Xe]6s^2 4f^{14} 5d^7$	136	878.0

TABLE A3.1 Ground-State Electron Configurations, Atomic Radii, and First Ionization Energies of the Elements *(Continued)*

Element	Symbol	Atomic Number Z	Ground-State Configuration	Atomic Radius (pm)	First Ionization Energy (kJ/mol)
platinum	Pt	78	$[Xe]6s^1 4f^{14} 5d^9$	139	868.4
gold	Au	79	$[Xe]6s^1 4f^{14} 5d^{10}$	144	890.1
mercury	Hg	80	$[Xe]6s^2 4f^{14} 5d^{10}$	151	1007.1
thallium	Tl	81	$[Xe]6s^2 4f^{14} 5d^{10} 6p^1$	170	589.4
lead	Pb	82	$[Xe]6s^2 4f^{14} 5d^{10} 6p^2$	154	715.6
bismuth	Bi	83	$[Xe]6s^2 4f^{14} 5d^{10} 6p^3$	150	703.3
polonium	Po	84	$[Xe]6s^2 4f^{14} 5d^{10} 6p^4$	167	812.1
astatine	At	85	$[Xe]6s^2 4f^{14} 5d^{10} 6p^5$	140	924.6
radon	Rn	86	$[Xe]6s^2 4f^{14} 5d^{10} 6p^6$	145	1037.1
francium	Fr	87	$[Rn]7s^1$	242	380
radium	Ra	88	$[Rn]7s^2$	211	509.3
actinium	Ac	89	$[Rn]7s^2 6d^1$	188	499
thorium	Th	90	$[Rn]7s^2 6d^2$	179	587
protactinium	Pa	91	$[Rn]7s^2 5f^2 6d^1$	163	568
uranium	U	92	$[Rn]7s^2 5f^3 6d^1$	156	587
neptunium	Np	93	$[Rn]7s^2 5f^4 6d^1$	155	597
plutonium	Pu	94	$[Rn]7s^2 5f^6$	159	585
americium	Am	95	$[Rn]7s^2 5f^7$	173	578
curium	Cm	96	$[Rn]7s^2 5f^7 6d^1$	174	581
berkelium	Bk	97	$[Rn]7s^2 5f^9$	170	601
californium	Cf	98	$[Rn]7s^2 5f^{10}$	186	608
einsteinium	Es	99	$[Rn]7s^2 5f^{11}$	186	619
fermium	Fm	100	$[Rn]7s^2 5f^{12}$	167	627
mendelevium	Md	101	$[Rn]7s^2 5f^{13}$	173	635
nobelium	No	102	$[Rn]7s^2 5f^{14}$	176	642
lawrencium	Lr	103	$[Rn]7s^2 5f^{14} 6d^1$	161	—
rutherfordium	Rf	104	$[Rn]7s^2 5f^{14} 6d^2$	157	—
dubnium	Db	105	$[Rn]7s^2 5f^{14} 6d^3$	149	—
seaborgium	Sg	106	$[Rn]7s^2 5f^{14} 6d^4$	143	—
bohrium	Bh	107	$[Rn]7s^2 5f^{14} 6d^5$	141	—
hassium	Hs	108	$[Rn]7s^2 5f^{14} 6d^6$	134	—
meitnerium	Mt	109	$[Rn]7s^2 5f^{14} 6d^7$	129	—
darmstadtium	Ds	110	$[Rn]7s^2 5f^{14} 6d^8$	128	—
roentgenium	Rg	111	$[Rn]7s^2 5f^{14} 6d^9$	121	—
copernicium	Cn	112	$[Rn]7s^2 5f^{14} 6d^{10}$	122	—
nihonium[a]	Nh	113	$[Rn]7s^2 5f^{14} 6d^{10} 7p^1$	136	—
flerovium	Fl	114	$[Rn]7s^2 5f^{14} 6d^{10} 7p^2$	143	—
moscovium[a]	Mc	115	$[Rn]7s^2 5f^{14} 6d^{10} 7p^3$	162	—
livermorium	Lv	116	$[Rn]7s^2 5f^{14} 6d^{10} 7p^4$	175	—
tennessine[a]	Ts	117	$[Rn]7s^2 5f^{14} 6d^{10} 7p^5$	165	—
oganesson[a]	Og	118	$[Rn]7s^2 5f^{14} 6d^{10} 7p^6$	157	—

[a]Names and symbols for these were recommended by the IUPAC in June 2016.

TABLE A3.2 Miscellaneous Physical Properties of the Naturally Occurring Elements[a]

Element	Symbol	Atomic Number	Physical State[b,c]	Density[d] (g/cm³)	Melting Point (°C)	Boiling Point (°C)
hydrogen	H	1	gas	0.000090	−259.14	−252.87
helium	He	2	gas	0.000179	<−272.2	−268.93
lithium	Li	3	solid	0.534	180.5	1347
beryllium	Be	4	solid	1.848	1283	2484
boron	B	5	solid	2.34	2300	3650
carbon	C	6	solid (gr)	1.9–2.3	~3350	sublimes
nitrogen	N	7	gas	0.00125	−210.00	−195.8
oxygen	O	8	gas	0.00143	−218.8	−182.95
fluorine	F	9	gas	0.00170	−219.62	−188.12
neon	Ne	10	gas	0.00090	−248.59	−246.08
sodium	Na	11	solid	0.971	97.72	883
magnesium	Mg	12	solid	1.738	650	1090
aluminum	Al	13	solid	2.6989	660.32	2467
silicon	Si	14	solid	2.33	1414	2355
phosphorus	P	15	solid (wh)	1.82	44.15	280
sulfur	S	16	solid	2.07	115.21	444.60
chlorine	Cl	17	gas	0.00321	−101.5	−34.04
argon	Ar	18	gas	0.00178	−189.3	−185.9
potassium	K	19	solid	0.862	63.28	759
calcium	Ca	20	solid	1.55	842	1484
scandium	Sc	21	solid	2.989	1541	2380
titanium	Ti	22	solid	4.54	1668	3287
vanadium	V	23	solid	6.11	1910	3407
chromium	Cr	24	solid	7.19	1857	2671
manganese	Mn	25	solid	7.3	1246	1962
iron	Fe	26	solid	7.874	1538	2750
cobalt	Co	27	solid	8.9	1495	2870
nickel	Ni	28	solid	8.902	1455	2730
copper	Cu	29	solid	8.96	1084.6	2562
zinc	Zn	30	solid	7.133	419.53	907
gallium	Ga	31	solid	5.904	29.76	2403
germanium	Ge	32	solid	5.323	938.25	2833
arsenic	As	33	solid (gy)	5.727	614	sublimes
selenium	Se	34	solid (gy)	4.79	221	685
bromine	Br	35	liquid	3.12	−7.2	58.78
krypton	Kr	36	gas	0.00373	−157.36	−153.22
rubidium	Rb	37	solid	1.532	39.31	688
strontium	Sr	38	solid	2.64	777	1382
yttrium	Y	39	solid	4.469	1526	3336
zirconium	Zr	40	solid	6.506	1855	4409
niobium	Nb	41	solid	8.57	2477	4744
molybdenum	Mo	42	solid	10.22	2623	4639

TABLE A3.2 Miscellaneous Physical Properties of the Naturally Occurring Elements[a] *(Continued)*

Element	Symbol	Atomic Number	Physical State[b,c]	Density[d] (g/cm^3)	Melting Point (°C)	Boiling Point (°C)
technetium	Tc	43	solid	11.50	2157	4538
ruthenium	Ru	44	solid	12.41	2334	3900
rhodium	Rh	45	solid	12.41	1964	3695
palladium	Pd	46	solid	12.02	1555	2963
silver	Ag	47	solid	10.50	961.78	2212
cadmium	Cd	48	solid	8.65	321.07	767
indium	In	49	solid	7.31	156.60	2072
tin	Sn	50	solid (wh)	7.31	231.9	2270
antimony	Sb	51	solid	6.691	630.63	1750
tellurium	Te	52	solid	6.24	449.5	998
iodine	I	53	solid	4.93	113.7	184.4
xenon	Xe	54	gas	0.00589	−111.75	−108.0
cesium	Cs	55	solid	1.873	28.44	671
barium	Ba	56	solid	3.5	727	1640
lanthanum	La	57	solid	6.145	920	3455
cerium	Ce	58	solid	6.770	799	3424
praseodymium	Pr	59	solid	6.773	931	3510
neodymium	Nd	60	solid	7.008	1016	3066
promethium	Pm	61	solid	7.264	1042	~3000
samarium	Sm	62	solid	7.520	1072	1790
europium	Eu	63	solid	5.244	822	1596
gadolinium	Gd	64	solid	7.901	1314	3264
terbium	Tb	65	solid	8.230	1359	3221
dysprosium	Dy	66	solid	8.551	1411	2561
holmium	Ho	67	solid	8.795	1472	2694
erbium	Er	68	solid	9.066	1529	2862
thulium	Tm	69	solid	9.321	1545	1946
ytterbium	Yb	70	solid	6.966	824	1194
lutetium	Lu	71	solid	9.841	1663	3393
hafnium	Hf	72	solid	13.31	2233	4603
tantalum	Ta	73	solid	16.654	3017	5458
tungsten	W	74	solid	19.3	3422	5660
rhenium	Re	75	solid	21.02	3186	5596
osmium	Os	76	solid	22.57	3033	5012
iridium	Ir	77	solid	22.42	2446	4130
platinum	Pt	78	solid	21.45	1768.4	3825
gold	Au	79	solid	19.3	1064.18	2856
mercury	Hg	80	liquid	13.546	−38.83	356.73
thallium	Tl	81	solid	11.85	304	1473
lead	Pb	82	solid	11.35	327.46	1749
bismuth	Bi	83	solid	9.747	271.4	1564

Continued on next page

TABLE A3.2 Miscellaneous Physical Properties of the Naturally Occurring Elements[a] *(Continued)*

Element	Symbol	Atomic Number	Physical State[b,c]	Density[d] (g/cm³)	Melting Point (°C)	Boiling Point (°C)
polonium	Po	84	solid	9.32	254	962
astatine	At	85	solid	unknown	302	337
radon	Rn	86	gas	0.00973	−71	−61.7
francium	Fr	87	solid	unknown	27	677
radium	Ra	88	solid	5	700	1737
actinium	Ac	89	solid	10.07	1051	~3200
thorium	Th	90	solid	11.72	1750	4788
protactinium	Pa	91	solid	15.37	1572	unknown
uranium	U	92	solid	19.05	1132	3818

[a]For relative atomic masses and alphabetical listing of the elements, see the flyleaf at the front of this volume.
[b]Normal state at 25°C and 1 atm.
[c]Allotropes: gr = graphite, gy = gray, wh = white.
[d]Liquids and solids at 25°C and 1 atm; gases at 0°C and 1 atm (STP).

TABLE A3.3 A Selection of Stable Isotopes[a]

Isotope ^{A}X	Natural Abundance (%)	Atomic Number Z	Neutron Number N	Mass Number A	Atomic Mass (amu)	Binding Energy per Nucleon (MeV)[b]
^{1}H	99.985	1	0	1	1.007825	—
^{2}H	0.015	1	1	2	2.014000	1.160
^{3}He	0.000137	2	1	3	3.016030	2.572
^{4}He	99.999863	2	2	4	4.002603	7.075
^{6}Li	7.5	3	3	6	6.015121	5.333
^{7}Li	92.5	3	4	7	7.016003	5.606
^{9}Be	100.0	4	5	9	9.012182	6.463
^{10}B	19.9	5	5	10	10.012937	6.475
^{11}B	80.1	5	6	11	11.009305	6.928
^{12}C	98.90	6	6	12	12.000000	7.680
^{13}C	1.10	6	7	13	13.003355	7.470
^{14}N	99.634	7	7	14	14.003074	7.476
^{15}N	0.366	7	8	15	15.000108	7.699
^{16}O	99.762	8	8	16	15.994915	7.976
^{17}O	0.038	8	9	17	16.999131	7.751
^{18}O	0.200	8	10	18	17.999160	7.767
^{19}F	100.0	9	10	19	18.998403	7.779
^{20}Ne	90.48	10	10	20	19.992435	8.032
^{21}Ne	0.27	10	11	21	20.993843	7.972
^{22}Ne	9.25	10	12	22	21.991383	8.081
^{23}Na	100.0	11	12	23	22.989770	8.112
^{24}Mg	78.99	12	12	24	23.985042	8.261
^{25}Mg	10.00	12	13	25	24.985837	8.223
^{26}Mg	11.01	12	14	26	25.982593	8.334

TABLE A3.3 A Selection of Stable Isotopes[a] *(Continued)*

Isotope ^{A}X	Natural Abundance (%)	Atomic Number Z	Neutron Number N	Mass Number A	Atomic Mass (amu)	Binding Energy per Nucleon (MeV)[b]
^{27}Al	100.0	13	14	27	26.981538	8.331
^{28}Si	92.23	14	14	28	27.976927	8.448
^{29}Si	4.67	14	15	29	28.976495	8.449
^{30}Si	3.10	14	16	30	29.973770	8.521
^{31}P	100.0	15	16	31	30.973761	8.481
^{32}S	95.02	16	16	32	31.972070	8.493
^{33}S	0.75	16	17	33	32.971456	8.498
^{34}S	4.21	16	18	34	33.967866	8.584
^{36}S	0.02	16	20	36	35.967080	8.575
^{35}Cl	75.77	17	18	35	34.968852	8.520
^{37}Cl	24.23	17	20	37	36.965903	8.570
^{36}Ar	0.337	18	18	36	35.967545	8.520
^{38}Ar	0.063	18	20	38	37.962732	8.614
^{40}Ar	99.600	18	22	40	39.962384	8.595
^{39}K	93.258	19	20	39	38.963707	8.557
^{41}K	6.730	19	22	41	40.961825	8.576
^{40}Ca	96.941	20	20	40	39.962591	8.551
^{42}Ca	0.647	20	22	42	41.958618	8.617
^{43}Ca	0.135	20	23	43	42.958766	8.601
^{44}Ca	2.086	20	24	44	43.955480	8.658
^{46}Ca	0.004	20	26	46	45.953689	8.669
^{48}Ca	0.187	20	28	48	47.952533	8.666
^{45}Sc	100.0	21	24	45	44.955910	8.619
^{46}Ti	8.0	22	24	46	45.952629	8.656
^{47}Ti	7.3	22	25	47	46.951764	8.661
^{48}Ti	73.8	22	26	48	47.947947	8.723
^{49}Ti	5.5	22	27	49	48.947871	8.711
^{50}Ti	5.4	22	28	50	49.944792	8.756
^{51}V	99.750	23	28	51	50.943962	8.742
^{50}Cr	4.345	24	26	50	49.946046	8.701
^{52}Cr	83.789	24	28	52	51.940509	8.776
^{53}Cr	9.501	24	29	53	52.940651	8.760
^{54}Cr	2.365	24	30	54	53.938882	8.778
^{55}Mn	100.0	25	30	55	54.938049	8.765
^{54}Fe	5.9	26	28	54	53.939612	8.736
^{56}Fe	91.72	26	30	56	55.934939	8.790
^{57}Fe	2.1	26	31	57	56.935396	8.770
^{58}Fe	0.28	26	32	58	57.933277	8.792
^{59}Co	100.0	27	32	59	58.933200	8.768
^{204}Pb	1.4	82	122	204	203.973020	7.880

Continued on next page

TABLE A3.3 A Selection of Stable Isotopes[a] *(Continued)*

Isotope ^{A}X	Natural Abundance (%)	Atomic Number Z	Neutron Number N	Mass Number A	Atomic Mass (amu)	Binding Energy per Nucleon (MeV)[b]
^{206}Pb	24.1	82	124	206	205.974440	7.875
^{207}Pb	22.1	82	125	207	206.975872	7.870
^{208}Pb	52.4	82	126	208	207.976627	7.868
^{209}Bi	100.0	83	126	209	208.980380	7.848

[a]Selection is complete through cobalt-59. Where natural abundances do not add to 100%, the differences are made up by radioactive isotopes with exceedingly long half-lives: potassium-40 (0.0117%, $t_{1/2} = 1.3 \times 10^9$ yr); vanadium-50 (0.250%, $t_{1/2} > 1.4 \times 10^{17}$ yr).
[b]1 MeV (mega electron-volt) = 1.6022×10^{-13} J.

TABLE A3.4 A Selection of Radioactive Isotopes

Isotope ^{A}X	Decay Mode[a]	Half-Life $t_{1/2}$	Atomic Number Z	Neutron Number N	Mass Number A	Atomic Mass (amu)	Binding Energy per Nucleon (MeV)[b]
^{3}H	β^-	12.3 yr	1	2	3	3.01605	2.827
^{8}Be	α	$\sim 7 \times 10^{-17}$ s	4	4	8	8.005305	7.062
^{14}C	β^-	5.7×10^3 yr	6	8	14	14.003241	7.520
^{22}Na	β^+	2.6 yr	11	11	22	21.994434	7.916
^{24}Na	β^-	15.0 hr	11	13	24	23.990961	8.064
^{32}P	β^-	14.3 d	15	17	32	31.973907	8.464
^{35}S	β^-	87.2 d	16	19	35	34.969031	8.538
^{59}Fe	β^-	44.5 d	26	33	59	58.934877	8.755
^{60}Co	β^-	5.3 yr	27	33	60	59.933819	8.747
^{90}Sr	β^-	29.1 yr	38	52	90	89.907738	8.696
^{99}Tc	β^-	2.1×10^5 yr	43	56	99	98.906524	8.611
^{109}Cd	EC	462 d	48	61	109	108.904953	8.539
^{125}I	EC	59.4 d	53	72	125	124.904620	8.450
^{131}I	β^-	8.04 d	53	78	131	130.906114	8.422
^{137}Cs	β^-	30.3 yr	55	82	137	136.907073	8.389
^{222}Rn	α	3.82 d	86	136	222	222.017570	7.695
^{226}Ra	α	1600 yr	88	138	226	226.025402	7.662
^{232}Th	α	1.4×10^{10} yr	90	142	232	232.038054	7.615
^{235}U	α	7.0×10^8 yr	92	143	235	235.043924	7.591
^{238}U	α	4.5×10^9 yr	92	146	238	238.050784	7.570
^{239}Pu	α	2.4×10^4 yr	94	145	239	239.052157	7.560

[a]Modes of decay include alpha emission (α), beta emission (β^-), positron emission (β^+), and electron capture (EC).
[b]1 MeV (mega electron-volt) = 1.6022×10^{-13} J.

Appendix 4

Chemical Bonds and Thermodynamic Data

TABLE A4.1 Average Lengths and Energies of Covalent Bonds

Atom	Bond	Bond Length (pm)	Bond Energy (kJ/mol)
H	H—H	75	436
	H—F	92	567
	H—Cl	127	431
	H—Br	141	366
	H—I	161	299
C	C—C	154	348
	C═C	134	614
	C≡C	120	839
	C—H	110	413
	C—N	147	293
	C═N	127	615
	C≡N	116	891
	C—O	143	358
	C═O	123	743[a]
	C≡O	113	1072
	C—F	133	485
	C—Cl	177	328
	C—Br	179	276
	C—I	215	238
N	N—N	147	163
	N═N	124	418
	N≡N	110	945
	N—H	104	391
	N—O	136	201
	N═O	122	607
	N≡O	106	678
O	O—O	148	146
	O═O	121	498
	O—H	96	463
S	S—O	151	265
	S═O	143	523
	S—S	204	266
	S—H	134	347
F	F—F	143	155
Cl	Cl—Cl	200	243
Br	Br—Br	228	193
I	I—I	266	151

[a]The bond energy of C═O in CO_2 is 799 kJ/mol.

TABLE A4.2 Critical Temperatures (T_c) and van der Waals Parameters (a, b) of Real Gases

Gas[a]	Molar Mass (g/mol)	T_c (K)	a ($L^2 \cdot atm/mol^2$)	b (L/mol)
H_2O	18.015	647.14	5.46	0.0305
Br_2	159.808	588	9.75	0.0591
CCl_3F	137.367	471.2	14.68	0.1111
Cl_2	70.906	416.9	6.343	0.0542
CO_2	44.010	304.14	3.59	0.0427
Kr	83.798	209.41	2.325	0.0396
CH_4	16.043	190.53	2.25	0.0428
O_2	31.999	154.59	1.36	0.0318
Ar	39.948	150.87	1.34	0.0322
F_2	37.997	144.13	1.171	0.0290
CO	28.010	132.91	1.45	0.0395
N_2	28.013	126.21	1.39	0.0391
H_2	2.016	32.97	0.244	0.0266
He	4.003	5.19	0.0341	0.0237

[a]Listed in descending order of critical temperature.

TABLE A4.3 Thermodynamic Properties at 25°C

Substance[a,b]	Molar Mass (g/mol)	ΔH_f° (kJ/mol)	S° [J/(mol · K)]	ΔG_f° (kJ/mol)
ELEMENTS AND MONATOMIC IONS				
$Ag^+(aq)$	107.87	105.6	72.7	77.1
Ag(*g*)	107.87	284.9	173.0	246.0
Ag(*s*)	107.87	0.0	42.6	0.0
$Al^{3+}(aq)$	26.982	−531	−321.7	−485
Al(*g*)	26.982	330.0	164.6	289.4
Al(*s*)	26.982	0.0	28.3	0.0
Al(ℓ)	26.982	10.6	39.6	−1.2
Ar(*g*)	39.948	0.0	154.8	0.0
Au(*g*)	196.97	366.1	180.5	326.3
Au(*s*)	196.97	0.0	47.4	0.0
B(*g*)	10.811	565.0	153.4	521.0
B(*s*)	10.811	0.0	5.9	0.0
$Ba^{2+}(aq)$	137.33	−537.6	9.6	−560.8
Ba(*g*)	137.33	180.0	170.2	146.0
Ba(*s*)	137.33	0.0	62.8	0.0
Be(*g*)	9.0122	324.0	136.3	286.6
Be(*s*)	9.0122	0.0	9.5	0.0
$Br^-(aq)$	79.904	−121.6	82.4	−104.0
Br(*g*)	79.904	111.9	175.0	82.4
$Br_2(g)$	159.808	30.9	245.5	3.1
$Br_2(\ell)$	159.808	0.0	152.2	0.0
C(*g*)	12.011	716.7	158.1	671.3
C(*s*, diamond)	12.011	1.9	2.4	2.9

TABLE A4.3 Thermodynamic Properties at 25°C *(Continued)*

Substance[a,b]	Molar Mass (g/mol)	ΔH°_f (kJ/mol)	S° [J/(mol · K)]	ΔG°_f (kJ/mol)
C(s, graphite)	12.011	0.0	5.7	0.0
Ca^{2+}(*aq*)	40.078	−542.8	−55.3	−553.6
Ca(*g*)	40.078	177.8	154.9	144.0
Ca(s)	40.078	0.0	41.6	0.0
Cl^{-}(*aq*)	35.453	−167.2	56.5	−131.2
Cl(*g*)	35.453	121.3	165.2	105.3
Cl_2(*g*)	70.906	0.0	223.0	0.0
Co^{2+}(*aq*)	58.933	−58.2	−113	−54.4
Co^{3+}(*aq*)	58.933	92	−305	134
Co(*g*)	58.933	424.7	179.5	380.3
Co(s)	58.933	0.0	30.0	0.0
Cr(*g*)	51.996	396.6	174.5	351.8
Cr(s)	51.996	0.0	23.8	0.0
Cs^{+}(*aq*)	132.91	−258.3	133.1	−292.0
Cs(*g*)	132.91	76.5	175.6	49.6
Cs(s)	132.91	0.0	85.2	0.0
Cu^{+}(*aq*)	63.546	71.7	40.6	50.0
Cu^{2+}(*aq*)	63.546	64.8	−99.6	65.5
Cu(*g*)	63.546	337.4	166.4	297.7
Cu(s)	63.546	0.0	33.2	0.0
F^{-}(*aq*)	18.998	−332.6	−13.8	−278.8
F(*g*)	18.998	79.4	158.8	62.3
F_2(*g*)	37.997	0.0	202.8	0.0
Fe^{2+}(*aq*)	55.845	−89.1	−137.7	−78.9
Fe^{3+}(*aq*)	55.845	−48.5	−315.9	−4.7
Fe(*g*)	55.845	416.3	180.5	370.7
Fe(s)	55.845	0.0	27.3	0.0
H^{+}(*aq*)	1.0079	0.0	0.0	0.0
H(*g*)	1.0079	218.0	114.7	203.3
H_2(*g*)	2.0158	0.0	130.6	0.0
He(*g*)	4.0026	0.0	126.2	0.0
Hg_2^{2+}(*aq*)	401.18	172.4	84.5	153.5
Hg^{2+}(*aq*)	200.59	171.1	−32.2	164.4
Hg(*g*)	200.59	61.4	175.0	31.8
Hg(ℓ)	200.59	0.0	75.9	0.0
I^{-}(*aq*)	126.90	−55.2	111.3	−51.6
I(*g*)	126.90	106.8	180.8	70.2
I_2(*g*)	253.81	62.4	260.7	19.3
I_2(s)	253.81	0.0	116.1	0.0
K^{+}(*aq*)	39.098	−252.4	102.5	−283.3
K(*g*)	39.098	89.0	160.3	60.5
K(s)	39.098	0.0	64.7	0.0
Li^{+}(*aq*)	6.941	−278.5	13.4	−293.3

Continued on next page

TABLE A4.3 Thermodynamic Properties at 25°C *(Continued)*

Substance[a,b]	Molar Mass (g/mol)	ΔH_f° (kJ/mol)	S° [J/(mol · K)]	ΔG_f° (kJ/mol)
Li(*g*)	6.941	159.3	138.8	126.6
Li^+(*g*)	6.941	685.7	133.0	648.5
Li(*s*)	6.941	0.0	29.1	0.0
Mg^{2+}(*aq*)	24.305	−466.9	−138.1	−454.8
Mg(*g*)	24.305	147.1	148.6	112.5
Mg(*s*)	24.305	0.0	32.7	0.0
Mn^{2+}(*aq*)	54.938	−220.8	−73.6	−228.1
Mn(*g*)	54.938	280.7	173.7	238.5
Mn(*s*)	54.938	0.0	32.0	0.0
N(*g*)	14.007	472.7	153.3	455.5
N_2(*g*)	28.013	0.0	191.5	0.0
Na^+(*aq*)	22.990	−240.1	59.0	−261.9
Na(*g*)	22.990	107.5	153.7	77.0
Na^+(*g*)	22.990	609.3	148.0	574.3
Na(*s*)	22.990	0.0	51.3	0.0
Ne(*g*)	20.180	0.0	146.3	0.0
Ni^{2+}(*aq*)	58.693	−54.0	−128.9	−45.6
Ni(*g*)	58.693	429.7	182.2	384.5
Ni(*s*)	58.693	0.0	29.9	0.0
O(*g*)	15.999	249.2	161.1	231.7
O_2(*g*)	31.999	0.0	205.0	0.0
O_3(*g*)	47.998	142.7	238.8	163.2
P(*g*)	30.974	314.6	163.1	278.3
P_4(*s*, red)	123.895	−17.6	22.8	−12.1
P_4(*s*, white)	123.895	0.0	41.1	0.0
Pb^{2+}(*aq*)	207.2	−1.7	10.5	−24.4
Pb(*g*)	207.2	195.2	162.2	175.4
Pb(*s*)	207.2	0.0	64.8	0.0
Rb^+(*aq*)	85.468	−251.2	121.5	−284.0
Rb(*g*)	85.468	80.9	170.1	53.1
Rb(*s*)	85.468	0.0	76.8	0.0
S(*g*)	32.065	277.2	167.8	236.7
S_8(*g*)	256.520	102.3	430.2	49.1
S_8(*s*)	256.520	0.0	32.1	0.0
Sc(*g*)	44.956	377.8	174.8	336.0
Sc(*s*)	44.956	0.0	34.6	0.0
Si(*g*)	28.086	450.0	168.0	405.5
Si(*s*)	28.086	0.0	18.8	0.0
Sn(*g*)	118.71	301.2	168.5	266.2
Sn(*s*, gray)	118.71	−2.1	44.1	0.1
Sn(*s*, white)	118.71	0.0	51.2	0.0
Sr^{2+}(*aq*)	87.62	−545.8	−32.6	−559.5
Sr(*g*)	87.62	164.4	164.6	130.9
Sr(*s*)	87.62	0.0	52.3	0.0

TABLE A4.3 Thermodynamic Properties at 25°C *(Continued)*

Substance[a,b]	Molar Mass (g/mol)	ΔH_f° (kJ/mol)	S° [J/(mol · K)]	ΔG_f° (kJ/mol)
Ti(*g*)	47.867	473.0	180.3	428.4
Ti(*s*)	47.867	0.0	30.7	0.0
V(*g*)	50.942	514.2	182.2	468.5
V(*s*)	50.942	0.0	28.9	0.0
W(*s*)	183.84	0.0	32.6	0.0
Zn^{2+}(*aq*)	65.38	−153.9	−112.1	−147.1
Zn(*g*)	65.38	130.4	161.0	94.8
Zn(*s*)	65.38	0.0	41.6	0.0
POLYATOMIC IONS				
CH_3COO^-(*aq*)	59.045	−486.0	86.6	−369.3
CO_3^{2-}(*aq*)	60.009	−677.1	−56.9	−527.8
$C_2O_4^{2-}$(*aq*)	88.020	−825.1	45.6	−673.9
CrO_4^{2-}(*aq*)	115.994	−881.2	50.2	−727.8
$Cr_2O_7^{2-}$(*aq*)	215.988	−1490.3	261.9	−1301.1
$HCOO^-$(*aq*)	45.018	−425.6	92	−351.0
HCO_3^-(*aq*)	61.017	−692.0	91.2	−586.8
HSO_4^-(*aq*)	97.072	−887.3	131.8	−755.9
MnO_4^-(*aq*)	118.936	−541.4	191.2	−447.2
NH_4^+(*aq*)	18.038	−132.5	113.4	−79.3
NO_3^-(*aq*)	62.005	−205.0	146.4	−108.7
OH^-(*aq*)	17.007	−230.0	−10.8	−157.2
PO_4^{3-}(*aq*)	94.971	−1277.4	−222	−1018.7
SO_4^{2-}(*aq*)	96.064	−909.3	20.1	−744.5
INORGANIC COMPOUNDS				
AgCl(*s*)	143.32	−127.1	96.2	−109.8
AgI(*s*)	234.77	−61.8	115.5	−66.2
$AgNO_3$(*s*)	169.87	−124.4	140.9	−33.4
Al_2O_3(*s*)	101.961	−1675.7	50.9	−1582.3
B_2H_6(*g*)	27.669	35.0	232.0	86.6
B_2O_3(*s*)	69.622	−1263.6	54.0	−1184.1
$BaCO_3$(*s*)	197.34	−1216.3	112.1	−1137.6
$BaSO_4$(*s*)	233.39	−1473.2	132.2	−1362.2
$CaCO_3$(*s*)	100.087	−1206.9	92.9	−1128.8
$CaCl_2$(*s*)	110.984	−795.4	108.4	−748.8
CaF_2(*s*)	78.075	−1228.0	68.5	−1175.6
CaO(*s*)	56.077	−634.9	38.1	−603.3
$Ca(OH)_2$(*s*)	74.093	−985.2	83.4	−897.5
$CaSO_4$(*s*)	136.142	−1434.5	106.5	−1322.0
CO(*g*)	28.010	−110.5	197.7	−137.2
CO_2(*g*)	44.010	−393.5	213.8	−394.4
CO_2(*aq*)	44.010	−412.9	121.3	−386.2
CS_2(*g*)	76.143	115.3	237.8	65.1
CS_2(*ℓ*)	76.143	87.9	151.0	63.6

Continued on next page

TABLE A4.3 Thermodynamic Properties at 25°C *(Continued)*

Substance[a,b]	Molar Mass (g/mol)	ΔH_f° (kJ/mol)	S° [J/(mol · K)]	ΔG_f° (kJ/mol)
$CsCl(s)$	168.358	−443.0	101.2	−414.6
$CuSO_4(s)$	159.610	−771.4	109.2	−662.2
$Cu_2S(s)$	159.16	−79.5	120.9	−86.2
$FeCl_2(s)$	126.750	−341.8	118.0	−302.3
$FeCl_3(s)$	162.203	−399.5	142.3	−334.0
$FeO(s)$	71.844	−271.9	60.8	−255.2
$Fe_2O_3(s)$	159.688	−824.2	87.4	−742.2
$HBr(g)$	80.912	−36.3	198.7	−53.4
$HCl(g)$	36.461	−92.3	186.9	−95.3
$HCN(g)$	27.02	135.1	201.81	124.7
$HF(g)$	20.006	−273.3	173.8	−275.4
$HI(g)$	127.912	26.5	206.6	1.7
$HNO_2(g)$	47.014	−79.5	254.1	−46.0
$HNO_3(g)$	63.013	−135.1	266.4	−74.7
$HNO_3(\ell)$	63.013	−174.1	155.6	−80.7
$HNO_3(aq)$	63.013	−206.6	146.0	−110.5
$HgCl_2(s)$	271.50	−224.3	146.0	−178.6
$Hg_2Cl_2(s)$	472.09	−265.4	191.6	−210.7
$H_2O(g)$	18.015	−241.8	188.8	−228.6
$H_2O(\ell)$	18.015	−285.8	69.9	−237.2
$H_2S(g)$	34.082	−20.17	205.6	−33.01
$H_2O_2(g)$	34.015	−136.3	232.7	−105.6
$H_2O_2(\ell)$	34.015	−187.8	109.6	−120.4
$H_2SO_4(\ell)$	98.079	−814.0	156.9	−690.0
$H_2SO_4(aq)$	98.079	−909.2	20.1	−744.5
$KBr(s)$	119.002	−393.8	95.9	−380.7
$KCl(s)$	74.551	−436.5	82.6	−408.5
$KHCO_3(s)$	100.115	−963.2	115.5	−863.6
$K_2CO_3(s)$	138.205	−1151.0	155.5	−1063.5
$LiBr(s)$	86.845	−351.2	74.3	−342.0
$LiCl(s)$	42.394	−408.6	59.3	−384.4
$Li_2CO_3(s)$	73.891	−1215.9	90.4	−1132.1
$MgCl_2(s)$	95.211	−641.3	89.6	−591.8
$Mg(OH)_2(s)$	58.320	−924.5	63.2	−833.5
$MgSO_4(s)$	120.369	−1284.9	91.6	−1170.6
$MnO_2(s)$	86.937	−520.0	53.1	−465.1
$CH_3COONa(s)$	82.034	−708.8	123.0	−607.2
$NaBr(s)$	102.894	−361.1	86.82	−349.0
$NaCl(s)$	58.443	−411.2	72.1	−384.2
$NaCl(g)$	58.443	−181.4	229.8	−201.3
$Na_2CO_3(s)$	105.989	−1130.7	135.0	−1044.4
$NaHCO_3(s)$	84.007	−950.8	101.7	−851.0
$NaNO_3(s)$	84.995	−467.9	116.5	−367.0
$NaOH(s)$	39.997	−425.6	64.5	−379.5

TABLE A4.3 Thermodynamic Properties at 25°C *(Continued)*

Substance[a,b]	Molar Mass (g/mol)	ΔH°_f (kJ/mol)	S° [J/(mol · K)]	ΔG°_f (kJ/mol)
$Na_2SO_4(s)$	142.043	−1387.1	149.6	−1270.2
$NF_3(g)$	71.002	−132.1	260.8	−90.6
$NH_3(aq)$	17.031	−80.3	111.3	−26.50
$NH_3(g)$	17.031	−46.1	192.5	−16.5
$NH_4Cl(s)$	53.491	−314.4	94.6	−203.0
$NH_4NO_3(s)$	80.043	−365.6	151.1	−183.9
$N_2H_4(g)$	32.045	95.35	238.5	159.4
$N_2H_4(\ell)$	32.045	50.63	121.52	149.3
$NiCl_2(s)$	129.60	−305.3	97.7	−259.0
$NiO(s)$	74.60	−239.7	38.0	−211.7
$NO(g)$	30.006	90.3	210.7	86.6
$NO_2(g)$	46.006	33.2	240.0	51.3
$N_2O(g)$	44.013	82.1	219.9	104.2
$N_2O_3(g)$	76.01	86.6	314.7	142.4
$N_2O_4(g)$	92.011	9.2	304.2	97.8
$NOCl(g)$	65.459	51.7	261.7	66.1
$PCl_3(g)$	137.33	−288.07	311.7	−269.6
$PCl_3(\ell)$	137.33	−319.6	217	−272.4
$PF_5(g)$	125.96	−1594.4	300.8	−1520.7
$PH_3(g)$	33.998	5.4	210.2	13.4
$PbCl_2(s)$	278.1	−359.4	136.0	−314.1
$PbSO_4(s)$	303.3	−920.0	148.5	−813.0
$SO_2(g)$	64.065	−296.8	248.2	−300.1
$SO_3(g)$	80.064	−395.7	256.8	−371.1
$ZnCl_2(s)$	136.30	−415.1	111.5	−369.4
$ZnO(s)$	81.37	−348.0	43.9	−318.2
$ZnSO_4(s)$	161.45	−982.8	110.5	−871.5
ORGANIC COMPOUNDS				
$CCl_4(g)$	153.823	−102.9	309.7	−60.6
$CCl_4(\ell)$	153.823	−135.4	216.4	−65.3
$CH_4(g)$	16.043	−74.8	186.2	−50.8
$CH_3COOH(g)$	60.053	−432.8	282.5	−374.5
$CH_3COOH(\ell)$	60.053	−484.5	159.8	−389.9
$CH_3OH(g)$	32.042	−200.7	239.9	−162.0
$CH_3OH(\ell)$	32.042	−238.7	126.8	−166.4
$HC{\equiv}CH(g)$	26.038	226.7	200.8	209.2
$CH_2{=}CH_2(g)$	28.054	52.4	219.5	68.1
$CH_3CH_3(g)$	30.070	−84.67	229.5	−32.9
$CH_3CH_2OH(g)$	46.069	−235.1	282.6	−168.6
$CH_3CH_2OH(\ell)$	46.069	−277.7	160.7	−174.9
$CH_3CHO(g)$	44.053	−166	266	−133.7
$CH_3CH_2CH_3(g)$	44.097	−103.8	269.9	−23.5
$CH_3(CH_2)_2CH_3(g)$	58.123	−125.6	310.0	−15.7

Continued on next page

TABLE A4.3 Thermodynamic Properties at 25°C *(Continued)*

Substance[a,b]	Molar Mass (g/mol)	ΔH_f° (kJ/mol)	S° [J/(mol · K)]	ΔG_f° (kJ/mol)
$CH_3(CH_2)_2CH_3(\ell)$	58.123	−147.6	231.0	−15.0
$CH_3COCH_3(\ell)$	58.079	−248.4	199.8	−155.6
$CH_3COCH_3(g)$	58.079	−217.1	295.3	−152.7
$CH_3(CH_2)_2CH_2OH(\ell)$	74.122	−327.3	225.8	
$(CH_3CH_2)_2O(\ell)$	74.122	−279.6	172.4	
$(CH_3CH_2)_2O(g)$	74.122	−252.1	342.7	
$(CH_3)_2C{=}C(CH_3)_2(\ell)$	84.161	66.6	362.6	−69.2
$(CH_3)_2NH(\ell)$	45.084	−43.9	182.3	
$(CH_3)_2NH(g)$	45.084	−18.5	273.1	
$(CH_3CH_2)_2NH(\ell)$	73.138	−103.3		
$(CH_3CH_2)_2NH(g)$	73.138	−71.4		
$(CH_3)_3N(\ell)$	59.111	−46.0	208.5	
$(CH_3)_3N(g)$	59.111	−23.6	287.1	
$(CH_3CH_2)_3N(\ell)$	101.191	−134.3		
$(CH_3CH_2)_3N(g)$	101.191	−95.8		
$C_6H_6(g)$	78.114	82.9	269.2	129.7
$C_6H_6(\ell)$	78.114	49.0	172.9	124.5
$C_6H_{12}O_6(s)$	180.158	−1274.4	212.1	−910.1
$CH_3(CH_2)_6CH_3(\ell)$	114.231	−249.9	361.1	6.4
$CH_3(CH_2)_6CH_3(g)$	114.231	−208.6	466.7	16.4
$C_{12}H_{22}O_{11}(s)$	342.300	−2221.7	360.2	−1543.8
$HCOOH(\ell)$	46.026	−424.7	129.0	−361.4

[a]Substances are arranged alphabetically by chemical formula within each class: (1) elements and monatomic ions; (2) polyatomic ions; (3) inorganic compounds (including CO and CO_2); (4) organic compounds (hydrocarbon-based).
[b]Symbols denote standard enthalpy of formation (ΔH_f°), standard third-law entropy (S°), and standard Gibbs free energy of formation (ΔG_f°). Entropies in aqueous solution are referred to $S^\circ[H^+(aq)] = 0$, not to absolute zero.

TABLE A4.4 Vapor Pressure of Water as a Function of Temperature

T (°C)	P (torr)
0.0	4.579
10.0	9.209
20.0	17.535
25.0	23.756
30.0	31.824
40.0	55.324
60.0	149.4
70.0	233.7
90.0	525.8
100	760.0
105	906.0

Appendix 5

Equilibrium Constants

TABLE A5.1 Ionization Constants of Selected Acids at 25°C

Acid	Step	Aqueous Equilibrium[a]	K_a	pK_a
acetic	1	$CH_3COOH(aq) + H_2O(\ell) \rightleftharpoons H_3O^+(aq) + CH_3COO^-(aq)$	1.76×10^{-5}	4.75
arsenic	1	$H_3AsO_4(aq) + H_2O(\ell) \rightleftharpoons H_3O^+(aq) + H_2AsO_4^-(aq)$	5.5×10^{-3}	2.26
	2	$H_2AsO_4^-(aq) + H_2O(\ell) \rightleftharpoons H_3O^+(aq) + HAsO_4^{2-}(aq)$	1.7×10^{-7}	6.77
	3	$HAsO_4^{2-}(aq) + H_2O(\ell) \rightleftharpoons H_3O^+(aq) + AsO_4^{3-}(aq)$	5.1×10^{-12}	11.29
ascorbic	1	$H_2C_6H_6O_6(aq) + H_2O(\ell) \rightleftharpoons H_3O^+(aq) + HC_6H_6O_6^-(aq)$	9.1×10^{-5}	4.04
	2	$HC_6H_6O_6^-(aq) + H_2O(\ell) \rightleftharpoons H_3O^+(aq) + C_6H_6O_6^{2-}(aq)$	5×10^{-12}	11.3
benzoic	1	$C_6H_5COOH(aq) + H_2O(\ell) \rightleftharpoons H_3O^+(aq) + C_6H_5COO^-(aq)$	6.25×10^{-5}	4.20
boric	1	$H_3BO_3(aq) + H_2O(\ell) \rightleftharpoons H_3O^+(aq) + H_2BO_3^-(aq)$	5.4×10^{-10}	9.27
	2	$H_2BO_3^-(aq) + H_2O(\ell) \rightleftharpoons H_3O^+(aq) + HBO_3^{2-}(aq)$	$<10^{-14}$	>14
bromoacetic	1	$CH_2BrCOOH(aq) + H_2O(\ell) \rightleftharpoons H_3O^+(aq) + CH_2BrCOO^-(aq)$	2.0×10^{-3}	2.70
butanoic	1	$CH_3CH_2CH_2COOH(aq) + H_2O(\ell) \rightleftharpoons H_3O^+(aq) + CH_3CH_2CH_2COO^-(aq)$	1.5×10^{-5}	4.82
carbonic	1	$H_2CO_3(aq) + H_2O(\ell) \rightleftharpoons H_3O^+(aq) + HCO_3^-(aq)$	4.3×10^{-7}	6.37
	2	$HCO_3^-(aq) + H_2O(\ell) \rightleftharpoons H_3O^+(aq) + CO_3^{2-}(aq)$	4.7×10^{-11}	10.33
chloric	1	$HClO_3(aq) + H_2O(\ell) \rightleftharpoons H_3O^+(aq) + ClO_3^-(aq)$	~1	~0
chloroacetic	1	$CH_2ClCOOH(aq) + H_2O(\ell) \rightleftharpoons H_3O^+(aq) + CH_2ClCOO^-(aq)$	1.4×10^{-3}	2.85
chlorous	1	$HClO_2(aq) + H_2O(\ell) \rightleftharpoons H_3O^+(aq) + ClO_2^-(aq)$	1.1×10^{-2}	1.96
citric	1	$CH_2(COOH)C(OH)(COOH)CH_2COOH(aq) + H_2O(\ell) \rightleftharpoons$ $H_3O^+(aq) + CH_2(COOH)C(OH)(COO^-)CH_2COOH(aq)$	7.4×10^{-4}	3.13
	2	$CH_2(COOH)C(OH)(COO^-)CH_2COOH(aq) + H_2O(\ell) \rightleftharpoons$ $H_3O^+(aq) + CH_2(COO^-)C(OH)(COO^-)CH_2COOH(aq)$	1.7×10^{-5}	4.77
	3	$CH_2(COO^-)C(OH)(COO^-)CH_2COOH(aq) + H_2O(\ell) \rightleftharpoons$ $H_3O^+(aq) + CH_2(COO^-)C(OH)(COO^-)CH_2COO^-(aq)$	4.0×10^{-7}	6.40
dichloroacetic	1	$CHCl_2COOH(aq) + H_2O(\ell) \rightleftharpoons H_3O^+(aq) + CHCl_2COO^-(aq)$	5.5×10^{-2}	1.26
ethanol	1	$CH_3CH_2OH(aq) + H_2O(\ell) \rightleftharpoons H_3O^+(aq) + CH_3CH_2O^-(aq)$	1.3×10^{-16}	15.9
fluoroacetic	1	$CH_2FCOOH(aq) + H_2O(\ell) \rightleftharpoons H_3O^+(aq) + CH_2FCOO^-(aq)$	2.6×10^{-3}	2.59
formic	1	$HCOOH(aq) + H_2O(\ell) \rightleftharpoons H_3O^+(aq) + HCOO^-(aq)$	1.77×10^{-4}	3.75
germanic	1	$H_2GeO_3(aq) + H_2O(\ell) \rightleftharpoons H_3O^+(aq) + HGeO_3^-(aq)$	9.8×10^{-10}	9.01
	2	$HGeO_3^-(aq) + H_2O(\ell) \rightleftharpoons H_3O^+(aq) + GeO_3^{2-}(aq)$	5×10^{-13}	12.3
hydr(o)azoic	1	$HN_3(aq) + H_2O(\ell) \rightleftharpoons H_3O^+(aq) + N_3^-(aq)$	1.9×10^{-5}	4.72
hydrobromic	1	$HBr(aq) + H_2O(\ell) \rightleftharpoons H_3O^+(aq) + Br^-(aq)$	$\gg$1 (strong)	<0
hydrochloric	1	$HCl(aq) + H_2O(\ell) \rightleftharpoons H_3O^+(aq) + Cl^-(aq)$	$\gg$1 (strong)	<0
hydrocyanic	1	$HCN(aq) + H_2O(\ell) \rightleftharpoons H_3O^+(aq) + CN^-(aq)$	6.2×10^{-10}	9.21
hydrofluoric	1	$HF(aq) + H_2O(\ell) \rightleftharpoons H_3O^+(aq) + F^-(aq)$	6.8×10^{-4}	3.17
hydr(o)iodic	1	$HI(aq) + H_2O(\ell) \rightleftharpoons H_3O^+(aq) + I^-(aq)$	$\gg$1 (strong)	<0
hydrosulfuric	1	$H_2S(aq) + H_2O(\ell) \rightleftharpoons H_3O^+(aq) + HS^-(aq)$	8.9×10^{-8}	7.05

Continued on next page

TABLE A5.1 Ionization Constants of Selected Acids at 25°C *(Continued)*

Acid	Step	Aqueous Equilibrium[a]	K_a	pK_a
	2	$HS^-(aq) + H_2O(\ell) \rightleftharpoons H_3O^+(aq) + S^{2-}(aq)$	$\sim 10^{-19}$	~19
hypobromous	1	$HBrO(aq) + H_2O(\ell) \rightleftharpoons H_3O^+(aq) + BrO^-(aq)$	2.3×10^{-9}	8.64
hypochlorous	1	$HClO(aq) + H_2O(\ell) \rightleftharpoons H_3O^+(aq) + ClO^-(aq)$	2.9×10^{-8}	7.54
hypoiodous	1	$HIO(aq) + H_2O(\ell) \rightleftharpoons H_3O^+(aq) + IO^-(aq)$	2.3×10^{-11}	10.64
iodic	1	$HIO_3(aq) + H_2O(\ell) \rightleftharpoons H_3O^+(aq) + IO_3^-(aq)$	1.7×10^{-1}	0.77
iodoacetic	1	$CH_2ICOOH(aq) + H_2O(\ell) \rightleftharpoons H_3O^+(aq) + CH_2ICOO^-(aq)$	7.6×10^{-4}	3.12
lactic	1	$CH_3CHOHCOOH(aq) + H_2O(\ell) \rightleftharpoons H_3O^+(aq) + CH_3CHOHCOO^-(aq)$	1.4×10^{-4}	3.85
maleic (*cis*-butenedioic)	1	$HOOCCH{=}CHCOOH(aq) + H_2O(\ell) \rightleftharpoons H_3O^+(aq) + HOOCCH{=}CHCOO^-(aq)$	1.2×10^{-2}	1.92
	2	$HOOCCH{=}CHCOO^-(aq) + H_2O(\ell) \rightleftharpoons H_3O^+(aq) + {}^-OOCCH{=}CHCOO^-(aq)$	4.7×10^{-7}	6.33
malonic	1	$HOOCCH_2COOH(aq) + H_2O(\ell) \rightleftharpoons H_3O^+(aq) + HOOCCH_2COO^-(aq)$	1.5×10^{-3}	2.82
	2	$HOOCCH_2COO^-(aq) + H_2O(\ell) \rightleftharpoons H_3O^+(aq) + {}^-OOCCH_2COO^-(aq)$	2.0×10^{-6}	5.70
nitric	1	$HNO_3(aq) + H_2O(\ell) \rightleftharpoons H_3O^+(aq) + NO_3^-(aq)$	$\gg$1 (strong)	<0
nitrous	1	$HNO_2(aq) + H_2O(\ell) \rightleftharpoons H_3O^+(aq) + NO_2^-(aq)$	4.0×10^{-4}	3.40
oxalic	1	$HOOCCOOH(aq) + H_2O(\ell) \rightleftharpoons H_3O^+(aq) + HOOCCOO^-(aq)$	5.9×10^{-2}	1.23
	2	$HOOCCOO^-(aq) + H_2O(\ell) \rightleftharpoons H_3O^+(aq) + {}^-OOCCOO^-(aq)$	6.4×10^{-5}	4.19
perchloric	1	$HClO_4(aq) + H_2O(\ell) \rightleftharpoons H_3O^+(aq) + ClO_4^-(aq)$	$\gg$1 (strong)	<0
periodic	1	$HIO_4(aq) + H_2O(\ell) \rightleftharpoons H_3O^+(aq) + IO_4^-(aq)$	2.3×10^{-2}	1.64
phenol	1	$C_6H_5OH(aq) + H_2O(\ell) \rightleftharpoons H_3O^+(aq) + C_6H_5O^-(aq)$	1.3×10^{-10}	9.89
phosphoric	1	$H_3PO_4(aq) + H_2O(\ell) \rightleftharpoons H_3O^+(aq) + H_2PO_4^-(aq)$	6.9×10^{-3}	2.16
	2	$H_2PO_4^-(aq) + H_2O(\ell) \rightleftharpoons H_3O^+(aq) + HPO_4^{2-}(aq)$	6.4×10^{-8}	7.19
	3	$HPO_4^{2-}(aq) + H_2O(\ell) \rightleftharpoons H_3O^+(aq) + PO_4^{3-}(aq)$	4.8×10^{-13}	12.32
propanoic	1	$CH_3CH_2COOH(aq) + H_2O(\ell) \rightleftharpoons H_3O^+(aq) + CH_3CH_2COO^-(aq)$	1.4×10^{-5}	4.85
pyruvic	1	$CH_3C(O)COOH(aq) + H_2O(\ell) \rightleftharpoons H_3O^+(aq) + CH_3C(O)COO^-(aq)$	2.8×10^{-3}	2.55
sulfuric	1	$H_2SO_4(aq) + H_2O(\ell) \rightleftharpoons H_3O^+(aq) + HSO_4^-(aq)$	$\gg$1 (strong)	<0
	2	$HSO_4^-(aq) + H_2O(\ell) \rightleftharpoons H_3O^+(aq) + SO_4^{2-}(aq)$	1.2×10^{-2}	1.92
sulfurous	1	$H_2SO_3(aq) + H_2O(\ell) \rightleftharpoons H_3O^+(aq) + HSO_3^-(aq)$	1.7×10^{-2}	1.77
	2	$HSO_3^-(aq) + H_2O(\ell) \rightleftharpoons H_3O^+(aq) + SO_3^{2-}(aq)$	6.2×10^{-8}	7.21
thiocyanic	1	$HSCN(aq) + H_2O(\ell) \rightleftharpoons H_3O^+(aq) + SCN^-(aq)$	$\gg$1 (strong)	<0
trichloroacetic	1	$CCl_3COOH(aq) + H_2O(\ell) \rightleftharpoons H_3O^+(aq) + CCl_3COO^-(aq)$	2.3×10^{-1}	0.64
trifluoroacetic	1	$CF_3COOH(aq) + H_2O(\ell) \rightleftharpoons H_3O^+(aq) + CF_3COO^-(aq)$	5.9×10^{-1}	0.23
water	1	$H_2O(aq) + H_2O(\ell) \rightleftharpoons H_3O^+(aq) + OH^-(aq)$	1.0×10^{-14}	14.00

[a]The formulas of most carboxylic acids are written in an RCOOH format to highlight their molecular structures.

TABLE A5.2 Acid Ionization Constants of Hydrated Metal Ions at 25°C

Free Ion	Hydrated Ion	K_a
Fe^{3+}	$Fe(H_2O)_6^{3+}$	3×10^{-3}
Sn^{2+}	$Sn(H_2O)_6^{2+}$	4×10^{-4}
Cr^{3+}	$Cr(H_2O)_6^{3+}$	1×10^{-4}
Al^{3+}	$Al(H_2O)_6^{3+}$	1×10^{-5}
Cu^{2+}	$Cu(H_2O)_6^{2+}$	3×10^{-8}
Pb^{2+}	$Pb(H_2O)_6^{2+}$	3×10^{-8}
Zn^{2+}	$Zn(H_2O)_6^{2+}$	1×10^{-9}
Co^{2+}	$Co(H_2O)_6^{2+}$	2×10^{-10}
Ni^{2+}	$Ni(H_2O)_6^{2+}$	1×10^{-10}

TABLE A5.3 Ionization Constants of Selected Bases at 25°C

Base	Aqueous Equilibrium	K_b	pK_b
ammonia	$NH_3(aq) + H_2O(\ell) \rightleftharpoons NH_4^+(aq) + OH^-(aq)$	1.76×10^{-5}	4.75
aniline	$C_6H_5NH_2(aq) + H_2O(\ell) \rightleftharpoons C_6H_5NH_3^+(aq) + OH^-(aq)$	4.0×10^{-10}	9.40
diethylamine	$(CH_3CH_2)_2NH(aq) + H_2O(\ell) \rightleftharpoons (CH_3CH_2)_2NH_2^+(aq) + OH^-(aq)$	8.6×10^{-4}	3.07
dimethylamine	$(CH_3)_2NH(aq) + H_2O(\ell) \rightleftharpoons (CH_3)_2NH_2^+(aq) + OH^-(aq)$	5.9×10^{-4}	3.23
ethylamine	$CH_3CH_2NH_2(aq) + H_2O(\ell) \rightleftharpoons CH_3CH_2NH_3^+(aq) + OH^-(aq)$	5.6×10^{-4}	3.25
methylamine	$CH_3NH_2(aq) + H_2O(\ell) \rightleftharpoons CH_3NH_3^+(aq) + OH^-(aq)$	4.4×10^{-4}	3.36
nicotine (1)	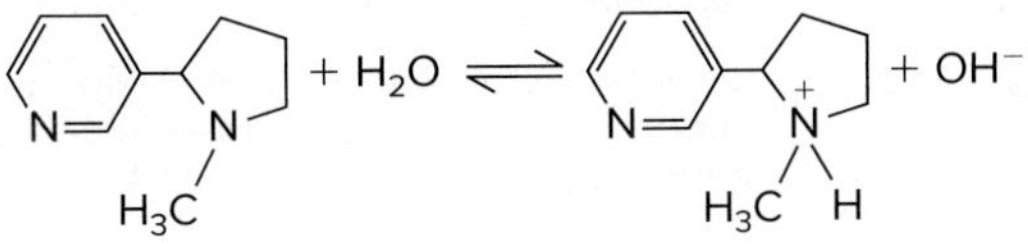 $+ H_2O \rightleftharpoons$ $+ OH^-$	1.0×10^{-6}	6.0
(2)	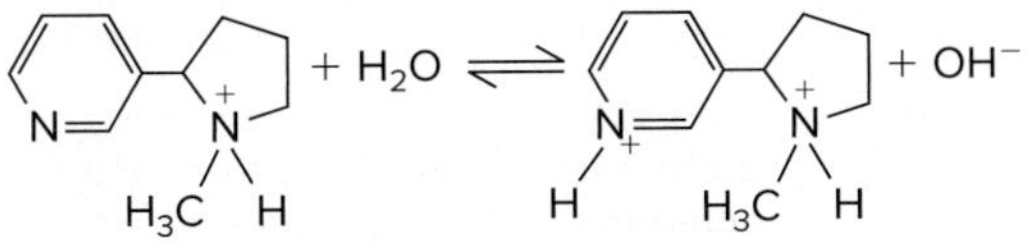 $+ H_2O \rightleftharpoons$ $+ OH^-$	1.3×10^{-11}	10.9
pyridine	$C_5H_5N(aq) + H_2O(\ell) \rightleftharpoons C_5H_5NH^+(aq) + OH^-(aq)$	1.7×10^{-9}	8.77
quinine (1)	$+ H_2O \rightleftharpoons$ $+ OH^-$	3.3×10^{-6}	5.48
(2)	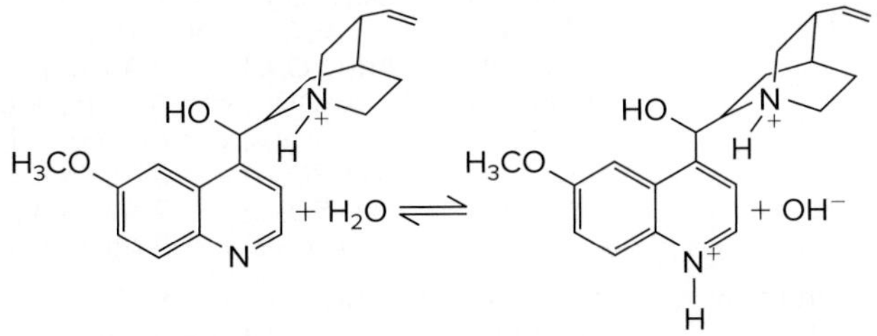 $+ H_2O \rightleftharpoons$ $+ OH^-$	1.4×10^{-10}	9.9
trimethylamine	$(CH_3)_3N(aq) + H_2O(\ell) \rightleftharpoons (CH_3)_3NH^+(aq) + OH^-(aq)$	6.46×10^{-5}	4.19
urea	$H_2NCONH_2(aq) + H_2O(\ell) \rightleftharpoons H_2NCONH_3^+(aq) + OH^-(aq)$	1.3×10^{-14}	13.9

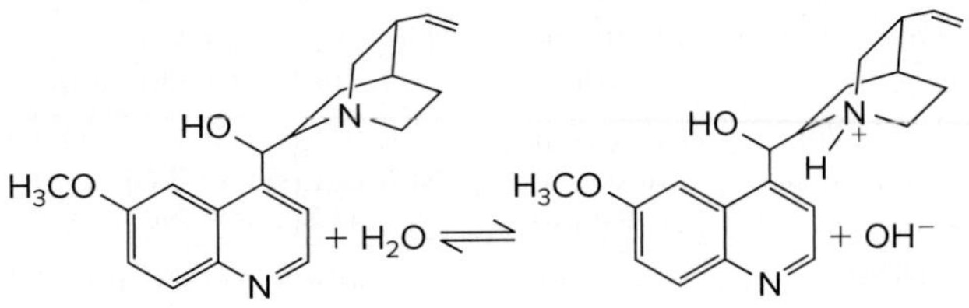

TABLE A5.4 Solubility-Product Constants at 25°C

Cation	Anion	Heterogeneous Equilibrium[a]	K_{sp}[b]
aluminum	hydroxide	$Al(OH)_3(s) \rightleftharpoons Al^{3+}(aq) + 3\ OH^-(aq)$	1.3×10^{-33}
	phosphate	$AlPO_4(s) \rightleftharpoons Al^{3+}(aq) + PO_4^{3-}(aq)$	9.84×10^{-21}
barium	carbonate	$BaCO_3(s) \rightleftharpoons Ba^{2+}(aq) + CO_3^{2-}(aq)$	2.58×10^{-9}
	fluoride	$BaF_2(s) \rightleftharpoons Ba^{2+}(aq) + 2\ F^-(aq)$	1.84×10^{-7}
	sulfate	$BaSO_4(s) \rightleftharpoons Ba^{2+}(aq) + SO_4^{2-}(aq)$	1.08×10^{-10}
calcium	carbonate	$CaCO_3(s) \rightleftharpoons Ca^{2+}(aq) + CO_3^{2-}(aq)$	2.8×10^{-9}
	fluoride	$CaF_2(s) \rightleftharpoons Ca^{2+}(aq) + 2\ F^-(aq)$	5.3×10^{-9}
	hydroxide	$Ca(OH)_2(s) \rightleftharpoons Ca^{2+}(aq) + 2\ OH^-(aq)$	5.5×10^{-6}
	phosphate	$Ca_3(PO_4)_2(s) \rightleftharpoons 3\ Ca^{2+}(aq) + 2\ PO_4^{3-}(aq)$	2.07×10^{-29}
	sulfate	$CaSO_4(s) \rightleftharpoons Ca^{2+}(aq) + SO_4^{2-}(aq)$	4.93×10^{-5}
cobalt(II)	carbonate	$CoCO_3(s) \rightleftharpoons Co^{2+}(aq) + CO_3^{2-}(aq)$	1.4×10^{-3}
	phosphate	$Co_3(PO_4)_2(s) \rightleftharpoons 3\ Co^{2+}(aq) + 2\ PO_4^{3-}(aq)$	2.05×10^{-7}
	sulfide	$CoS(s) \rightleftharpoons Co^{2+}(aq) + S^{2-}(aq)$	2.0×10^{-25}
copper(I)	bromide	$CuBr(s) \rightleftharpoons Cu^+(aq) + Br^-(aq)$	6.27×10^{-9}
	chloride	$CuCl(s) \rightleftharpoons Cu^+(aq) + Cl^-(aq)$	1.72×10^{-7}
	iodide	$CuI(s) \rightleftharpoons Cu^+(aq) + I^-(aq)$	1.27×10^{-12}
copper(II)	phosphate	$Cu_3(PO_4)_2(s) \rightleftharpoons 3\ Cu^{2+}(aq) + 2\ PO_4^{3-}(aq)$	1.4×10^{-37}
	hydroxide	$Cu(OH)_2(s) \rightleftharpoons Cu^{2+}(aq) + 2\ OH^-(aq)$	2.2×10^{-20}
iron(II)	carbonate	$FeCO_3(s) \rightleftharpoons Fe^{2+}(aq) + CO_3^{2-}(aq)$	3.13×10^{-11}
	fluoride	$FeF_2(s) \rightleftharpoons Fe^{2+}(aq) + 2\ F^-(aq)$	2.36×10^{-6}
	hydroxide	$Fe(OH)_2(s) \rightleftharpoons Fe^{2+}(aq) + 2\ OH^-(aq)$	4.87×10^{-17}
	sulfide	$FeS(s) \rightleftharpoons Fe^{2+}(aq) + S^{2-}(aq)$	6.3×10^{-18}
lead	bromide	$PbBr_2(s) \rightleftharpoons Pb^{2+}(aq) + 2\ Br^-(aq)$	6.60×10^{-6}
	carbonate	$PbCO_3(s) \rightleftharpoons Pb^{2+}(aq) + CO_3^{2-}(aq)$	7.4×10^{-14}
	chloride	$PbCl_2(s) \rightleftharpoons Pb^{2+}(aq) + 2\ Cl^-(aq)$	1.7×10^{-5}
	fluoride	$PbF_2(s) \rightleftharpoons Pb^{2+}(aq) + 2\ F^-(aq)$	3.3×10^{-8}
	iodide	$PbI_2(s) \rightleftharpoons Pb^{2+}(aq) + 2\ I^-(aq)$	9.8×10^{-9}
	sulfate	$PbSO_4(s) \rightleftharpoons Pb^{2+}(aq) + SO_4^{2-}(aq)$	2.53×10^{-8}
lithium	carbonate	$Li_2CO_3(s) \rightleftharpoons 2\ Li^+(aq) + CO_3^{2-}(aq)$	2.5×10^{-2}
magnesium	carbonate	$MgCO_3(s) \rightleftharpoons Mg^{2+}(aq) + CO_3^{2-}(aq)$	6.82×10^{-6}
	fluoride	$MgF_2(s) \rightleftharpoons Mg^{2+}(aq) + 2\ F^-(aq)$	5.16×10^{-11}
	hydroxide	$Mg(OH)_2(s) \rightleftharpoons Mg^{2+}(aq) + 2\ OH^-(aq)$	5.61×10^{-12}
manganese(II)	carbonate	$MnCO_3(s) \rightleftharpoons Mn^{2+}(aq) + CO_3^{2-}(aq)$	2.34×10^{-11}
	hydroxide	$Mn(OH)_2(s) \rightleftharpoons Mn^{2+}(aq) + 2\ OH^-(aq)$	1.9×10^{-13}
mercury(I)	bromide	$Hg_2Br_2(s) \rightleftharpoons Hg_2^{2+}(aq) + 2\ Br^-(aq)$	6.40×10^{-23}
	carbonate	$Hg_2CO_3(s) \rightleftharpoons Hg_2^{2+}(aq) + CO_3^{2-}(aq)$	3.6×10^{-17}
	chloride	$Hg_2Cl_2(s) \rightleftharpoons Hg_2^{2+}(aq) + 2\ Cl^-(aq)$	1.43×10^{-18}
	iodide	$Hg_2I_2(s) \rightleftharpoons Hg_2^{2+}(aq) + 2\ I^-(aq)$	5.2×10^{-29}
	sulfate	$Hg_2SO_4(s) \rightleftharpoons Hg_2^{2+}(aq) + SO_4^{2-}(aq)$	6.5×10^{-7}
mercury(II)	hydroxide	$Hg(OH)_2(s) \rightleftharpoons Hg^{2+}(aq) + 2\ OH^-(aq)$	3.2×10^{-26}
	iodide	$HgI_2(s) \rightleftharpoons Hg^{2+}(aq) + 2\ I^-(aq)$	2.9×10^{-29}
nickel(II)	carbonate	$NiCO_3(s) \rightleftharpoons Ni^{2+}(aq) + CO_3^{2-}(aq)$	1.42×10^{-7}
	phosphate	$Ni_3(PO_4)_2(s) \rightleftharpoons 3\ Ni^{2+}(aq) + 2\ PO_4^{3-}(aq)$	4.74×10^{-32}
	sulfide	$NiS(s) \rightleftharpoons Ni^{2+}(aq) + S^{2-}(aq)$	1×10^{-24}
silver	bromide	$AgBr(s) \rightleftharpoons Ag^+(aq) + Br^-(aq)$	5.35×10^{-13}
	carbonate	$Ag_2CO_3(s) \rightleftharpoons 2\ Ag^+(aq) + CO_3^{2-}(aq)$	8.46×10^{-12}
	chloride	$AgCl(s) \rightleftharpoons Ag^+(aq) + Cl^-(aq)$	1.77×10^{-10}
	chromate	$Ag_2CrO_4(s) \rightleftharpoons 2\ Ag^+(aq) + CrO_4^{2-}(aq)$	1.12×10^{-12}
	hydroxide	$AgOH(s) \rightleftharpoons Ag^+(aq) + OH^-(aq)$	2.0×10^{-8}
	iodide	$AgI(s) \rightleftharpoons Ag^+(aq) + I^-(aq)$	8.52×10^{-17}
	phosphate	$Ag_3PO_4(s) \rightleftharpoons 3\ Ag^+(aq) + PO_4^{3-}(aq)$	8.89×10^{-17}
	sulfate	$Ag_2SO_4(s) \rightleftharpoons 2\ Ag^+(aq) + SO_4^{2-}(aq)$	1.20×10^{-5}
	sulfide	$Ag_2S(s) \rightleftharpoons 2\ Ag^+(aq) + S^{2-}(aq)$	6.3×10^{-50}
strontium	carbonate	$SrCO_3(s) \rightleftharpoons Sr^{2+}(aq) + CO_3^{2-}(aq)$	5.60×10^{-10}
	fluoride	$SrF_2(s) \rightleftharpoons Sr^{2+}(aq) + 2\ F^-(aq)$	4.33×10^{-9}
	sulfate	$SrSO_4(s) \rightleftharpoons Sr^{2+}(aq) + SO_4^{2-}(aq)$	3.44×10^{-7}
zinc	carbonate	$ZnCO_3(s) \rightleftharpoons Zn^{2+}(aq) + CO_3^{2-}(aq)$	1.46×10^{-10}
	hydroxide	$Zn(OH)_2(s) \rightleftharpoons Zn^{2+}(aq) + 2\ OH^-(aq)$	3.0×10^{-17}

[a]Equilibrium is between solid phase and aqueous solution.

[b]From Dean, J. *Lange's Handbook of Chemistry* (The McGraw-Hill Companies, 1998).

TABLE A5.5 Formation Constants of Complex Ions at 25°C

Complex Ion	Aqueous Equilibrium	K_f
$[Ag(NH_3)_2]^+$	$Ag^+(aq) + 2\ NH_3(aq) \rightleftharpoons Ag(NH_3)_2^{\ +}(aq)$	1.7×10^7
$[AgCl_2]^-$	$Ag^+(aq) + 2\ Cl^-(aq) \rightleftharpoons AgCl_2^{\ -}(aq)$	2.5×10^5
$[Ag(CN)_2]^-$	$Ag^+(aq) + 2\ CN^-(aq) \rightleftharpoons Ag(CN)_2^{\ -}(aq)$	1.0×10^{21}
$[Ag(S_2O_3)_2]^{3-}$	$Ag^+(aq) + 2\ S_2O_3^{\ 2-}(aq) \rightleftharpoons Ag(S_2O_3)_2^{\ 3-}(aq)$	4.7×10^{13}
$[AlF_6]^{3-}$	$Al^{3+}(aq) + 6\ F^-(aq) \rightleftharpoons AlF_6^{\ 3-}(aq)$	4.0×10^{19}
$[Al(OH)_4]^-$	$Al^{3+}(aq) + 4\ OH^-(aq) \rightleftharpoons Al(OH)_4^{\ -}(aq)$	7.7×10^{33}
$[Au(CN)_2]^-$	$Au^+(aq) + 2\ CN^-(aq) \rightleftharpoons Au(CN)_2^{\ -}(aq)$	2.0×10^{38}
$[Co(NH_3)_6]^{2+}$	$Co^{2+}(aq) + 6\ NH_3(aq) \rightleftharpoons Co(NH_3)_6^{\ 2+}(aq)$	7.7×10^4
$[Co(NH_3)_6]^{3+}$	$Co^{3+}(aq) + 6\ NH_3(aq) \rightleftharpoons Co(NH_3)_6^{\ 3+}(aq)$	5.0×10^{31}
$[Co(en)_3]^{2+}$	$Co^{2+}(aq) + 3\ en(aq) \rightleftharpoons Co(en)_3^{\ 2+}(aq)$	8.7×10^{13}
$[Co(C_2O_4)_3]^{4-}$	$Co^{2+}(aq) + 3\ C_2O_4^{\ 2-}(aq) \rightleftharpoons Co(C_2O_4)_3^{\ 4-}(aq)$	4.5×10^6
$[Cu(NH_3)_4]^{2+}$	$Cu^{2+}(aq) + 4\ NH_3(aq) \rightleftharpoons Cu(NH_3)_4^{\ 2+}(aq)$	5.0×10^{13}
$[Cu(en)_2]^{2+}$	$Cu^{2+}(aq) + 2\ en(aq) \rightleftharpoons Cu(en)_2^{\ 2+}(aq)$	3.2×10^{19}
$[Cu(CN)_4]^{2-}$	$Cu^{2+}(aq) + 4\ CN^-(aq) \rightleftharpoons Cu(CN)_4^{\ 2-}(aq)$	1.0×10^{25}
$[Cu(C_2O_4)_2]^{2-}$	$Cu^{2+}(aq) + 2\ C_2O_4^{\ 2-}(aq) \rightleftharpoons Cu(C_2O_4)_2^{\ 2-}(aq)$	1.7×10^{10}
$[Fe(C_2O_4)_3]^{4-}$	$Fe^{2+}(aq) + 3\ C_2O_4^{\ 2-}(aq) \rightleftharpoons Fe(C_2O_4)_3^{\ 4-}(aq)$	6×10^6
$[Fe(C_2O_4)_3]^{3-}$	$Fe^{3+}(aq) + 3\ C_2O_4^{\ 2-}(aq) \rightleftharpoons Fe(C_2O_4)_3^{\ 3-}(aq)$	3.3×10^{20}
$[HgCl_4]^{2-}$	$Hg^{2+}(aq) + 4\ Cl^-(aq) \rightleftharpoons HgCl_4^{\ 2-}(aq)$	1.2×10^{15}
$[Ni(NH_3)_6]^{2+}$	$Ni^{2+}(aq) + 6\ NH_3(aq) \rightleftharpoons Ni(NH_3)_6^{\ 2+}(aq)$	5.5×10^8
$[PbCl_4]^{2-}$	$Pb^{2+}(aq) + 4\ Cl^-(aq) \rightleftharpoons PbCl_4^{\ 2-}(aq)$	2.5×10^1
$[Zn(NH_3)_4]^{2+}$	$Zn^{2+}(aq) + 4\ NH_3(aq) \rightleftharpoons Zn(NH_3)_4^{\ 2+}(aq)$	2.9×10^9
$[Zn(OH)_4]^{2-}$	$Zn^{2+}(aq) + 4\ OH^-(aq) \rightleftharpoons Zn(OH)_4^{\ 2-}(aq)$	2.8×10^{15}

Appendix 6

Standard Reduction Potentials

TABLE A6.1 Standard Reduction Potentials at 25°C

Half-Reaction	n	$E°$ (V)
$F_2(g) + 2\ e^- \rightarrow 2\ F^-(aq)$	2	2.866
$H_2N_2O_2(s) + 2\ H^+(aq) + 2\ e^- \rightarrow N_2(g) + 2\ H_2O(\ell)$	2	2.65
$O(g) + 2\ H^+(aq) + 2\ e^- \rightarrow H_2O(\ell)$	2	2.421
$Cu^{3+}(aq) + e^- \rightarrow Cu^{2+}(aq)$	1	2.4
$XeO_3(s) + 6\ H^+(aq) + 6\ e^- \rightarrow Xe(g) + 3\ H_2O(\ell)$	6	2.10
$O_3(g) + 2\ H^+(aq) + 2\ e^- \rightarrow O_2(g) + H_2O(\ell)$	2	2.076
$OH(g) + e^- \rightarrow OH^-(aq)$	1	2.02
$Co^{3+}(aq) + e^- \rightarrow Co^{2+}(aq)$	1	1.92
$H_2O_2(\ell) + 2\ H^+(aq) + 2\ e^- \rightarrow 2\ H_2O(\ell)$	2	1.776
$N_2O(g) + 2\ H^+(aq) + 2\ e^- \rightarrow N_2(g) + H_2O(\ell)$	2	1.766
$Ce(OH)^{3+}(aq) + H^+(aq) + e^- \rightarrow Ce^{3+}(aq) + H_2O(\ell)$	1	1.70
$Au^+(aq) + e^- \rightarrow Au(s)$	1	1.692
$PbO_2(s) + SO_4^{2-}(aq) + 4\ H^+(aq) + 2\ e^- \rightarrow PbSO_4(s) + 2\ H_2O(\ell)$	2	1.691
$PbO_2(s) + HSO_4^-(aq) + 3\ H^+(aq) + 2\ e^- \rightarrow PbSO_4(s) + 2\ H_2O(\ell)$	2	1.685
$MnO_4^-(aq) + 4\ H^+(aq) + 3\ e^- \rightarrow MnO_2(s) + 2\ H_2O(\ell)$	3	1.673
$NiO_2(s) + 4\ H^+(aq) + 2\ e^- \rightarrow Ni^{2+}(aq) + 2\ H_2O(\ell)$	2	1.678
$HClO(\ell) + H^+(aq) + e^- \rightarrow \frac{1}{2}\ Cl_2(g) + H_2O(aq)$	1	1.63
$Ce^{4+}(aq) + e^- \rightarrow Ce^{3+}(aq)$	1	1.61
$Mn^{3+}(aq) + e^- \rightarrow Mn^{2+}(aq)$	1	1.542
$MnO_4^-(aq) + 8\ H^+(aq) + 5\ e^- \rightarrow Mn^{2+}(aq) + 4\ H_2O(\ell)$	5	1.507
$BrO_3^-(aq) + 6\ H^+(aq) + 5\ e^- \rightarrow \frac{1}{2}\ Br_2(\ell) + 3\ H_2O(\ell)$	5	1.52
$ClO_3^-(aq) + 6\ H^+(aq) + 5\ e^- \rightarrow \frac{1}{2}\ Cl_2(g) + 3\ H_2O(\ell)$	5	1.47
$PbO_2(s) + 4\ H^+(aq) + 2\ e^- \rightarrow Pb^{2+}(aq) + 2\ H_2O(\ell)$	2	1.455
$Au^{3+}(aq) + 3\ e^- \rightarrow Au(s)$	3	1.40
$Cl_2(g) + 2\ e^- \rightarrow 2\ Cl^-(aq)$	2	1.358
$Cr_2O_7^{2-}(aq) + 14\ H^+(aq) + 6\ e^- \rightarrow 2\ Cr^{3+}(aq) + 7\ H_2O(\ell)$	6	1.33
$MnO_2(s) + 4\ H^+(aq) + 2\ e^- \rightarrow Mn^{2+}(aq) + 2\ H_2O(\ell)$	2	1.23
$O_2(g) + 4\ H^+(aq) + 4\ e^- \rightarrow 2\ H_2O(\ell)$	4	1.229
$IO_3^-(aq) + 6\ H^+(aq) + 5\ e^- \rightarrow \frac{1}{2}\ I_2(s) + 3\ H_2O(\ell)$	5	1.195
$IO_3^-(aq) + 6\ H^+(aq) + 6\ e^- \rightarrow I^-(aq) + 3\ H_2O(\ell)$	6	1.085
$Br_2(\ell) + 2\ e^- \rightarrow 2\ Br^-(aq)$	2	1.066
$HNO_2(\ell) + H^+(aq) + e^- \rightarrow NO(g) + H_2O(\ell)$	1	1.00
$VO_2^+(aq) + 2\ H^+(aq) + e^- \rightarrow VO^{2+}(aq) + H_2O(\ell)$	1	1.00
$NO_3^-(aq) + 4\ H^+(aq) + 3\ e^- \rightarrow NO(g) + 2\ H_2O(\ell)$	3	0.96
$2\ Hg^{2+}(aq) + 2\ e^- \rightarrow Hg_2^{2+}(aq)$	2	0.92

TABLE A6.1 Standard Reduction Potentials at 25°C *(Continued)*

Half-Reaction	*n*	*E*° (V)
$ClO^-(aq) + H_2O(\ell) + 2\ e^- \rightarrow Cl^-(aq) + 2\ OH^-(aq)$	2	0.89
$HO_2^-(aq) + H_2O(\ell) + 2\ e^- \rightarrow 3\ OH^-(aq)$	2	0.88
$Hg^{2+}(aq) + 2\ e^- \rightarrow Hg(\ell)$	2	0.851
$Ag^+(aq) + e^- \rightarrow Ag(s)$	1	0.800
$Hg_2^{2+}(aq) + 2\ e^- \rightarrow 2\ Hg(\ell)$	2	0.797
$Fe^{3+}(aq) + e^- \rightarrow Fe^{2+}(aq)$	1	0.770
$PtCl_4^{2-}(aq) + 2\ e^- \rightarrow Pt(s) + 4\ Cl^-(aq)$	2	0.73
$O_2(g) + 2\ H^+(aq) + 2\ e^- \rightarrow H_2O_2(\ell)$	2	0.68
$MnO_4^-(aq) + 2\ H_2O(\ell) + 3\ e^- \rightarrow MnO_2(s) + 4\ OH^-(aq)$	3	0.59
$H_3AsO_4(s) + 2\ H^+(aq) + 2\ e^- \rightarrow H_3AsO_3(aq) + H_2O(\ell)$	2	0.559
$I_2(s) + 2\ e^- \rightarrow 2\ I^-(aq)$	2	0.536
$Cu^+(aq) + e^- \rightarrow Cu(s)$	1	0.521
$2\ NiO(OH)(s) + 2\ H_2O(\ell) + 2\ e^- \rightarrow 2\ Ni(OH)_2(s) + 2\ OH^-(aq)$	2	0.52
$H_2SO_3(\ell) + 4\ H^+(aq) + 4\ e^- \rightarrow S(s) + 3\ H_2O(\ell)$	4	0.449
$Ag_2CrO_4(s) + 2\ e^- \rightarrow 2\ Ag(s) + CrO_4^{2-}(aq)$	2	0.447
$O_2(g) + 2\ H_2O(\ell) + 4\ e^- \rightarrow 4\ OH^-(aq)$	4	0.401
$Fe(CN)_6^{3-}(aq) + e^- \rightarrow Fe(CN)_6^{4-}(aq)$	1	0.36
$Ag_2O(s) + H_2O(\ell) + 2\ e^- \rightarrow 2\ Ag(s) + 2\ OH^-(aq)$	2	0.342
$Cu^{2+}(aq) + 2\ e^- \rightarrow Cu(s)$	2	0.342
$BiO^+(aq) + 2\ H^+(aq) + 3\ e^- \rightarrow Bi(s) + H_2O(\ell)$	3	0.32
$AgCl(s) + e^- \rightarrow Ag(s) + Cl^-(aq)$	1	0.222
$HSO_4^-(aq) + 3\ H^+(aq) + 2\ e^- \rightarrow H_2SO_3(\ell) + H_2O(\ell)$	2	0.17
$Sn^{4+}(aq) + 2\ e^- \rightarrow Sn^{2+}(aq)$	2	0.154
$Cu^{2+}(aq) + e^- \rightarrow Cu^+(aq)$	1	0.153
$2\ MnO_2(s) + H_2O(\ell) + 2\ e^- \rightarrow Mn_2O_3(s) + 2\ OH^-(aq)$	2	0.15
$S(s) + 2\ H^+(aq) + 2\ e^- \rightarrow H_2S(g)$	2	0.141
$HgO(s) + H_2O(\ell) + 2\ e^- \rightarrow Hg(\ell) + 2\ OH^-(aq)$	2	0.0977
$AgBr(s) + e^- \rightarrow Ag(s) + Br^-(aq)$	1	0.095
$Ag(S_2O_3)_2^{3-}(aq) + e^- \rightarrow Ag(s) + 2\ S_2O_3^{2-}(aq)$	1	0.01
$NO_3^-(aq) + H_2O(\ell) + 2\ e^- \rightarrow NO_2^-(aq) + 2\ OH^-(aq)$	2	0.01
$2\ H^+(aq) + 2\ e^- \rightarrow H_2(g)$	2	0.000
$Pb^{2+}(aq) + 2\ e^- \rightarrow Pb(s)$	2	−0.126
$CrO_4^{2-}(aq) + 4\ H_2O(\ell) + 3\ e^- \rightarrow Cr(OH)_3(s) + 5\ OH^-(aq)$	3	−0.13
$Sn^{2+}(aq) + 2\ e^- \rightarrow Sn(s)$	2	−0.136
$AgI(s) + e^- \rightarrow Ag(s) + I^-(aq)$	1	−0.152
$CuI(s) + e^- \rightarrow Cu(s) + I^-(aq)$	1	−0.185
$N_2(g) + 5\ H^+(aq) + 4\ e^- \rightarrow N_2H_5^+(aq)$	4	−0.23
$Ni^{2+}(aq) + 2\ e^- \rightarrow Ni(s)$	2	−0.257
$PbSO_4(s) + H^+(aq) + 2\ e^- \rightarrow Pb(s) + HSO_4^-(aq)$	2	−0.356
$Co^{2+}(aq) + 2\ e^- \rightarrow Co(s)$	2	−0.277
$Ag(CN)_2^-(aq) + e^- \rightarrow Ag(s) + 2\ CN^-(aq)$	1	−0.31
$Cd^{2+}(aq) + 2\ e^- \rightarrow Cd(s)$	2	−0.403
$Cr^{3+}(aq) + e^- \rightarrow Cr^{2+}(aq)$	1	−0.41

Continued on next page

TABLE A6.1 Standard Reduction Potentials at 25°C *(Continued)*

Half-Reaction	*n*	*E*° (V)
$Fe^{2+}(aq) + 2\ e^- \rightarrow Fe(s)$	2	−0.447
$2\ CO_2(g) + 2\ H^+(aq) + 2\ e^- \rightarrow H_2C_2O_4(s)$	2	−0.49
$Ni(OH)_2(s) + 2\ e^- \rightarrow Ni(s) + 2\ OH^-(aq)$	2	−0.72
$Cr^{3+}(aq) + 3\ e^- \rightarrow Cr(s)$	3	−0.74
$Zn^{2+}(aq) + 2\ e^- \rightarrow Zn(s)$	2	−0.762
$Cd(OH)_2(s) + 2\ e^- \rightarrow Cd(s) + 2\ OH^-(aq)$	2	−0.81
$2\ H_2O(\ell) + 2\ e^- \rightarrow H_2(g) + 2\ OH^-(aq)$	2	−0.828
$SO_4^{2-}(aq) + H_2O(\ell) + 2\ e^- \rightarrow SO_3^{2-}(aq) + 2\ OH^-(aq)$	2	−0.92
$N_2(g) + 4\ H_2O(\ell) + 4\ e^- \rightarrow 4\ OH^-(aq) + N_2H_4(\ell)$	4	−1.16
$Mn^{2+}(aq) + 2\ e^- \rightarrow Mn(s)$	2	−1.185
$Zn(OH)_2(s) + 2\ e^- \rightarrow Zn(s) + 2\ OH^-(aq)$	2	−1.249
$ZnO(s) + H_2O(\ell) + 2\ e^- \rightarrow Zn(s) + 2\ OH^-(aq)$	2	−1.25
$Al^{3+}(aq) + 3\ e^- \rightarrow Al(s)$	3	−1.662
$Mg^{2+}(aq) + 2\ e^- \rightarrow Mg(s)$	2	−2.37
$Na^+(aq) + e^- \rightarrow Na(s)$	1	−2.71
$Ca^{2+}(aq) + 2\ e^- \rightarrow Ca(s)$	2	−2.868
$Ba^{2+}(aq) + 2\ e^- \rightarrow Ba(s)$	2	−2.912
$K^+(aq) + e^- \rightarrow K(s)$	1	−2.95
$Li^+(aq) + e^- \rightarrow Li(s)$	1	−3.05

Appendix 7

Naming Organic Compounds

While organic chemistry was becoming established as a discipline within chemistry, many compounds were given trivial names that are still commonly used and recognized. We refer to many of these compounds by their nonsystematic names throughout this book, and their names and structures are listed in Table A7.1.

TABLE A7.1 Organic Compounds and Their Commonly Used Nonsystematic Names

Name	Formula	Structure
ethylene	C_2H_4	$H_2C{=}CH_2$
acetylene	C_2H_2	$HC{\equiv}CH$
benzene	C_6H_6	
toluene	$C_6H_5CH_3$	
ethyl alcohol	CH_3CH_2OH	$H_3C{-}CH_2{-}OH$
acetone	CH_3COCH_3	$H_3C{-}C({=}O){-}CH_3$
acetic acid	CH_3COOH	$H_3C{-}C({=}O){-}OH$
formaldehyde	CH_2O	$H{-}C({=}O){-}H$

The International Union of Pure and Applied Chemistry (IUPAC) has proposed a set of rules for the systematic naming of organic compounds. When naming compounds or drawing structures based on names, we need to keep in mind that the IUPAC system of nomenclature is based on two fundamental ideas: (1) the name of a compound must indicate how the carbon atoms in the skeleton are bonded together, and (2) the name must identify the location of any functional groups in the molecule.

Alkanes

Table A7.2 contains the prefixes used for carbon chains ranging in size from C_1 to C_{20} and gives the names for compounds consisting of unbranched chains. The name of a compound consists of a prefix identifying the number of carbons in the chain and a suffix defining the type of hydrocarbon. The suffix *-ane* indicates that the compounds are alkanes and that all carbon–carbon bonds are single bonds.

TABLE A7.2 Prefixes for Naming Carbon Chains

Prefix	Example	Name	Prefix	Example	Name
meth	CH_4	methane	undec	$C_{11}H_{24}$	undecane
eth	C_2H_6	ethane	dodec	$C_{12}H_{26}$	dodecane
pro	C_3H_8	propane	tridec	$C_{13}H_{28}$	tridecane
but	C_4H_{10}	butane	tetradec	$C_{14}H_{30}$	tetradecane
pent	C_5H_{12}	pentane	pentadec	$C_{15}H_{32}$	pentadecane
hex	C_6H_{14}	hexane	hexadec	$C_{16}H_{34}$	hexadecane
hept	C_7H_{16}	heptane	heptadec	$C_{17}H_{36}$	heptadecane
oct	C_8H_{18}	octane	octadec	$C_{18}H_{38}$	octadecane
non	C_9H_{20}	nonane	nonadec	$C_{19}H_{40}$	nonadecane
dec	$C_{10}H_{22}$	decane	eicos	$C_{20}H_{42}$	eicosane

Branched-Chain Alkanes

The alkane drawn here is used to illustrate each step in the naming rules:

$$\begin{array}{c} \qquad\qquad CH_3 \\ \qquad\qquad | \\ CH_3CH_2CHCH_2CHCHCH_2CH_2CH_3 \\ \quad | \qquad\quad | \\ \quad CH_3 \qquad CH_2CH_3 \end{array}$$

1. **Identify and name the longest continuous carbon chain.**

$$\begin{array}{c} \qquad\qquad CH_3 \\ \qquad\qquad | \\ \boxed{CH_3CH_2CHCH_2CHCHCH_2CH_2CH_3}\ \text{Nonane} \\ \quad | \qquad\quad | \\ \quad CH_3 \qquad CH_2CH_3 \end{array}$$

2. **Identify the groups attached to this chain and name them.** Names of substituent groups consist of the prefix from Table A7.2 that identifies the length of the group and the suffix *-yl* that identifies it as an alkyl group.

$$\begin{array}{c} \qquad\qquad \text{methyl-} \\ \qquad\qquad \boxed{CH_3} \\ \qquad\qquad | \\ CH_3CH_2CHCH_2CHCHCH_2CH_2CH_3 \\ \quad | \qquad\quad | \\ \quad \boxed{CH_3} \qquad \boxed{CH_2CH_3} \\ \quad \text{methyl-} \qquad \text{ethyl-} \end{array}$$

3. **Number the carbon atoms in the longest chain,** starting at the end nearest a substituent group. Doing this identifies the points of attachment of the alkyl groups with the lowest possible numbers.

$$\begin{array}{c} \qquad\qquad \text{methyl-} \\ \qquad\qquad \boxed{CH_3} \\ 1\ \ 2\ \ 3\ \ 4\ \ 5|\ \ 6\ \ 7\ \ 8\ \ 9 \\ CH_3CH_2CHCH_2CHCHCH_2CH_2CH_3 \\ \quad | \qquad\quad | \\ \quad \boxed{CH_3} \qquad \boxed{CH_2CH_3} \\ \quad \text{methyl-} \qquad \text{ethyl-} \end{array}$$

4. **Designate the location and identity of each substituent group with a number,** followed by a hyphen, and its name.

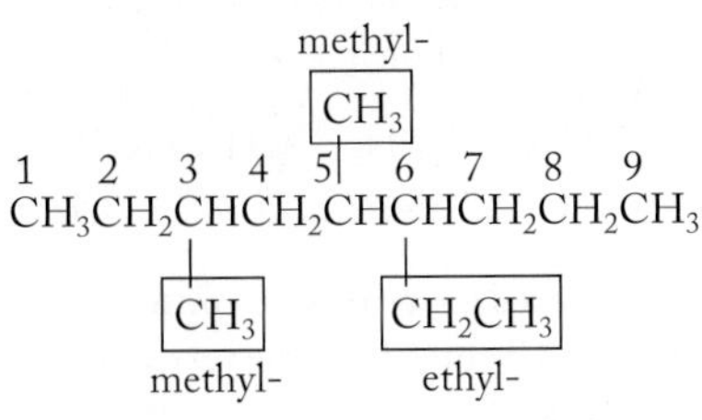

3-methyl-, 5-methyl-, 6-ethyl-

5. **Put together the complete name by listing the substituent groups in alphabetical order.** If more than one of a given type of substituent group is present, prefixes *di-*, *tri-*, *tetra-*, and so forth, are appended to the names, but these numerical prefixes are not considered when determining the alphabetical order. The name of the last substituent group is written together with the name identifying the longest carbon chain.

$$\begin{array}{c} CH_3 \\ | \\ CH_3CH_2\underset{\substack{|\\CH_3}}{CH}CH_2CH\underset{\substack{|\\CH_2CH_3}}{CH}CH_2CH_2CH_3 \end{array}$$

6-Ethyl-3,5-dimethylnonane

Cycloalkanes

The simplest examples of this class of compounds consist of one unsubstituted ring of carbon atoms. The IUPAC names of these compounds consist of the prefix *cyclo-* followed by the parent name from Table A7.2 to indicate the number of carbon atoms in the ring. As an illustration, the names, formulas, and line structures of the first three cycloalkanes in the homologous series are

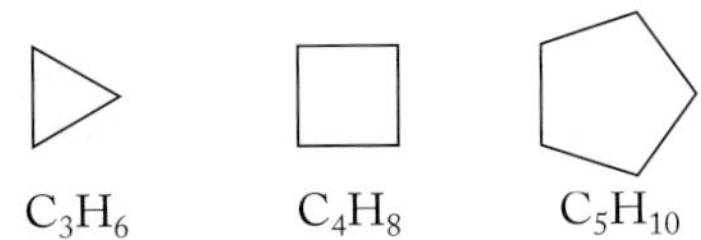

C_3H_6	C_4H_8	C_5H_{10}
Cyclopropane	Cyclobutane	Cyclopentane

Alkenes and Alkynes

Alkenes have carbon–carbon double bonds and alkynes have carbon–carbon triple bonds as functional groups. The names of these types of compounds consist of (1) a parent name that identifies the longest carbon chain that includes the double or triple bond, (2) a suffix that identifies the class of compound, and (3) names of any substituent groups attached to the longest carbon chain. The suffix *-ene* identifies an alkene; *-yne* identifies an alkyne.

The alkene and alkyne drawn here are used to illustrate each step in the naming rules:

$$CH_3\overset{\substack{CH_3\\|}}{C}HCH{=}CHCH_2CH_2CH_3 \qquad CH_3\overset{\substack{CH_3\\|}}{C}HC{\equiv}CCH_2CH_2CH_3$$

1. **To determine the parent name,** identify the longest chain that contains the unsaturation. Name the parent compound with the prefix that defines the number of carbons in that chain and the suffix that identifies the class of compound.

$$\underset{\text{Heptene}}{\boxed{CH_3\overset{\substack{CH_3\\|}}{C}HCH{=}CHCH_2CH_2CH_3}} \qquad \underset{\text{Heptyne}}{\boxed{CH_3\overset{\substack{CH_3\\|}}{C}HC{\equiv}CCH_2CH_2CH_3}}$$

2. **Number the parent chain from the end nearest the unsaturation so that the first carbon in the double or triple bond has the lowest number possible.** (If the unsaturation is in the middle of a chain, the location of any substituent group is used to determine where the

numbering starts.) The smaller of the two numbers identifying the carbon atoms involved in the unsaturation is used as the locator of the multiple bond.

$$\overset{\displaystyle CH_3}{\underset{1\ \ 2\ \ 3\qquad 4\ \ 5\ \ 6\ \ 7}{CH_3\overset{|}{C}HCH{=}CHCH_2CH_2CH_3}} \qquad \overset{\displaystyle CH_3}{\underset{1\ \ 2\ \ 3\quad 4\,5\ \ 6\ \ 7}{CH_3\overset{|}{C}HC{\equiv}CCH_2CH_2CH_3}}$$

3-Heptene 3-Heptyne

3. **Stereoisomers of alkenes are named by writing *cis*- or *trans*- before the number identifying the location of the double bond.** Chapter 19 in the text addresses naming stereoisomers.
4. **The rules for naming substituted alkanes are followed to name and locate any other groups on the chain.**

$$\overset{\displaystyle CH_3}{CH_3\overset{|}{C}HCH{=}CHCH_2CH_2CH_3} \qquad \overset{\displaystyle CH_3}{CH_3\overset{|}{C}HC{\equiv}CCH_2CH_2CH_3}$$

2-Methyl-3-heptene 2-Methyl-3-heptyne

Halogens attached to an alkane, alkene, or alkyne are named as fluoro- (F–), chloro- (Cl–), bromo- (Br–), or iodo- (I–) and are located by using the same numbering system described for alkyl groups.

Benzene Derivatives

Naming compounds containing substituted benzene rings is less systematic than naming hydrocarbons. Many compounds have common names that are incorporated into accepted names, but for simple substituted benzene rings, the following rules may be applied.

1. **For monosubstituted benzene rings,** a prefix identifying the group is appended to the parent name benzene:

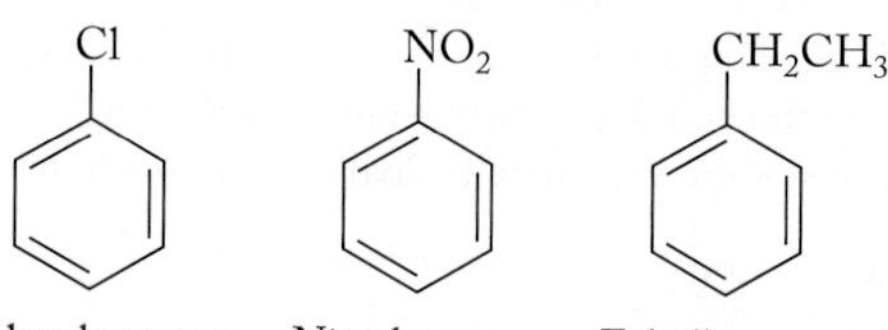

Chlorobenzene Nitrobenzene Ethylbenzene

2. **For disubstituted benzene rings,** three isomers are possible. The relative position of the substituent groups is indicated by numbers in IUPAC nomenclature, but the set of prefixes shown are very commonly used as well:

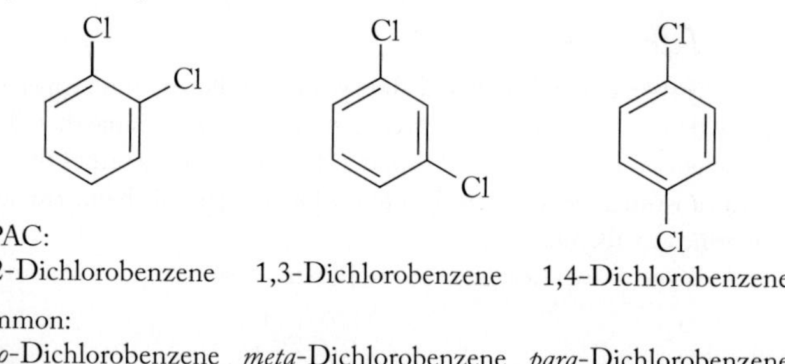

IUPAC:
1,2-Dichlorobenzene 1,3-Dichlorobenzene 1,4-Dichlorobenzene

Common:
ortho-Dichlorobenzene *meta*-Dichlorobenzene *para*-Dichlorobenzene
o-Dichlorobenzene *m*-Dichlorobenzene *p*-Dichlorobenzene

3. **When three or more groups are attached to a benzene ring,** the lowest possible numbers are assigned to locate the groups with respect to each other.

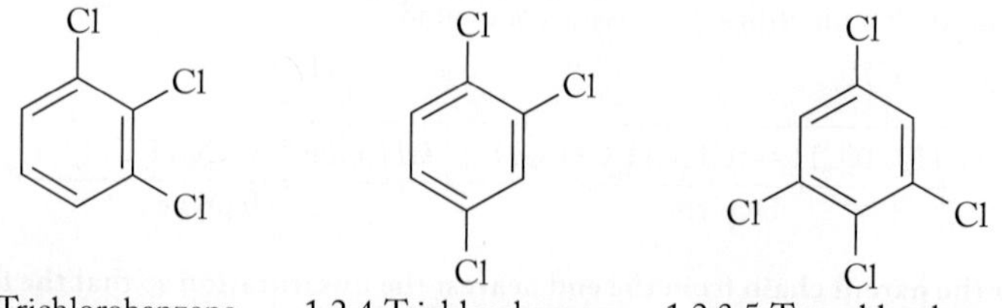

1,2,3-Trichlorobenzene 1,2,4-Trichlorobenzene 1,2,3,5-Tetrachlorobenzene

(*Note*: Not 1,3,4-trichlorobenzene; and not 1,3,4,5-tetrachlorobenzene.)

Hydrocarbons Containing Other Functional Groups

The same basic principles developed for naming alkanes apply to naming hydrocarbons with functional groups other than alkyl groups The name must identify the carbon skeleton, locate the functional group, and contain a suffix that defines the class of compound. The following examples give the suffixes for some common functional groups; when suffixes are used, they replace the final *-e* in the name of the parent alkane. Other functional groups may be identified by including the name of the class of compounds in the name of the molecule.

Alcohols: Suffix *-ol*

$CH_3CH_2CH_2OH$ — 1-Propanol

$CH_3CH(OH)CH_3$ — 2-Propanol

$CH_3CH_2CH_2CH(OH)CH_3$ — 2-Pentanol

Aldehydes: Suffix *-al*

HCHO ($H-C(=O)-H$) — IUPAC: Methanal; Common: Formaldehyde

$H_3C-C(=O)-H$ — IUPAC: Ethanal; Common: Acetaldehyde

Because the aldehyde group can only be on a terminal carbon, no number is necessary to locate it on the carbon chain.

Ketones: Suffix *-one* The location of the carbonyl is given by a number, and the chain is numbered so that the carbonyl carbon has the lowest possible value. Many ketones also have common names generated by identifying the hydrocarbon groups on both sides of the carbonyl group.

$H_3C-C(=O)-CH_3$ — IUPAC: Propan-2-one; Common: Acetone

$H_3C-C(=O)-CH_2CH_3$ — IUPAC: Butan-2-one; Common: Methyl ethyl ketone

Carboxylic Acids: Suffix *-oic acid* The carboxylic acid group is by definition carbon 1, so no number identifying its location is included in the name.

$H_3C-C(=O)-OH$ — IUPAC: Ethanoic acid; Common: Acetic acid

$H_3C-CH=CH-C(=O)-OH$ — IUPAC: *trans*-2-Butenoic acid

Salts of Carboxylic Acids Salts are named with the cation first, followed by the anion name of the acid from which *-ic acid* is dropped and the suffix *-ate* is added. The sodium salt of acetic acid is sodium acetate.

$H_3C-C(=O)-OH$ — Acetic acid

$H_3C-C(=O)-O^-$ — Acetate ion

$Na^+\ {}^-O-C(=O)-CH_3$ — Sodium acetate

Esters Esters are viewed as derivatives of carboxylic acids. They are named in a manner analogous to that of salts. The alkyl group comes first, followed by the name of the carboxylate anion.

$R-O-C(=O)-R$ — Alkyl (R) Carboxylate (O–C(=O)–R)

$CH_3CH_2-O-C(=O)-CH_3$ — Ethyl acetate

Amides Amides are also derivatives of carboxylic acids. They are named by replacing *-ic acid* (of the common names) or *-oic acid* of the IUPAC names with *-amide.*

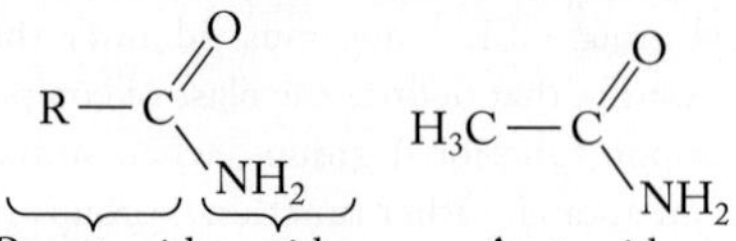

$$\underbrace{R-C}_{\text{Parent acid}}\underbrace{(=O)NH_2}_{\text{-amide}} \qquad H_3C-C(=O)NH_2$$

Parent acid -amide

Acetamide
(ethanamide)

Ethers Ethers are frequently named by naming the two groups attached to the oxygen and following those names by the word *ether.*

CH_3OCH_3 — Dimethyl ether

$CH_3CH_2OCH_2CH_3$ — Diethyl ether

Amines Aliphatic amines are usually named by listing the group or groups attached to the nitrogen and then appending *-amine* as a suffix. They may also be named by prefixing *amino-* to the name of the parent chain.

H_3C-NH_2 — Methylamine

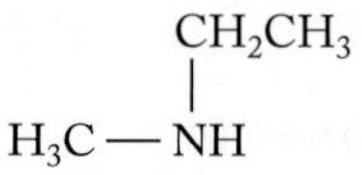

$H_3C-NH(CH_2CH_3)$ — Ethylmethylamine

$H_2NCH_2CH_2OH$ — 2-Aminoethanol

This brief summary will enable you to understand the names of organic compounds used in this book. IUPAC rules are much more extensive than this and can be applied to all varieties of carbon compounds, including those with multiple functional groups. It is important to recognize that the rules of systematic nomenclature do not necessarily lead to a unique name for each compound, but they do always lead to an unambiguous one. Furthermore, common names are still used frequently in organic chemistry because the systematic alternatives do not improve communication. Remember that the main purpose of chemical nomenclature is to identify a chemical species by means of written or spoken words. Anyone who reads or hears the name should be able to deduce the structure and thereby the identity of the compound.

Glossary

A

absolute zero (0 K) Zero point on Kelvin temperature scale; theoretically the lowest temperature possible. (Ch. 1)

accessible microstate A unique arrangement of the positions and momenta of the particles in a thermodynamic system. (Ch. 12)

accuracy Agreement between an experimental value and the true value. (Ch. 1)

activated complex A short-lived species formed in a chemical reaction. (Ch. 13)

activation energy (E_a) The minimum energy molecules need to react when they collide. (Ch. 13)

active site The location on an enzyme where a reactive substance binds. (Ch. 20)

addition polymer A macromolecule prepared by adding monomers to a growing polymer chain. (Ch. 19)

addition reaction A reaction in which two molecules couple together and form one product. (Ch. 19)

alcohol An organic compound whose molecular structure includes a hydroxyl group bonded to a carbon atom that is not bonded to any other functional group(s). (Ch. 6)

aldehyde An organic compound containing a carbonyl group bonded to one R group and one hydrogen; its general formula is RCHO. (Ch. 19)

alkali metal An element in group 1 of the periodic table. (Ch. 2)

alkaline earth metal An element in group 2 of the periodic table. (Ch. 2)

alkane A hydrocarbon in which each carbon atom is bonded to four other carbon or hydrogen atoms. (Ch. 6)

alkene A hydrocarbon containing one or more carbon–carbon double bonds. (Ch. 19)

alkyne A hydrocarbon containing one or more carbon–carbon triple bonds. (Ch. 19)

allotropes Different molecular forms of the same element, such as oxygen (O_2) and ozone (O_3). (Ch. 4)

alloy A blend of a host metal and one or more other elements, which may or may not be metals, that are added to change the properties of the host metal. (Ch. 18)

α-amino acid An amino acid in which the carboxylic acid and amine groups are both attached to the same (α) carbon atom. (Ch. 20)

α-carbon The carbon atom attached directly to a functional group. (Ch. 20)

alpha (α) decay A nuclear reaction in which an unstable nuclide spontaneously emits an α particle. (Ch. 21)

α helix A coil in a protein chain's secondary structure. (Ch. 20)

alpha (α) particle A radioactive emission with a charge of 2+ and a mass equivalent to that of a helium nucleus. (Ch. 2)

amide An organic compound in which the –OH of a carboxylic acid group is replaced by $-NH_2$, –NHR, or $-NR_2$, where R can be any organic group. (Ch. 19)

amine An organic compound that contains a group with the general formula RNH_2, R_2NH, or R_3N, where R is any organic subgroup. (Ch. 19)

amino acid A molecule that contains at least one amine group and one carboxylic acid group. (Ch. 20)

Amontons's law The pressure of a gas is proportional to its absolute temperature if its volume does not change. (Ch. 10)

amphiprotic Describes a substance that can behave as either a proton acceptor or a proton donor. (Ch. 8)

analyte The substance whose concentration is to be determined in a chemical analysis. (Ch. 8)

angular (also called *bent*) Molecular geometry about a central atom with a steric number of 3 and one lone pair or a steric number of 4 and two lone pairs. (Ch. 5)

angular momentum quantum number (ℓ) An integer having any value from 0 to $(n - 1)$ that defines the shape of an orbital. (Ch. 3)

anion A negatively charged ion. (Ch. 2)

anode An electrode at which an oxidation half-reaction (loss of electrons) takes place. (Ch. 17)

antibonding orbital A term in MO theory describing regions of electron density in a molecule that destabilize the molecule because they do not increase the electron density between nuclear centers. (Ch. 5)

antimatter Particles that are the charge opposites of normal subatomic particles. (Ch. 21)

aromatic compound A cyclic, planar compound with delocalized π electrons above and below the plane of the molecule. (Ch. 5)

Arrhenius acid A compound that produces H_3O^+ ions in aqueous solution. (Ch. 15)

Arrhenius base A compound that produces OH^- ions in aqueous solution. (Ch. 15)

Arrhenius equation Relates the rate constant of a reaction to absolute temperature (T), the activation energy of the reaction (E_a), and the frequency factor (A). (Ch. 13)

atom The smallest particle of an element that retains the chemical characteristics of the element. (Ch. 1)

atomic absorption spectra Characteristic patterns of dark lines produced when an external source of radiation passes through free gaseous atoms. (Ch. 3)

atomic emission spectra Characteristic patterns of bright lines produced when atoms are vaporized in high-temperature flames or electrical discharges. (Ch. 3)

atomic mass unit (amu) The unit used to express the relative masses of atoms and subatomic particles that is exactly 1/12 the mass of 1 atom of carbon with 6 protons and 6 neutrons in its nucleus. (Ch. 2)

atomic number (Z) The number of protons in the nucleus of an atom. (Ch. 2)

aufbau principle The concept of building up ground state atoms so that their electrons occupy the lowest energy orbitals available. (Ch. 3)

autoionization The process that produces equal and very small concentrations of H_3O^+ and OH^- ions in pure water. (Ch. 15)

average atomic mass The weighted average of the masses of all isotopes of an element, calculated by multiplying the natural abundance of each isotope by its mass in atomic mass units and then summing the products. (Ch. 2)

Avogadro's law The volume of a gas at constant temperature and pressure is proportional to the quantity (number of moles) of the gas. (Ch. 10)

Avogadro's number (N_A) The number of carbon atoms in exactly 12 grams of carbon-12; $N_A = 6.0221 \times 10^{23}$. It is the number of particles in 1 mole. (Ch. 2)

B

band gap (E_g) The energy gap between the valence and conduction bands. (Ch. 18)

band theory An extension of molecular orbital theory that describes bonding in solids. (Ch. 18)

barometer An instrument that measures atmospheric pressure. (Ch. 10)

becquerel (Bq) The SI unit of radioactivity; one becquerel equals one decay event per second. (Ch. 21)

belt of stability The region on the graph of number of neutrons versus number of protons that includes all stable nuclei. (Ch. 21)

bent (also called *angular*) Molecular geometry about a central atom with a steric number of 3 and one lone pair or a steric number of 4 and two lone pairs. (Ch. 5)

beta (β) decay The process by which a neutron in a neuron-rich nucleus decays into a proton and a β particle. (Ch. 21)

beta (β) particle A radioactive emission that is a high-energy electron. (Ch. 2)

β-pleated sheet A puckered two-dimensional array of protein strands held together by hydrogen bonds. (Ch. 20)

bimolecular step A step in a reaction mechanism involving a collision between two molecules. (Ch. 13)

binding energy (BE) The energy that holds the nucleons together in a nucleus. (Ch. 21)

biocatalysis The strategy of using enzymes to catalyze reactions on a large scale. (Ch. 20)

biomass The sum total of the mass of organic matter in any given ecological system. (Ch. 20)

biomolecule An organic molecule present naturally in a living system. (Ch. 20)

body-centered cubic (bcc) unit cell A cell with atoms at the eight corners of a cube and at the center of the cell. (Ch. 18)

boiling point elevation The decrease in the vapor pressure of the solution relative to that of pure solvent at all temperatures corresponds to an increase in the boiling point of the solution. (Ch. 11)

bomb calorimeter A constant-volume device used to measure the energy released during a combustion reaction. (Ch. 9)

bond angle The angle (in degrees) defined by lines joining the centers of two atoms to a third atom to which they are chemically bonded. (Ch. 5)

bond dipole Separation of electrical charge created when atoms with different electronegativities form a covalent bond. (Ch. 5)

bond energy The energy needed to break one mole of a particular covalent bond in the gas phase; also called *bond strength*. (Ch. 4)

bond length The distance between the nuclei of two atoms joined together in a bond. (Ch. 4)

bond order The number of bonds between atoms: 1 for a single bond, 2 for a double bond, and 3 for a triple bond. (Ch. 4)

bonding capacity The number of covalent bonds an atom forms to have an octet of electrons in its valence shell. (Ch. 4)

bonding orbital A term in MO theory describing regions of increased electron density between nuclear centers that serve to hold atoms together in molecules. (Ch. 5)

bonding pair A pair of electrons shared between two atoms. (Ch. 4)

Born–Haber cycle A series of steps with corresponding enthalpy changes that describes the formation of an ionic solid from its constituent elements. (Ch. 9)

Boyle's law The volume of a gas at constant temperature is inversely proportional to its pressure. (Ch. 10)

Bragg equation An equation that relates the angle of diffraction (2θ) of X-rays to the spacing (d) between the layers of ions or atoms in a crystal: $n\lambda = 2d \sin \theta$. (Ch. 18)

branch A side chain bonded to the main (longest) chain in an organic molecule. (Ch. 19)

breeder reactor A nuclear reactor in which fissionable material is produced during normal reactor operation. (Ch. 21)

Brønsted–Lowry acid A proton (H^+) donor. (Ch. 8)

Brønsted–Lowry base A proton (H^+) acceptor. (Ch. 8)

Brønsted–Lowry model Defines acids as H^+ ion donors and bases as H^+ ion acceptors. (Ch. 8)

buffer capacity The quantity of acid or base that a pH buffer can neutralize while keeping its pH within a desired range. (Ch. 16)

C

calorimeter A device used to measure the absorption or release of energy by a physical change or chemical process. (Ch. 9)

calorimeter constant ($C_{calorimeter}$) The heat capacity of a calorimeter. (Ch. 9)

calorimetry The experimental determination of the quantity of energy transferred during a physical change or chemical process. (Ch. 9)

capillary action The rise of a liquid in a narrow tube as a result of adhesive forces between the liquid and the tube and cohesive forces within the liquid. (Ch. 6)

carbohydrate An organic molecule with the generic formula $C_x(H_2O)_y$. (Ch. 20)

carbonyl group A functional group that consists of a carbon atom with a double bond to an oxygen atom. (Ch. 6)

carboxylic acid A compound containing the –COOH functional group. (Ch. 8)

catalyst A substance added to a reaction that increases the rate of the reaction but is not consumed in the process. (Ch. 13)

cathode An electrode at which a reduction half-reaction (gain of electrons) takes place. (Ch. 17)

cathode rays Streams of electrons emitted by the cathode in a partially evacuated tube. (Ch. 2)

cation A positively charged ion. (Ch. 2)

cell diagram Symbols that show how the components of an electrochemical cell are connected. (Ch. 17)

cell potential (E_{cell}) The electromotive force with which an electrochemical cell can pump electrons through an external circuit. (Ch. 17)

ceramic A solid inorganic compound or mixture that has been transformed into a harder, more heat-resistant material by heating. (Ch. 18)

chalcogen An element in group 16 of the periodic table. (Ch. 2)

chain reaction A self-sustaining series of fission reactions in which the neutrons released when nuclei split apart initiate additional fission events and sustain the reaction. (Ch. 21)

Charles's law The volume of a gas at constant pressure is directly proportional to its absolute temperature. (Ch. 10)

chelate effect The greater affinity of metal ions for polydentate ligands than for monodentate ligands. (Ch. 23)

chelation The interaction of a metal with a polydentate ligand (chelating agent); pairs of electrons on one molecule of the ligand occupy two or more coordination sites on the central metal. (Ch. 23)

chemical bond A force that holds two atoms or ions in a molecule or a compound together. (Ch. 1)

chemical energy Potential energy stored in chemical bonds. (Ch. 9)

chemical equation A description of the identities and proportions of the reactants and the products in a chemical reaction. (Ch. 7)

chemical equilibrium A dynamic process in which the concentrations of reactants and products remain constant over time and the rate of a reaction in the forward direction matches its rate in the reverse direction. (Ch. 14)

chemical formula A notation for representing the elemental composition of a pure substance using the symbols of the elements; subscripts indicate the relative number of atoms of each element in the substance. (Ch. 1)

chemical kinetics The study of the rates of change of concentrations of substances involved in chemical reactions. (Ch. 13)

chemical property A property of a substance that can be observed only by reacting the substance to form another substance. (Ch. 1)

chemical reaction The conversion of one or more substances into one or more different substances. (Ch. 1)

chemistry The study of the composition, structure, and properties of matter and of the energy consumed or given off when matter undergoes a change. (Ch. 1)

chiral Describes a molecule that is not superimposable on its mirror image. (Ch. 5)

chromatography A process involving a stationary and a mobile phase for separating a mixture of substances based on their different affinities for the two phases. (Ch. 1)

cis isomer (also called *Z isomer*) A molecule with two like groups (such as two R groups or two hydrogen atoms) on the same side of the molecule. (Ch. 19)

Clausius–Clapeyron equation Relates the vapor pressure of a substance at different temperatures to its heat of vaporization. (Ch. 11)

codon A three-nucleotide sequence that codes for a specific amino acid. (Ch. 20)

coenzyme An organic molecule that, like an enzyme, accelerates the rate of biochemical reactions. (Ch. 23)

colligative properties Characteristics of solutions that depend on the concentration and not the identity of particles dissolved in the solvent. (Ch. 11)

combination reaction A reaction in which two or more substances form a single product. (Ch. 7)

combined gas law The ratio PV/T for a given quantity of gas is a constant. (Ch. 10)

combustion A rapid reaction between fuel and oxygen that produces energy. (Ch. 7)

combustion analysis A laboratory procedure in which a substance is burned completely in oxygen to produce known compounds whose masses are used to determine the composition of the original material. (Ch. 7)

common-ion effect The shift in the position of an equilibrium caused by the addition of an ion taking part in the reaction. (Ch. 16)

complex ion An ionic species consisting of a metal ion bonded to one or more Lewis bases. (Ch. 16)

compound A substance composed of characteristic proportions of two or more elements chemically bonded together. (Ch. 1)

condensation polymer A macromolecule formed by the reaction of monomers, yielding a polymer and water or another small molecule as products of the reaction. (Ch. 19)

condensation reaction Two molecules combining to form a larger molecule and a small molecule (typically water). (Ch. 19)

conduction band An unoccupied band higher in energy than a valence band in which electrons are free to migrate. (Ch. 18)

conductor A material with partially filled valence bands or filled valence bands that overlap with empty conduction bands, leading to highly mobile electrons. (Ch. 18)

confidence interval A range of values that has a specified probability of containing the true value of a measurement. (Ch. 1)

conjugate acid The acid formed when a Brønsted–Lowry base accepts a H^+ ion. (Ch. 15)

conjugate acid–base pair A Brønsted–Lowry acid and base that differ from each other only by a H^+ ion: acid $\rightleftharpoons$ conjugate base + H^+. (Ch. 15)

conjugate base The base formed when a Brønsted–Lowry acid donates a H^+ ion. (Ch. 15)

constitutional isomer One of a set of compounds with the same molecular formula but different connections between the atoms in their molecules; also called *structural isomer.* (Ch. 6)

conversion factor Fraction in which the numerator is equivalent to the denominator but is expressed in different units, making the fraction equivalent to 1. (Ch. 1)

coordinate bond A covalent bond formed when one anion or molecule donates a pair of electrons to another ion or molecule. (Ch. 16)

coordination compound A compound made up of at least one complex ion. (Ch. 23)

coordination number The number of sites occupied by ligands around a metal ion in a complex. (Ch. 23)

copolymer A macromolecule formed from the chemical combination of two different monomers. (Ch. 19)

core electrons Electrons in the filled, inner shells in an atom or ion that are not involved in chemical reactions. (Ch. 3)

corrosion A process in which a metal deteriorates through spontaneous electrochemical reactions. (Ch. 17)

counter ion An ion whose charge balances the charge of a complex ion in a coordination compound. (Ch. 23)

covalent bond A bond created by two atoms sharing one or more pairs of electrons. (Ch. 4)

covalent network solid A solid consisting of atoms held together by extended arrays of covalent bonds. (Ch. 18)

critical mass The minimum quantity of fissionable material needed to sustain a chain reaction. (Ch. 21)

critical point A specific temperature and pressure at which the liquid and gas phases of a substance have the same density and are indistinguishable from each other. (Ch. 6)

crystal field splitting The separation of a set of *d* orbitals into subsets with different energies as a result of interactions between electrons in those orbitals and lone pairs of electrons in ligands. (Ch. 23)

crystal field splitting energy (Δ) The difference in energy between subsets of *d* orbitals split by interactions in a crystal field. (Ch. 23)

crystal lattice An ordered three-dimensional array of particles. (Ch. 4)

crystal structure An ordered arrangement in three-dimensional space of the particles (atoms, ions, or molecules) that make up a crystalline solid. (Ch. 18)

crystalline solid A solid made of an ordered array of atoms, ions, or molecules. (Ch. 18)

cubic closest-packed (ccp) A crystal structure composed of face-centered cubic unit cells and layers of particles having an *abcabc* . . . stacking pattern. (Ch. 18)

curie (Ci) Non-SI unit of radioactivity; 1 Ci = 3.70×10^{10} decay events per second. (Ch. 21)

cycloalkane A ring-containing alkane with the general formula C_nH_{2n}. (Ch. 19)

D

dalton (Da) A unit of mass equal to 1 atomic mass unit. (Ch. 2)

Dalton's law of partial pressures The total pressure of a mixture of gases is the sum of the partial pressures of all the gases in the mixture. (Ch. 10)

de Broglie equation Relates the wavelength of any moving object to its mass and its speed. (Ch. 3)

degree of ionization The ratio of the quantity of a substance that is ionized to the concentration of the substance before ionization; when expressed as a percentage, called *percent ionization.* (Ch. 15)

density (*d*) The ratio of the mass (m) of an object to its volume (V). (Ch. 1)

deposition The transformation of a gas (vapor) directly into a solid. (Ch. 1)

diamagnetic Describes a substance with no unpaired electrons that is weakly repelled by a magnetic field. (Ch. 5)

diffusion The spread of one substance (usually a gas or liquid) through another. (Ch. 10)

dilution The process of lowering the concentration of a solution by adding more solvent. (Ch. 8)

dipole moment (μ) A measure of the degree to which a molecule aligns itself in an applied electric field; a quantitative expression of the polarity of a molecule. (Ch. 5)

dipole–dipole interaction An attraction between regions of polar molecules that have partial charges of opposite sign. (Ch. 6)

dipole–induced dipole interaction An attraction between a polar molecule and the oppositely charged pole it causes in another molecule. (Ch. 6)

distillation A process using evaporation and condensation to separate a mixture of substances with different volatilities. (Ch. 1)

double bond A bond formed when two atoms share two pairs of electrons. (Ch. 4)

E

***E* isomer** (also called *trans isomer*) Molecule with two like groups (such as two R groups or two hydrogen atoms) on opposite sides of the molecule. (Ch. 19)

effective nuclear charge (Z_{eff}) The attraction toward the nucleus experienced by an electron in an atom; the positive charge on the nucleus is reduced by the extent to which other electrons in the atom shield the electron from the nucleus. (Ch. 3)

effusion The process by which a gas escapes from its container through a tiny hole into a region of lower pressure. (Ch. 10)

electrochemical cell An apparatus that converts chemical energy into electrical work or electrical work into chemical energy. (Ch. 17)

electrochemistry The branch of chemistry that examines the transformations between chemical and electrical energy. (Ch. 17)

electrode A solid electrical conductor that is used to make contact with a solution or other nonmetallic component of an electrical circuit. (Ch. 8)

electrolysis A process in which electrical energy is used to drive a nonspontaneous chemical reaction. (Ch. 17)

electrolyte A material that conducts electricity because it contains free ions; ionic solutions and molten salts are examples of electrolytes; a solute that produces ions in solution, which enables its solutions to conduct electricity. (Ch. 8)

electrolytic cell A device in which an external source of electrical energy does work on a chemical system, turning reactant(s) into higher-energy product(s). (Ch. 17)

electromagnetic radiation Any form of radiant energy in the electromagnetic spectrum. (Ch. 3)

electromagnetic spectrum A continuous range of radiant energy that includes gamma rays, X-rays, ultraviolet radiation, visible light, infrared radiation, microwaves, and radio waves. (Ch. 3)

electromotive force (emf) Also called voltage, the force pushing electrons through an electrical circuit. (Ch. 17)

electron A subatomic particle that has a relative charge of 1− and negligible mass. (Ch. 2)

electron affinity (EA) The energy change that occurs when 1 mole of electrons combines with 1 mole of atoms or ions in the gas phase. (Ch. 3)

electron capture A nuclear reaction in which a neutron-poor nucleus draws in one of its surrounding electrons, which transforms a proton in the nucleus into a neutron. (Ch. 21)

electron configuration The distribution of electrons among the orbitals of an atom or ion. (Ch. 3)

electron transition Movement of an electron between energy levels. (Ch. 3)

electronegativity A relative measure of the ability of an atom to attract electrons to itself within a bond. (Ch. 4)

electron-pair delocalization The spreading out of electron density over several atoms.

electron-pair geometry The three-dimensional arrangement of bonding pairs and lone pairs of electrons about a central atom. (Ch. 5)

electrostatic potential energy (E_{el}) The energy a charged particle has because of its position relative to another charged particle; it is directly proportional to the product of the charges of the particles and inversely proportional to the distance between them; also called *coulombic attraction.* (Ch. 4)

element A substance that cannot be separated into simpler substances by any chemical process. (Ch. 1)

elementary step A molecular-level view of a single process taking place in a chemical reaction. (Ch. 13)

empirical formula A chemical formula in which the subscripts represent the simplest whole-number ratio of the atoms or ions in a compound. (Ch. 1)

enantiomer One of a pair of optical isomers of a compound. (Ch. 5)

end point The point in a titration when a color change or other signal indicates that enough titrant has been added to react with all of the analyte in the sample. (Ch. 8)

endothermic process One in which energy—usually in the form of heat—flows from the surroundings into the system. (Ch. 9)

energy The capacity to do work (w). (Ch. 1)

energy profile A graph showing the changes in potential energy for a reaction as a function of the progress of the reaction from reactants to products. (Ch. 13)

enthalpy (*H*) A measure of the total energy of a system; the sum of the internal energy and the pressure–volume product of a system; $H = E + PV$. (Ch. 9)

enthalpy change (ΔH) The quantity of heat transferred into or out from a system during a chemical reaction or physical change at constant pressure. (Ch. 9)

enthalpy of fusion (ΔH_{fus}) The energy required to convert one mole of a solid substance at its melting point into the liquid state; also called the *heat of fusion.* (Ch. 9)

enthalpy of reaction (ΔH_{rxn}) The enthalpy change that accompanies a chemical reaction; also called the *heat of reaction.* (Ch. 9)

enthalpy of solution ($\Delta H_{solution}$) The overall change in enthalpy that occurs when a solute is dissolved in a solvent; also called the *heat of solution.* (Ch. 9)

enthalpy of vaporization (ΔH_{vap}) The energy required to convert one mole of a liquid substance at its boiling point into the vapor state; also called the *heat of vaporization*. (Ch. 9)

entropy (S) A measure of how dispersed the energy in a system is at a specific temperature. (Ch. 12)

enzyme A protein that catalyzes a reaction. (Ch. 20)

equilibrium constant (K) The value of the ratio of concentration (or partial pressure) terms in the equilibrium constant expression at a specific temperature. (Ch. 14)

equilibrium constant expression The ratio of the equilibrium concentrations or partial pressures of products to reactants, each term raised to a power equal to the coefficient of that substance in the balanced chemical equation for the reaction. (Ch. 14)

equilibrium (reversal) potential, E_{ion} An electrochemical potential that results from a concentration gradient of a particular ion on opposite sides of a cell membrane. (Ch. 22)

equivalence point The point in a titration when just enough titrant has been added to react with all of the analyte in the sample. (Ch. 8)

essential amino acid Any of the eight amino acids that make up peptides and proteins but are not synthesized in the human body and must be obtained through the food we eat. (Ch. 20)

essential element Element present in tissue, blood, or other bodily fluids that has a physiological function. (Ch. 22)

ester An organic compound in which the –OH of a carboxylic acid group is replaced by –OR, where R can be any organic group. (Ch. 19)

ether An organic compound that contains an oxygen atom with single bonds to two carbon atoms. (Ch. 6)

excited state Any energy state above the ground state. (Ch. 3)

exothermic process One in which energy—usually in the form of heat—flows from a system into its surroundings. (Ch. 9)

extensive property A property that varies with the amount of substance present. (Ch. 1)

F

face-centered cubic (fcc) unit cell An array of closest-packed particles that has eight of the particles at the corners of a cube and six of them at the centers of each face of the cube. (Ch. 18)

family (also called *group*) (of elements) All the elements in a column of the periodic table. (Ch. 2)

Faraday constant (F) The magnitude of electric charge in one mole of electrons; its value to three significant figures is 9.65×10^4 C/mol. (Ch. 17)

fat A solid triglyceride containing primarily saturated fatty acids. (Ch. 20)

filtration A process for separating solid particles from a liquid or gaseous sample by passing the sample through a porous material that retains the solid particles. (Ch. 1)

first law of thermodynamics The principle that the energy gained or lost by a system must equal the energy lost or gained by the surroundings. Energy can not be created or destroyed. (Ch. 9)

formal charge (FC) The value calculated for an atom in a molecule or polyatomic ion by determining the difference between the number of valence electrons in the free atom and the sum of the lone pair electrons plus half of the electrons in the atom's bonding pairs. (Ch. 4)

formation constant (K_f) An equilibrium constant describing the formation of a metal complex from a free metal ion and its ligands. (Ch. 16)

formation reaction A reaction in which one mole of a substance is formed from its component elements in their standard states. (Ch. 9)

formula mass The mass in amu of one formula unit of an ionic compound. (Ch. 2)

formula unit The smallest electrically neutral unit of an ionic compound. (Ch. 2)

fracking The process of injecting liquids at high pressures into rock formations to force open existing fissures in the formations for the purpose of extracting oil or gas. (Ch. 19)

fractional distillation A method of separating a mixture of compounds on the basis of their different boiling points. (Ch. 11)

Fraunhofer lines A set of dark lines in the otherwise continuous solar spectrum. (Ch. 3)

free energy A measure of the maximum amount of work a thermodynamic system can perform. (Ch. 12)

free radical An atom, ion, or molecule with unpaired electrons. (Ch. 4)

freezing point depression Solutions freeze at lower temperatures than their pure solvents. (Ch. 11)

frequency (ν) The number of crests of a wave that pass a stationary point of reference per second. (Ch. 3)

frequency factor (A) The product of the frequency of molecular collisions and a factor that expresses the probability that the orientation of the molecules is appropriate for a reaction to occur. (Ch. 13)

fuel cell A voltaic cell based on the oxidation of a continuously supplied fuel; the reaction is the equivalent of combustion, but chemical energy is converted directly into electrical energy. (Ch. 17)

fuel density The quantity of energy released during the complete combustion of a particular volume of a liquid fuel. (Ch. 9)

fuel value The quantity of energy released during the complete combustion of 1 g of a substance. (Ch. 9)

functional group A group of atoms in a molecule that is largely responsible for the physical and chemical properties of a molecular compound. (Ch. 6)

G

galvanic cell (also called *voltaic cell*) An electrochemical cell in which chemical energy is transformed into electrical work by a spontaneous redox reaction. (Ch. 17)

gas A phase of matter that has neither definite volume nor definite shape, and that expands to fill its container; also called *vapor*. (Ch. 1)

Geiger counter A portable device for determining nuclear radiation levels by measuring how much the radiation ionizes the gas in a sealed detector. (Ch. 21)

Gibbs free energy (G) The maximum amount of energy released by a process occurring at constant temperature and pressure that is available to do useful work. (Ch. 12)

glyceride Lipid consisting of esters formed between fatty acids and the alcohol glycerol. (Ch. 20)

glycolysis A series of reactions that converts glucose into pyruvate; a major anaerobic (no oxygen required) pathway for the metabolism of glucose in the cells of almost all living organisms. (Ch. 12)

glycosidic bond A C—O—C bond between sugar molecules. (Ch. 20)

Graham's law of effusion The rate of effusion of a gas is inversely proportional to the square root of its molar mass. (Ch. 10)

gray (Gy) The SI unit of absorbed radiation; 1 Gy = 1 J/kg of tissue. (Ch. 21)

green chemistry Laboratory practices that reduce or eliminate the use or generation of hazardous substances. (Ch. 6)

ground state The most stable, lowest energy state of a particle. (Ch. 3)

group (also called *family*) (of elements) All the elements in a column of the periodic table. (Ch. 2)
Grubbs' test A statistical test used to detect an outlier in a set of data. (Ch. 1)

H

half-life ($t_{1/2}$) The time interval during which the concentration of a reactant decreases by half in the course of a chemical reaction. (Ch. 13)
half-reaction One of the two halves of a redox reaction; one half-reaction is the oxidation component, and the other is the reduction component. (Ch. 17)
halogen An element in group 17 of the periodic table. (Ch. 2)
heat The transfer of energy from one object or place to another due to differences in the temperatures of the objects or places. (Ch. 1)
heat capacity (C_P) The energy required to raise the temperature of an object 1°C at constant pressure. (Ch. 9)
Heisenberg uncertainty principle The principle that one cannot simultaneously know the exact position and the exact momentum of an electron. (Ch. 3)
Henderson–Hasselbalch equation An equation used to calculate the pH of a solution in which the concentrations of acid and conjugate base are known. (Ch. 16)
Henry's law The concentration of a sparingly soluble gas in a liquid is proportional to the partial pressure of the gas. (Ch. 11)
hertz (Hz) The SI unit of frequency with units of reciprocal seconds: $1 \text{ Hz} = 1 \text{ s}^{-1} = 1$ cycle per second (cps). (Ch. 3)
Hess's law The principle that the heat of reaction (ΔH_{rxn}) for a process that is the sum of two or more reactions is equal to the sum of the ΔH_{rxn} values of the constituent reactions; also called *Hess's law of constant heat of summation*. (Ch. 9)
heteroatom Any atom other than carbon or hydrogen in an organic compound. (Ch. 19)
heterogeneous catalyst A catalyst in a different phase than the reactants. (Ch. 13)
heterogeneous equilibrium An equilibrium in which reactants and products are in more than one phase. (Ch. 14)
heterogeneous mixture A mixture in which the components are not distributed uniformly, so that the mixture contains distinct regions of different compositions. (Ch. 1)
heteropolymer A polymer made of three or more different monomer units. (Ch. 19)
hexagonal closest-packed (hcp) A crystal lattice in which the layers of atoms or ions have an *ababab* . . . stacking pattern. (Ch. 18)
hexagonal unit cell An array of nine closest-packed particles that are the repeating unit in a hexagonal closest-packed crystal. (Ch. 18)
homogeneous catalyst A catalyst in the same phase as the reactants. (Ch. 13)
homogeneous equilibrium An equilibrium in which reactants and products are in the same phase. (Ch. 14)
homogeneous mixture A mixture in which the components are distributed uniformly throughout and the composition and appearance is uniform. (Ch. 1)
homologous series A set of related organic compounds that differ from one another by the number of common subgroups, such as $-CH_2-$, in their molecular structures. (Ch. 19)
homopolymer A polymer composed of only one kind of monomer unit. (Ch. 19)
Hund's rule The lowest energy electron configuration of an atom has the maximum number of unpaired electrons, all of which have the same spin, in degenerate orbitals. (Ch. 3)
hybrid atomic orbital In valence bond theory, one of a set of equivalent orbitals about an atom created when specific atomic orbitals are mixed. (Ch. 5)
hybridization In valence bond theory, the mixing of atomic orbitals to generate new sets of orbitals that are then available to form covalent bonds with other atoms. (Ch. 5)
hydrocarbon An organic compound whose molecules contain only carbon and hydrogen atoms. (Ch. 6)
hydrogen bond The strongest dipole–dipole interaction, which occurs between a hydrogen atom bonded to a N, O, or F atom and another N, O, or F atom. (Ch. 6)
hydrogenation A chemical reaction between molecular hydrogen (H_2) and another substance. (Ch. 19)
hydronium ion (H_3O^+) An H^+ ion plus a water molecule, H_2O; the form in which the hydrogen ion is found in an aqueous solution. (Ch. 8)
hydrophilic Describes a "water-loving" or attractive interaction between a solute and water that increases water solubility. (Ch. 6)
hydrophobic Describes a "water-fearing" or repulsive interaction between a solute and water that decreases water solubility. (Ch. 6)
hydroxyl group A functional group that consists of an oxygen atom with a single bond to a hydrogen atom. (Ch. 6)
hypothesis A tentative and testable explanation for an observation or a series of observations. (Ch. 1)

I

ideal gas A gas whose behavior is predicted by the linear relations defined by the combined gas law. (Ch. 10)
ideal gas equation (also called the *ideal gas law*) The pressure, volume, number of moles, and temperature of an ideal gas are related by the equation $PV = nRT$, where R is the universal gas constant. (Ch. 10)
ideal gas law (also called *ideal gas equation*) The pressure, volume, number of moles, and temperature of an ideal gas are related by the equation $PV = nRT$, where R is the universal gas constant. (Ch. 10)
ideal solution One that obeys Raoult's law. (Ch. 11)
immiscible liquids Combinations of liquids that do not mix with, or dissolve in, each other. (Ch. 1)
inhibitor A compound that diminishes or destroys the ability of an enzyme to catalyze a reaction. (Ch. 20)
inner coordination sphere The ligands that are bound directly to a metal via coordinate bonds. (Ch. 23)
integrated rate law A mathematical expression that describes the change in concentration of a reactant in a chemical reaction with time. (Ch. 13)
intensive property A property that is independent of the amount of substance present. (Ch. 1)
intermediate A species produced in one step of a reaction and consumed in a subsequent step. (Ch. 13)
internal energy (E) The sum of all the kinetic and potential energies of all of the components of a system. (Ch. 9)
interstitial alloy An alloy in which the nonhost atoms occupy spaces between atoms of the host. (Ch. 18)
ion An atom or molecule that has a net positive or negative charge. (Ch. 1)
ion channel A group of helical proteins that penetrates cell membranes and allows selective transport of ions. (Ch. 22)
ion–dipole interaction An attractive force between an ion and a molecule that has a permanent dipole. (Ch. 6)
ion exchange A process in which one ion is displaced by another. (Ch. 8)

ion pair A cluster formed when a cation and an anion associate with each other in solution. (Ch. 11)

ion pump A system of membrane proteins that exchanges ions inside the cell with those in the intercellular fluid. (Ch. 22)

ionic bond A bond resulting from the electrostatic attraction of a cation for an anion. (Ch. 4)

ionic compound A compound that consists of a characteristic ratio of positive and negative ions. (Ch. 1)

ionic solid A solid consisting of monatomic or polyatomic ions held together by ionic bonds. (Ch. 18)

ionization energy (IE) The amount of energy needed to remove 1 mole of electrons from 1 mole of ground-state atoms or ions in the gas phase. (Ch. 3)

ionizing radiation High-energy products of radioactive decay that can ionize molecules. (Ch. 21)

isoelectronic Describes atoms or ions that have identical electron configurations. (Ch. 3)

isomer One of a group of compounds having the same chemical formula but different molecular structures. (Ch. 5)

isotopes Atoms of an element containing the same number of protons but different numbers of neutrons. (Ch. 2)

J

joule (J) The SI unit of energy, equivalent to 1 kg · (m/s)^2. (Ch. 1)

K

Kekulé structure A structure using lines to show all of the bonds in a covalently bonded molecule, but not showing lone pairs on the atoms. (Ch. 19)

ketone An organic compound that contains a carbonyl group bonded to two other carbon atoms. (Ch. 6)

kinetic energy (KE) The energy of an object in motion due to its mass (m) and its speed (u). (Ch. 1)

kinetic molecular theory (KMT) A model that explains the behavior of gases based on the motion of the particles that make them up. (Ch. 10)

L

lattice energy (U) The energy released when one mole of an ionic compound forms from its free ions in the gas phase. (Ch. 4)

law of conservation of energy The principle that energy cannot be created or destroyed but can be changed from one form to another. (Ch. 1)

law of conservation of mass The principle that the sum of the masses of the reactants in a chemical reaction is equal to the sum of the masses of the products. (Ch. 7)

law of constant composition The principle that all samples of a particular compound have the same elemental composition. (Ch. 1)

law of definite proportions The principle that compounds always contain the same proportions of their component elements; equivalent to the *law of constant composition*. (Ch. 1)

law of mass action The ratio of the concentrations or partial pressures of products to reactants at equilibrium has a characteristic value at a given temperature when each term is raised to a power equal to the coefficient of that substance in the balanced chemical equation for the reaction. (Ch. 14)

law of multiple proportions The principle that, when two masses of one element react with a given mass of another element to form two different compounds, the two masses of the first element have a ratio of two small whole numbers. (Ch. 1)

Le Châtelier's principle A system at equilibrium responds to a stress in such a way that it relieves that stress. (Ch. 14)

leveling effect The observation that all strong acids completely ionize in aqueous solutions, forming H_3O^+ ions; strong bases are likewise leveled in water and are completely converted into their conjugate acids and OH^- ions. (Ch. 15)

Lewis acid A substance that accepts a lone pair of electrons in a chemical reaction. (Ch. 16)

Lewis base A substance that donates a lone pair of electrons in a chemical reaction. (Ch. 16)

Lewis structure A two-dimensional representation of the bonds and lone pairs of valence electrons in an ionic or molecular compound. (Ch. 4)

Lewis symbol The chemical symbol for an element surrounded by one or more dots representing valence electrons; also called a *Lewis dot symbol.* (Ch. 4)

ligand A Lewis base bonded to the central metal ion of a complex ion. (Ch. 16)

limiting reactant A reactant that is consumed completely in a chemical reaction. The amount of product formed depends on the amount of the limiting reactant available. (Ch. 7)

linear (straight-chain) hydrocarbon A hydrocarbon in which the carbon atoms are bonded together in one continuous line. Linear alkane chains have a methyl group at each end of methylene groups connecting them. (Ch. 19)

lipid A class of hydrophobic, water-insoluble, oily organic compounds that are common structural materials in cells. (Ch. 20)

lipid bilayer A double layer of molecules whose polar head groups interact with water molecules and whose nonpolar tails interact with each other. (Ch. 20)

lipoproteins Soluble proteins that combine with and transport fat or other lipids in the blood plasma. (Ch. 20)

liquid A phase of matter that occupies a definite volume but flows to assume the shape of its container. (Ch. 1)

London dispersion force An intermolecular force between atoms or molecules caused by the presence of temporary dipoles in the molecules. (Ch. 6)

lone pair A pair of electrons that is not shared. (Ch. 4)

M

macrocyclic ligand A ring containing multiple electron-pair donors that bind to a metal ion. (Ch. 23)

magnetic quantum number (m_ℓ) Defines the orientation of an orbital in space; an integer that may have any value from $-\ell$ to $+\ell$, where ℓ is the angular momentum quantum number. (Ch. 3)

main group elements (also called *representative elements*) The elements in groups 1, 2, and 13 through 18 of the periodic table. (Ch. 2)

major essential element Essential element present in the body in average concentrations greater than 1 mg of element per gram of body mass. (Ch. 22)

manometer An instrument for measuring the pressure exerted by a gas. (Ch. 10)

mass (m) The property that defines the quantity of matter in an object. (Ch. 1)

mass action expression Equivalent to the equilibrium constant expression but applied to reaction mixtures that may or may not be at equilibrium. (Ch. 14)

mass defect (Δm) The difference between the mass of a stable nucleus and the masses of the individual nucleons that comprise it. (Ch. 21)
mass number (A) The number of nucleons in an atom. (Ch. 2)
mass spectrometer A device that separates, weighs, and counts ions based on their mass (m) to charge (z) ratio, m/z. (Ch. 2)
mass spectrum A graph of the data from a mass spectrometer, where m/z ratios of the deflected particles are plotted against the number of particles with a particular mass. Because the charge on the ions typically is 1+, $m/z = m/1 = m$, and the mass of the particle may be read directly from the m/z axis. (Ch. 2)
matter Anything that has mass and occupies space. (Ch. 1)
matter wave The wave associated with any moving particle. (Ch. 3)
mean (*arithmetic mean*) An average calculated by summing all of the values in a series and then dividing the sum by the number of values. (Ch. 1)
mean free path The average distance that a particle can travel through air or any gas before colliding with another particle. (Ch. 10)
membrane potential, E_{membrane} A weighted average of the equilibrium (reversal) potentials of the major ions based on their concentrations and abilities to pass through the membrane. (Ch. 22)
meniscus The concave or convex surface of a liquid. (Ch. 6)
messenger RNA (mRNA) The form of RNA that carries the code for synthesizing proteins from DNA to the site of protein synthesis in a cell. (Ch. 20)
metallic bond A bond consisting of the nuclei of metal atoms surrounded by a "sea" of shared electrons. (Ch. 4)
metalloids (also called *semimetals*) Elements that tend to have the physical properties of metals and the chemical properties of nonmetals. (Ch. 2)
metals Elements that are typically shiny, malleable, ductile solids that conduct heat and electricity well and tend to form positive ions. (Ch. 2)
meter (m) The standard unit of length, equivalent to 39.37 inches. (Ch. 1)
methanogenic bacteria Bacteria using simple organic compounds and hydrogen for energy; their respiration produces methane, carbon dioxide, and water, depending on the compounds they consume. (Ch. 19)
methyl group (CH_3–), a structural unit that can make only one bond. (Ch. 19)
methylene group (–CH_2–), a structural unit that can make two bonds. (Ch. 19)
miscible Liquids that are mutually soluble in any proportion. (Ch. 6)
mixture A combination of pure substances in variable proportions in which the individual substances retain their chemical identities and can be separated from one another by a physical process. (Ch. 1)
molality (m) Concentration expressed as the number of moles of solute per kilogram of solvent. (Ch. 11)
molar heat capacity ($c_{P,n}$) The energy required to raise the temperature of one mole of a substance by 1°C at constant pressure. (Ch. 9)
molar mass ($\mathcal{M}$) The mass of 1 mole of a substance. (Ch. 2)
molarity (M) The number of moles of solute per liter of solution: $M = n/V$; also called *molar concentration*. (Ch. 8)
mole (mol) An amount of a substance that contains Avogadro's number ($N_A = 6.0221 \times 10^{23}$) of particles (atoms, ions, molecules, or formula units). (Ch. 2)
mole fraction (x_i) The ratio of the number of moles of a particular component i in a mixture to the total number of moles in the mixture. (Ch. 10)
molecular equation A balanced equation that describes a reaction in solution in which all reactants and products are written as neutral compounds. (Ch. 8)
molecular formula A chemical formula that shows how many atoms of each element are in one molecule of a pure substance. (Ch. 1)
molecular geometry The three-dimensional arrangement of the atoms in a molecule. (Ch. 5)
molecular ion (M^+) An ion formed in a mass spectrometer when a molecule loses an electron after being bombarded with high-energy electrons. The molecular ion has a charge of 1+ and has essentially the same molecular mass as the molecule from which it came. (Ch. 2)
molecular mass The mass in amu of one molecule of a molecular compound. (Ch. 2)
molecular orbital A region of characteristic shape and energy where electrons in a molecule are located. (Ch. 5)
molecular orbital diagram In MO theory, an energy-level diagram showing the relative energies and electron occupancy of the molecular orbitals for a molecule. (Ch. 5)
molecular orbital (MO) theory A bonding theory based on the mixing of atomic orbitals of similar shapes and energies to form molecular orbitals that belong to the molecule as a whole. (Ch. 5)
molecular recognition The process by which molecules interact with other molecules to produce a biological effect. (Ch. 5)
molecular solid A solid formed by neutral, covalently bonded molecules held together by intermolecular attractive forces. (Ch. 18)
molecularity The number of ions, atoms, or molecules involved in an elementary step in a reaction. (Ch. 13)
molecule A collection of atoms chemically bonded together. (Ch. 1)
monodentate ligand A species that forms only a single coordinate bond to a metal ion in a complex. (Ch. 23)
monomer A small molecule that bonds with other monomers to form polymers. (Ch. 19)
monoprotic acid An acid that has one ionizable hydrogen atom per molecule. (Ch. 15)
monosaccharide A single-sugar unit and the simplest carbohydrate. (Ch. 20)

N

nanoparticle Approximately spherical sample of matter with dimensions less than 100 nanometers (1×10^{-7} m).
natural abundance The proportion of a particular isotope, usually expressed as a percentage, relative to all the isotopes of that element in a natural sample. (Ch. 2)
Nernst equation An equation relating the potential of a cell (or half-cell) reaction to its standard potential ($E°$) and to the concentrations of its reactants and products. (Ch. 17)
net ionic equation A balanced equation that describes the actual reaction taking place in solution; it is obtained by eliminating the spectator ions from the total ionic equation. (Ch. 8)
neutralization reaction A reaction that takes place when an acid reacts with a base and produces a solution of a salt in water. (Ch. 8)
neutron An electrically neutral (uncharged) subatomic particle with a mass number of 1. (Ch. 2)
neutron capture The absorption of a neutron by a nucleus. (Ch. 21)
noble gases The elements in group 18 of the periodic table. (Ch. 2)
nonelectrolyte A molecular substance that does not dissociate into ions when it dissolves in water. (Ch. 8)
nonessential element Element present in humans that has no known function. (Ch. 22)
nonmetals Elements with properties opposite those of metals, including poor conductivity of heat and electricity. (Ch. 2)
nonpolar covalent bond A bond characterized by an even distribution of charge; electrons in the bonds are shared equally by the two atoms. (Ch. 4)

nonspontaneous process A process that requires outside intervention to proceed. (Ch. 12)

n-type semiconductor A semiconductor containing an electron-rich dopant. (Ch. 18)

nuclear chemistry The study of reactions that involve changes in the nuclei of atoms. (Ch. 21)

nuclear fission A nuclear reaction in which the nucleus of an element splits into two lighter nuclei. The process is usually accompanied by the release of one or more neutrons and energy. (Ch. 21)

nuclear fusion A nuclear reaction in which subatomic particles or atomic nuclei collide with each other at very high speeds and fuse together, forming more massive nuclei and releasing energy. (Ch. 21)

nucleic acid One of a family of large molecules, which includes deoxyribonucleic acid (DNA) and ribonucleic acid (RNA), that contains the genetic blueprint of an organism and controls the production of proteins. (Ch. 20)

nucleon A proton or neutron in a nucleus. (Ch. 2)

nucleosynthesis The natural formation of nuclei as a result of fusion and other nuclear processes. (Ch. 21)

nucleotide A monomer unit from which nucleic acids are made. (Ch. 20)

nucleus (of an atom) The positively charged center of an atom that contains nearly all the atom's mass. (Ch. 2)

nuclide A specific isotope of an element. (Ch. 2)

O

octahedral Molecular geometry about a central atom with a steric number of 6 and no lone pairs of electrons, in which all six sites are equivalent. (Ch. 5)

octet rule Atoms of main group elements make bonds by gaining, losing, or sharing electrons to achieve a valence shell containing eight electrons, or four electron pairs. (Ch. 4)

oil A liquid triglyceride containing primarily unsaturated fatty acids. (Ch. 20)

orbital diagram A depiction of the arrangement of electrons in an atom or ion using boxes to represent orbitals. (Ch. 3)

orbitals Defined by the square of the wave function (ψ^2); regions in an atom where the probability of finding an electron is high. (Ch. 3)

ore Naturally occurring compounds or mixtures of compounds from which elements can be extracted. (Ch. 18)

organic chemistry The study of compounds containing C—C and/or C—H bonds. (Ch. 19)

organic compounds Compounds that contain carbon, hydrogen, and sometimes other elements, including oxygen, nitrogen, sulfur, and a halogen. (Ch. 6)

organometallic compound A molecule containing direct carbon–metal covalent bonds. (Ch. 23)

osmosis The flow of a fluid through a semipermeable membrane to balance the concentration of solutes in solutions on the two sides of the membrane. The flow of solvent molecules proceeds from the more dilute solution into the more concentrated one. (Ch. 11)

osmotic pressure (Π) The pressure applied across a semipermeable membrane to stop the flow of water from the compartment containing pure solvent or a less concentrated solution to the compartment containing a more concentrated solution. (Ch. 11)

outlier A data point that is distant from the other observations. (Ch. 1)

overall reaction order The sum of the exponents of the concentration terms in the rate law. (Ch. 13)

overlap A term in valence bond theory describing bonds arising from two orbitals on different atoms that occupy the same region of space. (Ch. 5)

oxidation A chemical change in which an element loses electrons; the oxidation number of the element increases. (Ch. 8)

oxidation number (O.N.) (also called *oxidation state*) A numerical value (which may be positive, negative, or zero) based on the number of electrons that an atom gains or loses when it forms an ion or that it shares when it forms a covalent bond with an atom of another element. (Ch. 8)

oxidation state (also called *oxidation number [O.N.]*) A numerical value (which may be positive, negative, or zero) based on the number of electrons that an atom gains or loses when it forms an ion, or that it shares when it forms a covalent bond with an atom of another element. (Ch. 8)

oxidizing agent A reactant that accepts electrons from another in a redox reaction, thereby oxidizing the other reactant; the oxidizing agent is reduced in the reaction. (Ch. 8)

oxoacid A compound composed of oxoanions bonded to H^+ ions. (Ch. 4)

oxoanion A polyatomic anion that contains at least one nonoxygen central atom bonded to one or more oxygen atoms. (Ch. 4)

P

packing efficiency The percentage of the total volume of a unit cell occupied by the spheres. (Ch. 18)

paramagnetic Describes a substance with unpaired electrons that is attracted to a magnetic field. (Ch. 5)

partial pressure The contribution to the total pressure made by a component in a mixture of gases. (Ch. 10)

Pauli exclusion principle Principle that states no two electrons in an atom can have the same set of four quantum numbers. (Ch. 3)

peptide A compound of two or more amino acids joined by peptide bonds. Small peptides containing up to 20 amino acids are *oligopeptides*; the term *polypeptide* is used for chains longer than 20 amino acids but shorter than proteins. (Ch. 20)

peptide bond The result of a condensation reaction between the carboxylic acid group of one amino acid and the amine group of another. (Ch. 20)

percent composition The composition of a compound expressed in terms of the percentage by mass of each element in the compound. (Ch. 7)

percent ionization The ratio of the quantity of a substance that is ionized to the concentration of the substance before ionization, expressed as a percentage. (Ch. 15)

percent yield The ratio, expressed as a percentage, of the actual yield of a chemical reaction to the theoretical yield. (Ch. 7)

period (of elements) All the elements in a row of the periodic table. (Ch. 2)

periodic table of the elements A chart of the elements in order of their atomic numbers and in a pattern based on their physical and chemical properties. (Ch. 2)

permanent dipole Permanent separation of electrical charge in a molecule due to unequal distributions of bonding and/or lone pairs of electrons. (Ch. 5)

pH The negative logarithm of the hydrogen ion concentration in an aqueous solution. (Ch. 15)

pH buffer A solution that resists changes in pH when acids or bases are added to it; typically a solution of a weak acid and its conjugate base. (Ch. 16)

pH indicator A water-soluble weak organic acid that changes color as it ionizes. (Ch. 16)

phase diagram A graphical representation of the dependence of the stabilities of the physical states of a substance on temperature and pressure. (Ch. 6)

phospholipid A molecule of glycerol with two fatty acid chains and one polar group containing a phosphate; phospholipids are major constituents of cell membranes. (Ch. 20)

phosphorylation A reaction resulting in the addition of a phosphate group to an organic molecule. (Ch. 12)

photochemical smog A mixture of gases formed in the lower atmosphere when sunlight interacts with compounds produced in internal combustion engines and with other pollutants. (Ch. 13)

photoelectric effect The release of electrons from a material as a result of electromagnetic radiation striking it. (Ch. 3)

photon A quantum of electromagnetic radiation. (Ch. 3)

photosynthesis A set of chemical reactions driven by the energy of sunlight that convert carbon dioxide and water into oxygen and a wide variety of molecules containing C, H, and O. (Ch. 7)

physical property A property of a substance that can be observed without changing the substance into another substance. (Ch. 1)

pi (π) bond A covalent bond in which electron density is greatest above and below the bonding axis. (Ch. 5)

pi (π) molecular orbital In MO theory, an orbital formed by the mixing of atomic orbitals oriented above and below, or in front of and behind, the bonding axis; electrons in π orbitals form π bonds. (Ch. 5)

Planck constant (h) The proportionality constant between the energy and frequency of electromagnetic radiation expressed in $E = h\nu$; $h = 6.626 \times 10^{-34}$ J · s. (Ch. 3)

pOH The negative logarithm of the hydroxide ion concentration in an aqueous solution. (Ch. 15)

polar covalent bond A bond resulting from unequal sharing of bonding pairs of electrons between atoms. (Ch. 4)

polarizability The relative ease with which the electron cloud in a molecule, ion, or atom can be distorted, inducing a temporary dipole. (Ch. 6)

polyatomic ion A charged group of atoms joined together by covalent bonds. (Ch. 4)

polydentate ligand A species that can form more than one coordinate bond per molecule. (Ch. 23)

polymer A very large molecule with high molar mass; the root word *meros* is Greek for "part" or "unit," so *polymer* literally means "many units." (Ch. 19)

polyprotic acid An acid that has two or more ionizable hydrogen atoms per molecule. (Ch. 15)

polysaccharide A polymer of monosaccharides. (Ch. 20)

porphyrin A type of tetradentate macrocyclic ligand. (Ch. 23)

positron A particle with the mass of an electron but with a positive charge. (Ch. 21)

positron emission The spontaneous emission of a positron from a neutron-poor nucleus. (Ch. 21)

potential energy (PE) The energy stored in an object because of its position or composition. (Ch. 1)

precipitate A solid product formed from a reaction in solution. (Ch. 8)

precision The extent to which repeated measurements of the same variable agree. (Ch. 1)

pressure (P) The ratio of a force to the surface area over which the force is applied. (Ch. 6)

pressure–volume (P–V) work The work associated with the expansion or compression of a gas. (Ch. 9)

primary (1°) structure The sequence in which the amino acid monomers occur in a protein chain. (Ch. 20)

principal quantum number (n) A positive integer describing the relative size and energy of an atomic orbital or group of orbitals in an atom. (Ch. 3)

product Substance formed as a result of a chemical reaction. (Ch. 7)

protein A biological polymer made of amino acids. (Ch. 20)

proton A subatomic particle in the nuclei of atoms that has a relative charge of 1+ and a mass number of 1. (Ch. 2)

pseudo first order A reaction in which all the reactants but one are present at such high concentrations that they do not decrease significantly during the course of the reaction, so that reaction rate is controlled by the concentration of the limiting reactant. (Ch. 13)

p-type semiconductor A semiconductor containing an electron-poor dopant. (Ch. 18)

Q

quantized Having values restricted to whole-number multiples of a specific base value. (Ch. 3)

quantum (plural *quanta*) The smallest discrete quantity of a particular form of energy. (Ch. 3)

quantum mechanics (also called *wave mechanics*) A mathematical description of the wavelike behavior of electrons and other particles. (Ch. 3)

quantum number One of four related numbers that specify the energy, shape, and orientation of orbitals in an atom and the spin orientation of electrons in the orbitals. (Ch. 3)

quantum theory A model of matter and energy based on the principle that energy is absorbed or emitted in discrete packets, or quanta.

quarks Elementary particles that combine to form neutrons and protons. (Ch. 21)

quaternary (4°) structure The larger structure functioning as a single unit that results when two or more proteins associate. (Ch. 20)

R

R A symbol in a general formula standing for an organic group that has one available bond; it is used to indicate the part of a molecule that is not the functional group of interest. (Ch. 19)

racemic mixture A sample containing equal amounts of both enantiomers of a compound. (Ch. 5)

radioactive decay The spontaneous disintegration of unstable particles accompanied by the release of radiation. (Ch. 21)

radioactivity The spontaneous emission of high-energy radiation and particles by materials. (Ch. 2)

radiocarbon dating A method for establishing the age of a carbon-containing object by measuring the amount of radioactive carbon-14 remaining in the object. (Ch. 21)

radiometric dating A method for determining the age of an object based on the quantity of a radioactive nuclide and/or the products of its decay that the object contains. (Ch. 21)

radionuclide A radioactive (unstable) nuclide. (Ch. 2)

random coil An irregular or rapidly changing part of the secondary structure of a protein. (Ch. 20)

Raoult's law The vapor pressure of a solution is the sum of the vapor pressures of the volatile components of the solution, which are each the product of the vapor pressure of the pure component and its mole fraction in the solution. (Ch. 11)

rate constant The proportionality constant that relates the rate of a reaction to the concentrations of reactants. (Ch. 13)

rate-determining step The slowest step in a multistep chemical reaction. (Ch. 13)

rate law An equation that defines the experimentally determined relation between the concentration of reactants in a chemical reaction and the rate of that reaction. (Ch. 13)

reactant Substance consumed during a chemical reaction. (Ch. 7)

reaction mechanism A set of steps that describes how a reaction occurs at the molecular level; the mechanism must be consistent with the rate law for the reaction. (Ch. 13)

reaction order An experimentally determined number defining the dependence of the reaction rate on the concentration of a reactant. (Ch. 13)

reaction quotient (Q) The numerical value of the mass action expression for any values of the concentrations (or partial pressures) of reactants and products; at equilibrium, $Q = K$. (Ch. 14)

reaction rate A measure of how rapidly a reaction occurs; it is related to rates of change in the concentrations of reactants and products over time. (Ch. 13)

reducing agent A reactant that donates electrons to another in a redox reaction, thereby reducing the other reactant; the reducing agent is oxidized in the reaction. (Ch. 8)

reduction A chemical change in which an element gains electrons; the oxidation number of the element decreases. (Ch. 8)

relative biological effectiveness (RBE) A factor that accounts for the differences in physical damage caused by different types of radiation. (Ch. 21)

replication The process by which one double-stranded DNA forms two new DNA molecules, each one containing one strand from the original molecule and one new strand. (Ch. 20)

representative elements (also called *main group elements*) The elements in groups 1, 2, and 13 through 18 of the periodic table. (Ch. 2)

resonance A characteristic of electron distributions when two or more equivalent Lewis structures can be drawn for one compound. (Ch. 4)

resonance stabilization The stability of a molecular structure due to the delocalization of its electrons. (Ch. 4)

resonance structure One of two or more Lewis structures with the same arrangement of atoms but different arrangements of bonding pairs of electrons. (Ch. 4)

reverse osmosis (RO) A purification process in which solvent is forced through semipermeable membranes, leaving dissolved impurities behind. (Ch. 11)

reversible process A process that happens so slowly that an incremental change can be reversed by another tiny change, restoring the original state of the system with no net flow of energy between the system and its surroundings. (Ch. 12)

root-mean-square speed (u_{rms}) The square root of the average of the squared speeds of all the particles in a population of gas particles; a particle possessing the average kinetic energy moves at this speed. (Ch. 10)

S

salt The product of a neutralization reaction; it is made up of the cation of the base in the reaction and the anion of the acid. (Ch. 8)

saturated hydrocarbon An alkane. (Ch. 19)

saturated solution A solution that contains the maximum concentration of a solute possible at a given temperature. (Ch. 8)

Schrödinger wave equation A description of how the electron matter wave varies with location and time around the nucleus of a hydrogen atom. (Ch. 3)

scientific law A concise and generally applicable statement of a fundamental scientific principle. (Ch. 1)

scientific method An approach to acquiring knowledge based on the observation of phenomena, the development of a testable hypothesis, and additional experiments that test the validity of the hypothesis. (Ch. 1)

scientific theory A concise explanation of widely observed phenomena that has been extensively tested. (Ch. 1)

scintillation counter An instrument that determines the level of radioactivity in samples by measuring the intensity of light emitted by phosphors in contact with the samples. (Ch. 21)

second law of thermodynamics The principle stating that the total entropy of the universe increases in any spontaneous process. (Ch. 12)

secondary (2°) structure The pattern of arrangement of segments of a protein chain. (Ch. 20)

seesaw Molecular geometry about a central atom with a steric number of 5 and one lone pair of electrons in an equatorial position; the atoms occupy two axial sites and two equatorial sites. (Ch. 5)

semiconductor A material with electrical conductivity between that of metals and insulators that can be chemically altered to increase its electrical conductivity. (Ch. 18)

semimetals (also called *metalloids*) Elements that tend to have the physical properties of metals and the chemical properties of nonmetals. (Ch. 2)

SI units A set of base and derived units used worldwide to express distances and quantities of matter and energy. (Ch. 1)

sievert (Sv) SI unit used to express the amount of biological damage caused by ionizing radiation. (Ch. 21)

sigma (σ) bond A covalent bond in which the highest electron density lies between the two atoms along the bond axis. (Ch. 5)

sigma (σ) molecular orbital In MO theory, the orbital that results in the highest electron density between the two bonded atoms. (Ch. 5)

significant figures All the certain digits in a measured value plus one estimated digit. The greater the number of significant figures, the greater the certainty with which the value is known. (Ch. 1)

simple cubic (sc) unit cell A cell with atoms only at the eight corners of a cube. (Ch. 18)

single bond A bond that results when two atoms share one pair of electrons. (Ch. 4)

solid A phase of matter that has a definite shape and volume. (Ch. 1)

solubility The maximum quantity of a substance that can dissolve in a given volume of solution. (Ch. 6)

solubility product, K_{sp} (also called *solubility-product constant*) An equilibrium constant that describes the formation of a saturated solution of a slightly soluble salt. (Ch. 16)

solubility-product constant (also called *solubility product, K_{sp}*) An equilibrium constant that describes the formation of a saturated solution of a slightly soluble salt. (Ch. 16)

solute Any component in a solution other than the solvent. A solution may contain one or more solutes. (Ch. 6)

solution Another name for a *homogeneous mixture*. Solutions are often liquids, but they may also be solids or gases. (Ch. 1)

solvent The component of a solution for which the largest number of moles is present. (Ch. 6)

***sp* hybrid orbitals** Two hybrid orbitals on opposite sides of the hybridized atom formed by mixing one *s* and one *p* orbital. (Ch. 5)

sp^2 hybrid orbitals Three hybrid orbitals in a trigonal planar orientation formed by mixing one *s* and two *p* orbitals. (Ch. 5)

sp^3 hybrid orbitals A set of four hybrid orbitals with a tetrahedral orientation produced by mixing one *s* and three *p* atomic orbitals. (Ch. 5)

sp^3d hybrid orbitals Five equivalent hybrid orbitals with lobes pointing toward the vertices of a trigonal bipyramid that form by mixing one *s* orbital, three *p* orbitals, and one *d* orbital from the same shell. (Ch. 5)

sp^3d^2 hybrid orbitals Six equivalent hybrid orbitals pointing toward the vertices of an octahedron that form by mixing one *s* orbital, three *p* orbitals, and two *d* orbitals from the same shell. (Ch. 5)

specific heat (c_P) The energy required to raise the temperature of 1 g of a substance by 1°C at constant pressure. (Ch. 9)

spectator ion An ion that is unchanged by a chemical reaction. (Ch. 8)

spectrochemical series A list of ligands rank-ordered by their ability to split the energies of the *d* orbitals of transition metal ions. (Ch. 23)

sphere of hydration The cluster of water molecules surrounding an ion in an aqueous solution. (Ch. 6)

spin quantum number (m_s) Either $+\frac{1}{2}$ or $-\frac{1}{2}$, indicating the spin orientation of an electron. (Ch. 3)

spontaneous process A process that proceeds without outside intervention. (Ch. 12)

square planar Molecular geometry about a central atom with a steric number of 6 and two lone pairs of electrons that occupy axial sites; the atoms occupy four equatorial positions. (Ch. 5)

square pyramidal Molecular geometry about a central atom with a steric number of 6 and one lone pair of electrons; as typically drawn, the atoms occupy four equatorial sites and one axial site. (Ch. 5)

standard atmosphere (atm) The average pressure at sea level on Earth. (Ch. 6)

standard cell potential (E°_{cell}) A measure of how forcefully an electrochemical cell, in which all reactants and products are in their standard states, can pump electrons through an external circuit. (Ch. 17)

standard conditions In thermodynamics: a pressure of 1 atm (~1 bar) and some specified temperature, assumed to be 25°C unless otherwise stated; for solutions, a concentration of 1 *M* is specified. (Ch. 9)

standard deviation (*s*) A measure of the amount of variation, or dispersion, in a set of related values. (Ch. 1)

standard enthalpy of formation (ΔH°_f) The enthalpy change that takes place at a constant pressure of 1 atm when one mole of a substance is formed from its constituent elements in their standard states; also called *standard heat of formation.* (Ch. 9)

standard enthalpy of reaction ($\Delta H^\circ_{\text{rxn}}$) The enthalpy change associated with a reaction that takes place under standard conditions; also called the *standard heat of reaction.* (Ch. 9)

standard free energy of formation (ΔG°_f) The change in free energy associated with the formation of one mole of a compound in its standard state from its elements in their standard states. (Ch. 12)

standard hydrogen electrode (SHE) A reference electrode based on the half-reaction $2\,H^+(aq) + 2\,e^- \rightarrow H_2(g)$ that produces a standard electrode potential defined to be 0.000 V. (Ch. 17)

standard molar entropy (S°) The absolute entropy of one mole of a substance in its standard state. (Ch. 12)

standard reduction potential (E°) The potential of a reduction half-reaction in which all reactants and products are in their standard states at 25°C. (Ch. 17)

standard solution A solution of accurately known concentration that is used in chemical analysis. (Ch. 8)

standard state The most stable form of a substance under 1 atm pressure and some specified temperature (usually 25.0°C). (Ch. 9)

standard temperature and pressure (STP) 0°C and 1 bar as defined by IUPAC; 0°C and 1 atm are commonly used in the United States. (Ch. 10)

state function A property of an entity based solely on its chemical or physical state or both but not on how it achieved that state. (Ch. 9)

stereoisomers Molecules with the same formula and bonding order, but with different spatial arrangements of their atoms. (Ch. 5)

steric number (SN) The sum of the number of atoms bonded to a central atom plus the number of lone pairs of electrons on the central atom. (Ch. 5)

stimulatory effect Increased activity, growth, or other biological response to a stimulant. (Ch. 22)

stoichiometry The mole ratios among the reactants and products in a chemical reaction. (Ch. 7)

strong acid An acid that completely dissociates into ions in aqueous solution. (Ch. 8)

strong base A base that completely dissociates into ions in aqueous solution. (Ch. 8)

strong electrolyte A substance that dissociates completely when it dissolves in water. (Ch. 8)

strong nuclear force The fundamental force of nature that keeps quarks together in subatomic particles and nucleons together in atomic nuclei. (Ch. 21)

structural formula A representation of a molecule that uses short lines between the symbols of elements to show chemical bonds between atoms. (Ch. 1)

subatomic particles The neutrons, protons, and electrons in an atom. (Ch. 2)

sublimation The transformation of a solid directly into a gas (vapor). (Ch. 1)

substitutional alloy An alloy in which atoms of the nonhost metal replace host atoms in the crystal lattice. (Ch. 18)

substrate The reactant that binds to the active site in an enzyme-catalyzed reaction. (Ch. 20)

supercritical fluid A substance in a state that is above the temperature and pressure at the critical point, where the liquid and vapor phases are indistinguishable. (Ch. 6)

supersaturated solution A solution that contains more than the maximum quantity of solute predicted to be soluble in a given volume of solution at a given temperature. (Ch. 8)

surface tension The ability of the surface of a liquid to resist an external force. (Ch. 6)

surroundings Everything in a thermochemical study that is not part of the system. (Ch. 9)

system The part of the universe that is the focus of a thermochemical study. (Ch. 9)

T

temporary dipole The separation of charge produced in an atom or molecule by a momentary uneven distribution of electrons; also called *induced dipole.* (Ch. 6)

termolecular step A step in a reaction mechanism involving a collision among three molecules. (Ch. 13)

tertiary (3°) structure The three-dimensional, biologically active structure of the protein that arises because of interactions between R groups on amino acids. (Ch. 20)

tetrahedral Molecular geometry about a central atom with a steric number of 4 and no lone pairs of electrons. (Ch. 5)

theoretical yield The maximum amount of product possible in a chemical reaction for given quantities of reactants; also called the *stoichiometric yield*. (Ch. 7)

thermal energy The portion of the total internal energy of a system that is proportional to its absolute temperature. (Ch. 9)

thermochemical equation The chemical equation of a reaction that includes the change in enthalpy that accompanies the reaction. (Ch. 9)

thermochemistry The study of the changes in energy that accompany chemical reactions. (Ch. 9)

thermodynamics The study of energy and its transformations. (Ch. 9)

third law of thermodynamics The principle stating that the entropy of a perfect crystal is zero at absolute zero. (Ch. 12)

threshold frequency (ν_0) The minimum frequency of light required to produce the photoelectric effect. (Ch. 3)

titrant The standard solution added to the sample in a titration. (Ch. 8)

titration An analytical method for determining the concentration of a solute in a sample by reacting the solute with a solution of known concentration. (Ch. 8)

total ionic equation A balanced equation that shows all species, including spectator ions, present in a reaction, either as molecular materials or separate ions. (Ch. 8)

trace essential element Essential element present in the body in average concentrations between 1 and 1000 μg of element per gram of body mass. (Ch. 22)

trans isomer (also called E *isomer*) A molecule with two like groups (such as two R groups or two hydrogen atoms) on opposite sides of the molecule. (Ch. 19)

transcription The process of copying the information in DNA to mRNA. (Ch. 20)

transfer RNA (tRNA) The form of RNA that delivers amino acids, one at a time, to polypeptide chains being assembled by the ribosome–mRNA complex. (Ch. 20)

transition metals The elements in groups 3 through 12 of the periodic table. (Ch. 2)

transition state A high-energy state between reactants and products in a chemical reaction. (Ch. 13)

translation The process of assembling proteins from the information encoded in mRNA. (Ch. 20)

tricarboxylic acid (TCA) cycle A series of reactions that continues the oxidation of pyruvate formed in glycolysis. (Ch. 20)

trigonal bipyramidal Molecular geometry about a central atom with a steric number of 5 and no lone pairs of electrons, in which three atoms occupy equatorial sites and two other atoms occupy axial sites above and below the equatorial plane. (Ch. 5)

trigonal planar Molecular geometry about a central atom with a steric number of 3 and no lone pairs of electrons. (Ch. 5)

trigonal pyramidal Molecular geometry about a central atom with a steric number of 4 and one lone pair of electrons. (Ch. 5)

triple bond A bond formed when two atoms share three pairs of electrons. (Ch. 4)

triple point The temperature and pressure where all three phases of a substance coexist. Freezing and melting, boiling and liquefaction, and sublimation and deposition all proceed at the same rate, so no net change takes place in the system. (Ch. 6)

T-shaped Molecular geometry about a central atom with a steric number of 5 and two lone pairs of electrons that occupy equatorial positions; the atoms occupy two axial sites and one equatorial site. (Ch. 5)

U

ultratrace essential element Essential element present in the body in average concentrations less than 1 μg of element per gram of body mass. (Ch. 22)

unimolecular step A step in a reaction mechanism involving only one molecule on the reactant side. (Ch. 13)

unit cell The basic repeating unit of the arrangement of atoms, ions, or molecules in a crystalline solid. (Ch. 18)

universal gas constant The constant R in the ideal gas equation; its value and units depend on the units used for the variables in the equation. (Ch. 10)

unsaturated hydrocarbon An alkene or alkyne. (Ch. 19)

unsaturated solution A solution that contains less than the maximum quantity of solute predicted to be soluble in a given volume of solution at a given temperature. (Ch. 8)

V

valence The capacity of the atoms of an element to form chemical bonds. (Ch. 4)

valence band A band of orbitals that are filled or partially filled by valence electrons. (Ch. 18)

valence bond theory A quantum mechanics–based theory of bonding that assumes covalent bonds form when half-filled orbitals on different atoms overlap or occupy the same region in space. (Ch. 5)

valence electrons Electrons in the outermost occupied shell of an atom having the most influence on the atom's chemical behavior. (Ch. 3)

valence shell The outermost occupied shell of an atom. (Ch. 3)

valence-shell electron-pair repulsion theory (VSEPR) A model predicting the arrangement of valence electron pairs around a central atom that minimizes their mutual repulsion to produce the lowest energy orientations. (Ch. 5)

van der Waals equation An equation describing how the pressure, volume, and temperature of a quantity of a real gas are related; it includes terms that account for the incompressibility of gas particles and interactions between them. (Ch. 10)

van der Waals force Any interaction between neutral atoms and molecules, including hydrogen bonds, other dipole–dipole interactions, and London dispersion forces; the term does not apply to interactions involving ions. (Ch. 10).

van 't Hoff factor (i) The ratio of the concentration of solute particles in a solution to the concentration of particles that would be there if the solute did not dissociate. (Ch. 11)

vapor pressure The pressure exerted by a gas in equilibrium with its liquid phase at a given temperature. (Ch. 11)

vinyl group The subgroup $CH_2{=}CH–$. (Ch. 19)

vinyl polymer One of the family of polymers formed from monomers containing the subgroup $CH_2{=}CH–$. (Ch. 19)

viscosity The measure of the resistance to flow of a fluid. (Ch. 6)

volatile Having a significant vapor pressure at a given temperature. (Ch. 11)

volatility A measure of how readily a substance vaporizes. (Ch. 1)

voltaic cell (also called *galvanic cell*) An electrochemical cell in which chemical energy is transformed into electrical work by a spontaneous redox reaction. (Ch. 17)

W

wave function (ψ) A solution to the Schrödinger wave equation describing how the matter wave of an electron varies in both time and location in an atom. (Ch. 3)

wave mechanics (also called *quantum mechanics*) A mathematical description of the wavelike behavior of electrons and other particles. (Ch. 3)

wavelength (λ) The distance from crest to crest or trough to trough on a wave. (Ch. 3)

weak acid An acid that only partially dissociates in aqueous solutions. (Ch. 8)

weak base A base that only partially dissociates or ionizes in aqueous solutions. (Ch. 8)

weak electrolyte A substance that only partly dissociates into ions when it dissolves in water. (Ch. 8)

weak-link rule The rule that the result of a calculation is known only as well as the least well-known value used in the calculation. (Ch. 1)

work (w) The exertion of a force (F) through a distance (d): $w = F \times d$. (Ch. 1)

work function (Φ) The amount of energy needed to dislodge an electron from the surface of a material. (Ch. 3)

X

X-ray diffraction (XRD) A technique for determining the arrangement of atoms or ions in a crystal by analyzing the pattern that results when X-rays are scattered after bombarding the crystal. (Ch. 18)

Z

***Z* isomer** (also called *cis isomer*) Molecule with two like groups (such as two R groups or two hydrogen atoms) on the same side of the molecule. (Ch. 19)

zeolites Natural crystalline minerals or synthetic materials consisting of three-dimensional networks of channels that contain sodium or other 1+ cations. (Ch. 8)

zwitterion A molecule that has both positively and negatively charged groups in its structure. (Ch. 20)

Answers

to Particulate Review, Concept Tests, and Practice Exercises

Chapter 1

Particulate Review

Image C is liquid nitrogen; Image A is dry ice (solid CO_2); Image B is dry ice after sublimation (gaseous CO_2).

Concept Tests

p. 9 Mass and volume are extensive; density and hardness are intensive.

p. 14 (a) Filtration; (b) chromatography; (c) distillation

p. 16 (a) Sublimation; (b) deposition

p. 18 1.49 or 49%

p. 20 CH_4O or CH_3OH

p. 25 3, 3, 4

p. 27 46.94 miles per day; the trip took 66,727 minutes, which is known to five significant figures: one more than the distance hiked (2175 miles). Therefore, distance is the weak link.

p. 33 The average deviation is 0.0076 mg/dL, which is less than the standard deviation because squaring deviation values in calculating a standard deviation gives more weight to higher values.

Practice Exercises

1.1. Nickel

1.2. (a) and (c) are physical; (b) and (d) are chemical

1.3. (a) Only oxygen gas

1.4. (a) Gas in the left box turns into a solid in the right box as a result of deposition; (b) sublimation

1.5. Statistics (a) and (e) are exact numbers; statistics (b), (c), and (d) have inherent uncertainty.

1.6. (a) 1.130 g/cm^3; (b) more dense

1.7. $T(\text{K})_{\text{low}} = 40$ K and $T(\text{K})_{\text{high}} = 396$ K; $T(°\text{F})_{\text{low}} = -387°\text{F}$ and $T(°\text{F})_{\text{high}} = 253°\text{F}$

1.8. 161 J

1.9. 93 cm^3/s and 5.7 in^3/s

1.10. Store A

1.11. Mean = 36.4 mg/L, standard deviation = 0.3 mg/L, 95% confidence interval = 36.4 ± 0.4 mg/L

1.12. The mean and standard deviation values are 193.3 ± 18.9 mg/dL. Testing 215 mg/dL as an outlier gives a calculated Z value of 1.14, which is slightly less than the reference Z values (1.155) for $n = 3$ from Table 1.7 at both the 95% and 99% confidence levels. Therefore, the 215 value should *not* be considered an outlier.

Chapter 2

Particulate Review

There are 4 atoms (including 2 oxygen atoms) in one molecule of H_2O_2. The ratio of H:O is 1:1 in H_2O_2, but 2:1 in H_2O.

Concept Tests

p. 50 Top plate

p. 52 An α particle would be attracted to, and combine with, an electron.

p. 56 The concept of atomic number was unknown in the mid-19th century.

p. 63 One box of tissues provides a convenient number (500) of them, just as one mole of a substance is a convenient number of particles because their mass is equal to the molar mass of the substance.

p. 67 One gram of silver (Ag) contains more atoms than 1 gram of gold because Ag atoms have less mass, so there are more of them per gram.

p. 73 The masses of the stable isotopes of Br (^{79}Br and ^{81}Br) and Cl (^{35}Cl and ^{37}Cl) both differ by 2 amu, so a molecular ion containing an atom of ^{13}C and either an atom of ^{79}Br or ^{35}Cl would not have the same m/z value as a molecular ion with no ^{13}C and an atom of either ^{81}Br or ^{37}Cl.

Practice Exercises

2.1. (a) $^{56}_{26}$Fe; (b) $^{15}_{7}$N; (c) $^{37}_{17}$Cl; (d) $^{39}_{19}$K

2.2.

	Symbol	Protons	Neutrons	Electrons
(a)	$^{40}Ca^{2+}$	20	20	18
(b)	$^{79}Br^-$	35	44	36
(c)	$^{35}Cl^-$	17	18	18
(d)	$^{23}Na^+$	11	12	10

2.3. (a) As, arsenic; (b) Mg, magnesium; (c) Hg, mercury

2.4. (a) 131.29; (b) 131.06

2.5. 37.0% ^{191}Ir; 63.0% ^{193}Ir

2.6. 1.5×10^{10} atoms

2.7. 180.15 g/mol

2.8. 5.25 μm

2.9. 0.0541 mole C or 3.26×10^{22} atoms C

2.10. 1.81×10^{-3} moles, 1.09×10^{21} molecules

2.11. The molecular mass of testosterone is 288.42 amu, and there is a prominent peak at m/z= 288 amu in the spectrum. Carbon-13 atoms in the molecular ions would contribute to the smaller peak at 289 amu.

2.12. (a) Subtracting the mass of the CH_3 group (15 amu) from the CH_3Cl molecular-ion peaks at 50 and 52 amu yields Cl isotopic masses of 35 and 37 amu. (b) The relative heights of the peaks are 100 and 32, which makes the natural abundances of the isotopes (100/132) or 76% ^{35}Cl and 24% ^{37}Cl. These values match those obtained from the HCl mass spectrum in Figure 2.25.

Chapter 3

Particulate Review

1. 79 protons, 118 neutrons, and 79 electrons
2. Image (a)
3. Image (b) includes a nucleus that was able to deflect an alpha particle.

Concept Tests

p. 89 UV rays have higher frequencies than IR radiation.
p. 90 (c)
p. 91 b and d
p. 99 c > a > b > d
p. 114 ns^2np^5
p. 114 (a) Ultraviolet radiation
p. 120 An electron configuration with a half-filled $4f$ subshell has lower energy.
p. 126 Because the valence electrons are farther from the nucleus and shielded from it by more inner-shell electrons
p. 127 The magnitude of IE and EA both increase with increasing Z across a row (except for group 18). EA values do not display clear trends within groups whereas IE values decrease with increasing Z.

Practice Exercises

3.1. $\lambda = 3.30$ m
3.2. $\lambda = 397$ nm $= 3.97 \times 10^{-7}$ m
3.3. $\lambda = 262$ nm
3.4. 375.0 nm; no, this wavelength is in the ultraviolet region.
3.5. Prediction: Less energy is required to remove an electron from the hydrogen atom in the $n = 3$ state versus that for a hydrogen atom in the $n = 1$ state. Calculated value: 2.420×10^{-19} J
3.6. $\lambda = 3.3 \times 10^{-10}$ m
3.7. $\Delta_x \geq 7 \times 10^{-11}$
3.8. For $n = 5$ there are $n^2 = 25$ orbitals: $5s$, $5p$, $5d$, $5f$, and $5g$.
3.9.

n	ℓ	m_ℓ	m_s
3	1	−1	$+\frac{1}{2}$
3	1	0	$+\frac{1}{2}$
3	1	1	$+\frac{1}{2}$

3.10. Co = $[Ar]3d^74s^2$
3.11. $K^+ = [Ar]$; $I^- = [Kr]4d^{10}5s^25p^6 = [Xe]$; $Ba^{2+} = [Xe]$; $S^{2-} = [Ne]3s^23p^6 = [Ar]$; $Al^{3+} = [Ne]$; K^+ and S^{2-} are isoelectronic with Ar.
3.12. $Mn^{3+} = [Ar]3d^4$; $Mn^{4+} = [Ar]3d^3$
3.13. (a) $Na^+ < F^- < Cl^-$; (b) $Al^{3+} < Mg^{2+} < P^{3-}$
3.14. Ne > Ca > Cs

Chapter 4

Particulate Review

- a. N_2, b. O_2, c. H_2O
- a. 7 + 7 = 14, b. 8 + 8 = 16, c. 8 + (2 × 1) = 10
- a. 5 + 5 = 10, b. 6 + 6 = 12, c. 6 + (2 × 1) = 8

Concept Tests

p. 145 a. RbCl < KCl < NaCl
b. KF < CaO < MgO
p. 152 Hydrofluoric acid
p. 154 Bonding Capacity = 18 − Group Number
p. 169 ⟷
:C≡O:
p. 172 −1
p. 178 P (an element in row 3) can expand its octet and form five bonds; N (in row 2) cannot.
p. 179 Neither vibration is infrared active because the bonds are not polar.

Practice Exercises

4.1. Should be more negative than KCl; -3.25×10^{-18} J
4.2. (a) $SrCl_2$; (b) MgO; (c) NaF; (d) $CaBr_2$
4.3. $MnCl_2$; MnO_2
4.4. (a) $Sr(NO_3)_2$; (b) $KHCO_3$; (c) $BaCrO_4$
4.5. (a) Calcium phosphate; (b) magnesium sulfite; (c) lithium nitrite; (d) sodium hypochlorite; (e) potassium permanganate
4.6. (a) Diarsenic pentoxide; (b) carbon monoxide; (c) nitrogen trichloride
4.7. (a) Hydrobromous acid; (b) bromous acid; (c) carbonic acid
4.8. Na^+ $[:\ddot{\underset{..}{O}}:]^{2-}$
4.9. H—C—H with H above and below C (CH₄)
4.10. :C̈l—P̈—C̈l: with :C̈l: below P
4.11. [H—N—H with H above and below N]$^+$
4.12. :Ö=C=Ö:
4.13.
$[:\ddot{N}=N=\ddot{N}:]^- \leftrightarrow [:N\equiv N-\ddot{\underset{..}{N}}:]^- \leftrightarrow [:\ddot{\underset{..}{N}}-N\equiv N:]^-$
$[:\ddot{O}=N=\ddot{O}:]^+ \leftrightarrow [:O\equiv N-\ddot{\underset{..}{O}}:]^+ \leftrightarrow [:\ddot{\underset{..}{O}}-N\equiv O:]^+$
4.14. H—Ö—N(=Ö)—Ö: ⟷ H—Ö—N(—Ö:)=Ö
No, the N—O bond on the left is a single bond and should be longer than the other two with a length that is between that of a single and a double N—O bond.
4.15. Be and Cl; No, $\Delta\chi = 1.5$, so the bond is polar covalent.
4.16. $[:\ddot{O}=N=\ddot{O}:]^+$

4.17.

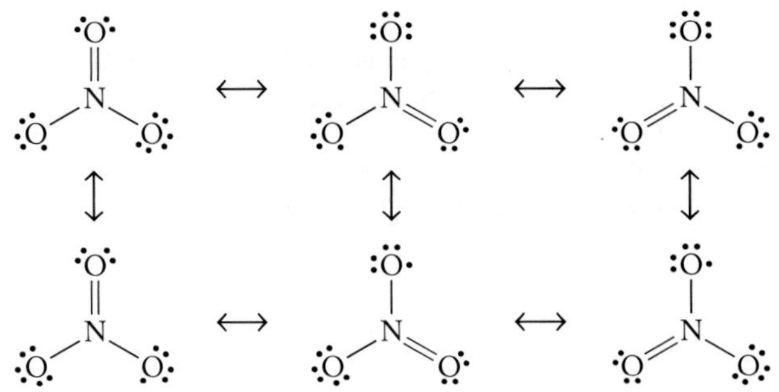

4.18.

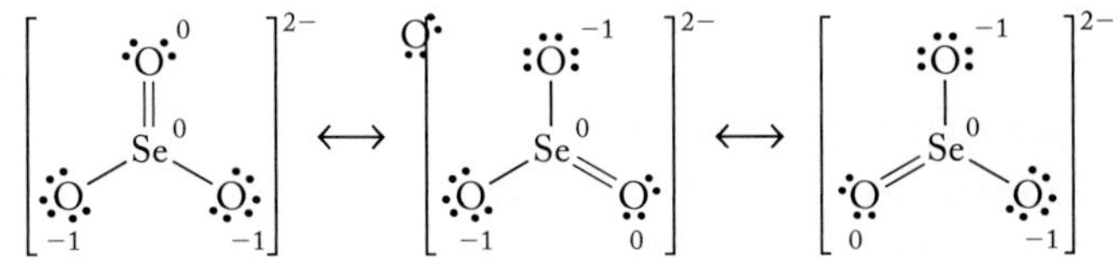

Chapter 5

Particulate Review

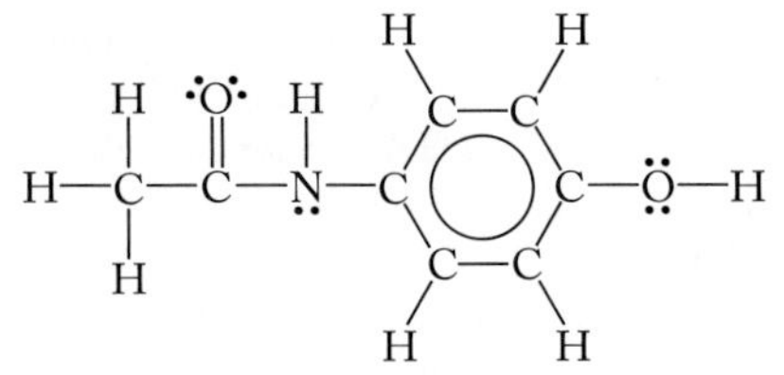

- Three pairs of delocalized π electrons.
- The O—H bond.

Concept Tests

p. 197 SN = 2 or 3 for less than an octet; SN = 4 for exactly an octet; SN = 5 or 6 for an expanded octet

p. 204 Slightly smaller bond angles in the seesaw molecular geometry

p. 208 Sulfur is less electronegative than oxygen.

p. 213 It has no unhybridized p orbitals with which to form π bonds.

p. 217 In structures (a) and (c) because the double bonds are conjugated

p. 223 If the muscarine came from mushrooms, it would be one enantiomer and its solution would be optically active. Because the solution did not rotate the plane of polarized light, the muscarine present was a racemic mixture, which had to be the product of a laboratory synthesis. The coroner concluded that the victim was poisoned by someone who had access to synthetic muscarine.

p. 229 (a) Be_2 and Ne_2; (b) B_2

p. 233 No, bond order is lower because one more electron is in an antibonding orbital and one fewer is in a bonding orbital.

Practice Exercises

5.1. Tetrahedral;

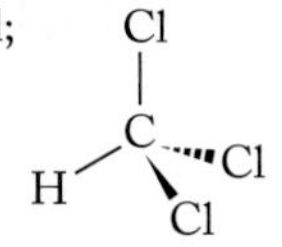

5.2. N_2O has the largest bond angle; NO_2^- has the smallest.

5.3. Tetrahedral molecular geometry with average bond angles ~109.5°C

5.4. No, because of its linear molecular geometry

5.5. CCl_4 and PH_3

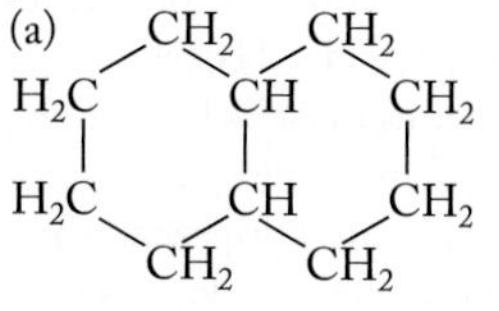

5.6. sp^3d^2

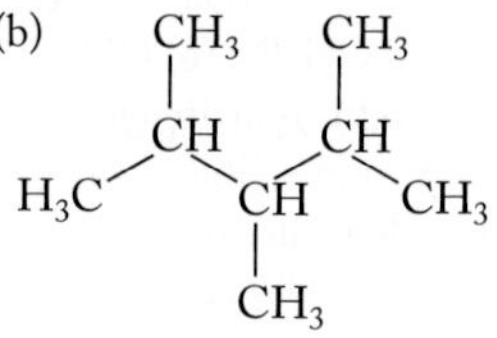

5.7.

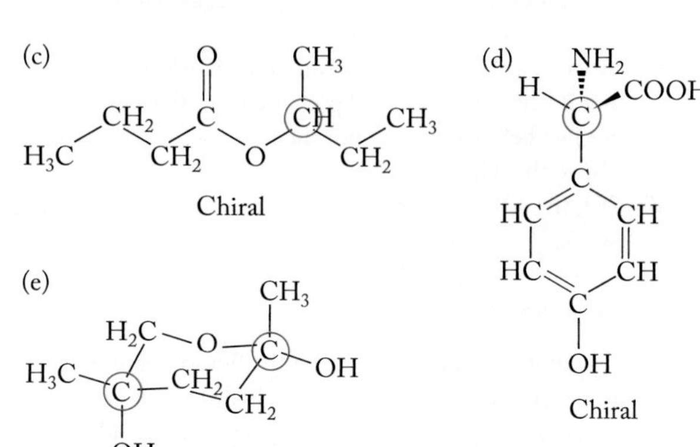

(c) Chiral

(d) Chiral

(e) Chiral

5.8. Bond order is 0.5.

5.9. Molecules experiencing an increase in bond order are B_2, C_2, and, if they existed, Be_2 and Ne_2.

5.10.

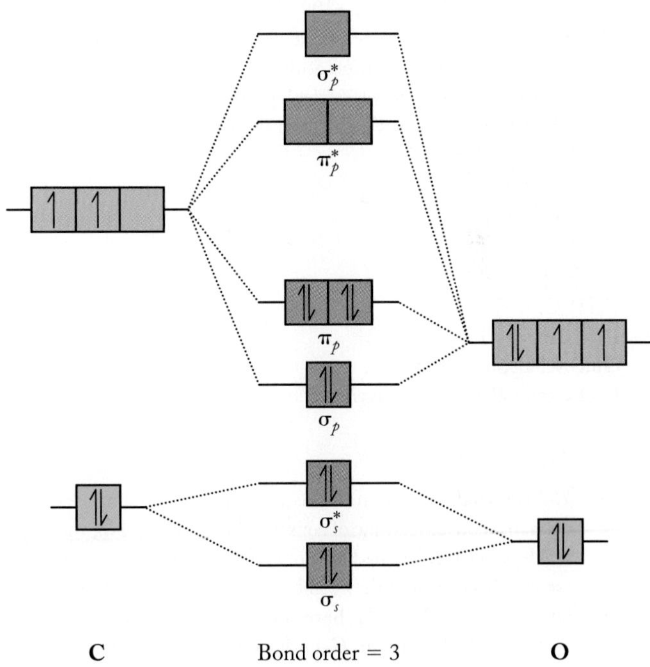

Chapter 6

Particulate Review

- H_2O
- CO_2
- O_3

Concept Tests

p. 249 CCl_4 is larger and experiences larger dispersion forces.
p. 251 (a) CH_3Br; (b) CH_3CH_2OH; (c) $(CH_3CH_2)_3N$
p. 253 To form hydrogen bonds, H atoms must be bonded to N, O, or F atoms. The H atoms in CH_3F are bonded to C atoms.
p. 254 Molecules of N_2H_4 can form H bonds with each other; C_2H_6 molecules cannot and experience only London dispersion forces.
p. 260 The solubilities of these alcohols decrease with increasing number of C atoms per molecule.
p. 263 A weighted wire can pass through a block of ice without melting it because downward pressure from the weights melts the ice just below the wire, but then the melt water refreezes just above the wire.
p. 267 No, the adhesive forces between mercury and glass are too weak.

Practice Exercises

6.1. To enter the vapor phase from the liquid phase, molecules of ethylene glycol must break more hydrogen bonds than molecules of isopropanol. Therefore, ethylene glycol has a higher boiling point.
6.2. In exactly 100 milliliters of overproof rum, there are 1.36 moles of water and 1.29 moles of ethanol, so water is the solvent.
6.3. The dipole–induced dipole interactions between molecules of water and He atoms are much weaker than those between molecules of water and N_2.
6.4. At 25 atm pressure and −100°C, the sample of CO_2 is solid. As the temperature is increased to about −50°C, the solid melts into a liquid; as the temperature is raised further (to about −20°C), liquid CO_2 vaporizes. As the pressure is raised from 25 to 100 atm at 50°C, CO_2 forms a supercritical fluid.

Chapter 7

Particulate Review

- $C_6H_{12}O_6$ and $C_3H_6O_3$
- 1 C : 2 H : 1 O in both
- Glucose: alcohol, ether
  Lactic acid: alcohol, carboxylic acid

Concept Tests

p. 290 Yes, the total mass of product equals the combined masses of the reactants consumed (law of conservation of mass). No, the masses of Mg and O_2 consumed are not equal because their molar masses are not the same and they combine 2 mol Mg : 1 mol O_2.
p. 293 (a) C_2H_4 and (b) $C_{20}H_{40}$ have the same empirical formula (CH_2); (c) C_2H_2 and (d) C_6H_6 have the same empirical formula (CH).
p. 295 Only (b) isopropanol (C_3H_8O)
p. 297 Molecular formula
p. 302 Combustion of propane in a gas grill or wood in a campfire
p. 303 (a) C is the limiting reactant. (b) 0.5 mol A and 1.5 mol B are left over.

Practice Exercises

7.1. $2\ C_4H_{10}(g) + 13\ O_2(g) \rightarrow 8\ CO_2(g) + 10\ H_2O(g)$
7.2. $P_4O_{10}(s) + 6\ H_2O\ (\ell) \rightarrow 4\ H_3PO_4(aq)$
7.3. 3.58 g O_2
7.4. Percent composition of tetracycline, $C_{22}H_{24}N_2O_8$: C = 59.45%; H = 5.443%; N = 6.305%; O = 28.80%
7.5. Cu_2S
7.6. $FeCr_2O_4$
7.7. Empirical formula: P_2O_5; molecular formula: P_4O_{10}
7.8. $C_8H_8O_3$
7.9. 22.1 g $NaBH_4$
7.10. The fuel–oxygen mixture is fuel-lean.
7.11. 75%

Chapter 8

Particulate Review

(a) could be either Fe^{3+} or Fe^{2+}; (c) is the nitrate ion
Two compounds that could be formed are iron(III) nitrate, $Fe(NO_3)_3$, and iron(II) nitrate, $Fe(NO_3)_2$.

Concept Tests

p. 322 b > a > c > d (b is the most concentrated solution)
p. 327 a < d < c < e < b (a is the most dilute solution)
p. 329 (a), (b), and (e) are electrolytes; (c) and (d) are nonelectrolytes
p. 332 $HEDTA^{3-} < H_2EDTA^{2-} < H_3EDTA^{-} < H_4EDTA$ (H_4EDTA is the strongest acid, $EDTA^{4-}$ is not an acid)
p. 336 Soluble: a. $CaCl_2$, b. K_3PO_4, d. Na_2S. Slightly soluble: c. $Fe(OH)_3$ and e. ZnO
p. 338 React aqueous lead(II) nitrate with the stoichiometric amount of aqueous potassium dichromate. Separate the solid lead chromate from the aqueous reaction mixture by filtration. Wash filtered solid to remove residual potassium nitrate and allow the solid to dry.
p. 342 Iron is oxidized.
p. 346 Reactions a and b are redox reactions. Reaction c is not. In reaction a, Br_2 is the oxidizing agent and Sn^{2+} is the reducing agent. In reaction b, F_2 is the oxidizing agent, and H_2O is the reducing agent.

Practice Exercises

8.1. 1.88 M $MgCl_2$
8.2. 0.109 M K^+
8.3. 4.48 g $NaCH_3CH(OH)CO_2$
8.4. $V_{initial} = 2.50$ mL
8.5. Molecular: $H_3PO_4(aq) + 3\ NaOH(aq) \rightarrow$
$3\ H_2O(\ell) + Na_3PO_4(aq)$
Total ionic: $H_3PO_4(aq) + 3\ Na^+(aq) + 3\ OH^-(aq) \rightarrow$
$3\ H_2O(\ell) + 3\ Na^+(aq) + PO_4^{3-}(aq)$
Net ionic: $H_3PO_4(aq) + 3\ OH^-(aq) \rightarrow 3\ H_2O(\ell), + PO_4^{3-}(aq)$
8.6. a. No precipitate forms.
b. Hg_2Cl_2 precipitates from the mixture. The net ionic equation is $Hg_2^{2+}(aq) + 2\ Cl^-(aq) \rightarrow Hg_2Cl_2(s)$.
8.7. 26.6 mmol/L SO_4^{2-}
8.8. 0.174 g HgS
8.9. (a) +4; (b) +1; (c) +5
8.10. Oxygen is reduced and is the oxidizing agent; SO_2 is oxidized and is the reducing agent.
8.11. $8\ H^+(aq) + Cr_2O_7^{2-}(aq) + I^-(aq) \rightarrow$
$2\ Cr^{3+}(aq) + IO_3^-(aq) + 4\ H_2O(\ell)$
8.12. $3\ HO_2^-(aq) + H_2O(\ell) + 2\ MnO_4^-(aq) \rightarrow$
$2\ MnO_2(s) + 3\ O_2(g) + 5\ OH^-(aq)$

8.13. The lemon juice is 0.292 $M\ C_6H_8O_7$; 100 mL of juice contains 5.62 g $C_6H_8O_7$.

Chapter 9

Particulate Review

Solid = (c); liquid = (a); gas = (b)

The physical change from (a) to (b) is evaporation, so energy must be added to overcome the hydrogen bonding between water molecules. The physical change from (a) to (c) is freezing, so energy would be released; the physical change from (c) to (a) is melting, so energy must be added.

Concept Tests

p. 376 a. A cooling pot of water with tight lid is a closed system; a cooling pot of water without a lid is an open system.

b. The pot of water without the lid cools faster because it can exchange matter with the surroundings and also exchanges energy faster.

p. 383 Water has the highest heat of vaporization in Table 9.2 because H_2O experiences the strongest intermolecular forces of attraction—namely, multiple hydrogen bonds with neighboring molecules.

p. 392 a. Less heat is transferred to the water, so calculated c_P will be too low.

b. Drops of hot water add more heat to the cool water, so the calculated c_P will be too high.

c. Heating other materials means less heat is transferred to the water, so calculated c_P will be too low.

d. Heat transfer from the system would lower the final temperature leading to a calculated c_P that is too low.

p. 393 Combustion of 2 mol of H_2 gives $\Delta H_{rxn} = -572$ kJ.

p. 413 Fuel densities (kJ/mL) of C_3 to C_{10} alkanes increase with molar mass even though fuel values (kJ/g) decrease because the densities of the alkanes increase with molar mass.

p. 416 Combustion of 1 g of glucose releases less energy than combustion of 1 g of sucrose because the carbon content of sucrose (144.12/342.30 or 42% C; glucose is 72.06/180.16 or 40% C).

Practice Exercises

9.1. a. The match is the system, $q < 0$, and the process is exothermic.

b. The wax is the system, $q < 0$, and the process is exothermic.

c. The dry ice (solid CO_2) is the system, $q > 0$, and the process (sublimation) is endothermic.

9.2. $w = 1.56 \times 10^7$ L · atm $= 1.58 \times 10^9$ J $= 1.58 \times 10^6$ kJ

9.3. $\Delta E = 32$ J

9.4. 1.13×10^3 g or 1.13 kg

9.5. 0.0°C

9.6. $\Delta H_{rxn} = 136$ kJ

9.7. 17.7 kJ

9.8. $2\ CH_4(g) + 3\ O_2(g) \rightarrow 2\ CO(g) + 4\ H_2O(g)$ $\quad H_{comp} = -1038$ kJ

$2\ CO(g) + O_2(g) \rightarrow 2\ CO_2(g)$ $\quad \Delta H_{comp} = -566$ kJ

$2\ CH_4(g) + 4\ O_2(g) \rightarrow 2\ CO_2(g) + 4\ H_2O(g)$ $\quad 2 \times \Delta H_{comp} = -1604$ kJ

$\Delta H_{comp} = -802$ kJ

9.9. a. $Ca(s) + C(s) + \frac{3}{2}\ O_2(g) \rightarrow CaCO_3(s)$

b. $2\ C(s) + 2\ H_2(g) + O_2(g) \rightarrow CH_3COOH(\ell)$

c. $K(s) + Mn(s) + 2\ O_2(g) \rightarrow KMnO_4(s)$

9.10. $\Delta H^\circ_{rxn} = -41.2$ kJ

9.11. $\Delta H_{rxn} = -36$ kJ

9.12. $U = -3792$ kJ

9.13. 1.1×10^3 L of kerosene per second

9.14. $C_{calorimeter} = 11.2$ kJ/°C

Chapter 10

Particulate Review

- $CH_4(g) + 2\ O_2(g) \rightarrow CO_2(g) + 2\ H_2O(g)$
- CH_4 is the limiting reactant.
- Seven moles, assuming four moles of water vapor are produced. If the water condenses, there are three moles of gas present: 1 mol O_2 and 2 mol CO_2.

Concept Tests

p. 433 (c)

p. 439 (c)

p. 444 (c)

p. 448 (c)

p. 455 (a) i; (b) ii

p. 460 Yes

p. 463 The van der Waals pressure correction factor a of SO_2 is nearly 2 times that for $CO_2(g)$ because bent, polar SO_2 experiences collectively greater van der Waals forces, including dipole–dipole interactions, than linear, nonpolar CO_2.

Practice Exercises

10.1. Ar

10.2. $u_{rms,He} = 1.36 \times 10^3$ m/s, or 2.65 times faster than N_2

10.3. $\Delta h = 127$ mmHg

10.4. 7.0 L

10.5. $V_2/V_1 = 1.61$

10.6. 38 psi

10.7. $V_2 = 2.5 \times 10^3$ L

10.8. (a) 9.41×10^5 g He; (b) 6.77×10^6 g air -9.4×10^5 g He $= 5.8 \times 10^6$ g lift

10.9. The O_2 balloon in air will sink to the floor.

10.10. $\mathcal{M} = 44.0$ g/mol, CO_2

10.11. 1.1×10^2 g NaN_3

10.12. $P_{O_2} = 5.24 \times 10^{-2}$ atm, outside air must be compressed 3.82 times

10.13. 2.2 mg H_2

10.14. $P_{ideal} = 252$ atm for tank of CH_4 ($d = 275$ g/L) at 25°C; $P_{real} = 212$ atm

Chapter 11

Particulate Review

Ethanol molecules interact via London dispersion forces, hydrogen bonds, and dipole–dipole interactions. Dimethyl ether molecules interact via London dispersion forces and dipole–dipole interactions. The strength of the hydrogen bonds between molecules of ethanol makes ethanol more likely to be a liquid and dimethyl ether a gas at room temperature.

Concept Tests

p. 485 Toward the KCl
p. 490 b < c < a < d
p. 491 Gasoline
p. 494 Dimethyl ether because its molecules should experience weaker dipole–dipole interactions in the liquid phase than those of acetone. Therefore, its vapor pressure should be higher than that of acetone in a 50:50 mixture.
p. 496 (a) Acetone and ethanol, because acetone molecules cannot form hydrogen bonds with each other but can with ethanol. Therefore, mixtures of the two should have lower vapor pressures than predicted by Raoult's law.
p. 498 The vapor pressure of a substance depends only on its identity, not on how much of it there is in the liquid phase, so vapor pressure is an intensive property.
p. 501 The solution is mostly water, which has a density of 1 kg/L.

Practice Exercises

11.1. 27.6 atm
11.2. 27 atm
11.3. 6.40×10^4 g/mol
11.4. $\Delta H_{vap} = 28.4$ kJ/mol
11.5. 1.6
11.6. $P_{solution} = 70$ torr
11.7. 0.840 m
11.8. 4.4 m
11.9. −6.9°C
11.10. $i = 1.10$
11.11. 4.6 g CO_2

Chapter 12

Particulate Review

The two phase changes are (a) freezing and (b) melting. Process (a) is spontaneous at −5°C. Process (a) is exothermic; process (b) is endothermic.

Concept Tests

p. 520 a, c, and d
p. 524 Prediction: (b) 10^6; $W = 4.47 \times 10^3$
p. 533 $\Delta S_{univ} > 0$
p. 538 The reaction is very slow at room temperature and pressure.
p. 539 No, $\Delta G°_{rxn}$ values for combustion of three C_8 alkanes are not the same because each compound has a different $\Delta G°_f$ value; reaction stoichiometry is the same for all three, so the other values in Equation 12.12 for calculating $\Delta G°_{rxn}$ do not change.
p. 540 a. Yes, it is spontaneous ($\Delta G < 0$).
b. No, spontaneity tells you nothing about the rate of reaction.

p. 541

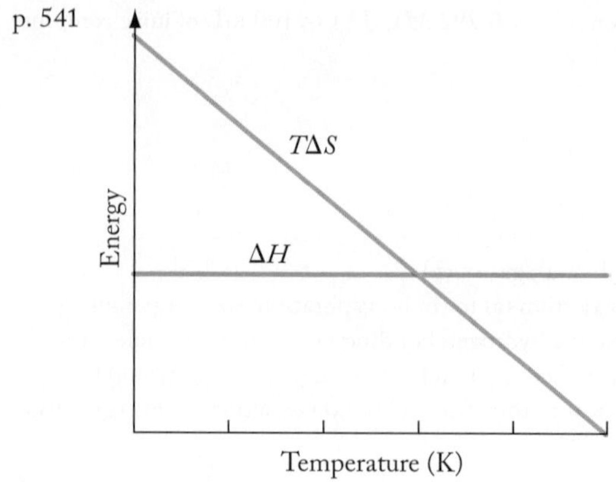

Practice Exercises

12.1. (a) Increase; (b) decrease; (c) increase
12.2. a > b > d > c (benzene has lowest $S°$)
12.3. Prediction: S_{sys} decreases; $\Delta S°_{rxn} = -242.6$ J/K
12.4. −1484.7 kJ
12.5. a. ΔS_{rxn} is expected to be negative.
b. $\Delta S°_{rxn} = -326.4$ J/K
c. $\Delta H°_{rxn} = -571.6$ kJ
d. The reaction is spontaneous at 298 K and 1 bar.
12.6. $\Delta G°_{rxn} = 142.2$ kJ
12.7. The reaction is spontaneous at high temperatures.
12.8. $\Delta G°_{rxn} = -196$ kJ

Chapter 13

Particulate Review

$N_2(g) + O_2(g) \rightarrow 2\ NO(g)$
(a) highest temperature, (b) lowest temperature
(a) faster moving molecules experience the greatest number of collisions

Concept Tests

p. 560 (a) Positive ($\Delta S°_{rxn} > 0$); (b) There are the same number of moles of gaseous reactants and products.
p. 567 (d)
p. 570 44 M/s
p. 583 Plot the four calculated k' values (y-axis) versus $[O_3]$ (x-axis); the slope of the curve = k.
p. 590 None of them are true.
p. 591 There are n activation energies (and n transition states).
p. 597 Similar rate laws suggest similar reaction mechanisms.

Practice Exercises

13.1. (a) [CO] decreases twice as rapidly as $[O_2]$.
(b) $\frac{\Delta[CO_2]}{\Delta t} = -2\frac{\Delta[O_2]}{\Delta t}$
13.2. $\frac{\Delta[A]}{\Delta t} = -4.0\ M\,s^{-1}$; $\frac{\Delta[B]}{\Delta t} = -2.5\ M\,s^{-1}$; $\frac{\Delta[D]}{\Delta t} = 3.0\ M\,s^{-1}$
13.3. $1.2 \times 10^{-6}\ M\,s^{-1}$
13.4. Rate = $k[NO][NO_3]$; $k = 1.57 \times 10^{10}\ M^{-1}\,s^{-1}$
13.5. The decomposition of H_2O_2 is first order in $[H_2O_2]$; $k = 8.30 \times 10^{-4}\ s^{-1}$

13.6. (a) $k = 4.33 \times 10^{-4}$ yr^{-1}
(b) 96%
(c) 1.06×10^4 yr
13.7. The reaction is second order in $[NO_2]$; $k = 0.751\ M^{-1}\ s^{-1}$
13.8. $t_{1/2} = 9.23 \times 10^{-3}$ s
13.9. The pseudo-first-order rate constant is $k' = 6.11 \times 10^{-4}\ \mu s^{-1}$. The second-order rate constant is $k = 7.2 \times 10^6\ M^{-1}\ \mu s^{-1}$ or $7.2 \times 10^{12}\ M^{-1}\ s^{-1}$.
13.10. $E_a = 6.9$ kJ/mol; $A = 1.17 \times 10^{10}\ M^{-1}\ s^{-1}$
13.11. 21 times larger
13.12. Rate $= k_{overall}[A]^2[B]^2$
13.13. No, none of the rate laws for the elementary steps matches the overall rate law.
13.14. Yes, NO_2 acts as a catalyst: it speeds up the reaction, but it's not, overall, consumed by it.

Chapter 14

Particulate Review

The reaction taking place in the system is $H_2O(g) + CO(g) \rightleftharpoons H_2(g) + CO_2(g)$. Addition of $CO(g)$ to the system increases collisions between H_2O molecules and CO molecules, which increases the rate of the forward reaction, and results in the formation of additional products. Removing $CO_2(g)$ from the system decreases collisions between H_2 molecules and CO_2 molecules, decreasing the rate of the reverse reaction, and results in the formation of additional products.

Concept Tests

p. 624 3.0
p. 626 b. $[CO_2] = [H_2] > [CO] = [H_2O]$
p. 637 Zone a in Figure 14.5 where $Q < K$
p. 641 There is no term for liquid water because its concentration is considered constant.
p. 642 None stay the same: the concentrations of CO_2, H_2O, and CO will all have decreased and $[H_2]$ will have increased.
p. 646 Increasing temperature would shift the reaction to the right. Decreasing temperature would shift the reaction to the left.
p. 654 (e) There is more B than A present.

Practice Exercises

14.1. $K_c = \dfrac{[CO][H_2]^3}{[CH_4][H_2O]}$ $\quad K_p = \dfrac{(P_{CO})(P_{H_2})^3}{(P_{CH_4})(P_{H_2O})}$
14.2. $K_c = [CH_3OH]/[CO][H_2]^2 = 2.9 \times 10^2$
14.3. $K_p = 2.7 \times 10^4$
14.4. $K_p = 1.4 \times 10^{-5}$
14.5. $K_c = 0.32$
14.6. $K_{c,overall} = 1.7 \times 10^2$
14.7. This reaction is not at equilibrium and proceeds to the right.
14.8. a. $K_p = \dfrac{(P_{CO})^2}{(P_{CO_2})}$

b. $K_p = \dfrac{(P_{CO})}{(P_{CO_2})(P_{H_2})}$

14.9. $K_c = 1.4 \times 10^{-9}$
14.10. a. When the reaction is cooled and water vapor condenses, one product is removed from the reaction mixture and the equilibrium shifts to the right, forming more SO_2.
b. When SO_2 gas dissolves in liquid water as it condenses, products are removed and the equilibrium shifts to the right, forming more products.
c. When O_2 is added, the concentration of one reactant increases and the equilibrium shifts to the right, forming more products.
14.11. Increasing the pressure shifts the equilibrium in the reaction to the products (i.e., NH_3), the side of the reaction that has fewer moles of gas.
14.12. The value of K for the endothermic reaction increases with increasing reaction temperature.
14.13. $P_{HI} = 0.156$ atm
14.14. $[NO_2] = 0.058\ M$ and $[N_2O_4] = 0.016\ M$ at equilibrium.
14.15. $K_p = 7.8 \times 10^2$
14.16. $K_{p,298\,K} = 2.96 \times 10^{-37}$ and $K_{p,2000\,K} = 9.2 \times 10^{-13}$

Chapter 15

Particulate Review

The covalent bond between the H atom and the Cl atom breaks in $HCl(g)$. The covalent bond that breaks in $CH_3COOH(\ell)$ is between the O atom and the H atom. In both cases, the H atoms form covalent bonds with water to form hydronium ions, $H_3O^+(aq)$. The intermolecular forces that form in both solutions are ion-dipole forces. The $HCl(aq)$ solution consists primarily of hydronium ions, solvated chlorine ions, and water molecules. The acetic acid solution consists primarily of CH_3COOH molecules and water molecules because CH_3COOH is a weak electrolyte.

Concept Tests

p. 677 Reduced concentrations of dissolved CO_2 in the blood shift both equilibria to the left, which lowers the concentration of H_3O^+ ions and makes the blood more basic.
p. 687 It should be weaker because the Cl atom draws electron density away from the $-NH_2$ group and the lone pair of electrons it needs to accept H^+ ions.
p. 691 13.77 = strongly basic; 10.03 = weakly basic; 7.00 = neutral; 4.37 = weakly acidic; 0.22 = strongly acidic
p. 693 The solution of the weak acid has a higher pH.
p. 695 An acid with a larger K_a value is more ionized than an acid with a smaller K_a value at the same concentration so its % ionization is larger.
p. 697 C, A
p. 705 Phosphoric acid is a weak acid, so the most abundant particles are molecules of phosphoric acid, H_3PO_4. The next most abundant particles are equal concentrations of $H_2PO_4^-$ and H_3O^+ ions. The products of the second and third ionization steps are much less abundant.
p. 707 The values of K_{a_2} and K_{a_3} for phosphoric acid are much smaller than its K_{a_1} value; so the second and third ionization steps should have no significant impact on the pH of 0.100 M H_3PO_4. For citric acid, because the value of K_{a_2} is close to that of K_{a_1}, there is likely to be an impact.

Practice Exercises

15.1. $HClO_2 > CH_2ClCOOH > C_6H_5COOH > HN_3 > HCN$
15.2. Trifluoroacetic > difluoroacetic > fluoroacetic > chloroacetic > bromoacetic acid. The first three acids are ranked by the

number of electronegative F atoms per molecule that withdraw electron density away from the carboxylic acid group and the resultant carboxylate anion; the last three are ranked by electronegativity of the halogen in the molecule from most electronegative (fluorine) to least (bromine).

15.3. H_2SO_4 is the strongest (more electronegative central atom and highest number of O atoms per molecule); H_2SeO_3 is the weakest.

15.4. Acetic acid and the acetate ion are one conjugate acid–base pair; H_2O and H_3O^+ are a conjugate base–acid pair.

$$\underset{\text{acid}}{CH_3COOH(aq)} + \underset{\text{base}}{H_2O(\ell)} \rightleftharpoons \underset{\text{conjugate base}}{CH_3COO^-(aq)} + \underset{\text{conjugate acid}}{H_3O^+(aq)}$$

15.5. $S^{2-} > ClO^- > CH_3COO^- > HSO_3^- > Br^-$.

15.6. Greatest percent increase: b. pH 7 → pH 4; greatest percent decrease: d. pH 5 → pH 9.

15.7. The pH at the start of the industrial revolution was 8.18 so the value of pH has dropped 0.11 pH units. The ocean has a higher $[H_3O^+]$ today and a lower pH.

15.8. $[H_3O^+] = 2.00 \times 10^{-12}\ M$; $[OH^-] = 5.01 \times 10^{-3}\ M$

15.9. pH = 2.23; percent ionization = 12%; $K_a = 7.9 \times 10^{-4}$

15.10. Percent ionization = 7.2%; $K_b = 5.6 \times 10^{-4}$

15.11. pH = 12.65

15.12. pH = 3.11

15.13. pH = 11.96

15.14. pH = 7.52

15.15. pH = 0.68; percent ionization = 5.4%, which is smaller than in a solution with half the concentration of acid (9.8%). This makes sense because the degree of ionization of a weak acid, such as HSO_4^-, tends to increase as acid concentration decreases.

15.16. pH = 5.60

15.17. $SO_4^{2-}(aq) + H_2O(\ell) \rightleftharpoons HSO_4^-(aq) + OH^-(aq)$

15.18. pH = 11.65

15.19. 0.77 pH units less acidic

Chapter 16

Particulate Review

- $CaCO_3$, Li_2O, KCl
- (a) lithium oxide; (b) potassium chloride; (c) calcium carbonate
- Li^+ and OH^- ions would be present in the highest concentration (2 *M*) because O^{2-} ions react with water forming OH^- ions.

Concept Tests

p. 735 Image (c)

p. 738 Images (d), (a), (c), (b)

p. 739 (b) is the weak acid at the start of the titration, (a) is the midpoint, and (c) is the equivalence point.

p. 741 There would be a smaller change in pH at the equivalence point with a weak base, so small that the volume corresponding to it would be difficult to detect precisely.

p. 742 pH = 10.64

p. 742 At the beginning = image (c), best buffer = image (b), equivalence point = image (a).

p. 745 It takes less titrant to reach the first equivalence point than the second because neutralization of the CO_3^{2-} during the first step produces more HCO_3^-, which adds to the HCO_3^- present initially in the sample that reacts in the second step, making the second plateau wider than the first.

p. 752 $Fe^{2+}(aq)$ is a weaker acid than $Fe^{3+}(aq)$ because ions with greater charges are stronger acids. The K_a of $Fe^{2+}(aq)$ should be smaller, so $Fe^{2+}(aq)$ would be below $Fe^{3+}(aq)$ in Table 16.1.

p. 753 $K_{sp} = 1.7 \times 10^{-6}$

p. 756 Nitric acid would increase solubility, sodium fluoride would decrease solubility, and hydrofluoric acid would have little effect.

Practice Exercises

16.1. pH = 3.24

16.2. pH = 10.16

16.3. Sulfurous acid ($pK_a = 1.77$)

16.4. Dimethylamine ($pK_b = 3.23$)

16.5. 0.61 kg sodium ascorbate; 0.34 kg ascorbic acid

16.6. Initial pH = 4.97; after addition of NaOH, pH = 5.07

16.7. After addition of OH^-, the pH of the 1.16 *M* buffer increased by 0.18 pH unit; the pH of the 0.58 *M* buffer increased by 0.38 pH unit.

16.8 Buffer B should experience the greater change because 50:50 acid–conjugate base buffers (such as buffer A) are generally the best at controlling pH. The pH of buffer A decreases by 0.087 pH units; the pH of buffer B decreases by 0.094 pH units.

16.9. pH = 11.39

16.10. pH = 4.20

16.11. It would be only if there were no bicarbonate initially in the sample. In that case $[CO_3^{2-}] = [HCO_3^-]$ at the midpoint because half of the carbonate originally in the sample has combined with hydrogen ions forming bicarbonate ions and $pH = pK_{a_2} + \log 1$.

16.12. 9.00 mL

16.13. The oxide ion from CaO is the Lewis base and CO_2 is the Lewis acid.

16.14. $[Ag^+] = 1.6 \times 10^{-8}\ M$

16.15. $S = 7.8 \times 10^{-7}\ M = 2.4 \times 10^{-5}$ g/100 mL

16.16. $S = 8.2 \times 10^{-4}\ M$

16.17. $S = 0.49\ M$

16.18. Yes

16.19. Yes, Ba^{2+} and Ca^{2+} ions in solution can be completely separated by selective precipitation with F^-.

Chapter 17

Particulate Review

- $4\ Fe(s) + 3\ O_2(g) \rightarrow 2\ Fe_2O_3(s)$
- Iron is oxidized; the oxygen in molecules of O_2 is reduced.
- Electrons are transferred from iron to oxygen.

Concept Tests

p. 778 No, because Zn and Cu have slightly different molar masses.

p. 781 Only a is spontaneous.

p. 784 The two E°_{cell} values are the same because cell potential is an intensive property of the cell reaction.

p. 786 $\Delta G_{cell} > 0$, $E_{cell} < 0$

p. 788 0.257 V

p. 793 (a) watt-minute; (b) kilowatt-hour; (c) kilowatt-hour

p. 803 H_2

p. 805 Negatively charged CO_3^{2-} ions would migrate toward anode.

Practice Exercises

17.1. $2\ NO_2^-(aq) + O_2(g) \rightarrow 2\ NO_3^-(aq)$

17.2. Oxidation: $HNO_2(aq) + H_2O(\ell) \rightarrow NO_3^-(aq) + 3\ H^+(aq) + 2\ e^-$
Reduction: $MnO_4^-(aq) + 8\ H^+(aq) + 5\ e^- \rightarrow Mn^{2+}(aq) + 4\ H_2O(\ell)$
Overall: $5\ HNO_2(aq) + 2\ MnO_4^-(aq) + H^+(aq) \rightarrow 5\ NO_3^-(aq) + 2\ Mn^{2+}(aq) + 3\ H_2O(\ell)$

17.3. The net ionic equation describing the redox reaction is $3\ Cu^{2+}(aq) + 2\ Al(s) \rightarrow 3\ Cu(s) + 2\ Al^{3+}(aq)$.
The cell diagram is
$Al(s) \mid Al^{3+}(aq) \parallel Cu^{2+}(aq) \mid Cu(s)$.

17.4. The net ionic equation is $Cd(s) + 2\ NiO(OH)(s) + 2\ H_2O(\ell) \rightarrow Cd(OH)_2(s) + 2\ Ni(OH)_2(s)$. $E^\circ_{cell} = 1.33$ V

17.5. $\Delta G_{cell} = -290$ kJ

17.6. $E_{cell} = 1.64$ V

17.7. $K = 1.8 \times 10^{62}$

17.8. 137 g Mg

17.9. 0.85 g Pb

Chapter 18

Particulate Review

- The potassium and iodide ions in potassium iodide are held together by ionic bonds. The platinum atoms are held together by metallic bonds.
- The K^+ and I^- experience electrostatic attractions in KI. The K^+ are repelled by other K^+ ions and the I^- ions experience electrostatic repulsions from other iodide ions.
- Neighboring platinum atoms experience electrostatic interactions between atom nuclei and the valence electrons on neighboring atoms.

Concept Tests

p. 822 For example:

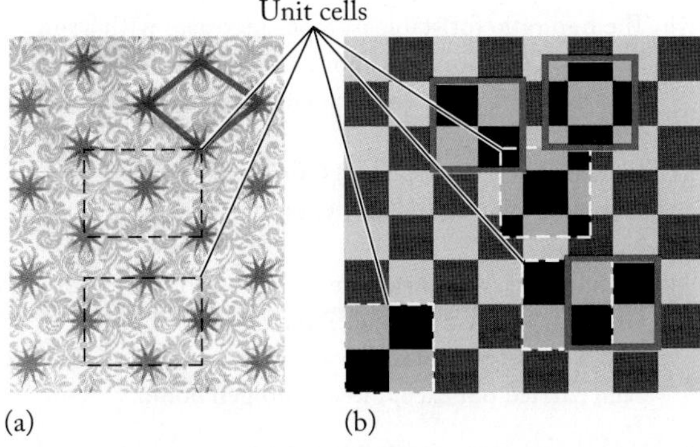

(a) (b)

p. 824 A crystal lattice extends indefinitely in three dimensions. A unit cell is the smallest repeating pattern in the arrangement of the particles in the lattice.

p. 832 Li/Al

p. 835 The filled valence band overlaps with the empty conduction band, allowing valence electrons to migrate into the conduction band where they are free to move when an electrical field is applied.

p. 837 Selenium

p. 840 Ice is a molecular solid in which the molecules are linked by a network of hydrogen bonds.

p. 845 $x = y = 1.2$ or $Mg_{1.2}Al_{1.2}(Si_2O_5)(OH)_4$

Practice Exercises

18.1. 128 pm

18.2. The density of silver is 10.57 g/cm^3, which is close to the 10.50 g/cm^3 value in Appendix 3; the density of gold is 19.41 g/cm^3, which is close to the 19.3 g/cm^3 value in Appendix 3.

18.3. $d = 18.5$ g/cm^3, which differs from the value in Appendix 3 (19.3 g/cm^3) by about 4%. The packing efficiency = 68.0%.

18.4. Gold forms substitutional alloys with both silver and copper.

18.5. 101 pm

18.6. 2.16 g/cm^3

18.7. $d = 412$ pm

Chapter 19

Particulate Review

H—C≡C—H

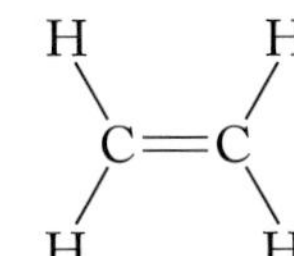

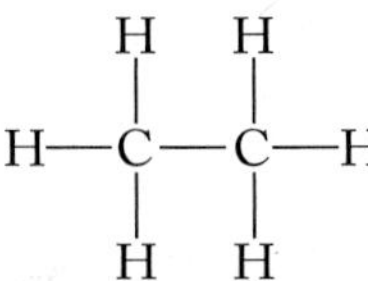

C_2H_2 has the strongest carbon–carbon bond. C_2H_6 has the longest carbon–carbon bond and contains only sigma bonds.

Concept Tests

p. 880 No. Both would consume 3 mol H_2 per mol of compound.

p. 884 (e)

p. 886 Dispersion forces

p. 891 The π electrons are localized in the double bonds.

p. 894 amphetamine = primary amine; Benadryl = tertiary amine; adrenaline = secondary amine; pargyline = tertiary amine.

p. 898 MTBE > diethyl ether > ethanol > methanol

p. 902 Zingerone: alcohol, ether; aromatic ring; carvone: carbon–carbon double bonds; cinnnamaldehyde: aromatic ring, carbon–carbon double bonds.

Practice Exercises

19.1. $CH_3(CH_2)_5CH_3$

19.2. (a) 7; (b) 9; (c) 8

19.3. (a) constitutional isomers; (b) different compounds; (c) constitutional isomers

19.4. (a) 4,5-diethyl-3-methyloctane; (b) 2,3,4,5,6,7-hexamethyloctane. The two compounds are unrelated.

19.5. Yes. Compound A reacts with 2 mol H_2 per mol; compound B reacts with 3 mol H_2 per mol. The product in both cases is heptane: $CH_3CH_2CH_2CH_2CH_2CH_2CH_3$.

19.6.

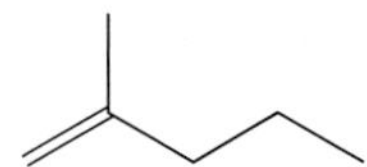
2-Methyl-1-pentene

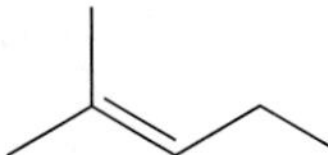
2-Methyl-2-pentene

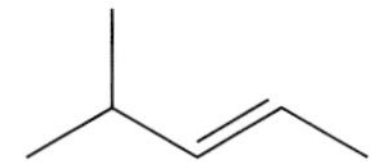
trans-4-Methyl-2-pentene

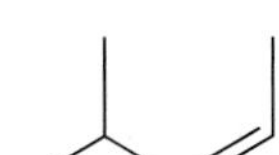
cis-4-Methyl-2-pentene

4-Methyl-1-pentene

19.7. H, CH_3, C=C, H, C=O, O, CH_3 $H_2C{=}C(CH_3)C(O)OCH_3$

19.8. Butane, by 3.8 kJ/mol.

19.9. Dipole–induced dipole interactions and London dispersion forces.

19.10. O, O, n

19.11. The polar fibers of cotton and polyester repel nonpolar grease and oil molecules while simultaneously wicking moisture away from the skin to cool it.

19.12. The two monomers are

HO, OH, O, O; NH_2, NH_2

(a)

And the repeat unit of the polymer is

H, N, O, O, N, H, n

(b)

Chapter 20

Particulate Review

Hydrogen bonds form between molecules of (b) and between molecules of (c). Molecule (b) contains an acidic hydrogen atom, which is highlighted in red in the formula of the compound: CH_3COOH. Molecule (c) has a primary amine ($-NH_2$) group, which can act as a Lewis base by donating the lone pair of electrons on its nitrogen atom to an electron-pair acceptor. Molecule (a) has two lone pairs of electrons on the carbonyl oxygen, so it can also act as a weak Lewis base.

Concept Tests

p. 929 sp^3 hybrid; ideal bond angles 109.5°

p. 937 Any with nonpolar R groups, e.g., leucine, isoleucine.

p. 938 Increased internal energy at high temperature overcomes the strength of the intermolecular forces responsible for a protein's secondary structure.

p. 945 (a)

p. 946 The breakdown of glucose by glycolysis produces acetyl-co A, which is a small molecule used in the biosynthesis of cholesterol.

p. 950 Combustion of olestra will produce more thermal energy per mole because it has more hydrocarbon chains per molecule.

p. 955 No. The direction in which the code is read matters.

Practice Exercises

20.1.

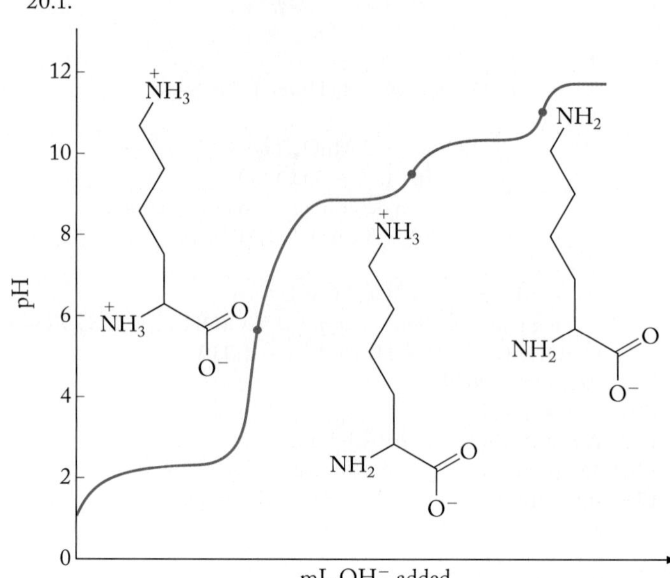

20.2. 6

20.3. 6

20.4. (a) GCCATAGGCTA (b) AATTCGGCGATC

Chapter 21

Particulate Review

From left to right, nuclide 1 has a mass number of 11 (5 protons + 6 neutrons) as does nuclide 2 (6 protons + 5 neutrons). Nuclide 3 has a mass number of 13 (6 protons + 7 neutrons). Nuclides 2 and 3 are isotopes.

Concept Tests

p. 974 The n–p ratio for stable isotopes increases with atomic number. For heavy isotopes it is about 1.5 to 1. Losing α particles increases this ratio, creating neutron-rich nuclides that undergo β decay.

p. 981 Increasing (C-14 depleted) CO_2 in the air by burning fossil fuels will reduce the $^{14}C/^{12}C$ ratio in the air and in plants.

p. 990 Neutron

p. 994 To fuse nuclei their coulombic repulsion must be overcome. Doing so means that nuclei must collide at very high velocities, which requires very high temperatures. Artificial fusion has not been carried out except in a hydrogen bomb.

Practice Exercises

21.1. $^{214}_{82}Pb \rightarrow {}^{214}_{83}Bi + {}^{0}_{-1}\beta$
$^{214}_{83}Bi \rightarrow {}^{210}_{81}Tl + {}^{4}_{2}He$

21.2. $^{28}_{13}Al \rightarrow {}^{28}_{14}Si + {}^{0}_{-1}\beta$ (β decay)
$^{18}_{9}F \rightarrow {}^{18}_{8}O + {}^{0}_{1}\beta$ (positron emission)
$^{18}_{9}F + {}^{0}_{-1}e \rightarrow {}^{18}_{8}O$ (electron capture)

21.3. 98.6%

21.4. 1.07:1

21.5. 7.572×10^7 kJ/mol

21.6. 1.753×10^9 kJ/mol

21.7. $A = 7.01 \times 10^9$ Bq; $A = 187$ mCi

21.8. 2.1×10^{11} X-rays

Chapter 22

Particulate Review

Sulfur atoms are in keratin: disulfide bonds are located between the α helices.

Phosphorus atoms are in the phospholipid bilayer: phosphorus atoms are in the phosphate units in the hydrophilic head groups.

Phosphorus atoms are in DNA: phosphorus atoms are in the phosphate units connecting the ribose units in the backbone.

Keratin has nitrogen atoms in its peptide linkages and in the amine groups on some of its side chains; phospholipids contain nitrogen atoms in the quarternary amine groups in the hydrophilic head group; DNA contains nitrogen atoms in the nucleotide bases.

Concept Tests

p. 1026 Nonspontaneous—they require energy to pump ions

p. 1029 $K > 1$ for the drug to be effective

p. 1032 Phosphates are less soluble because they have greater charge and interact more strongly with cations than do singly charged nitrate ions.

p. 1040 Essential elements are located toward the top of the periodic table. Nonessential elements are generally located toward the bottom. They have larger atomic numbers and are less abundant than the essential elements in the same group.

p. 1041 To avoid tissue damage, the nuclide should not emit α particles or high-energy β particles.

Practice Exercises

22.1. $E_{K^+} = -101$ mV; work to pump K^+ ions into the cell = 2.7 kJ/mol

22.2. pH = 1.19; the volume of $Mg(OH)_2$ solution required to neutralize the acid solution is 0.583 mL.

22.3. Nitrogen is reduced from 0 N = 0 in N_2 to −3 in NH_3.

22.4.

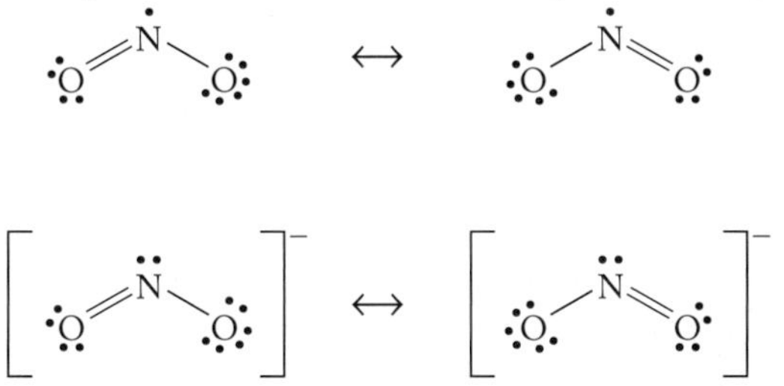

Both species have bent molecular geometry about the nitrogen atoms; the nitrogen atoms are sp^2 hybridized, and lone electrons or pairs of electrons are in sp^2 hybrid orbitals.

22.5. Bismuth-212: 2.0×10^{-6} mg; bismuth-213: 9.5×10^{-9} mg

Chapter 23

Particulate Review

The Lewis structures for ammonia (NH_3), borane (BH_3), and water (H_2O) are

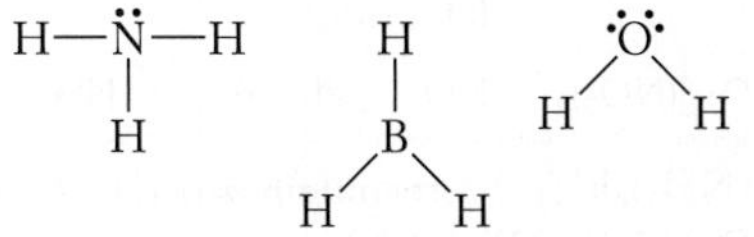

Ammonia has one lone pair on its central atom, and water has two lone pairs on its central atom; therefore, both can function as Lewis bases. Borane has no lone pairs on the central boron atom and an incomplete octet; it can function as a Lewis acid.

Concept Tests

p. 1054 CN^- ions occupy the inner coordination sphere of Fe^{3+}; Na^+ is the counter ion. $Na_3[Fe(CN)_6]$ would have about the same conductivity as $[Co(NH_3)_6]Cl_3$.

p. 1058 Tetrachloroplatinate(II)

p. 1068 Yellow-green

p. 1071 (a) CN^- is a stronger field ligand than pyridine. (b) Ru^{2+} ions are larger than Fe^{2+} and have a larger Δ_o; their 4*d* electrons interact more with ligand lone pairs than do the 3*d* electrons of Fe^{2+}.

p. 1074 No for square planar; yes for tetrahedral

p. 1075 Yes. Mn(IV) has three unpaired *d* electrons with only one possible distribution; there is no low-spin option. Mn(III) has four *d* electrons; depending on the strength of the ligands, it can be either high spin (four unpaired electrons) or low spin (two unpaired electrons).

p. 1077 H_2

p. 1082 A longer half-life means slower decay; it will take a longer time to collect a sufficient signal to get an image.

Practice Exercises

23.1. $[Ru(NH_3)_4Cl_2]^+$ is the complex ion; Ru is present as Ru^{3+}; chloride is the counter ion.

23.2.

		LIGANDS								
Compound	Counter ion	Formula	Name	Number	Prefix	Formula	Name	Number	Prefix	M^{n+}
$[Zn(NH_3)_4]Cl_2$	Cl^- (chloride)	NH_3	ammine	4	tetra-					2+
$[Co(NH_3)_4(H_2O)_2](NO_2)_2$	NO_2^- (nitrite)	NH_3	ammine	4	tetra-	H_2O	aqua	2	di-	2+

a. $[Zn(NH_3)_4]Cl_2$ = tetraamminezinc(II) chloride
b. $[Co(NH_3)_4(H_2O)_2](NO_2)_2$ = tetraamminediaquacobalt(II) nitrite

23.3. Four: three carboxylic acid groups and the O atom in the hydroxyl group

23.4. Ni^{3+} can have either a high-spin or a low-spin configuration; none of the ions is diamagnetic.

23.5.

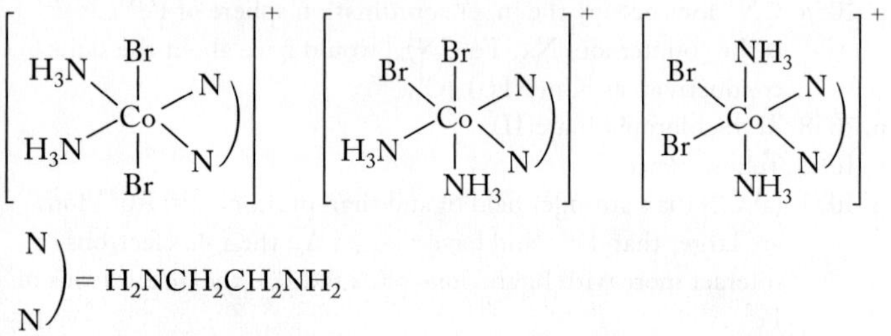

N–N = $H_2NCH_2CH_2NH_2$

cis-Diammine-*trans*-dibromo(ethylenediamine)cobalt(III)
cis-Diammine-*cis*-dibromo(ethylenediamine)cobalt(III)
trans-Diammine-*cis*-dibromo(ethylenediamine)cobalt(III)

23.6. 2.5×10^6

Answers

to Selected End-of-Chapter Questions and Problems

Chapter 1

1.1. (a) A pure compound in the gas phase; (b) A mixture of blue element atoms and red element atoms: blue atoms are in the gas phase, red spheres are in the solid or liquid phase.

1.3. (b)

1.5. CH_2O_2

1.7. A hypothesis is a tentative explanation of an observation or set of observations; a scientific theory is a concise explanation of a natural phenomenon that has been extensively tested and explains why certain phenomena are always observed.

1.9. Dalton's atomic theory explained the small, whole-number, mass ratios in his law of multiple proportions because the compounds contained small, whole-number ratios of atoms of different elements per molecule or formula unit.

1.11. Proust's law of definite proportions needed to have corroborating evidence to fully support it. At the time, experiments to prepare a compound of tin with oxygen gave varying compositions. The prepared compounds later turned out to be mixtures of two compounds of tin oxide.

1.13. Whereas *theory* in normal conversation means someone's idea or opinion that is open to speculation, a scientific theory is a concise and testable explanation of natural phenomena based on observation and experimentation that can accurately predict the results of future experiments.

1.15. A Snickers bar (b) and an uncooked hamburger (d) are heterogeneous mixtures.

1.17. Orange juice (d) and tomato juice (e) are heterogeneous mixtures.

1.19. Soil particles are not volatile, but water is. Therefore, distillation could be used to separate soil particles from water. However, filtration could also be used and would involve much less energy.

1.21. One chemical property of gold is its resistance to corrosion. Gold's physical properties include its density, color, melting temperature, and electrical and thermal conductivity.

1.23. We can distinguish between table sugar, water, and oxygen by examining their physical states (sugar is a solid, water is a liquid, and oxygen is a gas) and by their densities, melting points, and boiling points.

1.25. Density, melting point, thermal and electrical conductivity, and softness (a–d) are all physical properties while tarnishing and reaction with water (e and f) are both chemical properties.

1.27. Extensive properties will change with the size of the sample and therefore cannot be used to identify a substance.

1.29. Carbon dioxide is a nonflammable gas (chemical property) and it is denser than air (physical property). Therefore, the fire-extinguishing properties of CO_2 are due to both its physical and chemical properties.

1.31. In both liquid and ice, the water molecules touch each other. In ice, however, the water molecules are in a rigid hexagonal arrangement; in liquid, the molecules can move around each other and there is no long-range structure to their arrangement.

1.33. The particles in gases have the greatest freedom of motion; those in solids have the least.

1.35. The snow, instead of melting, sublimes to water vapor.

1.37. Energy is the ability to do work and must be expended to do work.

1.39. (a), (b), (c)

1.41. 13 times as much

1.43. SI units can be easily converted into a larger or smaller unit by multiplying or dividing by multiples of 10. English units are based on other number multiples and thus are more complicated to manipulate.

1.45. Scientists prefer the Celsius scale because it is based on the well-documented freezing and boiling points of a pure substance: water.

1.47. The absolute temperature scale (Kelvin scale) has no negative temperatures, and its zero value is placed at the lowest possible temperature.

1.49. 93.2%

1.51. 2.5 mi

1.53. 1330 Cal

1.55. 4.1×10^{13} km

1.57. 2.0 h

1.59. 4.03 m/s

1.61. 23 g

1.63. 19.0 mL

1.65. 26.5 g; 0.0265 kg

1.67. 58.0 cm^3

1.69. 73.8 mL

1.71. 5.1 g/cm^3

1.73. Yes

1.75. 109 g; 3.85 oz

1.77. (a), (c), (d), (f)

1.79. (a) 9.2×10^2; (b) 1.293×10^{20}; (c) 1.53×10^{-23}; (d) 3.73×10^{-6}

1.81. −269.0°C

1.83. 54.1°F; 285.4 K

1.85. −89.2°C; 183.9 K

1.87. The T_c for $YBa_2Cu_3O_7$ is already expressed in kelvins, $T_c = 93.0$ K. The T_c of Nb_3Ge converted to K is 23.2 K. The T_c of $HgBa_2CaCu_2O_6$ converted to K is 127.0 K. The superconductor with the highest T_c is $HgBa_2CaCu_2O_6$.

1.89. The Grubbs' test can be used to test for only one outlier in a data set.

1.91. Mean ± standard deviation (s) is slightly greater because the 95% confidence interval for 7 data points is

$$\text{Mean} \pm \left(\frac{t \cdot s}{\sqrt{n}}\right) = \pm \left(\frac{2.447 \cdot s}{2.646}\right) = \pm\, 0.9249\, s$$

1.93. (a) The mean and standard deviation values (all in μm) are 0.5106 ± 0.0047 for Manufacturer 1; 0.5134±0.0009 for

Manufacturer 2; 0.5012, 0.0008 for Manufacturer 3; (b) The 95% confidence intervals are 0.5106 ± 0.0058 for Manufacturer 1, 0.5134 ± 0.0011 for Manufacturer 2, and 0.5012 ± 0.0010 for Manufacturer 3. To three significant figures, the 95% confidence interval for Manufacturer 3's data includes 0.500; (c) Manufacturer 3 is both precise and accurate; Manufacturer 2 is precise but not accurate.

1.95. This point is an outlier.
1.97. 0.031 mg/L
1.99. Mixtures (a) and (b) react so that neither sodium nor chlorine is left over.
1.101. Day 11
1.103. 16 times more administered than prescribed
1.105. (a) Can't tell from the data given. All readings end in an even number which may mean that the minimum detectable change in temperature is 0.2°C; (b) It is not likely that any of them would give an accurate reading to the nearest 0.1°C.
1.107. 3.246×10^4 mi/h

Chapter 2

2.1. Purple (H, hydrogen)
2.3. Purple (H, hydrogen)
2.5. (a) Green (Au, gold); (b) Blue (Na, sodium); (c) Lilac (Cl, chlorine)
2.7. Red arrow, alpha; green arrow, beta
2.9. Dichloromethane: the differences in mass between the peaks near 50 amu and 85 amu correspond to the masses of stable Cl isotopes: 35 and 37 amu.
2.11. (a) Blue (K) and green (Ag); (b) Grey (Mg); (c) Yellow (Sc); (d) Purple (I); (e) Red (O)
2.13. The plum-pudding model could not explain the infrequent, large-angle deflections of alpha particles that Rutherford's students observed. However, these deflections could be explained by the particles' colliding with tiny, dense atomic nuclei that took up little of the volume of gold atoms, but that contained all of the positive charge and most of the atoms' mass (Rutherford's model).
2.15. The fact that cathode rays were deflected by a magnetic field indicated that the rays were streams of charged particles.
2.17. Through the alpha decay of uranium and other radioactive nuclides.
2.19. Absorption of an α particle by a gold atom would have transmuted the gold atom into an atom of thallium.
2.21. Greater than 1
2.23. Hydrogen (^{1}H)
2.25.

	Atom	Mass Number	Atomic Number = Number of Protons	Number of Neutrons = Mass Number − Atomic Number	Number of Electrons = Number of Protons
(a)	^{14}C	14	6	8	6
(b)	^{59}Fe	59	26	33	26
(c)	^{90}Sr	90	38	52	38
(d)	^{210}Pb	210	82	128	82

2.27. The neutron-to-proton ratio for each of these elements is (a) ^{4}He, 2/2 = 1.00, (b) ^{23}Na, 12/11 = 1.09, (c) ^{59}Co, 32/27 = 1.19, (d) ^{197}Au, 118/79 = 1.49. As the atomic number (Z) increases, the neutron-to-proton ratio increases.

2.29.

Symbol	^{23}Na	^{89}Y	^{118}Sn	^{197}Au
Number of protons	11	39	50	79
Number of neutrons	12	50	68	118
Number of electrons	11	39	50	79
Mass number	23	89	118	197

2.31.

Symbol	$^{37}Cl^-$	$^{23}Na^+$	$^{81}Br^-$	$^{226}Ra^{2+}$
Number of protons	17	11	35	88
Number of neutrons	20	12	46	138
Number of electrons	18	10	36	86
Mass number	37	23	81	226

2.33. Group 2, RO; group 3, R_2O_3; group 4, RO_2
2.35. Mendeleev based his groups on chemical reactivity. No compounds of the noble gases existed to indicate their presence in nature.
2.37. C, N, and O
2.39. (a) Palladium (Pd); (b) Rhodium (Rh); (c) Platinum (Pt)
2.41. Three (Na, Mg, and Al)
2.43. A *weighted average* takes into account the proportion of each value in the group of values to be averaged.
2.45. $(m_X + m_Y)/2$
2.47. Platinum must have other isotopes with masses greater than and less than 195. The abundances of the heavier and lighter isotopes must be roughly the same so that their weighted average atomic mass is 195.08 amu.
2.49. ^{40}Ar
2.51. (a) ^{11}B; (b) ^{7}Li; (c) ^{14}N
2.53. 63.55 amu
2.55. Yes
2.57. 47.948 amu
2.59. (a) CaF_2, 78.074 amu; (b) Na_2S, 78.045 amu; (c) Cr_2O_3, 151.989 amu
2.61. (a) 1; (b) 3; (c) 6; (d) 6
2.63. (e) CH_4 < (d) NH_3 < (a) CO < (c) CO_2 < (b) Cl_2
2.65. A dozen is too small a unit to express the very large number of atoms, ions, or molecules present in laboratory quantities such as a mole.
2.67. No
2.69. (a) 7.3×10^{-10} mol Ne; (b) 7.0×10^{-11} mol CH_4; (c) 4.2×10^{-12} mol O_3; (d) 8.1×10^{-15} mol NO_2
2.71. (a) 1 mol; (b) 2 mol; (c) 1 mol; (d) 3 mol
2.73. 10.3 g
2.75. (a) 0.125 mol; 7.53×10^{22} atoms; (b) 0.125 mol; 7.53×10^{22} atoms; (c) 0.250 mol; 1.51×10^{23} atoms; (d) 0.375 mol; 2.26×10^{23} atoms
2.77. (a) Both contain the same; (b) N_2O_4; (c) CO_2
2.79. (a) 3.00 mol; (b) 4.50 mol; (c) 1.50 mol
2.81. (a) 64.063 g/mol; (b) 47.997 g/mol; (c) 44.009 g/mol; (d) 108.009 g/mol
2.83. (a) 152.148 g/mol; (b) 164.203 g/mol; (c) 148.204 g/mol; (d) 132.161 g/mol
2.85. 41.63 mol
2.87. 0.25 mol; 10 g
2.89. (a) NO; (b) CO_2; (c) O_2
2.91. 0.752 mol

2.93. Diamond

2.95. The peak with the largest m/z value in many mass spectra is the molecular ion, M^+, peak whose mass is the molecular mass of the compound.

2.97. Yes

2.99. (a) 222 amu; (b) 296 amu; (c) 316 amu; (d) 450 amu

2.101. (a) The three peaks are due to the different combinations of isotopes in Cl_2: $^{35}Cl_2$ is at 70 amu, $^{35}Cl^{37}Cl$ is at 72 amu, and $^{37}Cl_2$ is at 74 amu; (b) The peak at 70 amu is much more intense than the peak at 74 amu because the ^{35}Cl is much more abundant than ^{37}Cl, so it is more likely that a molecule of Cl_2 will contain two ^{35}Cl isotopes than contain two ^{37}Cl isotopes.

2.103. The biggest peak at 34 amu is the molecular ion of $H_2{}^{32}S$ (nearly 95% of all S atoms are ^{32}S). The peaks at 33 and 32 amu are produced when electron bombardment of $H_2{}^{32}S$ molecules fragments them, forming ions of $H^{32}S$ and ^{32}S atoms. The source of the small peak at 36 amu is probably the molecular ion of $H_2{}^{34}S$ (the second most abundant S isotope). The even smaller peak at 35 amu is probably a combination of mostly $H^{34}S$ and a very little $H_2{}^{33}S$ (this isotope of sulfur has a natural abundance of less than 1%). Molecular and fragment ions of the even less abundant ^{35}S isotope were probably not detected.

2.105. 303 g/mol

2.107. (a) Electrons; (b) The negatively charged electrons were attracted to the positively charged plate and repelled by the negatively charged plate as they passed through the electric field. (c) If the polarities of the plates were switched, the electron would deflect in the opposite direction.

2.109. 10.00% ^{25}Mg and 11.01% ^{26}Mg

2.111. 1.2×10^{14}

2.113. 615 μg

2.115. (a) 0.7580 mol C; (b) 4.565×10^{23} atoms C

Chapter 3

3.1. (a) Purple (Na), red (Cr), and orange (Au); (b) Blue (Ne); (c) Orange (Au); (d) Red (Cr); (e) Blue (Ne) and green (Cl)

3.3. (a) Green (Cl); (b) Purple (Na), red (Cr), and yellow (Au)

3.5. Blue (Rb), green (Sr), and orange (Y)

3.7. (c)

3.9. (a)

3.11. (a) e; (b) a; (c) a

3.13. All these forms of light have perpendicular, oscillating electric and magnetic fields that travel together through space.

3.15. The lead shield protects the parts of our bodies that might be exposed to X-rays but are not being imaged. Lead is a dense metal and effective absorber of X-rays.

3.17. It still emits infrared radiation.

3.19. 5.490×10^{14} s^{-1}; yes

3.21. (a) 3.32 m; (b) 3.14 m; (c) 3.04 m; (d) 2.79 m

3.23. The frequency of the X-rays emitted from (a) copper is higher.

3.25. 8.317 min

3.27. The absorption spectrum consists of dark lines at wavelengths specific to that element. The emission spectrum has bright lines on a dark background. The wavelengths of the emission and absorption lines are the same.

3.29. Because each element shows distinctive and unique absorption and emission lines, the bright emission lines observed for the pure elements could be matched to the many dark absorption lines in the spectrum of sunlight. This approach can be used to deduce the sun's elemental composition.

3.31. The quantum is the smallest indivisible quantity of radiant energy that can be absorbed or emitted.

3.33. No

3.35. 6.62×10^{-19} J

3.37. (b) The door elevations have discrete values. There are none in between those on adjacent floors.

3.39. 7.71×10^{-19} J

3.41. No

3.43. Potassium; 8.04×10^5 m/s

3.45. 3.17×10^{18} photons/s

3.47. The Balmer equation is equivalent to the Rydberg if $n_1 = 2$.

3.49. It is the difference between n levels that determines emission energy.

3.51. (a)

3.53. No

3.55. At $n = 7$, the wavelength of the electron's transition ($n = 7$ to $n = 2$) is in the ultraviolet region.

3.57. 1875 nm; infrared

3.59. No, because for hydrogen the transition is in the ultraviolet region and as Z increases the wavelength will shorten further.

3.61. (a) No; (b) Yes

3.63. In the de Broglie equation, λ is the wavelength the particle of mass m exhibits as it travels at speed u, where h is Planck's constant. This equation states that (1) any moving particle has wavelike properties because a wavelength can be calculated through the equation, and (2) the wavelength of the particle is inversely related to its momentum (mass multiplied by velocity).

3.65. No

3.67. (b) and (c)

3.69. (a) 10.8 nm; (b) 0.180 nm; (c) 8.2×10^{-28} nm; (d) 3.7×10^{-54} nm

3.71. $\Delta x \geq 1.3 \times 10^{-13}$ m

3.73. The Bohr model orbit showed the quantized nature of the electron in the atom as a particle moving around the nucleus in concentric orbits. In quantum theory, an orbital is a region of space where the probability of finding the electron is high. The electron is not viewed as a particle but as a wave, and it is not confined to a clearly defined orbit; rather, we refer to the probability of the electron being at various locations around the nucleus.

3.75. Three: n, ℓ, and m_ℓ.

3.77. (a) 1; (b) 4; (c) 9; (d) 16; (e) 25

3.79. 3, 2, 1, 0

3.81. (a) 6*s*; (b) 3*d*; (c) 2*p*; (d) 5*g*

3.83. (a) 2; (b) 2; (c) 10; (d) 2

3.85. (b)

3.87. Degenerate orbitals have the same energy and are indistinguishable from each other.

3.89. Starting with the fourth row elements, the outermost (n = row number) *s* orbitals fill before the $(n - 1)$ *d* orbitals do. For example K and Ca atoms contain 4*s* electrons, but no 3*d* electrons.

3.91. (c) 3*s* < (a) 3*d* < (d) 4*p* < (b) 7*f*

3.93. Li^+: $1s^2$ or [He]; Ca: [Ar]$4s^2$; F^-: [He]$2s^22p^6$ or [Ne]; Mg^{2+}: [He]$2s^22p^6$ or [Ne]; Al^{3+}: [He]$2s^22p^6$ or [Ne]

3.95. K: [Ar]$4s^1$; K^+: [Ar]; Ba: [Xe]$6s^2$; Ti^{4+}: [Ne]$3s^23p^6$ or [Ar]; Ni: [Ar]$4s^23d^8$

3.97. (b) for Mn; (d) for Mn^{2+}

3.99. (a) 3; (b) 2; (c) 0; d) 0

3.101. Ti, two unpaired electrons

3.103. Cl^-, no unpaired electrons

3.105. (a) and (d)

3.107. $5p$, yes

3.109. Na atoms are larger than Cl atoms because the latter have six more positive charges in their nuclei, which pull the atom's electrons inward more strongly. However, Na^+ ions are much smaller than Na atoms because their $n = 3$ shell is empty. In addition, Cl^- ions are larger than Cl atoms because each ion has one more electron, which means greater electron-electron repulsion and an expanded cloud of electrons around the ion's nucleus.

3.111. (a) Al > P > Cl > Ar; (b) Sn > Ge > Si > C; (c) K > Na > Li > Li^+; (d) Cl^- > Cl > F > Ne

3.113. (a) As the atomic number increases down a group, electrons are added to higher n levels, leading to a decrease in effective nuclear charge and ionization energy.
(b) As the atomic number increases across a row, the effective nuclear charge increases. This means that the ionization energy increases across a period of elements.

3.115. Because less energy would be required to ionize a larger atom, the wavelength of light required would be longer for atoms of higher atomic number than for atoms of lower atomic number in the same group on the periodic table.

3.117. (a) I < Br < Cl < F; (b) Na < Li < Mg < Be; (c) N < O < F < Ne

3.119. No, sodium has a negative (favorable) electron affinity but is never found as an anion in nature. It is always a cation in salts such as NaCl and Na_2CO_3.

3.121. Electrons are added to shells that are farther form the nucleus with increasing Z. This means electron affinities become weaker, that is, less negative, which means their arithmetic values increase.

3.123. (a) $[Xe]6s^2$; (b) $n = 5$, $\ell = 2$, $m_\ell = -2, -1, 0, 1, 2$; (c) No; (d) Yes

3.125. (a), (c), and (d)

3.127. Yes. It is generally observed that as Z increases so does the IE_2. However, Ge's second IE_2 is lower than Ga's because to ionize the second electron in Ga, we need to remove an electron from a lower energy $4s$ orbital. Also, Br's IE_2 is lower than Se's because the electron pairing ($4p^4$) in one of the p orbitals for the Br^+ ion lowers its IE_2 slightly.

3.129. (a) Sn^{2+}: $[Kr]4d^{10}5s^2$; Sn^{4+}: $[Kr]4d^{10}$; Mg^{2+}: $[He]2s^22p^6$ or [Ne]; (b) Cadmium has the same electron configuration as Sn^{2+} and neon has the same electron configuration as Mg^{2+}; (c) Cd^{2+}

3.131. (a) Ne, 5.76; Ar, 6.76; (b) The outermost electron in argon is a $3p$ electron, which is mostly shielded by the electrons in the $n = 2$ level (10 electrons) and the $n = 1$ level (2 electrons), whereas the outermost electron in neon is a $2p$ electron, which is shielded only by the electrons in the $n = 1$ level (2 electrons).

3.133. Assume that the electrons in p orbitals behave like standing waves. Then the node between the two lobes could be a place where their waves have zero amplitude, but wave amplitudes in the lobes are not zero.

Chapter 4

4.1. (e) Al^{3+}

4.3. Only the middle structure is a resonance structure of the thiocyanate ion, SCN^-, because its middle atom is carbon.

4.5. (d) Cl is slightly more electronegative than Br, so the Cl atom should be redder than the Br atom.

4.7. (a) Oxygen is more electronegative than sulfur, so the O atoms at the ends of the molecule should have partial negative charges (depicted by their orange color) and the S atom in the middle should have a partial positive charge.

4.9. All three modes are infrared active.

4.11. The red curve represents the interaction between potassium and fluoride ions: its minimum is more negative and at a smaller distance between nuclei than the blue curve because of the smaller ionic radius of F^- compared to Cl^-.

4.13. The number of valence electrons equals the group number for groups 1–12. Once the d orbitals are filled, they are considered "core" electrons.

4.15. There are Coulombic forces of attraction between the ions of opposite charge in ionic compounds. In covalent compounds, chemical bonds are formed through electron sharing.

4.17. -6.94×10^{-18} J

4.19. (b) TiO_2

4.21. CsBr < KBr < $SrBr_2$

4.23. Roman numerals indicate the charge on the transition metal cation.

4.25. XO_2^{2-}

4.27. (a) NO_3, nitrogen trioxide; (b) N_2O_5, dinitrogen pentoxide; (c) N_2O_4, dinitrogen tetroxide; (d) NO_2, nitrogen dioxide; (e) N_2O_3, dinitrogen trioxide; (f) NO, nitrogen monoxide; (g) N_2O, dinitrogen monoxide; (h) N_4O, tetranitrogen monoxide

4.29. (a) Na_2S, sodium sulfide; (b) $SrCl_2$, strontium chloride; (c) Al_2O_3, aluminum oxide; (d) LiH, lithium hydride

4.31. (a) cobalt(II) oxide; (b) cobalt(III) oxide; (c) cobalt(IV) oxide

4.33. (a) BrO^-; (b) SO_4^{2-}; (c) IO_3^-; (d) NO_2^-

4.35. (a) nickel(II) carbonate; (b) sodium cyanide; (c) lithium hydrogen carbonate; (d) calcium hypochlorite

4.37. (a) hydrofluoric acid; (b) sulfurous acid; (c) H_3PO_4; (d) HNO_2

4.39. (a) K_2S; (b) K_2Se; (c) Rb_2SO_4; (d) $RbNO_2$; (e) $MgSO_4$

4.41. (a) manganese(II) sulfide; (b) vanadium(II) nitride; (c) chromium(III) sulfate; (d) cobalt(II) nitrate; (e) iron(III) oxide

4.43. (b) Na_2SO_3

4.45. In the diatomic molecule XY shown here

:X:Y:

Lewis counts 6 e⁻ in 3 lone pairs on both X and Y. He also counts the 2 e^- shared between X and Y separately (2 e^- for X and 2 e^- for Y). However, there are not 4 e^- being shared, only 2 e^-. It seems that the Lewis counting scheme counts the shared electrons twice.

4.47. For the H—O—H bonding pattern, the oxygen of the central atom forms bonds to the two hydrogen atoms. This uses 4 of the 8 e^- leaving 4 e^- left over for the 2 lone pairs. Each hydrogen atom has a duet of electrons, so the lone pairs reside on oxygen and form an octet on oxygen.

H—Ö—H

For H—H—O bonding, the two covalent bonds again use 4 of the 8 e^-, leaving 4 e^- for 2 lone pairs. If these are placed on the oxygen atom as shown here,

H—H—Ö:

oxygen does not complete its octet and the central hydrogen atom has 4 e^-, not a duet. This structure would violate the Lewis structure formalism.

4.49. K· ·Mg· ·P·

4.51. K^+ Al^{3+} $[:N:]^{3-}$ $[:I:]^-$

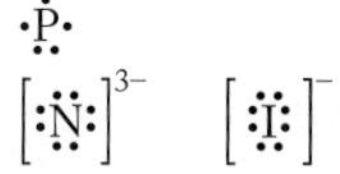

4.53. I^- and Ca^{2+}

4.55. (a) 10; (b) 8; (c) 8; (d) 10

4.57. (a) :C≡O:

(b) O=O

(c) $[:Cl—O:]^-$

(d) 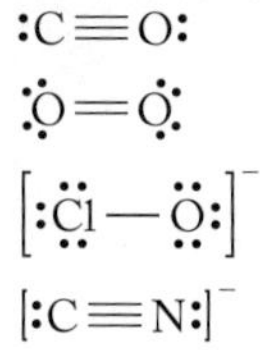 $[:C≡N:]^-$

4.59. (a) 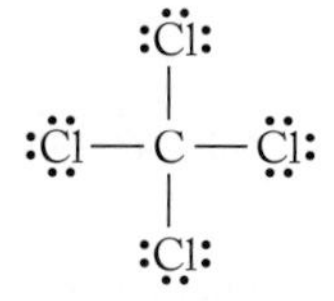 CCl_4: :Cl—C(Cl)(Cl)—Cl:

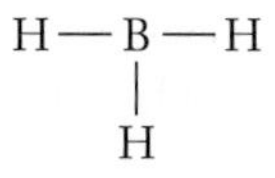

(b) H—B(H)—H

(c) :F—Si(F)(F)—F:

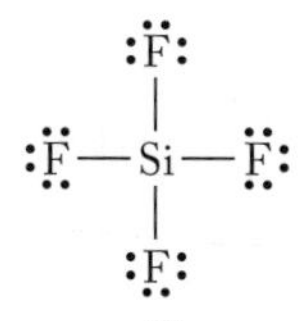

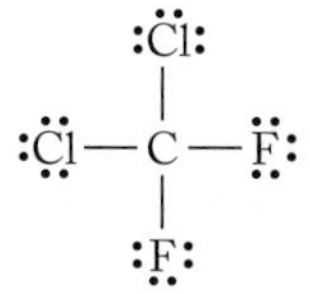

(d) 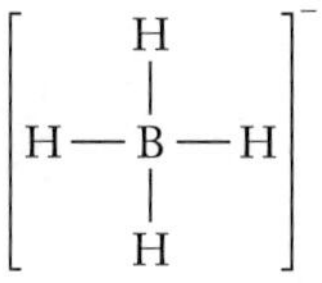 $[H—B(H)(H)—H]^-$

(e) 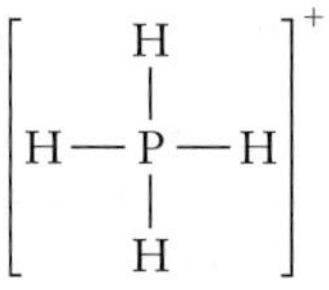 $[H—P(H)(H)—H]^+$

4.61. (a) 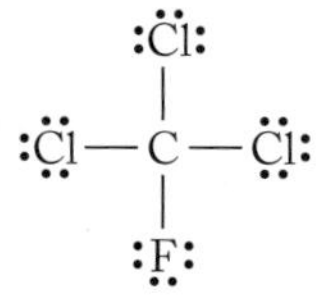 :Cl—C(Cl)(F)—Cl:

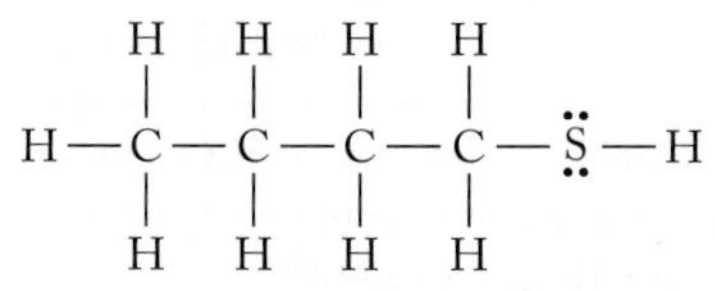

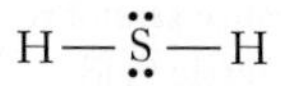

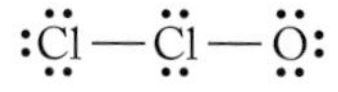

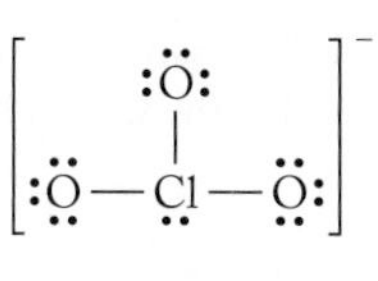

(b) :Cl—C(Cl)(F)—F:

(c) :F—C(Cl)(F)—F:

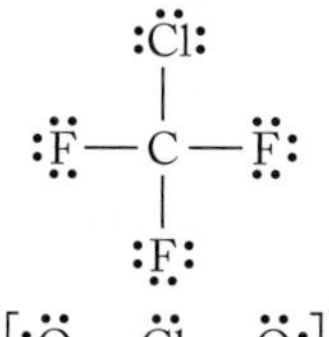

(d) 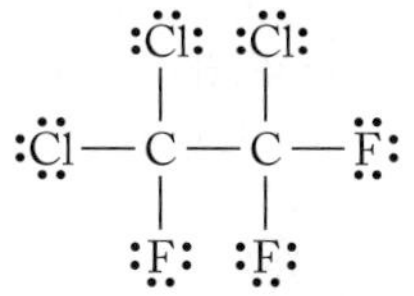 :Cl—C(Cl)(F)—C(Cl)(F)—F:

(e) 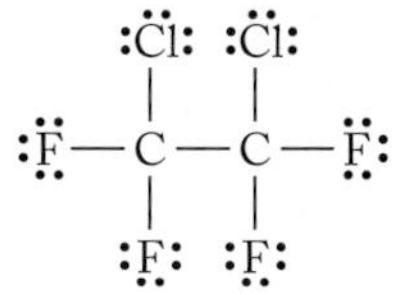 :F—C(Cl)(F)—C(Cl)(F)—F:

4.63. (a) 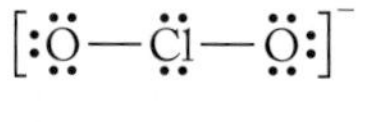 $[:O—Cl—O:]^-$

(b) 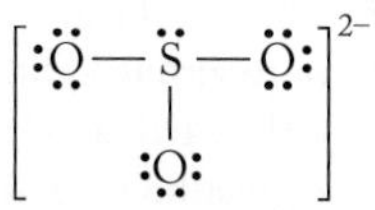 $[:O—S(O)—O:]^{2-}$

(c) 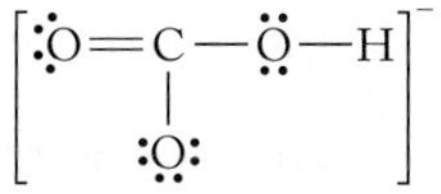 $[O=C(O)—O—H]^-$

4.65. H—C(H)(H)—C(H)(H)—C(H)(H)—C(H)(H)—S—H

H—S—H

4.67. :Cl—Cl—O:

$[:O—Cl(O)—O:]^-$

4.69. H—C(H)(H)—O—H

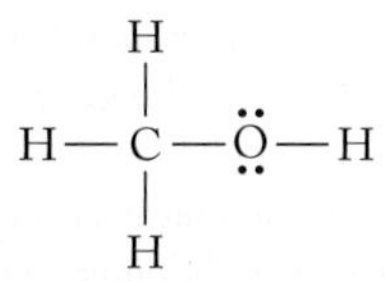

4.71. Resonance occurs when two or more valid Lewis structures may be drawn for a molecular species. The true structure of the species is a hybrid of the structures drawn.

4.73. Resonance is often the result of a single and a double bond to the same atom switching places within a structure or two adjacent double bonds becoming a single bond and a triple bond.

4.75. The resonance Lewis structures of NO_2 both contain a single and a double N—O bond. From a formal charge perspective, the structures are equivalent.

:O—N=O: (−1, +1, 0) ⟷ :O=N—O: (0, +1, −1)

The resonance forms of CO_2 show that one is dominant (the one in which all formal charges are zero) and so the other forms contribute little to the bonding in CO_2.

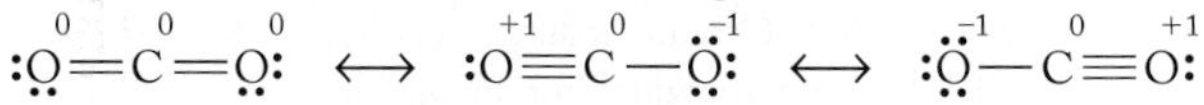

:O=C=O: (0, 0, 0) ⟷ :O≡C—O: (+1, 0, −1) ⟷ :O—C≡O: (−1, 0, +1)

4.77. Cyclobutadiene (H on each C): C=C / C=C ⟷ C—C with vertical double bonds

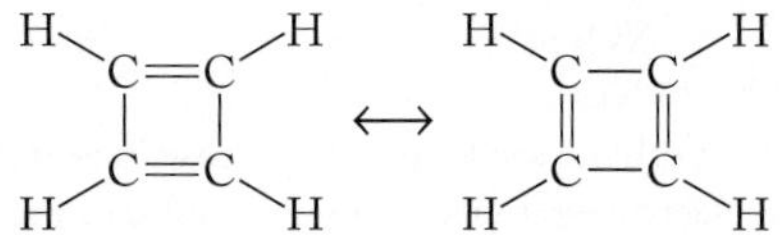

4.79. N_2O_2:

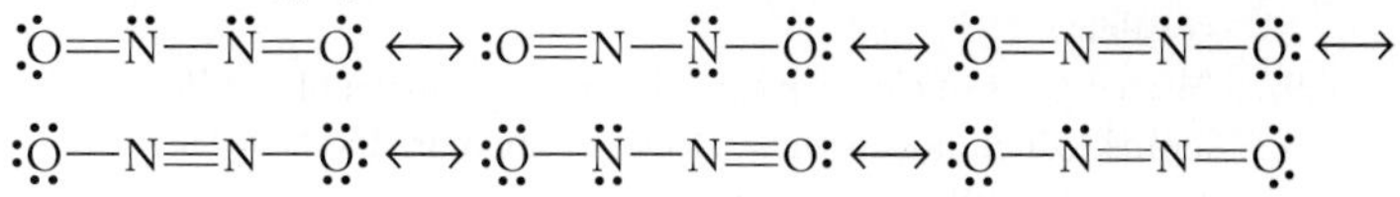

O=N—N=O ⟷ :O≡N—N—O: ⟷ O=N=N—O: ⟷

:O—N≡N—O: ⟷ :O—N—N≡O: ⟷ :O—N=N=O

N_2O_3:

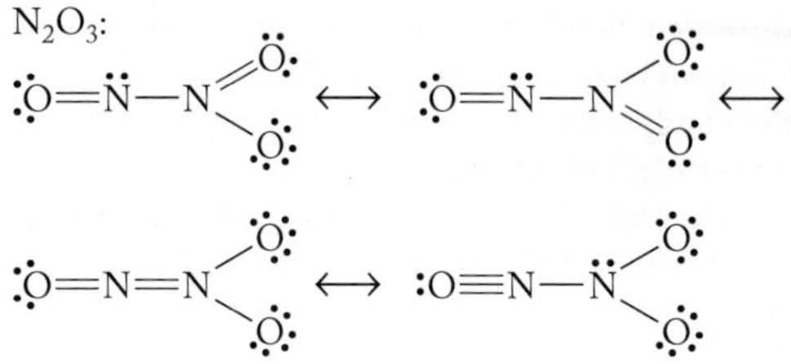

O=N—N(=O)—O ⟷ O=N—N(—O)=O ⟷

O=N=N(O)O ⟷ :O≡N—N(O)O

4.81.

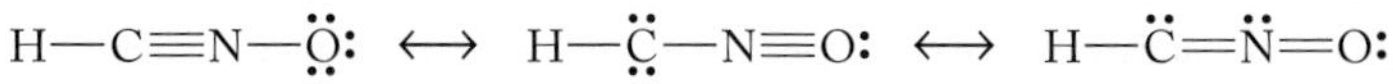

H—C≡N—O: ⟷ H—C—N≡O: ⟷ H—C=N=O:

4.83. $O_2N—O—NO_2$ resonance structures (four forms, N=O double bonds placed on different terminal O atoms)

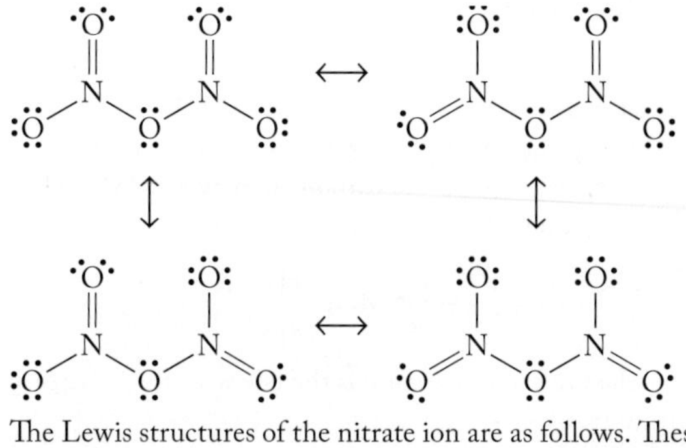

4.85. The Lewis structures of the nitrate ion are as follows. These show that the bond order in NO_3^- is 1.33 because of resonance.

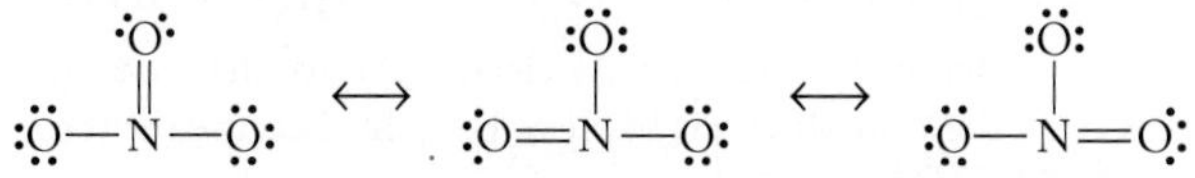

:O—N(=O)—O: ⟷ :O=N(O)—O: ⟷ :O—N(O)=O:

The Lewis structures of the nitrite ion are as follows. These show that the bond order in NO_2^- is 1.5 because of resonance.

$[O=N—O:]^-$ ⟷ $[:O—N=O]^-$

The lower bond order of the nitrogen–oxygen bonds in nitrate (NO_3^-) ions means that these bonds should longer than those in nitrite (NO_2^-) ions.

4.87. The nitrogen–oxygen bonds in N_2O_4 each have a bond order of 1.5, as evident in the blend of single and double N—O bonds in these four equivalent resonance structures:

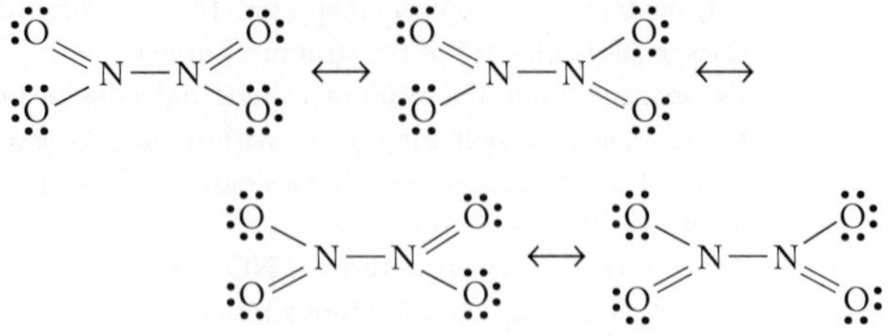

An N—O bond order of about 1.5 in N_2O makes sense if the left and middle resonance structures contribute the more to boding than the structure on the right, which has less favorable formal charges:

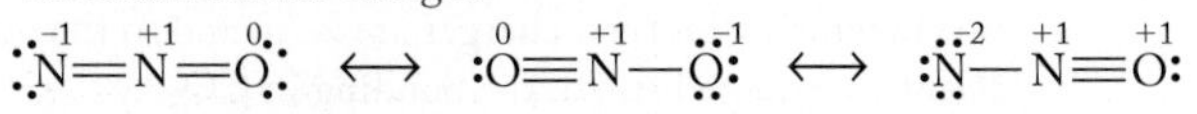

$\overset{-1}{N}=\overset{+1}{N}=\overset{0}{O} \leftrightarrow \overset{0}{O}\equiv\overset{+1}{N}-\overset{-1}{O} \leftrightarrow \overset{-2}{N}-\overset{+1}{N}\equiv\overset{+1}{O}$

Similar N—O bond orders in N_2O_4 and N_2O are consistent with their nearly equal bond lengths.

4.89. $NO^+ < NO_2^- < NO_3^-$

4.91. $NO_3^- < NO_2^- < NO^+$

4.93. C—C bonds in ethane are longer than those in acetylene.

4.95. If there is an electronegativity difference of 2.0 or greater, the bond between the atoms is ionic; below 2.0, the bond is covalent.

4.97. Shared pairs of electrons are closer to the nuclei of smaller atoms, which means these electrons are more likely to feel a stronger pull from the nucleus of a smaller atom, which contributes to higher electronegativity.

4.99. A polar covalent bond is one in which the electrons are shared, but not equally, by the bonded atoms.

4.101. The polar bonds and the atoms with the greater electronegativity (underlined) are C—<u>O</u> and <u>N</u>—H.

4.103. Binary compounds of (b) Al and Cl and (c) C and O have polar covalent bonds. The binary compound of (d) Ca and O has ionic bonds.

4.105. The structure likely to contribute the most to bonding is the one in which formal charges are closest to zero, and in which negative formal charges are on atoms of the most electronegative elements.

4.107. No

4.109. $\overset{0}{H}-\overset{+1}{N}\equiv\overset{-1}{C}$: $\quad \overset{0}{H}-\overset{0}{C}\equiv\overset{0}{N}$:

The formal charges are zero for all the atoms in HCN, whereas in HNC the carbon atom, with a lower electronegativity than N, has a −1 formal charge.

4.111. $(\overset{0}{H})_2\overset{+1}{N}=\overset{0}{C}=\overset{-1}{N} \leftrightarrow (\overset{0}{H})_2\overset{0}{N}-\overset{0}{C}\equiv\overset{0}{N}$

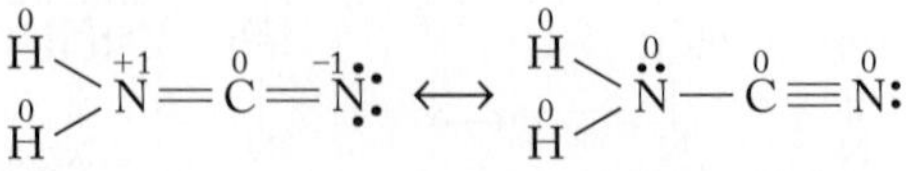

The preferred structure is the one with the C triple-bonded to N.

4.113. $\overset{-2}{N}-\overset{+2}{O}\equiv\overset{0}{N} \leftrightarrow \overset{-1}{N}=\overset{+2}{O}=\overset{-1}{N} \leftrightarrow \overset{0}{N}\equiv\overset{+2}{O}-\overset{-2}{N}$

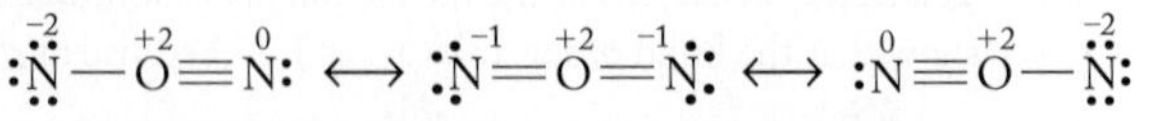

Because oxygen is more electronegative than nitrogen, none of these structures is likely to be stable because the formal charge on O is positive.

4.115. (a)

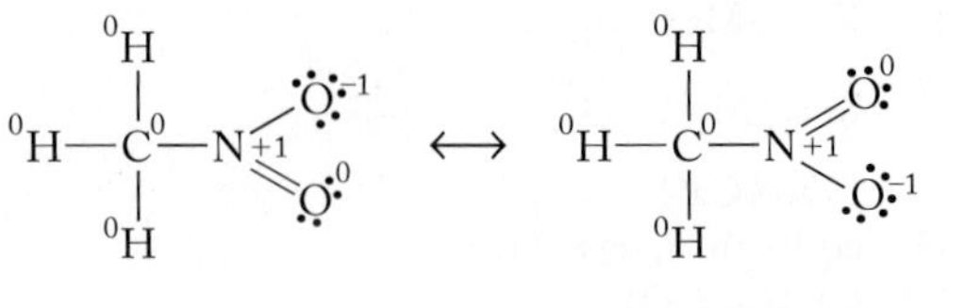

(b)

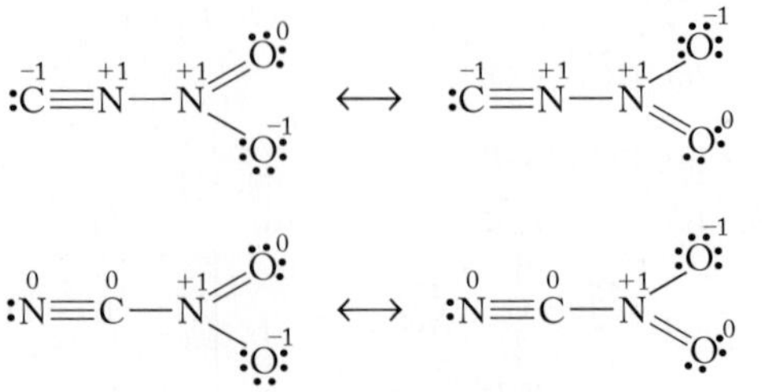

Formal charges are minimized in structures with a terminal N atom, so they are preferred; (c) No

4.117. Yes

4.119. Atoms of these 2nd-row elements cannot expand their octets because their valence shells contain only four orbitals.

4.121. (a) SF_6, (b) SF_5, and (c) SF_4

4.123. (a) 12; (b) 8; (c) 12; (d) 10

4.125.

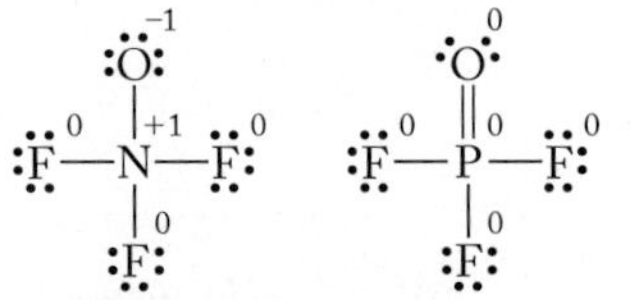

In POF_3 there is a double bond and no formal charges; in NOF_3 there are only single bonds and formal charges are present on N and O.

4.127.

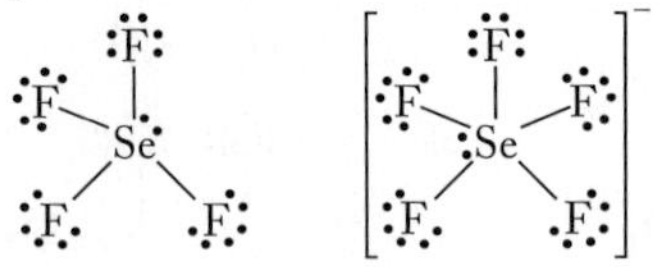

In both structures Se has more than 8 valence electrons.

4.129.

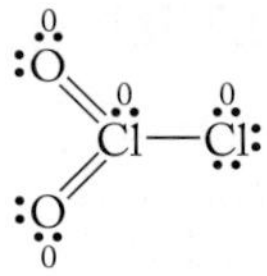

The central chlorine atom has an expanded octet.

4.131. (c) ClO_4, (d) ClO_3, and (e) ClO_2

4.133. (a) S; (b) N; (c) C; (d) O

4.135. (d)

4.137. Like the panes of glass in a greenhouse, the greenhouse gases in the atmosphere are transparent to visible light. Once the visible light warms the surface of the earth and is reemitted as infrared (lower energy) light, the greenhouse gases absorb the infrared light in the same way that the panes of glass do not allow the heat from inside the greenhouse to escape.

4.139. Both. The N—N and N—O bonds in N_2O undergo symmetrical and asymmetrical stretching: like CO_2 except that *both* stretching modes are IR active in N_2O.

4.141. CO does absorb IR radiation because stretching its linear bond gives rise to a fluctuating electric field.

4.143. Infrared radiation does not have enough energy to break covalent bonds.

4.145. The bond in CO is a triple bond, which means it is stronger and requires more energy to vibrate than the C—O bond in H_2CO, which is a double bond.

4.147. (a) ·Be· (b) ·Al· (c) ·C· (d) He:

4.149. $\overset{0}{S}=\overset{0}{C}=\overset{0}{S}$ $\overset{0}{S}=\overset{+2}{S}=\overset{-2}{C}$

Carbon is the central atom in the preferred structure.

4.151. $Cl_2C=O$ (C central, double bonded to O, single bonded to two Cl)

4.153.

(a)

$\overset{+1}{O}\equiv\overset{0}{C}-\overset{-1}{N}-\overset{-1}{N}-\overset{0}{C}\equiv\overset{+1}{O} \leftrightarrow \overset{0}{O}=\overset{0}{C}=\overset{0}{N}-\overset{0}{N}=\overset{0}{C}=\overset{0}{O} \leftrightarrow$

$\overset{0}{O}=\overset{-1}{C}-\overset{+1}{N}\equiv\overset{+1}{N}-\overset{-1}{C}=\overset{0}{O} \leftrightarrow \overset{-1}{O}-\overset{0}{C}\equiv\overset{+1}{N}-\overset{+1}{N}\equiv\overset{0}{C}-\overset{-1}{O} \leftrightarrow$

$\overset{0}{O}=\overset{0}{C}=\overset{0}{N}-\overset{+1}{N}\equiv\overset{0}{C}-\overset{-1}{O} \leftrightarrow \overset{-1}{O}-\overset{0}{C}\equiv\overset{+1}{N}-\overset{0}{N}=\overset{0}{C}=\overset{0}{O}$

(b)

$Br-N=O$

$\overset{+1}{O}\equiv\overset{0}{C}-\overset{-1}{N}=\overset{0}{N}-\overset{-1}{O} \leftrightarrow \overset{0}{O}=\overset{-1}{C}-\overset{0}{N}=\overset{+1}{N}=\overset{0}{O} \leftrightarrow \overset{-1}{O}-\overset{-1}{C}=\overset{0}{N}-\overset{+1}{N}\equiv\overset{+1}{O}$

(c)

$\overset{+1}{O}\equiv\overset{0}{C}-\overset{-1}{N}-\overset{0}{C}(=\overset{0}{O})-\overset{-1}{N}-\overset{0}{C}\equiv\overset{+1}{O} \leftrightarrow \overset{0}{O}=\overset{0}{C}=\overset{0}{N}-\overset{0}{C}(=\overset{0}{O})-\overset{0}{N}=\overset{0}{C}=\overset{0}{O} \leftrightarrow$

$\overset{0}{O}=\overset{0}{C}=\overset{0}{N}-\overset{0}{C}(-\overset{-1}{O})=\overset{+1}{N}=\overset{0}{C}=\overset{0}{O} \leftrightarrow \overset{-1}{O}-\overset{0}{C}\equiv\overset{+1}{N}-\overset{0}{C}(=\overset{0}{O})-\overset{+1}{N}\equiv\overset{0}{C}-\overset{-1}{O} \leftrightarrow$

$\overset{-1}{O}-\overset{0}{C}\equiv\overset{+1}{N}-\overset{0}{C}(=\overset{0}{O})-\overset{0}{N}=\overset{0}{C}=\overset{0}{O} \leftrightarrow \overset{0}{O}=\overset{0}{C}=\overset{0}{N}-\overset{0}{C}(=\overset{0}{O})-\overset{+1}{N}\equiv\overset{0}{C}-\overset{-1}{O} \leftrightarrow$

$\overset{0}{O}=\overset{0}{C}=\overset{+1}{N}=\overset{0}{C}(-\overset{-1}{O})-\overset{0}{N}=\overset{0}{C}=\overset{0}{O}$

4.155. $O=Cl(=O)(=O)-Cl(=O)(=O)=O$ $O=Cl(=O)(=O)-O-Cl(=O)(=O)$

4.157. (a) ·C≡N:; the more likely structure is the one that contains the C—C bond.

$\overset{0}{N}\equiv\overset{0}{C}-\overset{0}{C}\equiv\overset{0}{N}$ $\overset{-1}{C}\equiv\overset{+1}{N}-\overset{+1}{N}\equiv\overset{-1}{C}$

(b) The structure of oxalic acid is consistent with the structure predicted by formal charge analysis in Part a.

4.159. $\overset{0}{N}\equiv\overset{0}{C}-\overset{0}{S}F_3$ (all formal charges 0)

4.161. $[TeOF_5]^{2-}$: $\overset{-1}{O}$, $\overset{-1}{Te}$, five F

4.163. Ca^{2+} $[CO_3]^{2-}$ Mg^{2+} $2[:\ddot{O}H]^-$

4.165. (a), (b)

$\overset{0}{N}\equiv\overset{+1}{N}-\overset{0}{N}=\overset{-1}{N} \leftrightarrow \overset{-1}{N}=\overset{+1}{N}=\overset{+1}{N}=\overset{-1}{N} \leftrightarrow \overset{-1}{N}=\overset{0}{N}-\overset{+1}{N}\equiv\overset{0}{N}$

The middle structure has the most nonzero formal charges separated over three bond lengths, so this one is least preferred. The first and last resonance structures are preferred and are indistinguishable from each other.

(c) N_4 ring: $\overset{0}{N}-\overset{0}{N}$ / $\overset{0}{N}-\overset{0}{N}$ (with double bonds) $\leftrightarrow$ $\overset{0}{N}=\overset{0}{N}$ / $\overset{0}{N}=\overset{0}{N}$

All formal charges are zero.

4.167. (b) and (c)

4.169. (a), (b)

$[\overset{-2}{N}-\overset{0}{N}=\overset{0}{N}-\overset{+1}{N}\equiv\overset{0}{N}]^- \leftrightarrow [\overset{-2}{N}-\overset{0}{N}=\overset{+1}{N}=\overset{+1}{N}=\overset{-1}{N}]^- \leftrightarrow$

$[\overset{-2}{N}-\overset{+1}{N}\equiv\overset{+1}{N}-\overset{0}{N}=\overset{-1}{N}]^- \leftrightarrow [\overset{-1}{N}=\overset{0}{N}-\overset{+1}{N}\equiv\overset{+1}{N}-\overset{-2}{N}]^- \leftrightarrow$

$[\overset{0}{N}\equiv\overset{+1}{N}-\overset{0}{N}=\overset{0}{N}-\overset{-2}{N}]^- \leftrightarrow [\overset{-1}{N}=\overset{0}{N}-\overset{0}{N}=\overset{+1}{N}=\overset{-1}{N}]^- \leftrightarrow$

$[\overset{-1}{N}=\overset{+1}{N}=\overset{0}{N}-\overset{0}{N}=\overset{-1}{N}]^- \leftrightarrow [\overset{0}{N}\equiv\overset{+1}{N}-\overset{-1}{N}-\overset{0}{N}=\overset{-1}{N}]^- \leftrightarrow$

$[\overset{-1}{N}=\overset{0}{N}-\overset{-1}{N}-\overset{+1}{N}\equiv\overset{0}{N}]^-$

The structures that contribute most have the lowest formal charges (last four structures shown).

(c) N_3^- has the Lewis structures

$[\overset{-2}{N}-\overset{+1}{N}\equiv\overset{0}{N}]^- \leftrightarrow [\overset{-1}{N}=\overset{+1}{N}=\overset{-1}{N}]^- \leftrightarrow [\overset{0}{N}\equiv\overset{+1}{N}-\overset{-2}{N}]^-$

From these resonance structures we see that each bond is predicted to be of double-bond character in N_3^-. Therefore, in N_5^- there are two longer N–N bonds than in N_3^-. N_3^- has the higher average bond order.

4.171.

Electronegativity vs. First ionization energy (kJ/mol); $y = 0.0025x - 0.2912$

Using the equation for the best-fit line where x = the ionization energy of neon (2081 kJ/mol) gives a value of y (electronegativity) of neon: $y = 0.002(2081) - 0.2912 = 4.9$

4.173. (a) Isoelectronic means that the two species have the same number of electrons.

(b–d)

$[\overset{-2}{N}-\overset{+1}{N}\equiv\overset{+2}{F}]^+ \leftrightarrow [\overset{-1}{N}=\overset{+1}{N}=\overset{+1}{F}]^+ \leftrightarrow [\overset{0}{N}\equiv\overset{+1}{N}-\overset{0}{F}]^+$

The central nitrogen atom in all the resonance structures always carries a +1 formal charge. Of the three resonance forms, the one on the right should contribute the most to the actual bonding in the ion.

(e) Yes, the fluorine could be the central atom in the molecule, but this would place significant positive formal charge on the fluorine atom (the most electronegative element). These structures are unlikely:

$[:\ddot{N}^{-2}-F^{+3}\equiv N:^{0}]^{+} \longleftrightarrow [:\ddot{N}^{-1}=F^{+3}=\ddot{N}^{-1}]^{+} \longleftrightarrow [:N^{0}\equiv F^{+3}-\ddot{N}:^{-2}]^{+}$

4.175.

```
    H       H
    |       |
H — C — Ö — C — H
    |       |
    H       H
```

4.177.

```
    H   H   H   H
    |   |   |   |
H — C — C — C — C — H
    |   |   |   |
    H   H   H   H
```

Chapter 5

5.1. (a)

5.3. N_2F_2 and NCCN are planar; there are no delocalized π electrons in any of these molecules.

5.5. More

5.7. Yes, there is a carbon in the ring structure connected to four different groups, so the compound is chiral.

5.9. The axial F–Re–axial F bond angle is 180°. The axial F–Re–equatorial F angle is 90°. The equatorial F–Re–equatorial F bonds are all 72°.

5.11. Because the electrons take up most of the space in the atom and because the nucleus is located in the center of the electron cloud, the electron repel each other before the nuclei get close enough to repel each other.

5.13. Because they both have the same steric number around the central nitrogen atom.

5.15. The lone pair of electrons on the nitrogen atom in NH_3 gives it a steric number of 4; the boron atom in BH_3 has a steric number of 3.

5.17. The steric number of the carbon atom in CH_4 is 4, which means a tetrahedral molecular geometry and bond angles of 109.5°. The steric number of the carbon atom in CH_2O is 3, which means trigonal planar molecular geometry and bond angles near 120°.

5.19. The seesaw geometry has only two lone pair–bond pair interactions at 90° (compared to three in a trigonal pyramidal structure), so it has lower energy.

5.21. (c) H_2S < (a) NH_2Cl < (b) CCl_4

5.23. (a) and (c)

5.25. (a) Tetrahedral; (b) Trigonal pyramidal; (c) Bent; (d) Tetrahedral

5.27. (a) Tetrahedral; (b) Trigonal planar; (c) Bent; (d) Square pyramidal

5.29. (a) Tetrahedral; (b) Tetrahedral; (c) Trigonal planar; (d) Linear

5.31. O_3 and SO_2

5.33. SCN^- and CNO^-

5.35. S=S=O Bent

O=S=S=O Bent at each S atom

or

O=S(=O)=S Trigonal planar

5.37. $[XeF_5]^-$ (Xe bonded to five F atoms in a pentagon)

Pentagonal planar

5.39.

```
    H  O(0)  H—C—H
    |  ||    |
H — C — P — O — C — H
    |   |     |
    H  :F:  H—C—H
              |
              H
```

The geometry around the P atom in Sarin is tetrahedral.

5.41. A polar bond is only between two atoms in a molecule. Molecular polarity takes into account all the individual bond polarities and the geometry of the molecule. A polar molecule has a permanent, measurable dipole moment.

5.43. Yes

5.45. C—Cl

5.47. Polar molecules are (b) $CHCl_3$, (d) H_2S, and (e) SO_2. Nonpolar molecules are (a) CCl_4 and (c) CO_2.

5.49. All of the molecules (a–c) are polar.

5.51. (a) $CBrF_3$; (b) CHF_2Cl

5.53. $COCl_2$ < $COBr_2$ < COI_2

5.55. Atomic orbitals must have similar energies.

5.57. (a) sp^3 ; (b) sp^2; (c) sp^3; (d) sp; (e) sp

5.59. Both Lewis structures of N_2F_2 have sp^2 hybridized orbitals on N. Each F atom is sp^3 hybridized. In acetylene, C_2H_2, the carbon atoms are sp hybridized.

5.61.

CO_2	NO_2	O_3	ClO_2
sp	sp^2	sp^2	spd^2

5.63. $[O(0)=Cl(0)(=O(0))(=O(0))-O(-1)]^-$ ⟷ [and three others with a single bond to the left, top, and right-side O atoms.]

Tetrahedral molecular geometry. The three π bonds require that three of the p orbitals on Cl not be involved in the hybridization so that they can form π bonds. One option: Cl uses 3d orbitals in place of the p orbitals for sd^3 hybridization to form the 4 σ bonds to oxygen.

5.65. H—Ar—F: sp^3d hybridized

5.67. Yes

5.69.

```
       :F:
        |
[ :F — S — O: ]⁻
        |
       :F:
```

Molecular geometry = seesaw; sp^3d hybridized

5.71. Yes

5.73. Yes, in resonance structures the electron distribution is blurred across all the resonance forms, which, in essence, defines the delocalization of electrons.

5.75. One N atom has trigonal pyramidal geometry. The other N atom has trigonal planar geometry. No, the hybridization of both N atoms is not the same.

5.77. Both the S and N atoms have SN = 4, which means tetrahedral electron-pair geometry. The presence of a lone pair on N gives this atom trigonal pyramidal geometry and the nitrogen atom is sp^3 hybridized. The four bonds around sulfur means that it has tetrahedral molecular geometry. At first glance, it also means that the hybridization on S is sp^3. However, formal charge considerations argue for turning two of the lone pairs electrons from two O into two π bonds. This requires that two of the p orbitals on S not be involved in hybridization. Therefore, S must use two if its $3d$ orbitals in place of two $3p$ orbitals for spd^2 hybridization to form the four σ bonds to oxygen and nitrogen.

5.79.

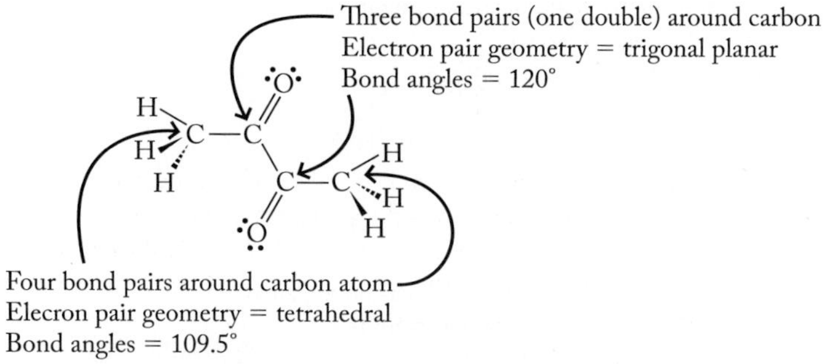

Three bond pairs (one double) around carbon
Electron pair geometry = trigonal planar
Bond angles = 120°

Four bond pairs around carbon atom
Elecron pair geometry = tetrahedral
Bond angles = 109.5°

5.81. (a), (c), and (d)

5.83. No, sp hybridized carbon is linear with only two bonds to each carbon and to be chiral the carbon must be bonded to four different groups.

5.85. Homogeneous

5.87. (a) and (c)

5.89. (a)

5.91.

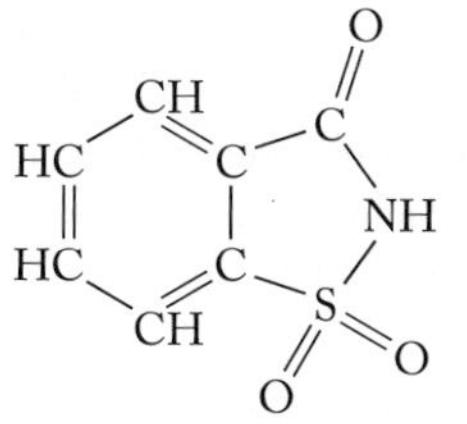

Saccharin

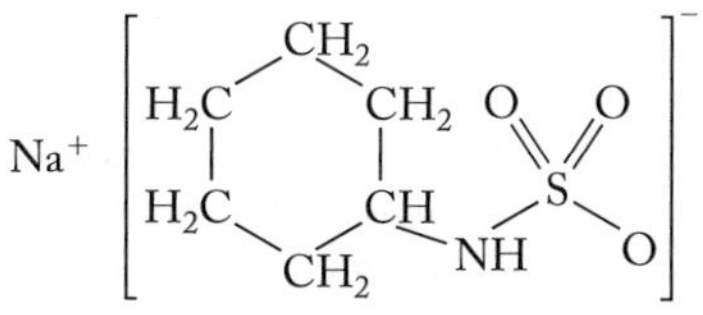

Sodium cyclamate

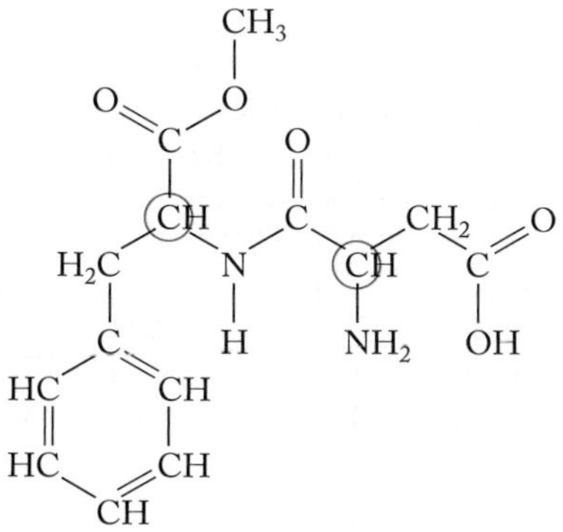

Aspartame

5.93.

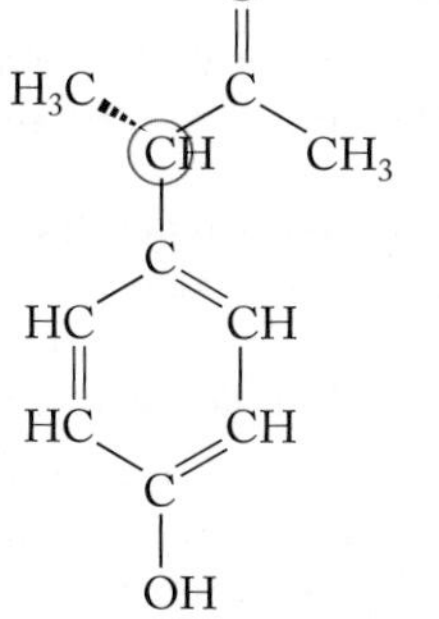

5.95. Molecular orbital theory

5.97. No

5.99. No. The overlap of 1s and 2s orbitals is not as efficient as 1s–1s or 2s–2s overlaps. The match in size and energy is poor.

5.101.

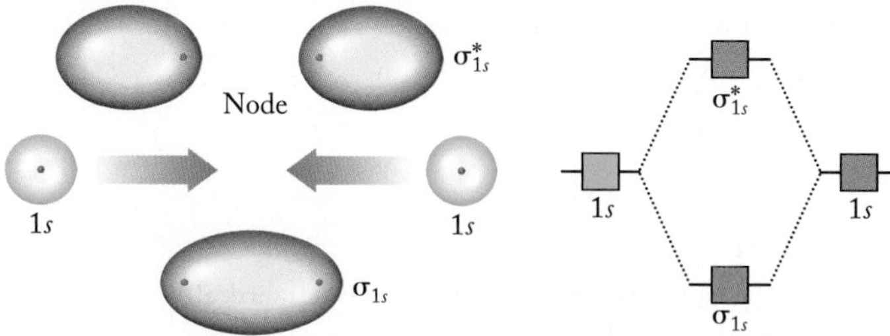

5.103. N_2^+ BO = 2.5; O_2^+ BO = 2.5; C_2^+ BO = 1.5; Br_2^{2-} BO = 0; All species with nonzero bond order (N_2^+, O_2^+, and C_2^+) may exist.

5.105. The paramagnetic species with one or more unpaired electrons are (a) N_2^+, (b) O_2^+, and (c) C_2^+.

5.107. The species with electrons in π^* orbitals are (b) N_2^{2-}, (c) O_2^{2-}, and (d) Br_2^{2-}.

5.109. (a) B_2 and (b) C_2

5.111. No

5.113.

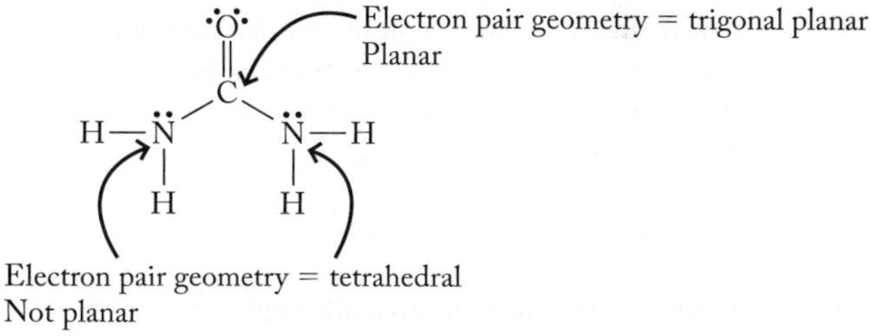

Electron pair geometry = trigonal planar
Planar

Electron pair geometry = tetrahedral
Not planar

While the geometry around the carbon atom is planar, the geometry around the nitrogen atoms is not; this molecule is not planar overall.

5.115.

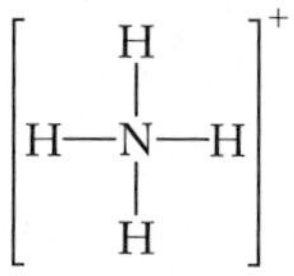

Tetrahedral

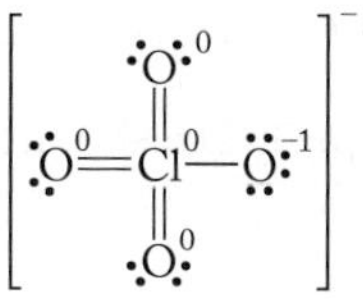

Tetrahedral

5.117.

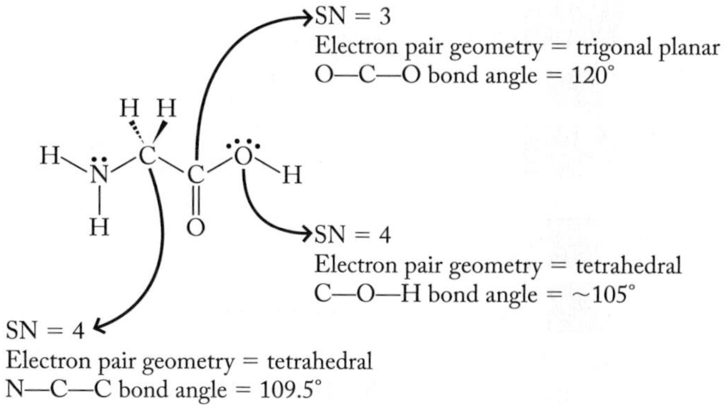

SN = 3
Electron pair geometry = trigonal planar
O—C—O bond angle = 120°

SN = 4
Electron pair geometry = tetrahedral
C—O—H bond angle = ~105°

SN = 4
Electron pair geometry = tetrahedral
N—C—C bond angle = 109.5°

5.119. (a) No, neither of the two molecules is linear; (b) Cl–O–Cl–O would have a permanent dipole.

5.121. (a) $[:\ddot{Cl}=\ddot{O}:]^+$

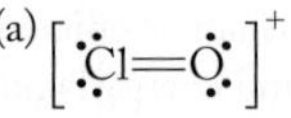

(b) BO = 2

5.123.

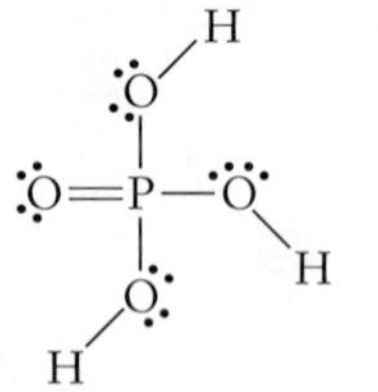

```
        H
       /
     :Ö
      |
:O=P—Ö:
      |     \
     :O:     H
     /
    H
```

The molecular geometry around P is tetrahedral.

5.125. (a)

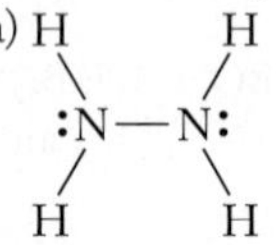

```
 H        H
  \      /
  :N—N:
  /      \
 H        H
```

(b) sp^3; (c) It is polar because the presence of a lone pair of electrons and two N—H bonds on each N atom assures that the molecule is asymmetrical.

5.127. (a), (b)

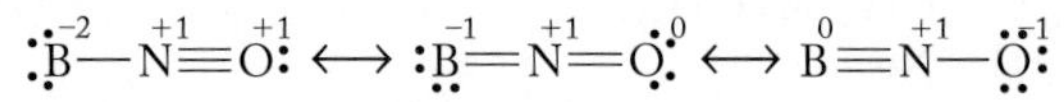

$:\ddot{B}^{-2}—N^{+1}≡O^{+1}: \longleftrightarrow :B^{-1}=N^{+1}=\ddot{O}^{0} \longleftrightarrow B^{0}≡N^{+1}—\ddot{O}^{-1}:$

The structure in which boron is triply bonded to nitrogen is the best description; the –1 formal charge is on the most electronegative atom. All these structures have boron with an incomplete octet; (c) Because SN = 2 around the central nitrogen atom, the molecular geometry is linear.

5.129. (a) and (b)

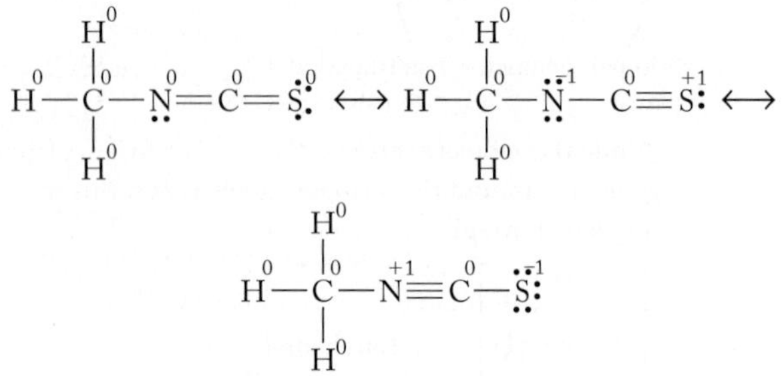

$H^0—C^0H^0_2—\ddot{N}^0=C^0=\ddot{S}^0 \longleftrightarrow H^0—C^0H^0_2—\ddot{N}^{-1}—C^0≡S^{+1}: \longleftrightarrow$

$H^0—C^0H^0_2—N^{+1}≡C^0—\ddot{S}^{-1}:$

The structure on the left is likely to contribute the most to bonding. The methyl (CH_3) carbon is tetrahedral; (c) The isothiocyanate (NCS) group is linear.

5.131. Bond Order = 0

5.133. N_2O_5 and N_2O_3; N_2O_2 depending on its actual structure

5.135. F_2^+

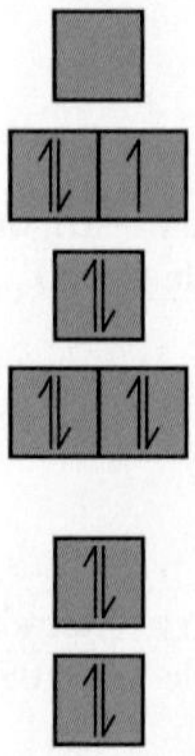

[ ]

[↿⇂][↿]

[↿⇂]

[↿⇂][↿⇂]

[↿⇂]

[↿⇂]

There are 3 more electrons in bonding orbitals than in antibonding orbitals, so the bond order is 3/2 = 1.5.

5.137.

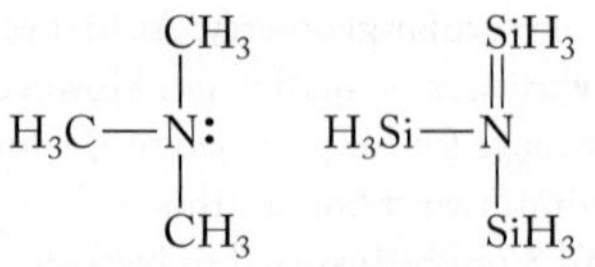

```
     CH3            SiH3
     |              ||
H3C—N:        H3Si—N
     |              |
     CH3            SiH3
```

With a lone pair on N in trimethylamine, the molecular geometry is trigonal pyramidal. If the lone pair on nitrogen double-bonds with the Si atom, SN = 3 and the molecular geometry of trisilylamine is trigonal planar. Silicon can form this double bond by expanding its octet since it has available 3*d* orbitals to do so.

5.139.

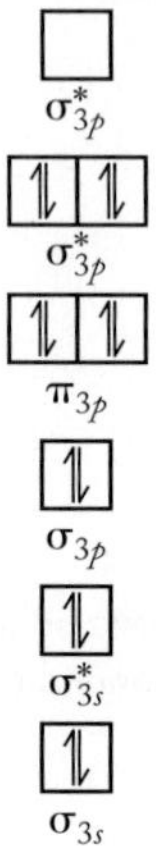

[ ] σ^*_{3p}

[↿⇂][↿⇂] π^*_{3p}

[↿⇂][↿⇂] π_{3p}

[↿⇂] σ_{3p}

[↿⇂] σ^*_{3s}

[↿⇂] σ_{3s}

BO = $\frac{1}{2}(8 - 6) = 1$

Because S_2^{2-} has no unpaired electrons, this species is diamagnetic.

Chapter 6

6.1. (a)

6.3. Ammonia should have the higher boiling point even though its molecules experience weaker London dispersion forces because its molecules form hydrogen bonds with each other and phosphine molecules do not.

6.5. (a) Ketone; (b) Alcohol; (c) Ether

6.7. Solid

6.9. Solid to liquid

6.11. No

6.13. Liquid to solid

6.15. Solid; dispersion forces

6.17. There is less molecular surface area on a branched alkane and, therefore, lower total intermolecular forces.

6.19. (a) C_2Cl_6; (b) C_3H_8; (c) CS_2

6.21. Fuel oil

6.23. The smaller dipole moment of CH_2Cl_2 is more than offset by the greater dispersion forces experienced by molecules of CH_2Cl_2 ($\mathcal{M}$ = 85 g/mol) compared to molecules of CH_2F_2 ($\mathcal{M}$ = 52 g/mol).

6.25. (a) CCl_4; (b) C_3H_8

6.27. Molecule (a) has an asymmetrical, trigonal pyramidal structure, which means that it has a permanent dipole and experiences dipole–dipole forces that nonpolar molecule (b) does not. However, molecule (b) experiences stronger dispersion forces than molecule (a) because (b) has a larger molecular mass and more electrons.

6.29. Water molecules are oriented so that one if not both of their hydrogen atoms are oriented toward the anion.

6.31. Because of the full positive or negative charge on an ion, ion–dipole interaction is stronger than the dipole–dipole interactions.

6.33. The partial positive charge on an H atom bonded to an electronegative O, N, or F means that H atoms and O, N, or F atoms on adjacent molecules experience dipole–dipole interactions. There particular interactions are called hydrogen bonds and are unusually strong because the small electron cloud around each H atoms allows neighboring O, N or F atoms to come very close to the center of the H atom, which leads to very strong attractions as predicted by Coulomb's law.
6.35. CH_3F molecules polar and have larger masses than molecules of nonpolar CH_4, which means CH_3F molecules experience dipole–dipole interactions (which CH_4 molecules don't) and they experience stronger dispersion forces.
6.37. Carbon is not electronegative enough to form the strongly polar bonds with H atoms that are needed to produce hydrogen bonds.
6.39. (b) CF_2Cl_2 and (d) $CFCl_3$
6.41. (a) Methanol and (d) acetic acid
6.43. Cl^- ions because they have the smallest ionic radii and can get closer to the H atoms of water molecules than can Br^- and I^- ions.
6.45. The solvent is the component present in the largest number of moles.
6.47. *Miscible* and *soluble* both describe a substances that dissolves in another. *Miscible*, however, refers to liquids that dissolve in each other in all proportions.
6.49. Hydrophilic substances dissolve in water. Hydrophobic substances do not.
6.51. The (a) compound is more soluble because it is polar.
6.53. (a) $CHCl_3$; (b) CH_3OH; (c) NaF; (d) BaF_2
6.55. All of the substances in each pair will be miscible.
6.57. (b) KI
6.59. (c) CH_3OCH_3
6.61. Sublimation is the change of state in which a solid turns into a gas. In evaporation a liquid becomes a gas.
6.63. Along an equilibrium line in a phase diagram, the two phases that border that line can coexist in equilibrium with each other.
6.65. (a) Solid phase; (b) Gas phase
6.67. Yes
6.69. About 310°C
6.73. −57°C
6.75. (a) Liquid; (b) Liquid; (c) Gas
6.77. Decrease the pressure to 0.006 atm and decrease the temperature to 0.01°C.
6.79. A needle floats on water but not on methanol because of the higher surface tension of water due to the larger number of hydrogen bonds per molecule that it forms.
6.81. The expansion of water in the pipes upon freezing may create sufficient pressure on the walls of the pipes to cause them to burst.
6.83. The cohesive forces in mercury are stronger than the adhesive forces of the mercury to the glass.
6.85. Molecules on the surface of a liquid interact with fewer molecules than those below in the bulk liquid. The area of the surface must expand to allow an object to penetrate it, which means turning bulk molecules into surface molecules, which means disrupting intermolecular interactions between them.
6.87. Water
6.89. Dispersion forces
6.91. Although the dispersion forces between methanol molecules are greater than those between water molecules, water can form more hydrogen bonds per molecule.
6.93. The greater dispersion forces of CH_2Cl_2 add to the dipole–dipole interactions to give stronger intermolecular forces between the CH_2Cl_2 molecules compared to those of CH_2F_2 molecules.
6.95. Increases
6.97. Contracts
6.99. (b)

Chapter 7

7.1. Empirical, NO_2; molecular, N_2O_4
7.3. (a) $X(g) + Y(g) \rightarrow XY(g)$; (b) $X(g) + Y(g) \rightarrow XY(s)$; (c) $X(g) + 2\,Y(g) \rightarrow XY_2(g)$; (d) $X_2(g) + Y_2(g) \rightarrow 2\,XY(g)$
7.5. (b) and (d)
7.7. 5.00 g
7.9. Reactant B
7.11. $2x$
7.13. 1.5 times more
7.15. Yes
7.17. 2 mol
7.19. (a) $N_2(g) + O_2(g) \rightarrow 2\,NO(g)$; (b) $2\,N_2(g) + O_2(g) \rightarrow 2\,N_2O(g)$; (c) $NO(g) + NO_3(g) \rightarrow 2\,NO_2(g)$; (d) $4\,NO(g) + O_2(g) + 2\,H_2O(\ell) \rightarrow 4\,HNO_2(\ell)$
7.21. (a) $N_2O_5(g) + Na(s) \rightarrow NaNO_3(s) + NO_2(g)$; (b) $N_2O_4(g) + H_2O(\ell) \rightarrow HNO_3(aq) + HNO_2(aq)$; (c) $3\,NO(g) \rightarrow N_2O(g) + NO_2(g)$
7.23. (a) $C_3H_8(g) + 5\,O_2(g) \rightarrow 3\,CO_2(g) + 4\,H_2O(g)$; (b) $2\,C_4H_{10}(g) + 13\,O_2(g) \rightarrow 8\,CO_2(g) + 10\,H_2O(g)$; (c) $2\,C_6H_6(g) + 15\,O_2(g) \rightarrow 12\,CO_2(g) + 6\,H_2O(g)$; (d) $2\,C_8H_{18}(g) + 25\,O_2(g) \rightarrow 16\,CO_2(g) + 18\,H_2O(g)$
7.25. (a) $C_2H_4(g) + 3\,O_2(g) \rightarrow 2\,CO_2(g) + 2\,H_2O(g)$; (b) $2\,C_3H_6(g) + 9\,O_2(g) \rightarrow 6\,CO_2(g) + 6\,H_2O(g)$; (c) $2\,C_4H_{10}(g) + 13\,O_2(g) \rightarrow 8\,CO_2(g) + 10\,H_2O(g)$; (d) $C_4H_8(g) + 6\,O_2(g) \rightarrow 4\,CO_2(g) + 4\,H_2O(g)$
7.27. (a) $2\,SO_2(g) + O_2(g) \rightarrow 2\,SO_3(g)$; (b) $2\,H_2S(g) + 3\,O_2(g) \rightarrow 2\,SO_2(g) + 2\,H_2O(g)$; (c) $16\,H_2S(g) + 8\,SO_2(g) \rightarrow 3\,S_8(s) + 16\,H_2O(g)$
7.29. Both equations give the same answer because the molar ratio of CO_2 to C_2H_6 is 2:1 in both.
7.31. (a) 4.5×10^{11} mol C; (b) 2.0×10^{10} kg CO_2
7.33. (a) $2\,NaHCO_3(s) \rightarrow CO_2(g) + H_2O(g) + Na_2CO_3(s)$; (b) 6.55 g CO_2
7.35. 1.17 kg
7.37. 29 g O_2
7.39. (a) 1.48 kg; (b) 1.11 kg
7.41. 346 g
7.43. An empirical formula shows the lowest whole-number ratio of atoms in a substance. A molecular formula shows the actual numbers of each kind of atom that compose one molecule of the substance.
7.45. No
7.47. (a) 74.19% Na, 25.81% O; (b) 57.48% Na, 40.00% O, 2.52% H; (c) 27.37% Na, 1.20% H, 14.30% C, 57.13% O; (d) 43.38% Na, 11.33% C, 45.28% O
7.49. Pyrene, $C_{16}H_{10}$
7.51. Al_2O_3

7.53. No
7.55. Ti_6Al_4V
7.57. (a) P_2O_5; (b) P_4O_{10}
7.59. $Mg_3Si_2H_4O_9$
7.61. $CuCl_2O_8$
7.63. The excess of oxygen is required in combustion analysis to ensure the complete reaction of the hydrogen and carbon to form water and carbon dioxide.
7.65. Yes, if the empirical formula is the same as the molecular formula.
7.67. $C_3H_6O_2$
7.69. Empirical, C_2H_3; molecular, $C_{20}H_{30}$
7.71. (c) Less than the sum of the masses of Fe and S.
7.73. Theoretical yield is the greatest amount of a product possible from a reaction and assumes that the reaction goes to completion. Percent yield is the actual yield divided by the theoretical yield and multiplied by 100.
7.75. The ratio of the reactants in the balanced chemical equation also is important. If A and B are necessary in equal molar ratios, then yes, with fewer moles of A, A would be the limiting reactant. However, if more than 1 mol of B is required for every mole of A and fewer moles of A than B are present, we could not unequivocally say that A is the limiting reactant.
7.77. 3 cups
7.79. 87% effective; 0.60 metric tons escaped
7.81. (a) $CHCl_3$; (b) 308 g; (c) 529 g
7.83. (a) $C(s) + H_2O(g) \rightarrow CO(g) + H_2(g)$; (b) 61%
7.85. (a) $C_6H_{12}O_6(aq) \rightarrow 2\ C_2H_5OH(\ell) + 2\ CO_2(g)$; (b) 77.1%
7.87. (a) Calcium triphosphate hydroxide; (b) 39.89%; (c) Decreases slightly
7.89. 99%
7.91. (a) 45 g; (b) 15 g; (c) 4.66 cm^3
7.93. (a) $a = 1$, $b = 3$, charge on U is 6+ (b) $c = 3$, $d = 8$, charge on U is 5.33+; (c) $x = 2$, $y = 2$, $z = 6$
7.95. (a) No; (b) $C_5H_{10}O_5(s) + 5\ O_2(g) \rightarrow 5\ CO_2(g) + 5\ H_2O(\ell)$
$2\ C_7H_{12}O_7(s) + 13\ O_2(g) \rightarrow 14\ CO_2(g) + 12\ H_2O(\ell)$
7.97. 0.26 g HCO_2H
7.99. (a) $3FeO(s) + H_2O(\ell) \rightarrow Fe_3O_4(s) + H_2(g)$;
(b) $12\ FeO(s) + 2\ H_2O(\ell) + CO_2(aq) \rightarrow 4\ Fe_3O_4(s) + CH_4(g)$
7.101. (a) 55 mol ethanol; (b) 110 mol CO_2
7.103. Re
7.105. 82.4%
7.107. (a) $2\ SO_2(g) + 2\ CaO(s) + O_2(g) \rightarrow 2\ CaSO_4(s)$; (b) 2.13 tons
7.109. Mg_2SiO_4

Chapter 8

8.1. Yellow
8.3. Blue, Na^+; green, SO_4^{2-}
8.5. From +2 to +4
8.7. (a) N_2O_5 > (b) N_2O_4 = (e) NO_2 > (d) N_2O_3 > (c) NO
8.9. (a) $H^+(aq) + HSO_4^-(aq) + Ba^{2+}(aq) + 2\ OH^-(aq) \rightarrow BaSO_4(s) + 2\ H_2O(\ell)$; (b) (b)
8.11. (a) 5.6 M $BaCl_2$; (b) 1.00 M Na_2CO_3; (c) 1.30 M $C_6H_{12}O_6$; (d) 5.92 M KNO_3
8.13. (a) 0.29 M; (b) 0.23 M; (c) 1.76 M; (d) 0.084 M
8.15. (a) 11.7 g NaCl; (b) 4.99 g $CuSO_4$; (c) 6.41 g CH_3OH
8.17. 2.72 g
8.19. (a) 1.9×10^{-2} mol; (b) 4.07×10^{-4} mol; (c) 82.72×10^{-4} mol; (d) 4.00×10^{-1} mol
8.21. Orchard sample: 3.4×10^{-4} mmol/L; Residential area sample: 5.6×10^{-5} mmol/L; After storm sample: 3.2×10^{-2} mmol/L
8.23. 1.2×10^{-3} M
8.25. (b) $AgNO_3$ and (c) $Fe(NO_3)_2 \cdot 6\ H_2O$
8.27. 4.57×10^{-2} M Mg^{2+}
8.29. (a) 1.81×10^{-2} M Na^+; (b) 2.7×10^{-1} mM LiCl; (c) 1.28×10^{-2} mM Zn^{2+}
8.31. 1.95 M
8.33. The concentration of the adult-strength medication is 1.8 mg/mL; 23 mL is needed to prepare the child-strength cough syrup.
8.35. Table salt produces Na^+ and Cl^- ions in solution when it dissolves in polar solvents, which makes its solutions electrically conductive. Sugar does not dissociate into ions and is not conductive.
8.37. Electrolytes produce charged particles (cations and anions) when they dissolve in polar solvents; nonelectrolytes do not produce ions when they dissolve.
8.39. In order of decreasing conductivity, 1.0 M Na_2SO_4 (c) > 1.2 M KCl (b) > 1.0 M NaCl (a) > 0.75 M LiCl (d).
8.41. (a) 0.025 M; (b) 0.050 M; (c) 0.075 M
8.43. Acids transfer H^+ ions to bases; in water they increase the concentration of H_3O^+ ions in solution.
8.45. Strong acids include HCl, HNO_3, $HClO_4$, H_2SO_4, HI, and HBr; most other common acids including CH_3COOH, HCOOH, HF, and H_3PO_4 are weak acids.
8.47. A base accepts H^+ ions from another species (an acid) in solution.
8.49. Strong bases include NaOH, KOH, CsOH, LiOH, RbOH, $Ba(OH)_2$, $Sr(OH)_2$, $Ca(OH)_2$; weak bases include NH_3, CH_3NH_2, C_5H_5N.
8.51. (a) The acid is H_2SO_4; the base is $Ca(OH)_2$.
Ionic equation: $H^+(aq) + HSO_4^{2-}(aq) + Ca^{2+}(aq) + 2\ OH^-(aq) \rightarrow 2\ H_2O(\ell) + SO_4^{2-}(aq) + Ca^{2+}(aq)$
Net ionic equation: $H^+(aq) + HSO_4^{2-}(aq) + 2\ OH^-(aq) \rightarrow 2\ H_2O(\ell) + SO_4^{2-}(aq) + Ca^{2+}(aq)$
(b) $PbCO_3$ is the base; sulfuric acid is the acid.
Ionic and net ionic equation: $PbCO_3(s) + H^+(aq) + HSO_4^{2-}(aq) \rightarrow PbSO_4(s) + CO_2(g) + H_2O(\ell)$
(c) $Ca(OH)_2$ is the base; CH_3COOH is the acid.
Ionic equation: $Ca(OH)_2(s) + 2\ CH_3COOH(aq) \rightarrow Ca^{2+}(aq) + 2\ CH_3COO^-(aq) + 2\ H_2O(\ell)$
Net ionic equation: $Ca(OH)_2(s) + 2\ CH_3COOH(aq) \rightarrow Ca^{2+}(aq) + 2\ CH_3COO^-(aq) + 2\ H_2O(\ell)$
8.53. (a) Molecular equation: $Mg(OH)_2(s) + H_2SO_4(aq) \rightarrow MgSO_4(aq) + 2\ H_2O(\ell)$
Net ionic equation: $Mg(OH)_2(s) + H^+(aq) + HSO_4^{2-}(aq) \rightarrow Mg^{2+}(aq) + 2\ H_2O(\ell) + SO_4^{2-}(aq)$
(b) Molecular equation: $MgCO_3(s) + 2\ HCl(aq) \rightarrow MgCl_2(aq) + H_2CO_3(aq)$
Net ionic equation: $MgCO_3(s) + 2\ H^+(aq) \rightarrow Mg^{2+}(aq) + H_2O(\ell) + CO_2(g)$
(c) Molecular equation: $NH_3(g) + HCl(g) \rightarrow NH_4Cl(s)$. This is also the net ionic equation.
8.55. $PbCO_3(s) + 2\ H^+(aq) \rightarrow Pb^{2+}(aq) + CO_2(g) + H_2O(\ell)$; $Pb(OH)_2(s) + 2\ H^+(aq) \rightarrow Pb^{2+}(aq) + 2\ H_2O(\ell)$
8.57. A saturated solution contains the maximum concentration of a solute. A supersaturated solution temporarily contains more than the maximum concentration of a solute at a given temperature.

8.59. Gaseous water vapor in the atmosphere is in a homogeneous solution; snow and rain "fall out" of this solution in the same way that a solid precipitates from a solution.

8.61. A saturated solution may not be a concentrated solution if the solute is only sparingly or slightly soluble in the solution. In that case, the solution is a saturated dilute solution.

8.63. (a) Barium sulfate, (e) lead hydroxide, and (f) calcium phosphate

8.65. (a) Balanced reaction: $Pb(NO_3)_2(aq) + Na_2SO_4(aq) \rightarrow PbSO_4(s) + 2\ NaNO_3(aq)$
Net ionic equation: $Pb^{2+}(aq) + SO_4^{2-}(aq) \rightarrow PbSO_4(s)$
(b) No precipitation reaction occurs.
(c) Balanced reaction: $FeCl_2(aq) + Na_2S(aq) \rightarrow FeS(s) + 2\ NaCl\ (aq)$
Net ionic equation: $Fe^{2+}(aq) + S^{2-}(aq) \rightarrow FeS(s)$
(d) Balanced reaction: $MgSO_4(aq) + BaCl_2(aq) \rightarrow MgCl_2(aq) + BaSO_4(s)$
Net ionic equation: $Ba^{2+}(aq) + SO_4^{2-}(aq) \rightarrow BaSO_4(s)$

8.67. $CaCO_3$

8.69. 2.11×10^{-2} g $MgCO_3$

8.71. 5.4×10^{-2} g O_2

8.73. Some other source

8.75. The number of electrons gained or lost is directly related to the decrease or increase, respectively, in oxidation number.

8.77. (a) −1; (b) +1; (c) −2; (d) −3

8.79. Silver

8.81. As n increases, the oxidation state of carbon increases.

8.83. (a) +1; (b) +5; (c) +7

8.85. No change

8.87. 0.25 mol

8.89. (a) Reactants: Si = +4, O = −2, 2/3 of Fe = +3, 1/3 of Fe = +2
Products: Fe = +2; Si = +4; O = −2 in Fe_2SiO_4 and 0 in O_2
Oxygen is oxidized (O^{2-} to O_2) and (some) iron is reduced (Fe^{3+} to Fe^{2+}).
(b) Reactants: Si = +4; Fe = 0; O = −2 in SiO_2 and 0 in O_2
Product: Fe = +2, Si = +4, O = −2
Iron is oxidized (Fe^0 to Fe^{2+}) and oxygen is reduced (O_2 to O^{2-}).
(c) Reactants: Fe = +2, H = +1, O = −2 in FeO and H_2O and 0 in O_2
Product: Fe = +3, O = −2, H = +1
Iron is oxidized (Fe^{2+} to Fe^{3+}) and oxygen is reduced (O_2 to O^{2-}).

8.91. (a) 0.25 mol; (b) 0.17 mol

8.93. $NH_4^+(aq) + 2\ O_2(g) \rightarrow NO_3^-(aq) + 2\ H^+(aq) + H_2O\ (\ell)$

8.95. $2\ Fe(OH)_2^+(aq) + Mn^{2+}(aq) \rightarrow 2\ Fe^{2+}(aq) + 2\ H_2O\ (\ell) + MnO_2(s)$

8.97. $2\ H_2O(\ell) + 4\ Ag(s) + 8\ CN^-(aq) + O_2(g) \rightarrow 4\ Ag(CN)_2^-(aq) + 4\ OH^-(aq)$

8.99. (a) $2\ ClO_3^-(aq) + SO_2(g) \rightarrow 2\ ClO_2(g) + SO_4^{2-}(aq)$
(b) $4\ H^+(aq) + 2\ ClO_3^-(aq) + 2\ Cl^-(aq) \rightarrow 2\ ClO_2(g) + 2\ H_2O(\ell) + Cl_2(g)$
(c) $2\ ClO_3^-(aq) + Cl_2(g) \rightarrow 2\ ClO_2(g) + 2\ Cl^-(aq) + O_2(g)$

8.101. (a) 5.00 mL; (b) 31.5 mL; (c) 215 mL

8.103. 500 mL

8.105. 556.0 mM, 19.23 g/kg

8.107. To deionize water, cations such as Na^+ and Ca^{2+} are exchanged for H^+ at cation-exchange sites. Anions such as Cl^- and SO_4^{2-} are exchanged for OH^- at the anion-exchange sites. The released ions (H^+ and OH^-) combine to form H_2O.

8.109. K^+ would act like Na^+ in removing the ions that make water hard, and its use avoids raising Na^+ ion concentrations which may be a problem for people with hypertension.

8.111. $7.98 \times 10^{-4}\ M\ SO_4^{2-}$

8.113. (a) 11.7 M; (b) 42.7 mL; (c) 1.72 kg

8.115. (a) $2\ OH^-(aq) + 2\ H_2O(\ell) + 3\ S_2O_4^{2-}(aq) + 2\ CrO_4^{2-}(aq) \rightarrow 6\ SO_3^{2-}(aq) + 2\ Cr(OH)_3(s)$; (b) Sulfur is oxidized; chromium is reduced; (c) Oxidizing agent = CrO_4^{2-}; reducing agent = $S_2O_4^{2-}$; (d) 38.7 g

8.117. (a) Balanced equation: $4\ Ag(s) + 2\ H_2S(g) + O_2(g) \rightarrow 2\ Ag_2S(s) + 2\ H_2O(\ell)$
(b) $3\ Ag_2S(s) + 3\ H_2O(\ell) + 3\ OH^-(aq) + 2\ Al(s) \rightarrow 6\ Ag(s) + 3\ HS^-(aq) + 2\ Al(OH)_3(s)$

8.119. (a) H_3PO_4; (b) H_2SeO_3; (c) H_3BO_3

8.121. $2\ H^+(aq) + ClO^-(aq) + 2\ I^-(aq) \rightarrow Cl^-(aq) + H_2O(\ell) + I_2(aq)$
$I_2(aq) + 2\ S_2O_3^{2-}(aq) \rightarrow 2\ I^-(aq) + S_4O_6^{2-}(aq)$

8.123. (a) $NaClO_4$, NH_4ClO_4; (b) 427 kg; (c) 2.82×10^{10} gal; (d) The MA lab

8.125. (a) $3\ CH_2O \rightarrow CO_2 + C_2H_5OH$; (b) $C_2H_5OH + O_2 \rightarrow CH_3COOH + H_2O$; (c) 0 in CH_2O and CH_3COOH; +4 in CO_2; −2 in C_2H_5OH; (d) 66.7 g acetic acid

8.127. (a) The first reaction is a redox reaction.
(b) $2\ H^+(aq) + SO_4^{2-}(aq) + CaCO_3(s) \rightarrow CaSO_4(s) + H_2O(\ell) + CO_2(g)$
(c) $SO_4^{2-}(aq) + CaCO_3(s) \rightarrow CaSO_4(s) + CO_3^{2-}(aq)$

8.129. (c) and (d)

Chapter 9

9.1. Internal energy increases.

9.3. Highest fuel value, (c) C_5H_{12}; lowest fuel value, (b) C_6H_{12}

9.5. q is negative, w is positive, ΔE is negative; 67% yield

9.7. (a) $CH_4(g) + H_2O(g) \rightarrow CO(g) + H_2(g)$; (b) $\Delta H^\circ_{rxn} = -206.1$ kJ/mol; (c) Heat flows out from the reaction mixture to the surroundings.

9.9. Energy makes work possible.

9.11. The value of a state function is independent of the path; only the initial and final values are important.

9.13. The particles attract each other.

9.15. Raise the temperature of the sample or decrease its volume through compression.

9.17. Energy is the ability to do work, so energy and work must have equivalent units. P–V work is equivalent to force × distance ($f{\cdot}d$) work because pressure is force (f) per unit area, which has units of d^2. Volume has units of d^3. Therefore, the units on P–V work are equivalent of $(f/d^2) \times d^3 = f{\cdot}d$.

9.19. (a) Exothermic; (b) Endothermic; (c) Exothermic

9.21. The heat of vaporization absorbed by the water raises its internal energy.

9.23. $w = -0.500\ \text{L} \cdot \text{atm} = -50.7\ \text{J}$

9.25. (a) $\Delta E = 50.0$ J; (b) $\Delta E = 6.3$ kJ; (c) $\Delta E = -1.23 \times 10^4$ kJ

9.27. $\Delta E = -276$ kJ

9.29. (b); w is negative

9.31. A change in enthalpy is the sum of the change of internal energy and the product of the system's pressure and change in volume.

9.33. When an exothermic reaction transfers heat to its surroundings, the sign of q (and ΔH) is negative.

9.35. Negative

9.37. Positive

9.39. Negative
9.41. Specific heat is heat capacity per gram of the substance and an intensive property of the substance. Heat capacity is an extensive property of an object.
9.43. To vaporize water all of the strong intermolecular hydrogen bonds between liquid water molecules must be broken, only a fraction of those between molecules of ice are broken when ice melts.
9.45. The line segments have slightly different slopes in each curve because the liquids they represent have different molar heat capacities.
9.47. Water's high heat capacity compared to air means that water carries away more energy from the engine for every Celsius degree rise in temperature, so water is a good choice to cool automobile engines.
9.49. q = 29.3 kJ
9.51.

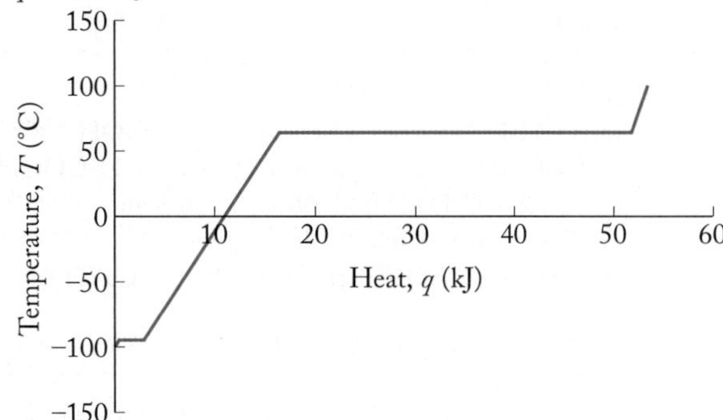

9.53. 95.4 g
9.55. −47.5°C
9.57. It is needed because the product of heat capacity and temperature change of calorimeter is needed to determine the quantity of heat transferred from or into the thermodynamic system inside the calorimeter.
9.59. Yes
9.61. $C_{calorimeter}$ = 8.044 kJ/°C
9.63. 5129 kJ/mol
9.65. 3.07°C
9.67. $\Delta H_{rxn} = -464$ kJ
9.69. In using Hess's law, we assume the enthalpy changes in each step of the overall reaction are conserved so that their sum is equal to the enthalpy change of the overall reaction.
9.71. $\Delta H^\circ_{rxn} = -297$ kJ/mol
9.73. $\Delta H^\circ_{rxn} = 28.0$ kJ/mol
9.75. $\Delta H^\circ_{rxn} = -103$ kJ
9.77. Reacting C(s) with a limited supply of $O_2(g)$ produces a mixture of products, not just CO.
9.79. No, because O_2 is more stable than O_3, which means ΔH°_f of $O_3 > \Delta H^\circ_f$ of O_2 (which is zero).
9.81. We must account for all the bonds that break and all the bonds that form in the reaction, which requires a balanced chemical equation.
9.83. Bond energies are based on bond formation in the gas phase from free atoms also in the gas phase.
9.85. (a) and (c)
9.87. $\Delta H^\circ_{rxn} = -164.9$ kJ
9.89. $NH_4NO_3(s) \rightarrow N_2O(g) + 2\,H_2O(g)$
$\Delta H^\circ_{rxn} = -35.9$ kJ
9.91. $\Delta H^\circ_{rxn} = -7198$ kJ
9.93. (a) $\Delta H_{rxn} = 852$ kJ; (b) $\Delta H_{rxn} = 90$ kJ; (c) $\Delta H_{rxn} = 96$ kJ
9.95. 554 kJ
9.97. $\Delta H^\circ_{rxn} = -610$ kJ
9.99. Strong ion–dipole interactions increase the likelihood that a heat of solution will be exothermic.
9.101. *Fuel value* refers to the thermal energy released per gram of fuel during complete combustion.
9.103. (a) CH_4; (b) H_2
9.105. −717 kJ/mol
9.107. 201 kg
9.109. (a) 48.99 kJ/g; (b) 4.90×10^4 kJ; (c) 5.97 g
9.111. The first reaction using hydrogen and chlorine gases is exothermic, so we should cool that reaction; the second reaction using sodium chloride and sulfuric acid is endothermic, so we should heat that reaction.
9.113. (a) $2\,NaOH(aq) + H_2SO_4(aq) \rightarrow 2\,H_2O(\ell) + Na_2SO_4(aq)$; (b) No; (c) $\Delta H^\circ_{rxn} = -57.1$ kJ/mol H_2O
9.115. 26.0°C
9.117. (a) $\Delta H^\circ_{rxn} = -1255.5$ kJ/mol; (b) 48.22 kJ/g
9.119. Exothermic
9.121. (a) $CH_3OH(g) + N_2(g) \rightarrow HCN(g) + NH_3(g) + \frac{1}{2}\,O_2(g)$; (b) A reactant; (c) 307 kJ
9.123. $\Delta H^\circ_{rxn} = -1400$ kJ
9.125. $\Delta H^\circ_{rxn} = 67.8$ kJ
9.127. $\Delta H^\circ_{rxn} = 31.5$ kJ
9.129. Hydrogen

Chapter 10

10.1. (c)
10.3. There is no effect assuming no change in temperature.
10.5. (a) The pressure will double; (b) The frequency of collisions will double; (c) The most probable speed does not change.
10.7. (a) a; (b) b; (c) b
10.9. (a)
10.11. Curve 1 is the SO_2 profile; propane matches curve 2.
10.13. (a) Red line; (b) Slope = 3.3×10^{-2} cm^3/K; $P = 2.5\times10^3$ atm
10.15. Line 2
10.17. Line 2
10.19. 1. (b); 2. (a); 3. (c)
10.21. The root-mean-square speed (u_{rms}) is the square root of the arithmetic mean of the squares of the speeds of all the particles in a gas sample. It is the speed of a particle that has the average kinetic energy of all the particles in the sample
10.23. No, only temperature affects the root-mean-square speed.
10.25. The rank order in terms of increasing root-mean-square speed is $N_2O_5 < N_2O_4 < NO_2 < NO$.
10.27. 32 g/mol
10.29. (a) CH_4: 681 m/s; C_2H_6: 497 m/s; C_3H_8: 411 m/s; (b) C_4H_{10}: 359 m/s; C_5H_{12}: 298 m/s; (c) Methane
10.31. 509 m/s
10.33. Br
10.35. Force is the product of the mass of an object and the acceleration due to gravity. Pressure is force exerted per unit of surface given area.
10.37. The ethanol barometer
10.39. A sharpened blade has a smaller area over which the force is distributed compared to a dull blade.
10.41. (a) 0.020 atm; (b) 0.739 atm
10.43. 914 g; 3.58×10^3 N/m^2 or 3.66×10^{-2} atm
10.45. (a) 814.6 mmHg; (b) 1.072 atm; (c) 1086 mbar
10.47. As temperature increases, the speed of gas particles increases, which means they collide with container walls with more force and more frequently, resulting in an increase in pressure.

10.49. The balloonist should decrease the temperature.
10.51. 1.45 atm
10.53. 6.79 L
10.55. 4.2 L
10.57. (a) 3.06 atm; (b) 21 m
10.59. Raising the temperature in b produces an 12.8% increase, which is greater than the 7.9% increase due to lowering pressure in a.
10.61. (a) No change; (b) Decrease of ¼ original volume; (c) Increase in volume by 17%
10.63. 144 L
10.65. 6.6 atm (assuming 7.1 atm is the actual pressure inside the tire and not a gauge pressure).
10.67. STP is defined as 1 atm and 0°C (273 K); $V = 22.4$ L
10.69. 0.67 mol
10.71. 2.36 atm
10.73. 4.19×10^4 g
10.75. 1.1 % more
10.77. (a) 1.00 mol; (b) Helium
10.79. No, because the densities of different gases are proportional to their molar masses. Only those with the same molar mass have the same density.
10.81. Density (a) increases with increasing pressure and (b) increases with decreasing temperature.
10.83. (a) 9.08 g/L; (b) In the basement, particularly if its source is underground.
10.85. (a) 2.60 g/L; (b) SO_2
10.87. (a) 28.0 g/mol; (b) CO
10.89. 394 g
10.91. 6.2×10^2 g
10.93. 3.78 L
10.95. The partial pressure of a gas is the pressure that a particular gas contributes to the total pressure of a mixture of gases.
10.97. 0.20
10.99. $P_{total} = 31$ atm
$P(N_2) = 22$ atm
$P(H_2) = 6.2$ atm
$P(CH_4) = 3.1$ atm
10.101. 0.0190 mol
10.103. (a) Greater than; (b) Less than; (c) Greater than
10.105. 67% more
10.107. 889 L
10.109. 13%
10.111. At low temperatures the gas particles move more slowly, and their collisions become inelastic; they stick together due to the weak attractive forces between them. The particles, therefore, do not act separately to contribute to the pressure in the container, and the pressure is lower than would be expected by the ideal gas law. Also, the gas particles take up real volume in the container and as the pressure increases the volume of the particles takes up a greater volume of the free space in the container. This has the effect of raising the pressure–volume product above what we would expect from the ideal gas law (in a plot of PV/RT versus P).
10.113. Because b is a measure of the volume that the gas particles occupy, b increases as the sizes of the particles increase.
10.115. Because Ar has more electrons and will experience stronger London dispersion forces between its atoms than He does.
10.117. H_2
10.119. (a) $P = 1590$ atm; (b) $P = 595$ atm
10.121. (a) 70.3 mL; (b) No, because when gases are compressed they heat up.
10.123. 0.25 m/s
10.125. 18 kg
10.127. 2.28 L
10.129. (a) CH_3COOH and $(CH_3)_3N$; (b) HCl and $(CH_3)_3N$; (c) No
10.131. $P_{H_2} = 110$ kPa
$P_{He} = 20$ kPa
$P_{CH_4} = 3$ kPa
10.133. 2.7 atm
10.135. (a) 38.4 L N_2; (b) 192 L CO_2; (c) 1.74 g/L
10.137. 70.0 g; 2.40 g more NaN_3 is needed at 10°C
10.139. 0.00781 mol O_2

Chapter 11

11.1. (a)
11.3. (a) NaCl; (b) $MgCl_2$; (c) $C_6H_{12}O_6$; (d) K_3PO_4
11.5. $bp_X \approx 20$°C and $bp_Y \approx 40$°C; Y has stronger intermolecular forces.
11.7. Solution A must be the more concentrated solution because solvent flows through the membrane from the less concentrated to the more concentrated side.
11.9. (a)
11.11. A semipermeable membrane is a boundary between two solutions through which some molecules may pass but others cannot. Usually, small molecules may pass through but large molecules are excluded.
11.13. Solvent flows across a semipermeable membrane from the more dilute solution side to the more concentrated solution side.
11.15. Reverse osmosis transfers solvent across a semipermeable membrane from a region of higher solute concentration to a region of lower solute concentration. Because reverse osmosis goes against the natural flow of solvent across the membrane, the key component needed is a pump to apply pressure to the more concentrated side of the membrane. Other components needed include a containment system, piping to introduce and remove the solutions, and a tough semipermeable membrane that can withstand the high pressures required.
11.17. A strong electrolyte completely dissociates in solution, which means a solution of it has greater osmotic pressure than a solution of a nonelectrolyte with the same concentration. For example, 1 M NaCl ($i = 2$) has twice the osmotic pressure as 1 M glucose ($i = 1$).
11.19. Solute ions of opposite charge attract each other in solution and can temporarily form ion pairs that reduce the total number of independent particles, which lowers the experimental value of the van't Hoff factor.
11.21. (a) From side A to side B; (b) from side B to side A; (c) from side A to side B
11.23. (a) 57.5 atm; (b) 0.682 atm; (c) 52.9 atm; (d) 46.5 atm
11.25. (a) 2.75×10^{-2} M; (b) 1.11×10^{-3} M; (c) 1.00×10^{-2} M
11.27. Line A (red)
11.29. 94.1 g/mol
11.31. 2.2 atm
11.33. When the average kinetic energy of the liquid molecules increases, more of the molecules can escape the liquid phase and enter the gas phase. More molecules in the gas phase increase the vapor pressure.
11.35. As intermolecular forces increase in strength, the vapor pressure decreases.

11.37. (a) CH_3CH_2OH < (b) CH_3OCH_3 < (c) $CH_3CH_2CH_3$
11.39. 41.0 kJ/mol
11.41. For isooctane, 0.105 atm or 80.0 torr; for tetramethylbutane, 0.0483 atm or 36.7 torr
11.43. The components of crude oil can be separated by fractional distillation, which uses differences in the volatility (boiling points) of the compounds.
11.45. C_5H_{12}
11.47. 60 torr
11.49. The vapor pressure of water in the compartment of pure water is greater than that in the seawater compartment, so water evaporates faster from the pure water compartment, but the rate of condensation of water vapor into the two compartments is the same. Over time this leads to the transfer of water from the pure water compartment to the seawater compartment, because seawater will always contain dissolved salts even after dilution and, therefore, evaporate more slowly than pure water.
11.51. Molarity is the moles of solute in one liter of solution. Molality is the moles of solute in one kilogram of solvent.
11.53. Because seawater contains a higher concentration of dissolved salt.
11.55. $\chi_{water} = 0.70$; $P_{soln} = 17$ torr
11.57. (a) 0.58 *m*; (b) 0.18 *m*; (c) 1.12 *m*
11.59. (a) 1.17 *m*; (b) 1.77 *m*; (c) 8.36 *m*; (d) 1.88 *m*
11.61. (a) 307 g; (b) 86.8 g; (c) 28.8 g
11.63. 6.5×10^{-5} *m* NH_3, 8.7×10^{-6} *m* NO_2^-, 2.195×10^{-2} *m* NO_3^-
11.65. 3.1°C
11.67. 2.52×10^{-2} *m*
11.69. −1.89°C
11.71. 0.5 *m* $CaCl_2$
11.73. 0.0400 *m* NH_4NO_3
11.75. (a) 0.06 *m* $FeCl_3$ < (b) 0.10 *m* $MgCl_2$ < (c) 0.20 *m* KCl
11.77. Molar mass = 194 g/mol; molecular formula = $C_8H_{10}N_4O_2$.
11.79. CO_2 reacts with water, forming H_2CO_3 which partially ionizes. This decreases the concentration of CO_2, allowing more of it to dissolve.
11.81. The first bubbles to form contain atmospheric gases (N_2, O_2, CO_2) that had dissolved in the water but became less soluble during heating.
11.83. 3.7×10^{-2} mol/(L · atm)
11.85. (a) 2.7×10^{-3} *M*; (b) 2.3×10^{-2} *M*
11.87. Yes
11.89. For 0.0935 *m* NH_4Cl, $i = 1.85$; for 0.0378 *m* $(NH_4)_2SO_4$, $i = 2.46$
11.91. Molar mass = 164 g/mol. The molecular formula of eugenol is $C_{10}H_{12}O_2$.
11.93. 4270 g/mol
11.95. (a) CO_2; (b) CO_2, unlike the other gases, reacts with water, forming H_3O^+ and HCO_3^- ions.

Chapter 12

12.1. (a) 3; (b) 6
12.3. Tire with blue gas on the right has greater pressure; tire with blue gas on the right has greater entropy.
12.5. No, the distribution of gases is not affected by gravity because the particles have sufficient energy to move throughout the container.
12.7. Spontaneous at low temperature
12.9. Condensation, freezing, and deposition
12.11. The sign is reversed.
12.13. Eight microstates; the most likely microstates have sums of +1 and −1.
12.15. 4.47×10^3
12.17. (b) < (a) < (c)
12.19. (d) $Cr(NO_3)_3$
12.21. (a) $S_8(g)$; (b) $S_2(g)$; (c) $O_3(g)$
12.23. Fullerenes
12.25. (a) $CH_4(g) < CF_4(g) < CCl_4(g)$; (b) $CH_2O(g) < CH_3CHO(g) < CH_3CH_2CHO(g)$; (c) $HF(g) < H_2O(g) < NH_3(g)$
12.27. ΔS_{sys} is positive; ΔS_{surr} is negative.
12.29. (a) and (b)
12.31. Decreases
12.33. ΔS_{surr} must be more positive than 66.0 J/K.
12.35. ΔS°_{rxn} is positive.
12.37. $\Delta S^\circ_{rxn} < 0$ because in precipitation reactions solid products have less entropy than the dispersed ions or molecules in the solution from which the precipitate forms.
12.39. (a) 24.9 J/K; (b) −146.4 J/K; (c) −73.2 J/K; (d) −175.8 J/K
12.41. 218.9 J/(mol · K)
12.43. When ΔG is positive, the reaction is nonspontaneous; when ΔG is negative, the reaction is spontaneous.
12.45. Because so many of them *are* spontaneous—their ΔH values either offset negative $T\Delta S$ values or reinforce positive ones.
12.47. ΔS is positive, ΔH is positive, ΔG is negative.
12.49. (a) and (d)
12.51. For NaBr, −18 kJ/mol; for NaI, −29 kJ/mol
12.53. $\Delta G^\circ_{rxn} = 91.4$ kJ
12.55. −35.3 kJ
12.57. −90.3 kJ
12.59. No, if ΔS_{rxn} were positive the exothermic reaction would be spontaneous at all temperatures.
12.61. 981.3 K
12.63. $\Delta H^\circ = 51.5$ kJ/mol; $\Delta S^\circ = 123.1$ J/mol · K; 418 K
12.65. (a) Spontaneous only at low temperature; (b) spontaneous only at low temperature; (c) spontaneous at all temperatures
12.67. (a) −29.2 kJ; (b) 29.5 kJ, not spontaneous
12.69. The spontaneous reaction must have a more negative or less positive free energy than the nonspontaneous reaction. Also, the two reactions must share an *intermediate*, which is produced in one of the reactions but consumed in the other.
12.71. ATP has the important function of energy storage and transfer. The cell must regenerate this important molecule during spontaneous reactions so that it is available to drive other reactions that are not spontaneous.
12.73. (a) $\Delta G^\circ_{rxn} = 142.2$ kJ; (b) $\Delta G^\circ_{rxn} = -28.6$ kJ; (c) $CH_4(g) + 2\,H_2O(g) \rightarrow CO_2(g) + 4\,H_2(g)$; (d) $\Delta G^\circ_{rxn} = 113.6$ kJ, nonspontaneous
12.75. $\Delta G^\circ_{rxn} = -51.5$ kJ
12.77. $\Delta S^\circ_{vap} = 83.5$ J/K · mol
12.79. $T = 618$ K
12.81. −630.4 kJ, spontaneous
12.83. $T_b \approx 294$ K or 21°C
12.85. 9.58 J/(mol · K)
12.87. 0.805 J/(mol · K)
12.89. There are more atoms in $CaCO_3$, so its S° is greater; 1099 K
12.91. (a) ΔH is negative, ΔS is positive; (b) No, because the reverse reaction would have positive ΔH and negative ΔS and, therefore, would never be spontaneous.

Chapter 13

13.1. $[N_2O]$ is represented by the green line and $[O_2]$ is represented by the red line.

13.3. (b)

13.5. (a) 1; (b) 5; (c) 2; (d) 4; (e) 3

13.7. (b)

13.9. (c)

13.11. (b)

13.13. Nitrogen (light blue)

13.15. Palladium (blue) and platinum (yellow-orange)

13.17. Ozone generation requires the presence of NO_2 in the atmosphere, which forms slowly from the NO in engine exhaust and mid-day sunlight.

13.19. Photodecomposition of NO_2 is key to restoring atmospheric NO concentrations. However, less radiation from the sun late in the day slows the decomposition process.

13.21. $\Delta H^\circ_{rxn} = -114.2$ kJ

13.23. (a) $2\ N_2(g) + O_2(g) \rightarrow 2\ N_2O(g)$;
(b) $2\ N_2(g) + 5\ O_2(g) \rightarrow 2\ N_2O_5(g)$

13.25. An average rate value describes the rate of a reaction over a defined time interval, whereas the instantaneous rate is the rate at a specific moment.

13.27. As the reaction proceeds, the concentrations of the reactants decrease. Because most reactions depend on collisions between reactant molecules, a decrease in reactant concentrations lowers the reaction rate.

13.29. (a) The rates are the same. (b) The rate of formation of NO_2^- and of H^+ is two-thirds the rate of consumption of O_2.
(c) The rate of consumption of NH_3 is two-thirds the rate of consumption of O_2.

13.31. (a) $2.85 \times 10^{-6}\ M \cdot s^{-1}$; (b) $-5.7 \times 10^{-6}\ M \cdot s^{-1}$

13.33. (a) Rate $= \dfrac{\Delta[CO_2]}{\Delta t} = -\dfrac{2}{3}\dfrac{\Delta[CO]}{\Delta t}$;
(b) Rate $= \dfrac{\Delta[COS]}{\Delta t} = -\dfrac{\Delta[SO_2]}{\Delta t}$;
(c) Rate $= \dfrac{\Delta[CO]}{\Delta t} = 3\dfrac{\Delta[SO_2]}{\Delta t}$

13.35. (a) $1.2 \times 10^7\ M$/s; (b) $2.9 \times 10^4\ M$/s

13.37. The average reaction rates based on $\Delta[O_3]/\Delta t$ are $1.4 \times 10^{-5}\ M/\mu s$ between 0 and 100 μs and $5.5 \times 10^{-6}\ M/\mu s$ between 200 and 300 μs

13.39. Plotting the concentration of ClO versus time:

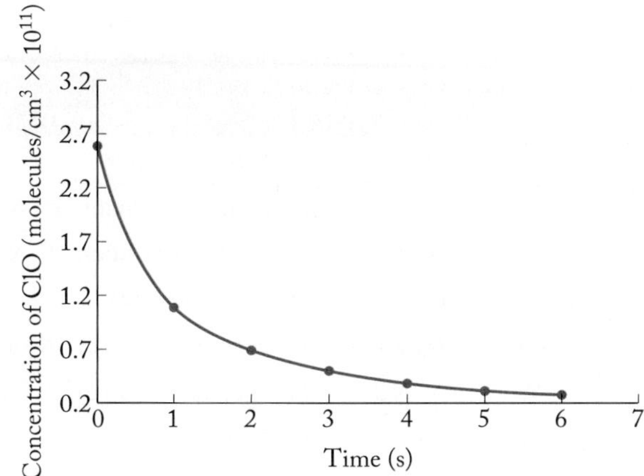

The instantaneous $\Delta[ClO]/\Delta t$ value at 1 s is -8.3×10^{10} molecules $cm^{-3}\ s^{-1}$. Assuming no Cl_2O_2 is present initially, the change in Cl_2O_2 concentration versus time is

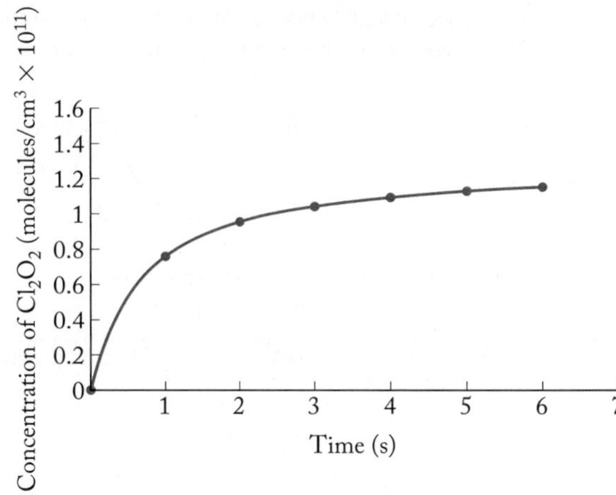

and the instantaneous $\Delta[Cl_2O_2]/\Delta t$ value at 1 s is 4.1×10^{10} molecules $cm^{-3}\ s^{-1}$.

13.41. No change

13.43. Yes

13.45. The half-life will be halved.

13.47. (a) First order in both A and B, and second order overall; (b) second order in A, first order in B, and third order overall; (c) first order in A, third order in B, and fourth order overall

13.49. (a) Rate $= k[O][NO_2]$; k units $= M^{-1}s^{-1}$;
(b) Rate $= k[NO]^2[Cl_2]$; k units $= M^{-2}s^{-1}$;
(c) Rate $= k[CHCl_3][Cl_2]^{1/2}$; k units $= M^{-1/2}s^{-1}$;
(d) Rate $= k[O_3]^2[O]^{-1}$; k units $= s^{-1}$

13.51. (a) Rate $= k[BrO]$; (b) Rate $= k[BrO]^2$; (c) Rate $= k[BrO]$;
(d) Rate $= k[BrO]^0 = k$

13.53. We need to determine the change in the rate when only [NO] or [ClO] is changed.

13.55. (a) Rate $= k[NO_2][O_3]$; (b) $4.9 \times 10^{-11}\ M$/s;
(c) $4.9 \times 10^{-11}\ M$/s; (d) the rate doubles.

13.57. (c)

13.59. Rate $= k[NO][NO_2]$

13.61. Rate $= k[ClO_2][OH^-]$; $k = 14\ M^{-1}s^{-1}$

13.63. Rate $= k[NO]^2[H_2]$; $k = 6.32\ M^{-2}s^{-1}$

13.65. $0.32\ \mu M^{-1}min^{-1}$

13.67. Rate $= k[NH_3]$, $k = 0.0030\ s^{-1}$

13.69. 0.051 M, 90%

13.71. (a) Rate $= k[N_2O]$; (b) 4

13.73. (a) Rate $= k\ N_{32_P}$ where N represents number of atoms;
(b) 0.0485 day^{-1}; (c) 14.3 days; (d) 95 days

13.75. (a) $k = 5.40 \times 10^{-12}\ cm^3\ molecules^{-1}\ s^{-1}$; (b) $t_{1/2} = 0.712$ s

13.77. Rate $= k[C_{12}H_{22}O_{11}][H_2O] = k'[C_{12}H_{22}O_{11}]$,
$k' = 5.19 \times 10^{-5}\ s^{-1}$

13.79. The larger the activation energy, the slower the reaction.

13.81. When the energy of the products is lower than the energy of the reactants

13.83. An increase in temperature increases the frequency and the kinetic energy at which the reactants collide, which speeds up the reaction. The order of the reaction is unaffected.

13.85. The reaction with the larger activation energy (150 kJ/mol)

13.87. $E_a = 17.1$ kJ/mol, $A = 1.002\ cm^3/(molecule \cdot s)$

13.89. (a) $E_a = 314$ kJ/mol; (b) $A = 5.03 \times 10^{10}$;
(c) $k = 1.06 \times 10^{-44}\ M^{-1/2}s^{-1}$

13.91. (a) $E_a = 39.1$ kJ/mol, $A = 1.27 \times 10^{12}\ M^{-1}\ s^{-1}$;
(b) $k = 5.85 \times 10^3\ M^{-1}s^{-1}$

13.93. No

13.95. No, because they have different rate laws

13.97. Pseudo-first-order kinetics occurs when one of the reactants is in sufficiently high concentration that its concentration does not change appreciably over the course of the reaction.

13.99.

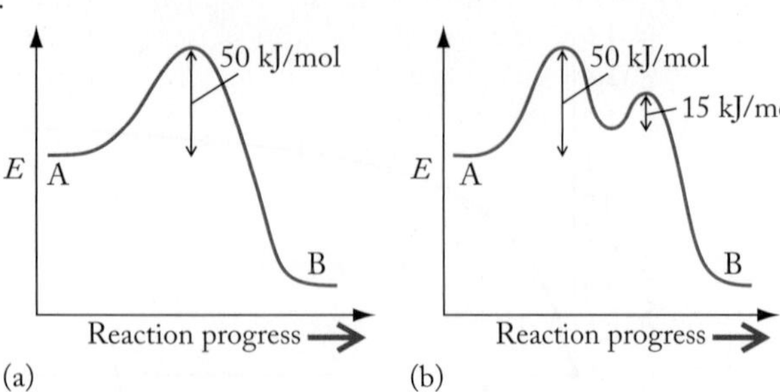

50 kJ/mol

E A

B

Reaction progress

(c)

13.101. (a) Rate = $k[SO_2Cl_2]$, unimolecular; (b) Rate = $k[NO_2][CO]$, bimolecular; (c) Rate = $k[NO_2]^2$, bimolecular

13.103. (a) $H_2O_2(aq) \rightarrow 2\ H_2O(aq) + O_2(g)$; (b) Rate = $k[H_2O_2][I^-]$; (c) I^-; (d) IO^-

13.105. The second step

13.107. The first step

13.109. Photochemical decomposition: a; thermal decomposition: b or c

13.111. Yes

13.113. Yes

13.115. The concentration of a homogeneous catalyst may not appear in the rate law if the catalyst is not involved in the rate-limiting step. However, if the catalyst is involved in the slowest step of the mechanism, it can appear in the rate law.

13.117. NO is the catalyst.

13.119. The reaction of O_3 with Cl has the larger rate constant.

13.121. When the concentration of a reactant (O_2 for the combustion reaction) increases, the rate of combustion increases.

13.123. The bodily reactions that use O_2 are slower at colder temperatures.

13.125. Yes, we could use other times, not just $t = 0$, as long as the rate of the reverse reaction is still much slower than the forward reaction.

13.127. In this plot $1/[X] - 1/[X]_0$ divided by $t - t_0$ is the slope of the line that corresponds to k, the reaction rate constant.

13.129. For an elementary step to take place, some involvement from a molecular or atomic species must occur. Therefore, there can be *no* independence (or zero order) of the reactants in an elementary step of a reaction mechanism.

13.131. The rate of consumption of O_3 is the same as the rate of formation of N_2O_5 and O_2 and one-half the rate of consumption of NO_2.

13.133. $k = 3.6 \times 10^{-4}\ s^{-1}$, Rate = $(3.6 \times 10^{-4}\ s^{-1})[N_2O_5]$

13.135. (a) Yes; (b) $E_a = 62.5$ kJ/mol; (c) Rate = $1.2 \times 10^{-12}\ M/s$; (d) at 10°C (283 K), $k = 21\ M^{-1}s^{-1}$; at 35°C (308 K), $k = 1.8 \times 10^2 \times M^{-1}s^{-1}$

13.137. (a) Rate = $k[Na(H_2O)_6^+]$; (b) neither

13.139. (a) Rate = $k[NO][ONOO^-]$, $k = 1.30 \times 10^{-3}\ M^{-1}s^{-1}$; (b)

$[\ddot{O}^{0}{=}\ddot{N}^{0}{-}\ddot{O}^{0}{-}\ddot{O}^{-1}]^- \longleftrightarrow [\ddot{O}^{-1}{-}\ddot{N}^{0}{=}\ddot{O}^{+1}{-}\ddot{O}^{-1}]^- \longleftrightarrow$

(preferred)

$[\ddot{O}^{-1}{-}\ddot{N}^{-1}{-}\ddot{O}^{+1}{=}\ddot{O}^{0}]^-$;

(c) −55 kJ

13.141. (a) Second order; (b) no

13.143. (a) Rate = $k[NH_2][NO]$; (b) $1.2 \times 10^9\ M^{-1}\ s^{-1}$

Chapter 14

14.1. Reaction C $\rightleftharpoons$ D has the larger k_f, the smaller k_r, and the larger K_c.

14.3. (a) A $\rightleftharpoons$ B; (b) 2.0

14.5. The reaction is endothermic. As temperature increases, K increases, indicating that more products form at higher temperatures.

14.7. No, because at 20 μs the concentrations of A and B are still changing.

14.9. (a)

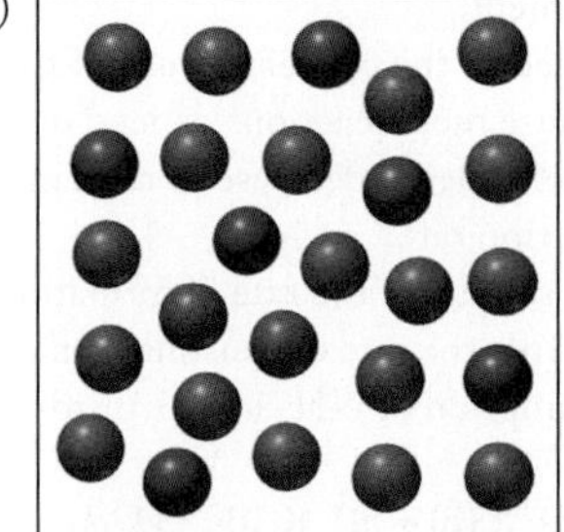

(b)

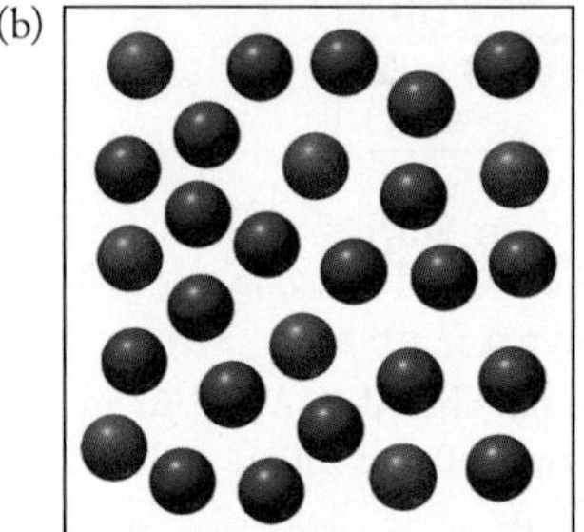

14.11. No

14.13. Greater than 1

14.15.

Molar Mass	Compound	How Present
28	$^{14}N_2$	Originally present
29	$^{15}N^{14}N$	From decomposition of $^{15}N^{14}NO$
30	$^{15}N_2$	From decomposition of $^{15}N_2O$
32	O_2	Originally present
44	$^{14}N_2O$	From combination of $^{14}N_2$ and O_2
45	$^{15}N^{14}NO$	From combination of $^{15}N^{14}N$ and O_2
46	$^{15}N_2O$	Originally present

14.17. 0.333

14.19. When the number of moles of gaseous products equals the number of moles of gaseous reactants and $\Delta n = 0$

14.21. (a) $K_c = \frac{[N_2O_4]}{[N_2][O_2]^2}$ and $K_p = \frac{(P_{N_2O_4})}{(P_{N_2})(P_{O_2})^2}$

(b) $K_c = \frac{[NO_2][N_2O]}{[NO]^3}$ and $K_p = \frac{(P_{NO_2})(P_{N_2O})}{(P_{NO})^3}$

(c) $K_c = \frac{[N_2]^2[O_2]}{[N_2O]^2}$ and $K_p = \frac{(P_{N_2})^2(P_{O_2})}{(P_{N_2O})^2}$

14.23. 0.50
14.25. 0.068
14.27. 0.50
14.29. 1.5
14.31. 780
14.33. 0.0583
14.35. (b) and (c)
14.37. 0.10
14.39. When scaling the coefficients of a reaction up or down the new value of the equilibrium constant is the first K raised to the power of the scaling constant.
14.41. 11.0

14.43. $K_{c,forward} = \frac{[NO_2]^2}{[NO][NO_3]}$; $K_{c,reverse} = \frac{[NO][NO_3]}{[NO_2]^2}$;

$K_{c,reverse} = \frac{1}{K_{c,forward}}$

14.45. $K_c = \frac{[SO_3]}{[SO_2][O_2]^{1/2}}$; $K_c' = \frac{[SO_3]^2}{[SO_2]^2[O_2]}$; $K_c' = (K_c)^2$

14.47. (a) 0.049; (b) 420; (c) 20
14.49. 7.4
14.51. The reaction quotient of a reaction mixture that may, or may not, have achieved chemical equilibrium, is the ratio of the concentrations (or partial pressures) of the products raised to their stoichiometric coefficients to the concentrations (or partial pressures of reactants raised to their stoichiometric coefficients.
14.53. The system is at equilibrium.
14.55. No, $Q < K$ so the reaction proceeds to the right to reach equilibrium.
14.57. Mixture a is at equilibrium.
14.59. $Q > K$, so the reaction will proceed to the left.
14.61. (a)
14.63. $K_c = [Cu^{2+}][S^{2-}]$
14.65. The quantities of the solids ($CaCO_3$ and CaO) in the reaction mixture may change, but their influence on the composition of the mixture at equilibrium does not – as long as both are present.
14.67. No
14.69. As the concentration of O_2 increases, the reaction shifts to the right and the CO on the hemoglobin is displaced.
14.71. According to Le Châtelier's principle, an increase in the partial pressure (or concentration) of O_2 above the water shifts the equilibrium to the right so that more oxygen becomes dissolved in the water. This is consistent with Henry's law.
14.73. (b) and (d)
14.75. (a) Increasing the concentration of the reactant O_3 shifts the equilibrium to the right, increasing the concentration of the product O_2. (b) Increasing the concentration of the product O_2 shifts the equilibrium to the left, increasing the concentration of the reactant O_3. (c) Decreasing the volume of the reaction to one-tenth its original volume shifts the equilibrium to the left, increasing the concentration of the reactant O_3.
14.77. The equilibrium shifts to the left.
14.79. (a)
14.81. When K is very small, the proportion of reactants that are transformed into products maybe so small that the "−x" values in their equilibrium concentration terms can be ignored. When there are no products (Z) present, the reaction must proceed to the right – at least a little – to achieve equilibrium.
14.83. (a) $P_{PCl_5} = 0.024$ atm, $P_{PCl_3} = 1.036$ atm, $P_{Cl_2} = 0.536$ atm; (b) The partial pressure of PCl_3 decreases and the partial pressure of PCl_5 increases.
14.85. $[H_2O] = [Cl_2O] = 3.76 \times 10^{-3}$ M; $[HOCl] = 1.13 \times 10^{-3}$ M
14.87. 9×10^5
14.89. $P_{CO} = 2.4$ atm, $P_{CO_2} = 3.8$ atm
14.91. (a) $P_{NO} = 0.272$ atm; $P_{NO_2} = 7.98 \times 10^{-3}$ atm; (b) $P_T = 0.416$ atm
14.93. $P_{O_2} = 0.17$ atm, $P_{N_2} = 0.75$ atm, $P_{NO} = 0.080$ atm
14.95. 5.75 M
14.97. $P_{CO} = P_{Cl_2} = 0.258$ atm, $P_{COCl_2} = 0.00680$ atm
14.99. $[CO] = [H_2O] = 0.031$ M, $[CO_2] = [H_2] = 0.069$ M
14.101. $[NO] = 0.027$ M, $[NO_2] = 0.277$ M, $[N_2O_3] = 0.098$ M
14.103. Yes
14.105. To the right
14.107. (c)
14.109. (a) −197.8 kJ; (b) 7.4×10^{24}; (c) −142 kJ/mol from K_P, −142.0 kJ/mol from Appendix 4
14.111. 2.96×10^{-20}
14.113. Exothermic
14.115. Exothermic
14.117. 1.3×10^{-31}
14.119. −115 kJ/mol
14.121. The reaction is endothermic: at 1500 K, $K_p = 5.5 \times 10^{-11}$; at 2500 K, $K_p = 4.0 \times 10^{-3}$; at 3000 K, $K_p = 0.40$. This reaction does not favor products even at very high temperature, so this is not a viable source of CO and is not a remedy to decrease CO_2 as a contributor to global warming. Also, the process produces poisonous CO gas.
14.123. 9×10^{-22} M
14.125. $K_{p,25°C} = 2.47 \times 10^{-70}$; $K_{p,500°C} = 1.96 \times 10^{12}$
14.127. $P_{SO_2} = 9.2 \times 10^{-74}$ atm

Chapter 15

15.1. Red line
15.3. The bar with the lowest percent ionization
15.5. Piperidine is the stronger base because the presence of oxygen (high electronegativity) on the morpholine withdraws electrons making the N atom less basic.
15.7. (a) basic; (b) the amine functional group
15.9. The ionizable protons are those bonded to the oxygen atoms.
15.11. HF is the Brønsted–Lowry acid; H_2O is the Brønsted–Lowry base.
15.13. H_2O is the Brønsted–Lowry acid; NH_3 is the Brønsted–Lowry base.
15.15. (a) HCl is the acid; NaOH is the base. (b) HCl is the acid; $MgCO_3$ is the base. (c) H_2SO_4 is the acid; NH_3 is the base.
15.17. NO_2^-; OCl^-; $H_2PO_4^-$; NH_2^-
15.19. H_2SO_4 is the conjugate acid; SO_4^{2-} is the conjugate base.
15.21. 0.65 M
15.23. 0.0410 M
15.25. Sulfur is more electronegative than selenium. The higher electronegativity of the sulfur atom stabilizes the anion HSO_4^- more than the anion $HSeO_4^-$.

15.27. (a) H_2SO_3; (b) H_2SeO_4

15.29. Because the pH function is a $-\log$ function, as $[H^+]$ increases, the value of $-\log[H^+]$ decreases.

15.31. Any strong acid at a concentration greater than 1.0 *M*

15.33.

```
    H   H                 H   H   H
    |   |                 |   |   |+
H — C — C — Ö:        H — C — C — Ö — H
    |   |                 |   |
    H   H                 H   H
```

15.35. (a) pH = 2.28, pOH = 11.72, acidic; (b) pH = 8.42, pOH = 5.58, basic; (c) pH = 5.14, pOH = 8.86, acidic; (d) pH = 0.00, pOH = 14.00, acidic

15.37. (a) 1.2×10^{-11} *M*; (b) 7.6×10^{-11} *M*; (c) 2.2×10^{-12} *M*; (d) 3.4×10^{-10} *M*

15.39. (a) pH = 0.810, pOH = 13.190; (b) pH = 2.301, pOH = 11.699; (c) pH = 2.38, pOH = 11.62; (d) pH = 1.81, pOH = 12.19

15.41. 7.003

15.43. (a) In order of decreasing $[H_3O]^+$: $HCl > HNO_2 > CH_3COOH > HClO$. (b) In order of increasing acid strength: $HClO < CH_3COOH < HNO_2 < HCl$.

15.45. HNO_3 ionizes completely in a 1.0 *M* aqueous solution. However, 1.0 *M* HNO_2 only partially ionizes. The greater number of ions in the HNO_3 solution makes it a better conductor of electricity.

15.47. $K_a = \dfrac{[H_3O^+][F^-]}{[HF]}$

15.49. (a) Water; (b) water

15.51. H_2O is the Brønsted–Lowry acid; CH_3NH_2 is the Brønsted–Lowry base.

15.53. 8.91×10^{-4}

15.55. 3.26%; $K_a = 1.37 \times 10^{-4}$

15.57. 2.52

15.59. 2.3 times

15.61. (a) Weaker; (b) 10.86; (c) 1.000×10^{-4} *M*

15.63. (a) 9.73; (b) 9.30

15.65. With each successive ionization, it becomes more difficult to remove H^+ from a species that is more negatively charged.

15.67. Ge is less electronegative than C.

15.69. 0.12

15.71. 2.80

15.73. 9.50

15.75. 10.27

15.77. F^- reacts with water to produce the weak acid HF and OH^- making the solution basic, but Cl^- does not react with water, and NaCl solutions are neutral.

15.79. Ammonium nitrate

15.81. The citric acid in the lemon juice neutralizes the volatile trimethylamine to make a nonvolatile dissolved salt.

15.83. 3.32

15.85. 7.35

15.87. (a) CH_3COOH and $HClO_4$; (b) $Ca(OH)_2$; (c) CH_3COOH and $HClO_4$; (d) $Ca(OH)_2$ and CH_3NH_2

15.89. Yes, all Arrhenius bases are Brønsted–Lowry bases because Arrhenius bases are all H^+ ion acceptors. However, the Arrhenius definition of bases refers to their behavior in aqueous solution, but Brønsted–Lowry bases can accept H^+ ions in the gas phase or in non-aqueous media.

15.91. Combustion of sulfur embedded in fossil fuels produces sulfur dioxide: $S(s) + O_2(g) \rightarrow SO_2(g)$;
The sulfur in SO_2 is further oxidized in the atmosphere in a series of reactions that can be summarized as follows:
$2\,SO_2(g) + O_2(g) \rightarrow 2\,SO_3(g)$
Sulfur trioxide combines with water vapor, forming sulfuric acid:
$SO_3(g) + H_2O(g) \rightarrow H_2SO_4(\ell)$, which dissolves in rain and falls to Earth as dilute sulfuric acid: $H_2SO_4(aq)$. Should this acid contact material made of marble, the reaction: $CaCO_3(s) + H_2SO_4(\ell) \rightarrow H_2O(\ell) + CaSO_4(aq) + CO_2(g)$. Slightly soluble calcium sulfate is washed away by the rain and the marble structures slowly dissolve.

15.93. 1.4×10^9 L

15.95. (a) Basic because Prozac has an amine group that will pick up a proton from water to form ammonium cations and OH^-; (b) HCl protonates the N atom of the amine group forming the hydrochloride salt; (c) because it is charged and water molecules form stronger ion–dipole forces around the molecule than the dipole–induced dipole forces between the neutral molecule and water

15.97. (a)

```
[          F          ]⁻
           |
           C:
        ⁄     ⁀
 F — C           C — F
     ‖           ‖
      C    —    C
     ⁄           ⁀
    F             F    ]
```
;

(b) because the presence of five very electronegative F atoms on the carbon ring stabilizes the anion formed when the proton is lost

15.99. (a) No; (b) acid = H_2SO_4, base = HNO_3, conjugate acid = H_3O^+, conjugate base = HSO_4^-

15.101. Most acidic: ii. 0.1 *M* H_2SO_4, most basic: v. 0.30 *M* Na_2SO_3

15.103. (a) 1.00×10^{-3} *M* NaOH; (b) 0.345 m*M* HBrO; (c) 45 m*M* $Ba(OH)_2$; (d) 1.6 m*M* C_6H_7N; (e) 252 m*M* KOH; (f) 105 m*M* CH_3NH_2; (g) 1.50×10^{-5} *M* pyridine; (h) 20 m*M* $C_7H_5O_2Cl$

15.105. 8.97

15.107. (a) $ClCH_2COO^-$, $K_b = 7.14 \times 10^{-12}$; (b) NH_3, $K_b = 1.76 \times 10^{-5}$; (c) CN^-, $K_b = 1.61 \times 10^{-5}$; (d) $CH_3CH_2O^-$, $K_b = 76.9$; $CH_3CH_2O^-$ is the strongest conjugate base

15.109. (a) $H^+(aq) + OH^-(aq) \rightarrow H_2O(\ell)$; acid = HNO_3, base = $Ca(OH)_2$; (b) $CO_3^{2-}(aq) + 2\,H^+(aq) \rightarrow CO_2(g) + H_2O(\ell)$; acid = H_2SO_4, base = Na_2SO_3; (c) $CH_3NH_2(aq) + H^+(aq) \rightarrow CH_3NH_3^+(aq)$; acid = HBr, base = CH_3NH_2; (d) $CH_3COOH(aq) + OH^-(aq) \rightarrow CH_3COO^-(aq) + H_2O(\ell)$; acid = CH_3COOH, base = $Mg(OH)_2$; (e) $CaO(s) + H_2O(\ell) \rightarrow Ca^{2+}(aq) + 2\,OH^-(aq)$; acid = H_2O, base = CaO; (f) $LiH(aq) + H_2O(\ell) \rightarrow Li^+(aq) + OH^-(aq) + H_2(g)$; acid = H_2O, base = LiH; (g) $H^+(aq) + OH^-(aq) \rightarrow H_2O(\ell)$; acid = H_2SO_4, base = $Ba(OH)_2$; (h) $SH^-(aq) + H^+(aq) \rightarrow H_2S(g)$; acid = HNO_3, base = NaSH

Chapter 16

16.1. The blue titration curve represents the titration of a 1 *M* solution of strong acid. The red titration curve represents the titration of a 1 *M* solution of weak acid.

16.3. The indicator with a pK_a of 9.0

16.5. The red titration curve represents the titration of Na_2CO_3; the blue titration curve represents the titration of $NaHCO_3$.

16.7. Ammonium chloride is dissolved in the yellow solution, sodium acetate is dissolved in the blue solution, and ammonium acetate is dissolved in the middle solution.

16.9. (a) Buffer; (b) the weak acid molecules of HCOOH will react with NaOH to produce more $HCOO^-$ and H_2O.
16.11. The presence of both the acid and its conjugate base means that the system can neutralize small additions of either acid or base with only slight changes in pH. The weak acid alone has little capacity to control pH against additions of acid.
16.13. Citric acid/citrate, hydrofluoric acid/fluoride, or iodoacetic acid/iodoacetate
16.15. Buffer capacity is the quantity of acid or base that a buffer can neutralize while keeping its pH within a desired range
16.17. Acetic acid/sodium acetate > formic acid/sodium formate > hydrofluoric acid/sodium fluoride
16.19. The should be no significant change in pH
16.21. 2.55
16.23. 6.88
16.25. 10:1
16.27. 9.26 g bromoacetic acid, 21.5 g sodium bromoacetate
16.29. 4.53 g dimethylammonium chloride, 4.26 g dimethylamine
16.31. 9.25
16.33. 53.66 mL
16.35. (a) 3.50; (b) 3.42
16.37. Yes, because at the equivalence point only ions that do not react with water are formed; the pH will always be neutral
16.39. (1) It changes color near the pH of the first equivalence point. (2) It is colorless at low pH, which means it will not obscure the color change of a second indicator added after the first equivalence point to detect the second equivalence point.
16.41. 4.44
16.43. After 10.0 mL of OH^- has been added, pH = 4.754; after 20.0 mL of OH^- has been added, pH = 8.750; after 30.0 mL of OH^- has been added, pH = 12.356.
16.45. pH = 2.92 after 2.50 mL; 3.40 after 5.00 mL; 3.88 after 7.50 mL, and 8.31 after 10.00 mL

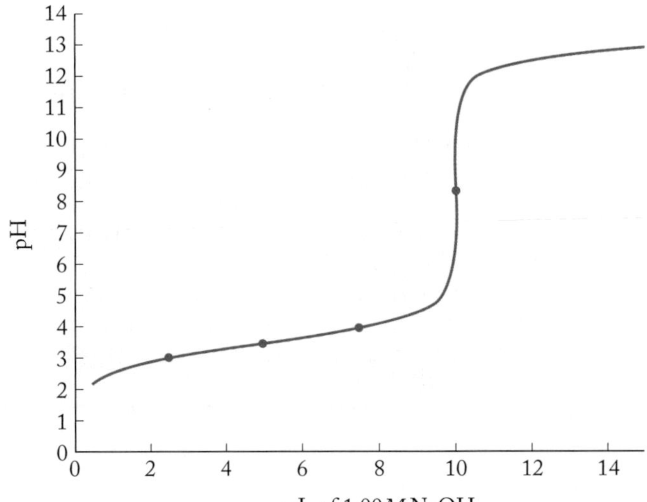

16.47. 25.0 mL
16.49. (a) 2.949 *M*; (b) no effect
16.51. (a) pH < 7; (b) pH > 7; (c) pH = 7
16.53. Two equivalence points: bromthymol blue for the first one and phenolphtalein for the second
16.55. No, because a substance can be a Lewis base if it can donate an electron pair but does not accept a proton; Cl^- is an example
16.57. Yes, because Brønsted–Lowry acids must be able to accept pairs of electrons so that the protons they donate can form bonds to Brønsted–Lowry bases
16.59. H—Ö:(—H) + H—Ö:(—H) → $[H—\ddot{O}—H(—H)]^+$ + $:\ddot{O}:(—H)^-$
16.61. H—Ö:(—H) + S(=Ö:)=O → H—Ö—S(=Ö:)—Ö:—H
16.63. $B(OH)_3$ + H_2O → $[B(OH)_4]^-$ + H^+
16.65. Water molecules
16.67. Because Ag^+ ions form a complex with ammonia, $Ag(NH_3)_2^+$, which is soluble
16.69. 1.25×10^{-10} *M*
16.71. 7.9×10^{-16} *M*
16.73. (b) and (d)
16.75. The solution will be more acidic.
16.77. In basic solution: $Cr(OH)_3(s) + OH^-(aq) \rightleftharpoons Cr(OH)_4^-(aq)$
In acidic solution: $Cr(OH)_3(s) + 3\ H^+(aq) \rightleftharpoons Cr^{3+}(aq) + 3\ H_2O(\ell)$
16.79. $Al(OH)_3$ reacts with OH^- in solution to form soluble $Al(OH)_4^-$. The other ions do not form this type of soluble complex ion.
16.81. 2.80
16.83. 1.80
16.85.

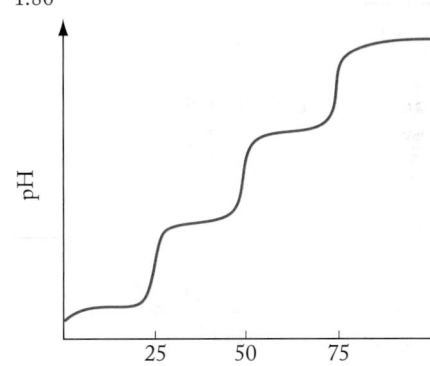

pH
25 50 75
Volume of 0.50 *M* NaOH (mL)

16.87. Molar solubility is the quantity (moles) of substance that dissolves in a liter of solution. The solubility product is an equilibrium constant whose value is based on the product of the concentrations of the ions released by an ionic compound in a saturated solution of the compound.
16.89. Sr^{2+}
16.91. Endothermic
16.93. Acidic substances react with the OH^- released on dissolution of hydroxyapatite. The equilibrium is shifted to the right, dissolving more hydroxyapatite.
16.95. 1.08×10^{-10}
16.97. $[Cu^+] = [Cl^-] = 4.1 \times 10^{-4}$ *M*
16.99. 1.00×10^{-5} g/mL
16.101. 10.15
16.103. (d)

16.105. Probably, because the product of the $[Sr^{2+}]$ and $[SO_4^{2-}]$ values exceeds the K_{sp} of $SrSO_4$.
16.107. Yes
16.109. (a) SO_4^{2-}; (b) 3.8×10^{-7} *M*
16.111. Subsequent additions of HCO_3^- react with water to form bicarbonate's conjugate acid (H_2CO_3) and its conjugate base (CO_3^{2-}) in the same proportions as the first addition, so pH does not change.
16.113. 6.6×10^{-16}
16.115. a. The acidity of the oceans increases (lower pH) because of increased ionization of the weak acid carbonic acid.

$$CO_2(g) \rightleftharpoons CO_2(aq)$$
$$CO_2(aq) + H_2O(\ell) \rightleftharpoons H_2CO_3(aq)$$
$$H_2CO_3(aq) \rightleftharpoons H^+(aq) + HCO_3^-(aq)$$

b. The acidity of the ocean will be about six times greater.
c. Because oyster shells are composed of $CaCO_3$, which may dissolve or not form in the first place as carbonic acid concentrations increase: $CaCO_3(s) + H_2CO_3(aq) \rightarrow Ca(HCO_3)_2(aq)$.

16.117. e. (50:50 mixture of carbonate and bicarbonate)
16.119. (a) Acetic acid; (b) hypobromous acid
16.121. (a) 1.6×10^{-4}; (b) 5×10^{-5}; (c) 1.3×10^{-11}
16.123. Formic acid, pH = 8.22; boric acid, pH = 9.94. Yes, phenolphthalein should work for both.
16.125. (a) 15.00 mL; (b) 1.77; (c) 7.21; (d) Na^+ and HSO_3^-; (e) Na^+ and SO_3^{2-}
16.127. (a) 4.68; (b) bromocresol green; (c) at 0.0 mL pH = 9.40; at 1.0 mL pH = 6.26; at 3.0 mL pH = 5.69; at 7.0 mL pH = 5.05; at 9.0 mL pH = 4.69; at 10.0 mL pH = 4.43; at 10.5 mL pH = 4.25; at 11.0 mL, pH = 3.96; at 11.5 mL pH = 3.16; at 11.8 mL pH = 2.43; at 12.5 mL pH = 1.79; at 15 mL pH = 1.26; at 20 mL pH = 0.93
(d)

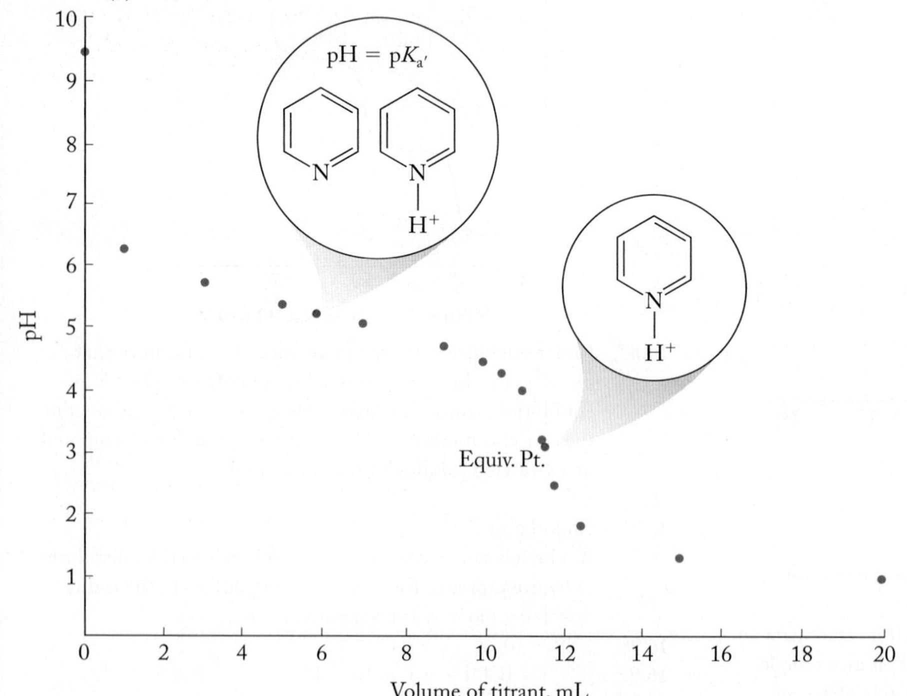

16.129. They have the same molar solubility but not the same solubility in g/mL.
16.131. Calcium hydroxide
16.133. For trypsin, carbonic acid/sodium bicarbonate; for pepsin, oxalic acid/sodium oxalate
16.135. Lowering CO_2 will lower the amount of CO_3^{2-} in the blood, which will lower the $CaCO_3$ in shells.

Chapter 17

17.1. Through careful layering, two stacked half-cells are created each consisting of a metal electrode in contact with a solution of its cations. Direct contact between the layered solutions obviates the need for a porous bridge to allow ions to migrate between them.
17.3. Ag is the cathode; Pt in the SHE is the anode; electrons flow from the SHE to the Ag electrode.
17.5. Blue line
17.7. (a) Left: $2\,H^+(aq) + 2\,e^- \rightarrow H_2(g)$ $E^\circ_{cathode} = 0.000$ V
Right: $2\,H_2O(\ell) \rightarrow O_2(g) + 4\,H^+(aq) + 4\,e^-$ $E^\circ_{anode} = 1.229$ V
(b) The ions it forms makes the solution more conductive
17.9. (a) anode: $V^{2+}(aq) \rightarrow V^{3+}(aq) + e^-$;
cathode: $VO_2^+(aq) + 2\,H^+(aq) + e^- \rightarrow VO^{2+}(aq) + H_2O(\ell)$;
(b) 1 electron is transferred as:
$VO_2^+(aq) + V^{2+}(aq) + 2\,H^+(aq) \rightarrow$
$VO^{2+}(aq) + V^{3+}(aq) + H_2O(\ell)$

(c)

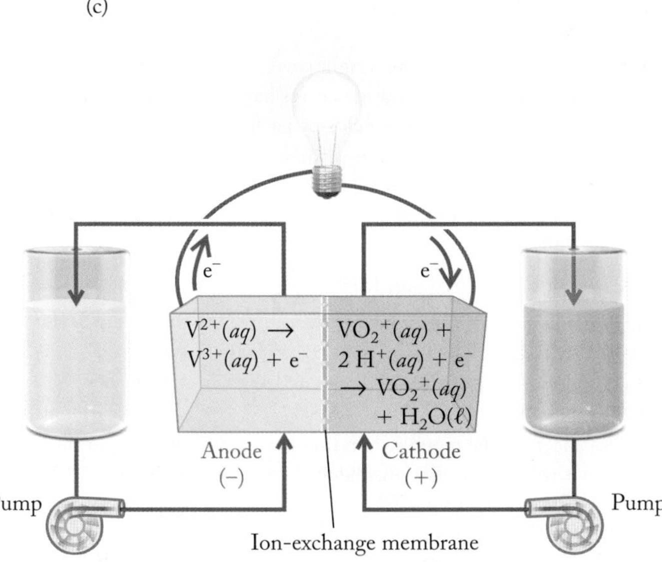

(d) As pH increases, E_{cell} decreases.

17.11. A half reaction is the oxidation or the reduction portion of a redox reaction

17.13. Ions migrate between the compartments in an electrochemical cell, not free electrons such as those that conduct electricity in wires.

17.15. (a) $Br_2(\ell) + 2\,e^- \rightarrow 2\,Br^-(aq)$, reduction;
(b) $Pb(s) + 2\,Cl^-(aq) \rightarrow PbCl_2(s) + 2\,e^-$, oxidation;
(c) $O_3(g) + 2\,H^+(aq) + 2\,e^- \rightarrow O_2(g) + H_2O(\ell)$, reduction;
(d) $H_2S(g) \rightarrow S(s) + 2\,H^+(aq) + 2\,e^-$, oxidation

17.17. $2\,Fe_3O_4(s) + H_2O(\ell) \rightarrow 3\,Fe_2O_3(s) + 2\,H^+(aq) + 2\,e^-$

17.19. (a) anode: $2\,Li(s) + S^{2-}(non\text{-}aq) \rightarrow Li_2S(s) + 2\,e^-$;
cathode: $FeS_2 + 4\,e^- \rightarrow Fe + 2\,S^{2-}(non\text{-}aq)$
(b) $4\,Li(s) + FeS_2(s) \rightarrow 2\,Li_2S(s) + Fe(s)$;
(c) 4

17.21. (a) Anode: $Zn(s) + 2OH^-(aq) \rightarrow Zn(OH)_2(s) + 2\,e^-$
Cathode: $NiO(OH)(s) + H_2O(\ell) + e^- \rightarrow Ni(OH)_2(s) + OH^-(aq)$
(b) $Zn(s) + 2\,NiO(OH)(s) + 2\,H_2O(\ell) \rightarrow Zn(OH)_2(s) + 2\,Ni(OH)_2(s)$
(c) $Zn(s) \mid Zn(OH)_2(aq) \parallel NiO(OH)(s) \mid Ni(OH)_2(s)$

17.23. The standard electrode potentials in Equation 17.1 are based on half reactions written as reductions:

$$E^\circ_{cell} = E^\circ_{reduction(cathode)} - E^\circ_{reduction(anode)}$$

However, we know that oxidation occurs at the anode, which means its reduction half-reaction actually runs in reverse. Reversing a half-reaction changes the sign of its standard potential. Therefore, the standard potential of the anode half reaction written as an oxidation is:

$$E^\circ_{oxidation(anode)} = -E^\circ_{reduction(anode)}$$

Substituting the above equality into Equation 17.1:

$$E^\circ_{cell} = E^\circ_{reduction(cathode)} + E^\circ_{oxidation(anode)}$$

17.25. (a) $Hg^{2+}(aq) + 2\,e^- \rightarrow Hg(\ell)$ with $Co(s) \rightarrow Co^{2+}(aq) + 2e^-$;
$E^\circ_{cell} = 1.128$ V;
(b) $Hg^{2+}(aq) + 2\,e^- \rightarrow Hg(\ell)$ with $Cu(s) \rightarrow Cu^{2+}(aq) + 2\,e^-$;
$E^\circ_{cell} = 0.509$ V

17.27. No

17.29. Less than 1.10 V

17.31. (a)
Anode: $Zn(s) \rightarrow Zn^{2+}(aq) + 2\,e^-$ $E^\circ_{anode} = -0.7618$ V
Cathode: $Hg^{2+}(aq) + 2\,e^- \rightarrow Hg(\ell)$ $E^\circ_{cathode} = 0.851$ V
$Zn(s) + Hg^{2+}(aq) \rightarrow Zn^{2}(aq) + Hg(\ell)$
$E^\circ_{cell} = E^\circ_{cathode} - E^\circ_{anode} = 1.613$ V

(b)
Anode: $Zn(s) + 2\,OH^-(aq) \rightarrow ZnO(s) + H_2O(\ell) + 2\,e^-$
$E^\circ_{anode} = -1.25$ V
Cathode: $Ag_2O(s) + H_2O(\ell) + 2\,e^- \rightarrow 2\,Ag(s) + 2\,OH^-(aq)$
$E^\circ_{cathode} = 0.342$ V
$Zn(s) + Ag_2(O)S \rightarrow ZnO(s) + 2\,Ag(s)$
$E^\circ_{cell} = E^\circ_{cathode} - E^\circ_{anode} = 1.59$ V

(c)
Anode: $2 \times [Ni(s) + 2\,OH^-(aq) \rightarrow Ni(OH)_2(s) + 2\,e^-]$
$E^\circ_{anode} = -0.72$ V
Cathode: $O_2(g) + 2\,H_2O(\ell) + 4\,e^- \rightarrow 4\,OH^-(aq)$
$E^\circ_{cathode} = 0.401$ V
$2\,Ni(s) + O_2(g) + 2\,H_2O(\ell) \rightarrow 2\,Ni(OH)_2(s)$
$E^\circ_{cell} = E^\circ_{cathode} - E^\circ_{anode} = 1.12$ V

17.33. Because the voltaic cell does work *on* its surroundings, which is defined as negative work

17.35. (a) $\Delta G^\circ = -34.5$ kJ; $\Delta E^\circ_{cell} = 0.358$ V
(b) $\Delta G^\circ = 2.9$ kJ; $\Delta E^\circ_{cell} = -0.030$ V

17.37. −290 kJ

17.39. (a) Ni is +2 and Cl is −1 in $NiCl_2$; Na is +1 and Cl is −1 in NaCl; (b) 2; (c) −509.4 kJ based on ΔG°_f values

17.41. (a) $\Delta E^\circ_{cell} = -0.478$ V; $\Delta G^\circ = 92.2$ kJ; (b) $\Delta E^\circ_{cell} = 0.548$ V; $\Delta G^\circ = -97.4$ kJ

17.43. It is an inert electrode, a surface upon which electron transfer can occur.

17.45. Voltage of a battery (a voltaic cell) is governed by the Nernst equation:

$$E_{cell} = E^\circ_{cell} - \frac{RT}{nF} \ln Q.$$

As a battery discharges, the value of Q increases and the value of E_{cell} decreases but not by very much because of the log relationship between Q and E_{cell} . It is not until nearly all of the reactants have been turned into products that the drop in E_{cell} becomes significant.

17.47. 1.27 V

17.49. 8.56×10^{19}

17.51. −0.414 V

17.53. $E_{cell} = 1.54$ V; E_{cell} will decrease.

17.55. (a) 0.62 V; (b) 0.61 V

17.57. (a) 0.349 V; (b) 1.02×10^{57}

17.59. c and f

17.61. Cell B

17.63. Cell F

17.65. In a voltaic cell, the electrons are produced at the anode so a negative (−) charge builds up there; in an electrolytic cell, electrons are being forced onto the cathode so that it builds up negative (−) charge. The flow of electrons in the outside circuit is reversed in an electrolytic cell compared to the flow in a voltaic cell.

17.67. Br_2

17.69. More negative

17.71. 6.8 g

17.73. +3

17.75. 18.0 minutes

17.77. (a) 5.78×10^{-3} L; (b) no, because Cl_2 and Br_2 would be produced.

17.79. -0.270 V

17.81. A hybrid vehicle uses a relatively inexpensive fuel (gasoline) in the internal combustion engine and has good fuel economy but still gives off emissions. A fuel-cell vehicle does not give off emissions (the reaction produces H_2O) but requires a more expensive and explosive fuel (hydrogen); moreover, current battery technologies incorporate materials that are still expensive and bulky.

17.83. Electric engines are more efficient by converting more of the energy into motion instead of losing it as heat.

17.85. a. $\overset{-4+1}{CH_4}(g) + \overset{+1\ -2}{H_2O}(g) \rightarrow \overset{+2\ -2}{CO}(g) + 3\ \overset{0}{H_2}(g)$

$\overset{+2-2}{CO}(g) + \overset{+1\ -2}{H_2O}(g) \rightarrow \overset{0}{H_2}(g) + \overset{+4-2}{CO_2}(g)$

b. For the reaction of CH_4 with H_2O, $\Delta G^\circ_{rxn} = 142.2$ kJ. For the reaction of CO with H_2O, $\Delta G^\circ_{rxn} = -28.6$ kJ. For the overall reaction, $\Delta G^\circ_{overall} = \Delta G^\circ_{rxn_1} + \Delta G^\circ_{rxn_2} = 113.6$ kJ.

17.87. All the values in Appendix 6 would increase by 0.8277 V.

17.89. (a) $Li \rightarrow Li^+ + e^-$; $2\ SO_2 + 2\ e^- \rightarrow S_2O_4^{2-}$; (b) 2

17.91. 0.0745 V

17.93. $E^\circ_{cell} = -\dfrac{RT\ln K}{nF}$

17.95. (a) -0.87 V; (b) Mo_3S_4: Mo = +2.67 (the average of 2 Mo^{3+} ions for every 1 Mo^{2+} ion); $MgMoS_4$: Mo = +2; (c) Mg^{2+} is added to the electrolyte to better carry the charge in the cell. This cation is produced at the anode and consumed at the cathode.

17.97. (a) In K_2MnF_6: K = +1, Mn = +4, F = −1; in SbF_5: Sb = +5, F = −1; in $KSbF_6$: K = +1, Sb = +5, F = −1; in MnF_3: Mn = +3, F = −1; in F_2: F = 0; this is a one-electron process; (b) −656 kJ; (c) 6.80 V; (d) too low; (e) in H_2: H = 0; in F_2: F = 0; in KF: K = +1, F = −1; in KHF_2: K = +1, H = +1, F = −1; this is a two-electron process.

17.99. (a) Mg; (b) U; (c) $E^\circ_{cathode} < -2.37$ V; (d) yes

17.101. (a) Cathode; (b) no, Mg^{2+} has a lower (less negative) reduction potential than Na^+; (c) no; (d) H_2 and Cl_2

Chapter 18

18.1. a, b and d are crystalline; c is amorphous.

18.3. The chemical formula is A_4B_4 or AB.

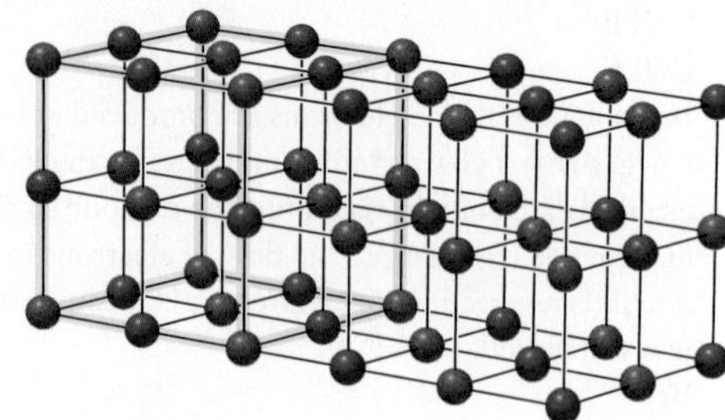

18.5. 3 A atoms, 1 B atom

18.7. The chemical formula is ABX_3.

18.9. 3.81 g/cm^3, cubic hole

18.11. b

18.13. (a) a.; (b) b; (c) c; (d) f

18.15. MgB_2

18.17. (a) Four large spheres, eight small spheres; (b) 57 pm

18.19. Cubic closest-packed structures have an *abcabc* stacking pattern and hexagonal closest-packed structures have an *abab* pattern.

18.21. Body-centered cubic

18.23. These structural forms are not allotropes because iron is not molecular.

18.25. For bcc $\ell = \dfrac{4r}{\sqrt{3}}$; for fcc $\ell = \dfrac{4r}{\sqrt{2}}$

18.27. 104.2 pm

18.29. 513 pm

18.31. c

18.33. No

18.35. Because the metallic bonding between Cu and Ag in the alloy is weaker due to a mismatch of their atomic sizes

18.37. Greater than

18.39. (a) XY_3; (b) YX_3

18.41. Octahedral holes

18.43. Substitutional alloy

18.45. Yes

18.47. (a) AB; (b) A_2B; (c) AB

18.49. One-fifth

18.51. Yes, the alloy is more dense.

18.53. Application of an electrical potential across a metal causes its mobile valence electrons to move away from the negative and toward the positive potential sources.

18.55. Metallic bonds are weaker than ionic bonds.

18.57. Yes, assuming it exists as H atoms in the solid phase

18.59. Groups 2 and 12

18.61. Phosphorus gives silicon a higher conductivity because it has one more valence electron making an n-type semiconductor.

18.63. Group 14

18.65. (a) n-type

(b)

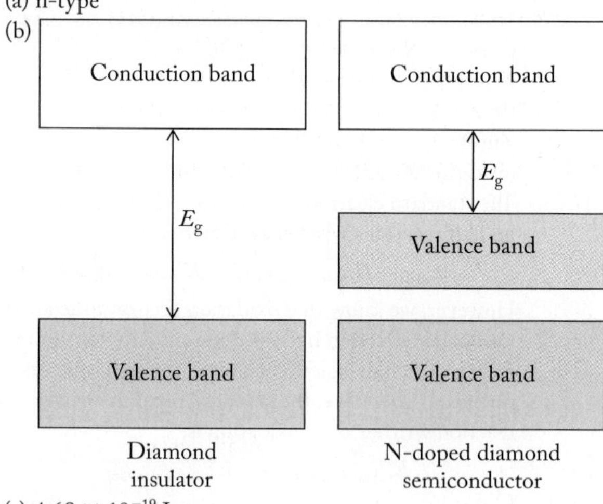

(c) 4.68×10^{-19} J

18.67. InN

18.69. Each S atom has a bent geometry due to sp^3 hybridization and bond angles of about 109.5°. Therefore, the ring is not flat.

18.71. The ring of BN atoms is flat due to sp^2 hybridization of the B and N atoms.

18.73. (a) 2 O^{2-} ions and 4 H^+ ions per unit cell; (b) $[:\ddot{\underset{..}{O}}:]^{2-}$ 2 H^+

18.75. 109.5°

18.77. K^+ is large and so does not fit well into the octahedral holes of the fcc lattice.

18.79. Because the radius of Cl^- is 181 pm and the radius of Cs^+ is 170 pm, their radii are very similar. The Cs^+ ion at the center of Figure P18.79 occupies the center of the cubic cell, so CsCl could be viewed as a body-centered cubic structure. The structure of CsCl could also be viewed as two interpenetrating arrays of simple cubic unit cells: on of Cs^+ ions and the other of Cl^- ions.

18.81. We might, but Cl^- ions are much larger than Na^+ ions, so they don't fit into the holes of an fcc unit cell of Na^+ ions. Instead, the unit cell would have to expand to accommodate the Cl^- ions, as the Ca^{2+} ions do in the fluorite structure.

18.83. Less than

18.85. $MgFe_2O_4$

18.87. UO_2

18.89. (a) Octahedral; (b) half

18.91. This rock salt arrangement is more dense than the sphalerite arrangement because in sphalerite the lattice of S^{2-} ions must expand to accommodate the Cd^{2+} ions.

18.93. (a)

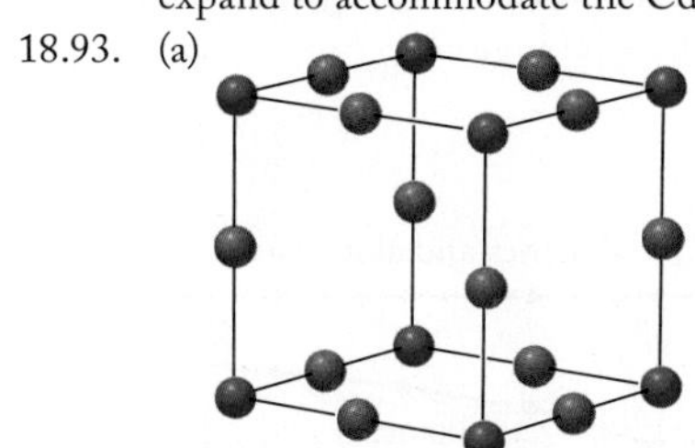

(b) 5.25 g/cm^3; (c) 79%

18.95. 421 pm

18.97. Ceramics are (b) thermal insulators. (a) Ductility, (c) electrical conductivity, and (d) malleability describe metals.

18.99. $Mg_3(Si_2O_5)(OH)_4$

18.101. $2\ KAlSi_3O_8(s) + 2\ H_2O(\ell) + CO_2(g) \rightarrow Al_2(Si_2O_5)(OH)_4(s) + 4\ SiO_2(s) + K_2CO_3(aq)$; this is not a redox reaction.

18.103. (a) $3\ CaAl_2Si_2O_8(s) \rightarrow Ca_3Al_2(SiO_4)_3(s) + 2\ Al_2SiO_5(s) + SiO_2(s)$ (b) In anorthite, the silicate anion is $Si_2O_8^{\ 8-}$. In grossular, the silicate anion is $SiO_4^{\ 4-}$. Kyanite consists of an irregular array of Al^{3+}, $SiO_4^{\ 4-}$, and O^{2-} ions.

18.105. Cubic holes can accommodate Ba^{2+}. Octahedral holes can accommodate Ti^{4+}.

18.107. An amorphous solid has no regular, repeating lattice to diffract X-rays.

18.109. X-rays have wavelengths of the order of the separation of atoms in crystals. Microwaves have wavelengths too long to be diffracted by crystal lattices.

18.111. If a crystallographer uses a shorter λ wavelength, the data set can be collected over a smaller scanning range.

18.113. Halite

18.115. The values of n are 2 ($\theta = 6.99°$) and 3 ($\theta = 10.62°$). The average lattice spacing is $d = 582$ pm and 580 pm, respectively.

18.117. $2\theta = 4.76°$

18.119. XYZ_3

18.121. 33.5%

18.123. (a) 139 pm; (b) 9.96 g/cm^3; (c) 6750 Mo atoms

18.125. (a) 7.53 g/cm^3; (b) 3.42 g/cm^3; (c) 3.36 g/cm^3

18.127. (a) AuZn; (b) CuZn, 4.58%; AgZn, 7.19%; AuZn, 7.19%; yes; (c) yes

18.129. (a) MS; (b) 2 M^{2+} and 2 S^{2-}; (c) large yellow spheres; (d) 4

18.131. Every other bcc unit cell of Cu atoms could have an Al atom in its center. That way half the bcc unit cells would have 2 Cu atoms each and half would have 1 Cu and 1 Al atom for a total of 3 Cu atoms and 1 Al atom.

Chapter 19

19.1. (a) One degree of unsaturation; (b) two degrees of unsaturation; (c) no degrees of unsaturation; (d) three degrees of unsaturation

19.3. Pine oil and oil of celery

19.5. b and d

19.7. Form a is the rigid polymer because its regular structure allows the chain to pack in a regular way; form b is the rubbery polymer because its random arrangement of side chains makes it more difficult to pack in a uniform manner.

19.9.

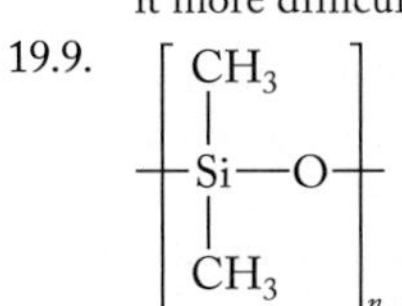

19.11. (a) For *cis*-polyisoprene:

For *trans*-polyisoprene:

(b) The more linear structure of the *trans* isomer allows stronger between-strand interactions, which explains the rigidity of materials made from it.

19.13. *sp* (triple bond and single bond or two double bonds), *sp*^{2} (double bond and two single bonds), *sp*^{3} (four single bonds)

19.15. No

19.17. Amine, alcohol, ether, aldehyde, ketone, carboxylic acid, ester, and amide

19.19. Sample A because of its higher average molar mass and stronger London dispersion intermolecular forces

19.21. Yes

19.23. *sp*^{3}

19.25. The structure of cyclohexane shows that C atoms are *sp*^{3} hybridized with bond angles of 109.5°. It cannot be a planar molecule.

19.27. No

19.29. No

19.31.

Pentane 2-Methylbutane 2,2-Dimethylpropane

19.33. (a) 2,3-dimethylhexane, (c and d) 2-methylheptane

19.35. (a) C_8H_{18}; (b) C_9H_{20}; (c) C_8H_{18}; (d) C_8H_{18}; (e) C_9H_{20}

19.37. Earliest to latest: $C_6H_{14} < C_9H_{20} < C_{12}H_{26} < C_{18}H_{38}$

19.39. Hexane

19.41. No

19.43. When the double bond is "terminal" (occurring at the end or beginning of the carbon chain), there are three like groups (H) so no cis and trans isomers are possible.

19.45. The C=C double bond outside of the ring does not show cis–trans isomerism because there are not two dissimilar groups on the terminal carbon atom. The C=C double bond in the ring of carbon atoms is cis in the structure of carvone. This bond cannot be trans or the ring of six carbon atoms would not be possible.

19.47. Ethylene has a C=C bond with which HBr is reactive, but polyethylene has only saturated C—C bonds that do not react with HBr.

19.49. −124 kJ
19.51. a is trans, E; b is cis, Z
19.53. 681.2 kJ, endothermic
19.55. −174.30 kJ
19.57.

19.59. Because they lack the conjugated C—C double bonds that give polyacetylene it conductivity.
19.61. In benzene, each C atom is sp^2 hybridized with bond angles of 120°. This geometry at each of the carbon atoms in the ring makes benzene a planar molecule.
19.63. Tetramethylbenzene has three constitutional isomers; pentamethylbenzene has no constitutional isomers.
19.65. Yes
19.67.

19.69. Fuel value for 1 mol benzene = 41.83 kJ/g; fuel value for 3 mol ethylene = 50.30 kJ/g; 1 mol benzene has a lower energy content than 3 mol ethylene.
19.71. Methylamine has a smaller nonpolar hydrocarbon chain compared to n-butylamine and so it is more soluble in water.
19.73.

HO, NH_2 1° amine, N 2° amine, H — Serotonin

NH_2 1° amine — Amphetamine

19.75. −138.7 kJ
19.77. a, b, and c.

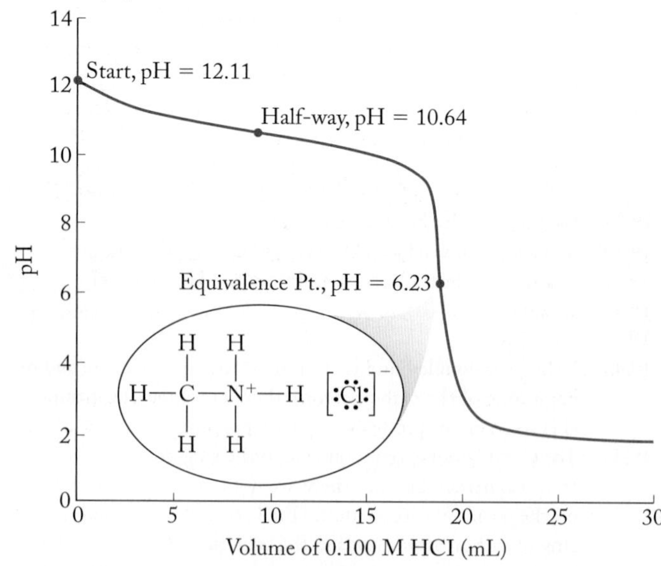

19.79. The more oxygenated the fuel, the lower the fuel value.
19.81. Ethers have lower boiling points than alcohols because they have weaker dipole–dipole forces compared to the alcohols, which have hydrogen bonding between the molecules.
19.83. Evaporation of ethanol from the skin is an endothermic process (phase change from liquid to vapor). The heat transfers from the skin to the ethanol so the skin feels cold.
19.85. a and d are alcohols, b and c are ethers; b < c < d < a
19.87. Fuel value for diethyl ether = 36.74 kJ/g; fuel value for n-butanol = 36.10 kJ/g; diethyl ether has a slightly higher fuel value.
19.89. Fuel value for methanol = 22.67 kJ/g; fuel value for ethanol = 29.67 kJ/g; yes, the answer supports the prediction made in Problem 19.80.
19.91. Both carboxylic acids and aldehydes have polar functional groups. Carboxylic acids, however, are more soluble in water because they form strong hydrogen bonds with water and partially ionize when they dissolve in it.
19.93. Yes
19.95. No
19.97. Structure a because all of the formal charges are zero
19.99. An amide includes a carbonyl (C═O) as part of its functional group in addition to the $-NH_2$ group.
19.101. (a), (b), and (d)
19.103. (b)
19.105.

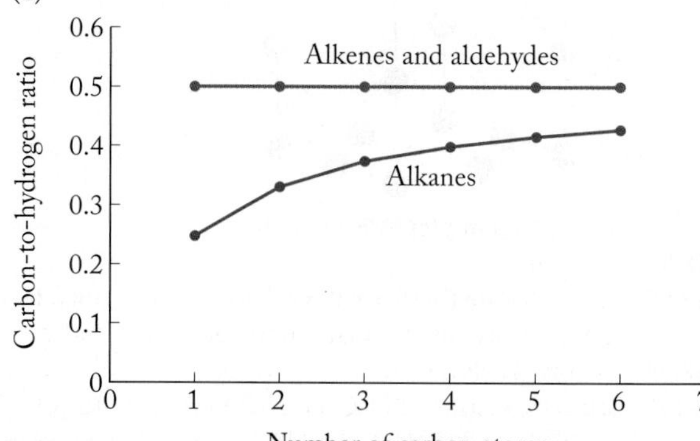

The plot of C:H ratio versus number of C atoms for aldehydes correlates exactly to that of alkenes and poorly to that of alkanes.
19.107. (a) Pineapples

Acetic acid + n-Butanol → ester + H_2O

(b) Bananas

2-Methylbutanoic acid + Ethanol → ester + H_2O

(c) Apples

Acetic acid + 3-Methylbutanol → ester + H_2O

19.109. Nicotine's highlighted N atom is in a tertiary amine group; valium's highlighted N atom is in an amide group.
19.111. Fuel value for formaldehyde = 19.00 kJ/g; fuel value for formic acid = 6.531 kJ/g; formaldehyde has a significantly higher fuel value than formic acid.
19.113. For reaction 1, $\Delta H^\circ_{rxn} = 17.5$ kJ; for reaction 2, $\Delta H^\circ_{rxn} = 312.1$ kJ

19.115. (a) 6; (b) 8; (c) 10

19.117. (a) Condensation; methanol; (b) because of the presence of the six-membered ring, Kodel might be better able to accept nonpolar organic dyes.

19.119. No; enantiomer and optically active can describe the same chiral molecule, but achiral cannot.

19.121. Yes. If R contains a chiral center, the cis or trans isomer of RCH=CHR would have optical isomers.

19.123. (a) No; (b) 12

19.125. None

19.127. 2.52 g methanol; 3.45 g carbon dioxide

19.129. 1,4-Dimethylbenzene has a symmetrical structure so it packs very efficiently in the solid state and, therefore, has a much higher melting point.

19.131. (a) Trans

(b)

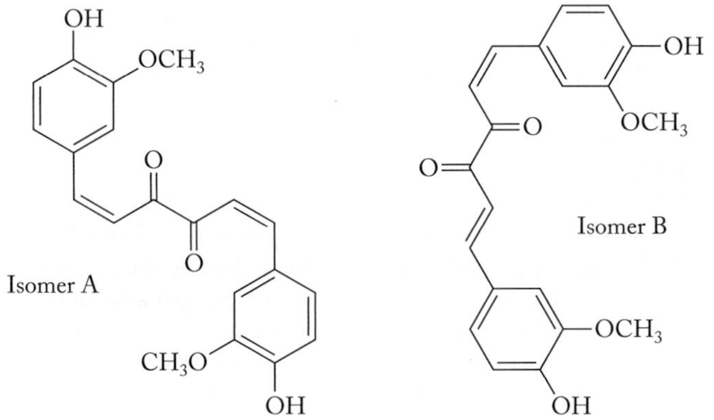

Isomer A

Isomer B

(c) The hybridizations on the carbon atoms in curcumin are sp^3 for the carbon of the $-OCH_3$ group and sp^2 for all other carbon atoms.

19.133.

For polymer a

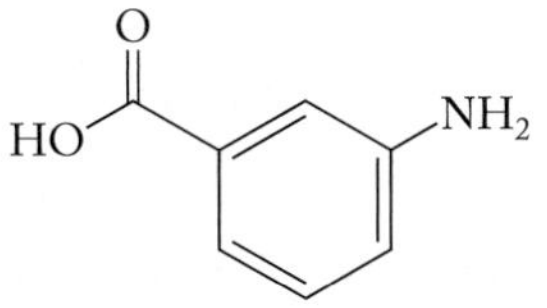

For polymer b

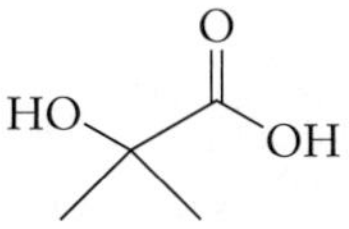

19.135.

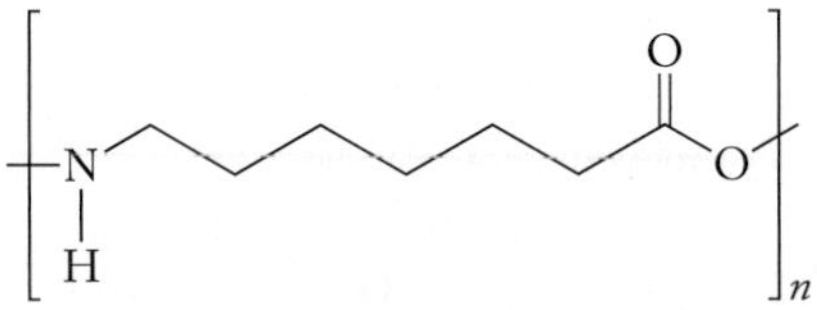

In this polymer there is one monomeric repeating unit with seven carbon atoms because it is prepared from the difunctional $H_2N(CH_2)_6COOH$ monomer. In nylon-6 the polymer also has a single monomeric unit but with six carbon atoms.

19.137. Cross-linking increases the strength, hardness, melting (softening) point, and chemical resistance of a polymer.

19.139.

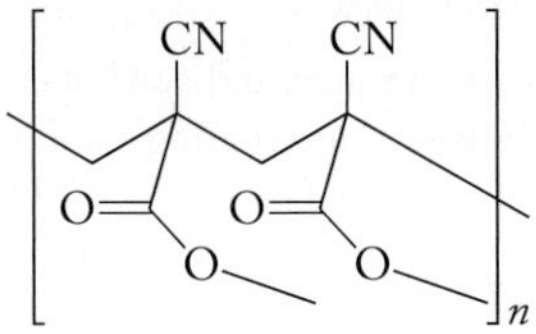

19.141. (a)

For piperine

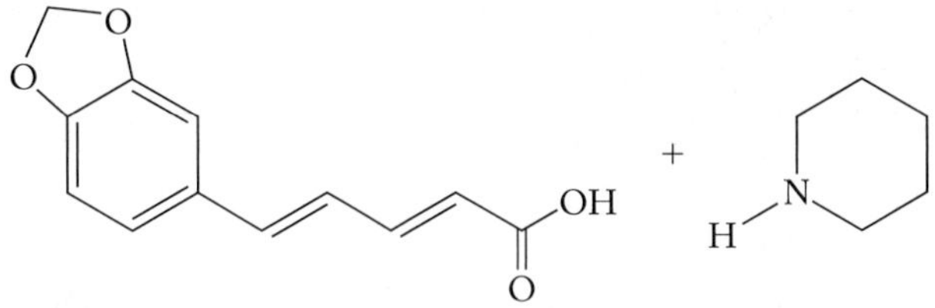

For capsaicin

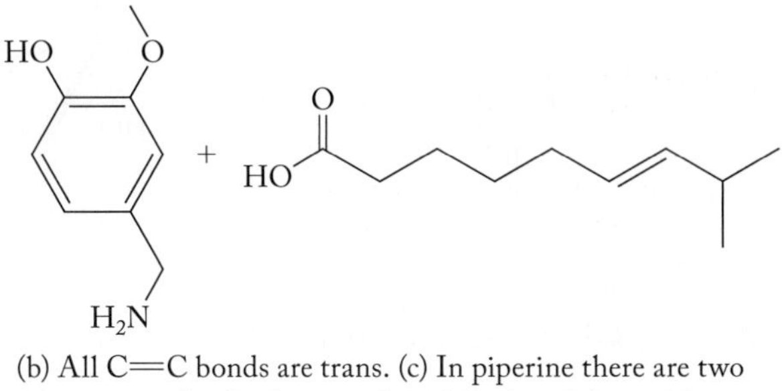

(b) All C=C bonds are trans. (c) In piperine there are two ether groups (in the five-membered ring) and the amide group. In capsaicin there is an ether group, an alcohol group, and the amide group.

Chapter 20

20.1.

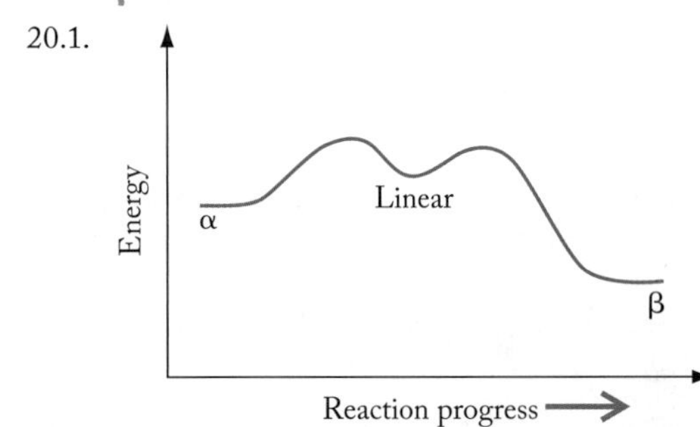

20.3. (a) Palmitic acid in (a), stearic acid in (b); (b) cis

20.5. Tyrosine, glycine, glycine, phenylalanine, and methionine

20.7. a and c

20.9. Sucrose. The difference in the structures is that in sucralose, three $-OH$ groups on sucrose have been replaced by Cl atoms. Being derived from sucrose implies that the sugar is natural, but the presence of Cl atoms on sugars is not natural.

20.11. Decreases

20.13. The "α" refers to the single carbon atom in amino acids to which both $-NH_2$ and $-COOH$ groups are bonded.

20.15. D- and L- refer to how the four groups on a chiral carbon are oriented.

20.17. a and c

20.19. Most amino acids are zwitterions at pH ≈ 7.4 because the amino group will be protonated and the carboxylic acid group will be deprotonated, giving $RC\overset{+}{N}H_3COO^-$.

20.21.

(a)

$H_2N-CH(CH_3)-C(=O)-NH-CH(CH_2OH)-C(=O)-OH$

$H_2N-CH(CH_2OH)-C(=O)-NH-CH(CH_3)-C(=O)-OH$

(b)

$H_2N-CH(CH_3)-C(=O)-NH-CH(CH_2C_6H_5)-C(=O)-OH$

$H_2N-CH(CH_2C_6H_5)-C(=O)-NH-CH(CH_3)-C(=O)-OH$

(c)

$H_2N-CH(CH_3)-C(=O)-NH-CH(CH(CH_3)_2)-C(=O)-OH$

$H_2N-CH(CH(CH_3)_2)-C(=O)-NH-CH(CH_3)-C(=O)-OH$

20.23. (a) Alanine + glycine; (b) leucine + leucine; (c) tyrosine + phenylalanine

20.25. NH_3

20.27. Primary structure

20.29. Tertiary structure

20.31. In this model the binding site changes slightly when the substrate binds with the active site so that it fits the transition state better for the reaction.

20.33. The substrate and the active site must fit each other; if the molecular lock and key do not match, then the reaction cannot be catalyzed at that site.

20.35. Lysine contains two amino groups, one of which is on a long carbon tail that can react with the carboxylic acid on the carbon tail of glutamic acid to form a bridge.

20.37. Starch has α-glycosidic bonds and cellulose has β-glycosidic bonds. Starch coils into granules and cellulose forms linear molecules.

20.39. No

20.41. Carbohydrates are fuels providing energy necessary for operation of organs and tissues, physical activity, and brain function.

20.43. To calculate the free-energy change for a two-step process we need only to sum the individual ΔG values for each reaction.

20.45.

β-Galactose ⇌ open-chain form ($_1$CHO, H–$_2$C–OH, OH–$_3$C–H, OH–$_4$C–H, H–$_5$C–OH, $_6CH_2OH$) ⇌ α-Galactose

20.47. c

20.49. a

20.51. b

20.53. −16.4 kJ

20.55. Saturated fatty acids have all C—C single bonds in their structure; unsaturated fatty acids have C=C double bonds.

20.57. Fatty acids have a high fuel value and eating sticks of butter affords Arctic explorers with more energy per gram of food compared to carbohydrates or proteins.

20.59. If the two fatty acids linked to the glycerol at C-1 and C-3 are different, then the triglyceride has a chiral center.

20.61. b and c

20.63.

(a) Glycerol with octanoic acid (C_7H_{15} chains)

(b) Glycerol with decanoic acid (C_9H_{19} chains)

(c) Glycerol with dodecanoic acid ($C_{11}H_{23}$ chains)

20.65. Phosphate group, a five-carbon sugar, and a nitrogen base; the backbone of DNA is composed of alternating sugar residues and phosphate groups.

20.67. Hydrogen bonds

20.69.

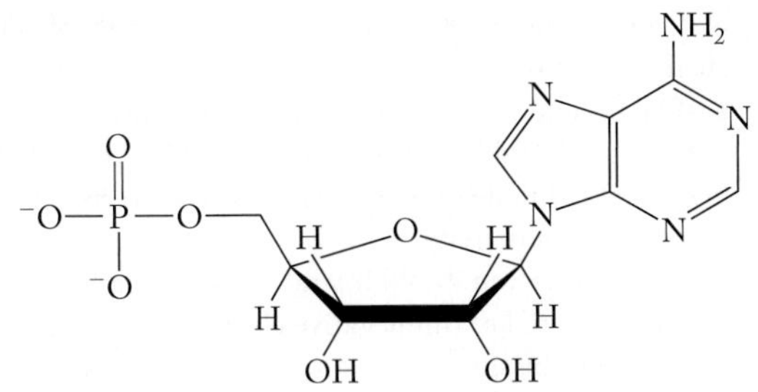

NH$_2$
N N
O
$^-$O—P—O— H O H N N
$^-$O H H
OH OH

20.71. A-G-C-C-A-T

20.73. (a) Sucrose; (b) esters; (c) $C_{15}H_{31}COOH$

20.75. (a) There is an extra $-CH_2-$ group in homocysteine's sulfur-containing side chain; (b) yes

20.77. Yes

20.79. (a) No; (b) when the $-NH_2$ group of glycine reacts with the $-COOH$ group of creatine:

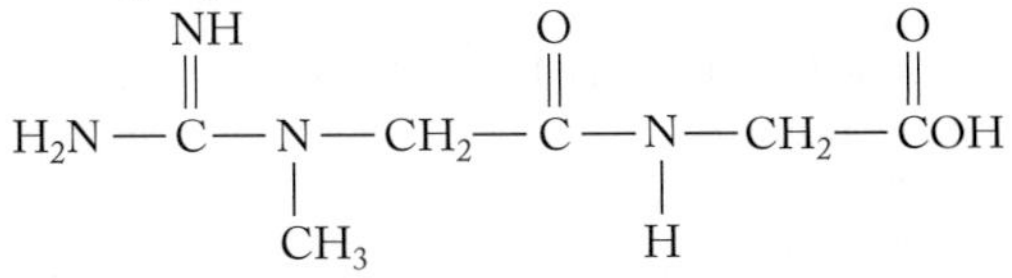

$$\mathrm{H_2N{-}\overset{NH}{\overset{\|}{C}}{-}\underset{CH_3}{\underset{|}{N}}{-}CH_2{-}\overset{O}{\overset{\|}{C}}{-}\underset{H}{\underset{|}{N}}{-}CH_2{-}\overset{O}{\overset{\|}{C}}OH}$$

When the $-NH_2$ group of creatine reacts with the $-COOH$ group of glycine:

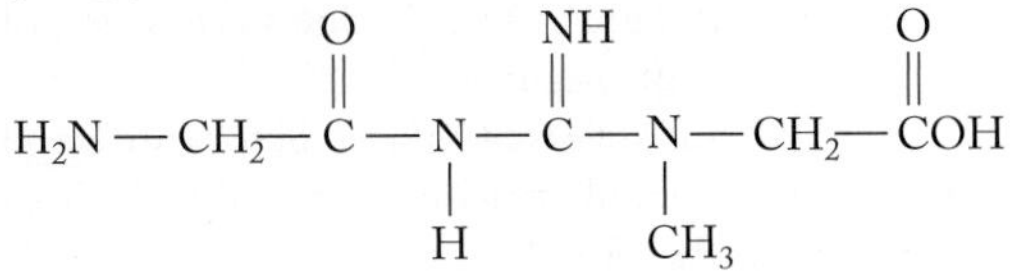

$$\mathrm{H_2N{-}CH_2{-}\overset{O}{\overset{\|}{C}}{-}\underset{H}{\underset{|}{N}}{-}\overset{NH}{\overset{\|}{C}}{-}\underset{CH_3}{\underset{|}{N}}{-}CH_2{-}\overset{O}{\overset{\|}{C}}OH}$$

20.81. Glutamic acid, cysteine, and glycine

20.83. Yes. Because there is no difference in the number of C—C, C—H, C=O, C—O, or N—H bonds between the two compounds, we expect on the basis of average bond energies that the fuel values of leucine and isoleucine should be identical. Isoleucine might have a lower fuel value because the CH_3 group is closer to the COOH and NH_2 groups, and this difference in shape must contribute to the slightly different fuel values.

Chapter 21

21.1. (a) b – synthesis of a supermassive nuclide; (b) d - fission

21.3. Red (radium)

21.5. a

21.7. Blue line (b)

21.9. Process 1 represents fission; process 2 represents fusion.

21.11. No, it penetrates much less the other two.

21.13. The ratio decreases.

21.15. Nuclides within the belt of stability (represented by the green dots in Figure 21.1) are stable. The nuclides represented by the orange dots in Figure 21.1 that are above the belt of stability are neutron-rich and tend to undergo β decay. The nuclides below the belt of stability are neutron-poor and tend to undergo positron emission or electron capture.

21.17. Alpha decay by these nuclides produces nuclides with even greater neutron-to-proton ratios, which means they are more likely to undergo β decay.

21.19. ^{64}Cu to ^{64}Zn is β decay; ^{64}Zn to ^{64}Ni is either positron emission or electron capture.

21.21. Greater than 1

21.23. (a) n/p = 1.24, β decay
(b) n/p = 1.23, positron emission or electron capture
(c) n/p = 1.56, alpha decay (see Figure 21.2)

21.25. (a) $^{192}_{77}Ir \rightarrow {}^{0}_{-1}\beta + {}^{192}_{78}Pt$; (b) $^{192}_{77}Ir + {}^{0}_{-1}e \rightarrow {}^{192}_{76}Os$

21.27. ^{56}Co and ^{44}Ti are both neutron-poor and should undergo either electron capture or positron emission.

21.29. Because only about 0.2% of ^{14}C initially in the sample is still there. This is too small a quantity to be determined accurately

21.31. If the rocks are less than 300,000 years old, less than 0.02% of the ^{40}K in the rocks has decayed, which is too small a change to be determined accurately

21.33. 12.5%

21.35. 646 h

21.37. 131 years

21.39. 35%

21.41. 85%

21.43. 36,640 years old

21.45. Nuclear reactions result in a loss of mass from reactants to products, which is released as energy, some of it in the form of heat.

21.47. (a) $^{11}_{6}C \rightarrow {}^{11}_{5}B + {}^{0}_{1}\beta$; (b) 1.176×10^{-11} J

21.49. For ^{35}Cl, 2.877×10^{10} kJ/mol; for ^{37}Cl, 3.110×10^{10} kJ/mol

21.51. $^{209}_{83}Bi + {}^{4}_{2}\alpha \rightarrow {}^{211}_{85}At + 2\,{}^{1}_{0}n$

21.53. (a) $2\,{}^{1}_{0}n$; (b) $^{1}_{0}n$; (c) $^{4}_{2}\alpha$; (d) $^{1}_{1}p$

21.55. (a) $^{0}_{-1}\beta$; (b) $^{122}_{53}I$; (c) $^{10}_{5}B$; (d) $^{68}_{30}Zn$

21.57. Electron = β particle = positron < proton < neutron < deuteron < α particle

21.59. Antihydrogen has the same mass as hydrogen but its nucleus has a negative charge with a positively charged electron. It contains the antiproton in the nucleus and a positron in place of the electron.

21.61. Fusion in the sun has more steps than did the primordial synthesis. Fusion in the sun produces beta particles along with alpha particles (helium) while primordial nucleosynthesis produced deuterium that was later used along with neutrons as fuel for the process.

21.63. Helium

21.65. From earlier generations of massive stars that exploded in supernovae

21.67. $\Delta E = 3.01 \times 10^{-10}$ J; $\lambda = 1.32 \times 10^{-15}$ m or 1.32×10^{-6} nm

21.69. (a) 4.37×10^{-12} J; (b) 6.80×10^{-12} J; (c) 2.69×10^{-12} J; (d) 1.60×10^{-12} J

21.71. 9.36×10^{-13} J

21.73. (a) ^{16}O; (b) ^{24}Mg; (c) ^{36}Ar

21.75. (a) ^{99}Tc; (b) ^{121}Sb; (c) ^{109}Cd

21.77. Control rods made of boron or cadmium are used to absorb the excess neutrons to control the rate of energy release.

21.79. The neutron-to-proton ratio for heavy nuclei is high and, when the nuclide undergoes fission to form smaller nuclides, it must emit neutrons because the fission products require a lower neutron-to-proton ratio for stability.

21.81. (a) $^{138}_{52}Te$; (b) $^{133}_{51}Sb$; (c) $^{143}_{55}Cs$

21.83. (a) $^{103}_{39}Y$; (b) $^{130}_{48}Cd$; (c) $^{138}_{52}Te$

21.85. Level of radioactivity is the number of decay events per unit of time. Dose is a measure of exposure to ionizing radiation over time.

21.87. Radon-222 is a gas. If inhaled, it may undergo α decay to solid polonium-218 while in the lungs, where it may continue to emit α radiation. Alpha radiation is one of the most damaging kinds of radiation when in contact with biological tissues. Exposure increases the risk of lung cancer.

21.89. 5 μSv = 5 μGy; 250 μJ

21.91. (a) $^{90}_{38}Sr \rightarrow {}^{0}_{-1}\beta + {}^{90}_{39}Y$; (b) 3.28×10^{8}; (c) strontium-90 is found in milk and not in other foods because it is chemically similar to calcium and milk is rich in calcium.

21.93. (a) 0.15 decay events/s; (b) 7.0×10^4

21.95. (a) The half-life should be long enough to effect treatment of the cancerous cells but not so long as to cause damage to healthy tissues.
(b) Because α radiation does not penetrate far beyond a tumor, α decay mode is best.
(c) Products should be nonradioactive, if possible, or have short half-lives and able to be flushed from the body by normal cellular and biological processes.

21.97. (a) Positron emission or electron capture; (b) positron emission or electron capture; (c) positron emission or electron capture

21.99. 74.3 days

21.101. Yes

21.103. (a) 79%; (b) 153 h

21.105. (a) $^{10}_{5}B + ^{1}_{0}n \rightarrow ^{7}_{3}Li + ^{4}_{2}\alpha$; (b) 4.43×10^{-13} J
(c) Alpha particles have a high RBE and they do not penetrate into healthy tissue if the radionuclide is placed inside a tumor.

21.107. (a) Besides releasing a large amount of energy to power the starship *Enterprise*, hydrogen is an abundant fuel in the universe and therefore could easily react with any antihydrogen produced.
(b) Antihydrogen would react with all matter so it would be difficult to contain except maybe with a magnetic or energy field.

21.109. Binding energy for $^{8}Be = 9.052 \times 10^{-12}$ J; binding energy for 2 atoms of $^{4}He = 9.067 \times 10^{-12}$ J

21.111. (a) Geiger counter; (b) 2877 y
(c) smoke detectors are safe to handle because the ^{241}Am is an α emitter and α particles do not travel more than a few inches in air and cannot penetrate the first layer of skin.

21.113. (a) $^{249}_{98}Cf + ^{48}_{20}Ca \rightarrow ^{294}_{118}Og + 3\ ^{1}_{0}n$; (b) $^{290}_{116}Lv$; (c) $^{286}_{114}Fl$; (d) $^{282}_{112}Cn$;
(e) because ^{294}Og is a member of the noble gas family, it should have chemical and physical properties similar to naturally occurring radon.

21.115. $\left[\mathrm{H{-}\ddot{O}\cdot}\ (\text{with H bonded below O})\right]^+$ $\left[\mathrm{H{-}\ddot{O}{-}H}\ (\text{with H bonded below O})\right]^+$ $\mathrm{H{-}\ddot{\underset{\cdot\cdot}{O}}\cdot}$

21.117. (a) $^{64}_{28}Ni + ^{124}_{50}Sn \rightarrow ^{188}_{78}Pt$; (b) $^{196}_{78}Pt$

21.119. 3.35: 1

21.121. (a) $^{40}_{19}K \rightarrow ^{40}_{18}Ar + ^{0}_{1}\beta$; (b) because the half-life of ^{40}K is so much longer than that of ^{14}C

21.123. 118 s

21.125. $^{12}_{5}B \rightarrow ^{8}_{3}Li + ^{4}_{2}He$; $^{12}_{5}B \rightarrow ^{12}_{6}C + ^{0}_{-1}\beta$

21.127. $^{12}_{6}C + ^{1}_{1}p \rightarrow ^{10}_{5}B + ^{3}_{2}He$; $^{14}_{7}N + ^{1}_{1}p \rightarrow ^{10}_{5}B + ^{5}_{3}Li$

21.129. (a) $^{104}_{44}Ru + ^{1}_{0}n \rightarrow ^{105}_{45}Rh + ^{0}_{-1}\beta$; (b) 29 h

Chapter 22

22.1. d

22.3. Pink, group 16

22.5. −6.7 kJ

22.7. Both are trigonal planar.

22.9. Phosphate

22.11. Essential elements have beneficial physiological effects whereas nonessential elements have no known function.

22.13. The amount present in the body

22.15. (a) 1.6 ppm; (b) 525 ppm; (c) 0.43 ppm

22.17. (a) oxygen; (b) oxygen; (c) carbon

22.19. Because the highest energy electrons in the atoms of groups 1 and 2 are in their valence shell *s* orbitals; the highest energy electrons in the atoms of groups 13–18 are in their valence shell *p* orbitals

22.21. Li_2O is an ionic solid with strong ion–ion interactions between its Li^+ and O^{2-} ions, whereas CO is a polar molecule with weaker dipole–dipole interactions between its molecules.

22.23. Electrical conductivity

22.25. Be^{2+} is smaller and would interact more strongly with binding sites with partial of full negative charges. Ca^{2+} ions might not even fit into the sites.

22.27. Potassium

22.29. $Li^+ < Al^{3+} < Mg^{2+} < Cl^-$

22.31. $K < Mg < S < F$

22.33. The ionization of a Cl^- ion is the reverse process of a Cl atom acquiring an electron.

22.35. Diffusion, ion channels, and ion pumps

22.37. The inner portion of the phospholipid bilayer is nonpolar.

22.39. K^+

22.41. $CaCO_3$ is less soluble than $CaSO_4$.

22.43. 0.56 atm

22.45. 0.072 V

22.47. 0.16 mol

22.49. 1.2×10^{-5} *M*

22.51. Cs^+ can interfere with K^+-dependent functions. It is also a β emitter and could cause radiation sickness and cancer.

22.53. ΔS positive; ΔG negative

22.55. Peroxide ions are easily reduced by a two-electron process to O^{2-} The standard reduction potential for hydrogen peroxide in acid is nearly 1.8 V

22.57. ^{137}Ba and a β particle $^{137}_{55}Cs \rightarrow ^{137}_{56}Ba + ^{0}_{-1}\beta$

22.59. 3.06

22.61. $K = \dfrac{[OH^-]}{[F^-]}$; the equilibrium lies to the right.

22.63. (a) More soluble; (b) 8.5×10^{-8} *M*; (c) bacteria's weak acids react with hydroxide ions in hydroxyapatite, which makes the compound more soluble.

22.65. 5.3×10^{-118}

22.67. H_2O is the most polar; H_2Te is the least polar.

22.69. Different decay products have different energies and different relative biological effectiveness (RBE) values

22.71. β particles cause less collateral tissue damage and are easier to detect because they penetrate through tissues better.

22.73. ^{213}Bi has a high neutron-to-proton ratio; ^{111}In has a low neutron-to-proton ratio.

22.75. Dipole–induced dipole

22.77. 41 h

22.79. $[:Bi{\equiv}O:]^+$

22.81. $Al(OH)_3(s) + 3\,H^+(aq) \rightarrow Al^{3+}(aq) + 3\,H_2O(\ell)$

22.83. Aluminum hydroxide on both a per mole and a per gram basis

22.85. 2.5 g $Mg(OH)_2$

Chapter 23

23.1. Chromium (green) and cobalt (yellow)

23.3. Zinc (blue)

23.5. 4

23.7. (a) $[Co(CN)_6]^{3-}$; (b) $[CoF_6]^{3-}$; (c) $[Co(NH_3)_6]^{3+}$

23.9. (a) Identical; (b) isomers; (c) identical

23.11. (c) is $CoCl_4{}^{2-}$ and (a) is $CoI_4{}^{2-}$

23.13. Molecules of H_2O

23.15. Molecules of H_2O

23.17. Na^+

23.19. $[Pt(NH_3)_6]Cl_4$, $[Pt(NH_3)_5Cl]Cl_3$, $[Pt(NH_3)_4Cl_2]Cl_2$, $[Pt(NH_3)_3Cl_3]Cl$, $[Pt(NH_3)_2Cl_4]$

23.21. Hexaammineplatinum(IV) chloride, pentaamminechloroplatinum(IV) chloride, tetraamminedichloroplatinum(IV) chloride, triamminetrichloroplatinum(IV) chloride, diamminetetrachloroplatinum(IV)

23.23. (a) Hexaamminechromium(III); (b) hexaaquacobalt(III); (c) pentaamminechloroiron(III)

23.25. (a) Tetrabromocolbaltate(II); (b) aquatrihydroxozincate(II); (c) pentacyanonickelate(II)

23.27. (a) bis(ethylenediamine)zinc(II) sulfate; (b) pentaammineaquanickel(II) chloride; (c) potassium hexacyanoferrate(II)

23.29. A sequestering agent is a multidentate ligand that separates metal ions from other substances so that they cannot react with them. Properties that make a sequestering agent effective include strong interactions between metal ions and ligands that lead to large complex formation constants.

23.31. As pH increases, the chelating ability increases because carboxylic acid groups ionize, and amino groups deprotonate, providing additional coordination sites.

23.33. When the transition metals bond to ligands, their *d* orbital energies split. If *d*-to-*d* electron transitions are possible, the compound is likely to be colored.

23.35. The lobes of the d_{xy} orbital are directed toward the four corners of the square formed by the ligands' lone pairs of electrons. Replusion between the lone pairs and electrons in the d_{xy} orbitals raise the electrons' energy.

23.37. The yellow solution contains (b) $Cr(NH_3)_6^{3+}$. The violet solution contains (a) $Cr(H_2O)_6^{3+}$.

23.39. Yellow, because the absorption band extends into the violet end of the visible spectrum

23.41. $NiCl_4^{2-}$

23.43. The magnitude of the crystal field splitting energy compared to the pairing energy of the electrons in a lower energy *d* orbital

23.45. Fe^{2+} has four unpaired electrons; Cu^{2+} has one unpaired electron; Co^{2+} has three unpaired electrons; Mn^{3+} has four unpaired electrons.

23.47. Cr^{3+}

23.49. (a) Mn^{4+} in MnO_2; 2 Mn^{3+} and 1 Mn^{2+} in Mn_3O_4; (b) both low-spin and high-spin configurations are possible in Mn_3O_4 (d^4 and d^5) but not in MnO_2 (d^3).

23.51. Paramagnetic

23.53. For an octahedral geometry *cis*- means that two ligands are side by side and have a 90° bond angle between them. Ligands that are *trans*- to each other have a 180° bond angle between them.

23.55. At least two different ligands

23.57. Yes

23.59. $[CuBr_2Cl_2]^{2-}$ (Br, Cl top; Cu; Br, Cl bottom) — Cis; $[CuBr_2Cl_2]^{2-}$ (Cl, Br top; Cu; Br, Cl bottom) — Trans

No, neither isomer is chiral.

23.61. (a) To catalyze reactions by lowering the activation energy; (b) no

23.63. Lowers the activation energy

23.65. Much greater than 1

23.67. Positive

23.69. 1.59×10^8 times

23.71. Zero order

23.73. To kill targeted disease cells such as cancer

23.75. Gamma radiation

23.77. They bind to DNA to prevent replication.

23.79. Paramagnetic

23.81. 126

23.83. [penicillamine] $= 9.98 \times 10^{-5}$ *M*, [cysteine] $= 1.00$ *M*

23.85. (a) One has the N_3 ligands *cis* and the other has the N_3 ligands *trans*; (b) *cis*-diaaminediazido *trans*-hydroxoplatinum(IV); *trans*-diaaminediazido *trans*-hydroxoplatinum(IV); (c) yes; (d)

23.87. (a) No; (b) diamagnetic

23.89. The oxidized (Co^{3+}) complex

23.91. As the sample is cooled, the Δ_o must increase and a transition from high spin to low spin occurs.

23.93. CN^- is a strong-field ligand with a very large Δ_o, which leaves d^6 Fe^{2+} with no unpaired electrons and diamagnetic. SCN^-, however, is a weaker (low field) ligand, which leaves d^6 Fe^{2+} with four unpaired electrons and paramagnetic.

23.95. Five unpaired electrons

Credits

Chapter 1

Pages 2–3: Gianni Dagli Orti/The Art Archive at Art Resource, NY; p.5: (inset) BM/SCIENCE PHOTO LIBRARY; p. 5: Geoff Tompkinson/ Getty Image; (b) shutterstock; (c) GIPhotoStock/Science Source; Erickson Photography/Getty Images; p. 6: (a) Photographer's Choice/Punchstock; p. 9: (a) iStockphoto; (b) Andy Clarke/Science Source; p. 10: 2009 Richard Megna/Fundamental Photograps; p. 12: (b) Aquacone courtesy Solar Solutions, Inc.; p. 13: Provided by the SeaWiFS Project, NASA/ Goddard Space Flight Center, and ORBIMAGE; (bottom left) 2009 Richard Megna/Fundamental Photographs; p. 14: (both) 2009 Richard Megna/Fundamental Photographs; (b) 2009 Richard Megna/ Fundamental Photographs; p. 16: (all) David Wrobel/Visuals Unlimited; (top left) Dorling Kindersley/Getty Images; (bottom right) Erik Page Photography/Getty Images; (middle right) Stefan Sollfors/Alamy Stock Photo; (bottom left) Podisu/Dreamstime; (middle left) Arenacreative/ Dreamstime; (freezer) Katherine Fawssett/Getty Images; (top right) Shutterstock; p. 17: Christian Charisius/DPA; p. 20: Dirk Wiersma/Science Photo Library/Science Source; p. 22: Photo courtesy of A&D Weighing, San Jose, CA. www.andweighing.com; p. 23: (top) Shutterstock; (center) Shutterstock; (bottom both) Photo courtesy of A&D Weighing, San Jose, CA. www.andweighing.com; p. 27: (bottom) Canadian Press Images; p. 42: Brian Vance/ Motor Trend; (left) Transtock/ Corbis RF; Wing Lun Leung/Alamy Stock Photo; (bottom) AP Photo/ Neal Hamberg; p. 44: Brian Hartshorn/Alamy; p. 45: (bottom) Shutterstock; (top) NASA/JHUAPL/SwRI.

Chapter 2

Pages 46–47: Sonja Pacho/Getty; p. 48: The Royal Institution, London/ Bridgeman Art Library; p. 50: De Agostini/Getty Images; p. 62: Seiden Allan/Getty Images/Perspectives; p. 63: Richard Megna/Fundamental Photographs; p. 65: (bottom) Vladimir Mucibabic/Dreamstime; (top) Rona Tuccillo; p. 67: Francois Lenoir/Zuma Press; p. 68: (a) AP Photo/ Erik S. Lesser; (b) Adrien Deneu /Morpho/Safran; p. 83: (left) Science & Society Picture Library/Getty Images; (right) Shutterstock.

Chapter 3

Pages 84–85: Hemis/Alamy Stock Photo; p. 87: (a) Digital Vision/Getty Images; (b) imagebroker/Jochen Tack/Newscom; (c) Andrew Holt/ Photographer's Choice/Getty Images; (d) Jason T. Bailey/UPI/Landov; (e) Tim Graham/Getty Images; p. 89: (top a) Getty Images/iStockphoto; (top b) Department of Physics, Imperial College London/Science Source; (bottom a) Richard Megna/Fundamental Photographs, NYC; (bottom b) Richard Megna/Fundamental Photographs, NYC; (bottom c) Richard Megna/Fundamental Photographs, NYC; p. 90: (all) © 1984 Richard Megna, Fundamental Photographs, NYC; p. 91: (bottom) Science Source/Science Source; p. 92: (top) James Leynse/Corbis via Getty Images; p. 94: (both) U.S. Army NVESD; p. 114: (a) David Taylor/ Science Source; (b) Kindersley/Getty Images; p. 121: Paul Sutherland/ Getty Images; p. 132: SPL/Science Source; p. 133: Michael Dalton/ Fundamental Photographs; p. 134: Blend Images/Alamy Stock Photo.

Chapter 4

Pages 140–141: Shutterstock; p. 145: David Wrobel/Visuals Unlimited, Inc.; p. 147: Charles D. Winters/Science Source; p. 148: (both) Martyn F. Chillmaid/Science Source; p. 162: W. Perry Conway/Getty Images; p. 167: Roman Samokhin/Dreamstime.com; p. 186: (left) Charles D Winters/Getty Images; (middle) Scott Eells/Bloomberg via Getty Images; (right) GIPhotoStock/Getty Images; p. 190: sbretz/Shutterstock.

Chapter 5

Pages 192–193: Image Source/Getty Images; p. 194: (a) ullstein bild via Getty Images; (b) Konstantin Gushcha/Getty Images; p. 197: (all) © 2007 Richard Megna/Fundamental Photographs; p. 219: (all) Vanderbilt Institute of Chemical; Biology/http://vanderbilt.edu/vicb/Discoveries Archives/reversing_polarity.html; p. 223: (a) Phil Degginger/www.color-pic.com; (b) Phil Dotson/Science Source; p. 224: Shutterstock; p. 229: © Yoav Levy/Phototake; p. 236: © Yoav Levy/Phototake; p. 240: MARK Garlick/SPL/Getty Images; p. 242: Shutterstock; p. 243: Shutterstock; p. 244: Thinkstock.

Chapter 6

Pages 246–247: Getty Images/iStockphoto; p. 251: © 2012 Richard Megna, Fundamental Photographs, NYC; p. 260: Goo Gone © is a premier brand of The Homax Group, Inc.; p. 262: ISRO/NASA/ JPL-Caltech; p. 263: (c) 1998 Richard Megna/Fundamental Photographs; p. 265: (top) Martin Shields/ Science Source; (bottom) Jeff Daly/Visuals Unlimited; p. 266: (top) Sinclair Stammers/Science Photo Library/Science Source; (bottom) Phil Degginger/www.color-pic.com; p. 267: (top) KARIM SAHIB/AFP/Getty Images; (b) iStockphoto.

Chapter 7

Pages 276–277: Emmanuel LATTES/Alamy Stock Photo; p. 282: (left) 1994 NYC Parks Photo Archive/Fundamental Photographs; (right) 1994 Kristen Brochmann - Fundamental Photographs; p. 284: Shutterstock; p. 291: Mark A. Schneider/Science Source; p. 293: Millard H. Sharp/ Science Source; p. 294: John Cancalosi/Getty Images.

Chapter 8

Pages 318–319: NASA/NOAA/GSFC/Suomi NPP/VIIRS/Norman Kuring; p. 323: Leigh Smith Images/Alamy; p. 325: Richard Megna/ Fundamental Photographs; p. 326: Richard Megna/Fundamental Photographs; p. 327: (top) Photo courtesy of A&D Weighing, San Jose, CA. www.andweighing.com; (bottom) © 2009 Richard Megna, Fundamental Photographs, NYC; p. 328: (a–e) © 2009 Richard Megna/Fundamental Photographs; p. 329: (both) © 1994 Richard Megna - Fundamental Photographs; p. 337: (both) Richard Megna/Fundamental Photographs; p. 341: (bottom left, all) © 1990 Richard Megna - Fundamental Photographs; (top right) Javier Trueba/MSF/Science Source; (bottom right) Richard Thom/Visuals Unlimited; p. 342: iStockphoto; p. 345: (all) Phil Degginger/www.color-pic.com; p. 347: NASA/JPL-Caltech; p. 348:

(both) Peticolas/Megna/Fundamental Photographs; p. 351: (bottom) Bill Ross/Getty Images; (top, both): Wetlands Field Manual/ Courtesy USDA; p. 352: (top, all) © 2009 Richard Megna, Fundamental Photographs, NYC; p. 354: (all) Richard Megna/Fundamental Photographs; p. 357: (top) Joel Arem/Science Source; p. 358: (a) © 2002 Richard Megna, Fundamental Photographs, NYC; (b) TUMS and the shape of the TUMS bottle are trademarks of GlaxoSmithKline. The image was provided courtesy of GlaxoSmithKline; p. 359: (top left) Johnbell/ Dreamstime.com; (bottom left) © 2009 Richard Megna - Fundamental Photographs; (right both) Peticolas/Megna/Fundamental Photographs; p. 365: P. Rona/NOAA; p. 367: (both) Courtesy Thomas Gilbert; p. 369: Joel Arem/Science Source.

Chapter 9

Pages 370–371: Shutterstock; p. 372: Richard and Ellen Thane/Getty Images; p. 373: Stocktrek Images, Inc./Alamy; p. 377: Shutterstock; p. 380: Newscom; p. 381: AP Photo; p. 384: Johner Images/Getty Images; p. 387: Sprokop | Dreamstime.com; p. 406: DigitalVues/Alamy RF; p. 407: (all) Tom Gilbert; p. 408: (a) Charles D. Winters/Science Photo Library/Science Source; (b) Charles D. Winters/Science Photo Library/Science Source; p. 414: (top) Shutterstock; (bottom) AP Photo; p. 416: (top) Granger Historical Picture Archive; (middle) Courtesy of Alcoa; (bottom) Courtesy of Alcoa; p. 422: (b) Charles D. Winters/Science Source; (h) Charles D. Winters/Science Source; p. 423: Shutterstock.

Chapter 10

Pages 430–431: Bill Stormont/Getty Images; p. 432: (b) British Antarctic Survey/Science Source/Science Source; p. 433: Predrag Vuckovic/Getty Images; p. 435: iStockphoto; p. 440: (a) Galen Rowell/Getty; (b) Hubert Stadler/Getty Images; (c) Bill Ross/Getty Images; p. 441: (a) Sam Ogden/ Science Source; p. 442: (bottom) Stan Pritchard/Alamy; p. 443: (bottom) Matt Anderson Photography/Getty Images; (top) LIBRARY OF CONGRESS/SPL/Getty Images; p. 452: © 2001 Richard Megna, Fundamental Photographs, NYC; p. 453: iStockphoto; p. 454: (a) Richard Anthony/Minden Pictures; (b): Thierry Orban/Sygma/Getty Images; p. 455: (all) © 2009 Richard Megna, Fundamental Photographs, NYC; p. 458: Art Directors & TRIP/Alamy; p. 460: © 2008 Richard Megna, Fundamental Photographs, NYC; p. 463: Fuse/Getty Images; p. 464: (top) Stephen B. Goodwin/Shutterstock; (bottom) Nicholas Burningham/ Dreamstime.com; p. 470: Kevin Arnold/Getty Images; p. 471: Jonathan Blair/Getty Images; p. 472: AFP/AFP/Getty Images; p. 473: F. Jack Jackson/Alamy; p. 475: Hugh Peterswald/Icon Sportswire DHE/ Hugh Peterswald/Icon Sportswire/Newscom; p. 476: National Museums, Scotland.

Chapter 11

Pages 478–479: JEAN-MARIE LIOT/AFP/Getty Images; p. 481: (center) David M. Phillips/Science Source; (left) David M. Phillips/ Science Source; (right): SPL/Science Source; p. 486: Courtesy of Katadyn; p. 489: W. W. Norton; p. 490: Shutterstock; p. 493: Martyn F. Chillmaid/ Science Source; p. 505: Phil Degginger/Color-Pic; p. 512: (right) Mark Bolton /Getty; (left) Shutterstock; (top right) OSH/Alamy.

Chapter 12

Pages 516–517: iStockphoto; p. 519: Phil Degginger/Alamy; p. 520: (both) © 1987 Richard Megna, Fundamental Photographs, NYC; p. 528: (a) Jon Stokes/Science Source; (b): Ken Lucas/Visuals Unlimited; p. 533: B. & C. Alexander/Science Source; p. 543: (a) Reinhard, H./picture alliance/Arco Images G/Newscom; (b) SCIMAT/Science Photo Library/ Science Source; p. 551: (b) John Patriquin/Portland Press Herald via Getty Images; (d) Richard Megna/Fundamental Photographs; (f) Richard Megna/Fundamental Photographs; (h) Getty.

Chapter 13

Pages 558–559: Csondy/Getty Images; p. 585: (b) Wikimedia Commons; p. 597: Image courtesy the TOMS science team & and the Scientific Visualization Studio, NASA GSFC; p. 599: (top) Clive Streeter/Getty Images; p. 605: (c, f, i) Image courtesy the TOMS science team & and the Scientific Visualization Studio, NASA GSFC; p. 608: Lawrence Migdale/Science Source; p. 614: Andrew Lambert Photography/Science Photo Library/Science Source.

Chapter 14

Pages 618–619: Phil Degginger/Alamy; p. 623: (bottom) © 1983 Chip Clark - Fundamental Photographs; (top) Roger Ressmeyer/Getty Images; p. 627: (both) Richard Megna/Fundamental Photographs, NYC; p. 638: Morev Valery/ITAR-TASS/Landov; p. 639: (a) Thermo Fisher Scientific; p. 663: (both) Richard Megna/Fundamental Photographs, NYC.

Chapter 15

Pages 674–675: Eduardo Jose Bernardino/Getty Images; p. 699: Shutterstock; p. 700: blickwinkel/Alamy Stock Photo; p. 710: Ed Endicott/ Alamy Stock Photo.

Chapter 16

Pages 722–723: Shutterstock; p. 736: (a–c) Courtesy Thomas Gilbert; p. 737: © 1994 Richard Megna/Fundamental Photographs, NYC; p. 744: Shutterstock; p. 749: Richard Megna, Fundamental Photographs, NYC; p. 752: David R. Frazier Photolibrary, Inc./Alamy Stock Photo; p. 753: Shutterstock; p. 754: (a, b) Charles D. Winters/Science Source; p. 762: Richard Megna, Fundamental Photographs, NYC; p. 763: (c, g) Richard Megna/Fundamental Photographs.

Chapter 17

Pages 770–771: dpa picture alliance/Alamy Live News; p. 770: (bottom) EggImages/Alamy Stock Photo; p. 773: Phil Degginger/www.color-pic.com; p. 779: (a) iStockphoto; p. 781: Mary Evans Picture Library/Alamy; p. 793: Paul Mogford/Alamy; p. 794: (a) Newscom; p. 796: Chris So/ The Toronto Star via ZUMA Wire; p. 799: (a) Getty; p. 802: AP Photo/ M. Spencer Green; p. 804: (a) Toyota Motor Sales, USA, Inc.; p. 807: Phil Degginger/www.color-pic.com; p. 809: (top) Alix/Science Source; (bottom) sciencephotos/Alamy; p. 810: (a) Science Source; (c) bagwold /Getty Images/iStockphoto;(g) Richard Megna/Fundamental Photographs; (i) Richard Megna/Fundamental Photographs; p. 815: (both) Courtesy of Thomas Gilbert.

Chapter 18

Pages 818–819: Rudi Van Starrex/Getty Images; p. 820: (a) Richard Megna/Fundamental Photographs, NYC; p. 821: (a) Dorling Kindersley/ Getty Images; (b) imagebroker/Alamy; p. 824: (top) Manfred Kage/Science Source; (bottom) Shutterstock; p. 827: David Parker/Science Source; p. 829: Kris Mercer/Alamy; p. 833: AlexeyKamenskiy/Getty Images; p. 834: Charles D. Winters/Science Source; p. 837: (a) yurazaga/Getty Images/

iStockphoto; (b) Shutterstock; p. 838: (a) Jon Stokes/Science Source; (b): Ken Lucas/Visuals Unlimited; (c) Andre Geim & Kostya Novoselov/Science Source; p. 840: (top) 1994 Richard Megna, Fundamental Photographs, NYC; (bottom) Charles D. Winters/Science Source; p. 841: (bottom) Raul Gonzalez Perez/Science Source; p. 842: Harry Taylor/Dorling Kindersley/Science Source; p. 844: (top) Getty Images/DeAgostini; (bottom): E. R. Degginger/Science Source; p. 845: Arthur Hill/Visuals Unlimited, Inc.; p. 846: Ashley Cooper/Alamy; p. 857: (bottom) Shutterstock; p. 858: Photo by VCG Wilson/Corbis via Getty Images.

Chapter 19

Pages 862–863: Chuck Myers/MCT via Getty Images; p. 878: Dr. Keith Wheeler/Science Source; p. 880: (a, b) Shutterstock; (c) Digital Vision/Jupiterimages; p. 887: (top) Lynnette Peizer/Alamy; (bottom) CANADA/Newscom; (a) Bsip/Photoshot/ZUMA; Press/Newscom; (bottom left) Dr. Ann Schmierer; (b) Manfred Gottschalk/Getty Images; p. 889: (bag) Emily Spence/Lexington Herald-Leader/MCT/Newscom; (carpet) Fotosearch; (rope) studiomode/Alamy; (chair) Elizabeth Whiting & Associates/Alamy Stock Photo; p. 892: (top) Polystyrene cup, Photo by Luiscarlosrubino; http://creativecommons.org/licenses/by-sa/3.0/; (center) Stephen Stickler/Getty Images; (bottom) © 2010 Richard Megna/Fundamental Photographs NYC; (a, b) Getty Images; (c) Shutterstock; (d) Douglas Peebles/Getty; p. 905: Barry Slaven/PMODE Photography; p. 906: Creatas/Jupiterimages; p. 908: Courtesy DuPont.

Chapter 20

Pages 926–927: Laguna Design/Getty Images; p. 929: Shutterstock; p. 936: (a) Dennis Kunkel Microscopy/Visuals Unlimited; (b) Omikron/Science Source; p. 938: (top) Courtesy of N.I.S.T.; (bottom) Chemical Desiupiterimagesgn/Science Photo Library/Science Source; p. 946: Biophoto Associates/Science Source; p. 956: (top) Roger Ressmeyer/Getty; (middle) NASA; (bottom) © W.R. Normak, courtesy USGS.

Chapter 21

Pages 968–969: LADA/Science Source; p. 981: Rob Blakers/Getty Images; p. 985: Corbis via Getty Images; p. 989: (top) NASA/HST/J. Morse/K. Davidson; (bottom) X-ray: NASA/CXC/ASU/J.Hester et al.; Optical: NASA/ESA/ASU/J.Hester & A.Loll; Infrared: NASA/JPL-Caltech/Univ. Minn./R.Gehrz; p. 993: (top, a) Astrid & Hanns-Frieder Michler/Science Photo Library/Science Source; (top b) Tom Tracey Photography/Alamy; (top c) Urenco; (bottom) Argonne National Laboratory, US DOE Office of Environmental Management; p. 995: (a) Iridiumphotographics/Getty Images; (b) Courtesy of Ecotest (www.ecotestgroup.com); (c) Courtesy of Ecotest (www.ecotestgroup.com); p. 999: (top) Photo by Igor Kostin/Sygma via Getty Images; (bottom, both) © 2006 T. A. Mousseau and A. P. Moller; p. 1000: Science Source; p. 1003: (a, b) Dr. Robert Friedland/Science Source; (left) © 1990 Richard Megna/Fundamental Photographs; (right) The American Weekly, 1926; p. 1005: Rob Blakers/Getty Images; (right) X-ray: NASA/CXC/ASU/J.Hester et al.; Optical: NASA/ESA/ASU/J.Hester & A.Loll; Infrared: NASA/JPL-Caltech/Univ. Minn./R.Gehrz; p. 1009: (top) Tony Camacho/Science Source; (middle): blickwinkel/Alamy; (bottom right) Pavel, et al; "Human Presence in the European Arctic Nearly 40,000 Years ago," Nature, 413:64,2001. Copyright ¬© 2001, Rights Managed by Nature Publishing Group; p. 1015: IPA USA/Karen L. King/Harvard via Sipa/Newscom.

Chapter 22

Pages 1016–1017: Claudio Ventrella/Getty Images; p. 1022: (both) E. R. Degginger/www.color-pic.com; p. 1023: (b) Don W. Fawcett/Science Source; p. 1035: (left) Dreamstime; (right) Tom Brakefield/Stockbyte/Getty Images; p. 1039: E. R. Degginger/www.color-pic.com; p. 1041: (both) SIU/Visuals Unlimited, Inc.; p. 1044: E. R. Degginger/www.color-pic.com.

Chapter 23

Pages 1050–1051: bbszabi/Getty Images/iStockphoto; p. 1053: (a,b) Photo by Albris, May 2011; http://creativecommons.org/licenses/by-sa/3.0/deed.en; p. 1055: Image copyright © The Metropolitan Museum of Art. Image source: Art Resource, NY; p. 1061: (a) Richard Megna, Fundamental Photographs; (b): Richard Megna, Fundamental Photographs; p. 1064: © 1997 Richard Megna, Fundamental Photographs, NYC; p. 1066: (middle) © 1990 Richard Megna, Fundamental Photographs, NYC; p. 1069: (d) Vaughan Fleming/Science Source; p. 1072: (a, b) E. R. Degginger/www.color-pic.com; p. 1081: (c) Zephry/Science Photo Library/Science Source; p. 1082: (a, b) From Small molecular gadolinium (III) complexes as MRI contrast agents for diagnostic imaging ‰ Chan Kannie Wai-Yan and Wong wing-Tak, Coordination Chemistry Reviews, Sept., 2007, Elsevier B.V., Copyright Clearance Center; p. 1088: (e) Richard Megna/Fundamental Photographs.

Index

Note: Material in figures or tables is indicated by *italic* page numbers. Footnotes are indicated by n after the page number.

A

B

C

F

G

H

I

J

K

L

O

P

Q

R

S

T

U

X

Y

Z

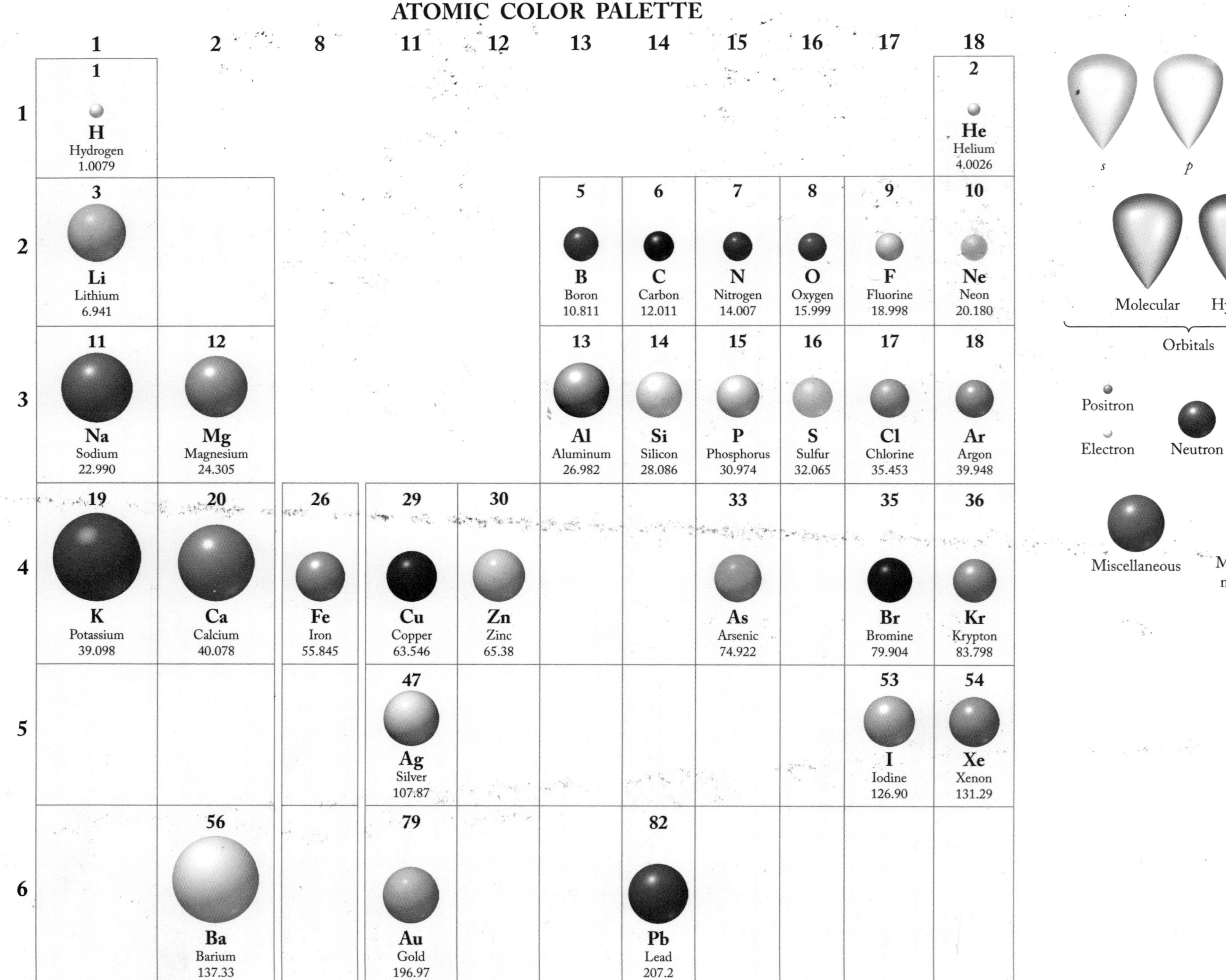

ATOMIC COLOR PALETTE
1
2
8
11
12
13
14
15
16
17
18
1
2
3
4
5
6
1 H Hydrogen 1.0079
2 He Helium 4.0026
3 Li Lithium 6.941
5 B Boron 10.811
6 C Carbon 12.011
7 N Nitrogen 14.007
8 O Oxygen 15.999
9 F Fluorine 18.998
10 Ne Neon 20.180
11 Na Sodium 22.990
12 Mg Magnesium 24.305
13 Al Aluminum 26.982
14 Si Silicon 28.086
15 P Phosphorus 30.974
16 S Sulfur 32.065
17 Cl Chlorine 35.453
18 Ar Argon 39.948
19 K Potassium 39.098
20 Ca Calcium 40.078
26 Fe Iron 55.845
29 Cu Copper 63.546
30 Zn Zinc 65.38
33 As Arsenic 74.922
35 Br Bromine 79.904
36 Kr Krypton 83.798
47 Ag Silver 107.87
53 I Iodine 126.90
54 Xe Xenon 131.29
56 Ba Barium 137.33
79 Au Gold 196.97
82 Pb Lead 207.2
s
p
d
Molecular
Hybrid
Orbitals
Positron
Electron
Neutron
Proton
Miscellaneous
Muted water molecule in solution